STRICKBERGER'S

EVOLUTION

THE INTEGRATION OF GENES, ORGANISMS AND POPULATIONS

FOURTH EDITION

BRIAN K. HALL
Dalhousie University

BENEDIKT HALLGRIMSSON
University of Calgary

JONES AND BARTLETT PUBLISHERS
Sudbury, Massachusetts
BOSTON TORONTO LONDON SINGAPORE

World Headquarters
Jones and Bartlett Publishers
40 Tall Pine Drive
Sudbury, MA 01776
978-443-5000
info@jbpub.com
www.jbpub.com

Jones and Bartlett Publishers Canada
6339 Ormindale Way
Mississauga, Ontario L5V 1J2
CANADA

Jones and Bartlett Publishers International
Barb House, Barb Mews
London W6 7PA
UK

Jones and Bartlett's books and products are available through most bookstores and online booksellers. To contact Jones and Bartlett Publishers directly, call 800-832-0034, fax 978-443-8000, or visit our website, www.jbpub.com.

Substantial discounts on bulk quantities of Jones and Bartlett's publications are available to corporations, professional associations, and other qualified organizations. For details and specific discount information, contact the special sales department at Jones and Bartlett via the above contact information or send an email to specialsales@jbpub.com.

Production Credits
Chief Executive Officer: Clayton Jones
Chief Operating Officer: Don W. Jones, Jr.
President, Higher Education and Professional Publishing: Robert W. Holland, Jr.
V.P., Design and Production: Anne Spencer
V.P., Manufacturing and Inventory Control: Therese Connell
V.P., Sales and Marketing: William J. Kane
Executive Editor, Science: Cathleen Sether
Acquisitions Editor, Science: Shoshanna Grossman
Managing Editor, Science: Dean W. DeChambeau
Associate Editor, Science: Molly Steinbach
Editorial Assistant: Briana Gardell
Senior Production Editor: Louis C. Bruno, Jr.
Production Assistant: Leah E. Corrigan
Senior Marketing Manager: Andrea DeFronzo
Text and Cover Design: Anne Spencer
Illustrations: Elizabeth Morales
Photo Research Manager/Photographer: Kimberly L. Potvin
Associate Photo Researcher: Christine McKeen
Composition: NK Graphics
Printing and Binding: Malloy
Cover Printing: Lehigh Press
Cover Image: © Matthew Oldfield Scubazoo/SPL/ Photo Researchers, Inc.

The animal pictured in the right-hand page corner throughout this book is an early tetrapod, a "four-footed" animal. Moving from water to land was a critical step in tetrapod evolution; fossil evidence of stem and early tetrapods indicates that this occurred about 365 million years ago (see Chapter 18). Hypothesized advantages to moving to land include avoiding predators and exploiting terrestrial food sources, such as insects. Flip the pages of this book from front to back to watch the animal "evolve."

About the cover: Pygmy seahorse (*Hippocampus bargibanti*) camouflaged on a sea fan coral (Muricella sp.). These seahorses have scaleless bodies and prehensile tails, which they may use to hold on to their gorgonian home. The coloration and texture of this tiny seahorse has evolved to match the color and shape of its host coral. It inhabits the tropical coral reefs of the western Pacific Ocean and reaches no more than 2.4 centimeters in length. Photographed in Pulau Kapalai, Sabah, Malaysia.

Library of Congress Cataloging-in-Publication Data
Hall, Brian Keith, 1941–
Strickberger's evolution / Brian K. Hall, Benedikt Hallgrímsson.
p. cm.
Rev. ed. of: Evolution / Monroe W. Strickberger. 3rd ed. c2000.
Includes index.
ISBN 978-0-7637-0066-9 (alk. paper)
1. Evolution (Biology) I. Hallgrímsson, Benedikt. II. Strickberger, Monroe W. Evolution. III. Title. IV. Title: Evolution.
QH366.2.S78 2008
576.8—dc22 2007008981
6048

Printed in the United States of America
11 10 09 08 07 10 9 8 7 6 5 4 3 2 1

Brief Contents

Contents

Preface

It has been a privilege to have been given the opportunity to revise and update *Evolution* by Monroe Strickberger, the most broadly based textbook on evolution. Available in three editions, the first in 1990, the third in 2000, *Evolution* has become a staple for undergraduate education in evolutionary biology. Over the last six years, however, the field has changed enormously. A major change is that embryonic development, which separates natural selection from its effect on genes, is no longer a black box; an entirely new field linking evolution with development has come into a mature phase of development. Another is that the sequencing of entire genomes and the emergence of bioinformatic tools for dealing with such vast quantities of information is beginning to allow us to ask questions at the level of biological systems rather than individual genes. Finally, the fossil record continues to expand and our understanding of it continues to mature, especially through paleobiology, the integration of paleontology with such biological disciplines as ecology, developmental biology and physiology. To the best of our ability, the fourth edition incorporates these changes in evolutionary biology and the broader biological sciences.

Essential to our understanding of evolution is that groups of organisms are bound together by their common inheritance; that the past has been long enough for inherited changes to accumulate; and perhaps most essential of all, that discoverable biological processes and natural relationships among organisms explain the reality of evolution. Although each of these aspects has been studied and discussed at various times in human history, only after the mid-nineteenth century when Charles Darwin developed his theory of evolution by natural selection did biological evolution become an acceptable scientific alternative to earlier explanations. This acceptance brought about enormous changes in how we view the world, understand our place in it, and explain natural phenomena.

Of special interest in historical sciences such as evolution is the emphasis on understanding a particular sequence of historical events that cannot always be understood by the application of general laws such as those of physics and chemistry. For example, subjects of historical sciences include events that led to our solar system, to the separation of South America from Africa and to the origin of humans, recognizing at the same time that these events are singular and do not apply to all stars, all continental separations or the evolution of all species. Thus, although sciences that deal with the past make use of general laws such as gravity, mechanics and biochemistry, *their aim is to discover the causes of diversity and uniqueness as much as to discover principles and laws that apply uniformly to all matter or all life.* Indeed, the ability to extrapolate from general principles *and* to explain the role of contingency is both central to the dynamic interdisciplinary nature of the study of evolution and makes it a robust and testable science. The properties of different hydrogen atoms, for example, can be explained on a common physical basis, whereas the properties of different organisms — the organization and function of their component parts — are explained within the context of their organismal histories, which include adaptations to specific lifestyles at particular times. *The central aim of evolutionary biology is to explain the origins and diversity of life.*

When historical conditions are repeated, and different organisms are subjected to similar selective evolutionary forces, some common features can be predicted; geographically widely separated populations and species, such as fish adapted to cave conditions, consistently show rudimentary eyes, enhanced development of chemosensory organs and loss of pigment, among other common attributes. Other evolutionary changes are contingent on past history, for example, the evolution of flight among reptiles and not among frogs or most mammals. In this text we both provide the evidence upon which general principles can be based and explore those aspects of organisms and biological processes involved in historical contingency. To this end, there is a logic in the way we have arranged the subject matter:

- beginning with the history of evolutionary ideas (Part 1);
- moving to the origins of the universe, Earth, molecules, cells, organisms and natural selection in Part 2;
- considering the genetic basis for inheritance, the nature of species, how species are grouped and the role of development in evolution in Part 3; before
- examining the major groups of organisms on Earth (Part 4);
- discussing how populations are maintained and how species arise (Part 5);
- finishing with how cultural, social and biological evolution interact in our own population, how humans have had an impact on the evolution of other species and how evolution has an impact on religion (Part 6).

PART 1 provides the historical, cultural and social framework leading up to Darwin's theory of the origin of species by means of natural selection. We approach the history of evolutionary biology by asking, "What happened?" "Who were the major players?" and "Why did it happen the way it did?"

PART 2 deals with events that occurred between 5 and 15 billion years ago (Bya is the abbreviation we use throughout) to create 100 billion galaxies, each with 100 billion or more stars, of which Earth's is one in a solar system that arose 4.6 Bya. Earth's origin, how molecules arose, whether the first molecules were DNA, RNA or protein, and how the first prokaryotic and eukaryotic cells arose 3.5 Bya and 2 Bya, respectively, are all discussed in the five chapters of Part 2.

PART 3 takes us into the realms of genes, cells and organisms as we consider those fundamental genetic, molecular, cellular and developmental features and processes that provide the basis for inheritance, variation and evolution — *descent with modification,* to use Darwin's terms. Now that we know that much of the genome is shared between organisms and has an evolutionary history that can be traced back to the origins of eukaryotes, much emphasis is being placed on gene regulation as pivotal to developmental and evolutionary change. Change the level or time of expression of a regulatory gene and you can change the phenotype. Selection on that new phenotype can result in the altered state of gene expression spreading through the population, leading to evolutionary change — descent with modification. The five chapters in Part 3 provide the background for you to appreciate how such modification occurs and how we recognize new species when they arise.

PART 4 contains seven chapters on the history of life in which we consider the diversity of organisms that now populate and have populated Earth. Emphasis is on how new groups of organisms arise and adapt to changing environments, and how organisms are related one to another. Much in these seven chapters is about transitions — from unicellular protozoans to multicellular plants and animals; from water to land in plants and animals; from invertebrates to vertebrates; from jawless to jawed vertebrates; from reptiles to birds.

PART 5 examines those genetic and other mechanisms operating to maintain populations as stable entities or to change populations and allow new species to arise.

PART 6 takes us into how cultural and biological evolution interact in our own species, how we now have the ability to influence the evolution of other species and how evolution and religion interact.

Although there is a logical arrangement for Chapters 1 to 26, the structure of the book allows you to begin with Part 6 or to read individual parts as units independent of the other parts.

Summaries are provided at the beginning of each of the six parts and at the beginning of each chapter. We suggest that you read the summary both before and after you read each chapter — before to see whether what is coming is familiar or not; after to be sure you have acquired the essence of the chapter (which you can check by looking over the discussion questions at the end of each chapter). **Boxes** are used to draw attention to particular topics, often ones that relate to more than one chapter. **Interviews** with active researchers, conducted in 1998–1999, and included in the third edition, have been updated for this edition by adding a Web site and a sample recent publication. We find these interviews valuable guides to where different fields were perceived to be moving. Comparing them with the text shows vividly how rapidly studies on evolution are progressing.

Much more information and analysis is available on all the events and concepts discussed than we could possibly have included in the book. Access to information in particular chapters is gathered together in two ways: as a list of **recommended reading** at the end of each chapter; and, at the end of the book in a **single list of literature cited** in each chapter and sources for the figures.

We have two major reasons for including references to the primary literature. One is intrinsic in the nature of the subject matter. Evolution is a science, and as such, you as the reader of this scientific textbook should not take our representations and interpretations of scientific research for granted. As you should be able to check the primary literature for yourself, we have cited enough of that primary literature to enable you to enter it with ease. Rejection or acceptance of a scientific hypothesis is based on whether data gathered to test the hypothesis refute it or not. Thus, the sequence of hominin, primate-like fossils extending from the far past to the present supports the hypothesis that humans have a primate origin (Chapter 20). Similarly, correspondence in the amino acid sequences of myoglobin and hemoglobin supports an evolutionary relationship between them (Chapter 12).

The second reason is that whether this book is used for an introductory or a more advanced class — and it is, in our view suitable for either — the primary literature cited provides an ideal basis for tutorials, discussions, essays and/or presentations.

Part I takes the publication in 1859 of *The Origin of Species by Means of Natural Selection* by Charles Darwin as the starting point. An oft-recommended version is the facsimile of the first edition published by Harvard University Press under the editorship of Ernst Mayr in 1964 (Darwin, 1964) and as an inexpensive paperback in 2005. Project Gutenberg Online Book Catalog has many of Darwin's books downloadable in reliable text form (http://www.gutenberg.org/).

An effective way to access Darwin's writings and commentaries on his work and influence, is through the *Norton Critical Edition* (Appleman, 2001), which contains an introduction by Philip Appleman, an evaluation of Darwin's life by Ernst Mayr, selections from Darwin's writing, reprints of eight essays on scientific thought before Darwin, and almost 100 commentaries on Darwin's influence on science, social thought, philosophy, ethics, religion and literature.

The following **encyclopedias, texts** and **collections of essays** are recommended, either because they cover evolution in its entirety or because they cover a major aspect of evolution.

Appleman, P, (ed.), 2001. *A Norton Critical Edition.* Darwin: Texts, Commentary, 3rd ed. W. W., New York.

Bowler, P. J., 2003. *Evolution: The History of an Idea,* 3rd ed. University of California Press, Berkeley, CA.

Browne, E. J., 1995. *Charles Darwin: Voyaging.* Alfred A. Knopf, New York.

Browne, E. J., 2002 *Charles Darwin: The Power of Place. Volume II of a Biography.* Alfred A. Knopf, New York.

Evolution, 2006. A Scientific American Reader. The University of Chicago Press, Chicago.

Hall, B. K., and W. M. Olson (eds.), 2003. *Keywords & Concepts in Evolutionary Developmental Biology.* Harvard University Press, Cambridge, MA.

Jones, S., R. D. Martin, and D. R. Pilbeam (eds.), 1996. *The Cambridge Encyclopedia of Human Evolution.* Cambridge University Press, Cambridge, England.

Keller, E. F., and E. A. Lloyd, 1992. *Keywords in Evolutionary Biology.* Harvard University Press, Cambridge, MA.

Knoll, A. H., 2003. *Life on a Young Planet: The First Three Billion Years of Evolution on Earth.* Princeton University Press, Princeton, NJ.

Niklas, K. J., 1997. *The Evolutionary Biology of Plants.* The University of Chicago Press, Chicago and London

Supplements to the Text

Jones and Bartlett offers an array of ancillaries to assist instructors and students in teaching and mastering the concepts in this text. Additional information and review copies of any of the following items are available through your Jones and Bartlett sales representative, or by going to http://www.jbpub.com/biology/.

For the Student

Developed by Dr. Bill Brindley of Utah State University exclusively for the fourth edition of *Strickberger's Evolution,* the companion Web site, http://biology.jbpub.com/evolution, offers a variety of resources to enhance understanding of evolution. The site contains a free on-line study guide with chapter outlines, quizzes and exercises to test comprehension and retention and an interactive glossary. This site also has links to other interesting and informative Web sites and seminal papers in the field of evolution.

For the Instructor

Compatible with Windows and Macintosh platforms, the Instructor's ToolKit—CD-ROM provides instructors with the following traditional ancillaries:

- The *Test Bank* of over 800 questions is available as MSWord or Rich Text Files.
- The *PowerPoint™ Image Bank* provides the illustrations, photographs and tables (to which Jones and Bartlett holds the copyright or has permission to reproduce digitally) inserted into PowerPoint slides. You can quickly and easily copy individual images or tables into your existing lecture slides.
- The *PowerPoint Lecture Outline Slides* presentation package provides lecture notes and images for each chapter of *Strickberger's Evolution.* Instructors with Microsoft PowerPoint software can customize the outlines, art and order of presentation.
- Extra *problems* and *exercises* to enhance your students' comprehension of and appreciation for the material in the text.

Acknowledgements

We are most grateful to Virginia Trimble for providing a new chapter (4) on cosmology and the origin of chemical elements and to June Hall for her contributions to Chapters 4 and 25. The finished textbook has benefited enormously from June Hall's keen editorial eye as she read through the entire manuscript at least twice. Sina Adl and Alastair Simpson provided many helpful comments on the topics of Chapters 6 to 8 and 12. Garland Allan, Gordon McOuat and Jan Sapp provided helpful comments on the comparative scientific and writing styles of Gregor Mendel and Charles Darwin. Mark Johnston provided insights into a number of aspects of fertilization in plants (Chapter 14); Marty Leonard did the same for clutch size in birds. For providing a place of solitude in which initial editing was done, BKH thanks Clive and Linda Bedford-Brown of Subiaco, Western Australia. Some of the figures were drafted by Lisa Budney and Tim Fedak; our thanks to you both. Lisa Budney also provided helpful insights into ways to present relationships, phylogenies and cladograms.

The following individuals, each a specialist in evolutionary biology, in a particular area of evolutionary biology and/or who teach evolution, took time from their academic schedules to comment on drafts of chapters or boxes (in parentheses below) for this edition. We appreciate your insights and forthright comments:

Ehab Abouheif, McGill University (13)
Sina Adl, Dalhousie University (8)
Joseph Beilawski, Dalhousie University (11, 12)
Wouter Bleeker, Geological Survey of Canada (4, 5)
Bill Brindley, Utah State University (9–16)
Lisa Budney, Dalhousie University (21–24, 26)
Richard Burian, Virginia Tech (1–3, 25)
Albert Buckelew, Bethany College (21–25)
M. Michael Cohen, Dalhousie University (20)
David Deamer, University of California Santa Cruz (6, 7)
Tamara Franz-Odendaal, Mount Saint Vincent University (13, 17)
Kenneth Gobalet, California State University Bakersfield (17–19)
Dalton Gossett, Louisiana State University, Shreveport (1–4, 19)
Rosemary Grant, Princeton University (23)
Michael Gray, Dalhousie University (8)
Robert Guralnick, University of Colorado, Boulder (16)
Jenna Hellack, University of Central Oklahoma (21–25)
Heather Jamniczky, University of Calgary (18)
Megan Johnson, University of Calgary (18)
Mark Johnston, Dalhousie University (14)
Hollie Knoll, University of Calgary (18)
Shigeru Kuratani, Riken Center for Developmental Biology, Japan (13, 17)
Robert Lee, Dalhousie University (11, 14)
Sally Leys, University of Alberta (Box 16-2)
Ian Mclaren, Dalhousie University (23)
Gordon McOuat, University of Kings College (1–3, 25)
Jeffrey Meldrum, Idaho State University (1–3)
Mary Murnik, Ferris State University (1–3)
Karl Niklas, Cornell University (14)
Glen Northcutt, University of California, San Diego (17)
John Olsen (5–8), Rhodes College
Wendy Olson, University of Northern Iowa (1–3, 25)
Leslie Orgel, the Salk Institute (6,7)
Louise Page, University of Victoria (16)
Keith Pecor, College of New Jersey (9–16)
David Piper, Atlantic Geological Survey (5)
David Polly, Indiana University, Bloomington (19)
Mark Regan, Griffiths University, Brisbane (11)
Andrew Roger, Dalhousie University (6,7)
Anthony Russell, University of Calgary (18)
Alastair Simpson, Dalhousie University (8)
Shiva Singh, University of Western Ontario (9, 10)
Eric Sniverly, University of Calgary (18)
George Spagna, Randolph-Macon College (4)
John Stone, McMaster University (Box 23-2)
Matt Vickaryous, University of Calgary (18)
Marvalee Wake, University of California Berkeley (18)
Alan Weiner, University of Washington (6, 7)
Ken Weiss, Pennsylvania State University (21, 22)
Polly Winsor, University of Toronto (11, 12)
Pat Wise, University of Calgary (18)
Nate Young, Stanford University (20)

The index is a vital part of any book and especially of a textbook. A colleague who prepared the index for a book that he had just finished writing was surprised to see a theme emerge in a set of entries, a theme he was not conscious of having developed in the book. We had the index "test-driven" by a number of colleagues in different areas of evolution. We asked them to check that the index provided the entries they would expect to find and in the places they would expect to find them. This valuable exercise resulted in additional entries and cross-references, both of which enhance the utility of the index. We thank Chris Corkett, Mark Johnston, Alan Pinder, Wendy Olson and Liz Welsh for their dry runs on the index.

Brian K. Hall
Dalhousie University

Benedikt Hallgrímsson
University of Calgary

October 2007

About the Authors

Brian Hall (left) and Benedikt Hallgrímsson (right) photographed outside Charles Darwin's home, Down House, in July 2006 after having participated in the symposium on *Tinkering: The Microevolution of Development,* held at the Novartis Foundation in London.

Brian Hall, born, raised and educated in Australia, has been associated with Dalhousie University in Halifax, Nova Scotia since 1968, most recently as a University Research Professor and George S. Campbell Professor of Biology, and since July 2007 as University Research Professor Emeritus. He was Killam Research Professor at Dalhousie University (1990–1995), Faculty of Science Killam Professor (1996–2001) and Canada Council for the Arts Killam Research Fellow (2003–2005).

Trained as an experimental embryologist, for the past 40 years he has undertaken research into vertebrate skeletal development and evolution and played a major role in integrating evolutionary and developmental biology into the discipline now known as *Evolutionary Developmental Biology (evo-devo);* he wrote the first evo-devo text book, published in 1990 and in a second edition in 1999 (Hall 1999a).

A fellow of the Royal Society of Canada and Foreign Honorary Member of the American Academy of Arts and Sciences, Dr. Hall has earned numerous awards for his research, teaching and writing, including the 2005 Killam Prize in Natural Sciences, one of the top scientific awards in Canada.

Benedikt Hallgrímsson was born in Reykjavík, Iceland, and completed his studies at the University of Alberta and the University of Chicago. A biological anthropologist and evolutionary biologist, he combines developmental genetics and bioinformatics with morphometrics to address the developmental basis and evolutionary significance of phenotypic variation and variability. His work has focused on humans and other primates as well as mouse models and has employed both experimental and comparative approaches to study the evolutionary developmental biology of variation. He is the editor-in-chief of *Evolutionary Biology,* a journal dedicated to the synthesis of ideas in evolutionary biology and related disciplines.

Based at the University of Calgary, Dr. Hallgrímsson teaches organismal biology and anatomy. There he has received several Gold Star Teaching Awards, a Letter of Excellence Lecturer Award and the McLeod Distinguished Achievement Award. He is featured on the University of Calgary "Great Teachers" website. From the American Association of Anatomists, he received the Basmajian/Williams and Wilkins Award for educational contributions in 2001.

Drs. Hall and Hallgrímsson have worked together over many years. Their latest completed collaboration before this text is the edited volume *Variation: A Central Concept in Biology* (2005), which addresses a concept of fundamental importance in evolutionary biology. The late Ernst Mayr concluded his foreword to the book with "In short, variation is an endless source of challenging questions."

Answering Life's Timeless Questions

Where do we come from? What are we? Where are we going?

—Translation of the title *D'où venons nous? Que sommes nous? Où allons nous?* one of Paul Gaugin's most famous works, perhaps his ultimate masterpiece, in which he depicts his vision of life's great questions.

There is grandeur in this view of life, with its several powers, having been originally breathed into a few forms or into one; and that, whilst this planet has gone cycling on according to the fixed laws of gravity, from so simple a beginning endless forms most beautiful and most wonderful have been, and are being evolved.

—The last sentence of *On the Origin of Species by Means of Natural Selection, or the Preservation of Favoured Races in the Struggle for Life,* by Charles, Darwin, 1859.

Darwin's alienation of the inside from the outside was an absolutely essential step in the development of modern biology.... The time has come when further progress in our understanding of nature requires that we reconsider the relationship between the outside and the inside, between organism and environment.

—From *The Triple Helix: Gene, Organism, and Environment* by Richard Lewontin, 2000.

PART

1

The Historical Framework

Evolution as Science

It has become common to acknowledge that our views of the world are strongly influenced by the culture in which we grow up. That is, different cultures place different emphases on how people perceive various events and relationships and on how they explain these perceptions. Thus, in one culture death is a recycling of a person's spirit into another organismic form; another culture believes death is a state of reward and punishment for an individual's behavior, while still another culture regards death as the end of a person's existence.

What we consider science is also culturally dependent in the sense that large differences can exist between cultures as to whether or how scientific concepts are applied to natural events. Explanations that many of us accept as scientific — analyses based on rational, understandable principles and laws — others do not necessarily accept, or accept to varying degrees. Thus, some people consider that human behavior and interactions can be explained by natural processes; others believe these are predestined actions produced by one or more godlike creators; and still others propose that events are determined by constellations of planets, stars and phases of the moon that have mysterious powers and properties.

In general, nonliving phenomena have been considered more acceptable to scientific analyses in Western European culture than matters that touch on life itself, and on human life in particular. For example, physics and chemistry were well established as sciences by the nineteenth century, whereas biology, especially evolutionary approaches, was the subject of vitalistic interpretations by biologists in the past and by various religious groups even to this day.

A philosophical criticism sometimes raised against evolution as a science is that evolutionary explanations (hypotheses) cannot be tested and supported in the same fashion as hypotheses in physics and chemistry. The claim is made that because evolutionary studies deal with *events that occurred in the past* — events that are generally impossible to repeat in a laboratory — evolutionary biology can never reach the status of such sciences as physics and chemistry. Some critics extend these arguments to paleontology, geology and astronomy, three fields of study also dealing with the past and with matters on such a large scale that they cannot be repeated experimentally in the laboratory. Another criticism mounted against evolution claims that many studies designed to demonstrate evolution cannot be properly evaluated by the scientific method.[1]

Rejection or acceptance of a scientific hypothesis is based on whether data gathered to test the hypothesis refute it or not. Hypotheses constructed so they can never be refuted ("falsified," according to the philosopher Karl Popper) are not considered scientific.[2] Thus, concepts that invisible angels are responsible for the birth or death of an organism, or that God created the universe, are not scientific. Any events that seem to conflict with such a concept can always be reinterpreted to support it. Some claim that because evolutionary concepts are historical, they appear

[1] *Ways of Knowing* by Pickstone (2001) contains a unified argument against such criticisms in an analysis of the development of the scientific method in science, technology and medicine, an analysis described by the late Roy Porter as, "the most exciting synthesis we now possess."

[2] A number of philosophers of science have noted that naïve falsification is not always sufficient to distinguish scientific from nonscientific theories. As their example: if we took Popper's definition seriously, then Newton's inverse square law would be utterly unscientific because it cannot be refuted.

irrefutable and unscientific because unknown past events might always be recruited to support a hypothesis.

We can, however, explain historical events as rationally as we explain other scientific events, provided that the explanations we use are consistent with observations. Thus, the sequence of primate-like hominin fossils from the past to the present supports the hypothesis that humans have a primate origin (see Chapter 20). Similarly, correspondence in the amino acid sequences of myoglobin and hemoglobin supports an evolutionary relationship between them (see Chapter 12). Because either hypothesis could be disproved by finding, for example, horse-like or birdlike hominid fossilized ancestors such as centaurs or gargoyles, or noting an absence of any amino acid sequence similarities between myoglobin and hemoglobin, such hypotheses bear the hallmarks of science without requiring laboratory repetition. So "historical" sciences concerned with the past — astronomy, geology, paleontology, and evolution — can make use of observations to refute or support proposed hypotheses.

Of special interest in historical sciences is the emphasis on **understanding a particular sequence of historical events** rather than on primarily discovering general laws such as those of physics and chemistry. Stephen J. Gould (1941–2002) promoted this view under the rubric of **contingency** and **historical constraint**, which is that past events condition present and future change. For example, topics of historical science include events that led to our solar system, to the separation of South America from Africa, and to the origin of humans, recognizing at the same time that many of these events are singular and do not apply to all stars, all continental separations, or the evolution of all species. Thus, although sciences that deal with the past make use of general laws such as those of gravity, mechanics and biochemistry, their aim is to discover the **causes of diversity and uniqueness *and* the principles and laws that apply uniformly to all matter or all life.** Indeed, the ability to extrapolate from processes that affect all organisms such as the existence of variation, mutation, changing environments, *and* to explain the role of past history, is both central to the dynamic interdisciplinary nature of the study of evolution, and makes it a robust and testable field of inquiry.

The ever-present influence of past evolutionary history is far more than a theoretical postulate. It can be demonstrated experimentally, as, for example, when replicated populations of the fruit fly, *Drosophila melanogaster,* are returned to a common ancestral environment and allowed to breed for 50 generations. These populations undergo reverse evolution to an ancestral state, but do not all revert to the same universal state. The changes depend on past evolutionary history (contingency) and on the particular character analyzed, a mosaic evolution in which characters evolve at different rates (Teotónio and Rose, 2000). This interesting experiment illustrates that we must assess "the influence of past evolutionary history" on a character-by-character basis — some features retain more of their past history than others.

The properties of different hydrogen atoms, for example, can be explained on a common physical basis, whereas the properties of different organisms — the organization and function of their component parts — can only be understood in the context of their history, which includes adaptations to specific lifestyles at particular times. There is an obvious distinction between attributes such as temperature that have "the same meaning for all physical systems," and a biological attribute such as fitness (see Chapter 22), which "although measured by a uniform method, is qualitatively different for every different organism" (Fisher, 1958). Nevertheless, when historical conditions are repeated, and different organisms are subjected to similar selective evolutionary forces, some common features can be predicted. Geographically widely separated populations and species of fish have adapted to cave conditions; they consistently show rudimentary eyes, enhanced development of chemosensory organs and loss of pigment among other common attributes.[3]

For our purposes, an appropriate way to deal with the history of evolutionary biology is to ask, "What happened?" "Who were the major players?" and "Why did it happen the way it did?" In seeking to answer these questions within historical, cultural and social contexts, Chapters 1 to 3 provide the background for discussions of the evidence for evolution, for analysis of evolution as a process continuing today and for discussion of the mechanisms driving evolution as a process.

[3] See Jeffery et al. (2000), Yamamoto et al. (2003) and Franz-Odendaal and Hall (2006) for studies on cavefish sensory systems.

1 Before Darwin

Chapter Summary

Many intellectual threads led to the modern theory of evolution, a theory that requires recognition that Earth is ancient, that there is a common inheritance within a biological group, and that natural events can be explained by discoverable natural laws. But it took a long time before these threads were woven into an evolutionary tapestry.

Plato's idealistic concept, that all natural phenomena are imperfect representations of the true essence of an ideal unseen world, was for centuries the prevailing philosophy in Western Europe. Following Platonic ideas, Aristotle suggested that not only were species immutable but that there was a hierarchical order of species from most imperfect to most perfect, a concept refined over the centuries as the "Great Chain of Being." In hindsight, this philosophy profoundly inhibited the development of evolutionary ideas because it maintained that the world of essences is perfect and all change is illusory. This unchanging order remained unquestioned until inexplicable gaps in the chain of nature prompted philosophers such as Gottfried Leibniz to propose that the universe was not perfect, only that it might go through successive intermediate stages on the way to perfection.

By the seventeenth and eighteenth centuries, new attention to animals and plants as well as far-flung explorations led to an increasing interest in classifying organisms within the natural chain. The Swedish founder of taxonomy, Carolus Linnaeus, revolutionized systematics by using the species as the basic unit and building a hierarchical system from species upward to larger taxonomic categories. The naturalist Buffon (Georges-Louis Leclerc, Comte Buffon) went farther, implying that the species is not just a category in classification, but rather the only natural grouping of historical and interbreeding entities. Buffon maintained that a species was a "real" but static unit. Jean-Baptiste Lamarck evaded this problem by proposing that species are arbitrary (not "real") and that there must be forms intermediate between species. So, Lamarck's view of a species was very different from Charles Darwin's. If species were arbitrary (Lamarck), then species never went extinct but instead evolved into other "species."

The ideas that organisms could arise from nonliving materials by spontaneous generation or that they did not change during their embryonic development but were "preformed" in their ancestors, provided perfectly rational explanations to questions posed at the time. Not until late in the nineteenth century was spontaneous generation finally disproved (many of the first Darwinians believed in it) and the idea established that organisms develop epigenetically, that is, by differentiating from undifferentiated tissues. At last, biological phenomena became amenable to rational explanation.

Discovery of a fossil record of life became a rich source of data for individuals trying to understand relationships between organisms. The discovery of (1) fossils of unknown types of organisms, (2) organisms similar to but not the same as organisms living in that locality, and (3) the apparently inappropriate location of some fossils, suggested that Earth's surface and the organisms on it had existed for a long time, and that organisms succeeded one another through time. These ideas conflicted, however, with the Judeo-Christian view of a recent origin, according to which fossil data were interpreted to accord with biblical catastrophes such as the Noachian flood or as "jokes of nature." Geologists asserted that fossil evidence was only explicable if Earth was indeed old and if forces of nature had shaped its surface. Changes on Earth's surface would have led to alterations in the organisms that lived on it, and these changes would be reflected in their fossil remains. Charles Lyell, a contemporary of Darwin, invalidated the idea that capricious catastrophic and miraculous events had influenced Earth's geological structure. Lyell developed the earlier principle of uniformitarianism, in which the same geological forces acted in the past as in the present. Extrapolating processes back in time helped establish the validity of a world that was both comprehensible and rational.

Biological evolution is concerned with inherited changes in populations of organisms over time leading to differences among them. **Individuals do not evolve,** in the sense that an individual exists only for one generation. Individuals within each generation, however, do respond to **natural selection.**[1] **Genes** within individuals (**genotypes**) in a population, which are passed down from generation to generation, and the features (**phenotypes**) of individuals in successive generations of organisms do evolve. Accumulation of heritable responses to selection of the phenotype, generation after generation, leads to evolution: Darwin's descent with modification (**Box 1-1**).

All organisms, no matter how we name, classify or arrange them on **The Tree of Life,** are bound together by four essential facts:

1. They share a common inheritance.
2. Their past has been long enough for inherited changes to accumulate.
3. The discoverable taxonomic relationships among organisms are the result of evolution.
4. Discoverable biological processes explain both how organisms arose and how they were modified through time by the process of evolution.

Although each of these aspects has been studied and discussed at various times in human history, only after Charles Darwin developed and published his theory in the mid-nineteenth century did biological evolution become an acceptable scientific (naturalistic) alternative to earlier explanations. Darwin's proposed mechanism, **natural selection,** however, was regarded by many to be of secondary importance to evolution. The acceptance that organisms could change over time brought about enormous shifts in the way we view the world and explain natural phenomena. We begin our exploration by looking at some of the earliest attempts to understand the world around us.

[1] **Selection** is the sum of the survival and fertility mechanisms that affect the reproductive success of genotypes (see Chapter 22).

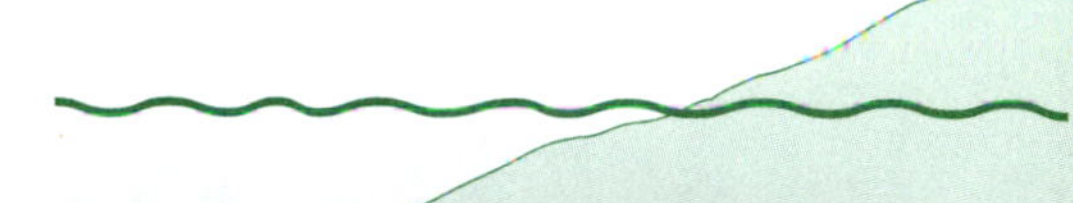

BOX 1-1 Evolution: An Overview of the Term and the Concept(s)

"Evolution. Development (of organism, design, argument, etc.); Theory of E. (that the embryo is not created by fecundation, but developed from a pre-existing form); origination of species by development from earliest forms" (*Concise Oxford Dictionary*, 5th ed., 1969).

The Word *Evolution*

As the definition above indicates, the word *evolution* has different meanings and the concept of evolution applies to a wide variety of human activity: the evolution of an argument; the evolution of the computer; the evolution of heart valves; evolutionary medicine.

The first definition — evolution as organismal development — reflects the original seventeenth century definition of the word, when evolution (from the Latin *evolutio,* unrolling) was defined as and used for the unfolding of the parts and organs of an embryo to reveal a preformed body plan. An example would be a caterpillar unfolding into a butterfly. Only in the nineteenth century did evolution come to mean transformation of a species or transformation of the features of organisms.[a]

Evolution as development can be traced to the Swiss botanist, physiologist, lawyer and poet Albrecht von Haller (1708–1777), who in 1774 used evolution to describe the development of the individual in the egg:

> But the theory of evolution proposed by Swammerdam and Malpighi prevails almost everywhere . . . Most of these men teach that there is in fact included in the egg a germ or perfect little human machine . . . And not a few of them say that all human bodies were created fully formed and folded up in the ovary of Eve and that these bodies are gradually distended by alimentary humor until they grow to the form and size of animals (Haller, 1774, cited from Adelmann, 1966, pp. 893–894).

Another Swiss lawyer, Charles Bonnet (1720–1793), further solidified evolution as preformation in his theory of encapsulation (*emboîtment*). He wrote that all members of all future generations are preformed within the egg: cotyledons within the seeds of plants; the insect imago inside the pupa; future aphids in the bodies of parthenogenetic female aphids, and so forth.[b]

Just nine years before Darwin published *On the Origin of Species,*[c] evolution was still being used for individual development. Here is a question from an examination held at Cambridge University in 1851, a question that presumes that species have not evolved: "Reviewing the whole fossil evidence, shew that it does not lead to a theory of natural development through a natural transmutation of species" (cited in Hall, 1999a). Even Darwin, who proposed a theory of evolution as descent with modification from generation to generation, only used the word evolution once, as the last word of his book.

> There is grandeur in this view of life, with its several powers, having been originally breathed by the Creator into a few forms or into one; and that, whilst this planet has gone cycling on according to the fixed law of gravity, from so simple a beginning endless forms most beautiful and most wonderful have been, and are being *evolved.*

It appears that Darwin had two concerns over the word evolution. One was its established use for the unfolding of development according to some preset plan. Another, but more difficult to establish, was Darwin's concern that he not be viewed as promoting the idea of **progress** (see Chapter 2 and Box 19-1).

Concept of Evolution as a Process

Perhaps not surprisingly in hindsight, and given the nature of the evidence they discovered, geologists were among the first to use the term *evolution* for the transformation of species and progressive change through geological time.

Robert Grant (1793–1874), who gave the name Porifera to the sponges, used the term evolution in 1826 for the gradual origin of

[a] See the essays in Hall and Olson (2003) for evaluations of the major concepts in evolution and development.

[b] Various topics (*emboîtment,* preformation, transformation, population genetics) and some of the individuals (Bonnet, Darwin and Lyell) introduced in this box are discussed in greater detail in the text.

[c] What is the appropriate (proper?) way to abbreviate the title of Darwin's book, the full title of which is, *On the Origin of Species by Means of Natural Selection or the Preservation of Favoured Races in the Struggle for Life?* Some abbreviate it to *On the Origin of Species,* others to *On the Origin,* yet others to *the Origin.* We use either *On the Origin of Species,* the mode preferred by Darwinian scholars and historians, or *The Origin.*

Idealism, Species and the Species Concept

Attempts to understand the world in a rational way — that is, by methods of thought and logic — began about the fifth century BC in Greece.

Plato (428–348 BC), the philosopher who along with Aristotle (384–322 BC) had the greatest impact on Western thought, suggested that the observable world (our experience) is no more than a shadowy reflection of underlying "ideals" that are true and eternal for all time. Most things, according to Plato, were originally in the form of such eternal ideals, a philosophy we now know as **idealism.** Plato and his successors assumed that only ideal generalizations are real; all else was merely a shadowy illusion. In Plato's famous parable in his dialogue, *The Republic,* humans deprived of philosophy are depicted as cave-confined prisoners facing a wall upon which are displayed their own shifting, distorted shadows as well as shadows of objects situated behind them that they cannot see directly. Their chains prevent them from turning their heads toward the light. As a result, the prisoners interpret the observed deformed, shadowy aberrations

invertebrate groups. Charles Lyell (1797–1875) used evolution in 1832 for gradual improvement associated with the transformation of aquatic to land-dwelling organisms: "the testacea of the ocean existed first, until some of them, by gradual evolution, were *improved* into those inhabiting the land." Even so, an argument can be made that both Grant and Lyell were using evolution in the sense of change during development. Not so for the engineer, journalist and writer Herbert Spencer (1820–1903), whom we associate with the origin of the term 'struggle for existence' and with Social Darwinism (the application of evolutionary theory to society). He used evolution in 1852 to mean **progression towards greater complexity,** heralding a century and a half-long controversy over whether evolution leads to progress, as Lyell, Spencer, and perhaps Thomas Huxley thought it did (Ruse, 1996; Shanahan, 2004). Julian Huxley penned one of the clearest statements of biological progress in 1947, a statement to which Darwin would not have objected: "Biological progress exists as a fact of nature external to man, and . . . consists basically of three factors — increase in control over the environment, increase in independence of the environment, and the capacity to continue further evolution in the same progressive direction."

Although many did not accept natural selection as the most important mechanism of evolution, most naturalists/biologists accepted that evolution had occurred. From the publication of *On the Origin of Species* in 1859 until 1900, evolution was the study of:

- **the origination and transformation of species** (one species of horse → another species of horse);
- **the transformation of major groups/lineages of organisms and the search for ancestors** (invertebrates → vertebrates; fish → tetrapods); and
- **the transformation of features** such as jaws, limbs, kidneys, nervous systems **within lineages of organisms.**

A new approach to evolution followed the rediscovery in 1900 of Gregor Mendel's experiments and the development of Mendelian genetics. Geneticists began to work with pure lines of organisms, with animals maintained in laboratories or plants in green houses, and with strains or cultivars that would have a hard time surviving in nature. The discovery of mutations — mostly of large effect, as only these were readily manifest in morphology and could be recognized and quantified — led to notions of large-scale evolution by saltation. This pitted geneticists against Darwinists, many of whom labeled geneticists as anti-Darwinian, as indeed many were. Not until the origination in 1908 of what became known as the Hardy-Weinberg law for calculating gene frequencies in populations under natural selection (see Chapter 21) and 1918 when R. A. Fisher published his paper, "The correlation between relatives on the supposition of Mendelian inheritance" (see Chapter 22), were doors opened that could reconcile Mendelism with Darwinism. During the 1930s, it led to the rise of population genetics, in which speciation was seen as resulting from genetic changes within a lineage as reflected in changes in gene frequency (Chapter 22). In the 1940s, the synthesis of population genetics, systematics and adaptive change forged what we know as the **Modern Synthesis of Evolution** or **neo-Darwinism.**

Although some thought otherwise, population genetics does not provide a complete theory of evolution. Now evolution is seen as **hierarchical,** operating on organisms on at least three levels:

- the *genetic level*, seen as substitution of alleles, changes in gene regulation and changes in gene networks;
- the *organismal level*, seen as individual variation and differential survival through adaptation and the evolution of new structures, functions and/or behaviors; and
- changes in *populations* of organisms, seen as the curtailment of gene flow between populations and the subsequent origin, radiation and adaptation of species (see Box 11-1, Hierarchy).

Natural selection acts because of the differential survival of individual organisms with particular features. The response to selection lies in the information content of the genome, information that can change because of mutation.

Because evolution acts at genetic, organismal and population levels, a definition ideally should reflect evolution at all three levels. In many respects, Darwin's concept of **descent with modification** remains an inclusive definition of (biological) evolution. Evolution is descent with modification, encompassing evolutionary change at genetic, organismal and/or population levels. How our views of evolution originated and have changed, and how evolution operates at the three levels of genes, organisms and populations are the major topics of this book. Rather, we should say, *is the major topic of this book* — an understanding and integration of all three levels are required to paint a complete picture of evolution.

as reality, while the actual unchanging humans and objects are the true "essences" or "forms." Any change would cause disharmony and so disrupt eternal ideals.

The Platonic goal for human society was to analyze experience in order to understand and strive for ideal perfection. Plato's writing had a goal that was founded in his belief in the centrality of beauty, truth and justice and the need to shape a society in which all could attain those goals. The concepts of perfect circles to explain the motions of the heavenly bodies (**Fig. 1-1**), perfect numbers such as 6 (1 + 2 + 3) and 10 (1 + 2 + 3 + 4), and the four "elements" (earth, water, fire, and air) to which all matter could be reduced were among the results of their search for perfection.[2]

To a large extent, idealism originates from the practice of abstracting concepts from experience. For example, to

[2] Variations on this theme were common. To the four elements Empedoceles (c. 490–430 BC) added two active principles: *love*, which binds elements together, and *hate*, which separates them. In respect to mystical numbers, Lorenz Oken (1779–1851), one of the German Natural Philosophers, proposed that the highest mathematical idea is zero, and God, or the "primal idea," is, therefore, zero.

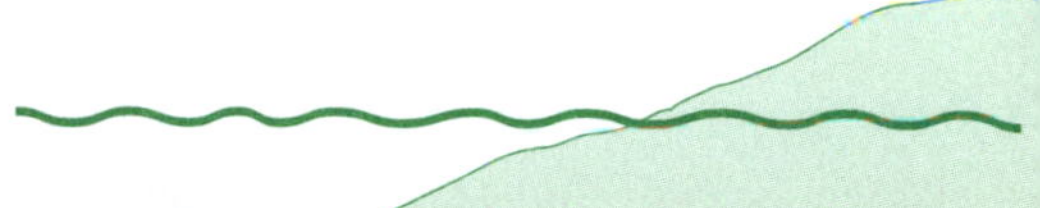

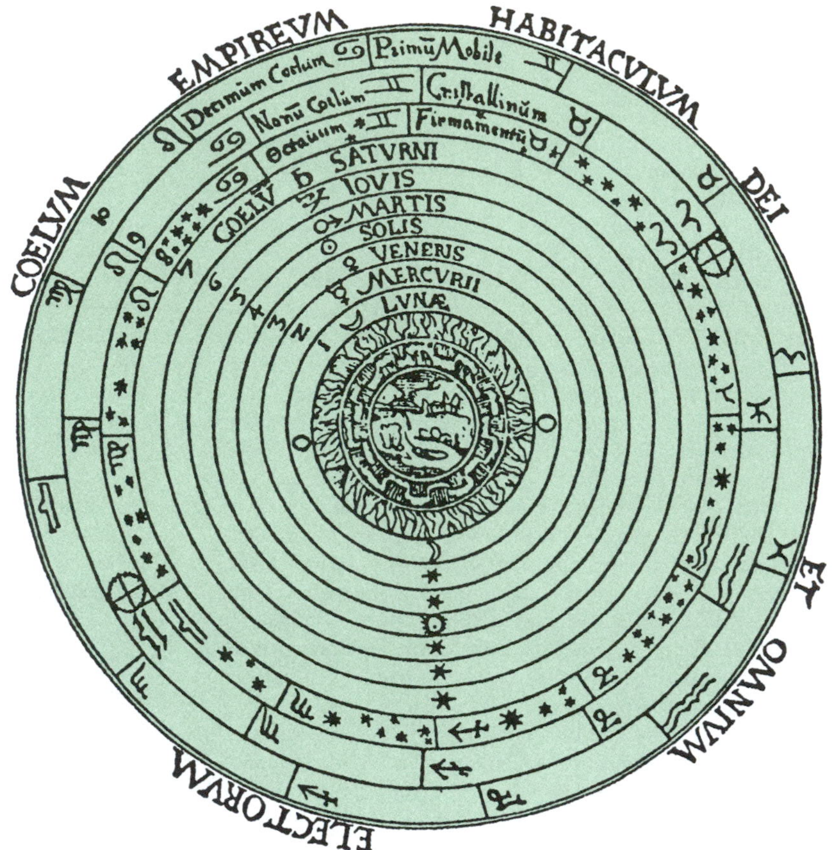

FIGURE 1-1 A medieval concept of the ten spheres of the universe with Earth and its four elements (earth, air, fire, water) at the center, according to Apian's *Cosmographia* (published 1539 in Antwerp). Surrounding Earth are transparent crystal spheres containing in succession the moon, Mercury, Venus, the sun, Mars, Jupiter, Saturn, the fixed stars, and spheres involved in the motion of the stars and of the entire universe ("Primu Mobile"). Beyond these spheres lies Heaven ("The Empire and Habitation of God and All the Elect").

think of "cat" in a way that includes all cats, rather than one particular animal of specific size and head shape, with claws, tail, fur, and so on. Abstraction allows us to generalize our experience, to differentiate between cat and tiger, to pet the cat and run from the tiger and to communicate these general concepts or universals to others through symbolic language. Despite these advantages, however, generalizations are not always reliable. Experiences can modify the generalizations; for example, not all cats or tigers are the same. The dilemma for **natural scientists**[3] has always been to recognize the reality of differences among members of a group and yet to recognize the reality of the group itself. Idealism offered practically no means of reconciling these two aspects of reality. No sooner do we conceive of some new generality than we discover further instances that may force us to modify our original concept.

Experience stresses continual change; generalization stresses stability. With the notable exception of Heraclitus (540–475 BC), few Greek thinkers tried to incorporate change into their philosophies. Heraclitus' philosophy disturbed that equilibrium, and we think, rightly so. Heraclitus maintained that all things are made of fire and so all things are constantly in motion or changing. The adages, "you cannot step into the same river twice" and "there is nothing permanent except change," capture his philosophy, as you will realize if you think about these two phrases for a moment or two. Biology is change, even when appearances are to the contrary. Maintaining a constant body temperature involves an enormous amount of change on second-by-second, minute-by-minute time scales; maintaining physiological stability *(homeostasis)* is really all about maintaining constancy in the face of change.

To Plato, the form of a structure could be understood from its function because function dictated form; the form of the universe derives from its function of goodness and harmony imposed by an external creator. Aristotle, whom

[3] We use the terms *natural scientist* and *natural historian* when referring to individuals working before the twentieth century, the term *scientist* having been introduced in 1830. The term *biology* was first used in 1800, independently by Lamarck and by the German naturalist, Gottfried Treviranus (1776–1837). The term *biologist*, which was introduced in the nineteenth century by William Whewell (1794–1866), did not come into general usage until the twentieth century. Although some studied zoology, others morphology and yet others botany or physiology, these specialized terms were not in general use. Nowadays, biology has become so specialized and fragmented that (1) one of the functions of a book such as this, (2) the integration of evolution and development as evolutionary developmental biology ("evo-devo"; Hall, 1999a; Hall and Olson, 2003), and (3) the call for an integrative biology, is to provide an integrative biology of the twenty-first century, an integration that was second nature to natural scientists in the nineteenth century.

many regard as the founder of biology (among other sciences), modified this notion to accommodate the embryonic development of organisms,[4] pointing out that the last stage of development — the adult form — explains the changes that occur in the immature forms. This type of explanation is called teleological (goal-oriented), where the adult represents the "telos," or final goal, of the embryo.

To many later thinkers, **teleology** became associated with Platonic processes by which advanced stages influenced and affected earlier stages. Because ideals implied conscious creation, it seemed as though organs and organisms were designed for some special purpose and that each species was created as an ideal in anticipation of its future use. Pliny the Elder (23–79 AD) carried this notion to the point of claiming that all species were created for the benefit of man, a concept laid out in the Jewish Testament. Some two hundred years later, Lucius Lactantius (c. 260–340 AD) wrote, "Why should anyone suppose that, in the contrivance of animals, God did not foresee what things were living, before giving life itself?" This view helped cast the teleological origin of species more permanently into the religious form it took in Christian Europe from the Middle Ages until Darwin, during which time natural science was inseparable from religion. Thus, the prominent thirteenth century Christian theologian Thomas Aquinas (c. 1225–1274) wrote in his *Summa Theologica*:

> Whatever lacks knowledge cannot move towards an end, unless it be directed by some being endowed with knowledge and intelligence; as the arrow is directed by the archer. *Therefore some intelligent being exists by whom all natural things are directed to their end*; and this being we call God (emphasis added).

Five centuries later, Linnaeus extended teleology even to science:

> If the Maker has furnished this globe, like a museum, with the most admirable proofs of his wisdom and power; if this splendid theater would be adorned in vain without a spectator; and if man the most perfect of all his works is alone capable of considering the wonderful economy of the whole; it follows that man is made for the purpose of studying the Creator's work that he may observe in them the evident marks of divine wisdom (Linnaeus, 1754, *Reflections on the Study of Nature*).

[4] *Embryology* was the term used for the study of embryonic development until the mid-twentieth century. In the early twentieth century, embryology was divided into descriptive and experimental and often taught as such in separate classes. In the mid 1950s, embryology was renamed *developmental biology*.

The Great Chain of Being

The idealistic concept of a species became strongly tied to its use in explaining the divine origin and design of nature. Plato had defined the species as representing the initial mold for all later replicates of that species: "The Deity wishing to make this world like the fairest and most perfect of intelligible beings, framed one visible living being containing within itself all other living beings of like nature." Aristotle expanded this view to a chain-like series of forms called the **Scale of Nature,** with each form representing a link in the progression from least perfect to most perfect (**Fig. 1-2**). This concept continued long into the history of European thought, merging with other ideas into the **Ladder of Nature** and the **Great Chain of Being** (Lovejoy, 1936).

Philosophically satisfying as it was, the concept of the Great Chain of Being did not necessarily put humans on the highest, or even near the highest, rung of the Ladder of Nature. Many who contemplated the innumerable steps between humans and perfection (God) felt the despair of occupying a relatively lowly position and only consoled themselves with the thought that there were even more lowly organisms. We insult someone if we call them a worm. Yet, despite its discomforts, the Great Chain of Being was accepted well into the eighteenth century.

In Germany, this notion was fostered by Johann Gottfried von Herder (1744–1803, a Protestant minister, philosopher and author), Johann von Goethe (1749–1832, the polymath who coined the term morphology), and others of the Natural Philosophy (*Naturphilosophie*) school, who tied the Great Chain of Being to an idealistic concept of biological forms. According to Goethe, the creation of each level of organisms was based on a fundamental plan: an **archetype** or ***Bauplan*** (pl. *Baupläne*). Goethe conceived the morphology of plants, for example, as founded on an *Urpflanze* (ancestral plant) that had only one main organ, the leaf, from which the stem, root and flower parts derived as variations (**Fig. 1-3a**). Similarly, the bones of the skull were modifications of the vertebrae of an *Urskeleton* (animal archetype) composed only of vertebrae. Ribs were modifications of vertebral processes (**Fig. 1-3b**).[5]

To most of its exponents, the Ladder of Nature had the comforting quality of stressing a precisely ordered regularity of relationships among organisms and could also be used to support and justify the prevailing social and political orders. As expressed by the literary Christian apologist Soame Jenyns (1757):

[5] *Archetypes* were taken up in a big way in the nineteenth century by Richard Owen; see Amundson (2007) for a reprinting of a classic paper by Owen on the essential nature of limbs. Hall (1994) and Bowler (2003) treat these topics further.

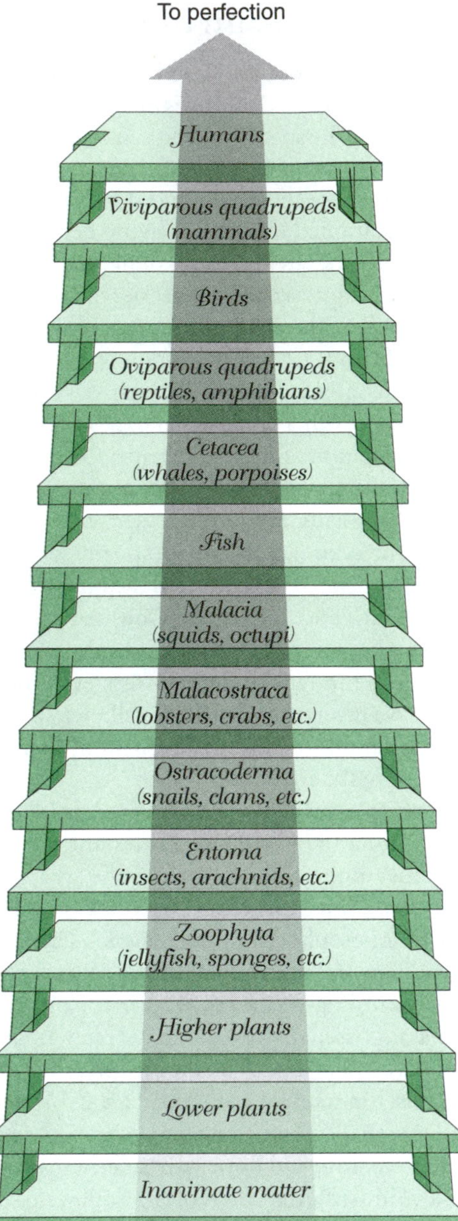

FIGURE 1-2 Aristotle's Scale of Nature (Adapted from descriptions in Guyénot, E., 1941. *Les Sciences de la Vie: L'Idee d'Evolution.* Albin Michel, Paris).

> The universe resembles a large and well-regulated family, in which all the officers and servants, and even the domestic animals, are subservient to each other in a proper subordination; each enjoys the privileges and perquisites peculiar to his place, and at the same time contributes, by that just subordination, to the magnificence and happiness of the whole.

Among the relatively few who disputed this concept was Voltaire (François-Marie Arouet; 1694–1778), who addressed the question of the many observed gaps between species, an observation that did not seem to be in accord with the expected innumerable steps in the continuous progression from imperfect to perfect. Voltaire proposed that although there were no living species to fill these gaps, such gaps were real, perhaps caused by the extinction of species, the concepts of adaptation and extinction having been developed by Lucretius as early as 55 BC. In this respect, Voltaire essentially echoed the writings of the philosophers René Descartes (1596–1650) and Leibniz (1646–1716).

Progress to Perfection

To Leibniz, the evolution of species was part of the perfection toward which the universe continually progressed. His philosophy represented a *major shift* from a perfectly created universe to one in the process of becoming perfect. Progress toward the perfection of species was expressed by natural historians such as Charles Bonnet (1720–1793), who maintained that the development[6] of any organism from its "seed" was an unfolding of a preconceived plan inherent in the seeds of previous generations. The notion of progress therefore fitted into a teleological framework: that it was "necessary" and directed toward some particular end.

As with so many changes in thought during the eighteenth century, these evolutionary rumblings were associated with major changes occurring in society following the Reformation that included the Enlightenment, the rise of empiricism and challenges to Papal authority (Chapter 3). The progressive weakening of feudalism, which had begun in the fourteenth century with the rise of commerce and the new power of the merchant classes, was accelerating because of rapid advances in technology and the Industrial Revolution. The old, rigid, land-based class structures were breaking up. Social institutions and the ideas expressed by many thinkers reflected these changes and became more mobile and flexible.

The Oxford historian John Roberts captured the essence of the Enlightenment as, "thousands of Europeans . . . felt, that they need no longer distrust the spread of knowledge; indeed, the idea that new knowledge was, in its social tendency, fundamentally progressive was another characteristic of 'Enlightenment'." He saw four changes as being particularly important in changing minds and attitudes, all of which had an impact on the reception of evolution:

- A new emphasis on and encouragement towards science and the manipulation of the natural world;
- A new skepticism that began to sap religious belief;
- A desire to want to be more humane, and "most important of all";
- "The growth of the idea that Progress was normal" (1985, p. 236).

The Great Chain of Being had important effects on plant and animal classification, which derived partly from the search

[6] Bonnet used the term evolution (*evolutio*, Latin) in its original meaning of unfolding during development (see Box 1-1).

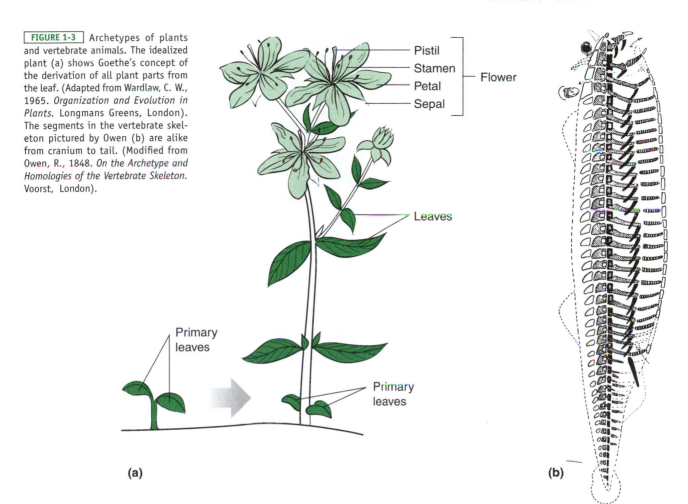

FIGURE 1-3 Archetypes of plants and vertebrate animals. The idealized plant (a) shows Goethe's concept of the derivation of all plant parts from the leaf. (Adapted from Wardlaw, C. W., 1965. *Organization and Evolution in Plants*. Longmans Greens, London). The segments in the vertebrate skeleton pictured by Owen (b) are alike from cranium to tail. (Modified from Owen, R., 1848. *On the Archetype and Homologies of the Vertebrate Skeleton*. Voorst, London).

for the multitude of organisms that many believed occupied all the rungs of the Ladder of Nature. There were proposals that even humans could be linked to other species through the "wildman of the woods" (the orangutan; **Fig. 1-4**). Other anthropologists thought the link between humans and animals was via a South African tribe, the Hottentots, whom Europeans believed to be almost indistinguishable in reasoning power from apes and monkeys. Despite the observed gaps between many species, all had been linked by a principle of continuity, expressed by Leibniz as, "Nature makes no leaps." Although not espousing the evolution of species as such, Emanuel Kant expressed this same idea as "the principle of affinity of all concepts, which requires continuous transition from every species to every other species by a gradual increase of diversity."

Thus, despite its idealistic nature, the Great Chain of Being led almost directly to the idea that the perfection of organisms may demand multiple intermediate stages. By the eighteenth century, the basic concept of **evolution** — the transformation of one species into another — required only the philosophical acceptance of actual change between the innumerable steps in the Great Chain of Being.

Classification and the Reality of Species

From the biological viewpoint, however, considerable difficulties remained concerning how species were to be defined, classified, distinguished one from the other and placed into groups that reflected their most significant features. Without a rational system of classification, evolutionary relationships between most species would have been impossible to establish (see Chapter 11). But recognition of the **biological importance** of species took considerable time.

In Europe during the Middle Ages, species were collected and described on the basis of their culinary or medical properties. The discovery of many new lands, floras, faunas and species of plants and animals, as a result of the expansion of worldwide exploration and trade in the sixteenth and seventeenth centuries, greatly increased the problems of classification and began to raise questions about relationships. Thomas Moufet (1553–1604), a prominent sixteenth century English entomologist, described grasshoppers and locusts:

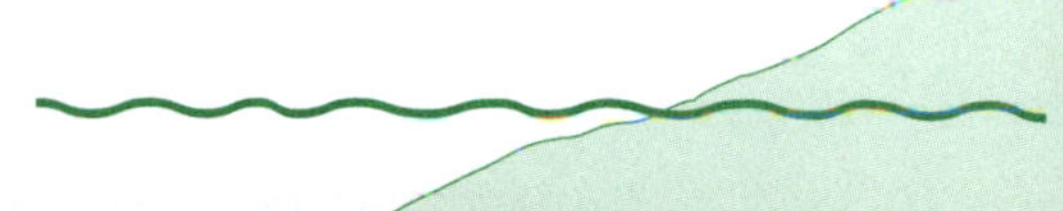

FIGURE 1-4 Presumed "missing links" between apes and humans in the Ladder of Nature. These individuals received binomial species designations, and Linnaeus tried to place them in his *Systema Naturae*. This figure is reproduced from an eighteenth century work by Linnaeus's student, C. E. Hoppius, who also noted the close similarity between humans and apes, "So near are some among the genera of Men and Apes as to structure of body: face, ears, mouth, teeth, hands, breasts; food imitation, gestures, especially in those species which walk erect and are properly called Anthropomorpha, so that marks sufficient for the genera are found with great difficulty." Social institutions, however, often greeted such proposed relationships with horror or derision: the 1770 suggestion by DeLisle de Sales that the orangutan was the human ancestor led to a prison sentence. (Reproduced from the original held by the Department of Special Collections of the University Libraries of Notre Dame.)

> Some are green, some black, some blue. Some fly with one pair of wings, others with more; those that have no wings they leap, those that cannot either fly or leap, they walk; some have longer shanks, some shorter. Some there are that sing, others are silent. And as there are many kinds of them in nature, so their names were almost infinite, which through the neglect of naturalists are grown out of use.

Nor were plants exempt from difficulties in classification. For example, Al-Dinawari's (820–895) *Book of Plants*, which was consulted through the Middle Ages, grouped plants according to at least two different systems: overwintering and growth.

> Plants are divided into three groups: in one, root and stem survive the winter; in the second the winter kills the stem, but the root survives and the plant develops anew from this surviving rootstock; in the third group both root and stem are killed by the winter, and the new plant develops from seeds scattered in the earth. All plants may also be arranged in three other groups: some rise without help in one stem, others rise also but need the help of some object to climb, whilst the plants of the third group do not rise above the soil, but creep along its surface and spread upon it.

Early attempts at classification usually followed Aristotle in postulating a broad category (for example, "substance"), and then creating subsidiary categories, each with its distinguishing elements (for example, body, animal), until an individual species could be placed into a particular subdivision. The method of classification devised by Carl Linnaeus (1707–1778)[7], the founder of systematics, represented a major advance. Beginning with a precise description of each species, Linnaeus grouped species closely related by their morphology into a category called a **genus** (plural genera). He then grouped related genera into **orders**, orders into classes, and established a system of **binomial nomenclature** in which each **species** name defines its membership in a genus and provides it with a unique species name and identity, for example, *Homo sapiens* (humans).

Designating the species as the basic unit of classification enabled Linnaeus to arrive at groupings far more "natural" in their interrelationships than many of the previously proposed artificial groups. In his scheme, species were separated or united into groups on the basis of fundamental structural and morphological features. To use a somewhat simplified

[7] Carl von Linné, usually known by the Latinized form of his name, Carolus Linnaeus, inherited his love of plants and their names from his father, Nils Ingemarsson Linnaeus, a Lutheran pastor.

example, one pre-Linnaean classification separated animals into those that can fly and those that cannot. Consequently, flying fish were considered to be hybrids between birds and fish. By ignoring categories based on lifestyle and confining his attention to a detailed description of the species itself, Linnaeus showed that a basic relationship of flying fish to other fish underlay the change in its fins that enables it to glide.[8] Therefore, except for those features shared by all vertebrate groups, there are no special birdlike structures in flying fish at all. Even though his system was idealistic — it treated species as ideal forms — Linnaeus's contribution to classification was an important step in allowing natural evolutionary relationships between organisms to be revealed.

Although late in life Linnaeus toyed with the concept of transitions between species, for much of his career he conceived of the species as a fixed entity. His concept was derived from the natural historian John Ray (1627–1705), who defined a species on the basis of its common descent. "The specific identity of the bull and the cow, of the man and the woman, originate from the fact that they are born of the same parents," wrote Ray. He attempted to separate different species on the basis of whether they could be traced to different ancestors. "A species is never born from the seed of another species." Thus, a species, with only rare exceptions, could never change, and its ultimate ancestor could only be divinely created. Linnaeus adopted this view but with the proviso that varieties within a species may show considerable non-heritable differences among themselves. Two examples from his work are his subdivision in 1758 of humans (*Homo sapiens*) into four races (Asiatic, American, European, African) and his designation in 1753 of the species *Beta vulgaris* for beets, whose cultivated varieties (spinach beet, chard, beetroot, fodder beet and sugar beet) were given varietal names, for example, *Beta vulgaris perennis* for sea beet. Linnaeus had second thoughts and in 1763 assigned the sea beet as a separate species, *Beta maritima*. Taxonomists now designate the sea beet as a subspecies *Beta vulgaris* subsp. *maritima* (L.). Why? Because it interbreeds with cultivated varieties of *Beta vulgaris*, from which it can be almost impossible to distinguish morphologically; variation within a single species can be considerable. Again, we see the idealism in Linnaeus' conception of species.

Under Linnaeus, the art of systematics developed rapidly as many species were described, mainly on the basis of their reproductive parts, and classified into groupings that are still valid today, despite the mix of artificial and natural categories.[9] Generally, however, classification was almost always based on appearance and not on observations of ancestry, because

TABLE 1-1 A few of the major figures whose contributions in the nineteenth century or earlier influenced evolutionary concepts

Author	Contribution
Linnaeus 1707–1778	*System Naturae* (1735–1758)
Buffon 1707–1788	*Natural History* (1749–1767)
Malthus 1766–1834	*An Essay on the Principle of Population* (1798)
Cuvier 1769–1832	*Lessons of Comparative Anatomy* (1805)
Lamarck 1774–1829	*Zoological Philosophy* (1809)
Lyell 1797–1875	*Principles of Geology* (1830–1833)
Darwin 1809–1882	*Voyage of the Beagle* (1837) *On the Origin of Species* (1859)
Gray 1810–1888	*Darwiniana* (1876)
Mendel 1822–1884	*Experiments in Plant Hybridization* (1866)
Wallace 1823–1913	Joint essays with Darwin (1858)
Huxley 1825–1895	*Collected Essays* (1893–1894)
Weismann 1834–1914	*Studies in the Theory of Descent* (1882)
Haeckel 1834–1919	*General Morphology of Organisms* (1866)
Bateson 1861–1926	*Materials for the Study of Variation* (1894)

the classifiers (taxonomists) usually described preserved specimens whose natural behavior and origins were often unknown. In accord with idealist concepts, each species was believed to possess a unique "essence" that determined all its specific characters. This **"essentialist"** or **"typological"** view of species[10] was reinforced by taxonomists who deposited "type" specimens in museums or herbaria to be used as the standards (types) for classifying further specimens.

Although Linnaeus placed special emphasis on the species as the practical unit of classification, Buffon (1707–1788; **Table 1-1**) codified the notion that species are the only biological units that have a natural existence ("*Les espéces sont les seuls êtres de la nature*"). Buffon introduced the idea that species distinctions should be made on the basis of whether there were reproductive barriers to crossbreeding between groups ("reproductive isolation"), evidenced by whether fertile or sterile hybrids were produced:

> We should regard two animals as belonging to the same species if, by means of copulation, they can perpetuate themselves and preserve the likeness of the species; and we should regard them as belonging to different species if they are incapable of producing progeny by the same means.

To Buffon, considerable variation could occur between individuals of a species, perhaps eventually even produc-

[8] See the chapters in Hall (2007a) for the development, transformation and evolution of fins and limbs.

[9] Less well known, Linnaeus spent much of his life attempting to organize the economy of Sweden according to scientific principles, to adapt crops such as rice and tea to grow in the Arctic tundra and to domesticate elk, buffalo and guinea pigs as farm animals.

[10] A number of historians of systematics (P. F. Stevens, 1994; Winsor, 2003, 2006; and Müller-Wille, 2003) make the case that Linnaeus and most other pre-Darwinian systematists were not essentialists.

ing completely new varieties, for example, different kinds of dogs. Despite such variation, a species itself remained permanently distinguished from other species, although at times Buffon seemed to indicate the possibility that a species could change significantly (and through degeneration[11]), as the last lines of the quotation below show):

> Not only the ass and the horse, but also man, the apes, the quadruped, and all the animals, might be regarded as constituting but a single family. . . . If it were admitted that the ass is of the family of the horse, and differs from the horse only because it has varied from the original form, one could equally well say that the ape is of the family of man, that he is degenerate man, that man and ape have a common origin; that, in fact, all the families, among plants as well as animals, have come from a single stock, and that all animals are descended from a single animal, from which have sprung in the course of time, as a result of progress or of degeneration, all the other races of animals (*Natural History*, 4th volume, 1753).

Despite this clear statement of an evolutionary view, however, Buffon rejected transformation in part because it was contrary to religion ("all animals have participated equally in the grace of direct creation").

Strangely enough, the eighteenth-century barrier to the acceptance of evolution seemed to rest mostly on the reality of species. If species were indeed real, they seemed inevitably fixed. How could new species arise? Buffon, who had proposed evolutionary events on cosmological and geological scales, established three basic arguments *against* biological evolution, arguments that were used by antievolutionists well into the nineteenth century:

- New species have not appeared during recorded history.
- Although mating between different species fails to produce offspring or results only in sterile hybrids, this mechanism could certainly not apply to mating between individuals of the same species. How could individuals of a single species be separated from others of the same kind and become transformed into a new species?
- Where are all the missing links between existing species if transformation from one to the other has taken place? Numerous missing links had been imagined (Fig. 1-4) but Buffon claimed that none had been found, despite Tyson's (1699) dissection and comparison of monkeys, orangutans, apes and humans as variations on a single type.

Because these arguments were not refuted until after Darwin, it is no surprise that one of the first serious pre-Darwinian proponents of biological evolution, Jean-Baptiste de Lamarck (1744–1829), proposed that one must do away with the concept of the fixity of species (species distinctions as artificial and arbitrary, although they may be helpful in classification) to establish the possibility of evolution. The observable gaps between species, genera, families, and so on were only apparent, not real; all intermediate forms existed someplace on Earth, although they were not necessarily easy to discover.

Thus, although Lamarck shared the concept of the Great Chain of Being that species do not become extinct, because of his conception of a species, he did not believe that species were separately created, proposing rather that they had evolved from each other. His branching classification of animals (**Fig. 1-5**) introduced a direct challenge to the venerable doctrine of a Scale of Nature, which goes in only one direction, from imperfect to perfect. "In my opinion, the animal scale begins with at least two separate branches and . . . along its course, several ramifications seem to bring it to an end in specific places."

As discussed in Chapter 2, the mechanisms Lamarck offered to account for these evolutionary changes were inadequate. However, even if Lamarck's explanations had seemed reasonable, a more serious impediment to evolutionary thought concerned the questions of life itself: Is continuity between the generations of a species necessary at all? What if species arise anew every generation?

Spontaneous Generation

Until perhaps the middle of the nineteenth century, the common belief was that although most large organisms reproduce by sexual means, smaller organisms could arise spontaneously from mud or organic matter. Some folklore suggested that, when they died, larger organisms decomposed into smaller ones. There were even legends that magical transitions could change an individual of one species into another, a human into a werewolf, for example. About 400 years ago the physician and chemist Johann van Helmont (1577–1644) offered a classic expression of spontaneous generation:

> If you press a piece of underwear soiled with sweat together with some wheat in an open mouth jar, after about 21 days the odor changes and the ferment, coming out of the underwear and penetrating through the husks of wheat, changes the wheat into mice. But what is more remarkable is that mice of both sexes emerge, and these mice successfully reproduce with mice born naturally from parents . . . But what is even more remarkable is that the mice which come out of the wheat and underwear are not small mice, not even miniature adults or aborted mice, but adult mice emerge!

Two serious and somewhat contradictory obstacles to the development of evolutionary concepts therefore prevailed almost simultaneously. The Linnaean contribution

[11] See Box 19-1 for further discussion of degeneration in parasites with respect to progress.

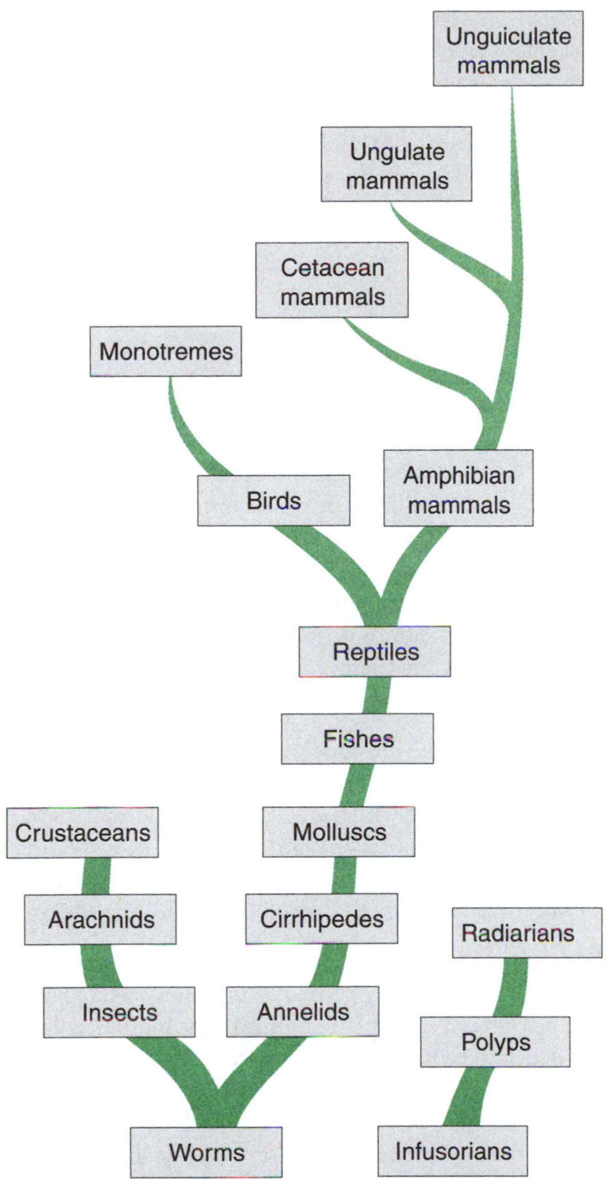

FIGURE 1-5 Evolutionary relationships among animals according to Lamarck.

of *species constancy* helped raise the question of the origin of species, but by insisting on *species fixity*, prevented consideration of any evolutionary transformations. Belief in **spontaneous generation**, in contrast, seemed contrary to species fixity, but at the same time cast doubt on any permanent continuity between organisms. If species could arise *de novo* at any time or be capriciously changed into another species, could there ever be a rational mechanism to explain their origin or the sequence of their appearance?

By the late seventeenth century, use of the experimental method had begun, and a number of experimentalists showed that, at least for insects, spontaneous generation does not occur. In 1668, Francesco Redi (1621–1697) demonstrated that maggots (larvae) arise only from eggs laid by flies, and flies arise only from maggots. If meat is protected so that adult flies cannot lay their eggs, maggots and flies are not produced. A year later, Jan Swammerdam (1637–1680) showed that the insect larvae found in plant galls arise from eggs laid by adult insects. Within a century, further experiments demonstrated that even the appearance of the microscopic "beasties" observed by Antony van Leeuwenhoek (1632–1723) in decaying or fermenting solutions and broth could be explained as originating from previously existing particles. The Abbé Spallanzani (1729–1799) heated various types of broth in sealed containers and observed no growth of tiny organisms. Only when the containers were open to airborne particles did organisms grow.

Preformation

Although the theory of spontaneous generation was not abandoned until the crucial experiments of the chemist Louis Pasteur (1822–1895) and the physician John Tyndall (1820–1893) in the nineteenth century, serious attempts to replace it with a theory of **preformationism** had begun much earlier (Farley, 1977). In the words of Swammerdam, preformation embodied the idea that "there is never generation in nature, only an increase in parts." That is, at conception, each embryonic organism is preformed as a perfect replica of the adult structure, which gradually enlarges through the nourishment provided by the egg and the environment. Some preformationists, now known as *ovists*, proposed that the miniature adult was contained within the maternal egg. Others (now known as *spermists* or animalculists) imagined that the adult in miniature was contained within the paternal seminal fluid (**Fig. 1-6**).

In its most extreme form, preformationism led to the ***emboîtment*** (encasement) theory, espoused by Bonnet and others, in which the initial member of a species encapsulates within it the preformed "germs" of all future generations; Eve's ovaries contained the entire preformed human species nested within like an infinite set of Russian dolls. Although preformation had the satisfying quality of explaining the many different plans of organismal growth and discounting the idea of spontaneous generation, it led once again to the fixity of species and brought the question of the origin of species back to an unknowable creation and/or creator.

Epigenesis

By the nineteenth century, improved experimental techniques and microscopic observations resulted in the replacement of preformationism[12] with the theory of **epigenesis**, according to which an embryo develops by gradually dif-

[12] The preformation story is much more complex than presented here. Today, we do not consider preformation versus epigenesis or that one side won out over the other. Each animal starts life as a preformed egg, so in this sense, preformation is alive and well. The egg nucleus, membranes and genes all are preformed. The way that these inherited structures are deployed and elaborated in development is epigenetic, as discussed further in Chapter 13.

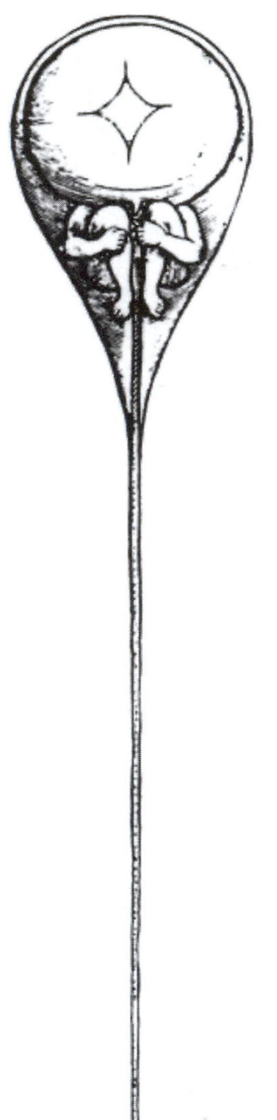

FIGURE 1-6 A fully formed human (homunculus) encased within the head of a sperm as imagined by Nicholas Hartsoeker (1694).

ferentiating undifferentiated tissues into organs that were not present at conception. At first, this was believed to occur because of nonphysical forces, such as the contribution of "form" by the seminal fluid (Aristotle), an "*aura seminalis*" (William Harvey, 1578–1657) or "*vis essentialis*" (Kaspar Wolff, 1733–1794).

Such explanations are **vitalistic**: They ascribe to living beings a vital force that cannot be explained by any physical or chemical principles. By the mid-third of the nineteenth century — coinciding with the studies of the comparative embryologist Karl von Baer (1792–1876) — the prevailing view of epigenesis had changed and biologists could accept differentiation and growth as being as natural and explainable a set of processes as any others. In addition, Friedrich Wohler's (1800–1882) 1828 biochemical synthesis of an organic compound (urea), the first such extraorganismal synthesis, showed there was no mystical essence in organic molecules that could not be explained by the laws of chemistry. Such ideas of rational biology helped cultivate the climate in which evolutionary concepts could develop further. A sample of some of those who contributed to evolving evolutionary concepts in the eighteenth and nineteenth centuries is outlined in Table 1-1.

Fossils

An essential basis for understanding evolutionary relationships between organisms of the past, and for appreciating their lengthy history, was the study of their **fossil** remains.

It had long been known that the fossilized bones of animals do not resemble extant species, and that strange seashells can be found in the most unlikely places, such as mountaintops. The ancient Greeks were aware of such fossils, and a number of ancient writers, including Herodotus (484–425 BC), suggested that they could be explained by changes in the positions of sea and land. To Aristotle, there was no question that these changes occurred over considerable periods of time:

> The whole vital process of the earth takes place so gradually and in periods of time which are so immense compared with the length of our life, that these changes are not observed; and before their course can be recorded from the beginning to end, whole nations perish and are destroyed (Aristotle, Treatise on *Meteorology*).

But as Christianity gained ascendancy in Europe, influential church authorities began to estimate Earth's age by the number of generations since Adam in the biblical book of Genesis, calculating Earth's origin as no earlier than perhaps 4000 to 7000 BC. Limited to such a relatively short period, fossils could hardly be ascribed to a long historical process. During the sixteenth and seventeenth centuries, although regarded as being "naturally formed," these *stones* (now known as fossils) were regarded as images of God's creation, placed on Earth for man's admiration and use but naturally formed by God (**Fig. 1-7**.)

The discovery of fossils in exposed riverbanks, mines, and on eroded surfaces posed a challenge to the concept of the Great Chain of Being. For example, Thomas Jefferson (1743–1826; agriculturalist, botanist, fossil hunter and third president of the United States) discovered the extinct clawed giant sloth *Megalonix jeffersoni* (**Fig. 1-8**), which he mistakenly thought was a giant lion that perhaps still existed in the unexplored areas of North America (Rudwick, 2005).

Did fossils indicate possible errors in the plan of nature, causing some species to become extinct? They were commonly called *lusi naturae*, or "jokes of nature." Were there gaps in the Ladder of Nature caused by the loss of

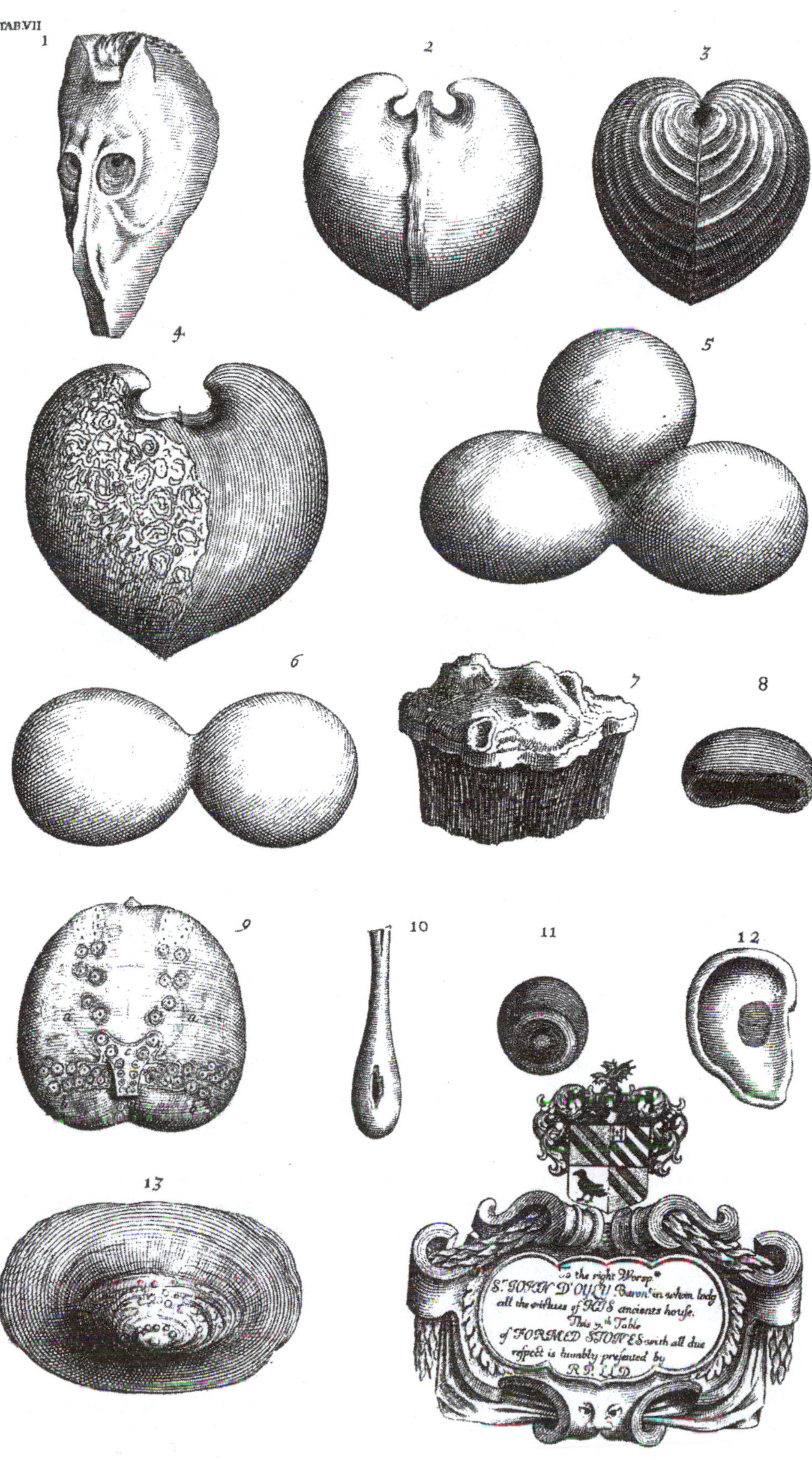

FIGURE 1-7 Plate 7 from *The Natural History of Oxfordshire* by Robert Plot (1676) illustrating naturally formed stones interpreted as representing parts of the human body, including: the heart (1-4) the holes in 1 being interpreted as the major artery taking blood away from the ventricles; the brain (9, showing the cerebellum and medulla oblongata); nerves (10, the olfactory nerve); the eye obscured by a cataract (11); and the external ear or pinna (12).

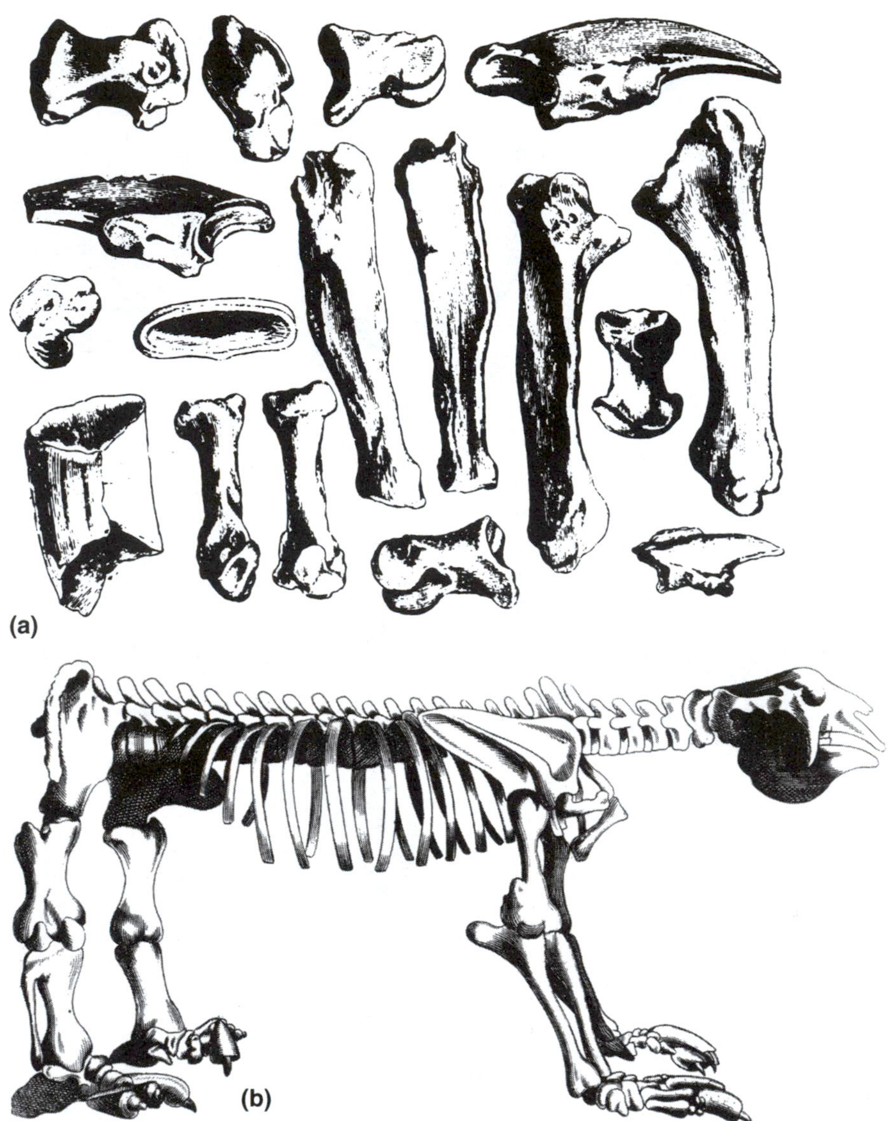

FIGURE 1-8 (a) Some of the bones of the extinct giant sloth, *Megalonix jeffersoni*, discovered in western Virginia in 1796. (b) Reconstruction by Cuvier of the skeleton of a similar extinct South American giant sloth, *Megatherium*. Both sloths were edentates, clawed mammals without cutting teeth. (From Greene, J. C., 1959. *The Death of Adam*. Iowa State University Press, Ames.)

these extinct species? Like many others who addressed these questions, Jefferson proposed that these species were not really extinct, only rare: "Such is the economy of nature, that no instance can be produced of her having permitted any one race of her animals to become extinct; of her having formed any link in her great works so weak to be broken." Other theories sought to explain fossils as caused by the Noachian flood described in Genesis or having purposely been implanted into Earth at the time of creation in order to test humanity's faith in religion.

Contrary arguments such as those proposing the reality of fossil species by the physician and naturalist Robert Hooke (1635–1703) and by the anatomist and geologist Nicolaus Steno (1638–1686) led to more naturalistic attempts to understand fossil origins. Such views helped place fossils in a historical sequence. When arranged by stratigraphic age, with deeper strata signifying older age than superimposed strata, older fossils showed greater differences from extant species than did later fossils (see Fig. 5-5), indicating changes over time (**Box 1-2** — Classification of Geological Strata).

Once the reality of fossils and of extinction was accepted, it was possible to conceive of a "law of succession" in which one form replaced another.

One of the commonly held theories during the late 1700s and early 1800s was **catastrophism**, popularized largely by followers of Georges Cuvier (1769–1832), one of the most gifted French comparative anatomists and the founder of paleontology (Table 1-1).[13] According to catastrophism, the sharp discontinuities in the geological record — stratifications of rocks, layering of fossils and transition from marine

[13] Weishampel and White (2003) provided the wonderful service of translating and reprinting important papers in the discovery of dinosaur fossils, including seven articles by Cuvier on crocodile fossils.

BOX 1-2 Classification of Geological Strata

THE SYSTEM OF GEOLOGICAL CLASSIFICATION adopted in the eighteenth and early nineteenth centuries, which was initially suggested by the mining specialist and "Father of Italian Geology," Giovanni Arduino (1714–1795), followed the practice of designating **primary rocks** as those without fossils. These were believed to date from the origin of Earth's crust and appeared as typical nonstratified, ore-bearing outcroppings in mountainous areas.

Geologists called stratified fossiliferous rocks, such as sandstone and limestone, **secondary.** Secondary strata contained ancient molluscan fossils such as ammonites and belemnites (see Chapter 15) as well as early fish and reptiles that differed considerably from present forms (see Chapters 17 and 18). Geologists believed **tertiary sedimentary rocks** to be derived from secondary strata by flooding, erosion, volcanic action and so on, and to contain ancient representatives of more recent forms such as mammals.

Quaternary rocks represented the glacial and alluvial deposits of relatively recent times. Because neither all mountains nor all strata are of the same age, these divisions were difficult to apply universally. All except tertiary and quaternary were abandoned. *Tertiary* came to mean the period of preglacial deposits corresponding to most of the Cenozoic period. *Quaternary* means the period dating from the Pleistocene ice age deposits to the present.

fossils to freshwater fossils — indicated sudden upheavals caused by catastrophes, glaciations, floods, and so on. Fossils were recognized as extinct species "whose place those which exist today have filled, perhaps to be themselves destroyed and replaced by others." To some upholders of the biblical account, catastrophism had the advantage of explaining at least some catastrophes as obvious departures from "natural" laws that could be ascribed to divine intervention. Some, such as the Swiss paleontologist, geologist, naturalist and founder of the Museum of Comparative Zoology at Harvard University Louis Agassiz (1807–1873), proposed that there may have been as many as 50 to 100 successive special divine creations. This approach justified the prior existence of fossil species *and* the biblical flood, and made it possible to conceive that all extant organisms arose within the time span the Judeo-Christian Bible provided, although preceded by many geological ages.

In contrast to Cuvier's catastrophist position, Lamarck proposed that geological discontinuities represented gradual changes in the environment and climate to which species were exposed. Through effects on organisms these changes led to species transformation. This **uniformitarian** concept, that the steady, uniform action of the forces of nature could account for Earth's features, had been foreshadowed by Buffon and others and was strongly developed in the work of the geologist James Hutton (1726–1797).[14] Later, Charles Lyell (1797–1875), a geologist and contemporary of Charles Darwin, offered the *uniformitarian reply to catastrophism* through the following arguments:

1. Sharp, catastrophic discontinuities are absent if geological strata are examined over widespread geographical areas. Any widely distributed stratum often shows considerable regularity in its structure and composition (see Box 1-2). Only in specific localities do rapid shifts seem to appear and then because of local changes.
2. Changes in the geological record arise from the action of erosive natural forces such as rain and wind as well as from volcanic up thrusts and flood deposits (Box 1-2). The laws of motion and gravity that govern natural events are constant through time. Therefore, past events are caused by the same forces that produce phenomena today although the extent of phenomena, such as volcanism, might have fluctuated in the past. This means that all natural causes for phenomena should be investigated before supernatural causes are used to explain them.
3. Earth must be very old for its many geological changes to have taken place by such gradual processes.

The frontispiece of Lyell's 1830 *Principles of Geology* is a portrait of the three remaining columns of the ruined "Temple of Serapis" in Pozzuoli, Italy, showing that they had been historically subjected to rise and fall in sea level. A 3-m section of these columns contains holes bored by molluscan bivalves, indicating that these columns were once partially submerged. Lyell used this portrait through 12 editions of his book as an example of gradual geological change. Thus, although uniformitarianism did not exclude sudden geological changes such as floods, volcanic eruptions and meteorite impacts — events that were of common or recorded knowledge — it led to the view that even such "catastrophes" could be naturally caused and rationally explained.

The transition from catastrophism to uniformitarianism had profound effects because it helped liberate scientific thinking from the concept of a static universe powered by capricious, unexplainable changes to one that is perpetually dynamic and more historically understandable. Charles

[14] In physics, Isaac Newton pointed out that, "we are to admit no more causes of natural things than such as are both true and sufficient to explain their appearance." The case has been made that evolution was (is) to biology as energy was (is) to physics, as each synthesizes previously separate domains to find deeper common principles. However, because the laws of physics pertaining to the physical interactions of matter relate to the transfer of energy, while biology deals with transform of information, biology cannot be reduced to physics.

Ernst Mayr

Birthday: July 5, 1904
Birthplace: Germany
Died: February 5, 2005[15]
"Late Professor Emeritus
Harvard University
Cambridge, Massachusetts."

What prompted your initial interest in evolution?

I was a born naturalist, roaming the fields and woods ever since I was a small boy. What most attracted me was the immense diversity of life. Why are there so many different kinds of species? Later on I asked how they originated. I looked for the solution on expeditions to New Guinea and the Solomon Islands in the 1920s.

What do you think has been most valuable or interesting among the discoveries you have made in science?

Science advances by discoveries and by the introduction of new concepts. It is to the latter that I made most of my contributions:

- The biological species concept
- The concept of sibling species
- The importance of geographic speciation
- Speciation through founder populations
- The origin of evolutionary novelties
- Species turnover on islands
- The relation between population size and evolution
- The holistic concept of the genotype
- Individuals and social groups are the targets of selection, not genes

My most interesting discovery was the great speedup of evolutionary rate in small populations isolated beyond the previous species borders.

What areas of research are you (or your laboratory) presently engaged in?

I am now exploring the philosophical consequences of the discovery of new evolutionary principles.

In which directions do you think future work in your field needs to be done?

The internal structure of the genotype, the workings of the central nervous system, and the interaction of species in the ecosystem are now the most exciting frontiers of biology.

What advice would you offer to students who are interested in a career in your field of evolution?

My advice to beginners is to become thoroughly familiar with one particular group of organisms in order to be able to test any new ideas or theories against the background of that set of solid facts. Pure speculation and model building without a solid factual basis rarely leads to sound advances.

---

[15] Ernst Mayr, one of the most influential evolutionary biologists of the twentieth century, was a staunch defender of the Modern Synthesis, a synthesis that he played a major role in creating. His last book entitled, *What Evolution Is*, published in 2001, is a thorough and thoughtful exploration of evolutionary biology. His contributions are celebrated in *Systematics and the Origin of Species: On Ernst Mayr's 100th Birthday*, which you can read on the Internet at *http://newton.nap.edu/catalog/11310.html#toc*.

EVOLUTION ON THE WEB biology.jbpub.com/book/evolution

Darwin first offered an acceptable explanation for historical changes among organisms and thereby helped tie all organisms together by a community of descent: **evolution.**

KEY TERMS

archetype
binomial nomenclature
Bauplan
catastrophism
emboîtment
epigenesis
fossil
Great Chain of Being
idealism
Ladder of Nature
preformationism
species
spontaneous generation
teleological
uniformitarian
vitalistic

DISCUSSION QUESTIONS

1. What is Platonic idealism?
2. Why did idealism become such an important approach to nature?
3. What is the connection between idealism and the description and classification of organisms?
4. Is the concept of the Great Chain of Being (Ladder of Nature) idealistic? Why or why not?
5. Can the concept of biological evolution be held by those who believe in idealism?
6. Why has freedom from cultural constraints and prejudices been more difficult for evolutionary studies than for physics or chemistry?

7. Discuss how the concept of the spontaneous generation of species contradicts the concept that each species is created individually.
8. What obstacles made it difficult for individuals to consider the reality of fossil species?
9. What differences exist between the concepts of catastrophism and uniformitarianism in their explanations for evolutionary changes?
10. Can uniformitarianism be defined to include occasional catastrophic changes?

EVOLUTION ON THE WEB

Explore evolution on the Internet! Visit the accompanying Web site for *Strickberger's Evolution*, Fourth Edition, at http://www.biology.jbpub.com/book/evolution for Web exercises and links relating to topics covered in this chapter.

RECOMMENDED READING Historical Aspects of Evolution

Appleman, P. (ed.), 2001. A Norton Critical Edition. *Darwin: Texts, Commentary*, 3d ed. Philip Appleman (ed.). W. W. Norton & Company, New York.

Bowler, P. J., 1996. *Life's Splendid Drama: Evolutionary Biology and the Reconstruction of Life's Ancestry, 1860–1940*. The University of Chicago Press, Chicago.

Bowler, P. J., 2003. *Evolution: The History of an Idea*, 3d ed. University of California Press, Berkeley, CA.

Bowler, P. J., 2005. Variation from Darwin to the Modern Synthesis. In *Variation: A Central Concept in Biology*, B. Hallgrímsson and B. K. Hall (eds.). Elsevier/Academic Press, New York, pp. 9–27.

Gould, S. J., 1977. *Ontogeny and Phylogeny*. Harvard University Press, Cambridge, MA.

Mayr, E., 1982. *The Growth of Biological Thought: Diversity, Evolution, and Inheritance*. Harvard University Press, Cambridge, MA.

Mayr, E., 2001. *What Evolution Is*. With a Foreword by Jared Diamond. Basic Books, New York.

Nyhart, L. K., 1995. *Biology Takes Form: Animal Morphology and the German Universities, 1800–1900*. The University of Chicago Press, Chicago.

Rudwick, M. J. S., 1985. *The Meaning of Fossils: Episodes in the History of Palaeontology*, 2d ed. The University of Chicago Press, Chicago.

Rudwick, M. J. S., 1995. *Scenes From Deep Time. Early Pictorial Representations of the Prehistoric World*. The University of Chicago Press, Chicago.

Rudwick, M. J. S., 2005. *Bursting the Limits of Time: The Reconstruction of Geohistory in the Age of Revolution*. The University of Chicago Press, Chicago.

Ruse, M., 1996. *From Monad to Man: The Concept of Progress in Evolutionary Biology*. Harvard University Press, Cambridge, MA.

Ruse, M., 2001. *The Evolution Wars: A Guide to the Debates*. Rutgers University Press, Rutgers, NJ.

2 Darwin and Natural Selection

Chapter Summary

The basic ideas essential to evolutionary theory were already present by the time Darwin made his famous voyage on the *Beagle*. Most important were the concepts that Earth is ancient, fossils represent the remains of extinct species, many species show close similarities, and organisms have descended from previously existing organisms. However, the mechanisms for evolutionary change and the agents that enabled populations of organisms to change over time, and new species to arise, remained unknown.

Information Darwin accumulated on his five-year voyage laid the foundation for the theory of evolution he later developed. A keen observer of geology and natural history, Darwin noted geological formations that gave evidence of historical transformation as well as peculiar geographical distributions of organisms and close similarities of species. Slowly, he came to realize that the only rational explanation for these phenomena must be that species could be transformed. At first, however, he could find no mechanism by which transformation might occur.

Darwin retained Lamarck's mechanism that structures survived or deteriorated through use or disuse and, furthermore, that traits so acquired could be inherited. Darwin also proposed another alternative, *natural selection*, which, given the state of knowledge of heredity, could be held along with Lamarck's mechanism of inheritance of acquired characters. From reading Malthus (see Chapter 1 and below), he derived the idea of an excess of progeny competing for limited resources. Competition provided Darwin with a process for changing the composition of a population. Organisms that had traits better suiting them to their environment (*adaptations*) would tend to reproduce more prolifically than others so that their traits would be passed on in higher proportions to future generations. Thus, populations could continually adjust, adapting to the environments in which they lived and to which they reacted. Populations with inadequate adaptations would become extinct. Limited environmental resources, eventually faced by all organisms, made reproductive success the ultimate judge of survival. By coincidence, the naturalist Alfred Russel Wallace simultaneously proposed a similar mechanism. The papers of both men were presented in 1858, and Darwin's *On the Origin of Species* was published a year later. The question of how species might change, however, remained unresolved. The theory of natural selection depended on inheritable variations on which selection could act. Neither Darwin nor Wallace knew how such variants might be produced. Not until the science of genetics developed was this difficulty resolved (but other difficulties were raised).

By the early nineteenth century, many of the basic concepts necessary to develop a scientific explanation for the diversity of life were in place:

- Geologists such as James Hutton (1726–1797) conceived *Earth's age to be in the range of millions of years.*
- Natural historians and many members of the general public accepted the reality of *extinct fossil species.*
- Systematists, comparative anatomists and embryologists noted the close *similarities among many different species.*
- Most if not all "men of science" were convinced that *organisms had descended through inheritance* from previously existing organisms.

The notion of a divine "common plan" to account for the relationships among species by supernatural acts of creation was only one step away from the materialist evolutionary notion that species relationships derive from their common ancestry: a change *from archetypes to ancestors.* Nevertheless, at least two important questions remained: What natural cause or mechanism could explain organismal change? What hereditary mechanism could enable new species to arise?

When Charles Darwin (1809–1882) answered the first question in 1859, he transformed natural history (biology as we know it today) into a science, the science of evolution or evolutionary biology. An acceptable answer to the second question had to wait for the twentieth century, although soon after Darwin, Gregor Mendel (1822–1884) provided the essential basis for understanding the material basis of biological inheritance (see Chapter 10).

FIGURE 2–1 Portrait of Charles Darwin in 1849 (aged 40), by T. H. Maguire. © National Library of Medicine.

Charles Darwin

Many biographies have been written of Charles Darwin (**Fig. 2-1**). He has been described as a man who defied his own social and religious background, not only by espousing a radical concept, but by becoming the instrument that made it acceptable to many of his compatriots. As described by Desmond and Moore (1991), the enigma was:

> How could an ambitious thirty-year-old gentleman open a secret notebook [in 1837] and, with a devil-may-care sweep, suggest that headless hermaphrodite mollusks were the ancestors of mankind? A squire's son, moreover, Cambridge-trained and once

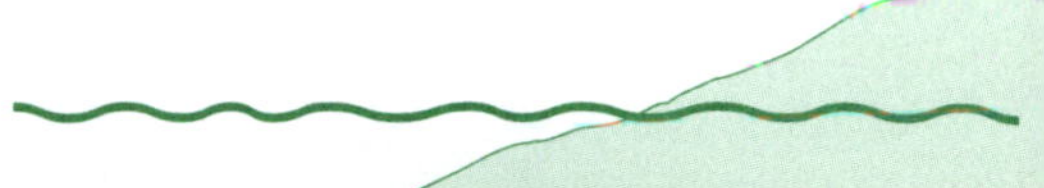

> destined for the cloth. A man whose whole family hated the "fierce & licentious" radical hooligans.[1]

In its barest outlines, Darwin's life is the history of a genial, curious, and intellectually creative man who was courageous yet fearful, living in a society undergoing considerable change. He was born to a British affluent, upper-middle-class family whose fortunes derived largely from his father, Robert Darwin (1766–1848), and paternal grandfather, Erasmus Darwin (1731–1802), both prosperous physicians. Erasmus Darwin was, in fact, an early thinker on and writer about evolution, but most of his contemporaries judged him wildly speculative in science. When searching for evolutionary explanations in the 1830s, Charles was aware of his grandfather's theories. Although some historians believe that evolutionary ideas, no matter what the source, had little effect on Charles' early development, interestingly, Herbert Spencer, who coined the phrase 'struggle for existence' (see Chapter 3), was much influenced by the views of Erasmus Darwin (Elliott, 2003).

At age 16, Charles left grammar school in Shrewsbury for Edinburgh University to study medicine. Surgical procedures at that time were dreadfully brutal, to the extent that, having witnessed two operations — one of them on a child — Darwin realized that he could never become a surgeon. He then went to Cambridge University with the intention of becoming a minister of the Church of England. However, his interests were not in academic or ministerial pursuits but in hunting, collecting, natural history, botany and geology. He despised formal classical education and was no more than a mediocre student. His father believed that Charles had betrayed the family trust of industrious professionalism and castigated him, "You care for nothing but shooting, dogs, and rat-catching, and you will be a disgrace to yourself and all your family."

In 1831, through the recommendation of John Henslow (1796–1861), Regius Professor of Botany at Cambridge University, and the intercession of an uncle, Josiah Wedgwood (1769–1843), Darwin was able to put off further study for the ministry and set off on his now famous voyage around the world on ***H.M.S. Beagle*** (**Fig. 2-2**). His post was ostensibly that of naturalist, a special unpaid position created by the British Admiralty on naval ships making broad geographical surveys. In fact, Darwin's primary role on the voyage was to serve as a dining companion to the Captain, Robert Fitzroy, to whose service his duties as naturalist were secondary. For this reason, it was of no importance to the British Admiralty that Darwin had not distinguished himself academically or even finished his studies at Cambridge. His qualifications as a gentleman, good shot and sportsman were quite sufficient.

[1] This is an interesting comment; some scholars consider that Erasmus Darwin (Charles' grandfather) was a bit of a fierce and licentious radical.

The *Beagle* voyage lasted approximately five years, during which time Darwin transformed himself from a casual amateur to a dedicated geologist and biologist. His letters to Henslow on many of the observations made during the voyage, along with his collections of plants, animals, fossils, and minerals, excited considerable scientific interest even before his return to England. Darwin's account of the voyage published later as his *Journal of Researches* and in an abridged popular version as *Voyage of the Beagle* remains one of the most perceptive chronicles of exploration in the nineteenth century.

On his return to England, a substantial income and inheritance enabled Darwin to forgo financial pursuits and dedicate himself entirely to biology. He married his cousin, Emma Wedgwood (1808–1896), and in 1842 settled near the village of Down in the Kent countryside, 16 miles (25.7 km)[2] from London, where he and his wife began to raise a large family.

For 40 years to the time of his death in 1882, Darwin lived at home, mostly as a semi-invalid subject to heart palpitations, rashes and gastric discomfort. The cause(s) of his disability is not known. Theories range from parasitic infection (trypanosomes that cause Chagas' disease) to heavy metal (arsenic) poisoning via some of the so-called cures of his time, to psychosomatic illness involving severe symptoms of panic disorder. Whatever the cause, his illness(es) isolated him — or was used by Darwin to isolate himself — from most of the world about him, except through letters and publications. Through his correspondence, however, Darwin was extraordinarily well connected socially and academically, corresponding prodigiously with naturalists, scholars and breeders. The resulting exchange of ideas was enormously important to his work. As Janet Browne (2006) argues, the relatively advanced development of the British postal system, which was founded in 1660, was a critical background element to Darwin's success. Darwin spent £20 (£1,000 [US$1,890] today) on stationery, stamps and newspapers in 1851 and £53 (£2,650 [US$5,000]) in 1877, when a postage stamp cost a penny.

Despite his physical discomforts, Darwin lived a harmonious life, probably because of his own warm personality and behavior and the sympathetic concern of his wife. It has often been said that Darwin was the perfect patient and his wife the perfect nurse.

Voyage of the *Beagle*

The five-year voyage on the *Beagle* (**Fig. 2-3**) enabled Darwin to observe and think about a wide range of organisms and geological formations. He collected birds, insects, spiders and plants in the Brazilian tropical forests. At Punta Alta

[2] See the conversion scale on the inside of the back cover.

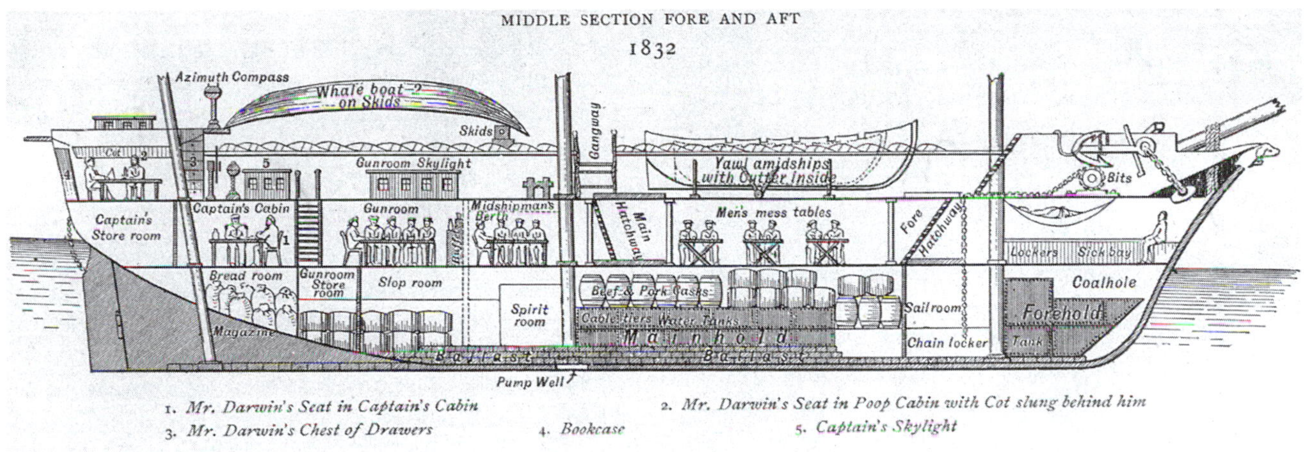

FIGURE 2-2 Top: H.M.S. (*His Majesty's Ship*) *Beagle* in the Strait of Magellan at the southern tip of South America. The ship was a 10-gun brig, 27-m long and weighing 240 tons. In 1831 (the year Darwin began his voyage), it had been refitted for circumnavigation in order to fix world longitudinal markings and to chart the coast of South America. (© SPL/Photo Researchers, Inc.) Bottom: Side and top elevations of the *Beagle*, based on a drawing by one of Darwin's shipmates showing the general plan of the ship and the cramped quarters occupied by the crew of about 70. (In an account of the ship's voyages, Thompson [1995] notes, "To say that the *Beagle* was extremely cramped, even given the expectations of the time, would be a supreme understatement. The ship was, after all, no longer than the distance between two bases on a baseball field," that distance being 90 feet (27.4 m). Darwin slept in the poop cabin at the stern of the ship, which he shared with two officers. This cabin also held a 10' × 6' (3 × 1.8 m) chart table and various chart lockers, as well as drawers for his equipment and specimens. He wrote, "I have just room to turn around and that is all." (© Mary Evans Picture Library/Alamy Images.)

on the coast of Argentina, he unearthed fossil bones of a 6-m-high giant sloth, *Megatherium;* the hippopotamus-like *Toxodon;* the giant armadillo *Glyptodon;* and other animals resembling present species, yet recognizably different. The primitiveness and wildness of the Tierra del Fuego Indians at the southern tip of South America impressed Darwin, as did the severity of their struggle for subsistence in a meager and unrelenting environment.

During the voyage, Darwin carried with him the first and then only published volume of Lyell's *Principles of Geology*, a gift from the captain of the *Beagle*, Robert Fitzroy, and assiduously noted the geological features of many terrains he covered.[3] To explain some of the geological uplifting pro-

[3] A classic geology textbook, *Earth Sciences*, is now in its ninth edition (Tarbuck and Lutgens, 2006). Although initially known as geology and an ancient science (Chapters 1 to 3, and see Rudwick, 2005 for a history of geohistory), most university departments of geology have been renamed departments of earth sciences to reflect the multi- and interdisciplinary approaches brought to bear on the study of Earth.

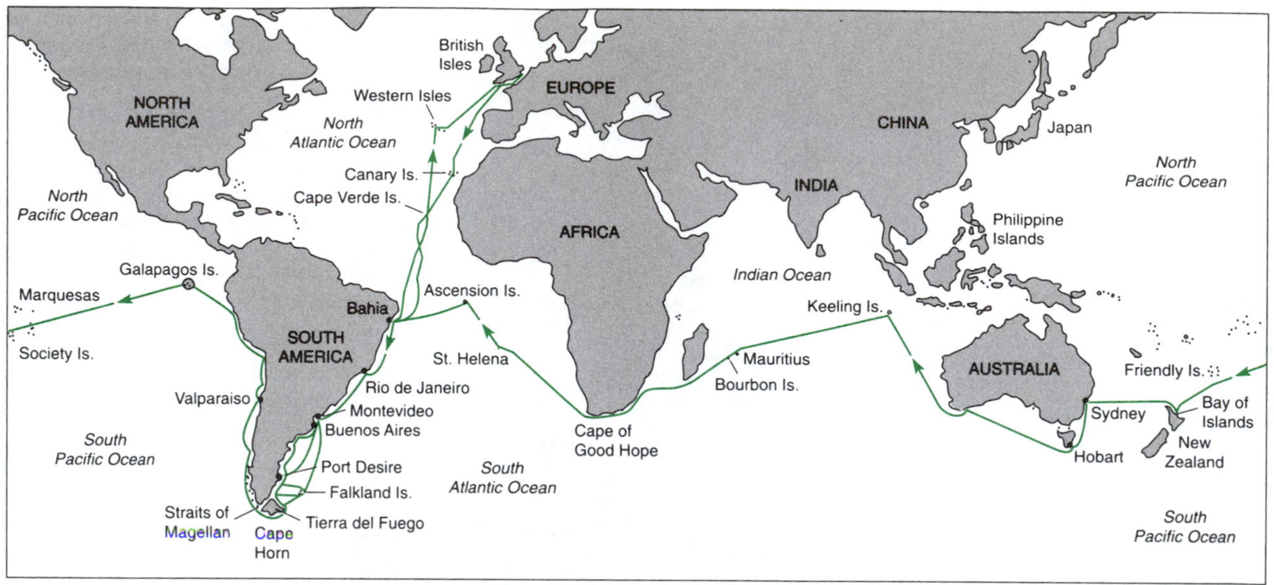

FIGURE 2-3 Route of the five-year voyage of the *Beagle*, beginning at Plymouth, England, in December 1831, and ending in Falmouth, England, in October 1836. Almost four years were spent in South America, including one month (September–October 1835) among the Galapagos Islands.

cesses that shaped the South American landscape, Darwin gathered evidence showing the distribution of marine shells at various places above sea level, the loss of pigment in the older shells found at higher elevations and the terracing of land by erosion as the land was elevated. At the Bay of Concepción on the coast of Chile, Darwin experienced a severe earthquake, which raised the level of the land in some places from two or three feet (0.6-0.9 m) to as much as ten feet (3 m) above sea level. This experience had a deep effect on Darwin, although it did not shake his belief in uniformitarianism or gradualism:

> A bad earthquake at once destroys our associations: the earth, the emblem of solidity, has moved beneath our feet like a thin crust over a fluid; one second of time has created in the mind a strange idea of insecurity, which hours of reflection would not have produced.

An experience that years later had great impact on Darwin's thinking about evolution was the month he spent in the bleak, lava-covered **Galapagos Islands** off the coast of Ecuador. Here, 800 km from the mainland, was a strange collection of organisms: giant tortoises, meter-long marine and land iguanas as well as many unusual plants, insects, lizards and seashells. As he had already noted on the mainland, different geographical localities, although possessing some environmentally similar habitats, were not always occupied by similar species. Darwin was particularly struck by the situation in the Galapagos, where insect-eating warblers and woodpeckers were absent but various species of finches, usually seed eating, assumed the insect-eating patterns of the missing species (**Fig. 2-4**). Also, the observation that each island appeared to have its own constellation of species raised the important question: What could account for this distribution of organisms? In Darwin's words:

> It is the circumstance that several of the islands possess their own species of the tortoise, mocking-thrush, finches, and numerous plants, these species having the same general habits, occupying analogous situations, and obviously filling the same place in the natural economy of this archipelago, that strikes me with wonder.

Did separate and different creations make one species in one place slightly different from another species in another place? Why? Darwin was still some years from asking, "How?"

Darwin's account of his 1831 to 1836 voyage on the *Beagle* was first published in 1838 and revised some years later. The ornithologist David Lack and various Darwin historians have pointed out that although Darwin's *Journal of Researches* expresses these and other evolutionary forethoughts, the significance of his observations on the Galapagos Islands and elsewhere did not become apparent to Darwin until after his return to England. This was especially true for the various Galapagos finches, which were first classified in England by John Gould (1804–1881), a British ornithologist whom Darwin met in 1837. Only then did Darwin manage to sort out his notes on which island was home to which species, and begin to see the significance of isolation (**Box 2-1**). Keep these ideas and dates in mind as you read on.

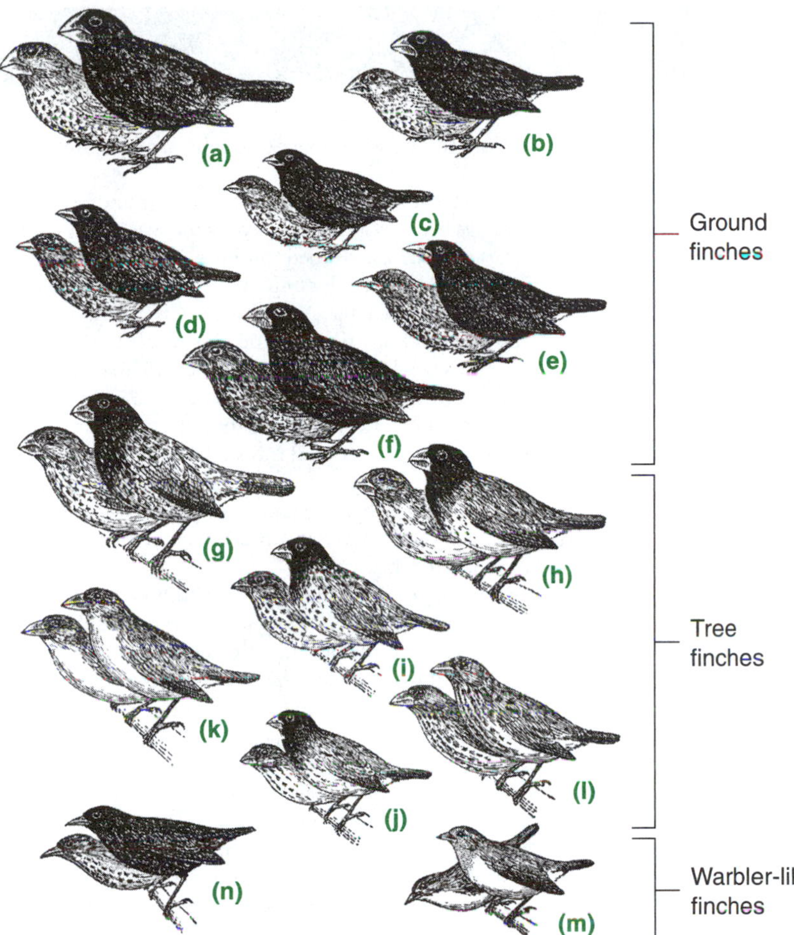

FIGURE 2-4 Species of finches (male on left, female on right; about 20 percent of actual size) that Darwin observed in the Galapagos Islands. (a) *Geospiza magnirostris* (large ground finch), (b) *G. fortis* (medium ground finch), (c) *G. fuliginosa* (small ground finch), (d) *G. difficilis* (sharp-beaked ground finch), (e) *G. scandens* (cactus ground finch), (f) *G. conirostris* (large cactus ground finch), (g) *Camarhynchus crassirostris* (vegetarian tree finch), (h) *C. psittacula* (large insectivorous tree finch), (i) *C. pauper* (large insectivorous tree finch on Charles Island), (j) *C. parvulus* (small insectivorous tree finch), (k) *C. pallidus* (woodpecker finch), (l) *C. heliobates* (mangrove finch), (m) *Certhidea olivacea* (warbler finch), and (n) *Pinaroloxias inornata* (Cocos finch). Evolutionary relationships among these finches are illustrated in Figure 3-5. (From *Darwin's Finches: An Essay on the General Biological Theory of Evolution*, 1947, by D. Lack. Reprinted by permission of Cambridge University Press.)

The *Beagle* voyage stirred Darwin, in 1837, to begin his first notebook on the transmutation of species. Darwin came to the view that only changes among species could reasonably explain the observation that present species resemble past species and that different species share similar structures. "The only cause of similarity in individuals we know of is relationship." The differences between the flora and fauna of different geographical areas, he thought, must have arisen because not all plants or animals are universally distributed. For the Galapagos Islands, for example, Darwin raised the question:

> Why on these small points of land, which within a late geological period must have been covered with ocean, which are formed of basaltic lava, and therefore differ in geological character from the American continent, and which are placed under a peculiar climate — why were their aboriginal inhabitants . . . created on American types of organization?

It seemed clear to Darwin that islands such as the Galapagos contained only those organisms able to reach them, and that evolution could transform only those species available. "Seeing this gradation and diversity of structure in one small, intimately-related group of birds, one might really fancy that from an original paucity of birds in this archipelago one species had been taken and modified for different ends."

The mechanism for the transformation of species, however, was by no means as obvious as was the reasonable assumption that such transformation had occurred. Why do species change? In seeking an answer over the next 20 years, Darwin explored a variety of theories. One of the most persistent, a theory that later had many adherents in France and the United States, was put forth by Lamarck.

BOX 2-1 Darwin's Finches

THE YOUNG CHARLES DARWIN was profoundly impressed with the diversity of wildlife he saw on the Galapagos Islands during his brief stay there. Whether his observations on these remote islands provided seeds for his later insights into the mechanisms of evolution or not, he certainly drew heavily on them for examples throughout *The Origin of Species*. Prominent among these was analysis of the finches he brought back from the *Beagle* expedition, work that was actually performed by John Gould, a prominent English ornithologist.

Gould pointed out to Darwin that a diverse assemblage of birds that Darwin had mistakenly assigned to diverse taxa was in fact a remarkably varied group of finches. Darwin had not recorded the island of provenance for the birds he collected on the Galapagos, but by examining the birds collected by other crew members who had kept such records, Gould showed that many of these varied finch species, which varied dramatically in body size and beak shape (**Fig. 2-5**), were confined to particular islands. Once he became aware of it, this important observation influenced Darwin's thinking because it implied to shim that a homogeneous population of finches must have colonized the islands after their volcanic origin from the ocean floor and then evolved into different forms on particular islands. In this way, the Galapagos finches and other varied groups that he encountered on these islands influenced Darwin's thinking on how geographic isolation relates to the formation of new species.

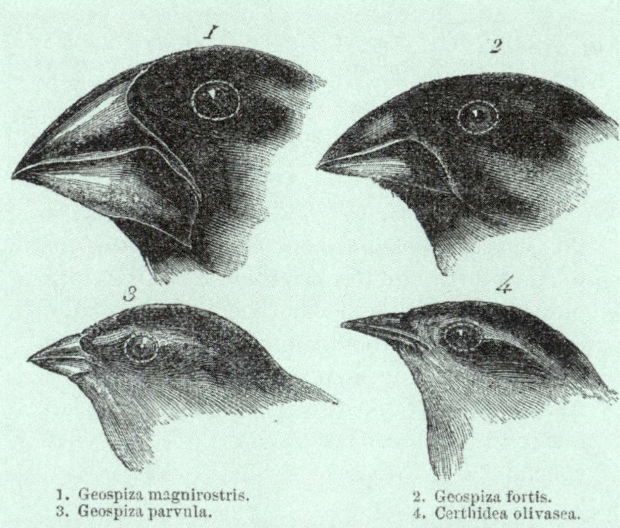

FIGURE 2-5 Drawings by Charles Darwin of four species of Darwin's finches showing variation in size and beak morphology. (©Photos.com.)

The importance of this group of birds to our understanding of evolution has been greatly enhanced by the work of Peter and Rosemary Grant who have studied Darwin's finches continuously since 1973. Their work has documented evolution in action by relating environmental changes resulting in evolutionary changes in both beak size and body size over the span of a few generations of birds. An early study by Grant and colleagues (1976) confirmed a prediction made by Leigh Van Valen that the degree of variation in important environmental factors should be reflected in the amount of variation in the morphological traits relevant to those factors. Thus, species encountering a food that varies greatly in size and hardness should show higher variation in beak dimensions. Within each species, individual birds choose food that is of the appropriate size and hardness for their beaks.

As their work progressed and the longitudinal datasets built up, the Grants showed that natural selection can produce evolutionary change very rapidly but that the direction of selection in particular populations can change frequently and unpredictably as environmental conditions fluctuate (Grant and Grant, 2002; **Fig. 2-6**). By documenting evolution in action in natural populations, the work of the Grants has made profound contributions to our understanding of the evolutionary process.

A question raised by the remarkable results of the Grants' study of evolution in Darwin's finches is how, in genetic and developmental terms, evolution can proceed so rapidly. If changes to beak size and shape required selection to alter the expression or function of many genes in some coordinated fashion, it is unlikely that evolutionary changes could happen as quickly as they clearly have over the past 30 years in the Galapagos finches.

An insight into this question has come from recent work by Abzhanov et al. on the developmental-genetic basis for variation in beak size and shape in Darwin's finches. Abzhanov and colleagues showed that remarkably, the amount and area of expression of a single gene, *Bmp-4*, is correlated with variation in beak size and shape across species of Darwin's finches (**Fig. 2-7**).

In a fascinating parallel, increased levels of *Bmp-4* in a cichlid fish from Lake Malawi are associated with biting and crushing feeding on algae, in comparison with lower levels of *Bmp-4* in another cichlid species from the same lake that feeds by suction feeding on plankton in the water column (Albertson et al., 2005).

Feeding is effected through jaws and teeth. Another study (Wise and Stock, 2006) shows that, although expression of *Bmp-4* is conserved in tooth-forming regions across three species of fish (zebrafish, Japanese medaka and Mexican tetra), *Bmp* genes are not expressed in

Lamarckian Inheritance

As discussed in Chapter 1, Erasmus Darwin and Jean-Baptiste de Lamarck were among the first to actively advocate evolution. Lamarck made the important leap from what appeared to others as species extinction — as evidenced by fossils — to proposing the *continuity of species* by gradual modification through time. By the early nineteenth century, most naturalists accepted the inheritance of acquired characters, the utility (adaptedness) of features and the concept of some internal force toward change.

To explain how modifications occurred and the exqui-

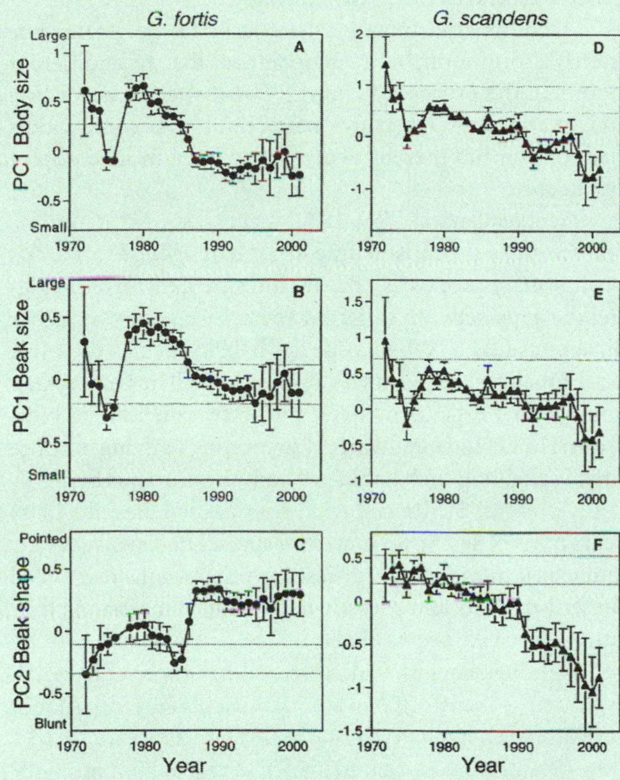

FIGURE 2-6 Morphological changes in adults of two species of Darwin's finches, *Geospiza fortis* (A to C) and *G. scandens* (D to F), from 1973–2001. The horizontal hashed lines represent the confidence intervals that define the null hypothesis of no evolutionary change. These results show both significant evolutionary change and also the fluctuating directions of change created by changing environmental conditions. (Reproduced from Grant, P. and Grant, R. 2002. *Science*. 296, 707–711. Reprinted with permission from AAAS.).

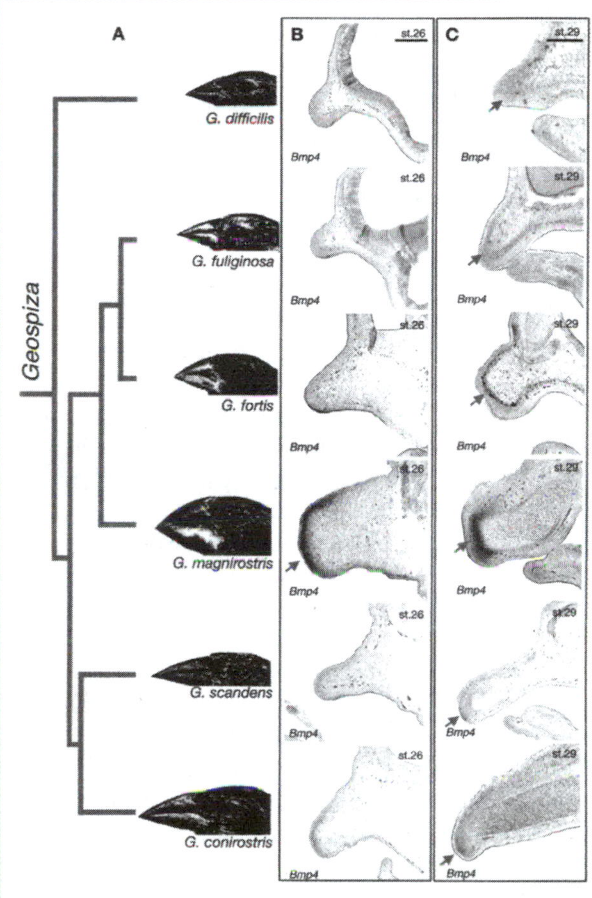

FIGURE 2-7 Relationship between variation in *Bmp4* expression and beak morphology in Darwin's finches. The histological sections shown are from the developing face (stages 26 and 29, facial prominences) from embryos of six species of Darwin's finches. Darkly stained areas indicate high levels of *Bmp4* expression. (Reproduced from Abzhanov, A., Protas, M., B. Rosemary Grant, B. R., Grant, P. R., and Tabin, C. J. 2004. *Bmp4* and morphological Variation of beaks in Darwin's finches. *Science*. **305**; 1462–1465. Reprinted with permission from AAAS).

toothless regions of the jaws, implicating *Bmp* in evolutionary tooth loss, and linking genetic regulation of teeth and jaws.

These results suggest that the rapid evolvability of the beak in Darwin's finches (and perhaps in many other lineages of birds), and jaw and tooth morphology associated with feeding in fish, is explained by the fact that a relatively simple developmental-genetic mechanism can produce a coordinated pattern of variation in beak size and shape. Other similar genetic and developmental mechanisms underlying speciation are explored in Chapter 24.

site relationships (**adaptations**) through which organisms exploited their environments, Lamarck is credited with proposing that:

- Variations among organisms originate because of response to the needs of the environment; and
- This ability to respond in a particular direction accounts for the adaptation of new features.

For example, Lamarck suggested that the long legs of water birds such as heron and egret have arisen through the following mechanism:

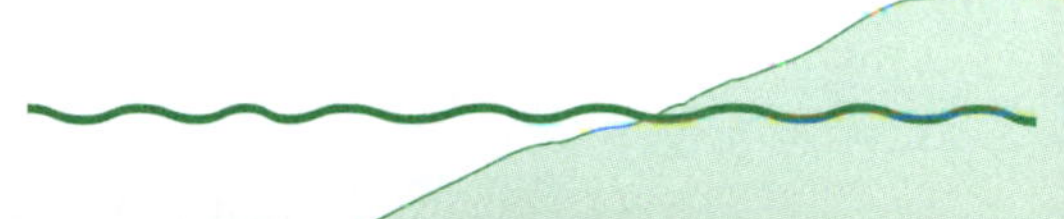

> We find . . . that the bird of the waterside which does not like swimming and yet is in need of going to the water's edge to secure its prey, is continually liable to sink in the mud. Now this bird tries to act in such a way that its body should not be immersed in the liquid, and hence makes its best efforts to stretch and lengthen its legs. The long-established habit acquired by this bird and all its race of continually stretching and lengthening its legs, results in the individuals of this race becoming raised as though on stilts, and gradually obtaining long bare legs, denuded of feathers up to the thighs and often higher still.

A central feature of this hypothesis was that, when confronted with new environments, organisms moving toward improvement learn new habits and can change appropriately by exercising an unknown, inner *perfecting principle.* This special power could sense the needs of the environment and respond by developing new traits in appropriate adaptive directions, mostly from simple to complex. The source for such directional orientation was not clear to Lamarck, who had no idea of a device comparable to natural selection that could lead to continued improvement of adaptive mechanisms.

At times Lamarck ascribed his belief in evolutionary "progress" to an inner vitalistic property of life *(feu éthéré,* ethereal fire). At other times he denied such supernatural causes. However, no matter whether their direction was caused by natural or supernatural events, the origin of organic changes and their transmission to further generations was believed by Lamarck to be aided by *two universal mechanisms in two natural processes* that he codified into two *Laws of Nature.* Both laws can be traced to the folklore of antiquity and were incorporated into *Zoonomia,* a popular work by Charles Darwin's grandfather, Erasmus Darwin:

> 1. **Principle of Use and Disuse.** In every animal that has not passed the limit of its development (like Linnaeus, Lamarck saw organisms as progressing through some internal principle), a more frequent and continuous use of any organ gradually strengthens, develops and enlarges that organ, and gives it a power proportional to the length of time it has been so used; while the permanent disuse of any organ imperceptibly weakens and deteriorates it, and progressively diminishes its functional capacity, until it finally disappears.
> 2. **Inheritance of Acquired Characters.** All of the acquisitions or losses wrought on individuals, through the influence of the environment in which their race has long been placed, and hence through the influence of the predominant use or permanent disuse of any organ, are passed on by reproduction to the new individuals that arise, provided that the acquired modifications are common to both sexes, or at least to the individuals producing the young.[4]

Some would list the drive toward perfection as a third process central to Lamarck's thinking.

According to Lamarck, what is called a species must be merely a continuum between organisms that are at different points in the process of change. Fossil species, according to Lamarck, were not truly extinct but had become modified in time and thereby evolved into later, more complex organisms.

Georges Cuvier (1769–1832) marshaled what at the time seemed to be the most telling argument *against* Lamarck's evolutionary proposals: that no intermediate forms bridging the gaps between different species were known, either alive or as fossils. When a species hybrid such as the mule occasionally occurred, it was always sterile. The Lamarckian concept that organisms strive for perfection seemed ludicrous: How could new habits of swimming or flying produce organs enabling such habits without these organs being already present? Furthermore, Cuvier argued, despite 4,000 years of recorded history, no new species had evolved. Cats and wheat found in the earliest Egyptian tombs resembled present-day cats and present-day wheat. Why assume they can evolve? Like begets like!

Although later proved incorrect, the processes Lamarck espoused — the effects of use and disuse and the inheritance of acquired characters — had the important advantage that they were uniformitarian in principle and did not immediately rely on supernatural or catastrophic events. By proposing a **materialistic explanation for evolution** (inheritance of acquired characters), an explanation that reflected the **close interactions known to occur between organisms and their environment,** Lamarck fostered a climate of opinion in which evolution could be understood in the same fashion as any other natural event.

Natural Selection

If we accept the basic idea that organisms can change through time and that use and disuse does affect organisms, but discard the Lamarckian explanations for how they change, the question then is: How do they change? Darwin elaborated an evolutionary mechanism — **natural selection** — but was not the first to think about selection in relation to living organisms.

[4] Although abandoned by practically all biologists today, a form of **Lamarckism** was adopted by the Soviet Union as official policy between 1948 and 1963 as a result of experiments by the Russian agronomist T. D. Lysenko and his supporters, who claimed to have shown inheritance of environmental adaptations in crops.

A Short History of the Concept of Natural Selection

Empedocles (c. 490–430 BC) suggested that life initially appeared in the form of parts and organs floating freely and combining together to form whole organisms. Those organisms that adapted to "some purpose" survived, those that did not, "perish and still perish." From this original selective act, Empedocles proposed, all present organisms stem. Aristotle disagreed with Empedocles' concept of randomness and selection with an argument often used since. The *Scale of Nature,* like any other teleological process, Aristotle said, arises through a fixed progression of steps from lowest to highest stages. There cannot, therefore, be anything arbitrary or random in this progression that would require selection (see Chapter 1).

In the eighteenth century, Buffon saw natural selection — and **artificial selection** on domesticated animals and plants by humans — as the agent responsible for the extinction of species. He did not, however, see natural selection as responsible for the generation of new species. Buffon believed that new species arose by spontaneous generation and that differences in the conditions under which spontaneous generation occurred caused differences between species.

Lamarck was a materialist, perhaps one of the last eighteenth century naturalists to seek systematic, comprehensive and materialistic explanations for natural phenomena, although he never completely abandoned the notion of a mystical drive, vital force or *feu éthéré.* In the Lamarckian view, environmental effects initiated variations that occurred only in an adaptive direction. There could be no extinction of "imperfect" or "defective" species because organisms could always adapt themselves to changing environments by inheriting acquired characteristics. To Lamarck, variations were not separate from evolution, and therefore they could not be random. Thus, selection was not needed for adaptive traits to appear.

In the early nineteenth century, a number of natural historians, including the physiologist William Wells (1757–1817) and the horticulturalist and evolutionary theorist Patrick Matthew (1790–1874), separated the origin of variations from the forces responsible for preserving them and used the principle of natural selection to explain changes within species. Because insufficient evidence was provided, their ideas seemed highly speculative. Furthermore, their works were recorded in obscure publications that did not come to the attention of later workers.

Malthus and Darwin

Many more individuals of each species are born than can possibly survive. Consequently, there is a frequently recurring struggle for existence. It follows that any being, if it varies however slightly in any manner profitable to itself under the complex and variable conditions of life, will have a better chance of surviving and thus be naturally selected. From the strong principle of inheritance, any selected variety will tend to propagate its new and modified form (Darwin).

Behind this simple explanation is a complex set of causative events, which Darwin spent most of his life investigating.

Darwin ascribed his notion that species tend to produce more members than resources can sustain — the primary population pressure that leads to competition and selection — not to biological literature but to the sociology of his time. Largely as a consequence of the rapid increase in the number of poor people resulting from the Industrial Revolution, a variety of social and economic problems became apparent in Victorian England. This tide of poverty had begun with the impoverishment of small handicrafts establishments and was continually fed by small farmers pushed from their lands (the "commons") by the Enclosure Acts.

One consequence, famously expressed by the Rev. Thomas Malthus (1766–1834), was that the fate of the poor is inescapable; their reproductive powers will always exhaust their means of subsistence. Food supplies, **Malthus** pointed out, at best can increase arithmetically (1 → 2 → 3 → 4 → 5 . . .) by the gradual accretion of land and improvement of agriculture. The number of people, however, will increase geometrically (2 → 4 → 8 → 16 . . .) because parents usually produce more than two children. Thus, famine, war and disease inevitably become major factors limiting population growth.

Malthus' 1798 essay on the principle of population is reprinted in Appleman (2001). What will amaze you when you read it, is not only that it is a bare 600 words long, but that Malthus applies his reasoning not only to human populations but to all animals and plants, as the excerpt below demonstrates:

> Assuming then my postulata as granted . . . *First, That food is necessary to the existence of man. Secondly, That the passion between the sexes is necessary, and will remain nearly in its present state* . . . , I say, that the power of population is indefinitely greater than the power in the earth to produce subsistence for man.
>
> Population, when unchecked, increases in a geometrical ratio. Subsistence increases only in an arithmetical ratio. A slight acquaintance with numbers will shew the immensity of the first power in comparison of the second.
>
> By that law of our nature which makes food necessary to the life of man, the effects of these two unequal powers must be kept equal. This implies a strong and constantly operating check on population from the difficulty of subsistence. This difficulty must fall somewhere and must necessarily be severely felt by a large portion of mankind.

> *Through the animal and vegetable kingdoms,* nature has scattered the seeds of life abroad with the most profuse and liberal hand. She has been comparatively sparing in the room and the nourishment necessary to rear them. The germs of existence contained in this spot of earth, with ample food, and ample room to expand in, would fill millions of worlds in the course of a few thousand years. *Necessity, that imperious all pervading law of nature, restrains them within the prescribed bounds. The race of plants and the race of animals shrink under this great restrictive law.* And the race of man cannot, by any efforts of reason, escape from it. *Among plants and animals its effects are waste of seed, sickness, and premature death. Among mankind, misery and vice* (Malthus, *An Essay on the Principle of Population,* 10th ed., 1803; emphases added).

The only hope that Malthus held out for the poor was self-restraint: delay marriage and refrain from sexual activity. All other solutions — the Poor Laws (welfare), redistribution of wealth, improvement of living conditions — were, in his view, inadequate. Such measures would stimulate a further increase in the number of poor people and again begin the cycle of famine, war and disease.

Like many others of the time, Darwin was deeply impressed by the Malthusian argument, although Malthus did not espouse evolution and Darwin only was influenced by Malthus after he had come to accept the possibility of the transformation of species. In fact, *Malthus believed that limiting population growth would prevent evolutionary change* because individuals who departed from the population norm would be more susceptible to extinction, an idea that had been spelled out in the 1830s and with which many biologists would have been comfortable. To Darwin, however, the importance of Malthus' theory lay in revealing the conflict between a population's limited natural resources and its continued reproductive pressure. In contrast to Malthus' proposals for alleviating the impact of population increase, Darwin pointed out that plants and animals had no such alternatives: "There can be no artificial increase in food, and no prudential restraint from marriage."

Darwin collected an enormous amount of evidence for artificial selection. From these data and his inquiries on breeding domesticated species, Darwin obtained clear evidence that selection (in this case, artificial selection) could have marked hereditary effects. Indeed, Darwin was amazed at the ease with which domesticated plants and animals could be changed and the small number of generations of selection it took to do so. He continually referred to artificial and natural selection throughout his writing.

Through the combination of:

- the ease of effecting change via artificial selection;
- the vastness of geological time;
- the application of the principle of uniformitarianism by which present-day processes could be extrapolated back in time; and
- the identification of a natural selector, the pressure of continuously limited resources,

and with brilliant intuition, Darwin saw that selection acted by choosing for reproduction those individuals or types with increased chances of survival, thereby changing the composition of the population.

Malthus therefore played an important role in Darwin's thinking out natural selection as a mechanism of evolutionary change. In his autobiography Darwin wrote:

> I soon perceived that selection was the keystone of man's success in making useful races of animals and plants. But how selection could be applied to organisms living in a state of nature remained for some time a mystery to me.
>
> In October 1838, that is, fifteen months after I had begun my systematic enquiry, I happened to read for amusement 'Malthus on Population' and being well prepared to appreciate the struggle for existence[5] which everywhere goes on from long-continued observation of the habits of animals and plants, it at once struck me that under these circumstances favourable variations would tend to be preserved, and unfavourable ones to be destroyed. The result of this would be the formation of new species. Here then I had at last got a theory by which to work; but I was so anxious to avoid prejudice, that I determined not for some time to write even the briefest sketch of it. In June 1842 I first allowed myself the satisfaction of writing a very brief abstract of my theory in pencil in 35 pages; and this was enlarged during the summer of 1844 into one of 230 pages, which I had fairly copied out and still possess.

Wallace and Darwin

Darwin delayed publishing his theory. In part this reflected his desire to amass a wealth of evidence that would overcome any opposition to the generality of his theory. His theory that natural selection was the mechanism of speciation also was incomplete until 1852. Only then did Darwin devise the *principle of divergence* — which he represented in the only

[5] It took Darwin some time to accept and use Herbert Spencer's term and metaphor *survival of the fittest* (See Box 22-1). In several letters, Wallace tried to convince Darwin to use the phrase, which he (Wallace) had adopted. Only after several editions of *On the Origin of Species* had been published did Darwin adopt the term for the processes that lead to the differential survival of organisms from one generation to the next. Darwin wrote that it might have been preferable had he used *struggle for reproduction* rather than struggle for existence; see Huxley (1947) for a discussion.

figure in *The Origin* — and which shows that as long as competition between subpopulations exists, subpopulations can begin to specialize and diverge to the point of speciation. With this principle, Darwin could maintain that *competition was the most powerful selective force; geographical isolation featured much less prominently.*

A third factor that made Darwin cautious of public reception to his theory was the book, *Vestiges of the Natural History of Creation.* It was published anonymously in 1844 but shown after the author's death to have been written by Robert Chambers, a member of a prominent Scottish publishing house. *Vestiges* elaborated the idea that all matter, inorganic and organic, evolved out of inorganic dust by the accumulation of accidental mutations caused somehow by changes in nutrition or environment. Although enormously popular for a time (there was much public discussion of the 14 editions in Britain) considerable religious and scientific denunciation focused on this work, which made Darwin fearful of exposing his own ideas to ridicule. At the time of *Vestiges* publication (1844), Darwin already had written a lengthy essay setting out his theory, an essay, which although complete, he set aside and never published. Reading *Vestiges* and the responses it elicited, convinced Darwin that he needed to amass as much evidence as possible before exposing his theory to public scrutiny (and possible ridicule).

But evolution was in the air. In 1858, Alfred Russel Wallace (1823–1913), a naturalist then collecting birds, insects and mammals in the islands of Southeast Asia, sent Darwin a paper ready for publication in which he described the theory of natural selection in the essential form Darwin had envisaged. Wallace also had puzzled about a mechanism for evolution and had read Malthus. Remarkably, Malthus performed the same function for Wallace as he had for Darwin 20 years earlier:

> At that time [February 1858] I was suffering from a rather severe attack of intermittent fever at Ternate in the Moluccas . . . and something led me to think of the positive checks described by Malthus in his '*Essay on Population,*' a work I had read several years before, and which had made a deep and permanent impression on my mind. These checks — war, disease, famine, and the like — must, it occurred to me, act on animals as well as on man. Then I thought of the enormously rapid multiplication of animals, causing these checks to be much more effective in them than in the case of man; and, while pondering vaguely on this fact there suddenly flashed upon me the idea of the survival of the fittest — that the individuals removed by these checks must be on the whole inferior to those that survived. In the two hours that elapsed before my ague fit was over I had thought out almost the whole of the theory, and the same evening I sketched the draft of my paper, and in the two succeeding evenings wrote it out in full, and sent it by the next post to Mr. Darwin (Introductory note to Chapter II of *Natural Selection and Tropical Nature,* revised edition, 1891).

To prevent Darwin losing his priority in applying natural selection to evolution, his friends Lyell and Hooker arranged for short papers on the topic by both authors to be published in 1858 in *The Journal of the Linnaean Society.*[6] Surprisingly, the two papers evoked little response from the scientific and nonscientific communities, indicating perhaps that the theory itself, without supporting evidence and without enlisting large-scale evolutionary phenomena, did not invite serious interest and did not threaten established views. Only with the publication of Darwin's expanded work with its huge weight of supporting evidence, *On the Origin of Species,* in November 1859, did the world take notice.[7]

The evolutionary principle expressed by Darwin and Wallace is briefly outlined in **Figure 2-8**. Note that the evolutionary process is a continual one; the achievement of an adaptation by individuals leads to enhanced reproductive ability relative to other individuals, followed by further competition for the limited resources and further natural selection. Because each evolutionary stage builds on the one before, the process spirals in the direction of improved adaptation for any particular environment.

At the base of the process is the continual introduction of new heritable variations on which selection can act. Darwin did not know the biological basis for heredity or its variations, and his arguments were weakest in these areas. At times, he proposed that environmental changes, a large increase in numbers, or some disturbance of the reproductive organs, might enhance variability among a population of individuals; at other times, he adopted the Lamarckian view of the inheritance of use and disuse (Bowler, 2005).

The fundamental problem for Darwin, as for others at the time, was the commonly accepted theory that heredity is mostly a blend of the heredity of both parents, much as diluting red paint with white produces pink. As Chapter 3 shows, the idea of *blending inheritance* confronted evolutionary theory with a serious enigma. If they are blended out by mating of their carriers with other members of the population, how can adaptive variations be preserved by

[6] Often we are left with the impression that Wallace and Darwin came up with identical theories. They did not. They differed over the use of Spencer's term *struggle for existence*, but more fundamentally, Wallace was unconvinced of the utility of artificial selection as a means to understand natural selection, a coupling that was essential for Darwin.

[7] Although crammed with evidence, including much data gleaned from the studies of others, Darwin provided almost no documentation of the sources of the evidence he amassed.

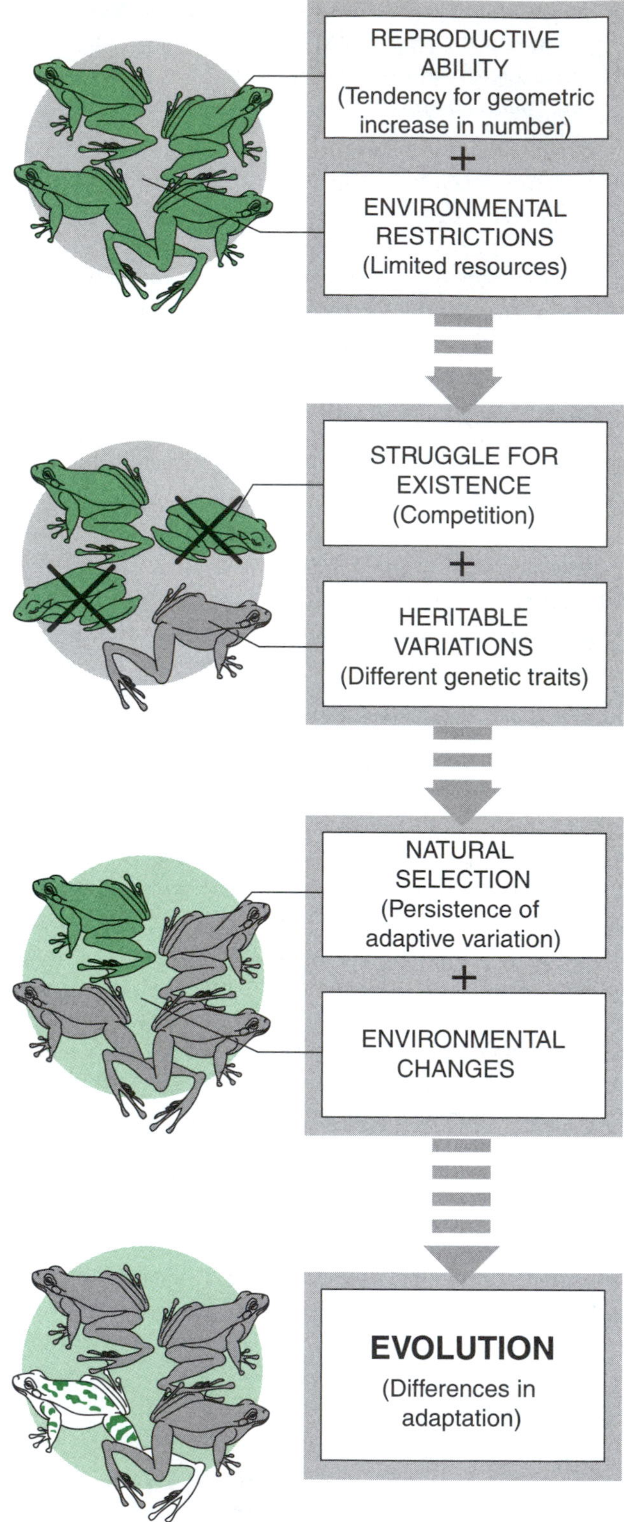

FIGURE 2-8 Schematic of the main conceptual arguments for evolution by natural selection given by Charles Darwin and Alfred Russel Wallace. For current versions of these arguments, see Chapters 10 and 21 to 24. (Adapted from a table in Wallace, A. R., 1889. *Darwinism: An Exposition of the Theory of Natural Selection with Some of Its Applications.* Macmillan, London.)

natural selection? Several views on variation and heredity are compared in **Table 2-1**.[8]

Instead of blending inheritance, some ten years after *The Origin* was published, Darwin reinstituted an old theory, **pangenesis**, according to which small particulate *gemmules* or *pangenes* derived from all the tissues of a parent are incorporated into the parental gametes. When fertilization occurs and parental gametes unite, these gemmules disperse to form the tissues of the offspring (**Fig. 2-9a**). By suggesting that changes can arise in the *frequencies* of particular gemmules but by postulating that the *structure* of gemmules can remain constant, pangenesis helped account for the presumed effects of use and disuse and for the observation that not all traits become blended (Vorzimmer, 1963).

There was no evidence for pangenesis, however, although August Weismann (1834–1914; see Table 1-1) believed he disproved it some years later, when he cut off the tails of 22 generations of mice and showed that tail length was not affected by the presumed loss of tail gemmules in each generation. Although proof against the inheritance of acquired characters, this is indirect proof, at best, against pangenesis. A better test was undertaken by Francis Galton, Darwin's cousin, who transfused blood between different breeds of rabbits without altering the nature of their offspring (Galton, 1871a,b). Darwin's response was that he had not proposed that gemmules circulated in the blood.

Nineteenth-century Lamarckians understood "acquired characters" to be somatic traits that can be reproduced each generation without instruction from germ-line tissue and that could be transmitted through somatic or germinal cells. Weismann devised the **germ plasm theory** of inheritance, in which only the reproductive tissues (testes and ovaries) transmit the heredity factors of the entire organism; changes that occur in nonreproductive somatic tissues are not transmitted (**Fig. 2-9b**).[9] Thus, changes in heredity cannot simply be explained by inheritance of acquired characters or by use and disuse.

With respect to heredity and inheritance, the rediscovery in 1900 of Mendel's work produced a generation of Mendelians, many of whom were opposed to Darwin, especially to his mechanisms of natural selection. Evolution meant something very different to them. It would be the 1930s and 1940s before many of the difficulties Darwin faced were resolved. The reception of Darwin's theory by his contemporaries is taken up in the next chapter.

---

[8] The chapters in Hallgrímsson and Hall (2005) summarize the central role of variation in biology.

[9] Weismann's theory of the *continuity of the germ plasm* is applied far too uncritically as if it applies to all animals. Separation of germ plasm and soma is true only for a minority of phyla within the animal kingdom and is not true in plants, fungi or protists. This means that gametes (animals) or a new individual (plants) could arise from any body cell.

TABLE 2-1 Comparison of views on variation and heredity

	Creationist	Lamarck	Darwin	Present Biology
What accounts for the similarity among many species?	The divine plan of creation (varied mythological purposes) produced the basic "kinds" of organisms.	Descent from a common ancestor.	Descent from a common ancestor.	Descent from a common ancestor.
What accounts for the origin of variations among members of a species?	Although they can be environmentally caused,[a] they are part of the divine plan of creation.	Environmentally caused.	At times: unknown causation. At times: environmental changes cause new variations, although the variations may be in any direction.	Heritable differences are caused by random changes (mutation) in the genetic material. Noninheritable differences are caused by the environment.
What accounts for the presence of particular organs and structures through time?	They were initially designed so by the creator. Many present creationists believe that organ defects, diseases, etc., are caused by the fall of humans from divine grace and/or intervention by a devil.	Use enhances the development of adaptive variations, and disuse eliminates nonadaptive ones.	Natural selection perpetuates only adaptive traits and eliminates nonadaptive traits. At times: use and disuse.	Primarily natural selection but other forces may be involved, as discussed in Chapter 22.
What accounts for the variation among species?	The separate creation of each species. Many present creationists believe that the original "kinds" of organisms were perfect, and variations leading to species differences have been degenerative.	Each species has responded to different environmental needs by developing new organs or discarding old ones.	At times: selective differences among species account for their changed inheritance. At times: differences in the use and disuse of particular organs has caused changed inheritance.	Changes occur in the genetic material of each species through the process of mutation and the various forces that change gene frequencies.
What accounts for the resemblance of organisms to their parents?	Mechanisms unknown, but acquired characters are inherited as part of the divine plan.[b]	Those characters acquired through use and disuse are inherited through a pangenesislike process.	At times: unknown. At times: pangenesis.	Transmission of genetic material through the germ plasm.

[a] See the story of Jacob and the sheep in Genesis 30: 37–39.
[b] "Visiting the iniquity of the fathers upon the children and children's children unto the third and fourth generation." (Exodus 34: 7.)

KEY TERMS

adaptations
artificial selection
blending inheritance
Galapagos Islands
germ plasm theory
H.M.S. Beagle
Inheritance of Acquired Characters
Lamarckism
Malthus
natural selection
pangenesis
Principle of Use and Disuse
variation

DISCUSSION QUESTIONS

1. How did his experiences in the Galapagos Islands affect Charles Darwin's thinking in searching for evolutionary explanations?
2. a. What were the Lamarckian explanations for evolutionary change?
   b. What was their influence on Darwin?
   c. Why are these explanations now unacceptable to most biologists?

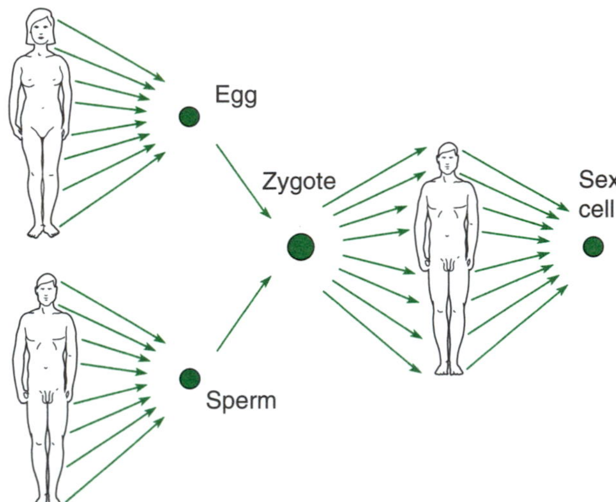

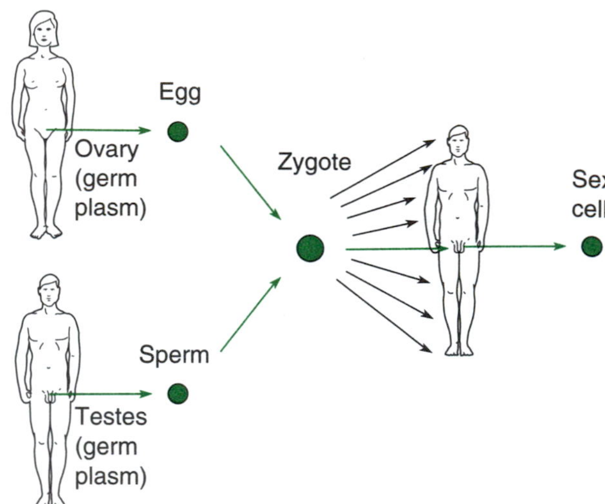

FIGURE 2-9 Comparison between (a) pangenesis and (b) germ plasm theories in the formation of a human. In pangenesis, all structures and organs throughout the body contribute copies of themselves to a sex cell. In the germ plasm theory, the plans for the entire body are contributed only by the sex organs. (From *Genetics*, third edition, by Monroe W. Strickberger. © 1985 by Monroe W. Strickberger. Reprinted by permission of Prentice Hall, Inc., Upper Saddle River, NJ.)

3. a. What is the concept of natural selection?
   b. From what sources did the concept of natural selection develop?
   c. What are the differences between Lamarckian explanations for evolution and the Darwin–Wallace concept of natural selection?
4. If Darwin's concept of natural selection explained how giraffes attained longer necks, by reaching higher branches during times of drought and intense competition, what explanation could he have offered to explain why sheep do not develop long necks?
5. What is the pangenesis theory, and why did Darwin espouse it?

EVOLUTION ON THE WEB

Explore evolution on the Internet! Visit the accompanying Web site for *Strickberger's Evolution,* Fourth Edition, at http://www.biology.jbpub.com/book/evolution for Web exercises and links relating to topics covered in this chapter.

RECOMMENDED READING of Books By and On Darwin

Browne, E. J., 1995. *Charles Darwin: Voyaging.* Alfred A. Knopf, New York.

Browne, E. J., 2002. *Charles Darwin: The Power of Place. Volume II of a Biography.* Alfred A. Knopf, New York.

Darwin, C., 1845. *The Voyage of the Beagle* (Originally published as *Journal of Researches,* it has now appeared in numerous editions).

Darwin, C., 1859. *On the Origin of Species by Means of Natural Selection or the Preservation of Favoured Races in the Struggle for Life.* Murray, London.

Darwin, C., 1964. *On the Origin of Species.* Facsimile of the First Edition with an introduction by E. Mayr. Harvard University Press, Cambridge, MA. [paperback version, 2005]

Darwin, F., 1887. *The Life and Letters of Charles Darwin.* Appleton, New York.

Desmond, A., and J. Moore, 1991. *Darwin.* Warner Books, New York.

Ospovat, D., 1981. *The Development of Darwin's Theory: Natural History, Natural Theology, and Natural Selection, 1838–1859.* Cambridge University Press, Cambridge, England.

Richards, R. J., 1987. *Darwin and the Emergence of Evolutionary Theories of Mind and Behavior.* The University of Chicago Press, Chicago.

Richards, R. J., 1992. *The Meaning of Evolution: The Morphological Construction and Ideological Reconstruction of Darwin's Theory.* The University of Chicago Press, Chicago.

Thompson, K. S., 1995. *HMS Beagle: The Story of Darwin's Ship.* W. W. Norton, New York.

3 Arguments and Evidence for Evolution

Chapter Summary

The publication of Charles Darwin's *On the Origin of Species* in 1859 aroused a variety of objections against his theory of evolution by natural selection. The prevalence of the idea that an individual's traits were a blend of those of the parents made it difficult to see how traits could be selected for; successive generations of mating with nonadapted individuals would dilute any trait. Darwin's insistence that evolutionary forces act on small and continuous variations led to criticism that selection could not recognize such slight variations and could, therefore, not lead to the formation of new species.

An additional problem was the appearance of new traits, which Darwin thought might form by preadaptation or the conversion of an existing structure to a new use. While selection might account for the evolution of one species into another, it remained difficult to understand how many species might arise from a single one. A branching pattern of evolution comes about through isolation of groups from one another, something that Darwin did not emphasize, but which others such as Gulick and Wagner did. Finally, given that geologists had yet to establish Earth's great antiquity, there did not seem to be sufficient time for evolution to occur. Nevertheless, Darwin's theory had the advantages of relying on explicable mechanisms, being supported by a massive amount of data, and successfully explaining many features of natural systems.

Intermediate forms observed by taxonomists could be explained using Darwin's theory. Evolutionary theory could also explain why organisms with similar characteristics occur geographically close to each other, while groups separated by geographical barriers have fewer characteristics in common. Structures found in different organisms but having an underlying similarity of plan could be explained by their relationship through a common ancestor (*homology*). Vestigial structures such as the rudiments of limb bones in snakes and whales would be the remnants of organs that were present, fully formed, in ancestors. Indeed, whale embryos produce hind limb buds that regress early in development. The similarities among vertebrate embryos during early developmental stages also suggested a common evolutionary past.

Fossil evidence was particularly important. Some fossil forms intermediate between extant organisms (such as *Archaeopteryx*, intermediate between reptiles and birds) were found. But horses provided the most complete and continual fossil record of evolution of a group. The horse evolved from a small, four-toed browsing animal to a large, single-toed creature with continuously growing teeth, adapted to chewing tough grasses. Finally, Darwin recognized that humans, in seeking to perpetuate desirable traits in domestic plants and animals, could select (over a relatively short period of evolutionary time) radical alterations in organisms, albeit artificially rather than naturally. The heavy accumulation of data provided support for Darwin's theory for evolution.

While the brief papers on natural selection by Darwin and Wallace, published in the *Journal of the Linnaean Society* in 1858, evoked little response from the scientific and the non-scientific communities, publication of *The Origin of Species* a year later had profound effects. The first printing sold out within hours of its release, and its contents initiated the first global public debates concerning science and the place and role of science in society.

Many found Darwin's detailed and elegant exposition of his theory, supported by 20 years of thought and documentation, impossible to overlook. Many recognized evolution as the cause for the diversity of species known to exist. Some accepted, to various degrees, that natural selection was an important if not primary mechanism for evolution. Thomas Huxley (1825–1895), who later became Darwin's main public defender, wrote, "... How extremely stupid of me not to have thought of that!" This oft-quoted phrase gives the impression that Huxley immediately captured the essence and significance of Darwin's theory. However, the first part of the sentence (which is not usually quoted) is, "My reflection when I first made myself master of the central idea of the *Origin* was, how incredibly stupid ..." He had to work to comprehend ("master") Darwin's ideas, which were revolutionary and counter to all prevailing thought, even to one as versed in the sciences as Huxley.[1]

[1] Huxley did not accept Darwin's theory uncritically, finding natural selection difficult to accept and disagreeing with Darwin on the reality of species. Thomas Huxley was a nineteenth-century polymath. One of the most eloquent statements of science and the scientific method was contained in a sermon Huxley delivered on 6 January 1866. "I cannot but think that the foundations of all natural knowledge were laid when the reason of man first came face to face with the facts of Nature; when the savage first learned that the fingers of one hand are fewer than those of both; that it is shorter to cross a stream than to head it; that a stone stops where it is unless it be moved, and that it drops from the hand which lets it go; that light and heat come and go with the sun; that sticks burn away in a fire; that plants and animals grow and die; that if he struck his fellow savage a blow he would make him angry, and perhaps get a blow in return, while if he offered him a fruit he would please him, and perhaps receive a fish in exchange. When men had acquired this much knowledge, the outlines, rude though they were, of mathematics, of physics, of chemistry, of biology, of moral, economical, and political science, were sketched" (cited from T. H. Huxley, 1910, pp. 48–49).

Scientific Objections

Objections came rapidly and in various forms, not only in attacks from the established churches but also as objections to specific aspects of his theory from Darwin's scientific colleagues. Some of these are discussed below, along with responses from Darwin, or, more usually, from one or more of his contemporary supporters, such as Huxley.

Blending Inheritance

Because the prevailing concept of inheritance was that maternal and paternal contributions blend in their offspring, a number of critics objected that new adaptations would be successively diluted with each generation of interbreeding. According to this argument, natural selection would be incapable of maintaining a trait for more than a few generations. Darwin made various replies, among them the five outlined below.

1. A beneficial trait could maintain itself if the individuals carrying the trait were isolated from the rest of the population. Darwin pointed to the familiar practice among animal breeders of isolating newly appearing "sports" and their offspring. This mechanism was commonly used to develop new stocks of domesticated animals or plants. Darwin placed major emphasis on artificial selection as engineered by man as evidence supporting his conception of natural selection.
2. Some traits are *prepotent* (dominant) and appear undiluted in later generations.
3. An adaptive trait does not appear only once in a population. Rather, such traits must arise fairly often; witness the large amount of variation in most populations. Because variation is common, went the argument, it would not dilute out as easily as it would if it was rare. Moreover, Darwin believed, some forms that carry a particular variation pass on to future generations the tendency for the same variation to arise again.

4. Natural selection both enhances the reproductive success of favorable variants and diminishes the reproductive success of unfavorable ones. Thus, the frequency of favorable variations increases when unfavorable ones die out, reducing the likelihood of diluting out favorable variations.
5. As explained in Chapter 2, Darwin developed the concept of pangenesis by gemmules to help explain the inheritance of traits — which he believed were affected by use and disuse — and to provide constancy for the determining agents of inheritance. There were presumably many gemmules for each particular trait, and their numbers could vary during passage from one generation to another. Gemmules could be lost but were not changed by "blending."

Variation

In *The Origin*, Darwin explicitly confined evolution by natural selection to small, continuous variations and (in the earlier editions of his book) excluded larger variations as not being useful. He adopted the dictum that "nature makes no leaps," although in his red notebook, Darwin convinced himself that, in some instances, differences between species were so great that they could only be achieved *per saltum* (by huge leaps). A number of objections followed almost immediately.

The first, raised by various critics, concerned the limits of **variability** on which selection could act. Except for monstrosities (which are highly abnormal or sterile), most observed variations involved only small changes and did not depart from the species pattern. How then could new species arise? Darwin replied that no limits really apply to variation because each stage in the evolution of a species entails further variation upon which selection acts. Darwin maintained that a succession of changes through time, rather than a single simultaneous set of changes, leads to species differences.

A second, more common objection was the difficulty of determining how selection would recognize each of the small modifications Darwin proposed. With many features, such as size, a small modification might hardly be enough to confer significant advantage to an organism, whereas a large modification might well be selected. Although Darwin could not successfully reply to this argument, he doggedly held to his concept of the gradual accretion of small modifications. The rediscovery of Mendel's work in 1900 and the findings of genetics (see Chapter 10) added considerable support for gradual change, although not for small modifications. As discussed in Chapter 9, Mendel's experiments involved larger scale characters.

We now know that many traits stem from small heritable changes ascribed to many different genes (*polygenes*), each with small effect. For example, size differences are often distributed in populations so that some large individuals possess many genes that lead to an increase in size while smaller individuals have relatively few such genes; could this be a cause and effect relationship? Although differences in size may stem from genetic differences of small effect, selection may act on accumulations of such differences in individuals of various sizes (see Chapters 10 and 22).

A further aspect of this same problem was the question of determining the initial adaptive level that a trait or organ would have to reach to be selected. If the trait already existed before selection acted on it, perhaps some quantitative expression of the trait would suffice for further evolution; a larger eye might function better than a smaller eye. But if the trait did not exist, or only barely existed, how could selection act on it? Many critics felt that the earliest incipient stages of complex organs such as the eye or brain would have no function at all and could hardly be selected. How could one possibly imagine an appropriate adaptive function in only one cell of an eye, brain or leg?

An evolutionary answer to this question of the origin of new traits came from the concept of **preadaptation**, a principle of which Darwin was aware.[2] According to preadaptation, a new organ need not arise *de novo* but may be present in an organism *but being used* for a purpose other than that for which it is later selected. For example, in his monograph on barnacles, Darwin suggested that the cementing mechanism by which present-day barnacles attach to their substrate is related to the cementing mechanism by which the barnacle oviduct coats its eggs to attach them to the substrate. Only after its earlier evolution in oviducts was this mechanism adapted for attaching the barnacle itself. Similarly, Darwin pointed out that the evolution of lungs from fish swim bladders illustrated, "that an organ originally constructed for one purpose, namely flotation, may be converted into one for a wholly different purpose, respiration." Among more recent evidence for this notion are the findings that optical neural pathways of blind cave animals, no longer needed for sight, may assume new olfactory and tactile functions. Positive selection for enhancement of sensory structures (taste buds and the lateral line) and a lack of selection to retain sight resulted in loss or reduction of the eyes to rudiments in cave-dwelling fish.[3]

Given continued environmental pressure and selection, such evolutionary intermediates are a commonly expected feature. A highly specialized organ like the vertebrate eye did not arise all at once but represents a succession of changes of a previous light-gathering organ and its ancillary tissues, which may have originally involved only a few cells. A turn-of-the-twentieth century illustration of one such progression is shown in **Box 3-1**. Even a one percent per generation change in eye anatomy produces a marked effect in a brief geological period. A flat patch of light-sensitive cells can change into a complex squid-like eye with a focused refractive lens in less

[2] See Chapter 19 for a discussion on preadaptation (exaptation).
[3] This example is discussed more fully in Box 13-2, and in Jeffery et al. (2000) and Franz-Odendaal and Hall (2006).

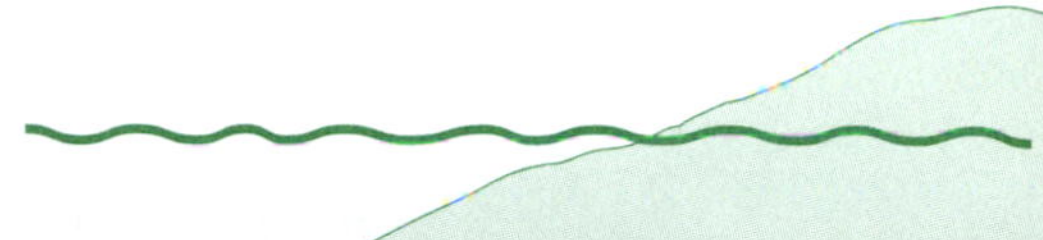

BOX 3-1 Eye Evolution

THE VARIOUS TYPES of mollusks in **Figure 3-1** show a wide range of light-gathering organs ("eyes") from the simplest (a) to the most complex (f) as depicted in 1900 (see Box 8-3 for a discussion of complexity in evolution).

(a) Perhaps the simplest form of light-gathering organ in the animal kingdom is a pigment spot with neural connections that light can stimulate. The spot could be comprised only of a small number of cells, theoretically of a single cell.

(b) Folding of pigment cells concentrates their light-gathering activity, providing improved light detection.

(c) A partly closed, water-filled cavity surrounded by pigment cells allows images to form on the pigmented layer as in a pinhole camera.

(d) A transparent fluid, secreted by the cells — and so a fluid extracellular matrix — rather than water forms a barrier that protects the pigmented layer (the retina) from injury.

Figure 3-1 Light-gathering organs in mollusks

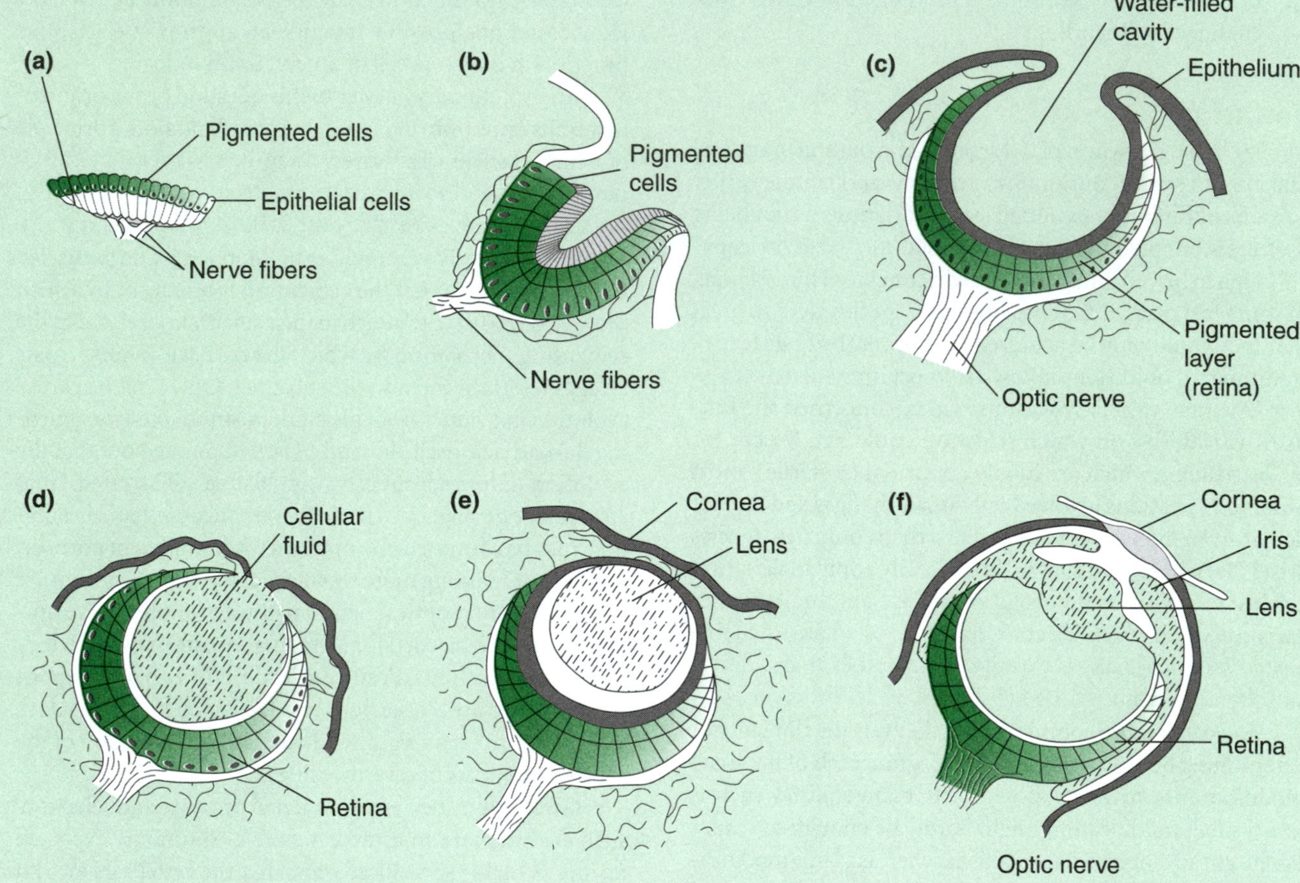

than 500,000 years (Box 3-1; see Land and Nilsson, 2002, for an excellent overview of the evolution of eyes).

Darwin's search for small modifications led him to place less emphasis on the many traits involving distinct steps and differences, such as different colors, presence and absence of structures and different numbers of structures. These large variations may be important for selection, as various individuals, including Huxley, suggested. Interestingly, large observable differences in plant traits were used by Mendel to develop the basic laws that explain inheritance (Chapter 10). Until the twentieth century, Darwinists did not resolve the problem of where, how, and to what extent variations originated, an issue that remained the most often attacked element in Darwin's theory.[4]

Geographical Isolation

Critics also singled out Darwin's strong emphasis on the transformation of a single species into another single species (phyletic evolution; see Chapter 11). They pointed out

[4] See the chapters in Hallgrímsson and Hall (2005) for discussion of phenotypic variation as central to evolutionary biology.

(e) A thin film or transparent 'skin' covers the entire eye apparatus, adding further protection. Some of the fluid extracellular matrix within the eye hardens into a convex lens that improves the focusing of light on the retina.[a]

(f) A complex eye, as seen in species of squid, which has an adjustable iris diaphragm and a focusing lens (adapted from Conn, 1900).

Interestingly, genetic studies indicate that a similar inherited factor (the *Pax-6* gene) regulates development of anterior sense organ patterns in invertebrates and vertebrates. Nevertheless, despite some common regulatory features, specific cellular pathways in embryonic eye development differ noticeably between squid (f) and vertebrates. Squid photoreceptor cells derive from the epidermis; vertebrate retinae derive from the central nervous system. As explained by the process of *convergent evolution*, the structural similarity of squid and vertebrate eyes does not come from an ancestral visual structure in a recent common ancestor of mollusks and vertebrates, but rather from convergent evolution as similar selective pressures led to similar organs that enhance visual acuity. Such morphological convergences may have arisen independently in numerous other animal lineages subject to similar selective visual pressures. What is shared is deep in metazoan ancestry: the ability to form light-gathering cells or organs.[b]

The series of images in **Figure 3-2** shows stages in eye evolution as depicted by a computerized model in which random changes in eye structure are followed by selection for visual acuity. Beginning with a light-sensitive middle layer of skin backed by pigment (a), successive selective steps for improved optical properties lead to a concave buckling that enhances light gathering (b–e), a focusing lens (f, g), and an eye with a flattened iris in which the focal length of the lens equals the distance between lens and retina (h) (Adapted from Nilsson, D.-E., and S. Pelger, 1994. A pessimistic estimate of the time required for an eye to evolve. *Proc. Roy. Soc. Lond.* (B), **256,** 53–58.)

[a] Now we know that this is a function of a four-member family of lens proteins, the crystallins (Table 12-2; Land and Nilsson, 2002; True and Carroll, 2002).

[b] For further discussions of the evolution of eyes and photoreceptor pigments, see Arendt (2003).

Figure 3-2 Stages in eye evolution

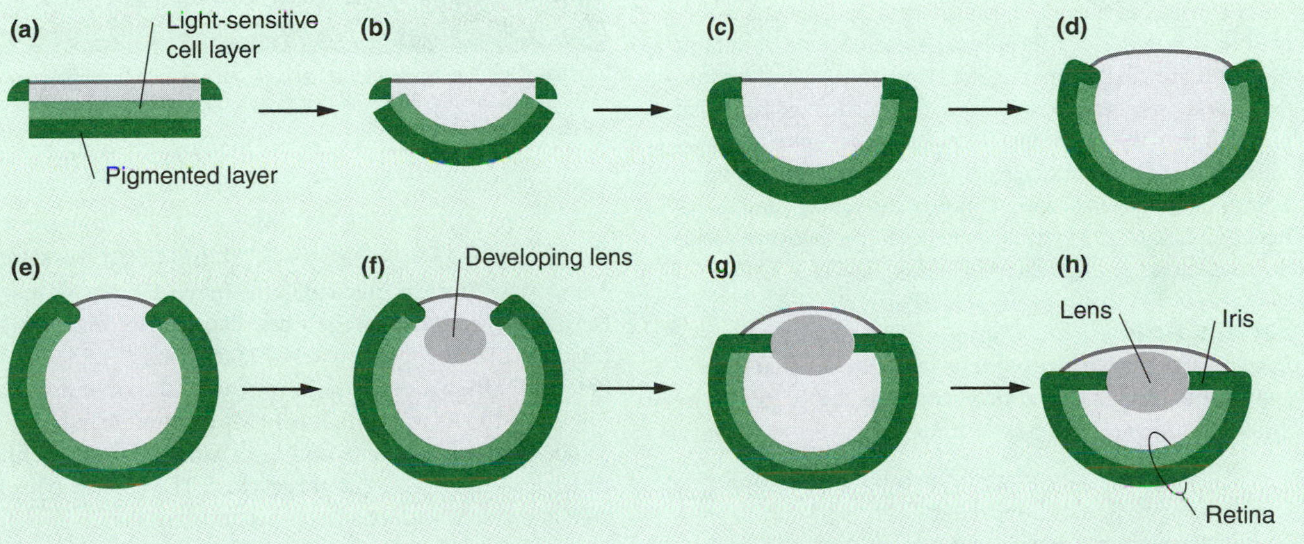

that, although Darwin's approach accounted for the evolution of a particular species in time, it did not easily account for the multiplication of species in geographical space. What explains the origin of many new species rather than the transformation of one old species? Furthermore, argued Moritz Wagner (1813–1887), among others (especially John Gulick, see below), Darwin did not even fully explain the evolution of a single species into a single new species. How could a new species possibly evolve in the same locality as its parents?

> Free crossing of a new variety with the old unaltered stock will always cause it to revert to the original type . . . Free crossing, as the artificial selection of animals and plants uncontestably teaches, not only renders the formation of new races impossible, but invariably destroys newly formed individual varieties.

Missing from Darwin's 1859 argument was a strong emphasis on the barriers that prevent exchange of hereditary material between different groups so that each such isolated group can follow its own evolutionary path.

Because Darwin did not emphasize isolation among groups as a primary cause for evolution, he also dismissed the notion that sterility among separately evolved groups might

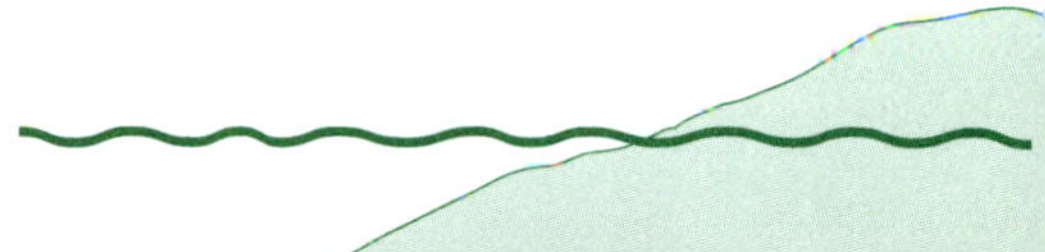

be beneficial. As Wallace (1889) showed, it would be advantageous for isolated populations, each with its special adaptations, to produce sterile hybrids when they meet, because sterility would allow each group to maintain its specific adaptations without dilution. Darwin insisted that sterility was primarily accidental. This view blocked him from explaining the almost universal prevalence of sterility among hybrids and from using this important isolating barrier to account for the divergent evolution of closely related species.

Although Darwin knew isolation could be important in helping a population evolve, he thought that it was more essential to establish that speciation could occur through time without isolation. It therefore remained the task of others to explore the role of isolation in forming species (see Chapter 24). Those others included John Gulick (1832–1933), a missionary and evolutionary biologist much influenced by Darwin. Although almost unknown today, during the late nineteenth and early twentieth centuries Gulick was one of the world's most well-known and influential evolutionary biologists. Gulick's collection (during the early 1850s) of over 200 species of land snails of the genus *Achatinella* from Oahu, Hawaii (**Fig. 3-3**),[5] analysis of their restricted geographical distributions — often to a single valley or region with a valley — and publication of his results beginning in 1872, provided what many considered the missing mechanism in Darwin's theory of evolution, namely **geographical isolation**, whereby one or more individuals becomes isolated from the rest of the population and their descendants subsequently diverge from that population to the extent that they become separate species. A century later, Ernst Mayr called this process the **founder effect,** acknowledging Gulick's origination of the concept and recognizing Gulick as, "the first author to develop a theory of evolution based on random variation" (Mayr, 1988a, p. 139).

FIGURE 3-3 Tree snails (*Achatinella mustalina*) on a leaf of a guava tree in the rainforest of Mount Tantalus on the island of Oahu. (Photo courtesy of Amy Tsuneyoshi.)

Earth's Age

Essential to Darwin's argument was a belief that Earth's age extended beyond anything ever proposed before. As he pointed out in an 1844 essay:

> The mind cannot grasp the full meaning of the term of a million or hundred million years, and cannot consequently add up and perceive the full effects of small successive variations accumulated during almost infinitely many generations.

This emphasis on evolution taking a long time ran counter to the time spans usually given at that time. The **heliocentric theory**, in which the sun was the center of the universe, tied the origin of Earth to the sun. Newton had calculated that a sphere the size of Earth would take about 50,000 years to cool down to its present temperature. Because even such a short period contradicted the 5,000 or so years of history allowed in the Judeo-Christian Bible, Newton piously rejected these calculations. Buffon, in contrast, calculated approximately 75,000 years (74,832 to be precise) for the age of Earth (in an unpublished manuscript, Buffon's estimate was as high as 10 My [Million Years]): "For 35,000 years our globe has only been a mass of heat and fire which no sensible being could get close. Then, for 15,000 or 20,000 years, its surface was only a universal sea."

In Darwin's time, William Thomson (Lord Kelvin, 1824–1907) reassessed the temperature gradients observed in mine shafts, the conductivity of rocks, and the presumed temperature and cooling rate of the sun to calculate the total age of Earth's crust at about 100 My but suggested that only the last 20 to 40 My could have been sufficiently cool for life to exist. Although similar to Phillips' (1860) calculation of 96 My (**Box 3-2**), in Kelvin's view, this number was still too small to account for the Darwinian evolution of organisms.

Tradition has it that Darwin had no answer to these various calculations, but as discussed in Box 3-2, he both did his own calculations and included them in the first and second edition of *The Origin.*

[5] You can read more about evolution in Hawaii on the Internet at http://www.nap.edu/catalog/10865.html#toc. See Hall (2006a,b) for Gulick's important contribution on geographical isolation.

BOX 3-2 Darwin and Earth's Age

Although in the traditional view, Darwin could not deal with the problem of the long geological periods of time needed for life to evolve, he did made some calculations based on rates of erosion that demonstrated the enormity of geological time. This estimation was not the 100 million years estimated by Kelvin, but an even longer geological history: 300 million years.

It is an interesting story. Darwin included his calculations in the first and second editions of *The Origin*, but removed them from later editions (over objections from a number of his scientific friends). The appropriate section from the second edition (emphases added) reads:

> I am tempted to give one other case, the well-known one of the denudation [erosion] of the Weald [the rolling countryside between the South and North Downs in Kent and adjacent counties in southern England], though it must be admitted that the denudation of the Weald has been a mere trifle, in comparison with that which has removed masses of our Palaeozoic strata, in parts ten thousand feet in thickness, as shown in Prof. Ramsay's masterly memoir on this subject. Yet it is an admirable lesson to stand on the North Downs and to look at the distant South Downs; for, remembering that at no great distance to the west the northern and southern escarpments meet and close, one can safely *picture to oneself the great dome of rocks which must have covered up the Weald within so limited a period as since the latter part of the Chalk formation.* The distance from the northern to the southern Downs is about 22 miles, and the thickness of the several formations is on an average about 1100 feet, as I am informed by Prof. Ramsay. But if, as some geologists suppose, a range of older rocks underlies the Weald, on the flanks of which the overlying sedimentary deposits might have accumulated in thinner masses than elsewhere, the above estimate would be erroneous; but this source of doubt probably would not greatly affect the estimate as applied to the western extremity of the district. If, then, we knew the rate at which the sea commonly wears away a line of cliff of any given height, we could measure the time requisite to have denuded the Weald. This, of course, cannot be done; but *we may, in order to form some crude notion on the subject, assume that the sea would eat into cliffs 500 feet in height at the rate of one inch in a century.* This will at first appear much too small an allowance; but it is the same as if we were to assume a cliff one yard in height to be eaten back along a whole line of coast at the rate of one yard in nearly every twenty-two years. I doubt whether any rock, even as soft as chalk, would yield at this rate excepting on the most exposed coasts; though no doubt the degradation of a lofty cliff would be more rapid from the breakage of the fallen fragments. On the other hand, I do not believe that any line of coast, ten or twenty miles in length, ever suffers degradation at the same time along its whole indented length; and we must remember that almost all strata contain harder layers or nodules, which from long resisting attrition form a breakwater at the base. Hence, under ordinary circumstances, I conclude that for a cliff 500 feet in height, *a denudation of one inch per century for the whole length would be an ample allowance. At this rate, on the above data, the denudation of the Weald must have required 306,662,400 years; or say three hundred million years.* But perhaps it would be safer to allow two or three inches per century, and this would reduce the number of years to 150 or 100 million years. So that it is not improbable that a longer period than 300 million years has elapsed since the latter part of the Secondary period *(Darwin, On the Origin of Species*, 2d ed., p. 285).

In 1860, in his presidential address to the Geological Society of London, the geologist, paleontologist and Reader in Geology at Oxford, John Phillips (1800–1874) provided his own estimate of Earth's age on the basis of *rates of sedimentation* by the Ganges River (Darwin had used *rates of erosion*) and the thickness of the known geological strata. On this basis, Phillips (1860) came up with an estimate of 96 My. Phillips, who did not support evolution, was quick to point out Darwin's "abuse of arithmetic."

Darwin's estimates of geological times caused him considerable anguish. How could a piece of the English countryside be three times older than Kelvin's estimate of the age of Earth itself, especially when Kelvin's calculations were based on physical rates (see text) and Darwin's on "back of the envelope" calculations? We see this in two letters written to his close friend and confidant, Charles Lyell. On January 10, 1860, responding to suggestions for changes to the first edition made by Lyell, Darwin wrote (emphasis added): ". . . It is perfectly true that I owe nearly all the corrections to you, and several verbal ones to you and others; I am heartily glad you approve of them, *as yet only two things have annoyed me*; those confounded millions of years (not that I think it is probably wrong), and my not having (by inadvertence) mentioned Wallace towards the close of the book in the summary, not that any one has noticed this to me." On November 20th of the same year, in a footnote to a letter to Lyell concerning changes to the next edition of his book, Darwin wrote, "The confounded Wealden Calculation to be struck out, and a note to be inserted to the effect that I am convinced of its inaccuracy from a review in the *Saturday Review*, and from Phillips, as I see in his Table of Contents[a] that he alludes to it."

Darwin never referred to the age of Earth again!

[a] The Table of Contents referred to by Darwin is the printed version of Phillips' lecture published in 1860, entitled, *Life on the Earth: Its Origin and Succession* (Phillips, 1860, reprinted 1980).

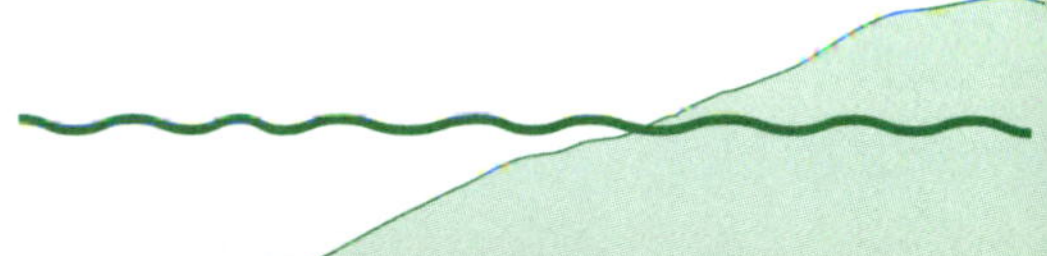

Support for Darwin

Although critics opposed Darwin, and although he and his supporters did not always have the information to answer each objection satisfactorily, Darwin's works made the evolution of species an acceptable concept. One important reason was that although Darwin presented many mechanisms — the struggle for existence, natural selection, divergence between species and the improvement of adaptations — each relied on natural processes and could be supported by observations. These natural processes were in strong contrast to previous, more speculative, evolutionary theories such as Lamarck's, which were tied to nonmaterial agents impossible to observe.

Another attractive feature of Darwin's theory was his expansion of the role of biology to include the study of relationships among all living creatures, including humans, formerly thought to be divine, and therefore separate. From anthropology to botany to paleontology to zoology, Darwinism opened new lines of thought and new areas of investigation. What are the relationships among different kinds of cells, different parts of cells, different flowers? How did the remarkable range of pollination mechanisms in orchids evolve? How did bone structures change? What are the steps in the evolution of circulatory systems, nervous systems? Why do some species mimic others? How did sterile insect castes such as worker bees evolve? What accounts for the geographical distribution of specific organisms? Although each topic demanded separate techniques and study, all sprang from an evolutionary source that rational, understandable mechanisms could explain. By offering an overall view of adaptation and evolution, Darwin provided a major step forward in binding all of biology together. Consequently, many lines of direct and indirect evidence began to accumulate in support of an evolutionary view. Some are outlined briefly in the sections that follow.

Sometimes we forget or perhaps are unaware of the scope of the evidence Darwin brought to bear on his theory of evolution by natural selection. The evidence is staggering for two reasons: its shear weight, and the range of topics and subjects Darwin used. Indeed, many of Darwin's lines of evidence were drawn from fields that now are sciences in their own right. A list is provided in **Table 3-1.** Keep this list in mind as you read through the sections that follow.

TABLE 3-1 A summary of the major lines of evidence used by Charles Darwin in presenting his theory of evolution by natural selection[a]

Line of Evidence	Chapter
Artificial selection	1
Comparative anatomy	14
Comparative embryology	14
Embryology	14
Fossils and the geological succession	11
Geographical distribution of organisms	12, 13
(the) Geological record	10
Homology	14
Hybridization	9
Instinct	8
Laws of variation	5
Morphology	14
Natural selection	4, 6, 7
(the) Struggle for existence	3
Survival of the fittest	4
Systematics	12, 13
Variation in nature	2
Variation under domestication	1
Vestiges and rudiments	14

[a] Chapters in current text in which each line of evidence is developed are indicated.

Geographical Distribution and Systematics

Although the evidence was indirect, after Darwin it seemed clear that the gradation of different organisms observed through attempts at classification, whether from simple to complex or from one type to another, could most reasonably be explained by evolutionary relationships (**Fig. 3-4**).

Many became aware that groups of organisms that are evolutionarily related are usually connected geographically. Large geographical barriers such as oceans and mountain ranges isolate groups from one another and led to considerable differences among the separated groups. Colonizers that transcend such barriers often become the ancestors of entirely new groups. This showed up most graphically in the wide evolutionary radiation of species that descended from finches reaching the Galapagos Islands. Beginning with what was probably an ordinary mainland finch, new kinds of finches evolved in the Galapagos that could function in habitats left vacant by other bird species (**Fig. 3-5**; see Chapter 23).

The name given to this process, **adaptive radiation**, signifies the rapid evolution of one or a few forms into many different species occupying a variety of habitats within a new geographical area. The radiation of marsupial mammals in Australia (**Fig. 3-6**) shows how protection from competition with placental mammals by the isolation of a continent can lead to an array of species with widely divergent functions,

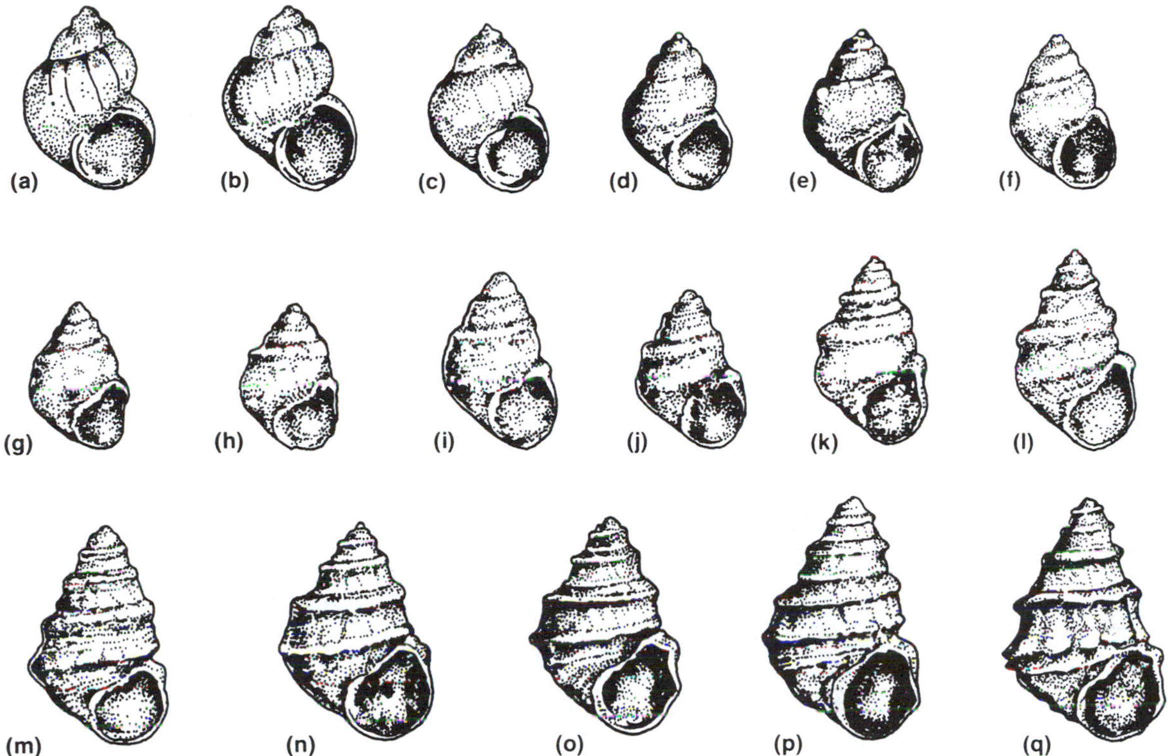

FIGURE 3-4 A nineteenth century illustration of obvious evolutionary relationships among fossil species of the mollusk *Paludina*, ranging from the oldest form *P. neumayri* (a), to the youngest *P. hoenesi* (q). To Darwin and many others, differences of this kind "blend into each other in an insensible series; and a series impresses the mind with the idea of an actual passage." (From Romanes, G. J., 1910. *Darwin, and After Darwin*. Open Court, Chicago.).

ranging from herbivores to carnivores, and paralleling placental mammals on other continents.[6]

Comparative Anatomy

Comparative anatomy is the study of comparative relationships among anatomical structures in the adults of different species. For some decades after Darwin, comparative anatomy and especially comparative (evolutionary) **embryology**— the study of comparative relationships among anatomical structures in the embryos of different species — became the most popular way to study evolution. Comparative anatomy followed the logic that organisms with shared structures derived from a common group of ancestors, whereas organisms with unlike structures represented divergent evolutionary pathways. A search for evolutionary relationships made it possible to trace, especially in vertebrates, many stepwise changes in bones, muscles, nerves, organs and blood vessels (**Fig. 3-7**). Such studies made clear that as each species and group of species evolved, previously inherited structures were modified in entirely new ways.

[6] Schulter (2000) provides an in-depth analysis of the links between ecology and adaptive radiation, while Selden and Nudds (2004) examine the evolution of ecosystems themselves.

Comparative anatomists followed the logic that organisms with shared structures had arisen from a common group of ancestors, whereas organisms with unlike structures represented divergent evolutionary pathways. Fascinating for Darwin and other comparative anatomists, and providing dramatic demonstrations of the non-randomness of natural selection — or at least of the results of natural selection — were examples of more or less distantly related groups of organisms attaining a similar morphology. These included marsupials and placental mammals (Fig. 3-6), penguins and auks, bats and birds, and so forth. Adaptive radiation, natural selection, homology and convergence formed a powerful tetrad explaining relationships between more and more distantly related organisms.

Homology

Derived from terminology introduced in the 1840s by the comparative anatomist Richard Owen (1804–1892), organs regarded as the same, though serving different functions, are **homologous**. For example, studies of similar bones and muscles — the humerus in the upper arm, radius and ulna in the lower arm, digits — showed that the forelimbs of widely different vertebrates were homologous as forelimbs, albeit with such different functions as walking, flying

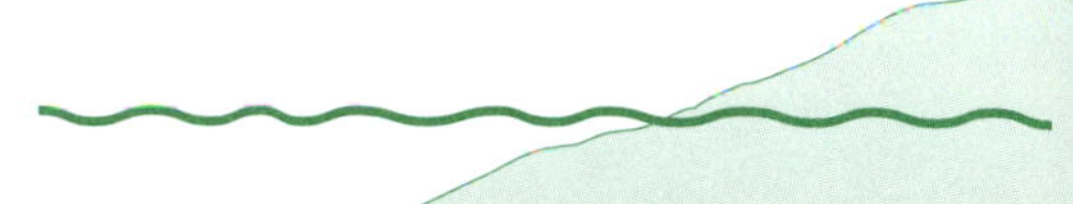

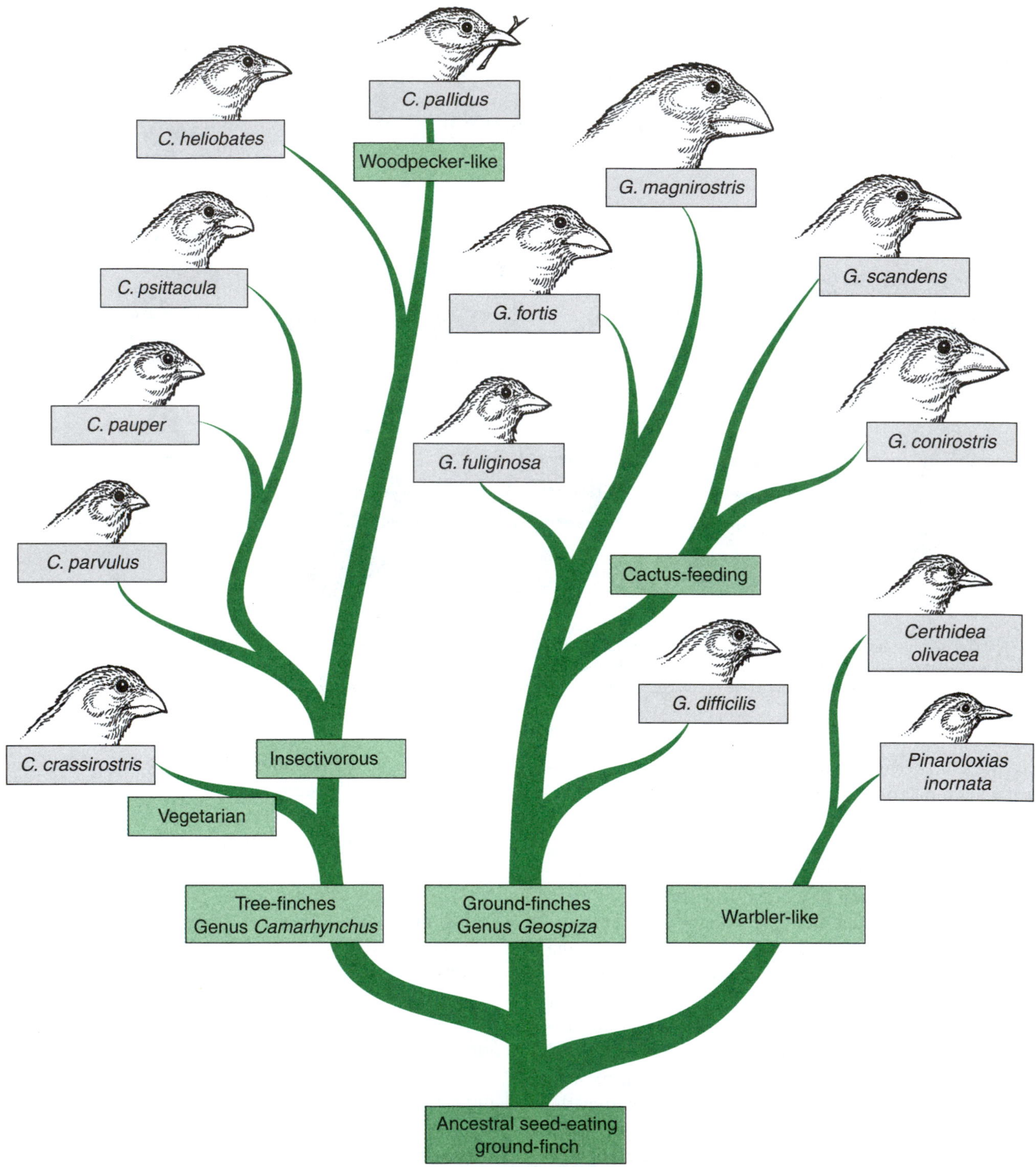

FIGURE 3-5 Evolutionary tree of Darwin's finches showing adaptations of the beaks of different species. A molecular study by Sato and coworkers (1999) supports some aspects of this phylogeny. Using comparisons among mitochondrial DNA sequences (see Chapter 12), the molecular study distinguishes between tree finches and ground finches, but shows that distinctions among members within each group have not yet been firmly established. This molecular study also indicates that these finches are all descended from a single species, now identified as a warbler-type, "dull-colored grassquit" (genus *Tiaris*). (Adapted from Lack, D., 1947. *Darwin's Finches.* Cambridge University Press, Cambridge, England.)

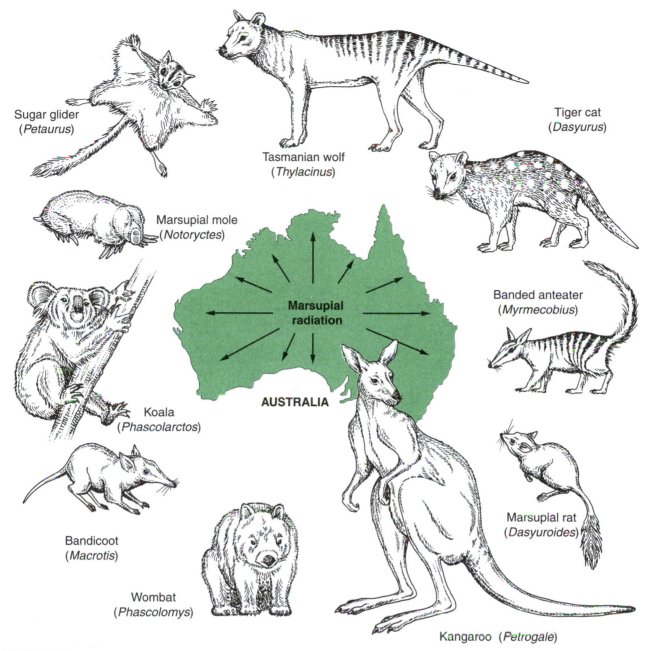

FIGURE 3-6 Adaptive radiation of Australian marsupial mammals. Many divergent forms evolved independently of but often in parallel to changes occurring among placental mammals on other continents. The striking similarity between some of these marsupial mammals and placental mammals arises because selection for survival in similar habitats can lead to similar adaptations as well as parallel or convergent evolution (see also Fig. 3-7).

and swimming (**Figs. 3-8** and **3-9**). In contrast, organs that perform the same function in different groups but do not share a similarity of structure are **analogous**. For example, the wings of bats, which are built around a bony skeleton, and the wings of insects, which are not, do not show a common underlying structural plan.

After Darwin, it was realized that homologues were found in organisms with a shared evolutionary history, in particular, those with a shared common ancestor. Analogues were found in animals without a common ancestor, although if we go far enough back in evolution all animals share a common ancestor. The evolution of different organisms, or parts of organisms, in similar directions — **convergent evolution (convergence)** — indicates that selection for similar features in different evolutionary lineages can, and often does, lead to functionally similar anatomical structures (**Fig. 3-10**).

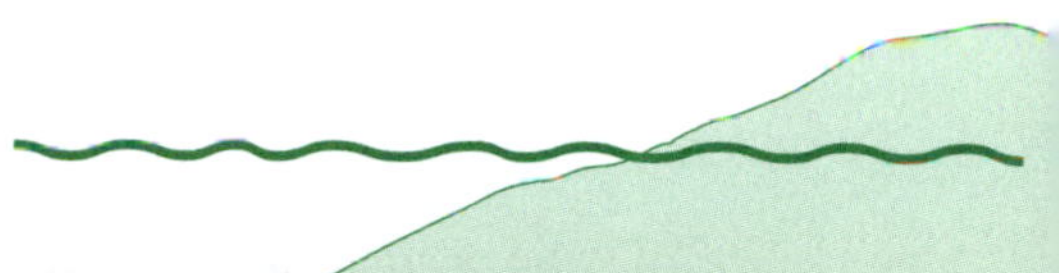

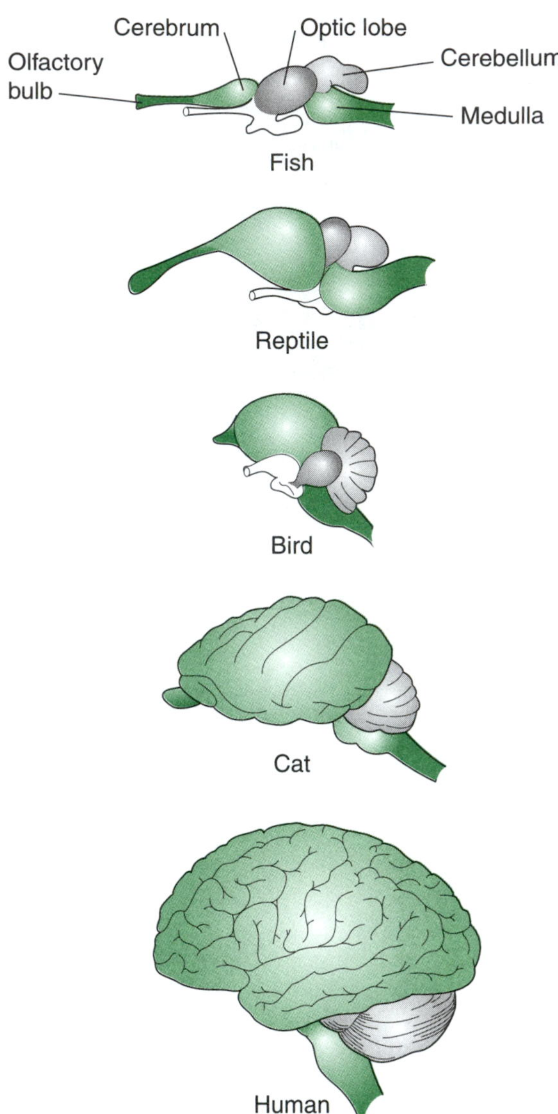

FIGURE 3-7 Redrawn nineteenth century diagrams of the brains of representative vertebrates. Homologous structures (labeled for the fish) are the cerebellum, cerebrum, medulla, olfactory bulbs and optic lobes. In humans, overgrowth of the cerebrum covers the olfactory bulbs and optic lobes. (Adapted from LeConte, A., 1888. *Evolution, Its Nature, Its Evidences, and Its Relation to Religious Thought.* Appleton, New York.)

Careful anatomical dissections and comparisons provided the criteria for constructing detailed **evolutionary trees**, an activity that preoccupied late nineteenth-century biologists as they sought evidence for Darwin's theory of the relatedness of all organisms and for descent with modification. Such comparisons were complicated by any change in design and/or function of organs within a group (the flippers of whales, for example) and by loss of design and/or function over time (the loss of limbs in snakes, for example). In the first attempt to redefine homology in an evolutionary context in 1870, Ray Lankester, an active and vociferous participant in tree building, summed up the prevailing view, which was of a single Tree of Life[7]:

> "In [the various] kinds of animals and plants [we see] simply the parts of one great genealogical tree, which have become detached and separated from one another in a thousand different degrees, through the operation of the great destroyer, Time . . ."

Vestiges

Of considerable interest to comparative anatomists were structures that seemed to have lost some or all of their ancestral functions. From an evolutionary viewpoint, biologists could explain such rudimentary or **vestigial organs** as arising because an organism adapting to a new environment usually carries along some previously evolved structures that are no longer necessary. According to the principles of natural selection, individuals that devote less energy to the specific elaboration and maintenance of such extraneous structures would be more reproductively successful than individuals that maintain them. Moreover, a structure that is no longer necessary might interfere with the functioning of new adaptations or even could give rise to an entirely new evolutionary function. Among new innovations are rudiments of bones of the reptilian lower jaw apparatus that evolved into the three[8] middle ear ossicles that carry sound to the inner ear in mammals (see Fig. 19-7), and the assumption of new olfactory and tactile functions by former optical neural pathways in blind cave animals.

As time went on, obsolete structures would tend to diminish, showing only traces of their former size and function. Examples of these are the rudiments of hind limb bones in the whale and snake species shown in **Figure 3-11**. Indeed, the earliest snakes had legs.[9] The presence of rudiments of limb skeletons indicates that early stages of limb development must occur in "limbless" vertebrates. Indeed they do, as demonstrated by the presence of hind limb buds early in development of whales and dolphins, which as adults lack hind limbs and transform the forelimbs into flippers (**Fig. 3-12**).

The presence of reduced eye stalks in blind, cave-dwelling crustaceans also speaks to an evolutionary process

[7] Also see Box 12-4 and Chapter 14 for recent efforts to construct (or reconstruct) the Tree of Life.

[8] It is not always appreciated that early reptiles had six to eight bones in their lower jaws. Three evolved into middle ear ossicles, others persisted as bones of the lower jaw in more derived reptiles, and yet other bones were lost.

[9] To speak of *snakes with legs* may seem to be an oxymoron. The direct ancestors of modern-day snakes were indeed limbed, snake evolution involving the loss of the limbs along with elongation of the body. At least four genera of limbed snakes are now known as fossils: *Haasiophis, Pachyrhachis, Eupodophis and Najash.*

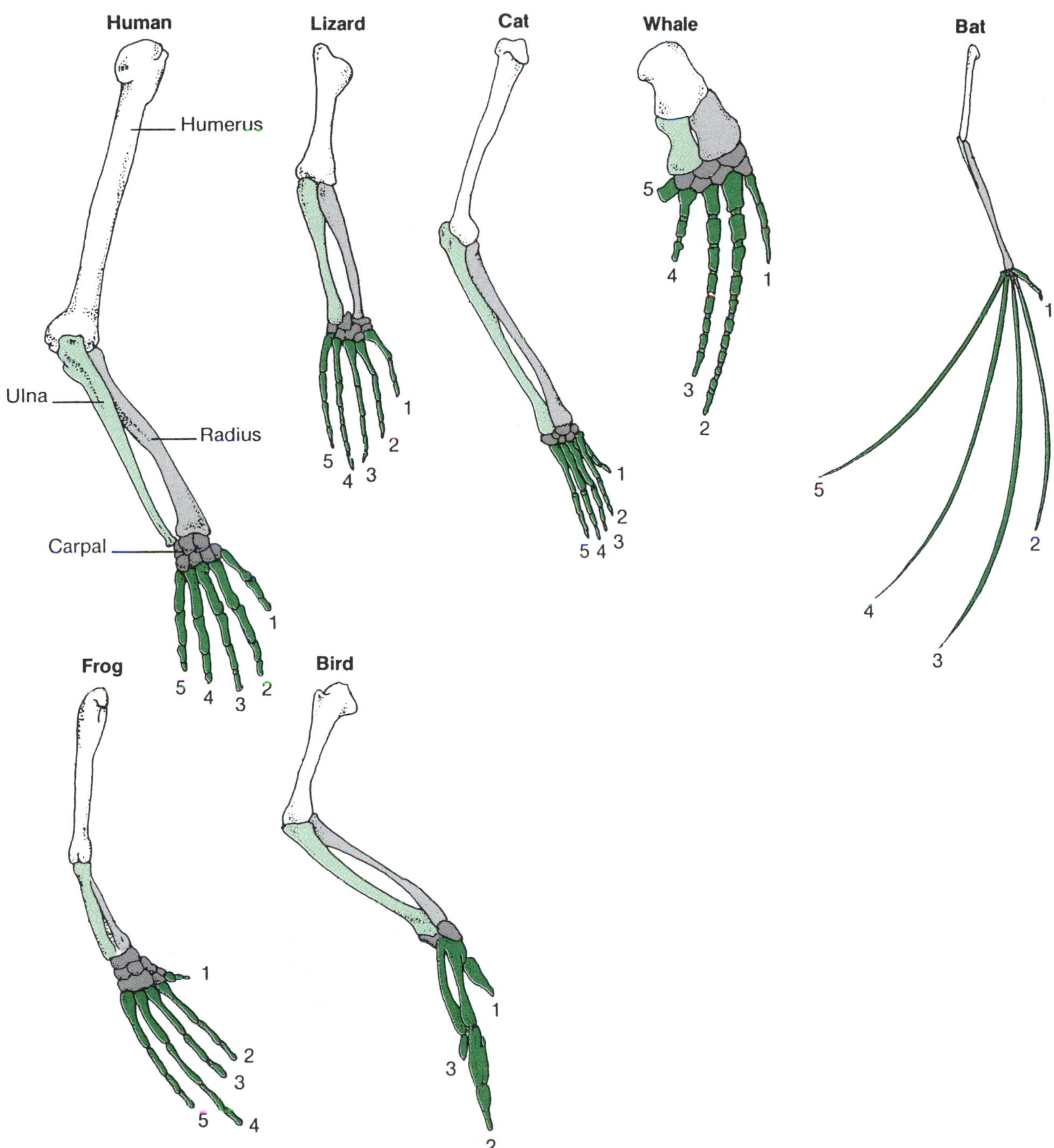

FIGURE 3-8 Skeletal structures of the forelimbs in various representative tetrapods (literally animals with four pods or feet) to show the homology among bones at different proximo-distal levels in the limbs (where proximal is near the body and distal farthest from the body), homology that is preserved despite changes in size, shape and function of the forelimbs during the course of evolution. Compare, for example, the single upper arm bone (the humerus in white) in each limb, or the distal digits in green. Even when digit number varies, homology of the digits remains (Hall, 2005a). As Darwin noted, "What can be more curious than that the hand of man formed for grasping, that of a mole, for digging, the leg of a horse, the paddle of a porpoise and the wing of a bat, should all be constructed on the same pattern and should include similar bones and in the same relative positions?"

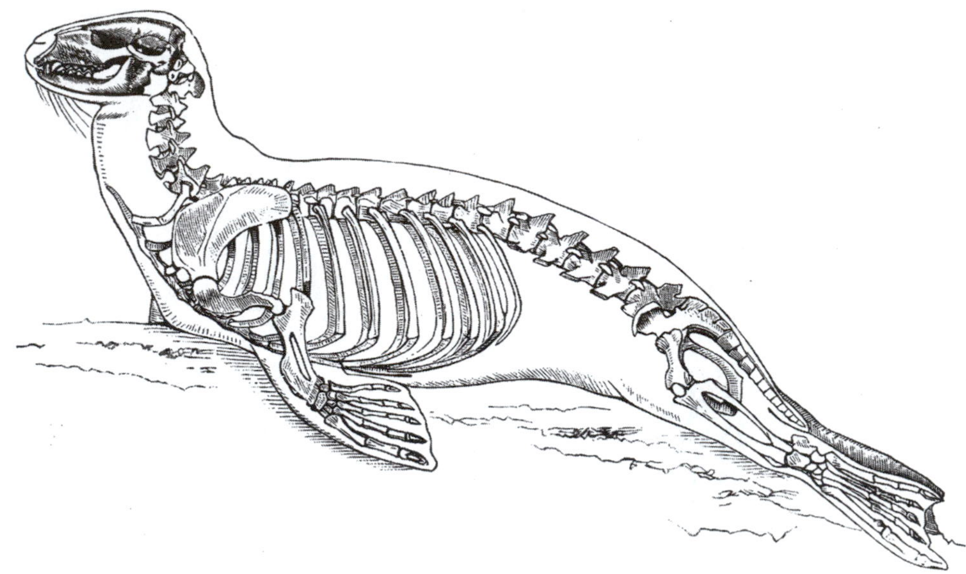

FIGURE 3-9 The skeleton of a harbor seal, *Phoca vitulina*, showing how readily the bones in the flippers can be homologized to the limb bones of other vertebrates (shown in Fig. 3-8). (From Romanes, G. J., 1910. *Darwin, and After Darwin*. Open Court, Chicago.)

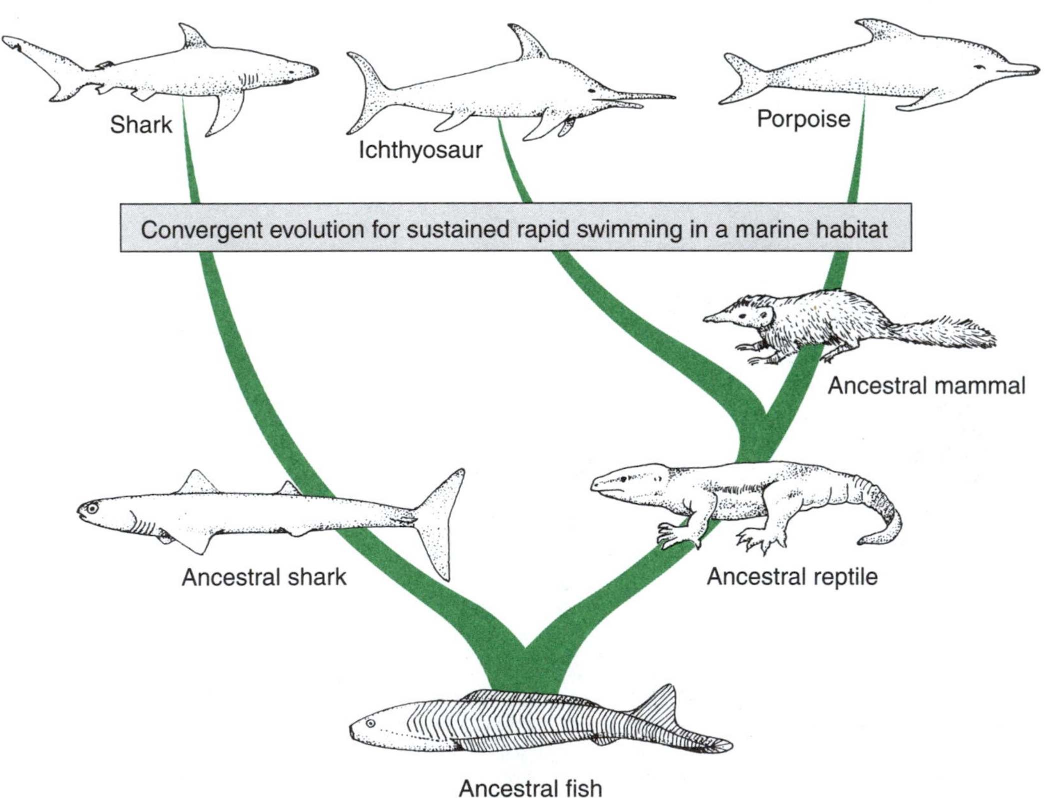

FIGURE 3-10 Convergent evolution in three marine predators with different ancestries — shark (cartilaginous fish), ichthyosaur (an extinct reptile), and porpoise (a placental mammal) — because similar adaptations for rapid movement through water were independently selected in each of the lineages.

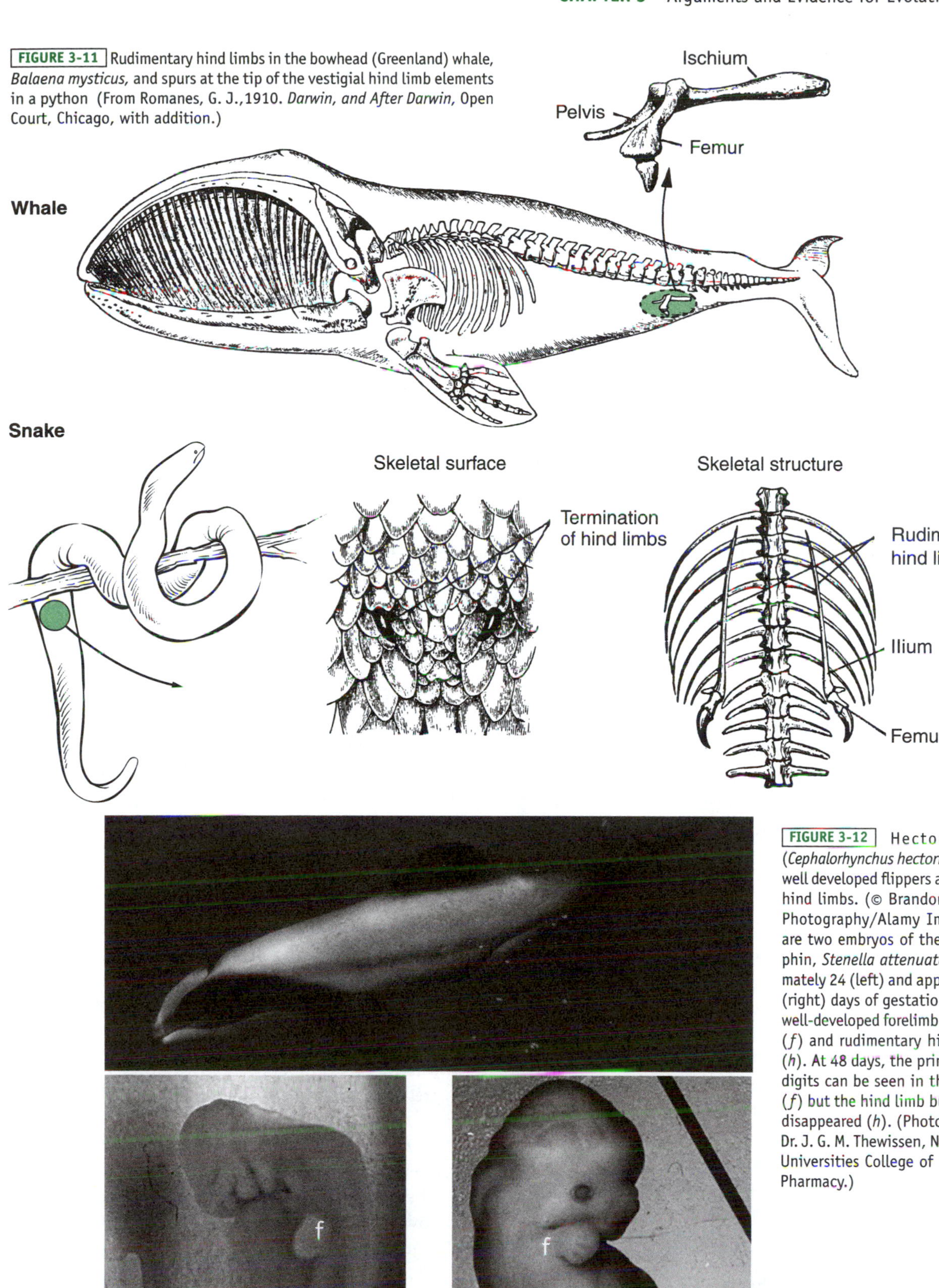

FIGURE 3-11 Rudimentary hind limbs in the bowhead (Greenland) whale, *Balaena mysticus,* and spurs at the tip of the vestigial hind limb elements in a python (From Romanes, G. J.,1910. *Darwin, and After Darwin,* Open Court, Chicago, with addition.)

FIGURE 3-12 Hector's dolphin (*Cephalorhynchus hectori*) showing the well developed flippers and absence of hind limbs. (© Brandon Cole Marine Photography/Alamy Images) Below are two embryos of the spotted dolphin, *Stenella attenuata,* at approximately 24 (left) and approximately 48 (right) days of gestation to show the well-developed forelimb (flipper) buds (*f*) and rudimentary hind limb buds (*h*). At 48 days, the primordia for the digits can be seen in the flipper bud (*f*) but the hind limb bud has almost disappeared (*h*). (Photos courtesy of Dr. J. G. M. Thewissen, Northeast Ohio Universities College of Medicine and Pharmacy.)

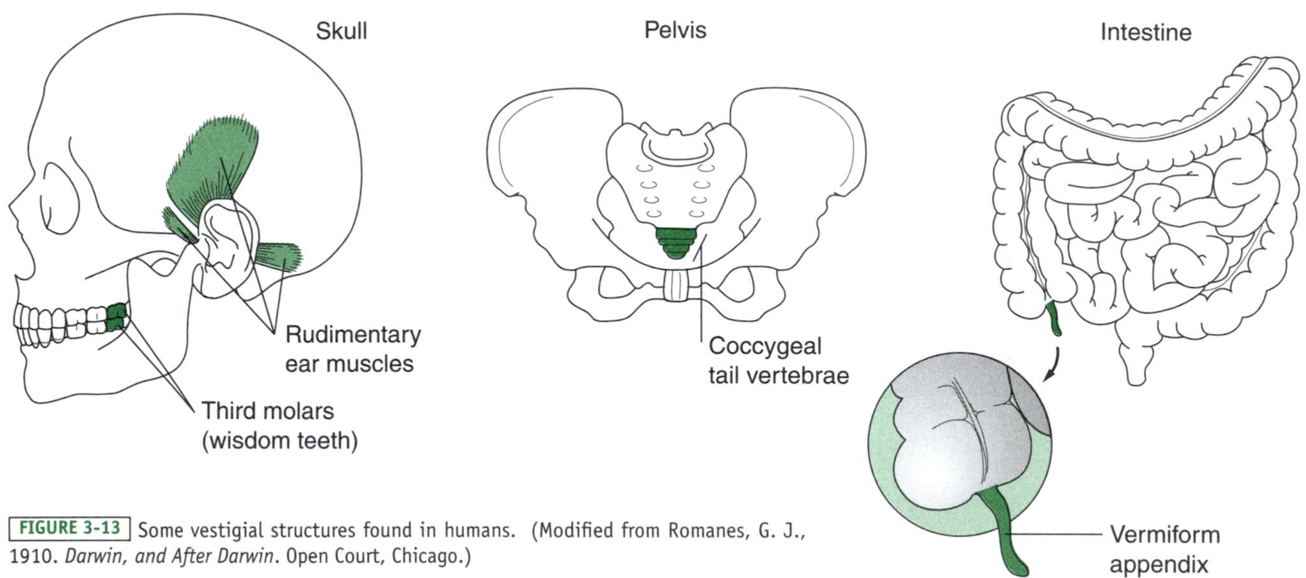

FIGURE 3-13 Some vestigial structures found in humans. (Modified from Romanes, G. J., 1910. *Darwin, and After Darwin*. Open Court, Chicago.)

by which obsolete structures gradually became rudimentary. Obvious vestigial organs in humans and the other great apes are reduced tail bones (*os coccyx*), remnants of a few tail muscles, the nictitating membrane of the eye, the appendix of the cecum, rudimentary body hair and wisdom teeth (**Fig. 3-13**). A number of vestigial structures in humans indicate not only the obsolescence of organs but a relationship to other mammals and primates. For example, muscles of the external ear and scalp common to many mammals are rudimentary and often nonfunctional in humans. Inflection of the feet for grasping, along with divergence of the great toe — a primate trait that is often expressed in human infants — degenerates in adults.

Embryology

Early in the nineteenth century, Karl von Baer (1792–1876) discovered the remarkable similarity among the embryos of vertebrates that are quite different from one another as adults. He generalized his findings into a "law": early embryos of related species bear more common features than do later, more specialized developmental stages. Darwin and others, especially Ernst Haeckel (1834–1919), built on von Baer's observations and considered the early stages of development more conservative or evolutionarily stable than adult stages (**Fig. 3-14a**, top row). von Baer's views were comparative and taxonomic but not evolutionary. Consequently, this *building on* represented a major shift to an evolutionary explanation. As Darwin stated:

> In two groups of animals, however much they may at present differ from each other in structure and habits, if they pass through the same or similar embryonic stages, we may feel assured that they have both descended from the same or nearly similar parents, and are therefore in that degree closely related. Thus, community in embryonic structure reveals community of descent.

Haeckel, who was a major proponent of evolution in Germany, further developed and popularized this concept into the **biogenetic law**:

> Ontogeny [development of the individual] is a short rapid recapitulation of phylogeny [the ancestral sequence] . . . The organic individual repeats during the swift brief course of its individual development the most important of the form — changes which its ancestors traversed during the slow protracted course of their paleontological evolution according to the laws of heredity and adaptation (1866).

To Haeckel, this meant that early stages of embryonic development recapitulate adult ancestral forms, that is, the tadpole of a living frog recapitulates a tailed ancestor. This biogenetic law has been disputed widely, with most biologists considering Haeckel's law an extreme oversimplification that is especially evident:

- when applying the concept to features that show progressive structural or functional changes during the life cycle;
- when a feature is present at one but not at another life history stage; or
- when a feature found in a group or lineage has no obvious connections with a feature(s) in another group/lineage (the turtle shell, for example).

Turtle shells continue to baffle us, for although we understand something of their development, which is from a bud analogous to a limb bud (**Fig. 3-15**), no potential ancestors of turtles have a shell. Consequently, the relationship of turtles to other reptiles remains unresolved (see Chapter 18).

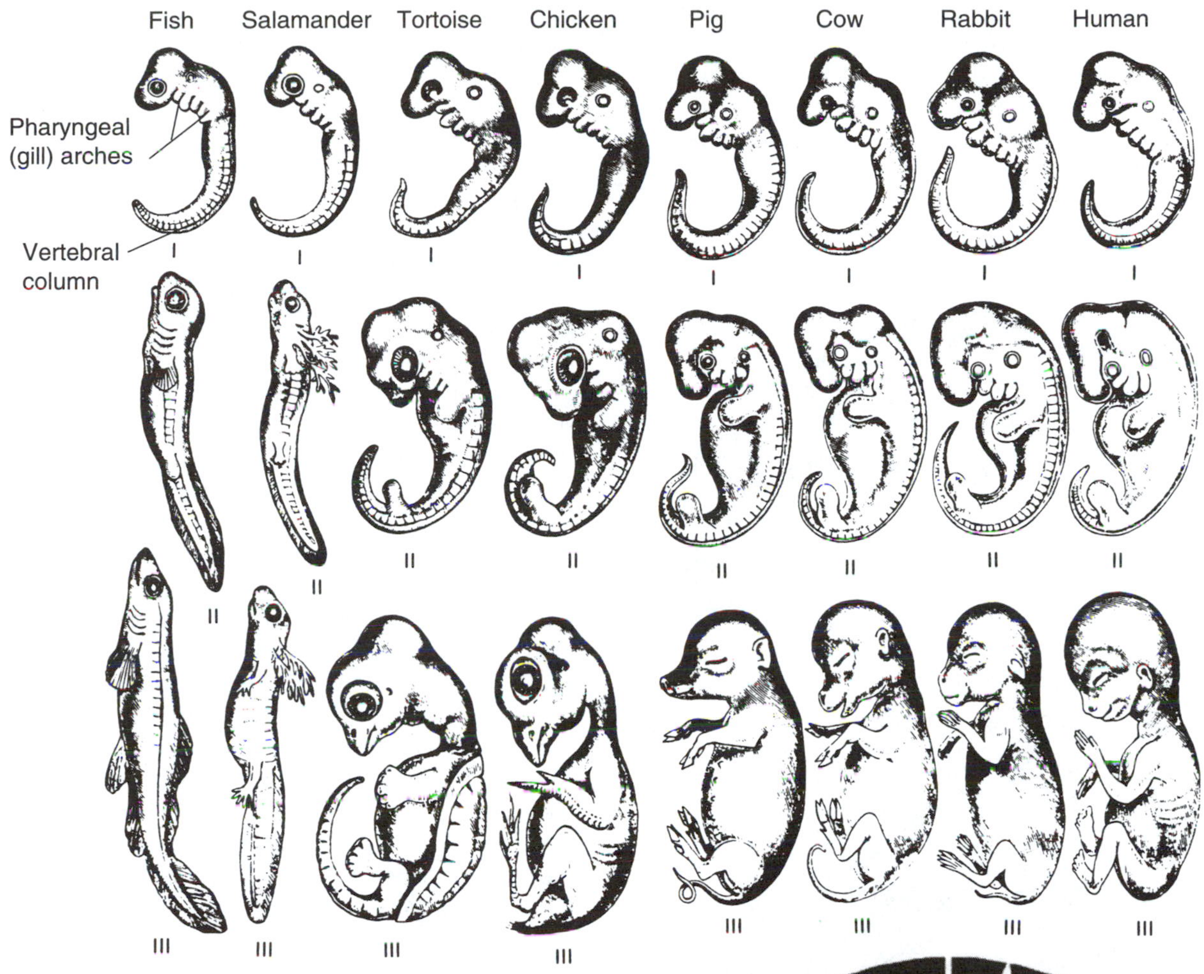

Fish Salamander Tortoise Chicken Pig Cow Rabbit Human

Pharyngeal (gill) arches

Vertebral column

I I I I I I I I

II II II II II II II II

III III III III III III III III

(a)

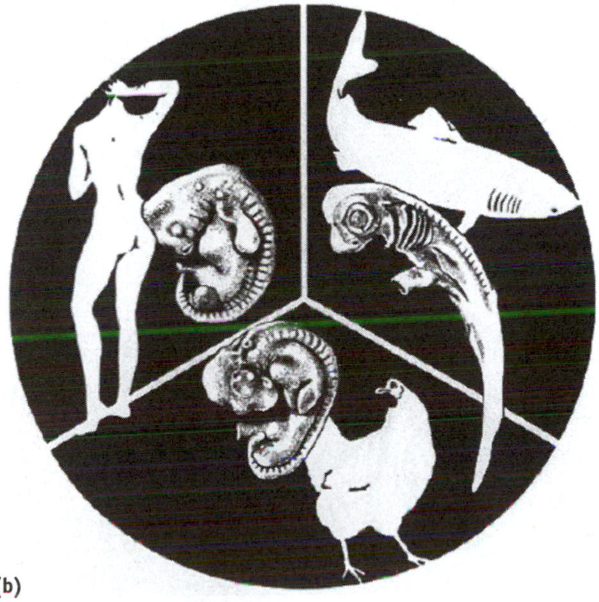

(b)

FIGURE 3-14 (a) Haeckel's classic illustration of different vertebrate embryos at comparable stages of development. Although Haeckel took some liberties in drawing these figures, the earlier stages are more similar to one another than later stages are to one another. Each embryo begins with a similar number of pharyngeal arches, paired eyes and internal structures such as a single midline heart (not shown). Some biologists call this a *phylotypic* stage (see also Fig. 13-9), representing the distinctive developmental substructure of the body plan shared by individuals in a particular phylum. In later stages of development, these and other structures are modified to yield the characteristic form of each subgroup. The embryos in the different groups have been scaled to the same approximate size so that comparisons can be made among them. (b) A hundred years later, we continue to recognize that early embryonic stages are more alike than are adults, depicted here as shark, chicken and human. (a from Romanes, G. J., 1910. *Darwin, and After Darwin*. Open Court, Chicago; b from de Beer, G. R., 1964. *Atlas of Evolution*. Nelson, London.)

According to current views, early stages of development recapitulate only early ancestral developmental stages. Organisms that share common descent make use of common underlying embryological patterns on which to build later but different adult patterns. Evidence from genetic, molecular and developmental biology and from the integration of evolutionary and developmental biology (evo-devo; see Chapter 13) provided strong support for this view.

The discovery that **juvenile** stages of ancestral organisms can be retained in the adult forms of their descendants — for

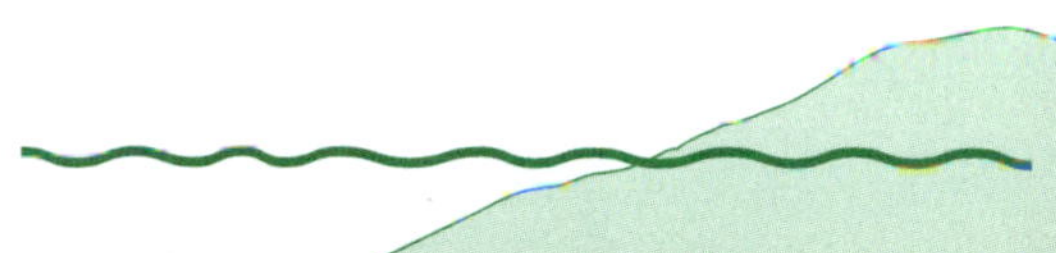

example, the preservation in adult humans of features found in juvenile apes (**neoteny**; see Gould, 1977, and Fig. 20-26) — directly contradicts the Haeckelian notion that descendants retain ancestral adult features. Embryological processes are more often retained in closely related species. Closely related organisms usually rely on common genetic agents able to produce characteristic developmental stages that have persisted for tens of millions of years (see Fig. 3-14). Examples of the evolutionary persistence of underlying patterns in human development include the pharyngeal (branchial) arches — often referred to (inappropriately) as "gill" arches (**Fig. 3-16**) — which serve as the embryonic basis for further development of head and throat structures, yet no functional gills ever form.

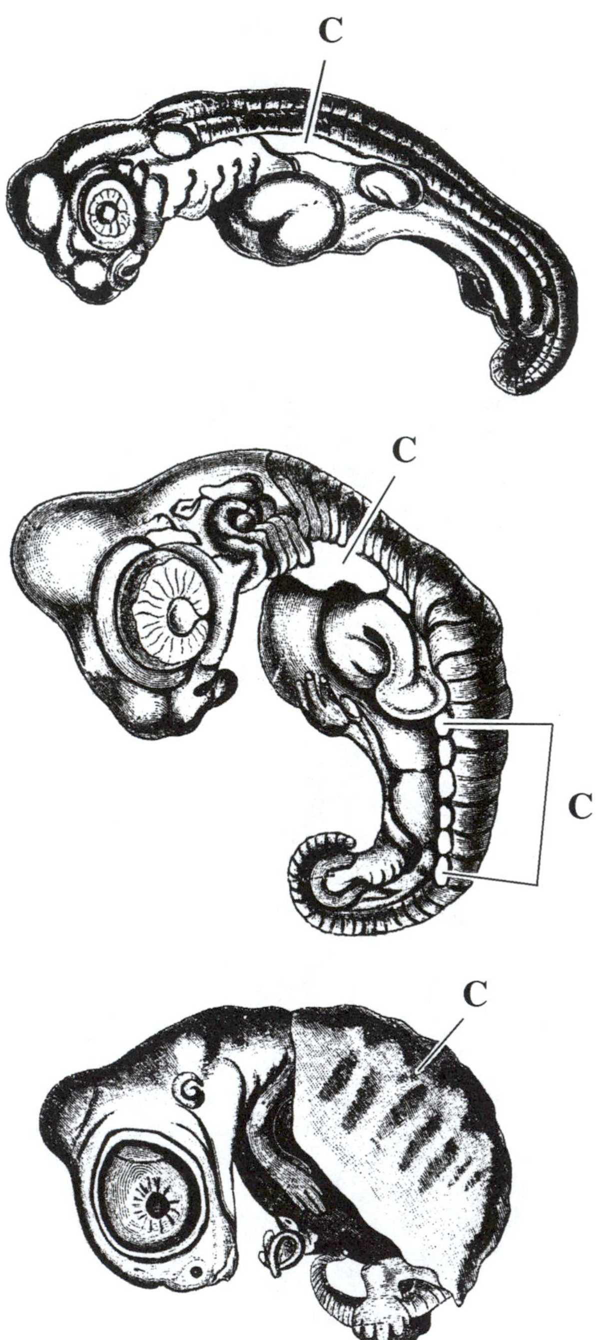

FIGURE 3-15 Three stages in embryonic development of the European turtle, *Chelonia midas*, showing the dorsal element of the shell or carapace (*C*) at the stage of initial outgrowth (top), when it has spread along the body axis (middle), and when it is more evidently recognizable as a shell (bottom). (Modified from Balfour, F. M., 1880–1881. *A Treatise on Comparative Embryology*. Two Volumes. Macmillan, London.)

Even the anatomical positions of some adult nerves, blood vessels and other structures are intelligible only in the light of their evolutionary developmental patterns. For example, as Figure 3-16a shows, each branch of the vagus nerve in fish runs through an arterial arch pierced by a gill slit. From studies in many vertebrates, we know that two of these vagal nerve branches eventually evolved in mammals for stimulating the larynx. The most anterior of these, called the *anterior laryngeal nerve,* loops around the third arterial arch (now the carotid artery), and a posterior nerve branch, called the *recurrent laryngeal nerve,* loops around the sixth arterial arch (Fig. 3-16b). In contrast to its function in fish, the left side of the sixth arterial arch in mammals is the ductus arteriosis, which is used embryonically to divert blood past the undeveloped lungs and directly to the body until birth, but which then atrophies and transforms into a pulmonary artery ligament. Because the ligamentous remnant of this old sixth arterial arch is close to the heart in adult mammals, to complete its circuit the left side of the recurrent laryngeal nerve must travel from the cranium to the thoracic cavity and back to the larynx. In mammals with long necks, such as giraffes, this extra loop is obviously much longer than it would be if nerve development were independent of evolutionary history.

All these embryonic stages and anomalies make sense to biologists only when we consider that humans and other terrestrial vertebrates had fishlike ancestors that provided them with some of their basic developmental patterns.

Fossils

Fossil remains are predominantly found in sedimentary rocks originally laid down by a succession of deposits in seas, lakes, riverbeds, deserts, and so on. Fossils occur in some areas but not in others (**Fig. 3-17**). Even in appropriate sedimentary environments, many dead organisms decompose before they fossilize or are destroyed by the erosion of sedimentary rocks even when they have fossilized (see Chapter 5).

Also, because the isolation of populations encourages and sustains their differences, we rarely find intermediate forms in the same place as the original forms. A complete evolutionary progression of fossils from most ancient to most recent has never been found in a single locality. Despite these difficulties, the search for fossils engendered considerable interest, because fossils provided hard evidence for evolution.[10]

[10] For analyses of the fossil record, see Fortey (2002), Milsom and Rigby (2003) and Benton (2004).

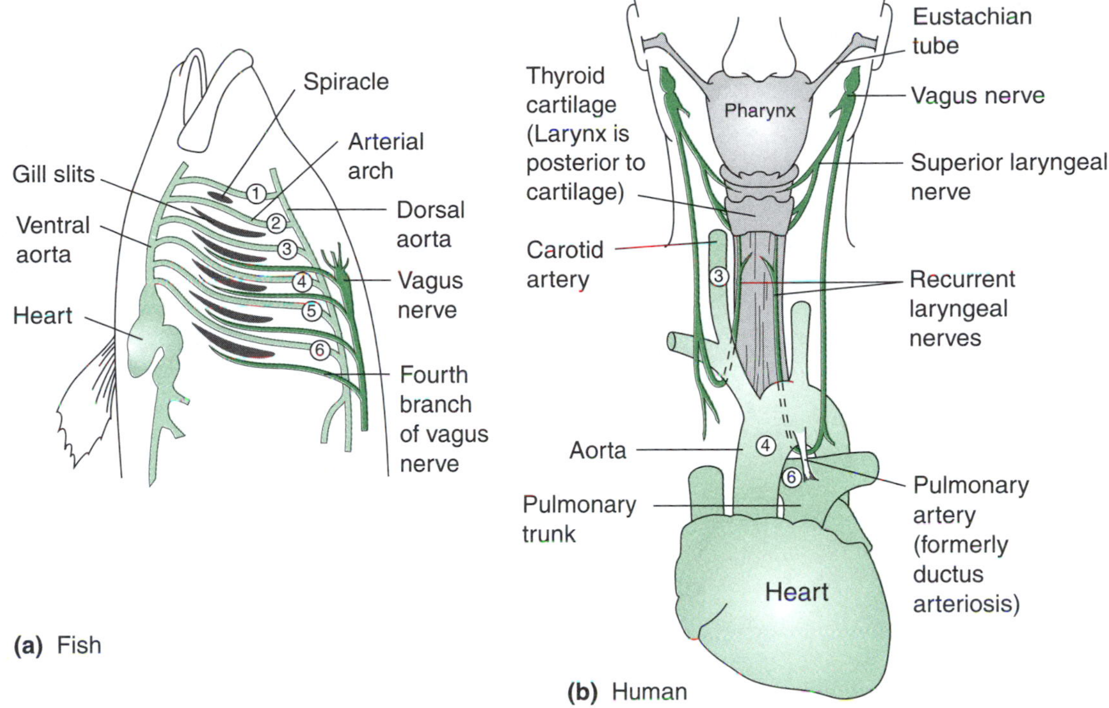

FIGURE 3-16 Schematic diagram showing the relationship between the vagus cranial nerve and the arterial arches in fish (*a*) and humans (*b*). Only the third, fourth, and part of the sixth arterial arches remain in placental mammals, the sixth acting only during fetal development to carry blood to the placenta. The fourth vagal nerve in mammals (the recurrent laryngeal nerve) loops around the sixth arterial arch just as it did in the original fishlike ancestor, but must now travel a greater distance because the remnant of the sixth arch is in the thorax.

In Darwin's day, the fossil record was sporadic, the result of serendipitous collecting and random finds. Nevertheless, during Darwin's lifetime a few paleontological findings came to light that strongly supported his theory of descent with modification. One was the discovery in 1861 of what had been proposed as a true "missing link," in this case, an animal that was interpreted as intermediate between reptiles and birds. As shown in **Figure 3-18**, this fossil, *Archaeopteryx*, had a number of *reptilian features*, including teeth and a tail of 21 vertebrae, but it also had a number of *birdlike features*, such as a wishbone and feathers (see Chapter 18). Huxley argued convincingly that *Archaeopteryx* was probably a "cousin" to the lineage running from reptiles (dinosaurs) to birds, and that such "primitive" forms were predictable consequences of evolution that helped prove the theory.

One year after the publication of *On the Origin of Species*, Richard Owen (1804–1892) described the earliest known horse-like fossil, first called *Hyracotherium*, now often referred to as eohippus (the dawn horse) although eohippus is a common, not a scientific name. *Hyracotherium* was some 50 cm high, weighed about 23 kg, had four toes on its pad-footed front legs and three on its hind legs (modern horses only have one toe on each foot), and simple teeth adapted for browsing on soft vegetation (**Figs. 3-19** and **3-20**). Later fossil finds indicate that *Hyracotherium* was present from North America to Europe as a number of herbivorous species, some no larger than an average-sized modern-day house cat.

In the approximately 60 My since *Hyracotherium* arose, horses have changed radically. They now run on hard ground, chew tough, silica-containing grasses, and show special adaptations for this particular environment. Their elongated legs are built for speed, bearing most of the limb muscles in the upper part of the legs, enabling a powerful, rapid swing. Today, horses are the only tetrapods with a single toe on each foot (Fig. 3-19). This arrangement, coupled with a special set of ligaments, provides horses with a pogo-stick-like springing action while running on hard ground. Their teeth also show exaggerated qualities adapted for chewing tough, abrasive grasses, their teeth being much longer than the teeth of other grazers. The molars and premolars are nearly identical in shape, long, and continue growing for the first eight years of life until the roots form.

By the 1870s, paleontologists such as Othniel C. Marsh (1831–1899) were able to use fossils of North American and European horses to present the first classic example of a stepwise evolutionary tree among vertebrates, showing various transitional stages (Figs. 3-19 and 3-20). Remarkably, we now know almost all the intermediate stages between eohippus (*Hyracotherium*) and the extant horse, *Equus*, including

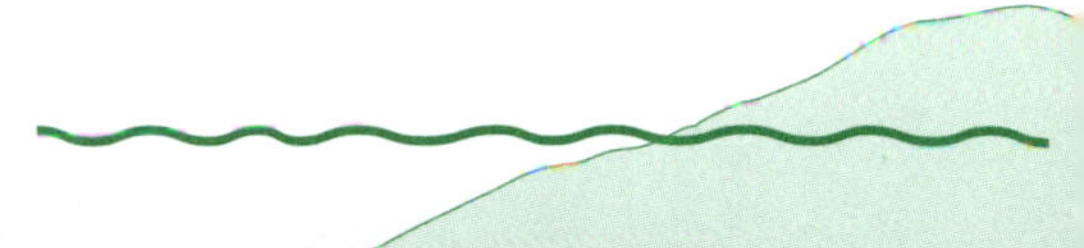

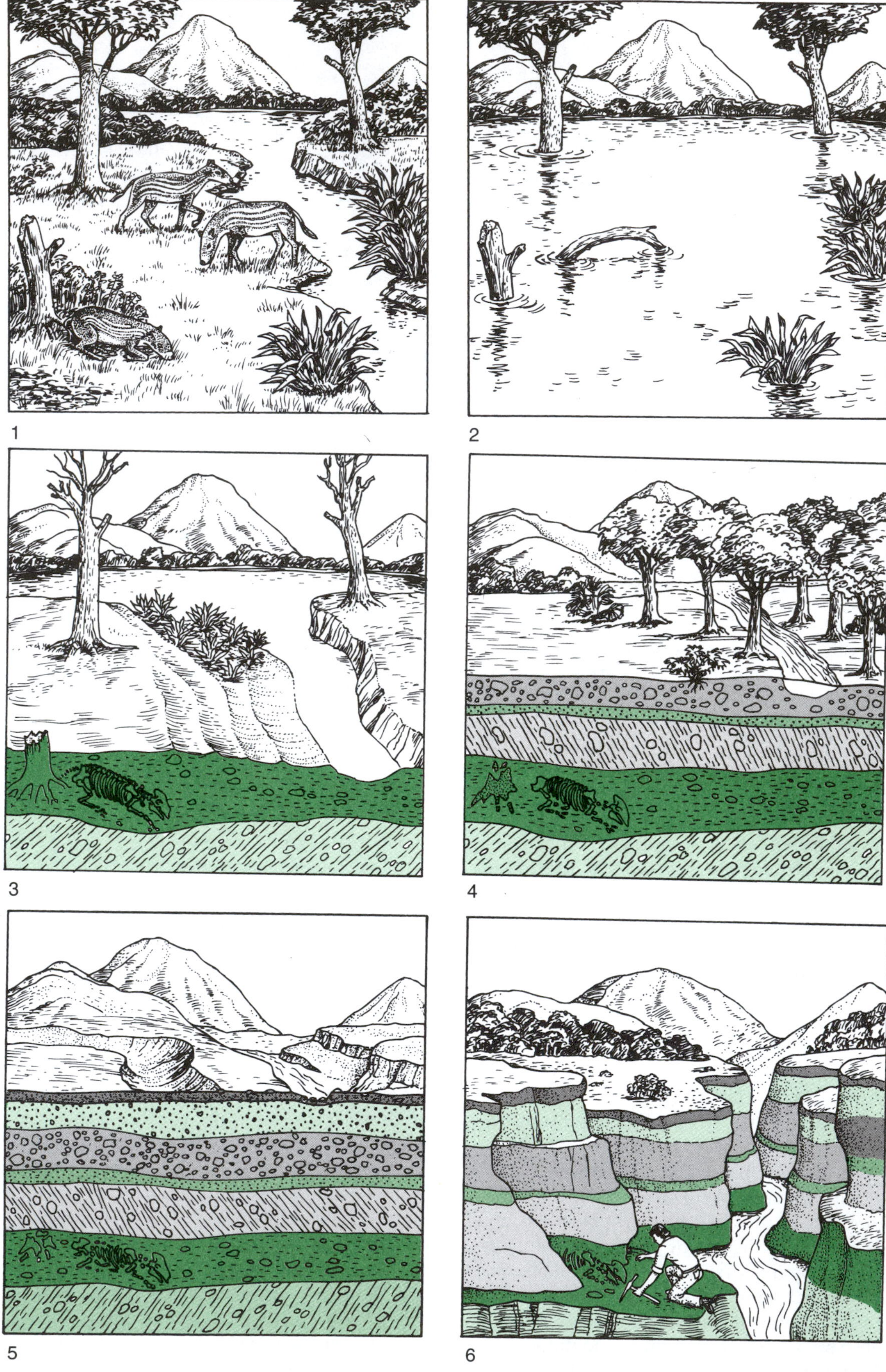

FIGURE 3-17 The process of fossilization in which an organism (in this case, an animal) dies in a watery environment that protects it from scavengers. Reduced oxygen levels in deeper water further resist deterioration. The remains are gradually silted over and eventually covered by successive layers of soil that compact into sedimentary rock. In time, because of erosion, the fossil surface may become exposed. (From Kardong, K. V., 2006. *Vertebrates: Comparative Anatomy, Function, Evolution,* 4th ed. McGraw-Hill, New York, © The McGraw-Hill Companies, Inc.)

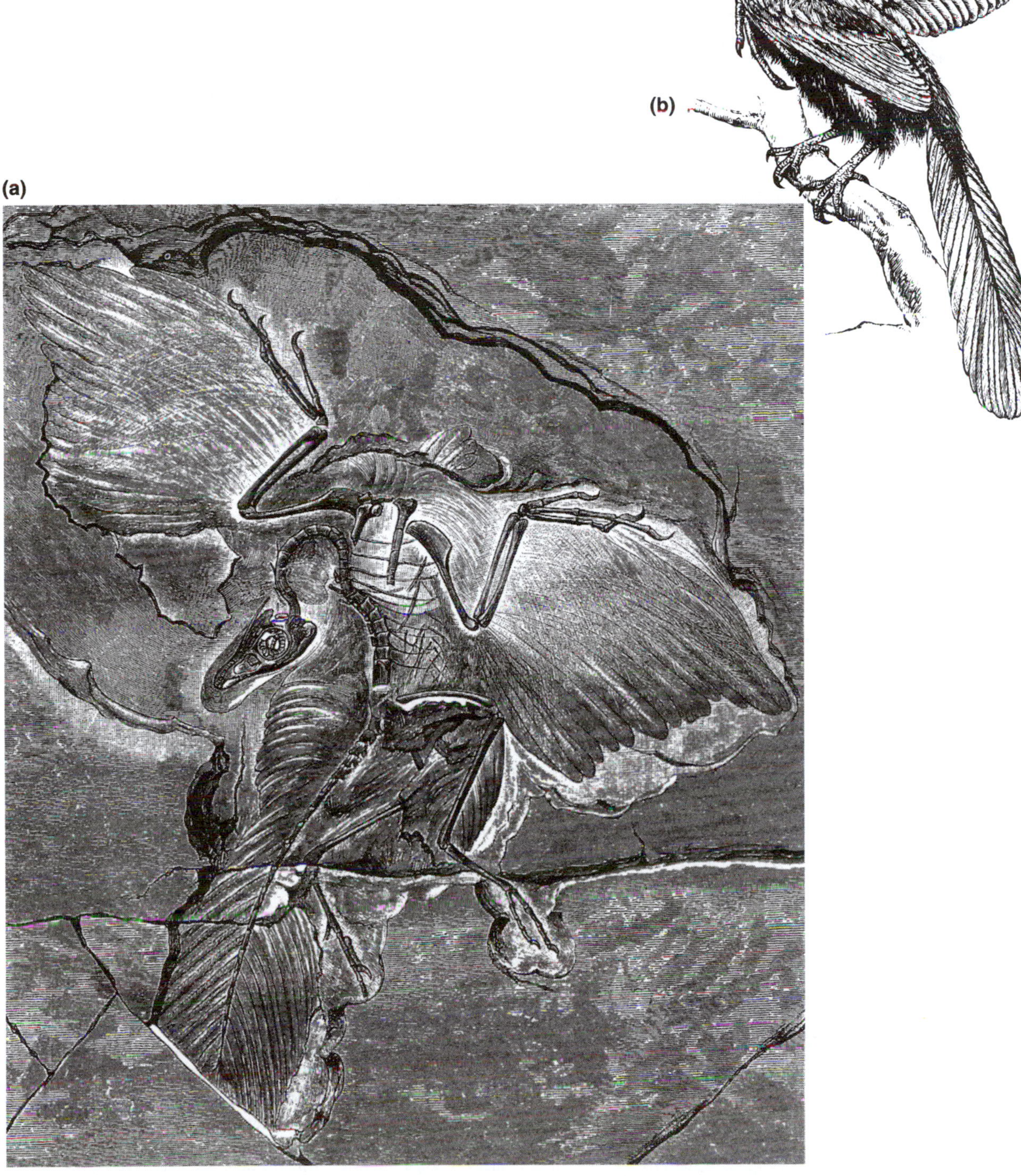

FIGURE 3-18 (*a*) The Berlin specimen of *Archaeopteryx* found in the Upper Jurassic limestones of Bavaria. (*b*) A nineteenth century restoration of what *Archaeopteryx* may have looked like. (From Romanes, G. J., 1910. *Darwin, and After Darwin*. Open Court, Chicago.)

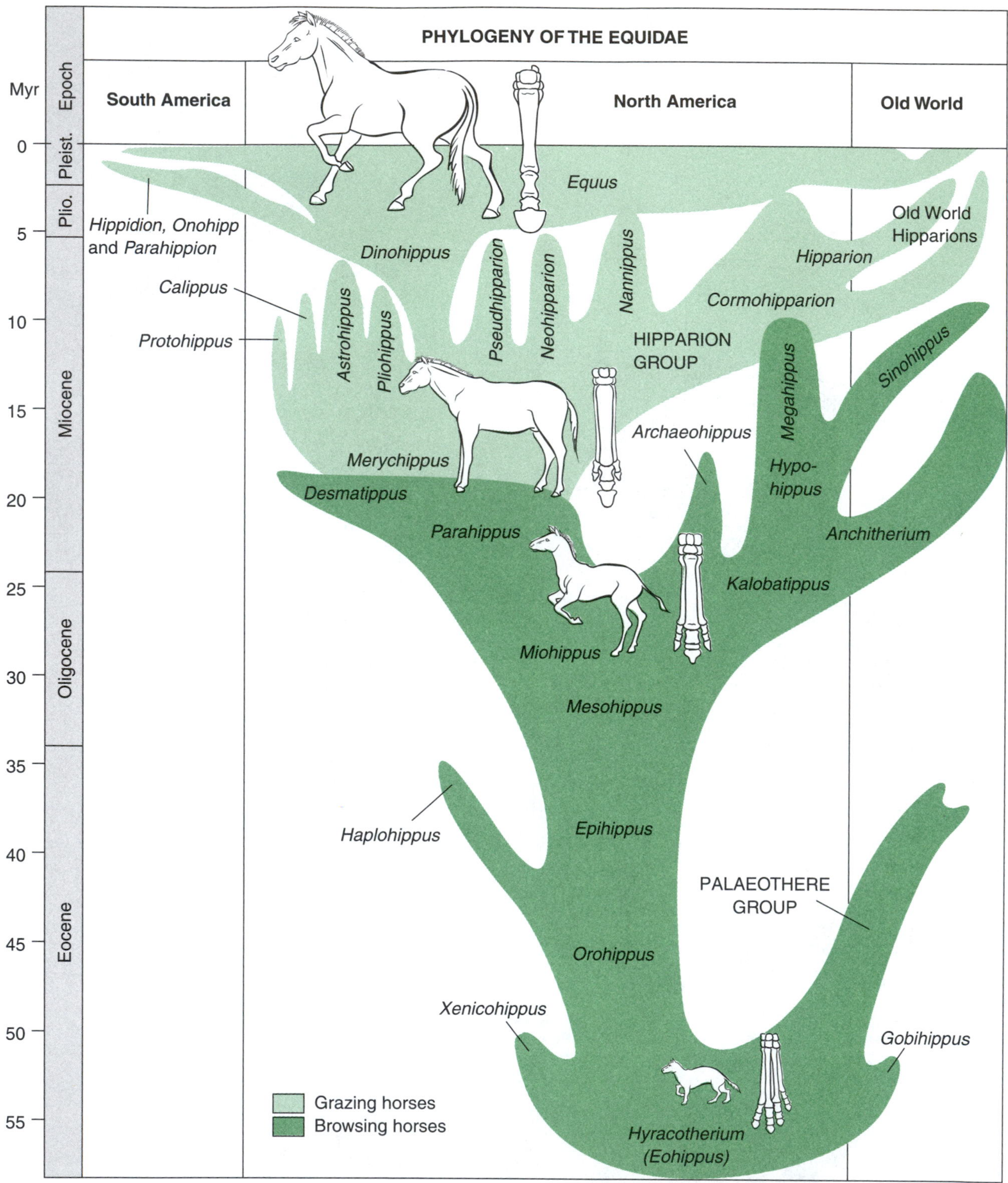

FIGURE 3-19 Evolutionary relationships among various lineages of horses, with emphasis on North American and Old World groups. Sample reconstruction of the digits ('toes') of the hind feet of some fossil horses and of the extant horse *Equus* are shown. The number of digits declined from four to one during evolution of the lineage. Using terms described in Chapter 11, horse lineages followed two different evolutionary patterns: cladogenetic (branching) and anagenetic (nonbranching). *Equus* died out in the Western Hemisphere but continued in Europe and Asia. (Adapted from MacFadden, B. J., 1992. *Fossil Horses: Systematics, Paleobiology, and Evolution of the Family Equidae.* Cambridge University Press, Cambridge, England, with additions.)

transitions from low- to high-crowned teeth, browsers to grazers, pad-footed to spring-footed and small- to large-brained. As shown in Figure 3-19, evolutionary changes among these forms did not proceed in a single direction (**orthogenesis**), being better represented as a "bushy" family tree. Horses evolved adaptations for their habitats in different ways; some lineages maintained distinct structures until they went extinct. Although they all occupied the same general area, they made use of different environmental resources (*resource partitioning*). Some species became browsers, feeding on shrubs and trees. Others remained grazers, feeding on grasses. Still others grazed and browsed, while reversion from grazing to browsing occurred in some Florida species (MacFadden et al., 1999). Although all had high-crowned molars, indicating a grazing ancestry, differences in feeding habits can be deduced from dental scratches caused by grazing and dental pits caused by browsing, and from differences in the carbon isotope ratios ($^{12}C/^{13}C$) of grasses and shrubs; different diets produced different $^{12}C/^{13}C$ ratios in teeth.

The rate of evolution for any particular trait among the various branches was not constant. Size, for example, underwent relatively few changes for the first 30 million and the last few million years of horse evolution. Even when evolution was proceeding rapidly, as it did during the Miocene, both small- and large-sized species evolved. No unidirectional orthogenetic trend in size applied to all lineages. This finely detailed phylogeny encompasses hundreds of fossil species and is one of the best illustrations of some of the realities and complexities of adaptive radiation in evolution.

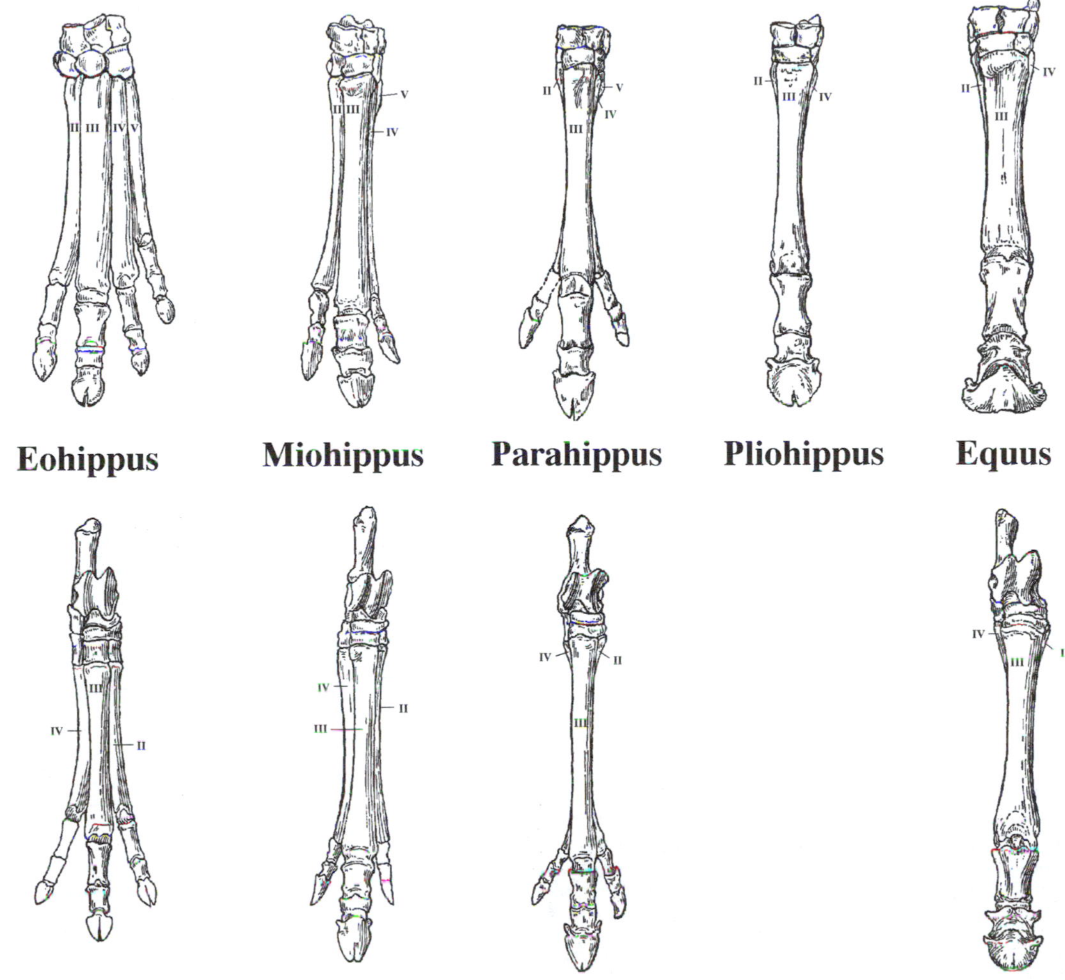

FIGURE 3-20 Reduction of the toes from four to one in both forelimbs (top) and hind limbs (bottom) in horses from the Eocene "dawn horse," *Eohippus* (*Hyracotherium*), to the modern horse genus *Equus*, which appeared in the Pleistocene and has persisted to today. Digit III is retained in all, digits II and IV are reduced to splint bones in *Parahippus* and *Equus* splint bones, while digits I and V (the outer digits) were lost as early as *Miohippus*. (Modified from Gregory, W. K., 1951. *Evolution Emerging. A Survey of Changing Patterns from Primevil Life to Man*. Two Volumes. The Macmillan Company, New York.)

Living Fossils

Interestingly, some ancient lineages have persisted with minimal morphological changes to the present day.

Occasionally, species are discovered that are so remarkably similar to organisms believed to have become extinct many ages ago that they are called **living fossils**. Sturgeons, lungfish, horseshoe crabs, *Lingula* (a brachiopod) and ginkgo trees are living fossils. For example, coelacanths are ancient lobe-finned fishes (**Fig. 3-21**) related to those that evolved into terrestrial vertebrates about 200 Mya (million years ago) (see Chapter 17). Although the fossil record of coelacanths (which began in the Devonian about 380 Mya) ended 80 to 100 Mya, fishermen find live coelacanths (*Latimeria chalumnae*) in deep waters off the eastern coast of South Africa. Similarly, an ancient form of segmental mollusk (*Neopilina*), believed extinct since the Devonian, has been found in deep-sea trenches off Costa Rica, Peru and southern California.

Although these findings support the validity of paleontological claims that fossils indicate the existence of real organisms, they also show the inadequacy of the paleontological record; disappearance of fossils of a particular type from the fossil record does not necessarily mean that this type immediately went extinct; "absence of evidence is not evidence of absence." However, aside from such rare "living relics," fossils in almost every instance differ from present-day forms, as more recent geological strata contain forms more like the present than those in the older strata. Fossils provide strong support for evolution.

Artificial Selection

To support his concept of evolutionary change, Darwin discussed a number of examples of evolution by selection, although the selection method involved human choice rather than natural events. In **artificial selection**, the breeder — whether of dogs, cats, pigeons, cattle, horses, peas, wheat — selects the parents deemed desirable for each generation and culls (destroys) the undesirable types. Because the selected parents may produce a variety of different offspring, the breeder can usually continue to select

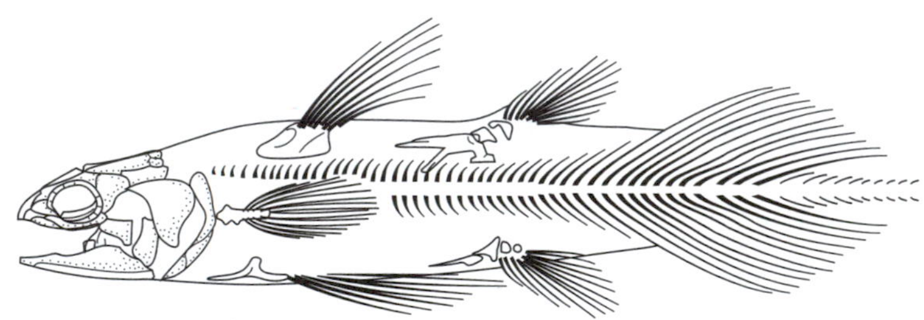

(a) Triassic (230 million years ago)

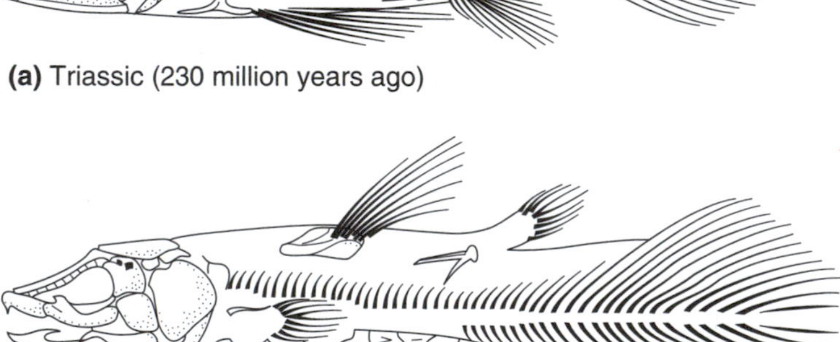

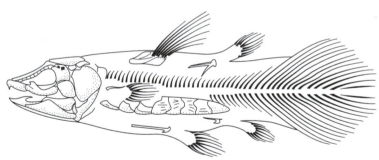

(b) Cretaceous (80 million years ago)

(c) Coelacanth

FIGURE 3-21 (a) A fossil lobe-finned coelacanth (*Laugia groenlandica*) from the Early Triassic, about 230 Mya. (b) *Macropoma mantelli* from the Late Cretaceous about 80 Mya. (From P. L. Forey, 1988. Golden jubilee for coelacanth *Latimeria chalumnae*. *Nature*, **336**, 727–732. Reprinted by permission). (c) The extant coelacanth (*Latimeria chalumnae*) found off the eastern coast of South Africa.

FIGURE 3-22 Hens from the control (unselected) breeding line (*left*) and from a line (*right*) in which selection for increase in shank length (the unfeathered portion of the leg) was carried out for 14 years, by the end of which time the length of the shank had increased by 13.5%. Note the correlated changes in the rest of the selected line, which is more than 13% taller than the chicken from the unselected line. The index of fitness (see Chapter 22) of the selected line dropped from 4.67 to 0.96 over the same time. (Reproduced from Lerner, I. M., 1954. *Genetic Homeostasis.* Oliver and Boyd, Edinburgh.)

in a particular direction until the traits in which he or she is interested are considerably different from their initial appearance (**Fig. 3-22**). For example, for thousands of years humans have selected dogs, which now range in size from the St. Bernard to the Chihuahua, and in features from greyhound to bulldog (**Fig. 3-23**). Fanciers have long bred pigeons, which now show a wide variety of beaks, shapes and feathers. The same is true for sheep, cattle and all the many agricultural species of plants and animals.

Artificial selection demonstrated to Darwin and his colleagues that continued selection was powerful enough to cause observable changes in almost any species. The Darwinian claim that natural selection for particular environments could accomplish even greater changes than artificial selection and lead to speciation therefore seemed reasonable, given the much longer periods of time in evolutionary history and the "unrelenting vigilance" of natural selection.

The greater efficacy of natural over artificial selection can be demonstrated experimentally, especially in an organism with a fast generation time, allowing many generations to be analyzed. When the efficacy of artificial versus natural selection on mycelial growth rate of the filamentous fungus, *Aspergillus nidulans*, was compared, the natural selection regime (weekly transfer of a random sample of spores to a fresh growth plate) led to faster adaptation (evolution) than did artificial selection (weekly transfer of the fastest growing portion of the fungus to a fresh growth plate).[11]

[11] See Schoustra *et al.* (2005) for the study using *Aspergillus*, Scheiner (2002) for the methodology involved in conducting artificial selection experiments, and see Wilner (2006) for Darwin's experiments with artificial selection.

One of the main reasons for the effectiveness of Darwin's book was his use of the words and assumptions of the day to destroy the words and assumptions of the day. Darwin argued by analogy, especially to natural selection from his own extensive research on artificial selection. His research gave him much of his primary data. He knew that in a few generations a pigeon breeder could produce a pigeon with a head so big it couldn't fly. Such "sports" and variations could be so different from their stock a few generations before that they would have been considered different species if found in the wild. Darwin had done the calculations for the denudation of the Weald (see Box 3-2), which gave more than enough time for evolution to occur. His theory is so firmly planted in data familiar to many that his theory had to be accepted. Not to do so would have required them to reject their own experiences and/or data.

Natural Selection

Darwin's evolutionary view of selection can be illustrated in simple form as follows.

Although chance events arise, evolution is primarily a historical process; what evolves depends on what has evolved before. Thus, no complex structure arises all at once by a lucky combination of events, but rather evolution builds new structures from old ones. For example, if one had a large bowl full of ten different letters (A, C, E, I, L, N, O, T, U, V) with each letter present in equal frequency (see Fig. 4-2), nine letters randomly drawn from this bowl can be arranged in many millions of ways. The chances of getting the exact word EVOLUTION from a random draw of nine letters is obviously small: $1/10^9 = 0.000000001$.

FIGURE 3-23 Heads of some different breeds of dogs produced by artificial breeding. Molecular evidence from mitochondrial DNA sequencing (see Chapter 12) shows that the domestic dog is more closely related to the gray wolf than to any other species of the canid family (such as the jackal and fox). This evidence also shows that domestication started independently on several occasions 10,000 to 15,000 years ago. (Modified from the American Kennel Club, *The Complete Dog Book,* 1985. Howell Book House, New York, p. 196.)

However, if we assume that a selection mechanism exists that will perpetuate certain adaptive combinations, the "evolutionary" attainment of the word EVOLUTION is far greater than on first appearance. Assume that E is the only letter that can survive by itself. The chance of drawing an E from the bowl is, on average, 1/10. Assume next that a V in combination with the already adaptive E has additional survival value. The chance of achieving this particularly adaptive combination EV is again 1/10. (Note that once E has been drawn and survives, the chance of getting the EV combination is the 1/10 chance of drawing V and not the 1/100 chance of drawing EV together.)

Similarly, if we assume that the next adaptive combination consists of adding an O to EV, and that further successive combinations are selected because of their enhanced relative fitness, the entire word, EVOLUTION, can eventually be selected with relatively high probability without the intervention of any agent other than the strictly opportunistic one of what is adaptive at each separate stage. Because there may be other adaptive combinations in this bowl full of letters, similar selective mechanisms can also lead to words such as EVOCATION and ELEVATION. Like EVOLUTION, the chances are extremely small for such words to arise by choosing nine letters at random in a single selective event, yet they can easily be produced by successive selection of adaptive combinations.

Complex organs such as the eye and the brain can arise by naturally selecting successively improved adaptations for preservation. The *variation* on which the choice is exercised is *random* (the letters, or mutations, are of different kinds, adaptive and nonadaptive), but the *structure* that is built over many generations of selection has been historically molded and created and *is not at all random.* Thus, in the words of the population geneticist R. A. Fisher (1930), "natural selection is a mechanism for generating an exceedingly high degree of improbability." Or to paraphrase Jacques Monod (1971), mutation provides the random noise from which selection draws out the nonrandom music. François Jacob (1977) referred to evolution as **bricolage (tinkering),** a term that rightly places the emphasis on modification of existing genes, and which has been employed effectively at levels above the gene.[12] All organisms have a history and evolution is about that history. There has been but one history of life; evolution is the working out of that history.

To Darwinians, all biology has had an accidental origin in the sense that hereditary variables arose at first randomly without purposeful foresight. Yet most, if not all, biological features that survived were adaptive in the sense that only adaptive combinations of these random variables could perpetuate themselves in the face of selection.

KEY TERMS

adaptive radiation
analogous
biogenetic law
convergence
geographical isolation
homologous
isolation
living fossils
neoteny
orthogenesis
preadaptation
variability
vestigial organs

DISCUSSION QUESTIONS

1. Why did the concept of blending inheritance conflict with the concept of natural selection, and how did Darwin attempt to deal with this problem?
2. Because Darwin believed that evolution was a gradual process resulting from the accumulation of only small differences over time, how did he seek to deal with the following objections?

[12] For tinkering, see Duboule and Wilkins (1988) and Goode (2007).

   a. How can new species arise when the variations that organisms produce are only small?
   b. How can small variations be recognized by natural selection?
   c. How can organs with new functions arise?
3. In what way did critics use the following points as objections against Darwin's theory, and how were these objections later resolved?
   a. The multiplication of species.
   b. Earth's age.
4. Nineteenth century support for Darwinism came from various sources and studies: systematics, geographical distribution of organisms, comparative anatomy, embryology, fossils and artificial selection among plants and animals. How did these areas of biology support Darwinism? Provide examples of this support. (For example, how were concepts such as adaptive radiation, homology, analogy, convergent evolution, vestigial structures, and living fossils used to support Darwinism?)
5. Which studies in Question 4 indicated that:
   a. Evolution always goes in a single direction.
   b. There is always an increase in the number of structures or organs in a lineage.
6. What are the differences between explaining vestigial structures by the Lamarckian principle of use and disuse, or by the Darwinian principle of natural selection?
7. What is Haeckel's biogenetic law, and how do biologists presently regard it?

EVOLUTION ON THE WEB

Explore evolution on the Internet! Visit the accompanying Web site for *Strickberger's Evolution*, Fourth Edition, at http://www.biology.jbpub.com/book/evolution for Web exercises and links relating to topics covered in this chapter.

RECOMMENDED READING

Appleman, P., (ed.), 2001. *A Norton Critical Edition. Darwin: Texts, Commentary*, 3d ed. W. W. Norton & Company, New York.

Bowler, P. J., 1996. *Life's Splendid Drama: Evolutionary Biology and the Reconstruction of Life's Ancestry, 1860–1940*. The University of Chicago Press, Chicago.

Fortey, R., 2002. *Fossils: The Key to the Past*. Natural History Museum, London.

Haeckel, E., 1905. *The Evolution of Man*. Translated from the 5th German ed., by J. McCabe. Watts, London.

Hall, B. K., (ed), 1994. *Homology: The Hierarchical Basis of Comparative Biology*. Academic Press, San Diego, CA. (paperback, 2001)

Land, M. F., and D-E. Nilsson, 2002. *Animal Eyes*. Oxford University Press, Oxford, England.

MacFadden, B. J., 1992. *Fossil Horses: Systematics, Paleobiology, and Evolution of the Family Equidae*. Cambridge University Press, Cambridge, England.

Mayr, E., 1988. *Toward a New Philosophy of Biology: Observations of an Evolutionist*. The Belknap Press of Harvard University Press, Cambridge, MA.

Milsom, C., and S. Rigby, 2003. *Fossils at a Glance*. Blackwell Science, Oxford, England.

Nyhart, L. K., 1995. *Biology Takes Form: Animal Morphology and the German Universities, 1800–1900*. The University of Chicago Press, Chicago.

Oldroyd, D. R., 1980. *Darwinian Impacts: An Introduction to the Darwinian Revolution*. Open University Press, Milton Keynes, Oxford, England.

Rudwick, M. J. S., 1985. *The Meaning of Fossils: Episodes in the History of Palaeontology*, 2d ed. The University of Chicago Press, Chicago.

Rudwick, M. J. S., 1995. *Scenes From Deep Time: Early Pictorial Representations of the Prehistoric World*. The University of Chicago Press, Chicago.

Rudwick, M. J. S., 2005. *Bursting the Limits of Time. The Reconstruction of Geohistory in the Age of Revolution*. The University of Chicago Press, Chicago.

Ruse, M., 1979. *The Darwinian Revolution*. The University of Chicago Press, Chicago.

Ruse, M., 1988. *But Is It Science? The Philosophical Question in the Creation/Evolution Controversy*. Prometheus Books, Buffalo, NY.

Schulter, D., 2000. *The Ecology of Adaptive Radiation*. Oxford University Press, Oxford, England.

Scott, E. C., 2000. Not (just) in Kansas anymore. *Science*, 288, 813–815.

Scott, E. C., 2005. *Evolution vs. Creationism: An Introduction*. University of California Press, Berkeley CA.

Scott, E. C., 2006. Creationism and evolution: It's the American way. *Cell*, **124**, 449–450.

Selden, P., and J. Nudds, 2004. *Evolution of Fossil Ecosystems*. Manson Publishing Ltd., London.

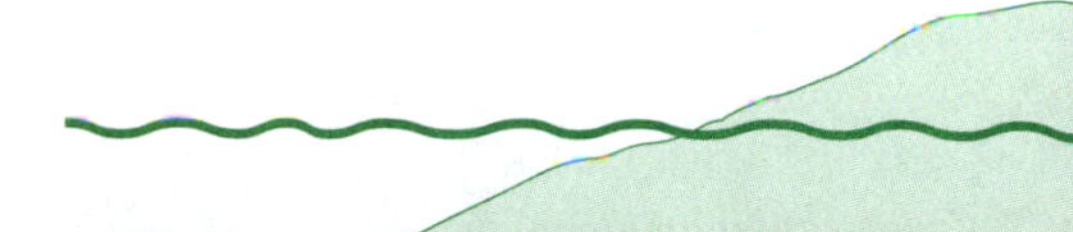

PART

2

Origins

The Enormity of Time

It is impossible to appreciate the vastness of the universe, or the enormity of time. How can we comprehend the **Big Bang,** that most fleeting of beginnings, occurring as it did at the first 10^{-35} of a second of the life of our universe? Or the series of events that occupied the billions of years that have since elapsed, or the outcome of those events: 100 billion galaxies, each with 100 billion or more stars? Without these almost unimaginable events, however, there would be no atoms, no molecules, no solar system, no Earth and thus no life as we know it.

The five chapters in Part II take us from the Big Bang 13.7 billion years ago (Bya) through the formation of atoms, stars, molecules, and Earth itself, to the origin of the first prokaryotic cells 3.5 Bya and eukaryotic cells 1.5 By later. They set the stage for the discussion of inheritance and development that follows in Part III.

Chapter 4 provides a brief sketch of current hypotheses about the early years of the universe, an area of science flooded with exciting new research findings, most of which are far beyond the scope of this book. In particular, it outlines the ways in which all the naturally occurring elements of the periodic table beyond hydrogen were manufactured (and are still being manufactured) in the fiery thermonuclear reactors we call stars.

Chapter 5 narrows our focus to the origin of the solar system and then, about 4.56 Bya, to the origin of Earth. It tracks the formation of Earth's atmosphere, rocks, continents and oceans, describing how conditions gradually became amenable to life, and ends with a discussion of ways in which movements of the gigantic plates that constitute the lithosphere have influenced the distribution of organisms over time.

But how did those organisms arise? To answer that question, we need to understand **how molecules arose** and how some of them became **organized into biological structures subject to natural selection.** In Chapter 6, we discuss (a) the nature of the major biological (organic) molecules; (b) experimental studies conducted in attempts to recreate the conditions under which molecules, membranes and protocells may first have arisen; and (c) the onset of the process of natural selection, triggering the evolution of further molecules. (Natural selection, and the way it operates on accumulating variation, resulting in the panoply of organisms that have ever existed, is, of course, the major topic of this book.)

According to the central dogma of biology, DNA is transcribed into mRNA, which is then translated into protein. But which of these molecules appeared first? Logic would dictate that they evolved in that order, with DNA leading the way. On the other hand, assuming that current function equals original function, the functions of these three classes of molecules today make it more logical to expect that proteins evolved first, a genetic code evolving later. Both of these possibilities have been extensively investigated, but as we show in Chapter 7, many recent findings are consistent with an initial **RNA world,** with RNA serving the dual functions required of a molecule capable of surviving and evolving: the ability to function as an enzyme — that is, to function as proteins do today — and the ability to replicate — that is, to function as DNA does today.

How did naked molecules go on to produce membranes that evolved into protocells and then cells? In Chapter 8, we consider how molecules and membranes resulted in the first (prokaryotic) cells 3.5 Bya, and how the evolution of prokaryotes led to the origin of the first **eukaryotic cells** (cells with nuclei) 2 Bya. Because cellular function depends on transforming external sources of material and energy into processes that take place within cells, the existence of cells implies metabolism. Chapter 8, thus, introduces metabolic pathways and their evolution, the biochemistry that drives cellular activity.

4 Origins of Cosmic Structures and Chemical Elements

Chapter Summary

Many lines of evidence converge to tell us that we live in a universe that is (1) expanding, (2) about 13.7 billion years old, (3) very large, possibly infinite, (4) made of both familiar and unfamiliar kinds of matter and energy, and (5) highly structured on scales all the way from planets and moons to clusters of galaxies.

Out of the early universe following the initial Big Bang came familiar kinds of energy and matter (light, neutrinos, hydrogen, helium), but also dark matter and dark energy, which do not interact with light, so that we know about them only from their gravitational effects. Very subtle density fluctuations in the matter developed under gravity into galaxies, stars, planets, clusters, quasars, indeed, all that we see today. Nuclear reactions in stars, particularly ones more massive than our own sun (which is a very typical, boring star, apart from being ours), eventually fused a small percentage of the hydrogen and helium into all the rest of the chemical elements, including the biologically important ones: carbon, oxygen, nitrogen, phosphorus and iron. After a few billion years of this chemical evolution, gas in our galaxy and others reached the stage where it was able to form the solar-type stars that might host planets.

The title of this chapter in earlier editions was "Origin of the Universe," but in fact astronomers have no idea whether our universe is the only one that has ever existed, or indeed why it began in the way it did, with a "**Big Bang.**" We do, however, know quite a lot about what happened during the Big Bang stage and afterwards, through an accumulation of a wide variety of observations and calculations using standard physics. The situation is a bit like tracing your own genealogy. Not being sure whether your very first ancestor was Adam or a primordial bacterium does not cast doubts upon your immediate parentage.

Cosmology — the study of the universe — is today an exciting whirlwind of research and discovery. Aided by the *Hubble Space Telescope*, satellites, space probes, and research on the ground, we are "seeing" further and further into the past, discovering planets orbiting far-distant stars and much more. Yet the more we learn, the more we realize how little we know. What is the nature of dark matter, dark energy and black holes? Is there life on other planets, and is ours the only universe that has ever existed?[1]

Steady State or Evolutionary Cosmology?

The term *Big Bang* began life as an insult coined by one of the theory's greatest opponents, the British astronomer and mathematician, Sir Fred Hoyle (1915–2001). In 1948, Hoyle proposed an alternative theory called the *Steady State Universe*, which was very elegant but in major disagreement with later observations and so is no longer under consideration.

A more formal title for the Big Bang hypothesis is **evolutionary universe**, for the hypothesis covers not only the beginnings of our universe but also its evolution. Evolution in astronomy and cosmology is very different from biological evolution (and much closer to the original meaning of the word, *unfolding*). In astronomy, evolution means changes with time of individual objects and systems, changes that are describable by the laws of physics (gravitation, electromagnetism, thermodynamics, quantum mechanics and so on). In biological evolution, contingency (chance) plays a major role; in cosmology, it almost certainly does not (see Chapter 1).

The universe is not only all the celestial objects you can think of — stars, planets, galaxies, black holes — but also the **space-time** that contains them, and several kinds of stuff you have probably never heard of. Cosmology means the study of the large-scale structure and evolution (in the cosmological sense) of the universe and, increasingly in recent years, contemplation of what other sorts of universes might be possible and even exist, though we presently may not be able to communicate with them.

What is the evidence for our current hypothesis about the universe? Below are just a few of many important observations consistent with the Big Bang theory:

- The universe is *expanding*, a fact first recognized by Edwin Hubble in 1929. Together with other evidence, the current rate of expansion implies an age for the universe of about 13.7 billion years (By).
- Indeed, the *oldest stars* turn out to be just a little younger than this, and consist almost entirely of hydrogen and helium.
- The *ratio of hydrogen to helium* in the oldest stars is 10:1 by the number of atoms, and 77:23 by mass, reflecting the density of ordinary matter (called **baryonic matter**) when the universe was young and hot.
- A sea of *microwave photons* — **the cosmic microwave background** (CMB) — fills the observable universe. Its temperature is close to 2.7 Kelvin[2], and it differs from one place to another in the sky by only a few parts in 100,000. These tiny fluctuations provide a map of the early universe.
- Comparison of the apparent brightness of differently aged examples of a kind of stellar explosion, a **Type Ia supernova**, leads us to conclude that **the rate of cosmic expansion is currently accelerating** and not slowing down, as you might expect.
- There is a **total absence** of any evidence **of edges, boundaries or complex geometry** in the universe, telling us that the total volume must be very much larger than the part we can observe, and quite possibly is infinite. The observable universe is set by the distance light can travel in 13.7 By, moving at 300,000 km/sec, slightly modified by the ongoing expansion.

Inflation

Despite its name, the Big Bang did not involve an explosion, nor did it begin at a particular place or expand *into* anything, being "better thought of as the simultaneous appearance of space everywhere."[3] **Table 4-1** outlines a recent view of what happened in the first 100 seconds after the birth of the universe. The speed of the processes, the extraordinarily high temperatures (and density), and especially the exponential increase in size of the universe (to about 10^{24} meters or one hundred million light years) during the epoch of inflation at around 10^{-35} of the first second, defy comprehension.

[1] For a primer on cosmology, visit http://map.gsfc.nasa.gov/m_uni.html, a NASA Web site devoted to the *Wilkinson Microwave Anisotropy Probe (WMAP)*. In particular, click on Universe, in the top bar, and then scroll down the page to access Cosmology 101.

[2] Kelvin signifies absolute zero, at which all molecular motion ceases. See the conversion scale on the inside back cover to convert Kelvin into °C or °F.

[3] NASA, 2006. Foundations of Big Bang Cosmology. Available at http://map.gsfc.nasa.gov/m_uni/uni_101bb2.html. Accessed March 2007.

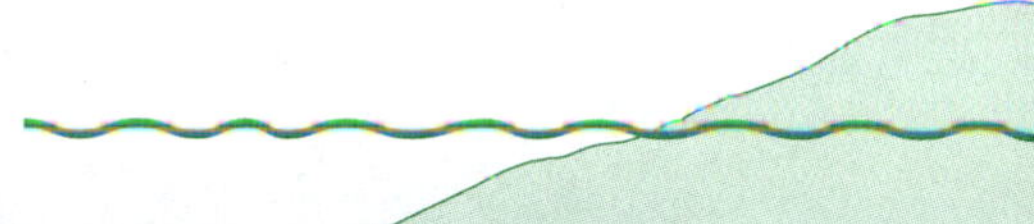

TABLE 4-1 The first 100 seconds in the life of the universe

Time (seconds)	Event	Temperature (°C)
0	Birth of the universe	
10^{-43}	Era of quantum gravity and exotic physics	10^{32}
10^{-35}	Universe expands exponentially	10^{28}
10^{-11}	Electromagnetic and weak forces differentiate; quarks and gluons emerge	10 quadrillion
0.1 microsecond		20 trillion
1 microsecond		6 trillion
10 microseconds	Quarks bound into protons and neutrons	2 trillion
100 seconds	Formation of helium and other elements from hydrogen	1 billion

Source: Riordan, M. and Zajc, W. A., 2006. The first few microseconds. *Scientific American.* **294,** 24–31.

The idea of inflation goes back 25 years and has been tested repeatedly as our observations of the universe have improved. So far, it has passed every test and so might almost be called a theory, in the same sense that biological evolution is a theory: a repeatedly tested idea that explains a very great deal. Yet how is it possible? How could the universe expand at speeds far, far in excess of the speed of light?

The answer is that space-time itself was inflating and the matter within it was merely carried along for the ride. Since that time, the universe has continued to expand, albeit at a slower pace (**Fig. 4-1**). This is perhaps "the most important fact ever discovered about our origins" (Lineweaver and Davis, 2005), because it is the basic tenet of the Big Bang hypothesis. A galaxy may appear to be moving away from us, when in fact it is the space-time between us that is expanding. Galaxies (and solar systems, and you) are held together by gravity, and so are scarcely affected by the expansion.

But how did we learn this?

One implication of Einstein's 1916 **General Theory of Relativity** was an **expanding universe**, but until 1929, when Hubble confirmed the hypothesis, even Einstein found such a notion absurd. To achieve the stationary universe he expected, Einstein added a **cosmological constant** to his equations, something he later came to regret (but see below).

Then in 1965, Arno Penzias (1933–) and Robert Wilson (1936–), two young radio astronomers working at the Bell Telephone Laboratories in Murray Hill, New Jersey, inadvertently discovered that the entire sky is permeated by a remarkably uniform sea of photons — radiation from the initial Big Bang — exactly as predicted in 1948 by George Gamow (1904–1968), building on earlier work in theoretical physics. The discovery of this cosmic microwave radiation (CMB) — which earned the two radio astronomers a Nobel Prize — has allowed us, among other things, to work out the age of the universe by the degree to which the photons reaching us have been **redshifted**, that is, stretched so that their wavelength moves towards the red end of the spectrum

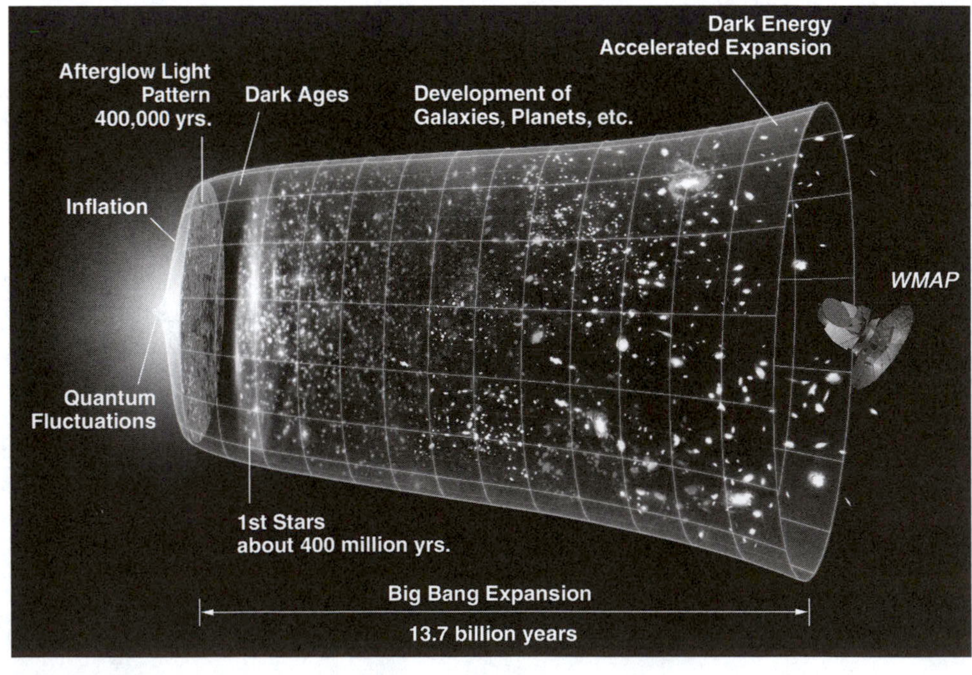

FIGURE 4-1 Time line of the universe, from the Big Bang to now, shown as a bell to simulate inflation. The initial inflation at 10^{-35} of the first second (left) was replaced by a far smaller rate of inflation that is currently accelerating as a result of dark energy (hence the bell's flare), implying that the universe will expand forever. *WMAP*, a probe operated by NASA, here represents the present. (Courtesy of NASA/WMAP Science Team.)

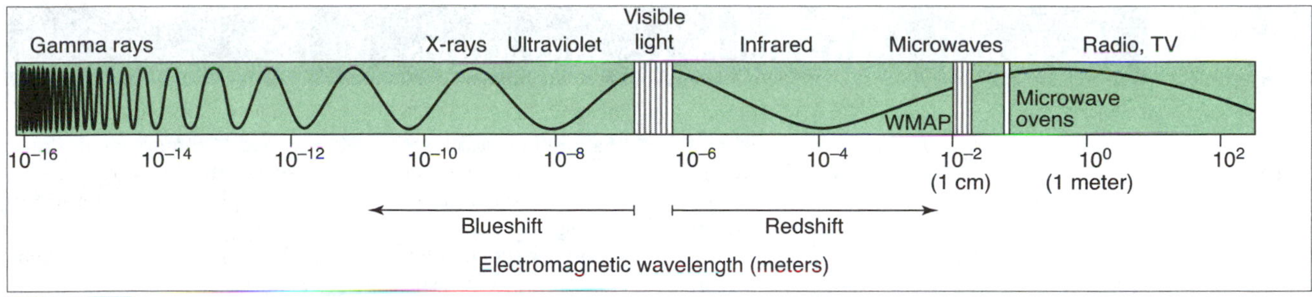

FIGURE 4-2 Electromagnetic waves vary in length, depending on their source. In this diagram, visible light appears as a spectral band in the center, with blue on the left and red on the right. The microwaves measured in space by the *WMAP* are light waves left over from the Big Bang that have been redshifted (that is, stretched far towards the red end of the spectrum) as a resut of the expansion of the universe over time.

(**Fig. 4-2**).[4] Photons are units of electromagnetic radiation, including light.

Astronomers use the **cosmological redshift** of light or radio waves or whatever (usually written as z) as their basic unit of time and distance; the older (and more distant) the object, the higher the number (z) and the faster the distance between the two objects is increasing. Some typical values of z are shown in **Table 4-2**.

A second crucial fact about the CMB — that it is *not totally uniform* — will be discussed later on. But first, we return to the beginnings of the universe, and the origin of cosmic microwave radiation.

TABLE 4-2 Selected redshift values (z) and the corresponding number of years after the Big Bang

Redshift Value (z)	Number of Years after the Big Bang
1100	380,000
40	75 million
20	150 million
6	1 billion
1	6 billion

The First 380,000 Years

If you look again at Table 4-1, you will see that by 10 microseconds after the birth of the universe, a great deal of complicated physics has already occurred, and protons and neutrons have formed. Recent experiments carried out at the Brookhaven National Laboratory (Long Island, New York) have given us a glimpse of what the universe may have looked like just before this stage:

> During those early moments, matter was an ultra hot, super dense brew of particles called quarks and gluons rushing hither and yon and crashing willy-nilly into one another. A sprinkling of electrons, photons and other light elementary particles seasoned the soup. This mixture had a temperature of trillions of degrees, more than 100,000 times hotter than the sun's core (Riordan and Zajc, 2006).

Out of this brew — a plasma rather than a gas — evolved everything that exists today: hundreds of billions of stars in more than 100 billion galaxies, every atom of which consists of relics of this era. Despite this, recent measurements show that ordinary matter (**baryons**[5]) accounts for only 4% of the total density of the universe (**Fig. 4-3**). Far more abundant are **dark matter** (22%), which exerts gravitational forces on itself and the baryons but neither emits nor absorbs light, and **dark energy** (74%), which neither emits nor absorbs light or other radiation and has a negative pressure, and so is causing the rate of expansion of the universe to increase. (**Radiation** in all forms and neutrinos make up only about 0.1% of the total.) The bulk of the universe, thus, comprises dark matter and dark energy, which we know about only through their gravitational effects.

Does this sound odd? Yes, sometimes even to physicists and astronomers. But branches of physics dealing with fundamental particles (not just protons and neutrons, but the quarks and leptons of which they are made) predict several kinds of particles — variously called axions, WIMPs, inos and lowest-mass supersymmetric partners — with the required properties for dark matter. Dark energy may well be equivalent to the cosmological constant in Einstein's original theory of general relativity. If such materials are indeed identified, we should be able to explain the workings of the universe with known physics, of which the most important aspects are the *four forces* (really two forces and two interac-

[4] Some nearby objects, including the Andromeda galaxy, show **blueshifts**; that is, they are approaching us as a result of local gravitational forces overcoming the general expansion of the universe.

[5] Baryon (Greek, *heavy*) is the collective name for those subatomic particles composed of three quarks, a type of subatomic particle. Baryons include protons and neutrons; baryonic matter includes all atoms.

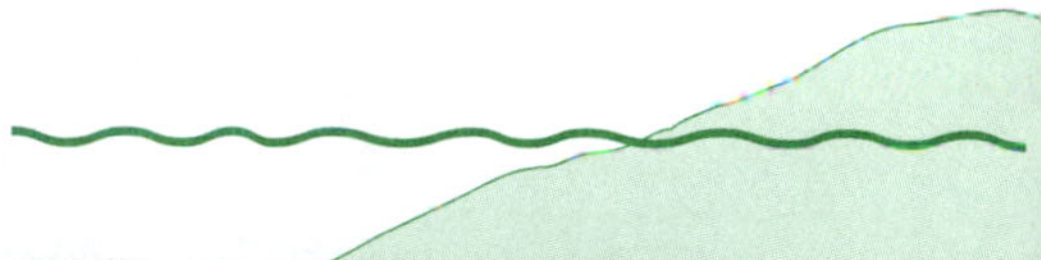

BOX 4-1 The Four Fundamental Forces/Interactions

FOUR FUNDAMENTAL FORCES act on matter in the universe:

- The **strong interaction,** which pulls particles of the atomic nucleus (protons, neutrons, etc.) together into densities of 1 billion tons per cubic inch. This force acts only over very small distances, no greater than 1 ten-trillionth of an inch.
- Another nuclear interaction, one a million times weaker than the strong interaction, is called the **weak interaction,** which is responsible for the manner in which neutrons can eject electrons and transform into protons during radioactive decay.
- The **electromagnetic force,** which is 100 to 1,000 times weaker than the strong interaction, binds electrons to nuclei-forming atoms but weakens with distance. "Weak" though it is, by linking atomic nuclei, electromagnetism introduces "chemistry" into the universe, advancing the complexity of matter to the molecular level.
- **Gravity** (gravitation) is a force that is 10^{38} times weaker than electromagnetism but can aggregate matter into structures and patterns. Gravity acts over long distances such as between earth and its moon, the sun and its planets, and the galaxy and its suns.

tions): gravity, electromagnetic force, the strong (nuclear) interaction and the weak nuclear interaction (**Box 4-1**).

But back to our story (see Figure 4-1). Despite the initial rapid cooling (see Table 4-1), for a long time extreme heat prevented electrons from combining with protons in the dense plasma to make neutral hydrogen atoms, and with helium nuclei to make helium atoms. Because free electrons scatter photons quite efficiently, the early universe would have appeared like a dense fog, much in the way that clouds are opaque. Once the temperature had cooled to around 3000 K, however, atoms could at last form. This period, known as the **epoch of recombination** (though the "re" part makes little sense), occurred at a redshift of 1100, about 380,000 years after the Big Bang. As a result, the universe became transparent, allowing photons emitted during the Big Bang to stream forth. This newly released radiation — detected today as the cosmic microwave background — provides us with a map of the universe as it was then.

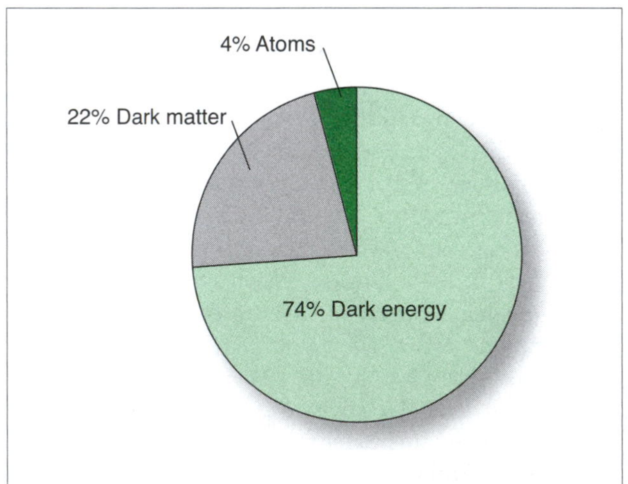

FIGURE 4-3 Content of the universe. A pie chart showing the proportions of atoms (baryons), dark matter, and dark energy in the universe. It is important to realize, however, that most of the universe is virtually empty. As discussed later in the text, the average density of the universe is only 10^{-29} g/cm^3 or about six atoms of hydrogen per cubic meter. Hydrogen is by far the most abundant element in the universe.

Cosmologists have only recently begun to decipher that map and to gain evidence about the origin of structure in the universe. Two satellites launched by NASA have provided much of the evidence.

In 1992, data from the *Cosmic Background Explorer (COBE)*, the first satellite devoted to cosmology, revealed minute fluctuations (parts per hundred thousand) in the temperature of the CMB, which now averages a very cold 2.725 K. These fluctuations (anisotropies) are believed by most cosmologists to reflect minute variations in the distribution of matter in the early universe, the seeds of future structure. The 2006 Nobel Prize in Physics went to John Mather (1946–) and George Smoot (1945–) for their measurements of the earliest moments of the universe using *COBE*.

A second satellite, the *Wilkinson Microwave Anisotropy Probe (WMAP)*, which was launched in 2001 and will continue transmitting data until at least 2009, has produced even more accurate data. Among other findings, it has:

- pegged the age of the universe at 13.7 billion years, with just a one percent margin of error;
- produced an intricate map or "fingerprint" of the CMB fluctuations;
- pinned down the time when the CMB streamed forth; and
- revealed that the first stars appeared about 200 My after the Big Bang.

First Lights and the Formation of Galaxies

After the epoch of recombination, ordinary matter in the universe consisted entirely of neutral hydrogen and helium, with a little bit of lithium. It was still hot, extremely dense and dark, because there were no stars to emit visible light. This period is known as the **dark ages.**[6]

Theories abound about the origin of structure in the early universe, with its tiny number of chemical elements and almost complete lack of the chemical processes we recognize today. Indeed, the end of the dark ages, turn-on of the first stars and other luminous sources, and growth of structure in the universe from subtle to obvious, all belong to theorists with extremely powerful computers. At some stage, however, dark matter particles, experiencing only gravity, began pulling themselves together into clumps. In due course, gravity would have attracted the neutral atoms of hydrogen and helium to the clumps, leading eventually to a vast cloud consisting of both forms of matter, in the center of which the first, massive **star** was born (Abel et al., 2002).

Several groups are pursuing such calculations independently and, happily, finding similar results. Dark matter assemblages of 10^6 to 10^9 *solar masses* (the astronomer's usual unit of mass for big things, equal to 2×10^{30} kg) have accreted enough gas for a few of them to form a few stars as early as redshifts near 40. Because the stars are made of pure hydrogen and helium, they are quite different from the stars around us, with masses of 10 to 1000 solar masses, lifetimes of only a million years and hot surfaces.[7]

Gravity, of course, continues to operate, and the dominant process from the earliest times we can see directly to the present is the interaction and merger of small **protogalaxies** into the full range of sizes and shapes of **galaxies** we see today. Gravity also brings the galaxies themselves together into clusters, which in turn can merge to make larger clusters and super clusters that are distributed in very complex arrangements.

Research carried out with the *Hubble Space Telescope* reveals that by 900 My after the Big Bang, there already existed many hundreds of galaxies bright enough for their light to reach us, meaning that the total number of galaxies at the time must have been far larger (Bouwens and Illingworth, 2006). Yet *Hubble* so far has been able to detect the light of only one similarly bright galaxy from 200 My earlier, giving us some idea of the speed with which galaxy merger must have proceeded. (Smaller galaxies had been around for a long time by then.)

Galaxies are given names based on their shapes: elliptical, spiral, barred spiral, irregular, dwarf spheroidal, and so forth. But from our point of view, the critical item is that each merger phase triggers additional star formation. Every galaxy, at some time in the past, has gone through, or will at some time in the future experience, a period when it contains some gas with a sufficient complement of carbon, oxygen, magnesium, iron, and so on, to form stars like the sun, many with planetary systems.

The merger process continues today. Our Milky Way, a spiral galaxy, is surrounded by more than a dozen much smaller galaxies of 10^6 to 10^9 solar masses, compared to 10^{12} for the Milky Way. Some of these are clearly in the process of being torn up and added in, with trails of gas and stars behind them. Others, including the best-known Large and Small Magellanic Clouds, will succumb to gravitational tides caused by the Milky Way over the next few billion years. Come back in 10 billion years, and you will probably find that we and the next-nearest large galaxy, the Andromeda galaxy (Messier 31), have come together into something like an elliptical galaxy.

But we're getting ahead of ourselves. In the next section we relate the life and death of the universe's nuclear fusion reactors, its chemical factories: stars. Without stars, there would be no elements other than hydrogen and helium, and we — and life, and Earth — could not exist. Indeed, we humans are largely made of stardust, dust from long-dead stars.

Formation of Stars and Chemical Elements

The structure and evolution of stars is one part of modern astrophysics that is in good condition. That is, we can write down a set of differential equations describing the physics inside stars, solve them, compare to the kinds of stars we observe and find a good 1:1 correlation. Stars shine with energy liberated in nuclear fusion reactions, and the products of all the reactions expected after several generations of stars comprise the full range of chemical elements and their isotopes.

In barest outline, the nuclear reactions work their way up the periodic table, beginning always with the fusion of hydrogen to helium, and continuing with helium fusion (normally called *helium burning*) and reactions among the carbon and oxygen nuclei that are its products. These extend to iron (element 26) and the elements adjacent to it in the periodic table. Only stars of more than about eight solar masses get beyond helium burning. Elements heavier than iron come from the addition of many neutrons to iron and its neighbors. Some of this happens in stars like the sun and those a bit bigger, and some occur only in the most massive stars, in both cases near the ends of their lives.

It is important, therefore, to recognize that the heavy elements in our Earth and sun were made in many earlier

[6] Our best hope for studying the Dark Ages would seem to be the 21-cm radio line, which neutral hydrogen can emit or absorb if it is at a slightly different temperature from the surrounding radiation bath. We readily observe this 21-cm radiation from the neutral gas in the spiral arms of our own galaxy and others (indeed, that is how we learned for sure that we live in a spiral), but the universe is expanding. It's hoped that a planned facility called LOFAR (*LO*w *F*requency *AR*ray) will allow us to search for such neutral hydrogen structure in a few years time. LOFAR will consist of thousands of antennae spread across the Netherlands and northern Germany.

[7] Other theories about first lights include the collapse of massive rotating stars to black holes, detectable as distant bursts of gamma rays, and accretion onto black holes at the centers of protogalaxies. Obviously there is still much to learn.

generations of stars, with the products accumulating gradually. Stars forming now have a bit more than the 1.7% of the heavy elements (mostly carbon, hydrogen and oxygen) in the solar system; the oldest stars have 0.01% or less. The very first generation of stars — massive, hot, and short-lived — behaved somewhat differently but they also produced some heavy elements, especially oxygen, neon, silicon, magnesium, and sulfur, and spewed them back out into the remaining gas, so that today all the stars we see have heavy elements.

We now outline eight steps in star formation:

1. Four hydrogens fuse to form helium with the release of energy as light and in other forms. All stars live on this for 90% or more of their lives. Two reaction sequences are possible, both identified by Hans Berthe in 1938–1939.

   In one (which must have come first), two protons meet, one turns into a neutron, and the resulting nucleus of deuterium goes around collecting things until it has made a helium-4 (^{4}He) nucleus with two protons and two neutrons.

   In the other reaction, carbon (or nitrogen or oxygen) acts as a catalyst, but the net result is still the assemblage of four hydrogens into one helium plus energy (some of which comes out in neutrinos in both cases). We have seen a type of neutrino known as a t neutrino coming from the sun, directly verifying that hydrogen fusion is going on now. The required temperature is about 10^7 K.
2. A star that has exhausted its core hydrogen fuel so that only the center is hot enough to burn becomes a **red giant,** in which hydrogen fusion continues in a shell around the core. We mention this as a separate stage only because *the fusion, which is always of the catalytic kind, is the source of nitrogen in the universe, and organisms need nitrogen for their proteins.*
3. The star core continues to contract and heat until it reaches 10^8 K, and helium begins to fuse. Carbon and oxygen are made in roughly equal amounts from three and four heliums, respectively, for carbon-12 (^{12}C) and oxygen-16 (^{16}O). A star like the sun becomes so bright in the later phases of helium fusion that its outer layers blow off, the reactions cease, and the remaining core of carbon and oxygen simply cools off very slowly.

   In case you are wondering what will happen to the solar system 5 By in the future, the sun will engulf the orbits of Mercury and Venus, and even Earth won't be quite safe, because the much brighter sun will heat us well above the boiling point of water. In the final, dying core stage (called a **white dwarf**) the sun becomes very faint, and we all freeze. This is the eventual fate of all single stars less than about eight solar masses when they are born, that is, about 99.9% of all stars.
4. More massive stars reach 10^9 K as their centers continue to contract. The main stages of fusion are then:
   - carbon burning (dominant product neon)
   - neon burning (dominant products oxygen, magnesium)
   - oxygen burning (dominant product silicon)
   - silicon burning (dominant products iron and adjacent elements)

   But during all of these, there are also stray protons, neutrons and helium nuclei running around and being captured, so that from carbon burning, for instance, you get not only the common neon-20 (^{20}Ne) but also its less common isotopes.
5. Something horrible is going to happen to that massive star soon, but before we follow it, let's go back and pick up two stray threads:
   a. Lithium, beryllium and boron are not made in this sequence, because they fuse at even lower temperatures than does hydrogen. They are destroyed in stars, but made in the interstellar gas when very fast-moving protons (called cosmic rays) hit atoms of carbon, hydrogen, and oxygen and break them apart.
   b. Glossed over in step 3 above is a short-lived stage with hydrogen and helium both fusing in thin shells around a carbon-oxygen core. Rising and falling gas streams bring the two shells into partial contact and minor branch reactions liberate neutrons. These are captured one by one, with occasional nuclear decays in between, sometimes up to 150 neutrons per nucleus. It sounds slow and painful, and is indeed called the *s*- or *slow process.* Some nuclides of all elements up to lead and bismuth are made and blown off with the star's outer layers before it becomes a white dwarf.
6. Meanwhile, our massive star has built up an iron core that is growing ever more massive as reactions continue. When it reaches a critical mass, the core collapses suddenly and catastrophically, blowing out most of the heavy elements built up at stage 4 above. Again there are many free neutrons, and they are captured rapidly, in the so-called **r- or rapid process**, to make nearly all the rest of the nuclides beyond iron, reaching to uranium, thorium, and plutonium. Perhaps we all would have been better off without the r-process!
7. We see the explosions, at a rate of a few per century per galaxy, and call them **supernovae** (of Type II). The collapsed cores are generally **neutron stars (1)** or **pulsars,** though occasionally black holes.
8. A still wider range of processes and reactions happen when two stars are born close enough together to interact when they expand to become red giants. Worth noting here is a rare, interesting case in which either the system evolves to leave two white dwarfs, or one white dwarf plus a red giant. Eventually the two stars

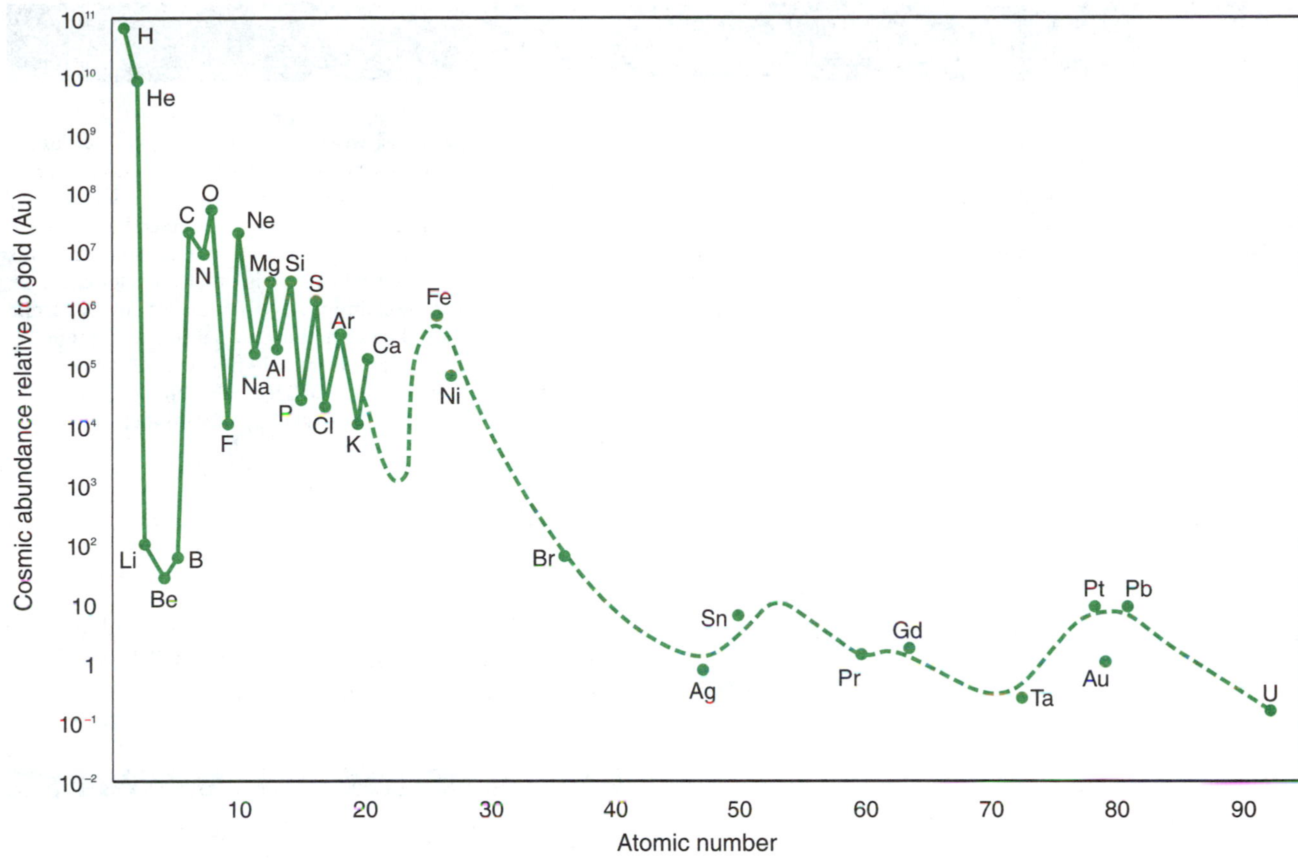

FIGURE 4-4 The relative abundance of stable elements as found in the solar system, other stars and galactic interstellar gas. The units are logarithmic (so that 10 is ten times larger than 9), meaning that hydrogen and helium together constitute close to 100% of the total number of atoms, and that iron at 6 is a million times more abundant than iodine (the rarest element essential to life) at 0. The patterns reflect both the intrinsic properties of nuclei (notice that those with even numbers of protons are always more common than the adjacent ones with odd numbers of protons) and the sequence of nuclear fusion reactions in the early universe and stars that have formed them. (Adapted from Jastrow, R. and M. H. Thompson, 1972. *Astronomy: Fundamentals and Frontiers.* John Wiley, New York.)

come together to make a white dwarf larger than the critical mass mentioned in step 4 above, and the whole thing explodes in a few seconds, producing large quantities of iron and its neighbors, but also carbon, sulfur, and so forth. These are the **Type Ia supernovae** used to study the time history of expansion of the universe. They are even rarer, by a factor of ten or so, than the core-collapse, Type II supernovae.

Galactic Chemical Evolution

The Milky Way and other large galaxies typically have old, metal-poor stars in their outskirts, and old, metal-rich stars at their center. Young stars and ongoing star formation are found largely in the disks of spirals, while the small irregular galaxies have had rather little star formation so far and consist of gas and metal-poor stars with a range of ages.

We can learn much about this process by studying both nearby galaxies and those at redshifts of 1 to 5, where the gradual enrichment was taking place. At least some galaxies at their centers accumulated 1% to 2% of heavy elements within the first billion years after the Big Bang, which is sufficient for planetary systems to form. We are currently aware of about 160 planets forming out of disks surrounding newborn stars through a process that requires dust to condense; you need elements beyond hydrogen and helium to make dust.

In later chapters, you will encounter descriptions of the chemical content of living creatures. Counting by atoms, the most common is hydrogen, which is mostly in water that is left over from the Big Bang. Next are oxygen and carbon, the products of helium burning, followed by nitrogen. Most of the other biologically important elements — potassium, phosphorus, chlorine, sodium, and so forth — come from the heavy-element burning of stage 4. The heaviest and most rare of the essential elements is iodine. Thus, not only are we made of stardust, we are made of rather common, ordinary stardust (**Fig. 4-4**).

BOX 4-2 Alternative Theories to Inflation

EVEN THOUGH ALL BUT HALF a dozen living astronomers and physicists agree that there was a Big Bang — a hot, dense, early stage, from which both radiation and matter remain — many alternatives to inflation have been proposed. Some make use of more than three spatial dimensions; others are variants such as *self-reproducing* and *external inflation*. These imply the existence of a large or infinite number of universes, each with its own space-time and contents, but with the potential of very different numbers for the amounts of the various kinds of matter and energy, different strengths for the forces, perhaps even a different number of forces, and so forth. In that case, generally called the **multiverse,** there is an extreme selection effect: only universes in which observers can arise will have entities like us who ask why their universe is the way it is.

In that case, we can ask whether our universe is a probable one, in the same way we can ask whether our star, the sun, and our galaxy, the **Milky Way,** are common sorts. (The answer for them is, yes.) The cosmological community is divided over whether the multiverse concept is really part of science. The critical issue is whether it is predicted by physics that can be tested in other ways, for example, at accelerators now at the planning stage, observations of the cosmic microwave background or something we haven't yet thought of. Astronomers have strong hopes for results from the European *Planck* satellite, which will analyze the CMB for clues about the formation of galaxies; the satellite is scheduled to fly later this decade.

Shape and Future of the Universe

The longevity of the universe and the way it has evolved depend on its contents. Much more matter, and the universe would have quickly collapsed back on itself; much less, and it would have expanded forever, but probably never have formed stars. Recent measurements tell us, however, that the average density of the universe is remarkably close to what is known as the **critical density**, the point at which neither of these things happen. This critical density — which is roughly 10^{-29} g/cm^3 or six atoms of hydrogen/m^3, meaning that most of the universe is virtually empty — determines the geometry of the universe (see Figs. 4-3 and 4-4). Above this value, and the universe would be *closed* and "positively curved like a sphere," below, and it is "*open*, negatively curved like the surface of a saddle." Around the critical density, however, "the universe is *flat* like a sheet of paper."[8] Indeed, observations by *WMAP* and other sources provide strong evidence that the universe is both flat and — as result of the negative pressure exerted by dark energy, which is causing the rate of expansion to increase — likely to expand forever (**Box 4-2**).

What lies ahead? The early universe is now thought to have comprised a small number of giant galaxies, each of which was a maelstrom of star formation, massive black holes, and so on. The more recent universe is more dispersed, with stars and black holes being formed in large numbers of smaller galaxies. Though this activity is considerably more impressive than previously believed, it is nevertheless far less violent than in earlier days. In effect, the universe is undergoing "*a cosmic downsizing* . . . a massive transition from the mighty to the meek" (Barger, 2005). Whatever the reasons for this trend, it's likely that all galaxies, including our Milky Way, will eventually run out of fuel, and that "the universe will darken, and its only contents will be the fossils of galaxies from its glorious past" (Barger).

[8] NASA, 2005. Is the Universe Infinite? Available at http://map.gsfc.nasa.gov/m_uni/uni_101shape.html. Accessed March 2007.

KEY TERMS

baryonic matter
Big Bang
cosmic microwave background
cosmological constant
cosmological redshift
cosmology
critical density
dark ages
dark energy
dark matter
epoch of recombination
evolutionary universe
expanding universe
expansion rate
galaxy
Milky Way
neutron stars (pulsars)
photon
protogalaxy
radiation
red giant
redshifted
space-time
steady state universe
superclusters
supernovae
white dwarf

DISCUSSION QUESTIONS

1. What evidence indicates that the universe is expanding?
2. Discuss the Big Bang, the epoch of recombination and the dark ages.
3. With regard to stars,
   a. How do stars originate?
   b. What sequence do they follow in their evolution?
   c. How do the various elements form in stars?
   d. What are red giants and white dwarfs?
4. What is likely to happen to our universe in the future? Why?

Virginia Trimble

Born: November 15, 1943

Birthplace: Los Angeles, California

Undergraduate degree: University of California–Los Angeles, 1964

Graduate degrees: M.S., California Institute of Technology, 1965; Ph.D., California Institute of Technology, 1968; M.A., University of Cambridge, England, 1969

Present position: Professor of Physics and Astronomy, University of California–Irvine, and Staff member, Las Cumbres Observatory, Goleta, CA.[9]

What prompted your initial interest in evolution?

My interest in biological evolution is that of an enthusiastic amateur, and dates back to reading, while I was still in grade school, a marvelous book called, *You and Heredity,* by Amram Scheinfeld. That stars and galaxies also evolve, as individuals and as populations, I did not discover until college. Some aspects of stellar and galactic structure and evolution that I continue to find fascinating are (1) that they can be described by exactly the same principles of physics — gravitation, electromagnetism, thermodynamics, and the rest — that we study in terrestrial laboratories; (2) that they all fit together to make a consistent pattern, in which changing populations of stars add up to make the galaxies we see here and now and long ago and far away; and (3) that if any of a number of things had been different (not much carbon built by helium fusion; only massive, short-lived stars formed, etc.) we could not be here to worry about scientific problems.

What do you think has been most valuable or interesting among the discoveries you have made in science?

My Ph.D. dissertation was a study of the Crab Nebula, remnant of a stellar explosion seen in the year 1054. I was able to show that the remnant indeed started expanding about then; that it is being "pushed on" by magnetic fields and high energy particles and so speeding up; and that the amount of matter in it is consistent with the evolution of the giant stars that we think ought to give rise to such supernova explosions. Other areas where I have published original papers include (1) the determination of some of the properties of white dwarfs — the stars left behind when small, long-lived stars like the sun die; (2) studies of populations of binary stars (gravitationally bound pairs) suggesting that their formation is the last stage in a more general problem of star formation; and (3) investigations of some of the short-lived and rare phases of stellar evolution, showing that they, too, fit into the pattern, though they are things our sun will never do.

What areas of research are you (or your laboratory) presently engaged in?

In recent years, I have focused increasingly on the field called scientometrics and on history of science (that is, in effect, on structure and evolution of the astronomical and physics communities, rather than astronomical objects). This has resulted in information about how productive different telescopes are, what kinds of careers astronomers and others of different ages can expect, and other items useful for the community in planning ahead.

In which directions do you think future work in your field needs to be done?

I think most people in the field would agree that the single most important unsolved problem in modern astrophysics is the formation of galaxies. The basic problem is for the matter to gather together into large, complex agglomerations, while leaving the cosmic microwave radiation (which also comes to us out of a hot, dense big bang) smooth throughout the observable universe. Understanding galaxy formation requires information about dark matter, particle physics, plasmas, and a number of other topics.

What advice would you offer to students who are interested in a career in your field of evolution?

Two generations ago, a wonderful woman astronomer named Cecilia Payne Gaposchkin said, "A woman should do astronomy only if nothing else will satisfy her; because nothing else is what she will get." At the present time, I think the main modification that should be made in this advice is to replace "A woman" by "you" and "astronomy" by "scientific research." That is, it applies to both genders, all races, and so forth, and to all of the sciences. You are unlikely to get rich, unlikely to become famous, unlikely to be understood by your family and friends. The rewards of finding out things that no one has ever known before (that is what is meant by "research") are enormous, but they are to be achieved only by exceedingly hard work, and do not bring much recognition or other secondary rewards.

[9] University of California—Irvine. Virginia Trimble (biography). Available at http://www.ps.uci.edu/physics/trimble.html. Accessed March 2007. Recent publication, text of the Klopsteg Lecture: Trimble, V., 2006. Early photons from the early universe. *New Astron. Revs* **50,** 844–849.

EVOLUTION ON THE WEB biology.jbpub.com/book/evolution

EVOLUTION ON THE WEB

Explore evolution on the Internet! Visit the accompanying Web site for *Strickberger's Evolution,* Fourth Edition, at http://www.biology.jbpub.com/book/evolution for Web exercises and links relating to topics covered in this chapter.

RECOMMENDED READING

Abel, T., G. L. Bryan, and M. L. Norman, 2002. The formation of the first star in the universe. *Science*, **295,** 93–98.

Barger, A., 2005. The midlife crisis of the cosmos. *Sci. Amer.*, **292**(1), 46–53.

Bouwens, R. J., and G. D. Illingworth, 2006. Rapid evolution of the most luminous galaxies during the first 900 million years. *Nature,* **443,**189–192.

Hawking, S. W., 2001. *The Universe in a Nutshell.* Bantam Books, Toronto.

Kaler, J. B. 2006. *The Cambridge Encyclopedia of Stars*. Cambridge University Press, Cambridge, UK.

National Aeronautics and Space Administration (NASA), 2007. *Cosmology: The Study of the Universe.* (Web site devoted to the Wilkinson Microwave Anisotropy Probe.) Available at http://map.gsfc.nasa.gov/m_uni.html. Accessed March 2007.

Kragh, H., 1996. *Cosmology and Controversy: The Historical Development of Two Theories of the Universe.* Princeton University Press, Princeton, NJ.

Lineweaver, C. H., and T. M. Davis, 2005. Misconceptions about the Big Bang. *Sci. Amer.,* **292**(3), 24–33.

Loeb, A., 2006. The dark ages of the universe. *Sci. Amer.,* **295**(5), 46–53.

Longair, M. S. 2006. *The Cosmic Century*. Cambridge University Press, Cambridge, UK.

Rees, M. J. 2001a. *Our Cosmic Habitat*. Princeton University Press, Princeton, NJ.

Rees, M. J. 2001b. *Before the Beginning.* Addison Wesley, Reading, MA.

Riordan, M., and W. A. Zajc, 2006. The first few microseconds. *Sci. Amer.,* **294**(5), 24–31.

Trimble, V. 2003. Cosmology: Man's place in the universe. *Amer. J. Physics* **79,** 1175–1183.

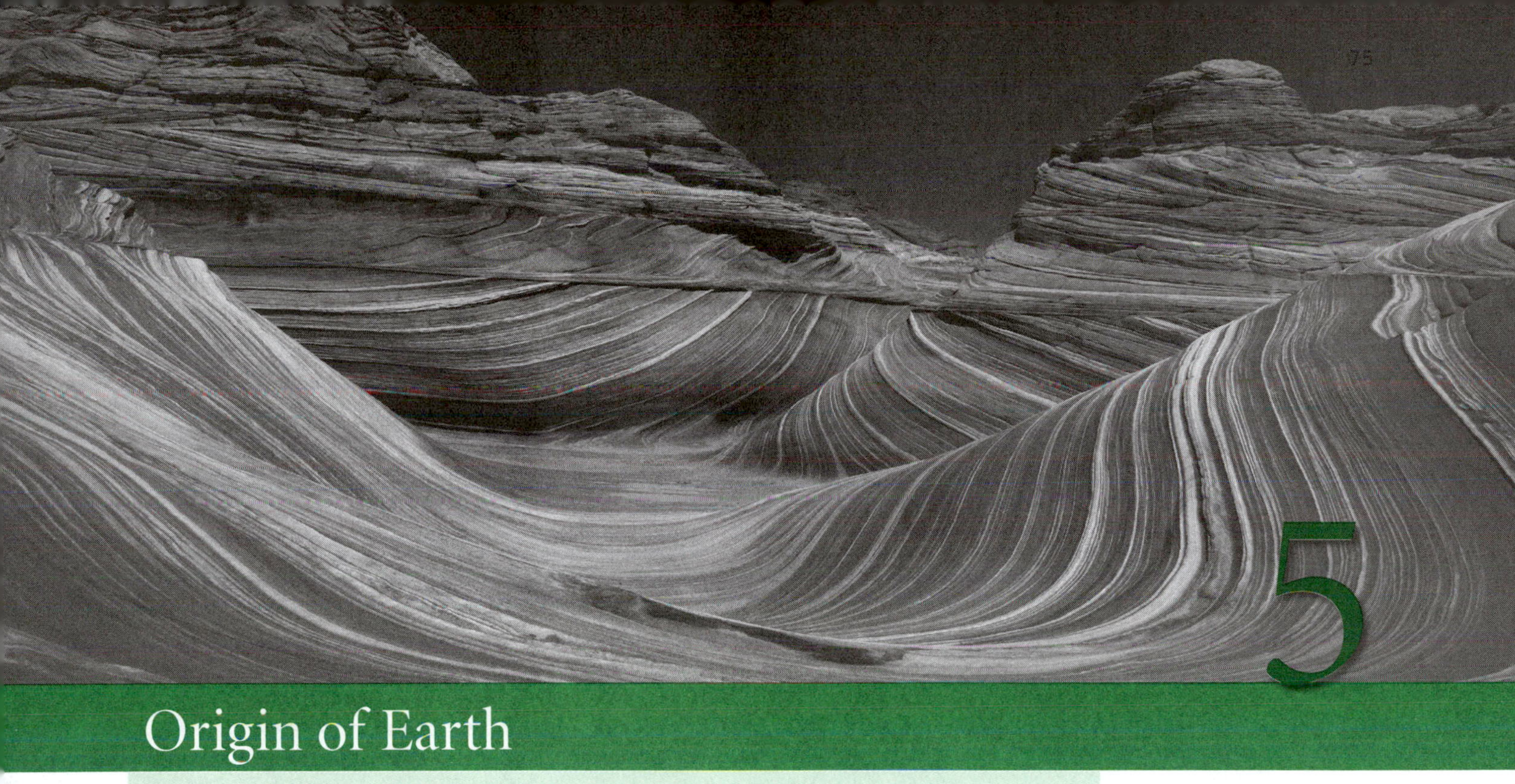

5 Origin of Earth

Chapter Summary

Our solar system is calculated to have originated by condensation from a rotating mass of gas and dust about 4.6 billion years ago (Bya). The central mass became the sun and peripheral masses became the planets. The four inner planets are rocky and Earth-like, and the four larger outer planets are made of gas. Earth's **atmosphere** at first was composed mostly of hydrogen and helium, but these elements soon were replaced by a secondary atmosphere comprised of gases such as ammonia, water vapor, methane, carbon dioxide and nitrogen, forming an atmosphere with little or no oxygen. The high proportion of oxygen in Earth's present atmosphere almost certainly owes its origin to the actions of photosynthetic organisms.

Earth's interior comprises several concentric layers differing in composition and physical properties. The **core** mostly consists of iron and nickel and is surrounded by a thick, partially ductile mantle. Blanketing the mantle is a relatively thin crust of igneous, sedimentary and metamorphic rocks, distributed in layers or strata according to their age. Different strata of sedimentary rocks can be recognized by the types of fossils they contain. Exact dating of strata had to await the discovery of radioactive isotopes of elements such as uranium whose disintegration rates could be measured. Using such techniques, the oldest rocks are estimated to be 4.1 By old.

Alfred Wegener's (1880–1930) proposal in 1912 that the present landmasses were once united as one huge continent, *Pangaea,* has garnered much support. Evidence includes the compatible profiles of continents, the similarity of rocks and fossils in previously conjoined areas, the direction of magnetism in rocks and the "youthful" structure of the ocean floor. Pangaea began to disintegrate about 225 million years ago (Mya) when one section, *Gondwana,* broke into massive blocks — South America/Africa and India/Antarctica/Australia —

leaving behind *Laurasia* (North America/Eurasia). Subsequent rifts and fusions created the continents we know today.

We can explain the history of the continents and many of Earth's features, such as earthquake belts, by the movement of at least eight giant *plates* comprising Earth's lithosphere. These plates can separate from, slide past, or converge on each other. We call such activity *plate tectonics*, and explain the distribution of many organisms on the basis of the historical fusion and separation of these huge landmasses. Most notable is the way in which the separation of part of Gondwana from the rest of Pangaea restricted monotreme mammals to Australia and marsupial mammals to Australia and South America. Elsewhere, placental mammals replaced monotremes and marsupials. Many other examples provide evidence for the effects of alterations of Earth's surface on the evolutionary patterns of life.

On February 14, 1990, after giving us views of Jupiter and Saturn a decade before, the *Voyager I* spacecraft, 6.4 billion km from Earth and on its way out of our solar system, turned its cameras back towards the sun to give us our first view of our solar system.

Origin of the Planets

Astronomers have proposed two main theories for the evolution of planets in our solar system. According to the **collision theory**, proposed in 1749 by Georges Buffon (Rudwick, 2005) and promoted in various forms by others until the early twentieth century, a star or comet either collided with our sun, throwing out debris that formed the planets and other bodies, or passed close enough to pull out, through gravity, the material that became the planets. The most serious of many difficulties with this theory is the extreme rarity of such events: Astronomers estimate collisions or near collisions between stars to have occurred in our galaxy only ten times in the last 5 By, whereas more than a billion stars in our galaxy have planets.

In one of its various forms, most astronomers today accept an alternative **condensation theory** or **nebular hypothesis**, first suggested by Immanuel Kant in 1755 and later elaborated by Pierre-Simon Laplace. In its modern form, the condensation theory posits that the huge mass of dust and gas out of which our solar system condensed began collapsing under its own gravity about 5 to 5.6 Bya, perhaps as the result of a nearby disturbance, such as a supernova, which is the spectacular explosion of a massive, dying star (see Chapter 4). The large condensing mass at the center of the cloud began heating up, eventually reaching thermonuclear reaction temperatures about 4.6 Bya and becoming the sun. The remaining material formed a whirling "accretion disk" around the sun in which smaller condensations grew as the material cooled (**Fig. 5-1**). At first, dust particles merely adhered to each other, but once the resulting bodies reached diameters of about a kilometer, gravity kicked in, and the process of planet formation was boosted as vast numbers of bodies, now called **planetesimals**, collided together and coalesced, eventually forming **protoplanets** and then **planets.** These peripheral masses remained tied to the solar orbit, although some gained subplanets or moons of their own (**Box 5-1**).

Recall from Chapter 4 that our sun is not a first generation star, and that the dust and gas out of which our solar system developed contained the full range of elements. The distribution of those elements as the planets formed, however, was far from uniform. The four inner planets (Mercury, Venus, Earth, Mars) are comprised largely of rocks and minerals, and are much denser than the four planets farthest from the sun (Jupiter, Saturn, Uranus, Neptune),[1] which are essentially huge balls of helium and hydrogen surrounded by rings and many satellites (moons). The inner planets, on the other hand, lack rings and have few satellites. This distribution reflects the temperature gradient associated with distance from the sun. Unlike the materials comprising the inner planets, hydrogen and helium solidify only at extremely low temperatures and were unable to form condensations so close to the sun. They were, therefore, unable to withstand the *solar wind*, a powerful, fast-moving stream of particles that swept clean the warmer inner areas of these gases and any remaining dust.

The sun itself has not escaped changes, both on astronomical time scales — the sun has become progressively warmer since its origin almost 5 Bya — and on more human time scales. Since the early 1700s, there has been an 11-year cycle in solar magnetic activity and therefore in sunspots. These cycles are not absolute. A 70-year period free of sunspots between 1645 and 1715 reduced solar power by 0.5%, which seems trivial, but was sufficient to plunge the more northern parts of the Northern Hemisphere into a Little Ice Age that lasted until between 1850 and 1900.[2]

[1] As of August 2006, tiny, distant Pluto is no longer recognized as a planet (see Box 5-1).

[2] In 1645, the inhabitants of the village of Chamonix, located at the foot of Mont Blanc in Switzerland, were concerned that the advancing glacier was overtaking and destroying their villages and farms. They called in the Bishop of Geneva, who performed an exorcism at the front of the advancing glacier. The glacier retreated only to advance again despite repeated exorcisms. These villages were facing the same Little Ice Age that iced up the North Atlantic and killed centuries-old orange groves in China.

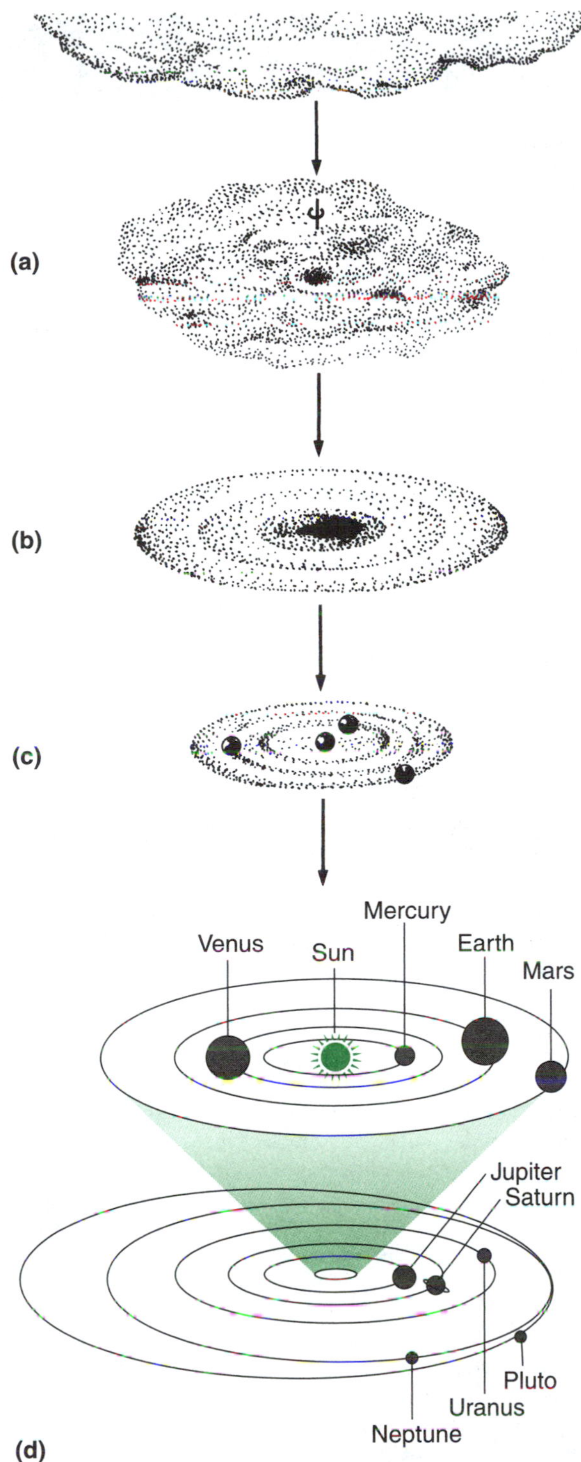

FIGURE 5-1 Stages during the condensation of the solar nebula into the solar planetary system. (a) Fragmentation of an interstellar cloud. (b) Contraction and flattening of the solar nebula. (c) Condensation of nebular material into meteorites and protoplanetary bodies. (d) Solidification of planets, with an indication of present orbits. An "asteroid belt" consisting of many thousands of bodies with sizes ranging up to about 1,000 kilometers in diameter, lies between Mars and Jupiter. Its fragmented nature is probably the result of the proximity of such a massive planet as Jupiter, which swept up or ejected objects that would ordinarily have aggregated into a planet of their own. Along with comets whose orbits cross that of Earth, the asteroid belt and its subsidiaries provide nearly all the 20,000 meteorites that annually enter Earth's gravitational field, most weighing in the range of 1 to 10 kilograms.

Scientists have long wondered if planets orbit stars other than our own. Because such stars are far too distant for us to see directly, astronomers have had to rely on indirect observation. By August 2006, about 200 extrasolar planets had been discovered, many of them parts of systems with more than one planet. The existence of other planetary solar systems underscores a long-standing question addressed in **Box 5-2**.

Origin of Earth's Atmosphere

Earth's **first atmosphere**, which consisted of hydrogen and helium, did not last. Extremely powerful winds from the early sun, the relatively low gravitational forces of Earth that were unable to hold the light gases, and heat generated both within Earth itself and by the sun all have been suggested as forces dissipating the first atmosphere.

A **secondary atmosphere** arose between 4.2 and 3.8 Bya, largely from volcanic out-gassing during a period of active volcanism, although comets have been suggested as another source of at least some of the gases, especially water vapor. The major constituents of the second atmosphere are thought to have been water vapor and carbon dioxide (CO_2); other gases present probably included hydrogen (H_2), carbon monoxide (CO), nitrogen (N_2), ammonia (NH_3), methane (CH_4), hydrochloric acid (HCl) and hydrogen sulfide (H_2S), most of which issue from volcanoes and hot springs today. Some scientists have suggested that the methane in the atmosphere was a byproduct of a group of single-celled microbes known as **methanogens**. As discussed in Chapter 8, recent studies provide evidence for methane and methanogens in rocks dated at 3.5 Mya, just a little later than the estimated dates for the accumulation of the first atmosphere (3.8 to 4.2 Mya). In such an atmosphere devoid of oxygen, this greenhouse gas could well have accumulated in quantities sufficient to act as a warming blanket, preventing the icy conditions that might have been expected, given that the sun then gave off only about 80% of its current heat (Kasting and Catling, 2003). Indeed, the heat reaching us from the sun has steadily increased since Earth's early days.

As time went on, Earth's surface temperature cooled and liquid water formed, allowing CO_2 to be absorbed by the oceans and to react with silicates to produce carbonates, reducing the amount of CO_2 in the atmosphere. A number of the noble gases such as neon, argon and xenon also may have been prevalent and should have persisted to this day in relatively high quantities because they are chemically inert.

BOX 5-1 Planets and Dwarf Planets

PLUTO, WHICH IS MAINLY composed of rock and ice, was recognized as the ninth planet in our solar system until August 24, 2006. On that day in Prague, Czechoslovakia, delegates at the closing ceremony of the general assembly of the 2006 International Astronomical Union (IAU) passed several resolutions. One of those was a definition of a planet as a "celestial body that (1) is in orbit around the Sun, (2) has sufficient mass for its self-gravity to overcome rigid body forces so that it assumes a hydrostatic equilibrium (nearly round) shape, and (3) has cleared the neighborhood around its orbit." Pluto fails the test because it does not dominate its neighborhood.

So if not a planet, what is Pluto? According to a second resolution passed in Prague, Pluto is one of three members of a new class of **dwarf planets**, a class so far restricted to our solar system; the other two are Ceres (previously an asteroid) and a dwarf planet given the temporary name of 2003 UB313. It is expected that many other small bodies will be added to the list.

- A **planet** is now a celestial object that orbits the sun, is sufficiently large to have become round due to the force of its own gravity, and dominates the neighborhood around its orbit, sweeping up asteroids, comets and other small objects.
- A **dwarf planet** fulfills the first two parts above, but fails the third.
- A third category of celestial objects recognized in Prague are **solar system bodies**, a category that includes asteroids, comets and moons.

Needless to say, some astronomers are unhappy that Pluto has been voted out of the planetary club. Indeed, a vote may seem a strange way for a scientific body to decide something as fundamental as the definition of a planet. What it reflects, of course, is the difficulty of naming objects that fall outside the standard definition of a category, or that have stopped their development at what appears to be an intermediate stage. Subspecies and varieties lie in the same relation to species as dwarf planets and solar system bodies do to planets and as protocells do to cells. At every level, evolution brings home to us the difficulties associated with dealing with process and pattern, and with origins and final form as we attempt to reconstruct historical events from one or more snapshots of the historical processes responsible for those events, processes that are ongoing.

BOX 5-2 Is there Life Elsewhere in the Universe?

OUR CONCEPT OF OURSELVES in relation to the universe we live in has changed radically, especially in this last century. Not only are we on a planet on the fringes of a galaxy containing more than 100 billion stars, we are in a universe containing billions of other galaxies. As we saw in Chapter 4, there are proposals among astrophysicists that our universe may be just one of many. Inevitably, this raises the question of whether the evolution of life has been repeated on planets of other stars, and how different from life on Earth such life might be. The evidence suggests that life on Earth depends on the following features:

- appropriate atomic elements and available reactive molecules;
- a sun of moderate size (between 0.8 and 1.5 solar masses) providing radiant energy for many hundreds of millions of years;
- a planet properly distant from its sun, following an orbit that eliminates extreme temperatures;
- a protective yet reactive atmosphere; and
- liquid water, a solvent that allows essential biochemical reactions.

Are these features unique to Earth, or do they exist elsewhere? Could life exist in the absence of one or more of the features? Among proposals for an alternative to our carbon-based, water-solvent biochemistry are silicon-based systems and an ammonia solvent.[a]

As mentioned in the text, the recognition of planets in other solar systems indicates that planet formation must be common in our galaxy and in most, or even all, others. Most astronomers feel that many features supporting life could have developed throughout the universe. Given the more than 100 billion stars in our galaxy, even a one percent chance for the origin of an Earthlike planet would provide more than a billion opportunities for the evolution of life.

Back in our own solar system, all the major types of environment that would have existed on Earth during the formation of life would have been expected on early Mars: hot springs, salt pools, rivers, lakes, volcanoes, and so forth. There would even have been tidal pools on Mars, albeit at a much reduced level because the tides were solar. The possible nonbiological sources of organic material (comets and other solar system bodies; see Chapter 6) would have supplied both planets. Perhaps the major unknown is whether Mars had Earthlike environments for a long enough time to allow for the origin of life. The possibility of life on Mars remains a controversial subject within the scientific community.

Two approaches are currently used to investigate the possibility of life elsewhere. First is the search for electromagnetic signals emitted by intelligent creatures. Such signals, used in radar or in radio and television communication, can carry considerable information through pulsed or modulated frequencies that move through space at the speed of light. The second method, discussed in the next chapter, is the identification of organic molecules in meteorites.

[a] Information about silicon or ammonia-based systems is available at http://nai.arc.nasa.gov/astrobio/feat_questions/silicon_life.cfm and http://www.daviddarling.info/encyclopedia/A/ammonialife.html. Accessed March 2007.

TABLE 5-1 Present composition of Earth's atmosphere[a]

Gas	Percent by Volume
Nitrogen (N_2)	78.09
Oxygen (O_2)	20.95
Argon (Ar)	0.93
Water vapor (H_2O)	Variable (up to 1.00)
Carbon dioxide (CO_2)	0.038
Neon (Ne)	0.002
Helium (He), methane (CH_4), carbon monoxide (CO), krypton (Kr), nitrous oxide (N_2O), hydrogen (H_2), ozone (O_3), xenon (Xe)	Less than 0.001

[a] See Table 5-2 for some facts and figures concerning Earth.

Their almost complete absence in the current atmosphere is so far unexplained.

In any case, Earth's second atmosphere was probably **reducing** because of the prevalence of hydrogen compounds capable of providing electrons to **oxidizing** agents capable of accepting them. Evidence exists in thick river deposits laid down in South Africa and other places around 3 Bya, and since then locked away from Earth's atmosphere. Such deposits include sand grains of sulfides of iron (FeS), lead (PbS) and zinc (ZnS), compounds that are highly unstable in the presence of oxygen. If oxygen were present in the atmosphere at the time these compounds formed, they would have oxidized to sulfates (for example, $FeSO_4$).

Oxygen

The present composition of Earth's atmosphere is outlined in **Table 5-1**. Importantly, oxygen makes up 21% by volume. Geochemists generally agree that the proportion of free oxygen in the atmosphere began to increase about 2.3 Bya. Driving this increase were cyanobacteria (see Chapter 12), the first organisms to leave behind fossils, beginning more than 3.5 Bya. Known to some as "the architects of the atmosphere," cyanobacteria are aquatic and photosynthetic. As discussed in Chapter 8, electron transfer in the photosynthetic process involves the removal of hydrogen ions from water molecules, producing free oxygen, which diffuses to the atmosphere. At first, free oxygen would have reacted with ammonia to form free nitrogen (N_2) and with surface materials such as iron. Eventually, after hundreds of millions of years, oxygen began accumulating in the atmosphere.

Two other results of this process would have been declining levels of CO_2 — consumed during photosynthesis — and the creation of an ozone layer through ultraviolet (UV) irradiation of oxygen high in the atmosphere. Life on land was not possible without the protection against UV light afforded by the ozone layer. These topics are elaborated in later chapters. In addition, oxygen had enormous impacts on life on Earth.

Origin of Earth's Structure and of the Moon

Earth today is a layered structure, its four major layers nested within each other (**Fig. 5-2**, **Table 5-2**). During the period when planetesimals coalesced to form our planet, collisions and radioactive decay of elements already present caused the temperature to increase substantially. Consequently, Earth became partially molten, allowing material to move around. Heavier elements (particularly iron) sank inwards, while lighter material gravitated towards the surface (**Fig 5-3**). Differentiation into layers is thought to have been well underway when the moon formed about 4.53 Bya, about 34 My after the date (4.56 Bya) that many believe marks Earth's origin.

The Moon

The generally accepted theory is that the **moon** formed when a planetesimal at least the size of Mars crashed into Earth at an oblique angle (called by some *the Big Splash*), a theory first proposed by Hartman and Davis in 1975. (Because the moon lacks iron, the impact must have happened *after* much of Earth's iron had sunk into its middle.) After impact, the core of the planetesimal would have merged with Earth, resulting in remelting of much of our planet. A major part of the planetesimal and material from Earth rebounded into space, clumped together, and remained gravitationally tied to Earth as its moon. This process was followed by intense meteorite bombardment of both Earth and the moon, ending about 3.8 Bya. The massive cratering on our side of the moon, marked by about 50 basins, each more than 300 km across, dates to this period.[3]

Events of this period also indicate that present life began no earlier, because such intense bombardments may well have sterilized Earth's surface (see Fig. 8-5). Nevertheless, such impacts probably brought to Earth large amounts of water, hydrogen, nitrogen, other elements and organic compounds (see Fig. 6-8) that could be used for biological purposes (Chyba et al., 1995). Wetherill (1995) suggested that, because large amounts of comet impact material were swept up by the giant planet Jupiter, there were far fewer catastrophic impact events on planets within the inner solar system after this period. The formation of Jupiter, therefore, allowed Earth's orbit and climate to stabilize. As Earth cooled, a crust began forming (see below) and eventually water condensed to form rain and finally the oceans.

[3] See Hartman and Davis (1975), Hartman (1997), and Canup (2004) for the origin of the moon.

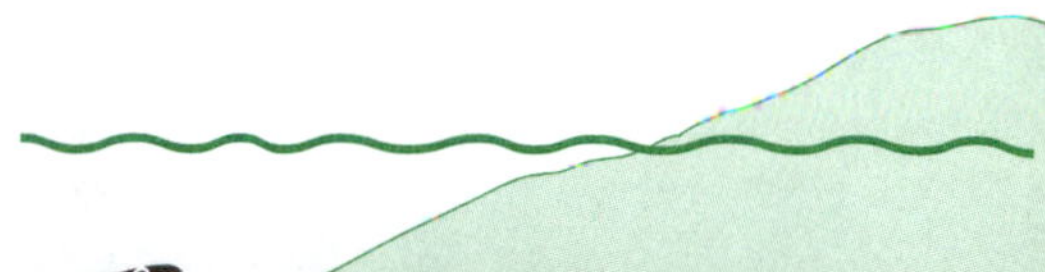

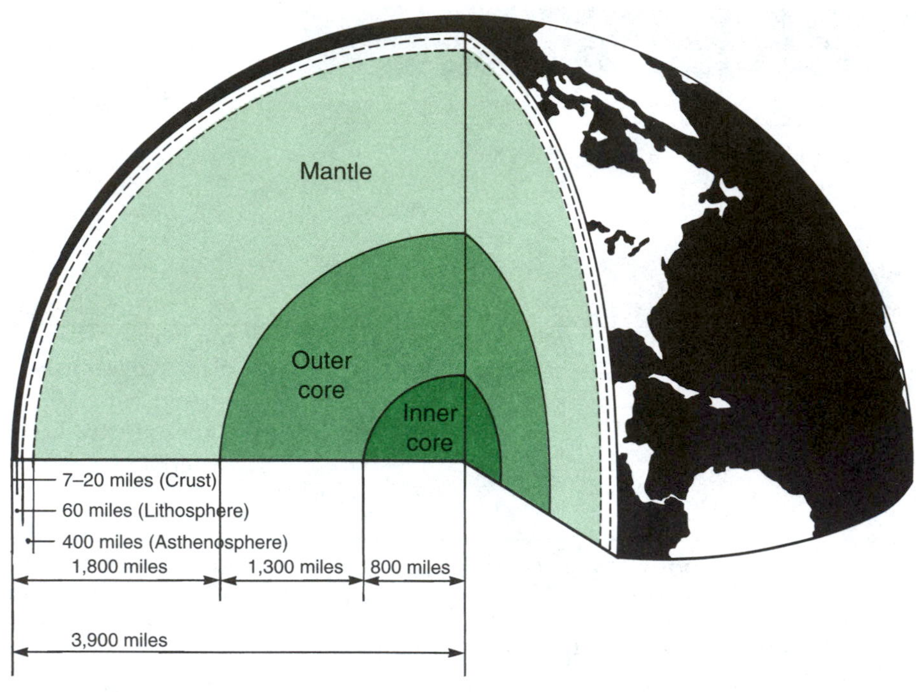

FIGURE 5-2 A section through Earth, which has a radius of 3,948 miles (6,357 km) at the poles and 3,960 miles (6,378 km) at the equator. The lithosphere consists of relatively rigid plates composed of the rocky crust plus a portion of the underlying mantle, reaching a depth of about 50 miles (80 km) at the oceanic basins and 60 to 90 miles (96 to 145 km) at the continents. Below the lithosphere is a more plastic, deformable material, the asthenosphere, which allows the lithospheric plates to move about. Further distinctions between these and other layers in the outer Earth are reviewed in Rogers, J. J. W., 1993. *A History of the Earth.* Cambridge University Press, Cambridge, England.

TABLE 5-2 Some basic facts and figures concerning Earth[a]

Diameter	12,756 km (7,926 miles)
Mass	5.974×10^{24} kg
Volume	1.08×10^{12} km^3
Day length	23.934 hrs
Earth year	365.256 days
Distance from the sun	147 to 152 million km (91 to 94 million miles)

[a] See Table 5-1 for the composition of Earth's atmosphere.

Earth's Core

Earthquakes have helped us unravel the mysteries of Earth's interior. Sensitive seismographs are used to detect **seismic waves,** whose paths and velocities depend on the composition, fluidity and thickness of the materials through which they travel. Combined with studies of Earth's magnetic, electric and gravity fields, the seismic information gleaned reveals a complex picture. At the center is the extremely hot **inner core,** a solid iron mass (with some nickel) about 1,287 km in radius surrounded by the **outer core,** a molten iron envelope mixed with sulfur or silicon about 2,090 km thick (see Fig. 5-2). Shifts in the outer core are believed responsible for changes in Earth's magnetic field.

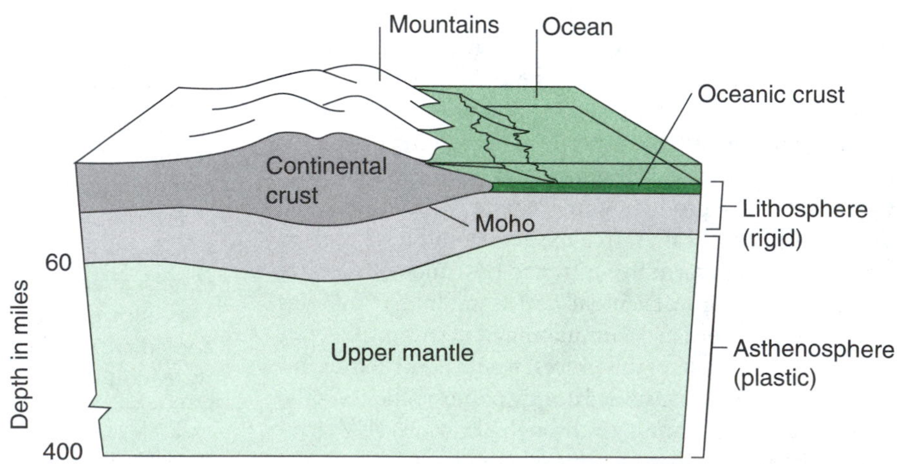

FIGURE 5-3 A section of Earth's crust, showing differences in thickness. Because the crust has less density than Earth's subsurface material, it floats on the mantle. Different thicknesses of the crust float at different levels. The thicker, more buoyant continental landmasses float higher than do the thinner, less buoyant oceanic basins. At the Moho discontinuity, a sharp change occurs in the velocity of certain seismic waves. Geologists consider this area to represent the boundary between crust and mantle.

Many of us may think of Earth's **magnetic field** as just that, an energy field that is magnetized. But it is much more than that. Earth is surrounded by or encased in an enormous magnetic bubble, the magnetosphere, which, along with our atmosphere, shields us from incoming radiation: x-rays, UV, and cosmic rays, and ionized particles. But the **magnetosphere** does more; it shields us from the electrically charged particles carried by solar winds, which pass Earth at speeds of 400 km/second.

Earth's Mantle and Crust

Surrounding the iron core is the **mantle**, a layer of rock about 2,900 km thick that comprises approximately four-fifths of Earth's volume (see Fig. 5-2). Because of radioactive decay, pressure and localized heating or cooling by convection, the mantle has experienced repeated melting and crystallization. Geologists now characterize it as a partly plastic or ductile structure whose density and temperature increase with nearness to the core; mantle physics is extremely complex.

Floating on the surface of the mantle is the **crust**, of which there are two types: the thicker (and mostly older) **continental crust**, and the thinner, younger **oceanic crust**. The crust plus the uppermost portion of the mantle are collectively known as the **lithosphere**, crust and mantle being separated by a discontinuity known as the **Moho** (Mohorovicic) **discontinuity**, made evident because of a sharp change in the velocity of earthquakes as they transit this layer (see Fig. 5-3 for details).[4] The lithosphere varies in thickness from 250 km or more under mountains to a few kilometers in mid-ocean ridges (see below), though the oceanic crust in general is of the order of several tens of kilometers thick.

We know most about the crust, in which we can distinguish *three fundamental types of rocks*:

1. **Igneous rocks** are formed when molten rock (magma) formed within the mantle cools and solidifies. If magma cools slowly, deep within Earth, the minerals in it crystallize out to form coarse-grained rocks such as *granite* (although granite can form in other ways). Magma may also be deposited directly on the surface in the form of *lava*, which cools quickly to form fine-grained rocks such as basalt.
2. **Sedimentary rocks.** Weathering of igneous (and other) rocks by water, wind, glaciers and chemical reactions produces particles that are transported and reformed into new arrangements along with the dust thrown up by volcanic activity. Thus, a stream may deposit its sediments at the bottom of a lake, and blowing sand may form dunes. With time, layers of considerable depth form. Should such layers harden, either through the pressure of other material above them or by chemical means, *sedimentary rocks* result. Sandstone (sand origin), shale (mud origin) and limestone (calcium carbonate) are examples of sedimentary rocks. Limestone may be formed from the remains of organisms such as corals, mollusks and other organisms that live in marine reefs and shallow seas, which incorporate calcium carbonate into their skeletons and deposit calcium carbonate into their habitat.
3. **Metamorphic rocks** have undergone significant changes because of heat, pressure, and/or chemical interactions. For example, marble is a metamorphic rock that was originally limestone; slate is a metamorphic rock that was once shale.

As shown in **Figure 5-4**, a rock cycle exists in which, given enough time, these three major types of rock transform from one to the other, although not necessarily in equal proportions. Geologists have determined that Earth's crust consists, by volume, of 65% igneous rocks, 8% sedimentary rocks and 27% metamorphic rocks. A layer of sedimentary rocks covers most of the surfaces of continental landmasses and much of the ocean floor.

Geological Dating

Beginning in the seventeenth and eighteenth centuries, it became known that the relative positions of different rocks could be used to determine their respective ages.

A Danish founder of geology and early proponent of the validity of fossils, Nicholas Steno (1638–1686) was among the first to establish the **law of superposition**: if a series of sedimentary rocks has not been overturned, the oldest layers or strata will be at the bottom of the series and the youngest layer at the top. More than a century later, the English geologist William Smith (1769–1839)[5] discovered how to identify different strata by the unique kinds of fossils within them, and published the first geological map in 1815. Using

[4] The magnitude of an earthquake is measured using one of two scales: the *Richter magnitude scale* (named after a Californian seismologist), which measures wave energy, and the *Mercalli intensity scale*, which measures the destructive effects of an earthquake. The Richter scale is logarithmic, from 1 to 10, and the Mercalli is numeric, from I to XII. At I on the scale, we feel no Earth movement, but at XII the ground moves in waves and almost everything is destroyed (see the Web site maintained by the U. S. Federal Emergency Management Agency [FEMA] at http://www.seismo.unr.edu/ftp/pub/louie/class/100/mercalli.html). The largest earthquake recorded — off the Chilean coast in 1960 — measured 9.5 on the Richter scale.

[5] Simon Winchester (2001) has written an engaging biography of William Smith and the geological map Smith published. A classic "geology" textbook, *Earth Sciences,* is now in its ninth edition (Tarbuck and Lutgens, 2006). Although initially known as geology and an ancient science (Chapters 1 to 3, and see Rudwick [2005] for a history of what he calls **geohistory**), most university departments of geology have been renamed departments of earth sciences to reflect the multi- and interdisciplinary approaches brought to bear on the study of Earth.

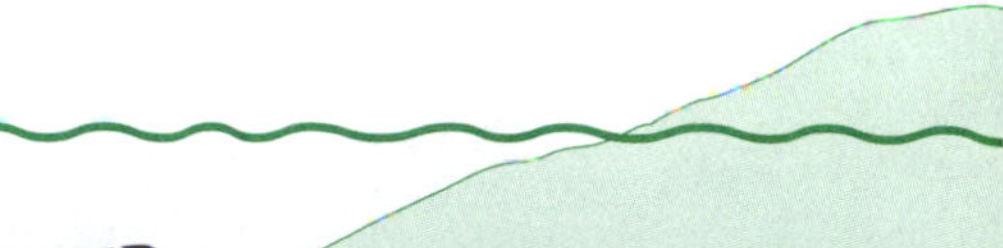

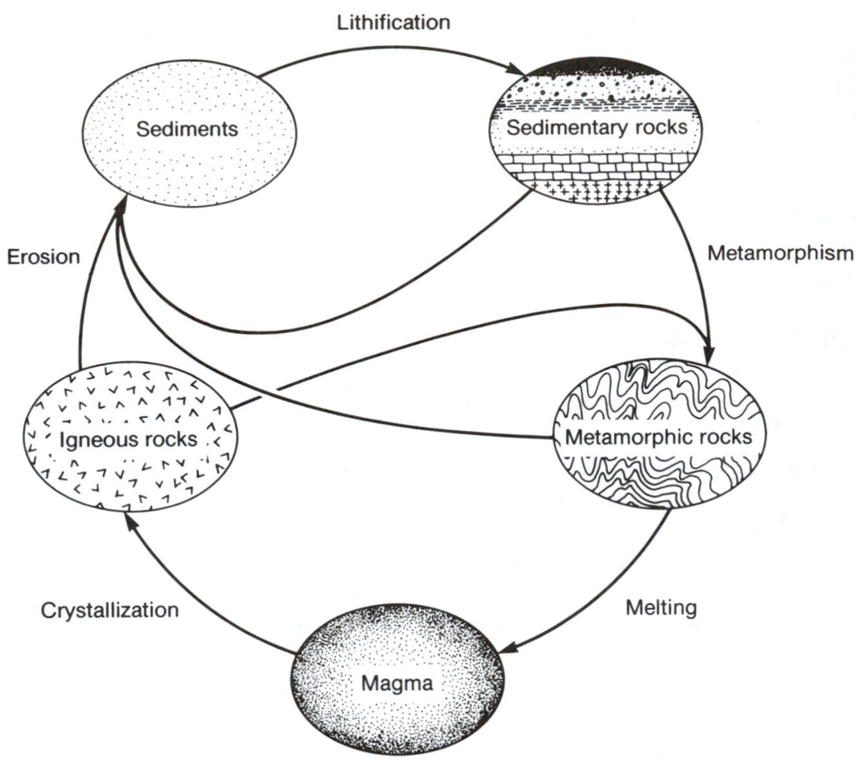

FIGURE 5-4 Diagrammatic representation of transitional events in the rock cycle. Some crustal rocks have recycled many times and others have persisted with little change from the initial formation of crustal rocks. Geologists estimate that about half of all crustal rocks have formed during the last 600 My (Adapted from Hawkesworth, C. J. and A. I. A. Kemp, 2006. Evolution of the continental crust. *Nature*, **443**, 811–817.)

changes in fossils, Smith also identified the Permo-Triassic boundary, a time of enormous extinction of life.

As Georges Cuvier and others showed, the relative ages of the fossils corresponded closely to the relative ages of the strata in which the fossils were found; fossils from the uppermost strata seemed more like extant organisms than fossils from lower strata (**Fig. 5-5**; Rudwick, 2005). Although relative dating by stratigraphic methods usually establishes a sequential relationship between different rocks and between different fossils, stratigraphy does not offer information on the lengths of time involved. Sediments do not deposit in identical thicknesses from time to time or place to place. Furthermore, large sections of the geological record in all localities have been worn away by weathering and erosion or destroyed by new rock formations and Earth movements. Nowhere does the geological record offer a complete sequence that we can trace continuously, year by year, to the present time.

These limitations notwithstanding, fossils became a primary means by which a particular *geological stratum* (*layer*) *or group of strata* (*system*) could be traced from one region to another. For example, the Cambrian system (named after a Welsh tribe by Adam Sedgwick in 1835) represents strata in which many marine invertebrate skeletons such as those of trilobites, brachiopods and mollusks first appear (see Chapter 15 for the Cambrian explosion of animal evolution). Cambrian strata exist on all continents and occupy the same relative positions; that is, they lie above Precambrian strata (identified, in part, by the absence of fossil shells) and below Ordovician and Silurian strata, which contain corals, echinoderms, small early fishes and so on.

Because they represent only a partial sampling of organisms — mostly those with shells, skeletons or hard parts deposited in appropriate sediments — fossils are infrequent in all geological strata (see Fig. 3-17). Soft-bodied organisms, which can be used to identify strata, are extremely rare in the fossil record. Furthermore, because they may have lived only in restricted habitats or areas, the same fossils are not always present in all the locations where a stratum is found. Nevertheless, a particular stratum can be identified because it contains at least some fossils characteristic of that period.

By these means, geologists have determined that significant numbers of hard-bodied organisms existed more than half a billion years ago. They call the time that has elapsed since then the **Phanerozoic Eon** (from *phanero*, visible, and *zoon*, life). An *eon* is the largest division of geologic time, consisting of two or more **eras**. The Phanerozoic consists of three major eras of geological strata, beginning with the Paleozoic. As shown in **Table 5-3**, each era contains a number of subsidiary systems or **periods**, often further subdivided into **epochs**. William Smith's nephew, John Phillips, who provided the first estimate of Earth's age based on rates of sedimentation (see Chapter 2), used changes in fossil assem-

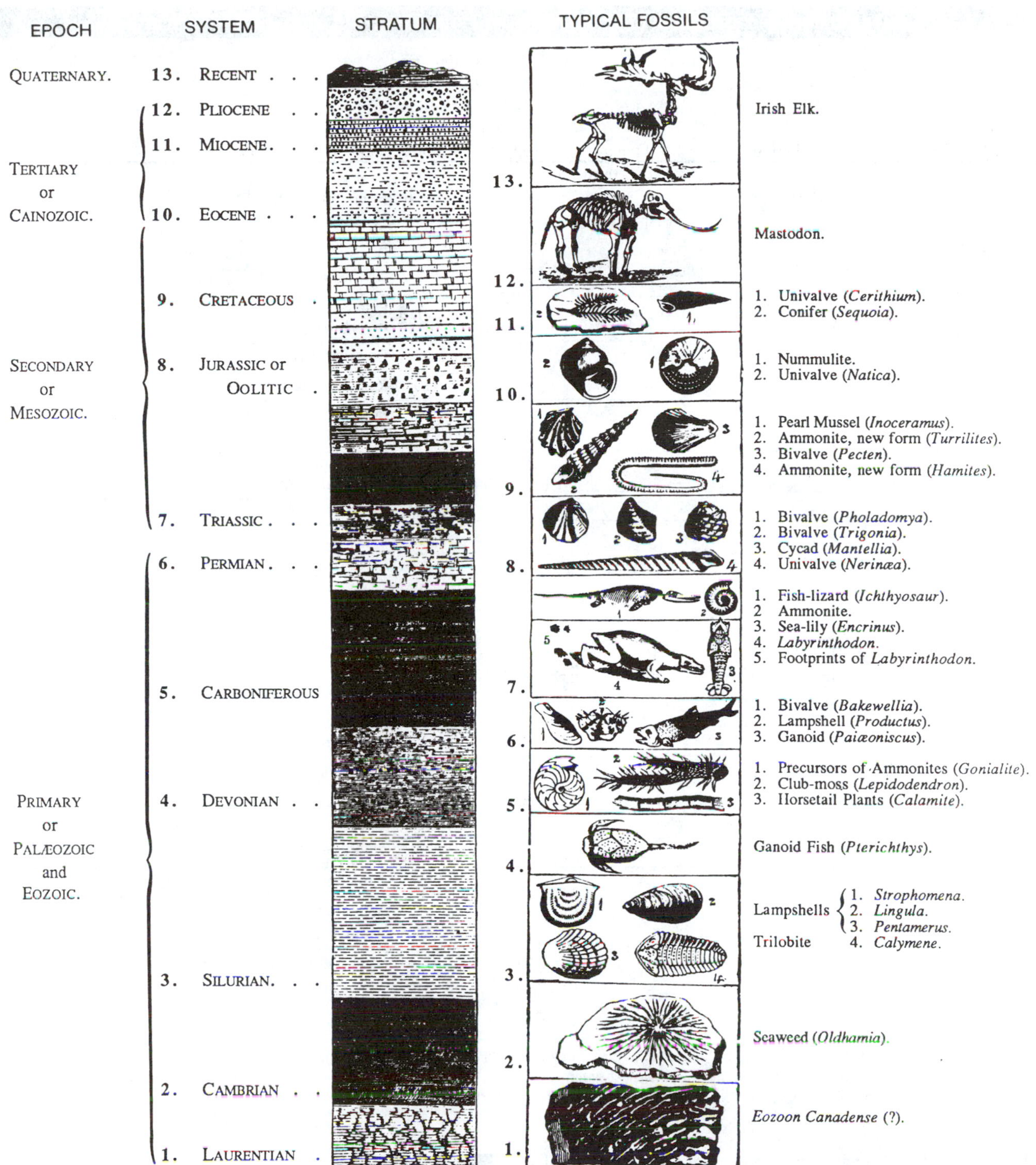

FIGURE 5-5 Nineteenth century illustration of a table of stratified rocks that classifies geological strata according to their relative age and shows some of the fossils associated with each period. (From Clodd, E., 1988. *Story of Creation*. Longmans Green, London.)

TABLE 5-3 Geological ages and associated organic events

Time Scale (eon)	Era	Period	Epoch	Millions of Years Before Present (approx.)	Duration in Millions of Years (approx.)	Some Major Organic Events
Phanerozoic	Cenozoic	Quaternary	Recent (last 5,000 years)	0.01	1.8	Appearance of humans
			Pleistocene	1.8		
		Tertiary	Pliocene	5.3	3.5	Dominance of mammals and birds
			Miocene	23.8	18.5	Proliferation of bony fishes (teleosts)
			Oligocene	34	10.2	Rise of modern groups of mammals and invertebrates
			Eocene	55	21	Dominance of flowering plants
			Paleocene	65	10	Radiation of primitive mammals
	Mesozoic	Cretaceous		142	77	First flowering plants Extinction of dinosaurs
		Jurassic		206	64	Rise of giant dinosaurs Appearance of first birds
		Triassic		248	42	Development of conifer plants
	Paleozoic	Permian		290	42	Proliferation of reptiles Extinction of many early forms (invertebrates)
		Carboniferous	Pennsylvanian	320	30	Appearance of early reptiles
			Mississippian	354	34	Development of amphibians and insects
		Devonian		417	63	Rise of fishes First land vertebrates
		Silurian		443	26	First land plants and land invertebrates
		Ordovician		495	52	Dominance of invertebrates First vertebrates
		Cambrian		545	40	Sharp increase in fossils of invertebrate phyla
Precambrian	Proterozoic	Upper		900	355	Appearance of multicellular organisms
		Middle		1,600	700	Appearance of eukaryotic cells
		Lower		2,500	900	Appearance of planktonic prokaryotes
	Archean			4,000–4,400	1,400	Appearance of sedimentary rocks, stromatolites and benthic prokaryotes
	Hadean			4,560	160–560	From the formation of Earth until first appearance of sedimentary rocks; no observable fossil organisms

**Notes:* Dates derived mostly from Gradstein, F. M., et al., 2004. *A Geological Time Scale 2004*. Cambridge University Press, Cambridge, England, and from Geologic Time Scale, obtainable from http://www.stratigraphy.org, a Web site maintained by the International Commission of Stratigraphy. See Box 1-2 for the origins of this scheme for the classification of geological strata.

blages to identify and name the three major eras of geological time, the Palaeozoic ("old life," dubbed the Age of Fishes), Mesozoic ("middle life," the Age of Reptiles) and Caenozoic ("new life," the Age of Mammals).[6]

[6] Evolution of fish, reptiles and mammals is treated in Chapters 17 to 20. (*Palaeozoic* and *Caenozoic* are the original English terms for these eras. The American spellings are *Paleozoic* and *Cenozoic*.)

Radiometric Dating

The atoms of a particular element all have the same number of protons, but the number of neutrons (and so the atomic weight of the atom) can vary. The different forms of an element are known as **isotopes**, some of which decay naturally over time, producing new isotopes and releasing energy. The decay of such radioactive isotopes is orderly, allowing geologists to date tiny fragments of rocks even bil-

TABLE 5-4 Radioactive isotopes used in dating

Parent Isotope (symbol)	Daughter Product	Half-life	Usable Range	Use for Dating
Samarium (Sm) 147	Neodymium (Nd) 147	110 billion years[a]	>1 billion years	Basalt, ancient meteorites
Rubidium (Rb) 87	Strontium (Sr) 87	49 billion years	>100 million years	Granites, igneous and metamorphic rocks
Thorium (Th) 232	Lead (Pb) 208	14 billion years	> 300 million years	Mineral crystals in crustal rocks
Uranium (U) 238	Lead (Pb) 206	4.5 billion years	>100 million years	Mineral crystals in crustal rocks
Potassium (K) 40	Argon (Ar) 40	1.3 billion years	>100 thousand years	Volcanic rocks
Uranium (U) 235	Lead (Pb) 207	0.7 billion years	>100 million years	Intrusions and mineral grains
Uranium (U) 234	Thorium (Th) 230	0.25 million years	>1 million years	Animal bones and teeth, corals
Carbon (C) 14	Nitrogen (N) 14	5,730 years	<50,000 years	Organic remains

[a] Two other naturally-occurring radioisotopes are Sm^{148}, with a half-life of 8,000 trillion years, and Sm^{149}, with a half-life of 10,000 trillion years.

lions of years old with considerable accuracy. **Table 5-4** is a list of radioactive isotopes used in dating. All methods of **radiometric dating** rely on three main factors:

- the ease with which many radioactive isotopes can be detected;
- the known stable products into which their atoms disintegrate; and
- the known rates at which this disintegration occurs.

For example, the radioactive isotope uranium 238 (^{238}U) is present in the mineral zircon, found in most igneous rocks. ^{238}U disintegrates to form the lead isotope ^{206}Pb at a rate that transforms half of the uranium into lead over a period (the **half-life**) of about 4.5 billion years (**Fig. 5-6**). After we make allowances for any lead not produced by uranium disintegration (^{204}Pb), and assuming these two isotopes have persisted equally, their relative amounts in a particular rock provide a fairly accurate dating method for older rocks. A somewhat simplified formula for this purpose is

$$t = 1/\lambda \ln (D/P + 1)$$

where t is time in years, λ is decay rate per year (1.537×10^{-10} for ^{238}U), ln is the natural logarithm (base e), D is the number of atoms of the daughter isotope in the sample and P is the number of atoms of the parent isotope in the sample. Thus, if the ratio of ^{206}Pb to ^{238}U in a sample was 0.360, we would calculate the age of the sample as:

$$t = 1/(1.537 \times 10^{-10}) \ln (1.360) = (6.508 \times 10^{9}) (.307)$$
$$= 1.998 \times 10^{9} \text{ years}$$

so approximately 2 By have elapsed since the ^{238}U was first incorporated in the sample. Dates determined in this fashion can be checked by the disintegration rates of other radioactive isotopes present in the same material, for example, the decay of ^{235}U to ^{207}Pb (half-life of about 0.7 By).

Carbon 14 (^{14}C), which disintegrates into nitrogen 14 (^{14}N) with a half-life of 5,730 years, is especially useful for dating organic material. Produced continuously in the upper atmosphere, ^{14}C eventually finds its way via photosynthesis into all living plants, and, hence, into animals in a fixed ratio to nonradioactive carbon. After the organism dies, the amount of ^{14}C slowly decreases, allowing time of death to be calculated. In this way, we've been able to date Egyptian mummies, mammoths, ancient trees and more.[7] Another method for dating young volcanic rocks (as well as ceramic artifacts) is to count the fission tracks that formed within them over time from the steady decay of uranium atoms.

So far, geologists have mostly applied radiometric dating methods to igneous rocks, but they can extend the dates to sedimentary rocks by comparing the relative positions of the two kinds of rock (**Fig. 5-7**). Thus, igneous rocks that coincide with the age of the Cambrian sediments are approximately 540 My old. Later sedimentary rocks, as shown in Table 5-4, can be dated fairly precisely up to the Recent period. Sedimentary rocks also can be dated quite accurately by dating layers of volcanic rock and ash in between the sedimentary strata. The first sedimentary rocks date from the Archean era, which lasted 1.4 By and ended some 2.5 By before the present, when weathering and erosive processes enabled the first sedimentary rocks to form.

When it comes to dating Earth itself, the oldest rock identified so far is a little older than 4 By, whereas estimates based on the combined isotope composition of lead in all Earth materials ($^{206}Pb/^{204}Pb$ ratio) point to an overall terrestrial age of about 4.56 By. This estimate accords with the ages of moon rocks brought to Earth by the *Apollo* lunar missions as well as with similar estimates made for meteorites that astronomers believe originated at the birth of the solar system.[8]

[7] A perceptive reviewer pointed out to us that the modern-day equivalent of ^{14}C is mitochondrial DNA (mtDNA) and the modern-day equivalent of longer-life isotopes is 5S rRNA, both of which are used to trace lineages.
[8] Norton (2002) and see the Web site maintained by the U. S. Geological Survey at http://www.usgs.gov/.

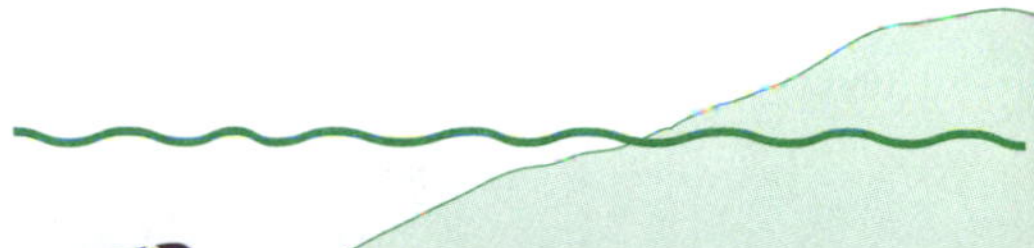

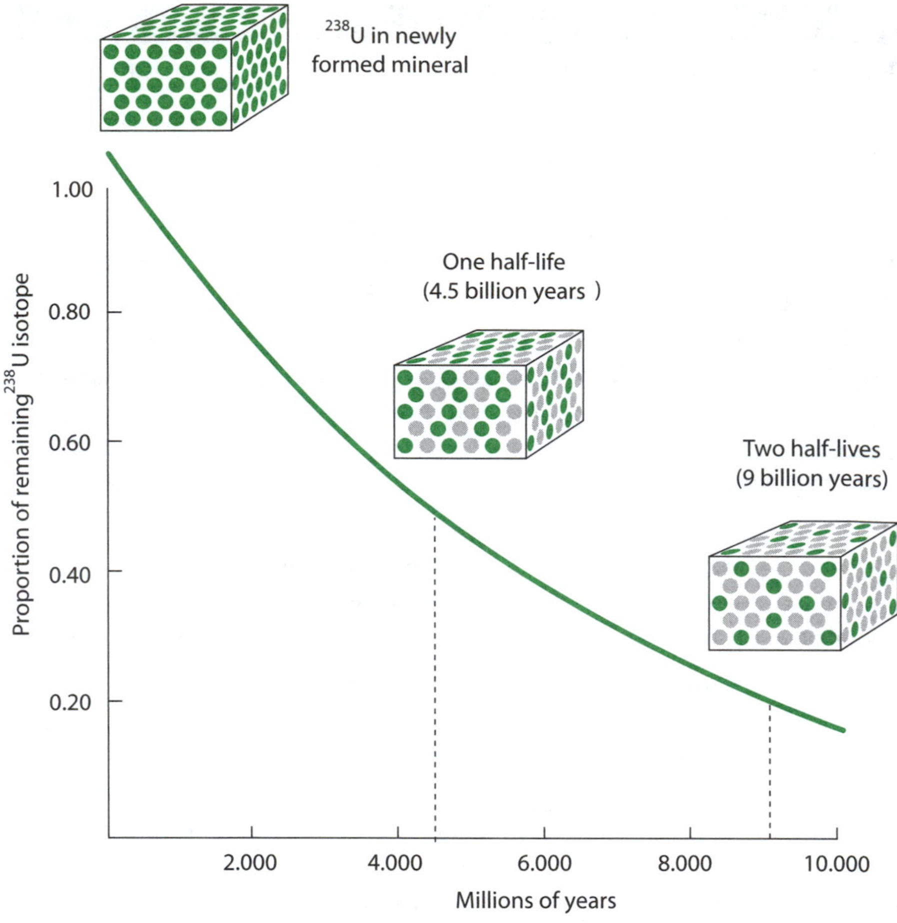

FIGURE 5-6 Theoretical relationship between duration of time in millions of years and the proportion of the original ^{238}U isotope that remains in a rock, given a half-life of about 4.5 By. Note that the line is curved, not straight, and never quite reaches zero, because each half-life period reduces the amount of ^{238}U by 50 percent and some of the original isotope will always remain if the initial amount is large. The fraction of ^{238}U that remains for any given period, x, is calculated as $(1/2)^y$ where $y = x/(4.5 \times 10^9)$.

Until recently, it had generally been accepted that Earth formed at low temperatures, but that radioactive heating and constant bombardment by meteorites turned the planet into a hellish place of roiling magma for much of the first 500 My and more, a period known as the **Hadean period.** In the early 2000s, dates as old as 4.4 By were confirmed for hundreds of zircon crystals found in northwestern Australia, pointing to a far earlier cooling of the planet than had been thought. To form, this type of zircon crystals requires liquid water and low temperatures. This active field of research continues to refine its techniques and extend our knowledge of Earth's earliest history.[9]

Origin of the Continents: Continental Drift

The notion that continents once existed but have been lost is an ancient one. The English essayist and master of inductive reasoning, Francis Bacon (1561–1626) proposed that a continent named **Atlantis** once existed in the middle of the Atlantic Ocean, but later sank beneath the surface.

Between 1912 and 1930, the meteorologist Alfred Wegener developed the concept that all the continents were at one time a single landmass, **Pangaea.** Wegener suggested that fissures occurred within this mass, the resulting fragments drifting apart to form the continents. He collected evidence for his theory of **continental drift**, but failed to come up with a convincing mechanism of how it worked.

For the next few decades, most geologists considered Wegener's theory little more than an imaginative fantasy until the evidence for continental drift became so overwhelming that they could no longer ignore it.[10] This evidence includes observations made of the fit between continents; similarity of rocks, fossils and ancient glaciation; paleomagnetism; and ocean-floor spreading.

Fit of the Continents

As shown in **Figure 5-8**, one of the most striking geographic correlations is the exact match between the east coast of South

[9] See Norton (2002) and the Web site maintained by the U. S. Geological Survey at http://www.usgs.gov/ for isotope dating and see Valley (2005) for information about zircon crystals.

[10] One of us (BKH), who took geology in the early 1960s, found the lecturer in structural geology advocating continental drift in his lectures and the Professor of Geology vehemently arguing against drift in his.

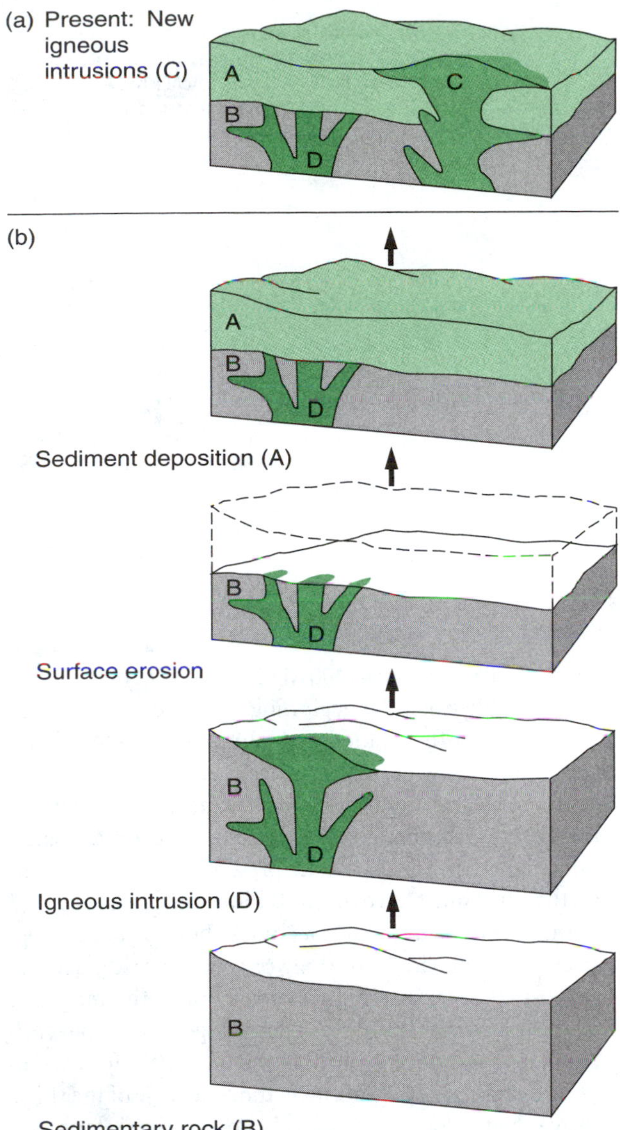

FIGURE 5-7 Classic use of relative and absolute dating in determining the ages of sedimentary and igneous rocks. (a) Diagram showing the observed relationships among four geological assemblies: two sedimentary layers (A and B) and two igneous intrusions or dykes (C and D). (b) Historical interpretation of the arrangement of these rocks based on the rules of superposition (younger sediments lie above older sediments) and crosscutting relationships (igneous rocks are younger than the rocks through which they cut across). According to these principles, B sediments are the oldest of rocks, and D represents a later igneous intrusion into B. Erosion then occurred, removing part of the intrusive igneous rock B, followed by the deposition of A sediments. The last geological event was a new igneous intrusion of rock C into both A and B. The age relationships are therefore B > D > A > C. If the absolute ages of the two intrusions, C and D, can be determined by radiometric dating techniques, the upper and lower limits for the age of the A sediments can be determined. Interbedded volcanic flows or ash deposits allow even more precise dating of sedimentary rocks.

America and the west coast of Africa. Not quite so obvious, but observable, is the match between the east coast of North America and the northwest coast of Africa. These and other geographical juxtapositions indicate that the continents at one time were either joined together or extremely close.

Similarity of Rocks, Fossils and Glaciations

A group of rock strata in India, called the **Gondwana system**, dates from the late Carboniferous to the early Cretaceous. Formations of extremely similar nature and composition exist in South Africa, South America, Antarctica, the Falkland Islands and Madagascar. As shown in **Figure 5-9**, associated with a few of these Gondwana formations are unique types of fossil plants (*Glossopteris*) and animals (*Mesosaurus*, *Lystrosaurus*, *Cynognathus*). Furthermore, all of the areas bearing Gondwana formations, along with Australia, were covered by the same glaciation event during a Paleozoic ice age. These land areas were much closer to the South Pole than their present locations and therefore glaciers could develop on them. To account for these observations, geologists proposed the existence of a massive southern continent, **Gondwana**, which included the areas that now carry the Gondwana formations and Australia.

Paleomagnetism

As new rocks arise from the cooling of magma, ferrous material within them (for example, magnetite, Fe_3O_4) magnetizes in a direction that depends on the location and strength of Earth's magnetic field at the time. Should this magnetic field change, the magnetic field of newly formed rocks would also be expected to change. Thus, we can study rocks from all eras and all continents for their fossilized magnetism, termed **paleomagnetism**, and deduce the direction and distance of Earth's magnetic poles relative to these rocks.

Although we would expect paleomagnetic studies to show slight shifts in the magnetic poles, it was most surprising to find that these poles had shifted over thousands of kilometers during past ages and that the *magnetic poles of different continents did not coincide for long periods of time.* For example, although magnetite deposits in recent igneous rocks from South America and Africa show the same magnetic orientation, this is not true for older Paleozoic rocks. As shown in **Figure 5-10a**, the magnetic poles derived from analyzing continental rocks that date between the Silurian and Permian indicate "seemingly" independent positions for each continent. Because different magnetic poles could not exist simultaneously, we can best explain these different polar wanderings as arising from the movement of continents relative to each other as well as from their movements relative to the poles.

As shown in **Figure 5-10b** the South American and African poles coincide during the Silurian-Permian if we

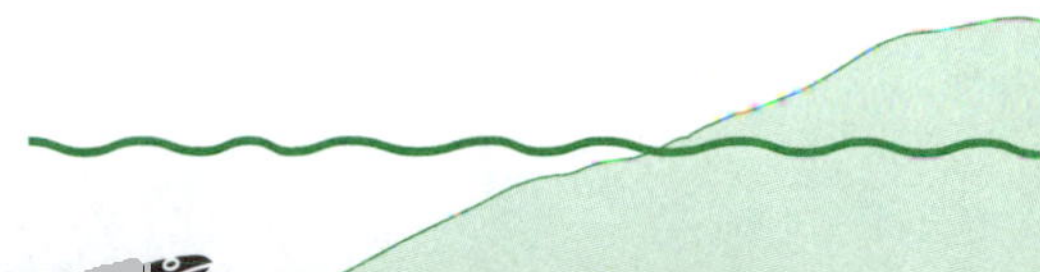

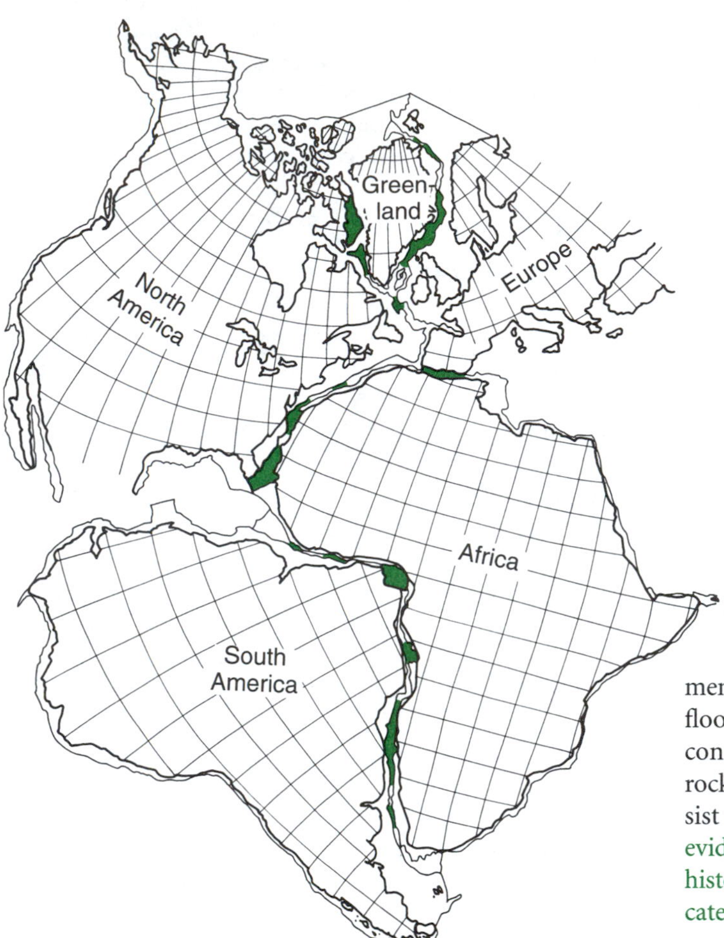

FIGURE 5-8 Matched fit between the offshore continental shelves at 500 fathoms (914 meters) deep on opposite sides of the Atlantic Ocean. (From Eicher, Don L., McAlester, A. L., *History of the Earth*, © 1980, p. 173. Adapted by permission of Pearson Education, Inc., Upper Saddle River, NJ.)

juxtapose the positions of these two continents. Both continents were united during the Paleozoic, so the magnetic orientation of magnetite in rocks formed during that period all pointed to the same geographic position for the South magnetic pole. As the continents separated in the Mesozoic, the "fossilized" magnetic orientations of these deposits now pointed to different South magnetic pole positions, giving rise to the anomalies shown in Figure 5-10a. What changed was not the position of the Paleozoic magnetic pole, but the geographic location of Paleozoic magnetized rocks.

Ocean Floor

Oceans, which have influenced Earth since its earliest history, have an enormous influence on Earth's climate; the heat capacity of the upper two meters of the world's oceans is equal to that of Earth's entire atmosphere. Knowing how oceans form is, therefore, an important component of understanding the evolution of Earth and the origins of life on Earth.

We know that the oceans are ancient, yet soon after sampling of the ocean floor began, geologists were most surprised to find that the ocean floor was relatively young, with sediments no older than 100 to 200 My. Fifty percent of the ocean floor is no older than the beginning of the Tertiary. Also, in contrast to the often folded and compressed sedimentary rocks of continental mountains, the oceanic mountains consist almost exclusively of igneous basalts. As considerable evidence shows that oceans have existed since early geological history, the relative youth of the present ocean basins indicates that they must have replaced older ocean floors.

Another unusual oceanic feature is the existence of magnetized belts that parallel the long **mid-oceanic ridges** found in almost all ocean basins. Measurement of the magnetic direction on both sides of such ocean ridges shows that each belt is paired with a belt of approximately equal width and of the same magnetic orientation on the other side of the ridge. However, belts adjacent to each other on the same slope usually magnetize differently. Radiometric dating shows that the youngest belts were closest to the crest of the ridge and the older belts located farther away. Changes in magnetic orientation between adjacent belts, known as **reversals**, are caused by a 180° reversal in the polarity of Earth's magnetic pole. This is because the degree of magnetism weakens with time, until, at some time, it reverses so that the south-pointing needle on a compass now points north. Using data from ocean floors and other sources, geologists have shown that the duration of a particular magnetic polarity, before it reverses, may vary from several thousand to 700,000 years. Reversals are irregular, occurring with frequencies between several hundred thousand and 50 My.

All these observations can best be explained if we assume that the mid-oceanic ridges represent fissures out of which new ocean floor emerges and spreads to either side. Molten rock spouting from the oceanic ridge magnetizes upon cooling and is displaced from the ridge by later-emerging

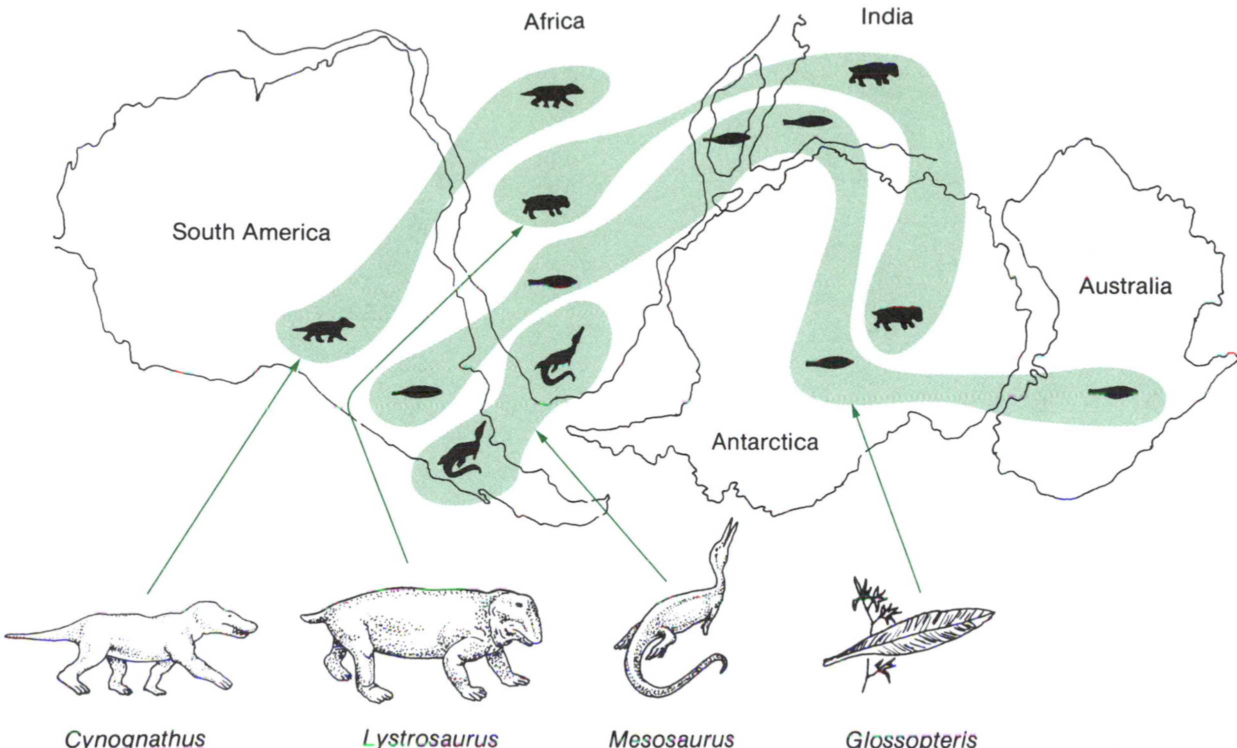

FIGURE 5-9 Distribution of various fossil plants and animals throughout Gondwanan continents. The presumed fit of the continental margins during the Permian–Triassic is also shown. *Cynognathus*, a carnivorous mammal-like reptile (therapsid, see Chapter 18) with a distinctive doglike skull, is found in Triassic-period deposits in South America and Africa. *Lystrosaurus*, another Triassic mammal-like reptile but larger than *Cynognathus* and probably herbivorous, had beaklike jaws and two large tusks. The genus *Mesosaurus* represents a fossil order of freshwater reptiles restricted to Permian deposits in Brazil and South Africa. This reptile was about 1.5 feet (0.5 m) in length with distinctive features of skull and limbs. *Glossopteris* was a fossil plant with many features similar to seed ferns (pteridosperms), bearing also large tongue-shaped leaves patterned with many reticulate veins. These fossil leaves appear in all the Gondwana formations and date to the Early Permian. (Adapted from Colbert, C. H., 1973. *Wandering Lands and Animals*. Hutchinson, London.)

material (**Fig. 5-11**). Just as trees lay down growth rings, the ocean floor retains its history in a series of parallel bands of rocks marked by magnetic fields prevailing at the time of their origin. **Sea-floor spreading** is not uniform, however. Its annual rates vary from about 1 cm in the North Atlantic Ridge to 3 cm in the South Atlantic Ridge — the latter being the rate of annual growth of our fingernails — to as much as 9 cm in some portions of the Eastern Pacific Ridge.

To and From Pangaea

One view of events that emerges from these studies is shown in **Figure 5-12.** It begins with a Devonian geography indicating separation between the Gondwana group of continents and a North American–Eurasian group called **Laurasia** (Fig. 5-12a). Most geologists now agree that by the end of the Paleozoic, these two major continental groups had united to form the giant landmass Pangaea (Fig. 5-12b,c). According to these reconstructions, Pangaea began to break up during the Triassic, about 225 Mya (Fig. 5-12d).

Fragmentation of Pangaea began when one oceanic rift developed between Western Gondwana (South America and Africa) and Eastern Gondwana (India, Antarctica, and Australia), and another separated Laurasia from Western Gondwana (Fig. 5-12e). By the Late Jurassic, sea-floor spreading began to separate North America from Africa, and by the Cretaceous to separate North America from Greenland and South America from Africa (Fig. 5-12f,g). The Indian subcontinent, moving independently from about the mid-Cretaceous on, continued northward from the Antarctic–Australian mass until it reached Southern Asia in the Cenozoic about 40 Mya. The Himalayan Mountains, in which mountain building is still going on, demonstrate the compressional forces exerted by the Indian–Asian collision.

In the Western Hemisphere, the rapid drift of South America away from Africa, which began about 100 Mya, led eventually to a reunion with North America some 4 or 5 Mya. In the Southern Hemisphere, New Zealand had drifted away from the Australian–Antarctican–South American landmass

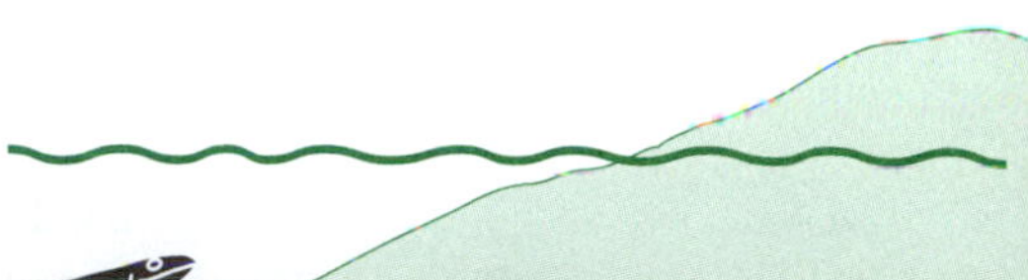

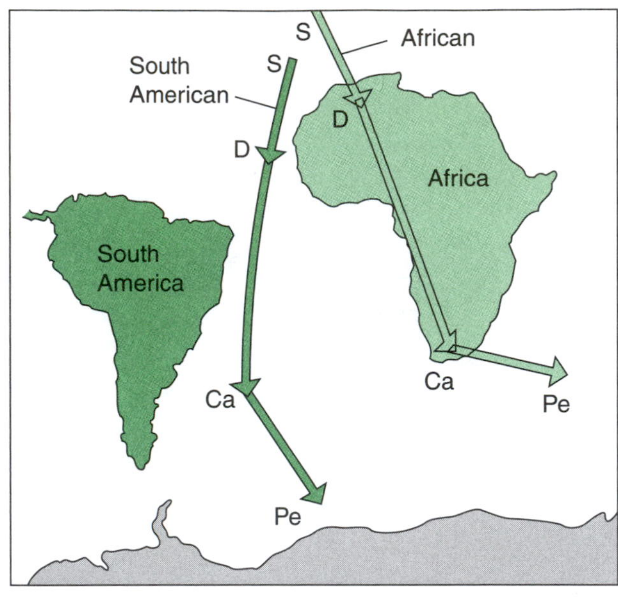

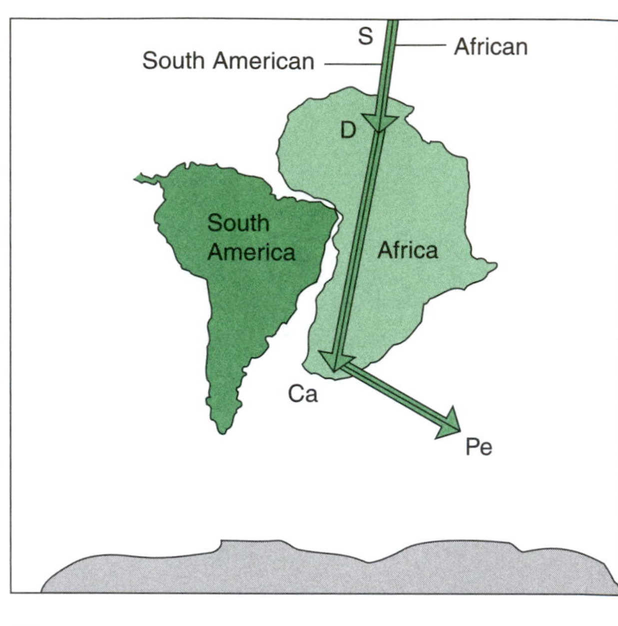

(a) (b)

FIGURE 5-10 Magnetic pole wanderings of South America and Africa. (a) The two continents in their present relative positions, showing paleomagnetically determined locations of the south magnetic pole for the Silurian (S), Devonian (D), Carboniferous (Ca) and Permian (Pe). The magnetic pole for these four periods is in a different location for each continent, an anomaly that would be difficult to explain if each continent had always occupied its present relative position. (b) This reconstruction demonstrates that the two seemingly independent polar pathways shown in (a) coincide when the two continents are fitted together (see Fig. 5-8). Paleomagnetic data indicate that the continents moved as a unit across the South Pole (see Fig. 5-12) through the Paleozoic and began to separate during the Mesozoic. (Adapted from Cox, C. B. and P. D. Moore, 2005. *Biogeography,* 7th ed. Blackwell, Oxford, England.)

before the end of the Cretaceous. The other Southern continents were joined until the beginning of the Tertiary about 65 Mya. However, by the Eocene, 20 My later, India had begun the northward journey that would eventually unite it with Asia.

The process of drifting and colliding continents extended even to Precambrian times. For example, geologists have shown that the magnetic poles of the North American continent and the Gondwana group shared a common pathway for more than a billion years during the Proterozoic (**Fig. 5-13**). Although these events indicate the existence of a giant continent containing most of Earth's surface, Northern Europe may have remained independent of North America until the middle of the Paleozoic. A number of Asian subsections did not unite with each other and with Europe until the Mesozoic (Ziegler et al., 1979).

Plate Tectonics

We now know that these varied and intricate continental movements are based on movements of gigantic plates (**tectonic plates**). Eight major plates and several minor ones have been identified, primarily on the basis of earthquake belts that accompany movements at the edge of the plates (**Fig. 5-14**).

In general, three major types of plate boundary events have been described:

1. Plates can **separate from each other** by the addition of new lava to their *adjoining boundary*. Such events occur in the oceanic ridges and account for sea-floor spreading and an increase in the size of some oceanic basins (Fig. 5-11).
2. Two adjoining plates can **slide past each other** *at a common boundary*, or *fault*, without any significant change in size. An example is the motion of the Pacific Plate carrying a section of western California past the North American Plate at the San Andreas Fault. The speed at which this is occurring (6 cm per year) will bring Los Angeles to the same latitude as San Francisco in about 10 My and to the Aleutian Islands near Alaska in about 60 My.
3. One plate can **move toward another**, causing a *convergent boundary*. When one such plate carries oceanic crust, the convergent event is often marked by the loss of plate material as the crustal mass plunges into the mantle. For example, the Pacific Plate in its motion northward meets a border of the North American Plate at the Aleutian Islands and **descends**[11] (is **subducted**) under this trench (**Fig. 5-15a**). The descending plate changes or deforms the mantle, which produces vol-

[11] Whether the Pacific Plate is being pushed or pulled remains unclear; hence, use of the more neutral term *descends*. Knowledge acquired from recent volcanic eruptions is providing new evidence on mechanisms of plate tectonics (Sigmundsson, 2006).

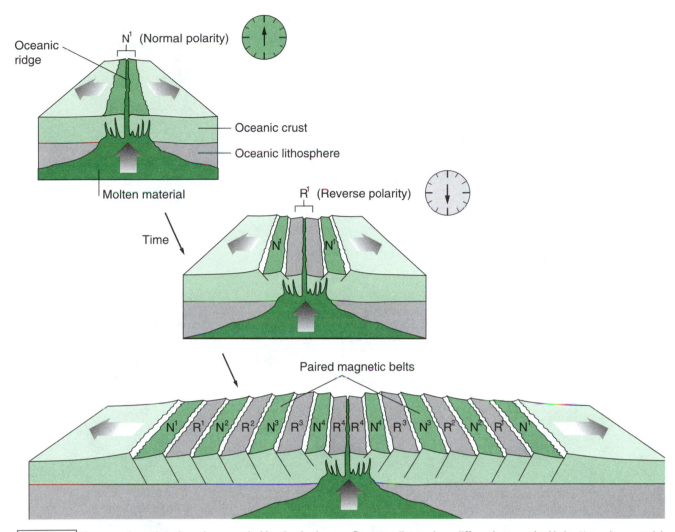

FIGURE 5-11 Diagrammatic sections through an oceanic ridge showing how sea-floor spreading produces differently magnetized belts. Hot molten material adds to the ridge from the mantle, falls away on both sides and magnetizes in the orientation of the prevailing magnetic field as it cools. With time, the magnetic field changes in strength and/or direction, and new material added to the ridge forms a pair of belts distinctly different from adjacent belts.

canic activity and mountain formation in the crustal region above.[12] Such processes account for some belts of volcanic islands (island arcs), such as those near the Aleutian and Java trenches, and also explain the origin of the Andes Mountains, which lie near the Chilean Trench, where the Pacific (Nazca) Plate moves beneath the westward-moving South American Plate (**Fig. 5-15b**). Once **subduction** has begun, a plate can continue to descend into the mantle, bringing oceanic sediments along with it until a continental mass meets the convergent boundary. Because rocks of the continents are lighter and thicker than the ocean floor, they cannot be forced far under another plate. As a consequence, mountains such as the Himalayas form through the foldings and pressures of these colliding landmasses (**Fig. 5-15c**).

Although we do not know the exact causes, **plate tectonics** more than any other geological theory has helped to explain the relative motion of continents since the Mesozoic, and to clarify the localization of Gondwana deposits, island arcs, earthquake belts, and so on. Before the Mesozoic, and certainly before the Paleozoic, mountain building and conti-

[12] Interestingly, in an 1838 meeting of London's Geological Society, Charles Darwin pointed to the relationship between volcanic action and mountain building: "The contemplation of volcanic phaenomena in South America has induced the author to infer, that the crust of the globe in Chile rests on a lake of molten stone, undergoing some slow but great change . . . that mountain building and volcanos are due to the same cause, and may be considered as mere subsidiary phaenomena, attendant on continental elevations; that continental elevations, and the action of volcanos, are phaenomena now in progress, caused by some slow but great change in Earth's interior; and, therefore, that it might be anticipated that the formation of mountain-chains is likewise in progress; and at a rate which may be judged of, by either actions, but most clearly by the growth of volcanoes."

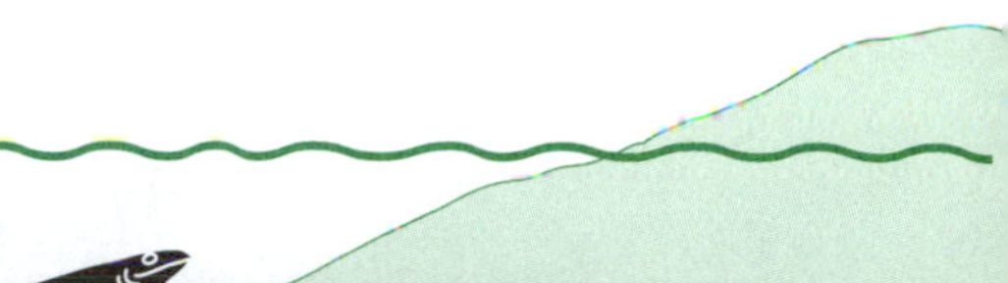

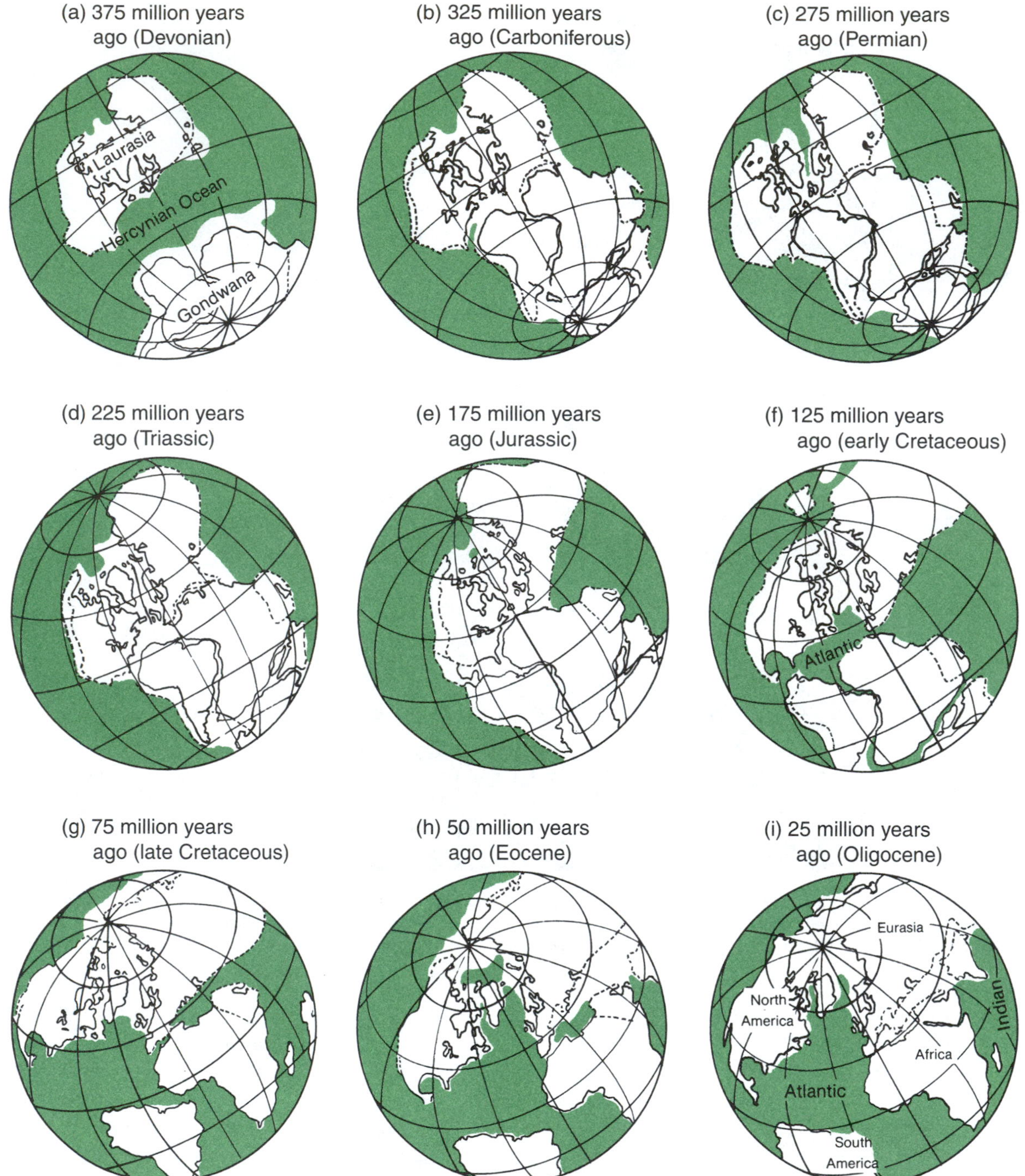

FIGURE 5-12 Some of the major geographical changes caused by continental drift (a–i) from the Devonian onward. (Irving E., 1977. Drift of the major continental blocks since the Devonian. *Nature* **270**, 304–309.)

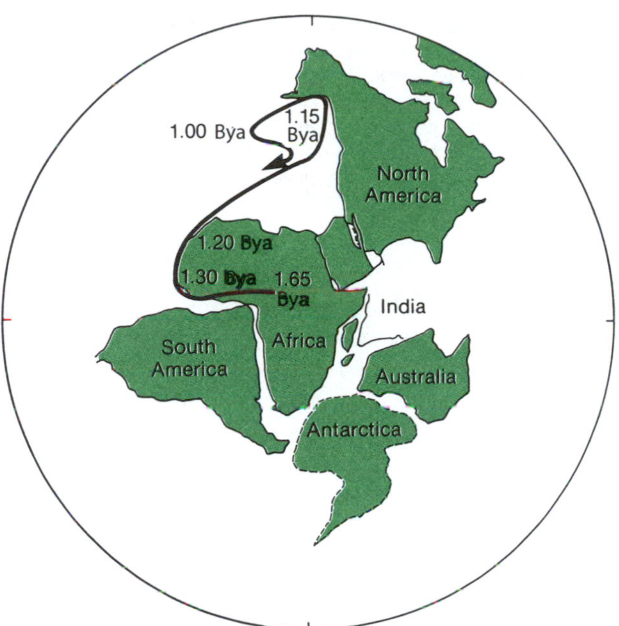

FIGURE 5-13 A possible reconstruction of the Gondwana–North American landmass for the approximate period 2.2 to 1.0 Bya. The heavy dark line shows the pathway of the magnetic pole determined for this landmass for given years (for example, the magnetic pole was at the northwest tip of Africa 1.2 Bya). The exact position of Antarctica is uncertain. Dalziel presents a more detailed scenario showing movements of the North American continents between 750 and 250 Mya. (Adapted from Piper, J. D. A. 1974. Proterozoic crustal distribution, mobile belts and apparent polar movements. *Nature*, **251**, 381–384.)

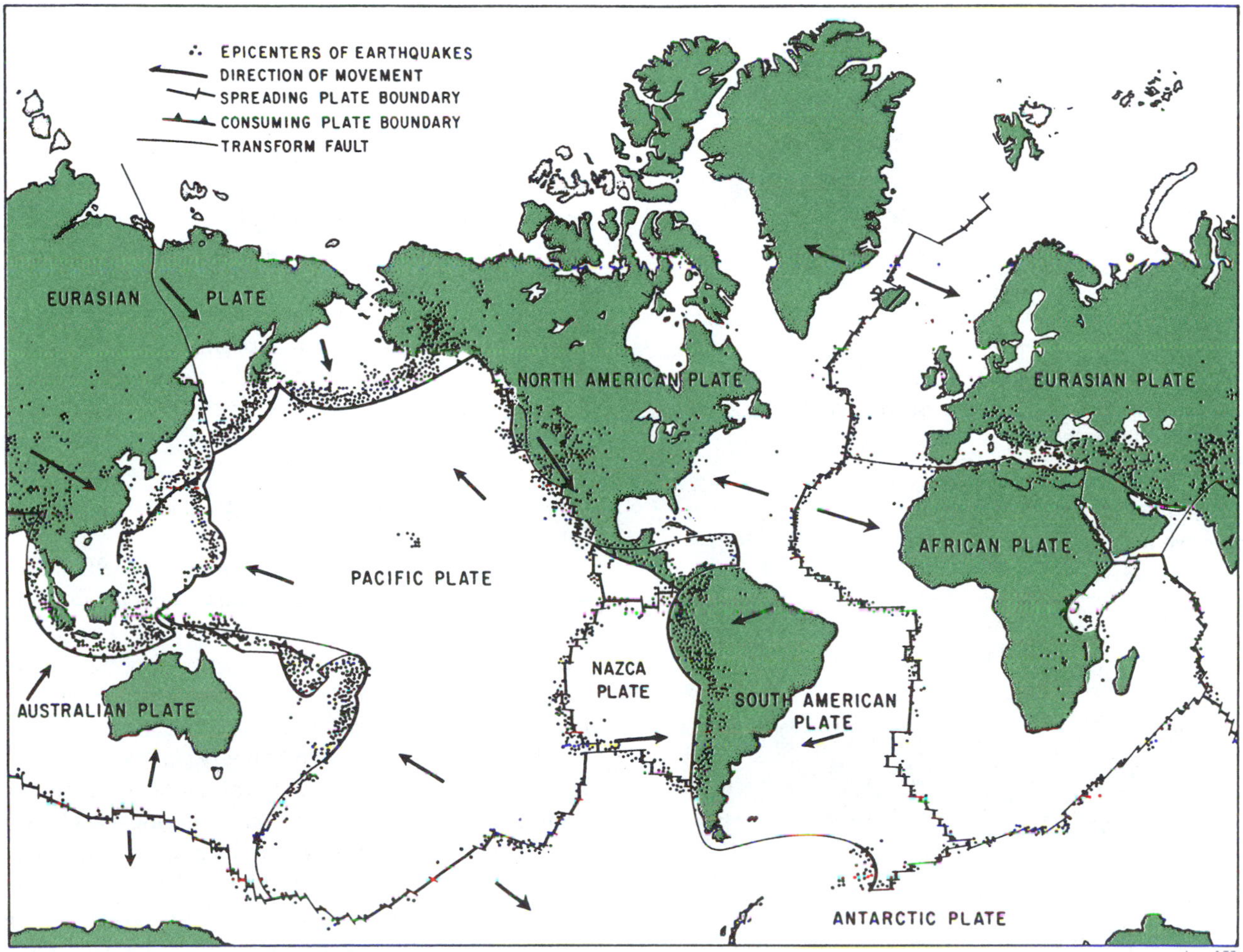

FIGURE 5-14 The major geological plates (and their boundaries) that account for many of the crustal movements. (From Cloud P., 1978. *Cosmos, Earth and Man: A Short History of the Universe.* Yale University Press, New Haven, CT, Fig. 11, p. 82. Reprinted by permission.)

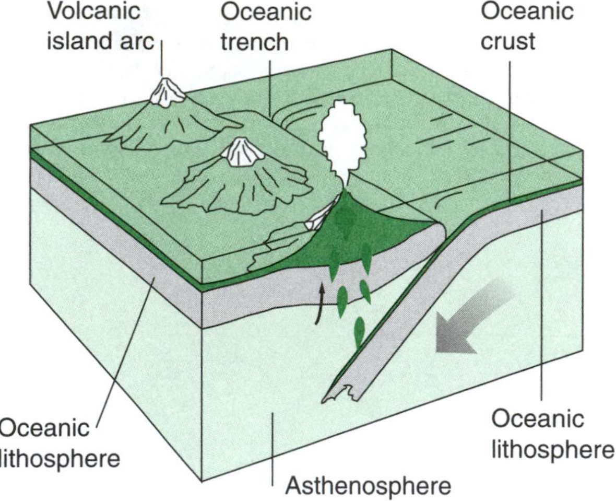

(a) Collision of two plates, both with oceanic lithosphere

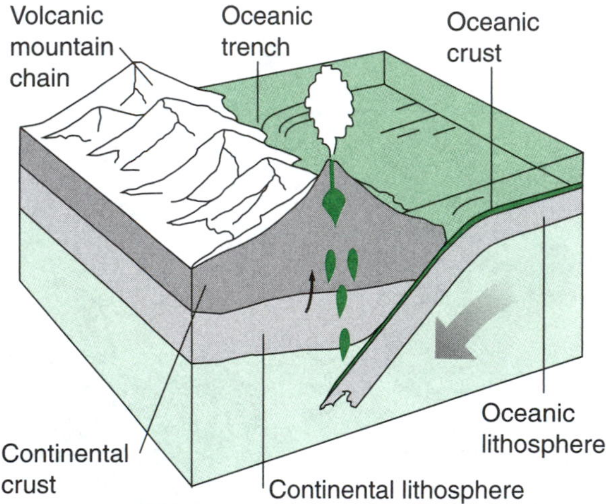

(b) Collision of two plates, one oceanic and one continental

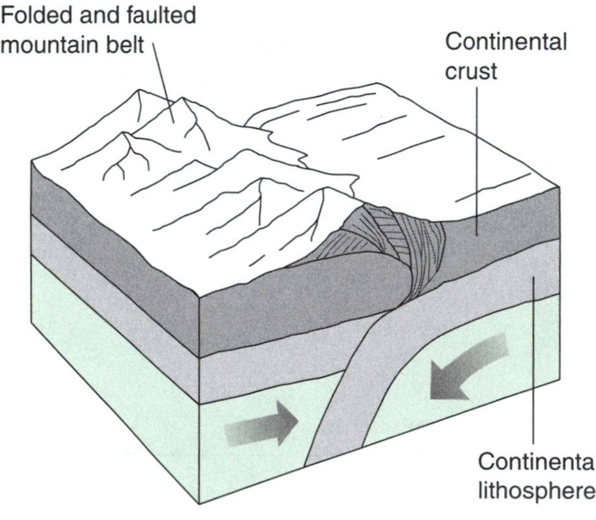

(c) Collision of two plates, both with continental lithosphere

FIGURE 5-15 Three types of convergent events (a–c) that can occur between lithospheric plates. Note that mountain building (c) can be accompanied by rare "overthrusting" events that invert the normal position of older and younger strata. Such events, however, are detectable by correlating the age of each stratum with its fossil remains, and by comparing stratigraphic contents and structures with their more widespread counterparts formed during more quiescent periods.

nental drift occurred, but how far back tectonic plates formed is unclear. Some researchers suggest that tectonic plates originated during the period after chemical differentiation of the mantle had begun and lighter materials (aluminum silicates) had surfaced. Others suggest that convection currents arising from heat produced by radioactivity during the Hadean period were four times larger than at present, and led to motion of the slag-like crust, beginning a tectonic-like process. However tectonic movements began, the separation and joining of landmasses from the Proterozoic onward had important biological effects, determining the distribution of organisms and so influencing their evolution.

Biological Consequences of Plate Tectonics

Plate tectonics has profound biological effects. It subjects moving landmasses to new climatic conditions and geographical relationships, resulting in their inhabitants being selected for different evolutionary adaptations.

Because of the breakup of landmasses, continental drift separates groups of organisms that were formerly associated, setting each such isolated species or group on its own evolutionary pathway. On the other hand, the joining of landmasses because of plate tectonics leads to competition among previously separated groups of plants and animals that had evolved unique adaptations in the interim. These organisms can then interact, leading to increased complexity for some groups, and extinction for others.

One of the most prominent examples of the effect of plate tectonics on the distribution and evolution of organisms is the unique collection of mammals found in Australia and South America. Although mammals are discussed in more detail in Chapter 19, we introduce them here because they illustrate the biological consequences of plate tectonics so well.

Mammals (hairy skin, mammary glands, special auditory skull bones; see Chapter 19) can be classified into three groups/lineages: **Prototheria** (monotremes) such as the duckbilled platypus, **Metatheria** (marsupials) such as kangaroos, and **Eutheria** (placental mammals) such as cats, dogs and humans (**Figure 5-16**). In most of the world today, only Eutherian mammals are present. Indeed, by the Pliocene, placentals had replaced monotremes and marsupials in all areas except Australia and South America.

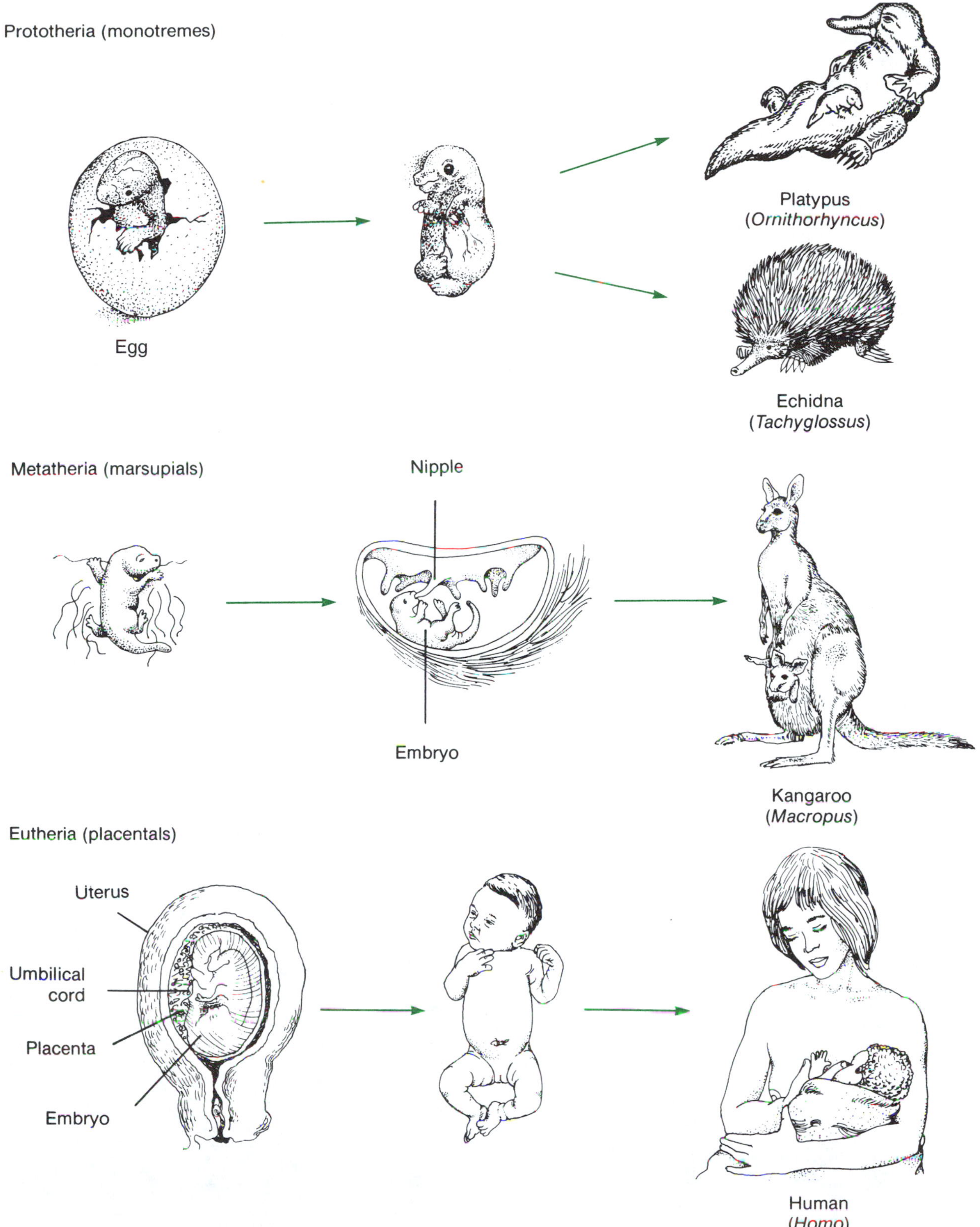

FIGURE 5-16 The three major lineages of existing mammals, showing early developmental stages on the left and adults on the right. All have hair, mammary glands and other features that distinguish them from reptiles, but they differ from each other in their prenursing development. Prototheria (a) lay eggs; the embryos hatch out to attach onto maternal abdominal hairs connected to mammary glands. In Metatheria (b), the embryo undergoes early development inside the egg *in utero;* the emerging offspring climbs into the maternal pouch and attaches itself to a nipple. Eutherians (c) maintain the fetus *in utero* until a relatively late stage of development.

Representatives of two families of monotremes and 13 families of marsupials are still found in Australia. With the exception of bats — Australia has two indigenous species — no placental mammals were present on the continent until the relatively recent introductions by humans of dogs, rabbits, and so on. The South American mammalian fauna seems to have been somewhat more derived than that of Australia and, until the Mid-Tertiary, included a number of placental families in addition to five families of marsupials. However, even the native South American placentals are regarded as generally "primitive," as evidenced by some of the mammals there today: armadillos, anteaters and tree sloths.

The picture of mammalian evolution that emerges from these studies is that early prototherians and metatherians entered southern parts of Pangaea by the Late Jurassic and Early Cretaceous (**Fig. 5-17a**). The subsequent rifting of Australia isolated its mammalian fauna from later competition, with more derived eutherians evolving in western Pangaea during the Late Cretaceous and Early Tertiary (Fig. 5-17b). In South America, metatherians and early eutherians had replaced prototherians by the Early and Mid-Tertiary, but by that time the South American continent had drifted considerably from Africa and separated from North America (Fig. 5-17c).

The evolution of mammals on the isolated island of South America was therefore largely independent of mammalian evolution elsewhere, until South America rejoined North America via the Panama Isthmus during the Pliocene (Fig. 5-17d). During the Oligocene, however, some island hopping combined perhaps with transport on floating debris (*rafting*) occurred as various monkeys and caviomorph rodents made their way to South America from Africa or North America.

By the Pliocene, however, considerable evolution toward more derived eutherian forms had occurred either in Africa or Laurasia (North America–Eurasia); most of the South American mammalian fauna showed far less change. When the Pleistocene began, massive invasions of northern euthe-

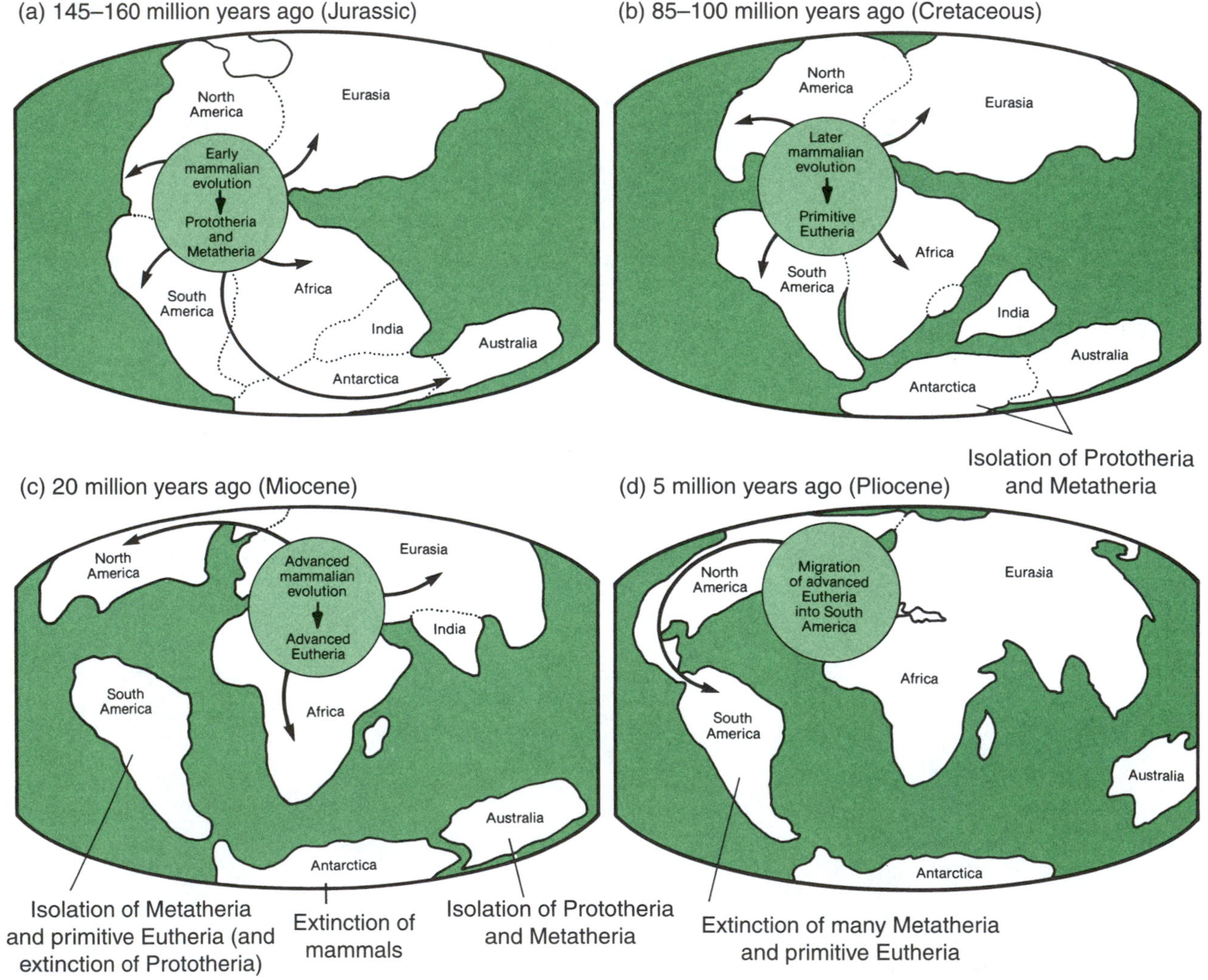

FIGURE 5–17 (a–d). Effect of continental drift on the dispersion and isolation of major lineages of mammals.

rians made their way south across the Central American land bridge, causing the rapid extinction of many South American mammalian families (see Chapter 19). Only rarely, as with opossums, did early South American mammals manage to successfully invade North America.

Supporting this view of plate tectonics and mammalian evolution is abundant fossil evidence of South American extinctions in the Pliocene and Pleistocene, and the absence of any eutherian fossils in Australia up to recent times. This view leads us to expect fossil remains of prototherians and metatherians in Upper Triassic deposits of other landmasses that separated from Pangaea concurrently with Australia, such as India and Antarctica. Discovery of such fossils would add considerably to the predictive value of the theory of plate tectonics.

An important lesson that emerges is the close connection between biological evolution and historical circumstances of all kinds. Evolutionary changes are allied with geological changes, which can cause environmental effects of many varieties and descriptions, in turn causing many kinds of organismal interactions and evolutionary consequences. This tie between the development of biological systems and the impact of diverse historical circumstances molded biology into a historical science.

KEY TERMS

atmosphere
collision theory
condensation theory
continental drift
core
crust
dwarf planets
Eutheria
geological dating
Gondwana
Hadean period
half-life
igneous rocks
law of superposition
lithosphere
magnetosphere
mantle
metamorphic rocks
Metatheria
nebular hypothesis
oxidizing
paleomagnetism
Pangaea
Phanerozoic Eon
planetesimals
planets
plate tectonics
protoplanets
Prototheria
radiometric dating
reducing
sea-floor spreading
sedimentary rocks
seismic waves
solar system bodies
subduction
tectonic plates

DISCUSSION QUESTIONS

1. What is the prevailing theory for the origin of the solar system?
2. Describe how Earth's atmosphere has changed over time.
3. Geological structures.
   a. What are the general compositions of the major types of rocks?
   b. How do rocks transform from one type to another?
4. Geological dating.
   a. How did scientists date geological strata in the nineteenth century?
   b. How do geologists perform radiometric dating?
   c. How can geologists use igneous rocks to date sedimentary rocks? (How do they use the rules of superposition and crosscutting relationships for this purpose?)
5. Geological time scales.
   a. What is the sequence of eras in the Phanerozoic Eon?
   b. What sequence of geological periods do scientists ascribe to each of these eras?
6. Plate tectonics.
   a. What evidence supports plate tectonics? (Explain terms such as continental fit, Gondwana, Pangaea, paleomagnetism and sea-floor spreading.)
   b. How do geologists delineate tectonic plates, and how do these plates interact?
   c. What major changes in the positions of continents occurred during the Mesozoic and Cenozoic eras? (Include changes in North America, Eurasia, Africa, South America, India, Antarctica and Australia.)
   d. How did plate tectonics affect the distribution and evolution of the major mammalian subgroups?

EVOLUTION ON THE WEB

Explore evolution on the Internet! Visit the accompanying Web site for *Strickberger's Evolution*, Fourth Edition, at http://www.biology.jbpub.com/book/evolution for Web exercises and links relating to topics covered in this chapter.

RECOMMENDED READING

Benton, M. J., 2003. *When Life Nearly Died: The Greatest Mass Extinction of All Time*. Thames and Hudson, London.

Condie, K. C., and R. E. Sloan, 1998. *Origin and Evolution of Earth: Principles of Historical Geology*. Prentice Hall, Upper Saddle River, NJ.

Cox, C. B., and P. D. Moore, 2005. *Biogeography: An Ecological and Evolutionary Approach*, 7th ed. Blackwell Publishing, Oxford, England.

Hallam, A., 2005. *Catastrophies and Lesser Calamities*. Oxford University Press, Oxford, England.

Redfern, M., 2003. *The Earth: A Very Short Introduction*. Oxford University Press, New York.

Rudwick, M. J. S., 2005. *Bursting the Limits of Time. The Reconstruction of Geohistory in the Age of Revolution*. The University of Chicago Press, Chicago.

Tarbuck, E. J., and F. K. Lutgens, 2006. *Earth Science*, 9th ed. Prentice Hall, Englewood Cliffs, NJ.

Taylor, S. R., 1998. *Destiny or Chance: Our Solar System Evolution and Its Place in the Cosmos*. Cambridge University Press, Cambridge, England.

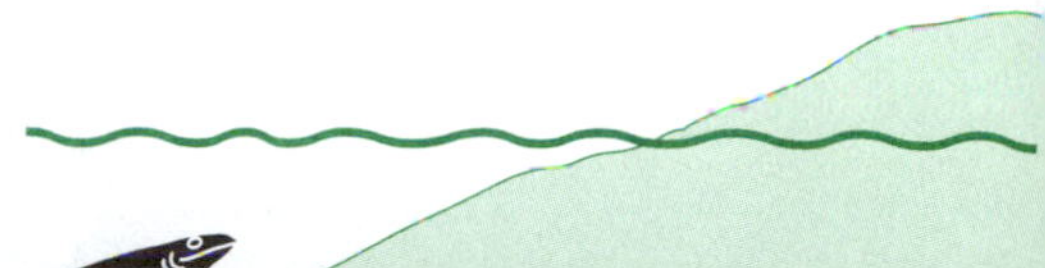

6 Molecules, Protocells and Natural Selection

Chapter Summary

For many years, even those who believed in the reality of evolution could not explain how cells, with their complex membranes and compartmentalized systems, could have arisen, even though chemical pathways that may have led to the formation of the most important molecules in cells had been identified.

How did molecules appear on Earth and become organized into biological structures subject to natural selection? In this chapter we discuss (1) the nature of the major biological (organic) molecules; (2) experimental studies conducted as attempts to recreate the conditions under which such molecules may first have arisen; and (3) how the origin of natural selection would have resulted in the evolution of more complex and efficient molecules. In Chapter 7, we discuss which of the three classes of molecules — DNA, RNA and proteins — evolved first to begin life on Earth.

Because of their catalytic functions as enzymes, proteins are crucial to organisms. Proteins are composed of specific linear sequences of amino acids linked together by peptide bonds. Their structure is specified by the sequence of nucleotides in nucleic acids, which act as information storage molecules in cells. The problem is to explain the origin of the information molecule (nucleic acid) and the active product (protein). If we assume that life arose on Earth by chemical means, the origin of complex organic molecules from simple substances present early in Earth's history has to be explained. The probability of the evolution of complex organic molecules is less daunting, if the first organic molecules were not as complex as they are at present and if chemical reactions were biased to produce such molecules. As discussed in the previous chapter, conditions on Earth several billion years ago probably offered a favorable environment: sufficient and continuous energy, availability of carbon and other important elements, abundance of water (an excellent solvent) and hydrogen and its compounds.

There is considerable experimental evidence that proteins can arise from a combination of simple compounds exposed to an external source of energy. A little over 50 years ago, Miller (1953) demonstrated experimentally that amino acids and other organic compounds could be synthesized from hydrogen, ammonia, methane and water in the presence of an energy source, although whether Earth's atmosphere was reducing and could allow such synthesis is now in doubt. Fox (1978) showed that amino acid mixtures subjected to high temperatures polymerize to form polymers with peptide-like properties, such as nonrandom amino acid frequencies and low-level catalytic activity. Amino acids and even nucleotides can also polymerize in certain types of clay. Whether such findings can be extrapolated to the conclusion that the original molecular world was a "protein world" is discussed in Chapter 7.

More recent attention has focused on how the original complex molecules became organized into cells. Again, experimental studies were initiated in attempts to create the basic elements of a cell, namely, a membrane-bounded system of molecules: a protocell. Intermolecular forces can bind macromolecules into protocells, which exhibit some features of living systems such as organization, selective permeability, and energy use, and which can be subject to natural selection. These protocells, like biological systems such as viruses, can self-assemble in nonrandom ways, suggesting that similar structures might have occurred on the pathway to living organisms. Those protocells that could best maintain themselves, perpetuated. Later, protocells that could reproduce, however inefficiently, would be acted on by natural selection, successively enhancing the probability of advantageous molecules, reactions and structures.

The problem of how life could have originated by ordinary chemical means long seemed insuperable. This is hardly surprising when we consider the **complexity** of even the simplest cell, surrounded as it is by a highly selective permeable membrane composed of lipids and proteins that regulate the kinds of substances that pass through (**Fig. 6-1**). Within the cell, the cytoplasm consists of a multitude of structures and substructures involved in the synthesis, storage and breakdown of a large variety of chemical compounds.

Comparing the cells in Figure 6-1, you are first struck by the *increasing complexity* of eukaryote cells (plant and animal shown) when compared with the eukaryote. But look more closely. There is an *amazing similarity* between the three cell types, reflecting **fundamental (deep) homologies**, some of which (such as the cell membrane and genes) are billions of years old. Others, as seen in the plant and animal cells (nuclei, ribosomes, Golgi apparatus, endoplasmic reticulum, cytoskeleton) had their origin in the earliest eukaryotes (see Chapter 8). Yet others (mitochondria, chloroplasts) have only a slightly more recent origin(s), but again, in their essential similarity, reflect common solutions to common problems as plants and animals emerged. A deep cellular (structural) homology underlies life as much as do deep genetic and molecular homologies. The two most fundamental and homologous molecules of life are *amino acids* from which proteins are constructed, and *nucleotides* from which DNA and RNA are constructed.

Amino Acids

Foremost among the metabolic agents that enable cells to function are many different **proteins** that catalyze and regulate practically all the chemical reactions within organisms. In basic structure, proteins consist of subunits called **amino acids**, with the following features[1]:

$$H_2N-\overset{\displaystyle H}{\underset{\displaystyle R}{\overset{|}{\underset{|}{C^*}}}}-\overset{\displaystyle O}{\overset{\|}{C}}-OH$$

- A central alpha carbon atom (C*) to which the following functional groups are attached.
- An amino group with a potential positive charge (NH_2^+).

[1] This chapter assumes some basic knowledge of chemistry. If you find the going heavy, we suggest reviewing unfamiliar material in a basic introductory chemistry textbook or on the Internet.

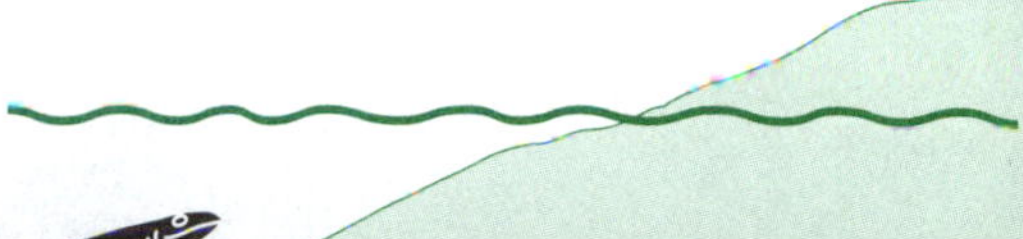

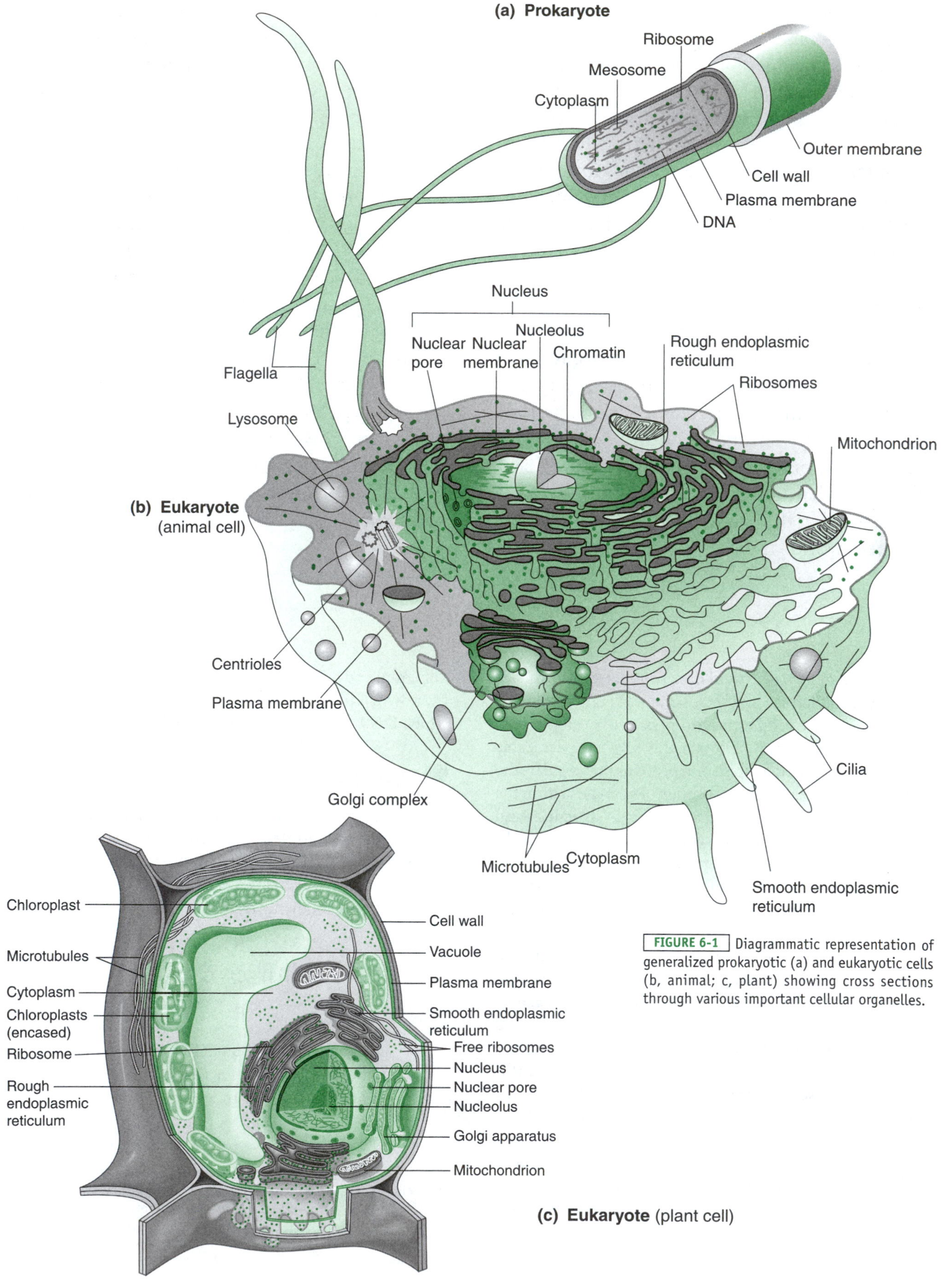

FIGURE 6-1 Diagrammatic representation of generalized prokaryotic (a) and eukaryotic cells (b, animal; c, plant) showing cross sections through various important cellular organelles.

(a) Amino acids containing one amino and one carboxyl group:

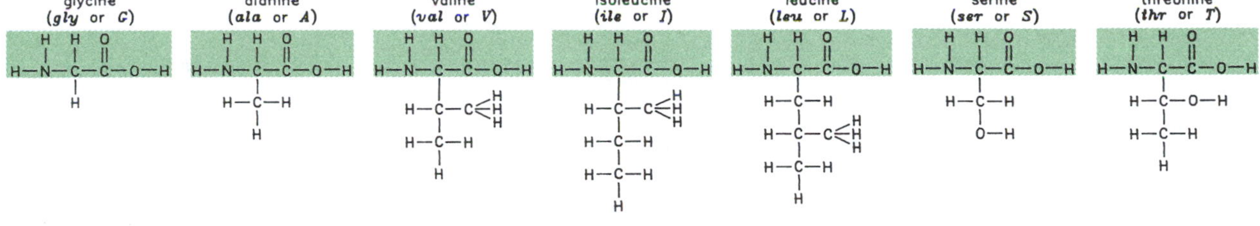

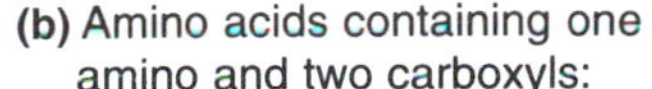

(b) Amino acids containing one amino and two carboxyls:

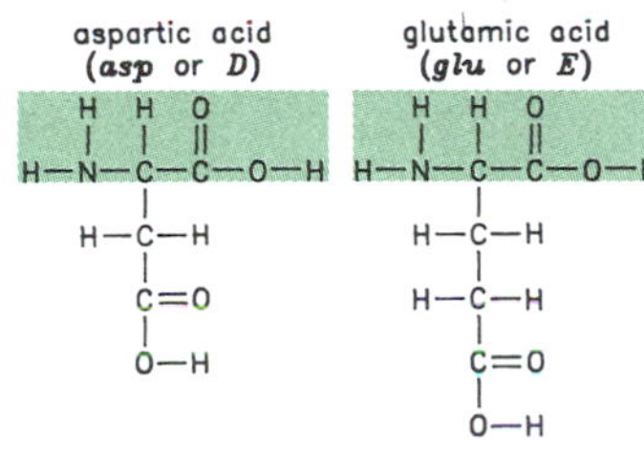

(c) Amides of dicarboxyl amino acids:

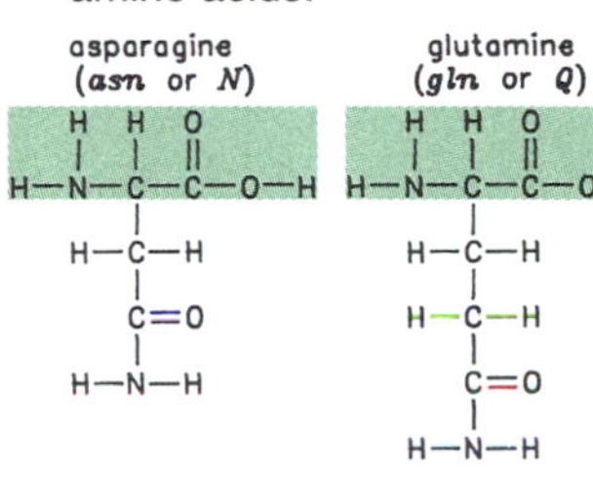

(d) Basic amino acids (additional NH groups):

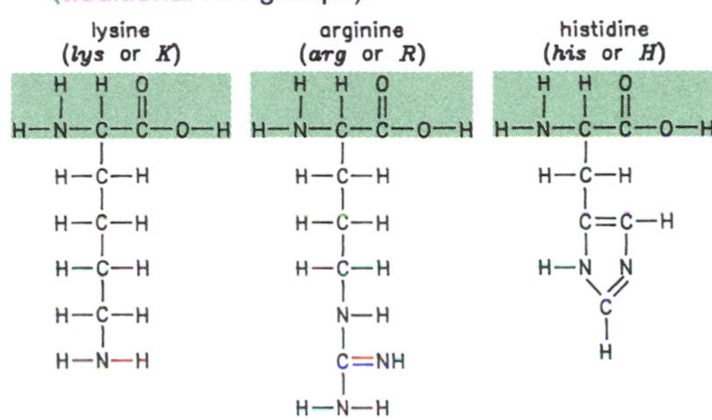

(e) Imino or cyclic amino acids:

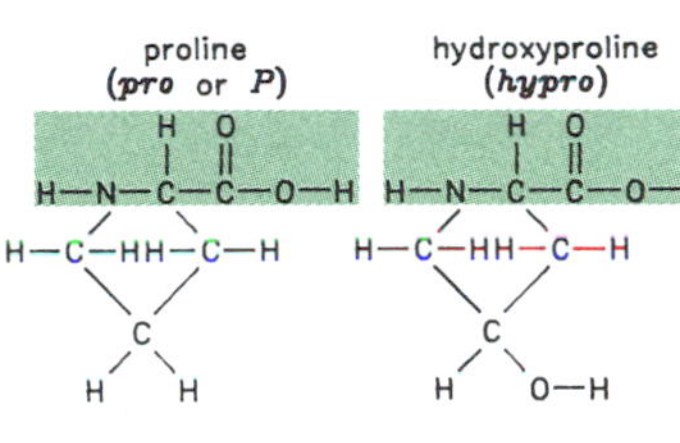

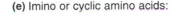

(f) Aromatic amino acids (containing benzyl ⬡ group):

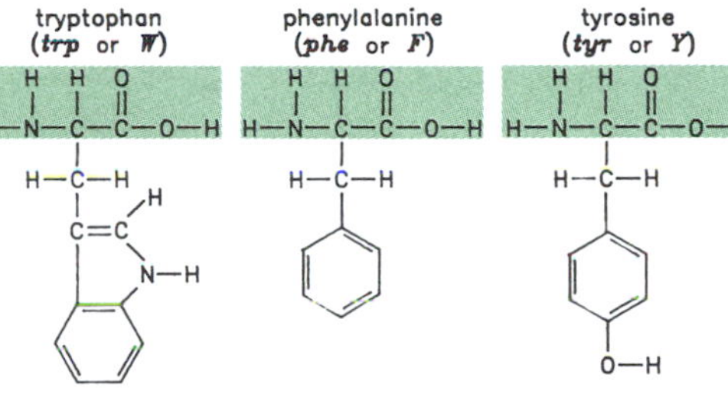

(g) Sulfur-containing amino acids:

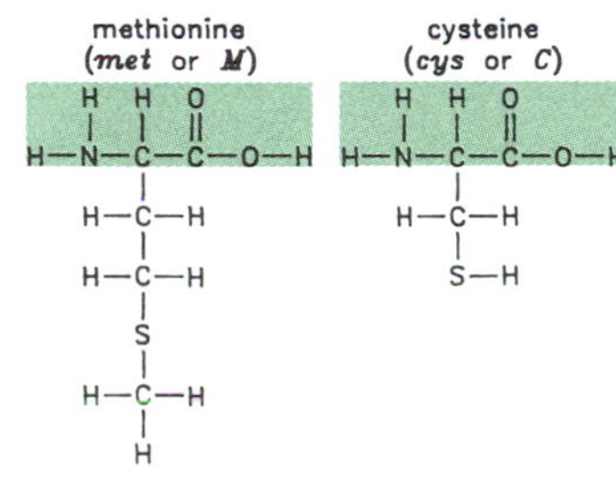

FIGURE 6-2 Structures of the 20 common amino acids found in proteins, arranged in six groups on the basis of numbers of amino and carboxyl groups (a, b), presence of amides (c) and types of amino acids: basic (d), imino or cyclic (e), aromatic (f), sulfur-containing (g). Alert readers will have noted that 21 not 20 amino acids are illustrated. The 21st, hydroxyproline, arises by hydroxylation of the amino acid proline, and so is a secondary or post-translationally modified amino acid.

- A carboxyl COOH group with a potential negative charge (COO^-).
- A hydrogen atom.
- An R side chain that varies in structure among the different amino acids (**Fig. 6-2**).

These amino acids link together by chemical bonds called **peptide linkages** into linear **polypeptide chains** that are the constituents of proteins (shown later in Figure 6-11). The highly specific structure of any protein molecule, whether it functions as an **enzyme** (catalyst) or for some other purpose, derives from the exact linear placement of the amino acids. Specific amino acid sequences enable polypeptide chains to fold into specific three-dimensional forms that confer individual properties on proteins. Most phenomena associated with life derive from the enzymatic and regulatory activities of long sequences of amino acids. Complexity, however, is not limited to proteins, since the amino acid sequences of proteins are determined by the nucleotide sequences in another group of basic molecules, nucleic acids.

Nucleic Acids

Nucleic acids are long-chained molecules composed of **nucleotide** subunits, each of which contains a pentose

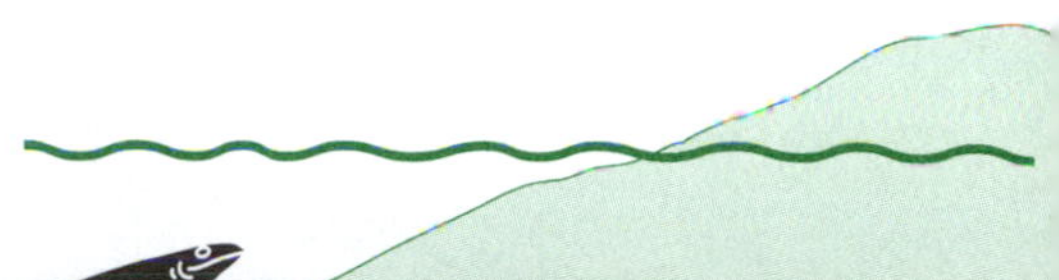

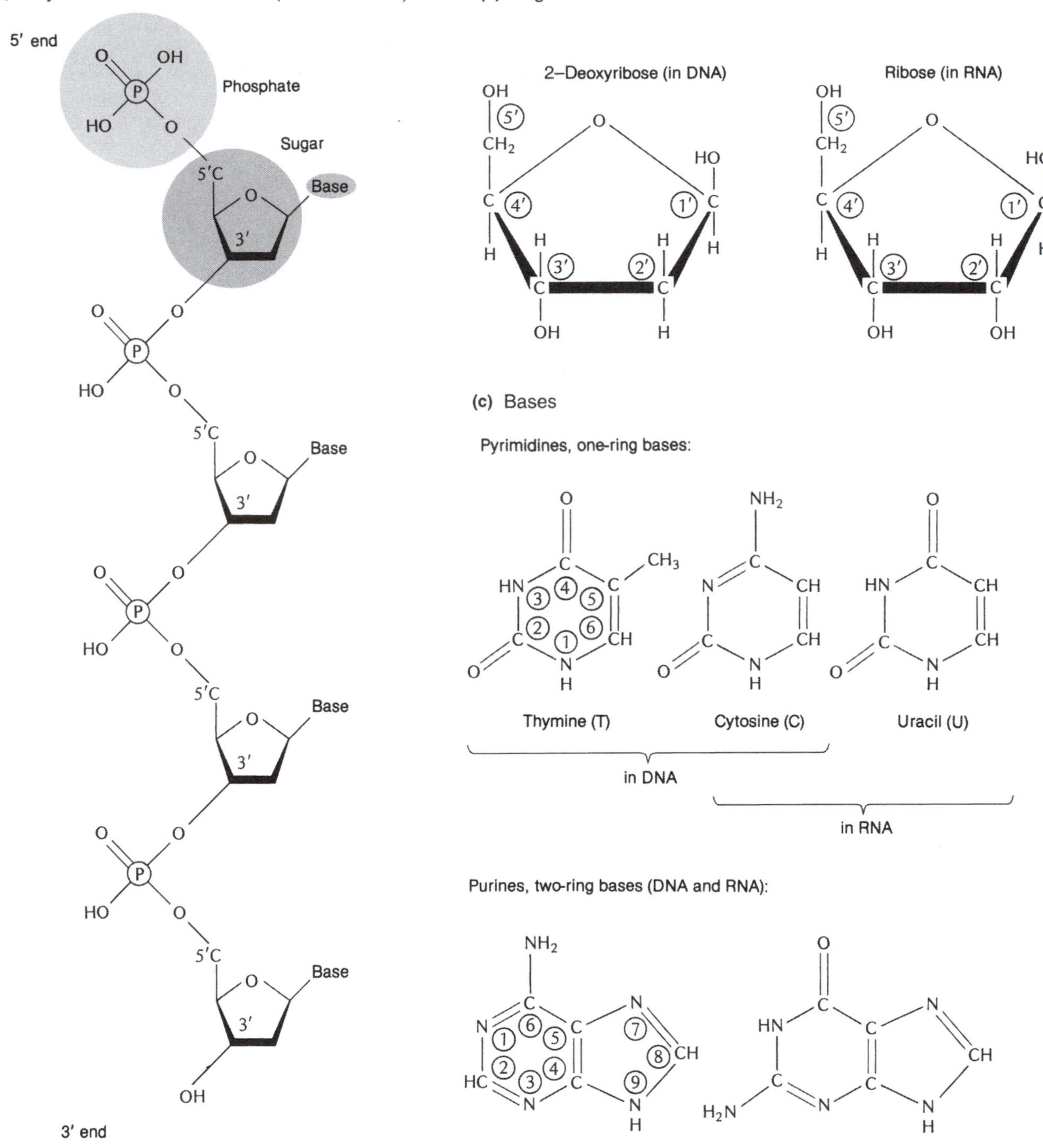

FIGURE 6-3 (a) General structure for DNA and RNA chains. The chains, or strands, may be of considerable length and are composed of a linear sequence of nucleotides. Each nucleotide consists of a phosphate group, a sugar, and a nitrogenous base, linked together in the manner shown. (b) Differences between the sugars found in DNA (deoxyribose) and RNA (ribose). (c) The basic kinds of nitrogenous bases found in DNA (T, C, A, G) and RNA (C, U, A, G).

(5-carbon) sugar, a monophosphate group and a nitrogenous base (**Fig. 6-3**).

The two kinds of sugar used in nucleic acids, ribose (hydroxylated at the 2′ carbon position) and deoxyribose sugar (lacks 2′ hydroxyl), provide the names for the two kinds of nucleic acids, **ribonucleic acid** (**RNA**) and **deoxyribonucleic acid** (**DNA**). In both these nucleic acids, the phosphate groups occupy the same position, tying the 3′

carbon of one sugar to the 5′ carbon of its neighbor via a phosphodiester bond. Connected to the 1′ carbon of each sugar is one of four kinds of nitrogenous heterocyclic bases, of which two are purines [A (*adenine*), and G (*guanine*), in both DNA and RNA] and two are pyrimidines [C (*cytosine*), and T (*thymine*), in DNA, and C (cytosine) and U (*uracil*), in RNA].

Because the complexity of proteins derives from the complexity of nucleic acids, you might think that the restriction of nucleic acid composition to only four different kinds of bases would limit the message-bearing capacity of these molecules to only four kinds of messages, but it doesn't. Nucleic acid molecules may be many thousands or millions of nucleotides long, with each message encoded by a unique linear sequence of nucleotides, endowing these molecules with the capacity to carry an immense variety of highly complex messages. For any one nucleotide position, four different messages are possible (A, G, C, or T); for two nucleotides in tandem, 4^2 or 16 different messages are possible (AA, AG, AC, AT, GG, GC, . . .); and so on; the rule being simply that for a linear sequence of *n* nucleotides, 4^n different possible messages can be encoded. Thus, a linear sequence of only 10 nucleotides can discriminate among more than one million (4^{10}) potentially different messages.

Replication and the Double Helix

This biochemical information helps explain the information-carrying role of nucleic acids but does not explain how they *replicate and transmit this information*. The model for nucleic acid replication derives from the now-familiar **double helix** structure deduced by James Watson and Francis Crick in 1953. **Table 6-1** shows a timeline of the major discoveries from Mendel's experiments showing that heredity is genetic (1900) to the establishment of the complete genetic code used in translating codons into amino acids (1966).[2]

The DNA double helix consists of two antiparallel strands coiled around each other in the form of a right-hand helix, with complementary pairing between purine bases on one strand and pyrimidine bases on the other (A-T, G-C). In the familiar B form of DNA, diagrammed in **Figure 6-4**, there are approximately 10 base pairs for each complete turn of the helix, and the bases are stacked almost perpendicularly to the helical axis. DNA takes various structural forms depending on relative humidity, salt concentration, and so on. Rosalind Franklin, an x-ray crystallographer whose photographs provided Watson and Crick with essential information in devising the model, gave the first two DNA forms investigated, crystalline (dry) and wet (hydrated), the names A and B, respectively. In the A form, each turn of the helix has 11 bases, and the bases are tilted 20° relative to the helical axis.

[2] Table 25-3 takes this list to the sequencing of the human genome in 2003.

TABLE 6-1 Some advances since 1900 leading to our understanding of genes, genomes and gene regulation

Year	Advance
1900	Mendel's 1866 experiments rediscovered and heredity seen as genetic
1903	The chromosomal theory of inheritance proposed
1911	Genetic linkage and X-chromosomes discovered
1913	The first genetic maps produced; recombination discovered
1941	The discovery that genes code for enzymes
1943	Evidence that DNA is the genetic material
1952	Proteins eliminated as basis of genes
1953	Proposal for the double helical structure of DNA
1958	Isolation of the first DNA replicating enzyme
1959	Recognition of the Down syndrome chromosome abnormality in humans
1960	Discovery of messenger RNA
1961	Nucleic acids hybridized
1962	Synthetic RNA used to unravel the genetic code; repressor and transcriptional control of genes discovered.
1966	Establishment of the complete genetic code used in translating codons into amino acids

Source: Based on data from Caskey, C. T. 1992. DNA-based medicine: Prevention and therapy. In *The Code of Codes: Scientific and Social Issues in the Human Genome Project,* Kevles, P. J. and L. Hood (eds.). Harvard University Press, Cambridge, MA, pp. 112–135; and Anderson, W. F. 1992. Human gene therapy. *Science,* **256**, 808–813; Anderson, W. F. 1998. Human gene therapy. *Nature,* **392** (suppl.), 25–30; and others.

The replicative power of the DNA double helix derives from the ability of each of the two strands to serve as a template for a newly complementary strand, so that two new double helices can form bearing nucleotide sequences that are identical to each other as well as to the parent molecule (**Fig. 6-5**). This unique quality of exact molecular replication, enabling similar messages to be transmitted from generation to generation, confers on nucleic acids their function as *genetic material.*

Transcription and Translation

Fundamental to our understanding of the relationship between genetic material and proteins is an important concept:

- the three-dimensional structure of a protein — its form, shape and subsequent function — is primarily determined by the linear sequence of amino acids, which in turn,

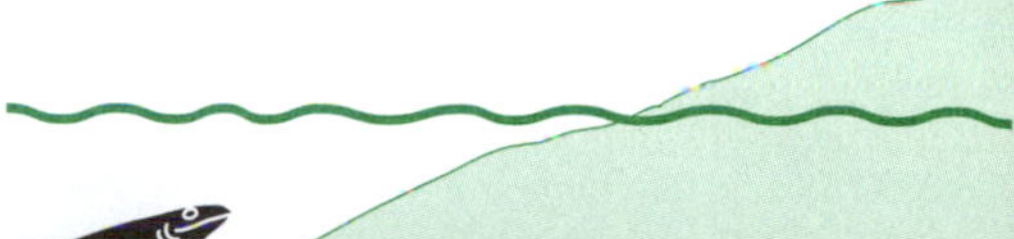

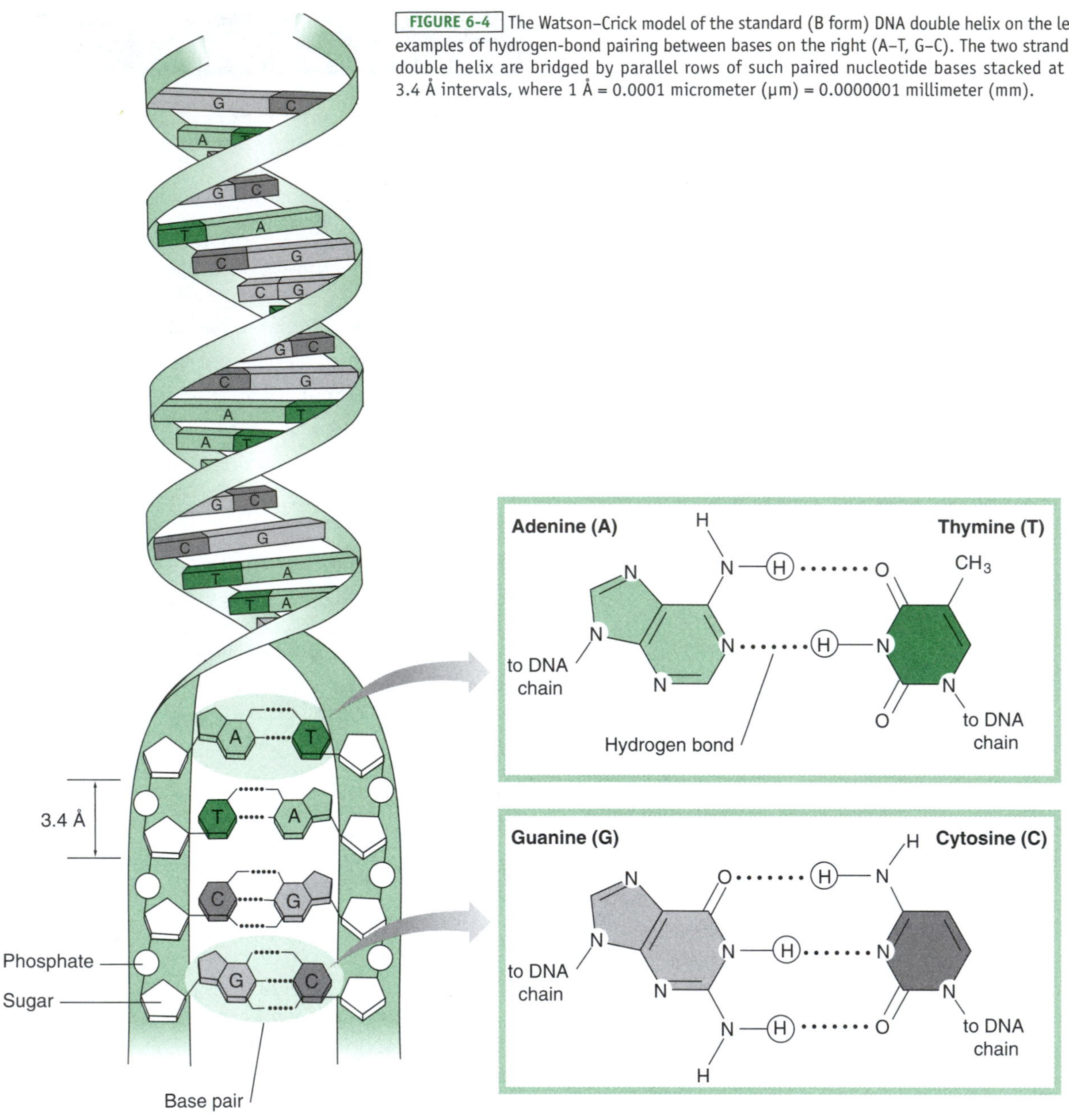

FIGURE 6-4 The Watson–Crick model of the standard (B form) DNA double helix on the left, with examples of hydrogen-bond pairing between bases on the right (A–T, G–C). The two strands of the double helix are bridged by parallel rows of such paired nucleotide bases stacked at regular 3.4 Å intervals, where 1 Å = 0.0001 micrometer (μm) = 0.0000001 millimeter (mm).

- derives from the linear sequence of bases in nucleic acids by means of a protein-synthesizing apparatus involving three different kinds of RNA.

In brief, and as diagrammed in Figure 6-5, the genetic material, through the process of **transcription**, produces a molecule of **messenger RNA** (**mRNA**) that is, base for base, a complement to the bases on one of the DNA strands. Through the mediation of ribosomes, which themselves consist of **ribosomal RNA** (**rRNA**) and protein, a sequence of bases in mRNA translates into a sequence of amino acids. This translation follows the **triplet rule** that a sequence of three mRNA bases designates one of the 20 different kinds of amino acids used in protein synthesis.

During translation, no physical material is inserted by the mRNA into the protein; *only information is transferred.* The mRNA only designates the linear position in which each amino acid is to be placed by special molecules of **transfer RNA** (tRNA) that bring the amino acids to the messenger (**Fig. 6-6**). With the aid of the ribosome and various enzymes, the amino acids connect in sequence through peptide linkages. A polypeptide chain forms in which the

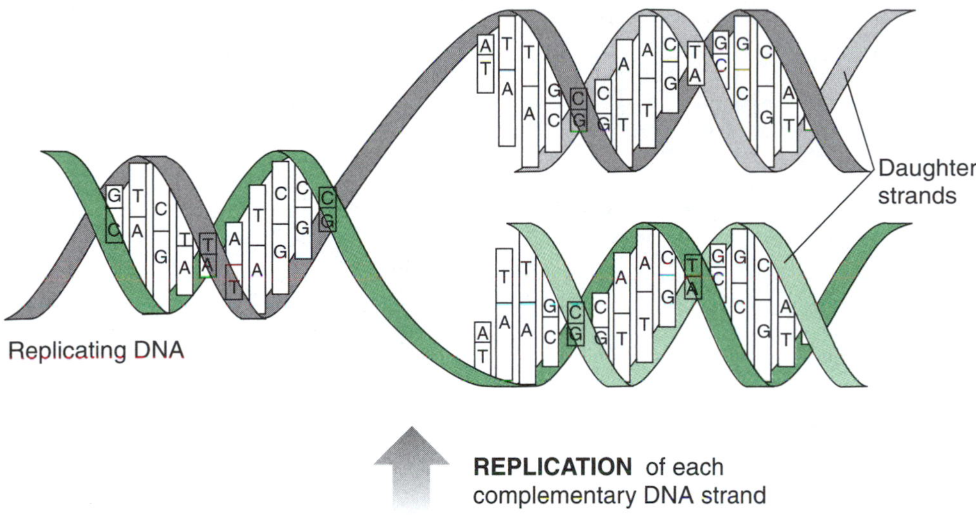

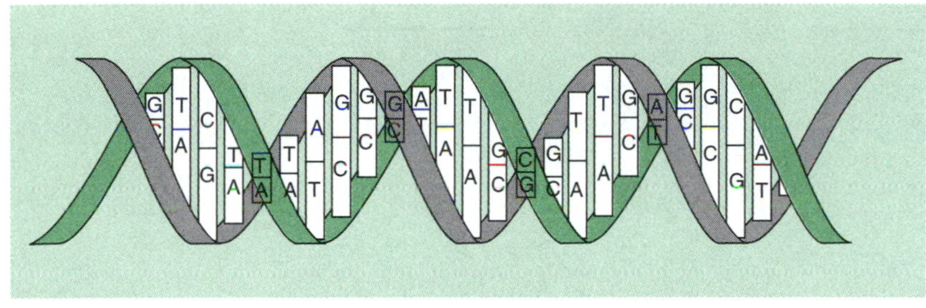

DNA

TRANSCRIPTION of a DNA template strand

G U C U U A U C C C A U U G C C A A C U G G C A

Messenger RNA

Codon

TRANSLATION of messenger RNA

Amino acid

Valine Leucine Serine Histidine Cysteine Glutamine Leucine Alanine

Polypeptide

COILING of polypeptide

Protein (myoglobin)

FIGURE 6-5 Diagrammatic illustration of how DNA replicates and how information is transferred from DNA to RNA to protein. In DNA replication, special proteins break the hydrogen bonds between paired bases, allowing the two strands to unwind. Each unwound strand then acts as a template producing a new complementary strand through base pairing catalyzed by a DNA polymerase enzyme. As a result, two double-stranded DNA molecules form, each an exact replica of the original parental double helix. In transcription, one of the two DNA strands (*dark gray color*) serves as a template upon which a molecule of messenger RNA is transcribed (*light gray color*). This messenger then serves in turn as a template on which a molecule of protein is translated. Note that the messenger RNA is exactly complementary to its DNA template and that a sequence of three nucleotides (a triplet codon) on the messenger specifies one amino acid (for example, 24 nucleotides = 8 amino acids). The amino acids in the illustrated polypeptide chain coil into a right handed α-helix that enables this particular molecule to fold into a special protein called myoglobin used in the cellular transport of oxygen. Other proteins, of course, have different amino acid sequences, which may form different kinds of polypeptide structures such as ß-pleated sheets and which may consist of two or more polypeptide chains.

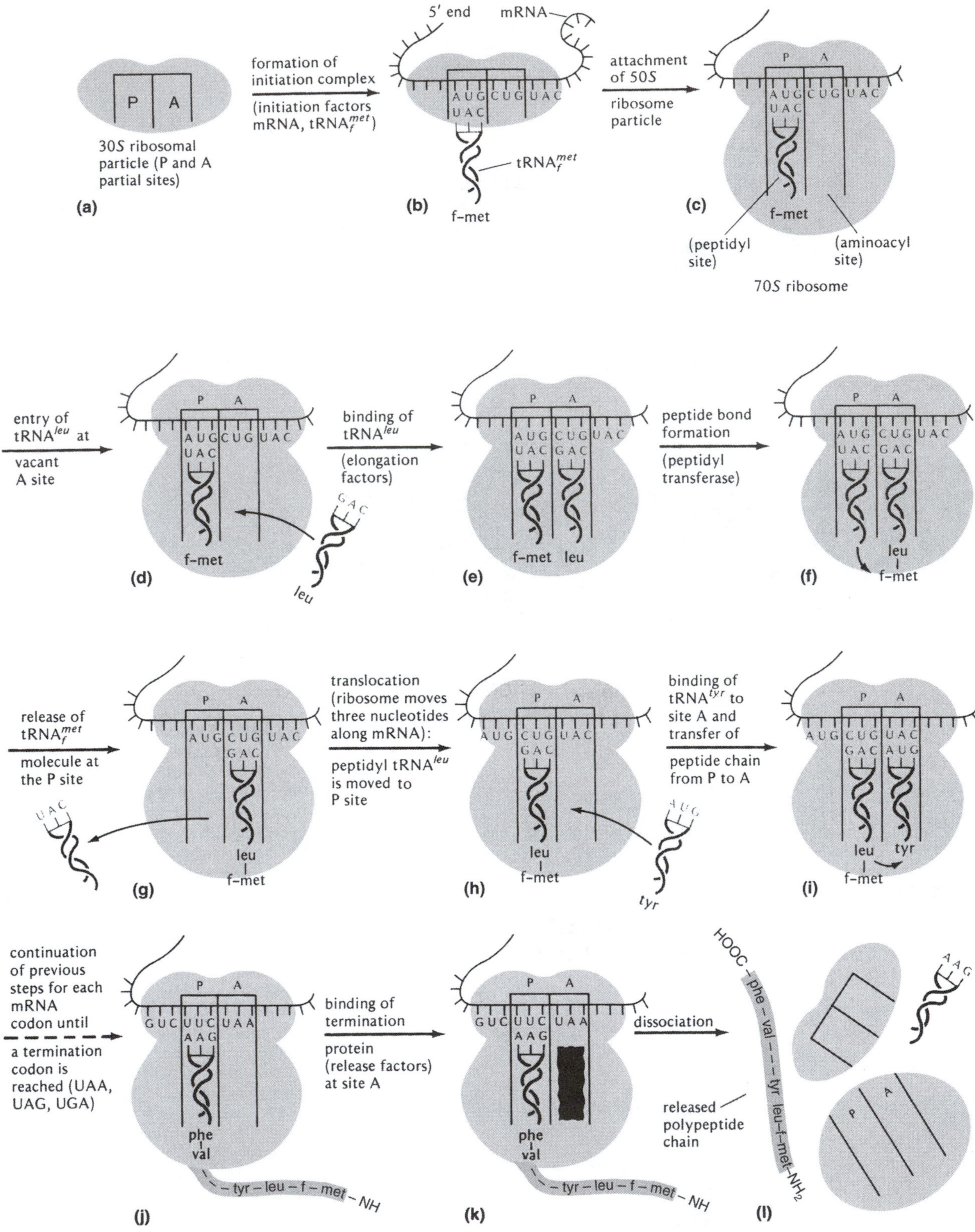
30S ribosomal particle (P and A partial sites)
(a)
formation of initiation complex
(initiation factors mRNA, tRNA$_f^{met}$)
5′ end
mRNA
tRNA$_f^{met}$
f-met
(b)
attachment of 50S
ribosome particle
(peptidyl site)
(aminoacyl site)
70S ribosome
(c)
entry of tRNA^{leu} at
vacant A site
(d)
binding of tRNA^{leu}
(elongation factors)
(e)
peptide bond formation
(peptidyl transferase)
(f)
release of tRNA$_f^{met}$
molecule at the P site
(g)
translocation (ribosome moves three nucleotides along mRNA):
peptidyl tRNA^{leu} is moved to P site
(h)
binding of tRNA^{tyr} to site A and transfer of
peptide chain from P to A
(i)
continuation of previous steps for each mRNA codon until
a termination codon is reached (UAA, UAG, UGA)
(j)
binding of termination
protein (release factors) at site A
(k)
dissociation
released polypeptide chain
HOOC – phe – val – – – tyr leu–f–met–NH$_2$
(l)

FIGURE 6-6 General scheme of protein synthesis in the bacterium *Escherichia coli.* The ribosome consists of two subunits designated by their rates of sedimentation in a centrifuge as 30*S* and 50*S*. Each subunit in turn consists of various proteins and molecules of ribosomal RNA. (a) The 30*S* ribosomal subunit bears two partial sites, peptidyl (P) and acceptor (A), which become functionally complete only when combined with the larger 50*S* subunit. (b) In the presence of mRNA and necessary protein initiation factors, only the special initiator transfer RNA, $tRNA_f^{met}$, bound to a formyl-methionine amino acid (f-met) occupies the partial P site opposite the mRNA initiation codon. (c) The 30*S* and 50*S* subunits join, accompanied by cleavage of the phosphate-bond energy donor, guanosine triphosphate (GTP), releasing guanosine diphosphate (GDP) and inorganic phosphate (P_i). (d) Once the 70S ribosome forms, the completed A site of the ribosome can be entered by aminoacyl-charged tRNAs [tRNAs that carry amino acids such as leucine (leu) and tyronsine (tyr)] whose anticodons can match the mRNA codon at that site. (e) Special protein elongation factors allow the binding of the appropriate aminoacyl-charged tRNA to the A site, and this step also accompanies cleavage of GTP to GDP and P_i. (f) The amino acid (or peptide) at the P site transfers to the amino acid at the A site through peptide bond formation (see Fig. 6-11a) catalyzed by a peptidyl transferase enzyme located on the 50*S* subunit. As a result, the lowermost amino acid in this diagram is at the amino (NH_2) end of the chain, and the uppermost amino acid is at the carboxyl (COOH) end. (g, h) Once the peptide bond forms, the tRNA molecule that has donated its amino acid or peptide to the A site releases from the P site. Simultaneously, the ribosome translocates along the mRNA molecule for a distance of three nucleotides, thereby placing the former A-site tRNA (with its newly elongated peptide chain) in the P site. An additional protein elongation factor is needed for this to occur, and the reaction GTP → GDP 1 P_i again provides energy. This translocation step allows a new mRNA codon to appear at the now vacant A site. (i) Transfer of the peptide chain from P-site tRNA to A-site tRNA again follows binding of aminoacyl-charged tRNA to the A site. (j) These steps are repeated as the ribosome moves along the mRNA molecule until it reaches a termination codon. (k) A protein release factor recognizes the termination codon at the A site and prevents further translocation of the ribosome along the mRNA molecule. (l) A termination reaction releases the peptide chain from the P site tRNA and expels the tRNA molecule from the ribosome. The ribosome separates from the mRNA strand and dissociates into 30*S* and 50*S* subunits. (From *Genetics*, 3d ed., by Monroe W. Strickberger. © 1985 by Monroe W. Strickberger. Reprinted by permission of Prentice Hall, Inc., Upper Saddle River, NJ.)

genetic material has designated the precise position of each component amino acid. DNA (or RNA in some viruses) provides the **genotype**, or **genetic endowment**, of an organism. [Defining a *gene*, the fundamental unit of the genotype, is not straight forward. How genes are defined in different specialties is laid out in **Box 6-1**.] The expression of this nucleic acid information, via transcription and/or translation, provides the various features of an organism, its **phenotype**.

The circular interdependency of all these events, such as the replication of nucleotides because of appropriate enzymes, and the determination of enzyme structure because of appropriate nucleotide sequences, points to certain difficulties in finding a reasonable explanation for the origin of life. Which of the many biochemical agents came first? How did they arise? How could they have functioned before an entire cellular structure evolved?

Theories of How Life Started

Creation Myths

The concept that life did not originally arise on Earth is embodied in almost all human **creation myths**. These myths usually presume the special creation of life on Earth by one or more superior, intelligent, and all-powerful beings who themselves possess attributes of life such as sensation, thought and purposive movement (see Chapter 3). This concept does not explain the source of the initial creator and therefore does not explain life's origin.

Spontaneous Generation

Another ancient concept claims that life can arise **spontaneously** (see Chapter 1), for example, the presumed origin of insects from sweat, crocodiles from mud or insect larvae

BOX 6-1 What is a Gene?

IT MAY SURPRISE YOU to find that there are several definitions for the basis unit of inheritance, the **gene**. In large part, this is because different specialists study the gene at different ways and at different levels. To embrace these approaches, a gene can be defined as a biological unit with one or more than one of the following properties, in what R. A. Wilson (2005) terms *a homeostatic property cluster*. To each property listed by Wilson, we have added the field(s) in which a particular property is sufficient to define a gene:

- a region of DNA whose activation leads to the formation of a discrete hereditary characteristic — developmental biology; developmental genetics.
- correspondence to a single protein or RNA — biochemistry.
- encompassing coding and noncoding segments — molecular biology; molecular genetics.
- located on chromosomes — cytology; genetics, especially medical genetics.
- reliably copied across generations in association with reproduction or transfer — evolution; genetics, especially population genetics.
- a physically localizable developmental resource — some philosophical approaches to development and evolution.

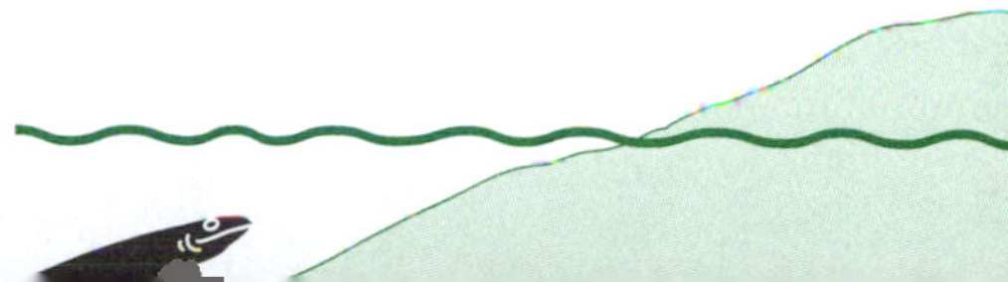

from inside sealed jars. Strangely enough, this view, which was popular in Europe throughout medieval times, was often held simultaneously with the view that life derives from a conscious creator. Pasteur and others put the idea of spontaneous generation as a mechanism for the origin of life to rest in the 1860s. Pasteur's evidence came incidentally from experiments he was undertaking to understand the fermentation process used in making beer and wine. Nothing grew when Pasteur sealed broths of beer in airtight glass vessels (Farley, 1977).

Life in All Matter

In another theory, life is somehow *ingrained in all matter* and the creation of matter no matter what the cause of the creation of life. Clearly, this philosophical notion, although ascribing life to natural events, *assumes a different definition of life* than defining organisms as living, and nonorganisms as nonliving, and so need be considered no longer. Or need it?

Viruses and **prions** function as living entities when associated with live organisms. A prion, believed to be the smallest agent of infection, is a particle composed of a hydrophobic protein. Prions contain neither DNA nor RNA. Can a virus or a prion be said to be alive when not within a live cell? Not if we follow the definition of life offered by Lynn Margulis (1998), namely that "the *units of life are cells*." Prions are misshapen proteins and the products of cells, but cannot function unless incorporated into a live cell, where they convert endogenous proteins into prions like themselves. Such attributes of prions suggest to some that prions (and viruses) lie on the edge of life (see Box 7-1 for information on prions as the cause of diseases).

Discussion of viruses and prions naturally leads us to seek a **definition of life**. R. A. Wilson (2005) provided a set of nine properties defining living agents. His list is so all encompassing, yet succinct, that we reproduce it as **Box 6-2**.

Extraterrestrial Origins of Life

Another theory, for which some scientists claim to have obtained evidence, is that life on Earth arose elsewhere, perhaps on a planetary body circling a distant star, and was then transported to Earth by radiation-resistant spores or other means. This theory overcomes the difficulty of seeking a chemical explanation for the origin of life on Earth but does not, of course, explain the origin of life elsewhere (see Box 5-1).

This notion, called **panspermia**, was fostered by the Nobel Prize-winning Swedish chemist Svante Arrhenius (1859–1927) in the early part of the last century and, perhaps surprisingly, still has some adherents. There is an Interstellar Panspermia Society — the Society for Life in Space (SOLIS) — whose aim essentially is reverse panspermia, the society being "dedicated to promote the future of life in space through directed panspermia. Payloads of selected microorganisms will start evolution in new solar systems."

What type of data could be used to support panspermia as a theory? Some proponents point to the discovery of a number of different organic compounds in carbon-containing meteorites (**carbonaceous chondrites**), ranging from carbohydrates to amino acids.

The way that molecules deflect light is known as **chirality**, from the Greek for *handedness*. Chirality is measured by the optical rotation of a beam of light shown through a molecule. Molecules either deflect light to the left (are *levorotary*) or to the right (*dextrorotary*). A mixture of equal forms will average out the optical activity and so show no chirality. The structural forms or isomers of amino acids in carbonaceous chondrites possess both right- and left-handedness (chirality) in approximately equal amounts, therefore comprising a **racemic mixture** that shows little if any optical activity. In contrast, amino acids of living forms generally show only one type of chirality, levorotary (**Fig. 6-7**).

Furthermore, a number of the amino acids found in meteorites do not appear in proteins. Along with other

BOX 6-2 What is Life?

FOLLOWING THE CRITERIA laid out by R. A. Wilson (2005), living agents:

- have parts that are heterogeneous and specialized;
- include a variety of internal mechanisms;
- contain diverse organic molecules, including nucleic acids and proteins;
- grow and develop;
- reproduce;
- repair themselves when damaged;
- have a metabolism;
- bear environmental adaptation; and
- construct the niches that they occupy.[a]

[a] Wilson extends this list to what he refers to as the *Tripartite View* of an organism. An organism is (1) a living agent (that is, has the nine properties outlined above), (2) belongs to a reproductive lineage, and (3) has minimal functional autonomy (2005, p. 59). As emphasized by Wilson, this definition applies to plants, animals, Archaea, Eubacteria, Protista, and Fungi.

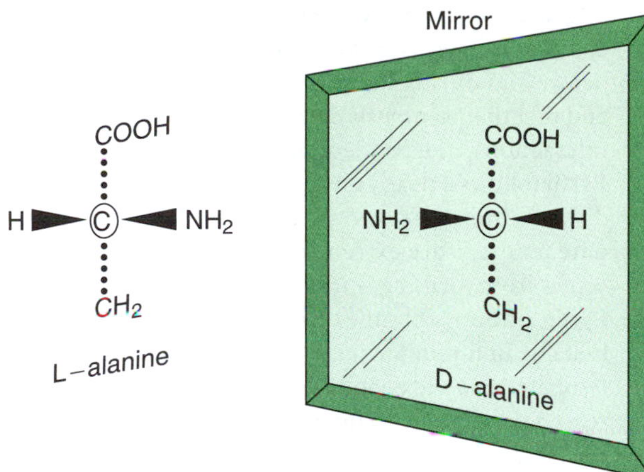

FIGURE 6-7 Structures of laevo (L)-form and dextro (D)-form amino acid. The dark, wedge-shaped bonds indicate that the attached NH_2 and H groups project above the plane of the paper. Note that although D-alanine is a mirror image of L-alanine, they are not identical; the two molecules cannot be superimposed on each other. Each stereoisomer has optical activity that can be measured by the extent to which it rotates polarized light. However, the two forms rotate light in opposite directions, and an equimolar mixture of the two is not optically active.

observations, these findings indicate that organic compounds could have formed through random chemical reactions in the meteorite itself, or in its parent body, rather than through ordered processes within organisms. However, the vastness of space would disperse molecules so widely that their chance of reaching Earth is infinitesimally small. Such misgivings, along with a rapid increase in our understanding of molecular biology and biochemistry, have influenced most scientists to concentrate their attention on a terrestrial origin of life.

Terrestrial Origin of Life

The difficulty in visualizing life originating on Earth is one of visualizing the molecular environment and events that occurred in a long-distant past. However, we can try to *deduce* the general nature of some of the original molecular events from present living structures and reactions and try to reconstruct such events experimentally under controlled conditions.

Likelihood and Complexity

The chances of most complex organic structures arising spontaneously are infinitesimally small. Even a small enzymatic sequence of 100 amino acids would have only one chance in 20^{100} ($= 10^{130}$) to arise randomly, because there are 20 different amino acids for each position in the sequence. Thus, if we randomly generated a new 100-amino-acid–long sequence each second, we could expect such a given enzyme to appear only once in 4×10^{122} years.

These long odds are formidable. Theorists often pointed out that according to the second law of thermodynamics, the energy in a system tends toward diffusion rather than concentration. That is, in an *isolated* system, in which events occur in the absence of outside sources of energy, conditions go from a more ordered to a less ordered state because **entropy** (disorder) increases rather than decreases. Thus, there appeared to be only a negative answer to the question posed by Louis Pasteur in the nineteenth century: "Can matter organize itself?" Many scientists today point to three important considerations:

1. The likelihood that primeval "living" organisms did not have many of the complexities associated with extant organisms (see Box 8-3).
2. The formation of organic molecules and subsequent organic structures was not the result of completely random events, although such events were nevertheless natural, in the sense that they followed chemical and physical laws.
3. Life is not an "isolated" system unable to maintain organized structures, but continually receives energy and materials from outside the organism that enables it to preserve "order" until death.

The first consideration that early organisms were less complex than those of today, is extremely important.

Perhaps the most **basic quality of life** we would recognize in even a "primitive" living organism is its ability to perform those reactions necessary for it to **survive and replicate.** The endless loop of metabolism and information transfer now embodied in the intricate relationship between proteins and nucleic acids need not always have been of the same complexity. As we shall see, protein-like compounds may arise in reaction mixtures of amino acids *without* the intervention from nucleic acids. Although formed randomly, these compounds may *function in a variety of enzymatic ways.* As RNA also can function as an enzyme under certain conditions, we have to keep an open mind on whether RNA or proteins came first (see Chapter 8).

The second and third considerations, that life did not arise from absolute chaos and was not isolated from external forces that decrease entropy, have been supported in many ways. This is because the evolution of our solar system offered a number of essential prerequisites that enabled the development and sustenance of life: a sun, chemical diversity, Earth's orbit and water.

Prerequisites for the Origin of the First Molecules

1. Our planet possesses a sun of moderate size that was on the main sequence of stellar evolution. This sun provided

TABLE 6-2 Present energy sources that probably were available for organic synthesis early in Earth's history

Source	Energy (calories/cm²/year)
Total solar radiation (all wavelengths)	260,000
Ultraviolet light wavelengths (in angstroms)	
Below 3,000	3,400
Below 2,500	563
Below 2,000	41
Below 1,500	1.7
Electrical discharges (lightning, corona discharges)	4
Shock waves (meteorite impacts, lightning bolt pressure waves)	1.1
Radioactivity (to depth of 1 km)	0.8
Volcanoes (heat)[a]	0.13
Cosmic rays	0.0015

[a] Not all volcanic eruptions release the same amount of energy. Of the four categories, *Hawaiian volcanoes* such as Mauna Loa release lava but are not explosive and do not expel fragments of rock. *Stumbling eruptions,* such as Mt. Storable, expel lava in frequent explosions accompanied by incandescent, luminous vapor. *Vulcan Ian eruptions,* the second most violent, erupt violently after the build up of gases that release the magma plug blocking the vent. Finally, *Paleyan eruptions* (named after Mt. Pelé in Martinique) expel rock, ash, lava and superheated gas high in the air (See *TIME Magazine,* 2006. Nature's Extremes. Inside the Great Natural Disasters that Shape Life on Earth. Time, Inc., New York).

Source: Reprinted by permission from Miller, S. L. and L. E. Orgel, 1974. *The Origins of Life on the Earth.* Prentice Hall, Englewood Cliffs, NJ.

a steady rate of emitted radiation over a long enough period of time for life to develop. As shown in **Table 6-2,** the amount of energy available from solar radiation seems to have always been far greater than any other source, although energy from other sources may have been important in initiating particular chemical reactions.

2. The presence of critical elements — H, O, C, N, S, P and Ca — must have provided considerable chemical diversity, enabling reactions to occur that were necessary to form organic molecules involving carbon. The important chemical attributes of carbon, its ability to maintain four covalent bonds and the tetrahedral arrangement of its outer electrons provided the opportunity for the formation of a large number of different kinds of stable molecules with considerable three-dimensional variety and complexity. Interestingly, the terrestrial presence of such molecules is not unique. Using spectroscopy, we can observe a variety of organic molecules in the dense interstellar clouds that give rise to stars and planets (**Fig. 6-8**). Such observations indicate that a number of compounds necessary for the origin of life were present before and during the formation of our solar system, and that their synthesis derived from **abiotic** chemical interactions, independent of living systems.
3. Earth follows a nearly circular orbit at a fairly uniform distance from the sun. Such an even orbit would eliminate temperature extremes preventing organic molecules from forming, surviving and functioning.
4. Large amounts of an excellent solvent, water, which is stable in liquid form over a relatively wide range of temperatures and enables acids and bases to ionize and react, were present. Water floats in its crystalline frozen form (ice), so that bodies of water containing organic matter may remain liquid under a surface of ice, rather than freezing because of the subsurface accumulation of ice. Geochemists now believe that water must have been present early in Earth's history in the cold planetesimal condensations and appeared in liquid form as soon as the lithosphere reached appropriate temperatures. Additional water has been continually emitted into the atmosphere through volcanic activity (which was greater in the past than at present). Since water causes crustal erosion, which leads to sedimentary rock formation, the presence of significant amounts of water must date back to the beginning of the geological record.
5. Resulting from the high cosmic abundance of hydrogen and the consequent abundance of its compounds, hydrogen-containing gases existed for a long initial period in Earth's history. Even though free hydrogen was probably lost early in Earth's history, outgassing from Earth's interior may have led to at least a partially reducing atmosphere in some or many localities in which hydrogen atoms could be donated to a variety of hydrogen-accepting elements, especially carbon. Such gases, along with the energy from solar and ultraviolet radiation, enabled the formation of a variety of organic molecules (for example, amino acids, sugars, fatty acids, purines and pyrimidines) that could, in turn, provide structure and energy for life's processes.

Theories about the composition of Earth's early atmosphere have varied over time and range from strongly reducing to mildly reducing to neutral and nonreducing. However, all theories agree that gases presently emitted from Earth's interior include hydrogen (H_2) and methane (CH_4). Furthermore, hydrothermal deep-sea vents, with water temperatures of 350°C, can produce large amounts of ammonia by combining nitrogen or its oxides with overheated water in the presence of iron and other mineral catalysts (Brandes et al., 1998; and see **Box 6-3**). There were thus appropriate energy sources, chemicals, temperature and solvent to lay the foundations of a "universal organic chemistry."

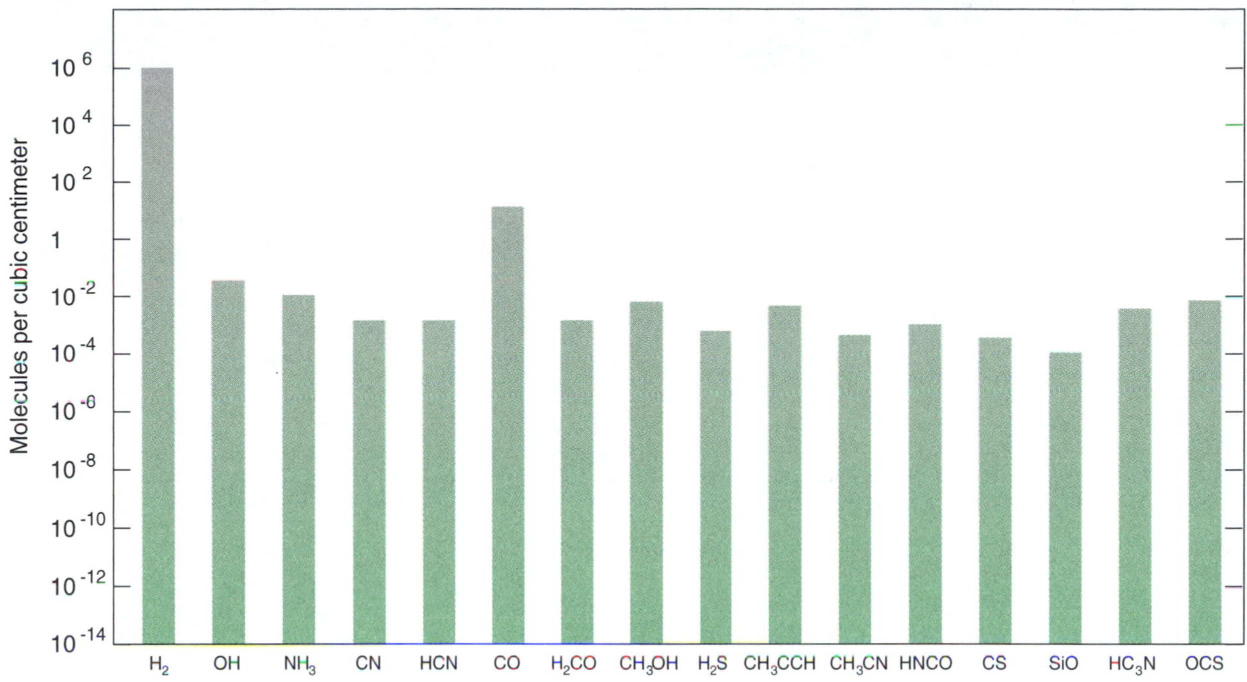

FIGURE 6-8 Densities of various molecules observed in molecular clouds within our galaxy. H_2 = hydrogen, OH = hydroxyl, NH_3 = ammonia, CN = cyanogen, HCN = hydrogen cyanide, CO = carbon monoxide, H_2CO = formaldehyde, CH_3OH = methyl alcohol, H_2S = hydrogen sulfide, CH_3CCH = methylacetylene, CH_3CN = methyl cyanide, HNCO = isocyanic acid, CS = carbon monosulfide, SiO = silicon monoxide, HC_3N = cyanoacetylene and OCS = carbonyl sulfide. (Adapted from Buhl, D., 1974. Galactic clouds of organic molecules. *Origins of Life*, **5**, 29–40.)

In order to investigate the kinds of reactions that could have taken place, researchers designed experiments and collected evidence from meteorites.

Evidence from Experiments and from Meteorites

In the 1920s, Aleksandr Oparin (1894–1980), a Russian biochemist, and J. B. S. Haldane (1892–1964), an English physiologist and geneticist, independently suggested that the early atmosphere was reducing and that organic compounds formed in such an atmosphere might be similar to those used by organisms. However, almost 30 years elapsed before anyone tested this hypothesis.

In 1953, Stanley Miller placed together in a glass apparatus, methane, ammonia and hydrogen gases, generated an electric spark in a large five-liter flask, and boiled water in a smaller flask to provide vapor to the spark and to circulate the gases (**Fig. 6-9**). Compounds formed by sparking were condensed, or recirculated if they were volatile. After a week of continuous electrical discharge, Miller used chromatography to analyze any products that may have accumulated in the aqueous phase. Remarkably, compounds had formed, including amino acids and other substances, such as urea, found in organisms (**Table 6-3**). Perhaps not surprisingly, all of the molecules had simple structures, but all were compounds known to be essential for life.

In addition to electrical discharges, other energy sources such as ß-rays, γ-rays, x-rays, thermal heating and ultraviolet light facilitated the production of amino acids and other organic compounds from simple gases. For example, a variety of organic compounds are produced when a heated mixture of carbon monoxide and hydrogen is passed over a catalyst. Adding ammonia produces purine and pyrimidine nucleotide bases that the Miller reactions do not produce.

Although it now is doubtful that the Earth's atmosphere was a reducing one that would have allowed these reactions to occur, these and subsequent similar experiments pointed strongly to the likelihood that the chemical environment that existed before the origin of life was not "chaos." Rather, Earth already had a significant amount of simple organic molecules that could participate in forming organisms. Moreover, we can see such compounds in interstellar clouds in our galaxy (see Fig. 6-8) and in various **carbonaceous meteorites** believed to represent material remaining in space from the original solar condensation 4.6 Bya.

One example, the **Murchison meteorite**, which fell in Australia in 1969, contains more than 80 kilograms of carbonaceous material, of which about one percent is organic

BOX 6-3 Origin of Molecules in Deep-Sea Hydrothermal Vents

DEEP-SEA HYDROTHERMAL VENTS, discovered in 1977 and where temperatures of 350°C are common, have emerged over the past decade as a plausible site for the origin of organic molecules, and perhaps for the first life on Earth. Evidence has accumulated for the production of the amino acid **alanine** and its assembly into short chain peptides in the following sequence:

> water and metallic ions → activated acetic acid, which with CO_2 → pyruvic acid, which with ammonia → alanine, which can interact with water to form small peptides.

In a scenario proposed by Huber and Wächtershäuser for the origin of the first molecules, metallic ions played the role now played by enzymes; iron pyrite is abundant in rocks of the ocean bottom. Acetic acid, one of the most simple acids composed of oxygen, hydrogen and carbon, is a primary initiator of many metabolic pathways in extant organisms, including the Krebs cycle discussed in Chapter 8 (especially Box 8-4). Experiments in 1998 provided laboratory-based evidence for the production of acetic acid at 100°C, but critics countered that such molecules could not survive such temperatures. Laboratory tests at a wider range of temperatures and at the pressures found at ocean depths produced ammonia that was stable to temperatures as high as 800°C. When combined with pyruvic acid, the amino acid alanine was produced and formed short peptides.[a]

[a] See Brandes et al. (1998) and Huber, C., and G. Wächtershäuser (1998) for these studies, and see Hazen (2005) for an overview of these and other studies on the evolutionary origin of molecules.

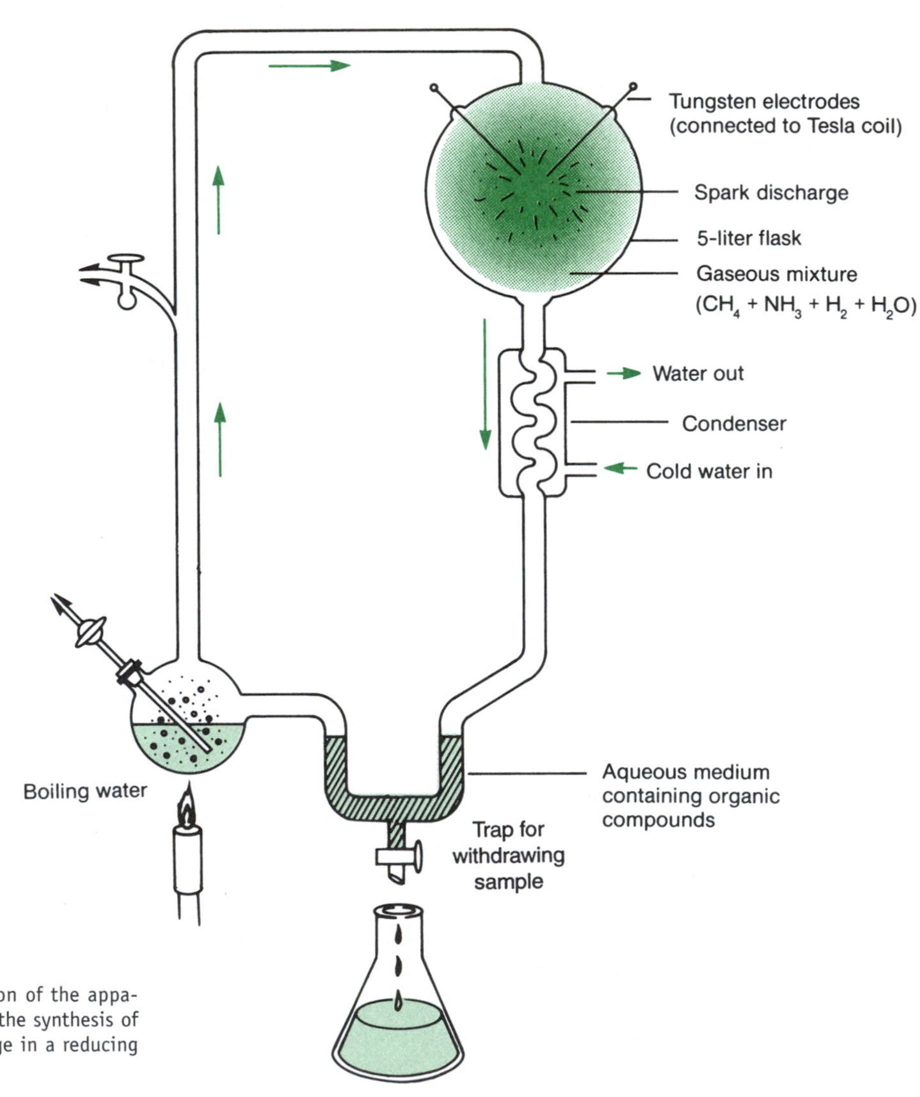

FIGURE 6-9 Diagrammatic representation of the apparatus Stanley Miller used to demonstrate the synthesis of organic compounds by electrical discharge in a reducing atmosphere.

TABLE 6-3 Yields of various organic compounds obtained from a mixture of water, hydrogen, methane, and ammonia exposed to electrical sparking[a]

Compound	Yield (%)[b]
Glycine	2.1
Glycolic acid	1.9
Sarcosine	0.25
Alanine	1.7
Lactic acid	1.6
N-methylalanine	0.07
α-amino-n-butyric acid	0.34
α-aminoisobutyric acid	0.007
α-hydroxybutyric acid	0.34
β-alanine	0.76
Succinic acid	0.27
Aspartic acid	0.024
Glutamic acid	0.051
Iminodiacetic acid	0.37
Iminoaceticpropionic acid	0.13
Formic acid	4.0
Acetic acid	0.51
Propionic acid	0.66
Urea	0.034
N-methyl urea	0.051
TOTAL	15.2

[a] These products represented only about 15% of the carbon that had been added to the apparatus. The remaining carbon products, mostly in the form of polymerized tar-like substances, were not analyzed.

[b] The percent yields are based on the amount of carbon that was added to the mixture as methane.

Source: Reprinted by permission from Miller, S. L. and L. E. Orgel, 1974. *The Origins of Life on the Earth*. Prentice Hall, Englewood Cliffs, NJ.

carbon (**Fig. 6-10**). Because a number of different laboratories analyzed the meteorite almost immediately after it landed and the results were consistent overall, researchers doubt that the protein-type amino acids could have originated from terrestrial contamination. Also remarkable is the fact that practically all the amino acids found in the meteorite, protein and non-protein, are similar to the amino acids produced by sparking mixtures in laboratory experiments (**Table 6-4**).

For evidence of prebiotic fatty acid synthesis, we can also look to carbonaceous chondrites that contain compounds of the kind synthesized in the early solar system. Interior portions of the Murchison meteorite, for example, contain fatty acids up to eight carbons long. Moreover, a portion of uncontaminated Murchison meteorite compounds can produce fatty-like structures *and* bounded vesicles that resemble membranes. Some researchers suggest that high temperatures and pressures found in hydrothermal plumes, common along mid-oceanic ridges, could have provided conditions appropriate for fatty acid synthesis on Earth. The chemicals found in these meteorites strongly indicate that laboratory experiments reflect actual chemical processes that occurred in the synthesis of prebiotic organic compounds early in Earth's history.[3]

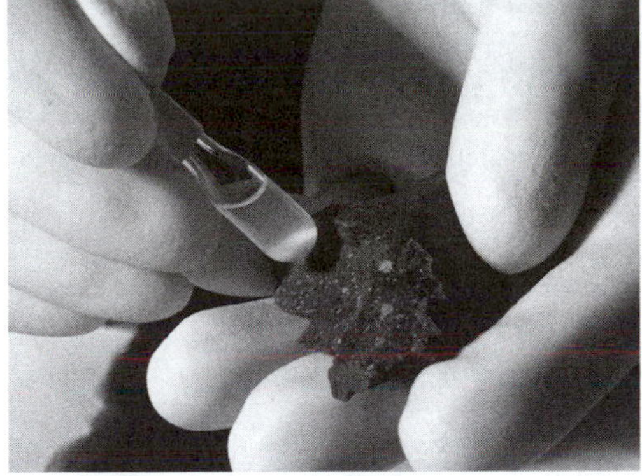

FIGURE 6-10 Researchers from the University of Chicago isolated individual components (shown in the test tube) of the Murchison meteorite that are unchanged since their condensation from material ejected by a star. Frictional heating produced the charred (black) crust as the meteorite passed through Earth's atmosphere; this meteorite crashed to Earth near Murchison, Victoria, Australia, in 1969. Geochemists have suggested that major meteorite impacts during the Hadean era contributed significant amounts of water and organic compounds used in origin-of-life reactions. (Photo courtesy of Argonne National Laboratory.)

The pathways required to synthesize five classes of organic molecules — aldehydes, sugars, purine and pyrimidine bases, fatty acids and pyrroles/porphyrin rings/coenzymes — under simple conditions such as may have existed early in Earth's evolution, are outlined in **Box 6-4**. An examination of these pathways shows that many of the basic organic molecules used in organisms form relatively easily in many reactions. The amounts per reaction are usually small, but the overall quantities of such substances may have been quite large if sufficient reactive compounds such as methane were present. Even if we limit reactions to available energy, significant concentrations of organic material could have been produced. One photon of ultraviolet radiation produces a quantum yield of about 1/100,000 to 1/1,000,000 of a simple organic molecule.

Of course, ultraviolet radiation and heat also decompose organic material, and such degradative effects may have been considerable. Nevertheless, once organic material formed, it would have had many opportunities to accumulate in relatively cool, protected localities, such as the fissures of rocks, oceanic depths and pools inaccessible to decomposition by ultraviolet rays.

[3] See Deamer (1986, 2003) for experiments with the Murchison meteorite, and see Ferris (1992) for the role of high temperatures and pressures.

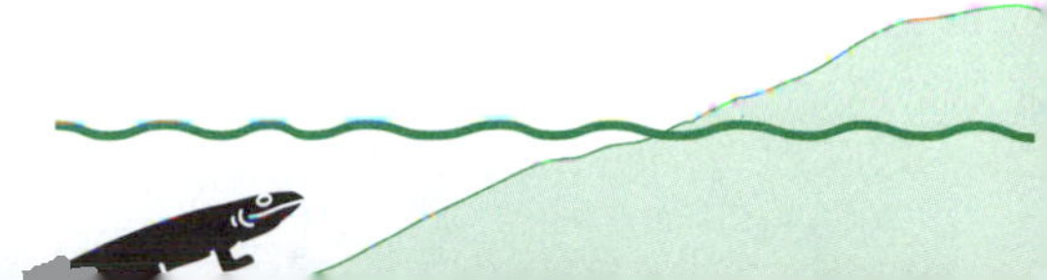

TABLE 6-4 Comparison between the relative abundances (asterisks) of amino acids in the Murchison meteorite and in electric discharge synthesis[a]

Amino Acid[b]	Murchison Meteorite	Electric Discharge
Glycine	****	****
Alanine	****	****
α-amino-n-butyric acid	***	****
α-aminoisobutyric acid	****	**
Valine	***	**
Norvaline	***	***
Isovaline	**	**
Proline	***	*
Pipecolic acid	*	*
Aspartic acid	***	***
Glutamic acid	***	**
β-alanine	**	**
β-amino-n-butyric acid	*	*
β-aminoisobutyric acid	*	*
γ-aminobutyric acid	*	**
Sarcosine	**	***
N-ethyglycine	**	***
N-methylalanine	**	**

[a] Analysts did not observe purine and pyrimidine compounds found in the nucleic acids of organisms in the meteorite, although they did note traces of nonbiological pyrimidines (for example, 4-hydroxypyrimidine).

[b] Strongly indicative of their nonbiological origin was the finding that the meteorite amino acids were optically inactive (racemic) mixtures of D and L forms, rather than consisting exclusively of the levorotary forms produced biologically on Earth.

Source: From *Cold Spring Harbor Symposia on Quantitative Biology,* Volume LII, 1987. Cold Spring Harbor Laboratory Press, New York.

Condensation and Polymerization

Given localized concentrations of amino acids, sugars, and other organic molecules, further chemical evolution would depend on the **polymerization** or **condensation** of these monomers into peptides, polysaccharides, and so on. Pathways for the production of five classes of organic molecules — aldehydes, sugars, purine and pyrimidine bases, fatty acids, and pyrroles, porphyrin rings and coenzymes — are outlined in Box 6-4.

As depicted in **Figure 6-11**, most polymerizations depend on the removal of water molecules from the monomers to be condensed. Amino acids, for example, are ordinarily bonded into peptides on cellular ribosomes through **phosphate-bond** energy (see Fig. 6-6). Outside the cell, the task is more difficult, but bonds can nevertheless form in aqueous media or under anhydrous conditions, and many condensing agents produce **peptide bonds** (**Box 6-5**).

Under anhydrous conditions, with no or few water molecules, heat can promote condensation and polymerization by causing extraction and loss of water molecules from chemical substrates even in the absence of specific condensing agents. One such reaction, accomplished by heat in the laboratory and by enzymes in organisms, is the formation of high-energy phosphate bonds from orthophosphate:

$$2\left[\mathrm{O{=}P(OH)_2{-}OH}\right] \longrightarrow \mathrm{O{=}P(OH){-}O{-}P(=O)(OH){-}O^-} + \mathrm{H_2O}$$

Orthophosphate → Pyrophosphate

High yields of inorganic pyrophosphate (PPi) also can be synthesized by the condensing agent cyanic acid (cyanate) reacting on precipitated hydroxyapatite [$Ca_{10}(PO_4)_6(OH)_2$], a major phosphate mineral. Such pyrophosphates can then be made available to form adenosine diphosphate (ADP) and **adenosine triphosphate** (ATP) in reactions that can be reversed by **hydrolysis** to yield energy and provide the primary means by which cells remove further water molecules during condensation reactions. Polyphosphate chains may have provided some of the first organismal energy sources with the adenosine component in ATP added later to act as a label that would allow enzymatic recognition.

Can any of the reactions outlined above and in Box 6-5 be accomplished in laboratory experiments that mimic conditions on the early Earth?

Production of Polymers in the Laboratory

In the 1950s, Sidney Fox and coworkers developed a technique in which heat could be used to produce **peptides** from dry mixtures of amino acids (Fox, 1984). Depending on the kinds of amino acids in the mixture, temperatures of 150° to 180°C could produce as much as a 40 percent yield of peptide-like products with molecular weights between 4,000 and 10,000 Daltons. Fox called these **polymers** *proteinoids* (also *thermal proteins*), and he and his group proposed that they bear protein-like features. According to their analyses, these polymers possess nonrandom proportions of amino acids; their compositions are not simply based on the frequency of the different amino acids in the initial mixture (**Table 6-5**).

They also suggest that, because some amino acids preferentially occupy the N- and C-terminals of the polymers, the positions of the amino acids in the polymer are not based on their overall frequencies in the chain. The nonrandomness of the polymer structure is supported by the

FIGURE 6-11 Examples of condensation reactions leading to the formation of peptides, polysaccharides, lipids and nucleic acids. In (d) the sugar unit is ribose and the nucleic acid produced is ribonucleic acid (RNA), whereas the sugar unit in deoxyribonucleic acid (DNA) lacks the oxygen atom at the 2′ carbon position. (Adapted from Calvin, M., 1969. *Chemical Evolution*. Oxford University Press, Oxford, England.)

(a) Proteins

$H_2N-CH(R_1)-CO-OH + H-NH-CH(R_2)-COOH \rightarrow H_2O + H_2N-CH(R_1)-CO-NH-CH(R_2)-COOH$

Amino acid 1 + Amino acid 2 → H_2O + Dipeptide

further condensations → Polypeptide

(b) Polysaccharides

Glucose + Glucose → H_2O + Maltose (disaccharide)

further condensations → Starch (polysaccharide)

(c) Lipids

$H-C(OH)(H)-C(OH)(H)-C(OH)(H)-O-H + HO-C(=O)-(CH_2)_n-CH_3 \rightarrow H_2O + H-C(OH)(H)-C(OH)(H)-C(OH)(H)-O-C(=O)-(CH_2)_n-CH_3$

Alcohol (glycerol) + Fatty acid → H_2O + Lipid

(d) Nucleic acids

Adenine (base) + Ribose (sugar) → H_2O + Adenosine (nucleoside)

Adenosine + Phosphate ($H-O-P(=O)(OH)-OH$) → H_2O + Adenylic acid (nucleotide)

2 Adenylic acid molecules —condensation→ Dinucleotide —further condensation→ RNA (nucleic acid)

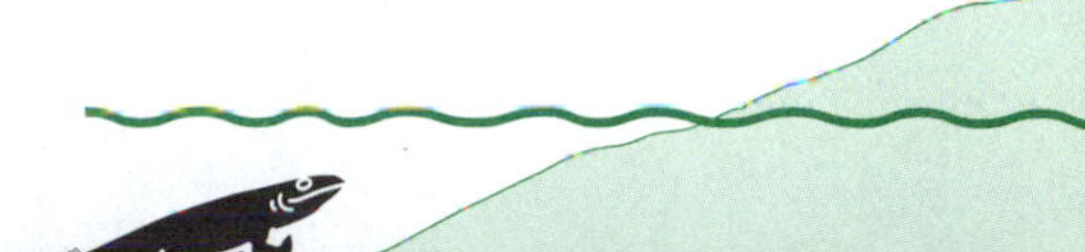

BOX 6-4 Pathways for the Production of Five Classes of Organic Molecules

FIVE POSSIBLE PATHWAYS as initiators of early organic molecules are outlined below.

Aldehydes

Amino acids synthesized under early Earth conditions may have arisen primarily from the formation of aldehydes, organic compounds formed by the oxidation of alcohols and that contain a CHO group as depicted below

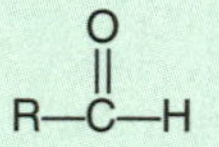

$$R{-}\overset{O}{\overset{\|}{C}}{-}H$$

where R may represent any group, which then interact with ammonia and cyanide compounds to produce ammonia, acetylene, hydrogen cyanide, formaldehyde or acetaldehyde. These reactive chemicals may have arisen from a variety of simple gases or from their further interactions:

Ammonia

$$N_2 + H_2 \rightarrow NH_3$$
$$N_2 + H_2O \rightarrow NH_3$$

Acetylene

$$CH_4 \longrightarrow C_2H_2\ (HC{\equiv}CH)$$

Hydrogen cyanide

$$CH_4 + N_2,\quad CH_4 + NH_4,\quad CO + NH_3,\quad C_2H_2 + N_2 \rightarrow HCN\ (HC{\equiv}N)$$

Formaldehyde

$$CH_4 + H_2O,\quad CH_4 + CO_2,\quad CO_2 + H_2,\quad CO_2 + H_2O \rightarrow HCHO\ (H\overset{O}{\overset{\|}{C}}H)$$

Acetaldehyde

$$CH_4 + H_2O \longrightarrow CH_3CHO\ (H_3C{-}\overset{O}{\overset{\|}{C}}H)$$

According to one of the possible pathways of amino acid synthesis (the **Strecker synthesis**), subsequent steps are as outlined below:

(1) Formation of aldimine from aldehyde + HN_3

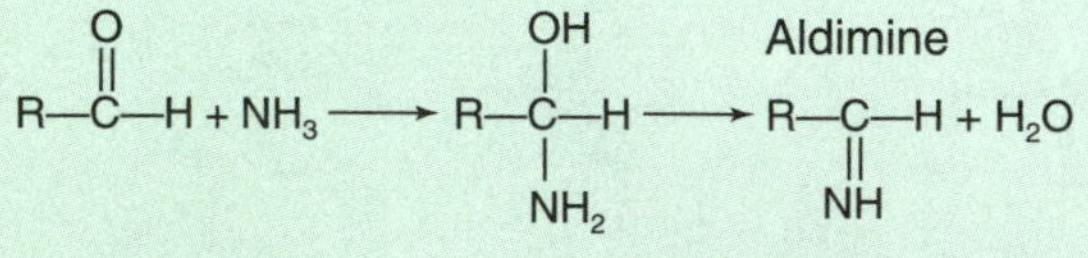

$$R{-}\overset{O}{\overset{\|}{C}}{-}H + NH_3 \longrightarrow R{-}\overset{OH}{\overset{|}{\underset{NH_2}{\underset{|}{C}}}}{-}H \longrightarrow R{-}\underset{NH}{\underset{\|}{C}}{-}H + H_2O \quad \text{Aldimine}$$

(2) Formation of aminonitrile from aldimine

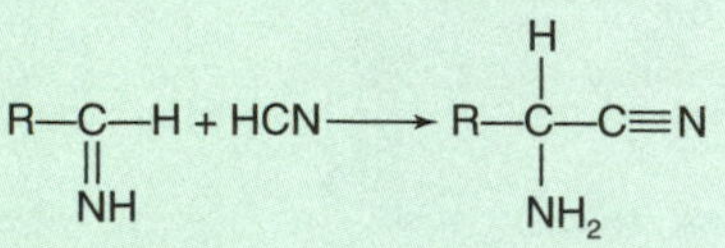

$$R{-}\underset{NH}{\underset{\|}{C}}{-}H + HCN \longrightarrow R{-}\overset{H}{\overset{|}{\underset{NH_2}{\underset{|}{C}}}}{-}C{\equiv}N \quad \text{Aminonitrile}$$

(3) Formation of aminoamide from aminonitrile

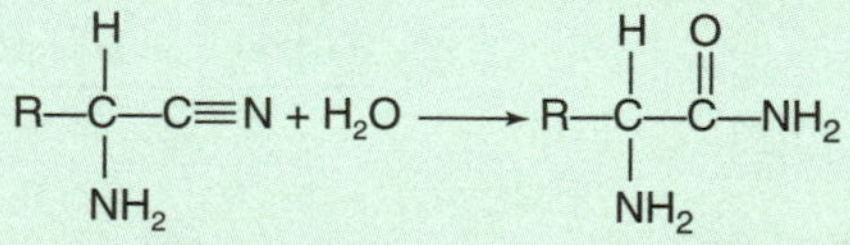

$$R{-}\overset{H}{\overset{|}{\underset{NH_2}{\underset{|}{C}}}}{-}C{\equiv}N + H_2O \longrightarrow R{-}\overset{H}{\overset{|}{\underset{NH_2}{\underset{|}{C}}}}{-}\overset{O}{\overset{\|}{C}}{-}NH_2 \quad \text{Aminoamide}$$

(4) Formation of α-amino acid from aminoamide

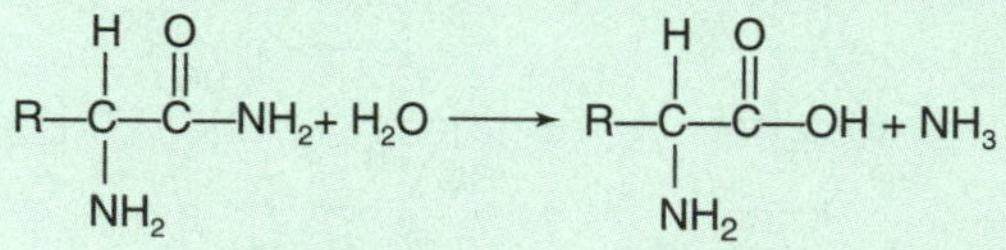

$$R{-}\overset{H}{\overset{|}{\underset{NH_2}{\underset{|}{C}}}}{-}\overset{O}{\overset{\|}{C}}{-}NH_2 + H_2O \longrightarrow R{-}\overset{H}{\overset{|}{\underset{NH_2}{\underset{|}{C}}}}{-}\overset{O}{\overset{\|}{C}}{-}OH + NH_3 \quad \alpha\text{-amino acid}$$

If R in the preceding reactions is a hydrogen atom, that is, if the initial molecule is formaldehyde (HCHO),

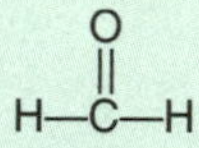

$$H{-}\overset{O}{\overset{\|}{C}}{-}H$$

the **result is *glycine*, one of the simplest of the 20 naturally occurring amino acids** (see Fig. 6-2), which are organic compounds containing both carboxyl (COOH) and amino (NH^2) groups and which function as the basic building blocks of proteins:

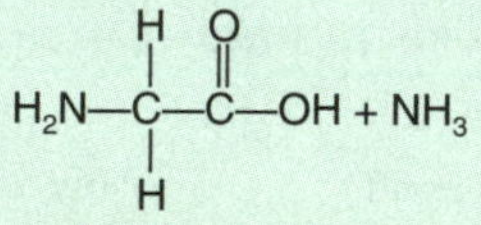

$$H_2N{-}\overset{H}{\overset{|}{\underset{H}{\underset{|}{C}}}}{-}\overset{O}{\overset{\|}{C}}{-}OH + NH_3$$

Glycine

Adding formaldehyde to glycine under alkaline conditions produces ***serine*, another amino acid** (see Fig. 6-2):

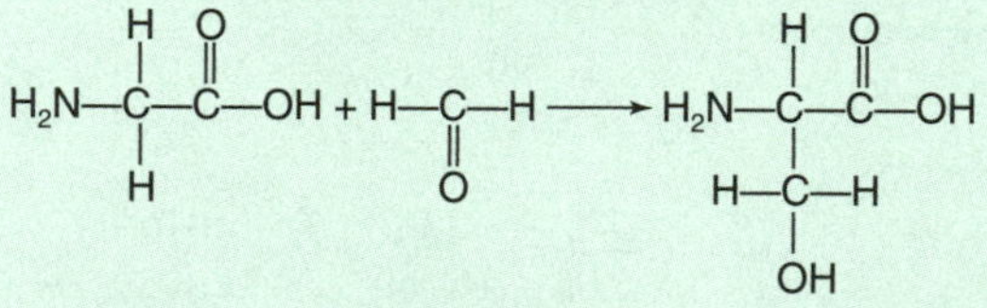

$$H_2N{-}\overset{H}{\overset{|}{\underset{H}{\underset{|}{C}}}}{-}\overset{O}{\overset{\|}{C}}{-}OH + H{-}\underset{O}{\underset{\|}{C}}{-}H \longrightarrow H_2N{-}\overset{H}{\overset{|}{\underset{H-C-H,\ OH}{\underset{|}{C}}}}{-}\overset{O}{\overset{\|}{C}}{-}OH$$

Glycine Formaldehyde Serine

All 20 different amino acids now used in protein synthesis have a similar structural pattern (see Fig. 6-2), although synthesized by different biochemical pathways.

Sugars

Among other basic organic molecules that could easily be synthesized under fairly simple conditions are **sugars**. In the *formose reaction*, small but significant yields of glucose and ribose are obtained from condensing formaldehyde:

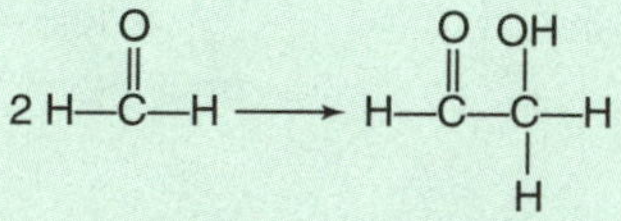

$$2\,H{-}\overset{O}{\overset{\|}{C}}{-}H \longrightarrow H{-}\overset{O}{\overset{\|}{C}}{-}\overset{OH}{\overset{|}{\underset{H}{\underset{|}{C}}}}{-}H$$

Formaldehyde Glycoalidehyde

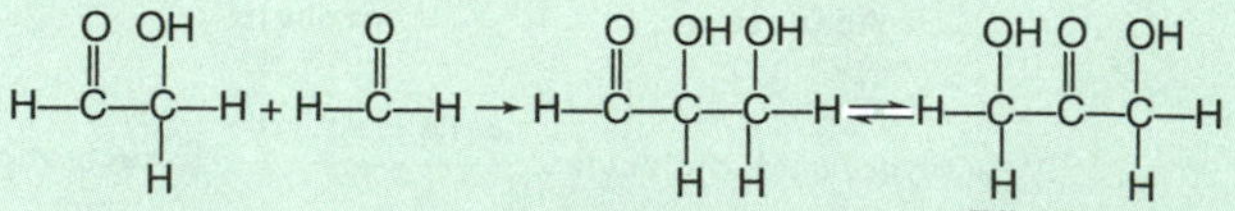

$$H{-}\overset{O}{\overset{\|}{C}}{-}\overset{OH}{\overset{|}{\underset{H}{\underset{|}{C}}}}{-}H + H{-}\overset{O}{\overset{\|}{C}}{-}H \rightarrow H{-}\overset{O}{\overset{\|}{C}}{-}\overset{OH}{\overset{|}{\underset{H}{\underset{|}{C}}}}{-}\overset{OH}{\overset{|}{\underset{H}{\underset{|}{C}}}}{-}H \rightleftharpoons H{-}\overset{OH}{\overset{|}{\underset{H}{\underset{|}{C}}}}{-}\overset{O}{\overset{\|}{C}}{-}\overset{OH}{\overset{|}{\underset{H}{\underset{|}{C}}}}{-}H$$

Glyceraldehyde Dihydroxyacetone

⋮

Aldose sugars Ketose sugars

D-ribose D-glucose

Among all possible six-carbon sugars that could have been used as the common metabolic fuel, the dominance of **glucose** is most probably related to its environmental stability. It may therefore have accumulated in greater quantities than other sugars, establishing its basic role in early organismal reactions. Gest and Schopf (1983) suggest that only "sugar-based cellular systems . . . provided the successful starting point for biochemical and cellular evolution."

Purine and Pyrimidine Bases

The **purine** and **pyrimidine** bases that are **essential components of nucleic acids** can be synthesized under conditions thought to simulate the prebiotic conditions of the early Earth. For example, heating aqueous solutions of ammonium cyanide (prepared by the reaction of HCN with NH_4OH) produces up to a 0.5 percent yield of *adenine*. Similarly, ultraviolet radiation acting on hydrogen cyanide solution produces a number of *purines*, including adenine and guanine (Chela-Flores et al., 2002).

Condensation reactions in forming adenine have now been studied in some detail. As one example, adenine can be produced from cyanoethylene. Reaction of cyanoacetylene with cyanates such as urea has produced the **pyrimidine cytosine:**

Cyanoacetylene Urea β-ureidoacrylonitrile

Cytosine

Similar synthetic procedures have been proposed for the two other pyrimidines, *uracil* and *thymine*. Orotic acid (which is a pyrimidine), whether produced *biologically* as a precursor to uracil or *abiologically* by polymerization of HCN, can be relatively efficiently decarboxylated to uracil by ultraviolet light. Because the reactions producing cytosine from cyanoacetylene and uracil from orotic acid are so alike, Ferris and Joshi (1978) suggested that early biological synthesis of uracil derivatives could easily have followed the pattern of abiological synthesis, but using catalytic enzymes rather than ultraviolet light (see Ferris, 2005a for further studies).

Fatty Acids

Now used in membranes and storage tissues of organisms, fatty acids are among other basic molecules that have been synthesized under high atmospheric pressures, with γ-rays as an energy source:

$$CO_2 + [H{-}CH{=}CH{-}H] \longrightarrow CH_3(CH_2)_nCOOH$$

Ethylene molecules Fatty acid

As discussed in the text, 8-carbon-long fatty acids have been isolated from carbonaceous chondrites.

Pyrroles, Porphyrin Rings and Coenzymes

Pyrroles, which are precursors of porphine-like compounds, can be synthesized in mixtures of CH_4, NH_3 and H_2O, and react with formaldehyde or benzaldehyde to form pyrrole structures. Although not considered to be a prebiotic sequence, oxidation then yields the **porphyrin rings** found in heme, chlorophyll and other pigments:

Pyrrole Formaldehyde

Oxidation

Porphinelike structure Porphyrin-type ring

The porphyrin structure has alternating double and single bonds that can *resonate* by assuming a variety of different configurations without changing the position of their constituent atoms. Such resonance confers stability on porphyrins, enabling them to hold extra electrons and to function as **electron acceptors (oxidation)** or **electron donors (reduction)**. Similar oxidative and reductive functions can be performed by nucleotide derivatives such as nicotinamide adenine dinucleotide (NAD), called **coenzymes** because they act in union with protein enzymes to catalyze a wide variety of chemical reactions.

BOX 6-5 Condensing Agents and the Formation of Peptide Bonds

COMPOUNDS IDENTIFIED AS condensing agents and that could have existed early in Earth's history include the six whose formulae are shown below:

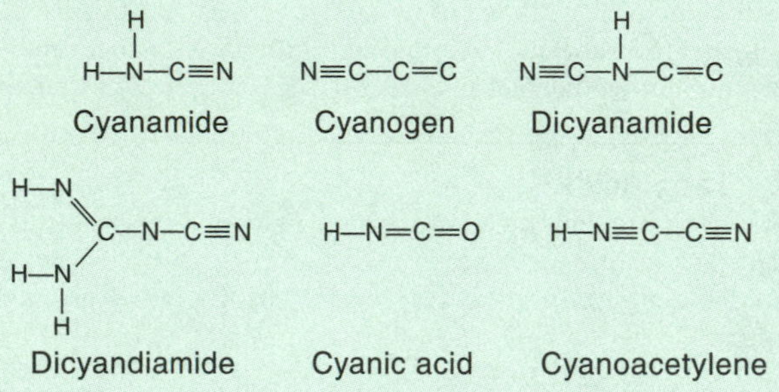

$H{-}N(H){-}C{\equiv}N$ Cyanamide

$N{\equiv}C{-}C{=}C$ Cyanogen

$N{\equiv}C{-}N(H){-}C{=}C$ Dicyanamide

$(H{-}N)(H{-}N(H))C{-}N{-}C{\equiv}N$ Dicyandiamide

$H{-}N{=}C{=}O$ Cyanic acid

$H{-}N{\equiv}C{-}C{\equiv}N$ Cyanoacetylene

Some, such as cyanamide, are crystalline and highly acidic. Others, such as cyanogen, are gaseous univalent radicals. In each of these molecules, the unsaturated cyano–carbon–nitrogen bonds enable the condensing agent to combine with water to produce a molecule of a type that became metabolically important with the origin of life, and to release energy during the hydration. For example,

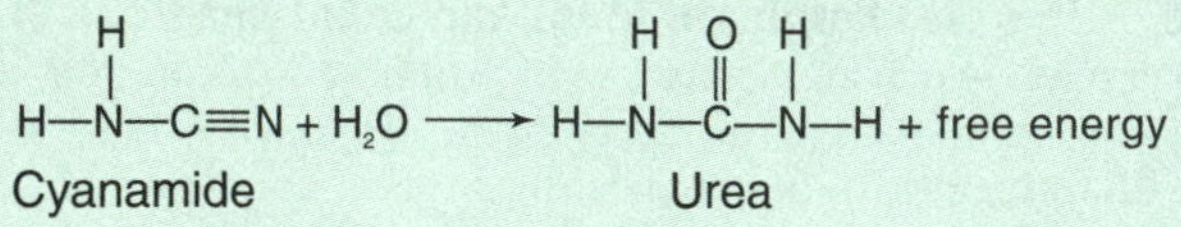

$$H{-}N(H){-}C{\equiv}N + H_2O \longrightarrow H{-}N(H){-}C({=}O){-}N(H){-}H + \text{free energy}$$

Cyanamide → Urea

Furthermore, hydrolysis of cyanamide can couple production of urea to the condensation of two amino acids into a dipeptide:

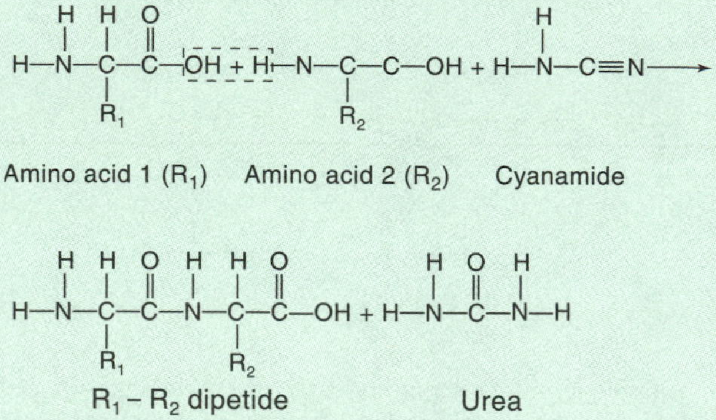

$$H{-}N(H){-}C(H)(R_1){-}C({=}O){-}OH + H{-}N{-}C(R_2){-}C{-}OH + H{-}N(H){-}C{\equiv}N \longrightarrow$$

Amino acid 1 (R_1) Amino acid 2 (R_2) Cyanamide

$$H{-}N(H){-}C(H)(R_1){-}C({=}O){-}N(H){-}C(H)(R_2){-}C({=}O){-}OH + H{-}N(H){-}C({=}O){-}N(H){-}H$$

R_1 - R_2 dipetide Urea

Because of their apparent preference for reacting with organic molecules carrying anions (for example, phosphate, HPO_4^-), many cyanic condensing agents produce peptide bonds between amino acids even in aqueous solutions. Some, such as cyanogen and cyanamides, also cause *nucleotides to form by the phosphorylation of adenosine, uridine, and cytosine.*

findings that they all show similar properties when tested by sedimentation rates, electrophoretic techniques, column fractionation and other measurements. Preferential interaction between amino acids in polymer formation seems to dictate their position and frequency and lead to some degree of uniformity in the kinds of molecules produced.

Not all the amino acid bonds formed in such polymers are of the usual peptide variety, nor do the shapes of these molecules follow the familiar α-helix of protein structure. Nevertheless, there seem to be enough peptide linkages to characterize them as proteins in many tests. Polymers give positive color tests with the same reagents that proteins do, their solubilities resemble proteins, they are precipitable with similar reagents and they have other protein-like traits (**Table 6-6**). It has been suggested that *some polymer reactions, combined into a particular sequence, may have served as the beginnings of later metabolic systems.* One sequence, beginning with decarboxylation of oxaloacetic acid, leads to alanine or to acetic acid and carbon dioxide (**Fig. 6-12**). Furthermore, some of these polymers show relatively sophisticated *hormonal activity* and can stimulate the production of melanin (pigment)-producing cells.

Possible Sites for the Origin of Molecules

Although researchers have investigated whether extensive thermal synthesis of proteins could occur in present natural surroundings, and although the conditions early in Earth's history are not known, they would have varied geographically. Surfaces near some volcanic regions, or upwelling from shallow marine hydrothermal plumes, may have maintained appropriate temperatures for the condensation of amino acids. Cooling rains or currents may then have dispersed such thermally produced polymers to places where further interactions could take place.

The unusual living conditions of many organisms discovered in the past two decades — *oceanic sea vents, boiling hot springs* — show that life can be maintained and may have arisen under the most stringent conditions. Thermophilic (heat-loving) organisms have been found at temperatures above the boiling point of water, some living a kilometer and a half below the surface. The superheated (350°C) plumes arising from hydrothermal vents produce large amounts of ammonia and can produce polymers as shown by the polymerization of glycine amino acids in a flow reactor system that simulates submarine hydrothermals. Indeed, as outlined in Box 6-3, a theory for the origin of life in deep-sea vents has gained considerable support. Volcanic sediments would have

TABLE 6-5 Amino acid compositions in molar percentages of two polymers compared to the initial reaction mixtures

	2:2:1 Polymer		2:2:3 Polymer	
Amino Acid	Initial Polymer Mixture	Polymer Product	Initial Polymer Mixture	Polymer Product
Aspartic acid	42.0	66.0	30.0	51.1
Glutamic acid	38.0	15.8	27.0	12.0
Alanine	1.25	2.36	2.72	5.46
Lysine	1.25	1.64	2.72	5.38
Semicystine	1.25	0.94	2.72	3.37
Glycine	1.25	1.32	2.72	2.79
Arginine	1.25	1.32	2.72	2.44
Histidine	1.25	0.95	2.72	2.03
Methionine	1.25	0.94	2.72	1.73
Tyrosine	1.25	0.94	2.72	1.66
Phenylalanine	1.25	1.84	2.72	1.48
Valine	1.25	0.85	2.72	1.16
Leucine	1.25	0.88	2.72	1.06
Isoleucine	1.25	0.86	2.72	0.90
Proline	1.25	0.28	2.72	0.59
Serine[a]	1.25	0.6	0.0	0.0
Threonine[a]	1.25	0.1	0.0	0.0

[a]Serine and threonine were omitted from the 2:2:3 polymer. Tryptophan was present in the 2:2:1 polymer.

Source: Reprinted by permission from S. W. Fox, K. Harada, K. R. Woods, C. R. Windsor, 1963. Amino acid compositions in proteinoids. *Arch. Biochem. Biophys.*, **102**, 439–445.

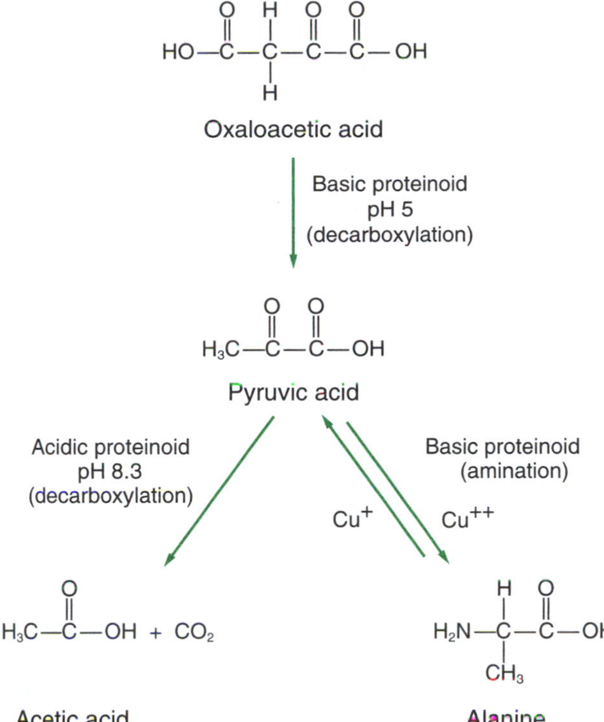

FIGURE 6-12 Some sequential reactions believed to be catalyzed by different proteinoids or proteinoid complexes. (Adapted from Fox, S. W., and K. Dose, 1972. *Molecular Evolution and the Origin of Life.* W. H. Freeman, San Francisco.)

produced sponge-like minerals (*zeolites*) that can retain and catalyze organic compounds. Proposals that much of Earth's early landmass was composed of volcanic islands, and that high global "greenhouse" temperatures persisted for more than a billion years because of high CO_2 atmospheric pressures, are consistent with such rigorous scenarios. Without the current "greenhouse effect," the average temperature on Earth would be of the order of 15°C lower than it is, an eventuality that, had

TABLE 6-6 Properties common to thermally produced polymers (proteinoids) and to biologically produced proteins

Qualitative amino acid composition	Some optical activity (for polymers of L-amino acids)
Range of quantitative amino acid composition (except serine and threonine)	Hypochromicity
Limited heterogeneity	Infrared absorption patterns
Range of molecular weights (4,000 to 10,000)	Recoverability of amino acids with mineral acid hydrolysis
Reaction in color tests (including biuret reaction)	Susceptibility to proteolytic enzymes
Inclusion of nonamino acid groups (iron, heme)	Various enzymelike properties
Range of solubilities	Inactivation of catalysis by heating in aqueous buffer
Lipid quality	Nutritive qualities
Salting-in and salting-out properties	Hormonal activity (melanocyte stimulation)
Precipitability by protein reagents	Tendency to assemble into microparticle systems

Source: Used with permission from Fox, S. W. and K. Dose, 1972. *Molecular Evolution and the Origin of Life.* W. H. Freeman, San Francisco.

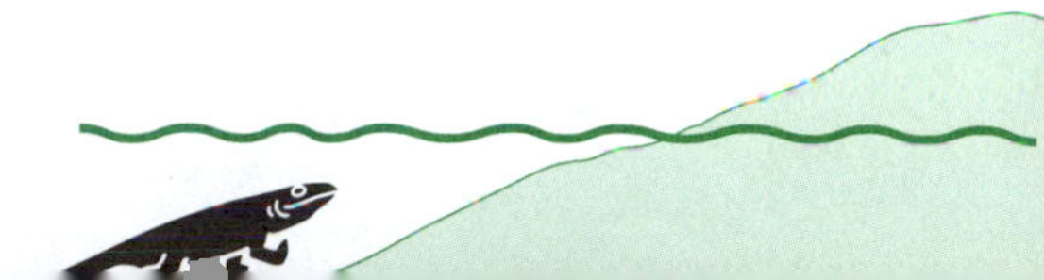

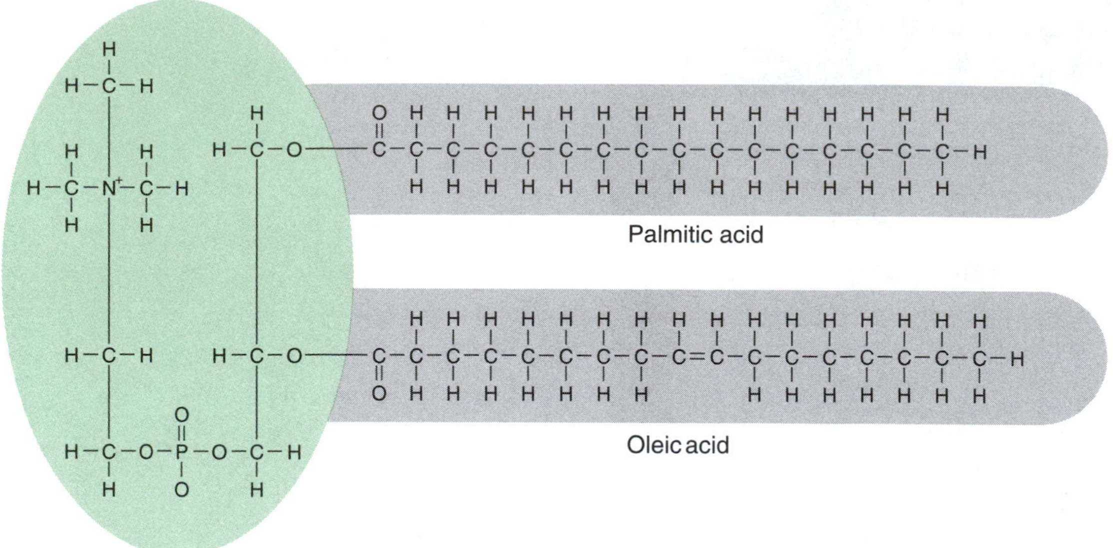

(a) Glycerophosphorylcholine

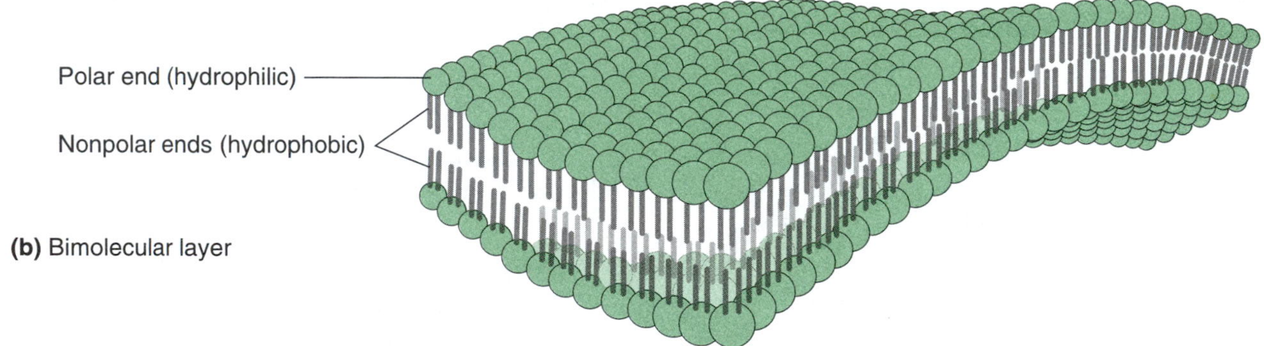

(b) Bimolecular layer

FIGURE 6-13 (a) A phospholipid molecule, lecithin. (b) A diagrammatic view of a bimolecular sheetlike double layer of phospholipid molecules that have self-assembled with their hydrophilic phosphate heads (*colored circles*) facing the water solvent, and their hydrophobic hydrocarbon tails facing each other. Such polar-nonpolar molecules are called *amphipathic* or *amphiphilic*, and characterize the plasma membrane that circumscribes the cell.

it occurred, would have taken the average global temperature in 2001 from 14.5°C to just below 0°C, not a happy prospect.[4]

Other researchers have proposed that *layered clays such as montmorillonite* served as polymerizing templates on which condensation occurred (see Chapter 5). Clays were used to show that phosphate-activated amino acids such as aminoacyl adenylates will condense to form high yields of polypeptide chains. The amino acid ends of the adenylates apparently penetrate the narrow layers of the clay, and the condensation reactions take place there. Nucleotides can be joined into 55-length chains on montmorillonite clays, and amino acids can polymerize into peptides on other mineral surfaces such as hydroxyapatite or illite (Ferris, 2005a,b).

The early availability of cyanamides, heat, clays and other condensing agents makes it highly probable that polypeptides, polysaccharides, lipids and perhaps even polynucleotides were present early in Earth's history, and could have been used for early organism-like reactions and structures.

[4] See Kerr (1997) for thermophiles, Imai et al. (1999) for hydrothermal simulation, Brandes et al. (1998) for ammonia production and J. V. Smith (1998) for zeolites.

Membranes: The First Structures

The presence of appropriate organic monomers and polymers is only a first step or series of steps in the origin of life. Processes of metabolism and function occur because the materials of which organisms consist are highly organized. How did such organization come about?

At its earliest, interactions among molecules must have led them to assume relative positions based on forces such as hydrogen bonding, ionization, solubility, adhesion and surface tension. Phospholipids, for example, are organic molecules with a phosphorus-containing polar group at one end and nonpolar fatty acid groups at the other (**Fig. 6-13a**). In water, which is a polar solvent, the polar ends of these molecules are oriented toward water (**hydrophilic**), while their nonpolar ends are oriented toward each other, away from water (**hydrophobic**). As a result, phospholipid membranous structures can form quickly, yielding *vesicles* composed of bimolecular layers in which the nonpolar surfaces of each of the two layers "dissolve" in each other (**Fig. 6-13b**). Carried further, such vesicles can encapsulate inclusions in tide pools, which undergo drying and wetting cycles (Monnard et al., 2002). The growth of fatty acid membranes can produce a gradient in pH across the **membrane.** ***Membranous droplets*** or **vesicles** composed of lipids, polypeptides, or other molecules, produced by the mechanical agitation of molecular films on liquid surfaces (**Fig. 6-14**) or even spontaneously, undoubtedly formed in great quantities.[5]

Protocells

Attainment of the **droplet level of organization** would have been an important step in the origin of life, not the least because, depending on its structure and permeability, the membrane surrounding the droplet can selectively permit some compounds to enter from the environment and others to exit from the droplet (**selective permeability**), functioning as a **protocell**, which we can define as a *self-assembled membrane-bounded system containing molecules.* This — along with the small size of protocells — allows different concentrations of the same compound to be maintained on either side of the membrane, facilitating reactions that could not have taken place at lower concentrations. Precipitation of such compounds would also be facilitated.

Small protocell size and local concentrations of molecules or compounds would have resulted in the establishment of concentration gradients, and later in the organization of compartments and substructures. The presence of a basic protein would result in a 100-fold increase in entrapment of nucleic acids into protocells, facilitating organization of nucleic acids into one compartment, the nucleus. As Jay and Gilbert (1987) point out, "Protein-mediated encapsulation creates high local concentrations of protein and nucleic acids within the vesicular volume . . . This would enhance the interaction of molecules with low affinities, potentiating the formation of aggregates with biological function."

Protocell membranes could incorporate "amphipathic" peptides — proteins that span phospholipid bilayers — and so channel ionic and molecular transfers. Once in place, some channels would have evolved into "proton pumps," enabling hydrogen ion (H^+) gradients for energy transport and accumulation (see Fig. 8-15). Because of their semipermeable membranes, protocells are not isolated, and

[5] See Chen and Szostak (2004) for the pH gradient, Deamer (1986, 2003) for studies with mechanical agitation and Hanczyc et al. (2003) for properties such as encapsulation growth and division.

(a)

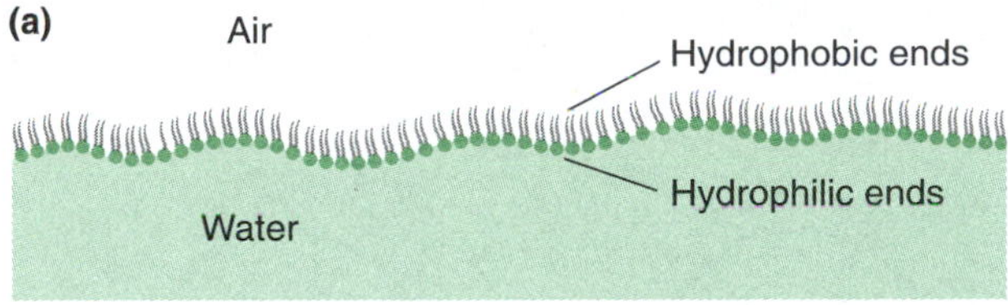

(b)

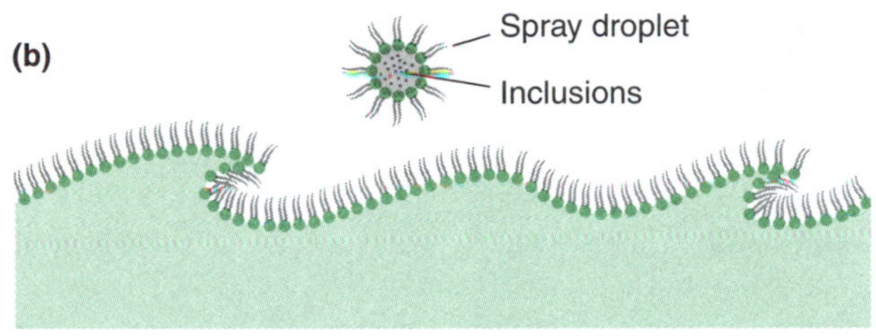

(c)

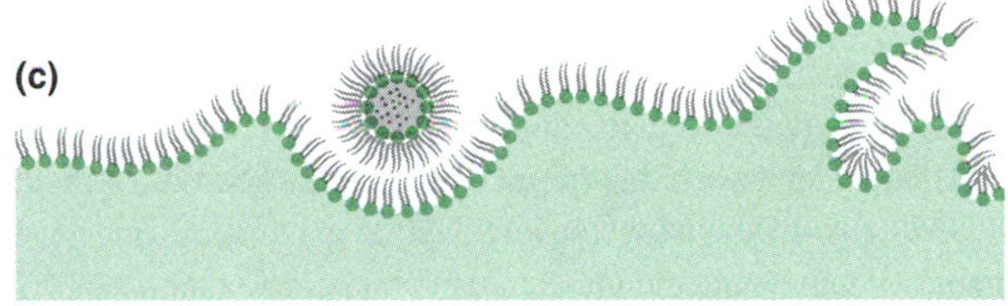

(d)

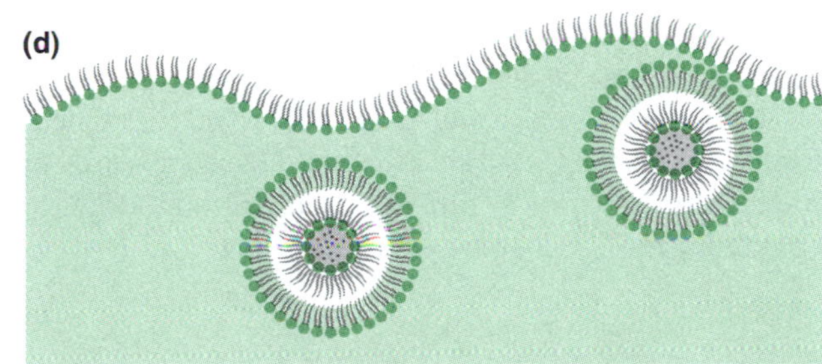

FIGURE 6-14 Effect of mechanical wave action on a surface film (e.g., "foaming") containing (a) molecules oriented with one end pointed away from water (hydrophobic) and the other end toward the water (hydrophilic). In organic molecules such as lipids, the hydrocarbon chain is hydrophobic and the carboxyl end is hydrophilic. Wave action (b and c) causes the formation of droplets and the bilayered vesicles shown in (d). Surface aggregates of peptide chains, nucleic acid sequences or their combinations may have been among the various vesicle inclusions. Appearance of cell-like double-membrane structures may have occurred through incorporation of one bilayered vesicle within another. Other accretions, composed at least partially of hydrophobic amino acids (see Table 7-3), may have persisted within a membrane's layer of hydrophobic tails, acting as selective channels to help transport materials, molecules and ions between exterior and interior.

can acquire external energy and matter to retain, and even enhance, their organizational and informational structures as long as they can continue to perform biological processes. Protocells preserved their organizational framework by partially separating themselves from the entropy or disorder in their surrounding environment. Although *entropy tends to increase* in the universe according to the second law of

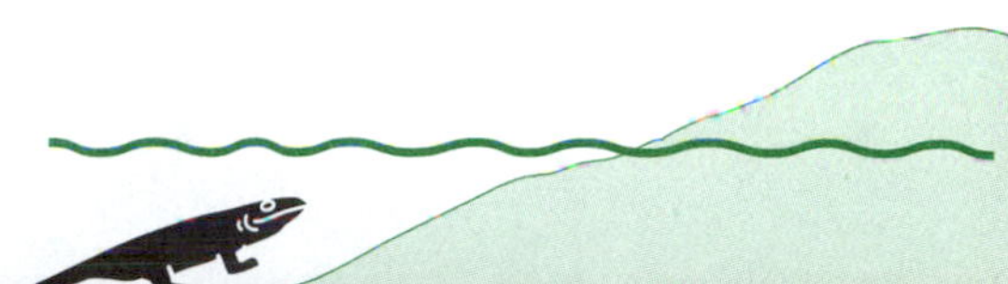

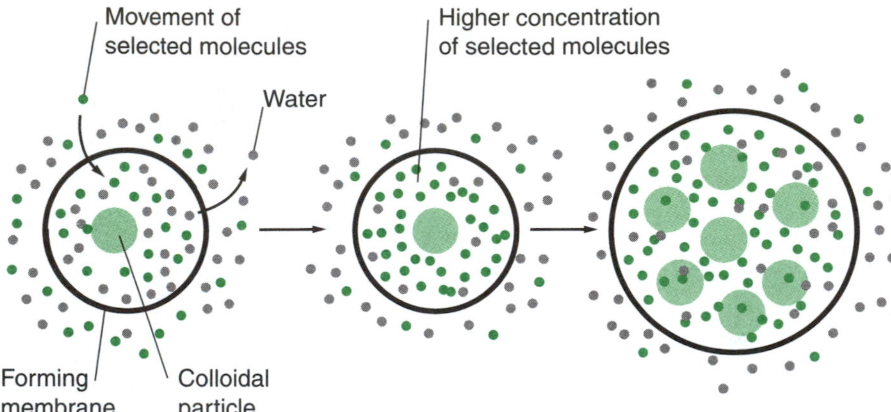

FIGURE 6-15 Formation of coacervates by the exclusion of water molecules (*dots*) from associated colloidal particles (*colored circles*). The intervening water molecules can be removed through dehydration (for example, increased salt concentration), or when colloidal particles are attracted to each other because they have opposite charges (for example, negatively charged gum arabic and positively charged gelatin), or because some colloids are basic (for example, histones) and others are acidic (for example, nucleic acids). (Modified from Kenyon, D. H. and G. Steinman, 1969. *Biochemical Predestination*. McGraw-Hill, New York.)

thermodynamics, *entropy can decrease* in protocells during their life spans.

For an early form to show the most essential "living" attributes, it would have had to be able to maintain its individuality, to grow and to divide. Interestingly, some authors ascribe such properties to bimolecular vesicles (for example, Morowitz, 1992, 2002), while at least two types of fairly simple laboratory-produced structures discussed in this chapter — Oparin's coacervates and Fox's microspheres — possess some of the basic prerequisites of protocells. Although these structures were created artificially, they point to the likelihood that nonbiological membrane enclosures (protocells) could have sustained reactive systems for at least short periods of time, and led to research on the experimental production of membrane-bound vesicles containing molecules: protocells.

Production of Coacervates in the Laboratory

Coacervates occur when dispersed colloidal particles separate spontaneously out of solution into droplets because of special conditions of acidity, temperature, and so on (**Fig. 6-15**). If there is more than one type of macromolecular particle in the colloid, complex coacervates can form that show a number of interesting properties:

- They have a simple but persistent organization.
- They can remain in solution for extended periods.
- They have the ability to increase in size.

Aleksandr Oparin, the first to draw serious attention to these droplets, developed artificial coacervate systems that could incorporate enzymes able to perform functions such as the synthesis and hydrolysis of starch (**Fig. 6-16**) and synthesize polynucleotides.

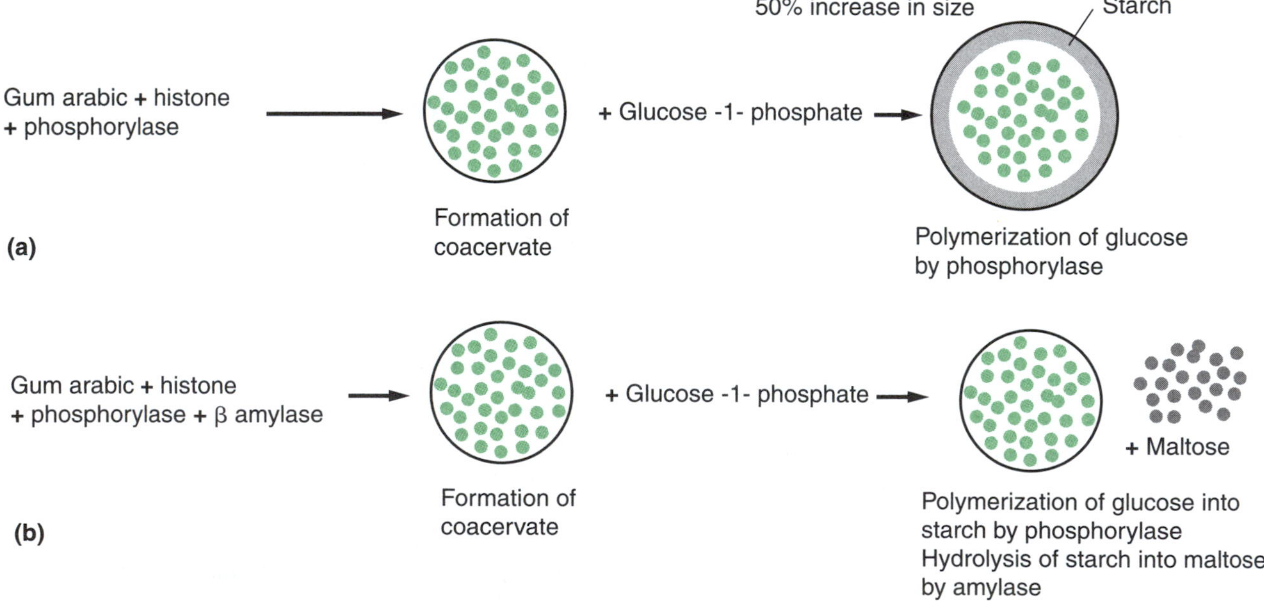

FIGURE 6-16 Synthesis and hydrolysis of starch in coacervate systems in which enzymes have been included in the droplets. In (a) the phosphorylase enzyme acts to polymerize phosphorylated glucose into starch, while in (b) the starch formed this way is hydrolyzed into maltose by the amylase enzyme.

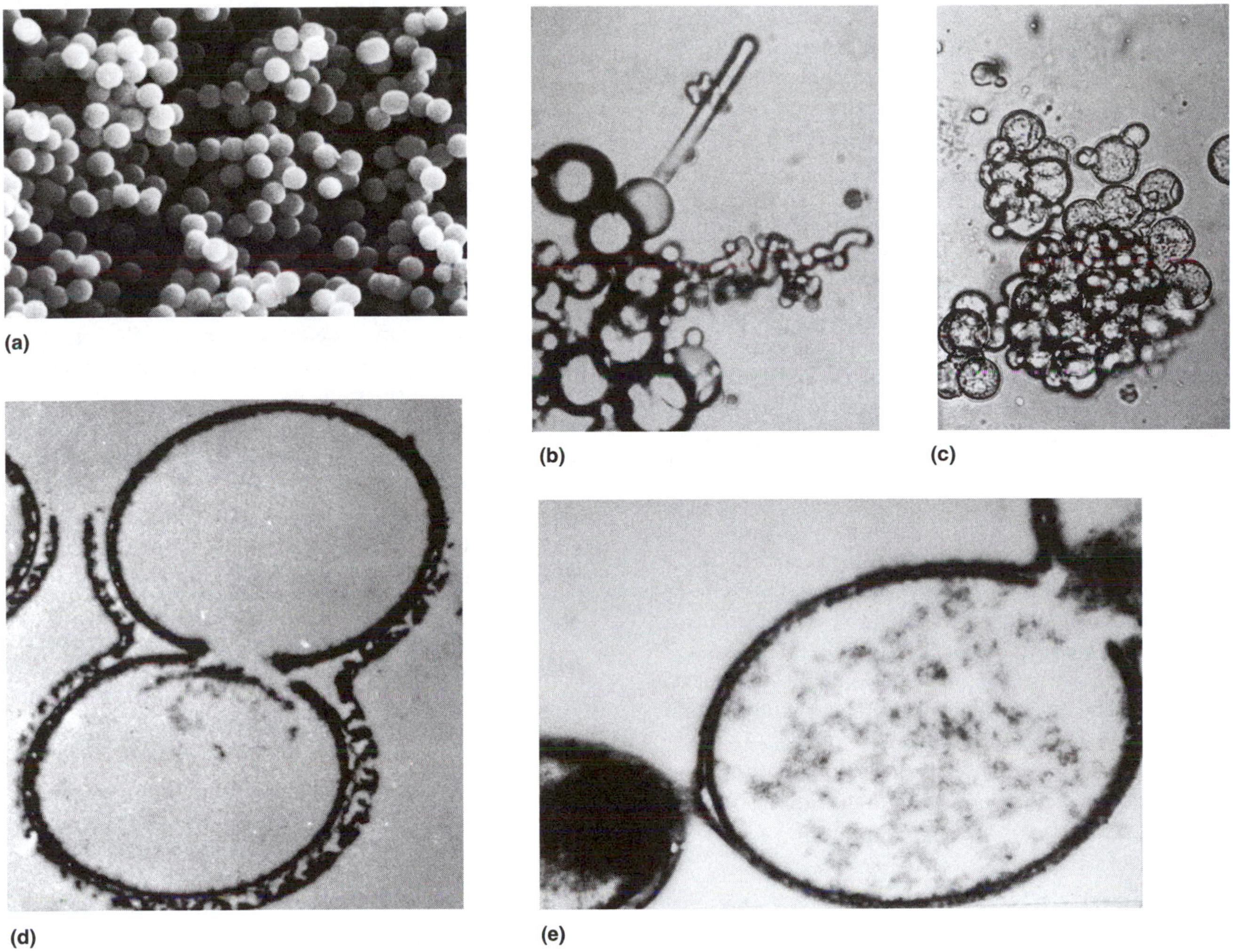

(a)

(b)

(c)

(d)

(e)

FIGURE 6-17 Microspheres produced in the laboratory. (a) Proteinoid microspheres visualized using scanning electron microscopy. (b, c) Proteinoid microspheres showing evidence of outgrowths and budding, visualized using light microscopy. (d, e) Proteinoid microspheres show the double-membrane structure and evidence of budding as seen with transmission electron microscopy. (© Science VU/Dr. Sidney Fox/Visuals Unlimited.)

In a coacervate system containing chlorophyll irradiated with visible light, Oparin and coworkers (Oparin, 1924) showed that there can be a constant inflow of reduced ascorbic acid and oxidized methylene red, which converts into a constant outflow of oxidized ascorbic acid and reduced methylene red. Chlorophyll picks up electrons from the ascorbic acid and supplies these for the methylene red reduction; this process is similar to common noncyclic photosynthesis in which water molecules supply electrons for reducing the coenzyme $NADP^+$ to NADPH (see Chapter 9).

Production of Microspheres in the Laboratory

Sidney Fox showed that 10^8 to 10^9 **microspheres**/gm polymer formed when thermally produced polymers were boiled in water and allowed to cool. The microspheres (essentially protocells) are uniform in size, stable, bounded by double membranes that appear somewhat cell-like and can undergo fission **and** budding (**Fig. 6-17**). Among the qualities of the microsphere indicating active internal processes are their selective absorption and diffusion of certain chemicals but not others; their **growth** in size and mass; and observations demonstrating osmosis, movement and rotation (see Fig. 6-17). Moreover, microspheres show the potential for transferring information, in that polymer particles pass through junctions between them.

The spontaneous **self-assembly** of macromolecules into vesicles and protocells indicates that the occurrence of similar entities early in Earth's history probably would not have been an unusual event. Such entities are not cells, of course, and considerable time may have elapsed before more elegant structures with more complex metabolic capabilities developed. Although not cells, their self-assembly and potential to undergo selection justifies the term and concept protocell.

Substantial evidence shows that the component materials of even more complex structures can self-assemble

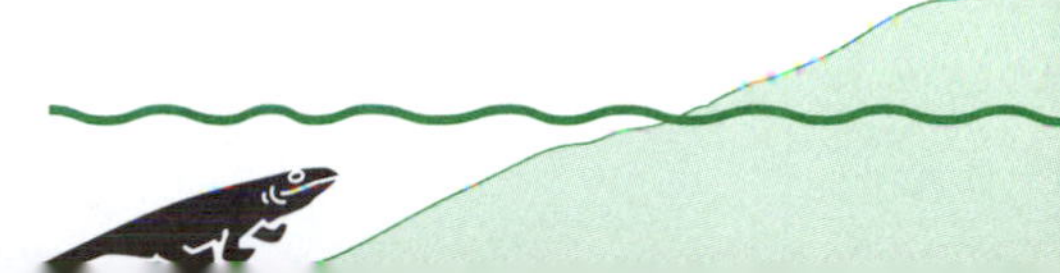

without the requirement of a prior pattern. One example, described in the section on origin of selection (below), is the creation of a virus from DNA. A second example is the naturally occurring **tobacco mosaic virus.**

Tobacco Mosaic Virus

The protein and nucleic acid components of tobacco mosaic virus spontaneously aggregate into the exact configuration needed to produce an active virus. A more complex virus such as T4 (see Fig. 13-1), containing many different kinds of protein, also has a significant number of steps in which self-assembly occurs. Nomura (1973) showed that even cellular organelles such as ribosomes can form by self-assembly from component materials. Each level of self-assembly, from monomers to polymers to protocells to various cellular organelles, may have derived from nonrandom events, in that certain combinations form more quickly and easily than others. In their most essential aspects, such as the precisely ordered monomer sequences found in proteins and nucleic acids, extant forms of life did not result from mere chemical attractions between component amino acids or nucleotides, although the positioning of monomers in these precisely ordered sequences probably had a nonrandom basis.

How can the nonrandom biological order of amino acid and nucleotide sequences arise from disorder? The answer lies in **selection.**

Origin of Selection

The earliest structures, whether protocells or other localized organizations, would have had one important evolutionary feature: they served as the first distinctive, multichemical **individuals** that could interact as units with their environment. Together with their various neighbors and progenies, such individuals would form a group or **population** on which **selection** could act.

The simplest form of cellular life that has been produced artificially had around 300 genes (see Box 6-1). Using readily available DNA, a "virus" containing 5,386 base pairs and capable of infecting and killing bacterial cells has been created in the laboratory. Of course, it was produced using the DNA-RNA-protein system that characterizes modern life and so even this simple life form must be much more complex that the first life form on which selection would have acted.

Natural selection (see Chapter 22) would have arisen when the following conditions were reached:

- A population of individuals existed.
- The properties of these individuals were governed by reactions in which they absorbed and transformed environmental material into their own material.
- Individuals differed in the efficiency with which these processes took place.
- Availability of materials and energy was limited so that not all types of individuals could form or survive.

Protocells incorporating such organization and metabolic properties allowing them to grow and divide — and hence to perpetuate themselves — would have increased most, either in relative frequency or in area occupied.

The mechanism that enabled the formation of protocells is a crucial issue in understanding both the origin and the nature of natural selection. If protocells could have originated only by self-replication, selection must always have operated on the same efficient basis as it does now, with advantageous traits rapidly transmitted to succeeding generations by fairly exact replicative mechanisms. However, if protocells initially formed only through acts of prevailing environmental chemistry, selection was probably not very efficient and would itself have undergone evolution from chemical nonreproductive selection to biological natural selection.

Early selection would have been confined to the survival of nonreproductive individuals that could wrest the most material from their environment and transform it for their own benefit with the least expenditure of energy. Although differences among such individuals could not be precisely transmitted, the fact that some such individuals survived and others did not would have affected the composition and further interactions of succeeding groups. Kauffman (1993) offers mathematical models showing that a "chaotic" system of interacting components can spontaneously become an ordered system on which selection can act.

Selection for What?

Selection bridged the gap between chemical evolution (changes in the composition of nonreproductive or poorly reproductive molecules, protocells, and so on) and biological evolution (changes in inherited differences among reproductive organisms). Selection is the only natural mechanism we know that can account for the changes among nonreproductive individuals that led them in the direction of organisms: "molecules" → "function" → "message." Although the events may be complex, the device is simple: **organisms that react to their environment with improved useful information replace those that lack such information.** The finding that a specific ribozyme can be selected from among trillions of random RNA molecules with only few selective steps in the laboratory demonstrates that selection operates on molecules in the absence of any higher levels of organization (Wilson and Szostak, 1999). But selection is not merely a passive agent that sifts the good from the bad, the adaptive from the nonadaptive.

Because of its historical continuity, selection enables a succession of adaptations that allow something entirely

J. William Schopf

Born: September 27, 1941
Birthplace: Urbana, Illinois
Undergraduate degree: Oberlin College (Ohio), 1963
Graduate degrees: A.M. Harvard University, 1965 (Biology) Ph.D. Harvard University, 1968
Present position: Professor of Paleobiology Director, Center for the Study of Evolution and the Origin of Life University of California–Los Angeles[6]

What prompted your initial interest in evolution?

My father was a paleobotanist and professor in the Department of Geological Sciences at Ohio State University. As a high school student, I was encouraged both by my parents and by my teachers to pursue a career in the natural sciences.

What do you think has been most valuable or interesting among the discoveries you have made in science?

Helping to establish and set up the interdisciplinary format of a new area of science, Precambrian paleobiology. Encouraging international activity in this area of science.

On the basis of direct fossil evidence, posing and attempting to answer new questions fundamental to science regarding, for example, the time, mode, and environmental and evolutionary impact of the origin of oxygen-producing photoautotrophy; the time of origin of eukaryotic cells and the evolutionary impact of eukaryotic sexuality; and the differences in both tempo and mode between Precambrian (prokaryotic) and Phanerozoic (eukaryotic) evolution.

Discovery of the oldest fossils now known — approximately 3,465 million-year-old cyanobacterium-like cellular filaments from the Apex chert of northwestern Western Australia.

What areas of research are you (or your laboratory) presently engaged in?

We are continuing work on the last two items just listed.

In which directions do you think future work in your field needs to be done?

Increased studies are needed of the evolutionary impact of major, long-term environmental change. For example, over the past approximately 4 BY, we need to know more about what changes occurred in day length, atmospheric composition, solar luminosity, and ambient surface temperature.

More studies are needed of the extent to which the concept of "molecular clocks" (revealed by the biochemistry of extant organisms) can be applied to determining the rate of ancient evolutionary change.

What advice would you offer to students who are interested in a career in your field of evolution?

Early in their undergraduate years, students should obtain a sound background in organic chemistry or biochemistry, biology (including molecular biology), and geology. Nature, in all its evolutionary splendor, is not compartmentalized into discrete disciplines, which means that major advances in understanding nature will come from soundly based interdisciplinary studies.

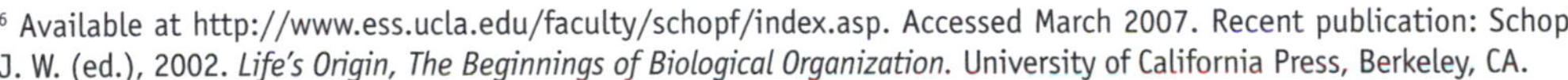

[6] Available at http://www.ess.ucla.edu/faculty/schopf/index.asp. Accessed March 2007. Recent publication: Schopf, J. W. (ed.), 2002. *Life's Origin, The Beginnings of Biological Organization.* University of California Press, Berkeley, CA.

EVOLUTION ON THE WEB biology.jbpub.com/book/evolution

new to accumulate. Selection has allowed possible biological organizations that would otherwise have been highly improbable. To use a previous example, a polypeptide chain consisting of a specific sequence of 100 amino acids has an extremely low probability (10^{-130}) of occurring spontaneously without selection. However, as explained in Chapter 7, if each step in the growth of the chain attains a selective advantage when the correct amino acid inserts, the probability of achieving a functionally advantageous polypeptide is almost immeasurably increased. For example, assuming that the 20 different amino acids are present in the surrounding medium in equal frequency, there is a probability of 1/20 that random chance will supply the correct, functionally advantageous amino acid to any position of the chain. Thus, $100 \times 1/20 = 5$ positions that will be correctly occupied by chance alone without selection. For the remaining 95 positions, selection will operate so that each position, once occupied by its correct amino acid, enables further selection to occur at the next succeeding position. This stepwise procedure would entail perhaps 20 trials to achieve the correct amino acid at one position, another 20 trials to achieve the correct amino acid at one position, and so on. In sum, a succession of only $95 \times 20 = 1{,}900$ trials may be necessary for selection to elicit a functional amino acid sequence for a chain 100 amino acids long, which is a probability of $1/1{,}900 \approx 2 \times 10^{-3}$ compared to the 10^{-130} probability in the absence of selection.

Once the game of life began, the evolutionary replacement of players bearing information — whether those players were coacervates, microspheres, protocells, genes, organisms, species or other entities — became inextricably bound to their ability to play the game further, an ability that selection has previously molded and that in turn is measured anew by selection. Some random replacement of the players certainly occurs by accident rather than by selection, but participation in life is selective by its very nature since the resources

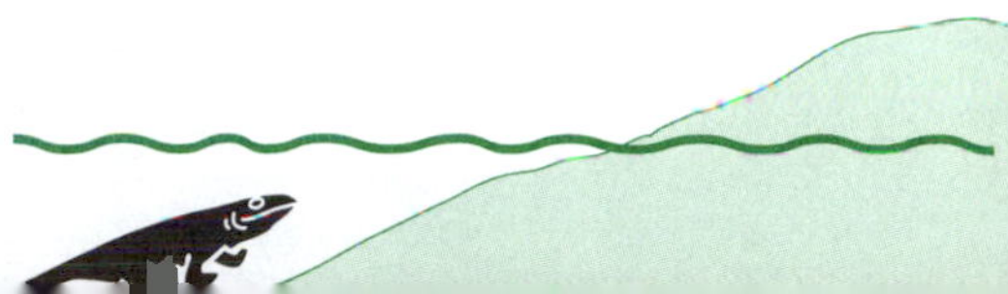

of life are always limited in one way or another (see Chapters 2 and 22).

Given reproductive expansion, limited resources, transmitted variation and environmental change — the organismal condition — selection characterizes living systems. As discussed in the next chapter, this ability has led to a coupling between two very different life processes — **function** (*proteins*) and **reproduction** (*nucleic acids*) — providing living forms with their relatively rapid evolutionary rates. How this relationship arose and evolved is the subject of the next chapter.

KEY TERMS

abiotic
adenosine triphosphate (ATP)
amino acids
carbonaceous chondrites
coacervates
coenzymes
condensation
double helix
enzyme
genotype
hydrolysis
individuals
membranes
membranous droplets
messenger RNA (mRNA)
microspheres
nucleic acids
nucleotide
peptide linkages
phenotype
phosphate bond
polymers
polymerization
polypeptide chains
population
protocell
purine
pyrimidine
ribosomal RNA (rRNA)
selective permeability
self-assembly
transfer RNA (tRNA)
vesicles

DISCUSSION QUESTIONS

1. DNA.
   a. What explains the information-carrying capacity of DNA?
   b. How does DNA replicate?
   c. How does information transfer from DNA to protein?
2. What is the difference between genotype and phenotype?
3. The terrestrial origin of life.
   a. What prevailing environmental conditions would have enabled the origin of life on Earth?
   b. What chemical compounds and reactions may explain the early origin of basic biological molecules? In what possible quantities?
   c. What reactions would have condensed such basic organic molecules into macromolecules?
4. Polymers.
   a. How are polymers formed?
   b. What biological properties do they possess?
5. Protocells.
   a. How do protocells permit increased control and organization of biological molecules, structures and reactions?
   b. Why does entropy decrease in such systems?
   c. What are coacervates, and what are their lifelike properties?
   d. What are microspheres, and what lifelike properties do they have?
6. Selection.
   a. What conditions permit selection to occur?
   b. Can selection occur among nonreproductive entities?
   c. Can selection produce biological compounds and structures that, most probably, would not have arisen spontaneously?

EVOLUTION ON THE WEB

Explore evolution on the Internet! Visit the accompanying Web site for *Strickberger's Evolution,* Fourth Edition, at http://www.biology.jbpub.com/book/evolution for Web exercises and links relating to topics covered in this chapter.

RECOMMENDED READING

Brandes, J. A., N. Z. Boctor, G. D. Cody, B. A. Cooper, R. Hazen, and H. S. Yoder, Jr., 1998. Abiotic nitrogen reduction on early Earth. *Nature*, 395, 365–367.

Chela-Flores, J., G. Lemarchand, and J. Oró (eds), 2002. *Astrobiology: Origins from the Big-Bang to Civilization.* Kluwer Academic Press, Dordrecht, The Netherlands.

Deamer, D. W., 2003. A giant step towards artificial life? *Trends Biotechnol.*, 23, 336–338.

Farley, J., 1977. *The Spontaneous Generation Controversy from Descartes to Oparin.* Johns Hopkins University Press, Baltimore.

Hazen, R. M., 2005. *Genesis. The Scientific Quest for Life's Origins.* Joseph Henry Press, Washington, DC.

Knoll, A. H., 2003. *Life on a Young Planet: The First Three Billion Years of Evolution on Earth.* Princeton University Press, Princeton, NJ.

Miller, S. L., 1992. The prebiotic synthesis of organic compounds as a step toward the origin of life. In *Major Events in the History of Life*, J. W. Schopf (ed.). Jones and Bartlett, Boston, pp. 1–28.

Morowitz, H. J., 2002. *The Emergence of Everything: How the World Became Complex.* Oxford University Press, New York.

TIME, 2006. *Nature's Extremes: Inside the Great Natural Disasters that Shape Life on Earth.* Time, Inc., New York.

Wilson, R. A. 2005. *Genes and the Agents of Life. The Individual in the Fragile Sciences*, Cambridge University Press, Cambridge, England.

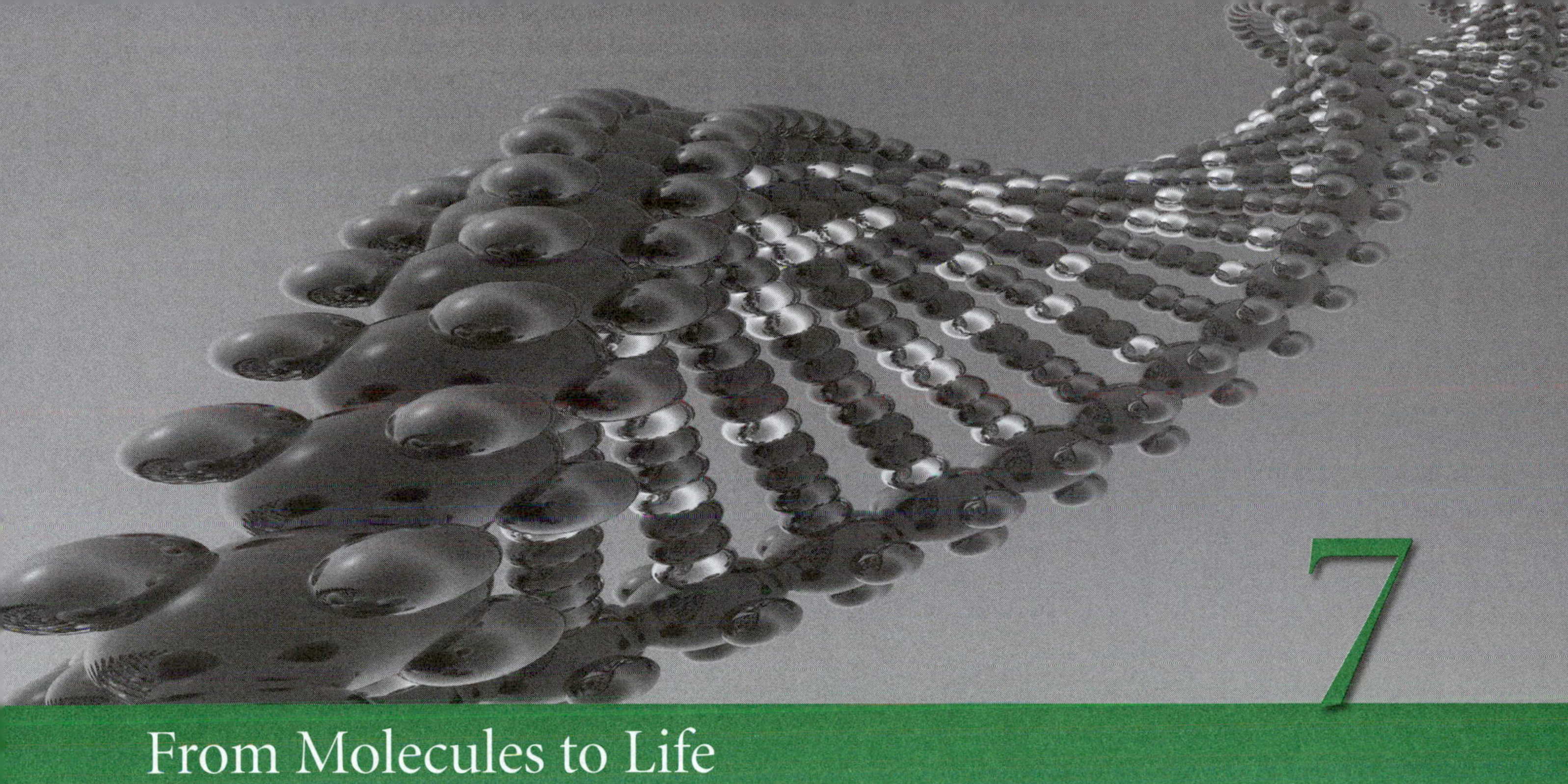

7 From Molecules to Life

Chapter Summary

RNA in the form of three-nucleotide units known as *codons* carries the genetically coded message from the storage molecule, DNA, to the protein synthesizing machinery. From an evolutionary perspective, the genetic code's most essential features are its universality and its redundancy or *degeneracy* (the use of several codons for the same amino acid). Nowadays, cells depend totally on protein catalysts for the production of proteins and nucleic acid templates. In this chapter, we discuss whether some types of nucleic acids appeared early in biotic history in an *RNA world,* or whether proteins were the earliest functioning polymers. Both types of molecules would have offered the advantage of catalytic activity. Their functions also may have extended to self-replication, a process that is more plausible for nucleic acids than for polypeptides. As there are reports that a few bacteria synthesize antibiotics using the polymerizing enzyme as a template, RNA may have had catalytic activity early in its evolution.

Because of the importance of lengthened molecular chains for structure and function, polymerization was likely one of the earliest cellular processes to evolve. If we assume the existence of some kind of template (perhaps an RNA ribozyme or a peptide/nucleic acid chimera) as well as a polymerizing enzyme, selection could have led to the origin of adaptor molecules carrying amino acids to the template, coenzymes for the use of high-energy molecules, a base-pairing system by which adaptor and nucleotide portions of the template could recognize one another, and some basic type of ribosome encompassing all these functions. Subsequently, storage and production functions separated into distinct molecules, a separate nucleic acid molecule (DNA) for information storage (a primitive genetic code) and other nucleic acid sequences acting as messengers, adaptors and regulators (mRNA, tRNA).

An early genetic code, even one composed of three nucleotides, probably provided information for fewer amino acids than now, and/or for more general classes of amino acids. The relative nonspecificity (redundancy) in the third nucleotide in each codon would have minimized errors in replicating the master code or in translating it into protein. The first and second nucleotides are less flexible and perhaps specified particular amino acids or discriminated among different groups of amino acids on the basis of such features as their hydrophobic or hydrophilic properties. Having two accurately translated nucleotides increased the number of amino acids that could be determined and at the same time regularized the primary structures of proteins. The code was constrained once 20 amino acids were specified and a large number and variety of proteins synthesized. At that time the code "froze," and it has remained virtually invariable in every cellular organism since.

Organismal function depends on transforming external sources of material and energy into processes that take place within those organisms. Critical to the evolution of life at the cellular level was development of chemical energy providers, such as adenosine triphosphate (ATP), that release small but significant amounts of phosphate-bond energy that the enzyme apparatus of the cell can control and localize to specific reactions. The elaboration of **organic catalysts** (**enzymes**) restricted chemical reactions to the most opportune times and place and increased the efficiency of the reactions.

Catalysts function by lowering the energy level necessary for a reaction, thereby increasing its frequency. Enzyme proteins add even more speed to this process by providing specific sites at which potential reacting molecules can be localized and manipulated to enhance the reaction. For example, inorganic ferric ion (Fe^{3+}) shows some catalytic activity in a variety of reactions (**Fig. 7-1a**). However, when such ions are incorporated into porphyrin molecules to form **heme** (Fig. 7-1b), the molecules are about a thousand times more effective than Fe^{3+} alone. If the protein component of the enzyme catalase is then added to the heme unit, catalytic efficiency increases by a further factor of one billion (Fig. 7-1c).

How did such proteins evolve and come to be coupled to a genetic code that could transfer the information necessary to produce them from generation to generation? We saw in the last chapter that various types of molecules associated with life can be produced experimentally and with comparative ease. But which class of molecules evolved first? Was original life an "RNA world" or a "protein world?" As you will see later (and may already have surmised), an initial "DNA world" is a far less plausible alternative.

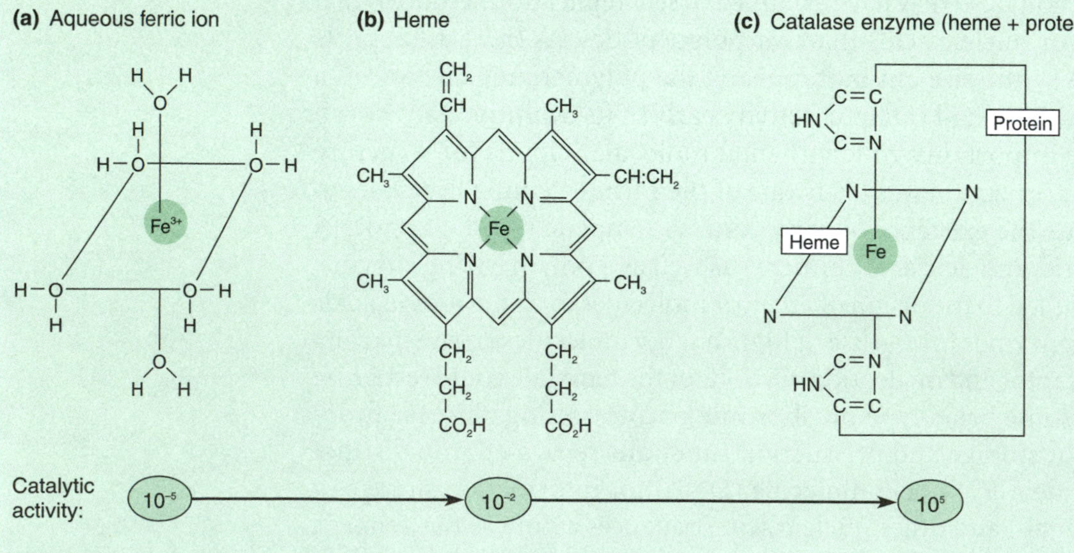

FIGURE 7-1 Change in catalytic activity for the reaction $2H_2O_2$ > $2H_2O + O_2$ when the iron atom is used by itself (a), or in different molecular combinations (b, c). The catalase protein in (c) provides an enzymatic advantage to the reaction because it binds rapidly to hydrogen peroxide molecules and distorts them so their decomposition proceeds at a lower "activation energy" than without the enzyme. (Adapted from Calvin, M., 1969. *Chemical Evolution.* Oxford University Press, Oxford, United Kingdom.)

An RNA World

In organisms alive today, and in those stretching back over a billion years, the amino acid sequences of the enzymes that serve as catalysts derive entirely from the nucleotide sequences in ribonucleic acid (RNA), which in turn derive from the nucleotide sequences of deoxyribonucleic acid (DNA), as discussed in Chapter 6. The chain of information transfer, known as the **central dogma:**

Replication (DNA → DNA), Transcription (DNA → RNA), Translation (RNA → protein)

depends on appropriate enzymes (**Fig. 7-2**). The universal interdependence of the three processes of the replication of DNA, transcription to RNA and translation of RNA to protein, poses the serious question of whether these functional and informational systems could have evolved independently of one other.

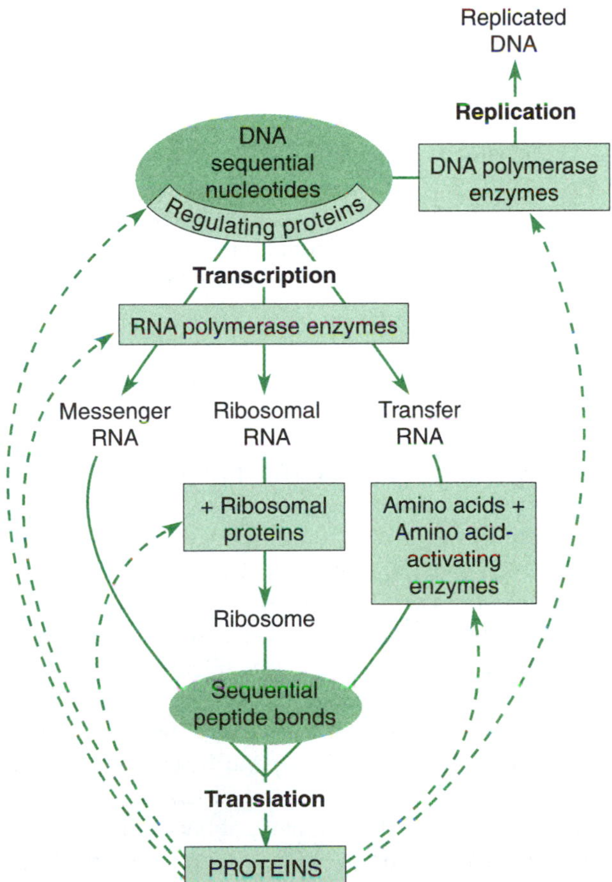

FIGURE 7-2 Schematic diagram showing the mutual dependence of information carried by nucleotide sequences and function governed by proteins. *Solid lines* indicate the general directions of information transfer, and *dashed lines* point to proteins that this process synthesizes. Clearly, the nucleotide sequence information determines the amino acid sequences of proteins, and proteins in turn regulate and catalyze the transfer of nucleotide information; one process could not have developed without the other.

It seems likely that proteins always had more functionally different forms than did nucleic acids. Part of this functional variety arises from the almost inexhaustible array of permuted amino acid sequences that proteins can achieve. For example, because 20 different amino acid alternatives exist for each position in a polypeptide, a sequence of five amino acids has 20^5 or more than three million possible arrangements. By contrast, a sequence of five nucleotides in a nucleic acid has only 4^5 or 1,024 possibilities. Moreover, many amino acids are quite different in structure (see Fig. 6-2), enabling them to interact in many different ways, with each other and with molecules such as water, metal ions, and various monomers and polymers. These differences confer an astronomical variety of possible three-dimensional configurations on a protein in contrast to the relatively more rigid shapes assumed by many nucleic acids.

Despite the enormous possible arrangements of amino acids, today, **protein synthesis** virtually always depends on RNA and DNA. One logical extension of this universal dogma is that nucleic acids must have arisen before proteins; the geneticist Hermann J. Muller long ago suggested that because the phenotype of an organism derives essentially from its genotype, this relationship must have existed in the past. That is, the genotype was first in the evolutionary sequence.

RNA First

In the early 1980s, three scientists (Leslie Orgel, Francis Crick and Carl Woese) independently proposed what is now known as the **RNA world** as the first stage in the evolution of life in which RNA catalyzed all molecules necessary for survival and replication. The self-replicating power of nucleic acids then enabled responses to selection to develop protein systems that would support further **self-replication.** Protein synthesis might then have evolved through specific amino acids directly interacting with specific nucleotide sequences or perhaps through indirect placement of amino acids into such sequences by adaptor molecules that brought amino acids to the nucleotide chain. The reasons for an initial RNA world rather than a protein world are discussed below. Despite its head start, RNA did not become the molecule of inheritance and form the genetic code, topics taken up toward the end of the chapter.

Supporters of the view that **nucleic acid replication** arose first argue that only a self-replicating system can provide the basis on which selection can build a cooperative functional unit: an organism. In the absence of self-replication, function presumably would quickly disappear and the "organism" could not maintain itself. Therefore, scientists have sought an **autocatalytic** process to explain the origin of a *"naked gene" that could replicate itself without the help of proteins.* The discovery that some RNA molecules possess catalytic properties even in the absence of proteins has

given some support to the possibility of autocatalytic nucleic acid replication. By using such **RNA catalysts** or **ribozymes** (enzymes made of RNA, rather than proteins), short RNA sequences can replicate without protein enzymes by forming templates for complementary RNA sequences (Loomis, 1988).

Increasing numbers of researchers have therefore proposed the early existence of an RNA world dependent on self-replicating RNA nucleotide sequences, although G. F. Joyce and coworkers (1987) pointed out that RNA replication would have been inhibited strongly in prebiological times because different stereoisomers in the ribose–sugar backbone would have prevented complementary base pairing. The consistent prebiological supply of ribose needed for such backbone formation has also been questioned; like many sugars, ribose is extremely unstable (Larralde et al., 1995). To overcome such difficulties, Joyce and coworkers (1987) suggested that early nucleic acid genetic material was not based on ribose-containing nucleotides but rather on ribose-like analogues, in which such pairing difficulties were minimized, and which were more easily synthesized than ribose. Such a system was demonstrated experimentally by Zielinski and Orgel (1987).[1]

The hypothesis that an early RNA world antedated cellular enzymatic proteins (see Hazen, 2005 and Penny, 2005 for excellent recent reviews) is now based on a variety of types of evidence that are consistent with such a theory:

- The discovery of enzymes made from RNA (*ribozymes*) that cut and rejoin preexisting strands of nucleotides and so replicate RNA.
- The "intron" portion of transcribed RNA in the protozoan ciliate *Tetrahymena* splices itself out of the RNA molecule and helps form a chemical bond between the protein-coding RNA sections ("exons") on either side.
- Some cellular RNA molecules catalyze reactions that include binding to ATP, the common energy transfer molecule, increasing the likelihood that, in an RNA world RNA could have replicated itself. The experimental studies of Sassanfar and Szostak (1993), on which this statement is based, were important in providing the first evidence that **RNA replication** could have occurred in a prebiotic world. From random oligonucleotides they isolated a catalyst that functioned to join oligonucleotides together. The catalyst obtained energy for the interactions from triphosphates, that is, an *RNA molecule can function as protein catalysts do today*.
- Researchers can design synthetic RNA molecules to perform precise catalytic reactions and, when replicated with ribonuclease, to become resistant to this enzyme, which normally degrades RNA.[2]
- Selection experiments in the laboratory have led to the evolution of RNAs into new kinds of molecules that show catalytic activities many orders of magnitude greater than in the initial mixture (Lehman and Joyce, 1993). Some such selected ribozymes can even combine a ribose sugar with a thiouracil base to make nucleotides at a rate more than ten million times greater than the uncatalyzed reaction (Unrau and Bartel, 1998).
- Some RNA molecules function as gene regulators by binding to nucleic acids and affecting gene expression.
- RNA "fragments" appear in coenzymes used in various metabolic reactions, for example, coenzyme A, nicotinamide adenine dinucleotide (NAD) and flavin adenine dinucleotide (FAD). Because other chemicals could have assumed the function of these RNA fragments, researchers interpret their presence as "historical," that is, a remnant of an earlier RNA world (Benner et al., 1993).
- The protein translation system that all cells use depends on a variety of RNAs: messenger, ribosomal and transfer RNA. In fact, peptide formation catalyzed by ribosomal enzymatic action, called *peptidyl transferase*, depends more on ribosomal RNA than on ribosomal proteins. Zhang and Cech (1997) demonstrated that even a selected noncellular ribozyme could perform this same amino acid-binding function. The RNA in ribosomes, rather than ribosomal proteins, may function to catalyze peptide bonds.

Templates, Catalysts and Mutations

The findings just discussed focus attention on RNA originally serving as a self-replicating molecule by acting as both template and catalyst, translating itself into a sequence of RNA nucleotides or ribozymes whose enzymatic activity included RNA replication. Initially, these would have been short nucleotide sequences. As the fidelity of replication increased in response to selection, longer RNA molecules could persist, even when their survival relative to other molecules was low. Similarly, superiority of survival influences length: the greater its relative superiority over other molecules, the longer the sequence of nucleotides that can be maintained for some given fidelity of replication. We can calculate that 90 percent fidelity of replication will conserve a molecule no longer than 12 nucleotides (at some given level of relative survival), whereas 95 percent fidelity of replication will conserve a molecule about twice as long. But fidelity of replication has the essential requirement that an

[1] See Deamer (2003), Orgel (2004) and Chen et al. (2005) for recent studies.

[2] See Hanczyc et al. (2003) for an update.

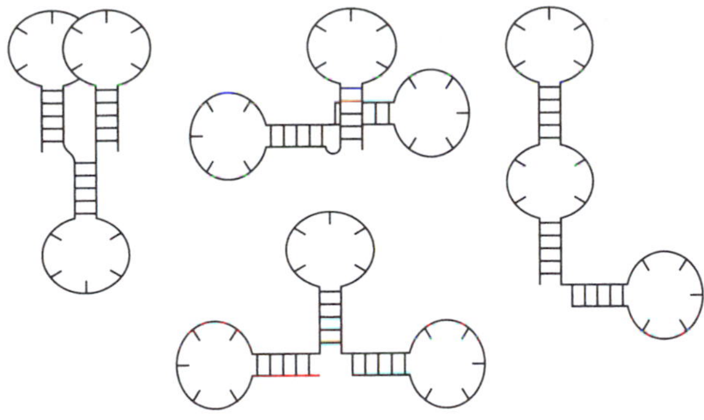

FIGURE 7-3 Some triple stem-loop structures that RNA molecules can assume. (From Joyce, G. F., and L. E. Orgel, 1993. Prospects for understanding the origin of the RNA world. In *The RNA World*, R. F. Gesteland and J. F. Atkins (eds.). © Cold Spring Harbor Laboratory Press, Cold Spring Harbor, New York, pp 1–25. Reprinted by permission.)

RNA molecule be long enough to act as a replicase enzyme. Joyce and Orgel (1993) suggest that the minimum length for RNA catalytic activity is "a triple stem loop containing 40 to 60 nucleotides" (**Fig. 7-3**). According to Fraser and coworkers (1995), even the smallest of cellular organisms (*Mycoplasma genitalium*) need a minimum of 90 different proteins for translation and about 30 for DNA replication. Other problems of RNA replication fidelity arise because known RNA polymerase enzymes lack the "proofreading" attributes of DNA polymerases, and so are subject to relatively **high mutation rates.** Although high mutation rates might seem an advantage in a rapidly evolving world, they could cause increasingly inefficient enzymes with each mutational generation. Moreover, RNA sequences, especially guanine nucleotides, coil back on themselves to produce tangled strands impossible to replicate without intervention of proteins. How could RNA sequences duplicate by RNA alone? Furthermore, synthesis of an enzymatic protein that can replicate RNA, even without proofreading, probably requires no less than 1,000 nucleotides, coding for at least 300 or more amino acids, which is a number too large for RNA itself to replicate without an already existing replicase enzyme.

According to one proposal, this dilemma —"enzymes need coding by long RNA sequences, and long RNA sequences need replicating by long enzymes" — may have been overcome by smaller functional polypeptide subunits, each coded by smaller, more easily replicated RNA sequences. In a system called a **hypercycle** (Eigen, 1992), these subunits enter into a symbiotic relationship conferring advantages to the overall system by successively coupling their individual effects, one subunit aiding replication of the other, forming a network that can produce a much longer, more accurately replicated, sequence. If such a system is compartmentalized, the success of the compartment ("cell") and the success of its symbiotic components become mutually dependent, a form of "group selection" discussed later (see Chapter 23). To these proposals, Maynard Smith and Szathmáry (1995) add other systems that could have accounted for an increase in genetic information.

Another avenue was opened up with the synthesis of *pyranosyl RNA* (pRNA), in which ribose has five carbon atoms rather than the four in RNA. pRNA forms double-stranded molecules that have fewer structural variations and do not twist around one another, providing a more efficient means of replication in a world without protein enzymes.

Why Is RNA still with us?

Given the evolution of proteins, why is RNA still necessary for translation? The simple answer would be that no successful alternative presented itself in the past as an intermediary between genetic material and protein translation. Through base pairing, RNA serves as a "messenger" to ribosomes (*messenger RNA*), and through codon–anticodon pairing it serves as an "adaptor" to transfer amino acids to polypeptides (*transfer RNA*). We do not fully understand its role in the ribosome itself, but judging from the complexity of ribosomal RNA, probably various types of nucleotide pairing and interaction are needed. For example, a nucleotide sequence in ribosomal RNA mediates base pairing between the codons and anticodons used in translation. Perhaps once pathways evolved using RNA for these or other purposes (e.g., intron splicing), substituting a different molecular mechanism would have entailed too many widespread changes, causing lethality (see "Frozen Accidents" later in this chapter). The same reasoning is used to explain retention of such fundamental features of organisms as cells producing extracellular matrices, conserved tissues, common body plans, often under the umbrella of concepts such as constraint, *Baupläne*, the zootype and phylotypic stages (see Chapter 13).

Molecules from Clays

Instead of nucleic acids, Cairns-Smith (1982) proposes that the earliest self-replicating unit may have consisted of organic, **clay-like silicate** crystals or layered clay mixtures (for example, *montmorillonite*). As crystals can grow by adding subunits into their highly ordered structures, we can view them as having some *self-replicatory powers*. According to this view, nucleic acids such as RNA could have incorporated into such an assembly, followed by the formation of peptides and the evolution of a protein-synthesizing system.

Although mineral surfaces are gaining more attention as a locale for biosynthetic reactions, their ability to develop a highly complex metabolic sequence extending from self-replicating RNA to protein synthesis is difficult to visualize. Some biologists, therefore, have emphasized the possibility that *proteins themselves* or *protein-nucleic acid combinations* were the first self-replicating systems. Some early peptides might have served as templates for the aggregation of nucle-

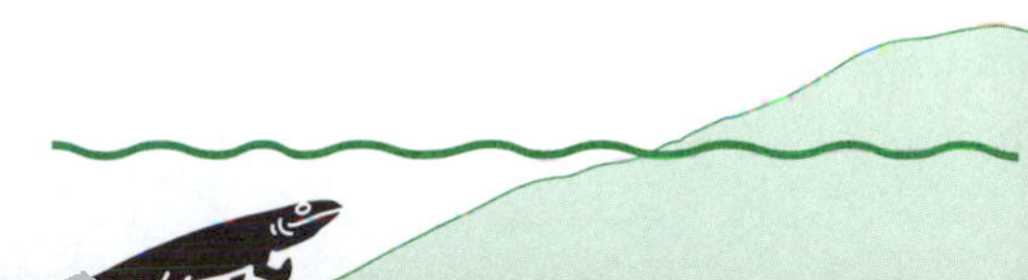

otides that may then have bonded together to form a precise mold for the replication of these same peptides.

This view offers a mechanism by which the replication of both peptide chains and nucleotide chains would have been interdependent from the start. Also, as one chain lengthens during further evolution, so does the other, each gradually improving its replicatory role until some of the present features of nucleic acid replication and protein synthesis evolve. As with naked genes, we still find it difficult to imagine the spontaneous origin of such a precisely organized nucleic acid template. Indeed, Ferracin (1981) incorporated these ideas into a theory that Darwinian selection did not apply in chemical evolution in the first protocells, but that random processes associated with self-assembly drove the first biogenetic processes.

Proteins First

Various past theories proposed that protein systems developed diverse functional properties *before* they coupled to nucleic acid replicative systems. A number of authors (for example, S. W. Fox, 1978) go further and claim that a protein catalytic system must have developed *before* a nucleic acid replicative system.

Many researchers, however, feel this view has a serious shortcoming: If proteins arose first and were used by the earliest cells or particles for functional purposes, how could they have replicated without a nucleic acid translational system? Could proteins alone have synthesized proteins? The chemical experiments under presumed prebiological conditions (discussed in Chapter 6) also show the difficulty of producing polymerized nucleic acids spontaneously, although long-chained polypeptides are produced in such experiments with relative ease.

If we consider only extant organisms, we have difficulties conceiving of protein synthesis independent of nucleic acids. No complementary relationships exist between amino acids as exist between nucleotides. That is, the precise **stereochemical fit** that occurs between the base pairs of complementary nucleotide chains (adenine–thymine, adenine–uracil, cytosine–guanine) and accounts for the replicative, transcriptional and translational properties of nucleic acids, nowhere echoes in a similar complementary stereochemical fit between amino acids in polypeptide chains.

Peptide Synthesis without RNA

Nevertheless, a process does exist in which **peptides** (short lengths of amino acids with functional properties) are made *de novo,* for, as Lipmann (1971) and others showed, a spore-forming bacterium, *Bacillus brevis*, produces at least two antibiotics, *gramicidin S* and *tyrocidin*, exclusively by enzymes in the absence of messenger RNA. Both of these antibiotic molecules are circular oligopeptides ten amino acids long, and both are synthesized by the sequential addition of amino acids. Should one amino acid be omitted, peptide synthesis ceases, indicating that the *enzyme involved functions as a precisely ordered template for the amino acid sequence*. An unfilled position on the template prevents bonding between amino acids on either side. Interestingly, such antibiotics often contain both dextro (D) and levorotatory (L) amino acids; peptides generated via mRNA are composed only of L amino acids.

Of special interest is the form in which peptides elongate among these antibiotics. Single amino acids bind to sulfhydryl (–SH) groups on the enzyme before they join the peptide chain and are then connected by the sequential removal of their sulfur (thiol) groups and the formation of peptide bonds. The chain maintains a thiol at its "head" end to furnish the connection for sequential growth. Lipmann called this process *head-growth polymerization* and pointed out its striking similarity to the polymerization of carbon groups during fatty acid synthesis (also polymerized by use of sulfhydryl bonds) and to the polymerization of amino acids during ribosomal peptide synthesis (polymerized by phosphate bonds).

Relating these findings to the evolutionary origins of proteins, it was suggested that these polymerizing similarities indicate a common underlying polymerization process that may have arisen early (but not necessarily before RNA) during chemical evolution. Heinen and Lauwers (1996) propose that thiol synthesis may have occurred, even in a non-reducing atmosphere. Of course, the enzymes *now* involved in antibiotic peptide synthesis are themselves synthesized via information transferred from genetic material, but the fact that proteins *can* produce proteins in these systems points to the possibility that repeatable copies of short-chained but functional peptides, 15- to 20-amino acids long, could have been produced in the past in the absence of nucleic acids.

In further support is a self-replicating peptide experiment that makes use of a 32-amino acid long helical peptide. This coiled structure serves as a template to bond 17- and 15-amino acid fragments, which become templates for further replication (Lee et al., 1996). Although dependent on laboratory conditions and components, such results imply that small α-helical catalytic structures can self-replicate once they have evolved. Matrices on which such reactions could have initially organized may have been the clays and zeolites mentioned above and in the previous chapter, or mineral surfaces such as iron sulfides discussed by Edwards (1998).

Perhaps also relevant to this view are the *prion diseases* discussed in **Box 7-1.**

Evolution of Protein Synthesis Machinery

Whatever the early composition of the templates used in condensing amino acids, the *polymerization process* itself must have had an early function for which selection occurred,

BOX 7-1 Prion Diseases

As discussed in Chapter 6 in relation to the definition of life, a **prion** is a particle *derived from a cell*, composed of a hydrophobic protein, but lacking DNA and RNA. Although prion protein sequences are genetically determined, the interactions that change their conformation can be transmitted and reproduced between cells and individuals without further genetic information. A prion is the smallest known agent of infection.

Only functional when incorporated into a living cell, prions convert cellular proteins into new prions. The prion protein, normally harmless, can undergo a pathogenic change in its three-dimensional shape, which converts other such proteins to a similar form. Through such "domino effect," a **prion disease** acting remarkably like a non-nucleic acid infectious agent, progressively develops as prion proteins increasingly change to pathogenic form (Prusiner, 1995).

By entering brain cells and taking over protein synthesis prions are thought to be responsible for neurological degeneration in humans (Creutzfeldt-Jakob disease) and for "mad cow" disease (bovine spongiform encephalopathy) in cattle. Prions also are thought to be the agents for *kuru* (laughing death) in the Fore tribe of New Guinea, whose custom was to eat the brains of dead relatives. Kuru, which is a form of transmissible spongiform encephalopathy, is characterized by loss of coordination, dementia, and, paradoxically, outbursts of laughter as the disease progresses; hence the name, laughing death.

because only in peptide form do amino acids attain their catalytic properties.

As time went on, selection for more efficient polymerization would have led to the production of more efficient templates, which were perhaps themselves polymerized by more efficient polymerases. Evolution of a template capable of producing an enzyme whose amino acid chain was long enough to polymerize the template itself, was a critical step in the transition to a self-replicating molecular life form. However this selection occurred (and RNA itself is a possibility), whether for successively lengthened templates or for improved enzymatic activity or through both processes (see, e.g., Szathmáry, 1989), the impact of increasing the amount of coded genetic material would have been profound. It seems reasonable that the circular, autocatalytic tautology of life — to make more of those substances that can interact with the environment to make more such substances — was firmly established during this early period.

Evolution of protein synthesis may have had its start in a basic polymerase enzyme that could replicate inefficiently on a template of polypeptides and other materials. The original force responsible for polymerizing organic molecules may have arisen because of hydrophobic interactions between various amino acids, interactions enabling the amino acids to separate from water. The most stable interactions that would dissipate hydrophobic forces and produce the lowest free energy level presumably lies in the folded globular organization of polymerized protein. Energy to attain these polymerizations may be derived from simultaneous degradation of other organic compounds. The coupling of degradation and polymerization would be followed by the selection of polymers that could catalyze their own polymerization.

Once peptide polymerization established itself, selection for its improvement would have led to an increased number of polymerase enzymes per protocellular unit, and improved template efficiency of these polymerases and their ability to increase the rapidity of amino acid condensation. Evolutionary refinements of peptide polymerization would have emphasized selection for other attributes, such as a code for polymer replication, metabolic pathways, and response to environmental substances, homeostasis and development. In later stages, genetic and translational functions sequestered to different parts of the cell, mRNA moving from its new site of transcription where genetic information was now stored, to the ribosome for translation. Three separate functional classes of nucleotide sequences eventually arose: **storage, messenger,** and **translational** (*ribosomal* and *transfer*) — all RNA, because this nucleic acid still fulfills two of these purposes in all organisms (messenger and adaptor) and fulfills genetic storage purposes in some viruses. The evolution of new kinds of ribosomes — no longer committed to the production of particular proteins — enabled the *same ribosome to translate different mRNAs.* This transferred the burden of regulating which proteins to make from ribosomes to the transcriptional process. That is, a particular protein could now be selected by regulating which mRNA molecules to transcribe from the stored genetic material, a process that eventually gave rise to sophisticated regulatory systems such as that shown in Figure 10-13.

Accompanying the innovations outlined in Figure 10-13 must have been changes from depending on only few enzymes and proteins, each with multiple functions and lower coupling specificities, to employing greater numbers with restricted functions and higher binding specificity. Such transitions would have followed an increase in the number of mRNA molecules by **gene duplication**. Mutations within duplicated genes and subsequent selection among them would allow individual gene products to diverge and evolve in new and more specific directions. Once gene numbers

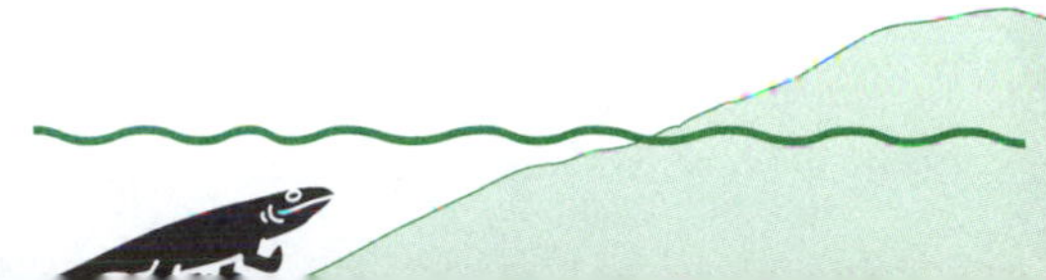

increased, their linkage into chromosomes would have been advantageous, ensuring their collective presence in each cell. Chromosomes also enable linked genes to replicate as a synchronized unit rather than as individual competitors.

Why DNA?

Because **RNA** was probably used initially for **information storage and protein translation,** difficulties in separating these two functions must have offered advantages to organisms that could use a different nucleic acid, **DNA**, for storage purposes. Enzymes that translate RNA into protein do not function with DNA, thus restricting the more uniformly structured double helix DNA exclusively to the storage of information and to transcribing one of its strands to form mRNA. The observation that deoxyribonucleotides are synthesized from ribonucleotides in cellular pathways supports the notion that DNA arose later in cellular evolution than RNA.

Some have proposed that DNA genetic material offered a molecule more easily protected against mutation than RNA because the 2′ hydroxyl group in the ribosome sugar of RNA causes it to undergo more rapid hydrolytic cleavage than the 2′ deoxyribose of DNA. Moreover, the 2′ hydroxyl group in RNA provides it with catalytic activity, whereas DNA is relatively inactive: *no DNA molecule acts as a catalyst.* In addition, the transition from RNA to DNA as the genetic material may have simply entailed special **reverse transcriptase** enzymes that can transcribe RNA sequences into DNA sequences. Such enzymes, perhaps originally only involved in RNA replication (they can function as RNA polymerases), could later have served to transfer genetic information from RNA to DNA.

A question raised is why both RNA and DNA are limited to four different bases, each matched with a single other base during complementary base pairing (adenine–uracil and guanine–cytosine in RNA, and adenine–thymine and guanine–cytosine in DNA). Among possible answers is the vulnerability of base pairing to mutation frequency. It seems likely that the greater the kinds of bases involved in genetic replication, the greater the chances for mismatching during the pairing process and the greater the loss of replication accuracy (Szathmáry, 1991). Because RNA most likely preceded DNA as genetic material, and because RNA polymerases, then as now, could not proofread or easily repair mismatching errors, high mutation frequency would have confined genetic material to short sequences, producing short, inefficient enzymes. To lengthen and improve enzyme function, genetic material would have had to increase in length and replication accuracy, a process that became dependent on restricting the number of different kinds of bases to a genetic alphabet of four code letters replicating via only two complementary base pairings.

Evolution of the Genetic Code

Information transfer between nucleic acids and proteins follows a **genetic code** that determines the placement of a particular type of amino acid within a protein from the placement of a particular trinucleotide sequence in mRNA. **Table 7-1** lists the terminology and characteristics of the code, and **Table 7-2** provides the code itself.

As for any other biological trait, the genetic code evolved from an earlier and presumably simpler form, although no ancestral codes have been discovered. Attempts at an **evolutionary reconstruction of the code** have relied on a detailed analysis of ten basic features that characterize the present code:

1. Messenger RNA molecules consist of only *four kinds of nucleotide bases*: adenine (A), guanine (G), uracil (U) and cytosine (C). These compose chains of varying lengths and varying sequences.
2. An mRNA codon that specifies a particular amino acid is a triplet consisting of a chain of *three nucleotides*.
3. The code is **nonoverlapping** (comma-less), so each codon *translates* in a continuous, uninterrupted sequence, three successive nucleotides at a time, from one end of an mRNA reading frame to the other.
4. *Reading frames* in messenger RNA generally begin with the codon AUG, and terminate at the stop codons UAA, UAG or UGA.
5. The codon sequence complements an *anticodon* sequence on the adaptor or transfer (tRNA) molecule (**Fig. 7-4**) that carries a particular amino acid to the mRNA codon.
6. All organisms share the *universal* coding dictionary outlined in Table 7-2, with some codon differences in mycoplasmas (bacteria lacking polysaccharide cell walls) and ciliate protistans. Mitochondrial organelles show a few codon differences from those used by cellular nuclei.
7. Ambiguities have not been found in the code; that is, *the same codon does not specify two or more different amino acids.*
8. With the exception of methionine and tryptophan, *more than one codon designates each amino acid.*
9. The pattern of code *redundancy is mostly in the third codon position*. For example, eight amino acids (including valine, threonine and alanine) use quartets of codons, each member of a quartet varying only at the third position.
10. When an amino acid uses only a *duet* (two) of the codons in a quartet, the third codon positions in the duet both are pyrimidines (U and C) or both purines (A and G), never one pyrimidine and one purine.

Redundancy

Explanations for some features of **code redundancy** have not been difficult to find. In part, redundancy derives from the

TABLE 7-1 Definitions of common terms used in describing the present genetic code

Term	Meaning
Code letter	Nucleotide, for example, A, U, G, C (in mRNA) or A, T, G, C (in DNA). Note that there are four code letters in each nucleic acid "alphabet," forming two kinds of complementary base pairs: A–U and G–C in RNA, and A–T and G–C in DNA.
Codon or code word	Sequence of nucleotides specifying an amino acid, e.g., the RNA codon for leucine = CUG (or GAC in DNA).
Anticodon	Sequence of nucleotides on transfer RNA that complements the codon, e.g., GAC = anticodon for leucine (see Fig. 7-6).
Genetic code or coding dictionary	Table of all the codons, each designating the specific amino acid into which it translates (Table 7-2).
Codon length or word size	Number of letters in a codon, e.g., three letters in a **triplet code** (these are the same as coding ratio in a nonoverlapping code).
Nonoverlapping code	Code in which only as many amino acids are coded as there are codons in end-to-end sequence, e.g., for a triplet code, UUUCCC = phenylalanine (UUU) + proline (CCC).
Redundant or degenerate	Presence of more than one codon for a particular amino acid, e.g., UUU, UUC = Code = phenylalanine. Twenty different amino acids are therefore coded by a total of more than 20 codons.
Synonymous codons	Different codons that specify the same amino acid in the redundant code, e.g., UUU = UUC = phenylalanine.
Ambiguous code	Circumstances when one codon can code for more than one amino acid, e.g., GGA = glycine, glutamic acid. No ambiguities exist in the present code although such ambiguities may have existed in the past.
Comma-less code	Absence of nucleotides (spacers) between codons, e.g., UUUCCC = two amino acids in triplet non-overlapping code.
Reading frame	Particular nucleotide sequence coding for a polypeptide that starts at a specific point and partitions into codons until it reaches the final codon of that sequence.
Frame shift mutation	Change in the reading frame because of the insertion or deletion of nucleotides in numbers, other than multiples of the codon length. This changes the previous partitioning of codons in the reading frame and causes a new sequence of codons to be read.
Sense word	Codon that specifies an amino acid normally present at that position in a protein.
Replacement mutation	Change in nucleotide sequence, either by deletion, insertion or substitution, resulting in the appearance of a codon that produces a different amino acid (**missense mutation**) in a particular protein, e.g., UUU (phenylalanine) mutates to UGU (cysteine).
Stop mutation	Mutation that results in a codon that does not produce an amino acid, e.g., UAG (also called a chain-terminating codon or **nonsense codon**).
Universal code	Use of the same genetic code in all organisms, e.g., UUU = phenylalanine in bacteria, mice, humans and tobacco (with some exceptions, e.g., mitochondria, see Table 7-2).

Source: Strickberger, M. W. 1985. *Genetics,* 3rd ed. Reprinted by permission of Prentice Hall, Inc., Upper Saddle River, NJ. © 1985 by Monroe W. Strickberger.

the presence of more than one kind of tRNA for a single amino acid. The tRNAs used for leucine, for example, may have the anticodons AAU, AAC and GAG. Crick explained that the redundancy pattern mostly attaches to the third codon position because of **wobble pairing** between certain bases of the tRNA anticodon and the mRNA codon in this position.

In Crick's **wobble hypothesis,** since proved correct, anticodons bearing inosine (I) at this third position can pair with either U, A or C; anticodons bearing G can pair with either of the pyrimidines U and C; and anticodons bearing U can pair with either of the purines A and G. Third codon position redundancy therefore points to the importance of the first two codon positions in specifying amino acids; these positions suffice to code for the eight amino acids that use codon quartets.

Why This Particular Code?

The striking **universality of the code** — its common features in all independent organisms and its restriction to the **same**

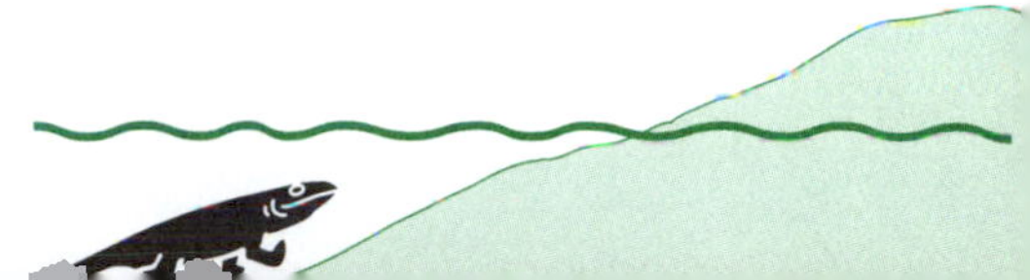

TABLE 7-2 Nucleotide sequences in mRNA codons specifying particular amino acids: The "standard" genetic code[a]

Codon	Amino acid	Codon	Amino acid	Codon	Amino acid	Codon	Amino acid
U U U	*phe*	U C U	*ser*	U A U	*tyr*	U G U	*cys*
U U C		U C C		U A C		U G C	
U U A	*leu*	U C A		U A A	STOP[b]	U G A	STOP[b]
U U G		U C G		U A G		U G G	*trp*
C U U	*leu*	C C U	*pro*	C A U	*his*	C G U	*arg*
C U C		C C C		C A C		C G C	
C U A		C C A		C A A	*gln*	C G A	
C U G		C C G		C A G		C G G	
A U U	*ile*	A C U	*thr*	A A U	*asn*	A G U	*ser*
A U C		A C C		A A C		A G C	
A U A		A C A		A A A	*lys*	A G A	*arg*
A U G[c]	*met*	A C G		A A G		A G G	
G U U	*val*	G C U	*ala*	G A U	*asp*	G G U	*gly*
G U C		G C C		G A C		G G C	
G U A		G C A		G A A	*glu*	G G A	
G U G		G C G		G A G		G G G	

[a]Rare exceptions to this code occur in various animal mitochondria in which AUA also specifies methionine, and AGA specifies serine or serves as a stop codon (Jukes and Osawa, 1991). Such mitochondrial changes seem to be in the direction of economizing in the kinds of transfer RNA produced in a small organelle that makes relatively few polypeptides. Some other exceptions are found in several organisms displayed in Fig. 7-7.

[b]These codons are also called *chain-terminating codons* or, in the past, *nonsense codons.*

[c]This is the common codon used to initiate protein synthesis.

20 amino acids — makes it reasonable to ask: Why this particular code? Because at least 1,070 possible different codes use 64 codons to code for 21 entities (20 amino acids + chain termination), either the exclusive use of this particular code must have derived from accidental causes, or some relationship between amino acids and their codons (or anticodons) must exclude large numbers of other possible codes — or perhaps both factors operated. Two different answers to this question have been offered:

1. Some authors suggested that the amino acids originally associated with their codons or anticodons because of stereochemical fitting or by **sharing other complementary characters** such as hydrophilic and hydrophobic properties. For example, some kind of pairing may have occurred between an amino acid such as phenylalanine and a codon such as UUC or its anticodon AAG. Others maintain that, despite considerable search, there are few, if any, examples of preferential affinity between amino acids and their codons or anticodons (Biro et al., 2003).
2. Theorists therefore have offered an alternative hypothesis that the universality of the code derives from the **survival of only one of the possible codes** that existed in the past; that is, early amino acid–codon relationships arose largely by chance rather than by restricted stereochemical pairing and may therefore have produced a number of different primordial genetic codes, each used by different groups. As time went on, however, only one group carrying a particular code continued evolving; the others became extinct. So, the code reached its present form after what has been called a frozen accident.

Frozen Accidents

In extant organisms (1) all proteins consist of L-amino acids, (2) all nucleic acids consist of nucleotides with D-sugars, and (3) DNA double helices coil in a right-handed rather than a left-handed direction.

Phenomena known as **frozen accidents** could help explain this universal composition and these universal optical rotations. Further evolution of the code was restricted (a frozen accident) because protein synthesis in its mature form precisely

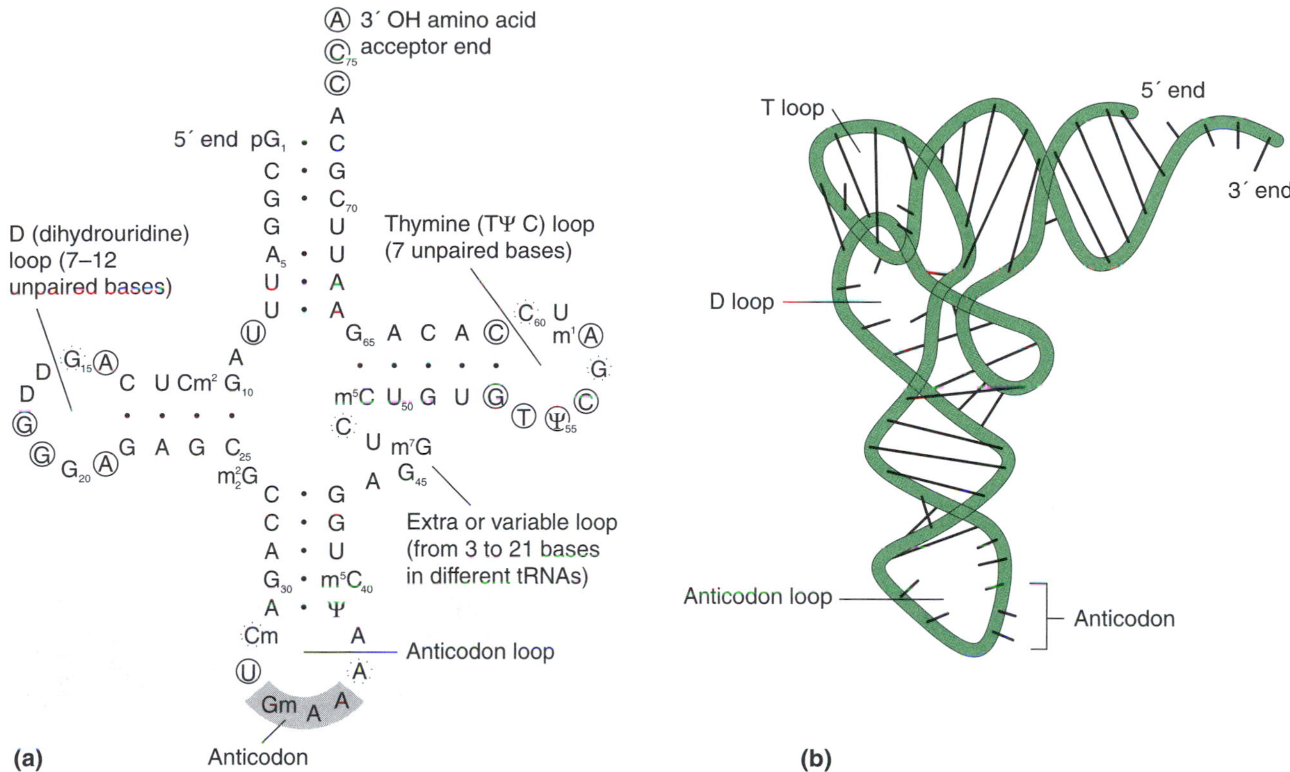

FIGURE 7-4 (a) Sequence of the 76 nucleotides in phenylalanine tRNA of yeast shown in the commonly portrayed two-dimensional cloverleaf form. Hundreds of different kinds of tRNA molecules sequenced in a variety of organisms can fit into this same cloverleaf pattern, offering a maximum of pairing (*dots*) between complementary bases. The four major tRNA loops are indicated, including the anticodon loop, which contains, in this case, the special sequence AAG that matches the mRNA codon for phenylalanine UUC (and also UUU because of wobble pairing; see also Figure 7-6). Transfer RNA molecules that are specific for other amino acids bear, of course, different anticodon sequences as well as different nucleotides in some of the other positions. Bases encircled with *solid lines* occupy the same positions in all the different tRNAs examined so far, whereas those encircled with *dots* are more variable, indicating base pair positions that are always occupied by either purines or pyrimidines. (b) Three-dimensional L-shaped structure showing the molecule's functional form as two distinct domains at right angles to each other: an acceptor domain for attaching specific amino acids and an anticodon domain for binding to specific mRNA codons. Some researchers propose that the acceptor domain evolved first, acting as a tag to attract RNA-replicating ribozymes in the RNA world (Maizels and Weiner, 1994). As evolution proceeded toward protein synthesis, some sequences within these tags served as "codons" that also coupled with ribozymes carrying amino acids. These coupling sequences duplicated in the anticodon domain, enabling them to bind to mRNA. Numbering of the bases, shown as subscripts, begins at the 5′ end of the nucleotide sequence. Unusual nucleotides found in tRNA include D = dihydrouridine, Ψ = pseudouridine, mX = methylated nucleotide and T = thymine.

positioned (and still positions) each particular amino acid in every long-chain polypeptide in which it is found. Any change in the genetic code for a widely used amino acid would significantly change many different proteins carrying that amino acid. As one example, if the code for phenylalanine (UUU) extended to include the serine codon (UCU), tRNA molecules carrying phenylalanine would insert into polypeptide positions formerly occupied by serine. Because these two amino acids differ considerably in structure and function, the result undoubtedly would be the death of the organism.

Before "freezing," the genetic code must itself have evolved to accompany some of the changes taking place in protein synthesis. As discussed earlier in the chapter, early proteins were probably much shorter than proteins today, would have consisted of fewer kinds of amino acids and were produced by a much less accurate translation mechanism, allowing for changes between some codons in response to selection. Furthermore, not all of the current amino acids would have been incorporated into the primeval code(s).

We believe the genetic code was triplet even during its beginnings, or perhaps a **doublet code** with single **nucleotide spacers.** Mechanical considerations support this view, because anything less than a triplet codon would not provide a stable pairing relationship between a tRNA anticodon and an mRNA codon. The trinucleotide width of a triplet anticodon also is thought to provide a minimal space, enabling tRNA molecules to lie close enough together for peptide bonding between their amino acids. Given a small group of amino acids coded by triplets, further evolution of the code probably would have proceeded under the three selective conditions outlined below.

1. *Nucleotide Substitution.* Nucleotide substitutions caused by errors in replication (mutations) should produce as few amino acid changes as possible. Selection would

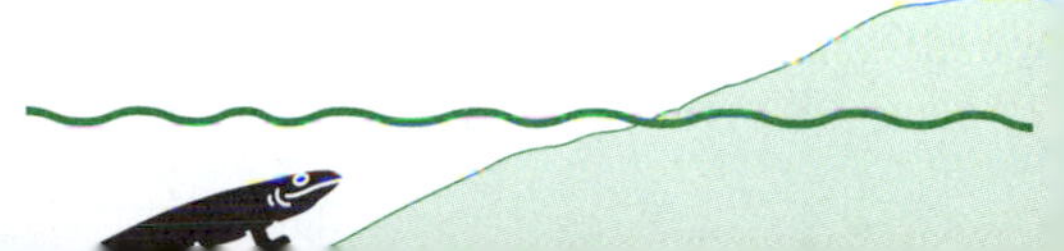

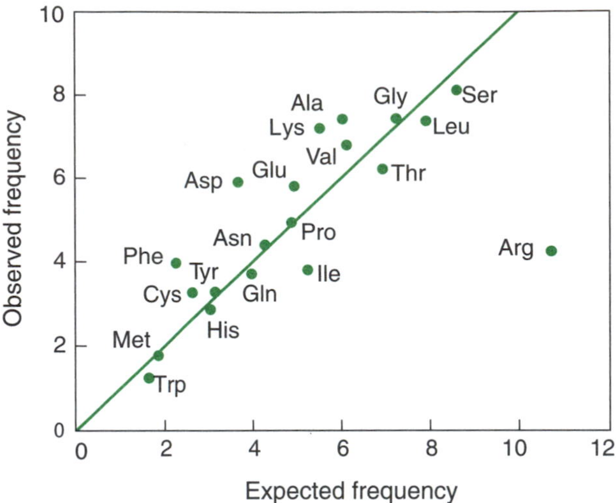

FIGURE 7-5 Comparison between observed and expected frequencies of amino acids at 5,492 positions in 53 different vertebrate polypeptides. If we exclude arginine, the correlation coefficient for these data is 0.89. (The expected frequency of each amino acid derives from a technique of calculating the nucleotide composition of the mRNA from the frequency of bases in the first two codon positions used by the various amino acids that compose the proteins: U = 0.220, A = 0.303, C = 0.217, G = 0.261. Randomized triplet codons for this nucleotide composition furnish the expected amino acids.) (Reprinted with permission from King, J. L. and T. H. Jukes, 1969. Non-Darwinian evolution: Random fixation of selectively neutral mutations. *Science*, **164**, 788–798. © 1969, American Association for the Advancement of Science.)

tend to expand the number of different codons used by an amino acid so that random base changes would still produce the same amino acid.

2. *Codon Frequency.* The number of different codons per amino acid generally should be proportional to the frequency in which the amino acid occurs in proteins. With respect to the numbers of codon assignments, a proportional relationship exists between the relative number of codons possessed by most amino acids and their frequencies in proteins (**Fig. 7-5**), although the source for this relationship is not clear. It may indicate a selective relationship (more codons are selected for the use of those amino acids that occur more frequently) or an accidental relationship (the overall composition of proteins derives from the frequency of amino acid codons), or both.
3. *Translation.* Errors occurring during the mRNA–tRNA translation process should lead to as little drastic protein changes as possible.

This third level of codon evolution, **selection for minimizing translational errors,** may have considerably affected codon assignments. For example, the prevailing redundancy at the third codon position is precisely what we would expect if this position were the one most involved in translational errors; that is, the various quartets (for example, GUU, GUC, GUA, GUG) and duets (for example, UUU, UUC) arise from the selective advantage of assigning to the same amino acid codons that could most easily be mistaken for one another. The next most error-prone translational event occurs at the first codon position. At this position, the code again shows some redundancy — for example, UUA (leucine) → CUA (leucine) — or is so constructed that an error may occasionally enable the substitution of an amino acid with related function, for example, UUA (leucine) GUA → (valine).

In general, the first codon position is considerably less error prone than the third, because of modifications of the 40th tRNA nucleotide immediately adjacent to this position. These modifications include the addition of methyl or even bulkier groups to nucleotide 40, preventing wobble at the first codon position. One such modified nucleotide is threonyl-6-adenine, which prevents the wobble pairing of U in the adjacent position with G in the first codon position.

TABLE 7-3 Classification of 18 amino acids according to the nucleotide found at their second codon positions and known hydration potentials of their side chains[a]

Amino Acid	Second Codon Letter	Hydration Potential (k/cal/mole)	
Gly	G	+2.39	**Most hydrophobic**
Leu	U	+2.28	↑
Ile	U	+2.15	
Val	U	+1.99	
Ala	C	+1.94	
Phe	U	−0.76	
Cys	G	−1.24	
Met	U	−1.48	
Thr	C	−4.88	
Ser	C(G)	−5.06	
Trp	G	−5.89	
Tyr	A	−6.11	
Gln	A	−9.38	
Lys	A	−9.52	
Asn	A	−9.68	
Glu	A	−10.19	
His	A	−10.23	↓
Asp	A	−19.92	**Most hydrophilic**

[a] Other proposals suggest that the first codon position specifies amino acids made through similar biosynthetic pathways, or distinguishes between small and large amino acids (see Maynard Smith and Szathmáry, 1995). *Source*: Data are from R.V. Wolfenden, P. M. Collis, and C. C. F. Southgate, 1979. Water, protein folding and the genetic code. *Science*, **206**, 575–577.

(a) An early code: 16 anticodons for perhaps 15 amino acids

Codon	Anticodon	Amino acid	Codon	Anticodon	Amino acid	Codon	Anticodon	Amino acid	Codon	Anticodon	Amino acid
UUU	AAU	phe?	UCU	AGU	ser	UAU	AUG	tyr	UGU	ACU	cys?
UUC			UCC			UAC			UGC		
UUA			UCA			UAA	STOP		UGA		
UUG			UCG			UAG			UGG		
CUU	GAU	leu	CCU	GGU	pro	CAU	GUU	his?	CGU	GCU	arg
CUC			CCC			CAC			CGC		
CUA			CCA			CAA			CGA		
CUG			CCG			CAG			CGG		
AUU	UAU	ile	ACU	UGU	thr	AAU	UUU	asn?	AGU	UCU	ser?
AUC			ACC			AAC			AGC		
AUA			ACA			AAA			AGA		
AUG			ACG			AAG			AGG		
GUU	CAU	val	GCU	CGU	ala	GAU	CUU	asp?	GGU	CCU	gly
GUC			GCC			GAC			GGC		
GUA			GCA			GAA			GGA		
GUG			GCG			GAG			GGG		

tRNA gene duplications and mutations → Evolution of new anticodons

(b) A later code: 31 anticodons for perhaps 18 amino acids

Codon	Anticodon	Amino acid	Codon	Anticodon	Amino acid	Codon	Anticodon	Amino acid	Codon	Anticodon	Amino acid
UUU	AAG	phe	UCU	AGG	ser	UAU	AUG	tyr	UGU	ACG	cys
UUC			UCC			UAC			UGC		
UUA	AAU	leu	UCA	AGU	ser	UAA	STOP		UGA	ACU	trp
UUG			UCG			UAG			UGG		
CUU	GAG	leu	CCU	GGG	pro	CAU	GUG	his	CGU	GCG	arg
CUC			CCC			CAC			CGC		
CUA	GAU	leu	CCA	GGU	pro	CAA	GUU	glu	CGA	GCU	arg
CUG			CCG			CAG			CGG		
AUU	UAG	ile	ACU	UGG	thr	AAU	UUG	asn	AGU	UCG	ser
AUC			ACC			AAC			AGC		
AUA	UAU	ile	ACA	UGU	thr	AAA	UUU	lys	AGA	UCU	arg
AUG			ACG			AAG			AGG		
GUU	CAG	val	GCU	CGG	ala	GAU	CUG	asp	GGU	CCG	gly
GUC			GCC			GAC			GGC		
GUA	CAU	val	GCA	CGU	ala	GAA	CUU	glu	GGA	CCU	gly
GUG			GCG			GAG			GGG		

(c) The modern code: 43 known anticodons for 20 amino acids

Further tRNA gene duplications and mutations → Evolution of new anticodons (and deletion of ACU anticodon to produce a UGA stop codon)

Codon	Anticodon	Amino acid	Codon	Anticodon	Amino acid	Codon	Anticodon	Amino acid	Codon	Anticodon	Amino acid
UUU	AAG	phe	UCU	AGI	ser	UAU	AUG	tyr	UGU	ACG	cys
UUC		phe	UCC		ser	UAC		tyr	UGC		cys
UUA	AAU	leu	UCA	AGU	ser	UAA	STOP		UGA	STOP	
UUG	AAC	leu	UCG	AGC	ser	UAG			UGG	ACC	trp
CUU	GAG	leu	CCU	GGI	pro	CAU	GUG	his	CGU	GCI	arg
CUC		leu	CCC		pro	CAC		his	CGC	GCG	arg
CUA	GAU	leu	CCA	GGU	pro	CAA	GUU	gln	CGA		arg
CUG	GAC	leu	CCG		pro	CAG	GUC	gln	CGG		arg
AUU	UAI	ile	ACU	UGI	thr	AAU	UUG	asn	AGU	UCG	ser
AUC	UAG	ile	ACC	UGG	thr	AAC		asn	AGC		ser
AUA	UAC*	ile	ACA	UGU	thr	AAA	UUU	lys	AGA	UCU	arg
AUG	UAC	met	ACG		thr	AAG	UUC	lys	AGG		arg
GUU	CAI	val	GCU	CGI	ala	GAU	CUG	asp	GGU	CCG	gly
GUC	CAG	val	GCC		ala	GAC		asp	GGC		gly
GUA	CAU	val	GCA	CGU	ala	GAA	CUU	glu	GGA	CCU	gly
GUG	CAC	val	GCG		ala	GAG	CUC	glu	GGG	CCC	gly

FIGURE 7-6 Some possible stages (a–c) in the evolution of the genetic code based on a scheme that Jukes suggested. The mRNA codons are at the left of each box. tRNA anticodons are in shaded capital letters to their right. The cytidine nucleotide in the UAC anticodon marked with an asterisk is acetylated, restricting this tRNA molecule to AUA (isoleucine) codons on mRNA. Osawa provides lists of known anticodons in eukaryotes, prokaryotes, chloroplasts and mitochondria. (From Strickberger, M. W. *Genetics*, 3rd ed. © 1985 by Monroe W. Strickberger. Reprinted by permission of Prentice Hall, Upper Saddle River, NJ.)

The second codon position, the least error prone during translation, may at one time have separated entire classes of amino acids with unique functions. This system would offer a selective advantage by ensuring that amino acids are rarely substituted between classes. A possible remnant of such a grouping may be the assignment of U to the second codon position for leucine, isoleucine and valine, all of which are **hydrophobic amino acids** that exist mostly in the interior of proteins, and assignment of A to the second codon position for glutamic acid, histidine, aspartic acid and other **hydrophilic amino acids** that commonly exist on the protein surface (**Table 7-3**).

In *summary*: The triplet genetic code may have initially coded for relatively few amino acids; or (based primarily on the second codon position), may distinguish among general classes of amino acids. As time went on, the first codon posi-

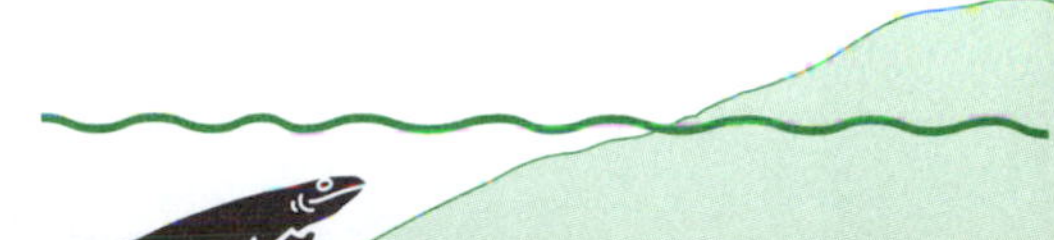

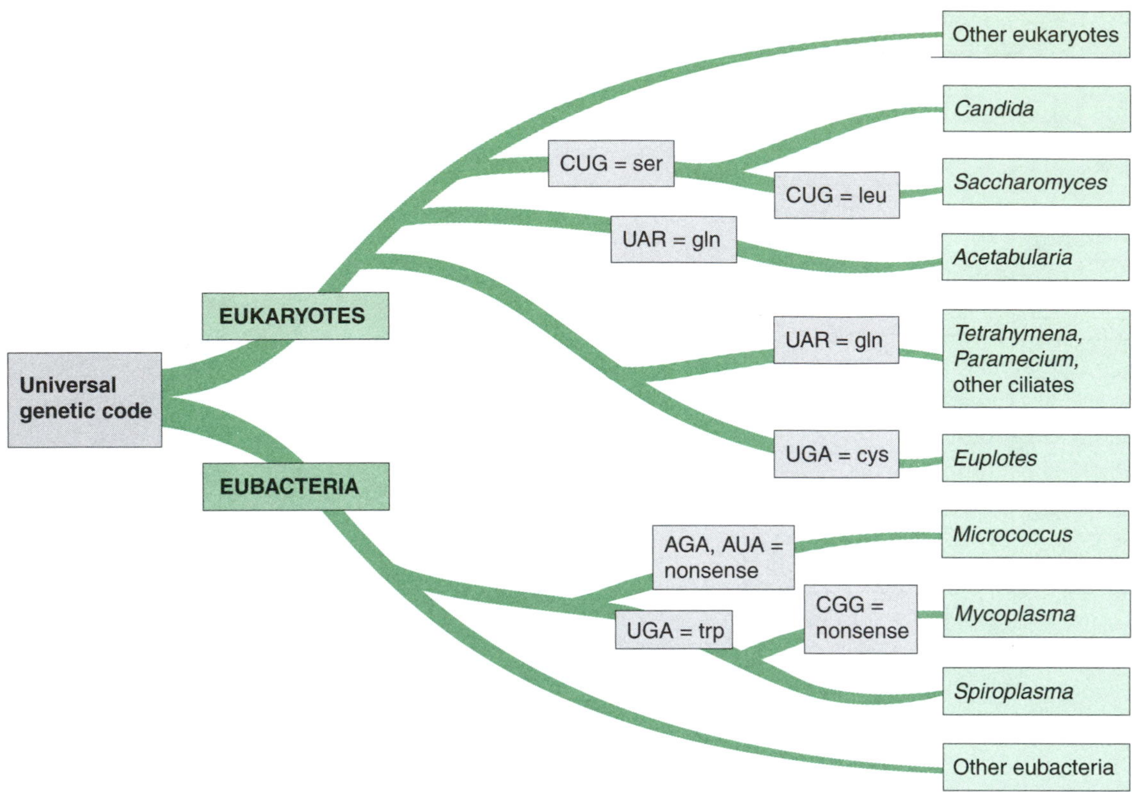

FIGURE 7-7 Changes in the universal genetic code found in nuclei of some organisms considered "primitive" among eukaryotes and eubacteria. R designates a purine nucleotide, for example, UAR = UAA or UAG. (Adapted from Osawa, S., 1995. *Evolution of the Genetic Code.* Oxford University Press, Oxford, United Kingdom.)

tion came into use for amino acid positioning because of the modifications that could be made to the immediately adjacent tRNA nucleotide. With the advent of two accurately translated codon positions, the genetic code could expand its repertoire of amino acids. Thus, until translation accuracy was firmly achieved, some shuffling of codons may have taken place between different amino acids, or entirely new amino acids may have entered the process, using some or all of the codons of older amino acids. The retention of third codon position redundancy or wobble may have economized on the number of tRNA molecules needed to translate amino acids such as valine, threonine and alanine.

Accordingly, Jukes proposed an early stage in the evolution of the genetic code somewhat like that shown in **Figure 7-6a** (see Jukes and Osawa, 1991). Each **codon quartet** or **codon family** of four codons specifies perhaps one amino acid, and each amino acid is represented by only one kind of tRNA molecule whose anticodon could pair with all four codons in the family. This type of code commonly appears in present-day mitochondrial organelles that, because of their small size and limited function, economize in the kinds of both tRNA and proteins they produce. Such extreme wobble limited this code to no more than 15 or 16 amino acids. One family or partial family of codons terminated translation because there were no amino acid-bearing tRNA molecules whose anticodons could pair with these stop codons.

In subsequent evolution (Fig. 7-6b,c), tRNA gene duplications enabled new anticodons to evolve, some of which mutated so that new amino acids activated them. By such means, the kinds of tRNA molecules could increase and new amino acids could add to the code. It appears that when 20 different amino acids had incorporated into the code, these ancestral organisms were producing a large-enough number of proteins that codon changes necessary to include any further amino acids would lead to widespread protein malfunction and widespread lethality. At that point, the code *froze*, limiting its codon assignments to the prevailing amino acids. The **universality of the genetic code** (with a few rare codon exceptions) indicates that only the ancestral bearers of this particular code successfully survived the early evolutionary period.

Genetic code exceptions found in mitochondria and in the nuclei of a few organisms (**Fig. 7-7**), indicate that some proteins in these entities tolerated some variation in amino acid composition or chain termination without ill effect. Osawa (1995) proposed that these exceptions are not ancient relics but are of recent origin, and that some codon changes may still be evolving in such organisms. He suggests that particular codons will be the subject of mutation or selection, thereby allowing unused codons to lose their former assignment and be "captured" to code for a new or different amino acid.

KEY TERMS

anticodon
autocatalytic
codon
degenerate code
DNA
doublet code
evolution of protein synthesis
frozen accident
genetic code
hydrophilic amino acids
hydrophobic amino acids
missense mutation
nonsense codon
nonoverlapping code
nucleic acid replication
organic catalysts
polymerization enzymes
protein synthesis
reverse transcriptase
ribozymes
RNA catalysts
RNA world
self-replication
sense word
stereochemical fit
stop mutation
synonymous codons
triplet code
universal genetic code

DISCUSSION QUESTIONS

1. What is the present relationship among cellular proteins, RNA and DNA?
2. What evidence would lead you to conclude that an RNA world came before proteins or DNA?
3. Nucleic acids, clays, proteins.
   a. Why have researchers suggested each of these substances as the earliest self-replicating genetic material or earliest cellular material?
   b. What problems have other researchers raised against these proposals?
4. Provide a scenario of how protein synthesis could have evolved.
5. Genetic code.
   a. What are the major features of the genetic code?
   b. How does wobble pairing account for redundancy of the code?
   c. What is a "frozen accident," and how would it explain the universality of the present genetic code?
   d. Why is the present genetic code believed to have derived from a code based on triplet codons rather than from a code based on doublets or singlets?
   e. What selective factors have researchers proposed that would have influenced the assignment of specific codons to specific amino acids?
   f. How could a smaller number of different amino acids specified by an earlier code have increased so that the present genetic code could specify a larger number (20)?

EVOLUTION ON THE WEB

Explore evolution on the Internet! Visit the accompanying Web site for *Strickberger's Evolution,* Fourth Edition, at http://www.biology.jbpub.com/book/evolution for Web exercises and links relating to topics covered in this chapter.

RECOMMENDED READING

Biro, C., B. Benyó, C. Sansom, Á. Szlávecz, G. Fördös, T. Micsik, and Z. Benyó, 2003. A common periodic table of codons and amino acids. *Biochem. Biophys. Res. Commun.*, **306**, 408–415.

Joyce, G. F., and L. E. Orgel, 1993. Prospects for understanding the origin of the RNA world. In *The RNA World*, R. F. Gesteland and J. F. Atkins (eds.). Cold Spring Harbor Laboratory Press, Cold Spring Harbor, New York, pp. 1–25.

Loomis, W. F., 1988. *Four Billion Years: An Essay on the Evolution of Genes and Organisms.* Sinauer, Sunderland, MA.

Maynard Smith, J., and E. Szathmáry, 1995. *The Major Transitions in Evolution.* Freeman, Oxford, United Kingdom.

Osawa, S., 1995. *Evolution of the Genetic Code.* Oxford University Press, Oxford, United Kingdom.

Pagel, M., (ed. in chief), 2002. *Encyclopedia of Evolution*, 2 volumes. Oxford University Press, New York.

Szathmáry, E., 1991. Four letters in the genetic alphabet: A frozen evolutionary optimum? *Proc. Roy. Soc. Lond. (B)*, **245**, 91–99.

Unrau, P. J., and D. P. Bartel, 1998. RNA-catalysed nucleotide synthesis. *Nature*, **395**, 260–263.

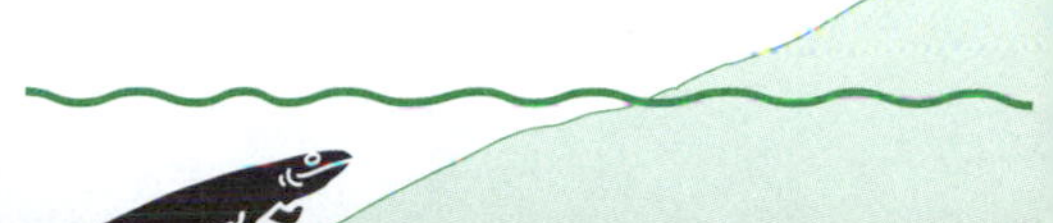

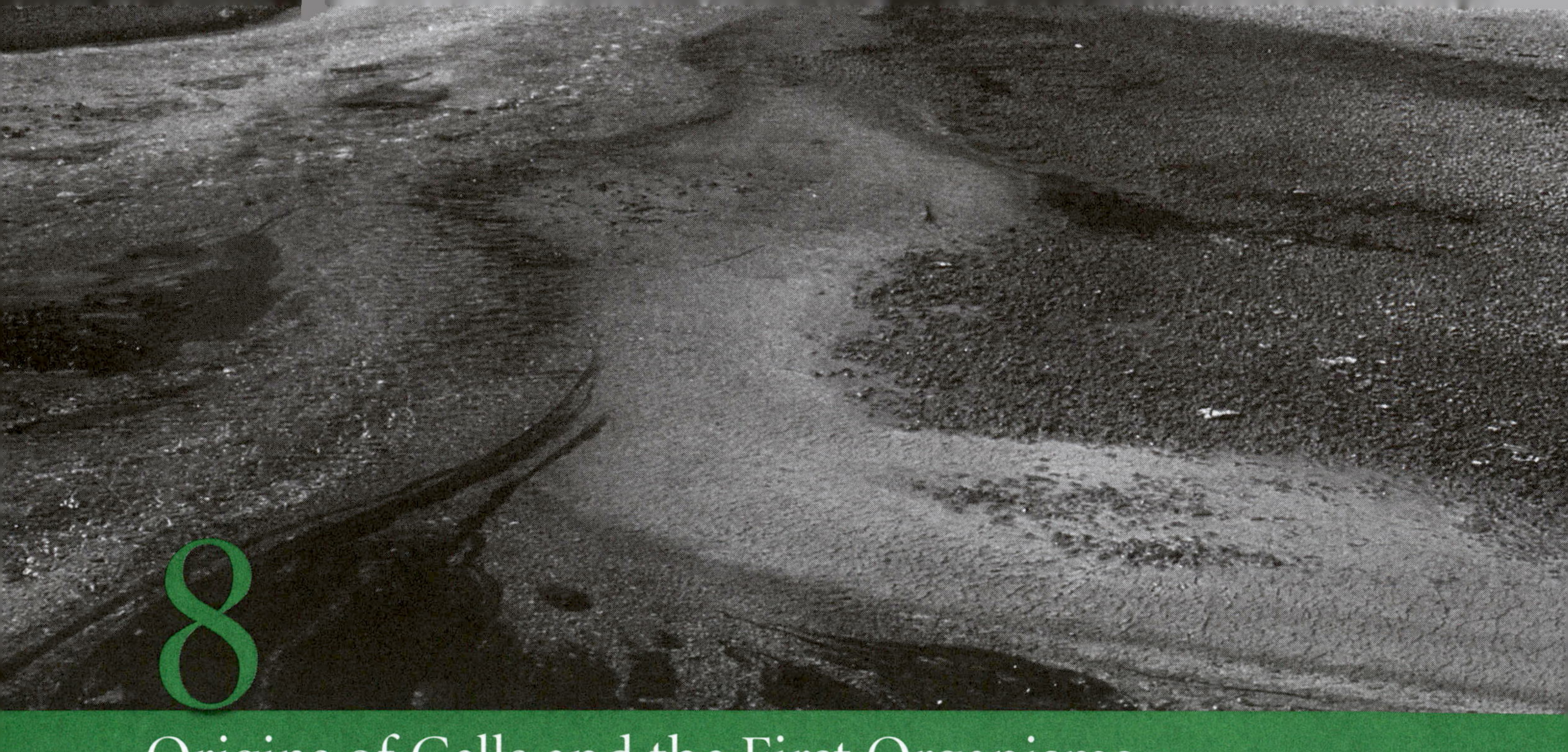

8 Origins of Cells and the First Organisms

Chapter Summary

Prokaryote-type cells arose about 3.5 Billion years ago (Bya). About two billion years later, some had diversified into eukaryotes. Extant prokaryotes, which include eubacteria and archaebacteria, are small (about 1 μm), contain no nuclear membrane, cytoskeleton or complex organelles and divide by binary fission. Eukaryotes are generally aerobic and more complex; they contain many organelles, internal membranes, a cytoskeleton and a microtubular apparatus for mitotic cell division. Eukaryotic genes, unlike those of eubacteria, often have nucleotide sequences within them (introns) that do not translate into peptide sequences. While similar in organismal complexity to eubacteria, archaebacteria share several core similarities with eukaryotic cells but perhaps all three types of cells arose from a single common ancestor, the progenote.

The features of mitochondria and chloroplasts suggest that these organelles in eukaryotes are remnants of two ancient symbiotic relationships. The large amount of DNA in eukaryotic cells led to its fission into several chromosomes and to the appearance of microtubule-dependent cell division. Because meiosis is almost universal in eukaryotes, meiosis must have arisen early. Somewhat later, perhaps just prior to the Phanerozoic, multicellular eukaryotes appeared and underwent several explosive radiations.

The prevalence of metabolic pathways common to many organisms suggests that comparative biochemistry would be a fruitful approach to investigate the evolution of metabolism. Anaerobic glycolysis is an almost universal pathway in which energy released as glucose degrades to pyruvic acid, and the reactions comprising it may illuminate its evolutionary past. Because of their sequential nature, metabolic pathways could not have arisen randomly but were "predetermined" by available preexisting compounds. Many pathways (and enzymes controlling individual reactions) survived because of their selective advantages, and commonly appear in living systems.

Eventually, high-energy carbon compounds became less plentiful and some organisms switched to reducing more copious carbon compounds, such as CO_2, by means of a membrane-bound oxidation–reduction system. A revolution occurred when some organisms began to reduce carbon by using light as an energy source: photosynthesis. Such reduction requires a source of electrons, for which substances such as H_2S were used. It later became advantageous to split water, which was plentiful. Oxygen, a by-product of this process, increased to the extent that it was exploited to enhance energy yields from glucose breakdown. A new pathway, the Krebs cycle, arose from anaerobic pathways, the entire process being contingent on oxygen. Electrons removed from oxidized compounds in the cycle transfer to a membrane-bound electron transport system for which oxygen acts as the ultimate electron acceptor molecule. Much more energy releases in this process than is released in anaerobic glycolysis.

Early Fossilized Cells

From what we can discern so far, the evolution of metabolic pathways (discussed below) followed a progression from:

- simple **anaerobic** systems dependent on energy sources in the primeval "soup"; to
- **autotrophic** systems capable of generating chemical phosphate bond energy from sunlight; to
- **aerobic** systems that derive energy from the transfer of electrons to oxygen.

Figure 12-20 shows a phylogeny of **prokaryotes,** based on the sequencing of amino acids in proteins and nucleotides in RNA as inferred from the DNA sequences of the genes that encode them.

An immediate word of caution is required concerning the use of the term *prokaryotes*, as it implies a single monophyletic group of unicellular organisms. As this chapter progresses, you will see a much more realistic division is of three kingdoms of single-cells organisms: **Eubacteria, Eukarya** and **Archaea** (**Box 8-1, Table 8-1** and **Fig. 8-1**; and see Pace, 2006 for an overview).

When these evolutionary steps took place is not known, but various cell-like objects appear in geological strata that date back as far as 3.5 Bya. These strata are mostly unmetamorphosed rocks called **cherts**, which are dark or black because of their high carbon content and also bear considerable silicon deposits. The oldest in the Warrawoona group of Western Australia (**Fig. 8-2**), like many other Archean cherts, are associated with layered organic deposits called **stromatolites**. Two fascinating finds, both from Western Australia, were reported in 2006, indicating that the hunt for Early Archaean life is in full swing and yielding important findings.

1. From analysis of silica dykes in greater than 3.5-By-old cherts of the Dresser Formation in Pilbara craton, Western Australia, Ueno and colleagues (2006) demonstrated *methane of microbial origin* within minute fluid inclusions. Given their careful distinction between abiotic and biotic sources of the methane, and the coexistence of one of the oldest microfossils — filamentous single ("prokaryote") cells — within the same geological unit, their study provides strong evidence for the origin of microbiological life more than 3.5 Bya. Ueno and colleagues further argued that microbial methane would have played an important role in regulating the climate 3.5 Bya.[1]
2. The second report is of *seven types of stromatolites* from a 10-km-long exposure of a shallow water marine, 3.4-By-old reef, the Strelley Pool Chert in Western Australia (Allwood et al., 2006). The clear impression is of stromatolites that formed a structured, extensive biological ecosystem.

Stromatolites

In their modern form, stromatolites consist of mats of microorganisms that trap various aqueous sediments, which cement together to form characteristic laminated structures shaped like giant knobs. As shown in **Figure 8-3**, these modern structures are remarkably similar to ancient stromatolites, which must also have arisen from the deposition of biological organisms. According to Golubic and Knoll (1993),

> Stromatolites . . . are initiated by the establishment of a thin mat of microbes on a sediment surface. As sediment particles accumulate on top of mats,

[1] See Ueno et al. (2001, 2006) for the methanogens, Garcia-Ruiz et al. (2003) for a contrary interpretation and see Kasting and Catling (2003) for the effect on climate.

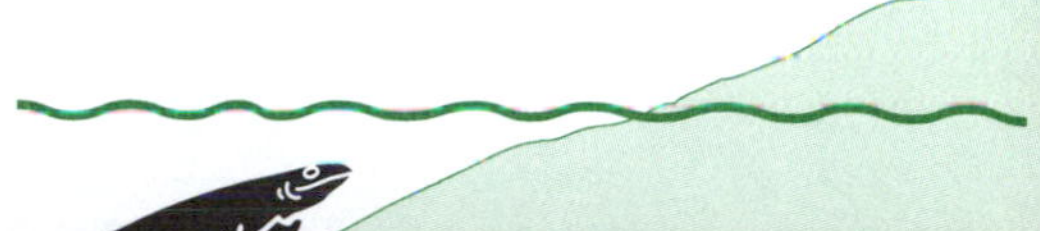

BOX 8-1 Kingdoms and Domains of Life

Two Kingdoms

FOR OVER TWO THOUSAND YEARS, life has been classified into **two kingdoms, Plantae** (L. *planta*, plant) and **Animalia** (L. *anima*, breath, life). Assignment to one kingdom or the other is based on structure, function, metabolism — plants use photosynthesis, animals do not — and locomotion (animals move from place to place, plants do not, other than during seed or spore dispersion).

Two Different Kingdoms

Evidence such as that provided in Chapter 6 and in the present chapter has been used to separate life into two broad domains: prokaryotes, which arose about 3.5 Bya, and eukaryotes, which arose about 1.5 Bya. Figure 6.1 shows representative prokaryotes and eukaryotes (a typical plant and an typical animal cell) but recall the caution at the beginning of the chapter that *prokaryotes are not monophyletic*. Extant unicells (prokaryotes), which include eubacteria and archaebacteria, are small, contain no nuclear membrane or complex organelles and divide by binary fission. Eukaryotes are generally aerobic and contain many organelles and a microtubular apparatus for mitotic cell division.[a] Eukaryotic genes contain introns, which prokaryote genes do not. Although similar in organismal complexity to one lineage of unicells, eubacteria, a second group of "prokaryotes," the archaebacteria, share several core similarities with eukaryotic cells (see text and *Three Domains* below).

Five Kingdoms

In 1959, things became more complex when Whittaker organized the 96 phyla of animals and divisions of plants, and prokaryotes and eukaryotes into five kingdoms, one prokaryote and four eukaryotes (**Table 8-1**).

[a] The division into prokaryotes and eukaryotes was made in 1941 by Stanier and van Niel. By 1962 they had established the characters listed as distinguishing prokaryotes from eukaryotes. See Stanier and van Niel (1941, 1962), Stanier (1970) and Margulis and Schwartz (1998).

TABLE 8-1 The Five Kingdoms of Life

Kingdom	Common Name(s)[a]
Prokaryotes (Monera)	bacteria (17)
Protista (Protoctista)	algae, protozoans, slime molds (27)
Fungi	fungi, mushrooms, molds, lichens (5)
Plantae	plants (mosses, ferns, cone-bearing and flowering land plants; 10)
Animalia	animals (worms, sea urchins, crabs, monkeys, and so forth; 37)[b]

[a] Based on Whittaker, R. H. 1959. On the broad classification of organisms. *Quart. Rev. Biol.*, **34**, 210–226. Protists, fungi, plants and animals are eukaryotes. Numbers of phyla (the highest grade within animals) or divisions (the highest grade within plants) are shown in parentheses; see also Margulis and Schwartz (1998) and Hall (1999a).

[b] Proposals for grouping animal phyla are discussed in Box 11-2 and in Chapter 15.

Five Kingdoms — the title of an influential book by Lynn Margulis and Karlene Schwartz, first published in 1982 and twice in updated editions — helped widely popularize this scheme. The situation was stable until the early 1980s when Woese (1981) and later Woese and colleagues (1990) found that rRNA sequences from an archaebacterium, representative of a lineage of prokaryotes, were sufficiently different from those of other prokaryotes that archaebacteria should be placed into a separate domain or division of life, which they termed **Archaea**.

Three Domains

One of the first organisms to have its complete genome sequenced was the methane-producing marine "bacterium," *Methanococcus jannaschii*, found at depths of 2,600 m where the pressure is over

they are trapped and bound into a coherent layer by the microorganisms. Microbially mediated precipitation of calcium carbonate can also contribute to sediment accumulation and stabilization . . . Through time, commonly, a laminated structure accretes, each lamina marking a former position of the living mat community . . . On the present-day Earth, filamentous cyanobacteria are predominant mat-builders, but coccoid cyanobacteria, other types of bacteria, and a variety of eukaryotic algae produce well-defined mats.

Extant and ancient forms differ in distribution. Modern stromatolites appear only in extremely inhospitable environments — salinities twice that of normal seawater and temperatures greater than 65°C — where they are protected from grazing metazoans such as snails and sea urchins. Ancient stromatolites were distributed more widely, presumably because such herbivores were absent. Many of the fossil organisms found in stromatolite deposits are remarkably similar to prokaryotes alive today (**Fig. 8-4**). In the words of Schopf, Hayes and Walter (1983):

1. Shallow water and intermittently exposed environments (and possibly, land surfaces and open oceanic waters) were habitable by unicellular organisms at least as early as 3.5 Bya.
2. Such organisms comprised finely laminated, multi-component, stromatolitic communities of the sediment-water interface, biocenoses where the principal surficial mat-building taxa were filamentous and photo-responsive, forms that may have been capable of phototactic, gliding motility.

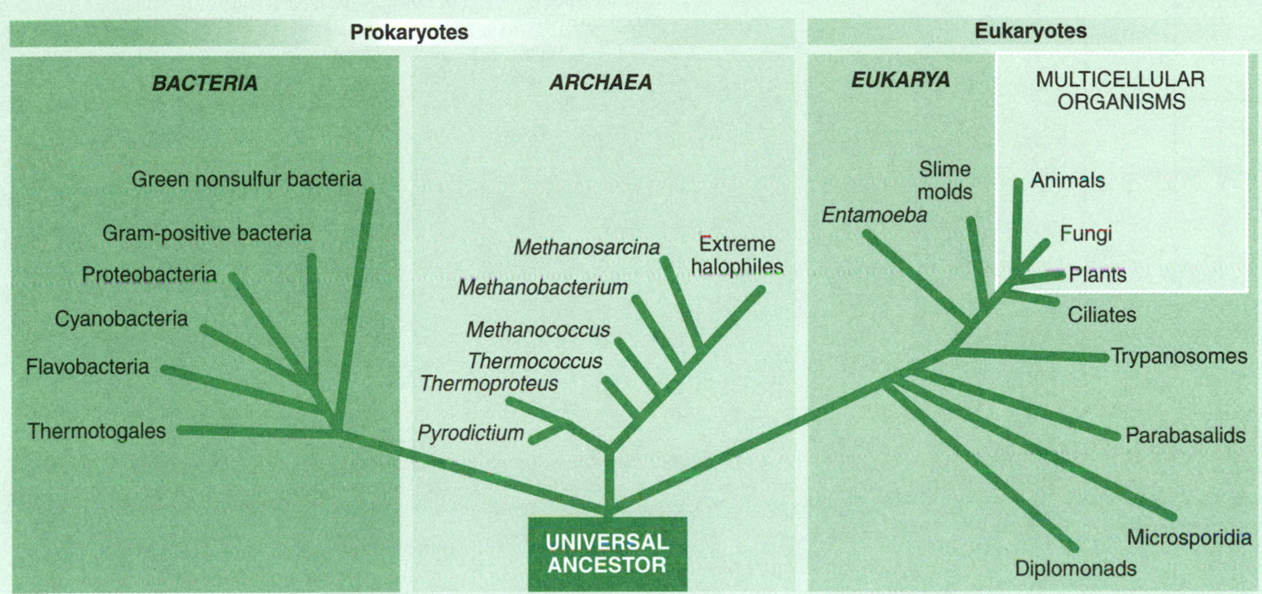

FIGURE 8-1 A simplified version of the three domains of life, Eubacteria, Archaea and Eukarya, showing some representative groups within the Eukarya.

200 atmospheres (Bult et al., 1996). Eubacteria do not contain histone proteins, but *M. jannaschii* has genes coding for histones. Furthermore, 56 percent of the genome of *M. jannaschii* is not found in other organisms, which is sufficient to place such organisms into a separate domain of life, the Archaea. With this and subsequent genetic analyses, **Eubacteria, Archaea** and **Eukarya** were recognized as three domains of life that cut across the traditional five kingdom and two divisions (prokaryotes and eukaryotes) scheme (Wheelis et al., 1992). By 'cut across' we mean that, for example, slime molds and ciliates, which were regarded as prokaryotes, now nest within the Eukarya[b] with plants and animals (**Fig. 8-1**). This scheme effectively divides Prokaryotes (and the kingdom Monera in the five kingdom system) into two groups, and increases them in rank and equivalence to eukaryotes. This action fits well with the current consensus that Archaea are more closely related to Eukaryotes than they are to Eubacteria.

More recent divisions of life into five or six super groups are outlined in Box 12-4.

[b] The first complete genome of a species from the Eukarya domain — the budding yeast, *Saccharomyces cerevisiae* — was sequenced by Clayton and colleagues in 1997.

3. Such communities probably included anaerobes and both autotrophic and heterotrophic microorganisms.

Schopf (1996) points out that aerobic cyanobacteria and nonaerobic bacteria can coexist in stromatolites by using different light-gathering pigments that enable them to occupy different subhabitats:

> The oxygen-producing cyanobacteria live in the uppermost layers of stromatolites, with the non-oxygen-producing photosynthesizers just beneath. Much of the light energy is absorbed by the cyanobacteria . . . but this does not snuff out the anoxygenic photosynthesis of the green sulfur and purple bacteria that live below because these more primitive photosynthetic anaerobes are literally able to see through the cyanobacterial layer — their pigments absorb light unused by the cyanobacteria above.

Using $^{13}C/^{12}C$ ratios we can surmise that many of these ancient cell-like fossils had a biological origin. These two carbon isotopes differ in respect to their participation in cellular metabolism. These differences lead to ratios that are unique compared to those found in nonbiological material. From such analyses, almost all the stromatolite deposits dating from 3.5 Bya and later have carbon isotope ratios similar to rocks from the Carboniferous and other strata in which living forms appear. Such biologically derived isotope carbon ratios have

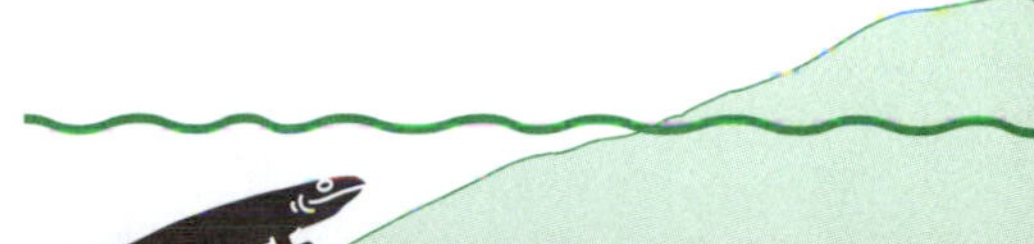

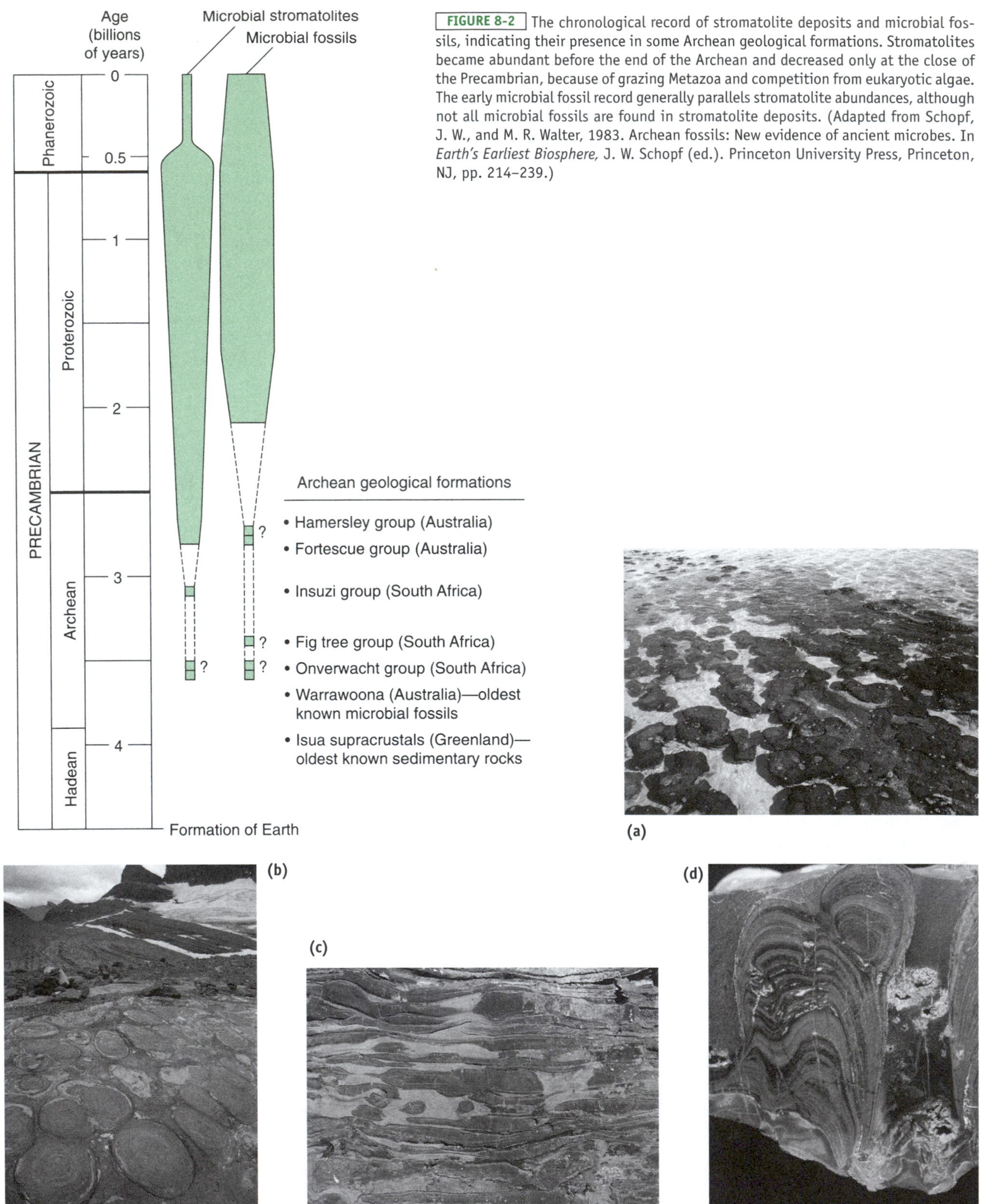

FIGURE 8-2 The chronological record of stromatolite deposits and microbial fossils, indicating their presence in some Archean geological formations. Stromatolites became abundant before the end of the Archean and decreased only at the close of the Precambrian, because of grazing Metazoa and competition from eukaryotic algae. The early microbial fossil record generally parallels stromatolite abundances, although not all microbial fossils are found in stromatolite deposits. (Adapted from Schopf, J. W., and M. R. Walter, 1983. Archean fossils: New evidence of ancient microbes. In *Earth's Earliest Biosphere,* J. W. Schopf (ed.). Princeton University Press, Princeton, NJ, pp. 214–239.)

FIGURE 8-3 Stromatolites, large, cushion-like masses, composed of layers of calcium carbonate secreted by cyanobacteria (blue-green algae), are both the oldest organisms known (up to 3.5 By old) and the longest lasting. Compare the 800-million-old but living stromatolites in Namibia (a) with the 1-to-2-By-old fossil stromatolites from the Helena Formation in Glacier National Park, Montana, USA (b). Cross sections of the Namibian (c) and fossil (d) stromatolites show the internal layers of calcium carbonate. (Photo [a] © Chung Ooi Tan/ShutterStock, Inc.; photo [b]© Marli Miller/Visuals Unlimited; and photos [c and d] © Sinclair Stammers/Photo Researchers, Inc.)

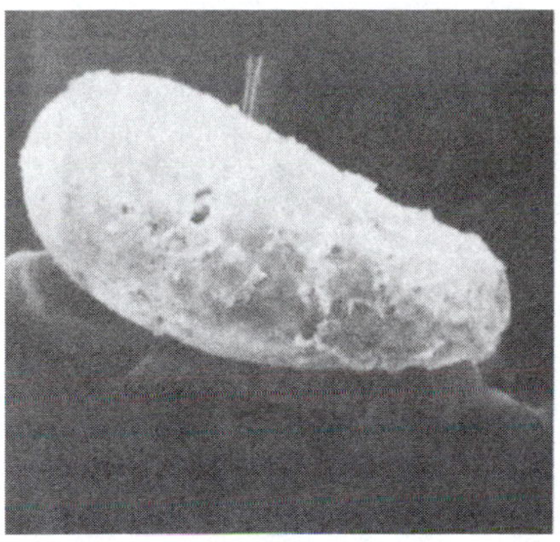
(a) Grand Canyon shales
(800 million years old)

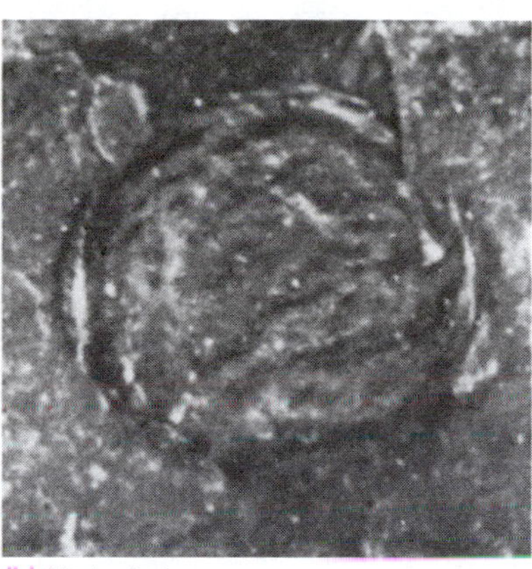
(b) Utah shales
(950 million years old)

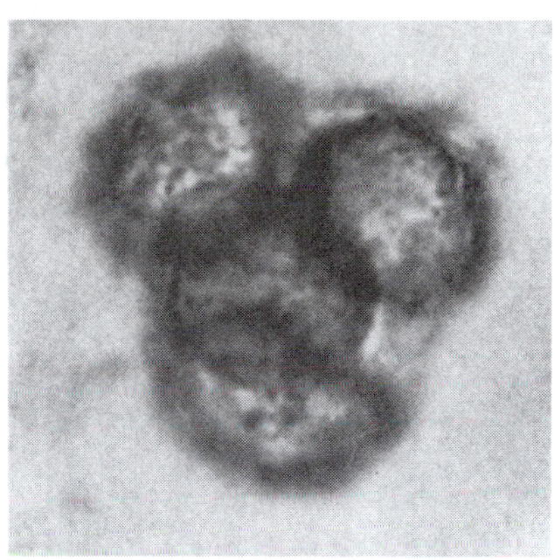
(c) Central Australian sediments
(850 million years old)

FIGURE 8-4 Microfossils of probable eukaryotic cells that date back to the Proterozoic. The cells in (a) and (b) are many times larger than any known prokaryotic cells and are considerably more complex. The group of cells in (c) is in a characteristic tetrahedral arrangement, suggesting they formed through either mitotic or meiotic eukaryotic cell division mechanisms. (From *The Evolution of the Earliest Cells* by J. W. Schopf, *Scientific American* 239, 1978. Reprinted by permission.)

been found in some sedimentary rocks of the Isua group, further supporting the concept that life on Earth existed more than 3.5 Bya (H. D. Holland, 1997).

Cloud (1974) suggested that **banded iron formations**, the oldest of which date to 3.76 Bya, also may have formed biologically. Internal sources of oxygen were responsible for the change from ferrous to ferric ion, and these sources were live protocyanobacteria that were splitting water molecules and releasing O_2:

$$\underset{\text{(ferrous oxide)}}{4\,FeO} + \underset{\text{(from photoautotrophs)}}{O_2} \rightarrow \underset{\text{(ferric oxide)}}{2\,Fe_2O_3}$$

The ferrous ion served as a "sink" for molecular oxygen generated during photosynthesis and would have protected the anaerobic metabolic systems of these early photoautotrophs. The oxidized iron bands may have resulted from episodic supplies of ferrous ion, which enabled growing anaerobic photoautotrophs to precipitate iron in the form of ferric oxide.

What were these early organisms like? We answer this question in the next sections by examining the nature of cells and genes of the two major types of organisms, **prokaryotes and eukaryotes** (remember that these are not monophyletic assemblages). Then we turn to the metabolic pathways found in these early cells.

Prokaryotes and Eukaryotes

Figure 8-5 summarizes some of the major biological and geological evolutionary events from the lowermost Hadean division to the Phanerozoic Eon, which originated with the Cambrian about 545 Mya. Aside from the origin of photosynthesis, perhaps the most significant biological change is the difference in *cellular complexity* that results in the "superkingdom" division between organisms classified as prokaryotes and eukaryotes.

1. *Prokaryotes*, a term used to include all bacteria, and used for one of two kingdoms of life (Table 8-1; see Box 8-1), now recognized as reflecting two major groups:
   - **Eubacteria** encompass the major forms of bacteria as well as the cyanobacteria, practically all possessing unique peptidoglycan or murein cell walls consisting of chains of sugars cross-linked with short peptides, some of which contain D-amino acids.
   - **Archaebacteria** use other materials for their cell walls and often live under more rigorous environmental conditions than eubacteria, such as hot sulfur springs and extreme salt concentrations.[2]
2. The term *eukaryote* embraces many single-celled **protistan** organisms, including photosynthetic and nonphotosynthetic algae; photosynthetic and nonphotosynthetic protozoans (some algae and protists can use both modes of metabolism) as well as multicellular plants (**Metaphyta**) animals (**Metazoa**) and fungi.[3] **Multicellularity** is a successful strategy that evolved many times (**Box 8-2**).

The most obvious difference between prokaryotes and eukaryotes is the absence of any internal membranous network such as a nuclear membrane or cytoskeleton in the generally smaller prokaryotic cells (0.5–10 μm) and its presence in generally larger eukaryote cells (10–100 μm). In addition, prokaryotes reproduce by **binary fission**, which does not involve the mitotic mechanisms seen in nearly all

[2] Some 10 percent of the unicells in the photic zone of the open ocean are archaebacteria, but because they are difficult to culture they are little understood (Alastair Simpson, personal communication). The third group of unicells, the Archaea, is discussed in Box 8-1.

[3] The term multicellular can be tricky. Fungi are no more multicellular than filamentous algae; some brown algae are multicellular; and many so-called fungi are actually yeasts.

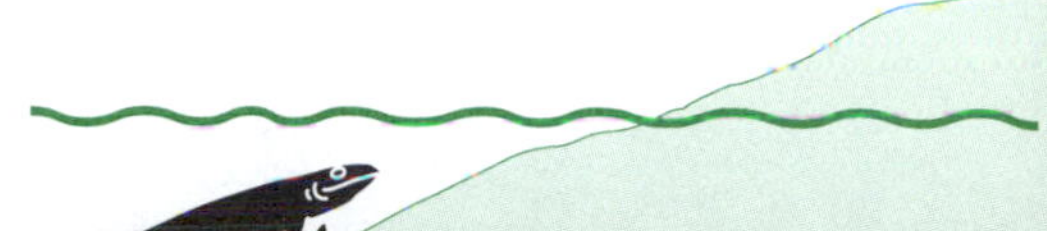

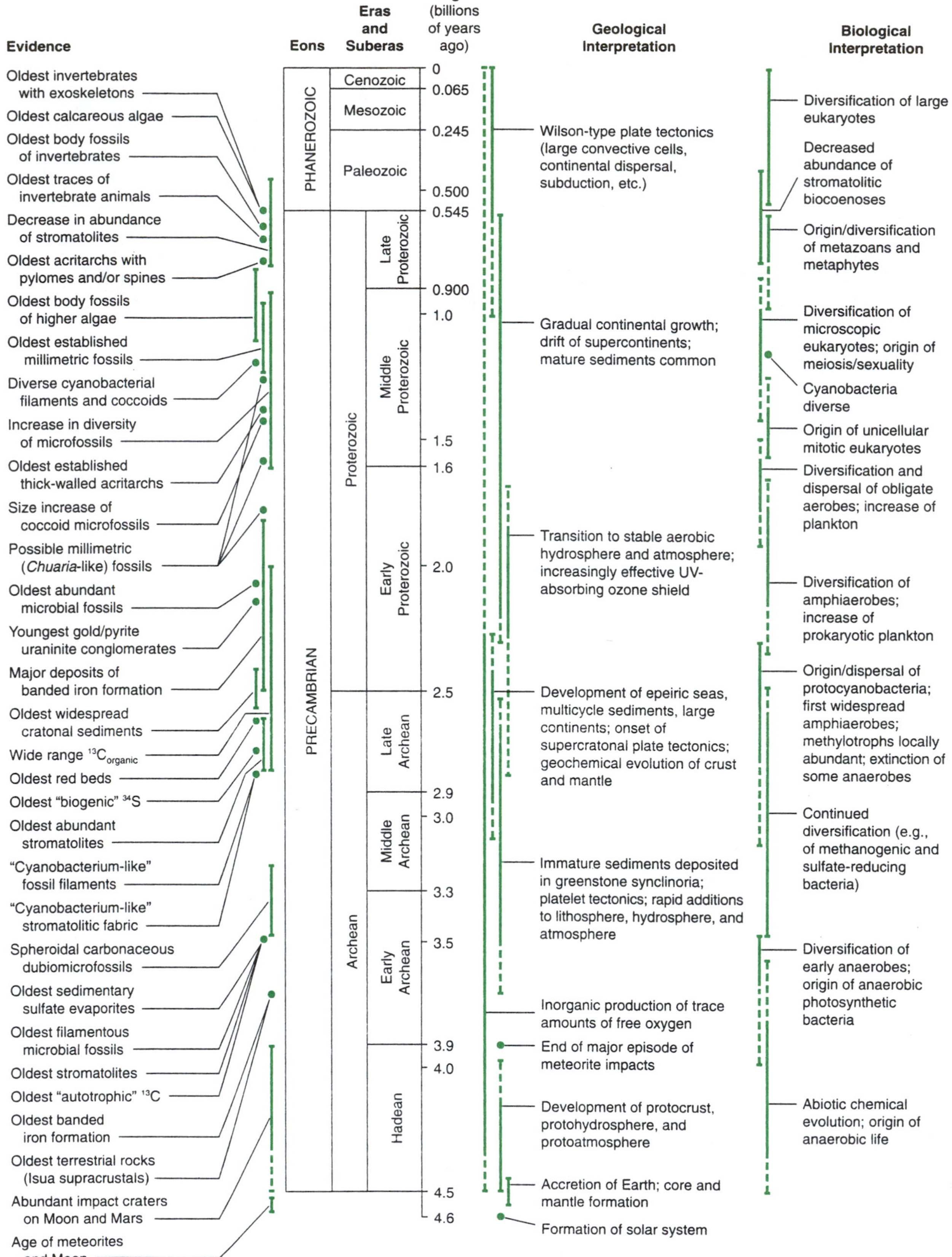

FIGURE 8-5 A summary of the evidence for Precambrian evolution along with the geological and biological interpretations of these observations. A series of major meteorite impacts occurred between 4.1 and 3.9 Bya as indicated in the moon rocks retrieved during the Apollo mission. Geologists believe these collisions produced enough heat to sterilize Earth's surface. As a result, some proponents believe that the origin of life most probably occurred somewhat later, perhaps 3.8 Bya, and according to S. L. Miller (1992), during an interval that may have been as short as 10 My or less. (Modified from J. William Schopf, ed., *Earth's Earliest Biosphere: Its Origin and Evolution*. © 1983 Princeton University Press.)

BOX 8-2 Multicellularity and Pluricellularity

INCREASE IN ORGANISMAL SIZE can have and usually does have a considerable effect on relative fitness. One way to achieve increase in size is to become **multicellular,** a term that we use to refer to individual organisms consisting of more than a single cell. Whether we examine genomic and cellular organization or phylogenetic relationships, we find that multicellularity has evolved numerous times.[a] Plants (metaphytes), animals (metazoans) and brown algae (Phaeophyta) are multicellular.

Conditions that resemble multicellularity, and which we term **pluricellularity,**[b] evolved independently in several lineages of unicellular organisms. Pluricellularity is used to embrace such multi-individual types of organization as *colony formation, filament formation* and *aggregation.* Why distinguish between multi- and pluricellularity? Because the former produces a single, many-celled organism, and the latter is the coming together of single-celled individuals, as in colony formation. Growth regulation, cell-cell recognition systems and modes of cooperation differ fundamentally in the two conditions. Different forms of pluricellularity evolved within bacteria; some aggregate, some are colonial, while others are filamentous. So, we distinguish multi- from pluricellularity.

Multicellularity evolved (independently) in several lineages of eukaryotes; for example, red and brown seaweeds have evolved multicellularity independently of true plants. Pluricellularity evolved on six occasions in the Volvocales (*Volvox*) and independently in other flagellated green algae, many of which are colonial or exist as multicellular individuals with complex life cycles. The smaller multicellular volvocales are comprised of between four and 64 cells embedded in a common extracellular matrix. The largest, such as *Volvox,* consist of thousands of cells. Evolution of pluri- or multicellularity if favored in lineages with a rigid cell wall within which multiple divisions can occur (Bell and Koufopanou, 1991; Michod, 1997).

The advantages of multicellularity over the unicellular condition are many and include:

1. Potential for increase in size beyond the limits set by the surface-to-volume of single cells;
2. Specialization into distinct cell types, each (or each compartment within the organism) with its own function, such as food gathering, reproduction or protection from predators; and
3. Enhanced dispersal, especially of immature stages.

Once distinct cell lineages arose, selection would act on those cell lineages, which are, in reality, the phenotype of each organism. A key innovation of multicellularity was the ability to regulate where and when transcription occurs within multicellular embryos or organisms. Within distinctive cell lineages genes became linked into networks or cascades, a different cascade for each cell lineage, furthering diversification. Multicellularity facilitated increasing complexity (see Box 8-3). Ravasz et al. (2002) provide support for tightly organized and coherent metabolic networks organized hierarchically across 43 organisms at very different levels of organization — a systems-level feature whose basic elements would have responded strongly to selection associated with the evolution of multicellularity.

The origin of multicellularity would have required single cells that reproduced by fission to separate into individual unicellular organisms, to develop mechanisms either to:

- **prevent the two cells from separating,** producing an organism whose cells would have had the *same genetic constitution,* or
- **facilitate aggregation with a cell(s) from another individual,** potentially producing an organism whose cells would have *different genetic constitutions.*

Some multicellular organisms such as *Volvox* arise by cells staying together. Others, such as myxobacteria (slime bacteria), myxomycetes (slime molds), dictyostelid amoebae, and some ciliated protozoa, arise by aggregation.

Although aggregation and failure to separate are different cellular mechanisms, they could have similar molecular bases. Aggregation and failure to separate reflect properties of cell membranes. Aggregation requires the evolution of cell adhesion molecules (CAMs) or ionic coupling. Failure to separate requires the development of a mechanism to keep coupled cells together, again via (CAMs) or ionic coupling. (CAMs) and substrate adhesion molecules (SAMs) are likely candidates for these roles.

[a] J. T. Bonner discusses the evolutionary importance of size in his latest (2006) book, *Why Size Matters: From Bacteria to Blue Whales.*

[b] Sina Adl (Dalhousie University) drew the fundamental differences between multi- and pluricellularity to our attention.

eukaryotes. Prokaryotes consequently lack organelles such as mitotic spindles and centrioles and do not have the histone proteins that structurally organize the relatively larger and more numerous eukaryotic chromosomes. Such comparisons and discussion of differences in cellular complexity between pro- and eukaryotes raise two obvious and absolutely central issues and questions: Can we measure or assess complexity, and has complexity increased during evolution? Both questions are addressed in **Box 8-3** and **Figure 8-6**.

Superkingdom and kingdom divisions are in flux. The traditional **prokaryote-eukaryote division of life** was subdivided by Woese (1998b) into a **tripartite classification of archaebacteria, eubacteria,** and **eukaryotes** (see Box 8-1 and **Fig. 8-7**). In fact, the numbers of kingdoms proposed ranges from two to 13, reflecting many shared features, at least some of which are the result of lateral gene transfer and/or capture of single-celled organisms and their transformation into organelles.

Organelles

Prokaryotes lack the membrane structures and **organelles** such as endoplasmic reticulum (usually associated with

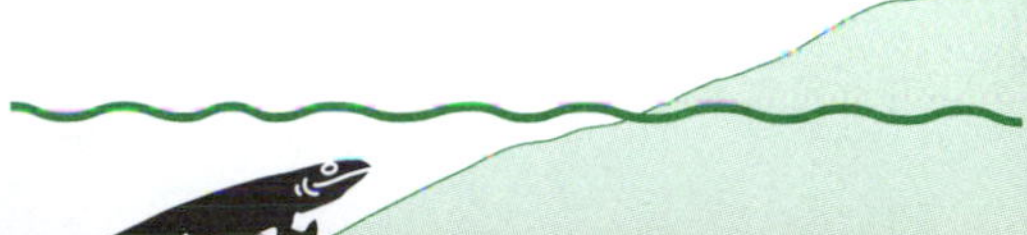

BOX 8-3 Complexity

THE EVOLUTION OF MULTICELLULAR from unicellular organisms raises the difficult issue of **complexity**. Are some organisms more complex than others? It seems self-evident that multicellular organisms (animals, plants, fungi) are more complex than unicellular organisms such as protists. But are some multicellular organisms more complex than others? Are we more complex than fungi? Are flies more complex than worms?

Evolutionary biologists shy away from the concept that evolution results in increasing complexity almost as much as they shy away from the concept of progress (see Chapter 1). To quote Szathmáry and Maynard Smith (1995):

> There is no theoretical reason to expect evolutionary lineages to increase in complexity with time, and no empirical evidence that they do so. Nevertheless, eukaryotic cells are more complex than prokaryotic cells, animals and plants are more complex than protists, and so on.

As the explanation, Szathmáry and Maynard Smith propose that "this increase in complexity may have been achieved as a result of a series of major evolutionary transitions. These involved changes in the way information is stored and transmitted."

Nevertheless, no agreed-upon criteria exist for: (1) defining the "information" used to measure complexity, or (2) identifying increasing complexity. Nor do all agree that complexity *has* increased throughout the evolution of life; the issue is as much philosophical as it is scientific. Criteria used to measure complexity include:

- genome size or the total number of genes in an organism;
- the number of genes that encode proteins;
- the number of *parts* or units in an organism (where parts might be segments, organs, tissues, and so forth);
- the number of cell types possessed by an organism;
- increased compartmentalization, specialization or subdivision of function between organisms being compared;
- the number of genes, gene networks or cell-to-cell interactions required to form the parts of an organism;
- the number of interactions between the parts of an organism (reflective of functional complexity or a high degree of integration; and
- the length and complexity of the minimal statement required to describe the organism, an approach that effectively combines the other eight.[a]

So which of these attributes of organisms allow us to answer "yes" to these questions, yes being the answer most biologists and laypersons would give if asked whether some organisms are more complex than others. Several of the criteria listed above, notably genome size, gene number, gene networks, and compartmentalization, are discussed in other chapters. Perhaps the most commonly used criterion, the number of cell types, is discussed below, as is a criterion, increase in organismal size, not on the list.

Increase in Cell Number

Increase in complexity is perhaps easiest to see during the embryonic development of an animal, when the individual progresses from a single-celled zygote to a multicellular organism with as many as several hundred different cell types and as many as the 100 trillion individual cells (10^{14}) found in an adult human. Perhaps as a consequence of parallels between development and evolution (see Chapters 3 and 13), the most commonly used metric of complexity is the number of different types of cells possessed by an organism, which as shown in Figure 8-6 and discussed in this box, has increased over the course of animal evolution. Estimates of cell numbers range from 6 to 12 in sponges and cnidarians, 20 to 30 in flatworms, 50 to 55 in mollusks, arthropods, annelids and echinoderms to as high as 200 to 400 in humans. James Valentine and his colleagues (1994) plotted the time of origin of metazoan body plans against number of cell types and found that the upper limits of complexity increased from the earliest metazoans, such as sponges, to the vertebrates, and that early rates of increase were

[a] See Cavalier-Smith (1985) and Greilhuber et al. (2005) for genome size; Maynard Smith and Szathmáry (1995) for numbers of genes; McShea (2005) and Szathmáry and Maynard Smith (1995) for numbers of parts; Bonner (1988, 2006), Valentine et al. (1994), Bell and Mooers (1997) and Vickaryous and Hall (2006a) for number of cell types; and McShea (2005), Larsen (1992) and Hall (1999a) for networks and interactions.

ribosomes in protein synthesis), Golgi apparatus (secretory bodies) and mitochondria (used in oxidative phosphorylation) found in eukaryotes. Such membrane-enclosed compartments allow eukaryotic cells to isolate enzymes for specific reaction sequences, confining transcription to the nucleus, translation to the cytoplasm, aerobic metabolism to mitochondria, and so forth.

Eukaryotes share mitotic and protein-synthesizing mechanisms and are almost all aerobic. Exceptions include some forms such as microsporidians and yeast that can function anaerobically (amphiaerobes) and some protists such as *Giardia*, *Trichomonas* and *Pelomyxa*, which have anaerobic energy metabolism systems but are probably descended from aerobic ancestors. Even yeasts can adapt to aerobic conditions, as demonstrated by two changes in the *FLO11* gene of the yeast *Saccharomyces* that increase cell surface hydrophobicity in those individuals that float on the surface as a buoyant film, providing an adaptation to access to oxygen at the surface. Individuals without the mutation are restricted to the oxygen-poor medium, demonstrating a mutational basis for adaptation to aerobic versus anaerobic conditions. The bases for the changes are a 111 nucleotide deletion in a repression region that increases gene expression, and a rearrangement in the central tandem repeat domain of the coding region that yields a more hydrophobic gene product (Fidalgo et al., 2006).

high in comparison to changes within the vertebrates (see Fig. 8-6). Comparisons also have been made across kingdoms.[b]

It is difficult to escape the conclusion that, as in development, complexity has increased during at least part of the evolutionary history of life on Earth, although some like McShea (2005) view increasing complexity as best evidenced by the increasing hierarchical developmental complexity associated with (perhaps causally) the origin of body plans in the early Phanerozoic.

Increase in Organismal Size

Embryonic development (and so increasing complexity) is accompanied by increasing size. Early stages of the evolution of life also were accompanied by an increase in organismal size. So, why has the increase in organismal size over evolutionary time, which clearly occurred and is known as **Cope's rule,**[c] not been used as a criterion of complexity? Because evolution is often for decrease in size, as seen in parasites in many different groups of animals, or in organisms that live between the sand grains at high tide levels on beaches. Different groups of organisms of the same size show different levels of complexity, when complexity is measured by criteria other than organism size. In future chapters, we will discuss examples where lineages decreased in size during their evolution (dinosaurs in Chapter 18), and further examples where small size enabled survival and so facilitated evolution (Mesozoic mammals; see Chapter 19).[d]

[b] See Bonner (1988) and Vickaryous and Hall (2006a) for estimates of numbers of cell types, and see Bell and Mooers (1997) for comparisons across kingdoms.

[c] See Hall (1999a) for Cope's rule. Using a combinatorial approach to Cope's rule, Moen (2006) compared body size evolution across six clades of extant turtles *and* through the turtle phylogenetic tree, without finding any evidence for Cope's rule. Indeed, Moen argues that no analysis of extant taxa has provided evidence for Cope's rule.

[d] See Bell and Mooers (1997) and Bonner (2006) for conflicting measures of complexity.

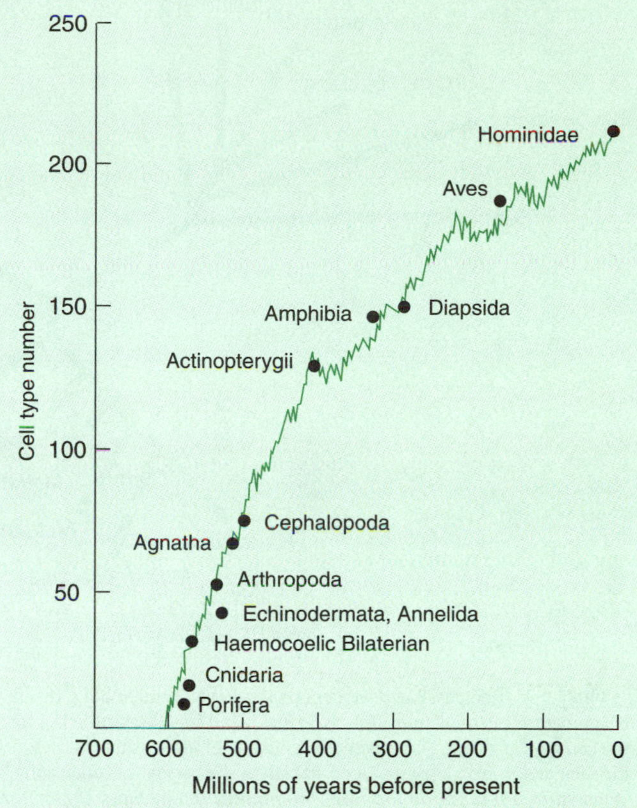

FIGURE 8-6 Estimated time of origin of various metazoans with their estimated somatic cell-type numbers. The marked increase from low to high is considered by many evolutionary biologists to reflect an increase in complexity. (Valentine, J. W., A. G. Collins, and C. P. Meyer, 1994. Morphological complexity increases in metazoans. *Paleobiology,* **20,** 131–142. Reprinted by permission.)

Gene Structure

Among further attributes of eukaryotes almost entirely absent from prokaryotes are **split genes** in which amino-acid–coding nucleotide sequences, separated by hundreds of bases in DNA, are combined in the final messenger RNA to translate into a single polypeptide product. The nucleotide sequences that code for amino acids in such polypeptides are called **exons** (*expressed sequences*). The intermediate noncoding nucleotide sequences are called **introns** (*intervening sequences*). These split genes transcribe their nucleotide sequences from DNA to RNA, but the RNA is processed so that the introns are removed and the exons spliced together (**Fig. 8-8**). Such processed mature mRNA molecules are transported through pores of the nuclear envelope to the cytoplasm where they are translated into polypeptides.

Split genes occur in almost all vertebrate protein-coding genes examined thus far, and in many similar genes in other eukaryotes. Originally each exon may have coded for a single polypeptide domain with a specific function. Because exon arrangement and intron removal are flexible, the exons coding for these polypeptide subunits act as *modules* (called *domains*), combining in various ways to form new genes. Single genes can produce different functional proteins by arranging their exons in several different ways through **alternate splicing** patterns, a mechanism now

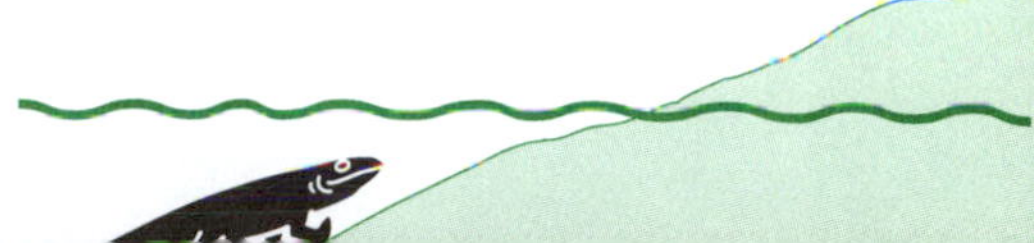

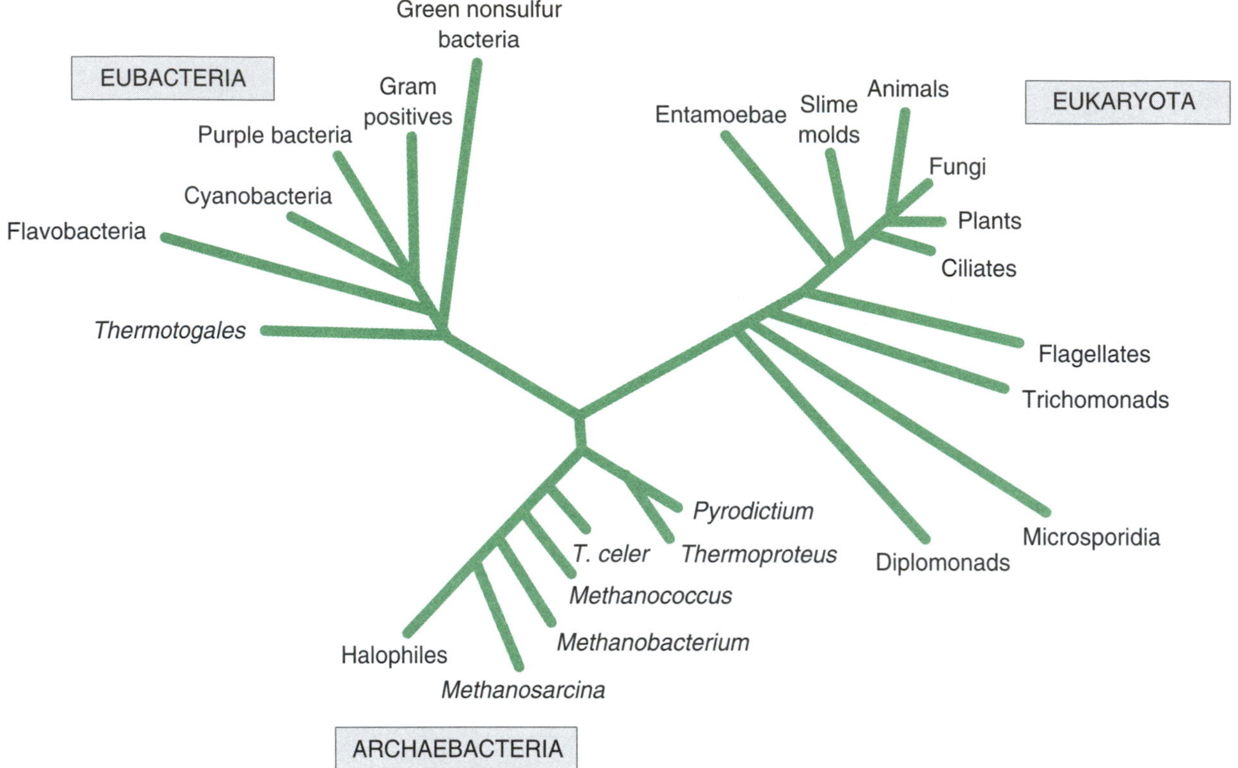

FIGURE 8-7 One possible phylogenetic tree for all cellular life ("the universal tree") based on nucleotide sequence similarities among some ribosomal components (small-subunit RNA, 16S rRNA). In this illustration, the tree is "unrooted," meaning that no source ("root") is indicated for the origin of the common ancestor. The tree shows a number of interesting features including the close tie between animals and fungi and the widely separate origins for some eukaryotic lineages (e.g., flagellates, entamoebae, diplomonads, ciliates, microsporidia) that have traditionally been classified in the single kingdom Protista. The diplomonads are represented by *Giardia,* a protozoan whose position as the earliest eukaryotic offshoot is supported by both nucleotide and amino acid sequence analyses but not by other analyses. Although *Giardia* was thought to lack mitochondria, mitochondrial organelles have now been found in *Giardia* and in *Entamoeba*. *Giardia* and *Entamoeba histolytica* possess nuclear genes that must have originated by nuclear transfer from a prokaryotic symbiont, which was largely lost from the cytoplasm in a process called *cryptic endosymbiosis* (Henze et al., 1995; Roger et al., 1998). (Adapted from Sogin, M. L., 1991. Early evolution and the origin of eukaryotes. *Current Opinion Genet. Devel.*, **1**, 457–463 and from Wheelis, M. L., O. Kandler, and C. R. Woese, 1992. On the nature of global classification. *Proc. Natl Acad. Sci. USA*, **89**, 2930–2934.)

known to play a major role in generating both protein and functional diversity in metazoans.[4]

Introns Early or Introns Late

W. F. Doolittle (1978) and others support an "*introns early*" hypothesis: cellular organisms ancestral to unicells (prokaryotes) and eukaryotes probably had intron–exon structures that were mostly abandoned in prokaryotic lines because selection streamlined DNA replication and improved transcription efficiency.

An "*introns late*" hypothesis suggests that introns first inserted into full-length eukaryotic genes *after* eukaryotes separated; that is, gene assembly did not depend on the initial use of introns to connect and shuffle modular exons. Among the evidence are tests showing lack of correspondence between exons and functional protein units, the demonstration that some introns are of recent origin in eukaryotes (Logsdon et al., 1998). A decision between the two hypotheses is yet to come. Indeed, both may be correct, for it has been proposed that as many as 40 percent of introns were "early" and the remainder "late" (De Souza et al., 1998).

How did late introns enter into eukaryotic genes? Why did they become so prevalent? Why are they maintained? Answers vary, but introns are mobile DNA sequences that can splice themselves out of, as well as into, specific "target sites," acting like mobile transposon-like elements (see Chapter 10). Because the insertion of such elements would add excess nucleotides to messenger RNA and disrupt normal gene expression, their survival would rely on splicing themselves out of RNA transcripts before translation; that is, by acting as introns. Other suggestions for intron function and persistence include a possible regulatory role; an

[4] Alternate splicing includes the formation of alternative 5′ or 3′ sites, retention of introns, mutually exclusive alternative exons, alternative promoters, or poly A sites establishing a gene regulatory code that lies between the transcriptional network and posttranscriptional and translational regulation (Blencowe, 2006).

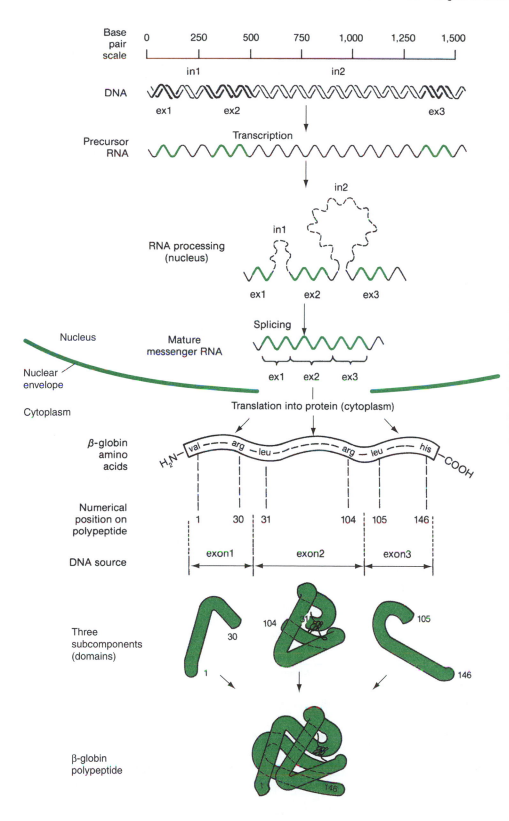

FIGURE 8-8 The intron-exon structure for nucleic acid sequences involved in the production of the human β-globin polypeptide chain, 146 amino acids long, one of the components of normal adult hemoglobin. The top portion of the diagram indicates the approximate 1,500 nucleotide-base-pairs in the β-globin DNA sequence and the structure of this sequence in terms of two introns (in1, in2) and three exons (ex1, ex2, ex3). Intron 1 is 130 nucleotides long and separates codons that will later translate into amino acids numbered 30 and 31 on the β-globin chain. Intron 2 is 850 nucleotides long and separates codons for β-globin amino acids numbered 104 and 105. Although the entire DNA sequence transcribes into mRNA, introns are precisely removed, by special RNA processing in the nucleus, and the exon sequences splice together and translate into a continuous β-globin amino acid sequence. At the bottom of this illustration, a single β-globin polypeptide chain has been diagrammatically split to indicate the three subcomponents (domains) respectively produced by the three exon nucleotide sequences. (Adapted from Strickberger, M. W., 1985. *Genetics,* 3rd ed. Macmillan, New York.)

increase in the number of genes from a few thousand or so in prokaryotes to tens of thousands in many eukaryotes may have depended on adding introns to the gene regulatory system.

In prokaryotes, protein products of other genes regulate gene expression (see, e.g., Fig. 10-13). Some propose that nucleotide sequences in eukaryotic introns also serve as regulatory agents. Because prokaryotes do not have nuclei,

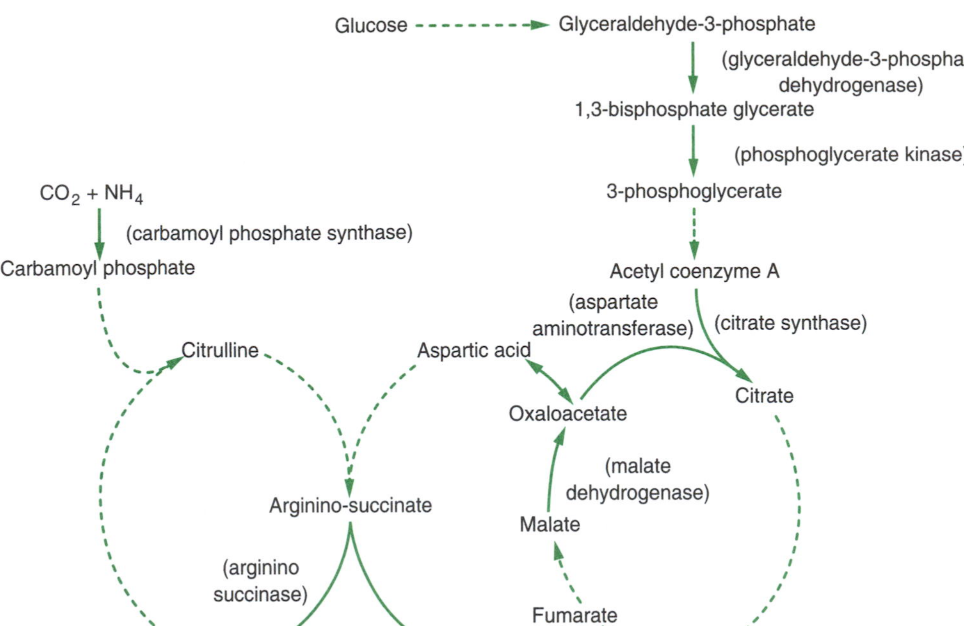

FIGURE 8-9 Some of the metabolic pathways presumed to exist in the most recent common ancestor to archaebacteria, eubacteria and eukaryotes, based on nucleotide and amino acid comparisons between enzymes (*in parentheses*) common to these groups. *Solid lines* indicate metabolic steps that were catalyzed by such reconstructed ancestral enzymes. *Broken lines* indicate steps presumed to have been present but whose enzymes have not yet been reconstructed. (Adapted from Benner, S. A., *et al,* 1993. Reading the palimpsest: Contemporary biochemical data and the RNA world. In *The RNA World*, Gesteland, R. F. and J. F. Atkins (eds.). Cold Spring Harbor Laboratory Press, Cold Spring Harbor, New York, pp. 27–70.)

and ribosome attachment with consequent protein synthesis occurs simultaneously with the transcription of messenger RNA, introns — should they be common in prokaryotes — would translate as part of messenger RNA, resulting in nonfunctional proteins. In eukaryotes, however, RNA processing is sequestered in the nucleus. Intronless messenger RNA translates in the cytoplasm, allowing the nuclear introns to function in a regulatory role without affecting messenger RNA translation. As discussed by Mattick (1994), this proposed intron-regulatory advantage may have enabled the evolution of more complex gene systems in eukaryotes and their advance to multicellularity (see Box 8-2).

Large numbers of introns found in some genes cast doubt on the presumed regulatory function. For example, the gene for dystrophin protein, whose absence or malfunction causes muscular dystrophy, has more than 65 introns totaling more than two million base pairs, whereas its exons total only 14,000 base pairs. Most of these introns seem excessive if their role is confined to regulating gene expression. *Transposable element* duplication has been used to explain such high intron numbers. As discussed in Chapter 10, because they replicate independently and more frequently than their host chromosomes, transposons can accumulate in large quantities as "junk DNA." Consistent with the introns-late hypothesis, frequent mobile transposon activity rather than regulation may account for the large number of introns in certain genes.

The Ancestral Organism

Progenote is a term for the hypothetical ancestral cellular form that gave rise to archaebacteria, eubacteria and eukaryotes. Woese and coworkers (1990) catalogued the presence and frequencies of various sequences in the 16S rRNA component of ribosomes in the three groups and showed that they have distinctive differences (see **Table 8-2** and **Figure 8-7**).[5]

Although accumulating molecular and other data seemed to show that archaebacteria occupied an intermediate evolutionary position between eubacteria and eukaryotes (see Figures 12-18 and 12-20 and Box 12-4), these data are not sufficient to determine the exact source of their common ancestor. Perhaps no one group derived from any other but all diverged from a common cellular ancestor (or from a community of early cells in which mutation rates were high and genes were being freely exchanged: a population of progenotes (Woese, 1998a). In support, all three of these groups have common basic attributes such as a similar mode of DNA replication (new nucleotides add to the 3′ end of the molecule), a common genetic code, a similar protein-synthesizing system, many similar metabolic pathways, a similar cell membrane structure consisting of a phospholipid bilayer and a similar mechanism of molecular transfer across membranes (**active transport**). **Figure 8-9** shows that this ancestral organism probably had enzymes involved in glycolysis, and had both the Krebs and urea cycles. The enzymatic fidelity needed by such processes indicates that DNA with its much lower mutation rate must have already replaced RNA as the cellular genetic material.

However, subsequent events remain unclear. Because of

[5] 16*S* rRNA molecules are essential for ribosomal protein synthesis and, like 5*S* rRNA, portions of their sequences are conserved across a wide range of organisms. The 16*S* rRNA molecule or prokaryotes is homologous to the 18*S* rRNA molecule of eukaryotes, both now being referred to as SSUrRNA (small subunit ribosomal RNA).

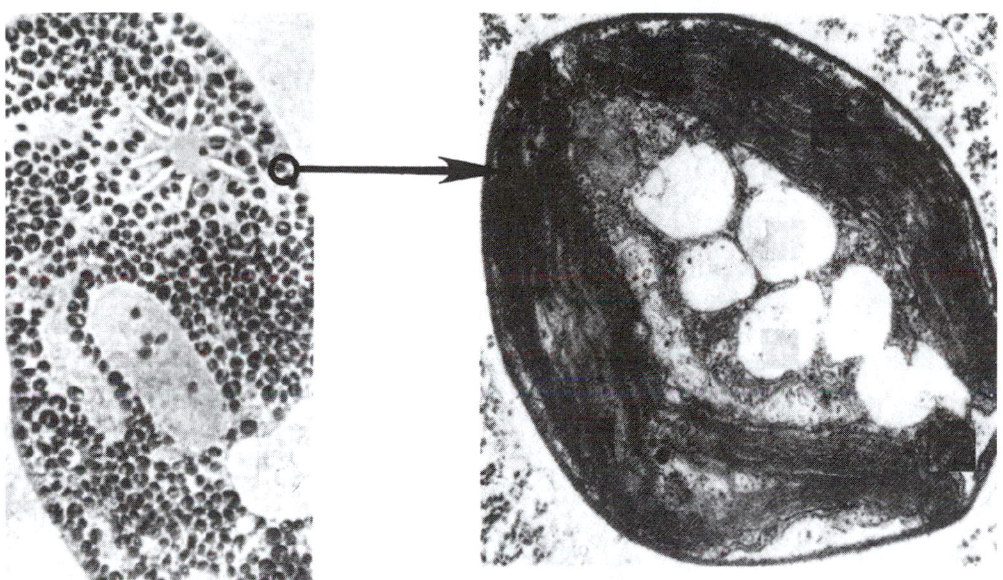

FIGURE 8-10 Symbiotic relationships between a eukaryote and its photosynthetic organelles. The protozoan ciliate *Paramecium bursaria* (left) harbors hundreds of symbiotic algae (right) that may be released from the cell and cultured independently. (From *Symbiosis in Cell Evolution*, by Lynn Margulis, 1993. New York: W. H. Freeman. Reprinted by permission.)

TABLE 8-2 Presence and frequencies of some oligonucleotide sequences found in the 16*S* ribosomal RNA component of eubacteria, archaebacteria and eukaryotes

	% Occurrence		
Sequence[a]	Eubacteria	Archaebacteria	Eukaryotes
CYUAAYACAUG	83	0	0
AYUAAG	1	62	100
ACUCCUACG	97	0	0
CCCUACG	0	97	0
ACNUCYANG	0	0	100
YYUAAAG	3	97	0
AUACCCYG	93	3	0
CAACCYUYR	91	0	0
CCCCG	0	100	0
UCCCUG	0	97	100
AUCACCUC	91	100	0

[a] R designates either purine (adenine or guanine), Y designates either pyrimidine (uracil or cytosine), N designates any of the four bases.

Source: Abridged from Woese, C. R. 1985. Why study evolutionary relationships among bacteria? In *Evolution of Prokaryotes*, Schleifer, K. H. and E. Stackebrandt (eds.). Academic Press, London, United Kingdom.

their ancient origins and the uncertainties caused by **horizontal gene transfer** — by which DNA from one organism incorporates into the DNA of another organism — many relationships are being questioned.[6] Connections among these three major lineages are often depicted as an "unrooted" tree (see Fig. 8-7) without designating a particular domain as the ancestral source. Although we do not know the "root," and are uncertain about some branches (Pennisi, 1998), we can discern general relationships; for example, some commonalities, particularly in organization of transcription and translation point to a closer connection between eukaryotes and archaebacteria than between either of them and eubacteria (Rowlands et al., 1994):

Eubacteria   Archaebacteria   Eukaryotes

Evolution of Eukaryotic Organelles

Information has been accumulating with respect to the origin of the mitochondria and chloroplast organelles found within eukaryotic cells. The most popular hypothesis, which first Max Taylor, then Stanier, Margulis and others offer, proposes that eukaryotic cells evolved by physically incorporating prokaryotic organisms into their cytoplasm. According to this theory, the ancestral anaerobic eukaryotic cell achieved the ability to perform **endocytosis,** to ingest supramolecular particles because of new surface membrane properties that included loss of the rigid cell wall. An internal cytoskeleton, composed in part of actin filaments and microtubules, helped maintain cell shape, and, in conjunction with molecular motors, allowed the cell to locomote and to manipulate its shape. This latter ability was probably essential for the development of phagocytosis, which is the endocytosis of "large" objects, such as prokaryotic cells.

[6] See the various studies by Woese for an archaebacterial origin, Poole *et al.* (1998) for prokaryotes from eukaryotes and Cavalier-Smith (1993, 2004) for eukaryotes from prokaryotes.

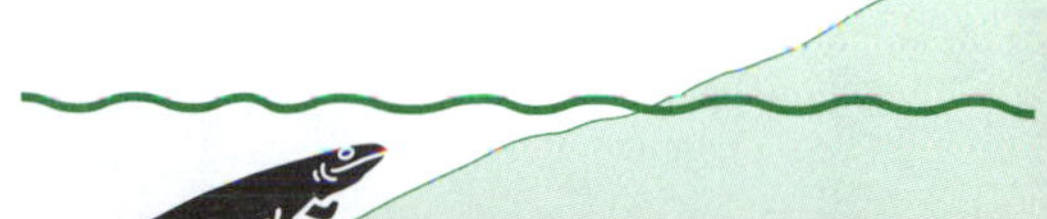

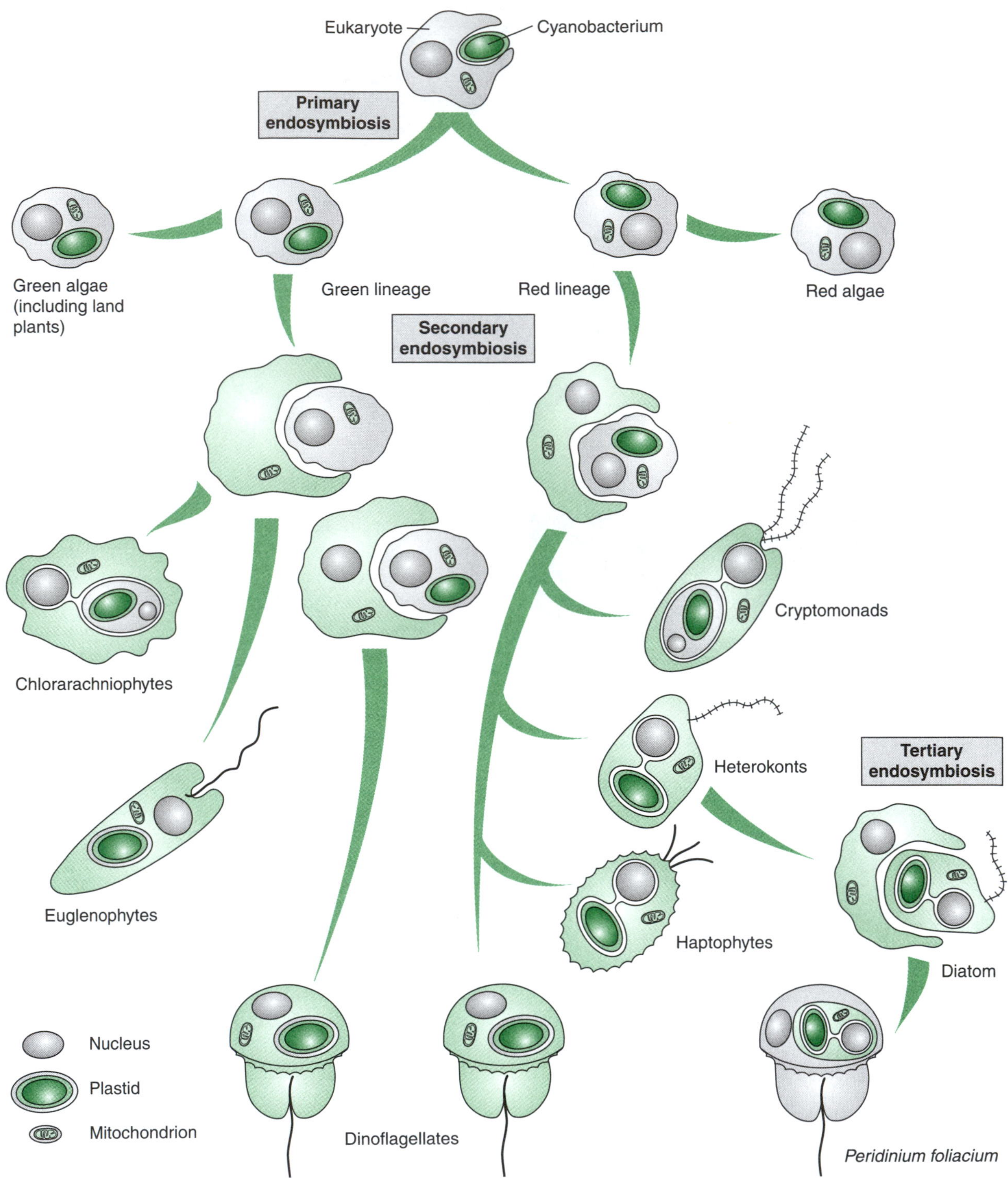

FIGURE 8-11 Primary, secondary and tertiary endosymbiosis and the origin of organelles in various lineages of eukaryotes including algae, land plants, dinoflagellates and cryptomonada, all of which are chimeras. As discussed in the text, horizontal gene transfer between lineages complicates reconstruction of a phylogenetic tree of organelle evolution. (Modified from Cracraft, J. and M. J. Donoghue (eds), 2004. *Assembling the Tree of Life*. Oxford University Press, Oxford, United Kingdom.)

Provided with a new flexible surface, some members of this eukaryotic lineage became active predators. Selection led to increased predatory abilities, such as an increase in cell size, as well as to other innovations that affect movement, capture of prey and digestion. Among the various prokaryotes on which this microbe fed were bacteria-like aerobes capable of **oxidative phosphorylation** and cyanobacteria capable of photosynthesis. At various times, such ingested photosynthetic and aerobic prokaryotes assumed mutually advantageous **symbiotic** relationships with their hosts, similar to those that even now frequently occur (**Fig. 8-10**).

According to one scheme, an early eukaryote lineage first established a symbiotic relationship with mitochondrion-like aerobic bacteria that enhanced eukaryotic metabolism and broadened eukaryotic predatory activity. Later, one or more lines of these new aerobic eukaryotes began similar symbiotic relationships with photosynthetic cyanobacteria, eventually evolving into the various eukaryotic algae and plants (**Fig. 8-11**), although the discovery that eukaryotic algae were engulfed by other eukaryotes (**secondary symbiosis**) complicates this interpretation to a considerable degree. To account for their evolutionary success, both kinds of symbiosis must have conferred selective advantages on their eukaryotic hosts, increasing their frequency compared to nonsymbiotic competitors.

Supporting the symbiotic origin of mitochondria and chloroplasts is the finding that both organelles have their own genetic material (DNA). Persistence of organelle DNA within host cytoplasm can lead to undesirable intracellular competition among different mutant organelle genomes, causing replacement of functional organelles by those less functional but more successfully reproductive. One of several ways this problem was resolved was by restricting organelle diversity through *uniparental inheritance;* that is, by transmitting mitochondria and chloroplasts through only a single parent, commonly the female.

Supporting this view is that the ribosomes of mitochondria and chloroplasts are more like those of prokaryotes than host organellar ribosomes in size, biochemistry, sensitivity to antibiotics and nucleotide sequences of ribosomal RNA components. Such similarities are obvious in the mitochondrion of a freshwater protozoan, *Reclinomonas,* which bears the largest collection of mitochondrial genes so far discovered, and is therefore interpreted by some as more basal than other such organelles that lost genes as they evolved in concert with more derived eukaryotes. The mitochondrial genome of *Reclinomonas americana* strain ATCC 50394 is 69,034 base pairs and codes for 97 genes. Most mitochondria are only a quarter this size and code for 18 to 30 genes. According to Andersson and coworkers (1998), many genes remaining in *Reclinomonas* mitochondria are strikingly similar to the unicell causing epidemic typhus, *Rickettsia prowazekii,* an obligate intracellular parasite. Apparently, an aerobic rickettsial ancestor, adapted to parasitizing eukaryotic cells, gave rise to mitochondria.

Evidence for the origins of mitochondria and chloroplasts from prokaryotic endosymbionts (*primary symbiosis*) is now strong (see Fig. 8-11). Evidence also now indicates that some eukaryotic algae probably acquired chloroplasts through secondary endosymbiosis from a eukaryote rather than from a prokaryote. Such secondary invasions by chloroplast-carrying eukaryotes may account for the retention of an algal-type cell within some protozoan parasites.[7]

Once incorporated in the eukaryotic cytoplasm, these symbiotic organelles transferred many of their genes to the eukaryotic nucleus. The genes that were lost, such as those for anaerobic glycolysis and amino acid biosynthesis, could be provided by host nuclei. Such transfers and deletions had the advantage of maintaining and replicating only two copies of a symbiotic gene in a diploid host nucleus instead of sustaining a separate gene copy within each of many cellular organelles. Because some cells carry enormous numbers of organelle genomes — more than 8,000 copies of the mitochondrial genome in some human cells and even more copies of chloroplast genomes in some plant cells — reductions in organelle gene number by deletion or nuclear incorporation must have been highly selected. Nuclear genes that code for symbiotic organelle proteins have been widely identified (Gillham, 1994). The proteins are made by the common cellular cytoplasmic translation process, and transported to the correct organelle position using special "signal" and "transit" peptides (coupled with signal peptides in those chloroplasts of secondary endosymbiotic origin).

Nuclei

As may be expected, the eukaryotic nucleus also has collected hypotheses suggesting its origin via endosymbiosis, but how strong is the evidence? The eukaryotic nuclear genome has a mosaic appearance, containing large numbers of genes that seem to be similar to genes from Archaea as well as large numbers of genes that are similar to Eubacterial genes.

Lake and Rivera (1994) suggest that nuclear membranes were derived from a captured prokaryotic cell that provided a portion of the eukaryotic genetic material. Using molecular sequencing techniques (see Chapter 12), Gupta and coworkers (1994) presented support for this view; significant similarities exist among endoplasmic reticular genes in the ancient eukaryote *Giardia* (see Fig. 8-7) and in genes of certain types of Gram-negative bacteria.[8] Because the eukaryotic double-membrane endoplasmic reticulum is continuous with the eukaryotic nuclear membrane (see

[7] See Gray (1993), Gray et al. (1999, 2001), Archibald and Keeling (2002) and Burger et al. (2004) for primary and secondary (or first and second wave) endosymbiosis.

[8] The genes analyzed for these studies produce a family of universally-conserved heat shock proteins called hsp 70 that, among other features, act as chaperones to help transfer proteins from one intracellular locality to another.

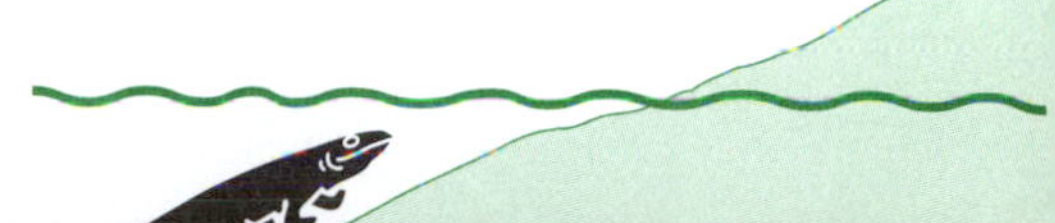

Fig. 7-1), the original eukaryote may have formed by fusion between an archaebacterium and a Gram-negative eubacterium. More specifically, an archaebacterium could have received energy and carbon through symbiosis with a bacterium that excreted hydrogen and carbon dioxide, an ancestor of the **hydrogenosome** organelle common in eukaryotes. A similar symbiosis is seen today in some eukaryotes that have organelles called hydrogenosomes, which share a common ancestry with mitochondria but have a radically different, and anaerobic energy metabolism. Other microbiologists explain the presence of both "eubacterial-like" and "archaeal-like" genes in the eukaryotic nucleus as the product of numerous lateral gene transfer events involving endosymbionts and prey organisms, rather than one dramatic cell fusion event.[9]

Metabolism

For cellular protein and nucleic acid synthesis to evolve (see Chapter 7) and for the organisms described above to arise, biochemical pathways must have been selected in which components of these and other polymers could be produced and chemical energy used. We see the results of such biochemical selection everywhere; cellular metabolism is highly organized in time and space so that each metabolic step in a sequence occurs in a highly repeatable order. Moreover, different metabolic sequences are often precisely coupled and regulated so that the products of one sequence (e.g., ATP) are available for use in other sequences.

Because there are no existing relics of ancient uncoordinated metabolic pathways, we cannot obtain direct evidence of precellular or early cellular metabolism. Two other types of evidence are available, however. One approach toward understanding metabolic evolution has been through comparative biochemistry, which has sought metabolic pathways or sequences within such pathways that different organisms share, considering such shared pathways as ancestral to these organisms. This approach is discussed below in the context of the evolution of metabolic pathways. Recovery of the *phylogenetic history of DNA sequences, including entire genes,* provides a second, powerful means for understanding the nature of the earliest life on Earth. This approach is discussed in Chapter 12. A third approach, based on reconstructing organismal relationships using molecular and morphological evidence, also is discussed in Chapter 12.

Given the anaerobic conditions known to have existed when life arose several billions of years ago, we begin with anaerobic metabolic pathways.

Anaerobic Metabolism

Anaerobic glycolysis, the breakdown of glucose in the absence of oxygen, is perhaps the most elemental metabolic pathway; all living organisms share various sections of this pathway.

This universality depends on the fact that all organisms derive their free energy from the chemical breakdown of monosaccharides. In **heterotrophic** organisms, monosaccharides or organic materials that can convert to monosaccharides derive from sources outside the organism. **Autotrophic** organisms make organic materials within themselves by reducing carbon dioxide.

In both types of organisms, glycolytic pathways may begin directly with glucose or with almost any organic material — sugars, fats or amino acids — that can be converted into glucose. The **Embden-Meyerhof glycolytic pathway** leads to pyruvic acid (pyruvate), providing a net yield of two *high-energy phosphate bonds in ATP, the basic currency for cellular chemical energy* (**Fig. 8-12**):

$$C_6H_{12}O_6 + 2\,ADP + 2\,P_i + 2\,NAD^+ \rightarrow$$
$$2\,C_3H_4O_3 + 2\,ATP + 2\,NADH + 2\,H^+ + 2\,H_2O$$

Only two of the reactions furnish ATP. The other steps in the pathway are preparatory to these primary reactions. The impression is of a pathway that is quite extended, even uneconomical. Some of these steps may partially recapitulate the succession of biochemical events that furnished energy to organisms in the past. For example, enough smaller molecules such as glyceraldehyde may have been available to allow relatively simple energy conversion in only a few steps. As these molecules depleted, organisms that could metabolize some of the more available larger molecules to the glyceraldehyde stage could continue to use the same, but now extended, pathway. By these means, succeeding organisms would need only to add or modify one or a few enzymes for each additional step as it occurred, rather than continually elaborate entirely new metabolic pathways.[10]

Retrograde *(Backward)* Evolution

Horowitz (1945) was the first to propose that the gradual depletion of a necessary molecule causes biochemical pathways to lengthen, enabling them to use an available related compound. Researchers have called this process retrograde or backward evolution and believe it accounts for many intermediate steps in the biochemical pathways that lead to the synthesis of compounds such as amino acids.

Let us assume, as outlined in **Figure 8-13**, that A is a product essential for cellular function, B is a molecular precursor of A and C is a molecular precursor of B. Obviously, there is no need for the organism to develop a pathway for the synthesis of A when A is present in the environment. But as A depletes, a considerable selective advantage occurs

---

[9] See Martin and Müller (1998) for hydrogenosomes and see Doolittle (1998) for lateral gene transfer.

[10] See Margulis et al. (2006) for a scenario for the evolution of eukaryote motility systems including the Embden-Meyerhof glycolytic pathway.

for catalysis of the available B precursor molecules into A (e.g., by action of enzyme 2). Similarly, as the supply of B is exhausted, selection occurs for the conversion of precursor C into B (e.g., by enzyme 1). A chain of metabolic reactions becomes established (. . . → C → B → A) that represents the evolution of enzyme 1 first, 2 second, 3 third, and so on, each new enzyme catalyzing a single step of a pathway that extends backward from a relatively complex molecule to its simpler precursors.

Rules of Change

Sequential pathways result from the chemical relatedness among compounds and the convertibility of one compound into another. This is most apparent in the simple stepwise changes noted in the glycolytic pathway. Biochemical pathways did not arise randomly, but followed rules of chemistry in pathways subject to selection. The survival of such pathways must have depended on their ability to cope with persistent chemical problems.

For example, although *glucose* may not have been the first energy-yielding compound, it is now a common sugar whose stability and ready availability in plants and animals led to the importance of *glycolysis* in virtually all organisms; each enzyme in the glycolytic pathway is found throughout multicellular organisms and in most single-celled organisms.

As an interesting testimonial to evolutionary economy, seven of the glycolytic enzymes also function in glucose biosynthesis, exactly reversing the direction of glycolysis. Furthermore, the amino acid sequences in many of these enzymes are remarkably conserved in organisms that have been evolving separately for at least a billion years. The amino acid sequence of a catalytically active site in the enzyme triosephosphate isomerase is found in organisms ranging from bacteria (*Escherichia coli*) to corn (*Zea mays*). (See Figure 7-2 for a listing of amino acids and their letter codes.)

E. coli	QGAAAFEGAVIAYEPVWAIGTGKSATPAQ
Yeast	EEVKDWTNVVVAYEPVWAIGTGLAATPED
Fish	DDVKDWSKVVLAYEPVWAIGTGKTASPQQ
Chicken	DNVKDWSKVVLAYEPVWAIGTGKTATPQQ
Rabbit	DNVKDWSKVVLAYEPVWAIGTGKTATPQQ
Corn	EKIKDWSNVVVATEPVWAIGTGKVATPAQ

Most likely, a single ancestral sequence existed for this catalytic purpose in the common progenitor of all these organisms, the b; it is highly improbable that so many similar amino acid sequences in so many different organisms arose by convergence. Continuous selection for the same enzymatic function essential in glycolysis in these various lineages preserved sequence similarity.

Other anaerobic pathways, probably also of ancient lineage, date back to the early anaerobic atmosphere. The Entner-Doudoroff pathway circumvents some early steps of the Embden-Meyerhof pathway by allowing a more direct conversion of glucose-6-phosphate to three-carbon compounds, such as pyruvate, but produces a net gain of one less ATP molecule per metabolized glucose:

$$\text{Glucose} + 2\,\text{NAD}^+ + \text{ADP} + \text{P}_i \rightarrow 2\ \text{pyruvate} + 2\,\text{NADPH} + \text{ATP}$$

Reduction–Oxidation

By whatever means anaerobic metabolism evolved, organisms that relied on it must eventually have faced a depletion of reduced carbon compounds to use as substrates. A change to an ability to use simpler and more readily available carbon sources such as CO_2 must have offered a considerable evolutionary advantage. To be effectively used, however, CO_2 must be reduced, or fixed, by a process that provides electrons and hydrogen ions. H_2, H_2S, NH_3 and others were probably among the reducing compounds available for this purpose. Early organisms and some taxa alive today (Vetter, 1991) probably broke down these inorganic compounds with the aid of catalysts (e.g., hydrogenase enzyme):

$$H_2(S) \xrightarrow{\textit{hydrogenase}} 2\,H^+ + 2\,e^- + (S)$$

furnishing electrons and hydrogen ions that could then be used for producing energy and hydrogenating carbon.[11]

The consensus is that many of these early reactions took place with the aid of membrane-bound enzymatic systems that allowed a successive chain of ***reduction–oxidations*** to occur as the electrons are picked up by acceptor molecules, which transform from the oxidized to the reduced form. These ***electron acceptors*** can then act as ***electron donors,*** transferring electrons farther down the chain until they reach the ultimate electron acceptor. Among electron carriers in these pathways are the iron-containing polypeptides, ferredoxins (**Fig. 8-14**), which are sufficiently small and widespread in nature to have been, along with porphyrins, among the first cellular oxidative–reductive agents.

During the process of electron transfer, which now involves other agents such as cytochromes, the transfer of protons across the cell membrane consumes energy released in specific oxidation–reduction steps, thereby creating a ***proton gradient*** (**Figure 8-15**). The potential energy available in this gradient can convert into chemical energy by **phosphorylating** ADP to ATP. The overall advantage of the Embden-Meyerhof pathway to the cell derives from the net formation of two high-energy phosphate bonds. Also indicated is the reduction of the pyridine nucleotide coenzyme,

[11] It has been proposed that the energy source for early life came through the formation of pyrite (FeS_2) from hydrogen sulfide by the reaction, $FeS + H_2S \rightarrow FeS_2 + 2H^+ + 2\,e^-$, providing effects similar to those just modeled by the hydrogenase enzyme. Subsequent organic reactions presumably took place on the pyrite surface and led to various basic metabolic pathways (see Box 6-3). Objections against such pyrite-forming reactions include the difficulty of developing a complex pathway on a pyrite surface. Nevertheless, the fact that mineral surfaces can enhance polymerization of both amino acids and nucleotides makes them valuable resources for exploring molecular evolution.

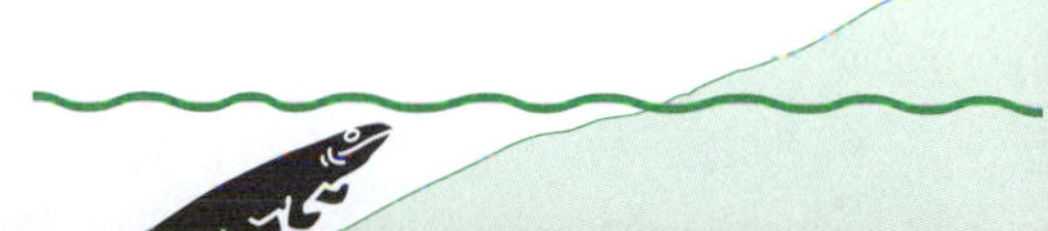

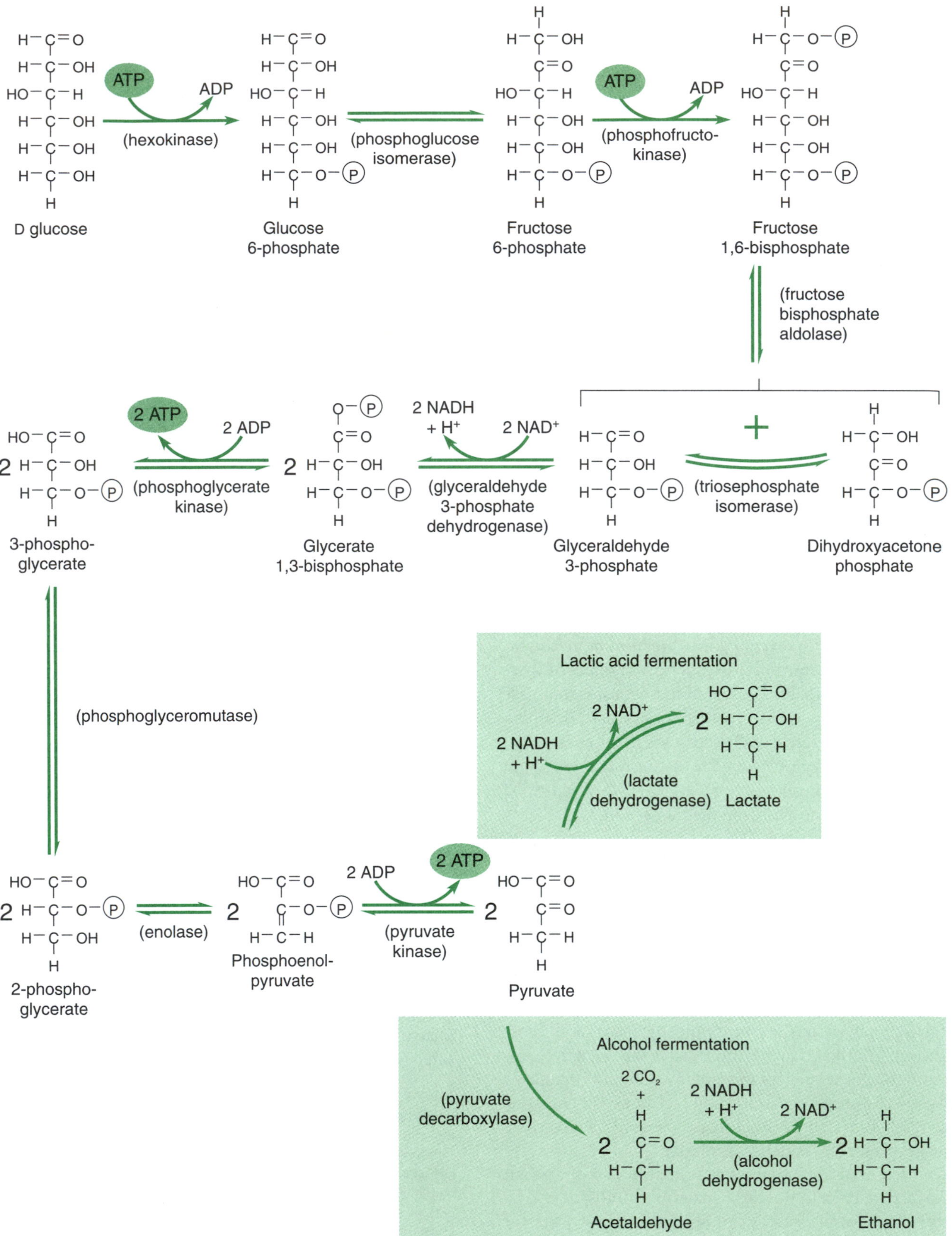

FIGURE 8-12 Steps in the anaerobic Embden-Meyerhof glycolytic pathway from glucose to pyruvate. Beginning with one molecule of glucose, the pathway degrades two ATP molecules to ADP but phosphorylates four ADP molecules to ATP. *Circled Ps* indicate phosphate groups (H_2PO_3), and *the specific enzyme for each reaction is in parentheses.*

NAD^+. The reduced form of this compound (NADH) can be oxidized to regenerate NAD^+ by reactions that donate hydrogens and electrons to form either lactic acid or ethanol:

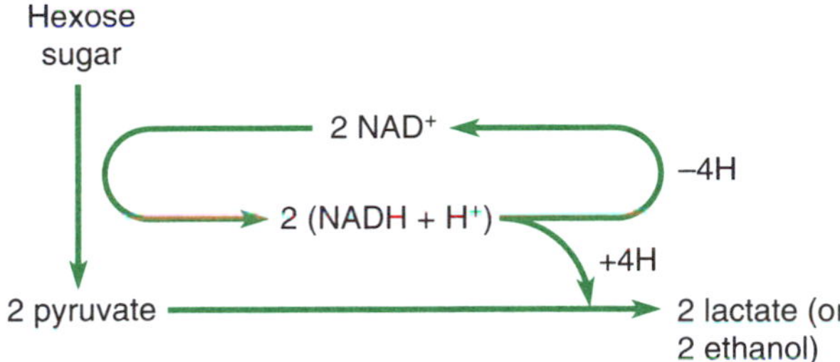

The pathway is more complex than illustrated here because the intermediate compounds formed during glycolysis can also function as substrates in synthesizing amino acids and nucleic acids. For example, 3-phosphoglycerate can serve as a substrate leading to the synthesis of serine, glycine or cysteine amino acids, or purine bases. The illustrations in this chapter show the various carboxylic acids in protonated form (HO–C=O or COOH) but are usually named as though they were unprotonated ($^{-}$O-C=O or COO^{-}), for example, pyruvic acid = pyruvate.

Photosynthesis

Despite advantages offered by electron transport systems in these early stages of evolution, reliance on chemical energy sources undoubtedly restricted organisms to those specific localities or conditions containing these organic and inorganic compounds. Perhaps the most important step toward *environmental independence* occurred when a mechanism evolved allowing light-absorbing porphyrins to move H^+ ions across membranes to generate ATP (photophosphorylation), and, with further selection, to act as photosensitive pigments.

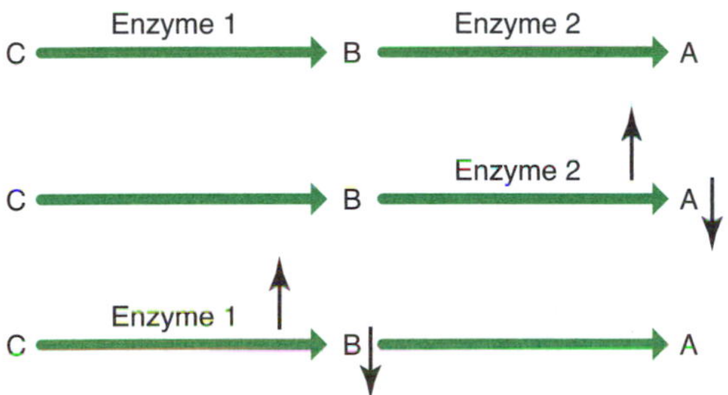

FIGURE 8-13 Evolution of a new metabolic pathway by modification of an existing pathway. Depletion of product A (downward arrow) favors selection for enzyme 2. Depletion of product B (downward arrow) favors selection of enzyme 1. See text for details.

Most photosynthetic mechanisms now depend on **chlorophyll,** a porphyrin derivative, although photosynthesis in various prokaryotes uses the transmembrane protein, bacteriorhodopsin as a light-driven **proton pump.** Woese and coworkers (1985) pointed out that five of the major eubacterial groups — Gram-positive, purple, green sulfur, green nonsulfur and cyanobacteria — possess photosynthetic species, and suggest that most, perhaps all extant eubacterial groups may have derived from a common photosynthetic ancestor. Very recent analysis of 15 complete cyanobacterial genomes by Mulkidjanian and colleagues (2006) is consistent with the origin of photosynthesis in anaerobic protocyanobacteria in the cyanobacterial lineage in response to selection pressure from UV light and depletion of electron receptors.

As with other porphyrins, chlorophyll has a number of resonance forms in which double and single bonds shift while the molecule remains rigid and stable (**Fig. 8-16**). This resonating structure enables chlorophyll to maintain light-absorbed energy as well as to transfer energy to similar molecules or receive it from pigments such as carotenoids, which absorb light energy at other wavelengths.

A photosynthetic pathway believed to be quite ancient is *cyclic photosynthesis*, in which solar energy acting on light-sensitive chlorophyll excites the molecule to a high-energy state, allowing an electron to pass on to other electron transfer agents (see Fig. 8-16, and Knoll, 2003). At an early evolutionary stage, this system must have bound to the membrane of a protocell, allowing photoactive energy to couple to existing membrane-contained systems that could phosphorylate ADP to ATP. ATP generated by this new, coupled system would be available for metabolic needs and enable growth to occur independently of environmental chemical energy sources such as glucose intake.

One such system probably originated as a membrane complex able to deacidify the cell interior by transporting protons (H^+) out of the cell using energy provided from ATP → ADP breakdown. Incorporation of anaerobic oxidation–reduction enzymes and electron transfer components into the cell membrane would have provided an opportunity for the evolution of a system that could also reverse the ATP →ADP reaction — an ATP-synthesizing system energized by re-entry of protons into the cell. The association of chlorophyll with such membrane components offered a pathway for powering proton gradient formation — the *proton pump* — by photoactivity.

Carbon Source

The source of carbon is as important to autotrophs as to heterotrophs. Some bacterial photophosphorylators such as purple nonsulfur bacteria in the genus *Rhodospirillum* depend, at least partly, on complex organic molecules for their carbon sources and also use organic substrates as elec-

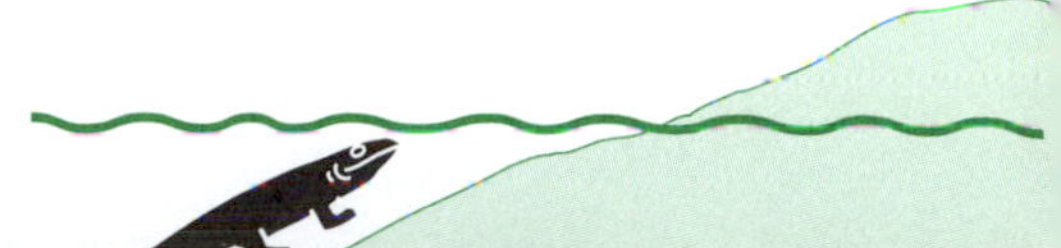

1 ... 28
A Ala-Phe-Val-Ile- Asn-Asp-Ser-Cys-Val-Ser-Cys-Gly-Ala-Cys-Ala-Gly-Glu-Cys-Pro-Val-Ser-Ala-Ile-Thr-Gin-Gly-Asp-Thr-
B Ala-Leu-Tyr-Ile- Thr-Glu-Glu-Cys-Thr-Tyr-Cys-Gly-Ala-Cys-Glu-Pro-Glu-Cys-Pro-Val-Thr-Ala-Ile-Ser-Ala-Gly-Asp-Asp-
C Ala-Leu-Met-Ile- Thr-Asp-Gin-Cys-Ala-Asn-Cys-Asn-Val-Cys-Gin-Pro-Glu-Cys-Pro-Asn-Gly-Ala-Ile-Ser-Gin-Gly-Asp-Glu-
D Pro-Ile-Gin- Val-Asp-Asn-Cys-Met-Ala-Cys-Gin-Ala-Cys-Ile-Asn-Glu-Cys-Pro-Val-Asp-Val-Phe-Gin-Met-Asp-Glu-Gin-

29 ... 55
A Gin-Phe-Val-Ile-Asp-Ala-Asp-Thr-Cys-Ile-Asp-Cys-Gly-Asn-Cys-Ala-Asn-Val-Cys-Pro-Val-Gly-Ala-Pro-Asn-Gin-Glu
B Ile-Tyr-Val-Ile-Asp-Ala-Asn-Thr-Cys-Asn-Glu-Cys-Ala Ala-Cys-Val-Ala-Val-Cys-Pro-Ala-Glu-Cys-Ile-Val-Gin-Gly (60)
Gly—Leu-Asp—Glu-Gin
C Thr-Tyr-Val-Ile-Glu-Pro-Ser-Leu-Cys-Thr-Glu-Cys-Val Asp-Cys-Val-Glu-Val-Cys-Pro-Ile-Lys-Asp-Pro-Ser-His-Glu-...-Gly (81)
Val-Cys
Gly-His-Tyr-Glu-Thr-Ser-Glu
D Gly-Asp-Lys-Ala-Val-Asn-Ile-Pro-Asn-Ser-Asn-Leu-Asp-Asp-Glu-Cys-Val-Glu-Ala-Ile-Gin-Ser-Cys-Pro-Ala-Ala-Ile-Arg-Ser (56)

(a)

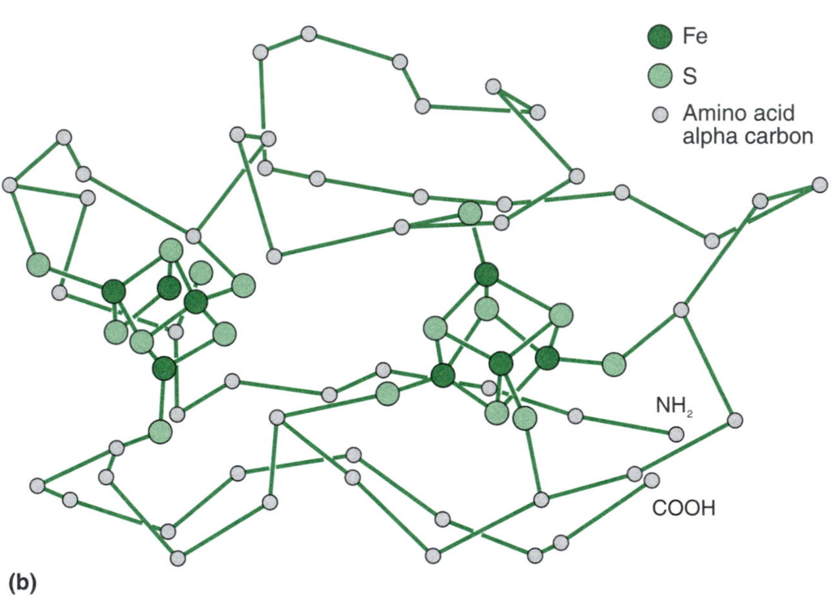

FIGURE 8-14 (a) Amino acid sequences of ferredoxin in four bacterial species: (A) *Clostridium butyricum*, an anaerobic fermenting bacterium; (B) *Chlorobium limicola*, a green photosynthetic bacterium; (C) *Chromatium vinosum*, a purple photosynthetic bacterium; (D) *Desulfovibrio gigas*, a sulfate-reducing bacterium. (From Hall, D. O. et al., 1977.) Note considerable similarity between the amino acid sequence in the first halves of these molecules (nos. 1 to 28) and the sequence in the second halves, an observation that initially prompted Eck and Dayhoff (1966) to suggest that this protein originated from a genetic duplication. (b) Folding of a ferredoxin molecule for *Peptococcus aerogenes* as revealed by x-ray analysis. Two identically distorted cubes, each containing four iron atoms, are held by sulfur bonds arising from the cysteine residues enclosed in shaded boxes in part (a). (Adapted from Adman, E. T., L. C. Sieker, and L. H. Jensen, 1973. The structure of a bacterial ferredoxin. *J. Biol. Chem.*, **248,** 3987–3996.)

tron donors, for example, the oxidation of succinic acid to fumaric acid. By contrast, others, such as purple sulfur bacteria in the genus *Chromatium* use CO_2 exclusively and can obtain electrons from inorganic material such as H_2S.

Although it is unclear which condition occurred first, the ability to use CO_2 as a source of carbon must have offered early photosynthesizers an important opportunity to expand their distribution. In fact, the **Calvin cycle**, the most common pathway for the reduction of CO_2, appears in practically all photosynthetic organisms. As shown in **Figure 8-17**, one carbon dioxide molecule incorporates for each turn of this cycle, and one molecule of ribulose bisphosphate regenerates for each CO_2 molecule incorporated. Six turns of the cycle are necessary to produce one glucose molecule. The overall reaction is:

$$6\,CO_2 + 18\,ATP + 12\,NADPH + 12\,H^+ \rightarrow \text{glucose} + 18\,ADP + 18\,P_i + 12\,NADP^+$$

Despite the advantage of using readily available CO_2 and easily obtainable photosynthetic energy, because they depended on compounds such as H_2S for hydrogen sources, the distribution of early photosynthesizers probably

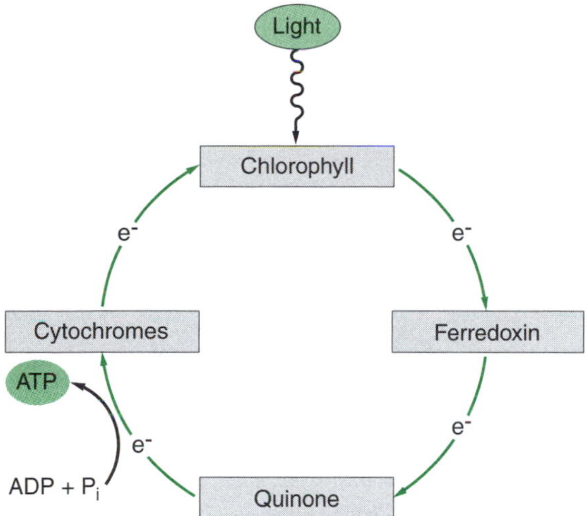

FIGURE 8-15 Diagrammatic view of a cyclic photophosphorylation pathway. Excited chlorophyll molecules force electrons to flow toward the more electronegative direction occupied by the electron acceptor, ferredoxin. Electrons flow back to the now positively charged chlorophyll through electron carriers that include quinones and various cytochrome pigment proteins. During the transfer of electrons toward electropositive components, a proton gradient forms across the membrane. This gradient is used to supply energy for phosphorylation of ADP to ATP (see Fig. 8-12).

remained restricted. Such dependence is found today among some bacterial photosynthesizers (purple and green sulfur bacteria) in which hydrogen sulfide provides the electrons for the hydrogenation of carbon:

$$2\,H_2S + CO_2 \xrightarrow{\text{light}} \underset{\text{carbohydrate}}{(CH_2O)} + H_2O + 2\,S$$

A primary revolution in organismal distribution occurred when photosynthetic mechanisms evolved that could derive their electrons from readily available water molecules. This process now involves two chlorophyll systems (**noncyclic photosynthesis**), resulting from lateral gene transfer between two kinds of photosynthetic bacteria, each possessing a somewhat different protein reaction center for transferring electrons. As shown in **Figure 8-18**, the source of electrons for the photosystem II chlorophyll component is the oxidation (dissociation) of water into electrons and H^+ ions and the release of molecular oxygen:

$$2\,H_2O \rightarrow 4e^- + 4H^+ + O_2\uparrow$$

The photosystems are localized within distinctively specialized photosynthetic membranes (*thylakoids*) (Fig. 8-18) found today in cyanobacteria and in the chloroplasts of eukaryotic algae and plants. The union of these two photosystems (I and II) is believed to have aided carbon dioxide reduction, especially with the advent of an aerobic atmosphere.

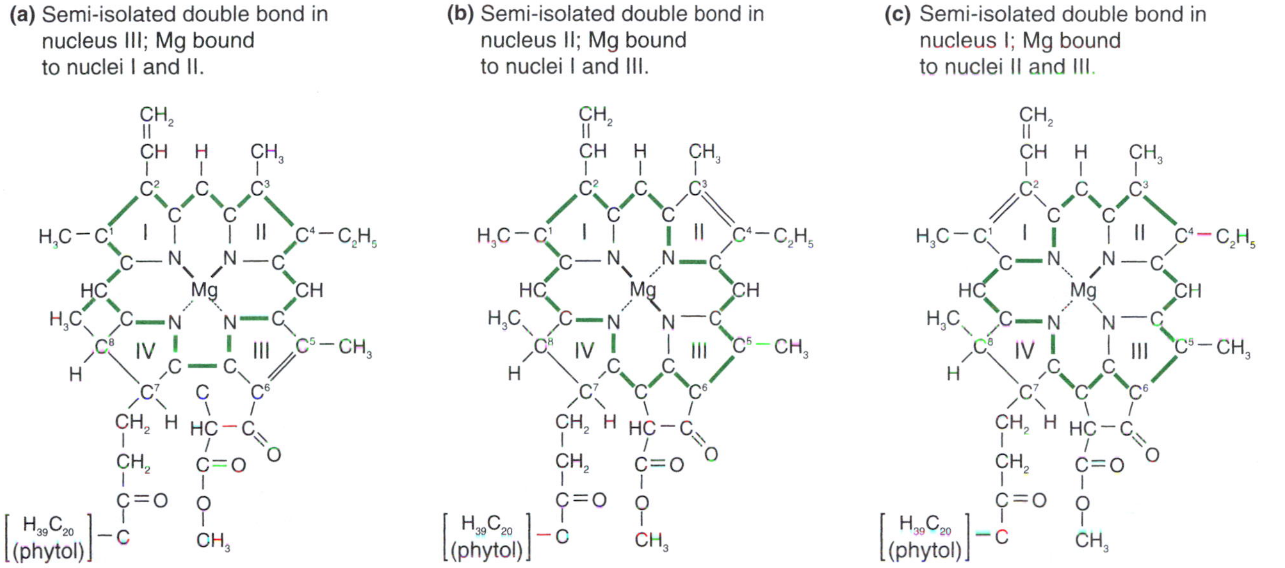

(a) Semi-isolated double bond in nucleus III; Mg bound to nuclei I and II.

(b) Semi-isolated double bond in nucleus II; Mg bound to nuclei I and III.

(c) Semi-isolated double bond in nucleus I; Mg bound to nuclei II and III.

FIGURE 8-16 Three resonance forms of chlorophyll *a*, showing stability of the molecule as its double and single bonds shift around the ring system (C^1–C^8) in various ways (*colored lines* in a–c). This ability to resonate enables chlorophyll to temporarily retain a high electron energy level resulting either from the excitation of electrons exposed to appropriate wavelengths of light or from electrons transferred to chlorophyll from other pigments such as carotenoids, phycobilins or flavines. Chlorophyll can also transmit such energy to other molecules used in photosynthetic reactions. (Adapted from Wald, G. 1974. Fitness in the universe: Choices and necessities. In *Cosmochemical Evolution and the Origins of Life*, J. Oró, S. L. Miller, C. Ponnamperuma, and R. S. Young (eds.). D. Reidel, Dordrecht, Netherlands, pp 7–27.)

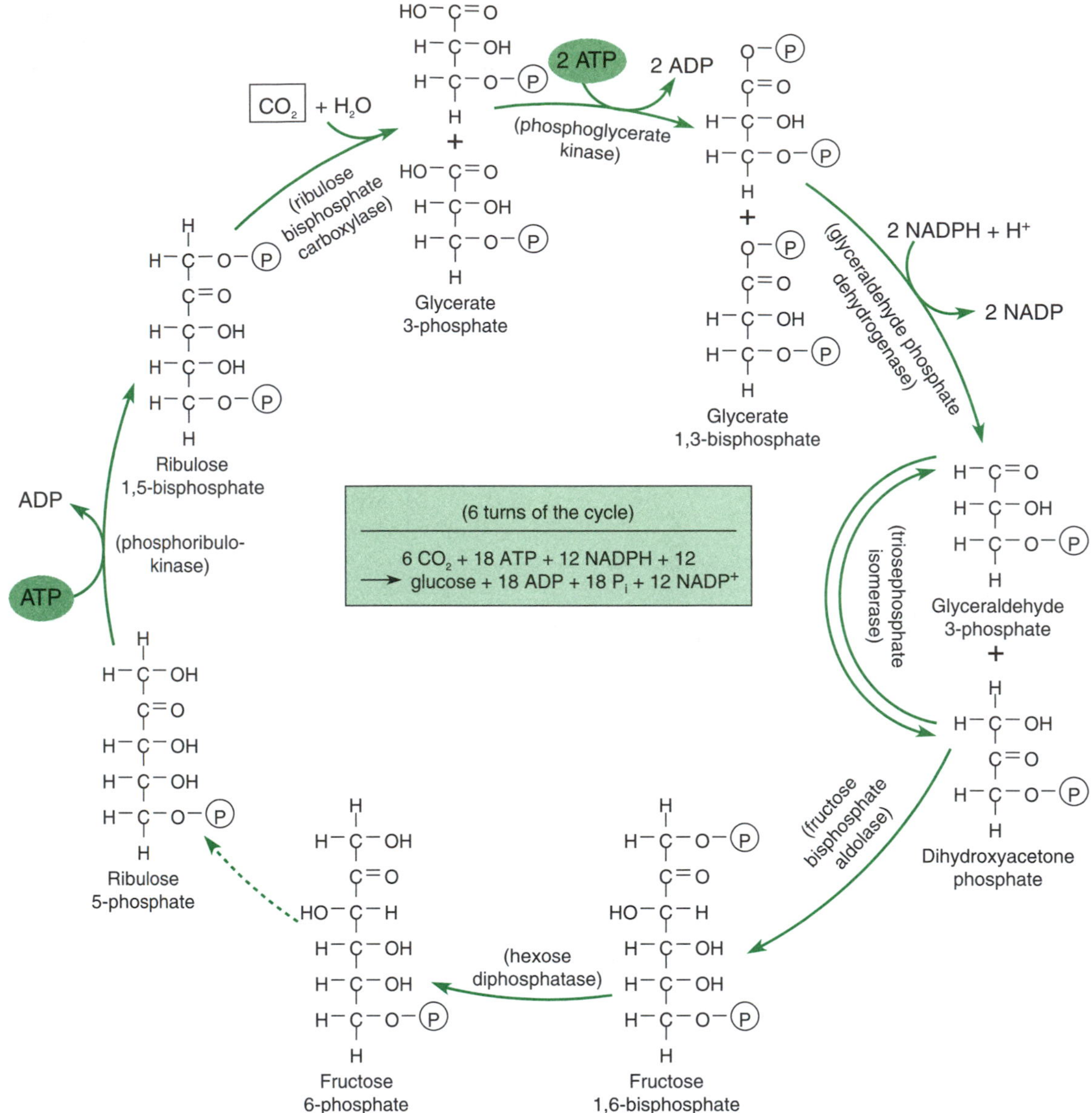

FIGURE 8-17 Simplified diagram of the Calvin cycle, the main metabolic pathway for carbon dioxide fixation in photosynthetic organisms. It relies on reducing power contributed by NADPH formed during photosynthetic reactions and on energy provided by ATP. ($NADP^+$ is the oxidized form of the coenzyme nicotinamide adenine dinucleotide phosphate, and NADPH the reduced form.) Between glyceraldehyde 3-phosphate and ribulose 5-phosphate in the pathway are a number of reactions, only two products of which (fructose 1, 6-bisphosphate and fructose 6-phosphate) are illustrated here.

Oxygen

The consequences of using water as an electron and hydrogen donor in photosynthesis were profound. Liberation of molecular oxygen began to produce an oxidizing, aerobic environment whose chemical effects were quite different from the previous relatively more reducing environment. We don't know the speed at which oxygen accumulated because of photosynthesis, but organisms that show up in 3-By-old South African Bulawayan limestone and gunflint strata seem to have been oxygen-generating *cyanobacteria*, as are the 3.5-By-old filamentous cells found at Warrawoona, Australia (**Fig. 8-19**).

Oxygen concentration in the atmosphere may have remained at one percent of the present level until about 2 Bya and gradually increased to its present 21 percent concentration with increased success of photosynthetic forms (see Chapter 5). Initiation of an oxygen atmosphere most

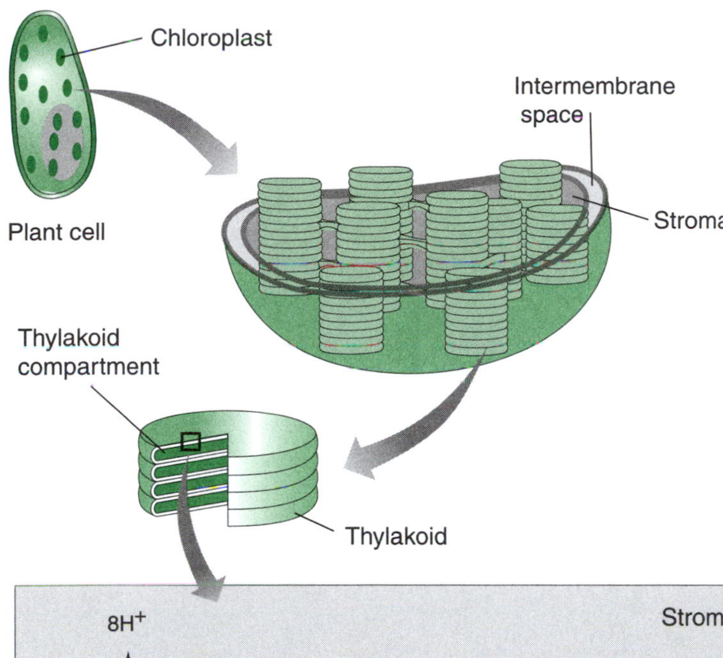

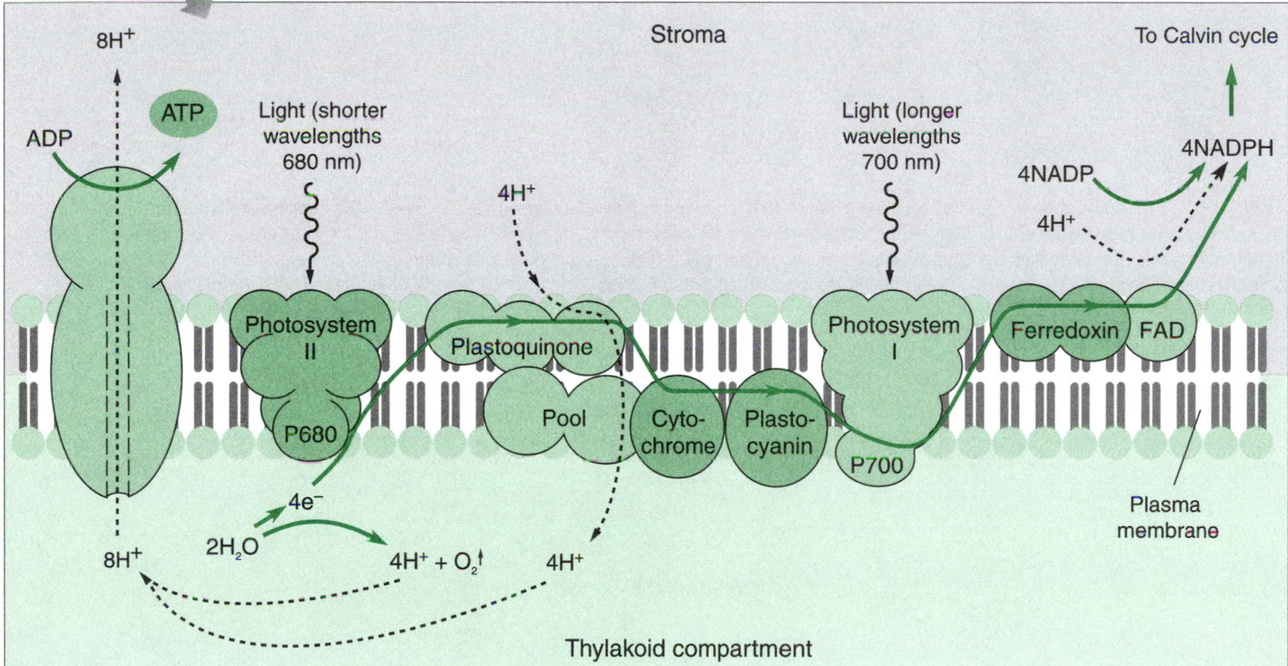

FIGURE 8-18 Diagram of noncyclic electron flow in which electrons obtained from water molecules transfer to the electron acceptor NADP+. Two photosystems, I and II, each sensitive to slightly different wavelengths of light, can be activated to high energy levels so that electrons pass along the chain. The transfer of electrons from photosystem II compensates for the loss of electrons to NADP+ by photosystem I. In turn, electrons derived from the photooxidation of water molecules replace photosystem II electrons. As in cyclic photophosphorylation, the flow of electrons produces a proton gradient that can supply energy for the phosphorylation of ADP to ATP. This appears more clearly in the thylakoid membrane structure where the ATP-synthesizing enzyme, ATPase, acts as a proton pump generating ATP by tapping energy from the gradient of H^+ ions flowing across the membrane into the thylakoid compartment. Also shown are the general locations of the two photosystem components and some of the electron transport chain proteins. An *unbroken line* indicates the presumed flow of electrons, and *dashed lines* indicate the flow of H^+ ions. (Membrane sequence from Wolfe, S. L., 1981. *Biology of the Cell*, 2d ed. Wadsworth, Belmont, CA. and Zubay, G., 1988. *Biochemistry*, 2d ed. Macmillan, New York; see also Barber , J., and B. Andersson, 1994. Revealing the blueprint of photosynthesis. *Nature*, **370**, 31–34.)

probably led to an increase in the number and kinds of organisms capable of utilizing aerobic metabolic pathways. By the Cambrian or somewhat earlier, oxygen levels had risen sufficiently to permit rapid evolution of large aerophilic multicellular organisms (see Chapter 15).

Because oxygen is a highly reactive element that can rapidly "burn" (oxidize) organic material, one immediate effect of an oxygen atmosphere was to increase the frequency of cellular **antioxidant compounds**, similar to vitamins E and K and the carotenes found today (**Fig. 8-20**). Interestingly, some extant organisms such as luminescent bacteria use what may have been an early mechanism to detoxify oxygen, namely, localizing oxidation to special reactions that emit fluorescent light. Oxidation in such organisms forms the peroxides that react with luciferase enzymes to produce organic acids and water, radiating light in the process.

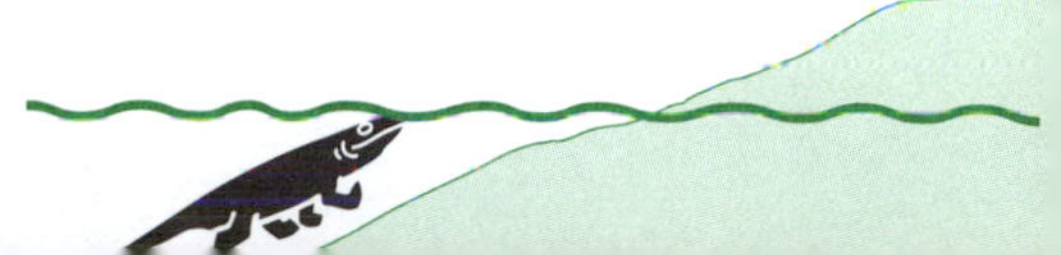

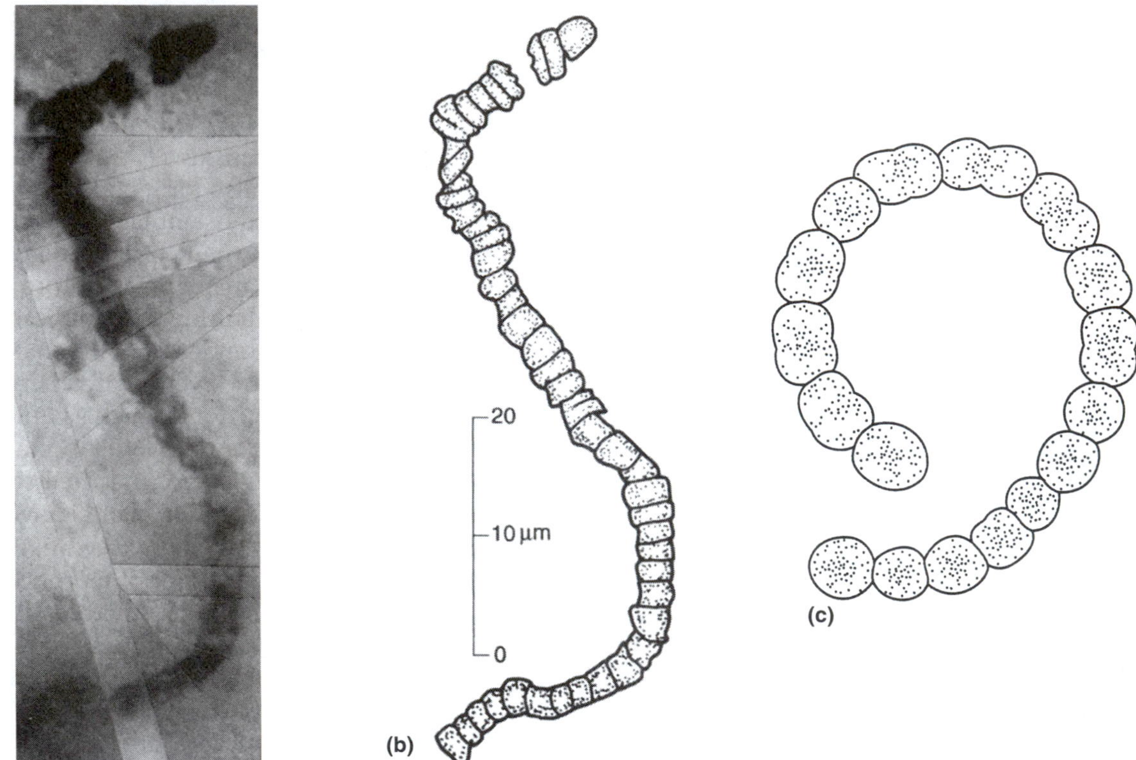

FIGURE 8-19 (a) Thin section of a filamentous unicellular fossil found in stromatolite chert in the 3.5-By-old Warrawoona formation in Western Australia. (b) Reconstruction of this fossil. (c) Diagram of a phase-contrast microphotograph of a filament of cells of an extant cyanobacterium (*Anabaena* sp.). Among other extant prokaryotes that appear similar to the microfossil in (a) are colorless sulfur-gliding bacteria (e.g., *Thioplaca schmidlei*) and green sulfur bacteria (*Oscillochloris chrysea*). (Reproduced from J. William Schopf. 1993. *Science.* 260:640–646. Reprinted with permission from AAAS.)

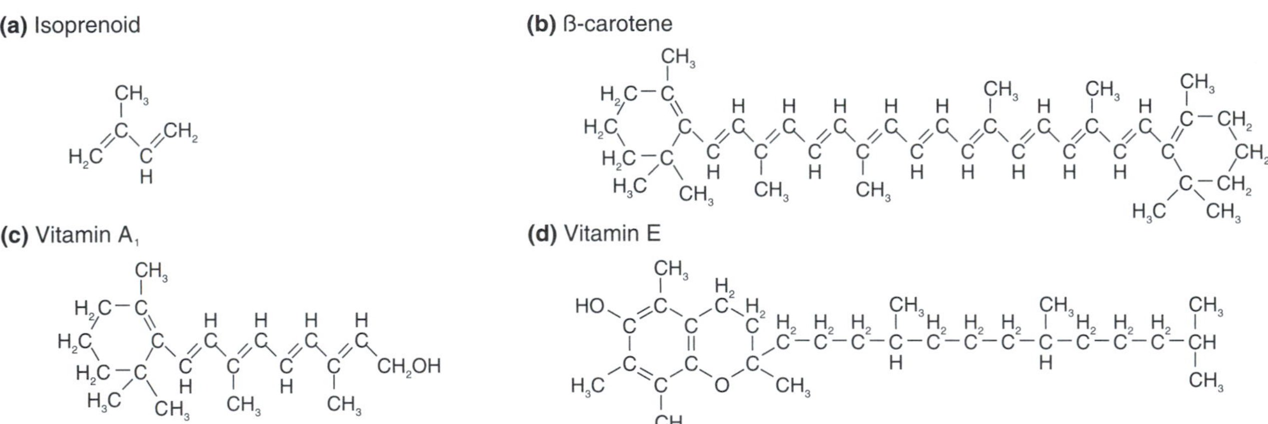

FIGURE 8-20 Examples of isoprenoid compounds. (a) The basic five-carbon isoprenoid unit. (b) β-carotene, a plant isoprenoid offering protection against oxidation in visible light. (c) Vitamin A_1 (retinol 1), a fat-soluble vitamin that exists only in animals, formed by the cleavage of carotenes. (d) Vitamin E (α-tocopherol), a plant antioxidant on which rodents are nutritionally dependent.

Aerobic Metabolism

Perhaps the most significant change in metabolism to accompany the new aerobic environment was the evolution of a respiratory pathway by which oxygen is used to produce much more energy from the breakdown of glucose than anaerobic glycolysis can produce (**Box 8-4**). Although tied to **aerobic metabolism**, the Krebs cycle does not immediately depend on molecular oxygen, and so may have originated from anaerobic pathways.

Although the Krebs cycle itself does not use molecular oxygen, its evolution and adoption by aerobic organisms seems based on a **membrane-bound electron transport system** where oxygen serves as the final electron acceptor in the chain: one more stage in the evolution of membrane-bound systems. In this electron transport process, the pyridine nucleotide coenzyme NAD^+ picks up electrons and associated protons (e^- and H^+) and transfers them into a respiratory chain consisting of various electron carriers. The final electron transfer to oxygen, the ultimate electron acceptor, produces water:

$$O_2 + 4\,e^- + 4\,H^+ \rightarrow 2\,H_2O$$

As diagrammed in **Figure 8-21**, such electron transfers occur along an electrical potential gradient that provides sufficient energy exchange at three coupling sites to allow an ADP molecule to be phosphorylated.

In summary, complete aerobic oxidation of a molecule of glucose, including **oxidative phosphorylation**, produces maximally about 38 molecules of ATP (compared to only 2 molecules of ATP formed by anaerobic glycolysis):

$$C_6H_{12}O_6 + 6\,H_2O + 6\,O_2 + 38\,ADP + 38\,P_i \rightarrow 6\,CO_2 + 12\,H_2O + 38\,ATP$$

If we consider that the hydrolysis of a mole of ATP (507 g) to ADP provides at least seven kilocalories, cellular oxidation of a mole of glucose (180 g) can produce at its theoretical maximum about $38 \times 7 = 266$ kilocalories. This amount — 39 percent of the 686 kilocalories produced by burning a mole of glucose in air — indicates that the efficiency of oxidative phosphorylation can be perhaps as high or higher than many human-designed mechanical energy conversion systems.

Ozone

In addition to its metabolic effects, the oxygen atmosphere enabled a stratospheric **ozone** (O_3) layer to form. Ozone, which is commonly formed by breakdown of diatomic molecular oxygen ($O_2 \rightarrow 2\,O$; $O + O_2 \rightarrow O_3$), screens out short-wave ultraviolet radiation from reaching Earth's surface, most medium-wave radiation (UVB), and almost all the more dangerous short-wave (UVC) radiation. Because absorption of such wavelengths by organic ring structures (e.g., nucleotides) can kill cells or organisms by deteriorating or modifying vital molecules, ozone screening must have been essential in enabling the expansion of life to ocean surfaces as well as to land. Ozone is also a biologically destructive oxidant in cells.

Nitric oxide (NO), however, impedes ozone's oxidation effect. Feelisch and Martin (1995) suggest that its early prevalence may have made it the first antioxidant. Once nitric oxide was incorporated into cellular metabolism, it could be used in other cellular activities such as vasodilation, endocrine secretion and neuronal cell communication. Unfortunately, nitric oxide is also a common atmospheric pollutant, produced by automobile exhausts and industrial plants, which reduces the protective ozone screen.

Evolution of Eukaryotes

Whatever the origin of their membranes, organelles and metabolism, by 1.5 Bya or even earlier, eukaryotic cells appeared and were fossilized. The approximate 2 By interval between the age of these fossils and the earliest prokaryotes at 3.5 Bya (see Fig. 8-5) may have been caused by the length of time needed to incorporate and coordinate the many profound changes in cell structure and function (*coadaptive mutations*). Gaining a foothold in a world dominated by prokaryotes may have been a protracted struggle. Because of their relatively large size and complexity, early protistans presumably had much more genetic material (DNA) than is normal among prokaryotes. The difficulty of manipulating and replicating one or more large DNA molecules having circular prokaryotic forms, each with only a single point (origin) of replication, might have led to the division of eukaryotic DNA into one or more linear chromosomes, each with multiple origins of replication: the more such origins, the more rapid the replication.

The near-universal presence of cytoskeletal microtubules in eukaryotes enables nuclear chromosomal division to supplant the usual prokaryotic method, which depends on separating dividing chromosomes by their individual attachment to a lengthening cell membrane. Instead, each of the relatively larger, more complex, and often more numerous eukaryotic chromosomes has a **centromere** (or **kinetochore**) that attaches to a microtubular network of sliding spindle fibers during division, allowing the daughter chromosomes to move to opposite poles. So far, we do not know the exact sequence of steps in the evolution of eukaryotic chromosome structure and cell division, although researchers (e.g., Cavalier-Smith, 1985) have offered hypotheses.

Very likely, once eukaryotic **mitosis** evolved, sex cell division or **meiosis** must have quickly followed, because meiosis and sexuality almost universally appear among the major eukaryotic taxonomic groups and practically all asexual multicellular eukaryotes seem derived from sexual forms. Some advantages of sexual reproduction, and the need for the meiotic cell divisions to reduce the number of chromosomes before cross-fertilization, are discussed in Chapters 10 and 13.

Although early eukaryotes were all single-celled organisms or simple filaments (see Fig. 8-4), their affinities to particular protistans is uncertain. They have been called acritarchs (Greek *akritos*, "undecided"). For the present, we can note that although some small multicellular forms may have evolved from single-celled protistans as long as one billion or more years ago, larger visible multicellular fossils with differentia-

BOX 8-4 The Krebs Cycle

The origin of this metabolic system involved the elaboration of a series of enzymes that transform pyruvic acid into an activated acetic acid group (acetyl-coenzyme A or one of its evolutionary precursors), carry these small acetyl groups along a special cycle in which they convert into carbon dioxide and lose their hydrogens:

$$CH_3COOH\text{–coenzyme A} + 2\,H_2O \rightarrow 2\,CO_2 + 8\,(e^- + H^+) + \text{coenzyme A}$$

This part of the pathway, variously called the **Krebs cycle,** *citric acid cycle*, or *tricarboxylic acid cycle* (**Fig. 8-22**), has an interesting self-catalytic feature (as does the Calvin cycle): The cycle continuously generates intermediate products necessary for the cycle to occur. Oxaloacetic acid, for example, combines with acetic acid to begin the cycle and regenerates from malic acid at the end of the cycle. One or a few molecules of oxaloacetate can therefore function continuously to permit an infinite number of acetate molecules to enter the cycle.

One proposed evolutionary sequence offered by Weitzman (1985) is of the cycle having evolved from an earlier stage in anaerobes in which it split into *two metabolically different arms*. One arm followed a reductive pathway:

Pyruvate → oxaloacetate → malate → fumarate → succinate → succinyl-coenzyme A

Coupled to reactions in the reductional side was the oxidation of NADH to NAD^+, necessary for replacing NAD^+ used in glycolysis, which, in turn, was necessary for ATP formation.

The other arm also began with an initial pyruvate but engaged in oxidative metabolism:

Pyruvate → acetyl-coenzyme A → citrate → *cis*-aconitate → *iso*citrate → α-oxoglutarate

Reactions in the oxidative side had the obvious function of providing reduced nucleotides (NADH, NADPH) for carbohydrate synthesis as well as providing compounds such as α-oxoglutarate, a precursor of glutamic acid. Both these sides of the reaction still function in some organisms, such as cyanobacteria and anaerobically grown *Escherichia coli.*

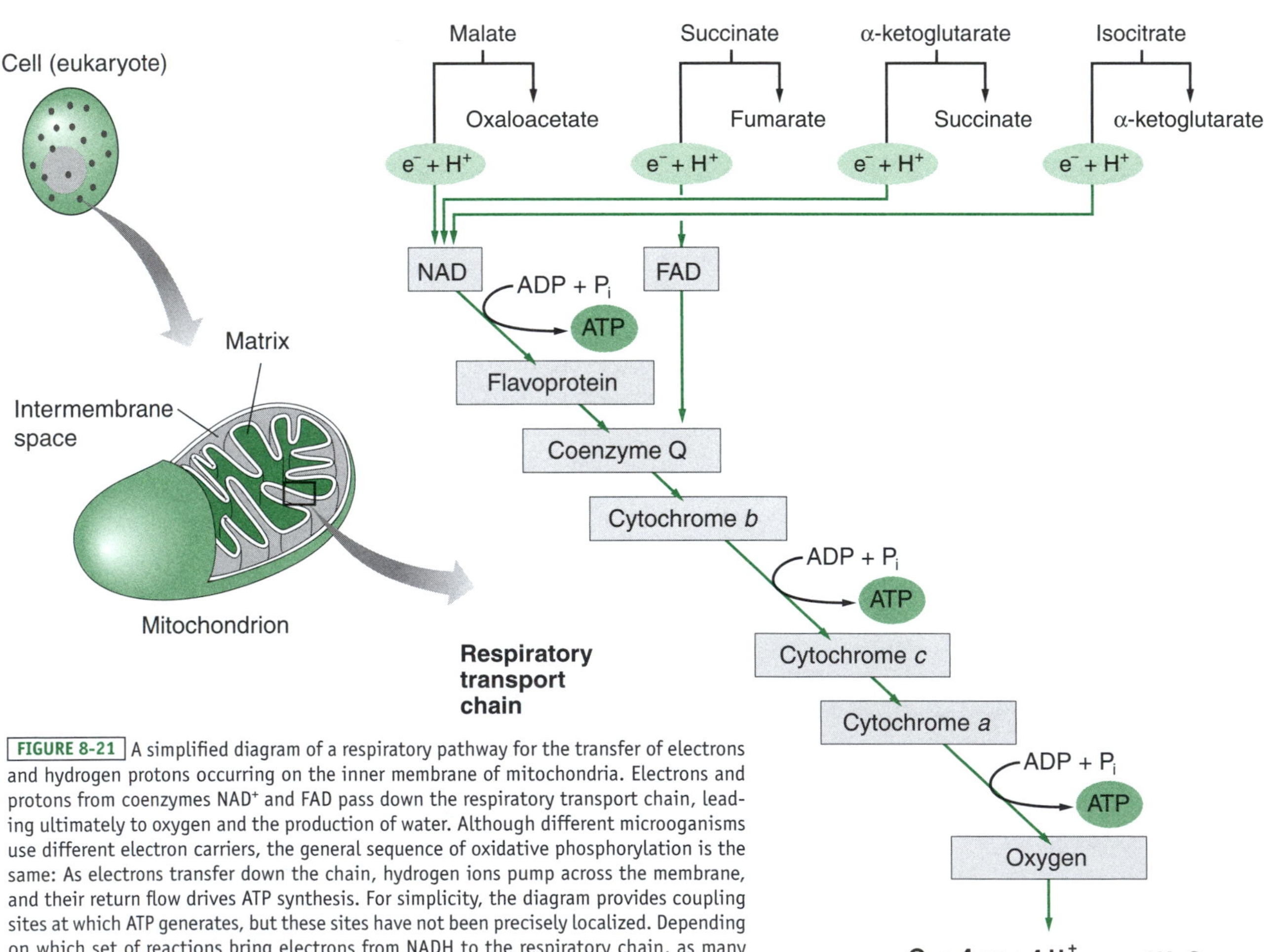

FIGURE 8-21 A simplified diagram of a respiratory pathway for the transfer of electrons and hydrogen protons occurring on the inner membrane of mitochondria. Electrons and protons from coenzymes NAD^+ and FAD pass down the respiratory transport chain, leading ultimately to oxygen and the production of water. Although different microoganisms use different electron carriers, the general sequence of oxidative phosphorylation is the same: As electrons transfer down the chain, hydrogen ions pump across the membrane, and their return flow drives ATP synthesis. For simplicity, the diagram provides coupling sites at which ATP generates, but these sites have not been precisely localized. Depending on which set of reactions bring electrons from NADH to the respiratory chain, as many as 38 ATPs can be formed during complete oxidation of glucose.

Pyruvate + Coenzyme A

(pyruvate dehydrogenase complex)

$2(e^- + H^+)$

CO_2 + Acetyl-Coenzyme A

H_2O

(citrate synthase)

Coenzyme A

Oxaloacetate

Citrate

(aconitase)

H_2O

Cis-aconitate

H_2O

(aconitase)

Isocitrate

(isocitrate dehydrogenase)

CO_2

$2 \, [e^- + H^+]$

α-oxoglutarate + Coenzyme A

(α-oxoglutarate dehydrogenase)

CO_2

$2 \, [e^- + H^+]$

Succinyl-Coenzyme A

(succinyl thiokinase)

ADP

ATP

Coenzyme A

Succinate

(succinate dehydrogenase)

$2 \, [e^- + H^+]$

Fumarate

H_2O

(fumarase)

Malate

(malate dehydrogenase)

$2 \, [e^- + H^+]$

$$CH_3COOH + 2\,H_2O \longrightarrow 2\,CO_2 + 8[e^- + H^+]$$

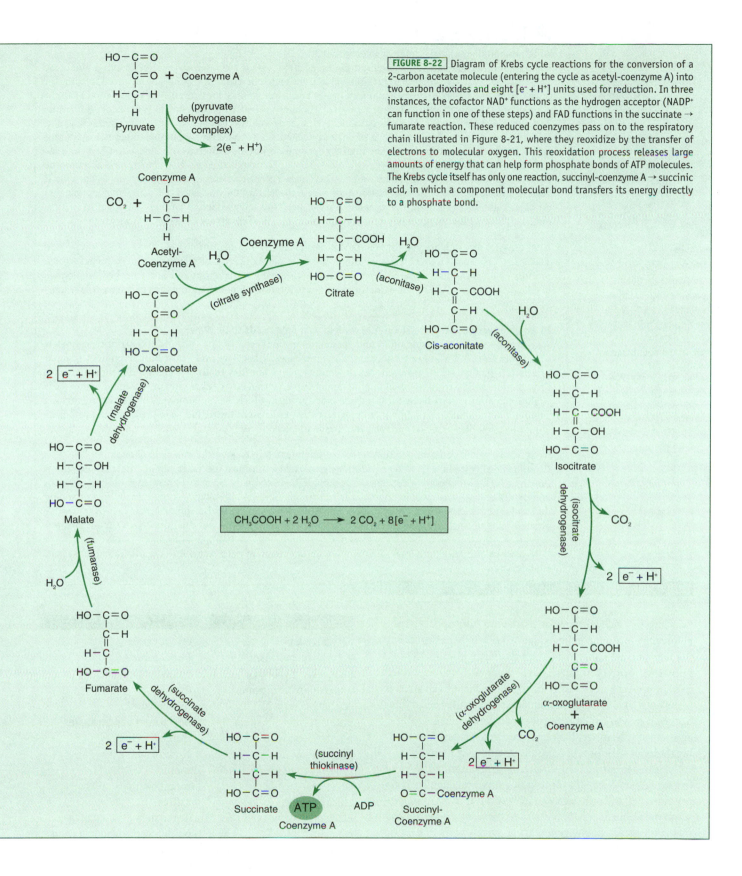

FIGURE 8-22 Diagram of Krebs cycle reactions for the conversion of a 2-carbon acetate molecule (entering the cycle as acetyl-coenzyme A) into two carbon dioxides and eight $[e^- + H^+]$ units used for reduction. In three instances, the cofactor NAD^+ functions as the hydrogen acceptor ($NADP^+$ can function in one of these steps) and FAD functions in the succinate → fumarate reaction. These reduced coenzymes pass on to the respiratory chain illustrated in Figure 8-21, where they reoxidize by the transfer of electrons to molecular oxygen. This reoxidation process releases large amounts of energy that can help form phosphate bonds of ATP molecules. The Krebs cycle itself has only one reaction, succinyl-coenzyme A → succinic acid, in which a component molecular bond transfers its energy directly to a phosphate bond.

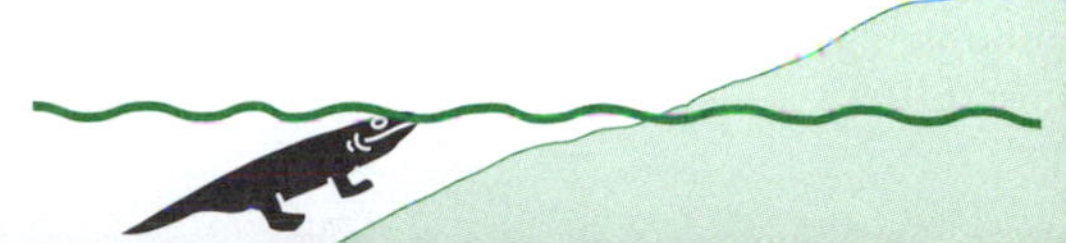

Lynn Margulis
Born: March 5, 1938
Birthplace: Chicago, Illinois
Undergraduate degree: University of Chicago, 1957
Graduate degree: Ph.D. University of California–Berkeley, 1965
Postdoctoral training: Brandeis University, Waltham, Massachusetts, 1964–1966
Present position: Distinguished University Professor Department of Geosciences University of Massachusetts at Amherst[12]

Lynn Margulis

What prompted your initial interest in evolution?

I have always loved nature and the diversity of life, especially the woods. Although my undergraduate training at the University of Chicago was primarily in the liberal arts, my interest in science was piqued by prescribed readings of original scientific papers. After graduation, the master's degree program at the University of Wisconsin at Madison afforded me the opportunity to do protist research. I became fascinated by studies of the genetics of cytoplasmic particles. These particles — chloroplasts, mitochondria, and motility organelles such as ciliary kinetosomes and mitotic centrioles — were not usually investigated by geneticists but questions about their inheritance patterns, origin, and evolution intrigued me. Although my doctoral work at the University of California at Berkeley centered on inducible mutations in *Euglena* — the loss of their photosynthetic plastids — the irreversibility of some of these mutations led me to the relationship of non-nuclear genetic systems to the history of eukaryotic cells. It became clear to me that symbiotic relationships were at the heart of the nucleated cell, whose major organelles probably had bacterial origins. The ideas and investigations that followed led to various papers and, in 1970, to publication of my first large major work on the problem, *Origin of Eukaryotic Cells* (Yale University Press).

What areas of research are you (or your laboratory) presently engaged in?

The area of my research that I feel most important right now concerns our discovery that spirochetes form motility associations and undergo developmental cycles that are analogous and may even be homologous to cilia development. Allied to this question is a strong interest in understanding the genetic basis of microtubular structures.

In which directions do you think future work in your field needs to be done?

A great need exists in my opinion to understand more about cell division (mitotic and meiotic) in smaller organisms. Protoctists (algae, protozoans, slime molds, etc.), bacteria, and yeasts have often been ignored or misclassified as "small" invertebrates or "lower" plants, but they really represent independent groups. Protists that evolved from bacterial symbioses are ancestral to all other eukaryotes. An essential part of their evolutionary importance lies in the fact that it is this group in which the fundamentals of mitosis and meiosis first appeared — patterns of growth and sexuality that radiated outward and were preserved in protoctists (single or few-celled protists) and their multicellular descendants.

What advice would you offer to students who are interested in a career in your field of evolution?

Paraphrasing Ernst Mayr, "In order to do any original science one must specialize, but in order to do evolution one must be far broader than a specialist." Although there are many aspects of knowledge evolutionists can and should utilize — natural history, chemistry, atmospheric science, geology — evolutionary understanding demands integration rather than separation. Above all, I believe evolutionists should learn as much as possible about live organisms, because through the windows of the living important clues can be seen about the past.

[12] Available at http://www.geo.umass.edu/faculty/margulis/, accessed March 2007. A recent publication is Margulis et al. (2006).

EVOLUTION ON THE WEB biology.jbpub.com/book/evolution

tion into several cell types or tissues are only apparent just before the beginning of the Phanerozoic. When the Cambrian, the first of the Phanerozoic periods, began about 545 Mya, numerous forms of multicellular invertebrates made a sudden marked appearance as forms that could fossilize. In an interval of perhaps 100 My or less, an explosive radiation of eukaryotes occurred in which a large number of animal lineages (phyla) appeared. Evolutionary biologists have long made understanding their derivations and relationships a major focus.

However, before we deal with these phenomena, we need to gain a clear concept of some basic genetic mechanisms that provide fuel and substance to evolution, that is, to understand *the origin of hereditary variation among organisms and its transmission among generations.* What kinds of genetic variability are there? What are their causes, and how are they transmitted? The next chapter briefly reviews this topic.

KEY TERMS

active transport
aerobic metabolism
anaerobic glycolysis
antioxidant compounds
Archaebacteria
autotrophic
banded iron formations
binary fission
Calvin cycle
cherts
chlorophyll
electron acceptors
electron donors
Embden-Meyerhof glycolytic pathway
endocytosis
Eubacteria
eukaryotes
exons
heterotroph
horizontal gene transfer
introns
Krebs cycle
meiosis
metabolism
mitosis
organelles
oxidative phosphorylation
ozone
phosphorylation

prokaryotes
proton gradient
proton pump
reduction–oxidation
split genes
stromatolites
symbiont

DISCUSSION QUESTIONS

1. How does Horowitz's hypothesis of retrograde evolution explain evolution of the Embden-Meyerhof glycolytic pathway?
2. What evidence exists for the evolutionary conservation of active sites of enzymes in such pathways?
3. By which means could carbon sources such as CO_2 reduce to cellular organic compounds in the early stages of evolution?
4. Electron transport systems and protein gradients.
   a. Why do researchers consider ferredoxin an early electron carrier?
   b. What roles do electron transport systems and protein gradients play in cellular energy production?
5. Photosynthesis.
   a. What advantage does chlorophyll offer in photosynthesis?
   b. How do cyclic and noncyclic photosynthesis differ? Which probably evolved first?
   c. What is the relationship between photosynthesis and the Calvin cycle?
   d. What effect did noncyclic photosynthesis have on the proportions of oxygen and ozone in the atmosphere?
   e. Why do cellular antioxidants offer a selective advantage, and what types are there?
6. Aerobic metabolism.
   a. According to Weitzman, how did the Krebs cycle evolve?
   b. What is the connection between the Krebs cycle and oxidative phosphorylation?
   c. How can one compare the efficiency of aerobic metabolism to anaerobic metabolism?
7. What accounts for the formation, structure and age of stromatolites?
8. Prokaryotes and eukaryotes.
   a. What are some major differences between prokaryotes and eukaryotes?
   b. Approximately how many years ago did each type of cell originate?
   c. What are split genes and how do they function?
   d. Is there support for the concept of a progenote? If so, then describe it.
   e. What hypotheses have researchers offered to explain the origin of eukaryotic organelles? Which hypothesis has gathered the most support, and what kinds of evidence have been used?
9. If genes can be transferred from symbiotic organelles to the eukaryotic nucleus, what might account for the persistence of genes in the eukaryotic mitochondrion?
10. How do researchers determine how many kingdoms or domains best reflect relationships among organisms?

EVOLUTION ON THE WEB

Explore evolution on the Internet! Visit the accompanying Web site for *Strickberger's Evolution,* Fourth Edition, at http://www.biology.jbpub.com/book/evolution for Web exercises and links relating to topics covered in this chapter.

RECOMMENDED READING

Adl, S. M., A. G. B. Simpson, M. A. Farmer, R. A. Andersen et al., 2005. The new higher level classification of Eukaryotes with emphasis on the taxonomy of protists. *J. Eukaryot. Microbiol.*, **52**, 399–451.

Allwood, A. C., M. R. Walter, B. S. Kamber, P. Marshall, and I. W. Burch, 2006. Stromatolite reef from the Early Archaean era of Australia. *Nature*, **441**, 714–718.

Bonner, J. T., 2006. *Why Size Matters: From Bacteria to Blue Whales.* Princeton University Press, Princeton, NJ.

Burger, G., M. W. Gray, and B. F. Lang, 2004. Mitochondrial genomes: Anything goes. *Trends Genet.,* **19**, 709–716.

Cavalier-Smith, T., 2002. The phagotrophic origin of eukaryotes and phylogenetic classification of Protozoa. *Int. J. Syst. Evol. Microbiol.*, **52**, 297–354.

Cavalier-Smith, T., 2003. Protist phylogeny and the high-level classification of Protozoa. *Eur. J. Protistol.*, **39**, 338–348.

Doolittle, W. F., 1999. Lateral genomics. *Trends Cell Biol.*, **9**, M5–M8.

Garcia-Ruiz, J. M., S. T. Hyde, A. M. Carnerup., A. G. Christy, et al., 2003. Self-assembled silica-carbonate structures and detection of ancient microfossils. *Science*, **302**, 1194–1197.

Gray, M. W., G. Burger, and B. F. Lang, 2001. The origin and early evolution of mitochondria. *Genome Biol.*, **2**, 1–5.

Keeling, P. J., G. Burger, D. G. Durnford, B. F. Lang, et al., 2005. The tree of eukaryotes. *Trends Ecol. Evol.*, **20**, 670–676.

Knoll, A. H., 2003. *Life on a Young Planet: The First Three Billion years of Evolution on Earth*. Princeton University Press, Princeton, NJ.

Lane, N., 2005. *Power, Sex, Suicide. Mitochondria and the Meaning of Life.* Oxford University Press, Oxford, United Kingdom.

Pagel, M. (ed. in chief), 2002. *Encyclopedia of Evolution.* 2 Volumes, Oxford University Press, New York.

Schopf, J. W., 1996. Metabolic memories of Earth's earliest biosphere. In *Evolution and the Molecular Revolution*, C. R. Marshall and J. W. Schopf (eds.). Jones and Bartlett, Sudbury, MA, pp. 73–107.

Simpson, A. G., 2003. Cytoskeletal organization, phylogenetic affinities and systematics in the contentious taxon Excavata (Eukaryota). *Int. J. Syst. Evol. Microbiol.*, **53**, 1759–1777.

Ueno, Y., K. Yamada, N. Yoshida, S. Maruyama, and Y. Isozaki, 2006. Evidence from fluid inclusions for microbial methanogenesis in the early Archaean era. *Nature*, **440**, 516–519.

Woese, C. R., O. Kandler, and M. L. Wheelis, 1990. Toward a natural system of organisms: Proposal for the domains Archaea, Bacteria and Eucarya. *Proc. Natl Acad. Sci. USA*, **87**, 4576–4579.

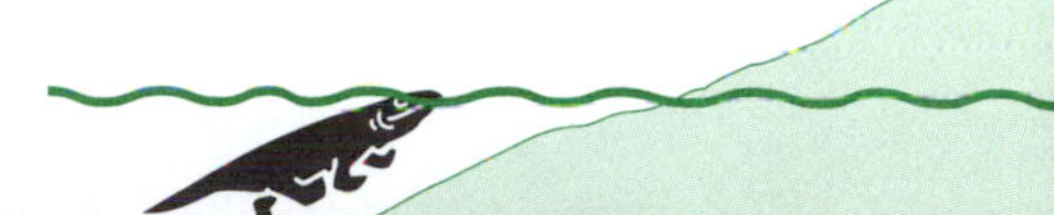

PART

3

The Organic Framework

Genes Cells and Development

The five chapters in Part 3 discuss what we call "The Organic Framework," by which we mean those fundamental genetic, molecular, cellular and developmental processes responsible for inheritance, variation and evolution.

Mechanisms of inheritance are discussed in Chapters 9 and 10. The fundamental biological processes of cell division, Mendelian genetics and sex determination are the topics of Chapter 9; chromosomes, mutation and variation the topics of Chapter 10. Together, these two chapters provide the background required to understand how genes are transferred from generation to generation (**inheritance**), how changes in genes (**mutations**) arise, and how those altered genes accumulate to provide the *genetic variation* upon which **selection** acts (evolution). Indeed, the production of variation is an absolutely essential part of the process of natural selection. It allows changes to occur within species and forms that we recognize as new species to arise.

In Chapter 11 we ask, "What is a species?" You might expect the answer to be simple — after all, we have been studying species for hundreds of years (see Chapter 1) — but it is not.

The concept of a species as a reproductively isolated group(s) or population(s) of individuals — the **biological species concept** we use for animals — cannot be applied universally. It cannot be used for fossil species, for which morphological characters are the primary and often the only way of identifying species, and it often breaks down when applied to plants. Furthermore, the concepts we erect for sexually reproducing organisms apply neither to asexually reproducing nor to unicellular organisms. Chapter 11 explains why this is so, and then goes on to discuss how we arrange species and higher groups or clades (such as classes or phyla) into hierarchies. Ways of doing this have changed over time, especially with the use of molecular criteria to create phylogenetic trees, the topic of Chapter 12.

Trees based on a single molecule (18S rRNA), on several molecules or on morphological characters may not coincide. This should not surprise us but it does; indeed, it alarms some biologists. But evolution is mosaic. "Parts" of an organism often evolve in different ways and in response to different selective pressures, not all of which will be reflected in a tree based on a single character.

Identifying corresponding "parts" of organisms should be easy, but is often extremely difficult, especially in situations when the morphological appearance of the parts has changed over evolutionary history. Classic examples are the skeletons of bony fish fins and of our limbs, the middle ear ossicles of mammals and the jawbones of reptiles, and the feathers of birds and the scales of reptiles. How we separate parts that *are similar* because of shared evolutionary history (**homology**) from parts that *look similar* because of parallel or convergent evolution (**homoplasy**) is an important topic discussed in Chapter 11.

The vast majority of *unicellular organisms* cannot be said to develop; they begin and end their lives as single cells. Exceptions are those that associate in some way, for example, by becoming colonial. Development, however, is a critical part of the life cycle of *multicellular organisms* — animals, plants and fungi — because the genotype does not make the phenotype. Genes operate only in context. That context is cells and, once the first embryonic cell division occurred, cells operated in a multicellular context. While the importance of this for embryonic development is appreciated, the importance of development for evolution has had a more checkered history, having been grasped by evolutionary embryologists (morphologists) with enthusiasm after Darwin published *On the Origin of Species,* dropped once Mendelian and population genetics took hold (see Chapter 2), and was not taken up again until the 1980s. In Chapter 13, we review the *important roles that changes in development play in mediating evolutionary change.* Developmental changes range from a mutation that influences one embryonic stage or one cell population, through altered interactions between cells and changes in the timing of developmental processes, to the impact of other individuals/species and environmental factors on development and therefore on evolution.

Now that we know that so much of the genome is shared between organisms and has an evolutionary history that can be traced to the origins of eukaryotes over 2 Bya, much more emphasis is being placed on **gene regulation** as pivotal to developmental and evolutionary change. Change the level or time of expression of a regulatory gene and you can change the phenotype. Selection on that new phenotype can result in the altered state of gene expression spreading through the population, leading to evolutionary change: descent with modification. Chapters 9 to 13 provide you with the background to appreciate how such modifications occur.

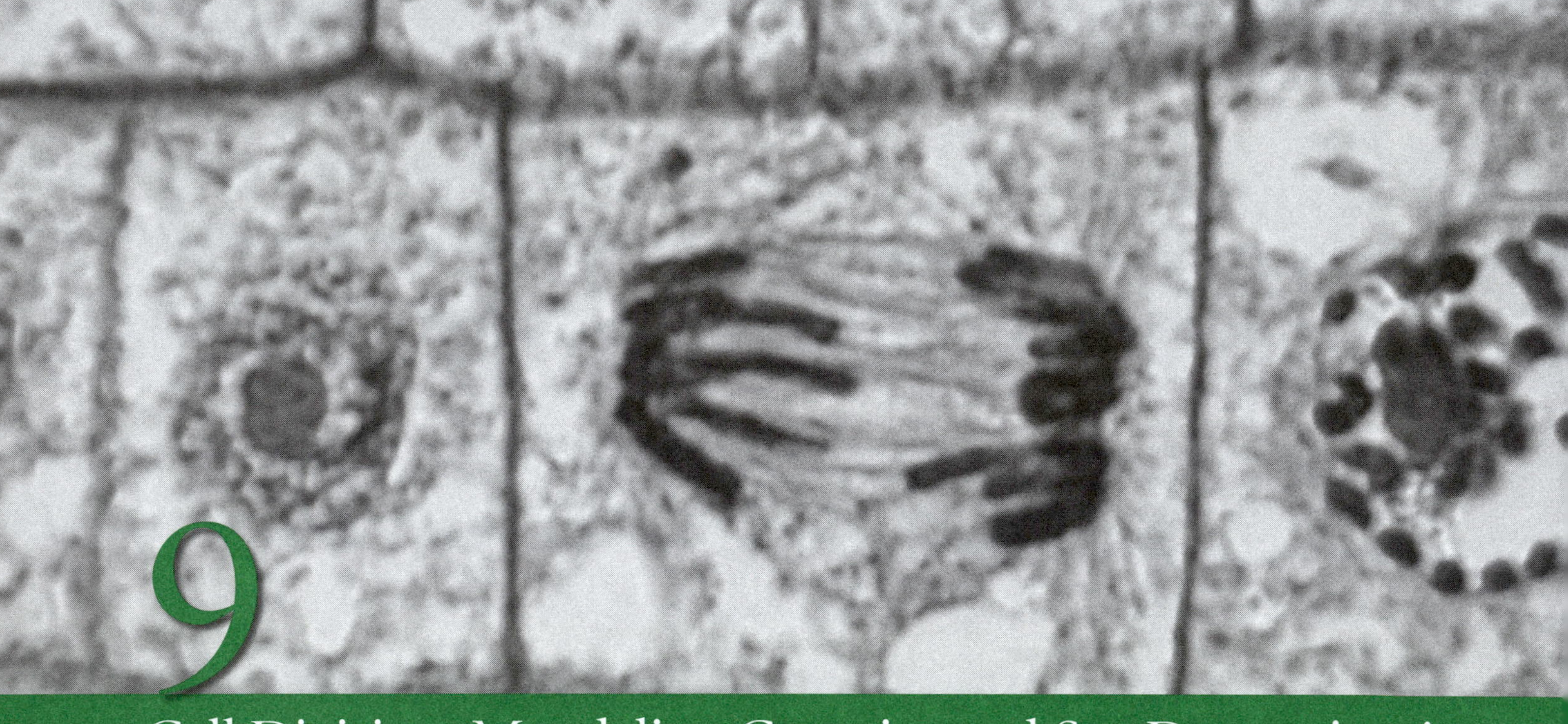

9 Cell Division, Mendelian Genetics and Sex Determination

Chapter Summary

Life depends on the fidelity with which organisms transmit genetic information to their offspring, but evolution cannot occur without genetic variability. Genetic traits, changed or unchanged, are transmitted from one generation to the next by cell division: binary fission in prokaryotes, mitosis (somatic cell division) or meiosis (gamete-producing division) in eukaryotes. In mitosis, all the daughter cells are genetically equivalent; in meiosis genetic variation is provided by recombination between homologous chromosomes, and by random assortment of chromosomes into the gametes.

In the nineteenth century, most biologists advocated (and believed in) blending inheritance. In the mid 1860s, Gregor Mendel broached two principles of inheritance that contradicted blending inheritance: (1) the principle of segregation — what we now know to be alleles of a single gene segregate from each other into the gametes as discrete entities — and (2) the principle of independent assortment — the independent segregation into gametes of the genes on different chromosomes. Although the phenotypic effects of some genes may blend with those of other genes, the genes themselves do not blend with each other.

Mendelian ratios result from segregation of the alleles of a gene on homologous chromosomes, independent assortment of different pairs of chromosomes, interaction of alleles of a gene (dominance, codominance), and interaction involving different genes (e.g., epistasis). Deviation from Mendelian ratios can occur because of linkage (the close location of two or more genes on the same chromosome), sex linkage (location of the gene on a sex chromosome), genes located on organelle chromosomes in mitochondria and chloroplasts and any biased transmission of alleles (segregation distortion) among others.

Among the genetic sources of variation are incomplete dominance or codominance of alleles, multiple alleles of a gene and the many instances when one gene locus affects the phenotypic expression of another (epistasis). In linkage, genes located on the same chromosome are constrained to varying degrees (recombination frequencies, linkage distances) to remain together during meiosis. In sex linkage, the heterogametic sex (XY) expresses all alleles lying on the X chromosome without regard to dominance relationships.

Organismal evolution relies on two fundamental aspects of genetics or biological inheritance: **constancy** and **variability**.

Constancy and Variability

Constancy resides in the observation that *like produces like,* and derives from the ability of nucleic acid macromolecules to replicate using a semiconservative mechanism (see Chapter 7). Constancy has the evolutionary significance that all life processes, from biochemistry to behavior, physiology to populations, energetics to ecology, development to decline, depend on the transmission of information from previous generations.

In contrast, *variability* resides in the observation that *like can produce unlike.* Replication of biological information is not always constant or exact because of changes (**mutations**) in the DNA of the genes. Mutations in genes within animal cells are lost when the individual dies. Mutations in genes within germ cells are passed on to the next generation and provide the basic foundation for genetic variation. The evolutionary significance of genetic variability is its potential to fuel evolution as organisms change to differ from their ancestors (see Chapter 10). Constancy and variability are indelibly intertwined in evolution through hereditary information (the **genotype**) that primarily determines organismal features (the **phenotype**). There is no one-to-one correspondence between genotype and phenotype.

As developmental and environmental signals influence (perhaps even determine) which genes, gene pathways and gene networks are expressed at which times/stages of the lifecycle, multiple phenotypes can arise from a single genotype. Examples are tadpole and frog, caterpillar and butterfly, and queen, soldier and worker ants. This chapter begins the exploration of **genotype-phenotype relationships** by reviewing the two fundamental concepts of the genetic system that enable organisms to change and yet, preserve their biological attributes: **genetic constancy** and **variation.**[1]

[1] For more in-depth discussions, see Hartl and Clark (1997), Hartl and Jones (1998), West-Eberhard (2003), Weiss and Buchanan (2004), and the chapters in Hall et al. (2004) and Hallgrímsson and Hall (2005).

Cell Division

The transmission of biological information is coordinated with cell division so that parental and daughter cells carry copies of the same information. Such cell division processes must have originated early in the history of life as a solution to the problem of how membrane-enclosed organisms could grow without enlarging themselves to the point of endangering their existence. Because biological information is coded as long-chain nucleic acid molecules, it is basic to cell division that the structures that bear these molecules — the **chromosomes** — must replicate and divide.

In **prokaryotes,** which have a single chromosome, the two products of chromosomal replication attach to different points on the cell membrane. As the cell elongates to form two daughter cells, the two chromosome products separate, each becoming enclosed in a separate daughter cell.

In **eukaryotes,** all of which have at least three chromosomes, more complex cell division processes occur, divided into **mitosis** (cell division of somatic or body tissues), and **meiosis** (division of the cells that produce gametes). In animals, we see a third type of cell division, **cleavage,** which is the division of the fertilized egg (**zygote**) to form the initial multicellular embryonic stage: the blastula.

The three types of cell division in animals differ at the cellular level with respect to:

- whether the cytoplasm is equally divided at each division;
- whether growth (synthesis of new cytoplasm) occurs before the next division; and
- whether the cells separate at each division (mitosis and meiosis) or remain attached (cleavage).

These three aspects and whether the cells have one or two sets of chromosomes (are haploid or diploid) are illustrated in **Figure 9-1.**

Mitosis

In mitosis, parental and daughter cells have exactly the same numbers and kinds of chromosomes. Mitotic cell division

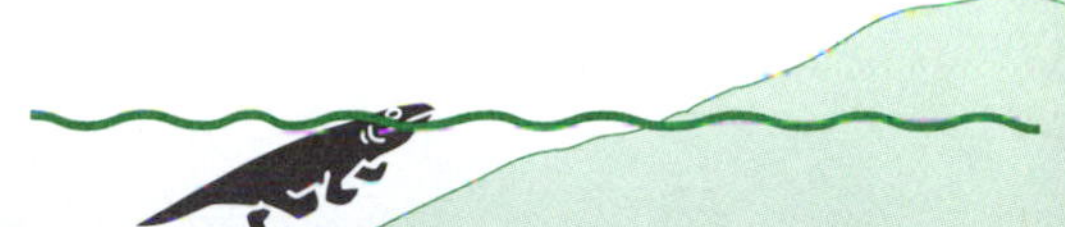

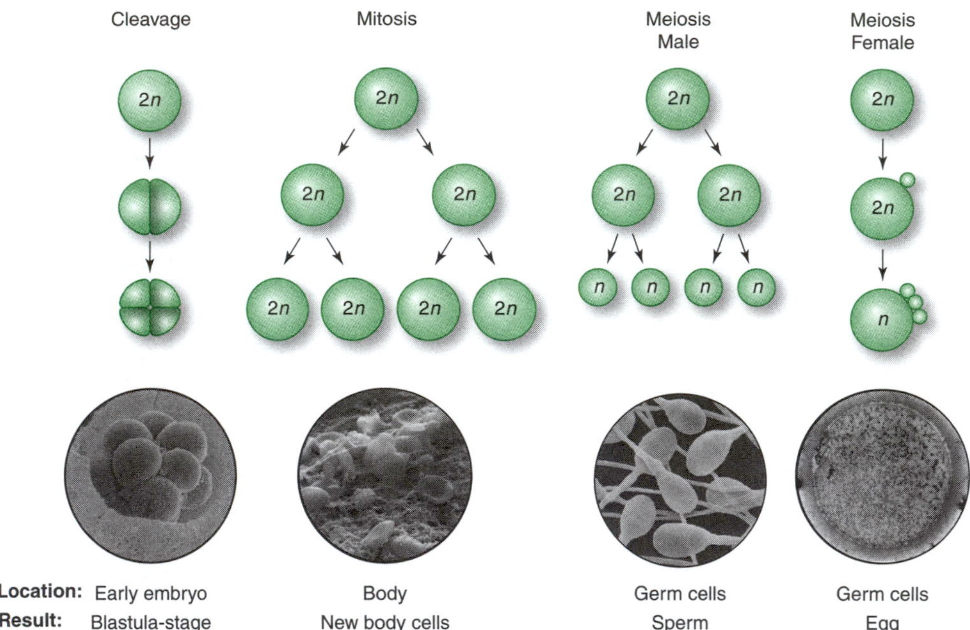

FIGURE 9-1 The essential differences between cleavage, mitosis and meiosis in animals, with emphasis on cell size and chromosome number. In *cleavage,* cell size is reduced at each division, most simply by equal division of cytoplasm as shown, but often by unequal distribution of cytoplasm between the two cells, which remain attached to form the multicellular embryo. All cells are diploid (shown as $2n$). In *mitosis,* cell size remains the same after each division because of synthesis of new cytoplasm (growth) after each wave of DNA synthesis and equal distribution of cytoplasm to each daughter cell, which may be identical or may differ (as when a stem cell divides to produce one stem cell and one cell that differentiates). All cells are diploid. Although the mechanism of chromosome reduction during *meiosis* is the same in germ cells of males and females, the fate of the cells differs. Equal distribution of cytoplasm in the male germ line results, in the example shown, in four sperm forming from a single cell after two divisions; there are many more divisions in real life. Unequal distribution of cytoplasm in the female germ line, in the example shown, results after the first division in one egg-forming cells and a much smaller polar body, and after two divisions in one future egg and three polar bodies. Again, there are many more divisions in real life. In both male and female germ lines, the reduction (meiotic) division reduces chromosome number from diploid ($2n$) to haploid (n). ([a] © Phototake/Alamy Images, [b] © Eye of Science/Photo Researchers, Inc., [c] © Phototake/Alamy Images, [d] © L. Zamboni, D. W. Fawcett/Visuals Unlimited.)

provides all the body cells of an organism with the same chromosome constitution, or **karyotype.** As illustrated in **Figure 9-2,** mitosis first becomes obvious during prophase when the gradually condensing chromosomes replicate, each chromosome then consisting of two **chromatids** connected at their centromeres. Chromatids arrange on a metaphase plate, in which the two members of a pair of chromatids connect to spindle fibers that go to opposite poles. The chromatids of a replicated chromosome separate in **anaphase** (see Fig. 9-2) and move to the two poles (telophase). Telophase is followed by cytokinesis, which partitions the cytoplasm into two daughter cells, each with a complete set of chromosomes identical to the parental cell. Mitosis provides the foundation for development and differentiation of a multicellular organism, including asexual reproduction where it occurs.

Meiosis

In meiosis, the cell division process used by eukaryotes engaged in sexual reproduction, the gametes that form contain only one of each pair of chromosomes. Meiosis is characterized by *one round* of DNA replication — resulting in every chromosome having two chromatids — and *two* cell divisions. The major feature of meiosis is the pairing of homologous chromosomes, allowing orderly separation of one of the two chromosomes to the daughter cells during the first cycle of meiotic division, reducing the chromosome number in half, each half containing only one homologue of each pair of chromosomes. This first cycle follows the second division, where the two chromatids of a replicated chromosome separate to yield two cells that develop as male or female gametes. A male gamete (sperm) fertilizes a female gamete (egg) to form the single-celled zygote. The mitochondria (and therefore mitochondrial genes) and most of the cytoplasm of the zygote come from the egg, but each gamete (egg and sperm) contributes an equal genetic element to the zygote nucleus. Two situations obtain:

1. If these events occur in an organism whose life cycle is primarily **diploid** (cells possessing two sets of chromosomes), mitosis replicates the number of chromosomes (the karyotype) in every somatic cell, until meiosis forms diploid gametes in the male and female gonadal tissues.
2. Each meiotic gamete has a **haploid** karyotype (only one set of chromosomes). If the zygote forms in an organism

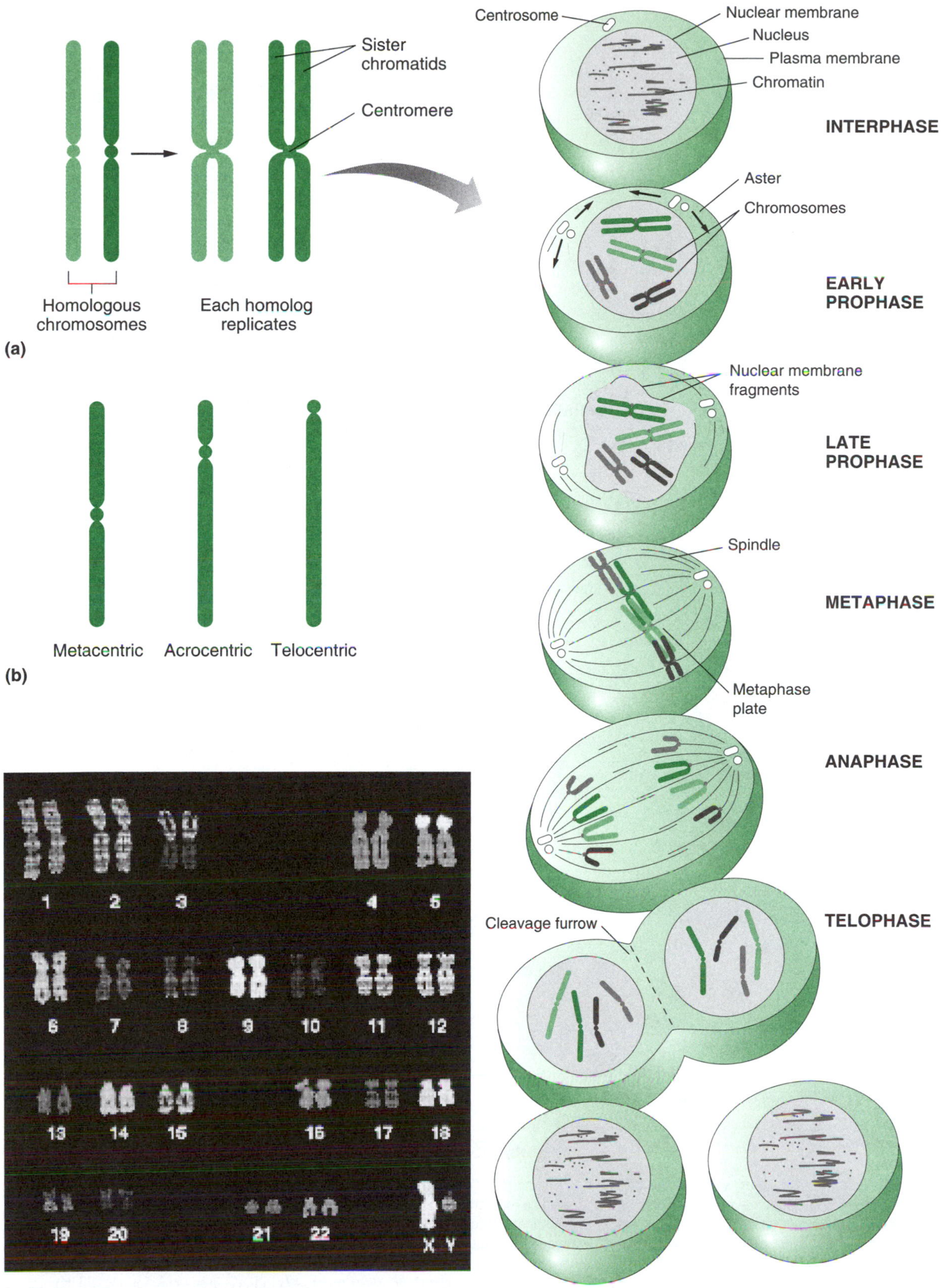

Sister chromatids

Centromere

Homologous chromosomes

Each homolog replicates

(a)

Metacentric

Acrocentric

Telocentric

(b)

1 2 3 4 5

6 7 8 9 10 11 12

13 14 15 16 17 18

19 20 21 22 X Y

(c) Human karyotype

Centrosome

Nuclear membrane

Nucleus

Plasma membrane

Chromatin

INTERPHASE

Aster

Chromosomes

EARLY PROPHASE

Nuclear membrane fragments

LATE PROPHASE

Spindle

METAPHASE

Metaphase plate

ANAPHASE

Cleavage furrow

TELOPHASE

(d) Mitosis

FIGURE 9-2 Diagrammatic presentation of various stages of mitosis in a somatic cell. During early, medium, and late prophase stages chromosomes (a–c) thicken and condense. Each chromosome with its attached replicate (a) lines up on a "metaphase" plate. During anaphase (c), the replicates (daughter chromosomes) separate, going to opposite poles. During telophase (c) a nuclear membrane reforms around each polar group of daughter chromosomes, and these chromosomes revert to the more extended interphase state. Division of the cytoplasm (cytokinesis) is also completed during this final mitotic stage. (Part [c] is from Hartl, D. L. and E. W. Jones, 1998. *Genetics: Principles and Analysis.* Jones and Bartlett, Sudbury, MA.)

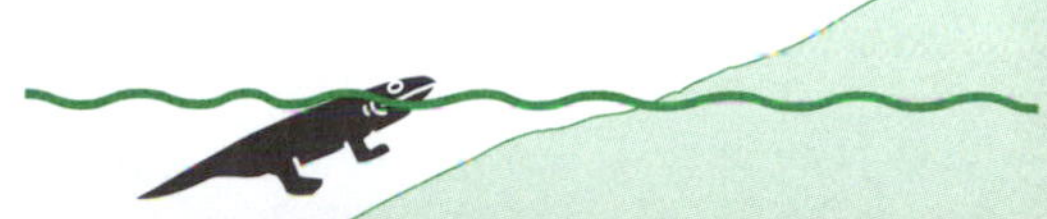

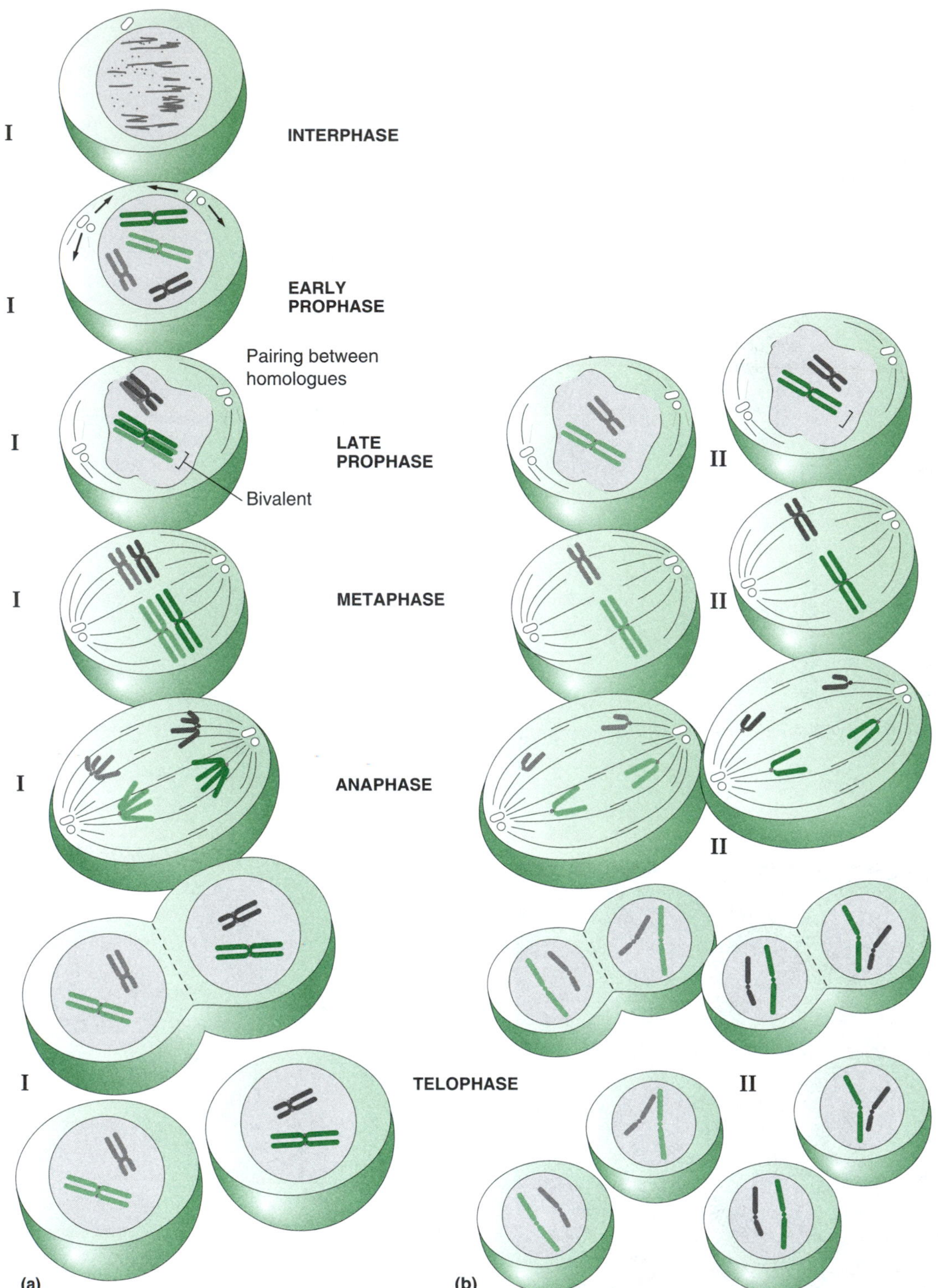

FIGURE 9-3 Principal stages of meiosis, showing the first (I) and second (II) meiotic divisions. (a) Pairing between homologous chromosomes occurs during prophase of meiosis I. By then, each chromosome has replicated to form two sister chromatids, connected at the centromere, which pair with the two sister chromatids of its homologue to form a bivalent pair of a homologous chromosomes with four chromatids. Although not shown in this illustration, some exchanges between non-sister chromatids can occur or have already occurred (see Fig. 9-4). At metaphase I, these paired groups arrange themselves on an equatorial plate preparatory to their separation. The two chromosomes of a pair have an equal probability of facing a given pole. At anaphase I, the two chromosomes of a homologous pair separate, each taking their chromatids with them to opposite poles. A telophase stage may follow, which involves some form of cytokinesis. In many organisms, this first meiotic division is a reduction division because it reduces the number of homologues in a nucleus from two to one. Thus, if homologous chromosomes A^1 and A^2 differ from each other, the resultant nuclei after meiosis I will not be identical, because each will have only one of these chromosomes. (b) The next meiotic division (II) that follows in such organisms separates the two chromatids of each chromosome, yielding two nuclei for each nucleus formed during division I. A diploid cell undergoing meiosis will, therefore, produce four haploid cells, each of which carries one representative of a homologous pair of chromosomes. The four haploid products of a sperm-producing cell can all function as gametes and produce sperm, while only one of the haploid products of an egg-producing cell functions as a gamete and produces an egg (see Figure 9-1).

whose life cycle is primarily haploid, the meiotic process produces nongametic cells carrying one of these four haploid sets of chromosomes.

In either case, whether the organism is haploid or diploid, meiosis, which generally follows the stages illustrated in **Figure 9-3,** is based on close homologous pairing between similar kinds of chromosomes called *homologues.* Pairing between homologues allows them to separate (disjoin) from each other at the end of the first meiotic division, giving each gamete a representative of each kind of chromosome. **Homologous pairing** ensures equal division between chromosomes that carry similar, although not necessarily identical, genes.

In most eukaryotes, each homologous chromosome replicates before the pairing process begins and, as in mitosis, produces two sister chromatids (see Fig. 9-3).[2] Probable causes (or at least an accompaniment) of pairing between homologous chromosomes are microscopically observed crosses (*chiasmata,* singular chiasma) between paired chromatids, called where transfers of genetic material takes place. Such events are usually distinctive of meiosis (**Fig. 9-4**). Mitotic chromatid replication and division occurs without formation of chiasmata.

Sources of Variation

Sexual reproduction and its accompanying meiotic divisions provide two important sources of variation, an obviously adaptive advantage in providing new genetic variation when environments are changing. First, because of the phenomenon of **recombination** or genetic exchange (also called **crossing over**), sections of homologous chromosomes can exchange material, thereby forming different linear arrays of nucleotides (see Fig. 9-4). As nucleotide sequences constitute *genes* (see Chapter 8), recombination can produce different combinations of genes along a chromosome, and different kinds of genes when crossovers occur within their nucleotide sequences. In addition, as the number of pairs of homologous chromosomes increases, the meiotic process assorts them into increasingly *numerous varieties* of chromosome combinations.

Two homologous pairs of chromosomes with genes *A* and *B* having alleles A^1 A^2 and B^1 B^2

$$(A^1 A^2, B^1 B^2)$$

segregating in a cross can yield four different kinds of gametes

[2] In some organisms such as dinoflagellates (flagellated marine, freshwater, and/or parasitic protists) whose phylogenetic position is in some respects midway between prokaryotes and eukaryotes (Morden and Sherwood, 2002), homologous chromosomes pair without chromatid replication. Their homologous chromosomes pair when two cells unite sexually, but then disjoin to opposite poles in a single meiotic division to form two haploid cells. Meiosis of this kind, consisting of only a single division, may have been a stage before eukaryotes were able to use chromatid replication for gene exchange, which necessitates two meiotic divisions.

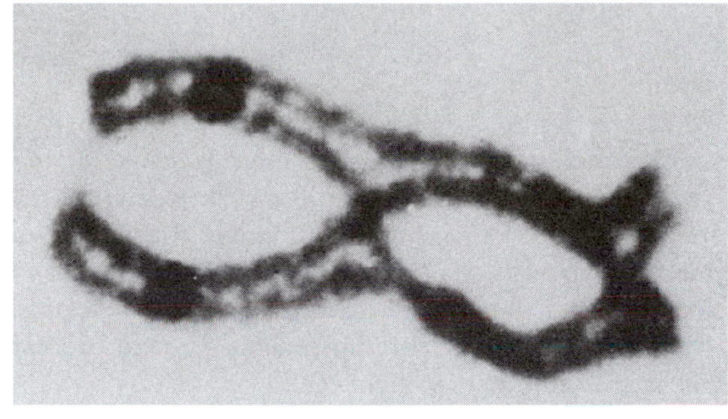

(a)

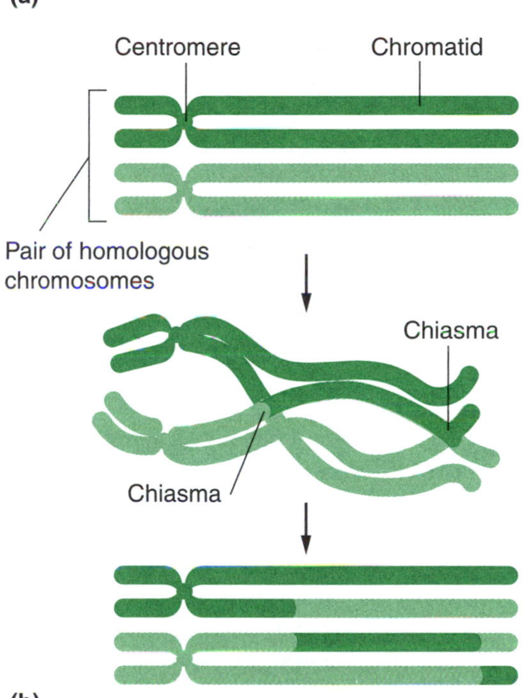

(b)

FIGURE 9-4 Diagrammatic interpretation of pairing and chiasma formation between homologous chromosomes in meiosis. By the prophase stage, each chromosome has replicated into two "sister" chromatids. Pairing between the two members of a homologous chromosome pair and their sister chromatids forms a "four-strand" bivalent showing physical crossovers (chiasma) between chromatids of different homologues. (a) A chiasma marks a point at which genetic material is exchanged. (b) Anaphase I of meiosis separates each chromosome member of a homologous pair, but the two sister chromatids of each chromosome do not separate from each other until anaphase II. (From Hartl, D. L. and E. W. Jones, 1998. *Genetics: Principles and Analysis.* Jones and Barlett, Sudbury, MA.)

$$(A^1B^1, A^1B^2, A^2B^1, A^2B^2).$$

Segregation of three loci (A, B, C), each with two alleles (1, 2), each located on three different chromosomes

$$(A^1 A^2, B^1 B^2, C^1 C^2)$$

can yield eight different gametic combinations

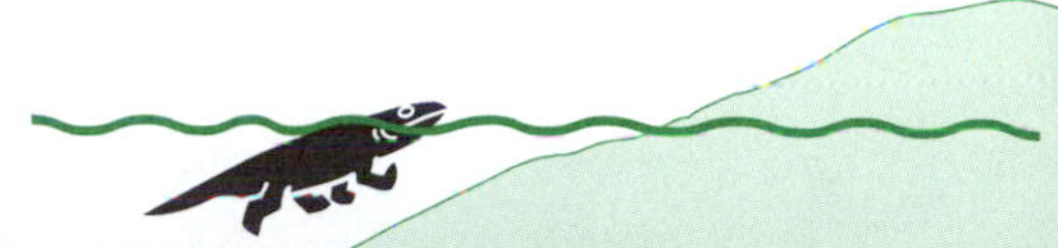

$$(A^1B^1C^1, A^1B^1C^2, A^1B^2C^1, A^1B^2C^2, A^2B^1C^1, A^2B^1C^2, A^2B^2C^1, A^2B^2C^2)$$

If we extend this to four pairs of chromosomes, 16 different gametic combinations are possible, the rule being that the number of possible different kinds of gametes equals 2^n, where n represents the number of pairs of chromosomes undergoing meiosis. In humans with normal karyotypes of 23 pairs of chromosomes, the different kinds of gametes that can form in this simple chromosome assortment process alone is 2^{-23}, or more than eight million varieties.

Mendelian Segregation and Assortment

Because the molecular basis of genetics long remained unknown, fundamental genetic principles primarily derived from observations on the transmission of more visible and obvious biological characteristics. In the early 1860s, Gregor Mendel in Brno, Czechoslovakia discovered the earliest genetic laws derived from such observations although biologists did not rediscover his work until the beginning of the twentieth century.

As described in Chapter 3, the pre-Mendelian view of heredity prevailing throughout the nineteenth century was that heredity followed a **blending inheritance** process in which offspring inherited a dilution, or blend, of parental characteristics derived primarily from their appearance (phenotype). Mendel's exceptional contribution was to demonstrate that organisms have a distinct hereditary system (now known as the genotype), which transmits biological characteristics through discrete units (now known as **genes**) that remain undiluted in the presence of other genes.

Mendel's Experiments

Mendel experimented with a number of characters in the pea plant, *Pisum sativum,* in which each character possesses two alternative appearances, or **traits,** such as smooth or wrinkled seeds, yellow or green seeds, tall or short plants. Mendel bred pea plants for many generations, observing the appearance and counting the numbers of each different trait (whether the seed coat was smooth or wrinkled, whether its color was yellow or green and whether the flowers were colored or white) among the individuals in every generation in a large number of plants.

Mendel spent two years (1856–1858) ensuring that his plants bred true. This was important, both for the characters he chose and because he was working with 37 varieties of peas from this one species, some of which were subspecies, giving his results an immediate link to classification and relationships. In case you want to repeat some of Mendel's experiments, the characters he found most distinctive and the easiest to distinguish are listed below:

seed shape — round(ish) or angular and wrinkled
seed color — pale yellow/bright yellow/orange-colored, or green
seed coat color — white or grey/grey brown/ leather brown; presence or absence of purple spots
mature seed pods — inflated or deeply constricted and less wrinkled
upright pod color — light to dark green or vividly yellow
flower position — along the main stem, or at the tip of the stem, and
stem length — 6 to 7 feet to 1 foot.

From the seven years (1856–1863) Mendel spent breeding peas, recording characters and calculating ratios, he developed *two fundamental principles of heredity: the principle of segregation, and the principle of independent assortment.*

Principle of Segregation

When two pure-breeding parental stocks that differ with respect to a character are crossed, the first filial (F_1) generation carries the determinants (genes) for each of these traits. One of the characters of pea plants used by Mendel was the texture ("the feel") of the seed's coat, its outer covering. The seeds were either smooth or wrinkled. Because the pea plants Mendel used were diploid, we now know that each individual plant had a pair of genes for seed coat texture as well as pairs of genes that govern other characters such as seed color and plant size.

In current terminology, the two members of a particular gene pair (smooth or wrinkled seeds, for example) are known as **alleles.** Geneticists use many types of symbols to distinguish alleles. For simplicity, we will label the smooth and wrinkled alleles as, respectively, *S* for smooth and *s* for wrinkled, the system used by Mendel. These alleles segregate in the gametes of the F_1 hybrid, which, if crossed, unite to form a second filial generation (F_2) in predictable proportions of 1 *SS*, 2 *Ss*, 1 *ss*, expressed as three smooth seeds and one wrinkled seed. This translates — which gives 3 smooth to 1 wrinkled seed in our example. Why a 3:1 ratio?

As shown in **Figure 9-5,** these F_2 proportions arise because the allele for one trait (*S*) has **dominant effects** over the allele (*s*), which is known as the **recessive allele,** when both are present in *Ss* individuals; that is, *SS and Ss* seeds are smooth. The only wrinkled seeds are those with the *ss* genotype, hence, the 3:1 ratio. Individual organisms that carry two different alleles for any particular character (*Ss*) are **heterozygotes.** Those that carry two identical alleles (*SS, ss*) are **homozygotes.** Mendel's experiments showed that the factors responsible for heredity (genes) are neither changed nor blended in the heterozygote but segregate from each other to be transmitted as discrete and constant particles between generations.

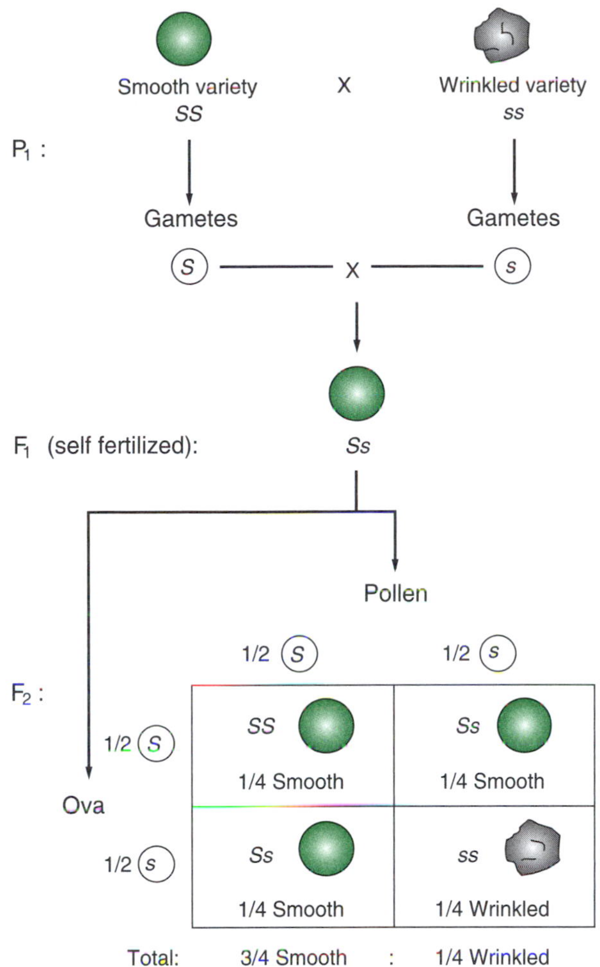

FIGURE 9-5 Explanation of Mendel's observed F_1 and F_2 results for the inheritance of seed texture in garden peas, based on the segregation of alleles *S* (dominant) and *s* (recessive).

Principle of Independent Assortment

When Mendel cross-pollinated plants that had two different characters (seed coat texture and seed color in our example), he demonstrated that the observed results could be predicted if one character had no effect on the segregation of the alleles of the other character. Genes for different characters segregated independently of each other. Now, we express this result as, "Mendel found that the results were predictable if different forms of genes (alleles) that determined *one character* (seed texture) have no effect on the segregation of alleles for the *other character* (seed color)." Alleles of different genes segregate independently of each other.

As shown in **Figure 9-6**, the F_1 of a cross differing in two such characters (smooth yellow × wrinkled green) shows only the dominant phenotypes (smooth and yellow seeds), but the F_2 proportions are predictably 9:3:3:1 — 9/16 smooth yellow, 3/16 smooth green, 3/16 wrinkled yellow and 1/16 wrinkled green. These ratios reflect independent association between seed texture alleles (*S* and *s*) and seed color alleles (*Y* and *y*), which means that the four kinds of gametes carrying one of the two alleles for each gene (*SY*, *Sy*, *sY* and *sy*) are produced in equal proportions, 1/4 each.

The cellular explanation for such **independent assortment** turned out to be the localization of the genes for each of the two characters on different pairs of chromosomes (**Fig. 9-7**). During meiosis, the two halves of a pair of homologous chromosomes move (sort themselves) toward opposite poles independently of any other pair of chromosome. As a result, the segregation of genes on one chromosome is independent of the segregation of genes on other chromosome(s).

Building on Mendel: Dominance Relations and Multiple Alleles

The particular phenotypic F_2 ratios Mendel observed in his diploid pea plants — 3:1 for one character (seed color), 9:3:3:1 for two characters (seed color and seed texture) — held true only as long as:

1. the phenotypic difference was due to a single gene;
2. the genes involved showed complete dominance and recessiveness; and
3. the gene only had two alleles for each character.

Incomplete Dominance and Codominance

Although Mendel published his results in 1866, they were unappreciated for 35 years when they were discovered, independently, by four individuals.[3]

Following the rediscovery in 1900 of Mendel's work and the principles of segregation of alleles and independent assortment, geneticists quickly realized that a single phenotypic difference could be due to a difference in more than one gene and that the dominance relationship between alleles could range widely. For example, the condition of **incomplete dominance** is expressed in the flower color of some plants in which the homozygote G^1G^1 produces red flowers, the homozygote G^2G^2 produces white flowers, and the heterozygote G^1G^2 produces an intermediate color, pink

[3] The structure of Mendel's paper is worthy of a short comment. Unlike the discursive literary "natural-history" style employed by Darwin with a minimum of analyses of data beyond the descriptive, Mendel's paper is in the plant hybridization tradition (Mendel was not the first to hybridize plants) with its data, analyses, ratios, citations of past work and more direct style. Indeed, the notation *A* for the dominant character, *a* for the recessive, and *Aa* for the hybrid are used throughout Mendel's paper. Darwin certainly did experiments in plant hybridization, reported some traits in 3:1 ratios, and used statistical methods in his analyses. Whether Darwin was one of the last of the "old-style" writers in biology and Mendel one of the first of the "new" requires attention from a suitably motivated literary historian. See McOuat (2001) for different styles of reasoning during this period, Sapp's analysis, *The Nine Lives of Gregor Mendel* on the Web site http://www.mendelweb.org/MWsapp.intro.html, and see other items on the Mendel Web site (http://www.mendelweb.org/; accessed March 29, 2007) for further information about Gregor Mendel.

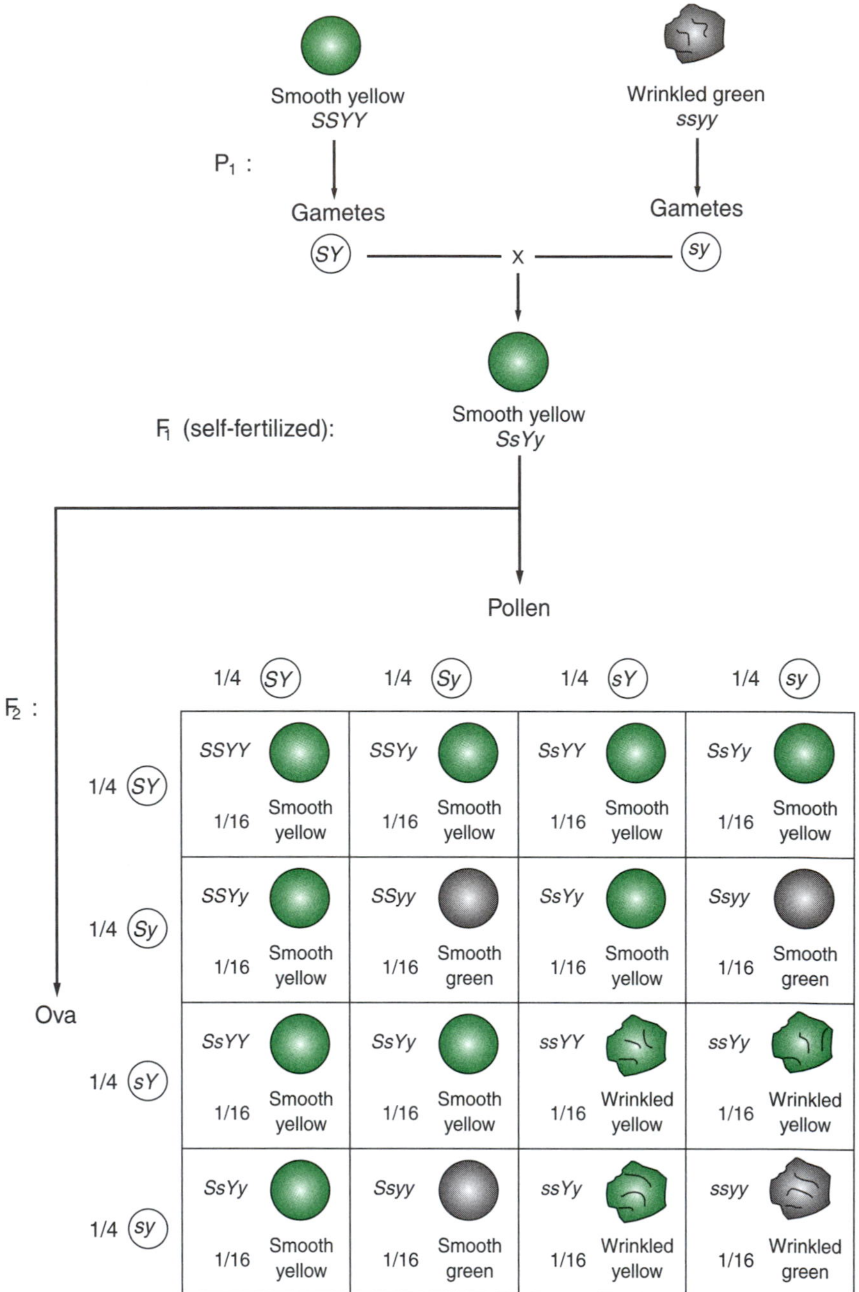

FIGURE 9-6 Explanation of Mendel's results for the segregation and assortment of alleles at two pairs of genes; *S* and *s* for seed texture, and *Y* and *y* for seed color. (From Strickberger, M. W., 1985. *Genetics,* 3rd ed. © 1985 by Monroe W. Strickberger. Reprinted by permission of Prentice Hall, Inc., Upper Saddle River, NJ.)

flowers. G^1 and G^2 are considered **codominant** when they have distinguishable effects in the heterozygote; that is, they each produce a uniquely recognizable substance so that the heterozygote G^1G^2 has both compounds.

Multiple Alleles

Furthermore, a gene determining a particular character could be represented by more than two alleles, $G^1, G^2, G^3, \ldots$ producing, for example, a different color or different compound in each different kind of homozygote, $G^1G^1, G^2G^2, G^3G^3, \ldots$. Further still, there may be different dominance relationships between the alleles in such a system so that G^1, for example, produces a codominant effect with G^2, but alleles G^1 and G^2 act as though they are each completely dominant over G^3. Such multiple allelic systems arise from the fact that a gene consists of a linear array of hundreds or thousands of nucleotides, and that an allelic difference may be caused by only a single nucleotide change at one or many positions along its length.

Epistasis

In addition, interactions between genes in different allelic systems may occur so that the expression of allele G^1, for

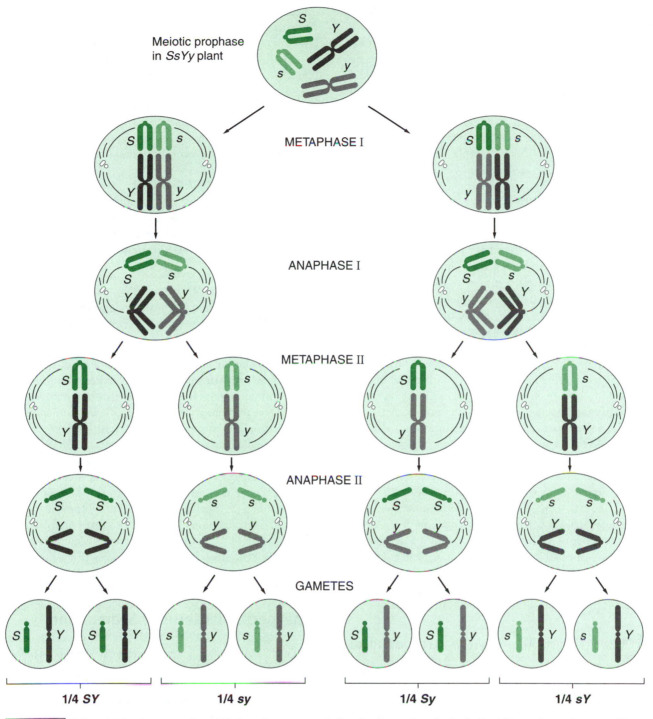

FIGURE 9-7 Explanation for the segregation and independent assortment of seed texture and seed color in Mendel's experiments in terms of factors or genes (*S, s,* and *Y, y*) localized on different chromosomes. Because of independent assortment, all four combinations of chromosomes in the gametes correspond to all four possible combinations of genetic factors. (Adapted from Strickberger, M. W., 1985. *Genetics,* 3rd ed. Macmillan, New York.)

example, changes because certain alleles are present at a different gene pair, such as H^1 or H^2. Among such interactions are those that **can modify the dominance relations of a particular allele** so that the effect of G^1, for example, is dominant over G^2 in certain genotypes at the H locus, but is recessive when the genotype at the H locus is different; that is, when the genetic background differs. Such interactions, when one or more gene pairs change qualitatively the phenotypic effect caused by other gene pairs, have been given the name **epistatic** interactions. Genes that change the action

of another gene(s) affecting quantitative traits are referred to as **modifier gene(s)**.

In general, wild type alleles are dominant in diploid organisms, presumably because they produce advantageous products in the presence of other alleles whose products may not be as advantageous or may be deleterious. On the molecular level, such wild type allelic products are mostly functional proteins such as enzymes. The products of mutant alleles often appear nonfunctional, partly functional or absent (null alleles). For example, the recessive allele causing Mendel's wrinkled peas prevents normal expression of an enzyme that makes branched starch molecules (Bhattacharyya et al., 1990).

Because so many allelic differences and different dominance and interaction effects are possible, the variation generated is far beyond that described previously for the independent assortment of chromosomes during meiosis. To take a simple example, a gene with four alleles (G^1, G^2, G^3, G^4) can produce 10 different diploid genotypes (G^1G^1, G^1G^2, G^1G^3, G^1G^4, G^2G^1, G^2G^2, and so on). A hundred such genes can produce 10^{100} genotypic combinations, a figure larger than the estimated number of protons and neutrons in the universe. Because any cellular organism carries more than 100 genes (which, incidentally, is the number of genes estimated to have been present in the original progenote), each with many possible alleles, the potential genetic variability in a single organism is greater numerically than any conceivable aspect of reality.

Viewing epistasis as a case of genetic interaction, and considering that increases in organismal complexity likely reflect genome organization, we can ask whether epistasis itself has evolved. From a recent study, the answer appears to be yes. In a large-scale analysis across multiple kingdoms of life, viruses exhibit fewer epistatic interactions (mutations have smaller effects than expected if epistasis is common) and eukaryotes more (mutations have greater effects than expected), findings that Sanjuán and Elena (2006) attribute to the compact versus more complex nature of the genomes of the two groups.

Exceptions to Mendelian Genetics

Although Mendel was not aware of the cause, his principles were based on the meiotic process of chromosome segregation and assortment, and the independent assortment of different pairs of chromosomes during gamete formation. Soon after Mendel's results were discovered, the association between Mendelian factors and meiotic chromosome distribution was recognized and provided the foundation for the new science of genetics.

Within a few years, many other studies in a variety of plants and animals echoed Mendelian ratios and their variations. Because eukaryotes share meiotic division processes, Mendelian genetics — the segregation of homologous chromosomes leading to the segregation of alleles, and the independent segregation of nonhomologous chromosomes (independent assortment) — seemed applicable to most, if not all, eukaryotic organisms with sexual reproduction (**Box 9-1**).

In the decades that followed two important deviations from Mendelian genetics became apparent: *extranuclear inheritance* and *aberrant patterns of the segregation of alleles.* Both circumvent conventional meiotic expectations and both have genetic and evolutionary significance.

Extranuclear Inheritance

The first unusual finding was that some traits do not follow a nuclear inheritance pattern but rather *transmit through egg cytoplasm.* Such **extranuclear inheritance** (also called **cytoplasmic** or **maternal inheritance**) arises because:

- cytoplasmic organelles such as mitochondria and chloroplasts have their own DNA genetic material, and
- the female parent deposits gene products (proteins, mRNAs) into the egg as it is being formed.

During fertilization of an egg, practically all the zygote cytoplasm including organelles is maternally derived. The sperm contributes primarily nuclear genetic material and, at most, a few mitochondria. In fact, it was through cytoplasmic inheritance that genes were first identified in mitochondria and chloroplasts, supporting proposals that these organelles entered eukaryotes by endosymbiosis. Mitochondria, chloroplasts, the nucleus and its membrane and the microstructure of the cytoplasm are all present *as pre-formed structures* in the zygote (indeed, in the egg) at the start of every generation. Such inherited structures provide evidence that we should not abandon entirely the notion of preformation (see Chapter 1). Eggs and all their contents are preformed inherited structures, as are patterns sometimes known as the "phenotype of the gene," for example, the process of methylation and imprinting discussed in Chapter 10.

Unexpected Patterns of Segregation

In a second departure from Mendelian genetics, some nuclear genes *do not segregate as expected in heterozygotes.* As mentioned previously, the four haploid products produced by the meiotic process in heterozygotes should lead to four gametes, two of which carry one allele and two the other (e.g., $Rr \rightarrow 2R{:}2r$). In a few cases, however, **segregation** is not normal, and the heterozygote for certain alleles produces more of one allele than the other (e.g., $Uu \rightarrow 3U{:}1u$). Such biased transmission, known as **segregation distortion** (or meiotic drive), is associated with a few rare genes, most notably *segregation distorter* in *Drosophila* and the *tailless* gene in mice. Although relatively uncommon, meiotic drive can lead to the persistence of a segregation distorter gene even though the gene may have deleterious effects.

In perhaps another deviation from Mendelian genetics — actually an elaboration rather than a departure — genes do not necessarily assort independently of each other if they are linked together on the same chromosome. These investigations began with inferences that some genes — **sex-linked genes** — are localized to the X (sex) chromosome. These inferences lead us into discussions of the genetics of sex determination (below), the evolution of sex-determining mechanisms (see Box 9-1) and sex linkage.

Sex Determination

Sexual reproduction involves differentiation into two different sexual forms in which reproduction occurs only as a result of union between sex cells (gametes) of different individuals to form an offspring's initial cell, the zygote (see Fig. 9-1). It has many interesting aspects, not the least of which is diversity. Indeed, do not presume that what occurs in humans or any other species is necessarily the model for all species. There is an amazing range of mechanisms for sex determination in organisms. Take *penis fencing*, for example.

Even though they are **hermaphrodite** (i.e., both male and females organs are found within each individual; see below) marine flatworms of the species *Pseudobiceros hancockanus* both hunt and fight for mates. Having found another individual with which to mate, these worms engage in penis fencing. Indeed, some individuals have two penises, giving them a decided advantage. Why bother, when you can fertilize yourself? Aside from the obvious disadvantage of inbreeding, it turns out that your sex depends on whether you are the first to pierce your opponent (mate) with your penis. The first to do so functions as a male, delivering a package of two-tailed sperm to the other individual, which, perforce, becomes the female and then invests resources in egg production.

Or take *temperature.*

The sex of alligators and quite a few other reptiles is set during embryonic development by the temperature at which the eggs develop. In some species, a higher temperature determines maleness while in other species, maleness is set at lower temperatures and femaleness at higher temperatures.

Or take fish that *change their sex* as adults.

We probably all know about the many chemicals that when released into the environment, lead to the appearance of female characteristics (feminization) in male animals. Not all will know that the majority of tropical reef fishes change sex during their lifetimes, often in a matter of hours, some switching from male to female, others from female to male, and some switching without inhibition of the other sex, becoming hermaphrodites. We assume the signals are hormonal but they await investigation, just as hormones play an important role in sex reversal in birds, although it requires experimental manipulation to change sex.

Male chickens have right and left testes of similar sizes. Females have left and right gonad rudiments of similar sizes, but only the left goes on to become a functioning ovary, the right remaining rudimentary. If the ovary is removed, the rudiment develops, sometimes into an ovary, sometimes into a testis and sometimes into an "ovotestis," which has features of both sexes. Unlike other vertebrates, hormones provide the trigger; treating males with estrogen also leads to the development of an "ovotestis."[4]

In the last example, honeybee males develop from unfertilized eggs and so are haploid. Females develop from fertilized eggs and so are diploid (*haplodiploidy;* see Chapter 16). But only one female becomes a queen and lays eggs. All other females help raise the queen's eggs and so contribute, albeit indirectly, to her reproductive success (kin selection; see Chapter 20). The queen is not merely a passive recipient of the contributions of these sterile workers; she releases a hormone that suppresses fertility in the workers (see Chapter 16).

Genetics of Sex Determination

As a rule, the two parental gametes that unite during fertilization are physically distinct from each other (sperm and egg) and form from different organs (testis and ovary). This division of labor, in which one sex was selected to produce a large stationary egg and the other to produce small mobile sperm, has the adaptive advantage of yielding a large zygote that provides an embryo with increased nutrition, yet still receives similar nuclear genetic contributions from each parent; recall maternal cytoplasmic contributions to the egg.

In most plants and many "lower" animals, the different sexual organs combine within single hermaphroditic or **monoecious** individuals, each capable of producing both types of sexual gametes. In contrast, some plant species (and many animals) consist of separate male and female (**dioecious**) individuals that produce pollen or ovules (sperm or eggs), but not both. The genetic and phenotypic variation produced by sexual reproduction appears essential for the long-term survival of those groups facing changing environments (see Box 9-1).

Sex Chromosomes

For many organisms, especially mammals, sex determination is associated with chromosomal differences between the two sexes, XX females and XY males, with the Y chromosome often smaller and mostly inactive, except for male determining and male-fertility genes. Inactive though it may be, the Y often serves an important role in meiosis, pairing with the X chromosome enabling the two chromosomes to separate during anaphase and go to opposite poles. The **heterogametic** XY individual can produce two kinds of gametes (X-bearing and Y-bearing) in equal proportions, accounting for an approximately equal sex ratio among offspring.

However, this is not the only sex chromosome system. Examples of other systems include XX females and XO

[4] See Davison (2006) and research on ovotestes throughout the animal kingdom.

BOX 9-1 Evolution of Sex-Determining Mechanisms

For chromosomal sex determination to occur, monoecy (hermaphroditism; male and female reproductive systems in single organism) must first change to dioecy (separate sexes) most probably fueled by the gain in genetic diversity from cross-fertilization (outbreeding) compared to its relative absence in self-fertilization (inbreeding). Given such benefit, there is selective advantage for a monoecious population to produce females that the male gametes of other members of the population can fertilize. As shown in the top sequence in **Figure 9-8,** in monoecious organisms the transition to such females requires little more than a mutation causing male sterility.

The presence of females, in turn, offers advantages to individuals that specialize in producing male gametes, which can also arise from monoecious organisms, but the mutations cause female sterility (Fig. 9-8, bottom sequence). The result, in this example, is a dioecious population bearing two kinds of chromosomes devoted to sex determination, the X distinguished by recessive male sterility, and the Y by dominant female sterility.

A cascade of consequences (10 of which are outlined below) follows from the two findings that individuals that are both male- and female-sterile cannot reproduce, and (see Fig. 9-8) that such individuals can arise from crossing over between sex-determining genes:

1. **Selection to prevent crossing over** promotes genes or chromosome arrangements that interfere with recombination between the X and Y, and helps preserve the tight linkage between genes on the Y chromosome important for the heterogametic (for example, male) sex. Thus, sexually antagonistic genes that benefit males but harm females, such as those used exclusively for male mating success, can be selected to localize on the Y chromosome, and are passed on only to Y-bearing offspring. To counter effects that can harm females, Rice (1998) proposed selection of X-linked and autosomal genes, which would enhance female fitness, resulting in an evolutionary see-saw of male-female "antagonistic coevolution."
2. Once X–Y recombination diminishes, its absence allows **harmful mutations on the Y chromosome to accumulate.** An example is a process known as *Muller's ratchet,* in which, because of sampling errors, populations more easily lose that class of individuals bearing the fewest harmful mutations, so that classes with increasing numbers of such mutations tend to increase with time. Whatever the cause, Rice (1994) experimentally demonstrated an increase in Y-chromosome deleterious mutations associated with lack of recombination.
3. To eliminate such deleterious mutant effects, **inactivation of the Y chromosome** (except for its sex-determining genes) **becomes selectively advantageous,** resulting in the common X–Y sex chromosome karyotype. We see chromosome inactivation when an entire chromosome or sections of it assume a state called **heterochromatin,** which stains differently under the microscope from normally active, unmodified sections, **euchromatin.** The Y chromosome is often heterochromatic

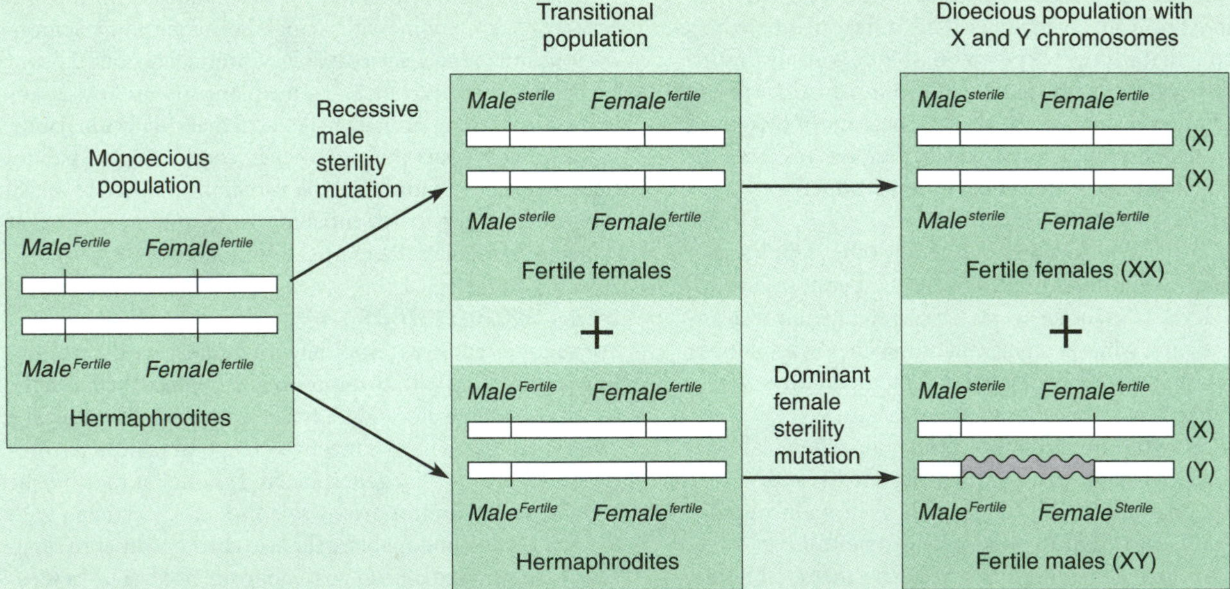

FIGURE 9-8 Sequence of nuclear mutational events leading to a transition from a monoecious to a dioecious population. Each horizontal bar represents a sex chromosome. In this example, the Y chromosome determines male sex development because of dominant *Male*^{Fertile} and *Female*^{Sterile} alleles, whereas the X chromosome carries recessives at these loci (*Male*^{sterile}, *Female*^{fertile}); the absence of the Y chromosome produces females. (Note that other mutational events, such as females produced by initially dominant *Male*^{Sterile} and *Female*^{Fertile} mutations, can lead to heterogametic ZW females and homogametic ZZ males.) The wavy section between Male and Female genes on the Y chromosome indicates a recombination deficient interval (caused either by close linkage or by crossover suppressors such as inversions) that interferes with crossovers that might otherwise lead to sterile individuals carrying *Male*^{sterile} and *Female*^{Sterile} alleles, or back to the hermaphroditic condition, *Male*^{Fertile} *Female*^{Fertile}. (Based on Charlesworth, B., 1996. The evolution of chromosomal sex determination and dosage compensation. *Current Biol.*, **6**, 149–162, with modifications.)

and the X euchromatic. Among other consequences of Y chromosome inactivation are increased numbers of transposable elements (see Chapter 10) and repetitive DNA sequences that can persist in chromosome sections not being actively selected for viability and where recombination cannot eject them.

4. As larger sections of the Y chromosome become inactive because of reduced recombination with the X, the **Y chromosome can decrease in size.** Selection no longer operates to preserve the Y chromosome, a feature common to many XX–XY species.

   Human Y chromosomes have been studied especially intensively. While human X chromosomes are some 165 Mb in size with around 1,000 genes, the Y chromosome is only 60 Mb with fewer genes. It is becoming apparent that the Y chromosome has been and continues to be subject to specific forces of selection, being subject to higher levels of mutation, deletion and insertion than any other genes in mammalian genomes. Consequently, mammalian Y chromosomes are regarded as undergoing systematic and ongoing degeneration to the point that **human Y chromosomes are close to extinction** (Graves, 2006). Other mammals have gone further down this road; two species of mole voles (*Ellobius*), and the Japanese spinous country ray *Tokudaia* have lost their Y chromosomes. Interestingly, and perhaps telling us something quite important, in flowering plants the mutation rate in pollen grains is also higher than in ovules.[a]

5. At some stage in this progression, the **Y chromosome has few if any functional genes** other than the sex-determining genes. If these are translocated elsewhere or if their functions are assumed by genes in another chromosome, the result can be an XX female/XO male species, in which sex determination depends on a *Drosophila*-like X/A ratio. In *Drosophila melanogaster,* sex determination is based on the X chromosome/autosome ratio. Individuals with a ratio of 1 (two X chromosomes and two sets of autosomes) develop as females. Individuals with a ratio of 0.5 (one X chromosome and two sets of autosomes) develop as males. A key regulatory gene involved in sex determination is *Sex lethal* (*Sxl*), which produces female development when turned on and male development when turned off. Turning *Sxl* on and off derives from the relationship among certain genes on the X chromosome and autosomes. An X chromosome/autosome ratio of 1 turns *Sxl* on while a ratio of 1/2 turns *Sxl* off.[b] Conversely, a **translocation** between the Y and an autosome produces a "neo-Y" chromosome, which now pairs in meiosis with its translocated autosome (X_2) as well as with the former X (X_1), leading to an X_1X_2Y male/$X_1X_1X_2X_2$ female sex chromosome constitution. **Figure 9-9** provides an example of *Drosophila* karyotypes showing a succession of X and Y chromosome changes.

6. In some cases, **only a single copy of a sex-determining gene is required** to embark on a pathway of sexually distinctive development. In mammals, the SRY gene on the Y chromosome serves such a purpose for male development. Other vertebrates, from fish to reptiles, do not share SRY's mammalian sex-chromosome locus, indicating that its mammalian role may have been derived from a special mammalian function.[c]

7. The difference in X (or Z) **chromosome gene dosage** between the heterogametic and homogametic sexes **can be compensated** for in various ways. One device is simply to ignore the problem if the X (or Z) chromosomes are small with few active genes; the ZW females in birds and butterflies generally produce half the amount of Z gene products of the ZZ males, a difference that does not seem to affect development.

   Two major systems of **dosage compensation** are known to function when X chromosomes are large, with many active genes. In *Drosophila,* the single X chromosome in the male is about twice as active as each of the two X chromosomes in the female, while in placental mammals, only one X chromosome of an XX female is functional in each cell. The result of both systems is therefore that an XX female has essentially the same level of X chromosome gene activity as an XY male. Some of the genes are involved in sex determination, of which many genes have now been identified, especially in the fruit fly *Drosophila* and the nematode *Caenorhabditis.*[d]

8. ***Haldane's rule,*** often used in describing defective hybridization between animal species, states, "When in the F_1 offspring of two different animal races one sex is absent, rare, or sterile, that sex is the heterozygous [heterogametic — XY, XO, ZW, or ZO] sex." Geneticists most often have ascribed the cause for this rule to sex-linked deleterious recessive alleles. Interspecific hybrids carrying autosomal alleles from one species that are incompatible with sex-linked recessive alleles from the other species show deleterious effects more readily in the heterogametic sex. X-linked deleterious recessive alleles in XY males, for example, are not masked by dominant alleles on the other X chromosome as they are in homogametic XX females. To this explanation, subsequent research has added two concepts: (1) that the sexual traits of males are produced by specific sets of genes that evolve rapidly, making some or many such males more sensitive to hybrid sterility, and (2) that sex-biased hybrid inferiority can act as an isolating mechanism capable of impeding gene flow.[e]

9. Although the heterogametic sex can theoretically produce equal numbers of male- and female-determining gametes, **equal sex ratios need not necessarily follow.** Some nuclear genes may affect the viability of male or female zygotes. Other genes may alter the ability of the X or Y chromosomes to

[a] See Whittle and Johnston (2003) for higher mutation rates in pollen than in ovules. Another aspect, little understood or studied, is the production in some species of two types of sperm or two types of pollen grains, leading, of course, to competition between sperm or pollen grains (Till-Bottraud et al., 2005).

[b] See Cline (1993) and Parkhurst and Meneely (1994) for sex determination in *Drosophila,* but note that this mechanism does not operate in all flies.

[c] The classic books by White (1973, 1978) and Bull (1983) provide many other examples of sex chromosome varieties that arise from diverse genetic and chromosomal changes.

[d] See Charlesworth (1996), Cline and Meyer (1996) and Lucchesi (1998) for dosage compensation.

[e] See Turelli (1998) for hybrid sterility, and see Wang (2003) for impeded gene flow.

BOX 9-1 Evolution of Sex-Determining Mechanisms (*continued*)

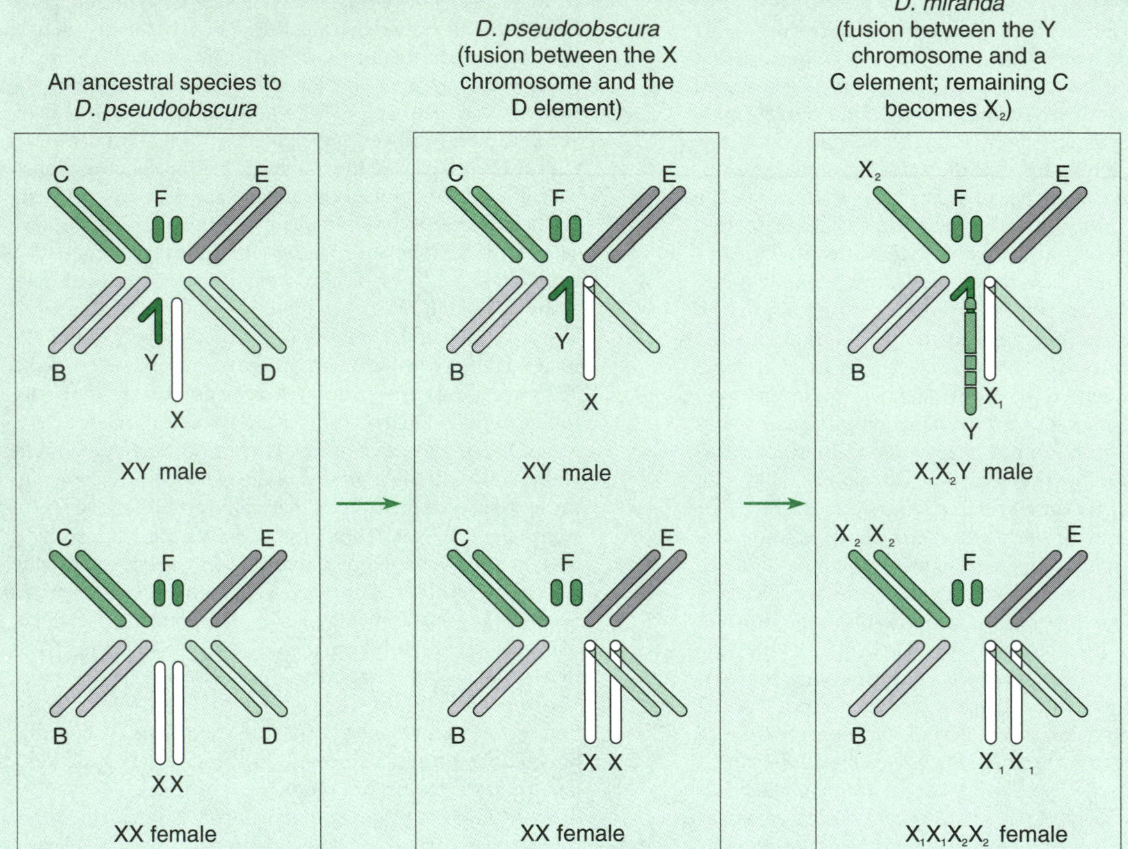

FIGURE 9-9 Diagrammatic view of sex chromosome changes in a lineage leading to *Drosophila miranda*. As shown here, the original chromosomal elements in the genus *Drosophila* are often designated by the six letters A to F, with the A element serving as the sex chromosome. The species on the left represents a hypothetical ancestor in which the common XY chromosome karyotype has been established. This is followed by fusion of the X with the D element in *D. pseudoobscura* and fusion of the Y with the C element in *D. miranda*. The sex chromosome constitution of *D. miranda* is therefore X_1X_2Y males and $X_1X_1X_2X_2$ females, each sex carrying three pairs of autosomes (elements B, E and F). (Adapted from Lucchesi, J. C.,1994. The evolution of heteromorphic sex chromosomes. *Bioessays*, **16,** 81–83.)

males, XX hermaphrodites and XO males, heterogametic (ZW) females and **homogametic** (ZZ) males, and XXXX females and XXY males. Surprisingly, because monotremes diverged from placental mammals 210 Mya (see Chapter 5), the platypus (*Ornithorhynchus anatinus*) has five X and five Y chromosomes, which form a translocation chain at meiosis, a system thought to be partly a carry-over from a more ancient reptilian or avian sex-determining system, and partly a modification of the mammalian X chromosome system.[5]

[5] See Scherer and Schmid, (2001) for the other sex chromosome systems, and see Grützner et al. (2004) for the platypus.

Autosomes and Sex Determination

The almost universal presence of sex chromosomes among sexual species does not necessarily mean these are the only chromosomes affecting sexual development.

Sex is a complex developmental character affected by numerous non–sex-chromosome (**autosomal**) genes. In *Drosophila*, for example, the ratio of X chromosomes to sets of autosomes (X/A) determines sex: In diploids (two sets of autosomes), females have an X/A ratio of 1 (2X/2A) and males have an X/A ratio of 1/2 (1X/2A). Ratios that differ from these two can produce sexual abnormalities; for example, in triploids (three sets of autosomes), an X/A ratio of 2/3 (2X/3A) produces "intersexes."

segregate normally during meiosis ("segregation distorters"). Either type of abnormality, whether in viability or gametic segregation, can affect male or female frequencies. Y-chromosome inactivation has the advantage of inactivating segregation distortion genes that would cause more Y- than X-carrying gametes, and therefore lead to more males than females.

One way of accounting for the prevalence of an equal male : female sex ratio in many species is to consider what happens when a particular sex is rare. Should a population have a scarcity of females (for example, a male:female sex ratio of 5:1), females, being more frequently mated, would on average produce more offspring than males, many of which cannot find mates. This would provide a reproductive advantage to genotypes that produce more females than males, and thus tend to correct the distorted sex ratio.

On the other hand, assuming that males are now relatively rare, the advantage would shift to genotypes that produce more males than females, again tending to correct the distorted sex ratio. At some point, the population will approach the stable sex ratio of 1:1, a value that no longer provides a benefit for a genotype to produce more of one sex than the other. Experimentally, it has been shown that populations of the platyfish *Xiphophorus maculatus*, composed of different sex ratio genotypes, evolve in this direction, as do those of *Drosophila*.[f]

[f] See Fisher (1958) for stable sex ratios, Basolo (1994) for platyfish and Carvalho and coworkers (1998) for *Drosophila*.

10. Interestingly, chromosomal sex determination **can transform into environmental sex determination** when the sex of one or both of the XY and XX karyotypes reverses because of sensitivity to agents such as temperature or hormones.

Various instances of sex reversal exist and might well lead to an environmental sex-determining system when the environment is "patchy" rather than uniform; that is, when parts of the environment influence one sex more than the other. For example, a greater food supply in some patches may lead to the production of larger females, in contrast to other, less nutritional patches that produce smaller males. Such differences can act as an important selective factor in establishing sex determination based on environmental sensitivity rather than on a genetic or chromosomal basis that does not distinguish among environments.

An example of environmental impact on sex ratio has been described in Seychelles warblers, *Acrocephalus sechellensi*, birds that commonly use their daughters as "helpers" in raising additional offspring. When food is plentiful, helper daughters increase their parents' reproductive success, producing broods with a female:male ratio of about 6:1. When food is scarce, such daughters hinder their parents' reproductive success by competing for the limited supply, and the female:male offspring ratio drops to about 1:3 (Komdeur et al., 1997).

In many cases, the potential for becoming either male or female exists at the time of fertilization, no matter what the sex chromosome complement. Thus, gene mutations such as *transformer* in *Drosophila* can turn XX zygotes into males, and the gene that causes testicular feminization syndrome in humans can turn XY zygotes into females. The function of sex chromosomes or the X/A ratio is to act as part of the "switch" mechanism that directs development into one of the sexes — male or female — the organism is capable of becoming.

Among the many questions these various findings raise are the following:

- How did sex determination become tied to sex chromosomes?
- What led to the different varieties of sex chromosome karyotypes?
- Because the heterogametic XY or XO (or ZW) sex carries only one copy of X (or Z) chromosome genes, and the homogametic XX (or ZZ) sex carries two copies, are there mechanisms that compensate for this difference in gene dosage and how did they evolve?
- When only one sex is sterile or nonviable among the offspring of a cross between two species, why is it almost always the heterogametic sex ("Haldane's rule")?
- Is there some advantage for the commonly observed male/female sex ratio of 1/1?
- How and why did some species abandon chromosomal sex determination and adopt environmental sex determination?

Various evolutionary biologists have offered answers to these questions, some of which are outlined in Box 9-1.[6]

Environmentally Induced Sex Determination

In contrast to what seem to be purely genetic mechanisms, **environmentally dependent sex-determining mechanisms** exist in many animals and in many of those plants with separate sexes.[7]

[6] See Maynard Smith (1978, 1988a), Bull (1983), Hodgkin (1992) and Charlesworth (1996) for seminal studies on the evolution and maintenance of sex.

[7] If defined as the formation of separate male and female organisms, sex determination, whether genetic or environmental, does not apply to most plants; some 90 percent of seed plants have both male and female gametes. Five percent of seed plant species have separate males and females (dioecy; see Chapter 14), and 5 percent are females or hermaphrodites (gynodioecy). Only a minority of dioecious species have sex chromosomes. Even in the absence of sex chromosomes, sex determination may be genetic (or it may

1. In the echiuroid sea worm, *Bonellia*, larvae that are free-swimming and settle on the sea bottom develop into females with 10- to 20-cm long body and a meter-long proboscis. Larvae that land on the proboscis of a female metamorphose into tiny, 1-mm-long males that lack digestive organs, existing as a parasite embedded within the genital ducts of the female, and producing virtually nothing but sperm. (See also the parasitic males of the bone-eating marine worm *Osedax* described in Box 11-2.)
2. Egg size appears to be the sex-determining mechanism in the sea worm *Dinophilus*, large eggs producing females and small eggs males.
3. As introduced earlier, in some reptiles, high egg-incubation temperatures produce mostly males (e.g., in a lizard, the red-headed rock agama, *Agama agama*), whereas warm temperatures in others produce mostly females (e.g., in the painted turtle *Chrysemys picta*).
4. In certain fish, social behavior can influence sex so that loss of socially dominant males from a group causes sex conversion of the dominant female in the group. In some coral reef fishes, sex changes seem to occur as a result of visual stimulation: Females become males when the surrounding fish in the group are relatively small.
5. In horsetails (*Equisetum*), female characteristics appear when the plant lives in good growth conditions, male characteristics when in poor conditions.
6. Some wasps deposit their eggs inside insect larvae. Female parasitoid wasps deposit eggs that will develop into females into larger host larvae. Large versus small host larvae seem to be recognized by the wasp as a relative property of the larvae available; as long as there is a size difference, wasps will differentiate between them, thereby enabling females to grow to larger sizes in the larger hosts and allowing those females to produce more eggs.

Sex Linkage

Early in the twentieth century, geneticists discovered that various genes could be localized to a particular chromosome associated with sex determination.

A prominent example is the "bleeder" disease, *hemophilia*, localized to the X chromosome of humans and other mammals, in which males are XY and females are XX. Males (the **heterogametic sex**) produce sperm that contain either X or Y chromosomes, whereas females (the **homogametic sex**) produce only X-bearing eggs. If the proportions of X- and Y-bearing sperm are equal, fertilization restores the two sexes in equal frequency. However, because males carry only a single X chromosome, and the Y chromosome is mostly inactive, alleles present on the X chromosome express their effects in males, although such alleles may be recessive when in females. Thus, a single hemophilia-producing allele on the X chromosome of a male with XY causes the classic hemophilia disease, whereas in XX females, two such alleles are necessary to cause hemophilia, a situation that rarely happens.

Ohno (1979) proposed that we expect any sex-linked gene found in one mammalian species to be sex linked in other mammals. Interestingly, many of the more than 100 sex-linked genes now identified in humans are also sex-linked in other mammals (**Table 9-1**). Conservation of the same genes on different mammalian X chromosomes indicates that a large part of this chromosome has persisted throughout mammalian evolution, at least 90 My. The evolution of sex-determining mechanisms and some of their effects are briefly discussed in Box 9-1.

be environmental; we just do not know for the vast majority of species). Interestingly, one of the most primitive of living plants, *Ginkgo biloba*, has sex chromosomes (Mark Johnston, personal communication).

Linkage and Recombination

The localization of certain genes to the X chromosome, which geneticists first demonstrated in the fruit fly, *Drosophila melanogaster*, provided the opportunity for establishing *distances between such genes*. Such determinations (**linkage** relationships) arise from crossing over or recombination events in which exchanges occur between the chromatids of paired homologous chromosomes. For example, in the experiment diagrammed in **Figure 9-10**, two *Drosophila* sex-linked genes, those involved in *white eyes* (w^+ and w) and *miniature wings* (m^+ and m), recombined with a frequency of about 38 percent to produce new chromosomal combinations. Because genetic crossing over does not normally occur in male *Drosophila*, and males only have a single X chromosome, these recombination events take place in females. Interestingly, certain crosses between *Drosophila* strains can produce hybrid anomalies in which considerable male recombination takes place. The intrusion of P transposable elements in some populations causes such events.

This frequency of recombination provided a measure of the linkage distance between genes on the same chromosome: the greater the recombination frequency, the greater the distance. Thus, because the recombination frequency between the genes for *white eyes* and *cut wings* was about half the frequency of that between *white* and *miniature*, geneticists could assume the linkage distance between *white* and *miniature* to be about twice that between *white* and *cut*. Such recombinations with sex-linked and non–sex-linked (autosomal) genes provided linkage maps such as shown in **Figs. 9-11** and **9-12**, indicating that chromosomes consist of linear arrays of genes, or **loci**, whose relative positions are additive (if three genes link in the order H-G-I, then the H–I linkage distance is the sum of H–G + G–I distances).

The relative position of different genes on chromosomes also can be established using a number of other more

TABLE 9-1 Some gene products and diseases linked to the X chromosome in humans and other mammals

X-Linked in Humans	X-Linked in Other Mammals
α-galactosidase deficiency	More than 25 species, including chimpanzee, gorilla, sheep, cattle, pig, rabbit, hamster, mouse, cat, dog, kangaroo
Anhidrotic ectodermal dysplasia	Cattle, dog, mouse
Bruton-type agammaglobulinemia	Mouse, cattle, horse
Copper transport deficiency	Mouse, hamster
Duchenne/Becker muscular dystrophy	Mouse, dog
Glucose-6-phosphate dehydrogenase deficiency	More than 30 species, including chimpanzee, gorilla, sheep, cattle, pig, horse, donkey, hare, hamster, cat, mouse, kangaroo, opossum
Hemophilia A (factor VIII deficiency)	Dog, cat, horse
Hemophilia B (factor IX deficiency)	Dog, cat, mouse
Hypoxanthine-guanine phosphoribosyl transferase (Lesch-Nyhan syndrome)	More than 25 species, including horse, hamster, dog, cat, mouse, cattle, gibbon, pig, rabbit, kangaroo
Ornithine transcarbamylase deficiency	Mouse, rat
Phosphoglycerate kinase	More than 30 species, including chimpanzee, gorilla, cattle, horse, hamster, mouse, kangaroo, opossum
Steroid sulfatase deficiency (ichthyosis)	Mouse, wood lemming
Testicular feminization syndrome	Cattle, dog, mouse, rat, chimpanzee
Vitamin D-resistant rickets	Mouse
Xg blood cell antigen	Gibbon

Source: Adapted from Strickberger, M. W. 1985. *Genetics,* 3d ed. Macmillan, New York, with modifications and additions. Further listings can be found in J. R. Miller (1990).

modern methods and approaches. These include segregation of gene-specific DNA markers, *in situ* hybridization of gene specific DNA on metaphase chromosomes and somatic cell hybridization involving deletions and chromosomal aberrations. The sequencing of the complete DNA sequences of many species, including humans, provides the most detailed relationship between genes on a chromosome. Unlike linkage relationships, which rely on percent recombination between two genes during meiosis, the latter methods provide the actual distance in base pairs between the two genes.

Using these methods, all human genes have been localized to specific chromosomes of the 23 human chromosomes. Moreover, as with similar sex-linked genes in mammals, genes localized to particular human chromosomes (identified by special banding patterns; see Fig. 10-7) can be predicted to lie on similar chromosomes in other mammals — a phenomenon known as **synteny**. In addition, as shown in Figure 9-11, linkage relationships among genes with similar functions in these mammals also persist. Elaborately detailed linkage maps such as shown in Figure 9-13 are available for bacteria and for viruses. The concept of synteny appears to hold in that closely related species appear to maintain the linkage relationship between genes, suggesting that during evolution, these relationships have been maintained.

All of these studies indicate the immense variability that can be generated through recombination; any chromosome may come to differ from its homologues by carrying a distinctive combination of alleles. In the next chapter, we examine how variation arises.

KEY TERMS

allele
anaphase
autosomal
chromatids
chromosomes
codominant
crossing over
cytoplasmic inheritance
diploid
dominant allele
epistasis
euchromatin
extranuclear inheritance
genotype
Haldane's rule
haploid
hermaphrodite
heterochromatin
heterogametic sex
heterozygotes
homogametic sex
homologous pairing
homozygotes
incomplete dominance
independent assortment
inversions

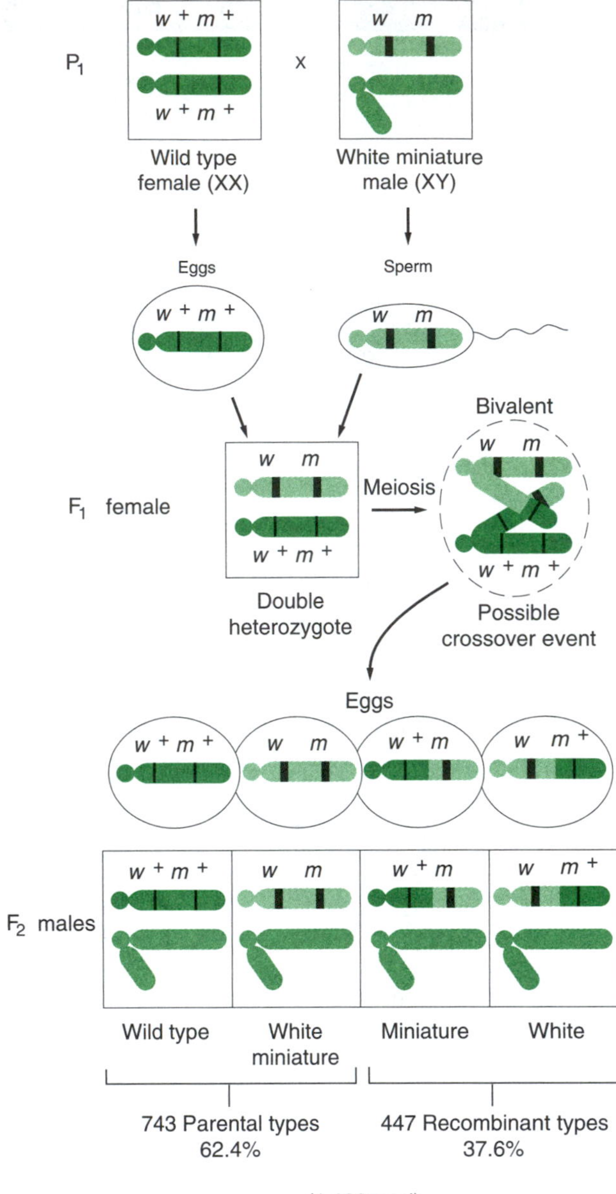

FIGURE 9-10 Recombination between the sex-linked genes *white* and *miniature* in *D. melanogaster* as evidenced by recombinant classes among F_2 males. Because the only sex chromosome the paternal parent contributes to male offspring is the Y, the meiotic events in the double-heterozygous F_1 female strictly determine the phenotypes of F_2 males in respect to *white* and *miniature*. As shown in this illustration, these meiotic events involve crossovers between chromatids at the four-chromatid, or bivalent, stage (see Fig. 9-3). The frequency of observed recombinants for any two given linked genes thus depends on the frequency in which such meiotic crossovers occur between the two genes. (Adapted from Strickberger, M. W., 1985. *Genetics,* 3rd ed. Macmillan, New York)

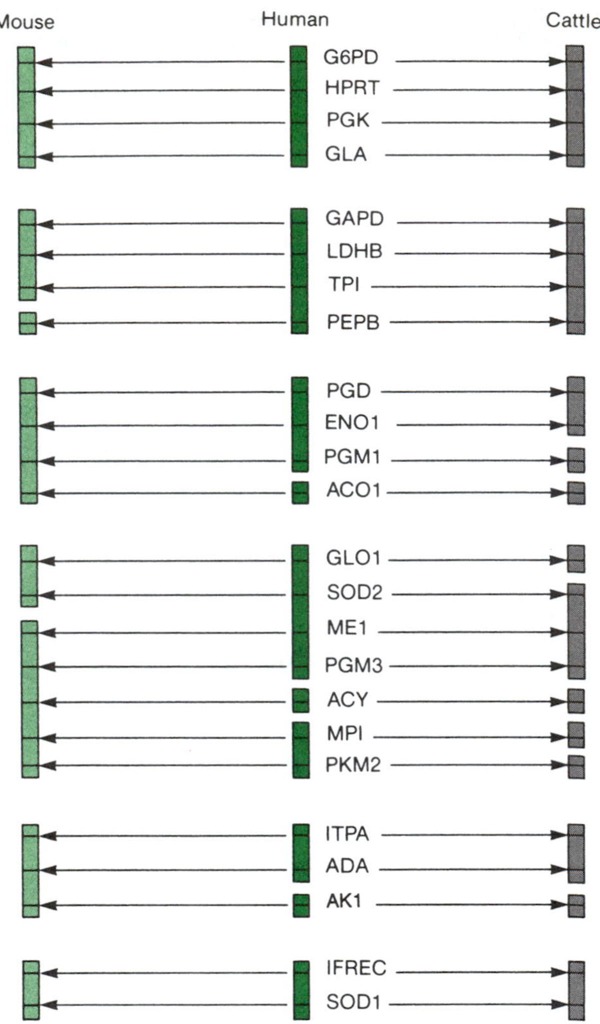

FIGURE 9-11 Linkage relationships between 24 human gene loci (*center column*) compared to those of similar genes in mice (*left*) and cattle (*right*). Each vertical rectangular block of genes indicates a chromosome in the respective species. For example, the top rectangles in each column give the relative positions of four genes on the X chromosome of each species. Note that the linkage order for many gene loci is the same in each of the three species, indicating that a large number of linkage relationships have been evolutionarily conserved. (Adapted from Womack, J. E., and Y. D. Moll, 1986. Gene map of the cow: Conservation of linkage with mouse and man. *J. Hered.,* **77,** 2–7.)

I (X)

0.0, 0.1, 0.3, 0.8, 1.0, 1.5, 3.0, 5.5, 6.9, 7.5, 13.7, 17.9, 18.9, 20.0, 21.0, 23.0, 27.7, 29.2, 32.8, 33.0, 36.0, 36.1, 36.9, 41.9, 43.0, 44.4, 44.5, 45.2, 51.5, 53.5, 56.3, 56.7, 57.0, 59.2, 59.4, 59.5, 62.5, 64.0, 64.8, 67.7

yellow body; *achaete* bristles; *Hairy-wing*; *scute* bristles; *silver* body; *suppressor of sable*; *minute* bristle (1); *Bld*; *deep orange* eyes; *prune* eyes; *zeste* eyes; *white* eyes; *facet* eyes; *Notch* wings; *echinus* eyes; *bifid* wing veins; *ruby* eyes; *crossveinless* wings; *shifted* wing veins; *carmine* eyes; *cut* wings; *singed* bristles; *ocelliless*; *lozenge* eyes; *Hexokinase* enzyme; *raspberry* eyes; *vermilion* eyes; *nod* meiosis; *miniature* wings; *furrowed* eyes; *wavy* wings; *sable* body; *garnet* eyes; *tiny* bristles; *narrow abdomen*; *scalloped* wings; *small-wing*; *sonless*; *forked* bristles; *Bar* eyes; *outstretched* wings; *Beadex* wings; *fused* wing veins; *carnation* eyes; *short-wing*; *maroonlike* eyes; *bobbed* bristles; centromere

II

0.0, 0.1, 0.3, 1.3, 4.0, 5.0, 6.0, 6.1, 7.6, 10.0, 11.0, 12.0, 13.0, 16.5, 22.0, 22.9, 31.0, 38.0, 41.0, 41.5, 44.0, 48.5, 50.5, 53.9, 54.5, 54.8, 55.0, 55.1, 55.3, 55.4, 55.9, 56.1, 57.5, 62.0, 67.0, 69.7, 71.1, 72.0, 75.5, 77.3, 81.0, 83.0, 91.5, 99.2, 100.5, 104.5, 107.0, 107.4, 108.0

net wing veins; *aristaless* antennae; *expanded* wings; *dachsous* wings; *Star* eyes; *Suppressor of Star*; *heldout* wings; *female-sterile*; *Enhancer of Star*; *Curly* wings; *Phosphoglycerokinase*; *Detached* vein; *echinoid*; *fat* body; *Gull* wings; *dumpy* wings; *clot* eye; *Sternopleural* bristles; *lysine* accumulation; *dachs* legs; *abnormal oocyte*; *Jammed* wings; *daughterless*; *abrupt* wing vein; *black* body; *Suppressor of Hairless*; *hook* bristles; *purple* eyes; *Bristle* short; *light* eyes; centromere; *Segregation distorter*; *straw* body; *thick* legs; *apterous* wings; *tarsi irregular*; *lightoid* eyes; *cinnabar* eyes; *engrailed* scutellum; *vestigial* wings; *waxy* wings; *comb-gap* bristles; *Lobe* eyes; *curved* wings; *Amylase* enzyme; *roof* wings; *abero*; *smooth* abdomen; *arc* wings; *plexus* wing veins; *brown* eyes; *speck* wing; *balloon* wings; *Minute* bristle (2) 33a

III

0.0, 0.2, 1.4, 5.4, 19.2, 20.0, 23.0, 26.0, 26.5, 35.5, 37.0, 40.4, 40.5, 41.4, 43.2, 44.0, 45.0, 46.0, 47.7, 47.9, 48.0, 48.5, 48.7, 49.7, 50.0, 51.3, 51.7, 52.0, 54.8, 57.4, 58.2, 58.5, 58.8, 59.0, 62.0, 63.1, 64.0, 66.2, 69.5, 70.7, 72.5, 75.7, 76.2, 88.0, 90.0, 91.1, 93.8, 95.5, 100.7, 101.0, 102.9, 104.3, 106.2

roughoid eyes; *veinlet* wings; *Roughened* eyes; *meiotic*; *javelin* bristle; *Moire* eyes; *Henna* eyes; *sepia* eyes; *hairy* body; *eye-gone*; *rotated* abdomen; *Dichaete* wings; *Lyra* wings; *Glued eyes*; *thread* arista; *scarlet* eyes; *transformed* sex; *Wrinkled* wings; centromere; *proboscipedia* mouth; *Antennapedia*; *pink* eyes; *tetraltera* wings; *blistery* wings; *maroon* eyes; *curled* wings; *tetrapter* halteres; *karmoisin* eyes; *rosy* eyes; *suppressor of Hairy-Wing*; *crossover suppressor* gene (*c*3*G*); *Stubble* bristle; *spineless* bristle; *bithorax*; *Roof* wings; *stripe* thorax; *glass* eyes; *kidney* eyes; *Delta* wing veins; *Hairless* bristle; *ebony* body; *detached* veins; *cardinal* eyes; *white ocelli*; *mahogany* eyes; *Prickly* bristles; *rough* eyes; *Beaded* wings; *suppressor of purple*; *claret* eyes; *Minute* bristle (3) 1; *Killer of prune*; *brevis* bristle; *Minute* bristle (3)g

IV

centromere; *cubitus-interruptus* vein
0.0 *Minute* bristle (4)
0.0-0.2 *abdomen rotatum*
0.2 *grooveless* scutellum
1.4 *bent* wings
2.0 *eyeless*
3.0 *shaven* bristles
4.0 *sparkling* eyes

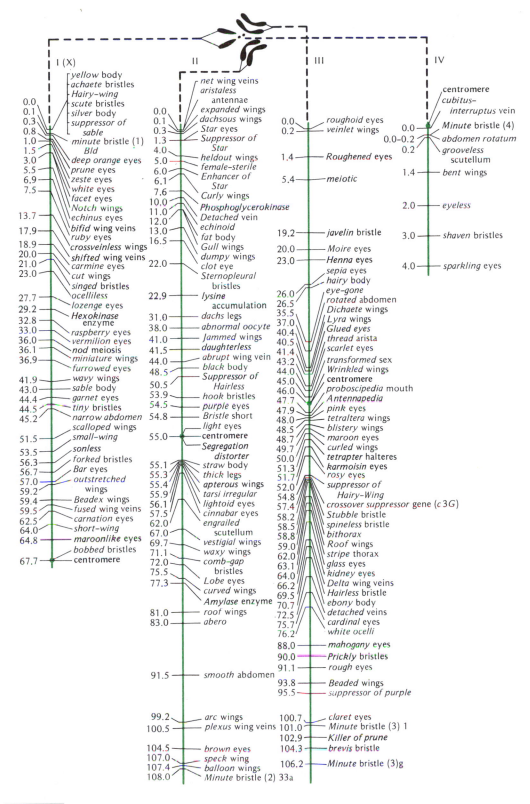

FIGURE 9-12 Linkage map of some of the important genes in the four chromosomes of *D. melanogaster*. Note that a variety of different genes may affect a specific character such as eye color, wing shape, and bristles, indicating that many steps exist in the development of a particular function, each step governed or capable of being modified by separate and different genes. (From Strickberger, M. W., 1985. *Genetics*, 3rd ed. © 1985 by Monroe W. Strickberger. Reprinted by permission of Prentice Hall, Inc., Upper Saddle River, NJ.)

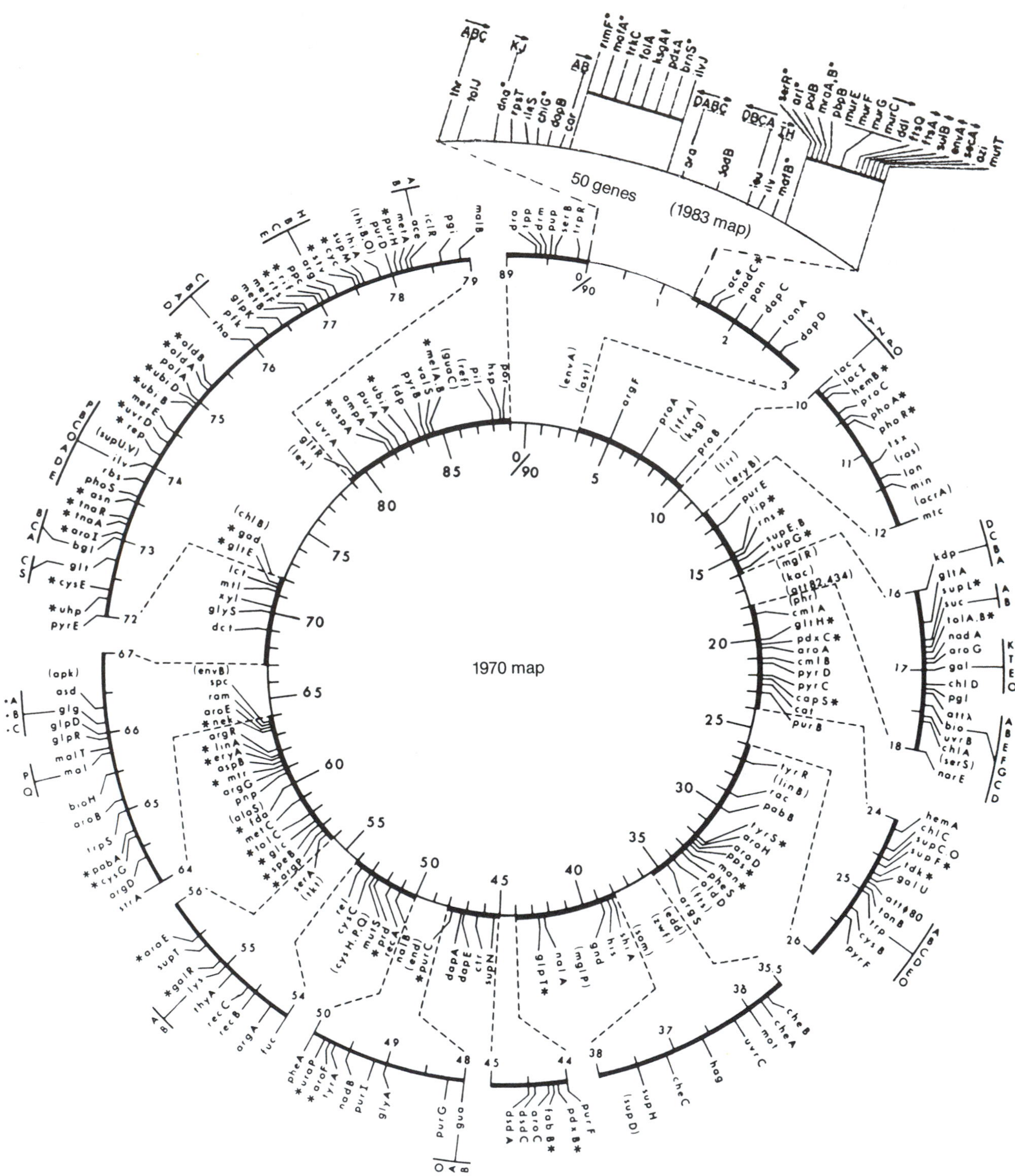

FIGURE 9-13 An early linkage map of an *E. coli* chromosome (Taylor, 1970). Geneticists have continually enlarged this map as they have discovered and localized new genes. For example, in the upper right corner is a small section, about one-fiftieth of the genome, to which 18 genes were localized in 1970 and 50 genes localized in 1983. (B. J. Bachmann, 1983. Linkage map of *E. coli* K-12 ed. 7 *Microbiological Reviews,* 47, 180–230. Reprinted by permission.)

linkage
locus
maternal inheritance
meiosis
metacentric
mitosis
modifier gene
mutations
phenotype
recessive allele
recombination
segregation
segregation distortion
sex linkage
synteny
telocentric
translocation
variation
variability
zygote

DISCUSSION QUESTIONS

1. Why are constancy and variability both essential in genetic material?
2. Cell division.
   a. What role do homologous chromosomes play in producing the difference between mitosis and meiosis?
   b. Which of these modes of cell division generates variability, and how is this accomplished?
3. How do the Mendelian principles of segregation and independent assortment differ?
4. How do dominance relations, multiple allelic systems and gene interactions generate variability?
5. What findings support Ohno's proposal that a sex-linked gene in one mammalian species will also be sex linked in other mammals?
6. Is Ohno's proposal also supported for genes that are not sex linked?

EVOLUTION ON THE WEB

Explore evolution on the Internet! Visit the accompanying Web site for *Strickberger's Evolution,* Fourth Edition, at http://www.biology.jbpub.com/book/evolution for Web exercises and links relating to topics covered in this chapter.

RECOMMENDED READING

Bull, J. J., 1983. *Evolution of Sex Determining Mechanisms.* Benjamin/Cummings, Menlo Park, CA.

Carvalho, A. B., M. C. Sampaio, F. R. Varandas, and L. B. Klaczko, 1998. An experimental demonstration of Fisher's principle: Evolution of sexual proportion by natural selection. *Genetics,* 148, 719–731.

Fisher, R. A., 1958. *The Genetical Theory of Natural Selection,* 2nd ed. Dover, New York.

Grützner, F., W. Rens, E. Tsend-Ayush, N. El-Mogharbel, P. C. O'Brien, and R. C. Jones, 2004. In the platypus a meiotic chain of ten sex chromosomes shares genes with the bird Z and mammal X chromosomes. *Nature,* 432, 913–917.

Haldane, J. B. S., 1922. Sex-ratio and unisexual sterility in hybrid animals. *Genetics,* 12, 101–109.

Hallgrímsson, B., and B. K. Hall (eds), 2003. *Variation: A Central Concept in Biology.* Elsevier/Academic Press, Burlington, MA.

Hartl, D. L., and E. W. Jones, 1998. *Genetics: Principles and Analysis.* Jones and Bartlett, Sudbury, MA.

Mendel, G., 1866. *Versuch über Pflanzenhybriden. Verh. Natur. Brünn,* 4, 3–47. (This is Mendel's classic paper, originally published in the Proceedings of the *Brünn Natural History Society.* It was translated into English by William Bateson in 1901 under the title *Experiments in Plant Hybridization.* Available at http://www.mendelweb.org/Mendel.html. Accessed March 29, 2007.

Ohno, S., 1979. *Major Sex Determining Genes.* Springer-Verlag, Berlin.

Pagel, M., (ed. in chief), 2002. *Encyclopedia of Evolution.* 2 volumes, Oxford University Press, New York.

Scherer, G., and M. Schmid, (eds), 2001. *Genes and Mechanisms in Vertebrate Sex Determination.* Birkhäuser Verlag, Basel.

Strickberger, M. W., 1985. *Genetics,* 3rd ed. Macmillan, New York.

Weiss, K. M., and A. V. Buchanan, 2004. *Genetics and the Logic of Evolution.* Wiley-Liss, Hoboken, NJ.

West-Eberhard, M. J., 2003. *Developmental Plasticity and Evolution.* Oxford University Press, Oxford, United Kingdom.

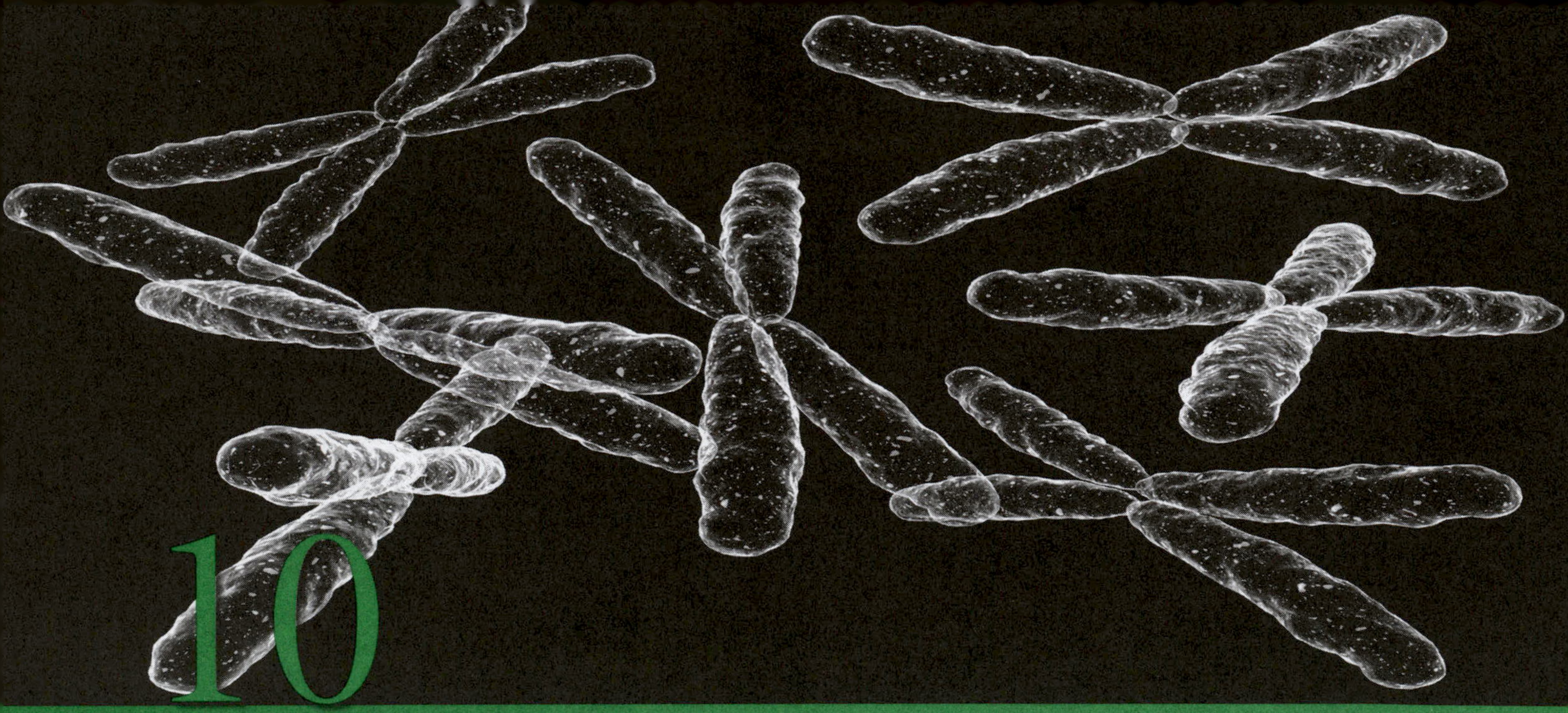

10 Chromosomes, Mutation, Gene Regulation and Variation

Chapter Summary

Although some biologists in the past held the view that **genes** could mutate in response to environmental demand, it is now clear that selection acts on existing variants. Most populations are polymorphic, with up to three-quarters of gene loci having more than one allele. With this great reservoir of variation, populations can respond to evolutionary pressures without new variants having to arise by mutation.

Variation is enhanced when there are alterations in chromosome number, either changes in entire sets (**euploidy**) or in individual chromosomes (**aneuploidy**), or when chromosomes undergo modifications to their structure such as deletions, duplications, inversions or translocations. Translocations can modify the size, composition or number of chromosomes. Small, localized chromosomal changes (mutations) are due to alterations in nucleotides or nucleotide sequences and can lead to changes in a gene or gene product. Regulatory mutations affect the systems controlling genetic activity. A classic example demonstrated mutations in the regulatory gene that governs the expression of genes coding for lactose-using enzymes in bacteria. These studies laid the foundation of our current understanding of gene regulation in prokaryotes and in eukaryotes. Mutations within a single gene are generally rare, and usually have a characteristic frequency. But the environment and special movable DNA sequences called *transposons* affect mutation rates. Continuous variation is often attributed to regions of chromosomes that contain blocks of genes (*quantitative trait* loci). The randomness of mutation, variability in mutation rates among genes and among species, and the ability of DNA to repair itself, all contribute to the maintenance of genetic polymorphism within individuals, providing a source of genetic variable that serves as the raw material for evolution.

Darwin's most fundamental insight was that the small differences among individuals within populations are the raw material for evolution. But before and during his time, most scholars, even those like Thomas Huxley (1825–1895) who supported Darwin, held the view that **saltations** (large discontinuous changes in phenotype) represented the level at which variation was relevant to evolution. Resolving the important issue of small or large differences at that time was not possible because the mechanisms of inheritance were not understood. For instance, Darwin and many of his contemporaries thought that offspring represented perfect or blended copies of their parents' parts. As Bowler (2005) points out, this *view sets variation against inheritance as antagonistic processes,* with variation representing deviations from the expected inherited resemblance between parents and offspring.

Darwin's concept of variation in relation to evolution focused on individuals rather than populations. Today, however, we recognize variation as a property of populations rather than of individuals. The current view of variation — that populations contain reservoirs of individual-level variation on which selection can act — finally emerged with the modern synthesis as the emerging field of genetics incorporated quantitative analyses of phenotypic variation. The shift in approach to variation from individuals to populations was most clearly articulated by Ernst Mayr, who "recognized the fundamental importance of population thinking in evolutionary biology."[1] Investigations into the existence, significance and causes of variation have thus been fundamental to our emerging understanding of the evolutionary process.

Variation and Variability

Gunter Wagner and colleagues (1997) have drawn an important distinction between **variation** and **variability.** Like "solubility" or "fragility," *variability refers to the tendency or propensity to exhibit variation rather than a particular instance of variation.* The study of variability, or of the determinants of patterns of variation, is important because variation is the raw material for natural selection.[2] It is therefore most surprising to find Mayr lamenting:

> It is amazing to what extent variation in natural populations has been neglected in the study of evolution. Amazing, because natural selection would be meaningless without variation. This conclusion gave me the idea to consider *the production of variation as a step in the process of natural selection* (Mayr, 2005, p. xviii; emphasis added).

Genetic variation and variability and evolutionary potential are two sides of the same evolutionary coin. The topic of this chapter — how variation arises at different levels — is both central to evolution and is the evolutionary mechanism that is least understood. Questions concerning variation and variability cross a range of sub-disciplines within evolutionary biology. Discussion of these issues, initiated in this chapter, pervades the book, but is especially addressed again in Chapters 13, and 21 to 23. The issues are set out as five unanswered questions in **Box 10-1.** We will refer to the central questions concerning variation and variability summarized in this box.

Variationin Chromosome Number

Numerical chromosome variations are of two major kinds: (1) changes in the number of entire sets of chromosomes, euploid variations, and (2) changes in the number of single chromosomes within a set, known as aneuploid variations. Because most sexually reproducing eukaryotes are diploids with two sets of chromosomes ($2n$), euploid variations may extend from the haploid or monoploid condition (n) to various levels of **polyploidy** ($3n, 4n \ldots$), as shown in **Table 10-1.** Such events may be caused by more than one sperm fertilizing an egg, or by cell division failures in which, for example, a diploid rather than a haploid gamete results.

In aneuploids (**Table 10-2**), a wide range of variations may result when chromosomes are either added to or subtracted from a normal set. Such events may occur because the meiotic disjunction between homologous chromosomes is abnormal (**nondisjunction**). Plants seem to be more tolerant of such chromosomal variations than animals, though aneuploids are found in both groups.

Polyploidy

The appearance of extra sets of chromosomes within a species (**autopolyploidy**) seems to be a common mode of evolution in many plant groups, including mosses, apples, pears, bananas, tomatoes and corn. Polyploidy originating from the interbreeding of different species (**allopolyploids**) appears in some plants, including wheat (see below). For example, a gamete from species *A* may fertilize species *B* to produce the diploid hybrid *AB,* which can undergo polyploidy to form the allotetraploid *AABB,* or other variations shown in **Figure 10-1.** Both auto- and allopolyploids can be produced after exposure to chemicals, such as colchicine, which break down spindle fiber microtubules and thereby interfere with chromosome segregation during cell division.

[1] Cited from the dedication in Hallgrímsson and Hall (2005, p. *ii*) for which Ernst Mayr wrote the foreword.

[2] The concept of *evolvability* attempts to understand the basis for why some structures/organisms change more than others over evolution. The "opposite" pattern is phylogenetic inertia, although inertia may have both positive and negative affects on selection (Blomberg and Garland, 2002). All three concepts may be investigated at either genotypic or phenotypic levels.

BOX 10-1 Central Questions Concerning Variation and Variability in Evolution

THE FIVE UNANSWERED QUESTIONS concerning the place of variation and variability in evolution are outlined below. Addressing each would make a wonderful class project. Each could provide the topic for a class devoted only to that question. Each would make a great term essay or take home exam. So, yes, these are important questions.

1. What is the relationship between *genetic (genotypic) variation,* which is transmitted from generation to generation, and *phenotypic variation,* the direct object of selection? This question is a major preoccupation of evolutionary developmental biology (see Chapter 13).[a]
2. What are the *ecological and developmental determinants* of the phenotypic variation among species? This includes such questions as how body size, geographic range, home range size, niche width, lifespan, environmental stress and population size and density affect the tendency of populations to exhibit phenotypic variation (see Chapter 23).[b]
3. How do *inbreeding* and *outbreeding* affect phenotypic variability? The population genetics of inbreeding is well understood, but the relationship(s) between inbreeding and phenotypic variability in natural populations is not and continues to be an active area of research (see Chapters 21 and 22).
4. What is the relationship between *selection intensity and phenotypic variability?* The effect on genetic variance of selection of various kinds and under different circumstances has, and continues to be, an active area of research in population genetics (see Chapters 21 and 22).
5. How do *mutations* influence variability? Mutations are a source of variation and so must increase phenotypic variance. Less obvious is the possibility that mutations may also affect variance through disruptive effects on development (see Chapter 13).

[a] A review of a recent book, *Genetics and the Logic of Evolution* (Weiss and Buchanan, 2004) that tackles this major question, begins:

> Addressing the problem of the genotype-phenotype map is arguably one of the main challenges in current biology. Its equally challenging corollary is to connect the evolutionary dynamics of genotypes with patterns of phenotypic evolution. The key phenomenon for grappling with these issues is biological development.

[b] In a review of a recent book, *Variation: A Central Concept in Biology,* edited by Hallgrímsson and Hall (2005) we find:

> Heritable phenotypic variation is essential for evolution by natural selection. It is generated by the more or less poorly understood intersection of genes, developmental processes and the environment. It is not surprising that the features of such a central concept [heritable phenotypic variation], including its production, regulation and measurement, remain current topics of research.

TABLE 10-1 Euploid variations, involving entire sets of chromosomes

Euploid Type	Number of Homologues Present for Each Chromosome	Example
Haploid or monoploid	One (1n)	A B C
Diploid	Two (2n)	AA BB CC
Polyploid	More than 2	
Triploid	Three (3n)	AAA BBB CCC
Tetraploid	Four (4n)	AAAA BBBB CCCC
Pentaploid	Five (5n)	AAAAA BBBBB CCCCC
Hexaploid	Six (6n)	AAAAAA BBBBBB CCCCCC
Heptaploid	Seven (7n)	AAAAAAA BBBBBBB CCCCCCC
Octaploid, etc.	Eight (8n), etc.	AAAAAAAA BBBBBBBB CCCCCCCC, etc.

Source: Adapted from Strickberger, M. W. 1985. *Genetics,* 3d ed. Macmillan, New York.

TABLE 10-2 Aneuploid variations, involving individual chromosomes within a diploid set

Type	Number of Chromosomes Present	Example
Disomic (normal diploid)	2n	AA BB CC
Monosomic	2n − 1	AA BB C
Nullisomic	2n − 2	AA BB
Polysomic	Extra chromosomes	
Trisomic	2n + 1	AA BB CCC
Double trisomic	2n + 1 + 1	AA BBB CCC
Tetrasomic	2n + 2	AA BB CCCC
Pentasomic	2n + 3	AA BB CCCCC
Hexasomic	2n + 4	AA BB CCCCCC
Septasomic	2n + 5	AA BB CCCCCCC
Octasomic, etc.	2n + 6, etc.	AA BB CCCCCCCC, etc.

Source: Adapted from Strickberger, M. W. 1985. *Genetics,* 3rd ed. Macmillan, New York.

The evolution of the wheat (*Triticum aestivum*)used to make flour for bread is an oft-cited example of chromosome duplication, linked as it is to the expansion of agriculture after the last ice age. The story goes as follows:

- At least 30,000 years ago, in the Fertile Crescent of southwest Asia, a natural hybrid formed between two grasses, a species of *Triticum* (wild einkhorn) and a type of goat grass, both of which had 14 chromosomes. Hunter-gatherers harvested the seeds of this new 28-chromosome plant (*wild emmer*) for millennia. Around 10,000 years ago, by which time the ice ages had ended, humans began cultivating the hybrid. By 9500 years ago, harvesting of the best plants resulted in the selection of desirable qualities had led to a new form of emmer, itself regarded as a species.
- Cultivated emmer was an important crop for 7000 years, spreading throughout the Near East and into Egypt. When it reached an area southwest of the Caspian Sea around 9000 years ago, however, cultivated emmer came in contact with a second 14-chromosome goat grass. The new hybrid, known as *spelt,* had 42 chromosomes. Neither emmer nor spelt was easy to harvest, but about 500 years later a fortuitous mutation changed the nature of the spike or ear, producing a shell that would allow threshing, but also meaning that the plant was no longer able to propagate naturally.
- Since then, this last species in the chain — bread wheat, *T. aestivum* — has undergone considerable change. For millennia, farmers sowed seed they had collected the previous year, creating in the process a multitude of "land races" adapted to local conditions. More recently, agronomists have bred improved varieties of wheat, enormously raising yields.
- About 7000 years ago, cultivated emmer evolved in a second direction when mutation again resulted in free-threshing grain, facilitating the spread of agriculture. Having evolved directly from spelt, *Triticum durum* (macaroni or durum wheat) also has 28 chromosomes.

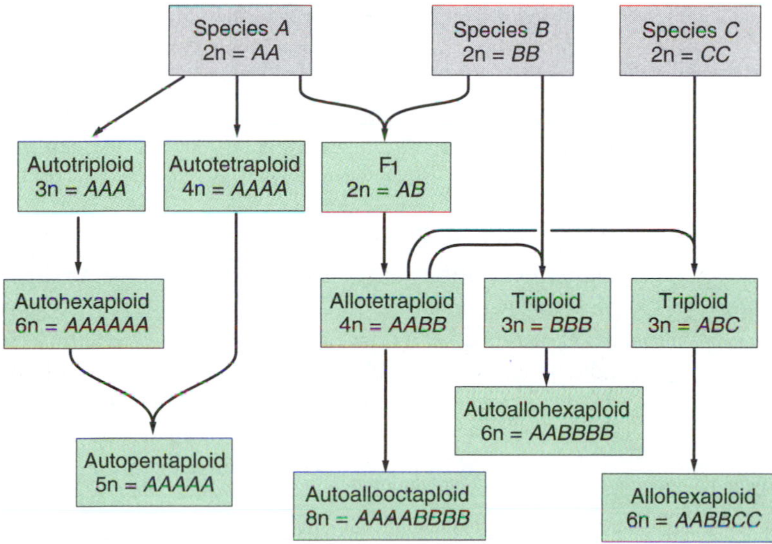

FIGURE 10-1 Terminologies used for different polyploids, and some of the pathways by which they originate. (Adapted from Strickberger, M. W., 1985. *Genetics,* 3rd ed. Prentice Hall, Upper Saddle River, NJ.)

The first *laboratory-created species* was also the product of a cross between two plants, in this case two species of tobacco: *Nicotiana tabacum* (a diploid with 48 chromosomes whose haploid gametic number, *n*, was 24) and *N. glutinosa* (24 chromosomes, $n = 12$). As shown in **Figure 10-2**, the hybrid was sterile. In such cases, sterility arises because an even division of chromosomes during meiosis depends on pairing between homologous chromosomes. These homologues are absent when the two cross-fertilizing species have evolved chromosomal differences between them. The hybrid tobacco plant was propagated by vegetative cuttings. Eventually, a chromosome-doubling event produced a fertile allopolyploid with the consequence that each chromosome had a pairing mate, and normal meiosis could take place. The resulting species, *N. digluta*, had 72 chromosomes and was fertile when self-crossed, indicating that all of its chromosomes had undergone homologous pairing. Many grocery stores carry the grain triticale. This, too, is a new species we have created using the principles of polyploidy to cross wheat with rye.

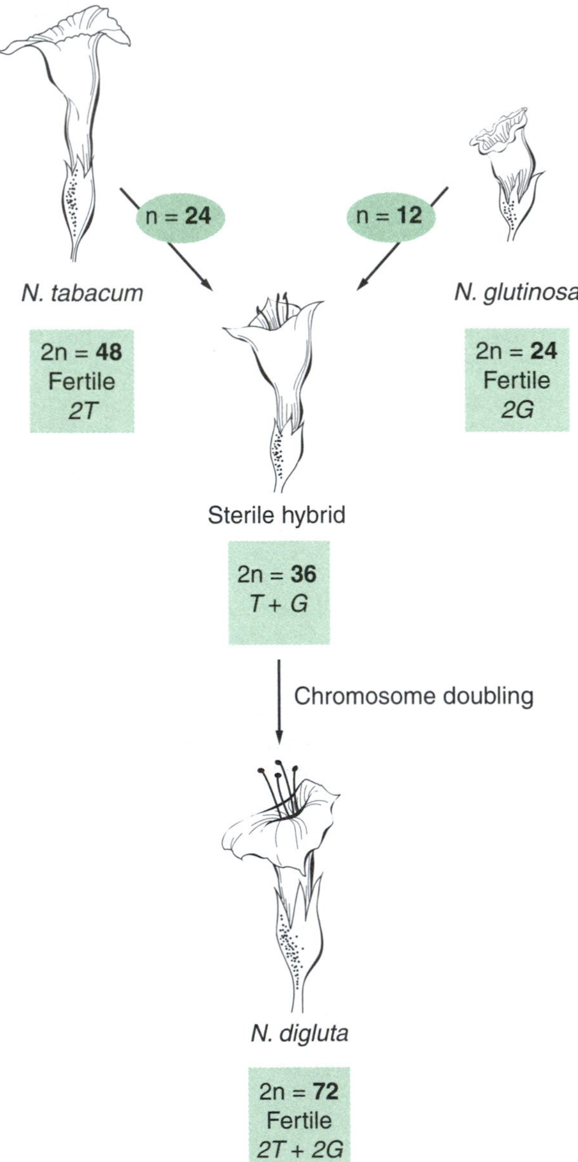

FIGURE 10-2 Flowers and karyotypes of *Nicotiana tabacum, N. glutinosa,* and their sterile diploid hybrid, and the new allotetraploid species formed by chromosome doubling in the sterile hybrid.

Polyploidy is much more rare in animals than in plants because most animals show much greater developmental sensitivity to even a small change in chromosome number. This sensitivity extends to chromosomal sex-determining mechanisms; animal polyploids face difficulties in maintaining the same proportions of X and Y chromosomes present in normal diploids. For example, polyploidy in a species in which males are XY and females XX could lead to establishment of XXYY males and XXXX females. The subsequent combinations of gametes produced by these individuals (XY sperm × XX eggs) might well produce XXXY individuals that are neither completely male nor completely female. Therefore, when animal polyploid species do occur, they usually employ some form of **asexual reproduction** such as **parthenogenesis** (development of eggs without fertilization).

Nevertheless, polyploidy has some advantages. In plants and animals, the extra polyploid chromosomes may act as multiple buffers in various physiological process or anatomical features, improving vigor or enabling polyploid individuals to face new conditions. Under such circumstances, the fecundity of a polyploid may be sufficient to replace diploids. Furthermore, the additional chromosomes may provide the opportunity to evolve new functions for the extra sets of genes, leading, among other features, to increased diversity and heterozygosity (Soltis and Soltis, 1995).

Phenotype of the Chromosome

A variety of changes in chromosome structure are diagrammed in **Figure 10-3.** Because these are changes in the appearance of the chromosome, it is reasonable to speak of them as the **phenotype of the chromosome,** just as we can speak of the **phenotype of the gene** when visualizing sites of methylation or imprinting, or the **phenotype of the individual** when comparing characters.

Deletions or Deficiencies

The terms **deletions** and deficiencies describe losses of chromosomal material (Fig. 10-3a). In general, the severity of a deletion depends on how extensive it is and which nucleotides or genes are missing. If functional genes are involved, deletions can be quite harmful in diploids and haploids but not necessarily harmful in polyploids or aneuploids, where such genes may be present in extra chromosomes.

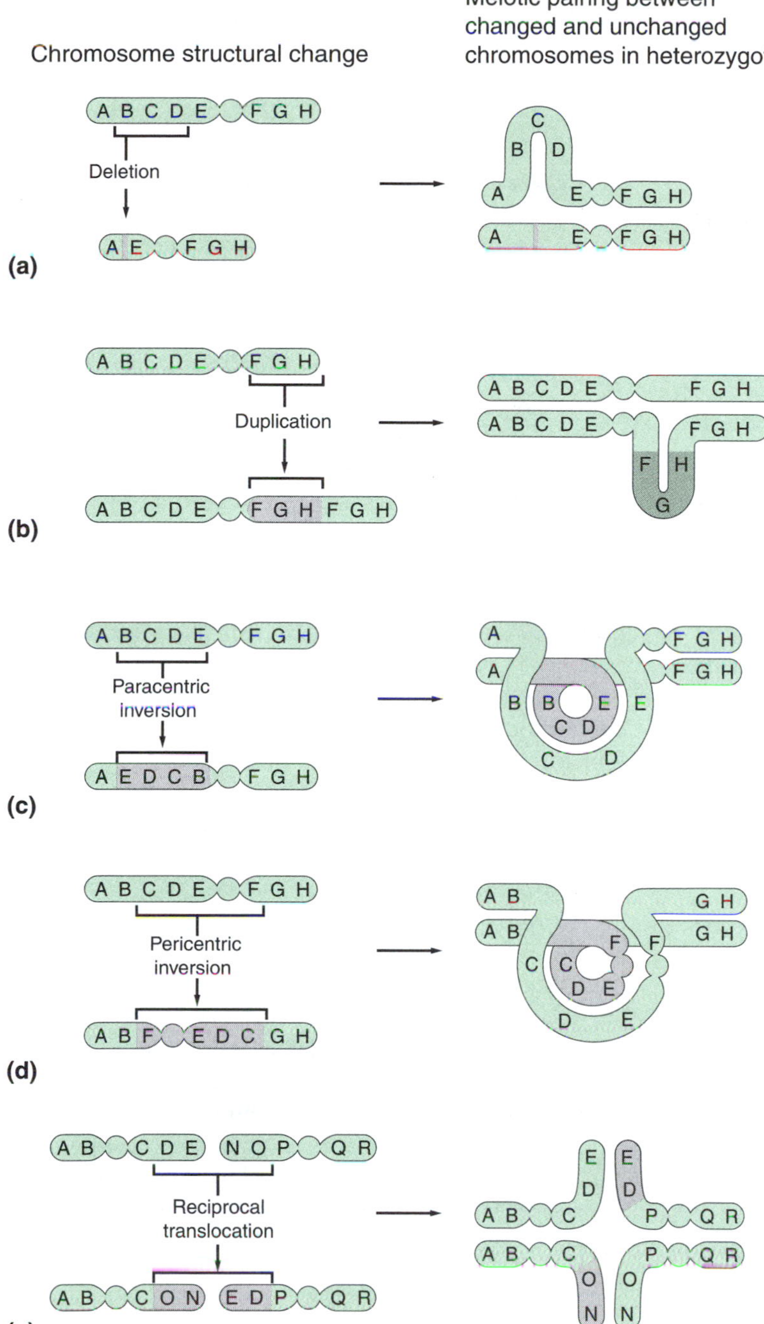

FIGURE 10-3 (a–e) Major kinds of structural chromosomal changes and their effects on meiotic pairing in heterozygotes carrying both changed and unchanged homologues. (For diagrammatic simplicity, the double chromatid structure of each chromosome is omitted, although meiotic pairing between two homologues involves four chromatids.)

Duplications

Duplications are either short or long segments of extra chromosome material originating from duplicated sequences within a genome (Fig. 10-3b).

Among the mechanisms producing duplication at the *chromosome level* is **unequal crossing over**, in which homologous pairing during recombination is slightly askew, producing a crossover product that contains extra chromosomal material (see Fig. 12-4). At the *level of the gene*, duplication (perhaps with rearrangement) of existing genes or parts of genes results in an additional copy or copies of the gene or part of the gene. Because numerous **gene families** of similar or identical genes occur in many species, duplications have been common during evolution.

Various eukaryotes produce the RNA components of ribosomes in clusters of long tandem arrays of duplicated genes. Other instances of duplication, such as the genes involved in producing different hemoglobin-type proteins

(see Figures 12-3 and 12-5), show that duplicated genes have evolved along different pathways, enabling them to perform different adaptive functions, and, as discussed in Chapter 12, to mediate evolutionary change.

Inversions

Inversions are reversals in chromosomal gene order that can sometimes be observed in the formation of loops during meiotic pairing in heterozygotes (Fig. 10-3c, d). Inversions generally lower the recombination frequency within the inverted sequence; crossing over within such sequences in heterozygotes can lead to chromosomal abnormalities. As a result, the genes included within an inversion tend to remain together as a nonrecombinant block, called by some, a *supergene.*

Two major kinds of inversions are known in eukaryotes, (1) *paracentric,* which do not include the centromere (Fig. 10-3c), and (2) *pericentric,* which include the centromere (Fig. 10-3d). Among the possible effects of pericentric inversions is a shift in the relative position of the centromere, as shown in Figure 10-3d. In some cases, this shift may be drastic, moving the centromere from the chromosomal center (metacentric position) to one end (acrocentric). Among deer mice, *Peromyscus,* there are species whose 48 chromosomes are entirely metacentric (for example, *P. collatus*) and species whose 48 chromosomes are almost entirely acrocentric (for example, *P. boylei*).

Translocations

The term **translocation** primarily refers to the transfer of material from one chromosome to a nonhomologous chromosome. When the exchange of such material is mutual, it can result in the kind of *reciprocal translocation* shown in Figure 10-3e.[3] Such translocations are recognizable by a cross-shaped configuration between translocated and nontranslocated chromosomes during meiotic pairing in the heterozygote. Because such meioses can produce gametes containing duplications and deficiencies, translocation heterozygotes are often sterile.

The fertile and viable gametes produced by translocation heterozygotes result primarily from alternate **segregation,** in which the translocated chromosomes segregate separately from the nontranslocated chromosomes, shown as the diagonal combinations of chromosomes in Figure 10-3e. Thus, genes in the translocated and nontranslocated chromosomes of such heterozygotes tend to be inherited as separate blocks, behaving as though all genes on each block were linked together.

Because translocations cause various sterility and fertility problems in heterozygotes, there is an advantage to individuals homozygous for translocations to be isolated from those homozygous for nontranslocated chromosomes. According to some plant geneticists, selection for separation between these groups can easily lead to speciation; examples among plants such as *Clarkia* indicate that such events have occurred. Some chromosomal changes could lower the reproductive success of hybrids between groups homozygous for different arrangements, although chromosomal speciation has been proposed as involved in the separation of humans from chimpanzees (see Chapter 20).[4]

In terms of their cytological effects, translocations can cause changes in the number and structure of chromosomes. For example, translocations may combine two different nonhomologous chromosomes into one larger chromosome (**Fig. 10-4a**), cause chromosomes to significantly change in shape (Fig. 10-4b), or cause fission events that increase the number of chromosomes (Fig. 10-4c). Thus, in some populations of European wild mice, *Mus musculus,* reductions occur from the standard number of 20 pairs of chromosomes to as few as 11 pairs. Among Asiatic muntjac deer, the Indian muntjac (*Muntiacus muntjac*) has only three pairs of large chromosomes, the black muntjac (*Muntiacus crinifrons*) has four pairs, and the Chinese muntjac (*Muntiacus reevesi*) 23 (**Fig. 10-5**). The relative amounts of DNA in the two species are about the same, the large muntjac chromosomes arising from successive translocations combining the smaller muntjac chromosomes.

Chromosomal conservatism, in contrast, may be a strong feature in other mammalian groups. For example, extant species of the 30-My-old lineage that includes Asian (Bactrian) and African (dromedary) camels as well as South American guanaco, vicuna, llama and alpaca, all have the same number of morphologically similar chromosomes, 74. Qumsiyeh (1994) proposes that the number of chromosome pairs reflects a species' adaptive capability: larger numbers allow increased genetic recombination, thereby producing more variability and environmental flexibility; smaller numbers allow genetic combinations to persist, leading to greater specializations for specific habitats.

Chromosomal Evolution in *Drosophila* and Primates

In those instances where we can identify linear sections of chromosomes through distinctive bandings, we can chart chromosomal evolution in great detail. In species of *Drosophila* and other similar insects, the chromosomes of salivary gland cells and other tissues have replicated many

[3] Note that the gametes produced by the translocation heterozygote illustrated in Figure 10-3e would have duplications and deficiencies if the two upper chromosomes went to one pole and the two lower chromosomes went to the other pole.

[4] See M. King (1993) for chromosome speciation and H. Lewis (1973) for *Clarkia*.

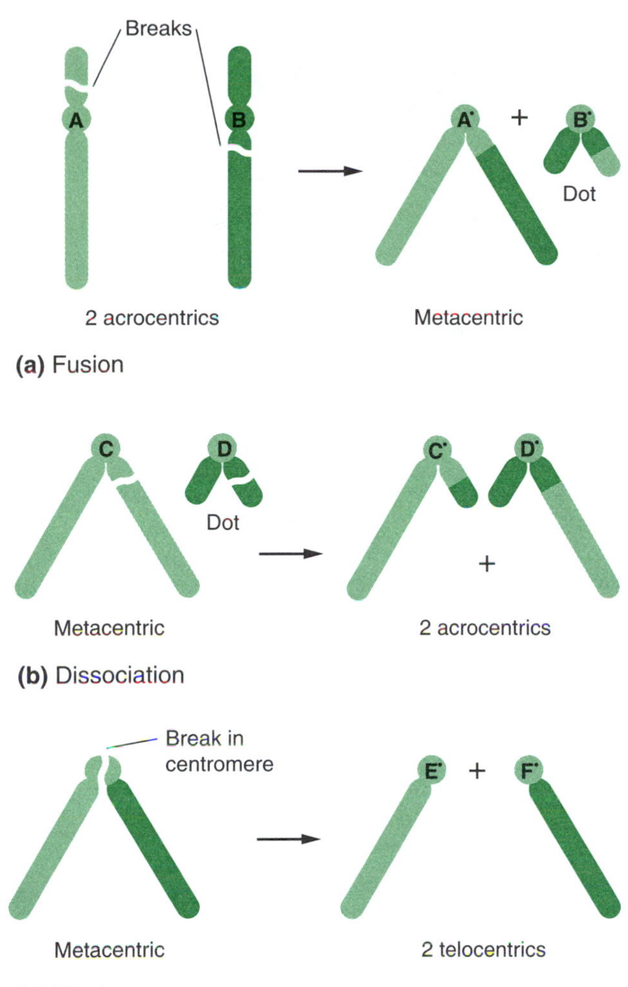

FIGURE 10-4 (a) Translocations that lead to fusion between the arms of two acrocentric nonhomologous chromosomes (A and B) to form one metacentric chromosome (A') and a "dot" chromosome (B') that carries only a small amount of chromosomal material. In diploids, loss of the dot chromosome will reduce the chromosome number by two, since homozygotes for the metacentric now carry the chromosome material formerly present in the two acrocentrics. (b) The reverse process of dissociation involves reciprocal translocations between the metacentric (C) and dot (D) chromosomes leading to C' and D' acrocentrics. (c) The fission mechanism proposed to explain the origin of presumed single-armed telocentric chromosomes (E', F') from two-armed metacentrics or acrocentrics. (From *Genetics,* 3rd ed by Monroe W. Strickberger. © 1985 by Monroe W. Strickberger. Reprinted by permission of Prentice Hall, Inc., Upper Saddle River, NJ.)

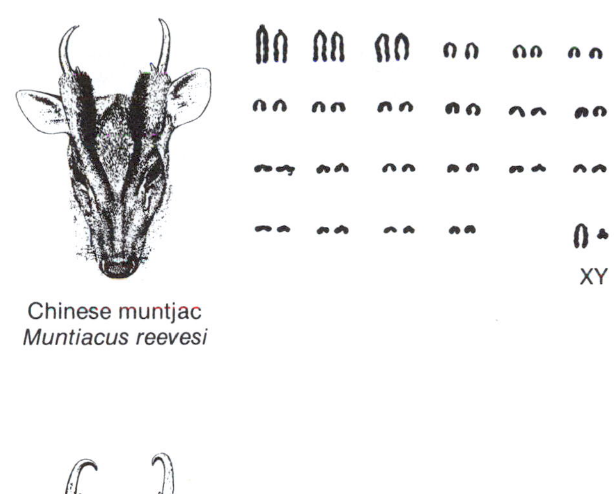

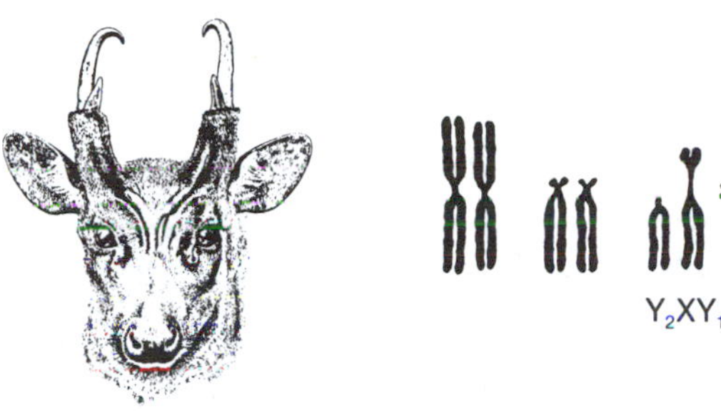

FIGURE 10-5 The Chinese and Indian muntjac deer (*Muntiacus reevesi, M. muntjac*) and their karyotypes. The Indian muntjacs, with two pairs of autosomes and three sex chromosomes, have the lowest known chromosome number of any mammal. (Adapted from Austin, C. R., and R. V. Short, 1976. *The Evolution of Reproduction.* Cambridge University Press, Cambridge, England.)

times, but each replicate remains closely apposed to the other replicates within the nucleus. The resulting **polytene chromosomes** are tremendously enlarged and show highly detailed banding configurations that enable even minor chromosomal changes to be traced. Geneticists have documented practically all the chromosomal changes in the evolution of hundreds of these fly species, a sample of which are shown in **Figure 10-6.**

Although polytene chromosomes are absent in many organisms, chromosomal staining techniques enable detailed comparisons, even between relatively small mammalian chromosomes. The G-banding technique illustrated in **Figure 10-7** allows a comparison of human, chimpanzee, gorilla and orangutan chromosomes. These bandings show that some chromosomes — nos. 6, 13, 19, 21, 22 and X — are almost identical in all four species, and that various arms or sections of other chromosomes are homologous throughout.

We can account for the changes that occurred during the evolution of these primates by the chromosomal variations just described. Thus, the difference in chromosome number between humans ($n = 23$) and apes ($n = 24$) derives from a fusion event that combined the two chimpanzee-type chromosomes indicated in Figure 10-7 to form the no. 2 human-type chromosome. DNA sequencing has provided the evolutionary odyssey of human chromosome no. 3 starting from its probable origin in an early rodent-like population perhaps 90 Mya (**Fig. 10-8**).

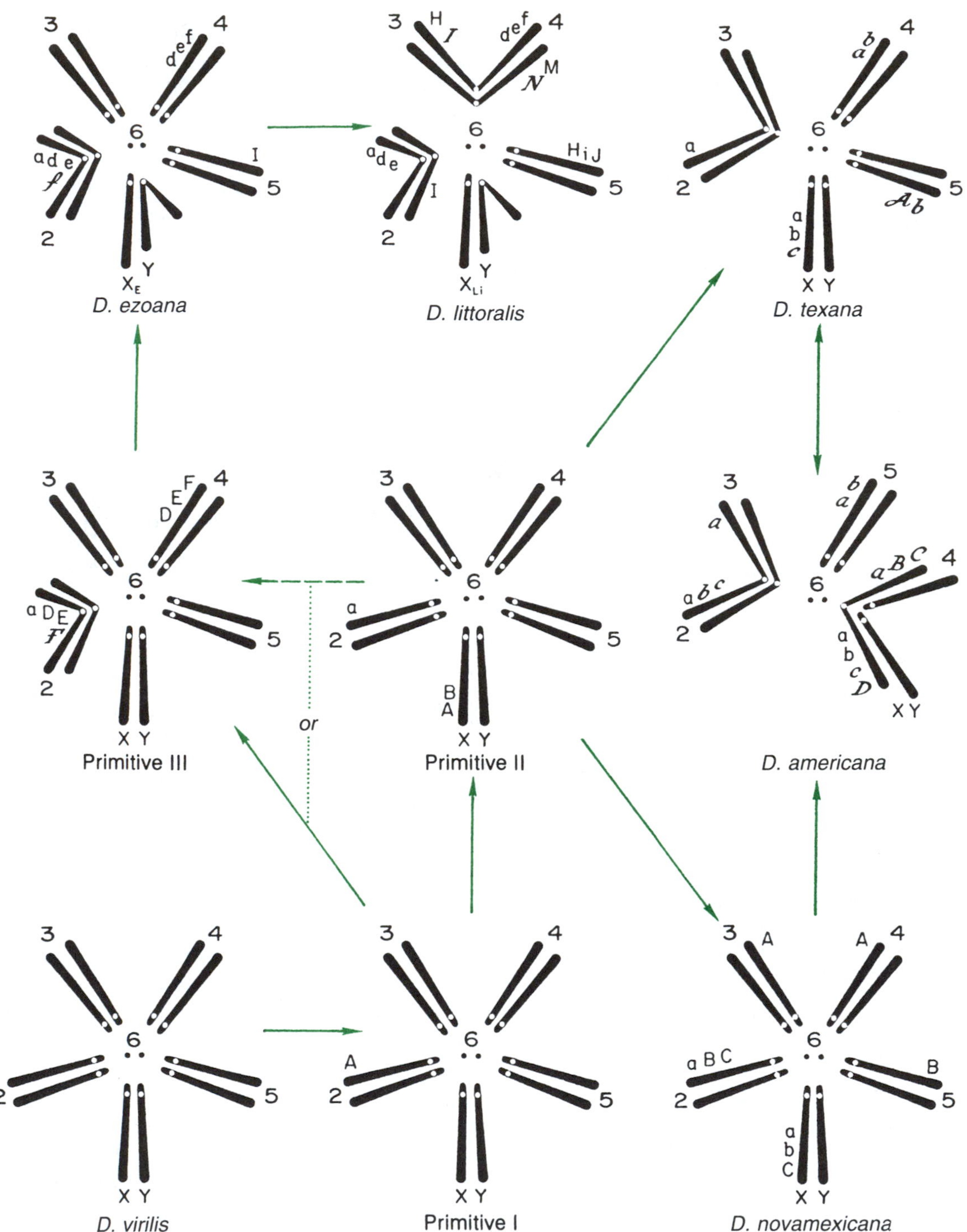

FIGURE 10-6 Paths of chromosomal evolution in some species of the *virilis* group of *Drosophila*. The chromosomes of what was probably the original karyotype of the genus *Drosophila* (*lower left*) are numbered from 1(X) to 6, and specific chromosomal banding arrangements are indicated by letters. (Adapted from Stone, W. S., 1962. The dominance of natural selection and the reality of superspecies (species groups) in the evolution of *Drosophila*. *Univ. of Texas Publ.*, **6205**, 507–537.)

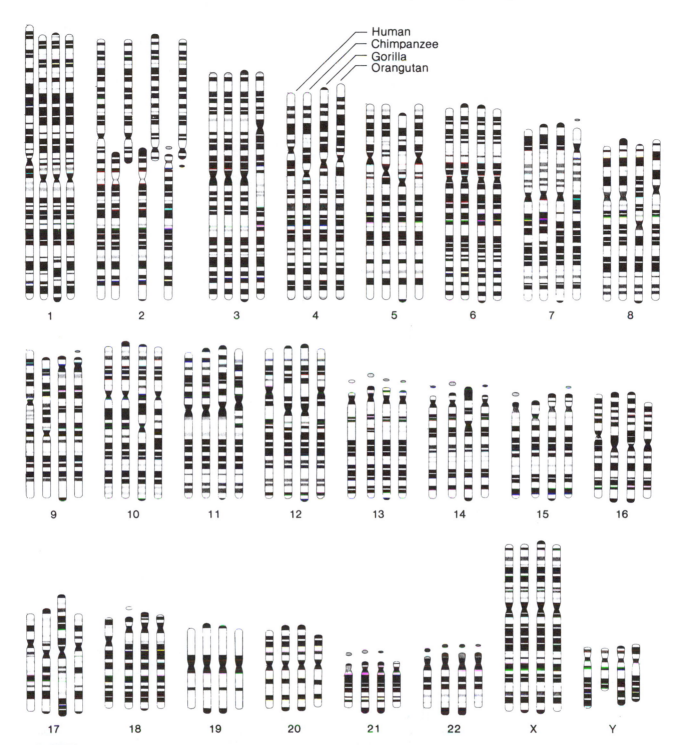

Human
Chimpanzee
Gorilla
Orangutan

1 2 3 4 5 6 7 8

9 10 11 12 13 14 15 16

17 18 19 20 21 22 X Y

FIGURE 10-7 Banding arrangements of the chromosomes of humans, chimpanzees, gorillas and orangutans, in respective order from left to right for each chromosome. Note that the 24 pairs of chromosomes in the great apes reduce to 23 pairs in humans (number 1 to 22 + XY) because two different chromosomes fuse into a single no. 2 human chromosome. This fusion, along with other changes (for example, inversions in chromosomes 1 and 18), must have occurred some time after the human line separated from a human-chimpanzee common ancestor. On the whole, these banding arrangements indicate that humans have a closer evolutionary relationship with chimpanzees than with gorillas and a more distant one with orangutans. (Reprinted with permission from *The Origin of Man: A Chromosomal Pictorial Legacy* by Yunis, J. J. © 1982 American Association for the Advancement of Science.)

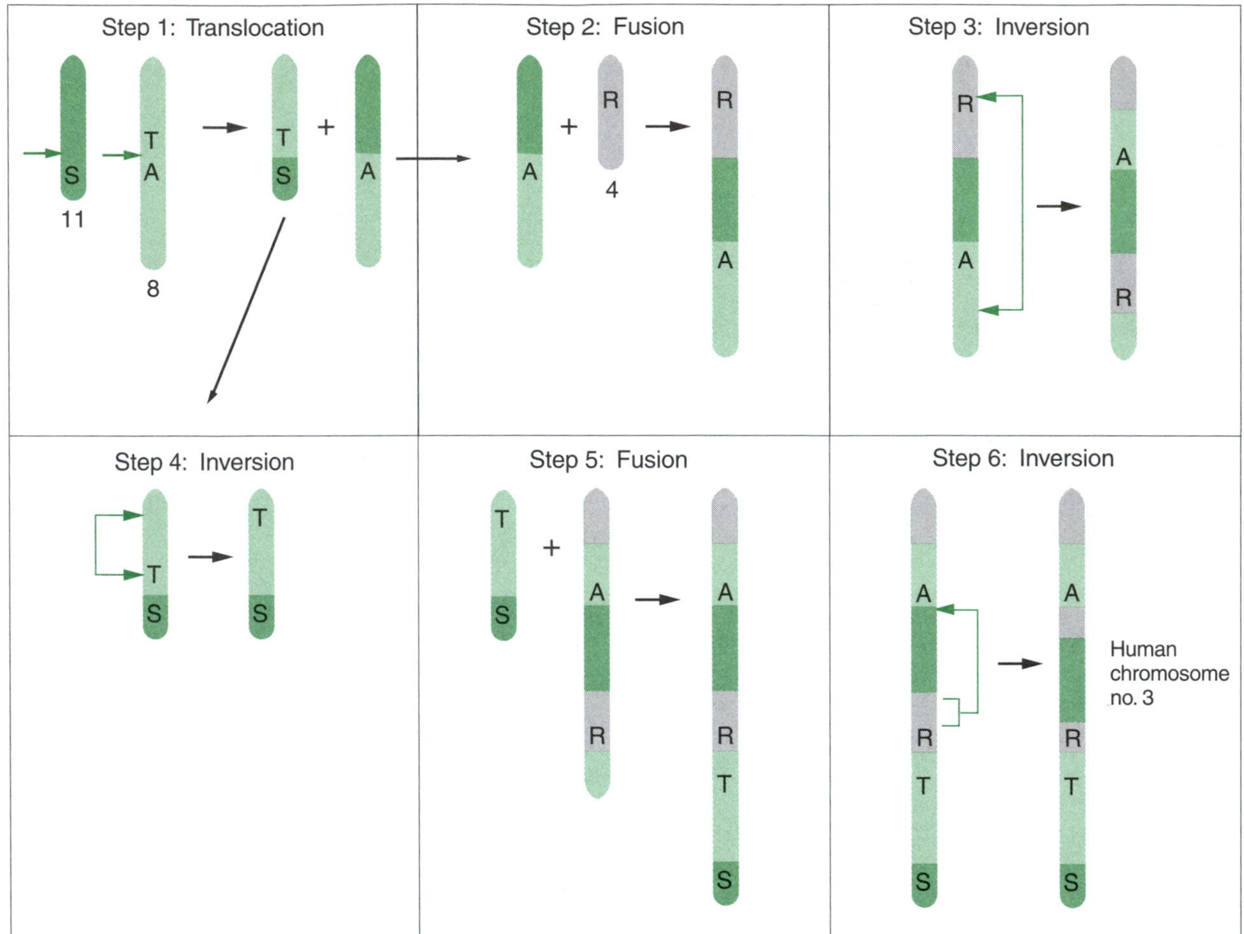

FIGURE 10-8 Six proposed major steps in the evolution of human chromosome no. 3, beginning some time during the Cretaceous with a translocation between the two rodent-like chromosomes nos. 8 and 11. This was followed by fusion of one translocation product to the no. 4 rodent chromosome. As shown, further steps involved lengthening of the chromosome through fusion (step 5) with the remaining translocation product of step 1, and two inversions. The letters A, R, S, and T, indicate four of the marker genes whose loci we know in rodents, humans, and other mammals: A = aminoacylase-1, R = rhodopsin, S = somatostatin, T = transferrin. For simplicity, we have omitted other known chromosomal changes from this illustration, including the insertion of small segments from rodent-like chromosomes 2, 14, 15, and 16. (Modified from Hino, O., J. R. Testa, K. H. Buetow, T. Taguchi, et al., 1993. Universal mapping probes and the origin of human chromosome 3. *Proc. Natl Acad. Sci. USA,* **90,** 730–734.)

Gene Mutations

Gene **mutations,** or point mutations, affect the nucleotide structure of the gene itself and so are not observable at the chromosomal level. Many gene mutations can be discovered either directly via nucleotide sequencing or indirectly from the effects mutations have on the amino acid sequences of proteins.

Among the various kinds of mutational changes at the molecular level are **base substitutions,** nucleotide changes that substitute one base for another. As shown in **Figure 10-9,** these changes are termed transitions when exchanges occur between purines (A ↔ G) or pyrimidines (T ↔ C) and transversions when purines exchange for pyrimidines or *vice versa* (A, G ↔ T, C).

Substitutions may occur spontaneously through copying errors caused, for example, by rare nucleotide base changes that enable complementary pairing between adenine and cytosine or guanine and thymine (**Fig. 10-10**). Other base substitutions may arise from the action of mutagenic agents such as nitrous acid or mustard gas. Also, in some instances, specific nucleotide sequences cause increased mutations among adjacent nucleotides. Such "hot spots of mutation" may act by coiling the DNA molecule in ways that influence DNA polymerase enzymes to produce replication errors. The replication accuracies of polymerase enzymes differ; some strains carry enzymes that are more prone to produce mutational errors than others.

Other mutational events at the molecular level include nucleotide deletions, duplications and insertions, which

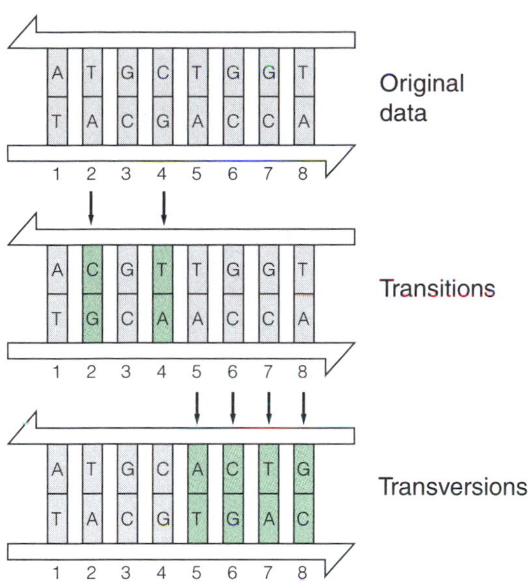

FIGURE 10-9 Examples of specific base pair changes in a section of double-stranded DNA.

can arise spontaneously within the cell or be evoked by externally applied mutagenic agents such as acridine dyes. Rearrangements of nucleotides — inversions (reversals in nucleotide order) or transpositions (movement of nucleotide sequences to new positions) — also occur.

In general, mutational effects caused by these mechanisms may be expressed at two levels of gene activity: (1) **changes within the gene product itself** (for example, in the amino acid constitution of a particular protein), and (2) **changes in the regulation of a gene product** (the gene product is not itself affected, but the timing of its appearance differs from normal).

Mutational effects that result in a changed gene product may arise because of nucleotide changes that cause:

- *missense mutations,* substitutions for one or more of the amino acids in a protein (**Fig. 10-11a, b**).
- *nonsense mutations,* changes that insert protein termination ("stop") codons in the middle of a gene sequence, thus causing premature termination of polypeptide chain synthesis (Fig. 10-11c).

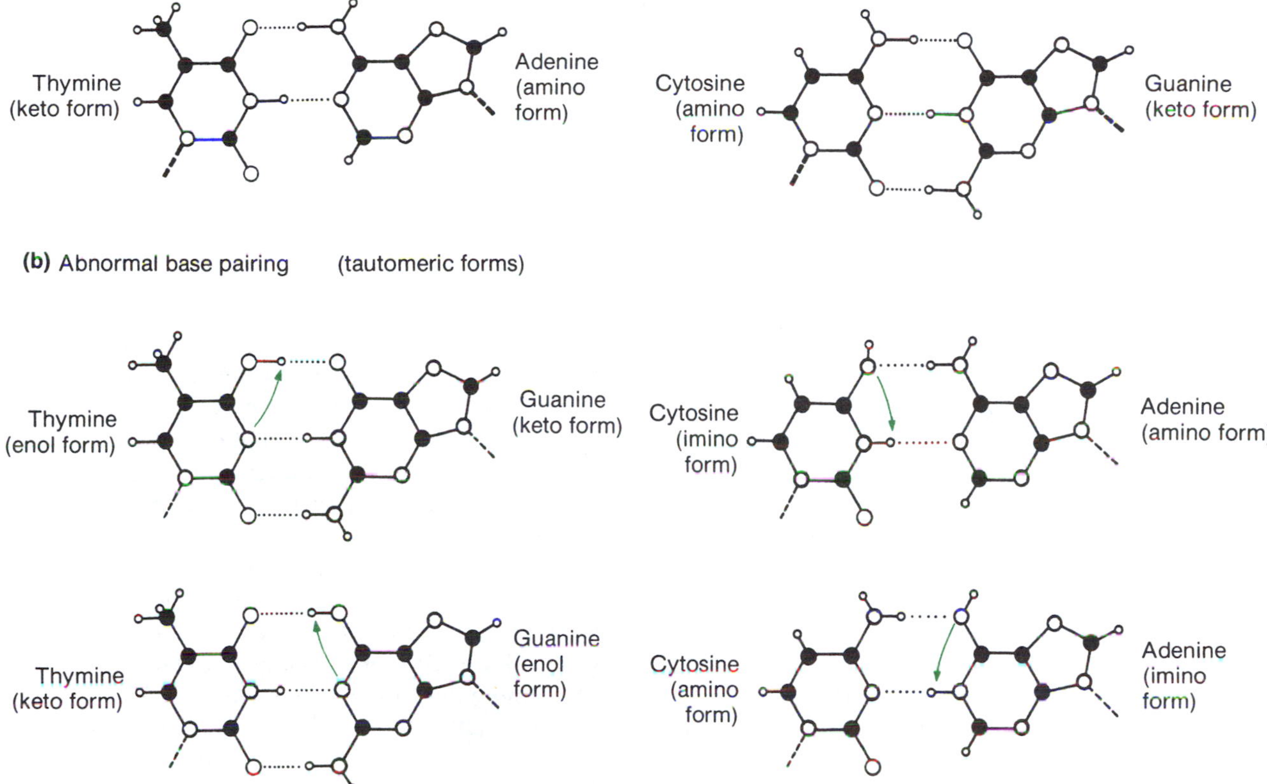

FIGURE 10-10 (a) Normal complementary pairing between nucleotide bases during DNA replication. (b) Modified base-pairing relationships that result from *tautomeric* molecular changes. Because of such changes, base substitutions can occur that produce, for example, transitions from T–A base pairs to C–G base pairs. (Adapted from Drake, J. W., 1970. *The Molecular Basis of Mutation.* Holden-Day, San Francisco.)

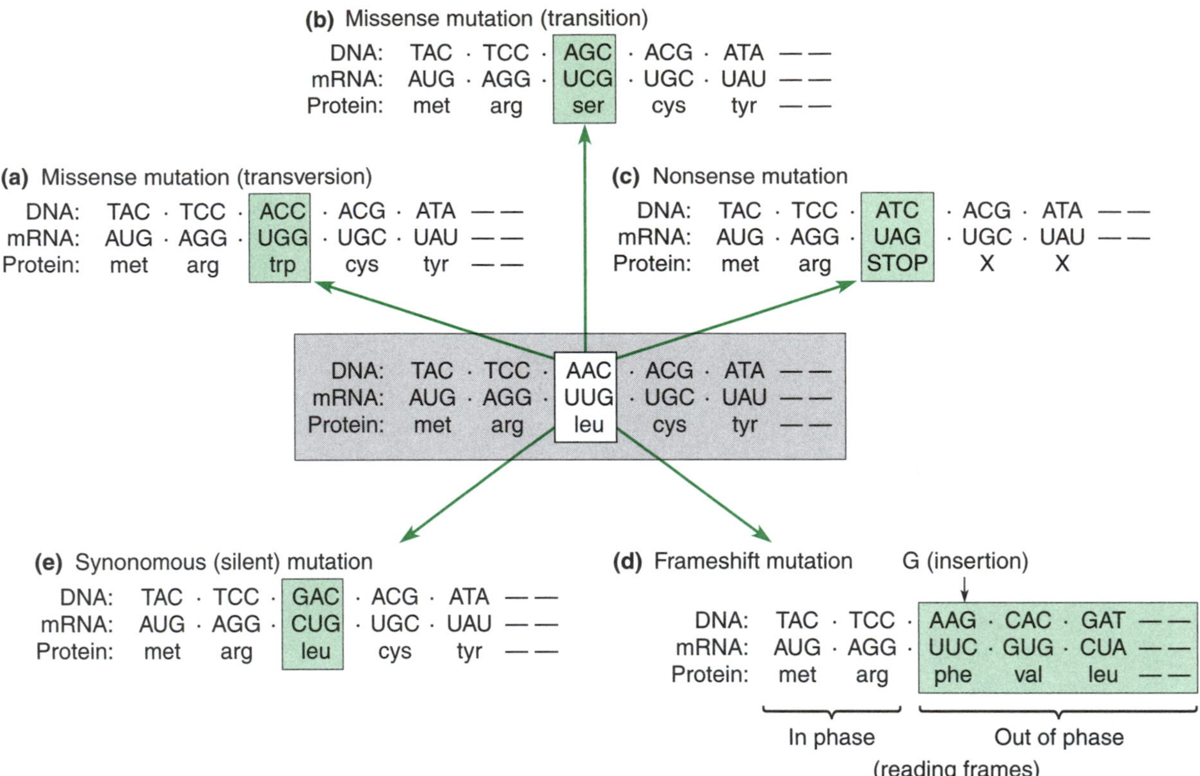

FIGURE 10-11 Different kinds of mutations produced by the indicated nucleotide changes. In all of these sequences, the AUG codon on messenger RNA initiates translation into amino acids. Note that despite mutations, translation proceeds unimpeded except in (c), which bears a stop codon that prevents further reading of messenger RNA. Translatable sequences of nucleotides (for example, *a, b, d, e*), free of stop codons, are called open reading frames.

- *frame shift mutations,* nucleotide insertions or deletions that modify the messenger RNA (mRNA) protein translation reading frame so that a new and different sequence of codons appears (Fig. 10-11d).
- *synonymous (or silent) mutations,* which, because practically all amino acids are coded by more than one kind of codon, can change an amino acid codon without producing an amino acid substitution (Fig. 10-11e).

This wide range of mutational possibilities has important consequences for organisms. Both too many and too few mutations can interfere with the effect that natural selection has on genetic variation, the basis for adaptation. Too many mutations will generate continued errors in organisms already selected for their environment. Too few mutations will reduce the opportunity for natural selection to lead to adaptations. Optimal mutation rates are therefore advantageous (Orzack and Sober, 2001).

Depending on its position, however, a single nucleotide mutation may have important consequences for the organism, even though it causes only a single amino acid substitution in a long-chain protein. A prominent example is the effect caused by the sickle cell mutation in humans, an allele that, in the United States, is almost entirely confined to Americans of African origin.

Normally, the adult hemoglobin molecule in human blood cells consists of four polypeptide globin chains, two αs and two βs, each about 140 amino acids long with its own specific sequence. However, in homozygotes for the sickle cell allele (Hb^S/Hb^S) all β-globin chains differ from normal βs at the no. 6 position because of a transversion that changed the glutamic acid codon GAA to the sickle cell valine codon GUA. As shown in **Figure 10-12,** the effects of this single genetic mutation are profound, causing a variety of phenotypic changes that often lead to death. (**Pleiotropy** is the name given to multiple phenotypic effects of a single gene.) **Sickle cell anemia** is known to kill, before the age of 20, more than 10 percent of African-Americans who are homozygotes for the allele, and kills even more individuals in Africa, where medical facilities are more limited. The high frequency of this gene in these populations is related to the selective advantage of sickle cell heterozygotes in regions in which malaria is endemic, a topic discussed in Chapter 22.

Normal
Mutation →
Sickle cell

Normal		Sickle cell
---GAA--- / ---CTT---	DNA (template strand)	---GTA--- / ---CAT---
---GAA---	Messenger RNA	---GUA---
--- glu ---	β Hemoglobin chain	--- val ---

Normal hemoglobin → Normal development

Abnormal hemoglobin → Deoxygenation of hemoglobin in tissues → Sickling of red blood cells

Clumping of cells and interference with blood circulation
Increased red blood cell destruction
Accumulation of sickle cells in spleen
Local failures in blood supply
Anemia
Breakdown of hemoglobin
Urogenital damage
Muscle and joint damage
Ischemia, necrosis
Brain damage
Heart damage
Dilation of heart
Bone marrow overactivity and increase in amount of marrow
Enlargement, then fibrosis of spleen
Bone damage
Lung damage
Accumulation of bilirubin

Stroke and paralysis
Delayed sexual maturity
Osteo-myelitis
Pneumonia
Weakness and lassitude
"Tower skull"
Gallstones

Impaired ability to fight infection
Kidney failure
Bone deformity
Pain crises
Heart failure
Decreased growth
Impaired mental function
Jaundice

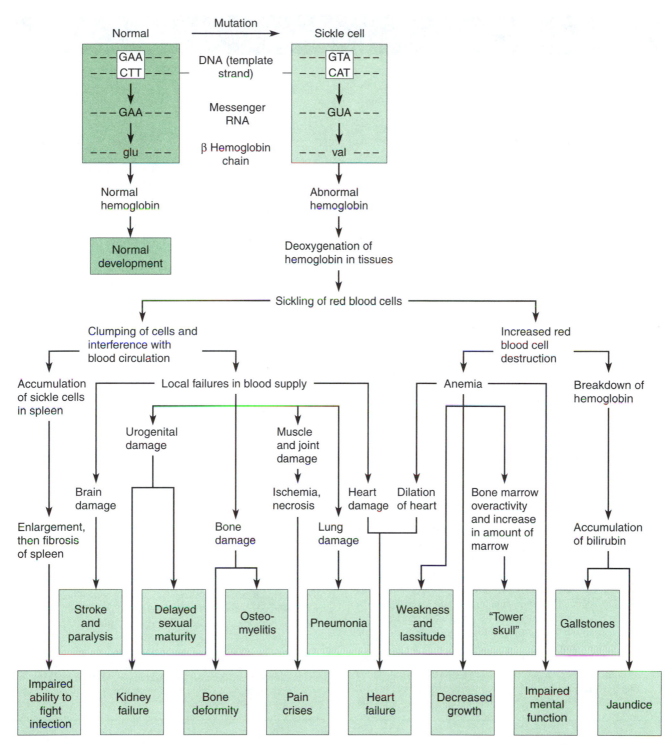

FIGURE 10-12 Varied (pleiotropic) effects of the sickle cell mutation, beginning with the transversion that changed a thymine nucleotide to an adenine nucleotide on the DNA template strand of the β-hemoglobin gene. The resultant GUA trinucleotide (triplet) coding sequence on messenger RNA translates into a valine amino acid instead of the normal glutamic acid, producing developmental consequences that can seriously affect sickle cell homozygotes. (From *Genetics,* 3rd ed. by Monroe W. Strickberger. © 1985 by Monroe W. Strickberger. Reprinted by permission of Prentice Hall, Inc., Upper Saddle River, NJ.)

BOX 10-2 Barbara McClintock and the Discovery of Transposons

THE HISTORY OF the discovery of **transposons** is a fascinating one, teaching us much about openness to new ideas in the scientific community.

The transposition between chromosomes of particular genetic elements, now known as transposons, and the regulatory role of the transported elements on gene expression (now gene regulation), were both discovered in the 1940s and 1950s by Barbara McClintock (1902–1992), a plant cytogeneticist who worked on maize (McClintock, 1950). If you know the more recent history, especially of gene regulation, there are two reasons why this will surprise you:

- The discovery of transposons is usually attributed to research on bacteria in the early 1970s, 30 years after McClintock's research.
- The knowledge that genes in prokaryotes are regulated dates back to Jacob and Monod's research on the *lac* operon in *E. coli* (Fig. 10-13), published in 1961, research for which they shared the Nobel Prize in 1965. The knowledge that eukaryote genes also are regulated, however, has an even more recent history, usually attributed to research on animals such as *Drosophila* and *Costridium elegans* (see Chapter 13).

McClintock's name should be known to every molecular biologist, and to an extent it is, though only occasionally do you find citations to her insightful work on gene regulation. In 1983, she was belatedly awarded the Nobel Prize in Physiology and Medicine for her discovery of genetic transposition. A biography published in the same year (Keller, 1983) also brought her stellar career to the forefront. Among her many other achievements, McClintock discovered genetic recombination by crossing-over during meiosis; produced the first genetic map for maize, enabling physical traits to be linked to specific regions of the chromosomes; and demonstrated the importance of telomeres and centromeres in conserving genetic information (see Chapter 9). Her essential findings on transposition, made between 1944 and 1953 from her studies of the basis of the mosaic color patterns of maize seeds (and the known unstable inheritance of this character), are outlined below.

1. McClintock identified two dominant and interacting gene loci, *Dissociator* (*Ds*) and *Activator* (*Ac*), discovering that *Ds* has an effect on nearby genes, but only when *Ac* is present. Surprisingly, *Ds* and *Ac* are able to change their positions on the chromosome.
2. After conducting generations of crosses, assessing kernel colors and carrying out painstaking cytogenetics, she concluded that the transposition of *Ds* was controlled by *Ac* and required chromosomal breakage. *Ds* normally suppresses a color gene, which is only activated after *Ds* moves. Because transposition is random, some cells produce pigment, others do not, resulting in the color mosaicism so characteristic of maize seeds.
3. By the late 1940s, McClintock had developed a theory in which these two mobile elements are controlling elements, as distinct from genes, and extended her theory of gene regulation to all multicellular organisms. But biology then was pre-DNA and certainly pre-molecular, meaning that the nature of genetic material remained a mystery. McClintock's genome was much too dynamic to fit the then-current view that genes provide preprogrammed instructions or a blueprint to cells.
4. In 1961, when the discovery of the *lac* operon was published, McClintock responded with an informed comparison of gene regulation in bacteria and maize. But only after the (re)discovery of transposons in the late 1960s did McClintock's discoveries on genetic transposition begin to receive the credit they deserved, although her pioneering work on gene regulation remains underappreciated to this day. *Dissociated*, *Activator*, and a third mobile element, *Suppressor-mutator* (*Spm*) also discovered by McClintock, are all now known to be transposons, which normally only move when the cell containing them is under stress (hence the link to chromosome breakage). *Ds* and *Ac* are now routinely used by molecular biologists to study gene regulation in plants.

Barbara McClintock was decades ahead of the field. But although her results were published in major journals (*Genetics*, *PNAS*)[a], and she spoke at meetings and trained and mentored students, her "colleagues" were so skeptical of her findings that in 1953 she stopped publishing on transposition, turning instead to the cytogenetic and ethnobotanical aspects of races of maize in South America.

She articulated her frustration in a letter to a geneticist at the University of Leeds who had invited her to participate in a workshop to be held in September 1973, an invitation she declined. Writing of "my attempts during the 1950s to convince geneticists that the action of genes had to be and was controlled." She continued, "It is now equally painful to recognize the fixity of assumptions that many persons hold on the nature of controlling elements in maize and the manners of their operation. *One must await the right time for conceptual change*" (emphasis added).

[a] The November 2006 issue of *Proceedings of the National Academy of Science U.S.A.* contains a special feature with eight papers on transposable elements, including an introduction (pp. 17600–17601) placing McClintock's research into context and summarizing the currently known types of transposable elements.

Gene Regulation

One of the more fascinating developments in evolutionary biology over the past decade has been the realization that mutations that affect the DNA sequence of genes and the amino acid composition of their protein products may not be as important as mutations that affect when, where and how much of a gene product is expressed in the organism. Although the idea of regulatory mutations dates back to the mid 1970s and earlier (**Box 10-2**), a combination of empirical, bioinformatic and conceptual advances in the past decade has brought regulation into the mainstream.[5]

[5] See King and Wilson (1975) and Jacob (1977) for seminal ideas on regulation, and Hall et al. (2004), S. B. Carroll (2005a,b) and Edwards et al. (2005) for recent approaches.

Gene Regulation in *Escherichia coli*

Viruses and prokaryotes have provided a great deal of information on regulatory mechanisms at the genetic level. The first study leading to the discovery of gene regulation was that by Jacob and Monod (1961) on the genes governing the production of enzymes involved in lactose sugar metabolism in the bacterium *Escherichia coli* (**Fig. 10-13a**), a fundamental discovery for which they and André Lwoff, the director of the Pasteur Institute where the research was conducted, received the 1965 Nobel Prize in Physiology and Medicine.

Because *E. coli* do not commonly encounter lactose, a **repressor protein** that occupies a specific regulatory operator site normally prevents the genes used in *lac* enzyme synthesis (**structural genes**) from being transcribed into mRNA. Molecular binding between the repressor and operator DNA prevents the RNA polymerase enzyme from attaching to its **promoter**, near which transcription normally begins. As a result, transcription of *lac* enzyme genes into mRNA is prevented (Fig. 10-13b). However, when bacteria encounter lactose sugar in the medium, some lactose molecules convert to a form called *allolactose*, which acts as an inducer and binds with the repressor. This combination releases the repressor from the operator site, allowing the RNA polymerase to transcribe the genes necessary to metabolize lactose (Fig. 10-13c).

Because of the complexity of regulatory systems, various kinds of mutations affect the quantity and timing of gene productivity. For example, some mutations in the *I* regulator gene that produces the *lac* repressor can prevent it from binding to the operator, thereby causing *lac* enzyme synthesis to occur, even in the absence of the inducer (Fig. 10–13d). Conversely, other *I* mutations produce repressor proteins that cannot bind to inducer molecules, thus persistently preventing *lac* enzyme synthesis by keeping the repressor attached to the operator site, even in the presence of the inducer (Fig. 10-13e). In addition, various mutations of DNA at the **promoter site** may either increase or decrease the rate of transcription by preferentially enhancing or diminishing the attachment of RNA polymerase enzymes. RNA polymerases also operate in eukaryotic cells, the 2006 Nobel Prize in Chemistry having been awarded to Roger Kornberg for elucidating the roles of RNA polymerases in eukaryotes.[6]

Levels of Gene Regulation in Eukaryotes

Several levels of regulation of gene expression operate in eukaryotes. The four outlined below are *cis-*, *trans-*, miRNA- and RNAi-regulation.

1. ***cis*-regulation:** One level involves special sites on DNA sequences, called CAAT and TATA boxes, to which special proteins attach so that transcription can occur (**Fig. 10-14**). The DNA double helix at some eukaryotic regulatory sites changes from a right-hand to a left-hand form, accompanied by a zigzag placement of phosphate groups (**Fig. 10-15**). Such sites are adjacent to the gene they regulate, interact with the promoter of those genes, and are known as *cis*-regulatory elements (**Fig. 10-16**). Changes in (and the evolution of) *cis*-regulation, either because modification of existing *cis*-regulatory elements alters binding sites for transcriptional activators or repressors (Fig. 10-16), or by duplication or co-option and expansion of function of an existing element, is now recognized as an important mechanism of genetic change leading to morphological change in evolution (see Chapter 13).
2. ***trans*-regulation:** At a second level are the *proteins that bind to these* cis-*regulatory elements.* Such proteins, called **transcription factors,** regulate the transcription of genes not in close proximity. Transcription factors may be transcribed on different chromosomes from the genes they regulate and are therefore referred to as *trans*-regulatory elements (Fig. 10-16).
3. **RNA interference (RNAi)** is the process of regulating gene transcription using *small interference* (si) RNA (siRNA). RNAis were demonstrated following experiments in which insertion into plants of extra copies of a gene that produces pigment, blocked the expression of any pigment. RNAis are abundant and highly conserved. A single RNAi can regulate many transcripts in an RNA interference pathway in which double stranded RNA is processed into 20 to 25 nucleotide siRNAs that assemble into RNA-induced silencing complexes (RISCs). siRNA strands guide RISCs to complementary RNA molecules, where they cleave and destroy the RNA; synthetic siRNAs can induce RNAi in mammalian cells. As several hundred RNAi classes are known, many associated with two of the most important cellular processes — division and differentiation — their regulatory role is enormous; the 2006 Nobel Prize in Physiology and Medicine went to U.S. researchers Andrew Fire and Craig Mello for their research on RNAi in animals.[7] *Drosophila* genomes contain more than 100 RNAis and human genomes may encode for more than 800. As RNAis regulate gene expression in many if not most developmental and physiological contexts, mutations that affect their activity are potential sources of evolutionarily significant variation.
4. **microRNA:** A fourth level is *post-transcriptional regulation* by microRNA (miRNA). miRNAs are short (18 to 25 nucleotide) sequences of non-coding single-stranded RNA that regulate the translation of proteins in plants

[6] See Landick (2006) for a description of Kornberg's research.

[7] See Zamore (2006) for a summary of their research. The first studies on RNAi were done in plants; see Hamilton and Baulcombe, 1999.

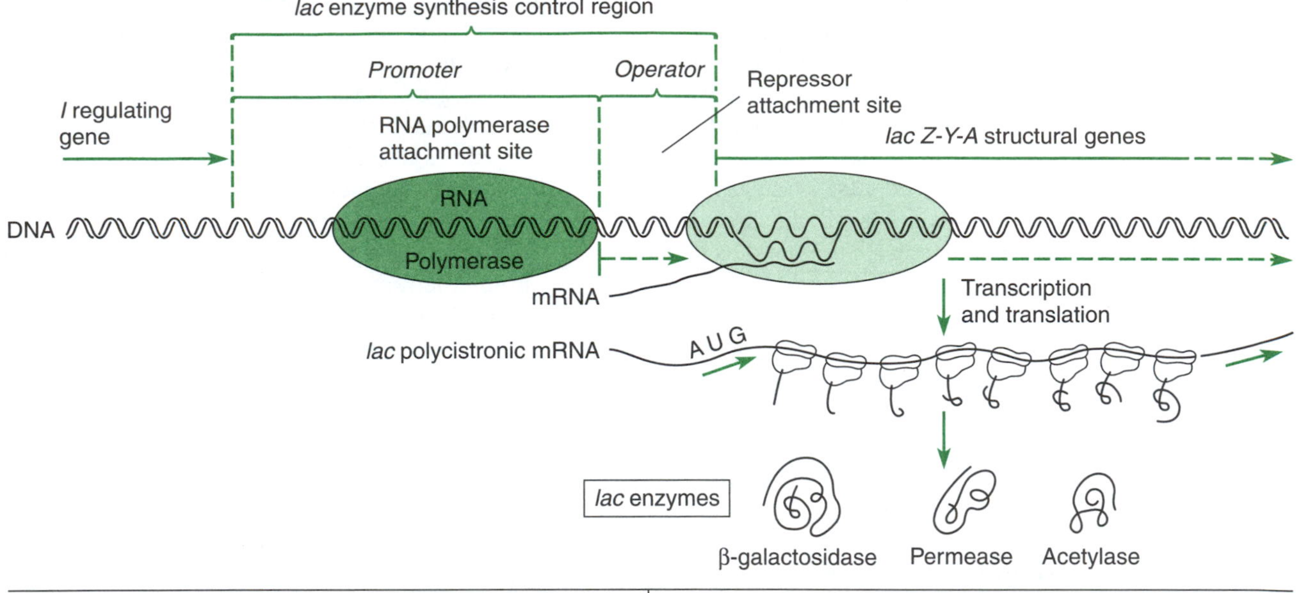
(a) Mode of lac enzyme synthesis in absence of repressor
lac enzyme synthesis control region
Promoter
Operator
Repressor attachment site
I regulating gene
RNA polymerase attachment site
lac Z-Y-A structural genes
DNA
RNA
Polymerase
mRNA
Transcription and translation
lac polycistronic mRNA
AUG
lac enzymes
β-galactosidase
Permease
Acetylase

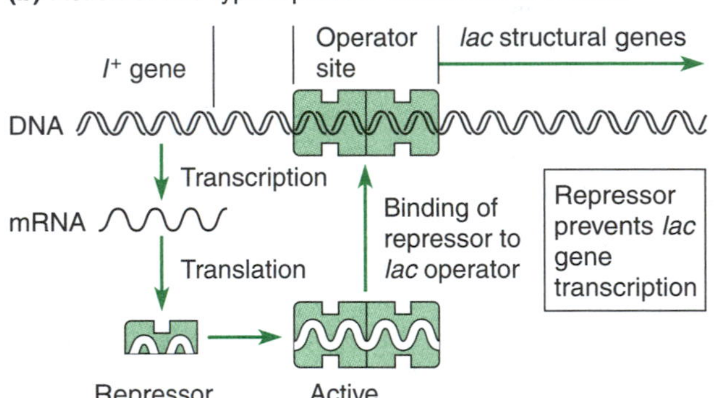
(b) Action of wild-type repressor in absence of inducer
I+ gene
Operator site
lac structural genes
DNA
Transcription
mRNA
Translation
Binding of repressor to lac operator
Repressor prevents lac gene transcription
Repressor polypeptide (monomer)
Active repressor (tetramer)

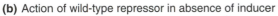

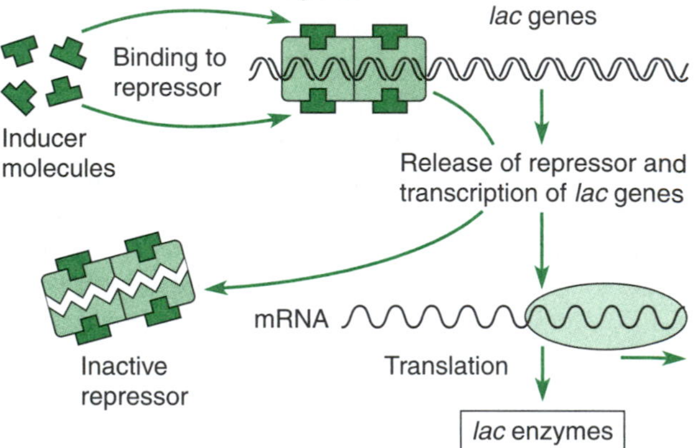
(c) Effect of inducer on repressor: induced lac enzyme synthesis
Operator site
lac genes
Binding to repressor
Inducer molecules
Release of repressor and transcription of lac genes
mRNA
Inactive repressor
Translation
lac enzymes

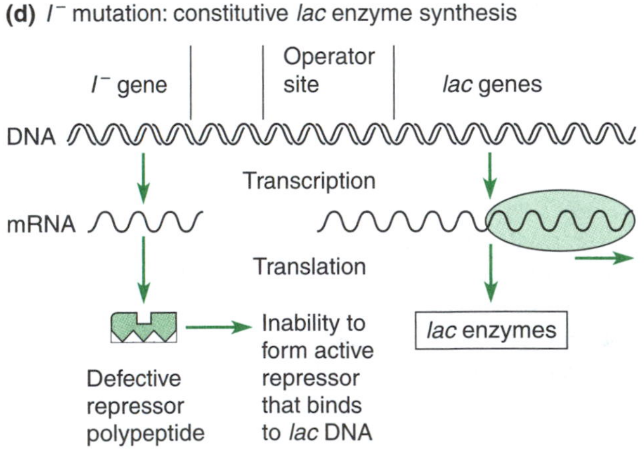
(d) I− mutation: constitutive lac enzyme synthesis
I− gene
Operator site
lac genes
DNA
Transcription
mRNA
Translation
Defective repressor polypeptide
Inability to form active repressor that binds to lac DNA
lac enzymes

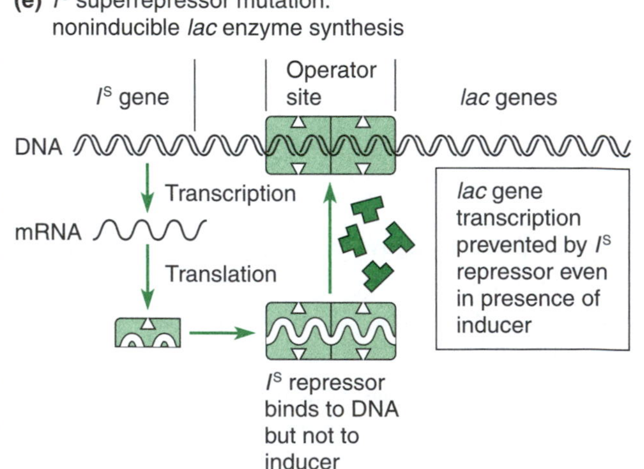
(e) IS superrepressor mutation: noninducible lac enzyme synthesis
IS gene
Operator site
lac genes
DNA
Transcription
mRNA
Translation
lac gene transcription prevented by IS repressor even in presence of inducer
IS repressor binds to DNA but not to inducer

FIGURE 10-13 General scheme of *lac* enzyme synthesis in *E. coli* and the effects of repressor function or dysfunction on this inducible system.

(a) The DNA region involved in controlling transcription of the *lac* structural genes *Z, Y,* and *A* consists of two major regulatory sites, the operator and promoter, each of which serves to bind specific proteins. In the absence of the repressor protein, RNA polymerase begins transcribing at the operator, which is also the site to which the repressor attaches. (The *lac* repressor itself is coded at a regulator locus, *I,* adjacent to the *lac* locus, but such proximity is not necessarily true for all systems controlled by repressor genes.) Transcription of the *lac* genes, coupled with translation, leads, as shown, to synthesis of the three *lac* enzymes.

(b) Transcription and translation of the I^+ normal gene produces a normal repressor protein that binds to the operator site of the *lac* locus, blocking the transcription of *lac* genes by RNA polymerase. This repressed state appears in normal *E. coli* cells that are not grown on a lactose medium.

(c) Transfer of cells to a lactose medium leads to the introduction of allolactose inducer molecules, which causes the repressor to dissociate from its DNA binding site on the operator. This allows transcription of the *lac* structural genes to proceed, followed by their translation and synthesis of *lac* enzymes. As shown diagrammatically, the repressor is a tetramer (a molecule composed of four polypeptide chains) that acts as a regulatory protein, signifying that it has more than one binding site: in this case, one for DNA and one for the inducer. Binding of the inducer to the repressor changes the form or steric configuration of the DNA binding site, making the repressor inactive.

(d) When an I^- mutation produces an inactive repressor that cannot bind preferentially to the *lac* operator, the repressor does not impede transcription of *lac* genes. The synthesis of *lac* enzymes thus proceeds "constitutively" in the absence of inducers, that is, when grown on a nonlactose medium.

(e) A super repressor mutation at the *I* locus causes the production of a *lac* repressor that no longer recognizes inducers but maintains its site for normal lac operator attachment. The result is repression of *lac* enzyme synthesis even in the presence of inducer. (From *Genetics,* 3rd ed, by Monroe W. Strickberger. © 1985 by Monroe W. Strickberger. Reprinted by permission of Prentice Hall, Inc., Upper Saddle River, NJ.)

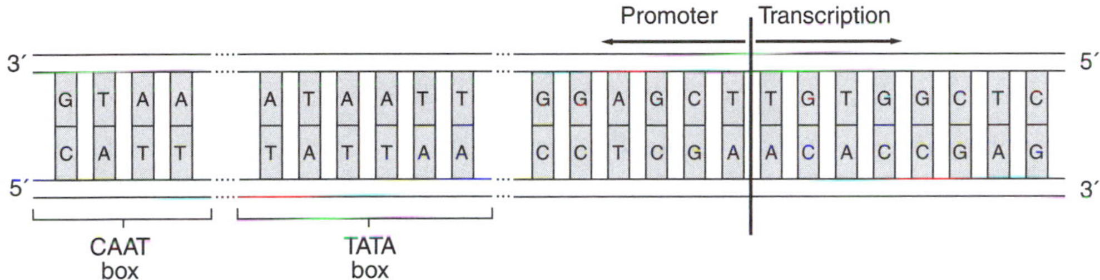

FIGURE 10-14 An eukaryotic nucleotide sequence regulating the gene that produces the thymidine kinase enzyme. The promoter region of this gene (used for attachment of RNA polymerase and other transcription-assisting proteins) contains two short sequences or boxes, called CAAT and TATA, which have also been found in promoter regions of other genes. Altering these sequences reduces the level of gene expression. In addition to promoters, other effects on transcription occur through "enhancers," sites that are often at some distance from promoters yet also affect RNA polymerase attachment. For transcription effects caused by extracellular signals (signal transduction), see Figure 13-3.

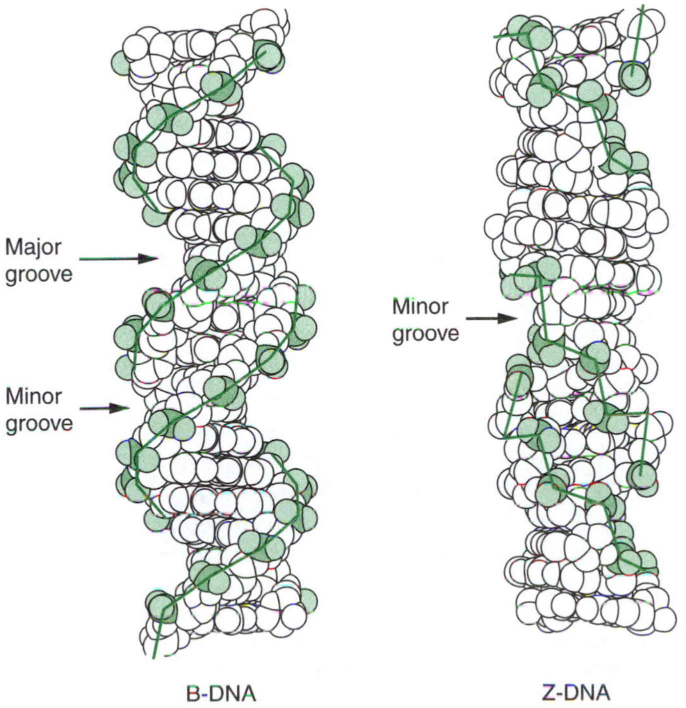

FIGURE 10-15 Space-filling models of B-DNA and Z-DNA double-helix molecules. The *lines* used to connect the phosphate groups (*color-shaded*) in each chain show the zigzag placement of phosphates in Z-DNA in contrast to the smoother curve of their relationship in B-DNA. The major and minor grooves in B-DNA differ in depth but do not extend to the central axis of the molecule, whereas the indicated Z-DNA groove penetrates the axis of the double helix. (From Rich, E., A. Nordheim, and A. H.-J. Wang. 1984. The chemistry and biology of left-handed Z-DNA. *Ann. Rev. Biochem.,* **53,** 791–846. Reproduced with permission from the *Annual Review of Biochemistry,* Volume 53, © 1984 by Annual Reviews Inc.)

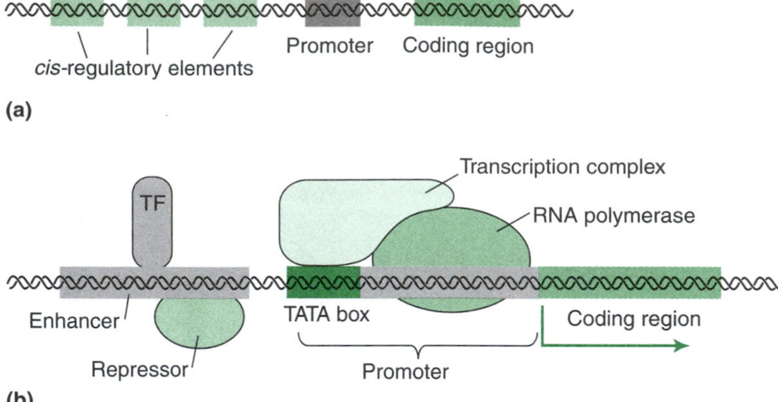

FIGURE 10-16 *cis*-regulatory elements and gene transcription in animals. (a) Location of three *cis*-regulatory elements upstream of the promoter and coding region. (b) Transcription factors (TF) in combination with coactivators (not shown) bind to *cis*-regulatory elements (as do repressors in combination with corepressors). The transcription factor(s) with RNA polymerase II (RNA) forms a transcription complex on the TATA box of the promoter, initiating transcription (arrow). (Modified from Carroll et al., 2005. *From DNA To Diversity,* 2nd ed., Blackwell Publishing, Malden, MA, 2005).

and animals by binding to matching target mRNAs leading to degradation of the mRNA itself. miRNAs are encoded by RNA genes that are transcribed from DNA. miRNAs do not produce proteins. Rather, the functional miRNA transcript binds to the 3′-untranslated region of one or more mRNAs, which are then destroyed. You can think of miRNA as killer RNAs. As there are hundreds of miRNAs and as each miRNA regulates a number of genes, the role of miRNAs in gene regulation is enormous.[8]

How Gene Regulation Works

Some mutations affect the amount or rate at which a gene product is produced, although the products themselves may be unaffected.

An example is the *thalassemias,* genetic diseases in which either α or β hemoglobin chains are not produced or are produced in diminished numbers. Although such mutations, like the sickle cell allele, often cause death of homozygotes, like the sickle cell allele they also offer protection against malarial parasites. Presumably, this special advantage of some thalassemia heterozygotes explains the frequency of these genes in human populations.

Another example is a mutation in a repressor element for the insulin-dependent growth factor II (*IGF2*) gene in pigs. Pigs carrying this mutation have a threefold increase in the production of IGF_2, resulting in a more muscular and leaner phenotype (Van Laere et al., 2003).

Other regulatory mutations affect **when** during development and **in which** anatomical locations a gene product is produced. Work by Sean Carroll and his colleagues, for instance, has shown how evolutionary changes in *Drosophila* wing pigmentation and male abdominal pigmentation patterns are produced by changes in *cis*-regulatory elements that determine the spatial distribution of expression of the *yellow* gene on the wing, and by gain of a **homeobox**-protein binding site in a *cis*-regulator of the *yellow* gene operating in abdominal cells. These changes resulted in five independent losses of wing spots within species of *Drosophila melanogaster,* and two independent gains with species of *D. obscura* (S. B. Carroll, 2005a,b; Jeong et al., 2006).

Because regulation plays an essential role in the timing and placement of all metabolic reactions, regulatory mutations can easily affect morphology and function of any organism, a topic that will be more thoroughly discussed in Chapter 13. The most dramatic examples of gene regulation involve mutations that alter the development of **entire parts** of organisms. For example, the *bithorax* locus in *D. melanogaster,* which specifies a particular segment, the thorax, has mutations that produce an extra set of wings by converting the abdominal segment into a thoracic segment and abdominal appendages (halteres) into thoracic appendages (wings) (**Fig. 10-17**)[9]. Changes that affect developmental growth coordinates (depicted in **Figure 10-18** as changes between the adults of two species of fish) may arise by regulatory

[8] Research into miRNA is at an active stage, with recent descriptions of 447 new miRNA genes in chimpanzee and human brains (Berezikov et al., 2006), evidence that human miRNAs respond to selection (Chen and Rajewsky, 2006), the discovery of a core set of miRNAs in protostomes and deuterostomes but not in sponges or cnidarians (Sempere et al., 2006, and see Chapter 16), and the suggestion that miRNA diversity between chick, mouse and two fish (Japanese medaka and zebrafish) relate to physiological differences between different groups of animals (Ason et al., 2006).

[9] Had this mutation leading to a second pair of wings occurred in nature, it may have forced entomologists to erect a special order (a higher level of taxon, see Chapter 11).

FIGURE 10-17 A four-winged *Drosophila,* caused by mutations at the *bithorax* locus. Normally, as in all dipteran insects, *Drosophila* has only a single pair of wings, which arise from the second of the three thoracic segments. (The second dipteran wing pair evolved into balancing organs, *halteres.*) As shown here, certain *bithorax* mutations cause the third thoracic segment to transform into a second thoracic segment complete with wings, a condition similar to the ancestral four-winged fly. These mutations, and others of this kind occur in genes that regulate development, a topic discussed in Chapter 13. (Figure provided by E. B. Lewis. Reprinted by permission.)

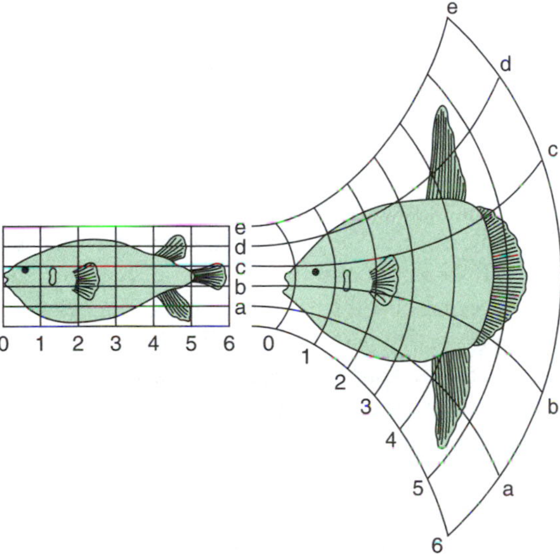

FIGURE 10-18 D'Arcy Thompson's demonstration of how completely different shapes of adults (in this case, fish) can be generated by simple developmental changes in geometric coordinates. Although shown as if one adult transforms into another, the changes affect development rates of differential growth. If the vertical coordinates of the puffer fish, *Diodon* (left), change into concentric circles, and its horizontal coordinates into hyperbolas, the resultant animal is shaped like the sunfish, *Orthagoriscus* (right). Such changes may be caused by a new or mutated developmental "morphogen" (see Chapter 13) that increases cell production as it moves along a gradient from anterior to posterior, and dorsally and ventrally from the midline. (Adapted from Thompson, D'A. W., 1942. *On Growth and Form,* 2d ed. Cambridge University Press, Cambridge, England.)

changes in patterning genes; recent evidence shows that the difference in size between wings and halteres is controlled by **Hox gene** regulation[10] of a diffusible product from the gene decapentaplegic protein (Crickmore and Mann, 2006).

The extent to which mutations in *cis*- and *trans*-acting regulatory elements or posttranscriptional regulation produce the phenotypic differences we see among individuals within species and between different species is becoming increasingly apparent. To quote one of the major researchers in the field:

> In my view, *cis*-regulatory information processing, and information processing at the gene regulatory network circuit level, are the real secret of animal development. Probably the appearance of genomic regulatory systems capable of information processing is what made animal evolution possible (Davidson, 2006, p. *x*).

Indeed, there are compelling reasons for the view that mutations in regulatory genes are of great importance in evolution (see Chapter 13), four of which are outlined below.

1. **The relatively high degree of gene sequence conservation among species.** Gene products tend to be fairly conservative across species in terms of both structure and function. This is contrary to what one would expect if structural and functional variation in genes and their protein products accounted for most organismal diversity.
2. We now know that there are **many fewer genes in metazoan genomes** than we had thought and that **most genes perform multiple developmental or physiological functions.** Mutations that affect the structure of a gene product are, therefore, rare; they tend to interfere with several, often unrelated, aspects of the phenotype. In contrast, mutations that influence regulation can affect the role of a gene in a particular part of the phenotype without interfering with other functions of the gene in other parts of the organisms or at other times during the life cycle. Variation in gene regulation helps explain how such a small number of genes can initiate the endless variety seen in nature.
3. Evolution by gene regulation helps explain why **evolutionary convergence is more common than expected by chance.** If complex developmental pathways can be tweaked by up- or down-regulating the expression of individual genes or proteins, or by turning on or off cascades of events that determine the fates of individual

[10] Homeobox-containing genes in vertebrates are known as Hox genes, Ho from homeodomain and x for vertebrate. Thus *Pax-6* is a vertebrate homologue of the *Drosophila paired rule* (*Pa*) gene, which is combined with x to make *Pax.* The 6 indicates that 5 other genes in the Pax family were known when *Pax-6* was named. Convention also dictates (although not all follow the convention) that homeobox or *Hox* genes are italicized (*antl*) but the gene protein is not, and that *Hox* genes in humans are italicized and capitalized (*HOX*).

parts, it is easy to see how changes in gene regulation could produce the repeated occurrence of similar (albeit rarely identical) evolutionary changes.

4. The importance of gene regulation may explain why **such a large proportion of many metazoan genomes consists of noncoding regions** that have, in the past, been regarded as "junk" DNA. Large amounts of noncoding DNA may be necessary for the intricate patterns of temporal, tissue, region and physiological state-specific gene regulation seen in multicellular organisms.

Even if gene regulation is an important or even the central level at which genomic evolutionary change occurs, this does not mean that changes in gene function produced by sequence variation in genes are not also important. Analyses of how mutations in Hox-gene function relate to morphological evolution in invertebrates show how changes in *cis*-regulation of Hox gene expression *and* sequence variation in the Hox genes themselves can interact to produce evolutionary change (Ronshaugen et al., 2002).

Continuous Variation

As indicated at the beginning of the chapter when introducing variation and variability, beginning with Darwin, many evolutionary biologists suggested that rather small heritable changes provide most of the variation on which natural selection acts. In Darwin's words from *On the Origin of Species:*

> Extremely slight modifications in the structure and habits of one species would often give it an advantage over others; and still further modifications of the same kind would often still further increase the advantage . . . Under nature, the slightest differences of structure or constitution may well turn the nicely balanced scale in the struggle for life, and so be preserved.

The relative importance for evolutionary change of small or large changes has been under active investigation ever since. As discussed in Chapter 21, this issue began to be clarified once genetic differences were understood in terms of gene frequencies in populations and the forces affecting them. Whatever the differences between genes, it was recognized that they furnish the basic elements of evolutionary change, and their frequencies and distributions provide means for identifying causes of evolutionary change.

Nevertheless, for many measurable traits such as size and yield, researchers such as medical or agricultural geneticists usually focused more narrowly on small changes or **continuous variation**. These are often seen for characters distributed in bell-shaped curves (normal distributions), for example, human height (**Fig. 10-19**). From an evolutionary view, it is clear that such small differences can accumulate through selection to give large quantitative differences. As illustrated in **Figure 10-20**, selecting for the presence or absence of white spotting in Dutch rabbits can lead to completely colored or completely white strains. The genetic causes for these changes are genes with small phenotypic effect — **polygenes,** and **quantitative trait loci** (QTLs) — that can produce normal distributions when they assort independently.

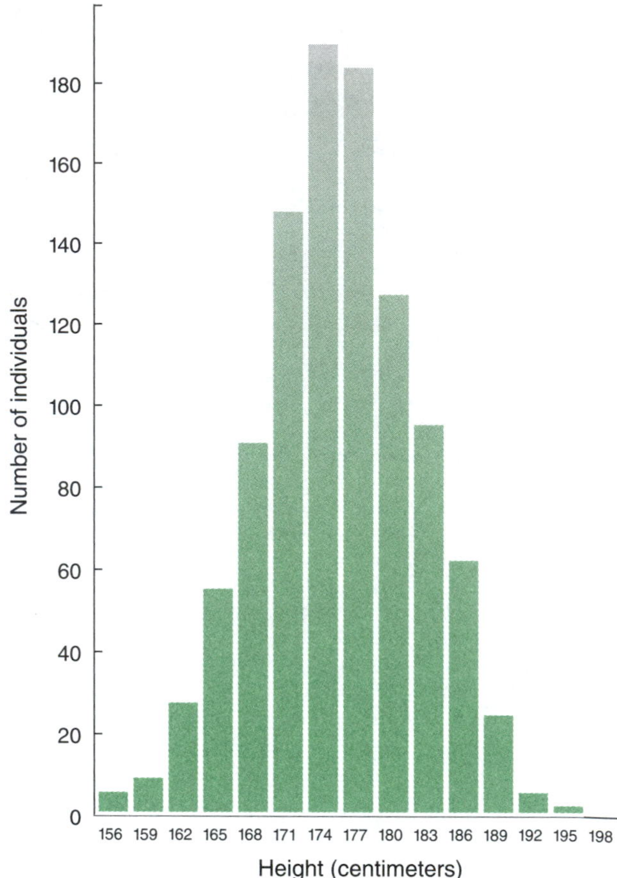

FIGURE 10-19 Distribution of the heights of 1,000 Harvard College students aged 18 to 25. (Adapted from Castle, W. E., 1932. *Genetics and Eugenics,* 4th ed. Harvard University Press, Cambridge, MA.)

A cross between strains of wheat heterozygous for three pairs of genes, each with two alleles — one colored, one white — produces the phenotypes shown in **Figure 10-21,** which range from all colored to all white. Experiments and analyses of this kind demonstrate that we can explain selection for quantitative characters on the basis of the segregation and assortment of simple Mendelian genes whose individual small and discontinuous effects may add up to large continuous phenotypic differences.

Quantitative Trait Loci

The term *quantitative trait loci* (*QTL*) is shorthand for all of the genes (alleles) in a particular region of a chromosome that affect a quantitative aspect of the phenotype, such as size or shape, or processes such as growth and morphogenesis that influence size and shape. In the past, those undertak-

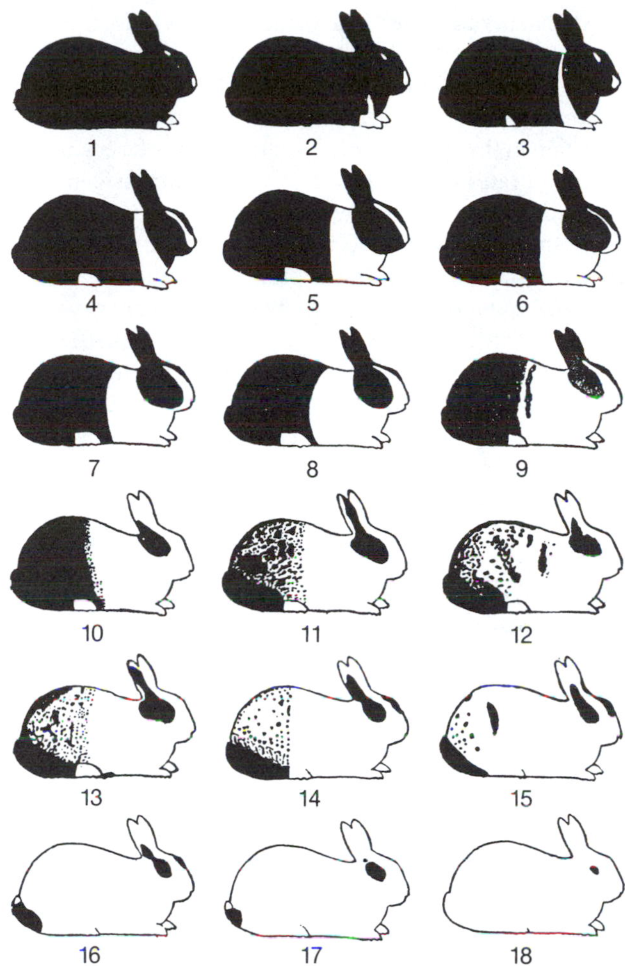

FIGURE 10-20 White spotting in Dutch rabbits, ranging from almost no spotting (grade 1) to complete spotting (grade 18). Selection experiments on animals with intermediate spotting (for example, grades 7 to 12) showed that spotting can be increased or decreased. (Reprinted by permission of the publisher from *Genetics and Eugenics: A Text-book for Students of Biology and a Reference Book of Animal and Plant Breeders* by W. E. Castle, Cambridge, Massachusetts: Harvard University Press, Copyright © 1916, 1920, 1924, 1930 by the President and Fellows of Harvard College.)

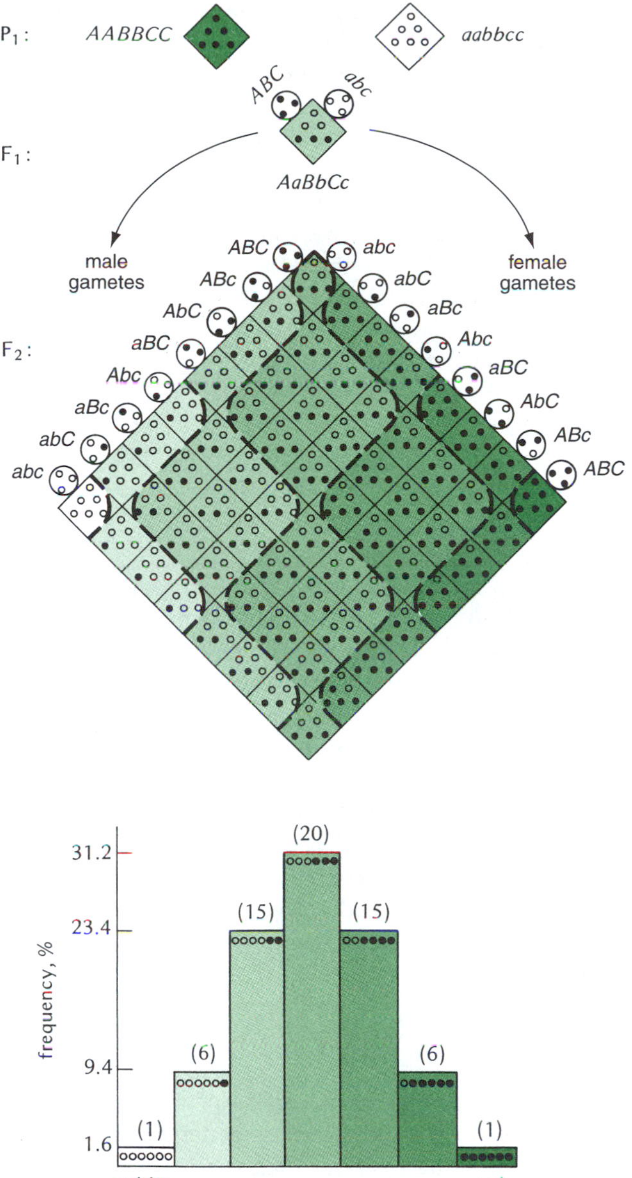

FIGURE 10-21 Results of crosses between two strains of wheat differing in three gene pairs that determine grain color. Each gene pair assorts independently of the others, and the alleles at each gene pair lack dominance, so that *Aa*, for example, has a color intermediate between *AA* and *aa*. The F_1, carrying three color (*ABC*) and three noncolor alleles (*abc*) is therefore intermediate in color to the parental stocks, and the F_2 produces a range of colors in the frequencies shown in the histogram. (From *Genetics*, 3rd ed, by Monroe W. Strickberger. © 1985 by Monroe W. Strickberger. Reprinted by permission of Prentice Hall, Inc., Upper Saddle River, NJ.)

ing QTL analyses assumed that the cumulative actions of many genes of small effect were responsible for the QTL effect. More recent analyses in *Drosophila*, mice and fish have revealed that a small number of genes within the QTL may contribute disproportionately to the QTL effect. Quantitative inheritance, of course, does not exclude factors with large effect: highly analyzed characters such as *Drosophila* bristle number are the result of genes with variable influence.[11]

Detecting the numbers and chromosomal positions of QTLs has become possible because of the availability of DNA sequences to which such genes may be linked. Differences among DNA sequences of the kinds discussed later in the chapter — short tandem repeat polymorphisms (STRPs), and restriction fragment length polymorphisms — serve as molecularly identifiable gene markers. These markers, located at specific linkage positions, can be correlated with specific measurements of a trait, and thus indicate numbers and approximate positions of quantitative loci.

[11] See Mackay (1996, 2004) and Mackay and Lyman (2005) for QTL analyses in *Drosophila*.

QTL analysis enables us to isolate suites of genes acting on particular parts of the phenotype at particular stages in ontogeny and to determine their relative effects with respect to one another. For instance, in selected lines of rats, four QTLs influence tail growth and body weight. One set has a substantial effect on both traits and acts throughout development. A second QTL has a large effect late in development but only a small effect early. A third shows the reverse pattern, while a fourth has minor effects at one or two discrete stages (Cheverud, 2004).

A second example comes from natural populations of the threespine stickleback, *Gasterosteus aculeatus,* in lakes in British Columbia. Sticklebacks show variation in the length of the dorsal spines, the number of dermal plates that make up the body armor and the number of gill rakers, each of which is influenced by QTLs. One QTL explains much of the variation in plate number, and two QTLs explain 66 percent of the variation in gill raker number.[12]

Environmental effects such as external conditions (available food, light, moisture, and so forth) also play a role, affecting size and yield. Sorting out these various influences and weighing the importance of genes, and the environmental effects of different genes in a particular environment and of individual genes in different environments, involves sophisticated statistical analysis, worthy of separately dedicated texts.[13] One common measure emerging from such studies is **heritability,** signifying the extent to which genetic differences affect a character; that is, the degree to which a character can be modified by selection. The various kinds of selection discussed in Chapter 22 depend on heritable variation.

Mutation Rates

Considering everything discussed so far, we see that the opportunity for all kinds of mutations derives from various sources and may have various phenotypic effects. For newly arisen mutations, these effects will most likely be harmful; prevailing phenotypes are generally well adapted for their particular environments, and most changes are unlikely to enhance that adaptation.

Mutations and, therefore, mutation rates, are properties of genes or, more strictly, of alleles and so should be *expressed* in terms of genes and not in terms of features of the phenotype. However, new mutations are usually *detected* because we observe their harmful effects on the phenotype, some of which are listed in **Table 10-3.** An average mutation rate therefore is a function of rate at which an allele (strictly a pair of nucleotides) will change each time the DNA is replicated; that is, each time a cell divides. On this basis, the *average mutation rate/nucleotide/replication is one in 10 billion (10^{-10}).* In practice, mutation rate/nucleotide pair varies within eukaryote genomes. When expressed in terms of the rate of mutation in the entire genome in the lifetime of a eukaryote, mutation rates vary from 0.1 to 100 mutations/genome/generation. Expressed in terms of mutations in functional genes, the mutation rate is closer to 0.003 mutations per generation, a similar rate to that in prokaryote genomes.

The mutation rates shown by these data are generally low, of the order of about one mutation per 100,000 copies of a gene. In organisms such as humans, carrying an estimated 25,000 genes per haploid genome, this means that each sperm and egg may well carry less than one newly arisen mutation or an average of 0.4 such mutations in a diploid fertilized zygote. However, if we extend our search for new mutations to larger numbers of genes and many possible nucleotide changes, human mutation rates are higher still.

In a major study of human mutation rates, Eyre-Walker and Keightley (1999) compared amino acid compositions of 46 different proteins between humans and chimpanzees to discover changes that had occurred during the time these lineages diverged. Based on a predicted genome size of 60,000 genes (now known to be less than half that), they calculated an overall rate of 4.2 amino acid-changing "missense" mutations each generation, of which at least 1.6 (38%) were eliminated by natural selection because they were deleterious. We circumvent many deleterious effects by medical treatment and a highly improved environment. Cumulative amounts of such mutations, however, may still be harmful, and their impact may eventually significantly affect our health and lifestyles (see Chapter 25; see also Crow, 1997).

However mutations are induced, mutation rates are not necessarily constant. Among the causes that can modify mutation rates are genes for polymerase enzymes that replicate DNA. Some alleles of these genes can increase mutation rates many fold. Other alleles act as antimutators and decrease mutation rates. Mutation rates, like other essential traits, seem mostly selected for optimum values, balancing on the delicate adaptive line that stands between retaining prevailing adaptive features yet allowing new ones to occur.

External causes may considerably affect mutation rates, including, surprisingly, infectious elements that can be transmitted from other individuals. Because they release nuclease enzymes that attack host DNA, viruses such as herpes simplex, rubella (German measles) and chicken pox can cause breaks and deletions in chromosomes.

DNA Repair Mechanisms

Important factors that act to correct nuclear damage from these and other influences, including ultraviolet radiation, are a variety of *DNA repair mechanisms.* These enzyme

---

[12] See Peichel et al. (2001) and Bell et al. (2004) for studies on sticklebacks.

[13] See the textbooks by Falconer and Mackay (1996), Lynch and Walsh (1998) and Roff (1997).

TABLE 10-3 Spontaneous mutation rates at specific loci for various organisms

Organism	Trait	Mutation/100,000 gametes[a]
DNA virus[b] : T4 bacteriophage	Rapid lysis ($r^+ \rightarrow r$)	7.0
	New host range ($h^+ \rightarrow h$)	0.001
Bacteria: *E. coli*	Streptomycin resistance	0.00004
	Phage T1 resistance	0.003
	Leucine independence	0.00007
	Arginine independence	0.0004
	Arabinose dependence	0.2
Salmonella typhimurium	Threonine resistance	0.41
	Histidine dependence	0.2
	Tryptophan independence	0.005
Fungus: *Neurospora crassa*	Adenine independence	0.0008 to 0.029
	Inositol independence	0.001 to 0.010
Insect: *D. melanogaster*	y^+ to *yellow*	12.0
	bw^+ to *brown*	3.0
	e^+ to *ebony*	2.0
	ey^+ to *eyeless*	6.0
Plant: corn (*Zea mays*)	*Sh* to *shrunken*	0.12
	C to *colorless*	0.23
	Su to *sugary*	0.24
	Pr to *purple*	1.10
	I to *i*	10.60
Rodent: *Mus musculus*	a^+ to *nonagouti*	2.97
	b^+ to *brown*	0.39
	c^+ to *albino*	1.02
	d^+ to *dilute*	1.25
Primate: *Homo sapiens*	Achondroplasia	0.6 to 1.3
	Aniridia	0.3 to 0.5
	Dystrophia myotonica	0.8 to 1.1
	Epiloia	0.4 to 1.0
	Huntington chorea	0.5
	Intestinal polyposis	1.3
	Neurofibromatosis	5.0 to 10.0
	Retinoblastoma	0.5 to 1.2

[a] Analysts base mutation rate estimates in viruses, bacteria, and fungi on particle or cell counts rather than on gametes.

[b] Mutation rates for RNA viruses are generally much higher than for DNA viruses, reaching in some cases one mutation per genome per replication (Drake, 1993). Reasons for these high RNA viral mutation rates include the absence of proofreading functions in RNA polymerase enzymes and the lack of repair mechanisms such as those used to correct DNA base pair mismatches.

Source: Adapted from Strickberger, M. W. 1985. *Genetics,* 3d ed. Macmillan, New York.

systems can excise DNA molecular distortions and replace mutant nucleotide sequences by inserting normal sequences from specially defined DNA strands such as those introduced during recombination. Although mostly known from *E. coli,* DNA repair mechanisms appear to exist in practically all cells, indicating that all forms of life have faced common problems of DNA damage.

A remarkable way in which some organisms accumulate mutations without experiencing their immediate effects is to bind their gene products with heat shock proteins, such as heat shock protein-90, which normally chaperone and protect other proteins. When such protective heat shock proteins are disabled, the assemblage of masked mutational products can cause significant developmental changes, perhaps leading to new evolutionary opportunities (Rutherford and Lindquist, 1998).

According to some views, improved DNA replication and repair mechanisms, especially those accompanied by diploidy and sexual reproduction, were required for the transition from small prokaryotic genomes to much larger eukaryotic

genomes. No matter how repair mechanisms arose, their value persists, because the inability to repair DNA damage can be lethal. In humans, such deficiencies appear in genetic diseases such as *xeroderma pigmentosum,* which often causes death because it increases the incidence of cancer.

On the other hand, defective DNA repair systems can be as important as polymerase enzymes in providing a wide array of mutations, some of which enable an organism to face new environmental challenges. Among such examples are strains of *E. coli* and *Salmonella enterica,* in which faulty DNA repair systems caused increased mutation rates (LeClerc et al., 1996), allowing some mutants to circumvent antibiotic challenges, including increased bacterial infectivity and pathogenicity, and causing several serious epidemics of food-related illness.

Transposons, Horizontal Gene Transfer and Selfish DNA

Transposons

Other sources of mutational change in prokaryotes and eukaryotes are **transposons,** nucleotide sequences that promote their own transposition among different genetic loci. The history of the discovery of transposons says much about the difficulty of gaining acceptance for ideas ahead of their time (see Box 10-2).

The transposon produces special transposase enzymes that allow copies of the transposon to insert into various target sites. For example, the IS1 transposon illustrated in **Figure 10-22** makes staggered cuts at each side of a nine-nucleotide base pair sequence, and a copy of IS1 inserts within the gap these cuts produce. Depending on where transposons insert, mutations of all kinds may arise, marked by target site repeats in which similar sequences of nucleotide bases appear at each end of the insertion but in inverted order.

Horizontal Gene Transfer

Inverted repeats were used to detect transposable elements in various species. Such repeats indicate that a transposon can pick up DNA sequences and transfer them to other DNA locations by recombination (see Mizuuchi, 1992). Antibiotic resistance genes, for example, can be transmitted between bacterial strains by small, *circular DNA particles* called **plasmids** that have received transposon insertions.

Such instances indicate that some hereditary traits carried by transposons may have passed horizontally (laterally) between individuals of the same generation — **horizontal or lateral gene transfer** — rather than through normal "vertical" (gametic) transmission between generations. For hori-

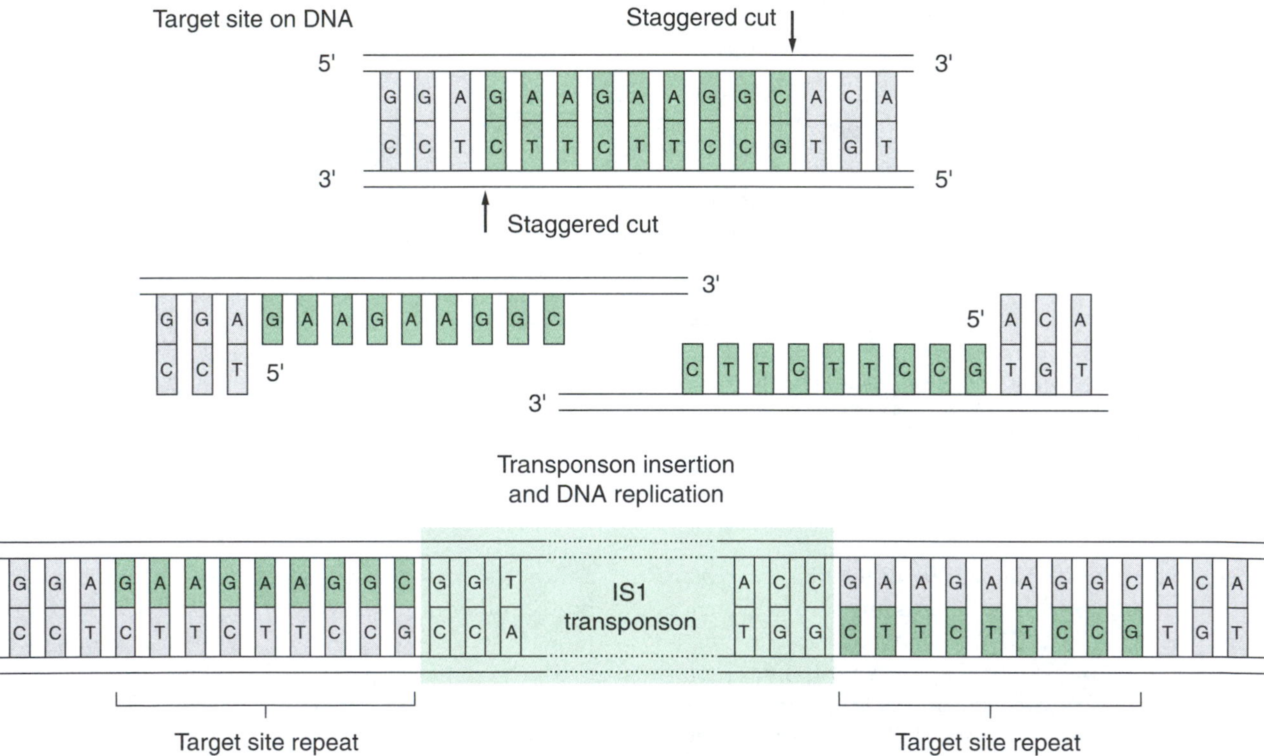

FIGURE 10-22 Mode of insertion of the IS1 transposon. The transposon recognizes the nine-base pair DNA sequence at the top of the diagram as a target site and cleaves it at the indicated *arrows.* IS1 inserts into the resulting gap, and DNA sequences are synthesized complementary to the former single-strand sections of the target site. This process produces identical but inverted nine-base pair repeats at each end of the transposon. (Adapted from Strickberger, M. W., 1985. *Genetics,* 3rd ed. Prentice Hall, Upper Saddle River, NJ)

zontal gene transfer to occur, DNA comprising a complete gene has to move with the aid of a virus or plasmid into another cell and survive in that cell (and, of course, cells have mechanisms to eliminate foreign molecules). If the gene survives, it has to incorporate into the host genome and provide a sufficient advantage to the host (or be sufficiently neutral) that it survives and can become subject to selection. Successful transfer is most common between closely related unicellular organisms (see below). Once established in their new hosts, horizontally transferred genes can undergo further mutation, leading to entirely new characteristics. Thus, as an example of rapid adaptive evolution, various pathogenic bacteria have widened their spectrum of resistance to many different antibiotic drugs in a decade or less.[14]

Over longer evolutionary periods, a large but difficult to determine number of horizontal gene transfers are thought to have occurred in some bacterial lines. According to Lawrence and Ochman (1998), *E. coli* must have experienced more than 200 such events from the time this prokaryote diverged from the *Salmonella* lineage 100 Mya, almost 20 percent of the *E. coli* genome having been obtained by horizontal gene transfer. You can see that horizontal gene transfer obscures phylogenetic relationships when otherwise distantly related organisms share a gene or sequence obtained through horizontal gene transfer rather than through common ancestry.

How then do we reconstruct such transfer events? By using enormously large data sets (see Chapter 12). In an analysis of over 220,000 proteins from the completely sequenced genomes of 144 prokaryotes (15 phyla), Beiko et al. (2005) were able to infer patterns of gene sharing (see Chapter 11). Using an approach that is independent of phylogenetic relationships, Dagan and Martin (2007) concluded that among 57,670 gene families from 190 genomes, at least two thirds and possibly all of the gene families have been subjected to horizontal gene transfer at sometime in their evolutionary history. Add horizontal gene transfer, the origin of organelles (and their genes) by endosymbiosis, and genome fusion, and we have three major means by which lineages of organisms acquire new genetic information.

Horizontal transfer of one family of transposons (*P* transposons or *P* elements) also has been demonstrated in a few eukaryotic organisms. As one example from the fruit flies, the *P* transposon, originally a transposable element common to *Drosophila willistoni*, first spread to *D. melanogaster* about 50 years ago and now appears in all wild populations of that species. The likelihood that the original *willistoni-melanogaster* event involved horizontal transfer is supported by the finding that the two species differ significantly in their gene nucleotide sequences, indicating a separation of about 20 My, yet their *P* transposons differ by only a single nucleotide (Daniels et al., 1990). More recent research has shown that *P* elements have repeatedly crossed species barriers in the fruit fly genus *Drosophila* and also spread to related genera, potentially confusing delineations of species based on molecular data.

Although discriminating between horizontal and vertical transfer can be difficult, other transposon elements (such as *Hobo*) also show evidence of horizontal transfer. The phenomenon may be more frequent than most geneticists anticipated. Evidence has now accumulated for transposon-mediated horizontal gene transfer between animals, between the mitochondrial genes of plants, and, most recently, between nuclear genes of millet and rice.[15] As Kidwell (1994) and others point out, one important advantage transposons gain by horizontal transfer is circumventing the barriers of reproductive isolation among species, and escaping inevitable extinction in vertical lineages when a species dies out.

Surprisingly, although we might expect a large increase in transposon numbers within a genome because of their simple transfer mechanisms, this is not always so. For example, the *Drosophila* genome carries only about 30 to 50 copies of the *P* element and a similar number of *copia* transposons. Regulatory agents within these transposons control their number and thereby limit their mutagenic effects, a feature that may have been selected to ensure survival of their hosts, and therefore their own survival.

Restricted numbers, however, do not extend to all transposons. In primates, for example, a 300 base-pair sequence with transposon-like features (*Alu*) is present in perhaps more than one million copies per human diploid cell. Smaller repetitive sequences of the type discussed in Chapter 12 are widely prevalent in various eukaryotes. For example, only three percent of human DNA codes for proteins, whereas the remainder consisting of noncoding sequences comprising many kinds of repetitive elements (see Table 12-4). What explains the widespread distribution and persistence of what appears to be extraneous DNA?

Selfish DNA

According to some molecular biologists, many transposable elements and other forms of repeated sequences contribute little, if any, function to their host cells.[16] Because the DNA replication process cannot discriminate between functional and nonfunctional sequences, it replicates any introduced DNA sequence. Transposon DNA and repeated sequences may therefore perpetuate parasitically as either "junk" or **"selfish" DNA.** Other explanations propose that some such sequences may function as essential elements in regulating gene activity, help maintain the structure and integrity of

[14] See Davies (1994) for resistance to antibiotic drugs. Several chapters in Hey et al. (2005) discuss current perspectives on horizontal gene transfer.

[15] See Cummings (1994) for horizontal and vertical transmission, and Silva and Kidwell (2000) and Diao et al. (2006), respectively, for transmission between animal species and between plant species.

[16] Orgel and Crick (1980) and Doolittle and Sapienza (1980) pioneered this view.

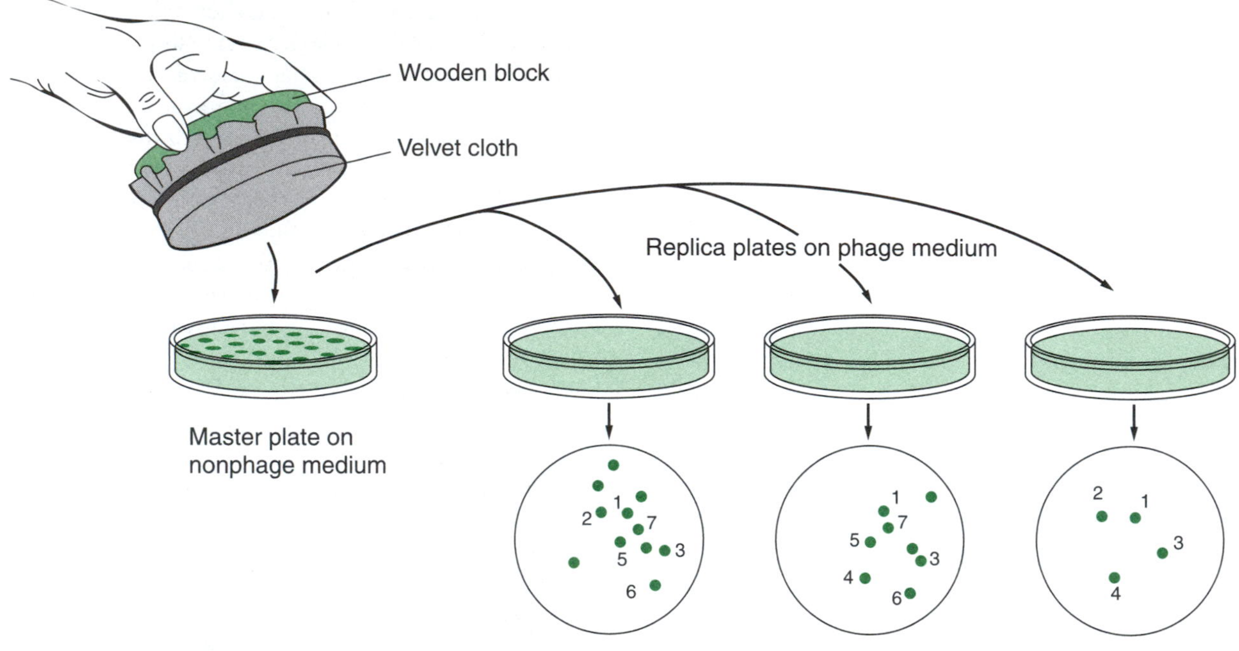

FIGURE 10-23 Replica-plating technique used to test the location of clones of *E. coli* resistant to T1 bacteriophage (virus). The test begins with a master plate that shows diffuse bacterial growth of phage-sensitive *E. coli* on a nonphage medium. Replicas are made by pressing a velvet covered wooden block against the master plate, pressing this, oriented in the same direction, to the surface of Petri dishes containing culture medium mixed with phage T1. One master plate has sufficient bacteria to start colonies on a number of replica plates. The replica plates show occurrence of resistant colonies (for example, 1, 3, 4) in identical locations, indicating that resistance to phage T1 must have been present at each of these positions in the master plate. (Adapted from Strickberger, M. W., 1985. *Genetics,* 3rd ed. Prentice Hall, Upper Saddle River, NJ; adapted from Lederberg, J., and E. M. Lederberg, 1952. Replica plating and indirect selection of bacterial mutants. *J. Bact.*, **63,** 399–406.)

the chromosome, serve as origins of DNA replication, or act as genes that occasionally provide new adaptive mutations.[17] No agreement exists on how to weigh the selfishness or unselfishness of such sequences, and some repeated sequences fulfill different roles.

Randomness of Mutation

Until the 1950s, the accepted view among bacteriologists was that bacteria had a unique "plastic heredity" in which appropriate mutations arise as an immediate response to the needs of the environment.

This concept seemed to be supported by observations in which bacteria exposed to some virus or antibiotic would quickly develop a resistant form. The explanation offered was that mutations do not originate on a random basis before exposure to some selective agent (**preadaptive** mutations), but rather that appropriate mutations arise only after bacteria have encountered a selective agent (**postadaptive** mutations). As with Lamarck, the postadaptive mutation concept had to rely on some unknown agency that allowed the environment to cause the appearance of new adaptive hereditary factors instead of employing Darwin's natural selection of hereditary factors already present.

J. and E. Lederberg (1952) performed an important test that evaluated these pre- and postadaptive models, using a novel technique called replica-plating. They transferred samples from a "lawn" of bacteria growing on a Petri dish (master plate) to other Petri dishes (replica plates) that contained selective media such as viruses (bacteriophages) or antibiotics (streptomycin). These transfers were made in a manner enabling bacteria derived from specific clones on the master plate to be localized to the same positions on the replica plates (**Fig. 10-23**). The finding that resistant bacteria occupied identical positions on various replica plates indicated that they arose from the same clone on the master plate, a clone that must have been present before exposure to the selective medium on the replica plates. In other words, these mutants shared a common preadaptive origin on the master plate and did not arise postadaptively because of a Lamarckian response to a stimulus by the selective medium.

[17] Having discussed that mutations are not adaptive in their own right, but only in response to selection, the use of the term "adaptive mutation" may seem odd. And indeed, the term is used in the context of adaptation to prolonged selection, which is the sense in which Kugelberg and colleagues (2006) use it in studies on mutations in the *lac* operon of bacteria, where selection favors gene amplifications that make sequence changes more probable by adding additional targets for selection to act upon; that is, mutations building on mutations.

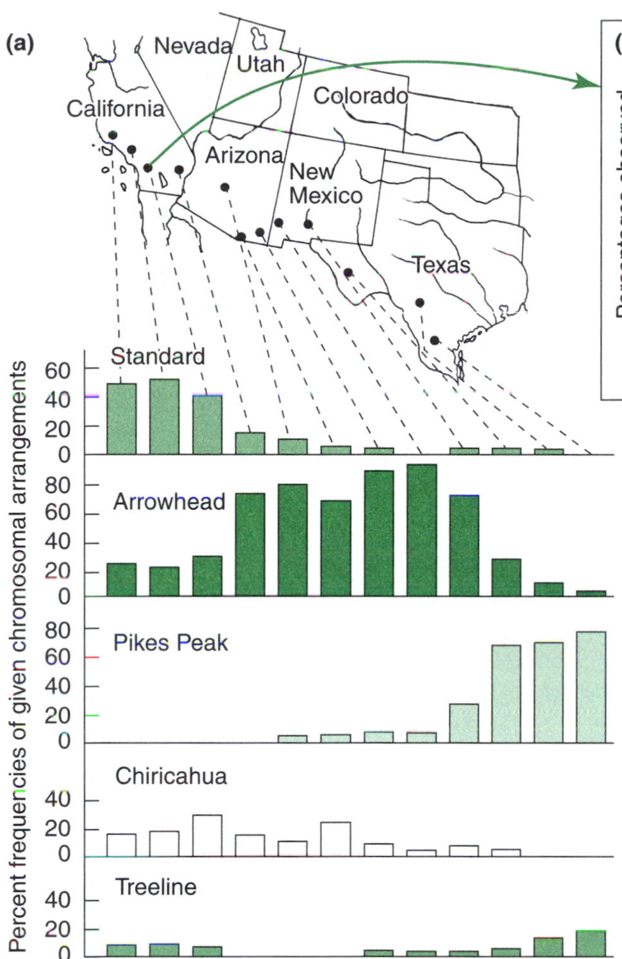

FIGURE 10-24 (a) Frequencies of five different third chromosome gene arrangements in *D. pseudoobscura* in 12 localities on an east–west transect along the United States–Mexican border. Each kind of arrangement consists of inversions observable as a specific system of chromosomal banding patterns. (Adapted from Dobzhansky, Th., 1944. Chromosomal races in *Drosophila pseudoobscura* and *D. persimilis. Carnegie Inst. Wash. Publ.* No. 554, Washington, DC, pp. 47–144.) (b) Percentages of different third chromosomal arrangements in *D. pseudoobscura* found at different months during the year in one of these localities, Mount San Jacinto, California. Each of these chromosome arrangements maintains a specific gene combination that enables adaptive interactions between component alleles (epistasis). Because inversions inhibit recombination, such allelic combinations can be maintained without disruption. (Adapted from Dobzhansky, Th., 1947. A directional change in the genetic constitution of a natural population of *Drosophila pseudoobscura. Heredity,* 1, 53–64.)

Mutations, however, are random only in the sense that particular mutations have not been shown to arise in response to environmental needs. Mutations are not random in any other sense. Mutations, for instance, are *not random with respect to function.* Mutations that decrease fitness by interfering with function are much more common than those that increase fitness. Similarly, mutations are *not random with respect to their effect on protein structure.* In most organisms, mutations that have no effect on the structure and function of proteins are more common than those that do. Different *types of mutations vary in their probability of occurrence and are thus not random.* For base pair substitutions, for example, transitions are more common than transversions even though there are twice as many possible transversions. Finally, mutation rates in genes on the Y chromosome are higher than on X or autosomal chromosomes (see Box 9-1).

Genetic Polymorphism: The Widespread Nature of Variation

Evolutionary potential and genetic variability are two sides of the same evolutionary coin.

New mutations that have an immediate beneficial effect on the organism seem generally to be quite rare, although some mutations are either neutral in their effect or harmful only when they occur in relatively rare homozygotes. Such neutral and deleterious but recessive mutations can therefore often accumulate in a population without any immediately serious deleterious effects, thereby furnishing a reservoir of genetic variability.

Genetic variation, expressed in a population by the existence of two or more genetically distinct forms — **polymorphism** — may include the maintenance of different kinds of chromosomal abnormalities such as inversions, translocations and extra chromosomes. In *D. pseudoobscura,* for example, populations in different localities in the western United States are polymorphic for a wide variety of gene arrangements on the third chromosome (**Fig. 10-24a**). The

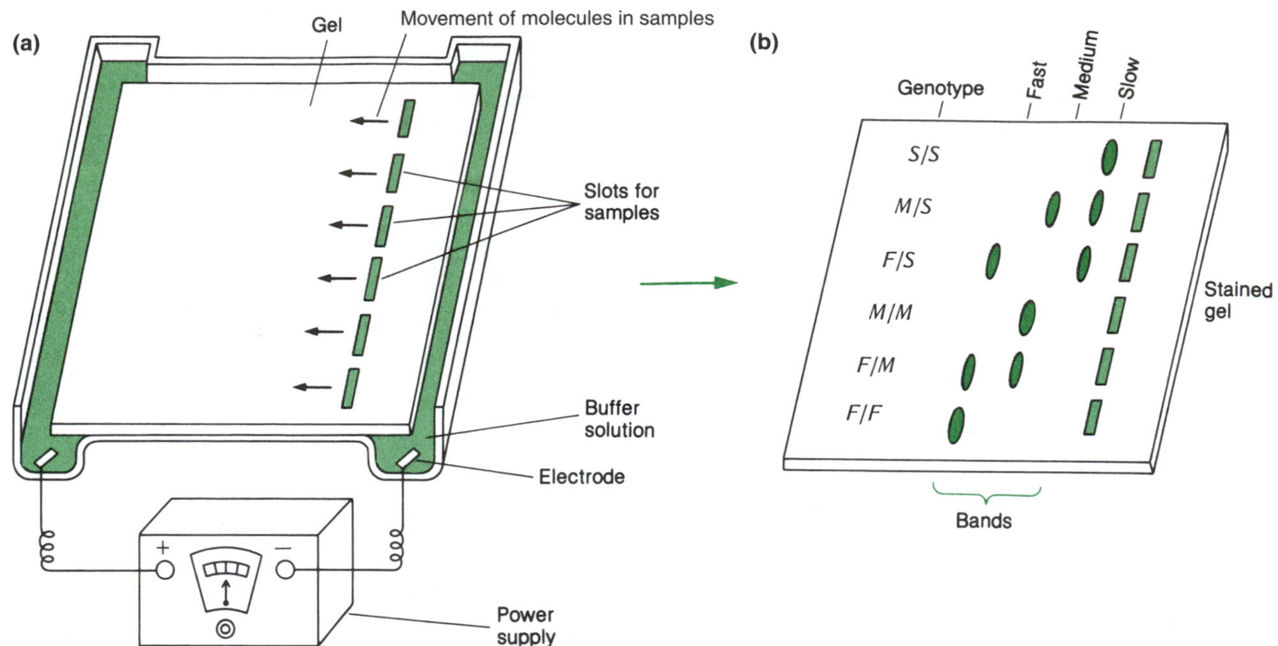

FIGURE 10-25 (a) A general scheme for electrophoresis, using a gel (starch or polyacrylamide) in which samples are placed along a row and subjected to an electric current carried through an aqueous buffered solution. Depending on their size and electrical charge, molecules in the samples separate, going toward either the negative or positive pole. The position they occupy on the electrical gradient can be identified as bands when the gel is treated with agents that can assay molecular or enzymatic activity or is exposed to ultraviolet light. (b) Treating the gel with dyes sensitive to a specific enzyme activity shows that the enzyme on this gel has three different forms, each migrating at a distinct rate (slow, medium, and fast) toward the positive pole. Because each of these three enzymatic forms is produced by a single allele of the gene for this protein, *S, M, F,* it is often called an *allozyme.* Thus, an individual may possess one of six genotypes, either homozygous (*S/S, M/M, F/F*) or heterozygous (*M/S, F/S, F/M*), each different genotype producing an identifiable electrophoretic pattern of allozymes. (Some terminologies use the more general term isozyme for any distinct electrophoretic form of a protein, whether its uniqueness arises from genetic or nongenetic causes.) (Adapted from Strickberger, M. W., 1985. *Genetics,* 3rd ed. Prentice Hall, Upper Saddle River, NJ.)

frequencies of such arrangements may change seasonally (Fig. 10-24b), indicating not only that chromosomal polymorphism is generally adaptive in this species, but also that certain polymorphic variations are preferentially adaptive in helping populations adjust to their specific environment at specific times.

Variation in chromosome number between similar species can associate with adaptive features. According to Nevo and coworkers (1994), stressful ecological conditions such as periodic aridity and other unpredictable hardships appear to correlate with chromosome numbers in Israeli and Turkish populations of the mole rat *Spalax.* They suggest that the chromosomal increase caused by dissociation or fission (Fig. 10-4b,c) leads to increased genetic diversity by increasing the numbers of different chromosome combinations, so that species in such localities can specialize for a rise in opportune ecological variation.

On the DNA level, diversity can be identified by molecular techniques (see Chapter 12), which determine particular sequences or the number of repeating nucleotide strings such as CACA, CACACA . . . CA_n. These latter repeats, known as **short tandem repeat polymorphisms (STRPs),** are distributed throughout the genome. Differences in number of repeats among the polymorphisms serve as gene markers for particular chromosomal locations.

On the gene level, geneticists can discern the magnitude of polymorphism by techniques that make allelic differences visible. Among such studies are electrophoretic methods that measure the mobility of a protein in an electric field, distinguishing slight variations in conformation and electric charge (**Fig. 10-25**). Because each different electrophoretic form of a protein usually indicates a different amino acid sequence (and therefore a different nucleotide sequence in the gene that produced it), each signifies an allelic difference, or **allozyme.** Applying this technique to natural populations, beginning in 1966 with the studies of Harris, and Lewontin and Hubby, led to the surprising result that populations maintain considerably more genetic variability than previously estimated. As shown in **Table 10-4,** a large number of species display allozymic differences at an average of about one quarter of all loci tested, indicating that the chance of an individual being heterozygous for any particular tested locus is more than seven percent.

Because it is estimated that only perhaps one third of amino acid changes in proteins are detectable by electrophoretic techniques, these observed values should probably

TABLE 10-4 Estimates of genetic variability found in natural populations, based on electrophoretic studies

Organisms Tested	Number of Species Examined	Average Number of Loci (proteins) Studies/Species	Proportion of Polymorphic Loci	Heterozygosity per Locus
Plants	15	18	.259	.071
Invertebrates				
Various groups except insects	27	25	.399	.100
Various insects except Drosophilidae	23	18	.329	.074
Drosophila species	43	22	.431	.140
Vertebrates				
Fish	51	22	.152	.051
Amphibia	13	22	.269	.079
Reptiles	17	23	.219	.047
Birds	7	21	.150	.047
Mammals (except primates)	43	26	.148	.036
Primates				
Humans	1	71	.28	.067
Chimpanzees	1	43	.05	—
Macaque monkeys	1	29	.10	.014
Totals and Averages	242	23	.263	.074

Source: From Strickberger, M.W., 1985. *Genetics,* 3rd ed. Macmillan, New York; data from Nevo, E., 1978. Genetic variation in natural populations: Patterns and theory. *Theor. Pop. Biol.,* **13,** 121–177.

be tripled. About two thirds to three quarters of all loci in many species are polymorphic, and the average individual may be heterozygous for as much as one quarter to one third of all its loci. This means that in *Drosophila* species, with an estimated 2,000 gene loci, an individual can be a heterozygote for about 500 genes or more; and in humans, with an estimated 25,000 gene loci, individuals may be heterozygous for as many as 6,250 genes. Britten (1986) for example, estimates that of the three billion nucleotides in the haploid human genome, one human differs from another at an average of about five million sites. Clearly, such past accumulations provide a much greater amount of genetic variability than do the relatively few new mutations that arise each generation.

Genetic polymorphisms allow many populations to confront new environmental challenges with a large variety of mutations, some of which may be advantageous in the new environment. For example, the exposure of insect populations to DDT (dichloro-diphenyl-trichloroethane) pesticides has caused a widespread increase in the frequency of various DDT-resistant mechanisms such as:

- an increase in lipid content that allows the fat-soluble DDT to separate from other parts of the organism;
- enzymes that break down DDT into relatively less toxic products;
- a reduced toxic response of the nervous system to DDT;
- changes in the permeability of the insect cuticle to DDT absorption; or
- a behavioral response that reduces contact with DDT.

It is, therefore, not surprising that insecticide resistance is associated with numerous genes. For example, the genes responsible for DDT resistance in *Drosophila* are located on all major chromosomes, each gene acting as a polygene with a small incremental effect (**Fig. 10-26**). Similar antipesticide selective events are common; Roush and McKenzie (1987) list more than 400 such cases.

Because variation can be studied at a variety of levels ranging from molecular genetics to macroevolutionary dynamics, other areas of active study have not been discussed

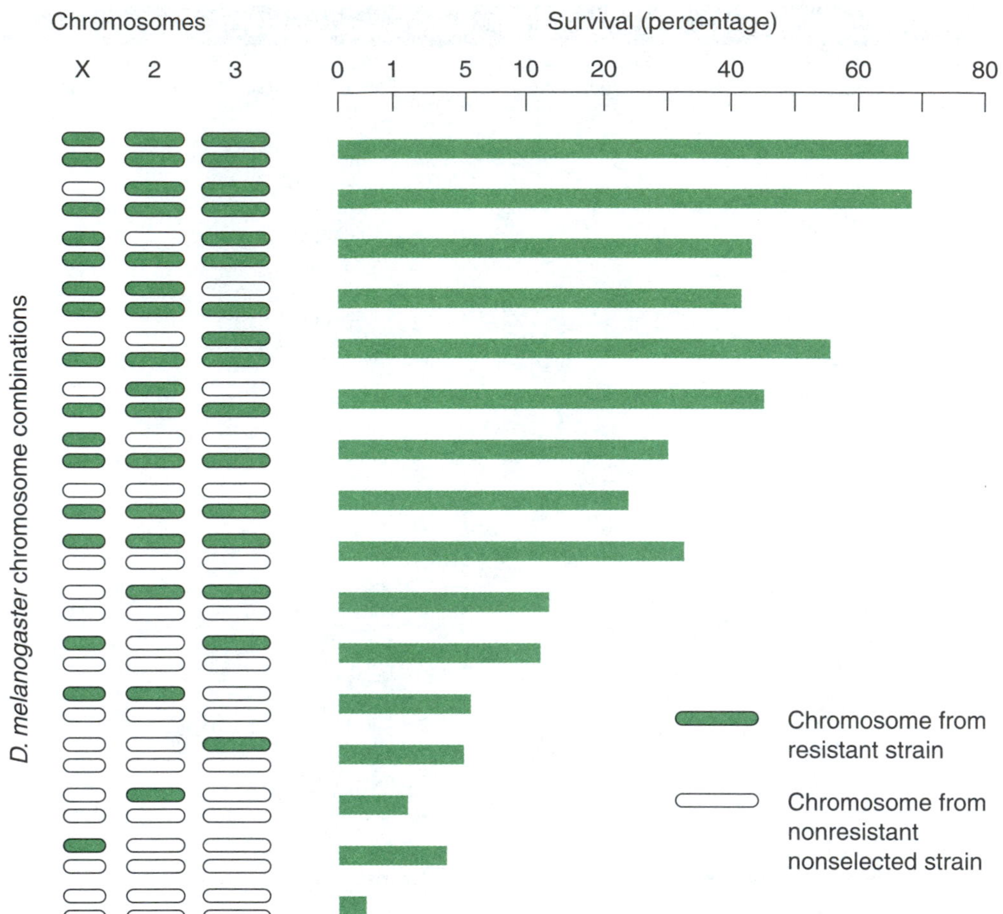

FIGURE 10-26 Percent survival of 16 different types of *D. melanogaster* flies exposed to a uniform dose of the insecticide DDT. Each type of fly carries a set of chromosomes derived from DDT-resistant and DDT-nonresistant strains. Obviously DDT resistance increases with the increased number of chromosomes from resistant strains. (Adapted from Crow, J. F., 1957. Genetics of insect resistance to chemicals. *Ann. Rev. Entomol.*, **2**, 227–246.)

in this chapter. They are outlined in Box 10-1. Although we have come a long way since Darwin in terms of understanding the theoretical importance of variation and variability, we still have a lot to learn about what causes patterns of phenotypic variation in nature and how variation is transmitted between levels in the biological hierarchy. Chapters 13, 21 and 22 address these important issues further. Before we tackle these issues, however, we need to revisit the thorny issue of how we determine the relationships between the organisms and/or genes we want to study. This we do in the next chapter.

KEY TERMS

allele
allopolyploids
allozyme
aneuploidy
asexual reproduction
autopolyploidy
base substitutions
continuous variation
deletions
DNA repair mechanisms
duplications
euploidy
genes
gene families
heritability
horizontal (lateral) gene transmission
inversions
open reading frames
orthogenesis
paracentric inversion
parthenogenesis
pericentric inversion
phenotype
plasmids
pleiotropy
point mutations
polygenes
polymorphism
polyploidy
polytene chromosomes
preadaptive
promoter site
quantitative trait loci (QTLs)
quantitative variation
reciprocal translocation
repetitive DNA
repressor protein
selfish DNA
short tandem repeat polymorphisms (STRPs)
sickle cell anemia
structural genes
translocation
unequal crossing over
variation

DISCUSSION QUESTIONS

1. What are the major forms of variation in chromosome number? Types of euploidy? Types of aneuploidy?
2. How can a new allopolyploid species arise rapidly in nature or be created in the laboratory?
3. What are the major forms of variation in chromosome structure?
4. What structural chromosome changes can lead to changes in chromosomal number?
5. What kinds of chromosomal changes have occurred among some hominoids (humans and great apes)?
6. What are the different types of mutational changes on the nucleotide level, and how do they arise?
7. What is pleiotropy? Give an example.
8. How are the genes that produce the enzymes used for lactose metabolism regulated in prokaryotes? (Explain the terms *inducers, repressors, operators* and *promoters.*)
9. What are some examples of gene regulation in eukaryotes?
10. How can multiple factors or polygenes explain continuous quantitative variation?
11. Can gene mutation rates account for the prevalence of gene mutations in populations?
12. Transposons.
    a. How do transposons generate mutations?
    b. Why do transposons persist in a genome?
13. How can one discriminate whether mutations arise randomly without regard to their adaptive or nonadaptive effects (preadaptively) or whether they arise as adaptive responses to specific selective environments (postadaptively)?
14. Polymorphism.
    a. What examples are there of chromosomal and gene polymorphisms?
    b. What evolutionary value do such polymorphisms offer?
15. Write a five-page essay answering each of the questions asked in Box 10-1?

EVOLUTION ON THE WEB

Explore evolution on the Internet! Visit the accompanying Web site for *Strickberger's Evolution,* Fourth Edition, at http://www.biology.jbpub.com/book/evolution for Web exercises and links relating to topics covered in this chapter.

RECOMMENDED READING

Carroll, S. B., 2005. *Endless Forms Most Beautiful.* W. W. Norton & Co., New York.

Carroll, S. B., J. K. Grenier, and S. D. Weatherbee, 2005. *From DNA to Diversity. Molecular Genetics and the Evolution of Animal Design,* 2nd ed. Blackwell Publishing, Malden, MA.

Falconer, D. S., and T. F. C. Mackay, 1996. *Introduction to Quantitative Genetics,* 4th ed. Longman, Harlow, Essex, UK.

Graves, J. A. M., 2006. Sex chromosome specialization and degeneration in mammals. *Cell,* **124,** 901–914.

Hey, J., W. M. Fitch, and F. J. Ayala (eds), 2005. *Systematics and the Origin of Species on Ernst Mayr's 100th Anniversary.* The National Academies Press, Washington, DC.

Jacob, F., 1977. Evolution as tinkering. *Science,* **196,** 1161–1166.

Jacob, F., and J. Monod, 1961. Genetic regulatory mechanisms in the synthesis of proteins. *J. Mol. Biol.* **3,** 318–356.

King, M. C., and A. C. Wilson, 1975. Evolution at two levels: Molecular similarities and biological differences between humans and chimpanzees. *Science,* **188,** 107–116.

Mackay, T. F. C., 2004. The genetic architecture of quantitative traits: Lessons from *Drosophila. Curr. Opin. Genet. Dev.,* **14,** 253–257.

Pagel, M., (ed. in chief), 2002. *Encyclopedia of Evolution.* 2 Volumes, Oxford University Press, New York.

Roff, D. A., 1997. *Evolutionary Quantitative Genetics.* Chapman and Hall, New York.

11 Species, Phylogeny and Classification

Chapter Summary

Before Darwin, classification involved observations of similarities and differences among organisms, without regard to their origins. Since Darwin's time, however, taxonomists have sought to construct a system that would reflect evolutionary relationships. Traditionally, the main goals of evolutionary taxonomy were to use morphology to recognize and describe species, the basic unit of classification, and to order them into as realistic a phylogenetic scheme as possible.

In many instances, morphological classification is a reliable and practical alternative for taxa with many species. Morphological classification often corresponds with more "modern" approaches but sometimes differences have to be rationally resolved. With the advent of molecular phylogeny, the traditional morphological approach is often replaced by new techniques and philosophies. Because of patterns of evolution arising from parallelism or convergence, molecular techniques often provide results that can only be deciphered reasonably with interpretations by a taxonomist skilled in the biology of the group. The need for such validation underscores the continuing need to train taxonomic specialists in all groups of organisms.

Grande (2004) proposes that we view morphology and morphological variation in three ways: *taxonomic* (differences between taxa), *ontogenetic* (differences between growth/developmental stages of an individual) and *individual* (within or between individuals in the same species. The ability of populations to interbreed and produce viable offspring is the primary criterion for determining species boundaries using the *biological species concept.* However, because of the enormous number of species, their geographic dispersal, and because most species are extinct, it is impossible to differentiate many species by these criteria.

Furthermore, the biological species concept cannot be applied to fossils and only rarely to plants.

The evolutionary species concept represents an effort to resolve these difficulties by defining species based on their evolutionary isolation from each other. Ideally, this method uses morphological, genetic, molecular, behavioral and ecological variables, although, like other species concepts, it does not resolve all the problems intrinsic to species taxonomy, because not all traits evolve at the same rate or in the same sequence (mosaic evolution).

In the past, it was easier to identify a species than to reconstruct a phylogeny. Phenotypes may be alike morphologically because: (1) of common origin (*homology*), (2) of similar evolutionary patterns arising separately in different lines from not-too-distant ancestors (*parallelism*), or (3) of the development of similar characteristics in groups originating from completely different ancestors (*convergence*). Because it can be difficult to distinguish between "not-too-distant" and "completely different," parallelism and convergence are commonly included in the term **homoplasy**.

Traditional classifications use hierarchical schemes based on natural criteria to order organisms into taxa. But such taxa are not necessarily monophyletic (coming from a common ancestor), and their designations are contested. Recent approaches are beginning to resolve these problems by establishing methods that identify monophyletic groups, but recognizing homoplasy remained a problem. Cladistics recognizes phylogeny as a series of dichotomous branches, each branch point giving rise to taxa of equal rank. Multiple radiations (multiple branch points) create problems for cladistic methods. The availability of molecular data, our ability to create trees using only molecular data, and the increasing interest in groups of organisms in which morphological variation is minimal (much of the Archaea, for example) have brought molecular taxonomy/phylogeny to the fore and revolutionized many traditional classifications, including the number of kingdoms into which organisms can be grouped.

The chapters in Part 4 seek to outline, in a general way, the probable course of events that took place in the evolution of various groups of organisms. Much of this effort is based on enumerating and comparing similarities and differences in morphology and/or function among and between organisms. This is the traditional province of **classification** and **systematics**. The terms *systematics, classification* and *taxonomy* are often used interchangeably, although some taxonomists (such as G. G. Simpson) saw systematics as a much broader field: the study of the diversity of organisms and all their comparative and evolutionary relationships, including such topics as comparative anatomy, comparative ecology, comparative physiology and comparative biochemistry. In his recent book entitled, *Genes and the Agents of Life,* Robert Wilson (2005) captured the essence and the role of taxonomy:

> Taxonomies are simply categories and kinds, and they exist throughout the biological sciences . . . Taxonomic practices in biology range from carbon dating and anatomical reconstruction of fossils in Paleozoology to molecular phylogenetic analysis of both microorganism and plant lineages, and range up and down various biological hierarchies . . . Taxonomic categories in genetics and developmental biology include cells, genes, tissues, organs, bones, cytoskeletal structures, membranes, pathways, modules, molecules, and various biological systems (nervous, digestive, immune, respiratory). . . . In ecology they include organisms, demes, adaptive niches, ecosystems, communities, symbiotic pairs, and predatory-prey systems (p. 31).

The present chapter and Chapter 12 both deal with classification and relationships and should be read as if one. The present chapter tells you about the various species concepts, about the nature of determining relationships and phylogeny and of the need to ensure that only homologous characters are compared when establishing relationships or creating a phylogeny. The emphasis is on the basic concepts that have a history, which we outline. Most of the characters used as examples are morphological, behavioral or ecological. Chapter 12 takes us from these characters to the (increasing) use of molecular data to infer relationships, phylogeny and ancestry. These are not either/or approaches. They are treated in separate chapters because how to recognize and relate species is so enormous and so central to the study of evolution.

Classification and Phylogenetic Relationships

As we saw in Chapter 1, natural historians formulated techniques of classifying organisms much earlier than they accepted the concept that their similarities and differences arose from evolutionary causes. It is, in fact, fairly easy to classify organisms in ways that distinguish among them but do not reveal their common origins. For example, we can classify fish and whales in one group, flies and birds in another, frogs and crocodiles in a third and squirrels and monkeys in a fourth.

With the advent of the Darwinian revolution, an additional consideration entered into the approach of at least some systematists: whether they could or should use classification to reflect evolutionary relationships. There were strong indications that many organisms, grouped together because they possessed a large number of similar features, could also be said to *descend from a common ancestor.* But such determinations revealed only one aspect of evolutionary classification. Another aspect was to discern *lines of descent* among groups that perhaps shared only a few features. As Darwin put it, "Our classification will come to be, as far as they can be so made, genealogies . . . we have to discover and trace the many diverging lines of descent in our natural genealogies, by characters of any kind which have long been inherited."

One reason for the trouble paleontologists have in determining phylogenetic relationships using morphological data is the difficulty in finding an unbroken line of ancestors connecting different groups. The fossil record may be fairly complete for some groups such as horses, but far less complete for birds, whales, insects, early flowering plants and many other organisms. Inadequacies of the fossil record arise from a number of causes:

- Few bodies are fossilized, and those that are usually are localized to a former aqueous region such as a river, lake or ocean shoreline (see Fig. 3-12). It has been estimated that the remains of individuals from only 10 or 15 species that died on a tropical river bank can be identified out of the 10,000 or so species that would have lived in the area. The proportion of preserved species is greater in some ocean environments such as limestone reefs, but even then is no greater than 1 to 2 percent.
- Strong winds, heavy wave action or other potent forces can disturb the formation of sedimentary layers.
- Even after sedimentary layers form, various geological events may erode or move them about, causing discontinuities in the record.
- Only a small proportion of fossil-bearing sedimentary rocks are accessible.
- Finally, because we know that many extant species are separated by seasonal or behavioral barriers, we cannot necessarily be confident that a species identified in the fossil record was not, in fact, also subdivided into subspecies by seasonal or behavioral barriers with consequent lack of reproductive interaction across the whole species range.

Despite these difficulties, fossils offer an important advantage of showing the historical order in which different phylogenetic characters were acquired or lost. And of course, they are the chief and often the only record available to paleontologists.

Neither taxonomic knowledge nor approaches are stagnant. Consequently, classifications and relationships between groups of organisms change over time. As shown in **Figure 11-1,** entomologists have offered five different schemes for the same groups of insects. The single genus *Rubus* (blackberries, raspberries, loganberries) has been divided into 500 species by one botanist, 200 by another and 25 by a third. While this means that you have to run to keep up with the latest work, it undercuts neither the validity nor the importance of taxonomic analysis. Taxonomic identification, comparison and classification are vital in such areas as health, environment, industry and conservation. Is that tick a carrier of lyme disease? Will this invading fish species outcompete local species? Will this beetle destroy a forest industry? Taxonomy underpins the answers to these questions.

Species Concepts

Taxonomic problems are at least twofold:

- How to recognize the basic unit of classification, the **species,** and associate this unit, if possible, with a fundamental evolutionary unit.
- How to order species into systems that will connect them into a reasonably accurate **phylogenetic** scheme.

Morphological Species

An old adage for a species definition — "a species is what a competent taxonomist says is a species" — illustrates the specialized knowledge required to identify species and the need to continue to train such specialists. Among the variety of species definitions that have been offered, taxonomists have generally used morphological criteria, because this is how most individuals have been compared. Thus, in a classic textbook, Davis and Heywood (1963) define **morphological species** as, *"assemblages of individuals with morphological features in common and separable from other such assemblages by correlated morphological discontinuities in a number of features."* Davis and Heywood are experts in their field.[1]

[1]See the papers in R. A. Wilson (1999) for interdisciplinary approaches to species definition, and see R. A. Wilson (2005) for an excellent discussion of species concepts.

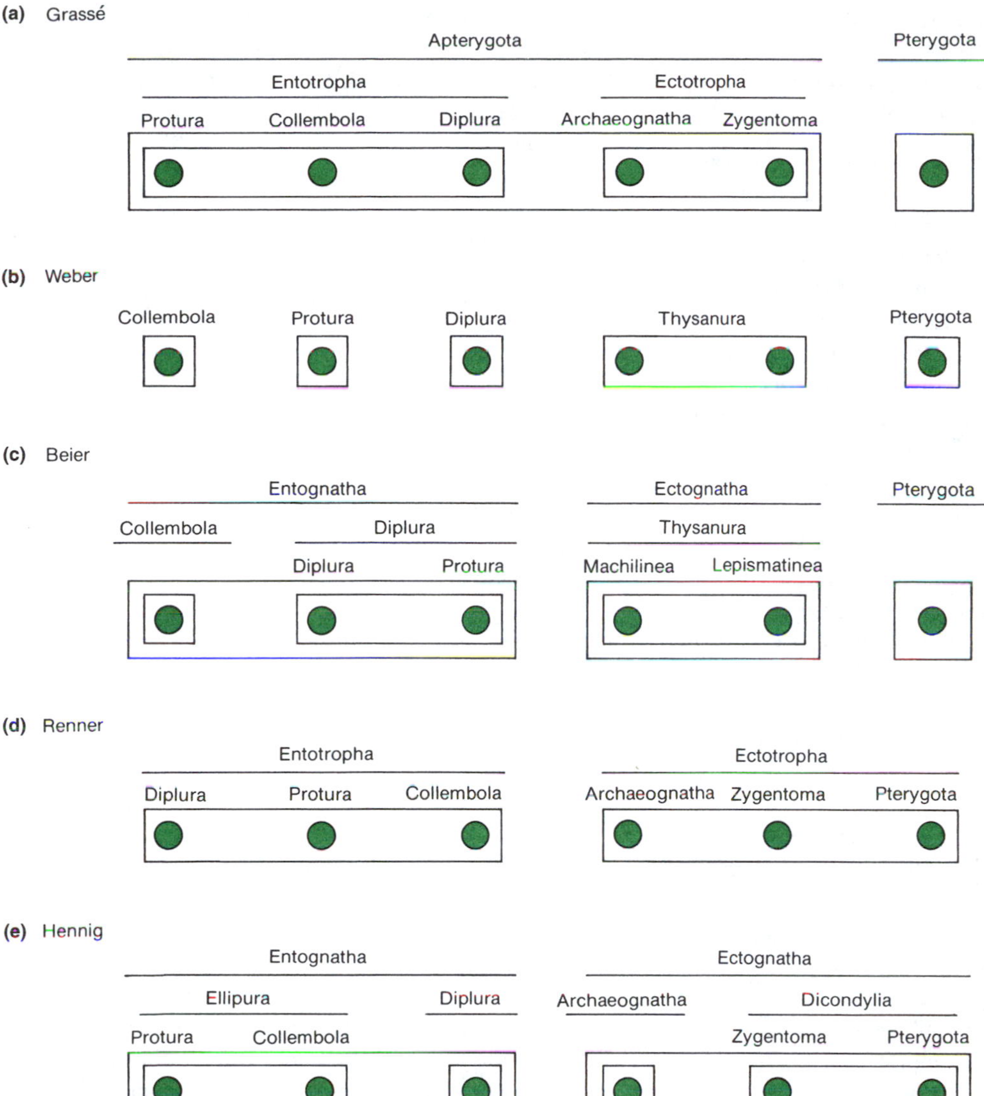

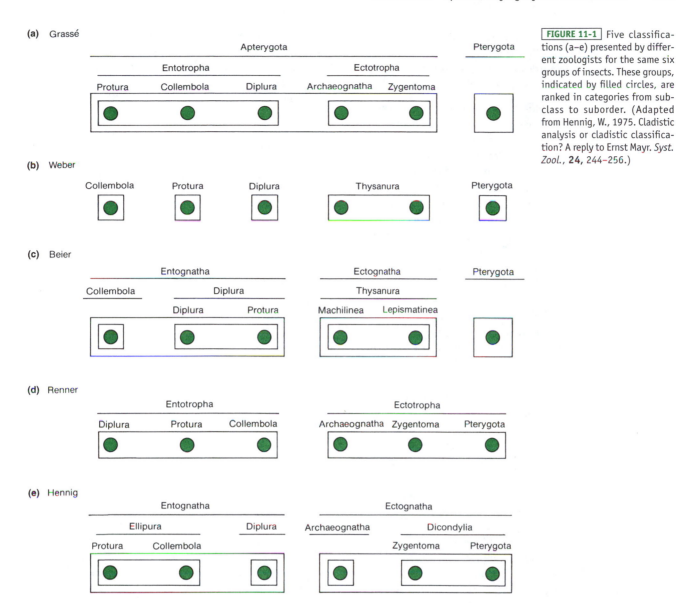

FIGURE 11-1 Five classifications (a–e) presented by different zoologists for the same six groups of insects. These groups, indicated by filled circles, are ranked in categories from subclass to suborder. (Adapted from Hennig, W., 1975. Cladistic analysis or cladistic classification? A reply to Ernst Mayr. *Syst. Zool.*, **24**, 244–256.)

As it has evolved, taxonomy has sought increasingly objective ways to identify characters. Some taxonomists proposed numerical methods in which taxonomic distinctions depend on the degree of the statistical correlation for as large a number of characters as possible. The presumption is that a high statistical correlation among individuals for a large number of such characters indicates their membership in the same species or groups. Low correlation points to their separation into different species or groups. This method, **numerical** or *phenetic* **taxonomy**, which largely formalized some of the processes taxonomists used intuitively but added a degree of quantitative numerical consistency, has been supplanted by other approaches.

Whether the approach is traditionally morphological or more quantitatively and statistically numerical, both demand some degree of subjective judgment; that is, they both exemplify arbitrary choice, or, as John Locke commented in *An Essay Concerning Human Understanding* (1689, http://plato.stanford.edu/entries/substance/), "the boundaries of the species, whereby men sort them, are made by men." To Darwin, the attempt to find "the essence of the term species" was "a vain search."

Biological Species

Many biologists adopt a **biological species concept.** Derived from Buffon and others (see Chapter 1), this concept defines a species as a sexually interbreeding or potentially interbreeding group of individuals normally separated from other species by the absence of genetic exchange, that is, by **reproductive isolation.** As an obvious advantage of this definition, species distinctions can be objectively tested by two relatively simple criteria:

- Do individuals in populations in the same locality normally interbreed?
- Should cross-fertilization occur, do embryos develop and are the hybrid progeny viable and fertile?

If the answer to either question is no, we consider the evaluated populations as species separated by reproductive isolation barriers (see Chapter 24).

Instead of defining a species by negative attributes — for example, inability to cross-fertilize with other species — some evolutionary biologists propose a more positive approach, emphasizing *inclusive* qualities that tie species members together. Paterson (1985) used the *recognition concept,* by which species members are defined by their "common fertilization system"; Van Valen (1976) asserts that a species occupies a unique *adaptive* [ecological] *zone;* Templeton (1989) defines a species as possessing *intrinsic cohesion mechanisms* by essentially adding to Van Valen's definition of a uniform ecological niche the ability to transfer genetic material between species members. Positive attributes are not necessarily distinctive, however; two species that produce sterile hybrids may still have a common fertilization system, and two species may occupy the same adaptive zone. In addition, ecological and biogeographical divergence can occur not only between but also within species.[2]

Applying the biological species concept has allowed taxonomists to make species distinctions between similar-appearing populations that could not be separated on the basis of the usual morphological taxonomic criteria. The fruit fly species *Drosophila pseudoobscura* and *D. persimilis* are called **sibling species** because they are almost identical in appearance and do not normally cross-fertilize. This is also true for some leafy-stemmed sibling species in the phlox family, *Gilia tricolor* and *G. angelensis.*

Applying the biological species concept also has allowed taxonomists to unify different groups into single species that, based on morphological and geographical criteria, had been separated into distinct species. One example is the union of several species of North American sparrows into a single **polytypic species**, the song sparrow, *Passarella melodia,* consisting of multiple geographic subspecies. Similarly, groups of yarrows (*Achillea* sp.), plants that show distinct ecological adaptations restricting their growth to particular environments (see Figs. 24-1 and 24-2), are generally identified as varieties rather than species because they could exchange genes if they came into contact.

But a caution — Basing species distinctions mainly on reproductive isolation led some to conclude that the actual proportion of species that do not fit well into the biological species concept may be quite small, even among plants. A study made of 838 named plant species in the Concord, Massachusetts, area showed that 93 percent could be distinguished according to the biological species concept. Despite these advantages, attempts to apply the biological species concept universally face difficulties, five of which are outlined below. These are not merely arcane issues of concern only to taxonomists. They apply to any attempt to explain evolution by a single mechanism. Indeed, if these problems did not exist, we would not need textbooks on evolutionary biology. Here are the five difficulties.

1. Although it may be possible to observe reproductive barriers between groups found in the same locality (*sympatric populations;* see Chapter 24), many practical limitations block crosses between groups that are ordinarily separate (*allopatric populations*).
2. Even if we cross allopatric populations, we have to make subjective decisions. For example, results from interbreeding experiments may range from no genetic exchange at all for certain attempted crosses to a fairly large degree of genetic exchange for others (**Fig. 11-2**). At which point on this *scale of interbreeding values* shall we separate species?

   Interestingly, even when genetic exchange is completely uninhibited between some allopatric populations, it may still be appropriate to classify them as separate species because they do not hybridize in nature. One well-known example is two widely separate populations of trees occupying similar habitats, one found in China (the Chinese catalpa, *Catalpa ovata*) and the other, the Indian bean tree (*C. bignoides*) in the eastern United States. Although they can cross with each other to produce hybrids that are as viable and fertile as the parents, these populations have been separate for many millions of years, and botanists have therefore generally agreed to continue their separate species identifications.
3. We cannot test *fossil populations* to see if they can interbreed. Fossil species will always have to be defined on a different basis.
4. The biological species concept *only applies to sexually reproducing animals* and not to plants. In animals with asexual reproduction, no matter whether reproduction occurs by fission or by parthenogenesis, each clone of individuals is essentially genetically isolated from every other clone, yet few, if any, biologists would consider describing each clone as a separate species.[3]
5. Plasmid-mediated *horizontal gene transmission* between different species (see Chapter 10), although once assumed to be rare and restricted to prokaryotes and

[2] For a brief discussion and evaluation of these various species concepts, see Endler (1986) and R. A. Wilson (1999). For an insightful analysis of the validity and reality of biological species, see Lee (2003.)

[3] Ernst Mayr (1987) referred to asexual groups as *paraspecies,* Michael Ghiselin (1987) called them *pseudospecies.*

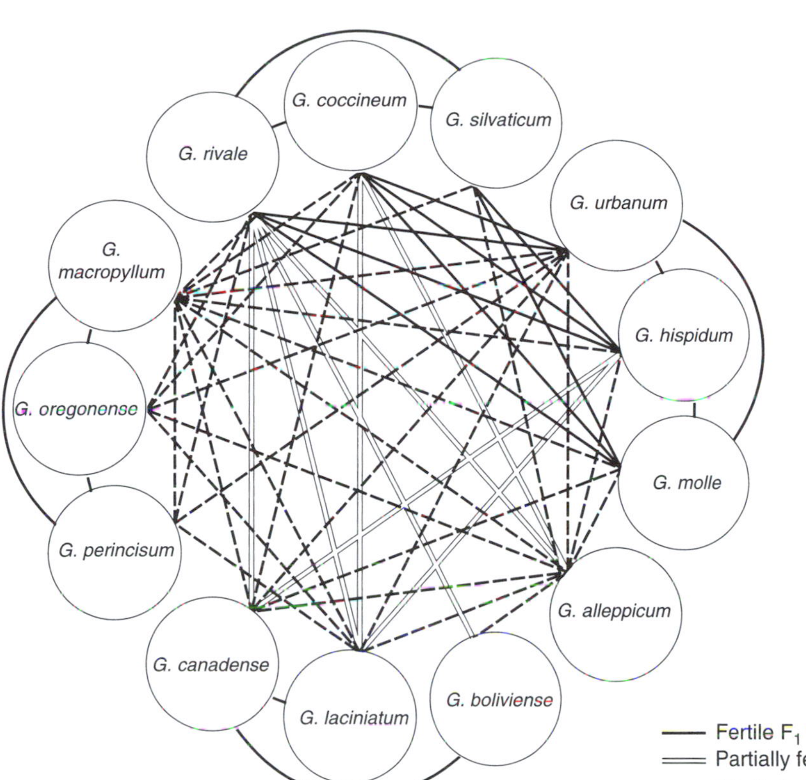

FIGURE 11-2 Differences in fertility observed for F_1 hybrids derived from crosses between 13 species of the plant genus *Geum*, a perennial herb. (Reprinted by permission from Gajewski, W. 1959, Evolution in the Genus Geum, *Evolution*, **13,** 378–388)

to the distant past, can transcend reproductive barriers and cast doubt on species (and even clonal) distinctions. For example, many bacterial clones and species groups (such as *Escherichia*, *Salmonella* and *Shigella*) have similar genes and gene sequences, indicating that horizontal gene transfer has occurred among supposedly unrelated or distantly related groups. Viruses (bacteriophages) that infect these bacteria demonstrate "access, by a horizontal exchange, to a large common gene pool" (Hendrix et al., 1999). Horizontal transfer also has been described among plants and among animals (see Chapter 10).

An analysis of more than 220,000 proteins from the completely sequenced genomes of 144 prokaryotes, representing 15 phyla, identified what Beiko et al. (2005) called "highways of gene sharing" between two types of lineages: closely related taxa on the one hand, and distantly related taxa that inhabit similar environments on the other. The ability of Beiko and colleagues to discriminate genes that are not or rarely subject to horizontal transfer (16*S* rRNA, genes that code for cell wall, cell divisions and ribosomal proteins) from those more subject to transfer (aminoacyl-tRNA synthetases) provides hope that we may be able to erect more accurate relationships by concentrating on genes not subject to horizontal gene transfer. Nevertheless, horizontal gene transfer poses major problems for attempting to recreate a Tree of Life for unicellular organisms (see Box 12-4). Indeed, because so much genetic exchange occurred among prokaryotes — perhaps as much as a third of their genomes (see Chapter 12) — the very notion of a prokaryote Tree of Life may be an entirely inappropriate way to represent prokaryote evolution.

Evolutionary Species

To circumvent some of these problems and to explicitly take evolution into account, various paleontologists, taxonomist and evolutionary biologists proposed an *evolutionary species concept* in which species are defined in terms of differences that are not dependent on sexual isolation but rather on their "evolutionary" isolation, of which sexual isolation is only one aspect. In Simpson's (1961) words, "an evolutionary species is a lineage (an ancestor-descendant sequence of populations) evolving separately from others and with its own unitary evolutionary role and tendencies." Here, for the first time, a species concept incorporated change over time (evolution) rather than static features: species as an evolutionary entity. Such an approach laid the groundwork for considering those changes resulting from competition and interaction among species.

One problem with an evolutionary species definition arises. Because speciation is treated as an evolutionary process, defining the stage when groups of organisms have

reached complete separation is subjective. This is why defining species as individuals and a species as an individual, clearly separated from other individuals, can be a gross oversimplification.[4] Nevertheless, the evolutionary species concept justifies using ecological, behavioral, genetic and morphological evidence to help judge evolutionary separation or distance.

In summary, the difficulties in species taxonomy are, to a large extent — and perhaps almost entirely and inevitably — inherent in the process of speciation itself. Although it may seem obvious, it is often not fully appreciated that the differences among populations that make some of them hard to classify as varieties, subspecies, or species arise from the fact that they underwent evolutionary changes that differed in intensity, sequence, duration and time during the history of life, place and circumstances (see Chapters 22 to 24). Phenotypic and genotypic distinctions do not evolve uniformly among groups; comparisons among them show different degrees of change in various characters. In sexual forms, the overall result of such differences is different degrees of reproductive isolation and morphological distinctiveness. In asexual forms, evolutionary distinctions are reflected in differences other than reproductive isolation.

In both sexual and asexual cases, members of a species share a community of descent that explains many of their common features, yet these individuals cannot be described as members of fixed and unchanging entities. While members of a species are identified by their similarity, their relationship derives from their shared history, which rarely, other than in clones, results in all individuals in a species being identical. A species name indicates *singularity*, but the individual members of a species display **variation;** not all individuals are identical to the type specimen. Therefore, classification and evolution emphasize different aspects of organisms. Although an evolutionary explanation does not always make the taxonomist's task easier, it does offer an explanation for some difficulties in determining species.

Phylogeny

Given that species evolve from existing species, the question of the **origin of species** is basically the question of the **origin of new species.**

Darwin mostly devoted himself to explaining how, under natural selection, a single species can change through time. **Figure 11-3** illustrates such an example of successional changes within a single lineage, called **phyletic evolution.** However, an important concern is species multiplication: how to explain the splits and divisions within an ancestral line that result in the emergence of more than one species,

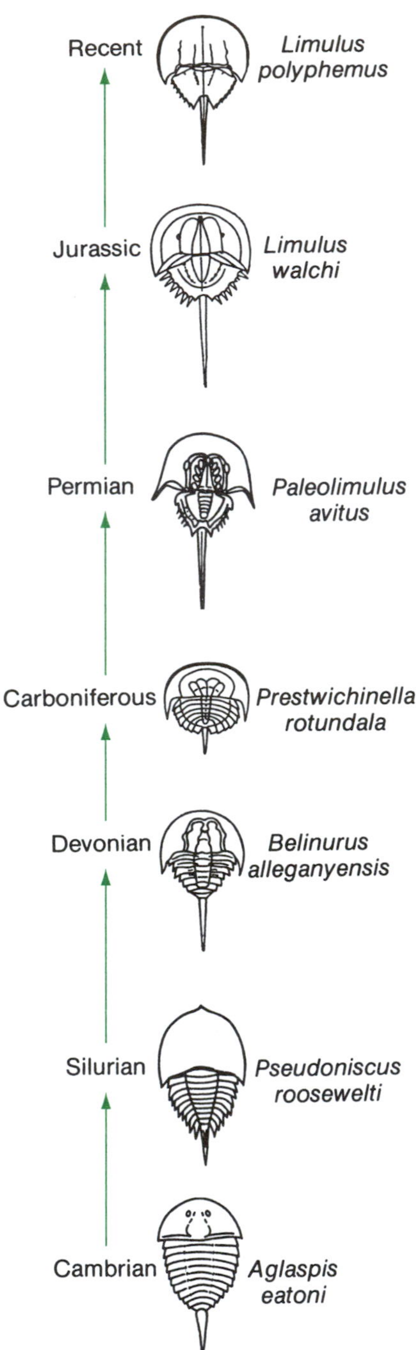

FIGURE 11-3 Phyletic evolution in merostome arthropods (horseshoe crabs), indicating relatively small phenotypic changes over long periods of time. That evolution proceeds only in such a linear direction is, of course, a diagrammatic illusion. As in other such sequences (see Fig. 3-14), there were variations within and between each of these stages. Orthogenesis, the now discarded concept that evolution occurs linearly without selection, was quite popular among paleontologists before geneticists developed an explanation of how variation was generated and maintained (see Chapters 10 and 21). (Adapted from Newell, N. D., 1959. The nature of fossil record. *Proc. Amer. Phil. Soc.*, **103,** 264–285.)

[4] See Ruse (1987) for an extended discussion on whether species are individuals, classes or populations.

Neofelis *Felis* *Puma* *Lynx* *Acinonyx* *Panthera*

Smilodon *Homotherium*

PLEISTOCENE

Dinofelis

Machairodus

PLIOCENE

Metailurus

MIOCENE

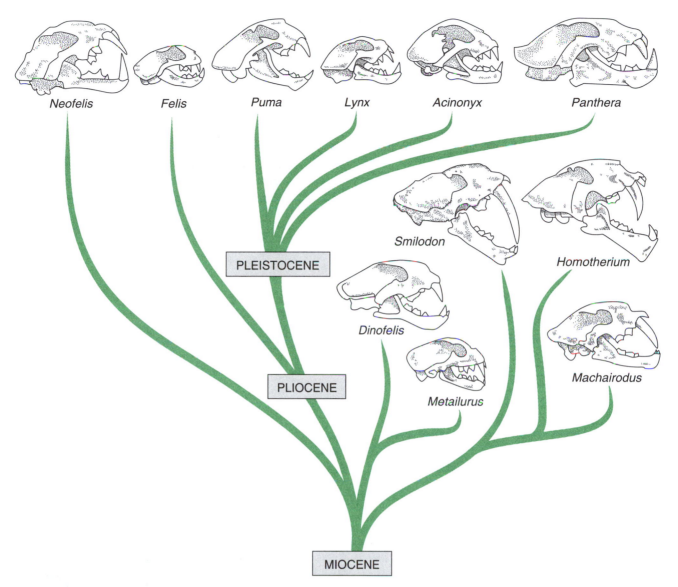

FIGURE 11-4 Phylogenetic branching (cladogenesis) of the cat family beginning with the origin of cats in the Miocene about 20 Mya. The six groups along the *top row* are extant, the other five groups on the *lower right* are extinct. (Modified from Benton, M. J., 1991. *The Rise of the Mammals.* Apple Press, London.)

which is a cluster of species or a **clade**. This pattern, known as **phylogenetic branching** or **cladogenesis** (**Fig. 11-4**), was popularized in the 1860s by Haeckel in his paintings of phylogenetic trees. The idea of cladogenesis is attributed to Lamarck (see Fig. 1-5) but the physician and naturalist Vitalioan Donati (1717–1762), who made important contributions to botany, mineralogy, geology, climatology and the study of earthquakes, may have been the first to suggest phylogenetic branching.[5]

Viewed through time, as in the evolution of horses (see Fig. 3-14), reduction in the number of toes coupled with speciation appears to have proceeded in a straightforward, linear manner. However, as explained previously, establishing a phylogenetic tree is often difficult because the common ancestors of different groups of organisms are long extinct and the fossil record usually insufficient to directly read a phylogeny from it. Nevertheless, as we look back over half a billion years for evidence of the history of life on Earth, a great deal of our knowledge and understanding of evolutionary history is based on fossils. Even though the fossil record is incomplete, it provides a key proof and clarification of evolution. The absence of complete fossil information does place considerable emphasis on constructing phylogenies by comparisons between known organisms, whether extant or fossil, a process that comes with its own difficulties and constraints.

[5] Professor Mark Ragan, University of Queensland, drew Donati to our attention. Donati described nature as a "tissue of many threads that communicate, relate, and unite with one another."

In general, the more a group of species shares common inherited attributes, the more likely their descent from a common ancestor. Taxonomists, therefore, enlist all available heritable characteristics in making comparisons between species: morphological, embryological, behavioral, physiological, biochemical, genetic and chromosomal. Nevertheless, even when species share similar features, the genotypic basis for this may derive from different evolutionary causes, each of which has a specific and time-honored name (**Fig. 11-5**):

- **Homology.** The same feature occurs in different species because it derives from a common ancestor that bore the same characteristic. For example, the many similar features in the forelimb skeleton of vertebrates indicate derivation from a common vertebrate ancestor (see Fig. 3-6). However, because organismal "features" are not only morphological or structural but also physiological, developmental and genetic, what is homologous at one level may not be so at another. Homologous genes may direct development of nonhomologous structures and nonhomologous genes may direct development of homologous structures (see Chapter 13).
- **Parallelism.** A similar feature occurs in different species, but their immediate common ancestor was different. For example, anteater-like features have appeared in several lines of mammals, each of which descended from nonanteater mammalian groups (**Fig. 11-6**).
- **Convergence.** A similar feature arose independently in different species whose ancestral lineages had been separated for a considerable time. We have seen an example of convergence in the similarity of marine hydrodynamic forms among the widely separated fish, reptile and mammalian classes of vertebrates (see Fig. 3-7). As in parallelism, convergence derives from the exposure of different lineages to similar environmen-

Homology: two species bearing the same phenotype caused by common ancestry for the same genotype

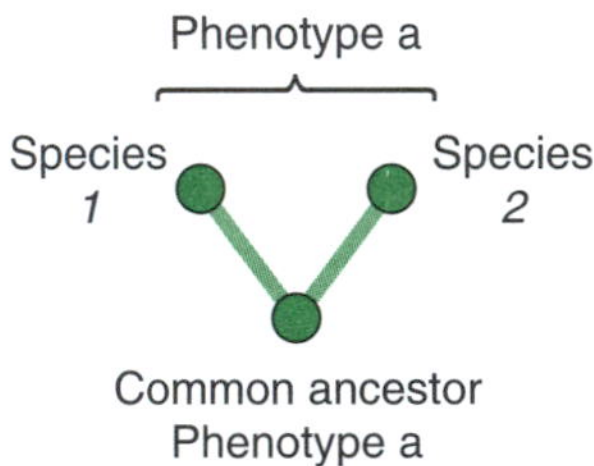

Parallelism: two species with the same phenotype descended from a common ancestor with a different phenotype and genotype

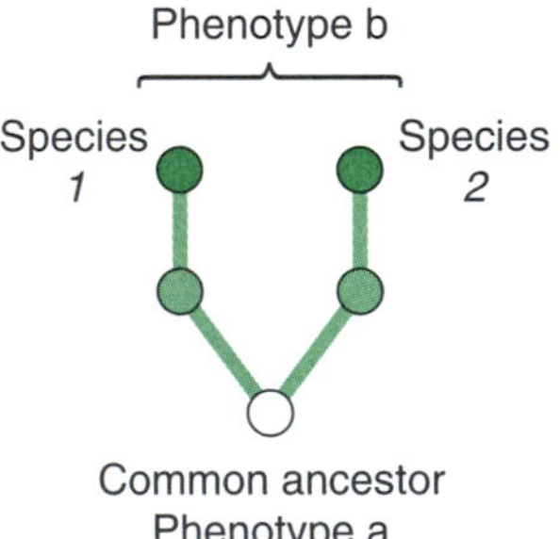

Convergence: two species with the same phenotype whose common ancestor is very far in the distant past

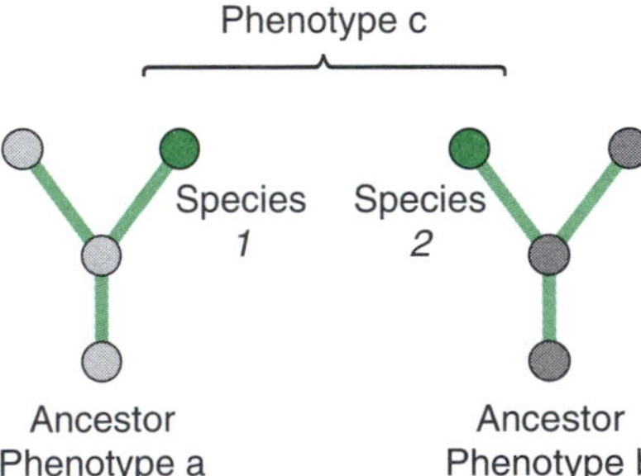

FIGURE 11-5 Homology, parallelism and convergence diagrammed for two species, labeled 1 and 2, that are phenotypically similar for the state in which a particular character appears (for example, long, short, round, colored). Note that the distinction between parallelism and convergence may be subjective, because there are no rules that specify how far in the past one can establish a common ancestor for parallel evolution, and even convergent lineages have common, albeit distant, ancestors.

Prototheria (*Echidna*)

Metatheria (*Myrmecobius*)

Eutheria (*Myrmecophaga*)

FIGURE 11-6 Similar phenotypic features among anteaters (long snout and tongue, powerful claws) that evolved independently within each of the three major groups of extant mammals. Depending on how we estimate their distance from a common ancestor, these forms can be considered either parallel (sharing mammalian ancestry) or convergent (descended from different mammalian groups).

> tal factors evincing similar selective forces. Common adaptive features thus can be attained in each group through independent genetic changes.

Although the difference between parallelism and convergence is mostly based on time to the last common ancestor, this distinction can seem arbitrary, and both terms are now often united in the single term: *homoplasy.*

Homology

Because it is absolutely fundamental to any comparison between two biological objects, homology is discussed in some depth. For most biologists, *homology* is the term for a feature of the phenotype in taxa that share a common ancestry, even though there may be little apparent similarity in the features in different taxa.

Valuable though it may seem, and important as it is, establishing homologous relationships is not a simple task, mostly because parallelism and convergence (homoplasy) can lead to false assumptions that the organisms share a recent common ancestor, and because the genetic and/or developmental basis for homology is not always identical. Given the chain of organismal events that extends from information to function:

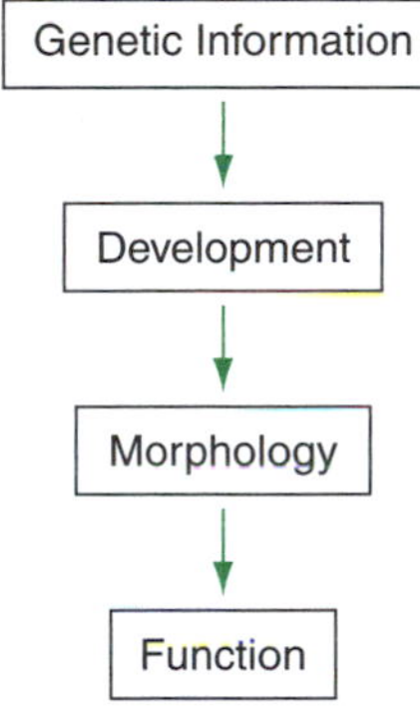

it seems that the more distant a factor is from its informational source in this sequence, the greater the impact of environmental nonhereditary agents and, therefore, the greater the opportunity for convergent influences to select for a similar form in unrelated organisms.

The term **serial** ("iterative") **homology** is used for similarities among parts of the same individual or organism, for example, different types of vertebrae (cervical, thoracic, caudal) or hemoglobin molecules (α, β, γ and so on). Serial homology often reflects **duplication** of a gene responsible for producing or affecting a particular structure. Such duplicates originally may have had similar features, but may evolve differently from each other, a point made by Darwin (although not in the context of genes).

> We already have seen that parts many times repeated are eminently liable to vary in number and structure; consequently, it is quite probable that natural selection, during a long continued course of modification, should have seized on a certain number of primordially similar elements, many times repeated, and have adapted them to diverse purposes. By their functional divergence, duplications can offer an important evolutionary advantage compared to new genes formed with entirely novel and untried nucleotide sequences.

Whether homology is determined at molecular or other levels of the phenotype, convergent events can occur (see Box 12-3). Moreover, organisms are organized hierarchically, with information building upon the level(s) below (see Box 8-1). New "emergent" properties arise at each level and, just as importantly, cannot be predicted from the properties of the level below (**Box 11-1**). If evolution is constrained, as, for example, by stabilizing selection (see Chapters 13 and 22), we expect little evolutionary change in the genetic or developmental processes from which homologous features are constructed.

Features that are homologous, however, need not arise via homologous developmental pathways; the lenses that develop in the eyes of embryonic urodeles and the lenses that form by regeneration following removal of the lens from an adult urodele are homologous as lenses, even though one arises from undifferentiated ectodermal cells and the other by the dedifferentiation of pigmented cells of the iris (Hall, 1999a). Features of the phenotype not homologous as structures (limbs and genitalia in tetrapods) nevertheless share genes or gene pathways that are homologous; homology exists at one level (genes) but not at another (structure). Therefore, it is critically important to specify the level at which homology is designated. In the examples just used, we specify homology of urodele lenses, homology of the genetic pathway, and nonhomology of the limbs and genitalia.[6]

Homology, Partial Homology and Percent Similarity

As molecular sequences became available and were compared, the term homology began to be applied to sequence similarity. To enable comparisons of genes between different organisms and to differentiate a gene duplicated within a species from a gene(s) shared with other species, Fitch and Margoliash (1967) created the terms paralogous and orthologous genes:

- **Orthologous genes** (orthologues) are *genes duplicated in different species.*
- **Paralogous genes** (paralogues) are *genes duplicated within a species;* that is, they are extra copies of the gene (see Chapter 12).

[6] For discussions of homology at different levels of biological organization, see P. C. J. Donoghue (2002), Hall (1995, 2003a, 2006c) and Hall and Olson (2003). See Hillis (1994) for discussion of molecular homology.

BOX 11-1 Hierarchy

One of the most, if not *the* most important aspects of complexity is **increase in hierarchical organization,** reflected in the criteria outlined in Box 8-3. Simon (1962) called this essential hierarchical property "the architecture of complexity," emphasizing that hierarchical systems evolve faster than non-hierarchical ones of similar size.

Perhaps the most well recognized hierarchy of life is that from:

genes → molecules → organelles → cells → tissues → organs → organisms → populations → species

Sandra Mitchell, a philosopher of science, recognizes three categories of complexity, reflecting on a number of hierarchical organizations:

- *constitutive* (compositional) complexity: organisms are comprised of numerous parts;
- *dynamic* complexity: developmental and evolutionary processes generate organisms; and
- *evolved* complexity: similar adaptive challenges have resulted in a diversity of forms of organisms, not a single form (Mitchell, 2003).

A hierarchical approach to understanding evolution has emerged and consolidated over the past 40 years, replacing or running parallel with the reductionist approach, which proposes that explanations for events on one level of complexity can and should be reduced to (deduced from) explanations on a more basic level.[a]

Why does evolutionary biology not reduce to a reductionist approach? As discussed in the introduction to Part 1, we require a hierarchy of mechanisms to explain emerging properties and increasing complexity precisely because properties at one level are insufficient to explain properties at higher level(s). Biology is not physics. Here are a few examples of the principle of properties or explanations at one level not extending to other levels at different levels in the hierarchy (the levels are in italics):

- Although the way in which a repressor *molecule* functions in a cell depends on its *atomic structure,* the repressor function the molecule shares with other regulators is a specific molecular property, and would be difficult, if not impossible, to predict as an atomic property.
- Glycolysis is a property shared by a number of metabolic *pathways* but not a property of any particular *molecule.*
- Dominance is a property shared by some *genes* but not a property of a *nucleotide.*
- Sexual behavior is a property shared by some *organisms* but not a property of *ovaries* and *testes.*
- Population density is a property of a *group* of organisms, but not of any *single* organism.

As evident from the concept and from these examples, a hierarchy extends across many levels, from atoms to molecules to cells to tissues to organs to individuals to populations to species to cultures, each with specific functional properties. François Jacob (1977) referred to "this hierarchy of successive integrations, characterized by restrictions and by the appearance of new properties

Only comparisons between orthologous genes reveal phylogenetic relationships.

Unlike the qualitative use of homology (two features are or are not homologous), early studies comparing molecules often referred to percent similarity or percent homology. Two sequences would be spoken of as 70 percent similar and so homologous, while other molecules were 50 percent similar and so not homologous. But homology is an either/or state. Comparisons were made with pregnancy; just as you cannot be partly pregnant, you cannot be partly homologous. An influential letter published by Reeck and colleagues in *Cell* in 1987, written by a veritable who's who of molecular evolution, took a strong stance against the use of homology for partial sequence similarity, and also emphasized that even a high degree of similarity is no guarantee that the two molecules did not evolve in parallel, rather than from a common ancestral molecule. Their stance was taken to heart. Any notion of partial homology has disappeared.

Homology and Constructing Phylogenetic Trees

However determined, homology is the basis for establishing phylogenetic lineages; the sharing of any character because of common descent signifies a closer relationship between organisms than any other cause for such similarity. Convergences, therefore, should be minimized so that phylogenies relate taxa using ancestrally derived homologies. In general, the observation that some shared characters between species may derive from distant ancestors and others from more recent ancestors provide a means for creating a phylogenetic tree.

A hypothetical tree constructed from information on four characters is shown in **Figure 11-7.** The eight descendants (species *H–O*) share a character (4) derived from a distant common ancestor (*A*). Although this **basal character** unites all eight species, detailed branching of the tree depends on information from less basal and more derived characters present in more immediate ancestors. Thus, species *H–K* are related to a more recent common ancestor *B* because they share character 2 in addition to character 4. Similarly, species *L–O* are related to the more recent common ancestor *C* by sharing character 3. Using these characters, species *H* and *I* can be derived from an even more recent common ancestor *D* because character 1 is shared. Similarly, *J* and *K* derive from *E* (shared character 3), *L* and *M* from *F* (shared character 2), and *N* and *O* from *G* (shared character 1).

at each level. . . . Each system at a given level uses as ingredients some systems of the simpler level, but some only."

Because of their multicomponent subunit structure, hierarchies are more stable and more easily constructed than a nonhierarchical structure in which all parts must be simultaneously assembled and in which the absence of any single component would interfere with any kind of assembly at all. For example, a cell composed of multiple subsystems is less vulnerable to accident and more easily synthesized than a similar compartment that has no subsystems but can function only when all of its many chemical components match perfectly and aggregate simultaneously. It is difficult to visualize a nonhierarchical system that maintains complex functional properties.

Hierarchical systems aid and stabilize evolution, enabling organisms to incorporate new functional properties and to avoid cataclysmic fragmentation should any minor single component be defective because of change or substitution. A marked change in selection from competitiveness to cooperation among cells exemplifies such benefits in the transition from unicellular to multicellular organisms. Similar cooperative changes occur in the transition from solitary individuals into social groups. Components of complex biological organizations can be modified by selection away from independence and toward cooperation.

Because hierarchical properties seem novel and "emergent," some have questioned whether hierarchical levels are understandable and open to analysis based on cause and effect, or whether their complexity makes them opaque to common scientific methods. However, hierarchical levels are explainable. To use the four examples introduced above, we can analyze:

- glycolysis biochemically (see Fig. 9-1);
- dominance enzymatically (see Chapter 10);
- sexual behavior genetically (see Fig. 24-6); and
- population density ecologically (see Chapter 23),

even though the terms used in the explanation are not interchangeable between levels.

These statements notwithstanding, aspects of biological hierarchies do pose problems. Because hierarchical properties confer distinct evolutionary values, some evolutionary biologists have proposed that each level is a special unit of selection: molecular selection, tissue selection, organ selection, species selection. Recognize that hierarchical organization is neither deterministic nor the automatic running of a genetic program. Hierarchical properties are not independent of environmental influence; environmental interactions modify gene and phenotypic expression. Temperature, light, density, competition, and so forth alter organismal properties.

[a] See Mitchell (2003) and Newman (2003) for overviews of hierarchy in biology, Jablonski (2000) for hierarchical approaches to micro- and macroevolution, and the chapters in Hall (1994) for a hierarchical approach to homology.

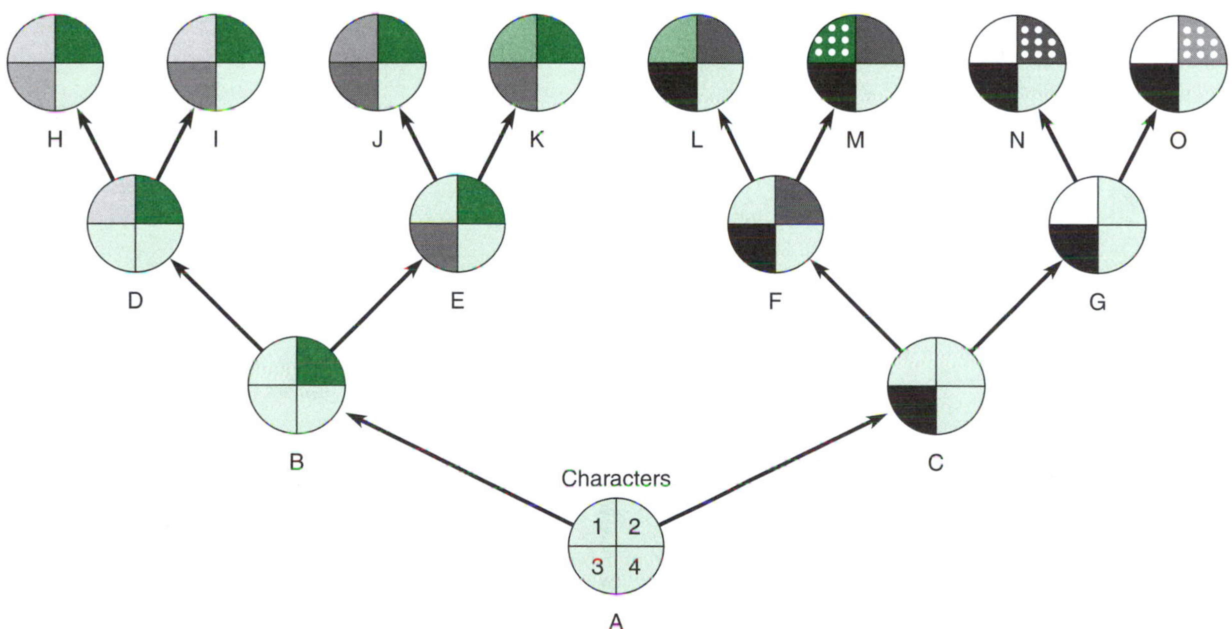

FIGURE 11-7 A simplified phylogenetic tree based on tracing four characters among 15 species (8 extant, 7 ancestral), each bearing variations of these characters ("character states") that allow connections to be established. As the text explains, characters that species share indicate evolutionary relationship; for example, they all share character 4, species *H–K* share character 2, species *L–O* share character 3, species *H–I* also share 1, and so forth. The shared characters indicate common descent from species in which these variations first arose: ancestral species *A*, *B–C*, *D–G*. Note, however, that a character may be shared because of parallelism or convergence rather than because of common ancestry. In this illustration, species *K* and *L* share character 1, although they derived from different lineages (ancestors *E* and *F*) in respect to characters 2 and 3.

Convergence and Parallelism

When similar organisms are exposed to similar environments in different localities, evolution can produce strikingly convergent or parallel features, whether a structure, function or behavior.

The wings of birds and bats are homologous as limbs with digits because they share an ancestor that possessed limbs with digits. Neither bat nor bird wings are homologous to insect wings as no common ancestor has a feature from which both types of wings can be derived. The evolution of wings in insects and in vertebrates is an example of convergence.

Losos and coworkers (1998) showed that anole lizards that colonized four different Caribbean islands diverged from their ancestors and evolved independently into phenotypically similar species occupying similar ecological zones on each island. In these cases, evolution seems to repeat itself fairly exactly, given similar genetic backgrounds and similar environmentally selective events. Identical behavioral breeding patterns found in widely separated groups of fiddler crabs also seem to arise from such comparable conditions (Sturmbauer et al., 1996). Nevertheless, *identical* convergences producing *indistinguishable phenotypes* are rarely expected in organisms that do not share a recent common ancestor.

Consequently, when determining phylogenetic relationships between organisms, it is critically important that we be able to separate homologies from convergence or parallelism, that is, from homoplasy. In reality, such distinctions are not easy to make. For example, we can define mammalian fossils by various criteria — lower jaw consisting of a single bone (the dentary), a dentary-squamosal jaw joint, and a middle ear with three ossicles (see Chapter 19) — and we can trace mammalian ancestry back to an ancient group of reptiles, therapsids. Using these criteria, some paleontologists claim that different groups among the therapsids may have independently evolved to the mammalian level, or grade of organization and function, because of **parallel evolution** (Kermack and Kermack, 1984; see also Miao, 1991), the implication being that mammalian features shared by some lines of early mammals did not arise from common shared ancestry (and so are not homologous) but represent parallelism. We use the term **polyphyletic evolution** for such cases in which different sets of organisms arrived independently at a particular grade of organization, and contrast **polyphyletic** with **monophyletic evolution**, in which sets of organisms derive from only a single ancestral population (**Fig. 11-8**). Over the past decade, we have come to recognize that homoplasy is common in evolution, and that the amount of homoplasy detected tells us something important about the plasticity of genetic and developmental systems and about their ability to response in similar ways when similar selective pressures are acting (see Chapter 13).

Phylogenetic Trees

The fundamental difference is that **phylogeny** is something that happened and classification is an arrangement of its results. Although phylogeny cannot be observed as such over periods long enough to be really significant, it existed as a sequence of factual events among real things (organisms) and in a philosophical or logical sense it is objective or realistic in nature. Classification is not. It is an artifice with no objective reality. It arises and exists only in the minds of its devisers, learners and users (G. G. Simpson, 1980).

Ideally, the most informative phylogenetic picture of a particular population of organisms would be a portion of a multilimbed tree that has branched connections to all extant and ancestral populations, and that indicates, through these connections, its degree of relationship to all other populations (**Fig. 11-9**).

Because as many as 10 to 30 million species of organisms may exist — some named, some not — and undoubtedly some hundreds of millions of species existed in the past, mostly unknown, a complete phylogenetic tree is impossible

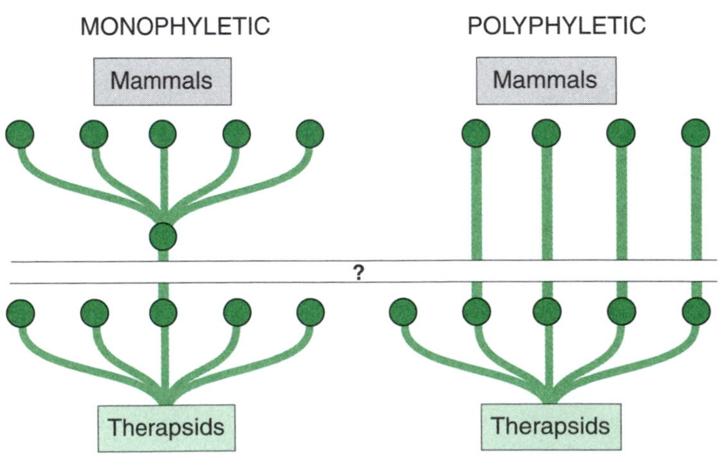

FIGURE 11-8 Monophyletic and polyphyletic schemes used to explain the evolution of mammals from therapsid reptiles. In monophyletic evolution, only a single therapsid group served as ancestor to the mammalian radiation. Two or more therapsid groups gave rise to mammals in the polyphyletic scheme.

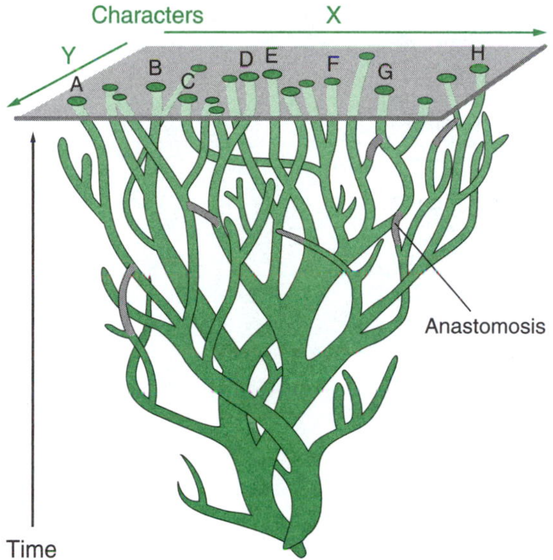

FIGURE 11-9 A phylogenetic tree of related populations shown as continuous branches undergoing evolutionary changes through time. Some populations have become extinct, while others have merged ("reticulate" evolution) or diverged to produce new and different forms. If we consider time as the vertical (*Z*) axis in this illustration, distances along the *X* and *Y* axes might indicate measurements of different genetic traits or groups of traits. Thus, the differences between some populations, such as A and H at the present time level (*top of illustration*), may be sufficiently great to warrant separate taxonomic designations, whereas others (for example, D and E) may not yet be taxonomically distinct. Note also that convergences between two separate lineages (for example, B and C) in respect to the measured traits can conceal their evolutionary separation. (Adapted from Levin, D. A., (ed.), 1979. *Hybridization: An Evolutionary Perspective.* Dowden, Hutchinson and Ross, Stroudsburg, PA.)

to create, nor could we even hope to describe or compute differences among the many possible trees. For 50 extant taxa alone, 3×10^{76} possible different phylogenies (rooted bifurcating trees) would connect them to a common ancestor, exceeding the estimated number of protons and neutrons in the universe (2.4×10^{70}). Nevertheless, a picture of some sort is desirable, for which other biologists rely on taxonomists.

Traditional classification, as in Linnaeus's system, did not depend on evolutionary criteria but on what seemed to be "natural" (see Chapter 1). The attempt to use *natural relationships* to organize the many groups of organisms being discovered into simpler but fewer categories, prompted hierarchical classifications in which biologists placed an organism not only in a particular species but also in ranked categories that included other species (the genus), other genera (the family, the order), and so forth. We still use this system today, although much extended since the time of Linnaeus; for example, his idea of natural differs from ours.[7]

Taxonomists call each unit of classification, whether it be a particular species, genus, order, or whatever **taxon** (plural, **taxa**) and give it a distinctive name. They arrange taxa in nested categories so that a taxon in a "higher" category includes one or more taxa in "lower" categories. Some of the categories and taxa often used in classifying humans (*Homo sapiens*) and fruit flies (*Drosophila melanogaster*) are shown in **Figure 11-10.**[8]

This mode of classification offers a simple scheme for identifying and cataloging large numbers of species. For example, we can use phyla such as Arthropoda and Chordata to distinguish many animals, and use mammalian orders such as Primates and Rodentia to distinguish many mammals. (**Box 11-2** provides additional information on phyla.) From an evolutionary point of view, there is much to commend this classification, because there are close relationships between species within taxa such as dipterans (flies) and primates. To a significant extent, these classification schemes reflect an underlying phylogenetic pattern in which each taxon "seems" to have originated from the one in which it is included. We use the word "seems" because taxa above the level of species are human constructs with no evolutionary potential. A better way to express this type of classification is taxa nested within other taxa.

Furthermore, each traditionally recognized group is not always monophyletic (derived from a single common ancestor), and so, in this sense, these classification schemes do not reflect evolutionary history. Mammals, as explained previously (see Chapter 5), may have had a polyphyletic origin. Also, some invertebrate biologists have concluded that the arthropod grade of organization, characterized by an exoskeleton with jointed appendages, was achieved by several segmented wormlike organisms undergoing parallel evolution; some gave rise to the ancient trilobites, others served as ancestors to groups such as crustaceans and insects (see Chapter 16). Traditional monophyly, assumed for some smaller taxonomic groups such as the mammalian order of insectivores, has been refuted (see Chapter 19). Major plant taxa — bryophytes, tracheophytes, gymnosperms and angiosperms — also may have had multiple origins, that is, be polyphyletic (see Chapter 14), further weakening any correlation between traditional classification and phylogeny.[9]

As a further serious problem, classification at this level, like species concepts, has subjective elements. For example, different taxonomists working with organisms such as arthropods or angiosperms can disagree about which groups

[7] See Mayr (1981), Kingsland (1995) and Winsor (2003, 2006) for discussions of what Linnaeus and we consider natural types.

[8] Some classifications introduce additional categories to those shown in Figure 11-10, either by adding new terms (for example, *tribe*, a rank between family and genus), or by adding the prefixes *super-*, *sub-*, and *infra-*, to the given categories. Thus, the class Mammalia is usually included within the subphylum Vertebrata and the order Primates in the infraclass Eutheria.

[9] See discussions in Thomas and Spicer (1987), Stewart and Rothwell (1993) and Niklas (1997) for plant evolution.

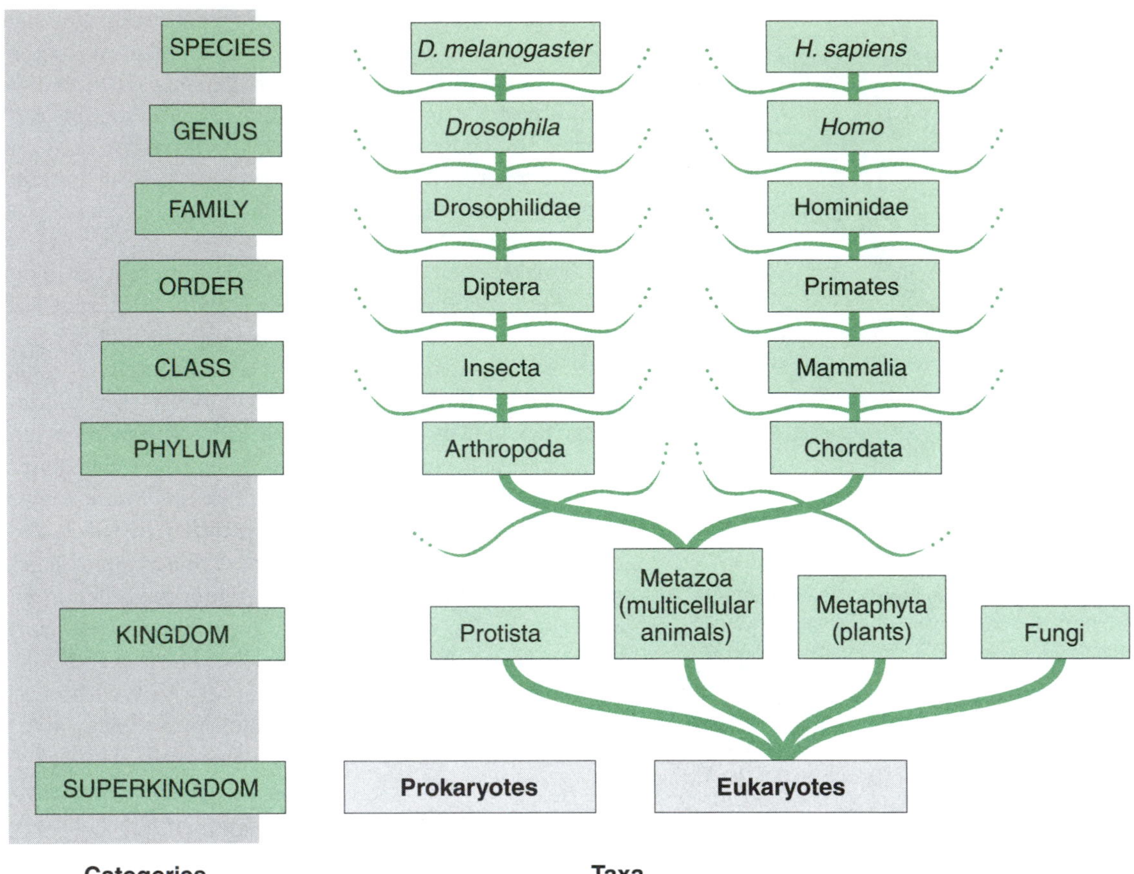

FIGURE 11-10 A classification of fruit flies and humans. Among other classification schemes, Wheelis and colleagues (1992) substitute three "domains" — Archaea, Bacteria, and Eucarya — for the prokaryote and eukaryote "super kingdoms" illustrated here (see also Fig. 8-7). As outlined in Box 8-1, revisions in the "kingdom" category also have been suggested, with numbers ranging from two ("plants" and "animals") to as many as 13. A classification system first proposed by Whittaker (1959; and see Margulis, 1998) uses five major kingdoms: Monera (prokaryotes), Protista (unicellular eukaryotes such as protozoa, algae, slime molds and some other groups), and the multicellular Fungi, Plantae and Animalia. However, the wide separation among many prokaryote and eukaryote branches shown in Figure 8-7 indicates the inadequacy of usual "kingdom" classifications when evaluated on a molecular level: plants, for example, are less separated from fungi or from animals than are microsporidia from diplomonad protistans.

are to be classified into which families. Some taxonomists split the various catlike mammals into 28 genera, while others combine them into one or two genera. Different interpretations exist about whether to assign a particular group to one category or to a higher (more inclusive) one. For example, some invertebrate biologists classify nematodes as a class within the phylum Aschelminthes, while others place them in an independent phylum. The difficulty of classifying organisms above the species level has led to a variety of proposals designed to make classification more objective. One, *phenetics*, arose from numerical taxonomy. The other, **cladistics**, represents a radical departure from previous approaches. Both resolve the subjective aspects of assigning taxa to higher taxa, but both sacrifice information on ancestor-descendant relationships and suffer from our ability to detect homoplasy. Phenetic classification is no longer used. Cladistics, on the other hand, has had a major impact on classification and therefore on evolutionary biology.

Cladistics

Cladistics (also **phylogenetic systematics**) is an approach to classification that relies on branching. The original proponent of this method was Willi Hennig (1913–1976), whose pivotal role in founding cladistics is recognized in the name of a professional association of cladisticians, the Willi Hennig Society.[10] In this system, every significant evolutionary step marks a dichotomous branching that produces two genetically separated **sister taxa** equal to each other in rank. Because the ranking of such sister groups is below the rank of the parental group that gave rise to them, the hierarchy or ranking of groups derives logically from their genealogical position.

[10] The Willi Hennig Society (http://www.cladistics.org/) publishes the journal *Cladistics*, which keeps alive Hennig's name and pivotal role in founding cladistics.

BOX 11-2 Phyla

A PHYLUM (division, in plants), the taxonomic category immediately below the kingdom in traditional classification schemes, is a group of animals (plants) that share a common plan and have a closer evolutionary relationship to one another than they do to animals (plants) in another phylum (division).

The approximately two million named species of animals are arranged into some 37 phyla, examples of most of which are discussed in Chapters 16 to 19. The phylum Arthropoda (arthropods) has more than a million species, Nematoda (roundworms) half a million. The phylum Placozoa (from the Latin for flat animal) consists of a single marine species, *Trichoplax adhaerens,* 300 μm "long" (perhaps we should say short or across as it has no obvious symmetry) with four cell types, 12 chromosomes and a low DNA content (1,000 Daltons). Gene number is not known (see Chapter 16). The phylum Cycliophora (from the Latin for small wheel-bearing) also consists of a single microscopic marine species, *Symbion pandora,* which spends its life attached to the mouth parts and appendages of the Norway lobster *Nephrops norvegicus.*[a]

Although some thirty phyla have been recognized for well over a century and some for a century and a half, six have been recognized and erected over the past 70 years, the latest in 1995. With the year they were recognized in parentheses, these are:

- Pogonophora or "beard worms" with some 100 species, all marine (1937);
- Archaeocyatha, a phylum of extinct sponge-like reef-dwelling animals restricted to the Cambrian (1955);
- Gnathostomulida, 100 species of microscopic (0.5–1.0 mm) marine "jaw worms" (1969);
- Loricifera, 10 species of marine, sediment dwelling "brush heads" (1983);
- Vestimentifera, six species of benthic the red tube worms up to 3 m in length (1985); and
- Cycliophora (1995), mentioned above.

Needless to say, investigations on the status of some of these organisms continue. Some specialists classify archaeocyathans as sponges; others have united vestimentiferans with pogonophorans as the family Siboglinidae within annelid worms, robbing both of phylum status.

[a] See Budd (2003), Hall (1999a), Margulis and Schwartz (1998) and Valentine (2004) for approaches to how to distinguish phyla and for descriptions of the phyla.

For example, as birds and crocodiles derive from a common reptilian stem ancestor (Hedges, 1994) cladistics consider them to be sister groups of equal rank within Archosauromorpha, which in turn ranks lower than Reptilia, within which it nests (see Chapter 17). Cladistics does not follow the usual classification of ranking birds (Aves) separate from reptiles (Reptilia). In cladistic analysis, groups that do not include all the descendants of a common ancestor — for example, the class Reptilia does not include their mammalian or avian offshoots — are **paraphyletic** and are not valid taxa. Similarly, cladistic schemes exclude polyphyletic groups such as Arthropoda and others that consist of convergent or parallel lineages. A group of taxa or a clade can achieve taxonomic status only if it is strictly monophyletic and there are no ambiguities; the group must include its common ancestor and *all* of the common ancestor's descendants. Birds, which are related to dinosaurs, "are not only descended from dinosaurs, they are dinosaurs (and reptiles)" (Padian and Chiappe, 1998; **Box 11-3**).

Many who do not use cladistic approaches claim that any partition of taxa into groups, whether paraphyletic, polyphyletic or monophyletic, is done for the convenience of taxonomists, and not because one partition is more "real" than another. Although there is continuity between an ancestor and its descendants, and although taxonomy must take into account evolutionary relationships, schemes for assembling descendants into groups often seem mostly a "mental" construct.

Cladistics stresses the separation of ancestral (**plesiomorphic**) characters from newly derived (**apomorphic**) characters, and emphasize the latter to establish phylogenies. The sharing of derived characters (**synapomorphy**) dictates the phylogeny. For example, how species A, B, and C relate can be determined by noting which share a newly derived character. Thus, specific character X shared by species A and C indicates a closer relationship than to species B; that is, A and C branched off together from a common ancestor, whereas species B branched off separately (**Fig. 11-11a**). This emphasis on using shared-derived characters provides more precise branching information than phenetic classification,

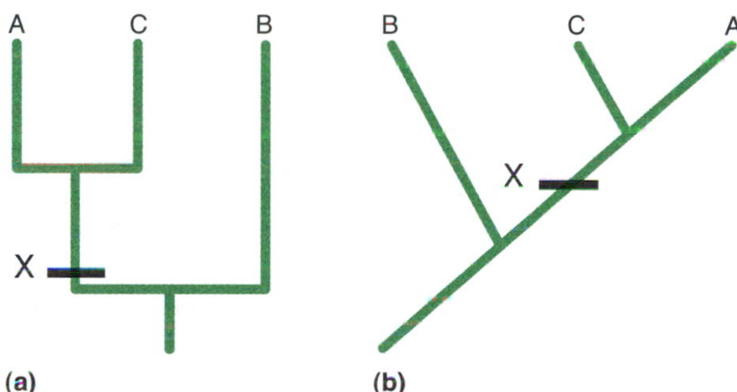

FIGURE 11-11 Relationships between three species, two of which share a newly derived character (X), as depicted using a cladistic approach (a), which does not make any statement of ancestry, and as depicted in a more traditional tree (b) in which species B is ancestral to species A and species C. Note that in both trees species A and C are more closely related to one another than either is to species B.

BOX 11-3 Phylogenetic Species

The urge to define taxa strictly according to their cladistic relationships has led some who use cladistics to replace the biological species with a phylogenetic species concept. Cracraft (1983) proposes that a phylogenetic species is a monophyletic group composed of "the smallest diagnosable cluster of individual organisms within which there is a parental pattern of ancestry and descent." The "diagnosable" element in this definition is based on similar, if not identical, characters and measurements used in numerical taxonomy, and leads to similar problems of defining such characters and evaluating such measurements. In addition, emphasis on the "smallest diagnosable cluster" can result in declaring varieties and even smaller genetically-distinct groups as separate species. Many consider this an unwarranted and unrealistic multiplication of species.[a]

Not all agree that cladograms and evolutionary relationships should correspond. To some, the goal of systematics is to discover and delineate the patterns of nature, a goal that taxonomists can best achieve by concentrating on shared-derived characters of organisms, however they evolved. According to this view, evolutionary preconceptions divert attention from the essential core of systematics, which is "to clearly define taxa through the characters used in cladograms and [to] place these taxa into a "natural order . . . basically independent of evolutionary theory" (Rieppel, 1984).

Because cladograms depend on distinguishing derived, ancestral and convergent characters — all of which arise from evolutionary phenomena — the proposed "independence" of cladograms from evolution seems unreal. For many who use cladistics, Ghiselin (1984) argues, cladograms "are nested sets of characters, not descriptions of the underlying historical order." A cladistic taxon cannot evolve from another taxon of the same category, but must be "nested" within a taxon of higher category. Thus, Mammalia is not a major vertebrate group evolved from the Reptilia, but a branch of Therapsida, a branch within Synapsida (see Chapters 18 and 19), which branches from higher taxa that nest successively into amphibian-like taxa, fish-like taxa, and so forth. There is a danger in this approach that phylogenies become ends in themselves rather than data sets and hypotheses of inferred relationships.

[a] See Talbot and Shields (1996) for a study based on the paraphyletic origin of brown bears, and Knowles (2004) for an analysis of the field of "statistical phylogeography," the study of the processes that regulate the distribution of lineages of organisms.

which offers only general estimates of similarity but no such direct method of generating branching dichotomies. And, it contains no suggestion as to which species is ancestral (Fig. 11-11a with 11-11b).

When this approach is extended to many taxa using many characters, fairly complex phylogenies are possible. To select one hypothesis of relationships from the set of possible hypotheses, sophisticated statistical methodology is used to infer the most likely tree (e.g., Felsenstein, 1982, 2004). Another popular technique is to choose the phylogenetic tree that minimizes the number of changes necessary to explain its evolutionary history (the **parsimony method**).

To use the simplified example shown in **Figure 11-12**, if one phylogenetic tree requires that a particular character evolved on more separate occasions than it would have had to evolve in another tree, the latter tree is preferred because evolution is assumed to occur by simple binary branching. Cladogram a in Figure 11-12 is preferred over cladogram b, because a requires two evolutionary events to explain the data, whereas b requires three. Cladogram a is more parsimonious than cladogram b. Clearly, parsimony is not a statistical method. Consequently, a more extensively used method is **maximum likelihood estimation**, a statistical method based on probability distribution developed by R. A. Fisher in the early nineteen twenties. This method is based on maximizing the most likely distribution for a set or sets of data, and can handle enormous numbers of data points.

Because of its formal terminology, precise rules, and strict genealogical consistency in assigning branching points and patterns, cladistic classification combined with maximum likelihood analysis has become highly popular (see Box 11-3). This popularity has brought terms such as "sister group," "synapomorphy," "paraphyly," and "homoplasy" into common use. Cladistics has brought a number of issues to the fore, some of which are outlined below. As with the issues raised by the biological species concept, this set of six issues reflects our incomplete knowledge of mechanisms of evolution although not the fact of evolution[11]:

1. Because cladistics does not assign a completely new taxonomic designation to each branch, if one branch remains identical to the previous ancestral population (**Fig. 11-13**), many clades are left unnamed.
2. As lineages also can be "reticulate," dichotomous branching oversimplifies the evolutionary process. Bear in mind that cladistics does not attempt to recover phylogenies directly, but rather informs phylogenetics, each tree being a hypothesis of relationships. The analogy is often made between the branching patterns seen in a tree versus those in a shrub. A lineage can connect genetically with related lineages through

[11] For cladistic classification, see, Forey et al. (1992), and Minelli (1993). For problems raised, Ghiselin (1984), Panchen (1992), and the historical account by Hull (1988).

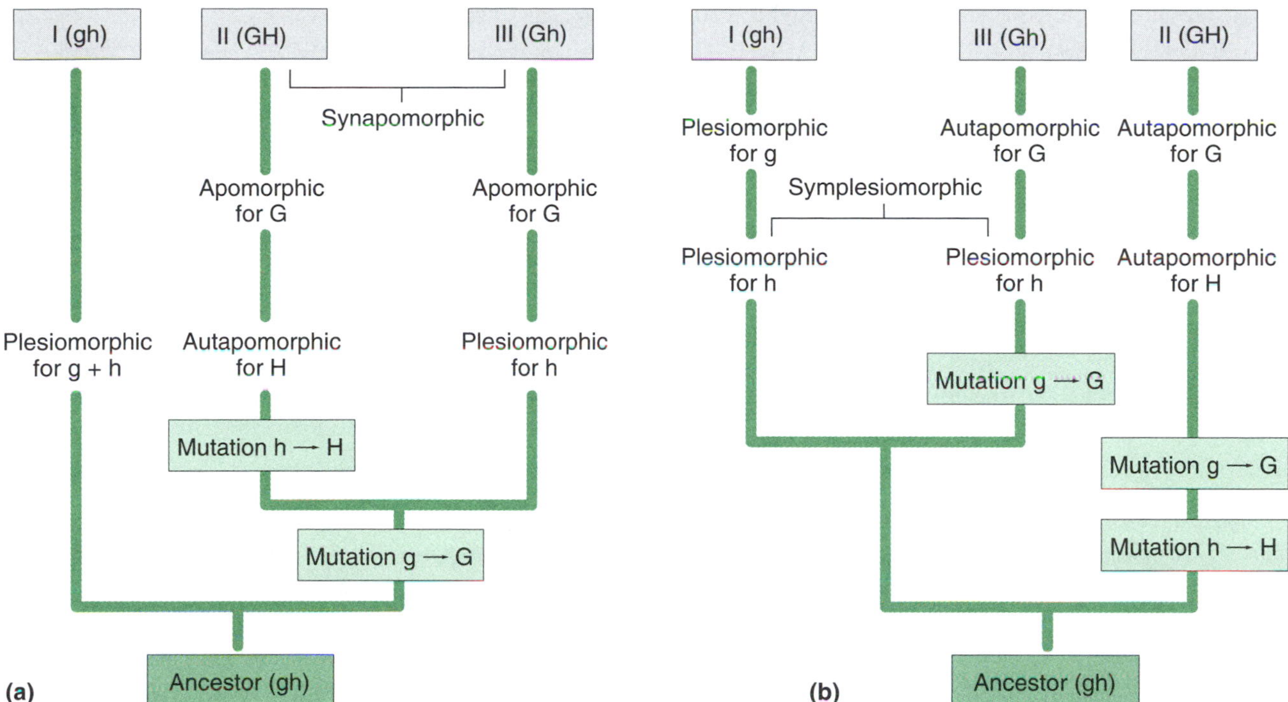

FIGURE 11-12 Cladograms for variations of two characters g and h in hypothetical taxa I, II, and III and their presumed common ancestor. Taxon I has the ancestral (plesiomorphic) characters g and h, taxon II has derived (apomorphic) characters G and H, and taxon III is apomorphic for G and plesiomorphic for h. As shown, two alternative cladograms are possible. Each step is assumed, for simplicity, to arise by a single mutation. In (a) the assumption is made that character G shared by taxa II and III evolved only once by mutation of g → G, and character H evolved in the lineage of taxon II also by a single mutation of h → H. Thus, taxa II and III share the derived G character (synapomorphy) and are sister taxa, while taxon I, in turn, is the sister group of (II + III). In (b) the assumption is made that I and III are sister taxa, because they share ancestral character h (**symplesiomorphy**), and taxon II is the sister group of (I + III). Note, however, for the (b) tree to be accepted two mutations of g → G are necessary (as well as one of h → H). Thus, by the principle of parsimony, the (a) cladogram based on synapomorphy, requiring a total of only two mutations is preferable to the (b) cladogram based on symplesiomorphy, which requires three mutations. Derived character states that are unique rather than shared with other lineages are called **autapomorphies** and do not indicate relationship.

hybridization (see Chapter 24), convergent evolution, or through "horizontal gene transmission" from even more distant lineages. Such phylogenies are networks rather than simple dichotomous branching trees.

3. Cladistics informs us about the relationships of taxa, not about ancestry. In cladistic terminology, a parental taxon that produces a new taxon becomes the sister group of its offspring, essentially annulling its ancestral relationship. This leads to considerable difficulties in acknowledging ancestral taxa, as you will see in Chapters 14 to 20, in which we attempt to use both cladistic and other approaches to reveal organismal relationships and phylogeny. Some molecular systematists suggest that because fossils are relatively rare and often morphologically incomplete, we should ignore the fossil record and base cladograms entirely on extant forms. Such cladograms arbitrarily dismiss the hard evidence for evolution and can overlook phylogenetic sequences.
4. Although phyletic changes in a species can occur through time (see Figures 3-4 and 11-3), cladistic clas-

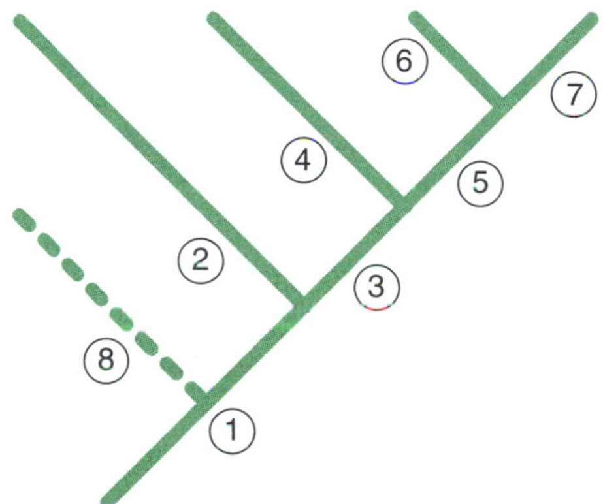

FIGURE 11-13 Illustration of the kind of diagram commonly used in cladistic classification (called a *Hennigian comb*), where each lineage that extends from the node of a branched fork represents a distinct taxon. Thus, populations 1, 3, 5 and 7 are given different taxonomic designations although they may not be recognizably different genetically. If a new lineage of organisms (*indicated by the dotted line 8*) is discovered to have split from population 1, cladistic classification changes the taxonomic designation for 1 so that it differs before and after this split. Such proceedings are based on the concept that any split produces two new sister groups of equal rank, neither of which is ancestral to the other, and the taxon that gave rise to these sister groups has become extinct.

sification used the same group name throughout evolution as long as a group had produced no discernible branches. This is changing as supraspecific taxon names are increasingly being defined on the basis of ancestry rather than on the possession of a suite of characters, providing a better reflection of phyletic evolution.

5. Some early cladistic studies assumed that new branches occur at a fairly constant rate; an older group has necessarily branched into a larger number of taxa than a younger group. This view seems supported by the burst of new major multicellular animal phyla more than 500 Mya during the Cambrian (see Chapter 14), phyla that subsequently radiated into diverse subcategories, recognized as classes and orders. However, groups can branch extensively over relatively short periods, as evidenced by the large numbers of Hawaiian Drosophilidae (more than 800 species) which descended from perhaps one or two ancestral species that arrived in the islands no earlier than 30 or so Mya (see Fig. 22-15). Similarly, more than 40 species of Hawaiian honeycreepers, sporting many differently shaped beaks for extracting floral nectar, descended from a single species that immigrated to these islands early in their history.
6. Phylogenies that include only synapomorphic characters can omit evolutionary information when a cladistic tree does not agree with evolutionary changes in genetic and/or developmental mechanisms. For example, cladistic phylogenies that associate groups because they are synapomorphic for lens crystalline proteins overlook the production of these proteins by nonhomologous genes (see Table 12-2), unless the genes are used as characters.

In summary, from the various critiques of all these taxonomic systems we can see that, to many biologists, the *desirable goals of classification* include:

- the arrangement of groups into a pattern that accurately reflects their evolutionary relationships (phylogeny) and/or
- the placement of groups into a reference system so that their major features are easily and efficiently described and identified (*information storage and retrieval*).

No single classification system reviewed so far fully accomplishes either of these purposes. Traditional morphological classification and numerical taxonomy simplify the placement of organisms into a classification scheme but run the risk of ignoring their evolutionary relationships. Cladistic classification offers hypotheses of relationships that can be tested as new data arise.

The need for flexibility and change in classification has become apparent as new molecular information challenges formerly accepted phylogenies and as cladistic analyses suggest new relationships (see Chapter 12). For example, and as discussed in Chapter 18, birds are flying reptiles. For a long time, cetaceans — whales, dolphins, porpoises — were classified as a separate mammalian group. More recent analyses of fossil, morphological and molecular data (12*S* and 16S ribosomal RNA sequences) suggest that cetaceans are *closely related* to the extant artiodactyls (even-toed ungulates), a group that includes camels, pigs, hippos and ruminants. Early whale fossils morphologically similar to artiodactyls have been found; they share a modified ankle joint (Bejder and Hall, 2002). Current evidence from mitochondrial, nuclear and chromosomal DNA shows that cetaceans arose deep *within* the artiodactyls. Cetaceans and artiodactyls are monophyletic.

Also complicating strict genealogies, especially for unicellular organisms and for plants, are horizontal gene transfers between species of distinctly different lineages, producing "chimeras." Studies that are now allowing us to begin to compare what we may term "lineage-intrinsic" and "horizontal-transfer aspects" of organisms, when combined with sophisticated statistical analyses and methods of making inferences (for example, maximum likelihood or Bayesian inference[12]), permit us glimpses of the past history of life, although we have some way to go. An ideal biological classification system may be as elusive as fitting all organisms into a single ideal species concept or into a single (universal) tree of life (see Box 12-4).

KEY TERMS

apomorphic
autapomorphies
biological species concept
clade
cladistics
cladogenesis
classification
convergence
evolutionary species concept
grade
homology
homoplasy
maximum likelihood estimation
monophyletic evolution
numerical taxonomy
parallel evolution
paraphyletic
parsimony method
phylogenetic branching
phylogenetic systematics
phylogeny
plesiomorphic
polyphyletic
symplesiomorphy
polytypic species
reproductive isolation
serial homology
sibling species
sister taxa
species
synapomorphy
systematics
taxon

[12] Bayesian inference takes its name from the Reverend Thomas Bayes (1702–1761). Bayes' theorem solved the problem in probability theory of estimating the parameters of a distribution (say age at death) based on a prior distribution and data from an observed distribution. "Spam" filters use Bayesian analysis to assess the probability that an incoming message is junk and so should be routed to the spam folder or, ideally, eliminated entirely. Because the validity of the prior distribution cannot be assessed statistically, Bayesian analysis is controversial; see Gelman et al. (2003) for a detailed analysis.

DISCUSSION QUESTIONS

1. Define systematics and phylogeny. Are these concepts connected? (Explain.)
2. Why can't the fossil record provide phylogenies for all organisms?
3. What difficulties do we encounter in defining a species in a manner that allows us to apply this definition universally? (Discuss such species concepts as morphological species, numerical species, biological species and evolutionary species.)
4. Which proposals for different numbers of super kingdoms (two, three, or more) would you support? (See Figure 11-10.)
5. Would you agree with Darwin that the attempt to find "the essence of the term species" is a "vain search"?
6. How do parallel and convergent evolutionary events affect phylogenetic determinations?
7. Would you include instances of polyphyletic evolution in phylogenies? Why or why not?
8. What difficulties arise in applying traditional morphological, phenetic and cladistic classifications to devise phylogenies?

EVOLUTION ON THE WEB

Explore evolution on the Internet! Visit the accompanying Web site for *Strickberger's Evolution,* Fourth Edition, at http://www.biology.jbpub.com/book/evolution for Web exercises and links relating to topics covered in this chapter.

RECOMMENDED READING

Ax, P., 1987. *The Phylogenetic System: The Systematization of Organisms on the Basis of Their Phylogenesis.* Wiley, Chichester, UK.

Beiko, R. G., T. J. Harlow, and M. A. Ragan, 2005. Highways of gene sharing in prokaryotes. *Proc. Natl Acad. Sci. USA,* 102, 14332–14337.

Felsenstein, J., 2004. *Inferring Phylogenies.* Sinauer Associates, Sunderland, MA.

Hall, B. K., (ed.), 1994. *Homology: The Hierarchical Basis of Comparative Biology.* Academic Press, San Diego, CA. [paperback issued 2001]

Hall, B. K., 1999. *Evolutionary Developmental Biology,* 2nd ed. Kluwer Academic Publishers, Dordrecht, Netherlands.

Hall, B. K., and W. M. Olson (eds.), 2003. *Keywords & Concepts in Evolutionary Developmental Biology.* Harvard University Press, Cambridge, MA.

Kingsland, S. E., 1995. *Modeling Nature: Episodes in the History of Population Ecology,* 2nd ed. The University of Chicago Press, Chicago.

Lee, M. S. Y., 2003. Species concepts and species reality: Salvaging a Linnean rank. *J. Evol. Biol.*, 16, 179–188.

Mayr, E., 1982. *The Growth of Biological Thought. Diversity, Evolution, and Inheritance.* Harvard University Press, Cambridge, MA.

Valentine, J. W., 2004. *On the Origin of Phyla.* The University of Chicago Press, Chicago.

Wilson, R. A. (ed.), 1999. *Species. New Interdisciplinary Essays.* MIT Press, Cambridge, MA.

12 Genes and Phylogenetic Relationships

Chapter Summary

In searching for methods to determine phylogenetic relationships, many techniques now allow molecular comparisons among organisms, even among those without uncommon morphological features.

Amino acid sequences in individual proteins can be informative about the evolution of individual genes. For example, amino acid sequencing indicates that myoglobin and the many different forms of hemoglobin arose from one ancestral globin gene that subsequently underwent a series of duplications. Such events were common, with many of the duplicated genes evolving along different lines and carrying out disparate functions.

Biologists can determine phylogenetic distances by comparing amino acid changes in orthologous proteins in different species and calculating the number of mutations necessary to convert one amino acid to another. To construct a realistic phylogenetic tree, however, we need information from many proteins. Techniques involving measurement of DNA levels generally show that the quantity of DNA is not always proportional to the taxonomic status of the group. Analyses of the types of DNA found in organisms indicate that much DNA is repetitive, sometimes highly so. By measuring the degree to which DNAs from different organisms hybridize, we can detect homologies that can help determine phylogenetic lineages and date the divergence points of various groups.

Restriction enzymes, which cleave DNA into fragments at particular sites, allow comparisons among DNAs of different species. The most informative of these techniques is a comparison of exact nucleotide sequences of DNAs from different sources. For example, data from 5*S* ribosomal RNA have been used to compose a phylogenetic tree in which nucleotide changes serve as a measure of evolutionary distance. Large datasets of RNA and DNA sequences, including mitochondrial genes and microsatellites are being used to construct and refine phylogenies. One can also extract DNA from dead or fossil organisms, although such DNA, referred to as "ancient DNA" is much more difficult to analyze.

Some have assumed that mutations are incorporated into genotypes at a fairly regular rate (the molecular clock), and that the rate of molecular change is constant over time for some genes. However, no single clock applies universally; different parts of the genome have different clocks, as do different taxonomic groups.

Major evolutionary change, even speciation, may occur because of changes in regulatory genes, rather than in genes coding for protein structure. Consequently, amino acid sequences for many proteins may be almost identical in different groups, but the phenotypes may vary considerably.

Molecular evolution can be studied by inducing mutations in microorganisms before subjecting the organisms to selection pressures in the laboratory. Under these conditions, evolutionary changes, such as enzyme adaptation to a new substrate or the conversion of a protein to a new function, have resulted. Evolution of viral genomes and even smaller nucleotide sequences has been demonstrated under molecular selection pressures.

Efforts to overcome the phylogenetic problems outlined in Chapter 11 include using molecular information rather than placing exclusive reliance on morphological characters. On the molecular level, we can obtain information by comparing sequences of nucleotides in various DNA and RNA molecules as well as by comparing sequences of amino acids (and their molecular configurations) in different proteins.

A molecular approach has the advantage that evolutionary changes marked by substituting amino acids and nucleotides can be measured and compared on a unit scale of amino acids or nucleotides no matter how much the organisms differ in phenotypic features such as morphology, behavior, ecology or physiology. Molecular comparisons transcend barriers among organisms whose relationships cannot be evaluated by traditional experimental techniques such as genetic exchange. We can compare an amino acid sequence in a particular protein among an array of organisms as different as humans and bacteria (**Fig. 12-1**).

This close tie to molecular biology has transformed evolution from a "theoretical" explanation of historical events to an observable and continuous link among all life forms. Organisms, past and present, all can be bound together through the genetics of evolution by innumerable threads of inherited molecular sequences and their variations. Such molecular records can encompass all basic genetic information transferred between organisms, affecting all attributes, from transcribed to nontranscribed nucleotide sequences.

Proteins and Phylogenetic Relationships

The first generation of comparative molecular methods employed immunological techniques in which antibodies produced in a particular host (usually a rabbit) against proteins (*antigens*) of one species were measured for their

Bacteria	P L F D F A Y Q G F A R G – L E E D A E G L R A F A A M H K E L I V A S S Y S K N F G L Y N E R V G
Yeast	A L F D T A Y Q G F A T G D L D K D A Y A V R X X L S T V S P V F V C Q S F A K N A G M Y G E R V G
Alfalfa	P F F D S A Y Q G F A S G S L D A D A Q P V R L F V A D G G E L L V A Q S Y A K N M G L Y G E R V G
Chicken	P F F D S A Y Q G F A S G S L D K D A W A V R Y F V S E G F E L F C A Q S F S K N F G L Y N E R V G
Rat	P F F D S A Y Q G F A S G D L E K D A W A I R Y F V S E G F E L F C P Q S F S K N F G L Y N E R V G
Horse	P F F D S A Y Q G F A S G N L D R D A W A V R Y F V S E G F E L F C A Q S F S K N F G L Y N E R V G
Pig	P F F D S A Y Q G F A S G N L E K D A W A I R Y F V S E G F E L F C A Q S F S K N F G L Y N E R V G
Human	P F F D S A Y Q G F A S G N L E R D A W A I R Y F V S E G F E F F C A Q S F S K N F G L Y N E R V G

FIGURE 12-1 A comparison of eight organisms for a 50-amino-acid–long sequence of the enzyme aspartate transaminase. For the amino acid abbreviations, see Figure 6-2 or Table 12-3. (Adapted from Benner, S. A., M. A. Cohen, G. H. Gonnet, D. B. Berkowitz, and K. P. Johnsson, 1993. Reading the palimpsest: Contemporary biochemical data and the RNA world. In *The RNA World*, R. F. Gesteland and J. F. Atkins (eds.). Cold Spring Harbor Laboratory Press, Cold Spring Harbor, New York, pp. 27–70.)

activity against proteins of other species. For example, if antibodies against species A precipitate much of the protein in species C but little of the protein in species B, we presume the A and C proteins have more similar molecular configurations (antigenic components) and are more evolutionarily alike (smaller antigenic distance) than those of A and B.

Using a variety of antibodies, immunological data can be analyzed by various mathematical rules (*algorithms*) to construct a phylogenetic tree that best correlates the antigenic distance between species and the length of time since they shared a common ancestor. Successive comparisons are made between all possible combinations of species until the entire phylogenetic tree is obtained. An early tree for anthropoids (humans, apes, monkeys) is shown in **Figure 12-2** (see also Chapter 20). In contrast to traditional taxonomy of the time, this technique shows that humans (*Homo*), gorillas (*Gorilla*), and chimpanzees (*Pan*) are antigenically closer to each other than to the Asian orangutan (*Pongo*), and so

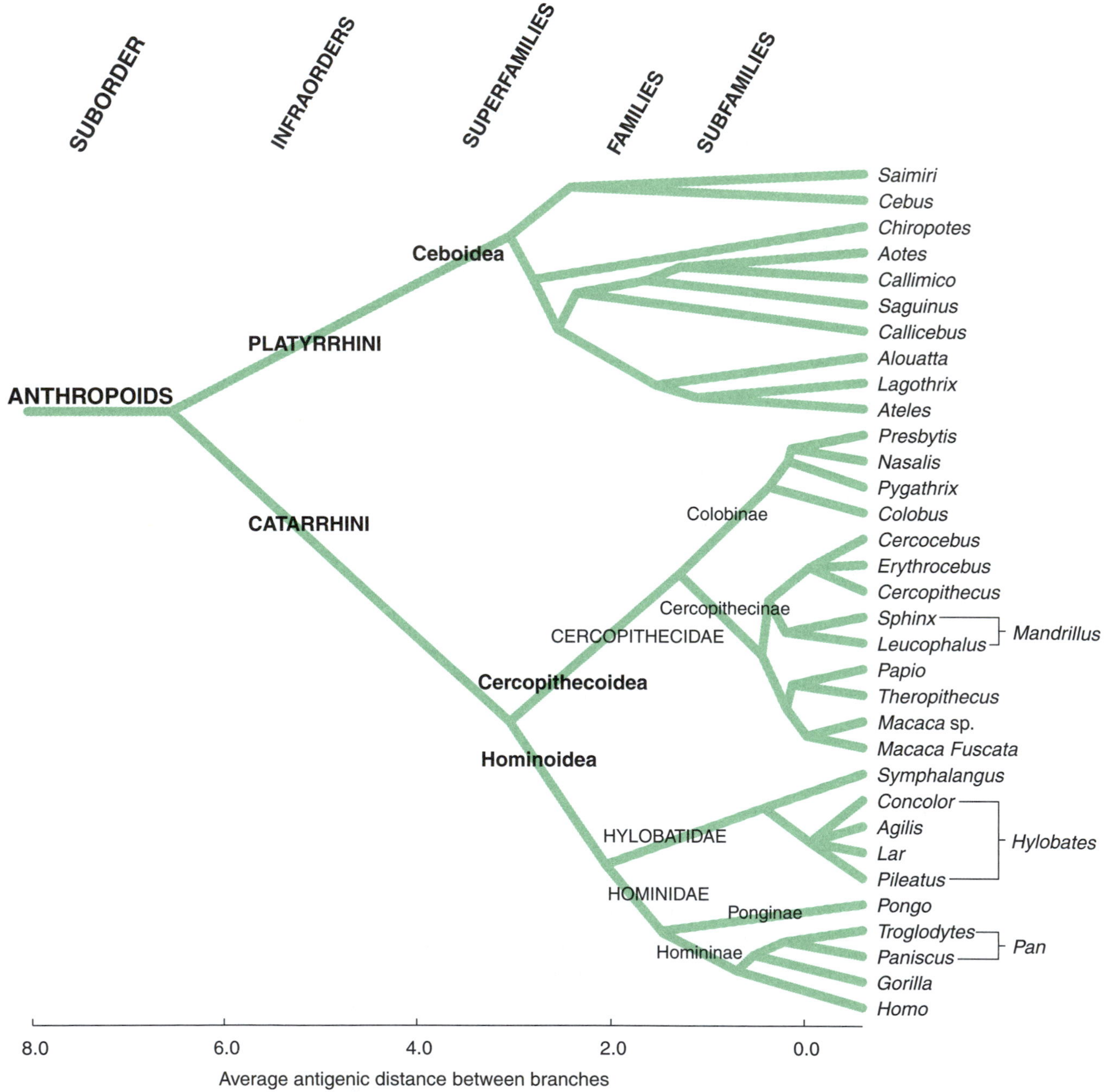

FIGURE 12–2 A phylogenetic tree for anthropoids based on immunological distances derived from blood serum antigens. (Adapted from Dene, H.T., M. Goodman, and W. Prychodko, 1976. Immunodiffusion evidence on the phylogeny of primates. In *Molecular Anthropology*, M. Goodman and R. E. Tashian (eds.). Plenum Press, New York, pp. 171–195.)

we can separate the first three genera from the orangutans and place them in a distinct group.

Other immunological approaches such as microcomplement fixation, involve measuring antibodies produced against specific proteins found in blood serum (albumin and transferrins) or enzymes such as lysozyme. Antigenic distance, detected by measuring the amount of antigen–antibody reactions, provides data generally supporting phylogenies obtained by other methods.

Amino Acid Sequences: Hemoglobin Phylogeny

A more direct approach to molecular phylogeny involves sequencing the amino acids in proteins by biochemical methods and comparing such sequences for the same protein in different species. Hemoglobin was among the first proteins to yield its amino acid sequence and remains one of the most investigated of all proteins (see Chapter 10).

The basic unit of **hemoglobin** consists of an iron-containing porphyrin (heme) that reversibly can bind oxygen attached to a globin polypeptide chain usually no less than 140 amino acids long. The demonstration that hemoglobin-like molecules appear in a wide range of organisms, from invertebrates to vertebrates, and even in plants, fungi and bacteria (Hardison, 1996) indicates their origin far back in the history of life. In vertebrates, hemoglobins are usually the primary protein of red blood cells, making them relatively easy to isolate and purify in large amounts.

As explained in Chapter 10, red blood cell hemoglobin of normal human adults is a four-chain molecule or tetramer, consisting of two pairs of polypeptide chains, one pair bearing the α sequence and the other pair mostly bearing the β sequence ($\alpha_2\beta_2$). Some adult hemoglobin uses δ chains instead of βs ($\alpha_2\delta_2$). A common form of embryonic hemoglobin has two α and two γ chains ($\alpha_2\gamma_2$). Other types of hemoglobin chains are known (for example, ε), and hemoglobin-like molecules such as myoglobin appear in other tissues. The findings that a species can possess different kinds of globin molecules and that each molecule can differ among different species, point to two major kinds of globin evolution:

- Different kinds of globin chains arose during evolution, producing the variety carried by a particular vertebrate.
- Each particular globin chain followed its own evolutionary path, which led to changes in its amino acid sequence in different species (α chains are different in different species, as are β chains, and so on).

As an example of the first kind, the amino acid sequences for five different human globin chains are shown in **Table 12-1.** All are variations on a single homologous theme. The chains are all about the same length; they possess identical amino acids at a significant number of positions; and all functioning human globin genes share a similar exon/intron structure: three exons separated by two introns (see Fig. 9-14). The three-dimensional structures that amino acid sequences dictate are similar, leading to their similar physiological functions. In addition, the genes for the β, γ and δ are closely linked on chromosome 11, although the gene for the α chain lies on a different chromosome, number 16.

Gene Duplication and Divergence

Observations of multiple hemoglobin chains, especially their sequence similarities, strongly suggest that, rather than arising from different genes that accidentally converged in sequence and function, all the different globin chains arose as **gene duplications** of an original globin-type gene. The temporal order in which the duplications occurred can be discerned from amino acid differences, on the basis that the greater the amino acid differences between any two chains, the further back in time their common ancestor. We can order a number of events from the following three observations:

1. The myoglobin chain differs most from all others because it has distinctive amino acids at more than 100 sites.
2. The α chain of hemoglobin differs from β at 77 sites.
3. The β chain differs from δ at 39 sites but differs from δ at only 10 sites.

The gene for myoglobin must have formed from an early duplication, which was followed by a later duplication that separated α and β genes. Because they differ least, the separation between β and γ chains derives from a fairly recent duplication.

Figure 12-3 portrays the genetic phylogeny of the five globins in terms of the numbers of nucleotides necessary to account for the amino acid differences, along with the chronological periods in which we presume each duplication occurred. Duplication events led to the early coexistence of myoglobin with an α-like chain, the former assuming (or maintaining) an intracellular function, the latter assuming a circulatory function.

When a duplication of the α-like gene further evolved into a β-like gene, the advantage of having two pairs of different chains in a tetramer hemoglobin molecule must have been sufficiently great to preserve tetramer organization in the circulating blood of most vertebrates. After the β-like gene formed, a translocation separated it from α and transferred it to a different chromosome. Duplications occurred in the β-like gene, eventually yielding β, δ and γ genes.

It is now clear that such gene duplications are not unusual and can easily arise from an **unequal crossing over**

TABLE 12-1 Amino acid sequences for human myoglobin and four human hemoglobin chains (α, β, γ and δ)

	1		2	3	4	5	6	7	8	9	10	11	12	13	14	15	16	17	18	19
Myoglobin	G	—	L	S	D	G	E	W	Q	L	V	L	N	V	W	G	K	V	E	A
α chain	V	—	L	S	P	A	D	K	T	N	V	K	A	A	W	G	K	V	G	A
β chain	V	H	L	T	P	E	E	K	S	A	V	T	A	L	W	G	K	V	—	—
γ chain	G	H	F	T	E	E	D	K	A	T	I	T	S	L	W	G	K	V	—	—
δ chain	V	H	L	T	P	E	E	K	T	A	V	N	A	L	W	G	K	V	—	—
	20	21	22	23	24	25	26	27	28	29	30	31	32	33	34	35	36	37	38	39
Myoglobin	D	I	P	G	H	G	Q	E	V	L	I	R	L	F	K	G	H	P	E	T
α chain	H	A	G	E	Y	G	A	E	A	L	E	R	M	F	L	S	F	P	T	T
β chain	N	V	D	E	V	G	G	E	A	L	G	R	L	L	V	V	Y	P	W	T
γ chain	N	V	E	D	A	G	G	E	T	L	G	R	L	L	V	V	Y	P	W	T
δ chain	N	V	D	A	V	G	G	R	A	L	G	R	L	L	V	V	Y	P	W	T
	40	41	42	43	44	45	46	47	48	49	50	51	52	53	54	55	56	57	58	
Myoglobin	L	E	K	F	D	K	F	K	H	L	K	S	E	D	E	M	K	A	S	
α chain	K	T	Y	F	P	H	F	—	D	L	S	H	—	—	—	—	—	G	S	
β chain	Q	R	F	F	E	S	F	G	D	L	S	T	P	D	A	V	M	G	N	
γ chain	Q	R	F	F	D	S	F	G	N	L	S	S	A	S	A	I	M	G	N	
δ chain	Q	R	F	F	E	S	F	G	D	L	S	S	P	D	A	V	M	G	N	
	59	60	61	62	63	64	65	66	67	68	69	70	71	72	73	74	75	76	77	
Myoglobin	E	D	L	K	K	H	G	A	T	V	L	T	A	L	G	G	I	L	K	
α chain	A	Q	V	K	G	H	G	K	K	V	A	D	A	L	T	N	A	V	A	
β chain	P	K	V	K	A	H	G	K	K	V	L	G	A	F	S	D	G	L	A	
γ chain	P	K	V	K	A	H	G	K	K	V	L	T	S	L	G	D	A	I	K	
δ chain	P	K	V	K	A	H	G	K	K	V	L	G	A	F	S	D	G	L	A	
	78	79	80	81	82	83	84	85	86	87	88	89	90	91	92	93	94	95	96	
Myoglobin	K	K	G	H	H	E	A	E	I	K	P	L	A	Q	S	H	A	T	K	
α chain	H	V	D	D	M	P	N	A	L	S	A	L	S	D	L	H	A	H	K	
β chain	H	L	D	N	L	K	G	T	F	A	T	L	S	E	L	H	C	D	K	
γ chain	H	L	D	D	L	K	G	T	F	A	Q	L	S	E	L	H	C	D	K	
δ chain	H	L	D	N	L	K	G	T	F	S	Q	L	S	E	L	H	C	D	K	
	97	98	99	100		102		104		106		108		110		112		114		
Myoglobin	H	K	I	P	V	K	Y	L	E	F	I	S	E	C	I	I	Q	V	L	
α chain	L	R	V	D	P	V	N	F	K	L	L	S	H	C	L	L	V	T	L	
β chain	L	H	V	D	P	E	N	F	R	L	L	G	N	V	L	V	C	V	L	
γ chain	L	H	V	D	P	E	N	F	K	L	L	G	N	V	L	V	T	V	L	
δ chain	L	H	V	D	P	E	N	F	R	L	L	G	N	V	L	V	C	V	L	
	116		118		120		122		124		126		128		130		132		134	
Myoglobin	Q	S	K	H	P	G	D	F	G	A	D	A	Q	G	A	M	N	K	A	
α chain	A	A	H	L	P	A	E	F	T	P	A	V	H	A	S	L	D	K	F	
β chain	A	H	H	F	G	K	E	F	T	P	P	V	Q	A	A	Y	Q	K	V	
γ chain	A	I	H	F	G	K	E	F	T	P	E	V	Q	A	S	W	Q	K	M	
δ chain	A	R	N	F	G	K	E	F	T	P	Q	M	Q	A	A	Y	Q	K	V	
		136		138		140		142		144		146		148		150		152		
Myoglobin	L	E	L	F	R	K	D	M	A	S	N	Y	K	E	L	G	F	Q	G	
α chain	L	A	S	V	S	T	V	L	T	S	K	Y	R							
β chain	V	A	G	V	A	N	A	L	A	H	K	Y	H							
γ chain	V	T	G	V	A	S	A	L	S	S	R	Y	H							
δ chain	V	A	G	V	A	N	A	L	A	H	K	Y	H							

Note: The amino acids are abbreviated by single capital letters as shown in Figure 6-2 and Table 12-3. The chains are aligned with the 153 amino acids in myoglobin, and the boxes indicate identical amino acids found in all chains at the designated numbered positions.

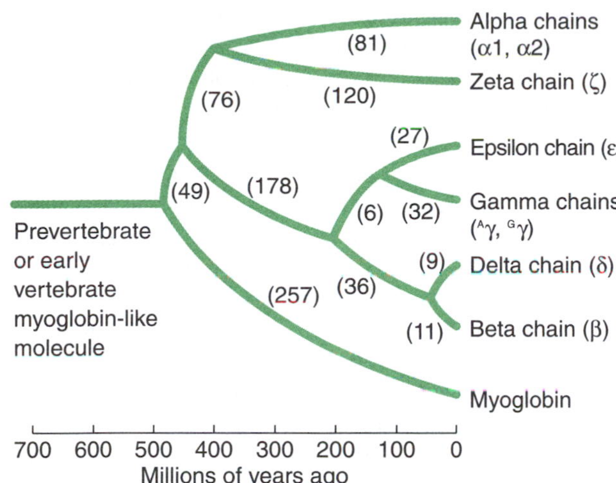

FIGURE 12-3 Phylogenetic relationships between globin-type proteins found in humans showing the estimated times when they diverged from each other. The estimated number of nucleotide replacements necessary to cause the observed amino acid changes in each branch of the lineage is given in parentheses. (From *Genetics Third Edition* by Monroe W. Strickberger. © 1985 by Monroe W. Strickberger. Reprinted by permission of Prentice Hall, Inc., Upper Saddle River, NJ.)

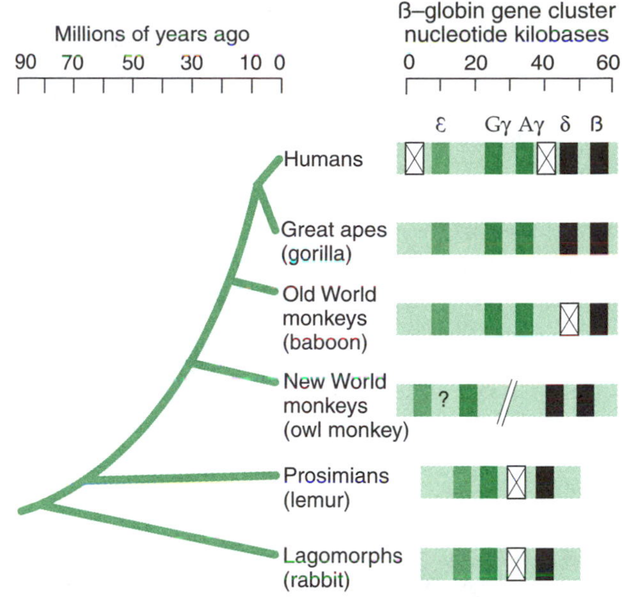

FIGURE 12-5 The clustered organization of the β-globin-type genes in five different primates and in the rabbit (*right*), along with a proposed evolutionary tree (*left*). The genes in each of these clusters are linked together on the same chromosome; for example, the human β-globin cluster is localized to a span of 60,000 nucleotides on chromosome 11. Each gene, denoted by a small rectangle, is transcribed from left to right, with the genes responsible for embryonic and fetal development on the left (*lighter shading*) and the genes that produce adult β-globins on the right (*darker shading*). Genes marked with crosses indicate pseudogenes, duplicates that have become nonfunctional for various reasons. Nadeau and Sankoff (1997) estimate that deleterious mutations in gene duplicates responsible for causing pseudogenes were as likely to occur among early vertebrates as were mutations allowing duplicates to diverge functionally. (From Strickberger, M. W., 1985. *Genetics*, 3rd ed. Macmillan, New York; adapted from Jeffreys, A. J., S. Harris, P. A. Barrie, D. Wood, A. Blanchetot, and S. M. Adams, 1983. Evolution of gene families: The globin genes. In *Evolution from Molecules to Men*, D. S. Bendall (ed.). Cambridge University Press, Cambridge, England, pp. 175–195.)

event that produces a recombinant product possessing increased chromosome material (**Fig. 12-4**). Unequal crossing over may cause gene fusion by eliminating chromosomal material between two formerly separate genes. The union in fungi between the A and B components of the tryptophan synthetase enzyme — which normally are separate in bacteria — appears to have occurred through such fusion events. The human α gene, for example, is known to have two side-by-side (tandem) duplicates, α1 and α2, and the β **gene cluster** (also called a **gene family** or **multigene family**) consists of a sequence of seven such genes (**Fig. 12-5**).

In some cases, such as globin chains, serine proteases (chymotrypsin, trypsin), and lactate dehydrogenases, duplicated genes have preserved similar although not identical functions. In other cases, duplications retained as **pseudogenes** lost function completely because of mutations that prevent transcription or translation. Most important from an evolutionary view are duplicated genes that have evolved in completely different

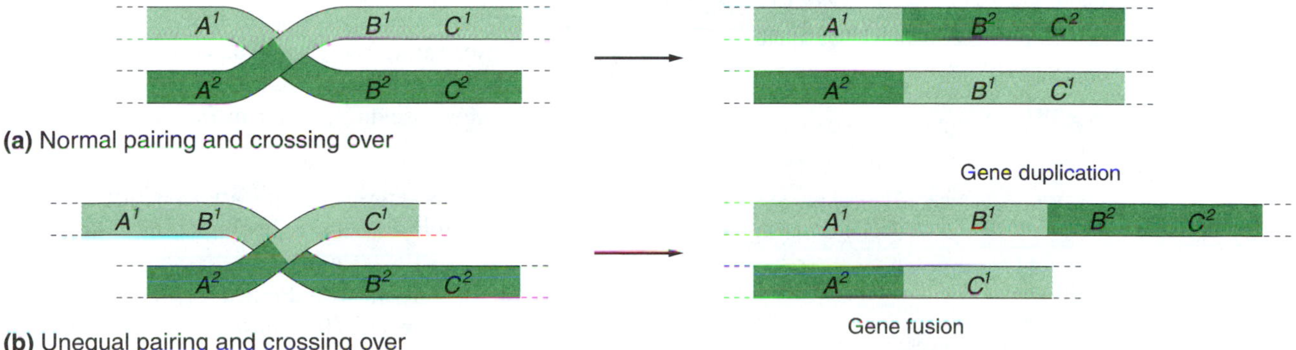

FIGURE 12-4 The results of equal and unequal crossing over for three gene segments on a chromosome. (a) When pairing between homologous sections on two chromosomes is equal, the crossover products have the same amounts of chromosomal material (e.g., they both have an *A*, *B*, and *C* gene segment). (b) When pairing between the two chromosomes is unequal, one of the crossover products carries a gene duplication (the *B* gene segment in this illustration), and the other product shows a fusion between gene segments (*A–C*) that were formerly separated by the intervening *B* segment.

functional directions, although they share enough common amino acid sequences to indicate their relationship.

As examples, the nerve growth factor protein that enhances the outgrowth of neural cells from sympathetic and sensory ganglia has amino acid sequences similar to insulin, and may share some functional similarities with insulin. The α-lactalbumin protein, which is part of an enzyme used in the synthesis of lactose in mammalian milk, has an amino acid sequence remarkably similar to that of the lysozyme enzyme found in tears, an enzyme that degrades the mucopolysaccharides of bacterial cell walls. This similarity is further reflected in evidence that both proteins are the products of tissues (mammary, tear duct) that were at one time probably sebaceous glands, and both use sugar molecules as their substrates.

The use of an enzyme for more than one adaptive purpose is common for transparent *crystallin proteins* in eye structures. Although at least ten such lens crystallins are found among organisms ranging from vertebrates to mollusks, each appears related to a functional enzyme such as alcohol dehydrogenase, α-enolase, and glutathione S-transferase (**Table 12-2**). In fact, ε-crystallin and lactate dehydrogenase enzyme in crocodiles and some birds derive from a single identical gene product (Piatagorsky and Wistow, 1989), indicating that a molecule or structure may be preadapted for a new function (lens crystallin) yet retain its former enzymatic catalytic function. Such cases, in which the same gene product performs two entirely different functions, have been termed *gene sharing* by Piatagorsky and Wistow.

How long it takes for duplications to diverge in function depends, of course, on how many amino acid substitutions are necessary, but also on changes occurring in other genes and in the adaptation/evolution of the organism itself. A most striking example of rapidity is a single amino acid mutation that converts lactate dehydrogenase (LDH) to malate dehydrogenase (MDH). LDH is used in glycolysis for the lactate-pyruvate reaction (see Fig. 8-12), MDH in the Krebs cycle for the malate–oxaloacetate reaction (see Fig. 8-22). This change can be engineered by simply substituting arginine for glutamine at the 102nd polypeptide position.

TABLE 12-2 A sample of enzymes used as lens crystallins in various taxa

Source	Crystallin Type	Enzyme
Cephalopods	S	glutathione S-transferases
Birds, crocodiles	ε	lactate dehydrogenase B
Cavies, camels	ξ	alcohol dehydrogenases
Rabbits, hares	λ	coenzyme A derivative dehydrogenases
Many species	τ	α-enolase
Mammalian cornea	BCP54	aldehyde dehydrogenase III

Abridged from Wistow, G. J., 1993. Identification of lens crystallin: A model system for gene recruitment. *Methods in Enzymol.*, **224,** 563–575.

Other adaptive conversions undoubtedly involve more amino acid changes, the histories of some of which are being inferred through "paleomolecular biochemistry," which traces past amino acid substitutions by noting the effects of purposely changing amino acids at particular sites using a technique called *site-directed mutagenesis* (Golding and Dean, 1998).

Duplications exist within genes themselves, as shown by three homologous amino acid regions (polypeptide domains) within the γ heavy chain of the immunoglobulin G antibody. In the human haptoglobin α-2 blood serum protein, a segment of 59 amino acids (positions 13 to 71) almost exactly repeat an adjacent segment (positions 72 to 130). Repeated amino acid sequences are seen also within the *ferredoxin* protein (which is used as an electron carrier in various biochemical processes; see Fig. 8-14), in the glutamate dehydrogenase enzyme, and in various other proteins.

The small size of ferredoxin, its limited sampling of amino acids, and the simple positioning of its iron atoms make it one of the most basic electron transport agents in cells. These features, along with its apparent duplicated structure, suggest that ferredoxin may owe its origin to an even earlier peptide formed on one of the then-primitive templates. (Ferredoxin's base amino acid sequence before duplication is only about 28 amino acids.)

Despite these examples, ancestral homologies based on similarities between amino acid sequences are not always obvious, a finding that prompted development of more sensitive methods. Because comparisons between short sequences can be misleading (similar sequences only three or four amino acids long may often arise in nonrelated proteins) the most sensitive methods depend on comparing fairly long amino acid sequences. Methods that align molecular sequences overcome difficulties caused by gaps or insertions.[1]

Phylogenies involving a large number of organisms come from the second aspect of protein evolution: comparisons between amino acid sequences in different species. To clarify such comparisons, Fitch and Margoliash (1967) suggested that we use different terms for two major kinds of homologous genes:

- **paralogous genes**, for duplications within a species (for example, α, β and γ hemoglobins in humans), and
- **orthologous genes**, for duplication of the same gene in different species (for example, α hemoglobins in horses and humans).

[1] See Nei (1987), McClure et al. (1994) and Goldman (1998) for the methods developed.

An issue raised by the existence of gene duplications, especially whole genome duplications, is whether such events facilitate the origin of novel organs, organisms, and/or radiation of clades. While the origins of land plants and of teleosts do correlate with genome duplication, the origins of birds and mammals do not, so gene duplication is not a requirement for the origin of new clades, although duplication of paralogous coincident with radiation is more likely to be established as a consequence of the selection associated with the radiation. The remainder of the chapter is devoted to exploring the fact that comparisons between orthologous genes are now an important way to determine the phylogenetic relationships of the genes themselves, and of the species that carry them.

Many Trees, One Phylogeny

The basic technique for detecting orthologous amino acid sequences provides precise phylogenetic positioning because it begins with an alignment of amino acid sequences from the same known protein in different species. Once aligned, the minimum number of mutations necessary to transform one amino acid in one sequence to that of a different amino acid in the same position in the other sequence can provide an estimate of mutational distances, such as those shown in **Table 12-3.**

Choosing among the number of relationships generated by such data involves invoking the parsimony principle or deriving maximum likelihood estimates (see Chapter 11 and Fig. 11-12). With **parsimony,** the more likely hypothesis generally involves fewer assumptions than a less likely hypothesis. "Unparsimonious" explanations are not necessarily false, but *in the absence of further information,* choices are made on the basis of parsimony. **Maximum likelihood estimates** indicate the most likely distribution for a data set.

By using increased numbers of species as sources of mutational data, we can establish a **phylogenetic tree** by various numerical and algorithmic methods. Important as these methods are, they remain beyond the scope of this book, involving sophisticated mathematical techniques in choos-

TABLE 12-3 Matrix of minimum number of nucleotide substitutions necessary to convert a codon for one amino acid (rows) into a codon for another amino acid (columns)

	Abbreviations		Minimum Number of Nucleotide Substitutions																			
Amino Acid	3-letter	1-letter	A	C	D	E	F	G	H	I	K	L	M	N	P	Q	R	S	T	V	W	Y
Alanine	Ala	A	0	2	1	1	2	1	2	2	2	2	2	2	1	2	2	1	1	1	2	2
Cysteine	Cys	C	2	0	2	3	1	1	2	2	3	2	3	2	2	3	1	1	2	2	1	1
Aspartic acid	Asp	D	1	2	0	1	2	1	1	2	2	2	3	1	2	2	2	2	2	1	3	1
Glutamic acid	Glu	E	1	3	1	0	3	1	2	2	1	2	2	2	2	1	2	2	2	1	2	2
Phenylalanine	Phe	F	2	1	2	3	0	2	2	1	3	1	2	2	2	3	2	1	2	1	2	1
Glycine	Gly	G	1	1	1	1	2	0	2	2	2	2	2	2	2	2	1	1	2	1	1	2
Histidine	His	H	2	2	1	2	2	2	0	2	2	1	3	1	1	1	1	2	2	2	3	1
Isoleucine	Ile	I	2	2	2	2	1	2	2	0	1	1	1	1	2	2	1	1	1	1	3	2
Lysine	Lys	K	2	3	2	1	3	2	2	1	0	2	1	1	2	1	1	2	1	2	2	2
Leucine	Leu	L	2	2	2	2	1	2	1	1	2	0	1	2	1	1	1	1	2	1	1	2
Methionine	Met	M	2	3	3	2	2	2	3	1	1	1	0	2	2	2	1	2	1	1	2	3
Asparagine	Asn	N	2	2	1	2	2	2	1	1	1	2	2	0	2	2	2	1	1	2	3	1
Proline	Pro	P	1	2	2	2	2	2	1	2	2	1	2	2	0	1	1	1	1	2	2	2
Glutamine	Gln	Q	2	3	2	1	3	2	1	2	1	1	2	2	1	0	1	2	2	2	2	2
Arginine	Arg	R	2	1	2	2	2	1	1	1	1	1	1	2	1	1	0	1	1	2	1	2
Serine	Ser	S	1	1	2	2	1	1	2	1	2	1	2	1	1	2	1	0	1	2	1	1
Threonine	Thr	T	1	2	2	2	2	2	2	1	1	2	1	1	1	2	1	1	0	2	2	2
Valine	Val	V	1	2	1	1	1	1	2	1	2	1	1	2	2	2	2	2	2	0	2	2
Tryptophan	Trp	W	2	1	3	2	2	1	3	3	2	1	2	3	2	2	1	1	2	2	0	2
Tyrosine	Tyr	Y	2	1	1	2	1	2	1	2	2	2	3	1	2	2	2	1	2	2	2	0

Source: Adapted from Fitch W. M., and E. Margoliash, 1967. Construction of phylogenetic trees. *Science,* **155,** 279–284.

ing among many phylogenetic possibilities that increase many-fold with increasing number of taxa.[2]

Many possible trees can be generated but there is only one true phylogeny. The taxa sampled therefore may not reflect the true phylogeny. They are hypotheses of phylogeny, hypotheses that must be tested using accepted and generalizable criteria. Statisticians have searched for methods to gauge how much confidence to place on proposed phylogenies. At present, a common method is bootstrapping, which gives the proportion (percentage) of acceptable trees in which a branch point ("node" or "clade") appears when data are repeatedly sampled and replaced.

For example, ten amino acid differences used in constructing species relationships can be resampled 100 times so that some differences are omitted and some appear more than once. Each resampling generates a tree in which a particular branch may or may not appear. The bootstrap value is the frequency in which the same branch appears. Using this method, the phylogenetic tree of cichlid fishes (shown later in Figure 12-19) indicates that, in 99 percent of 2,000 repeated samples of the data, the six Malawi species derive from a common branch.

Bootstrapping values are not infallible predictors but are limited by how well the data represent the phylogeny. In a study by Maley and Marshall (1998), changing the species used to represent arthropods from brine shrimp to spider changed the arthropod relationship to mollusks and echinoderms. Aside from other complications likely to confuse any tree — such as homoplasy and horizontal gene transfer (see Chapter 11) — high bootstrapping values increase confidence but do not guarantee a true phylogeny.

Nevertheless, parsimony, maximum likelihood estimates, and other techniques to offer reliable choices when taxonomic comparisons between related groups make use of many differences, such as **nucleotide** or amino acid changes comparably aligned in lengthy sequences (Russo et al., 1996). One such phylogeny for the α-hemoglobin sequence in 29 different vertebrates, based on analyzing a total of 630 nucleotide replacements, is shown in **Figure 12-6.**

Because hemoglobin phylogenies exclude invertebrates, plants and fungi, molecular taxonomists also compare more ubiquitous proteins that are present in widely varied groups of organisms. One such protein, cytochrome *c*, contains heme but generally functions in aerobic organisms as part of the respiratory electron transport chain (see Fig. 8-21). A phylogenetic tree based on 53 different amino acid sequences of cytochrome *c* is shown in **Figure 12-7.**

We can hardly expect phylogenies derived from studies of one or two proteins always to reflect true evolutionary history or to discriminate accurately even between different species. The lack of differences between human and chimpanzee α-hemoglobins is one example of such lack of discrimination. Nevertheless, it is remarkable how closely the information obtained from a limited sample of proteins out of many thousands approximates the phylogenetic relationships obtained from extensive studies in comparative anatomy and paleontology.

Convergent Molecular Evolution

The persistence of similarity in a protein's amino acid sequence through many branches of a long lineage is related to persistence of selection for the same function (see Figure 12-1).

Interestingly, such selection may also be strong enough to produce similar amino acid sequences in different branches for a protein that may have originally served a different function: **convergent molecular evolution.** For example, artiodactyl ruminants (cows) and langur monkeys, alone among mammals in fermenting vegetable matter in a foregut, are the only mammals that bear the same five amino acid changes in the lysozyme enzyme that breaks down cell walls of fermenting bacteria. Some of these lysozyme changes occur even in a foregut-fermenting bird, the hoatzin *Opisthocomus hoazin.* In each case, these common protein modifications must have arisen independently. Among other molecular **convergences** is a mammalian protein that binds to blood vessel fibrins and increases the risk of coronary dysfunction and cerebral stroke. Although most mammals lack this protein, apolipoprotein A1, it is found in humans and hedgehogs, having arisen independently from distinctively separate duplications of a plasminogen gene.[3]

Perhaps an even more distant molecular convergence — that between fungi and animals (and thus crossing kingdom boundaries) — are the highly similar cell wall proteins of certain dipteran insects (*Chironomus*) and filamentous fungi (*Trichoderma*). Although Rey and coworkers (1998) ascribe this similarity to common usage of a 39-base–long repeating nucleotide sequence, independent emergence of the same protein in both groups indicates selection for convergent molecular function.[4]

For a more faithful reflection of evolutionary history, it is preferable (indeed, it is essential) to *identify homologies and reduce the effect of convergences* by combining studies from many different proteins, allowing comparisons among many

[2] See Felsenstein (1988, 2004), Nei (1987) and Swofford et al. (1996) for numerical approaches. For example, although about 100 different phylogenies are possible for five taxa, there are more than 30 million possible phylogenies for 10 taxa and more than 200 billion billion (8.2×10^{21}) for 20 taxa. On reaching 50 taxa, more phylogenetic possibilities exist than estimated atoms in the universe (10^{74}). Considering more than small sections of a tree at a time can become mathematically cumbersome and impractical.

[3] See Stewart and Wilson (1987) and Kornegay et al. (1994) for the fermentation studies, and Lawn et al. (1997) for the plasminogen gene.

[4] M. S. Y. Lee (1999) reviews other cases of molecular convergence.

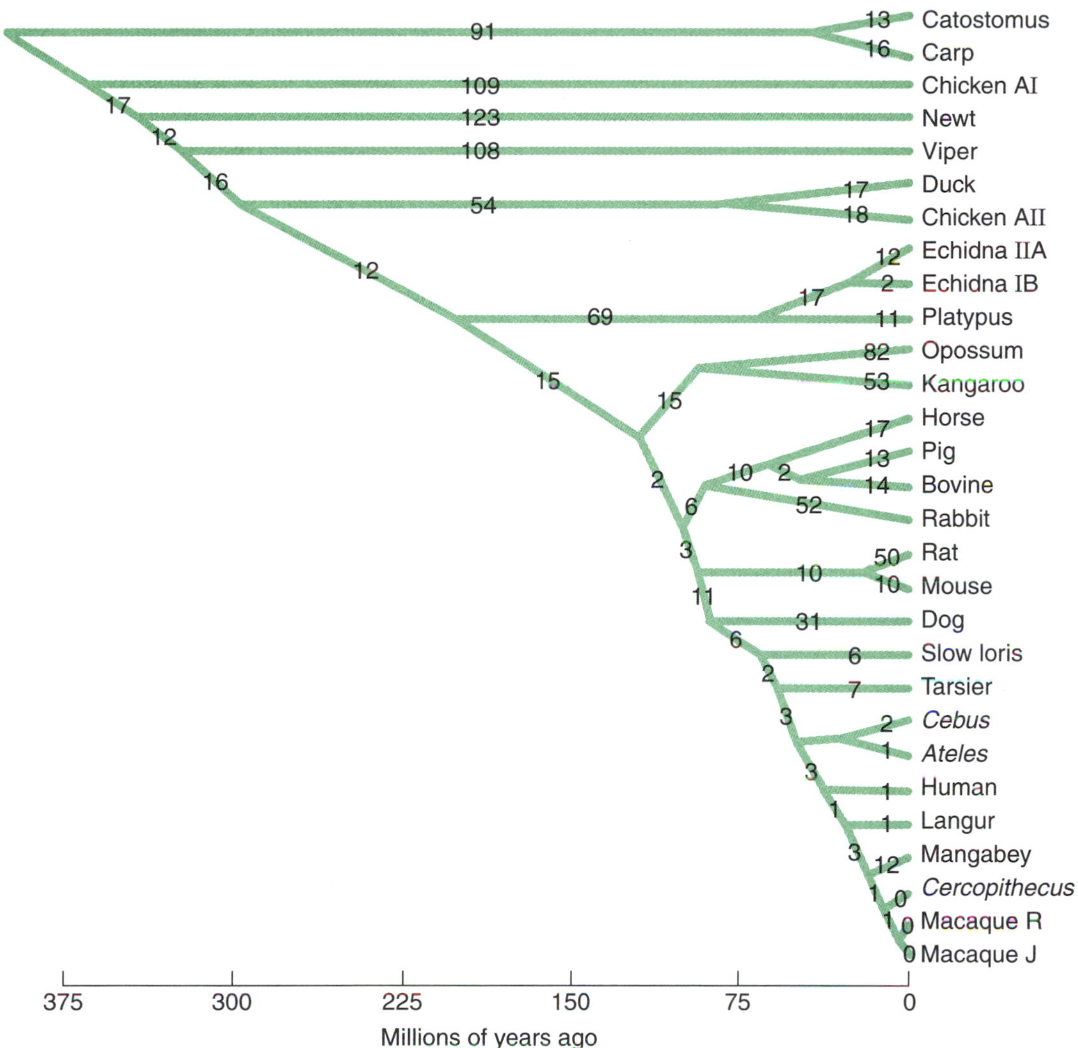

FIGURE 12-6 A phylogeny of α-hemoglobin chains in a variety of vertebrate species, determined by calculating the number of nucleotide replacements that account for the number of observed amino acid substitutions presumed to have occurred during each evolutionary interval. The horizontal scale, given in millions of years, is based on paleontological estimates for the age of the common ancestor of each branch. (Adapted from Goodman, M., 1976. Toward a genealogical description of the primates. In *Molecular Anthropology*, M. Goodman and R. E. Tashian (eds.). Plenum, New York, pp. 321–353.)

different amino acid positions. Although such procedures present difficulties, some attempts have been made. **Figure 12-8** diagrams the results of one study. Another example, involving proteins and nucleic acids, is discussed later.

DNA and Its Repetitive Sequences

The first self-replicating organisms are thought to have had no more than a small amount of genetic material, enough to maintain those relatively few functions necessary for their preservation in a mostly noncompetitive environment.

As time passed and these organisms faced more challenges, increased amounts of genetic material provided greater selective advantages by increasing the number of functions and their regulation. Competition among organisms for successful adaptation to their environments became dependent on the numbers and kinds of genes they had, and evolutionary changes have proceeded on both levels. One technique for evaluating some consequences of these evolutionary factors is to measure the amounts of genetic material in different organisms (**Box 12-1**). A further approach toward analyzing DNA in greater detail has been to determine differences among the kinds of DNA present in a single organism.

These studies began with Britten's (1986) discovery that DNA sheared to specific sizes and separated into single strands would reassociate into double-strand molecules at rates based on the nature of their nucleotide sequences. For example, a single strand of DNA bearing a sequence

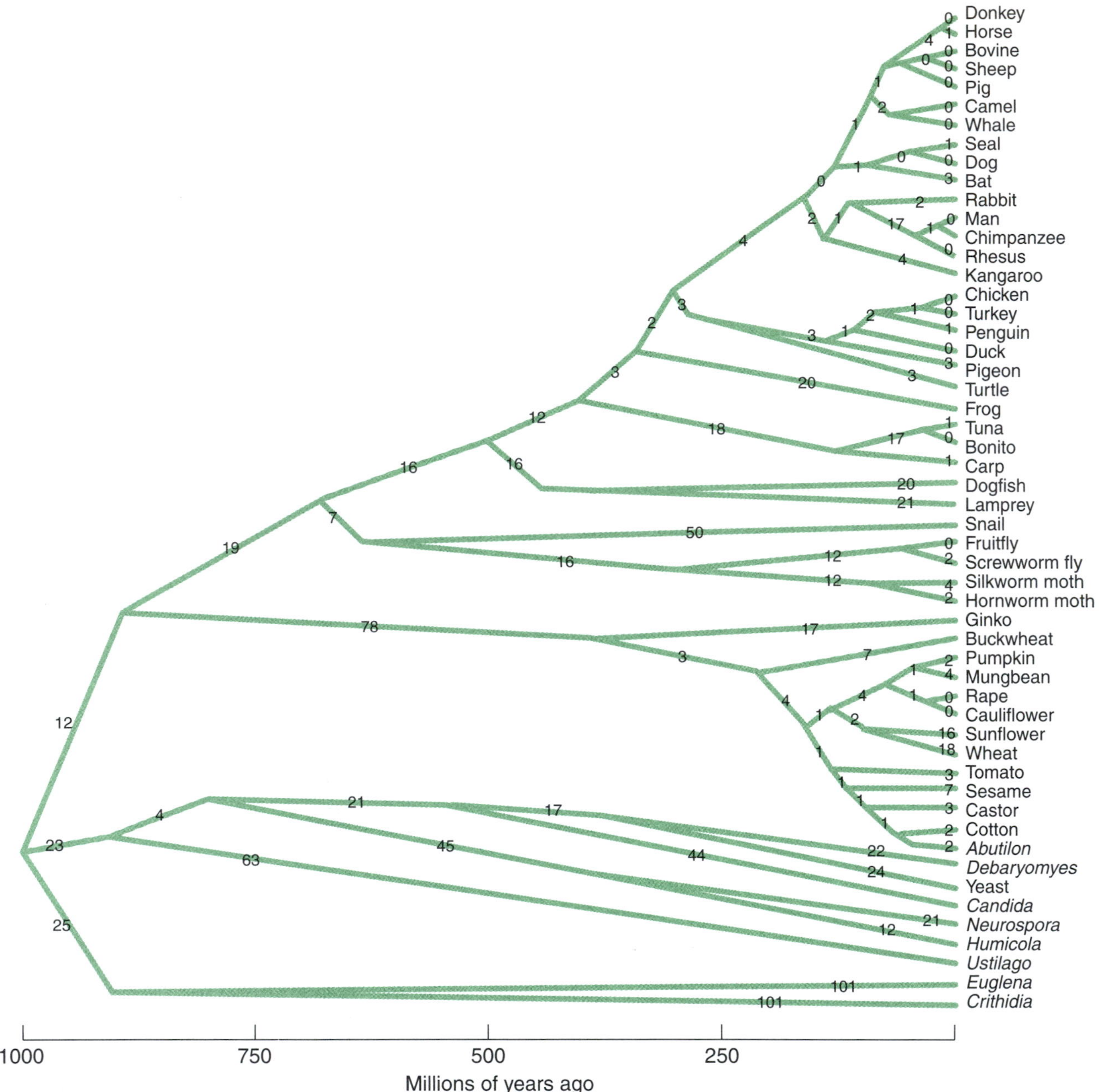

FIGURE 12-7 A cytochrome *c* phylogeny showing relationships among plants, fungi and a variety of animals. The estimated ages of branching points appear on the horizontal scale, and the estimated number of nucleotide replacements necessary for the evolution of each branch appears within each interval. When the number of nucleotide replacements exceeds five, some undetectable back or parallel mutations have occurred, and the number is augmented proportionally. (Adapted from Goodman, M., 1976. Toward a genealogical description of the primates. In *Molecular Anthropology,* M. Goodman and R. E. Tashian (eds.). Plenum, New York, pp. 321–353.)

repeated many times over throughout a genome would find a complementary "mate" and form a double-strand molecule much more rapidly than would a strand bearing a single sequence. We quantify the reassociation rate by measuring optical changes that occur in the transition from single- to double-strand DNA. By these means, Britten and coworkers classified DNA as either repetitive or unique, referring to sequences that occurred frequently and those that occurred only as single copies.[5] As shown in **Table 12-4,** a significant fraction of DNA in tested eukaryotic organisms is repetitive, some sequences being present in 200 copies or less and others repeated more than one million times.

The lengths of these **repetitive DNA** sequences may

[5] See Britten and Kohne (1968), Britten and Davidson (1971) and Britten (1986, 1998) for repetitive sequences.

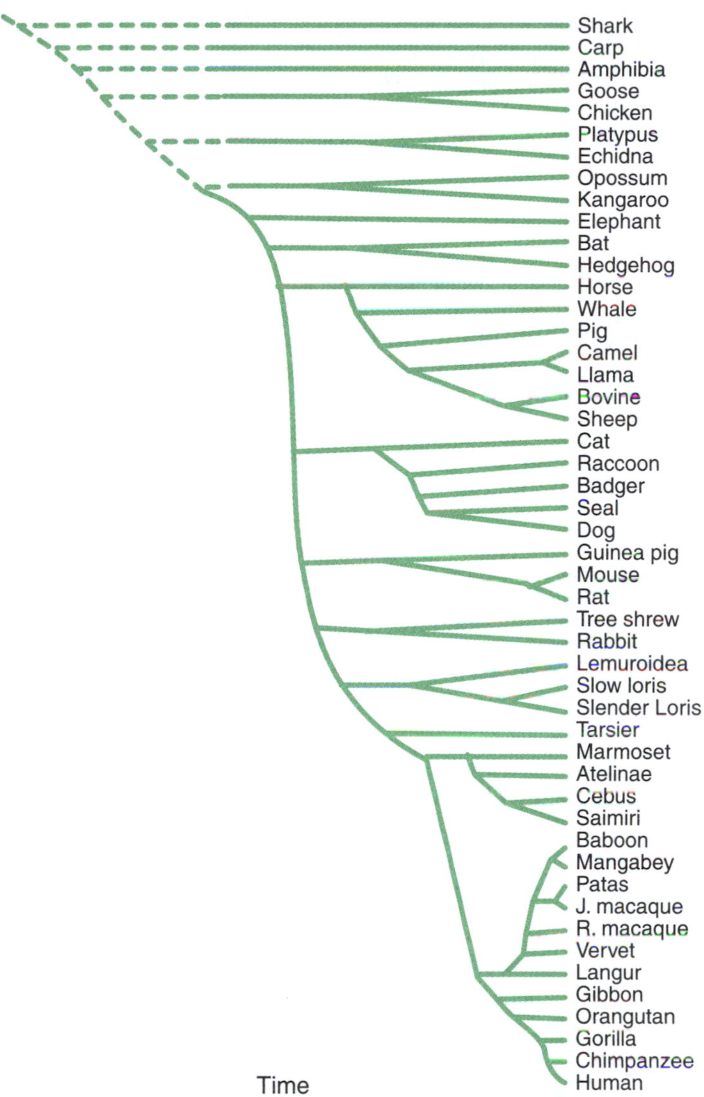

FIGURE 12-8 A phylogenetic tree for 49 vertebrate taxa derived from aligning the amino acid sequences in up to seven different polypeptide chains: α-hemoglobin, β-hemoglobin, myoglobin, lens α-crystallin A, fibrinopeptide A, fibrinopeptide B, and cytochrome *c*. (Adapted from Goodman, M., A. E. Romero-Herrera, H. Dene, J. Czelusniak, and R. E. Tashian, 1982. Amino acid sequence evidence on the phylogeny of primates and other eutherians. In *Macromolecular Sequences in Systematic and Evolutionary Biology,* M. Goodman (ed.). Plenum, New York, pp. 115–191.)

vary from less than 100 to more than 2,000 nucleotides. With the exception of *Drosophila,* the shorter-length repetitive sequences seem to be interspersed among sequences of nonrepetitive unique DNA, although the function of this arrangement is unknown. These interspersed sequences act as regulatory genes similar to prokaryotic operators and promoters (Chapter 10). Others suggested that the sequences are involved in packaging long, transcribed strands of heterogeneous nuclear RNA (HnRNA) so they can be further processed into messenger RNA. The fact that many eukaryotic mRNA molecules are shorter than their parental HnRNA molecules certainly indicates that some, or even many, DNA nucleotides in a gene are not translated into protein: some are at the gene termini, while *introns* (see Fig. 9-13 and Chapter 8) are distributed throughout eukaryotic protein-coding genes.

Among the long-length repetitive DNA sequences, some undoubtedly code for multigene families such as the many duplicate, tandemly arranged ribosomal RNA genes, transfer RNA genes, and histone genes.

One source accounting for sequence homogeneity is an error-correcting mechanism called *gene conversion* that is used in some multigene families such as those coding for chromosome structural histone proteins. This mechanism involves matching the nucleotides between DNA molecules, thereby enabling two or more gene sequences to correspond. Among agents that account for gene similarity in a multigene family, gene conversion occupies a prominent position and helps explain why such genes appear to evolve in unison (**concerted evolution**).

The function of the more repetitive but shorter nucleotide sequences in satellite DNA (recognized by its separation from the main portion of DNA after centrifugation) seems

BOX 12-1 Quantitative DNA Measurements

DNA Content

In a classic study, Mirsky and Ris (1951) measured the amounts of DNA in different organisms. Such measurements are now available for well over 1,000 species. The techniques include:

- Measuring the amount of stained DNA in cells (Feulgen staining).
- Isolating DNA chemically and deriving an average DNA amount from a known number of cells.
- Indirectly estimating DNA content from chromosomal size or nuclear volume.

Based on results from these techniques, **Figure 12-9** shows the observed range of cellular DNA content in numbers of nucleotides (1 picogram = 10^{-12} grams = 2.01×10^9 nucleotides).

Interestingly, these data indicate a relationship between the amount of DNA and general evolutionary status, going from small amounts in viruses and bacteria to relatively large amounts in more derived eukaryotes. However, this progression is not uniform. Some of the least-derived representatives of various groups (for example, ferns, lungfishes and salamanders) show relatively large amounts of DNA; the lungfish *Protopterus* has more than 40 times the amount of DNA in humans.

Comparative amounts of DNA are not much of a clue to complexity or to phylogenetic relationship, especially since the cells of different organisms (such as mammals and gymnosperms) have exactly the same DNA content, whereas obviously related species among amphibians and other groups vary widely in DNA content. Inconsistencies between genome size and phenotypic complexity in multicellular organisms have been called the *C-value paradox,* where C-value designates genome size in terms of numbers or weight of DNA base pairs (see Box 8-3).[a]

Nevertheless, some cell and molecular biologists have attempted to derive generalities from these data. Specialized species within some groups have less DNA than do more generalized species. For example, "more generalized" fish species such as salmon and cod have cellular DNA contents ranging from 1.2 to 4.4 picograms, whereas the DNA of "specialized forms" such as sea horses and anglerfish ranges from 0.45 to 0.80 picograms. At the same time, however, various species of algae, protozoa, ferns and amphibia are quite "specialized" yet contain relatively large amounts of DNA, although the basis on which we can differentiate generalized from specialized taxa is not clear.

DNA Amount and Chromosome Number

The relationship between **DNA amount and chromosome number** is another feature whose consistency varies: DNA content correlates well with chromosome number in plants and fishes but poorly in mammals. Even mammalian phenotypes that seem much alike, such as Chinese and Indian muntjac deer (*Muntiacus reevesi* and *M. muntjac*), may have similar amounts of DNA distributed in chromosomes that differ widely in number (see Fig. 10-5). In general, knowledge of DNA amount alone is insufficient to derive the fine textural patterns of evolutionary history. A more complete analysis is necessary.

Gene Numbers

Rather than measuring the nucleotide number in different organisms, another approach toward obtaining information on evolutionary status has been to **compare gene numbers.**

In recent years, the pace of DNA sequencing has increased dramatically due to advances in sequencing technology and bioinformatics. As of June 2006, the complete genomes of 476 different organisms have been sequenced, at least to the level of high-quality drafts. Of these, 46 are members of the Archaea, 319 are bacteria and 41 are eukaryotes (see Box 8-1). For some groups whole genome sequences are available for more than one species, seven species of *Drosophila,* for example, enabling comparisons at the speciation level of evolution. These numbers will expand dramatically in coming years; we are aware of 2,037 ongoing genome projects, of which 608 target eukaryotic genomes. The result will be good information about gene numbers in an increasing number of species from the major divisions of the Tree of Life. Surprisingly, the number of genes is lower than most evolutionary biologists expected. Table 12-6 shows actual or estimated gene numbers for selected species for which this information is known. Organization of the Tree of Life into five or six supergroups is discussed in Box 12-4.

Although there are still too few organisms with known gene numbers to draw binding conclusions, Bird (1995) proposes that gene number differences among groups (Table 12-6) are caused by differences in mechanisms that restrict inefficient gene production ("noise reduction"). Thus, he suggests that prokaryotes cannot contain more than a few thousand genes on average because bacterial transcriptional controls (see Fig. 10-13) become inefficient in the presence of greater numbers of genes, and would therefore cause some genes to be "turned on" even when they are not functionally necessary.

According to Bird, eukaryotes were able to circumvent such "noise" by a nuclear membrane that separates transcription from protein translation, allowing only translatable messenger RNA sequences to filter into the cytoplasm, and by tightly folding the DNA of functionally unnecessary genes into nontranscribable conformations, using nucleosomes and their histones. To these transcription-repressing mechanisms, Bird claims that vertebrates added DNA cytosine methylation, formerly used mostly to suppress genomic parasites such as transposons. Each major step in reducing wasteful transcription and translation of genes whose products were not required for immediate purposes resulted in organisms possessing more genes that could be more appropriately expressed. However, it is not known whether such a mechanism was a cause or accompaniment of improved systems of replication and mutation repair.

---

[a] See Cavalier-Smith (1985) and Greilhuber et al. (2005) for an excellent introduction and recent overview respectively, Gregory (2002) for genome size in birds, Bennett and Leitch (2005) for genome size in plants, and Bennett et al. (2003) for a comparison of genome size in *Caenorhabditis* (100 Mb), *Drosophila* (175 Mb), and *Arabidopsis* (157 Mb).

10^3 10^4 10^5 10^6 10^7 10^8 10^9 10^{10} 10^{11} 10^{12} 10^{13}

Dipnoans, primitive amphibia and primitive vascular plants

Mostly intermediate to highly advanced vertebrates and vascular plants

Complex multicellular forms, no vertebrates (except teleostei) or vascular plants

Single-cell or primitive multicellular forms

Viruses and bacteria

Psilopsida
Dipnoa (lungfishes)
Sphenopsida (horsetails)
Apoda (caecilians)
Urodela (salamanders)
Pteropsida (ferns)
Gymnospermae
Prototheria (monotreme mammals)
Lycopsida (club-mosses)
Eutheria (placental mammals)
Chondrostei (fishes)
Metatheria (marsupial mammals)
Chondrichthyes (sharks, rays, chimaeras)
Reptilia (reptiles)
Annelida (segmented worms)
Holostei (fishes)
Agnatha (lampreys and hagfishes)
Anura (frogs and toads)
Angiospermae
Aves (birds)
Echinodermata
Cephalochordata (amphioxus)
Mollusca
Bryophyta (mosses and liverworts)
Nematoda (roundworms)
Teleostei (fishes)
Coelenterata
Urochordata (tunicates)
Arthropoda
Protozoa
Porifera (sponges)
Algae (except blue-greens)
Fungi
Bacteria
Double-stranded DNA viruses
RNA and single-stranded DNA viruses

10^3 10^4 10^5 10^6 10^7 10^8 10^9 10^{10} 10^{11} 10^{12} 10^{13}

Numbers of Nucleotides

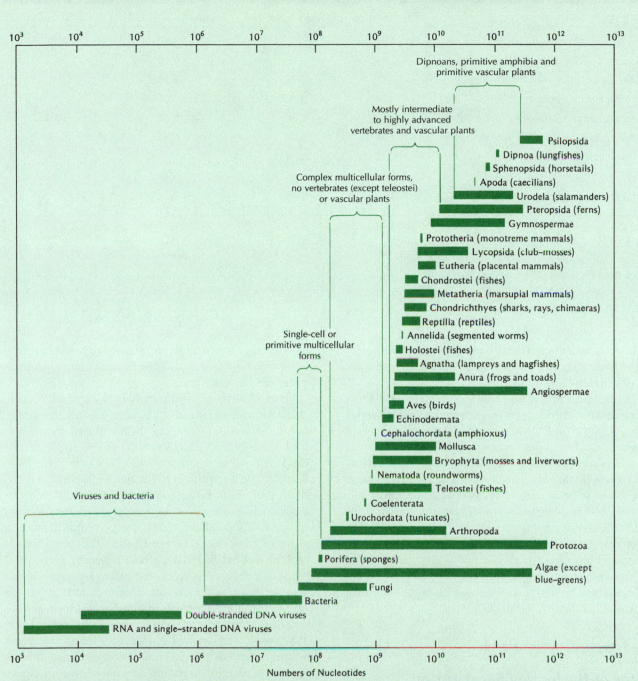

FIGURE 12-9 A comparison of the numbers of nucleotides in the genetic material of different types of organisms. Each bar in the illustration represents the range of nucleotide numbers found among the species sampled in a designated group. The nucleotide values given are largely derived from estimates of the weight of nucleic acid in the haploid complement of an organism according to the formula, 1 gram (g) of nucleic acid = 2.0×10^{21} nucleotides. Thus, the DNA in the diploid complement of human chromosomes (6.4×10^{-12} g) contains $(2.0 \times 10^{21}) \times (6.4 \times 10^{-12}) = 12.8 \times 10^9$ nucleotides, or 6.4×10^9 nucleotide pairs of double-stranded DNA. This equals a length of more than 2 meters (2.9×10^6 nucleotide pairs $\times$ 1 mm). (From Strickberger, M. W., 1985. *Genetics,* 3rd ed. Macmillan, New York; adapted from Sparrow, A. H., H. J. Price, and A. G. Underbrink, 1972. A survey of DNA content per cell and per chromosome of prokaryotic and eukaryotic organisms: Some evolutionary considerations. *Brookhaven Symp. Biol.,* **23,** 451–493)

TABLE 12-4 Estimates of the frequencies of nonrepetitive DNA sequences and three classes of repetitive DNA sequences in various genomes

Organism	Nonrepetitive[a]	Partially repetitive[b]	Intermediate repetitive[c]	Highly repetitive[d]
Chlamydomonas reinhardtii (alga)	0.70		0.30	
Physarum polycephalum (fungus)	0.58		0.42	
Ascaris lumbricoides (nematode)	0.77		0.23	
Drosophila melanogaster (fruit fly)	0.78	0.15	0.07	
Strongylocentrotus purpuratus (sea urchin)	0.38	0.25	0.34	0.03
Xenopus laevis (South African clawed toad)	0.54	0.06	0.37	0.03
Gallus domesticus (chicken)	0.70	0.24		0.06
Bos taurus (domestic cow)	0.55		0.38	0.05
Homo sapiens (human)	0.64	0.13	0.12	0.10

[a] Single copy
[b] to about 200 copies/genome
[c] 250 to 60,000 copies/genome
[d] 70,000 to 1,000,000 or more copies/genome

Source: Data are from Straus, N. A., 1976. Repeated DNA in eukaryotes. In *Handbook of Genetics,* vol. **5**, R. C. King (ed.). Plenum Press, New York, pp. 3–29.

more obscure, although these have often been localized to distinctively staining chromosome sections and centromere regions (**heterochromatin**).

In the house mouse *Mus musculus,* a **satellite DNA** that comprises about 10 percent of the genome interestingly shows no homology to the DNA of related rodents such as rats, field mice and hamsters. In contrast, a number of satellite sequences are widely conserved in groups such as crustaceans (crabs) and insects (*Drosophila*). Such studies and others indicate that at least some satellite DNA can arise quite rapidly during the evolution of a species by adding many copies of a new DNA sequence or amplifying ancestral DNA sequences. The different kinds of satellite DNA, their different amounts, and their possible different origins signify different functions, or perhaps no function at all. As discussed in Chapter 10, various biologists have been tempted to consider some or many such sequences as forms of "selfish DNA."[6]

Nucleic Acid Phylogenies

Rates of Nucleotide Substitution

To estimate the extent of homology between nucleic acids of different sources, molecular biologists can measure the degree to which homologous nucleotide sequences in different single strands pair up to form double-strand sections. In one technique, DNA molecules extracted from two organisms, X and Y, one of which is radioactively labeled, are dissociated into single strands and allowed to reassociate into X–Y hybrid double-strands by incubating them together at appropriate temperatures.

If the DNA from X and Y are 100 percent similar (often, but incorrectly referred to as 100% homologous), that is, they have no nucleotide differences, the melting temperature at which the hybrid X–Y DNA dissociates into single strands (**Fig. 12-10**) will be the same as either X–X or Y–Y. However, should X and Y sequences differ, the X–Y hybrid will dissociate more easily because of nucleotide mismatching; its stability reduces and its melting temperature lowers. Various experiments show that for each one percent difference in nucleotide composition between X and Y, the thermal stability of the X–Y hybrid DNA molecule lowers by about 1°C.

Such techniques enable comparisons among perhaps a billion or more nucleotides simultaneously and can provide considerably more information than usually obtained from comparing a few characters at a time. Sibley and Ahlquist (1987) produced detailed phylogenies for all the major groups (orders) of birds, a task that once seemed difficult or impossible. Primate phylogenies based on such DNA–DNA comparisons can specify relationships that formerly seemed obscure, such as human–chimpanzee–gorilla (**Fig. 12-11**). Assuming from paleontological evidence that the divergence between the lineages of Old World monkeys and apes–humans occurred about 33 Mya (Chapter 20), and observing an approximate 7.7°C change in thermal stability between these groups and their common ancestor, there is an average of 1°C change for each 33/7.7 = 4.3 million-year interval. Thus, the lower scale of Figure 12-11 provides

[6] See Doolittle and Sapienza (1980) Orgel and Crick (1980) and Charlesworth et al. (1994) for selfish DNA.

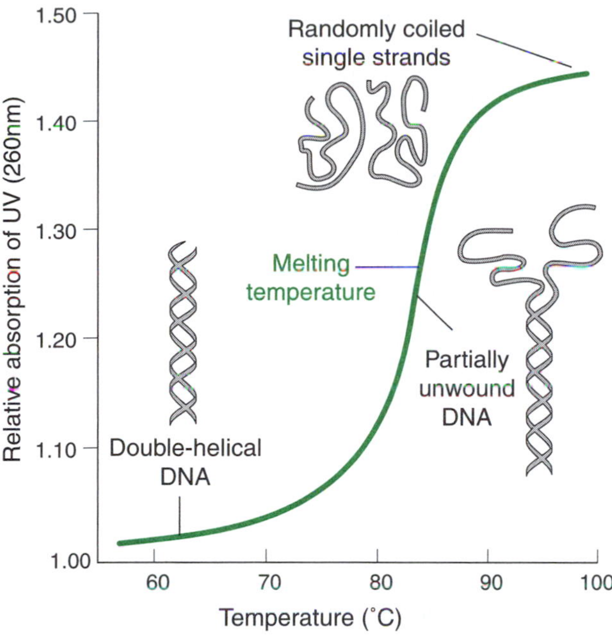

FIGURE 12-10 Melting temperature curve for DNA of T4 bacteriophage showing the marked change in ultraviolet absorption that occurs at approximately 84°C when about half the DNA has changed from double-helix to single-strand form. (From *Genetics Third Edition* by Monroe W. Strickberger. Copyright © 1985 by Monroe W. Strickberger. Reprinted by permission of Prentice Hall, Inc., Upper Saddle River, NJ.)

estimates of the dates at which these various primate taxa diverged.

These techniques make possible comparisons among nucleotide substitution rates obtained from **DNA–DNA hybridization** data and substitution rates from amino acid changes in known proteins. When such comparisons are undertaken, as they were by Thorpe (1982), for example, the rate of change in DNA seems more rapid than the rate of change from most proteins except for fibrinopeptides, which accumulate many changes over relatively short periods of time (**Fig. 12-12**).[7] Therefore, considerable portions of these tested DNA sequences do not code for essential proteins and may code for no proteins at all. Such sequences, perhaps largely "junk" or "selfish" DNA, accumulate changes more rapidly than genes that code for stringently selected proteins. The proportion of such noncoding junk DNA in many eukaryotes can be quite large. R. Nowak (2006), for example, estimates 97 percent of human DNA codes neither for proteins nor for functional RNA sequences. Among these presumed extraneous DNA sequences, he includes

[7] Rapid accumulation of changes in fibrinopeptides has been explained in relation to their function. Fibrinopeptides are the sections of fibrinogen molecules that are removed during formation of blood clots. Consequently, most amino acid changes in these sections have relatively little effect on fibrinogen function.

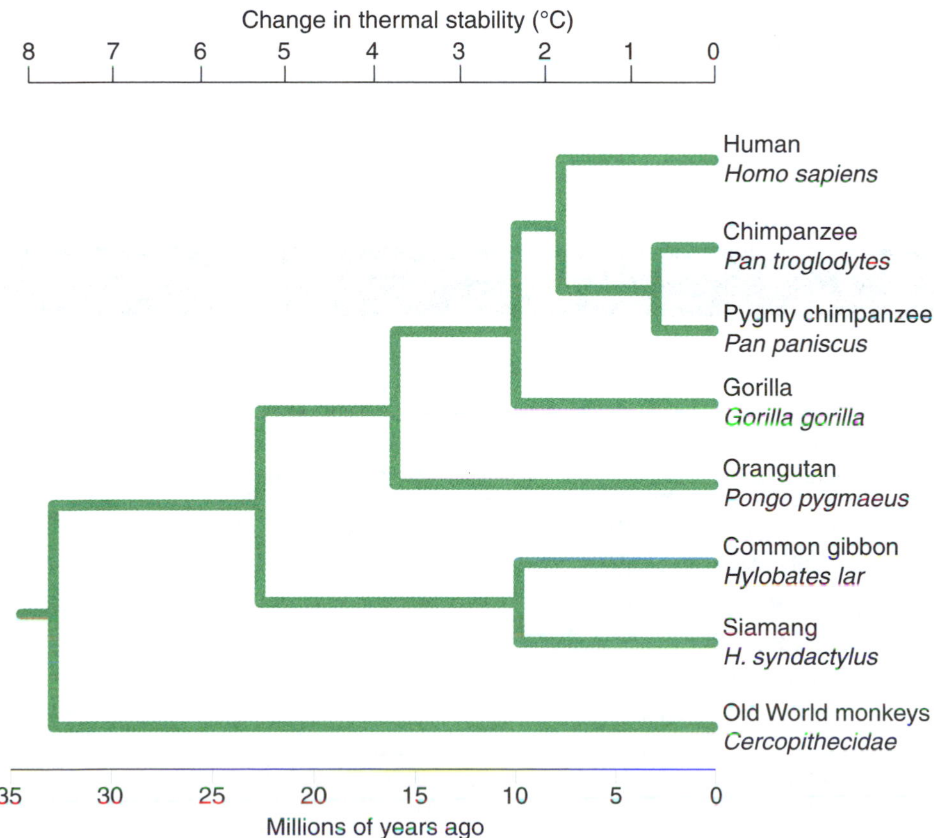

FIGURE 12-11 A phylogenetic tree and dates of divergence for humans, apes, and Old World monkeys based on DNA–DNA hybridization studies. According to Templeton (1986) it is difficult to distinguish between this illustrated phylogeny and one in which chimpanzees and gorillas are placed together in a lineage separate from humans, a view that some morphological comparisons seem to support (P. Andrews, 1987). In contrast studies using methods other than DNA-DNA hybridization support Sibley and Ahlquist (1984). (Adapted from Sibley, C. G., and J. E. Ahlquist, 1984. The phylogeny of primates as indicated by DNA-DNA hybridization. *J. Mol. Evol.*, **20,** 2–15.)

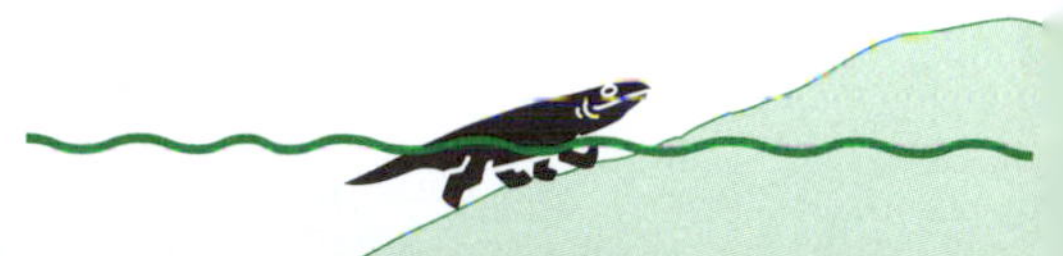

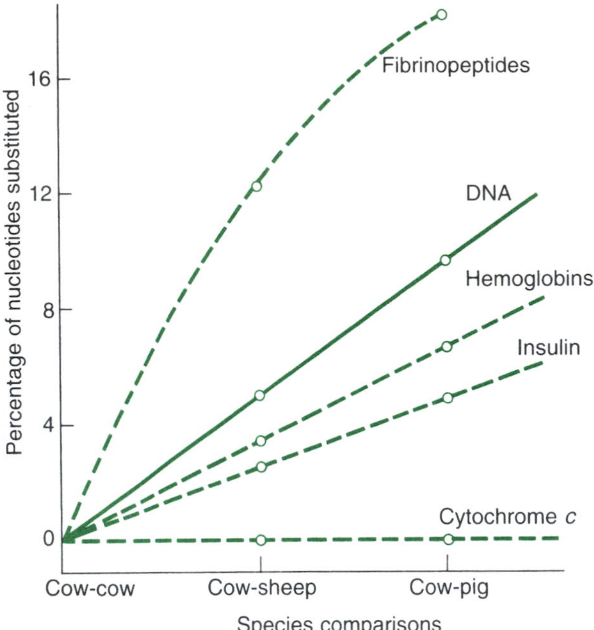

FIGURE 12-12 Nucleotide differences observed among three species of artiodactyls, using the DNA hybridization technique described in the text (*solid line*) and estimates of nucleotide substitutions derived from amino acid analysis of various proteins (*dashed lines*). (Adapted from McCarthy, B. J., and M. N. Farquhar, 1972. The rate of change of DNA in evolution. *Brookhaven Symp. Boyl.*, **23**, 1–41.)

introns (see Fig. 9-13), repetitive sequences such as satellites, microsatellites, and "interspersed elements" such as the *Alu* sequence. Thus, although we would expect the number of mutations to increase over time, we also expect that the number of mutations allowing amino acid substitutions correlates strongly with protein function.

The relationship between protein function and mutational change is supported by comparing the numbers of fixed mutations in amino acid codons incorporated into organisms. The more distant the taxonomic relationship between the listed organisms, the more evolutionary time elapsed from their common ancestor, and the greater the number of synonymous ("silent") mutations that do not cause amino acid substitutions (**Table 12-5**).

In contrast, the number of fixed amino acid substitution mutations follows a pattern based on function, from greater numbers of acceptable changes in some proteins to fewer changes in others, irrespective of evolutionary time. For example, the function of histone 3 is to bind and fold DNA molecules in identical fashion in all eukaryotic organisms, providing a common basic chromosome structure enabling common transcriptional processes as well as common chromosome replication mechanisms. Because of strong selection for such uniformity, histone 3 shows practically no amino acid replacements compared to hemoglobin or cytochrome *c*. Because the mutation process is random, as indicated by the fairly constant rate at which silent mutations occur in these proteins, we assume that a highly critical selective process is the primary agent restricting or permitting particular amino acid replacements.

Restriction Fragment Length Polymorphisms

One important approach to comparative DNA analysis is to use **restriction enzymes** that recognize specific short nucleotide sequences and cleave the molecule at these sites. For example, the enzyme *Eco*RI, isolated from *E. coli*, recognizes the GAATTC/CTTAAG sequence in double-strand DNA and

TABLE 12-5 Frequency comparisons of synonymous (silent) and replacement (amino acid substitution) mutations in homologous proteins from various organisms

			Changes per 100 Codons per 100 My		
Homologous Protein	Organism Compared	Relationship	Approximate Time (My) to Common Ancestor	Synonymous Mutations (silent changes)	Replacement Mutations (amino acid substitutions)
β hemoglobin	Rabbit:mouse	Same class (mammal), different order (lagomorph:rodent)	80	34	25
Cytochrome *c*	Rodent:chicken	Same phylum (chordate), different class (mammal:bird)	250	43	8
Histone 3	Sea urchin:trout	Different phylum (echinoderm:chordate)	650	50	0.8

Source: Modified from a table in Jukes, T. H., 1996. How did the molecular revolution start? What makes evolution happen? In *Evolution and the Molecular Revolution*, C. R. Marshall and J. W. Schopf (eds.). Jones and Bartlett, Sudbury, MA, pp. 31–52, with additions.

produces cleavage products at the points between G and A on both strands indicated by the arrows:

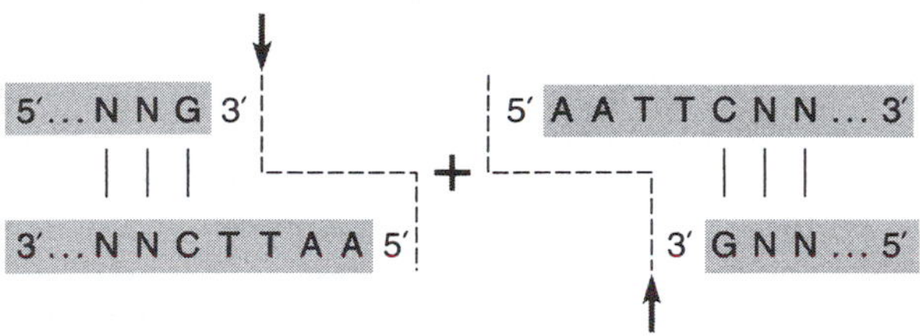

Because DNA molecules can differ from each other in nucleotide sequence, and therefore differ in the number and placement of sites recognized by *Eco*RI, each particular kind of DNA will have fragments of characteristic length when subjected to the enzyme (**Box 12-2** and **Table 12-6**). Also, as there are different kinds of restriction enzymes, many of which recognize target sites different from those recognized by other such enzymes, a DNA molecule subjected to a battery of different enzymes will produce cleavage products specific to that particular kind of DNA. Scoring differences in these inherited fragmentation patterns between individuals (**restriction fragment length polymorphisms**, also called **RFLPs**) enables estimates of genetic variation in populations (Nei, 1987), with each pattern designated as a **haplotype.**

The distinct sites at which restriction enzymes fragment DNA also provide restriction site maps for DNA sequences. In one example analyzed by S. D. Ferris and coworkers (1981), mitochondrial DNA from humans and apes was subjected to 19 different restriction enzymes, which cleaved approximately 50 sites in each mitochondrial chromosome, allowing a detailed comparison of target site sequences among the five species (**Fig. 12-14**). In accord with previously determined phylogenies, humans share more such sites with chimpanzees and gorillas than with orangutans and gibbons (Smouse and Li, 1987).

Nucleotide Sequence Comparisons and Homologies

A more precise method of phylogenetic determination is to **compare known nucleotide sequences from different organisms** rather than to infer relationships from hybridization studies or restriction enzyme maps. This procedure offers advantages in comparing changes between protein-coding and noncoding DNA sequences and in determining the extent of synonymous and nonsynonymous nucleotide substitutions in the amino acid coding regions. Data are available on over 90 billion DNA sequences (base pairs) from over 200,000 organisms ranging from viruses to eukaryotes (**Fig. 12-15**). They are accessible from GenBank (http://www.ncbi.nlm.nih.gov/Genbank/). Complete genome sequences, now known for 900 organisms, also are accessible in GenBank.

From an evolutionary view, sequencing technology advances and the availability of entire organismal genomes have opened a flood of molecular information, offering opportunities for a wide range of comparative genetic research. For example, important evolutionary features obtained from initial comparisons of sequenced prokaryotic genomes, include:

- extensive horizontal gene transfer between genomes;
- considerable amounts of gene duplication — as high as 25 percent in the *Bacillus subtilis* genome;
- greater similarity of archaebacterial protein sequences to eubacterial proteins than to eukaryotic proteins;
- proteins used in replication, transcription and translation show a reverse relationship: greater similarity between archaebacteria and eukaryotes;
- as much as 50 percent or more of genes in some genomes are "orphans" with no known function; and
- based on the 480 genes in *Mycoplasma genitalium,* that number, or an even smaller number may represent the minimal set of genes necessary for cellular life.

The latest genome sequence, published as this chapter was being written (November 2006), is that of the purple sea urchin, *Strongylocentrotus purpuratus,* a member of a group (echinoderms) more closely related to vertebrates than are any other invertebrates (see Chapter 17).

As discussed in **Box 12-3,** advances in technology, especially invention of polymerase chain reaction (PCR; **Fig. 12-16**), enable sequence information to be obtained from the DNA of some fossil organisms, the latest (again, November 2006), being the finding that Neanderthals and humans share genomes that are 98 percent similar (see Chapter 20).

Extensive PCR analysis of specific genes from extant organisms has now been undertaken, producing previously unsuspected relationships. One example is the phylogeny derived from comparing DNA sequences of *ß-globin gene clusters in various primates* (see Fig. 12-5). Some other examples come from nucleic acid structures that are more widely distributed, such as 5*S* rRNA, a component of the larger of the two ribosomal subunits that function in ribosome binding of the various transfer RNA molecules that carry the different amino acids. This cornerstone of the basic protein-synthesizing apparatus, once evolved, appears difficult, if not impossible, to change, and has been conserved evolutionarily in all organisms. The secondary structure of 5*S* rRNA seems universally the same (**Fig. 12-17**), a feature that enables all the various 5*S* rRNAs to be aligned for every nucleotide position. When such alignments are effected, differences among 5S rRNAs can be used to generate a phylogenetic tree in which nucleotide changes measure evolutionary distance.

As depicted in **Figure 12-18,** this tree illustrates a divergence between early prokaryotes and early eukaryotes at a time perhaps 50 percent earlier than the divergence between

BOX 12-2 Molecular Evolution in the Test Tube

THE PHENOMENAL GROWTH of molecular information in biology has sparked various attempts to demonstrate evolutionary processes in the laboratory, where detailed analysis of successive changes is more possible than in nature. Because this approach usually demands considerable biochemical analyses and rapid generation times as well as rigid control over genetic and environmental conditions, most of these studies have been undertaken using microbial organisms.

A common technique of *test tube evolution* is to subject a strain of bacteria to a new carbon source (e.g., xylitol) or nitrogen source (e.g., butyramide) that the cells cannot metabolize properly. When such cells are exposed to a mutagenic agent, mutations increase in frequency, and some adaptive mutations may arise. Natural selection then proceeds "directionally" toward a new goal by enhancing survival of those bacterial strains with enhanced metabolic efficiency for the new, demanding environment. Evolutionary changes of this kind often occur through the adaptation of enzymes that were initially inefficient on the new substrate, because they were used primarily for other purposes, but which were available to be adapted. Various experimenters have noted a variety of such adaptational changes:

1. Synthesis of the inefficient enzyme may become constitutive through a regulatory mutation (see Fig. 10-13d), increasing the amount of this enzyme in the presence of the new substrate.
2. A regulatory mutation may enable the new substrate to induce synthesis of the inefficient enzyme.
3. Mutations may occur that enable the substrate to enter the cell more easily.
4. A gene duplication may occur that enables increased production of the inefficient enzyme.
5. Mutations may occur in the enzyme's structural gene enabling the formerly inefficient enzyme to metabolize the new substrate more efficiently.

An experiment on protein from *E. coli* showed a striking demonstration of enzymatic evolution. Instead of trying to adapt bacteria to an unusual artificial medium, a strain of bacteria carrying a deletion of the β-galactosidase Z gene (see Fig. 10-13a) was cultured in lactose medium containing a dye as indicator. In the absence of β-galactosidase, lactose does not hydrolyze — recognized by red bacterial colonies — in contrast to the white color of lactose-using colonies. Within one month of growth on a lactose-containing medium, the *Z*-deficient strain gave rise to white colonies that could use lactose, although inefficiently. Further growth and selection among these new lactose-using cells gave rise to a more efficient strain. On lactose medium unsupplemented with other sugars, the final selected strain of bacterial cells, called *evolved* β*-galactosidase* (*ebg*), could form colonies as rapidly as could wild type *E. coli.*

Various tests showed that the *ebg* strain had evolved a lactose-hydrolyzing enzyme (EBG) completely different from β-galactosidase. This new enzyme had a larger molecular weight, different immunological properties and different ionic sensitivities. Interestingly, lactose can regulate the enzyme's appearance employing regulatory mutations similar to those that control β-galactosidase. After nucleotide sequencing of the genes involved, Stokes and Hall (1985) concluded that the EBG system is a remnant of an ancient duplication of the *E. coli lac* enzyme system. In sum, these experiments demonstrate that a protein with only vague affinities for a particular function can assume that function with remarkable efficiency by a stepwise evolutionary process of mutation and selection (Hartl and Clark, 1997).

On the nucleotide level, Mills and coworkers (1973) narrowed down test-tube evolutionary experiments to some small self-replicating molecules. They began with the RNA nucleic acid of a Qβ virus that was about 4,220 nucleotides long, a molecule that could replicate in test tubes after adding a replicase enzyme and various chemical components. Successive transfers of only the earliest replicating molecules to new cultures caused selection for rapid replication.

Under these conditions successful Qβ molecules only need retain those sequences that enable them to be recognized by the replicase enzyme in the culture; that is, they no longer need genes that formerly coded for what are now unnecessary proteins: the coat protein and the replicase enzyme. Selection was further intensified by placing fitness advantages on those RNA molecules that replicated only when a single such molecule was present in an entire culture. This single-stranded molecule (or plus strand) must rapidly attract a replicase enzyme to form a complement (or minus) strand that forms a new plus strand, and so on. These techniques selected short, independently replicating RNA molecules, including one type, called *midivariant,* only about 220 nucleotides long (**Fig. 12-13**).

When these experiments are reversed — that is, when mixtures begin without any Qβ RNA sequences at all — the Qβ replicase splices together nucleotides on its own without a template (Fig. 12-13). Moreover, evolution in such mixtures can produce a variety of *de novo* RNA sequences capable of adapting to different environmental conditions, including sequences that by accretion converge evolutionarily to reach that optimal self-replicating *midivariant* length of 220 nucleotides. Natural selection operates as it does elsewhere, establishing reproductively successful genotypes through a series of successively adaptive stages.

These and other experiments led Eigen and coworkers (1981) to conclude that the number of nucleotides (information content) in an RNA strand determines the frequency at which mutant sequences arise as well as the number of reproductive cycles necessary for the selection of an optimal mutant with a high reproductive rate. The prevailing (wild type) genotype achieves stability when its selective advantage is great enough to overcome the error rate of replication for that nucleotide sequence. If the error rate is too high ("mutational meltdown"), the genotype loses adaptive information. If the error rate is too low, the capacity for further adaptation declines, and a lineage may more easily become extinct compared to lineages with more optimal mutation rates.

Such factors govern the nucleotide length of a molecule in these experiments. For example, because RNA polymerases do not replicate nucleotides as accurately as DNA polymerases, the RNA molecules in single-stranded RNA viruses are usually no longer than 10^4 nucleotides, a value that can be calculated theoretically from error rates and selective advantages. (Of course, genomes need to achieve longer lengths in coding for enzymes that need longer sequences. One method of overcoming the separation and competitiveness of small nucleotide sequences is a hypercycle in which these subunits join into a mutual symbiotic group, each subunit enhancing the replication of the next.)

The advantages of test tube evolution in its ability to rapidly create entirely new molecules that can perform new biological roles have attracted many experimenters. For example, RNA molecules half the size of the Qβ sequence in Figure 12-13 evolved with their own signals for test-tube replication (Breaker and Joyce, 1994). By allowing selection to occur in a continuous self-evolving system, Wright and Joyce (1997) improved the efficiency of an RNA catalyst ("ribozyme") to 14,000 times its initial value in only 52 hours.

"Sex" can be introduced into test-tube evolution. Stemmer (1994) broke genetic sequences into subsidiary fragments and allowed them to recombine in various ways using the polymerase chain reaction technique (see Fig. 12-15). These new recombined molecular sequences were thousands of times more effective in their selected function (antibiotic resistance) than the sequences produced in the absence of recombination, testifying to an important advantage of sexual reproduction (see also Chapter 23), albeit confined to the test tube. Experimental molecular evolution is becoming a highly promising field for understanding how changes on the molecular level can occur, for grasping their biochemical and evolutionary significance, and for using their effects in producing new functional molecules.

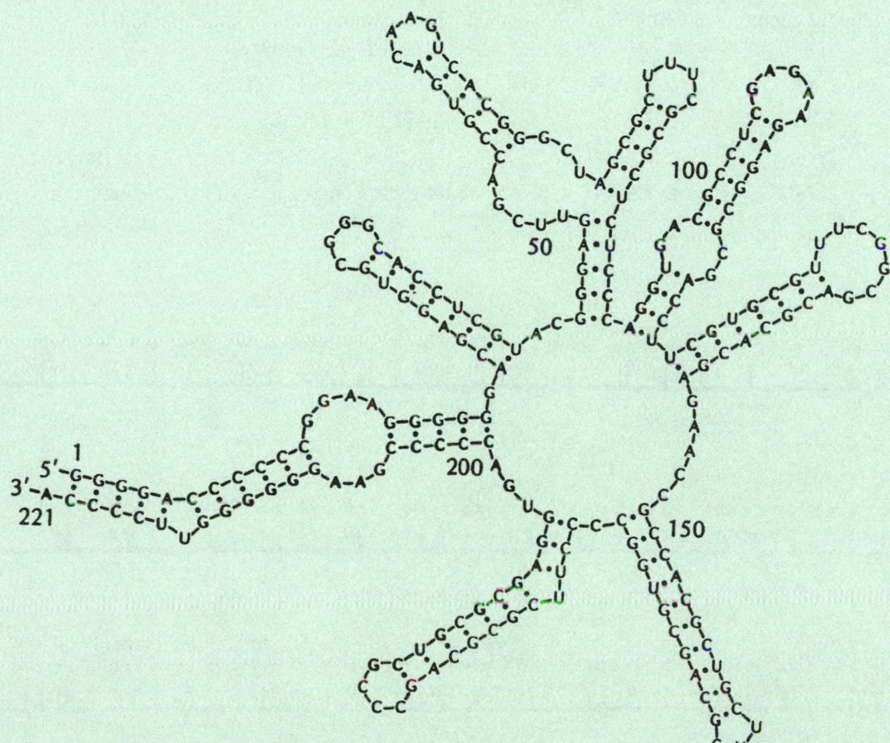

FIGURE 12-13 Nucleotide sequence and secondary structure of the smallest RNA molecule that can replicate independently in a mixture containing the replicase enzyme of Qß virus. (Adapted from Miele, E. A., D. R. Mills, and F. R. Kramer, 1983. Autocatalytic replication of a recombinant RNA. *J. Mol. Biol.*, **171**, 281–295.)

TABLE 12-6 Numbers of nucleotides and estimated numbers of genes for 18 taxa

Species	Number of Nucleotides	Estimated Number of Genes	Reference
Yersinia pestis (bacterium that causes bubonic and pneumonic plague)	4.5 Mb	3,956	Chain et al. (2006)
Escherichia coli (bacterium)	5 Mb	3,470	Chen and Szostak (2006)
Saccharomyces cerevisiae (yeast)	12 Mb	6,604	The Yeast Genome Directory (1997)
Cryptococcus neoformans (yeast)	20 Mb	6,500	Loftus et al. (2005)
Caenorhabditis elegans (nematode)	97 Mb	19,000	The *C. elegans* Sequencing Consortium (1998)
Arabidopsis thaliana (flowering plant from the mustard family)	115 Mb	25,498	*Arabidopsis* Genome Initiative (2000)
Drosophila melanogaster (fruit fly)	137 Mb	14,100	Adams et al. (2000)
Ciona intestinalis (ascidian)	160 Mb	15,852	Dehal et al. (2002)
Anopheles gambiae (mosquito)	278 Mb	13,683	Holt et al. (2002)
Takifugu rubripes (Fugu)	365 Mb	c. 30,000	Aparicio et al. (2002)
Tetraodon nigroviridis (freshwater pufferfish)	370 Mb	27,918	Jaillon et al. (2004)
Oryza sativa (rice)	466 Mb	c. 50,000	Yu et al. (2002)
Bombyx mori (silkworm)	530 Mb		Mita et al. (2004)
Danio rerio (zebrafish)	1.7 Gb	30,000	http://www.genome.org/cgi/content/full/11/11/1958
Mus musculus (house mouse)	2.4 Gb	c. 30,000 (24,174)	Waterston et al. (2002)
Rattus norvegius (Norway rat)	2.7 Gb	c. 30,000 (21,166)	Gibbs et al. (2004)
Homo sapiens (humans)	2.9 Gb	c. 30,000 (26,966)	Lander et al. (2001); Venter et al. (2001)
Pan troglodytes (common chimpanzee)	2.9 Gb	c. 30,000	The chimpanzee analysis and sequencing consortium (2005)

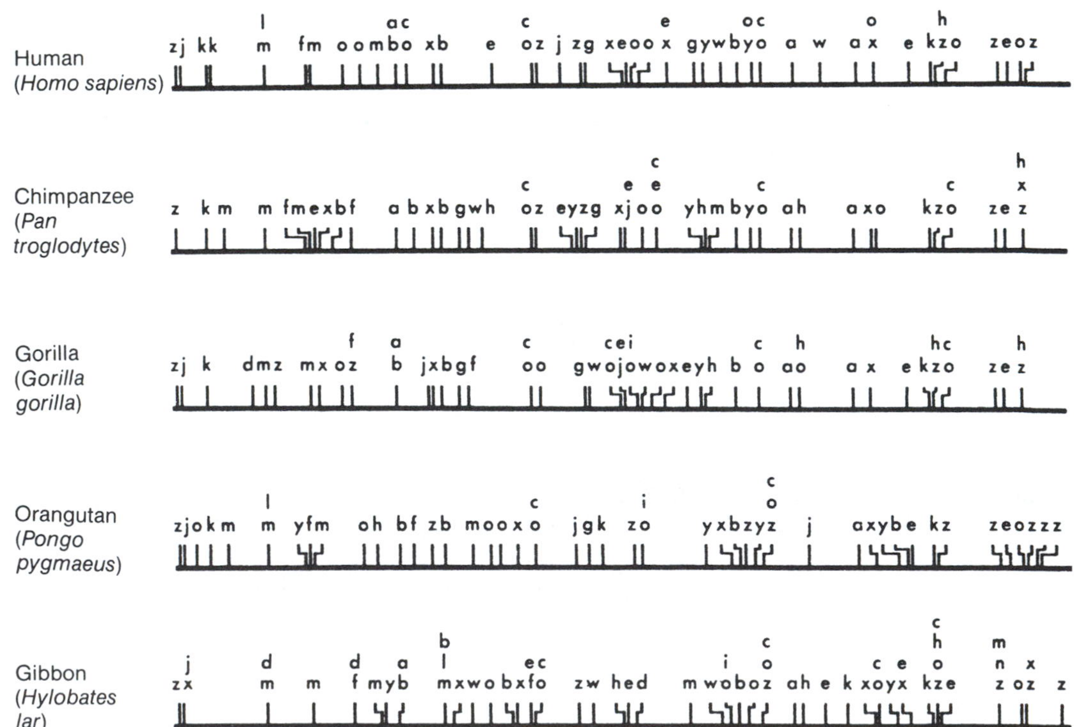

FIGURE 12–14 Cleavage maps of mitochondrial DNA from humans and four other primate species, derived from the use of 19 restriction enzymes. Cleavage sites for each enzyme are designated by small letters: a, EcoRI; b, HindIII; c, HpaI; d, BglII; e, XbaI; f, BamHI; g, PstI; h, PvuII; i, SalI; j, Sac I; k, KpnI; l, XhoI; m, AvaI; n, SmaI; o, HincII; w, BstEII; x, BclI; y, BglI; and z, FnuDII. The position at the left of each map is the replication origin of the mitochondrial chromosome. (Adapted from Ferris, S. D., A. C. Wilson, and W. M. Brown, 1981. Evolutionary tree for apes and humans based on cleavage maps of mitochondrial DNA. *Proc. Natl Acad. Sci. USA*, **78**, 2432–2436.)

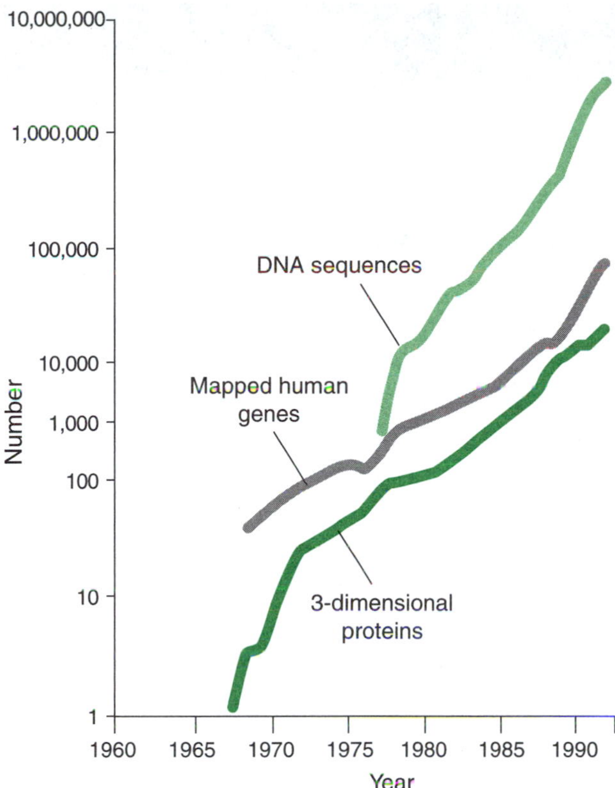

FIGURE 12-15 A growth chart of biomolecular information obtained during the last few decades. More than 2.5 million different DNA sequences (90 million base pairs as of August, 2005) from more than 205,000 organisms are now available in the GenBank facility, an amount doubling every 18 to 24 months. 30,000 human genes (3 billion base pairs) are now mapped, and some 37,000 three-dimensional protein structures are recorded in the Protein Data Bank. (From Boguski, M. S., 1998. Bioinformatics — A new era. In *Trends Guide in Bioinformatics*. Elsevier Trends Journals, Haywards Heath, West Sussex, UK, pp. 1–3.)

fungi (for example, yeast) and plants and animals (see also Gouy and Li, 1989). Because biologists estimate the latter divergence to have occurred about 1.2 Bya, the earlier prokaryote–eukaryote separation may well have occurred 1.8 Bya, a finding supported by the existence of fossil eukaryotic-type cells of Proterozic age (Chapter 9).

Because 5*S* rRNA molecules have not been found in animal mitochondria and some experimenters consider them too small for obtaining complex phylogenies, other ribosomal RNA sequences are used for broad-range nucleotide comparisons. For example, Figure 12-17 offers a phylogeny of eukaryotes based on sequence analysis of a larger (18*S*) ribosomal RNA component, and Figure 8-7 ("the universal tree") extends this analysis in differentiating among eubacteria, archaebacteria and eukaryotes.

Another study of this kind by M. W. Gray and coworkers (1984) uses mitochondrial and chloroplast RNAs to support the endosymbiotic theory of organelle evolution (Chapter 9) by showing that we can trace both these organelles to a eubacterial origin. Interestingly, this study also proposes that animal and fungal mitochondria originated from nonphotosynthetic aerobic bacteria, whereas plant mitochondria originated separately from cyanobacteria. In a later study (1989), Gray's group suggested that *plant mitochondria may be mosaics resulting from two different symbiotic events.* That mitochondria derive from a bacterial origin possessing a genetic code shared by all other organisms (the universal code) supports the concept that amino acid associations with mitochondrial codons changed after the universal code was established.

Mitochondrial DNA

In instances where biologists compare a narrower spectrum of organisms, such as a group of vertebrates, mammals or humans, they commonly use **mitochondrial DNA** sequencing (see Avise, 1994).

Among its advantages, mitochondrial DNA is easily isolated and evolves at a sufficiently rapid rate to allow recognition of distinctions and similarities among organisms that have only recently diverged. In one example, mitochondrial DNA sequence analysis indicates that the many morphologically diverse species of cichlid fishes in East Africa have a common monophyletic origin, but their similarities in different lakes are caused by parallelism or convergence rather than common ancestry (**Fig. 12-19**). An obvious lesson emerges from these and other molecular studies: Molecular and morphology data need not produce similar trees. Convergence produced similar morphologies in various aschelminth groups; molecular findings confirm their different ancestries.[8]

In addition to procedures using enzymatic alleles, nucleic acid techniques are being used to analyze relationships within species or between populations, for example, determining the number of genetically isolated populations of ocean-dwelling fish or mammal species. As discussed in Chapter 20, mitochondrial DNA marker studies are illuminating relationships among the various groups of humans.

Microsatellites

Other types of DNA analysis use **microsatellites** — tandem repeats of short nucleotide sequences such as cytosine–adenine–adenine (CAACAACAA . . .) — which may vary between individuals in the numbers of repeats at some particular chromosomal position.

Through techniques involving the isolation and analysis of microsatellites, molecular evolutionary biologists can distinguish an individual with ten CAA repeats at a particular locus, for example, from individuals with more or fewer

[8] See Meyer et al. (1990) and Goldschmidt (1996) for speciation in cichlids, and see Winnepenninckx et al. (1995) and Chapter 16 for aschelminths.

BOX 12-3 Ancient DNA

DNA CAN BE EXTRACTED from dead or fossil organisms. Such DNA is referred to as "ancient DNA."

After death, DNA degrades rapidly, initially because of enzymes present within cells and later due to oxidation, hydrolysis and radiation. Indeed, DNA becomes chopped into smaller and smaller fragments with increasing time post-mortem. Most ancient DNA studies analyze mitochondrial DNA because it is much more abundant than nuclear DNA and so tends to be better preserved. A second major problem is contamination. Because so little DNA with sufficient fragment length is present in ancient DNA samples, it is easy to contaminate the samples with DNA from other sources. Elaborate protocols have been devised to deal with these technical difficulties.

The first successful extraction and analysis of DNA sequences from dead (ancient) organisms occurred in 1984, with DNA from muscle tissue of a 140-year-old museum specimen of the quagga, a now extinct member of the horse family. Higuchi and coworkers (1984) analyzed 229 nucleotide base pairs of mitochondrial DNA, finding 12 base substitutions, causing only two amino acid replacements, when compared to a corresponding mitochondrial DNA sequence from zebra (*Equus zebra*). These data indicated common ancestry with the horse family as well as little (if any) modification of the DNA sequences after the quagga died.

There were, however, serious limitations. First, only relatively short DNA sequences could be extracted, none longer than 100 to 200 nucleotides. The same was true for DNA analyses done of *Alu* repeat sequences in Egyptian mummies.[a] Second, and most importantly, it proved difficult to obtain sufficient numbers and kinds of these short ancient DNA sequences using bacterial cloning to generate a DNA library; the carriers had difficulty cloning such small sequences.

Polymerase Chain Reaction

In the middle 1980s, Kary Mullis invented **polymerase chain reaction (PCR),** which amplifies even small traces of DNA with great success. By 1990, thousands of laboratories were using PCR for many purposes, including analysis of ancient DNA (Mullis et al., 1994).

Briefly, PCR is a test-tube process that involves placing special small "primer" sequences at each end of a target DNA sequence (for example, ancient DNA), and subjecting these to many replicating cycles (see Fig. 12-16). Each such cycle doubles the number of target sequences so that, after 30 to 40 cycles, there are many millions of replicates of the original ancient DNA sequence. By obtaining such large amounts of DNA, experimenters can achieve nucleotide sequencing of the target with accuracy and confidence.

The PCR method has revolutionized molecular genetics by permitting accurate nucleotide sequencing of any extracted nucleic acid from whatever tissues are available, be they living, museum or fossil. DNA sequences have been analyzed by this method from many fossil organisms: kangaroo rats, the marsupial (Tasmanian) wolf, amber-embedded insects, plants, and vertebrates, human remains from the Arctic zones and from peat bogs in Florida, New Zealand flightless birds (ratites), fossilized plant material and seeds and even fungal spores.[b]

Finding Ancient DNA

Because of fossil deterioration, ancient DNA appears in relatively few relics, and even when found, is often highly fragmented and modified. An additional problem comes from contamination, either from the DNA of bacteria, parasites, and symbionts that infected the original host, or from the DNA of later organisms that helped decompose the body, or even from the DNA of organisms present in the laboratory. Fortunately, there are means of detecting most such intrusions. The consequences for RNA are even more drastic: because RNA has a single strand, it lacks the molecular stability and protection a double-strand structure offers.

Despite these limitations, research in ancient DNA is growing rapidly, and is helping to answer archaeological controversies and to resolve problematic phylogenetic relationships. Ancient DNA from human bones on Easter Island indicates that its settlers were Polynesians from other Pacific islands, and not the South American Indians that the Danish explorer Thor Heyerdahl suggested. Also, Cooper and coworkers' 1992 study of flightless New Zealand birds shows that kiwis and extinct moas were much more distantly related than previously proposed, and that kiwis are more closely related to the Australian emus and cassowaries.

A most interesting discovery is the distinctive "ancient" mitochondrial DNA isolated from a Neanderthal skeleton 30,000 to 100,000 years old. In answer to persistent questions on the relationship between Neanderthals and humans (see Chapter 20), Krings and coworkers (1997) show that the Neanderthal sequence is significantly more different than would be expected if it were a sample of normal human variation. According to their findings, the Neanderthal lineage diverged from humans about 600,000 years ago, and "went extinct without contributing mitochondrial DNA to modern humans."

Although the results of these and other ongoing investigations are encouraging for evolutionary understanding, their oft-popularized potential to help us recreate older life forms — placing dinosaur DNA into frog's eggs to recreate dinosaurs — is limited or nonexistent. The difficulty of finding appropriate ancient DNA to analyze, and the highly fragmented form in which such DNA is found, makes it unlikely that we can resurrect entire ancient organisms from ancient DNA. More important, as discussed especially in the following chapter, the contribution of the maternal genome to animal eggs, and environment-genome interactions play essential roles in the proper read-out of DNA.

---

[a] See Pääbo (1993, 2003) for DNA extracted from Egyptian mummies. Research is underway to analyze the DNA of Tutankhamen, the boy king who, more than 30,000 years ago, ruled Egypt from the age of eight to seventeen.

[b] See the collection of papers edited by Herrmann and Hummel (1994) for analyses of DNA from a variety of extinct organisms.

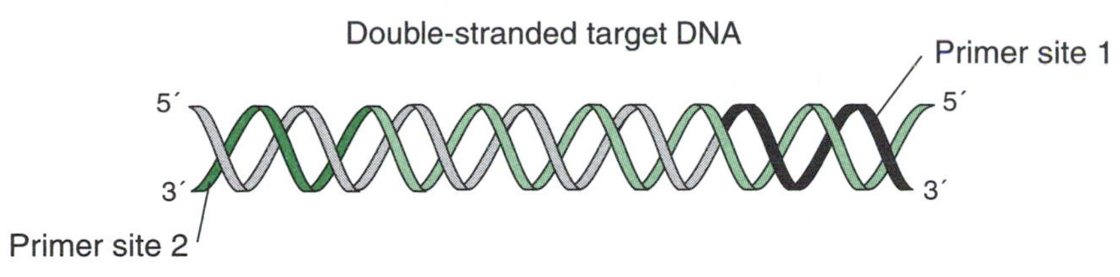

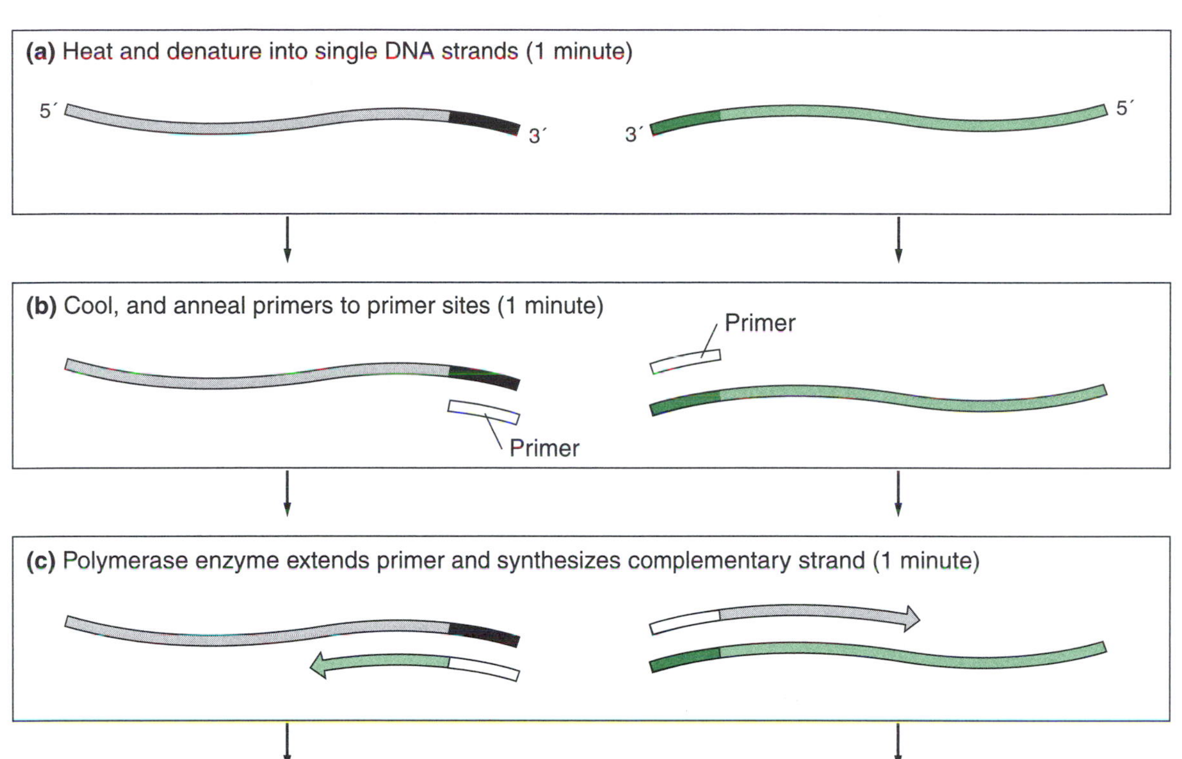

(d) Two replicates of target DNA double strands, each with the same primer sites as the initial target DNA

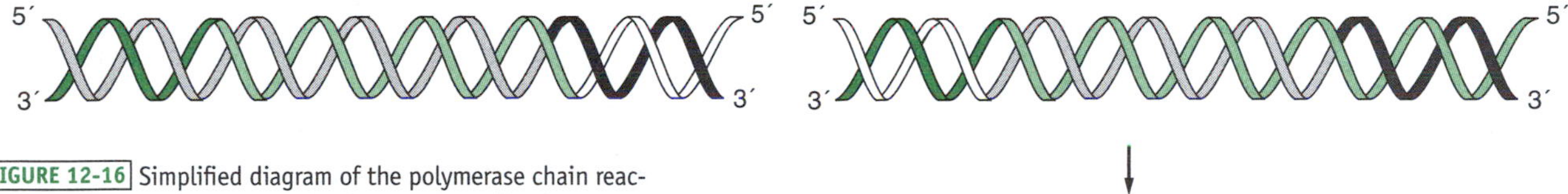

(e) Repeat steps a,b,c
Four replicates of target DNA double strands

(f) Repeat steps a,b,c
Eight replicates of target DNA double strands

(g) Multiple repeats of steps a,b,c
MULTIPLE COPIES OF TARGET DNA SEQUENCE

FIGURE 12-16 Simplified diagram of the polymerase chain reaction (PCR) technique showing basic steps in replicating a DNA target sequence that may be as long as 2,000 nucleotide base pairs. Short primer sites at each end of the target are identified (a), and oligonucleotide sequences synthesized (usually about 20 base pairs long) that can pair with the primer sites when "melting" the DNA into single strands (b). A heat-resistant DNA polymerase enzyme is used to extend nucleotide synthesis from the primers along each complementary strand (c), forming two double-stranded replicates of the original target DNA sequence (d). By alternately heating and cooling the mixture, each cycle (a–c) exponentially replicates the DNA target (e–g) so that an original sequence can potentially be amplified more than one million times ($225 = 4 \times 10^6$) in 25 cycles. The duration for steps a–c may vary in different experimental protocols, depending on lengths and compositions of primers and target DNA segments.

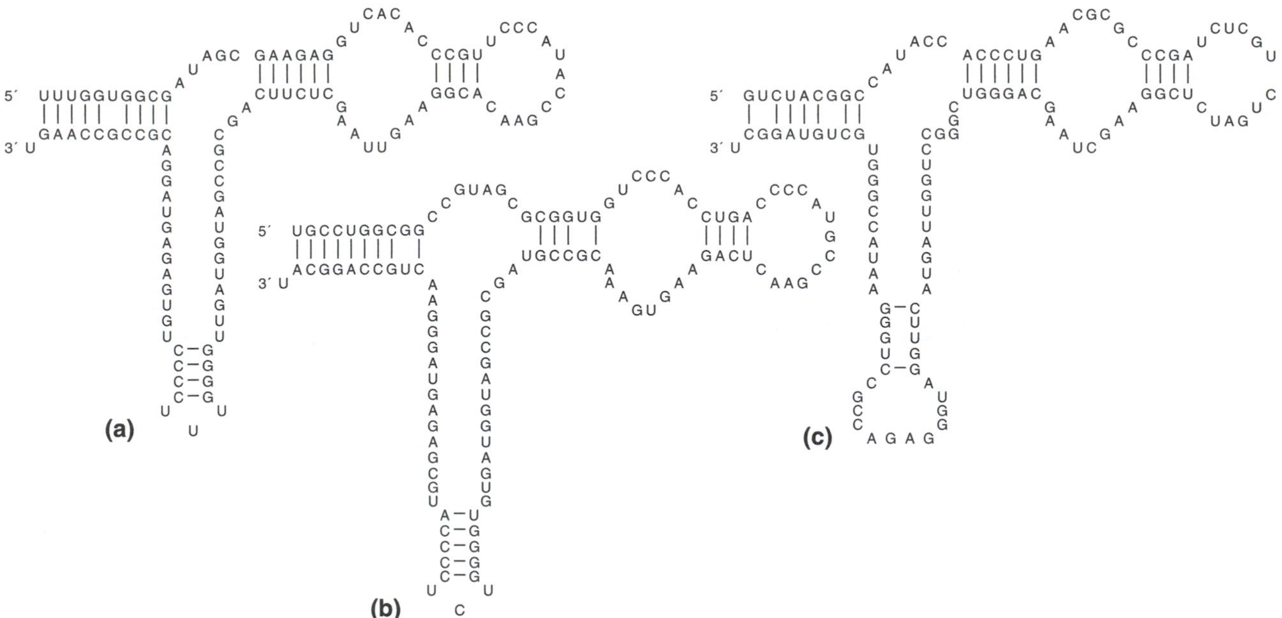

FIGURE 12-17 Models of the secondary structures of 5S ribosomal RNA molecules from three different organisms: two from bacteria (a) *Escherichia coli,* (b) *Bacillus subtilis,* and the third from humans (c). (Adapted from Hori, H., and S. Osawa, 1979. Evolutionary change in RNA secondary structure and a phylogenetic tree of 54 5S RNA species. *Proc. Natl Acad. Sci. USA,* 76, 381–385.)

repeats. Because there are many microsatellite loci (humans have at least 50,000 such loci per haploid genome), and they mutate at a relatively high rate compared to protein-coding sequences, opportunities for tracking individual differences and relationships abound. In general, molecular data from all sources can be analyzed by various elegant mathematical techniques to infer phylogenetic relationships.[9]

Combined Nucleic Acid–Amino Acid Phylogenies

Barnabas and coworkers (1982) were among the first to offer a comprehensive **Tree of Life** that takes into account nucleotide sequences from 5S rRNA and amino acid sequences from ferredoxin and the *c*-type cytochromes (diagrammed in **Figure 12-20,** and see **Box 12-4** for two current versions of the Eukaryote Tree of Life). Because ferredoxin is a primitive iron-containing protein used in a number of basic oxidative–reductive pathways, the doubling event that occurred early in its evolution provides a baseline for the phylogenetic tree. Organisms whose ferredoxins most resemble the inferred sequence of the primitive duplicated molecule are anaerobic and heterotrophic bacteria such as *Clostridium.* They later diverged, developing into anaerobic photosynthetic bacteria such as *Chromatium* and into the main line of aerobic respiratory organisms.

The cytochrome *c* analysis that Barnabas and coworkers (1982) used shows a phylogeny in which the eukaryotic sequences are most similar to the cytochrome *c*2 sequences of the nonsulfur purple photosynthetic bacteria (Rhodospirillaceae). Because cytochrome *c* functions exclusively in the eukaryotic mitochondrion (although coded by DNA in the nucleus), most likely it is the mitochondrion organelle itself (rather than the eukaryotic cell) that derives from this bacterial line. Studies on 5*S* rRNA support this view (see Figure 12-20), showing that eukaryotes diverged from an earlier prokaryotic form and not from the later *Rhodopseudomonas* line. Apparently, the gene for cytochrome *c* incorporated into the eukaryotic nucleus after the inclusion of the mitochondrion organelle into the eukaryotic cell.

Similarly, the evidence of strong homologies between cyanobacteria and plant chloroplasts for their cytochrome and ferredoxin sequences indicates that eukaryotic plant cells must have incorporated a chloroplast organelle that may at one time have been a prokaryotic cyanobacterium. This phylogeny offers strong support for the symbiotic theory of organelle function (see Chapter 9). Such endosymbiotic events occurred more than once, with some eukaryotic algae receiving their chloroplasts through secondary transfer from other eukaryotes.[10]

In any case, symbiosis was not only cytoplasmic but extended to **nuclear incorporation of organelle genes.** Because chloroplasts synthesize only a small portion of the

[9] See Queller et al. (1993) for microsatellite analysis, and Swofford et al. (1996) for techniques of phylogenetic analysis.

[10] See M. W. Gray et al. (1989, 1999, 2001), Douglas and Turner (1991), Palmer (2003), Burger et al. (2004), Keeling et al. (2005) and Margulis et al. (2006) for studies on endosymbiosis.

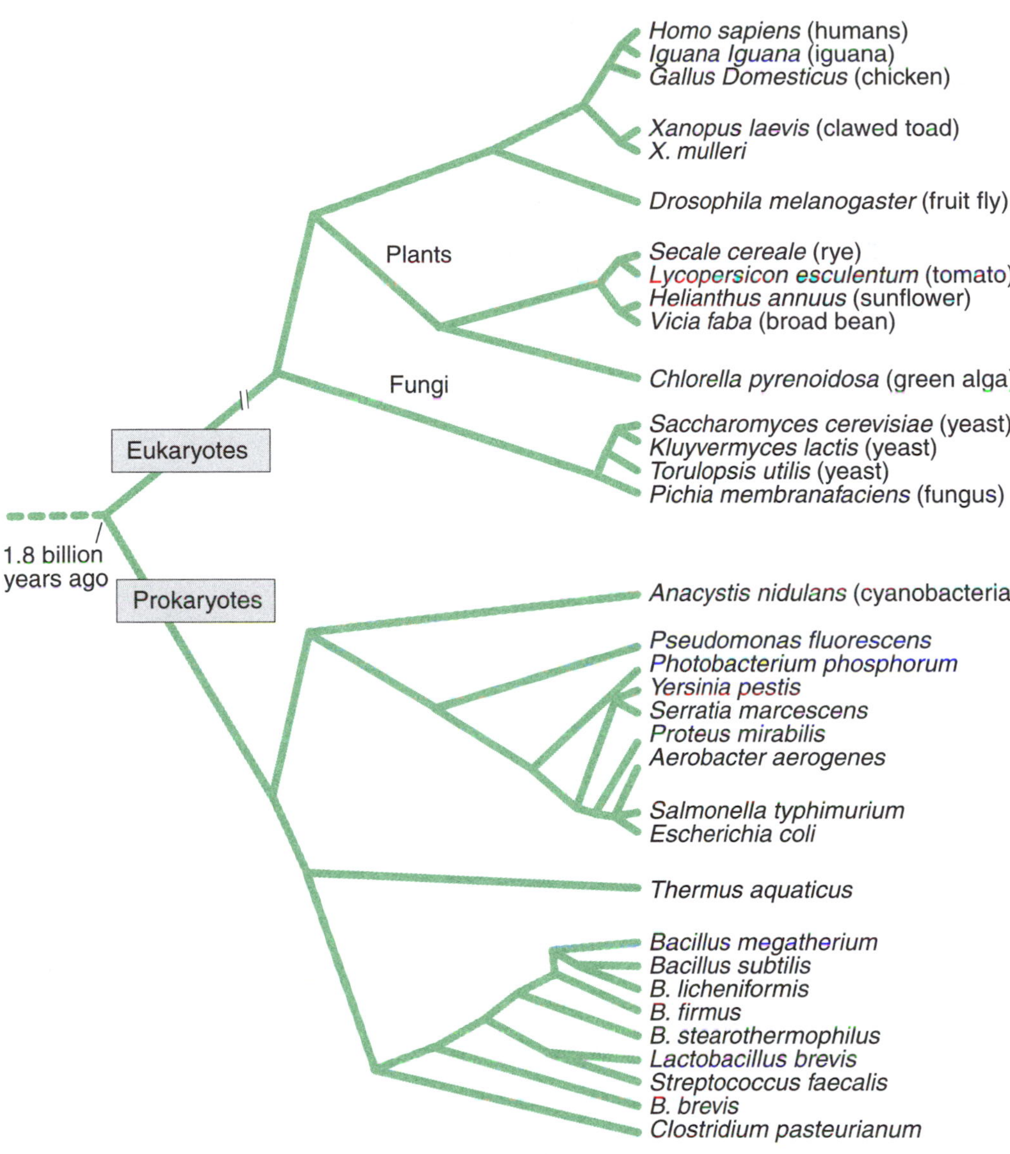

FIGURE 12-18 A phylogenetic tree that best explains the data gathered from comparing 5*S* rRNA nucleotide sequences among many different species. The separation between eukaryotes and prokaryotes appears to be close to about 2 Bya. Other studies comparing amino acid sequences from 57 prokaryotic and eukaryotic enzymes indicate that the two groups shared a common ancestor about 2 Bya (Feng et al., 1997). Among additional problems is that such estimates leave in question the kinds of complex cells found in stromatolites that appear much earlier in the fossil record (Chapter 9). Were these cells variations on prokaryote-eukaryote themes or were they entirely different? As discussed in the text, applying a single molecular clock to all data causes difficulties. *Source:* Hori, H. and S. Osawa, 1979, Evolutionary change in 5*S* RNA secondary structure and a phylogenic tree of 54 5s RNA species, *Proc. Natl Acad. Sci. USA*, **76:** 381–385. Reprinted by permission.

proteins they use, geneticists have often pointed out that many of the genes introduced by the original cyanobacterial endosymbionts were transferred to the nucleus; their protein products are now reimported into the chloroplast. Among the evidence is the finding that the chloroplast enzyme, glyceraldehyde-3-phosphate dehydrogenase (GAPDH) used in the Calvin cycle (see Fig. 8-17), is made by a nuclear gene whose nucleotide sequence is similar to one of the three GAPDH genes in the cyanobacterium *Anabaena variabilis.* Even more telling is W. Martin and coworkers' 1993 finding that another *Anabaena* GAPDH gene has a similar nucleotide sequence to the GAPDH gene used in glycolysis by all plants, animals and fungi (see Fig. 8-17). Thus, prokaryotic endosymbionts may have transferred to eukaryotes a number of genes other than those used in mitochondria and chloroplasts. Eukaryotes, therefore, are **chimeras,** consisting partially of various prokaryotic constituents.

Rates of Molecular Change: Molecular Clocks

Inherent in all phylogenies is the concept that evolutionary differences between organisms arise from mutational differences and, generally, the greater the number of mutational differences between organisms, the greater their evolutionary distance between those same organisms.

In some of the phylogenies considered so far (e.g., Figure 12-11), mutational differences have been used to provide evolutionary time scales; that is, they have assumed that muta-

Lake Tanganyika

Lake Malawi

Bathybates

Ramphochromis

Lobochilotes

Placidochromis

Tropheus

Pseudotropheus

Cyphotilapia

Cyrtocara

Julidochromis

Melanochromis

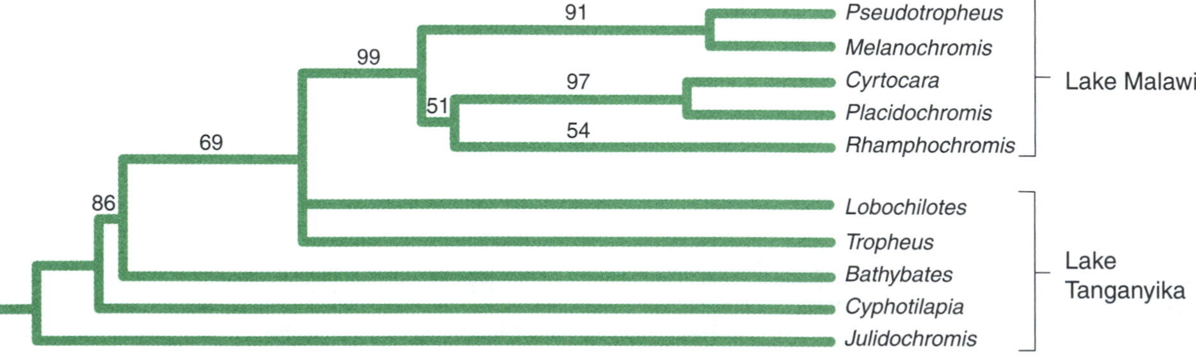

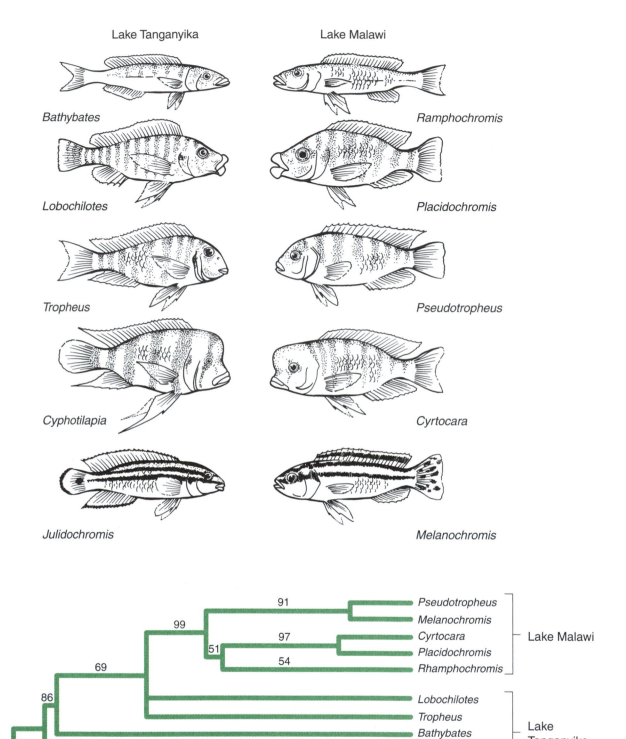

FIGURE 12-19 (Top): Phenotypic comparisons among some of the cichlid species from 12 genera found in Lake Tanganyika and Lake Malawi. (Bottom): Phylogenetic tree showing the separate origins of these species, indicating that the similarities among them are convergent rather than homologous. Numbers represent percent bootstrap values that are over 50 percent based on 2,000 samples of the data. Note that these species are only a small sample of the thousand or so different cichlid species found in various East African lakes and rivers. Having originated no later than about 3 to 4 Mya, these cichlids represent a prime example of explosive radiation among vertebrates, mostly because variations in a specialized pharyngeal jaw apparatus enabled individual groups to specialize on different prey, and because their breeding, sheltering and feeding behaviors can restrict them to extremely localized habitats (Meyer, 1990; Meyer et al., 1990). Perhaps the most dramatic of such speciation events occurred in Lake Victoria, the youngest of the East African lakes. It is known to have dried completely during the last Ice Age 12,400 years ago, yet it has produced about 500 new cichlid species since that time (Galis and Metz, 1998). (Adapted from Kocher, T. D., J. A. Conroy, K. R. McKaye, and J. R. Stauffer, 1993. Similar morphologies of cichlid fish in Lake Tanganyika and Lake Malawi are due to convergence. *Mol. Phylog. Evol.*, **2**, 158–165.)

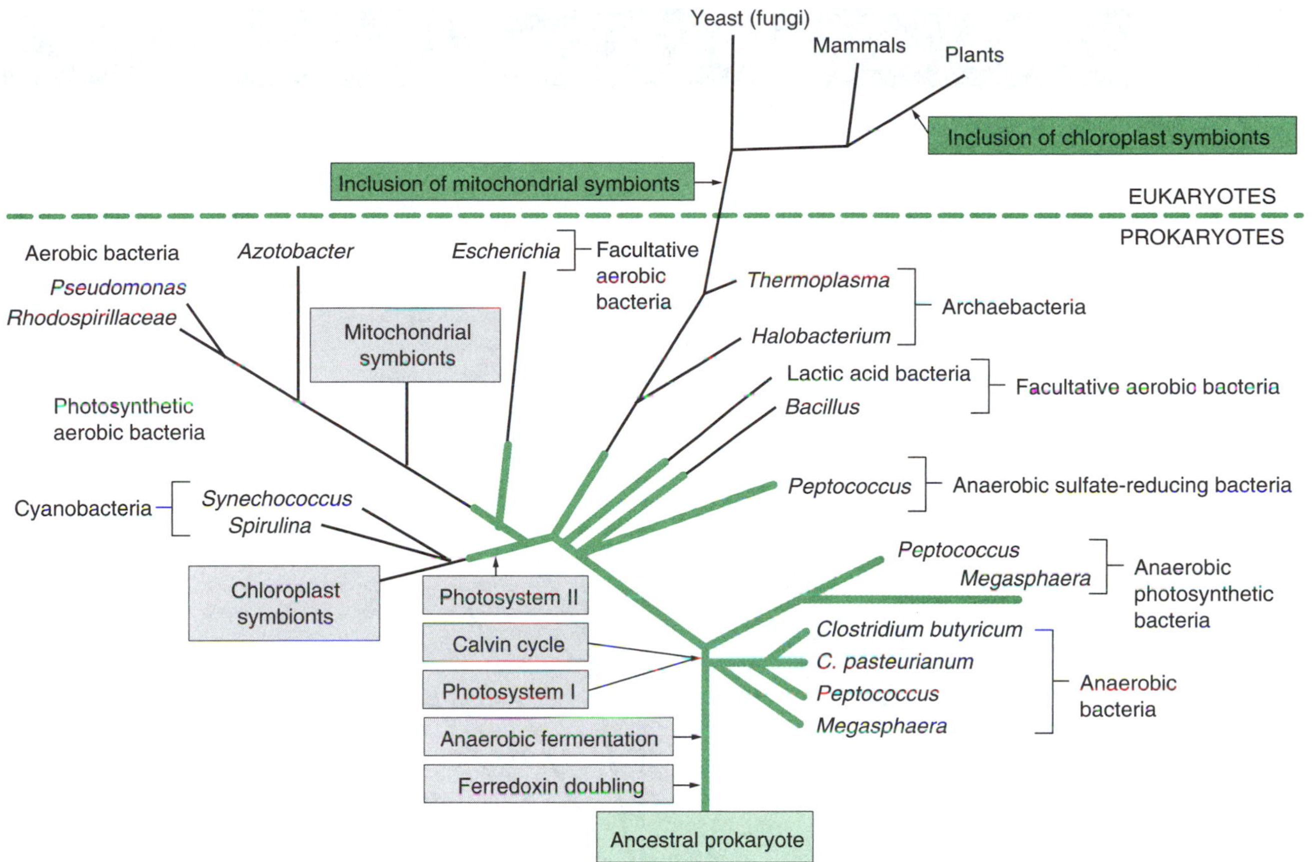

FIGURE 12-20 A composite evolutionary tree based on sequence analyses of ferredoxin, *c*-type cytochromes, and 5S ribosomal RNAs. The colored lines indicate segments of the tree dependent on anaerobic metabolism, and the thin black lines indicate groups using aerobic respiration. Not all branching relations have been clearly resolved. Note that photosynthesis evolved fairly early in prokaryotic phylogeny, is presumed to have been independently lost in a number of derived lines such as *Bacillus* and the eukaryotes, but was regained later in eukaryotic plants that ingested chloroplasts from a lineage close to the cyanobacteria. According to this tree, early eukaryotic-type cells were facultatively aerobic, because they stemmed from a line close to *Escherichia* and *Bacillus.* Only later did they replace these pathways with a more derived aerobic system introduced by the mitochondrial symbiosis. The dashed line across the illustration separates eukaryotes (above) from prokaryotes (below). (Adapted from Barnabas, J., R. M. Schwartz, and M. O. Dayhoff, 1982. Evolution of major metabolic innovations in the Precambrian. *Origins of Life,* **12,** 81–91.).

tions are incorporated (or fixed) at fairly regular rates over time, and that the degree of mutational distance for a phylogenetic interval correlates with the length of time in which such phylogenetic evolution took place. In other words, regarding changes in a specific gene, this assumption suggests that a **molecular clock** reflects the rates at which many mutations become fixed. Because fixation of these mutations mostly depends on the clock rather than on their adaptive or selective value, Kimura and other geneticists proposed that mutations are primarily neutral in their effects, the **neutral theory of evolution,** discussed in Chapter 23. The theoretical foundations of the neutral theory — and therefore of the concept of molecular clocks — comes from population genetics rather than from molecular phylogenetics (see the section on the cost of evolution and the neutralist position in Chapter 23). However, because molecular clocks have been used extensively in establishing phylogenies using molecular data, we include a discussion of them in this chapter.

One prominent finding that supports the concept of a molecular clock is the constant number of differences in amino acid sequence for the same hemoglobin chain derived from different vertebrates. Specifically, if we look at comparisons with the shark sequence for α-hemoglobin, other vertebrates differ from it by similar numbers of amino acid changes: carp 85, salamander 84, chicken 83, mouse 79, and human 79. This finding indicates that although considerable morphological changes have occurred in these different lineages over a 400 million-year period, constant rates of mutation may have been occurring for at least some proteins.

Even more obvious clocklike effects are seen in data showing increasing numbers of mutational differences between pairs of organisms separated by increasing time spans (**Table 12-7**). Thus, the amount of β-hemoglobin chain differences between human and monkey lineages — which separated from a common ancestor about 33 Mya (a in Table 12-7) — increases more than three times if we compare dif-

BOX 12-4 Five or Six Supergroups in the Eukaryote Tree of Life

In Box 8-1 (Kingdoms and Domains of Life), we discussed how increasing understanding of relationships between organisms has been reflected over time in the division of life by taxonomists into two broad divisions (prokaryotes and eukaryotes), two kingdoms (plants and animals), five kingdoms (prokaryotes, protista, fungi, plants and animals), and three domains (Bacteria, Archaea and Eukarya).

Over the last two decades, increasing concern has been voiced that the four eukaryotic kingdoms recognized in the five-kingdom system outlined in Box 8-1, do not accurately reflect evolutionary relationships. For example, the Kingdom Protista includes organisms that are closely related to animals (choanoflagellates), organisms closely related to plants (e.g., red algae) and other major lineages that are extremely distantly related (e.g., ciliates and slime molds, both of which are protists, are more distantly related to each other than animals are to Fungi; see Figs. 8-7 and **12-21**).

Molecular and evolutionary biologists have continued to find

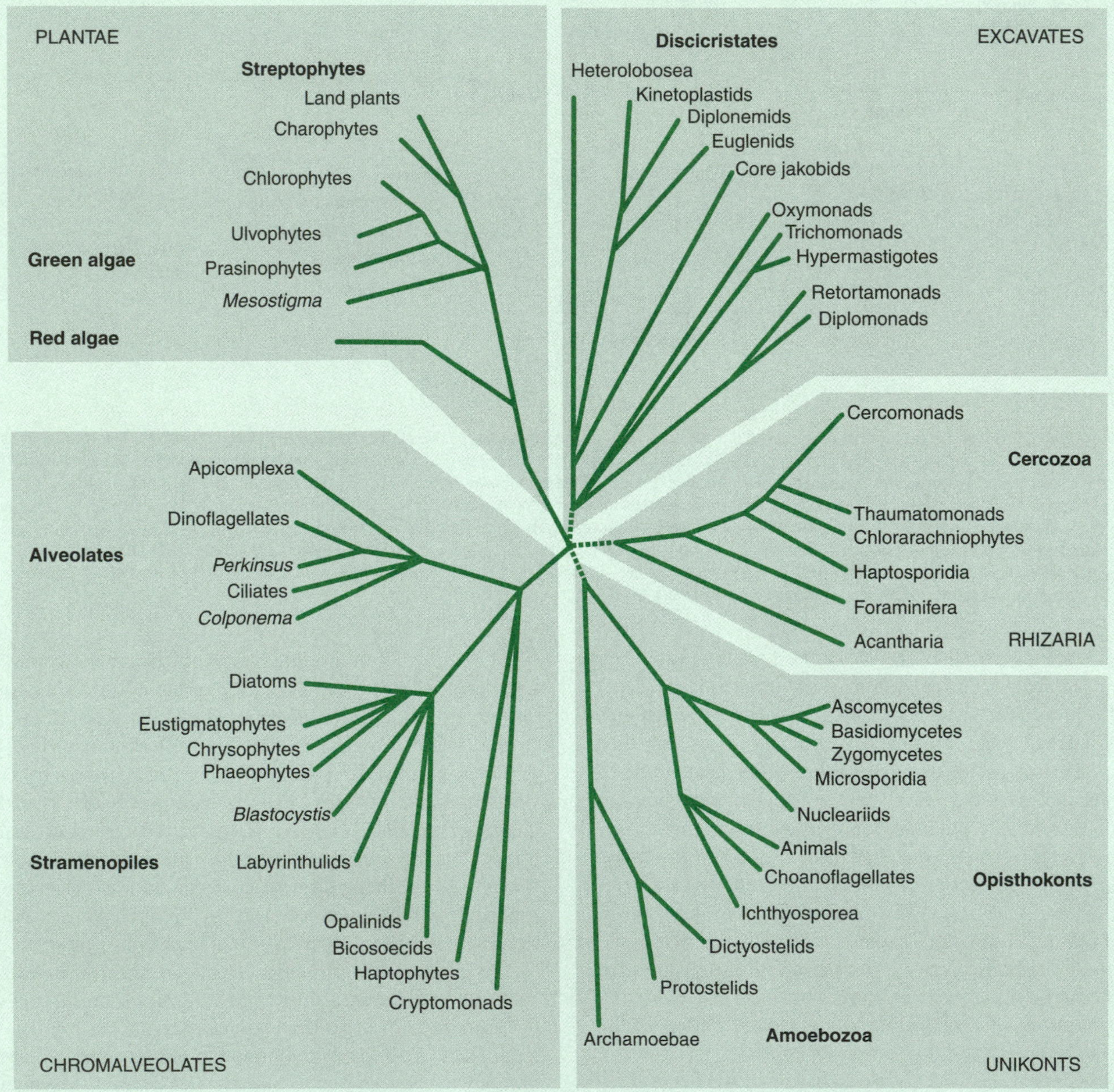

FIGURE 12-21 The eukaryotic Tree of Life as five supergroups. Relationships are unresolved. (Modified from Keeling, P. J., G. Burger, D. G. Durnford, B. F. Lang, R. W. Lee, R. E. Pearlman, A. J. Roger, and M. W. Gray, 2005. The tree of eukaryotes. *Trends Ecol. Evol.*, **20**, 670–676.).

new types of data and more and more organisms to analyze, to approximate better what is now called the **Tree of Life** (TOL) or sometimes the Universal Tree of Life (UTOL). As introduced in Chapter 10 and as discussed at the end of this box, the concept of a UTOL is severely compromised by horizontal gene transfer among prokaryotes. Because horizontal gene transfer is not a major factor in eukaryote evolution, re-creation of an eukaryote tree of life is a much more realistic project. Indeed, recent accounts of the diversity of eukaryote-based molecular data, internal cell morphology and cell biological considerations, have led to the replacement of the concept of four eukaryotic kingdoms with a concept of **five or six supergroups.** Although relationships between the supergroups are not resolved and although it is unclear how they branched off the TOL (see Fig. 12-21), affinity between members within a group is greater than affinity between the five supergroups.

Outlined below are two variations of the eukaryote TOL, one with five and the other with six supergroups.

Five Eukaryote Supergroups

- **Archaeplastida** (also known as **Plantae**). A supergroup that includes red and green algae and land plants; characterized by chloroplastids that arose by primary endosymbiosis and whose ancestors are thought to have been the first photosynthetic organisms.
- **Excavata.** Organisms that previously were members of the Protista. Commonly known excavates include *Giardia,* which causes the intestinal illness giardiasis (beaver fever, traveler's tummy), *Trypanosoma brucei,* which causes sleeping sickness, and *Trichomonas,* which causes trichomoniasis. Excavates are not united by any single morphological or molecular feature possessed by all members. Rather, ultrastructural or molecular features unite overlapping subsets of the 10 groups within this supergroup. Consequently, there is considerable controversy about this supergroup (**Fig. 12-22** and Keeling et al. (2005) for a recent analysis.

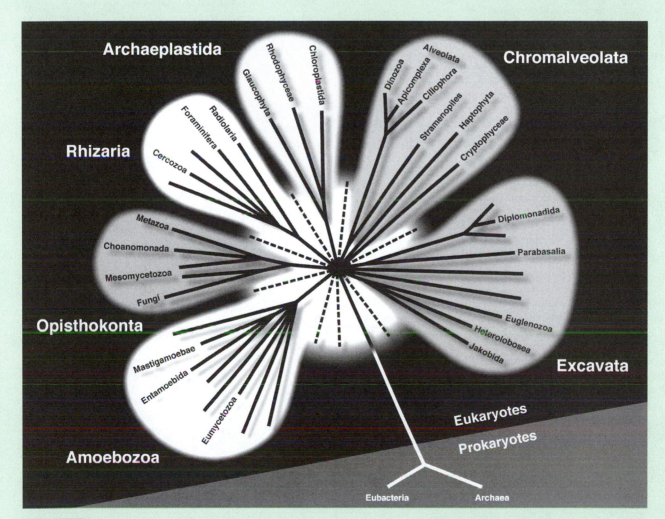

FIGURE 12-22 The eukaryotic Tree of Life as six supergroups. Relationships are unresolved. (Reproduced from Adl, Sina M., Alastair, G. B., et al. 2005. 52(5):399–451. Copyright © 2005 Blackwell Publishing. www.blackwell-synergy.com. Courtesy of Dr. Sina M. Adl.)

BOX 12-4 Five or Six Supergroups in the Eukaryote Tree of Life (*continued*)

- **Chromalveolates.** A supergroup of some 23 previous groups, including various types of algae (kelp, dinoflagellates, diatoms) that possess chloroplasts acquired by secondary endosymbioses as well as some important nonphotosynthetic groups, notably ciliates and the apicomplexan parasites (e.g., *Plasmodium,* which causes one form of malaria[a]). One subgroup, alveolates (Ciliates, apicomplexa and the dinoflagellates) is especially well supported through phylogenies of nuclear genes, but the monophyly of Chromalveolates as a whole is currently controversial.
- **Rhizaria.** A group of eukaryotic organisms recognized by Cavalier-Smith (2002, 2003) on the basis of molecular data alone. For some, this demonstrates the power of molecular approaches to the Tree of Life, but for others for whom groups should share some aspect of their morphology, it is disturbing. Many, but not all, Rhizaria are heterotrophic cells that use fine pseudopodia to capture prey such as prokaryotes and other eukaryotes. Perhaps the most well known rhizarians are the 4,000+ species of foraminiferans and radiolarians, examples of which Ernst Haeckel drew in such wonderful and glorious detail (**Fig. 12-23**).
- **Unikonta.** Unites two major groups: opisthokonts and amoebozoans (Cavalier-Smith, 2002). Opisthokonts alone include animals, fungi, some amoebae and some parasitic protists as well as choanoflagellates, that is, organisms traditionally placed within three of Whittaker's five kingdoms (see Table 8-1). Amoebozoans include slime molds and many types of more typical amoebae. Organisms comprising unikonts are a surprising collection — animals, fungi, slime molds, some amoebae and some parasitic protists — that are organisms traditionally placed within four of the five kingdoms in Table 8-1.

No aspect of the search for the Tree of Life stands in more stark contrast to the original phylogenetic trees drawn by Haeckel than the placement of animals as one of the five major groups of unikonts, and as the sister group to choanoflagellates. Haeckel's tree had man (not humans) at the top, and Haeckel would have placed Germans at the top of the human tree, with Bismarck as the top German.

Six Eukaryote Supergroups

A subgroup from within the *International Society of Protistologists,* after extensive consultation with specialists from many taxonomic disciplines (phycologists, mycologists, parasitologists, etc.), proposed six supergroups to embrace the eukaryotes (Fig. 12–21). The difference between the five- and six-supergroup schemes is that the latter treat opisthokonts and Amoebozoa as two separate supergroups, rather than grouping them as unikonts; that is, they recognize an additional monophyletic lineage within the protists (Adl et al., 2005). Briefly, the six supergroups are:

- **Archaeplastida.** Red and green algae, plants, Glaucophyta (unicellular flagellates with cyanobacterial-like plastids), Charophyta, *Mesostigma.*
- **Excavata.** Oxymonads, diplomonads, jakobids, and heterotrophic flagellates (some).
- **Chromalveolata.** Ciliates, dinoflagellates, Apicomplexa (collectively the Alveolata), brown algae, diatoms, zoosporic fungi (many), opalinids, Haptophyta, Cryptophyceae.
- **Rhizaria.** Foraminiferans, radiolarians (most), Cercozoans (most).
- **Opisthokonta.** Animals, fungi, sponges, choanoflagellates, Mesomycetozoa (parasitic protists), Mesozoa, *Trichoplax.*
- **Amoebozoa.** Amoebae (most), slime molds, testate amoebae (many), amoeboflagellates (some) and several species lacking mitochondria.

Both five- and six-supergroup classifications recognize the origin of the three multicellular groups — animals, plants, fungi — from monophyletic lineages of protists: animals and fungi within the Opisthokonta; plants from within the Charophyta; a multicellular group within the Archaeplastida (A. G. B. Simpson, 2003; Simpson and Roger, 2004).

In both schemes, plants (Plantae) have been expanded (and renamed Archaeplastida), to include several monophyletic protist groups as close allies of plants and of red and green algae. Both schemes reflect accumulating information on phylogenetic relationships, and the dynamic state of classification and of nomenclature, which, after all, should reflect our knowledge of organismal relationships, origins and evolution.

Nevertheless, in the current state of our knowledge, no matter how many sets and different types of molecular data we obtain, and no matter how sophisticated the ultrastructural analyses being applied to organisms previously ignored, difficulties remain, including determining:

- the rate of evolution of specific genes;
- whether rates of gene evolution have varied with time;
- the degree of sophistication of the phylogenetic methods employed and of the statistical approaches used to determine significant results;
- the role of horizontal gene transfer, which was common among most prokaryotes, means that creating a TOL representing prokaryote ancestry as ancestor-descendant relationships is fraught with major problems and, therefore, so is creating a universal tree of life (W. F. Doolittle, 1999, 2000; Doolittle et al., 2003); and
- accommodating the fact that most organisms do not separate their germ lines from somatic cell lines, which means that evolutionary change can potentially come from changes in different cell lines in different generations (Hall, 1999a; Extavour and Akam, 2003).

FIGURE 12-23 Part of the diversity of forms of foraminiferans drawn by Ernst Haeckel. (From Haeckel, E., 1904. *Kunstformen der Natur.*)

ferences between humans and artiodactyls (b), more than five times if we compare marsupials and placental mammals (c), and more than twelve times if we compare sharks and bony vertebrates (g in Table 12-7).

If molecular clocks exist, two consequences can be expected:

1. The lines of descent leading from a common ancestor to all contemporary descendants should have similar rates of fixed mutations because they have experienced similar durations.
2. The proportional rate of fixation that occurs in one gene relative to the rates of fixation in other genes should stay the same throughout any line of descent.

In a classic study, Fitch and Langley (1976) tested attributes of the molecular clock hypothesis for seven proteins whose amino acid sequences they examined in 18 vertebrate taxa. Using commonly accepted dates of divergence for the various common ancestors of these taxa to "calibrate" their molecular clock, they obtained temporal lengths for each separate line of descent. This allowed them to compare the number of nucleotide substitutions that occurred within a given time period for all proteins together and for each protein individually. The results showed that the rate at which all proteins have changed together varies significantly among the branches in the different lines of descent, indicating that molecular changes are not uniform over these geological periods.

Moreover, we cannot simply explain these differences in rates of protein change as arising from different generation times in different lines of descent; the rate at which individual proteins changed relative to other proteins differs significantly within single branches. If changes in generation time caused molecular rate changes, we would expect all proteins to behave similarly within any particular branch, and their individual relative rates to remain unchanged. This analysis indicates that the ticking of the molecular clock in each of these seven proteins is not constant in each branch of this phylogeny, whether scored with respect to time or generation.

Nevertheless, when we average the nucleotide substitutions over all seven proteins for each branching point in the phylogeny (rather than summing them or considering them individually), we find a marked uniformity in the rate of molecular change over time.

As Figure 12-23a shows, a mammalian phylogeny derived from the mutational distance data provides an average number of nucleotide substitutions at each branching point that corresponds with a linear relationship to time of divergence (see Fig. 12-23b); that is, this procedure "calibrates" an evolutionary molecular clock for these proteins, linking change to time.[11]

[11] Müller and Reisz (2005) used a similar approach to rates of vertebrate evolution when they used four calibration dates from the fossil record: the split between lungfish and tetrapods 408 to 419 Mya, between birds and crocodiles 243 to 251 Mya, between birds and lizards 252 to 257 Mya, between alligators and caimans 66 to 71 Mya.

TABLE 12-7 Evidence for progressive increases in amino acid substitutions over time (the molecular clock) for vertebrate β-hemoglobin chains

Organisms Compared	Amino Acid Changes per 100 Codons	Approximate Time (millions of years) to Common Ancestor
(a) Human/monkey	5	30
(b) Human/cattle	18	90
(c) Marsupial/placental mammal	27	130
(d) Bird/mammal	32	250
(e) Amphibian/amniote vertebrate	49	320
(f) Teleost fish/tetrapod vertebrate	50	400
(g) Shark/bony vertebrate	65	500

Source: Abridged and modified from Jukes, T. H., 1996. How did the molecular revolution start? What makes evolution happen? In *Evolution and the Molecular Revolution,* C. R. Marshall and J. W. Schopf (eds.). Jones and Bartlett, Sudbury, MA, pp. 31–52.

Given this linear correlation, we can use the following calculations to derive the overall rate at which nucleotide substitutions occur that lead to amino acid changes. Because an average of 98.17 nucleotide substitutions occurs at the farthest point of the linear slope (see number 16 in Figure 12-23) for a time

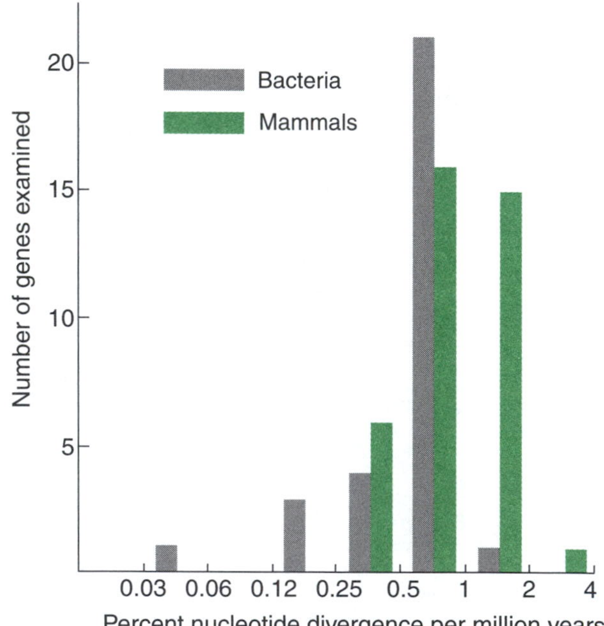

FIGURE 12-24 Rates of nucleotide substitutions per My at synonymous codon sites in 30 bacterial and 38 mammalian genes, according to A. C. Wilson et al. (Wilson, A. C., H. Ochman, and E. M. Prager, 1987. Molecular time scale for evolution. *Trends in Genet.,* **3,** 241–247.)

period of 120 My, and a total of 1,734 nucleotide positions in the seven proteins (578 codons $\times$ 3 nucleotides) exist, the rate of nucleotide change over 120 My is 98.17/1734 = 0.057; that is, about 6 out of 100 nucleotides caused amino acid substitutions during this interval. The annual rate of amino-acid–changing nucleotide substitutions in this lineage is therefore $0.057/(120 \times 10^6) = 0.47 \times 10^{-9}$.

The need to average changes among different genes to obtain an annual rate of nucleotide substitutions indicates that no single molecular clock applies to every nucleotide sequence. Variation in selection intensity in different parts of the genome, fixing mutations at different rates, is the most probable reason. In addition, as Britten (1986) points out, significant differences appear between taxonomic groups in nucleotide substitutions that have neutral effects on the phenotype, such as synonymous codon changes, for example, UUU (phenylalanine) → UUC (phenylalanine).

The data so far show **two different rates** at which such substitutions have incorporated: a *slow rate* of divergence for humans, apes and birds, and a *faster rate* for rodents and nonhuman primates, *Drosophila*, and sea urchins. Britten offers as a possible reason for slower rates of divergence, the lower mutation rate that would result from improved DNA repair systems, which he thinks favors evolution of increased parental investment in offspring, as seen in greater postnatal care, reflecting reduced maturity at birth. In contrast, A. C. Wilson and coworkers (1974) propose that the differences in evolutionary rates among taxonomic groups that Britten cites are probably exceptional. Instead, they suggest that an examination of many different genes in bacteria and mammals shows fairly *similar rates of nucleotide substitutions at synonymous codon sites* (**Fig. 12-24**). Riley (1989) disputed this view for such sites in *Drosophila* species, perhaps indicating taxon-specific rates operate. Although evidence may exist for a molecular clock in various lineages, it does not tick at the same rate in all taxonomic groups, nor for all genes or all proteins.[12]

Reasons offered for such variations include differences in selection intensity among genes and among groups, differences in DNA repair efficiency, different mutagenic experiences, different metabolic rates and different nucleotide generation times. Homologous genes coding for proteins with different properties — decarboxylases with different substrate specificities, for example — vary considerably in the rate of nucleotide substitutions. Sáenz-de-Miera and Ayala (2004) compared three decarboxylase enzymes in animals (in which they function as neurotransmitters) with two in plants (where they are involved in secondary metabolism). Depending on the enzyme and comparison, nucleotide substitutions per site ($\times 10^{-10}$) varied enormously:

- paralogous genes (dopa-decarboxylase and histidine decarboxylase) in the same mammalian lineage: 4.13 versus 1.95;
- orthologous genes between dipteran flies and mammals: 7.62 versus 4.13; and
- dopa-decarboxylase in *Drosophila*: 3.7 to 54.9 nucleotide substitutions per site ($\times 10^{-10}$),

results that are inconsistent with a uniform molecular clock, perhaps even with the notion of a molecular clock at all.

As a second example, Buggiotti and Primmer (2006), who compared the evolution of avian and mammalian growth hormone genes, found that the mammalian gene evolved in rapid bursts while the avian gene evolved at a more even pace. The bursts, which increased gene evolution 25- to 50-fold, are not randomly distributed through mammalian evolution but are associated with two lineages (ruminants and primates). Within the avian lineages analyzed, substitutions/amino acid site/year ranged between 0.56×10^{-9} and 0.80×10^{-9}. This can be compared with the mammalian gene in which the background ("non-burst") rate is 0.21 to 0.28×10^{-9} but the rapid rate is 5.6 to 10.8×10^{-9}. Unlike many studies of the timing of molecular evolution, these evolutionary biologists interpret the different rates in the context of the differing functional roles of growth hormone in controlling mammalian and avian growth that is, as adaptive, not as the running out of a molecular clock, an interpretation reinforced by their finding that functionally important sites in the avian gene are under positive selection.

Regulatory Genes and Their Evolutionary Consequences

One frequent observation that has emerged from *comparing different organisms on the molecular level is just how many organisms share the same kinds of proteins* (see Chapter 13).

For example, whether organisms are prokaryotes or eukaryotes, they share similar enzymes involved in basic biochemical processes such as glycolysis, amino acid synthesis, DNA replication and protein synthesis. When we examine them closely, the distinctive structural features of different organisms within any group, such as vertebrates, seem less dependent on differences among the kinds of proteins organisms have than on how they organize and regulate various common proteins such as actin, myosin, collagen and albumin.

In comparing the anatomy of cat with dog, or cat with mouse, differences primarily seem to depend on the extent and location of their common tissues (e.g., bone, muscle, nerve). This control over the quantity and placement of tissues arises from regulatory events during development that can easily be modified by mutational changes in the eukaryotic counterparts of prokaryotic regulator genes dis-

[12] See Li (1993), Gibbs and Dugaiczyk (1994), and Sáenz-de-Miera and Ayala (2004) for rates of ticking of evolutionary clocks.

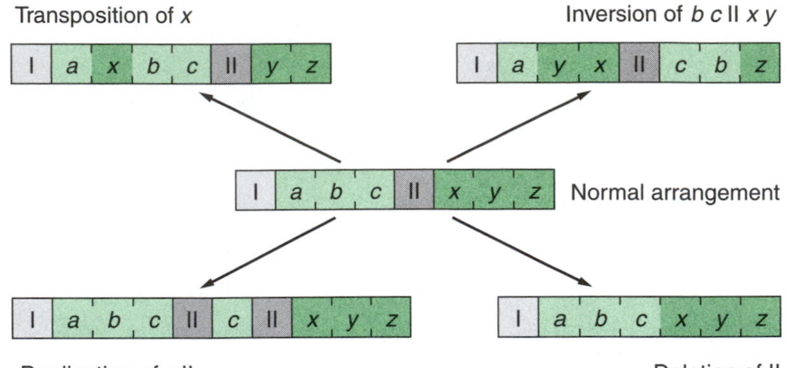

FIGURE 12-25 Some of the rearrangements that can occur in a gene sequence containing two sets of structural genes (*a, b, c* and *x, y, z*), each set controlled by separate regulator genes (*I, II*) and each change capable of causing striking mutant effects. For example, the deletion of regulator gene *II* places control of structural genes *x, y,* and *z* under regulator gene *I,* and the indicated inversion reverses the previous regulatory controls over genes *b, c* and *x, y*. (Adapted from Strickberger, M. W., 1985. *Genetics,* 3rd ed. Macmillan, New York; adapted from Wilson.)

cussed in Chapter 10 and later more fully in Chapter 13. As **Figure 12-25** shows, simple gene rearrangements such as deficiencies, duplications, inversions, and transpositions (translocations) can markedly change regulatory control over gene function. Because regulation is so important (see Chapter 10), changes in **gene regulation** may be responsible for major changes in evolution, that is, "new bottles for old wine" (see Chapter 13).

A. C. Wilson and coworkers (1974) based one study on comparisons between frogs and placental mammals. Early frogs appear in Triassic deposits of about 200 Mya, whereas placental mammals do not appear in the fossil record until some time during the Cretaceous, about 90 Mya. Despite their more ancient fossil history and their numerous arrays of species, frogs have undergone few phenotypic changes in evolution compared to the enormous adaptive radiation of placental mammals.

The more than 3,000 species of frogs look so much alike (**Fig. 12-26a**) that herpetologists have consigned them to a single clade, the order Anura. The 4,300 species of placental mammals (of which about 2,000 are rodents) diverged widely and usually have been classified into about 18 orders, ranging from bats to primates to whales (**Fig. 12-26b**). Because the amino acid sequences in proteins of both groups seem to have evolved at approximately the same rate, Wilson et al. suggested that the phenotypic similarity among frogs indicates that relatively few regulatory mutations have established themselves in frogs compared to mammals.

The importance of regulatory changes seems obvious in the sharp phenotypic contrast among related species such as humans and African apes. These two groups differ enough (brain size, facial structure, bipedal locomotion, and so on) for many anthropologists to place them in different taxonomic families (Hominidae, Pongidae), yet the composition of their structural proteins is strikingly similar: They have almost identical myoglobin and hemoglobin chains, cytochrome *c* proteins, and even fibrinopeptides (see Table 20-3). In fact, comparisons for any given protein between these two groups show an average of more than 98 percent identity in amino acid sequence (King and Wilson, 1975), so it is perhaps no surprise that on a much larger genome scale Neanderthals and humans share 98 percent of their genes.

These and other examples show that regulatory mutations can play a larger role in the morphological and functional differentiation of species than many structural gene changes. Some have used this observation to support the notion that new species (which might be sufficiently different from existing species to be classified into new genera or new orders) can arise from regulatory changes that have a large phenotypic effect, producing major taxonomic divisions over relatively short periods of time;[13] that is, a population undergoing only small gradual changes may persist that way for long periods until pronounced regulatory changes are incorporated that allow one or more of its small isolated groups to evolve rapidly into a higher taxonomic category.

Because some of the fossil data for vertebrates, mollusks, and other animals show such periodic bursts of evolutionary activity, paleontologists such as Eldredge and Gould (1972) gave the name **punctuated equilibrium** to what they saw as a process of long-term uniformity of fossil populations punctuated by geologically short periods of rapid speciation. Others argue that because the fossil record has gaps, punctuated equilibria are only apparent, not real; what seems to be rapid in geological time may involve many thousands of generations.

Although this issue has generated a great deal of discussion, at times heated, many arguments and counterarguments (see the discussion in Chapter 24), evolutionary studies demonstrate that gradual and rapid changes occur

[13] The term **macroevolution** is sometimes used for this view, but as originally proposed, macroevolution is evolution at or above the species level, and **microevolution** is evolution within a species, without reference to whether the mechanisms are the same at each level. See Chapter 24 for further discussion on this point.

Walter M. Fitch

Born: May 21, 1929
Birthplace: San Diego, California
Undergraduate degree: University of California–Berkeley
Graduate degree: Ph.D. University of California–Berkeley, 1958
Postdoctoral training: Stanford University, 1959–1961; University College, London, 1961–1962
Present position: Professor of Ecology and Evolutionary Biology School of Biological Sciences, University of California–Irvine[14]

What prompted your initial interest in evolution?
It is startling to be told that one might be related by ancestry to a fish, a fly, a worm, a plant, and a mushroom. To find evidence for or against such a possibility seemed like a wonderfully stimulating way to spend one's life.

What do you think has been most valuable or interesting among the discoveries you have made in science?
My most exciting experience was developing a method to analyze sequences of amino acids in proteins, then applying that method to 20 cytochromes *c* and seeing produced, from one small protein, a wonderful tree spanning most of the eukaryotic kingdom with considerable accuracy. (This was published in a 1967 paper in *Science* with E. Margoliash.)

What areas of research are you (or your laboratory) presently engaged in?
I am currently developing improved methods for

Multiple-sequence alignment
Reconstructing molecular trees
Assigning events (including gene conversions) to trees
Detecting and accounting for reticulate evolution (networks rather than trees)
Predicting the future course of human influenza evolution

I am currently applying such methods in many areas, but especially to viruses (flu, HIV, and vesicular stomatitis). I love the molecular clock problem, too.

In which directions do you think future work in your field needs to be done?
See my list above. Every time a result appears ambiguous, one should ask whether the problem is in the data or in the method. And every time you say "the method" you have just recognized a worthwhile problem that, if you solve it, permits you to be the first to apply a new and/or more powerful technique to many areas and discover new things as well as answer other people's questions.

What advice would you offer to students who are interested in a career in your field of evolution?
You can't know everything but in evolution, it pays to be broadly rather than narrowly trained. New insights frequently arise when different concepts come together to provide an illuminating spark of understanding, and that flash comes more readily by crossing disciplinary boundaries. And the more problems you understand, the more likely your observations will suggest a solution. In Pasteur's words, "Chance favors the prepared mind." The corollary to that is "Treasure [and understand] your exceptions."

[14] A biography is available at http://ecoevo.bio.uci.edu/Faculty/Fitch/Fitch.html. Accessed April 2007. Recent publication: Fitch, W. M., R. M. Bush, C. A. Bender, K. Subbarao, and N. J. Cox, 2000. On predicting the evolution of human influenza A. *J. Heredity,* **91,** 183–185

EVOLUTION ON THE WEB biology.jbpub.com/book/evolution

during evolution. What we have not yet resolved is the relative importance of these changes in explaining speciation and the evolution of higher taxonomic categories. On the phenotypic level, some speciation events certainly seem to be gradual; sibling species of *Drosophila,* for example, look much alike, whereas other related species that are probably no older than some *Drosophila* sibling species (for example, chimpanzees and humans) look different. Among questions raised by these observations are:

- How many genetic differences does it take to make a species?
- Do differences in morphology between species correlate with the number of genetic differences?

Although these questions are still difficult to answer (see also Chapter 24), molecular studies are now providing important information. For example, DNA mapping of male sterility factors in *Drosophila simulans-mauritania* hybrids indicates there may be more than 100 genes involved in reproductive isolation between these morphological and molecularly similar sibling species (Davis and Wu, 1996). We don't yet know whether such high numbers of genes explain reproductive isolation between other closely related species, but such findings certainly indicate that some speciation events depend more on genetic interaction between many loci (epistasis) expressed in hybrid sterility, than on morphological or developmental novelty in one or few loci. In support are findings of Omland (1997) of coupling between the amounts of morphological and molecular evolution in some groups.

The prospect for constructing phylogenies for lineages whose relationships long seemed obscure are now attainable because of improved nucleotide sequencing techniques and a rapidly increasing fund of molecular data. Ribosomal RNA

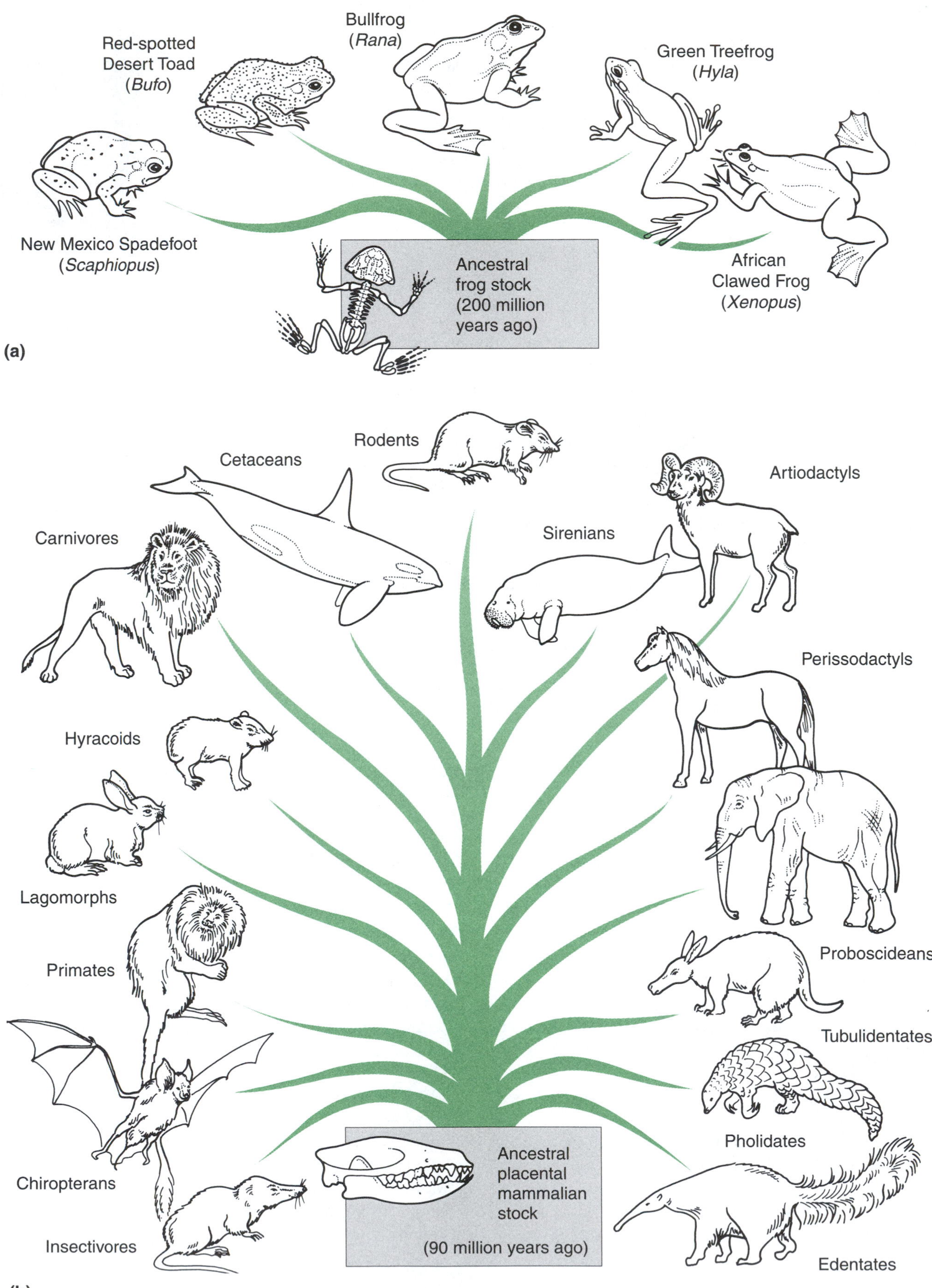
Bullfrog
(Rana)
Red-spotted
Desert Toad
(Bufo)
Green Treefrog
(Hyla)
New Mexico Spadefoot
(Scaphiopus)
Ancestral
frog stock
(200 million
years ago)
African
Clawed Frog
(Xenopus)
(a)
Rodents
Cetaceans
Artiodactyls
Carnivores
Sirenians
Perissodactyls
Hyracoids
Lagomorphs
Primates
Proboscideans
Tubulidentates
Pholidates
Chiropterans
Ancestral
placental
mammalian
stock
(90 million years ago)
Insectivores
Edentates
(b)

FIGURE 12-26 (a) Major forms of extant frogs exhibit considerable similarity in basic body forms. The skeleton of a Triassic anuran dating back to about 200 Mya is also shown. Although frogs are similar in adult phenotype, their development can differ considerably among species, producing immature forms ranging from tadpoles to froglets, in media ranging from water to brood pouches. (b) Representatives of 15 extant clades (orders) of placental mammals illustrating diversification in body form over 90 My of evolution. The trees do not indicate phylogenetic relationships (Adapted from Strickberger, M. W., 1985. *Genetics*, 3d ed. Macmillan, New York.)

sequence analysis has defined a new major group among prokaryotes, the Archaebacteria (see Fig. 8-7), and is being used to delineate relationships among eukaryotic phyla (see, for example, Fig. 15-7). Investigators have even extended the use of sequencing comparisons to studying molecular evolution on the laboratory level (see Box 12-2).

KEY TERMS

cloning
concerted evolution
convergences
DNA–DNA hybridization
gene cluster
gene conversion
gene duplications
gene family
gene regulation
haplotype
hemoglobin
heterochromatin
macroevolution
maximum likelihood estimates
microevolution
microsatellites
molecular clock
molecular cloning
multigene family
nucleotide
orthologous genes
paralogous genes
parsimony
phylogenetic tree
polymerase chain reaction (PCR)
pseudogenes
punctuated equilibrium
repetitive DNA
restriction enzymes
restriction fragment length polymorphisms (RFLPs)
satellite DNA
unequal crossing over

DISCUSSION QUESTIONS

1. What are the advantages in using proteins and nucleic acids to determine phylogenies?
2. How are polymerase chain reactions used in making phylogenetic determinations?
3. What accounts for the presence of paralogous and orthologous genes? Provide examples of their evolution.
4. Does the relative amounts of DNA among organisms reflect their phylogenetic positions? Explain.
5. What evidence, pro and con, has been provided on the validity of phylogenies based on DNA–DNA hybridizations?
6. In a choice between DNA sequences that change slowly during evolution and sequences that change rapidly, which provide a better estimate for establishing a phylogeny among closely related species and why?
7. Why have 5S rRNAs been used for nucleic acid phylogenies?
8. How do the findings of Barnabas and coworkers (see Fig. 12-19) support the concept that eukaryotic mitochondria and chloroplasts originated from symbiosis with prokaryotes?
9. How do researchers determine how many super kingdoms best reflect relationships among organisms? Do you think it will be possible to produce a true Tree of Life?
10. What evidence, pro and con, has been gathered on the validity of the molecular clock?
11. Gene regulation.
    a. What are regulatory gene changes?
    b. How do such changes arise?
    c. Why are such changes presumed to have greater evolutionary consequences than changes in structural genes?
12. Enzymatic evolution.
    a. How can you select for enzymatic changes in the laboratory?
    b. What kinds of adaptive enzymatic changes are observed?
    c. How did the EBG enzyme strain of *E. coli* evolve?
13. How did laboratory selection reduce the genome length of the Qβ RNA virus?

EVOLUTION ON THE WEB

Explore evolution on the Internet! Visit the accompanying Web site for *Strickberger's Evolution,* Fourth Edition, at http://www.biology.jbpub.com/book/evolution for Web exercises and links relating to topics covered in this chapter.

RECOMMENDED READING

Adams M. D., S. E. Celniker, R. A. Holt, C. A. Evans, et al., 2000. The genome sequence of *Drosophila melanogaster. Science,* **287,** 2185–2195.

Aparicio S., J. Chapman, E. Stupka, N. Putnam, et al., 2002. Whole-genome shotgun assembly and analysis of the genome of *Fugu rubripes. Science,* **297,** 1301–1310.

Arabidopsis Genome Initiative, 2000. Analysis of the genome sequence of the flowering plant *Arabidopsis thaliana. Nature,* **408,** 796–815.

Burger, G., M. W. Gray, and B. F. Lang, 2004. Mitochondrial genomes: anything goes. *Trends Genet.,* **19,** 709–716.

Feng, D.-F., G. Cho, and R. F. Doolittle, 1997. Determining divergence times with a protein clock: Update and reevaluation. *Proc. Natl Acad. Sci. U. S. A.*, **94,** 13028–13033.

Herrmann, B., and S. Hummel (eds.), 1994. *Ancient DNA*. Springer-Verlag, New York.

Krings, M., A. Stone, R. W. Schmitz, H. Kainitzki, M. Stoneking, and S. Pääbo, 1997. Neanderthal DNA sequences and the origin of modern humans. *Cell,* **90,** 19–30.

The *C. elegans* Sequencing Consortium, 1998. Genome sequence of the nematode *C. elegans:* a platform for investigating biology. *Science,* **282,** 2012–2018.

Venter J. C., M. D. Adams, E. W. Myers, P. W. Li, et al., 2001. The sequence of the human genome. *Science,* **291,** 1304–1351.

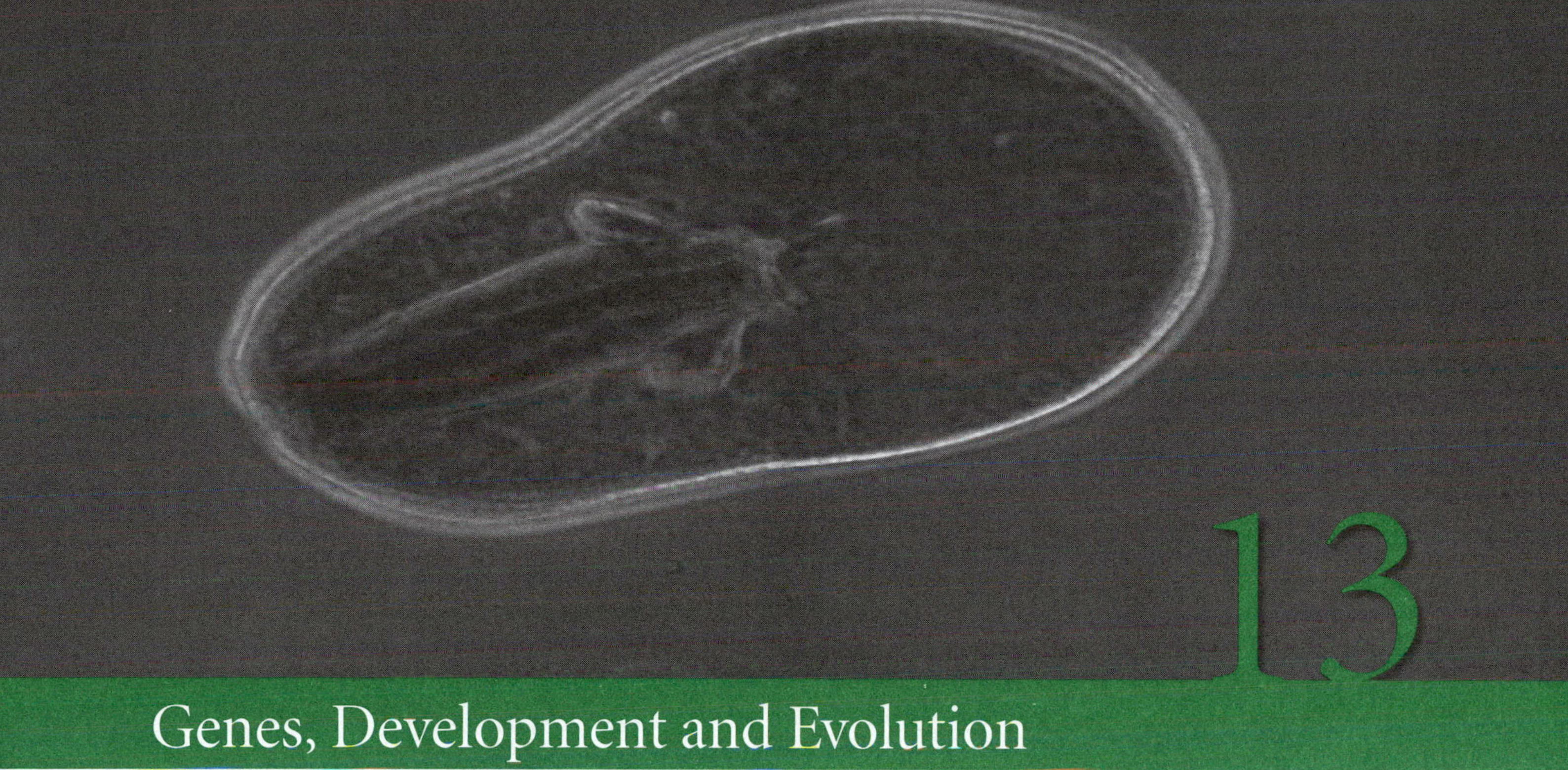

13 Genes, Development and Evolution

Chapter Summary

Molecular analysis of viruses and prokaryotes has provided many details of differential control of genes. For example, we know that interaction with their environment leads to negative and positive controls on viral growth, bacterial metabolism and sporulation. Development — the progression from egg to adult in multicellular organisms — is in large part a consequence of differential gene and cellular activity, both of which are regulated in time and space.

In eukaryotes, analyses of development, which focused initially on morphology, and then on cell-cell, cell-extracellular matrix and tissue-tissue interactions, have been extended by the discovery of the roles of genes in development. New emphasis has been placed on identifying genes and their role in the differences that accumulate between embryological stages and how such genetic effects are brought about. At least ten common themes have emerged in the molecular genetics of development.

Changes in cell structure and function during development derive from changes in the *presence or activity* of cellular proteins produced by gene transcription. Which genes are transcribed into messenger RNA, which transcripts are translated into proteins, and which proteins are activated, are the result of interactions that often begin with signals that initiate different genetic regulatory pathways or networks. As the developmental positions of cells and tissues change in time and space, they become subject to new sets of signals that lead to further developmental consequences. Furthermore, cell lineages (modules) that share common cellular histories are organized into specific tissues that perform specific functions. Genetic changes can modify each developmental stage by affecting regulatory agents and processes, from signal reception to transcription and translation, a process that François Jacob named *tinkering.*

Because of genetic variability, many kinds of developmental genetic changes occur, including those that shift a gene's expression from its normal anatomical position and cause misplacement of organs from expected body segments (**homeotic mutations**). Developmental novelties can arise from changes in the timing or position of gene expression (heterochrony and heterotopy), changes in growth rate (allometry) and/or changes in other regulated events. Increasing regulatory complexity enables increasing developmental novelty.

As with any other trait in which genes and their alleles have been selected for their functional (adaptive) value, an organism's development is the outcome of an historical evolutionary process. Whether new mutations can incorporate successfully into an organism's development depends on how they interact with existing developmental processes. Such interactions channel development and both place limits on, and create opportunities for, evolutionary change. Among the evolutionary effects that act upon development when organisms face a variety of different stimuli is the selection of genotypes that enhance the production of an array of phenotypic variants or the maintenance of a single adaptive phenotype.

The term **differentiation** is often used to describe developmental processes that cause structural or functional distinction among parts of an organism. Developmental biologists often restrict the term to the process(es) of cells becoming different from one another and/or becoming different from earlier stages in their own lineage: **differentiation as cell differentiation or specialization.** Thus, we speak of four stages in the differentiation of cartilage cells (*chondro*, cartilage) — prechondroblasts, chondroblasts, chondrocytes and hypertrophic chondrocytes — or we compare muscle cells with chondroblasts, especially in situations where both cells types can arise from a common stem cell.

In single-celled organisms, differentiation effects, by definition, only one cell or parts of a cell. In multicellular organisms, different cells differentiate from one another so that cells, tissues and organs come to differ from each other. Differentiation in multicellular organisms arises as a result of the appearance of gene products differing in quality or quantity from those in other parts of the organism, either because of *intrinsic maternal control* in a cell lineage, after *epigenetic* interactions with cells or cell products (growth factors, hormones), or in response to environmental cues such as temperature, pH, prey or predator density, to name but a few.[1]

Eukaryote Development

Morphology is a term for the shape of an organism or for a feature of an organism. **Morphogenesis** is the set of processes whereby that shape is generated. Despite the enormous range of morphologies, the number of morphogenetic processes is limited.

Interest in morphogenesis goes back at least as far as Aristotle, who studied embryonic development of the chick in great detail. Aristotle noted that chick embryos, which appear almost structureless early in their development, give rise to complex differentiated organisms later in development. The fact that development produces an exact replica of the parents was long held to result from the transmission of adult structures in miniature form. As discussed in Chapter 1, this **preformationist** doctrine was replaced in the eighteenth and nineteenth centuries by the **epigenetic** view that adult structures are absent from the early embryo but appear progressively during embryonic development.

The sequential pattern of development — during which cells and organs go through successive stages of differentiation — is reminiscent of the sequential pattern of organismal evolution itself; that is, organisms and their organs do not appear *de novo*, but rather result from a sequential process in which each successive change produces an additional and incremental feature (see Figure 3-1 for animals and Figure 14-10 for plants). In multicellular development, organs begin with only a few types of cells, which undergo a series of successive differentiations as

[1] For intrinsic and epigenetic control of development, see Hall (1999a), Pigliucci (2002) and Hall et al. (2004).

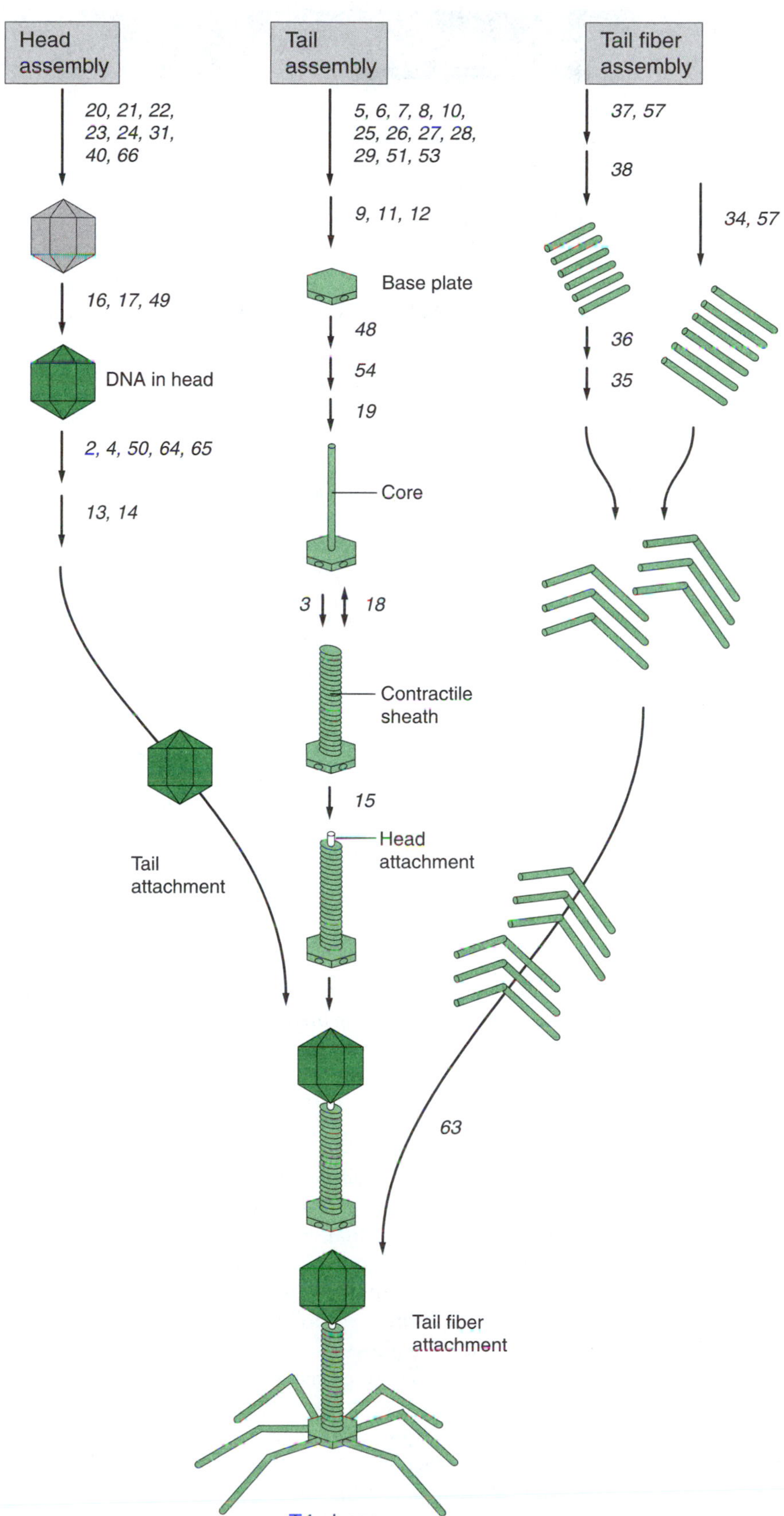

FIGURE 13-1 Sequence of T4 phage development and the genes involved in the various morphological steps. Although many of the genes (indicated by numbers) produce proteins or polypeptides directly incorporated into the T4 assembly, the products of other genes (e.g., *38, 57, 63*) are not incorporated but are still necessary for the assembly. (Modified from Wood, W. B., 1980. Bacteriophage T4 morphogenesis as a model for assembly of subcellular structure. *Q. Rev. Biol.*, **55**, 353–367.)

BOX 13-1 Viruses and Bacteria: Replication and Morphogenesis

Morphogenesis of the T4 Bacteriophage

Bacteriophages are viruses that infect and devour bacteria. The T4 bacteriophage infects the common bacterium *Escherichia coli* (often referred to simply as *E. coli*).

Each change in morphology in the production and assembly of T4 bacteriophages is under genetic control (Fig. 13-1). Phage DNA — T4 has 168,800 base pairs of DNA — uses bacterial host DNA-dependent RNA polymerase to synthesize phage mRNA, which translates on host ribosomes to produce a number of "early" proteins, many of which are necessary for the subsequent synthesis of phage DNA. Within five to seven minutes after infection, these early enzymes lead to the formation of a pool of "vegetative" phage DNA fibrils.

After the synthesis of early viral proteins ceases, a number of "late" proteins appear, including an inducing protein that acts as a scaffold to position the *head protein* molecules that form the viral head capsule or shell. As T4 phage DNA enters this viral shell, scaffold protein is destroyed. DNA packaging is completed when a "headful" of DNA has been enclosed. Other late proteins include those involved in the various tail structures as well as the lysozyme used to rupture the host cell walls. Altogether, a completed phage particle is composed of 30 to 40 different components, each genetically produced in a sequence of coordinated morphogenetic interactions.

Replication of the λ bacteriophage

Although viruses may seem passive parasitical replicators, some possess genes that have been evolutionarily selected so that they do or do not replicate. One example occurs in the lambda (λ) bacteriophage, which can remain quiescent within its host in *temperate form,* or can replicate and destroy its host by *lysis,* depending on the state of its bacterial host, and a sequence of gene actions that produce host-sensitive proteins. One gene, the gene for the λ cI protein (λ repressor protein) normally prevents viral replication by acting as a **repressor** that inhibits viral genes necessary for replication. Host conditions such as stress or ultraviolet radiation cause cleavage of the cI protein, leading to lysis (Ptashne, 1992).

Sporulation in Bacteria

Because microbial organisms offer advantages in dissecting and analyzing biochemical details of various developmental stages and their causative genetic elements, the molecular basis of even fairly complex morphological changes such as bacterial sporulation is well understood. For example, the decision to form a spore in some bacteria, such as *Bacillus subtilis,* is usually caused by a limitation in available nutrients, especially carbon and nitrogen. As shown in **Figure 13-2,** the new pathway consists of a succession of enzymatic steps that enable the cell to change morphologically. Some bacterial enzymes increase in amount, others decrease, and still others appear that are specific to sporulation — an exact pattern that is under precise genetic transcriptional control. New forms of RNA polymerase are produced through a change in component proteins that allow new sets of genes to be transcribed into mRNA. The spores that result neither replicate nor show metabolic activity, and yet may survive for hundreds of years under adverse conditions. When exposed to an appropriate growth environment, spores germinate and resume normal morphology and function for vegetative growth and replication.

they divide and increase, rather than collectively undergoing a single spontaneous differentiation into their final individual forms.

Development and Evolution

In the decades after Darwin's *Origin* was published, many embryologists began to explore evolutionary areas, emphasizing development as an evolving entity that could be studied by comparing embryological relationships. They provided a large literature on developmental changes in vertebrate embryos, patterns of the sutures of molluskan shells and other comparative data. Out of these works came emendations of Haeckel's "biogenetic law" (see Chapter 3), leading to the theory that ontogeny (development of an individual) can reflect an ancestral sequence of developmental processes, and that embryological stages can be used as morphological traits to evaluate relationships between some organisms.

Although these investigations offered considerable descriptive information, they left three unanswered questions:

1. What factors are responsible for *changes from egg to adult*?
2. How are such changes *transmitted between generations*?
3. What forces enable different developmental changes to become established in different generations and different groups?

By 1950, well before molecular genetics blossomed, a few geneticists (listed below; see literature cited for individual studies) were investigating the impact of genes on developmental processes in a variety of normal and mutant organisms, including:

- genes controlling rates of physiological processes (Richard Goldschmidt), research that led directly to the concept of heterochrony discussed later in the chapter;

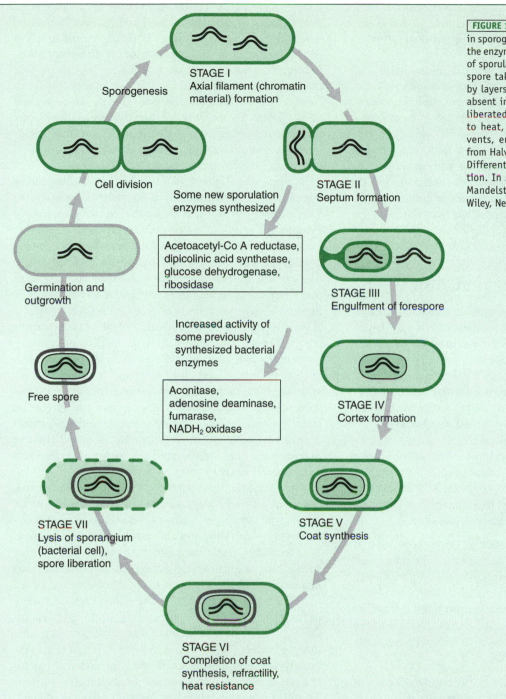

STAGE I
Axial filament (chromatin material) formation

STAGE II
Septum formation

STAGE IIII
Engulfment of forespore

STAGE IV
Cortex formation

STAGE V
Coat synthesis

STAGE VI
Completion of coat synthesis, refractility, heat resistance

STAGE VII
Lysis of sporangium (bacterial cell), spore liberation

Free spore

Germination and outgrowth

Cell division

Sporogenesis

Some new sporulation enzymes synthesized

Acetoacetyl-Co A reductase, dipicolinic acid synthetase, glucose dehydrogenase, ribosidase

Increased activity of some previously synthesized bacterial enzymes

Aconitase, adenosine deaminase, fumarase, $NADH_2$ oxidase

FIGURE 13-2 Sequence of morphological steps in sporogenesis in *Bacillus subtilis* and some of the enzymatic events involved. From the onset of sporulation to the production of a finished spore takes between 6 and 8 hours. Coated by layers of special proteinaceous materials absent in vegetative cells, the final spore is liberated by cell lysis. Spores are resistant to heat, ultraviolet radiation, organic solvents, enzymes and desiccation. (Adapted from Halvorson, H., and J. Szulmajster, 1973. Differentiation: Sporogenesis and germination. In *Biochemistry of Bacterial Growth*, J. Mandelstam and K. McQuillen (eds.). John Wiley, New York, pp. 494–516.)

- genes underlying different kinds of mutant phenotypes (Hermann Muller) and metabolic diseases in humans (Archibald Garrod);
- the genetic basis of pigment development in flowers (H. Onslow, R. Scott-Moncrieff) and in the eyes of *Drosophila* (George W. Beadle and Boris Ephrussi);
- vertebral development in normal and mutant mice (L. C. Dunn, S. Gluecksohn-Schoenheimer; Hans Grüneberg);
- the basis of pleiotropy in *Drosophila* (Ernst Hadorn); and
- metabolic pathways in *Neurospora* (George W. Beadle and E. L. Tatum).

Despite these efforts and the broad research that followed, many morphologists insisted that the gap between genotype and phenotype was great enough to be "unbridgeable," and that genetics could offer little information to explain development.

Based on research begun by Richard Goldschmidt, L. C. Dunn, and George Beadle and Boris Ephrussi and continued by many others, the science used to answer the three questions outlined above shifted from embryology to developmental genetics. Within the last two decades, such genetic investigations have accelerated rapidly with the phenomenal expansion of biomolecular techniques leading to the integration of molecular, developmental and evolutionary studies into what is now known as evolutionary developmental biology (or evo-devo), the origins of which are found in the evolutionary embryology that followed the publication of *The Origin* in the mid-nineteenth century.[2]

Regulation: Genes in Context

Much of our initial understanding of gene regulation came from analyses of how genes function in bacteria and viruses (see Chapter 10). One elaborate system, in which gene activity is responsible for each change in morphology, is seen in the production and assembly of *T4 bacteriophages* (**Figure 13-1**). Bacteria provided the first information on mechanisms of gene regulation, and on how genes regulated aspects of bacterial life history such as sporulation (**Box 13-1**).

A classic molecular example of how the activity of a bacterial cell can be regulated to perform differently under one condition (presence of lactose sugar) rather than another (absence of lactose) is illustrated in Figure 10-13. Production of the β-galactosidase enzyme that metabolizes lactose is regulated by a regulatory protein with two sites:

1. A site that binds to a particular DNA sequence that prevents (represses) transcription of the messenger RNA, which would ordinarily be used to synthesize β-galactosidase.
2. A site that binds to the inducer (produced by lactose) that changes the stereochemical conformation of the regulatory protein, causing it to vacate its normally repressive position on DNA.

For β-galactosidase synthesis, the regulatory protein acts as a repressor to prevent synthesis, unless lactose is present. Activator regulatory proteins, on the other hand, act oppositely. An inducer or specific condition empowers the activator to bind to a DNA sequence (*promoter;* see Fig. 10-14), enabling or enhancing messenger RNA transcription. Genetic control, negative or positive, is precisely tuned to allow specific functional shifts.

Soon it became clear to molecular geneticists that **regulatory processes also control eukaryotic gene expression,** both through negative (repressor) and positive (activator) mechanisms. How a cell differentiates in form and function depends on which of its genes are available for transcription. In turn, transcription depends on stimuli — "signals" — that may come from other cells by diffusion, direct contact or from the environment. Such signals, which can be organic or inorganic, may enter the cell directly through the plasma membrane or, commonly in development, attach to membrane receptors that amplify the signal within the cytoplasm, providing a mechanism of **signal transduction.**

In general, as **Figure 13-3** illustrates, signal transduction begins with an extracellular molecule that binds (ligates) to a cell-surface receptor specific for that *ligand.* A cascade of cytoplasmic reactions follows which can activate specific cytoplasmic transcriptional precursors, either by removing their inhibitors or by allowing them to associate functionally. These proteins, in turn, form multiprotein nuclear complexes that stimulate or repress transcription of specific genes by binding to gene promoters and enhancers. The intricate relationship between a ligand and its cell surface receptor is a good example of **coevolution,** in which changes in one element (ligand) select for changes in the other (receptor). Such signaling systems show extensive homology across various phyla, even when the features they are involved in producing are not homologous. Insect and vertebrate limb development is one example. As a second, Ras protein determines genital structures (vulva) in *Caenorhabditis,* eye development in *Drosophila* and cellular proliferation and differentiation in mammals.

Furthermore, because development is based on dynamic processes, a cell's responsiveness to signals changes with its history; the same signal pathway can express or inhibit different genes depending on a cell's position in time and space. This flexibility enables an economy of signal pathways for a multitude of purposes. Signaling pathways, such as that illustrated in Figure 13-3 and many others, are conserved in a wide variety of metazoans and regulate a wide variety of functions. For example, depending on the extracellular signal, phosphorylating enzymes such as mitogen-activated protein kinases (MAPKs) can activate different transcriptional proteins by following different routes from cell membrane to nucleus. In plants, common signals such as ethylene act in transduction pathways affecting a variety of functions, including seed germination, fruit ripening and cell development.[3]

Choices among binding positions on the DNA molecule allow choices of gene expression; there are more than 20 regulating DNA sites for the actin protein gene (used

[2] For the history of evo-devo, see Bowler (1996), and Hall (1999a, 2006d), Hall and Olson (2003) and Nyhart (1995).

[3] See Davidson (2006), S.B. Caroll et al. (2005), M. Freeman (1998), Gerhart and Kirschner (1997), and Wilkins (2002) for conservation of signal transduction

for cytoskeletal and muscle filaments) in the sea urchin *Strongylocentrotus purpuratus.* Some activate while others repress transcription. Interactions between regulatory proteins also can occur, converting activators to repressors and *vice versa.*

Regulatory processes affect not only transcription but also act at many other basic molecular levels:

- **Changes in chromatin** can affect DNA segments or even entire chromosomes. Changes include methylation, imprinting or those that cause *Drosophila* X-chromosome inactivation (see Chapter 10). Methylation of cytosine is an important means of silencing and regulating gene function. The first DNA "methylation map" of an entire genome, that of the flowering plant *Arabidopsis thaliana,* which was published in September 2006, indicates that methylation is found within transcribed regions in over one third of expressed genes (such genes are constitutively active), but only in 5 percent of promoter regions of expressed genes, whose expression is much more tissue-specific, indicating the role of methylation in gene regulation (Zhang et al., 2006).
- **Post-transcriptional modification** of messenger RNA can occur through different splicing patterns that produce different mature mRNAs from the same precursor molecule, or directly modify mRNA nucleotides (*RNA editing*) by transitions (C → U), deletions or insertions. Modified intron splicing during the evolution of domesticated rice caused a single base change in the *Waxy* gene, leading to less waxy protein (Hirano et al., 1998).
- **Translation** of mRNA into protein can be regulated by altering the rate of **mRNA degradation,** or by binding proteins or complementary RNA sequences to the mRNA molecule to prevent translation (*antisense control*).
- Post-translational modification of proteins can occur through **different splicing patterns** that remove amino acid sequences, or by chemically modifying amino acid residues by adding acetyl, sugar or phosphate side-chains (acetylation, glycosylation, phosphorylation).

Regulatory pathways and networks (**gene regulation**) and their specific activity in time and place, play a major role in accounting for the variety of different kinds of cells in multicellular organisms. The emerging picture is of all levels of gene activity, from DNA replication onwards, as exquisitely and sensitively controlled and regulated. Selective processes acting on these genes over time guide the form, function and direction of development.

Genetic Control of Embryonic Pattern

In multicellular eukaryotes, obvious and distinctive features include their axial polarity (usually anterior-posterior, left and right), appendages (usually paired) and the spatial arrangement of tissues and organs. Not only do body parts differ from one another — head from tail, arms from legs — they occupy positions within individuals and between different organisms that are generally uniform. Such precise and repeatable structures must be preceded by a precisely ordered pattern of development directed by a series of processes that create and maintain each cell's special relationships to other cells.

Patterns are present within the egg. Differences, such as those imposed by the unequal deposition of maternal substances in the cytoplasm or by external factors such as products from adjacent (nurse) cells, even gravity, help initiate localized differences in development. Such influences are part of the **epigenetic control of gene regulation in development.** The developmental differences make use of molecules that act as morphogenetic agents (morphogens) enabling different cellular responses. The *bicoid* gene in *Drosophila,* for example, helps determine the anterior–posterior pattern of the egg by producing a protein whose concentration follows a gradually decreasing gradient along the length of the egg. Should *bicoid* be defective, a headless embryo forms with tail structures at both ends. The gene *gurken* acts similarly by producing a messenger RNA that helps determine dorso–ventral differentiation (**Fig. 13-4**).

Figure 13-5 depicts how some cells or tissues obtain information on their position, information corresponding to such gradients. This positional information, in turn, can influence subsequent activity so that different groups of cells follow different developmental paths, such as moving to new locations and/or causing the production of substances (inducers) that set up developmental gradients of their own. Developmental biologists have suggested that each new set of positional coordinates activates special selector genes that cause cell lineages to commit themselves to a particular developmental direction in particular compartments. Subsequent exposure to further positional influence can activate a different battery of selector genes leading to a further commitment, and so on.

The uniformity of the developmental pattern in different individuals of the same species lies in the fact that the succession of regulatory events, or pattern history, is identical or similar. If the pattern changes, by an alteration in the type of inducer or gradient produced by a tissue or by a change in the ability of a tissue to recognize positional information, the phenotype can also change. For example, the *Drosophila* mutation *Antennapedia* results in a leg develop-

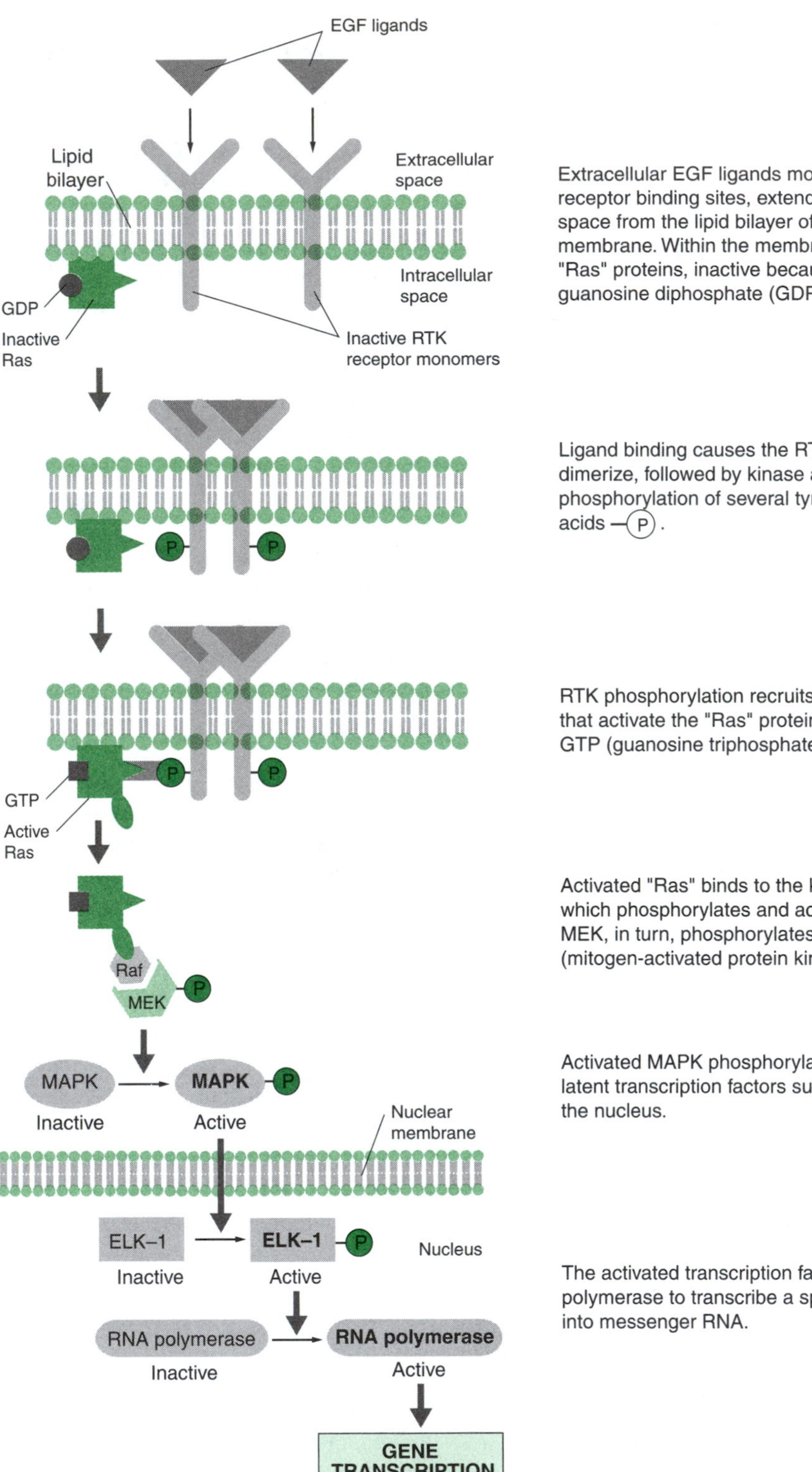

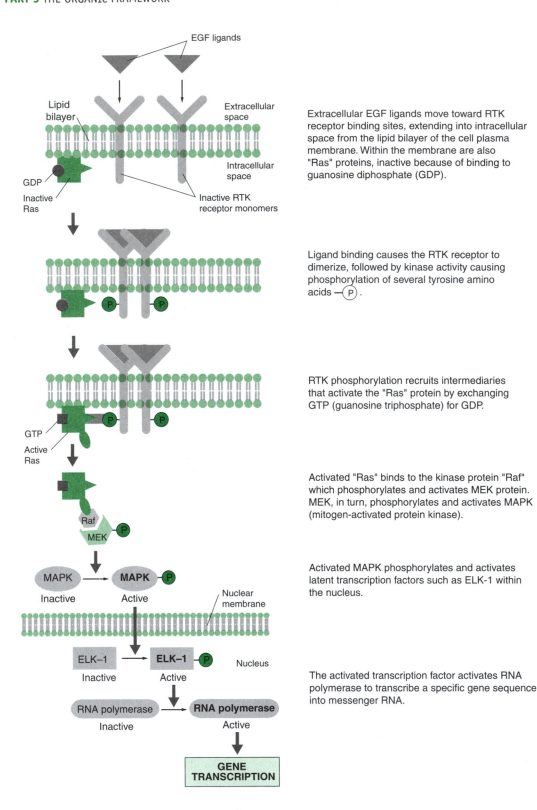

FIGURE 13-3 Diagrammatic illustration of signal transduction in which an extracellular epidermal growth factor (EGF) binds to a transmembrane receptor (RTK) with tyrosine kinase activity that activates reactions in a chain of cellular proteins, leading to messenger RNA transcription. When inactive, "G proteins" such as Ras are commonly complexed with the guanine nucleotide GDP, and are activated when GDP is phosphorylated to GTP. (After their activation and subsequent function, G proteins normally return to inactivity by phosphatase hydrolysis of GTP to GDP.) In some cases, targeted proteins in such cascades produce "second messengers," such as cyclic adenosine monophosphate (cAMP) or cyclic guanosine monophosphate (cGMP), that can affect glucose metabolism, fat storage, and cellular responses such as aggregation and secretion.

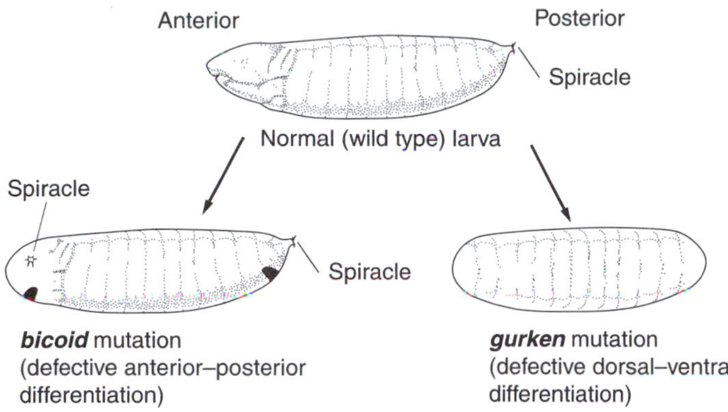

FIGURE 13-4 Effects of two *Drosophila melanogaster* mutations on larval differentiation. Mutations in the *bicoid* gene interfere with development of anterior–posterior asymmetry, producing a headless individual; *gurken* mutations produce a ventralized larva lacking normal dorsal tissues. Interestingly, an understanding of these morphological effects allows further manipulations. Thus, the *bicoid* gene sequence that determines anterior tissue localization can be transferred to a gene called *oskar* that normally transmits germinal tissue to the posterior pole of the egg, in which case gonads arise anteriorly rather than posteriorly (From Ephrussi, A., and R. Lehmann, 1992. Induction of germ cell formation by *oskar*. *Nature*, **358,** 387–392.).

FIGURE 13-5 Diagrammatic representation of how some cells or tissues located along the major body axes assume positional information during development. A gradient along one axis provides cells or tissues A, B and C with information as to their relative anterior–posterior positions. (For simplicity, only three anterior–posterior tissue blocks are shown. In normal development each of these blocks may be further differentiated "horizontally" into three or more anterior–posterior subdivisions, so there may be a total of nine or more blocks of differentiated tissue extending along the anterior–posterior axis.) Further growth and differentiation confers dorsal–ventral information upon clones of each A, B and C tissue labeled 1 and 2. Subsequent cell divisions of these tissues provide the α and β subclones with information as to their proximal–distal positions. As a result of their geographical position and of the activity of special "selector" genes — of which homeotic genes are a major example — some or many cells of tissue A, for example, may come to possess a specific set of developmental responses so that they develop differently from tissues B or C, or from other subclones within A. A simple gradient model in which different morphogen concentrations produce different cellular colors is called the "French flag": high levels "blue," intermediate levels "white," and low levels "red." (From Strickberger, M. W., 1985. *Genetics*, 3d ed. Macmillan, New York.)

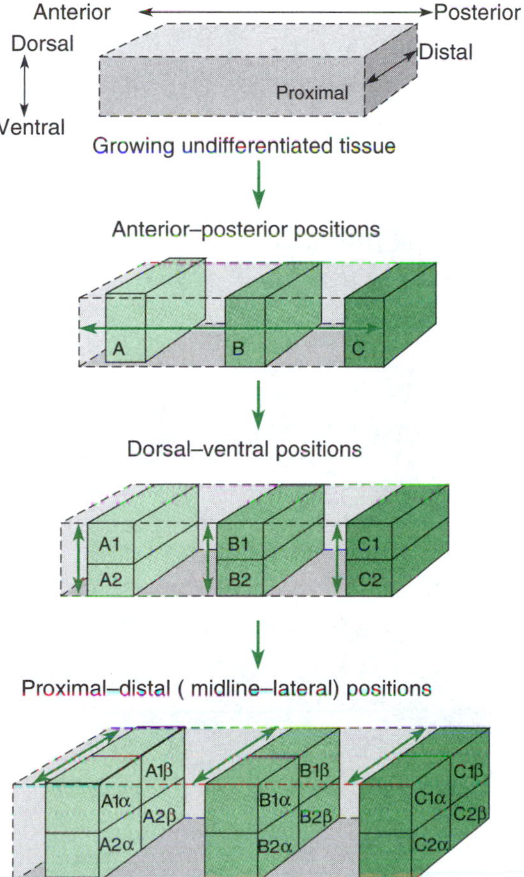

ing from the imaginal disc that normally forms an antenna. Such mutations are called **homeotic** because they change a particular organ in a particular body region to resemble an organ normally found in a different region along the body axis (**homeosis**). The correspondence between antennal parts and leg parts indicates that both structures bear the same positional information, but that cells carrying the *Antennapedia* mutation interpret these positions to produce leg rather than antennal tissues (**Fig. 13-6**).

The function of the normal *Antennapedia* gene product, *Ant+*, is as a selector gene whose activity is required for normal leg development in the thoracic region of the body. The *Antennapedia* mutation causes this gene to malfunction so that it is also active (gain of function) in the anterior head region, repressing genes, which if active, would result in antenna formation (Casares and Mann, 1998). A cluster of *Antennapedia*-linked loci, called the *antennapedia* complex, possesses similar regulatory functions.

Like the *Drosophila antennapedia* complex, which consists of six genes, the *bithorax* complex investigated by E. B. Lewis (and which contains three genes) also causes homeotic effects, and consists of at least a dozen genes that control the fates of various structures from the posterior part of the second thoracic segment to the tip of the abdomen (see Chapter 16 for a discussion of segmentation). Thus, should expression of the normal *Ultrabithorax* gene fail in the anterior abdomen, this region develops as though it was thoracic rather than abdominal (see Figure 10-17).

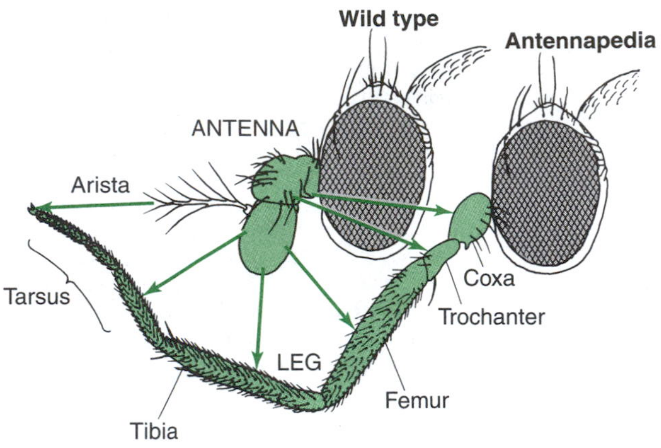

FIGURE 13-6 Positional correspondence between antennal sections and leg sections produced in *Drosophila melanogaster* carrying the homeotic *Antennapedia* mutation. The mutational substitution of leg tissues for antenna tissues follows the arrows (*color*) so that distal cells, for example, interpret their position as tarsal segments, and proximal cells as coxa or trochanter.

Because both the *bithorax* and *antennapedia* complexes confer specific identities on regions of the *Drosophila* body and on substructures within these regions, genes in both complexes are most likely to have arisen by duplication from a common ancestral gene that initially specified cell differentiation of uniform body regions or segments. Duplications of this primitive gene or gene complex, and subsequent divergent mutation of each duplicate, permitted different segments to undergo different developmental pathways, yet each duplicate gene complex still maintained sequences necessary to produce or interact with the basic underlying segmentation process. As evolution proceeded from uniform segments in the insect ancestor to the more complex segments found in dipteran flies such as *Drosophila* (**Fig. 13-7**), the number of regulatory genes successively increased, enabling specific structures in each region to be controlled. Gene duplication in *Drosophila,* other arthropods, and also in vertebrates, has played a major role in bursts of diversification during animal evolution (Chapters 16 and 17).

FIGURE 13-7 Proposed evolutionary progression to account for the segmental organization found in insects; see Chapter 16 for segmentation. Body segments were, at first, relatively uniform, becoming more complex as different lineages evolved. In dipteran insects (*bottom figure*) the segments not only bear complex structures but, except for some abdominal segments, also differ from each other considerably. For example, the three thoracic segments of *Drosophila* each have a pair of legs, but only the second thoracic segment has wings, and the third has the haltere or balancing organs. The three thoracic segments are numbered consecutively from the head boundary as T1, T2 and T3; and the eight recognized abdominal segments in *Drosophila* are numbered consecutively from the thoracic boundary. (Adapted from Strickberger, M. W., 1985. *Genetics,* 3d ed. Macmillan, New York.)

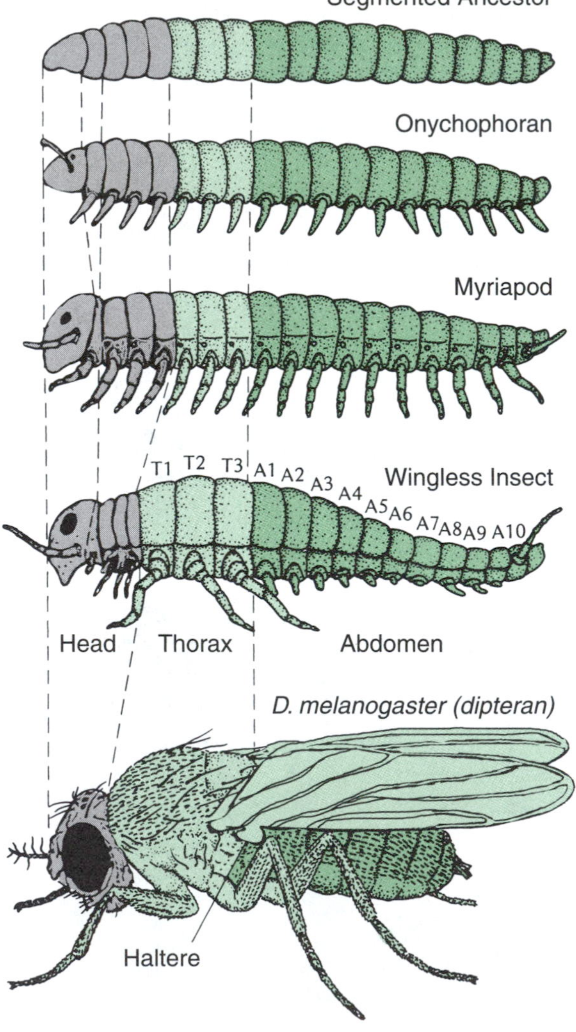

Homeobox Genes and Conserved Embryonic Stages

In support of evolutionary duplication and differentiation of developmental genes is the finding by McGinnis and coworkers (1984) that a family of related DNA sequences called **homeoboxes** is found in various locations in the *Drosophila* genome, including loci within the *bithorax* and *antennapedia* complexes.

Each homeobox codes for a polypeptide sequence about 60 amino acids long called a **homeodomain.** The homeodomain, in turn, is part of a transcription factor that binds to DNA, thereby regulating messenger RNA production. Genes regulated by these homeoboxes affect cell positioning and differentiation in *Drosophila,* in most other animals and in plants (Kappen et al., 1993; Niklas, 1997). Surprisingly, the linkage order of these homeobox-containing genes in the *Drosophila antennapedia–bithorax* clusters accords with their phenotypic expression along the anterior–posterior axis of the animal (**Fig. 13-8**). Equivalent (orthologous) homeodomains with equivalent linkage orders of homeobox (*Hox*) genes occur across the animal kingdom (**Figs 13-8** and **13–9**), indicating that positional information in different lineages has a common evolutionary origin.[4]

Furthermore, a particular homeobox protein can perform a similar function in different organisms; for example, much or all of the function of the *Drosophila* homeobox gene *Antennapedia* can be assumed by a protein produced by a homologous *Hox-b* gene in mice. Segmentation stripes, initiated by the *engrailed* gene in *Drosophila melanogaster* are produced by *engrailed* homologues in other insects, crustaceans, annelids and vertebrates. If we consider anterior–posterior organization, although expressed in different structures, a common *homeobox* gene feature relating arthropods and vertebrates, Geoffroy Saint-Hilaire's attempt to homologize vertebrates and invertebrates, made more than 150 years ago, was perceptive: "every animal lives within [arthropods] or without [vertebrates] its vertebral column." We now also know that *Dhox* genes pattern the developing intestinal tract along an A-P axis and that another class of genes, *distal-less* (*Dlx*), patterns vertebrate head development along the dorso-ventral axis (see Chapter 17). In metazoan lineages, Zhang and Nei (1996) suggest that homeobox genes began as two linked loci, each locus conferring a distinctive role in anterior or posterior development. These genes duplicated further, eventually generating the linked clusters of homeobox genes in triploblastic animals.[5]

Homologous gene performance of similar functions is also involved in the growth of neuronal axons in nematodes and vertebrates, and in muscle development in nematodes, *Drosophila* and mice. The development of anterior sense organs (e.g., eyes) and the central nervous system in organisms as different as humans, fish, mollusks, nematodes and insects, shows evolutionary relationship through a common regulatory *Pax-6* gene sequence, indicating that such homologies originated in the Precambrian triploblastic ancestor of invertebrates and vertebrates. Also, homeobox proteins expressed during segmentation in *Drosophila* function in forming non-segmental patterns in the nematode *Caenorhabditis elegans*. Appendages in animal phyla, from protostomes to deuterostomes, also make use of homologous genes that regulate body wall outgrowths along a proximal-distal axis.[6]

Several major conclusions come from studies of comparative gene structure and function:

- These important animal developmental (regulatory) genes all share a common, highly conserved, and evolutionarily ancient role as **transcription factors.**
- **What has changed with evolution is context;** the specific function of these conserved regulators varies from cell lineage to cell lineage, tissue to tissue, and organ to organ as well as from time to time during both development and evolution
- Conserved developmental genes are also found in flowering plants and fungi, both of which produce homeodomain proteins much like those of metazoans. Such proteins must have functioned early in eukaryotic history, and probably arose from duplication events before plants, fungi and metazoans diverged (Bharathan et al., 1997).

A common genetic evolutionary origin suggests conservation of basic developmental pathways such as those that establish or implement the A-P axis. Slack and coworkers (1993) took this further when they proposed that all animals use a common genetic developmental system, the **zootype,** which governs the basic spatial arrangement of their tissues. These common processes, established by their common genetic ancestry and retained because of common selective developmental value, characterize the kingdom Animalia. Furthermore, "the zootype is expressed most clearly at a particular stage of embryonic development." This stage, called the **phylotypic stage** is illustrated in **Figure 13-10** for each metazoan lineage in which the zootype's developmental genes are conserved and expressed. Such a stage may coincide with the time in development when interactions between parts (genes, cells, epithelia and mesenchyme) is most critical, and so has been conserved through stabilizing selection.

Given the deep conservation of developmental genes, and the stability that arises from their conserved functions, how is variation in development and descent with modification mediated at the genetic level? **Gene regulation** has emerged as a central mechanism explaining developmental and evolutionary change. In a seminal paper credited with directing many researchers to enter molecular biology, François Jacob (1977) saw the importance of regulation (which he named *tinkering*) in both develop-

[4] As indicated in footnote #10 in Chapter 10, and here as a reminder, convention dictates that homeobox genes in vertebrates are called ***Hox* genes, *Ho*** from homeodomain and ***x*** for vertebrate. Thus, *Pax-6* is a vertebrate homologue of the *Drosophila paired rule* (*Pa*) gene. Convention also dictates (although not all follow the convention) that homeobox or *Hox* genes are italicized (*antl*) but the gene protein is not, and that *Hox* genes in humans are italicized and capitalized (*HOX*).

[5] See Finnerty and Martindale (1998) and Martindale and Kourakis (1999) for further studies on homeobox gene clusters in triploblasts.

[6] See Land and Nilsson (2002) and Nilsson (1996) for triploblast ancestors, Hunter and Kenyon (1995) for segmentation and Panganiban et al. (1997) for appendages.

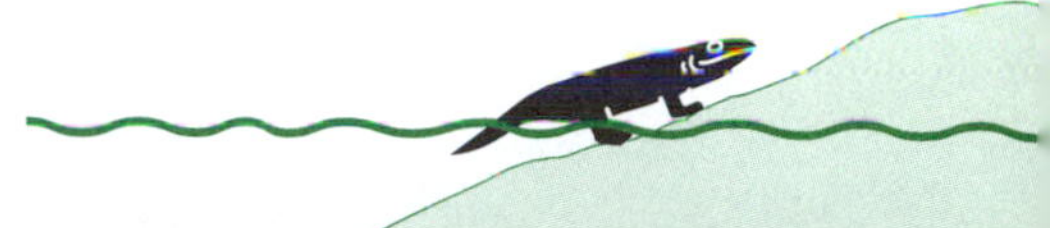

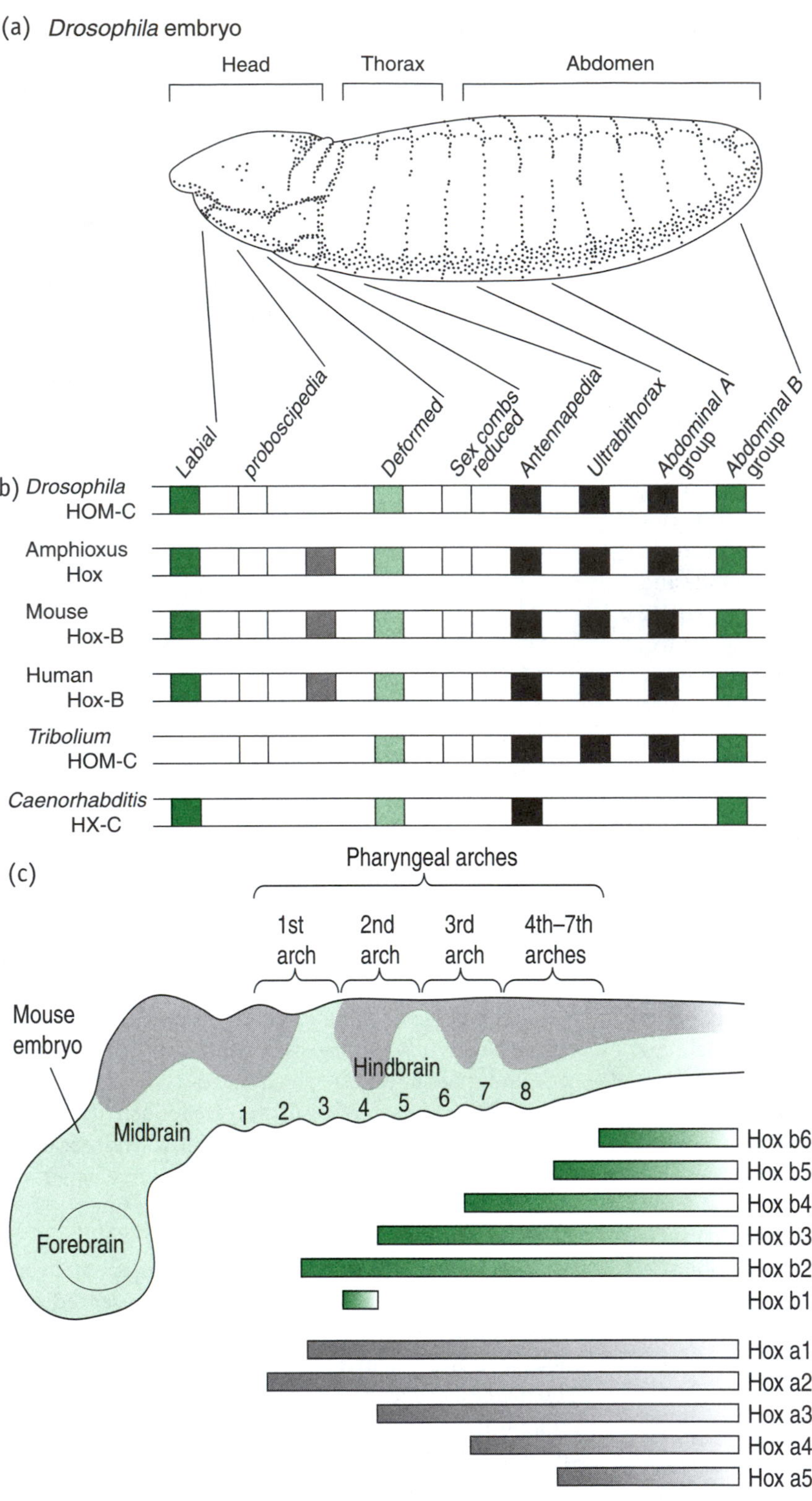

FIGURE 13-8 Homeobox gene relationships in representative animals. Almost all the genes illustrated here show a highly conserved relationship between their position along the chromosome (shown in b) and developmental function along the anterior–posterior axis (shown in a and c). (a) Regions of action of homeobox genes in *Drosophila melanogaster* along the anterior–posterior axis of the embryo with labial acting most anteriorly and Abdominal B group genes acting most posteriorly. (b) Clusters of homeobox (Hox) genes from *Drosophila,* amphioxus, the mouse, humans, a beetle (*Tribolium castaneum*), and a nematode (*Caenorhabditis elegans*). Homologous (orthologous) genes are aligned vertically. (c) Expression of Hoxa and Hoxb genes in the anterior region of a mouse embryo. The left hand end of each gray bar marks the extent of anterior (rostral) expression of each gene, which determine important boundaries such as the position of phar-yngeal arches 1 to 7, divisions 1 to 8 within the hindbrain, and boundaries between populations of migrating neural crest cells (*black*) fore, forebrain; mid, midbrain. (Adapted from various sources with additions and modifications.)

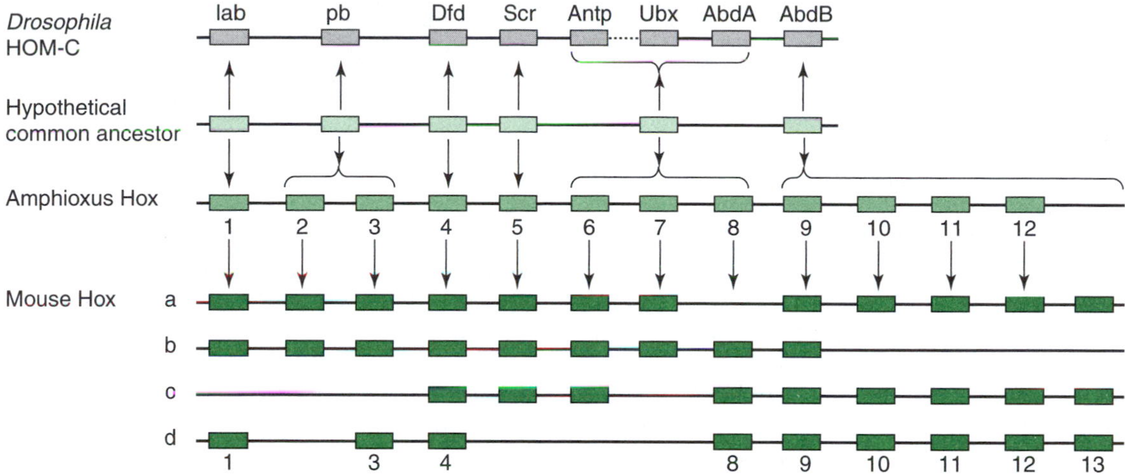

FIGURE 13-9 Relationships between homeotic gene clusters in *Drosophila,* Hox genes in amphioxus and mice, and homeotic genes in the hypothetical common ancestor of arthropods and chordates. Expansion of the six gene clusters in the common ancestor gave rise to eight clusters in *Drosophila* and to 13 in mice. (Adapted from Hall, B. K., 1999a. *Evolutionary Developmental Biology,* 2nd ed. Kluwer Academic Publishers, Dordrecht, The Netherlands, modified from Holland, P. W. H., and J. Garcia-Fernàndez, 1996. Hox genes and chordate evolution. *Devel. Biol.,* **173,** 382–395.)

ment and evolution and as the ultimate target of natural selection.

> Natural selection does not work as an engineer works. It works like a tinkerer — a tinkerer who does not know exactly what [s]he is going to produce.

> What makes one [organism] different from another is a change in the time of expression and in the relative amounts of gene products rather than the small differences observed in the structure of these products. It is a matter of regulation rather than of structure. (Jacob, 1977, p. 1163)

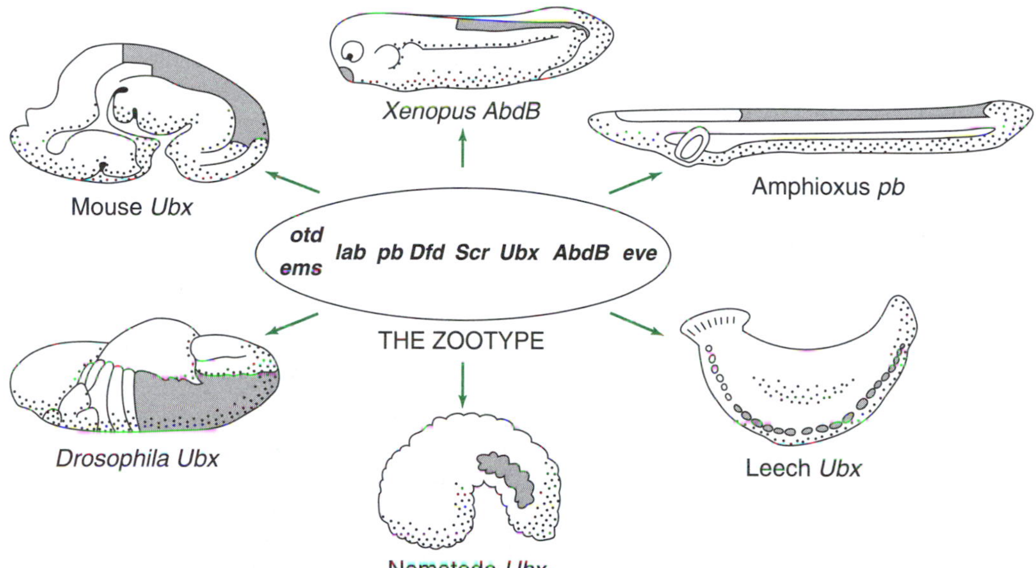

FIGURE 13-10 Diagrammatic view of the proposal by Slack and coworkers (1993) that a "zootype" (*center*) reflects the spatial development of different multicellular animals in their phylotypic stage. The genes in the zootype are designated by standard *Drosophila* abbreviations, and the expression of particular zootype genes in different animals is *shaded*. (Abbreviations: *otd, orthodenticle; ems, empty spiracles; lab, labial; pb, proboscipedia; Dfd, Deformed; Scr, Sex combs reduced; Ubx; Ultrabithorax; AbdA, Antennapedia–Abdominal A group; AbdB = Abdominal B1; eve = even-skipped*.) The phylotypic stage characterizes each phylum's distinctive development, and is called the pharyngula in vertebrates (see Fig. 3-14). Homeobox genes can be used in many different body plans. However, in addition to homeobox genes, it takes many genetic interactions to produce such basic body plans. (Adapted from Slack, J. M. W., P. W. H. Holland, and C. E. Graham, 1993. The zootype and the phylotypic stage. *Nature,* **361,** 490–492.)

Regulatory Genes and Evolutionary Change

Even developmental patterns that are imbedded in deep evolutionary time, such as segmentation, may shift toward new directions under changed selective conditions. As discussed in Chapter 16, arthropod segments differ extensively in number and function (see also Figure 13-7), and segmentation has been significantly reduced or else has disappeared in mollusks. Homeobox-containing genes in metazoans as diverse as nematodes, arthropods and vertebrates indicate that *orthologous (homologous) regulatory molecules* can be recruited for use in *entirely different developmental pathways,* leading to morphologies nonhomologous to those in other phyla (Hall, 1994, 2003a).

Because different organismal features can be initiated by homologous genes, and because similar "convergent" features can be caused by nonhomologous genes, "homology" can be an elusive concept when its genetic basis is ignored and its usage restricted to morphology and function. Distinctions between homology and convergence should be made genetically and developmentally.[7] In the words of Abouheif and coworkers (1997):

> Homology is a powerful concept. In order to use it consistently when making comparisons across taxa, features should be termed homologous if, and only if, they share a common evolutionary origin. Other criteria, particularly those based on functional similarity, can be misleading. Homology is a hypothesis about the evolutionary origins of a trait, and gene expression data can be an extremely valuable source of evidence supporting homology of a morphological feature, although they cannot be the sole criteria. Any hypothesis of morphological homology based on gene expression data should include: (1) a robust phylogeny of the taxa; (2) a reconstructed evolutionary history of the genes whose expression is being compared; (3) extensive taxonomic sampling, including a broad range of evolutionary informative species; and (4) a detailed understanding of comparative anatomy and embryology. Further, we should regard proposed homologies as falsifiable, and test the possibility that overtly similar gene expression patterns might be due to convergence or recruitment, rather than common ancestry.

Some homeobox genes are expressed in diverse vertebrate body structures such as the neural crest, limb buds and gill arches. Among insects, even when homeobox genes are organized into identical clusters, they can still lead to different developmental mechanisms underlying segmentation: "short-germ-band" insects (e.g, beetles) in which abdominal segments are added sequentially from one end, and "long-germ-band" insects (e.g, *Drosophila*) in which all segments form simultaneously (Raff, 1996).

New functions for such regulatory genes involve interactions with modified receptors and cofactors. For example, a transcriptional protein that regulates function A by attaching to a particular RNA polymerase promoter (see Fig. 10-14) may be recruited to regulate function B by combining with a new transcriptional cofactor that attaches to a different promoter. Or a promoter binding site that stimulates transcription in one species can change, repressing gene function in another species (Singh et al., 1998). Similarly, an enhancer DNA region that influences availability of promoter sites to the polymerase enzyme, and thus affects transcription, can mutate, causing a change in regulatory gene activity (Belting et al., 1998).

Developmental innovations can appear by employing old regulatory genes in new roles, rather than waiting for appropriate regulators to form entirely *de novo.* Some (if not many) major phenotypic changes derive from large effects in small numbers of genes rather than an accumulation of small effects from large numbers of genes; hence, the growing importance of analysis of quantitative trait loci (QTL) (see Chapter 10).[8]

Averof and Patel (1997) provide a prime example of how a change in regulatory gene function can produce a major evolutionary impact. The homeobox genes *Ubx* and *abdA* (Fig. 13-8), whose expressions initiate legs in crustacean thoracic segments, are expressed more posteriorly in more derived arthropods. With this delayed expression, anterior thoracic segments become more head-like, that is, produce head-like maxillary appendages with larger and more muscular thoracic features. These new limbs ("maxillipeds") can function for both feeding and walking. Among other findings are homologies between regulatory genes that pattern crustacean gill development with genes also involved in insect wing development, indicating a potential regulatory pathway leading from gills to wings (Averof and Cohen, 1997).

One example of conserved gene regulation, associated with what has long been regarded as a major evolutionary change, is specification of the dorso–ventral axis in vertebrates and of the ventro–dorsal axis in arthropods. In arthropods, as in annelids, the circulatory system is dorsal and the nerve cord ventral, whereas these positions are reversed in vertebrates (see Fig. 17-2). It came as a considerable surprise to discover that these fundamentally different patterns — dorsal dominance in vertebrates, ventral dominance in arthropods — result from the actions of the same

[7] For the concept of homology as hierarchical, see Koonin et al. (1996), Wray and Abouheif (1998) and Hall (2003, 2006c).

[8] See Albertson et al. (2000) and Cheverud (2004) for application of QTL analysis to developmental changes associated with evolutionary change.

two pairs of homologous genes. One pair — *decapentaplegic* (*dpp*) and *bone morphogenetic protein* (*bmp*) — produces products that dorsalize development in arthropods (*dpp* as shown in *Drosophila*) and ventralize development in vertebrates (*bmp* as shown in zebrafish and *Xenopus*). The other pair — *short order gastrulation* (*sog*) and *chordin* (*ch*) — have the opposite actions, *sog* specifying ventral in arthropods and *ch* specifying dorsal in vertebrates.[9]

Gene Networks and Modularity

Regulatory changes of all kinds have been among the key agents of organismal evolution. Like a child's Lego set, which produces differently shaped structures by rearranging the modular blocks, developmental evolution uses regulatory changes to produce a variety of new functions by rearranging constituent activity, such as changing the signals, pathways and targets of signal transduction. Such changes are inherent in the way development works. Akam (1998), for example, points out that the regulatory effect of the homeobox gene *Ultrabithorax,* in differentiating the normal *Drosophila* hind wing structure (the haltere), impacts on signals and pathways involving at least thirty target genes. Variation in such genes has been documented in both animals and plants.[10]

As regulatory pathways extend, change and interact, organismal integration and complexity both increase. This is one of the key reasons proposed for the observations that major changes in embryonic development occurred early in multicellular organismal evolution, explaining why such major changes cannot occur in extant organisms.

Similar gene networks can be used by different organs in the same individual, yet respond differentially to selection. An example is a gene network regulated by *hedgehog* (*Hh*), *Pax-6* and the homeobox gene *Prox1* (*prospero-related homeobox 1*), a network used in the development of sensory organs in various organisms, but we will discuss its operation in two forms (morphs)[11] of a fish, the Mexican tetra *Astyanax mexicanus.* One morph found in surface pools (the surface morph) is sighted, while the other morph, found in caves, is blind. Blind cavefish have reduced and vestigial eyes and no pigment but have expanded the taste buds and the sensory system found in a line that runs along the body (the lateral line system).

The gene network initiated by *Hh*/*Pax-6* controls development of the eye sensory modules. Over-expressing *Hh* in surface fish results in diminished eye development but increased numbers of taste buds; we don't know about the lateral line system. We do know that a homeobox gene, *Prox1,* expressed in all three modules, is the upstream regulator of the pathways (**Fig. 13-11**). So, we have three modules (eyes, taste buds, lateral line) that overlap in sharing *Prox1* and *Hh* as upstream regulators. Selection could operate on the adult cavefish phenotype, on the sensory modules or on the common gene network, but in all scenarios the modularity of the gene network with *Prox1* upstream and of the sensory organs (Fig. 13-11) enables cavefish to expand their taste buds and lateral lines while diminishing their eyes (Franz-Odendaal and Hall, 2006).

Gene networks are one example of the modularity of development (**Box 13-2**). As the cavefish example illustrates, a gene network may operate in several tissues, organs, or even regions of an organ, change in one, but not in the others. Similarly, the cells from which individual elements are constructed, for example, the cells that make the humerus or a finger bone, are modules at the cellular level, usually termed **condensations.** The behavior of cells within a condensation, including the gene and gene networks operating, is integrated, and differs from activity in cells outside the condensation. In the condensation for a cartilage, mRNA for cartilage-type collagen (type II collagen) is upregulated but mRNA for myosin is not. In a condensation for a muscle, mRNA for myosin is upregulated, but mRNA for type II collagen is not.

An example of modularity at both cellular and genetic levels is illustrated in **Figures 13-12** and **13-13.** What appears to be a single bone (the dentary) making up the lower jaw of mammals, is comprised of six modules, each of which arises from a separate population of cells (Fig. 13-12), and each of which is subject to independent genetic control, as shown by the effects of knocking out single genes, when one module can be prevented from developing or diminished in size, while other modules develop normally (Fig. 13-13). An example of the result of selection on such modules is shown in the extremely reduced dentary of a marsupial, the Western Australian honey possum *Tarsipes rostratus* (see Fig. 19-5), which feed by licking the nectar from flowers. Comparison of the honey possum with the dentary of the mouse or of a vole (see Figs. 13-12 and 19-6) shows the extent of reduction in the honey possum. You can readily see how, both in development and evolution, modularity provides a means for independent yet integrated development of embryos and for mosaic evolution in response to selection (Box 13-2).[12]

[9] See Hall (1999a,b) and Jenner (2005) for discussions of dorso-ventral and ventro-dorsal axes, and see Arendt and Nübler-Jung (1994, 1999) and Ferguson (1996) for *bmp, sog* and *chordin.*

[10] For the importance of networks in development and in evolution, see Moriyama and Powell (1996), Purugganan and Suddith (1998), Wilkins (2002) and Davidson (2006).

[11] See Figure 17-27 for two morphs of the pharyngeal jaws and teeth in cichlid fishes.

[12] For the development and application of the concept of modularity to development and evolution, see Atchley and Hall (1991), Gass and Bolker (2003), Hall (2003b), Schlosser and Wagner (2004), Emlen et al. (2005) and Franz-Odendaal and Hall (2006). For its application to mammalian evolution, see Kemp (2007).

(a)

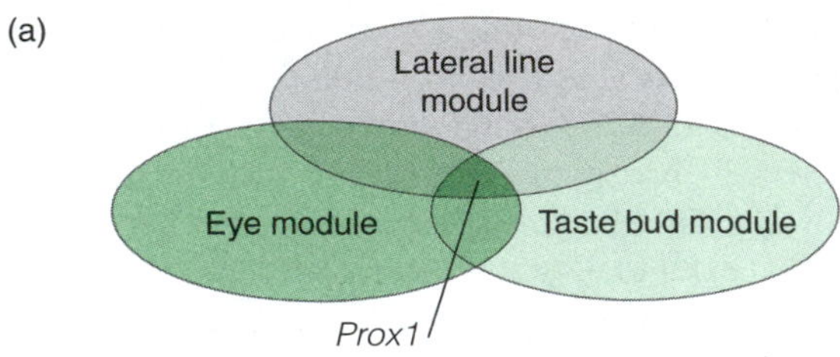

(b)

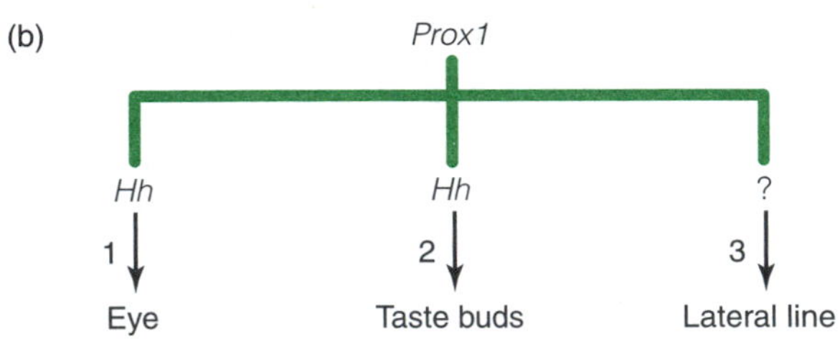

(c)

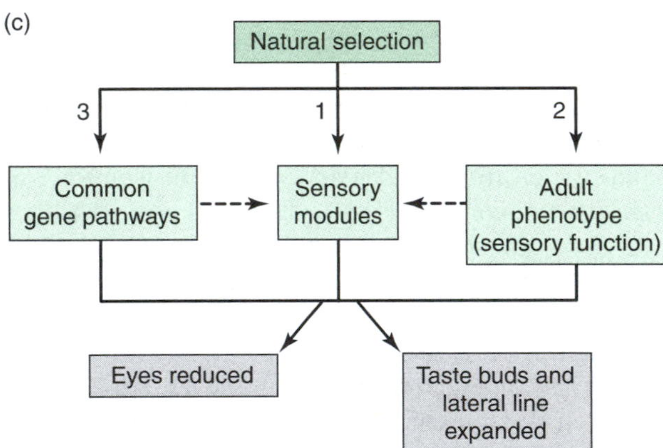

FIGURE 13-11 (a) Three modules (lateral line, taste buds, eyes) in Mexican tetra *Astyanax mexicanus* share an upstream control (*Prox1*) for gene pathways. (b) *Prox1* signals through hedgehog (*Hh*) in two of the pathways; the equivalent gene for lateral lines modules is not known. (c) The common upstream genetic control shown in (b) operates whether selection is on the adult phenotype, development (sensory modules) or common gene pathway, resulting in expansion of the lateral line and taste bud modules and reduction in the eye module. (Modified from Franz-Odendaal, T. A., and B. K. Hall, 2006. Modularity and sense organs in the blind cavefish, *Astyanax mexicanus*. *Evol. & Devel.* **8,** 94–100.)

Constraint and Selection

Among mechanisms proposed to regulate or circumscribe the activity of single genes or of gene networks are developmental interactions, canalization and constraints that either maintain a constant phenotype or that lead to variation and developmental plasticity (see Figure 13-15 later). The concept of **constraint** is the topic of the remainder of the chapter. In order to understand evolutionary conservation or change of phenotypes, the processes that underlie constraint must be integrated with the role of natural selection, a topic also discussed below.

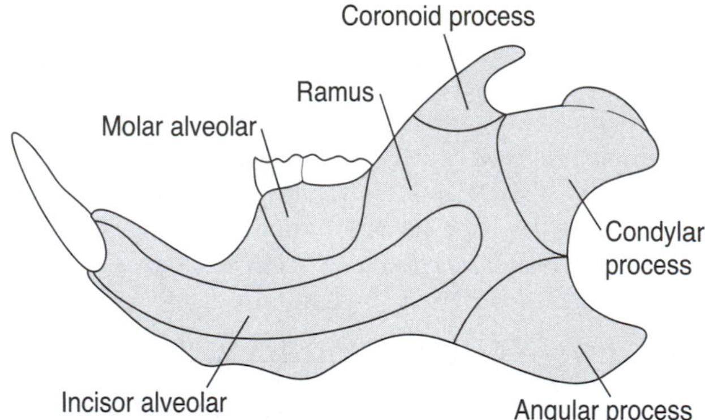

FIGURE 13-12 The single dentary bone of the rodent lower jaw is comprised of six modules that form a unified single, but complex element; see Figure 19-6 for the dentary in a meadow vole. The molar- and incisor-alveolar units are derived from the cell population that forms the molar and incisor teeth, respectively. (Note that rodents lack other teeth such as canines or premolars.) The ramal population, a separate cell population not associated with tooth formation, forms the body of the dentary. The coronoid, condylar and angular processes each derive in part from the ramal population and in part from a cartilage-forming population of cells at the tip of each process.

The existence of a developmental system with its regulated pathways raises the question of whether such a system acts as a constraint that channels, limits or enhances structural and functional variation, and thereby molds and directs future evolution. For example, one can ask why insects have six legs and land vertebrates four, when almost any number of legs, or even none at all, can suffice for locomotion (see Chapter 16). What role, if any, do constraints play in determining the number of legs in insects and in vertebrates, and how does such a mechanism relate to natural selection?

The answer is not entirely simple, although constraint must place limits on the range of forms that natural selection can bring forth. Schwenk (1995) separates "internal" genetic and **developmental constraints** that limit character expression from "external" **selective constraints** that affect lineages. These two types of constraint can and probably often do overlap — selection that affects a phylogenetic trend operates on constituent characters, while character constraints affect evolutionary direction — further emphasizing that we should not place constraint against selection, but rather see **the reciprocal interaction of constraint and selection.**[13]

Limb abnormalities of all kinds arise in both insects and vertebrates, indicating that limb number can change, but because such mutations generate major bodily disruptions, they are often lethal early in development or at birth. Although many developmental changes are possible, certain of these changes, especially those of large degree are mostly lethal; their effects cannot be successfully integrated with other developmental stages.

[13] See Gibson and Wagner (2000) and Schwenk and Wagner (2003) for further analyses of constraint.

BOX 13-2 Modularity

IN RECENT YEARS, MODULARITY[a] has emerged as an organizing principle at all levels from gene networks, through cell populations, to organ primordia. Modularity is being applied in developmental genetics, developmental biology, evolution and in evolutionary-developmental biology. Because modules subdivide what would otherwise be large biological units, and because modules are subject to selection, the concept of modularity is fundamental to understanding how the phenotype arises; modules link the genotype with the phenotype.

Modules have three central features:

- a distinctive fate;
- internal integration; and
- the ability to interact with other modules.

As is homology (indeed as is the organization of life), modularity is a hierarchical concept. A limb bud is a developmental module in the embryo that forms either a wing or a leg, both of which are modular components of the adult (Franz-Odendaal and Hall, 2006). Anterior and posterior limb buds share some modules at the level of gene networks (those that result in the differentiation of cartilage or muscle), but differ in other networks (those that specify that a limb will be anterior or posterior, for example).

Modularity is not merely a theoretical construct; modules can be tested for. If a region of an early vertebrate embryo from where the forelimbs are known to develop is grafted to another region of the embryonic body, it still forms forelimbs. If a gene network for forelimb is activated in another region of the embryo and initiates a forelimb from that region, the gene network is *both acting as a module and conferring a higher modular property onto cells in the new location.*

Modules are subject to natural selection and evolve; indeed they involve in much the same ways that new genes evolve, which are:

- by *duplication,* with one of the duplicates acquiring a new function;
- by *dissociation,* in which a module separates either in space (homeotic mutations, for example) or in time (heterochrony, for example), and acquires a new function; or
- by *co-option,* the incorporation of one module into another (Raff, 1996).

It will be clear that these are not mutually exclusive processes.[b]

[a] This Box is modified with the permission of the other authors from a column by Budney et al. (2005) in *The Palaeontological Association Newsletter* (*60,* 10–12).

[b] See Gass and Bolker (2003) for an overview of modularity and see Franz-Odendaal and Hall (2006) and the text of this chapter for an example of the hierarchical nature of modularity.

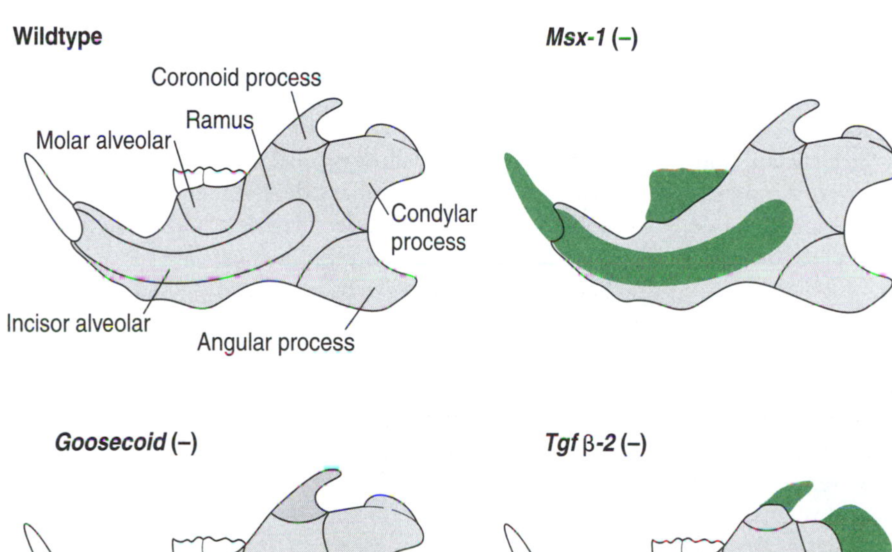

FIGURE 13-13 The modules in the single dentary bone of the rodent lower jaw are subject to independent genetic control as illustrated by the effect on dentary development of knocking out individual genes. The wild type shows six modules: (see Figure 13-12). Modules that fail to develop or that are under-developed in the knockout mice are shaded in the other three images. Lack of the gene *Msx-1* — shown as *Msx-1* (–) — prevents the alveolar units from developing; the other modules develop normally. Lack of *goosecoid* results in smaller coronoid and angular units but does not affect the growth or shape of the other modules. Lack of *TGF*β*-2* results in underdevelopment of all three posterior processes. For details see Hall (2003).

The need for integration necessitates developmental constraints whose limits, according to Raff and coworkers (1991), depend on the extent of interaction among the metabolic pathways, cells, tissues, organs and other components of a developing system. Development of those embryonic stages with greater numbers of interactions — such as the zootype or phylotypic stage — is more subject to constraint (less variation is permitted) than is development of other stages with fewer interactions, an end result that could be reached by constraint, stabilizing selection and/or by differential survival. The designation of stages as phylotypic (see Figs. 3-14 and 13-10) recognizes an embryonic stage where the resultants of constraint, selection and differential survival are expressed as a reduction in the amount of variation permitted, a bottleneck if you will.

An anatomical example of developmental constraint is the position of the recurrent laryngeal nerve in mammalian vertebrates. Because the sixth arterial arch in vertebrates forms early in embryonic development, it governs the distance traveled by the later-appearing recurrent laryngeal nerve (see Fig. 3-16). This distance is greatly increased in mammals, in which the arch has become displaced to the thorax and the nerve must double back a distance that may be longer than 1.8 m in giraffes; that is, arterial arch development rather than adaptive selection appears responsible for the nerve's length and positioning. Also relevant as an example of constraint, is the constant number of seven cervical vertebrae in practically all mammals, whether short-necked (whales) or long-necked (giraffes).[14]

Other examples, such as neuronal connections in animals and root connections in plants, indicate the presence of similar genetic systems producing arrays of potential phenotypic variants. These systems generate complex physiological, morphological and behavioral patterns that enable organisms to attune their responses to particular environmental stimuli by enabling a choice of gene action among different inherited possibilities. Such stimuli initiate genetic systems that respond to environmental signals, whether organic or inorganic, as in signal transduction (see, for example, Figure 13-3). An excellent example of such control is **phenotypic plasticity.**

Phenotypic Plasticity

Genes provide the necessary historical information used by cells as they differentiate. Cells then interact with each other and with their environment to assume their ultimate phenotypic shape and relationship (see Figure 13-15). Such interactions may include development of complex patterns from limited instructions. For example, each of the millions of different antibodies that can be produced by the mammalian immune system is not individually coded by the genome. Instead, they are produced by a developmental system that can produce almost any antibody to interact with almost any environmental antigen by selecting among an array of newly generated nucleotide sequences. Inherited mutations in the immune system, and not environmental antigens, are responsible for producing the variety of potential antibody types and numbers. What is evolutionarily adaptive, therefore, is not the appearance of one or more specific antibodies, but a genotype selected to confer the potentiality to make them.[15]

There is a logical tendency to think "one genotype–one phenotype." The human genotype produces a human phenotype; a worm genotype produces a worm. Of course, the relationship between genotype and phenotype is not as simple as this. A butterfly genotype produces both caterpillar and adult butterfly: two phenotypes. A frog genotype produces tadpole and adult frog, again two phenotypes. An ant genotype produces queen, workers and soldiers — three phenotypes — and does so in every generation (i.e, is genetically controlled), although the numbers of workers or soldiers in a population is modified by environmental factors. Polymorphisms within a species are constitutive polymorphisms.

In other situations, a single genotype can produce more than one phenotype, but will do so only in response to an external signal, a signal that may be environmental, diet or predator-induced and that is not necessarily present in every generation. If individuals are not exposed to the environmental signal in a given generation, the environmentally induced phenotype will not appear, but the ability to respond to the signal is passed on to the next generation. We refer to this facultative ability of a single genotype to produce more than one phenotype as phenotypic plasticity.

The large number of examples testifies to the ubiquity and utility of phenotypic plasticity, a topic that has been the subject of many studies in many fields, especially ecology and life history theory.[16] Indeed, *life history theory,* which is normally studied by those trained in ecology, has become of major interest to evolutionary, developmental and molecular biologists interested in understanding how phenotypes arise. Some instances of phenotypic plasticity are so well known that we take them as given, never giving a thought to how a single genome could produce such diverse phenotypes as tadpoles and frogs, larvae and adults of many invertebrates, and workers, soldiers and queen social insects. Other examples that appear more subtle (in that they are often more difficult to understand) include:

[14] See Galis (1999) and Galis et al. (2001) for this and further examples.

[15] See Danilova (2006) for an overview of the evolution of the immune system.

[16] See Stearns (1989), Hall et al. (2004), Pigliucci (2001), Piglucci and Preston (2004) and Sultan and Stearns (2005) for recent and extensive treatments of phenotypic plasticity from various viewpoints.

1. In the presence of predators (dragonfly larvae), tadpoles of the Australian brown striped marsh frog, *Limnodynastes peroni*, extend the development of one feature by increasing tail height and muscle mass, a feature that responds to selection to expand survival.
2. When food supply is restricted or population numbers high, some anuran and urodele tadpoles display modified development; they become cannibalistic, and develop much larger heads and more massive jaws than their noncannibalistic brothers and sisters, that they eat.
3. Rotifers produce an extra set of spines. The water flea *Daphnia* (a microscopic crustacean), produces enlarged spines and an enlarged helmet when eggs develop in the presence of a predator(s); these spines and enlarged helmet prevent the *Daphnia* morphs from being eaten by the predator(s) (**Fig. 13-14a,b**).
4. The North American moth *Nemoria arizonaria*, in which the caterpillars take on a morphology depending on whether they hatch out as *twig morphs* on oak trees when catkins are absent (summer) or *as catkin morphs* when catkins are present (spring), a seasonal polymorphism that is triggered by the levels of tannin in their diet.
5. Threespine sticklebacks *Gasterosteus aculeatus*, in which one, two, or three of the bones of the pelvic girdle are reduced in response to a combination of a *biotic factor* — absence of predatory fish — and an *abiotic factor* — the level of dissolved calcium in the lake (**Fig. 13-14c**).[17]

From these examples we see that phenotypic plasticity embraces interactions that occur:

- *within an individual* (thyroid hormone and transformation of the tadpole into a frog);
- *between individuals* of the same or different species (e.g., a chemical signal released by a predator, bacteria-squid interactions); and/or
- in response to diet or abiotic components of the environment (e.g., dissolved Ca^{++} levels).

Heredity and Developmental Constraint

To repeat their functional roles each generation, developmental factors and their interactions must have an underlying hereditary basis. As with any other trait, heredity ties organismal characters to evolutionary processes, with selection as a primary influence affecting transition and survival (see Chapter 22).

The notochord is maintained in vertebrate embryos, not because it serves the skeletal supporting function it served 500 Mya but because it induces ectodermal cells to produce nerve cells and then interacts with those cells to pattern the spinal cord. Selection for the developmental integration of notochord and central nervous system maintains the notochord.

The superfluous length of the mammalian recurrent laryngeal nerve (introduced earlier as seemingly embedded in mammals like a useless vestige), is a by-product of a more basic genetically determined developmental system that was subject to selection in the past, and whose conservative features are still selected today. Laryngeal nerve function and its circuitous positioning seem determined by separate genetic pathways, each maintained by selection: one for neurological activity and another for skeletal structure. The unusual location of the sixth arterial arch in the mammalian thorax is caused by selection for lengthening of the neck in mammals.

Developmental constraints are tied to past evolutionary events and contingencies, some more deep-seated than others. What appears as a "lethal" mutation may be the inability of a genetic variant to interact successfully with previously evolved systems that enable birth and survival. In broad perspective, a mutation constrained in one organism need not be constrained in organisms that occupy other habitats, live in other circumstances, and/or differ because of different selective histories. Indeed, even in inbred strains of mice, mutant genes have different effects depending on the genetic background of the specific inbred strain. Antecedent selective histories and intricately stabilized networks of developmental genetic interactions channel subsequent evolution. Compatibility of new phenotypic features with established genetic functions serves as a powerful selective mechanism.

The genetic basis for developmental constraint may not be apparent because a phenotypic effect may be at the end of a long chain of interactions. For example, although a gene provides a primary effect in terms of a structural or regulatory product (see, for example, Fig. 10-13), it may also have a second order, third order or even more distant effect because its product undergoes a series of interactions with different genetic and environmental factors. Of serious consequence are genes with pleiotropic effects (**pleiotropy**) influencing many different developmental aspects of an organism, such as the sickle cell mutation and its normal counterpart (see Fig. 10-12).

Although absence of an appropriate adaptive mutation can be declared a constraint, even more so is **selection,** which by taking adaptation in one direction constrains it in other directions by making "adaptive" mutations unadaptive for entirely different functions. Selection for excellence in swimming will impact on traits that enable excellence in running. Similarly, selection for excellence in running sacrifices excellence in flying, and so forth. Organismal (biological) adaptation that can successfully face all possible eventualities is

[17] For further discussion of these examples and for further examples, see Hall (1999a), Hall et al. (2004), Pigliucci (2001), Pigliucci and Preston (2004), Kraft et al. (2006) and Shapiro et al. (2007).

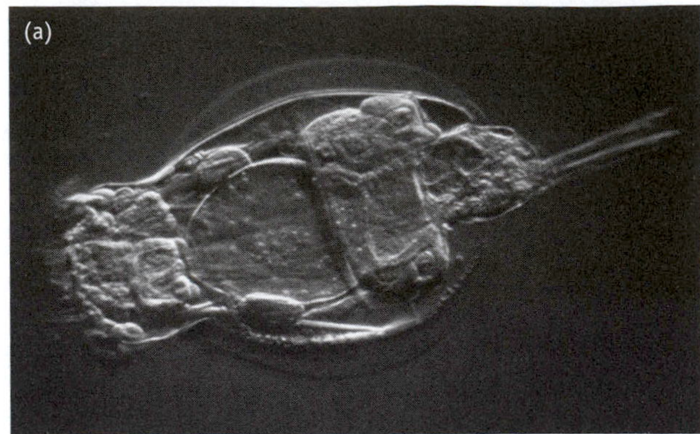

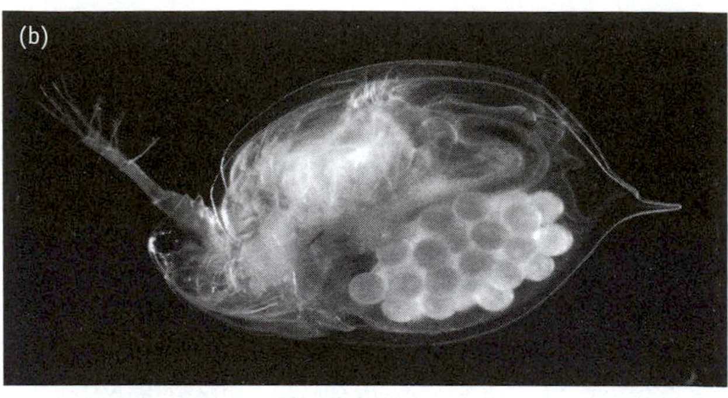

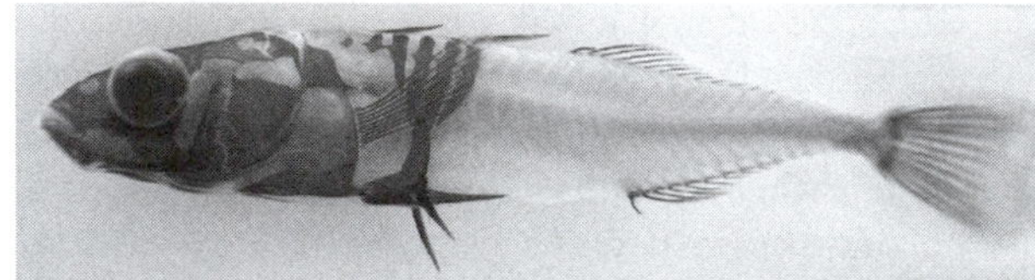

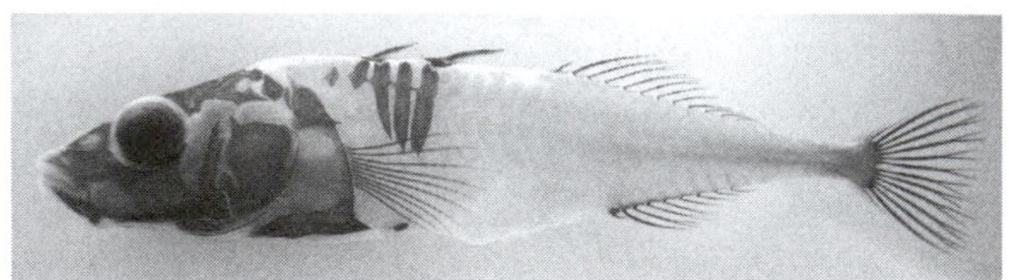

FIGURE 13-14 Three examples of phenotypic plasticity in response to chemical signals released by predators (see text for details). (a) Additional spines on a rotifer (© Tom E. Adams/Visuals Unlimited). (b) The helmet on a water flea, *Daphnia* (© Nigel Cattlin/Visuals Unlimited). (c) Reduction in elements of the pelvic girdle (top to bottom) in the threespine sticklebacks *Gasterosteus aculeatus.* (Reproduced from *Trends in Ecology & Evolution,* vol. 19 (9), Foster, Susan A. and Baker, John A., Evolution in parallell..., pp. 456–459, copyright 2004, with permission from Elsevier. Photo courtesy of Dr. William A. Cresko, University of Oregon.)

unattainable in a single lineage, simply because organisms do not possess unlimited internal resources to continually enlarge one feature without affecting others (Nijhout and Emlen, 1998). Nor can genes wholly adaptive to every possible eventuality interact successfully in a single organism. Selection toward becoming a "master of one trade" prevents an organism from becoming a "jack of all trades," and *vice versa.*

Because adaptive limitations arise from selective history, we call such channeling *adaptive constraint.* Selection thus can act as a phenotypic constraint on two levels:

1. a new trait may not be selected for in the future, and/or
2. may not have been selected for in the past.

Nevertheless, adaptive history indicates that constraints are not impassible barriers: Organismal components restricted in structure by a particular condition may evolve in new directions under different conditions. For example, given sufficient time and new opportunities, vertebrate lineages that evolved from fish into crawling terrestrial forms, evolved into still-different swimming marine forms (ichthyosaurs) and even into entirely novel winged aerial forms (birds and bats). Similarly, some mammals escaped the constancy of seven cervical vertebrae — three-toed sloths have eight or nine, manatees have six — and some reptiles had many more. The Cretaceous plesiosaur *Elasmosaurus* had as many as 76. A developmental constraint need not prevent evolutionary change. Constraint limits variation but does not prevent change, providing a further instance of the importance of understanding exactly how the limits to variation are set.

As we might expect, environmental conditions can also be restrictive, limiting adaptation. Organisms must cope with physical laws that govern their selective environment, including hydrodynamic laws involved in selection for aquatic speed, and aerodynamic laws involved in aerial flight and in pollen/seed dispersal. Environments can cause extinction when the requirements for survival exceed the developmental limits of the organism. From such a broader perspective, adaptive constraints may be imposed by other groups with which organisms interact. For example, the absence of new metazoan body plans after the Cambrian radiation could well have been caused by the restricted access of new phyla to "saturated" ecological niches already occupied by highly adapted preexisting phyla (see Fig. 15-1). Also, competition with successful resident phyla even in new "unsaturated" environments could have limited the development of body plans to variations of older forms rather than make room for distinctively new forms. Where species could enter relatively unoccupied environments, however, opportunistic adaptations and morphological diversity are much more likely, as seen in adaptive radiation (e.g, see Fig. 3-4).

Although not always obvious, the genetic dependency of adaptive constraints is also evidenced by the many regulatory genes that affect development. To these, we can add the genetic mechanisms of canalization and developmental **homeostasis,** which account for uniform expression of traits or patterns of development despite changes in environment or genotype.[18]

According to terminology introduced early in the development of Mendelian genetics and used later by Richard Goldschmidt, Ivan Schmalhausen and others, there are **reaction norms** of individual *genotypes* that allow differential phenotypic expression in response to environmental change. A reaction norm is detected experimentally by raising individuals in different environments such as different temperatures or densities of predators, measuring the trait of interest (size of an appendage, survival, and so forth) under each set of conditions and plotting the phenotype against the environmental variable. A reaction norm, therefore, is an index of organismal-environmental interaction on the one hand, and of flexibility/plasticity-canalization/constraint on the other.

A single line — a single reaction norm indicates a tightly canalized response. A spread of reaction norms indicates greater plasticity of responses. Human blood types (for example, *ABO* and *MN*) seem to be relatively unaffected by environmental changes and, once genotypically determined, persist unchanged throughout life. Other genotypes, such as those causing diabetes, can produce phenotypes quite sensitive to such environmental influences as diet or insulin levels.[19]

Schlichting and Pigliucci (1998) claim that reaction norms apply to *phenotypes* encountering different developmental environments; that is, selection among genotypes must occur in order for developmental plasticity to produce adaptive phenotypes, a concept strongly related to Waddington's theory of **canalization** (see below). Among Waddington's canalization experiments were demonstrations in *Drosophila* that phenotypically uniform expression of the normal *Ultrabithorax* gene in the face of environmental stress depended on other "background" genes. The genetic basis for such constant expression became apparent when Gibson and Hogness (1996) identified specific genetic loci that support *Ultrabithorax* transcriptional stability. Waddington (1957) also pointed out that traits elicited by environmental stimuli, such as crossveinless wings in *Drosophila,* can be genetically incorporated so they appear developmentally without the environmental stimulus, a process he termed *genetic assimilation.* This is not a Lamarckian process of direct instruction by the environment, but occurs because of selection for genotypes capable of such response. The close fit between organismal flexibility and environmental change is a product of underlying genetic components. Suzuki and Nijhout (2006) demonstrated this beautifully using color polymorphism in the hornworm *Manduca quinquemaculata,* a trait sensitive to environmental signals, and which results from modification of the regulation of juvenile hormone. A polymorphic line initiated by heat shock and then subjected to selection, produced the environmentally sensitive color morph in the absence of heat shock. The environmentally induced morphology had been genetically assimilated, arising, after selection, in the absence of the environmental signal.

Canalization

The concept of canalization has a fascinating history in embryology, developmental biology, genetics and in the study of evolution.

Waddington (1942) defined canalization as:

> Canalization or developmental canalization is the presence during embryonic development of paths or channels that result in the production of one or only few stable phenotypes *despite any genetic, epigenetic or environmental influences that would otherwise lead to deviations in development* (emphasis added).

In 1966, Waddington and Robertson reported the results of experiments in which developmental canalization was selected for. If canalization can be selected for, and given that selection is on the phenotype, there must be canalized traits and canalizing selection:

> . . . **canalizing selection** limits the expression of a trait, presumably by eliminating genotypes that would broaden phenotypic responsiveness to the environment and so broaden expression of the character, which could occur by eliminating genotypes that would permit an individual to respond to *any genetic, epigenetic or environmental influences that would lead to greater variability in the phenotypic expression of the trait* (emphasis added).

[18] For canalization, see Lerner (1954), Waddington (1962), For developmental homeostasis, see Hall (1999a, 2005b), Wilkins (2003) and Flatt (2005) .

[19] See Schlichting and Pigliucci (1998) and Sultan and Stearns (2005) for analyses of reaction norms and phenotypic plasticity.

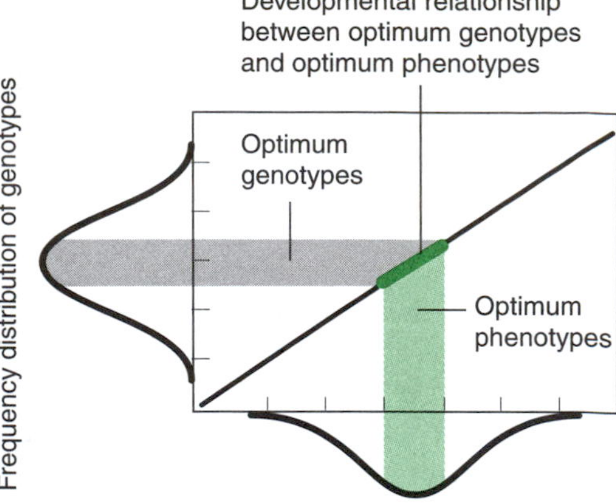

(a) Before canalizing selection

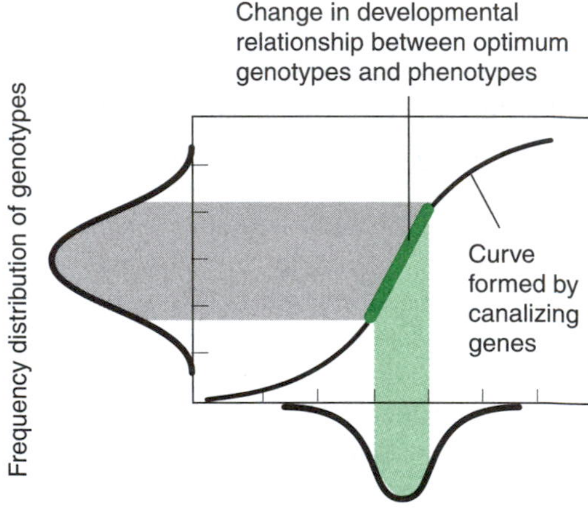

(b) Selection for canalization

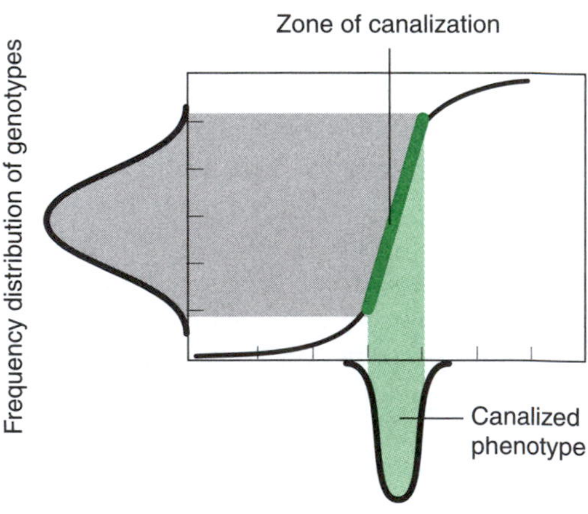

(c) After canalizing selection is completed

FIGURE 13-15 Sequence of selection for a canalized phenotype, or "zone of canalization." The *diagonal*, running from lower left to upper right, represents the developmental relationship between genotype (*vertical axis*) and phenotype (*horizontal axis*); the steeper this developmental curve, the greater the number of different genotypes that produce the same phenotype. (a) At this stage the developmental curve is relatively flat, and only a small section of the genotypic distribution produces the optimum phenotype. (b and c) As selection for canalization proceeds, the developmental curve assumes more of an "S" shape, and larger portions of genotypic variants produce the optimum phenotype. Thus, selection occurs for genotypes that can produce the same phenotype in spite of their variability. The same canalization process explains selection for genes that can preserve phenotypic constancy in the face of environmental change. (Adapted from Strickberger, M. W., 1985. *Genetics,* 3d ed. Macmillan, New York.)

The essence of canalization is, therefore, that not all developmental or evolutionary outcomes are possible. The evidence for canalization outside the laboratory comes from restricted patterns of variation observed in nature when we expect a much greater variety of patterns. Examples occur at all levels:

- *organismal:* all organisms within major lineages of bivalves occupy only a restricted amount of morphospace (see Fig. 15-5);
- *organ:* only six types of variation are found in the aortic arches of a sample of 3,000 rabbits, two patterns accounting for 97 percent of the variability;
- *gene:* variation in the functional region of a gene is more tightly canalized than are the nonfunctional elements.

Figure 13-15 diagrams a scheme devised by Rendel (1967) showing how canalization is selected in a population so that varied genotypes produce a particular optimum phenotype. Thus, in a *Drosophila* population selected for a change in bristle number caused by the *scute* gene mutation, selection can occur on two levels:

1. **Directional Selection.** Selection for change in the primary expression of the major bristle-determining genotype in a mutant *scute* stock, from an average of two to an average of almost four bristles, four being the number in wild type flies.
2. **Stabilizing Selection.** Selection for change in the zone of canalization so that even varied genotypes produce the same desired number of bristles (see Fig. 22-8).

A developmental change can be channeled (constrained) to modify an existing feature or produce a new feature in response to altered conditions by the selection of modifying genes that affect the trait's zone of canalization. Selection not only affects a trait's variability, it can significantly change the trait's pattern of variability, evolving an entirely new range of phenotypes.

Recently, evolutionary biologists have distinguished *genetic* from *environmental canalization* in approaches that emphasize the hierarchical nature of canalization as a process and the need to tease out the genetic from the environmental component(s).[20]

Even under selective change, however, constraints persist. Because organismal features neither arise nor exist independently, and because any phenotype is a product of interaction both with other characters and with the environment, constraint of some sort affects all characters. We know that genetically correlated, mutually interdependent traits cannot change easily without disrupting other traits, dampening the speed of their response to selection. Although hard to trace, particular constraints must be identified developmentally and placed in historical and phylogenetic contexts.

Heterochrony and Allometry

Despite their prevalence, some constraints are bypassed, allowing developmental innovations to occur, when opportune mutations (usually of regulatory genes) successfully integrate into a genetic network. Among such events are genetic changes in positional information, such as the homeotic mutations, discussed above, which cause structures to be placed, repeated or omitted in body segments.

Other innovations can be caused by genetic changes in timing, when a developmental stage or event shifts in sequence relative to other stages or events (**heterochrony**). The precocious onset of an adult stage — for example, sexual maturity in immature tunicate larvae — is one such instance, as is the retention of tadpole features in adult Mexican axolotl (*Ambystoma mexicanum*), the classic example of neoteny, a form of heterochrony. In sea urchins, a genetic change that affects a seemingly simple but fundamental trait, such as increase in egg size, can also lead to changes that radically modify larval development, up to and including elimination of the larval stage (Raff, 1992).

Among other mechanisms are differences in developmental growth rates between organs — **allometry** — because allometry can vary between species. For example, as shown in Figure 20-40, head and leg growth in humans follows a distinctive allometric path when compared to other primates. Niklas (2003) summarizes evidence for the importance of allometry in the evolution of the land plants. Webster et al. (2004) show how the echoes of evolutionary body size — whether ancestral groups were large or small — can be detected in such important life history aspects of carnivore evolution as age of maturity, litter size (both number of offspring per litter and size of offspring), maternal age at which litters are born, interval between litters, and so forth. This does not mean that body size will necessarily correlate with or be a predictor of speciation; body size does not correlate with taxonomic diversity in the Metazoa (Orme et al., 2002).

Development has long emphasized ontogenetic changes in features and organismal design. Evolution has long emphasized phylogenetic changes in fitness based on hereditary variation. Both are limited perspectives, neither providing an inclusive explanation or foundation for the relationship among form, variation and fitness (Hallgrímsson and Hall, 2005). Biological evolution expressed through changes in factors such as morphology, physiology and behavior, arises from developmental changes. Now joining forces as evolutionary developmental biology, the big challenge is to gain a full understanding of the relationships between mutation, gene regulation, developmental processes, variation and selection (see Chapters 10, 22 and 23).

[20] See Stearns et al. (1995), Gibson and Wagner (2000), Debat and David (2001), Hallgrímsson et al. (2002), and Wilkins (2003) for canalization.

KEY TERMS

allometry	homeosis
canalization	homeostasis
coevolution	homeotic mutations
constraint	*hox* genes
developmental constraints	morphogenesis
differentiation	phylotypic stage
epigenetic	preformationist
gene regulation	reaction norms
heterochrony	repressor
homeoboxes	signal transduction
homeodomain	zootype

DISCUSSION QUESTIONS

1. Discuss the following statements:
   a. "Heritable developmental effects are produced by environmental changes."
   b. "Heritable developmental effects respond to environmental changes."
2. Would you expect changes at early or late stages of development to have a greater effect on development? To be more "constrained"? Explain.
3. Can you suggest ways that a signal transduction pathway can change from transcribing one gene to transcribing another, using the same extracellular signal?
4. Would you expect to find the same kinds of constraint on development before and after a mass extinction? Explain.
5. How does an *Antennapedia* mutant act to modify a head segment or an *Ultrabithorax* mutant act to generate a second thoracic segment?
6. Explain the kinds of information needed to designate a developmental change as heterochronic.

EVOLUTION ON THE WEB

Explore evolution on the Internet! Visit the accompanying Web site for *Strickberger's Evolution,* Fourth Edition, at www.biology.jbpub.com/book/evolution for Web exercises and links relating to topics covered in this chapter.

RECOMMENDED READING

Albertson, R. C., J. T. Streelman, and T. D. Kocher, 2000. The beak of the fish: genetic basis of adaptive shape differences among Lake Malawi cichlid fishes. Assessing morphological differences in an adaptive trait: a landmark-based morphometric approach. *J. Exp. Zool.,* **289,** 385–403.

Carroll, S. B., 2005. *Endless Forms Most Beautiful.* W. W. Norton & Co., New York.

Carroll, S. B., J. K. Grenier, and S. D. Weatherbee, 2005. *From DNA to Diversity: Molecular genetics and the Evolution of Animal Design.* 2nd ed. Blackwell Publishing, Malden, MA.

Davidson, E. H., 2006. *The Regulatory Genome: Gene Regulatory Networks in Development and Evolution.* Elsevier/Academic Press, Burlington, MA.

Emlen, D. J., J. Hunt, and L. W. Simmons, 2005. Evolution of male- and sexual-dimorphism in the expression of beetle horns: Modularity, lability and developmental constraint. *Am. Nat.,* **166,** S42–S68.

Gould, S. J., 1977. *Ontogeny and Phylogeny.* Harvard University Press, Cambridge, MA.

Hall, B. K., 1999a. *Evolutionary Developmental Biology,* 2nd ed. Kluwer Academic Publishers, Netherlands. [Japanese translation, Kosakusha, Tokyo, 2001.]

Hall, B. K., and W. M. Olson (eds.), 2003. *Keywords and Concepts in Evolutionary Developmental Biology.* Harvard University Press, Cambridge, MA.

Hallgrímsson, B., and B. K. Hall (eds.), 2005. *Variation: A Central Concept in Biology.* Elsevier/Academic Press, New York.

Land, M. F., and D.-E. Nilsson, 2002. *Animal Eyes.* Oxford University Press, Oxford, UK.

Niklas, K. J., 1997. *The Evolutionary Biology of Plants.* The University of Chicago Press, Chicago and London.

Pagel, M., (ed. in chief), 2002. *Encyclopedia of Evolution,* 2 Volumes. Oxford University Press, New York.

Pigliucci, M., and K. Preston (eds.), 2004. *Phenotypic Integration: Studying the Ecology and Evolution of Complex Phenotypes.* Oxford University Press, Oxford, UK.

Raff, R. A., 1996. *The Shape of Life: Genes, Development, and the Evolution of Animal Form.* University of Chicago Press, Chicago.

West-Eberhard, M. J., 2003. *Developmental Plasticity and Evolution.* Oxford University Press, Oxford, UK.

Wilkins, A. S., 2002. *The Evolution of Developmental Pathways.* Sinauer Assoc., Sunderland, MA.

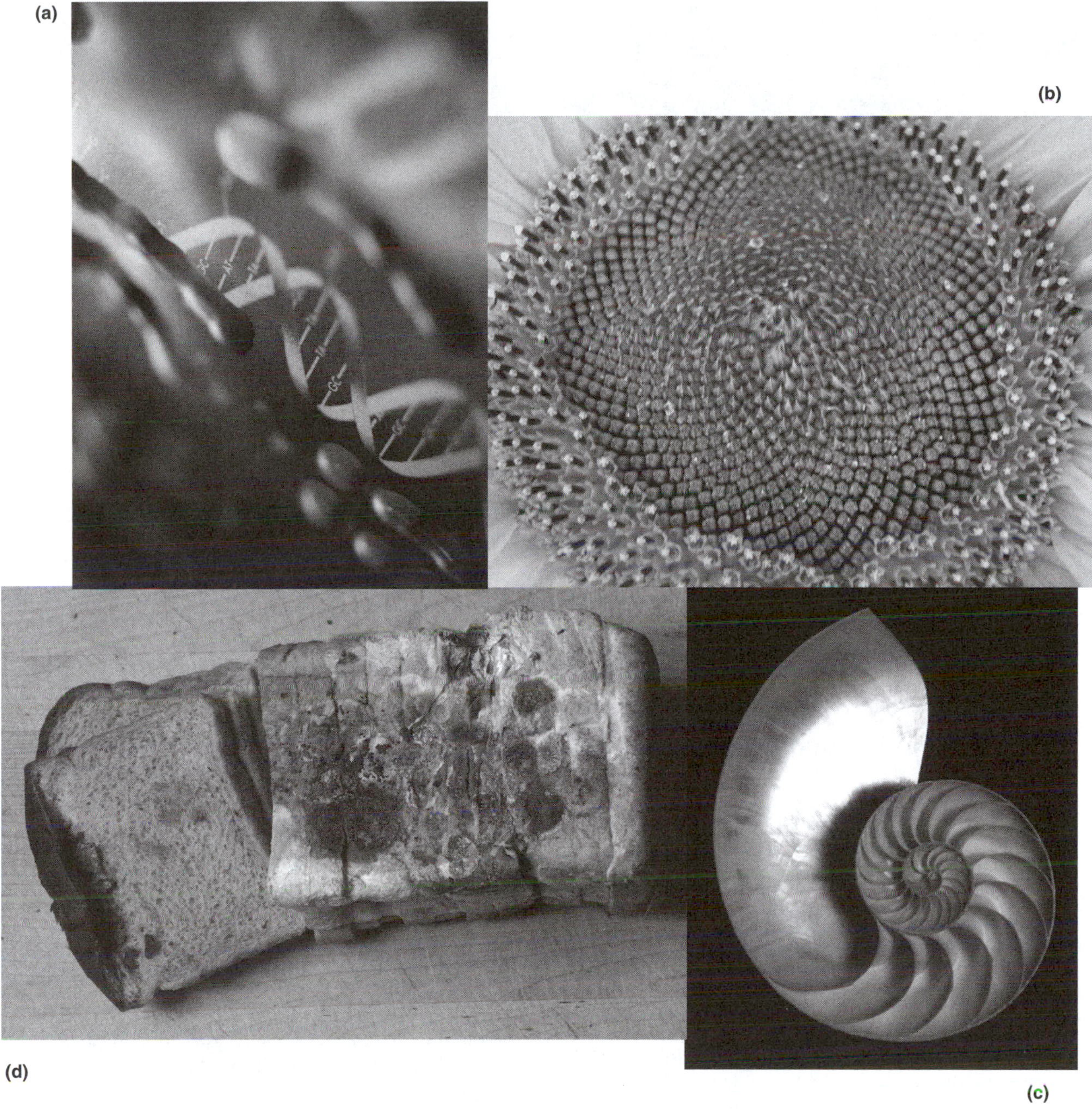

As we move from discussing genes, cells, and development — the organic framework underpinning evolution — to the diversity of organisms the "have been, and are being evolved," we move from the world of DNA (a) to the organisms described, again by Charles Darwin, as those "endless forms most beautiful and most wonderful," forms represented here by the incredible geometric symmetry and subtlety of the flower of a sunflower (b), by the record of development and growth left behind in the chambers of the shell of a nautilus (c), and, yes, even in the patterned growth of a mold on a forgotten loaf of bread (d).

PART

4

The Organisms

The Diversity of Life

Having examined the **origin of life** in Chapters 6 to 8, and the **organic framework** for evolution in Chapters 9 to 13, we turn in Chapters 14 to 20 to the **history of life** and consider both the organisms that populated Earth in the past and those alive today. Our approach represents a compromise at several levels.

We are accustomed to consider organisms in such commonly known and understood categories as animals (vertebrates and invertebrates), plants, microbes, fungi or protists. But phylogenetic analyses carried out over the past two decades demonstrate convincingly that these are not natural categories; a recent classification scheme subsumes the term plants (Plantae; see Box 12-4). So, should we approach organisms on the basis of names long since used for them, or should we assemble them according to the latest Tree of Life? We try, we hope successfully, to do both.

As one example typical of many, *reptiles* is a name for a group of animals that does not include birds, even though birds are derived from reptiles. In systematic classification, reptiles as the name for a group, does not include descendants of reptiles (for example, birds), which have features not found in reptiles (feathers). In cladistic terminology, reptiles would be a paraphyletic group and so not a valid taxon, paraphyly referring to a taxonomic group or lineage that includes some descendants of a single common ancestor, but not all descendants (in this example, birds are not included). Cladistically, Reptilia includes all reptiles, their descendants (birds) and their common ancestor, and so "birds" are flying dinosaurs. Consequently, in Chapter 18, we integrate more historical (classic) studies and terms with recent cladistic analyses to convey both "what is a reptile" and "why the old term reptiles is not a monophyletic."

The old name "mammal-like reptiles," used for those reptiles described as having given rise to mammals, is no longer considered appropriate for three major reasons:

1. The organisms (therapsids) comprised multiple independent lines, only one or a few of which gave rise to mammals.
2. The term confuses a crown group (the culmination of an evolutionary lineage) with a stem group (a lineage that gave rise to another group).
3. The name "mammal-like-reptiles" makes it sound as if some reptiles were "trying to become" mammals and so were "mammal-like."

The latter problem also has led to the replacement of the term *preadaptation* with *exaptation*.

To meet all these ends, Chapters 14 to 20 are organized around familiar groups: plants and fungi in Chapter 14; invertebrates in Chapter 16; chordates and vertebrates in Chapters 17 to 20. Within these chapters, we discuss relationships and changing terminology in a way that, we trust, allows you, the reader, to appreciate the diversity of familiar groups and their relationships to other groups. Chapter 16, for example, describes the diagnostic features of the major groups of invertebrates, but also provides a framework for our understanding of how the many invertebrate groups relate to one another and might have evolved. Chapter 20 describes the diagnostic features of the major groups of primates, but also provides a framework of our understanding of how humans diversified from other primates. To emphasize origins, two chapters explicitly deal with ancestry: Chapter 15, the origins of multicellular organisms, and Chapter 17, the origin of the vertebrates.

Within the space available, we have provided information about the major groups of organisms at multiple levels: genes, development, morphology, function, interaction with other organisms and interactions with the environment (ecology). A topic of long-standing interest to biologists — the transition of vertebrates from water to land — is discussed in Chapter 18, which also treats the important topic of extinctions, especially those mass extinctions of many different types of organisms that occurred at several times in the history of life. Given our curiosity about ourselves, the evolution of behavior in primates is discussed in some detail in Chapter 20, following an overview of mammals and mammalian evolution in Chapter 19.

14 Evolution of Plants and Fungi

Chapter Summary

The more than 500,000 species of land plants share certain features with green algae and are known to have arisen from organisms similar to some extant green algae. Among the characteristics shared by land plants and green algae are starch storage, cell plate formation during cytokinesis and the use of chlorophylls *a* and *b* in photosynthesis. Other land plant characteristics that may have been present in their algal or alga-like ancestors are alternation of diploid and haploid generations and meiotic spore production. Algal or alga-like progenitors evolved into multicellular organisms in which cells became specialized but lost individual motility.

According to many biologists, the increased variation among offspring resulting from meiosis, entrenched reduction division in the reproductive processes of most eukaryotes. Meiosis also may have served to reestablish the prevailing haploid condition in early, sexually reproducing organisms, advantages of sexual reproduction including both increased variation and increased efficiency of selection. Because of the advantages of having two sets of genetic information, the initially brief diploid stage has considerably lengthened in many forms.

The evolutionary origins of bryophytes — early land plants — are obscure. Bryophytes live in moist areas, have simple water- and food-distributing tissues and exhibit a distinctive life cycle in which the diploid generation, the **sporophyte,** parasitizes the haploid plant, the **gametophyte.** In land plants, however, the products of meiosis are resistant spores that germinate into the haploid gametophyte and fertilization of gametes produced by the gametophyte yields the diploid sporophyte. Although the gametophyte generation has persisted in all land plants, the sporophyte has become dominant.

The earliest vascular plants (ancestors of club mosses, horsetails and ferns) were leafless and rootless. Evolutionary trends in these plants included the development of two types of **spores** (heterospory), the formation of large leaves (megaphylls) and the differentiation of increasingly elaborate vascular tissues. Plants also developed techniques to minimize water loss when they became terrestrial.

When embryos gained independence from water and became enclosed in seeds, the gametophyte was greatly reduced. Plants, in the form of gymnosperms (which bear naked seeds) and angiosperms (with covered seeds), became fully terrestrial. Angiosperms have double fertilization in which they produce a triploid nutritional endosperm as well as a diploid embryo, and elaborate flower structures that attract pollinators.

We know comparatively little of angiosperm origins, although fossil pollen appears in Early Cretaceous deposits. Two theories proposed are that angiosperms developed as woody shrubs in climatically unstable highland areas or as lowland tropical plants. Because of their special characteristics, particularly reproductive features — extreme truncation of gametogenesis, efficient pollination and protected seeds — angiosperms became dominant in many environments.

Fungi — an ancient but actively evolving group — are unicellular, pluricellular or multicellular saprophytic or parasitic organisms. No longer regarded as simple plants without chlorophyll, fungi are now recognized as the third major grade of multicellular organisms (although some are unicellular), a sister group of animals and plants and one of the five kingdoms in Whittaker's proposal. The lack of chlorophyll is reflected in fungal life styles; fungi obtain nutrition either from live hosts by parasitism or saprophytically from dead organic matter. They may have derived from chemotropic protistans or from obligate parasites, some of which became saprophytic.

Interest in the origin of land plants has a long history. The proposition that the photosynthetic eukaryotic organisms that were ancestral to the land plants were similar to some extant green algae is an old idea, predating Darwin's publication of *On the Origin of Species.*

On the basis of current knowledge, the likely sequence in the early evolution of land plants is:

1. Unicellular organisms, which had incorporated new genes through horizontal gene transfer (see Chapter 10) and organelles by endosymbiosis, transformed into multicellular photosynthetic organisms through retention of cell division products.
2. These first small multicellular "plants" captured additional genes by horizontal gene transfer.
3. By these means these organisms increased in size, and in organelle and genome complexity, enabling their component cells to compartmentalize and specialize.
4. With further evolution, organisms in which most or all cells could reproduce the entire body transformed to organisms in which most cells were somatic and only a few were reproductive.

Before we learned more about the ancestry of plants, some steps in the evolutionary sequence from green algae to land plants were thought to be echoed in a series of taxa that extends from *Chlamydomonas* to *Volvox* (**Fig. 14-1;** see Box 8–2, Multicellularity and Pluricellularity). The taxa shown in Figure 14-1, however, are highly derived living forms and not ancestors; as discussed in Chapter 11 and in the introduction to Part IV, a crown taxon cannot be an ancestor to another taxon. Furthermore, the ancestor of land plants is now known to lie within the charophycean-like algae; *Volvox,* in contrast, is a chlorophycean alga (see below).

Based on morphological (especially ultrastructural) features, the ancestor of land plants is now thought to have been a single-celled flagellated green alga somewhat like the green flagellated alga *Mesostigma viride,* which itself evolved from eukaryotic cells that had been invaded by prokaryotic chloroplast-like symbionts (see Box 8-1, Kingdoms and Domains of Life). Identification of a land plant multigene family in this alga supports *Mesostigma* as the nearest relative of land plants (Nedelcu et al., 2006). Subsequently — exactly when is unclear — green flagellated algae became sessile by losing flagellar motility, initiating a lineage that produced the vascular plants. The accumulation of evidence for multiple invasions of land by aquatic algae lends credence to this scenario, which has been enriched and expanded by studies using 18*S* rRNA and the chloroplast gene *rbcL,* and the sequencing of nuclear and mitochondrial genomes. The resulting molecular phylogenies establish two major lineages with a single common ancestor:

- chlorophyte green algae as one lineage, and
- charophyte algae plus embryophyte land plants as the other.

Analysis of 18*S* rRNA sequences provides strong support for stoneworts (Charales) as a sister group to land plants.

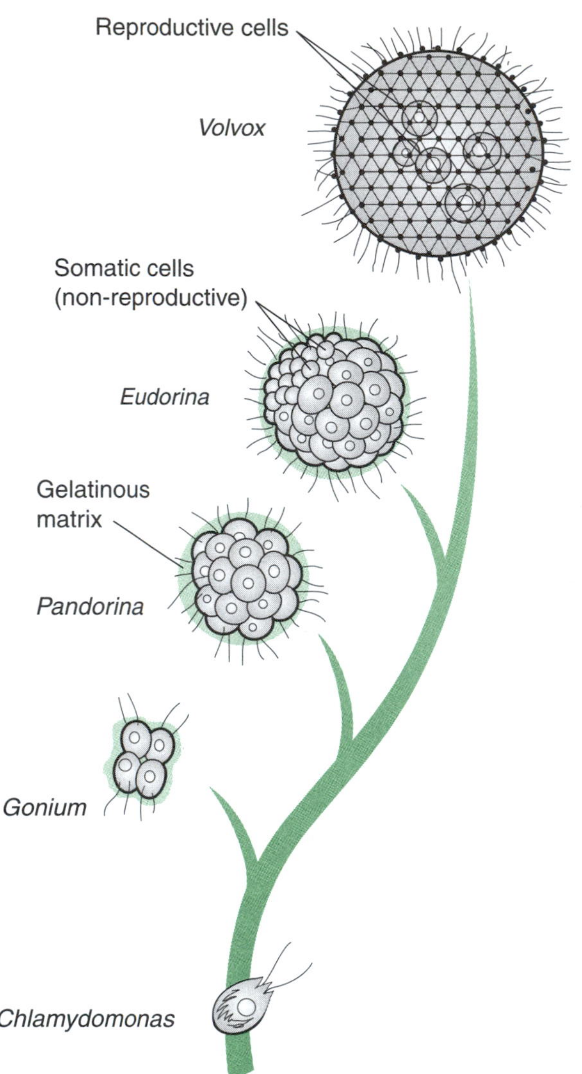

FIGURE 14-1 The sequence, classically presented as a phylogenetic sequence, to illustrate the origin of some multicellular algal aggregates such as *Volvox,* but note that this is a sequence of derived forms not an ancestor-descendant tree.

A study using a mitochondrial, a nuclear, and two plastid genes, and another study based on the entire mitochondrial genome of the common stonewort, *Chara vulgaris,* support the conclusion from the 18*S* rRNA data. Data from analysis of the chlorophyll gene *rbcL* support either Charales alone, or Charales plus a group of microscopic filamentous algae (Coleochaetales) as the sister group. Although identification of the nearest relative of the land plants remains incompletely resolved, the U.S. National Science Foundation-funded program, "Assembling the Tree of Life" (http://atol.sdsc.edu/; accessed April 23, 2007), holds the promise of enhancing enormously our understanding of plant origins and relationships, as is true for our understanding of the lineage relationships of all multicellular organisms.[1]

Terrestrial Algae

Having undergone hundreds of millions of years of evolution, extant sessile forms of algae are no longer representative of the ancient progenitors of land plants, though they have retained some features of those progenitors.[2]

Many **algae** grow on soil or as epiphytes on trees, and, like motile aquatic forms, may be either uni- or multicellular (see Box 8-2). Some land-dwelling algae such as *Coleochaete* bear morphological similarities to *Parka,* a fossil plant more than 400 My old, dating from the Late Silurian to Early Devonian (**Fig. 14-2**). The possibility that *Coleochaete* (a genus of approximately five species of green algae epiphytic on aquatic plants) may represent a prototype of extant land plants is based on biochemical similarities and their use of

[1]See Delwiche et al. (2002) for establishment of the two major lineages, Louis and McCourt (2004) for discussion of the gene studies and see Rokas et al. (2003, 2005) for the Tree of Life project. The ability of some green algae to thrive under extreme conditions such as baking desert environments, salt concentrations greater than 10 percent or at a pH below 2, opens other avenues to the exploration of algal-plant relationships.

[2]The same reasoning is used when we examine any group that contains extant taxa, some of which have more basal features than other taxa. Extant ascidians and hemichordates are used in this way when the evolutionary origin of vertebrates is evaluated (Chapter 17).

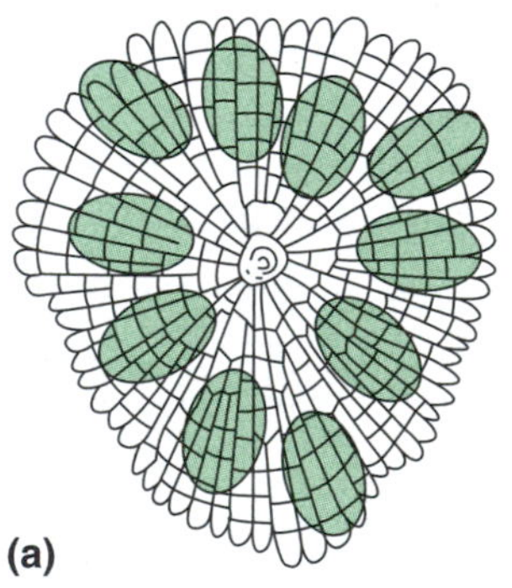

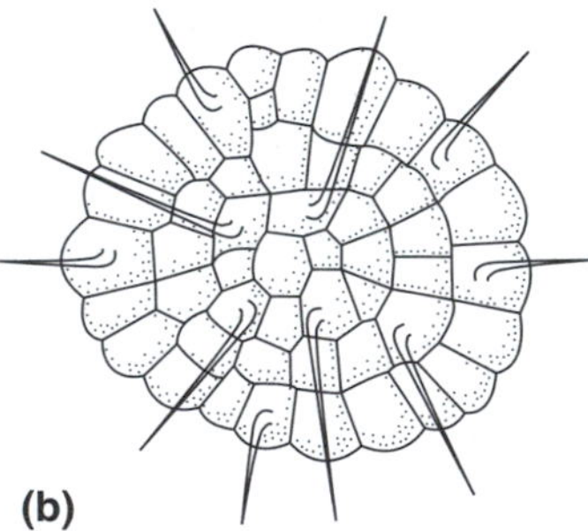

FIGURE 14-2 (a) Reconstruction of a 400-million-year-old fossil plant, *Parka decipiens.* (b) Plant body of the extant green alga, *Coleochaete.* (From Taylor, T.N., and E.L. Taylor, 1993. *The Biology and Evolution of Fossil Plants.* Reprinted by permission of Prentice Hall.)

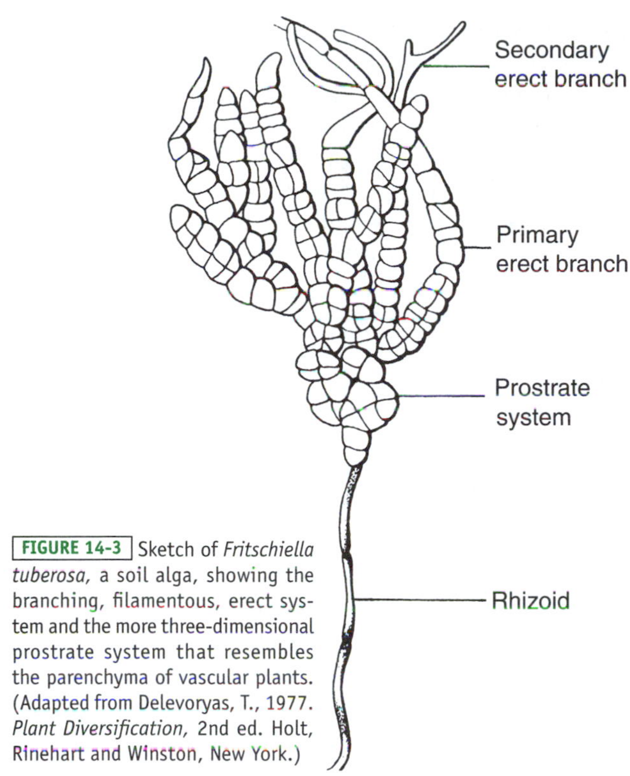

FIGURE 14-3 Sketch of *Fritschiella tuberosa,* a soil alga, showing the branching, filamentous, erect system and the more three-dimensional prostrate system that resembles the parenchyma of vascular plants. (Adapted from Delevoryas, T., 1977. *Plant Diversification,* 2nd ed. Holt, Rinehart and Winston, New York.)

a cell plate during cell division rather than cytoplasmic constriction or cell furrowing. *Coleochaete* is regarded by many botanists as the closest extant algal relative to both stoneworts (Characeae or Charales, which includes some of the largest green algae) and vascular plants.[3]

Another green alga, perhaps farther from land plant ancestry than *Coleochaete* but bearing other terrestrial adaptations, is *Fritschiella tuberosa* (Stewart and Rothwell, 1993), a terrestrial species whose rhizoids penetrate the ground and that maintains branched, prostrate and erect multicellular filaments (**Fig. 14-3**). Some algologists report that the *Fritschiella* life cycle alternates between haploid gametophyte and diploid sporophyte phases, both of which are common to some other green algae and to all land plants (**Fig. 14-4**);

[3] See Delwiche et al. (1989, 2002) and L. E. Graham (1993) for *Coleochaete* sp. as a prototype of extant land plants.

FIGURE 14-4 The mode in which haploid and diploid generations alternate in some green aquatic algae (e.g, *Ulva*), showing similar-appearing (isomorphic) gametophytes (n) and sporophytes (2n). In land plants, the sporophyte embryo is retained and nourished within the gametophyte tissue and matures into a different form from the gametophyte (heteromorphic alternation of generations). In bryophytes, as explained in the text, the sporophyte remains dependent on the gametophyte, whereas the sporophyte is independent in vascular plants (Tracheophyta).

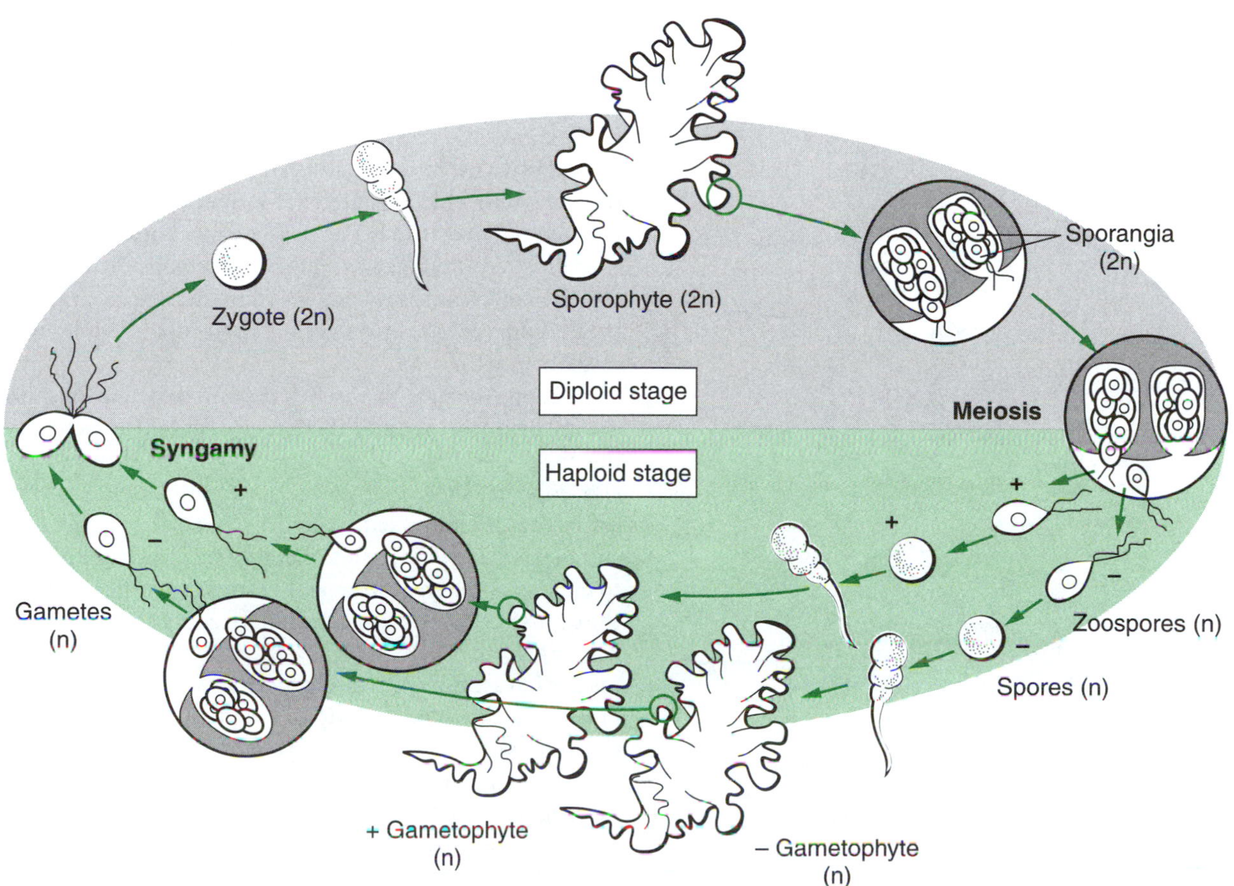

that is, although both phases are multicellular and grow through regular mitotic cell division, *the alga changes from diploid to haploid through a meiotic reduction division in the sporophyte.* Haploid spores produced by the sporophyte develop into the gametophyte phase, which produces sexual gametes mitotically. These unite, in turn, to form the diploid zygote and subsequent sporophyte.

Green algae and land plants share other similarities. Green algae store their carbohydrate reserves as starch. Many green algae also have rigid, cellulose-reinforced cell walls. Recent studies have been providing strong evidence that both algal and flowering plant cell walls arose in endosymbiotic and horizontal gene transfer from Eubacteria to a common ancestor of algae and plants (Niklas, 2004). In addition, green algae and vascular plants use similar types of chlorophyll (*a* and *b*) and carotenoids (α and β). A number of green algae, such as sea lettuce (*Ulva* sp.) and razor algae (*Caulerpa* sp., a genus that includes the highly invasive "killer seaweed," *C. taxifolia*), have membranous forms that simulate some vascular plants yet show their evolutionary ancestry by passing through an alga-like, filamentous stage (**Fig. 14-5**).

Perhaps the most significant aspect of adaptation to land was prevention of water loss through cell-surface evaporation. In those algae where dehydration (**desiccation**) is a problem, two major mechanisms of coping with this difficulty evolved. One mechanism used in chlorophyte algae such as *Trentepohlia* is to confine cellular growth to aquatic conditions and to become dormant under dry conditions. The absence of large, watery vacuoles in *Trentepohlia* cells, a feature shared with air-dispersed spores of other plants, enables such cells to suffer relatively little change in shape and volume during dehydration compared to cells with vacuoles.

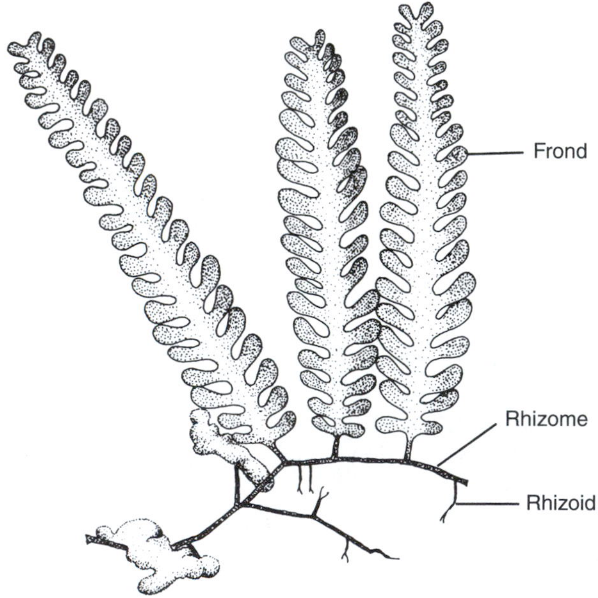

FIGURE 14-5 *Caulerpa*, a green alga with leaf-like forms. (Adapted from Delevoryas, T., 1977. *Plant Diversification*, 2nd ed. Holt, Rinehart and Winston, New York.)

In contrast, some terrestrial algal species with water-filled vacuolated cells, including *Cladophorella* and *Fritschiella*, maintain a waxy cuticle on their airborne parts to retard water loss. The vacuoles, in turn, enable cells to continue their metabolic activity — as though under a constant marine environment — and provide mechanical rigidity (turgor), which prevents cellular collapse. The large volume that vacuoles occupy forces the cytoplasm into a relatively thin sheet along the perimeter of the cell, maximizing available photosynthetic surface.

Origins of Land Plants

In a number of important respects, some shallow-water or mud-dwelling green algae seem eminently adapted to begin the journey to land. The lineage that made this transition is still unclear, although as discussed above, most botanists incline toward a land plant ancestry from a lineage that gave rise to *Coleochaete.* Nevertheless, as L. E. Graham (1993) suggests, algal land invasion was not a singular event: "there have probably been multiple colonizations of the terrestrial environment by green algae." Perhaps only one or two of these gave rise to vascular plants.

We do not know when the algal journey to land began, but botanists surmise it was post-Cambrian. A fall in sea level during Ordovician glaciations would have exposed aquatic plants in shoreline communities to selection for resistance to desiccation. Other environmental conditions contributing to land plant evolution include an increase in atmospheric oxygen levels, which fostered the formation of highly oxygenated polymers such as cutin (a waxy **cuticle** material used in waterproofing) and lignin (a stiffening polymer used for mechanical support and water-conducting tissues) and produced an ozone screen against harmful ultraviolet rays (see Chapter 8).

Among the algae themselves, evolutionary relationships are still being sorted out, although most botanists recognize that the golden-brown algae (Chrysophyta) and brown algae (Phaeophyta) show relatively derived features, especially in respect to differentiated structures. Both have *planktonic* (motile) and *benthic* (nonmotile) forms, the latter attaching to the sea floor in shallow areas. In some brown algae such as the genus *Fucus*, the most common large alga on the Pacific coast of North America (**Fig. 14-6**), cell division is localized (as it is in vascular plants) to a specific *meristematic growth area* below the elongating tip of the plant, and to differentiated organs that produce sperm and eggs. Furthermore, the relatively complex body tissues include specialized conducting cells that function similarly to the sieve tubes in the phloem of vascular plants, providing an example of conver-

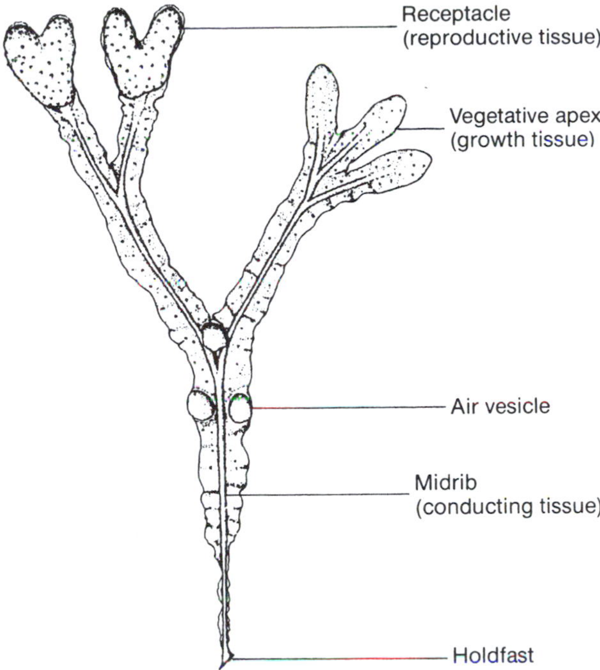

FIGURE 14-6 The habit of *Fucus vesiculosis*, a brown alga that commonly grows in the intertidal zone. (Adapted from Bold, H. C., C. J. Alexopoulos, and T. Delevoryas, 1980. *Morphology of Plants and Fungi*, 4th ed. Harper & Row, New York.)

gent evolution. Among factors they share with other algae, Chrysophyta and Phaeophyta lack the waterproof cuticles that prevent desiccation of land plants. Nevertheless, botanists do not consider these algae ancestral to land plants because their pigmentation (chlorophyll *c*, fucoxanthins, and so on) and storage products are so different.

Reproductive organs also distinguish algae from land plants. "True" land plants — photosynthetic eukaryotes marked by the ability to survive and sexually reproduce on land (Niklas, 1997) — are called ***embryophytes.*** Ranging from simple *bryophytes* to complex *angiosperms*, embryophytes are characterized by reproductive structures consisting of one or more multicellular layers that help protect and develop gametes. These surrounding "sterile" cells also provide embryonic nutrients for the egg. By contrast, algal reproductive structures are less complex, and gamete development lacks such multicellular enclosures.

Bryophytes

Botanists traditionally classify what are regarded as the simplest land plants — liverworts, hornworts and mosses — into a single clade, **bryophytes,** although some botanists restrict this name to the mosses, and classify liverworts and hornworts as Hepatophytes.

Bryophytes have features common to land plants: multicellular reproductive structures, a cuticle protecting their aerial parts, and epidermal pores (*stomata*) that permit the transfer of carbon dioxide, water vapor and oxygen between their tissues and the atmosphere. Some bryophytes have food and water transport tissues, although these do not seem as efficient as the phloem and xylem of vascular plants. Limitations in food and water transport restrict bryophytes to small size. They live mostly in moist environments where they can transport water along their surfaces. In arctic or arid environments, they usually suspend growth until the warm, moist season begins.

Among characteristics shared by all bryophytes is **alternation of generations,** in which the haploid gametophyte generation is free-living and the diploid sporophyte generation remains parasitically attached. In liverworts such as *Marchantia* (**Fig. 14-7**) and *Sphaerocarpos*, the sporophyte is relatively undifferentiated. It is considerably more complex in the hornwort *Anthoceros*.

Although we have few clues to the direct ancestry of bryophytes or even the phylogenetic relationships between their major groups, many botanists support the thesis that these plants have an algal origin because of spore-forming and gamete-forming tissues in various algae and the similarity between the filamentous growth pattern of some green algae and the branching filamentous protonema stage of many mosses. The moist environment of most bryophytes and their dependence on water for fertilization also point to an aquatic origin.

According to one proposed evolutionary sequence, popular in the past, algae evolved into bryophytes, which evolved into vascular plants (e.g, ferns). Today, most botanists support the view that bryophytes and vascular plants differ notably and may each have had an independent algal origin and therefore independent evolutionary histories. One molecular phylogeny suggests that bryophytes themselves are polyphyletic, different groups having arisen independently. Because the sporophyte generation of bryophytes depends for nutrients and support on the gametophyte — the sporophyte of vascular plants is completely independent — distinct algal origins for bryophytes and vascular plants seem even more likely.

The earliest unequivocal appearance of bryophytes in the fossil record is Devonian for liverworts and Carboniferous for mosses, whereas recognizable fossils of vascular land plants appear in earlier Silurian strata. A recent comprehensive phylogenetic analysis of liverworts, mosses, hornworts and vascular plants, drawing on three data sets (nuclear and chloroplast genes and structural characters), provides strong support for liverworts as the sister group to all other land plants and hornworts as the sister taxon to vascular plants (Qiu et al., 2006). An important implication of these data is that bryophytes were important in two major transitions,

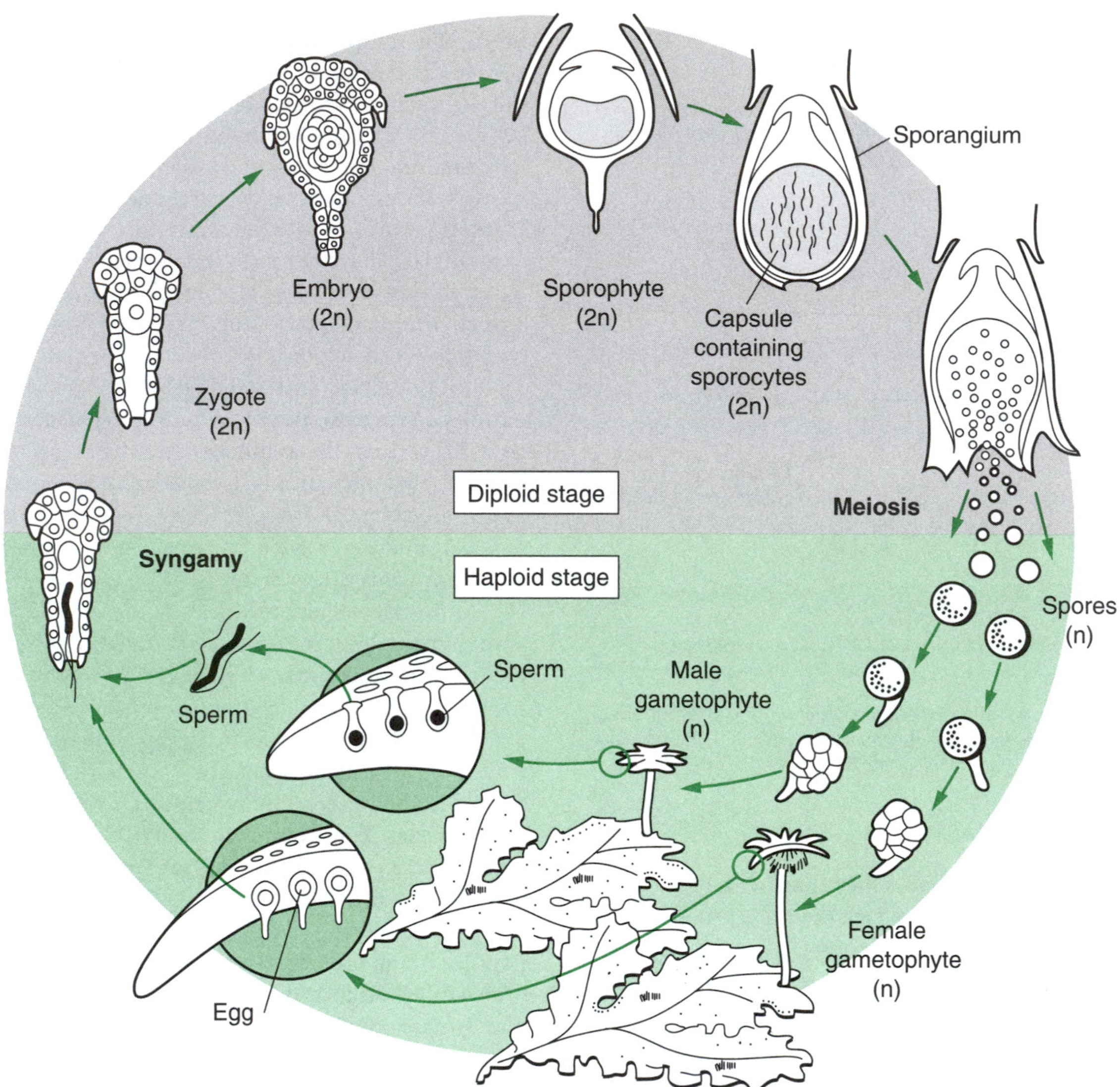

FIGURE 14-7 Life cycle of a liverwort species in the genus *Marchantia*. The gametophyte is a flattened, chlorophyll-bearing thallus with rootlike rhizoids on its undersurface. The spherical sporophyte grows on the tissue of the gametophyte. Each of its interior sporocytes divides meiotically to form a tetrad composed of four haploid spores. Germination of the spore leads to the gametophyte, and the cycle continues as shown. In gametophytes, the *antheridium* is the sperm-producing tissue, the *archegonium* the egg-producing tissue.

the transition from water to land, and the transition from a haploid gametophyte-dominated life cycle.

Whatever their relationship(s), so-called "simpler organisms" with seemingly only marginal or intermediate adaptations for a terrestrial existence persist despite the presence of so-called "complex organisms" with more derived adaptations. Nonvascular plants did not become extinct because of the evolution of vascular plants, and nonseed plants such as ferns still survive in the presence of seed plants; there are approximately 22,000 named species of bryophytes and 10,000 species of ferns.

Sexual Reproduction, Meiosis and Alternation of Generations

Sexual Reproduction

As indicated in Chapter 9, an important consequence of sexual reproduction is provision of the variation that enables a population to produce a wide array of genotypes and so enhance the efficiency of selection. This variation arises because when two parents contribute chromosomes to an offspring, the chromosomes reshuffle in the offspring's sexual meiotic tissues to produce chromosomal and genetic combinations different from those in either parent. Gametes containing these new combinations can combine to form the zygotes of the next generation, which have new genetic combinations. Chromosomes reshuffle in the following generation, and the process continues (see Chapter 9).

For populations continually interacting with changing environments, **sexual reproduction** is an obvious advantage in producing new combinations of alleles, some of which may be adaptive and permit the lineage to survive. Lineages without the variability introduced by such reproduction can more easily become extinct under changing circumstances. Data from the incidence of asexuality point to the likelihood that asexual species have higher extinction rates than sexual species; that is, asexuality only rarely appears as a prevailing character of large taxonomic groups, being more common among smaller groups. Occasional asexual species are found as members of larger, more inclusive groups that reproduce sexually. The scarcity of asexual families or higher groups indicates that asexuality rarely survives long enough to become the predominant character of a lineage.

In a long-standing population that continually experiences the *same* environment, genotypes eminently adapted to that environment evolve, and most, if not all, new genetic combinations have lower relative fitness than the parental generation. Under such constant circumstances, the advantages of having two sexes are not apparent. In fact, *many plants have abandoned sexual reproduction* and replaced it with asexual methods such as the spread of vegetative somatic tissues or parthenogenesis (reproduction through unfertilized eggs — seen also in some animals). Although some asexual plants are found over wide geographical ranges, their success is often restricted to specific environments or to conditions that severely limit cross-fertilization, because only small, inbred populations can survive. However, because environmental conditions are *not often constant,* eukaryotes generally use meiotic forms of sexual reproduction for at least part of their life cycle, and asexual groups rarely survive over long evolutionary periods.

Evolution of Meiosis

We do not know precisely when meiosis originated, although it must have appeared in conjunction with, or soon after, the evolution of sexual fertilization. Some evidence supports the proposal that the haploid–diploid cycle preceded the origin of sexual reproduction and survived in some asexual protists because it helped to eliminate mutations during the haploid stage while providing the advantages of increased functional genetic material or other benefits during the diploid stage. According to this hypothesis, sexual union could only occur after the meiotic reduction mechanism evolved to facilitate a regularized transition from diploid to haploid stages (Kondrashov, 1994).

In the absence of a mechanism to reduce chromosome numbers in gametes, sexual union leads to a doubling of nuclei and, hence, chromosome numbers. With each succeeding sexual generation, chromosome numbers would increase almost exponentially, forming large, unwieldy nuclei with difficulties in functioning and in coordination. A meiotic mechanism reducing the gametic chromosome number to half would have had selective value. At what stage of the life cycle of early organisms would meiosis have occurred? The answer to this is again conjectural, but the following scenario seems reasonable.

Because a doubling of chromosome number in somatic cells would not have been immediately advantageous to the earliest haploid organisms, meiosis probably took place immediately after fertilization in the diploid zygote cell itself. These sexual organisms would have immediately regained their haploid condition without the intervention of an extended diploid state, a situation similar to that found in algae such as *Chlamydomonas* and fungi such as *Neurospora.*

However, there are advantages to lengthening the diploid stage, not the least being that such cells may have two kinds of genetic information, one from each parent, enabling a single organism to use different developmental pathways in responding to different environmental conditions. In addition, the two alleles of a gene in a diploid may each produce different products that can buffer each other to ensure developmental uniformity in any particular environment (**heterozygote advantage**; see Chapter 22). Diploidy provides the opportunity for dominant genetic relationships that mask the effect of deleterious recessive alleles and yet, at the same time, allow a population to evolve further by retaining recessive alleles that may be advantageous under future conditions. Bearing pairs of homologous chromosomes also may allow one member of a pair to act during recombination as a template in repairing damage in the other. It can also be argued that when meiosis takes place immediately after fertilization the gametes have relatively limited genetic variability because only one reduction division has occurred. For example, a single diploid cell with three pairs of chromosomes (or three pairs of alleles), A^1, A^2, B^1, B^2 and C^1, C^2, might produce four haploid gametes from a meiotic division that are of constitutions $A^1B^1C^2$, $A^1B^1C^2$, $A^2B^2C^1$ and $A^2B^2C^1$.

Alternation of Generations

In contrast to a single diploid cell, a multicellular diploid organism can produce a greater variety of gametes because numerous kinds of reduction divisions can take place in a large number of parental cells undergoing meiosis.

Some meiotic products of such an organism could produce $A^1B^1C^2$ and $A^2B^2C^1$ gametes, others could produce $A^1B^2C^1$

and $A^2B^1C^2$ gametes or $A^2B^1C^1$ and $A^1B^2C^2$, and so forth. Along with other advantages, a population of organisms whose diploid meiotic tissues are multicellular would produce greater genetic variability among offspring and therefore have greater potential for evolutionary change than a population containing a similar number of organisms in which the diploid meiotic stage is unicellular. (For further hypotheses to explain the origin and persistence of two sexes see Box 9-1, The Evolution of Sex-Determining Mechanisms)

Persistence of haploidy as a major life cycle stage in some organisms may be related to the speed at which haploids eliminate deleterious alleles unprotected by the diploid stage. According to Mable and Otto (1998), such advantages favor haploids "if (i) sex is rare, (ii) recombination is rare, (iii) selfing is common, or (iv) assortative mating is common." However, they also point out that:

> Once a certain ploidy level has become dominant within a taxonomic group, it may be difficult to expand the alternate ploidy phase, either because the necessary mutations simply do not arise or because individuals with atypical ploidy levels are unable to develop normally . . . [A]n organism may evolve developmental pathways that depend on having the appropriate ploidy level.

In animals, the lengthened diploid stage became the dominant feature of the life cycle; the haploid stage is mostly restricted to the gametes themselves. In land plants and some green algae, the lengthened diploid stage, or sporophyte, also produces meiotic products as in animals, but these are spores rather than gametes. The meiotically produced spores develop into haploid gametophytes, which only later produce gametes by mitosis.

The sporophyte–gametophyte alternation of generations in plants (**Fig. 14-8**) has long been puzzling. Aside from the advantages of maintaining a diploid state, the sporophyte produces dispersible, encapsulated spores that resist desiccation. By contrast, land plant and animal gametes are relatively unprotected. The vulnerability of sexual gametes to terrestrial conditions may account for the persistence of the gametophyte stage in plants, a stage that easily disappears in animals. The features of plant sporophytes enabling resistance to desiccation and conquest of the land differ from the features of the gametophyte required for the aqueous transfer of plant gametes. The evolution of plant sporophytes that do not depend on the gametophyte reflects both the advantages of diploidy and the needs of spore production. Persistence of the gametophyte stage reflects the advantages that accrue to immobile gamete-producing individuals growing in proximity to each other in water.

A further animal–plant difference lies in how cells of the **germ line** separate from other tissues. Extensive migration of cells during the development of some animals enables germ plasm to localize in specific reproductive organs: ovaries and testes. Plant cells, by contrast, mostly maintain their relative positions during development, and form gametes only by transforming somatic cells in different vegetative regions. Reproduction in those animals with a separate germ line attains improved genetic stability by restricting germ plasm to a single tissue that has selective value in avoiding the somatically derived mutations that are so common to plant germ cells.

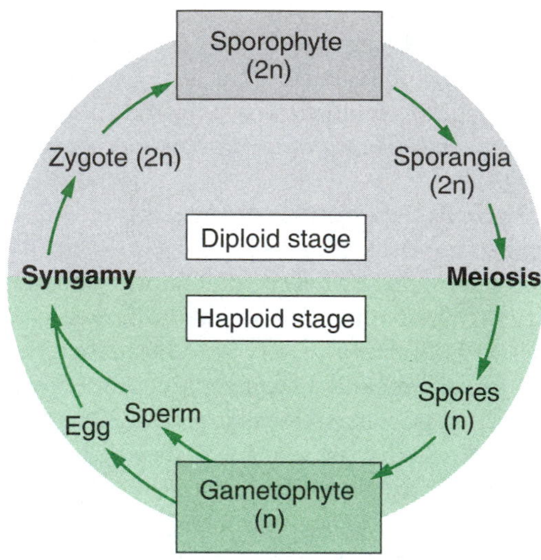

FIGURE 14-8 Alternation of gametophyte and sporophyte generations in the plant life cycle (see also Fig. 14-4). In some plants, the gametophyte is unisexual, either male or female. In others, the gametophyte is bisexual or hermaphroditic, producing male and female tissues.

Similar or Different

Traditionally, botanists concentrated discussions about the alternation of generations on whether the two generations were initially similar or different. According to the antithetic, or interpolation theory of sporophyte origin, all early plants were gametophytes that produced diploid zygotes, some of which underwent a period of delayed meiosis, yet kept dividing mitotically. These parasitic diploid tissues were the initial rudimentary sporophytes and, thus, differed functionally and morphologically from the parental gametophyte. With the colonization of land, an increasing proportion of sporophyte tissues assumed vegetative purposes such as photosynthesis, and the sporophyte became independent.

A contrasting homologous or transformation theory postulates that the sporophyte showed little initial difference from the gametophyte, because both share the same genetic constitution derived from the same organism and should therefore have shared the same patterns of growth. Supporting the homologous theory is the similarity between sporophyte and gametophyte in many algae, as well as structural similarities between the stems in the terrestrial fern-like and extant genus *Psilotum* and in some basal ferns such as

Stromatopteris. There also are basic developmental similarities: Sporophyte tissue can arise in some gametophytes without the intervention of gametic fertilization (*apogamy*), while some sporophytes may produce gametophytes without spore formation (*apospory*). Some proponents of the homologous theory suggest that land plant evolution eventually led to morphological divergence between sporophyte and gametophyte, the former becoming erect, the latter more prostrate.

Early Vascular Plants

We have not resolved the issue of sporophyte origin, although some gametophytes may have been among the early land plants. On the whole, land plant evolution is best characterized by increased importance of the sporophyte. In fact, the vascular tissues of vascular land plants are found only in sporophytes, although land plants may have arisen from charophycean algae that lacked a sporophyte (Gensel and Edwards, 2001). A small, algal, nonsporophytic ancestor would explain the lack of fossil evidence connecting algae and vascular plants.

Whatever the origin of vascular plants, the fossil record shows their rapid evolutionary radiation after their appearance in the Silurian more than 400 Mya.[4] By the Mid to Late Devonian the evolution of the cambium permitted plants to increase in size by an order of magnitude over those earlier in the Devonian. By the end of the Devonian some 75 to 100 My later, forests containing woody trees of relatively great variety had been established. These successful land plants were **vascular,** bearing conductive tissue (**xylem**) that enables water to reach the erect parts of the plant, associated with tissue (**phloem**) that enables food to be distributed. Botanists have often given them the name **Tracheophyta** because they have tracheids, fluid-conducting tubes impregnated with an organic substance (e.g, lignin) that also provides mechanical support for erect growth.

The earliest of the Silurian vascular fossils include a number of simple plants with leafless stems classified in the genus *Cooksonia* (**Fig. 14-9a**), some of which may have had terminal spore-bearing organs (**sporangia**). Some Chinese forms had tracheids. Together with other leafless and rootless fossil plants, these bear a superficial resemblance to plants in the extant genus *Psilotum* (Fig. 14-9c). According to some, such early plants were ancestors of multibranched plants that rapidly evolved into plants such as the herbaceous Devonian genus *Psilophyton* (Fig. 14-9b).

The first leafless plants also appeared in the Devonian, differing from *Cooksonia* types in carrying their sporangia laterally along branches rather than terminally. Botanists have long presumed that these plants — *Zosterophyllum* is an example (**Fig. 14-10a**) — were the ancestors of the early **club mosses** or **lycopods,** such as *Asteroxylon* (Fig. 14-10b), which in turn led to arborescent lycopods such as *Lepidodendron* (Fig. 14-10c), so abundant during the Carboniferous. Herbaceous lycopods have persisted until today.

The Devonian also saw the origin of sphenopsids (horsetails), plants with segmented stems and whorled leaves and branches. Horsetails were common until the Mesozoic, contributing huge trees to Carboniferous coal forests (**Fig. 14-11a**). Now only the genus *Equisetum* (Fig. 14-11b) remains, consisting of a group of about 25 herbaceous species.

Ferns

In terms of evolutionary persistence, an enduring lineage among these early spore-bearing plants is the **ferns** (Pterophyta), now numbering about 10,000 species, in which sporangia are carried directly on the fronds (**Fig. 14-12**).

From the outset, ferns included small forms and large tree ferns (**Fig. 14-13**). Ferns were the first plants to exploit the use of large, prominent leaves (megaphylls) in contrast to the smaller leaves (microphylls) used by lycopods and sphenopsids. We do not know the origin of either leaf type, although botanists have offered various theories, the two major ones being the telome and enation theories.[5]

In the **telome theory,** as Walter Zimmermann (1952) developed it half a century ago, basic thin branches, called *telomes,* evolved in two major directions:

- the first toward greater complexity and vascularization, leading eventually to the leaves and branches of ferns and vascular plants;
- the second regressing toward a single unbranched form, leading eventually to bryophytes such as the hornwort *Anthoceros.*

According to this hypothesis, leaves (essentially macrophylls) originated from small flat branches lying in the same plane. As shown in **Figure 14-14a,** webs formed between such flattened branches could have produced leaflike structures. The telome theory is based on relatively simple changes in fundamental developmental processes, and as such was one of the first theories of plant evolution to integrate development and evolution. In a reanalysis of the theory, Stein and Boyer (2006) retain the essential elements of Zimmermann's theory, augmented with now known rules of developmental dynamics in plants and an emphasis on multiple rather than single developmental trajectories that play out over evolutionary time.

A different proposal, the **enation** (extension) **theory** (**Fig. 14-14b**), hypothesizes that the earliest leaves (essentially microphylls) arose from small flaps or extensions of

[4] See Gensel and Edwards (2001) and Kenrick and Davis (2004) for the radiation of vascular plants.

[5] Although the evolutionary origin(s) of leaf types remains obscure, the hormonal and genetic basis for leaf shape is being revealed in studies such as that by Hay et al. (2006), in which transcription factors such as KNOX establish a gradient of the plant hormone auxin, which regulates leaf shape; KNOX, the *knotted1*-like homeobox family of genes, are active in apical meristems where they regulate leaf and stem development.

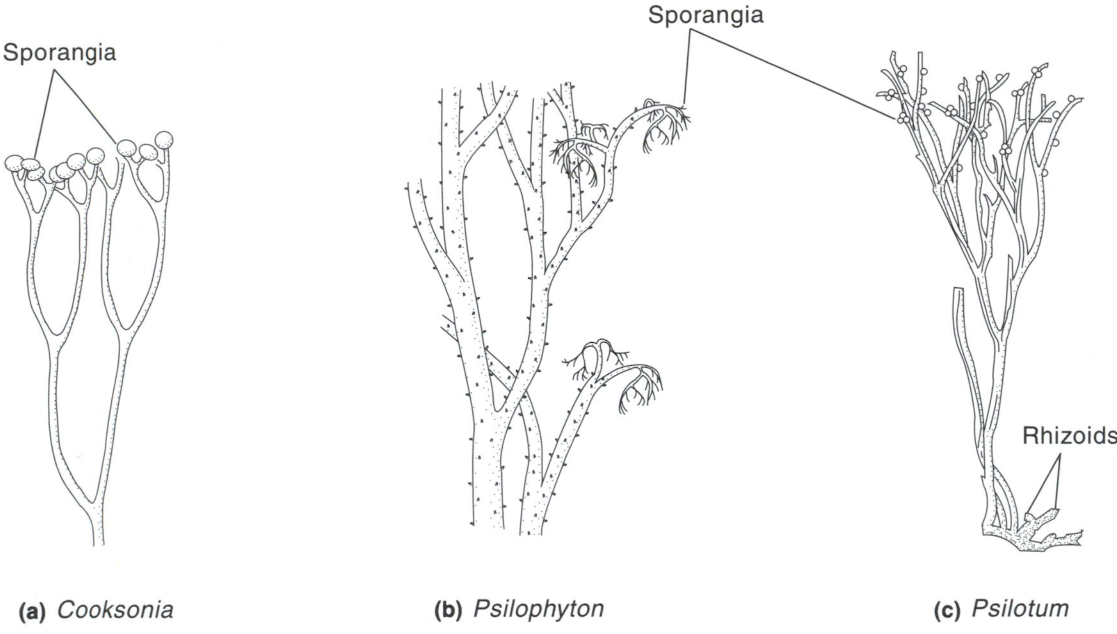

FIGURE 14-9 Reconstructions of the sporophytes of two early fossil plants and an extant representative of the Psilopsida. (a) A diminutive Upper Silurian plant, *Cooksonia caledonica,* about 25 mm or so high, which had no distinctive leaves or roots but showed naked, dichotomously branched axes with terminal sporangia. Similar fossils were long known from Devonian rocks in Rhynie, Scotland, and have been classified together as *Rhyniophyta,* or Rhynia-type plants. (b) *Psilophyton princeps,* section of a spiny, leafless fossil plant that first appears less than 10 My after *Cooksonia.* It had a main stem axis with lateral branches terminating in sporangia and a vascular structure that seems to have been larger than in *Cooksonia,* so *Psilophyton* may have grown taller. (c) The extant plant *Psilotum nudum* has a number of features that resemble the fossil forms: simple stems, nondiscernible leaves and absence of a root system. It is not clear whether this plant is a "fossil" or a secondary descendant of a more derived form (Stewart and Rothwell, 1993). (a and b adapted from Taylor, T. N., and E. L. Taylor, 1993. *The Biology and Evolution of Fossil Plants.* Prentice Hall, Englewood Cliffs, NJ; c adapted from Bold, H. C., C. J. Alexopoulos, and T. Delevoryas, 1980. *Morphology of Plants and Fungi,* 4th ed. Harper & Row, New York.)

FIGURE 14-10 Reconstructions of several Devonian plants. (a) *Zosterophyllum myretonianum* (about 10.5 cm). (b) *Asteroxylon mackiei* (about 0.6 m tall). (c) *Lepidodendron* species (about 46 m tall). The scars along the upper stem of *Lepidodendron* are leaf cushions, where the long, filamentous leaves attached during earlier growth. The heavy, pendulous cones carry the sporangia. (a and b adapted from Foster, A. S., and E. M. Gifford, Jr., 1974. *Comparative Morphology of Vascular Plants,* 2nd ed. Freeman, San Francisco; c adapted from Stewart, W. N., and G. Rothwell, 1993. *Paleobotany and the Evolution of Plants,* 2nd ed. Cambridge University Press, Cambridge, UK.)

(a) (b) (c)

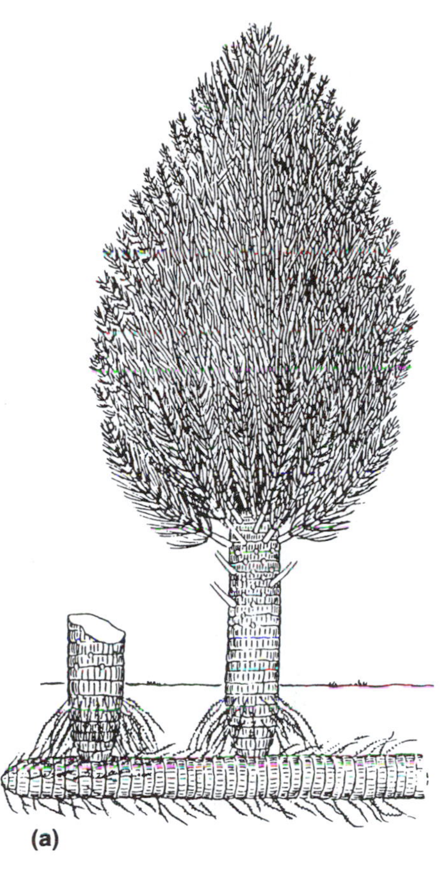

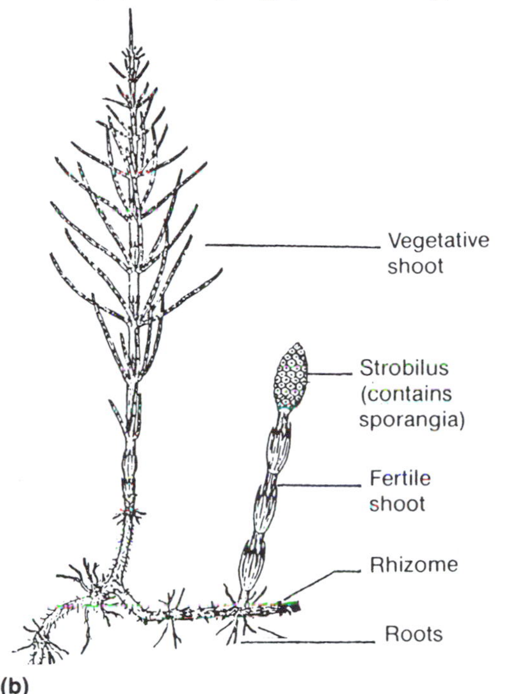

FIGURE 14-11 Ancient and extant representatives of Sphenopsida. (a) Reconstruction of *Calamites,* a common tree during the Carboniferous, that reached heights of 27 to 30 m with trunks 0.6 m thick. (Adapted from Foster, A. S., and E. M. Gifford, Jr., 1974. *Comparative Morphology of Vascular Plants,* 2nd ed. Freeman, San Francisco.) (b) Extant *Equisetum arvense,* showing vegetative and fertile shoots. Its reproductive system is homosporous: The strobili bear sporangia that produce spores alike in size. (Adapted from Bold, H. C., C. J. Alexopoulos, and T. Delevoryas, 1980. *Morphology of Plants and Fungi,* 4th ed. Harper & Row, New York.)

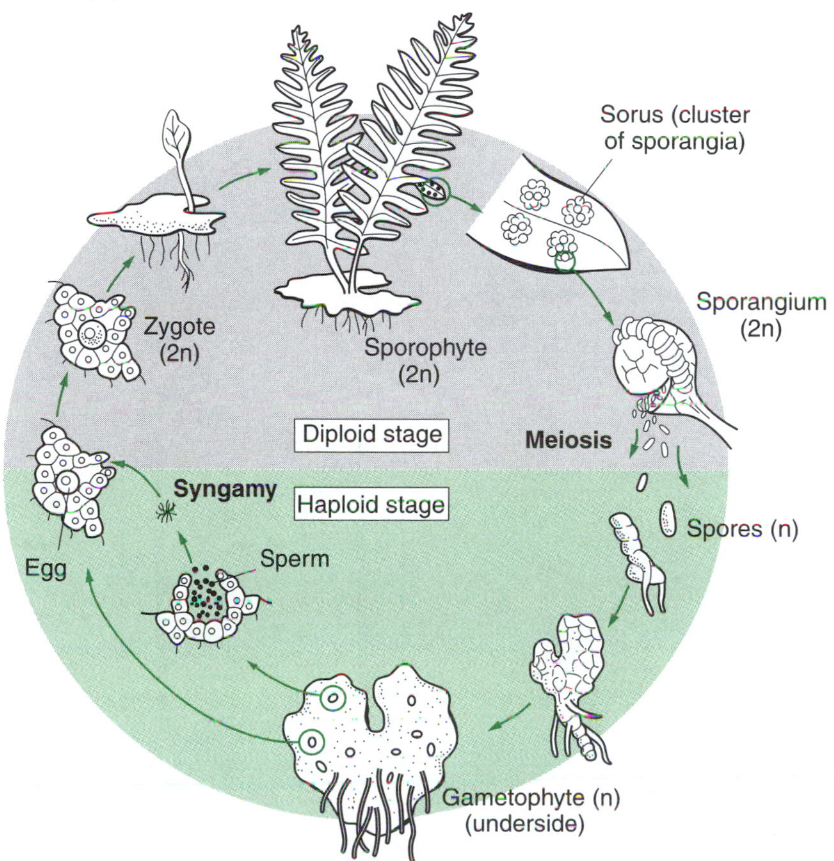

FIGURE 14-12 Life cycle of a fern, the common polypod, *Polypodium vulgare.*

FIGURE 14-13 Reconstruction of the Carboniferous tree fern *Psaronius,* about 7.6 m tall. Leaf scars left by earlier fronds that have fallen away are visible near the top of the trunk. The surrounding adventitious roots, which increase in thickness toward the base, cause the trunk's long pyramidal shape. The root structure suggests that these trees grew in swampy habitats. Both mitochondrial gene and morphological evolution in trees ferns has been so slow over the past 200 My that tree ferns can be regarded as molecular living fossils (Soltis et al., 2002). (Adapted from Foster, A. S., and E. M. Gifford, Jr., 1974. *Comparative Morphology of Vascular Plants,* 2nd ed. Freeman, San Francisco.)

tissue along the stem, somewhat like microphylls in the fossil lycopod *Asteroxylon.* Only later did these leaves vascularize.

We do not know which theory is correct; the telome theory works best for the evolution of megaphylls and the enation theory for the evolution of microphylls.

From phylogenetic reconstructions of the history of land plants, we do now know, that whatever the developmental mechanisms, **leaves evolved multiple times,** reflecting three independent evolutions within each of gametophyte and sporophyte generations:

- twice in liverworts and once in the line that gave rise to the mosses, that is, three times within the gametophyte generation, and
- once in the lycopods or club mosses, once in ferns and related forms and once in seed plants, that is, three times in the sporophyte generation.[6]

[6] See Niklas (1997) and Friedman et al. (2004) for the evolution of leaves and other plant organs.

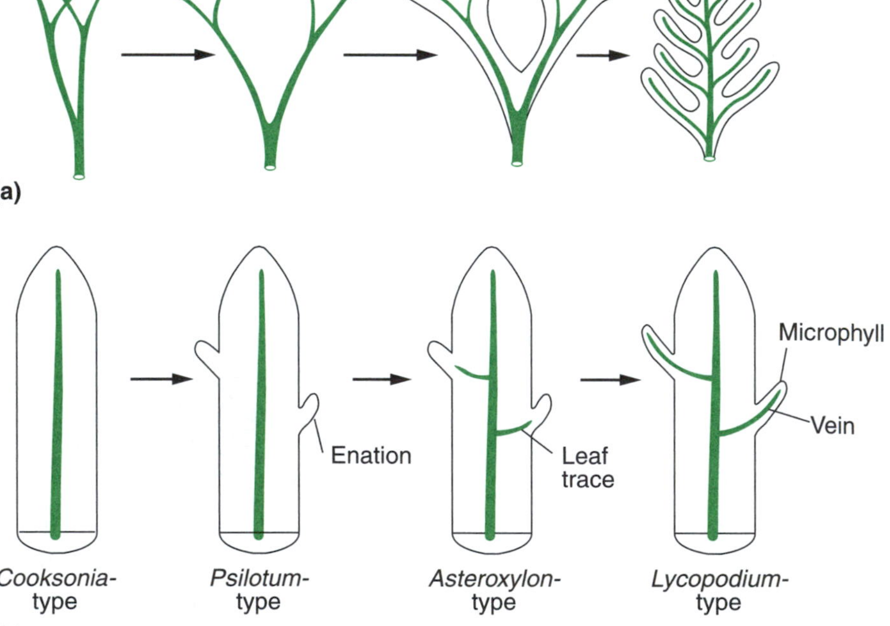

FIGURE 14-14 Diagrammatic representations of two theories explaining the origin of leaves. (a) Flattening (planation) of a branch system according to the telome theory, followed by webbing between the branches to form a flat, veined megaphyll. (b) Evolution of microphylls according to the enation theory. (Adapted from Bold, H. C., C. J. Alexopoulos, and T. Delevoryas, 1980. *Morphology of Plants and Fungi,* 4th ed. Harper & Row, New York.)

Of course, evolutionary changes affected almost every aspect of these early plants; for example, beginning with the first spore-bearing plants, there is a progressive change from **homospory,** in which all spores are alike, to **heterospory,** in which sporophytes produce both large-diameter **megaspores** (200 μm) and smaller-diameter **microspores.**

By the Late Devonian, the heterosporous lines had evolved *megaspores* more than 2,000 μm in diameter, an increase that resulted from a reduced number of cells in the megasporangium, so that it produced only a single tetrad of spores, of which three abort and one enlarges (**Fig. 14-15**). As botanists have often pointed out, we can view such a huge megaspore as the bearer of a female gametophyte with enhanced nutritional resources that, on fertilization, will provide stored food for the developing sporophyte embryo. Megaspores represent one small step in the evolution of **seeds,** complex structures consisting of the remains of the megaspore and tissues from three generations: the parent diploid, the parent haploid and the embryo.

Complexities of stem structures involved in conduction, support and storage also originated from a primitive form, the **protostele,** in which these vascular tissues took, at first, simple forms. Later in evolution they divided into various lobed and concentric arrangements (**Fig. 14-16**). The structural and functional advantages provided by many of these innovations bolstered the vertical growth of plants and permitted large trees and shrubs of all types to evolve.

By the Carboniferous, lush and extensive forests had developed in vast swamps along the east coast of North America and similar coastal regions of Europe and North Africa. The absence of annual growth rings in many tree trunks of this period indicates that the climate was mostly tropical and growth rapid. In this environment, rapid submergence in the swamp of many fallen trees and shrubs inhibited their decay, making them immune to attack by all but anaerobic bacteria. As water levels fluctuated, successive generations of swamps formed and submerged, and thick layers of organic strata were compressed into peat. Further sedimentation and compression led to the escape of volatile hydrocarbons, allowing enormously thick and extensive coal seams to form.

From Swamps to Uplands

Successful as they were, Carboniferous spore-bearing plants were limited to a moisture-laden environment because the motile male gamete depended on aqueous transmission to the female gametophyte. To extend their range onto dry land, plants had to evolve an enclosed desiccation-protected gametophyte in which cross-fertilization could occur by nonaqueous devices such as wind dispersal. Similarly, a considerable advantage attended the evolution of protected sporophyte embryos whose distribution could be independent of their parental gametophytes. In essence, size reduction of the gametophyte and the evolution of easily dispersible **pollen** (male gametophytes) and **seeds** (sporophyte embryos) helped further the conquest of dry land.

We don't yet know how early vascular plants evolved into pollen-producing, seed-bearing plants. According to the fossil record, **gymnosperms** (naked seeds) appear in notable frequency during the Carboniferous, eventually giving rise to representatives that include ginkgos and cycads as well as conifers such as pines, cedars and sequoias — about 750 different species. **Angiosperms** (covered seeds), now the dominant land plants accounting for about 220,000 (or more than 80 percent) of all plant species, have no identifiable fossils earlier than the Cretaceous. We have not yet discovered transitional fossil forms leading directly to either gymnosperms or angiosperms. Fossils such as *Archaeopteris,* dating to the Devonian, represent the earliest known seed-bearing plants.

As shown in **Figure 14-17,** *Archaeopteris* was a tall tree resembling a conifer with a crown of leafy branches. Botanists propose that its stem bore a number of features in common with gymnosperms, although it produced free spores rather than "naked" seeds. One proposal is that by combining pteridophytic, sporulating reproductive modes with more derived anatomical structures such as large trunks, Paleozoic plants of this kind (**progymnosperms**) gave rise to the gymnosperms that became so successful during the relatively dry Mesozoic (**Fig. 14-18**).

Other fossil groups, perhaps also arising from the progymnosperms, were fernlike plants that bore seeds rather than spores (**Fig. 14-19**). These **seed ferns** (Pteridospermales) show a variety of seed forms that suggest a progression in the method by which the seed integument enclosed the female gametophyte (**Fig. 14-20**). However, from the phylogeny in **Figure 14-21,** we don't know whether seeds originated in only one lineage of progymnosperms (monophyletic origin) or more than one lineage (polyphyletic origin). In any event, such evolution may have occurred early: Fossil seeds date back to the Late Devonian, 350 Mya.

Angiosperms

Angiosperms were the last major plant group to evolve, making their first appearance in the Early Cretaceous, but quickly radiating so that by the Late Cretaceous they were both abundant and varied.[7] Among other features, they share **flower** structures that enable insects or birds to pollinate many of them, and also bear seeds that are often adapted to dispersal by other animals.

The adaptive advantage of pollination by animals is the simple one of ensuring cross-fertilization with other members of the same species by using a relatively small amount

[7] The angiosperm phylogeny group maintains a Web site that contains the latest version (AGP II, 2003) of angiosperm classification (http://www.f-lohmueller.de/botany/apg/apg_ii.htm; accessed April 23, 2007).

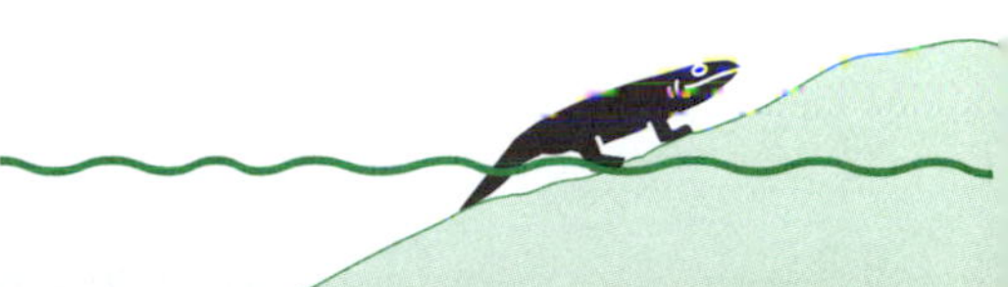

FIGURE 14-15 Proposed stages in the evolution of megaspores that develop into egg-bearing gametophytes. As shown in a reconstruction of a Devonian gymnosperm seed (*Archaeosperma arnoldii*) on the right, the selective advantage of this trend was the development of large eggs providing increased embryonic nutrition. (Adapted from Niklas, K. J., 1997. *The Evolutionary Biology of Plants.* University of Chicago Press, Chicago.)

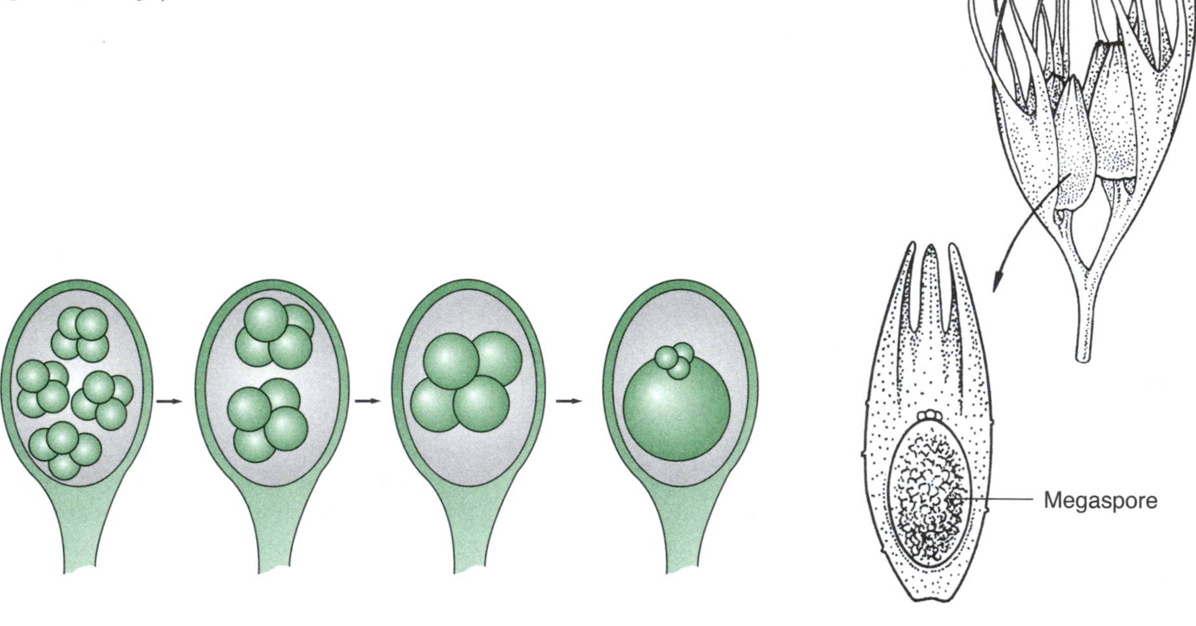

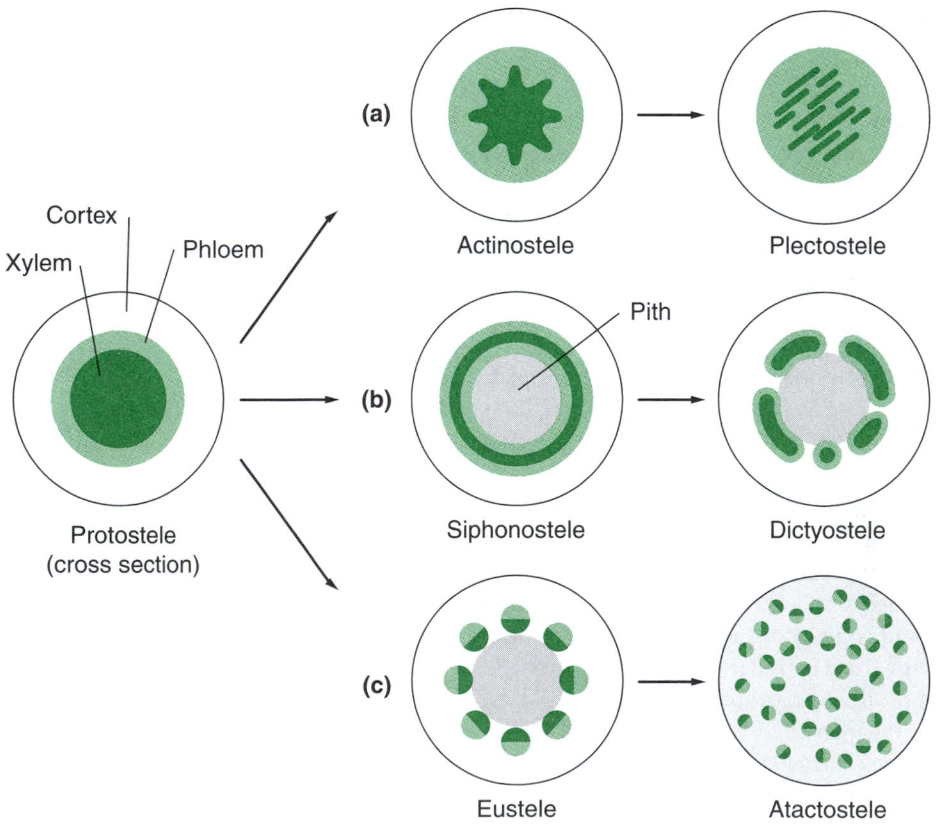

FIGURE 14-16 Proposed evolutionary relationships among some of the vascular cylinders (steles) found in plants. The various protosteles (a) are considered primitive and occurred in Rhynia-type plants; (b) siphonosteles and dictyosteles characterize many ferns; (c) various seed plants have eusteles; and some of the complex atactosteles appear in flowering plants.

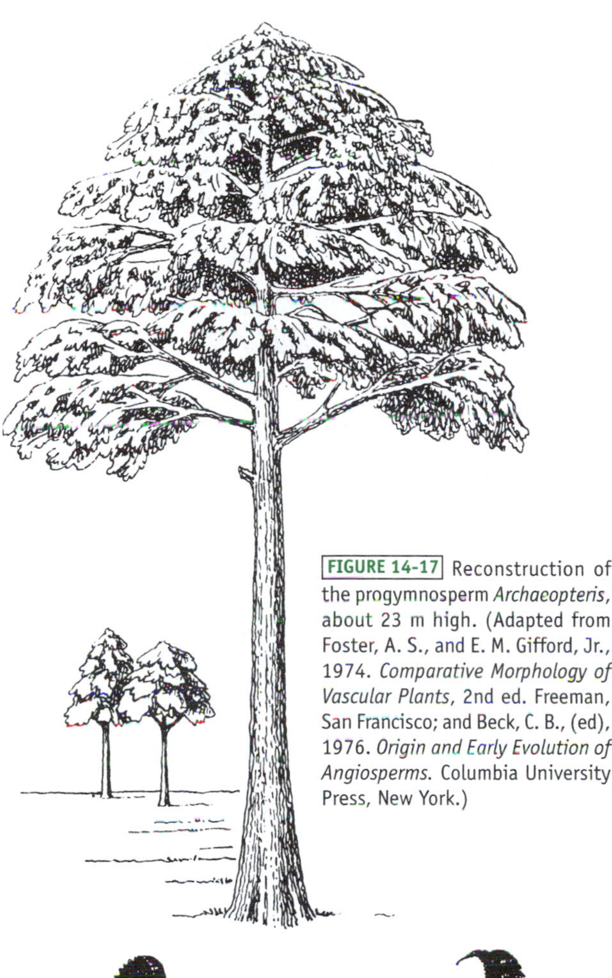

FIGURE 14-17 Reconstruction of the progymnosperm *Archaeopteris*, about 23 m high. (Adapted from Foster, A. S., and E. M. Gifford, Jr., 1974. *Comparative Morphology of Vascular Plants*, 2nd ed. Freeman, San Francisco; and Beck, C. B., (ed), 1976. *Origin and Early Evolution of Angiosperms*. Columbia University Press, New York.)

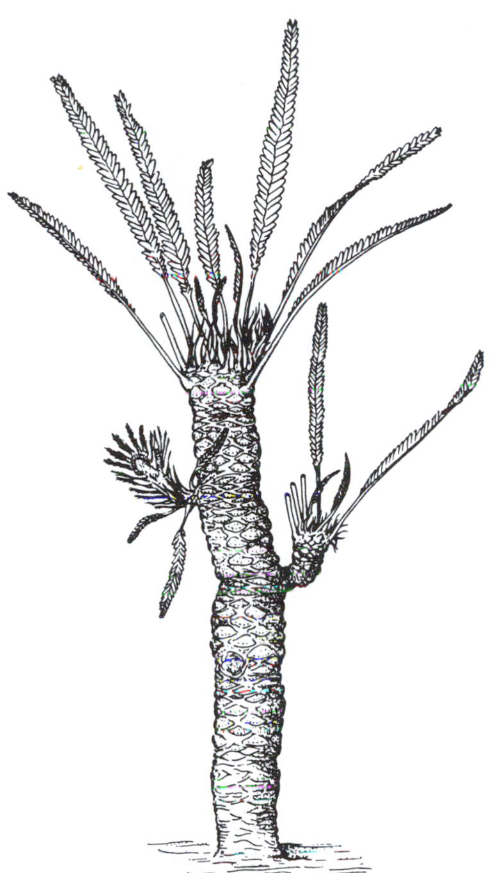

FIGURE 14-18 Reconstruction of a cycad gymnosperm, *Williamsonia sewardiana*, from Jurassic rocks in India. The cycads were abundant contemporaries of the dinosaurs, and hence the Jurassic period is also known as the Age of the Cycads. Like the dinosaurs, most cycads became extinct, although 100 species of cycads exist, most in the tropics. Another "living fossil" is the ginkgo, *Ginkgo biloba*, sole remnant of the Ginkgopsida, also common during the Mesozoic. (From Andrews, H. N. Jr., 1961. *Studies in Paleobotany*. John Wiley, New York.)

of pollen, compared to the large amounts of pollen necessary in random wind pollination as found, for example, in all gymnosperms. As a result, angiosperm flowers, derived from leaves modified into petals, sepals and related structures, are among the most intricate and attractive organs plants have developed (**Fig. 14-22**). Size, color and odor differences attract specific animal pollinators, an advantage that can spread so rapidly that some closely related plants have evolved flowers that can discriminate among pollinators, while pollinators have *coevolved* mechanisms to feed on specific flowers (**Fig. 14-23** and **Box 14-1**). A recent combined molecular and morphological analysis of bees, of which there are more than 16,000 extant species, indicates that the earliest mid-Cretaceous lineages included host-plant

FIGURE 14-19 Reconstruction of a seed fern, *Medullossa*, about 3.6 to 4.5 m tall. (Adapted from Andrews, H.N., 1951. *Studies in Paleobotany*, Wiley, New York.)

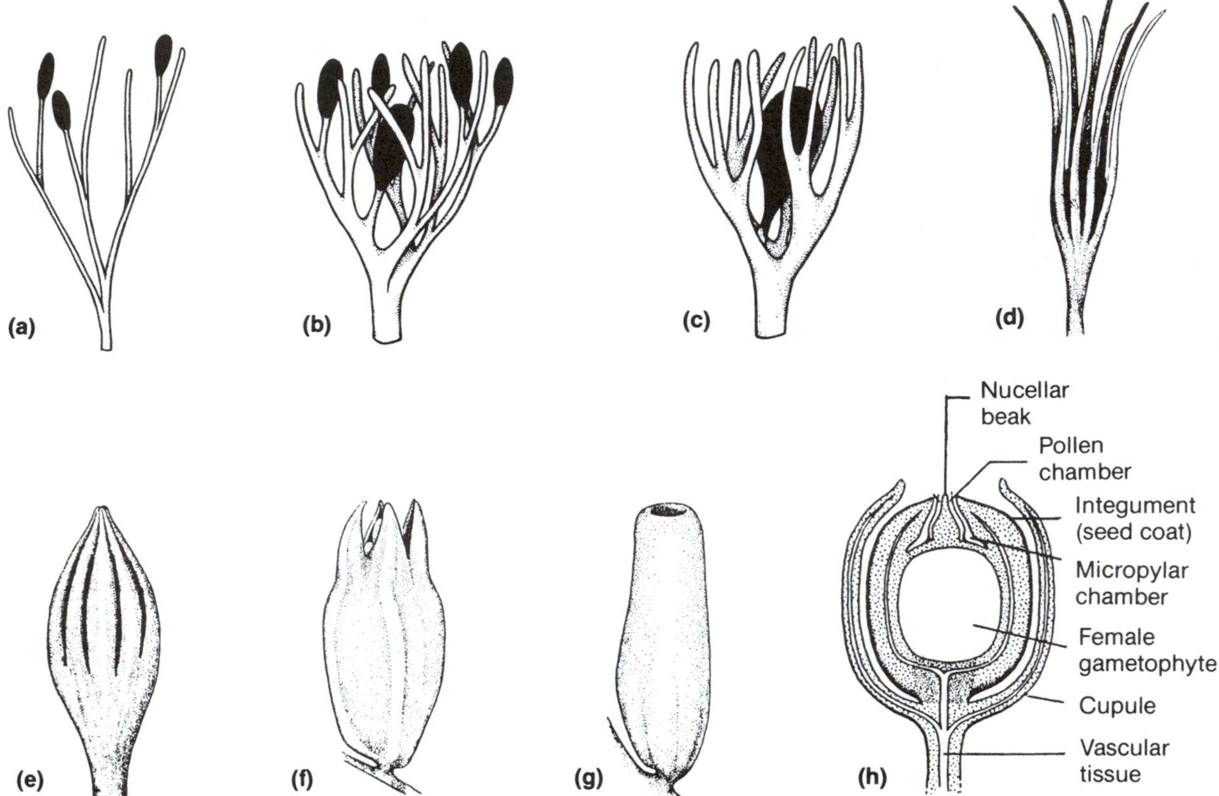

FIGURE 14-20 (a–g) One scenario for the evolution of the pteridosperm seed. The exposed sporangium (which produces megasporocytes) is gradually enclosed, enabling the female gametophyte (produced by the megaspore) to be completely protected within sporophyte tissue (nucellus). Fertilization takes place when a pollen tube (part of the male gametophyte) grows through the micropyle, enabling sperm to reach the female gametophytic egg. The complete seed that envelops the zygote and developing embryo is coated with an integument produced by the parental sporophyte. (d–g) Seeds of fossil pteridosperms. (h) Section of a pteridosperm ovule. (Adapted from Foster, A. S., and E. M. Gifford, Jr., 1974. *Comparative Morphology of Vascular Plants,* 2nd ed. Freeman, San Francisco.)

specialists, supporting host-plant specificity as an ancient feature in angiosperm evolution.[8]

Self-Sterility and Double Fertilization

On a basic genetic level, an extremely important mechanism for preventing self-fertilization was the development of **self-incompatibility** (or **self-sterility**) **alleles,** so that pollen bearing the same allele as an ovule could not grow on the female style. In self-incompatibility in the gametophyte generation, a haploid pollen grain carrying the self-sterility allele S^1 will not grow well on a female style carrying $S^1 S^3$, but can successfully fertilize a plant carrying $S^2 S^3$ or $S^3 S^4$, and so on. In "sporophytic" self-incompatibility, an S^1 pollen from an S^1S^2 sporophyte cannot fertilize any plant bearing either the S^1 or the S^2 allele. The consequences of these systems were avoidance of close inbreeding and protection of the genetic variation produced by sexual reproduction.

[8] See Dudareva and Pichersky (2006) for the role of floral scent in coevolution and Danforth et al., (2006) for the bee phylogeny.

Also distinctive in angiosperms is **double fertilization,** in which two gametic nuclei of the pollen tube fertilize the female gametophyte, one producing the diploid (2n) embryonic nucleus, the other often producing a polyploid (commonly triploid, 3n) **endosperm** nucleus used for embryonic nutrition. Maturation of the fertilized ovule leads to an angiosperm seed that has two integuments rather than the single integument found in gymnosperms. As the name angiosperm ("seed vessel") implies, these seeds often have covers that are fleshy, fruity tissues, adhesive burs, feathery parachutes or other devices that allow either animals or the elements to disperse them.

Dispersal ability is only one of the selective forces acting on seeds. Others include:

- the need for seed coats to protect seeds against predators and the elements;
- the necessity of adequate food storage for embryonic development; and

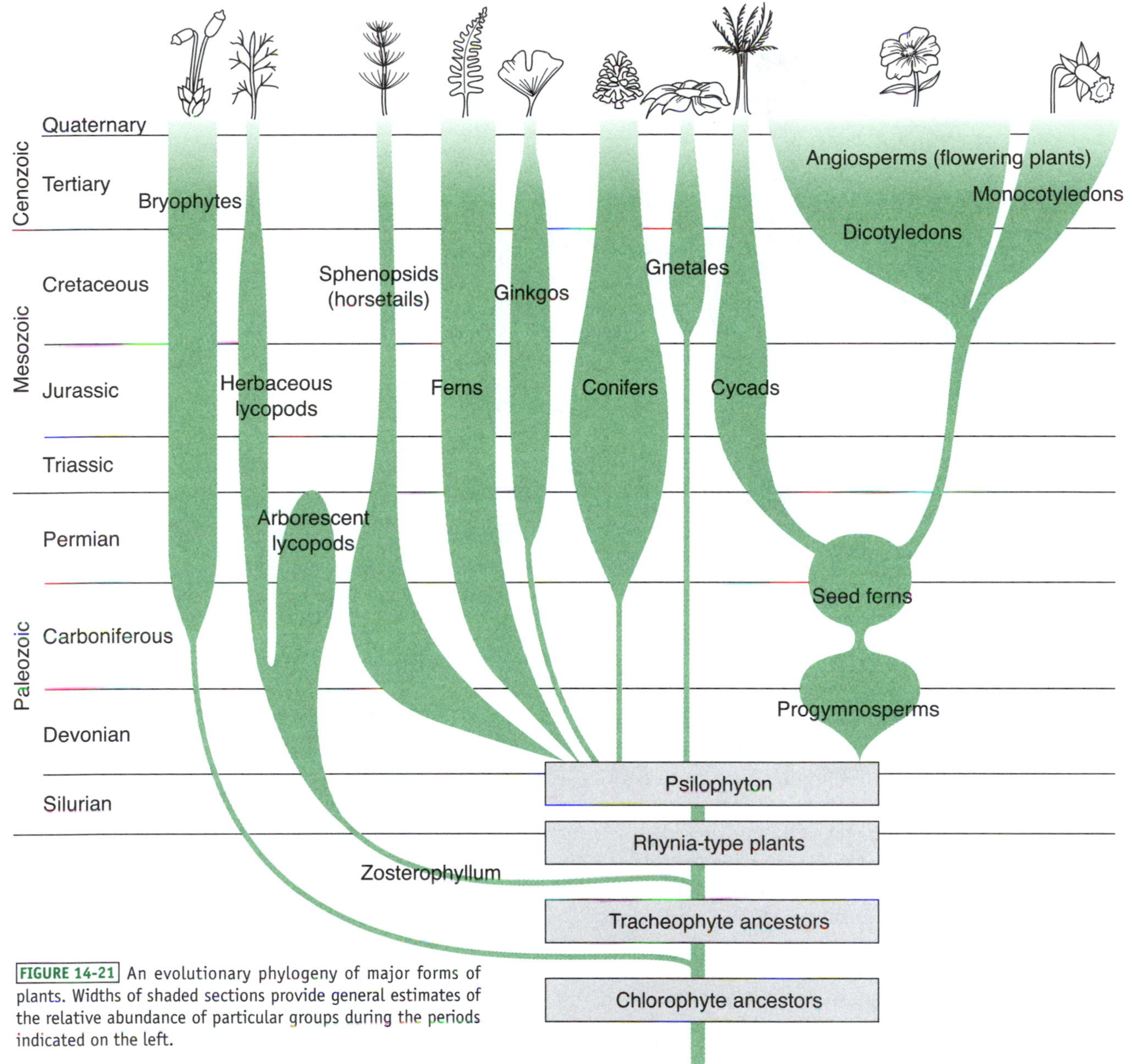

FIGURE 14-21 An evolutionary phylogeny of major forms of plants. Widths of shaded sections provide general estimates of the relative abundance of particular groups during the periods indicated on the left.

- programming of seed germination to coincide with the developmental period and with favorable environmental conditions.

Factors such as these lead to specific anatomical and physiological adaptations. In addition, some forces, such as selection for wide dispersal, put a premium on small size and large numbers. Others, such as selection for vigorous competitive embryos, emphasize large seeds and smaller numbers (discussed in Chapter 23 as *r*- and **K-selection**). On the whole, the size and number of seeds a particular species produces is a compromise between these various factors and the physical limitations of the environment.

Evolution of Angiosperms

A century ago, Darwin called angiosperm origin "an abominable mystery," a mystery that remains unsolved today. Unresolved are both the source and time of angiosperm origin, with estimates ranging from Early Permian to Late Carboniferous (Savard et al., 1994), to a date corresponding with the earliest angiosperm fossils in the later Mesozoic/Early Cretaceous.

Because most plant fossils discovered to date are associated with wet lowland areas where organic decomposition could be inhibited by silt and mud, those paleobotanists who propose an earlier Mesozoic or Paleozoic origin for angio-

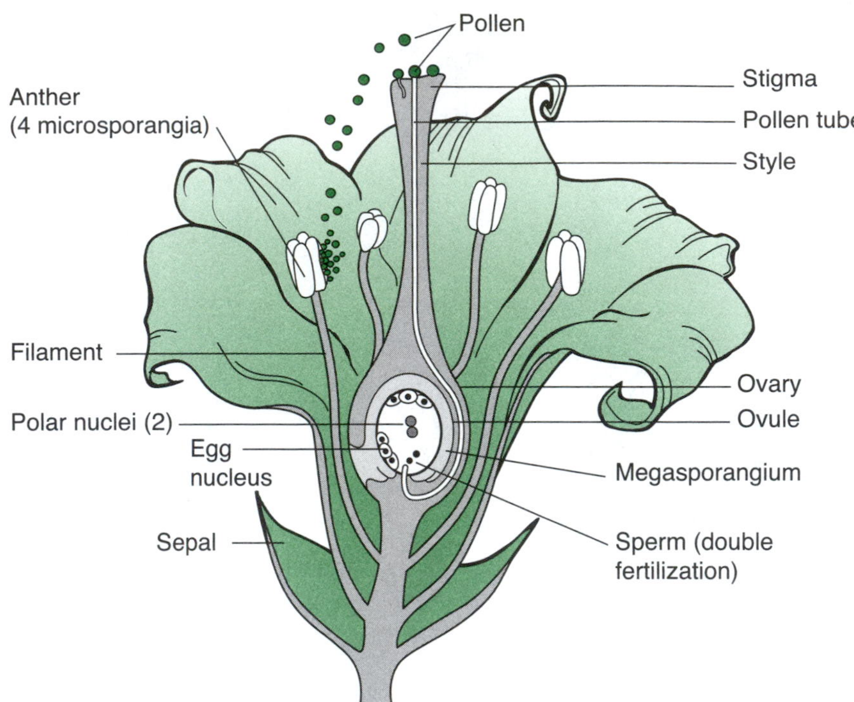

FIGURE 14-22 Diagram of a generalized angiosperm flower after fertilization. The female gametophyte is within the ovule; the male gametophyte is the pollen tube. The male gametophyte produces two sperm nuclei. One fertilizes the egg nucleus to produce the zygote (2n). The other fertilizes the two polar nuclei to produce the endosperm (3n). The petals are usually the more colorful part of the flower, and are organized in a whorl called the *corolla.*

sperms suggest that plants first arose in upland mountainous areas where fossilization is rare and fossil deposits rarely persist because of active erosion. Justification for this view lies in the finding that the earliest angiosperm fossil leaves already show considerable differentiation, as though preceded by a lengthy evolutionary period (**Fig. 14-25**). Discovery of Late Jurassic fossil insects with mouthparts adapted to flower pollination (Ren, 1998) lends support to pre-Cretaceous angiosperm evolution.

On the other hand, the earliest fossil pollen that we can confidently ascribe to angiosperms — single-furrowed (monocolpate) grains — is from the Early Cretaceous. New pollen types appear soon after. Rapid angiosperm evolution during the Early Cretaceous (rather than a much earlier unobserved origin) has been proposed to account for the variety of fossil forms found in this period. In a now classic study, Stebbins (1974) drew on summaries of a large amount of information to hypothesize that ancestral angiosperms were shrubs with spirally arranged, simple leaves and woody tissues formed from a single vascular cylinder.[9]

[9] Stebbins is recognized as the biologist who incorporated plants into the modern synthesis. *Variation and Evolution in Plants and Microorganisms: Toward a New Synthesis 50 Years After Stebbins* (2000), available free on http://www.nap.edu/openbook/0309070996/html/index.html (accessed April 23, 2007) commemorates his classic analysis.

Semiarid Mountains or Tropical Lowlands?

According to Stebbins (1974), the ecological impact that led to the evolution of angiosperms was alternation of dry and wet seasons, conditions that would have selected for rapid gametophyte and embryonic development. Rainy periods followed by calms after storms also provide an opportune time for flowering and insect pollination, as well as promoting selection for such protective seed structures as closed carpels.

It is not clear where such conditions might have appeared, but Stebbins suggests that semiarid mountainous regions with annual droughts would have offered early evolutionary opportunities for angiosperms, similar perhaps to those inferred from the rapid evolutionary rates observed among angiosperms that now inhabit mountainous regions in South Africa, Ethiopia, Ecuador and Mexico. This thesis is amply borne out by a recent analysis by Hughes and Eastwood (2006) of 81 species of lupins (genus *Lupinus*) endemic to the Andes. This study demonstrates rapid speciation, 2.5 to 3.5 species per one million year — the fastest speciation documented for any plants — on a continental scale, and in response to ecosystem diversification following uplifting of the Andean mountains.

Data collected by Stebbins (1974) along the U.S. Pacific coast for more than 8,000 plant species belonging to more than 800 genera show that some ecologically specialized

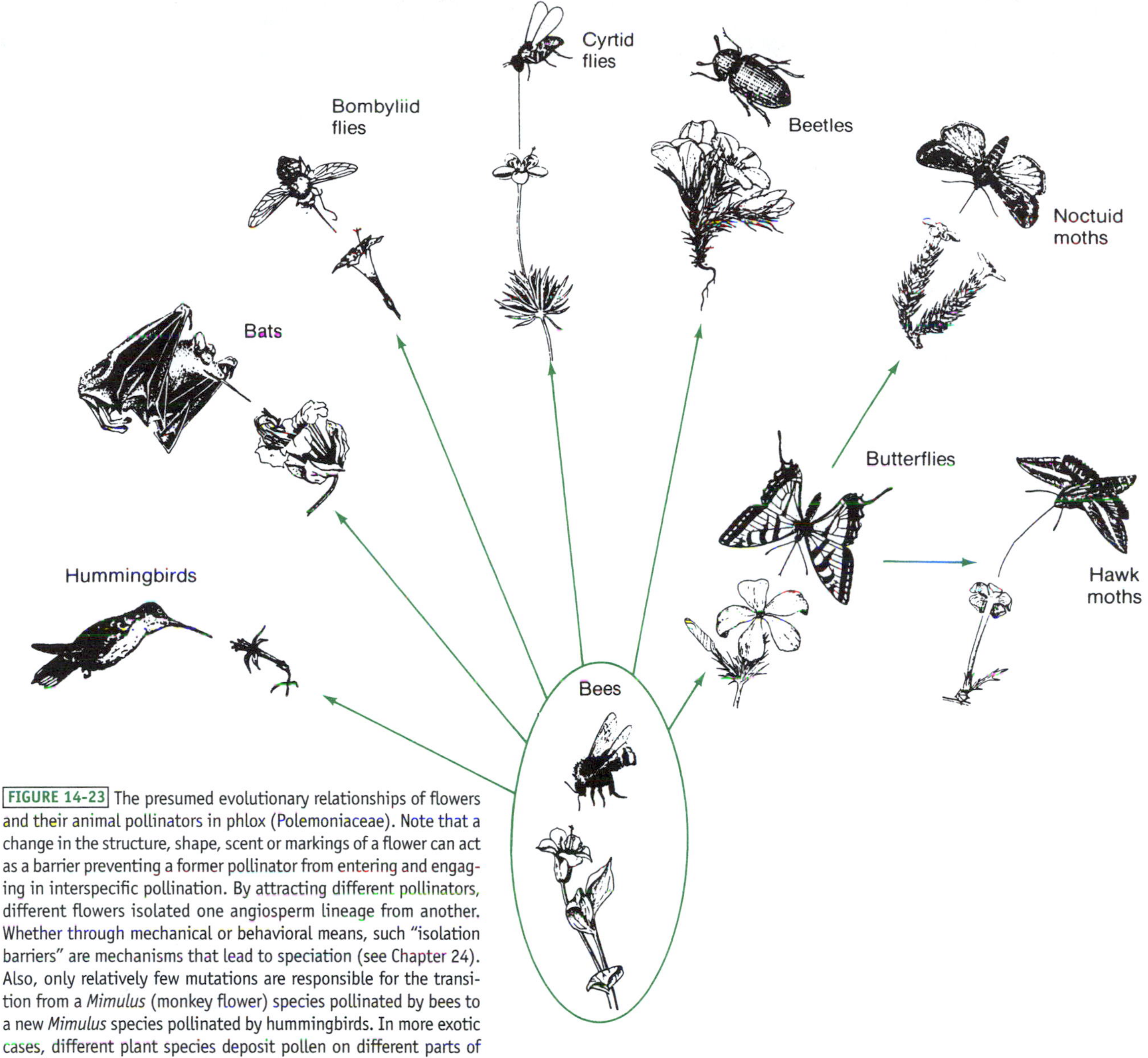

FIGURE 14-23 The presumed evolutionary relationships of flowers and their animal pollinators in phlox (Polemoniaceae). Note that a change in the structure, shape, scent or markings of a flower can act as a barrier preventing a former pollinator from entering and engaging in interspecific pollination. By attracting different pollinators, different flowers isolated one angiosperm lineage from another. Whether through mechanical or behavioral means, such "isolation barriers" are mechanisms that lead to speciation (see Chapter 24). Also, only relatively few mutations are responsible for the transition from a *Mimulus* (monkey flower) species pollinated by bees to a new *Mimulus* species pollinated by hummingbirds. In more exotic cases, different plant species deposit pollen on different parts of the pollinator's body, allowing the same animal species to pollinate different plant species. Underlying these relationships are mutual benefits: plants make as much use of animals as animals do of plants. (Adapted from Grant, V., and K. A. Grant, 1965. *Flower Pollination in the Phlox Family*. Columbia University Press, New York.)

regions such as alpine areas, deserts, lakes, swamps and bogs have fewer plant species per genus than do more ecologically variable habitats such as fields, meadows and open woods. Numerically, no more than four or five species exist in each genus in ecologically specialized regions, compared to about 10 species per genus in ecologically variable regions. The latter provides greater evolutionary opportunities than the former.

Stebbins calls tropical plant communities "museums, . . . plant communities that have suffered the least disturbance during the past 50 to 100 million years and so have preserved the highest proportion of archaic forms in an essentially unchanged condition." Tropical communities mostly represent geographical depositories for ancient plant groups rather than sites of origin. This scenario proposes that angiosperm diversity in tropical flora reflects dispersion into the tropics rather than origin from the tropics. It has been pointed out that tropical lowland conditions with their large insect populations are more favorable for plants that depend on insect pollination than on wind pollination. Because tropical climates expanded significantly during the Cretaceous, Stebbins suggests that this environment enabled the early angiosperms to disperse and become dominant.

BOX 14-1 Coevolution of Plants and Insects

COEVOLUTION BETWEEN FLOWERS AND POLLINATORS was evident to Darwin, who, being aware of the diversity of strange flowers among orchids (**Fig. 14-24**), postulated that even the most unusual orchid flower known to him would have a matching pollinator, although no pollinator was then known. The orchid Darwin discussed was the Madagascar star (Christmas) orchid, *Angraecum sesquipedale,* the nectary of which is at the base of a corolla tube 25 cm long. Darwin's prediction that a pollinator with a similarly long tongue must exist was borne out years later with the discovery of a giant hawk moth — Morgan's Sphinx moth, *Xanthopan morgani praedicta* — bearing an appropriately long (30 cm) tongue, longer than its body. In a pattern of coevolution, some hawk moths have been selected for longer tongues to pollinate star orchids selected for longer corolla tubes to be pollinated by a specific hawk moth (L. A. Nilsson, 1998).

An astonishing parallel to the coevolution of the hawkmoth and Malagasy star orchid appeared as this edition was being completed, not an example involving coevolution of another moth and orchid, but in this case coevolution of the tongue of the nectar bat, *Anoura fistulata,* and the length of the corolla of *Centropogon nigricans* in the cloud forests of the Ecuadorian Andes Mountains. Close relatives of the nectar bat — species in the same genus — have tongues that can extend to 37 to 39 mm. The tongue of the nectar bat can extend to 85 mm, 150 percent of its body length, a feat second only to the chameleon. And, when the tongue is not in use, the nectar bat stores it inside the thorax in a special tube (Muchhala, 2006).

Animal pollinators undoubtedly affected the sexual organization of angiosperm flowers, because it is to the advantage of a plant to contribute its pollen to a mobile animal pollinator that would at the same time bring pollen from another plant to fertilize its own ovules. Consequently, flowers would be selected in which pollen transfer and fertilization occurred in a single visit. The flowers of early angiosperms relying on insect pollination were probably bisexual, in contrast to their wind-pollinated ancestors, which would mostly have used unisexual flowers to help prevent self-fertilization between pollen and ovules of the same flower.

The fact that some groups of flowering plants have species that are wind pollinated indicates that evolutionary reversals can occur from one form of pollination to the other. In the evolutionary history of figs (*Ficus* sp.), such reversals may well have occurred more than once, the first reversal being from insect pollination to wind pollination in an ancestral group, the nettle family Urticaceae (the stinging-nettle, *Urtica dioica,* is one example), followed by a subsequent reversal to insect pollination in members of the Moraceae, in which the genus *Ficus* is almost exclusively pollinated by species of chalcid wasps.

A dramatic example of a novel mechanism of self-fertilization came to light in 2006 with the discovery that the pink-flowered orchid, *Holcoglossum amesianum,* rotates the anther of its bisexual flower through a full circle and against gravity to place pollen grains on its own stigma and so fertilize itself. *H. amesianum* grows on tree trunks at 1,200 to 2,000 m elevation in Yunnan province, China, where wind and potential insect-pollinators are scarce during the dry months of February to April when the orchid flowers.[a]

[a] See Liu et al. (2006) for the study of the pink-flowering orchid and see Alcock (2006) for a beautifully illustrated analysis of reproductive adaptations in orchids.

Ancestry

Place of origin is only one aspect of the "abominable mystery" of angiosperms. The other, of course, is their evolutionary origin(s).

One approach to angiosperm phylogeny has been to search for groups that have structures like those seen in extant "primitive" angiosperms such as magnolias. Botanists once hypothesized that the magnolia flower is strikingly similar to the axial grouping of sporangia-bearing structures (strobili) of gymnosperms, cycads and an extinct genus, *Bennettitales.* Closer analysis of fossil seeds in combination with molecular analyses, especially of MADS box genes (see below), shows that *Bennettitales* is a basal seed plant, with a sister-group relationship to cycads.

Small-flowered diminutive herbaceous plants also have been proposed as the ancestral angiosperm type, based on a fossil described by Taylor and Hickey (1990), who suggest that:

> . . . the lack of pre-Albian [pre-Early Cretaceous] fossil angiosperm wood is due to their diminutive habit and that the failure to recognize protoangiosperm fossils results from their diminutive size and an incorrect search image.

Excluding other possible progenitors, botanical evidence indicates that the most likely candidates for angiosperm ancestors are among the seed-ferns (pteridosperms). Some suggest that the unique characteristics of angiosperms indicate a monophyletic origin because it is doubtful that such characteristics could have arisen independently in different lineages or even that they arose more than once in the same lineage.[10]

Among these characteristics, reproductive mechanisms stand out. Mitosis is reduced to only two cell divisions between

[10] See Thomas and Spicer (1987) and Crepet (2000) for seed-fern ancestry.

FIGURE 14-24 A diversity of orchid flowers as drawn by Ernst Haeckel. (From *Kunstformen der Natur* 1904).

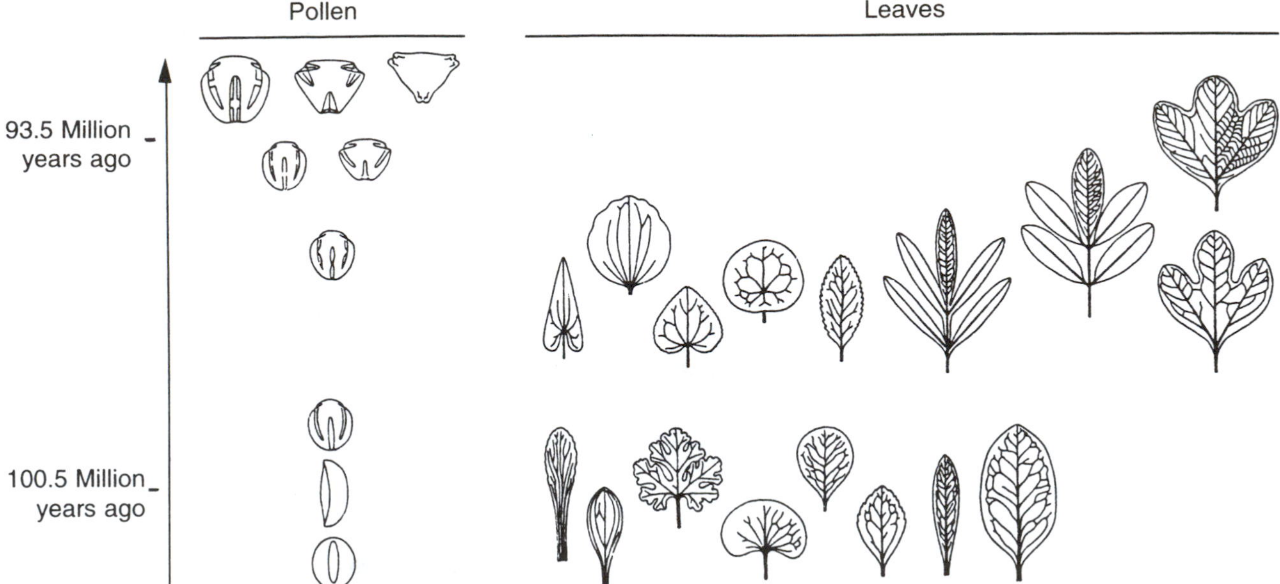

FIGURE 14-25 Appearance of angiosperm pollen and leaves at different times in the Potomac Group of the mideastern United States Cretaceous. The earliest pollens possess single furrows, but pollen grains became more sculpted as time went on. Leaves show a hierarchy of vein complexities: primary veins, secondary veins, intercostal veins, and so on. (Adapted from Doyle, J. A., and L. J. Hickey, 1976. Pollen and leaves from the mid-Cretaceous Potomac Group and their bearing on early angiosperm evolution. In *Origin and Early Evolution of Angiosperms,* C. B. Beck (ed.). Columbia University Press, New York, pp. 139-206.)

formation of the haploid microspore and production of the male gamete, and to only three cell divisions between the megaspore and the multinucleate embryo sac. In addition, angiosperms alone use double fertilization to simultaneously produce the diploid sporophyte zygote and the commonly triploid nutritionally supporting endosperm.

Characteristics such as xylem vessels, the sepals and petals of flowers and other traits (Stewart and Rothwell, 1993) are not universal among angiosperms, and may be considered similar to structures in gymnosperms and other vascular plants. Some botanists propose that the combinations of characteristics that allow particular plants to be classified as angiosperms may have arisen in more than one ancestral group, and that consequently angiosperms may have had a polyphyletic origin.

Some progress has been made recently with more fine-grained analyses of fossil seeds, leaves and flowers, the discovery of new fossils and molecular analyses of MADS box[11] and chloroplast genes, although the evidence has not led to a single theory of angiosperm origins. Two extant taxa are now regarded as at the base of the angiosperm phylogenetic tree: water lilies (*Nymphaea* sp.) and a more basal small shrub, *Amborella trichopoda,* found only on New Caledonia in the South Pacific. *Amborella* lacks a vascular system for conducting water and has other primitive features. Flowers found in 130 My old sediments resemble those of both genera and are consistent with the basal nature of both forms. However, rather than giving us a more detailed knowledge of angiosperm ancestors, *Amborella* and *Nymphaea* distance the angiosperms from other seed plants; the gaps in the record now go even further back. Indeed, new analyses of fossil leaves now lead us to question whether monocots diverged from dicots early in land plant evolution (see below).

However this matter is resolved, angiosperm advantages in rapid gametogenesis, biparental contributions to the endosperm, improved and often specialized pollination and fruity seed coverings, certainly enabled angiosperms to radiate into widely different ecological habitats and become dominant in many of them. Angiosperms, with their protected and nutritionally endowed seeds developed forms adapted to dry climates, wet climates and various types of terrain. Some reinvaded the sea, others became parasitic, and some — such as sundews and Venus flytraps — are even carnivorous!

Not surprisingly, these bountiful adaptations reflect *convergence* as selection under similar environmental conditions produced similar plant phenotypes in different lineages residing in different geographical localities. Prominent examples are some New World cacti and African euphorbs, both occupying desert environments and both highly similar in appearance, with sharp spines or thorns to dissipate heat and to guard their succulent, water-laden stems (**Fig. 14-26**). Such evolutionary convergences, like those of animals (see

[11] The MADS box in plants, fungi and animals, like the homeobox gene, is a DNA sequence shared by a family of genes whose proteins bind to DNA, acting as transcription factors functioning at various stages and locations in plant

FIGURE 14-26 Convergent evolution between representative desert species of American Cactaceae (*left*) and African Euphorbiaceae (*right*). (From Niklas, *The Evolutionary Biology of Plants,* 1997. Reprinted by permission of the University of Chicago Press.)

Fig. 3-7), derive some similarities by modifying different genetic pathways: the cactus spine is a modified leaf and the euphorb thorn a modified branch.

Figure 14-27 depicts a cross section of a hypothetical evolutionary tree with the various clades (orders) of angiosperms arranged as branches around an ancestral complex that served as the primeval trunk. The figure is drawn so that groups close to the ancestral complex are more basal in respect to early angiosperm characteristics than are those farther away. Although much remains unknown, such a tree gives us some idea of the successful radiation of angiosperms, their remarkable evolutionary plasticity, and some of the phylogenetic relationships among them. **Figure 14-28** shows the latest phylogenetic relationship among 24 of the extant families of plants.

Fungi

Fungi first appeared in the fossil record in the Silurian, along with the first vascular plants.

Fungi have cell walls and produce spores, features that led early biologists to include them within the plant kingdom, often defining fungi as "simple plants without chlorophyll." Because of their many specific attributes, this classification has changed in recent years, and biologists now place fungi in their own kingdom. According to cur-

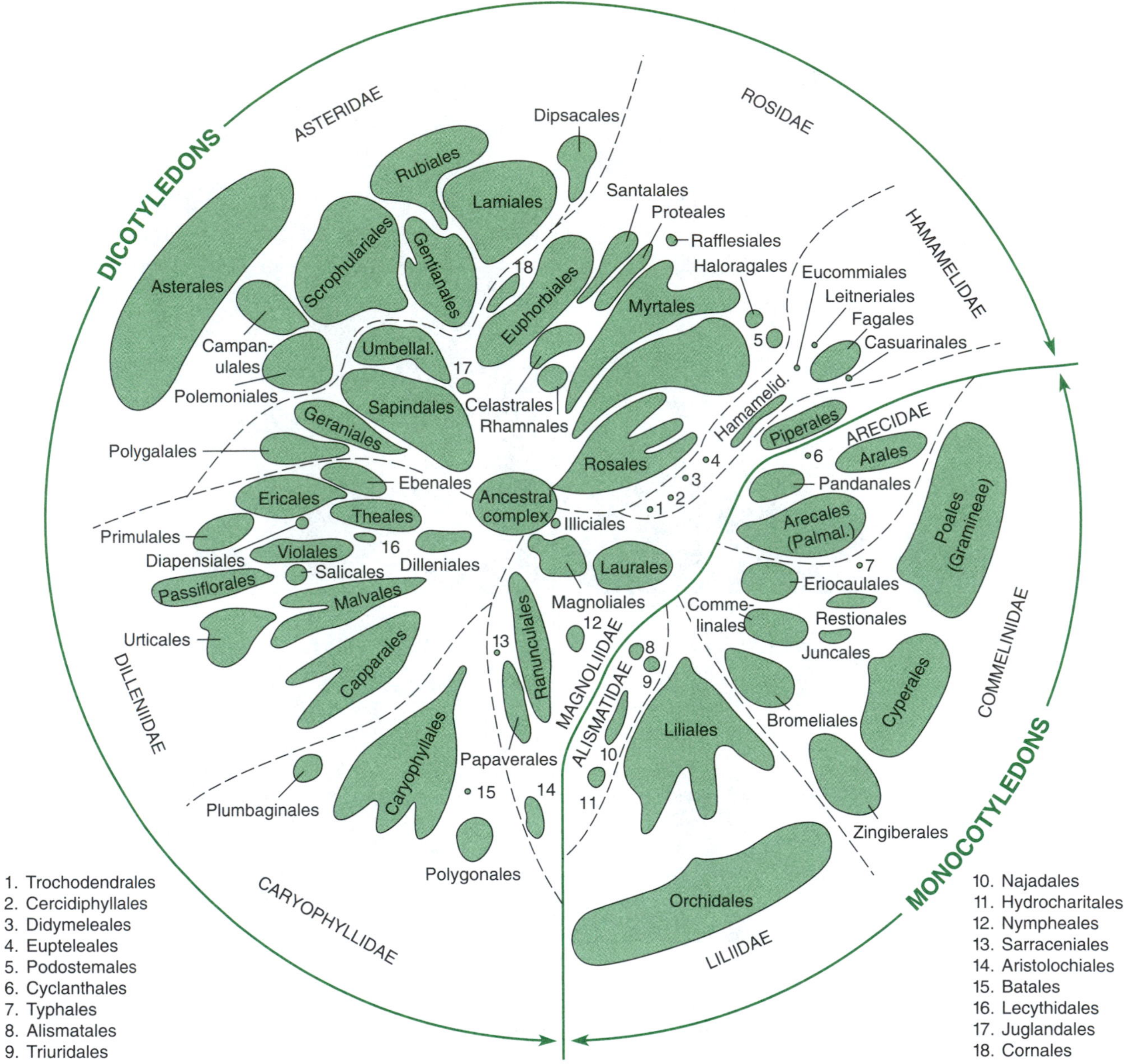

FIGURE 14-27 Proposed evolutionary relationships among major groups of angiosperms, according to Stebbins (1974). Many botanists presume that the **dicotyledons** (dicots; two embryonic leaves) evolved before the **monocotyledons** (monocots; one embryonic leaf). The phylogenetic relationships between subclasses and between orders are shown by their physical proximities in the illustration (e.g, the Hamamelidae may derive from the Magnoliidae). Within each subclass, each order occupies an area indicative of its relative population size. (Adapted from Stebbins, G. L., 1974. *Flowering Plants: Evolution Above the Species Level.* Harvard University Press, Cambridge, MA.)

rently accepted molecular phylogenies, fungi are the "sister group" of multicellular animals (see Figures 9-16 and 11-10). Nuclearid amoebae are the likely unicellular sister group to fungi, although see the discussion in Chapter 8 (especially Boxes 8-1 and 8-2) for the nature of multicellularity in fungi and other groups.[12]

[12] See Wainright et al. (1993) and Baldauf and Palmer (1993) for fungi as the sister group of animals and see Medina (2005) for nuclearid amoebae.

Some of the 120,000 fungal species such as yeasts are unicellular. Others have vegetative stages that are mostly in the form of branched multicellular or multinuclear filaments called **hyphae,** which aggregate into a mass called the **mycelium.** Restricted ecologically by their growth form and absence of chlorophyll, fungi are heterotrophic, deriving their nutrition **parasitically** (*live hosts*) or **saprophytically** (*dead organic material*). The parasitic–saprophytic transition occurred a number of times in fungal evolution and in both directions.

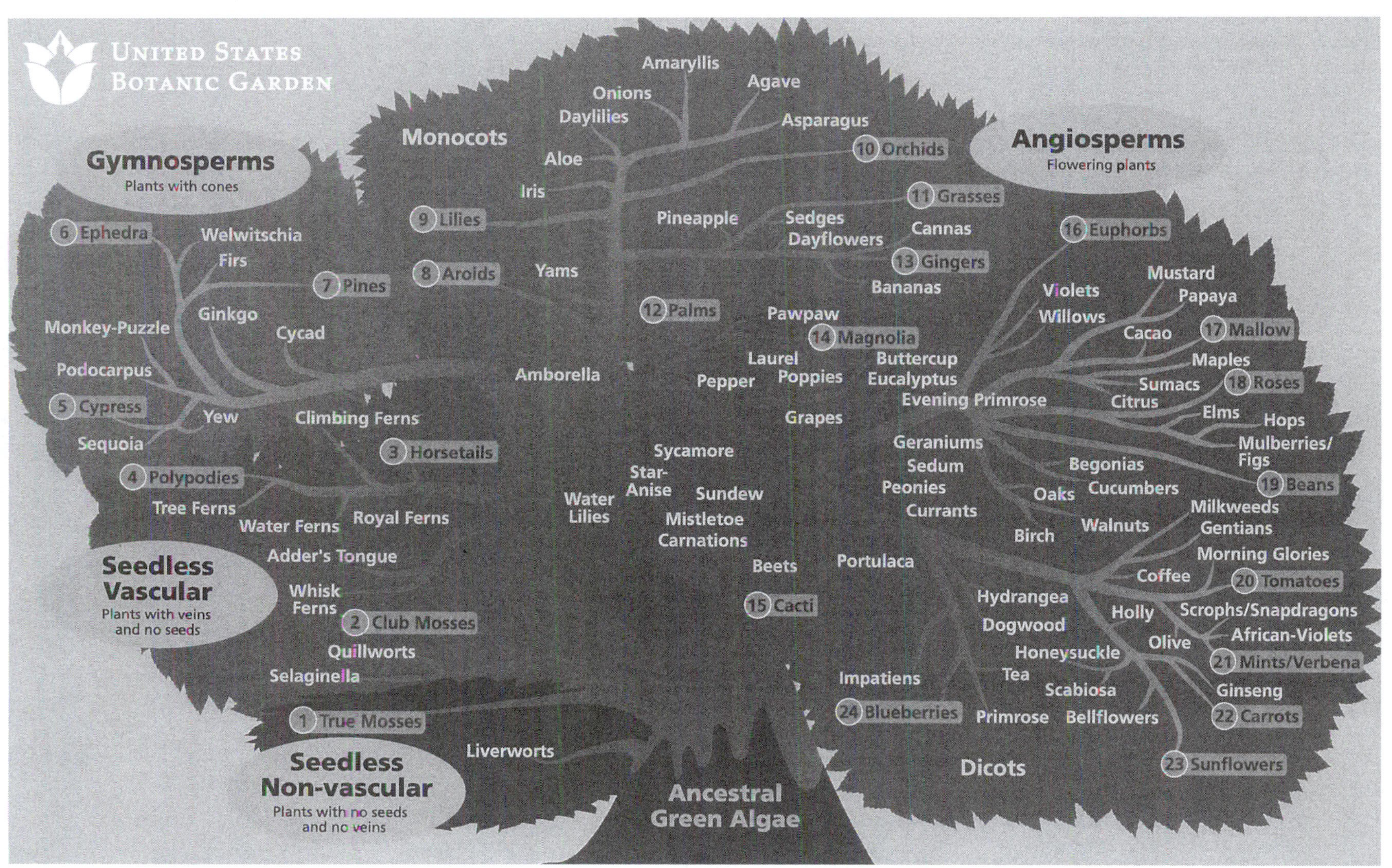

FIGURE 14-28 A phylogenetic tree for 24 families of plants, including vascular and nonvascular plants, gymnosperms and flowering plants (angiosperms). (Courtesy of U.S. Botanic Garden.)

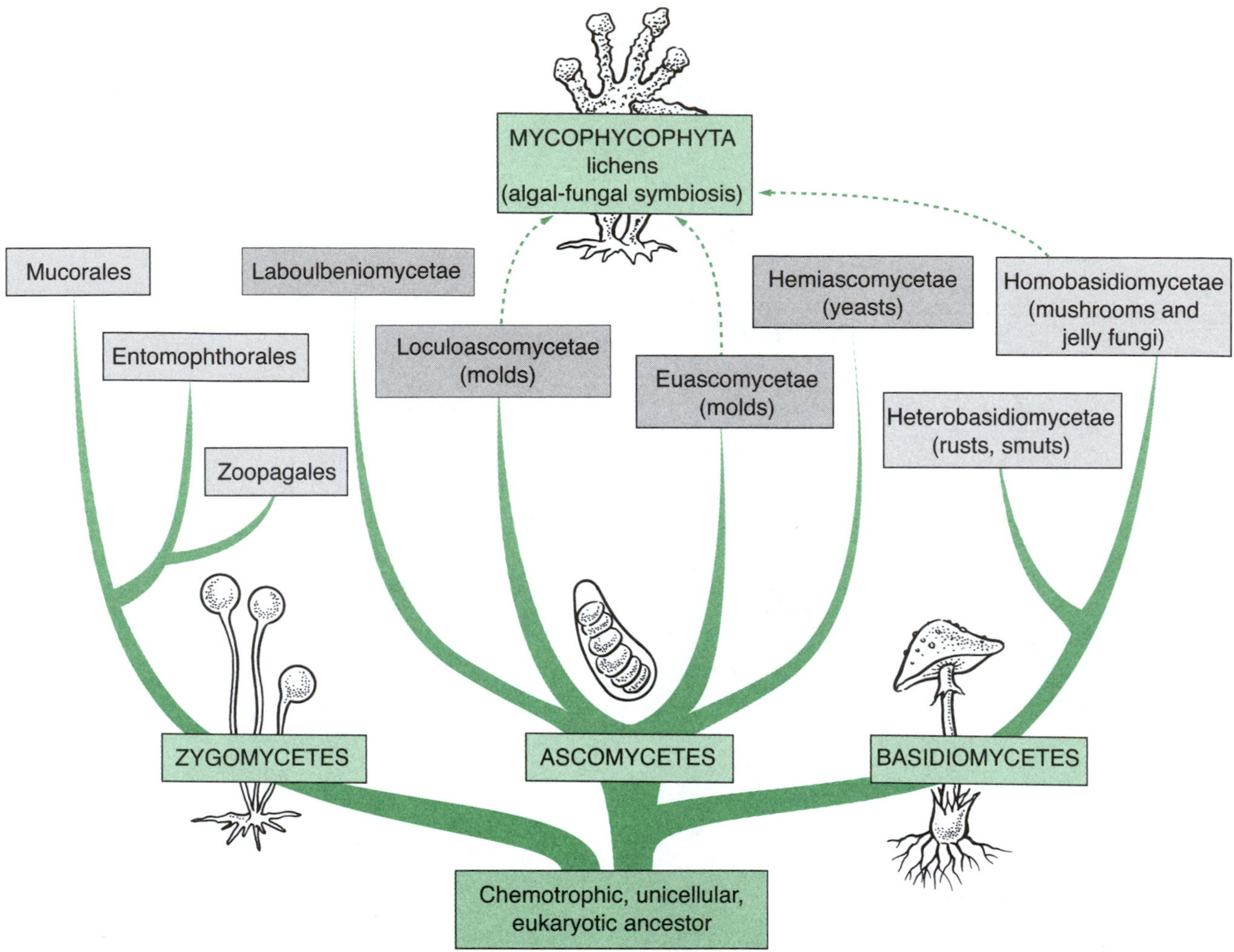

FIGURE 14-29 Phylogenetic relationships among the nonflagellated fungi, showing three subdivisions, Zygomycetes, Ascomycetes and Basidiomycetes. A fourth subdivision, Deuteromycetes (fungi imperfecti), consists of about 25,000 species and includes some common fungi such as *Penicillium*. Evidence suggests that some deuteromycetes evolved from Ascomycetes, others from Basidiomycetes. In addition to these nonflagellated fungi are flagellated forms, and the separately classified slime molds. As the text mentions, molecular studies are beginning to resolve fungal phylogenies. (Adapted from Margulis, L., and K. V. Schwartz, 1998. *Five Kingdoms: An Illustrated Guide to the Phyla of Life on Earth,* 3rd ed. W. H. Freeman, New York.)

Fungi can digest cellulose. Most classifications separate fungi from the various slime molds that have amoeboid stages (such as Myxomycetes, which form plasmodial, acellular aggregates and the cellular Acrasiales) but regard fungi and cellular slime molds as having a shared ancestry.

In one suggestion, early fungi shared a common ancestry with chemotrophic, flagellated cells using inorganic sulfur or nitrogen for energy. Those chemotrophs that gave rise to fungi evolved saprophytic forms that depended on organic materials synthesized by previous organisms. The fact that fungal cells are eukaryotic led to the proposal that some of the intermediate steps occurred through protozoan-like forms. Adopting a parasitic existence on live hosts would have offered the opportunity for early aquatic fungi (perhaps related to extant forms, the water moles) to resist desiccation in the host tissues of land plants and evolve subsequently into forms that use aerially dispersed spores (**Fig. 14-29**).[13]

According to some interpretations of the fossil record, the success of fungi can be traced back to the Precambrian (see earlier in this chapter), and fungi had evolved into most of their presently observed forms by the end of the Paleozoic (**Fig. 14-30**). Although outwardly they have not changed much, parasitic fungi are still actively evolving on the gene

[13] A recent comprehensive analysis of six gene regions in 200 species showed that the spore flagellum used in swimming may have been lost four times independently during evolution of the fungi, each of these losses being accompanied by the evolution of new mechanisms of spore dispersal (James et al., 2006). In this new phylogeny, basidiomycetes and ascomycetes are united as the *Dikarya,* in which part of the life cycle consists of cells with two nuclei.

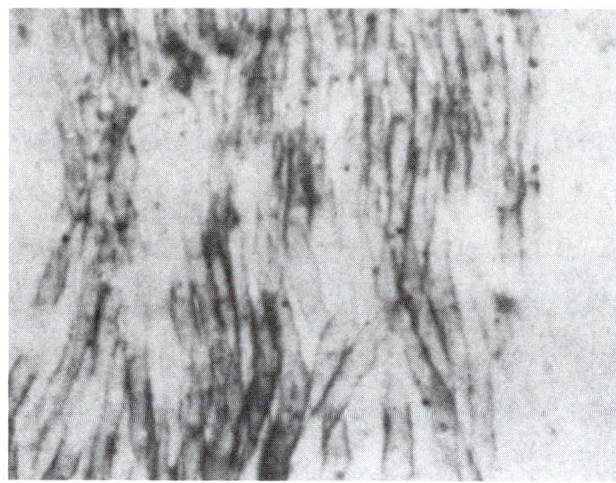
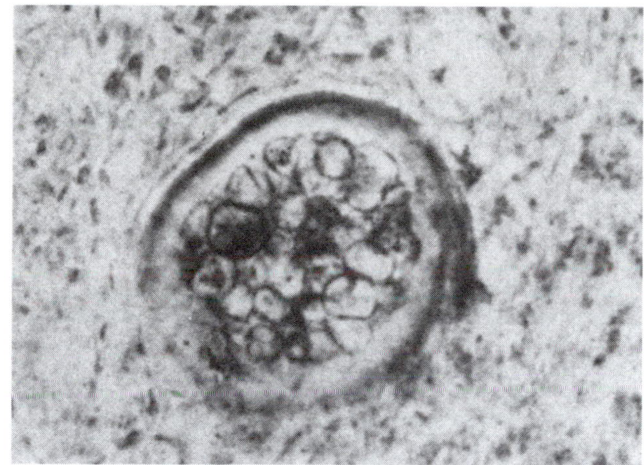

FIGURE 14-30 Fossils of different parts of a Devonian fungus, *Palaeomyces gordonii.* (*left*) Hyphae. (*right*) Spore. (From *Paleobotany,* 1981 by T.N. Taylor. Reprinted by permission of The McGraw-Hill Book Company.)

level, in what has been called an *arms race* between host and parasite. Each new genetic variant of a host that confers resistance against a fungal parasite is often overcome by selection for increased frequency of a fungal genetic variant that increases host susceptibility.

Before Darwin, botanists often suggested that fungi were a form of algae, and grouped them with algae into a single division, **Thallophyta,** their common characteristics being branched, threadlike filaments and motile, alga-like zoospores. With the appearance of *On the Origin of Species,* botanists sought the ancestry of fungi among the algae, especially the red algae, because the fungal lineage lost the algal chloroplasts responsible for its former photosynthetic mode of nutrition and consequently became exclusively parasitic or saprophytic. Because the resemblances between fungi and algae (presence of cell walls, nonmotile habit) may have arisen from convergence, other hypotheses have put forth. The weight of evidence is that fungi shared a choanoflagellate ancestor with animals and are the sister group to animals (see Box 12-4). Fungi as more closely related to animals than to plants is just one of the traditional concepts being overturned as we expand our understanding of the origin of animals, which is the subject of the next chapter.

KEY TERMS

algae
alternation of generations
angiosperms
benthic
bryophytes
club mosses
cuticle
desiccation
dicotyledons
double fertilization
embryophytes
enation theory
endosperm
ferns (lycopods)
flowers
fungi
gametophyte
gymnosperms
heterospory
homospory
hyphae
lycopods
megaspores
microspores
monocotyledons
mycelium
parasitism
pollen
progymnosperms
protostele
saprophytic
seed ferns
seeds
self-incompatibility alleles
self-sterility alleles
sporangia
spores
sporophyte
telome theory
Tracheophyta
vascular plants

DISCUSSION QUESTIONS

1. On what bases have biologists proposed the origin of land plants from algae?
2. Sexual reproduction and meiosis.
   a. What advantages does meiosis offer to sexual organisms?
   b. How does multicellularity increase the genetic variability produced by meiosis?
3. Alternation of generations.
   a. What explanations have botanists offered for the gametophyte–sporophyte alternation of generations in plants?
   b. What can account for the persistence of the multicellular gametophyte stage in plants and its absence in animals?
4. What were the major evolutionary changes among the earliest land plants in respect to their leaves, spores and vascular structures?
5. How does botanical evidence suggest that seeds originated?
6. Angiosperms.
   a. What are the unique characteristics of angiosperms, and what adaptive advantages do they offer?
   b. What proposals have been made concerning angiosperm origins?

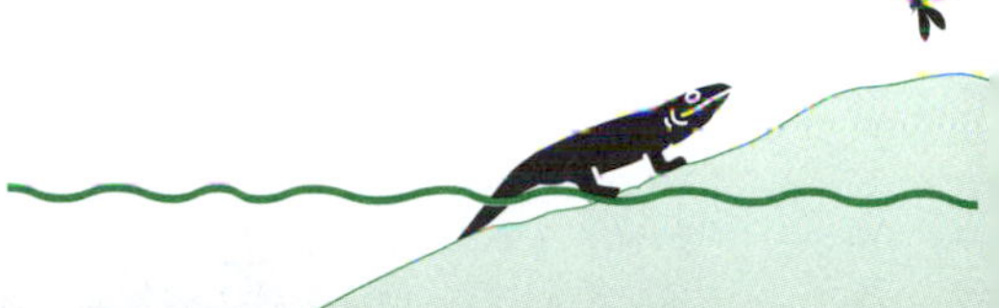

7. What proposals have been offered to explain the origin and evolution of fungi?

EVOLUTION ON THE WEB

Explore evolution on the Internet! Visit the accompanying Web site for *Strickberger's Evolution*, Fourth Edition, at www.biology.Jbpub.com/book/evolution for Web exercises and links relating to topics covered in this chapter.

RECOMMENDED READING

Alcock, J., 2006. *An Enthusiasm for Orchids: Sex and Deception in Plant Evolution.* Oxford University Press, Oxford, UK.

Crepet, W. L., 2000. Progress in understanding angiosperm history, success, and relationships: Darwin's abominably "perplexing phenomenon." *Proc. Natl. Acad. Sci. U.S.A.*, **97**, 12939-12941.

Darwin, C., 1862. *On the Various Contrivances by Which British and Foreign Orchids Are Fertilized by Insects, and on the Good Effects of Intercrossing.* Murray, London, UK.

Gensel, P. G. and D. Edwards, (eds), 2001. *Plants Invade the Land: Evolutionary and Environmental Perspectives.* Columbia University Press, New York.

Hughes, C. and R. Eastwood, 2006. Island radiation on a continental scale: Exceptional rates of plant diversification after uplift of the Andes. *Proc. Natl. Acad. Sci. U.S.A.*, **103**, 10334-10339.

Kenrick, P. and P. Davis, 2004. *Fossil Plants.* Natural History Museum, London.

Louis, L. A. and R. M. McCourt, 2004. Green algae and the origin of land plants. *Am. J. Bot.*, **9**, 1535-1556.

Margulis, L. and K. V. Schwartz, 1998. *Five Kingdoms. An Illustrated Guide to the Phyla of Life on Earth,* 3rd ed. Freeman, New York.

Mauseth, J. D., 1998. Botany: *An Introduction to Plant Biology,* 2nd ed. Jones and Bartlett, Sudbury, MA.

Niklas, K. J., 1997. *The Evolutionary Biology of Plants.* University of Chicago Press, Chicago.

Stewart, W. N. and G. Rothwell, 1993. *Paleobotany and the Evolution of Plants,* 2nd ed. Cambridge University Press, Cambridge, UK.

Thompson, J. N., 2005. *The Geographic Mosaic of Coevolution.* The University of Chicago Press, Chicago.

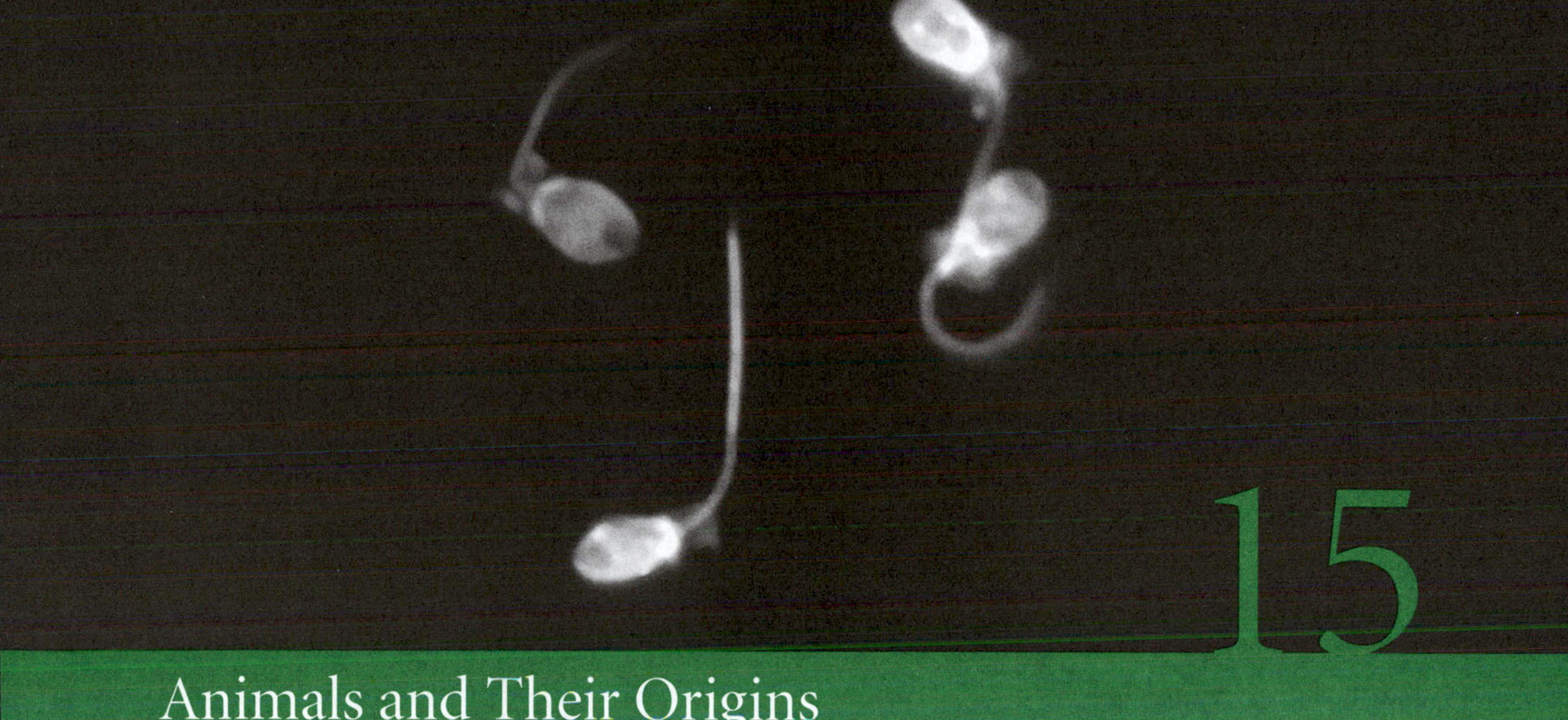

15

Animals and Their Origins

Chapter Summary

A fascinating but enigmatic Precambrian assemblage of organisms, known as the Ediacaran fauna, consisted of flattened, often frond-like fossils lacking any obvious organs or appendages, mouths or anuses, but some attaining heights (lengths?) of two meters. Variously assigned to separate domains, the Vendozoa or Rangeomorpha, they formed complex ecosystems but disappeared early in the Cambrian. Any relationship to metazoans is unresolved and may be non-existent.

About 545 Mya in the Early Cambrian, an explosive radiation of multicellular organisms appeared in the fossil record. For protection, organisms developed the exoskeletons found as Cambrian fossils, although we have evidence that their soft-bodied precursors arose much earlier. The origin of body plans, phyla and of these organisms, of which the Burgess Shale fauna is a prime example, is an area of active research. The appearance of so many "fully formed" body plans raises the question of their ancestors and why metazoans radiated and diversified so rapidly in the Early Cambrian. Explanations offered include a rise in atmospheric oxygen coincident with the evolution of geological features, the earlier existence of a "snowball earth" that prevented organisms from arising or surviving, a diversification of gene regulation and the evolution of predator-prey interactions as heterotrophic "croppers" opened new niches by preying on the autotrophic population.

The dramatic appearance of metazoans a billion years or more after prokaryotes arose raises obvious questions: What is the relationship between metazoans and prokaryotes? Did metazoans have a protistan origin? What are the consequences and advantages of multicellularity? By what means and however many times it arose, multicellularity permitted cellular specialization and more efficient food gathering as small ciliated metazoans became larger and bilaterally symmetrical.

Among several hypothesized pathways by which heterotrophic protistans could have become multicellular animals, the oldest two theories, both of which depend on our understanding of metazoan embryonic development, are (1) a proposal by Haeckel (1874) that flagellated colonial protistans became **bilaterally symmetrical** and developed an invagination of "digestive" cells, producing a gastrula-like structure, the gastraea; and (2) the theory that a solid ball of cells, the planula, differentiated into interior digestive cells and exterior locomotory cells, as occurs in some metazoan larvae. As discussed in Chapter 13, many genes operating in a wide range of segmented and nonsegmented animals share homologous nucleotide sequences, similar linkage orders, and even similar developmental targets. Genetic analysis of comparative embryonic development is providing a deep understanding of the types of developmental changes that occur, how such changes function and which changes may be selected and transmitted during evolution.

Although we can trace unicellular eukaryotic fossils back about 1.6 to 1.8 By or so, and there are claims of eukaryotic algae dating from 2.1 Bya or even older, diverse and more "complex" multicellular forms appeared only about 1 Bya.[1]

Multicellular Organisms Arise

We still don't know what changes led to the appearance of multicellular animal eukaryotes (**metazoans**), but by the beginning of the Cambrian (545 Mya) many differently skeletonized organisms were present. Within a relatively short geological time span, an explosive radiation of multicellular eukaryotes marked the emergence of a large number of distinctive and different animal *Baupläne* or body plans (**Figures 15-1** to **15-3**). However, another completely different, diverse and enigmatic assemblage of organisms, the **Ediacaran fauna** (**Fig. 15-4**) existed earlier in the Precambrian and into the Early Cambrian, although any relationship between Ediacaran organisms and metazoans has been hard to uncover and may be nonexistent (**Box 15-1**).

This chapter summarizes some of the explanations proposed for these events, although many uncertainties persist, both regarding the organisms themselves and the environment and climatic conditions in which they lived. We begin with the canonical Cambrian fauna, the fossils of the Burgess Shale.

Burgess Shale Fauna

As illustrated in Figure 15-1, many different metazoan body plans are known from the start of the Cambrian Era. Not all these body plans survived to the end of the Cambrian, the presumption being that extinction removed less successful forms. The most well-known formation in which Cambrian organisms are preserved is in a limestone reef, 160 m deep and more than 20 km long, in the Burgess Shale in Yoho National Park, British Columbia, Canada. The animals of the Burgess Shale were neither an isolated evolutionary experiment nor unrepresentative of the general situation in the Early to Middle Cambrian. At least 12 other sites contain animals with an equivalent range of body plans. These include faunas in Pennsylvania, north Greenland, China, Spain, Poland and the Chengjiang fauna in China.[2]

Although discovered in 1909 by Charles Doolittle Walcott (1850–1927), Secretary of the Smithsonian Institution in Washington, DC, and described by him between 1911 and 1926, it was not until an expedition in the late 1960s that the Burgess Shale revealed its true story, a story with a cast of 124 genera and 140 species. Just over a third of the genera are arthropods (especially trilobites), but sponges, brachiopods, polychaete worms, echinoderms, cnidarians and mollusks are all represented. So well known is the Burgess Shale fauna that the scientific names of some are known outside paleontology; *Marrella splendens* and *Canadapsis perfecta* (arthropods) and *Hallucigenia sparsa* (Fig. 15-3) may be familiar to you. Remarkably, for some species we have enormous numbers of specimens; at least 15,000 specimens of *Marrella splendens* are known. A species named *Pikaia gracilens* may represent the first chordate, the lineage to which we as vertebrates belong. The possibility of early Cambrian chordates is reinforced by the discovery of two potential chordate ancestors, *Cathaymyrus diadexus* and *Yunnanozoon lividum*, in the Early Cambrian Chengjiang

[1] See Mendelson (1993), Han and Runnegar (1992), and Knoll (1992) for 1- to 2-By-old fossils.

[2] For the Burgess Shale, see Conway Morris (1989, 1998a), Briggs et al. (1994), Gould (1989) and Hall (1999a). For the Chengjiang fauna in China, see Hou et al. (2004).

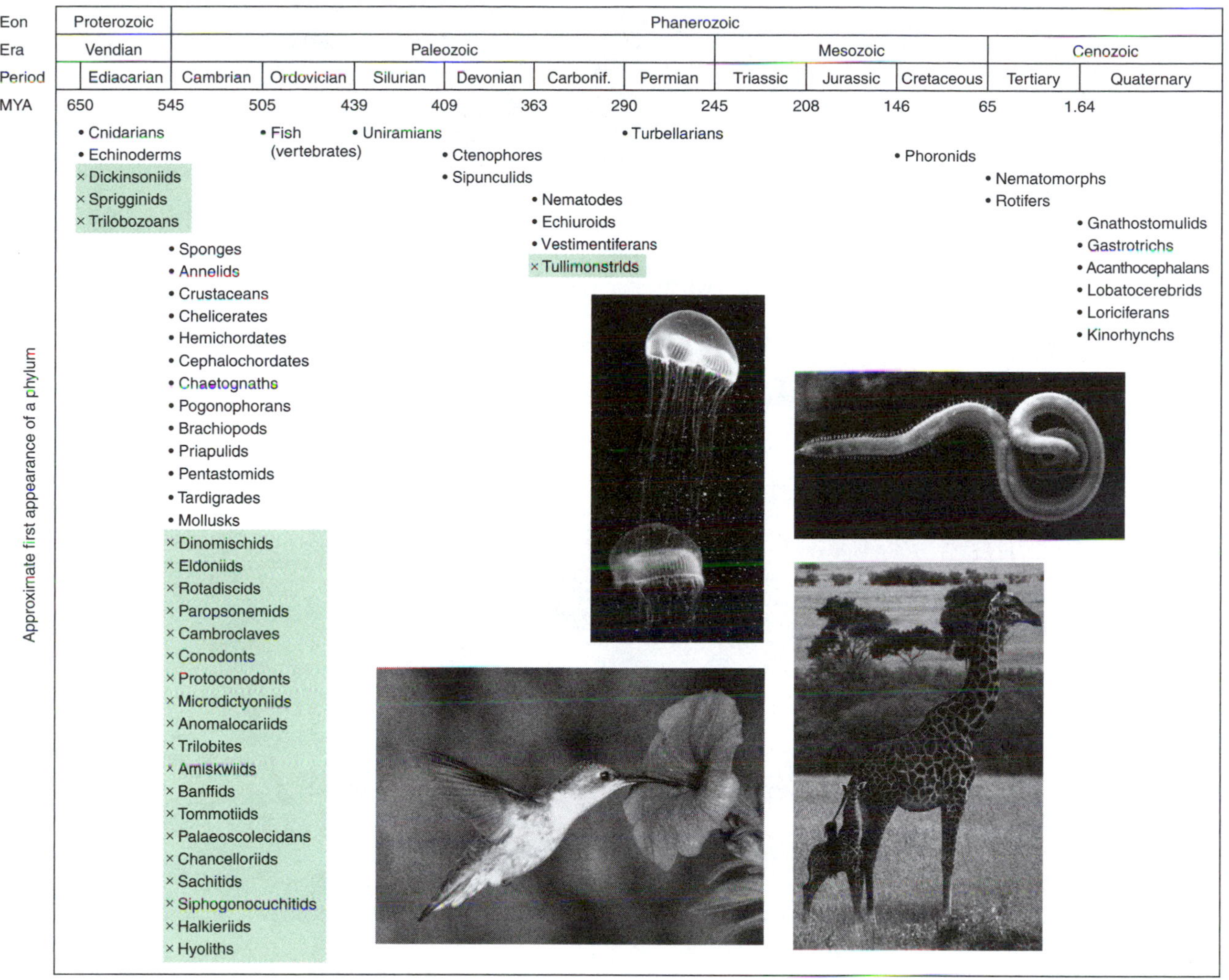

FIGURE 15-1 Approximate times at which various major metazoan lineages first appear in the fossil record. Unshaded groups marked with filled-in circles have extant descendants, although the original species representing these groups are long extinct. Shaded groups marked with **x**s represent extinct clades (orders or phyla) that have no known surviving descendants. (Based on data derived from Conway Morris, S., 1993. The fossil record and the early evolution of the metazoa. *Nature,* **361**, 219-225, and Conway Morris, S., 1998. *The Crucible of Creation: The Burgess Shale and the Rise of Animals*. Oxford University Press, Oxford, England.)

formation in Yunnan Province, China, which is some 10 My older than the Burgess Shale.

Assigning organisms to clades such as phyla at a time when distinct body plans that could be regarded as diagnostic of phyla were emerging or had only just emerged, presupposes:

- that organisms had already evolved the characters that define the range of present-day phyla;
- that morphological gaps separate phyla to the extent that we can assign an individual to one phylum and not to another; and
- that stem groups should be sought in earlier forms.

The paleontologist and paleobiologist David Raup (1966) developed an approach to plot the morphologies of organisms in three dimensions in what he termed **morphospace,** represented as a cube with different features of the organisms along each axis (**Fig. 15-6**). If all morphologies were possible, the cube — morphospace — would be filled. Occupancy by known forms of only a portion of morphospace would indicate that only a subset of morphologies was possible, that is, adapted and survived. Raup's analysis of shelled invertebrates (bivalves, brachiopods, cephalopods and gastropods) showed that their known morphologies cluster in one region of morphospace (Fig. 15-6). The obvious conclusion is that not all morphologies are possible. Unoccupied morphospace could represent impossible morphologies, nonadaptive morphologies, constraints on morphology and/or gaps between groups of organisms. Hall (2002) summarizes this approach.

A further difficulty when comparing Cambrian with extant species lies in the very basis on which we recognize

FIGURE 15-2 A variety of Cambrian animals including some exotic forms from the Middle Cambrian Burgess Shale of British Columbia, Canada. Many are arthropods, including some trilobite-like animals (*Naraoia*). *Aysheaia,* seen feeding on a sponge, is considered in the ancestral arthropod lineage, related to onychophora (velvet worms). *Pikaia,* with its putative notochord and fish-like zig-zag muscle arrangement, the first fossil found of a chordate-like animal, may be a vertebrate ancestor, provided that these features stand up to close study; no detailed study has yet been published. The probable lifestyles of these and other Burgess Shale animals are described by Conway Morris (1998a). (From Erwin, D., J. Valentine, and D. Jablonski, 1997. "The origin of animal body plans." *The American Scientist,* 85, 126–137. Reprinted with permission of D. W. Miller.)

a species. Paleontological species based on morphology are neither biological species — which are based on reproductive isolation — nor are they evolutionary species (a series of ancestral and descendant populations; Chapter 11).[3]

[3] See G. G. Simpson (1961), Conway Morris (1989), Hall (1999a), and Levinton (2001) for discussions of species types.

Conway Morris was quite pessimistic about our ability to resolve these problems:

> Here (in the phylum) is the quintessence of biological essentialism, a concept that is almost inextricably linked with that of the body plan. For most of the Phanerozoic, the status of phyla (and classes) seems to be effectively immutable, the one fixed point in

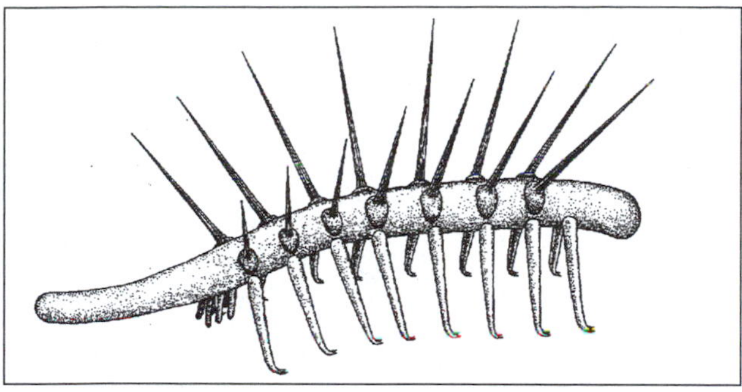

FIGURE 15-3 A reconstruction of *Hallucigenia sparsa*, from the Burgess Shale, thought to be related to the velvet worms (onychophorans) or an early arthropod. (Adapted from Ramsköld, L. and X.-G. Hou, 1991. New early Cambrian animal and onychophoran affinities of enigmatic metazoans. *Nature*, **351**, 225–228.)

> the endless process of taxonomic reassignments and reclassifications. However, for the Cambrian radiations, such precepts begin to fail and our schemes of classification lose relevance (1989, p. 345).

A large number of body plans are represented in the Early Cambrian. Why did so many animal body plans appear in the Early Cambrian and why have so few evolved since?

A Cambrian "Explosion"?

The diversity of Cambrian organisms suggests an adaptive radiation in which many new ecological opportunities were made available for organisms with the capacity to evolve in diverse ways (Fig. 15-2). As these environments would quickly have become saturated, extinctions would have occurred; as in other periods, both selective and accidental factors would have contributed to extinction or survival and adaptation.[4]

Increasing competition for available resources among so many groups with diverse yet overlapping characters must have led to selection among them. Accidental factors such as environmental calamities had serious impacts on groups whose numbers, habitats or lifestyles made them vulnerable to extinction, with "survival of the fittest" and "survival of the luckiest" both operating. Many biologists suggest that multicellular lineages that survived the Cambrian radiated in so many ways and generated such a wide range of body plans, that few later-emerging phyla could successfully compete with them; body plans that came first limited opportunities for those that came later. The absence of fossils of some groups, such as platyhelminths — a basic multicellular form — can be explained by their lack of a fossilizable epidermal cuticle.

[4] See Conway Morris (1998a,b), Schulter (2000) and Selden and Nudds (2004) for the Cambrian biota.

Although ancestral connections are unclear, some or many of the skeletonized forms that appeared during the Cambrian represent new adaptive radiations of Precambrian forms. Evolutionary molecular clocks, based on a considerable amount of nucleotide sequence data, indicate that divergences among major Cambrian lineages had already begun about 700 Mya (Ayala et al., 1998; Valentine, 2004). To more ancient soft-bodied organisms, modifications during the Cambrian added hard parts and mineralized skeletons that provided leverage for evolving muscles, support for body organs, enclosures for gills and filtering systems and protective shells and spines. As is true for many other Cambrian conditions, we do not know which factors initiated tissue mineralization. Perhaps the Cambrian marked a warming trend that allowed mineralization. Or perhaps the atmospheric accumulation of oxygen through photosynthesis reached sufficient levels to permit mineralization and to energize oxygen-dependent synthesis of metazoan connective tissue proteins (collagens).

Explanations of the Cambrian Explosion

Oxygen

Forty years ago it was suggested that metazoan evolution — which is dependent on aerobic metabolism — required an oxygen atmosphere that had reached perhaps one percent of current atmospheric oxygen. Now we know that free oxygen produced by algae was present for a billion years or more before the Cambrian.

According to Knoll (1992), even if Proterozoic oxygen pressure was low, some evidence exists for major increases in oxygen during the Precambrian that, in conjunction with new and more sophisticated controls of embryonic development, may have led to the evolutionary burst of large multicellular metazoans. Ohno (1996) proposed that animals capable of exploiting Early Cambrian oxygen possessed a battery of common genes including those for lysyloxidase, which uses oxygen to crosslink collagen; hemoglobin, a conveyor of molecular oxygen; and homeobox proteins that orient body plans along a directional axis (see Chapter 13).

Certainly, the ability of oxygen to form a protective blanket of ozone (see Chapter 5) facilitated the rapid expansion and radiation of multicellular animals in shallow waters and on varied surfaces, providing the opportunity for skeletal and soft tissue adaptations.

Fenchel and Finlay (1994) suggest that because anaerobic metabolism has a low energy yield and therefore low biomass production (10 percent growth efficiency), the evolution of large organisms required an aerobic environment (although recall that some rangeomorphs attained lengths of two meters). Anaerobic metabolism limits the food chain to perhaps only two steps, from bacteria, which consume organic matter, to single-celled eukaryotes, which consume

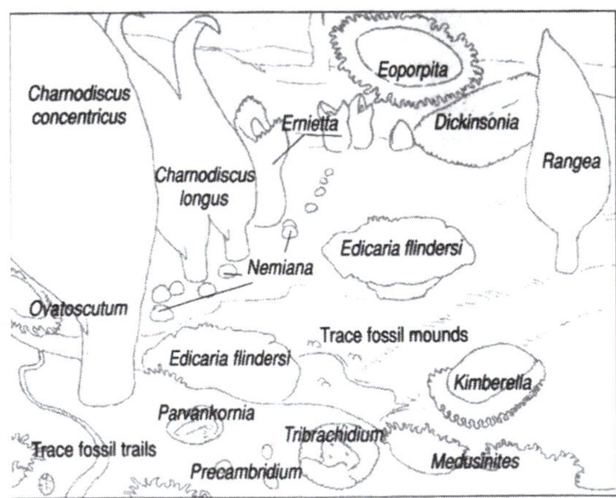

FIGURE 15-4 A panorama of soft-bodied animals found in Ediacaran tidal flat deposits of South Australia and 10 or so other places throughout the globe, most occurring approximately 545 to 650 Mya, just prior to the Cambrian. Some of these fossils (*Edicaria, Charniodiscus*) resemble cnidarians such as jellyfish and sea pens. Others bear likenesses to segmented annelid-like animals (*Dickinsonia*) and perhaps to echinoderms (*Tribachidium*) or mollusks (*Kimberella*). (From Erwin, D., J. Valentine, and D. Jablonski, 1997. "The origin of animal body plans." *The American Scientist*, 85, 126–137. Reprinted with permission of D. W. Miller.)

bacteria. Oxygen environments, in contrast, are about four times as efficient in energy yield, producing sufficient biomass to allow more levels in the food chain, each with increasing morphological size, although with fewer individuals at higher levels. Although physiologists normally regard metabolic rate as not scaling one to one with body mass (M), but scaling to the 3/4 power ($M^{0.75}$), scaling varies between pelagic and marine organisms and can vary during the life history, scaling closer to one to one in larvae, but at $M^{0.75}$ in adults; that is, like any other character, metabolism and metabolic rate is subject to evolutionary change (Glazier, 2006).

BOX 15-1 The Ediacaran Fauna

WHATEVER THE CAUSES for the proliferation of Cambrian shelled animals, the soft-bodied fossils found in the Precambrian Ediacaran strata of South Australia (known as the **Ediacaran fauna**) and more recently in deposits at other locations around the globe, have transformed our understanding of metazoan origins. The fossils at Mistaken Point, Newfoundland, are preserved beneath a thick layer of volcanic ash that has been dated using U-Pb at 565 ± 3 My, making these the oldest Ediacaran fauna, and providing amazing insights into a Precambrian ecosystem.

As depicted in Figure 15-4, some Ediacaran fossils resemble cnidarians such as jellyfish and sea pens. Others bear likenesses to segmented annelids and, more distantly, to mollusks and echinoderms. The conclusion that Ediacaran fossils represent various lineages of organisms is reinforced by the reconstruction of *Kimberella* from the Ediacaran fauna of the White Sea in Russia. A blastula-like organism, *Kimberella* may be a common ancestor of mollusks and allied invertebrates.[a]

Because paleobiologists have not unambiguously detected any of the basic anatomical features of later animals — eyes, mouths, anuses, intestinal tracts or locomotory appendages — in any Ediacaran fossil, the relationship of these organisms to Cambrian metazoans remains obscure. Glaessner (1984) proposed that the organisms were early representatives of known metazoan phyla. Seilacher (1989, 1992) placed them in a distinctive group, the *Vendozoa,* which he and others regard as unrelated to any metazoan. The absence of features usually associated with prey capture and a digestive tract has led to suggestions that many of these Ediacaran animals may have depended on photosynthetic or other types of symbionts. Retallack (1994) suggests most were really lichens — symbiotic associations between fungi and algae — flattened by compaction in quartzite deposits. Others have disputed this view.

Perhaps the most appealing thesis today comes from Guy Narbonne (2004), who invoked repeated structures (*modularity*) based on fractal repeats of a frond structure in his analysis of early Ediacaran organisms, a major group of which he has classified as **rangeomorphs** (**Fig. 15-5**). Narbonne is cautious when it comes to saying whether rangeomorphs were metazoans or an alternative (and earlier) evolutionary experiment in multicellularity. Rangeomorphs were organized into complex ecological communities with more than one trophic level. One species, *Charnia wardi,* found in the Drook Formation near Calvert, Newfoundland, grew to lengths (heights?) of two meters, and is the oldest representative of this fauna, dating as it does to 570 to 575 Mya.[b]

Whatever their affinities, their morphological differences indicate that a variety of Ediacaran organisms, which already had diverse evolutionary histories, existed right up to the Cambrian boundary. Although some survived into the Cambrian, they became extinct. Why? We don't know, although their lack of armor may have made them easy prey for the new mobile Cambrian predators?

[a] See Erwin et al. (1997) and Erwin and Davidson (2002) for *Kimberella*.
[b] For rangeomorphs and other Ediacaran fauna, see Clapham et al. (2003), Narbonne and Gehling (2003), Brasier and Antcliffe (2004) and Narbonne (2004).

FIGURE 15-5 A rangeomorphs frondlet to show the size (the scale bar is 0.25 cm) and the repetitive branching pattern. (Reproduced from Guy Narbonne, 2004. *Science* **305**: 1141–1144. Reprinted with permission from AAAS.)

Geological Conditions

A change in oxygen concentration is *only one of many* possible causes for the Cambrian explosion.

A formerly popular theory held that most signs of life disappeared in strata earlier than the Phanerozoic because of the heat and pressure involved in geological processes such as mountain building. A further notion was that life evolved mostly in fresh water. Fossils would thus be absent in Precambrian sediments, which are primarily marine. However, because there is considerable evidence of prokaryotic *and* eukaryotic Precambrian organisms, the Cambrian discontinuity seems real and not merely the result of geological metamorphism or imperfect fossilization. If the Precambrian ancestors of later metazoan phyla seem absent, it may be because they were quite small, no larger than a few millimeters, and, therefore, did not fossilize well or at all.

Among other physical causes offered for the Cambrian explosion are changes in the shape, extent and latitude of shorelines — the result of plate tectonics — profoundly transforming both climate and environment (see Chapter 6). The sea level changes that accompany glaciation undoubt-

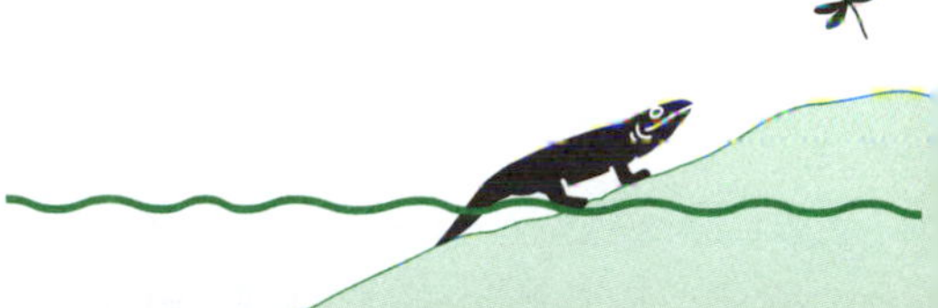

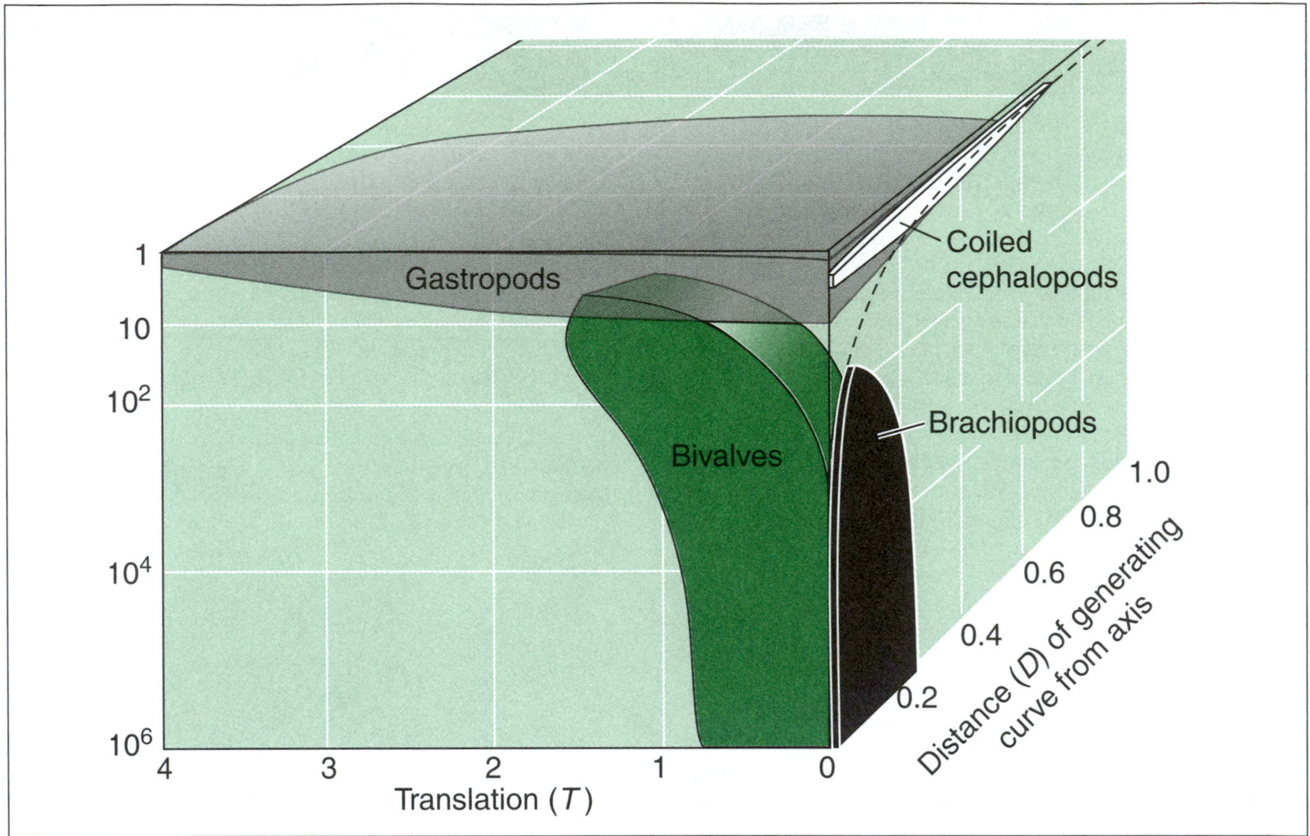

FIGURE 15-6 Only a portion of theoretically available morphospace (which is represented as a three-dimensional space) is occupied by known shell morphologies of bivalves, gastropods, brachiopods and coiled cephalopods. The axes represent three measures of shell morphology/growth. (From Hall, B.K., 1999a. *Evolutionary Developmental Biology,* 2nd ed. Kluwer Academic Publishers, Dordrecht, The Netherlands; modified from Raup, D. M., 1966. Geometric analysis of shell coiling: General problems. *J. Paleontol.*, **40,** 1178–1190.)

edly would have had similar effects, especially if the "snowball Earth theory" is correct (**Box 15-2**). Some proposals even suggest tidal effects caused by the moon. We cannot exclude any of these ideas with certainty, and each may have had some influence on metazoan evolution.[5]

Genes and Species Numbers

Biologists commonly suggest that the evolution of eukaryotic **sexual genetic exchange and/or regulatory genes** that control multicellular development (see Chapters 10 and 13) could have sparked the diversity and anatomical complexity that fueled the Cambrian explosion. To these proposals we can add the hypothesis that because there were few Precambrian multicellular species, their exponential increase could only become notable after reaching a threshold number at the start of the Cambrian. Again, no demonstrable support exists for these proposals — in fact, we can claim that both sexual reproduction and developmental regulatory elements existed long before the Cambrian — nor is there any way of ascertaining a "threshold number" of species.

Cropping

A further biological cause, suggested by Stanley (1973) depends on the principle of **cropping.**

In cropping, predators feed on the most abundant prey species, thereby reducing their numbers and allowing other species to use resources formerly monopolized by the dominant prey. For example, cropping of a field once restricted to a single dominant plant species soon opens many niches for other plant species, which now can grow in an area from which they were formerly excluded. The evolutionary value of cropping extends to predators through a feedback cycle. Diversification of prey species leads in turn to diversification of predator species. Prey-predator adaptations would have quickly escalated into what has been characterized as a persistent co-evolving arms race: successive rounds of selection for adaptive predator responses to their preys' protective devices, followed in turn by adaptations by the prey to their predators' aggressive devices. Each new environmental interaction would have allowed selection for new adaptive features and survival strategies, promoting diversity.

[5] See Valentine et al. (1991) and Valentine (2004) for elaboration of these points.

Once multicellular heterotrophic **herbivores** appeared, only a few evolutionary steps would have been required for the evolution of **carnivores** that feed on herbivores, and then carnivores that feed on other carnivores, evidenced in the explosive adaptive radiations in the Cambrian. In offering varying degrees of protection from predation, hard exoskeletons therefore may have provided a common function to a wide variety of Cambrian organisms. The fact that eukaryotes by this time had evolved sexual reproduction enhanced their ability to diversify into new functions and newly available habitats.

The Precambrian fossil record indicates the widespread existence of cyanobacterial stromatolites (see Chapter 8) growing under conditions unhampered by limitations other than those of available light and nutrients. As Stanley (1973) said, "We can envision an all-producer Precambrian world that was generally saturated with [autotrophic] producers and biologically monotonous." In the absence of cropping, new algal prokaryotic species and even new algal eukaryotes must have been preempted from occupying these areas. Prokaryotic algae do not exist in acid conditions — perhaps because their chlorophyll molecules are relatively unprotected in their sites on the plasma membrane — suggesting that such environmentally restricted areas may have provided the first opportunity for colonization by eukaryotic algae.

Although attractive, much of Stanley's hypothesis depends on the appearance of effective heterotrophic croppers immediately before the Cambrian, a view that other workers in this field do not share. Heterotrophic nonphotosynthetic organisms were no novelty in the Precambrian, and may have originated far back in the origin-of-life period. Yet, the discovery of Early Cambrian filter-bearing crustaceans capable of feeding on even the smallest planktonic autotrophs may be indicative of the sudden appearance of efficient "cropping" organisms that could have fueled the Cambrian expansion (Butterfield, 1994). A single source for this expansion is not yet generally accepted, and Cambrian events may have arisen from a combination of conditions.

From Unicellularity to Metazoans

The basic body plans derived from Precambrian metazoan ancestors were present as fossils by the Middle Cambrian. They included some exotic forms that did not survive much further. What were the first steps in multicellular animal evolution?

As discussed in Box 8-1 (Kingdoms and Domains of Life), the five-kingdom schemes classify all *unicellular eukaryotes* into the kingdom **Protista,** which includes (1) *protozoans,* which ingest their food directly, (2) *photosynthetic algae,* and (3) some *saprophytic fungi* (see Chapter 14). Protists form a large assemblage, covering more than 100,000 species; their location as one of five or six super kingdoms in the latest Tree of Life is discussed in Box 12-4.

Evolutionarily, protistan ancestry is old, dating back about 1.6 to 1.8 By. Their relationship to metazoans is commonly accepted, and has received molecular support from nucleotide analysis of 16S rRNA sequences,[6] which indicates that the closest extant metazoan relative is a single-celled choanoflagellate (collared sponge-like zooflagellate; see Chapter 16). Thus, aside from their diverse forms and the many ways in which they affect other forms of life, protistans are the essential link between early progenote cells and all multicellular eukaryotes.

As discussed in Chapter 8, biologists now accept that endosymbiotic events provided protistans with mitochondria, chloroplasts and perhaps other constituents. In addition to these organelles, protistans share many features with other eukaryotes, including a nuclear membrane as well as cilia and flagella, both of whose substructures are organized into nine pairs of microtubules circling two microtubules in the axial center (9 + 2 arrangement).

Because not all photosynthetic protistans contain the same light-gathering pigments, it has been suggested that photosynthetic prokaryotes invaded eukaryotic cells more than once. Although the number of endosymbiotic events accounting for chloroplast invasion into eukaryotes is still in question, most agree that the chloroplast source was cyanobacterial. Delwiche and coworkers (1995) offer data for a monophyletic cyanobacterial origin based on sequence analysis of a gene (*tufA*) coding for an elongation factor used in protein synthesis.

Some of these endosymbiotic events resulted in protistan algae completely reliant on autotrophic nutrition. Other events produced protistans, such as *Euglena,* that alternated from autotrophic to heterotrophic nutrition. The lack of chloroplasts, whether caused by their initial absence or later loss, gave rise to the large diversity of protistan heterotrophs, or protozoans.

One proposed scheme for protistan radiation is offered in **Figure 15-7**, which begins with a protoflagellate that has established endosymbiotic relations with a mitochondrion. Its nutrition is heterotrophic, based on phagocytosis through its naked cell membrane (see also Cavalier-Smith, 1993, 2004). Cell division among these early protistans then evolved in two main directions: (1) retention of the nuclear membrane during mitosis, somewhat similar to chromosomal division in prokaryotes (*closed division*), and (2) mitotic division accompanied by breakdown of the nuclear envelope (*open division*). Because phylogenetic proposals are hypotheses of relationships, research into the protozoan-

[6] 16*S* rRNA molecules are essential for ribosomal protein synthesis, and, like 5*S* rRNA, portions of their sequences are conserved across a wide range of organisms (Wainright et al., 1993).

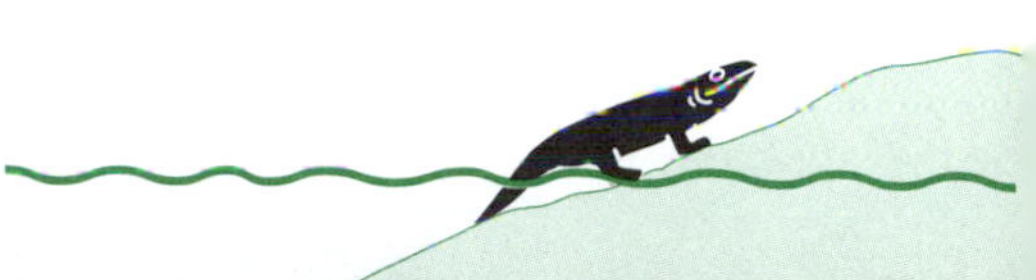

BOX 15-2 Snowball Earth

EVIDENCE HAS BEEN ACCUMULATING since the mid-1960s that Earth was enveloped in a blanket of ice as much as 1 km thick about 600 Mya, that this frigid state continued for 10 million years, and that during this time the average temperature hovered around –40°C.[a] Even more surprising, at least two and as many as four such episodes may have occurred between 750 and 600 Mya. Yet despite the mounting geological evidence, biologists and paleobiologists continued to resist the hypothesis — dubbed **"snowball Earth"** in the early 1990s — because it was difficult to imagine how life could possibly have survived such conditions. Here is how the hypothesis developed.

Until 1964, when geologist Brian Harland reported the presence of glacial deposits dating from 600 Mya in the baking deserts of Namibia, it was not known that tropical regions had ever experienced the ice sheets characteristic of ice ages. Reporting similar deposits on every continent, he was the first to suggest a long period of intense cold.[b]

Mathematical models developed by Mikhail Budyko in the 1970s showed how, under certain conditions, runaway freezing of the whole planet was almost inevitable. But with temperatures as low as –50°C, how could Earth have escaped the trap? The story goes as follows:

- The intense heat of volcanoes allowed them to penetrate the ice and thus to continue emitting CO_2, which as we know is a greenhouse gas.
- As time went by, CO_2 levels increased in the atmosphere because there was no rain to wash CO_2 from the atmosphere.
- Eventually, the CO_2 in the atmosphere reached such high levels — hypothesized to be as much as 10% or about 260 times current levels — that the greenhouse effect kicked in, elevating temperatures to levels where the ice began to melt.
- According to this scenario, the ice would have disappeared in as short a time as a few thousand years.[c]

When you consider that global climate change today is being driven by quite small increases in CO_2 levels — from less than 0.03% before the Industrial Revolution to 0.038% now — just imagine the impact of 10% CO_2. Runaway global warming would be the only logical outcome. And that's how geologists explain another puzzle in Namibia: the presence, immediately above the glacial deposits, of thick layers of carbonaceous rock that bear all the marks of having been deposited very quickly — and which form only in water. Again, this finding was repeated around the globe. We continue the story:

- As temperatures continued to soar — from the coldest conditions ever experienced on Earth to the hottest, at 50°C — evaporation from the oceans resulted in torrential rains that may have lasted for centuries, washing the CO_2 from the air.
- CO_2 dissolved in water forms a weak acid, carbonic acid. Under these conditions, prodigious amounts of carbonic acid would have bathed the globe, interacting with particles of rock eroded

---

metazoan transition benefits from new approaches, many of which have been applied to the evolutionary consequences of multicellularity.

Consequences of Multicellularity

The change from unicellularity to multicellularity occurred a number of times, giving rise to different lineages of organisms. For example, the evolution of multicellularity led to such unrelated groups as filamentous cyanobacteria, slime bacteria (myxobacteria), aggregating amoebae (for example, dictyostelids), algae (brown, red and green) and colonial ciliated protozoans (*Zoothamnium*), in addition to the plants and fungi discussed in Chapter 14.

All hypotheses of the evolutionary origin of metazoans assume the advantages of multicellularity (see Box 8-2, Multicellularity and Pluricellularity). A multicellular organism's food-gathering surface increases by extending its cells to places that it could not have reached were it small and unicellular. This increase ensures a more stable food supply to all its cells even where food distribution is uneven, and allows multicellular organisms to attack and digest larger particles of food by secreting greater quantities of digestive enzymes than single cells can secrete.

Aiding such development are signaling systems (see Chapter 13) that direct cells to move, aggregate, divide and specialize into different tissues (*division of labor*). Multicellularity, which is based on increased gene numbers and regulatory pathways, provides morphological and functional innovations that broaden the scope of protection, dispersion, food gathering, reproduction, excretion and other functions. Recent studies by Nichols and colleagues (2006) demonstrate that major classes of metazoan cell adhesion gene families coding for cell surface receptors and extracellular matrix proteins are present in the sponge *Oscarella carmela,* so their origin, which must have predated the divergence of sponges, is old. Although in the past, biologists used the diversity of metazoan features to indicate the likelihood of multiple (polyphyletic) origins for metazoans, molecular evidence has gone in the opposite direction, pointing to protistan choanoflagellates as a potential metazoan ancestor (**Fig. 15-8**; see Box 12-4).

Increase in Complexity?

The issue of whether complexity increases during the course of evolution was discussed in Box 8-3. Differences between different groups of metazoans associated with the origin

by the torrential rains. The resulting carbonaceous deposits formed a cap on the glacial remains. Eventually, conditions would have returned to more usual temperatures and CO_2 levels.

Which still leaves the question: How did life survive the cold? After all, even the warmest parts of the planet would have been covered in ice at least tens of meters thick. Two lines of evidence provide clues:

- Research carried out in Antarctica has revealed abundant life in water under ice tens of meters thick. The answer lies in the speed at which the ice formed — very slowly. At slow rates of formation the end result is ice of amazing clarity, ice that allows light rays to penetrate to considerable depths.
- Life has also been revealed flourishing in extreme conditions: around deep-sea hydrothermal vents, in and under polar snow, and in hot springs, for instance. These three habitats would all have existed on a snowball Earth.

Important in this discussion is the time period involved: just shortly (in geological terms) before the first complex large organisms appear in the fossil record. Also important is the extraordinary assemblage of new body plans that these fossils represent. Some last points to ponder:

- Given that there is no doubt that most life would have become extinct under the ice and snow, and that any surviving remnants would have lived in scattered and isolated niches, could the rapid reversal of conditions have stimulated a great burst of evolution as the formerly isolated (and by now evolutionarily distinct) populations met?[d]
- Could a series of such snowball Earths over the 150-million-year period in question explain the long delay before the flowering of new body plans?
- Is it possible that the ice acted as an "evolutionary filter," driving to extinction most of the evolutionary experiments that emerged in the inter-snowball period(s) and leaving only a few to emerge?

These and other fascinating unanswered questions await the results of ongoing research. One final note: An even earlier snowball Earth, dating from about 2.2 Bya and hypothesized on the basis of glacial deposits found in South Africa, also would have had profound impacts on life.[e] As in the Namibian case, these deposits were laid down in areas that paleomagnetic evidence confirms were tropical at the time.

[a] See Hoffman and Schrag (2000) for a review.
[b] See Harland (1964) and Harland and Rudwick (1964) for the initial hypothesis of snowball Earth.
[c] See Kirschvink (1992) and Kirschvink et al. (2000) for increasing CO_2 levels and melting.
[d] See Narbonne and Gehling (2003), and Kopp et al. (2005) for recent analyses of organismal survival and climate change at even earlier times, perhaps as far back as 2.2 By.
[e] See Evans et al. (1997) for snowball Earth 2.2 Bya.

of body plans provide perhaps the clearest indication of increase in complexity with evolution (McShea, 2005). Many of these features are associated with lifestyle; early *pelagic* metazoans swimming above the sea bottom mainly by ciliary motion became benthic animals crawling along the ocean floor and feeding on accumulated detritus.

A number of evolutionary steps would inevitably accompany a benthic existence. The scattered distribution of food sources would give a selective advantage to organisms that could eat more food more rapidly, leading to an increase in size and to evolution of a mouth and gut that would permit selective digestion (Bonner, 2006). Ciliary motion, by its nature slow and cumbersome for a large animal, would (as discussed in Box 16-3, Evolutionary Solutions to Problems of Locomotion) give way to leechlike and appendage-driven locomotion using circular and longitudinal muscles.

The increased success and proliferation of bottom feeders would open a niche for carnivorous animals that would emphasize speed of locomotion and development of a grasping mouth or other prehensile organs. Selection for increasingly efficient predation would lead to selection for better means of defense and escape in the prey, resulting in an arms race. Competition among all the different varieties of prey and predators would lead to an explosive evolutionary radiation, generating a large variety of morphological forms and adaptive strategies.

Some of the key features of metazoan evolution are shown in Figure 15-8, where they are plotted onto a metazoan phylogeny derived from studies comparing ribosomal RNA sequences (16S and 18S rRNA) among different phyla. The studies identify features that have arisen at progressively later stages of evolution, features that allow us to group related organisms together. Some such as collagen are basic characters (synapomorphies) of metazoans and so appear at the base of the tree. The next feature shown in Figure 15-8 is the **gastrula,** which is a universal animal embryonic stage in which a cup formed of two tissue layers (**ectoderm** and **endoderm**) encloses an inner digestive cavity called the archenteron (see Figs 15-11 and 15-12 later). The gastrula is found in all those invertebrates that develop from two germ layers (ectoderm and endoderm) and in those that possess the third embryonic germ layer, the **mesoderm** (**Fig. 15-8**).

Characters that evolved in metazoans with mesoderm (see Figs 15-11 and 15-12) include a blood vascular system

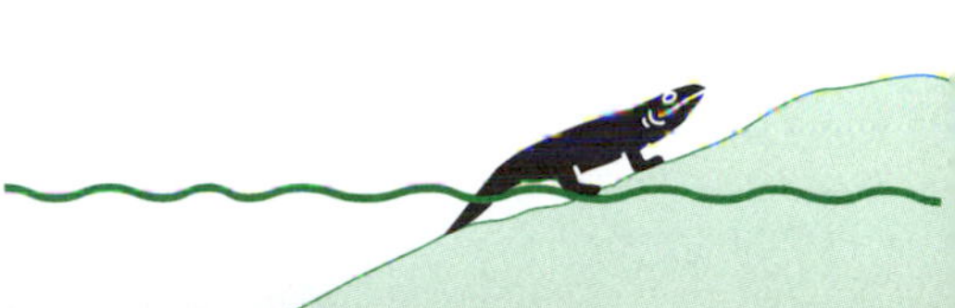

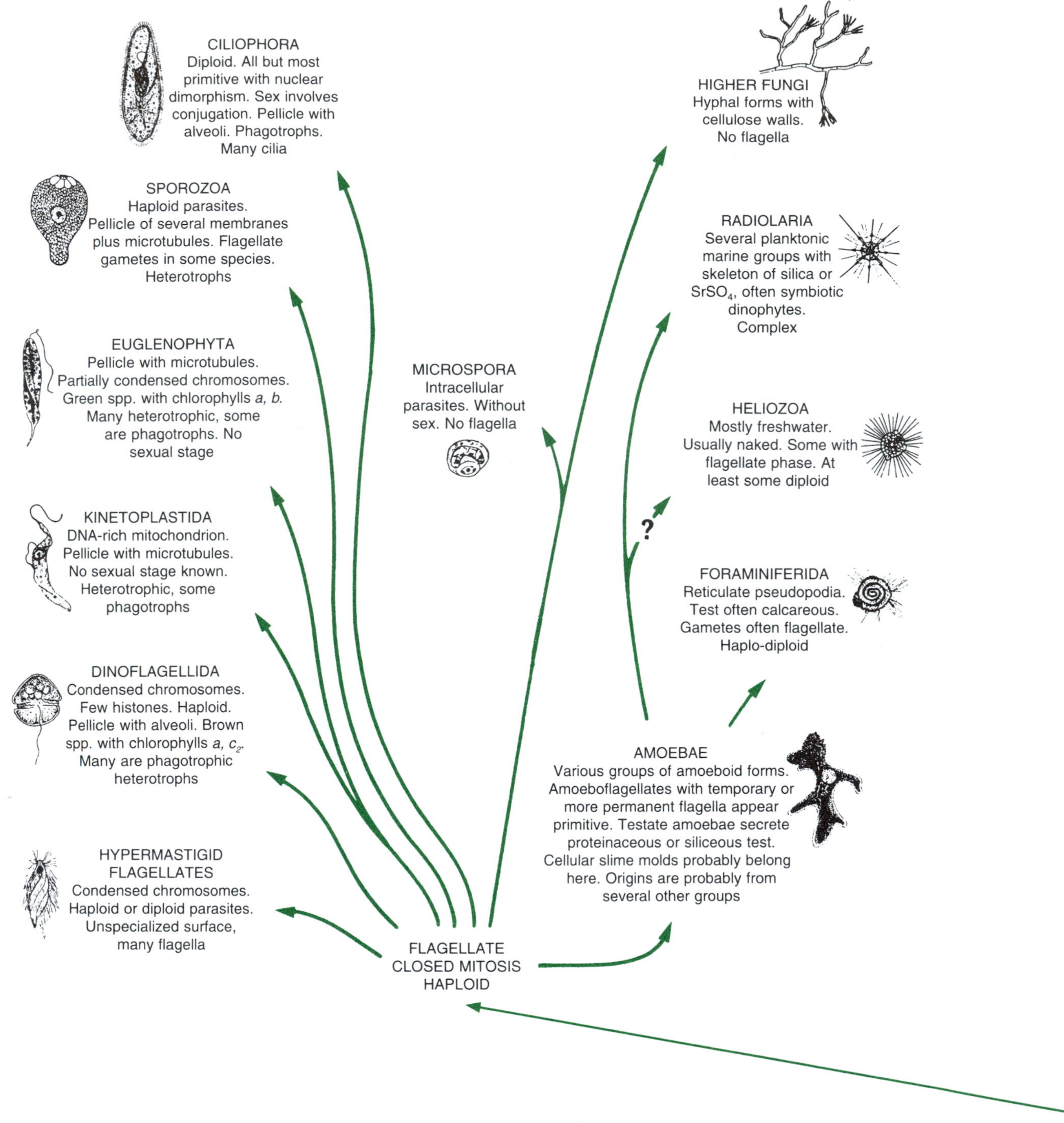
CILIOPHORA
Diploid. All but most primitive with nuclear dimorphism. Sex involves conjugation. Pellicle with alveoli. Phagotrophs. Many cilia
SPOROZOA
Haploid parasites. Pellicle of several membranes plus microtubules. Flagellate gametes in some species. Heterotrophs
EUGLENOPHYTA
Pellicle with microtubules. Partially condensed chromosomes. Green spp. with chlorophylls a, b. Many heterotrophic, some are phagotrophs. No sexual stage
KINETOPLASTIDA
DNA-rich mitochondrion. Pellicle with microtubules. No sexual stage known. Heterotrophic, some phagotrophs
DINOFLAGELLIDA
Condensed chromosomes. Few histones. Haploid. Pellicle with alveoli. Brown spp. with chlorophylls a, c_2. Many are phagotrophic heterotrophs
HYPERMASTIGID FLAGELLATES
Condensed chromosomes. Haploid or diploid parasites. Unspecialized surface, many flagella
MICROSPORA
Intracellular parasites. Without sex. No flagella
HIGHER FUNGI
Hyphal forms with cellulose walls. No flagella
RADIOLARIA
Several planktonic marine groups with skeleton of silica or $SrSO_4$, often symbiotic dinophytes. Complex
HELIOZOA
Mostly freshwater. Usually naked. Some with flagellate phase. At least some diploid
?
FORAMINIFERIDA
Reticulate pseudopodia. Test often calcareous. Gametes often flagellate. Haplo-diploid
AMOEBAE
Various groups of amoeboid forms. Amoeboflagellates with temporary or more permanent flagella appear primitive. Testate amoebae secrete proteinaceous or siliceous test. Cellular slime molds probably belong here. Origins are probably from several other groups
FLAGELLATE
CLOSED MITOSIS
HAPLOID

FIGURE 15-7 An early phylogeny for various protistan lineages leading to metazoans, fungi and land plants. How to decide between the many different classifications and the many possible phylogenetic relationships offered for those protistans has long been a subject of active research. Molecular data derived from nucleotide sequencing of ribosomal RNA and amino acid sequencing of various proteins are now helping resolve classifications as well as determining more precise relationships between protistans, metazoans and fungi (see, e.g, Figure 9-16). (Adapted from Sleigh., M. A., 1979. Radiation of the eukaryote Protista. In *The Origin of Major Invertebrate Groups.* M. R. House (ed.). Academic Press, London, pp. 23–54.)

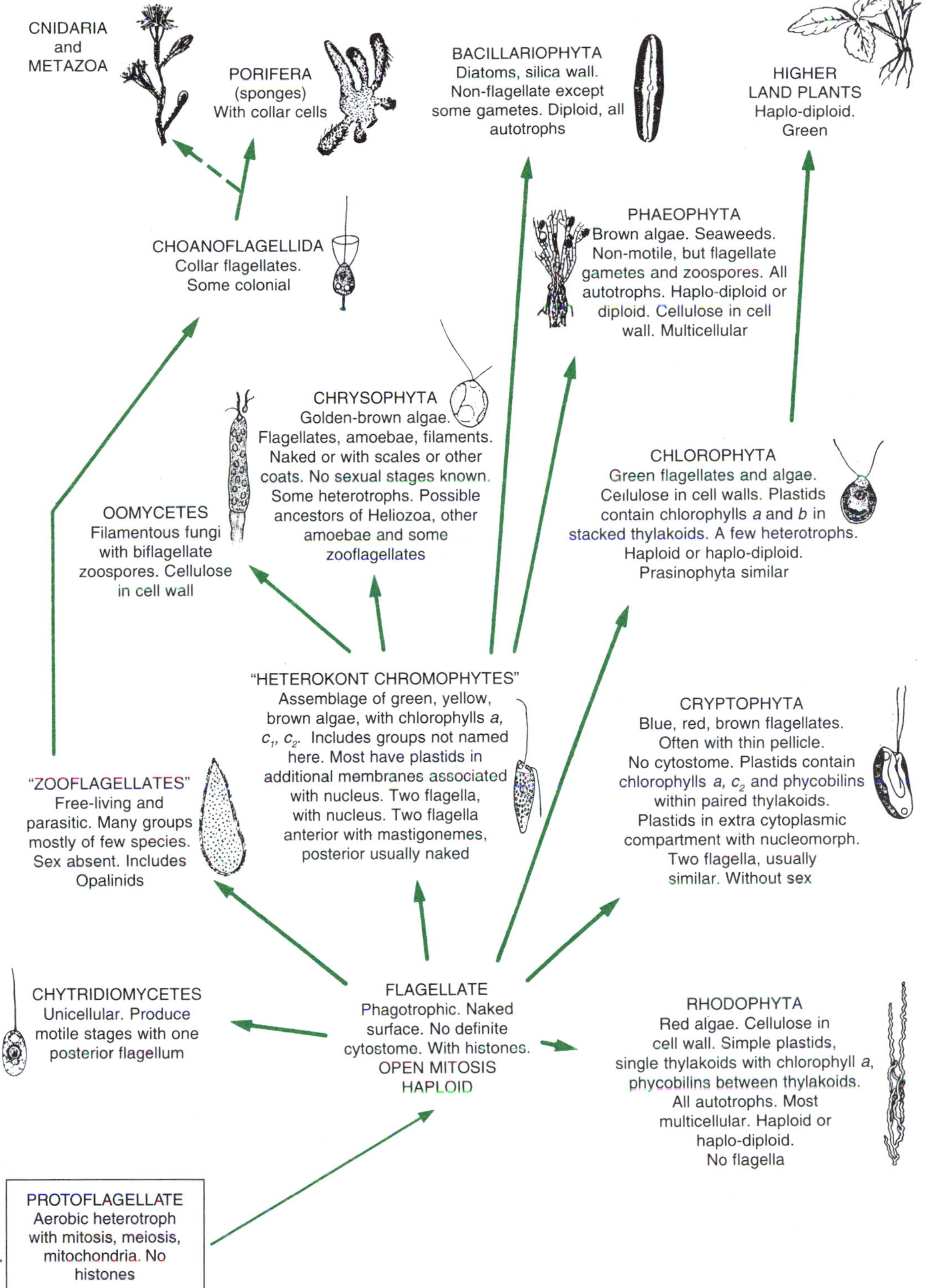
CNIDARIA
and
METAZOA
PORIFERA
(sponges)
With collar cells
BACILLARIOPHYTA
Diatoms, silica wall.
Non-flagellate except
some gametes. Diploid, all
autotrophs
HIGHER
LAND PLANTS
Haplo-diploid.
Green
CHOANOFLAGELLIDA
Collar flagellates.
Some colonial
PHAEOPHYTA
Brown algae. Seaweeds.
Non-motile, but flagellate
gametes and zoospores. All
autotrophs. Haplo-diploid or
diploid. Cellulose in cell
wall. Multicellular
CHRYSOPHYTA
Golden-brown algae.
Flagellates, amoebae, filaments.
Naked or with scales or other
coats. No sexual stages known.
Some heterotrophs. Possible
ancestors of Heliozoa, other
amoebae and some
zooflagellates
CHLOROPHYTA
Green flagellates and algae.
Cellulose in cell walls. Plastids
contain chlorophylls a and b in
stacked thylakoids. A few heterotrophs.
Haploid or haplo-diploid.
Prasinophyta similar
OOMYCETES
Filamentous fungi
with biflagellate
zoospores. Cellulose
in cell wall
"HETEROKONT CHROMOPHYTES"
Assemblage of green, yellow,
brown algae, with chlorophylls a,
c1, c2. Includes groups not named
here. Most have plastids in
additional membranes associated
with nucleus. Two flagella,
with nucleus. Two flagella
anterior with mastigonemes,
posterior usually naked
CRYPTOPHYTA
Blue, red, brown flagellates.
Often with thin pellicle.
No cytostome. Plastids contain
chlorophylls a, c2 and phycobilins
within paired thylakoids.
Plastids in extra cytoplasmic
compartment with nucleomorph.
Two flagella, usually
similar. Without sex
"ZOOFLAGELLATES"
Free-living and
parasitic. Many groups
mostly of few species.
Sex absent. Includes
Opalinids
CHYTRIDIOMYCETES
Unicellular. Produce
motile stages with one
posterior flagellum
FLAGELLATE
Phagotrophic. Naked
surface. No definite
cytostome. With histones.
OPEN MITOSIS
HAPLOID
RHODOPHYTA
Red algae. Cellulose in
cell wall. Simple plastids,
single thylakoids with chlorophyll a,
phycobilins between thylakoids.
All autotrophs. Most
multicellular. Haploid or
haplo-diploid.
No flagella
PROTOFLAGELLATE
Aerobic heterotroph
with mitosis, meiosis,
mitochondria. No
histones

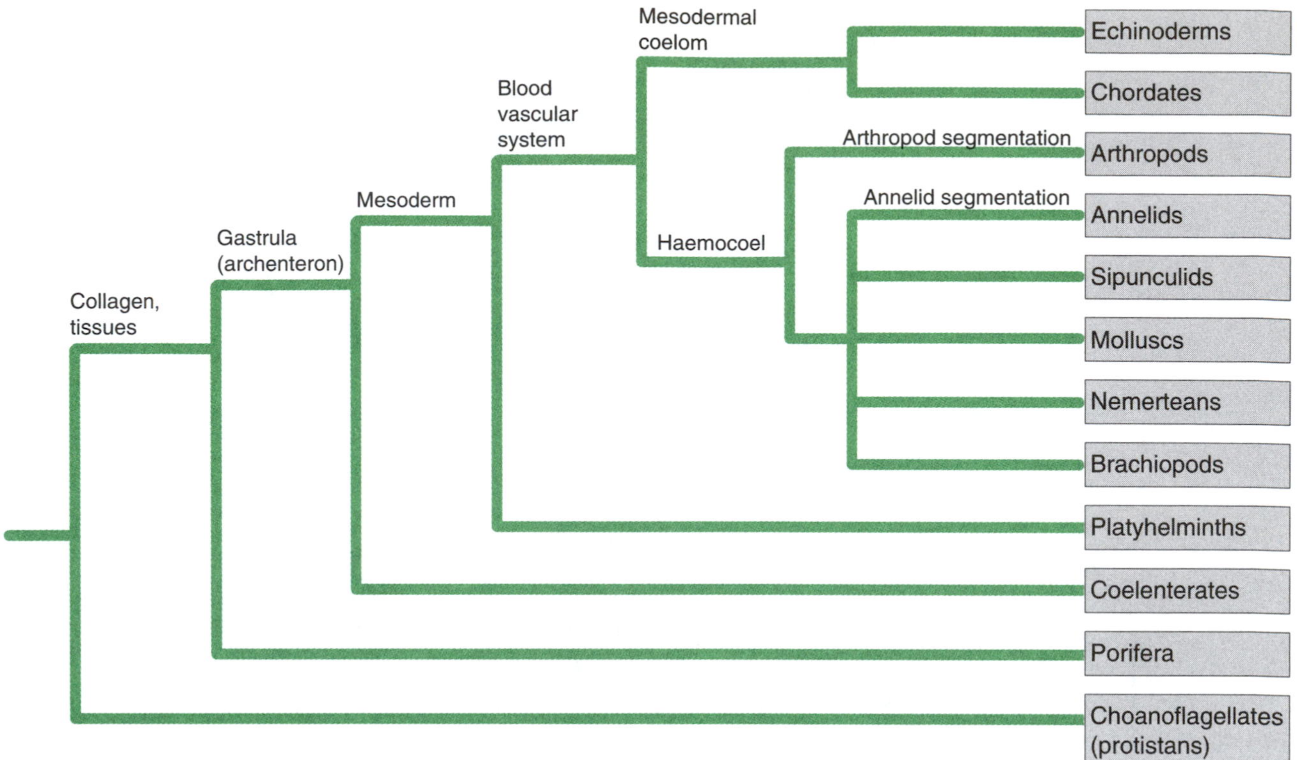

FIGURE 15-8 A metazoan phylogeny derived from a number of studies comparing ribosomal RNA sequences (16S as well as 18S components) among different phyla. These RNA molecules are essential for ribosomal protein synthesis, and, like 5S RNA, portions of their sequences are conserved across a very wide range of organisms. The time during metazoan evolution when features arose are shown. As described by Valentine et al. (1994), some of these innovative features correlate with changes in organismal complexity. Briefly: collagen is a distinctive fibrous protein used in animal connective tissue; the gastrula is the embryonic stage in which a cup formed of two tissue layers (ectoderm and endoderm) encloses an inner primitively digestive cavity, the archenteron; mesoderm is a third germ layer; blood vascular system is the vessels to transport blood fluids; mesodermal coelom is a fluid-filled cavity formed in mesoderm tissue; hemocoel is a fluid-filled cavity derived from the hollow interior (blastocoele) of the early embryonic blastula; annelid-type segmentation is where segments arise in a sequence along the anterior–posterior axis of the embryo; arthropod-type segmentation is where segments arising by successive splitting or doubling of primary units. We do not know the geological dates when these features appeared, but many, or even most, may have originated in the Precambrian. Although not fossil evidence, Bromham et al. (1998) support this view with molecular data showing that metazoans most probably diversified during an extended Precambrian period (see also Fig. 16-1). (Adapted from Valentine, J. W., 1994. Late Precambrian bilaterians: Grades and clades. *Proc. Nat. Acad. Sci.*, **91**, 6751–6757.)

and the elaboration of an internal cavity, either as a **hemocoel** by transformation of the interior cavity of the embryonic blastula, or as a **coelom** within the mesoderm. The last features identified in Figure 15-8 relate to subdivision of the body along the anterior–posterior axis of the embryo into compartments (**segments**) that arise in sequence along the anterior–posterior axis of the embryo, primarily in mesoderm. Although **metameric segmentation** occurs in both annelids and arthropods, it operates on different principles and is hypothesized to have evolved independently in the two lineages (Valentine, 2004; Valentine et al., 1994).

In Chapter 16 we discuss how elaboration of an internal body cavity, whether coelomic or not, enables us to group organisms into major types such as coelomate (having a coelom), acoelomate (lacking a coelom), or **pseudocoelomate** (having a false coelom or hemocoel).[7] We also discuss the effect of segmentation on body organs and on mode of locomotion.

As all these features are present in animals found in the Early Cambrian Burgess Shale, many or even most of them may have originated in the Precambrian. Bromham and coworkers (1998) support this view, which is based on morphological evidence, especially from embryonic development, with molecular data showing that metazoan phyla must have diversified during an extended period during the Precambrian (see also Figure 16-1).

Although past biologists used the diversity of metazoan features to indicate the likelihood of multiple (polyphyletic)

[7] See Freeman and Martindale (2002), and Martindale et al. (2002, 2004) for recent and insightful analyses on mesoderm and coelom development.

origins for metazoan multicellularity, molecular evidence has gone in the opposite direction, pointing to a close relationship between metazoans and protistan choanoflagellates (see Box 12-4; Fig. 15-8). Theories on how a unicellular or multicellular organism such as a choanoflagellate could have initiated more complex multicellularity as seen in metazoans are discussed in the next section.

Developmental Mechanisms

Although we recognize that all metazoans had a common ancestry, one of the most difficult questions in biology remains: How did unicellular ancestors transform into metazoan descendants?

Major difficulties confront those who seek the nature of the ancestor for any taxon. Many approaches involved searching for *features in extant taxa* (ideally taxa that are basal or least derived in the phylogenetic tree) that *are thought to have been present in the ancestor of that taxon and of related taxa.* But even the most basal member of any extant group (e.g., annelids) shows derived features that preclude using it as a surrogate for the ancestral state of, for example, the arthropods.

Research into metazoan origins has been active for more than a century, leading to various proposals, some receiving more attention than others. Three candidates for a unifying theory, all based on embryological evidence, are briefly described below. The rationale in all three has been to apply to evolutionary mechanisms our knowledge of animal embryonic development, in which the single celled zygote transforms into the **blastula** (the initial multicellular embryo), which then develops an internal cavity, as it becomes a gastrula. By now you should recognize this extrapolation of current mechanisms back in time as a familiar conceptual approach to evolutionary origins. It is the approach used in experiments on the origin of molecules and of the genetic code (see Chapters 6 and 7), and to test theories of how cells and cell division arose (see Chapter 8).

Cellularization of a Multinucleate Protozoan

A hypothesis first suggested in the middle of the nineteenth century, and that Hadzi (1963) developed further, proposes that some early multinucleate protozoans, bilaterally organized along the anterior–posterior axis, gave rise to the first "flatworms," organisms similar to platyhelminths (**Fig. 15-9**). Hadzi chose flatworms as a surrogate for the earliest metazoan ancestor because they lack a coelom (they are acoels or acoelomate; see Fig. 15-8), and do not show complete cellularization of their digestive tissues, or gut, nor of any other body cavity (Figs 15-8c and 16-6a, and see Chapter 16). Further basic features that seem to link them to protozoans are their small size (1 to 2 mm long), ciliated epidermis, ventral mouth, absence of excretory organs, intracellular

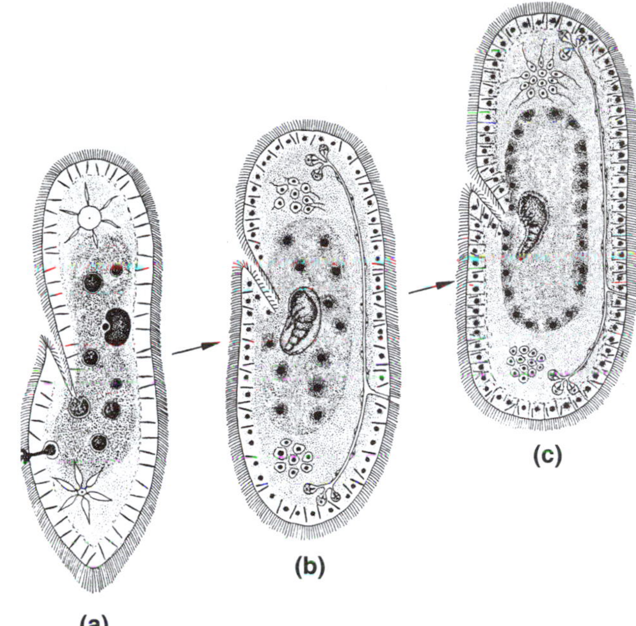

FIGURE 15-9 A hypothetical transformation of a ciliated paramecium-like protozoan (a) into a basic turbellarian metazoan (c). (Adapted from Hadzi, J., 1963. *The Evolution of the Metazoa.* Macmillan, New York.)

digestion and relatively little differentiation; for example, the cords of eggs and sperm lying side by side rather than being organized into ovaries or testes.

Because one fairly common mode of acoelan nutrition — dependence on internal symbiotic algae — is highly specialized, some zoologists disagreed that such animals were necessarily the most primitive bilaterally symmetrical metazoans (often referred to as *bilaterians*). Most invertebrate zoologists consider cnidarians (coelenterates) and ctenophores, which are **diploblastic** — that is, built from two germ layers, ecto- and endoderm — more basic and evolutionarily older than flatworms, which are **triploblastic,** that is, built from three germ layers, ecto- and endoderm plus mesoderm (see Figs. 15-8 and 15-12).

An attempt to salvage at least part of the hypothesis of protozoan-acoelomate evolution was the proposal that cnidarians may have originated from protozoans independently and that cnidarians and flatworms are therefore not directly related. The study by Field and coworkers (1988), comparing various regions of 18*S* rRNA molecules in a wide variety of eukaryotes (**Fig. 15-10**), seemed to support such a polyphyletic origin of metazoans, but see Figure 16-1. Their sequence comparisons show that cnidarians, fungi and plants derive from a distinctly separate protistan origin from other metazoans. However, even if we consider that all metazoans have a monophyletic origin (Fig. 15-8), acoelan flatworms may well represent basal bilateral triploblastic animals. According

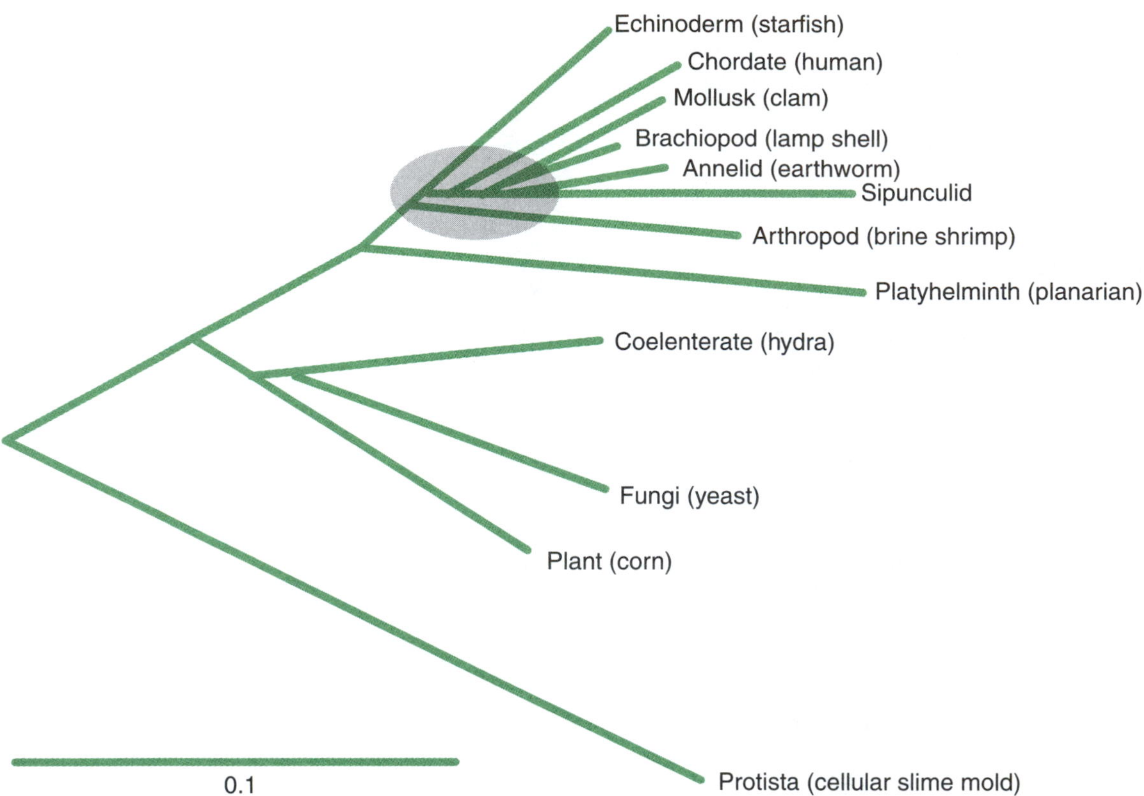

FIGURE 15-10 An evolutionary tree based on comparisons between 18S rRNA sequences from the different groups shown as determined in 1988; compare with Fig. 16-1, a 2002 tree. As indicated, the platyhelminths and later eucoelomate groups appear to have diverged from a protistan ancestry separate from the protistan ancestry of cnidarians, fungi and plants. Unfortunately, the data cannot resolve the exact branching order within the shaded ellipse, although it seems clear that these different groups all radiated relatively rapidly from what may well have been a common coelomate ancestor that was also segmented. The scale bar at the lower left represents an evolutionary distance calculated as 0.1 substitutions for each nucleotide position in the sequences compared (a total of more than 800 nucleotides). (Adapted from Field, K. G., G. J. Olsen, D. J. Lane, et al. 1988. Molecular phylogeny of the animal kingdom. *Science,* **239,** 748–753; Figs. 2 and 5.)

to molecular studies, acoelans, more than any other group, stand at the base of all triploblastic metazoans (see Chapter 16 for further discussion).[8] It is therefore difficult to accept Hadzi's proposal that an organism similar to a flatworm arose directly from a unicellular protozoan.

Gastrulation of a Colonial Protozoan

In the 1870s, Haeckel proposed that hollow-balled colonies of flagellated protozoans, not unlike the extant alga *Volvox* (see Figure 14-1) but lacking chloroplasts, developed an anterior–posterior axis. Ciliary action swept food particles in this basic blastula (a stage Haeckel proposed was recapitulated in the blastula embryonic stage of many metazoans) toward its posterior pole, and cells at that end specialized for digestive functions.

Haeckel and his followers claimed these digestive cells invaginated through a circular **blastopore** into the hollow interior of the organism to form an internal digestive tract or **archenteron** (**Fig. 15-11**). They proposed that this new bilayered, cuplike organism with ectoderm on the outside and endoderm on the inside was similar to one of the highly conserved developmental stages in some extant metazoa, the gastrula. The **Gastraea hypothesis** suggests that the primitive (basic) nature of sponges and cnidarians lies in their persistence at this diploblastic gastrula level (Figs. 15-11 and **15-12**).

A major difficulty with the Gastraea hypothesis is the finding that gastrulation by invagination as shown in Figure 15-11 is not common in the embryological development of many metazoans. Even in hydrozoan cnidarians, which Haeckel presumed exemplified the gastrula stage of evolution, endodermal tissues are formed by ectodermal cells that wander in from an intact epithelial surface, rather than by a cuplike folding process. To the extent that developmental

[8] See Wainright et al. (1993) and Borchiellini et al. (1998) for monophyly, and see Ruiz-Trillo et al. (1999) for the position of acoelans.

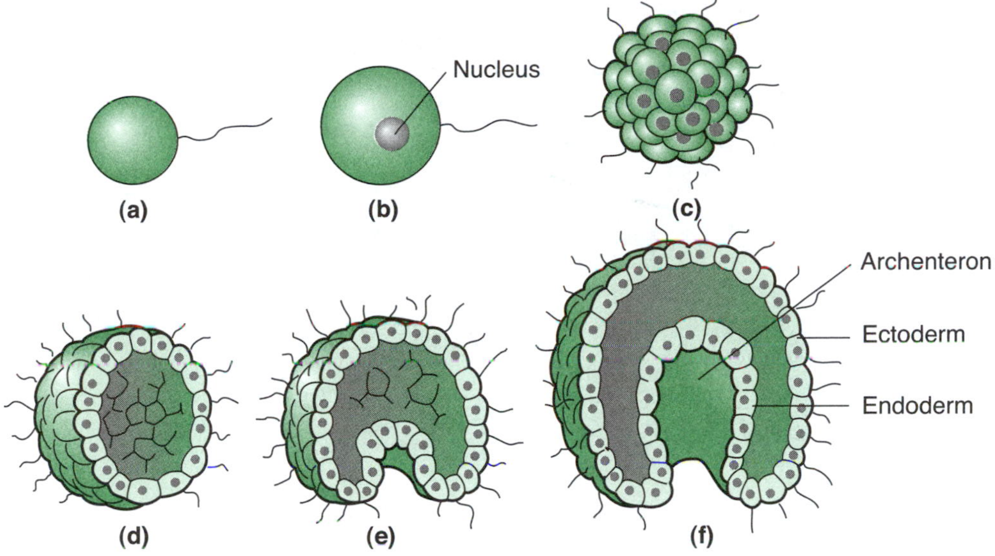

FIGURE 15-11 Stages in the evolution of a multicellular organism according to Haeckel (1874). The monerula (a) has no nucleus; the cytula (b) is nucleated; the morula in Haeckel's Gastraea hypothesis (c) is a compacted solid ball of cells; the blastula (d) is a hollow, single-layered cellular sphere; and the gastrula (e, f) is a bilayered organism with an exterior opening. (Adapted from Kerkut, G. A., 1960. *Implications of Evolution.* Pergamon Press, Oxford, England, based on Haeckel.)

FIGURE 15-12 Transformation (*solid arrows*) of a simple blastula (*left*) into a gastrula shown in longitudinal section (*middle column*) and in cross section (*right hand column*) to compare the developmental sequence in a diploblastic embryo *(upper row)* with that in a triploblastic embryo *(bottom row)*. These sequences also serve to show the phylogenetic transformation (*dashed arrow*) from a diploblastic to a triploblastic level of organization. Endoderm (*color*) invaginates to form a hollow gut, the archenteron (A). Ectoderm (*gray*) surrounds the outside of the embryo/organism. Mesoderm — represented as mesenchymal cells (*black*) — forms the third germ layer in triploblastic embryos/organisms. (Modified from Hall, B. K., 1999a. *Evolutionary Developmental Biology,* 2nd ed. Kluwer Academic Publishers, Dordrecht, The Netherlands.

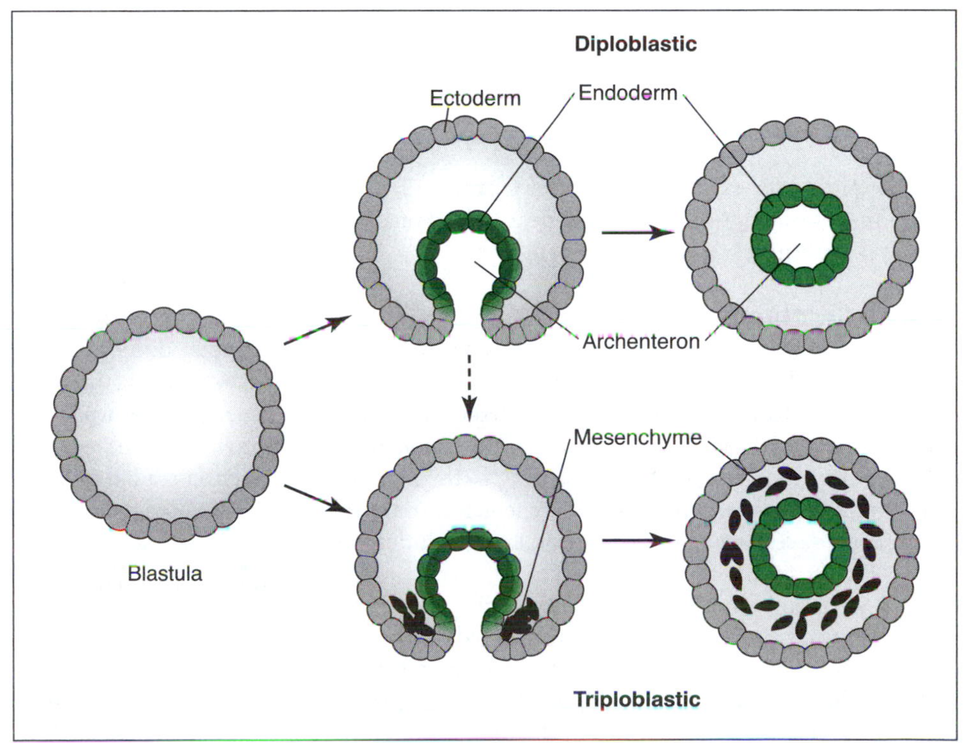

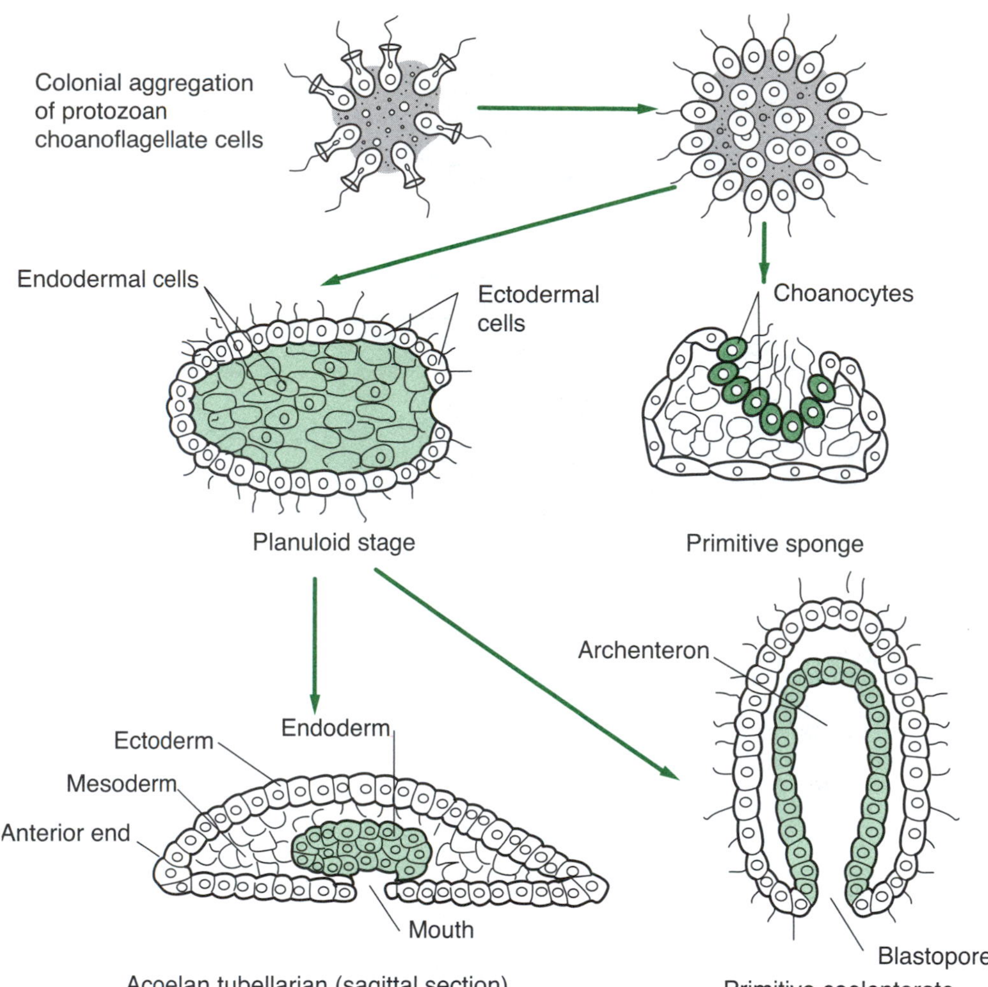

FIGURE 15-13 Illustration of one version of the planula hypothesis, beginning with a colonial choanoflagellate that evolved into planuloid organisms. According to this hypothesis, these, in turn, led to sponges, cnidarians and bilateral creeping turbellarians. (Adapted from Glaessner, M. F., 1984. *The Dawn of Animal Life*. Cambridge University Press, Cambridge, UK; from Ivanov.)

patterns are conserved in evolution (see Chapter 13), gastrulation by invagination does not appear to be a retained ancestral pattern among such cnidarians.

Planula Hypothesis

A third and more well supported hypothesis states that the blastula was followed not by gastrulation but by the formation of a solid ball of cells (a *planula*) in which the ectodermal cells became specialized for locomotion and the endodermal cells for digestion (**Fig. 15-13**). According to the **planula hypothesis,** the formation of a hollow archenteron and open blastopore would have occurred during later evolutionary stages (Fig. 15-13, upper right).

As Metschnikoff (1884) and others in the nineteenth century showed, many metazoans have no need of a mouth and digestive tube because digestion is phagocytic and intracellular. The finding of planula-type larvae in metazoans such as sponges and cnidarians and the observation that various groups among the platyhelminths have a solid gut filled with endodermal cells indicates that planula-type organisms would have been viable.

A proposal by Collins (1998) based on extensive analysis of 18*S* ribosomal RNA sequence suggests that planula-type larvae produced by ancestral cnidarians became reproductive *before* reaching the adult stage. This developmental change, called **paedomorphosis** (the incorporation of adult features into immature stages), enabled these larvae to depart from cnidarian radial development, elaborate mesodermal tissue and become bilaterally symmetrical (see Fig. 16-1).

Concluding Comments

Discovery of early fossils gives us some hope of getting closer to ancestral conditions, although even the Cambrian fauna is not basal, being composed of derived (crown) taxa. Early Cambrian organisms can be placed into many clades at the phylum-level,

James W. Valentine

Birthday: November 10, 1926
Birthplace: Los Angeles, CA
Undergraduate degree: Phillips University, Enid, Oklahoma, 1951
Graduate degrees: M.A. University of California at Los Angeles, 1954; Ph.D. University of California at Los Angeles, 1958
Present position: Professor of Integrated Biology Emeritus and Curator, Museum of Palentology University of California at Berkeley[9]

What prompted your initial interest in evolution?
I was a World War II G.I. who had not planned on going to college but who took advantage of the G.I. Bill for Education, choosing geology rather blindly out of a long list of majors available at first enrollment. Paleontology was the most interesting course to me, and on reading *Tempo and Mode in Evolution* by George Gaylord Simpson, I realized that the fossil record could be used as evidence in formulating hypotheses about evolutionary processes. Combined with data from genetics, zoology, botany and other life sciences, it was clear that paleontology could make important contributions to evolutionary theory.

What do you think has been most valuable or interesting among the discoveries you have made in science?
The most fun has been not so much to make discoveries as to pursue an approach: to attempt to treat the fossil record as an ecological theater, to use G. Evelyn Hutchinson's metaphor. There has been lots of pleasure in looking for and finding clues that reveal the processes that produced the evolutionary play.

What areas of research are you (or your laboratory) presently engaged in?
I am now most interested in the "Cambrian explosion," when the remains of animals with the body plans of many living phyla first appeared during a 9- to 10-million year period beginning about 530 Mya. This is one of the most spectacular events recorded by fossils, and must have involved extensive early evolution of metazoan developmental systems, but is so unique and so remote in time that it has been difficult to interpret. Dating of those ancient fossil assemblages by geophysical laboratories with new, extremely accurate techniques, and an integration of the fossil evidence with findings from molecular biology laboratories, has begun to clarify the fine structure of the evolutionary events leading to the explosion. In my laboratory we combine observations of living organisms with molecular studies to model the sorts of evolutionary pathways that were required to produce such a broad array of body plans.

In which directions do you think future work in your field needs to be done?
Evolutionary hypotheses generated by the fossil record need to be tested by the tools of molecular biology; these seemingly disparate fields have much to offer each other.

What advice would you offer to students who are interested in a career in your field of evolution?
Aside from the obvious need to master the basics of paleobiology and appropriate ancillary subjects, there are three bits of advice that seem particularly valuable. One is to question authority, a generally good idea in any area. A second is to contrast notions prevalent in different fields, because transfer of approaches from one field to another can lead to most creative results, and because contradictions between fields can lead to new insights as well. And finally, as you become interested in research, do as Peter Medawar has suggested and attack the most important problem you think you can solve, or at least advance.

[9] A list of published articles is available on http://www.ucmp.berkeley.edu/people/jwv/jwv.html. Accessed April 24, 2007. Recent publication: Valentine, J. W. 2004. *On the Origin of Phyla*. The University of Chicago Press, Chicago.

EVOLUTION ON THE WEB biology.jbpub.com/book/evolution

indicating that their origins were Precambrian. Furthermore, the discovery of even earlier wormlike burrows provides physical evidence for some of the features of Precambrian animals (Seilacher et al., 1998; Valentine, 2004).

Cladistic approaches (see Chapter 11) are advocated and used by many to establish relationships among metazoans to enable more derived features to be separated from more basal features (see Jenner, 2005, 2006, for recent and provocative analyses). Phylogenetic trees based on molecular characters provide a means of establishing phylogeny independent of morphology, which can then be mapped onto the tree (Fig. 15-8). Reconstruction of ancestral states using a combination of morphological and molecular characters provides further scenarios that can be tested as new data arise.

Two of the most informative approaches are those that reveal the genetic and cellular basis of development (including segmentation) in different metazoans, either through analysis of gene expression or by the use of natural or induced mutations. As discussed in Chapter 13, specific genes exercise control over segmental patterns, segmental borders and the ability of segmental tissues to differentiate into particular structures such as wings and legs. Moreover, genes that govern these and other developmental patterns in a wide range of segmented and nonsegmented animals share homologous nucleotide sequences, similar linkage orders and even similar developmental targets. Genetic analysis of development is providing a deep understanding of the types of developmental changes that occur, how such changes function, and which changes may be selected and transmitted during evolution (see Chapter 13). The four chapters that follow explore evolutionary changes that have taken place in animals over the past 600 or more million years.

KEY TERMS

archenteron
benthic
bilaterally symmetry
blastopore
blastula
Cambrian explosion
carnivores
coelom
cropping
diploblastic
ectoderm
Ediacaran fauna
endoderm
gastrula
Gastraea hypothesis
herbivores
mesoderm
metazoans
pelagic
planula hypothesis
Protista
pseudocoelomate
segments
triploblastic

DISCUSSION QUESTIONS

1. How do paleobiologists use the principle of "cropping" to explain the proliferation of multicellular animals during the time around the Cambrian?
2. Would you classify all Precambrian multicellular organisms into the same phyla used for classifying organisms from the Cambrian onward?
3. Metazoan origins.
   a. What are the advantages of animal multicellularity?
   b. What is the evidence, pro and con, for the various hypotheses of metazoan origin?
4. How would the development of a coelomic cavity have influenced the evolution of locomotion in wormlike organisms?
5. What advantages does segmentation offer to metazoans?
6. Why don't we regard extant taxa as ancestors of other extant taxa?
7. Are recent molecular studies on metazoan relationships consistent with any of the three theories of metazoan origins based on developmental mechanisms?

EVOLUTION ON THE WEB

Explore evolution on the Internet! Visit the accompanying Web site for *Strickberger's Evolution*, Fourth Edition, at http://www.biology.jbpub.com/book/evolution for Web exercises and links relating to topics covered in this chapter.

RECOMMENDED READING

Ayala, F. J., A. Rzhetsky, and F. J. Ayala, 1998. Origin of the metazoan phyla: Molecular clocks confirm paleontological estimates. *Proc. Natl. Acad. Sci. U.S.A.*, **95**, 606–611.

Bonner, J. T., 2006. *Why Size Matters: From Bacteria to Blue Whales.* Princeton University Press, Princeton, NJ.

Brusca, R., and G. Brusca, 2003. *Invertebrates,* 2nd ed. Sinauer Associates, Inc., Sunderland, MA.

Cavalier-Smith, T., 2004. Only six kingdoms of life. *Proc. Roy. Soc. Lond. B. Sci.* **271**, 1251–1262.

Clapham, M. E., G. M. Narbonne, and J. G. Gehling, 2003, Paleoecology of the oldest-known animal communities: Ediacaran assemblages at Mistaken Point, Newfoundland. *Paleobiology* **29**, 527–544.

Erwin, D. H. and E. H. Davidson, 2002. The last common bilaterian ancestor. *Development,* **129**, 3021–3032.

Hall, B. K., 1999a. *Evolutionary Developmental Biology,* 2nd ed. Kluwer Academic Publishers, Dordrecht, Netherlands.

Hoffman, P. F., A. J. Kaufman, G. P. Halverson, and D. P. Schrag, 1998. A Neoproterozoic snowball Earth. *Science,* **281**, 1342–1346.

Hou, X-G., R. J. Aldridge, J. Bergström, D. J. Siveter, and X-H. Feng, 2004. *The Cambrian Fossils of Chengjiang, China: The Flowering of Early Animal Life.* Blackwell Science, Oxford, UK.

Levinton, J., 2001. *Genetics, Paleontology, and Macroevolution.* 2nd ed. Cambridge University Press, Cambridge, England.

Nielsen, C., 2001. *Animal Evolution: Interrelationships of the Living Phyla,* 2nd ed. Oxford University Press, Oxford, UK.

Schulter, D., 2000. *The Ecology of Adaptive Radiation.* Oxford University Press, Oxford, UK.

Valentine, J. W., 2004. *On the Origin on Phyla.* The University of Chicago Press, Chicago.

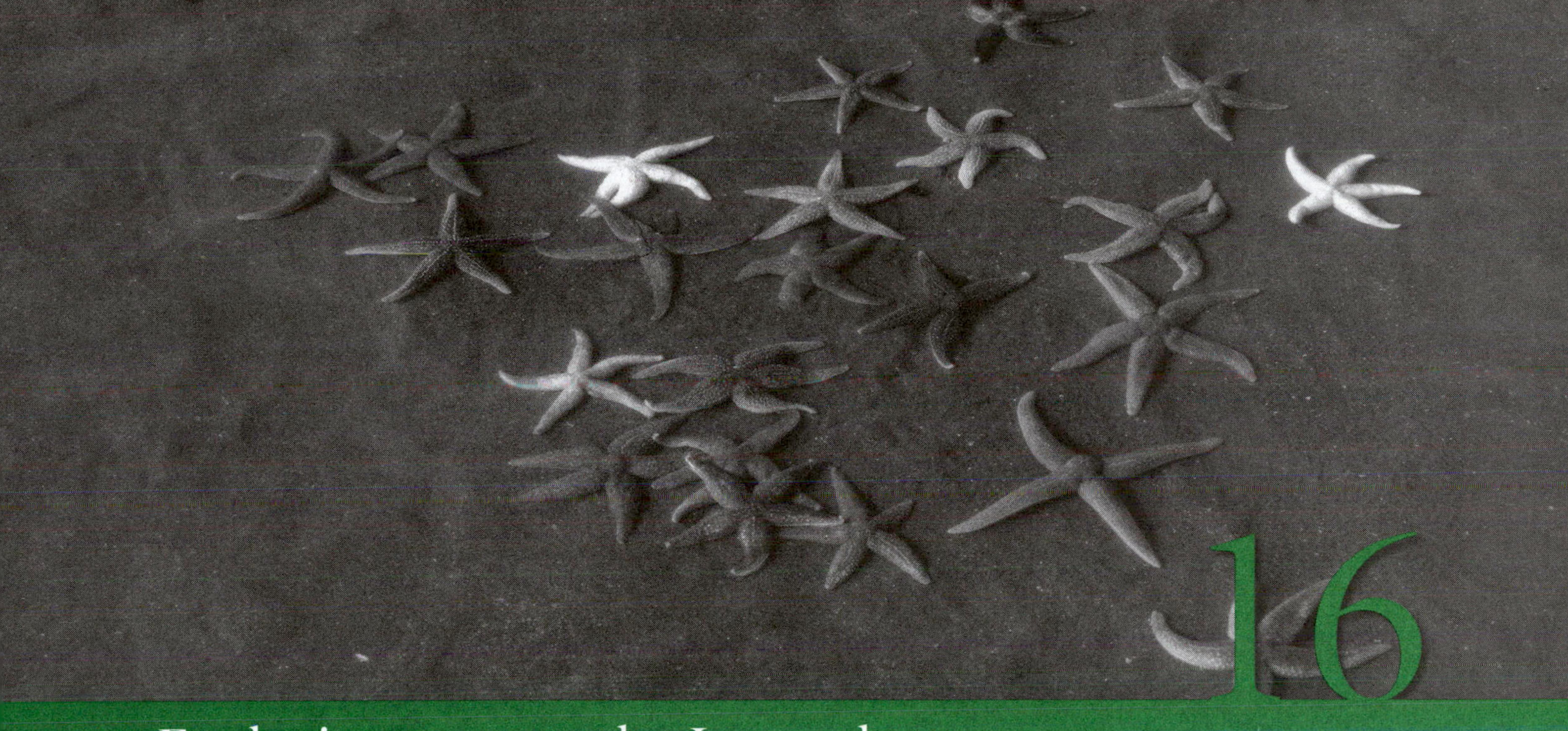

Evolution among the Invertebrates

Chapter Summary

Metazoans evolved many complex body plans that, when combined with embryological features, form the basis for invertebrate taxonomy. A number of themes, reflecting solutions to common problems, are seen in these body plans: the presence or absence of tissues or organs, body symmetry, segmentation, the presence of an internal body cavity and its mode of origin and evolution.

Porifera (sponges), Mesozoa and Placozoa are the least complex multicellular animals, with a variety of cell types, outer and inner layers on either side of a gelatinous mesohyl, but according to traditional views, no true tissues or organs.

Next in what we recognize as complexity are organisms with radial symmetry, the Cnidaria and Ctenophora, each with a digestive cavity and more obvious tissue organization than sponges. Characteristic of cnidarians are the planula larva, specialized stinging cells or nematocysts, and two body forms, the polyp and the medusa, both of which can occur in the life cycle of a single species. In other invertebrates, a third layer — the mesoderm — permitted the development of complex organ systems and facilitated an increase in size and the appearance of bilateral symmetry. Platyhelminths (flatworms) have complex reproductive systems as well as simple osmoregulatory, nervous and digestive systems.

Although Aschelminthes is no longer regarded as monophyletic, many organisms subsumed under this umbrella share characteristics such as a partially mesodermal body cavity (the pseudocoele), an exterior cuticle and a digestive tract with mouth and anus. Layers of circular and longitudinal muscle acting against the fluid-filled cavity provide a hydraulic skeleton.

All other invertebrates have a mesoderm-lined body cavity, the coelom or body cavity, whose evolutionary origin(s) is (are) uncertain, and that acts in many invertebrates as a hydraulic skeleton. Some animals have a body divided into segments (metameres) in which many organ systems serially repeat, although few animals are completely metameric, presumably because a set of organs had to be produced for each segment. The evolutionary source(s) of segmentation remains unknown; advantages range from ease of organ replacement if segments are lost to enhanced flexibility in swimming. Further adaptations, especially in hard-bodied forms, led to changes in body plan, including modification of the coelom and segments in most animal clades.

Zoologists divide coelomates into two major groups mainly on embryological criteria. Deuterostomes are characterized by development of the blastopore into the anus, a coelom formed by outpocketing of the mesoderm, and radial, indeterminate cleavage. In contrast, in protostomes, the blastopore becomes the mouth, the coelom forms within the mesoderm, and cleavage is spiral and determinate. Certain coelomates show some degree of metamerism, and some capture food by means of ciliated arms called lophophores. Two major lineages of protostomes are now recognized — those that molt (arthropods, nematodes and others), the Ecdysoza, and those with a tentacle (mollusks, annelids and others), Lophotrochozoa.

Mollusks are a diverse assemblage of mostly shelled coelomates specialized for creeping on a muscular foot and covered by a fold of the body wall, the mantle. These structures have been much modified in the different mollusks; cephalopods have lost the foot and most of the shell and become specialists at propulsive swimming. Molecular evidence suggests that mollusks, annelids and arthropods may have arisen from the same ancestral lineage.

In annelid worms, septa divide the coelom into compartments, a segmentation that allows peristalsis and an active burrowing lifestyle. Polychaetes have developed lateral appendages, or parapodia, for swimming, but their soft bodies are not effective foils for limb muscles.

In arthropods, the exoskeleton provides a firm attachment for muscles, with concomitant reduction of the coelom and in internal segmentation. Arthropods evolved into many diversified and enormously successful lineages — insects in particular — and have invaded most terrestrial habitats, although their exoskeleton and dependence on tracheal gaseous exchange limit their size.

Echinoderms are deuterostomes with a mesodermal skeleton and a water vascular system for movement and circulation. Although they have retained many of the characters they inherited from a free-living bilateral ancestry — rudimentary nervous system, isosmolarity with seawater, bilaterally symmetrical larvae — as adults they are radially symmetrical with pentamerous organization.

Ninety-nine percent or more of all extant Metazoan species are **invertebrates**, a polyphyletic assemblage that includes jellyfish, worms, squid, starfish, shrimp, flies and myriad others. The term *invertebrate* has been used on numerous occasions in the text without comment. A comment is called for at this stage.

We all know what invertebrates are; they are animals that lack vertebrae, that is, they are not vertebrates. This was seen as an entirely appropriate way to name organisms in the mid-nineteenth century. Nowadays, this term, along with terms such as *worms, fish(es), amphibians, reptiles,* can usefully be used to refer to animals in a way that will immediately conjure up an image of the animal in the mind of the reader, **as long as we realize that the terms are not being used for organisms that comprise monophyletic lineages.** Indeed, some would say that the terms should be abandoned for this very reason, although major textbooks such as those by Brusca and Brusca (2003) and Ruppert et al. (2004) are simply named *Invertebrates* and *Invertebrate Zoology,* respectively. We also have retained invertebrates as a collective term recognized by many, but remember this is not a term for a group of organisms that make up a monophyletic lineage.

Features and Origins

Some of the basic features of animals (metazoans) were outlined in Chapter 15 and plotted against a molecular phylogeny in Figure 15-8. Therefore, you should now be familiar with terms such as germ layers, ectoderm, endoderm, mesoderm, coelom and segmentation; each will be discussed further as we examine invertebrates in this chapter.

All animals are metazoans and all metazoans are animals. *Bilateria* is a collective name for those animals (1) with bilateral symmetry, (2) that develop from embryos with three germ layers and (3) in most cases, possess a body cavity derived from the coelom. Such a broad grouping encompasses the majority of animals, except for those animals whose embryos have only two germ layers, both those with tissues (cnidarians and ctenophores) and those without (placozoans, mesozoans and sponges). Bilateria includes echinoderms, which, although radially symmetrical as adults, are bilaterally symmetrical as larvae. All bilaterians have a concentration of sensory organs at the anterior end, defining a head.

Aside from parasitic invertebrates, which tended to become smaller and less differentiated during evolution, free-living invertebrates became more complex (as defined in Box 8-3, Complexity), able to adapt, and survive and thrive in a wider and wider range of environments. Such adaptations involve many organs that act in concert and form part of the architectural framework, or ***Baupläne*** (**body plan**), the origins of which were discussed in the previous chapter. Differences in body plans provide the basis for organizing extant invertebrates into around 36 clades (phyla), mostly on the basis of shared developmental and morphological features. We say "around" 36 because different experts recognize different numbers of phyla (see Box 8-2, Kingdoms and Domains of Life).[1]

As introduced in Chapter 15 and as discussed below, the lack of a **coelom** separates Platyhelminthes from many other animals. Presence or absence of a coelom reflects the elaboration of an internal body cavity and, subsequently, subdivision of that cavity, associated in many lineages with **segmentation** of the entire body. Similarly, the radial organization of the Cnidaria separates them from bilaterally organized animals (Bilateria), **bilateral** (left-right) **symmetry** reflecting a major change associated with and in part responsible for changing modes of locomotion and life in different habitats. On the embryological level, as described later, the fate of the blastopore (whether it becomes mouth or anus) separates protostomes and deuterostomes (**super phyla** in Linnean classification). Analysis of the elaboration and diversification of these features is a major aim of this chapter. Furthermore, these features have been used to classify metazoans. A well-recognized classification is shown in **Table 16-1.**

[1]For approaches to determining the number of phyla, see Hall (1999a), Nielsen (2003) and Valentine (2004).

From an evolutionary point of view, these and other morphological and developmental differences are evolutionary changes reflecting the ancestry of these organisms. **Figure 16-1** provides only one of a sampling of proposals offered to show evolutionary relationships. As shown in the lower half of this figure, the ancestral lineages that gave rise to the different phyla, date to Precambrian times, perhaps as early as 750 Mya. However, sharp morphological distinctions among these lineages did not appear, or may not have even developed, until the Cambrian (see Chapter 15).

As in many other instances, the lack of success in discovering intermediate forms between one clade (phylum) and another stems from the fact that many transitional events occurred in Precambrian times among small, soft-bodied organisms that fossilized poorly, if at all (Fig. 16-1). Moreover, different clades represent different major types of organization, members of each phylum often using a new mechanism for obtaining food, and having distinct metabolic needs, reproductive modes, and so on. Their origin would have been confined to small populations undergoing genotypic and morphological changes, with little or no chance that they would become fossilized.

As they arise, newer forms are subject to considerable competitive disadvantage. The scarcity of most transitional forms, either among extant species — where we might not expect to find them for reasons outlined in Chapter 11 — or in the fossil record, reflects their typical short-term survival and helps explain the relatively wide evolutionary gaps between major lineages that occupy different adaptive zones.

Some workers have made claims of worm-like burrows dating from 1.1 Bya. Over the past decade eggs and metazoan embryos have been discovered in Early Cambrian strata. While not numerous, these specimens have already given us a glimpse of cnidarian development 530 Mya, and of segments developing in what may be a protostome embryo. The absence of any larval stages as fossils strongly suggests that early metazoans had direct development as opposed to indirect development with a larval stage, although this issue is not resolved.[2]

Molecular studies based on differently calibrated evolutionary "clocks" provide data suggesting that protostomes and deuterostomes diverged some 750 Mya. In 1997, Aguinaldo and coworkers divided protostomes into **Ecdysoza,** which molt (e.g, arthropods, priapulids, nematodes), and **Lophotrochozoa,** which have tentacles (e.g, mollusks, annelids, brachiopods; see Box 16-1).

In Collins' (1998) scheme, a cnidarian-type planula larva (see Chapter 15) bearing a layer of mesodermal tissue became prematurely adult (*paedomorphosis*), giving rise to

[2] See Bengtson and Yue (1997) and Conway Morris (1998c) for embryos from the Early Cambrian, and see Dong et al. (2004) for embryos from the Middle and Late Cambrian.

TABLE 16-1 A scheme for differentiating metazoans according to morphological and developmental criteria

Criterion	Phylum	Number of Species
I. Differentiated tissues and organs poorly defined or absent		
	Porifera	>8,000
	Placozoa	1
	Mesozoa[a]	50
II. Differentiated tissues and organs		
A. *Radially symmetrical*	Cnidaria	10,000
B. *Bilaterally symmetrical*[b]	Ctenophora	100
1. Acoelomates	Platyhelminthes	24,000
	Gnathostomulida	100
2. Pseudocoelomates[c]	Gastrotricha	500
	Rotifera	> 1,800
	Acanthocephala	1,150
	Nematoda[d]	>200,000
	Nematomorpha	325
	Kinorhyncha	150
	Priapulida	9
	Loricifera	10
3. Uncertain affinity		
4. Coelomates		
a. Protostomes		
i. With lophophore (tentacled food gathering crown)[e]	Bryozoa	5,000
	Entoprocta	150
	Phoronida	14
	Brachiopoda	350
ii. Without lophophore		
(a) Nonmetameric or pseudometameric organization		
	Mollusca	70,000
	Sipunculida	300
	Nemertea	1,150
	Tardigrada	>500
	Pentastomida	100
(b) Metameric organization		
	Annelida	14,000
	Pogonophora	100
	Echiura	150
	Onychophora	>10
	Arthropoda	1,113,000
b. Deuterostomes	Echinodermata	6,000
	Hemichordata, Pterobranchia, Enteropneusta	92
	Chordata	52,000

[a] Although this classification of phyla is common, some zoologists disagree (see Box 8-1, Kingdoms and Domains of Life). Some minor clades considered "enigmatic" are not included, and the status of some others is not fully resolved (see Nielsen, 2001). One proposal suggests that mesozoans be divided into two separate phyla (see Fig. 16-6).

[b] Based on 18S ribosomal RNA sequence analysis and the discovery of triploblastic homeobox-containing (*Hox*) genes, the Myxozoa, formerly considered protistan, has now been placed among the metazoans, as parasitic hydrozoans.

[c] Some invertebrate biologists group these as Aschelminthes.

[d] Some estimates place the number of species of nematodes at 500,000, most of which are undescribed and unnamed.

[e] See Box 16-1 for a proposal to unite all protostomes into two monophyletic super groups, the Lophotrochozoa and the Ecdysozoa.

Source: Adapted from Lutz, P. E., 1986. *Invertebrate Zoology.* Addison-Wesley, Reading, MA. Species numbers are from Hall (1999a) and Ruppert et al. (2004).

FIGURE 16-1 Major metazoan phyla arranged according to one phylogenetic scheme, along with an illustrated sample species from each. Colored lines represent fossil lineages that current evidence indicates extended, in most cases, to the Early Cambrian. In older strata are found Ediacaran fossils and some others of questionable association (see Figs 15-4 and 15-5). The striking divergence among lineages, already obvious in the Cambrian, indicates considerable Precambrian history with possible relationships diagrammed in the lower part of the figure. (Linear distances connecting Precambrian lineages are not scaled chronologically.) (Adapted from Fortey, R., 2002. *Fossils: The Key to the Past.* Natural History Museum, London, UK.)

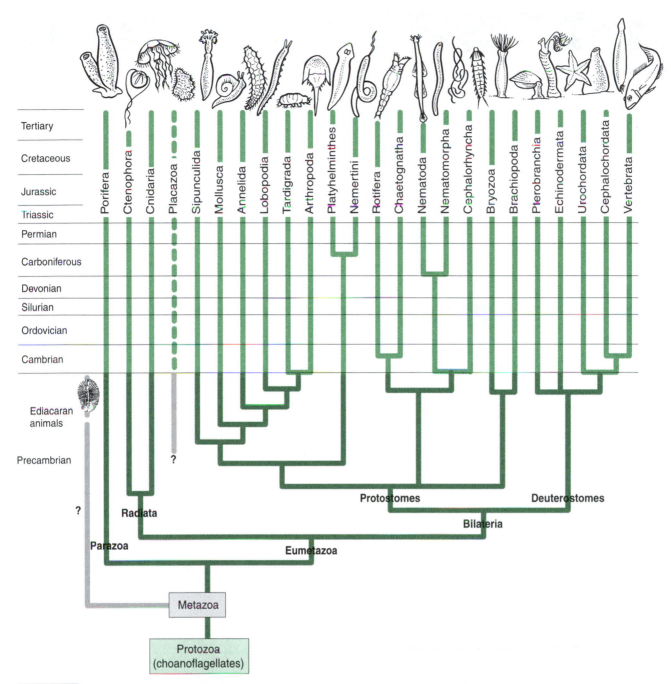

new triploblastic organisms. Supporting this view are findings using 18S ribosomal RNA sequences, suggesting that acoelan flatworms, usually classified within platyhelminths, are even more ancient than platyhelminths, and represent the earliest triploblastic (three germ layers) metazoans. Valentine (1997), who offers a phylogeny based on sequence analysis of the small subunits of ribosomal RNA and on patterns of embryonic cleavage having evolved a minimal

BOX 16-1 Ecdysozoa and Lophotrochozoa

In 1997, on the basis of data from 18S rRNA, Aguinaldo and colleagues (1997) proposed that *protostomes* could be fitted into just two monophyletic supergroups, the Ecdysozoa, which includes arthropods, rotifers, nematodes, onychophorans and tardigrades, and Lophotrochozoa, which includes annelids, mollusks and brachiopods.

Ecdysozoans possess a *cuticular* skeleton and undergo metamorphosis (*ecdysis*), the process by which the cuticle is shed under the control of steroid hormones, the *ecdysteroids* — hence the name for the supergroup. Many ecdysozoans possess separate sexes, with males depositing sperm inside the bodies of the females.

Lophotrochozoans possess a *trochophore larva* (**Figs. 16-2 and 16-3**) and a feeding structure known as a *lophophore* (although not all lophotrochozoans have both a trochophore larval stage and a lophophore as adults). They reproduce by releasing games into an aquatic environment.[a]

It took some time for this scheme to be accepted. In part this reflects differing interpretations of skilled taxonomists. Using morphological evidence from extant and extinct organisms, Nielsen (2003) proposed to deal with difficulties relating to the placement of the arthropods, by regarding Ecdysozoa as the sister group to annelids, but acknowledged that molecular data are inconclusive on this grouping. In part, it reflected other genetic analyses that failed to support the two supergroups. However, further and large-scale analyses — for example, an analysis by Hervé and colleagues (2005) involving 35 species, 146 genes and 35,371 positions — strongly support a single origin for (monophyly of) the two supergroups, which now are generally accepted.

[a] See Brusca and Brusca (2003), Halanych (2004) and Ruppert et al. (2004) for details.

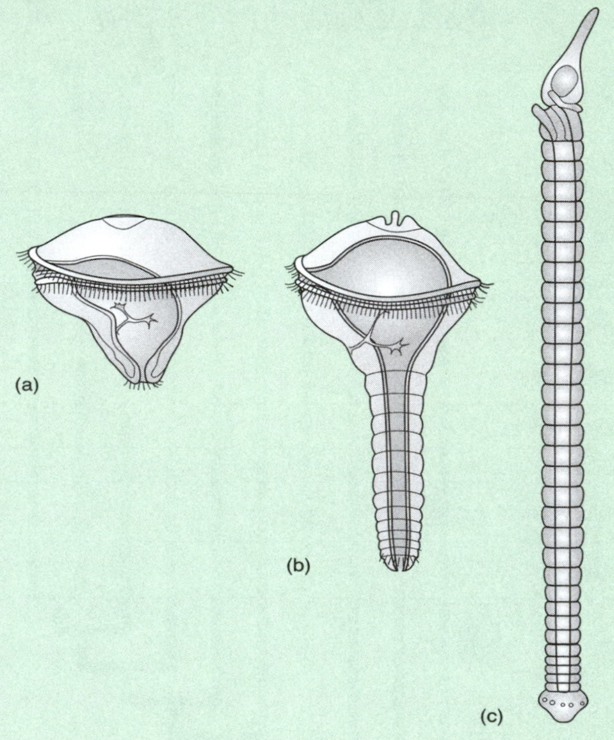

FIGURE 16-2 Three stages in the life history of a paedomorphic marine annelid, *Polygordius* sp. (a) shows a typical trochophore larva. (b) A larval stage with 12 segments in addition to the specialized anterior and posterior segments. (c) The adult.

number of times, proposes that radial cleavage preceded spiral cleavage (**Fig. 16-4**), making the deuterostome lineage ancestral to protostomes.[3]

From these and other studies, we can see that the clarification of metazoan relationships is an ongoing process as more molecular, genetic and developmental information becomes available. Molecular studies, based on comparing sequences from the billion or more base pairs in each metazoan genome, will help clarify and refine our understanding of many of these relationships, while new paleontological studies inform us on morphology and date their origins more precisely.[4]

Our approach is to provide an overview of the major features that distinguish several levels of increasing complexity among the invertebrates. **Germ layers,** presence of an internal cavity (**coelom**) and **segmentation,** all of which were introduced in Chapter 15, are three of the major features whose evolution we will follow, beginning with those organisms that have *only two of the germ layers, ecto- and endoderm* (see Fig. 15-12), moving on to discussing organisms with a simple internal cavity (digestive tract) and the "simplest" organisms with *three germ layers but no body cavity* (coelom), before turning to those animals *with a body cavity* (either a false coelom or a true coelom), with or without segmentation, with some emphasis on arthropods and echinoderms.

[3] See Sidow and Thomas (1994), Nielsen (2003), and Collins (1998) for deuterostomes-protostome-placozoan relationships, Ruiz-Trillo and coworkers (1999) for the 18S ribosomal RNA analysis, and Valentine (1997) for the RNA and cleavage pattern analyses.

[4] For detailed analyses of metazoan relationships, see Conway Morris (1998a), Adoutte et al. (1999), Nielsen (2001), Brusca and Brusca (2003), Ruppert et al. (2004), Jenner (2006), and Seidel and Schmid (2006)

Animals Composed of Cells and Prototissues

Sponges

Sponges (Porifera) are among the most simply constructed metazoans. Their body plan (**Fig. 16-5**) enables them to extract food particles from water currents and digest them intracel-

FIGURE 16-3 A selection of bryozoans drawn by Ernst Haeckel. Some, such as numbers 6 and 10 show the lophotrochophore particularly well. (From Haeckel, E., 1904. *Kunstformen der Natur.*)

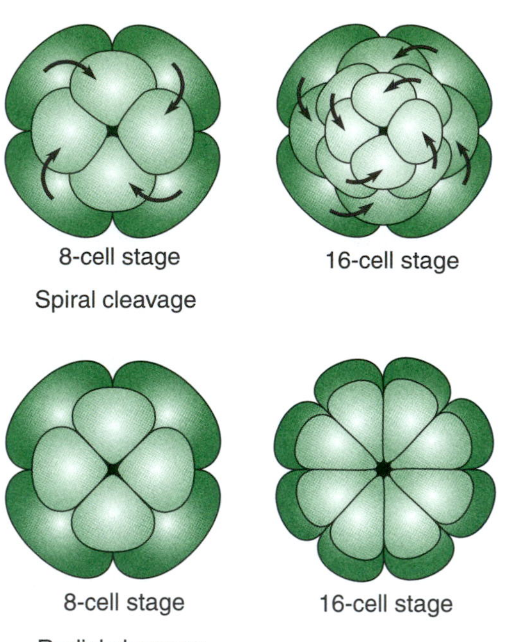

FIGURE 16-4 Modes of spiral and radial cleavage at 8- and 16-cell stages. According to Valentine (1997), cleavage patterns affect traits such as mesodermal tissue origin and coelomic development.

lularly. Within sponges, currents are generated by collared, flagellated cells (**choanocytes**) whose flagella move water out through a large exhalant body opening, thereby drawing water in through small inhalant pores. To filter sufficient water for food and to expel the effluent far enough so that it does not flow back, many tiny choanocyte flagellae combine to produce a forceful exhalant current that exceeds more than 15 cm a second. For this system to function efficiently, the spongiform body must stand erect, which is accomplished by networks of collagenous fibers and skeletons made of small spicules consisting of calcium or silicon compounds.

Although various groups differ in body organization and kinds of spicules, sponges consist of 8 or 10 different cell types, none of which are organized into tissues or organs of the types found in more derived metazoans (although recent analysis is challenging the view that sponges lack tissues (**Box 16-2**). Sponges have no organized digestive organs, muscular tissue or nervous network. Function is localized mostly in specific cells, with coordinated movements based on direct cellular contacts that usually extend no further than a small area. Moreover, cell determination is often quite flexible.

The most generalized sponge cell, the **archaeocyte,** is a large amoeboid cell that can differentiate into all the other cell types, many of which can dedifferentiate into archaeocytes. Archaeocytes therefore are stem cells. A classic experiment shows that a sponge strained through a sieve reaggregates into a complete organism. Repeat the experiment with two sponges of different colors, mix the cells together and watch as they reaggregate in a species-specific manner.[5]

Morphological differences exist between the external and internal layers of sponges, and a mostly gelatinous middle layer (*mesohyl*) that carries archaeocytes and a variety of other cells. Archaeocytes are widely used during asexual reproduction, being incorporated into buds and fragments or into small, hardy, spore-like spheres coated with spongin, called *gemmules.* Sexual reproduction based on meiosis also takes place, producing radially symmetrical, free-swimming larvae that provide the principal means of dispersal.

Taxonomists have used the simplicity of sponges, compared to other metazoans, to place them in their own metazoan phylum, Porifera, or subkingdom, Parazoa. Their archaic features include the *absence of* various structures found in triploblastic organisms: a distinctive mouth, tissues and organs, tightly bound cellular sheets (epithelia) and distinguishable anterior and posterior ends. Also unusual are the choanocyte cells, which strongly resemble choanoflagellate protozoans (see Fig. 15-13), and the presence of other cell types normally absent in other animals. Some have suggested that sponges bear some affinity to Archaeocyathids, an extinct phylum (see Box 11-2, Phyla). At least from the Cambrian onward, their simple body plan has successfully generated water currents for feeding, and they remain successful today, with about 8,000 species distributed widely from freshwater to marine areas and from shallow regions to great depths.

Placozoa

Morphologically allied to other metazoans are two phyla of multicellular animals, Placozoa and Mesozoa. **Placozoa** consists of only a single known species, *Trichoplax adhaerens,* a flattened, free-living marine organism only a few millimeters in diameter (**Fig. 16-6a**) and composed of an upper and lower layer of flagellated epithelial cells enclosing a sheet of loose, fibrous mesenchymal cells that some consider mesodermal tissue. Its body can assume irregular shapes as it creeps along the substratum like an oversized amoeba, enveloping food particles and digesting them through its lower surface. Asexual reproduction occurs by fission and budding, but sexually produced eggs have been found in the "mesenchymal" inner layer. Although we know little about this simple metazoan, it appears to be a remnant of an early metazoan offshoot.

[5] All you need to do this experiment is get a small piece of sponge (yellow and red sponges are easy to obtain from pet stores), some salt water and a piece of cheesecloth or other material with a fine mesh such as nylon stockings.

Osculum
Spicules
Choanocytes
Porocyte
Dermal pore
Water flow
Atrium
Mesohyl
Archaeocyte
Pinacoderm

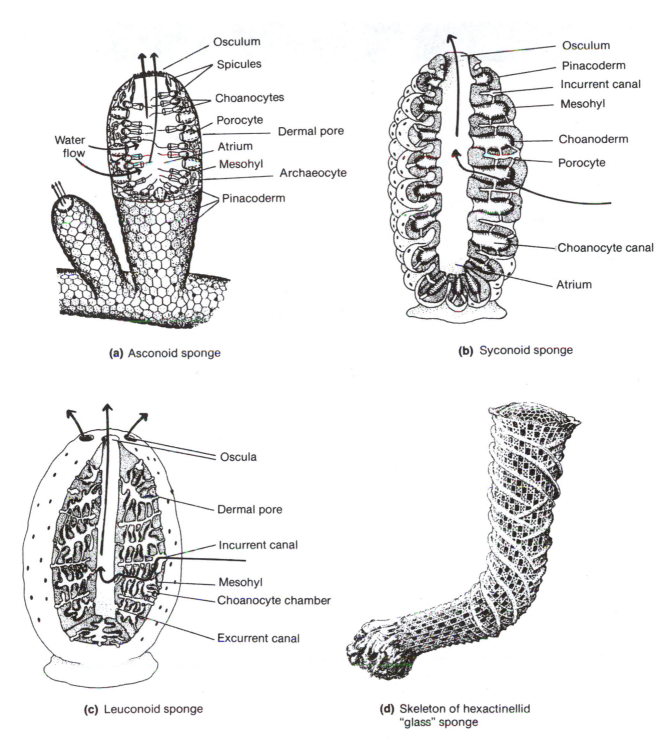

(a) Asconoid sponge

Osculum
Pinacoderm
Incurrent canal
Mesohyl
Choanoderm
Porocyte
Choanocyte canal
Atrium

(b) Syconoid sponge

Oscula
Dermal pore
Incurrent canal
Mesohyl
Choanocyte chamber
Excurrent canal

(c) Leuconoid sponge

(d) Skeleton of hexactinellid "glass" sponge

FIGURE 16-5 Models of the three major types of sponge morphology: (a) asconoid, (b) syconoid and (c) leuconoid. All three forms are found in the Calcarea; Demospongiae and Sclerospongiae include only the leuconoid type. Because the fourth lineage, the Hexactinellida, or glass sponges, has a distinctive skeletal framework (d), does not have the same type of pinacoderm found in other sponges, and lacks cell wall separation between many of their cells (syncytium), some taxonomists have proposed they be placed in a separate phylum. (From Bergquist, P. R., 1985. Poriferan relationships. In *The Origin and Relationships of Lower Invertebrates,* S. Conway Morris, J. D. George, R. Gibson, and H. M. Platt (eds.). Clarendon Press, Oxford, England, pp. 14–27.)

BOX 16-2 Do Sponges Have Tissues?

IN THE TEXT WE RECITE the traditional story that sponges lack tissues, a story that is as old as the invention of histology in the mid-nineteenth century. Underlying this statement is the unwritten presumption that sponges have not "evolved enough" to produce tissues, or organs that are built from tissues. You may have noticed, however, that in the text we state that sponges have no "true" tissues or organs. Why the qualification? The qualification reflects two approaches to sponges.

The first recalls the classification of all animals other than vertebrates as invertebrates because they lack a vertebral column. Sponges lack the types of tissues found in triploblastic animals, and so we regard them as tissue-less in comparison (Cole and Hall, 2004). This leaves unanswered the question of what a **prototissue** would look like. Sponges are epithelial and diploblastic, with an outer ectoderm and an inner endoderm. They lack mesoderm (if by mesoderm we mean a layer that arises during gastrulation to occupy the position between ectoderm and endoderm and that gives rise to muscle and other structures) but do possess a gelatinous middle layer between the epithelial layers, the *mesohyl,* that contains crawling cells — leaving unanswered the question of what a protomesoderm would look like. So, while it may be true that sponges lack the tissues of triploblastic organisms, sponges are not mere gelatinous bags. They have structural integrity and functional specialization, both roles played by tissues in triploblasts.

The second qualification is more of a question. If sponges lack tissues, do they have an organization that does for them what tissues do for other animals? Sally Leys from the University of Alberta[a] argues that the basic requirements of a tissue are threefold:

1. The cells are polarized.
2. They are joined by sealing junctions.
3. They sit on a basement membrane, which they are presumed to have deposited.

Indeed, she reports that epithelial cells in calcareous sponges are polarized, and that epithelia in demosponges are comprised of flat 'pinacocytes,' superficially similar to the pavement of cells in gill epithelia in fish, have an outer coating of mucous, and are joined by sealing junctions composed of *protocadherins* rather than the *cadherins* found in triploblasts, which is very much what one might expect of prototissues. This specialized epithelium has a specialized function — it prevents water from diffusing into the sponge — in distinction from the porocytes, which allow water to enter. So we have two distinct cell types in a layer that is specialized to regulate water flow into the animal.

Of the more than 20 different types of collagen molecules, type IV collagen is the collagen found in epithelial extracellular matrices (basement membranes) of triploblastic animals. Sponges have type IV collagen where basement membranes are distinct and also in the collagenous mesohyl,[b] an important step in differentiating one type of cellular organization from another.

The major classes of metazoan cell adhesion gene families coding for cell surface receptors and extracellular matrix proteins (Wnt, Tgf-β, receptor tyrosine kinase, Notch, Hedgehog) are present in the sponge *Oscarella carmela* (Nichols et al., 2006), indicating much greater potential for cellular interactions and cell signalling in sponges than previously imagined. So, it certainly could be argued that sponges have structural and regional specializations that function as do tissues in other animals and that these prototissues display some of the features used by triploblastic organisms to make tissues.

[a] Based on personal communication.

[b] See Boute et al. (1996) and Boury-Esnault et al. (2003) for type IV collagen.

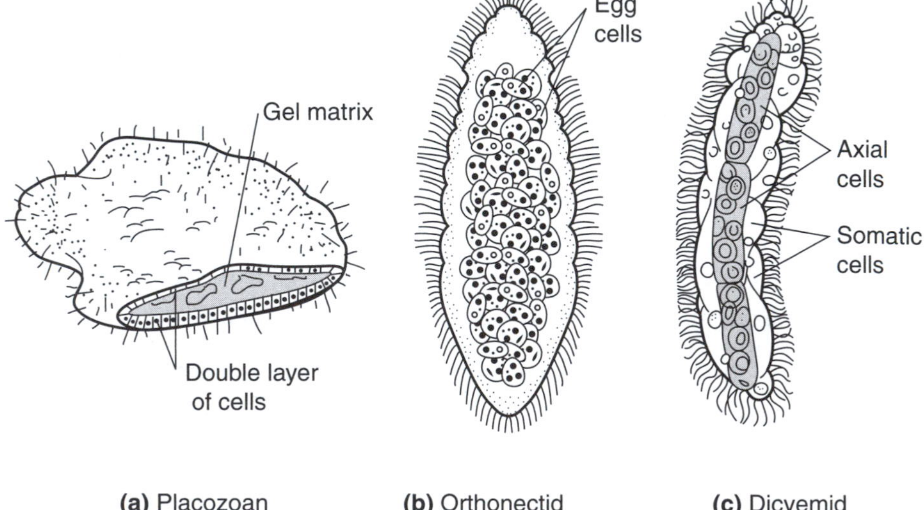

FIGURE 16-6 (a) *Trichoplax adhaerans,* the only known species in the Placozoa. Although orthonectids (b) and dicyemids (c) have been placed in the phylum Mesozoa, 18S ribosomal RNA sequence data indicate distinct origins. Pawlowski and coworkers (1996) propose classifying them into separate phyla.

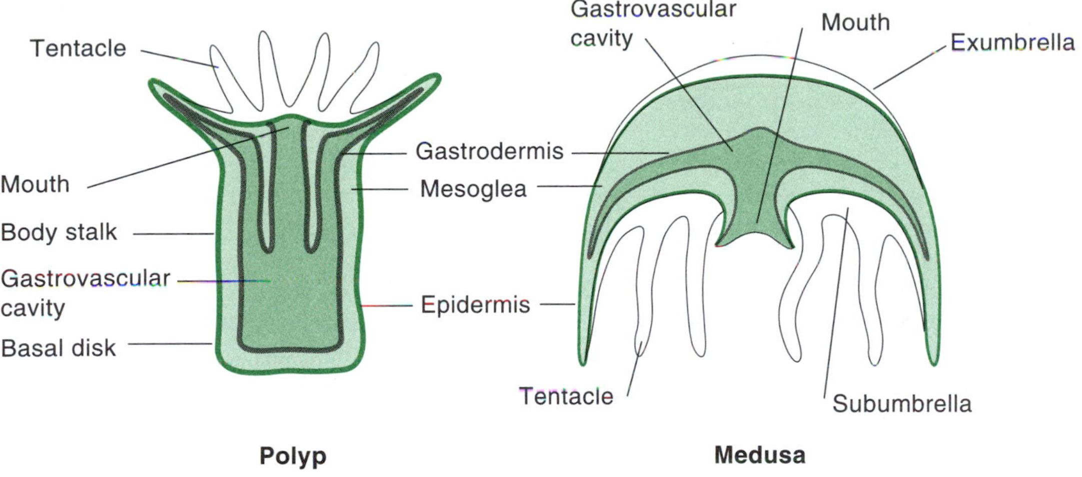

FIGURE 16-7 The two primary body forms found in Cnidaria. Note that these shapes are essentially inversions of each other, except that the intermediate mesogleal layer is usually thicker in the medusa form. (Adapted from Barnes, R. D., 1980. *Invertebrate Zoology,* 4th ed. Saunders, Philadelphia.)

Mesozoans

Also of a simple body plan **mesozoans** (**Mesozoa**; Fig. 16-6b, c), of which we know about 50 species, all parasites in or on marine invertebrates. These too are quite small but have more complex life cycles than do placozoans, involving male and female differentiation and various larval stages. Because of their parasitism, zoologists have proposed that mesozoans are degenerate platyhelminths that abandoned a free-living lifestyle, reducing and simplifying their tissues in the process. Most biologists, however, still classify them into either one or two separate phyla that may have become parasitic early in their evolution.

Relationships among Placozoa, Mesozoa and other metazoans are not completely resolved. Using an extensive 18S ribosomal RNA sequence analysis, however, Collins (1998) provided evidence that placozoans (Fig. 16-1) are the closest relatives of bilaterians (protostomes and deuterostomes).

Animals Based on Two Layers and Possessing Tissues

Cnidaria and Ctenophora

Cnidaria (corals, jellyfish, hydrozoans) and **Ctenophora** (comb jellies) show a major change in metazoan organization with possession of a mouth and a specialized gastrovascular digestive cavity (coelenteron). These expandable organs allow these organisms to ingest much larger food particles than sponges can filter, permitting them to break down entire prey organisms extracellularly before absorbing them intracellularly.

Tissue organization is more evident in cnidarians and ctenophores than in sponges (see Box 16-2), with a distinctively organized outer epidermis and inner gastrodermis, homologous with the ectodermal and endodermal layers, respectively, of triploblastic metazoans. Cnidarians have both epidermal and gastrodermal muscular tissue whose activity they coordinate with simple nerve nets that enable various body movements for locomotion and food capture. Ctenophores possess muscle but eight powerful rows of cilia (**Fig. 16-7**) provide the principal means of locomotion. The middle layer, the **mesoglea**, is primarily gelatinous; the other two layers carry on most body functions. No specific cells or tissues devoted exclusively to circulatory, respiratory or excretory functions are found in these animals.[6]

Both cnidarians and ctenophores are *radially organized* so that almost any plane through the central oral axis of the animal cuts it into two approximate mirror-image halves; the two long tentacles of ctenophores, one on each side of the body, provide a degree of bilateral symmetry to an otherwise radially symmetrical organization. Individuals in both clades are soft-bodied and use flexible tentacles to bring food to their extendable oral cavity. This soft-bodied structure facilitated varied changes in shape. Various shallow-water cnidarians, including coral reef builders, harbor symbiotic algae (zooxanthellae) that provide nutrition through photosynthesis.

Cnidarians use both sexual and asexual modes of reproduction. Gamete formation, when it occurs — not all can reproduce sexually — leads to a fertilized egg that develops into a solid, externally ciliated ball of cells, the **planula,** which was introduced in Chapter 15 when discussing the planula hypothesis of metazoan origins. Further development of

[6] See Seidel and Schmid (2006) for a recent analysis of tissue organization in cnidarians in relation to their common ancestry with bilaterians from an "Urtriploblast", and see Sempere et al. (2006) for the discovery of a core set of miRNAs in protostomes and deuterostomes but not in sponges or cnidarians, consistent with a role for miRNAs in the origin of body organs.

the planula varies in different groups, but the planula is a universal feature of sexual reproduction in cnidarians and in one genus of ctenophores.

Cnidarians comprise over 10,000 described species, ctenophores about 80. Present-day cnidarians are carnivorous, although methods of food capture vary between the two. In the more common of the two, the cnidarians, the tentacular epidermis (and often sections of the gastrodermis) is armed with specialized cells, *cnidocytes,*[7] containing miniature stinging, harpoon-like organelles called **nematocysts** that immobilize prey, which is then brought into the gastric cavity. This cnidarian feature, used for offense and defense, dates back to the their early history and may have originated from glandular secretory cells; there are no obviously homologous protozoan nematocysts. However it arose, this unusually effective mode of food capture certainly helps account for cnidarian evolutionary persistence.

Polyp and Medusa

Many cnidarians undergo developmental changes that produce one of two body forms, the *polyp* and the *medusa* (Fig. 16-7). The **polyp** is mostly a stationary (sessile) form with a tubular body, tentacles surrounding a mouth at the oral end, and a basal disk at the aboral end by which it attaches to the substratum. The **medusa,** in contrast, is usually a free-swimming form resembling an inverted umbrella-shaped polyp. Its concave undersurface bears a centrally located mouth surrounded by tentacles that hang down from the umbrella's margin. You can see the close relationship between these two body forms in some cnidarians in the transformation from medusa to polyp, during which the medusal stage attaches to a solid substrate. Generally, the polyp functions for stationary food gathering and the medusa for dispersion. Medusae are entirely absent in the Anthozoa (sea anemones and corals), polyps are absent in some of the Scyphozoa (jellyfish) and inconspicuous in others, while both forms can occur in some species of Hydrozoa (*Hydra* and its various solitary and colonial derivatives) and Cubozoa (sea wasps).

Although cnidarians and ctenophores differ in a number of traits, many invertebrate biologists consider them closely related phylogenetically. Consequently, and using data available to them at the time, zoologists have offered different views on the evolutionary relationships between cnidarians and ctenophores. Indeed, on the basis of their radial symmetry and two-layered level of organization, the two groups were long classified under one phylum, Coelenterata. (The term coelenterate is now mostly applied to members of the Cnidaria.) Shared radial symmetry also is used in some schemes to group the two under the superphylum Radiata, as distinct from the superphylum Bilateria, as shown in Figure 16-1.

Cnidarians show considerable morphological similarity to fossils of some Precambrian organisms. Evidence available now enables us to conclude that a Precambrian lineage capable of producing larval polyps and adult medusa gave rise to the various cnidarians and probably to ctenophores (**Fig. 16-8**).

Homeobox Genes

Homeobox genes play major roles in the triploblastic animals (see Fig. 13-9) discussed in the next section.

Bilaterality, in which the body has right and left sides, is an immediate consequence of a dorso–ventral, anterior–posterior anatomy and leads to opportunities for organizational complexity. All bilaterally symmetrical triploblastic animals use the same three sets of regulatory genes to determine the anterior–posterior axis, the gene sets being Hox and ParaHox genes and the genes *even-skipped* (*evx*) and *empty-spiracles* (*emx*). The conclusion that these genes and this patterning mechanism were present in the last common ancestor of all bilaterally symmetrical animals seems inescapable. Surprisingly, cnidarians have become an unexpected source of insights into metazoan evolution, for it now appears that cnidarians and ctenophores diverged before the origin of bilaterally symmetrical animals.

Homeobox genes have been observed both in cnidarians and in placozoans, two lineages of diploblastic animals. The cnidarian *Nematostella vectensis* has seven representatives of Hox and ParaHox genes in its genome, fewer genes than found in triploblastic animals but clustered on one chromosome as in triploblasts. Cnidarians lack bilateral symmetry, but the adults have an anterior–posterior axis (the *oral–aboral axis*). Both the oral–aboral of cnidarians and the *anterior–posterior* (A-P) axis of the bilaterally symmetrical organisms (Bilateria) are patterned by Hox genes, *Hydra* having six Hox-like genes, two of which determine the oral pole. Consequently, the oral–aboral axis is considered homologous to the A-P axis, implying either that cnidarians are diploblastic, as long thought, or that they had a triploblastic ancestor and secondarily lost genes during their evolution. The discovery that *Nematostella* has bilateral symmetry early in its development, only becoming radially symmetrical later, reinforces the view that cnidarians and triploblasts shared a common ancestor.[8]

[7] Ctenophores lack cnidocytes, employing special adhesive cells (*collocytes*) to capture their food.

[8] See Finnerty (2001), Schierwater and Kuhn (1998) and Darling et al. (2005) for cnidarians and placozoans, and Bode (2001) for *Hydra*. A recent characterization of the full Hox/ParaHox genes in *Nematostella vectensis* and *Hydra magnipapillata* by Chourrout et al. (2006) suggests that the ancestral protoHox cluster consisted of two anterior genes and is consistent with posterior genes in Hox and ParaHox clusters appearing independently in cnidarians and bilaterians.

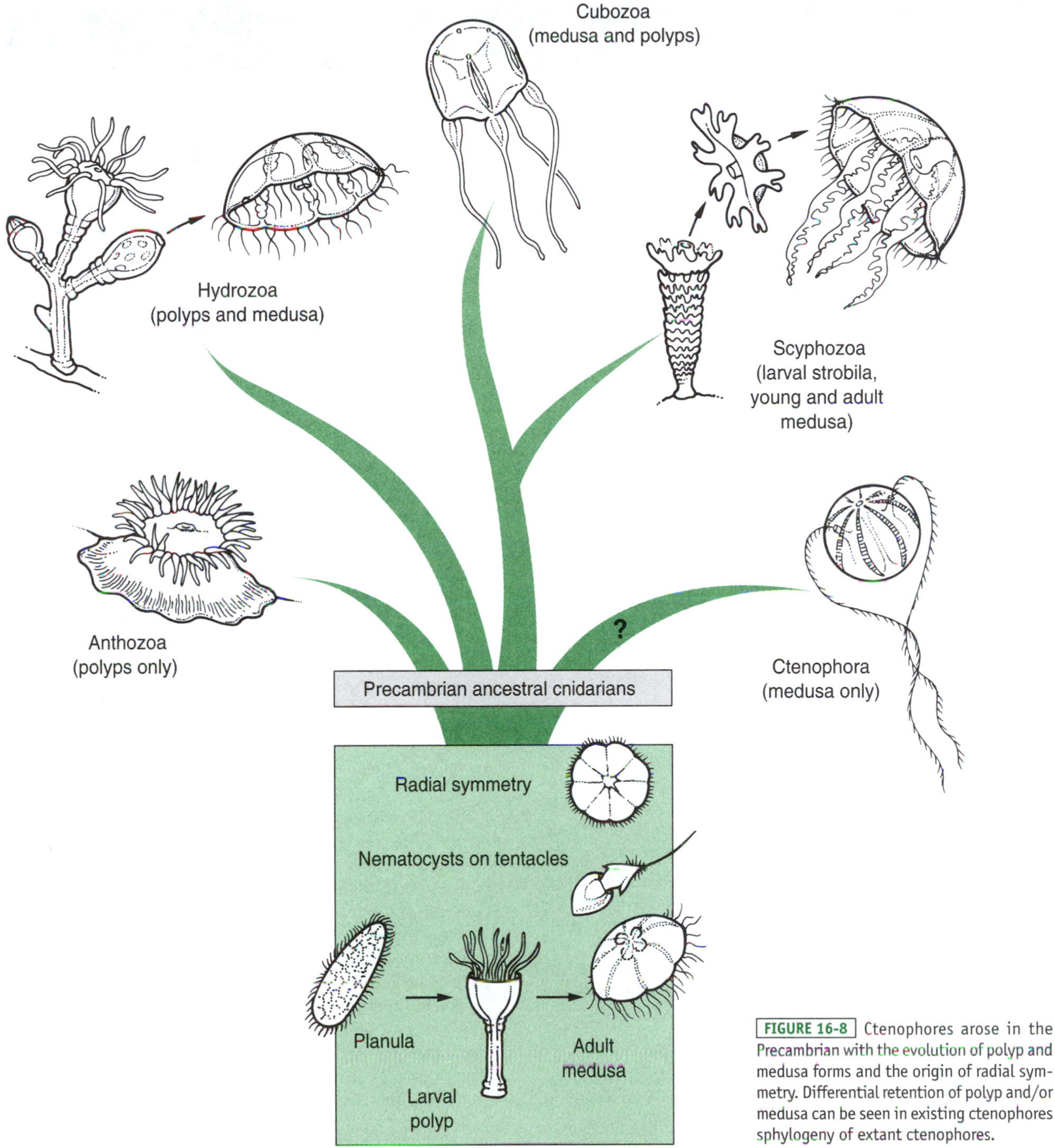

FIGURE 16-8 Ctenophores arose in the Precambrian with the evolution of polyp and medusa forms and the origin of radial symmetry. Differential retention of polyp and/or medusa can be seen in existing ctenophores sphylogeny of extant ctenophores.

Animals Based on Three Germ Layers but No Coelom

The transformation of the **diploblastic** animals with only two embryonic cell layers (ectoderm and endoderm), discussed above, to **triploblastic** animals that also possess mesoderm (see Figs. 15-11 and 15-12), can be correlated with (and indeed initiated) an increase in the number and organizational complexity of cells types in all three germ layers. These changes enabled a wide range of novel tissues and organs to arise, examples including bundles of circular and longitudinal muscles, excretory organs, circulatory channels and tissues and complex reproductive systems. Size increased and active locomotion became possible (**Box 16-3**). The evolution of an A-P orientation or polarity aided food gathering and provided various animal lineages with bilateral ("left–right") symmetry.

BOX 16-3 Evolutionary Solutions to Problems of Locomotion

THE TASK OF OBTAINING FOOD for animals is inextricably bound with a variety of adaptations: sensory, locomotory, ingestatory and others that support and enhance fulfillment of this primary need. Differences in locomotory behavior have provided functional and evolutionary biologists with opportunities to examine some **basic concepts of adaptive change.**

On the *unicellular level,* small size enables locomotion through relatively simple ciliary, flagellar, pseudopodial or even Brownian motion. Once the *metazoan grade* of organization was reached, locomotory cells faced the problem of moving relatively large masses in concerted activity. The earliest of metazoan animals, perhaps planula-like organisms, probably moved by ciliary activity not unlike the motion of small acoelan turbellarians. *Acoelans,* many no larger than a millimeter, use ciliary cells on their ventral surface for creeping.

Motion in a directional fashion confers an anterior–posterior orientation on the animal, making the anterior portion more concerned with those adaptations necessary for sensing and confronting the environment being entered. Bilaterality, the distinction between right and left sides, an immediate consequence of a dorso–ventral, anterior–posterior anatomy, leads to opportunities for organizational complexity, but necessitates changes in locomotion. Cilia can only generate relatively small forces **(Fig. 16-9)**. Therefore, as animals grow larger, the use of cilia alone limits more rapid locomotion and a range of other methods are employed, all of which depend on organized muscular tissues.

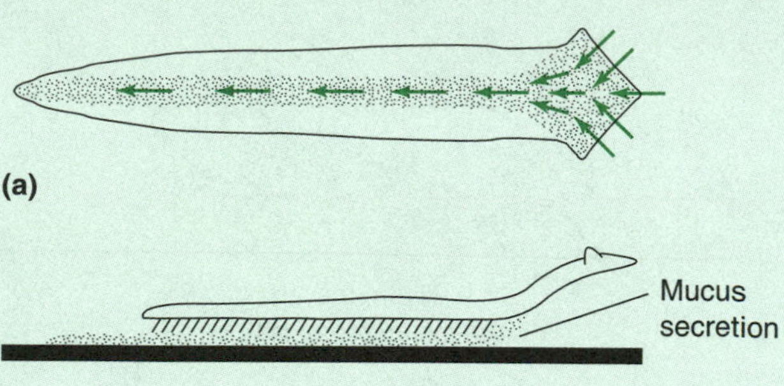

FIGURE 16-9 Ciliary distribution and locomotion in a platyhelminth turbellarian flatworm, *Planaria.* (a) Ventral surface of *Planaria* showing the direction of ciliary beats. (b) Mode of creeping, by means of ciliary beating on a mucus secretion deposited on the substratum. The efficiency of such ciliary creeping depends on the relatively small size and flattening of the turbellarian body to present as large a ventral surface as possible. In platyhelminths with larger and more circular dimensions, ciliary creeping is mostly, if not entirely, abandoned. (Adapted from Clark, R. B., 1979. Radiation of the metazoa. In *The Origin of the Major Invertebrate Groups,* M. R. House (ed.). Academic Press, London, UK, pp. 55–102; and from Clark, R. B., 1964. *Dynamics in Metazoan Evolution.* Clarendon Press, Oxford, England.)

Circular and Longitudinal Muscles

In its simplest form, tissue organization in *triploblastic animals* takes the shape of a worm, which can be defined as a long, flexible tube of constant volume enclosed by a muscular body wall. Coordinated activity results from organization of the muscle tissue into two major groups: *circular muscles* whose contraction reduce the diameter of the animal and increases its length, and *longitudinal muscles* that contract with opposite effect by reducing length and increasing diameter. The opposed activity of these muscle tissues means essentially that for one type of muscle to extend, the other must contract. Limitations on the extent of movement depend on the size of the muscles, their locations and attachments as well as the degree to which the body wall can distort.

Pedal Locomotion

In many platyhelminth turbellarians that have reached sizes much larger than acoelans, locomotory movements are almost entirely transmitted through a pedal longitudinal foot. *Pedal locomotory* waves arise by contraction and relaxation of those ventral longitudinal muscles that contact the surface. This creeping mode of locomotion enables animals whose bodies are mostly solidly filled with cells to locomote.

Peristalsis

As fluid accumulates in either cells or sinuses within the wormlike body, the effects of muscular changes are transmitted through greater distances, leading to improved locomotion. For example, although a coelom is lacking in ribbon worms (Nemertea), some have a gelatinous parenchyma, allowing the effects of muscular contractions to transfer more easily than through solid tissue. Undulatory swimming movements can now occur, produced by contraction of longitudinal muscles on opposite sides of the body. Furthermore, in some nemerteans we see the early signs of peristaltic movements, which become an important feature of animals possessing a true fluid-filled coelom. Essentially, the *coelomate* condition of a continuous body cavity provides a fluid skeleton that eliminates cellular barriers to hydrostatic pressure. Such a feature permits the effects of contractions in one part of the body to be transferred immediately to other parts.

Platyhelminths and Other Acoelomates

Platyhelminths (*flatworms*) represent one of the earliest evolutionarily successful stages in triploblastic organization, comprising at present some 20,000 species in a single phylum, **Platyhelminthes,** which is almost certainly polyphyletic.

The flattened flatworm body contains muscular tissue, an extensive hermaphroditic reproductive system, relatively simple osmoregulatory organs (*protonephridia*), but no circulatory system for gas exchange. Morphologically significant is their anterior–posterior organization with nervous and sensory structures concentrated at the cephalic end. The absence of any internal (coelomic) body cavity delimits

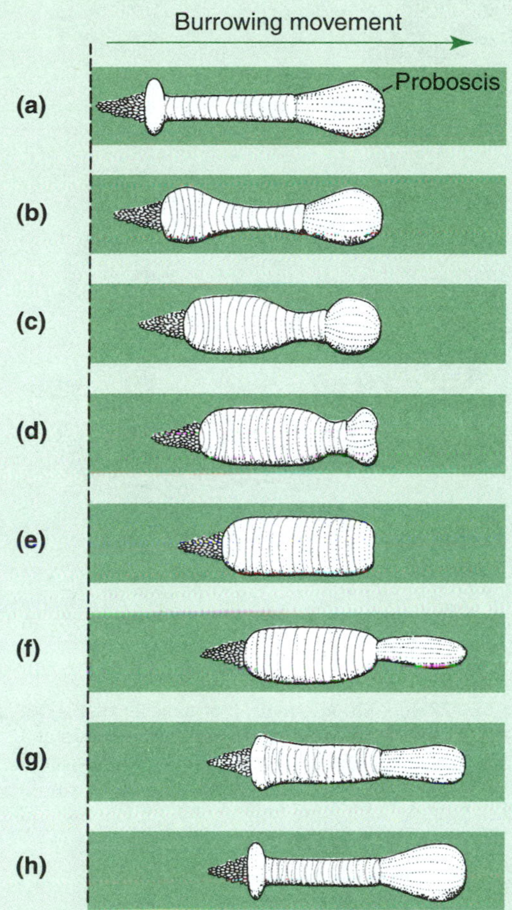

FIGURE 16-10 Stages in the burrowing activity of a priapulid "worm," *Priapulus,* beginning with (a) lengthwise extension of its body and enlargement of its proboscis, which serves as an anchor allowing the animal to move anteriorly (to the right). The animal contracts by increasing in diameter (b–e), enclosing the proboscis. In stages (f–h) the proboscis extends again, and the animal elongates to repeat the cycle. See Webster et al. (2006) for analysis of the slowly evolving mitochondrial genome of *Priapulus.* (Adapted from Clark, R. B., 1979. Radiation of the metazoa. In *The Origin of the Major Invertebrate Groups,* M. R. House (ed.). Academic Press, London, pp. 55–102.)

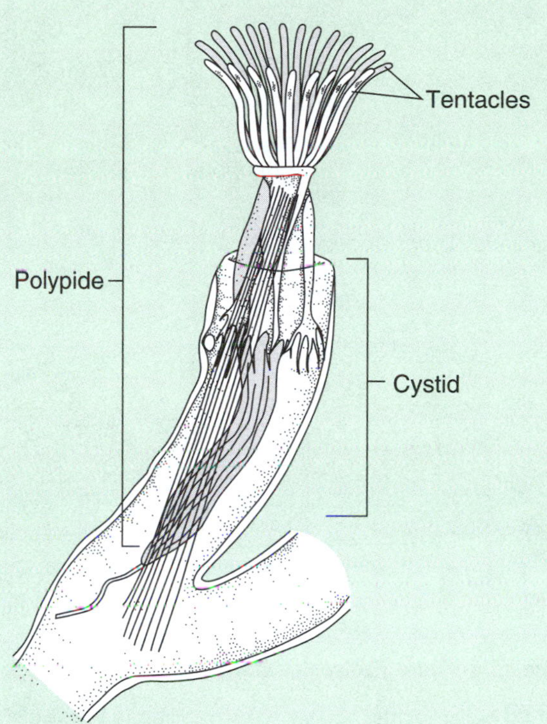

FIGURE 16-11 The ectoproct, *Fredericella sultana,* with everted polypide. (Adapted from Clark, R. B., 1979. Radiation of the metazoa. In *The Origin of the Major Invertebrate Groups,* M. R. House (ed.). Academic Press, London, pp. 55–102.)

Peristaltic motion — a wave of circular muscle contraction followed by a wave of longitudinal muscle contraction — can generate much larger forces than is possible in acoelomates because the entire coelomic hydrostatic skeleton and all the body wall musculature are involved. Peristalsis enhances the relative fitness of a burrowing animal by enabling it to use the entire circumference of the body in thrusting through the substrate. Furthermore, a fluid-filled coelom allows the rapid eversion of a proboscis or lophophore by hydrostatic pressure alone, as you can see in the rapid burrowing movements of priapulids (**Fig. 16-10**) and the extension and withdrawal of the tentacular polypide in ectoprocts (**Fig. 16-11**).

flatworms and other **acoelomates** from the remainder of the triploblastic animals, all of which, as discussed in the next section, have a coelomic cavity in one form or another.

In most turbellarians, including the free-living flatworms, the mouth serves as both entrance and exit for the digestive organ. In acoelan turbellarians, digestion is carried out by a communal cell mass, lacking a cavity **(Fig. 16-12a)** In other turbellarians digestion is carried out in one or more blind sacs (Fig. 16-12b), similar but not homologous to coelenterate digestive cavities. In all individuals, tissue fills spaces between the internal organs and the body wall. Turbellarian locomotion is normally by ciliary movement and/or ventral (pedal) muscular creeping (Fig. 16-9). The other two major groups of platyhelminths are entirely parasitic: Trematodes

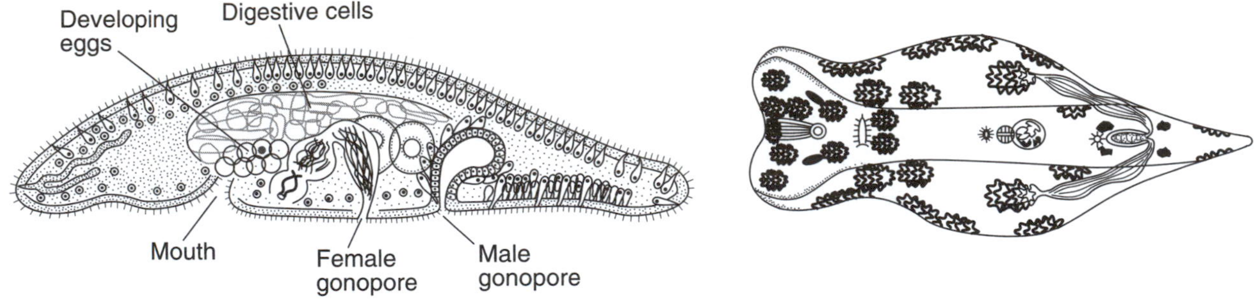

(a) Turbellaria: Order Acoela

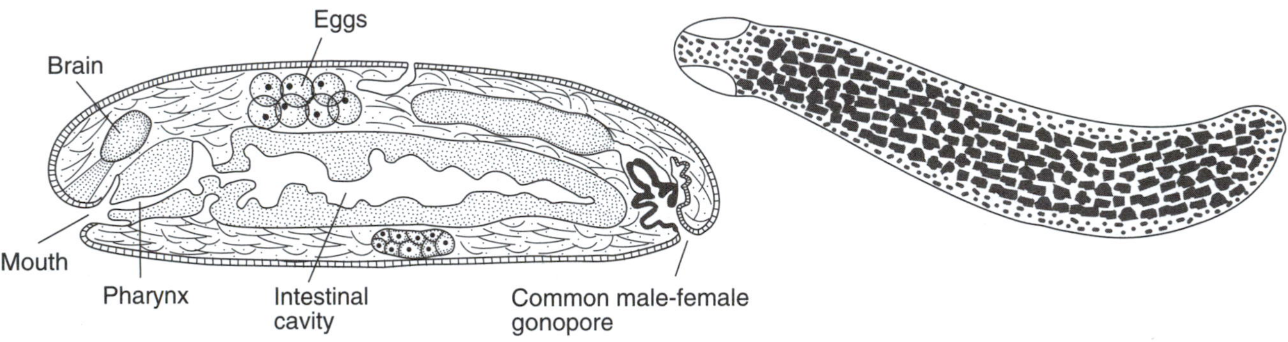

(b) Turbellaria: Order Prolecithophora

FIGURE 16-12 Median sagittal sections (*left side*) and dorsal views (*right side*) of genera from two different lineages (orders) of turbellarians. (a) An acoelan, *Convoluta*, showing the syncytial mass of digestive cells. (Adapted from Barnes, R. D., 1980. *Invertebrate Zoology*, 4th ed. Saunders, Philadelphia.) (b) The freshwater turbellarian *Hydrolimax* with muscular pharynx and gut cavity. (Adapted from Hyman, L., 1951. *The Invertebrates: Platyhelminthes and Rhynchocoela, the Acoelomate Bilateria*. McGraw-Hill, New York.)

(flukes), like turbellarians, have a mouth and digestive cavity, whereas Cestodes (tapeworms) depend entirely on absorbing host nutrients through the body wall.

Parasitism

Once relatively large potential host organisms evolved, **parasitism** became a successful way of life for many platyhelminths. Flatworm parasitic adaptations involve devices that fasten onto host tissues, such as hooks and suckers, as well as the reduction or loss of sensory and digestive organs no longer needed for a dependent existence. Most important, in the continuous "arms race" between parasite and host, parasites evolved integuments that protect them against host enzymes and against the antibodies that hosts in turn evolved as protection against parasites When the parasite does not occur over the full geographical range of the host, parasite adaptation is inversely proportional to the fraction of the host's geographical range occupied by the parasite, with at least two thresholds of overlap that influence the stability of allele frequency and adaptedness of the parasite. Where the ranges overlap completely, the expectation is coevolution of host and parasite, creating a "coevolutionary hot spot."[9]

Life Cycle

We know that as a parasite evolves, many of its organs become simplified or disappear altogether.[10] This is not the case in the life cycle of the parasite, which may become enormously complex and involve multiple intermediate hosts. Some tapeworms, for example, may pass through a number of intermediate hosts ranging from arthropods to fish before the adult stage develops in the primary mammalian host. Such developmental and life history networks must often have taken advantage of opportunities that other evolution-

[9] See Nuismer et al. (2003) for the range studies, and see Poulin (2006) for the evolutionary ecology of parasites.

[10] Dollo's law states than a lineage cannot regain a structure lost earlier in the evolution of the lineage, or if the structures re-evolves, it will not be in the same form as the ancestral feature. Although parasites have long been regarded as exemplars of *Dollo's law*, recent studies are beginning to consider that organ loss associated with parasitism may neither be a one-way street nor an evolutionary dead end (Cruickshank and Paterson, 2006).

ary events provided. Hyman (1951) suggested that some members of one lineage of trematodes (Digenea) originally only infected mollusks. After fish and other vertebrates evolved, many digenetic trematodes invaded these newer groups but retained mollusks as the intermediate host.

The adaptive advantage of such life cycle complexity is that a parasite can build up population numbers in intermediate hosts, enhancing its chance of infecting a primary host. In addition, spreading its early stages among intermediate hosts conserves the resources of the primary host, enabling the adult parasite to remain productive for relatively long periods. For example, some adult *Schistosoma* trematodes, the causative agent of the widespread tropical disease *schistosomiasis,* may live for 30 years. Some human tapeworms are active for 20 years or more.

Reproduction

Characteristic of these parasites are their enormous reproductive powers. Asexual reproductive stages often supplement sexual reproduction so that a single adult in some species can produce hundreds of thousands, if not millions, of offspring. Various trematodes and cestodes seem to devote almost their entire anatomy and physiology to reproduction. Because their offspring have extremely low survival rates — a result of the many chance factors and hazards in parasite distribution and infection — such features have been long selected. Limited survival opportunities explain why usually only a minority of an appropriate host species is infected by a particular parasite at any one time.

Because turbellarians possess the least derived characters, parasitic platyhelminths are considered to have been derived from a turbellarian ancestor. We do not know whether the ancestral turbellarian was an acoelan (and therefore lacked a gut) or had a simple gut of the type shown in Figure 16-12b. Molecular studies by Ruiz-Trillo and coworkers (1999) suggest that acoelans stand close to the evolutionary base of all bilateral triploblastic metazoans and should be separated from platyhelminths as a distinctive lineage.

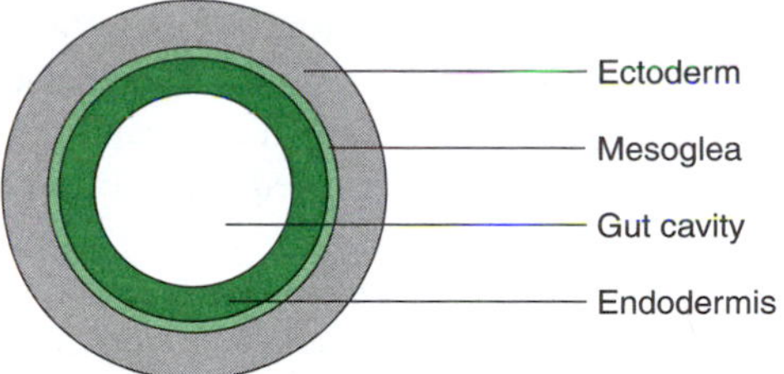

(a) Diploblastic coelenterate

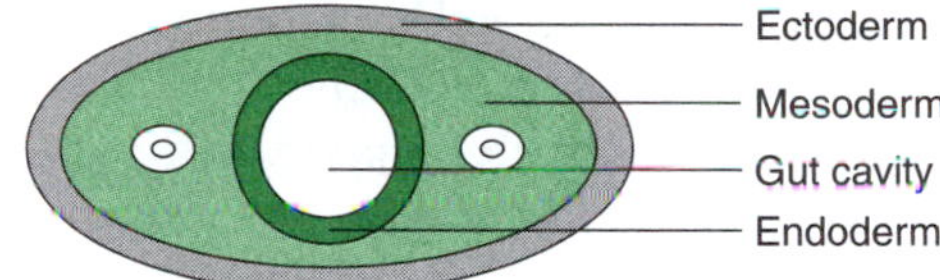

(b) Triploblastic acoelomate

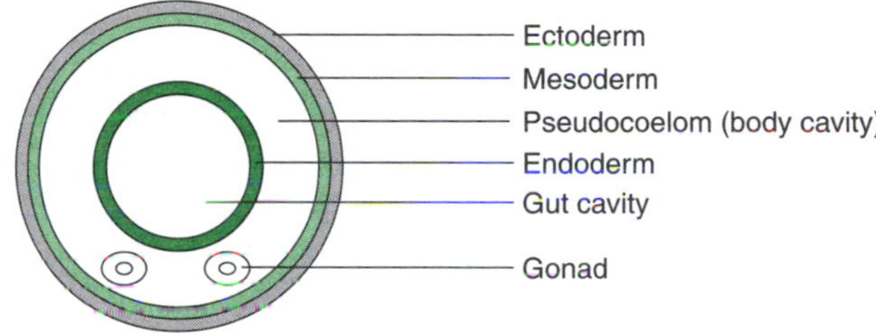

(c) Pseudocoelomate

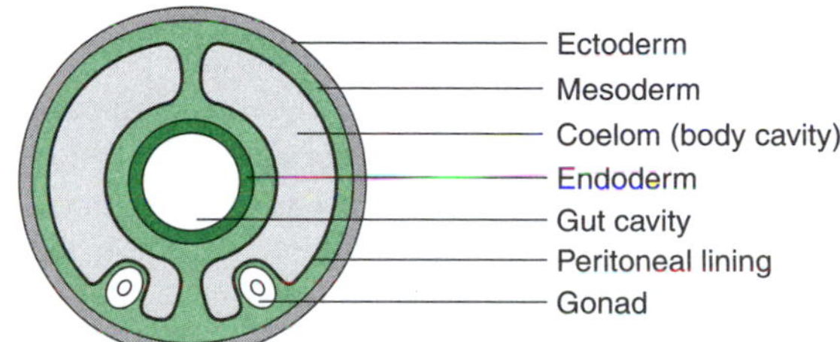

(d) Eucoelomate

FIGURE 16-13 Diagrammatic illustrations of general kinds of metazoan body cavities. (a) Diploblastic body plan in cnidarians such as *Hydra.* (b) Triploblastic plan in which the coelom is absent (acoelomate), as it appears in platyhelminths. (c) Pseudocoelomate plan in various aschelminths in which the body cavity is only partially lined with mesoderm. (d) Eucoelomate body plan in arthropods, annelids, chordates and echinoderms.

Animals Based on Three Germ Layers and a Coelom

The Coelom

Introduced in Chapter 15, the coelom, an internal cavity formed within mesoderm (as Haeckel defined it in 1872), was one of the most successful early metazoan adaptations. It is lined with an epithelium that often contains testes or ovaries and has ducts to the exterior used to transmit gametes or waste products. As shown in **Figure 16-13**, the term *coelomate* is used for two types of animals:

- **pseudocoelomates** (*false coelomates*), in which the body cavity (also called the hemocoel) derives from the persistence of the cavity of the blastula (the blastocoele), and is only partially lined with mesoderm; and
- **eucoelomates** (*true coelomates*), in which the coelom arises as a cavity within mesodermal tissue and which therefore is completely lined with mesoderm.

In pseudocoelomate and eucoelomate organisms, the body cavity is filled with fluid, enabling it, among other functions, to act as a hydrostatic skeleton that can transmit pressure from one part to another. Thus, the efficiency of peristaltic motions in coelomate animals is considerably greater than in non-coelomates; waves of circular and lon-

gitudinal muscle contraction can be transmitted more easily through the hydrostatic skeleton. This dynamic flow enables undulatory swimming movements as well as greater efficiency in burrowing resulting in faster speed of capture and speed of escape.

Hypotheses concerning the evolutionary origin of the coelom — indeed, whether the coelom may have originated more than once — include its beginnings from outpocketings from the gastric cavity, or from intercellular fluid-filled cavities within mesodermal structures. Both of these modes of development (enterocoely and schizocoely, respectively) occur in coelomates, supporting hypotheses of multiple origins of the coelom. In whatever manner it originated, the coelom conferred an important advantage in some lineages by providing a mechanical hydrostatic function. As these lineages evolved and diverged, some made use of the coelom in unsegmented form (e.g, priapulids). Others adopted various segmental organizations (see Table 16-1) that had profound effects on their future evolution.[11]

Pseudocoelomates (Aschelminthes)

Judging from the large numbers of species with a false coelom, the pseudocoelomate condition and its various adaptations represented a major evolutionary advance. Whatever its nature, a fluid-filled tube provided with circular and longitudinal muscles offered significant advantages both as a hydrostatic organ for locomotion and for carrying metabolites, wastes and gases throughout the body.

Although their form and structure vary considerably, metazoans with a distinctive, fluid-filled body cavity include lineages that biologists often group together under the name **Aschelminthes** (**Table 16-2**). The body cavity of these pseudocoelomates characteristically encloses a thin-walled digestive cavity that lacks the peritoneal linings, muscles or supporting mesenteries that are found in all coelomates (Fig. 16-13). Many have an epidermal cuticle, adhesive organs, constant cell numbers (*eutely*) and a digestive tract with mouth, anus and muscular pharynx that pumps food into the flaccid gut cavity.

The most common and perhaps most representative of these animals, **nematodes,** have a tubular shape maintained by high internal pressures that distend the animal to the extent permitted by its thick cuticle (almost like an overstuffed sausage). Overall changes in length are slight because nematodes are almost always fully extended. Not highly adapted for burrowing — which demands peristaltic activity — they are undulatory swimmers and coilers, using antagonistic longitudinal muscle contractions. The well-known nematode *Caenorhabditis elegans* has become a commonly used laboratory (model) organism whose complete genome has been sequenced (see Chapter 12).

Some zoologists have proposed that aschelminths share enough features to unite them into a single clade. However, most zoologists find phylogenetic relationships among the aschelminths difficult to discern, and their distinctions seem great enough to justify classifying them separately. Among them, gastrotrichs are considered the most basic animals; they are aquatic and ventrally ciliated.

Although some of these animals are related, pseudocoelomate features of the others may be the result of **convergence**, which would make the aschelminth assemblage *polyphyletic*, having arisen from one or more ancestral lineages (Winnepenninckx et al., 1995). For example, many zoologists suggest that gastrotrichs, nematodes and nematomorphs share a common heritage in their derivation from a single lineage of acoelan turbellarians, with the four remaining body forms derived independently from other acoelans. There are also suggestions that some aschelminths originally had true coelomic structures but lost them because of severe size reduction.

Coelomates: Protostomes and Deuterostomes

In terms of known numbers of species, distribution of habitats and total mass, the so-called *true coelomates* are among the most successful of all metazoans.

We classify coelomates into 16 to 18 clades (phyla), each characterized to varying degrees by a coelom surrounded by mesodermal tissues (Fig. 16-13). Further features are commonly used to organize coelomates into two major groups, mostly distinguished by *the embryonic location of the mouth:*

- In **protostomes** the mouth develops at or near the blastopore, the opening of the blastula through which the endoderm invaginates to form the gut (see Fig. 15-13).
- In **deuterostomes**, the blastopore develops into the anus, and the mouth develops elsewhere and secondarily (see also Chapter 15).

Although the fossil record is and always will be incomplete for small soft-bodied organisms, most coelomates are known to be ancient, being of Precambrian origin (see Fig. 16-1). Certainly by the Cambrian, the divergence of protostomes from deuterostomes was complete. The Paleozoic marked

[11] See R. B. Clark (1979), Martindale et al. (2002, 2004) and Valentine (2004) for the evolution of the coelom. Recent sequencing of the complete mitochondrial genome and 42 nuclear genes of the priapulid *Priapulus caudatus* indicate very slow evolution of mitochondrial, sufficiently slow that Webster and colleagues (2006) think that priapulids may constitute living fossils to provide insights into what the ecdysozoan genome was like in the Cambrian.

TABLE 16-2 Some characteristics of Pseudocoelomates

Phylum[a]	Approximate Number of Described Species	Adult of a Sample Species	Lifestyles	Habitat	Features
Nematoda (roundworms)	12,000		Both free-living and parasitic	Marine, fresh-water, and soil	Complex flexible cuticle; lack flagella or cilia; tubular excretory system; mostly dioecious sexual reproduction
Nematomorpha (horsehair worms)	230		Adults free-living, but larvae parasitic in arthropods	Mostly fresh-water and damp soil	Thick, flexible cuticle; digestive tract absent in adults; dioecious
Gastrotricha (gastrotrichs)	400		All free-living	Marine and freshwater	Ciliated ventral surface; pseudocoel is diminished or absent; mostly hermaphroditic, but some parthenogenetic reproduction
Rotifera (wheel animals)	1,800		Mostly free-living; some sessile and colonial forms; a few are parasitic	Mostly fresh-water; others in marine habitats and in bryophytes	Ciliated crown; grinding pharynx (mastax); sexually dioecious, but some parthenogenesis
Acanthocephala (spiny-headed worms)	500		Parasitic	Larval stages in arthropods; vertebrates are final hosts	Both the retractable proboscis and body wall covered with short spines; digestive tract absent; dioecious
Kinorhyncha (kinorhynchs)	100		All free-living	Burrow in marine sediments	Segmented cuticle; lack external cilia; have large movable spines on trunk; dioecious
Loricifera	1		Free-living	Marine sediments	Spiny head; telescoping mouth; abdomen enveloped by plates (lorica); two oarlike tail appendages in larvae for swimming and climbing; dioecious

[a]Some authors include an eighth phylum, Priapulida in this group, related to kinorhynchs and loriciferans.

the firm establishment of most metazoan body plans.[12] Early embryonic cleavage patterns often differ between the two, with **spiral cleavage** common in protostomes and **radial cleavage** in deuterostomes (see Fig. 16-4). Spiral cleavage maintains the initial spatial relationships between cells as they cleave so that any differences that develop between regions of the early embryo are maintained in protostomes but not by the radial cleavage seen in deuterostomes.[13]

Despite its advantages, a large coelom means that sustained peristaltic movement involves the entire musculature. Unsegmented coelomic worms such as sipunculids and echiuroids cannot localize their hydrostatic pressures and as a result are generally slow moving and/or relatively sedentary. Selection for **subdivision of the coelom** seems to have taken metazoans in at least two major directions:

- toward **oligomerous** animals, which have three main coelomic areas, the most anterior being an unpaired pocket, the protocoele, followed by a pair of mesocoeles and a posterior pair of metacoeles. Hemichordates such as *Balanoglossus* (the acorn worm), pterobranchs (**Fig. 16-14**) and echinoderms all have three coelomic sacs to various degrees;
- toward animals with **metameric segmentation,** which have multiple body wall divisions, allowing a smooth transition of the peristaltic wave. The sustained, efficient burrowing by oligochaete annelids such as earthworms is a direct consequence of their numerous segments (**Fig. 16-15**).

Segmentation of the entire body by **metamerism** (discussed below) allows the localized establishment of pressure gradients and overcomes generalized coelomic pressures, which affect the entire musculature simultaneously. An important caution is necessary here.

The bodies of organisms in a variety of lineages may appear to be segmented, but the segmentation is not metameric. Tapeworms are the outstanding example, but the "segments" that comprise their long bodies consist almost entirely of reproductive organs and the vast numbers of eggs they produce. In others, external rings running around part or all of the body do not involve a coelom; any segmentation is superficial and more apparent than real.

Segmentation

Metameric or **serial segmentation** of the body along an anterior–posterior axis is found in many metazoans, including annelids, arthropods and chordates (see Chapter 17).

Annelids, whose name means *little ring,* display the most complete evidence of segmentation (illustrated in Fig. 16-2). The whole body, with the exception of the specialized head and tail (pygidium), is comprised of segments. Segmentation is also obvious in most arthropods, although in many the segments are highly specialized and there is a trend toward fusion and/or loss of segments. In vertebrates (see Chapter 17), segmentation is most apparent during embryonic development, but can be seen in the ribs, vertebrae and surrounding muscles of adults. (The role of homeobox genes in establishing segmentation in arthropod and vertebrate embryos is discussed in Chapter 13.)

In annelids and arthropods, organ systems such as nephridia, gonads and nerve ganglia often repeat within each *segment,* or *metamere,* and the segments are commonly marked by constrictions of the body wall musculature and

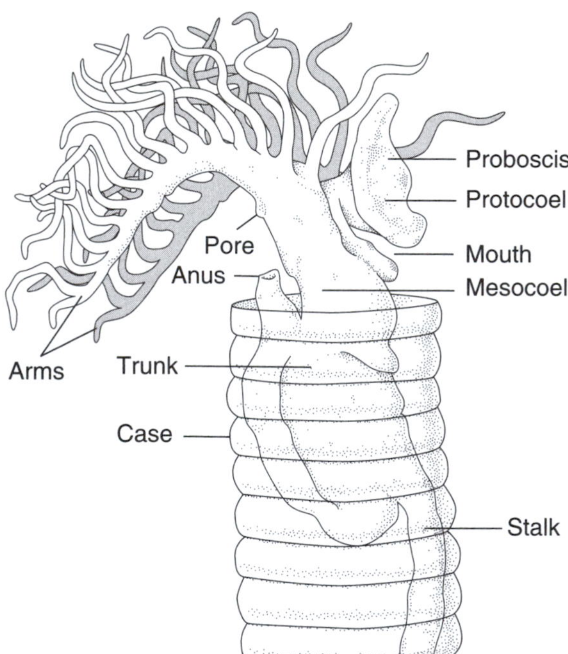

FIGURE 16-14 *Rhabdopleura,* a genus of colony formers (pterobranchs) that zoologists consider related evolutionarily to chordates and place in the Hemichordata, which may be polyphyletic. These individuals have three coelomic regions: the proboscis or cephalic shield contains the protocoel; the collar has a pair of mesocoeles that extend into the tentacle-bearing arms; and the trunk contains the paired metacoel cavities. (Adapted from Borradale, L. A., F. A. Potts, L. E. S. Eastham and J. T. Saunders, 1959. *The Invertebrata,* 3rd ed., revised by G. A. Kerkut, Cambridge University Press, Cambridge, England.)

[12] Because a discussion of each of these phyla represented by these body plans is beyond the scope of this book, only some general evolutionary trends are discussed. For further information, refer to textbooks by Nielsen (2001), Brusca and Brusca (2003), Ruppert et al. (2004) and Valentine (2004) as well as various collections of articles, such as those edited by Conway Morris et al. (1985) and by Jenner (2006).

[13] As a further consequence, protostome development is mostly *determinate* (*mosaic*) because regions of the egg differ; embryonic cells descended from particular egg regions are committed to their fate at early stages and cannot develop into a complete animal when separated from other cells. In most deuterostomes, eggs are more homogeneous, and development is *indeterminate* (*regulative*). If isolated, individual early-cleavage cells generally cannot develop into complete organisms.

by the repetition of coelomic cavities. No extant animals have identical segments throughout — the head and anal segments differ from other metameres in all known cases (Fig. 16-2).

Once segmentation appeared, further locomotory adaptations such as parapodia in polychaete annelids rapidly evolved (**Fig. 16-16**). Nevertheless, despite its locomotory advantages, segmentation has drawbacks. Separated as it is from its neighbor by a septum, each segment must have its own set of organs, such as nerve ganglia, nephridia, gonads and musculature. Consequently, the numbers and kinds of segments as well as the coelomic cavities inevitably become modified or reduced with changes in habit or function.

The presence of metameric segmentation in annelids and arthropods was seen as important evidence for an annelid ancestor for arthropods. The recent discovery of genes and regulatory agents involved in metazoan axial organization, however, shows that the mechanisms responsible for metameric segmentation are common to all metazoans, though expressed in different ways in different lineages.

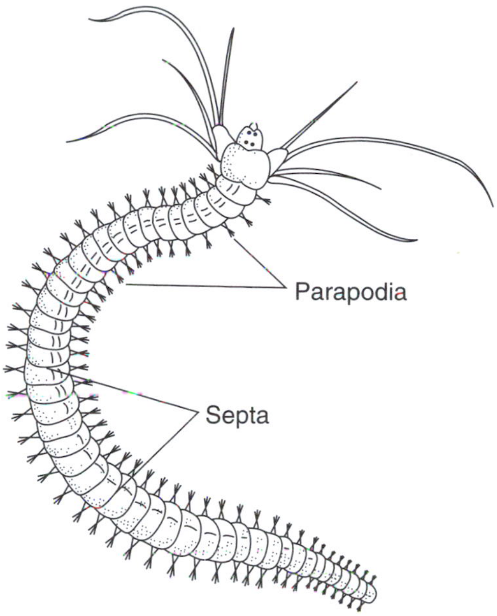

FIGURE 16-16 One of the free-swimming nereid polychaete annelids, *Platynereis*. (Adapted from Smith, J. E., (ed.), 1971. *The Invertebrate Panorama*. Universe Books, New York.)

Supergroups

Homologous developmental genes such as *engrailed*, which specifies compartmental distinctions within segments, function both in protostomes and in deuterostomes (see Chapter 13). Therefore, embryonic segmentation is believed to trace to a common ancestry of protostomes and deuterostomes. Both protostomes and deuterostomes include organisms that are primarily sessile and feed mainly by capturing food particles with ciliated tentacular arms called lophophores, a feature used by Aguinaldo and colleagues in 1997 to create the super group **Lophotrochozoa**, which includes brachiopods and polychaete annelids. Lophotrochozoans share posterior *Abd-B*-like Hox genes. A second super group, **Ecdysozoa**, was erected for nematodes, arthropods and other

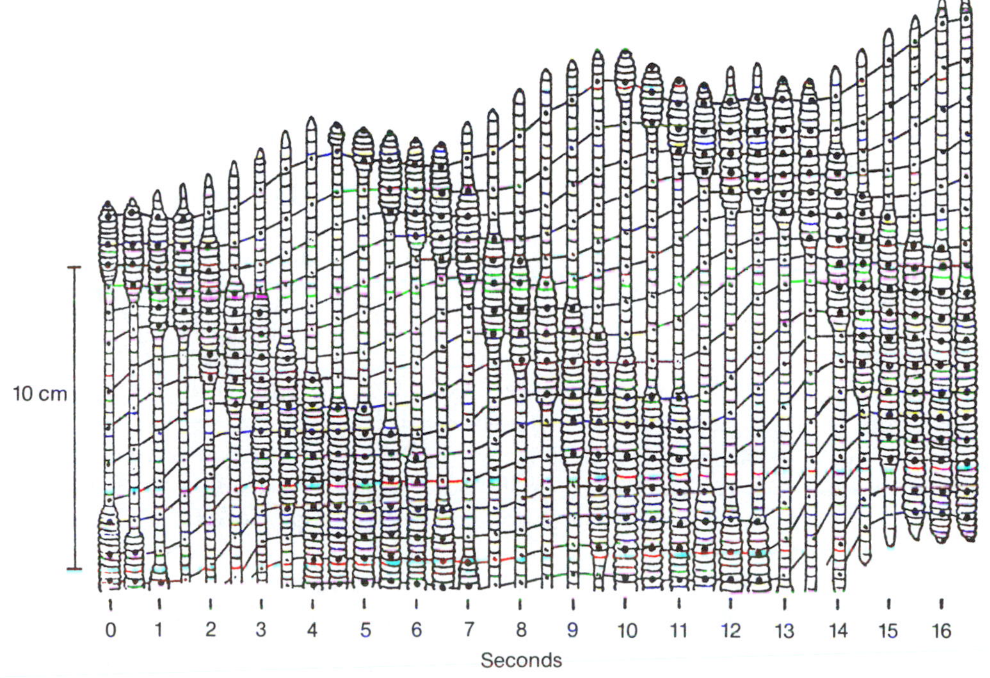

FIGURE 16-15 Peristaltic motion of burrowing earthworms as observed by changes in their segmental diameters. Contracting longitudinal muscles in a group of segments widens these segments and enables the animal to wedge against the sides of the burrow. Segments behind this group are pulled up by further longitudinal muscle contraction, shortening the body and increasing the number of segments in the "anchor." Some of the widened segments in the anchor undergo circular muscle contraction, elongating the body and extending these segments in a forward direction. These elongated segments contract and widen in turn, and the peristaltic cycle repeats. Connecting lines indicate the relative motion of particular segments. (Adapted from Clark, R. B., 1979. Radiation of the metazoa. In *The Origin of the Major Invertebrate Groups*, M. R. House (ed.). Academic Press, London, pp. 55–102; from Gray and Lissmann.)

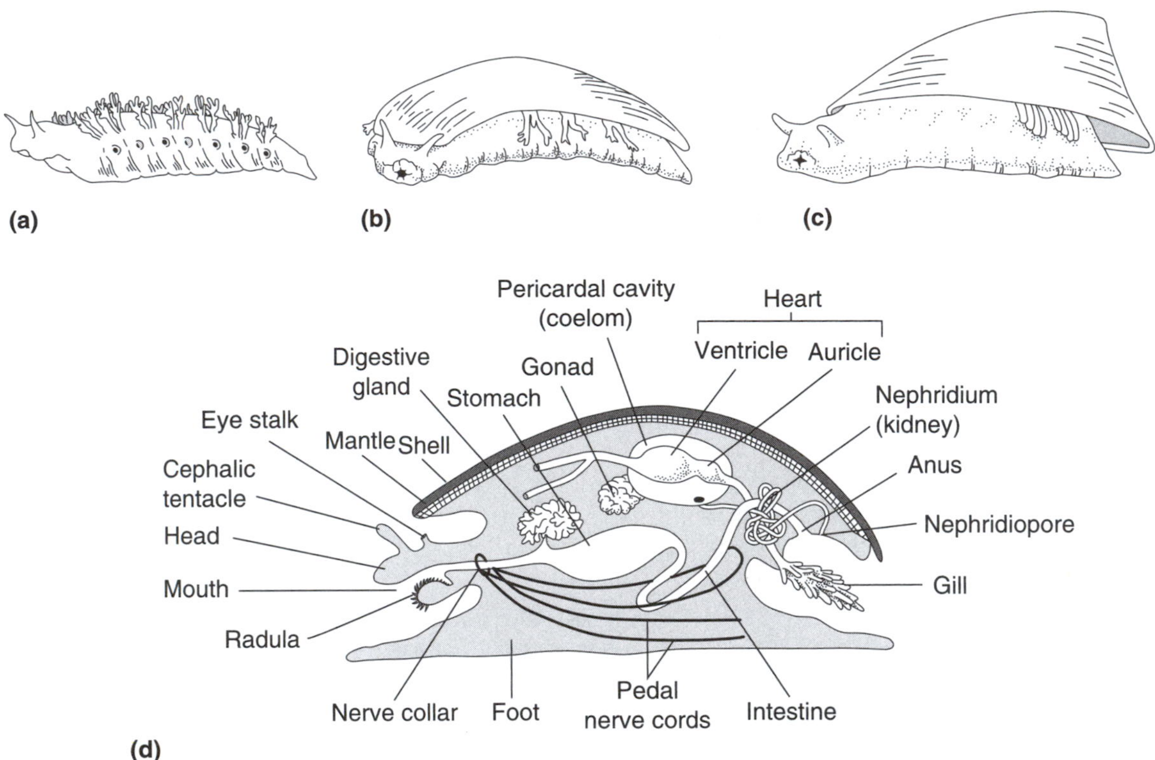

FIGURE 16-17 Hypothetical stages in the early evolution of mollusks. (a) A worm-like benthic creeper that may have had repeated sets of external gills and excretory pores. (b) Evolution of a protective calcified dorsal shield. (c) Development of the shield into a shell that moves forward to cover the head. (d) A lateral view of a later evolutionary stage showing the basic body plan and organs of a hypothetical ancient mollusk. (Adapted from Lutz, P. E., 1986. *Invertebrate Zoology*. Addison-Wesley, Reading, MA; and from Solem, G. A., 1974. *The Shell Makers*. John Wiley, New York.)

lineages that share a cuticular skeleton and metamorphosis (see Box 16-1). The common ancestor of lophotrochozoans and ecydosoans is thought to have had 8 to 10 Hox genes (see Fig. 13-9), indicative of a major expansion and diversification of Hox genes before bilateral clades arose and radiated (de Rosa et al., 1999; Martindale and Kourakis, 1999).

We now discuss examples of three protostome lineages — mollusks, annelids and arthropods, and one deuterostome — echinoderms — before turning, in Chapter 17, to the largest group of deuterostomes, the chordates, including vertebrates.

Mollusks

Mollusca, a clade (phylum) that now numbers over 50,000 species, were perhaps among the earliest of metazoan herbivores, with a body plan based on creeping over shallow marine substrates (**Fig. 16-17**). One of their distinctive features is a **radula**, a rasp-like organ bearing chitinous teeth that unrolls from the mouth and scrapes algae from rocks. Some mollusks abandoned this herbivorous tool because of new feeding habits, but its vestiges remain.

Because of their early herbivorous habits, many **mollusks** have long intestinal tracts, dorsally located so that the ventral creeping surface, the foot, remains free. Protecting and providing respiration for this visceral bulk is a hard shell and a set of active gills enclosed in a body fold, the **mantle.** Various mollusks using this initial architecture have evolved subsidiary changes in structures that are often recognizably molluskan but differ considerably from one another (**Fig. 16-18**). A few are described below.

Gastropods

Gastropods ("stomach-feet") include snails, whelks and limpets, often with a cone-shaped shell that was straight at the outset of gastropod evolution and served not merely as a dorsal shield but also as a protected retreat to house the entire animal. As they evolved into larger forms, spirally coiled shells were selected in various lines. The mantle bearing the gills rotated from facing posteriorly to facing anteriorly (**torsion**) with an accompanying rotation of the viscera. One hypothesis of the function of torsion is that it provided room for the head to be withdrawn and for water to enter the mantle cavity frontally rather than posteriorly. Interestingly, species of opisthobranchs (sea slugs), evolved *detorsion,* with a tendency towards reducing the remaining gill and forming new bilateral gills and bilateral symmetry.

As a consequence of torsion, the products of excretory organs enclosed within the shell could foul the respiratory

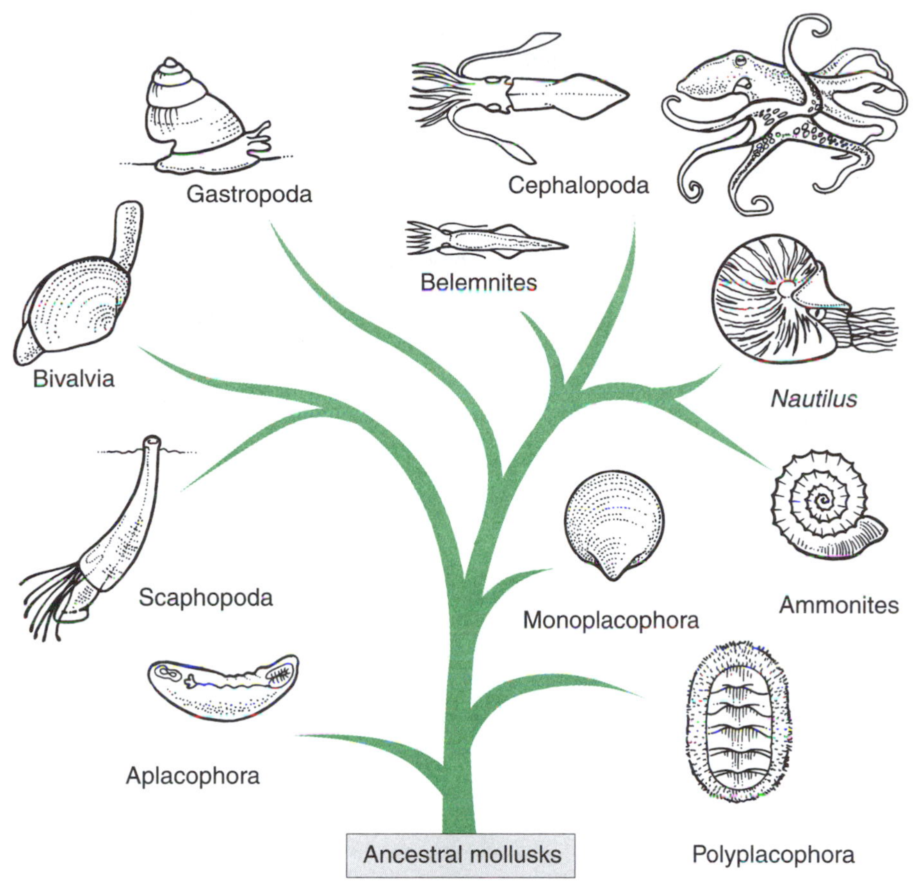

FIGURE 16-18 One phylogeny proposed for the various lineages (classes) of mollusks. Shelled mollusks are found in many fossil strata, some of which (the existing *Nautilus*) can be traced back to the Cambrian. Similar cephalopods with coiled external shells, the ammonites, became extinct at the end of the Cretaceous. The belemnites, marked by internal shells, also became extinct at that time and are allied with extant squids and octopoid forms. Comparisons of other molluskan phylogenies, morphological and molecular, may be found in Winnepenninckx et al. (1996).

gills. In keyhole limpets, the solution is a small hole in the mantle and shell immediately above the anus, positioned so the feces can be voided instead of passed down the gills. A more common solution was to eliminate one member of the original pair of gills so that incoming water circulates through the remaining gill and is propelled outward along the other side, which now contains the openings of the anal and kidney duct.

Bivalves

Bivalvia (also called Pelecypods or "hatchet feet"), such as clams and mussels have hinged shells flattened from side to side. The gills are much larger than in gastropods; most **bivalves** use ciliary tracts on their gills to carry captured food particles to the mouth. Because these particles are relatively small and dispersed, bivalves no longer need the radula and collect food in a sedentary fashion. The head is greatly reduced (sensory orientation is minimal) and the foot either reduced or converted into a burrowing tool. In some bivalves evolution has led to completely separate inhalant and exhalant water currents.

Cephalopods

Cephalopods ("head-feet") include the most mobile of all mollusks, squid and octopi. The cephalopod head, bearing the largest and most complex invertebrate brain, now occupies the main locomotory position, formerly occupied by the foot, and the foot has transformed into a ring of tentacles around the head. The mantle is heavily muscularized and draws water around the sides, squirting it out like a syringe through a tube-like opening, the **siphon.** Cephalopods can aim the siphon in any direction, moving by *jet propulsion* in the opposite direction. This style of life represents a transition from relative passivity to rapid mobile locomotion and aggression and was successful through the Mesozoic, as the large numbers of fossil ammonites and belemnites bear witness.

As a concomitant to hunting, the nervous system and vision developed extensively. The eyes form images that may be as clear as those of the evolutionarily convergent eyes of vertebrates (see Box 3-1, Eye Evolution). Extant cephalopods, with the exception of *Nautilus* and the horn squid *Spirula*, show reduced shell size and complexity; in squid the shell is reduced to a chitinous plate and functions as an internal skeleton. In octopi, only tiny shell remnants remain.

Relationships Among the Mollusks

Other mollusk lineages include two that show some metameric organization, chitons, or polyplacophorans (multiplated shell), and the monoplacophorans (single flat shell).

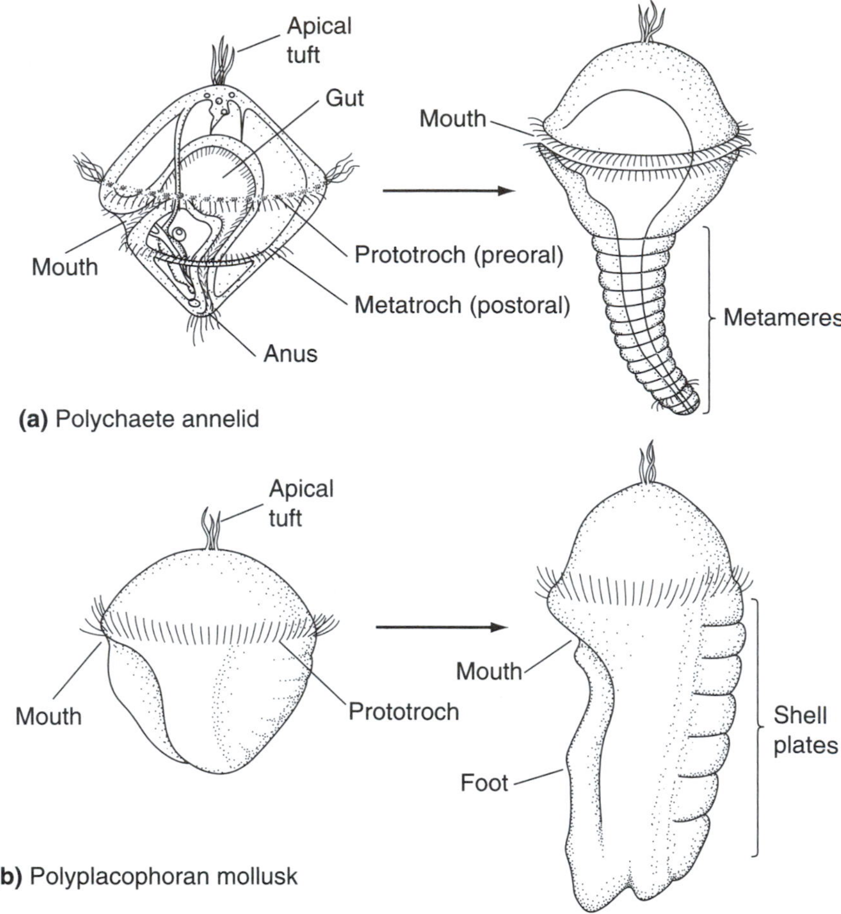

FIGURE 16-19 Trochophore larval stages in an annelid (a) and a mollusk (b). (Adapted from Lutz, P. E., 1986. *Invertebrate Zoology*. Addison-Wesley, Reading, MA; and from Barnes, R. D., 1980. *Invertebrate Zoology*, 4th ed. Saunders, Philadelphia.)

It was thought that the latter had died out at the end of the Devonian, but in the 1950s a Danish expedition discovered a living member, *Neopilina.* Many of its organs, such as gills, muscles and nervous system, show serial repetition, indicating that mollusks and annelids are probably closely related. Supporting this view is the occurrence of a **trochophore larva** in species of molluscs and annelids (**Figs. 16-2** and **16-19**).

Coelomate metamerism was discussed above. Many experts have pointed out that serial repetition of parts in mollusks is not necessarily the same as segmentation in annelids, although both lineages may have shared a common coelomate ancestor. Others carry this possibility further and suggest that the presumed molluskan coelom (its pericardial cavity) originated independently of the annelid-type coelom. Since no other evidence of a mollusk coelom exists, they propose that mollusks arose from benthic acoelomate animals that may have been similar to turbellarians and nemerteans. How this question will be resolved is unclear, although molecular studies (see Fig. 15-8) suggest that the ancestral lineage from which coelomic metameric arthropods are derived also probably gave rise to coelomic metameric annelids and mollusks.

Annelids

An assemblage of about 15,000 species, **annelids** have soft, worm-like bodies with various numbers of segmented coelomic compartments separated by transverse septa. Capping their anterior and posterior ends are specialized structures, the *prostomium* and *pygidium* (Fig. 16-2).

Although unsegmented coelomate worms are burrowers, their habits are relatively sedentary. None engage in continuous burrowing. The sausage-shaped Sipunculida (*peanut worms*) and Echiura (*proboscis worms*) use peristalsis to move slowly through marine substrata. Only segmented coelomates such as annelids localize hydrostatic pressures to specific segments, primarily because of selection to sustain active burrowing.[14]

Of the three annelid lineages (Oligochaeta, Hirudinea and Polychaeta (**Fig. 16-20**), the first represents an offshoot of what was probably an early annelid benthic stock. **Oligochaetes** include the familiar earthworms, which use

[14] A recent issue of the journal *Integrative and Comparative Biology* (2006, 46, 481–568) contains seven papers on annelid structure, development, evolution and relationships.

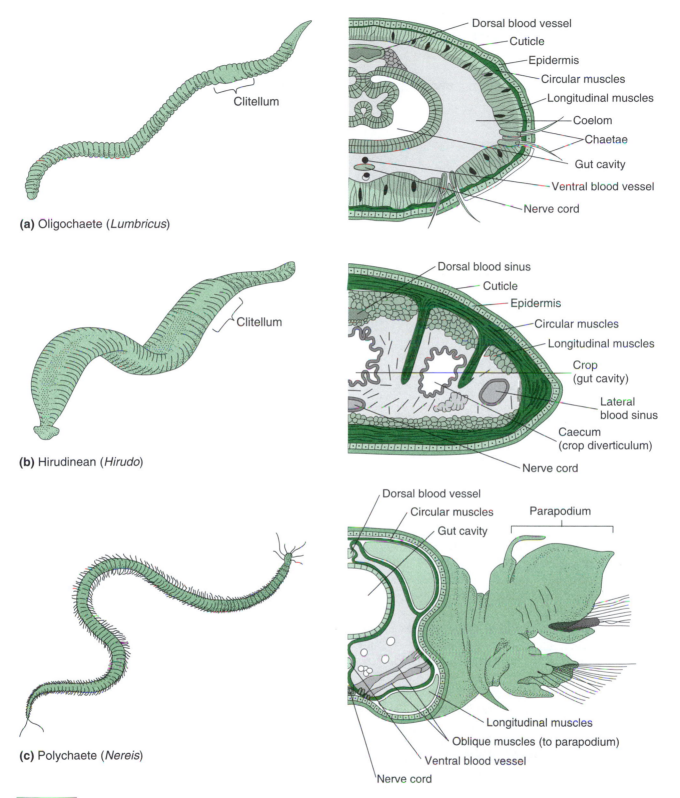

(a) Oligochaete (*Lumbricus*)

(b) Hirudinean (*Hirudo*)

(c) Polychaete (*Nereis*)

FIGURE 16-20 Sample genera in the three major lineages of annelids. On the *left* are entire animals with their anterior ends toward upper right. On the *right* side are diagrammatic transverse cross sections of a segment in each animal.

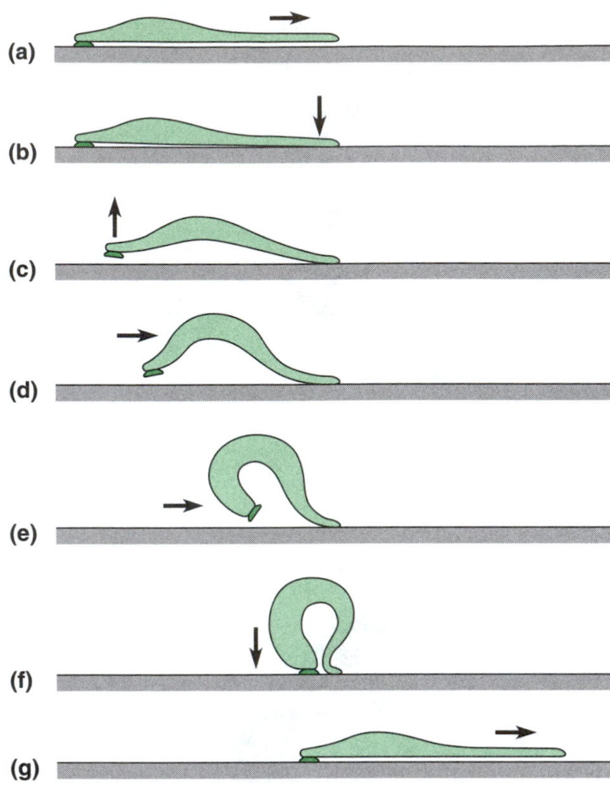

FIGURE 16-21 Stages in the locomotion of a leech, *Helobdella stagnalis*. (a) Posterior sucker (*left*) fixes to the substrate, and anterior end (*right*) extends in direction of arrow. (b) Anterior sucker is fixed. (c–e) Posterior sucker releases, and animal contracts. (f) Posterior sucker fixes immediately behind the anterior sucker, and the cycle begins again (g). (Adapted from Clark, R. B., 1964. *Dynamics in Metazoan Evolution*. Clarendon Press, Oxford, England.)

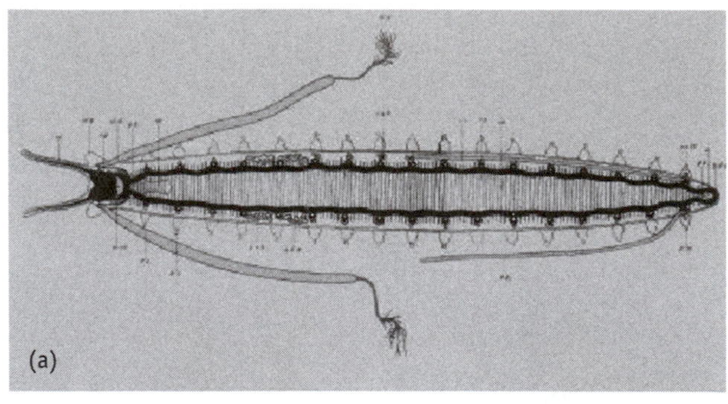

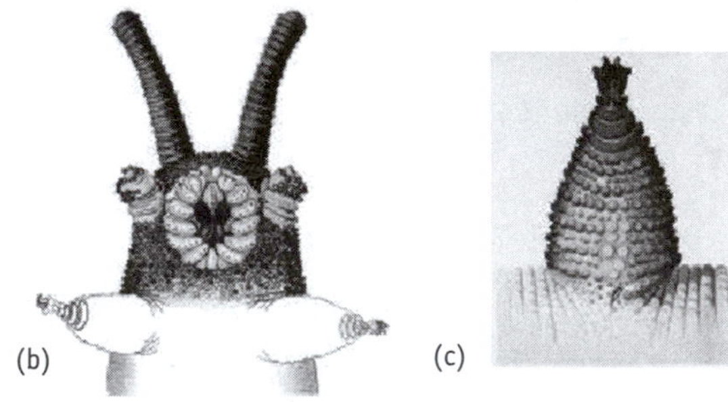

FIGURE 16-22 Three views of the velvet worm (onychophoran), *Peripatus capensis*, showing (a) the nervous and sensory systems, (b) details of the head, and (c) details of one of the clawed appendages. The detailed observation required to produce such images is breathtaking. (Balfor, F. M., 1883. The anatomy and development of *Peripatus capensis*. [Edited by H. N. Moseley and A. Sedgwick]. *Q. J. Microsc. Sci.*, **23,** 213–259.)

spine-like chaetae (or setae) for traction during the continuous burrowing that so enhances soil fertility.

Allied to oligochaetes are Hirudinea (*leeches*). Both earthworms and leeches are hermaphroditic and have a glandular organ called the **clitellum** that usually covers five to ten segments near the anterior end. The clitellum secretes a mucous coat that helps bind two copulating animals together and also produces a "cocoon" for depositing fertilized eggs. Because this organ seems homologous, some consider oligochaetes and hirudineans as subgroups within a "Clitellata" grade. Leeches, however, abandoned burrowing for predation or blood-sucking parasitism, and their locomotory habits are mostly based on "looping," in which they contract the entire circular and longitudinal musculature as a unit and don't use peristalsis at all (**Fig. 16-21**) (see Box 16-3).

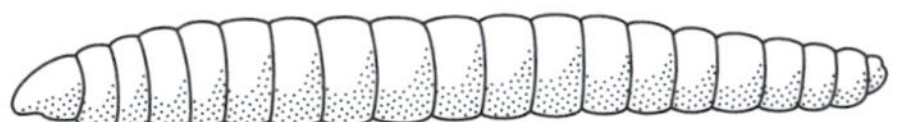

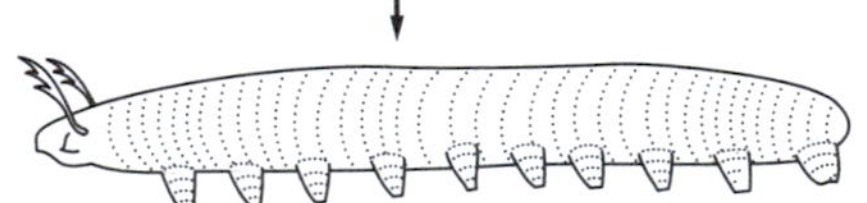

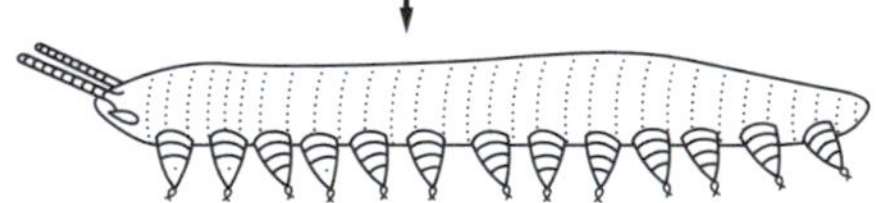

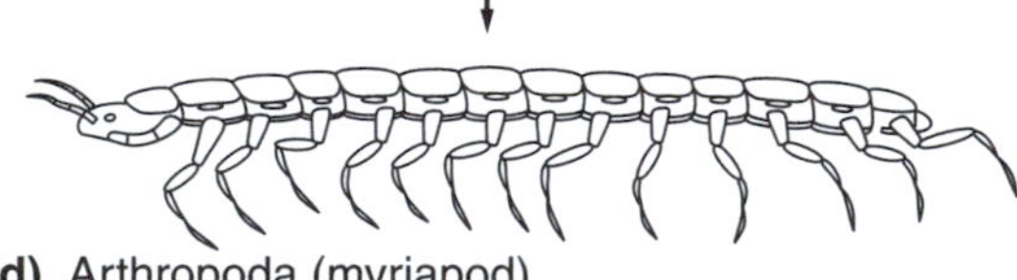

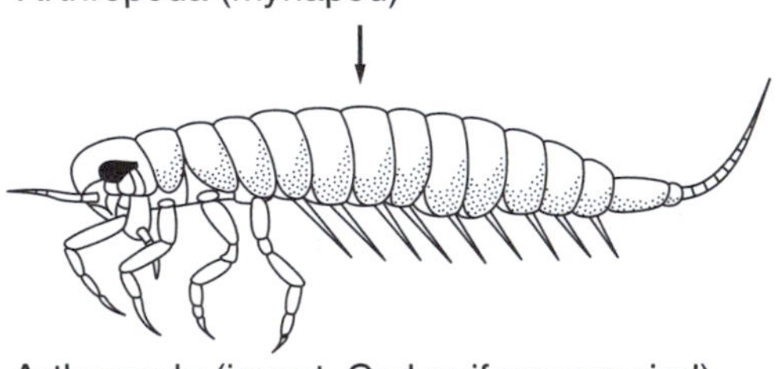

FIGURE 16-23 Evolutionary progression from a segmented worm-like ancestor to arthropods through hypothetical intermediate forms that may have been similar to those found among the Onychophora (see also Figure 13-7). (Adapted from Clarke, K. U., 1973. *The Biology of Arthropods*. Arnold, London.)

By contrast to oligochaetes and leeches, many free-living **polychaetes** such as the marine worm *Nereis* evolved lateral appendages or **parapodia** (Fig. 16-16, see also Figure 14-16), an adaptation for mobility in loose dispersed material. Parapodial animals would have had increased opportunities to feed directly on the bottom rather than hide below it.

Arthropods

In **arthropods**, a tough external layer, the cuticle, which is a complex exoskeleton composed of chitin, proteins, lipids, waxes and cementing substances, provides an exoskeleton with fixed hinges for the jointed appendages. Except in some arthropod larvae that are soft-bodied burrowers, this hardened skeleton eliminates the need for a hydrostatic scaffold, and the coelom contains only the excretory organs.

Although internal septa disappeared, various degrees of segmentation remain, indicating an ancestry that may have progressed through animals similar to onychophorans (*"velvet worms"*; **Figs 16-22** and **16-23**). Molecular studies cited previously (see Fig. 15-8) suggest that segmentation in arthropods and annelids are convergences rather than shared-derived characters.[15]

Exoskeleton

Jointed appendages, a hardened, cuticular exoskeletons with inner projections for muscle attachments, and new kinds of cephalic structures were among the many features providing opportunities for arthropod evolutionary radiation. Today, 80 percent of all animal species are arthropods (mostly insects), found in almost every conceivable ecological habitat.

Because phylogenetic relationships among major lineages are not fully resolved, zoologists classify the diversity of arthropod species in various ways. One common classification system, shown in **Table 16-3,** divides arthropods into four clades, a view supported by the molecular studies of Ballard and coworkers (1992), who also include onychophorans in Arthropoda. Homologous genes for leg development point to a close relationship between at least two arthropod lineages, crustaceans and insects (Panganiban et al., 1997).

A different view, initiated by Manton (1977) and others, is to classify each of the four groups as a separate phylum because each is likely to have originated from a different annelid-like ancestor, probably in Precambrian times; that is, arthropods evolved *polyphyletically.* This view, based on development and morphology, points to embryonic distinctions, differences in leg structure, and differences in metameric segments. Opposed to these conclusions are claims that such differences represent new features derived from a monophyletic origin. In support of monophyly are molecular studies showing that Hox gene expressions affecting head segmentation are the same in chelicerates (spiders, ticks, scorpions), myriapods (millipedes), crustaceans (crabs, shrimp, lobsters) and insects. Some relationships seem uncertain, and firmer conclusions await further sequencing.[16]

Whatever their origins, the armored and articulated arthropod body plan gave them almost immediate success. During the Cambrian, trilobites and other arthropods became dominant marine animals in size, mobility and predatory powers; some fossil eurypterids were almost 3 m long. In the later Paleozoic, these groups declined because of competition with other arthropods and invertebrates as well as with vertebrates (see Chapter 17).

Invasion of the Land: Insects

Arthropod invasion of the land resulted in the evolution of terrestrial arachnids; some terrestrial crustaceans; **insects,** millipedes, centipedes and their relatives known as *uniramians* because all share the feature that their legs consist of a single segmented lobe.

Preadaptations for a terrestrial existence included a hardened cuticle that, along with waxy waterproofing, could act as a barrier to desiccation; ability to burrow into shoreline sand and soil, enabling a terrestrial foothold; and protected gills that, with minor changes, could act as lungs. Among arthropods, *insects* underwent what is arguably the most explosive radiation of any metazoan lineage, occupying practically all of the many varied (and some of the most extreme) terrestrial habitats (**Figs.** 16-23 and **16-24;** see Grimaldi and Engel, 2005 for insect evolution). Entomologists have described approximately three-quarters of a million insect species. Perhaps more than twice that number remain undescribed, many in the tropics.

The modular organization of insects seems to derive from the facility with which regulatory changes, such as homeotic mutations (see Chapter 13), can modify individual segments in a serially repetitive structure and provide a wide spectrum of adaptations. Used in this fashion, different modular replacements and substitutions lead to mosaic evolution or tinkering, a much more rapid tempo of evolution than attempting to fabricate a completely novel architecture from scratch.[17]

Marden and Kramer (1994) suggest that insect wings evolved in aquatic forms that used gill plates for rowing and skimming across water. As evidence of a water-skimming origin, they point to stoneflies, which flap their wings along

[15] A recent issue of the journal *Development, Genes and Evolution* (2006, 216, 355–465) contains ten papers on various aspects of arthropod development, evolution and relationships.

[16] See Regier and Shultz (1997) and Damen et al (1998) for studies on conservation of Hox gene functioning in arthropods.

[17] See Whiting and Wheeler (1994) for the adaptations, and see Duboule and Wilkins (1998), Goode (2007), West-Eberhard (2003) and Wilkins (2002) for tinkering.

TABLE 16-3 Some characteristics of the four major lineages of Arthropoda

Subphylum	Approximate Number of Existing Species	Adult of a Sample Species	Habitat	Pairs of Antennae	Pairs of Legs	Characteristics and Features
Trilobitomorpha			Marine	1	Many	Body with variable numbers of segments organized into three longitudinal lobes; biramous (two-lobed) legs
Trilobites	Extinct					
Chelicerata			Marine	0	5	Anterior appendages are chelicerae (pincers or fangs); uniramous (one-lobed) legs; book gills, waxy cuticle, and muscular pumping pharynx (arachnids)
Merostomes (horse-shoe crabs; eurypterids–extinct)	4					
Arachnids (scorpions, spiders, ticks, mites and others)	73,400		Mostly terrestrial	0	4	
Crustacea						
Crustaceans (crabs, lobsters, shrimp, copepods, isopods, barnacles and others)	40,000		Mostly marine, some freshwater and terrestrial	2	Many	Biramous legs; carapace (dorsal cover); paired compound eyes; larvae when present are of nauplius type.
Uniramia						
Myriapods (centipedes, millipedes)	10,500		Terrestrial	1	Many	Uniramous legs; 2 pairs of maxillae; Malphigian tubule excretory system; paired compound eyes; complete metamorphosis through pupal stages (holometabolous and hemimetabolous insects); respiration via air tubules (trachea)
Insects (flies, beetles, bugs, bees, locusts and so on)	750,000		Terrestrial aerial, and aquatic	1	3	

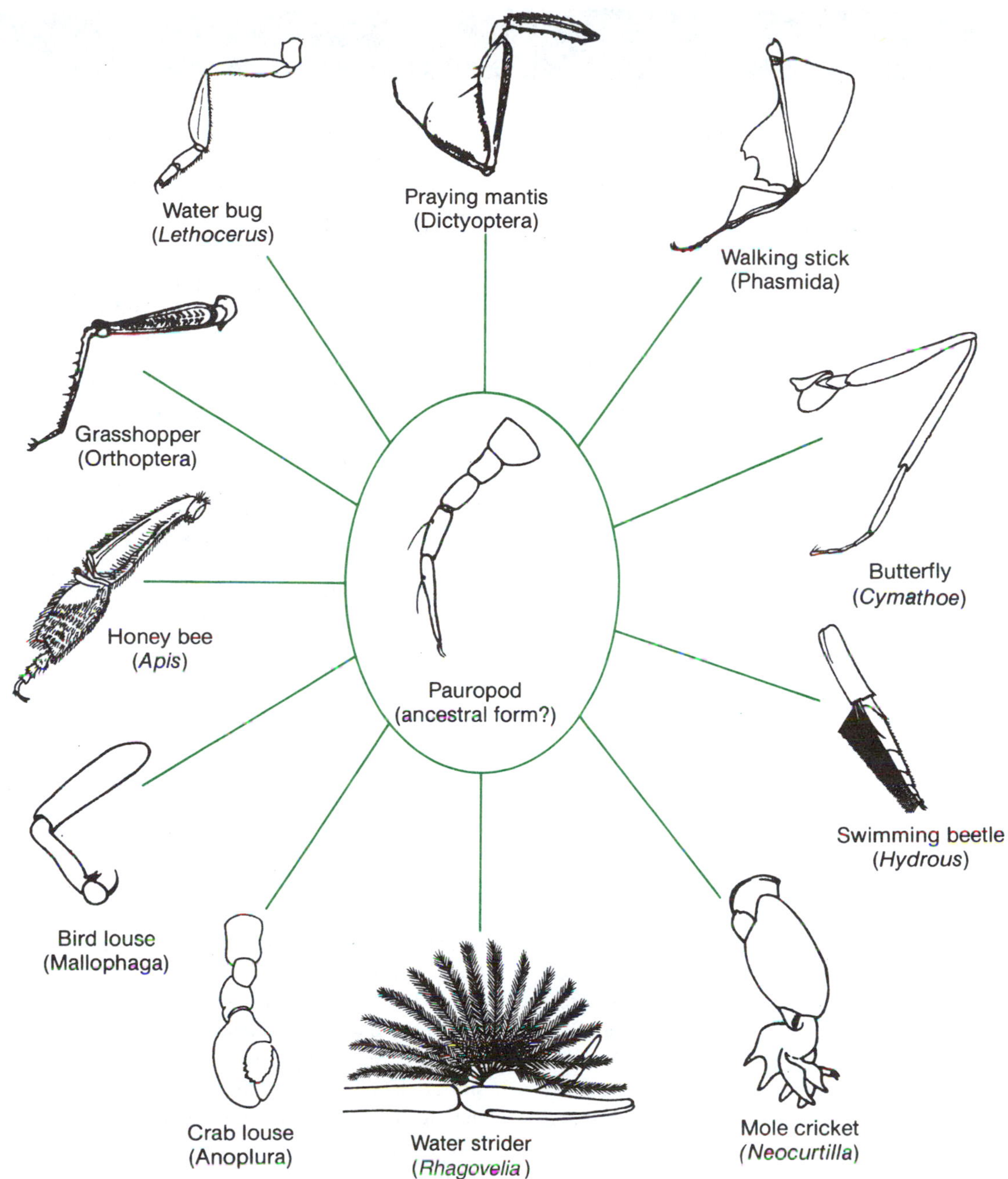

FIGURE 16-24 Legs in different insect lineages. (Adapted from Fox, R. M., and J. W. Fox, 1964. *Introduction to Comparative Entomology*. Reinhold, New York.)

the water surface to gain aerodynamic thrust without aerodynamic "lift." Once evolved for such purpose, stronger and larger water-skimming "wings" could undergo selection for aerial flight. The genetic basis for transforming gill appendages into wings may have come from changes in regulatory genes used in developing gill structures from biramous (branched) crustacean legs. Some genes, such as crustacean homologues of the *Drosophila* gene *engrailed,* regulate the development of anterior–posterior compartments, while others, such as homologues of *apterous,* initiate a dorsal–ventral pattern (Averof and Cohen, 1997).

Evolution of Insect Larvae

Other advantages accrue to insects that possess **specialized larval stages** living and feeding differently than adult forms. With this divided lifestyle, especially those with winged adults, insects can finely partition their resources and exploit a wide range of environments and diets. Beetles, with more than

BOX 16-4 Kin Selection and Haplodiploidy in Social Insects

In termites, both sexes derive from fertilized eggs, but in hymenopterans males develop from unfertilized (haploid) eggs and females from fertilized (diploid) eggs. This hymenopteran system of **haplodiploidy** is especially prone to evolve sociality since, as depicted in **Figure 16-25,** the diploid female offspring of a queen share more genes with their sisters (75%) than they share with their own daughters (50%).

Genetic relationship leads to a conflict of interest for a colony's sex ratio. Being more related to sisters, it is to the workers' genetic benefit that the colony supports more females and fewer distantly related males. Queens, by contrast, being related equally to sons and daughters, benefit genetically when males and females are produced in equal numbers. Because workers control egg-to-adult development, fratricide can be expected at various stages (Seger, 1996). When the queen mates more than once, genetic relationships between female workers decline, but a worker is still more closely related to her mother's female offspring than to her sister's female offspring, which accounts for the destruction of workers' eggs by other workers. The genetic relationship between workers and male offspring follows a somewhat different pattern, depending on whether queens mate once or more than once.

Although the haplodiploidy hymenopteran system may produce different sex ratios based on various environmental factors, in the parasitoid wasp *Lariophagus distinguendus,* which parasitizes the larvae of common granary weevils, *Sitophilus granaries,* the female lays differently sexed eggs in response to the size of the wheat grain containing the larval host (Charnov, 1982). If the wheat grain is relatively large, the wasp inserts a single fertilized female egg into the weevil larva; if the grain is relatively small, an unfertilized male egg is injected. This difference can be traced to the difference in resources needed for the fertility of male and female offspring: A larval host in a larger grain enables female offspring to be more fecund because the larger grain supplies more resources, whereas a larval host in a smaller grain still supplies enough resources to enable a male offspring to produce a large number of viable sperm.

According to Hamilton (1964), haplodiploidy encourages **kin selection** (a name coined by John Maynard Smith), in which it is to the genetic advantage of females to invest their energy in raising sisters (sisters are more closely related to them)than in producing daughters, who are more distantly related. In evolutionary terms, this means that the **altruistic behavior** of sterile female workers in helping raise more sterile worker progeny for their mother (the queen) is a phenomenon we would expect to arise frequently in Hymenoptera. The independent evolution of altruism in three different hymenopteran lineages — at least once in ants, eight times in bees, and twice in wasps **(Table 16-4)** — is consistent with Hamilton's proposal.

Although the benefits of sociality seem obvious, not all haplodiploid hymenopteran species are social, and evolutionary biologists are still not sure of its specific causes. They do agree that among likely socializing factors are opportunities for communal nesting, joint protection against predators and parasites, and shared foraging for food. Cooperative behaviors implementing such traits would have had high selective value among genetically related individuals, conferring special advantages on the group as a whole (*group selection*), even overcoming social problems caused by increased infective disease. (See Boute et al. (1996) and Boury-Esnault et al. (2003) for type IV collagen.)

300,000 described species, have an additional adaptation in the extra-thick exoskeletal "armor," the cuticle that protects even their vulnerable wings against predators and parasites.

Once such highly adaptive traits appeared, arthropods could enter and thrive in almost any small vacant terrestrial niche. Among metazoans occupying terrestrial insect-sized niches, insects triumphed by arriving first and adapting first.

Social Organization Among Insects

The only social organization other than that seen in vertebrates (if we exclude colonial organization) occurs in insects.

Social organization entails a division of labor among different members of a group, a phenomenon far beyond the kinds of simple aggregation exemplified by swarms of migrating locusts or the cooperative "tents" that some caterpillars construct. Social insect societies include termites (Isoptera) and various ants, bees and wasps (Hymenoptera).

In these socialized insects, different morphological types (castes) or age groups assume different social functions. Each colony, for example, usually has only one fertile queen engaged in egg production, one or more (sometimes many more) fertile males for egg fertilization, and one or more classes of sterile workers exclusively engaged in food gathering, cooperative brood care, nest maintenance and defense of the colony. In worker honeybees, age-related divisions of labor primarily tie younger bees to caretaking and maintenance, and older bees to food foraging and defense. The completion of the sequencing of the honeybee (*Apis mellifera*) genome has already revealed differences in gene expression associated with castes and foraging.[18]

These behavioral differences correlate with differences in juvenile hormone, younger bees having lower levels than older bees. Hormones provide one of the mechanisms linking and integrating environmental changes to gene action and to changes in the phenotype, as is especially evident

[18] See The Honeybee Genome Sequencing Consortium (2006) for the genome sequence. The one example of caste formation known from outside social invertebrates occurs in the naked mole-rat *Heterocephalus glaber,* in which complete sexual dimorphism occurs between reproductive and helper females, a dimorphism that is initiated after reproduction by a lengthening of the lumbar vertebrae (O'Riain et al., 2000).

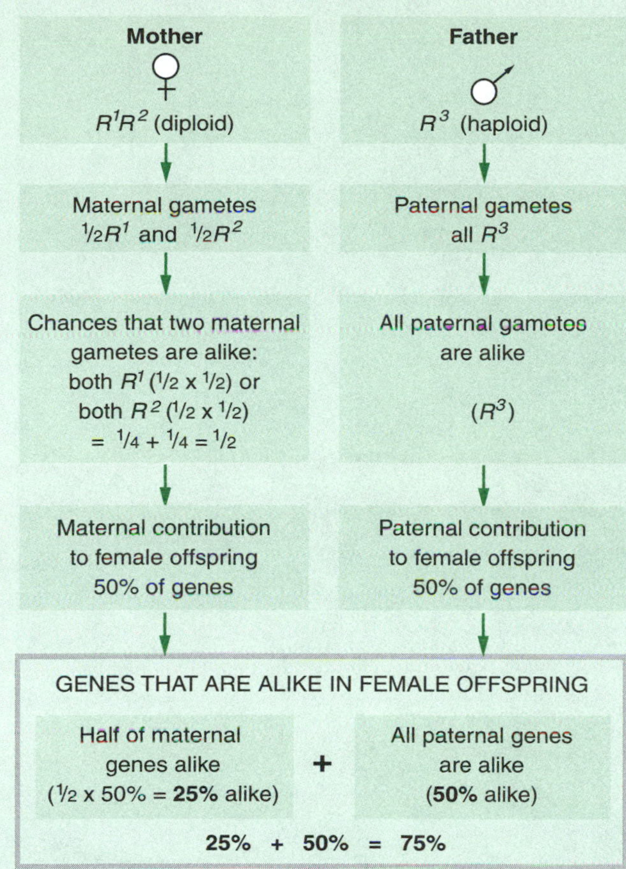

FIGURE 16-25 Schematic illustration of the effect of hymenopteran haplodiploidy on the relationship between sisters ("workers") derived from the same mother ("queen") and father, using three alleles of a hypothetical gene, *R*. Fifty percent of the genes shared by such workers come from their haploid father, who gives the same set of genes to all his daughters. All these female offspring share half their remaining genes (0.50 × 0.50 or 1/4 of their total complement) because genes in different haploid eggs produced by their diploid mother have the probability of being 50 percent alike. The total frequency of shared genes among workers is therefore 0.50 + (0.50 × 0.50) = 0.75. In contrast, these workers (R^1R^3 or R^2R^3) share only 50 percent of their genes with their queen mother (R^1R^2), since queen and workers have different fathers (for example, R^2 and R^3 males), who contribute 50 percent of all female genes. This helps account for workers' reliance on their mother to produce their more closely related sisters than engage in producing their own more distantly related daughters or allow their sisters to produce even more distantly related female offspring. (Adapted from Ratnieks, F. L W., and P. K. Visscher, 1989. Worker policing in the honeybee. *Nature,* 342, 796–797.)

TABLE 16-4 One proposed sequence of behavioral changes leading to the evolution of sociability and caste divisions in wasps

1. Female stings prey, then lays egg.
2. Female stings prey, places it in a convenient niche, then lays egg.
3. Female stings prey, constructs a nest on the spot, then lays egg.
4. Female builds a nest, stings prey, transports it to nest, then lays egg.
5. Female builds a nest, stings and transports a prey item, lays egg, then mass-provisions egg with several more prey, added before egg hatches.
6. As in (5) but prey items are progressively provided as the larva grows.
7. As in (6) but progressive provisioning occurs from the start.
8. In addition to provisioning in a nest, female macerates prey items and feeds the pieces directly to the larva.
9. The founding female is long-lived, so that offspring remain with her in the nest. Offspring add cells and lay eggs of their own.
10. Small colony of cooperating females exchange liquid food.
11. Behavioral differences between a dominant queen caste and a subordinate worker caste appear; unfertilized workers may still lay male (haploid) eggs.
12. Larvae are fed differentially; queen and workers that result are physically distinct, but intermediates remain common.
13. Worker caste is physically strongly differentiated, and intermediates are rare or absent.

Source: Hölldobler, B., and E. O. Wilson, 1990. *The Ants.* Harvard University Press, Cambridge, MA; after Evans, H. E., 1958. The evolution of social life in wasps. *Proc. 10th Int. Congr. Entemol.*, **2**, 449–457.

in organisms such as insects with one or more phases in the life cycle. Chemical influences on development and behavior also differentiate fertile queens and sterile workers. The queen produces "pheromones" that suppress fertility in workers, a suppression that can be overcome by various environmental influences such as special foods (in honeybees, royal jelly). Circadian rhythm (usually on a 23- to 24-hour cycle) varies with age because of different expression of a gene homologous to the *Drosophila* gene *period,* which is turned on in foragers, imparting diurnal activity,

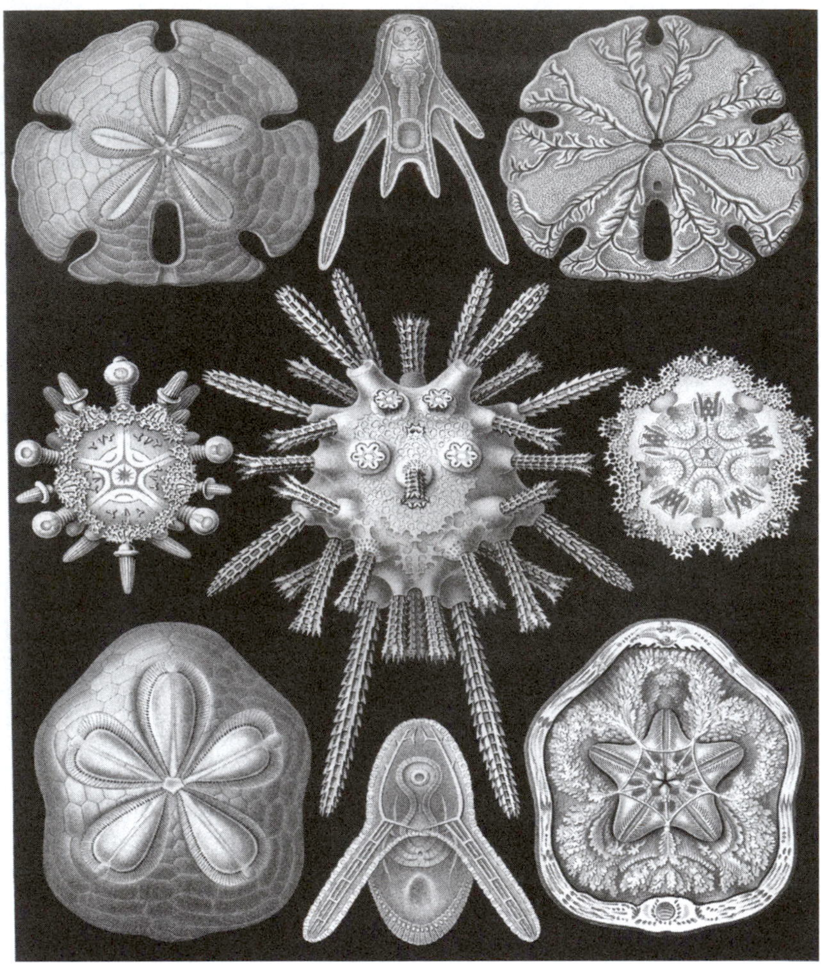

FIGURE 16-26 A diversity of echinoids (sand dollars) and Pluteus larvae (middle, top and bottom) as drawn by Ernst Haeckel. (From Haeckel, E., 1904. *Kunstformen der Natur.*)

and turned off in younger bees, which perform brood care in an irregular daily pattern.[19]

The development of different sexes or life forms from fertilized and unfertilized eggs and kin selection characterize some forms of social insects (**Box 16-4**).

In summary, insect success extends to social and non-social forms and has prevailed in almost every terrestrial environment since at least the Carboniferous. Apart from insects and vertebrates, relatively few lineages of invertebrates have colonized the land as successfully, and those that did — nematodes, earthworms and gastropods, for example — are often confined to special humid or soil-like conditions. Nevertheless, it is interesting to note that although some early insects reached relatively large sizes — some Carboniferous dragonflies measured 0.6 m between wing tips — they have tended to remain small, especially by comparison with most vertebrate species. There are limits to the volume of tissue in which insect tracheal tubules can effectively exchange gases. An even more limiting factor may be that the insect exoskeleton would have had to become much heavier and more unwieldy as insects became larger.

[19] For studies on environmental influences, see the chapters in Hall and Olson (2003), Hall et al. (2004), and see Flatt et al. (2005), Abouheif (2004) and Emlen et al., (2005). See Robinson (1998) for circadian rhythms in *Drosophila.*

Echinoderms

Echinoderms (Echinodermata) are the second largest group of deuterostomes after chordates (see Chapter 17), containing at present about 7,000 species. These "spiny-skinned" invertebrates are mostly marine bottom-dwellers that have three major distinguishing features:

- a mesodermal skeleton of small calcite plates or spicules (often fused in sea urchins) lying just below the epidermis (**Fig. 16-26**);

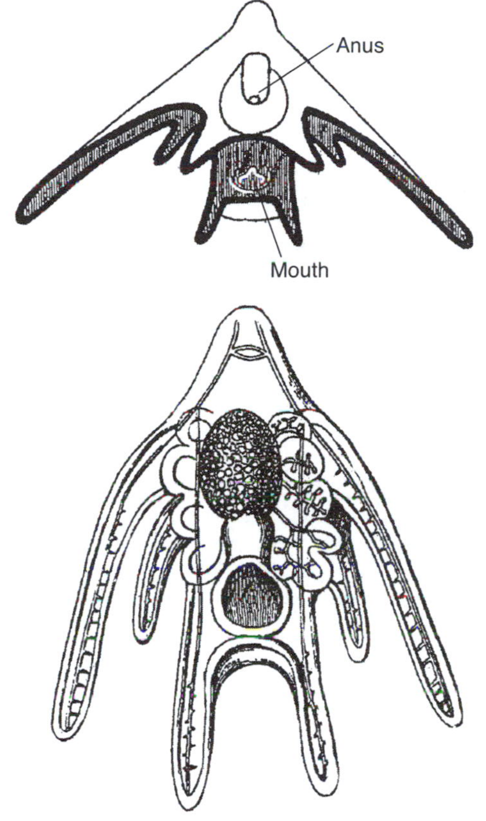

FIGURE 16-27 Representations of Pluteus larvae from brittle stars (echinoderms). Anus (An) and mouth (M) are identified.

- bilateral symmetry in Pluteus larvae (**Figs.** 16-26 and **16-27**), followed mostly by a five-rayed (**pentameral**) **symmetry** of the adult; and
- a water vascular system, derived from the coelom, in which tube feet used for locomotion and feeding connect to internal canals of circulating seawater.

All extant and extinct clades (classes) of echinoderms are known from various Paleozoic strata, indicating that this lineage evolved considerably before they appeared in the fossil record.

Given that their larval forms are pelagic and bilateral, it has long been assumed that echinoderms arose from a free-living, deuterostome, bilaterally symmetrical ancestor.[20]

The coelomic pouches eventually subdivided to include the specialized water vascular system, the perivisceral coelom surrounding the gut, and various sinuses used for circulatory purposes. Presumably some echinoderm features, such as the osmotic similarity between their coelomic fluids and seawater and their simple nervous system, date back to this early stage. Echinoderms lack excretory or osmoregulatory organs and have no anterior specialization of nerve cells or ganglia that we could call a brain.

According to Arenas-Mena and coworkers (1998), echinoderm larval development may hold a clue to the evolution of some or many bilateral metazoans. Only a small number of homeobox genes are expressed in larvae, the remainder functioning in adult development. They propose that this pattern is basic for bilateral animals, which evolved from simple larval-type organisms — needing only few regulatory genes — to more complex adult body-plan stages utilizing batteries of homeobox regulators.

Following an early free-living existence, an ancestral echinoderm probably adopted a sessile mode of life attached to the substrate, with oral surfaces upward and surrounded by food-gathering lophophore-like tentacles. Many Paleozoic echinoderms, including the eocrinoids, blastoids and others, show the radial organization that accompanies such sessile habits. Interestingly, although they lack head structures, they retain the same Hox genes used for developing anterior regions in other deuterostomes (Martinez et al., 1999). Again, regulatory genes can be channeled in different groups to perform different functions.

The genome of an echinoderm, the purple sea urchin, *Strongylocentrotus purpuratus,* the collaborative effort of 240 researchers from 70 institutions worldwide, was published as this section was being completed.[21] It contains 23,300 genes, of which 10,000 have been studied. One third of these genes are found in humans, but others only in birds or mammals among the vertebrates. Although echinoderms lack vision and smell, the purple sea urchin genome has 1,000 genes involved in vision or smell in other metazoans, a number highly unlikely to have arisen by convergence (especially as the sensory function on which selection would have acted are missing), but indicative of shared origins. Echinoderms have innate immune systems, as do humans, but ours plays a small role in comparison with induced immunity. The sea urchin has ten times more genes devoted to innate immunity than do we, opening fascinating possibilities for understanding the evolution of immunity.

In answer to the question, "Why did echinoderm radial symmetry assume a pentamerous form?" we don't yet know the answer from the genome sequence, but zoologists have offered at least two hypotheses. One claims that an organ-

[20] Investigations into genes known to be asymmetrically expressed in chordates have revealed asymmetrical expression in a sea urchin and in a starfish, but the expression is inverted in the echinoderms in comparison to chordates; *Pitx,* a gene expressed on the left in chordates, is expressed on the right in the echinoderm larvae, suggesting inversion of the left-right axis (Hibino et al., 2006).

[21] The Sea Urchin Genome Sequencing Project Consortium (2006). Also see the special issue "The sea urchin genome" of the journal *Developmental Biology* (2006), 300(1), which contains 36 papers based on the Sea Urchin Genome Project.

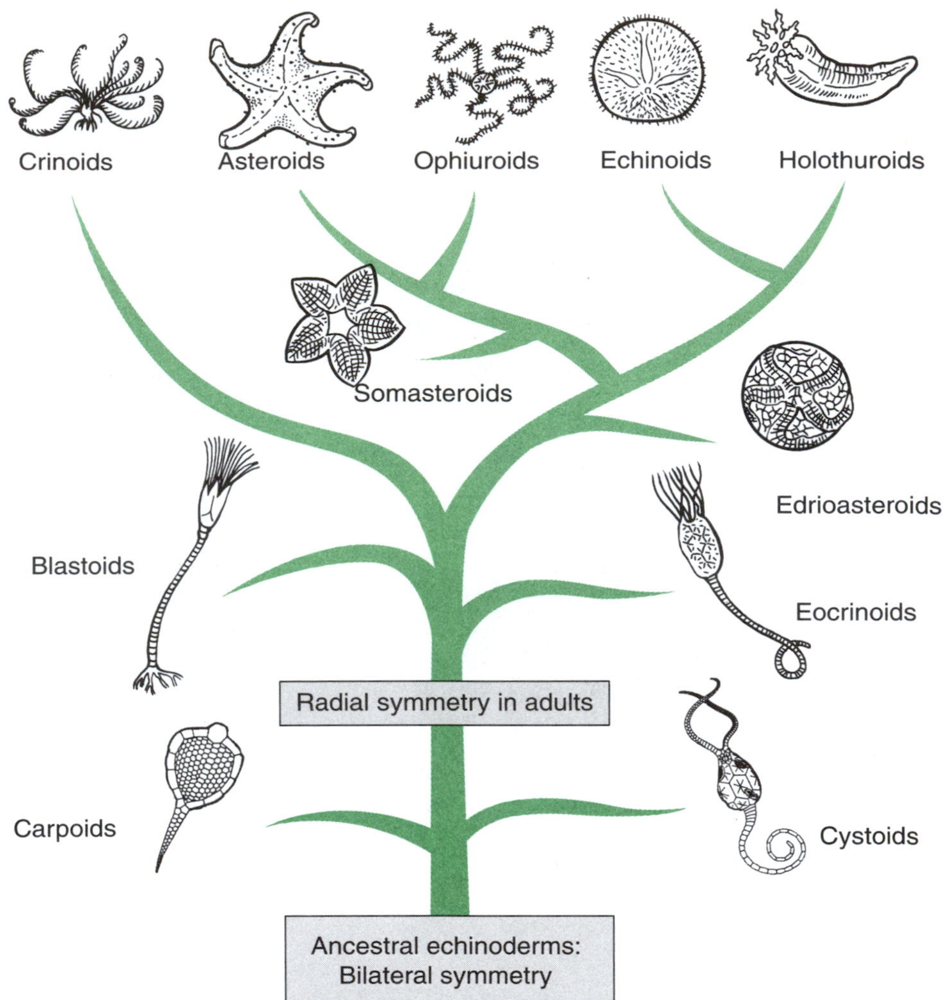

FIGURE 16-28 A phylogeny of echinoderm lineages arising from an ancestor that was radially symmetrical. The five groups at the *top* are extant; those *below* are extinct. (See Brusca and Brusca, 2003, and Smith, 1997, for echinoderm larval evolution.

ism with a five-tentacle lophophore (the pentactula) was the echinoderm ancestor, and its pentameral radial symmetry persisted in later forms. According to the second hypothesis, pentameral organization ensured that torsional strains that would tend to cleave a suture between plates surrounding an essential central area of the animal would not easily transmit to a suture that was immediately opposite; that is, in an animal with peripheral plates, a structure such as

is less likely to break apart than

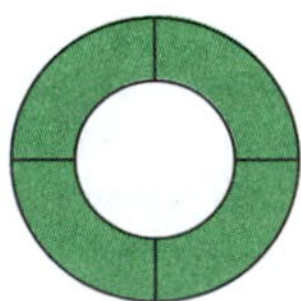

While echinoderm ancestry remains hidden, more recent work is consistent with a sessile ancestor that did not have pentameral symmetry. Once established, however, their calcite skeleton meant that echinoderms left an excellent fossil record, sufficient to provide tentative relationships among most major echinoderm lineages. **Figure 16-28** illustrates one possible phylogeny.

KEY TERMS

altruistic behavior
annelids
archaeocyte
arthropods
Aschelminthes
bivalves
bauplan (body plan)
cephalopods
choanocytes
clitellum
Cnidaria
coelenterate
deuterostomes
diploblastic
echinoderms
gastropods
haplodiploidy
insects
invertebrates
kin selection

Lophotrocozoa
mantle (mollusc)
medusa
mesozoans
mollusks
nematocysts
nematodes
oligochaetes
parapodia
parasitism
pentameral symmetry
Placozoa
planula
Platyhelminthes
polychaetes
polyp
protostomes
pseudocoelomates
radial cleavage
radula
siphon
social organization
spiral cleavage
sponges
superphylum
torsion
triploblastic
trochophore larva

DISCUSSION QUESTIONS

1. Should we use the criteria for classifying metazoan phyla (Table 16-1) to determine their phylogenetic relationships? Why or why not?
2. Would you consider sponges more related to early metazoans than to cnidarians (coelenterates)? Explain.
3. How could the complex life cycle of various platyhelminth parasites have evolved?
4. Would you unite the various pseudocoelomate groups shown in Table 16-2 into a single phylum, Aschelminthes? Why or why not?
5. What major features do zoologists use for separating protostome from deuterostome coelomates?
6. What are some changes in foot and shell that occurred in the evolution of major molluskan groups?
7. Would you consider mollusks to have originated from a metameric coelomate? Why or why not?
8. What explanations can you offer to account for the general absence of transitional forms between the major animal phyla?
9. What prevented annelids with appendages (e.g, polychaetes) from radiating into as many evolutionary niches as arthropods?
10. Why is haplodiploid reproduction (haploid egg males, diploid egg females) important in the evolution of socially organized insects?
11. Considering their widespread evolutionary radiation, why are insects generally small in size compared to vertebrates?
12. What factors can account for the pentameral symmetry of echinoderms?
13. To which, if any, invertebrate phyla can we ascribe polyphyletic origins? Explain.
14. How would the development of a coelomic cavity have influenced the evolution of locomotion in wormlike organisms?

EVOLUTION ON THE WEB

Explore evolution on the Internet! Visit the accompanying Web site for *Strickberger's Evolution*, Fourth Edition, at http://www.biology.jbpub.com/book/evolution for Web exercises and links relating to topics covered in this chapter.

RECOMMENDED READING

Brusca, R. C., and G. J. Brusca, 2003. *Invertebrates,* 2nd ed. Sinauer Associates, Sunderland, MA.

Carroll, S. B., 2005. *Endless Forms Most Beautiful.* W. W. Norton, New York.

Carroll, S. B., J. K. Grenier, and S. D. Weatherbee, 2005. *From DNA to Diversity: Molecular Genetics and the Evolution of Animal Design,* 2nd ed. Blackwell Publishing, Malden, MA.

Grimaldi, D., and M. S. Engel, 2005. *Evolution of the Insects.* Cambridge University Press, Cambridge, UK.

Hall, B. K., and W. M. Olson (eds), 2003. *Keywords and Concepts in Evolutionary Developmental Biology.* Harvard University Press, Cambridge, MA.

Hervé, P., N. Lartillot, and H. Brinkmann, 2005. Multigene analyses of bilaterian animals corroborate the monophyly of Ecdysozoa, Lophotrochozoa, and Protostomia. *Mol. Biol. Evol.*, 22, 1246–1253.

Nielsen, C., 2001. *Animal Evolution: Interrelationships of the Living Phyla,* 2nd ed. Oxford University Press, Oxford, UK.

Raff, R. A., 1996. *The Shape of Life: Genes, Development, and the Evolution of Animal Form.* University of Chicago Press, Chicago.

Ruppert, E. E., R. S. Fox, and R. D. Barnes, 2004. *Invertebrate Zoology: A Functional Evolutionary Approach,* 7th ed. Brooks Cole, Thomson Learning, Belmont, CA.

Valentine, J. W., 2004. *On the Origin on Phyla.* The University of Chicago Press, Chicago.

17 Chordate and Vertebrate Origins

Chapter Summary

Vertebrates have several shared-derived chordate features such as pharyngeal clefts, a notochord, an internal skeleton derived from mesoderm and neural crest, a tail and a hollow dorsal nerve cord. Chordate ancestry has been sought among annelids, arthropods, and, more plausibly, from a common ancestor with echinoderms, given their embryological similarities.

Some extant chordates are thought to more closely resemble the vertebrate ancestor than do others. Although clearly not ancestral to vertebrates — no extant group can be an ancestor of another extant group — cephalochordates are related and have a swimming and filter feeding lifestyle, thought to be characteristic of putative protovertebrates. Many larval urochordates have internal gill slits, a notochord and a dorsal nerve cord, structures that are lost in the sessile adults. If such larval types underwent paedomorphosis and reproduced, they might have evolved into free-swimming chordates and, eventually, vertebrates.

Until conodonts were recognized to be vertebrates, jawless "fishes" were always regarded as the most ancient fossil vertebrates. Many were covered with dermal bony plates and scales. They had a muscular pharynx, which they at first used for filter feeding and pumping water but that later evolved into a predatory organ. Lampreys and hagfishes are extant jawless vertebrates.

The evolution of jaws was a major evolutionary advance, resulting in gnathostomes. (jawed vertebrates). Jaws have long been thought to have evolved from pharyngeal arches used for filter feeding, an adaptation that enabled fish to become effective carnivores. The earliest jawed vertebrates, acanthodians and placoderms, died out to be replaced by cartilaginous fishes (Chondrichthyes) and bony fishes with gas bladders or "lungs" (Osteichthyes). Both lineages have elaborate and highly evolved fins, including a caudal fin, dorsal and anal fins for stability and paired pectoral and pelvic fins. Interestingly, cartilaginous fishes appear later in the fossil record than bony fishes. In many lineages, bone offers many advantages over cartilage, particularly as a supporting tissue.

Bony fishes are divided into ray-finned fish (Actinopterygii) and lobe-finned fish (Sarcopterygii). Living members of ray-finned fishes are sturgeon, gar and teleosts. Sarcopterygians retain the osteichthyan pineal eye, lungs of more ancient actinopterygians, developed unique fin supports and often had internal nostrils. If assessed by numbers of species, teleosts are the most successful vertebrates, with 24,000 species, reflecting changes in skull, fin and scales structures associated with feeding and locomotory adaptations.

The lungs of ancient actinopterygian were modified into swim bladders. Lungs arose in forms inhabiting freshwater environments (where oxygen content is low) and with colonization of the land. Extant lungfishes (Dipnoi) and coelacanths — the latter discovered last century, and previously thought extinct — are living fossils. Also among sarcopterygian fossils are osteolepiforms and panderichthyids, which are on or close to the lineage that gave rise to the first land vertebrates, tetrapods.

Separating all vertebrates from all invertebrates as another group reflects an exclusive quality we humans confer on organisms that resemble us compared to organisms that don't. To add insult to injury, we define all nonvertebrates on the basis of the absence of a fundamental vertebrate feature, the vertebral column. Vertebrates are chordates, represented by about 47,200 species that range in size from minuscule fishes and amphibians to giant whales, and that have invaded a wide variety of habitats from oceanic depths to high altitudes.

What features characterize **chordates**? **Figure 17-1** shows a composite of representatives of two chordates, a fish and a human. Although widely separated along the chordate and vertebrate spectra, the figure portrays some of the special morphological features (synapomorphies, homologies) of chordates:

- A paired series of **pharyngeal (branchial) clefts** that lead outward from the pharynx. Commonly present in embryos as transient features, they persist as gill clefts in the adults of most vertebrates (see Fig. 3-10).
- An **internal skeleton**, oriented along the anterior–posterior (rostral-caudal, head-tail) axis, that develops around a flexible rod, the **notochord.** Although a column of cartilaginous or bony **vertebrae** augment and even replace the notochord in the adults of many groups, the notochord gives the group the name **Chordata.**
- A single **hollow nerve cord** that runs **dorsal** to the notochord.
- A **tail** that extend beyond the anus in embryos of all groups (*postanal tail*) although not in all adults.

FIGURE 17-1 Composite drawing of a human and a fish, showing various vertebrate characteristics. (Reproduced from Ohno, S., 1996. The notion of the Cambrian pananimalia genome. *Proc. Acad. Sci.*, **93**, 8475–8478.)

Hypotheses of the Origin of Vertebrates

Since the nineteenth century, zoologists have investigated vertebrate origins using morphological, embryological, and, more recently, molecular comparative methods. We examine three hypotheses: *origins among annelids/arthropods, among echinoderms, or among cephalochordates, urochordates and hemichordates.*

An Annelid–Arthropod Origin

Among early proposals were suggestions for an *annelid* or *arthropod* origin based on some general similarities between the two lineages (**Table 17-1a**). However, as shown in **Figure 17-2**, such a transition meant positioning the ventral annelid/arthropod nerve cord dorsally to attain a dorsal vertebrate nerve cord and reversing the direction of blood flow; the animal had to be turned upside down and its mouth relocated from top to bottom. Because such morphological manipulations were radical, because they involved changing adult annelids or arthropods rather than changing embryos, and because of other major differences between them (Table 17-1b), this hypothesis was abandoned, although it has resurfaced with the accumulation of molecular data.

We now know that among genes affecting dorsal–ventral development in arthropods and vertebrates are two homologous pairs.

- One pair — *decapentaplegic* (*Dpp*) in *Drosophila* and *bone morphogenetic protein* (*bmp*) in vertebrates — produces proteins that dorsalize arthropods and ventralize vertebrates.
- The other pair — *short order gastrulation* (*sog*) in *Drosophila* and *chordin* in vertebrates — ventralizes arthropods and dorsalizes vertebrates.

The finding that the genes in each pair share common ancestry, and that they elicit complementary dorsal–ventral developmental activity, suggests that a common arthropod–vertebrate ancestor gave rise to both body plans, and that one is the inversion of the other. This most probably occurred in a small, soft-bodied Precambrian lineage that left few or no fossils, making such transitions difficult to visualize. So, while vertebrates arose neither from annelids nor from arthropods, these three lineages share a

TABLE 17-1 Summary of the main morphological types of evidence for and against an annelid/arthropod origin for vertebrates

	Annelids/Arthropods	Vertebrates
(a) Evidence consistent with an annelid origin		
Bilateral symmetry	Yes	Yes
Presence of coelom	Yes	Yes
Metameric organization	Yes	Yes
Terminal growth	Yes	Yes
Dorsal and ventral longitudinal blood vessels	Yes	Yes
Anterior "brain"	Yes	Yes
(b) Evidence not consistent with an annelid origin		
Complete segmentation through the body wall	Yes	No
Coelom embryology	Schizocoelous	Enterocoelous
Position of nerve cord	Ventral	Dorsal
Skeleton[a]	External	Internal
Pharyngeal clefts	No	Yes
Flow of dorsal blood vessel	Anteriorly	Posteriorly
Flow of ventral blood vessel	Posteriorly	Anteriorly
Fate of blastopore	Mouth (protostome)	Anus (deuterostome)

[a] See Hall (2005a) for further information on skeletal evolution in both vertebrates and invertebrates.
Source: Adapted from Neal, H. V., and H. W. Rand, 1939. *Comparative Anatomy*. Blakison, Philadelphia.

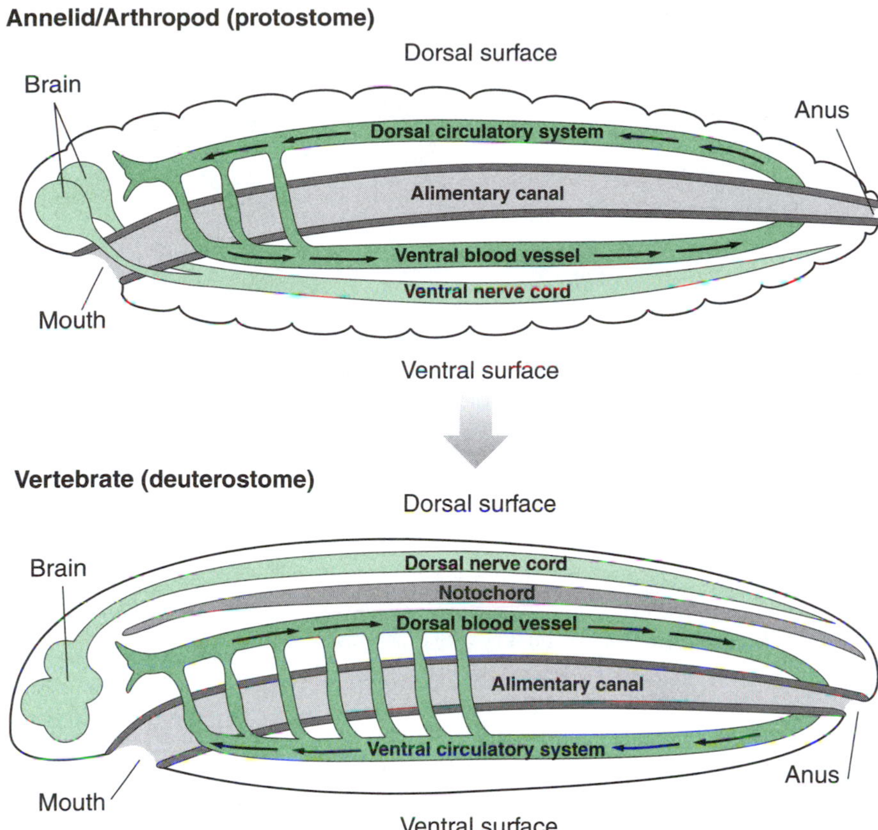

FIGURE 17-2 Diagrammatic illustration of proposed annelid/arthropod–vertebrate homologies that could account for a protostome–vertebrate transformation if the animal was turned upside down. Should such a reversal occur, the ventral nerve cord would be situated dorsally, and blood would flow anteriorly in the ventral aorta and posteriorly in the dorsal aorta. Counterarguments against this hypothesis are that such reversal would place a new vertebral mouth on the newly ventralized surface, while a notochord would have to be inserted as a new structure between the alimentary canal and the nerve cord. Such drastic manipulations found little support among zoologists. How would an organism feed and move while these major reconstructions were going on?

deeper connection, sometimes referred to as **deep homology**, than anticipated by even the most ardent nineteenth century enthusiast of an annelid–arthropod origin of vertebrates.[1]

Echinoderm Affinities

Probably the most popular hypothesis today is a shared ancestry for echinoderms and chordates. This concept rests, for the most part, on a variety of traits — many discovered in the latter half of the nineteenth century — that biologists have long presumed to be strongly conservative and use to distinguish deuterostomes from protostomes, discussed in Chapter 16. The most recent manifestation of these conservative features follows the analysis by Aguinaldo and colleagues (1997) uniting echinoderms, chordates and hemichordates as deuterostomes, to the exclusion of protostomes as Ecdysozoa and Lophotrochozoa (see Box 16-1). Traits considered to unite echinoderms and chordates include:

1. In deuterostomes, the **blastopore** produces the adult anus, whereas in protostomes the blastopore becomes the mouth.

2. *Radial cleavage* of early zygotic cells (blastomeres) rather than the *spiral cleavage* that seems the rule in the protostomes (see Fig. 16-4). Furthermore, because isolated blastomeres of echinoderms and amphibian vertebrates develop into normal embryos, zoologists consider their developmental process *indeterminate*, in contrast to the incomplete embryos produced by isolated blastomeres undergoing spiral cleavage (*determinate development;* see Chapter 16).

3. A feature little subject to evolutionary change is the origin of the **coelom** from the wall of the developing gut in both echinoderms and chordates in contrast to splitting within the mesoderm in annelids (see Chapter 16). However, in **hemichordates,** the sister group to echinoderms (see below), the coelom arises by both modes, depending on the species examined. Consequently, this feature is too plastic, and/or reflects convergent evolution, to be used in the either/or manner in which it was used in the past.

4. The skeleton was proposed to originate embryonically from mesodermal tissue in vertebrates and echinoderms but from ectodermal tissue in annelids. Or so was the view until the discovery of the **neural crest** (an ectodermal derivative; **Fig. 17-3**) as a source of

[1] For the molecular studies see DeRobertis and Sasai (1996), Arendt and Nübler-Jung (1994, 1999), Ferguson (1996), and Gerhart (2000). For nineteenth century views, see Nyhart (1995) and Bowler (1996).

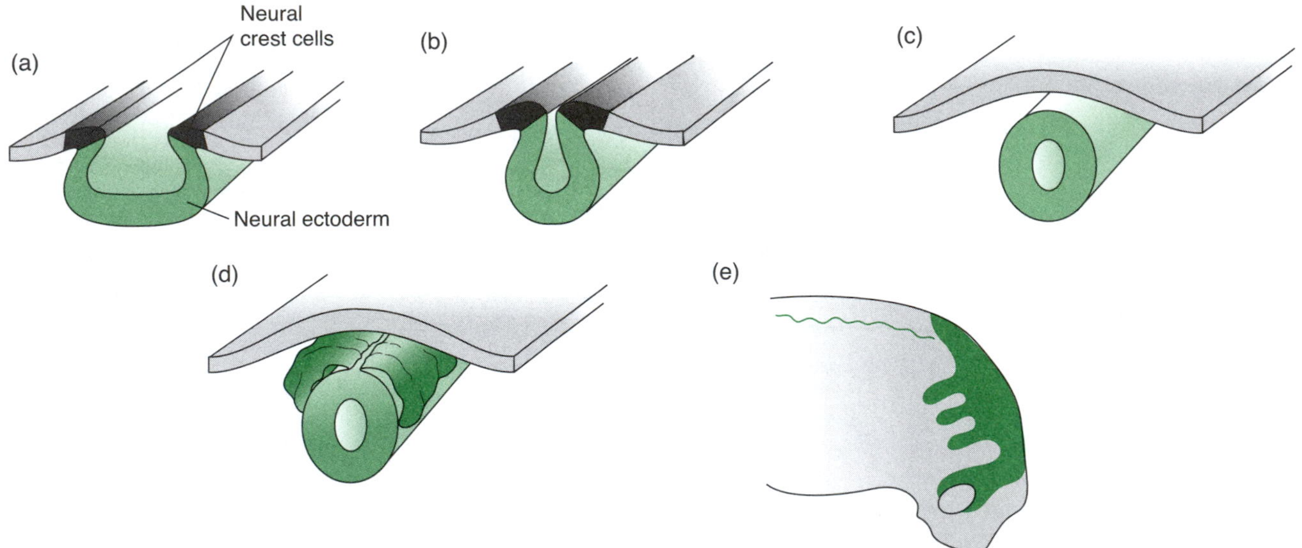

FIGURE 17-3 (a, b) A diagrammatic cross section through a vertebrate embryo at the time of initial formation of the neural tube from neural ectoderm (black). The crests of the neural folds contain neural crest cells (black). (c–e) As the neural tube closes over and the neural folds close to form the neural tube (c), neural cells begin to migrate from the neural tube (d). Subsequent migration (e) is of distinct populations of neural crest cells that surround the eye and move into individual branchial arches.

skeleton-forming cells in vertebrates, complicated this otherwise nice dichotomy of mesodermal versus ectodermal skeletons. The presumed common mesodermal origin of the skeleton in echinoderms and vertebrates is now difficult to accept since this structure differs so greatly in composition and pattern between them.[2]

5. Echinoderm larvae are of the auricularia type as seen in holothurians (sea cucumbers), or variations of it, such as the pluteus larvae (see Fig. 16-26). By contrast, annelids produce trochophore larvae. Because the acorn worm, *Balanoglossus* (**Fig. 17-4a**), has an auricularia-type larva (*tornaria*) and as other lines of evidence suggest that this animal is a chordate (hemichordate), it follows that echinoderms and vertebrates are closely allied (Fig. 17-4c, d). Homology with mammalian genes of three of the Hox gene clusters in the hemichordate *Saccoglossus kowalewski* is also consistent with a close relationship of hemichordates to chordates; **Box 17-1**).

Although many biologists accept some of this evidence for echinoderm–chordate affinity, others have raised interpretations against almost every point. For example, the blastopore does not always develop consistently in a particular clade: In some so-called protostome annelids, arthropods, brachiopods and mollusks, the blastopore closes completely and the mouth forms elsewhere. Cleavage patterns are also not well defined: other than in some crustaceans, arthropods do not show spiral cleavage, nor is cleavage consistently determinate in mollusks (Valentine, 1997).

Larval similarity between *Balanoglossus* and some echinoderms may be the least questionable of chordate–echinoderm affinities, but even that idea is difficult because it is based on accepting *Balanoglossus* as a chordate-like invertebrate — a point that has been argued ever since 1886 when Bateson proposed that the *Balanoglossus* preoral gut diverticulum was a notochord (Fig. 17-4b). *Balanoglossus* does share features with chordates, including pharyngeal clefts, and molecular similarities such as shared expression of *Pax1/9* in the endoderm of the pharyngeal clefts and of homeobox-containing (Hox) genes used in spatial organization.[3]

Many biologists now emphasize nucleotide sequence comparisons for determining phylogenetic relationships. Such studies show a close relationship between echinoderms and chordates. Keeping in mind their large differences in body plan — one bilaterally linear (chordates) and the other pentamerally radial (echinoderms, although the larvae are bilaterally symmetrical) — we may still accept a common ancestry and recognize that it must have been far in the past, certainly Precambrian. Once we reach the Cambrian, the search for chordate ancestors becomes much more encouraging. Recently discovered fossils represent, in many ways,

[2] See Northcutt and Gans (1983), and Northcutt (2005) for what is often referred to as "the new head hypothesis;" M. M. Smith and Hall (1990, 1993) and Hall (1999b, 2000a,b) for the neural crest and skeletal development/evolution, and **Box 17-2** and Stone and Hall (2004) for the neural crest. The discovery that the neural crest forms much of the craniofacial skeleton, and the prominent involvement of ectodermal tissues in forming enamel in vertebrate teeth and shark scales, present further challenges to the presumed restriction of vertebrate skeletal formation to mesodermal tissue.

[3] See Pendleton et al. (1993) and Ogasawara et al. (1999) for these studies in *Balanoglossus*.

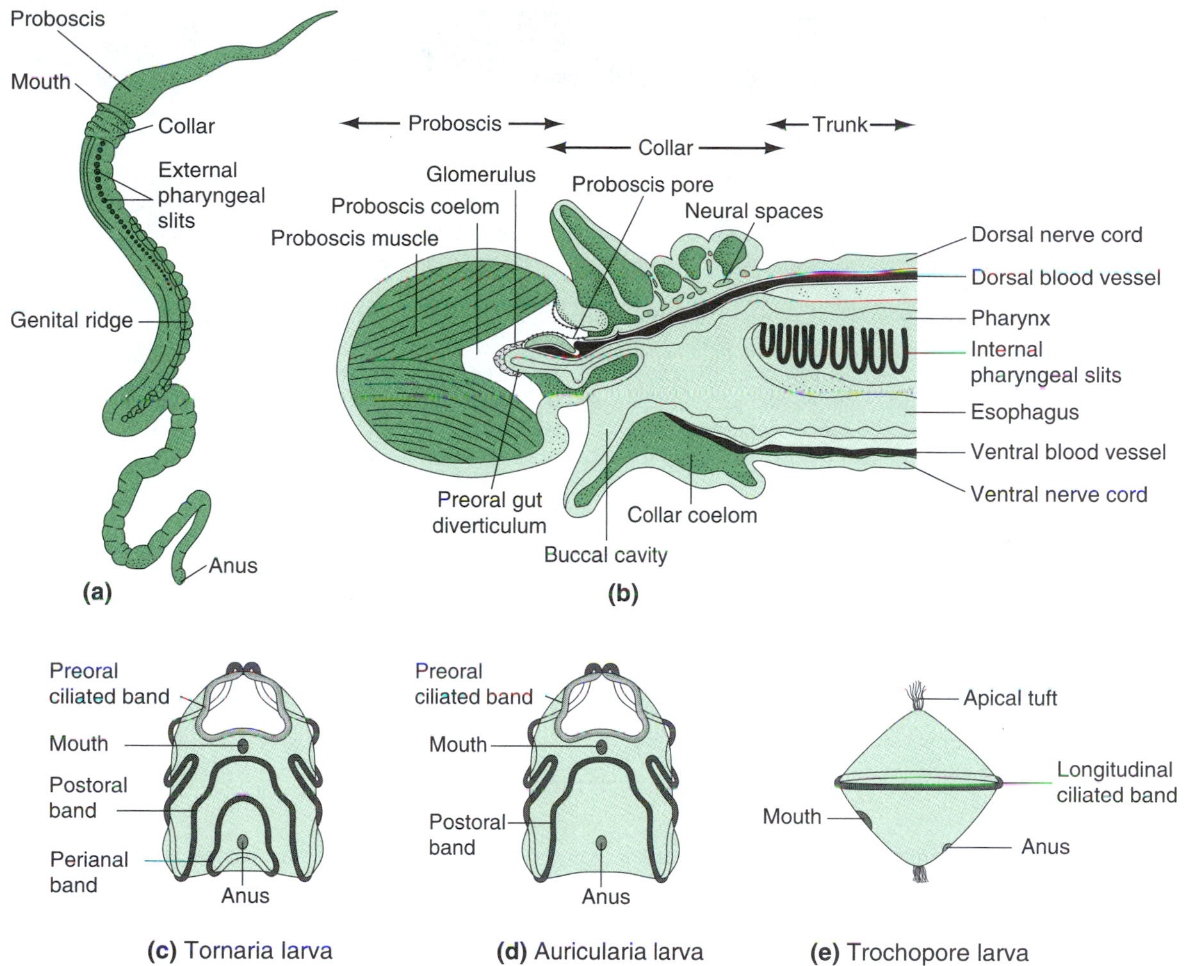

FIGURE 17-4 (a) Adult acorn worm of the genus *Balanoglossus*. (b) Internal structure of the anterior region of a *Balanoglossus* species. Some zoologists considered the pre-oral gut diverticulum to be homologous with the notochord, a view no longer held. (c) The Tornaria larva of *Balanoglossus*. (d) The auricularia larva of echinoderm holothurians. The major difference between (c) and (d) is the presence or absence of a perianal ciliated band. (e) The trochophore larva found in annelids and mollusks (see also Figures 16-2 and 16-19).

the culmination of research carried out over a century and a half by zoologists into small marine animals — cephalochordates and urochordates — that were thought to bear some relationship to vertebrates. The likelihood of a marine and estuarine origin of early vertebrates is strengthened by the observation that cephalochordates, urochordates, and hemichordates (whose relationships are discussed below) are all marine fauna, found mostly in shallow waters.

Cephalochordates

In addition to vertebrates, biologists usually classify several other clades within the Chordata, all of which share features that prompted zoologists to suggest what early chordates were like. Among these the most vertebrate-like are **cephalochordates** such as the marine lancelet, *Branchiostoma*, known by the common name of **amphioxus.**

These 25- to 50-mm–long marine animals swim by contracting metamerically organized muscles (myotomes) placed alongside a semi-rigid notochord that runs from tip to tail (**Figs. 17-5** and **17-6**). In early chordates, whose supportive skeletons were most likely similar to the notochords of tunicate larvae (**Figure 17-7**), muscles would have inserted directly onto the notochord. This was followed by segmentation of the dorsal mesoderm (but not of the notochord) by vertical septa at repeated intervals (as seen in the ammocoete larva of the lamprey *Petromyzon*). Such organization would provide increased mechanical advantage for longitudinal muscles.

Adult cephalochordates are mostly sedentary, burrowing in the sea bottom and exposing their anterior end into the waters above to filter feed. They have a large pharyngeal "branchial basket," penetrated by up to 200 perforations, that filters water drawn through the mouth by ciliary currents

BOX 17-1 Hox Genes and Vertebrate Origins

THE RECOGNITION OF THE importance of Hox genes (see Chapters 12, 13, 16) marks a fascinating episode in the history of the search for chordate and vertebrate origins.[a]

Vertebrate homeobox (Hox) genes with sequence homology to such *Drosophila* genes as *Ultrabithorax* and *Antennapedia* are a series of transcription factors organized as homeobox clusters on four chromosomes in vertebrates. As in *Drosophila*, the order of Hox genes within a cluster is paralleled by an anterior–posterior sequence of expression. The *patterning role* of Hox genes is demonstrated by findings, in mice, that knocking-out or knocking-in a Hox gene to eliminate or enhance its function is followed by transformation of skull, vertebral or other features into a more anterior element in the sequence, ***a transformation known as homeotic.*** Regenerating amphibian limbs can be duplicated following manipulation of Hox genes. In some species of frogs, an amputated tail can be made to duplicate, homeotically, the posterior portion of the body and to regenerate a limb complete with pelvic girdle and not a tail (Müller et al., 1996). Finally, conservation of the roles of vertebrate and *Drosophila* genes has been demonstrated by research showing that, for example, after being transfected into *Drosophila*, the mouse Hoxb-6 gene elicits leg formation in the place of antennae in *Drosophila*.

The number of Hox clusters varies among vertebrates: four clusters of 39 genes in mice, three clusters in lampreys, and up to seven in teleost fishes. Duplication of the genome at the origin of the chordates is the most likely current explanation for the four clusters; duplication sets up the possibility of future structural and functional divergence and *specialization of function among copies of the gene.* Four possibilities, which are not mutually exclusive, could explain this evolutionary change in gene function.

Two involve change in gene number, either in the number of Hox gene clusters or the number of genes per cluster. Duplication of *Hox* clusters before the teleosts arose — perhaps associated with duplication of large portions of chromosomes or entire chromosomes — would take the number from four to eight. This, coupled with subsequent loss of one cluster, would explain the seven clusters found in zebra fish.

The *other two possibilities involve altered function* by modification of the function of individual Hox genes or by increasing the complexity of interaction between gene networks, either of which could come about by alteration in the upstream and/or downstream regulation of a Hox gene(s).

[a] See Pendleton et al. (1993), Akam (1995) and Carroll et al. (2005) for discussion of the roles of Hox genes.

and passes extracted food (primarily algae) along a mucosal strand into the digestive tract.

The cephalochordate notochord, pharyngeal clefts, dorsal nerve cord, metamerically organized myotomes, posterior direction of blood flow in the dorsal vessels and anterior direction in the ventral vessels, as well as vertebrate-like organs such as the thyroid, indicate a close relationship between cephalochordates and vertebrates. It seems reasonable to propose that cephalochordates, although not ancestral to other vertebrates, represent a mode of life — swimming and **filter feeding** — that all early chordates shared. At least two lines of evidence support this view.

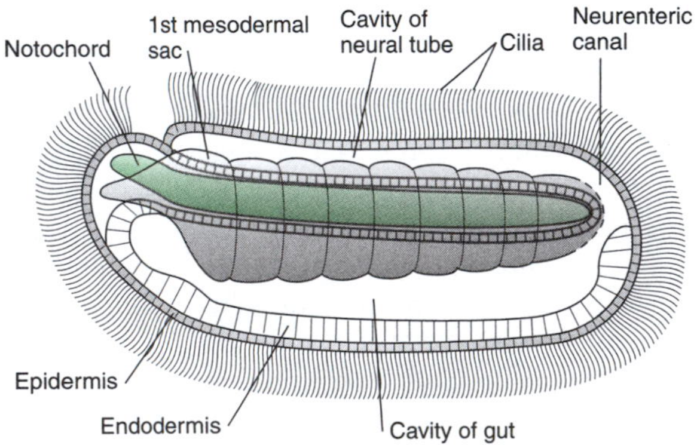

FIGURE 17-5 A diagram of a young ciliated amphioxus larva showing the notochord extending from anterior on the left to the posterior.

One is the observation that lampreys, a group of jawless, boneless animals, but vertebrates nevertheless, have an **ammocoete larva** that is a swimming filter feeder remarkably similar to cephalochordates. The earliest chordates were undoubtedly *filter feeders,* using a muscular pharynx to suck water into the pharynx.

Recent fossil finds in the 530-million-year-old Chengjiang Formation in Yunnan Province, China, allow us to see what early cephalochordates looked like and how they were constructed. Two genera with pharyngeal clefts, myomeres and a well-developed dorsal fin containing skeletal elements — *Myllokunmingia fengjiaoa,* and *Haikouichthys ercaicunensis* — with many similarities to extant hagfishes, have been described from the same fossil beds. Three species are noteworthy:

1. A single, 22-mm–long specimen of *Cathaymyrus diadexus,* described in 1996. *Cathaymyrus* has pharyngeal clefts, a pharynx, and segmented muscle blocks (myomeres), all of which are chordate features. A putative notochord also has been identified.

2. The discovery of *Yunnanozoon lividum,* first described as a chordate in 1995 has been reinterpreted as a hemichordate. Thirteen pairs of symmetrically arranged gonads and tiny (1-mm–long) teeth in the pharynx are distinctive features, as is the metameric nature of the muscles.

3. Further discoveries of hundreds of specimens of animals up to 40 mm in length and assigned to one or two species in the genus, *Haikouella.* More derived than *Yunnanozoon, H. lanceolata* and *H. jianshanensis* have a notochord, six pharyngeal arches, heart, pharyngeal teeth and what has been interpreted as a tail with a fin.[4]

Some morphologists have pointed out that cephalochordates are not very similar to purported ancestral chordates because their notochord does not end at a "brain" but rather extends to the anterior tip of the head. Thus, cephalochordates might not be the nearest extant relatives of vertebrates. That honor may go to the **tunicates,** although older phylogenetic analyses using morphological characters and 18*S* rRNA placed cephalochordates as the closest sister group to the vertebrates.[5]

However, other homologies are being described, including:

- an eye spot and regions of the amphioxus brain that are homologous to vertebrate sensory organs and to divisions of the vertebrate brain;
- anterior–posterior organization/expression and regulation of Hox genes by retinoic acid are shared with vertebrates;
- homologous body segments recognized by homologous *Hox, Pax, engrailed* and *distalless* genes (**Fig. 17-8**);
- the hepatic diverticulum, a homologue of the vertebrate liver; and
- the 'liver' enzyme plasminogen in amphioxus.[6]

[4] See Shu et al. (1996) for *Cathaymyrus diadexus,* and see J.-Y. Chen et al. (1995), Shu and colleagues (1996, 2003a) and Mallatt et al. (2003) for *Yunnanozoon.*

[5] See Delsuc et al. (2006) for tunicates as ancestors, and see Wada and Satoh (1994) for the 18S rRNA studies.

Amphioxus has a single *Hox gene cluster containing at least 12 genes, known as AmphiHox-1* to *-12* (**Fig. 17-9**). The single cluster is consistent with duplication of the ancestral single *Hox* cluster after vertebrates diverged from cephalochordates; that is, the condition in amphioxus is closer to that in the common chordate ancestor, although not ancestral (Fig. 17-9). Gene duplication continued after the separation of amphioxus from other chordates; the gene *Brachyury* is duplicated in amphioxus but not in the vertebrates. (Gene duplication in vertebrates is not restricted to *Hox* genes but is seen in other regulatory genes — *engrailed, Msx, Otx, MyoD,* the globin gene family, and insulin-like growth factor — a pattern suggesting duplication of the entire genome with subsequent selective losses in particular lineages.)

Urochordates

According to many zoologists, a group of small marine filter feeding urochordates demonstrate how a sessile filter feeder may have evolved into an active swimmer.

[6] See Lacalli (2004) and Wicht and Lacalli (2005) for the eye spot and brain; L. Z. Holland et al. (2004) and Benito-Gutiérrez (2006) for shared genes; and Liang and Zhang (2006) for liver.

BOX 17-2 Latent Homology

You may wonder how we can speak of precursors of jaws in animals (jawless vertebrates) that have not evolved jaws. Developmentally, the question is how do we identify a mandibular arch in organisms (lampreys) that have not evolved mandibles? The evidence can be morphological, developmental, and/or molecular.[a]

The ability to recognize and identify such features as mandibular arches in jawless vertebrates (this chapter), precursors of middle ear ossicles in "mammal-like–reptiles" (see Chapter 18), prototissues in sponges (see Box 16-2), or a protoneural crest in ascidians and amphioxus, goes to the heart of the concept of homology (see Chapter 2), especially latent homology. Latent homology allows us to equate features even when they are not morphologically the same or even similar. Actinopterygian and reptilian lower jaw bones are morphologically distinct from mammalian middle ear ossicles, but because we can trace the transformation of the former to the latter in the fossil record and in the development of middle ear ossicles in marsupials (see Chapter 19), we can identify certain of the lower jaw bones as homologues of middle ear ossicles. The reptilian bones are not middle ear ossicles but are their evolutionary precursors (see Hall, 2003a, 2006c and Stone and Hall, 2004).

If the mandibular arch arose from the first pharyngeal arch of jawless vertebrates, *then* the *first pharyngeal arch of jawless vertebrates is a latent homologue of the mandibular arch of jawed vertebrates,* and the skeleton of the first pharyngeal arch is a homologue of the skeleton of the mandible. Such a theory has prevailed since the nineteenth century.

If, however, the mandible did not arise from a pharyngeal arch — whether or not the most anterior (first) arch — *then,* despite their strong similarity, *mandibular and pharyngeal arches are not homologous,* and even though they share many genetic, developmental and structural features, vertebrate jaws would have been an evolutionary novelty.

[a] See a text on vertebrate anatomy such as Liem et al. (2001) or Kardong (2006) for the morphological evidence, which has been available for 140 years; see Kuratani et al. (2001) for developmental and molecular evidence; and see Stone and Hall (2004) for latent homology as exemplified by the neural crest and chordate origins.

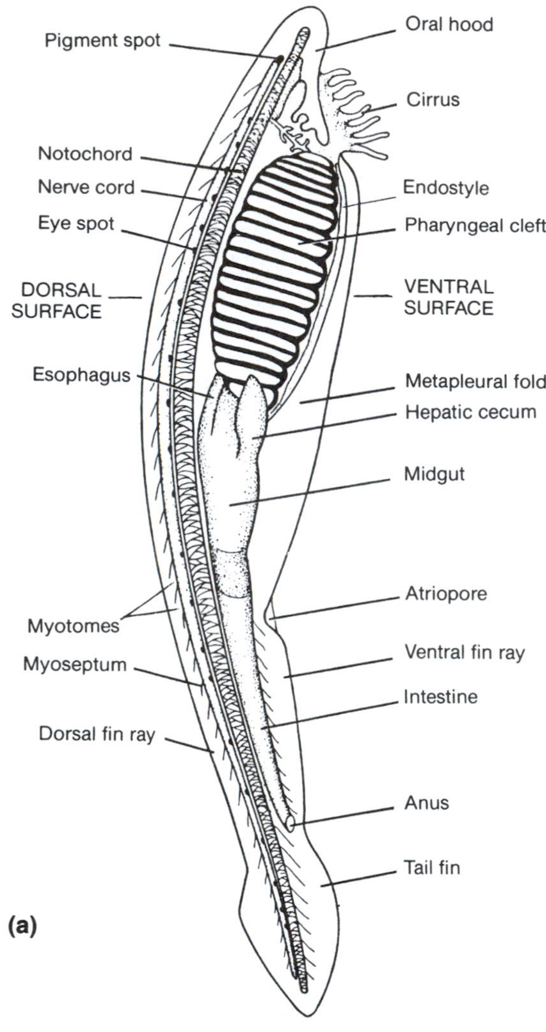

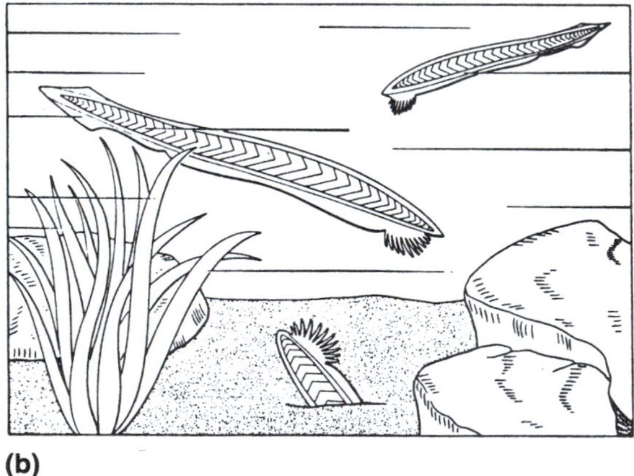

FIGURE 17-6 Anatomical features (a) and habitat (b) of the cephalochordate *Branchiostoma*.

Among the **urochordates** is a subgroup of ascidians (also called *tunicates* or *sea squirts*), which are sessile as adults, feeding by means of a large pharyngeal basket through which they draw and filter water, which they expel through many clefts **(Fig. 17-10).** In addition to pharyngeal clefts, ascidian larvae have a notochord and dorsal nerve cord (see Fig. 17-7a and d). Larvae are distributed widely, settle on suitable substrates and metamorphose into a sessile adult form (Fig. 17-7b and c).

Ascidians possess such fundamental chordate features of embryonic development as:

- basic organ formation (Fig. 17-7d);
- a notochord;
- induction of the neural plate by the notochord;
- induction of sensory pigment cells by the developing dorsal neural (Satoh, 1994);
- more than one set of Hox genes and their regulation by retinoic acid; and
- Pax genes (*HrPax37*) that pattern the neural tube, where *Hr* stands for the species *Halocynthia roretzi* and 37 for a homologue of the chordate genes *Pax-3* and *Pax-7* (Wada et al., 1996).

The clear implication is either that ascidians are on the ancestral line leading to vertebrates, or the common ancestor of urochordates and vertebrates already shared these genes and gene functions, and that *evidence of ancestry should be sought in larval and not in adult ascidians.*

Larval Evolution: Paedomorphosis and Neoteny

In the early 1920s, Walter Garstang developed the concept, quickly accepted by many biologists, that ascidian metamorphosis could have been increasingly delayed over successive generations until the swimming larval form itself became sexually mature. This process, termed **paedomorphosis** (*shaping like a child*), involved the incorporation of adult sexual features into earlier immature stages. Garstang hypothesized that some urochordates such as the Larvacea arose by paedomorphosis and can now be considered as sexually mature larvae. Garstang (1951) provided a poetic description of paedomorphosis in the Mexican axolotl (*Ambystoma mexicanum*):

Ambystoma's a giant newt who rears in swampy waters,
As other newts are wont to do, a lot of fishy daughters:
These Axolotls, having gills, pursue a life aquatic,
But, when they should transform to newts, are naughty and erratic.
They change upon compulsion, if the water grows too foul,
For then they have to use their lungs, and go ashore to prowl:

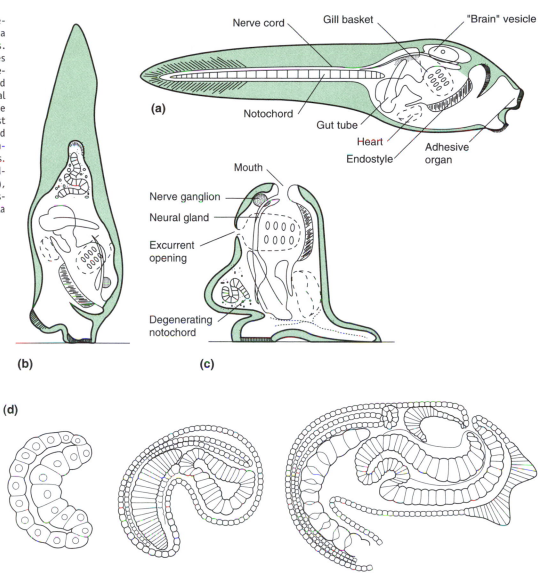

FIGURE 17-7 (a) Free-swimming ascidian larva before metamorphosis. (b) As the larva attaches to a substrate by its anterior suckers, the notochord degenerates and its internal structures rotate. (c) Stage at which the tail is almost completely resorbed and the body parts are assuming their adult positions. (d) Three stages, including (from left to right), the early embryonic blastula, gastrula, and neurula (organ-forming stage).

But when a lake's attractive, nicely aired, and full of food,
They cling to youth perpetual, and rear a tadpole brood.[7]

Extending Garstang's proposal, biologists suggested that early free-swimming chordates passed through this same process of paedomorphosis in descending from ancestors that had chordate-like larval stages; that is, the swimming, filter-feeding larvae of some prechordate animals replaced their adult sessile forms, perhaps because they could better follow and search out new food supplies as well as escape predators. Selection for preserving larval characteristics led to precocious sexuality that bypassed the sessile adult, yielding mobile mature "chordates."

For those who interpreted the evidence for the similarity between the tornaria larva of the hemichordate *Balanoglossus* and the auricularia larva of echinoderms as stemming from a phylogenetic relationship, Garstang's proposal had the added attraction of explaining that the transition between the two lineages occurred through a larval form rather than through changes in the considerably different adult forms. He visualized echinoderms evolving into chordates by developing dorsal neural folds from ciliated bands of the auricularia and adding pharyngeal clefts and a notochord. The hypothesis did not explain the origin of the latter structures, nor did it make clear how this unusual larva reached the adult sexual stage. Importantly, Garstang's theory turned Haeckelian recapitulation on its head. It is

[7] From Garstang (1951), *Larval Forms and other Zoological Verses.* Blackwell Publishers, Oxford, UK.

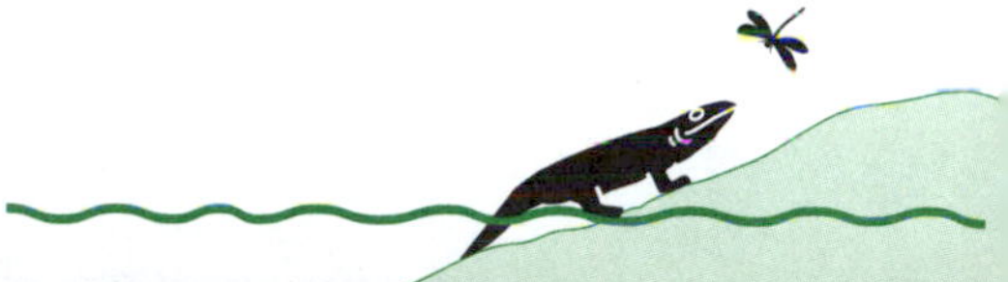

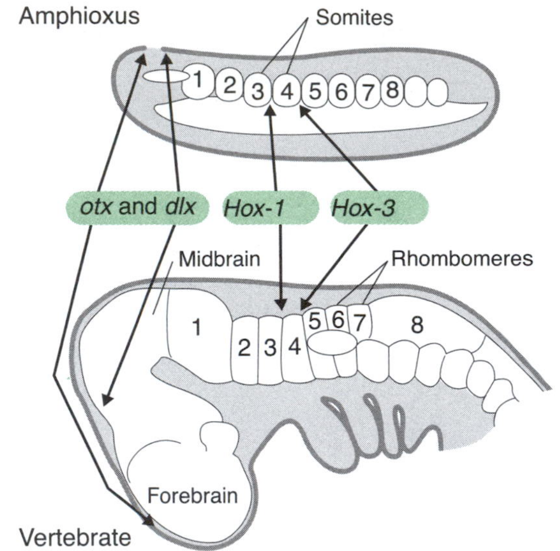

FIGURE 17-8 Expression pattern of Hox genes in amphioxus and a generalized vertebrate embryo. Numbered segments represent somites (trunk segments) in amphioxus embryos corresponding to rhombomeres (nervous system segments) in vertebrate embryos. Hox genes are expressed in similar locations in both embryos, taken by some as indicating that these are homologous structures and that amphioxus has a forebrain. Homology between vertebrate and amphioxus Hox gene clusters is shown in Figure 13-8. (Adapted from Stokes, M. D., and N. D. Holland, 1998. The lancelet. *Amer. Sci.*, **86**, 552–560.)

not that ontogeny recapitulates phylogeny (Haeckel), rather, ontogeny creates phylogeny (Garstang).

In general, the relationship of vertebrates to cephalochordates and urochordates makes it likely that the earliest chordates were actively swimming filter feeders, and paedomorphosis may well explain their origin. This process and its auxiliary, **neoteny** (retaining some immature morphological traits into adult stages), are not unusual events; for example, some salamanders such as the mudpuppy and axolotl reproduce as gilled, immature forms, as described in the poem above.[8]

Finding the Larval Ancestor

Because hemichordates such as the pterobranch *Rhabdopleura* (see Figure 16-14) use a lophophore (group of ciliated tentacles) for food capture, biologists have suggested that the chordate lineage can be extend back to a lophophorate-type ancestry that chordates may have shared with phoronids, brachiopods and ectoprocts as well as with the crinoid-like ancestors of the echinoderms. A proposed phylogeny of this type, extending from lophophore ancestor to vertebrates, appears in **Figure 17-11.**

However neat these solutions may seem, keep in mind that ancestral branching patterns extending back to Precambrian times are not easily settled. For example, related as urochordates and other chordates may seem morphologically, they are not so on all molecular levels analyzed: sequence analysis using the gene for 18*S* ribosomal RNA indicates a distant relationship, whereas the gene for the

[8] Evolutionary and developmental biologists use the terms *paedomorphosis* and *neoteny,* and other terms referring to evolutionary changes in developmental rates (*heterochrony*), in different ways. For example, some define paedomorphosis as the *result of neoteny;* that is, when the developmental rate of non-reproductive compared to reproductive tissues slows down (neoteny), the result is a full-sized adult with juvenile features that is sexually mature (paedomorphosis). Among other forms of paedomorphosis and heterochrony is *progenesis,* used to describe an increased rate of sexual development leading to early sexual maturity in an adult that remains quite small in size, such as the tiny parasitic *Bonellia* males; see McKinney and McNamara (1991), and Zelditch (2001) for types of heterochrony.

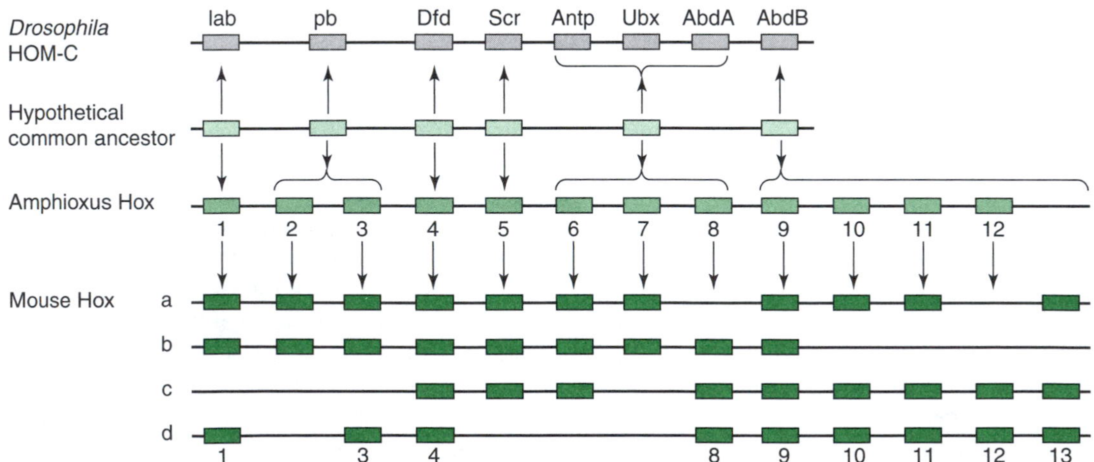

FIGURE 17-9 The Hox gene clusters in amphioxus (*Branchiostoma*) shown in relation to the six ancestral Hox genes, the eight genes in *Drosophila* and *Hoxa-Hoxd* in the mouse genome. (Adapted from Hall, B. K., (ed), 1999a. *Evolutionary Developmental Biology,* Second Edition. Kluwer Academic Publishers, Dordrecht, The Netherlands; modified from Holland P. W. H., and J. Garcia-Fernàndez, 1996. Hox genes and chordate evolution. *Dev. Biol.*, **173,** 382–395.)

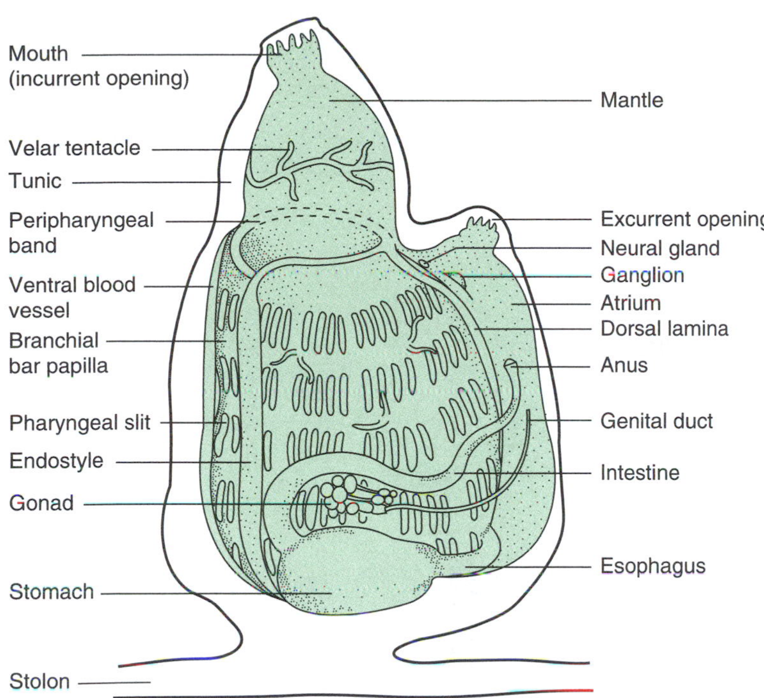

FIGURE 17-10 Internal anatomical features of an adult colonial ascidian (*Perophora*).

actin muscle protein indicates a close relationship. Biologists offer other hypotheses, such as chordate origin from free-living animals rather than from neotenous forms of sessile urochordates (Wada and Satoh, 1994).

Nevertheless, it seems clear to many zoologists that larvae of some urochordates reflect a basal chordate body plan: a slotted pharyngeal basket, and swimming by means of muscles attached to an undulating flexible notochord. Among the developmental homologies that support this view are:

1. The *Brachyury* gene, which is active during differentiation of the notochord in both mammals and ascidian larvae. The ascidian *Ciona intestinalis* regulates *Brachyury* in the notochord using a regulatory element related to the vertebrate *Notch* signaling pathway. A role for *Brachyury* in notochord formation therefore predates the origin of the vertebrates.[9] As the cephalochordate amphioxus has two copies of *Brachyury* and vertebrates only one, gene duplication continued after the separation of amphioxus from the vertebrates.
2. Shared genes involved in specification of cranial sensory placodes in vertebrates and of the filter-feeding organs in the ascidian *Ciona intestinalis.*[10]
3. Conservation of genes within the nervous system, for example, Pax genes (*Pax2/5/8*) in a region of the brain in *Ciona* homologous to the midbrain-hindbrain boundary in vertebrates, consistent with a pattern that can be traced to amphioxus [11]
4. Conservation of the transcription factors *CiPhox2* and *CiTbx20* in *Ciona* motor neurons that arise from a region of the hindbrain equivalent to the vertebrate hindbrain (Dufour et al., 2006).

The genome of *Ciona intestinalis* has now been sequenced, opening up vast avenues along which to explore ascidian-vertebrate relationships (Dehal et al., 2002). The genome contains 16,000 genes, which is about half the number found in vertebrates. The difference is that *Ciona* has only one copy of many genes, where vertebrates have two. Around 80 percent of the genes in *Ciona* are found in vertebrates, including genes involved in specification and/or formation of the brain and spinal cord; sense organs such as the eyes; the heart; and the thyroid gland and immune system. All of these systems have their homologues in *Ciona.*

But neither Ciona nor any other ascidian is a vertebrate ancestor. These are specialized organisms, with their own

[9] Detection of a homologue of *Brachyury* in the secondary mesenchyme of the sea urchin *Hemicentrotus pulcherrimus* indicates that the gene had an even more ancient role, predating the origin of the chordates.

[10] See Northcutt (2004) and Mazet and Shimeld (2005) for specification of placodes.

[11] See Lacalli (2006) and Rosa-Molinar and Pritz (2004) for Pax genes. Not all comparative biologists support this point. Wada et al. (1998) localized *Pax2/5/8* to brain regions that they interpreted as fore-, mid- and hindbrain. L. Z. and N. D. Holland (1998) localized *AmphiHox 1,2, Otx, Dll* and *Engrailed* to regions identified as fore- and hindbrain.

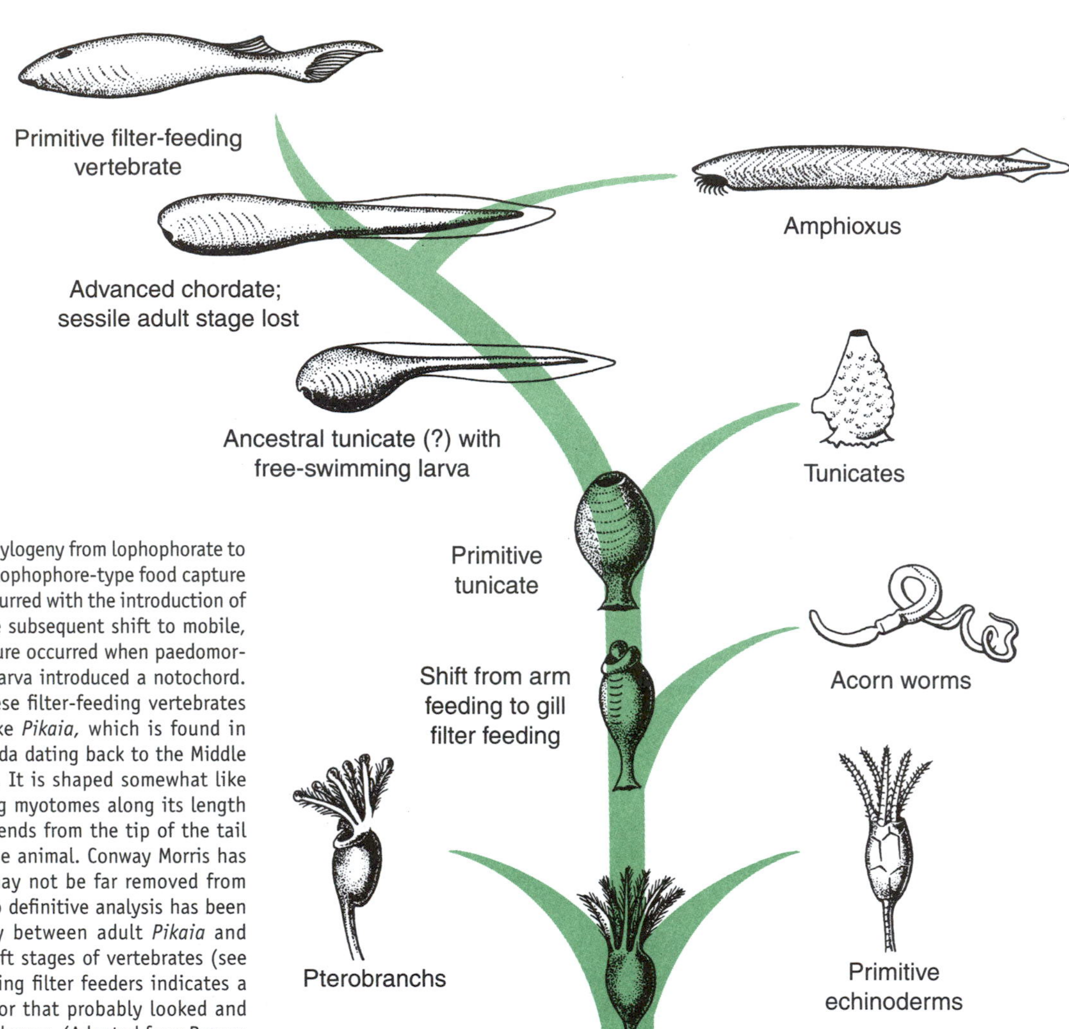

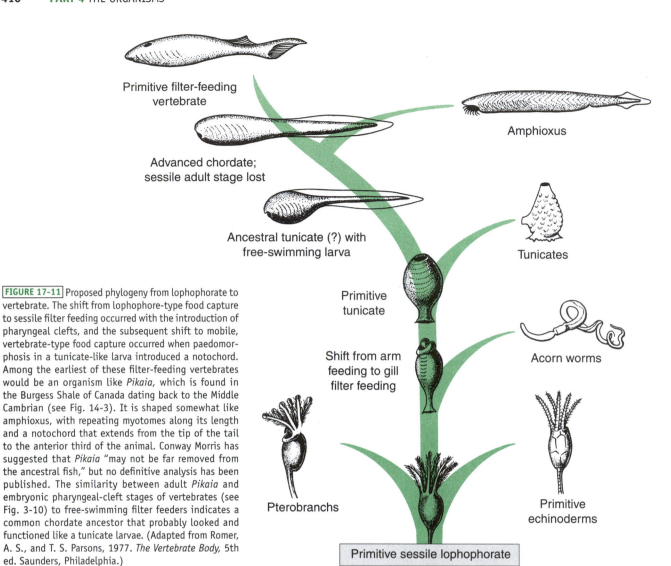

FIGURE 17-11 Proposed phylogeny from lophophorate to vertebrate. The shift from lophophore-type food capture to sessile filter feeding occurred with the introduction of pharyngeal clefts, and the subsequent shift to mobile, vertebrate-type food capture occurred when paedomorphosis in a tunicate-like larva introduced a notochord. Among the earliest of these filter-feeding vertebrates would be an organism like *Pikaia*, which is found in the Burgess Shale of Canada dating back to the Middle Cambrian (see Fig. 14-3). It is shaped somewhat like amphioxus, with repeating myotomes along its length and a notochord that extends from the tip of the tail to the anterior third of the animal. Conway Morris has suggested that *Pikaia* "may not be far removed from the ancestral fish," but no definitive analysis has been published. The similarity between adult *Pikaia* and embryonic pharyngeal-cleft stages of vertebrates (see Fig. 3-10) to free-swimming filter feeders indicates a common chordate ancestor that probably looked and functioned like a tunicate larvae. (Adapted from Romer, A. S., and T. S. Parsons, 1977. *The Vertebrate Body*, 5th ed. Saunders, Philadelphia.)

evolutionary history, which, although it began in a stem group from which vertebrates arose, has evolved independently. So it was a surprise when Dehal and colleagues (2002) discovered that *Ciona* has genes for the synthesis of cellulose (of which the external tunic is composed) and that these genes, which are not found in animals, resemble those found in bacteria and in fungi. Darwin would be overjoyed to find that ascidians and vertebrates did indeed descend from a common ancestor. As he wrote in *The Descent of Man* (1871), "one branch retrograding in development and producing the present class of Ascidians, the other rising to the crown and summit of the animal kingdom by giving birth to the Vertebrata."

As we pass from the deep unknowns of chordate origins in the Precambrian, fossils begin to accumulate, enough to offer a more detailed vertebrate history. What we know paleontologically is now reviewed.

Conodonts

Conodonts are soft-bodied, elongate creatures, 80 mm long, that captured prey and ground it into food using arrays of sharp, bony *conodont elements* thought to have been located in a pharynx (**Fig. 17-13**). Conodonts are almost entirely known from conodont elements, tooth-like structures unassociated with any conodont organism until recently, but common in various Paleozoic strata. Although the tissues that comprise conodont elements have been studied, whether these tissues include bone and dentine (vertebrate tissues) is not resolved. Most current experts place conodonts within the chordates, many considering them as vertebrates.[12]

[12] See Donoghue (2000) and Donoghue et al. (2002) for conodont elements and for the conodont animal.

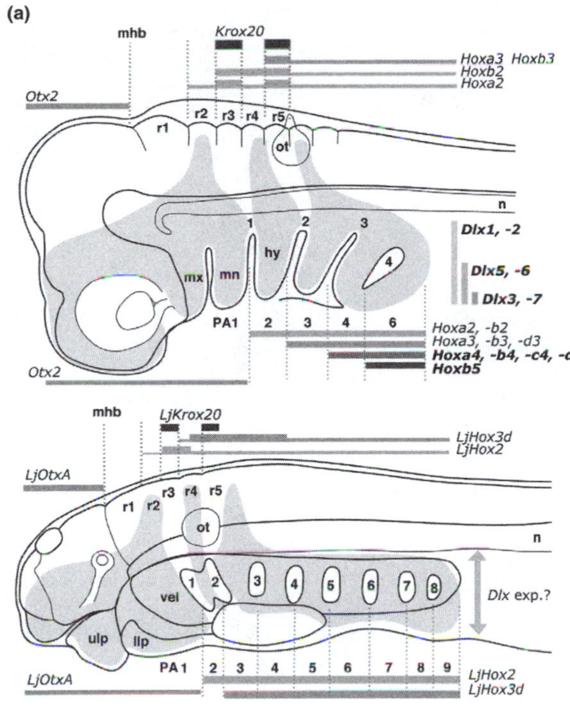

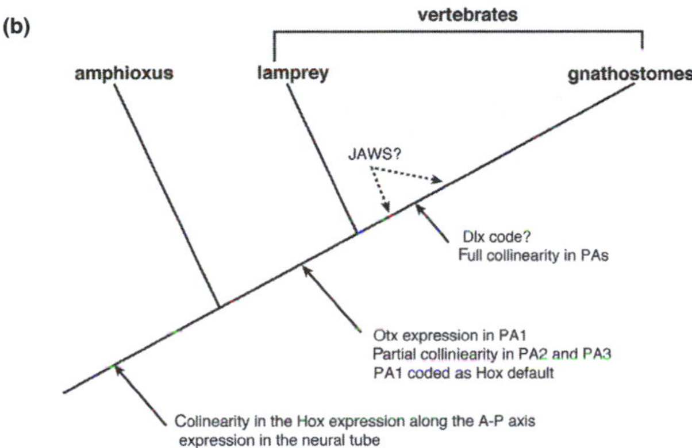

FIGURE 17-12 Gene expression patterns and the origin of jaws (a). A simplified representation of the *Hox* codes in a generalized gnathostome embryo (*above*), and a lamprey larva (*below*). Gnathostome-specific genes are in bold, pathways of migration of neural crest cells in gray. In the gnathostome, sets of *Hox* transcripts are nested in the pharyngeal arches (PA1-6). The mandibular arch (mn, PA1) is characterized as Hox gene-free and by the expression of the Otx gene *Otx2*. The rostral (anterior) border of the *Hox*-free zone, and the rostral expression boundary of the Hox genes in pharyngeal arches 2 and 3 are the same in gnathostomes and in the lamprey, implying a deep homology. A Dlx gene code (*Dlx1-7*) establishes the dorso–ventral pattern of each pharyngeal arch in gnathostomes (above) but has not been detected in lampreys (Dlx exp.?). Other abbreviations are: hy, hyoid arch; llp, lower lip; mhb, midbrain-hindbrain boundary; mn, mandibular process; mx, maxillary process; n, notochord; ot, otic vesicle; pa1, pharyngeal arch 1; r1-5, rhombomeres 1 to 5 in the hindbrain; ulp, upper lip; vel, velum; 1-8; 2-8; pharyngeal arches 1 to 8 and 2 to 8. (b) shows a scenario of evolutionary changes in regulation of the expression of the Otx, Dlx and Hox genes. (Courtesy of Shigeru Kuratani, Center for Developmental Biology.).

Fossil Jawless Vertebrates

Teeth and armored skin plates (dermal bones) of **jawless vertebrates** appear in marine deposits of the Late Cambrian and extend into the Late Devonian. Although often grouped as agnathans or Agnatha **(Fig. 17-14)**, it is important to realize that Agnatha is not a monophyletic lineage. The term Agnatha embraces a variety of jawless forms whose relationships to one another remain unclear (Donoghue et al., 2002).

Heterostracans, early jawless vertebrates, were small, 20 to 30 cm long, and encased in large dermal plates anteriorly and smaller plates or scales posteriorly (Fig. 17-14). In addition to two laterally placed eyes, heterostracans had a mid-dorsal opening for a median eye or pineal organ. Various forms are known, some bottom feeders, mostly with ventrally placed mouths; other heterostracans, with mouths on the dorsal surface, may have fed on plankton.

Heterostracans and an associated lineage about half their size, coelolepids (meaning hollow-scaled; Fig. 17-14b), had two nasal openings, distinguishing them from other jawless vertebrates with single nostrils, such as heavily armored osteostracans (bony shells; Fig. 17-14c) and lightly armored but more maneuverable "shieldless" anaspids (Fig. 17-14d), both of which appeared in the mid-Silurian and became extinct during the Devonian.

It has been suggested that all four of these jawless groups, often collectively called **ostracoderms** (bony skins; Figs. 17-14 and **17-15**), used a large, muscular pharynx to suck up food-laden water more rapidly and in much larger amounts than the ciliary activity of invertebrate lophophorates and filter feeders could achieve. In a seminal paper, Northcutt and Gans (1983) proposed that although the early vertebrate pharynx may have originally been used for pumping water and filter feeding, it evolved into an active predatory organ in adult ostracoderms. According to them, jawless vertebrates used the pharynx for scooping up small soft-bodied or lightly armored bottom-dwelling animals.

Through more effective filter feeding or active benthic predation, pharyngeal adaptations were an important factor in the early success of ostracoderms, which established themselves throughout Late Cambrian and Early Ordovician marine environments and attained widespread Devonian distribution in fresh- or mixed fresh- and saltwater areas in North America and Europe.

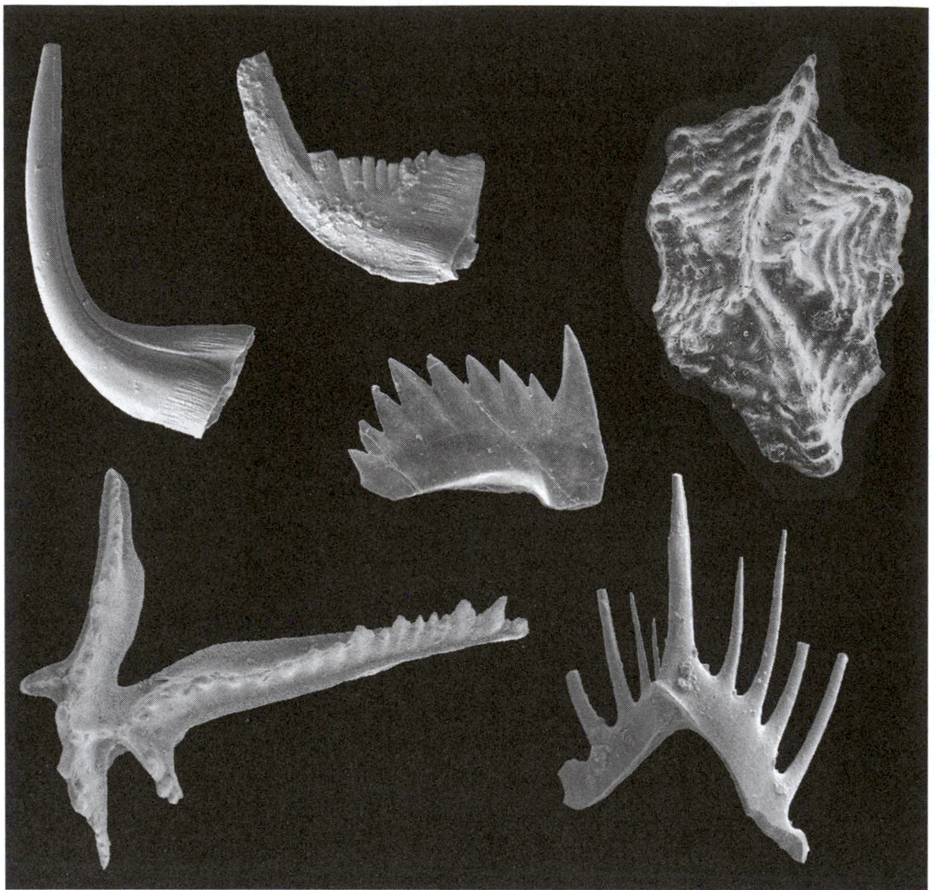

FIGURE 17-13 Some of the diversity of conodont elements. (Photo courtesy of Dr. Richard J. Aldridge, F. W. Bennett Professor of Geology, University of Leicester.)

Extant Jawless Vertebrates

Although ostracoderms are not found in the fossil record after the Devonian, several agnathan lineages persisted. The extent to which extant agnathans (**cyclostomes**) are related to one another or to the extinct jawless vertebrates continues as an active area of research, not only in paleontology, but also in molecular and developmental biology (**Fig. 17-16**).

As shown in Figure 17-14 panels e and f, there are two extant groups of round-mouthed jawless vertebrates: *lampreys* (Petromyzontiformes) and *hagfishes* (Myxiniformes). Both now occupy widespread but specialized ecological niches, either as ectoparasites (lampreys) or as burrowing detritus feeders and scavengers (hagfishes). Paleontologists have proposed that the fossil anaspid *Jamoytius* may be in the direct line of descent of ostracoderms, lampreys and perhaps hagfishes as well, although it is too derived to be a stem taxon. As noted earlier, two Cambrian hagfish are now known, *Myllokunmingia fengjiaoa* and *Haikouichthys ercaicunensis,* both considered stem craniates. A fossil lamprey, *Mesomyzon mengae,* from 125-My-old freshwater deposits in Inner Mongolia, demonstrates that lampreys invaded fresh water no later than the Early Cretaceous. A recent fossil find, *Priscomyzon riniensis,* a marine-estuarine lamprey from the Late Devonian of South Africa indicates that features associated with modern lampreys (an oral disc, circumoral teeth) arose much earlier than previously thought.[13]

New jawed and finned vertebrates arose during the mid-Devonian using hard bony tissue for structure and defense. Although cartilage can be softer than bone it appears as the major structural tissue in some later lineages such as sharks. Bone — an organic protein matrix mineralized with hydroxyapatite — turned out to be a highly adaptive tissue for vertebrates, providing a reservoir for Ca^{++} ions and for phosphate that was resistant to acid breakdown that would have accompanied increased metabolic rates (**Table 17-2**). Paleontologists suggest that the development of defensive bony plates was an essential element enabling early vertebrates to withstand predation by the voracious and widely distributed scorpion-like *eurypterids* — that is, until the vertebrates themselves evolved into important predators.[14]

---

[13] See Shu et al. (2003b) for the Cambrian hagfish, and Chang et al. (2006) and Gess et al. (2006) for the fossil lampreys.

[14] See R. L. Carroll (1988), Hall (1999a) and Benton (2004) for elaborations on this theme.

AGNATHA

Paired nostrils

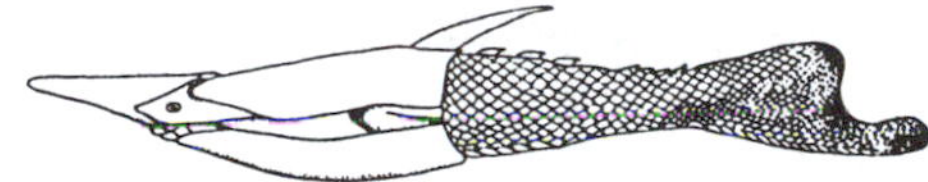

(a) Heterostracan (*Pteraspis*)

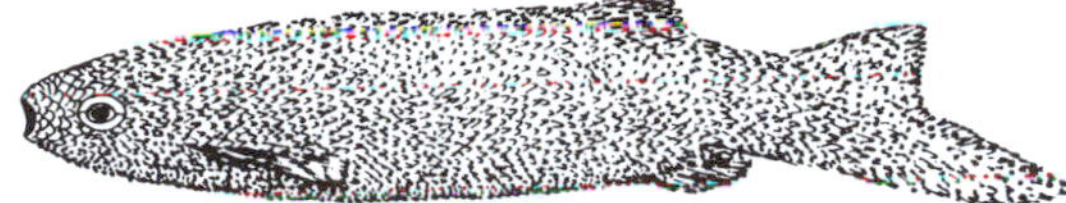

(b) Coelolepid (*Phlebolepis*)

Single nostril

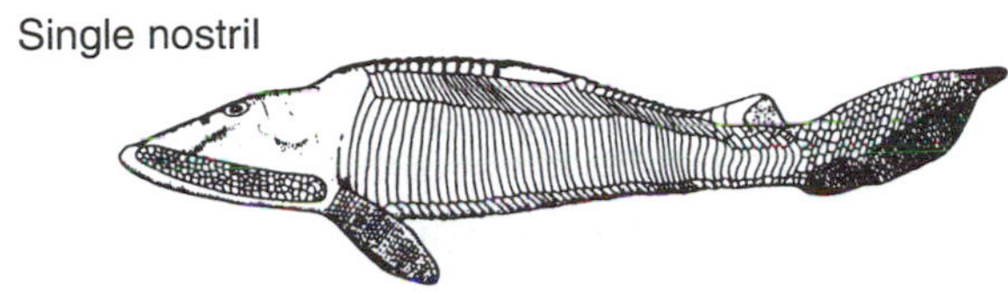

(c) Osteostracan (*Hemicyclaspis*)

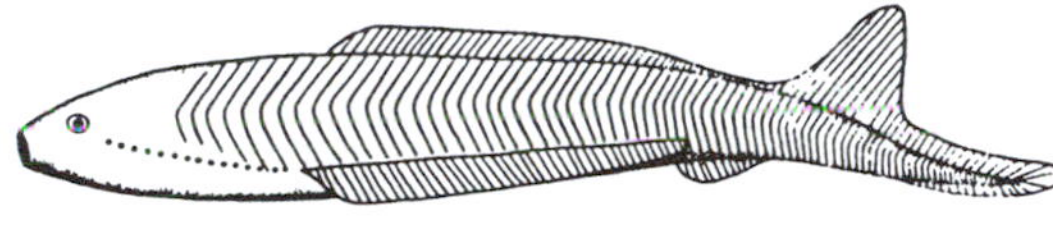

(d) Anaspid (*Jamoytius*)

Living forms (single nostril)

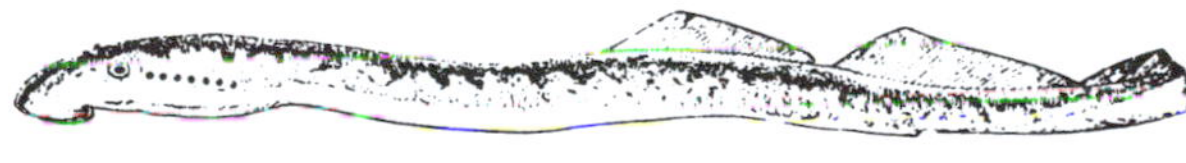

(e) Lamprey (*Petromyzon*)

(f) Hagfish (*Bdellostoma*)

FIGURE 17-14 Fossil (a–d) and extant (e, f) jawless vertebrates (Agnatha). Note that tail structures differ: the axial supporting element may be in the upper lobe (heterocercal) or in the lower lobe (hypocercal). Extant agnathans — *cyclostomes* — have round, sucking mouths and lack paired fin structures and dermal bones. Because of other basal features, which include a lens-less eye and simplified hindbrain, it has been suggested that hagfishes represent the most ancient of all extant vertebrate lineages.

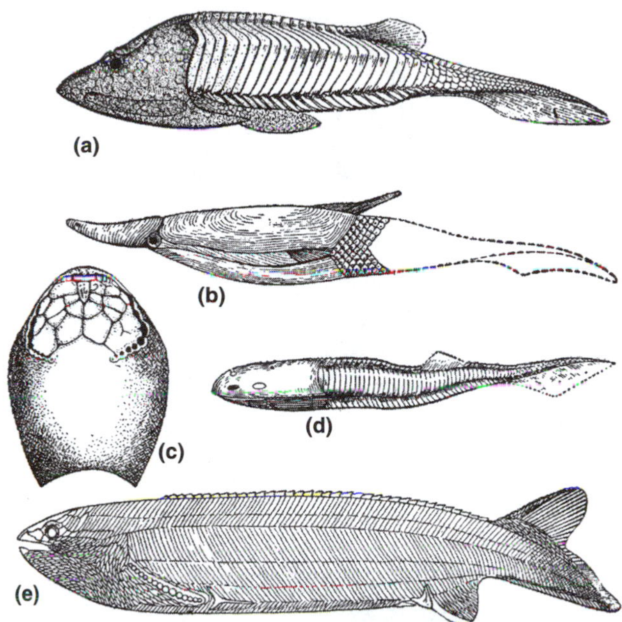

FIGURE 17-15 The diversity of ostracoderms in the Upper Silurian and Devonian is shown by these four taxa. (a) *Pterolepis;* (b, c) *Tremataspis* (c is a dorsal view of the head shield); (d) *Pteraspis;* (e) *Cephalaspis.* (Adapted from Gregory, W. K., 1929. *Our Face from Fish to Man. A Portrait Gallery of Our Ancient Ancestors and Kinfolk Together with a Concise History of Our Best Features.* G. P. Putnam's Sons, London, UK.)

Evolution of Jawed Vertebrates (Gnathostomes)

The first **jawed** fossil **vertebrates,** which appeared during the Silurian, can be assigned to one of two lineages, acanthodians and placoderms.

Acanthodians (spiny sharks; **Figs. 17-18 to 17-20**) are represented by *Climatius,* 80 mm long and characterized by both paired and unpaired spiny fins. Although the earliest known forms are marine, for the most part, acanthodians were freshwater animals found in rivers, lakes and swamps, many surviving up to Late Paleozoic times.

Placoderms (plate-skinned), which appeared toward the end of the Silurian, flourished during the Devonian, but became extinct. Some were bottom dwellers, such as the skate-like rhenanids (Fig. 17-19a) and the antiarchs with their stilt-like jointed pectoral appendages (Fig. 17-19b). Others were predators of gigantic proportions; the arthrodire *Dunkleosteus* was more than 9 meters long (Fig. 17-19c).

Jaws and the further development of paired fins are significant features in acanthodians and placoderms. Jaws revolutionized the way of life for these early vertebrates by offering them new food resources previously excluded because of the limitations of filter feeding and sucking. Fishes extended carnivorous behavior to all sizes of prey by grab-

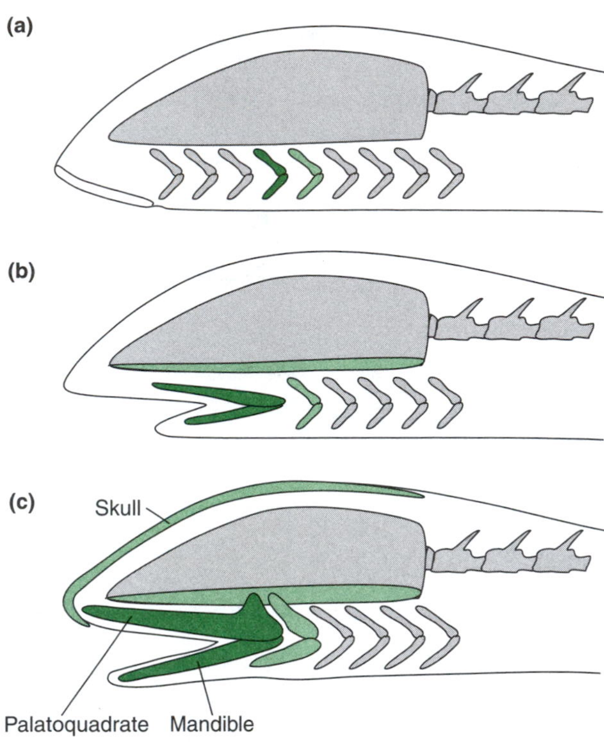

FIGURE 17-16 Diagrammatic representations in side view of the proposed transformation of pharyngeal arches of an agnathan (a; shown as open black Vs below the brain and anterior to the vertebrae) into the upper and lower jaws of a gnathostome (b), the upper jaw containing the palatoquadrate, the lower jaw the mandible. With further evolution the jaws became braced by the skeleton of the hyoid arch (c).

TABLE 17-2 Properties of bone that facilitated vertebrate evolution[a]

1. Excess *calcium ions,* diffusing through the skin and gills, can be deposited in the skin as an osmotically inert substance, conserving energy that organisms would otherwise expend in excreting these ions.
2. Tissue deposits of calcium and phosphate provide a *metabolic reserve* that can be mobilized when needed by partial bone decalcification. Cartilage can also calcify, be mobilized, and be used for this purpose.
3. Calcium phosphate tissues such as dentin and enamel can crystallize near electrosensory organs — for example, "lateral line" systems that detect electrical currents emitted by prey — insulating them from internal electrical body currents, improving their directional resolution. Underlying dermal bone structures would *mechanically stabilize* the position of these organs (Northcutt and Gans, 1982 and Northcutt and Northcutt, 2004).
4. The elaboration of bone structures became important in providing defensive *dermal armor.*
5. Hardened tooth surfaces evolved for *grasping and masticating food.*
6. The evolution of *ossified internal skeletons* offered rigid *supporting structures* for muscle and organ attachment, far stronger than the notochord.

[a] See Hall (2005a) for further information on skeletal evolution in both vertebrates and invertebrates.

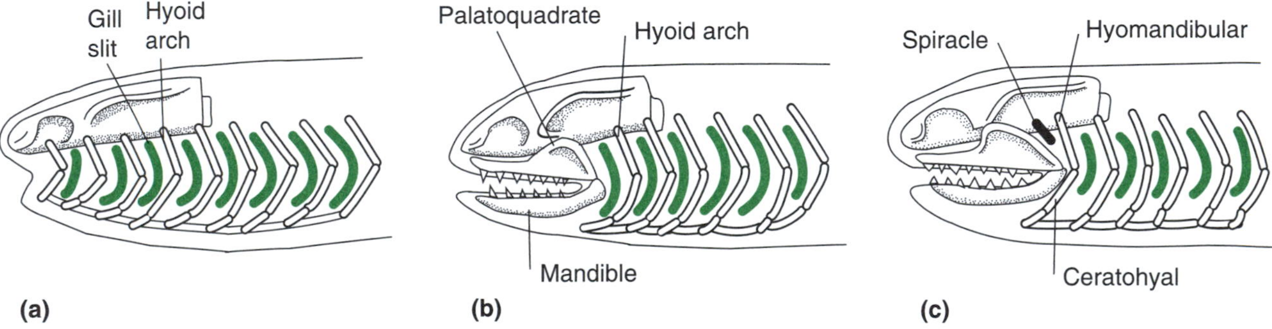

FIGURE 17-17 Stages in the evolution of jaws as derived from one of the anterior gill arches. (a) The jawless condition. (b) The conversion of an anterior gill arch into jaws. (c) Incorporation of bones from the hyoid arch to support the hinge of the jaw. In this progression, the gill cleft anterior to the hyoid arch reduced to the spiracle. In further evolutionary steps, the ceratohyal and hyomandibular become, respectively, the articular and quadrate bones in the jaw joint of amphibians and reptiles. In mammals this joint is replaced by a squamosal-dentary hinge. Selection for hearing led to the conversion of the articular/quadrate connection into malleus/incus ossicles (see Fig. 19-7). (Adapted from Romer, A. S., and T. S. Parsons, 1977. *The Vertebrate Body,* 5th ed. Saunders, Philadelphia.)

bing, tearing and chewing. Flattened, opposed teeth crushed hard or armored food materials (such as mollusks) that were formerly inaccessible. The teeth that provided the earliest jaws with their cutting function evolved either from skin "denticles" that initially served as armor plating in these early vertebrates or from cutting edges on the jaws themselves, as

FIGURE 17-18 The acanthodian *Climatius* showing the broad-based spiny fins running mid-dorsally and (in two rows) ventrally. Small armored scales completely covered these fins. (Adapted from Colbert, E. H., and M. Morales, 1991. *Evolution of the Vertebrates: A History of the Backboned Animals Through Time,* 3rd ed. Wiley, New York.)

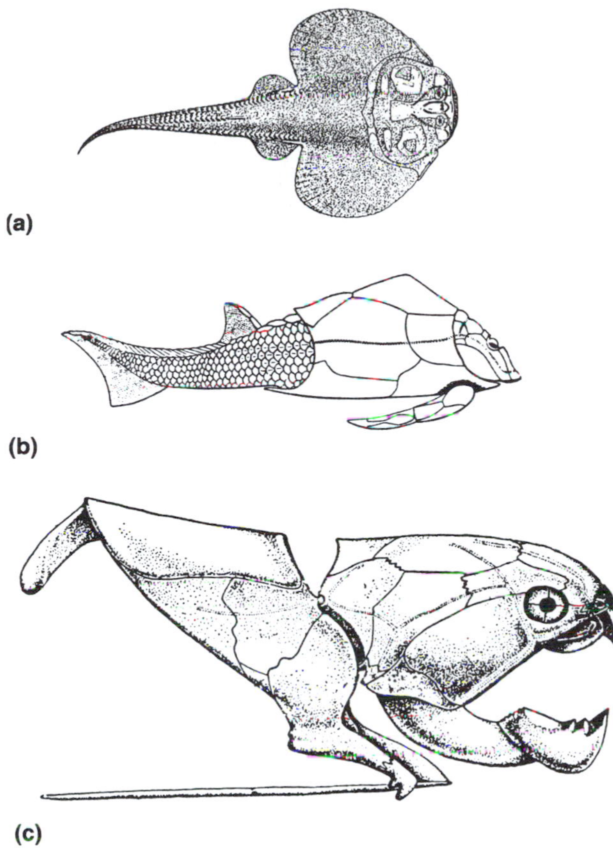

FIGURE 17-19 Placoderms. (a) The skate-like rhenanid (*Gemuendina*) with large lateral fins. (b) An antiarch (*Pterichthyodes*) showing the scaled posterior portion and the heavily armored anterior trunk and head regions. The pectoral "fins" of antiarchs were encased in bony plates that may have enabled them to crawl along the sea bottom. (c) The 3-m-long anterior bony shield of the arthrodire *Dunkleosteus*. The animal was about 9 m long. (Adapted from Romer, A. S., 1968. *The Procession of Life*. World Publishing, Cleveland, OH.)

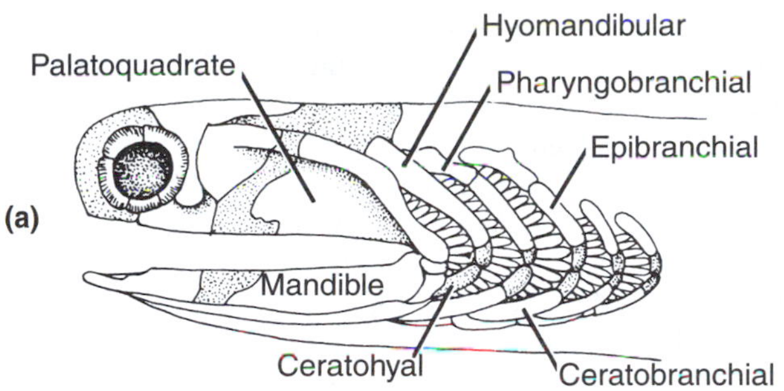

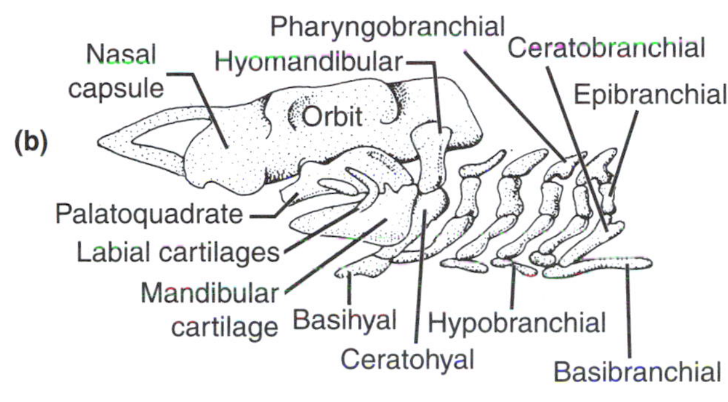

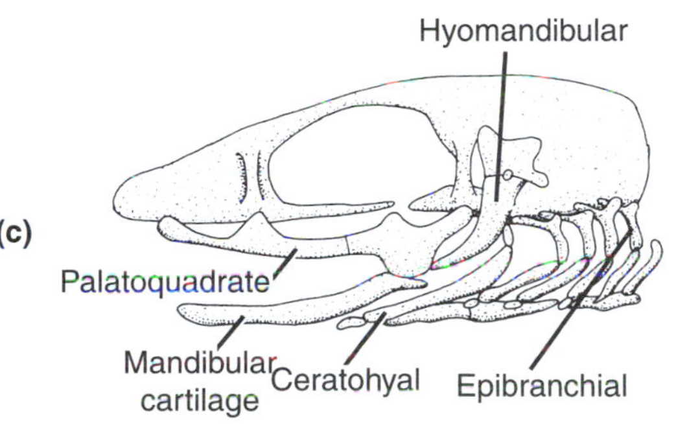

FIGURE 17-20 Braincases along with gill and mouth structures (visceral arches or skeletons) in a fossil acanthodian (a), a shark (b), and a bony teleost embryo (c). (Adapted from Romer A. S., and T. S. Parsons, 1977. *The Vertebrate Body*, 5th ed. Saunders, Philadelphia.)

in some of the placoderms. In addition, jaws facilitated both intra- and interspecific defensive and aggressive behaviors and offered these fish greater opportunities to manipulate the environment in building nests or grasping mates.

Because the jaw structures are so similar, the major extant forms of fish must have derived their jaws from a common ancestor. Two major hypotheses for the origin of jaws have been proposed, the first by transformation of a pharyngeal gill arch, the second proposing a non-gill-arch origin.

Pharyngeal Gill Arch Origin of Jaws

We do not know the intermediate steps between jawless and jawed vertebrates from the fossil record. In one long-held hypothesis, jaws evolved by the transformation of **pharyngeal arches** previously used for filter feeding and perhaps for respiration (see Figs. 17-16 and 17-17). An important issue raised by this hypothesis is how we recognize pharyngeal arches in organisms that lack jaws as the evolutionary precursors of jaws in gnathostomes. This issue, often termed *latent homology*, is explored in Box 17-2 and by Stone and Hall (2004) in relation to the neural crest and vertebrate origins.

Gill arches are paired on each side and supported by V-shaped hinged structures whose apices point backward. Possible fates of the anterior pair of gill arches include (1) disappearance, (2) incorporation into the base of the cranium, or (3) formation of one or more of the mouth structures. According to a hypothesis that had its origins in the nineteenth century, the first gill arch posterior to these anterior pairs changed so that the upper part of the hinge became the upper jaw, or *palatoquadrate*, and the lower part became the *mandible* (see Figs. 17-17 and 17-20). Posterior

to these structures, the *hyoid* arch was incorporated into the complex by contributing its dorsal portion, the hyomandibular to anchor the hinge of the jaw to the braincase.

Among evidence supporting this view are the arch-like appearance of the palatoquadrate and mandible in acanthodians and in later sharks and bony fishes (see Fig. 17-20), and the origin of these elements (and of the lower jaws) from the same (mid brain or mesencephalic) embryonic neural crest cells (see Fig. 17-3). In addition, one branch of the trigeminal cranial nerve in sharks runs down to the lower jaw, while another runs anteriorly to the upper jaw, as though a more anterior pharyngeal cleft had been present at one time **(Fig. 17-21).** The presence of cranial nerves anterior to the trigeminal indicates that there once were pharyngeal arches anterior to those involved in jaw formation. All of these features support the homology of mandibular and pharyngeal arches.[15]

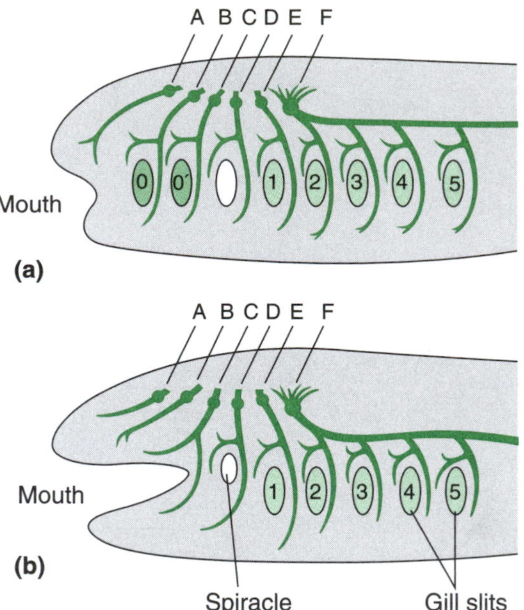

FIGURE 17-21 Diagrams of the dorsal root cranial nerves that innervated the gill arches (a) in a hypothetical early fish, and (b) in a later jawed fish such as a shark. The numbers 0 and 0′ represent gill clefts lost with the acquisition of jaws, spiracle represents the pharyngeal cleft later used as the spiracle in sharks, and 1 to 5 are gill clefts posterior to the spiracle. A to F are cranial nerves. A represents the terminal cranial nerve found in numerous vertebrates that may have innervated the anteriormost member of the gill series. B represents a nerve that combines with the trigeminal nerve in mammals. C is the trigeminal nerve that innervates the upper and lower jaws in all extant vertebrates. D, E, and F represent, respectively, the facial, glossopharyngeal, and vagus nerves. (See Fig. 3-16 for a comparison of innervations of the vagus nerve in fish and in mammals.) (Adapted from Romer, A. S., and T. S. Parsons, 1977. *The Vertebrate Body,* 5th ed. Saunders, Philadelphia.)

Non-Pharyngeal Gill Arch Origin of Jaws

For want of a better term, we refer to the alternate theory as the **non-pharyngeal arch origin of jaws** to emphasize that, whatever form variations of this theory take, all agree that jaws are not transformed pharyngeal arches, meaning, of course, that jaws would not be homologous with pharyngeal arches (see Box 17-2). This theory of the evolutionary origin of jaws proposes that no clear sign of a gill arch remnant anterior to the mouth appears in either extinct or extant vertebrates. Accordingly, the palatoquadrate and mandible may never have served as gill supports, and the mouth and its supporting structures were always separate from the pharynx.[16]

Support for this hypothesis comes, in part, from reanalysis of cranial development, and the distribution and function of Hox genes associated with head development in one or two species of lampreys (see Fig. 17-12), whose phylogenetic position is shown in **Figure 17-22.** A brief summary of the major lines of evidence follows.

1. **Cells anterior to where the mandibular arch lies in gnathostomes** form different structures in lampreys and gnathostomes. Lamprey mandibular arches form the lower lip — which resembles, but may not be homologous to, the lower jaws of gnathostome — and a specialized lamprey structure known as the *velum,* an organ that pumps water into the pharynx. Both of these findings question the homology of the oral regions in lampreys and gnathostomes. Lack of homology would mean that all or parts of the jaws arose as evolutionary novelties (Kuratani, 2004). In ammocoete larvae, the premandibular ectomesenchyme, which forms what is termed the prechordal cranium in gnathostomes, forms the upper lip, which is part of the oral apparatus.
2. **Different subsets of cells** form oral structures in lampreys and gnathostomes. These cells differentiate after undergoing inductive interactions with different embryonic epithelia in different regions of the head, but using homologous sets of genes (*Fgf8, Dlx1: Bmp2/4, Msx*) for the interaction. Such data strongly suggest that **heterotopy** and **heterochrony** (see Chapter 13) played an important role in the origin of the jaws (Kuratani, 2004; Long et al., 2006), provide an elegant demonstration of how the same genes can be utilized when deployed in different regions, but on their own are insufficient to demonstrate the "non-pharyngeal arch" hypothesis.
3. **Hox-free cells.** Both the jaws of gnathostomes and the equivalent portion of lamprey heads form from cells

[15] See Hall (1999b, 2000a) for overviews of the neural crest in development and evolution, and see Kuratani (2004) for fate mapping. Recent studies demonstrate that the neural crest has an important and previously unknown role in fore- and midbrain development (Creuzet et al., 2006).
[16] See R. L. Carroll (1988), Mallatt et al. (2003) and Kuratani (2004), for discussion of the origins of jaws.

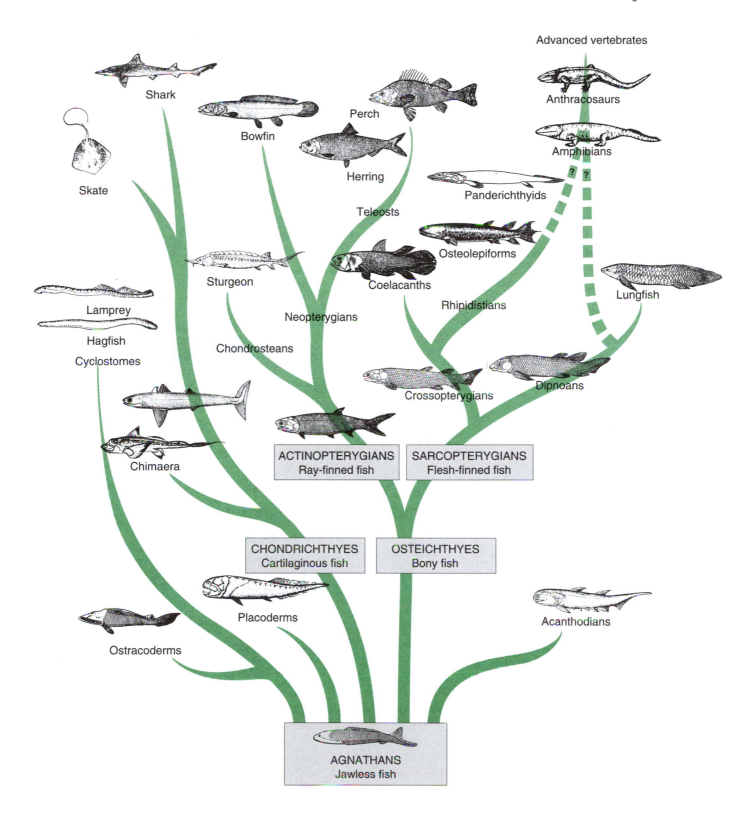

FIGURE 17-22 A traditional representation of the phylogenetic relationships among different lines of "fishes." The ancestor of each given lineage is not necessarily the species illustrated, but a species with similar characters. On the molecular level, Rasmussen and Arnason (1999) question whether sharks and skates originated earlier than the Mesozoic, and whether "advanced vertebrate" ancestry is osteichthyan, from within an even more ancient group of jawed fishes, as portrayed by paleontologists.

in the mandibular pharyngeal arch that *do not express* Hox genes (but do express *Otx* genes; see Figs 17-8 and 17-12), suggesting that a "Hox-free zone" was present in the common ancestor of lampreys and gnathostomes, and/or that absence of Hox-gene expression is not sufficient to explain the origin of the jaws.[17] An important consequence of this interpretation is the greater attention now devoted to the role played by Hox genes in initiating fundamental patterns along the A-P axis.

4. Recent studies in embryos of the river lamprey *Petromyzon marinus* show that SoxE genes are essential for development of the posterior pharyngeal arches but not for the mandibular arch, which either is evidence for the *non-homology of anterior and posterior pharyngeal arches,* or establishes that SoxE genes functioned in the common ancestor of lampreys and gnathostomes and underwent independent evolution in lampreys and gnathostomes (McCauley and Bronner Fraser, 2006).

Despite 150 years of research and recent exciting data from a variety of sources, it is still not easy to distinguish between these two hypotheses for the origin of the jaws. Research that integrates molecular and developmental approaches to understand evolution is in an exciting phase, however, and additional data and interpretations will continue to appear.

Evolution of Cartilaginous and Bony Fishes

Entirely new forms of fishes began to appear during the Devonian. Improved swimming efficiency resulting from increased neural and muscular coordination and progressively streamlined body forms marked their evolution. Among these were **cartilaginous fishes (Chondrichthyes)** and **bony fishes (Osteichthyes;** Figure 17-22).

Chondrichthyans — sharks, skates and rays — are cartilaginous, with cartilage that may be heavily calcified. Bone is found in these cartilaginous fishes, although only in minute amounts in vertebrae or at the bases of the teeth (Hall, 2005a). Almost all cartilaginous fish have tails that turn up at the end (*heterocercal*), a feature also common in early bony fish. Common ancestry of extant sharks, skates and rays can be traced to the Jurassic but the morphologically bizarre chimaeras have ancestral Carboniferous forms, indicating that chondrichthyans may not be monophyletic.

Osteichthyans, which are characterized by paired fins, a bony composition of skull, jaws, gill cover (opercule), scales,

[17] Gnathostomes use distalless (Dlx) genes to pattern each pharyngeal arch in the dorsal-to-ventral direction (Fig. 17-12). Dlx genes are also expressed in lamprey pharyngeal arches (Neidert et al., 2001), although, as with point two above, such data on their own are insufficient to demonstrate the "non-pharyngeal arch" hypothesis.

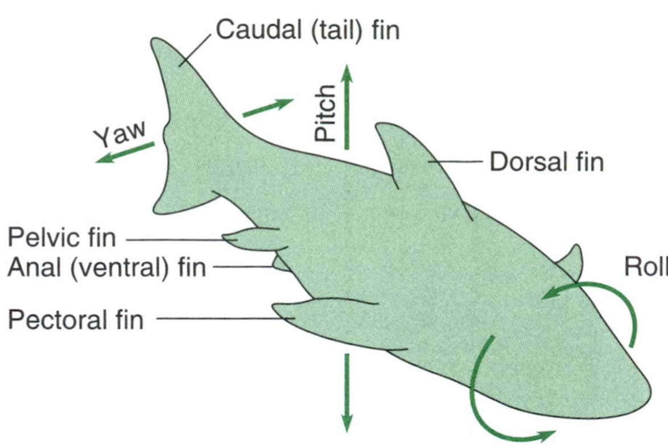

FIGURE 17-23 Generalized design of an extant fish that enables it to cleave through the water rapidly with considerable control and to cope efficiently with disturbing forces such as yaw, pitch and roll. (Adapted from Waterman, A. J., B. E. Frye, K. Johansen, A. G. Kluge, M. L. Ross, C. R. Noback, I. D. Olsen, and G. R. Zug, 1971. *Chordate Structure and Function.* Macmillan, New York.)

vertebrae and ribs, may not be monophyletic. In contrast to cartilaginous fishes, oste-ichthyans have as an ancestral feature a lung that became the **swim bladder** in derived forms. Quite a lot is known about fin evolution, some of which is outlined below.

Fins

Compared to acanthodians and placoderms, cartilaginous and bony fishes show new uses and arrangements of fins including:

- a caudal (tail) fin used for propulsory motion;
- dorsal and anal fins that act as keels to prevent rolling and side-slipping; and
- mobile, **paired** pectoral and pelvic **fins** that provide vertical controls, "brakes," and "bilge rudders" and that have been extensively modified in different fishes (**Fig. 17-23**).

As just two examples, the bases of the paired fins in the Devonian shark-like *Cladoselache* are quite wide, limiting movement (**Fig. 17-24**), but fin bases become narrower and more mobile in later forms. Portions of the pelvic fins in male sharks are modified as reproductive organs or claspers. As shown in **Figure 17-25,** the pelvic fins in some bony fishes are modified as suckers (Hall, 2007a).

A long-standing theory of the origin of fins is the "fin-fold" theory, which was developed independently in the mid-1800s by three zoologists. *There are really two theories, one for the median unpaired fins, and another for the paired pelvic and pectoral fins.* The former states that the median unpaired fins (the dorsal, caudal and anal fins) are derived from a continuous fold of skin originally present along the dorsal and ventral midline of the body. We know this to be the case during embryonic development, when a continu-

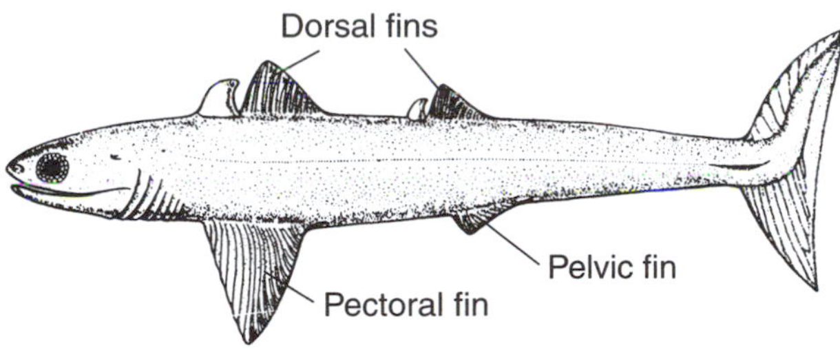

FIGURE 17-24 *Cladoselache,* a late Devonian shark-like fish ranging from 0.45 to 1.2 m long. It had paired pectoral and pelvic fins as well as fins on the mid-dorsal line, each supported by a broadened row of unconnected rods of cartilage. Some studies question whether *Cladoselache* was ancestral to lineages of sharks and skates, and suggest instead a more recent Mesozoic origin (Rasmussen and Arnason, 1999). (Adapted from Romer, A. S., and T. S. Parsons, 1977. *The Vertebrate Body,* 5th ed. Saunders, Philadelphia.)

ous fold that runs along the dorsal midline, around the tail bud, and along the ventral midline, becomes subdivided into separate dorsal, ventral and anal fins.

The second fin-fold theory holds that paired pectoral and pelvic fins arose in evolution from folds that ran beside the midline, also being later subdivided into separate pelvic and pectoral fins. No paleontological evidence supports this view, nor is such a continuous lateral fold found in the embryos of any extant fishes. Indeed, we now know that pelvic and pectoral fins — which develop separately, as do their homologues, the limbs of tetrapods — evolved separately and that the pelvic fins were the first to arise (Hall, 2005a, 2007a). Any connection between the median and paired fins has been elusive, although a recent gene expression study in the river lamprey lampreys *Petromyzon marinus* and in the catshark *Scyliorhinus canicula* has shown that *Hoxd* and *Tbx18* are both expressed in paired and midline fin primordia, suggesting to Freitas and colleagues (2006) that paired fins co-opted a gene network originating in median fins.[18]

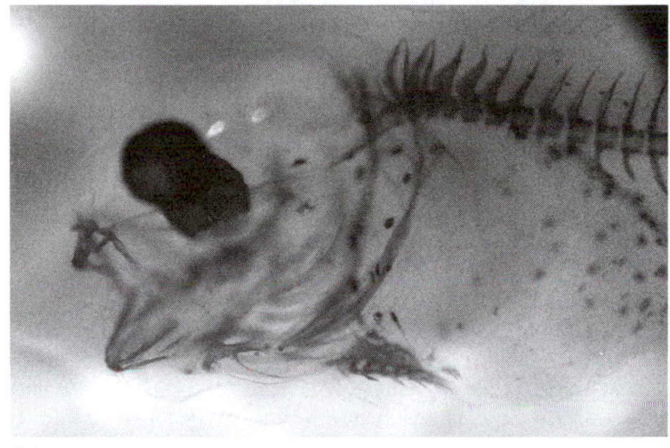

(a)

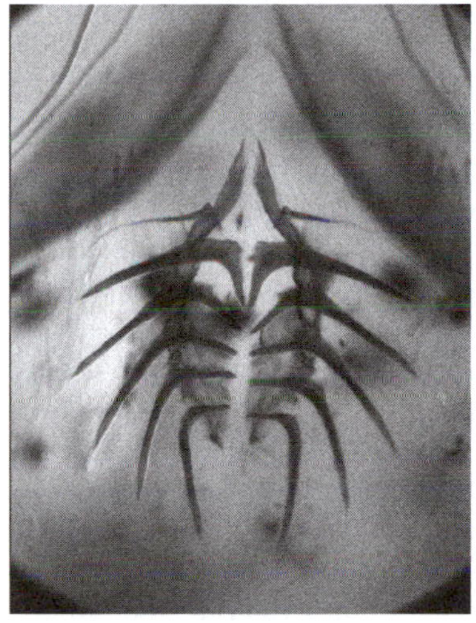

(b)

FIGURE 17-25 Pelvic suckers in the a 5-m–long inquiline snailfish *Liparis inquilinus* shown (a) in relation to the gills in lateral view, and (b) in ventral view to show the left and right halves of the sucker composed of fin rays. (Images by Lisa Budney.)

Bony Fish

Because land vertebrates evolved through lineages of osteichthyans, paleontologists have paid much attention to them. The Middle Devonian strata in which they first appear indicate their presence and major evolution in shallow waters. Such environments would have included stagnating lakes and slow-moving rivers poor in oxygen because of high temperatures or because blooms of algae and microorganisms had consumed much of the oxygen. This would have placed selective advantages on developing lungs as specialized respiratory tissues to help oxygenate blood in oxygen-depleted waters. Consistent with the presence of lungs in ancestral actinopterygians, early marine fish entering shallow waters would have carried sac-like air or swim bladders (**"lungs"**), previously used for buoyancy, but that could evolve into lungs.

However lungs arose, whether for buoyancy or respiration, Farmer (1997) suggests their primary value was to increase oxygen supply to the heart, enabling air-breathing fish to increase muscular activity even in nonhypoxic environments. J. B. Graham (1994) proposes that pulmonary air breathing evolved "as many as 67 times." Lungs and swim bladders may therefore have reversed functions because of selection in different conditions, each structure serving as a preadaptation for the other.

On the basis of fin structure and other features, the first bony fishes of the Devonian are identifiable as **ray-finned fish** or actinopterygians, and **lobe-finned fish** or sarcopterygians (see Fig. 17-22).

[18] *Tbx18* is a member of a family of T-box transcription factors, members of which are involved in specification of boundaries and regions within embryos.

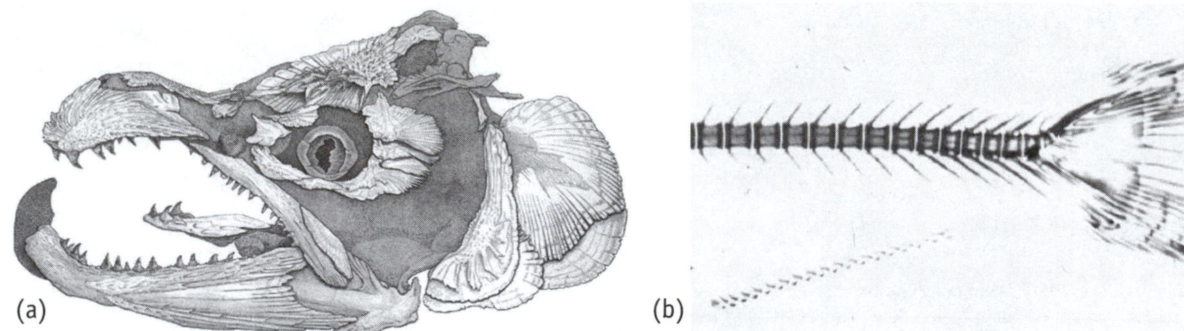

FIGURE 17-26 (a) The skull of a male Atlantic salmon *Salmon salar* illustrating the extensive and elaborate nature of the bones. Note the bone on the tongue and the teeth on that bone. The structure (gray) at the tip of the lower jaw is the kype that develops in male Atlantic salmon as they migrate to spawn. (From Tchernavin, V, 1937. Skulls of salmon and trout. A brief study of their differences and breeding changes. *Salmon and Trout Magazine,* **88,** 235–242.) (b) Specialization of the tail skeleton as seen in the Alewife, *Alosa pseudoharengus,* a member of the herring and sardine family, with duplication of the skeleton in the bilobed tail. (Photo courtesy of Brian Hall, created by T. Miyake)

Ray-Finned Fish (Actinopterygians)

Actinopterygians are ray-finned fish in which the fins are supported by parallel bony rays whose movements are controlled almost entirely by muscles within the body wall. All derive from a basic ancestral form that had paired pectoral and pelvic fins as well as a single dorsal fin balanced by a single anal fin on the ventral surface. The latest studies indicate that fin development in ray-finned and cartilaginous fishes both share conserved and ancient gene networks based on expression and regulation of the gene *sonic hedgehog* (Dahn et al., 2007).

Various lineages of ray-finned fishes evolved, differing in types of scales, tail structure, jaw structure, position of the fins and degree of ossification of the skeleton. The most basal actinopterygians (**Chondrostei**) include some extant species, including sturgeons (*Acipenser* spp.) and paddlefishes (*Polyodon spathula*). More complex forms, the **Neopterygii** ("new fins') include the bowfin (*Amia*) and gar (*Lepisosteus*), as well as most derived *Teleostei,* characterized by a highly ossified skeleton, thin scales, and fin specializations (**Figs.** 17-22 and **17-26**). **Teleosts** have expanded in abundance and diversity from the Cretaceous onward until they now number more than 24,000 species in 40 orders. Teleost diversity probably results from their capacity to exploit a wide variety of environments as a result of the plasticity of their feeding apparatus (see, for example, Fig. 12-18 and **Box 17-3**) as well as new fin and scale locomotory adaptations.

Lobe-Finned Fishes (Sarcopterygians)

Sarcopterygians,[19] lobe-finned ("flesh-finned") fishes support the fin with small individual bones arranged either along the fin axis or in rows parallel to the body. Muscles within the fin itself mostly control its movements. Early lobe-finned fishes had two dorsal fins. Some had internal nostrils that functioned in air breathing.

Among sarcopterygian descendants, one group persists in the form of African, Australian and South American **lungfishes** or **dipnoans (Fig. 17-29a).** During dry seasons, when their pools stagnate, the African and South American fishes encyst in mud. Openings in their burrows allow them to breathe via a pair of vascularized lungs. Australian lungfish, considered less derived, behave differently because they cannot survive out of the water. They come to the surface during the dry season and use their single lung to breathe air. The present distribution of lungfishes in Africa, South America and Australia can be explained by the proximity of these continents during the Mesozoic in Gondwanaland (see Chapter 5).

Because the origin of land vertebrates (**tetrapods;** see Chapter 18) was from one or more sarcopterygians, their phylogenetic relationships are of considerable interest. The traditional classification — which can lump organisms together even though they do not have a single phylogenetic origin — divides sarcopterygians into two subgroups, the dipnoans described above, and the *polyphyletic* **crossopterygians.** Crossopterygians are further divided into **rhipidistians** (Fig. 17-29b and c) and **coelacanths,** which can be traced back to the Early Devonian. Rhipidistians were primarily freshwater fishes; coelacanths were marine. As mentioned earlier, coelacanth fossils appear as far back as the middle Devonian, and it had been assumed that they became extinct sometime during the Cretaceous. This view prevailed until a living coelacanth, *Latimeria chalumnae,* was found in 1938 (see Fig. 3-21). Since then, many fish of this species have been caught in the Indian Ocean between Africa and Madagascar, and a separate species as far east as Indonesia. According to Zardoya and coworkers (1998), coelacanths, generally considered crossopterygian descendants, are further from the tetrapod lineage than are lung-

[19] The term *sarcopterygian* is used in two ways today. One (introduced above) is as one of the two clades (subclasses) of bony fishes. The second, used in cladistic terminology, includes the lobe-finned fish (coelacanths, lungfishes) of the original subclass and all their descendants, that is, all tetrapods (see Maisey, 1996). Mammals are sarcopterygians in this cladistic sense. Here, the term is used in its original sense as applying to fishes.

BOX 17-3 Pharyngeal Jaws in Cichlid Fishes

An especially nice example of the capacity to exploit a wide variety of environments because of flexibility of the feeding apparatus is seen in the evolution of a **second set of jaws** in a number of lineages of bony fishes **(Fig. 17-27).** Such evolutionary plasticity and evolvability has enabled fish such as the cichlids of the East African lakes to speciate rapidly and to exploit many microhabitats and diets within the lakes. The combination of developmental and morphological plasticity, ability to adjust the phenotype in rapid response to environmental changes, short generation time, an abundance of potential diet items, territoriality and aggressive defense of nests, make cichlids a "textbook" example of rapid evolutionary change, emphasizing the role of phenotypic plasticity in speciation.

Lakes Victoria, Malawi and Tanganyika in Africa contain 300, 200, and 125 endemic species of cichlids, respectively, all of which appear to have arisen rapidly and by sympatric speciation (see Chapter 24). Lake Victoria is no more than 750,000 years old, yet contains 300 species found nowhere else. Indeed, if Lake Victoria dried up between 12,500 and 14,000 years ago, as geological evidence suggests, then these species have arisen amazingly rapidly. If recent findings for the South American cichlid genus *Apistogramma* hold true for African cichlids, then these species numbers may be considerably underestimated; Ready et al. (2006) demonstrated that individuals from the same populations of *Apistogramma caetei* that *differ only in color* and so have been regarded as the *same species,* fail to interbreed because of female mate choice, that is, they behave as good biological species.

Morphological specialization in cichlids is especially evident in the feeding apparatus, reflecting ecological specialization in adaptation to a diversity of diets (feeding on plankton, detritus, insects, insect larvae, other fish and fish embryos; grazing on algae; crushing snails; T. Goldschmidt, 1996).

A specialization in cichlids, especially in those that crush snails, is the development of pharyngeal jaws (Fig. 17-27), which, as their name suggests, arise from a more posterior (5th to 7th) pharyngeal arch than do the lower or mandibular jaws (which arise from arch 1). Preformed in cartilage, both upper and lower cartilaginous elements ossify. Pharyngeal teeth develop on dermal bones that fuse to the ossifying cartilages. Pharyngeal jaws and teeth, which are used to process food, enable the 'true' jaws to be adapted for food capture, facilitating diversification, which combined with the other attributes of cichlids, facilitated speciation. As **Figure 17-28** shows, the pharyngeal jaw of cichlids consist of a massive crushing plate, with morphological specializations that relate to whether the food eaten is soft or hard.

Karel Liem (1974) discussed the evolution of pharyngeal jaws as an evolutionary avalanche ("a specialized, highly integrated innovation") with three successive changes:

- a shift in the insertion of the fourth levator externi muscles; and
- fusion of the left and right fifth ceratobranchial to form the lower pharyngeal jaws;
- the development of a mobile joint between the upper pharyngeal jaws and the base of the cranium.

Given that we know that cells of adjacent bones or cartilages readily respond to mechanical stimulation (Hall, 2005a), the formation of a new joint does not require any exotic mechanisms, and could have occurred quite rapidly, both in developmental and in evolutionary time. As Liem (1974) described it, "The conversion of the preexisting elements into a new and significantly improved cichlid adaptive complex of high selective value may have evolved in rapid steps under influence of strong selection pressure acting on the minor reconstruction of the genotype which is involved in evolutionary changes of the pertinent ontogenetic mechanism."

Sadly, just as preexisting elements transformed rapidly into the "significantly improved cichlid adaptive complex," the decline in Lake Victoria over the past four decades has been the most rapid and perhaps most drastic of any lake on Earth. The deadly combination of over-fishing, introduction of exotic species (especially the Nile tilapia *Oreochromis niloticus* and the Nile perch *Lates niloticus*), pollution and poor management of the surround land, all have depleted oxygen levels, choking the lake with algae to the extent that more than half of the indigenous cichlid species are now extinct. By 1982, the Nile perch represented 80 percent of the biomass in the lake, while algal blooms (including toxic blue-green algae) and proliferation of water hyacinth block channels and further reduce light penetration into the light. Cichlids are not the only species at risk; 30 million people depend on Lake Victoria to make a living, and ultimately to survive. A crater of speciation has become a crater of doom.

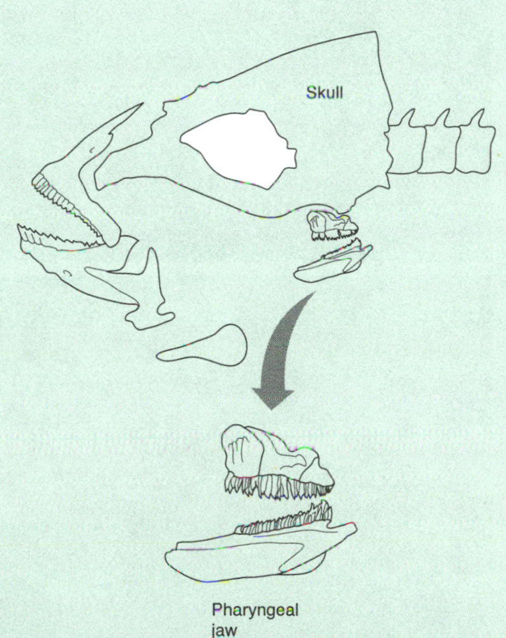

FIGURE 17-27 Head of a cichlid fish showing the location of the upper and lower pharyngeal jaws between the lower jaw and the ventral surface of the skull. The image below shows the pharyngeal jaws and their teeth at higher magnification (see also Figure 7-29).

BOX 17-3 Pharyngeal Jaws in Cichlid Fishes (*cont.*)

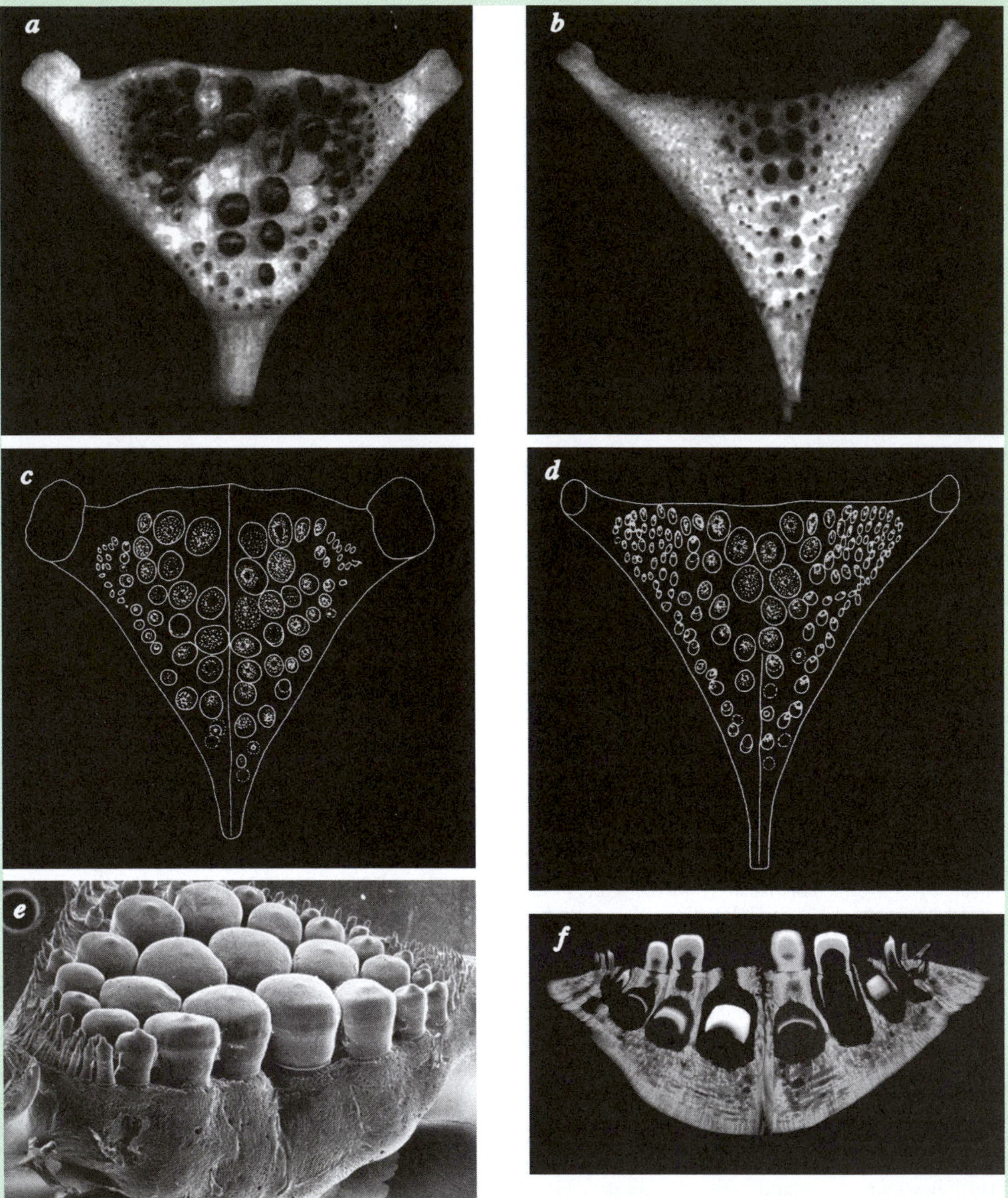

FIGURE 17-28 Lower pharyngeal jaws and teeth of cichlid fishes. The Midas cichlid *Amphilophus citrinellus* is polymorphic, one form with pharyngeal jaws with molariform teeth (a), the other with papilliform teeth (b), reflecting the different diets of the two morphs. The East African cichlid Allauad's haplo, *Astatoreochromis alluaudi,* also displays two morphs, one molariform (c) associated with hard food and a second papilliform (d) associated with soft diet, both of which can be modulated by changing the diet (a–d are shown as viewed from inside the mouth). e and f show molariform teeth as seen with scanning electron microscopy (e, Midas cichlid) and in an x-ray of a section through the pharyngeal jaws (f, Allauad's haplo). (Photos a, b, and e courtesy of Dr. Axel Meyer, University of Konstanz, and photos c, d, and f, courtesy of Dr. Ann Huysseune, Ghent University.)

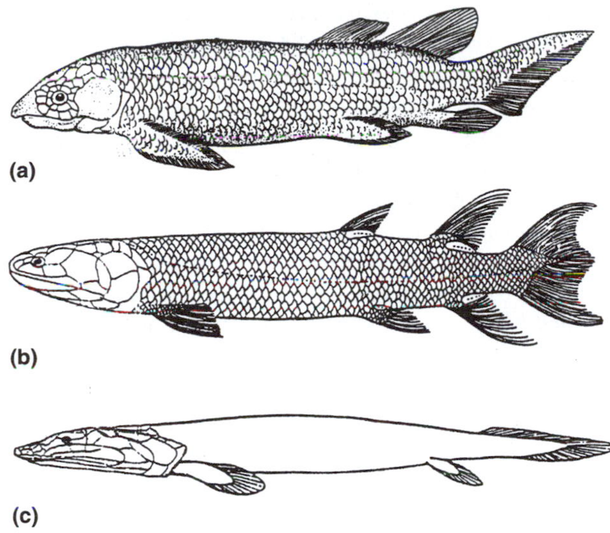

FIGURE 17-29 Reconstruction of three lobe-finned Devonian fishes of lineages long considered as possibly ancestral to early land-living tetrapods. (a) The dipnoan lungfish *Dipterus*. (b) and (c) Rhipidistians, *Eusthenopteron* (an osteolepiform) and *Panderichthys*.

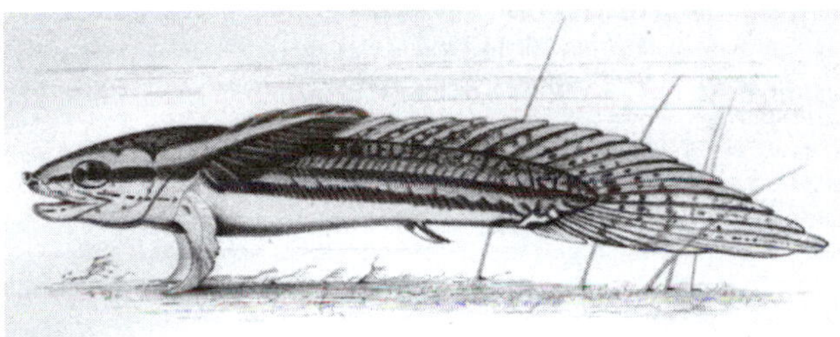

FIGURE 17-30 A 3-cm-long larva of the African lungfish *Protopterus*, drawn from life by John Budgett in 1901, the first zoologist to obtain lungfish embryos and larvae. Note the prominent external gills and the expansive tail both used in respiration.

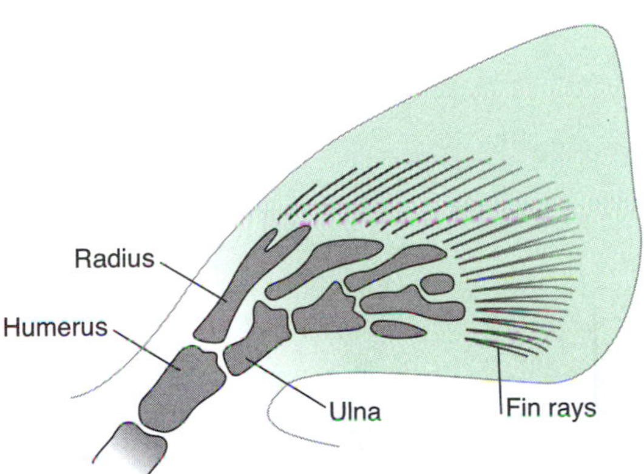

FIGURE 17-31 A representation of a transitional stage between the fore fin and anterior limb. Elements closest to the body can be identified as homologous to the humerus (H), radius (R), and Ulna (U) of tetrapod limbs but fin rays (FR) are still present. (Adapted from Hall, B. K., 1999a. *Evolutionary Developmental Biology*, 2nd ed. Kluwer Academic Publishers, Dordrecht, The Netherlands; modified from Shubin, N., 1995. The evolution of paired fins and the origin of tetrapod limbs. *Evol. Biol.* **28**, 39–86.)

fishes. A larval African lungfish, with features very similar to those of urodele amphibians, including external gills and a prominent tail is shown in **Fig. 17-30.**

Among the rhipidistians, two of the lineages that evolved — **panderichthyids** and **osteolepiforms** (Fig. 17-29) — show sufficiently strong similarities to the early tetrapods in their internal nostril openings, skull and limb structures (**Fig. 17-31**) that they are regarded as the **stem group** *stem tetrapods*. As we discuss in the next chapter, fossil finds and molecular sequence analyses of extant groups have added enormously to our understanding of these transitional forms.

KEY TERMS

acanthodians
actinopterygians
ammocoete larva
amphioxus
bony fishes
cartilaginous fishes
cephalochordates
Chondrichthyes
chordates
coelacanths
crossopterygians
cyclostomes
deuterostomes
dipnoans
dorsal hollow nerve cord
filter feeding
gill arches
hemichordates
heterochrony
heterotopy
jawed vertebrates
jawless vertebrates
lungfishes
lungs
muscular pharynx
neoteny
neural crest
notochord
Osteichthyes
osteolepiforms
ostracoderms
paedomorphosis
paired fins
panderichthyids
pharyngeal (branchial) clefts
pharyngeal arches
placoderms
protostomes
rhipidistians
sarcopterygians
swim bladder
tail
teleosts
tetrapods
tunicates
urochordates
vertebrae

DISCUSSION QUESTIONS

1. What features distinguish chordates from other phyla?
2. What evidence, pro and con, have paleontologists offered for the origin of chordates from (a) annelids or arthropods, (b) ascidians?
3. How have recent fossil discoveries contributed to our understanding of the origin of early vertebrates?
4. In the evolution of fish, what advantages can we ascribe to
    a. a muscular pharynx.
    b. bone.

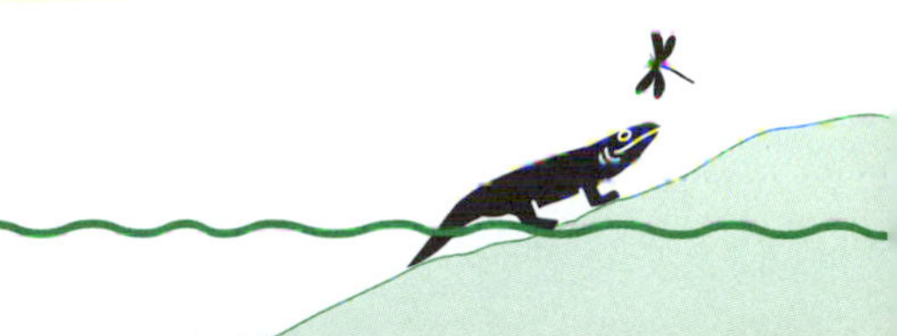

c. jaws.
d. fins.
e. lungs.

5. What hypotheses have paleontologists, developmental and molecular biologists offered to explain the evolution of jaws? Do the data from these three fields lead to similar conclusions?
6. What are the major groups and proposed phylogenetic relationships among the lobe-finned fishes (sarcopterygians)?

EVOLUTION ON THE WEB

Explore evolution on the Internet! Visit the accompanying Web site for *Strickberger's Evolution*, Fourth Edition, at http://www.biology.jbpub.com/book/evolution for Web exercises and links relating to topics covered in this chapter.

RECOMMENDED READING

Aguinaldo, A. M. A., J. M. Turbeville, L. S. Linford, et al., 1997. Evidence for a clade of nematodes, arthropods and other moulting animals. *Nature,* **387**, 489–493.

Arendt, D., and K. Nübler-Jung, 1994. Inversion of dorsoventral axis? *Nature,* **371**, 26.

Benton, M. J., 2004. *Vertebrate Palaeontology,* 3rd ed. Blackwell, Oxford, UK.

Carroll, R. L., 1988. *Vertebrate Paleontology and Evolution.* Freeman, New York.

Dehal, P., Y. Satou, R. K. Campbell, J. Chapman et al., 2002. The draft genome of *ciona intestinalis:* Insights into chordate and vertebrate origins. *Science,* **298**, 2157–2167.

Goldschmidt, T., 1996. *Darwin's Dreampond. Drama in Lake Victoria.* MIT Press, Cambridge, MA.

Hall, B. K., 1999b. *The Neural Crest in Development and Evolution.* Springer-Verlag, New York.

Hall, B. K., 2005a. *Bones and Cartilage: Developmental and Evolutionary Skeletal Biology.* Elsevier/Academic, London, UK.

Hall, B. K., (ed.), 2007a. *Fins into Limbs: Development, Evolution, and Transformation.* The University of Chicago, Chicago.

Kardong, K.V., 2006. *Vertebrates: Comparative Anatomy, Function, Evolution,* 4th ed. McGraw-Hill, New York.

Liem, K. F., W. E. Bemis, W. F. Walker, Jr., and L. Grande, 2001. *Functional Anatomy of the Vertebrates. An Evolutionary Perspective,* 3rd ed. Harcourt College Publishers, Fort Worth TX.

Maisey, J. G., 1996. *Discovering Fossil Fish.* Henry Holt, New York.

Northcutt, R. G., and C. Gans, 1983. The genesis of neural crest and epidermal placodes: a reinterpretation of vertebrate origins. *Q. Rev. Biol.,* **58**, 1–28.

Ready, J. S., I. Sampaio, H. Schneider, et al., 2006. Color forms of Amazonian cichlid fish represent reproductively isolated species. *J. Evol. Biol.,* **19**, 1139–1148.

Zelditch, M. L., (ed.), 2001. *Beyond Heterochrony. The Evolution of Development.* Wiley-Liss, New York.

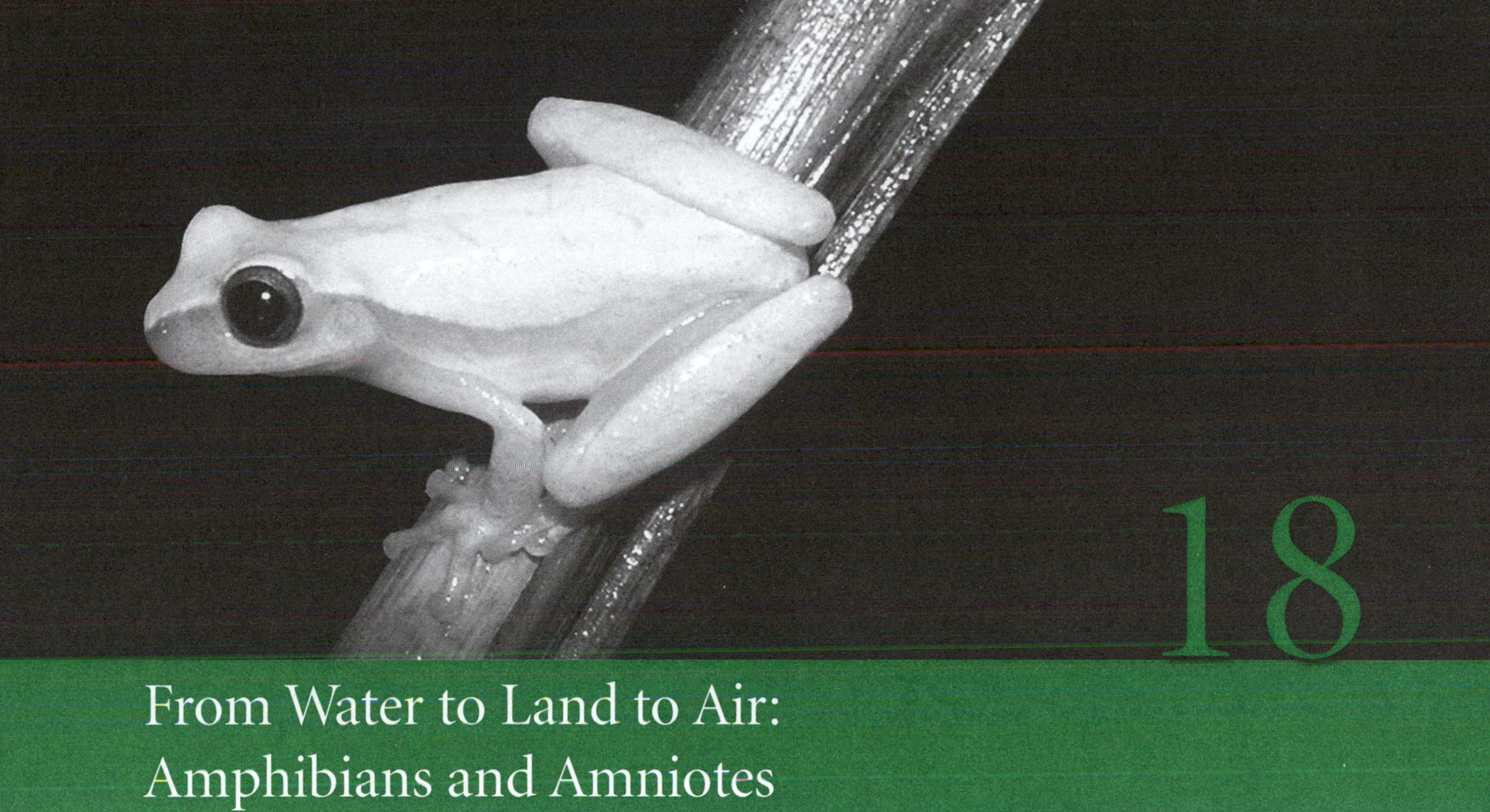

18

From Water to Land to Air: Amphibians and Amniotes

Chapter Summary

With their lungs and fleshy fins, the more derived sarcopterygian fish (stem tetrapods) were well placed to respond to selective pressures associated first with spending increasing periods of time out of the water and then with becoming terrestrial, invading the land in response to competition and/or predators in their swampy habitats. Seven stem tetrapods and early aquatic/semiterrestrial tetrapods from the Devonian provide us with considerable knowledge of the changes associated with leaving the water and becoming terrestrial, especially changes in the limbs, vertebral column and pectoral and pelvic girdles. The earliest terrestrial tetrapods, "amphibians," arose from these aquatic forms, retaining an aquatic larva, the tadpole.

The first amniotes arose from small, lizard-like anamniote tetrapods. Two ancestral amniote lineages are recognized, sauropsids and synapsids. The evolution of the amniotic egg liberated amniotes from dependency on water and was preceded by terrestrial egg laying and internal fertilization. Openings (fenestrae) in the temporal region of the skull — a trait found in amphibians, enhanced in amniotes, and often used by paleontologists to classify fossil amniotes — provided room for expansion and anchorage for enhanced jaw muscles association with food capture and processing on land.

The first amniotes (reptiles in traditional terminology) diversified into many terrestrial habitats to become the dominant tetrapods throughout the Mesozoic. Therapsids, ancestors of mammals, appear to have had temperature-regulating mechanisms and may have been the first endotherms. Specialization of limb bones facilitated bipedalism in sauropods, increasing their aerobic activity. The most famous archosaurs — the dinosaurs — successfully radiated into almost every terrestrial habitat, some becoming so enormous as to approach the upper limit for the size of a terrestrial animal. Considerable evidence from bone and integument structure, posture and biogeographical distribution, demonstrates that some dinosaurs were endothermic. By the end of the Cretaceous, however, most dinosaurs along with other large marine forms and various invertebrates had died out as the result of several mass extinctions, each of which may have been precipitated by the impact of a large meteor.

Adaptations for sustained flight appeared independently in pterosaurs and birds. Pterosaurs developed hollow bones and flight membranes (patagia) between trunk and forelimbs, and in some cases, lost the tail and teeth, leaving the jaw as a beak. Birds arose from bipedal, ground-dwelling or arboreal carnivorous dinosaurs. Feathers are generally considered as homologues of, and derived from, scales, and likely evolved as an insulating mechanism and/or as devices for display. Birds show altered bone structures and an enlarged brain, some returning to a flightless condition. On the paleontological level, we knew little of evolutionary relationships among birds until recent discoveries of fossil birds in China and elsewhere. Molecular techniques offer new phylogenetic information about the evolution of tetrapods, including birds.

We begin by reiterating an important point about classification, terminology and phylogenetic relationships made in the introduction to this section of the book (Part 4, The Organisms).

We are accustomed to consider vertebrates in commonly known and understood categories such as fish, amphibians and reptiles. But phylogenetic analyses carried out over the past couple of decades demonstrate quite convincingly that these are not natural categories, if by natural we mean that each constitutes a group, all of whose members can be traced to a single ancestral lineage, that is, a group that is monophyletic. The recent elimination of birds as a group and the recognition of "birds" as flying dinosaurs, while fully justified, illustrates the shoals on the course we are navigating.

As we saw in the last chapter, "fish" is a name for a polyphyletic group of animals that share many features, but a single evolutionary origin is not one of them. We thus refer to older terms and classifications as we integrate the more historical (classic) studies with recent cladistic analyses to convey both "what is a fish" and "why fish are not a monophyletic lineage." The first tetrapods (amphibians) arose from among one group of fishes (sarcopterygians). How do we distinguish the more derived sarcopterygian fish that gave rise to tetrapods? Because the sarcopterygian genus *Panderichthys* is the most derived of these fishes, some refer to the even more derived taxa as "postpanderichthyid stem tetrapods," sufficient of which are now known that we can begin to sort out their relationships.

The old term, "mammal-like reptiles," once used for those reptiles recognized as having given rise to mammals, is no longer considered appropriate because:

- these organisms (therapsids) consist of multiple independent lines, only one or a few of which gave rise to mammals;
- the term confuses a **crown group** (the culmination of an evolutionary lineage) with a **stem group** (a lineage that gave rise to another group); and
- the term makes it sound as if some reptiles were trying to become mammals and so were mammal-like.

Problems of terminology and relationships are in as much flux for tetrapods as they are in the unicellular world (see Chapter 8), despite the century and a half head start of vertebrate morphology (which is mostly macroscopic) over the study of unicells (which is mostly microscopic and/or molecular). So what we have tried to do, especially in this chapter, is to acquaint you with terms and relationships that will enable you to navigate both the older and the more recent literature. Sometimes we do this with the suggestion that you compare two figures, and sometimes by discussing one approach with comments on the other. With this background, let's begin where we left off at the end of the last chapter, with sarcopterygian fishes.

Water to Land

Three lineages of sarcopterygian fishes introduced at the end of the last chapter and in Figure 17-29 — panderichthyids, osteolepiforms and dipnoans — have been regarded as ancestral (stem) tetrapods, that is, as a stem group. These fish had functioning lungs and two pairs of bone-strengthened muscular fins (**Fig. 18-1**), with which they

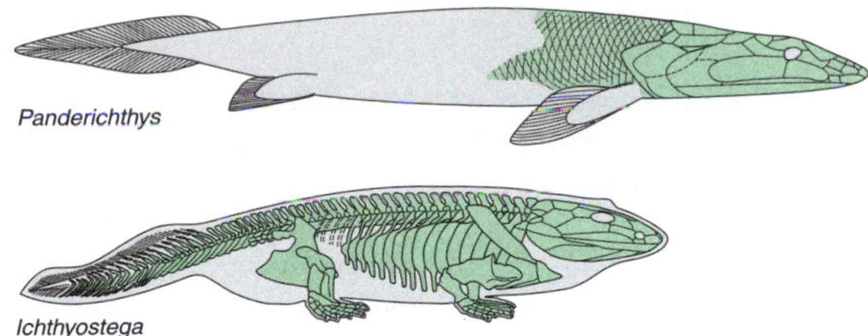

FIGURE 18-1 Reconstruction of a derived sarcopterygian fish *Panderichthys* (top) with muscular fins and of a partly aquatic–partly terrestrial stem tetrapod *Ichthyostega* (bottom).

could move their bodies and support themselves on the bottom and, we presume, during short excursions onto land.

In hindsight, it seems that many of these fishes needed relatively few further changes to attain at least a partial terrestrial existence (Figs. 18-1 and **18-2**). Features that characterize **tetrapods** — limbs instead of fins, lungs instead of gills — have been identified in sarcopterygian fishes that were still aquatic. The transition from aquatic, through partly aquatic/partly terrestrial, to terrestrial was almost certainly made by several lineages independently. Of course, we cannot call them fish if they have lungs and limbs and so we refer to them as stem tetrapods. Some decades ago we might have said "tetrapod-like fish," but that term implies a crown group and not a stem group.

Why Leave the Water/ Move to the Land?

We once described the move to land as fish abandoning their shallow-water habitats for a terrestrial life, implying a much more rapid shift than seems to have occurred. This is especially true now that we realize that limb formation began in aquatic forms that would have used their limbs for moving along the bottom (see Figs. 17-30 and 17-31). There is a big difference, however, between a limb that will allow you to "walk" along the bottom and one that will support your weight on land. These fish made forays onto land, a behavior that selection would have reinforced as changes resulting in stronger limbs, and *vice versa*.

One classic hypothesis for the water to land transition, popularized by Romer (1968) but with few adherents today, postulated that when the shallow, hypoxic habitats of ancient sarcopterygians dried up or stagnated, forms with lungs that were breathing atmospheric oxygen may have searched for new pools of water and survived on land for short periods of time. This was a perfectly reasonable hypothesis 40 years ago, associated as it was with the presumption that the *first tetrapods were terrestrial*.[1] At the time, most vertebrate biologists/paleontologists agreed with Romer. Because we know much more about the nature of the environment as a consequence of reanalysis of Early Devonian sediments, and because we have many new fossils, we now know that the **early tetrapods were aquatic**, inhabiting fresh water. Therefore, the first tetrapod features (limbs, lungs) evolved in aquatic environment, being adapted only later for more permanent life on land.

Another hypothesis, more commonly accepted now, suggests that these aquatic forms moved onto land — to what was then a moist tropical climate — because of pressures from increasing numbers of predators (probably other fish) and from competition for space, food and breeding sites in warm, swampy habitats. This is consistent with short forays out of the water. Combined ecological, environmental and climatic changes provided powerful selective advantages

[1] We mention Romer's hypothesis because it shows how hypotheses depend on the interpretation of the evidence available when the hypothesis is constructed. There are no right or wrong hypotheses, only more or less plausible hypotheses, testable or untestable hypotheses.

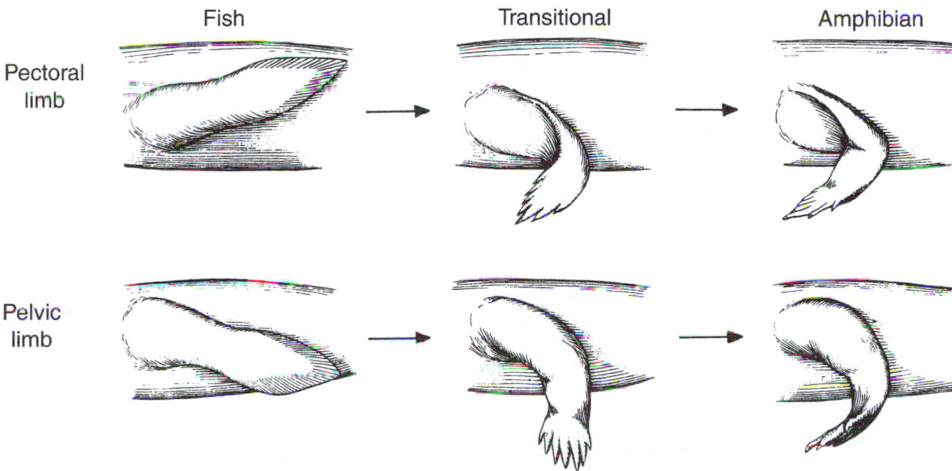

FIGURE 18-2 Shifts in the positions of pectoral and pelvic limbs associated with the transition from fish to tetrapod. (Adapted from Romer, A. S., and T. S. Parsons, 1977. *The Vertebrate Body*, 5th ed. Saunders, Philadelphia.)

to lineages spending progressively longer periods of time on land. There was certainly food available on land: Arthropods, among other invertebrates, had already made the terrestrial transition, and throughout the Devonian, land plants were establishing themselves in increasing number and variety. Further selection would have operated on many levels to enhance breathing, eliminate carbon dioxide, increase resistance to desiccation, increase mobility of the neck and enhance further transformations. You can see such changes in extant fish, some bony fish such as mudskippers, climbing perch and walking catfish have developed terrestrial adaptations, even to the point of climbing trees and capturing aerial food.

Origin of Tetrapods

Paleontological evidence leads us to conclude that land vertebrates are descendants of panderichthyids and osteolepiforms (rhipidistians) lobe-finned fishes and not from lungfishes or coelacanths, which are now seen as more like cousins of tetrapods. The transition from fish to four-legged tetrapod occurred by the end of the Devonian, about 354 to 360 Mya, during a relatively short geological interval of no more than 15 or 20 My (see Figs. 17-28 and 18-1).[2]

For many years we had fossils of only two major genera of Devonian tetrapods, *Acanthostega* and *Ichthyostega* (**Fig. 18-3**), but since then *Ventastega, Hynerpeton* and *Tulerpeton* have been discovered and described. All are Late Devonian in age, and all have well-developed tetrapod features, as summarized in **Box 18-1.**

The relationship among these late Devonian tetrapods is shown in Figure 18-8. *Tulerpeton* is the most derived and *Ventastega* the least. All are found within the last 6 My of the Devonian (354 to 360 Mya). None are found in the succeeding period, the Carboniferous, by which time five groups of more derived tetrapods had evolved, some 328 to 348 Mya.

A comment (indeed, a caution) on the nature of the fossil record is in order at this time, especially in relation to gaps in the fossil record.

It has been known for 40 years that few fossils of terrestrial vertebrates are known from the period between 360 and 345 Mya. This 15 My interval, named *"Romer's gap,"* in honor of Alfred Sherwood Romer (1894–1973) who first pointed it out, has now been shown to be a period with few fossils of terrestrial arthropods, although deposits immediately before and after Romer's gap contain many fossils. Ward and colleagues (2006) have now shown that a drop in O_2 levels coincides with Romer's gap and that "terrestrialization" of vertebrates and of arthropods occurred in two phases: a 65-My-period between 425 and 360 Mya when smaller forms became terrestrial, and a 55-My-period between 360 and 385 Mya when larger forms became terrestrial. These two periods bracket Romer's gap.

As shown in Figure 18-8, two more basal genera are known, *Panderichthys,* from 374 Mya, and *Elginerpeton,* from 367 Mya. *Panderichthys* and *Elginerpeton* represent the oldest links we have between tetrapods and sarcopterygian fishes. Because both genera have fins and not limbs, we cannot regard them as tetrapods. They are **stem tetrapods.**

To enable you to appreciate the gradual evolution of tetrapod features, the features of seven Devonian tetrapods and stem tetrapods are outlined in Box 18-1. We suggest you check this box before reading on.

Tetrapod Adaptations and Diversification

Early tetrapods flourished from the Carboniferous through the Triassic, producing a diversity of forms from the alligator-like *Eryops* (**Fig. 18-9**) to the smaller, dorsally armored *Cacops.*

The body (centrum) of each vertebrae arises from two elements, one anterior to the other, known, respectively, as the **intercentrum** and **pleurocentrum** (pl., intercentra and pleu-

[2] See R. L. Carroll (1995), Clack (2002), Carroll and Holmes (2007) and Coates and Ruta (2007) for overviews of tetrapod origins.

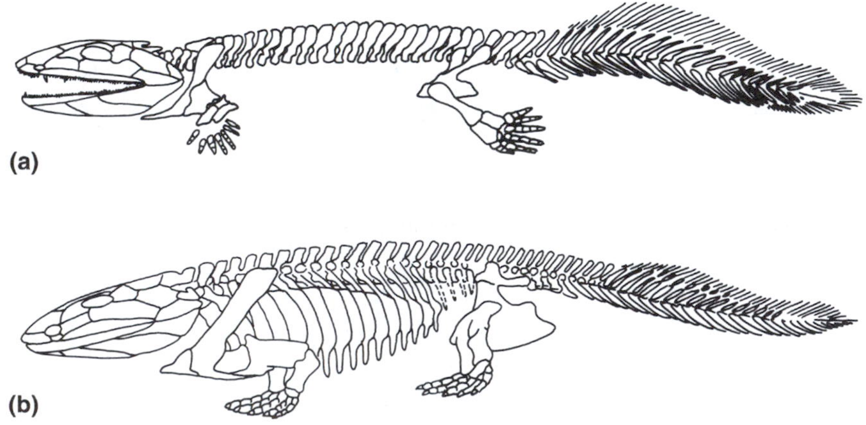

FIGURE 18-3 Two Late Devonian tetrapod skeletons (not to scale). (a) *Acanthostega,* a 60-cm–long Late Devonian tetrapod that retained fishlike internal gills and led an aquatic life, using its limbs in water rather than on land (see Coates and Clack [1991], Clack [2002] and Coates and Ruta [2007]). (b) The stout-limbed 1.5-m–long *Ichthyostega,* which maintained many fishlike features including scales, dorsal tail fin and a hydrodynamic shape, probably indicating a shallow-water environment.

rocentra). A neural arch forms dorsal to the centrum. By the Triassic, organisms appear in which the intercentrum alone became the main vertebral central element. Among this new group, called *stereospondyls,* are various forms with flattened heads, including one short-faced, armored type (*Gerrothorax*) that had evidently returned to an aquatic existence by maintaining gills in the adult stage (*paedomorphosis*).

We know almost nothing of the soft-body structures of these early tetrapods and have no obvious clue to how they prevented desiccation, although, unlike extant forms, many had a well-developed exoskeleton that could have reduced water loss. Like many extant amphibians, they spent a considerable part of their life in fresh water, with adults probably never wandering far from water. Their many anatomical post-aquatic innovations indicate that they and their successors were evolving adaptive solutions to at least some of the mechanical difficulties that face land-dwelling vertebrates, doing this with organs, such as limbs, which had first appeared while their ancestors were aquatic.

One early problem was *the need* to *prevent the impact of terrestrial locomotion from being transmitted to the braincase.* This was accomplished by freeing the limbs from the skull. Osteolepiform pectoral fins connect to the back of the skull, forming an inflexible "cage" around the internal organs. Such a system works well in an aquatic organism using gills for respiration, but not in a terrestrial organism using lungs, which would be compressed by such a solid cage.

A second problem in a terrestrial environment, therefore, was *the need to prevent the internal organs from being compressed as a result of the functioning of the limbs and limb girdles.* As illustrated in **Figure 18-10** these difficulties were solved by the limbs changing their relationships to the pectoral and pelvic girdles (which were present in ancestral fishes). The pelvic girdle fused with the vertebral column at the sacrum, while the pectoral girdle became less dependent on the vertebral column, which became a "suspension bridge" absorbing the forces associated with terrestrial motion, enabling the head to turn and lift independently of the body by placing it on a forward cantilever of cervical vertebrae. Selection for terrestrial skeletal rigidity had various consequences, especially strengthened vertebral elements and increasing contact between adjacent vertebrae (**Fig. 18-11**).

One group of Carboniferous tetrapods, **anthracosaurs,** a polyphyletic assemblage, had become extinct by the end of the Permian. Another lineage produced stout-legged terrestrial animals such as *Seymouria* (**Fig. 18-12a,** see Fig. 18-9). Until gilled seymourian larvae were discovered, *Seymouria* and its closest relatives (the Seymouriamorpha) were thought to be amniotes, providing a dramatic demonstration of how gradual the transitions were (**Fig. 18-13**).

As we are now approaching the geological period when the first amphibians appear, you may want to consult the phylogeny presented in Figure 18-13.

Nonamniote Tetrapods

Early **amphibians** all had new vertebral and skeletal structures. One group, often assembled under the name *microsaurs,* had extremely variable body proportions and vertebral structures. Although they have been proposed as possible amniote ancestors, R. L. Carroll (1997) and others have pointed out their many peculiar specializations as evidence against this hypothesis. As shown in Figure 18-13, the phylogenetic position of microsaurs remains unresolved, although they nest within Amphibia, as do the limbless and snakelike aïstopods and nectridians, with their peculiar hornlike projections of the rear skull bones (Fig. 18-9). These small aquatic forms flourished during the Carboniferous, but disappeared by the close of the Permian (Fig. 18-9).

Amphibia

Zoologists classify living amphibians as **Lissamphibia** (Fig. 18-9), separating them into three groups: *caecilians* (legless, wormlike burrowers), *anurans* (frogs and toads) and *urodeles* (newts and salamanders). In contrast to the scaled skins of early amphibians, extant lissamphibians have a permeable, glandular skin that permits considerable water and gaseous exchange (though small scales are embedded in scale pockets in the skin of some caecilians).

Although a permeable skin limits their life history to moist or aqueous surroundings, lissamphibians have achieved a variety of remarkable adaptations. Some frogs, including the spadefoot toads, *Scaphiopus* and *Pelobates* sp. and the water-holding frog, *Cyclorana* sp., are desert inhabitants capable of surviving long periods of drought in underground burrows by drawing moisture from the soil or by encapsulating their water-soaked bodies in relatively impermeable membranes (Kley and Kearney, 2007). In some caecilians such as "rubber eels" (*Typhlonectes*), offspring at birth can be almost half the maternal length, having fed on nutrient substance supplied by the oviduct.

Lissamphibians also show a wide range of reproductive patterns: from laying their eggs in water and the formation of gilled larvae; viviparous production of well-developed offspring; suppression of the larval (tadpole) stage and direct development. The eggs of some frog species develop conventionally into free-swimming tadpoles, while other species raise their tadpoles or froglets in parental brood pouches. Direct developers abandon the tadpole stage altogether, accelerate the development of the limbs, which are initiated while the embryo is still laying down its basic body plan (**Fig. 18-14**) and hatch as small frogs. Direct development

BOX 18-1 Devonian Stem Tetrapods and Early Tetrapods

LATE DEVONIAN STEM TETRAPODS all have well-developed tetrapod features:

1. Their skulls contain many dermal bones. The pattern of the frontal bone of panderichthyids and Devonian tetrapods are similar (**Fig. 18-4**). They retain a remnant of a preopercular bone, although the gill cover has been lost, reflecting reduction in aquatic respiration.
2. Fins are now limbs (Fig. 18-3). Bones of the limbs and their supporting pectoral and pelvic girdles are homologous in early tetrapods (**Fig 18-5**, see Fig. 17-30).
3. The teeth of osteolepiforms and Devonian tetrapods show similar complex labyrinthine folding of the pulp cavity (**Fig. 18-6**).
4. The sensory lateral line system of osteolepiforms, which extended across the skull, is considered homologous to a similar pattern of sensory canals embedded in the tetrapod skull (Fig. 18-4; R. G. and M. S. Northcutt, 2004).
5. Early tetrapods possessed a tail with dorsal fin rays showing obvious fish-like ancestry (compare Fig. 17-29 with Figs. 18-1 and 18-3).
6. In contrast, early tetrapod vertebrae changed relatively little from the vertebral structure of osteolepiform rhipidistians (**Fig. 18-7**).

To illustrate the incremental accumulation of tetrapod characters and the mosaic evolution that characterized the early stages in tetrapod evolution, the major features of seven Devonian tetrapods and stem tetrapods are outlined below (**Fig. 18-8**). Emphasis is placed on those characters typical of derived sarcopterygian fishes and those that came to characterize the first tetrapods.

Derived Sarcopterygians

Panderichthys

Panderichthys is the most derived sarcopterygian fish, the fish that possesses the greatest number of early tetrapod characters along with the most sarcopterygian characters. On the basis of current evidence, it is quite appropriate to regard *Panderichthys* as a link between sarcopterygians and tetrapods, precisely the type of taxon sought as a missing link.

The two known species are 90 to 130 cm in length (see Figs. 17-27 and 18-2). The many skull bones are typical of sarcopterygian fishes. The long and flattened head, nature of the tail and absence of dorsal and anal fins are all characteristic of tetrapods rather than of fishes. The vertebrae are more tetrapod-like, which, along with the presence of an elongated humerus in the forelimbs, reflects adaptations to short periods of terrestrial locomotion. The presence of fin rays in both fore- and hind limbs and the absence of digits (see Fig 17-27) demonstrate mosaic evolution within the limbs in a species whose environment was likely fresh water, tidal or estuarine.

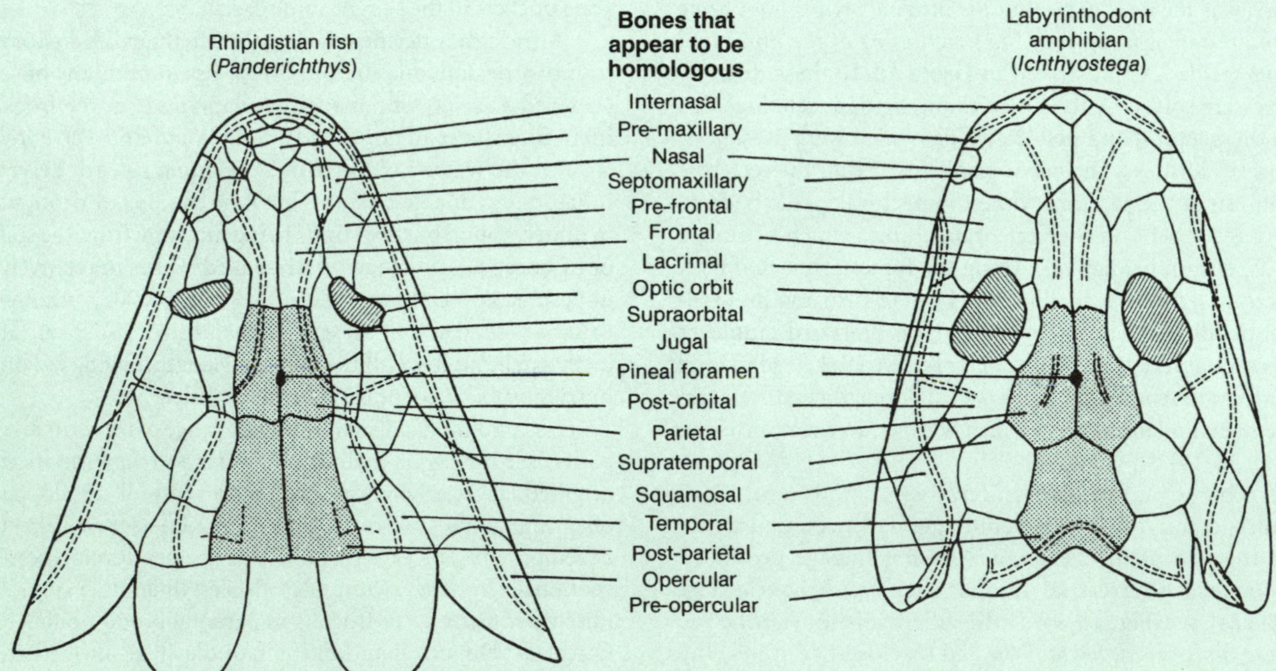

FIGURE 18-4 Dorsal views of the skull bones in skulls of an osteolepiform and early amphibian compared in terms of likely homologous structures. Dashed lines indicate sensory canals. (Adapted from Duellman, W. E., and L. Trueb, 1986. *Biology of Amphibians*. McGraw-Hill, New York.)

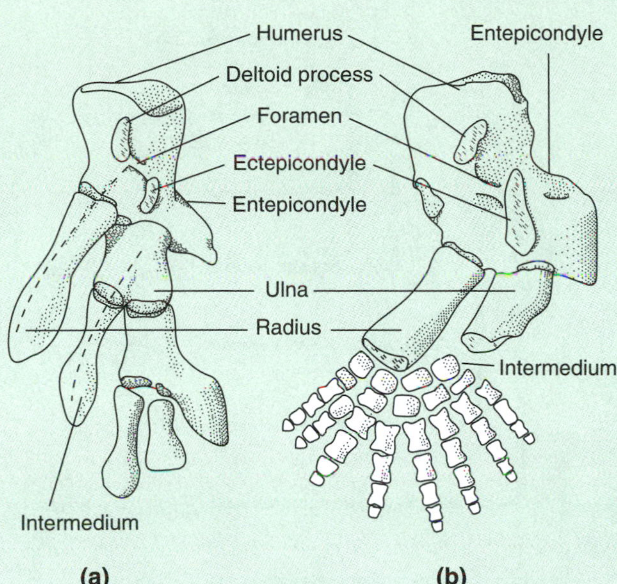

FIGURE 18-5 Comparison between bones in the pectoral fin of an osteolepiform, *Eusthenopteron* (a) and those in the forelimb of the Late Devonian tetrapod, *Acanthostega* (b). As shown, Devonian tetrapods had more than the five digits that characterize many extant tetrapods. *Acanthostega* had eight digits on each forelimb, *Ichthyostega* had seven on each hind limb. Later pentadactyl limbs may have evolved from the loss of supernumerary digits. (Adapted from Coates, M. I., and J. A. Clack, 1990. Polydactyly in the earliest known tetrapod limbs. *Nature,* **347**, 66–69.)

Tiktaalik

Another sarcopterygian, *Tiktaalik roseae,* recently discovered in Nunavut, Canada, and some 375 My old, has features — lungs, a small gill, wrists, changes in the middle ear and vertebrae (including a mobile neck) — that place it phylogenetically between *Panderichthys* and *Acanthostega* (Fig. 18-8).

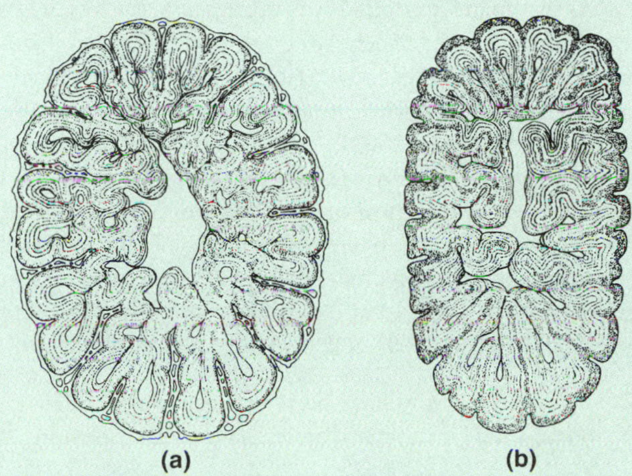

FIGURE 18-6 Cross sections of teeth from an osteolepiform, *Polyplocodus* (a), and a Devonian tetrapod, *Benthosuchus* (b).

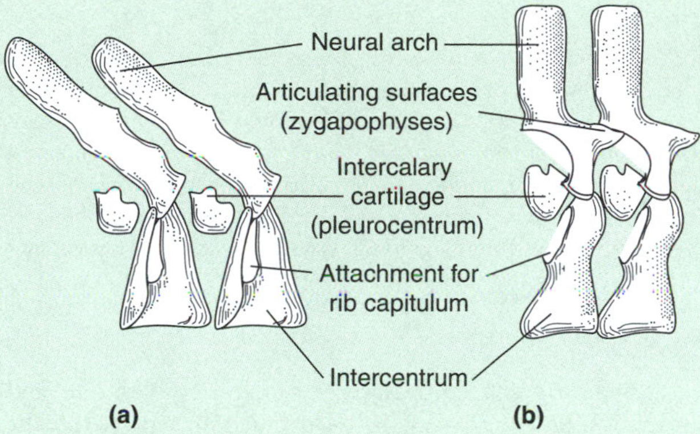

FIGURE 18-7 Lateral view of vertebrae from an osteolepiform, *Eusthenopteron* (a), and from the Late Devonian tetrapod *Ichthyostega* (b). (Adapted from Romer, A. S., and T. S. Parsons, 1977. *The Vertebrate Body,* 5th ed. Saunders, Philadelphia.)

Elginerpeton

Elginerpeton pancheni has a closely related form *Obruchevichthys.* Estimated as 1.5 m long, the jaws, girdles and skull are all a mosaic of sarcopterygian and tetrapod characters. The absence of fossilized appendages (fins/limbs?) makes placement of *Elginerpeton* difficult. Is it a derived sarcoptyergian or a postpanderichthyid stem tetrapod? Its long body says sarcopterygian, but without the fins/limbs we cannot say.

■ Postpanderichthyid Stem Tetrapods

Acanthostega

Acanthostega has had more influence on our ideas of early tetrapods than any other species. *A. gunneri,* up to 60 cm in length, is known from abundant fossils in eastern Greenland (Figs. 18-3 and 18-5). *Acanthostega* is the oldest stem tetrapod with digits. Presence of digits, absence of wrist or ankle bones and the nature of the hip joint in *Acanthostega* have allowed paleontologists to conclude that limb evolution was initiated in aquatic forms still living parts of their lives in an aquatic environment. Internal gills and a lateral line system (sarcopterygian features) are dramatic demonstrations that early tetrapods were aquatic, and of the mosaic evolution of major organ systems: limbs, lungs and sensory systems. Of even more dramatic consequence is the presence of *eight digits* on the forelimbs and *at least eight* on the hind limbs of *Acanthostega* (Fig. 18-5). Five digits per limb, long considered the basal tetrapod condition, can no longer be maintained.

Ventastega

Ventastega uvronica, known from a skull and isolated pieces of postcranial skeleton from near shore sediments in Latvia, is generally similar to *Acanthostega.*

(*continued*)

BOX 18-1 Devonian Stem Tetrapods and Early Tetrapods (*continued*)

Ichthyostega

Multiple specimens from Eastern Greenland, some up to 1.5 m long, are now interpreted as a single species, *Ichthyostega stensioei* (Figs. 18-1 and 18-8). A number of features place *Ichthyostega* as a more derived stem tetrapod than *Acanthostega* or *Ventastega:* seven digits on the hind limb (Fig. 18-5), broadly flanged and overlapping ribs, specializations in the ear not seen in other early tetrapods and tetrapod-like vertebrae (Fig. 18-7). Whether *Ichthyostega* possessed gills is not known.

Tulerpeton

Tulerpeton curtum, from Russia, has six digits on both fore- and hind limbs. As *Tulerpeton* is Late Devonian, it appears that the five-digit condition was not established until the Carboniferous. Pectoral girdle and forelimbs, especially the humerus, are more tetrapod-like than those of *Ichthyostega or Acanthostega.*

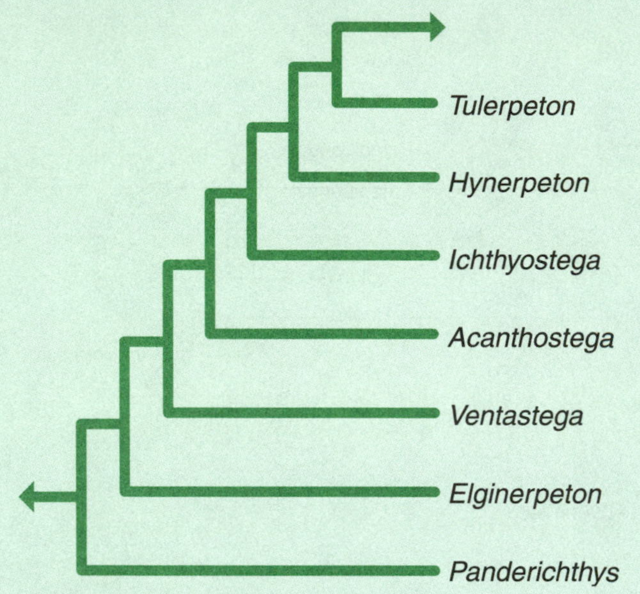

FIGURE 18-8 A cladogram showing the relationships between postpanderichthyid stem tetrapods, of which *Panderichthys* is the most basal and *Tulerpeton* the most derived. (Modified from Coates, M. I. and M. Ruta, 2007. Skeletal changes in the transition from fins to limbs. In *Fins into Limbs: Evolution, Development and Transformation,* B. K. Hall (ed). The University of Chicago Press, Chicago, pp. 15–38.)

also occurs in salamanders and in caecilians. Direct developers have either eliminated metamorphosis from the life cycle or complete metamorphosis before they hatch (experts differ on this point), a quite amazing example of an evolutionary change.

Many distinctive features and specializations (apomorphies) unite lissamphibians:

- Most lissamphibians have teeth in which a zone of fibrous tissue separates the calcified base and crown.
- They show a marked reduction of the amount of bone that contributes to the cranial skeleton.
- Compared to most early fossil tetrapods, the forelimbs of extant anurans and urodeles have four digits rather than five.[3]
- Anurans and urodeles have two auditory ossicles, the stapes and operculum, rather than only the stapes, as in earlier tetrapods.

Despite this list, resolving the ancestry of lissamphibians remains a problem; relationships within extant amphibian are much more certain.[4]

As shown in Figure 18-13, all the nearest relatives of living amphibians are extinct. Although some Mesozoic frogs and urodele fossils are known (**Fig. 18-15**), they are already so differentiated that we cannot easily trace their ancestries (Carroll and Holmes, 2007).

Like many such phylogenetic gaps, a lack of intermediates raises the question of whether amphibians have a monophyletic or a polyphyletic origin. According to some, lissamphibians are polyphyletic; frogs derived from within temnospondyls, urodeles and caecilians having had separate ancestries (see Fig. 18-9). R. L. Carroll (1997) points out,

> There seems to have been a succession of radiations among [amphibian] assemblages with common

[3] Until the early 1990s, the basal number of digits conserved throughout tetrapod evolution was understood to be five (pentadactyly). The discovery by Coates and Clack (1990) that the earliest tetrapods had at least seven or eight and as many as eleven digits (Box 18-1) totally overturned our century-old ideas on pentadactyly as basal. The early tetrapods are sometimes referred to as *polydactylous,* meaning having more than the normal number of digits. However, possession of more than five digits is the basal condition, subsequent vertebrates display hypodactyly (reduction in digit number).

[4] Relationships within extant amphibians are summarized in the Amphibian *Tree of Life (Frost et al., 2006).* Available at http://digitallibrary.amnh.org/dspace/handle/2246/5781. Accessed May 2, 2007.

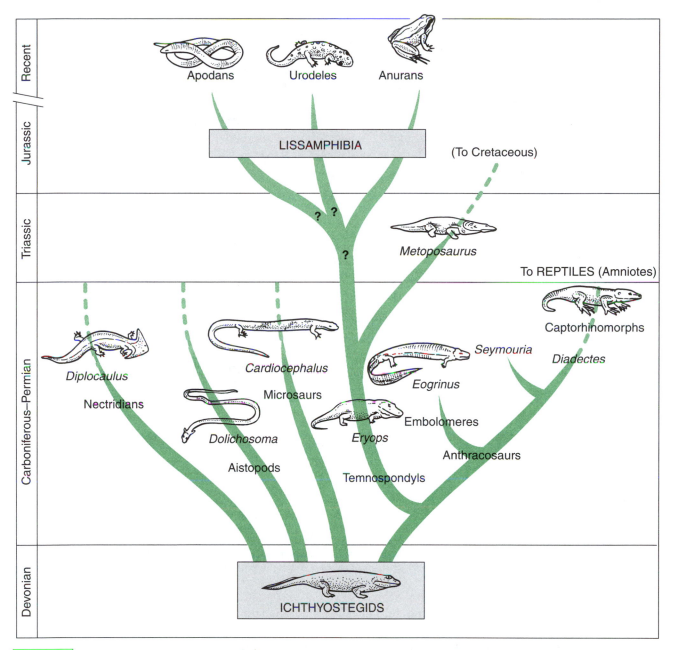

FIGURE 18-9 One phylogeny proposed for some of the primary lineages leading to lissamphibians. Note that many forms became extinct by the end of the Permian. We cannot yet trace any present-day amphibian (Lissamphibia) directly to any of these Paleozoic forms. (For more detailed phylogenies, see Benton, 1997.)

anatomical patterns, including the Paleozoic "labyrinthodonts" and "lepospondyls," the early Mesozoic "stereospondyls," and the Mesozoic and Cenozoic "lissamphibians," but none of these groups can be established as being monophyletic.

There is, however, some support for lissamphibian monophyly in the discovery of Early Jurassic caecilians showing some similarities to frogs and urodeles, although this could be due to convergence.

In the past, we tended to speculate on evolutionary origins by telling what are now regarded as "just-so" stories, after Rudyard Kipling's famous tales. The Russian evolutionary morphologist Ivan Schmalhausen (1884–1963) suggested that some if not most lissamphibians evolved in isolated, mountainous ponds and streams that protected them because the environment was cold and relatively inhospitable to other early tetrapods. Even today, amphibians are more frequent than lizards and snakes in some cooler areas. Whether this scenario represents reality or not, when con-

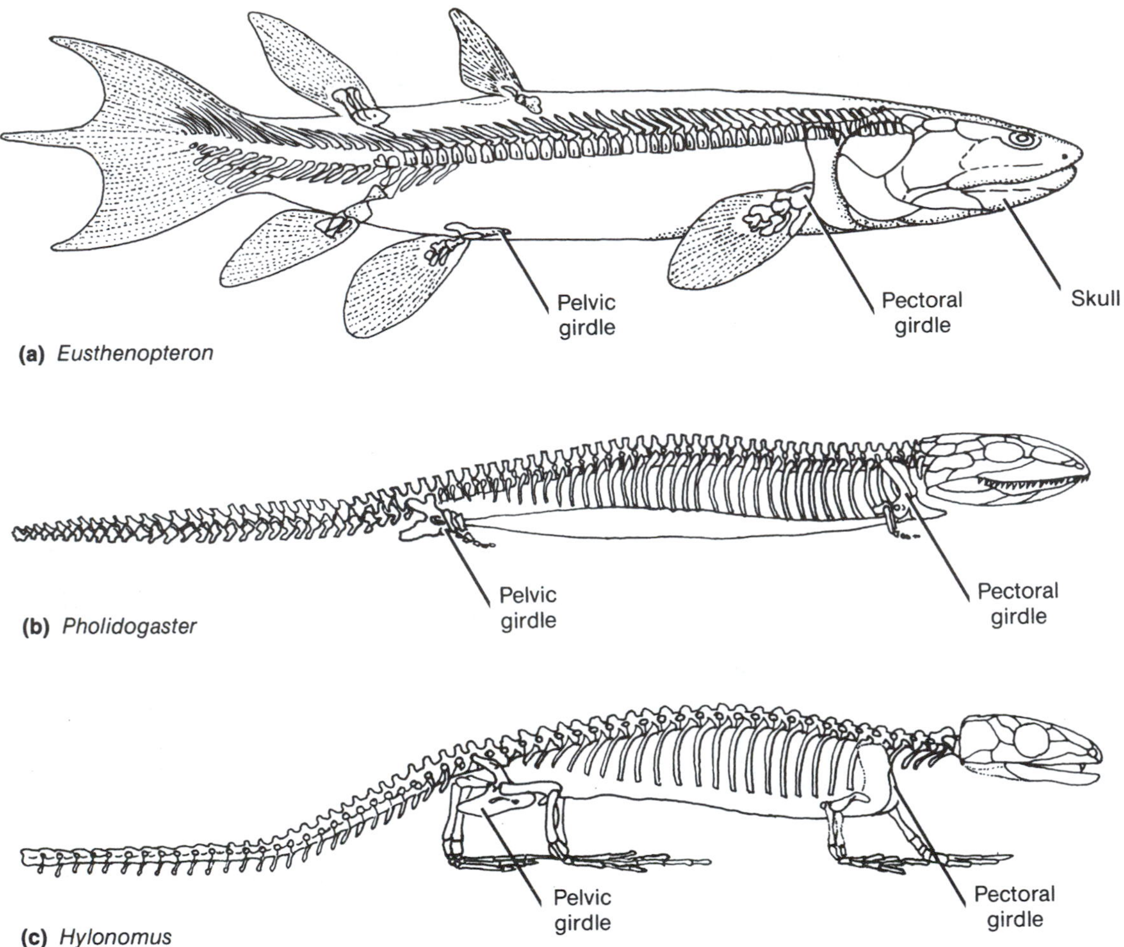

FIGURE 18-10 Attachments of the pectoral and pelvic girdles in an osteolepiform (a), an early amphibian (b) and an early reptile (c). In the change from swimming to walking, the pectoral girdle lost its direct attachment to the skull while the pelvic girdle became firmly attached to sacral ribs of the vertebral column. As Benton (1997) points out, "A tetrapod is rather like a wheelbarrow, since the main driving forces in walking come from the hind limbs, and the sacrum and pelvis had to [sic] become rigid to allow more effective transmission of thrust." (Panels a and b are adapted from Romer, A. S., and T. S. Parsons, 1977. *The Vertebrate Body,* 5th ed. Saunders, Philadelphia; (c) adapted from Carroll, R. L., 1991. The origin of reptiles. *In Origins of the Higher Groups of Tetrapods: Controversy and Consensus,* H.-P. Schultze and L. Trueb (eds.). Cornell University Press, Ithaca, New York, pp. 331–353.)

sidered in terms of their continued persistence for more than 200 My and the significant numbers of extant species (about 4,000), lissamphibians are an extraordinarily successful lineage. Nevertheless, many amphibians went extinct; their adaptations made them vulnerable to the changing environmental conditions in which tetrapods that evolved amniotic eggs could better survive. Those animals we know as **amniotes,** one lineage of which gave rise to sauropsids (including avian and nonavian reptiles), another to synapsids, including mammals.

The First Amniotes

If a single feature unites amniotes as a group it is the possession of an **amniotic egg.** Although many extant amphibians lay their eggs on land, a major evolutionary innovation during the Carboniferous was the evolution of an egg enclosing and protecting the embryo, especially when encased within a shell (**Fig. 18-16**). We call the vertebrates in which this innovation first appeared amniotes. Many systematists classify all amniote derivatives — *reptiles, birds* (either as a separate group or nested within reptiles) *and mammals* — into a single group, **Amniota** (Fig. 18-13).[5]

Although all vertebrate embryos have a yolk sac membrane that is continuous with the wall of the gut, amniote embryos produce an additional membrane, continuous with the embryonic body wall, that folds around it to yield an

[5] In addition to these taxonomic qualifications, cladistic systematists (see Chapter 11) do not accept this classification of Reptilia because it is paraphyletic. It does not include birds and mammals, descended from sauropods and synapsids, respectively.

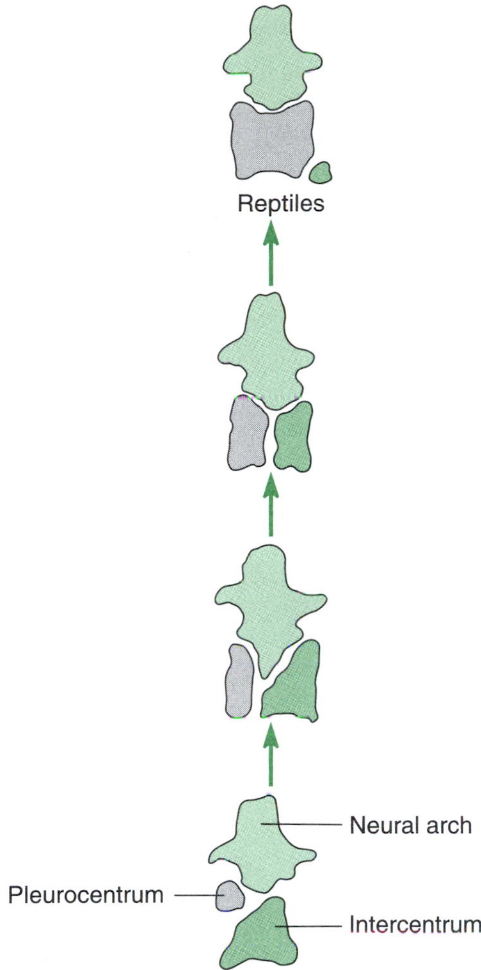

FIGURE 18-11 The evolution of vertebral central elements (lateral views) beginning with an early tetrapod (bottom) with an enlarged intercentrum (*brown*) and proceeding through stages found in anthracosaurs and early reptiles (top). In reptiles, birds and mammals, the entire vertebral central element consists of the expanded pleurocentrum, shown in *gray*. (Adapted from Romer, A. S., and T. S. Parsons, 1977. *The Vertebrate Body,* 5th ed. Saunders, Philadelphia.)

outer chorion and an inner amnion (Fig. 18-16). The amniotic cavity prevents adhesions by isolating the embryo from direct contact with the shell and helps provide protection against temperature fluctuations. In addition, amniote eggs have a sac (the **allantois**) that grows out of the embryonic hindgut and covers the inner surface of the chorion to form a highly vascularized **chorioallantoic membrane**, which acts as a respiratory organ for inward diffusion of oxygen through the permeable shell (placenta in other amniotes) and for outward passage of carbon dioxide. This respiratory function allows amniote embryos to achieve a large size. Nitrogenous wastes of amniote embryos are deposited into the allantoic cavity in relatively insoluble, nontoxic precipitates such as uric acid.

With the evolution of a shelled amniotic egg, amniotes were freed from reproductive dependence on an aqueous environment and are considered fully terrestrial. We see this diversification in the large numbers of amniote fossils that appear from the late Paleozoic onward to the stage when sauropsids (including dinosaurs), representing in numbers, size and mass, the major land vertebrates throughout the Mesozoic (Tanke and Carpenter, 2001).

The first amniote fossils appear in Pennsylvanian deposits of the Carboniferous. Paleontologists suggest they may have evolved from earlier anthracosaurs (Fig. 18-9), perhaps during the late Mississippian. R. L. Carroll (1988) points out the unusual nature of their fossilization, which illustrates the importance of accidental factors in such processes:

> These fossils are not found in normal coal-swamp deposits, such as those from which the majority of Carboniferous tetrapods have been found, but rather within the upright stumps of the giant lycopod *Sigillaria.* These trees grew in areas that were subject to periodic flooding, which resulted in the burial of the trees in several meters of sediments. The trees died and the central portion rotted out, but the bark was stronger and retained the cylindrical shape of the stump. After the withdrawal of the water, animals living on the newly developed land surface would occasionally fall into the hollow stumps. Eventually they died and were covered with sediments and fossilized.

Seymouria and most of the tetrapods discussed in the previous section were carnivorous, probably depending on a diet of invertebrates, fish or other tetrapods. *Diadectes* (see Fig. 18-12b), a member of a Late Carboniferous–Early Permian group, was one of the earliest, if not the first, tetrapod herbivore (Fig. 18-13). It had a massive skeleton with heavy

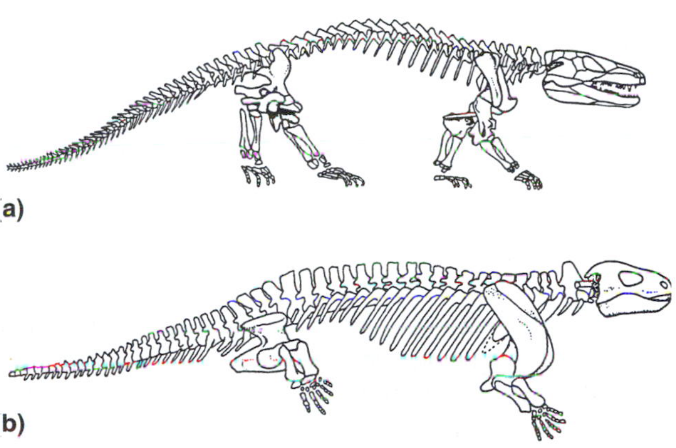

FIGURE 18-12 Skeletons of two Early Permian tetrapods with reptile-like features. *Seymouria* (a), about 50 cm long. *Diadectes* (b), about 2.4 to 2.7 m long. As noted in the text, the discovery of seymouriomorphs gilled larvae removes this group from the Reptilia. (Adapted from Benton, M. J., 1997. *Vertebrate Paleontology,* 2nd ed. HarperCollins, London, UK.)

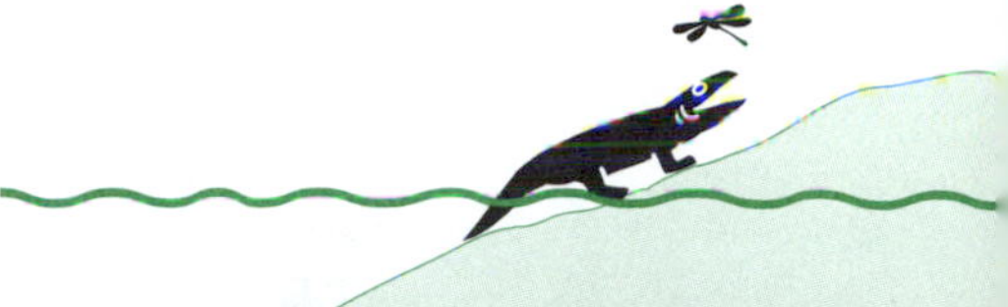

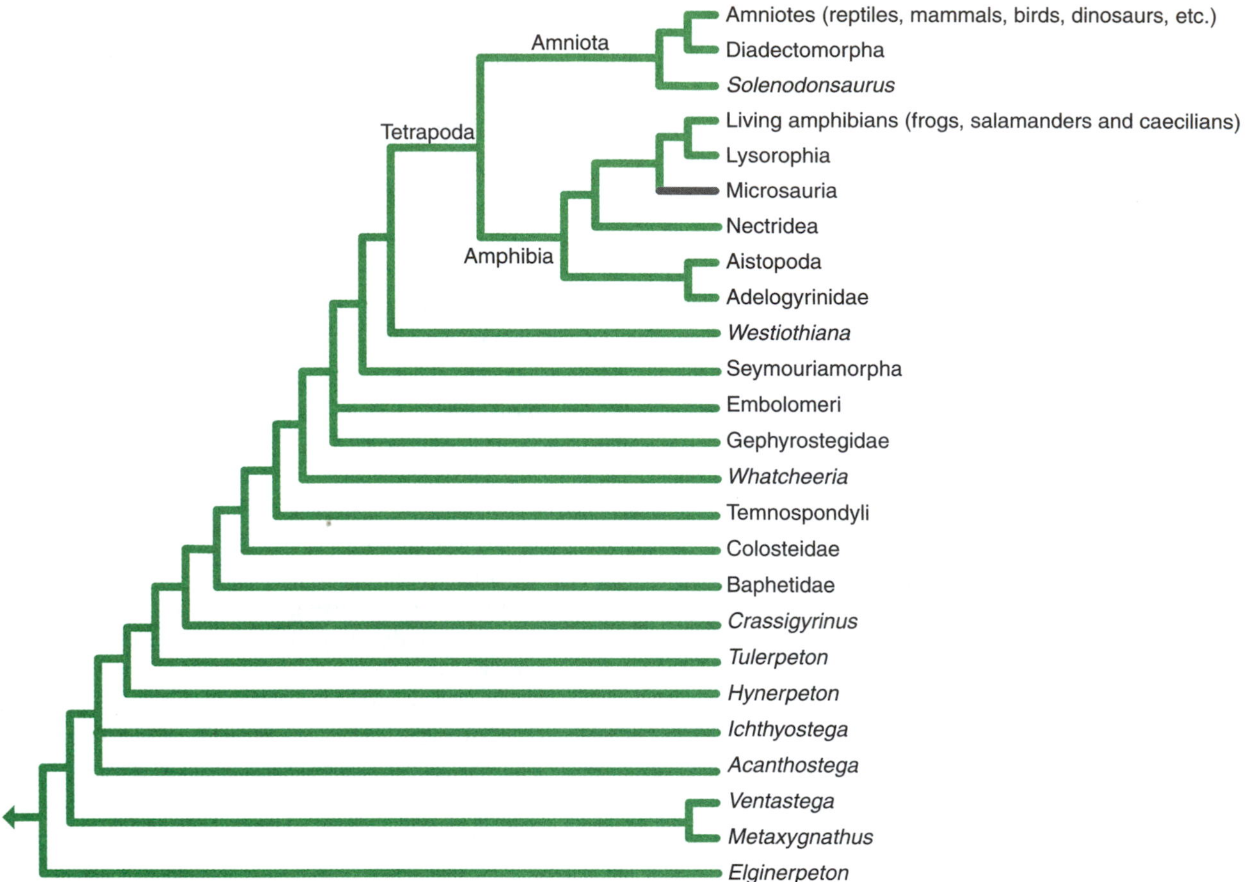

FIGURE 18-13 Phylogenetic relationships among stem tetrapods and tetrapods leading to Amphibia and Amniota. All lineages except living amphibians and Amniota are extinct.

vertebrae, probably necessary to support its considerable weight. The fact that *Diadectes* and other diadectomorphs have features similar to those seen in both amphibians *and* the first amniotes suggested that this group was more closely related to the amniote ancestor than was *Seymouria* and its relatives (seymouriamorphs), or *Diadectes* was a member of a sister-group to Amniota. Diadectomorphs now are recognized as the closest sister taxon to amniotes (Fig. 18-13).

Amniota

By the beginning of the Permian, new amniote groups that we would recognize as sauropsids and synapsids (**Fig. 18-17**) appear side by side with a great variety of amphibians. Outlined below are five of the many features that distinguish amniotes from extant amphibians:

1. *Eggs and embryonic membranes:* Amniotes produce protected, shelled eggs encircled by membranes that have no counterpart in the gel-coated eggs of amphibians.
2. *Skull and skeletal differences.* Sauropsid skulls have one occipital condyle compared to two in extant amphibians, and the reptilian sacrum incorporates at least two vertebrae, compared to one in amphibians.
3. *Heart.* Amphibians have a single ventricle. The sauropsid ventricle is partially divided, the left side sending oxygenated blood to the carotid artery, the right side sending venous blood to the pulmonary artery (**Fig. 18-18**). Birds, crocodilians and mammals have a four-chambered heart.
4. *Epidermis.* Amphibians have a soft, moist epidermis that permits some degree of gaseous and aqueous exchange, whereas the more heavily cornified reptilian epidermis acts as a barrier to both water loss and gas exchange.
5. *Gonadic ducts and excretion.* In many male amphibians a single excretory duct system serves as a common duct for gonads and kidneys, whereas sauropsids have separate ducts for each of these systems (**Fig. 18-19**). Sauropsids can also concentrate the nitrogenous prod-

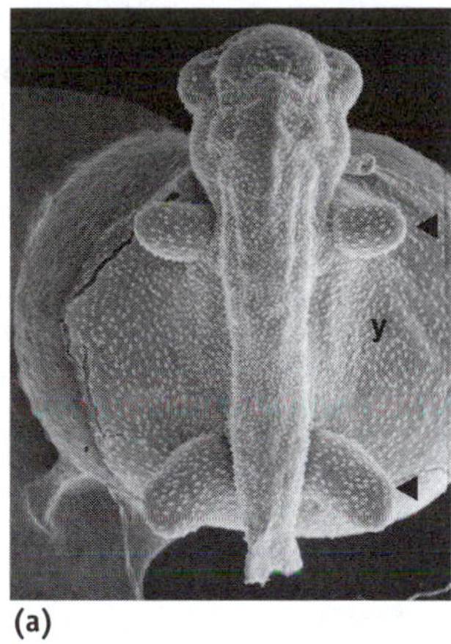

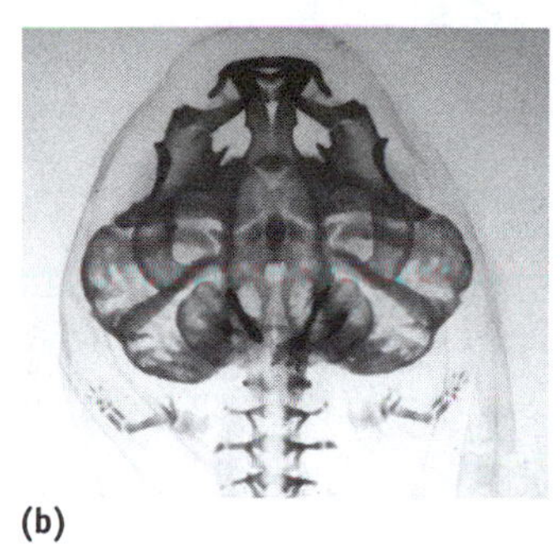

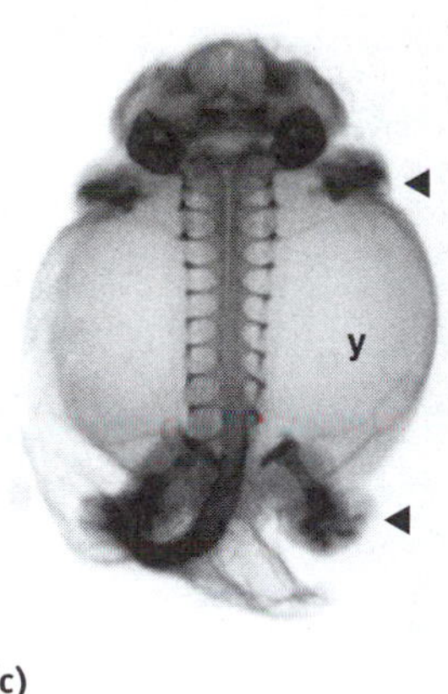

FIGURE 18-14 Images (a) and (c) are of an early embryonic stage of the direct-developing frog, *Eleutherodactylous coqui* from Puerto Rico.[6] Initiation of limb bud development (arrows) is enormously advanced, buds being present when the embryo is still attached to a large yolk sac (y) as seen in the scanning electron micrograph (a) and, on the right (c), in an embryo in which the early skeleton has been visualized with an antibody against "cartilage-type" collagen; skeletal elements are already laid down. Compare this precocious limb development of frogs and toads with a tadpole that metamorphoses into an adult, shown in (b) by the tadpole of the Oriental fire-bellied toad, *Bombina orientalis*, with fully developed head and gill skeletons and early forelimb limb development at this early stage in metamorphosis, much later than in coqui. (Photo a, courtesy of Dr. David Moury, Associate Professor of Biology at Texas A&M University–Corpus Chrisit; photos b and c, courtesy of Dr. James Hanken, Museum of Comparative Zoology, Harvard University.)

ucts of excretion (urea and uric acid) and do not need a large flow of water to remove them. In many amphibians, the urine is dilute and may contain considerable quantities of ammonia.

Of all the traits that allowed new environments to be exploited, the amniotic egg has been regarded as the most significant. With the exception of some viviparous forms, the moisture-dependent amphibian egg is an important element in the need to maintain ties to water or moist environments. Amniotes, in contrast, can lay their eggs in a large variety of terrestrial environments, amniotic fluid providing the aqueous environment. For most herpetologists, the intricacy of the amniotic egg and associated evolutionary changes — internal fertilization, laying eggs on land,[7] loss of gilled larvae — suggests that the transition to amniotes

[6] What may be the smallest tetrapod is another direct developing frog in the genus *Eleutherodactylus, E. iberia*, which is only 10 mm long (snout to vent).

[7] Accumulating information about dinosaur eggs includes evidence for eggs being laid in nests or buried in the substrate, incubated by the parents, and some very reasonable indications of the species whose members laid the eggs (Chiappe and Dingus, 2001; Deeming, 2006).

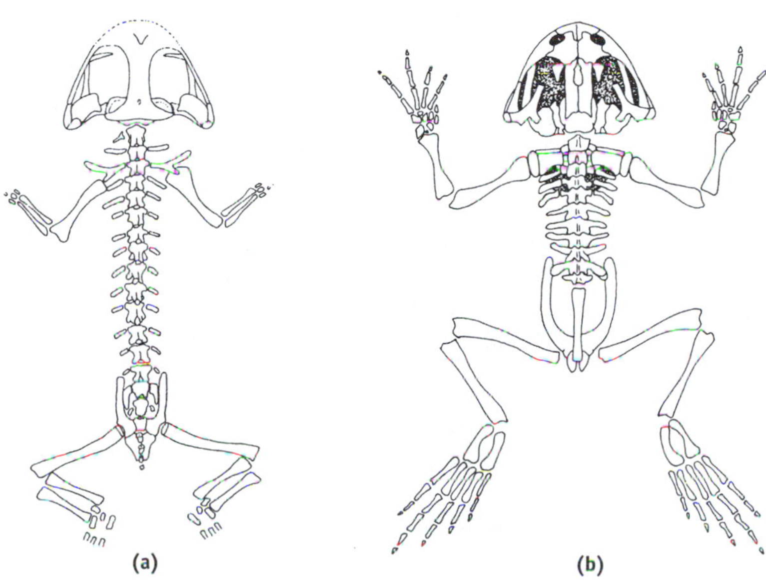

FIGURE 18-15 Skeletal reconstructions of two frog-like Mesozoic fossils: *Triadobatrachus* (a), found in Late Triassic deposits, and the early Jurassic *Neobatrachus* (b). Urodeles have a fossil record dating back to the Jurassic, as do caecilians, indicating that these three groups may have originated during the Triassic or even earlier.

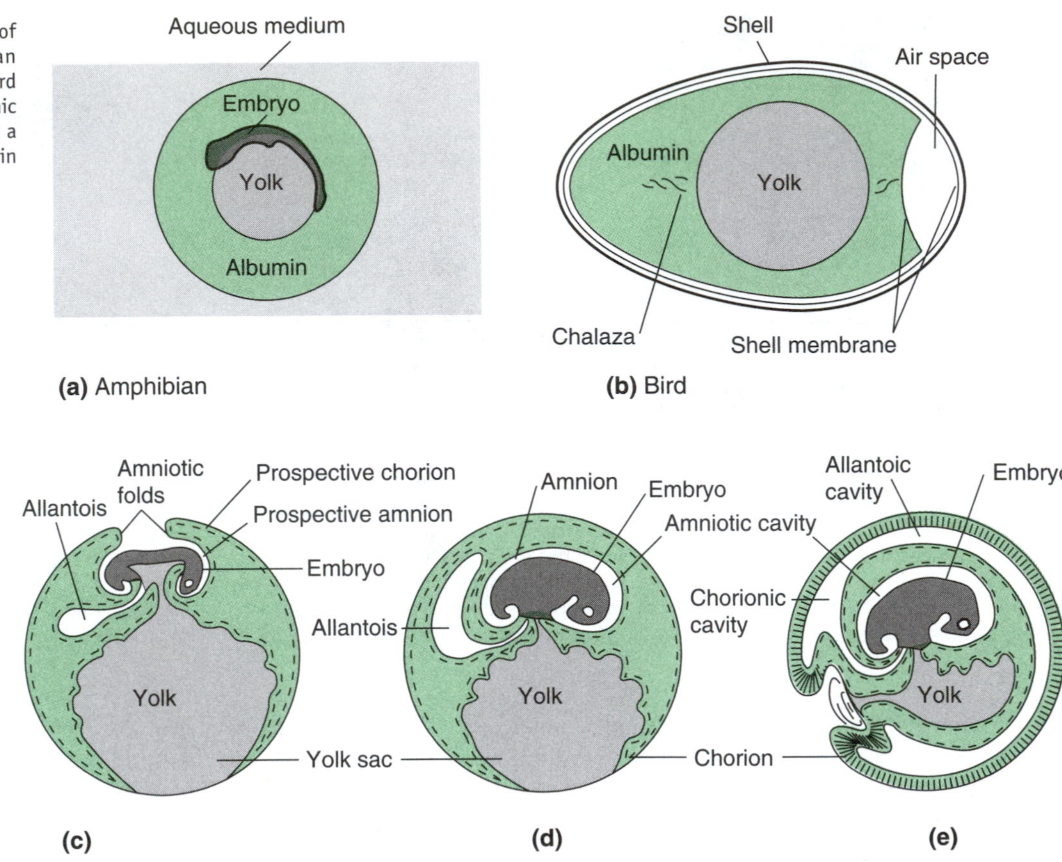

FIGURE 18-16 Diagrams of a generalized amphibian egg (a), an amniotic bird egg (b), and the embryonic membranes formed during a few developmental stages in avian development (c–e).

did not occur more than once, and that the amniote grade of evolution is therefore most probably monophyletic.

Skulan (2000) has made the interesting case that we need to revisit the necessity of evolving an amniotic egg for the expansion and diversification of tetrapods. One line of evidence is environmental, namely, his contention that conditions on land are mild compared with those experienced in water, especially if the soil in which eggs are laid retains moisture. Another is twofold and phylogenetic: Amniotes are no less likely than anurans and urodeles to retain primitive patterns of reproduction; the primitiveness of an amphibian type egg over the amniote egg is not so easily demonstrated. A third is that life on land provided opportunities for the evolution of new reproductive strategies rather than a constraint, raising the interesting possibility that tetrapods laid eggs on land *before* they became fully terrestrial.

Paleontologists seeking criteria for the origin of amniotes concentrate on skeletal characteristics such as the structure of the palate (e.g., the disappearance of the large fangs on the palatal bones), and reduction in size of the bones that roof the posterior portion of the skull. Other changes include increased heterogeneity of reptilian teeth compared to the uniformly-shaped amphibian teeth, changes in the proportions and degree of ossification in the pectoral and pelvic girdles and the evolution of a bone — the *astragalus* — in the ankle, probably associated with selection for support and locomotion in less aquatic habitats.

Anapsids/Diapsids and Synapsids

A longstanding classification of Amniota is based on openings in the temporal region of the skull behind the eye. These openings are called ***fenestrae,*** from the Latin *fenestra,* a window. Three types of reptiles have been recognized on the basis of skull morphology — **anapsids** with no opening, **synapsids** with one and **diapsids** with two (**Fig. 18-20**). We now realize that this classification reflects morphology but not lineage relationships of the forms with these morphologies.

The classification system in **Table 18-1** reflects this tripartite subdivision, although placement of groups such as mesosaurs, placodonts, turtles (**Fig. 18-21**) and ichthyosaurs (**Fig. 18-22**) are not completely resolved; some place turtles in diapsids, for example, but the novelty of the shell (see Fig. 3-15) and the absence of any potential ancestors with even the vestige of a shell, continues to frustrate herpetologists. Note that the scheme summarized in Table 18-1 includes stem reptiles. A more recent phylogenetic approach to the Amniota recognizes that *synapsids branched off early as a sister taxon to a combined anapsid/diapsid lineage,* the

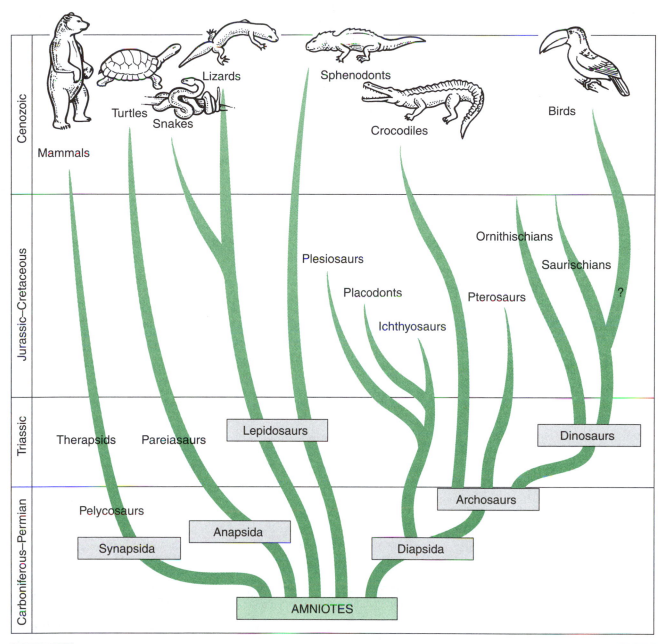

FIGURE 18-17 General evolutionary scheme showing relationships among the major reptilian groups, beginning with their origin in the Paleozoic. As indicated, further major evolutionary events include the origin of mammals from synapsids and the origin of birds from a lineage that also may have given rise to dinosaurs. Conflicting with traditional phylogeny, mitochondrial DNA analysis by Zardoya and Meyer (1998) places turtles as a diapsid lineage rather than anapsid, suggesting that diapsid skull fenestration was lost during turtle evolution.

sauropsids. We use the anapsid/diapsid and synapsid terminology in the discussion that follows as a transition to more recent approaches to amniote evolution.

The *unfenestrated condition* defines anapsids, which have a relatively rigid skull that restricts expansion of the temporalis muscle (Fig. 18-20a). The skull has a solid roof, and the temporalis muscles that close the jaw run between the inside (medial) surface of the lower jaw and the braincase, within the outer bony layer of the skull. Classically, the **Anapsida** comprised stem reptiles, mesosaurs and turtles. Given the uncertainty of the placement of mesosaurs and turtles, and the separation of stem lineages in more recent phylogenetic approaches, current approaches omit anapsids as a grade and recognize anapsids/diapsids as sauropsids and synapsids as the second lineage.

The evolution of openings in the skull enabled jaw muscles to increase in size, exert a stronger bite and effect more efficient food capture. The bony edges of fenestral openings

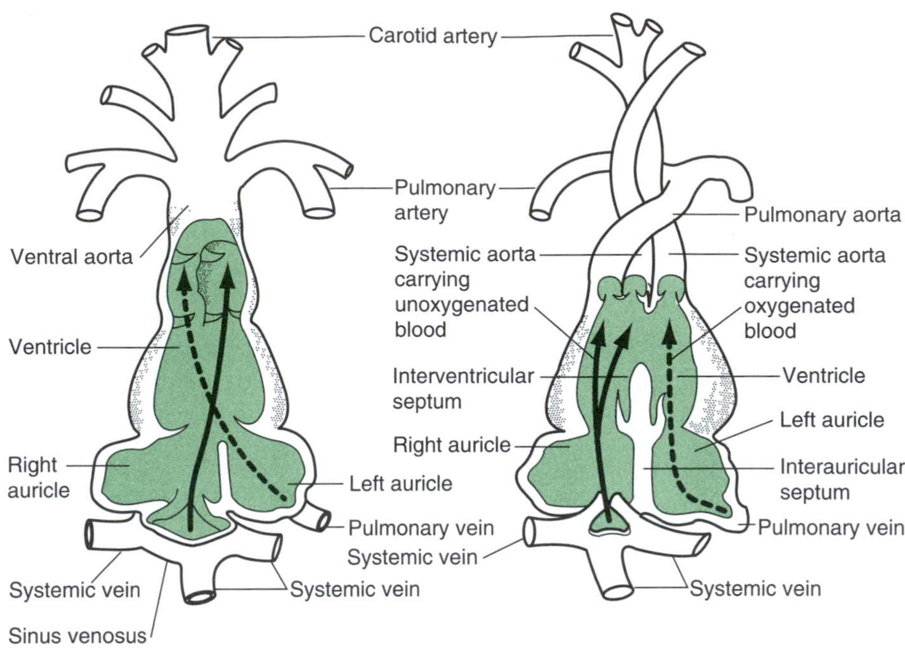

FIGURE 18-18 Ventral views of amphibian (a) and reptile (b) hearts, showing the chambers and blood flow (*arrows*). (Adapted from Stahl, B. J., 1974. *Vertebrate History: Problems in Evolution*. McGraw-Hill, New York.)

serve as a much stronger anchorage for jaw muscles than do the internal surfaces of the skull bones. Amniotes with single temporal openings — synapsids (Fig. 18-20b) — diverged early from a combined anapsid–diapsid lineage, the sauropsids. Phylogenetic terminology (Kemp, 2005, 2006), recognizes a single clade **Synapsida** for all synapsid lineages and their descendants, including mammals and their ancestors (**Table 18-2;** see Fig. 18-17), and **Sauropsida** for all avian and nonavian reptiles. The tripartite scheme and the phylogenetic approach, in which monophyletic clades and inclusive lineages are recognized, are summarized and compared in Table 18-2. If you scan this table for differences and similarities you will readily see the principles underlying the phylogenetic classification.

Fenestrae above and below a bar formed by joining the postorbital and squamosal bones characterize diapsids (Fig. 18-15c), a group that is further partitioned into Lepidosauromorpha, which includes snakes, lizards and sphenodonts, and Archosauromorpha, which includes crocodiles and dinosaurs (Fig. 18-20, Table 18-1, and see below).

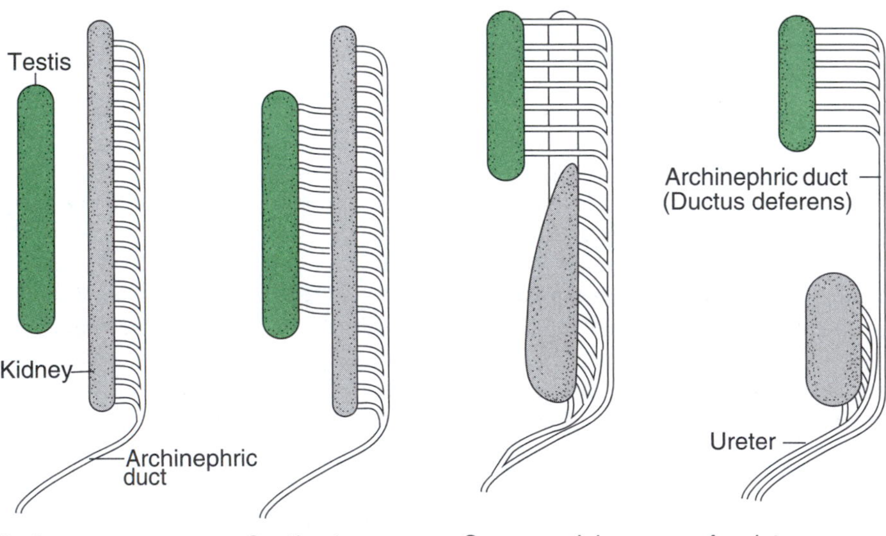

FIGURE 18-19 Male urinary and genital duct systems in various vertebrate groups. In hagfishes and lampreys, the gonad is not connected to the urinary system. In early-jawed fishes (gnathostomes), such as sturgeon and garpike, the testis has multiple connections to the kidney. Many sharks and urodeles show replacement of the anterior portion of the kidney by testicular ducts that drain directly into the urinary duct system. In more derived amniotes, the archinephric duct serves as the gonadic duct, and a single ureter is used for kidney drainage in both sexes. (Adapted from Romer, A. S., and T. S. Parsons, 1977. *The Vertebrate Body*, 5th ed. Saunders, Philadelphia.)

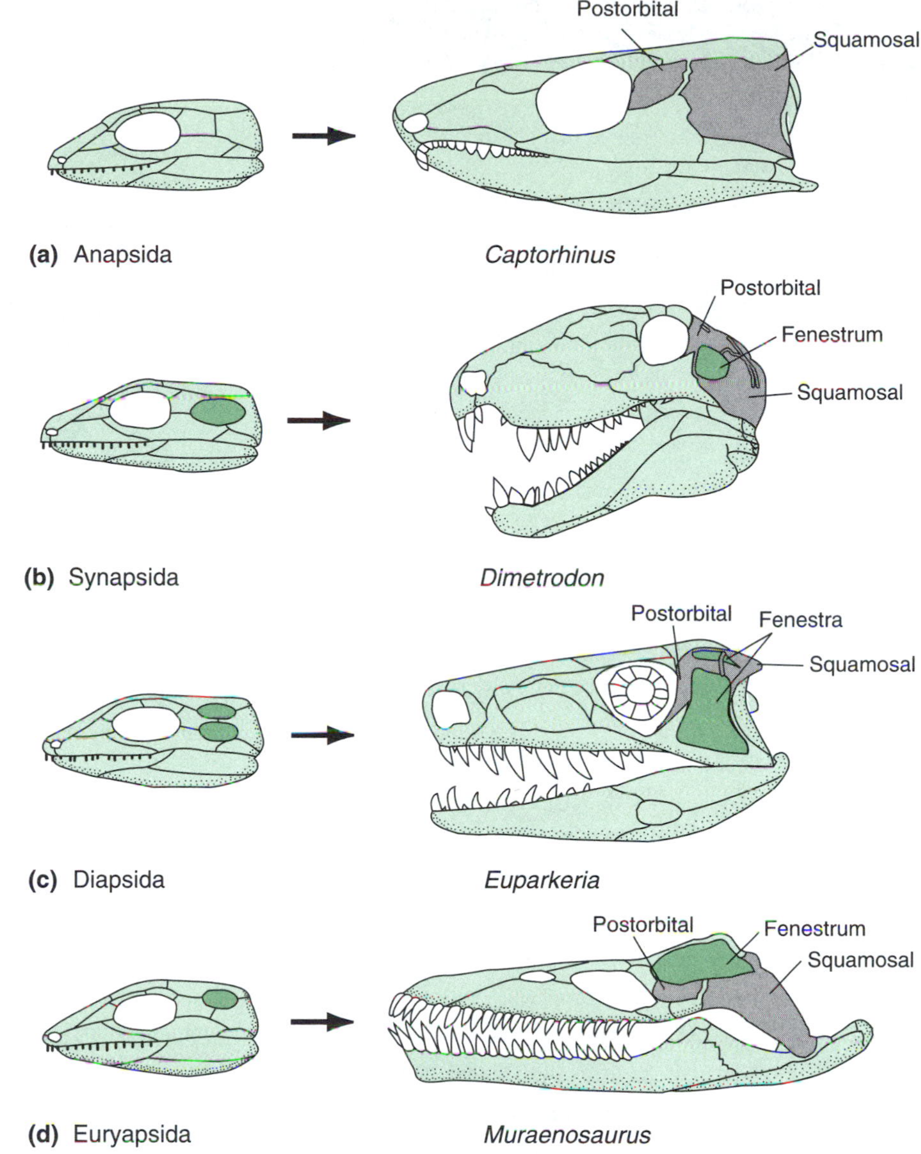

FIGURE 18-20 Schematic diagrams illustrating various kinds of reptilian postorbital temporal openings (left) and fossil skulls with their fenestrae (right). The euryapsid pattern (d), also called parapsid, is found in groups such as ichthyosaurs, nothosaurs and plesiosaurs (see Fig. 18-22). According to R. L. Carroll (1988), euryapsids derive from early diapsids, and their fenestral pattern evolved by loss of the temporal bar beneath the lower fenestra, accompanied by thickening of the post-orbital and squamosal bones. (Adapted from Colbert, E. H., and M. Morales, 1991. *Evolution of the Vertebrates: A History of the Backboned Animals Through Time,* 4th ed., John Wiley, New York.)

The earliest diapsids are found in Upper Pennsylvanian deposits.

Diversification and Adaptation

Figure 18-17 depicts a general scheme of some major amniote phylogenetic relationships and indicates approximately when various lineages became extinct.

Early synapsids, small, slender animals about 0.3 to 0.6 m long, appeared at a time in the Carboniferous when insects were evolving terrestrial forms (Grimaldi and Engel, 2005). The ecological relationship between insects and amniotes is obscure. Paleontologists have suggested that early synapsids functioned primarily as insectivores in the terrestrial food chain. Adaptation for a terrestrial existence seems also to have been shared by other synapsids, the *pelycosaurs* (Fig. 18-17), although pelycosaurs are not monophyletic but rather comprise at least three independently evolved lineages.

Early sauropsids, which specialized on nonaquatic food, were either herbivores or predators on various upland insectivorous and herbivorous forms, such as the carnivorous therapsids. The rapid radiation of these early forms meant that a wide variety of environments had been exploited by the end of the Carboniferous. Mesosaurs, basal sauropsids with long, toothy jaws, lived in aquatic habitats (see Fig. 5-9). Among terrestrial forms were some of the larger pelycosaurs. Others were more aquatic, preying on fishes and amphibians.

The end of the Permian and beginning of the Triassic mark a decline in amphibian lineages and a striking diversification of **therapsids**. By the Jurassic, almost all the major lineages had emerged. From the Jurassic onward, sauropsid

TABLE 18-1 Historical classification system for Reptilia

Subclass Anapsida
Order Captorhinida (Cotylosauria): stem reptiles
Order Mesosauria: mesosaurs (aquatic fresh water reptiles)
Order Testudinata: turtles
Subclass Synapsida
Order Pelycosauria: pelycosaurs (includes "sail-backed" reptiles)
Order Therapsida: mammal-like reptiles
Subclass Diapsida
Infraclass Lepidosauramorpha
Order Eosuchia: early diapsids
Superorder Lepidosauria
Order Sphenodontida: sphenodontids
Order Squamata: lizards and snakes
Superorder Sauropterygia: marine Mesozoic reptiles
Order Nothosauria: nothosaurs
Order Plesiosauria: plesiosaurs
Order Placodontia: placodonts
Superorder Ichthyopterygia: ichthyosaurs
Infraclass Archosauromorpha
Order Prolacertiformes: protorosaurs
Order Trilophosauria: trilophosaurids
Order Rhynchosauria: rhynchosaurs
Superorder Archosauria
Order Thecodontia: early (Triassic) archosaurs
Order Crocodylia: crocodiles and alligators
Order Pterosauria: flying reptiles
Superorder Dinosauria
Order Saurischia: lizard-hipped dinosaurs
Order Ornithischia: bird-hipped dinosaurs

Note: For other reptile classification systems, see Benton (1997). The origin of some groups, such as turtles, remains unresolved.
Source: Adapted from Carroll, R. L., 1988. *Vertebrate Paleontology and Evolution.* Freeman, New York.

adaptations enabled widespread dispersion to many habitats, including aquatic. The next 100 My or so was a veritable age of dinosaurs, pterosaurs and marine sauropsids, an age that lasted until the end of the Cretaceous, when most of the large-bodied forms disappeared, leaving lizards, turtles, crocodiles, birds and the New Zealand tuatara (*Sphenodon*).

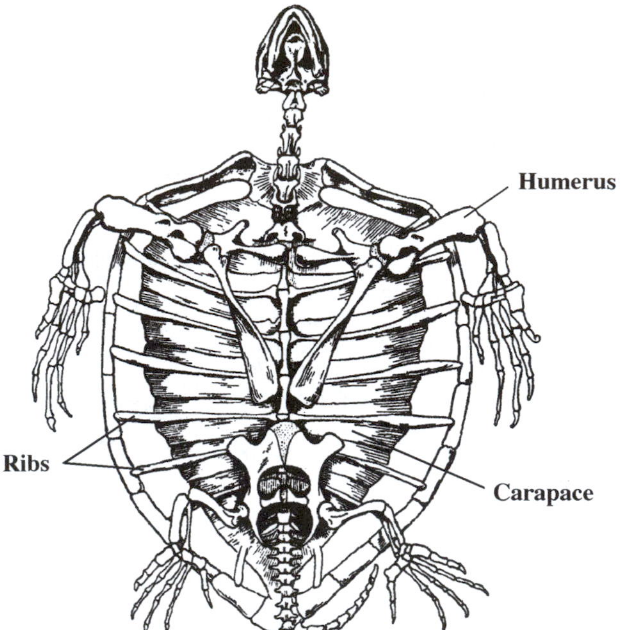

FIGURE 18-21 The skeleton of a turtle (*Chelydra*) as seen from the ventral surface showing the relationship between the ribs, limbs and shell. (From Hall, B. K., 1999a. *Evolutionary Developmental Biology,* 2nd ed. Kluwer Academic, Dordrecht, The Netherlands.)

Although many explanations have been offered for the remarkable drama of sauropsidian radiation and decline, only a sample of these hypotheses can be discussed here.

To indicate that they already possessed features found in mammals (and not in other reptiles), therapsids have been referred to as *mammal-like reptiles* (see Chapter 19). There are very good reasons to consider this term as inappropriate, not the least of which is that it refers to a paraphyletic assemblage, not all of whom were ancestors to mammals. The term does convey the fact that some reptiles developed features *later found in mammals,* while others did not, but "mammal-like reptiles" implies foresight in the evolutionary process. *Tetrapod-like fish* would be a similar term, indicating those fish that developed characters *later found in tetrapods.* The disadvantages of both terms are that (1) the group referred to in neither case (reptiles or fish) is monophyletic, and (2) not all mammal-like reptiles or tetrapod-like fish provided ancestors for mammals or tetrapods respectively.

Behavioral Evolution

Synapsids such as *Dimetrodon* (Fig. 18-17) are basal members of the line that gave rise to therapsids, which, in turn, gave rise to mammals (see Chapter 19).

Dimetrodon was a specialized animal with extremely long neural spines that paleontologists interpret as supporting a membranous dorsal "sail" used in regulating body temperature (**Fig. 18-23**). A sail of this kind, well supplied with blood vessels, would have enabled an animal that had cooled

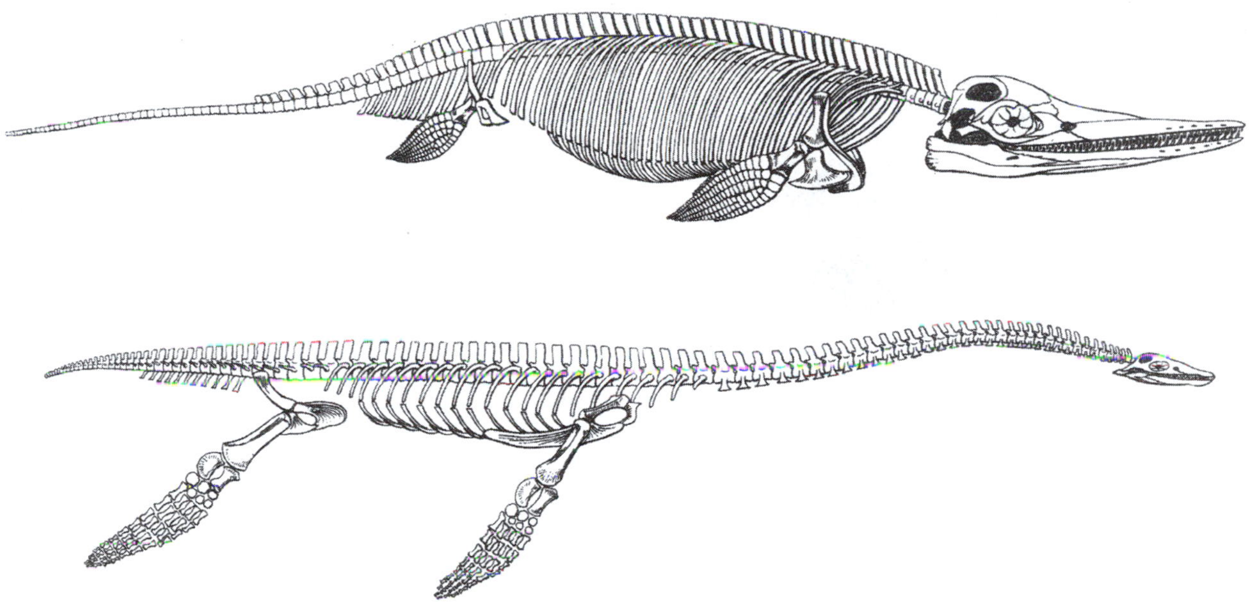

FIGURE 18-22 Reconstructions of the skeletons of an ichthyosaur (top) and a plesiosaur as depicted by Conybeare in 1824. Compare the differences in the limbs and cervical vertebrae between the two.

off at night to resume an optimum metabolic temperature soon after daybreak by placing its body perpendicular to the sun's rays. The animal would have accomplished further heating and cooling by increasing or decreasing blood flow into this large, heat-exchanging dorsal surface.[8]

Dimetrodon may also have cooled off during warm periods by moving into the shade or orienting itself parallel to the sun's rays. Because continuous enzymatic activity in muscle and other tissues depends on maintaining optimum body temperatures, selection for such mechanisms would have enabled early synapsids (including therapsids) to engage in longer periods of active predation or escape.

Paleontologists have concluded that therapsids used **temperature-regulating mechanisms** other than dorsal sails. Here we need to distinguish between "warm-blooded" and "cold-blooded" (*homeothermy* and *poikilothermy*), and the ability to maintain a constant body temperature (*endothermy*) or not (*ectothermy*). Some of these new forms may have been the first **endothermic** vertebrates;

[8] Another early amniote (*Edaphosaurus*), known only from fragments from the late Carboniferous and Early Permian, also had a sail along the dorsal side, but in this species the sail was supported by extensions from the vertebrae, reinforced by crossbars of bone. At 3.2 m (11 ft) long and weighing 300 kg (660 lbs) these were impressive inhabitants of swampy areas.

TABLE 18-2 Comparison of two classification schemes for reptiles[a]

Traditional Classification	Phylogenetic Classification
Amniota	Amniota
Reptilia	Synapsids (reptiles, birds and mammals)
Anapsids (stem reptiles, mesosaurs, turtles)	Reptilia
Synapsids (pelycosaurs, therapsids)	Testudines (turtles, tortoises)
Diapsids	Diapsids
Archosaurs (crocodilians, dinosaurs, pterosaurs)	Archosaurs (crocodilians, dinosaurs, pterosaurs and birds)
Lepidosaurs (lizards and snakes)	Lepidosaurs (lizards and snakes)

[a] See text for details.

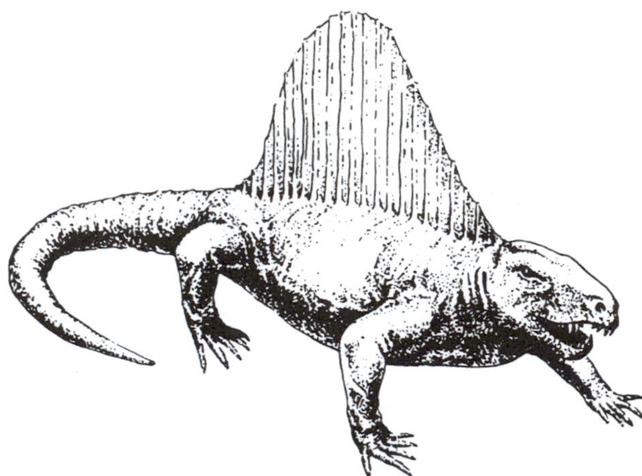

FIGURE 18-23 Reconstruction of the early Permian synapsid, *Dimetrodon*. (Adapted from Romer, A. S., 1968. *The Procession of Life*. World Publishing, Cleveland, OH.)

that is, they maintained optimum body temperatures through internal metabolic heat-producing reactions instead of being subject to more variable external **ectothermic** influences. Higher metabolic rates in endothermic animals not only raise body temperatures, but also necessitate higher levels of oxygen use, greater numbers of mitochondria and increased vascularization. In turn, increased aerobic metabolism supports more sustained activity and greater stamina than ectotherms can achieve; ectotherms become rapidly exhausted because they rely mostly on anaerobic metabolism.

Because an organism must consume much more food to provide energy for high metabolic rates, the cost of endothermy is relatively high. For a given body weight, an endothermic mammal or bird needs five to ten times more energy to maintain an elevated body temperature than a comparably sized ectothermic reptile or amphibian. Nevertheless, despite these energetic limitations, ectotherms are capable of short bursts of activity through anaerobic metabolism. Thus, ectotherms can survive well under conditions that stress low energy expenditure, and may even compete successfully with endotherms when predatory pursuit or escape requires only short distances. Lizards, for example, have prospered in many environments where they can move quickly in and out of protected surroundings — aided often by limb modifications, including reduction in the number of digits — and now number about 4,800 species. Similarly, snakes (2,500 species), about 25 percent of which are highly venomous, can use energy bursts for both offense and defense.[9]

Endotherms, of which therapsids may have been the earliest forms, adapted especially well to environments in which activity must be sustained. Their locomotory adaptations moved the legs to beneath the body, close to the median plane, thus diminishing side-to-side bending while running and facilitating movements of the ribs required for breathing. Lateral undulation associated with rapid locomotion, was dealt with differently by the most derived sauropsids and synapsids: Many dinosaurs became bipedal; mammals flex the trunk dorso-ventrally.[10]

Therapsids succeeded well, both as herbivores and carnivores (**Fig. 18-24**), some reaching 3 to 3.6 m in length. Surprisingly, although many therapsids lasted into the Mesozoic, their numbers significantly diminished before the end of the Triassic. Perhaps the warm, constant climate of the Mesozoic allowed many ectothermic sauropsids to maintain stable high body temperatures on less food intake, thus reducing the advantages of therapsid temperature-regulating abilities.

Diapsid archosaurs, characterized the remainder of the Mesozoic from the Late Triassic to the close of the Cretaceous (Fig. 18-17). Many other groups — turtles, lizards and snakes — certainly persisted and continued to evolve, but it was from among the archosaurs that the dinosaurs appeared, some of which dwarfed any other land vertebrate.

Early Archosaurs

As indicated above, two major groups of diapsids are recognized, lepidosauromorphs and archosauromorphs. They can be distinguished by many traits, the most important of which is the **archosaur** ankle-and-foot structure, which facilitated an upright posture. By the end of the Permian and the beginning of the Triassic, evolution had produced a variety of archosaurs, including bipedal forms in which the forelimbs were shorter than the hind limbs and the teeth were set in sockets (thecodont dentition).

The origination and persistence of archosaur **bipedalism** is associated with selection for faster running speed and large size, and resolves the problem of extant amphibians and nonavian reptiles that they cannot run and breathe at the same time. Bipeds can breathe while they run. Consequently, even small early Triassic archosaurs show the effects of selection toward bipedalism. For example, *Euparkeria*, an Early Triassic archosaur, was a carnivorous form about 0.6 m long, lightly built with hollow bones. Because the hind limbs were about 1.5 times the length of the forelimbs (**Fig. 18-25**), *Euparkeria* may have been capable of bipedal locomotion. *Euparkeria* lies within the lineage leading to bipedal dinosaurs.

[9] See McNab (2002) for an analysis of physiological ecological energetics in vertebrates and see Shapiro et al. (2007) for digit reduction in reptiles.

[10] See Alexander (1991). Alligators aid respiratory flow by flexing their bodies dorso-ventrally while moving, a mechanism known as pelvic aspiration (Farmer and Carrier, 2000; Seymour et al., 2004).

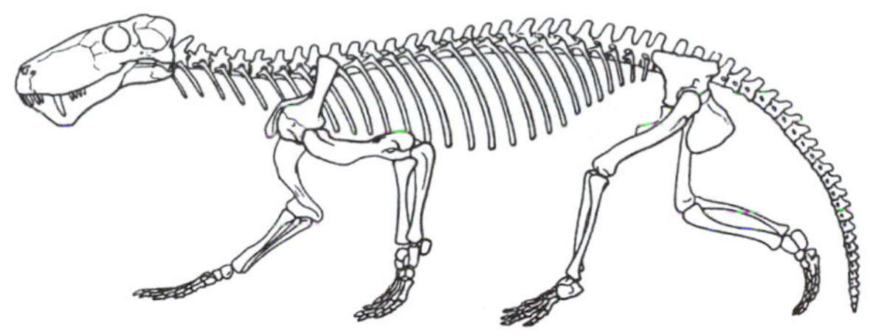

FIGURE 18-24 Skeleton and reconstruction of a carnivorous therapsid (*Lycaenops*) from the late Permian, showing a number of mammal-like features. It was about the size of a wolf, with large upper canines functioning for piercing and tearing. (Adapted from Colbert, E. H., and M. Morales, 1991. *Evolution of the Vertebrates: A History of the Backboned Animals Through Time,* 4th ed., John Wiley, New York, and from Romer, A. S., and T. S. Parsons, 1977. *The Vertebrate Body,* 5th ed. Saunders, Philadelphia.)

By the middle of the Triassic, bipedal innovations had developed further in a number of archosaur lines. In some bipedal forms, selection for increased length of stride by the hind legs resulted in the tibia being almost as long as the femur, an anatomical modification for speedy running; the tibia–femur ratio of a racehorse is about 0.9. Other lines, such as crocodiles, preserved a four-footed gait, although new information on crocodiles demonstrates that a four-footed gait is not obligatory — crocodiles can run on two legs.

By the end of the Triassic, many tetrapod groups, including many early archosaurs and synapsids, were extinct. According to Benton (2004), these Triassic extinctions, caused or accompanied by major climatic changes, enabled dinosaurs to enter a variety of vacant ecological niches, accounting for their extensive radiation during both the Jurassic and the Triassic.

Dinosaurs

The name Dinosauria was erected by Richard Owen in the nineteenth century for what are now recognized as two major lineages, Saurischia (sauropods) and Ornithischia. All **dinosaurs** share elongate hips, an S-shaped neck and an offset first digit, all of which reflect their early role as bipedal predators.[11]

A primary difference in the structure of the pelvic girdle appeared early in dinosaurian evolutionary history. **Saurischia** (lizard-hipped), the lineage that includes birds, retain the original archosaur three-part structure in which the pubis extends anteriorly and the ischium posteriorly (**Fig. 18-26a**). **Ornithischia** (bird-hipped), all of which were herbivorous and all of which went extinct, had a pelvis with the pubis usually parallel to the ischium (Fig. 18-26b). The two lineages show their bipedal ancestry in shorter forelimbs than hind limbs, although some dinosaurs returned to a quadrupedal stance with lengthened forelimbs.

As shown in **Figure 18-27,** diversification among dinosaurs proceeded throughout the Mesozoic. Among saurischians, and on the basis of body proportions, the carnivorous **theropods** (four or fewer toes on the hind feet) were among the most prevalent reptiles. By the Cretaceous, bipedal theropods had radiated widely, evolving into forms such as the small, ostrich-like *Ornithomimus,* small- to medium-sized forms such as *Deinonychus* ("terrible claw"), and large theropods such as *Tyrannosaurus,* which may have weighed 6 to 8 tons and stood 5 m above the ground.

The increased body size in some lines of theropods paralleled the increase in body size in some of their herbivorous prey; larger size offers protection to prey, but results in selection, in turn, for increase in predator size. On one hand, this cycle of selection for body size elicited the large theropods, and on the other, large herbivorous ornithischian and even larger herbivorous sauropods (five toes on the hind legs)

[11] Weishampel and White (2003) contains 48 seminal papers on dinosaurs published between 1676 and 1906, including Owen's 1842 paper, in which the term *dinosaur* was first used. See R. L. Carroll (1988), Currie et al. (2004), Weishampel et al. (2004) and Curry-Rogers and Wilson (2005) for recent analyses of dinosaur evolution and paleobiology.

FIGURE 18-25 Reconstruction of *Euparkeria,* an early Triassic archosaur, 0.6 to 0.9 m long, with a short trunk counterbalanced by a heavy, muscular tail that might have enabled it to run bipedally.

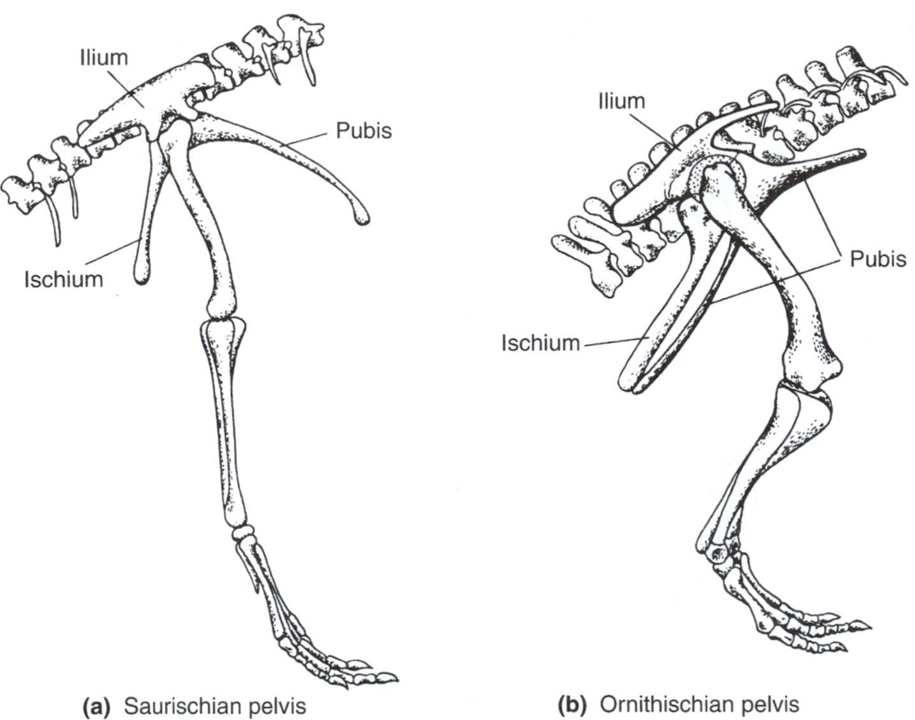

FIGURE 18-26 The two general types of pelves found in dinosaurs. In early reptiles, the pelvis is a solid, plate-like structure from which the femur projects horizontally (parallel to the ground). With the evolution of bipedalism and a vertical femur, greater leverage for moving the hind legs arose by attaching the limb muscles to fore and aft extensions of the pelvis. In birds, the posterior extension of the pubis helps support the ischium, but the anterior pubis is not well developed. (Adapted from Romer, A. S., 1968. *The Procession of Life.* World Publishing, Cleveland, OH.)

such as *Apatosaurus* (formerly *Brontosaurus*), which were 21 to 24 m long, and weighed 18 tons or more. Others were even more impressive (**Table 18-3**).[12]

Dinosaurs, in the broad sense of the Dinosauria, did increase in size as they evolved, that is, they obeyed *Cope's rule* (see Chapter 22). Although every schoolchild knows that dinosaurs were enormous in size, here the results stand on firmer footing, being grounded on 65 phylogenetically independent comparisons. On average, later genera were 25 percent larger than the earlier genera with which they were compared (Hone and colleagues, 2005). However, the trend is not universal; while small-bodied lineages increase in size, lineages that started out as large-bodied tended to decrease in size over time.

Decrease in size (often in only a few generations) is often seen in lineages that become isolated on islands. When we know that the lineage evolved from a larger-bodied form we speak of the process as **insular dwarfism**, a fine example of which was published as this chapter was being revised. Martin Sander and colleagues (2006) studied eleven individuals of a new, dwarf species of sauropod, *Europasaurus holgeri.* Using bone histology to gauge rates of growth of the long bones, they demonstrated that evolution of the smaller island species came about through a decrease in growth rate compared to the growth rates measured in the larger-bodied and ancestral mainland species.

A classic but long outdated hypothesis held that many giant sauropods were aquatic because only water could have buoyed up their immense weights. Opposed to this hypothesis are findings of a terrestrial existence: The fossilized stomach contents in one animal indicate a diet of woody, leafy material. Cretaceous trackways found in Texas and Argentina indicate that sauropods traveled in herds. Biomechanically, a skeleton with many air spaces enabled these animals to support their mass on land. At least some of the long-necked sauropods were ground feeders, gathering food from low-lying plants or from shallow lakes and ponds. Their large size, powerful tails and social organization gave adults considerable protection against all but the largest predators.[13]

Lineages of the exclusively herbivorous ornithischians were not as large as some lines of saurischians, but

[12] See Curry-Rogers and Wilson (2005) for sauropod evolution and paleobiology.

[13] See Wedel (2003) for biomechanics of air spaces, Stevens and Parrish (1999) for sauropod habitats and Senter (2007) for the proposal that the elongated necks of sauropods evolved in response to sexual selection rather than for feeding above the ground.

many were nevertheless quite large, reaching lengths of 9 m and weights of 5 tons or more. Among ornithischians are the duck-billed *hadrosaurs,* the armored *stegosaurs* (**Fig. 18-28**) and *ankylosaurs,* and the horned *ceratopsians,* some of which developed considerable variation and specialization (Carpenter, 2001). Hadrosaurs, which exhibited some of the greatest diversity in terms of skull morphology, with a large variety of crested forms (**Fig. 18-29**), may have comprised as 60 percent or more of the terrestrial vertebrate biomass in many places by the end of the Cretaceous.

Endothermy Versus Ectothermy

Dinosaurs were enormously successful throughout much of the Mesozoic; no other reptiles or mammals approached

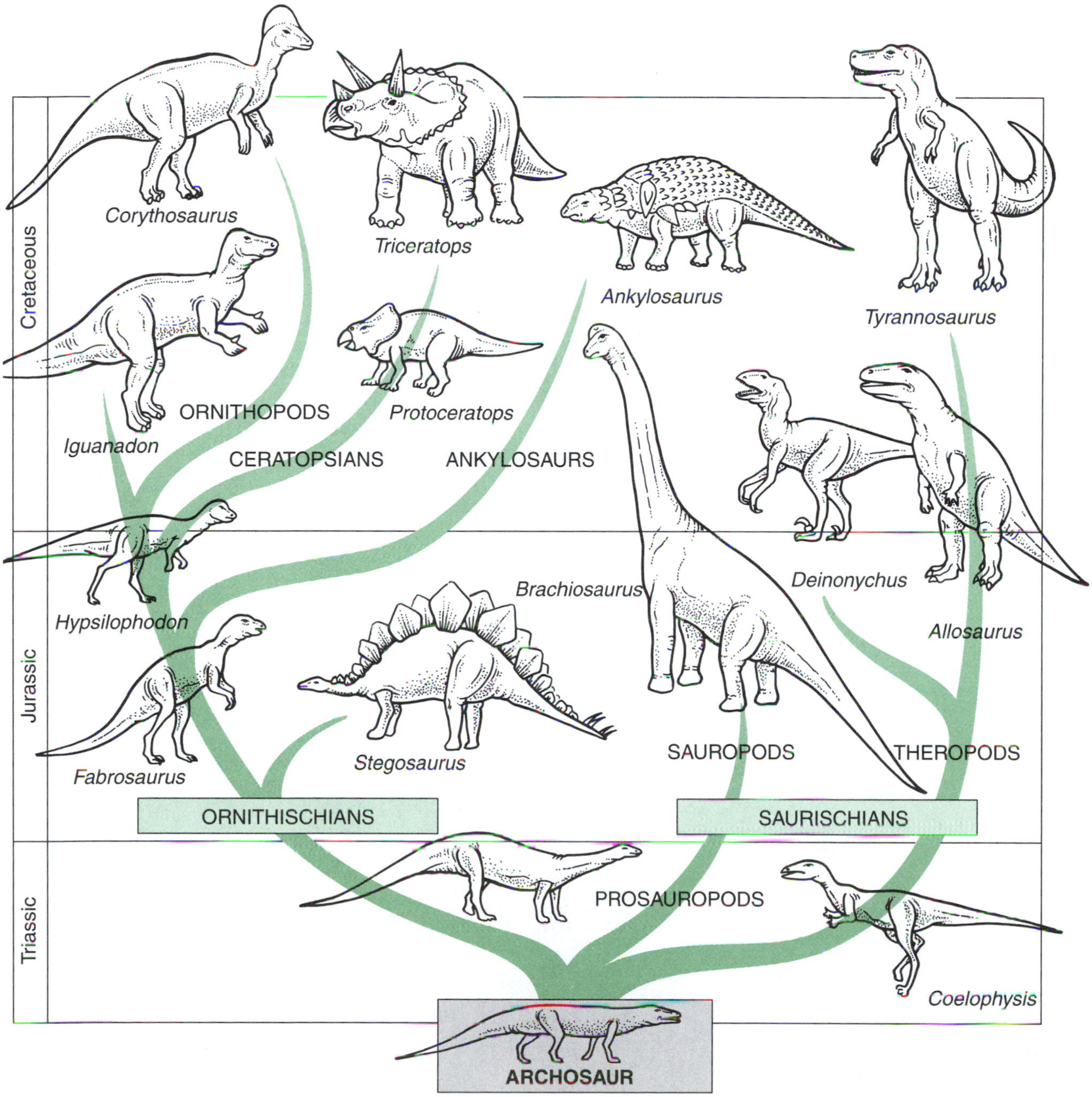

FIGURE 18-27 Dinosaur phylogeny from the Triassic to the end of the Cretaceous. (Adapted from Colbert, E. H., and M. Morales, 1991. *Evolution of the Vertebrates: A History of the Backboned Animals Through Time,* 4th ed., John Wiley, New York.) An outstanding review of all major dinosaur groups is the volume *The Dinosauria,* edited by Weishampel et al. (2004).

TABLE 18-3 The largest sauropod dinosaurs discovered to date

Species	Length (m)	Estimated Body Weight (tons)
Herbivorous[a]		
Argentinasaurus	37	80–100
Turiasaurus reiodevensis	38	40–48
Apatosaurus (formerly *Brontosaurus*)	21–24	>18
Carnivorous		
Giganotosaurus carolinii	14	6
Tyrannosaurus rex	12–15	6

[a] By comparison, herbivorous ornithischian dinosaurs reached lengths of 9 m and weights of 5 tons or more.

their size or prominence in ecosystems until after the Cretaceous.[14]

Some paleontologists, including Ostrom (1974) and Bakker (1986), have suggested that dinosaur success stemmed at least partially from the high metabolic rates provided by endothermy and the consequent high levels of activity, a topic introduced earlier in the section titled *Behavioral Evolution*. According to their proposal, a large, sluggish, ectothermic amniote, whether herbivore or carnivore, could hardly have competed successfully with endothermic therapsids and their mammalian descendants. Furthermore, if dinosaurs were ectothermic, why did endothermic mammals remain small and insignificant throughout the Mesozoic?

A number of lines of evidence have been marshaled in support of dinosaur endothermy. No single line of evidence is compelling; indeed, counter-arguments have been raised to the interpretation of each of the five considered below.

1. The fully erect *posture* of dinosaurs, with their hind limbs extending vertically downward beneath the body — as seen in earlier crocodilians — also appears among endotherms (mammals and birds). All extant ectotherms (nonavian reptiles and amphibians) have a sprawling gait. The speed and agility that accompanies an erect posture would have had its source and sustenance in high metabolic activity. Alternatively, erect posture may be the only stance a large, heavy, terrestrial animal can assume without unduly bending its supporting limbs, as seen in the sprawling limb posture of extant opossums.
2. The microscopic *structure of dinosaur bone* shows a high density of blood-carrying (Haversian) canals similar to those in mammals. Extant ectotherms show few such canals. But Haversian bone structure is also related to growth rate, body size and other factors rather than endothermy; some ectothermic sauropsids, such as turtles, show Haversian canals, while some small mammals and birds do not.
3. The *distribution of dinosaur fossils* during the Cretaceous extended from as far south as Antarctica to areas in Canada that must have been close to the Arctic Circle.

[14] See *Mesozoic Vertebrate Life* by Tanke and Carpenter (2001) and *The Cretaceous World* by Skelton (2003).

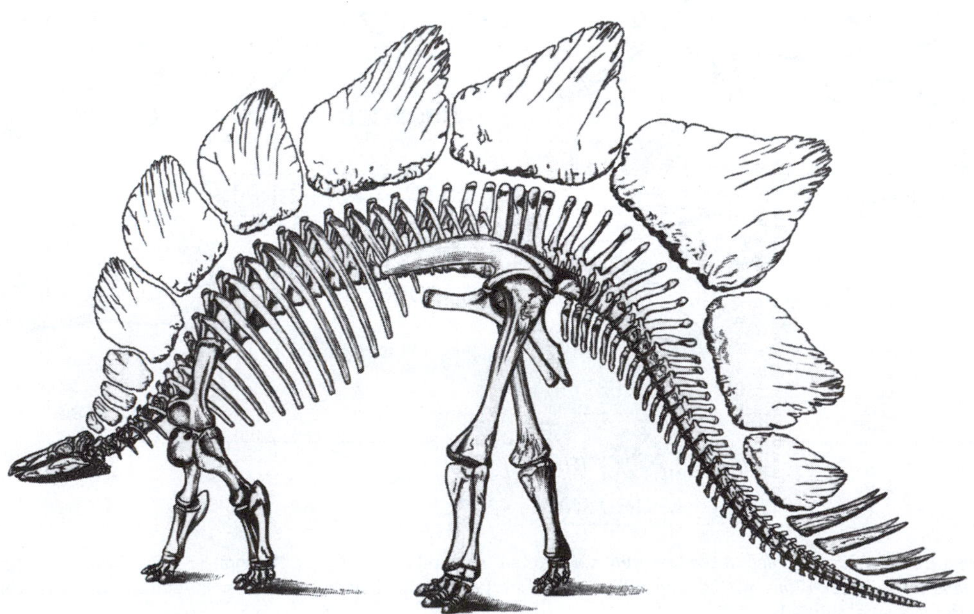

FIGURE 18-28 A restoration of *Stegosaurus ungulatus* as figured by March in 1891.

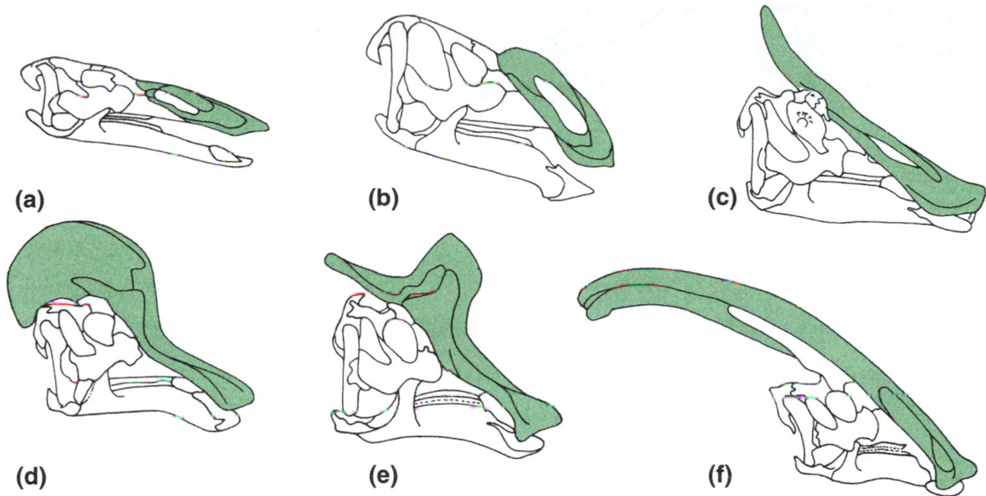

FIGURE 18-29 Skulls of various duck-billed hadrosaurs of the Late Cretaceous: (a) *Anatosaurus;* (b) *Kritosaurus;* (c) *Saurolophus;* (d) *Corythosaurus;* (e) *Lambeosaurus;* (f) *Parasaurolophus.* Behind a flattened, toothless beak lay rows of teeth (as many as 700 per jaw ramus) apparently used to grind tough vegetable matter. As shown, the nasal and premaxillary bones (*shaded*) assumed unusual shapes in different groups, producing long loops of nasal passages for purposes that are still undetermined. Among suggestions for their use have been enhancing the sense of smell, visual signals for species recognition, male weapons or shields during mating competition and resonators for amplifying sound. (From Romer, A. S., *Vertebrate Paleontology,* 3rd ed. 1966. Reprinted by permission of the University of Chicago.)

Over time, plate tectonics (see Chapter 5) had moved the latter areas further north, where even during the Cretaceous they would have had periods of continuous darkness. Assuming a cool climate in such localities, these dinosaurs, like extant endotherms, would have been able to supply their own body heat. Dinosaur radiation to northern latitudes, however, is not sufficient evidence for endothermy. Cretaceous climates in general were warmer than they are now, plate tectonics may since have moved their ancient habitats northward, and northern deposits also contain fossils of turtles, presumed to be ectothermic.

4. The *origin of birds* such as *Archaeopteryx* from small carnivorous dinosaurs, reinforced by finds of feathered dinosaurs in China, carries the corollary that primitive feather-like structures may have insulated ancestral dinosaurs against loss of body heat. Others argue that feathers were originally used for flight, while other scenarios have been proposed involving aerodynamical modeling. Discovery of *feathered dinosaurs* is consistent with insulation, although we should be aware that these dinosaur feathers are well-developed feathers, not protofeathers, which may have served a different purpose(s).[15]
5. The *predator-to-prey ratio* of carnivorous-to-herbivorous dinosaurs in some fossil deposits has been estimated by Bakker (1986) as of the order of 3:100. This ratio is similar to that for extant endothermic predators, which need a large prey population to support their high-energy metabolism. By contrast, ectothermic predators with lower metabolic requirements can exist on a prey population ten times smaller, 3:10. It is difficult, however, to discriminate between large ectotherms and endotherms on predator-to-prey ratios alone; both types may require similar amounts of food, and there may be a collecting bias. In general, incomplete fossilization makes it difficult, if not impossible, to determine the relative abundance of different species in a community and adds to the uncertainty of specifying predator-to-prey ratios.[16]

Temperature stability in extant ectothermic sauropsids is a function of size: the larger the animal, the more stable its body temperature. If dinosaurs were ectothermic (*gigantothermy*), large size alone may have both affected their success and imposed limitations. As discussed previously, large ectotherms in a warm climate would have been able to preserve body heat and perhaps attain fairly high rates of metabolism and activity without paying the high cost of endothermy, but at the cost of a substantial increase in basal metabolic rate. Given such dependence on high body temperatures, smaller dinosaurs, subject to greater temperature fluctuations because of their size and lack of insulation, would not have been as successful.

[15] See Feduccia (1985, 1999a), Tarsitano et al. (2000), Currie et al. (2004) and the section *Feathers and the Origin of Birds* later in this chapter for various scenarios for the original function of feathers.

[16] For endothermy, see Chinsamy and Hillenius (2004), for crocodiles, Seymour et al. (2004) and for *T. rex,* Wedel (2003).

TABLE 18-4 Number of genera identified during the interval that began 20 million years before the extinctions at the end of the Cretaceous and ended with the beginning of the Cenozoic, compared with the number that existed 10 million years later[a]

	Before Extinction	After Extinction	% of Genera After Extinction
Fresh water organisms			
Cartilaginous fishes	4	2	
Bony fishes[b]	11	7	
Amphibians	9	10	
Reptiles[c]	12	16	
	36	35	97
Terrestrial organisms (including fresh water organisms)			
Vascular plants	100	90	
Snails	16	18	
Bivalves	0	7	
Cartilaginous fishes	4	2	
Bony fishes[b]	11	7	
Amphibians	9	10	
Reptiles[c]	54	24	
Mammals	22	25	
	216	183	85
Floating marine microorganisms			
Acritarchs	28	10	
Coccoliths	43	4	
Dinoflagellates	57	43	
Diatoms	10	10	
Radiolarians	63	63	
Foraminifers	18	3	
Ostracods	79	40	
	298	173	58
Bottom-dwelling marine organisms			
Calcareous algae	41	35	
Sponges	261	81	*(continued)*

If dinosaurs were primarily ectothermic, even large size would not have protected them against the more variable climate that inaugurated the Cenozoic. Given its reduced ability to change body temperature rapidly, a large ectotherm would have had considerable difficulty losing heat during a hot summer and gaining enough heat during winter's prolonged cold periods. Endothermic mammals were able to survive the end of the Cretaceous and increase during the Cenozoic because activity depended less on external temperatures. Larger dinosaurs, such as *Tyrannosaurus rex*, did not survive. Small dinosaurs did, including some with feathers.

The Late Cretaceous Extinctions

No large dinosaurs, indeed no land vertebrate larger than 23 kg (50 lbs), survived the Cretaceous. Other terrestrial and marine organisms also became extinct; paleontologists estimate that more than half of all animal species, classified into the various groups given in Table 18-1, became extinct during a relatively short geological period at the end of the Cretaceous (**Table 18-4**). Jablonski (2002) has shown that for many groups, species or generic diversity was no higher 5 to 10 millions of years after extinction than before.

Many possible causes have been proposed to account for such a wide spectrum of **extinction**. A once popular hypothesis, now abandoned, sought to explain the extinction of dinosaurs by *internal rather than external causes.* Just as individuals are born, grow old and die, this hypothesis suggested that species and other taxonomic categories follow a similar life history driven by internal factors (**orthogenesis**) that cause evolution to proceed in a direction unrelated to selection and adaptation. Evidence used to support orthogenesis was the appearance of bizarre and what appeared

TABLE 18-4 Continued

	Before Extinction	After Extinction	% of Genera After Extinction
Bottom-dwelling marine organisms (cont'd)			
Foraminiferans	95	93	
Corals	87	31	
Bryozoans	337	204	
Brachiopods	28	22	
Snails	300	150	
Bivalves	399	193	
Barnacles	32	24	
Malacostracans	69	52	
Sea lilies	100	30	
Echinoids	190	69	
Asteroids	37	28	
	1,976	1,012	51
Swimming marine organisms			
Ammonites	34	0	
Nautiloids	10	7	
Belemnites	4	0	
Cartilaginous fishes	70	50	
Bony fishes[b]	185	39	
Reptiles[c]	29	3	
	332	99	30
Totals	2,858	1,502	52

[a] The record for terrestrial organisms is limited to North America but is global for marine organisms.
[b] Actinopterygii plus Sarcopterygii.
[c] Here reptiles is used in the Linnean sense.
Source: From Russell. Reproduced with permission from the *Annual Review of Earth and Planetary Sciences,* Volume 7, © 1979 by Annual Reviews Inc.

to be 'nonadaptive' characters. Some long considered the large, seemingly clumsy 3.3-m-wide antlers of the Irish elk *Megaloceros* (see Fig. 24-8) to be a cause or corollary of its senescence and extinction.

Various types of dinosaurs, however, persisted and continued to adapt to changing circumstances for more than 100 My. Even toward the end of the Cretaceous, new groups, such as the ceratopsians dinosaurs, were appearing. No evidence indicates that any biological mechanisms other than inability to cope with new environmental or competitive challenges caused the extinction and replacement of dinosaurs or any organism, although, as we shall now see, the cause of the Late Cretaceous extinctions may not have been biological.

There is no shortage of more plausible hypotheses than orthogenesis for these mass extinctions: intense volcanic activity, epidemics, changes in plant composition, shifting continental profiles, elevated carbon dioxide levels (greenhouse effect), changes in sea level or ocean salinity, high doses of ultraviolet radiation, dust clouds caused by collisions with comets or asteroids and ionizing radiation from supernova explosions or other sources.

The most generally accepted theory, collision with an asteroid (**Collision Theory**) gathered considerable support in the 1980s as a result of the discovery of iridium deposits in strata marking the Cretaceous–Tertiary boundary. Iridium is a rare earth element often found in meteorites. Evidence linking iridium to meteors was first obtained when geologists examined an area of destruction of all trees within a 14.5 km radius of a small area of the remote Tunguska region of Russia, and flattening of those trees within a 40-km radius. High levels of iridium in the soil were interpreted as meteoritic in origin, the result of a 30 m-diameter meteor that exploded 8 km above the ground in 1908. The worldwide presence of iridium in Cretaceous–Tertiary boundary strata, along with high-impact particles (glass-like spherules and

BOX 18-2 Extinctions and Extraterrestrial Impacts

Our ability to project present-day mechanisms back into the past does not mean that we follow a philosophy *restricted* to gradual change, such as the one Lyell expounded to explain all events on Earth (see Chapter 1).

Astronomers have demonstrated that the universe is a violent place. It had an abrupt birth. Its stars and galaxies were born in the midst of violent interactions. Its elements were created from the debris of many violent episodes, and violent impacts still occur — even in our small solar system. Whether from impacts, volcanism or plate tectonics (see Chapter 5), we recognize occasional catastrophes in Earth's history that had major effects on geology and life.[a]

As a result of the *Apollo* space program, we know that a series of heavy extraterrestrial impacts battered the moon about 4 Bya, indicating that similar events must have occurred on Earth's surface. The sterilizing heat such impacts generated, whether caused by asteroids, comets or collisions with even larger planet-like bodies, may be characteristic of the period after which forms of life arose (see Fig. 8-5). Although their frequency has greatly diminished from the Hadean age, impacts persisted, and geologists have identified more than 100 craters on Earth. **Table 18-5** relates the size of the crater to a general estimate of the size of the celestial body causing it, and estimates the time between such events.

Depending on their size, objects causing such impacts may have enormous environmental effects over months to years. The impact crater throws large numbers of particles into the atmosphere, producing dust clouds that interfere with photosynthesis, causing collapse of the food chain in various localities. Depending on whether the impact is on land or at sea, it can also cause large climatic temperature changes ranging from an immediate "winter" (dust) followed by a high-temperature "greenhouse effect" (water vapor). Heat generated by the object entering the atmosphere and the heated material it ejects on impact can spark raging forest fires and generate nitrous oxides that seed acid rains that destroy vegetation and marine organisms. Alvarez and Asaro (1992) estimate that the impact of a 10-km-diameter body would produce an explosion equal to that of 100 million hydrogen bombs, a force that is impossible to imagine (and perhaps to believe).[b]

Given such colossal effects, can we discern a relationship between extraterrestrial impacts and mass extinctions?

As far as we can tell from fossil data, there have been five major mass extinctions since the Cambrian (**Table 18-6**), each marked by the relatively abrupt disappearance of at least 75 percent of marine animal species. (Because marine environments are the major sites of sediment deposit, aquatic organisms more commonly fossilize than do terrestrial ones, and paleontologists consider estimates of their species frequency more reliable.) According to Raup and Sepkoski (1986), these large-scale extinctions are not isolated events but allied to other extinctions, perhaps caused by a series of impacts from extraterrestrial bodies that occurred throughout the Phanerozoic.[c] Mass extinctions have recurred at regular intervals (**Fig. 18-30**); Raup (1991) estimates that extinction events on the order of the type in Table 18-6 occur on average about every

TABLE 18-5 Size of meteorite impact craters, estimates of the size of the celestial body causing them, and the time between such events

Meteorite Diameter (km)	Crater Diameter (km)	Average Time (My) Between Impacts
10 km	>150	100
	>100	50
	>50	12.5
	>30	1.2
	>20	0.4
1 km	>10	0.11

Source: Raup, D. M., 1991. *Extinction: Bad Genes or Bad Luck?* Norton, New York.

TABLE 18-6 Details of the five major mass extinction since the Cambrian[a]

Extinction Period	Approximate Date (Mya)	Estimated % of Marine Animal Extinctions: Genera	Species
Late Ordovician	440	61	85
Late Devonian	365	55	82
Late Permian	245	84	96
Late Triassic	208	50	76
Late Cretaceous	65	50	76

[a] Each event was marked by the relatively abrupt disappearance of at least 75 percent of marine animal species.
Source: Raup, D. M., and J. J. Sepkoski, Jr., 1986. Periodic extinction of families and genera. *Science,* **231**, 833–835.

shocked, fractured quartz) strongly indicated collision with an extraterrestrial body (**Box 18-2**).[17]

But many paleontologists demonstrated that dinosaurs and other animals had already declined in numbers or disappeared *before* these impact layers were deposited. Contrary to the immediate effects of an extraterrestrial impact, dinosaur extinction may have taken a million years

[17] See Alvarez (1983) and Alvarez and Asaro (1993) for the iridium studies, and see Benton (2003), Macleod (2004), Hallam (2005) and Erwin (2006) for discussions of mass extinctions.

100 My. As we might expect, extinctions with less effect are more common, but the average interval between such events is still long enough that most marine invertebrate species could face extinction from such an event. For example, an event that eliminates five percent of species occurs on average only once every million years, and if the life spans of marine invertebrate species is of the order of 4 My, all marine invertebrate species would have experienced such an event.

Despite the extensive environmental effects of such impacts, good evidence for a close association with a foreign body impact exists *only* in the case of the Late Cretaceous extinction. As the text indicates, an anomalous iridium-rich layer at the Mesozoic-Cenozoic boundary, now identified at more than 100 different localities, lends credence to a large impact at the end of the Cretaceous (Alvarez and Asaro, 1992). Such an explosion would have carried the observed iridium and high-velocity particles — spherules, and shocked quartz — to the top of the atmosphere and spread them worldwide.

Among the best candidates for a crater large enough to produce such Late Cretaceous global effects is the Chicxulub crater off the coast of the Yucatán Peninsula in Mexico. This crater has an outer diameter of about 195 km and "records one of the largest collisions in the inner solar system since the end of the early period of heavy bombardment almost four billion years ago . . . Earth probably has not experienced another impact of this magnitude since the development of multicellular life approximately a billion years ago" (Sharpton et al., 1993).

Does this impact account for all the Late Cretaceous extinctions? Paleogeologists have proposed volcanic eruptions that, when explosive, may deposit iridium globally in atmospheric dust and ash (and even when nonexplosive may affect world climate) as a primary or auxiliary cause for these extinctions. An enormous outpouring of nearly one million cubic miles of volcanic lava covering one-third of India (the "Deccan Traps") during the Late Cretaceous extinctions would support this hypothesis, although current evidence indicates that the meteor fell between eruptions of plumes of deep mantle material.

As to proposals that extinctions other than the Late Cretaceous relate to impacts, no firm supporting evidence has appeared. Iridium deposits in strata associated with other extinctions are not great enough to assume extraterrestrial impact, while strata in the Late Permian extinction, the greatest of all extinctions, show almost no iridium. Nevertheless, the likelihood that the Permian extinction occurred over a period of only 10,000 to 165,000 years makes it as catastrophic as a conspicuous impact.[d]

Knowledge of the cause(s) for mass extinctions and the precise manner in which they manifested their effects awaits more exact information. Extinctions certainly had serious evolutionary consequences by producing a "major restructuring of the biosphere wherein some successful groups are eliminated, allowing previously minor groups to expand and diversify" (Raup 1994). In our unpredictable solar system, such events may occur again. Of the many asteroids that cross Earth's orbit at various times, perhaps a thousand or more could have an impact equal to the asteroid that may have caused the extinction of the dinosaurs.

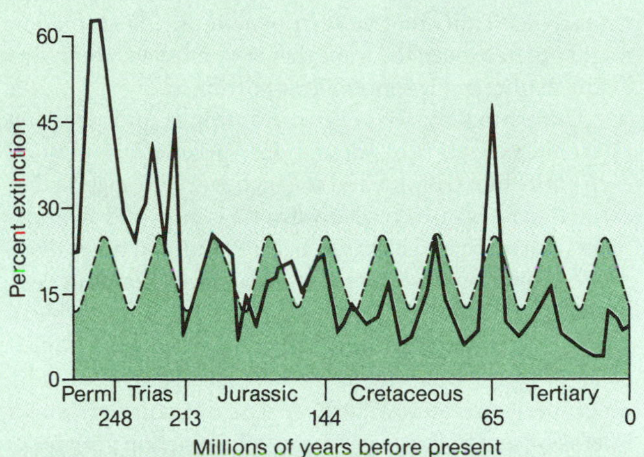

FIGURE 18-30 *Heavy solid lines* show the percentage of marine animal extinctions during the approximate 260-My interval from the Permian to the present. *Dashed lines* and *shaded areas* are based on a periodicity of 26 My. (Adapted from Fox, W. T., 1987. Harmonic analysis of periodic extinctions. *Paleobiology*, **13**, 257–271.)

[a] See Macleod (2004), Hallam (2005) and Erwin (2006) for the role of catastrophes in mass extinctions.

[b] For comparison, the eruption of Mount St. Helens in Washington State (1980) was equivalent to 350 megatons; Krakatoa (1883) 5,250 megatons; the eruptions that produced the Yellowstone Caldera in the Arizona, 875,000 megatons.

[c] Raup and Sepkoski (1986) point out that mass extinctions of families and genera occur with a periodicity of about 26 My. W. T. Fox's 1987 analysis of their data supports this view (Fig. 18-30). Although Fox finds the relationship between the extinctions and 26-My periods statistically significant, others question the basis for the periodicity. No clear cause for this periodicity has been found and so the Raup-Sepkoski hypothesis remains unresolved.

[d] See Bowring et al. (1998) for the duration of the Permian extinction, and see Erwin (1993, 2006) for analyses of other extinction events.

or more, and according to some claims, may have extended into the Paleocene, although no unequivocal nonavian dinosaur bones are known after the Mesozoic. Some paleontologists proposed a combination of stressful environments and an extraterrestrial impact. For others, the question of how to decide whether extinction was gradual or catastrophic for various vertebrates remains unanswered (Dingus and Rowe, 1998).

Some species or subpopulations of species did survive in the midst of extinction, either because certain traits they

possessed were beneficial or because of "luck"; many mammalian species survived the Cretaceous extinctions, as did many birds. Small size, and perhaps efficient endothermy, may have been crucial for terrestrial survival in the Late Cretaceous. Traits that were of benefit during extinction might not have been the traits that were most advantageous before extinction. Evolution is opportunistic.

Unfortunately, we are contributing at an increasing rate to the extinction of populations, species, communities, even entire biota. Pimm and colleagues (1995) suggest that extinction rates in even widely diverse groups and environments can reach and have reached 100 to 1,000 times those of pre-human levels. Although some contest these numbers, such appraisals, even if reduced by half, portend a bleak future for many species living in or near human-occupied areas. Humans not only usurp natural resources used by other species for our own use, we destroy natural resources by erosion and pollution — agents of extinction perhaps as powerful as global climate change and the impact of extraterrestrial bodies (Diamond, 2005).

Flight

Escape from extinction is a hallmark of biological survival, and may at times depend on the ability to move rapidly from threatened environments. Flight provided a number of advantages: rapid escape from terrestrial predators and menacing conditions; access to feeding and breeding grounds that would otherwise be difficult or impossible to reach; and relatively swift transit between localities.

Although forms capable of gliding for short distances arose in various vertebrate lineages including fishes, known adaptations for sustained ascending flights have appeared only three times in tetrapod evolution: pterosaurs, birds and bats. The two forms derived from reptiles — pterosaurs and birds — differ in respect to mechanisms and accompanying adaptations. In pterosaurs a *flight membrane* (*patagium*) was formed by a thin fold of skin stretched between the trunk and elongated fourth finger of each hand. In birds, the flying surface consists of many *stiff wing feathers* that project posteriorly from the wing. In general, by the time of their first fossil appearance, both pterosaurs and birds were well adapted to gliding and probably to sustained flying.[18]

Pterosaurs

Although early **pterosaurs** appear in the Late Triassic and birds in the Jurassic, the earlier evolution of these two types left no obvious record, probably because they started out as small arboreal creatures in highland habitats where the chance of fossilization was low (Buffetaut and Mazin, 2003). Nevertheless, their skeletal features provide a number of clear homologies with earlier reptiles (**Fig. 18-31**). One phylogeny suggests that pterosaurs originated from early Triassic bipedal archosaurs (**Fig. 18-32,** left side).

A well-described Jurassic pterosaur, *Rhamphorhynchus,* (Fig. 18-32a), was about 0.6 m long with a typical diapsid archosaurian skull and an additional antorbital fenestrum. The tail was long with a small, rudder-like flap of skin at the end. The bones of *Rhamphorhynchus* were light and hollow, and its elongated jaws were armed with strong, pointed teeth. According to Padian (1985), the sternum and its accessory bones provided sufficient surface for attachment of large flight muscles like those of flying birds. Padian also demonstrated that the wing membrane did not cover the hind limb, and the legs were, therefore, free for bipedal locomotion, somewhat like extant large birds. In contrast, others support a more traditional notion that pterosaur motion on land was quite awkward, based on a sprawled quadrupedal stance.

Whichever concept is correct, maximally efficient terrestrial locomotion may not have been essential: their fossil locations indicate that pterosaurs roosted near large lakes and coastal areas, perhaps on offshore islands in trees or protected cliffs, and hunted for fish. Flight did not protect them from extinction, and they too, like the nonavian dinosaurs, did not survive the Cretaceous. Instead, their cousins, the birds, became the most widely distributed group of Cenozoic flyers, while lepidosaurs and turtles thrived on the ground.

Pterodactyls

By the late Jurassic, rhamphorhynchoid success had given rise to another group, the pterodactyloids (Fig. 18-32b), which continued well into the Cretaceous. The tail is much reduced, a special bony element anchors the shoulder girdle to a number of vertebrae and the teeth tend to be reduced in size and number (leading eventually to a long, toothless beak). In *Pteranodon* a long skull crest extends behind the orbit.

Size differences were pronounced. The largest fossil pterodactyloid, a Late Cretaceous form, *Quetzalcoatlus,* found in Texas, has a wingspan that may have reached 12 m. These animals must have been successful gliders, probably using sea or land thermals for lift, and may have been capable of powered flapping for short distances, although how they did so remains a mystery; pterodactyloids are much larger than the largest flying birds either known or based on biomechanical considerations; see the end of the chapter for flightless birds, which can be very large.

[18] Caution is required in contrasting or comparing pterosaurs *as a group* with birds or bats *as groups.* Dyke et al. (2006) identified at least two features — forelimb to hind limb ratios, and hind wing membranes — that vary within pterosaurs. Only one of these features resembles the patterns in bats.

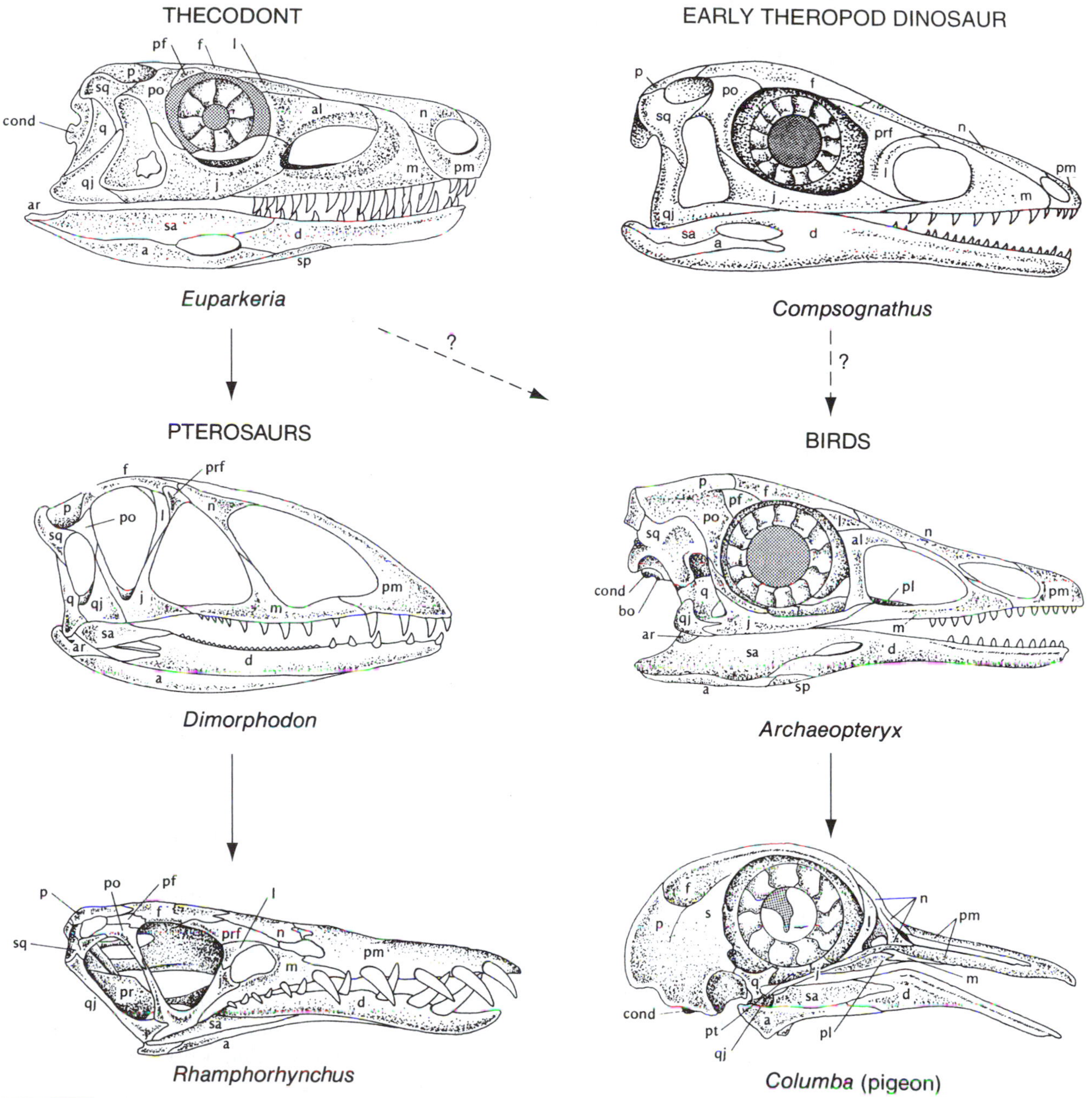

THECODONT

Euparkeria

EARLY THEROPOD DINOSAUR

Compsognathus

PTEROSAURS

Dimorphodon

BIRDS

Archaeopteryx

Rhamphorhynchus

Columba (pigeon)

FIGURE 18-31 Similarities in skull structures among early bipedal archosaurs such as *Euparkeria* and the pterosaur and bird lineages that paleontological evidence show may be derived from them. Abbreviations: *a,* angular; *al,* adlacrimal; *ar,* articular; *bo,* basioccipital; *cond,* occipital condyle; *d,* dentary; *f,* frontal; *j,* jugal; *l,* lacrimal; *m,* maxilla; *n,* nasal; *p,* parietal; *pf,* postfrontal; *pl,* palatine; *pm,* premaxilla; *po,* postorbital; *pr,* prootic; *prf,* prefrontal; *pt,* pterygoid; *q,* quadrate; *qj,* quadratojugal; *sa,* surangular; *sp,* splenial; *sq,* squamosal. (Adapted from Stahl, J. B., 1974. *Vertebrate History: Problems in Evolution.* McGraw Hill, New York; and Romer, A. S., 1996. *Vertebrate Paleontology,* 3rd ed. University of Chicago Press, Chicago; and Heilmann, G., 1927. *The Origins of Birds.* Appleton, New York.)

Nonavian Dinosaurs: Birds

Based on cladistic classification of lineages, birds are nested within dinosaurs. Consequently, not all dinosaurs went extinct; **birds** are living dinosaurs.

Three lineages of extant birds are recognized (**Fig. 18-33**):

- Palaeognathae (ostriches, emus, tinamous)
- Neognathae (penguins, flamingos, petrels, albatrosses, hummingbirds, swifts)
- Galloanserae (ducks, geese, quail, pheasants),

all within Neoaves, the lineage that included most modern birds.

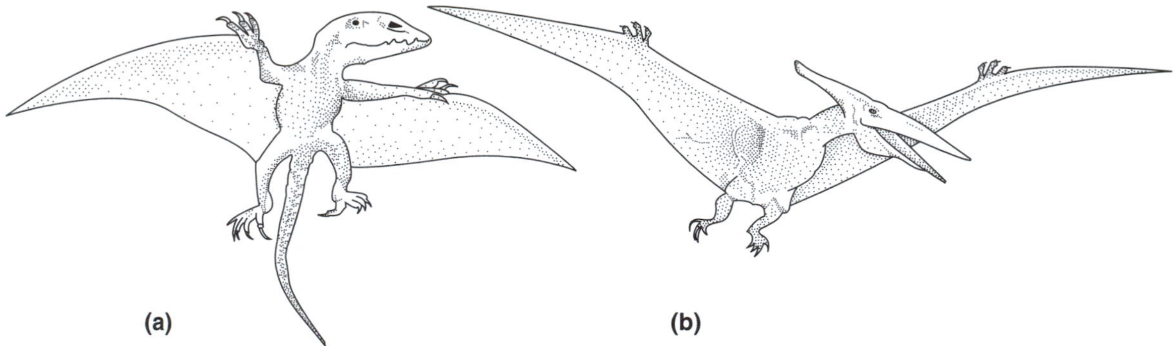

FIGURE 18-32 Comparison between rhamphorhynchoids (a) and pterodactyloids (b). Rhamphorhynchoids were tailed with wing spans of 0.3 to 2.1 m. Pterodactyloids had very short tails, and with wing spans from 15 cm to as much as 12 m. Other interpretations of pterosaur fossils suggest that, although narrow at the tips, the wings broadened at the trunk to attach from neck to ankles (Unwin and Bakhurina, 1994). This would have made pterosaur mobility on land no better than a bat's, a view supported by their sprawling quadrupedal posture (From J. M. Clark, et al., 1998. Foot pressure in a primitive pterosaur. *Nature* 391, 886–889).

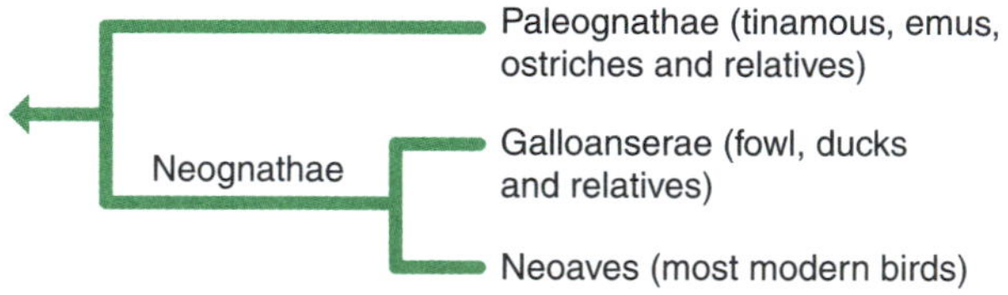

Paleognathae (tinamous, emus, ostriches and relatives)
Neognathae
Galloanserae (fowl, ducks and relatives)
Neoaves (most modern birds)

FIGURE 18-33 The phylogenetic relationship between the three major groups of birds. (*Source:* Mindell, D. P., and Brown J. W. 2005. Tree of Life Web Project, available at http://www.tolweb.org/neornithes. Accessed May 2, 2007.)

Extant Birds

Palaeognathous and neognathous birds can be differentiated morphologically on the basis of the nature of the palate and how the palate joins with the skull, the pelvis (fused or not), and sternum (keeled or not) and the detailed structures of the middle ear. Several molecular studies have begun to unravel the relationships among the 35 or so avian clades (orders), while recent discoveries of *feathered dinosaurs* have caused great excitement, not only among paleontologists but for all those who are fascinated by dinosaurs.

Systematists currently classify birds into 35 major clades (orders) subdivided into about 200 lineages (families). The distinctions among these groups are subtle external traits. Among the extant 9,600 species of birds are a variety of feeding and locomotory adaptations, ranging from flesh eating to nectar feeding, and from rapid running to gliding, swooping, diving and swimming (**Fig. 18-34**). Although some workers have proposed evolutionary changes that may have given rise to some of these adaptations, the absence of fossil intermediates makes relationships among extant avian groups uncertain, especially among those in arboreal habitats. A molecular approach to this problem (see Chapter 12) is providing a mass of information that should clarify the currently unresolved relationships of birds and other tetrapods.[19]

[19] See Sibley and Ahlquist (1990) for molecular studies on bird phylogeny, Cracraft et al. (2004) for a recent phylogeny of birds and see Box 12-2, Ancient DNA, for analysis of molecules preserved in fossils.

Fossil Birds

Perhaps the most famous bird fossil is ***Archaeopteryx.*** The first bird-like fossils known, classified as *Archaeopteryx*, were found in Upper Jurassic limestone deposits in Bavaria. Of the known seven skeletons, four are nearly complete, some of them with flight feathers of the wings and tail in their natural positions (**Figs. 18-35 and 18-36b**).

The similarity between *Archaeopteryx* and dinosaurs is so striking that paleontologists would have classified *Archaeopteryx* as a dinosaur had it lacked feathers (Fig. 18-36a). Birds are now nested within dinosaurs as *archosauromorphs.*[20] Most striking are their *similarities to small, carnivorous dinosaurs* (Fig. 18-36) whose teeth, separate clawed fingers, long bony tail and dozens of other features indicate their status as a link between earlier theropods and extant birds.[21]

Archaeopteryx is not a "**missing link**"; not only has it been found, it is too advanced to have been the first

[20] Richard Owen's 1864 paper describing the Solenhofen specimen of *Archaeopteryx* is reproduced in Weishampel and White (2003). See Heilmann (1927), Zhou and Zhang (2002), and Gatesy and Middleton (2007) for the evolution of adaptations for flying.

[21] From the late 1920s until the 1970s, absence of a wishbone (clavicle) from theropod dinosaurs was used as conclusive evidence that birds could not have arisen from dinosaurs, even though a mountain of other evidence supported a dinosaur origin (Heilmann, 1927). Beginning in the 1970s, this view was overturned as clavicles were identified in dinosaurs. Now, a theropod dinosaurian origin of birds — indeed, that birds are reptiles (flying dinosaurs) — holds sway, even though the homology of avian and reptilian clavicles remains open (Vickaryous and Hall, 2006b).

bird (see below). Within a short geological period — by the Early Cretaceous — a range of aquatic and shore birds had appeared, indicative of rapid speciation and habitat exploitation. Some of these fossils represent groups such as flamingos, loons, cormorants and sandpipers, although others, including *Hesperornis,* still retained teeth. Lineages that can be placed into most of the recognized extant orders of birds appear to have originated about 60 to 90 Mya.[22]

The relative position of birds is shown in **Figure 18-37**. *Archaeopteryx* is shown as a basal bird. *Confuciusornis* is known from 120-My-old (lower Cretaceous) deposits in China. Its beak was toothless, as in modern birds, indicating that teeth were lost early in avian evolution. Large claws on the forelimbs indicate that the several species known retain the claws seen on *Archaeopteryx.* The *Enantiornithines* (opposite birds), represented by *Nanantius eos,* comprised a lineage that had many of the features of extant birds — feet that allowed them to perch and well-developed flight — but did not survive the mass extinction 65 Mya.

[22] See Ostrom (1991) for the theropod-bird link, and see Currie et al. (2004) and Zhou and Zhang (2002) for fossil birds.

The oldest fossil bird with the greatest resemblance to modern birds is a loon-like shorebird, *Gansus yumenensis,* described on the basis of five specimens found in Cretaceous deposits in China. *Gansus,* and the oldest (already flightless) penguin, *Waimanu,* from the Paleocene on New Zealand, 60 to 62 Mya, when combined with analysis of mitochondrial DNA, have opened new windows on the origin and diversification of birds that we could not have imagined even a few years ago.[23]

Feathers and the Origin of Birds

Among its various consequences, our failure until recently to find any pre-*Archaeopteryx* fossils with preserved soft tissue kept alive the mystery of when and how feathers first evolved, and left the origin of avian flight unresolved. Indeed, recent fossil discoveries have us asking whether we should only call a species a bird if it has feathers.

- Were feathers primarily an adaptation for *insulating* the presumed endothermic reptilian ancestors of

[23] See You et al. (2006) for *Gansus yumenensis,* and see Slack et al. (2006) for the mitochondrial DNA study.

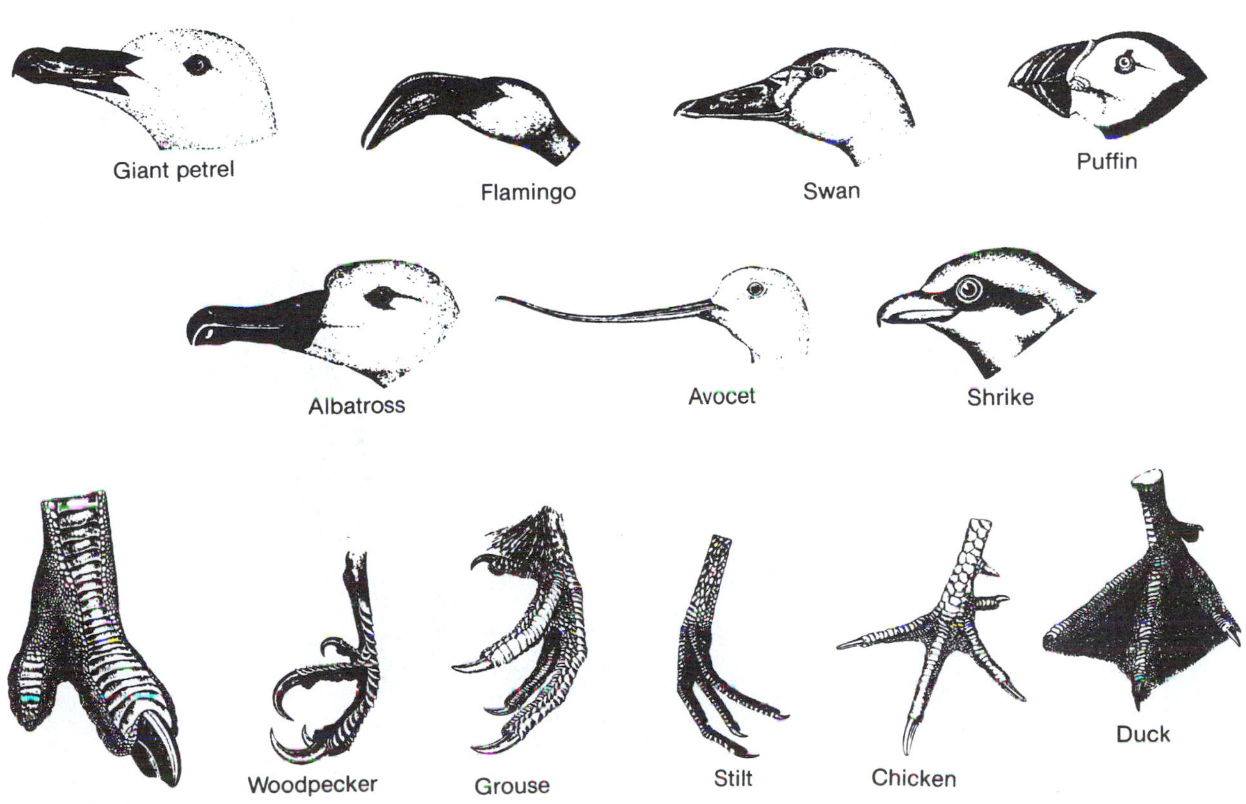

FIGURE 18-34 Some of the many adaptations of bird beaks (top) and feet (bottom). (Adapted from Feduccia, A., 1980. *The Age of Birds.* Harvard University Press, Cambridge, MA.)

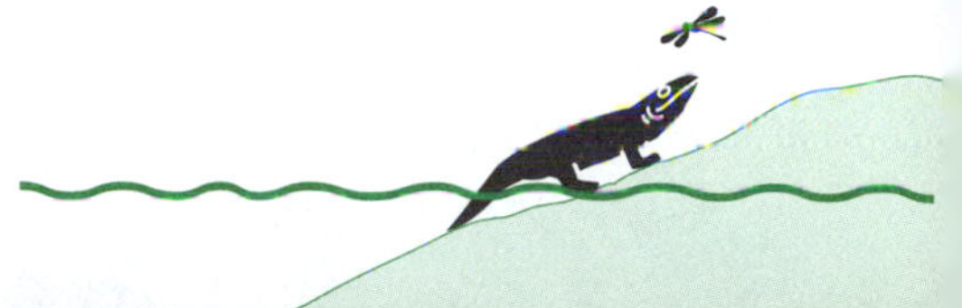

birds, or were they primarily associated with *flight* and only secondarily with insulating qualities? Feathered dinosaur fossils suggest — perhaps even demonstrate — an insulation/display function, but these are already well-developed feathers, not protofeathers, about whose origin we know little.

- Were ancestral birds originally *arboreal* reptiles that used their developing wings to glide from branch to branch, or were they **cursorial** (i.e., with legs adapted for running), ground-dwelling creatures, whose feathers formed planing surfaces enabling them to increase their speed?

Different interpretations extend to *Archaeopteryx* itself, although paleontologists generally agree that *Archaeopteryx* represents a fairly late stage in a long history of bird evolution. The primary feathers of *Archaeopteryx* are remarkably similar in vane structure to the primary (flight) feathers of today's flying birds, meaning that *Archae-*

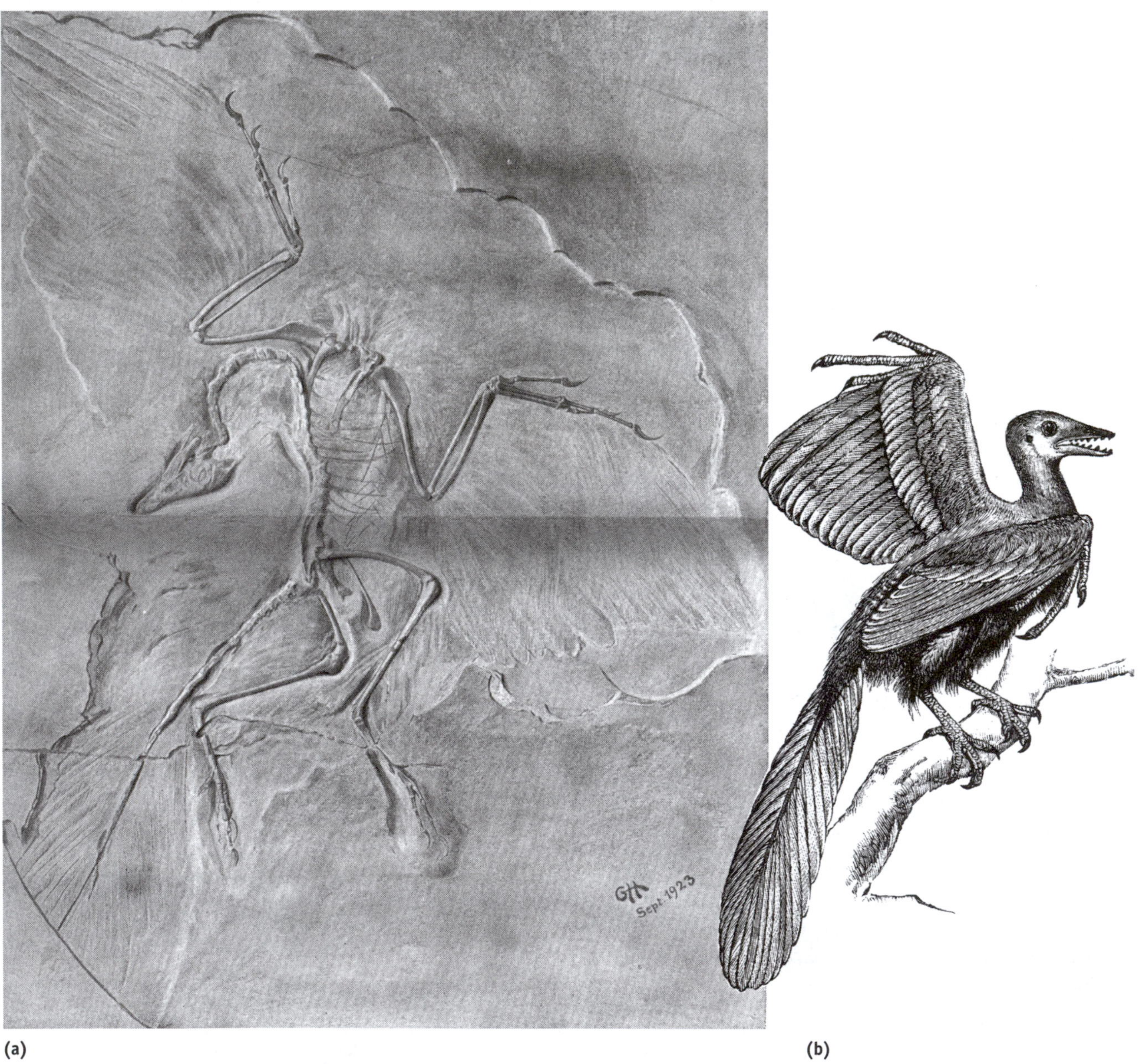

FIGURE 18-35 (a) *Archaeopteryx* as drawn by Gerhard Heilmann in 1923. Compare the accuracy of this drawing with the lithograph of the same specimen in Figure 3-18a, both of which show the flight features along the wings and tail. (b) Heilmann's reconstruction of what *Archaeopteryx* may have looked like in real life. (a and b from Heilmann, G., 1927. *The Origin of Birds.* Appleton, New York. Reprinted 1972, Dover Books, New York.)

opteryx could fly. Nonflying birds have feathers with structures similar to those found in nonavian carnivorous dinosaurs.

Ostrom (1974) presented the view that feathers evolved primarily as a means of controlling *heat loss* in some endothermic dinosaurs. These feathers, especially on the forelimbs, could have helped capture prey such as insects. A feathered "insect net"[24] of this type, in combination with accompanying muscular adaptations such as enlarged pectoral muscles, could serve as an incipient **wing**.

[24] Anthony Russell (University of Calgary) pointed out to us, that if pervious to air, such a "net" could not send a pressure wave ahead of it and so would not be effective as a "fly trap."

The report in 1998 by P.-J. Chen and coworkers of a mane of feathers down the back of a *Compsognathus*-like fossil caused immediate discussion. (See Figure 18-37 for the phylogenetic position of *Compsognathus*, a basal theropod.) A mane of feathers would provide strong support for Ostrom's notion of a feathered-dinosaur origin of birds. Since then, species of "feathered dinosaurs" belonging to 14 genera have been discovered. The most basal of these, *Sinosauropteryx* (Fig. 18-37) from the Jurassic-Cretaceous boundary, some 150 to 120 Mya, had hollow, feather-like structures on its entire body. *Protarchaeopteryx* and *Caudipteryx*, both some 135 to 121 My old, had feathers that resembled more closely the flight feathers of birds (Currie et al., 2004).

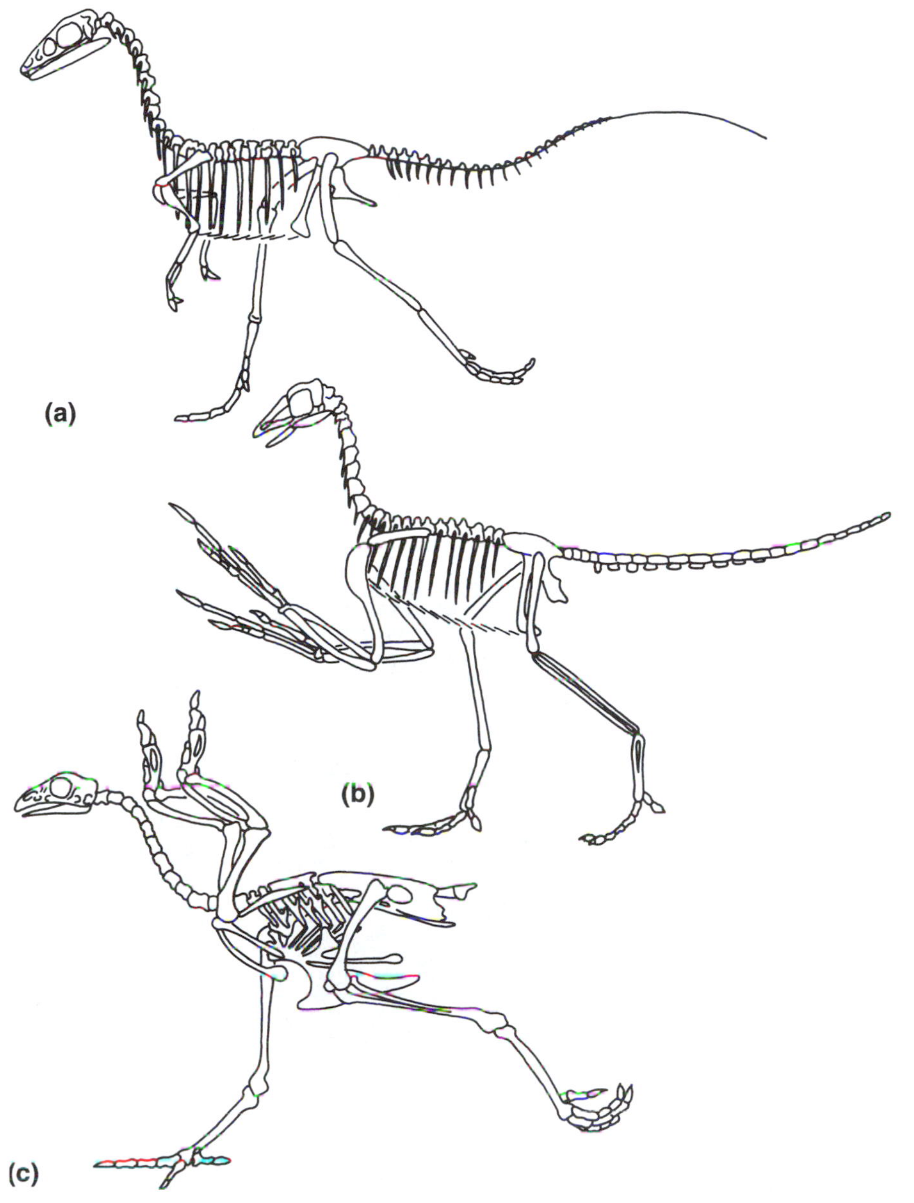

FIGURE 18-36 Skeletons of (a) *Compsognathus,* a dinosaur, (b) *Archaeopteryx,* a fossil bird, and (c) *Gallus,* a chicken. Although *Archaeopteryx* had proportionately longer arms and hands than *Compsognathus,* birds and dinosaurs share many common skeletal features. Birds underwent marked structural changes: the pelvis and sacrum coalesced into a single structure; the sternum enlarged with the expansion of flight muscles; the hand bones fused; and the long, bony tail diminished. [Panels (a) and (b) reprinted with permission of Cambridge University Press from *Patterns and Processes of Vertebrate Evolution,* Robert Lynn Carroll, © 1997. Panel (c) from Dingus, L., and T. Rowe, 1998. *The Mistaken Extinction: Dinosaur Evolution and the Origin of Birds.* W. H. Freeman, New York.]

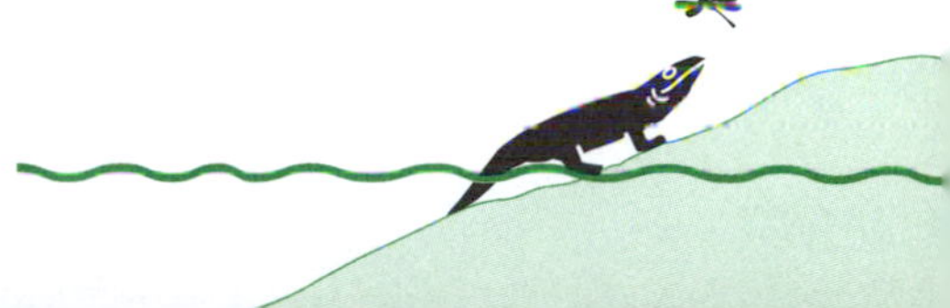

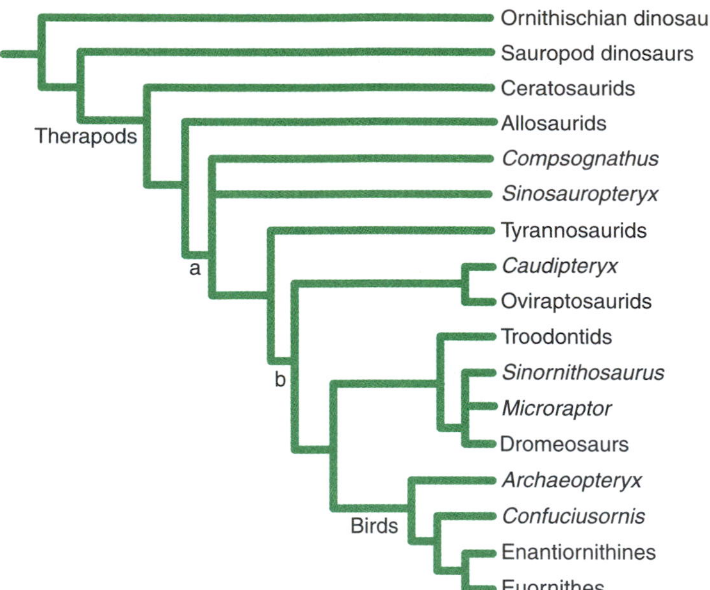

FIGURE 18-37 The phylogenetic relationships between birds and sauropod and ornithischian dinosaurs. Panel a indicates presence of simple feathers; panel b shows the presence of more complex (essentially modern) feathers.

Amazing as these animals are, to everyone's amazement, *Microraptor* had feathers on *both* fore- and hind limbs, strongly suggesting that this four-winged theropod could have glided through the Cretaceous forests (Xu et al., 2000). At 40 cm long, including a 25-cm-long tail, and thus close to the size of *Archaeopteryx*, it certainly was small and thus sufficiently light. The hypothetical missing "Proavian" drawn by Gerhard Heilmann in 1923, who also reconstructed the skeleton (**Fig. 18-38**), was prescient indeed.

The second issue listed above, whether the immediate ancestors of birds were arboreal gliders or exclusively ground-dwelling, has generated much interest and discussion. Burgers and Chiappe (1999) calculated that "[A] running *Archaeopteryx* . . . could have achieved the velocity necessary to become airborne by flapping feathered wings." In contrast, Feduccia (1999b) claims that a ground-dwelling origin of avian flight is "a near biophysical impossibility," and long-feathered gliding forms must have developed in trees.

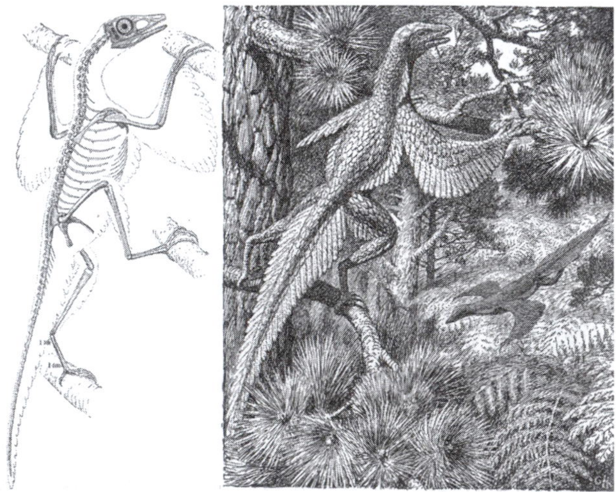

FIGURE 18-38 The hypothetical "Proavian," as drawn by Gerhard Heilmann with a skeletal "reconstruction" also by Heilmann, influenced the search for avian ancestors for 50 years. (From Heilmann, G., 1927. *The Origin of Birds.* Appleton, New York. Reprinted 1972, Dover Books, New York.)

Young extant birds have been shown to use flapping motion to maintain traction when running up an oblique or vertical slope, activity that has been used to help explain the adaptive value (relative fitness) of "half a wing." These words appeared in a challenge to Darwin's theory of evolution posed by St. George Mivart (1871), who questioned the adaptive value of intermediate forms when he asked, "What use is half a wing?" Dial's experiments could provide the answer if the flapping motion is a primitive character. Only if a primitive character, would the proposal provide an evolutionary explanation for the *origin* of "half-wings."[25]

Once past its initial stages, bird evolution proceeded rapidly from the Cretaceous through the beginning of the Tertiary. Cretaceous birds show an enlarged brain, fused skull bones and reduced temporal fenestrae. The greatly enlarged sternum in some forms may indicate the attachment of powerful flight muscles. Other skeletal changes included a fused pelvis and sacrum. Although bird fossils are never plentiful, there are fossils in sufficient numbers and kinds of Eocene, Oligocene and Cretaceous deposits in China to indicate that almost all the major clades of birds

[25] See Dial (2003) and Dial et al. (2006) for these studies on "flapping" locomotion.

Born: April 8, 1956
Birthplace: Aberdeen, Scotland
Undergraduate degree: University of Aberdeen (Zoology)
Graduate degree: Ph.D. University of Newcastle-Upon-Tyne, 1981
Postdoctoral training: University of Oxford 1982–1983
Present position: Professor of Vertebrate Paleontology Department of Earth Sciences University of Bristol, Bristol, England[26]

Michael J. Benton

What prompted your initial interest in evolution?

I first got into paleontology when I was seven or eight. I was given a small color book, *The Golden Guide to Dinosaurs* by Zim and Shaffer, and I was hooked. Then, I read books about Darwin, and I became fascinated by the interdisciplinary nature of the study of evolution. It's no different now from Darwin's day: zoologists, botanists, paleontologists, ecologists, experimental biologists and philosophers all have important contributions to make.

What do you think has been most valuable or interesting among the discoveries you have made in science?

One of the first efforts at a cladogram of basal diapsid reptiles, and the discovery that the split between lepidosauromorphs (the lizard group) and archosauromorphs (the bird-crocodile group) goes deep in time. The demonstration that the fossil record is not as bad as some people have suggested, as assessed by quantitative comparisons of phylogenies and stratigraphies.

What areas of research are you (or your laboratory) presently engaged in?

Determination of long-term patterns of the diversification of life, and assessment of whether they follow equilibrium or non-equilibrium patterns.Tests of the quality of cladograms: how well do they reconstruct phylogeny? Excavations at a huge dinosaur bone bed in the Mid Cretaceous of Tunisia in North Africa.

In which directions do you think future work in your field needs to be done?

We are living through an exciting time in the study of evolutionary patterns, which began about 1970. We now have two pretty well independent methods for reconstructing phylogeny (patterns of evolution), cladistics and molecular phylogenies. Biologists and paleontologists working in this area are real pioneers. Old ideas can be tested, and some dramatic new discoveries have been made about patterns of the one great evolutionary Tree of Life. This is original "one-off" enterprise, and in centuries to come, people will look back to the time from 1970 to 2010, when the outlines of the evolution of life were pinned down in a testable way.

What advice would you offer to students who are interested in a career in your field of evolution?

Students who wish to make original contributions to the growing field of phylogeny reconstruction and macroevolution must master a broad field of knowledge in biology and geology. Luckily, much of the work is reported in excellent, readable, popular books. Students must then read really widely in the current professional journals to be really up-to-date. They need enthusiasm and excitement, and there's no harm in dreaming about dinosaurs, huge asteroid impacts and the vastness of geological time. But it's important to master the necessary quantitative approaches, and to adopt a rigorous questioning approach.

[26] A list of publications is available at http://palaeo.gly.bris.ac.uk/personnel/benton/Benton.html. Accessed May 2, 2007. Recent publication: Benton, M. J., 2004. *Vertebrate Palaeontology,* 3rd ed., Blackwell Publishers, Oxford, UK.

EVOLUTION ON THE WEB biology.jbpub.com/book/evolution

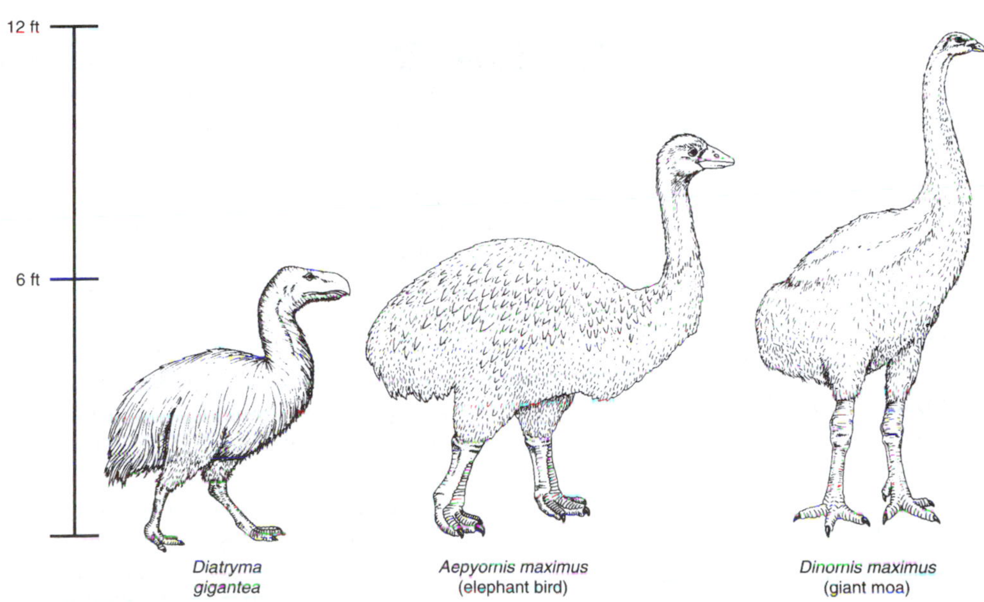

FIGURE 18-39 Reconstructions of some extinct large flightless birds showing their relative sizes. *Diatryma* was an early Cenozoic bird, the others date from the much later Pleistocene. (Adapted from Feduccia, A., 1980. *The Age of Birds.* Harvard University Press, Cambridge, MA.)

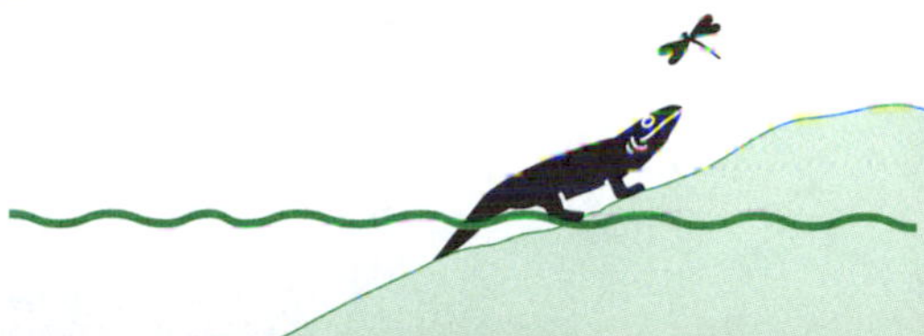

(recognized as orders) evolved by then, with some lineages perhaps tracing back to the Late Cretaceous.[27]

Flightlessness

One dramatic adaptive opportunity opened up by the extinction of the dinosaurs was the large number of vacant terrestrial niches into which various large flightless ground birds evolved. Giant forms, such as the 2.1-m-tall *Diatryma* (**Fig. 18-39**) and others that may have reached a height of 3 m or more, were widely distributed until they became extinct later in the Cenozoic. Few flightless birds now survive. Ostriches of Africa, emus of Australia, rheas of South America and the smaller flightless species such as kiwis and island rails are mostly confined to diminishing habitats.[28]

The evolution of flightlessness seems to have involved changes in the direction of selection, caused either by an absence of predation or as a response to marine habitats. In the first case, in protected or island habitats where major carnivorous forms were absent, local birds could evolve to dominate the terrestrial food chain. Once assuming such roles, selection for reduction in former flight structures would have reduced their energy expenditure.

In marine habitats, many birds "fly underwater," the major difference between flight in air and in water being the density of the medium and not the dynamics/mechanics of this form of locomotion. Some marine birds such as penguins and steamer ducks that spend little time flying in air responded to selection for wing modifications that enhanced underwater propulsion but reduced flight ability.

[27] Gee (2001) collected accounts of Chinese fossils published in the journal *Nature*. See Feduccia (1995) and Dingus and Rowe (1998) for major groups of fossil birds.

[28] Murray and Vickers-Rich (2004) present a most readable analysis of a group of flightless birds, the "thunder birds" (Dromornithidae) from Australia, which includes perhaps the largest bird ever, *Dromornis stirtoni*, estimated to have weighed 570 kg.

KEY TERMS

allantois
Amniota
amniotic egg
amphibians
Anapsida
anthracosaurs
archosaurs
Archaeopteryx
bipedalism
birds
cursorial
Diapsida
dinosaurs
ectothermic
endothermic
extinction
fenestrae
missing link
Ornithischia
orthogenesis
pterosaurs
pterodactyloids
rhamphorhyncoids
Saurischia
Synapsida
tetrapod
therapsids
theropods
wing

DISCUSSION QUESTIONS

1. What explanations have paleontologists offered to account for the transformation of sarcopterygian fishes and stem tetrapods to a land-based existence?
2. What evidence indicates that Devonian tetrapods, such as *Ichthyostega*, were related to rhipidistian (sarcopterygian) fishes?
3. What evidence indicates that tetrapods evolved their limbs in water?
4. What proposals have been made for the origin(s) of amphibians (Lissamphibia)?
5. What advantages did the amniotic egg offer to reptiles, and what preceding stages were necessary?
6. What factors can account for the radiation and extinction of therapsid reptiles?
7. What are the advantages and disadvantages of endothermy and ectothermy?
8. What selective factors can account for
   a. The large size of many dinosaurs?
   b. Dinosaur bipedalism, especially in theropods?
9. What is the evidence, pro and con, for dinosaur endothermy?
10. What proposals have paleontologists offered to explain the mass extinctions at the close of the Cretaceous?
11. What is the relationship between reptilian pterosaurs and birds?
12. What evidence do paleobiologists use to support the proposal that early birds were arboreal? Cursorial?

EVOLUTION ON THE WEB

Explore evolution on the Internet! Visit the accompanying Web site for *Strickberger's Evolution*, Fourth Edition, at http://www.biology.jbpub.com/book/evolution for Web exercises and links relating to topics covered in this chapter.

RECOMMENDED READING

Benton, M. J., 2003. *When Life Nearly Died: The Greatest Mass Extinction of All Time.* Thames and Hudson, London.

Benton, M. J., 2004. *Vertebrate Palaeontology,* 3rd ed. Blackwell, Oxford, England.

Carpenter, K. (ed), 2001. *The Armored Dinosaurs.* Indiana University Press, Bloomington, IN.

Carroll, R. L., 1988. *Vertebrate Paleontology and Evolution.* Freeman, New York.

Carroll, R. L., 1997. *Patterns and Processes of Vertebrate Evolution.* Cambridge University Press, Cambridge, England.

Chiappe, L. M., and L. Dingus, 2001. *Walking on Eggs: Discovering the Astonishing Secrets of the World of Dinosaurs.* Little Brown, Boston.

Clack, J. A., 2002. *Gaining Ground: The Origin and Evolution of Tetrapods.* Indiana University Press, Bloomington. IN.

Curry Rogers, K. A., and J. A. Wilson (eds), 2005. *The Sauropods: Evolution and Paleobiology.* University of California Press, Berkeley, CA.

Erwin, D. H., 1993. *The Great Paleozoic Crisis: Life and Death in the Permian.* Columbia University Press, New York.

Erwin, D. H., 2006. *Extinction: How Life on Earth Nearly Ended 250 Million Years Ago.* Princeton University Press, Princeton, NJ.

Frost, D. R., T. Grant, J. Faivovich, R. H. Bain et al., 2006. The amphibian tree of life. *Bull. Am. Mus. Nat. Hist.*, **297**, 1297.

Hall, B. K., 2005a. *Bones and Cartilage: Developmental and Evolutionary Skeletal Biology.* Elsevier/Academic Press, London.

Hall, B. K. (ed), 2007a. *Fins Into Limbs: Evolution, Development, and Transformation.* The University of Chicago Press, Chicago.

Hallam, A., 2005. *Catastrophes and Lesser Calamities.* Oxford University Press, Oxford, England.

Maisey, J. G., 1996. *Discovering Fossil Fish.* Henry Holt, New York.

Schluter, D., 2000. *The Ecology of Adaptive Radiation.* Oxford University Press, Oxford, England.

Skelton, P., (ed.), 2003. *The Cretaceous World.* The Open University, Milton Keynes, England.

Weishampel, D. B., P. Dodson, and H. Osmólska (eds.), 2004. *The Dinosauria,* 2nd ed. University of California Press, Berkeley, CA.

Weishampel, D. B., and N. M. White, (eds), 2003. *The Dinosaur Papers 1676–1906.* Smithsonian Books, Washington and London.

Zhou, Z., and F. Zhang, (eds), 2002. *Proceedings of the 5th Symposium of the Society of Avian Paleontology and Evolution, Beijing, 1–4 June, 2000.* Science Press, Beijing, China.

19 Evolution of Mammals

Chapter Summary

Many traits distinguish mammals — the most-derived synapsids — from their amniote predecessors. Among these are mammary glands, live birth, a four-chambered heart, a diaphragm, a distinct division between thoracic and lumbar regions of the body axis, skeletal changes in the skull and head and adaptations for homeothermy. Therapsids, which paleontologists consider to be the closest relatives of mammals, were probably endothermic, had cusped teeth with which they could grind food material and a secondary palate that kept food from blocking the nasal openings.

Most mammals shed their teeth once; their amniote ancestors continuously replaced their teeth. As the structure and function of teeth changed, so did the jaws and the muscles of mastication. The articular and quadrate bones articulating the jaw with the skull became the ossicles of the middle ear, which amplify and transmit sound vibrations to the inner ear, allowing the evolution of keen hearing in these small animals.

The earliest known mammals, morganucodontids, are a fossil lineage dating from the Triassic, with tricuspid molars and a precise occlusion of upper and lower cheek teeth. From them probably came the Prototheria or egg-laying mammals, the extant representatives of which are the platypus and the echidna. A later branching gave rise to therians with more elaborately cusped molars and jaws capable of grinding motions enabling them to use new food sources. As a response to archosaurian predation during the Mesozoic, most mammals were probably nocturnal with excellent sensory organs.

Response to selection by therians, which had small eggs, led to viviparity, maternal protection and rapid fetal development. In the marsupial branch, the egg is retained for most of the brief embryonic period, and the fetus emerges from the oviduct at an immature stage to be nourished by mammary glands. Placental mammals have lost the eggshell membranes, and the fetus develops for a much longer time inside the uterus where the placenta provides nourishment, oxygen and waste removal. Most probably, marsupials have persisted along with placental mammals because marsupials expend relatively little energy on their undeveloped newborn offspring.

After the extinction of the dinosaurs at the end of the Cretaceous, mammals diversified into many habitats, making use of new adaptations such as specialized limbs. Marsupials isolated in South America and Australia by continental drift radiated widely. In South America, they diversified as omnivorous taxa (mostly possums) in many different niches. When South and North America united in the Pliocene, many of the marsupials there became extinct and invading North American placental mammals diversified rapidly and took their place.

The most derived synapsids are the animals we know as mammals. A number of principles that apply to the evolution of all animals emerge from an examination of mammalian evolution:

- We cannot predict the continued evolutionary success of a particular group on the basis of its dominance at a particular earlier time.
- Crucial evolutionary advantages may accrue to groups that already have characteristics adaptable to new circumstances.
- Long-term evolutionary replacement among groups is not predictable.
- New modes of biological organization can enhance group survival.
- New levels of organization occur because of coordinated changes in many traits over long intervals of time.
- Once such new levels have been attained, widespread radiation often begins.
- Extinction is common if not inevitable because of constraints on the ability of a species to adapt to large or rapid environmental changes.

Mammalian Characteristics

Mammals derive their name from the **mammary glands** used to suckle the young after birth. Mammalian distinctiveness, however, extends to many other anatomical, physiological and behavioral traits that evolved throughout much of the Mesozoic. In addition to mammary glands, the *soft-body features* of mammals are:

- **live birth** (except for monotremes; Griffiths, 1978);
- body temperature control (**endothermy**) augmented with adaptations such as a hairy covering to control heat loss and sweat glands to enhance evaporation and cooling;
- a **diaphragm** to facilitate the inspiration of oxygen and expiration of carbon dioxide, both necessary for high metabolic activity;
- a **four-chambered heart** that completely separates oxygenated arterial blood from nonoxygenated venous blood; and
- enhanced brain function, derived from expansion of the **neocortex.**

Among the *skeletal differences* that set mammals apart from other vertebrates are:

- a **double occipital condyle** at the rear of the skull that articulates with the first of seven cervical vertebrae;
- division of the dorsal vertebral column into thoracic and lumbar regions;
- a **mandible** consisting of a single bone (the dentary) with a condyle that articulates with the squamosal bone of the skull;
- transformation of the quadrate and articular bones (formerly used for jaw articulation) into the incus and malleus **middle ear ossicles** used for sound transmission;
- a **bony secondary palate** separating the nasal passages from the mouth;
- a single nasal opening in the skull;
- a scapula spine;
- a relatively large braincase; and

FIGURE 19-1 Reconstruction of *Cynognathus,* a carnivorous cynodont therapsid of the early Triassic, about 1.2 m long. (Adapted from Colbert, E. H., and M. Morales, 1991. *Evolution of the Vertebrates: A History of the Backboned Animals Through Time,* 4th ed., John Wiley, New York.)

- greater differentiation among teeth (**heterodont dentition**), characterized largely by multirooted cheek teeth (molars and premolars) with multicusped crowns.

Because we find mammalian soft-body features difficult to trace, our hypotheses on the origin of mammals from egg-laying amniotes mainly derive from studies of fossilized skeletal materials and from molecular phylogeny. The most prominent candidates for mammalian ancestors lie among extinct groups of **synapsids** that initially appeared in the Carboniferous (see Chapter 18). By the Permian these pelycosaurs (see Fig. 18-23) had evolved into a variety of therapsid forms adapted primarily to a terrestrial existence (see Fig. 18-24). A number of paleontologists consider it likely that by the Early Triassic some therapsids had become the first of the vertebrate endotherms and may have had other mammalian soft-body features.

Skeletally, therapsids were distinct from other amniotes with features that, in hindsight, presage their evolution into mammals; for example, the sprawled ancestral amniote stance had changed in some of the dog-like **cynodont therapsids** to a more vertical placement of limbs, with the knees pointing forward and elbows backward (**Fig. 19-1**). This elevated the body from the ground, allowed ventilation of the lungs to be synchronized with locomotion by dorso-ventral flexion of the spine, and enhanced mobility by enabling direct "fore and aft" leg motion. Although the jaw articulation was still between the quadrate and articular, as in sauropsids, the later therapsid mandible was almost entirely composed of the **dentary bone** as in mammals (**Figs. 19-2** and **19-3**).

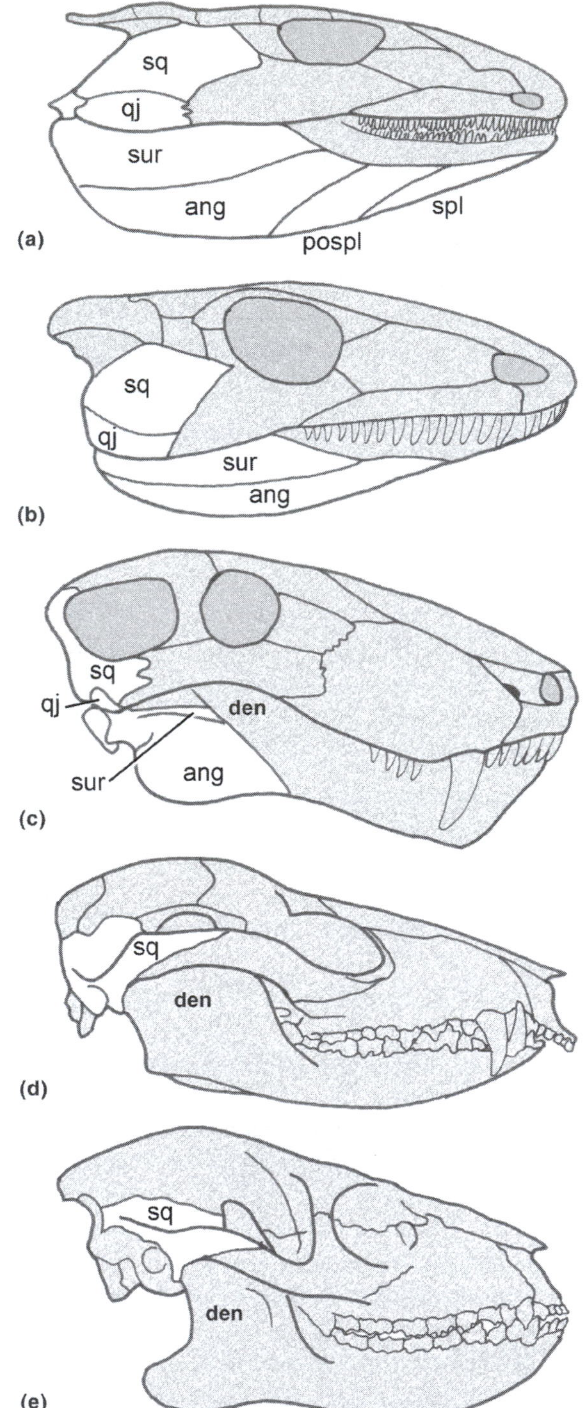

FIGURE 19-2 Changes in the skull associated with the evolution of the mammalian jaw joint as depicted in a classic text by Gregory (1929). Bones of the lower jaw, the quadratojugal (qj) and the squamosal (sq) are unshaded. (a) An early labyrinthodont (*Palaeoherpeton*) from the Lower Carboniferous with multiple bones in the lower jaw and articulation involving the surangular, quadratojugal and squamosal. (b) *Seymouria,* a basal (cotylosaurian) reptile from the Permian-Carboniferous with loss of anterior elements of the lower jaw. We now recognize that *Seymouria* is not on the line from therapsids to mammals but represents an independent lineage. As such, it demonstrates the close parallelism in jaw evolution at that time. (c) A therapsid, *Scymnognathus,* from the Permian with reduced surangular and involvement of the angular in the jaw articulation. (d) *Eodelphis,* an early marsupial from the Upper Cretaceous with a single bone (the dentary) in the lower jaw and involvement of the squamosal in the jaw articulation. Note the distinct posterior processes on the dentary. (e) An early Eocene primate (*Notharctus*) with a single bone (the dentary) in the lower jaw articulating with the squamosal. Ang, angular; den, dentary; pospl, postsplenial; spl, splenial; sur, surangular (From Hall, B. K. 2005a. *Bones and Cartilage: Developmental and Evolutionary Skeletal Biology.* Elsevier Academic Press, London, UK; substantially modified from Gregory, 1929).

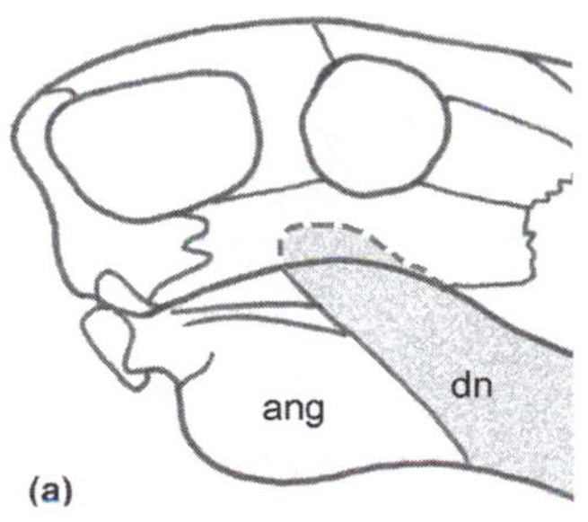

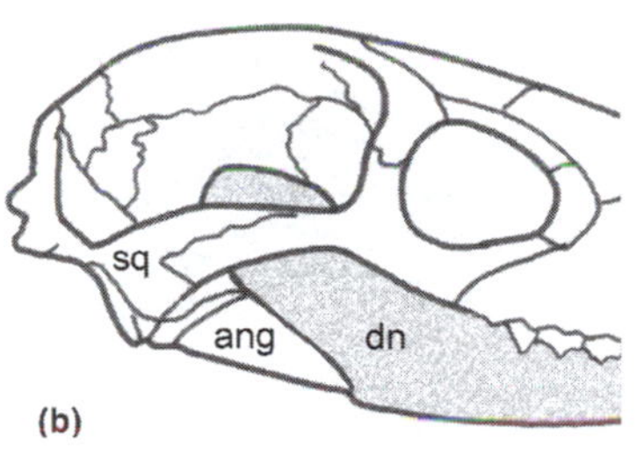

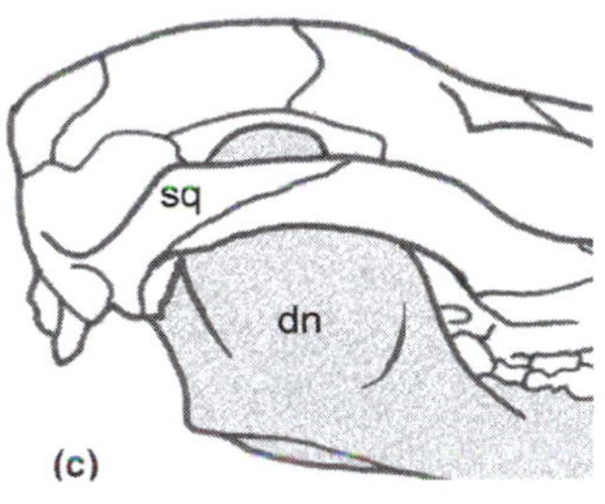

FIGURE 19-3 Changes in the dentary and angular bones associated with development of the mammalian jaw articulation. Elaboration and upward growth of the dentary (dn) and reduction of the angular (ang) are shown in early and late therapsids *Scymnognathus* (a) and *Ictidopsis* (b), and in *Thylacinus,* an early marsupial (c). (From Hall, B. K., 2005a. *Bones and Cartilage: Developmental and Evolutionary Skeletal Biology*. Elsevier Academic Press, London, UK; substantially modified from Gregory, 1929).

Teeth

Among significant mammal-like features in therapsids was the development of differently specialized (**heterodont**) teeth: incisors, canines and cheek teeth.

Therapsids used the various cusp-like surfaces on the crowns of the molar and premolar "cheek" teeth to cut and break food into small particles, rather than to gulp large chunks or swallow whole prey. Because continuous breathing is essential for the metabolic needs of mammals, retaining food orally while chewing led to a variety of innovative changes. In most amphibians and terrestrial sauropsids, the nasal openings are in the anterior portion of the mouth, with the consequence that breathing can be interrupted temporarily without ill effect while the mouth is full of food. Mammals, in contrast, depending more on constant aerobic respiration, would asphyxiate if the food bolus blocked inspired air long enough to chew a mouthful. Selection for an extended secondary palate in synapsids enabled air to be carried to and from a position beyond the mouth near the trachea (**Fig. 19-4**).

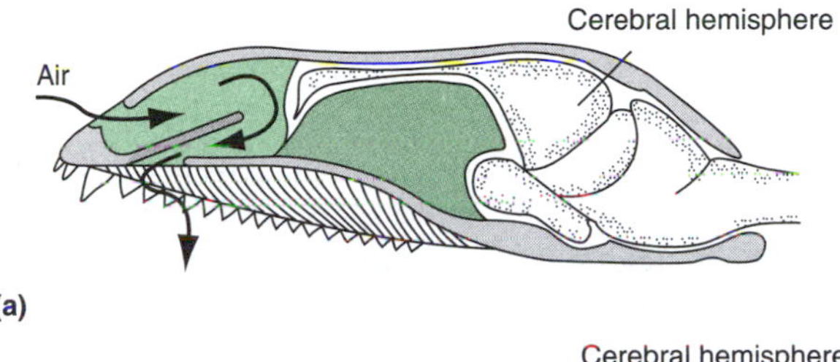

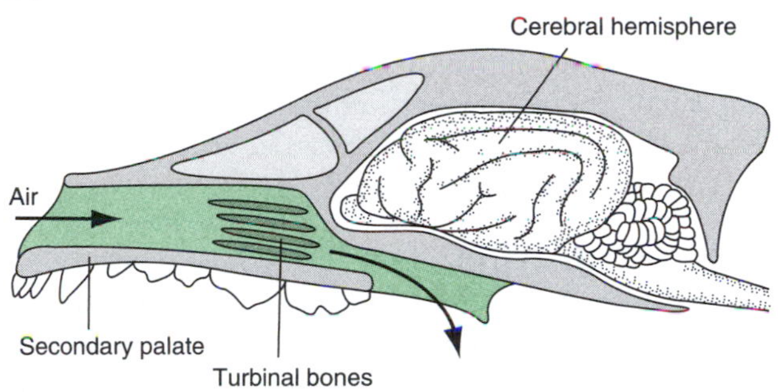

FIGURE 19-4 Air pathways in reptiles (a) and mammals (b), showing the long mammalian secondary palate that separates air entering the pharynx from food being retained in the mouth. Mucous membranes that cover the mammalian turbinal bones warm and moisten the entering air. (Adapted from Romer A. S., 1968. *The Procession of Life.* World Publishing, Cleveland, OH.)

According to Hillenius (1994), two therapsid lineages from the Late Permian, including the cynodonts, had nasal turbinate bony ridges covered by membranes. He points out that such features used to reduce "desiccation associated with rapid and continuous pulmonary ventilation" are confined to endothermic mammals and birds, and are absent in other ectotherms.

Considerable evolution also occurred in tooth replacement. In general, the addition and replacement of teeth is closely associated with relationships between the size of the teeth and the size of the skull during growth. There is considerable value for a growing animal whose head and mouth are enlarging to keep pace with the increased size of its food by increasing the number of large teeth.

In newborn nonavian reptiles, teeth along the jaw margins are small in relation to the size of the animal, and are shed and replaced by larger teeth in an alternating pattern. Because a fully mature nonavian reptile may have a skull ten times longer than when it began tooth replacement, this alternating cycle ensures that teeth of appropriate size are always present. In mammals, such a continuous replacement of alternate teeth would interfere with the precise fit between the upper and lower teeth necessary for effective chewing. Only one set, the **deciduous teeth** or "milk teeth," is replaced in mammals, which include incisors and canines as well as the postcanine deciduous premolars and molars (cheek teeth). In immature mammals, the deciduous premolars perform the chewing function. Permanent premolars replace them when the more posterior adult molars have emerged.

The necessity for only a single replacement of mammalian teeth probably derives from the relatively large size of mammalian newborns, which survive without teeth by suckling at mammary glands until their heads are even larger

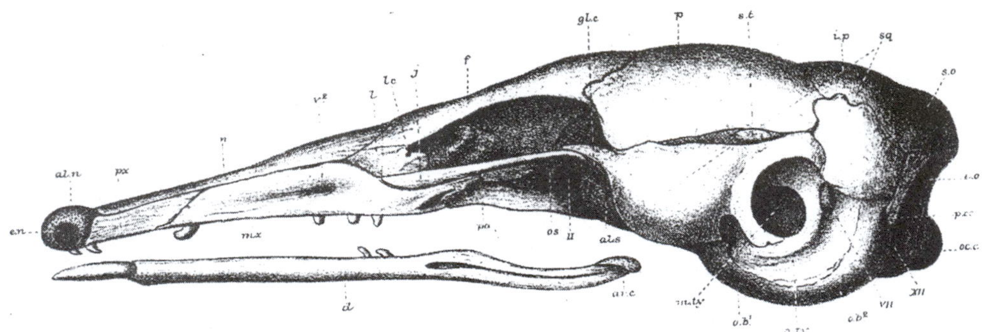

FIGURE 19-5 The Western Australian honey possum, *Tarsipes rostratus,* has perhaps the most reduced lower jaw and teeth of any mammal, the teeth consisting of peg-like molars and a forward-projecting incisor in a slender and much reduced dentary bone. (Reproduced from Parker, W. K. 1890. On the skull of *Tarsipes rostratus. Stud. Mus. Zool. Univ. Coll. Dundee,* **1890,** 79–83 + plate.).

than at birth. By the time the deciduous teeth have erupted, and weaned young mammals are subsisting on adult food, their skulls have reached about 80 percent of adult size. With relatively little more skull growth, permanent teeth replace their deciduous set. As shown in **Figure 19-5** we see an example of extreme reduction of the teeth and supporting dentary bone in the Western Australian honey possum, *Tarsipes rostratus,* which feeds by licking nectar from flowers. Such a mode of feeding exerts minimal forces on the jaws, resulting in the splint-like dentary, reduced peg-like molars and a forward-projecting incisor.[1]

Jaws and Hearing

The precise fit between upper and lower mammalian cheek teeth results from selection for chewing activity and correlates with changes in jaw muscles and tooth shape. In addition to the relatively limited grasping and puncturing functions of sauropsid jaws, synapsid evolution has emphasized shearing, grinding and crushing activity in premolar and molar regions. Masseter and temporalis muscles enable mammals to bite down with considerable force; the internal pterygoid muscle moves the jaw from side to side during chewing, and buccinator and tongue muscles move the food within the mouth. The rearrangement and change in mammalian jaw muscles, along with lengthening of the coronoid and angular processes of the dentary (**Fig. 19-6**), likely reduced strain at the jaw articulation. Correlated with these changes was the loss of what were posterior bony elements of the jaw, the articular and quadrate, which assumed auditory functions as middle ear ossicles with the stapes (columella, hyomandibula).

A major adaptive pressure for these changes — enhanced hearing to capture prey and escape predators at night — must have been one of the primary selective forces acting on early tetrapods. This was especially true for early mammals, which survived the Mesozoic "reptilian tyranny" by adapting to a **nocturnal** environment that required heightened auditory and olfactory perception. A possible evolutionary sequence for these events, diagrammed in **Figure 19-7,** unfolds as set out below.

The **tympanic membrane,** which functions as a taut, drum-like receptor for airborne sound, may well have been lacking in early land vertebrates and even perhaps in early synapsids, which relied mainly on ground-transmitted vibrations (Fig. 19-7a). In these therapsid ancestors, sound was mostly transmitted from ground to inner ear through bone, via contact of the relatively thick stapes with both the quadrate and the articular in the lower jaw (Fig. 19-7b).

Kermack and Mussett (1983) argued that the evolution of a tympanum in therapsids would have allowed hearing

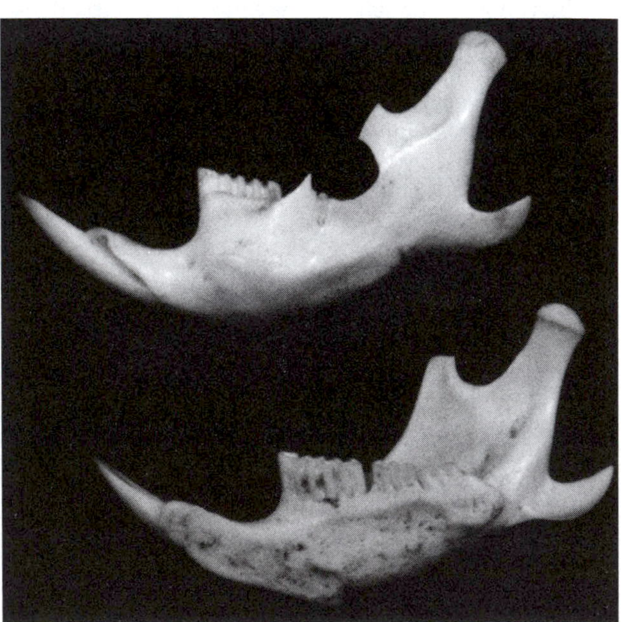

FIGURE 19-6 The dentary, the single bone of the mammalian lower jaw, from the Magdalen Island subspecies of the meadow vole, *Microtus pennsylvanicus magdalensis,* shown in lateral (upper) and medial (lower) views. Note the three prominent bony processes at the posterior end of the dentary (right). (Image provided by Brian K. Hall.)

[1] The honey possum has the lowest weight at birth (<5 mg) and the longest sperm (360 mm) of any mammal.

for airborne sounds but would nevertheless have been inefficient in detecting a wide range of frequencies because of the relatively large mass and immobility of the bones between the tympanic membrane and inner ear (Fig. 19-7c). Because the difficulty in reducing the size of the articular and quadrate bones in these lineages derives from their use as the jaw hinge, one solution was to use other bones as the jaw joint. The articular and quadrate diminished in size and became adapted for sound conduction.

This process is apparent in some Triassic therapsids such as *Scymnognathus* and *Diarthrognatus*, which shows *a new mammal-like jaw articulation* involving the dentary bone in the lower jaw and the squamosal bone in the upper jaw, *in addition to the ancestral articular–quadrate joint.* In early mammals such as *Morganucodon* and *Eodelphis*, a marsupial, the articular and quadrate had reduced even further, although they still form part of the jaw hinge (Fig. 19-7d). Apparently only in the Jurassic were the quadrate and articular freed from their function in jaw articulation and incorporated into the mammalian middle ear as small ossicles, the incus (anvil) and malleus (hammer), respectively (Fig. 19-7e, f). Supporting this view are embryological studies showing that the quadrate and articular in mammalian fetuses (especially clear in marsupials and monotremes)

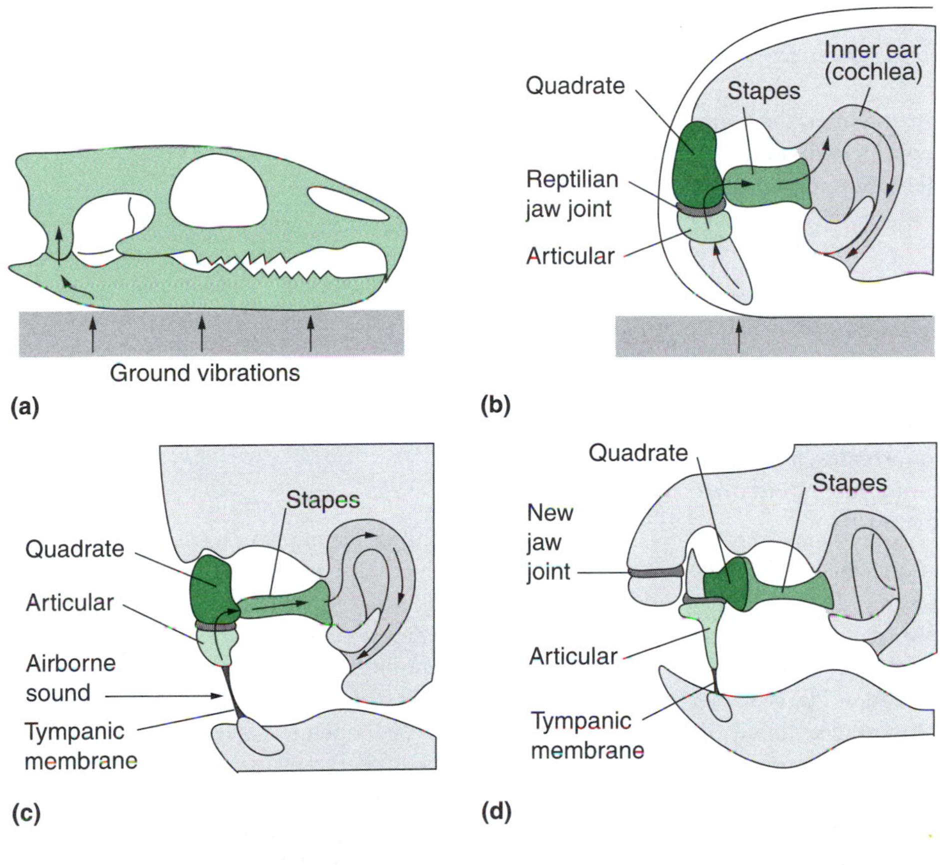

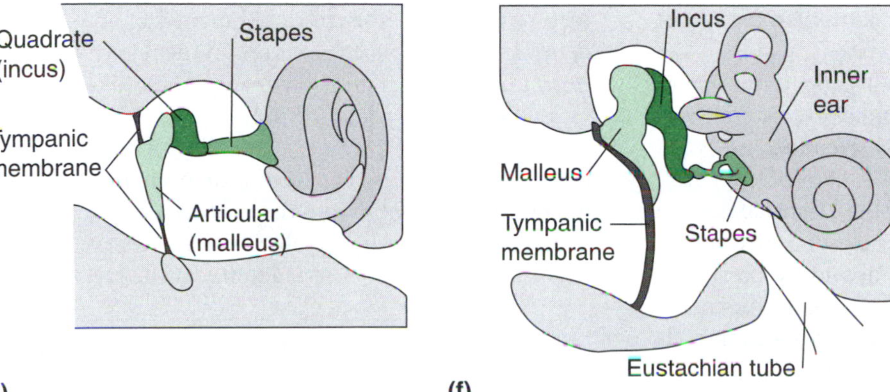

FIGURE 19-7 Proposed stages in the evolution of the ear apparatus, beginning with a land tetrapod that picks up ground vibrations through bone conduction (a, b). In synapsid lineages that lead to the mammal-like therapsids (c), a tympanic membrane picks up airborne sound and transmits it to the articular and quadrate bones of the jaw hinge and into the stapes that connects to the inner ear. As therapsid evolution proceeds, a new mammalian jaw (squamosal-dentary) joint evolves in response to selection for molar chewing abilities. In *Morganucodon*, an early Triassic mammal (d), both jaw joints are present, although the articular–quadrate–stapes bones are diminished in size. In late Triassic mammals (e), the squamosal–dentary joint has become the only jaw hinge, and the articular–quadrate–stapes bones are now entirely involved in hearing. The diagram in (f) presents a more anatomical view of the shape and positioning of these bones in the ear of an extant mammal. (Adapted from Kermack, K. A., and F. Mussett, 1983. The ear in mammal-like reptiles and early mammals. *Acta Palaeontolgica Polonica,* **28,** 147–158.)

first occupy an ancestral position on the side of the jaw and later transform into functional ear ossicles.

Early Mammals

Early mammalian fossils are still scarce, so we still don't know precisely *when and where* therapsids graded into mammals, although new fossil discoveries and molecular phylogenies help us pinpoint the likely ancestor-descendant sequence. Early mammals seem to have been small, about the size of mice or rats. With some exceptions, complete mammalian skeletons that have been discovered date no earlier than the Early Cretaceous although one Late Jurassic skeleton is known. As a result, mammalogists and paleontologists have derived possible evolutionary lineages among earlier Mesozoic mammals almost entirely from fossil teeth and jaws.

Having defined the features of mammals at the beginning of the chapter, we should now note, perhaps even caution, that there are **two competing definitions of the group Mammalia:**

- One is based on a synapomorphy described above, namely, the origin of the dentary-squamosal joint. Under this definition, **morganucodonts** (see below) are the earliest mammals.
- The second definition, which is more restrictive, makes mammals the clade whose first member is the last common ancestor of monotremes, marsupials and placentals. Under this definition, a more derived group would be the earliest mammals.

In this chapter we use the first definition, which is the more long-standing and familiar one (see Kemp, 2007).

The earliest mammalian departure from the therapsid line is found in a geographically widespread group of Late Triassic–Early Jurassic fossils, morganucodontids (**Figs. 19-8** and **19-9**). For the first time, the post canine teeth differentiate into premolars and molars, with only the last premolars showing evidence of tooth replacement; the molars show the mammalian pattern of permanence. Among other distinctive morganucodont traits is precise occlusion between upper and lower jaws, producing a consistent pattern of molar wear facets (Fig. 19-9).

In general, the structure of a morganucodont molar was three cusps aligned along the anterior–posterior axis of the tooth, an arrangement called **triconodont.** Fossil triconodonts with patterns similar to or derived from these appear throughout the remainder of the Mesozoic, and are among the variety of groups traditionally classified as protherians (subclass **Prototheria**). As mentioned previously (see Fig. 5-16), extant prototherian lines are the Australian and New Guinean **monotremes**, egg-laying mammals now represented by the grub- and shrimp-eating platypus

FIGURE 19-8 Proposed skeletal and full-body reconstructions of a Late Triassic–Early Jurassic mammal, the morganucodontid *Megazostrodon*, which was about 10 cm long and weighed approximately 30 g. (Adapted from Crompton, A. W., C. R. Taylor, and J. A. Jagger, 1978. Evolution of homeothermy in mammals. *Nature*, **272**, 333–336.)

(*Ornithorhyncus*) and ant-eating echidnas (*Tachyglossus* and *Zaglossus*). Although their fossil record is meager, some of their features may reflect those of their more numerous Mesozoic prototherian ancestors.

For example, monotremes retain a cloaca (in addition to its egg laying function) as a common chamber for both the rectal and urogenital openings and retain a number of ancestral amniote skull characters including an interclavicle in the pectoral girdle. The absence of teeth in monotremes is a rare specialization that helps make their phylogeny difficult to determine, but that can be explained by an evolutionary history confined either to mud burrowing or ant eating. Persistence of such specializations in the relatively isolated Australian continent probably enabled these Mesozoic relics to survive for such a long period.

Discoveries of a variety of fossil teeth indicate that by the end of the Triassic an elaboration occurred, separating the morganucodontids and their subsequent prototherian lineages from a new group called **therians,** marked by more sophisticated molars.[2] Therian molars, called **tribosphenic**

[2] See Hunter and Jernvall (1995) and Jernvall et al. (1996) for the evolution of mammalian tooth morphology.

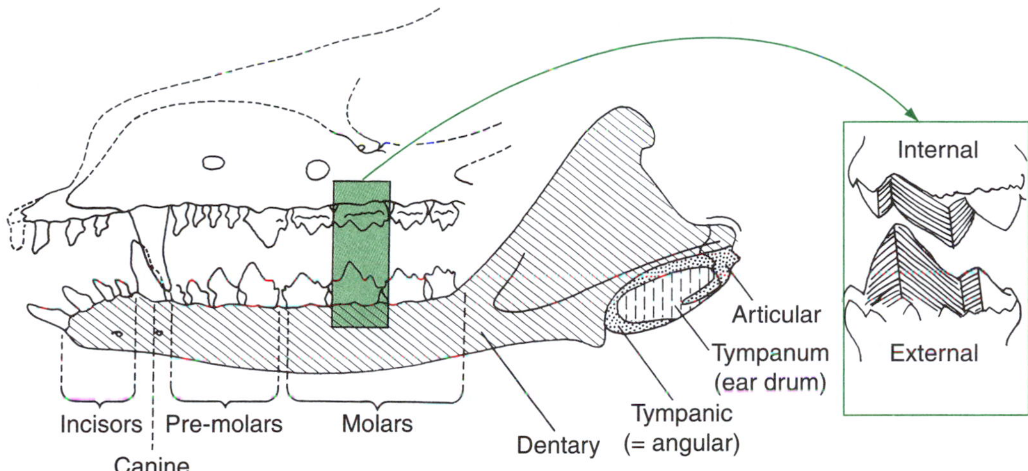

FIGURE 19-9 Lateral view of the 25-mm-long jaws of a late Triassic morganucodontid, *Morganucodon*. The molar teeth occlude more precisely than in any reptile, and, as can be seen in the inset, show matching wear facets between the internal surface of the upper molars and the external surface of the lower molars. To illustrate the internal surface of the upper molar, the tooth is drawn as though it were transparent. (Adapted from Crompton, A. W., and F. A. Jenkins, Jr., 1979. Origin of mammals. In *Mesozoic Mammals: The First Two-Thirds of Mammalian History*. J. A. Lillegraven, Z. Kielan-Jaworowska, and W. A. Clemens (eds.). University of California Press, Berkeley, pp. 59–73.)

or **tritubercular,** differ from prototherian types in having a triangular arrangement of cusps in the upper molars, one of which (the *protocone*) fitted closely into a lower molar basin (*talonid*), much like a pestle into a mortar (**Fig. 19-10a, b**). The crushing action of cusp-to-basin was supplemented by shearing and cutting surfaces that change in number from three to six as therian lineages evolved (Fig. 19-10c–g). These functions were further enhanced by evolution of a narrower lower jaw suspended in a sling of muscle that enabled side-to-side grinding action, a feature present in earlier mammals. These oral-pulverizing mechanisms probably accompanied new dietary opportunities as well as enhanced digestive processing.

A possible forerunner of the tribosphenic molar is found among therapsid contemporaries of the morganucodonts, called *kuehneotherids,* which some regard as stem mammals (Fig. 19-10c). Some paleontologists used this evidence to suggest a diphyletic or polyphyletic origin of mammals from different therapsid stocks; that is, more than one line of mammal-like therapsids — distinguished from other groups by cusped molar teeth — gave rise separately to morganucodonts, kuehnotheriids and perhaps to some other early mammalian lineages. Important braincase similarities between therian and prototherian mammals, however, prompted other paleontologists to adopt a monophyletic position. A firm decision between these views depends on tracing the complexity of early mammalian evolution among the various groups that fall under the broad category of cynodont therapsids. Kemp did that, proposing a *correlated progression* of the evolution of mammalian characters, leading to increased levels of metabolic activity and the ability to regulate body temperature (homeostasis) in response to a savannah-like environment that was seasonally arid. The Mid-Permian geographical continuity between this environment and more temperate regions facilitated the explosive radiation of therapsids that took place at that time.[3]

From discoveries made this century in the Yixian Formation in China, eight new species of the Early Cretaceous mammals are now known. Two species of triconodonts, *Repenomamus giganticus* and *R. robustus,* the largest Cretaceous mammals discovered so far (1 m long/12–14 kg, 0.5 m long/9–12 kg, respectively), were preyed upon by dinosaurs, and, judging from a young dinosaur preserved in the stomach of one specimen, ate young dinosaurs, in marked contrast to our previous image of Early Cretaceous mammals as small nocturnal insectivores (Hu et al., 2005). *Eomaia scansoria* is known from an almost complete skeleton and preserved fur. Tiny (10 cm long, 20–25 g body weight) and arboreal, its features have been interpreted as basal to the lineage leading to placental mammals. *Sinodelphys szalayi,* now the oldest marsupial fossil known, is tree-dwelling, insectivorous and probably an early opossum.

Habitats of Early Mammals

Even with many gaps in the fossil evidence, we can reconstruct some aspects of early mammalian lives and habitats.

Alterations in dentition and mastication in small Mesozoic mammals helped maintain constant body temperatures by promoting rapid food processing. Because of their endothermy, these early mammals may have been pri-

[3] See Kermack and Kermack (1984) for lineages of therapsids and see Kemp (2005, 2006, 2007) for correlated progression.

marily *nocturnal and insectivorous,* functioning in the cool of the evening when ectothermic sauropsid predators were inactive. In support of this nocturnal role, early mammalian brains were three or four times larger than those of even the most advanced therapsids (**Box 19-1**). We can attribute a significant portion of this increase to selection for additional neural connections that provided enhanced auditory (and perhaps also olfactory and visual) acuity associated with a nocturnal habitat.[4] Increased specialization of the mammalian auditory apparatus, partly accomplished by freeing the articular and quadrate bones from the jaw, enabling their transformation into middle-ear ossicles, is perhaps further evidence of selection for sensory ability in a nocturnal, light-diminished habitat. A nocturnal mammalian ancestry is supported by finding that the retinas of many extant mammals, such as insectivores, are rich in rod photoreceptors sensitive to dim light, in contrast to the daylight-adapted retinae of lizards and birds that are almost entirely composed of cone photoreceptors.

Marsupials and Placentals

Dental changes in early mammals, probably prompted by a nocturnal way of life, seem to have continued throughout the Jurassic; **Figure 19-11** illustrates such changes in fossil molars. In the Early Cretaceous, the first tribosphenic molars

[4] See Jerison (1973), Northcutt (2001) and Kielan-Jaworowska et al. (2004).

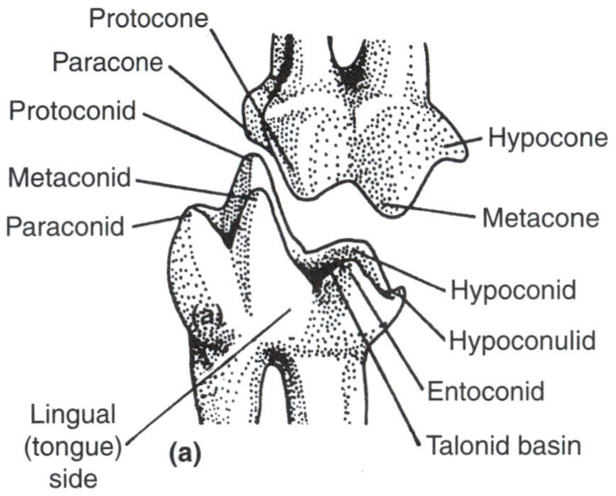

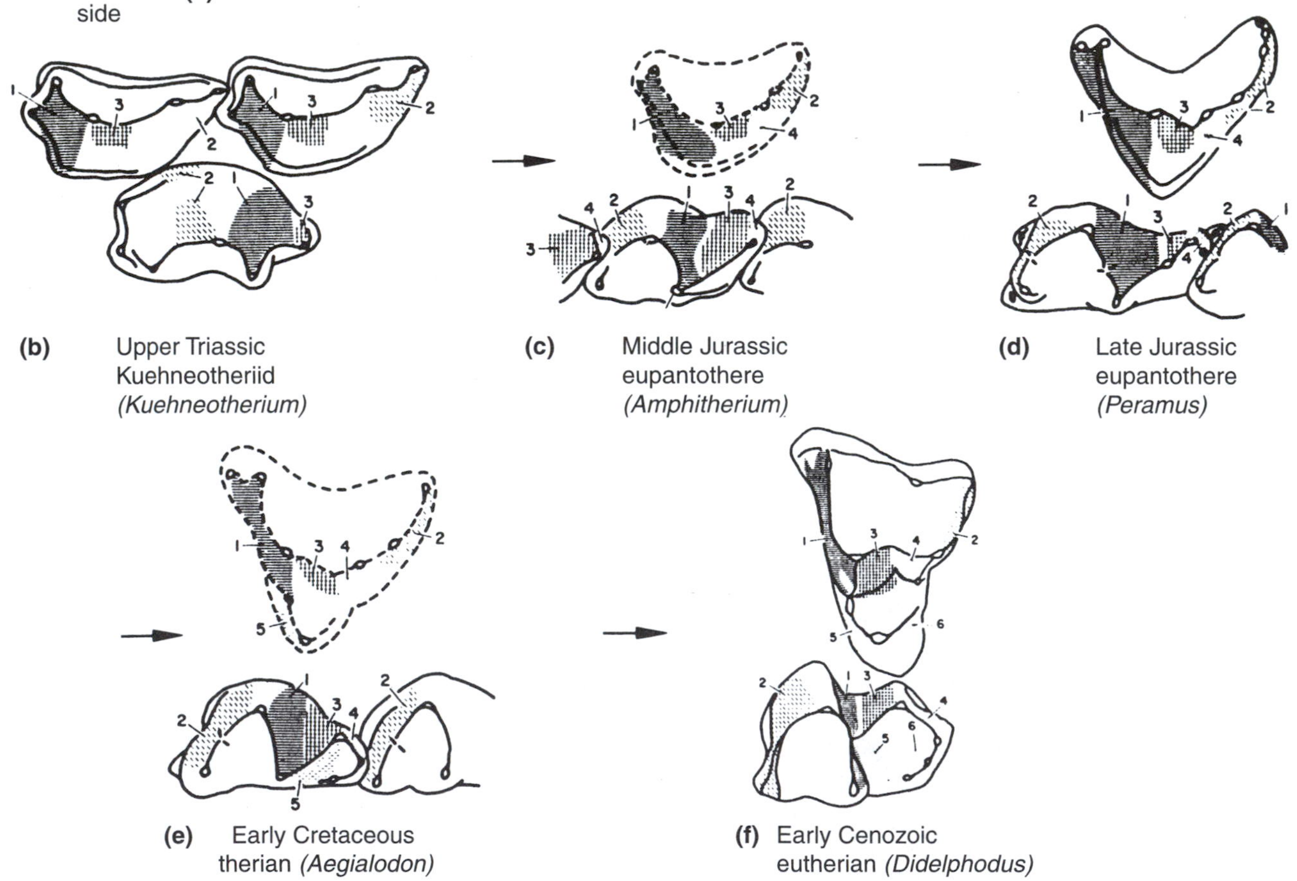

FIGURE 19-10 (a) Generalized upper and lower tribosphenic molars, based on those of an extant therian, the opossum *Didelphis,* oriented with the anterior of the animal to the left. (b–f) Crown (occlusal) views of upper molars (above) and lower molars (below) with matching wear facets (numbers) shaded alike, representing general stages in the evolution of the therian tribosphenic molar (only lower molars are known for fossils c and e).

BOX 19-1 Progress

ALTHOUGH ONE CAN CLAIM that time provides the arrow that orients the direction of evolutionary change and that an evolutionary trend represents **progress** (see Chapter 1), there is more than one direction and there are many lines of progress.

Some may consider the criterion for progress to be increasing morphological complexity exemplified by increasing number of cell types (see Box 8-3, Fig. 8-6), but many evolutionary lineages show no such tendency. Those parasites that lose organs that consume needless energy can certainly replace those who retain such structures. In these cases, *simplification and reduction rather than complexity* often direct parasitic "progress." Even if we restrict the term progress to increased complexity, its measurement is still unresolved, although most biologists would agree that more complex organisms are alive today than were alive 3.5 billion years ago(Bya).[a]

If there is a common evolutionary thread that runs through organisms and their many different lifestyles, it is the opportunism that became embedded in the earliest of their ancestors. We can call the results of this opportunism "progress" but in view of its semantic ambiguities and contradictions, and socially judgmental overtones, it is questionable whether this term or other value judgments help us understand evolution. In Darwin's words, ". . . natural selection, or the survival of the fittest, does not necessarily include progressive development — it only takes advantage of such variations as are beneficial to each creature under its complex relations of life." Biological evolution tracks opportunistic pathways, and is blind to destinations other than survival.

[a] See Valentine et al. (1994), McShea (1996, 2005), Valentine (2003, 2004) and Vickaryous and Hall (2006a) for various approaches to progress.

appear in parallel in two groups, the southern hemisphere australosphenidans (*Ausktribosphons*), which are related to monotremes, and the northern boreosphenidans (*Montanalestes*), which are related to the two major therian lineages that emerged in relative paleontological abundance: marsupials in North America and placentals in North America and in Asia.

In contrast to this paleontological grouping, Janke and coworkers (1997), using nucleotide and amino acid sequencing, argue against combining marsupials and placentals into a therian subclass, proposing instead a monotreme/marsupial sister group that separated from placentals in the Early to Mid Cretaceous, about 130 Mya. According to their molecular clock, the monotreme–marsupial divergence took place only 15 My later at 115 Mya. However, other molecular time scales place marsupial divergence from eutherians even further into the Mesozoic, about 170 Mya (Kumar and Hedges, 1998).

Fossils of marsupials and placentals differ in skull and tooth structure, with marsupials showing:

- a relatively small braincase;
- deciduous teeth reduced to only posterior premolars, which, in turn, often markedly differ from the anterior molars;
- a relatively large number of incisors: eight or more in the complete upper jaw;
- distinctive arrangements of molar cusps and ridges; and
- pelvic epipubic bones in some postcranial Cretaceous skeletons.

Reproduction

A major difference between marsupial and placental mammals lies in their modes of reproduction. Among various prevailing hypotheses, we offer the following scenario for changes in reproductive mode during mammalian evolution.

Early mammals, distinguished by small size, endothermy and heterodont dentition, most likely laid small eggs. Under such circumstances, a significant selective advantage would have accrued to animals that could raise their young past the immature stages of hatching. **Lactation** by mammary glands was apparently one successful solution to the problem. Monotremes are presumed to be a relic of this stage.

Given a system that provided maternal care, protection and nourishment, selection could have taken these animals in the direction of smaller eggs and more rapid development of the fetus before hatching. Endothermy would have facilitated such evolution as hatched offspring could be kept close to the maternal body at a temperature that was optimal for enzymatic activity. At some point, **viviparity** (viviparous reproduction) replaced **oviparity**; it would probably take only a few additional steps for hatching to occur within the oviduct. The embryo could be nourished on maternal fluids within a portion of the oviduct that would eventually become the uterus. Although we have no way of proving this hypothesis, current evidence suggests that mammalian viviparity was restricted to marsupials and placentals, with Hox genes (*HoxA-10* and *HoxA-11*) acquiring new roles enabling the uterus to permit embryo implantation (**Box 19-2**; and see Lynch and Wagner, 2006).

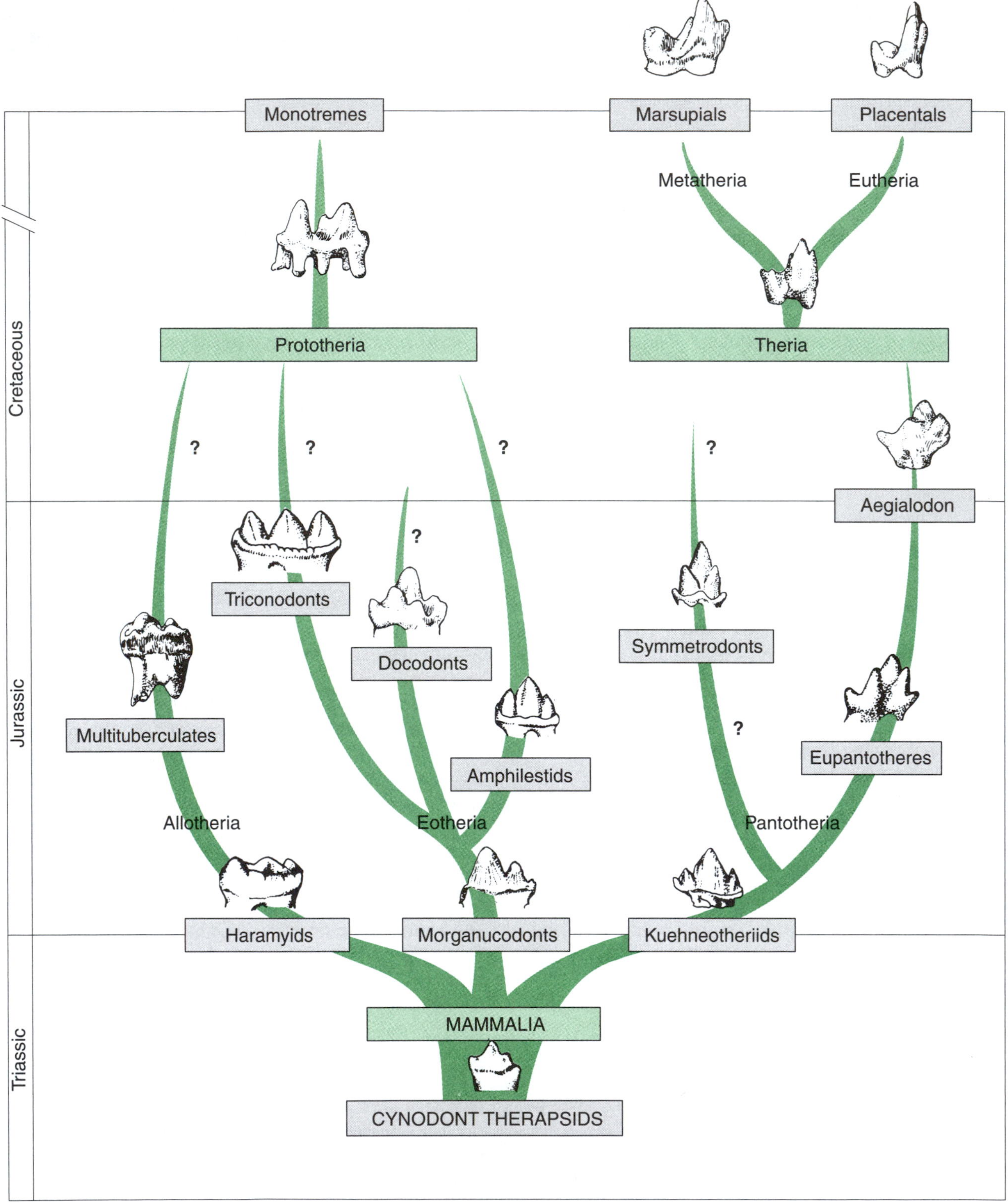

FIGURE 19-11 Relationships of stem mammals and of mammals from the Triassic to the Cenozoic based on evolutionary changes in the molar teeth. A complete triconodont skeleton found in China by Q. Ji and coworkers (1999) indicates that the triconodont tooth structure (and features such as the pectoral girdle) evolved at the end of the Jurassic in different mammalian lines through convergence, indicated here as question marks. (The Prototheria–Theria division is based on that in Carroll, R. L., 1988. *Vertebrate Paleontology and Evolution.* Freeman, New York.)

BOX 19-2 Marsupial and Placental Embryonic Development

IN MARSUPIALS, A THIN, permeable eggshell surrounds the embryo for most of the pregnancy, which ranges between 11 and 38 days depending on the species. During this time, marsupial embryos receive nourishment from both egg nutrients and uterus, but because of the short pregnancy, emerge from the vagina in highly immature form. It may be two to three months after birth before marsupial offspring are capable of independent locomotion.

In placentals, a shelled embryonic stage is no longer discernible and gestation (and so pregnancy) is much longer. Emerging offspring are larger than those of marsupials, far more developed and often capable of independent locomotion shortly after birth. The difference in marsupial–placental gestational periods may derive from differences in maternal immunity response to the fetus: The marsupial fetus is not protected against the maternal immune system, and must therefore abandon the uterus soon after egg hatching, before maternal leukocyte invasion can damage it. In contrast, the trophoblastic membranes surrounding the placental fetus ordinarily prevent exchange between maternal and fetal tissues and act as barriers that help keep the fetus from being immunologically rejected.

Once immunologically protected uterine retention of the fetus could be prolonged by incorporating maternal and fetal membranes into the eutherian placenta. Such a placenta, sustained by various endocrine secretions, nourishes the fetus, provides oxygen requirements for rapid developmental growth and acts as a waste removal system. Compared to any other mode of reproduction, uterine development probably confers greater protection to the embryonic organism during its most vulnerable stages. Also, because mother–child attachments continue past birth through mammary feeding, the stage is set for prolonged family relationships and enhanced learning.

Significantly, profound as placental advantages may have been, they did not eliminate marsupials. Marsupials have at least as long a history as placentals, are prevalent in Australia, include forms such as possums — which compete successfully with placentals in various placental-dominated localities — and number about 270 extant species. One important reason for the persistence of marsupials is their relatively minor reproductive investment: Their birth size is so small that they can abandon offspring soon after birth without great maternal loss. Placentals, in contrast, commit much greater resources to early reproductive stages; a pregnancy often continues in the face of serious maternal sacrifice. Thus, marsupial reproduction can more easily adjust to environmental conditions; reproduction and nursing continue when conditions are advantageous, and marsupials incur little expense in discarding their minuscule newborn offspring when conditions turn poor. Placentals take greater reproductive risks, because their commitment to their offspring is greater (pregnancy being often difficult to interrupt) and involves considerable cost.

The Mesozoic Experience: Emerging Rules of Evolution

As from all evolutionary history, we learn a number of lessons from the evolution of mammals during the Mesozoic. Eight lessons — perhaps even rules — of evolution emerge from the Mesozoic experience.

1. **Dominance and long-term success.** Dominance of a particular group at a particular time is not necessarily a measure of its long-term evolutionary success. This lesson has repeatedly been proven true: various therapsid groups replaced each other, dinosaur groups replaced therapsids and later dinosaur groups replaced earlier ones.
2. **Preadaptation.** The transition from one group to the other offers a second lesson, which is the importance of what was once referred to as **preadaptation.** Not all amniote lines evolved into dinosaurs or into therapsids, nor did all therapsid lines evolve into mammals. Those that did possessed features acquired during their evolutionary histories that, even if used for a different purpose, facilitated their response to the selective pressures that led to the evolution of mammals. For example, dinosaur bipedalism and mammalian endothermy trace, respectively, to the beginning of a bipedal stance in some early archosaurs and to the beginning of endothermy among some therapsids. Although most if not all characters are, or have been, adaptive, the inability of organisms to anticipate future evolutionary needs often makes it a matter of rare chance which of these characters will facilitate further changes. Because the term *preadaptation* has the connotation of a feature anticipating a future need, some evolutionary biologists today use the term *exaptation* or *exapted* introduced by Gould and Vrba in 1982.

3. **Unpredictability.** The third lesson offered is the unpredictability of long-term evolutionary succession. All of the many kinds of biological and environmental changes involved in evolution, although individually understandable in terms of cause and effect, act largely at random when we view them together over long (or even relatively short) periods of evolutionary time. Evolution is tied to historical contingencies. For example, could one have predicted which lines of early mammals would provide descendants that would survive into the Cenozoic 150 My distant from the Triassic, or even last for another 70 My through the Jurassic?
4. **NewModes of Organization.** Out of the hazards that await any particular lineage, another lesson emerges: New modes of biological organization can enhance opportunities for survival. True, not all Triassic lines of mammals survived into the Cenozoic, but some did, and many carried with them adaptations in temperature regulation, reproductive mode, nursing care, sensory perception, brain development, blood circulation, oxygen use, locomotion, dentition and so forth. Undoubtedly because of many or all of these biological innovations, along with their small size, some mammals made safe passage through the Late Cretaceous extinctions that destroyed the dinosaurs (see Box 18-2, Extinctions and Extraterrestrial Impacts).
5. **NewLevels of Organization.** The next lesson derives from the complexity of major biological adaptations (Box 19-1). Evolution of a new level of organization, as seen in mammals, is marked by coordinated changes in many different traits often occurring over a considerable period of time. For example, although we can characterize a mammal by one or another of its specific traits, the trait itself, such as dentition or reproductive mode, results from a number of successive mutations, each of which must coordinate with its entire genetic architecture.

Most (if not all) characters cannot evolve independently to maximize only a single adaptive function, but coevolve with other characters so that the many possible developmental interactions between them do not decrease fitness. The transition from ancestral amniote to placental mammal may well have taken 75 to 100 My because a wide range of **random** mutations had to be integrated into evolving organisms.

Also, the complexity of this integrative process makes it unlikely that many different lines would have continued to undergo the same succession of identical genetic changes. And of course, no two lines are identical. Each line starts from a different position. Different lines evolve similarly, as observed in therapsid–early mammalian transitions, because of parallel or convergent evolution as similar characters evolved through different genetic events. Furthermore, the "sweepstakes" nature of evolution makes it unlikely that even such parallel evolution could have continued in a variety of therapsid lines throughout the Mesozoic, with all such different lines attaining all the same placental mammalian features in the Cretaceous.

6. **Radiation.** Once a new adaptive innovation appears, or adaptive organization reaches a new grade, **opportunities for widespread radiation** can follow. Mammalian endothermy opened a nocturnal niche, just as mammalian dentition opened a dietary niche that few other tetrapods fully exploited. The diversity produced by such radiations is not "passive," but arises from selective acts on the mutational variation with which organisms confront environmental differences.
7. **Coevolutionand Coextinction.** A seventh lesson of the Mesozoic is that the survival or extinction of any group or lineage may be closely connected to the survival or extinction of other groups or lineages. Individuals of a particular group depend on the existence of entire constellations of associated organisms. Replacement of therapsids by dinosaurs was, for example, a mass phenomenon, involving many lineages. Also, as already mentioned and discussed later, a significant cause for mammalian radiation during the Cenozoic was the many new functional roles and habitats the extinction of many nonavian dinosaurs made available to mammals.
8. **Constraint.** Perhaps a final lesson, which like the others is not confined to the Mesozoic, is the high prospect of extinction, of the loss or replacement of lineages because of genetic constraint; a group's genomes cannot adapt to every environmental change it may possibly encounter. Although selection among genotypes can lead to adaptation in one or more directions, it constrains adaptation in others; selection for swimming sacrifices adaptations for running, and so forth. Even when adaptations appear to be "ingeniously provident or fortuitous" — such as those found in some desert animals and plants that can delay development during long unfavorable periods until opportune circumstances arise — extinction still occurs because some stressful physical or biological environmental impacts can only be circumvented by adaptations that would exceed the limit of species tolerance and developmental ability.

To put this last lesson somewhat differently, extinction (death) can result because environmental impacts that threaten organisms can occur more quickly than the adaptational changes necessary for a species to respond successfully. Short-term adaptations do not necessarily confer long-term advantage. Extinction is caused both by "bad luck" and by "bad genes."

The Cenozoic: The Age of Mammals and the Northern Continents

Although the Mesozoic presents a spectacular display of mammalian evolution, the full flowering of mammalian radiation was in the Cenozoic, which we discuss as an example of diversification and radiation in response to changing environments.

Extinction of the nonavian dinosaurs removed many mammalian Mesozoic constraints: Mammals invaded herbivorous and carnivorous niches formerly closed to them and became active **diurnally** as well as nocturnally. New mammalian lifestyles began to appear during the Paleocene, but observable morphological differences accumulated slowly.

In addition to dinosaur extinction opening new adaptive niches, a significant stimulus for mammalian radiation was the breakup of the large Pangaea landmass that began in the Mesozoic, and the movements of tectonic plates that continued throughout the Cenozoic, establishing new continents with their varying connections and separations (see Fig. 5-12), and dispersing and isolated major mammalian groups (see Fig. 5-17).

To these land movements with their marked effects on climate, environment and regionalization, we can add the uplifting of mountain systems that took place from the Cretaceous onward leading to chains such as the Rockies, Andes, Alps and Himalaya; the submersions and regressions of shallow seas; and the delineations of new shorelines. Changes in vegetation, especially the diversification and radiation of angiosperms, took place during these periods leading to new landscapes of grasslands, savannas and

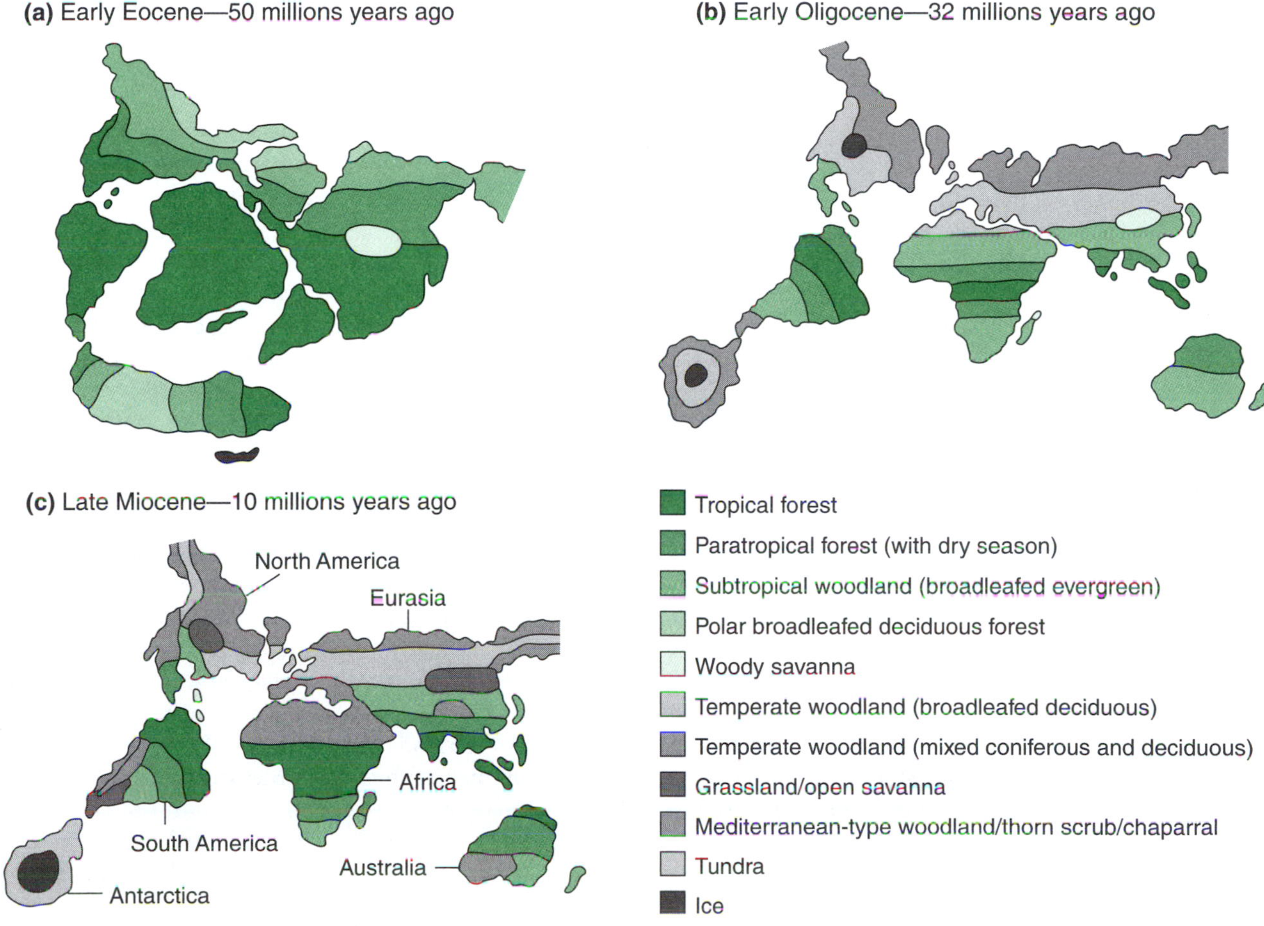

FIGURE 19-12 Positions of continental landmasses at three stages during the Cenozoic, showing the distribution of different kinds of vegetation during the (a) Early Eocene, (b) Early Oligocene, and (c) Late Miocene. (Adapted from Janis, C. M., 1993. Tertiary mammal evolution in the context of changing climates, vegetation, and tectonic events. *Ann. Rev. Ecol. Syst.*, **24**, 467–500.)

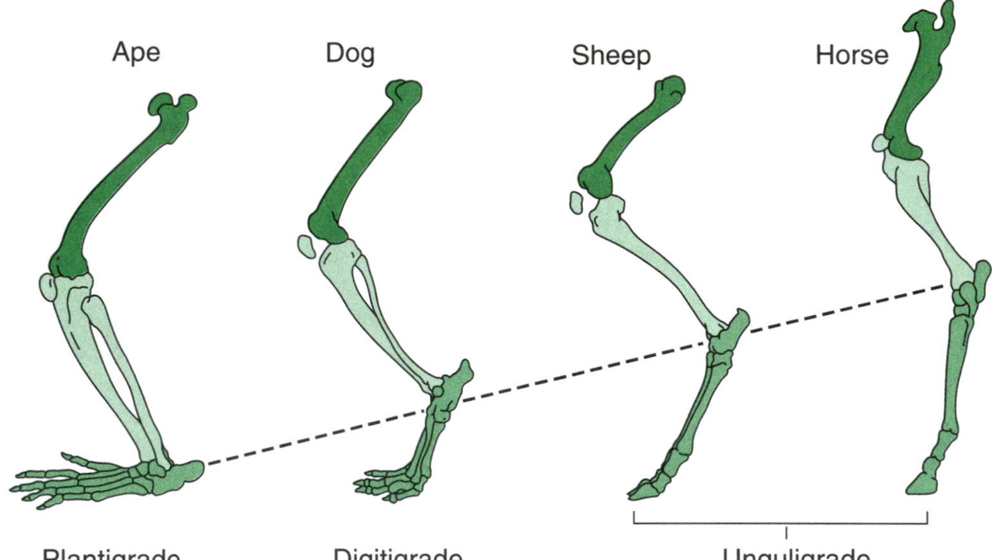

FIGURE 19-13 A series of modifications in the mammalian hind limb associated with increased running speeds (*from left to right*). These changes include positioning of the foot to run on the toe tips, lengthening of the foot bones (*indicated by height of the diagonal line from the ground*), and fusion of the tibia and fibula (also the forelimb radius and ulna) to prevent rotation of the foot during running. (Adapted from Savage, R. J. G., and M. R. Long, 1986. *Mammalian Evolution*. British Museum, London.

forests (**Fig. 19-12**). These novelties and modifications shaped new and different habitats, affecting mammalian adaptation, variation and distribution.

By the Middle and Late Paleocene, evidence for enormous amounts of evolutionary change appears in a few major centers, especially in North America, but also in Europe and Asia. In quantitative terms mammalian radiation was extraordinary: from about 21 diverse mammalian families at the end of the Cretaceous, 37 by the Early Paleocene, and 86 by the Late Paleocene (Benton, 1997). A greater partitioning of the environment also took place: North American fossil faunas that contained about 20 to 30 mammalian species in the Late Cretaceous had 50 to 60 species by the Middle Paleocene.

If we continue onward to the Middle and Late Eocene, 20 to 30 My after the Cretaceous, mammalian skeletal adaptations for creeping, running, digging, swimming, flying and climbing had morphologically differentiated many major fossil groups, from early whales to bats (**Figure 19-13**, and see Polly, 2007). For example, these and further changes led to modifications in the lower limbs of some terrestrially mobile animals from a flat-footed stance to running on the digits or on the tips of the toes A reduction in the number of digits and lengthening of the limbs and foot bones (**Fig. 19-14**) accompanied increased speed in groups such as horses and other hoofed animals.

Brain size, a mark of the ability to integrate sensory and motor information, increased relative to body size in both mammalian prey and predators as part of a continuous competition in which predators were selected for greater skill in capturing prey and prey for greater skill in avoiding predators (**Fig. 19-15**).

The relative brain size estimate in Figure 19-15 is given as the encephalization quotient (EQ), calculated as the ratio of actual brain weight to the brain weight expected for an animal of the same body size. (The expected brain weight for a broad sample of mammals, according to Jerison, 1973, is $0.12 \times$ body weight in grams$^{.67}$).[5] Although there is no exact correlation between brain size and intelligence, a large difference in EQ probably denotes a significant difference in mental capacity. An animal with an EQ of 0.5 possesses a brain that is half the size of an "average" extant mammal of that body weight, most likely indicating fewer intellectual powers and less complex behavior. By contrast, an EQ of 2.0 would signify twice the expected brain size and probably greater than expected mental ability.

Prey often invest more resources in defense than do predators in offense. As Dawkins (1986) pointed out, the struggle involves different needs "the rabbit runs faster than the fox, because the rabbit is running for his life, while the fox is only running for his dinner." It is to the advantage of rabbits to concentrate major resources on speed and evasion, whereas it is to the advantage of foxes to also look for different prey and use strategies other than running.

Such evolutionary changes as well as many others led to the replacement of most of the Mesozoic and Early Cenozoic mammalian forms, a process that continued throughout the Tertiary. By the end of the Pliocene, about 2 Mya, many mammalian lines had evolved, such as horses, deer, bats, insectivores, elephants, rodents, carnivores and early primates (**Fig. 19-16**). Two examples of morphological evolution are illustrated in **Figures 19-17** and **19-18.**

[5] Encephalization quotients for hominins are given in Table 20-6.

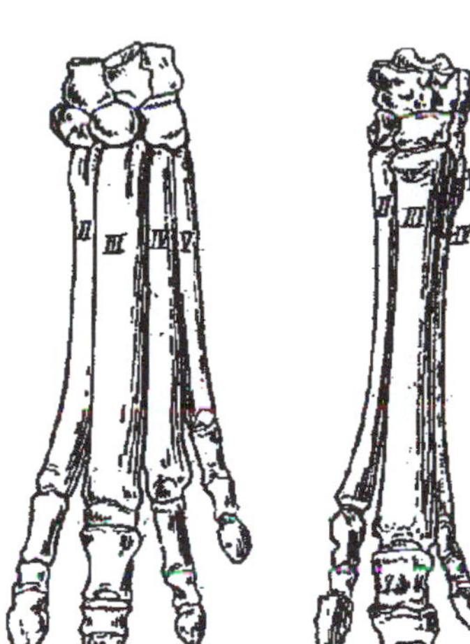

FIGURE 19-14 Reduction of the toes from four to one in the forelimbs in horses from the Eocene "dawn horse' eohippus (*Hyracotherium,* left) to the modern horse genus *Equus* (right), which appeared in the Pleistocene and has persisted to today. Digit III is retained in all (see also Fig. 19-13). Digits I and V (the outer digits) were lost as early as *Miohippus* (second from left). Digits II and IV are reduced to splint bones. See also Figures 3-19 and 3-20. (Modified from Gregory, W. K. 1951. *Evolution Emerging. A Survey of Changing Patterns from Primeval Life to Man.* Two Volumes. The Macmillan Company, New York.)

Notable as these morphological distinctions are, discerning their evolutionary ties and divergences has been difficult, although more accessible now with increasing molecular information. The results of an analysis of mitochondrial and nuclear gene sequences in a large number of mammals are shown in **Figure 19-19.** Their findings support some previously mentioned relationships such as between whales and artiodactyls, but also propose a *polyphyletic origin of Insectivora,* separating golden moles and tenerecs from moles and shrews. Some common insectivoran morphologies are thus ascribed to convergence (homoplasy) and not to homology.[6]

Glaciation

Mammalian diversity continued through the Pleistocene, an epoch that marked the appearance of many mammals in what we consider their current forms. Climatically, the Pleistocene also marks a period of at least seven glaciations, called the *Ice Ages,* which at times covered one third of Earth's surface.

Woolly mammoths and woolly rhinoceroses made their appearances in the northern continents during this interval, along with giant deer, giant cattle and large cave bears.

[6] See De Jong (1998) and Stanhope et al. (1998) for the genetic analysis, Bejder and Hall (2002) for an overview of the relationship of whales to artiodactyls, and see Figure 3-12 for hind limb loss in whales.

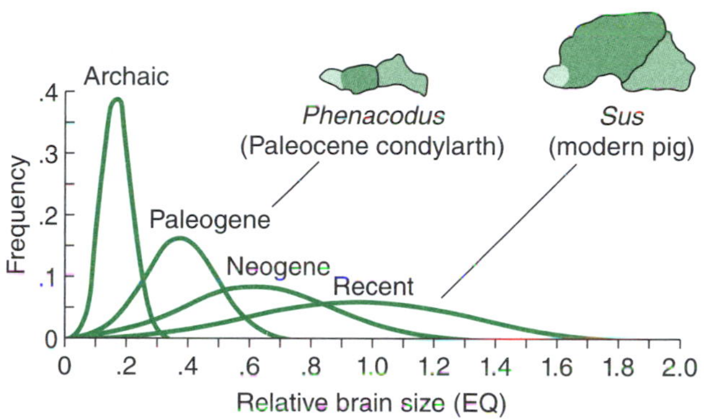

(a) Ungulates (prey)

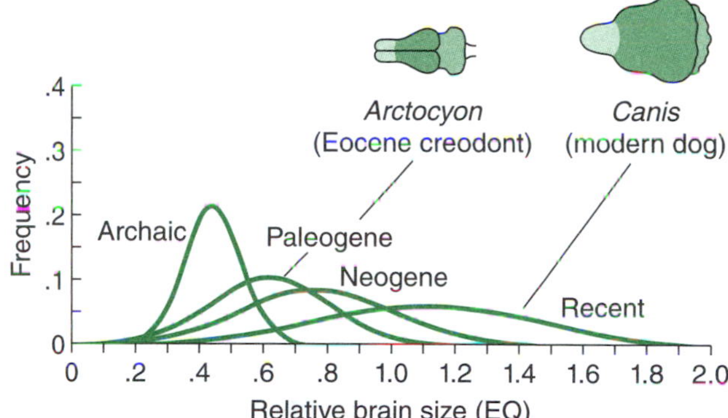

(b) Carnivores (predators)

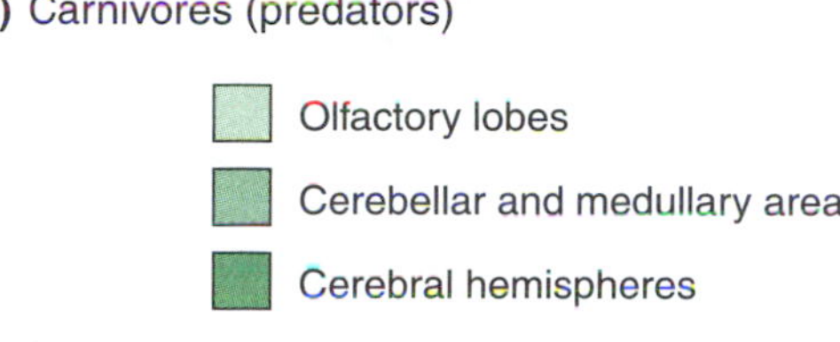

FIGURE 19-15 Normal distributions curves of relative brain sizes (encephalization quotient – EQ) for mammalian ungulates (a) and carnivores (b) during the Cenozoic showing brain outlines of some similar body-sized species. (Cerebral hemispheres are indicated by *dark shading,* olfactory lobes are *light,* and cerebellar and medullary areas are *medium.*) See text for details. (EQ distributions from Jerison, H. J., 1973. *Evolution of the Brain and Intelligence.* Academic Press, New York.)

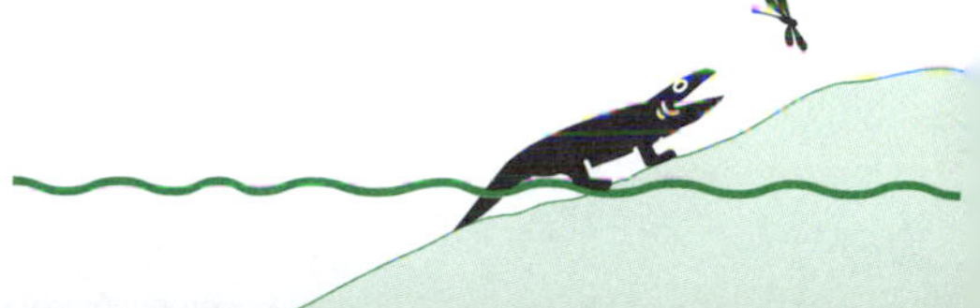

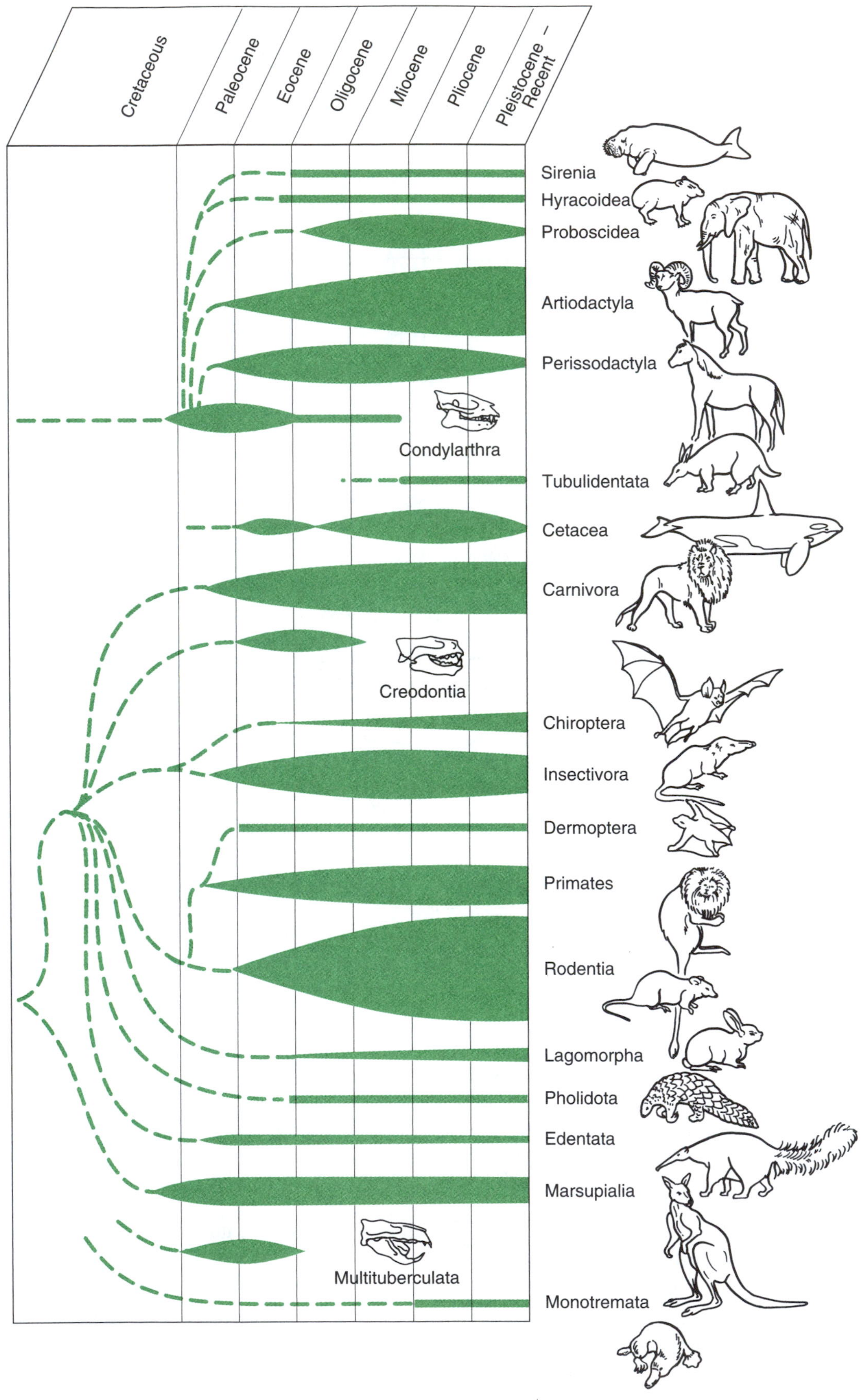
Cretaceous
Paleocene
Eocene
Oligocene
Miocene
Pliocene
Pleistocene – Recent
Sirenia
Hyracoidea
Proboscidea
Artiodactyla
Perissodactyla
Condylarthra
Tubulidentata
Cetacea
Carnivora
Creodontia
Chiroptera
Insectivora
Dermoptera
Primates
Rodentia
Lagomorpha
Pholidota
Edentata
Marsupialia
Multituberculata
Monotremata

FIGURE 19-16 Radiation patterns of mammalian clades beginning with the Cretaceous, including three extinct groups (multituberculates, condylarths and creodonts). Widths of *shaded bars* indicate rough estimates of relative fossil abundances at various times. Exact phylogenetic relationships among many of these mammalian orders (*dotted lines*) are still not resolved to the satisfaction of all paleontologists (Benton, 1988; Novacek et al., 1988), and molecular phylogenies, such as illustrated in Figure 19-19, show unexpected relationships, such as between elephants and some insectivores. Estimates of dates of divergence also differ between the fossil record and molecular studies: paleontological findings point to major mammalian radiations at or near the 65-million-year-old Cretaceous–Tertiary boundary, whereas some molecular divergence times are more than double (Bromham et al., 1998, Foote et al., 1999). Such differences are yet to be resolved. (Data are from Gingerich, P. D., 1998. Vertebrates and evolution. *Evolution,* **52,** 289–291.)

Interestingly, these large mammals, in addition to horses, camels, ground sloths and various other groups, all became extinct in North America about 11,000 years ago. Among approximately 79 mammalian species weighing more than 45 kg (100 lbs), 57 (72%) became extinct at that time. Limited extinction occurred in Europe.

Among possible explanations for these Late Pleistocene events is the hypothesis that climatic advantages for large animals deteriorated rapidly as the ice sheets retreated, causing their extinction. Another explanation is the predatory role of humans: stone-age hunters who entered formerly glaciated areas of North America and Europe slaughtered ("over killed") these large mammals. Whether or not some of our ancestors played this role, our present role as agents of extinction has unfortunately grown from incidental to flagrant (see Chapter 25).

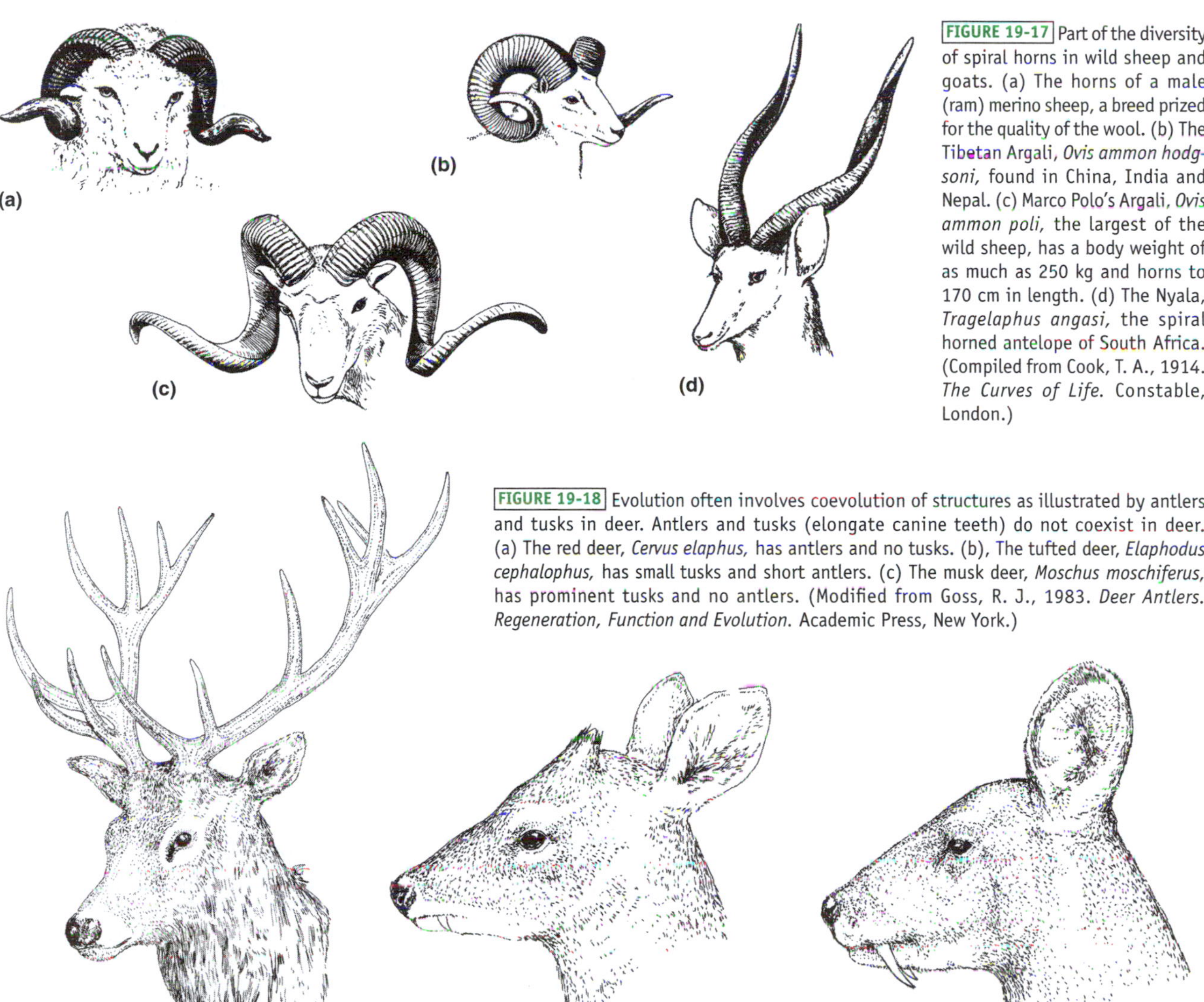

FIGURE 19-17 Part of the diversity of spiral horns in wild sheep and goats. (a) The horns of a male (ram) merino sheep, a breed prized for the quality of the wool. (b) The Tibetan Argali, *Ovis ammon hodgsoni,* found in China, India and Nepal. (c) Marco Polo's Argali, *Ovis ammon poli,* the largest of the wild sheep, has a body weight of as much as 250 kg and horns to 170 cm in length. (d) The Nyala, *Tragelaphus angasi,* the spiral horned antelope of South Africa. (Compiled from Cook, T. A., 1914. *The Curves of Life.* Constable, London.)

FIGURE 19-18 Evolution often involves coevolution of structures as illustrated by antlers and tusks in deer. Antlers and tusks (elongate canine teeth) do not coexist in deer. (a) The red deer, *Cervus elaphus,* has antlers and no tusks. (b), The tufted deer, *Elaphodus cephalophus,* has small tusks and short antlers. (c) The musk deer, *Moschus moschiferus,* has prominent tusks and no antlers. (Modified from Goss, R. J., 1983. *Deer Antlers. Regeneration, Function and Evolution.* Academic Press, New York.)

Two Island Continents: Australia and South America

Many new adaptive radiations occurred among placentals, although marsupials on two continents, Australia and South America, experienced significant evolutionary changes (see also discussion in Chapter 3).

Some reasons for the marsupial radiation derive from the isolation of the two southern continents landmasses because of continental drift during the Late Cretaceous and Early Cenozoic (see Chapter 5). Although paleontologists have offered various scenarios, most are consistent with marsupials originating in North America during the mid-Cretaceous, and, along with a couple of early placental groups, migrating down an arc of Central American islands into South America before the end of that period.

From South America, marsupials dispersed into an Antarctican continent that was considerably warmer than at present and, unaccompanied by placentals, reached Australia during the Early Eocene, about 50 Mya. By mid-Eocene, perhaps five My later, Australia separated from Antarctica and began its northern journey toward Asia, carrying along its isolated marsupial population. A colony of North American marsupials also reached Europe during the Early Eocene, while short-lived lineages made their way to Asia and Africa, probably through a North Atlantic–

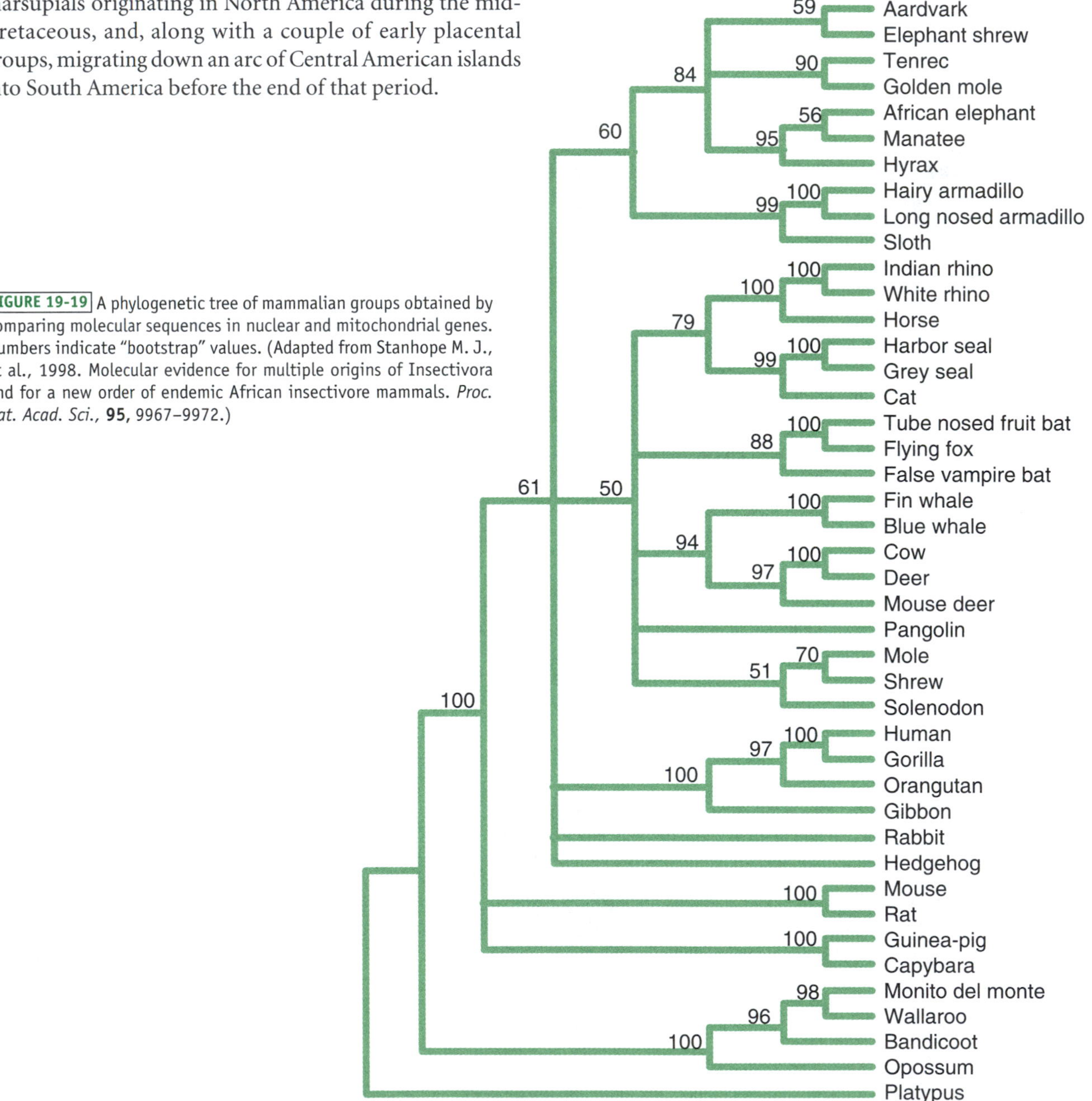

FIGURE 19-19 A phylogenetic tree of mammalian groups obtained by comparing molecular sequences in nuclear and mitochondrial genes. Numbers indicate "bootstrap" values. (Adapted from Stanhope M. J., et al., 1998. Molecular evidence for multiple origins of Insectivora and for a new order of endemic African insectivore mammals. *Proc. Nat. Acad. Sci.*, **95**, 9967–9972.)

Greenland–Europe connection. During the Miocene, marsupial populations of both northern continents became extinct. Only when Pliocene events re-established a North American–South American land bridge did marsupials reinvade North America from the south. New Zealand apparently separated from the Gondwana continents even earlier, probably during the Cretaceous. Only native marsupial mammals and bats had been found there until recently when a mouse-size mammal was discovered in Miocene deposits in New Zealand, the age of which is consistent with this mammal having been present before the divergence of marsupial from placental mammals (Worthy et al., 2006).

Paleontologists ascribe the success and diversity of herbivorous and carnivorous marsupials during their Australian radiation (see Fig. 3-4) to the absence of placental rivals. As a result, by mid-Miocene times, at least 15 families of marsupials existed. Many of these were browsers that probably fed on temperate rain forest vegetation, along with at least two groups that were carnivorous.

Monotremes, the only other mammalian clade present in Australia, were probably too specialized to offer much competition. Fossil evidence for monotremes is poor, although Pleistocene deposits show monotremes (both platypus and echidna genera) in Australia at that time, and paleochronologists have given a middle Miocene date to teeth that may have belonged to a platypus-type animal. A few Cretaceous platypus-like fragments have also been reported, indicating monotremes probably reached or arose in Australia by the Late Jurassic–Early Cretaceous.

By the Late Miocene, drier conditions led to an expansion of grasslands, followed by the evolution of many different kinds of grazing kangaroos, including one Pleistocene species whose adults were 3 m high. The invasion of Australia by humans, both during the Pleistocene and more recently, was accompanied by other placentals, including dogs, rabbits, sheep and rodents. Given their vulnerability to placental competition, many extant Australian marsupial groups will probably not survive without protection.

In South America, marsupial radiation followed a different pattern as placental herbivores and edentates channeled marsupials into carnivorous and insectivorous niches. These animals ranged from many species of opossum-like didelphids to jumping, gnawing and dog-like forms. Paleobiological evidence suggests that some of the latter, members of the borhyaenid family, are ancestral to *Thylacosmilus,* the marsupial saber-toothed "tiger" of the South American Pliocene (**Fig. 19-20a**), a carnivore strikingly similar to the large placental saber-toothed cat, *Smilodon,* of the North American Pleistocene (Fig. 19-20b). The borhyaenids also show marked similarities to the Australian marsupial family of thylacines that included the Tasmanian wolf (**Fig. 19-21**). Evolutionary *convergence or parallelism,* caused by response to selection by divergent groups of organisms, produced similar structures in the genetically different placentals and marsupials even on different continents.

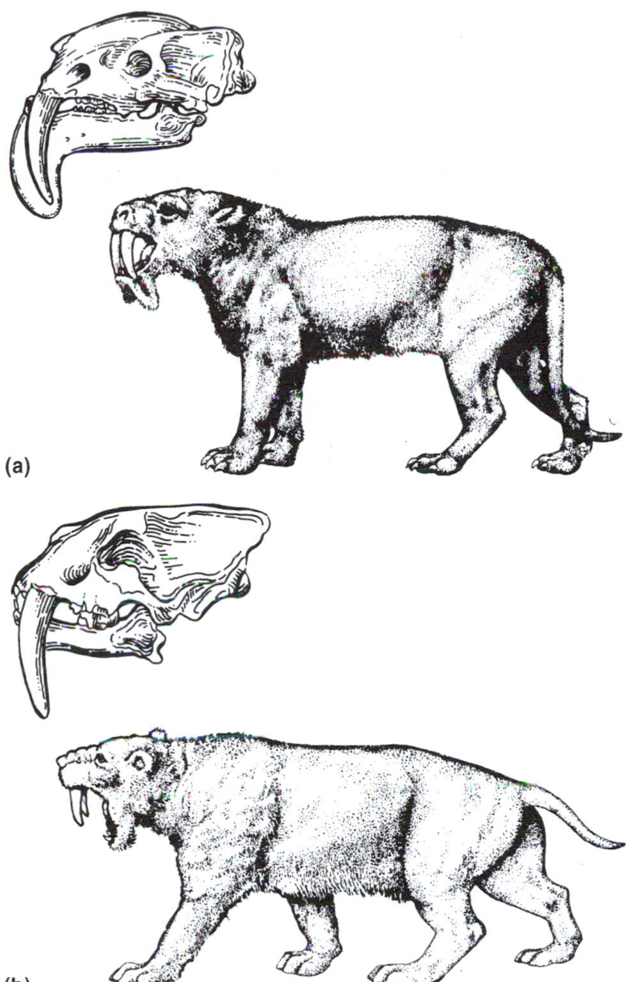

FIGURE 19-20 Reconstructions and skulls of (a) *Thylacosmilus,* a saber-toothed marsupial carnivore from Pliocene deposits in South America; and (b) *Smilodon,* a saber-toothed placental carnivore found in late Pleistocene deposits in North America.

The South American placentals, although beginning only with some ungulates and xenarthrans ("strange-jointed"), radiated perhaps even more rapidly than did marsupials on that isolated continent. By the Early Eocene, within 15 to 20 My of their initial Late Cretaceous colonization, placentals comprised 75 to 100 new genera, which we divide into about 15 families. The xenarthrans (also called **edentates** because of their reduced or suppressed dentition) produced a strange bestiary of armadillos, glyptodonts, sloths and anteaters (**Fig. 19-22a**). Also radiating widely were the mostly hoofed ungulates, which paleontologists

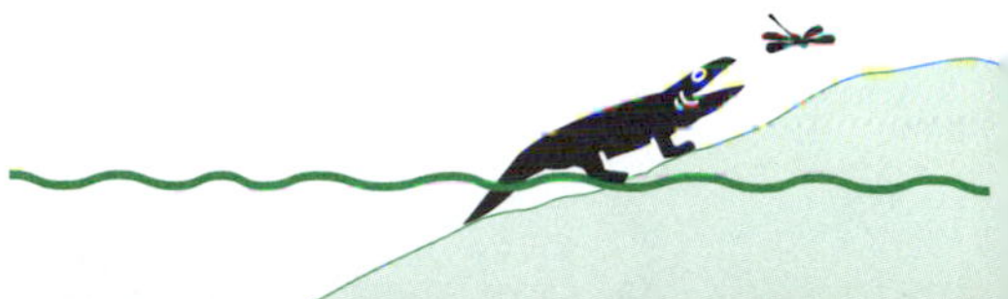

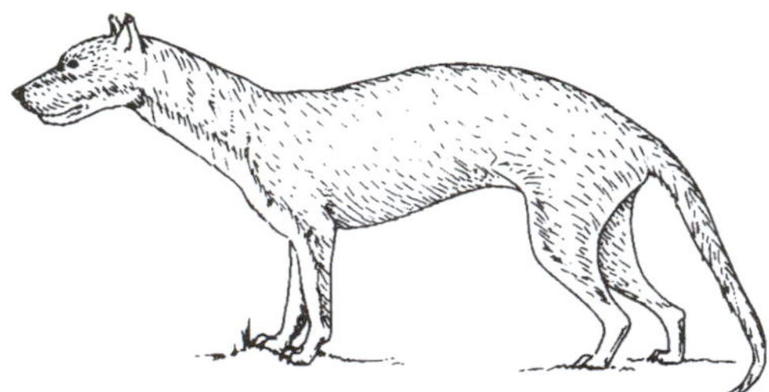

(a) Borhyaenid marsupial (Miocene, Argentina)

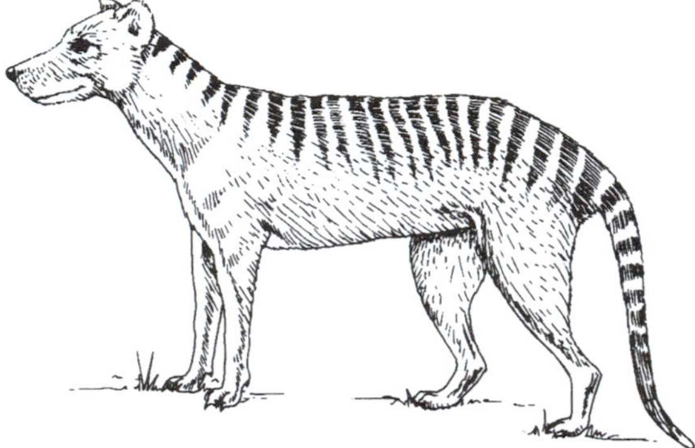

(b) Marsupial Tasmanian wolf (Tasmania, Australia)

(c) Placental wolf (North America)

FIGURE 19-21 (a) *Prothylacynus patagonicus,* a borhyaenid marsupial from the early Miocene in southern Argentina. (b) *Thylacinus cynocephalus,* the recently extinct marsupial Tasmanian wolf. (c) *Canis lupus,* the placental North American wolf. (From Marshall, L. Q., 1980. Marsupial paleogeography. In L. L. Jacob (ed.). *Aspects of Vertebrate History,* Museum of Northern Arizona Press, Flagstaff, pp. 345–386.)

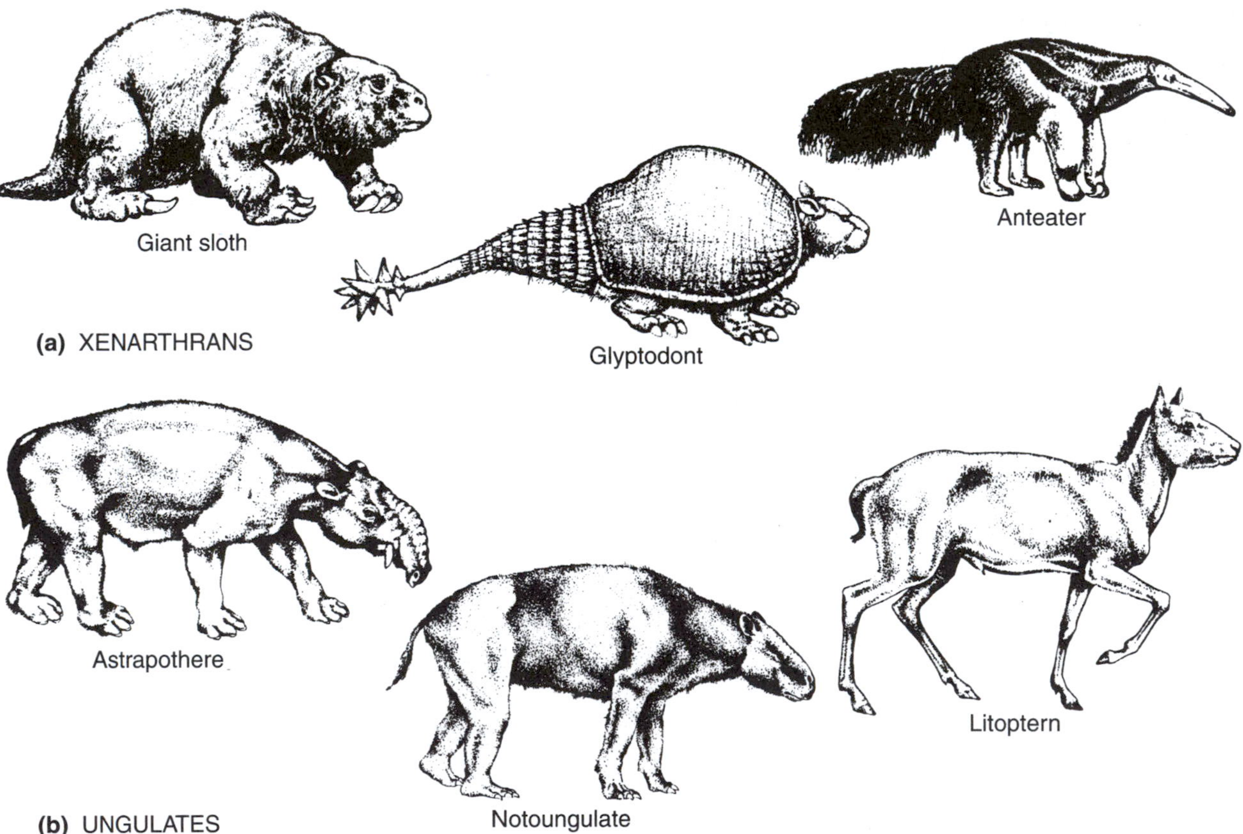

FIGURE 19-22 Reconstructions of some of the (a) xenarthrans (edentates) and (b) ungulates produced by the South American placental mammalian radiation. Flynn and Wyss (1998) point out that the marsupial immigration from South America to Australia led us to expect an Australian presence of other South American groups such as xenarthrans and ungulates. Their absence in Australia is unexplained. (Adapted from Steel, R., and A. P. Harvey, 1979. *The Encyclopaedia of Prehistoric Life.* Mitchell-Beazley, London.)

consider to have originated from an ancestral **condylarth** (Fig. 19-22b). Again, convergent or parallel evolution produced striking similarities: by the Early Miocene some South American litopterns (see Fig. 19-22b), apparently selected for grazing and rapid running, were remarkably similar to the one-toed horses that first developed about 20 My later in North America.

In the Oligocene, a similar rapid radiation began among the rodents and primates that had reached South America from Africa, probably by "island hopping" along the island chains on the oceanic ridges cast up in the South Atlantic Ocean. Rodents produced a great diversity of caviomorphs (cavies) distinguished by special jaw muscle attachments. Primates, confined mostly to tropical areas, produced the wide array of New World monkeys in the superfamily Ceboidea (see Hartwig, 2002 for the primate fossil record).

When South America was next united with North America, 30 to 35 My later in the Pliocene, an extensive interchange between the mammals of these two continents followed. Many South American groups, including marsupial carnivores and many placental ungulates, became extinct, at least partially because of competition with more derived North American placentals. Some of the successful invading North American groups also diversified quite rapidly. As elsewhere, extinction and radiation in South America seemed to go hand in hand, testifying again to the basic opportunism of evolutionary change.

KEY TERMS

bony secondary palate
condylarth
cynodont therapsids
deciduous teeth
dentary bone
diaphragm
diurnally
double occipital condyle
edentates
encephalization quotient (EQ)
endothermy
four-chambered heart
heterodont
lactation
live birth
mammary glands
mandible
marsupials
middle ear ossicles
monotremes
morganucodontids
neocortex
nocturnal
oviparity
placentals
Prototheria
quadrate-squamosal jaw joint
synapsids
triconodont
tympanic membrane
uterus
viviparity

DISCUSSION QUESTIONS

1. What major features distinguish mammals from other vertebrates?
2. Why do paleontologists consider mammals to have had an ancestry among the therapsid reptiles?
3. How did selection for endothermy and continuous metabolic activity affect the mammalian palate? Mammalian dentition?
4. What evolutionary stages can account for the transformation of the posterior elements of the reptilian jaw (articular and quadrate bones) into the mammalian ear ossicles?
5. How have biologists used changes in fossil teeth in constructing hypotheses about early mammalian evolution?
6. What lifestyle have paleontologists proposed for Mesozoic mammals, and what evidence supports this view?
7. How can we explain the evolution of major differences in reproductive modes among monotremes, marsupials and placentals?
8. In terms of the directions taken by evolution (such as radiation patterns, long-term predictability, new levels of organization, and group interactions and replacements), what are some lessons that we can learn from reptilian and mammalian evolution during the Mesozoic?
9. How can we explain the evolution and distribution of major groups of both fossil and extant mammals in South America and Australia?

EVOLUTION ON THE WEB

Explore evolution on the Internet! Visit the accompanying Web site for *Strickberger's Evolution,* Fourth Edition, at http://www.biology.jbpub.com/book/evolution for Web exercises and links relating to topics covered in this chapter.

RECOMMENDED READING

Archer, M., and G. Clayton (eds.), 1984. *Vertebrate Zoogeography and Evolution in Australia.* Hesperian Press, Carlisle, Australia.

Benton, M. J., 2004. *Vertebrate Palaeontology.* 3rd ed. Blackwell Publishers, Oxford, England.

Carroll, R. L., 1988. *Vertebrate Paleontology and Evolution.* Freeman, New York.

Griffiths, M., 1978. *The Biology of the Monotremes.* Academic Press, New York.

Hartwig, W. C. (ed.), 2002. *The Primate Fossil Record.* Cambridge University Press, Cambridge, England.

Jones, S., R. D. Martin, and D. R. Pilbeam, (eds), 1996. *The Cambridge Encyclopedia of Human Evolution.* Cambridge University Press, Cambridge, England.

Milner, R., 1990. *The Encyclopedia of Evolution. Humanity's Search for Its Origins.* Facts On File, New York.

Kemp, T. S., 2005. *The Origin and Evolution of Mammals.* Oxford University Press, Oxford and New York.

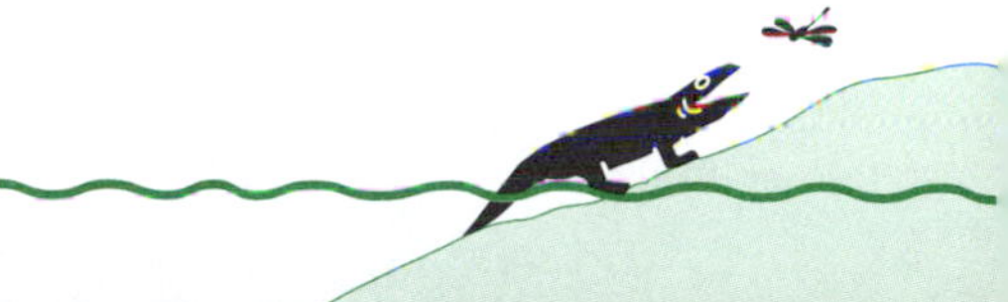

Kielan-Jaworowska, Z., R. L. Cifelli, and Z-.X. Luo, 2004. *Mammals from the Age of Dinosaurs: Origins, Evolution, and Structure.* Columbia University Press, New York.

Novacek, M. J., A. R. Wyss, and M. C. McKenna, 1988. The major groups of eutherian mammals. In *The Phylogeny and Classification of the Tetrapods,* vol. 2, M. J. Benton (ed.). Oxford University Press, Oxford, England, pp. 31–71.

Valentine, J. W., 2004. *On the Origin of Phyla.* The University of Chicago Press, Chicago.

Vickaryous, M., and B. K. Hall, 2006a. Human cell type diversity, evolution development classification with special reference to cells derived from the neural crest. *Biol. Rev. Camb. Philos. Soc.*, **81**, 425–455.

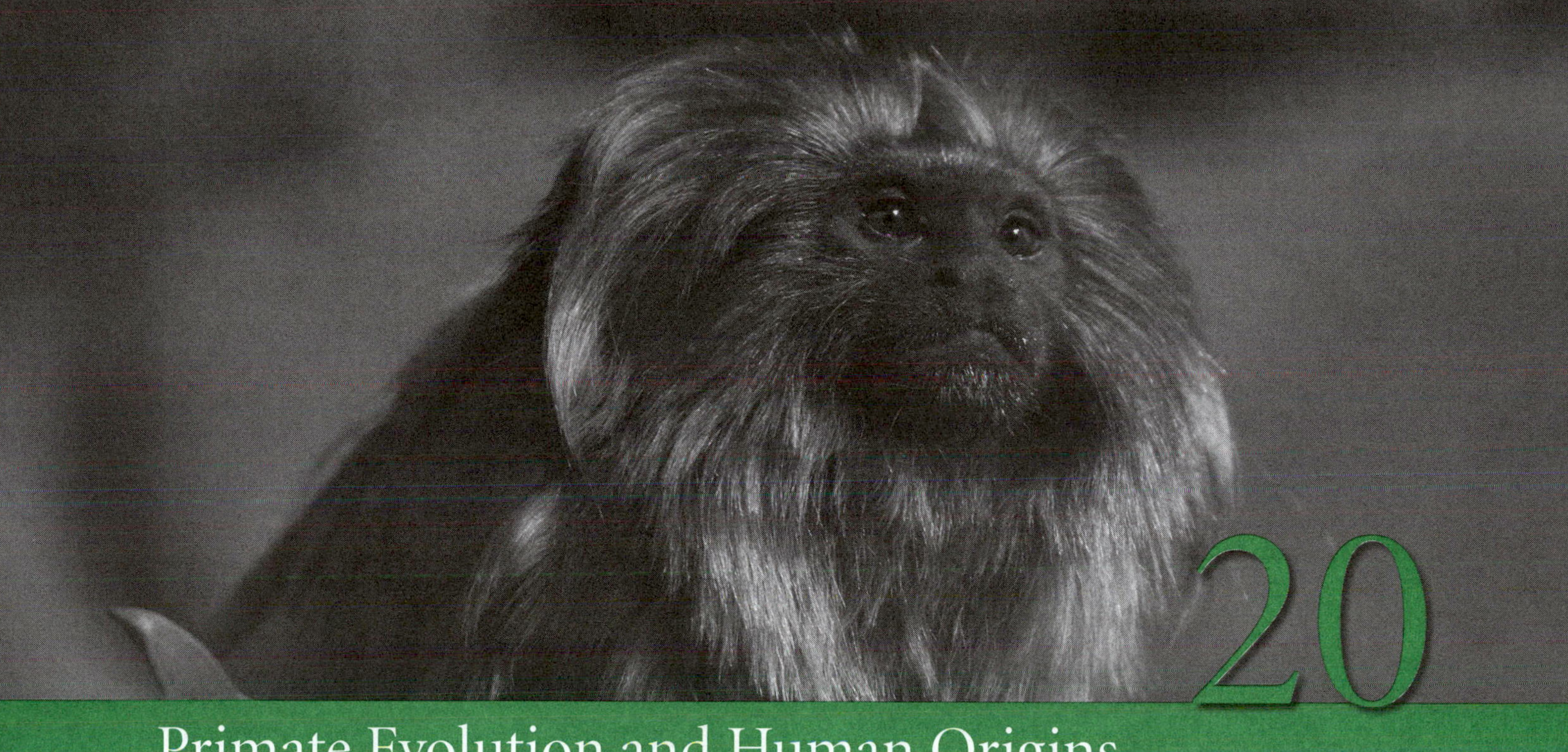

20 Primate Evolution and Human Origins

Chapter Summary

Primates share a number of features indicating an arboreal past as well as more recent adaptations such as large brain size. Primates raise their few offspring during an extended childhood, a time when the young learn and receive care from their parents. Claws and other early primate features characterize prosimians — lemurs, lorises and tarsiers — many of which are nocturnal and have a retinal tapetum that permits vision in dim light. Anthropoids — the Old and New World monkeys, apes and humans — are usually diurnal and have forward-directed eyes and larger brains than prosimians. Among hominoids — gibbons, orangutans, chimpanzees, gorillas and humans — humans are the most adapted for bipedal locomotion.

Primates originated about 85 Mya in the Late Cretaceous. The divergence of primates into Strepsirrhini (lemurs and lorises) and Haplorhini (tarsiers and anthropoids) occurred soon after. Molecular data suggest that within the anthropoids, apes diverged from monkeys in the early Oligocene (29 to 34 Mya), and that hominins and chimpanzees diverged more recently than 6.3 Mya.

Three African fossil species — *Sahelanthropus tchadensis, Orrorin tugensis* and *Ardipithecus kadabba* — compete for the designation of the earliest hominin. These three forms, which date from around the period of the human-chimpanzee divergence, show a mix of hominin features including evidence that suggests bipedal locomotion. They were followed by a succession of more human-like species in the genus *Australopithecus,* whose postcranial fossils show conclusive evidence of habitual bipedal locomotion, specializations of the teeth and jaws related to diet and slightly increased brain size.

Comparative as well as fossil evidence suggests that the last common ancestor of humans and chimpanzees resembled chimpanzees more than humans, was similar in body size to chimpanzees, and was very likely a knuckle-walking quadruped. Many explanations have been suggested for the evolution of bipedalism including advantages in foraging, evading predators, carrying provisions and using tools and weapons. Changes in the pelvis, lower limb, vertebral column, feet and skull accompanied the evolution of upright posture.

The earliest fossils of the genus *Homo* are about 2 My old and have brains larger than those of the australopithecines. Four species of *Homo* were present in Africa between 2.4 and 1.6 Mya. Of these, *H. ergaster* gave rise to *H. erectus,* the dominant species of hominin for the next million years. *H. erectus* may have persisted in Southeast Asia until quite late. Fossils that appear in the record around 500,000 years ago have often been loosely lumped as "archaic *H. sapiens.*" Increasingly, these specimens are being classified into distinct species such as *H. heidelbergensis, H. helmei* and *H. rhodesiensis.*

Anatomically modern humans, which first appear in the record 160,000 years ago in Ethiopia, originated in Africa and spread from there to Europe, Asia and eventually the New World, although there is debate about the extent of gene flow that may have occurred from local populations to the expanding modern human population. *H. neanderthalensis* (Neanderthals) arose in Europe probably from *H. heidelbergensis* about 250,000 years ago, persisted in Spain and Gibraltar as late as 24,000 years ago, and so may have coexisted with anatomically modern humans in parts of Europe.

Primates can transmit signals to other individuals through any of the sensory mechanisms: scent, grooming behavior, gestures and facial expressions, displays and vocalization. The use of symbols through language is obviously a defining characteristic of modern humans, but the origins of language in human evolution are poorly understood, although development of the tongue, soft palate and larynx allowed anatomically modern humans to produce a variety of sounds not possible in other primates.

Chimpanzees have some of the conceptual basis for linguistic skills but are unable to use spoken language because of anatomical differences in the structure of their vocal tract. Similarly, studies of chimpanzee behavior reveal many traits once thought unique to humans, such as tool use and modification, planning, organized aggression and complex social networks. What long appeared to be unbridgeable gaps between apes and humans are now being bridged by discoveries that show that many traits formerly considered exclusively human are also present in apes, but to lesser degrees.

Primates, the mammalian group that includes humans, have a number of adaptations indicating an arboreal (tree-living) ancestry (**Table 20-1**). Although not every feature listed in Table 20-1 characterizes every primate, all extant primates have enough of these features to distinguish them from other arboreal mammals such as shrews, squirrels and raccoons. These adaptations include:

1. ability to move the four limbs in various directions;
2. grasping power of the hands and feet;
3. slip-resistant cutaneous ridges (*dermatoglyphs*) on the ventral pads of these extremities, which also contain specialized tactile-sensitive organs (Meissner's corpuscles);
4. retention of the clavicle (collar bone) to support the pectoral girdle in positioning the forelimb; and
5. flexibility of the vertebral column allowing twisting and turning.

In addition to highly developed brains, anthropoid primates — monkeys, apes and humans — have a relatively long postnatal growth period accompanied by considerable parental care for a relatively small number of offspring. The selective value of this suite of traits arises from the limited number of offspring that can be born successfully and carried by highly mobile mammals, along with the long-dependent learning period that enabled them to cope with many complex environmental and social variables.

Primate Classification

There are presently about 230 primate species, which primatologists have traditionally classified into two suborders, prosimians and anthropoids (**Table 20-2** and **Figure 20-1**, top). Recent molecular evidence strongly supports dividing extant primates into one lineage comprised of lemurs and lorises and another comprised of anthropoids (monkeys and apes) and tarsiers. Lemurs and lorises comprise the suborder **Strepsirrhini** — moist, dog-like muzzle between nose and lip — while anthropoids and tarsiers comprise the suborder **Haplorhini** — dry skin or fur between nose and lip. Older accounts will usually include tarsiers in the Strepsirrhini (prosimians), but their closer relationship to anthropoids would make that suborder a paraphyletic group.

Strepsirrhines generally retain a greater number of earlier mammalian features (e.g., claws, long snout, lateral-facing eyes) than do anthropoid primates. With the exception of large-bodied prosimians in Madagascar, an island that separated from Africa during the late Cretaceous, before anthropoids evolved, prosimians are small bodied and nocturnal.

Haplorhines, which include monkeys, apes and humans, are mostly larger than prosimians and are generally diurnal rather than nocturnal. Compared to prosimians, anthropoids have more of the primate features enumerated previously — a shortened face, forward-directed eyes and a larger, more complex brain.

Lemurs

Lemurs appear exclusively in Madagascar. In this relatively protected and isolated location, they produced a range of small and large species that often parallel the role of forest monkeys on the mainland. But lemurs are less derived than monkeys, having a longer snout and a moist philtrum between nose and upper lip that accentuates their sense of smell. They also have a special toilet claw on the second toe, thick fur, sensitive facial hairs (vibrissae) and a dental comb formed by the nearly horizontal (procumbent) orientation of the lower incisors and canines, and that is used for grooming and feeding. Because lemurs have a *"tapetum lucidum"* (retinal layer that reflects incoming light back through the retina), some remain exclusively nocturnal (mouse and dwarf lemurs) while others are active during the dim, crepuscular light of late dusk and early dawn as well as diurnally (true lemurs).

TABLE 20-1 Traits and tendencies found in primate groups

- Independent mobility of the digits
- An opposable first digit in both hands and feet (thumb, big toe)
- Replacement of claws by nails to support the digital pads on the last phalanx of each finger and toe
- Teeth and digestive tract adapted to an omnivorous diet
- A semi-erect posture that enables hand manipulation and provides an optimal position preparatory to leaping
- Center of gravity positioned close to the hind limbs
- Well-developed hand–eye motor coordination
- Optical adaptations that include overlap of the visual fields to gain precise three-dimensional information on the location of food objects and tree branches
- An eye completely (anthropoids) or fractionally (prosimians) encased by bone (bony orbits)
- Shortening of the face accompanied by reduction of the snout
- Diminution of the olfactory apparatus in diurnal forms
- Compared with practically all other mammals, a large and complex brain in relation to body size

TABLE 20-2 Traditional systematics classification of extant subgroups of primates, with common names for some members in each group

- Suborder Prosimii
  - Superfamily Lemuroidea (lemurs)*
  - Superfamily Lorisoidea (lorises, galagos — bush babies)*
  - Superfamily Tarsioidea (tarsiers)
- Suborder Anthropoidea
  - Infraorder Platyrrhini (New World)
    - Superfamily Ceboidea
      - Family Callitrichidae (marmosets, tamarins)
      - Family Cebidae (capuchins, howler monkeys, spider monkeys)
  - Infraorder Catarrhini (Old World)
    - Superfamily Cercopithecoidea
      - Family Cercopithecidae (macaques, baboons, guenons, vervet monkeys)
      - Family Colobidae (langurs, colobines)
    - Superfamily Hominoidea
      - Family Hylobatidae (gibbons, siamangs)
      - Family Pongidae (orangutans, gorillas, chimpanzees and bonobos)
      - Family Hominidae (humans)

*These constitute the Strepsirrhini. The other taxa constitute the Haplorhini.

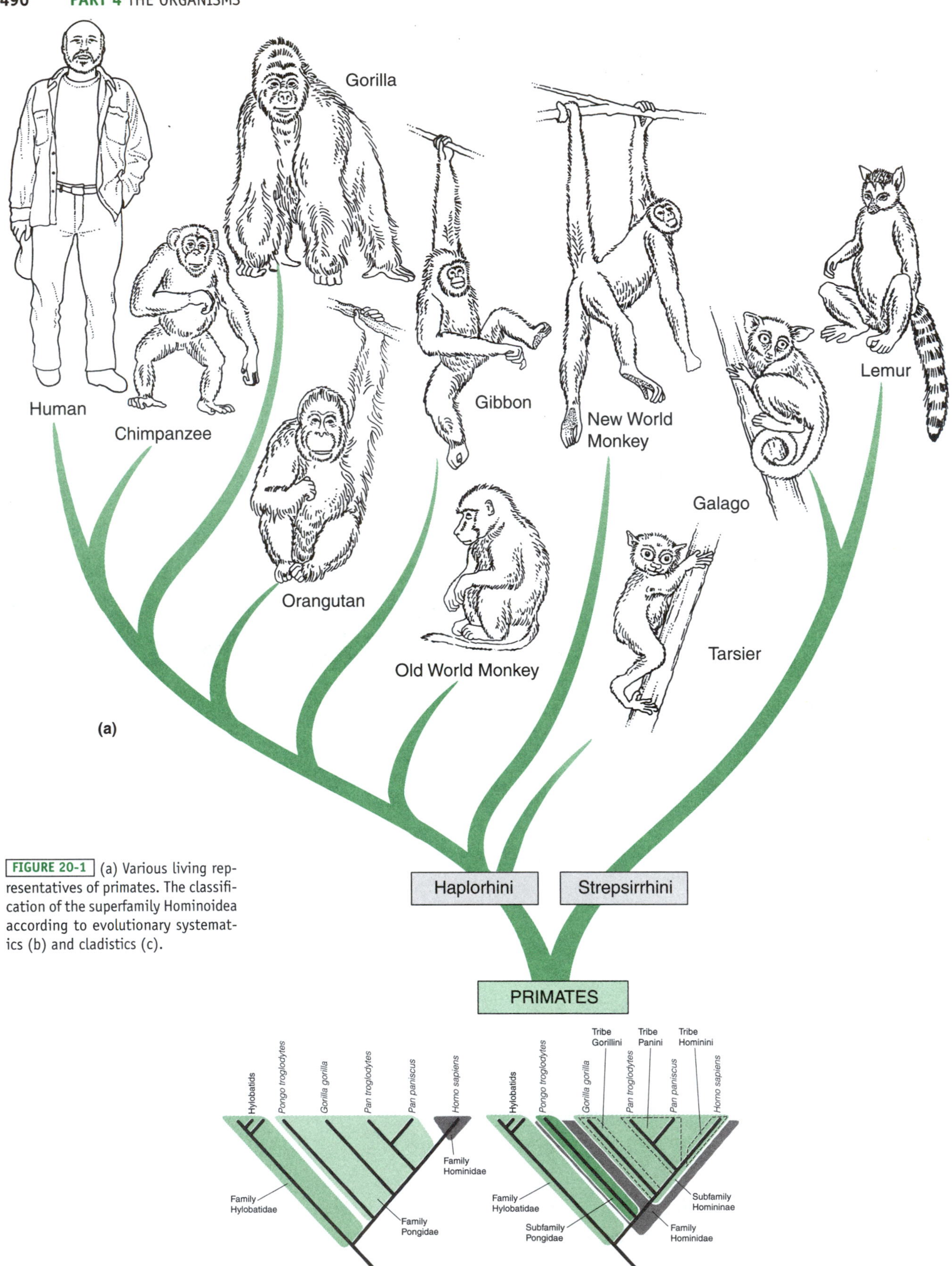

FIGURE 20-1 (a) Various living representatives of primates. The classification of the superfamily Hominoidea according to evolutionary systematics (b) and cladistics (c).

Lorises

Lorises are found in forests of Africa (pottos and galagos, or bush babies) and Southeast Asia (slender and slow lorises). The snout is shorter than in lemurs, and the relatively large eyes face forward, indicating adaptation to a larger forebrain and perhaps also to increased emphasis on visual predation. These and other adaptations, including a retinal tapetum (*tapetum lucidum*), permit either nocturnal or crepuscular activity. Like lemurs, lorises have dental tooth combs, a toilet claw and a moist philtrum.

Tarsiers

Tarsiers, nocturnal Southeast Asian primates, are more closely related to anthropoids (monkeys and apes) than to lemurs and lorises. Although they have two toilet claws on each foot and enormous eyes relative to head size, like anthropoids they lack the retinal tapetum that characterizes lemurs and lorises. Tarsiers also show anthropoid characteristics in the replacement of the moist philtrum by a dry, furry space between nose and lip, as well as upright lower incisors and a partially closed bony orbit around the eyes. Restricted to tarsiers is tibial-fibula fusion in the lower leg, an adaptation that enables them to make single leaps as long as 2 m.

Platyrrhines

Platyrrhines are the *New World monkeys* found in Central and South America. All are arboreal and characterized by broad noses with widely spaced nostrils facing laterally and three premolars on each side of the jaw. *Marmosets* and *tamarins,* small animals with claws on all digits except the big toe, comprise one family, Callitrichidae. Species in the other family, Cebidae, have nails instead of claws, and some have prehensile tails.

Catarrhines

Catarrhines include two superfamilies, the Cercopithecoidea (*Old World monkeys*) and Hominoidea (*apes and humans;* **Fig. 20-1b**). Catarrhines share narrowly spaced nostrils facing downward and a dental formula of 2.1.2.3 (two incisors, one canine, two premolars and three molars on each side of the centerline in both upper and lower jaws). Catarrhine monkeys are mostly larger than New World monkeys, lack prehensile tails, and include terrestrial (baboons, mandrills, vervets, patas monkeys and some macaques) as well as arboreal forms.

Hominoids

This catarrhine superfamily of apes and humans is distinguished from other catarrhines by a number of adaptations to **brachiation** (arm-hanging and -swinging) arboreal locomotion along with different degrees of adaptation to a ground-dwelling existence. The body proportions of the extant apes including humans are shown in **Figure 20-2.**

Perhaps because arboreal hominoids adapted to holding onto overhead tree branches, their posture is more erect than that of monkeys. As other aids in brachiation, the arms and shoulders are more flexible, the wrists and elbows more limber, and the vertebral column shorter and stiffer. Other attributes that distinguish hominoids from monkeys are a broader and larger pelvis to support more vertical weight;

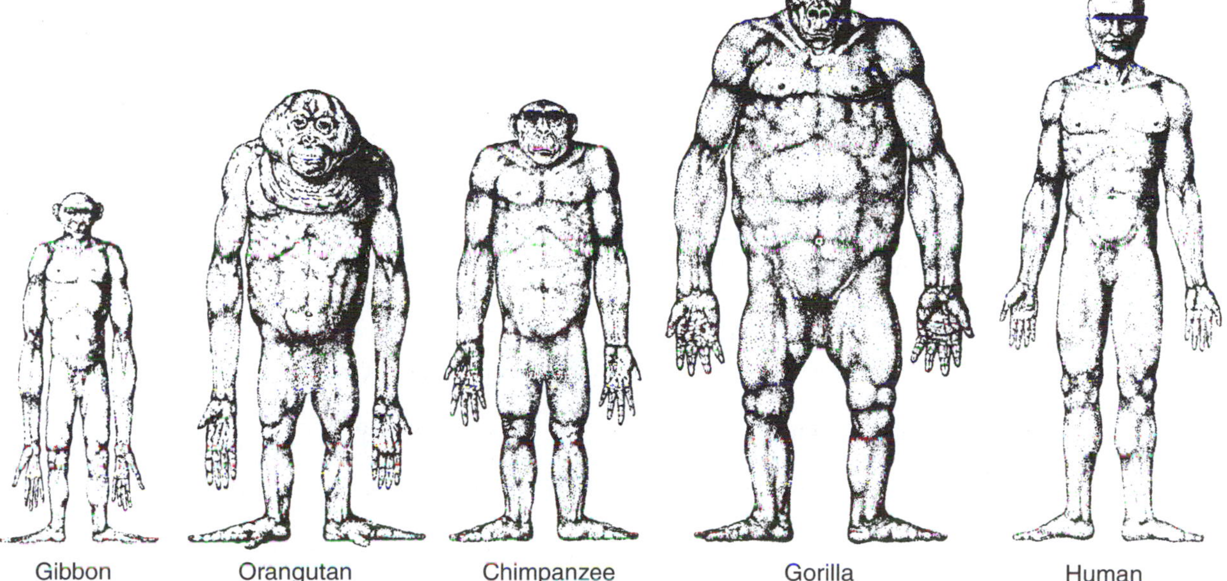

FIGURE 20-2 Body contours and proportions of adult male apes and humans with all hair removed, drawn to the same scale. (From Schultz, A.H., 1969. *The Life of Primates.* Universe Books, New York.)

visceral attachments and arrangements that provide more vertical support for the stomach, intestines and liver; loss of the tail; five-cusped lower molars rather than the four cusps in monkeys; a broad but shallow thorax because of the change to a more vertical posture; and scapulae (shoulder blades) placed dorsally on the thorax to position the shoulder joint so the arms can be extended laterally. Other hominoid features include a larger body size compared to Old World monkeys, and a life history with greater emphasis on extended postnatal development and complex social interactions.

Hylobatid Apes

Perhaps the least derived extant hominoids are **gibbons** and **siamangs,** an almost entirely arboreal group of small-bodied apes confined to Southeast Asia. They share with many Old World monkeys a relatively small size (none are more than 11 kg or 25 lbs) and ischial callosities (cornified sitting pads fused to the ischial bones). Compared to other hominoids, they are superb acrobatic brachiators, swinging with elongated arms through the trees, their legs often folded beneath them, only resorting to climbing or walking during traveling when brachiation is not possible.

Hylobatids exhibit only minimal sexual dimorphism in body size. Both sexes engage in a unique and spectacular form of arboreal locomotion known as **richochetal brachiation.** Swinging from arboreal supports by their upper limbs, they use specialized muscular adaptations to propel themselves across sizeable gaps in the forest canopy (Tuttle, 1972). Molecular data indicate that hylobatids and the lineage leading to the pongid apes and humans diverged 15 to 18 Mya.

Orangutans

Orangutans are large apes — some males may weigh more than 90 kg — restricted to Borneo and Sumatra. Orangutans are highly sexually dimorphic with males twice as heavy as females. With the exception of adult males, they are mainly arboreal. On the ground, they move mostly quadrupedally with clenched fists to support the upper torso. In the trees, orangutans move about in a manner best characterized as *versatile climbing,* using both hands and feet to grasp available supports and often bending branches to help bridge gaps in the canopy. Orangutans will occasionally brachiate, but much more slowly and what appears to us to be more cautiously than do hylobatids, and only over short distances. Like chimpanzees and gorillas, they lack ischial callosities (sitting pads). Molecular data indicate that orangutans and the lineage leading to African apes and humans diverged about 13 Mya.

Common Chimpanzees and Bonobos

There are two species of **chimpanzee,** *Pan troglodytes* or the common chimpanzee and *Pan paniscus,* the **bonobo.** Both species are found in equatorial Africa with bonobos confined to central Zaire.

Common chimpanzee males average 45 kg in weight, females about 40 kg, although these figures vary across the geographic range. Bonobo males average 40 kg in weight and females 32 kg so they are only slightly smaller than common chimpanzees on average. Bonobos are often called "pygmy chimpanzees" but this is misleading, not only because the difference in size has been overstated but, more important, because they are a distinct species and not a dwarfed form of the common chimpanzee.

Bonobos are more gracile than common chimpanzees and have less protruding faces. Both species live in fairly large groups of 50 to 120 individuals with temporary subgroups within the larger groups. Both have complex social structures that have long been the subject of detailed study because of the behavioral and phylogenetic closeness to humans.

Chimpanzees move around mostly on the ground but also spend a great deal of time in the trees feeding, sleeping and grooming. Like gorillas, they travel terrestrially by *"knuckle-walking,"* using friction pads on the middle phalanges of nonthumb digits as forelimb support. Their diet mostly consists of fruit, but they do eat termites and capture and eat young baboons, monkeys and occasionally even young chimpanzees. Compared to all other primates except humans, chimpanzees show a remarkably wide array of expressions, postures and gestures.

Gorillas

Gorillas, the largest apes — some males may weigh 227 kg (500 lbs) or more — inhabit equatorial Africa in two main distributions: lowland gorillas west of the Congo basin, and mountain gorillas eastward. Adult males make their sleeping nests on the ground and rarely climb in trees. Their social groups, usually fewer than 10 individuals, organize around a single dominant male ("silverback"), with other adult males occasionally present. Gorillas are almost entirely terrestrial although juveniles spend some time in trees. Like chimpanzees, gorillas are quadrupedal knuckle-walkers. They are less active than chimpanzees and have a diet that is almost entirely herbivorous.

Human–Ape Comparisons

Compared to other hominoids, humans (**hominins**) present the greatest number of adaptations to bipedal terrestrial locomotion. The hind limbs are longer relative to their forelimbs than in any of the apes (see Fig. 20-2), and their hands, freed from supporting the body, provide the most refined of manipulatory controls. Additional specific human anatomical traits include a relatively large brain and small retruded face, shorter canines, less body hair, and many cranial, dental, skeletal and other features.

TABLE 20-3 Amino acid differences between chimpanzees and humans for nine proteins

Protein	Number of Amino Acids	Human–Chimpanzee Differences
Hemoglobins		
α chain	141	0
β chain	146	0
$^{G}\gamma$ chain	146	0
$^{A}\gamma$ chain	146	0
δ chain	146	1
Myoglobin	153	1
Cytochrome *c*	104	0
Fibrinopeptides A and B	30	0
Carbonic anhydrase I	25	3
Totals	1,037	5

Source: From Diamond, J.M., 1995. The evolution of human creativity. *In Creative Evolution?!*, J.H. Campbell and J.W. Schopf (eds.), Sudbury, MA: Jones and Bartlett Publishers, www.jbpub.com. Reprinted with permission.

Despite these differences, a large number of molecular similarities exist. For example, comparisons of protein and DNA sequences for chimpanzees and humans show more than 96 percent identity, indicating that molecular differences are no more than, and often much less than, those between other related species (**Table 20-3**).

Anatomically, if we compare bone for bone, muscle for muscle, organ for organ, humans strikingly resemble apes, although differing in proportions (Fig. 20-2). You can see this similarity between apes and humans in some of their motions and postures:

- Because their arms extend laterally in brachiator fashion and their elbows and shoulders are remarkably mobile, both scratch the back of their head from the side, rather than from the front as do monkeys.
- As you can see from **Figure 20-3**, positioning of the limbs can be remarkably alike; both support the chin with their hands and cross their legs. Even some facial expressions seem similar (see Figure 20-34).
- In walking, the heel touches the ground first; in monkeys the metatarsals touch first.
- The knuckle-supporting stance of crouching American football players is the conventional stance for ground movement in chimpanzees and gorillas.

The Primate Fossil Record up to the Appearance of Hominins

According to the fossil record, many Mesozoic mammals were much like extant tree shrews (**Fig. 20-4a**; see also Fig. 19-11), adapted to an insectivorous lifestyle that encompassed the forest floor, trees and shrubs. Primates originated from one of these lineages of Cretaceous mammals. According to molecular evidence primates diverged in the Late Cretaceous about 85 Mya although the details of primate origins remain fairly obscure. Part of the difficulty lies in the relative absence of fossilized forest animals in sedimentary rocks. Generally, the bodies of arboreal animals disarticulate completely soon after they reach the forest floor, and only rare chance events wash their skeletons into rivers,

FIGURE 20-3 A chimpanzee and a movie actress (Dorothy Lamour) resting on a 1938 movie set of the motion picture *Jungle Love.* (Adapted from a photograph in Mann, W. M., 1938. Monkey folk. *Nat. Geograph.*, **73**, 615–655.)

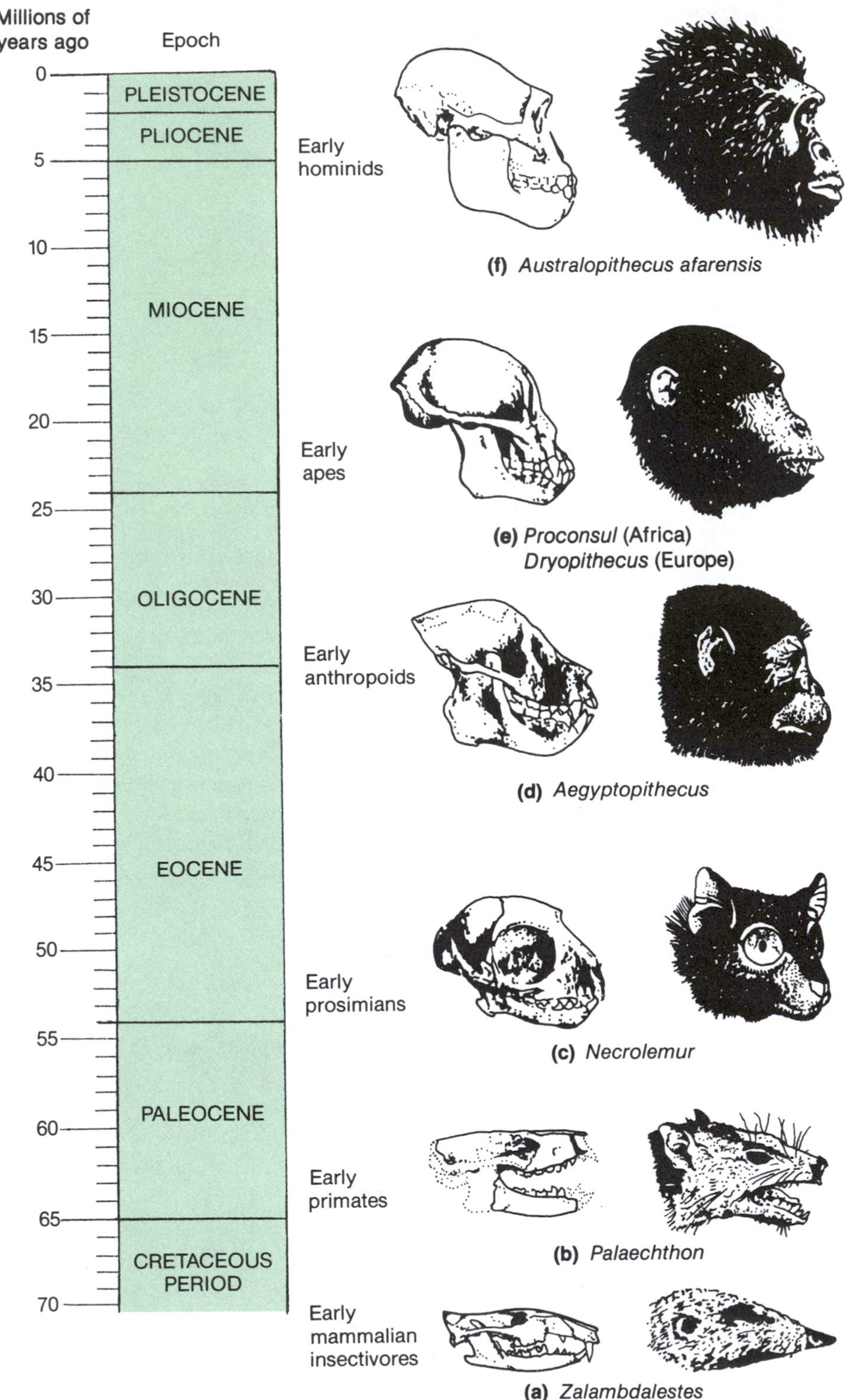

FIGURE 20-4 Beginning from an early Mesozoic insectivore (a) are samples from branches of the primate evolutionary tree (b–f) showing the chronology, skulls and reconstructions of some fossil species.

lakes or marine sediments where they can more easily fossilize. The earliest known primate, *Purgatorius,* known from teeth and jaw fragments, is represented by two species from the late Cretaceous and Early Paleocene (**Fig. 20-5**).

Several hypotheses have been proposed to explain the origin of primates. All focus on factors such as an arboreal lifestyle, predation and selective pressures from evolutionary changes in plants. The currently favored model is the *visual*

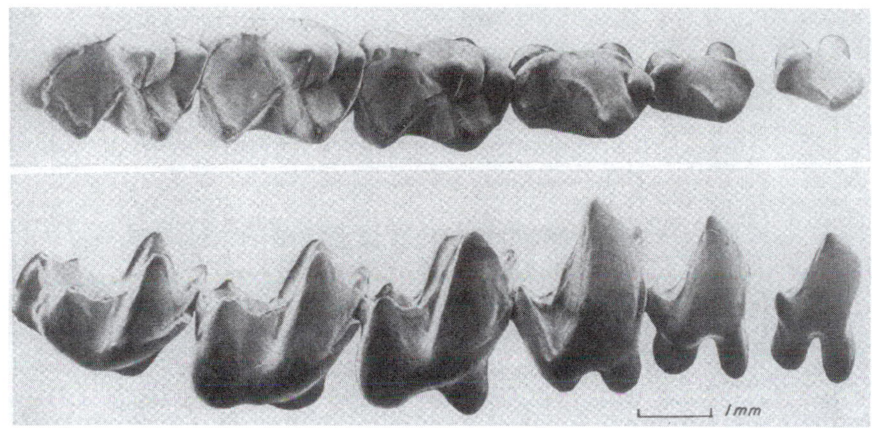

FIGURE 20-5 The dentition of *Purgatorius unio*. (Reproduced from William A. Clemens, *Science* **184:** 903–905. Reprinted with permission from AAAS. Courtesy of William A. Clemens, University of California, Berkeley.)

predation hypothesis, originally proposed by Cartmill (1974), who argued that convergence of the orbits, binocular vision and grasping hands and feet are adaptations to feeding on insects on small terminal branches.

Paleocene Primates: Plesiadapiformes

By the Middle Paleocene, deposits contained well-represented and abundant early archaic primates, classified as Plesiadapiformes. These animals have auditory regions and dentitions that clearly establish them as primates. Although they are primate-like in many respects, this group is also very specialized in that they possessed rodent-like incisors with a large gap or diastema separating the incisors from the more posterior teeth. Thus, the origins of subsequent groups of primates are probably not found among this group.

The earliest nonplesiadapid primate is *Altiatlasius,* from the late Paleocene (ca. 55 to 60 Mya). Molecular evidence suggests, though, that the divergence of the two major groups of living primates, strepsirrhines and haplorhines, occurred earlier than this, in the Late Cretaceous about 77 Mya. The first fossil primates that clearly belong to these two extant lineages, however, don't appear until the Eocene, around 55 Mya. Although a long gap, this is not unusual. The fossil record represents some time periods better than others and divergence dates are always older than the earliest known fossil representatives of that divergence.

Eocene Primates: Adapiforms and Omomyiforms

Eocene primates are mainly represented by two groups, adapiforms and omomyiforms. Some have proposed that adapids (**Fig. 20-6a**) contain the ancestors of lemurs and lorises, although this idea is not consistent with more recent fossil data. Much more accepted is the idea that the omomyids (Fig. 20-6b) contain the ancestral species for tarsiers and anthropoids. By the Middle Eocene, we find fossil species closely related to the ancestors of modern tarsiers and anthropoids (**Fig. 20-7**).

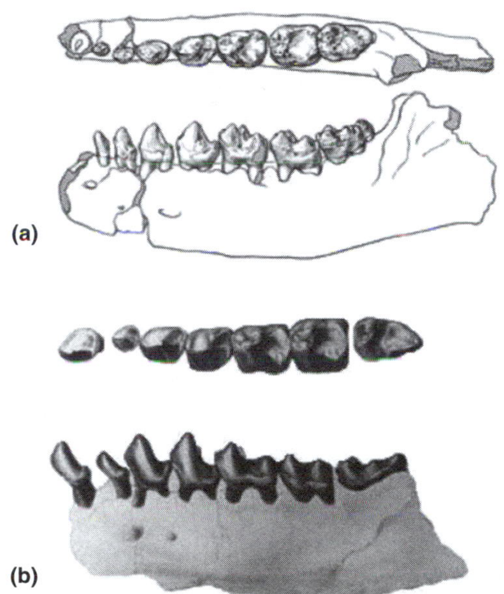

FIGURE 20-6 Examples of the two main lineages of Eocene primates. (a) is a representative of the Adapiform *Cantius torresi,* and (b) is a fossil specimen of the Omomyiform *Teilhardina americana.* (From Miller, E. R., Gunnell, G. F., and Martin, R.D. 2005. Deep time and the search for anthropoid origins. *Am. J. Phys. Anthropol. Suppl.* **41,** 60–95.).

Based on molecular evidence anthropoid primates diverge into Old and New World monkeys (Catarrhines and Platyrrhines) about 35 to 40 Mya. The earliest Platyrrhines lived in Africa or Asia, but modern representatives are confined to South and Central America. The earliest platyrrhine fossils in South America date to around 26 Mya. There is much debate about whether New World monkeys arrived in South America by island hopping the 1,000 km distance in the Early Oligocene or by crossing via Antarctica.

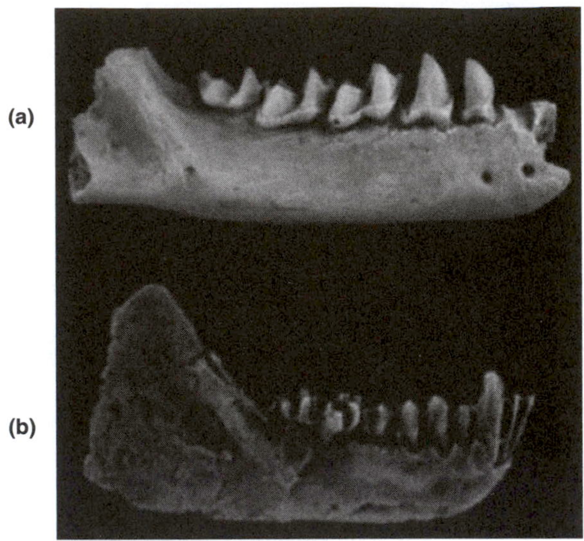

FIGURE 20-7 Middle Eocene representatives of tarsiers and anthropoids. (a) is the tarsiiformid, *Xanthorhysis tabrumi,* while (b) is the eosimid, *Eosimias centennicus.* (From Miller, E. R., Gunnell, G. F., and Martin, R.D. 2005. Deep time and the search for anthropoid origins. *Am. J. Phys. Anthropol. Suppl.* **41,** 60–95.)

Because there are no Cenozoic prosimian fossils in South America or Cenozoic anthropoid fossils in North America, the platyrrhines probably did not originate in either continent. Rather, the New World monkeys may well have come from Africa, which, because of plate tectonics, lay closer to South America during the Early and Mid Cenozoic than it does today (see Fig. 6-12). During this time, the two continents were spanned by one or more chains of islands that later submerged, some of which might have served as intermediate stations for the platyrrhine journey.

Miocene Primates: Hominoids and Old World Monkeys

Catarrhines, in turn, diverge into apes (Hominoidea) and *Old World monkeys* (Cerceopithecoidea). By molecular evidence, this split occurred between 29 and 35 Mya. Along the catarrhine trajectory and close to this divergence is *Aegyptopithecus,* (Fig. 20-4d), an early anthropoid fossil in the Fayum Province of Egypt, dating to the Oligocene, 30 Mya. The cercopithecoid branch is represented by fossil species such as *Victoriapithecus,* an abundant quadrupedal monkey from the Middle Miocene of East Africa (**Fig. 20-8**). While *Victoriapithecus* is clearly an Old World monkey, it does not share derived features with either of the two extant groups, colobines or cercophithecines.

Although the split with cercopithecoids occurred earlier, the earliest undoubted hominoids date from the Early Miocene (about 20 Mya). The best known of these is the African genus *Proconsul* (Fig. 20-4e). Along with the genera *Afropithecus* and *Kenyapithecus, Proconsul* lacks several derived features that all hominoids share and thus represents

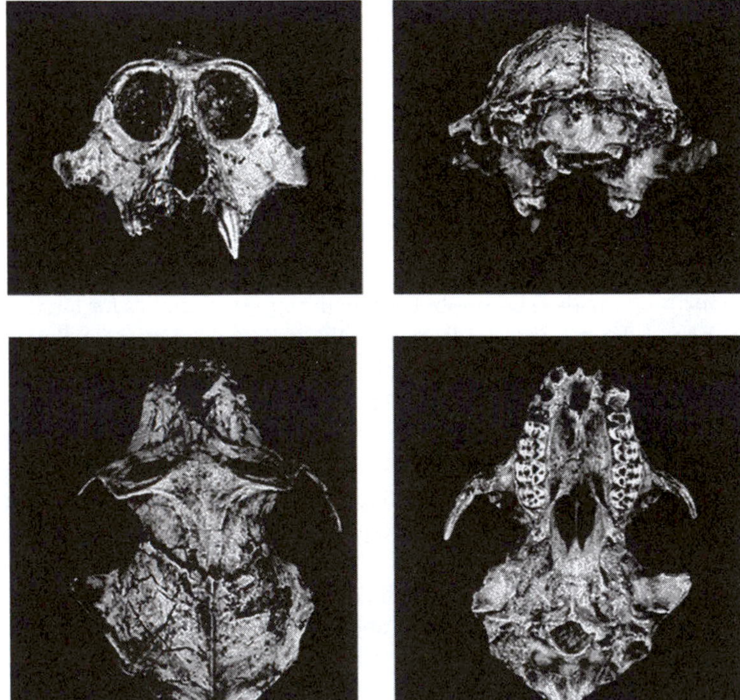

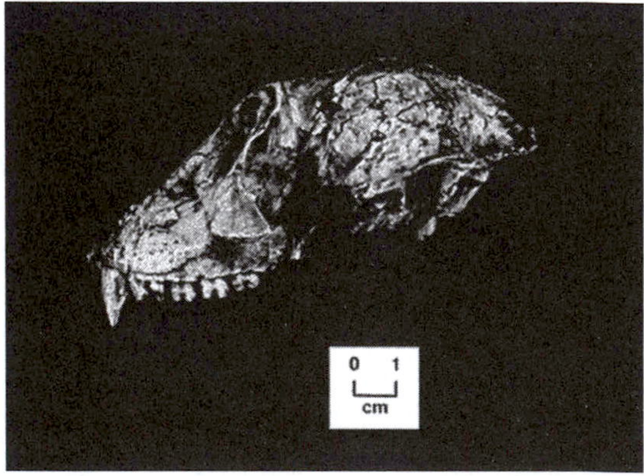

FIGURE 20-8 (a–e) Nearly complete skull of *Victoriapithecius,* an early Old World monkey from East Africa dated at 14 to 16 Mya. (From Benefit, B.R., and M. L. McCrossin, 1997. Earliest known Old World monkey skull. *Nature* **388,** 368–371.)

a lineage that branched off before the split of the great and lesser apes.

The earliest known hominoid, however, is *Morotopithecus* dated at 20.6 Mya (**Fig. 20-9**). This relatively large-bodied ape shares many features with the great apes and so is thought by some to be the earliest known representative of that lineage. This is problematic, however. According to molecular data, the great apes and lesser apes are estimated to have diverged about 18 Mya, which is later than the dating of *Morotopithecus* specimens. This either means that the great apes and *Morotopithecus* evolved these similarities in parallel, or that many features thought to be derived for the great apes are actually primitive to all hominoids, and that the lesser apes represent the more derived condition. These alternatives remain to be tested as our data set of Miocene and Late Oligocene hominoids grows.

Compared to their relatively lower numbers today, ape-like forms were more common during the Early and Middle Miocene than monkeys. They were particularly diverse and abundant in Europe and Asia where the great ape lineage is represented by genera such as *Drypoithecus, Oreopithecus, Sivapithecus* and *Lufengpithecus*. Some of these are clearly linked to the lineages of extant ape genera. *Sivapithecus* (**Fig. 20-10**), for instance, is certainly the sister genus or even direct ancestor of orangutans. *Dryopithecus* and *Lufengpithecus* are also assigned to the Pongidae, the same family as the extant great apes, while others, such as *Oreopithecus* represent distinct families within Hominoidea now extinct.

The relative abundance and diversity of apes and monkeys was reversed during the Late Miocene, when monkeys became much more numerous and widespread than apes. It has been suggested that dietary changes in Old World monkeys during the Miocene enabled them to compete successfully against many of the arboreal apes, perhaps by developing the ability to eat and digest fruits before they ripened enough for the hominoids. The more rapid reproductive rate of monkeys may have allowed them to compete successfully with apes as well as to radiate into new habitats such as savannas. With the exception of protohominin lineages most apes became restricted to wet forest habitats.

One important consequence of monkeys replacing apes in various arboreal habitats was that selective pressure among a few ape (and some monkey) species for ground-dwelling adaptations increased. One of these late Miocene ground-dwelling ape species is ancestral to humans.

FIGURE 20-9 Cranial and postcranial specimens of *Morotopithecus*, the earliest definitely assigned hominoid. (Reproduced from the *Journal of Human Evolution*, vol. 46(2), Young, N. M., and MacLatchy, L., The Phylogenetic position of *Morotopithecus*, pp. 163–184, copyright 2004, with permission from Elsevier. Photo courtesy of Dr. Nathan M. Young, Harvard University.)

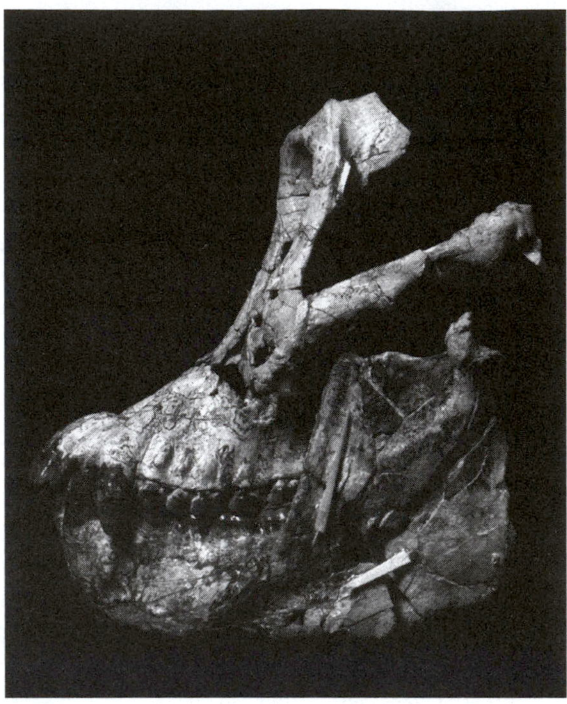
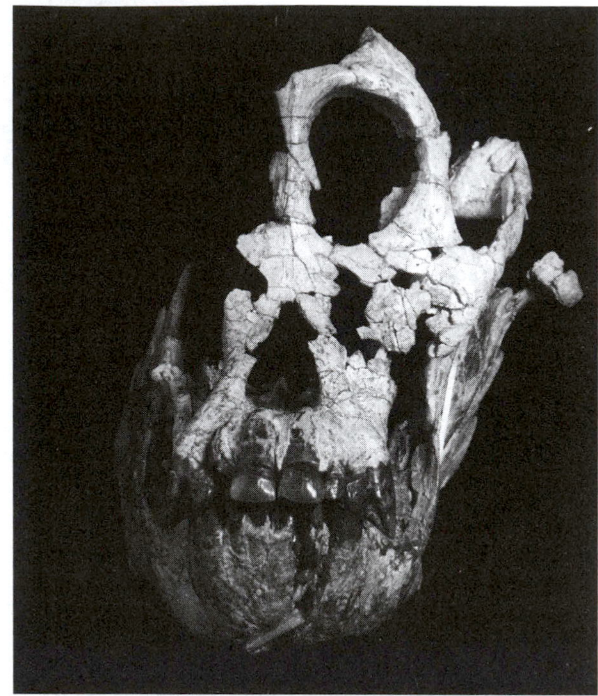

FIGURE 20-10 A beautifully preserved *Sivapithecus* specimen (GSP 15000) that shows the very distinctive and orangutan-like morphology of this genus. (© Steve Ward/Anthro Photo.)

Origins of the Hominini

Humans are more closely related to the African apes than to orangutans and even more closely related to chimpanzees than to gorillas. According to recent molecular dating, the divergence between the lineages leading to humans and chimpanzees occurred about 6.3 Mya, and so the earliest species in the tribe Hominini, to which humans belong (**Box 20-1**), would have appeared shortly afterwards.

BOX 20-1 Hominids and Hominins: Cladistics and the Construction of Our Family Tree

As EXPLAINED in Chapter 11, phylogenetic systematics or cladistics has increasingly influenced our views on how to organize the diversity of life into systematic categories. In cladistics, all taxonomic groups include all descendents from the origin of that group and cannot have multiple descendents. Proponents of this view argue that it creates internally consistent, nonarbitrary classifications.

Many long-standing taxonomic groups are **paraphyletic** in that they do not include all descendents. A famous example of this is the Class Reptilia, which by cladistic classification includes both mammals and birds. Amongst our closest relatives, the family Pongidae is one such paraphyletic group in that it excludes humans (Fig. 20-1).

As it has become increasingly clear that humans and the African Apes (particularly chimpanzees) are closely related and as the diversity of fossil species related both humans and the great apes has increased, many paleoanthropologists have felt that the traditional classification of humans and our closest relatives is an inadequate and inconsistent representation of our place in nature. For this reason, a cladistic classification of hominoids has gained favor in recent years (see Fig. 20-1 b, c). A consequence of this change is that humans and the great apes are now collectively referred to as **"hominids."** The reason for this is that the various taxonomic levels are assigned specific endings. Families all end in ". . . *ideae*," so if humans and the great apes belong to the family *hominidae*, they are all hominids. The term *hominids* was previously used exclusively for humans and the fossil species most closely related to humans. In the cladistic classification, humans and our closest extinct relatives belong to the "tribe" *hominini.* Tribe is a taxonomic level between subfamily and genus; thus, when paleoanthropologists in the recent literature use the term **"hominin,"** they mean the same thing as "hominid" in older literature. At stake is more than just nomenclature. This change reflects our changing view on how we classify the diversity of life and our own place in nature.

Three African fossil species — *Sahelanthropus tchadensis, Orrorin tugensis,* and *Ardipithecus kadabba* — compete for the designation of earliest hominin. These three forms, which were added to the known hominin fossil record within the past decade, date from around the period of the human–chimpanzee divergence and show a mix of hominin features including evidence that suggests bipedal locomotion. Before these forms were discovered, the period surrounding the divergence of hominins and panins (chimpanzees) was very much a black box. Now, however, we have a glimpse of the morphological changes that accompanied the origin of the human and chimpanzee lineages. Perhaps the most surprising finding thus far is the diversity of forms that appeared around the divergence. As with later periods of hominin evolution, our initial view of a single lineage leading progressively toward the human form is clearly an anthropocentric oversimplification.

Sahelanthropus tchadensis

Sahelanthropus tchadensis (**Fig. 20-11**) dates to 6.5 to 7.4 Mya, which places it slightly before the current molecular estimate for human–chimpanzee divergence. Specimens of this recently discovered species share several derived features with known hominins: a shorter, less prognathic face, a wide interorbital pillar and enlarged supraorbital tori. A recent quantitative morphological analysis, however, places *Sahelanthropus* closer to known hominins than to chimpanzees. Whether this species represents the earliest hominin or not, it is clearly very close to the divergence of the human and chimpanzee lineages.

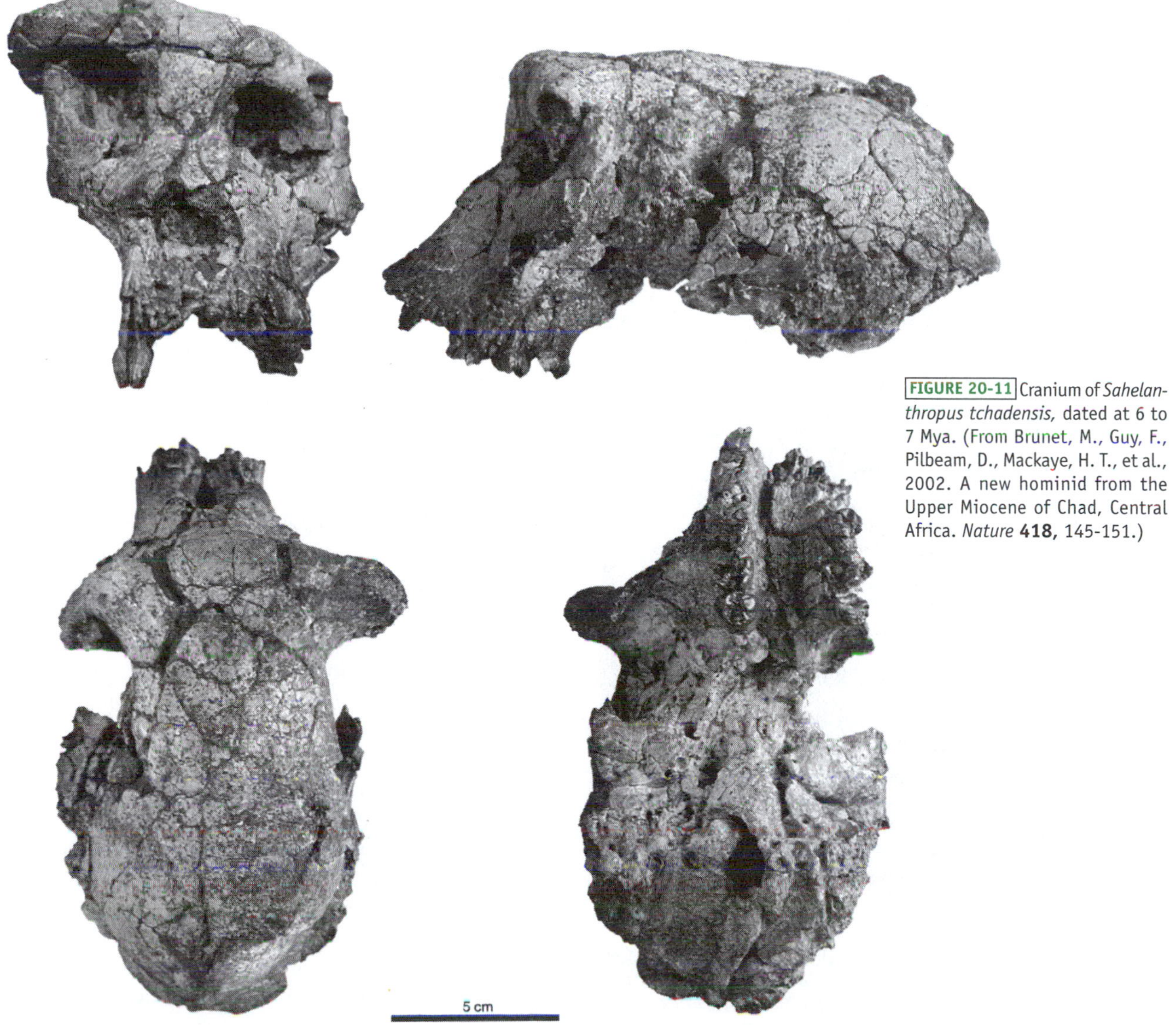

FIGURE 20-11 Cranium of *Sahelanthropus tchadensis,* dated at 6 to 7 Mya. (From Brunet, M., Guy, F., Pilbeam, D., Mackaye, H. T., et al., 2002. A new hominid from the Upper Miocene of Chad, Central Africa. *Nature* **418,** 145-151.)

Orrorin tugensis

Another candidate for the earliest hominin, *Orrorin tugensis* (**Fig. 20-12**), is represented by lower jaw fragments and postcranial fragments including three femora that show clear indications that it was a habitual biped when on the ground. The upper limb fragments also show arboreal traits shared with chimpanzees, indicating that this species may have spent a fair amount of time in trees.

Ardipithecus kababba

Slightly younger, at 5.6 to 5.8 Mya, *Ardipithecus kababba* is the third contender for the title of basal hominin (**Fig. 20-13**). Like *Orrorin,* this species is relatively well represented by post-cranial remains but less well by cranial specimens. It was clearly a bipedal hominin and shares derived postcranial and dental features with later Australopithecines along with more primitive features such as fairly large canine teeth. While *Ardipithecus* is clearly a hominin, there is debate about its relationship to later hominins including the lineage leading to *Homo.*

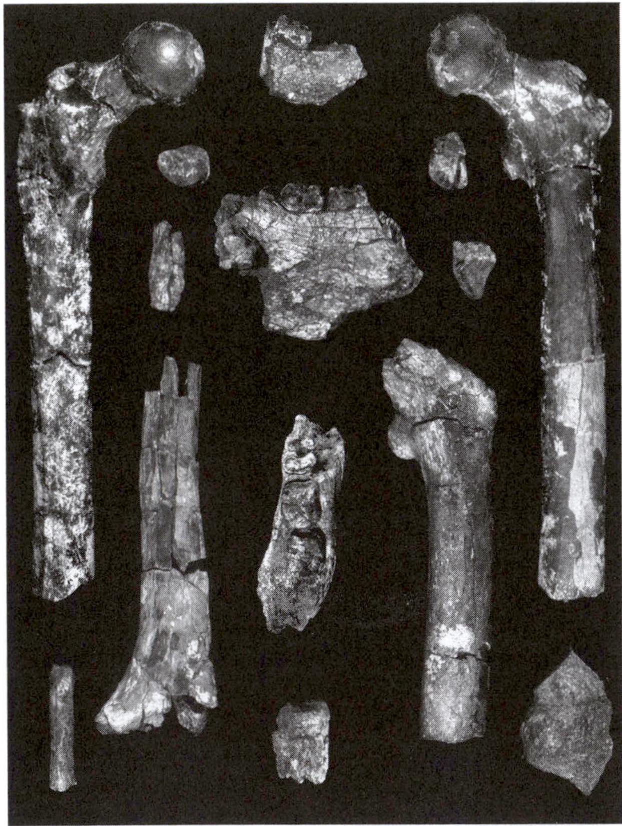

FIGURE 20-12 Specimens of *Orrorin tugensis* dated at c. 5.8 Mya. (From Senut, B., Pickford, M., Gommery, D., Mein, P., et al., 2001. First hominid from the Miocene. (Lukeino formation, Kenya). *C. R. Acad. Sci. Paris, Sciences de la Terre et des planètes / Earth and Planetary Sciences,* **332**, 137–144.)

Factors in Human Origins

The evolutionary origins of humans have been the focus of much speculation and theorizing ever since the publication of *The Origin of Species* and Darwin's subsequent book, *The Descent of Man.* The models proposed are difficult to test and a variety of possible scenarios are consistent with the available data. As our knowledge of the fossil record has grown, some possibilities that were previously plausible have been eliminated. For example, Darwin's suggestion that bipedalism, increased brain size and tool-use were intimately related can be eliminated because we know that bipedal locomotion evolved far earlier than either significantly increased brain size or tool use. This offers hope that as the record and our understanding of our closest relatives grows we will continue to narrow the range of possible explanations for human origins. In this section, we discuss several *factors that may have been involved in the origin of the human lineage.*

Bipedalism

Although paleoanthropologists have established **bipedalism** as a long-standing feature in hominin lineages, and although the questions of why and how bipedalism originated have been discussed for over a century, the origin of bipedalism remains a matter of controversy (Henke and Tattersall, 2007). We now know that bipedalism evolved very early in hominin evolution and that it represents a *key innovation* leading to the suite of adaptations that we associate with humans. Day (1986b) offers three arenas in which selective pressures might have enhanced bipedalism, each of which may have been influenced by the others. The three are improvements in acquisition of food, avoiding predators and reproductive success.

Improved Food Acquisition

Beginning with the Late Miocene and extending through Pliocene–Pleistocene times, there were periodic decreases in global temperature, marked by the onset of ice-sheet formation in Antarctica and the northern hemisphere. As ice locked up water, various terrestrial areas became relatively dry, and open environments such as woodlands and grasslands replaced rain forests in many tropical regions.

Early hominins lived in a patchy environment of mixed woodland and savanna (relatively dry grassland and bush land with occasional trees) that provided seasonal food supplies. Their habitat dictated an omnivorous diet, demanding relatively more time devoted to searching for food over longer distances than would be required in a more localized, homogeneous environment. An upright stance and bipedal striding would have enhanced long-distance foraging by enabling them to carry food gathered in different places.

Tanner (1987) proposed that food gathering was originally a female function, prompted primarily by food sharing with their offspring, leading to the invention and use of food-gathering tools. Unfortunately, we don't know what kinds of food, plant or animal, were acquired and carried in these early journeys. A primary diet of dispersed plant foods would accord with the ape-like teeth of early hominins. Yet some workers suggest that early hominins relied heavily on scavenging carcasses from migratory herds of ungulates, and that bipedalism became important for both terrestrial locomotion and in helping them to carry immature offspring.

Bipedalism may have arisen as a byproduct of adaptations that reduced forelimb involvement in quadrupedal support and movement. As hands became increasingly specialized for grasping, manipulating and carrying food, tools and offspring, selection occurred for an upright stance and for transferring locomotion to hind limbs.

Improved Predator Avoidance

Bipedalism enhances height and so improved a hominin's ability to see over tall grasses and obstructions and to wade in deeper water to pursue game or seek protection from predators. Day (1986b) points out that the ability to climb trees would have helped them to escape predators and to increase their field of view, benefits when it came to detecting danger and surveying their surroundings. The curved hand and foot bones and relatively long arms of early australopithecines and early *Homo* (*H. habilis*) point to persistent tree-climbing abilities.

Improved Reproductive Success

Lovejoy (1981) proposed that bipedalism enabled adult males to carry food manually to their females and offspring, who could remain sequestered in a single locality, the **home base.**

This mode of provisioning reduced the need for females to be continuously mobile in foraging for themselves and their attached offspring thereby offering three important advantages:

- a relatively stable home base that provided more constant social relationships and perhaps closer mother–infant relationships that improved infant survival;
- reduced infant injuries because infants no longer were attached to a continuously mobile mother; and
- reduction in the spacing between births by allowing parents to care for more offspring successfully.

Sexual Bonding

Although anthropologists still debate the extent to which the proposals above represent historical events, some with considerable vehemence — for example, see the collection edited by Kinzey (1987) — Foley (1987) points out that, "evolution is as much about reproductive strategy as foraging behavior." It would certainly seem that a survival strategy dependent on bipedalism and a home base requires other adaptations and preadaptations. For example, **sexual bonding** between males and females can motivate male foragers to continue to provision their family group because of their ties to particular females and can extend male involvement into helping parent their offspring, assuming they have good or reasonable assurance of paternity.

Certainly, one important element that encourages human sexual bonding and year-round copulation is the absence of seasonal estrus cycles marked by specific, externally recognizable signals ("concealed ovulation"). Human secondary sexual characteristics that persist from puberty onward stimulate continued interest of both sexes in sexual bonding. These traits include the relatively large penis in males, the enlarged mammary glands and increased subcutaneous fat deposits in females (e.g., buttocks) as well as the hair and apocrine scent glands displayed in both sexes. Some of these sexual features are associated with the bipedal stance, and point to an evolutionary process that may have begun among australopithecines but developed more fully in *Homo.*

Bipedalism fosters the use of manual weapons such as stick wielding and stone throwing, which extends the reach of hominins beyond the teeth, claws and other defenses of animal competitors, predators and prey. Bipedalism also allows hominins to carry tools and weapons from place to place as well as to move offspring from one camp to another or from one food resource to another.

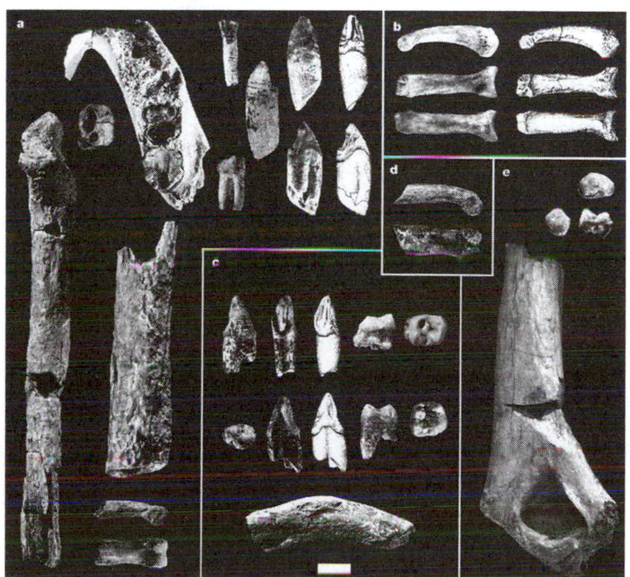

FIGURE 20-13 Specimens of *Ardipithecus kadabba* dated at 5.2 to 5.8 Mya. (From Haile-Selassie, Y., 2001. Late Miocene hominids from the Middle Awash, Ethiopia. *Nature,* **412,** 178–181.)

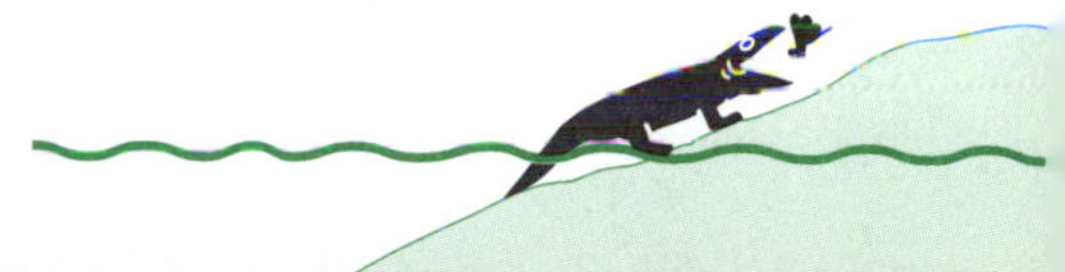

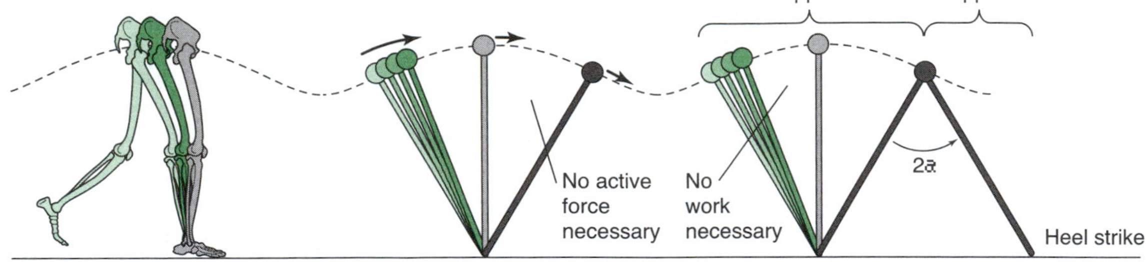

FIGURE 20-14 Schematic of the inverted pendulum model of human walking. In this model, very little energy is used for the pendular mechanics portion of walking. Active work is performed to make adjustments at various points during the stride and forward propulsion is provided by contraction of the calf muscles which plantarflex the foot before toe-off. (From Kuo A. D., J. M. Donelan, and A. Ruina, 2005. Energetic consequences of walking like an inverted pendulum: step-to-step transitions. *Exerc. Sport Sci. Rev.*, **33**, 88–97.).

Anatomical Basis for Bipedalism

Several key anatomical features are associated with bipedal locomotion. Paleoanthropologists use the presence or degree of expression of these features in hominin fossils to assess their positional behavior (locomotion and movement). These features comprise a functionally interrelated suite, the most important of which are outlined below:

1. *Increased relative length of the lower limb.* Walking is modeled as an inverted pendulum. At each stride, the center of mass of the trunk is shifted forward. At that point, one of the lower limbs swings largely passively like a pendulum and the body falls forward on to the foot of the swung limb. Increased limb length both increases the speed of walking and also the efficiency of this pendulum-like gait (**Fig. 20-14**).
2. *The valgus knee.* During walking, the body must be balanced on one foot during the support phase of the stride. To minimize lateral excursions of the center of mass during the stride, the feet thus need to be as close to the midline as possible. For this reason, the human femur angles towards the midline. This is associated with some distinctive features, shown in **Figure 20-15.**
3. *Lumbar curvature.* A key change allowing habitually upright posture in hominins was the evolution of an elongated and curved lumbar spine (**Fig. 20-16**). In fact, although the shape of the hominin pelvis is significantly altered in humans (**Fig. 20-17**) its orientation is actually remarkably similar to that of chimpanzees and other quadrupeds. A functional cost of the lumbar curve is greater instability of the lower lumbar vertebrae and thus lower back pain.
4. *Adducted hallux.* In human bipedalism during the stance phase of the stride, weight is transferred from the heel to the ball of the foot and finally to the hallux (big toe; **Fig. 20-18**). An adducted hallux with shortened but more robust phalanges is thus an important adaptation for walking and running bipedally.
5. *Enlarged heel and stable ankle.* As shown in Figure 20-18, the calcaneus provides the lever arm for the large muscles that plantarflex the foot; thus, an elongate and robust calcaneus is considered an adaptation for hominin bipedalism. Along with this are changes to the joint surfaces and ligaments of the ankle that stabilize the ankle and restrict its mobility. In contrast, a highly mobile ankle is important for climbing.
6. *A shortened and laterally rotated ilium.* As shown in **Figure 20-19**, three large muscles of the hip, the *gluteus maximus, medius* and *minimus* originate on the dorsal surface of the ilium. In chimpanzees and other quadrupeds the dorsal surface of the ilium is directed dorsally and the *gluteus medius* and *minimus* act as extensors of the femur. The *gluteus maximus* is smaller and acts as an abductor of the femur. In hominins, the *gluteus maximus* is positioned to be the major extensor of the thigh. That's why this muscle is by far the largest of the three in humans. On the other hand, the *gluteus medius* and *minimus* become abductors of the femur. Their function, however, is to balance the upright body on the planted limb during the stance phase of the stride. This is known as the *human pelvic tilt mechanism.*

Although chimpanzees are capable of walking bipedally, their mechanics of bipedalism differs dramatically from humans because chimpanzees do not have the adaptations listed above. As their center of gravity falls considerably in front of the pelvis (**Fig. 20-20**), chimpanzees cannot stand completely vertical and walk without leaning their trunk backwards. Instead, they walk hunched over. Because their *gluteus medius* and *minimus* function as extensors and not abductors and because of their short inflexible lumbar spine, chimpanzees must swing their body to one side to be balanced on the planted limb. For this reason, they swing from side to side when walking (Fig. 20-20).

How Similar Are Chimpanzees to Humans Genetically?

Because chimpanzees are our closest relatives, much effort has gone into comparing humans and chimpanzees in terms of behavior, morphology and genetics. A striking conclusion that emerges from these comparisons is that while the behavioral and morphological differences are obvious and significant, the genetic differences are more subtle.

It is now relatively accepted that the date of the **human–chimpanzee divergence** is around 6 Mya. This estimate is based on the alignment of some 20 million base pairs in a study of five primate species performed by Patterson and colleagues (2006). Fossil evidence places the divergence at 7 Mya. This million years' difference could reflect inadequacies in one or other dating methods, mosaic evolution, and/or interbreeding until the two groups became genetically distinct. The entire chimpanzee genome has recently been sequenced and a comparison of this initial sequence to the human genome sequence reveals 96% similarity at the base-pair level (The Chimpanzee Sequencing and Analysis Consortium, 2005). An average chimpanzee and an average human will thus differ by four base pairs out of every 100. By comparison, randomly chosen humans will be similar at 99.9% or will differ only in one base pair out of 1,000.

Although much work remains to be done on what these differences mean in functional terms, research to date shows remarkable similarity in function and sequence for the genes that have been studied. The current hypothesis, therefore, is that most of the differences between chimpanzees and humans relate not to differences in function in individual genes but rather in where, when and how much of these genes are expressed during development and later life. In

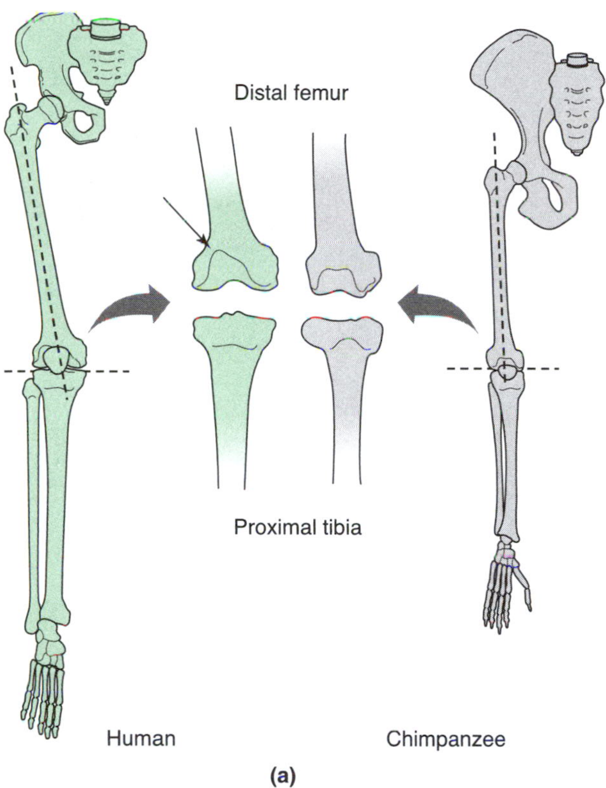

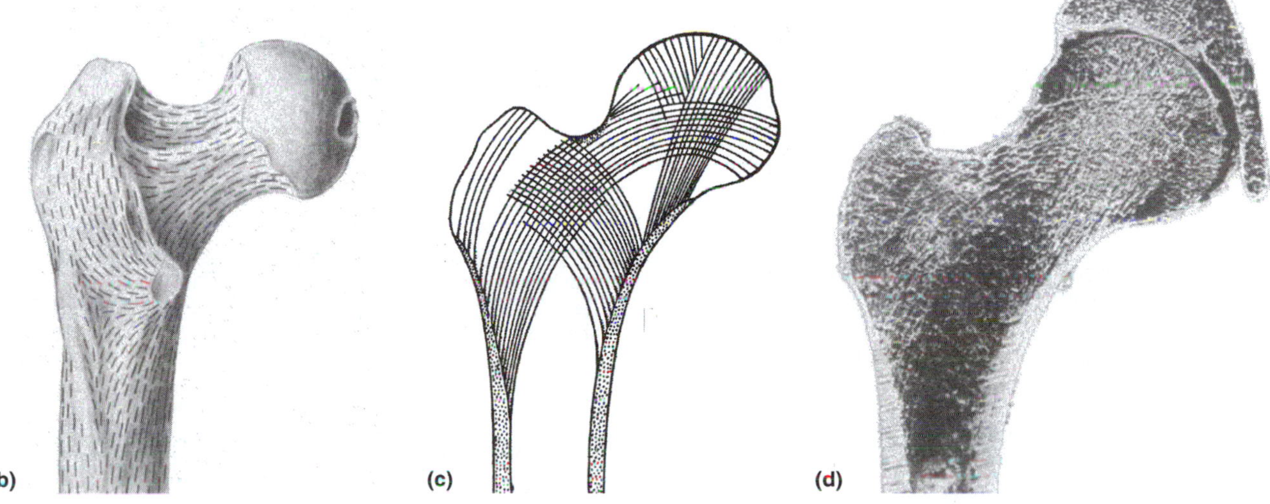

FIGURE 20-15 (a) A comparison between the angulation of the femur in humans and chimpanzees with enlarged views of the distal femur and proximal tibia. The medial angle of the femur in humans results in distinctive features on the distal femur such as the larger medial epicondyle and the lateral expansion of the patellar groove (*arrow*). These and many other features can be used to infer hominin locomotor patterns. (b–d) The orientation of bony trabeculae, which demonstrates the orientation of the major lines of force on the head of the human femur. (a, modified from Aiello, L., and C. Dean, 1990. *An Introduction to Human Evolutionary Anatomy*. Academic Press, London. b–d, from Hall, B. K., 2005a. *Bones and Cartilage: Developmental and Evolutionary Skeletal Biology*. Elsevier Academic Press, London, UK, modified from Murray, 1936).

other words, most of the differences between chimpanzees and humans may involve gene regulation rather than the genes themselves. One should appreciate that the incredibly complex nature of development means that small differences in gene function or pattern of gene expression can produce large changes in phenotype. Many of the important differences between humans and chimpanzees is probably accounted for by only a small proportion of the roughly 4% difference in base pair sequence.

Variation in gene regulation can be studied using **microarray analysis** (see Chapter 12). Analysis of an array of complementary DNAs (cDNAs) from the livers of humans (*Homo sapiens*), chimpanzees (*Pan troglodytes*), orangutans (*Pongo pygmaeus*) and rhesus macaques (*Macaca mulatta*) identified two classes of genes, one that was shared by the four species — indicating *stabilizing selection* over the 70 My since they diverged — and a second class (in which transcription factors were in excess), whose expression differed significantly in humans from the other three species, indicative of *directional selection.* The implication is evolution of

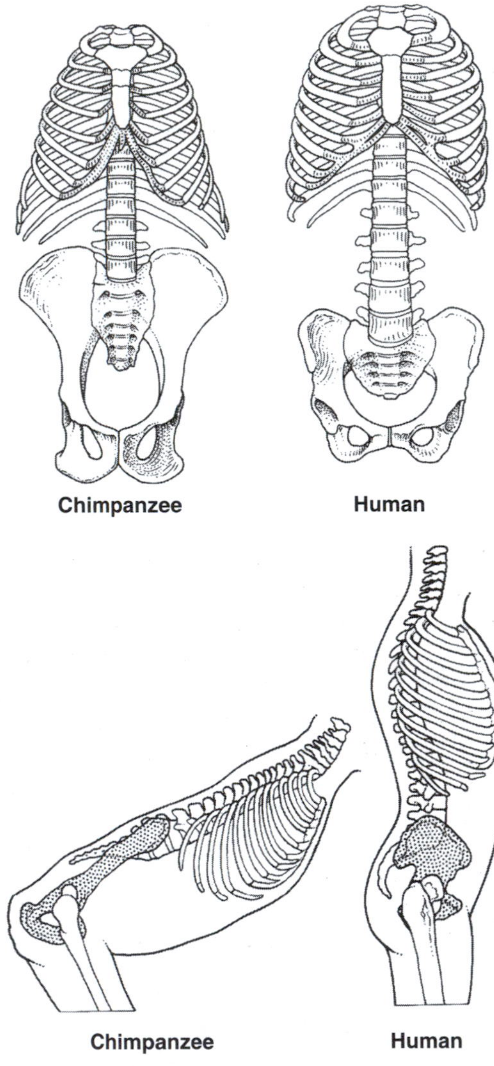

FIGURE 20-16 Comparison of the trunk skeleton in humans and chimpanzees. Note that in humans, the lumbar region of the spine is much longer and is also curved. The human pelvis is much shorter but is only slightly altered in terms of overall orientation. (From Aiello, L., and C. Dean, 1990. *An Introduction to Human Evolutionary Anatomy*. Academic Press, London, UK.)

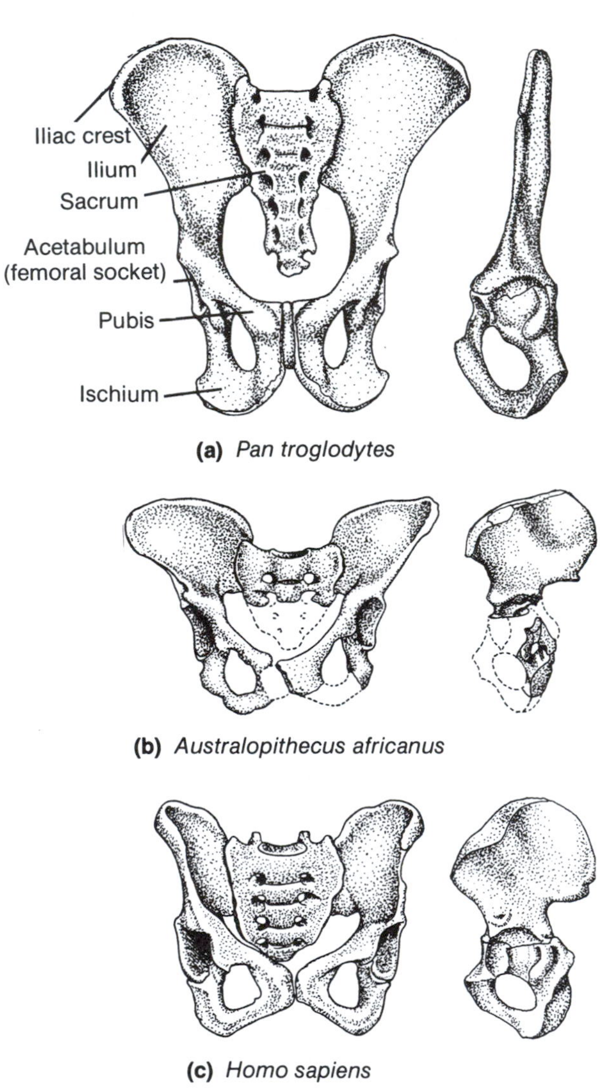

FIGURE 20-17 Comparison between the pelves of a chimpanzee (a), an australopithecine (b), and a modern human (c). Frontal views are on the left, lateral views on the right. Note that the distance between left and right acetabuli in *A. africanus* is less than in *H. sapiens.* This increase in interacetabular distance in humans may have resulted from selection for a relatively large birth canal to permit the passage of newborn infants with crania larger than those of australopithecine newborns (Rosenberg and Trevathan, 1996; see also Box 20-5). (Adapted from Le Gros Clark, W. E., 1978. *The Fossil Evidence for Human Evolution,* 3rd ed. University of Chicago Press, Chicago.)

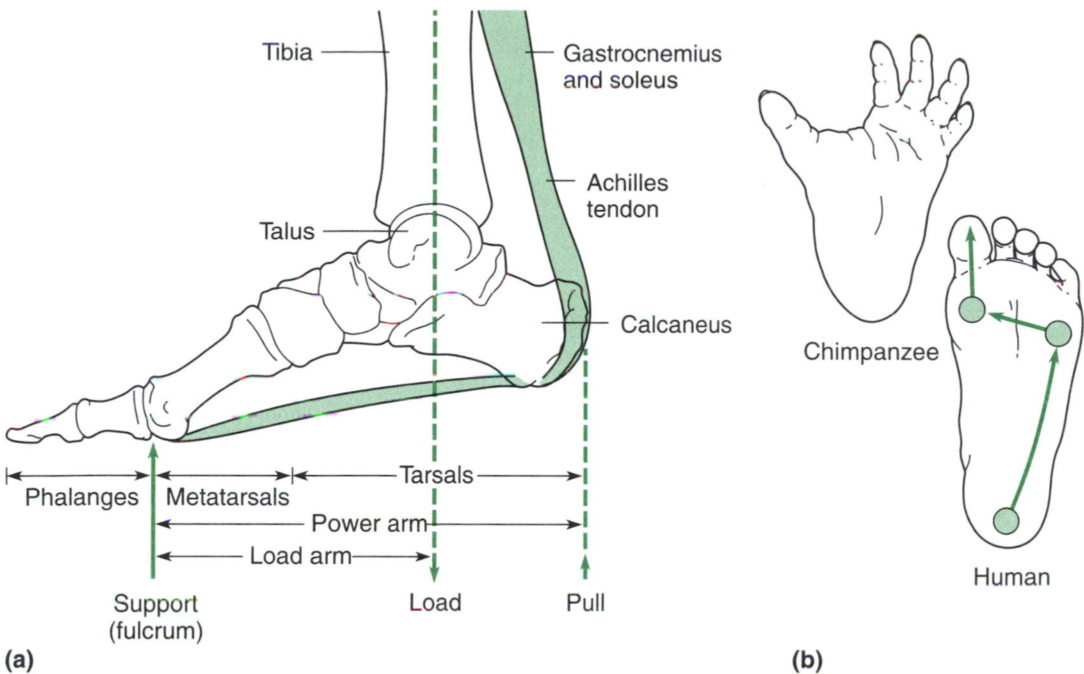

FIGURE 20-18 (a) Leverages in the human foot that provide propulsive forces for walking. Contraction of the gastrocnemius and soleus muscles in the calf pulls the Achilles tendon attached to the heel (calcaneus). This produces the power (power arm) required to transfer the weight load (load arm) to the metatarsals. (b) Plantar view of chimpanzee and human feet. In the chimpanzee, the first metatarsal and its accompanying phalanx (big toe) are at a marked angle to the other metatarsals, the phalanges are relatively long and curved and the foot can be used for grasping. In humans the phalanges are reduced, and the more robust first metatarsal and big toe are parallel to the others, thus enhancing the ability to walk or run directly forward. Note that although the human foot is narrow, it acts as a tripod on which weight is stably distributed (*indicated by the three circles*); the *arrows* show how weight transfers between these three centers to the "push off" on the big toe. (Adapted from Campbell, B. G. 1998. *Human Evolution: An Introduction to Man's Adaptations.* Aldine de Gruyter, New York.)

transcriptional regulation in the divergence of humans from the other great apes (Gilad et al., 2006).[1]

Australopithecines

In 1925, anthropologist Raymond Dart reported an early hominin fossil from a lime quarry at Taung in the Cape Province of South Africa.

Ascribed to a new genus, *Australopithecus* (southern ape), the fossil consisted of the front part of the skull and most of the lower jaw (**Fig. 20-21b**) of a six-year-old (the "Taung child"), which Dart named *Australopithecus africanus.* All the deciduous teeth were present as well as the first of the permanent replacement molars. Although these teeth were generally larger than those of humans, they showed features in the multicusped nature of the anterior milk molar, which is single-cusped in apes, seen in humans. Also, the lunate sulcus (the anterior border of the visual area in the brain) in the endocranial cast was further back than its usual position in apes, more like that of humans. Some interpretations of the Taung skull indicated that the adult brain volume would have been 450 cm^3 (midway between chimpanzee and gorilla) and that the adult body was smaller than that of chimpanzees, with weights between 18 and 32 kg (40 to 70 lbs).

Because of the Piltdown forgery (**Box 20-2**), it was more than 20 years after the Taung discovery before most anthropologists began to accept the human-like nature of the australopithecine fossils. By then, anthropologists had found many such fossils in other sites in Africa, such as the adult australopithecine skulls at *Sterkfontein*, not far from Taung (Fig. 20-21d–g). The thickly enameled teeth of these fossils indicated heavy tooth usage that wore the teeth flat before the dentin was exposed (perhaps also indicating a longer life span than apes). The Sterkfontein fossils also showed that, although the australopithecine premolars and molars were larger than those of humans, the canines were smaller than those of apes and no longer projected above the tooth row. Paleontologists set the date for these fossils at about 2.5 to 3 Mya.

Robinson (1972) and others showed post-cranial australopithecine skeletal material, such as the pelvis and vertebrae, to be human-like, with a distinct lumbar curvature of the spine indicating erect posture (**Fig. 20-22**). As in hu-

[1] Reinforcing regulatory control in human divergence is a study by Edwards et al. (2005) of 1,400 conserved noncoding but regulatory elements shared by the human genome and the genome of a teleost fish, fugu (*Takifugu rubipres*).

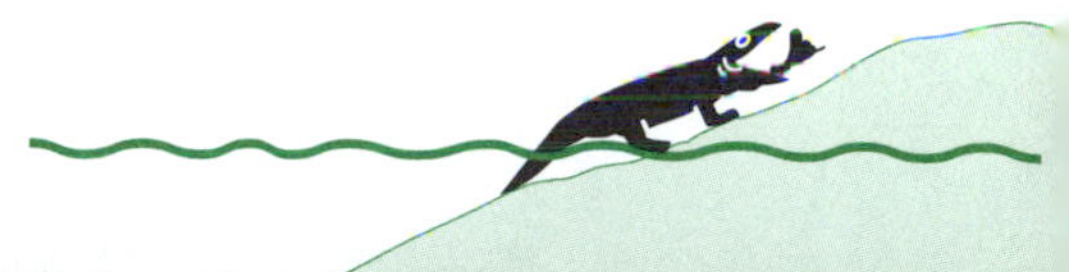

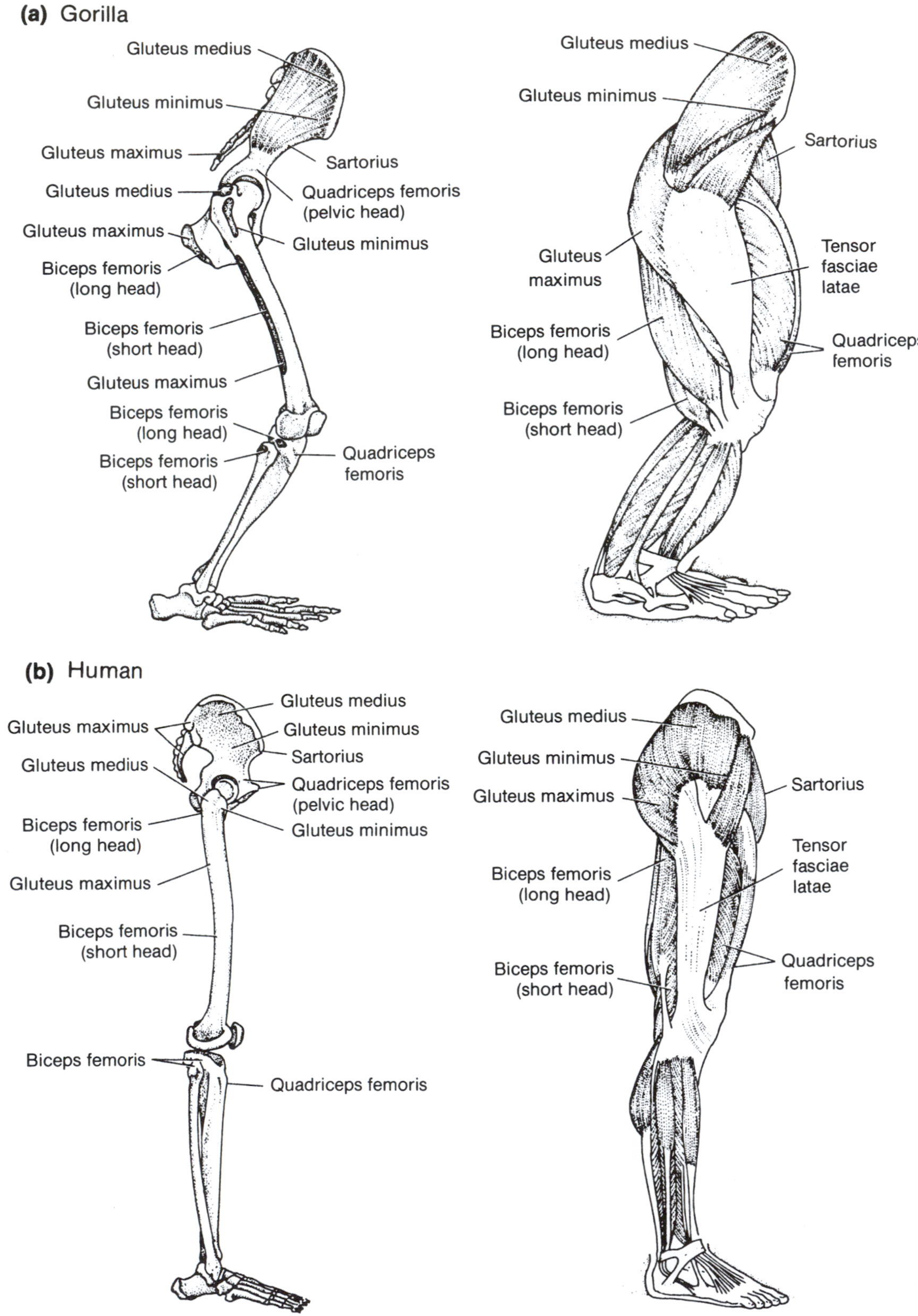

FIGURE 20-19 Bones of the hind limbs (*left side*) showing origins and insertions of the major muscles (*right side*) of a gorilla (a) and human (b). In both primates the muscles that cause the femur to swing forward in relation to the pelvis (flexion) are the sartorius and quadriceps. The broadened human pelvis provides leverage (pulling power) for these muscles when the individual is standing erect. Gorillas can only gain such leverage in the bent position shown. (See also Lovejoy, 1988.) (From *Paleoanthropology,* 1980 by M.H. Wolpoff. © The McGraw-Hill Companies, Inc.)

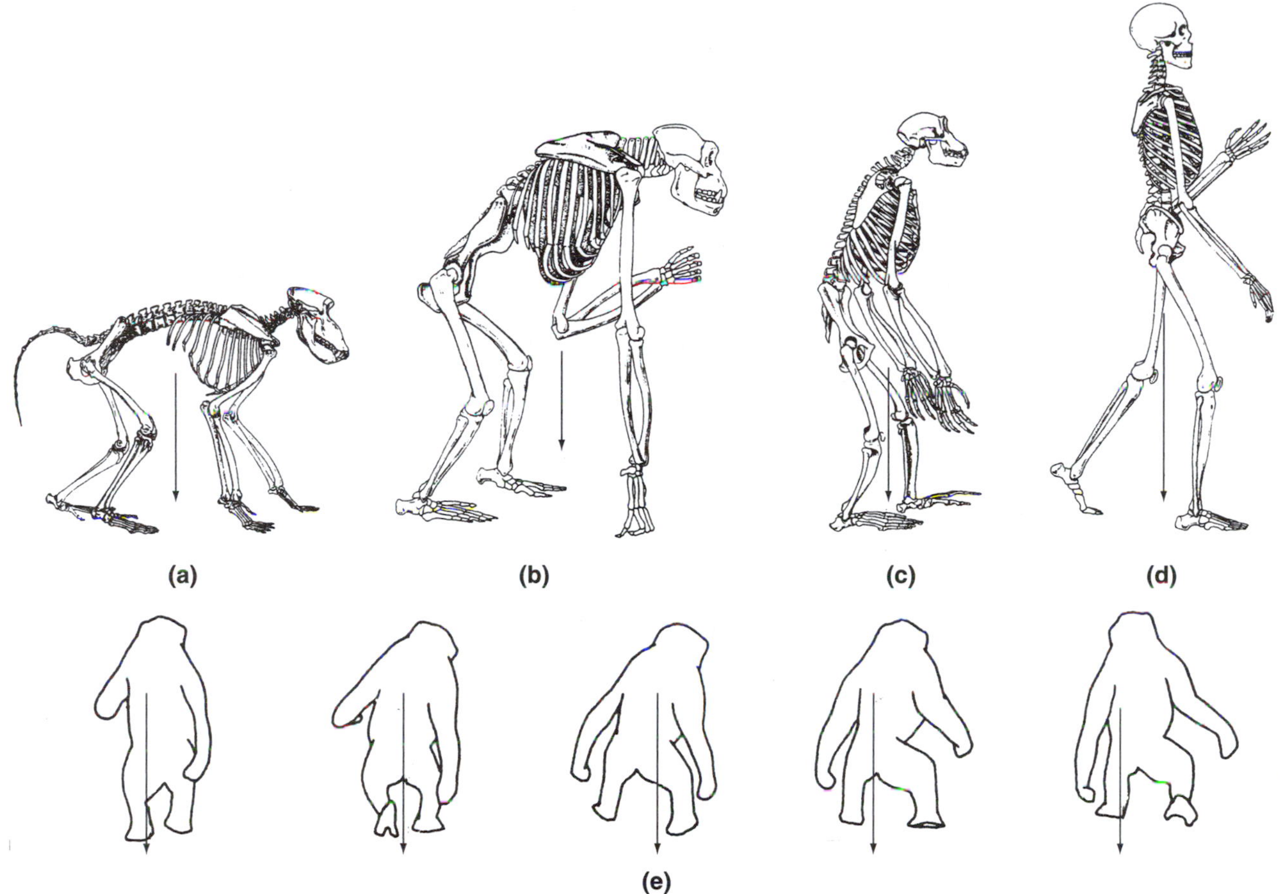

FIGURE 20-20 Centers of gravity (*arrows*) in the common stances of four ground-dwelling primates: baboon (a), gorilla (b), chimpanzee (c) and human (d). When humans walk bipedally the center of gravity remains in approximately the same vertical plane as the skeletal axis, the spinal lumbar curve helping to center upper body weight in line with hip and knee. Because their center of gravity lies in front of the hip, apes walking bipedally perform strenuous muscular activity to keep from falling forward — like humans trying to walk in a forwardly tilted or crouched position. Also, the abductor muscles between femur and hip used for lateral stability in human walking are used as extensors in apes, so a standing ape lifting its left leg, for example, will tilt toward the left unless it bends its trunk over the right leg to regain balance. A walking bipedal ape (e) thus tilts from side to side, causing shifts in the center of gravity that must be corrected by muscular exertion.

mans, vertical weight was transmitted through the outer condyle of the knee, and *Australopithecus* could walk bipedally, although this may not have been the exclusive mode of locomotion (Kingdon, 2004). Studies on australopithecine balancing organs (bony semicircular canals in the ear) suggest that bipedalism may have accompanied aerial climbing (Spoor et al., 1994).

Other Sites, Other Finds

Uncovered in other South African sites was a somewhat larger hominin having a mature weight of 36 or more kg (80 lbs), called *Australopithecus robustus* (Fig. 20-21e). In addition to larger size, this group — often represented as an offshoot of *A. africanus* — had significantly larger teeth and jaws, reflecting a different, perhaps more herbivorous, diet (e.g., seeds, nuts and tubers) with emphasis on more powerful grinding. The brain, too, was larger than that of *A. africanus,* with a volume of about 550 cm^3, but this may have been primarily associated with increased relative body size rather than increased intelligence.

In East Africa a Pliocene australopithecine, *A. boisei* (Fig. 20-21f), reflects selection for even larger molars than *A. robustus.* The largest of the australopithecines, it shows various cranial adaptations for powerful masticating jaw muscles that indicate a diet of tough plant food such as seeds and fruits with hard husks and pods.

Interestingly, an early robust australopithecine (*Australopithecus aethiopicus*) found in 2.5-My-old deposits at Lake Turkana, Kenya, indicates that these massively built australopithecines may have evolved separately but in parallel with the *africanus–robustus* lineage. If we follow both these lineages forward from the Pliocene into the Pleistocene less than 2 Mya, it seems that the molars were becoming larger while the face was shortening and becoming more

vertical and humanlike. Their common skeletal similarities, according to McCollum (1999), are the result of convergent evolution caused by "developmental by-products of dental size and proportion," rather than by shared cladistic synapomorphies. The presence of convergent/parallel changes in two different lineages is important, indicating as it does common selective forces and parallel responses to those forces.

Of an earlier age than the above groups are a series of fossils found in East African sites at Laetoli, Tanzania and the Afar (Hadar) region of Ethiopia. Anthropologists include these fossils, spanning an interval from about 3.9 to 3 Mya, among the australopithecines under the species name *A. afarensis.* Figure 20-21g shows a reconstruction of the face of this early hominin species, showing the heavy brow ridges, low forehead and projecting (prognathous) mouth. Despite some such similarities, *A. afarensis* displays a large number of important cranial, dental and skeletal differences from apes; for example, although the canines are larger in *afarensis* males than in females, this sexual dimorphism is much less pronounced than in apes or early Miocene pongid-like fossils. Selection resulted in modification of tooth positioning, enamel thickness, and the resulting wear facets to facilitate greater transverse jaw movements that retain cutting functions and optimize side-to-side grinding.

Perhaps the earliest australopithecine fossils known to date are relics of an even earlier Ethiopian species named *Ardipithecus ramidus,* dating between 4.3 and 4.5 Mya (White et al., 1995). The species shows hominin-like reduced

(a) Chimpanzee (*Pan troglodytes*)

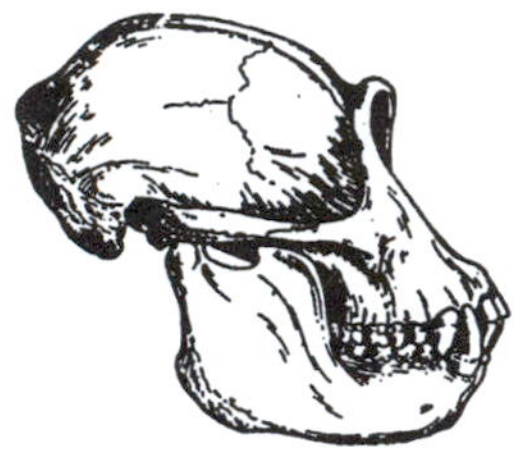
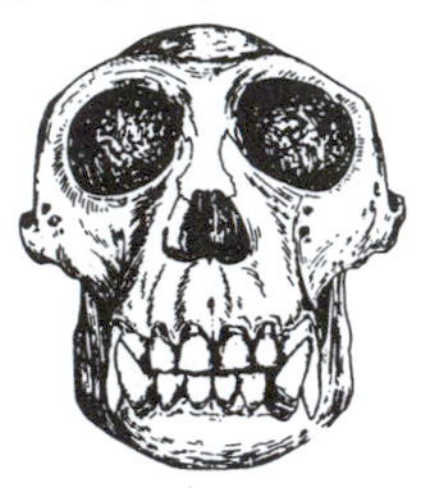

(b) Taung child (juvenile *A. africanus*)

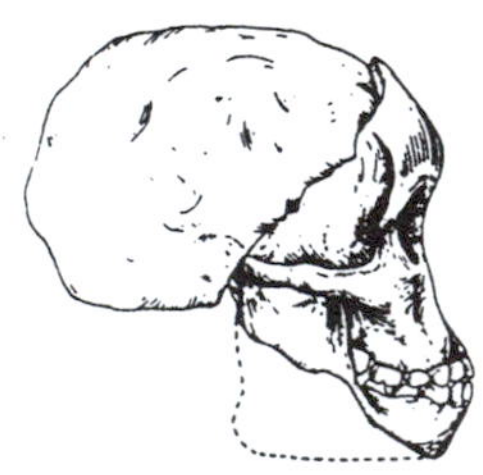

(c) "Piltdown man"

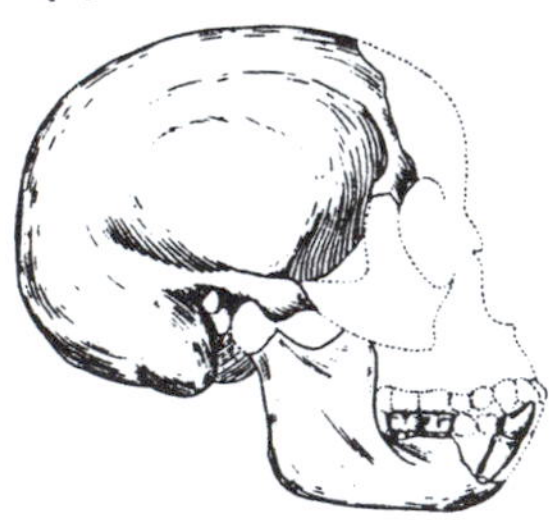
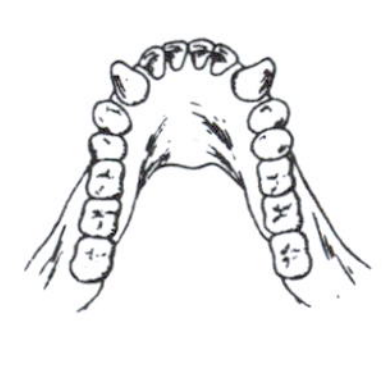

(d) Adult *A. africanus*

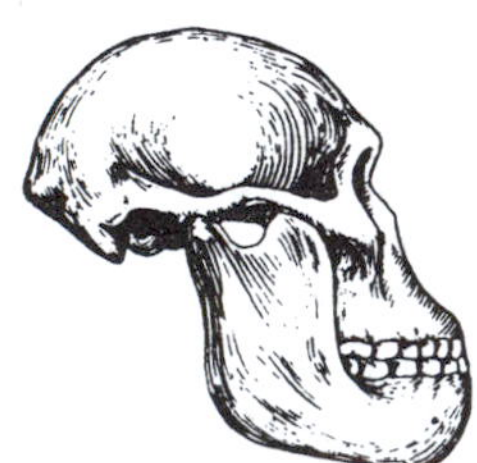

(e) *A. robustus*

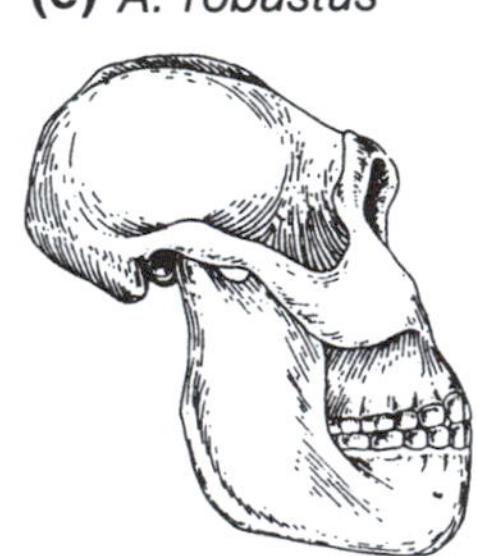
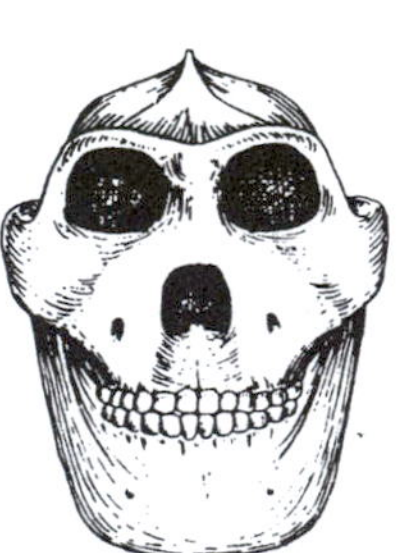

(f) *A. boisei*

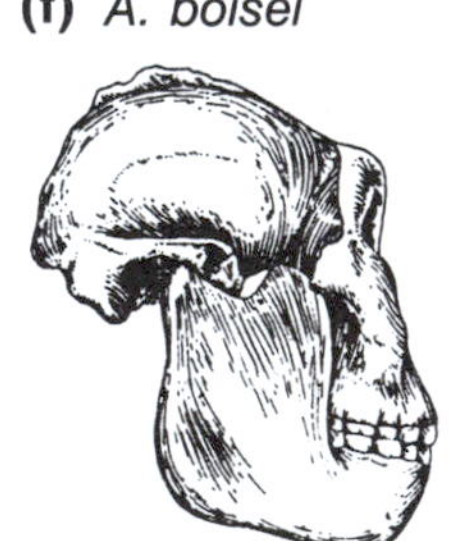
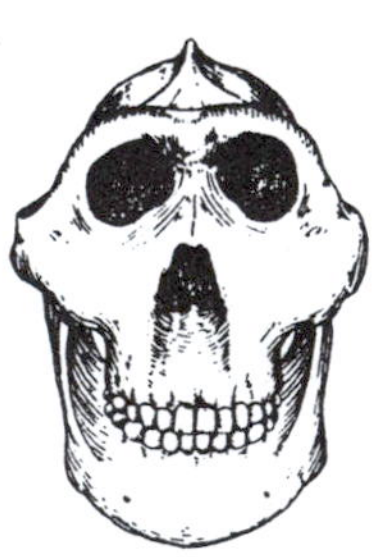

(g) *A. afarensis*

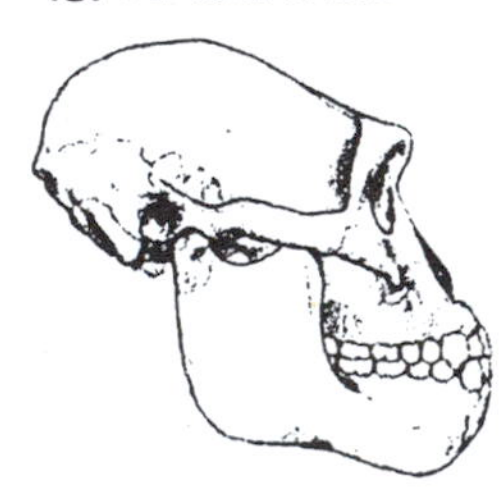

FIGURE 20-21 Skulls of a chimpanzee (a) and fossil hominins described in the text, including cranial and lower jaw fragments of the Piltdown forgery (c). (Adapted from Johanson, D., and M. Edey, 1981. *Lucy: The Beginnings of Mankind.* Simon and Schuster, New York.)

BOX 20-2 Piltdown Man

AT THE TIME of the discovery of the early hominin fossil at Taung in South Africa, most anthropologists interpreted the existing evidence to indicate that early humans had large braincases and ape-like jaws, with large canines. Evidence for this assumption came from what was interpreted as a fossil cranium and lower jaw found in 1912 at Piltdown, England, that showed such features (Fig. 20-21c). Many anthropologists accepted the Piltdown fossil as valid for about 40 years until a number showed that the entire skull was a hoax: The teeth had been artificially ground down, the cranium was of a different age than the jaw, artificial pigmentation had colored the bones and the molar teeth had long roots like those of apes. Moreover, the associated animal fossils at the Piltdown site had a large accumulation of radioactive salts whose origin could be traced to a site in Tunisia. **Piltdown man,** therefore, turned out to be a combination of a human cranium and the lower jaw of a female orangutan, a hoax perpetrated by someone who knew enough to destroy all obvious signs of the pseudofossil's true origin by removing the jaw joint and modifying other features.[a]

Spencer (1990) argues that the Piltdown forgery was perpetrated by Charles Dawson — the principal "discoverer" of the Piltdown fossils — in conspiracy with Arthur Keith, who was a leading British anatomist and physical anthropologist. Their motivations were presumably self-centered: Dawson's to become a Fellow of the Royal Society (FRS), and Keith's to promote his view of the antiquity of a large human brain. Although other historians of science suggest other suspects, no conclusive evidence has appeared identifying the perpetrator(s). Whoever perpetrated this hoax, the result for a time was the preservation of false views with false facts. The events that followed showed that false facts can be challenged in science and false views replaced.

[a] See Weiner (1955), Spencer (1990), Millar (1998) and Russell (2003) for the story of Piltdown man.

sexual dimorphism for canine teeth as well as indications of an upright stance based on forward positioning of the foramen magnum, the aperture through which spinal nerves enter the base of the skull. Most *A. ramidus* fossils are teeth that indicate relationships with both *A. afarensis* and extant great apes, especially chimpanzees. As its discoverers point out, "*A. ramidus* is the most apelike hominin ancestor known." More than any other hominin-like fossil, *A. ramidus* deserves to be called the "missing link" between hominins and apes.[2]

Because postcranial fossils of *A. ramidus* have not been found, questions about size and type of locomotion remain unanswered. However, *A. afarensis,* among early hominins, has provided some relevant information. For example, the fairly complete skeleton of an *afarensis* female ("*Lucy*") found at Afar in Ethiopia (Fig. 20-22a) shows a small, muscularly powerful body, perhaps only 1 to 1.2 m in height, with relatively longer arms than modern humans but presenting a habitual bipedal stance and some form of bipedal locomotion. In support of such early bipedalism are footprints dated to about 3.7 Mya, preserved under a layer of volcanic ash at Laetoli. These prints of two individuals who walked along the same path for a distance of more than 21 m are of distinctly bipedal hominins, demonstrating that bipedalism preceded many other hominin adaptations such as increased brain size.

[2] Although *Ardipithecus ramidus* may be the most deserving "missing link," it is not the most famous. That prize goes to *Pithecanthropus erectus* (now *Homo erectus*), whose features were predicted by Ernst Haeckel in 1868. Haeckel even gave it a name, *P. alali,* before it was discovered in Java by Eugène Dubois in 1891. This specimen was the center of discussion, confrontation and intrigue. Dubois's perseverance and truculence during two decades (1900–1920) of withdrawal from social contact — during which he refused to allow others to see the specimens — polarized the scientific community, and, paradoxically, helped to found paleoanthropology (Theunissen, 1981; Shipman, 2001; Shipman and Storm, 2002).

Because all australopithecine skeletal reconstructions starting with *afarensis* show the bipedal stance (Fig. 20-22), the later human genus, *Homo,* must have evolved from a population within this group. What is not yet clear are the exact lineages within these early hominin species:

- Was *ramidus* a bipedal australopithecine, or a tree-climbing member of some other genus?
- Was *A. afarensis* an early form of *A. africanus*?
- Was *A. afarensis* so specialized in the direction of increased food grinding that its descendants could only have been heavy-jawed australopithecines such as *A. robustus* or *A. boisei*?
- Or do the fossils we ascribe to *A. africanus* represent two species, one allied to the "robust" australopithecines and the other ancestral to the *Homo* lineage?

In any case, there is little question that *A. afarensis* stands at or near the base of hominin phylogeny, which indicates that we should regard the australopithecines as a group in which considerable evolutionary change was occurring, exemplifying rapid adaptive radiation of bipedal tropical apes.

Origins of *Homo*

The genus ***Homo*** first appears in the fossil record in East Africa around 2.4 Mya. While similar in size to members of

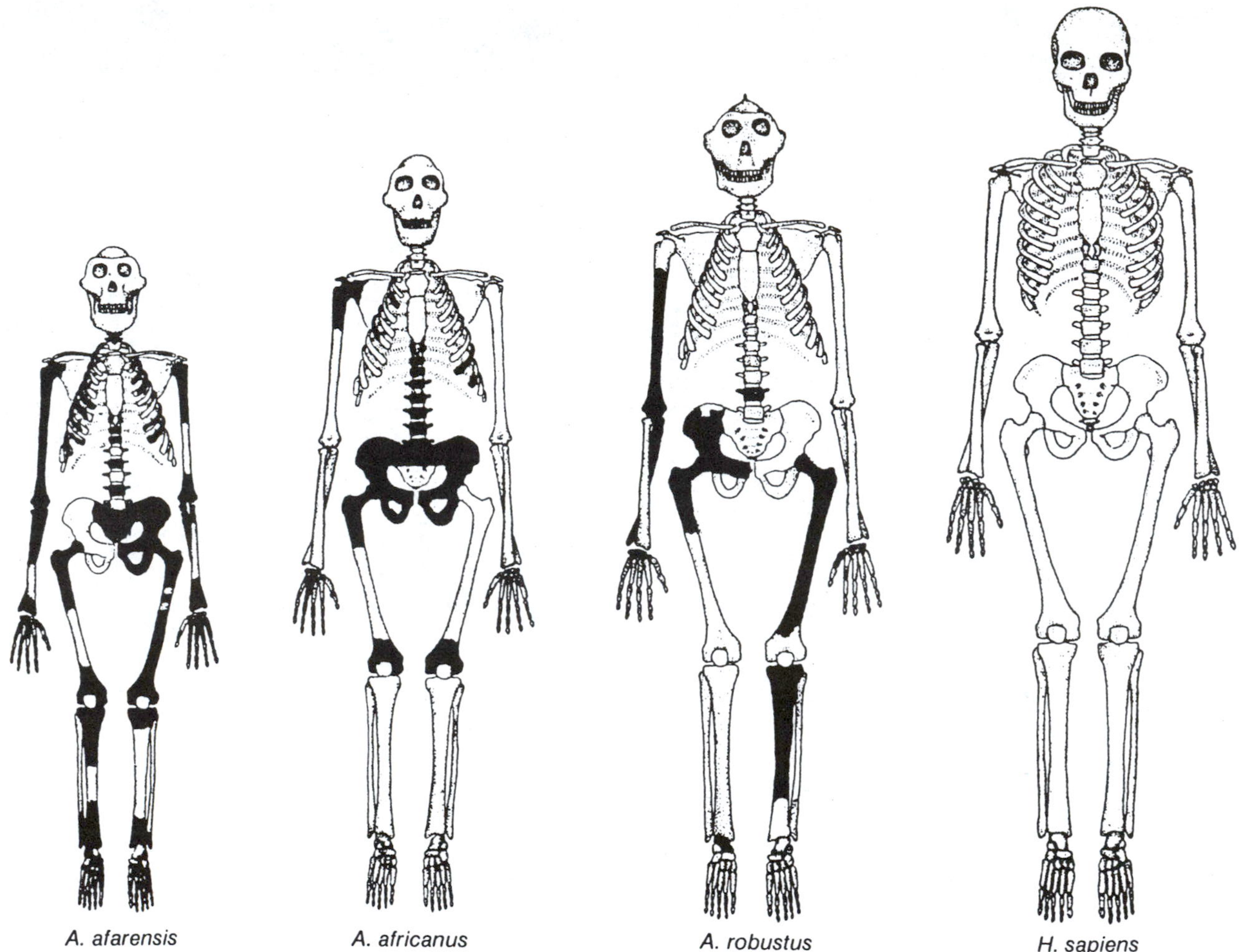

FIGURE 20-22 Skeletons of three australopithecines and a modern human. Black portions indicate fossil finds for each skeleton. In all these hominids, the pelvis is relatively shallow and rounded, and the femurs tilt toward the midline, indicating that the pelvis is supporting the trunk, and that body weight is transmitted directly downward through the hip and knees when the individual stands erect. Among other australopithecine human-like traits are their forelimb proportions, relatively short toes and presence of a large heel bone (calcaneous). (From *Lucy: The Beginnings of Humankind* by Donald Johanson & Maitland Edey. New York, Simon & Schuster, Inc., 1981. © 1981 by Donald C. Johanson and Maitland A. Edey.)

the genus *Australopithecus,* the earliest specimens assigned to *Homo* are distinguished by a suite of characters including increased cranial capacities (600 to 700 cm^3) and a reduction in the size of the molar teeth. Many workers recognize two species of early *Homo. H. rudolfensis* appears in East Africa as early as 2.4 to 1.8 Mya while *H. habilis* is found in a similar geographic range from 2.0 to 1.6 Mya. Examples of both of these early species are shown in **Figure 20-23**.

Associated with the appearance of *Homo* are artifacts indicating that these new hominins were engaged in making regularly patterned **stone tools,** products of the stone industry named *Oldowan* (**Fig. 20-24a** and **Box 20-3**). These tools were used for processing plant materials as well as in hunting and butchering animals (including small reptiles, rodents, pigs and antelopes) and perhaps also in scavenging carcasses of animals as large as elephants. Some tools date back 2.5 My, indicating that the earliest members of the genus *Homo* engaged in scavenging and hunting. Australopithecine species had probably earlier embarked on using simple stone tools, bones and sticks. This is likely because chimpanzees engage in simple tool use and modification.

Homo erectus

About 1.9 Mya, somewhat after the *H. habilis/rudolfensis* fossil period, new groups of hominin fossils appear that are taller than their predecessors, reaching 1.7 m or more.

In Africa, the species *H. ergaster* arose from either *H. habilis* or *H. rudolfensis. H. ergaster* had significantly larger brain size (750 to 1100 cm^3) and body size than earlier *Homo.* Associated with *H. ergaster* are occasional signs of the use of fire and considerable use of stone tools including large hand axes of the *Acheulean* type (Fig. 20-24b). Circumstantial evidence indicates that *H. ergaster* groups engaged in hunting large animals, but this hypothesis is very difficult to test and contrasts with the alternative hypothesis of scavenging. In contrast to both earlier *Homo* and later *H. erectus, H. ergaster* individuals appear to have had tall, lean builds. This is evident in the most complete *H. ergaster* fossil, the juvenile male Nariokotome skeleton that is dated at 1.6 Mya (**Fig. 20-25**, and see Walker and Leakey, 1993).

In 1891 in Trinil, Java, Eugene Dubois discovered the first of the fossils that are now assigned to the species *H. erectus* (Fig. 20-23b). These individuals are distinguished from *H. ergaster* by thicker skulls, heavier brow ridges and a distinctive sulcus behind the brow ridges. Like *H. ergaster,* they had smaller teeth and larger brain volumes (750 to greater than 1,200 cm^3) than early *Homo. H. erectus* specimens have been found in Africa, China and Europe. As with *H. ergaster,* there is circumstantial evidence that *H. erectus* groups engaged in hunting large animals. In addition, they very likely scavenged for meat and collected plant foods. *H. erectus* is associated with the *Achelean* stone industry, which remained remarkably consistent and conservative for over a million years (Box 20-3).

Homo erectus became geographically widespread and existed for a very long period of time.

If we knew them all, the various distinctions among different *H. erectus* groups over time might well be enough to mark off new species. These distinctions, however, encompass anatomical and behavioral traits that we cannot always see on the fossil level, and therefore we don't know when separations or transitions occurred. Were there parallelisms? Were there convergences? Obviously so. Common selective factors may well have caused adaptational similarities in different groups, clouding distinctions between *H. erectus* and *Homo sapiens,* our own species. Some insist there is too much variation among *H. erectus* fossils and too little distinction from later forms to define it as a species. "*H. erectus* is but an early version of *H. sapiens*" (Wolpoff and Caspari, 1997).

For example, hominin fossils found near the Solo River in Java, dated to less than 250,000 years ago, show brain volumes averaging 1,100 to 1,200 cm^3, significantly larger than those of middle Pleistocene *H. erectus* fossils from the same area; yet they are like older fossils in respect to their prominent brow ridges and some other features. Fossils from Swanscombe in England and Steinheim in Germany, dated to about 200,000 years ago, also show such increased brain volumes as well as anatomical traits intermediate between *H. erectus* and *H. sapiens.* Most paleoanthropologists now recognize these more modern forms as a new species, *H. heidelbergensis,* and many have concluded that this species arose from *H. ergaster* and not from the widespread *H. erectus.* It now appears likely that *H. erectus* persisted in parts of the world long after *H. heidelbergensis* arose and even later. *Homo erectus* may have persisted in Southeast Asia (Java) as late as 50,000 years ago. Startlingly, recent finds from the island of Flores in Indonesia reveal what some interpret as a relict population of dwarfed *H. erectus* that persisted as recently as 18,000 years ago (**Fig. 20-26**). This conclusion is controversial, however, as others have argued that the well preserved small bodied Flores individual is more modern but shows signs of pathological dwarfing. It is clear, however,

(a) *Homo habilis*

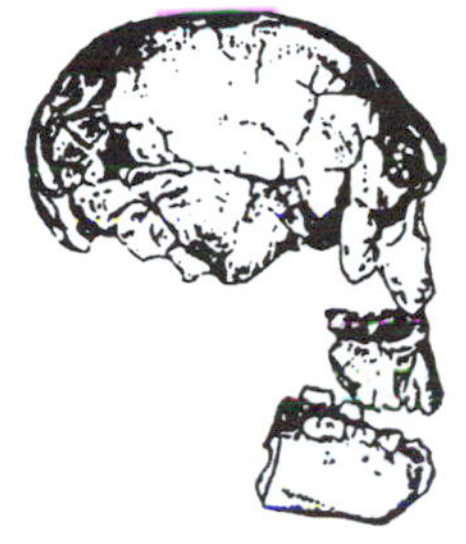

(b) *Homo erectus*

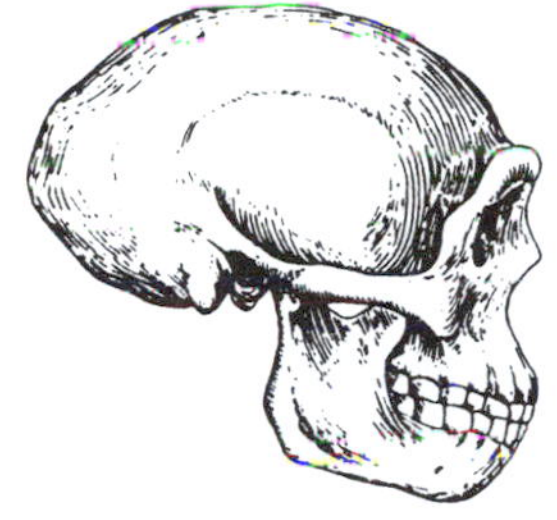
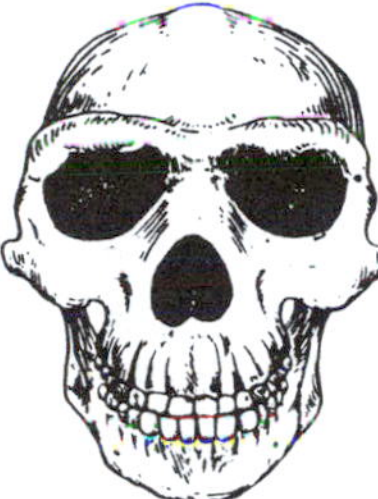

(c) *Homo sapiens neanderthalensis*

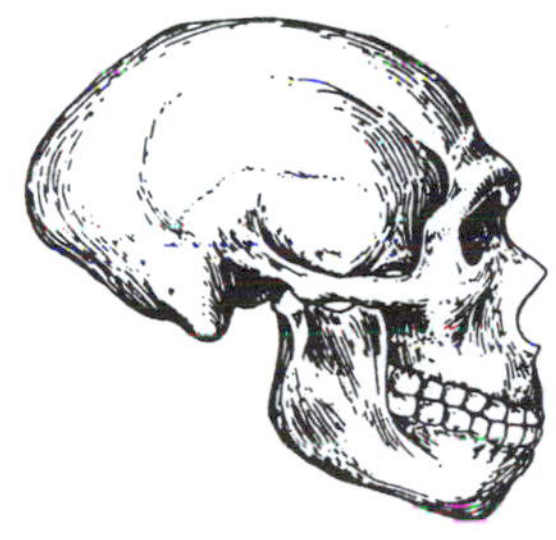
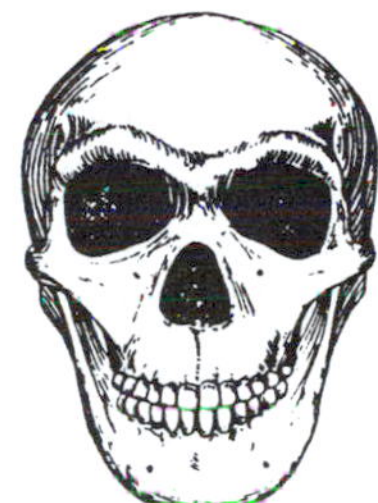

(d) *Homo sapiens sapiens*

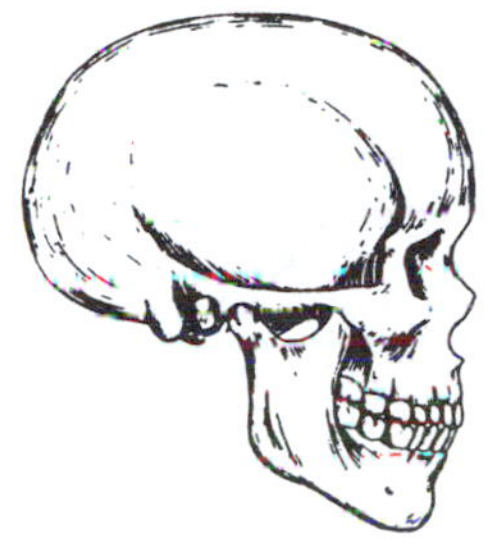
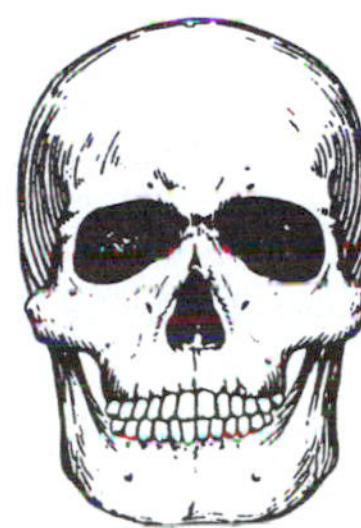

FIGURE 20-23 (a–d) Sample cranial specimens of species in the genus *Homo.*

BOX 20-3 Hominin Tools and Tool-Age Cultures

ANTHROPOLOGISTS CHARACTERIZE four major eras of tool making in hominin evolution (**Fig. 20-24**). The earliest tools appear almost 2.5 Mya. The latest period, the *Magdalenian,* ended some 12,000 years ago and is associated with the first rock paintings.

Oldowan tools, the oldest hominin tools (Fig. 20-24a), which first appear 2.4 Mya, are characterized by chipping stones to create a cutting edge to break up animal carcasses, break open marrow cavities of bones and strip the fibers from tough plants. A single blow is sufficient to create a sharp-edged tool. Oldowan tools are associated with fossils of *Homo habilis.*

Acheulean tools characterize a period about 1.5 Mya and are associated with fossils of *Homo ergaster* and *Homo erectus.* Acheulean tools were chipped from both sides with multiple strikes to create a two-sided cutting edge (Fig. 20-24b). Leaf-shaped tools ("hand axes") appear about 1 Mya.

By 500,000 years ago, Acheulean tool making had spread to Europe in association with *Homo heidelbergensis.* Acheulean tools were found in Europe as recently as 200,000 years ago.

Mousterian tools appear some 200,000 years ago and last until around 40,000 years ago. In Europe they are associated with *Homo neanderthalensis,* in other parts of the world with Neanderthals and with *Homo sapiens.* Characterized by several phases of manufacture, Mousterian tools of various sizes and shapes could be reworked, either to sharpen or to reshape them. They include spear points for hunting and scrapers to clean animal skins (Fig. 20-24c).

Upper Paleolithic tools were the dominant tool forms in Asia and Africa from 90,000 years in Africa (40,000 years in Asia) to 12,000 years ago. Regional differences attest to the sophistication of these tools, including sewing needles and fish hooks. Several periods are recognized:

- The *Aurignacian* period (40,000 to 28,000 years ago) associated with *Homo sapiens* (Cro-Magnon) and *Homo neanderthalensis* (Fig. 20-24d).
- A more temporally restricted *Châtelperronian* period from 40,000 to 34,000 years ago associated with *Homo neanderthalensis* in Europe.
- The *Gravettian* period, from 28,000 to 22,000 years ago, includes ivory beads and carved female Venus figurines.
- The brief *Solutrean* period, from 22,000 to 19,000 years ago, characterized by tools that were shaped by heating and cooling.
- The *Magdalenian* period, from 18,000 to 12,000 years ago, includes the first barbed and multibarbed harpoons, fine arrow heads and wooden tools (spear throwers made from bone, antlers or wood) and is associated with the first rock paintings representing elements of the natural world.

that the present-day condition of only one species of extant *Homo* is rather unusual in the history of our genus.[3]

Many paleoanthropologists have come recently to the view that *H. sapiens* evolved from *H. heidelbergensis.* Molecular evidence suggests that the ancestral population, which gave rise to *H. sapiens,* separated from *H. heidelbergensis* some 200,000 years ago and that this happened in Africa (**Box 20-4**). The paleontological evidence also indicates separation in Africa as Africa is where the earliest modern human fossils are found. Debate continues, however, on whether the molecular evidence excludes or allows the possibility of low amounts of gene flow between this ancestral population (or populations) and other populations of *H. heidelbergensis.* One thing that does appear to be fairly clear is that another species of humans also originated from *H. heidelbergensis* in Europe. This species, *Homo neanderthalensis* has stirred the imagination of scholars and the public alike since the discovery of the first remains of these ancient humans in a cave in the Neander valley, Germany, in 1856.

Neanderthals

Neanderthals (*Homo neanderthalensis*) are somewhat shorter than modern humans and show distinctive brow ridges, large jaws, small chins, robust skeletons with bowed femoral shafts, a large and rounded terminal phalange on each digit and other anatomical features not seen in humans (**Fig. 20-29,** see Fig. 20-23c). Brain size was some 10 percent larger than modern humans. Neanderthals were socially and behaviorally quite 'advanced' in many respects. They were skillful hunters of large animals such as the cave bear and mammoth, produced many complex stone tools, and apparently performed ritualistic social ceremonies, including placing flowers in the graves of their dead. Neanderthals inhabited Europe and Western Asia between 300,000 and 30,000 years ago. Many, although not all, anthropologists believe that Neanderthals deserve as full a membership in *H. sapiens* as higher-skulled groups; that is, *H. sapiens sapiens* (Fig. 20-23d), which began to replace Neanderthals in various parts of the world about 40,000 years ago.[4]

Little is known about the abrupt **disappearance of the Neanderthals;** some of their populations may have died out and others may have merged into the new dominant forms.

[3] Discovery of this dwarf or pygmy population is so recent and the findings so unexpected that analysis and interpretation is ongoing at a fast pace, especially the question of whether they represent a new species of *Homo.* A recent analysis of 140 cranial features led to the conclusion that these individuals represent an early pygmy population with signs in the one complete skull (see Fig. 20-26) present of the developmental anomaly, microcephaly (small brains and, therefore, small skulls). In this study (Jacob et al., 2006), the population is assigned to *H. sapiens,* as it is by R. D. Martin et al. (2006) and by Richards (2006), the former by comparison with microcephalic *Homo sapiens,* the latter invoking modification of the growth hormone-insulin-like growth factor 1 axis and mutation of the *MCPH* (autosomal recessive primary microcephaly) gene family, which is causally related to microcephaly, to explain their diminutive size.

[4] See the papers in Harvati and Harrison (2006) for a recent reanalysis of Neanderthals.

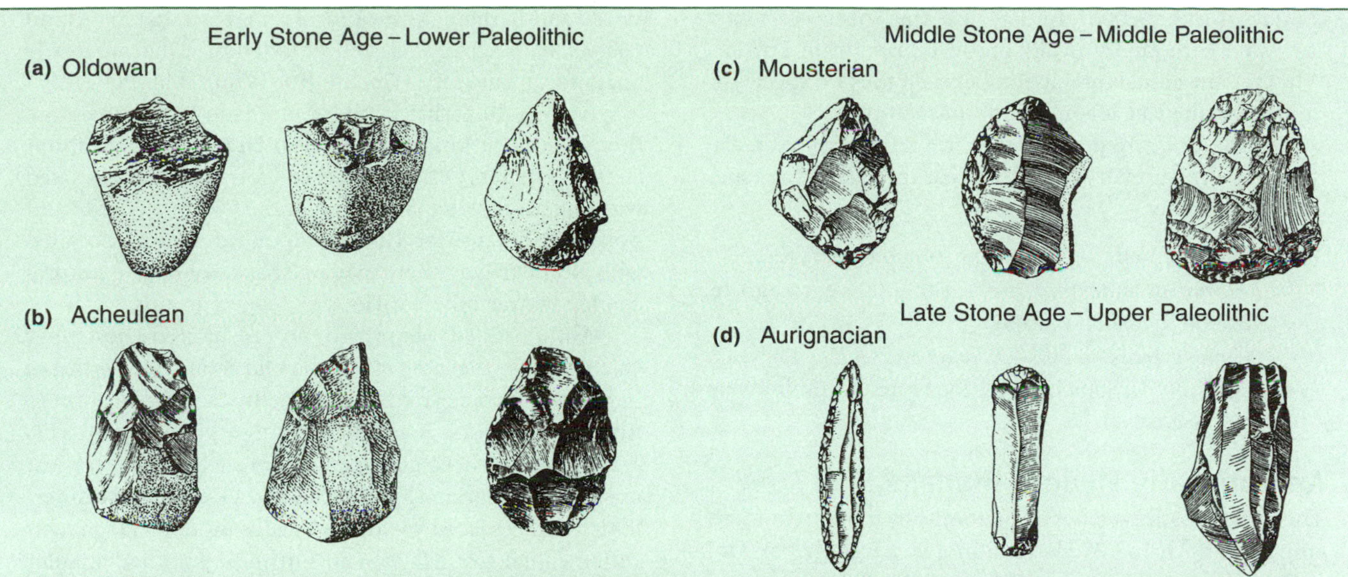

FIGURE 20-24 Stages in stone tool development beginning with the Oldowan stone industry (a), now known to date back at least 2.5 My (Semaw et al., 1997) and to have persisted for more than 1 My. Some paleoanthropologists (Wood, 1997) suggest that although Oldowan tools are traditionally ascribed to *Homo habilis,* they may have been used by earlier hominins. In any case, the cleavers and hand axes of the Acheulean stone industry (b) appear abruptly about 1.5 Mya, and are generally assigned to *H. erectus.* Later stages were associated with other groups: (c) Mousterian (*H. sapiens neanderthalensis*), (d) Aurignacian and Upper Paleolithic (*H. sapiens sapiens*). These industries and cultures constitute the anthropologists' Paleolithic Stone Age. Following this period are the Mesolithic and Neolithic ages, the latter beginning about 10,000 years ago and marked by polished stone tools, pottery, domesticated animals, cultivated plants and woven cloth.

One hypothesis suggests that Neanderthals represent a separate offshoot of the human line, differing from both the *H. sapiens* groups that preceded it and those that followed (Stringer and Gamble, 1993). In support of this view are findings that recovered sequences of Neanderthal mitochondrial DNA are outside the limits of normal *H. sapiens* variability. According to Krings and coworkers (1997), this evidence indicates that "Neanderthal mtDNA and the human ancestral mtDNA gene pool have evolved as separate entities for a substantial period of time and gives no support to the notion that Neanderthals should have contributed mtDNA to the modern human gene pool." It has been pointed out that the extraordinarily large Neanderthal face resulted from biological innovations that permitted strong biting forces to be exerted on their front teeth, which were much larger with deeper roots than in other *H. sapiens* groups. The fact that their incisors and canines often show heavy wear indicates that Neanderthals may have used these teeth for processing tough foods or hides, or both. The expanded nasal chamber in their large face may have served as a radiator, warming and humidifying inspired air in the dry, cold, glacial climates that many European Neanderthals inhabited.[5]

Recent excavations at Devil's tower in Gibraltar show that Neanderthals inhabited parts of Europe as late as 28,000 years ago and possibly as late as 24,000 years ago

[5] A study of 13 recent human populations across diverse geographical and climatic regions shows that shape of the facial portion of the skeleton does have a weak association with climate change (Harvati and Weaver, 2006).

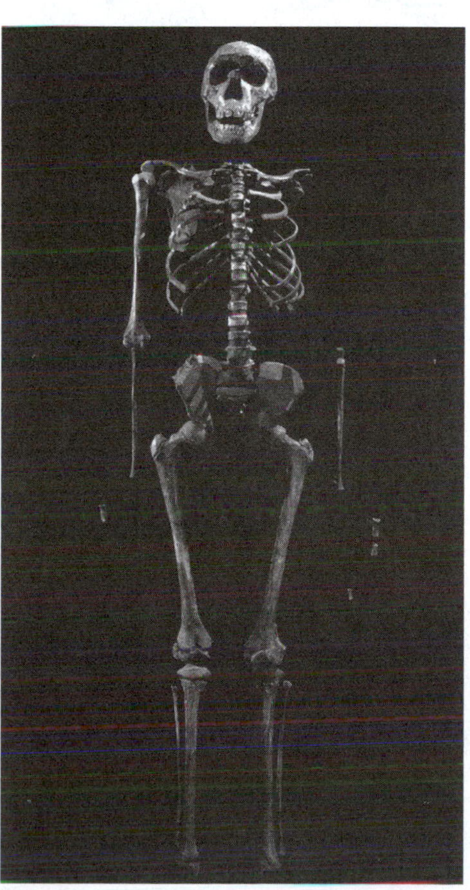

FIGURE 20-25 The Nariokotome skeleton. This juvenile male assigned to *Homo ergaster* is dated to 1.6 Mya. (© Kenneth Garrett.)

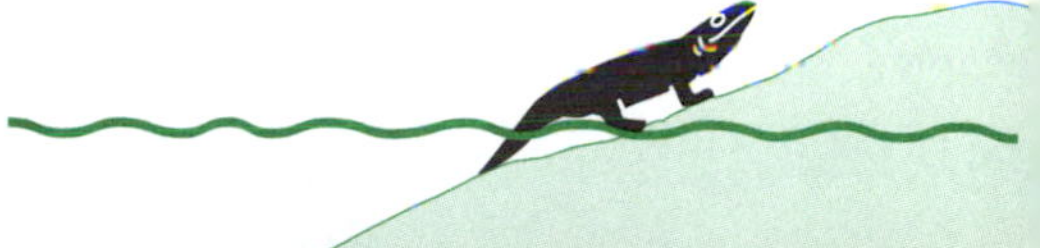

(Finlayson et al., 2006). In that case, this species certainly coexisted with anatomically modern humans in Europe. Whether any actual interaction between the two species of humans took place is a matter of speculation.

With our current knowledge it is reasonable to make the following summary concerning coexistence of three species of hominins:

- *Homo sapiens* — 500,000 years ago to the present
- *Homo neanderthalensis*— 500,000 years ago to 24,000 to 30,000 years ago
- *Homo erectus* — 500,000 years ago to 250,000 years ago (and perhaps 12,000 if the Flores hominins were *Homo erectus*).

Anatomically Modern Humans

The earliest fossil evidence of **anatomically modern humans** comes from Africa.[6] At Herto, Ethiopia, a team led by Tim White and Berhane Asfaw found remains dated to around 160,000 years ago of at least three individuals that are clearly anatomically modern (**Fig. 20-30**) (White et al., 2003).

Among the earlier fossils of more modern humans are those found in Mount Carmel in Israel, dated to about 90,000 years ago. Other transitional forms are all associated with the Mousterian Stone Age industry (Fig. 20-24c and Box 20-3). Because the Mousterian Culture is also associated with Neanderthals, there may well have been some cultural overlap between these various *H. sapiens* groups.

About 35,000 years ago, an era named the Upper Paleolithic — the last part of the Old Stone Age — began in Europe, characterized by new methods of flaking flint to form stone tools. A marked change in human fossils accompanied these cultures, among which the earliest was the Aurignacian (Fig. 20-24d and Box 20-3). Neanderthals were replaced by anatomically modern *H. sapiens* (often called *Cro-Magnon* in Europe), who had smaller brow ridges, higher skull vaults and smaller and less prognathic faces.

Behaviorally, the evolving human lifestyle followed a pattern of increasing technological sophistication and

[6] Throughout the book, we have tried to avoid use of terms such as *modern* and *advanced.* Here, use of the terms seems reasonable as a way of differentiating extant humans such as ourselves from earlier humans, and it is in this context that we use terms such as *anatomically modern humans, modern humans,* or *advanced culture.*

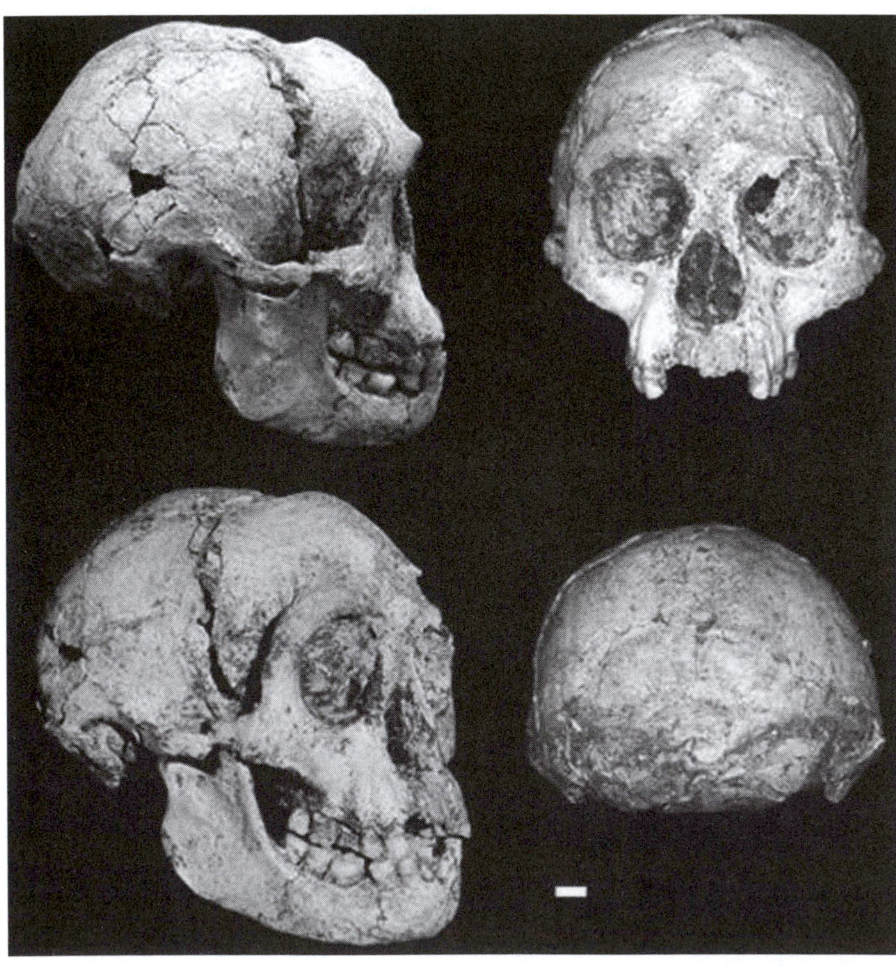

FIGURE 20-26 The skull of the most complete hominin from Flores. This individual, and the associated remains of other individuals dated at 18,000 before present, may represent a dwarfed relict population of *Homo erectus* (see Footnote 3). The figured individual is estimated based on associated postcrania to have stood a little over 1 m in height, had long arms in relation to the length of the legs, unusual shoulders and a chinless mandible. (Reproduced from Brown, P., Sutikna, T., Morwood, M. J., Soejono, R. P., Jatmiko, T., Saptomo, E. W. and Due, R. A., 2004. A new small-bodied hominin from the Late Pleistocene of Flores. *Nature,* 431, 1055–1061.)

BOX 20-4 Did *Homo sapiens* Arise in One or in Many Places?

There are two main views of the origin of modern humans from a *Homo erectus* ancestor (**Fig. 20-27**):

- The single-origin hypothesis, also called the *Out of Africa* or *Noah's Ark* model, proposes the origin of *Homo sapiens* in a single locality, Africa, followed by dispersal to other continents (Fig. 20-27a).
- The multiple-origin hypothesis, also called the *Candelabra* model because of its shape, proposes the parallel origin of *Homo sapiens* in different unconnected localities (Fig. 20-27b).

Among the information considered important in deciding between these alternatives is the date of the last common ancestor of modern humans. If this individual existed 1 My or more ago, such a date might well coincide with one of the dispersals of *Homo erectus* from Africa. It would indicate that modern humans found in different continents are the present end points of evolutionary lineages that each began with *H. erectus* in these geographically separated localities — support for the multiple origin hypothesis (see Henke and Tattersall, 2007, especially Volume 3). If the last common *Homo sapiens* ancestor was much more recent — for example, only 100,000 to 500,000 years old — the dispersal of *Homo sapiens* occurred after a 1- or 2-My-old dispersal of *Homo erectus*. This would indicate that populations of *Homo sapiens* entered localities where *H. erectus* had already been established, and eventually replaced these earlier hominins — supporting the single-origin hypothesis.

Because Cann and coworkers' (1987) original proposal of an approximate 200,000-year-old common mitochondrial DNA (mtDNA) ancestor to modern humans was widely challenged, molecular anthropologists undertook many new studies. Some of these calibrated the rate of mitochondrial nucleotide substitution using a 4- to 6-My-old date for sequence divergence between humans and chimpanzees (see Chapter 12). Others used a 60,000-year-old date for sequence divergence among Papuans, based on the time they first colonized Papua New Guinea (Stoneking, 1993). As **Table 20-4** shows, almost all these studies support the relatively young 200,000-year-old single-origin date. Even high upper confidence limits of about 500,000 years are still too recent to fit the 1-My-old or more dispersal age expected in the multiple-origin hypothesis or in its multiregional variation.

Mitochondrial DNA

Among other support for the single-origin hypothesis are four discoveries from studies of mtDNA:

- All of the nonAfrican mtDNA sequences are *variants of the African sequence.* If the nonAfrican mtDNA had been derived from resident nonAfrican populations, much more nonAfrican mtDNA variability would be expected. But no nonAfrican mtDNA types are known.
- *Most mtDNA sequence variability is found among African populations,* again suggesting that these are the oldest mtDNA pop-

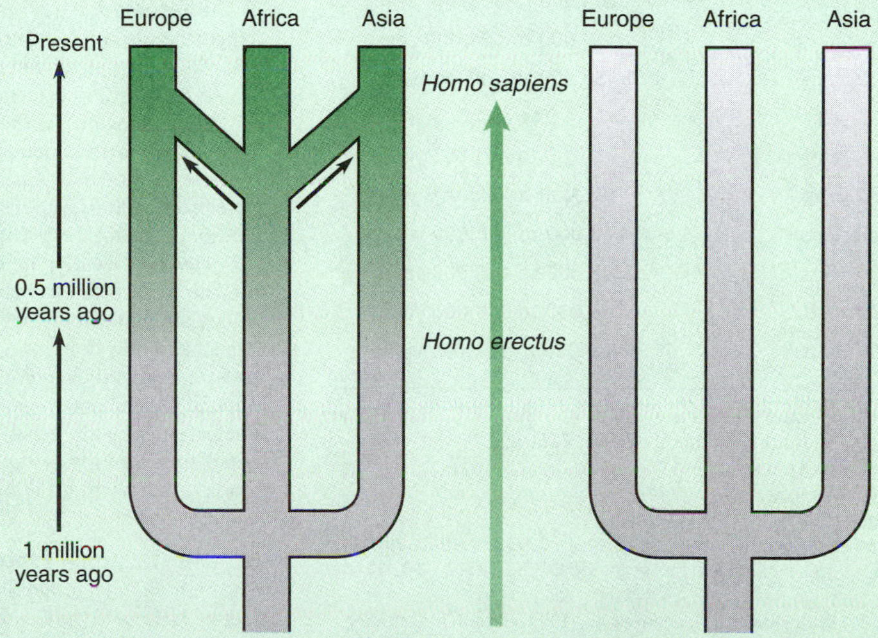

FIGURE 20-27 Diagrammatic representation of two models for the origin of *Homo sapiens*. In (a) humans (colored sections) originated in one locality (Africa) and migrated to other continents where they replaced relict *H. erectus* populations (gray sections) that had entered these continents 1 My or more ago. In (b) humans originated in different localities independently of other such groups. According to some proponents of multiple origins (the "multiregional" model; see Wolpoff, 1989), some genes exchanged between continental populations via cross-migration, enabling all these various evolving groups to reach the same *H. sapiens* grade.

BOX 20-4 Did *Homo sapiens* Arise in One or in Many Places? (*continued*)

ulations among modern humans and that the non-African populations are their derivatives.

- Because the multiple-origin hypothesis proposes that all populations evolved in parallel over long periods, we would *expect* them to have *similar amounts of variability,* a conclusion that the data *contradict.*
- Stoneking (1993) points out that the *most common mtDNA ancestor is most likely older* than the age at which the population bearing this ancestor diverged. For example, a 200,000-year-old date for an mtDNA ancestor means an even more recent date for the populations derived from this ancestor.

Y Chromosome

Despite its advantages, mtDNA acts as a single genetic unit whose genealogy may not necessarily coincide with other genetic units.

Do nonmitochondrial genes trace back to continents other than Africa? Other studies have sought information from nuclear genes. A prominent example is the small human **Y chromosome,** inherited exclusively through the male line, a counterpart to maternally transmitted mitochondria.

TABLE 20-4 Estimates of the age of the common human mtDNA ancestor with 95 percent confidence intervals for these estimates

Age of Common Ancestor	95 Percent Confidence Interval
280,000 years	180,000 to 380,000 years[a]
400,000 years	200,000 to 600,000 years
207,000 years	110,000 to 504,000 years
160,000 years	80,000 to 480,000 years
211,000 years	0 to 433,000 years[b]
101,000 years	0 to 205,000 years[c]
213,000 years	102,000 to 389,000 years
133,000 years	63,000 to 356,000 years[d]
137,000 years	63,000 to 416,000 years[d]
195,000 years	85,000 to 349,000 years[e]
143,000 years	125,000 to 161,000 years[f]

[a] The first nine studies are referenced in Stoneking, M. 1993. DNA and recent human evolution. *Evol. Anthropol.,* **2,** 60–71.
[b] Control region sequences.
[c] Coding sequences.
[d] Control region sequences: 2 methods.
[e] Ruvolo, M., S. Zehr, M. von Dornum, et al., 1993. Mitochondrial COII sequences and modern human origins. *Mol. Bol. Evol.,* **10,** 1115–1135.
[f] Horai, S., K. Hayasaka, R. Kondo, et al., 1995. Recent African origin of modern humans revealed by complete sequences of hominid mitochondrial DNAs. *Proc. Nat. Acad. Sci. USA,* **92,** 532–536.

The Y chromosome has a nonrecombining portion in which mutations, like those in mitochondria, can be used to establish phylogenetic and chronological trees. One such study by Hammer and coworkers (1998) surveyed more than 1,500 individuals from all continents, and traced all Y chromosomes to a common ancestor living about 150,000 years ago. Interestingly, like *Mitochondrial Eve,* this *Y-chromosome Adam* was of African origin, but African populations also received Y chromosomes returning from Asia. (As with mitochondria, we should keep in mind that our Y chromosome ancestor may have provided only a small part of our genome, which contains 22 autosomal chromosome pairs plus the X.)

Autosomal Genes

Broadening the data even further are studies that cover **autosomal genes.**

Nei and Roychoudhury (1993), for example, calculated genetic distances between 26 human populations for 29 different nuclear genes (see Fig. 24-3). They also suggest a single African origin for *Homo sapiens* with subsequent widespread geographic divergence (**Fig. 20-28**). Mountain and coworkers (1992) present similar findings. Polymorphism for Alu chromosomal DNA sequences specific to humans, support a recent African ancestry (Batzer et al., 1994).[a]

Anatomy

Whether mitochondrial, Y-chromosomal or autosomal, the genetic and molecular data do not convince all paleontologists.

A major line of evidence some offer to support multiple origins of *Homo sapiens* is the continuity of anatomical features in Chinese and Australian humanoid fossils. Some fossil characters in these localities, such as brow ridges and cranial size, seem to have progressed from *H. erectus*-like to *H. sapiens*-like in a fairly continuous sequence, pointing to their independent origin.

However, if each population evolved independently, how did they become so similar? Humans may differ in color and other minor attributes, but they all share basic *Homo sapiens* traits. Where and how did they acquire these similar traits? In answer, paleontologists such as Wolpoff (1989) proposed that *H. sapiens* populations did not evolve in complete isolation from each other but exchanged some genes. That is, *interpopulation gene flow* enabled the evolving groups to reach the same *Homo sapiens* grade — a new multiple-origin model given the name *multiregional evolution.*

Multiregional Evolution

The multiregional proposal, an apparent *compromise* between the Out of Africa and the Candelabra models, is also contested.

It has been argued that gene flow alone does not explain the necessary transitions, and that quantitative tests on fossil morphology also contradict multiple-origins. On the other hand, Templeton (1997) used genetics to make a strong case for multi regionalism. He calls it a "trellis" model in which geographically mobile ancestral humans exchanged genes between populations to form a genetic trellis as they dispersed from their African origin. There was movement both in and out of Africa, but because gene exchange depends on proximity, genetic differences between distant populations increased ("isolation by distance").

Although much molecular evidence seems to support the single-origin hypothesis, there are exceptions (e.g., see Ayala et al., 1994), and the data still generate different interpretations for how or whether a single origin occurred. According to Harris and Hey (1999), sequence analysis of an X-chromosome gene indicates a major genetic division between African and non-African populations about 200,000 years ago. Because this date is earlier than the presumed origin of "anatomically modern" humans of 100,000 to 130,000 years ago, such data appear to support the hypothesis that modern humans originated in different geographic localities after their initial African/non-African separation.

To *summarize:* Anthropologists and primatologists are continually obtaining and evaluating further molecular evidence that may eventually prove one or the other of these hypotheses. Such procedures, however, depend more on intricate statistical analyses and population genetics (Mountain, 1998) that, for some paleontologists, lack the realism of actual fossils. For the future, both strategic hominin discoveries and increasing nuclear gene information will help decide the matter. Some workers are willing to accept a somewhat intermediate position: "An African origin, with some mixing of populations, appears to be the most likely possibility" (Jorde et al., 1998).

[a] Alu sequences of DNA, some 300 base-pairs long, are a class of repetitive DNA elements known as short interspersed nuclear elements (SINEs). The human genome contains more than one million Alu sequences. Alu sequences are a conspicuous feature of exons in the human genome that are not found in other mammalian genomes (Zhang and Chasin, 2006).

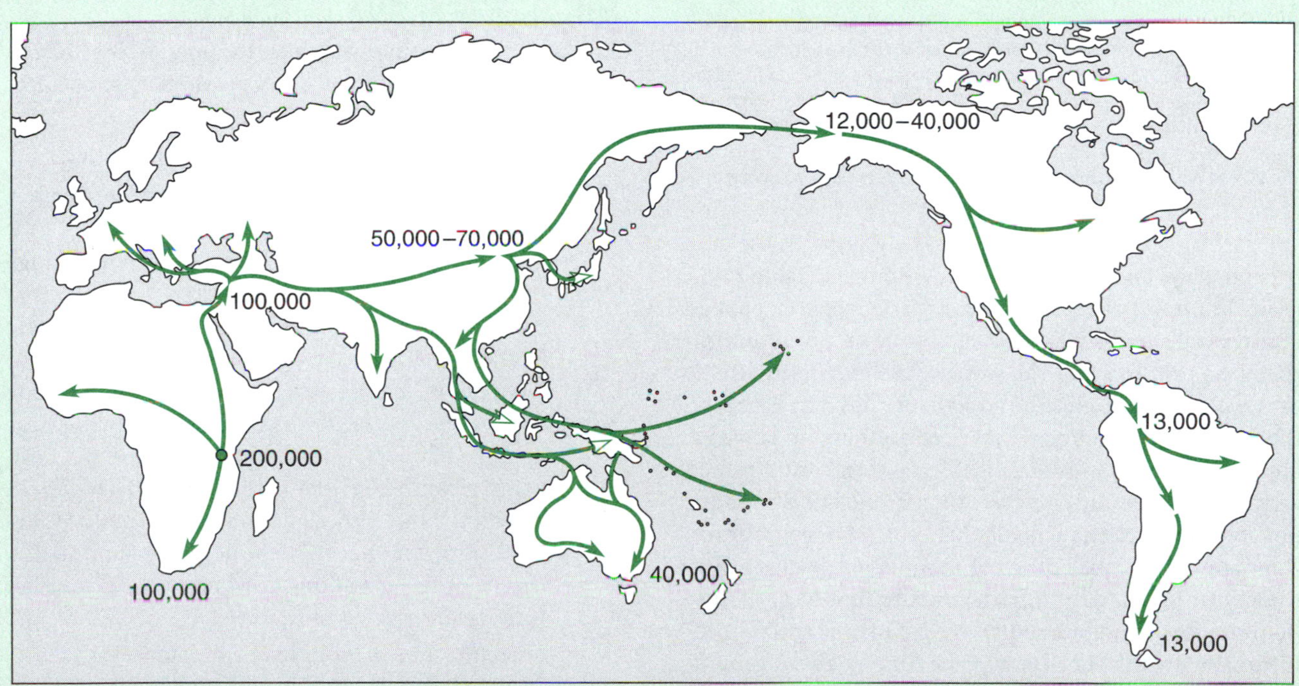

FIGURE 20-28 One proposed scenario for the geographic distribution of *Homo sapiens* from their African origin. The numbers derive from a study of genetic distances between 26 human populations and represent estimated dates (thousands of years ago) at which these populations reached their various destinations. According to Cavalli-Sforza and coworkers (1992), these migrations correlate highly with many major patterns of linguistic evolution. Surprisingly, some Native American genes may owe their origin to a European/Asia Minor population, indicating a long migration from Europe to North America that picked up Central or East Asian genes along the way (Brown et al., 1998). (Adapted from Nei, M., and A. K. Roychoudhury, 1993. Evolutionary relationships of human populations on a global scale. *Mol. Biol. Evol.*, **10**, 927–943.)

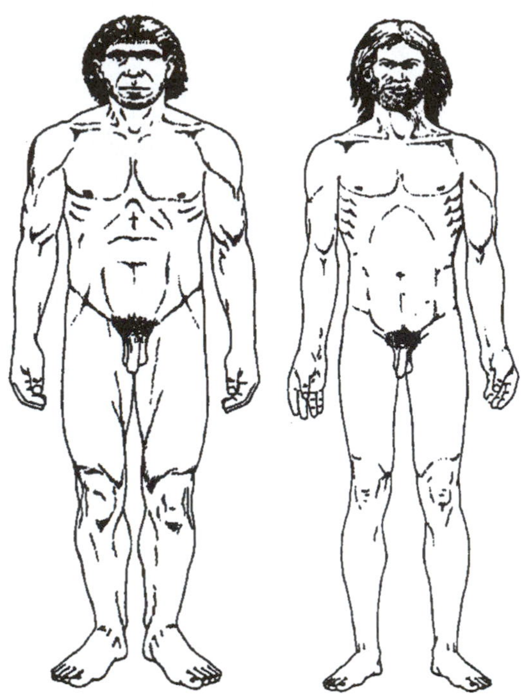

FIGURE 20-29 Reconstructions of the generalized Neanderthal male body form (left) and the Cro-Magnon (right). According to some anthropologists, the stockier body form of the Neanderthal came from selection for adaptation to colder climates (see also Fig. 24-5). Increased muscularity went along with their stockier build, giving Neanderthals a body weight about 30 percent greater than average modern humans (Ruff et al., 1997). (Reprinted with permission from *Pour La Science: Pour La Science,* **64,** 2/1983.)

expanded use and control of the environment (**Table 20-5**). Many examples of representational art appeared, painted on cave walls and sculpted in clay or bone. Some anthropologists propose that the success and enhanced artistic expression of anatomically modern humans may have been the result of developing social organization and language ability. In essence, "anatomically modern" humans had arrived in Europe and elsewhere. **Figure 20-31** illustrates one possible phylogenetic scheme that broadly traces relationships among various known hominin fossil groups beginning with forms dating back more than 4 My. Because hominin fossil finds are spotty, we don't know specific evolutionary events that took place among these groups, although many of the fossils are transitional. For example, were there two separate australopithecine lineages, as Figure 20-31 shows, or should we combine *A. robustus* and *A. boisei* in a single lineage? Also, because of marked pelvic distinctions and/or presumed limited language abilities, some physical anthropologists suggest considering Neanderthals as a separate species, *H. neanderthalensis,* rather than as a subspecies of *H. sapiens.*

Although anthropologists do not question the primitiveness of *Ardipithecus ramidus,* some are unsure of its position at the root of the hominin tree. Also at issue are other species names given to fossils placed along these lineages by various anthropologists. For example, *H. rudolfensis* is considered by some as antedating or contemporary with *H. habilis; H. ergaster* bifurcating into *H. erectus* and *H. sapiens,* and *H. heidelbergensis* ancestral to *H. neanderthalensis.* Many derive *H. sapiens* and *H. neanderthalensis* directly from *H. ergaster* while *H. erectus* as a lineage persists until late Pleistocene. Other proposed names include *H. rhodesiensis* for an African *H. erectus* derivative, and *H. antecessor* for Spanish fossils that may have been ancestral to both *H. sapiens* and *H. neanderthalensis.* Taking all the many hominin variations into account, it is clear that, "Instead of a ladder with humans at the pinnacle, there is a bush with humans as one little twig" (Foley, 1995). A major area of concern is the last twig of this phylogeny — the origin of humans. This topic, discussed further in Box 20-4, gained considerable attention with a 1987 publication by Cann and coworkers indicating that all modern human **mitochondrial DNA (mtDNA)** sequences probably originated in Africa between 140,000 and 290,000 years ago.

Mitochondrial DNA (mtDNA)

Among various possible molecular techniques (see Chapter 12), the use of primate mtDNA for evolutionary studies had a number of important advantages:

- mtDNA is a *circular molecule,* 16,569 base pairs long, the complete nucleotide sequence of which is known (**Fig. 20-32**).
- mtDNA is *inherited* primarily, if not entirely, through the *maternal lineage* as a sequestered extranuclear haploid unit (male sperm do not ordinarily transmit their cytoplasm to the egg during fertilization, and those few male mitochondria that do enter are soon diluted out in successive cell divisions by the large numbers of oocyte mitochondria) and does not ordinarily recombine either with nuclear DNA or with other mtDNA. (Although some [for example, Wallis, 1999] propose that mitochondrial recombination occurs, others question such findings and consider such instances as, at most, rare.)
- Therefore, unless modified by mutation (usually by single nucleotide substitutions), an mtDNA molecule *remains unchanged from one generation to the next,* and is homogeneous within an individual.
- Because there are about 10^{16} molecules of mtDNA per individual and up to thousands of copies per cell, we can *more easily isolate* mtDNA from human tissues than nuclear DNA genes, which have only two copies per cell.
- In contrast to nuclei, mitochondria *lack repair enzymes,* and *mutations can accumulate up to 10 times*

faster than nuclear DNA mutations. Such rapid evolution enables comparisons between groups that would be more difficult to differentiate using slower evolving and more complex nuclear DNA sequences.

- Assuming that most mitochondrial DNA changes have little effect on viability, and that mutations accumulate at a fairly constant rate, differences between mitochondrial DNA sequences can act as a molecular clock, marking the time taken for these DNAs to diverge. A mitochondrial gene "tree" can be used as a chronological "tree," depending on how accurately the molecular clock is calibrated (see Chapter 12), taking into account variability between taxa (Strauss, 1999).
- Most important, DNA sequencing techniques (discussed in Chapter 12) allow us to trace mtDNA differences among individuals, *establishing branching pathways* that help determine their evolutionary relationships.

As shown in **Figure 20-33**, Cann and coworkers (1987) proposed an evolutionary tree of human mtDNA that began with a single ancestral sequence in Africa (Fig. 20-33a). This led to nine major descendant sequences (Fig. 20-33b–j) that were subsequently dispersed to populations in Africa and other geographical regions in which further branching occurred. Because we can trace any geographical group outside Africa to more than one unique mtDNA branch, females carrying their particular mtDNA sequences made many colonizations of each area. For example, Cann and coworkers suggested that mitochondrial genomes in their sample of New Guinea Highlanders had seven different maternal origins, most from Asia and the remainder probably from Australia.

The timing of these migratory events is based on estimations of the rate at which mtDNA sequences diverge, using measurements of differences between mtDNA sequences whose common ancestry can be approximately dated. According to Cann and coworkers, this divergence rate was most likely between 2 and 4 percent nucleotide change per My in vertebrates, giving the dates shown in Figure 20-33. These results supported the view that early forms of *H. sapiens* were present in Africa between 100,000 and 200,000 years ago and radiated outward to different localities and differentiated into various groups, replacing indigenous *H. erectus.*[7]

Novel and interesting as these data are, it did not take long for statistical difficulties to arise. It was pointed out that the phylogenetic tree Cann and coworkers (1987) offered was only one of many possible trees, some of which could better explain the data. They also challenged the date Cann and coworkers proposed for an African mtDNA origin; later work provided new dates, some of which are shown in Box 20-4. Although discussion continues, Cann and coworkers' general conclusion for a single African origin of *H. sapiens* has been largely substantiated by fossil finds such as those at Klasies River Mouth, South Africa and Herto in Ethiopia.

Hunting Hominins

Many anthropologists want to understand how past environments affected and selected among various human traits, and how humans, in turn, affected and selected their environments.

As discussed, the introduction of *bipedalism* and the *home base*, however they arose, profoundly affected the

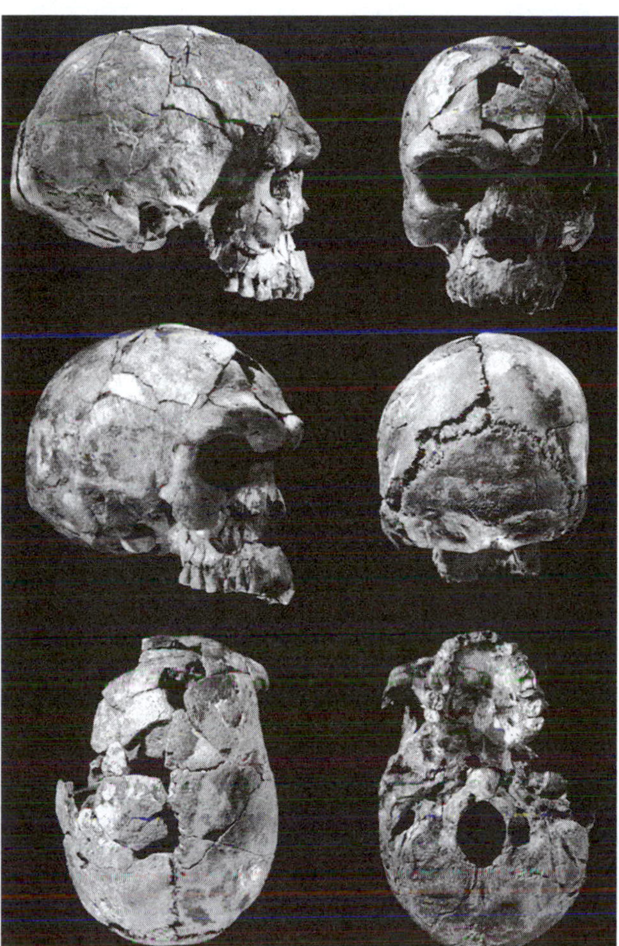

FIGURE 20-30 The cranium of one of three individuals from Herto, Ethiopia, dated at 160,000 to 154,000 years ago. These individuals are the earliest known anatomically modern humans. (From White, T. D., B. Aspaw, D. DeGusta, H. Gilbert, et al., 2003. Pleistocene *Homo sapiens* from Middle Awash, Ethiopia. *Nature,* **423,** 743–747.)

[7] Recent analyses of hominin genomes are complicating the mitochondrial DNA story, in great part because hominin gene loci are turning out to have widely varied evolutionary histories; see Pääbo (2003) and Garrigan and Hammer (2006) for overviews of these recent data.

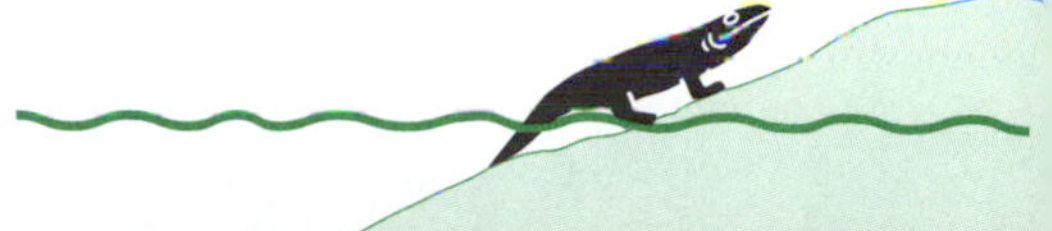

TABLE 20-5 Brief inferences about adaptations, behavior and ecological factors in hominin evolution

Lineage[a]	Approximate Time (Years ago)	Adaptations, Behavior and Habitats	Fossil and Archaeological Evidence
Hominin ancestors	8 to 5 My	Relatively large-bodied apes distributed in Central and Eastern Africa across forest–woodland mosaics	No fossil evidence yet, but when found, expected to be a group or groups ancestral to humans and chimpanzees
Australopithecines	4 to 2 My	Bipedal on the ground, occasionally arboreal Open savanna, and mosaic grasslands and woodland habitats Fibrous plant diet that may also have included meat[b]	Extensive fossils in Eastern and Southern Africa Large teeth and jaws
Homo habilis	Pliocene–Pleistocene 2 to 1.5 My	Improved bipedalism Tools to procure and process food Habitats in drier areas indicating larger home ranges Scavenging and active animal hunting	Skeletal changes and increase in brain size Early stone tools
Homo erectus	Early–Mid Pleistocene 1.5 to 0.5 My	Entry into new habitats and geographical zones Definite preconception of tool form Manipulation of fire Increased level of activity and skeletal stress	Fossils found in formerly unoccupied areas of Africa, and outside Africa Development of a stone tool industry Archaeological hearths Increased cranial and postcranial development
"Archaic" *Homo sapiens*	Mid Pleistocene 500 to 150 thousand years	Geographical divergence and ecological adaptations More complex tools	Old World distribution with some distinct regional morphologies Bifacial axes: Acheulean–Mousterian stone tool industries
H. sapiens neanderthalensis	Late Pleistocene 150 to 35 thousand years	Large and robust individuals More social complexity and development of ritual Increasingly sophisticated tools	Massive cranial and postcranial development Intentional burial of the dead Increased number of stone-tool types
H. sapiens sapiens	Late Pleistocene to Present	Decreased levels of activity and skeletal stress Expansion of technology Development of complex cultures Increase in population size	Appearance of "anatomically modern" humans From Upper Paleolithic (Aurignacian) stone tools to satellite communication Beginnings and expansion of agriculture

[a] "Lineage" designates the name commonly given to a major group found in the specified period. As the text indicates, other names have been used for fossil groups in these periods (for example, *H. ergaster, H. rhodesiensis, H. heidelbergensis*). Various groups also overlapped.

[b] New findings of 2.5-My-old hominins in Ethiopia suggest that behavioral changes associated with lithic (stone tool) technology and enhanced carnivory (butchered antelopes, horses, and other animals) may have been coincident with the emergence of the *Homo* clade that arose from *Australopithecus afarensis* in East Africa (de Heinzelin et al., 1999).

Modified from tables in Foley, R. 1996. The adaptive legacy of human evolution: A search for the environment of evolutionary adaptedness. *Evol. Anthropol.*, **4**, 194–203; and Potts, R., 1992. The hominid way of life. In *The Cambridge Encyclopedia of Human Evolution*, S. Jones, R. Martin, and D. Pilbeam (eds.). Cambridge University Press, Cambridge, England, pp. 325–334.

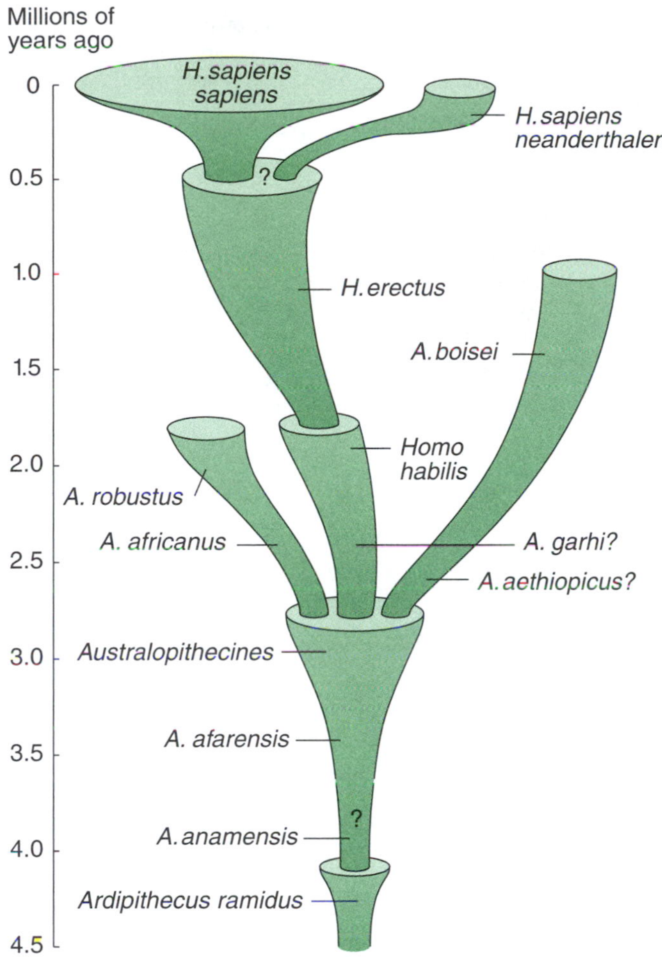

FIGURE 20-31 One possible phylogenetic scheme depicting evolutionary relationships among presently known hominid groups. This three-pronged tree is based on findings of an *A. boisei* form (*Australopithecus aethiopicus?*) that indicate the boisei line was evolving independently and parallel with the *A. africanus–A. robustus* line (McCollom, 1999), and that another australopithecine lineage, the Ethiopian *A. garhi*, is the "candidate ancestor for early *Homo*" (Asfaw et al., 1999). Some schemes combine *A. robustus* and *A. boisei* into a single lineage and suggest only two major branches leading from *A. africanus* — one to the robust australopithecines and the other to *Homo*. As might be expected, categorizing transitional fossils as species can be difficult in the absence of firmly accepted criteria. Genera designations are also difficult. For example, Wood and Collard (1999) argue that the genus *Homo* is not monophyletic because *H. habilis* and *rudolfensis* are more closely related to australopithecines than they are to the *H. sapiens* ancestor shared by other *Homo* species. Obvious differences in the fossil record indicate some groups can be distinguished because of their specific evolutionary positions and roles. How many such groups there are, and how to distinguish them, of course relates to the extent of fossil research and available collections, and to achieving some agreement among paleoanthropologists. (Adapted from Day, M. H., 1986a. *Guide to Fossil Man*, 4th Edition. Cassell, London, UK.)

kinds of environments early hominins could exploit. They could now move from forest to savanna with greater ease than ever before and cover much larger areas in search of food. Food resources in the savanna, however, differ from those in the forest and selected for a host of new behaviors and adaptations.

Boesch-Achermann and Boesch (1994) suggest that forest chimpanzees are much more versatile in behavior, tool use, hunting and cooperative food sharing than chimpanzees in savanna–woodlands. Because the chimpanzee–human phylogenetic relationship is so close, and the forest environment so demanding and selective, they propose that early hominin evolution was primarily associated with forest experiences.

The relatively low rainfall in the savanna provided fewer high-quality plant foods than in the forests and made the distribution of such resources patchy, that is, present in some places and not in others. These resource irregularities, combined with plant seasonality, would have initiated response to further selection for increased hominin mobility, broad dietary habits and flexible strategies in searching for food. Since the savanna grasslands also supported various herbivores, including migratory herds of large mammals, a selective advantage for meat eating may have appeared fairly early, including strategies for both avoiding and successfully competing with the large predators that preyed on these mammals.

An important change in the lifestyle of these hominins would have been an increase in the relative amount of meat in their diet. **Animal hunting** is not a novel trait in primates; many investigators have recorded instances of baboon and chimpanzee groups engaged in purposeful hunting of animals smaller than themselves.[8] Among primates, humans are the greatest **meat eaters,** a dietary habit that undoubtedly varied in degree at different times and places but became established early in human history. It therefore seems likely that even in their forest habitats, early hominins had become meat eaters, at least to some minor degree.

An increase in meat consumption would have offered early hominins many advantages:

- Meat is a rich source of essential amino acids used in the synthesis of proteins such as lysine, tryptophan and histidine.
- Meat provides more calories per unit weight than most plant foods.

[8] See Harding and Teleki (1981) and Goodall (1986). Goodall lists more than 200 observed incidents where Gombe chimpanzees caught and/or ate colobus monkeys near Lake Tanganyika, in addition to cannibalism and the capture and consumption of many other mammals. According to Teleki, the chimpanzee kill rate in Gombe is about 225 to 300 mammals a year, which agrees "with the kill rates of some large carnivores."

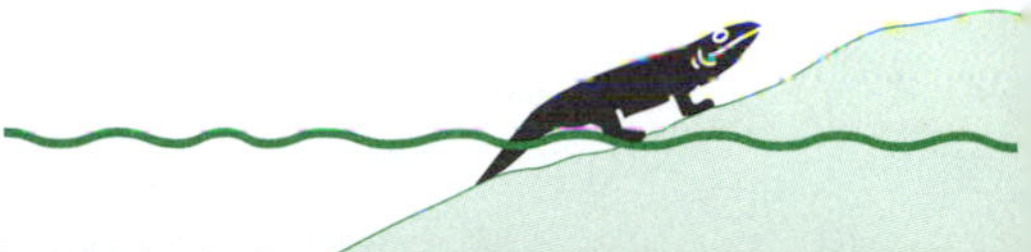

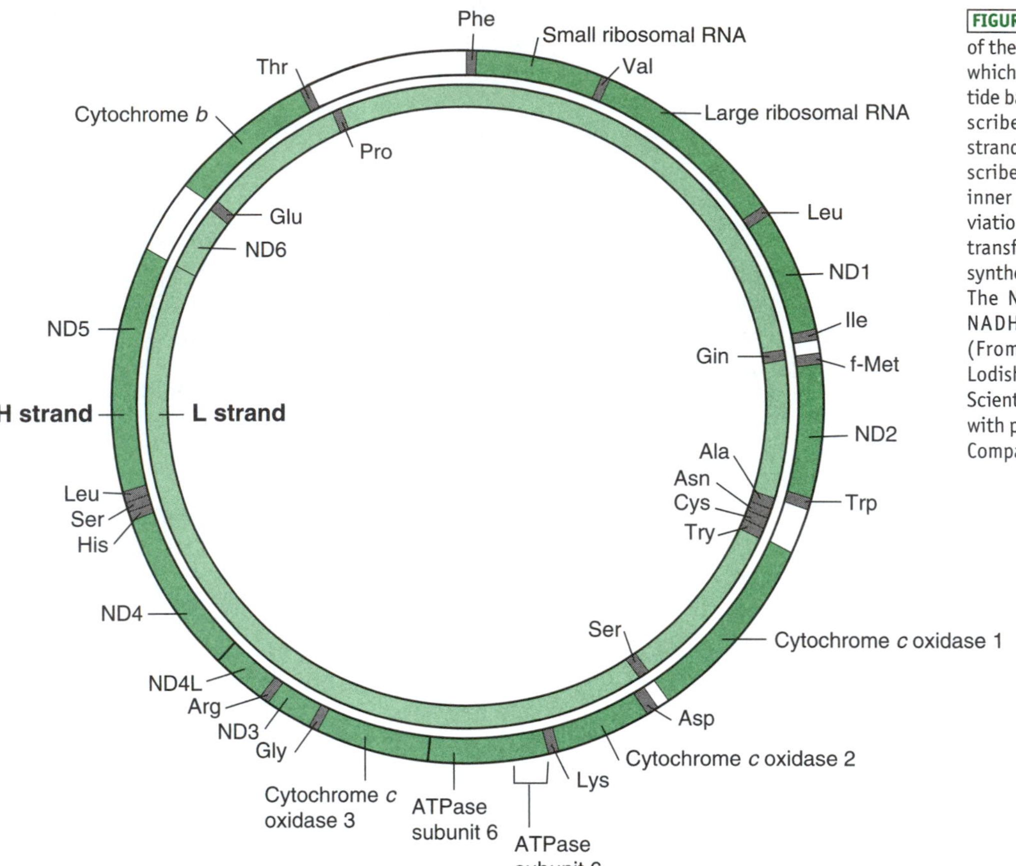

FIGURE 20-32 Genetic organization of the human mitochondrial genome, which is comprised of 16,569 nucleotide base pairs, with most genes transcribed clockwise along the outer strand (H), and the remainder transcribed counterclockwise along the inner strand (L). Amino acid abbreviations (e.g., Phe, Cal, Leu) refer to transfer RNA genes used in protein synthesis within the mitochondrion. The ND prefix refers to genes for NADH dehydrogenase subunits. (From: *Molecular Cell Biology* by Lodish *et al.* © 1986, 1990, 1996 by Scientific American Books, Inc. Used with permission by W.H. Freeman and Company.)

- Meat is either packaged (small animals) or can be modified by cutting and tearing (large animals) into units easy to transport to a home base.
- Killing only one large animal feeds a group of individuals, often for more than one day.
- Meat remains available in dry seasons when plant food diminishes.
- As an added food source in a marginal environment, meat would have helped provide the additional energy to develop and sustain a larger brain.

Hunter-Gatherer Societies

Although there are big differences, we can gain some idea of early hominin lifestyles from **hunter-gatherer societies** that continue today in places such as Central Africa (*Mbuti Pygmies*), South Africa (*Kalahari Bushmen*) and Australia (*Aborigines*). These groups consist of social communes or bands where males are usually the hunters and females the plant gatherers. Because their omnivorous diet depends on highly variable and often seasonal plant and animal food sources, each band moves several times a year over fairly wide ranges to different home bases or settlements.

Among some groups of Kalahari Bushmen, females gather about 60 percent of the diet in the form of vegetables and fruit, and male hunters bring in about 40 percent in the form of animal game. Because they usually share food, Bushmen waste little, and maximize the chances that all band members will receive some (Silberbauer, 1981).

We have no information on the proportions of plant and animal foods in the diets of ancient hominins, especially as many plants that such hominins may have used have not fossilized or are poorly preserved. Nevertheless, bone accumulations that appear with traces of hominin activities at East African sites indicate that hominin scavenging and hunting were most likely an important part of the lifestyle of various groups by the early Pleistocene, about 2 Mya, if not earlier.

Cooperative Hunting

So, although we may not know exactly when hunting began in human history, it was a significant industry for a long enough period — in many societies, up to the agricultural (Neolithic) revolution about 15,000 to 10,000 years ago — to have seriously influenced human behavior. Even if we grant that plant food was at least as important as meat in early

Africa
Asia
Asia Australia New Guinea Europe
Australia Europe
Africa Asia Australia Europe
Africa Asia Australia New Guinea Europe
Africa Asia Australia New Guinea Europe
Africa Asia New Guinea Europe
Africa Europe
Africa Asia Australia New Guinea Europe

(j) (62–125)
(i) (65–130)
(h) (68–130)
(g) (70–140)
(f) (75–150)
(e) (85–170)
(d) (98–195)
(c) (108–215)
(b) (112–225)
(a) (143–285)

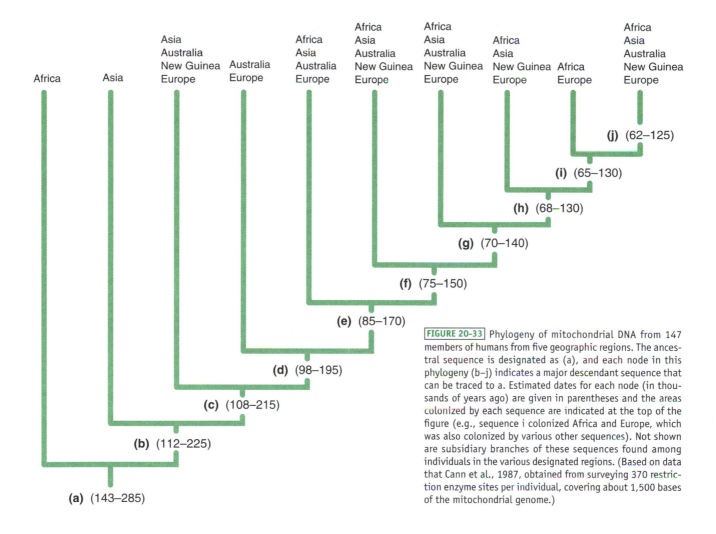

FIGURE 20-33 Phylogeny of mitochondrial DNA from 147 members of humans from five geographic regions. The ancestral sequence is designated as (a), and each node in this phylogeny (b–j) indicates a major descendant sequence that can be traced to a. Estimated dates for each node (in thousands of years ago) are given in parentheses and the areas colonized by each sequence are indicated at the top of the figure (e.g., sequence i colonized Africa and Europe, which was also colonized by various other sequences). Not shown are subsidiary branches of these sequences found among individuals in the various designated regions. (Based on data that Cann et al., 1987, obtained from surveying 370 restriction enzyme sites per individual, covering about 1,500 bases of the mitochondrial genome.)

human history (Tanner, 1987), it seems reasonable to ask: What selective forces and effects did hunting generate? As Tooby and DeVore (1987) stress, hunting "would elegantly and economically explain a number of the unusual aspects of hominin evolution."

First of all, successful medium and large game hunting requires *active cooperation among hunters.* We see this even in groups of foraging chimpanzees (usually two to five males), which will tree a monkey and then cut off its escape by assuming strategic positions around it. Hominin hunters, empowered with simple weapons such as wooden spears, clubs and hand axes, used such techniques and others, including tracking; stalking; and chasing game into cul-de-sacs and swamps, over cliffs, into ambushes, or by continuing the chase until the animal tired. With **cooperative hunting,** humans could bring down larger animals than could single hunters alone.

Second, cooperative hunting and the killing of large animals emphasized increased social cohesion during both hunting and the food sharing that followed. Transfer of information in successful hunts became vital, performing a necessary function in many later social interactions of the entire group. Improved communication became especially advantageous as individuals took more complex roles in planning, hunting, helping, food gathering, food sharing, infant care, child training and other vital activities.

Third, successful hunting *emphasized perceiving and retaining information on migratory pathways, watering sites, and home base settlements,* whose geographical positions extended over home ranges (regions habitually occupied by a group) greater than those occupied by most other primates or carnivores (Foley, 1987). Hominin hunters had to mentally dissect their experiences and observations into component geographical and ecological features, prey behaviors, weather effects, and seasonal changes, and store and synthesize this information into communicable mental maps that enabled prediction, planning and modification. A genetic basis for the selection and evolution of visual-spatial reasoning can be inferred in humans with Williams syndrome, who lack such abilities because of a defective gene.

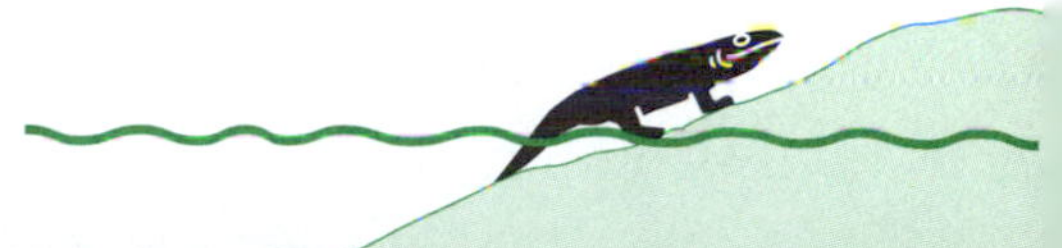

Fourth, hominin hunting involved stresses that fostered increased locomotory adaptations such as persistence in the chase (humans can continue jogging for distances that are generally longer than many large animals can continue running), maneuverability in the kill and long-distance traveling to or between home bases while carrying heavy burdens. Bramble and Lieberman (2004) recently presented anatomical and physiological evidence that adaptations for long-distance endurance running characterize the genus *Homo* from *Homo erectus* on. Various writers have also pointed out that the need for increased diffusion of metabolic heat during these pursuits would have selected for the loss of body hair and increased numbers of sweat glands, features that among primates are restricted to humans.

Fifth, the *technological skills* necessary for a clawless, canine-less hominin to capture and butcher large prey promoted the *making of a variety of snares, weapons and tools,* including stone implements that date back at least to *H. habilis* (Fig. 20-24). Such technologies, especially evident in fossil tools, involve shaping material according to some preconceived notion of what it should look like after the process is completed. These tool making skills involve not only manual dexterity, hand–eye coordination, and considerable concentration, but also the ability to plan and visualize an object that is not apparent in the raw material from which it is created. The artisan must conceptualize the final form of a tool in its three dimensions, and implement such concepts by mastering a series of techniques. These included finding and recognizing appropriate, workable stones in outcrops that were often widely dispersed, carrying these stones back to a base, and shaping them into tools by a sequence of precise strokes. The toolmakers also had to supplement the considerable mental abilities they used in tool making with social and communicatory abilities, in order to transmit such skills to other individuals who could continue the industry.

Finally, hunting placed further social emphasis on the home base to enable food exchange among foraging subgroups, particularly when the food supply was irregular, as it often is in hunting. A home base has value for nursing and pregnant females who could not always or easily cover the long distances necessary for large-scale hunting. The home base would have become a center for food sharing, shelter, hunting preparation, sexual bonding, childcare and other social exercises in which communication skills tied all members together.

However, as with the origin of bipedalism, some anthropologists have argued against the role of hunting among human ancestors, especially because we don't know the extent to which early human groups hunted (Harding and Teleki, 1981). Nevertheless, from what we can surmise from present-day hunting-gathering groups, and even from individuals in more technologically advanced societies who engage in hunting as a sport, the practice of hunting, whatever its role, was reinforced in various emotional ways: by the pleasures of seeking out and subduing prey; by the satisfactions of mastering the physical skills necessary for efficient aiming, throwing and grappling; and by mastering the intellectual skills used in devising cunning offensive and defensive strategies. These behaviors arise early in human development, especially in play among juveniles and adolescents. Their perfection in adults has been socially approved and rewarded in every known historical culture. The fact that most human societies no longer need hunting for food has not lessened interest in these behaviors; athletes and "warriors" who develop such skills are often greatly esteemed.

Communication

Communication is the means through which a stimulus from one individual can trigger a response in others. Methods include signals transmitted through any of the sensory channels: scent, touch, vision and sound. Practically all animals that interact with each other use one or more communicatory methods, but they are especially well developed in social animals where information is essential in providing cues to other individuals about factors such as food sources, predator encounters, territorial boundaries, sexual readiness, social ranking (dominance) and emotional states. We can find examples of these throughout primates.

For instance, various prosimians and monkeys use urine or scents emitted by special glands to mark trails and territorial boundaries. They also commonly use **olfactory cues** to attract sexual partners and signal the onset of ovulation. Such communication has its counterpart in humans, who emit odors from their axillary and genital regions. Although in Western culture people now generally wash off or disguise these odors by deodorants and perfumes, tests have shown that many can use such body scents to distinguish between the two sexes as well as among individuals.

Tactile communication assumes its most common primate form in grooming, or fur cleaning with fingers, lips and teeth. It is one of the most obvious and frequent kinds of interaction in many mammalian groups and seems to serve as the main social cement that binds pairs of individuals or group members together. Chimpanzees supplement grooming with other tactile behaviors such as holding hands, patting, embracing and kissing. Because humans have relatively little fur, they don't engage in the traditional form of primate grooming, and not surprisingly they mostly confine tactile social reassurances to other tactile behaviors.

Primate **visual signals** include physical gestures or anatomical displays such as the postures, genital swellings and colorations used to signal sexual receptivity. As shown in **Figure 20-34**, facial expressions may be quite varied and are easily visible in hairless faces. Some of these expressions, such as the glare and scream call (Fig. 20-34a and c), signal threat messages throughout the primates and mark an aggressive attitude even in human cultures (Fig. 20-34b and d). A fascinating recent study examined facial expressions in individual humans who have been blind since birth and found that expressions expressing sadness, anger and concentrated thinking were more similar to those expressions in related family members than in unrelated individuals, which is consistent with a genetic component in facial expressions (Peleg et al., 2006).

Compared to visual displays, vocalizations have the advantage that they leave the hands and body free for other activities. Oral sounds also have the advantage of providing feedback by allowing the vocalizer to hear his or her own vocalizations and thereby evaluate and control them while (or perhaps even before) uttering them. Moreover, although sound fades rapidly, it can be transmitted over long or short distances, in all directions, even around obstacles that would interfere with visual communication.

In general and as sound leaves no record, most primates usually send short, simple messages, denoting, for example, predator alarms or territorial calls. Yet some primate vocalizations have a variety of gradations, each providing a subtle meaning. Thus, Japanese macaques use particular variations of the "coo" sound in specific situations, such as females contacting their young, dominants contacting subordinates, females in estrus and when a male is separated from the group.

Other primate vocalizations not only reflect the emotional state of the vocalizer but also direct attention to specific external events. For example, vervet monkeys give three different alarm calls, each designating a specific kind of predator:

- A "rraup," uttered upon detecting a hawk, prompts the troop to look up and then seek cover in lower branches.
- "Chirps," uttered on seeing a mammalian predator, prompt the troop to ascend to the forest canopy;
- A "chutter," which denotes the presence of a snake, may prompt the troop to adopt aggressive positions on the ground.

Each of these calls is **symbolic** in the sense that it denotes an object that has no direct relationship to the call itself (e.g., a chirp is not a leopard) and we may consider the calls a primate preadaptation (exaptation) for human communication in which the sounds of words do not correspond to their meanings.

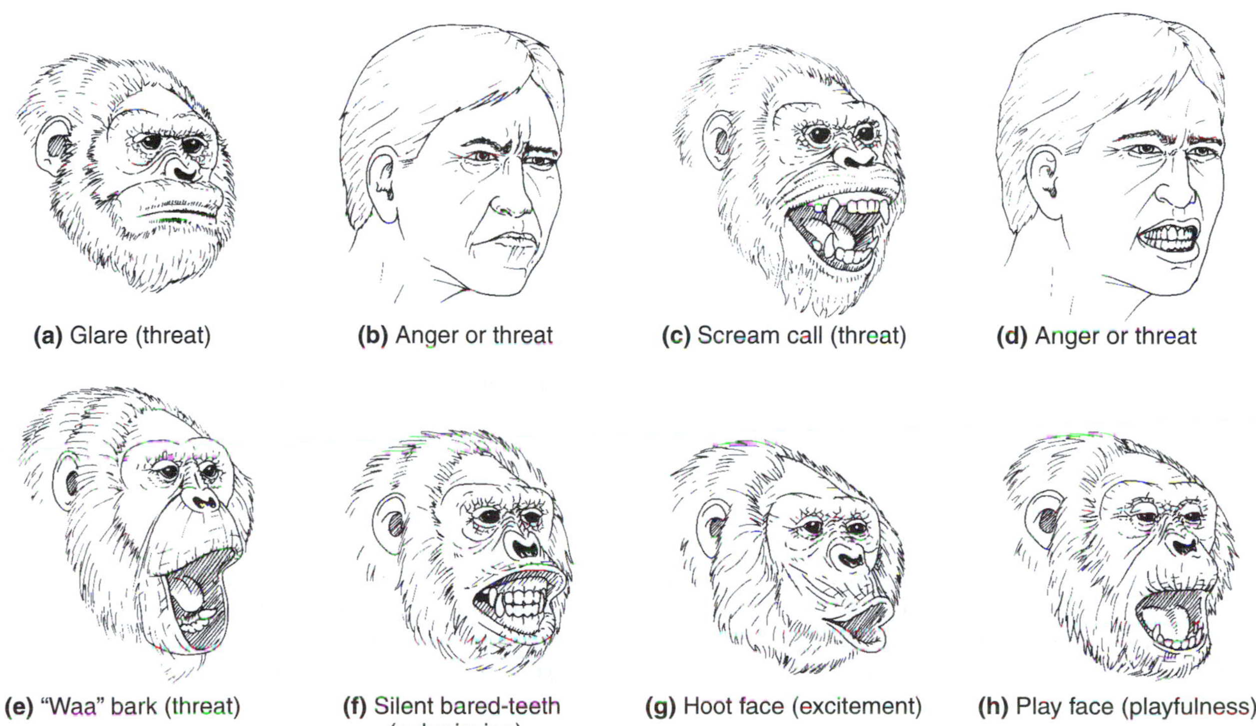

FIGURE 20-34 A small sampling of chimpanzee facial expressions indicating various emotional states, along with two seemingly related human expressions in (b) and (d). (Adapted from Chevalier-Skolnikoff, S. 1973. Facial expression of emotion in non-human primates. In *Darwin and Facial Expression*. P. Ekman (ed.) Academic Press, New York.)

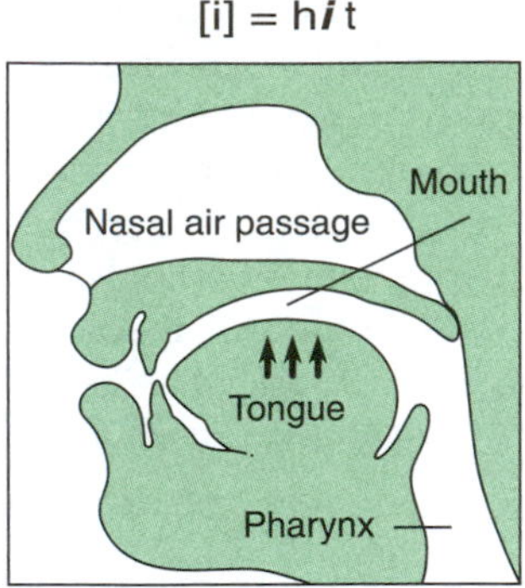

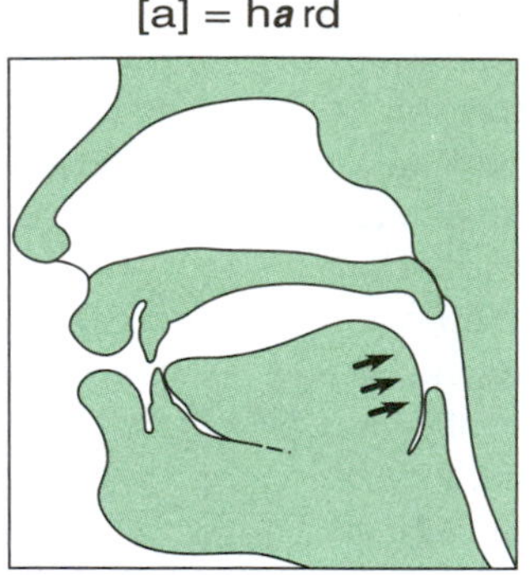

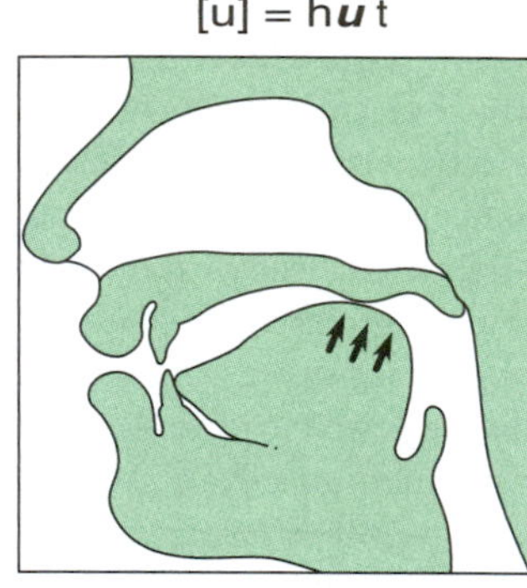

FIGURE 20-35 Diagrammatic views of how adult humans produce three vowel sounds by positioning the tongue (*arrows*) in different parts of the oral airway. Note that a sharp bend (formed by the hard palate above the mouth and rear wall of the pharynx) partitions this airway into right-angled mouth and pharyngeal sections that are essential for these vowel sounds. By contrast, the vocal airways of chimpanzees and newborn human infants are shorter and primarily in the shape of a slightly curved tube, making these vowels much less distinct. (Adapted from Aiello, L., and C. Dean, 1990. *An Introduction to Human Evolutionary Anatomy*. Academic Press, London, UK.)

Speech

The most symbolic of primate vocalizations is human **speech** and **language.** Here, we have introduced new characteristics through a wide range of different-sounding syllables that we string together in various ways to form a vocabulary of different meanings (words). Compared to a sequence of sounds limited to single tones, human speech provides a rapid means of communication. For example, we can interpret a sequence of dots and dashes, as in the Morse code, at a rate often much less than 50 words a minute, whereas we can often easily understand a sequence of spoken syllables delivered at 150 words a minute.

Analysis of the evolution of human speech has long occupied the minds of anatomists, physical anthropologists and behavioral biologists, from classic studies in comparative anatomy — in which the vocal tracts of newborn humans and chimps were regarded as equivalent to those of Neanderthals — to the advances that come from integrating approaches from anatomy, archaeology, neurology, psychology and musicology to argue for the essential role of rhythm in the evolution of human language, bipedal locomotion and musical ability.[9]

As in other mammalian vocalizations, the **larynx,** in the upper part of the tracheal tube, provides the basis for speech. Its origin is not connected with sound but stems from early air-breathing fish. These early vertebrates, like the lungfish of today, opened a valve in the floor of their pharynx to help swallow air into their lungs when they were out of the water, and closed the valve when in the water. As selection for air breathing continued in terrestrial vertebrates, this laryngeal valve developed fibers and cartilages that more precisely controlled laryngeal dilation and closure and allowed the animal to breathe more air when necessary. Like so many other evolutionary features, the ability of the larynx to generate oral sounds is an example of using an existing organ for a different purpose.

In producing sound, the larynx acts like a woodwind reed, controlling vocal pitch by opening and closing rapidly so that expired air from the lungs is interrupted to form puffs: the greater the frequency of puff formation, the higher the pitch. However, to produce the vowels of human-like speech, laryngeal puffs must pass through a tube-like airway (the pharynx) whose length and shape determine the eventual frequency patterns emitted and thus the quality of the different vowel sounds (**Fig. 20-35**).

All terrestrial, air-breathing animals that produce oral sounds, from frogs to mammals, use this basic mode of **vocalization** — *a laryngeal-like output and a supralaryngeal "filter."* In addition, neural auditory units in these animals seem to react with maximum sensitivity to specific ranges of frequencies. These specific neural sensitivities allow bullfrogs, for example, to respond to the mating calls of their own species and not to those of others. Humans seem to have special neural brain circuits that can perceive and identify various categories of sound combinations and thus distinguish between different kinds of spoken syllables, an ability that evolved from primates with more limited powers of distinction (Lieberman, 1991).

Figure 20-36a is a diagram of the adult human upper respiratory tract, with its sound-producing airway, which begins at the larynx and proceeds through the pharynx and mouth. Note, compared to that of the chimpanzee (Fig. 20-36b) the adult human mandible extends forward for a relatively shorter distance. Among the consequences of reduced mandibular size (which results in crowding of the teeth) and lower positioning of the human larynx in the vocal airway are the thickening and rounding of the tongue to form the anterior wall of the pharynx. In chimpanzees (and newborn

[9] See Lieberman (1991) for the anatomical studies, and see Mithen (2006) for the role of rhythm.

human infants), the pharyngeal section of the oral tract is shorter, and the epiglottis (used to cover the trachea during food intake) overlaps the soft palate. As a result, the tongue is isolated from the pharynx and respiration normally proceeds through the nasal cavity. Nonhuman primates and newborn humans (Fig. 20-36c) can drink and breathe simultaneously because the nasal respiratory pathway to the lungs can remain open while liquid passes around it into the esophagus; thus, although there is a better separation between breathing and swallowing in nonhuman primates and newborn human infants, they cannot produce adult human speech sounds by manipulating the tongue in the pharyngeal cavity.

Because speaking depends so much on soft, nonfossilizable tissues, we find it hard to uncover the phylogenetic history of speech. We can, however, make important correlations between skull structure and the positioning of the larynx, size of the tongue and length of the pharynx. According to such studies, the vocal tract of australopithecines (Fig. 20-36d) was no different from that of nonhuman

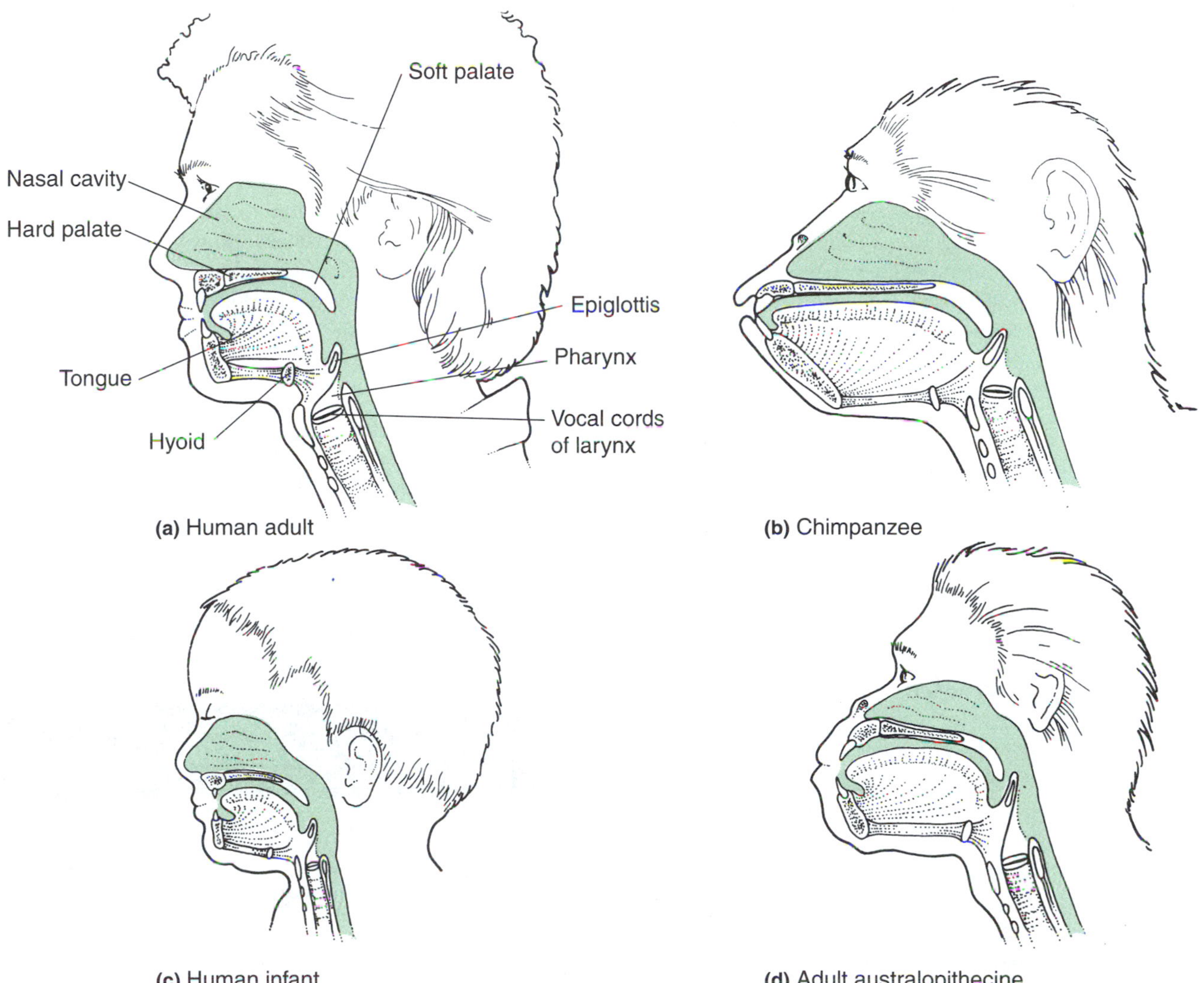

FIGURE 20-36 Upper respiratory systems of an adult human (a), chimpanzee (b), human infant (c), and adult australopithecine (d) showing important structures associated with vocalization. The pharynx is much longer in human adults than in chimpanzees because the larynx is displaced downward in the neck, and the bulging tongue formed by shortening of the mandible (note also the lower position of the hyoid bone and epiglottis) now forms the anterior wall of the pharynx. As a result, humans can enunciate vowels and syllables more clearly by positioning the tongue in both mouth and pharynx. Newborn human infants (c) show the same overlap between epiglottis and soft palate as nonhuman primates (b) but the pharynx lengthens considerably during infancy and childhood, transforming humans from obligate nose breathers as infants to the adult condition of voluntary mouth breathers. In (d) is a reconstruction of the presumed upper respiratory system of an adult australopithecine. As in nonhuman primates, the epiglottis overlaps the soft palate, the back of the tongue does not reach the pharynx and the larynx is relatively high in the vocal tract. (Based on Conroy, G. C., G. B. Weber, H. Seidler, P. V. Tobias, et al., 1998. Endocranial capacity in an early hominid cranium from Sterkfontein, South Africa. *Science,* **280,** 1730–1731; after Lieberman, P., 1984. *The Biology and Evolution of Language.* Harvard University Press, Cambridge, MA, and Lieberman, P., 1991. *Uniquely Human: The Evolution of Speech, Thought, and Selfless Behavior.* Harvard University Press, Cambridge MA.)

primates. This was probably also true for most, if not all, of the lineages classified as *H. erectus*.

Lieberman (1991) and others suggest that Neanderthals had a chimpanzee-like vocal tract that would have considerably limited its speech patterns. Sounds dependent on pharyngeal shape and control, such as [i], [a], [u], [k] and [g], would have been considerably distorted. Like chimpanzees and newborn human infants, Neanderthals would have nasalized speech because their nasal cavity could not be closed off from the pharynx. If this view is correct, a primary reason for the divergence of Neanderthals from humans may have been differences in phonetic ability and consequent differences in the kinds of languages that each employed and the level of social and technological organization that could be achieved.

A contrasting view is based on the discovery of a 60,000-year-old Neanderthal skeleton with an intact hyoid bone used for laryngeal muscle control. According to Arensburg and coworkers (1989), this crucial bone shows little difference from our own hyoid bone. They suggest that it signifies Neanderthal capability for speech similar to human speech. Because we do not know the actual position of the larynx and other soft tissues in this fossil or in earlier ones, we have no evidence for the extent of Neanderthal vocalization.

Language

In contrast to Darwin, who stated in *The Descent of Man* that it would be "impossible to fix on any definite point when the term 'man' ought to be used," Max Müller (1834–1898), a linguist, laid down the challenge that "language is our Rubicon, and no brute [ape] will dare cross it." Müller's barrier could only be overcome by teaching apes a human language. Various trials made throughout the early twentieth century were unsuccessful, although one is left with the impression that apes often understood far more than they could communicate.

Among the first serious long-term attempts to bridge the language gap was the undertaking by Keith and Catherine Hayes (1951), who raised an infant chimpanzee (Viki) in a normal human environment and tried to teach her to articulate human speech. The attempt was a failure, and after six years, Viki could not utter more than four distinguishable words: "Papa," "Mama," "cup," and "up." As already pointed out, because its vocal tract is relatively short and apes seem unable to control its shape, a chimpanzee can only produce a quite limited range of sounds. This is not surprising, because chimpanzees in their natural habitats are usually silent except when aroused.

A different and more successful approach was begun in 1965 by the Gardners, who raised a chimpanzee (Washoe) from the age of ten months in an environment in which its human caretakers used *American Sign Language*, which does not involve speech. By the time Washoe was five years old, she had learned to use at least 132 different signs covering a variety of names, actions, modifiers and functions. Her sentences, however, rarely extended to more than one or two words or their repetitions, indicating that her language abilities stopped at about the level of a two-year-old human child. This has also been true for other chimpanzees taught American Sign Language since Washoe.

Despite their limitations, chimpanzees have linguistic abilities, although on a more elementary level than humans. For example, as shown in **Table 20-6**, chimpanzees trained in American Sign Language by the Gardners, Roger Fouts and others, as well as Koko the gorilla, trained by Francine Patterson, can use various appropriate and imaginative combinations of signs to signify items not in their vocabulary. Chimpanzees can use words (or signs) as arbitrary symbols for real objects and actions, to deceive, gain information and to comment on the world about them. Chimpanzees create combinations of words for objects that are not in their vocabulary, refer to activities or objects distant in time or place (**displacement;** one of the important features of language), understand word sequences that use the logic of cause and effect, classify newly presented objects into groups (e.g., fruit, tool), use Arabic numerals to count and learn signs from other chimpanzees.

Even some simple word orders can be grasped by chimpanzees — Roger Fouts observed that Lucy could distinguish between "Roger tickle Lucy" and "Lucy tickle Roger." Chimpanzee communications include deceptions, as when chimpanzee Booee asked for a tickle from his chimpanzee companion, Bruno, while obviously trying to get to the raisins that Bruno possessed.

TABLE 20-6 Sign sequences (word combinations) created by chimpanzees and gorillas for items not in their ordinary vocabulary

Item	Sign Sequence
Onion, radish	"Cry fruit," "cry hurt fruit"
Watermelon	"Candy fruit," "drink fruit"
Alka Seltzer	"Listen drink"
Cigarette lighter	"Metal hot"
Ring	"Finger bracelet"
Swan	"Water bird"
Ostrich	"Giraffe bird"
Brazil nut	"Rock berry"
Hateful objects	"Dirty . . ."[a]

[a] Probably signifying fecal, for example, "dirty leash," "dirty monkey," "dirty Roger."

A somewhat more abstract two-dimensional language that David Premack (1971) taught to chimpanzees involves metal-backed plastic tokens of arbitrary shape (lexigrams) to represent words that are arranged in vertical sequence on a magnetized board. Duane Rumbaugh (1977) further developed this approach through the use of computer-connected keys embossed with lexigrams that, when pressed, light up sequentially on a screen above the keyboard console. Lana, one of the computer-trained chimpanzees, demonstrated that she could use this system to ask "intelligent" questions. For example, having been taught the symbol "name of," she used it to elicit the unknown name of the object ("box") that contained a desirable item, candy. First she asked, "Tim give Lana name of this," and, when given the name, called for the "box." Perhaps a greater linguistic feat was that of Sarah, trained in Premack's plastic token language, who used it to understand "if–then" causal relationships: for example, "If Sarah take apple — then Mary give chocolate Sarah; if Sarah take banana — then Mary no give chocolate Sarah."

Tool Making and Use

According to many observers apes have gone beyond merely perceiving events and objects to conceiving how they occur and how they are related (Savage-Rumbaugh and Rumbaugh, 1993).

In their natural habitat, these accomplishments are obvious in chimpanzee tool making and tool use, such as the way they employ twigs and vines in fishing for termites. *Termite fishing* involves being aware of when to fish (October and November), locating the sealed termite tunnels, often importing the necessary supply of tools from distances as far away as 0.8 km, shaping some of the tools by removing leaves, biting off the ends of the tools to achieve an optimum length, inserting the tool with a proper twisting motion required to follow the curves of the termite tunnel, vibrating it gently to bait the termite soldiers and retracting it carefully to avoid tearing off the termites. Learning these tasks takes years, and "even" an anthropologist who studied the technique for months was no better at it than a chimpanzee novice.

Similar demanding techniques used by nut-cracking chimpanzees involve finding a properly sized stone or hardwood club to be used as a "hammer," choosing a well-shaped tree root as an "anvil," and precisely positioning each nut on the anvil, keeping in mind that nuts from different species must be positioned differently. The hammer must be gripped in its most effective position, swung with the proper force, and aimed so that it hits the nut in exact locations to extract the maximum amount of nutmeat. Although young chimpanzees generally learn this technique from their mothers by imitation, Boesch (1993) recounts incidents in which mothers actively intervened in their offspring's unsuccessful nut-cracking attempts by taking the hammer, positioning the nut, and demonstrating the proper technique: active teaching.

Lieberman (1991) and others suggest that the "rules" involved in sequencing such motor-controlled operations are preadaptations for language, which must follow **sequencing rules,** because a sentence is a sequence of phrases made up of components such as noun, verb, adjective and subject. The ability to devise and follow such rules lies in the association centers of the brain. To enable a word (e.g., an auditory stimulus) to stand as a symbol for an object or action (e.g., a visual stimulus), a neural associative center must enable such connections to be made. Such **cross-modal associations,** as they are called, are present in many mammals; they understand and can act on some of the words spoken to them. In apes and even monkeys, cross modality extends to the ability to correlate objects they cannot see but can feel to objects they can see. Tests of apes, for example, show that even a glimpse of the photograph of an object allows them to choose correctly by feeling for the object among different objects hidden from view.

We do not yet know which neurological areas of the brain facilitate cross-modal associations and matching in apes, but neurological evidence shows one human center to be the *angular gyrus,* shown in **Figure 20-37**. Near this region, on the parietal lobe of the left cerebral hemisphere, is *Wernicke's area,* concerned with formulating and comprehending intelligent speech; people who have lesions in this area emit less information in mostly wordy babble. Patients with lesions in *Broca's area* on the left frontal lobe have difficulty speaking; this region coordinates vocal muscular movements.

We can reasonably conclude that some of the intellectual attributes of humans can be ascribed to evolutionary events that occurred early enough for humans and chimpanzees to share them. Both groups are composed of individuals who continually interact in complex and changing social patterns, and this was most likely true for their common ancestors as well. Because **self-awareness** and language depend almost entirely on social interactions — isolated young chimpanzees do not develop self-awareness, and isolated human children do not develop language — an environment of complex interpersonal relations as well as one that presented continually varying challenges in food gathering and capture, played an essential role in the evolution of these traits.

We can reasonably say that many flexible and complex human behavioral traits are adaptations to conditions in which early groups lived, and in turn, contributed to these conditions. Once started, such a reciprocal process could have accelerated rapidly, leading to increased behavioral complexities that emphasized selection for communication (language), reasoning and creative attributes, all contributing to the evolution of the brain (**Box 20-5**).

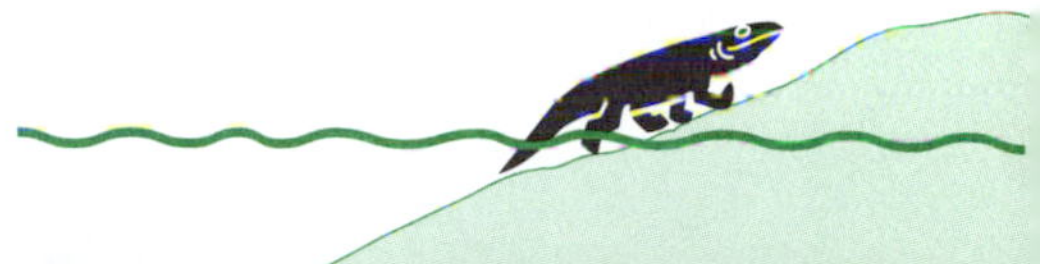

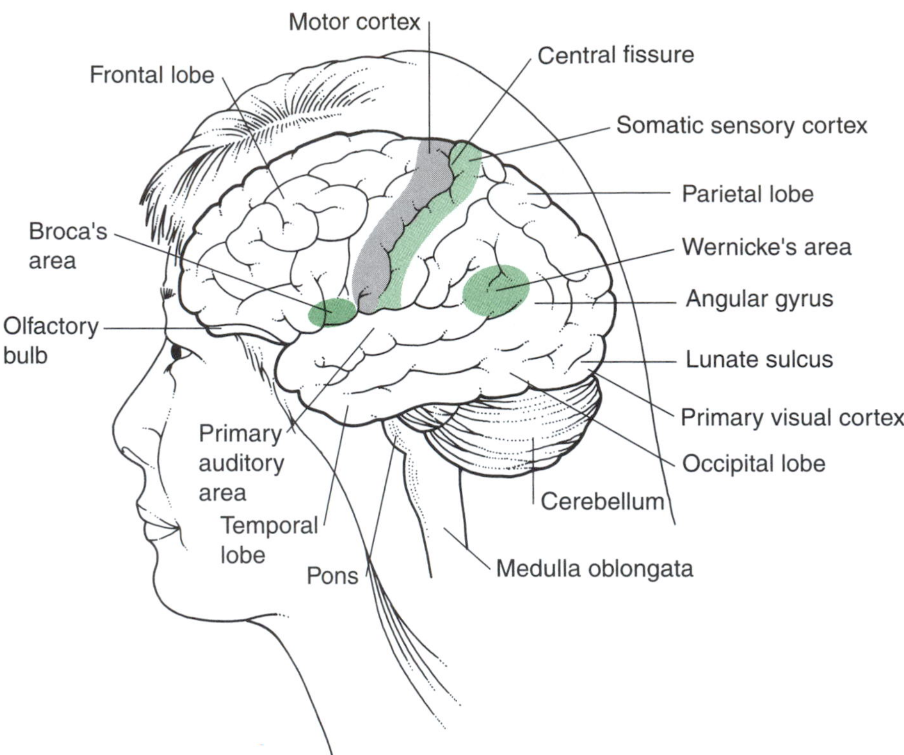

FIGURE 20-37 Some anatomical landmarks (e.g., lunate sulcus, central fissure) and some centers associated with functional specializations on the human left cerebral hemisphere. Regions identified with speech include Broca's area in the left frontal lobe and Wernicke's area in the left temporal lobe.

Altruism and Kin Selection

The importance of social interactions in developing behavioral and communication skills is seen throughout primate groups in the panoply of calls, grimaces, gestures and activities used to indicate social positions (dominance, subordination, group affiliation), needs (food, sex, reassurance) and changes in any of these areas (new social positions, alliances, sexual states or dietary interests). Such behaviors range from transmitting only information on themselves as individuals to actions that may immediately affect the survival of other group members.

Information affecting the survival of other group members is most obvious, for example, when a monkey encounters a leopard and reacts with a loud scream, signaling nearby listeners to take refuge. Although this warning signal may call the predator's attention to the screamer and diminish its own chances for survival, the effect can help preserve its relatives or compatriots.

Population geneticists beginning with Haldane in 1932 and Wright in 1949 (see Chapter 23) suggested that there were genetic advantages in such **altruistic behavior,** in which individuals may go so far as to endanger their own genetic future for those who carry closely related genotypes. In 1964, Hamilton popularized this cooperative process under the name **kin selection,** and provided formulae by which some of its benefits could be evaluated (Hamilton, 1964a, b) As Maynard Smith (1983) pointed out, "the main reason for thinking that kin selection has been an important mechanism in the evolution of cooperation is that most animal societies are in fact composed of relatives."

Reciprocal Altruism

Two decades after Hamilton's proposals, Trivers (1985) introduced a concept of altruism that seemed to have special applicability to human social behavior.

Trivers' theory of **reciprocal altruism** suggested that altruism can become established in a group where the frequency of interaction among individuals is high and the life span sufficiently long to enable recipients of altruistic acts to return favors to the altruists. Because even slight expenditures of altruistic energy (such as throwing a life preserver to a drowning individual) may have significant benefits to the altruist when it is reciprocated by the previous beneficiary or another group member, the benefits to individuals who partake in reciprocal altruism can far outweigh the costs. Frequent interaction and exchange of roles ("sometimes an altruist, sometimes a beneficiary") are necessary in order to recognize "cheaters" early on — individuals who would otherwise continually try to act as beneficiaries and exploit the altruists. Emphasis is placed on precise accounting and balancing of exchanges among individuals.

Dominance Relations

The other side of this social coin is where **dominance relations** hold sway, and cheating by one or more dominant individuals replaces cooperative relationships. Under such circumstances, reciprocal altruism would be absent or tend to diminish except in those areas in which social alliances are forged — "You help me dominate X, and I'll help you dominate Y" — or where dominance is excluded (chimpanzees who capture game are considered its proprietors whether they are dominant or subordinate, and the entire troop may then congregate for handouts.

However, even in human societies opposed to overt dominance relations, cheating may be practiced where it can remain undetected, or where local conditions may allow it to occur (e.g., smuggling). In Trivers (1985) words, humans differ "in the degree of altruism they show and in the conditions under which they will cheat." Even in societies where cheaters are considered to be "criminals," social and economic exploitation can be practiced and even institutionalized by those who use and manipulate a community's social resources for purposes other than social needs.

Morals and Emotion

By expecting altruistic behavior from other community members and by rejecting cheaters, through either punishment or exile, moral sentiments of approval and disapproval are developed and enhanced in such cooperative groups. As pointed out by Trivers (1985), the maintenance of such systems is supported by introducing or reinforcing a variety of **emotional traits** that are named for their expression in humans:

- *Friendship:* The emotional bonds established among individuals who behave altruistically toward one another.
- *Moral indignation and resentment:* The feelings of injustice and hostility toward cheaters and oppressors that can lead to retribution against them.
- *Gratitude:* The emotional responses of recipients to what they perceive as altruistic acts.
- *Sympathy, kindness, and generosity:* Emotional motivations that help individuals perform altruistic acts.
- *Guilt and repentance:* Emotions engendered as a result of active cheating or its contemplation that either prevent cheating from happening or lead to reparation and thereby help prevent the rupture of social bonds.

Although these emotions are not uniformly felt or expressed by all individuals under all circumstances, like other feelings such as love and security, they help preserve social groupings based on intricate reciprocal relationships. In *The Origin,* Darwin emphasized the value of social interactions on the evolution of instinctive "moral faculties," pointing out that these led to "love" of praise and "dread" of blame.

It seems likely that human morality and its accompanying sentiments, such as fairness and justice, did not emerge full blown from human thought but derive from an evolutionary history that traces back to early hominin groups or even earlier. The emotions just listed (as well as others such as empathy and remorse) led to abilities we call "moral judgment" and "moral conscience," enabling humans to conceive and incorporate "ethics" of right and wrong.

As Goodall (1986) and others have shown, there are hints of the existence of some of these emotions in other primates, in addition to well-expressed emotions such as fear. Because neurologists have shown that certain social-interactive behaviors ("processing of emotion") in different patients can be localized to the same section of the brain, these traits must have some genetic basis, and therefore, an evolutionary history. Rational decision-making has been traced to the prefrontal cortex in the forepart of the brain, and fearful conduct and depressed behavior appear connected to the amygdala and hypothalamus, in the center of the brain. Even our perception of emotions in others is localized to particular brain regions. Such views are supported by twin studies, in which identical twins raised apart demonstrate more shared emotional traits than genetically different fraternal twins raised apart. The genetically shared behaviors among identical twins extend to traits such as empathy and altruism.[10]

From what we can see of their effects, the *origin of emotions* is certainly no mystery; emotions help individuals perform tasks necessary for their own preservation or for that of the group. The *evolution of emotions* is also not a mystery: As a preadaptation, an emotion can be used for one purpose and later serve or be modified for another purpose. The strength of emotional ties is reflected in emotional responses when ties are broken. The closer the tie, the greater the grief. Such significant behavioral effects raise the question of the manner and extent to which evolutionarily derived emotions and strategies ("evolutionary psychology") guide our present social conduct. The topic concerned with this issue, *sociobiology,* has engendered considerable discussion and argument. Some of its aspects are discussed in Chapters 24 and 25.

[10] See Damasio et al. (1994) for evolutionary history and see Phillips et al. (1997) for localization to brain regions.

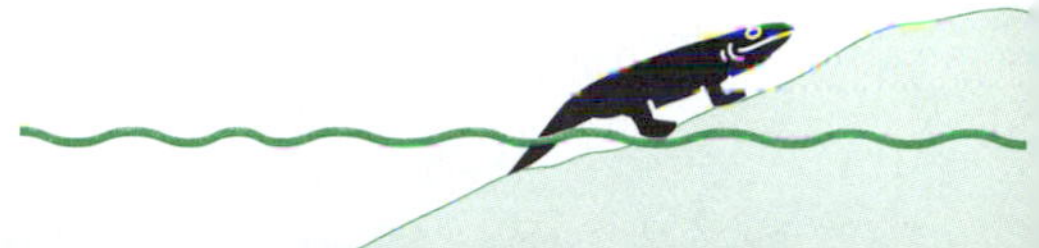

BOX 20-5 Evolution of the Human Brain

MANY WOULD CLAIM, perhaps rightfully so, that the most distinctive and interesting character in human evolution is our brain. This organ separates humans from other primates in both behavior and communication, and gives humans what we assume to be unique insight into their own thought processes, a feature commonly called **consciousness.**

Because of the brain's value and importance, scientists of all kinds, from anthropologists to psychologists to zoologists, have offered explanations to account for its evolution. For the most part, these hypotheses break down into two general types:

- The evolution of a large brain accommodated the increased number of nerve cells necessary for increased processing capacity.
- The evolution of more complex substructures and intricate neuronal circuits facilitated refinement of neural function.

These two broad concepts are not mutually exclusive because a brain can be both large and complex, and the human brain must have achieved its present state because of both types of evolution. With respect to brain size, we have already pointed to the marked increase in brain size among mammalian ungulates and carnivores during the Cenozoic (see Fig. 19-15). The data in **Table 20-7** carry this forward to primates by comparing average brain volumes in fossil hominins, humans, chimpanzees and gorillas.

As the data in the last column of Table 20-6 show, relative brain size in extant hominins is just about triple that of early australopithecines, and is triple that of extant great apes; the different proportions of the brains are shown in **Figure 20-38**. Most of the increase in brain volume is largely associated with increased area and thickness of the cerebral cortex (**Fig. 20-39**). In chimpanzees, as in other apes, the postnatal skull development primarily enhances mastication, leading to massive jaws and angular cranial projections to accommodate large muscular attachments and is marked by pronounced brow ridges. By contrast, greater emphasis in modern humans is placed on postnatal brain growth, vertical balance of the skull and much less on biting power. Selection for these traits leads to retention of primate infantile features such as small jaws and teeth, a round-domed cranium, nonprotuberant brow ridges and a long neck. Such *neotenous* human characters are deemed attractive in most cultural concepts of facial beauty and female sexual appeal, usually correlated with youthful adolescent images. Other human characteristics, such as the large size of the cerebral cortex and long legs, are obviously not neotenous but represent enhanced development of adult primate features — what some call peramorphosis. In general, because human features seem to be a mosaic of many growth patterns, some neotenous, some peramorphic, and some entirely unrelated to differences in developmental timing (heterochrony), no single descriptive pattern appears sufficient to explain human development. The brains of apes show specializations in the cerebellum. Specialization of human brains is seen in a disproportionate increase in the neocortex and increased connections between cerebellum and cortex. Lay

TABLE 20-7 Relative brain sizes in some hominin species

Species	Dates (My before present)	Estimated Body Weight (kg)[a]	Average Brain Volume (cc^3)	Relative Brain Size (EQ)[b]
Homo sapiens	0.4 to present	54	1350	5.8
Late *Homo erectus*	0.5 to 0.3	58	980	4.0
Early *Homo erectus*	1.8 to 1.5	55	804	3.3
Homo habilis	2.4 to 1.6	42	597	3.1
Australopithecus robustus	1.8 to 1.0	36	502	2.9
Australopithecus boisei	2.1 to 1.3	42	488	2.6
Australopithecus aethiopicus	2.7 to 2.3	?	399	?
Australopithecus africanus[c]	3 to 2.3	36	420	2.5
Australopithecus afarensis	4 to 2.8	37	384	2.2
Chimpanzee (*Pan troglodytes*)	Present	45	395	2.0
Gorilla (*Gorilla gorilla*)	Present	105	505	1.7

[a] An average based on combining male and female body weights given by McHenry (1994).
[b] Relative brain size is presented as the encephalization quotient (EQ) calculated as the ratio of actual brain weight or volume to the weight or volume expected for a mammal of that body size (see Fig. 19-15). McHenry calculated expected brain volume in these data as 0.0589 x (species body weight in grams)$^{0.76}$.
[c] Conroy and coworkers (1998) reported an endocranial capacity of 515 cm^3 in a particular *A. africanus* skull and suggested that other measurements may need to be reevaluated.
Source: Modified from McHenry, H. M., 1994. Tempo and mode in human evolution. *Proc. Natl Acad. Sci. USA,* **91,** 6780–6786.

FIGURE 20-38 Casts taken of the inner cranial surfaces (endocranial casts) of a chimpanzee (a), three fossil hominins (b–d), and *Homo sapiens* (e), showing a lateral view (*right side*) of each brain, with frontal lobes on the right. Most of the increase in brain size in the apes is caused by an increase in height, a trend that continued onward to *Homo sapiens,* in which expansion of the cerebral cortex also is seen (Adapted from Aiello, L., and C. Dean, 1990. *An Introduction to Human Evolutionary Anatomy*. Academic Press, London, UK; after Holloway.)

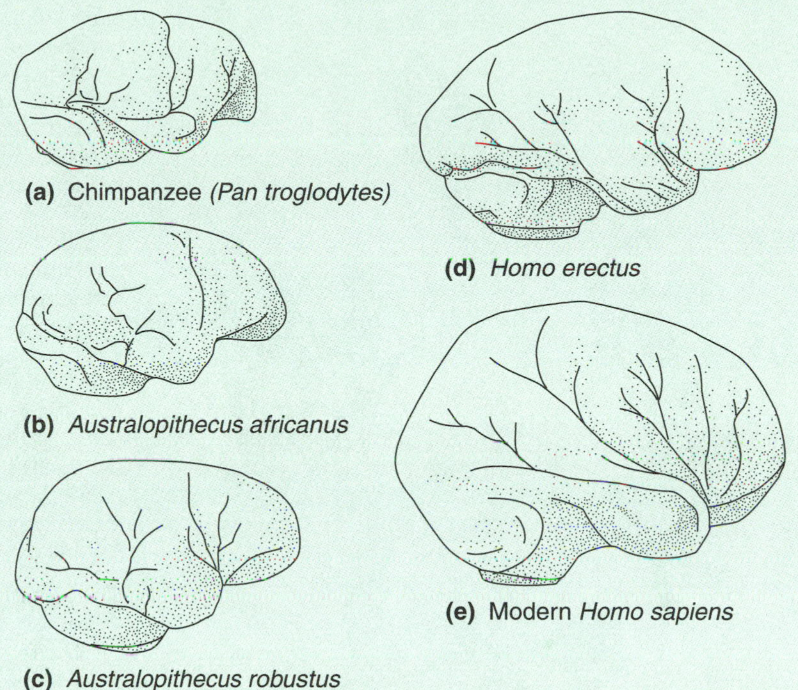

out the cortex of a rat's brain and it would cover an area the size of a postage stamp. In contrast, the cortex of a monkey brain is postcard size, of a chimpanzee is one page, while the human cortex would cover four pages. Like some other traits that also increased disproportionately relative to body size (e.g., shorter arms and longer legs relative to apes), brain size increase in hominins is an example of **allometry:** a difference in the rate at which a particular feature grows during development or evolution relative to the growth rate of other structures (**Fig. 20-40,** and see Chapter 13). Recent molecular studies have identified 49 regions of the genome (some of which are expressed in the neocortex) that are conserved in many mammals but that diverged rapidly since humans separated from chimpanzees (Pollard et al., 2006). Surprisingly, these are not protein-coding regions of the genome but may regulate genes that control neural development.

To achieve this relatively large brain size without an accompanying large body size, human brain development follows a different pattern than other mammals or even other primates. In most mammals and primates, brain growth is rapid relative to body growth during the fetal stages, but this rate diminishes after birth. In humans, prenatal brain growth is also quite rapid relative to body growth but this rate does not significantly diminish until infants are past one year of age (Martin, 1990). In that first year of postnatal growth, human brain weight almost triples from about 300 to about 900 grams, and then follows the usual primate brain–body growth rate, reaching its full size of about 1,350 grams in adulthood 15 to 20 years.

This emphasis on *rapid early postnatal brain growth* essentially extends the human gestation period from nine to 21 months by adding 12 months of extrauterine development (Portmann, 1990). Infant dependency, marked by helplessness and vulnerability, is a corollary of the advantages of increased brain size and must have had considerable response to selection, enhancing those social interactions required to support such dependency, interactions that provide the learning experiences that help stimulate postnatal brain development. According to some anthropologists, prolonged human infancy with its increased brain development has led to cranial *neoteny,* the retention of juvenile facial features in sexually mature adults.

Instead of explaining extended human childhood caused by a larger brain and the need for a longer learning period, Hawkes and coworkers (1998) explain lengthened immaturity resulting from increased female longevity, that is, the appearance of long-lived postmenopausal (nonfertile) grandmothers. They suggest these older females could supply food to the offspring of their childbearing daughters, enabling children to be weaned earlier, thus extending the juvenile learning period between infancy and sexual maturity. By contrast, others propose that the interval between menopause and death — about 30 to 40 years in humans — allows a female to help her last-born child reach independence. Different views on causes of the lengthened human childhood are not yet resolved.

The large cranial volume of newborns — a cost of brain growth *in utero* — helps explain the difficulties faced by human females in giving birth (Figs. 20-17 and 20-41) and the common need for social support and obstetrical assistance. Moreover, the energy requirements needed to sustain such a large mass of metabolically active neural tissue are significant.[a] Although the adult

BOX 20-5 Evolution of the Human Brain (*continued*)

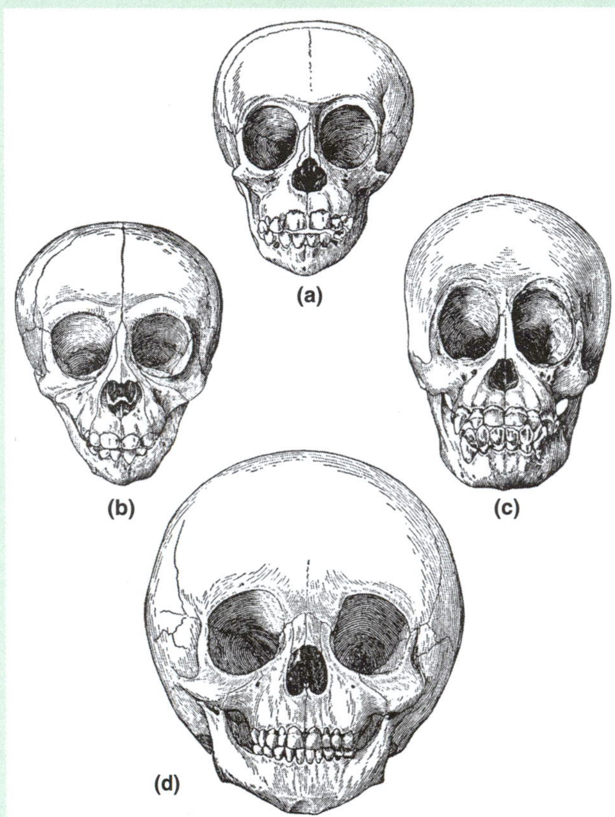

FIGURE 20-39 Skulls of juvenile chimpanzee, *Pan troglodytes* (a), gorilla, *Gorilla gorilla* (b), orangutan, *Pongo pygmaeus* (c), and human, *Homo sapiens* (d) showing the difference in proportions. (Modified from Gregory, W. K., 1929. *Our Face from Fish to Man. A Portrait Gallery of our Ancient Ancestors and Kinfolk together with a Concise History of Our Best Features.* G. P. Putnam's Sons, London, UK.)

human brain represents only two percent of total body weight, it can consume as much as 20 percent of the energy budget.

What beneficial effect of increased brain size would counter its obvious costs? Although commonly assumed, current data have generally not supported the proposal that increased brain size always correlates with increased intelligence (however measured). Including extreme examples, brain volumes can reach more than 1,500 cc^3 in some individuals diagnosed as mentally challenged, and less than 1,000 cc^3 in some "geniuses." Nor do we know of any correlation between overall brain volume among humans and particular behaviors and skills.

The obvious progression in hominin cranial size over the last 4 My must signify some crucial changes in mental capacity, especially because it is so expansive anatomically and metabolically. Among the questions raised are: What advantageous mental functions could have been selected to explain this increase in brain size? Can we identify their cerebral locations? How did these functions evolve? Although paleoneurology is a young science, we can list a few primary mental functions that most changed during the last few million years.

■ Linguistic Abilities

Aspects of language, such as grammatical structure (**syntax**) and vocabulary, seem associated with neural circuits in the prefrontal cerebral cortex that lie somewhat forward of Broca's area. The disproportionate enlargement of the prefrontal cortex is an obvious feature of *Homo sapiens* (**Fig. 20-41**, and see Rosa-Molinar and Pritz, 2004).

■ Technical Aptitudes

The suggestion that tool use and oral language may share a common neurological root is not far distant from the proposal that auditory and gestural languages are neurologically related.

Both language and manual tool manipulation are sequential processes. Because the appearance of stone tools coincides with the appearance of Broca's area in hominoids, both may have begun their evolutionary maturation together. Linkage between the two functions is further indicated by their common localization in the left or dominant hemisphere of right-handed individuals.

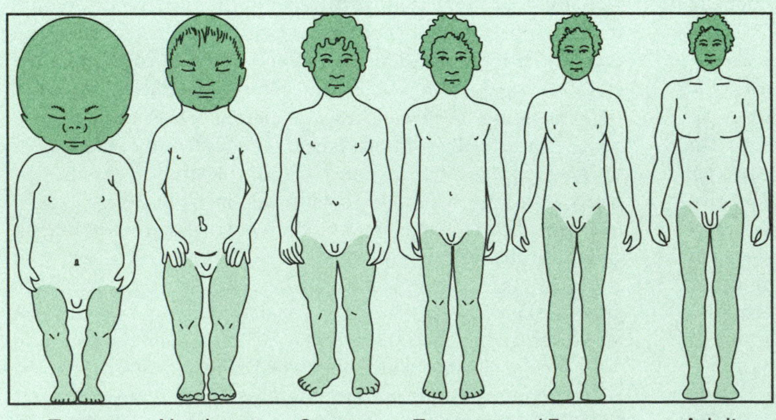

FIGURE 20-40 Different relative growth rates of human body parts — allometry. In proportion to the increase in body size that occurs during normal growth, the head (30% shaded) grows less rapidly and the legs more rapidly (10% shaded). Head and leg growth in other primates follow a less pronounced allometric pattern. (Based on Beck, F., D. B. Moffatt, and J. B. Lloyd. 1973. *Human Embryology and Genetics.* Blackwell, Oxford, England.)

For those left-handed people whose manual control center lies in the right hemisphere, language control is often found there as well.

This "lateralization" of the brain into left or right dominant hemispheres has evolutionary antecedents in the many organisms in which an important behavioral pattern locates in one hemisphere rather than the other. Song production in many birds is predominantly associated with the left hemisphere, as is the auditory perception of "coo" signals in Japanese macaque monkeys. Right-hemisphere specializations in humans also appear, especially spatial perceptions and other nonverbal traits such as rhythm and musical abilities.[b] Similarly, spatial mapping that rodents use for maze running is localized to the right hemisphere. Hemisphere dominance may therefore enable specialization by facilitating close interconnection of neural circuits devoted to a particular function.

Capacities for Social Interactions

Does a human who must exercise considerable mental flexibility in making appropriate social choices need much more brainpower than when performing solitary functions? Ridley (1996) points out that social complexities in vampire bats are responsible for the largest cerebral neocortex of any bat species. For survival, these bats need to recognize and keep track of those neighbors who will provide them with blood meals (by regurgitation) when their own hunt has been unsuccessful: that is, to discriminate between individuals, helping those who helped them and rejecting those who did not (*reciprocal altriusm*).

Language Evolution

Once vocalization evolved further development of language would have been stimulated by mutual ties with developing social in-

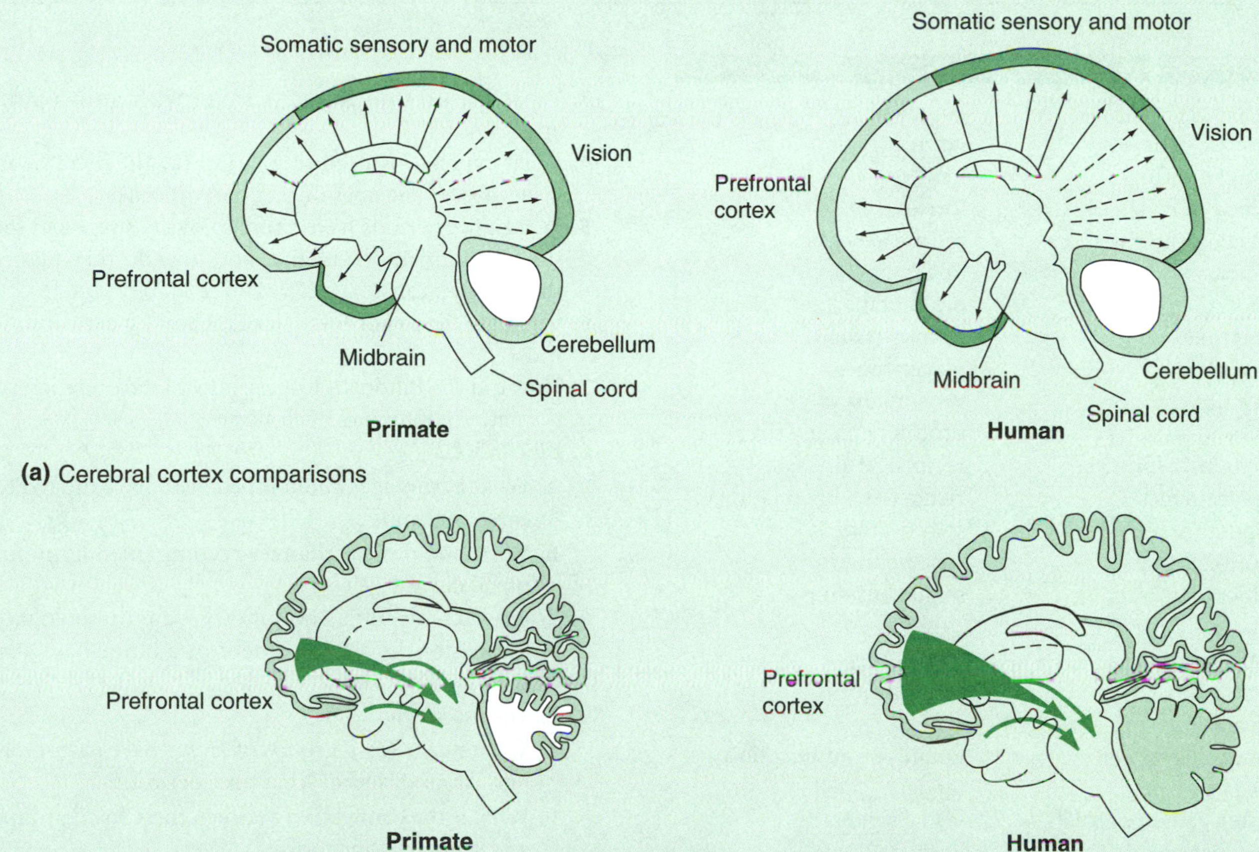

FIGURE 20-41 Comparisons between cortical areas in a typical primate (*left side*) and those in a human with an enlarged cerebrum (*right side*). Most of the disproportionate cortical enlargement in humans is of the prefrontal cortex (a), which produces, among other consequences, a greater effect on the midbrain limbic system, which governs many emotional responses (b). According to Deacon (1990), the prefrontal cortex is more than twice as large as one would predict for a comparably sized ape brain (Modified from Deacon, T. W., 1990. Rethinking mammalian brain evolution. *Am. Zool.*, **30**, 629–705.)

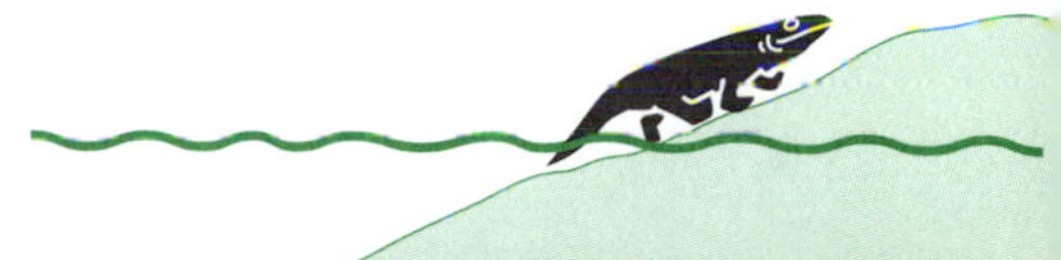

BOX 20-5 Evolution of the Human Brain (*continued*)

teractions. Given the verbal means for symbolically representing and comprehending experience, language facility could embark on an evolutionary trajectory of its own. Although the means by which such changes were acquired are still unknown, it would seem that, like other evolved complex traits (e.g., vision; see Fig. 3-1), the evolution of syntax involved a succession of selective stages. Any basic language improvement would have had selective value.

From all accounts, the primary functions of the cerebrum is to analyze and select information received from sensory nerves and coordinate it with stored information (i.e., memory), providing connections for the resulting impulses to be further stored or transmitted to other parts of the brain such as motor nerves.

The increased neural connections led to expansion of this elaborately structured cerebral organ, enabling development of those traits of the primate brain that are accentuated in humans. Deacon (1990) emphasized that "the correlated reorganization of underlying neural circuitry" may well account for the disproportionate enlargement of the human prefrontal lobes, because these cortical areas are involved in a variety of processes that include organizing sequential activity (Fig. 20-41).

[a] See Parker (1990), Martin (1990) and Bonner (2006) for the metabolic cost of large brains.

[b] A compelling case is made for a deep connection between the biology of music and the biology of language by Fitch (2005), based in part on examples of convergent evolution of the learning of complex songs by birds, whales and seals. Fitch reviews models for the function of music going back to Darwin's model of the evolution of a song-like musical **protolanguage**, which Fitch finds consistent with much of the evidence on the evolution of language and of music.

KEY TERMS

allometry
altruistic behavior
animal hunting
anthropoids
australopithecine
bipedalism
bonobo
catarrhines
chimpanzee
communication
consciousness
cooperative hunting
displacement
dominance relations
emotional traits
endocranial casts
gibbons
gorillas
home base
hominins
hominoids
Homo
hunter-gatherer societies
kin selection
language
larynx
lemurs
lorises
meat eaters
olfactory cues
orangutans
Piltdown man
platyrrhines
prosimians
protolanguage
reciprocal altruism
richochetal brachiation
self-awareness
sequencing rules
sexual bonding
speech
stone tools
symbolic calls
syntax
tactile communication
tarsiers
visual signals
vocalization

DISCUSSION QUESTIONS

1. What major features distinguish primates from other mammals?
2. What features characterize the major subgroups of primates?
3. What features do humans share with anthropoid apes?
4. What evidence is used to account for the increase in ground-dwelling apes during the Miocene?
5. What time periods have primatologists suggested for the ape–human divergence, and how do they justify these suggestions?
6. What are the major fossil lineages among early hominins?
7. How did the Piltdown forgery affect interpretation of the pattern of human evolution?
8. Bipedalism.
   a. What advantages could bipedalism have offered to early hominins?
   b. What anatomical changes accompanied hominin bipedalism?
9. What effects on human evolution have anthropologists proposed for the establishment of a home base and sexual bonding?
10. The genus *Homo*.
    a. What major fossil groups of *Homo* have paleontologists found, and in what time periods?
    b. What is the connection between these fossil groups and types of stone tools?
    c. What does the fossil evidence suggest about the origins of modern humans?
    d. How have geneticists used mitochondrial DNA to trace the origin and migration of *Homo sapiens?*
    e. Assuming that a common ancestral Y-chromosome sequence was found, would you expect the individual carrying that original sequence (the

"Y-chromosome Adam") to be a contemporary of the "Mitochondrial Eve"? Why or why not?

11. Hunting.
   a. What advantages does meat eating offer a primate?
   b. What social and behavioral characteristics are of selective value in hunting societies?

12. Communication.
   a. What modes of communication do primates use?
   b. What advantages does vocalization confer over other modes of communication?
   c. How do symbolic and nonsymbolic communication differ?

13. Speech.
   a. What anatomical features distinguish the human vocal tract from that of other primates?
   b. How do these features affect vocalization?

14. Language.
   a. What techniques have people used to communicate with apes?
   b. What kinds of linguistic abilities have apes demonstrated?

15. How can primates demonstrate self-recognition and self-awareness in nonverbal form?

16. What selective factors lead to altruistic behavior? To reciprocal altruism?

17. What types of emotions do societies dependent on complex and reciprocal social relationships develop or enhance?

18. What primate and human characteristics, physical and mental, would you consider to have an origin that can be ascribed to preadaptation? Explain.

19. Because humans had an ancestry among ancient apelike primates, would you agree that we can briefly describe humans as intelligent, naked, bipedal apes? Why or why not?

EVOLUTION ON THE WEB

Explore evolution on the Internet! Visit the accompanying Web site for *Strickberger's Evolution,* Fourth Edition, at http://www.biology.jbpub.com/book/evolution for Web exercises and links relating to topics covered in this chapter.

RECOMMENDED READING

Batzer, M. A., M. Stoneking, M. Alegria-Hartman, H. Bazan, et al., 1994. African origin of human-specific polymorphic Alu insertions. *Proc. Natl. Acad. Sci. USA,* **91,** 12288–12292.

Bickerton, D., 1995. *Language and Human Behavior.* University of Washington Press, Seattle.

Carlson, B., 2005. *Human Embryology and Developmental Biology,* 3rd ed. Mosby, Philadelphia, PA.

Cinque, G., 1999. *Adverbs and Functional Heads: A Cross-Linguistic Approach.* Oxford University Press, Oxford, UK.

Fitch, W. T., 2005. The evolution of music in comparative perspective. *Ann. N.Y. Acad. Sci.* **1060,** 29–49.

Fleagle, J. G., 1988. *Primate Adaptation and Evolution.* Academic Press, San Diego, CA.

Foley, R., 1995. *Humans Before Humanity.* Blackwell Publishers, Oxford, UK.

Gilad, Y., A. Oshlack, G. K. Smyth, T. P. Speed, and K. P. White, 2006. Expression profiling in primates reveals a rapid evolution of human transcription factors. *Nature,* **440,** 242–245.

Hammer, M. F., T. Karafet, A. Rasanayagam, E. T. Wood, et al., 1998. Out of Africa and back again: Nested cladistic analysis of human Y chromosome variation. *Mol. Biol. Evol.,* **15,** 427–441.

Harvati, K., and T. Harrison (eds.), 2006. *Neanderthals Revisited: New Approaches and Perspectives.* Springer, Netherlands.

Henke, W., and I. Tattersall (eds.), 2007. *Handbook of Paleoanthropology. Vol. 1: Principles, Methods and Approaches. Vol 2: Primate Evolution and Human Origins. Vol 3: Phylogeny of Hominines.* Springer-Verlag, New York.

Howells, W., 1993. *Getting Here: The Story of Human Evolution.* Compass Press, Washington, DC.

Jones, S., R. Martin, and D. Pilbeam (eds.), 1992. *The Cambridge Encyclopedia of Human Evolution.* Cambridge University Press, Cambridge, UK.

Kingdon, J., 2004. *Lowly Origin: Where, When, and Why Our Ancestors First Stood Up.* Princeton University Press, Princeton, NJ.

Mithen, S., 2006. *The Singing Neanderthals: The Origins of Music, Language, Mind, and Body.* Harvard University Press, Cambridge, MA.

Moore, K. L., and T. V. N. Persaud (eds.), 2003. *The Developing Human,* 7th ed. W. B. Saunders, New York.

Russell, M., 2003. *Piltdown Man: the Secret Life of Charles Dawson & the World's Greatest Archaeological Hoax.* Tempus Publishing, Gloucestershire, UK.

Shipman, P., 2001. *The Man Who Found the Missing Link: The Extraordinary Life of Eugene Dubois.* Simon & Schuster, New York.

Spencer, F., 1990. *Piltdown: A Scientific Forgery.* Oxford University Press, Oxford, UK.

Wood, B, 2005. *Human Evolution. A Very Short Introduction.* Oxford University Press, Oxford, UK.

PART

5

Populations and Speciation

Changes Within and between Species

Having examined organisms and the various lineages into which organisms fit as we continue to construct the Tree of Life (Part 4), we turn to populations and to the process — speciation — that has its origins in the natural selection of variations arising in individual organisms, generation after generation.

Gene frequencies are central to any evolutionary analysis of populations. By the 1930s, it was recognized that evolution could be represented as change in gene frequencies in populations, resulting from the processes of mutation, selection, and genetic drift. Indeed, evolution consolidated (some say hardened) into what we know as the neo-Darwinian or modern synthesis that changes in gene frequencies lead to differences among populations, species, and, finally, higher clades.

Nowadays, we have multiple concepts of what a gene is. For the analysis of gene frequencies in populations, genes are represented as alleles, so gene frequency equals allele frequency and a gene pool equals all the alleles in the population as measured by the Hardy–Weinberg equilibrium (principle; Chapter 21). The conditions under which alleles remain in equilibrium — **conservation of allele frequencies** — are also discussed in Chapter 21.

Change in **allele frequencies** is the topic of Chapter 22, in which we consider the reciprocal and interactive roles played by mutation, selection and genetic drift, as we did in Chapter 21 in the context of conservation of allele frequencies. How alleles with deleterious effects decline and those that improve fitness persist and accumulate is examined, as are the different types of natural selection under which genetic variation accumulates, the major types being stabilizing, directional, disruptive and diversifying.

As we discussed explicitly in Chapter 13, much of the current excitement about evolution comes from our ever-expanding knowledge of genomics and development, fields that are enhancing our understanding of mechanisms of evolution. Nevertheless, at the foundation of evolution lie interactions of individuals with their environment and with other individuals, interactions that we can only understand by analyzing the structure and interactions of populations, which is the topic of Chapter 23. Much past analysis of population structure focused on measuring population size, then attempting to understand how population sizes vary, and to discover the conditions that limit population growth. Two broad sets of conditions are summarized under the headings ***r*- and K-selection,** shorthand for increasing the numbers of offspring (*r*) or increasing the selective advantages of a smaller number of offspring (K), respectively. A third is the **genetic load,** the effect of deleterious alleles on a population. These three concepts begin to integrate ecological and genetic aspects of population structure. Competition and predation, discussed in Chapter 23, are two further important factors determining population structure.

Speciation occurs within populations and so, having discussed how populations are maintained in Chapter 23, we move in Chapter 24 to discuss the conditions under which populations transform into new species. Those conditions include whether a species is broken up into groups that are isolated geographically, genetically, reproductively, and/or behaviorally, all of which impede gene flow, which is a necessary precondition for speciation.

Chapter 23 on the structure and maintenance of populations, and Chapter 24 on how new species arise from existing populations should be read together. Although much of the current research in evolution involves genomics and development, interactions between individual organisms and of organisms with their environment occur in natural populations. Consequently, understanding population structure, dynamics and growth is fundamental to understanding long-term evolutionary change. In this regard, we know most about two aspects of populations: (1) how they are maintained, decline or grow, especially in the context of ecological studies of the carrying capacity (K) of the environment, and (2) how individuals from different species interact through competition and predation. Geneticists emphasize the amounts and kinds of genetic variation in populations; the large amounts of genetic variability present in practically all natural populations allow genetic evolutionary changes to occur. Some genetic variants may be neutral at certain times or under certain conditions but have selective value when the environment or genetic background changes, providing a basis for changes that can lead to speciation.

All members of a species share a common gene pool, although populations may vary genetically from one another. In Chapter 24 we discuss (1) those mechanisms of adaptation, behavioral strategies and sexual competition that maintain species and (2) those mechanisms of isolation, hybridization, bottlenecks and founder effects that can lead to speciation. Speciation occurs when genetic exchange among isolated populations is impeded, either by the slow accumulation of genetic differences or when a portion of the population, perhaps only a few individuals, is isolated — the founder effect.

21 Populations, Gene Frequencies and Equilibrium

Chapter Summary

In *On the Origin of Species by Means of Natural Selection* and its subsequent editions, Charles Darwin proposed that natural selection operates on small, continuous hereditary variations. Francis Galton and others accepted evolution, but maintained that variations are sharp and discontinuous. The controversy resolved when it was shown that several genes, each with small effect, can have a large effect when they mutually influence expression of a single phenotypic trait. By the 1930s, it became accepted through genetics that (1) evolution is a population phenomenon, which (2) we can represent as a change in gene (now allele) frequencies because of the action of various natural forces such as selection and genetic drift, and (3) that these changes can lead to differences among populations, species and higher clades. This population view of evolution became known as neo-Darwinism or the modern synthesis.

In population genetics, we are dealing with genes as alleles and gene frequencies as allele frequencies. Allele frequencies and the gene pool are two major attributes of a population, that is, a group of potentially interbreeding organisms. *Allele frequency* is the frequency of individual alleles of a gene in a population. A *gene pool* consists of all alleles in the population and therefore represents all the variation available in the current generation, and, setting mutation aside, the variation than can contribute to the next generation.

According to the Hardy–Weinberg principle, in a random mating population, allele frequencies are conserved unless external forces act on them and the equilibrium of genotype frequencies (e.g., $p^2 + 2pq + q^2$, with respect to two alleles at a locus) derive from the gene frequencies.

Many more alleles become available following a recombination event or if two or more pairs of genes assort independently. The more gene pairs, the longer it takes to achieve overall genotypic equilibrium. When linkage occurs, the higher the frequency of recombination between *linked genes* — that is, between, nucleotide positions on a chromosome (another way to define a gene) — the shorter the time needed to reach equilibrium. When genes are linked on the X chromosome, gene frequencies at equilibrium will be equal in both sexes, though this may take a number of generations if frequencies between the two sexes differ initially.

In natural populations, we can determine genotype frequencies if no allele is dominant. Such observations generally show that Hardy–Weinberg equilibrium has been achieved. With the gene sequence data now available, however, we realize (1) that this statement effectively only applies to single sites between genes that are quite far apart on the chromosome, and (2) that the notion of equilibrium (as is also true when we emphasize the mean) takes the emphasis away from the variation present in the population and available for selection. Two populations with the same Hardy–Weinberg equilibria may, and almost inevitably will, show different responses to selection.

If one of the two alleles is dominant, we can compute allele frequencies by assuming Hardy–Weinberg equilibrium and using the frequency of homozygous recessive individuals as q^2 in the genotypic equilibrium formula. When we do this for various recessive conditions present in low frequency, the frequency of heterozygous "carriers" is surprisingly high (because $2pq$ is much greater than q^2). In the case of multiple alleles, we can calculate genotype frequencies using a multinomial expansion of the Hardy–Weinberg equation. Most studies of this type indicate that gene pools are quite stable and remain at equilibrium unless selection or other conditions interfere, but such measures do not take into account the large numbers of alleles that are present at many loci. Inbreeding does not affect allele frequencies but does increase homozygosity, allowing relatively rare recessive alleles to be expressed. If these alleles are harmful, inbreeding depression may result.

Development of new methods of gene mapping (association mapping) and their application at the population level, enables genes associated with human genes to be mapped and then identified (the HapMap Project), and has the potential to reveal the extent of genetic variation in natural populations, leading us to the position where we will be able to create the elusive genotype-phenotype map.

At the center of Darwin's evolution theory was the concept that small inherited changes provide the **continuous variation** on which natural selection acts, each species representing a unique accumulation of small changes: "Species are only strongly marked varieties with the intermediate gradations lost."

Soon after publication of *On the Origin of Species,* Darwin's cousin Francis Galton became convinced that a mathematical approach to heredity showed that evolution must have proceeded in sharp, discontinuous steps. By 1871, Galton had already disproved Darwin's pangenesis hypothesis (see Chapter 2) to his own satisfaction by showing that transfusing blood between rabbit strains had no effect on heredity.

As in August Weismann's later germ plasm theory, Galton suggested that instead of somatically acquired "gemmules," the hereditary material was passed on between generations with little or no change. It seemed to Galton that parents who deviate significantly from the average for some continuous quantitative trait (such as height) tend to produce offspring who are closer to the average than the parents were (Galton's law of regression). He surmised that continuous variation was not the agency that leads to the origin of new species but that nonblending, **discontinuous variation** ("sports") provided the abrupt changes between species. Variation and variability (the propensity to vary) both are discussed in Chapter 10.

Mutationists and Selectionists: Continuous or Discontinuous Variation

By the end of the nineteenth century, two schools of thought had established themselves in England, one based on **continuous variation**, the other on **discontinuous variation.**

The former are usually now known as **mutationists** or **Mendelians,** the latter as *selectionists.* Supporters of the importance of continuous variation in evolution included the mathematically oriented *biometricians,* W. F. R. Weldon and Karl Pearson. In opposition to them were William Bateson and his supporters.

When Mendel's 1866 paper on the genetics of the common garden pea was rediscovered in 1900, mutationists and selectionists became further polarized. Bateson and the mutationists proposed that most hereditary characteristics were discontinuous and could be explained by the segregation of Mendelian factors. Weldon, Pearson and the selectionists insisted that most characteristics were continuous and that Mendelian factors were only involved in exceptional traits. To Mendelians such as De Vries and others, evolution could only be effective if selection operated on large mutations of the kind that produces varieties and species in the evening primrose, *Oenothera.* The biometricians allied themselves with Darwin's original concept: selection acting on small differences was the primary mechanism for evolutionary change.

Supporting the mutationist position were Wilhelm Johannsen's experiments, from which he concluded that selection was ineffective in quantitatively changing the size of beans descended from homozygous pure (inbred) lines. Furthermore, even when selection was practiced on beans descended from crosses between different pure lines, size differences among their descendants seemed to show relatively little change from the range of values initially observed in the F_2 generation of the cross. It seemed that marked changes in the size of beans could only come from mutations with large effect (**macromutations**), rather than from selection among the small differences observed in Johannsen's experiments.

Various biologists extended these views to propose that new species can arise in only one or a few mutational steps driven perhaps by mutation pressure in a particular, even nonadaptive direction, a hypothesis known as **saltation,** according to which, the slow, plodding process of Darwinian selection is not an essential factor in evolution.

Variations on this theme have been proposed a number of times. For example, those who consider selection merely a passive "sieve" that acts only to remove the "unfit" but does little to create the "fit," generally substitute mutation as the primary creative force in evolution. This creativity was often presumed to be caused either by a single mutation of major effect that led directly to a new species (macromutation) or by a succession of somewhat smaller changes that influenced development in a particular evolutionary direction (**orthogenesis**). As discussed later, some paleontologists used macromutation, in conjunction with saltation concepts, to explain the unevenness of the fossil record and the presumed origin of new groups. Orthogenesis via mutation also had its adherents.

Castle and Phillips (1914), in contrast, demonstrated that selection could lead to entirely new coat color patterns in hooded rats: some selected lines had "less pigment than any known type other than albino," whereas others were "so extensively pigmented that they would readily pass for the 'Irish type' which has white on the belly only." Although these studies identified no specific genes with quantitative effect, similar results with other organisms (see Fig. 10-20) did point to the likelihood that selection could act on small continuous characters to produce marked changes in phenotype.

Although the rift between mutationist and selectionist explanations of the basic mechanisms of evolution remained until the 1920s and 1930s, the gap was being narrowed by further experimental work on **quantitative characters** such as height or weight. For example, it was shown experimentally that a number of different gene pairs (multiple factors) may work in concert to affect a single quantitative character so that a wide array of possible genotypes can occur, each with a different phenotype. Alleles that segregate in typical Mendelian patterns were found to be responsible for many observed distributions of continuous traits (see Fig. 10-21).

In the end, no real difficulty arises in providing a Mendelian interpretation of Darwinian selection for quantitative traits; the most basic evolutionary changes can be explained without introducing mutations with large effect. The emphasis on evolution through small continuous characters does not mean that alleles with large effects on phenotype are always unadaptive. As Orr and Coyne (1992) pointed out, "mutations of large effect clearly play a substantial role in animal and plant breeding." Natural populations also show such effects. For example, insecticide resistance is often caused by only a few enabling mutations, while polymorphism for large and small beak size in the African seed-eating finch *Pyrenestes* results from segregation of only two alleles at a single locus (T. B. Smith, 1993).

Some adaptations do involve genes with major phenotypic effects, but these effects, taken singly, are not really those envisioned by macromutationists; that is, they do not create a complex organ or new species in a single stroke. Alleles with large effect and which cause considerable morphological differences can be selected, as supported by studies on the evolution of cultivated maize from its wild ancestor, teosinte. Changes in only about five genes are responsible, although these two plants are quite different in appearance (**Fig. 21-1;** Doebley et al., 1995). Similarly, in *Mimulus* (monkey flower), Bradshaw and coworkers (1995) showed that several mutational changes in flower structure (red flower color, long beak-shaped corolla tube, protruding anthers and stigma) can account for a shift from pollination by bees to hummingbirds, and thereby help give rise to a new species, *Mimulus cardinalis.* Continuous and discontinuous genetic variation exists, has a mutational basis and can be modified in response to selection on the phenotype.

FIGURE 21-1 (a) Teosinte (*Zea mays parviglumis*), the wild ancestor of cultivated maize, showing the mature plant and a kernel-bearing ear, and (b) a mature plant and ear of its descendant, maize (*Zea mays mays*). Although strikingly different in plant and ear architecture, these two forms differ in relatively few genes.

Three factors fueled the mutationist-selectionist fire: (1) The introduction by Kimura in 1979 of the concept that mutations could be neutral in effect (see Chapter 23), (2) identification of polymorphism at enzyme and other protein loci and (3) recognition of the importance of genetic drift, as illustrated by the following example from molecular biology/molecular evolution.

The guanine-cytosine (G-C) content of DNA is characteristic of the genomes of different organisms and consequently has been used as a molecular character in taxonomic analyses. *Plasmodium falciparum,* one of the protists that causes malaria, has a G-C content of 20% (the other 80% being the adenine-thymine or A-T content). The G-C content of the watercress *Arabidopsis thaliana* is 36%, of the yeast *Saccharomyces cerevisiae* 38% and of the bacterium *Streptomyces* 72%. Within a gene, exons tend to have higher G-C contents than do introns, while G-C bonds are more resistant to conditions such as high temperature than are A-T bonds. You can see the potential for differing interpretations, selectionists claiming that this feature is adaptive and so was selected for, mutationists that it reflects a mutational bias. Because the randomness of variation at the phenotypic level is not the same as variation in G-C content at the level of the gene, and because selection acts on the phenotype but mutation acts on the genotype, mutation "is the driving force [for evolution] at both the genic [genetic] and the phenotypic levels" (Nei, 2005).

Population Gene Frequencies

A population is a group of organisms belonging to one species (conspecific) occupying a more or less well-defined geographical region and exhibiting reproductive continuity from generation to generation.

Concurrent with the disputes between selectionists and mutationists, geneticists were proposing important new concepts emphasizing that populations rather than individuals should be the important evolutionary focus, and so geneticists had to pay attention to population allele frequencies, rather than only to whether a gene was present or absent.

It became clear that the collection of gametes a population contributes to the next generation can be considered as a giant gene pool from which offspring draw their various genotypic combinations at random. In the absence of selection and other factors that could change allele frequencies, these frequencies tend to be conserved in accordance with the **Hardy–Weinberg equilibrium.**

During the 1920s, Ronald Fisher, Sewall Wright and John (J. B. S.) Haldane developed in considerable detail a mathematical approach toward considering **evolution as change in gene (allele) frequencies.** Various studies dealt with the effects of inbreeding, the evolution of dominance and the effects of selection on allele frequencies as well as the effects of mutation, migration and genetic drift. In the early 1930s, these studies culminated in a variety of papers and books that laid the foundations for **population genetics,** the study of allele frequencies and their changes.[1]

Along with these mathematical models, natural populations were discovered to contain considerable genetic variation on which selection could act. All of these studies helped to establish the concept that populations have the variation necessary to explain evolutionary genetic change through space and time, whereas an individual is extremely limited in these dimensions (**Table 21-1**). Differences among

[1] See Provine (1986) and Crow (1988a) for the best histories of population genetics.

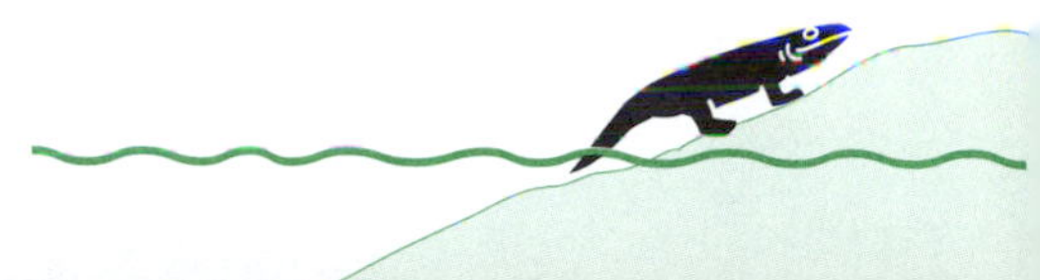

TABLE 21-1 Characteristics of individuals compared to those of populations

Characteristic	Individual	Population
Life span	One generation	Many generations
Spatial continuity	Limited	Extensive
Genetic characteristics	Genotype	Allele frequencies
Genetic variability	None	Considerable
Evolutionary characteristics	No changes, because an individual has only one genotype and is limited to a single generation	Can evolve (change in allele frequency), because evolution occurs between generations

a population's genotypes enable different reproductive rates among them, whereas an individual's genotype is constant from birth to death; that is, evolutionary changes depend on differentiation among genotypes: populations evolve, not individuals.

The neo-Darwinian or Modern Synthesis

This emphasis on the genetics of populations helped transform evolutionary theory into the form known as the **neo-Darwinian** (or **modern**) **synthesis.** At the base of this synthesis is the concept that mutations occur randomly and furnish the fuel for evolution by introducing genetic variation. We can define evolution as an ongoing process in which random mutation introduces genes (alleles[2]) whose frequencies change through time, with natural selection usually considered as the most important, although not the only, cause for such change; among other factors are migration and genetic drift, discussed in Chapter 22. In contrast to other biological disciplines that emphasized static typology, genetics offered the advantage of understanding and accentuating the *transmission, persistence, and modification of inherited variation* — the elements that enable evolution to occur. Therefore:

Biologists reject teleology out of hand, because if selection can only screen existing variation it cannot have the foresight to build today what will be needed in the future. It can only move tomorrow's generation closer to today's environment, based on yesterday's mutations (Weiss, 2002, pp. 4–5).

The accumulation of allele-frequency differences, *by whatever means,* eventually leads to more pronounced differences among populations in different localities. We say "by whatever means" because we now know that genetic variation consists of much more than changes in the frequencies of alleles. All of the following discoveries show how genetic variation can increase and/or be maintained:

- most genes have more than the basic Mendelian situation of one dominant and one recessive allele (see Chapter 10);
- genes and even entire genomes can be and have been duplicated (see Chapter 12);
- duplicated genes can change their function (see Chapter 13);
- exons can be shuffled between genes (see Chapter 7);
- repeat sequences are scattered through most genomes, often comprising the bulk of the DNA in the genome (see Chapter 7);
- transposable elements introduce novel variation (see Box 10-2); and
- the regulation of genes can alter, allowing more than one product to be produced from a single "gene" (see Chapters 10 and 13).

If we look beyond the gene as a sequence of DNA, we find another source of genetic variation in two other discoveries: (1) **alternate splicing** can produce multiple mRNAs from a single gene, and (2) **RNAi** and **post-translational modification** both increase the number of products a given gene can provide (see Chapter 10).

When gene exchange between populations can no longer occur because of reproductive barriers, separate species become established. Essentially, this approach gives a **genetic slant to the biological species concept** (see Chapter 11) in two ways:

[2] Alleles, rather than gene frequencies are the basic genetic unit calculated in population genetics. To the change in allele frequency resulting from mutation we now can add change in gene frequency because of gene duplication and exon shuffling. Furthermore, transposable elements and repeats provide sources of additional genetic variation, as does RNA interference (RNAi), differential splicing of mRNA, *cis*-regulation and post-translational modification. In Chapters 21 and 22, the emphasis is on (and examples are drawn from) the classic analysis of gene frequencies in populations in which each gene is assumed to have only two alleles, one of which (*A*) is dominant over the other (*a*).

1. By conceiving of a species as a population of individuals bearing distinctive genetic variation, separated from other species by biological mechanisms that prevent gene exchange. Now that we are further removed from the Mendelian perspective of two alleles per locus, one dominant and one recessive, we must remember that all individuals differ genetically, in part, because of multiple alleles at single loci. This does not make the above statement incorrect, but it does mean that the situation in any population is much more complex and varied than the statement implies.
2. By indicating how the mechanisms of mutation and selection, whose operation results in new varieties and species, are sufficiently general to enable us to arrange species of all organisms into higher taxa. This approach led to an important realization: whichever groups occupy a particular environment, whether designated as genotypes or higher taxa, selective interactions between them becomes an essential factor in their evolution.

Variation and Selection

From the discussion above, we can see that the neo-Darwinian synthesis helped explain how **mutation led to variation and how selection led to adaptation and "design."**

Although a major mechanism of speciation, mutation is not the only mechanism. Importantly, some of the others — genetic drift (see Chapter 22) or chromosomal changes that introduce mating barriers, for example — do not require adaptive changes in the phenotype as the initiating process. The synthesis proposed that allele frequencies and quantitative changes in allele frequencies in populations become qualitative changes through time as organisms interact with their environments, transforming some groups into varieties or subspecies and some varieties/subspecies into species (Bowler, 2005). To briefly summarize this process:

Mutations or genetic drift introduce variation that is subjected to selection resulting in adaptive changes in the phenotype and, as the process continues, to organismal diversity. The modern synthesis provides a conceptual sequence that helps explain the chain of events from genes to organisms to communities, and back again.

One can claim that the neo-Darwinian synthesis, although influential in providing a common evolutionary genetic theme for fields as diverse as embryology, systematics and paleontology, *did little more than introduce a genetic basis for Darwin's fundamental concepts of variation and divergence.* That is, although Darwin was unaware of the source, measurement and extent of variation, he understood that a species maintains sufficient variation that it can evolve varieties and subspecies that diverge from each other because of selection in new and different environmental conditions (Ospovat, 1981). This gradual divergence — the end result of variation and selection — produced in time the striking abundance of species and their many complex interactions. The genetics of neo-Darwinism made selection and variation scientifically understandable and helped reinforce a search for the genetics of evolutionary form and adaptation: *developmental genetics.* Molecular techniques introduced into such studies have demonstrated many of the intricate connections among genes and morphological patterns (Chapter 13).

Questions Provoked by the Modern Synthesis

By providing the general ideological framework in which to understand this continuum, the neo-Darwinian synthesis helped to motivate many evolutionary biologists to ask and answer more detailed questions as to how and why particular evolutionary events and adaptations occurred. Such questions began at basic levels:

- What are the allele frequencies in populations?
- How do allele frequencies change?
- Why do they change?
- What genetic differences separate varieties? Species?
- At what rates do these differences arise?
- What historical phenomena can account for these differences?

Perhaps the most prominent of such studies was a series on *Drosophila* species by Theodosius Dobzhansky (1938–1976) and coworkers, called the *Genetics of Natural Populations.* By the 1950s, these papers were widely influential in:

> Comprising a model of how genetic variation in natural populations could be studied [and] included observations of temporal variation and stability in polymorphism, estimates of migration and effective population size, evidence for the existence of selective differences in nature, and the creation of laboratory model populations in which selection could be demonstrated and estimated (Lewontin, 1997).

The neo-Darwinian synthesis popularized in books by Dobzhansky (1937) and Julian Huxley (1942), had a huge influence on biology, partly because of its fundamentally materialistic approach, and because it eliminated Lamarckian concepts and theories such as saltation and orthogenesis. The influence of population genetics, the essential component of this synthesis, extends to many other fields, including demography, ecology, epidemiology, plant and animal breeding and other areas in which gene variation and distribution affect the relationships and life patterns of organisms.

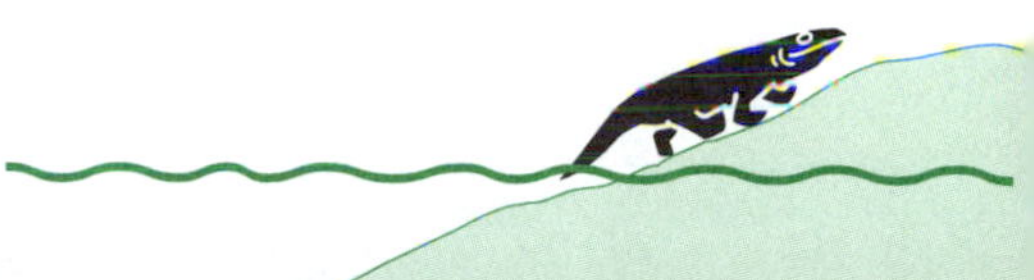

Populations and Allele Frequencies

Geneticists usually *define* a population as a group of sexually interbreeding or potentially interbreeding individuals, providing their specialist "take" on the definition given above.

Mendelian laws apply to the transmission of genes among such diploid individuals, which Wright called a **Mendelian population,** but do not apply to haploid, asexually reproducing populations. The size of the interbreeding population may vary, but is usually taken to be a local group (also called a **deme**), each member of which has an equal chance of mating with any other member of the opposite sex. Structural features can reduce the variation (and potential variation) within a population — a river that individuals cannot cross, or a landslip — by reducing the effective deme size, which therefore is much smaller than the population defined geographically or ecologically. In such situations, selection may take demes in different directions. Most theory and experiments emphasize populations of diploid organisms. The discussions that follow deal mostly with such cases.

Allele Frequency and the Gene Pool

Whether diploid or haploid, populations have two important attributes, allele frequencies and a gene pool. **Allele frequencies** are simply the proportion of the different alleles of the genes in a population. To obtain these proportions, we count the total number of organisms with various genotypes in the population and estimate the relative frequencies of the alleles involved. Except for gametes and occasional mutations, the genetic complement of all cells in a multicellular organism is the same. We may therefore adopt the convention that a haploid organism has only one gene at any one locus, a diploid has two, a triploid three, and so on.

To give a simple example, we can assume that the difference between humans who can and cannot taste the chemical phenylthiocarbamide resides in a single gene difference between two alleles, *T* and *t*. Because the allele for tasting, *T*, is dominant over *t*, two genotypes (homozygous *TT* and heterozygous *Tt*) represent tasters. The nontasters are *tt*. A population of 200 individuals composed of 90 *TT*, 60 *Tt*, and 50 *tt* will therefore have a total of 400 alleles at this locus. As shown in **Table 21-2a,** 240 of these are *T* (a frequency of 0.60) and 160 are *t* (a frequency of 0.40). We can calculate the same allele frequencies from the frequencies of the three genotypes, according to the formula, frequency of an allele = frequency of homozygotes for that allele + 1/2 frequency of heterozygotes, each of which contains one such allele out of two (Table 21-2b).

The **gene pool** is the sum total of alleles in the reproductive gametes of a population. It can be considered as a gametic pool from which samples are drawn at random to form the zygotes of the next generation. The genetic relationship between an entire generation and the subsequent generation is similar to the genetic relationship between a parent and its offspring. Because the frequencies of alleles in the new generation will depend to some degree at least on their frequencies in the old, we might say that allele frequencies rather than genes are inherited in populations.

In what form can we express and analyze these allele frequency relationships between generations?

One of the first attempts at using the concept of allele frequencies occurred in the dispute between the biometricians and Mendelians. Some argued that dominant alleles, no matter what their initial frequency, would be expected to reach a stable equilibrium frequency of three dominant individuals to one recessive, because this was the Mendelian segregation pattern for these alleles. That such ratios were not observed for low-frequency dominant alleles such as *brachydactyly* (short fingers) was offered as evidence that populations did not follow Mendelian rules, and that allele frequencies could be ignored in evolutionary studies.

Although this interpretation was widely accepted at first, Geoffrey H. Hardy (1877–1947) in England and Wilhelm Weinberg (1862–1937) in Germany independently disproved them. They demonstrated that allele frequencies do not depend upon dominance or recessiveness but remain essentially unchanged from one generation to the next under certain conditions. Such conservation of allele frequencies is briefly discussed in this chapter. Chapter 22 deals with forces that can change allele frequencies.[3]

Conservation of Allele Frequencies

The principle discovered by Hardy and Weinberg is illustrated using the tasting example previously mentioned.

For example, let us place on an island a group of children of the genotypes 0.45 *TT*, 0.30 Tt, 0.25 *tt*, where allele frequencies are 0.60 *T* and 0.40 *t*. Let us assume that the number of individuals in this newly formed population is large and that tasting or nontasting has no effect on survival (viability), fertility or attraction between the sexes.

As these children mature, they mate at random with individuals of the opposite sex. We can predict mating between any two genotypes solely on the basis of the genotypic frequencies in the population. As shown in **Table 21-3,**

[3] See Crow (1988a) for a history of the roles of Hardy and Weinberg. Various books — Falconer and Mackay (1996), Hartl and Clark (1997), Hedrick (2000), and Gillespie (2004) — treat more fully the basic theoretical principles of population genetics and provide more complete formula derivations and extensive examples.

TABLE 21-2 Techniques for obtaining allele frequencies for a diploid population of 200 individuals

a. Using numerical gene counts (there are 400 genes in 200 diploid individuals):

T	=	180 (in *TT*)	+	60 (in *Tt*)	=	240/400	=	.60
t	=	100 (in *tt*)	+	60 (in *Tt*)	=	160/400	=	.40
						Total		1.00

b. Using genotype frequencies:

T	=	.45 *TT*	+	1/2 (.30 *Tt*)	=	.45	+	.15	=	.60
t	=	.25 *tt*	+	1/2 (.30 *Tt*)	=	.25	+	.15	=	.40
								Total		1.00

Note: These individuals are of the following types:
90 *TT* + 60 *Tt* + 50 *tt* = 200 individuals
.45 *TT* + .30 *Tt* + .25 *tt* = 1.00 (genotypes)

nine different types of mating can occur, of which three mating are reciprocals of others (e.g., *TT* × *tt* = *tt* × *TT*). In all, these six different mating combinations will produce offspring in the ratios shown.

Note that although **random mating** has altered the frequencies of genotypes, allele frequencies among the offspring have not changed. For *T*, the offspring allele frequency is equal to 0.36 + 1/2(0.48) = 0.60, and the frequency of *t* is 0.16 + 1/2(0.48) = 0.40, exactly the same as before. Under these conditions, no matter what the initial frequencies of the three genotypes, the allele frequencies of the next generation will be the same as those of the parental generation. For example, if the founding population of this island contained 0.40 *TT*, 0.40 *Tt*, and 0.20 *tt*, the allele frequency for *T* would be 0.40 + 1/2(0.40) and 0.20 + 1/2(0.40) for *t*, the same as before. However, as shown in **Table 21-4,** despite the different initial genotype frequencies, offspring are again produced in the ratio 0.36 *TT*/0.48 *TT*/0.16 *tt*, or an allele frequency of 0.60 *T*/0.40 *t*. Two important conclusions follow:

TABLE 21-3 Allele frequencies produced by random mating among individuals in a population having the frequencies given in Table 21-2

		Males		
		TT = .45	*Tt* = .30	*tt* = .25
Females	*TT* = .45	.2025 ①	.1350 ④	.1125 ⑦
	Tt = .30	.1350 ②	.0900 ⑤	.0750 ⑧
	tt = .25	.1125 ③	.0750 ⑥	.0625 ⑨

Parents				Offspring		
Matings of Genotypes	Box Number(s) from Above		Mating Frequency	*TT*	*Tt*	*tt*
TT × *TT*	①	=	.2025	.2025		
TT × *Tt*	② + ④	=	.2700	.1350	.1350	
TT × *tt*	③ + ⑦	=	.2250		.2250	
Tt × *Tt*	⑤	=	.0900	.0225	.0450	.0225
Tt × *tt*	⑥ + ⑧	=	.1500		.0750	.0750
tt × *tt*	⑨	=	.0625			.0625
			1.0000	.3600	.4800	.1600

Allele frequencies among offspring:

T	=	.36 *TT*	+	1/2 (.48 *Tt*)	=	.45 + .15	=	.60
t	=	.16 *tt*	+	1/2 (.48 *Tt*)	=	.25 + .15	=	.40
						Total		1.00

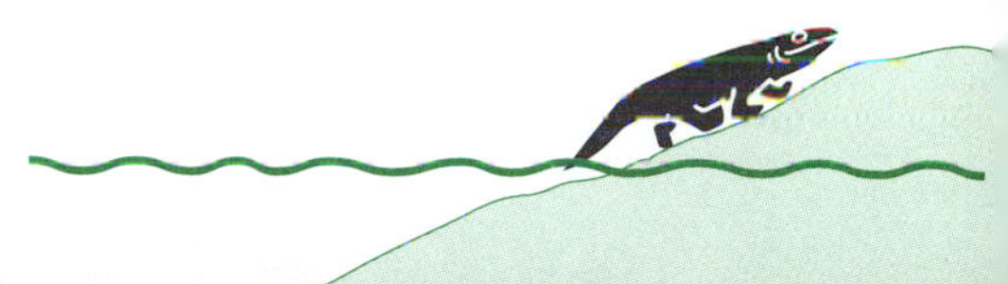

1. Under conditions of random mating (**panmixia**) in a large population where all genotypes are equally viable, allele frequencies of a particular generation depend on the allele frequencies of the previous generation and not on the genotype frequencies.
2. The frequencies of different genotypes produced through random mating depend only on the allele frequencies.

Both of these points mean that by confining our attention to alleles rather than genotypes, we can predict allele and genotype frequencies in future generations, providing outside forces are not acting to change their frequency and that random mating occurs between all genotypes. To continue our previous illustration, we may predict that under these conditions the initial allele frequencies in taster/nontaster populations will not change in the next or succeeding generations. Furthermore, after the first generation, the genotype frequencies will remain stable, that is, at **equilibrium.**

The Hardy–Weinberg Equilibrium

The theory describing this genotypic equilibrium, based on stable allele frequencies and random mating, is known as the **Hardy–Weinberg Equilibrium (principle** or **law)** and has served as the founding theorem of population genetics. Perhaps its main contribution to evolutionary thought lies in demonstrating that genetic differences in a randomly breeding population tend to remain constant unless acted on by external forces, a point contrary to the pre-Mendelian concept that heredity involves a blending of traits that become more dilute with each generation of interbreeding. **Figure 21-2** outlines major assumptions and steps in the Hardy–Weinberg equilibrium.

The general relationship between allele and genotype frequencies can be described in algebraic terms using the Hardy–Weinberg principle as follows: If p is the frequency of an allele at a given locus in a panmictic population (e.g., *T*) and q the frequency of another allele at that locus (e.g., *t*), so that $p + q = 1$ (i.e., there are no other alleles), then the equilibrium frequencies of the genotypes are given by the terms $p^2(TT)$, $2pq(Tt)$, and $q^2(tt)$. If the allele frequencies of

TABLE 21-4 Allele frequencies produced by random mating among individuals in a population that has genotypic frequencies of 0.40 *TT*, 0.40 *Tt* and 0.20 *tt* (allele frequencies: 0.60 *T*, 0.40 *t*)

		Males: *TT* = .40	Males: *Tt* = .40	Males: *tt* = .20
Females	*TT* = .40	.1600 (1)	.1600 (4)	.0800 (7)
Females	*Tt* = .40	.1600 (2)	.1600 (5)	.0800 (8)
Females	*tt* = .20	.0800 (3)	.0800 (6)	.0400 (9)

Parents: Matings of Genotypes	Box Number(s) from Above		Mating Frequency	Offspring: *TT*	*Tt*	*tt*
TT × *TT*	(1)	=	.1600	.1600		
TT × *Tt*	(2) + (4)	=	.3200	.1600	.1600	
TT × *tt*	(3) + (7)	=	.1600		.1600	
Tt × *Tt*	(5)	=	.1600	.0400	.0800	.0400
Tt × *tt*	(6) + (8)	=	.1600		.0800	.0800
tt × *tt*	(9)	=	.0400			.0400
			1.0000	.3600	.4800	.1600

Allele frequencies among offspring:

T = .36 *TT* + 1/2 (.48 *Tt*) = .45 + .15 = .60
t = .16 *tt* + 1/2 (.48 *Tt*) = .25 + .15 = .40
Total 1.00

T and t are $p = 0.6$ and $q = 0.4$, respectively, the equilibrium frequencies will be

$$(0.6)^2(TT) + 2(0.6)(0.4)(Tt) + (0.4)^2(tt) \\ = 0.36\ TT + 0.48\ Tt + 0.16\ tt$$

You can visualize this relationship by drawing a checkerboard in which the genotype frequencies stem from random union between alleles that are in the frequencies of p and q (**Fig. 21-3**). The same results derive from the **binomial expansion** $(p + q)^2 = p^2 + 2pq + q^2$. Therefore, with any given

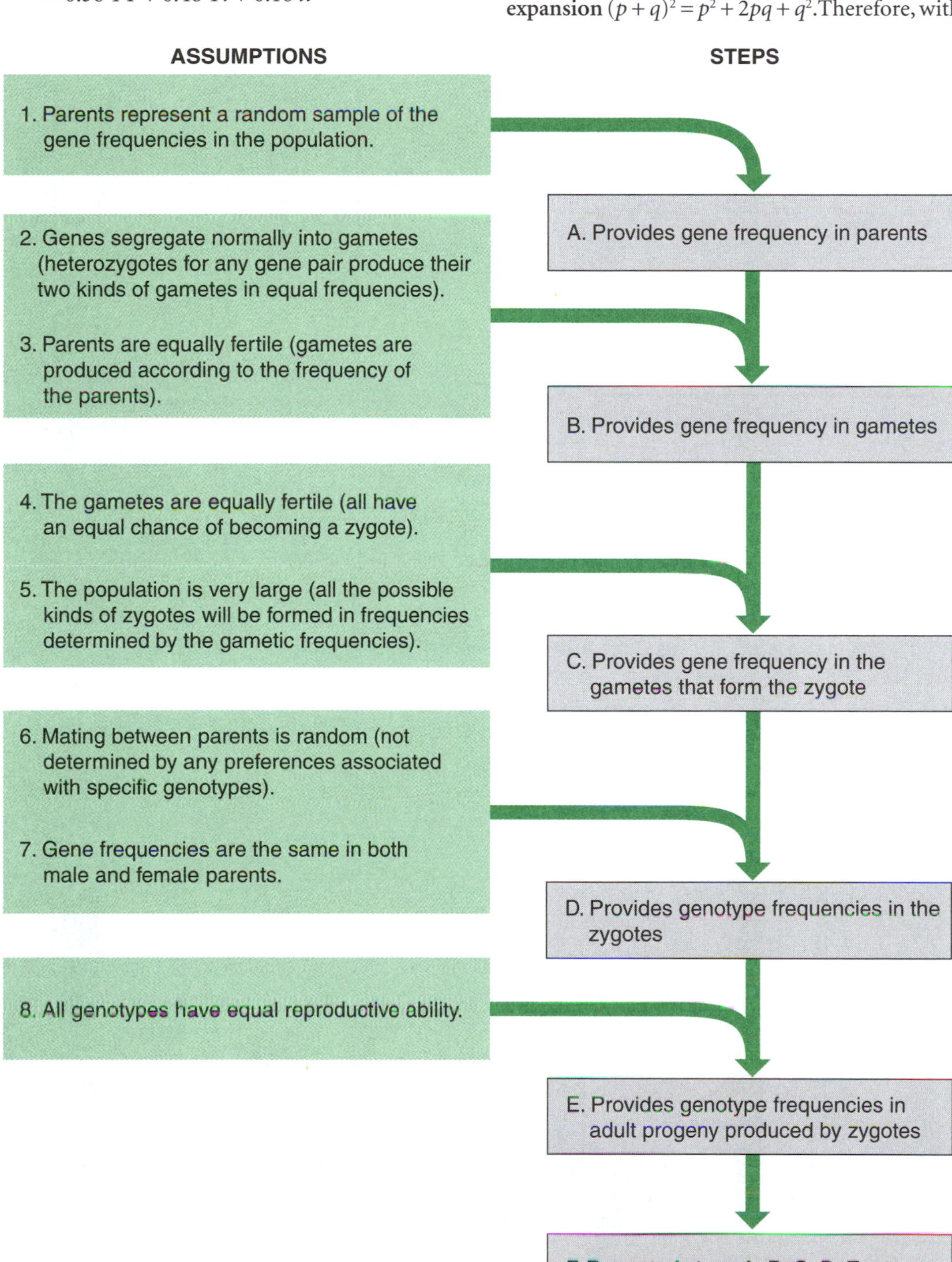

ASSUMPTIONS
STEPS
1. Parents represent a random sample of the gene frequencies in the population.
A. Provides gene frequency in parents
2. Genes segregate normally into gametes (heterozygotes for any gene pair produce their two kinds of gametes in equal frequencies).
3. Parents are equally fertile (gametes are produced according to the frequency of the parents).
B. Provides gene frequency in gametes
4. The gametes are equally fertile (all have an equal chance of becoming a zygote).
5. The population is very large (all the possible kinds of zygotes will be formed in frequencies determined by the gametic frequencies).
C. Provides gene frequency in the gametes that form the zygote
6. Mating between parents is random (not determined by any preferences associated with specific genotypes).
7. Gene frequencies are the same in both male and female parents.
D. Provides genotype frequencies in the zygotes
8. All genotypes have equal reproductive ability.
E. Provides genotype frequencies in adult progeny produced by zygotes
F. Repeat of steps A, B, C, D, E, etc.

FIGURE 21-2 Assumptions and steps in the Hardy–Weinberg equilibrium. (Adapted from Falconer, D. S., and T. F. C Mackay, 1996. *Introduction to Quantitative Genetics*, 4th ed. Longman, Harlow, Essex, UK.)

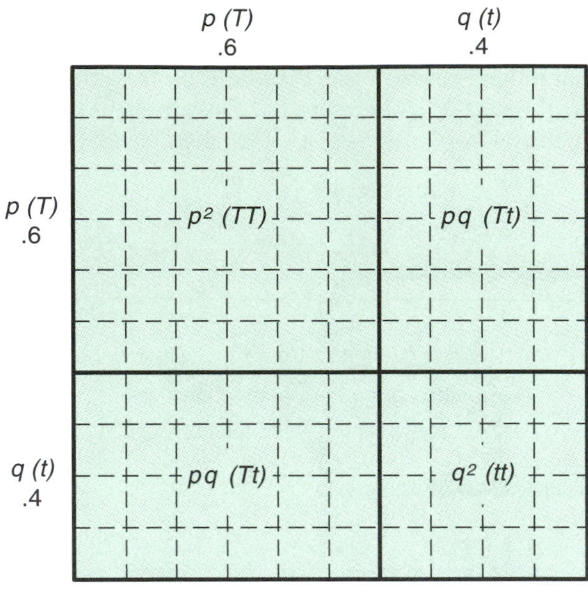

FIGURE 21-3 Genotypic frequencies generated under conditions of random mating for two alleles, T and t, at a locus when their respective frequencies are $p = .6$ and $q = .4$. Equilibrium genotypic frequencies are therefore .36 *TT*, .48 *Tt*, and .16 *tt*.

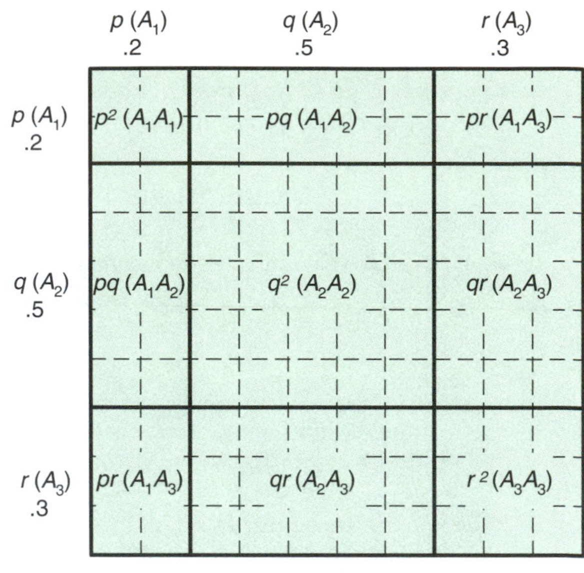

FIGURE 21-5 Genotypic frequencies generated under conditions of random mating when there are three alleles, A_1, A_2, and A_3, present at a locus. For purposes of illustration, the respective frequencies of these alleles are given as $p = .2$, $q = .5$ and $r = .3$. Equilibrium genotypic frequencies are therefore .04 A_1A_1, .20 A_1A_2, .12 A_1A_3, .25 A_2A_2, .30 A_2A_3 and .09 A_3A_3.

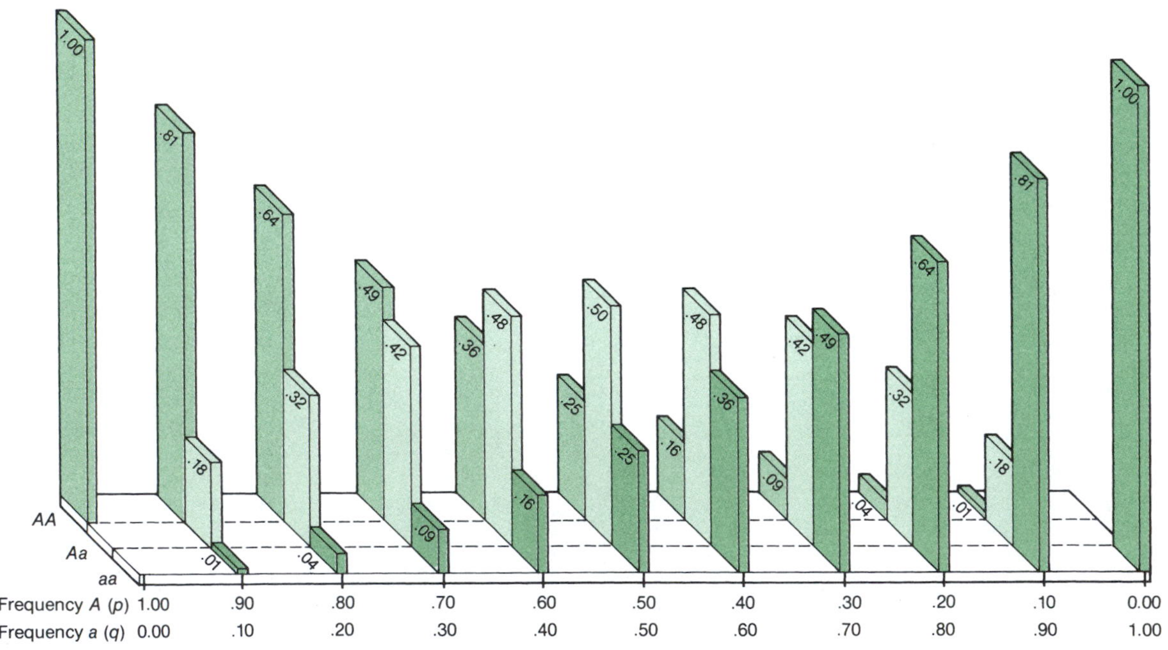

FIGURE 21-4 Genotypic frequencies at Hardy–Weinberg equilibrium for a variety of allele frequencies of A (p) and a (q). (Adapted from Wallace, B., 1970. *Genetic Load: Its Biological and Conceptual Aspects.* Prentice Hall, Englewood Cliffs, NJ.)

p and q and random mating between genotypes, one generation of a population in which generations do not overlap is enough to establish equilibrium for the frequencies of alleles and genotypes. Once established, the equilibrium will persist until the allele frequencies are changed.

Figure 21-4 shows the genotypic frequencies at Hardy–Weinberg equilibrium for a two-allele locus, where the frequency of each allele ranges from 0 to 1. Note that the frequency of heterozygotes never exceeds 0.50 but is significantly higher than the frequency of homozygotes for a rare allele (e.g., when a is 0.1, aa is 0.01 but Aa is 0.18).

When there are more than two alleles at a locus (a common situation), we must consider each allele frequency as an element in a multinomial expansion. For example, if there are only three possible alleles at a locus, A_1, A_2, and A_3, with respective frequencies p, q, and r, so that $p + q + r = 1$, the trinomial expansion $(p + q + r)^2$ determines the genotypic equilibrium frequencies. The six genotypic values are

$$p^2 A_1A_1 + 2pq\,A_1A_2 + 2pr\,A_1A_3 + q^2 A_2A_2 + 2qr\,A_2A_3 + r^2 A_3A_3$$

Because each haploid gamete contains only a single allele for any one gene locus, zygotic combinations will depend only on the frequency of each allele (**Fig. 21-5**) and, as when there are only two alleles, equilibrium is established in a single generation of random mating.

Attaining Equilibrium at Two or More Loci

Establishment of equilibrium in one generation holds true as long as we consider each single gene locus separately without being concerned about what is happening at other gene loci. If, however, we consider the products of two independently assorting gene pair differences simultaneously — for instance, *Aa* and *Bb* — the number of possible genotypes increases to 3^2 (*AABB, AABb, AaBB, AaBb,* and so on). As expected, more terms are now involved in the multinomial expansion, so that if we call p, q, r and s the allele frequencies of A, a, B and b, respectively, the equilibrium ratios of their genotypes are expressed as $(pr + ps + qr + qs)^2$, or p^2r^2 *AABB*, $2p^2rs$ *AABb*, $2p^2s^2$ *AAbb*, $2pqr^2$ *AaBB*, . . ., q^2s^2 *aabb*.

This equilibrium formula depends on the terms pr, ps, qr, qs, the equilibrium frequencies of the gametes *AB*, *Ab*, *aB* and *ab*, respectively. Once the gametic frequencies reach these equilibrium values, the equilibrium genotypic frequencies will be reached. The problem of attaining equilibrium resolves itself to the time it takes for the gametic frequencies to reach these values. If we begin only with heterozygotes ($AaBb \times AaBb$) in which the frequencies of all alleles are the same (i.e., $p = q = r = s = 0.5$), then all four types of gametes (*AB, Ab, aB, ab*) are immediately produced at equilibrium frequencies (0.25), and genotypic equilibrium is reached within one generation.

However, an initial population of heterozygotes is the only condition in which equilibrium is reached so rapidly. To take an extreme case, if we begin with the genotypes *AABB* and *aabb,* only two types of gametes are produced (*AB* and *ab*) and, since many genotypes are missing (e.g., *AAbb* and *aaBB*), equilibrium for all genotypes cannot be reached in the next generation. In general, we may ask two questions:

1. What are the expected equilibrium frequencies of gametes?
2. How rapidly are these frequencies achieved?

To deal with these questions, we can characterize *AB* and *ab* gametes as *nonrecombinant* in this case or *in coupling,* and *Ab* and *aB* gametes as *recombinant* in this case or *in repulsion.* However defined, both gametic types carry the same alleles (*A, a, B, b*), meaning that the frequency of each allele in one type (e.g., repulsion) is equal to its frequency in the other (e.g., coupling). We would expect the products of the frequencies of both types of gametes to be equal at equilibrium, $(AB) \times (ab) = (Ab) \times (aB)$.

For example, if the frequencies of A and B are each 0.6, and the frequencies of a and b are each 0.4, at equilibrium $(0.36)(0.16) = (0.24)(0.24)$, or both products equal 0.0576. If the coupling and repulsion products in the initial population differ, this difference represents the change in gametic frequencies that must occur for equilibrium values to be reached. If we call this difference disequilibrium, or d, and it is positive, so that coupling minus repulsion > 0, for example, $(AB)(ab) - (Ab)(aB) = +d$, at equilibrium this fraction will have been added to each of the coupling gametes and subtracted from each of the repulsion gametes. If d is negative, the reverse operation will occur. In both cases, disequilibrium will have diminished to zero, and equilibrium will have been established.

Until the final gametic ratios are reached, half the difference from equilibrium reduces each generation, so that within four to five generations more than 90 percent of this difference from equilibrium frequency has been attained by all gametes, and less than 10 percent of disequilibrium value remains. **Table 21-5** shows how to calculate d and how the changes in gametic frequencies occur until equilibrium is attained. For three gene pairs, the speed of approach to equilibrium is even further diminished, becoming slower still as more gene pairs are involved.

Linkage Disequilibrium

The classic concept of equilibrium allele frequencies is just that, classic. It is correct for situations in which a gene consists of one or a few alleles. However, in most populations

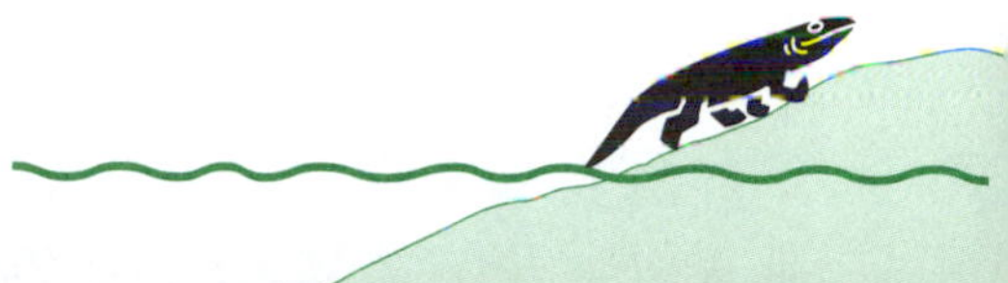

TABLE 21-5 Calculation of *d* and the equilibrium frequencies of gametes for a population in which the frequencies of two unlinked gene pairs *Aa* and *Bb* are *A* = *b* = 0.6 and *a* = *B* = 0.4, and the initial genotypic frequencies are *AABB* = *AAbb* = *aabb* = 0.30 and *aaBB* = 0.10

EQUILIBRIUM FREQUENCY OF GAMETES

Initial Population	Gametes: Type	Gametes: Initial Frequency	Gametes: Equilibrium Frequency	
30% *AABB*	*AB*	.3	.32 − *d*	(.6 x .4 = .24)
30% *AAbb*	*Ab*	.3	.31 + *d*	(.6 x .6 = .36)
30% *aaBB*	*aB*	.1	.11 + *d*	(.4 x .4 = .16)
10% *aabb*	*ab*	.3	.32 + *d*	(.4 x .6 = .24)

$d = (AB)(ab) - (Ab)(aB) = (.3)(.3) - (.3)(.1) = .06$

ATTAINMENT OF EQUILIBRIUM
WHEN RECOMBINATION IS UNHINDERED BY LINKAGE

Generation	Amount Added (*AB*, *ab*) or Subtracted (*Ab*, *aB*)	Proportion of Disequilibrium Remaining	Gametes: *AB*	Gametes: *Ab*	Gametes: *aB*	Gametes: *ab*
1		1.0*d*	.3	.3	.1	.3
2	.5*d*	.5*d*	.27	.33	.130	.27
3	.75*d*	.25*d*	.255	.345	.145	.255
4	.875*d*	.125*d*	.2475	.3525	.1525	.2475
5	.9375*d*	.0625*d*	.24375	.35625	.15625	.24375
•	•	•	•	•	•	•
•	•	•	•	•	•	•
•	•	•	•	•	•	•
Equilibrium	*d*	0*d*	.24	.36	.16	.24

Source: Adapted from Strickberger M. W., 1985. *Genetics*, 3rd ed. Macmillan, New York.

of most species, simple allelic equilibrium is unlikely to be present. There is a substantial element of stocasticity when multiple alleles are involved, which means that the pattern of allele frequency in a population or subpopulation at any one time depends on chance aspects of mutation, genetic recombination (including any recombination hot- or cold-spots), any mating preferences in finite populations or subpopulations (deviations from random mating), selection, population structure, and so forth.

As you might expect, and even in the simplified example used so far, linkage between two loci complicates reaching equilibrium because the chances that all the different types of dihybrid gametes will be found depend on crossover frequencies between the two loci. The closer the linkage, the longer it will take for the frequency of coupling gametes to equal the frequency of repulsion gametes. In other words, such **linkage disequilibrium** depends on recombination frequency, and lower recombination frequencies between linked loci accordingly delay the attainment of equilibrium (**Fig. 21-6**). Once equilibrium is attained, we have no way of distinguishing linked or unlinked genes except through tests for departures from independent assortment (recognizing, however, that even genes on different chromosomes may show linkage disequilibrium if the chromosomes do not assort independently).

Despite these theoretical considerations, and for various reasons, not all gametes of linked loci in natural populations reach equilibrium frequencies. For example, some linkage disequilibrium appears for alleles between which recombination is extremely rare. There are cases in which linkage disequilibrium persists because certain linked allele combinations seem beneficial so that reproductive success of alleles at one locus depends upon particular alleles at another locus (**epistasis**). The third chromosome gene arrangements com-

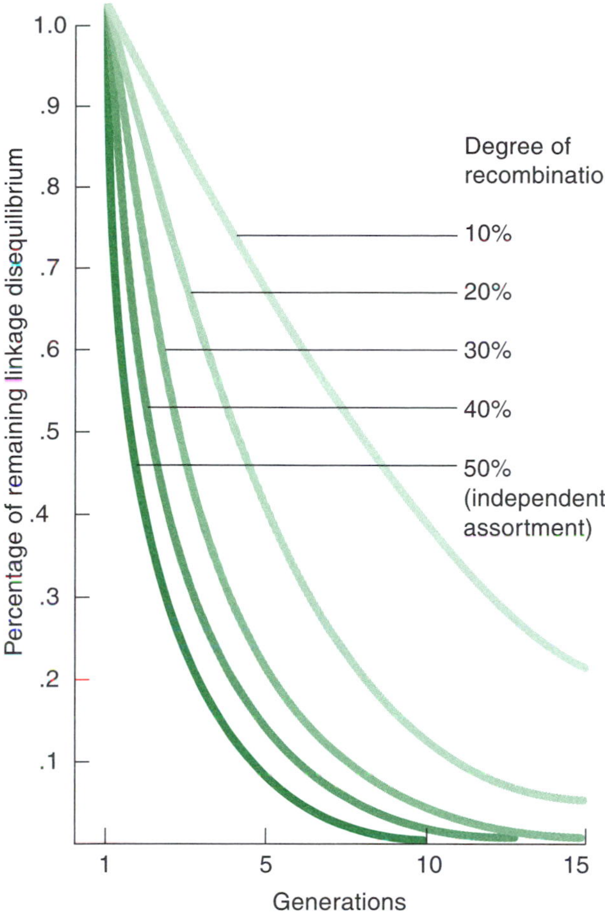

FIGURE 21-6 The proportion of linkage disequilibrium that remains in various generations (starting from an initial value of 1.0) when different degrees of recombination occur between two loci. (Adapted from Strickberger M. W., 1985. *Genetics,* 3rd ed. Macmillan, New York.)

mon in *Drosophila pseudoobscura* thus represent linked groups of genes that are advantageous under particular environmental conditions. Because these genes are included within inversions that restrict recombination, their linkage can be preserved for relatively long periods of time, thereby forming **coadapted gene complexes.** Other linkage effects are possible when a "neutral" allele at one locus, which itself has little or no discernible influence on reproductive success, is closely linked to an allele that has an adaptive effect. Because of linkage disequilibrium, the neutral allele will increase its frequency in conjunction with the linked advantageous allele through this type of *"hitchhiking."*

Association Mapping and the HapMap Project

Gene mapping, locating the position on individual chromosomes of genes that contribute to the phenotype, has become especially important in locating genes associated with diseases and syndromes in humans. Knowing the location of the gene facilitates identification of alleles of the gene associated with such diseases as Tay-Sachs. Genotyping individuals allows those alleles to be detected. Sequencing the allele allows its function to be explored, drugs to be produced and/or cures sought.

A long-standing method to map genes is linkage analysis of *individuals* with a known family history or pedigree (see the section on linkage disequilibrium in this chapter). Genotyping marker genes that lie near an allele related to or thought to be related to a particular disease has allowed many genes affecting many diseases or syndromes to be identified.

A more recent approach to gene mapping is based on associations between genes at the *population* level. Association or disequilibrium mapping takes advantage of the knowledge that mutated alleles associated with a disease generally appear against a particular genetic (haplotype) background. For example, some strains of mice are much more susceptible to cleft lip and/or cleft palate than are other strains; the prevalence of congenital anomalies varies by ethnicity in humans (see Table 24-1). Association mapping is essentially mapping by linkage analysis but at the population level, allowing (1) a much more complete genetic analysis of the past history of the allele, (2) much more fine-scale mapping of the allele, and (3) rare alleles or alleles with small effects to be detected more readily than if mapped to an individual genome.[4]

The international HapMap Project, a consortium of researchers from six countries, was established to map and catalogue genetic variation in groups of humans from different regions of the world. This enormously useful database will then be used by other researchers interested in identifying genes associated with diseases affecting humans, with the long-term aim of developing drugs or other modes of treatment of those diseases, and identifying individual responses to drugs, treatment regimes or environmental agents. Cancer therapy is already targeting the presence of diverse cell populations in individuals with the "same" cancer, thereby designing treatment on an individual cell population basis. Now this really is intelligent design. Initial screening of genetic differences between individuals from four ethnic populations has begun.

You will have realized that association mapping provides a powerful method for analyzing natural genetic variation in populations other than humans, especially for species in which the entire genome has been sequenced and/or for which the entire genome has been mapped, both of which are the case for mouse ear water cress, *Arabidopsis thaliana,* the model vascular plant used in genomic and developmental studies. Of course, any subdivision of populations com-

[4] See Hästbacka et al. (1992) for association mapping, and see Graham and Thompson (1998) and Morton (2005) for its use in mapping rare alleles.

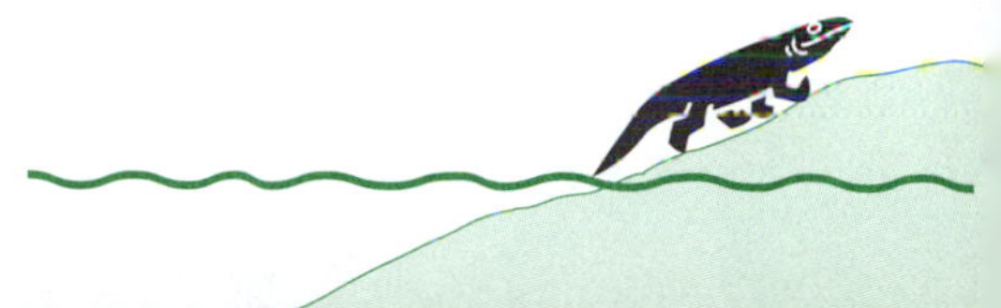

plicates such analyses, as was found to be true for *Arabidopsis*. Nevertheless, Aranzana and colleague (2005) were able to identify the major genes for all the phenotypes they tested, increasing the likelihood that methods of association mapping will continue to evolve, enhancing its role in the analysis of natural variation, and thereby, of course, increasing our understanding of how the genotype maps to the phenotype.

Sex Linkage

For sex-linked genes, the number of possible genotypes is increased because of the difference in the number of sex chromosomes between the homogametic and heterogametic sex. If females are chromosomally XX and males XY, five genotypes can occur for a sex-linked pair of alleles *A* and *a*, three in females (*AA, Aa, aa*) and two in males (*A* and *a*). If we assign the frequencies *p* and *q* to *A* and *a*, respectively, the equilibrium genotypic values in females are the same as for an autosomal gene, p^2 *AA*, 2*pq Aa*, and q^2 *aa*, but are expressed directly in hemizygous males as *p A* and *q a* genotypes. At equilibrium the sex-linked allele frequencies are the same in both sexes, although the genotypes differ.

Assuming all genotypes are equally viable for the sex-linked gene, a difference in allele frequencies between males and females indicates that the population is not at equilibrium. For example, in a population with the proportions 0.20 *A*/0.80 *a* in males and 0.20 *AA*/0.60 *Aa*/0.20 *aa* in females, the frequency of *A* is 0.2 in males and 0.5 in females. We can calculate equilibrium frequencies of all five genotypes by considering that because there is only one X chromosome in males and two in females, the average frequency of a sex-linked gene in a breeding population with equal numbers of males and females is the sum of one-third of its frequency in males plus two-thirds of its frequency in females, or $p = 1/3\ (p_{\text{males}}) + 2/3\ (p_{\text{females}}) = (p_{\text{males}} + 2p_{\text{females}})/3$. In the present example, this translates into an *A* (*p*) frequency of $[0.2 + 2(0.5)]/3 = (1.2)/3 = 0.4$, and an *a* (*q*) frequency of 0.6. The equilibrium genotypic values expected are 0.16 *AA*/0.48 *Aa*/0.36 *aa* in females, and 0.4 *A*/0.6 *a* in males.

In contrast to single autosomal loci with two alleles, however, equilibrium values for these genotypes will not be reached in a single generation. Because males inherit their X chromosomes only from their mothers, the frequency of a sex-linked gene among them is the same as its maternal frequency, whereas the frequency of the gene among daughters is an average of paternal and maternal frequencies because they each inherit one paternal and one maternal X chromosome. Therefore, if the females in a founding population had a frequency of *A* equal to 0.5, but the males had an *A* frequency of only 0.2, the daughters would have an *A* frequency of $(0.2 + 0.5)/2 = 0.35$, while their brothers would have the 0.5 frequency of their mothers.

Thus, in the first generation of random mating the daughters will not reach the *A* equilibrium value of 0.4 and the sons will exceed it. In the second generation, the difference from equilibrium values will diminish, but this time the sons will not achieve equilibrium ($A = 0.35$) and the daughters will exceed it [$A = (0.35 + 0.5)/2 = 0.425$]. As **Figure 21-7** shows, each succeeding generation will show a similar reversal, but nevertheless move toward achieving a successively closer approximation to the final equilibrium values.

Equilibria in Natural Populations

In natural populations, we can reliably estimate allele frequencies where we can score all segregants of a gene at a single locus. Using these values, we can easily compare observed genotype frequencies to their expected equilibrium values.

To take a simple example, codominance at the *MN* blood group locus (using only two alleles, *M* and *N*) enabled Boyd (1950) in a pioneering study to classify 104 American Ute Indians into genotype frequencies of 0.59 *MN*, 0.34 *MN* and 0.07 *NN*. Because the allele frequencies are 0.59 + 0.17 = 0.76 for *M*, and 0.07 + 0.17 = 0.24 for *N*, the expected genotype frequencies are $(0.76)^2 = 0.58$ for *MM*, $2(0.76)(0.24) = 0.36$ for *MN* and $(0.24)^2 = 0.06$ for *NN*. The close correlation between observed and expected genotypic values indicates that this population has reached Hardy–Weinberg equilibrium for these two alleles.

When we know the genotypes of mated couples in the population, we can test the assumption of random mating. Under random mating, the frequencies of the different mating combinations should depend only on the frequencies of their genotypes. A set of data collected by Matsunaga and Itoh (1958) — the blood types of 741 couples or 1,482 individuals in a Japanese town — showed genotypic frequencies of 0.274 *MM*, 0.502 *MN* and 0.224 *NN*. (Because the allele frequencies are 0.525 *M* and 0.475 *N*, the expected equilibrium genotypic frequencies are 0.276 *MM*, 0.499 *MN* and 0.225 *NN*, again indicating equilibrium for this locus.) **Table 21-6** demonstrates random mating among these individuals, with the number of observed matings of different combinations given in the last column. To the left of this column, the expected mating combination frequencies are calculated on the basis of the allele frequencies, *p* for *M* and *q* for *N*. For example, because the frequency of the genotype *MM* is p^2 at equilibrium, the random mating combination *MM* × MM should be p^4 when we calculate the frequencies of all of the expected mating combinations this way; comparing observed with expected frequencies agrees with the assumption of random mating.

A more common case in natural populations is when the effect of one allele at a locus is completely dominant over another so that we cannot phenotypically distinguish the heterozygous genotype (e.g., *Aa*) from the homozygous dominant (e.g., *AA*). Under such circumstances, we cannot obtain allele frequencies directly, as in codominance, because we don't know two of the genotypic frequencies (*AA*, *Aa*) but instead must rely on the distinctive recessive homozygote (*aa*), the genotypic frequency of which coincides with its phenotypic frequency; that is, if we assume that such a population has reached Hardy–Weinberg equilibrium (p^2 *AA*, $2pq$ *Aa*, q^2 *aa*), the recessive homozygotes are present in a frequency q^2 equal to the square of the recessive allele frequency, q. If q^2 is 0.49, then q is $\sqrt{0.49} = 0.70$, and the frequency of the dominant allele p is $1 - q$, or 0.30. The homozygous dominants, therefore, have the frequency $p^2 = (0.30)^2 = 0.09$ and the heterozygotes have the frequency $2pq = 2(0.30)(0.70) = 0.42$.

As one consequence of this analysis, when recessive phenotypes are rare, it is surprisingly common to find that the carrier heterozygotes, phenotypically disguised as dominants, are present in relatively high frequency. *Albinism*, for example, affects only about 1 in 20,000 humans in some populations, or $q^2 = 1/20{,}000 = 0.00005$. The allele frequency, q, of the albino gene is therefore 0.007, and the frequency, p, of the nonalbino allele is 0.993. The frequency of heterozygous albino carriers is therefore $2(0.993)(0.007) = 0.014$ or approximately 1 in 70 individuals. Thus, there are $0.014/0.00005 = 280$ times as many heterozygotes for this trait as there are homozygotes. Similarly high proportions of carriers of other recessive traits point to the difficulty of eliminating rare harmful recessives, because such deleterious alleles are carried mostly in the unexpressed heterozygous condition (**Table 21-7**).

When more than two alleles are present at a locus, the Hardy–Weinberg equilibrium is based on a multinomial expansion such as that described on page 548. For that example of three alleles (A_1, A_2, A_3), we expect six genotypes,

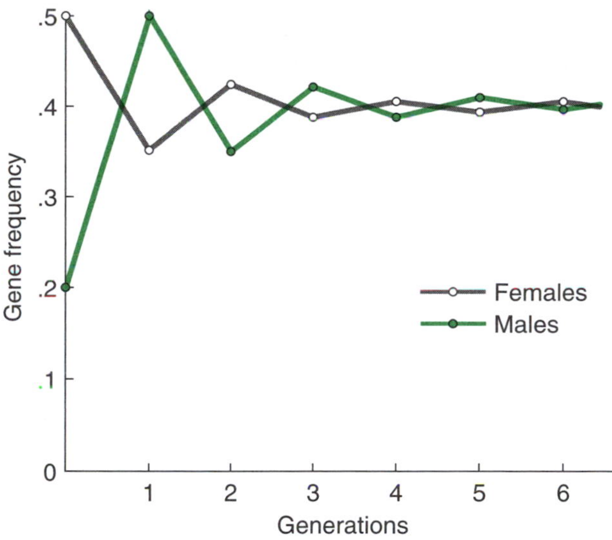

FIGURE 21-7 Frequencies of the sex-linked gene A in males and females in successive generations under conditions of random mating when the initial frequency of A is .2 in males and .5 in females.

and we can calculate the frequency of each allele (p, q, r) from the following equations:

$$p = 2\,\frac{(A_1A_1) + (A_1A_2) + (A_1A_3)}{2N}$$

$$q = 2\,\frac{(A_2A_2) + (A_1A_2) + (A_2A_3)}{2N}$$

$$r = 2\,\frac{(A_3A_3) + (A_1A_3) + (A_2A_3)}{2N}$$

TABLE 21-6 Comparison of mating combinations expected according to random mating and those observed in 741 couples by Matsunaga and Itoh, 1958 ($p = .525$, $q = .475$)

Mating Combination			Expected Frequency			Expected Number Observed (Frequency x 741)	Observed Number
MM x *MM*	$(p^2)(p^2)$	=	p^4	=	.0760	56.3	58
MM x *MN*	$2 \times (p^2)(2pq)$	=	$4p^3q$	=	.2749	203.7	202
MM x *NN*	$2 \times (p^2)(q^2)$	=	$2p^2q^2$	=	.1244	92.2	88
MN x *MN*	$(2pq)(2pq)$	=	$4p^2q^2$	=	.2487	184.3	190
MN x *NN*	$2 \times (2pq)(q^2)$	=	$4pq^3$	=	.2251	166.8	162
NN x *NM*	$(q^2)(q^2)$	=	q^4	=	.0509	37.7	41
Total					1.0000	741	741

Source: Adapted from Strickberger, M. W., 1985. *Genetics*, 3rd ed. Macmillan, New York.

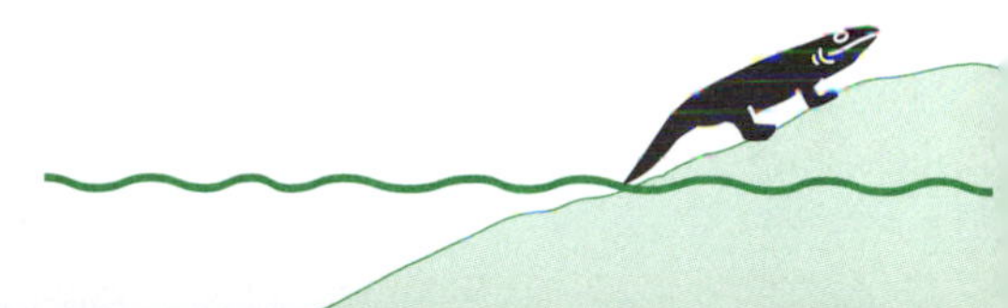

TABLE 21-7 Genotype frequencies for some human diseases caused by recessive alleles

Disease	Population	Gene Frequency (q)	Frequency of Homozygotes (q^2)	Frequency of Heterozygous Carriers ($2pq$)	Ratio of Heterozygous Carriers to Homozygotes ($2pq/q^2 = 2p/q$)
Achromatopsia	Pingelap (Caroline Islands)	.22	1 in 20	1 in 2.8	7/1
Sickle cell anemia	Africa (some areas)	.2	1 in 25	1 in 3	8/1
Albinism	Panama (San Blas Indians)	.09	1 in 132	1 in 6	21/1
Ellis-van Creveld syndrome	Old Order Amish	.07	1 in 200	1 in 8	26/1
Sickle cell anemia	African-Americans	.04	1 in 625	1 in 13	48/1
Cystic fibrosis	European-Americans	.032	1 in 1000	1 in 16	60/1
Tay-Sachs disease	Ashkenazi Jews	.018	1 in 3000	1 in 28	108/1
Albinism	Norway	.010	1 in 10,000	1 in 50	198/1
Phenylketonuria	United States	.0063	1 in 25,000	1 in 80	314/1
Cystinuria	England	.005	1 in 40,000	1 in 100	400/1
Galactosemia	United States	.0032	1 in 100,000	1 in 159	630/1
Alkaptonuria	England	.001	1 in 1,000,000	1 in 500	2000/1

Source: From *Genetics* Third Edition by Monroe W. Strickberger. Copyright © 1985 by Monroe W. Strickberger. Reprinted by permission of Prentice Hall, Inc., Upper Saddle River, NJ.

where A_1A_1, A_1A_2, A_1A_3, and so on refer to the numbers of genotypes in each category and N refers to the total number of individuals scored.

A system of this type is seen in human populations bearing different forms of the red blood cell enzyme acid phosphatase, which can be scored into six different phenotypes (AA, BB, CC, AB, BC or AC), as determined by all possible combinations of the alleles *A*, *B* and *C* at a single locus. As **Table 21-8** shows, investigations of a Brazilian population indicate that the observed phenotypic frequencies of the acid phosphatase combinations conform closely to those the Hardy–Weinberg equilibrium predicts. Although exceptions arise, conformities to the Hardy–Weinberg equilibrium seem quite common for autosomal and sex-linked genes and appear in a variety of sexually outbreeding organisms. In general, as noted earlier, these studies emphasize that population gene and genotype frequencies do not change without cause.

TABLE 21-8 Comparison of observed acid phosphatase phenotypes and those expected according to Hardy–Weinberg equilibrium in a sample of 369 Brazilian individuals

Phenotypes	AA	BB	CC	AB	AC	BC
Observed	15	220	0	111	4	19
Expected	14.4	219.9	0.4	112.2	4.4	17.7

Source: From Lai, L., S. Nevo, and A. G. Steinberg, 1964. Acid phosphatases of human red cells: Predicted phenotype conforms to a genetic hypothesis. *Science*, **145**, 1187–1188.

Inbreeding

Nonrandom mating is one set of conditions interfering with the Hardy–Weinberg equilibrium.

An important example occurs when related individuals of similar genotype mate preferentially with each other in **inbreeding.** (An extreme form of inbreeding, **self-fertilization,** when two gametes of a single individual unite to form a fertile zygote, is discussed later in the chapter.) Although the effect of inbreeding will not change the overall allele frequency, it will lead to an excess of homozygous genotypes. Inbreeding will thus cause a rare recessive allele to appear in greater homozygous frequency than under random mating, offering increased opportunity for selection to act on rare recessives.

Inbreeding Coefficient

We usually quantify inbreeding by an **inbreeding coefficient,** F, which measures the probability that the two alleles of a gene in a diploid zygote are identical; that is, descended from a single ancestral allele. For example, if we begin with a heterozygous diploid, A^1A^2, normal Mendelian segregation will confer a 1/2 probability that each F_1 offspring will receive the same A^1 allele. The allele transmitted by an F_1 individual to its offspring, in turn, has a 1/2 chance of being the same as the ancestral allele. This means that if two such F_1 offspring mate, the chances that the two alleles in one of their offspring (an F_2) are identical by descent (A^1A^1), is $1/2 \times 1/2 \times 1/2$, or the inbreeding coefficient is 1/8.

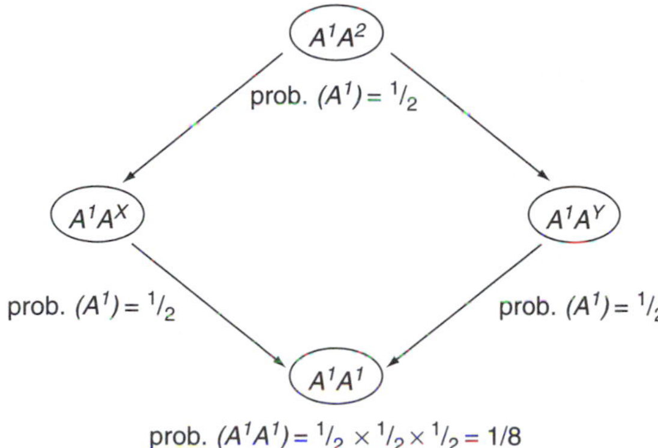

Having identical alleles, of course, means homozygosity, and F can range from one (complete homozygosity) to zero (complete **heterozygosity**). Of the inbred proportion measured by F, some will be AA and some aa, the frequencies of each depending on their respective population allele frequencies p and q. Thus, inbreeding will produce $pF\ AA$ and $qF\ aa$ genotypes. In addition to these, however, the remaining individuals in this population $(1–F)$ will bear genotypes with frequencies determined according to the Hardy–Weinberg equilibrium of $p^2\ AA$, $2pq\ Aa$ and $q^2\ aa$. The three genotypes will have the following frequencies:

$$\begin{aligned} AA &= -p^2(1-F) + pF = p^2 - p^2F \pm pF \\ &= p^2 + pF(1-p) = p^2 + pqF \end{aligned}$$

$$Aa = 2pq(1-F) = 2pq - 2pqF$$

$$\begin{aligned} aa &= -q^2(1-F) + qF = q^2 - q^2F + qF \\ &= q^2 + qF(1-q) = q^2 + pqF \end{aligned}$$

It is now easy to see that the increase in the frequency of each homozygote type by a factor of pqF flows from an equivalent fall in the heterozygote frequency $(-2pqF)$. Note also that this reduction in **heterozygotes** affects the allele frequencies p and q equally, so that only the genotypic frequencies change. When inbreeding is absent, $F = 0$, and the preceding equations reduce to the Hardy–Weinberg frequencies $p^2\ AA$, $2pq\ Aa$ and $q^2\ aa$. When inbreeding is complete, $F = 1$, $2pq - 2pqF = 0$ and the only remaining genotypes are pAA and qaa.

Self-Fertilization

Inbreeding is greatest under self-fertilization (e.g., *hermaphrodites*), where F is equal to 0.5 in the first generation and approaches 1 within four or five generations. As shown in **Figure 21-8**, any other mating scheme slows the rate of inbreeding. Similarly, as seen in **Figure 21-9**, population size can affect inbreeding, because the smaller the size, the greater the opportunity for related individuals to mate. *Genetic assortative mating* is the name given to systems such as brother–sister mating and first cousin mating in which individuals mate on the basis of their genetic relationship.

Phenotypic similarity, however, may also cause preferential mating; in many human societies mates are chosen who share characteristics such as height, color, facial form (especially symmetry according to recent studies), muscular build and intelligence. Homozygosity can increase in such *phenotypic assortative matings*, but only for those loci involved in the trait(s) on which the preferred matings are based. This is in contrast to the *genetic assortative mating* of inbreeding, which tends to increase homozygosity at all loci.

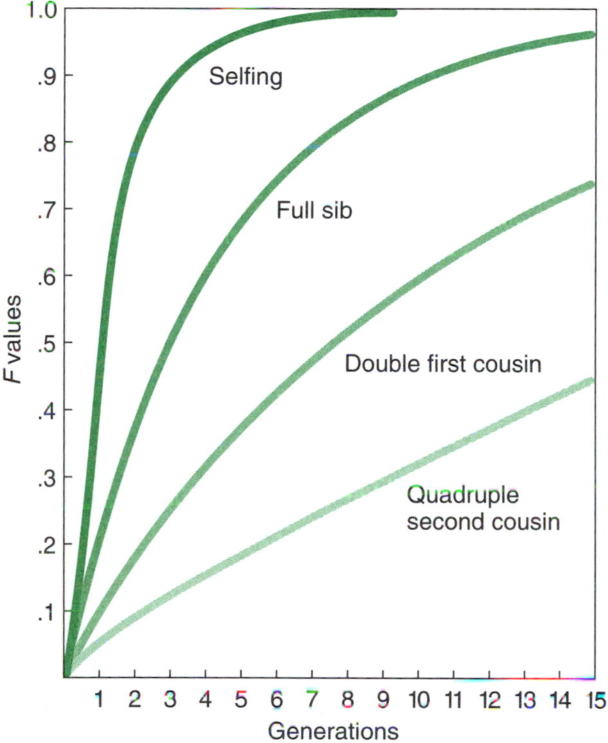

FIGURE 21-8 Inbreeding coefficients at generations 1 to 15 for four different systems of inbreeding (pedigrees given in Strickberger, 1985). You can obtain or derive formulas for calculating F in other inbreeding systems from Wright's (1921) basic work in this field; see the biography by Provine (1986) for Wright's life and contributions and for the best history of population genetics.

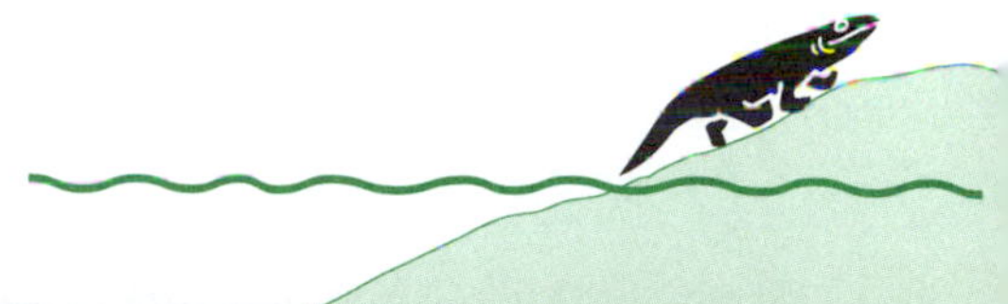

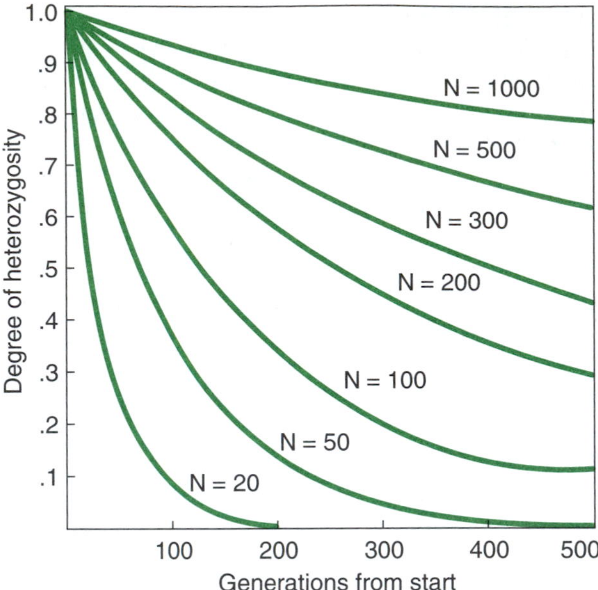

FIGURE 21-9 Degrees of heterozygosity (heterozygote frequency) remaining in populations of different sizes after given generations of random union between gametes. Calculations are based on 1.00 as the initial degree of heterozygosity (or when $F = 0$). (Adapted from Strickberger M. W., 1985. *Genetics,* 3rd ed. Macmillan, New York.)

Disassortative mating, a further type of mating practice, occurs when individuals of unlike genotype or phenotype form mating pairs, thereby preventing inbreeding and helping to maintain heterozygote frequency (heterozygosity). Various examples of such systems include alleles in plants that cause sterility of male gametes when they attempt to fertilize ova of the same genotype (*self-sterility alleles*).

Inbreeding Depression

Although some degree of inbreeding occurs in most outbreeding populations, significant amounts of inbreeding can cause **inbreeding depression** in which rare deleterious recessives may appear with increased homozygous frequency. If a recessive disease with genotype aa occurs with frequency q^2 in a random outbred population, its frequency will increase by pqF in an inbred population as just derived. The ratio of inbred to outbred frequency for the homozygous recessive will therefore be,

$$\frac{q^2 + pqF}{q^2} = \frac{q(q + pF)}{q^2} = \frac{q + pF}{q}$$

Obviously, if q is large and F is small, the inbreeding increment pF will be relatively small, and the increased frequency of homozygous recessives will hardly be noticeable. However, if q is rare and p is large, then pF provides a notable increase in recessives even when F is fairly small. For example, if q is 0.5, first cousin mating ($F = 0.0625$) will produce an inbred-to-out bred ratio of homozygotes of,

$$\frac{q + pF}{q} = \frac{0.5 + (0.5)(0.0625)}{0.5} = \frac{0.53125}{0.5} = 1.06$$

However, if q is 0.005, this ratio increases to 0.067/0.005 = 13.4. When $q = 0.0005$, the increase of homozygotes because of first cousin mating is 0.0630/0.0005, or 126 times that of randomly bred populations.

Should homozygous recessives have a quantitative effect on one or more traits, inbreeding would cause the measured values of these traits to tend in the direction of recessive values. Thus, in various outbred populations such as corn, inbreeding depression can reduce height, yield and other characters. Inbreeding depression, however, is not a universal phenomenon in all species, certainly not in many species that are normally self-fertilized and have eliminated most or all of their deleterious recessives. On the whole, ample evidence shows that most normally cross-fertilizing species deteriorate on consistent inbreeding, leading even to extinction (Saccheri et al., 1998), although some strains may escape because they carry relatively few deleterious recessive alleles.

KEY TERMS

allele frequencies
binomial expansion
coadapted gene complexes
continuous variation
deme
discontinuous variation
equilibrium
gene frequencies
gene pool
Hardy–Weinberg Equilibrium (principle or law)
heterozygosity
heterozygotes
inbreeding
inbreeding coefficient
inbreeding depression
linkage disequilibrium
macromutations
Mendelian population
Mendelians
modern synthesis
neo-Darwinian
nonrandom mating
panmixia
population genetics
quantitative characters
random mating
saltation
self-fertilization

DISCUSSION QUESTIONS

1. Continuous versus discontinuous variation.
   a. What controversy arose between Mendelians and biometricians?
   b. What controversy arose between mutationists and selectionists?
   c. What role did "pure lines," "macromutations," "saltations," and "quantitative characters" play in these arguments?
   d. How were these various issues resolved?
2. What are the elements of the neo-Darwinian (modern) synthesis?

3. How do we determine allele frequencies in a diploid population when we can identify the frequencies of all genotypes?
4. Hardy–Weinberg equilibrium (principle).
   a. Under what conditions are allele frequencies conserved?
   b. How do geneticists derive genotype frequencies according to the Hardy-Weinberg principle?
5. Equilibrium between genes (alleles) at two or more loci.
   a. Why does a population rarely, if ever, attain multilocus equilibrium in a single generation?
   b. How do linkage and recombination frequencies affect such attainment of equilibrium?
6. How do an autosomal locus and a sex-linked locus differ in reaching genotypic equilibrium?
7. How can we test the assumption of random mating in the Hardy–Weinberg principle for a particular gene segregating in a natural population?
8. How can we derive gene and genotype frequencies in a diploid population when we only know the frequency of recessive homozygotes?
9. What is the relationship between (a) the frequency of a recessive allele and (b) the ratio of heterozygous carriers to homozygotes for that allele?
10. How and why does inbreeding affect the frequency of homozygous genotypes?

EVOLUTION ON THE WEB

Explore evolution on the Web! Visit the accompanying Web site for *Strickberger's Evolution*, Fourth Edition, at http://www.biology.jbpub.com/book/evolution for Web exercises and links relating to topics covered in this chapter.

RECOMMENDED READING

Crow, J. F., 1988a. Eighty years ago: The beginnings of population genetics. *Genetics*, **119**, 473–476.

Dobzhansky, Th., 1937, 1941, 1951. *Genetics and the Origin of Species* (three eds.). Columbia University Press, New York.

Dobzhansky, Th., 2003. *Dobzhansky's Genetics of Natural Populations I–XLIII,* R. C. Lewontin, J. A. Moore, W. B. Provine, and B. Wallace (eds.). Columbia University Press, New York. [This volume is a collection of the 43 papers in Dobzhansky's influential series on *Drosophila* population genetics, with various coworkers, along with introductory articles by Provine and Lewontin.]

Falconer, D. S., and T. F. C. Mackay, 1996. *Introduction to Quantitative Genetics,* 4th ed. Longman, London, UK.

Gillespie, J. H., 2004. *Population Genetics: A Concise Guide,* 2nd ed. Johns Hopkins University Press, Baltimore, MD.

Hartl, D. L., and A. G. Clark, 1997. *Principles of Population Genetics,* 3rd ed. Sinauer Associates, Sunderland, MA.

Hedrick, P. W., 2000. *Genetics of Populations* 2nd ed. Jones and Bartlett, Sudbury, MA.

Pagel, M. (ed. in chief), 2002. *Encyclopedia of Evolution.* 2 Volumes, Oxford University Press, New York.

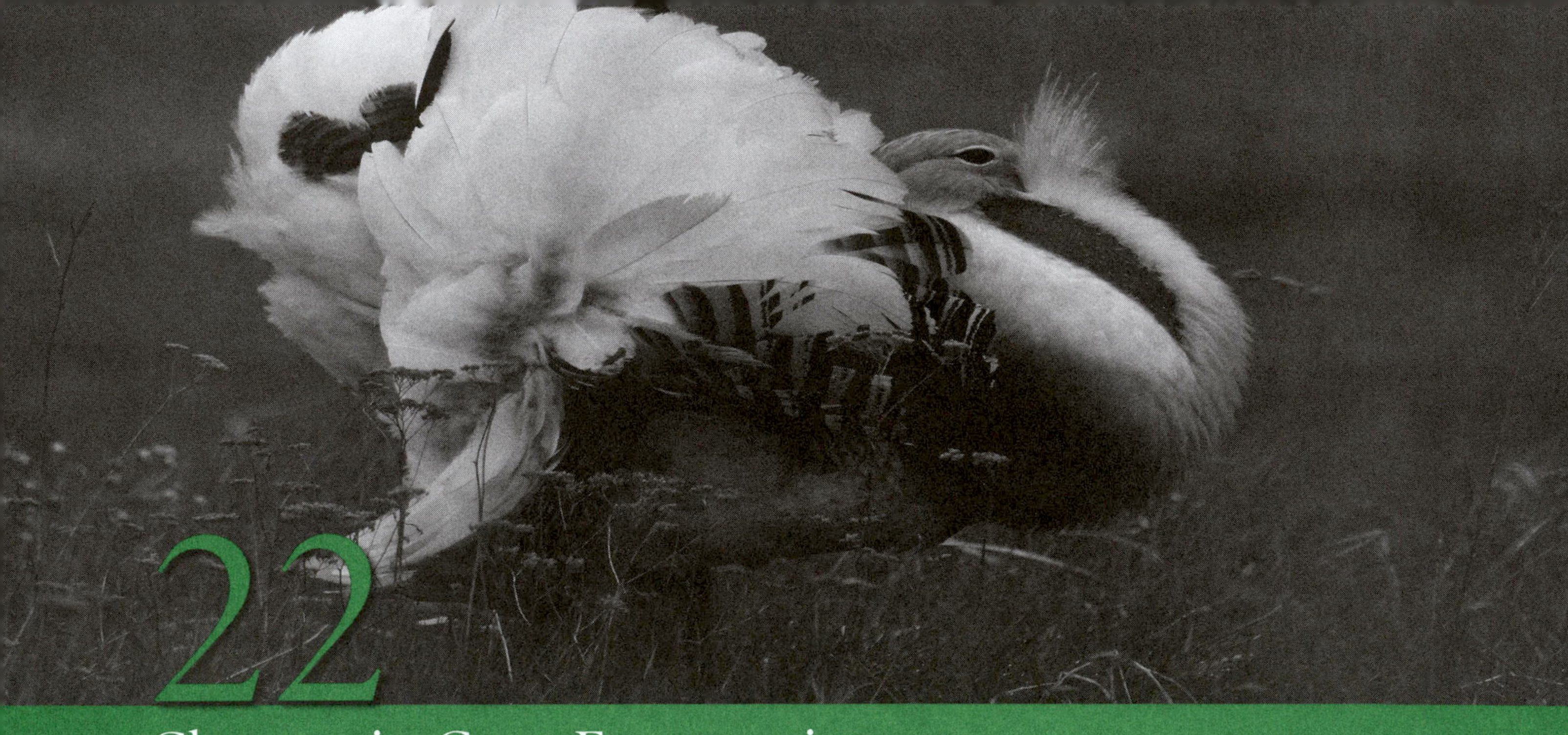

22 Changes in Gene Frequencies

Chapter Summary

To effect evolutionary change, new alleles arising because of mutation must persist in populations. Although most evolutionary models are of infinite alleles and/or unique mutations, we have simplified the discussion and examples to the classic Mendelian case of a gene with a dominant allele *A* and a recessive allele *a,* with mutation affecting *a* so that its frequency increases. In reality, codominance of complex and interacting loci is much more usual than the situation of a single dominant and single recessive allele acting alone.

Because mutational equilibrium is rarely attained through mutation alone, other mechanisms must influence the frequency of alleles introduced by mutation. Selection is a major mechanism. If certain alleles increase the reproductive success (i.e., fitness) of the carrier, they tend to become more frequent. Selection acts on the fitness differences among different traits. With systematic and persistent selection on the phenotype from generation to generation allele frequencies change. Weak selection, which is more common, only slowly leads to change in allele frequencies, often no more effectively than genetic drift.

Alleles with deleterious effects will decline in frequency, the rate of decline depending on the allele frequency, the amount of genetic drift in the population, and interaction or linkage with other genes or gene complexes. For example, deleterious recessive alleles decline rapidly from high frequencies but only slowly from low frequencies. In any case, selection may eventually result in the removal of the deleterious allele from the population or persistence of the allele (equilibrium), depending on factors such as new mutations.

Selection can preserve genetic variation (polymorphism) if the heterozygous genotype is more fit than either homozygote, as in sickle cell anemia, or when the frequency of an allele affects fitness. Allele frequency also, may vary in different environments, as in the case of the British peppered moth, where lighter forms are more plentiful in unpolluted areas and melanic forms in industrial regions.

Selection can operate on a population in several ways: (1) if phenotypes far from the norm are less fit (stabilizing selection) or where selection is proportional to the deviation from the mean, (2) through rapid elimination of deleterious mutations, (3) if an extreme phenotype has adaptive value (directional selection), (4) if different phenotypes do better in particular environments (disruptive selection), and/or (5) by heterozygote advantage, as occurs in the immune and olfactory systems (diversifying selection). For selection to occur, genetic variation must be present; the greater the genetic variability, the greater the chance for increasing fitness. Because both mutation and selection affect allele frequencies, mutation may maintain even deleterious alleles in a population, although selection will remove many individuals bearing them. Genetic drift also influences gene frequencies between populations.

Mutation, selection, drift and migration can lead to fixation of certain alleles and loss of others, or to equilibrium, fixation or loss being the fates of most alleles. Gene frequency can change more rapidly if a new population begins with a small sample ("founders") in which the gene frequency varies from that of the original and larger population. Drift and other random factors are often more important than mutation and selection in establishing unpredictable population variability.

For populations to evolve — that is, to change their **gene frequencies** — mutation must first introduce the nucleotide differences from which such changes arise.

The mere appearance of new genes (alleles), however, is no guarantee that they will persist or prevail over others. For example, no certainty exists that a newly mutated gene such as *a* (e.g., $A \times a$) will be transmitted to the next generation, because its carrier (e.g., *Aa*) may or may not survive and may or may not mate. Even if the *Aa* mutant carrier does mate ($Aa \times AA$), the chance of transmission declines because a significant proportion (40%) of mating in most stable populations produces families with no surviving offspring (*a* is lost) or only one offspring (*a* has a 50% chance of being lost during meiosis). Larger families also may lose gene *a*, since, even when, for example, the $Aa \times AA$ family produces two offspring the mutant gene has a 25 percent chance (0.5×0.5) of not being transmitted to either of them.

Fisher (1930) calculated that, because of its random loss in families of such different sizes, the chance that a newly mutated gene may be eliminated within one generation is more than 33 percent. By the time 30 generations have passed, the probability of elimination rises to almost 95 percent. To explain the persistence of many mutations and their increase in frequency, we must look elsewhere than the original mutational event.

Mutation Rates

One factor that affects allele frequency is the frequency of mutation. Although most situations involve infinite alleles and/or unique mutations, it is customary to simplify the discussion and examples to the classic Mendelian case of a gene with a dominant allele *A* and a recessive allele *a* with mutation affecting *a* so that its frequency increases. In reality, codominance and linkage of complex and interacting loci are much more usual than the situation of a single dominant and single recessive allele acting alone.

Most mutations do not involve the simple mutation of one allele to another, but accumulation in the population of new, often unique mutations that affect nucleotide sequences. In a population whose structure is stable demographically, an equilibrium level of heterozygosity will be reached, reflecting nucleotide diversity of the original and the new alleles.[1] Sequence clades can be recognized when gene sequences are analyzed. A nice example is an analysis of the control regions of mitochondrial genes from 191 extant domestic horses, and from ancient DNA from archaeological sites and Late Pleistocene deposits, which revealed widespread integration of matrilines (mother-daughter lineages) over long periods of horse evolution (Vila et al., 2001). So bear in mind that the simplified allele situation is used to illustrate the principles that operate, and not the known complexity of allele interactions.

If allele *A* continually mutates to *a*, the chances improve that *a* will increase in frequency with each generation. Given a long enough time and a persistent **mutation rate** in a population of constant size, *a* can eventually replace *A*. Of course, the mutation rate does not always occur in only one

[1] See Hartl and Clark (1997), Hedrick (2000) or Gillespie (2004) for more extended treatments and discussions.

direction, especially at loci with multiple alleles. For example, if u is the mutation rate of A to a, the allele a may mutate back to A with frequency v. We can estimate these effects quantitatively by calling the initial frequencies of alleles A and a, p_0 and q_0, respectively, and noting that a single generation of mutation will produce a frequency of A equal to $p_0 + vq_0$ and a frequency of a equal to $q_0 + up_0$.

If we now confine our attention to only one of the alleles, a, clearly it has gained the fraction up_0 (new a alleles) but lost the fraction vq_0 (new A alleles). In other words, the change in the frequency of a, which we call delta q (Δq), can be expressed as ($\Delta q = up_0 - vq_0$. Thus, if p was relatively large and q small, Δq would be large and q would increase rapidly; when q became larger and p became smaller, Δq would diminish. The point at which Δq is zero — that is, the point where there is no further change and p and q are balanced in relation to their mutation frequencies — we call the **mutational equilibrium** (frequency $a = \hat{q}$, or "q *hat*"): $\Delta q = 0 = up - vq$, or $up = vq$ at $\hat{q}$. However, because there are only two alleles, A and a, $p = 1 - q$, which leads to:

$$\begin{aligned} up &= vq \\ u(1-q) &= vq \\ u &= uq + vq = q(u+v) \\ \hat{q} &= \frac{u}{u+v} \end{aligned}$$

The same procedure applied to the frequency of A gives $\hat{p} = v/(u+v)$, so that $\hat{p}/\hat{q} = [v/(u+v)]/[u/(u+v)] = v/u$. Thus, when the mutation rates are equal ($u = v$) the **equilibrium gene frequencies** $\hat{p}$ and $\hat{q}$ will be equal. If the mutation rates differ, so will the equilibrium frequencies. For example, if $u = 0.00005$ and $v = 0.00003$, the equilibrium frequency $\hat{q}$ equals $5/8 = 0.625$ and $\hat{p} = 3/8 = 0.375$. However, the rate at which mutation reaches this equilibrium frequency is usually quite slow and we can derive it by calculus methods from Δq as,

$$(u+v)n = \ln\,[(q_0 - \hat{q})/(q_n - \hat{q})]$$

where n is the number of generations required to reach a frequency q_n when starting with a frequency q_0. For the example just considered, the number of generations necessary for q to increase from a frequency of one-eighth to three-eighths is,

$$\begin{aligned} (.00008)n &= \ln \frac{.125 - .625}{.375 - .625} \\ &= \ln 2.00 = .69315 \end{aligned}$$

$$n = \frac{0.69315}{.00008} = 8{,}664 \text{ generations}$$

Thus, based on the usually observed mutation rates of 5×10^{-5} or less (Table 10-3), the shift toward equilibrium is slow, so slow that mutational equilibrium is rarely if ever reached, especially because mutation rates are not constant.

Selection and Fitness

As a rule, attainment of mutational equilibrium is not the sole cause for existing gene frequencies. A more efficient mechanism that helps explain how gene frequencies change is the effect of **selection,** the "*scrutinizing process*" proposed by Darwin.

Some dispute whether selection should be considered primarily phenotypic or genotypic. We follow the view that although evolution is marked by phenotypic changes, the transmission of such changes between generations (evolution) is genotypic, although the genotype need not be only that of the individual. Genes of the parent (maternal gene effects) and epigenetic influences from the gene products of other organisms or from the environment (see Chapter 13) are also modes of transmission of change between generations; genotypic inheritance is both direct and indirect.

- Selection is the sum of the survival and fertility mechanisms acting on the phenotype that affect the reproductive success of genotypes.
- The extent to which a genotype contributes to the offspring of the next generation relative to other genotypes in a given environment is the **relative fitness** of the genotype.

To the extent that phenotypic differences are due to genetic variation, phenotypes *do* differ in viability and fertility and so influence the frequencies of their genotypes. If individuals carrying gene A are more successful in producing viable and fertile offspring than individuals carrying its allele a, and sufficient numbers of advantageous genotypes have arisen to overcome their loss by chance, frequency of the A allele will tend to increase relative to a.

The genetic consequences of selection on a particular trait in a population are confined to fitness differences among the different genotypes that affect that particular trait. When the selective process operates, gene frequencies tend to change among generations, unless the population has reached a genetic equilibrium, as described later in this chapter and in **Box 22-1.** For simplicity, the dynamics of selection and **fitness** are calculated in terms of their relative contribution within closed populations, rather than in terms of a population exchanging genes with or in competition with other populations. Such situations require much more sophisticated (and more controversial) models.

In their simplest form, fitness and selection are measured by the number of descendants produced by one

BOX 22-1 Selection and the Survival of the Fittest

CRITICS OFTEN CLAIMED THAT **"survival of the fittest"** is a circular, tautological or unprovable statement that cannot be challenged because it defines those that survive as fittest, and the fittest as those that survive. Some philosophers propose that the criterion of survival of the fittest be abandoned. Others suggest it has practical value. Natural selection on a phenotypic character can be a *cause* for change in gene frequency (evolution) but is *not the same as a change* in gene frequency. Selection can change the types of phenotypes that survive in a single generation. Most genes require multiple generations of selection in the same direction before their proportions change (see text).

For example, the frequency of a harmful recessive allele may remain stable despite selection against homozygotes for that gene if there is heterozygote advantage. In other words, the relative fitness of a genotype may not be the only reason for its survival, and the statement that evolutionary theory merely proposes the *survival of the fittest* is misleading and incorrect. In fact, Fisher begins his classic 1930 treatise *The Genetical Theory of Natural Selection* with the statement "Natural Selection is not Evolution." If we define evolution as hereditary changes over time, selection, although important, *is not the only process that can cause such changes*; other evolutionary mechanisms considered in Chapter 2 — mutation, migration and genetic drift — also affect gene frequencies. **Moreover, selection is a process, survival is the product of that process.** Neither natural selection nor its colloquial expression, "survival of the fittest," is a tautology.

An example of interplay of factors affecting fitness is that fitness is often frequency-dependent, especially in situations where a variant allele is either common or rare; maintaining a rare variant increased existing variation. Analysis of *frequency-dependent fitness* or *adaptive dynamics* originated in game theory so it is not for the mathematically faint-of-heart.[a] Mutations that would modify the uniform phenotypes are said to "invade" the population. After invasion, the fitness of the mutations is the invasion fitness from which, and by analogy to Wright's adaptive landscape, an invasion fitness gradient and invasion fitness landscape can be constructed.

Frequency-dependent fitness works best when applied to populations of asexual individuals with identical phenotypes. Although this may seem a severe limitation, the method has the advantage of allowing single evolutionary events, such as the origin and spread of a single mutation to be modeled and such evolutionarily single strategies compared with evolutionary stable strategies. Understanding frequency-dependent fitness therefore has the potential to inform us about branching points in evolution; modeling the feasibility of, and understanding the basis of, sympatric speciation is one obvious application.

[a] See Geritz et al. (1998) for a technical paper on the origins of adaptive dynamics and see Waxman and Gavrilets (2005) for an introduction based on answering 20 questions about adaptive dynamics.

genotype relative to those produced by another. For example, if individuals of genotype *A* produce an average of 100 offspring that reach full reproductive maturity while genotype *B* individuals produce only 90 in the same environment, the relative fitness of *B* relative to *A* is reduced by 10 offspring, or the fraction 10/100 = .1. If we designate the relative fitness of a genotype as W and the selective force acting to reduce its relative fitness as s (the **selection coefficient**), we can say that $W - 1$ and $s = 0$ for *A* in the preceding example, and $W = .9$ and $s = .1$ for *B*. The relationship between W and s for a particular genotype is $W = 1 - s$, or $s = 1 - W$.

Selection may elicit changes in either the haploid (gametic) or diploid (zygotic) stage or both, depending on which stage gene expression influences survival or fertility (**Fig. 22-1**). In any of these stages, selection may be obvious or subtle, its effects ranging from complete lethality or sterility ($s = 1$) to only slight reductions in relative fitness (e.g., $s = 0.01$). When selection occurs among haploids, there is no difference between dominant and recessive alleles, because their carriers phenotypically express both kinds of alleles.

Thus, as we might expect, the effect of selection on haploids is much more rapid and direct than on diploids; del-

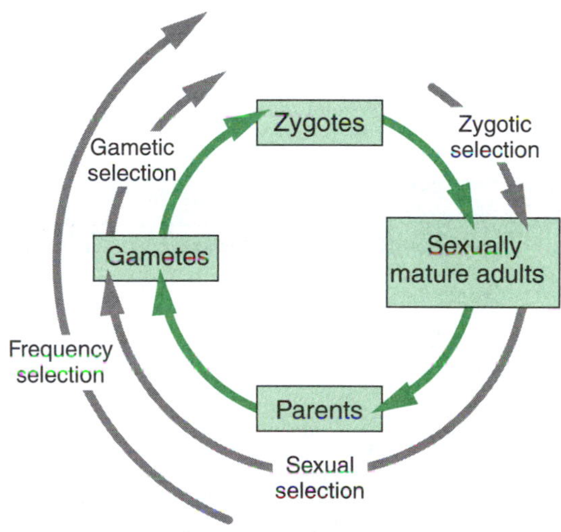

FIGURE 22-1 Simplified diagram of selection acting on the life stages of an organism during a single generation, from zygote to zygote ("egg to progeny"). In addition to sperm and pollen competition, sexual selection includes choice of sexual partners. Other complexities occur in organisms such as mammals, where fitnesses overlap between generations because a juvenile's fitness depends not only on its own attributes but also on receiving parental care — on the fitness of individuals in the preceding generation. (Adapted from Christiansen, F. B., 1984. The definition and measurement of fitness. In *Evolutionary Ecology*, B. Shorrocks (ed.). Blackwell, Oxford, England, pp. 65–79.)

eterious recessive alleles cannot be hidden from selection among heterozygotes as they are in diploids. **Table 22-1** provides estimates of the number of generations necessary to change the frequency of deleterious alleles in haploids under a variety of selective conditions. Note that in contrast to diploids (see Table 22-3), haploids completely eliminate a *deleterious lethal allele* ($s = 1$) in one generation. Even lower selection coefficients result in relatively rapid allele frequency changes.

In most animals and plants, selection takes place primarily on the diploid or zygotic and post-zygotic stages. In diploids, however, there are three possible genotypes for a single gene difference (for example, *AA, Aa, aa*), so that the effectiveness of selection depends, among other things, on the degree of **dominance**. **Table 22-2** shows calculation of the change in allele frequency (Δq) of *a* for one generation when complete dominance exists and selection occurs only against the recessive *aa*. **Table 22-3** summarizes the number of generations necessary to change such deleterious recessive allele frequencies when we project selection over periods of time and for various selection coefficients. Note that the initial change in allele frequency from .99 to .10 is relatively rapid for selection coefficients from $s = 1$ to $s = .10$. Further reductions in frequency are considerably slower; to reduce the allele frequency below .01 may take thousands of generations, even when the selection coefficient is relatively high. As indicated in Chapter 21, the reason for the relative insensitivity of rare recessives to selection is that most recessive alleles are present in heterozygotes where they are protected from selection: The more rarely an allele appears in a population, the more frequently it occurs in heterozygotes compared to homozygotes (see Table 21-7).

The selective situation can change so that the dominant allele is selected against and the recessive accumulates. Selection will be more effective, because the *deleterious dominant allele* is subject to selection in all genotypes in which it occurs. For example, should a dominant allele become lethal, its frequency falls to zero in a single generation. However, as the selection coefficient against the dominant allele decreases, replacement by the recessive is considerably slower. For the general case shown in **Table 22-4,** selection against a dominant allele of frequency *p* results in a change of

$$-sp(1-p)^2/[1-sp(2-p)]$$

Note that if *s* is small, the denominator is close to 1, and Δp is effectively equal to $-sp(1-p)^2$. Because $1-p$ is q and p is $1-q$, this means that Δp is now $-sq^2(1-q)$, or Δp is identical to Δq for a deleterious recessive at low selection coefficients. Under these conditions, we may apply the values of Table 22-3 in reverse order; that is, for a selection coefficient of .10 against the dominant allele (in favor of the recessive), 90,023 generations are necessary to increase the frequency of the recessive from .0001 to .001, or to reduce the frequency of the dominant from .9999 to .9990. Subsequent changes in frequency are more rapid as the recessive homozygotes become more frequent.

When dominance of the advantageous allele is incomplete, heterozygotes will show the effect of a deleterious allele because the heterozygous phenotype is at least partially harmful. If dominance is completely absent and the heterozygote has a phenotype exactly intermediate between the two homozygotes, its selection coefficient will be exactly half that in the deleterious homozygotes $[(1)+(1-2s)]/2 = (1/2)+(-s) = 1-s$. As Table 22-4 shows,

TABLE 22-1 Number of generations required for given frequency changes (q_0 to q_n) of a deleterious allele under different selection coefficients in haploids

Change in Gene Frequency		Number of Generations for Different *s* Value						
From (q_0)	To (q_n)	$s = 1$	$s = .80$	$s = .50$	$s = .20$	$s = .10$	$s = .01$	$s = .001$
.99	.75	↓	4	7	17	35	350	3,496
.75	.50	↓	1	2	5	11	110	1,099
.50	.25	↓	1	2	5	11	110	1,099
.25	.10	1	1	2	5	11	110	1,099
.10	.01	↓	3	5	12	24	240	2,398
.01	.001	↓	3	5	12	23	231	2,312
.001	.0001	↓	3	5	12	23	230	2,303

TABLE 22-2 Calculation of Δq for a deleterious recessive allele (a)

	AA	Aa	aa	Total	Frequency of a
Initial frequency	p^2	$2pq$	q^2	1	q
Relative fitness	1	1	$1 - s$		
Frequency after selection	p^2	$2pq$	$q^2(1 - s)$	$p^2 + 2pq + q^2 - sq^2 = 1 - sq^2$	
Relative frequency after selection	$\frac{p^2}{1 - sq^2}$	$\frac{2pq}{1 - sq^2}$	$\frac{q^2(1 - s)}{1 - sq^2}$		$\frac{pq + q^2(1 - s)}{1 - sq^2} = \frac{pq + q^2 - sq^2}{1 - sq^2} = \frac{q(1 - sq)}{1 - sq^2}$

Δq = relative frequency of a after selection — initial frequency of a

$$\Delta q = \frac{q^2(1 - sq)}{1 - sq^2} - q = \frac{q(1 - sq)}{1 - sq^2} - \frac{q(1 - sq^2)}{1 - sq^2} = \frac{q - sq^2 - q + sq^3}{1 - sq^2} = \frac{-sq^2 + sq^3}{1 - sq^2} = \frac{-sq^2(1 - q)}{1 - sq^2}$$

Source: Adapted from Strickberger, M. W.,1985. *Genetics,* 3rd ed. Macmillan, New York.

TABLE 22-3 Number of generations required for a given change in frequency (q_0 to q_n) of a deleterious recessive allele in diploids under different selection coefficients

Change in Gene Frequency		Number of Generations to Attain Given Gene and Genotype Frequencies for Different s Values					
From (q_0)	To (q_n)	$s = 1$ (lethal)	$s = .80$	$s = .50$	$s = .20$	$s = .10$	$s = .001$
.99	.90	1 (spanning .99 to .50)	3	5	13	25	250
.90	.75		2	3	7	13	132
.75	.50		2	3	9	18	176
.50	.25	2	4	6	15	31	310
.25	.10	6	9	14	35	71	710
.10	.01	90	115	185	462	924	9,240
.01	.001	900	1,128	1,805	4,512	9,023	90,231
.001	.0001	9,000	11,515	18,005	45,011	90,023	900,230

Source: From *Genetics* Third Edition by Monroe W. Strickberger. Copyright © 1985 by Monroe W. Strickberger. Reprinted by permission of Prentice Hall, Inc., Upper Saddle River, NJ.

the resultant change in allele frequency in one generation $[-sq(1-q)]/[1-2sq]$ is almost identical to that for gametic selection $[-sq(1-q)]/[1-sq]$.

In other words, the absence of dominance uncovers deleterious alleles and makes all of them available for selection, allowing rapid changes in allele frequencies mostly on the order of those observed in Table 22-1. The effectiveness of selection therefore strongly depends on the degree to which the heterozygote expresses the deleterious allele. Because population genetics indicates that most recessive alleles have some heterozygous expression, selection efficiency for or against them falls between the extremes of slow progress for complete dominance and rapid progress for absence of dominance.

Heterozygous Advantage

The examples of selection just considered always go in one direction, toward elimination of the deleterious allele and establishment or **fixation** of the other allele.

As long as the selection coefficient does not change, equilibrium between the two alleles is impossible without

TABLE 22-4 Single-generation changes in allele frequency for diploid genotypes subject to given selection coefficients under different conditions of dominance

	Relative Fitness for Genotype Frequencies Initially in Hardy-Weinberg Equilibrium			
Dominance Relations for the Three Given Genotypes	*AA* p^2	*Aa* $2pq$	*aa* q^2	Change in Gene Frequency[a]
Complete dominance: selection against the recessive allele	1	1	$1 - s$	$\frac{-sq^2(1-q)}{1 - sq^2}$
Complete dominance: selection against the dominant allele	$1 - s$	$1 - 2s$	1	$\frac{-sp(1-p)^2}{1 - sp(2-p)}$
Absence of dominance: selection against the *a* allele also occurs in the heterozygote	1	$1 - s$	$1 - 2s$	$\frac{-sq(1-q)}{1 - 2sq}$
Overdominance: selection against both homozygotes	$1 - s$	1	$1 - t$	$\frac{pq(ps - qt)}{1 - p^2s - q^2t}$

[a] As mentioned in Chapter 21, mathematical derivations for various formulas have been omitted for simplicity but can be found in many population genetics textbooks.

new mutations. Various conditions, however, permit the establishment of an equilibrium through which both alleles may remain indefinitely within the population. One condition, **overdominance**, occurs when the heterozygote has superior reproductive fitness to both homozygotes.

The superiority of the heterozygote, often called **heterosis** or **hybrid vigor**, may show itself in enhanced fitness characters such as longevity, fecundity and resistance to disease. An oft-cited example of heterosis is the dramatic increase in agricultural yield of hybrid corn achieved by crossing selected inbred lines. Beginning with an average yield of about 25 bushels per acre in the 1920s, hybrid corn enabled increases to as much as 140 bushels per acre (Crow, 1998b). However, agricultural geneticists are still investigating whether such hybrid vigor arises from the superiority of the heterozygote for particular gene differences (overdominance) or from other causes such as the introduction of dominant alleles at particular loci that were formerly homozygous for deleterious recessives.

When Δq is zero, equilibrium has been reached and the allele frequency will not change further. Note that three possible conditions will cause the numerator $[pq(ps-qt)]$ to be equal to zero and therefore Δq to equal zero. Under the first two conditions, when either p or q are zero, neither allele will be present in the population at the same time, and balance, or equilibrium, will be absent. The third condition occurs when $ps = qt$, so that the numerator of Δq is $pq(0) = 0$. When this happens, the following relationships can be derived:

$$ps = qt$$

Add qs to both sides of preceding equation			Add pt to both sides of preceding equation		
$ps + qs$	$=$	$qt + qs$	$ps + pt$	$=$	$qt + pt$
$s(p + q)$	$=$	$q(s + t)$	$p(s + t)$	$=$	$t(p + q)$

(Now, since $p + q = 1$)

$$q = \frac{s}{s + t} \qquad p = \frac{t}{s + t}$$

It is easy to see that if s and t are constant values, both p and q will reach a stable equilibrium; if q departs from the equilibrium value, then selection pressure will force it back into equilibrium with p. With positive Δq, allele frequency (q) increases, with negative Δq, q decreases (the negative or positive sign of Δq depending on whether q is above or below its equilibrium value). For example (**Fig. 22-2**), when $s = .2$ and $t = .3$, the equilibrium value for q is $s/(s + t) = .2/(.2 + .3) = .4$. As you can see from Figure 22-2, the effect of such heterozygote superiority is to drive the frequencies of the two alleles in the population to a stable equilibrium at $q = .4$.

Not all equilibria are permanent or stable, however. Unstable equilibria occur if any disturbance of equilibrium frequencies causes the frequency of one of the alleles to continue moving away from equilibrium. One such unstable equilibrium occurs when selection acts against the heterozygote at a gene locus with two alleles. If both homozygotes

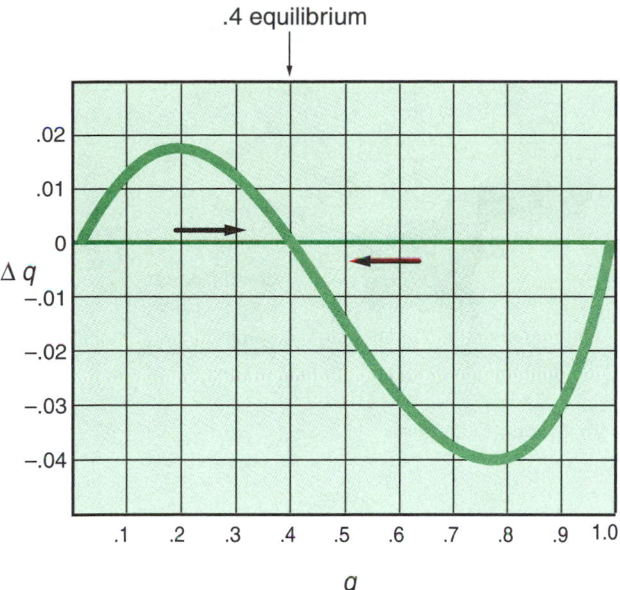

FIGURE 22-2 Change in the frequency (Δq) of allele *a* when the genotypic relative fitness are *AA* = 0.8, *Aa* = 1.0, *aa* = .7, and population size is infinite. These values provide a stable balanced polymorphism (Δq = 0) at q = .4. Values of q below .4 cause Δq to be positive, which increases q, whereas values of q above .4 cause Δq to be negative, which decreases q. Should one allele be accidentally eliminated — that is, q = 0 or 1 — then Δq is of course zero, but polymorphism is lost. (Adapted from Li, W. H., 1993. So what about the molecular clock hypothesis? *Curr. Opin. Genet. Devel.* **3**, 896–901.)

have equal relative fitness and the heterozygote is inferior, equilibrium will arise only when the frequency of each of the two alleles exactly equals .5. At this value the alleles are perfectly balanced. Equal amounts of each of the two are being removed in the heterozygote, and genotype frequencies are .25 *AA*, .50 *Aa* and .25 *aa*. However, any slight departure from these frequencies will cause the less frequent allele to have proportionally more of its alleles in heterozygotes than the more frequent allele. For example, if the gametic frequency of *A* rose accidentally to .6 and that of *a* fell to .4, then the genotypic frequencies under random mating are .36 *AA*, .48 *Aa*, and .16 *aa*, and the heterozygotes now contain a greater proportion of the *a* alleles than they do of the *A* alleles (.24/.16 > .24/.36). Thus, if the heterozygotes were lethal, the *A* frequency becomes .36/(.36 + .16) = .69, and *a* becomes .16/(.36 + .16) = .31. In the next generation, continued lethality of the heterozygotes would lead to an increase of the frequency of *A* to .83, while the frequency of *a* falls to .17. Within a relatively short time, the *A* allele would go to fixation and the *a* allele would be eliminated.

Selection and Polymorphism

The persistence of different genotypes through heterozygote superiority is an example of **balanced polymorphism**, a term coined by E. B. Ford to describe the preservation of genetic variability through selection.

We consider a gene locus polymorphic if at least two alleles are present, with a frequency of at least one percent for the second most frequent allele. Although selection coefficients are difficult to measure in natural populations, such polymorphisms are ubiquitous in practically all populations examined, on both chromosomal (see Fig. 10-35) and gene levels (see Table 10-4). One prominent example of polymorphism caused by overdominance is the sickle cell gene in humans, where heterozygotes (Hb^A/Hb^S) survive the malarial parasite more successfully than either normal (Hb^A/Hb^A) or sickle cell homozygotes (Hb^S/Hb^S). As shown in **Figure 22-3**, this gene — and others that appear to offer protection against malaria — persists in notable frequencies in geographical areas where malaria is endemic. An example of a balanced polymorphism that has been detected and described at the phenotypic level is selection of cryptic moth prey by blue jays (*Cyanocitta cristata*), where jays capture and consume more of the abundant than of the rare prey types, switching their capture strategy in response to changes in abundance of prey types.

In laboratory populations, where we can more easily control and measure genetic variance, many experiments achieve balanced polymorphism by some sort of overdominance. A balanced polymorphism that is found in natural populations of *Drosophila* has been amenable to laboratory analysis. In *Drosophila pseudoobscura*, for example, the frequencies of the Standard (ST) and Chiricahua (CH) third-chromosome arrangements come to a stable equilibrium when flies carrying these arrangements are placed together in a population cage kept continuously for a year or longer (**Fig. 22-4**). The superiority of the heterozygote can be seen in the relative fitness values calculated for the various third-chromosome combinations: *ST*/*ST* = .90, *ST*/*CH* = 1.00, *CH*/*CH* = .41.

In fact, we can calculate that even lethal recessive alleles may remain in a population if they confer only a small heterozygous advantage. For example, allele *a*, lethal in *aa* homozygous condition but providing a one percent advantage to the *Aa* heterozygote compared to the *AA* homozygote, would reach a frequency of approximately one percent at equilibrium:

	Genotypes		
	AA	*Aa*	*aa*
Relative fitness	.99	1.00	0
Selection coefficient	$s = .01$	0	$t = 1$

$$\hat{q} \text{ (frequency of } a) = \frac{s}{s+t} = \frac{.01}{1.01} = .0099$$

A further laboratory demonstration is when two types of *Escherichia coli* are co-cultured in glucose-limited mini-

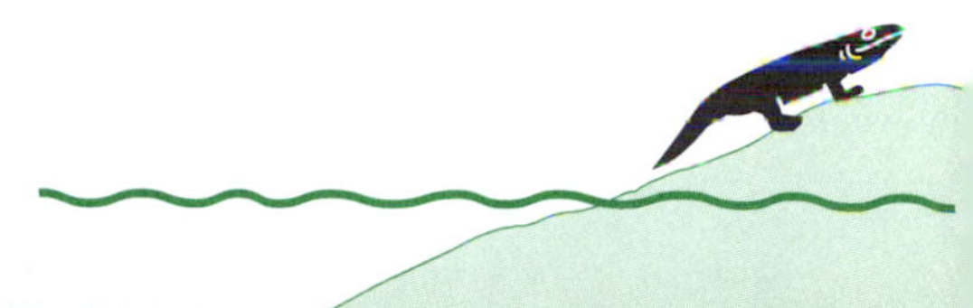

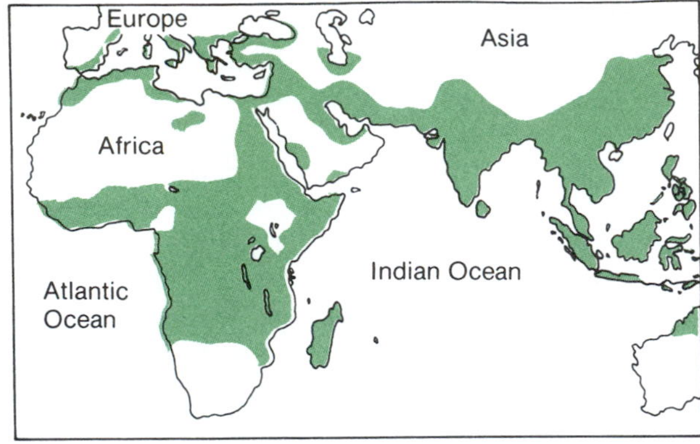

(a) Falciparum malaria

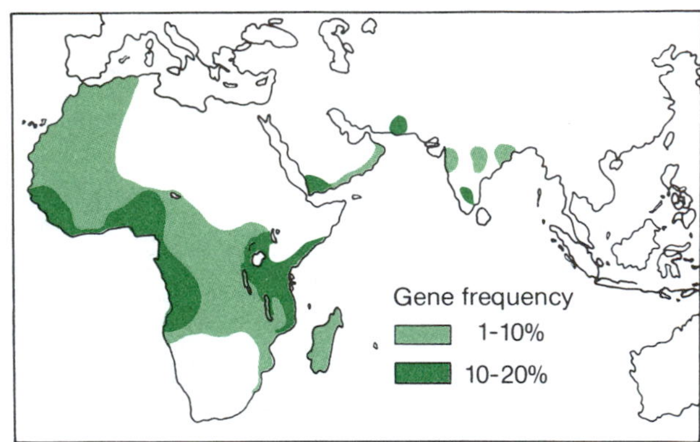

(b) Sickle cell anemia

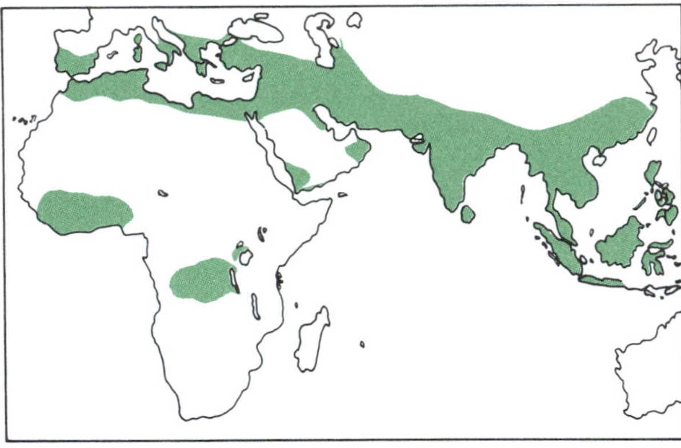
(c) Thalassemia

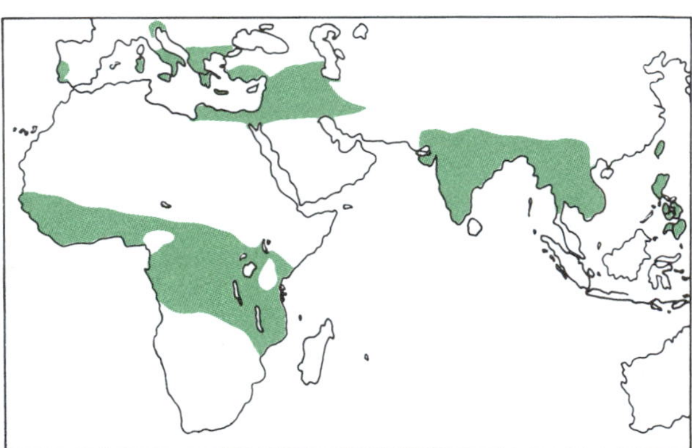
(d) G-6-pd deficiency

FIGURE 22-3 Relationship between the geographic distributions of malaria and genes that confer resistance against the disease. (a) Distribution of falciparum malaria in Eurasia and Africa before 1930. (b) Distribution of the gene for sickle cell anemia (Hb^S). (c) Distribution of the gene for β-thalassemia. (d) Distribution of the sex-linked gene for glucose-6-phosphate dehydrogenase deficiency in males in frequencies above 2 percent. (Adapted from Strickberger, M. W., 1985. *Genetics,* 3rd ed. Macmillan, NY.)

mal medium, the relative frequency of each type is frequency dependent, each type having an advantage when rare, thereby maintaining a balanced polymorphism in the population.

Mimicry

Other conditions responsible for polymorphism may include a change in selection coefficients so that genes detrimental at one time are advantageous at another.

Also, selection against a gene may depend on its frequency and may change or even be reversed when the gene is at low frequency, before it can be eliminated. An example of such **frequency dependent selection** is **Batesian mimicry**, in which palatable species that mimic distasteful models are protected against predators. In general, the more frequent the mimic and the less frequent its model, the greater the chances that predators will attack the mimic; conversely, the less frequent the mimic compared to the model, the greater the chances that the mimic will be protected. Mimicry among insects evolved early and, as you might expect if the mimic is effective, has been remarkably stable. A leaf-mimicking stick insect, *Eophyllium messelensis,* from the Eocene, 47 Mya, closely resembles extant insect mimics (Wedmann et al., 2007). Furthermore, this insect mimicked the leaves from several species of plants — myrtle and laurel trees, and alfalfa — indicating that phenotypic plasticity was established in the Eocene.

As shown in **Figure 22-5,** mimicry also occurs when a palatable mimic imitates a conspicuous warning coloration or pattern shared by two or more different unpalatable species. Mimicry between different unpalatable species (**Müllerian mimicry**) benefits both species by enabling predators to learn a single warning pattern that applies to all these potential but distasteful prey. Evolution of such warning patterns is probably again a matter of frequency dependence. When rare, conspicuous warning patterns on

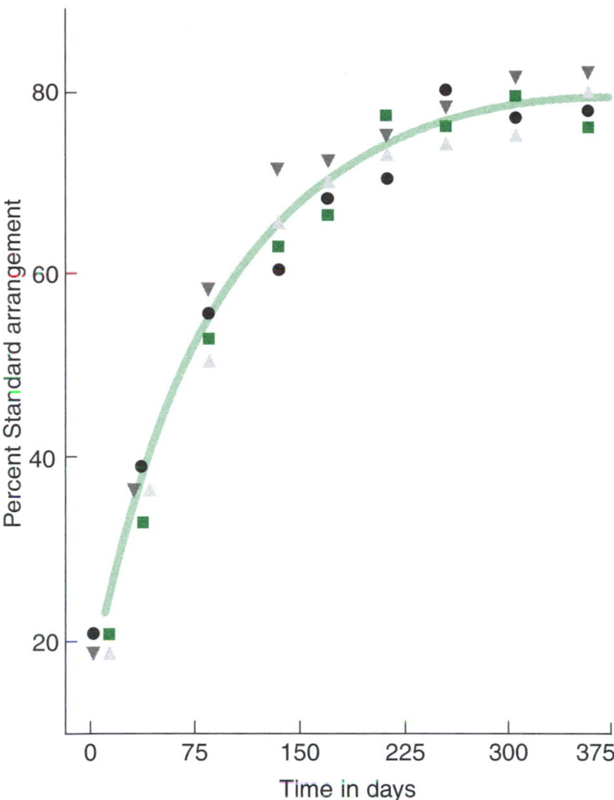

FIGURE 22-4 Results of four *Drosophila pseudoobscura* population cage experiments in which two third-chromosome arrangements are competing, Standard (ST) and Chiricahua (CH). Each population is denoted as a *circle, square* or *triangle,* and was begun with 20 percent ST and 80 percent CH, reaching equilibrium values of 80 to 85 percent ST after approximately one year. The solid curve represents the frequencies of the ST arrangement expected according to the relative fitness ST/ST = 0.90, ST/CH = 1.00, CH/CH = 0.41. (Adapted from Dobzhansky, Th., and O. Pavlovsky, 1953. Indeterminate outcome of certain experiments on *Drosophila* populations. *Evolution,* **7,** 198–210.)

unpalatable individuals probably offer little protection because predators have few chances to learn their distastefulness. Distinctive patterns, however, offer greater protection to unpalatable individuals when they are at higher densities, as shown by Sword in experiments with grasshoppers.

Self-Sterility

In plants, **self-sterility genes** that prevent fertilization between closely related individuals are also frequency-dependent.

For example, a haploid pollen grain carrying a self-sterility allele, S^1, will not grow well on a diploid female style carrying the same allele, such as S^1S^2, but can successfully fertilize a plant carrying S^2S^3 or S^3S^4. Once an allele (e.g., S^1) becomes common, its frequency is reduced by the many sterile mating combinations to which it is now exposed. Rare alleles, in contrast, will successfully fertilize almost every female plant they meet until they, too, become common. Thus, because of frequency dependence, self-sterility systems comprised of considerable numbers of alleles can become established, reaching, for example, 200 alleles or more in red clover.

Polymorphism and Industrial Melanism

Polymorphism may also be established when selection coefficients vary from one environment to another. A population sufficiently widespread to occupy many environments may maintain a variety of genotypes, each of which is superior in a particular habitat.

A prominent example is the polymorphism associated with the phenomenon known as **industrial melanism.** Certain moths and butterflies show increased proportions of dark-colored compared to melanic forms, usually caused by the increased frequency of a melanic form (morph) in

(a) Batesian mimicry

Danais plexippus
monarch butterfly
(unpalatable model)

Limenitis archippus
viceroy butterfly
(palatable mimic)

(b) Müllerian mimicry

Heliconius eucrate
(unpalatable)

Lycorea halia
(unpalatable)

FIGURE 22-5 Mimicry in different species of butterflies. (a) Batesian mimicry by a North American species, in which the more palatable viceroy butterfly (*right*) mimics the more unpalatable monarch (*left*). Resemblance between two South American unpalatable species in (b) provides a common warning pattern to predators and helps protect both prey species (Müllerian mimicry). Whether it is Batesian or Müllerian, mimicry is one of the most obvious examples of convergent evolution.

industrial areas where air pollution darkens vegetation because of coal smoke deposits (Majerus, 1998).

In the English industrial city of Birmingham, Kettlewell (1973) and others sought to explain the selective advantage of melanic morphs by releasing known numbers of light and melanic forms of the British peppered moth *Biston betularia* and recapturing a significantly greater proportion of melanic forms. Their data suggested that sooty areas offer greater protection to melanic forms than do light-colored forms; more of the former survive to be recaptured. The relative fitness of the melanic types may lie, at least partly, in their ability to remain concealed from bird predators on darkened twigs or tree trunks (**Fig. 22-6**). In nonindustrial areas, in contrast, trees covered with normal gray lichens offer decided advantages to the light-colored moths.

Whether because of environmental camouflage or unknown factors, English *B. betularia* populations show various degrees of polymorphism, ranging from high frequencies of the melanic form in industrial areas to almost zero in many rural areas. Based on dates of specimens collected by amateurs and those in museum collections, and assuming one generation per year, we can estimate that it took about 40 generations during the nineteenth century for the frequency of nonmelanic phenotypes of *B. betularia* to decrease in some industrial areas from about 98 percent to 5 or 6 percent. Using such data, we can arrive at an approximate selection coefficient for industrial melanism in this moth by noting that the nonmelanic phenotypes are homozygotes (frequency q^2) for the recessive nonmelanic allele (frequency q), and therefore q was reduced during this 40-generation interval from $\sqrt{.98} = .99$ to $\sqrt{.06} = .25$. As indicated in Table 22-3, it would take about 44 generations (13 + 7 + 9 + 15) to reduce q from .99 to .25 when the selection coefficient is .20. In other words, the selection coefficient against a nonmelanic gene in some of these industrial areas was fairly intense, about .20 or somewhat greater.

Passage of clean air legislation in Britain in 1956 reduced industrial smoke and sulfur dioxide in many formerly polluted areas. This reduction in pollution is now correlated with altered selection as the frequency of melanic forms of *B. betularia* and other insects has declined dramatically.[2] Spatial and temporal organization of the environment may significantly affect the extent to which a population will rely on genetic polymorphism as an adaptive strategy. Coarse-grained environments, in which different individuals in a population are exposed to different experiences, promote greater genetic polymorphism than fine-grained environments, in which all individuals experience the environmental differences. Hartl and Clark (1997) discuss other mechanisms that maintain polymorphism.

[2] See Brakefield (1998) and Majerus (1998) for melanism, and see Hooper (2002) for the history of research on the peppered moth.

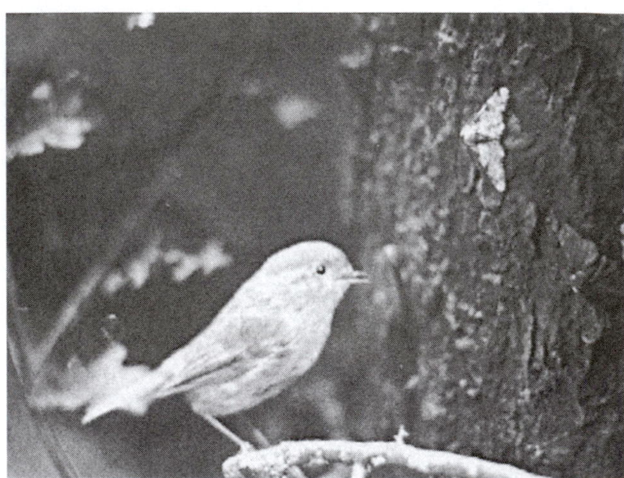

FIGURE 22-6 Light-colored and dark-colored tree trunks, each with a melanic and nonmelanic *Biston betularia* moth placed on a tree to show the contrast between them. The light-colored trunks derive their appearance from lichens, a symbiotic association between fungi and algae in which fungi receive products of algal metabolism and algae are protected from desiccation by fungal tissue. Although trees are commonly shown as resting sites for these moths, their actual resting habits are unknown. Some evolutionary biologists suggest caution in crediting color camouflage as the full explanation for *Biston betularia* polymorphism (Sargent et al., 1998).(© Science VU/Visuals Unlimited.)

Kinds of Selection

When selection has occurred for particular conditions over long periods of time, most populations achieve phenotypes that are optimally adapted to their surroundings; many phenotypes cluster around some value at which fitness is highest. We expect individuals that depart from these optimum phenotypes to show less fitness than those closer to the optimal values.

In a classic study on sparrows that survived a storm, Bumpus (1899) showed that measurements taken on eight of nine different characteristics clustered around intermediate phenotypic values, while sparrows killed by the storm showed much greater variability. In Bumpus' terms, "it is quite as dangerous to be conspicuously above a certain standard of organic excellence as it is to be conspicuously below the standard." Many studies on a variety of organisms, including snails, lizards, ducks and chickens, have since supported this view.

Among other characteristics, measurements of the birth weights of newborn human babies show selection for optimum values. As you can see from **Figure 22-7,** most survivors cluster around a birth weight of eight pounds (3.6 kg). Those departing from this value have fewer chances for survival. This reduction in frequency of extreme phenotypes has been called **stabilizing selection,** because it signifies selection for an intermediate stable value (**Fig. 22-8a**). Because mutation continually introduces departures from optimum character values, this mode of selection acts genetically to inhibit or reduce variation.[3]

However, not all character selection is stabilizing. Selection may well result in an extreme phenotype as organisms move in one or the other direction of a phenotypic distribution (Fig. 22-8b). Animal and plant breeders, who select for extremes of yield, productivity, resistance to disease, and so forth, commonly practice such **directional selection** (**Fig. 22-9**). Its role in evolution is especially important when the environment of a population is changing and when one or only few phenotypes happen to be adapted for new conditions.

Selection, whether stabilizing or directional, may act in a constant fashion if the selective environment is uniform. However, when conditions are changing, a population may be subjected to divergent or cyclically changing (oscillating) environments to which different genotypes among its members are most suited (Gibbs and Grant, 1987). Such selection is **disruptive,** because it establishes different optima within a population (Fig. 22-8c).

Because environmental conditions can be quite changeable, these different types of selection do not remain separate, but combine in different ways. For example, disruptive selection may be followed by directional selection, which may yield to stabilizing selection. The genetic means through which these forms of selection are expressed varies, from genes with large effect to polygenes with smaller effect, some causing simple developmental changes and others more complex canalizing processes (see Fig. 13-14).

[3] Data on human birth weight, especially low birth weight (<2,500 g [5.4 lbs]), coupled with the suggestion that low birth weight was an expression of a syndrome and not the lower end of the normal distribution of variation in birth weight, prompted research into the developmental origins of human health and disease, using an evolutionary approach; see Gluckman and Hanson (2005) for a recent analysis, and see the papers in Orzack and Sober (2001) for overviews of adaptation in the context of optimality.

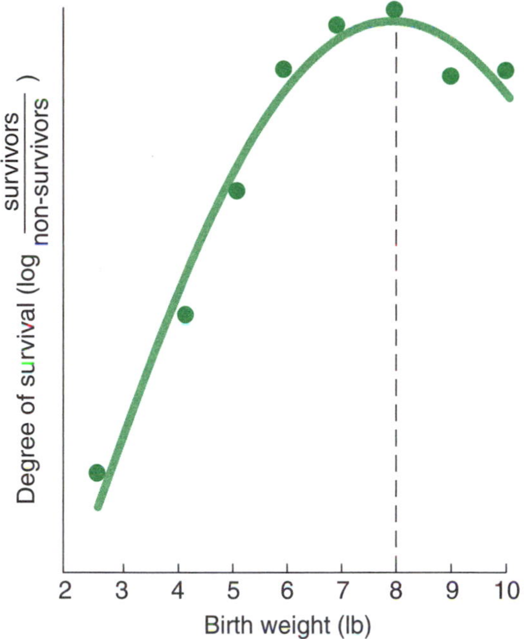

FIGURE 22-7 Relationship between birth weight and survival in female births in a London obstetric hospital between 1934 and 1946. Of 6,693 births, there were 6,419 survivors one month later, or a mortality rate of 4.1%. Only 25 died out of 1,689 born in "optimum" 8-pound (3.6 kg) class, representing 9.1% (25/274) of all deaths. Thus, 90.9% (249/274) of deaths occured among the nonoptimal classes, indicating that selection against nonoptimal phenotypes causes a high proportion of the deaths between birth and 1 month of age. Interestingly, the numerically largest class (7-pound, 3.2 kg) had a higher mortality rate (49/2,570 or 1.9%) than did the optimum class (25/1,689 or 1.5%). (Adapted from Karn, M.N., and L.S. Penrose, 1951. Birth weight and gestation time in relation to maternity age, parity, and infant survival. *Ann. Eugenics,* **161,** 147-164).

Variation and Variability

To ultimately affect evolution, selection (however it occurs) must change the frequencies of alleles or genotypes affecting fitness. This means that genetic variability must be present; lines that are homozygously uniform offer no opportunity for selection to produce any noticeable evolutionary change. Fisher formulated this principle mathematically as a fundamental theorem that essentially states, "*The greater the genetic variability upon which selection for fitness may act, the greater the expected improvement in fitness.*"

As a consequence of Fisher's theorem (1930), we expect populations long subjected to selection (the vast majority of populations) to have little remaining variability for genes affecting fitness; selection would have diminished such variability. The continued existence of selection therefore implies that variation itself is subject to selection, and so the propensity to vary (variability) is an important attribute of

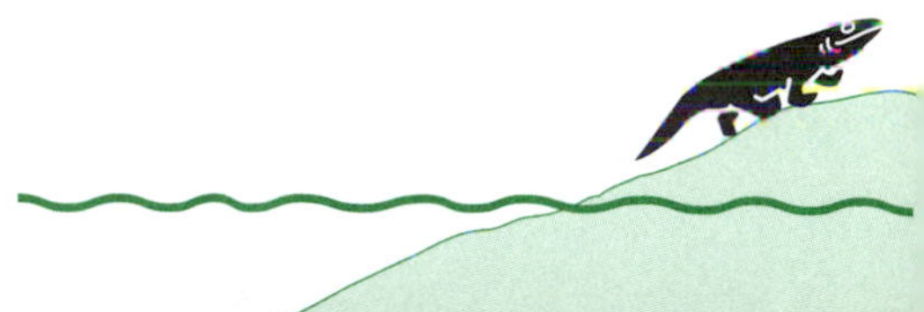

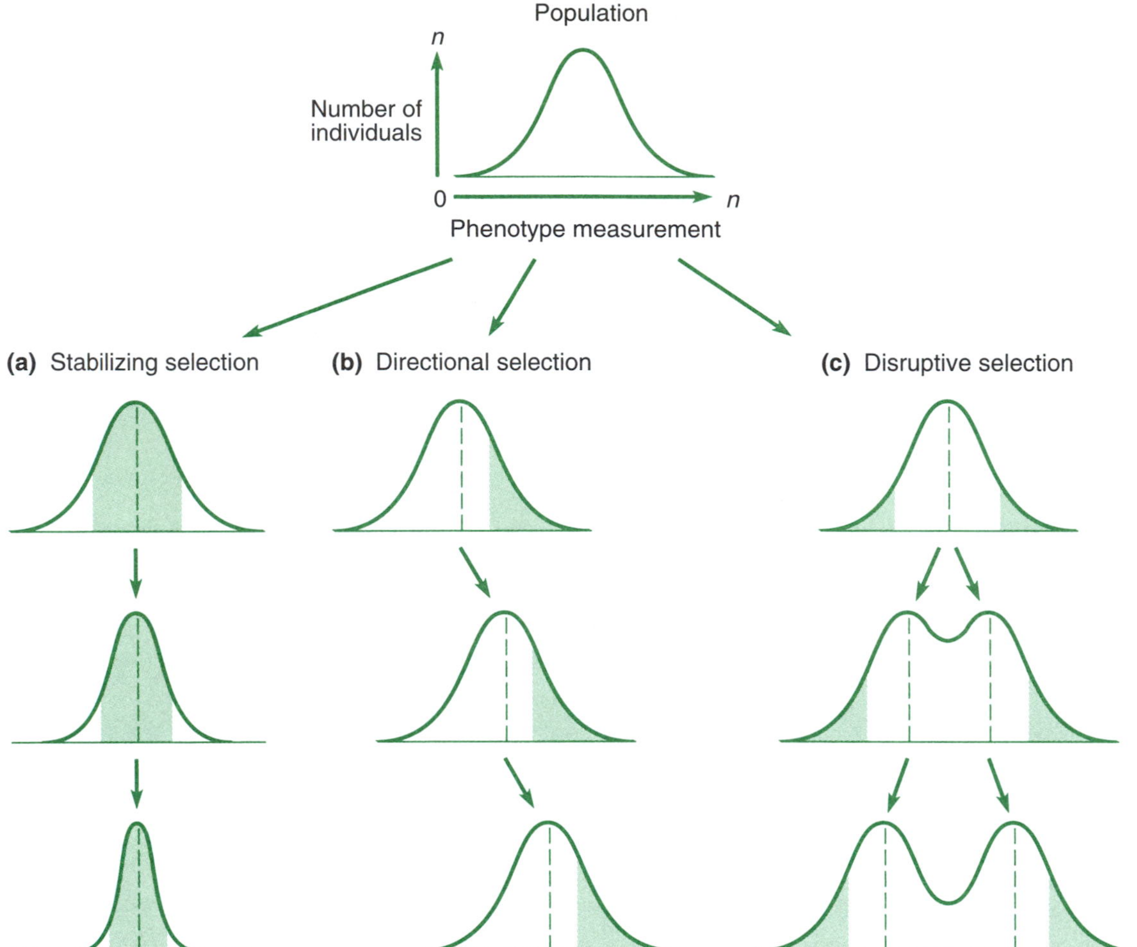

FIGURE 22-8 (a–c) Three basic modes of selection and their effects on the mean (*dashed lines*) and variation of a normally distributed quantitative character. The horizontal axis of each bell-shaped curve represents measurements of a quantitative character (e.g., from low on the left end to high on the right end), and the vertical axis represents the number of individuals found at each measurement. Shaded areas represent the individuals selected as parents of the next generation.

organisms.[4] Continuous changes in the environment affect formerly unselected genes, which now offer new opportunities for altering fitness. Such environmental changes include changes in resources, supplies, waste products and predator and parasite populations.

We do know that variability for genes affecting fitness can persist despite continued selection as, for example, when allele differences are retained through devices discussed in the previous section (e.g., frequency dependence). In general, however, populations tend to change genetically in directions that enhance fitness for their environment; Endler (1986) lists more than 160 cases in natural populations where selection has been demonstrated. Such findings indicate that genetic variability affecting fitness must have provided the baseline on which selection acts. Given persistent and recurrent genetic variability in factors such as differential mortality, differential fecundity and differential mating success, gene frequency change caused by selection is a constant feature of most or all populations facing changing environments. Interaction with the environment and other species is continuous. Van Valen (1976) proposed that species compete with each other for resources, so that an advantage, or increase in fitness, for one species represents a deterioration in the environment of others. He points out that species survival is in accord with the remark made by the Red Queen, whom Alice meets in Lewis Carroll's *Through the Looking Glass:* "Here, you see, it takes all the running you can do to keep in the same place."

According to this **Red Queen hypothesis**, each species continually faces new selective challenges because of

[4] See the papers in Hallgrímsson and Hall (2005) for analyses of the many factors affecting phenotypic variation and variability, and for proposal to further our understanding of these factors and how they interact.

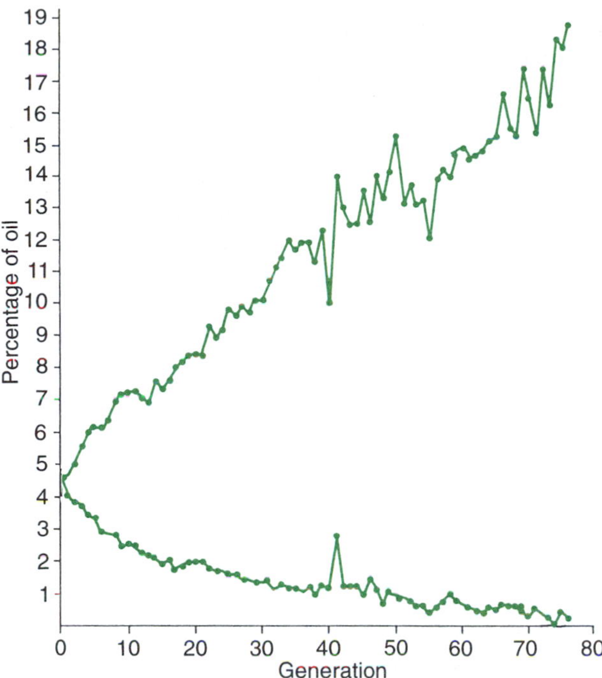

FIGURE 22-9 Results of selection for high and low oil content in corn kernels in an experiment begun in 1896 at the University of Illinois and continuing to the present. Selection for high oil content still continues to yield increases, whereas the effect of selection for low oil content has tapered off on reaching the 0 percent lower limit. (Adapted from Dudley, J. W., 1977. Seventy-six generations of selection for oil and protein percentages in maize. In *Proceedings of the International Conference on Quantitative Genetics*, E. Pollack, O. Kempthorne, and T. B. Bailey, Jr. (eds.). Iowa State University Press, Ames, pp. 459–473.)

environmental changes often associated with variations in the fitness of its interacting populations. Species must constantly confront and overcome recurring threats to fitness in order to survive. Nature perpetuates a cyclical process where adaptations in any one organism continually elicit selection for adaptations in others. Sooner or later, species face an "arms race" with a changing biological environment. Or as Darwin stated in *On the Origin of Species*, "If some of these many species become modified and improved, others will have to be improved in a corresponding degree or they will be exterminated."

The long-term consequence of the Red Queen's reign is to increase the competitive fitness of each interacting population, a theory that has gained support from experiments with RNA viruses (Clarke et al., 1994). The adaptations resulting from this process will, of course, vary between organisms, but we can discern patterns of increasing complexity over time. These changes include, for example, a steady increase in genome size from prokaryotes to eukaryotes, from about 10^6 to 10^9 or more nucleotide base pairs (see Fig. 12-9), and among vertebrates, a marked increase in relative brain mass from fish to reptiles to mammals, including humans (see Fig. 3-7).

According to Valentine and coworkers (1994), an additional manifestation of such pattern is increasing morphological **complexity** among metazoans as measured by their estimated number of different somatic cell types. As shown in Figure 8-5, this number has increased at an average rate of about one cell type per three million years, starting with the Precambrian–Cambrian, with no evidence of any downward trend. Vickaryous and Hall (2006a) place the number of cell types in humans at 411, of which 35 percent are nerve cells. Note, however, that the increased cell-type numbers that provide new adaptational opportunities do not imply replacement of all organisms that have fewer cell types (see also Box 8-3, Complexity). The race for survival goes on at all levels of morphological complexity, with survivors at each level, yet with seeming pressure to generate new levels of interaction.

Equilibrium Between Mutation and Selection

For convenience, we have considered changes in gene frequency to be caused by either mutation *or* selection acting separately. In nature, however, mutation and selection are ongoing processes and both factors influence gene frequency values. Predictions on the basis of one factor alone are misleading. Changing population size, lack of persistent selection over many generations, alterations in the intensity of selection and variation in genetic drift over time, are just some of the factors that make single factor analyses an unrealistic representation of the situation in nature. However, to determine the *mutation-selection equilibrium frequency*, we assume idealized conditions of equilibrium demography of population, selection, drift, etc.

Change in allele frequency per generation for a deleterious recessive a with frequency q is equal to a loss of $sq^2(1-q)/(1-sq^2)$. If s is small, we can consider the denominator 1, and the loss in frequency is then $sq^2(1-q)$. The frequency of newly mutated a alleles, however, is equal to the mutation rate (u) of $A \rightarrow a$ multiplied by the A frequency, which is $1-q$. Thus, the loss of a alleles through selection is exactly balanced by the gain of newly mutated a alleles when

$$sq^2(1-q) = u(1-q)$$

$$sq^2 = u$$

$$q^2 = \frac{u}{s}$$

$$q = \sqrt{\frac{u}{s}}$$

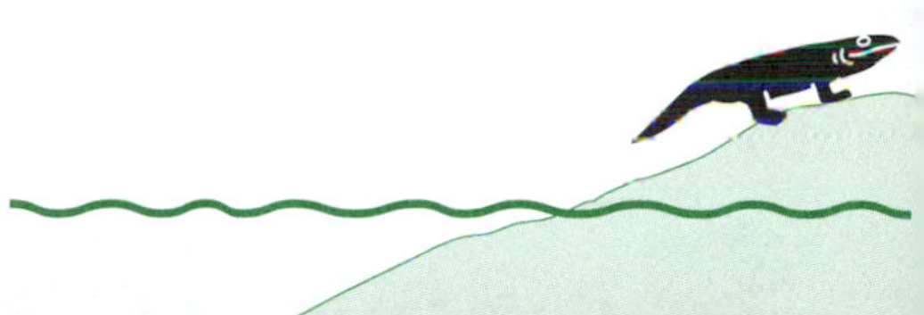

The equilibrium frequency of a mutant allele in a population is thus a function of mutation frequency and the selection coefficient. As you can see in the hydraulic model of this relationship in **Figure 22-10,** when the mutation rate increases, the equilibrium allele frequency also increases,

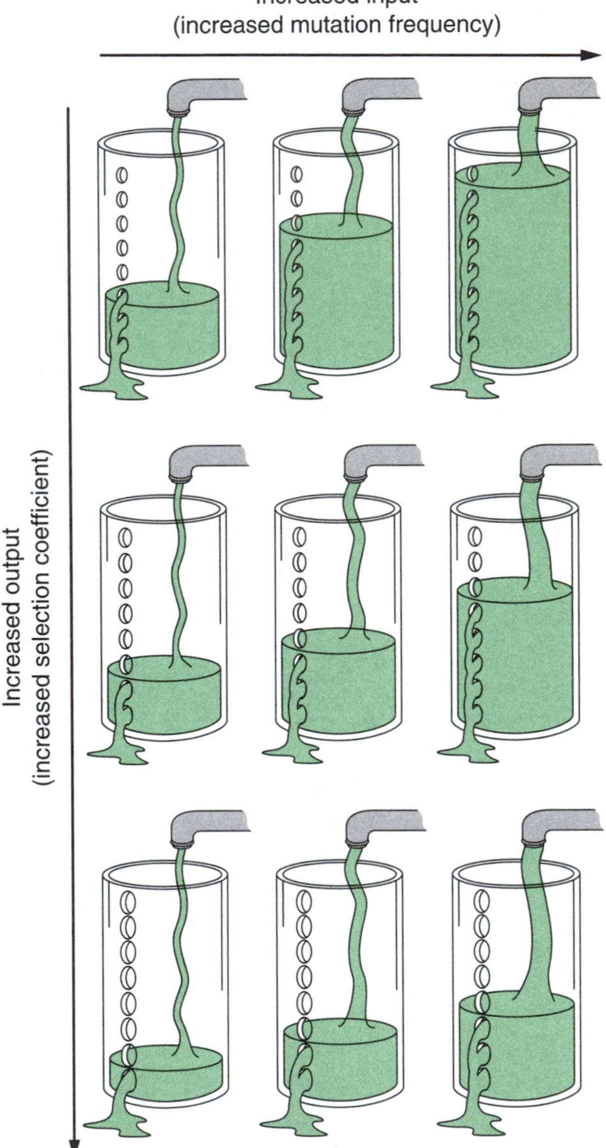

FIGURE 22-10 Hydraulic model of mutation-selection equilibrium. Each container is analogous to a population in which the water level represents the equilibrium frequency of a gene. As the water input (mutation frequency) increases, the standing water level (equilibrium gene frequency) increases. When the overflow holes are small (small selection coefficient) the water levels are higher for the same input (mutation frequency) than when the overflow holes are larger (large selection coefficient). (Adapted from Stern, C., 1973. *Principles of Human Genetics,* 3rd ed. W. H. Freeman, San Francisco.)

but the equilibrium frequency decreases when the selection coefficient increases.

For a deleterious dominant allele, similar algebraic manipulations point to an equilibrium frequency of about u/s, a value almost identical to the equilibrium frequency of alleles that lack dominance. Because such dominant or partially dominant alleles are of considerable disadvantage to heterozygotes, Fisher (1930) proposed that their deleterious effect diminishes in most organisms by selection of **modifier genes (alleles)** at other loci that change the degree of dominance. For example, mutant alleles at a particular locus *A* (for example, A^1, A^2, A^3 . . .) may act as partial dominants in the presence of the wild type allele A^1. Because these mutant alleles are mostly deleterious, modifier alleles at other loci (for example, B^1, B^2 . . . , or C^1, C^2 . . . , etc.) that increase the dominance of A^+ will be selected until the effects of mutations at the *A* locus are relatively recessive.

The successful selection for dominant and recessive modifiers that Ford (1940) demonstrated in the currant moth *Abraxas grossulariata* provides evidence for Fisher's view. In this moth a single gene, *lutea,* in homozygous condition, produces yellow instead of the normal white ground color but has an intermediate effect as a heterozygote. After four generations of selecting moths for greater and lesser expression of the *lutea* phenotype in the heterozygote, Ford obtained two distinct strains. In one, *lutea* acted almost as a complete dominant. In the other, *lutea* acted almost as a complete recessive. In each strain special modifiers had been chosen, some enhancing and some detracting from the dominance of this particular gene.

Instead of modifiers, Haldane (1939) suggested that special wild type alleles are selected (e.g., A^{X+}, A^{Y+}, A^{Z+}) that act as dominants in the presence of a mutant allele (e.g., A^1, A^2, A^3, . . .). Orr (1997) presents the view that, because haploid algae (*Chlamydomonas*) artificially transformed into diploids display dominance despite the lack of modifiers or alleles selected for this purpose, selection has little to do with dominance.

Whatever the initial cause for dominance, modifiers or alleles may affect dominance once it appears. Dominance modifiers and alleles occur in two species of cotton, *Gossypium barbadense* and *G. hirsutum,* in which certain alleles show simple dominance when variants of the same species cross or are crossed. In contrast, interspecific crosses show the effect of many modifying genes on these traits as well as differences in the degree of dominance of particular alleles. Despite these dominance-producing mechanisms, many harmful genes are still not completely recessive and seem to have some effect in heterozygous condition. Thus, in natural populations the equilibrium frequencies of harmful genes are higher than for dominants but lower than for recessives.

Migration

Mutation is not the only mechanism by which new alleles enter a population.[5]

A population may receive alleles by **migration** (also called **gene flow**) from a nearby population that maintains an entirely different allele frequency. When this occurs, at least three factors are important to the recipient population:

1. the difference in frequencies between the two populations;
2. the proportion of migrant genes that are incorporated each generation; and
3. the pattern of gene flow, whether occurring once or continually over time.

If we designate q_0 as the initial gene frequency in the recipient, or hybrid, population, Q as the frequency of the same allele in the migrant population and m as the proportion of newly introduced genes each generation, the allele frequency in the hybrid population will suffer a loss of q_0 equal to mq_0 and a gain of Q equal to mQ. Over n generations of migration, when the gene frequency of the hybrid population becomes q_n, one can calculate that the relationship between these factors will reach:

$$q_n - Q = (1-m)^n(q_0 - Q)$$

$$\text{or } (1-m)^n = \frac{q_n - Q}{q_0 - Q}$$

For populations where this equation can be applied, we must know four of these factors to calculate the fifth.

A classic example, which precedes our current extensive knowledge of DNA sequences, is blood group gene frequencies in African American and European Americans, two populations between which gene exchange has occurred. (**Box 22-2**). From the data and calculations in Box 22–2, you will see that on the basis of gene frequencies for blood group genes, the African-American population is about 70 to 80 percent African and 20 to 30 percent European-American.

An example of a more recent approach to estimating genetic exchange between African and European Americans is an analysis using nine autosomal DNA markers of the contribution from Europeans to 10 populations of African

[5] Alleles and genes are used interchangeably in this section.

BOX 22-2 Blood Group Gene Frequencies in Human Populations

Because of intermarriage among human populations, the frequencies of genes for particular blood groups can be calculated with some precision. The example used here is of African American and European Americans, two populations between which gene exchange has occurred over considerable numbers of generations. In general, although some genes from African Americans undoubtedly enter the European-American population, the latter population is so large that this introduction makes little difference to allele frequencies. In contrast, the African-American population is much smaller and has remained isolated from its African origin for two or more centuries. On this basis, one population can be considered as the gene donor or migrant population (Q) and other population as the recipient population (q_n). Although there is gene flow in both directions between European-American and African-American populations, for this example only, we take European Americans as the gene donors and African Americans as the recipient populations. First we present the classic study and then more recent analyses.

To obtain the original frequency of one of the Rh blood group alleles, R^0, in the African-American population (q_0), data of present East Africans was used, assuming that these data reflect the original allele frequencies of 200 to 300 years ago. Among the East Africans, R^0 showed a frequency of 0.630, indicating that the frequency of this gene had fallen in African Americans to its present frequency of 0.446. This fall in frequency could be ascribed to interbreeding with the European-American population, where the frequency of R^0 is about 0.028, much lower than among African Americans. According to Glass and Li (1953), this reduction had begun at the time of the initial introduction of Africans into the American colonies 300 years ago and had continued throughout 10 generations when the study was completed. Substituting these values into the preceding formula, we obtain,

$$(1-m)^{10} = \frac{q_{10} - Q}{q_0 - Q} = \frac{.446 - .028}{.630 - .028} = .694$$

$$1 - m^{10} = \sqrt{694} = .964$$

$$m = .036$$

This value of m means that, excluding all other causes such as mutation, 36 alleles per 1,000, or 3.6 percent of alleles in the African-American population, entered from the European-American population each generation. Because $1 - m$ represents the proportion of nonintroduced alleles, $(1-m)^{10} = .694$ is the proportion of alleles that have remained of African origin over the 10-generation period. Supported by somewhat similar estimates in more recent studies, blood group allele frequencies indicate that the African-American population is genetically about 70 to 80 percent African and 20 to 30 percent European American, with some differences between Southern and Northern African-Americans. See the text for percentage African ancestry based on a different approach and a different set of genes.

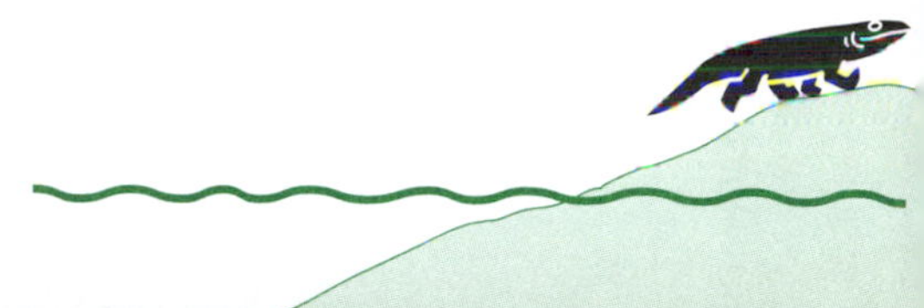

descent in the United States (Parra et al., 1998). Contributions from European ancestry ranged from 7 to 23 percent among the populations; that is, these populations are between 77 and 93 percent of African descent, statistics consistent with the results obtained from blood group genes. Combining their approach with data on mtDNA and Y Alu polymorphic markers enabled Parra and colleagues to identify a sex-biased flow, the contribution from European males being substantially greater than that from European females.

Where we lack exact information on gene frequency exchanges between populations — and this includes most populations — considerable discussion has flourished about the importance of migration. Migration can hinder local evolutionary changes by infusing genes from populations that are not adapted to local conditions. For example, some populations of mammals that live on dark, formerly volcanic lava flows have dark fur when they are isolated from neighboring populations that live on lighter colored backgrounds, but do not have dark fur when they receive immigrants from the lighter-colored surroundings. In contrast, populations of the checker spot butterfly, *Euphydryas editha,* can show no phenotypic changes, whether or not they are subject to migration from phenotypically different populations. In the absence of genetic information, these issues are difficult to resolve, but clearly, there is taxon-specificity.

Genetic Drift

The three forces we have considered up to now — mutation, selection and migration — share one important quality; they usually act directionally to change gene frequencies progressively from one value to another. Unopposed, these forces fix one allele and eliminate all others; when balanced, they can lead to equilibrium between two or more alleles. However, in addition to these directional forces are changes that have no predictable constancy from generation to generation. **Genetic drift,** one of the most important of such random or nondirectional forces, arises from variable sampling of the gene pool of each generation.

For practical reasons, we cannot always analyze mutation, selection, migration and drift in the same population/study. A recent reminder that these four processes act in concert comes from a set of simulations of interaction between genetic drift and gene flow under conditions where the population is assumed to be fragmented. The traditional assumption that any connections between subpopulations in the species range *inhibit* local adaptation was not borne out; low migration rates improved fitness in marginal populations, reflecting a complex interplay between migration rate, mutation rate, subpopulation size and species range (Alleaume-Benharira et al., 2006).

Genetic drift arises from variable sampling of the gene pool each generation. This is apparent if we consider that, in the absence of directional forces to change allele frequencies, there is always a strong likelihood of obtaining a good sample of the alleles of the previous generation as long as the number of parents in a population is consistently large. However, because real populations are limited in size, genetic drift will cause gene frequency changes due to sampling variation. For example, if only a few parents are chosen to begin a new generation, such a small sample of alleles may deviate widely from the allele frequency in the previous generation.

The extent of the deviation for all sizes of populations can be measured mathematically by the standard deviation of a proportion $\sigma = \sqrt{pq/N}$, where p is the frequency of one allele, q of the other, and N the number of genes sampled. For diploid parents, each carrying two alleles, $\sigma = \sqrt{pq/2N}$, where N is the number of actual parents. For example, if we begin with a large diploid population, where $p = q = .5$, and continue this population each generation by using 5,000 parents, then $\sigma = \sqrt{(.5)(.5)/10{,}000} = \sqrt{.000025} = .005$. The values of such populations will fluctuate mostly around .5 ± .005, or between .495 and .505. A choice of only two parents as founders will produce a standard deviation of $\sqrt{(.5)(.5)/4} = \sqrt{.625} = .25$, or values of .50 ± .25 (from .25 to .75).

In other words, sampling accidents because of smaller population size can easily yield allele frequencies that depart considerably from the initial .5 values in a single generation. If the population remained small and the next generation began with either of these extremes — that is, a frequency of .25 or .75 for a particular allele — in the following generation the frequency of that allele may fall to almost zero ($.25 \pm \sqrt{(.25)(.75)/4} = .25 \pm .22$; a range of .03 to .47) or increase almost to 1 ($.75 \pm \sqrt{(.75)(.25)/4} = .75 \pm .22$, a range of .53 to .97). Should such small populations continue each generation, the likelihood increases that one or more will eventually reach fixation for one of the alleles. The proportion of such populations that attain fixation (i.e., the **rate of fixation**) will eventually reach $1/2N$. Obviously, if N is large, fixation proceeds slowly, but even large populations can show some degree of drift, as diagrammed in **Figure 22-11.**

This reliance of drift on population number emphasizes the importance of what is called **effective population size** (N_e). Effective population size in the usual sense is the size of a totally random mating population that would generate the same amount of variation as observed in a real population, with all of its structure, etc., or the same amount of divergence per generation over time. In this sense, effective population size is a parameter that calibrates evolutionary change over time. Of course, selection also changes effective population size, which differs from the observed population size because not all members of a population are necessarily parents and because parentage also can be limited by a reduced number of one of the sexes. For example, if out of

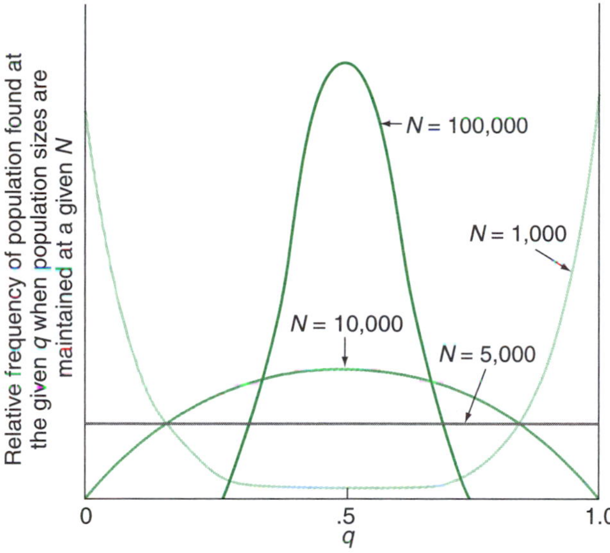

FIGURE 22-11 Distribution of equilibrium gene frequencies for populations of different sizes when selection is zero and a small amount of migration occurs into each population (m = .0001) from a population with a gene frequency of q = .5. Despite this migration, populations of sizes N = 1,000 and N = 5,000 show a considerable amount of genetic drift, many reaching elimination (q = 0) or fixation (q = 1). Only populations of relatively large sizes (N = *10,000*, N = 100,000) maintain the initial gene frequency q = .5 in appreciable proportions. (Adapted from Wright, S. 1951. The genetic structure of populations. *Ann. Eugenics,* **15,** 323–354.)

a total population of 1,000, three males mated to 300 females produced the next generation, the effective population size is more than six but still less than 303. Wright (1978) expressed the relationship as $N_e = 4N_fN_m/(N_f + N_m)$, where N_f is the number of parental females and N_m the number of parental males. In the preceding case, N_e would be $4(300)(3)/303 = 11$. Inequalities in numbers of offspring among different parents will also reduce the effective population size.

Wright proposed that genetic drift can be quite important in changing allele frequencies among populations when their effective sizes are small. Among the observations illustrating this concept are those by Buri (1956), who set up 107 separate lines of *Drosophila melanogaster,* each line carrying two alleles at the *brown* locus (*bw* and bw^{75}) at initially equal frequencies of 50 percent. Buri continued the lines for 19 generations by randomly selecting 8 males and 8 females as parents from each preceding generation ($N = 16 = 32$ brown

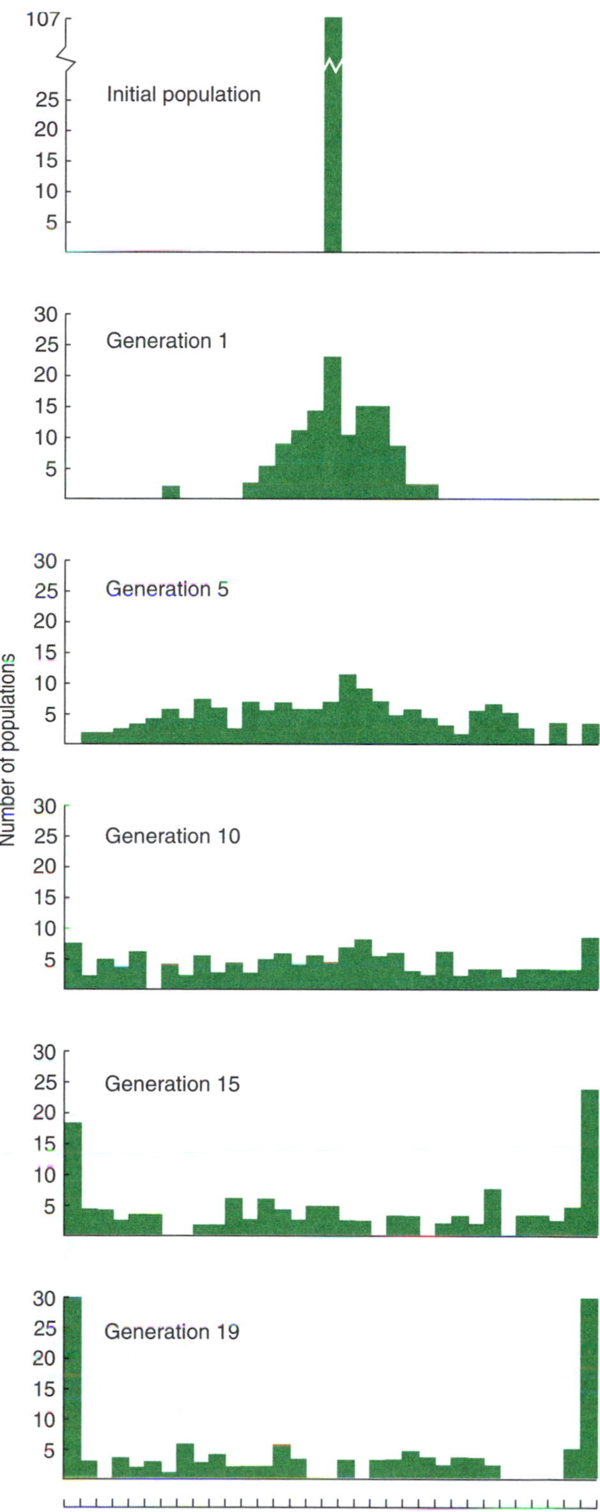

FIGURE 22-12 Distributions of the numbers of bw^{75} alleles in 107 lines of *D. melanogaster,* each with an initial frequency of 0.5 bw^{75}. Buri continued the lines for 19 generations, using 16 parents to start each generation (32 alleles at the brown locus), and the number of bw^{75} alleles found are given for the various lines. Note that by generation 19, the bw^{75} allele had been eliminated from 30 of these lines (0 alleles) and had been fixed in 28 of these lines (32 alleles). (Data are from Buri's series I cultures. Buri, P., 1956. Gene frequency in small populations of mutant *Drosophila. Evolution,* **10**, 367–402.)

alleles) and scoring the frequency of the two different brown alleles.

As shown in **Figure 22-12**, by the first generation of Buri's experiment, a number of populations already showed departures from the original 50 percent bw^{75} frequency. Genetic drift continued to increase successively so that by generation 19 more than half the 107 populations reached fixation for either the *bw* or bw^{75} alleles. Note that despite these genetic differences, when all populations are combined, the average frequency of *brown* alleles remains at about 0.5. Genetic drift therefore increases variation between populations, but on the average, not in any particular direction. These results also indicate that, because of genetic drift, selection in small populations, unless intense, may have little or no effect on a deleterious allele frequency such as bw^{75}.

The Founder Effect

Although the persistence of small population size over many generations is a cause of genetic drift, occasional size reductions for only one or a few generations may have pronounced effects on gene frequencies and future evolution.

At the extreme of such reductions, which Mayr called the **founder effect** — based on concepts developed by John Gulick in the 19th century, and discussed in Chapter 3 — a population may occasionally send forth only a few founders to begin a new population. Whatever genes or chromosome arrangements these founders take with them, detrimental or beneficial, all stand a chance of becoming established in the new population because of this sudden sampling accident (**Fig. 22-13**).

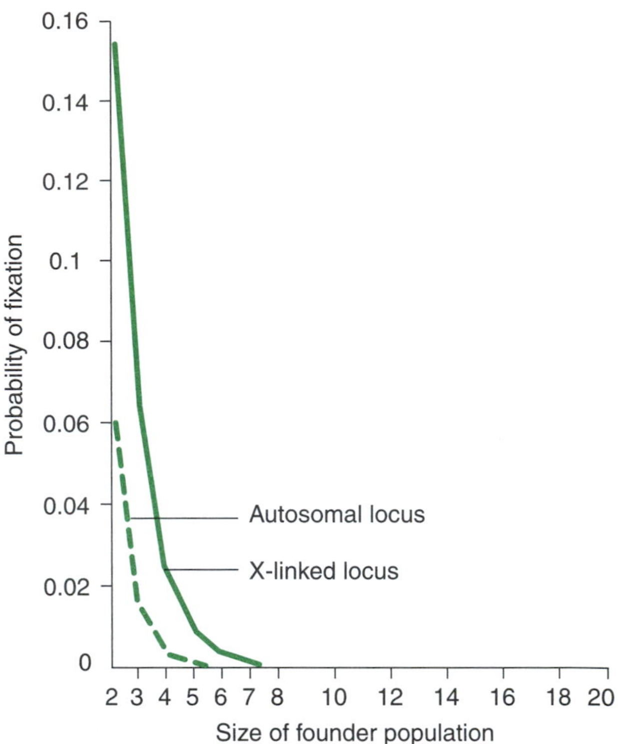

FIGURE 22-13 Theoretical probability of fixation (frequency = 1) of an allele that begins with a frequency of 0.5 in different founder population sizes. The dashed line is for an autosomal locus. The solid line is for an X-linked locus in a founder population consisting of twice as many males as females. In both cases, the smaller the founder population, the greater the chances that the allele will be fixed, but even so, note that the chances for fixation are not especially great. Less common alleles may have a greater "founder effect" on developmental processes, but their low frequency further reduces their chances for incorporation into a founder population. (Adapted from Templeton, A. R., 1996. Experimental evidence for the genetic transilience model of speciation. *Evolution*, **50**, 909–915.)

By carefully analyzing salivary chromosome banding patterns, Carson (1986, 1992) showed that the more than 100 native Hawaiian "picture-winged" *Drosophila* species were derived from founder events in which each island was settled by relatively few individuals whose descendants evolved into different species. For example, the 41 species endemic to the Maui island complex (**Fig. 22-14**) derive from only 12 founders — ten from Oahu and two from Kauai — with each single founder providing unique chromosome arrangements that we can trace in the descendant species.

Bottlenecks

There are by now numerous examples in many organisms, including humans, of unique gene frequencies that seem best explained by such founder events, or **bottleneck effects.** It seems likely that at least some populations began with only a few individuals carrying genotypes that may have differed greatly in frequency from their parental populations. Certainly the relatively high incidences of genes such as those listed in Table 21-7 — achromatopsia among the Pingelapese, and Ellis-van Creveld syndrome (polydactylous dwarfism) among the Lancaster County Amish — are difficult to explain except as founding accidents, because they seem to confer no advantage on either their homozygous or heterozygous carriers. The same conclusion is true for chromosomal translocations, which are usually selected against because they can cause sterility in heterozygotes, yet are nevertheless common features in the evolution of many mammalian lines.

Bottleneck effects may counter the effects of previous selection for a short period of time, an interval during which mutations that previously were advantageous may be lost and deleterious mutations may be fixed. However, it is difficult to imagine that any genetic trait that affects the armamentarium with which organisms face their environment can long continue to escape selective environmental pressures. Through nonselective genetic changes, a bottleneck

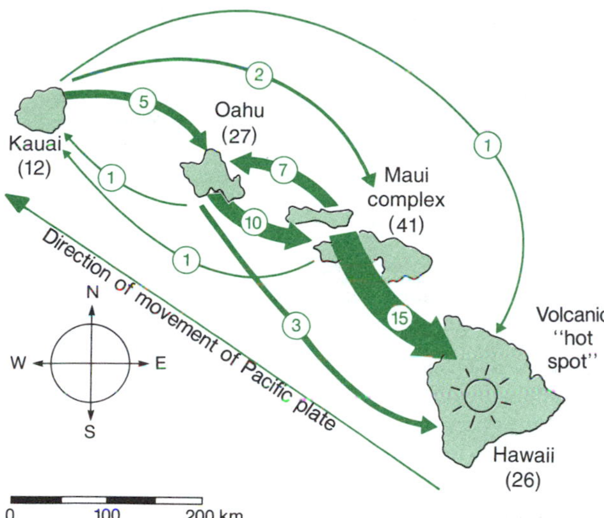

FIGURE 22-14 Colonization pattern showing the founder events that Carson proposed to explain the origin of native "picture-winged" *Drosophila* species found on the Hawaiian Islands. The width of each *arrow* is proportional to the number of founders (*circled*), and the number of *Drosophila* species now present on each island is given in *parentheses*. Each successful founder is presumed to have been a fertilized female, usually from a geologically older island. The oldest Hawaiian island is Kauai (about 5.6 My old), and 10 of the 12 species it has are considered the most ancient *Drosophila* elements in the islands. The youngest island, Hawaii, has been colonized entirely by founders from the older islands. Altogether the Drosophilidae family has more than 800 native Hawaiian species with an ancestry that dates back to founding episodes on islands even older than Kauai, some 30 or so Mya. Note that tectonic events formed these islands: a localized "hot spot" in Earth's mantle, lying under the large island of Hawaii, pierces the lithosphere and produces volcanic eruptions that form a succession of islands as the Pacific plate moves northwestward. In time various older islands erode; first becoming atolls, then seamounts (submerged volcanoes), which form a series that extends from Hawaii to Midway to a point near the far-western Aleutians. (Adapted from Carson, H. L., 1992. Inversions in Hawaiian *Drosophila*. In Drosophila *Inversion Polymorpism,* C. B. Krimbas and J. R. Powell (eds.). CRC Press, Boca Raton, FL, pp. 407–439.)

may cast the evolution of a population in a new direction, but from all we understand of evolution, this direction could not remain nonadaptive without extinction.

In general, the consequences of founding events on gene frequencies are largely unpredictable compared to variability estimates (i.e., σ) that can be arrived at when population size is constant. Other factors that make the variability of gene frequencies unpredictable are:

- unique historical events such as a change in the direction or intensity of selection because of a radical change in environment;
- a mutation subject to unusually high selection;
- a rare hybridization event with another variety; or
- an unusual swamping of a population by mass immigration.

The effects of such factors on evolutionary changes may be quite important, although genetic data for such events remain difficult to obtain.

KEY TERMS

balanced polymorphism
Batesian mimicry
bottleneck effects
directional selection
disruptive selection
effective population size
equilibrium gene frequencies
fitness
fixation
founder effect
frequency-dependent selection
gene flow
gene frequency
genetic drift
heterosis
heterozygous advantage
hybrid vigor
industrial melanism
migration
modifier genes
Müllerian mimicry
mutation rates
mutational equilibrium
overdominance
polymorphism
rate of fixation
Red Queen hypothesis
relative fitness
selection
selection coefficient
self-sterility genes
stabilizing selection

DISCUSSION QUESTIONS

1. Why does the origin of an allele by a single mutational event rarely lead to its persistence in a population?
2. How can we calculate mutational equilibrium, and why does a population rarely attain it?
3. Fitness.
   a. How would you define fitness? Would you equate fitness with survival? Why or why not?
   b. How would you measure fitness?
4. Selection.
   a. How are selection and fitness related?
   b. Why is selection generally more effective against an allele in haploids than in diploids? Would you say that this effectiveness confers an advantage on diploidy? Why or why not?
   c. Why is selection generally less effective in diploids against rare deleterious recessive alleles than against common deleterious recessive alleles?
   d. Does the action of artificial selection exclude natural selection? Explain.
5. Heterozygote superiority.
   a. Why does heterozygote superiority (overdominance) lead to allele frequency equilibria?
   b. How can we calculate such equilibria?
   c. Can allele frequency reach equilibrium when an allele is lethal to homozygotes?

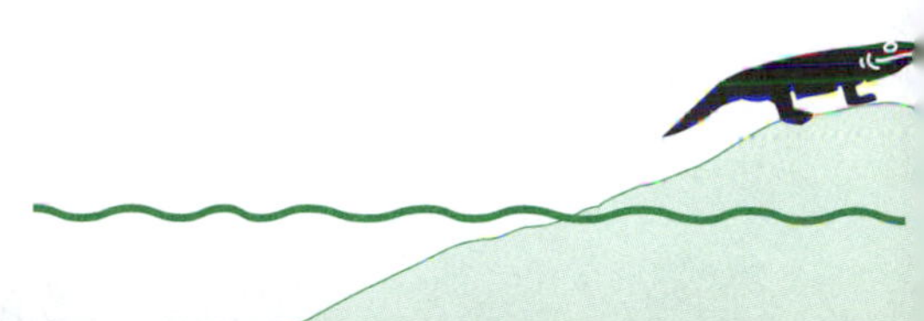

6. What selective conditions can explain balanced polymorphisms and the persistence of harmful genes in populations?
7. For selection on a particular quantitative character, what are the consequences if that selection is stabilizing, directional, or disruptive?
8. Why would you or would you not expect the effects of directional selection on a character to continue indefinitely?
9. Because selective success for increased fitness depends on genetic variability, could an increase in fitness occur in the absence of new mutation? Explain.
10. Is the population that always has the least remaining variability the most fit population in a stable environment? Explain.
11. Could a species evolve to the point where it could escape selection? Why or why not?
12. How is equilibrium allele frequency determined when both mutation rate and selection are acting simultaneously?
13. What hypotheses may explain the evolution of dominance at a particular locus?
14. How can we calculate the effect of migration on gene frequency?
15. How can we calculate the effect of genetic drift on gene frequency?
16. Will the long-term effect of genetic drift differ when population size is large compared to when it is small? Why or why not?
17. How would you support the hypothesis that a small number of founders ("founder effect," "bottleneck effect") can cause a radical change in genotype in a new population?
18. How, and under what conditions, would you rank (a) mutation, (b) selection, (c) migration, (d) genetic drift and (e) founder effect, as forces that cause rapid changes in gene frequencies?

EVOLUTION ON THE WEB

Explore evolution on the Internet! Visit the accompanying Web site for *Strickberger's Evolution,* Fourth Edition at http://www.biology.jbpub.com/book/evolution for Web exercises and links relating to topics covered in this chapter.

RECOMMENDED READING

Alleaume-Benharia, M., I. R. Pen, and O. Ronce, 2006. Geographical patterns of adaptation within a species' range: interactions between drift and gene flow. *J. Evol. Biol.,* **19**, 203–215.

Bonner, J. T., 1988. *The Evolution of Complexity by Means of Natural Selection.* Princeton University Press, Princeton, NJ.

Brower, L. P. (ed.), 1988. *Mimicry and the Evolutionary Process.* University of Chicago Press, Chicago.

Endler, J. A., 1986. *Natural Selection in the Wild.* Princeton University Press, Princeton, NJ.

Fisher, R. A., 1930. *The Genetical Theory of Natural Selection.* Clarendon Press, Oxford, UK. (2nd ed, 1958, Dover, New York.)

Gillespie, J. H., 2004. *Population Genetics: A Concise Guide,* 2nd ed. Johns Hopkins University Press, Baltimore, MD.

Hallgrímsson, B., and B. K. Hall (eds), 2003. *Variation: A Central Concept in Biology.* Elsevier/Academic Press, Burlington, MA.

Hartl, D. L., and A. G. Clark, 1997. *Principles of Population Genetics,* 3rd ed. Sinauer Associates, Sunderland, MA.

Hedrick, P. W., 2000. *Genetics of Populations,* 2nd ed. Jones and Bartlett, Sudbury, MA.

Kettlewell, H. B. D., 1973. *The Evolution of Melanism.* Clarendon Press, Oxford, UK.

Majerus, M. E. N., 1998. *Melanism: Evolution in Action.* Oxford University Press, Oxford, UK.

Maynard Smith J., and E. Szathmáry, 1995. The *Major Transitions in Evolution.* Freeman Press, Oxford, UK.

Mayr, E., 1942. *Systematics and the Origin of Species.* Columbia University Press, New York.

Orzack, S. H., and E. Sober. (eds), 2001. *Adaptation and Optimality.* Cambridge University Press, Cambridge.

Pagel, M. (ed. in chief), 2002. *Encyclopedia of Evolution,* 2 volumes. Oxford University Press, New York.

Provine, W. B., 1986. *Sewall Wright and Evolutionary Biology.* The University of Chicago Press, Chicago.

Ruse, M., 1996. *From Monad to Man: The Concept of Progress in Evolutionary Biology.* Harvard University Press, Cambridge, MA.

23 Structure and Maintenance of Populations

Chapter Summary

Much of the current excitement about evolution comes from our ever-expanding knowledge of genomics and development, fields that are enhancing our understanding of mechanisms of evolution. But, at the *foundation* of evolution lie interactions of individuals with their environment and with other individuals, interactions that we can only understand by analysis of the structures and interactions of populations, which are the topic of this chapter.

The structures of populations are so complex that quantitative evaluation of their evolutionary potential is difficult. Growth, however, is one characteristic that can be measured. Unlimited population growth becomes exponential, but as the environment imposes restrictions, a population will tend to stabilize at a size called the carrying capacity, or *K*. Strategies such as *r*-selection that increase the numbers of offspring, or K-selection that increase the selective advantages of a smaller number of offspring, often only one, enable populations to approach the carrying capacity of the environment.

Although some genetic variability is advantageous, the immediate effect of many alleles or genetic combinations may not be. The extent to which these detrimental genotypes affect a population is the genetic load. Both mutation and heterozygote superiority contribute to genetic load by perpetuating deleterious recessive alleles. Although this polymorphism reduces the frequency of optimum genotypes, it may increase the adaptiveness of the lineage at some future time. Because we cannot explain all genetic polymorphisms as resulting from heterozygote superiority, some have proposed that most alleles are neutral in effect and incur no genetic load. Under these circumstances, polymorphisms remain in the gene pool, not because of selection but because of mutation and genetic drift (see Chapter 22). Selectionists, however, argue that (1) relationships between ecological conditions and polymorphism, (2) similarities in allozyme frequencies in different species, (3) the association between polymorphism and enzyme function, and (4) the higher frequency of polymorphism in noncoding DNA sequences, cannot be entirely explained by random forces but must be due to selection pressures.

Sexual reproduction, recombination of linked genes and mutation can (but need not) all produce genetic combinations that increase fitness. When such advantageous genotypes occur, we say they occupy adaptive peaks of varying value according to their *degree of fitness.* However, to reach higher adaptive peaks, a population must often pass through less fit combinations (adaptive valleys), a process that, among other factors, involves genetic drift, according to Sewall Wright.

Although the issue is still unresolved, in some instances, for example, in a population with sexual reproduction and altruistic social behaviors, selection seems to occur for the benefit of the group, even though it may harm an individual's own genetic future, although theoretically, the average fitness of an individual should go up in a population where altruism is maintained.

Population interactions such as competition and predation may result in the selection of particular traits. Competition results in niche and character distinctions between groups — in some cases, eliminating one group entirely — while predation has complex effects on the size and structure of the populations of both prey and predator.

In nature, the structure and relationships of populations depart from many of the ideal conditions that would make their evolutionary behavior simple to understand. Populations are neither of constant size, uniformly distributed in space, nor always of the same mating pattern and are subject to variable conditions of mutation, migration and selection. Because most genes are comprised of multiple alleles, there are usually more than three diploid genotypes for any one locus, and through developmental interactions, the fitness these genotypes confer depends on genes at other loci.

Moreover, the environments in which populations evolve are usually changing. In terrestrial environments, changes include chemical composition, moisture, temperature, pressure and sunlight/shade. In aquatic environments, changes include pH, dissolved chemicals, currents, temperature, pressure and amount of light. In both environments, changes include such elements of the biological environment as prey, predators, parasites, hosts and competitors.

The relationship of a population to various ecological factors is not merely that of a passive recipient. A population often modifies its physical and biological environment in ways that can diminish or enhance its own resources and those of other populations. Because of all these interactions, it is no wonder that populations "must continue running in order to keep in the same place" (the Red Queen hypothesis; see Chapter 22). It is also not surprising that the structure of populations is difficult to predict mathematically. Nevertheless, mathematical models do allow us to apply information from populations under one set of conditions to other populations and other sets of conditions. Various populations share common features in their response to such factors as inbreeding, selection, mutation, migration and genetic drift. A popular approach is to measure various characteristics present in natural and experimental populations and then mathematically derive some broad evolutionary concepts about the structure of populations in general. This chapter briefly surveys such attempts at ecological and genetic levels.

Ecological Aspects of Population Growth

One area of mathematical modeling deals with the evolutionary potential of the reproductive power of populations. In the evolutionary process, the capacity for reproduction is always counterpoised against selection. In its simplest form, as in rapidly growing, asexual, unicellular species, the early stages of population growth in an environment well supplied with resources can occur geometrically ($2 \rightarrow 4 \rightarrow 8 \rightarrow 16 \rightarrow 32 \rightarrow 64$, and so forth) in each generation, closely following an exponential growth curve of the type shown in **Figure 23-1a.** Such a pattern of population growth would be limitless if reproduction rate was unchecked and space and resources limitless.

Rates

Quantitatively, the rate of numerical change (ΔN) is equal to the difference between birth (b) and death (d) rates multiplied by the number of individuals (N),

$$\Delta N = (b - d)N$$

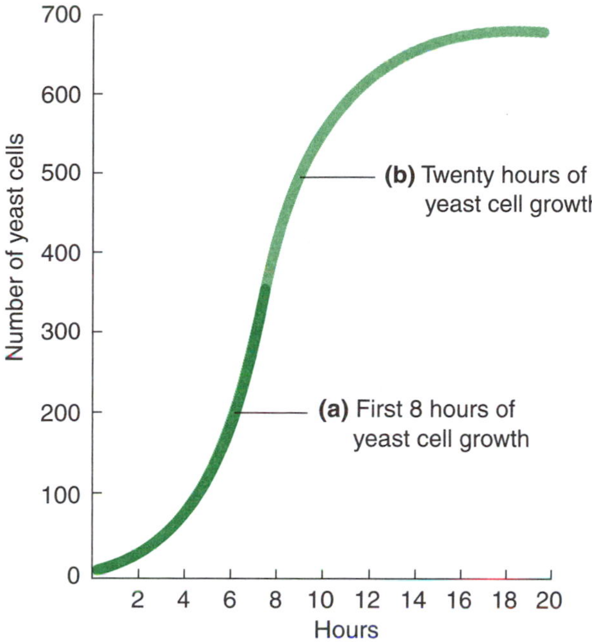

FIGURE 23-1 Numbers of yeast cells (*Saccharomyces cerevisiae*) in a defined volume of culture medium for two growth periods, beginning with approximately 10 cells per volume. (a) Geometric growth during the first 8 hours ($\Delta N = rN$, where $r = 0.5535$). (b) Sigmoidal growth curve approximating the logistic relationship [$\Delta N = rN \times (K - N)/K$, where $r = 0.5535$ and $K = 665$] for the 20-hour growth period. (Data are from Carlson, T., 1913. Über Geschwinddigkeit und grösze der Hefevermehrung in Würze. *Biochemische Zietschrift,* **57,** 313–334.)

Thus, when birth rate exceeds death rate, ΔN is positive and population size increases; equal birth and death rates yield $\Delta N = 0$ and an unchanged N, while a death rate higher than the birth rate yields negative ΔN and decreasing population size. Ignoring other causes in this simple illustration, $b - d$ is a primary factor determining population size. We can condense it as the rate of increase (r) so that $\Delta N = rN$. In environments where a population is free of those factors that limit its growth, the population attains what is called an **intrinsic rate of natural increase** (r_m).

Environmental resources and space are not limitless, of course, nor are individuals in a population unaffected by the waste products and toxins neighbors produce. Consequently, population growth is not limitless; population size eventually stabilizes at some near-constant value or may even "crash" to some low number. Malthus's popularization of the idea that war, famine and disease held in check exponential growth of human population led Darwin and Wallace independently to the concepts of the struggle for existence and natural selection (see Chapter 2). An important point, perhaps even a cautionary note, is to remember that natural selection happens within a generation and affects individuals of that generation. Response to natural selection happens between generations, and only for those features or aspects of features that are heritable.

To evaluate such impacts on a population we need to determine how many individuals the environment can support. This value, K (the **carrying capacity**) is determined as follows, using the example in Figure 23-1b, which shows how the size of a yeast population grown under a particular set of conditions levels off from its early nearly exponential direction to a plateau of about 665 individuals. The smooth S-shaped curve that results can be described mathematically as the modification of ΔN by a factor $(K - N)/K$ in the formula:

$$\Delta N = rN \times (K - N)/K \, (\mathrm{dN/dT})$$

where K = carrying capacity, so that when N is small this factor is essentially 1, and population growth is almost exponential (or $\Delta N = rN$ as before). However, as N increases in value closer to K, the $(K - N)/K$ factor becomes a fractional quantity closer to zero. Eventually, N is large enough to equal K so that $\Delta N = rN \times 0 = 0$, and, theoretically, population size no longer changes but stabilizes at the value K.

The relationship $\Delta N = rN(K - N)/K$ is known as the **logistic growth curve.**[1] In reality, populations rarely follow such smooth growth curves.

Population Parameters

As noted above, many parameters can effect fluctuations in numbers within a population (**Fig. 23-2**). We attribute such fluctuations to factors intrinsic to individuals (growth, metabolic rate), intrinsic to populations (subpopulations, differential mortality among age or spatial groups) or to the impact of environmental agents.

Density independent agents are those, such as climate, that are independent of population size and crowding. On the other hand, **density dependent** agents depend more on crowding. Examples of density-dependent agents include the effects of metabolically produced toxins, depletion of resources and intrapopulation aggression, all of which may serve to stabilize populations. Although obvious examples arise where one or both density factors influence population size, their quantitative effects are often difficult to measure, and ecologists continue to investigate their relative importance.

Alongside the effect of environmental agents, *the age structure of a population* is a most important determinant of the reproductive power of a population. If reproduction is associated with a particular age cohort in the population, the chances of surviving to that age and the number of off-

[1] Pierre Verhulst (1804–1849) discovered the original equation for logistic growth in 1838. You can find discussions of its derivation, along with possible applications, in the population and evolutionary ecology textbooks by C. W. Fox et al. (2001), Krebs (2001), Ricklefs and Miller (2000), Rockwood (2006) and Rose and Mueller (2006). See G. E. Hutchinson (1978) for an historical account.

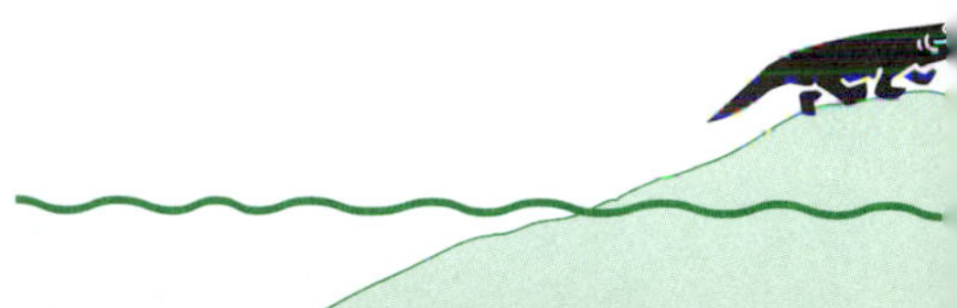

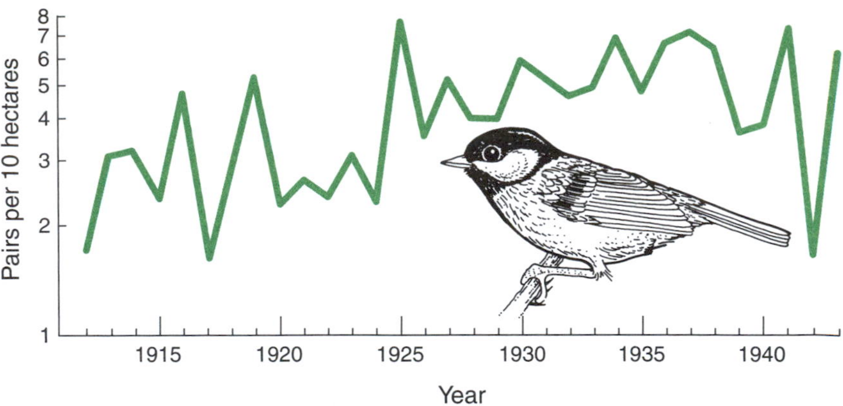

FIGURE 23-2 Fluctuations in the population size of the great tit (*Parus major*) in Holland between 1912 and 1943. (Adapted from Begon, M., and M. Mortimer, 1986. *Population Ecology*, 2nd ed., Blackwell, Oxford, UK.)

spring that such individuals produce are essential attributes represented by:

$$R_0 = \Sigma\, l_x m_x$$

1. The first of these, **survivorship,** we measure by a factor l_x, which represents that proportion of individuals who survive from age 0 to age *x* out of a group (cohort) that were all born during the same period.
2. The second, **fecundity,** we represent by m_x, the average number of offspring produced by an individual of age *x*.
3. The **net reproductive rate** of the entire population (R_0) is the sum of all its individuals multiplied by their fecundity at each age *x*.

This relationship between survival and fecundity can be elaborated in various ways, but a population that is stationary[2] will have a net reproductive rate (or replacement rate) of one, as illustrated in **Table 23-1.** Populations with R_0 less than one (e.g., decreased survival of age, class 3 in Table 23-1) and lacking compensating increases in the size and fecundity of other reproductive age classes, will decrease in size. Populations with R_0 greater than one will increase.

Older age classes such as class 6 in Table 23–1 can be reproductively barren. They have the lowest survival rate of all classes, although reproductive senescence may be rare in natural populations. In the past, population biologists considered the aging (**senescence**) and death of populations to have been selected as traits that benefit a population by removing individuals that might compete for its resources but no longer contribute to its reproductive success; that is, selection occurred among populations — **group selection,** discussed later in this chapter — by adding fitness to populations in which nonreproductive individuals died out because of the aging process. Although population benefits of this kind probably exist, such evolutionary explanations are presently out of favor, some instead suggesting that senescence evolved by mechanisms that allow alleles with deleterious effects on older, nonreproductive individuals to spread through a population:

1. because they are either neutral or advantageous to reproductive individuals in the same population, or
2. because somatic mutations that affect longevity accumulate as individuals grow older, leading to increased additive genetic variance in components of fitness (K. A. Hughes, 1995).

Both approaches reflect the concept that natural selection favors genotypes that confer survival to reproductive age, with little or no selection for genotypes to survive longer. As Dawkins points out, "we inherit whatever it takes to be young, but not necessarily whatever it takes to be old." However, we do know that at least some populations maintain genetic variability for aging, and so we continue to search for genes involved in senescence. For example, longevity in *Drosophila melanogaster* can decline when experimenters select individuals to reproduce early in the life cycle, or longevity can increase by selecting individuals to reproduce later in the life cycle. In one experiment in which individuals were selected for rapid growth rates, the decline in female fecundity (an estimate of senescence) was caused by increased homozygosity for deleterious recessive alleles in populations (*r*-selection, see discussion below).

Reproductive Strategies: *r*- and *K*-selection

Whatever the causes for differences in age distribution, the strategies used in reproduction differ among organisms. Many annual plants and insects, for example, breed only once during their lifetimes (**semelparous**). Many perennial plants and vertebrates breed repeatedly (**iteroparous**), although not necessarily annually. The number of offspring a reproductive female produces at any one time varies sig-

[2] The term *stationary* is used here, as in demographics, rather than the term *stable in size,* because a growing population can have a stable age distribution.

TABLE 23-1 Age structure of a hypothetically stable population with seven discrete age classes, ranging from non-reproductive juveniles (class 0, *lightly shaded*) to reproductive adults (classes 1 to 5, *darkly shaded*) to non-reproductive senescents (class 6, *lightly shaded*), illustrating the calculation of net reproductive rate (R_0)

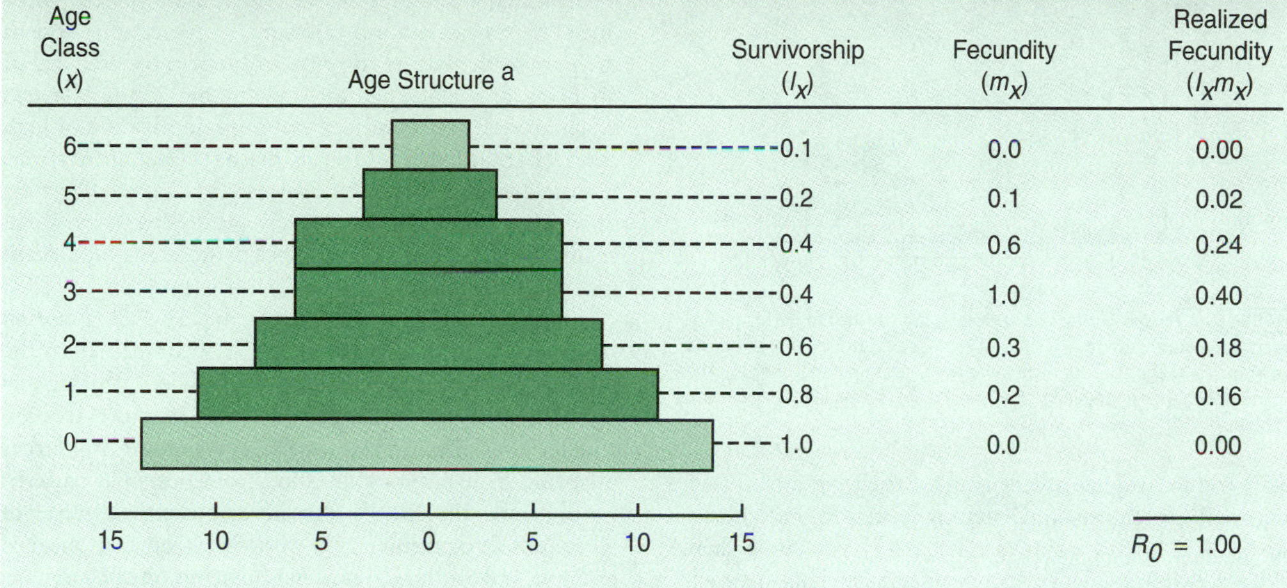

Age Class (x)	Age Structure [a]	Survivorship (l_x)	Fecundity (m_x)	Realized Fecundity (l_xm_x)
6		0.1	0.0	0.00
5		0.2	0.1	0.02
4		0.4	0.6	0.24
3		0.4	1.0	0.40
2		0.6	0.3	0.18
1		0.8	0.2	0.16
0		1.0	0.0	0.00
				R_0 = 1.00

[a] Frequency of each age class in a stable population is related to *lx* (Hutchinson, 1978) and can be noted here as the percentage graphed symmetrically on either side of the zero midline, for example, class 0 = 14.5 + 14.5 = 29.0 percent.

Source: Adapted from Pianka, E. R., 1988. *Evolutionary Ecology,* 4th ed., Harper & Row, New York.

nificantly, ranging from a single offspring in many larger mammals to the millions of eggs a codfish spawns. Furthermore, because an individual may die before reaching reproductive age, the sooner reproduction begins, the greater the chances of producing offspring. Thus, in some organisms, reproductive stages follow soon after hatching. Other organisms — more dependent for survival on reaching larger size, obtaining greater experience or needing more parental care — delay reproduction until the second half of the life cycle.

The alternatives of producing many offspring with little parental care or few offspring with greater prenatal investment in eggs and greater parental care are among properties often ascribed to differences between what are called ***r*- and *K*-selection** strategies.

Organisms such as bacteria and plant weeds that display rapid population growth in the face of wildly fluctuating environmental challenges and opportunities generally have a rapid rate of increase (r) at low population densities. Individuals in such populations are r-selected (**Fig. 23-3**). As stated by Darwin, "A large number of eggs is of some importance to species dependant upon a fluctuating amount of food, for it allows them rapidly to increase in numbers."

Other organisms, such as large vertebrates, face more uniform or predictable environments with population sizes that are close to the environmental carrying capacity (K). In these K-selected organisms (Fig. 23-3), there is density-dependent competition for food, nesting space and other resources, providing selective advantages that increase efficiency in resource use and ensure that offspring are raised to the stage when they themselves can compete. **Table 23-2** compares some characteristics associated with r- and K-selection. Mueller et al. (1991) demonstrated evolution of different competitive abilities and different rates of population growth under these different selective models in experimental *Drosophila* populations. Note, however, that r- and K-selections are not strict alternatives; some populations may *compromise between the two strategies,* **clutch size** in birds being a prime example.

Clutch size is the term used by ornithologists and avian ecologists to describe the number of eggs laid by a nesting pair of birds. Hatching success is the term used for the number of young that hatch from those eggs. (Although these terms are not used when referring to mammals, both numbers of eggs fertilized and numbers of offspring born are important adaptive factors in mammals, including humans.)

David Lack (1947) showed that small clutch sizes in birds may produce too few offspring for the individual to sustain through feeding and parental care. Large clutch sizes

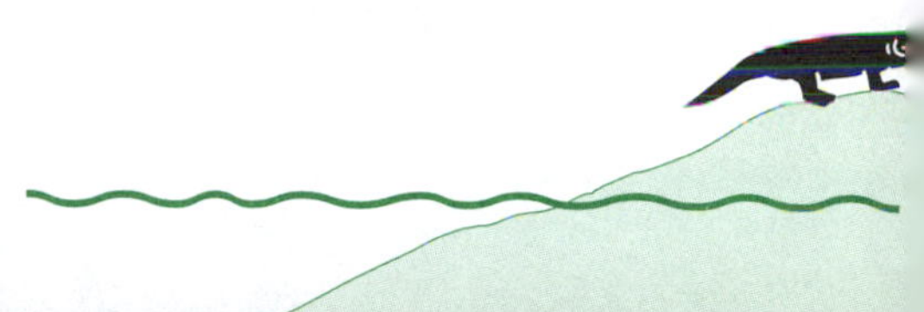

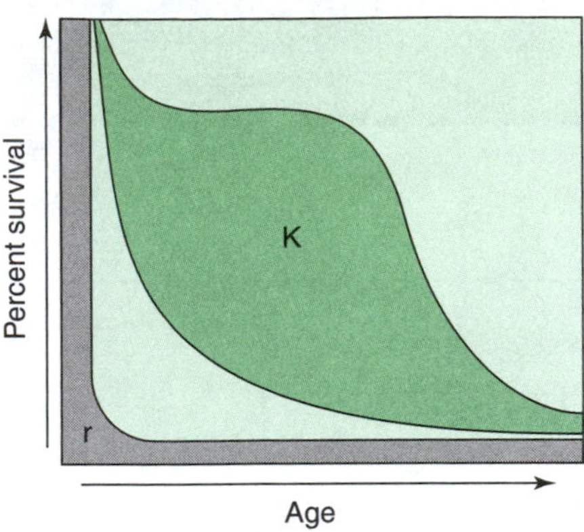

FIGURE 23-3 r- and K-selection strategies are contrasted in this plot of percentage survival against time. With r-selection, most individuals die at an early age. The survival curve for K-strategists falls within the darker green range bounded by slowly and more slowly declining survival with age.

may produce too many offspring for their parents to feed. "Optimal" clutch sizes and hatching success for individuals, like optimal birth weights (see Fig. 22-7), rest somewhere between extremes, dependent on factors such as the availability of food and parental longevity, factors that vary depending on whether reproductive success in one season or reproductive success between species is being compared; what is "optimal" depends on the criteria being used.

In birds, a plentiful food supply in a particular season is positively correlated with large clutch size and high hatching success (**Table 23-3**). Long-lived birds (which are often also large-bodied) tend to lay smaller clutches. Numbers of eggs produced by species of birds in food-rich environments is higher (K-selection) than numbers produced by species in food-poor areas (*r*-selection). Age of the parent incubating the eggs influences the survival of offspring, as recently demonstrated in herring gulls (*Larus argentatus*). Under conditions of food deprivation or high rates of infant or childhood diseases, the ability of a mature human female to produce eight or nine offspring, of which an average of two survive, is of selective value in ensuring that her lineage will persist in the face of high infant mortality.[3]

In Welsh populations of the rough periwinkle *Littorina rudis*, for example, larger adult size is accompanied by the production of increased numbers of smaller offspring in a boulder environment, while periwinkles in crevices are smaller as adults and produce relatively fewer but larger offspring. Thus, although *r*- and *K*-selection offer patterns for understanding life histories in various populations, not all groups fit conveniently into such categorizations. For example, and contrary to expectation, selection of *Drosophila* for faster development (which is not the same as selection for higher population density) leads to correlated evolution

[3] See Gluckman and Hanson (2005) for human birth weight, and Martin et al. (2000), Bogdanova et al. (2007) for the herring gull study, and Blondel et al. (2006) for factors impacting on clutch size and prenatal care in birds.

TABLE 23-2 Characteristics often associated with *r*- and *K*-selection

Characteristic	*r*-selection	*K*-selection
Climate	Variable or unpredictable	Fairly constant or predictable
Diversity of resources and habitats	Usually broad	Relatively narrow
Causes for mortality	Often catastrophic and density independent	Mostly density dependent
Survivorship	High mortality at younger stages, with high survivorship at later stages	Either constant rate of mortality at most stages, or little mortality until a certain stage is reached
Competitive interactions	Variable, mostly weak	Usually strong
Length of life	Relatively short, usually less than one year	Longer, usually more than one year
Selection pressure for	1. Rapid development	1. Slower development
	2. Rapid increase in numbers	2. Greater competitive ability
	3. Early reproduction	3. Delayed reproduction
	4. Small body size	4. Larger body size
	5. Semelparity	5. Iteroparity
	6. Many small offspring	6. Fewer and larger offspring
	7. Increased productivity (quantity)	7. Increased efficiency (quantity)

Source: Adapted from Pianka, E. R., 1988. *Evolutionary Ecology*, 4th ed., Harper & Row, New York.

TABLE 23-3 Characteristics of blue tit (*Cyanistes caeruleus*) populations from one mainland and two deciduous habitats in the Mediterranean

Trait	Mainland	Island 1	Island 2
Population density (pairs/hectare)	1.0	1.28	0.35
Caterpillar abundance (mg/m^2/day)	23	493	–
Body weight (g)			
Males	11.2	9.9	9.4
Females	10.7	9.7	9.4
Clutch size	9.8	8.5	7.2
Number of fledglings	7.5	7.3	5.0
Fledgling weight (g)	10.7	10.4	9.3

Source: Blondel, J., D. W. Thomas, A. Charmantier, P. Perret, et al., 2006. A thirty-year study of phenotypic and genetic variation of blue tits in Mediterranean habitat mosaics. *BioScience*, **56**, 661–673.

of decreased competitive ability, not higher competitive ability as *K*-selection predicts.

In general, the ecological approach briefly reviewed in this section and in a later one ("Group Interaction") emphasizes how populations respond to their environment in terms of their numbers and distribution. In the overwhelming majority of these studies, it is extremely difficult to discover and incorporate genetic information — for example, allele frequency estimates — that would explain how such changes correlate with the evolutionary mechanisms of mutation, selection and genetic drift discussed in Chapters 21 and 22.

Genetic Load and Genetic Death

Through theoretical advances and development of new methods of computation such as bioinformatics, population ecology can now be combined with population genetics to further our understanding of the bases of the structures of, and interactions within and between, populations (e.g., see Roughgarden, 1996).

In contrast to the ecological approach with its emphasis on population distributions numbers, growth rates and life histories, geneticists have placed more emphasis on the amounts and kinds of genetic variability present in populations and on uniting observations on natural populations with models of their genetic structure. Such studies received special impetus from the 1920s and 1930s onward. Most interesting in these studies was the demonstration that large amounts of genetic variability exist in practically all natural populations. As noted earlier (e.g., Figure 10-35 and Table 10-4), considerable polymorphism shows up on chromosomal and gene levels, a variability that allows genetic evolutionary changes to proceed.

However, despite the numerous advantages of genetic variability, many genes maintained in natural population, either in certain combinations or in homozygous condition, may compromise the survivability and fecundity of their carriers. As shown in Figure 22-7, selection can account for a significant loss of nonoptimal individuals even in long-standing populations. Thus, most, if not all, populations are genetically "imperfect" — "non-optimal" if you prefer.

The extent to which a population departs from a perfect genetic constitution is called its **genetic load,** and is marked by the loss of some individuals through **genetic death.** Genetic death is not necessarily an actual death before reproductive age but can be expressed through sterility, inability to find a mate or by any means that reduces reproductive ability relative to the optimum genotype(s). We therefore phrase estimates of these values in terms of the proportion of individuals eliminated by selection. For example, if a gene is deleterious in a homozygous condition, the homozygote frequency before and after selection is as shown in the following table:

	Genotypes		
	AA	*Aa*	*aa*
Frequency at fertilization	p^2	$2pq$	q^2
Relative fitness	1	1	$1 - s$
Frequency after selection	p^2	$2pq$	$q^2 - sq^2$

The loss in frequency of individuals or incurred genetic load equals sq^2, where s is the selection coefficient of *aa*. If N individuals were in the population before selection, genetic load has eliminated sq^2N.

However, this value of sq^2 also equals the mutation rate (u) at equilibrium for $A \rightarrow a$, which means that the genetic

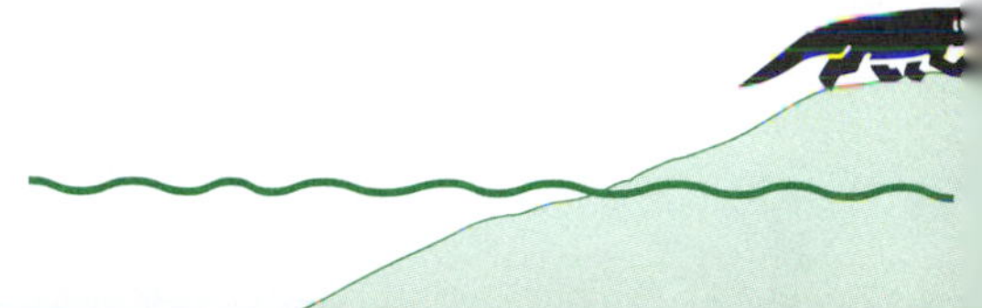

load that a deleterious homozygous recessive causes is equal to its mutation rate (but only in a genetically stable system). Haldane (1939) pointed out an important feature of this relationship; when the mutation rate is constant, it will make little difference to the genetic load whether s is small or large. As you can see in each single column (constant mutation rate) of the hydraulic model in Figure 22-10, if s is small, q will be large at equilibrium, and if s is large, q will be small.

High selection coefficients eliminate the gene more rapidly (low q), while low selection coefficients allow the gene to stay longer in the population (high q). In either case, the genetic load remains $sq^2 = u$ and the total number of genetic deaths remains at sq^2N. Insofar as mutation produces deleterious recessives, any increase in their mutation rate causes a corresponding increase in genetic load and thus in genetic deaths.[4]

According to Crow (1992), any factor that produces differences in fitness among genotypes can create a genetic load. Crow and others devised techniques to evaluate their relative importance. The **mutational load,** just discussed, is only one essential factor responsible for genetic load. Another is the **segregational** or **balanced load,** restricted to those instances in which a heterozygous genotype is superior to both types of homozygotes. For a gene with two alleles, the segregational load amounts to p^2s for the *AA* homozygotes plus q^2t for the *aa* homozygotes, because s and t are the selection coefficients against these homozygotes, respectively, when the heterozygote is overdominant (see Table 22-4). If we substitute the equilibrium frequencies of p and q under those circumstances into $p^2s + q^2t$ we obtain

$$\left(\frac{t}{s+t}\right)^2 s + \left(\frac{s}{s+t}\right)^2 t = \frac{st^2 + ts^2}{(s+t)^2} = \frac{st(s+t)}{(s+t)^2} = \frac{st}{s+t}$$

Thus, if s and t are both about .1, the segregational load will be .01/.2, or .05. This value, which is considerably higher than most mutation rates, demonstrates the increased genetic load that we expect segregation to cause in randomly breeding populations, compared with the load mutation causes ($sq^2 = u$).

Although most or all populations carry genetic loads of one kind or another — including even **recombinational loads** that can break up adaptive combinations of linked genes by crossovers — we have not determined the relative values of each type of load. Crow (1993) pointed out that, "The total deleterious mutation rate remains unknown in any animal except *Drosophila.*" From the viewpoint of evolution, no species ever reaches "genetic perfection" with its absence of genetic death. Environments change with time and the advantages of different genotypes change accordingly. On the one hand, it is conceivable that a population so perfectly adjusted to its environment — with little or no genetic load (no variability) — may become extinct within a short period because of rapid environmental change. A population with a relatively large genetic load may encounter a new environment in which formerly deleterious alleles endow it with adaptations that benefit survival.

Lacking a genetic load may harm a population more than having one — a point made previously in relation to optimal mutation rates. Therefore, although we can measure the genetic load in terms of departure from the optimum genotype, the evolutionary value of a particular optimum genotype may be limited. The optimum genotype may change in time or place, or even differ in the same place if, for example, there is a division of labor between different genotypes, as for example, between males and females.

Cost of Evolution and the Neutralist Position

Whatever type of load a population bears, natural selection adds to the load as some alleles are selected and others not. Because allele replacement is ongoing and pervasive, we can reasonably ask how many individuals a population must lose (in terms of genetic death) to replace a single allele by selection alone. If a dominant mutation arises that has greater selective value than the more frequent recessive, we have seen that the number of genetic deaths will be sq^2N for one generation of selection, where N is the number of individuals in the population.

For complete replacement of a deleterious allele, Haldane calculated that the total number of genetic deaths is determined by a factor D ($D = \ln p_0$) that is based primarily on the initial frequency of the new (p) allele, multiplied by the population number N. For a new but rare dominant allele, D is about 10, so the cost of evolution (DN) for eliminating the deleterious recessive allele is about 10 times the average number of individuals in a single generation. This cost, of course, is spread over many generations, the rapidity of gene replacement depending on the selection coefficient. If the rare new allele is recessive, DN increases to about $100N$, because homozygotes are extremely rare.

When s is large, gene replacement is more rapid than when s is small, but the population now faces the danger of extinction; its numbers may be too few to ensure mating partners of survival in an accident. In the extreme case, when $s = 1$, replacement may occur in a single generation if the allele is dominant, in which case recessive homozygotes sur-

[4] Because of their composition, particular DNA sequences — for example, guanine-cytosine rich sequences that replicate early during DNA synthesis — may mutate at different rates and in different directions from other sequences. As a result, such sequences can bias nucleotide genomic patterns yet still produce a minimal mutational load (Holmquist and Filipski, 1994)

vive. However, replacement would be fairly likely to occur if the new allele is present in low frequency, since few homozygous recessives will be available for survival.

According to Haldane (1939), many selected alleles are intermediate in dominance. He proposed an average death value of $30N$ for the replacement of a single allele and suggested that a population is capable of sacrificing about one-tenth of its reproductive powers for such selective purposes; that is, a gene substitution can occur at a rate $1/10 \times 30N$, or every 300 generations.

However, not all gene substitutions may be caused by selection. Kimura (1979) proposed that the rate of evolution at the molecular level is far more rapid than Haldane suggested. In vertebrates, for example, one can calculate that many hemoglobin protein amino acids are replaced at a rate of approximately one amino acid change per 10^7 years, yet the total amount of DNA is certainly more than necessary to code for 10^7 amino acids. So we would expect each vertebrate species to undergo at least one complete amino acid substitution per year if we assume that overall amino acid substitution rates are about the same for all proteins. If we accept Haldane's value of about $30N$ for the cost of a single gene substitution by selection and estimate an average of about three years per vertebrate generation, each vertebrate population must continually expend an enormous number of genetic deaths to maintain its size and escape extinction. In the present example, the population would have to devote 90 times its number in each generation ($3 \times 30N$) if selection were the primary cause for gene frequency changes.

Because of this presumably high cost of selection, it was proposed that most amino acid changes are neutral in effect. As selection does not act on **neutral mutations,** the fixation of such alleles incurs no genetic load and depends only on their mutation rates and on genetic drift. On an individual basis, the time needed to fix any particular neutral allele will depend on population size. Genetic drift in small populations greatly speeds up the fixation of these alleles, although drift does not change their probability of fixation. Kimura and Ohta (1972) calculated that the average number of generations required to fix a new neutral mutation is approximately four times the number of parents in the population in each generation.

This neutralist hypothesis (also called by some **non-Darwinian evolution** because of its dependence on mutation and not selection),[5] seems supported by the extensive degrees of enzyme and protein polymorphisms, indicating that allele differences persist at many thousands of loci in many species. We could of course claim that allele differences can persist in selectively advantageous heterozygotes without fixation and, because they will never be fixed, we can exclude the high cost of evolution that Haldane has shown necessary for gene substitution. However, maintaining polymorphism by selection may itself entail an enormous and intolerable genetic load.

If only the superiority of heterozygotes maintained polymorphism at a single locus, the segregational load would be $st/(s + t)$, as explained previously. Should the cause for this segregational load depend on the lethality of the two homozygotes (**balanced genetic load** or balanced lethals), the selection coefficients s and t are then both 1, the load is 1/2, and the remaining fitness of the population is $1 - 1/2 = 1/2$.

For two pairs of balanced lethal genes acting independently of each other (e.g., *AA* and *aa* are lethal, as are *BB* and *bb* and only the *AaBb* heterozygotes survive), the fitness of the population reduces to $(1 - 1/2)^2 = 1/4$. In general, no matter what the value of the selection coefficients, the average fitness of a population bearing a balanced or segregational load is $[1 - st/(s + t)]^n$, where n designates the number of gene pairs at which heterozygote superiority is being maintained. Kimura and Crow (1964) calculated that the fitness of such a population approximately equals $e^{-\Sigma L}$ where e (2.718) is the base of natural logarithms and ΣL designates the sum of individual loads for each gene pair involved.

Not surprisingly, this load can be quite large, even if the selection coefficients acting against homozygotes are small, as long as many gene pairs are involved in maintaining superior heterozygotes. For example, if superior heterozygotes are being maintained at 100 loci, each bearing a genetic load of .01 (e.g., $s = .02$, $t = .02$; $st/(s + t) = .0004/.04 = .01$), the average fitness of the population falls to about $e^{-100(.01)} = e^{-1} = 0.37$. For 500 gene pairs acting similarly the fitness is approximately $e^{-1} = .007$, and it is $e^{-10} = .00005$ for 1,000 such gene pairs. Thus, and even with only a small selective advantage for the heterozygote at each gene, to maintain polymorphism at 1,000 loci only one out of 20,000 offspring would survive. For every female to produce two surviving offspring (and thereby allow the population to survive), each such female would have to produce about 40,000 young for this selective purpose alone. Most polymorphic systems therefore must consist of neutral mutations, according to Kimura and his followers, who support this interpretation with observations of high frequencies of selectively neutral mutations in a rapidly mutating strain of *Escherichia coli.*

Among further evidence used by neutralists is the presumed constancy of amino acid substitution rates in particular proteins (the molecular clock; see Chapter 12). Kimura and Ohta (1972), for example, note that the same number of changes in the α-hemoglobin chain occurs, relative to amino acids in β-hemoglobin, whether the α-chain comes from the same species or from a different spe-

[5] Strictly selection does not act on mutations, whether neutral or not, but acts on the phenotypes that result from the incorporation of mutations. In this sense, Kimura's theory can be said not to depend on selection.

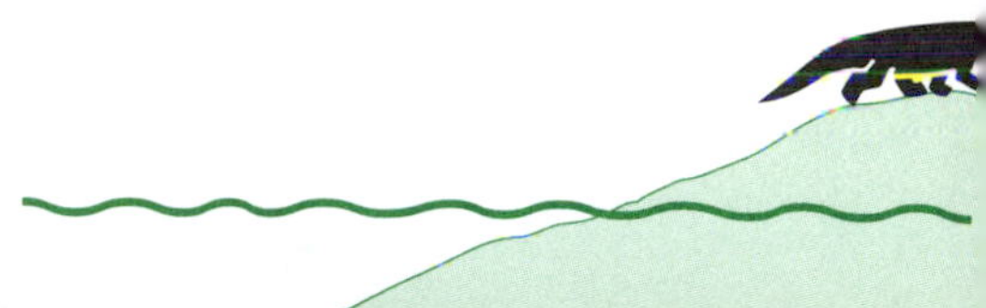

cies. Compared to the human β-chain, the human α-chain shows 75 differences, the horse α-chain 77 and the carp α-chain 77. Kimura and Ohta ask: Why should the α-hemoglobin chains of humans, horses and fish, each with a different selective history, have diverged from the human β-chain at exactly the same rate? According to the neutralists, we can most easily explain this uniformity by a common rate of neutral mutation and drift, rather than common selective conditions. As striking as this evidence is, the regularity of the molecular clock is unclear. There are certainly exceptions to constant evolutionary rates in some proteins, and different rates of nucleotide substitution appear in comparisons among different proteins (**Table 23-4**).[6]

The Selectionist Position

To counter the neutralist view, selectionists proposed various selection schemes to explain the persistence of many polymorphisms that would confer minimal genetic loads. One is **frequency-dependent selection,** which entails genetic loads only when the frequency of a relatively rare selected allele is changing, but produces no genetic load when the allele has reached equilibrium. Because polymorphisms exist at thousands of loci, it is questionable whether there are also thousands of individual frequency-dependent mechanisms in the environment.

It has been suggested that selection in a natural population lumps the effects of many individual genotypes into two main groups, the fit and the unfit. A "threshold" number of polymorphic loci in a population separates these two groups: heterozygotes for more gene loci than the threshold number show no increased heterotic effect, while heterozygotes for fewer alleles than the threshold are presumably all equally deleterious. The threshold thus acts as a form of **truncation selection** to eliminate or truncate an entire class of phenotypes, whatever their genotypes may be.

Although a threshold of this kind may shift, depending on the environmental stresses on the population (B. Wallace, 1970) the genetic loads such populations incur may be relatively small; because of lumping, differences between each of the many genotypes above or below the threshold do not add to the load. The likelihood that genes are not individually replaced in evolution but can be selected in linkage blocks, or as functional groups, supports the view that considerable interaction between genes may produce threshold effects; the possibility of such interactions indicates that we may erroneously calculate theoretical genetic loads if we assume that each locus acts independently of all others. Although the threshold concept is attractive to some population geneticists, we have little direct evidence that thresholds exist.

At present, selectionists emphasize the four aspects described below.

Protein Polymorphisms and Ecological Conditions

Selectionists presented proposals that a strong correlation between particular alleles and particular environmental conditions might indicate that selection is maintaining polymorphisms for protein variations, known as **allozymes.** One well-known example is the relationship between the gene for sickle cell anemia and malaria (see Fig. 22-3), where the heterozygote is superior in fitness to both homozygotes. A

[6] The erratic nature of the molecular clock and other "unsettled" issues of the neutral theory, are reviewed by Takahata (1996).

TABLE 23-4 Comparisons of evolutionary rates for two proteins

		Rate[a]	
Comparisons Between Groups	Time of Divergence in Millions of Years Ago	Superoxide Dismutase	Glycerol-3-Phosphate Dehydrogenase
Drosophila groups	45 ± 10	16.6	0.9
Drosophila subgenera	55 ± 10	16.2	1.1
Drosophila genera	60 ± 10	17.8	2.7
Mammals	70 ± 15	17.2	5.3
Drosophila families	100 ± 20	15.9	4.7
Animal phyla	650 ± 100	5.3	4.2
Multicellular kingdoms	1,100 ± 200	3.3	4.0

[a] Rate is measured as amino acid replacements per year $\times 10^{-10}$, after correction for multiple replacements. Both genes compare the same species in each group.

Source: From Ayala, F. J., 1999. Molecular clock mirages. *BioEssays,* **21,** 71–75.

further example of overdominance is an alcohol dehydrogenase gene in yeast. In such cases, however, maintaining polymorphisms by selection entails a significant expense in the loss of homozygotes.

A system that may have milder selective effects is a polymorphism in a freshwater fish, the desert sucker *Catostomus clarki.* The distribution of two alleles of an esterase enzyme seems to follow a temperature cline along the Colorado River basin. The homozygote for the most frequent allele in more southern (and warmer) latitudes produces an esterase enzyme that becomes more active as temperature increases. The allele that is more frequent in northern (and colder) latitudes forms an enzyme that is more active as temperature decreases. The heterozygote for the two alleles forms an enzyme that is most active at intermediate temperatures.

In *D. melanogaster,* allozymes produced by the alcohol dehydrogenase locus (*Adh*) show correlations between their frequencies and environmental temperatures. The stability of different *Adh* allozymes under high temperatures relates to the fitness they confer on flies by enabling them to survive high alcohol concentrations in the food medium. Selection acting on such allozyme differences can help explain some of the gene frequency changes observed along various north–south geographical gradients.

Although other such correlations between alleles and particular environments appear in plants and animals, in many species no obvious correlation exists. It is hard to prove that such correlations are always necessarily causal and may not be accidental. Neutralists have argued that some enzyme loci being scored for polymorphism may be strongly linked to a gene locus at which selection is operating, and that any protein polymorphism observed is only the effect of such linkage disequilibrium, or **hitchhiking**.

Nonrandom Allele Frequencies in Enzyme Polymorphisms

Geneticists have pointed out a number of enzyme polymorphisms, the advantages for which we do not know, but whose frequencies we find difficult to explain on a purely random basis. An early example of this kind was Prakash and coworkers' (1969) observation that populations of *Drosophila pseudoobscura* ranging from California to Texas show remarkably similar allozyme frequencies for a number of proteins. Although some have suggested that such similarities arise because of migration among different populations rather than through common selective factors, migration cannot explain similarities in gene frequencies among different species.

Genetically isolated species of the *D. willistoni* group share common frequencies for the alleles of many different enzymes. Furthermore, the pattern of similarity between them is not constant, and the data seem to show that different species share common selective factors for some enzymes but not for others. On a broad scale, coordinating genetics and ecology, Nevo (1978) surveyed 35 species from Israel, including insects, mollusks, vertebrates and plants, and concluded that the amount of allozyme polymorphism in a species correlates to factors such as life habit and climate.

Perhaps even more striking is Milkman's finding that *E. coli* clones isolated from the intestinal tracts of animals as diverse as lizards and humans, and from localities as widespread as New Guinea and Iowa, share common allozyme frequencies. For each of five different enzymes, Milkman (1973) found that one particular electrophoretic band was frequent in almost all samples. Because other allozymes of these proteins exist, the finding of such a narrow distribution of allozymes is difficult to explain on any basis other than selection. An alternate explanation would be that mutational and recombinational events in *EcoRI* may have been rare enough that the genotypes of the initial founding populations persisted over long periods of time.

Selection also operates to cause convergent evolution, in which the same functions arise in different taxonomic groups by different evolutionary pathways (see the example illustrated in Figure 3-7). On the molecular level, Yokoyama and Yokoyama (1990) used DNA sequences to demonstrate such convergences in fish and humans, which independently evolved red visual pigments from green visual pigments. Also, highly conserved gene products, such as histone proteins, which are alike in plants and animals, indicate that selection can actively reject any major change in an essential function.

Enzyme Function and Degree of Polymorphism

The function of an enzyme may influence the degree of polymorphism at its locus, enzymes falling into two classes: Those involved in restricted pathways of energy metabolism such as glycolysis, and those that can use a variety of substrates. Enzymes with more restricted substrates show significantly less polymorphism than enzymes with varying substrates. It was further proposed that enzymes involved in regulating metabolic pathways are generally more polymorphic than enzymes that are not primarily regulatory. Although we do not know the biochemical causes that sustain these differences in polymorphism, they are not random, indicating that many of the polymorphic alleles are not neutral.

Polymorphisms for DNA Coding Sequences

In general, nucleotide sequence analyses show that polymorphisms are significantly greater in those DNA sequences that do not determine amino acid sequences than in those DNA sequences that transcribe and translate into amino acids.

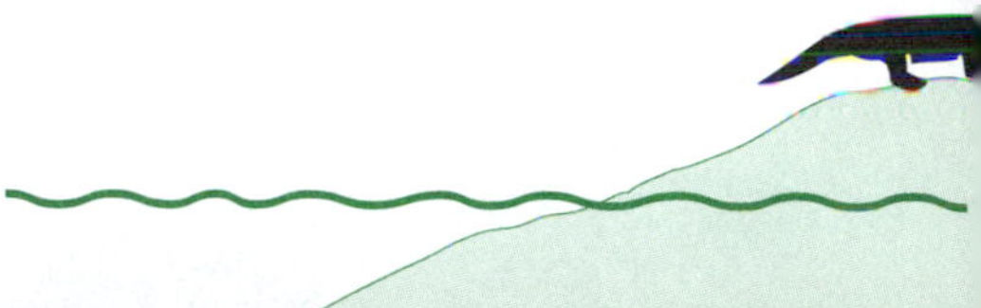

This finding suggests that selection reduces the variability in amino acid coding regions because such sequences have a greater effect on the phenotype than do noncoding regions. Or, put differently, those random "neutral" mutations mostly responsible for variability in noncoding regions, do not determine the variability that remains in amino acid coding regions. Thus, we would expect selection to operate more strongly on the first two amino acid codon positions; these are more involved in amino acid determination than is the "wobbly" third-codon position, which is essentially responsible for degeneracy of the genetic code (see Chapter 8).

Similarly, we would expect more polymorphism in introns (intervening sequences, see Chapter 9), which do not code for amino acids, than in exons (expressed sequences), which do. Both these expectations are fulfilled for the *D. melanogaster* alcohol dehydrogenase gene, which has six percent polymorphism among introns, seven percent polymorphism among third-codon positions but almost no polymorphism in exons or in the first two codon positions. Because random events can hardly explain such pronounced differences in polymorphism, these findings strongly suggest that selection must be the discriminating agent that determines which nucleotide base substitutions will become established in functionally different DNA sequences.

We can also see the active role of selection when we compare *duplicate genes* in the process of becoming functionally different. When comparing amino acid coding nucleotide sequences between these duplicates, we expect greater differences for their functionally strategic sections than for coding sequences for sections that are not becoming functionally differentiated. When Ngai and coworkers (1993) studied duplicate genes involved in producing olfactory receptor proteins in catfish, they found greater differences among these duplicate genes for a particular amino acid coding region of the gene than for other regions, indicating active selection for functional divergence rather than neutral mutation.

Functional divergence also accounts for mutations in sex-related genes — sex determination, mating behavior, spermatogenesis, egg fertilization — that occur more frequently in amino acid determining codons than in synonymous codons, which do not cause amino acid substitutions (see Chapter 12). Such findings in a wide variety of organisms, including mollusks, arthropods and mammals (e.g., see Civetta and Singh, 1998a), indicate that changes in sex-gene amino acid sequences were actively selected as species diverged.

A further test, also derived from DNA sequencing, is based on the concept that, according to the neutral theory, we expect most mutations in a specific gene to be neutral in whatever species they occur; that is, if we make interspecific and intraspecific comparisons for a particular locus, the degree of nucleotide divergence leading to amino acid differences between two related species should correlate with presumably neutral variation in other traits, such as the degree of nucleotide divergence (polymorphism) within each species. When Eanes and coworkers (1993) tested this neutral mutation model by comparing sequences in 44 copies of the glucose-6-phosphate dehydrogenase (*G6pd*) gene derived from two related *Drosophila* species, *D. melanogaster* and *D. simulans,* they found much more interspecific amino acid divergence (21 replacement differences) than we would expect from the degree of polymorphism within species and from the degree of interspecific synonymous mutational differences that do not cause amino acid changes. Their greater frequency indicates that these amino acid changes result from much more stringent selection than we would expect if they had experienced little selection; that is, if they had been neutral in effect.

In *summary,* selective forces unquestionably affect polymorphism and help maintain it under some circumstances. However, we do not know the number of genes on which selection operates at any one time, the linkage relationship between these genes, the kinds of selection that operate or the size of the selection coefficients (Hey, 1999). We cannot exclude neutral mutation and genetic drift as important causes for some, or even many, polymorphisms. Some genetic variants may be neutral at certain times or under certain conditions but have selective value when the environment or genetic background changes.

Some Genetic Attributes of Populations

As shown in Table 21-1, populations have unique evolutionary characteristics. Many genetic factors that affect the evolution of populations may act differently from our expectations if we consider individuals alone. Among these factors, we discuss sex, sexual reproduction, mutation and linkage.

Sex and Sexual Reproduction

We are dealing with two separate topics in this section, although both are intimately associated. One is the **evolution and possession of two sexes** (genders) in a population, male and female. The other is the **sexual reproduction** that accompanies the evolution of male and female genders. Sex may be of little value or even a disadvantage to an individual but can be a distinct advantage to a population.

Although the relationship between the *evolution of sex* and the *evolution of recombination* has generated considerable interest, we remain unsure of how or whether the two are connected. Despite close to two-dozen proposals, why sexual reproduction and recombination persist in natural populations remains a mystery. We expect natural selection to preserve associations between alleles that benefit the fitness of an individual. Sexual reproduction creates new association; recombination breaks up those alleles. Furthermore, the rate of reproduction would increase if females reproduced parthenogenetically; such females would leave twice as many genes to the next generation as females reproducing sexually. So, why sex?

With respect to *evolutionary origins,* one proposal is that sex (here considered as meiosis) originated as a way of overcoming DNA damage by using recombinational DNA repair mechanisms. The genetic variation that resulted was only an accidental by-product. However, according to Maynard Smith (1988a), even if recombination had a DNA repair origin, one can still argue that the genetic variability "by-product" became the primary sexual mainstay. One can also contend that DNA repair mechanisms are not essential for sexual reproduction because they are also quite efficient in asexual organisms including viruses, bacteria and even long-standing asexual eukaryotes such as bdelloid rotifers.

Of the many hypotheses to explain the persistence of sexual reproduction and (as a consequence) recombination, removing deleterious mutations is perhaps the most reasonable. In the cladoceran (water flea) *Daphnia pulex,* which can modulate its mode of reproduction from sexual to asexual, more mutations accumulate with asexual than with sexual reproduction (Paland and Lynch, 2006). It also has been proposed that sex is advantageous to organisms that exist in variable environments.[7]

Females who reproduce parthenogenetically accomplish the work of the usual two sexes more efficiently, because such females spare the expense of producing a superfluous male sex. Sexual reproduction is also often costly to individuals in terms of the risks associated with finding the opposite sex, fertilizing gametes in exposed circumstances and the extra resources expended during intrasexual competition. Predation and parasitic infection are, for many species, common perils of courtship and copulation. Furthermore, the recombination events that accompany sexual reproduction can rearrange and reassort genes — the previous combinations of which had already achieved high fitness — thereby lowering individual fitness.

Various population geneticists point to populations rather than individuals in seeking a cause for sexual reproduction. Among the most popular proposals, sexual crossing in a population allows single individuals to incorporate different beneficial mutations from other members through mating and recombination. Without sex, combinations of such beneficial mutations are more difficult to achieve. As shown in **Figure 23-4,** potentially advantageous mutations such as A^X, B^X and C^X may arise in an asexual population but remain in separate individual lineages. Only if the asexual population becomes extremely large (which may take considerable time) is there a reasonable chance that a second such mutation will occur in a lineage that contains a previous advantageous mutation. By contrast, by mating between the separate lineages that carry the mutations, individuals within a sexual population may rapidly incorporate mutations soon after they occur.

Furthermore, just as mating and recombination in sexual populations enable relatively rapid combinations of newly arisen mutations, they also enable relatively rapid combinations of already existing mutations. In both cases, the advantages achieved include additional variability allowing a population to persist in a changing environment. For example, parasites usually evolve rapidly to counter resistance in their hosts and hosts usually evolve rapidly to counter infectivity of their parasites. This "arms race" places a premium on rapid evolution through sexual recombination. Asexual populations, in contrast, must depend mostly on the variability they already have — a static condition that can limit their opportunities for rapid change. It is no surprise that experimental competition between sexual and asexual strains of yeast can evolve in favor of the former.[8]

In addition to an increased rate of adaptation, a further advantage of sexual populations, as Muller pointed out, is their relative ease in eliminating deleterious mutations via recombination rather than waiting for chance back mutations to occur. This position is based on the concept that at equilibrium values of mutation–selection, individuals will be carrying varying numbers of deleterious mutations (*del*) at many loci (e.g., $A^{del}B^+C^+D^+ \ldots$, $A^+B^{del}C^+D^+ \ldots$, $A^{del}B^+C^{del}D^+ \ldots$, $A^{del}B^{del}C^+D^{del} \ldots$) but the class of individuals carrying a total of zero deleterious mutations (e.g., $A^+B^+C^+D^+ \ldots$) will be quite small. If sampling error (genetic drift) should lose this zero class in an asexual population, only back mutation from a class carrying a single deleterious mutation could reconstitute it (e.g., $A^{del}B^+C^+D^+ \ldots \times A^+B^+C^+D^+ \ldots$; and $A^+B^{del}C^+D^+ \ldots \times A^+B^+C^+D^+ \ldots$).

Similarly, if the single-mutation class is lost through drift, its reconstitution depends on back mutation from a class carrying two deleterious mutations. Because of drift, therefore, classes with increasing numbers of deleterious mutations tend to replace those with fewer such numbers, a phenomenon called **Muller's ratchet.** As time goes on, it will become more difficult to eliminate deleterious mutations in asexual populations, and their genetic load will increase (for a confirming experiment, see Andersson and

[7] Various contributors in the collections edited by Stearns (1989), Michod and Levin (1988), and West-Eberhard (2003) discuss other aspects of this topic, including the proposal that parasitic elements, such as transposons and plasmids, initiated or promoted sexual fusion as a mechanism to infect other cells.

[8] See Maynard Smith (1978) and Hamilton et al. (1990) for the arms race and existing variability associated with asexual reproduction, and see Birdsell and Wills (1996) for competition studies.

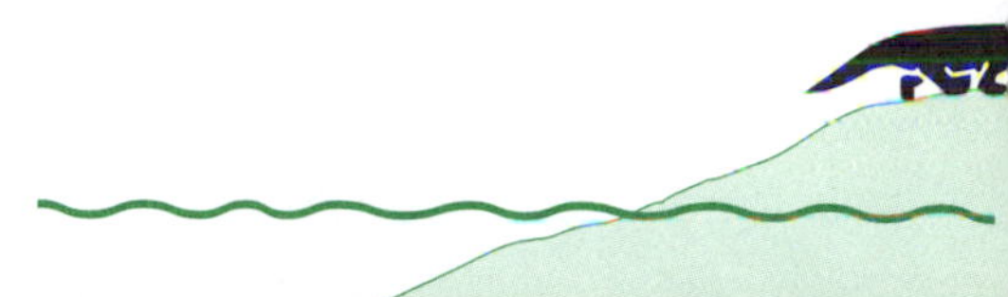

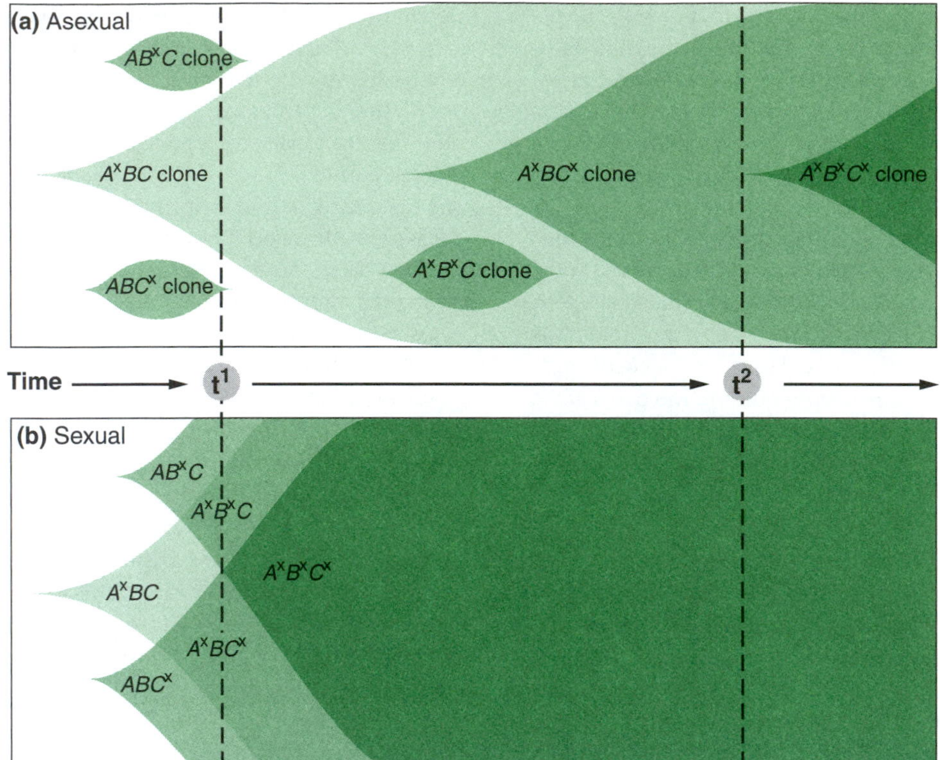

FIGURE 23-4 Muller's model of the difference between asexual and sexual populations in the speed at which they incorporate combinations of advantageous mutant alleles and lose deleterious alleles. In this illustration, we assume both kinds of populations begin with an initial state where three loci are fixed for the *A*, *B* and *C* alleles. We call the advantageous alleles that arise at these loci A^x, B^x and C^x, with individuals carrying A^x being more fit than those carrying B^x or C^x. Among the other combinations, we presume A^xBC^x to be more fit than A^xB^xC or AB^xC^x, while the individuals carrying all three mutant alleles, $A^xB^xC^x$, are the most fit of all. (a) In asexual clonal populations, each clone is independent of the others, so attaining the most fit genotype must await successive beneficial mutational events in a single clone ($A^xBC \rightarrow A^xBC^x \rightarrow A^xB^xC^x$, time t^2), while the least fit clones become extinct. (b) In large sexual populations, beneficial mutations need not occur successively in a single clone, and the $A^xB^xC^x$ genotype can be achieved relatively rapidly (*time* t^1) through recombination among individuals carrying the various advantageous alleles without the need of further mutational events. Also, as Peck pointed out, when deleterious mutations are common, beneficial mutations have a better chance of becoming established in sexual than in asexual populations because of the asexual requirement that beneficial mutations can only persist in clones that are already positively selected, whereas such rigid clonal selection does not govern sexual populations. (Adapted from Muller, H. J., 1932. Some genetic aspects of sex. *Amer. Nat.*, **66**, 118–138.)

Hughes, 1996). On the other hand, Muller's ratchet will not operate in sexual populations because the zero mutation class can easily be reconstituted through recombination (e.g., $A^{del}B^+C^+D^+ \ldots \times A^+B^{del}C^+D^+ \ldots \rightarrow A^+B^+C^+D^+ \ldots$).

From such an effect on fitness, we would expect recombination to be as subject to selection as any other character influencing reproductive success. Experiments with *Drosophila* show that high and low crossover rates can be selected, supporting this view and demonstrating a range of recombinational variability. Observed genomic recombination rates probably evolved as compromises between two contrasting effects:

- A beneficial effect that assembles specific alleles of different loci adapted for specific environments (Fig. 23-4), also disrupting disadvantageous gene combinations by reducing linkage between them (**linkage disequilibrium;** see Fig. 21-6).
- A deleterious effect that can break up advantageous combinations, whether they are composed of genes that interact with each other (**epistasis**) or of genes that act independently. Only mechanisms that inhibit recombination in specific areas, such as inversion systems in *Drosophila* (see Fig. 10-35), can prevent such loss. In this sense, lack of recombination provides an advantage to asexual populations that can persist long enough to accumulate strings of potentially advantageous mutations. (Note, however, that lack of recombination also enables an increase in linkage disequilibrium of deleterious alleles and will cause a loss of such asexual clones.)

Mutation

A further important attribute of populations is **mutation.** One might expect that because mutation is mostly random, as many new adaptive mutations will arise as deleterious ones. It would, therefore, appear as though evolution merely awaits the occurrence of superior adaptive individuals before it proceeds. Although such views have been expressed (W. H. Li, 1997), it is likely that the role of new mutations is not often immediately significant.

A population that has long been established in a particular environment will have many genes adapted for prevailing conditions. New mutations that arise, if not neutral in effect, will rarely be better, and likely will be worse, than the genes already present — a consequence not much different from the serious damage we can expect when a random change is introduced into any intricately organized and integrated system, such as computer wiring. Complex organisms are developmentally constrained by their evolutionary history, so that advantageous mutant effects are generally confined to few of the many intricate and sensitive developmental processes. For phylogenetically crucial gene products, conserved molecular sequences are commonplace, and viable changes occur only rarely. For example, although plants and animals are separated by more than one billion years of evolution, there are only two differences between them in the 100 amino acids of histone 4, a eukaryotic protein that binds and folds DNA.

In contrast, a change in environmental conditions may have a more important evolutionary effect; many alleles formerly in low frequency may suddenly have high relative fitness. We see this in:

- rapid genetic changes in many insect populations exposed to pesticides such as dieldrin and DDT (**Fig. 23-5**), where resistant alleles appear on all major chromosomes (see Fig. 10-37);
- large increases of melanic allele frequencies in populations of the peppered moth *Biston betularia* in industrialized regions;
- increased frequencies of resistant genes in some plant populations exposed to herbicides and metallic toxins; and in
- genes that modify red blood cell physiology offering protection against malaria in the human populations shown in Figure 22-3.

In all cases, many individuals die, leaving those (and it may be those few) individuals with the alleles for resistance, melanism or protection, to pass their alleles to the next generation. In essence, these individuals become the founders of the next generation.

Nevertheless, new mutations, such as those for new enzymatic functions derived from duplications, can occasionally be adaptive and lead to divergence. Whether old or new, mutations supply the variation upon which selection acts, and which selection incorporates into evolutionary change.

Linkage

Selection among individual mutant genes on separate chromosomes, however, is not the only method by which genetic progress is achieved.

Recombination between linked alleles may markedly affect the response to selection. To illustrate this point, assume that each of four loci (*Aa, Bb, Cc, Dd*) influence a character quantitatively, and that all capital letter alleles have a positive ("plus") effect on the character and all small letter alleles have a negative ("minus") effect. If the phenotypic optimum is an intermediate one, as it is for many characters, the genotype would benefit if it were also intermediate: *AaBbCcDd.*

One way of achieving such optimum genotypes is for tight linkage between these four loci, in the fashion of *AbCd* on one homologue and *aBcD* on the other. Thus, any combination of chromosomes will always have four plus genes and four minus genes, yet the population retains the variability of all the different alleles. Note, however, that such a linkage group must have three crossovers to form chromosomes containing all plus or all minus genes. If selection changes from an intermediate phenotype to an extreme phenotype (either all plus or all minus), some time may elapse before the appropriate crossovers can furnish the most adaptive combinations. Recombination rates may dictate the progress of selection by connecting as well as disrupting gene combinations affecting fitness.

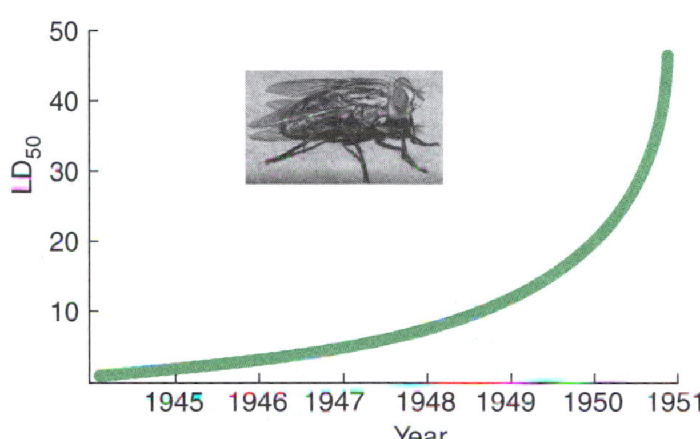

FIGURE 23-5 Resistance to DDT in houseflies, collected from Illinois farms, measured in terms of the lethal dose necessary to kill 50 percent of the flies (LD_{50}) against time since the introduction of DDT. (Adapted from Strickberger, M. W., 1985. *Genetics,* 3d ed. Macmillan, New York; data are from Decker, G. E., and W. N. Bruce, 1952. House fly resistance to chemicals. *Amer. J. Trop. Med. Hygiene,* **1,** 395–403.)

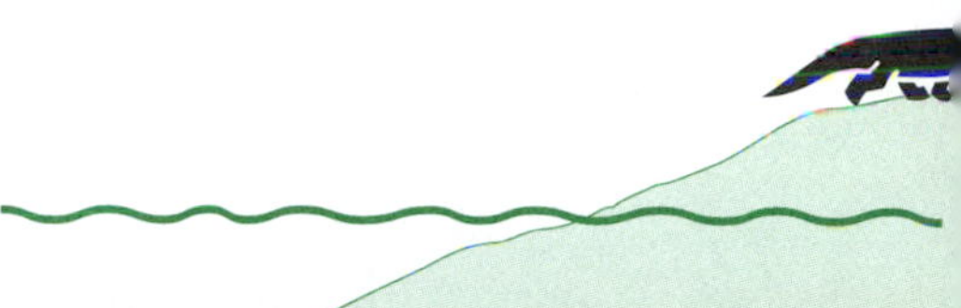

The Adaptive Landscape

In general, because more than one locus affects fitness in a population, increased fitness can evolve in many ways.

For simplicity, we can consider a population containing only homozygous genotypes, where the same four loci just mentioned affect a character, so that equal numbers of capital letter and small letter alleles determine the optimum phenotype. A variety of six optimal genotypes is then possible, *AABBccdd, AAbbCCdd, aaBBccDD,* and so forth. According to Wright (1963, and earlier publications), we may consider each of these genotypes to occupy an **adaptive peak** *in the* **adaptive landscape,** a position of high fitness associated with a specific environment. As long as no other factors change the fitness of these genotypes, each of these six peaks is of equal height, and a population consisting entirely of any one of these genotypes would achieve maximum fitness for this phenotype.

When fitness involves more than four loci with more than two alleles at any locus, the number of adaptive peaks increases astronomically. A locus with only four alleles has 10 diploid gene combinations; 100 loci with four alleles each have a total of 10^{100} gene combinations. Even limited to this relatively small number of loci, the number of possible combinations far exceeds the number of individuals in any species and even the estimated number of protons and neutrons in the universe (2.4×10^{70}). Thus, even if only a small portion of these gene combinations is adaptive, there are more possible adaptive peaks than a species can occupy at any one time.

Populations and Individual Genotypes

Again, however, we must differentiate between populations and individuals. A potentially high adaptive peak for a population need not coincide with a high selective peak for a genotype within the population. This discrepancy arises because the selective values of genotypes are based on competition with other genotypes but may not indicate their effect on the population. Haldane, Wright and others pointed out that **altruists** (see Chapter 24) who sacrifice themselves for the benefit of shared genotypes may have low selective value as individuals, although a population bearing such altruistic genotypes may have higher reproductive values than one without them. Sterile castes among insects that forgo reproduction represent an example of altruism of this type. Situations in which the altruist is not disadvantaged also have been proposed. When alarmed, many species of antelope "bounce" up and down with the joints of the legs locked and the legs in an extended posture. This behavior, called stotting or pronking, warns other individuals of present danger, but, it has been argued, also tells the predator that the individual 'pronker' is aware of his/her presence, and so may not be an easy target for predation.

Conversely, those organisms that are parasitic on social host species such as ants or termites (social parasites), and that increase their frequency at the expense of other genotypes in a population that may have high individual selective value, although they depress the reproductive fitness of the population as a whole. Illustrations of the latter type are alleles that modify segregation ratios so that the gametes produced by heterozygotes carrying these alleles consist mostly of such distorters rather than normal nondistorter alleles; see **segregation distorters** and **meiotic drive** in Chapter 9. The frequency of such segregation distorters tends to increase in a population even though some distorters are associated with deleterious or even lethal phenotypic effects such as *tailless* alleles in mice (Silver, 1993). In contrast to common expectations that selection for adaptation increases survival, meiotic drive that increases a deleterious genotype's reproductive success can lead to its extinction. Such extremely "selfish" genes and devices can only exist as transient rarities.

Adaptive Landscapes and Shifting Balance

For simplicity, assume that the present example of four gene pairs does not involve the complications just outlined. Even so, the concept of many adaptive peaks with uniform height departs from real conditions. It is likely that the effects of each of these four pairs of genes differs considerably, and we can assign relative fitness (adaptive) values to the effects of the A and B genes on the basis that the greater the number of capital letter genes in these two pairs, the greater the fitness. When we combine them with the previous relative fitness values, we can construct an *adaptive landscape* of peaks, as in **Figure 23-6**, showing the adaptive heights of different possible genotypes.

Note that now one peak superior to all others (*AABBccdd*) arises in a landscape of intermediate peaks, each surrounded by relatively inferior genotypes. A population may increase in fitness during evolution but nevertheless reach an intermediate peak that is not necessarily the most adaptive. To move from peak to peak until a population finds the highest one demands that the population travel through inferior genotypes that occupy the lesser **adaptive valleys** of this adaptive landscape. Arrows in Figure 23–6 indicate such reductions in fitness, showing the general route a population at *aabbCCDD* might take to reach the highest peak at *AABBccdd.*

At *least two nonadaptive stages* appear in this illustration, at which the population will suffer. If the landscape shown in Figure 23-6 remains stable, arrows show one possible path in the progress of a population from a lower adaptive peak at *aabbCCDD* to the highest adaptive peak at

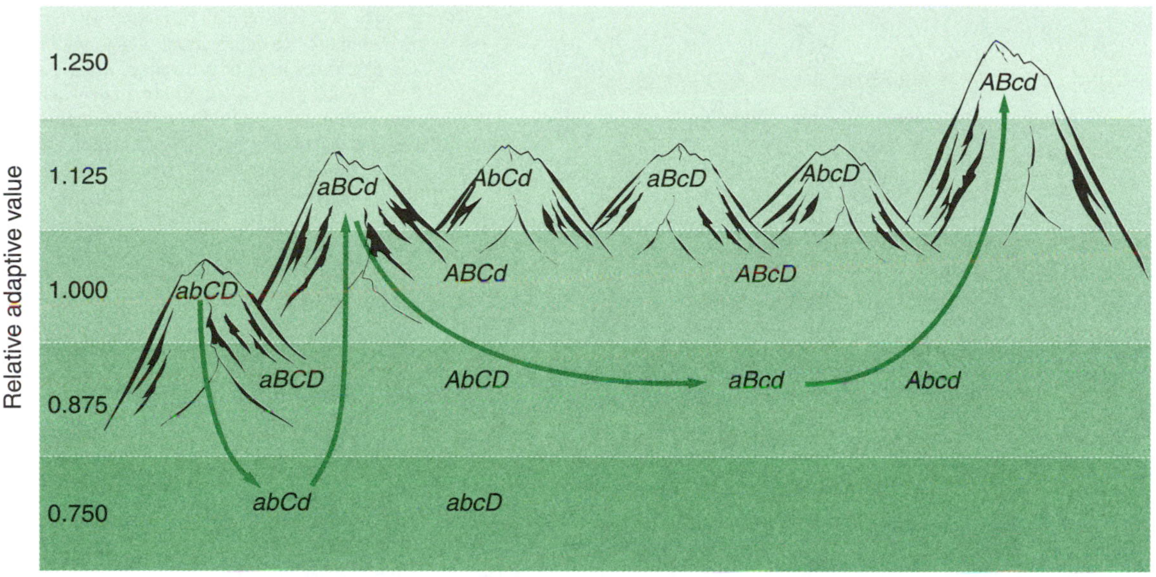

FIGURE 23-6 Adaptive landscape for homozygous genotypes at four loci in which six genotypes homozygous for capital letter alleles at two loci (*AAbbccDD, AAbbCCdd, aaBBccDD*, and so on) attain relatively high relative fitness (adaptive) values (peaks). Further differences among them are caused by the fitness of different alleles at the Aa and Bb loci so that genotypes bearing more capital letter alleles at these loci have higher relative fitness values than those that do not (e.g., *aaBBCCdd, aabbCCDD, AABBccdd, AAbbccDD*). (Adapted from Wright, S., 1963. Genic interaction. In *Methodology in Mammalian Genetics*, W. J. Burdette (ed.). Holden-Day, San Francisco, pp. 159–192.)

AABBccdd. Movement of a population from peak to peak depends on various factors, especially on the size of the population; small populations are more subject to genetic drift, so their frequencies vary more easily than in large populations. If selection is weak, and a population is not held firmly to a particular peak, movement across a "valley" becomes easier.

We also should recognize that peaks can be transitory, making the adaptive landscape change like a rubber sheet with adaptive "bumps" arising at different places at different times, reminiscent of the "epigenetic developmental landscape" C. H. Waddington (1957) offered in his classic book, *The Strategy of the Genes* (**Fig. 23-7**). Thus, as the environment changes, the relative fitness of genotypes change, and as new alleles are introduced through mutation or migration, interaction with other alleles (epistasis) changes their relative fitness. What is a "valley" at one time need not be a valley at another, and a population's position on the landscape can fluctuate accordingly.

Once such an adaptive landscape has evolved, further evolution will depend on the origin of a new selective environment and the creation of new adaptive peaks. However, if conditions are not changing rapidly, the same set of adaptive peaks may remain for long periods of time. A population on one adaptive peak can no longer reach a higher peak without going through a nonadaptive valley. Because constant selection can hardly occur for nonadaptation, it seems reasonable to ask: How can a population located on a relatively low adaptive peak evolve so that it occupies the highest or near-highest peaks on the adaptive landscape? To answer this problem, Wright proposed that many populations break into small groups of subpopulations. These local populations (known as **demes**) are small enough to differ genetically through the nonselective process of genetic drift but are not so widely separate as to completely prevent gene exchange and the introduction of new genetic variability. The adaptive landscape is therefore occupied by a network of demes, some at higher peaks than others. Selection takes place not only between genotypes competing within demes but also between demes competing within a general environment. Wright called the kaleidoscopic pattern of evolutionary forces acting on these demes the **shifting balance process**, shown in its simplest form in **Box 23-1.**

By contrast, Fisher (1958) suggested that most populations are large and fairly homogeneous, and that selection tests each new allele independently in competition with all other alleles in the population. This large population primarily increases its fitness by small, incremental ("additive") selective steps rather than major genetic drift.

According to Wright's scheme (**Table 23-5**), subdivided populations have many evolutionary advantages over a

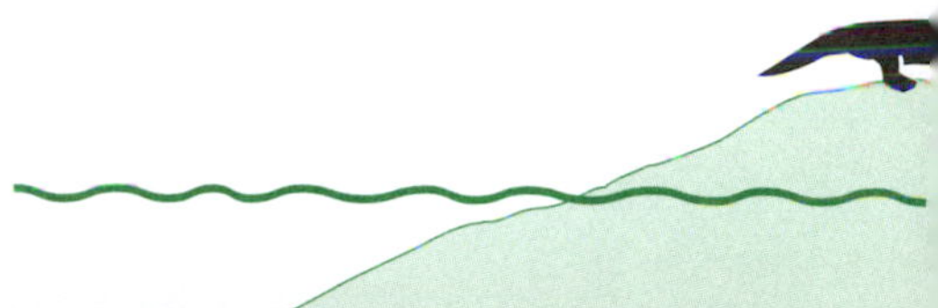

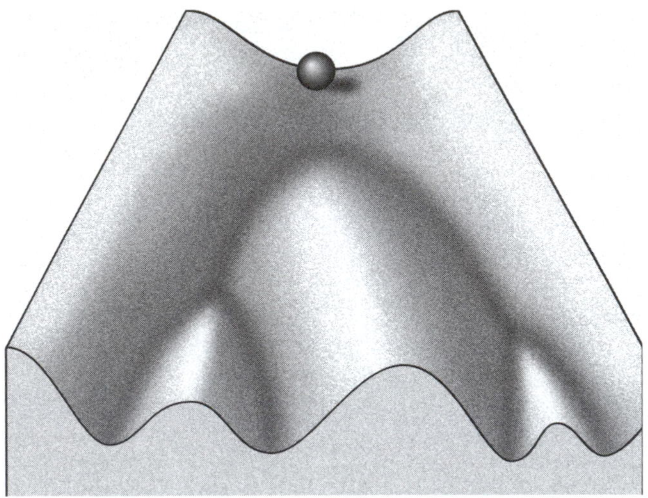

FIGURE 23-7 The epigenetic landscape, the analogy used by Waddington to depict a cell or embryo in development as a ball at the head of two valleys which each bifurcate. Once the cell or embryo has entered one valley the steepness of the valley wall (selection and the dynamics of development) keeps the cell or embryo in the valley. Once shifted into another branch, the cell or embryo stabilizes along the new trajectory.

single, large, homogeneous population, a pattern that helps explain how evolution occurs in most sexually interbreeding species. Selection, acting on the products of gene interaction, causes the close fit between organismal adaptation and environment, and adaptational change is influenced by population substructure in which forces such as genetic drift operate.

Investigation into the views of Fisher and Wright continues. Coyne and coworkers (1997) present strong arguments against Wright's shifting balance theory, pointing to

BOX 23-1 The Shifting Balance Theory Proposed by Sewall Wright

The following seven statements provide an outline of the shifting balance theory of population proposed by Sewall Wright in 1931.[a]

1. *Genetic drift,* acting on polymorphism and heterozygosity at various loci, allows a number of demes to change their allele frequencies and move across nonadaptive valleys to different parts of the adaptive landscape by developing new fitness values. (Wright maintained that gene interaction — epistasis — was a major component of variation in fitness; "Genes favorable in one combination, are, for example, extremely likely to be unfavorable in another.")
2. *Selection* pushes some of these demes up the nearest available adaptive peak by changing allele frequencies even further, that is, by making some loci homozygous or nearly so.
3. *Polymorphism* retained at other loci, or variability introduced through migration and mutation, provides further opportunity for genetic drift to trigger movement across the adaptive landscape, eventually enabling a population to occupy still higher adaptive peaks.
4. Because *population subdivision* impedes genetic recombination between demes, epistatic gene combinations producing novel advantageous interactions can persist (see review by Fenster et al., 1997). When we add the influence of genetic drift, demes can respond uniquely to selective pressure and evolve in distinctive and diverse directions.
5. A deme that has attained a high adaptive peak tends to *displace other demes* at lower peaks by expanding in size or dispersing outward and changing the genetic structure of other demes through migration.
6. *Environmental change* such as a stream of seismic earthquakes, can act on populations, continually producing new adaptive landscapes surfaced with new slopes and adaptive peaks. Channeling selection in new directions encourages populations to continually shift their genetic structures. As P. R. and B. R. Grant point out in a study of selective changes in Darwin's finches on the Galapagos Islands, "The population tracks a moving peak in an adaptive landscape under environmental fluctuations, and there is more than one individual fitness optimum within the range of phenotypes in the population."
7. Because selection, time, and genetic accident are needed to achieve the most optimum genotypes, while genetic loads and environmental contingencies can oppose such optimal achievement, *highest adaptive peaks are potential and not necessarily realized.* Populations can, at best, trail behind the summits of oncoming adaptive landscapes.

[a] Much of Sewall Wright's long career was spent providing evidence, explaining and defending his theory. See Coyne et al. (1997) for a review of the theory and its impact on evolution, genetics and animal breeding.

TABLE 23-5 Comparison of evolutionary processes in a single homogeneous population and in a subdivided population

	Homogeneous Population	Population Subdivided into Demes
What is selected	A gene, differing from its alleles in net selective value	Different allele frequency
Source of variation	Gene mutation	Genetic drift between demes and selection toward new adaptive peaks
Process of selection	Selection among individuals	Selection among demes
Evolution under static conditions	Progress restricted to a single peak	Continued shifts as new adaptive peaks are encountered
Evolution under conditions	Progress up nearest adaptive peak	Selection between *changing* different demes for occupancy of all available peaks

Source: From *Methodology in Mammalian Genetics* by Wright and Burdette, editors. Reprinted by permission of Holden-Day, Inc.

its dependence on unsupported assumptions of genetic drift, selection and gene flow. In contrast, Wade and Goodnight (1998) present examples of deme structure and intergroup selection that supports Wright's theory. Until further evidence appears, perhaps both Fisherian and Wrightian populations should be assumed to exist, large and homogeneous during one period and subdivided during another. Also, selection may have multiple objects, acting on genes with individual phenotypic effects and on genes with epistatic effect. Perhaps, as Wright (1988) stated, different aspects of populations demand different theoretical approaches:

> It is to be noted that the mathematical theories developed by Kimura, Fisher, Haldane, and myself dealt with four very different situations. Kimura's "neutral" theory dealt with the exceedingly slow accumulations of neutral biochemical changes from accidents of sampling in the species as a whole. Fisher's "fundamental theorem of natural selection" was concerned with the total combined effects of alleles at multiple loci under the assumption of panmixia in the species as a whole. He recognized that it was an exceedingly slow process. Haldane gave the most exhaustive mathematical treatment of the case in which the effects of a pair of alleles are independent of the rest of the genome. He included the important case of "altruistic" genes, ones contributing to the fitness of the group at the expense of the individual. I attempted to account for occasional exceedingly rapid evolution on the basis of intergroup selection (differential diffusion) among small local populations that have differentiated at random, mainly by accidents of sampling (i.e., by local inbreeding), exceptions to the panmixia postulated by Fisher.
>
> Wright concludes, "All four are valid."

Group Selection

One important consequence of Wright's shifting balance theory was to emphasize differences in survival or extinction among populations rather than only among individuals.

Selection among individuals in a population is a conservative force that pushes the population up a single adaptive peak. Selection among populations (accompanied by genetic drift) leads to occupation of higher adaptive peaks and replacement, extinction or colonization of populations at lower adaptive peaks (**Table 23-6**).

To some extent, we have already considered selection among groups in discussing the advantages of sexual reproduction. Because sexual reproduction may involve hazards as well as significant expenditure of resources, individuals often incur considerable disadvantages. Benefits in the evolution of sexual reproduction therefore seem most likely related to variability conferred on the sexual population as a whole. The truism of Table 21-1 repeats itself. "Populations evolve, not individuals."

Another factor causing differences among populations that can lead to their differential reproductive success is *mutation rate.* As outlined in Table 12-3, mutation rates that change nucleotide replication fidelity also change prospects for adaptation; high rates generate continual errors and break down adaptations; absent or low rates inhibit or pre-

TABLE 23-6 A comparison between individual and species selection

Process	Microevolution	Macroevolution
Unit of selection	Individual	Species
Source of variation	Mutation, genetic recombination	Speciation
Type of selection	Natural	Species
Character selected for	Survival against death	Survival against extinction
Evolutionary consequence	Rate of reproduction	Rate of speciation

Source: Modified from Stanley, S. M., 1979. *Macroevolution: Process and Product.* W. H. Freeman, San Francisco.

vent adaptation to new environments. Achievement of optimal mutation rates may be a matter of selecting among different lineages/populations for replication fidelity of their replicases and polymerases.

Kin Selection

A further example where population selection seems to operate on a level different from individual selection is **kin selection** used by population biologists to explain social behavior in some lineages of hymenopteran insects — ants, bees and wasps. As noted in Box 16-4 (Kin Selection and Haplodiploidy in Social Insects), these insects have haploid males and diploid females ("haplodiploidy") so that all females (sisters) derived from a single pair of parents are more closely related than are mothers to their own daughters. The genotype of their group (or kin) is therefore benefited by female workers who sacrifice their own reproductive ability and rely instead on their mother's reproductive ability by helping raise sisters rather than producing their own daughters. Advantages of kin selection are also seen in the alarm calls of vervet monkeys (see Chapter 20) and Belding's ground squirrels. Although these cries enhance the caller's danger, they provide the individual with indirect benefits by helping its genetic relatives.

Reciprocal Altruism

We refer to the situation in which individuals cooperate only with those who also cooperate as **reciprocal altruism.** Cooperative defense roles taken on by group members — lions, primates, cattle, birds, and so forth — involve shared genetic survival. Even fish aggregate in schools, the schools offering increased defense against predators compared to isolated individuals (**Fig. 23-8**).

A similar type of selection probably occurs in distasteful prey species in which mutant individuals with an aposematic[9] warning pattern arise. The warning pattern of the mutant makes them more susceptible to predation, but through their mortality they protect related genotypes that carry the same pattern. In the flour beetle *Tribolium confusum* egg-eating cannibalism by larvae declines in groups in which larvae only feed on genetically related eggs. This altruistic behavior of refraining from cannibalism is selected because it enhances the survival of related individuals that would be considered prey in the absence of altruism.

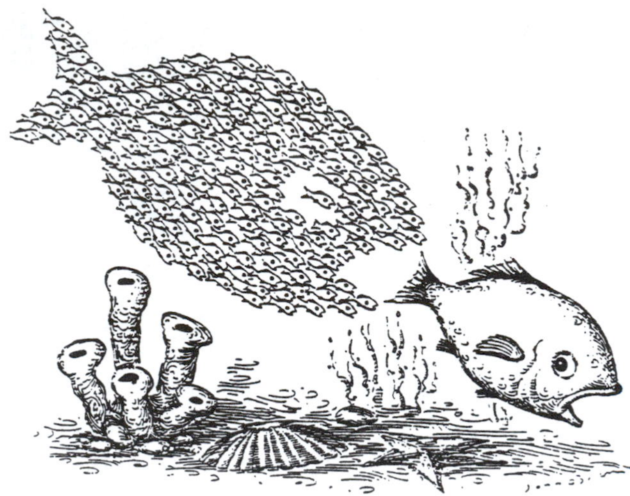

FIGURE 23-8 Group power. (© The New Yorker Collection 1991, John O'Brien from cartoonbank.com. All rights reserved.)

Social interactions can produce results that are positive or negative for both the individuals involved in the interactions, ranging from a mutual benefit to both individuals, to a positive advantage to the performer but not to the recipient, to the negative effects of altruistic sacrifice to the performer but advantage to the other(s) individual(s). As Figure 23-8 shows, the fitness effect of altruism on the altruist may be negative compared to selfishness, but it will have a positive effect on others in the population.

On the genetic level, an altruist really benefits genes it shares with compatriots, although possibly sacrificing its own. Thus, individuals in a socially interacting group containing many altruists are better off (achieve higher fitness) because the "effect on other" is more positive than in a group with fewer altruists. So, the success of a group depends on *a group property* (the frequency of individuals expressing certain behaviors) rather than on characters confined to only single individuals. As Darwin (1871) put it for humans:

[9] *Aposematic* is the term for the warning color markings that make a poisonous or distasteful animal conspicuous and recognizable to a predator. In some species, all individuals are aposematic — yellow and black striped wasps, or brightly colored poisonous frogs and snakes. In the example in the text, aposematic coloration appears occasionally as a result of mutation. See R. A. Wilson (2005) for a summary of current thinking on altruism.

> It must not be forgotten that although a high standard of morality gives but a slight or no advantage to each individual man and his children over the other men of the same tribe, yet that an increase in the number of well-endowed men and an advancement in the standard of morality will certainly give an immense advantage to one tribe over another.

Evolutionary biologists have long considered the extent to which group selection occurs, or whether it occurs at all. Among the arguments against group selection, it has been claimed that we can explain many so-called group adaptations as arising from selection among individuals. Others showed mathematically that selection among groups necessitates high extinction rates and practically no gene flow. Nevertheless, most would agree that group selection is at least theoretically possible, and from the examples just given and others, group selection is receiving much more attention and approval than in the past. Some have even broadened the discussion to consider "hierarchies" of selection that include competition between species (**species selection**) or accidental factors such as catastrophes that causes survival differences among them (**species sorting**). Other mass interactions, such as susceptibility or resistance to parasitic infection, also can involve group survival or extinction. The manner in which natural selection can be evaluated — reproductive success or survival — need not be confined to individuals.[10]

Although the issue is still unresolved, in some instances, for example in a population with sexual reproduction and altruistic social behaviors, selection seems to occur for the benefit of the group, even though it may harm an individual's own genetic future (although theoretically, the average fitness of an individual should go up in a population where altruism is maintained). Such situations remind us that fitness can be affected by behavior; through courtship display and mating strategies, feeding and care of offspring by individuals other than the parents (as occurs in birds), altruism, behavioral mimicry, schooling or swarming and aggression, being some examples.

Group Interaction

Natural communities of organisms are assemblages of species or groups, each interacting with others in various ways.

In terms of survival, growth or fecundity, a slight increase in the numbers of one particular group may cause an increase (+) in numbers of another group, a decrease (−) or have no discernible effect (0). We classify group interactions between two groups according to the terminology of **Table 23-7,** with categories ranging from neutral interaction (0, 0) to mutual gain (+, +), predation (+, −), and competitive inhibition (−, −).

Ecologists and naturalists have observed examples and variations of all such interactions. Ecologists have proposed an extensive variety of models to explain their mechanisms and effects, usually under the headings of population and community ecology, fields for which enormous numbers of studies exist.[11] A challenge not easily approached at the population level is the integration of the evolution and ecology of populations with developmental changes in individuals, changes that can be lifelong and affect all life history stages. One approach is outlined in **Box 23-2.**

Although ecological interactions have significant implications for any species, unraveling their exact evolutionary effects is difficult (Brooks and McLennan, 1991). We briefly review a few aspects of two of these interactions, **competition** and **predation,** which, in their various forms, often decide which species will be members of a localized community of organisms.

Competition

Competition arises when two groups depend on the same limited environmental resource(s) so that each group

[10] See G. C. Williams (1966, 1992), See D. S. Wilson (1992), L. C. Stevens et al. (1995), R. A. Wilson (2005) and Sober and Wilson (1998) for group selection; Gould (2002) and R. A. Wilson (2005) for species selection, and Vrba (1989) for species sorting.

[11] See the introductory ecology texts by Putman (1993), Begon et al. (1996), Krebs (2001) and Rockwood (2006).

TABLE 23-7 Interaction effects that can occur between two populations

Type	Effect on Species A	Effect on Species B	Nature of Interaction
Neutralism	0	0	Neither population affects the other.
Commensalism	0	1	Species A (e.g., the host) is not affected but species B (the commensal) benefits from the relationship.
Amensalism	0	2	Species A is not affected but species B is inhibited.
Mutualism	1	1	Both species benefit (e.g., Müllerian mimicry).
Predation or parasitism	1	2	Species A (predator or parasite) benefits at expense of species B (prey or host).
Competition	2	2	Each species inhibits the other.

Source: Adapted from Pianka, E. R., 1988. *Evolutionary Ecology,* 4th ed., Harper & Row, New York.

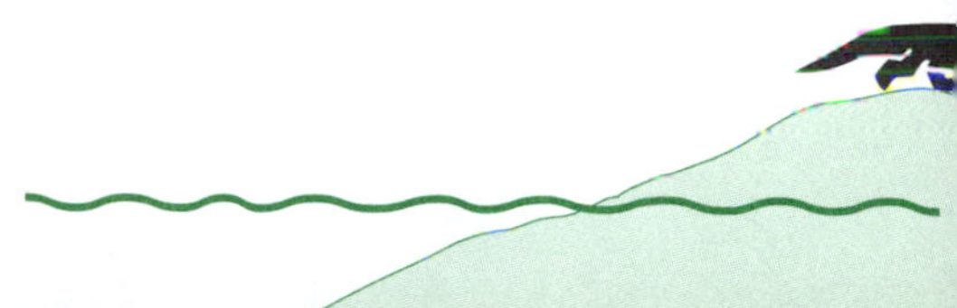

BOX 23-2 Integrating Evolution, Ecology and Development

Just as evolution and development have come together to form the integrated approach known as **evolutionary developmental biology (evo-devo)**, individuals have begun to explore how ecology can be added to the mix as *eco-evo-devo*. One strategy for approaching and potentially testing integrated evolutionary-ecological-developmental hypotheses, utilizes a refined character classification developed by Brooks and McLennan (1991) to establish four classes of features reflecting similarity or differences:

1. form (similar or dissimilar),
2. congruence (homology or homoplasy),
3. function (adaptive, nonadaptive), and
4. development (comparable and non-comparable).

The resulting 2^4 matrix provides the 16-class system depicted in **Table 23-8** and to which 16 known evolutionary categories can be mapped.

Two examples from Table 23-8 — combinations 5 and 10 — illustrate one approach, which is to use the matrix to assign a hypothesis of evolutionary mechanism to the phenotypic feature:

1. Two features have similar form but are homoplastic (that is, the similarity is superficial), adaptive, and show comparable development, representing an example of **parallelism**; feathers and insect bristles constitute one example.
2. Two features have dissimilar form but are homologous (that is, the similarity reflects recent ancestry), adaptive, and show non-comparable development, representing an example of **homeotic transformation**; transformation of an eye to an antenna in *Drosophila* constitute one example.

Conversely, the hypothesis that a feature arose by, for example, stabilizing selection, would be approached first by asking whether the features meets the four criteria set out in combination 1 in Table 23-8.[a] Try testing the approach using two features you want to compare.

[a] Other examples and a fuller development of the approach may be found in Stone and Hall (2006). See also Hall (2003a, 2006c) for background to the approach.

TABLE 23-8 A 16-class system proposed to help formulate and test hypotheses that combine evolutionary, ecological, development and adaptive approaches to phenotypic change during evolution[a]

Form	Congruence	Function	Development	Example
1 similar	Homologous	Adaptive	comparable[b]	Stabilizing selection
2 similar	Homologous	Adaptive	Noncomparable	Directional selection
3 similar	Homologous	Nonadaptive[c]	Comparable	Evolutionary novelty
4 similar	Homologous	Nonadaptive	Noncomparable	Evolution of development
5 similar	Homoplastic	Adaptive	Comparable	Parallelism (adaptive)
6 similar	Homoplastic	Adaptive	Noncomparable	Analogy
7 similar	Homoplastic	Nonadaptive	Comparable	Constraint
8 similar	Homoplastic	Nonadaptive	Noncomparable	Parallelism (nonadaptive)
9 dissimilar	Homologous	Adaptive	Comparable	Developmental flexibility
10 dissimilar	Homologous	Adaptive	Noncomparable	Homeotic transformation
11 dissimilar	Homologous	Nonadaptive	Comparable	Phenotypic plasticity
12 dissimilar	Homologous	Nonadaptive	Noncomparable	Genomic evolution
13 dissimilar	Homoplastic	Adaptive	Comparable	Regulatory genes
14 dissimilar	Homoplastic	Adaptive	Noncomparable	Regulatory gene network
15 dissimilar	Homoplastic	Nonadaptive	Comparable	Tissue inductions
16 dissimilar	Homoplastic	Nonadaptive	Nonconformational	Superficial similarity

[a] *Form* represents the phenotype, *congruence* represents evolutionary history, while *function* represents integration of the feature with the environment (*ecology*).
[b] Comparable is used to mean comparable development but can be used to indicate comparable genetic control.
[c] Exaptations and nonadaptive are other terms that reflect a mismatch between feature and environment.

Adapted from Stone and Hall (2006) with some change of terminology.

causes a demonstrable reduction in the other's numbers. Such reduced availability of common resources often has important ecological or behavioral consequences. For example, protective territorial mechanisms that inhibit competitors' use of such resources have evolved in some groups. These devices may include growth-inhibiting chemicals — for example, toxins such as the creosote some plants produce — or aggressive encounters — for example, fighting in many vertebrates — often responsible for species dispersal.

Resource Partitioning

Competition often leads to *ecological diversity.* It can be to the advantage of competing groups to minimize the harmful effects of direct competition by using different aspects of their common environmental resources. Among the many examples of such **resource partitioning** is the one illustrated in **Figure 23-9** for five species of warblers, each using different parts of their spruce tree habitat. Should such resource partitioning be disrupted, and habitats overlap, we would expect competition to lower the fitness of competing groups. A study by T. E. Martin (1996) supports this notion; nest predation increases when nesting sites of different species overlap.

Character Displacement

Character displacement is a further evolutionary response to competition, in which measurable phenotypic differences accompany resource partitioning among coexisting groups. One prominent example occurs among "Darwin's finches" in the Galapagos Islands where coexisting species show large differences in bill sizes, enabling each species to feed on differently sized seeds. In contrast, species isolated on different islands possess intermediate bill sizes enabling them to feed without partitioning seed resources. For example, one beak dimension measures about 8 mm for *Geospiza fuliginosa* and 12 mm for *G. fortis* on islands where both species exist, whereas it measures about 10 mm for each species on islands where they exist separately (**Fig. 23-10**). The advantages of close and continuous observations of natural populations are demonstrated by the ongoing studies on Darwin's finches of the Galapagos Islands by Peter and Rosemary Grant in which a natural process beginning with competitive interaction and character displacement leading to evolutionary change in one of the participating species was documented.

Competitive Exclusion

The many definitions of **niche** essentially propose that a species' niche includes all the environmental resources used

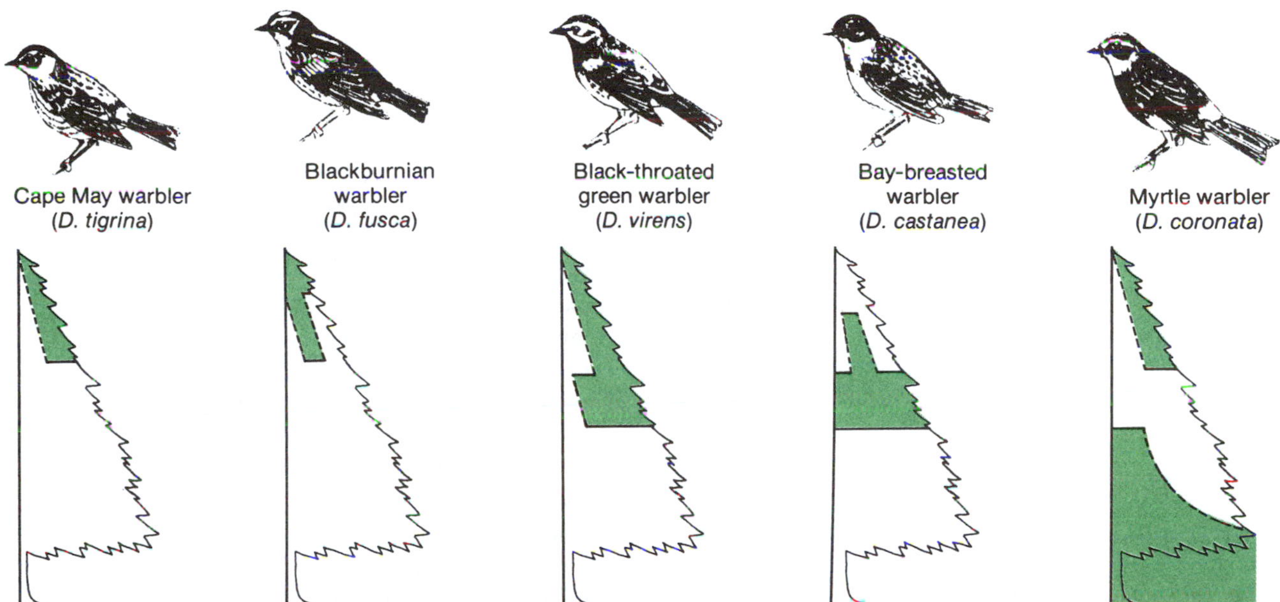

FIGURE 23-9 Most common feeding zones (shaded) in spruce trees for five species of northeastern U.S. warblers of the genus *Dendroica,* based on the number of birds observed. The Cape May warbler may be quite rare unless there is a large outbreak of insects. The Myrtle Warbler is the most abundant (and the least specialized) warbler in most coniferous forests. The more common warblers — Blackburnian, Bay-Breasted, and Black-Throated Green — are different enough in feeding zone preferences to explain their coexistence. Should such resource partitioning be disrupted, and overlap occur, we expect competition to lower the fitness of competing groups. A study by T. E. Martin (1996) supports this notion, showing that nest predation increases when nesting sites overlap between different species. (Adapted from Krebs, C. J., 2001. *Ecology: The Experimental Analysis of Distribution and Abundance.* 5th ed. Benjamin Cummings, San Francisco, CA.; based on MacArthur, R. H., 1958. Population ecology of some warblers of northeastern coniferous forests. *Ecology,* **39,** 599–619.)

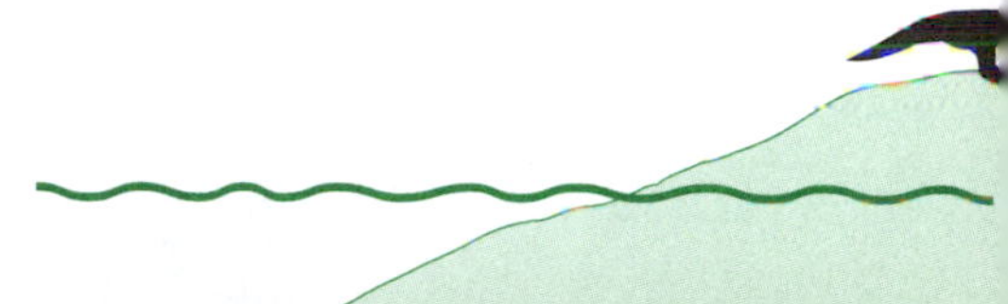

by the species, as well as the strategies the species applies in exploiting those resources (Odling-Smee et al., 2003). When competition is not checked by partitioning or fluctuation of resources and when two competing species use exactly the same resources in the same environment, they occupy the same niche.

In laboratory experiments, one species commonly dies out if two species occupy the same niche (**Fig. 23-11**), a finding that supports the **principle of competitive exclusion,** which states that two species cannot continue to coexist in the same environment if they use it in the same way. This principle, also called *Gause's axiom or Gause's law,* was foreshadowed by Darwin's statement in *On the Origin of Species:*

> Owing to the high geometrical rate of increase of all organic beings, each area is already fully stocked with inhabitants; and it follows from this, that as the favored forms increase in number, so generally will the less favored decrease and become rare.

Evolutionary biologists have seriously questioned whether competitive exclusion alone accounts for the differences observed among coexisting species — invoking for example, character displacement — or for the finding that closely related and potentially competitive species often occupy different habitats. Morphological differences among related coexisting species may have evolved in the past in places where these species did not compete, while related species occupying different habitats may not have diverged because of competition but because of different food preferences, nesting sites, and so on.

In fact, one can argue that much more coexistence appears among related species (e.g., species found in oceanic plankton) than would be expected if species were randomly distributed.

Predation

Predation is an important factor diminishing exclusion among competing, coexisting species; predators can reduce the ability of any single dominant species to reach its full potential carrying capacity, thus making room for other competitors (see also Cropping, Chapter 14). The predator entirely or partially consumes its prey, thus affecting the numbers of those organisms it feeds on.

Predation may be exercised in various ways on plants and animals, including overt attack and consumption of prey or **parasitism** (infestation and impairment of host tissues). The intimate dependence of predators on their prey often leads to a *coupling of their relative abundances;* an increase in numbers of prey is followed by an increase in predators that, in turn, can reduce prey that can then reduce predators. As shown in **Figure 23-12,** cyclic oscillations in population numbers may become a pattern, especially in the relationships of predators confined to a single species of prey.

Such simplicities are not the rule. Some predators don't cause a crash in abundance of prey but like the predation of wolves on caribou herds, prey mostly on the young or those weakened by age or disease and which have little reproductive value. Moreover, predator numbers may be buffered when more than one species of prey is being exploited, thereby reducing large oscillations in predator population size and spreading the effects of predation so that prey population sizes also remain fairly stable. Compensatory or additive mortality may be seen when a new predator is introduced into an ecosystem, examples including introduction of wolves or lions into populations of mule deer (*Odocoileus hemionus*), where the lions but not the wolves add to the mortality rate of the mule deer.

Some species in aquatic communities are held in balance because of chemical interactions between predator and prey. This fascinating phenomenon, a form of *phenotypic plasticity* (see Chapter 13), has been studied especially in interactions between species of predatory and prey rotifers, and between the water flea *Daphnia pulex* and its predator, the larvae of a midge. In both situations, the predator releases a chemical that acts on some *but not all* the eggs of the prey species (and it may be eggs of more than one prey species) so that the individuals that develop from those eggs produce an additional set of spines (rotifers) or longer spines than otherwise would be present (water flea), so that they are too big to be eaten by the predator. Phenotypic polymorphism therefore cycles dependent on predator levels, leading to such changes being named *cyclomorphosis.*[12]

An example of the role that phenotypic plasticity plays in allowing a species to adjust to changing environments (including food abundance) over time is the accumulation of data over 30 years on both phenotypic and genetic variation in a small passerine bird, the blue tit *Cyanistes caeruleus* in the Mediterranean. Comparisons have been made between deciduous and evergreen forest environments, both on the mainland and on islands. Extensive data are available on population structure, food abundance, breeding performance and morphology. Phenotypic plasticity emerged as a function of the distances over which birds dispersed, short-distance dispersal resulting in local specializations, longer-distance dispersals resulting in phenotypic plasticity, which, when local selection regimes opposed gene flow, led to local adaptations out of proportion to the geographical separation between the populations. Table 23-3 summarizes some of the habitat-specific differences recorded on the mainland and on two of the deciduous island sites. Blondel and colleagues (2006) distinguish deciduous oak and evergreen oak types but find maladapted populations in which, for example, breeding time does not coincide with peak food (cater-

[12] See Stearns (1989), Hall (1999a), Hall et al. (2004) for these and other examples of cyclomorphosis.

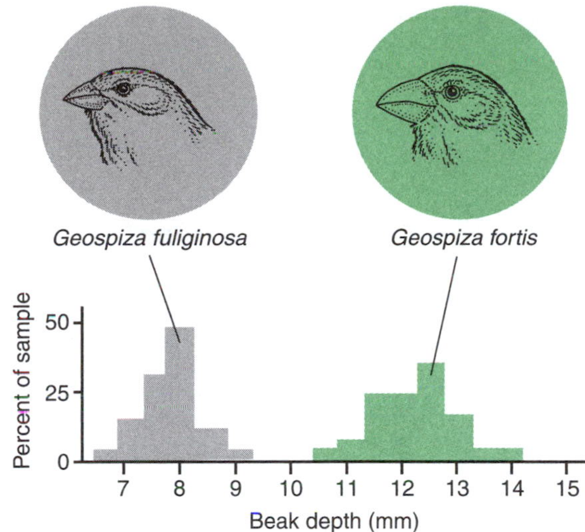

(a) Two different species on the same island (Abingdon, Bindloe, James, Jervis Isls)

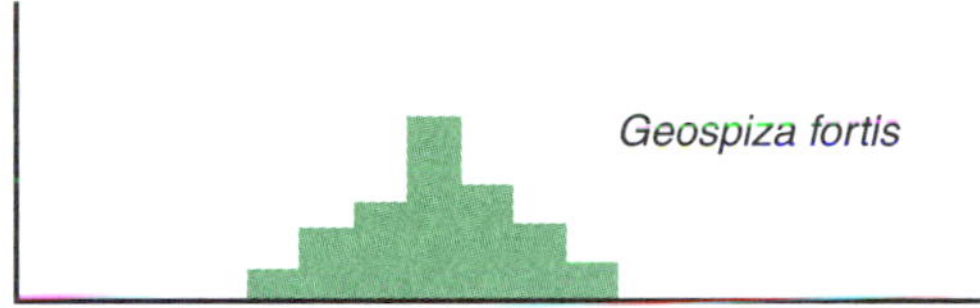

(b) A single species on an island (Daphne)

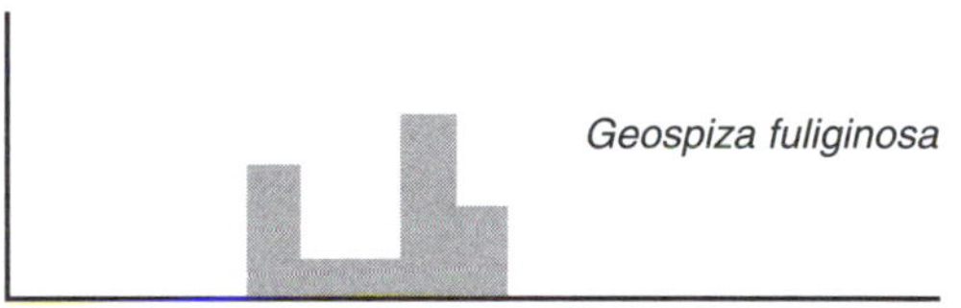

(c) A single species on an island (Crossman)

FIGURE 23-10 Character displacement among Darwin's finches in the Galapagos Islands. Coexisting species on four islands (a) show large differences in bill sizes, enabling each species to feed on different sized seeds. However, when either species exists alone on different islands (b, c), it possesses intermediate bill sizes (about 10 mm) enabling it to feed without partitioning seed resources. (Adapted from Givnish, T. J., 1997. Adaptive radiation and molecular systematics: Issues and approaches. In *Molecular Evolution and Adaptive Radiation*, T. J. Givnish and K. J. Sytsma (eds.). Cambridge University Press, Cambridge, England, pp. 1–54, based on Grant, P. R., 1986. *Ecology and Evolution of Darwin's Finches*. Princeton University Press, Princeton, NJ.)

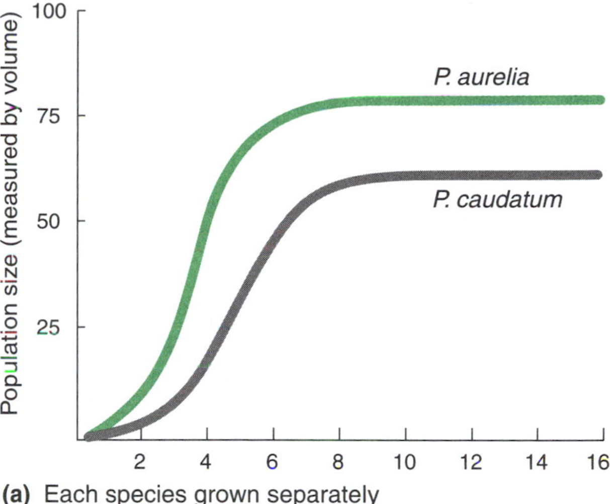

(a) Each species grown separately

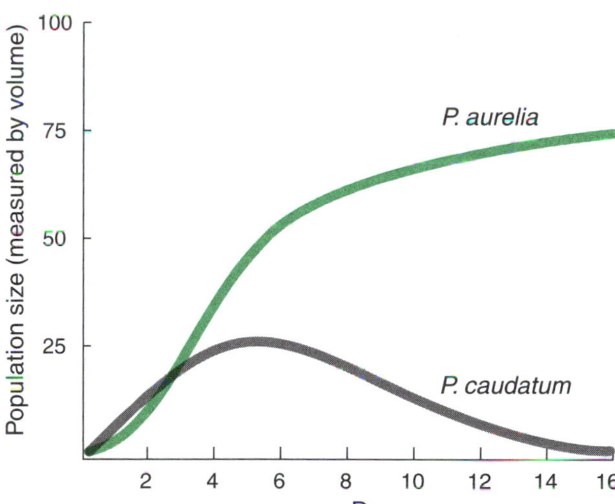

(b) The two species grown together

FIGURE 23-11 Growth of two species of *Paramecium* (a) in separate cultures, and (b) in mixed cultures. Although *P. aurelia* generally replaces *P. caudatum* as shown, in some mixed culture conditions *P. caudatum* multiplies faster than *P. aurelia*. Apparently *P. caudatum* is more sensitive to metabolic pollutants than is *P. aurelia*, so that removing such pollutants encourages *P. caudatum* population growth. The results of competitive interactions may be quite sensitive to external environmental factors, and, of course, will be more complex than in the mixed than in the separate cultures (Adapted from Gause, G. F., 1934. *The Struggle for Existence*. Williams & Wilkins, Baltimore.)

pillar) supply, concluding that, "wherever habitat patchiness is a mixture of deciduous and evergreen patches, the resulting reaction norm includes local specializations, phenotypic plasticity, or local maladaptation, depending on the size of habitat patches relative to the average dispersal range of the birds."

A second example is phenotypic plasticity in body forms in the pumpkinseed sunfish *Lepomis gibbosus* in which littoral and pelagic ecomorphs diverged from one another in response to predation by walleye (*Sander vitreus*). Januszkiewicz and Robinson (2007) conclude that this diversification is driven selection related to predation rather than to use of available resources by the two ecomorphs; that is, selection for predator-induced polymorphism drives divergence.

Although various mathematical models have been offered to generalize the intricacies of predator–prey rela-

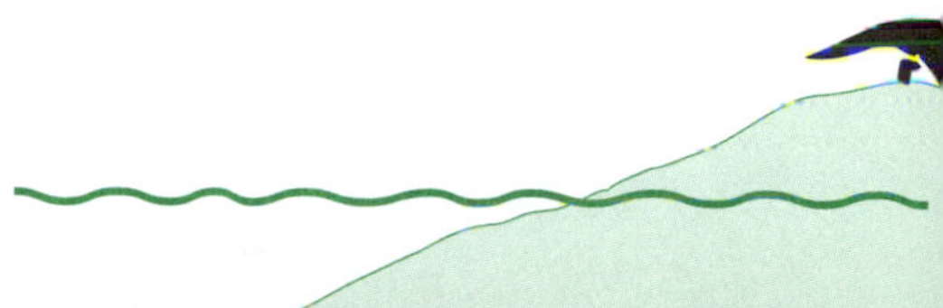

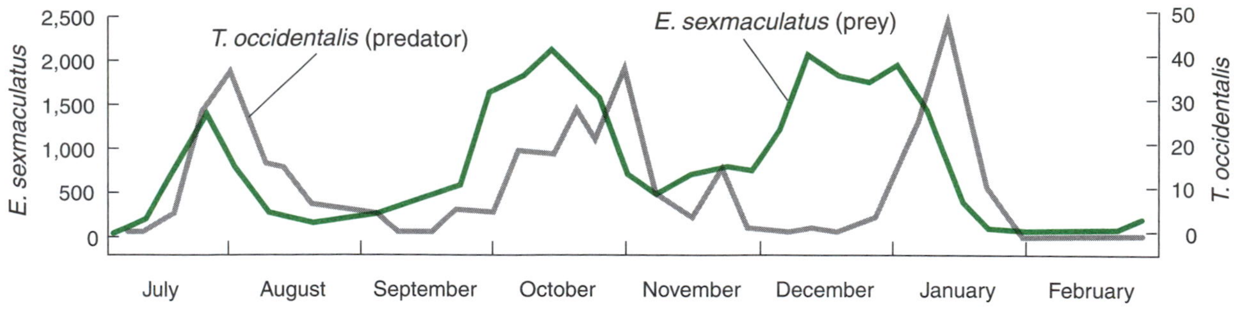

FIGURE 23-12 Three cycles of oscillating population numbers for two species of mites in a defined, controlled environment, one species (*Typhlodromus occidentalis*) being the predator and the herbivorous species (*Eotetranychus sexmaculatus*) being the prey. For each cycle, as numbers of prey increase, predator numbers follow, causing a "crash" in the prey population, followed by a crash in the predator population. (Adapted from Pianka, E. R., 1988. *Evolutionary Ecology,* 4th ed. Harper & Row, New York, from Huffaker C. B., 1958. Experimental studies on predation: Dispersion factors and predator-prey oscillations. *Hilgardia,* **27,** 343–383.)

tionships, there are no simple universal explanations. As Begon and Mortimer (1986) note:

> Predators and prey do not normally exist as simple, two-species systems. To understand the abundance patterns exhibited by two interacting species, these must be viewed in realistic multi-species context . . . Before multi-species systems are even considered, we must abandon our expectation of universal prey–predator oscillations, and look instead, much more closely, at the ways in which predators and their prey interact in practice.

Such interactions, whether of competition or predation, can be unique (i.e., seen only in one combination of species), dependent, for example, on the past evolutionary history of the populations involved, spatial limitations, climatic conditions, soil nutrients and the effects of other species in the community. Each interaction must be disentangled from others and explored separately — a difficult task, but one that has engendered considerable interest and effort among population biologists. Among such examples is evidence that different combinations of environmental factors can produce extensive differences in populations of Canadian snowshoe hares (Krebs et al., 1995), indeed, more differences than we might have expected. Either reducing predation or increasing the food supply caused a two- or threefold increase in numbers. Combining these factors caused an elevenfold increase. Obviously, different interactions occur (perhaps a large reduction in physiological stress) when predation is low and food is plentiful than when food is plentiful and predators are prevalent, or when predation is low and food is meager.

Coevolution

Group interactions have various consequences and coevolution is common. The intimate ecological relationships among many species derive from coevolutionary events in which adaptive changes in one species follow adaptive changes in others. What begin as casual interactions between different species develop into obligatory coevolving associations. There has been, and still is, a great deal of interdependence among many different life forms, and we cannot fully understand them except in an evolutionary framework.

We know that evolutionary changes in one species can prompt evolutionary changes in a species with which it interacts ecologically. Paleontological evidence points strongly to competition between herbivorous and carnivorous mammals during which size and speed seemed to increase sequentially in various members of both groups (see Chapter 19). Similar evolutionary progressions in respect to size, speed and protective devices occurred among the dinosaurs (see Chapter 18), have been a consistent trend in many prey–predator groups, and are a requirement when **coevolution** involves pollination of a plant by an insect (see Box 14-1, Coevolution of Plants and Insects) or parasitism.

Notable examples of coevolution in extant species are those between parasites or pathogens and their hosts. Twenty-seven genes in the flax plant *Linum usitatissium* confer resistance against a fungal rust pathogen. The pathogen, in turn, has a similar number of genes allowing it to overcome resistance these host genes confer. In such cases and in others, one can reasonably claim that an increased frequency of a resistant mutation in the host will be followed by selection for an increased frequency of one or more mutant genes in the parasite or pathogen that overcomes resistance (J. N. Thompson, 2005).

Not unexpectedly, coevolutionary events between host and parasite can be complex and may reduce the virulence of the parasite. One prominent example concerns a myxoma virus imported from South America to Australia to control the phenomenal population growth of the European rabbit *Oryctolagus cuniculus.* Although the virus caused only a mild disease in native South American cottontail rabbits *Sylvilagus brasilensis,* it acted as a highly lethal pathogen among the Australian rabbits, being transmitted primarily by mosqui-

Name: B. Rosemary Grant

Born: October 8, 1936

Undergraduate degree: University of Edinburgh, Scotland, 1960

Graduate degree: Ph.D., University of Uppsala, Sweden, 1985

Present position: Senior Research Scholar (Professor rank), Department of Ecology and Evolutionary Biology, Princeton University, New Jersey[13]

B. Rosemary Grant

What prompted your initial interest in evolution?

As a youngster growing up in The Lake District in England, I roamed the woods and fells and became fascinated with the diversity of organisms and particularly the differences between individuals of the same species. Why are no two individuals alike, and to what extent are these differences inherited are questions I discussed first with my parents and much later with professors at the University of Edinburgh.

What do you think has been most valuable or interesting among the discoveries you have made in science?

With my husband Peter Grant and colleagues I have carried out intensive studies on populations of finches in the Galapagos Islands for more than 25 years. The highlights of this study have been:

Establishing the heritability of morphological traits and the evolutionary responses to natural selection over short periods of time in natural populations.

Finding the extent to which genetic, ecological, and behavioral factors interact and the bearing this has on the formation of species.

Determining the causes and consequences of low levels of hybridization for evolution and the speciation process.

An exciting finding has been the role a culturally transmitted learned trait, imprinting on song, can play in evolution and the early stages of speciation.

These long-term studies have enabled us to interpret the evolutionary dynamics of natural populations living in climatically variable environments.

What areas of research are you (or your laboratory) presently engaged in?

Continuing our long-term study of individually marked birds in the Galapagos under the altered ecological conditions caused by the unprecedented severe El Niño of 1997–98.

In collaboration with two post-doctoral fellows, Ken Petren and Lukas Keller, we are using molecular genetic techniques in a study of phylogeny, paternity, hybridization, and inbreeding in Darwin's finches.

Investigating the consequences of inbreeding when populations fluctuate under extreme climatic conditions.

Exploring implications of our work for conservation where human induced fragmentation of the environment produces increased incidences of inbreeding and hybridization.

In which directions do you think future work in your field needs to be done?

Understanding the developmental and genetic basis for the modification of traits.

Investigating the interaction between culturally transmitted learned traits and genetic variation.

Establishing the connection between patterns of evolution in the past and evolutionary dynamics in contemporary time.

What advice would you offer to students who are interested in a career in your field of evolution?

It helps to know one organism or assemblage of organisms in depth, but at the same time to remain broadly interested and widely read. In this way, intuitions arise and can be explored in a thoroughly understood system. It is important to appreciate that there are many routes to the same objective. A diversity of approaches by different people can work synergistically and lead to a fuller understanding of both the problem and its solution. Most discoveries are the work of many people, not one, and by fostering these individualistic roles we can extend our knowledge.

[13] For a biography, visit http://www.eeb.princeton.edu/FACULTY/Grant_R/Grant_BR.html. Accessed May 17, 2007. Recent publication: Grant, P. R., and B. R. Grant, 2002. Unpredictable evolution in a 30 year study of Darwin's finches. *Science*, **296**, 701–711.

EVOLUTION ON THE WEB biology.jbpub.com/book/evolution

toes. Viral-caused lethality in 1950 to 1951 was as high as 99 percent among infected rabbits. It seemed as though the Australian rabbit population eventually would be eradicated by the virus or persist only at low numbers.

This expectation was not fulfilled, however. Although some Australian rabbits became increasingly resistant to the virus, the virus itself became less virulent. Because mosquitoes feed only on live rabbits, the rate at which they transmit the virus falls if the virus kills its immediate host too quickly. Highly virulent, rapidly replicating strains of virus were therefore selected against, whereas strains with reduced virulence were selected for — they allow infected rabbits to live long enough for the virus to spread more easily. Interestingly, this host–pathogen relationship is not static;

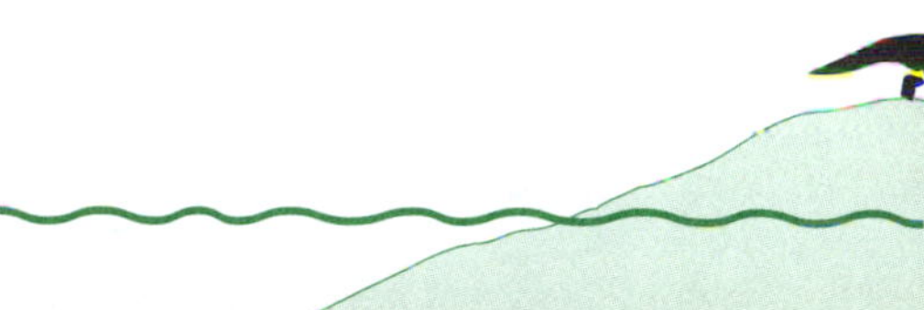

as rabbit resistance to the virus increases, increased viral virulence can become a selected trait. Thus, the relationship between virus and rabbit in Australia may eventually emulate the relationship in South America, in which a virulent virus has only a partially harmful effect on its native host. Such evolutionary outcomes are not unusual. Some small DNA viruses are genetically stable compared to rapidly evolving RNA viruses, and can persist and coevolve with their hosts causing little disease.[14]

Reduced parasitic virulence, however, is not always a successful option for competing parasitic strains. As we have seen, parasitic infection and parasitic virulence are positively correlated, indicating that success among competing parasitic strains depends on the ability to replicate rapidly, and that rapid parasitic replication is a major factor in causing virulence. Viruses that incorporate into host chromosomes and are transmitted between generations (**vertical transmission**) are called temperate, and have been selected for reduced virulence because they depend on host reproduction for survival. On the other hand, viruses that enter organisms exclusively through infection from other hosts (**horizontal transmission;** sometimes known as horizontal transfer) are selected to increase infectivity by replicating rapidly and, thus, almost always cause host destruction. Different infective opportunities come into play in selecting viral virulence. The sexually transmitted human immunodeficiency virus (HIV) has increased its virulence in populations that have greater sexual promiscuity (Ewald, 1994; Mindell, 2006).

The many instances of specificity between an insect species and its host plant are usually presented as examples of coevolution. Many may represent coevolution, some do not. The ability to establish the timing of the evolution of a lineage using molecular phylogenies and to calibrate that phylogeny against the fossil record raises some warning signs. Lopez-Vaamonde and colleagues (2006) used nuclear DNA to construct a molecular phylogeny of 100 species of leaf-mining moths in the genus *Phyllonorycter* and outgroups to obtain estimates of the time of origin of the moth species. Comparison with the distributional ages in the fossil record of the host plants of these moths led to the conclusion that the main radiation of this genus (27.3 to 50.8 Mya) took place *after* the radiation of their host plants, which was 84 to 90 Mya. Of course, it could not have taken place before the host plants radiated (unless the diet of earlier species differed). But the moth radiation appears *not* to be an example of coevolution (in the strict sense of that word) but of delayed colonization of host plants: "follow-along evolution."

Although competition, predation and parasitism reflect the popular concept of "nature red in tooth and claw," instances of cooperation and mutualism certainly modify its impact. Among these examples are cooperative relationships discussed earlier under "group selection" and "cyclomorphosis" as well as many instances of *symbiotic relationships,* including those between cellular organelles and their eukaryotic hosts (see Chapter 9), algae and fungi (see Fig. 22-7), and cellulose digesting protozoans in termite intestinal tracts. Long periods of coevolution made many such associations obligatory, but others such as between some pollinators and plants may be facultative, where neither species relies on the other for survival. Others have been forced upon species by our methods of cultivation; the dwarf palm *Chamaerops humilis* has been forced into a mutualism (in this case, forced pollination) by being cultivated in botanical gardens with the weevil *Derelomus chamaeropsis.* Depending on how one defines the concept, many other examples of coevolution abound, including mimics that evolve in step with the evolution of their models, ants that cospeciate with fungi they "farm" for food, competing species that evolve changes between them to reduce competition (e.g., character displacement), and many others.[15]

KEY TERMS

adaptive landscape
adaptive peak
adaptive valleys
allozymes
altruists
balanced genetic load
balanced load
carrying capacity
character displacement
coevolution
competition
demes
density dependent
density independent
fecundity
group selection
hitchhiking
intrinsic rate of natural increase
iteroparous
kin selection
logistic growth curve
meiotic drive
Muller's ratchet
mutational load
net reproductive rate
neutral mutations
parasitism
predation
principle of competitive exclusion
r- and *K*-selection
resource partitioning
segregation distorters
segregational load
semelparous
senescence
shifting balance process
species selection
species sorting
survivorship

[14] See Shadan and Villerreal (1993) and Fenner and Fantini (1999) for analyses.

[14] See Waser and Ollerton (2006) for pollinators and plants; Dufaÿ and Anstett (2004) for forced mutualism; Hinkle et al. (1994) for ant farming and J. N. Thompson (2005) for coevolution in general.

DISCUSSION QUESTIONS

1. What advantages does mathematical modeling offer evolutionary studies?
2. What is the relationship between the growth rate of a population and its carrying capacity? Or growth rate and the age structure of the population?
3. How do *r*- and *K*-selection strategies differ?
4. Genetic loads.
   a. How do geneticists calculate mutational and segregational (balanced) genetic loads?
   b. How do their effects differ?
   c. What changes in these genetic loads would you expect because of increased inbreeding?
5. In explaining the prevalence of genetic polymorphism, what evidence has been used to support the importance of neutral mutation? Of selection?
6. What are the comparative evolutionary advantages of sexual and asexual populations?
7. Does the survival of a population exposed to a new environment primarily depend on new adaptive mutations? Explain.
8. How can linkage and recombination affect adaptation?
9. Can genes that affect the survival of their individual carriers act differently for the survival of the population?
10. According to Wright, how can populations evolve to occupy high adaptive peaks when some genotypic combinations necessary to achieve these peaks are nonadaptive?
11. Why do populations not occupy all the adaptive peaks that are theoretically available?
12. Provide evidence, pro and con, for the concept of group selection.
13. Would you extend the notion of "hierarchies" of selection to include selection among genera, families, orders, classes and phyla? Why or why not?
14. What are the various consequences of species competition? Predation? What information would you deem necessary to declare these consequences predictable?
15. Provide examples in which interaction among different species affects their evolutionary direction.

EVOLUTION ON THE WEB

Explore evolution on the Internet! Visit the accompanying Web site for *Strickberger's Evolution*, Fourth Edition, at http://www.biology.jbpub.com/book/evolution for Web exercises and links relating to topics covered in this chapter.

RECOMMENDED READING

Begon, M., M. Mortimer, and D. J. Thompson, 1996. *Population Ecology: A Unified Study of Animals and Plants.* 3rd ed. Blackwell Science Ltd., Oxford, UK.

Bell, G., 1989. *Sex and Death in Protozoa: The History of Obsession.* Cambridge University Press, Cambridge, UK.

Benton, M. J., and F. J. Ayala, 2003. Dating the tree of life. *Science,* **300,** 1698–1700.

Ewald, P. W., 1994. *Evolution of Infectious Disease.* Oxford University Press, Oxford, UK.

Fox, C. W., D. A. Roff, and D. J. Fairbairn (eds.), 2001. *Evolutionary Ecology. Concepts and Case Studies.* Oxford University Press, New York.

Grant, P. R., 1986. *Ecology and Evolution of Darwin's Finches.* Princeton University Press, Princeton, NJ. [Reissued in 1999 with a Foreword by J. Weiner.]

Hartl, D. L., and A. G. Clark, 1997. *Principles of Population Genetics,* 3rd ed., Sinauer Associates, Sunderland, MA.

Krebs, C. J., 2001. *Ecology: The Experimental Analysis of Distribution and Abundance,* 5th ed., Benjamin Cummings, San Francisco, CA.

Lopez-Vaamonde, C., N. Wikström, C. Labandeira, H. C. J. Godfray, et al., 2006. Fossil-calibrated molecular phylogenies reveal that leaf-mining moths radiated millions of years after their host plants. *J. Evol. Biol.,* **19,** 1314–1326.

Maynard Smith, J., 1978. *The Evolution of Sex.* Cambridge University Press, Cambridge, UK.

Maynard Smith, J., 1998. *Evolutionary Genetics.* 2nd ed., Oxford University Press, Oxford, UK.

Mindell, D. P., 2006. *The Evolving World: Evolution in Everyday Life.* Harvard University Press, Cambridge, MA.

Odling-Smee, F. J., K. N. Laland, and M. W. Feldman, 2003. *Niche Construction. The Neglected Process in Evolution.* Princeton University Press, Princeton, NJ.

Pagel, M. (ed. in chief), 2002. *Encyclopedia of Evolution,* 2 Volumes. Oxford University Press, New York.

Ricklefs, R. E., and G. L. Miller, 2000. *Ecology,* 4th ed., W.H. Freeman & Co., New York.

Rockwood, L. L., 2006. *Introduction to Population Ecology.* Blackwell, Malden, MA.

Rose, M. R., and L. D. Mueller, 2006. *Evolution and Ecology of the Organism.* Prentice-Hall, Upper Saddle River, NJ.

Sober, E., and D. S. Wilson, 1998. *Unto Others: The Evolution and Psychology of Unselfish Behavior.* Harvard University Press, Cambridge, MA.

Thompson, J. N., 2005. *The Geographic Mosaic of Coevolution.* The University of Chicago Press, Chicago.

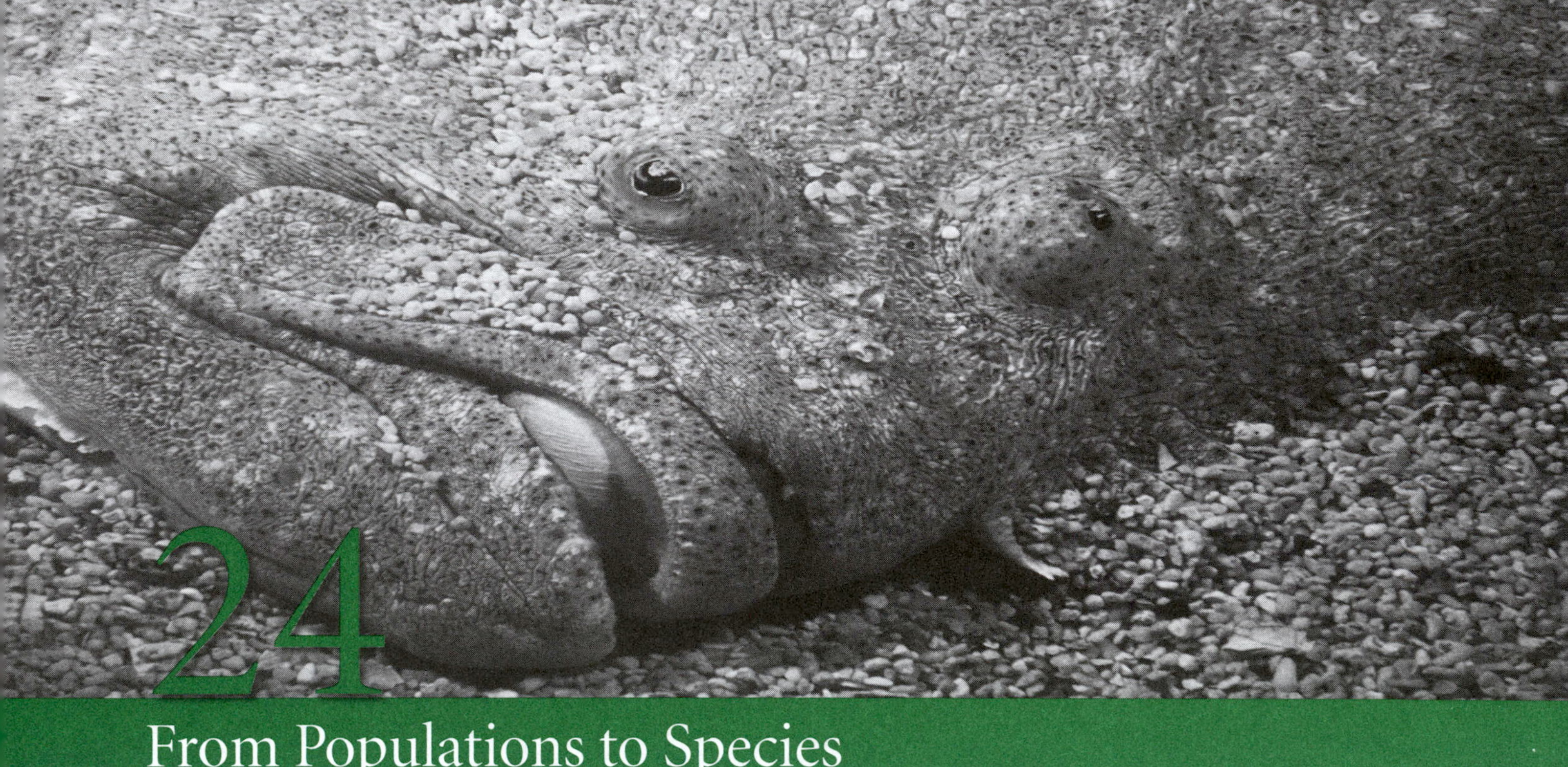

24 From Populations to Species

Chapter Summary

In this chapter, we discuss aspects of population structure as they relate to those mechanisms that maintain species (adaptations, behavioral strategies, sexual competition), those associated with speciation (isolation, hybridization, bottlenecks and founder effects), and rates of evolutionary change, especially rates of speciation.

All members of a species share a common gene pool, although populations within a species may vary genetically from one another. If the gene frequencies of these populations are sufficiently distinct, they are known as subspecies, varieties, subpopulations or geographical groups. Humans, at least, show so much polymorphism that their subpopulations cannot be distinguished by the presence or absence of certain alleles but only by variations in the frequencies of a number of alleles. At least some morphological and behavioral differences are adaptations to dissimilar environments. Learned and innate behaviors have a genetic component, although this component is more apparent in tropisms and instincts than in learned behaviors.

New species form when genetic exchange among isolated populations is impeded. Reproductive isolating mechanisms that provide the barriers for genetic exchange are of various kinds. Behavioral, seasonal and mechanical *prezygotic* mechanisms obstruct zygote formation. *Postzygotic* mechanisms result in nonviable or sterile offspring. According to some proposals, more of these mechanisms should develop in sympatric populations than in allopatric ones, which are geographically isolated from each other.

New species may form in allopatric groups by the slow accumulation of genetic differences, or when a portion of the population, perhaps only a few individuals, is isolated (founder effect). Speciation is presumed to occur in sympatric populations in response to a variety of selection pressures. Such populations become distinct because of persistent preferential mating.

The rate at which new taxa form is specific to particular lineages or taxa; evolutionary rates differ even within phylogenetic groups, in part because of variations in selection pressures on population at different times. Whether or not microevolutionary forces inducing change within species are identical to macroevolutionary forces generating new species remains under investigation within evolutionary biology.

The interbreeding nature of a sexual species serves as an important cohesive force holding a species together and maintaining a common gene pool. At the same time, a species may consist of many individual populations with various degrees of interbreeding. In this chapter we are concerned with:

- **mechanisms that maintain species:** those aspects of population structure that relate to adaptations, behavioral strategies, sexual competition and rates of evolutionary change;
- **mechanisms of speciation:** those aspects of population structure that relate to isolation, hybridization, bottlenecks and founder effects; and
- **rates of speciation:** those aspects of population structure that inform whether speciation is fast or slow, continuous or discontinuous, gradual or punctuated.

We also discuss whether speciation is an extension of microevolution or driven by separate processes, an issue introduced in Chapter 12.

Maintenance of Species

Epperson (1995) showed that an entirely heterozygous model population (e.g., all individuals *Aa*) may produce irregularly dispersed clusters of genotypes (*AA*, *Aa*, *Aa*) if individuals persistently mate with their neighbors. We expect widely separated populations to have less opportunity to share gene pools than those closer together, separating a species into various genetically diverse geographical subunits. Because the forces acting on these subunits may change among localities, observable differences among populations occur and are maintained.

A transect across central California shows that populations of the yarrow plant, *Achillea*, differ significantly in such traits as height and growing season (**Fig. 24-1**). We see the adaptive nature of such traits in the different responses of populations from different localities. Coastal plants are weak when grown at higher altitudes; high-altitude forms grow poorly at lower altitudes (**Fig. 24-2**). The adaptive features of many such plant populations are the genetic response of a population to a particular ecological habitat.

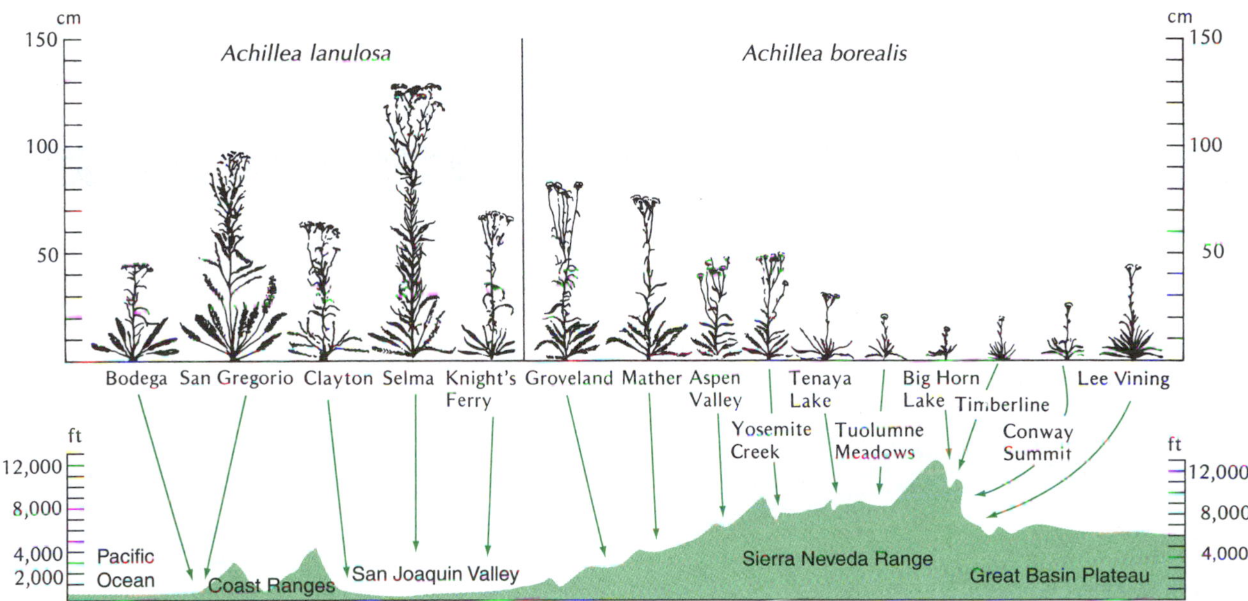

FIGURE 24-1 Representative plants from different populations of the common yarrow *Achillea* gathered from designated localities along a transect across central California and grown in a garden at Stanford, California. The fact that these populations, grown in a uniform environment, differ in size, leaf shape and other characteristics indicates that genetic differences have evolved among them. (Adapted from Clausen, J., D. D. Keck, and W. M. Hiesey, 1948. Experimental studies on the nature of species. III. Experimental responses of climatic races of *Achillea*. *Carnegie Inst. Wash. Publ. No. 581,* 1–219.)

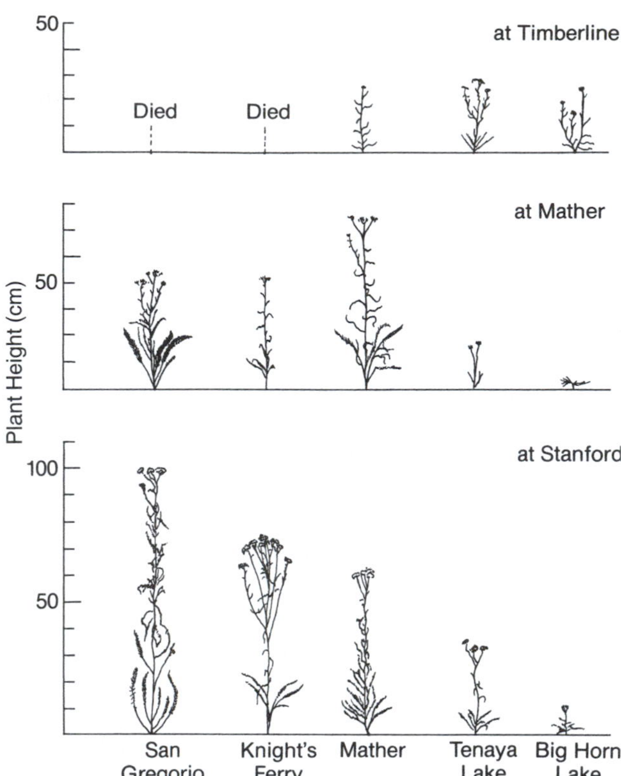

FIGURE 24-2 Responses of clones from representative *Achillea* plants originating from five localities in California and grown at three different altitudes: sea level (Stanford), 1,400 m (Mather) and 3,050 m (Timberline). (Adapted from Clausen, J., D. D. Keck, and W. M. Hiesey, 1948. Experimental studies on the nature of species. III. Experimental responses of climatic races of *Achillea. Carnegie Inst. Wash. Publ. No. 581,* 1–219.)

Many instances of changes between localities over various time intervals are known where allele frequencies can be scored. As shown in Figure 10-24, the frequencies of third-chromosome arrangements in *Drosophila pseudoobscura* differ notably in a range of environments across the American Southwest and also seasonally. Further genetic changes in this species extend over longer periods of time, as seen in the significant increase in the frequency of one arrangement (Pikes Peak) in many California populations from almost zero to as high as 10 percent over a 17-year period. As a second example, populations of the British peppered moth, *Biston betularia,* show significant changes in melanism frequency over even longer periods, all associated with specific localities and environments.

Groups

Groups (subspecies, varieties, subpopulations) are populations of the same species that differ markedly from each other. Terms specifying population differences vary. Taxonomists often use *subspecies, variety,* or even *subvariety* rather than *group* to designate taxonomically distinct groups within species.

The term race (racial) has been used in the past, as have tribe and ethnicity (**Table 24-1**). We have several reasons for not using these terms when referring to subgroups within a species. *Race* implies categories with strict, stable boundaries; has the connotation that races do not exchange genes, have distinct genes, and/or are "pure"; and has been used as a slur against individuals from other groups (racism). We follow Cavalli-Sforza:

> ". . . the psychological taxonomy of European races which became popular at the beginning of this [the 20th] century forms one of the most ludicrous confusions among customs, culture and genetics — in short, a perfect example of "scientific racism" (1994, p. 267).

The term *group* has no such connotations, and is in keeping with the fact that populations exchange alleles. We use group to designate any population that can be differen-

TABLE 24-1 Prevalence of birth defects per 10,000 live births in relation to ethnicity[a]

	Ethnicity		
Birth Defect	Non-Hispanic White	Non-Hispanic Black	Hispanic
Anencephaly	1.00	0.87	1.53
Spina bifida	1.00	0.96	1.19
Atrioventricular septal defect	1.00	1.17	0.92
Cleft palate	1.00	0.66	0.73
Cleft lip (± cleft palate)	1.00	0.56	0.98
Down syndrome (trisomy 21)	1.00	0.77	1.12
Trisomy 18	1.00	1.30	0.94

[a] Based on a sample of 2.4 million live births. Prevalence is given as the ratio using non-Hispanic whites as 1.00.

Based on data in Canfield, M. A., et al., 2006. National estimates and race/ethnicity-specific variation of selected birth defects in the United States, 1999–2001. *Birth Defects Res. (A) Clin. Mol. Teratol.*, **76**, 747–756.

tiated from another on the basis of its unique gene frequencies. Groups share the possibility of participating in the gene pool of the entire species, although they are sufficiently phylogenetically separated to exhibit specific gene frequencies. The distinction among groups is not absolute; they may differ in the relative frequency of a particular gene, but such differences do not prohibit gene exchange. An example is the prevalence of birth defects among three groups of humans, classified by Canfield and colleagues (2006) as non-Hispanic white, non-Hispanic black and Hispanic. They evaluated 21 birth defects from data collected from 6.24 million births between 1999 and 2001 in 11 U.S. states. As the sample of data in **Table 24-1** shows, prevalence of birth defects can be related to ethnicity. Intriguingly, differences between groups in defects known to have a chromosomal origin (and therefore to have a high heritability), for example trisomy 21, show lower differences between groups than do syndromes such as anencephaly or spina bifida, which have a more complex etiology involving environmental factors (e.g., intake of folic acid in the case of spina bifida).

Phylogeography

Study of the evolutionary processes regulating the geographic distributions of lineages/groups using reconstruction of genealogies of individual genes, groups of genes or populations is now known as **phylogeography**. Rather than sampling a single population, as is often the case in standard population genetics analyses, phylogeography involves analysis of different populations within a species. Consequently, phylogeographic approaches can reveal information about past patterns of migration (see Chapter 22), population bottlenecks and founder effects (this chapter) and current subdivisions of species into subpopulations, which can indicate incipient speciation. Phylogeography enables us to uncover and take into account the important issue raised in the introduction to Part I: *the role of history and of contingency in evolution.*

Criteria for evaluating differences among populations of a species are based on differences in gene frequencies. When these differences are plentiful, involve many genes, and it is appropriate to consider populations as separate entities, we may categorize them broadly as groups. At times, observable morphological differences accompany differences between groups, as among some human populations. At other times, observed differences extend only to gene or chromosomal differences such as those between Texas and California populations of *D. pseudoobscura* (see Fig. 10-24). We discuss two examples: (1) blood types in humans, and (2) the geographical distribution of strains of wheat.

Human Blood Types

For genes whose allele frequencies we can detect and score, gene frequencies rather than the presence or absence of particular genes help to discern geographical distinction between groups. **Table 24-2** shows a comparison of frequencies for a variety of gene systems in three major human groups, in practically all of which knowledge of a particular genotype alone is not, by itself, sufficient to indicate to which group an individual belongs. An individual of O blood type who is also Rh positive may belong to any of the groups if we were to look only at these genes; also recall the data on frequencies of genes for human blood presented in Box 22-2.

It is interesting to note that differences among human populations have not reached the point where one population is fixed for one allele at a particular locus and another population fixed for a different allele; when a population shows fixation for one allele, other populations are polymorphic for that allele. To calculate genetic divergences under such circumstances, Nei (1987) proposed the procedure shown in **Table 24-3.** When we apply such an index of **genetic distance** (*D*) to human populations (**Fig. 24-3**), we discern five major human groups:

1. **African** includes various tribes and groups that are (or were) indigenous to or can trace their ancestry to southern Africa.
2. **Caucasian** includes a variety of European populations (or individuals who can trace their ancestry to such populations), ranging from the Lapps of Scandinavia to the Mediterranean peoples of southern Europe and North Africa and as far east as India.
3. **Greater Asian** includes Mongoloid peoples, Polynesians and Micronesians.
4. **Amerindian** includes North American Inuit, American Indians and other indigenous peoples of North and South America.
5. **Australoid** groups native to Australia and Papua (Australopapuans).

Using the procedure outlined in Table 24-3, Nei and Roychoudhury (1993) estimated that, when averaged over many loci, genetic distance in humans (*D*) accumulates at some constant rate, roughly 3.75 My per unit of *D*. Thus, for the 85 loci that allowed comparisons, the estimated times when these groups diverged from each other are:

- Caucasian from Asian (*D*, 0.019), 41,000 ± 15,000 years ago
- Caucasian from African (*D*, 0.032), 113,000 ± 34,000 years ago
- African from Asian (*D*, 0.047), 116,000 ± 34,000 years ago

The oldest divergences are between Africans and other human groups, a view that other findings, mentioned in Chapter 20, seem to support. Also striking in these data are the consistently high levels of variability for the many genes

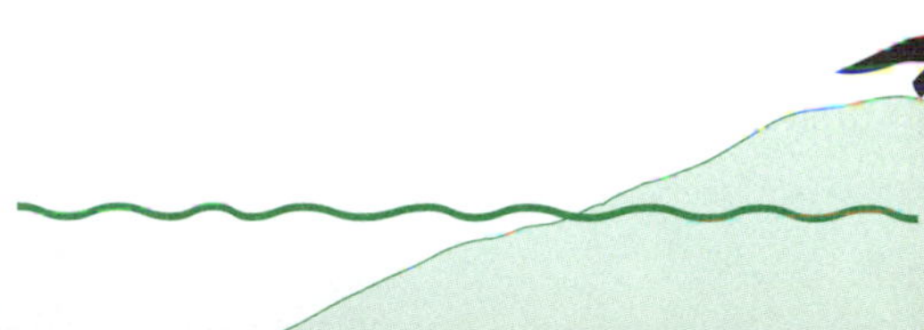

TABLE 24-2 Frequencies of alleles in samples of individuals taken from three broadly-defined human groups

Gene Locus	Allele	Caucasians	Africans	Mongols
Proteins				
Acid protein	Pa^1	0.21	0.14	0.42
	Pa^0	0.79	.86	.58
Adenylate cyclase	AK^1	.96	.99	1.00
	AK^2	.04	.01	–
Esterase D	ESD^1	.89	.97	.66
	ESD^2	.11	.03	.34
Glyoxylase I	GLO^1	.44	.26	.09
	GLO^2	.56	.74	.91
Haptoglobin-α	Hp^1	.43	.51	.24
	Hp^2	.57	.49	.76
Blood Groups				
ABO	*A*	.24	.19	.27
	B	.06	.16	.17
	0	.70	.65	.56
Duffy	Fy^a	.41	.06	.90
	Fy^b	.59	.94	.10
MN	*M*	.54	.58	.53
	N	.46	.42	.47
Rh (simplified to	Rh^+	.62	.70	.95
2 alleles)	Rh^-	.38	.30	.05

Source: Adapted from Strickberger, M. W., 1985. *Genetics,* 3d ed. Macmillan, New York. Data from Nei, M., and A. K. Roychoudhury, 1993. Evolutionary relationships of human populations on a global scale. *Mol. Biol. Evol.,* **10**, 927–943.

examined. According to Nei and Roychoudhury, the proportion of loci that were polymorphic in the three groups ranged from 45 to 52 percent for proteins and from 34 to 56 percent for blood groups. The average frequency of heterozygotes per locus ranged from 13 to 16 percent for proteins and from 11 to 20 percent for blood groups.

The considerable genetic variability in human groups shows the fallacy of concepts such as "pure" races. Members of such a group are not genetically pure in the sense of sharing a uniform genetic identity, nor does genetic uniformity even apply to members of the same family. Templeton (1997) pointed out that about 84 percent of the genetic variability among humans comes from differences among individuals and populations of the same group, while only 16 percent comes from differences among groups. And recall the studies discussed in Chapter 22 in which exchange of genes between groups of humans was shown to be of the order of 20 to 30 percent.

Geographical Spread of Asian Common Wheat

Strains of the Asian common wheat *Triticum aestivum* represent adaptations of a single species to a large number of environments, growing conditions and artificial selection pressures across Asia from the Ukraine and Turkey in the west to Japan in the east. Using the argument that the Tibetan plateau and large areas of desert in west China form effective geographical barriers to migration, three routes for the spread of wheat along trade routes to East Asia have been proposed:

- from Turkey to Sichuan China spread along an ancient Myanmar route;
- along the Silk Road; and
- populations related genetically to the lineage from the Silk Road introduction that were spread from the coastal area of China and Korea (**Fig. 24-4**).

TABLE 24-3 Example of the Nei procedure in calculating indices of genetic identity (I) and genetic distance (*D*) between Caucasians and Africans for two of the loci given in Table 24-2

Locus	Allele	Caucasians	Africans
Acid protein	Pa^1	$p_1 = 0.21$	$p_2 = 0.14$
	Pa^2	$q_1 = 0.79$	$q_2 = 0.86$
ABO blood group	*A*	$p_1 = 0.24$	$p_2 = 0.19$
	B	$q_1 = 0.06$	$q_2 = 0.16$
	O	$r_1 = 0.70$	$r_2 = 0.65$

$$I = \frac{\text{arithmetic mean of the products of allele frequencies}}{\text{geometric mean of the homozygote frequencies}}$$

$$I = \frac{[(p_1 \times p_2) + (q_1 \times q_2) + (r_1 \times r_2)]/\text{number of loci}}{\sqrt{[(p_1)^2 + (q_1)^2 + (r_1)^2] \times [(p_2)^2 + (q_2)^2 + (r_2)^2]/(\text{number of loci})^2}}$$

$$I = \frac{[(.21 \times .14) + (.79 \times .86) + (.24 \times .19) + (.06 \times .16) + (.70 \times .65)]/2}{\sqrt{[.21^2 + .79^2 + .24^2 + .06^2 + .70^2] \times [.14^2 + .86^2 + .19^2 + .16^2 + .65^2]/(2)^2}}$$

$$I = \frac{1.2190/2}{\sqrt{(1.2194)(1.2434)/4}} = \frac{0.6095}{\sqrt{1.5162/4}} = \frac{0.6095}{0.6157} = .9899$$

D (genetic distance) $= -\ln I = -\ln .9899 = .0101$

Note: I can range from zero (no similar alleles between the two populations) to one (complete similarity of alleles and frequencies). When $I = 1$, genetic dissimilarity (D) is zero. When I approaches zero, D can increase to large values, indicating that the alleles at some or many loci have been replaced one or more times.

Source: From Strickberger, M. W., 1985. *Genetics,* 3rd ed., Macmillan, New York.

This interpretation based on geographical barriers and trade routes developed and maintained by humans is supported by two studies based on the distribution of one or two genes, respectively.

Distribution of the strains of Asian common wheat was recently analyzed by Ghimire and colleagues (2005) in a multilevel study using isozymes of two enzymes, peroxidase and esterase. An amazing total of 648 "races" of wheat were used in the analysis (Fig. 24-4). Computations of genetic distances between these races revealed 33 populations that could be grouped into six clusters originated from three lineages (Fig. 24-4). This study revealed both genetic and geographical differentiation of Asian wheat and allowed the possible sources of origin and routes of migration of populations to be verified; phylogeography at its best.

Adaptational Patterns

As we have seen, forces producing differences within populations of a single species are often adaptive; at least some gene frequency changes are the response of a population to the selective forces operating within a particular environment, although it has been argued that some features of the phenotype are inevitable consequences of how a structure forms and so not themselves adaptive; they are spandrels (**Box 24-1**).

Climate, terrain, prey and predators can evince specific adaptations to differentiate a population, as many remarkable examples of camouflage and mimicry attest. In fact, a number of "rules" generalize the adaptive response of populations to certain ecological and geographical conditions. As with many other generalizations, exceptions to these **ecogeographical rules** exist, but the rules point to the importance of environmental selection in exacting parallel-convergent evolutionary changes in different species.

Among the best known of these climatic rules is **Bergmann's rule,** which relates body size in endothermic vertebrates to average environmental temperature. Bergmann's rule states that individuals of species in cooler climates tend to be larger than those in warmer climates. This relationship derives primarily because bodies with larger volumes have proportionately less exposed surface areas than bodies with smaller volumes. Because heat loss relates to surface area, larger bodies can retain heat more efficiently in cooler climates, whereas smaller bodies can eliminate heat more efficiently in warmer climates (**Fig. 24-5**).

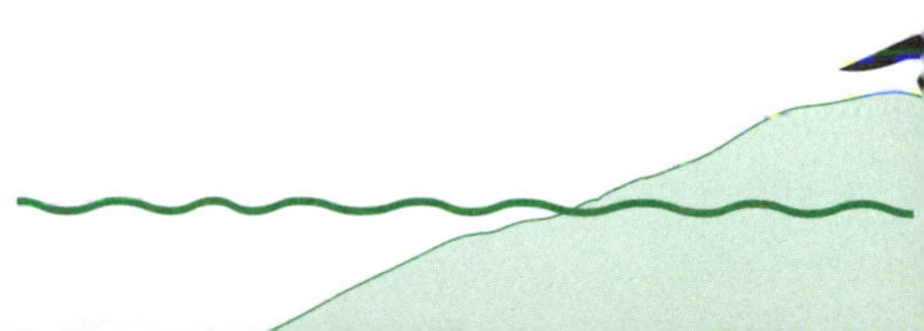

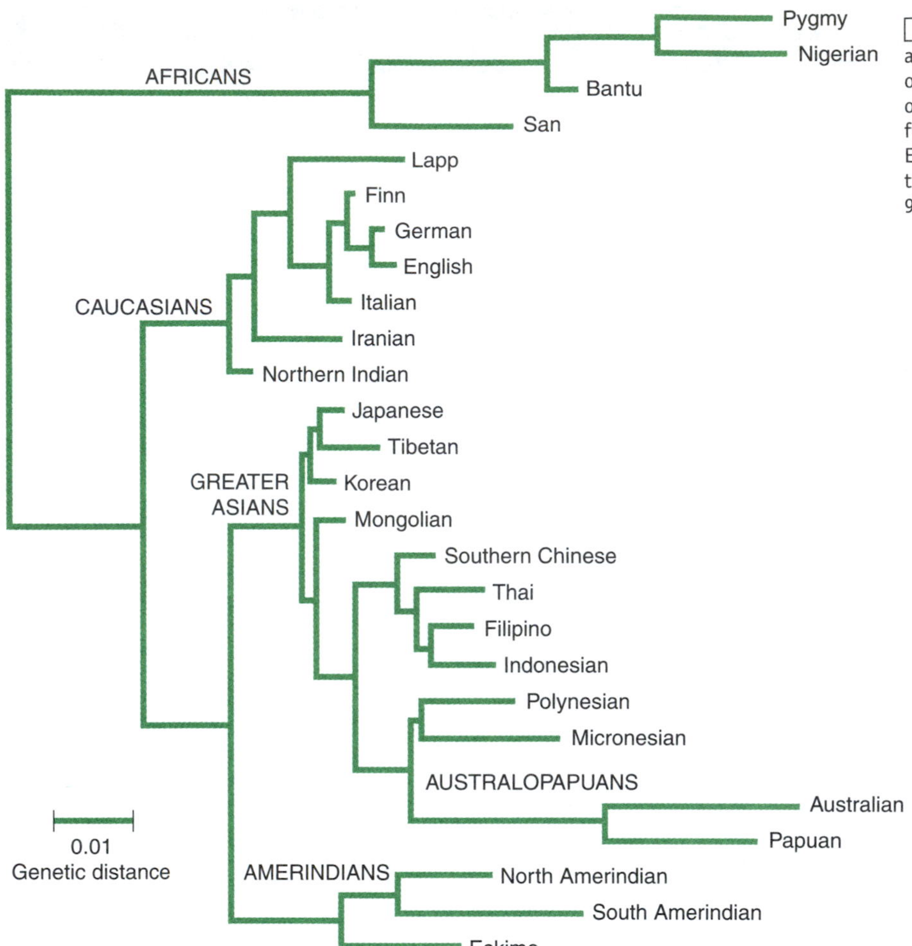

FIGURE 24-3 Phylogenetic relationships among 26 human populations based on using one of the genetic distance measurements (D_A) on 29 genetic loci bearing 121 alleles. (Adapted from Nei, M., and A. K. Roychoudhury, 1993. Evolutionary relationships of human populations on a global scale. *Mol. Biol. Evol.*, **10**, 927–943.)

Comparisons made among many north-south populations of terrestrial and marine birds and mammals corroborate the rule. For example, body size of bushy-tailed woodrats (*Neotoma cinerea*) closely follows climatic fluctuations from the time of the last glacial period about 25,000 years ago, with larger size in colder periods and smaller size in warmer periods (F. A. Smith et al., 1995). In some human groups, we can apply the rule by noting the ratio of body weight to body surface, comparing, for example, thick-chested, short-limbed Inuit groups to slender, long-limbed Nilotic African tribes (Fig. 24-5). An extension of Bergmann's rule is **Allen's rule,** which states that protruding body parts (tail, ears) are shorter in cooler climates than they are in warmer climates. Indeed, organisms of the same species (e.g., human groups) are shorter and stockier in colder climates than in tropical, a combined result of Bergmann's and Allen's rules. Some invertebrate taxa also show evolutionary patterns that conform to Bergmann's rule, for example, within lineage and genus-wide increases in body size in deep-sea ostracods of the genus *Poseidonamicus* that track climate change during the Cenozoic (Hunt and Roy, 2006).

Other provisional ecogeographical regularities are given by **Gloger's rule** (individuals are more heavily pigmented in warm, humid areas than in cool, dry areas; **Fig. 24-6**) and **Rapoport's rule** (species adapted to cooler climates are distributed along a wider range of latitudes than species adapted to warmer climates) as well as rules that apply primarily to insects, reptiles and amphibians (Mayr 1963; Jablonski, 2006).

Behavioral Adaptations and Strategies

In addition to morphological adaptations, organisms relate to their surroundings by assuming various motions and positions in escaping from predators, pursuing prey, interacting with conspecifics, and so on — all of which usually come under the terms behavioral adaptation or **behavior.**

Behavioral Adaptations

Behavioral adaptations — for example, migration to remain within a particular ecosystem as it shifts in response to climate change — can occur in the absence of any morphological changes. It appears as if the species are not adapting,

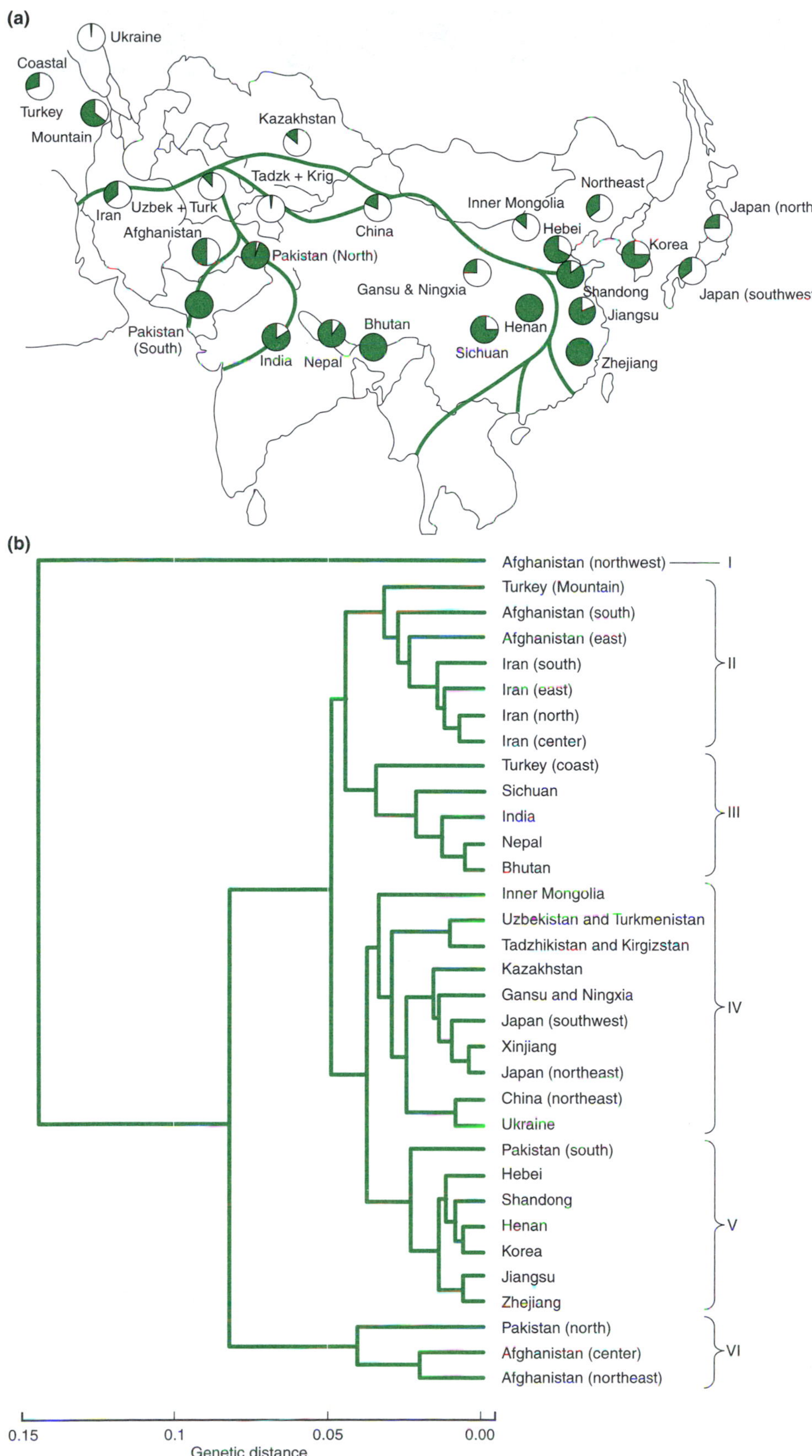

FIGURE 24-4 Phylogeographic analysis of Asian common wheat (*Triticum aestivum*) based on isozyme analysis. (a) The collection sites of the populations used in the study range across all of Asia. The major trade route (the "Silk Road" is shown in green. (b) The 33 populations analyzed were resolved into six clusters (I–VI) derived from three lineages. (Modified from Ghimire, S. K., Y. Akashi, C. Maitani, M. Nakanishi, and K. Kato, 2005. Genetic diversity and geographical differentiation in Asian common wheat (*Triticum aestivum* L.), revealed by the analysis of peroxidase and esterase isozymes. *Breeding Sci.*, **55**, 75–185.).

BOX 24-1 Adaptations and Spandrels

EVALUATING WHAT IS OR IS NOT an adaptation can be controversial. A feature's underlying functional and selective values are not always obvious, especially because they often derive from unknown past events. Some evolutionary biologists even question the validity of searching for functional values, whether current or retrospective, as many characters may be nonfunctional by-products of processes unrelated to the feature itself.

Spandrels are the curved triangular spaces between arches supporting the domed roves of cathedrals. In a widely cited paper, Gould and Lewontin (1979) claim that the spandrels in the Venetian Church of San Marco are, like many organic structures, nonfunctional artifacts.[a] Spandrels and various organismal features are really nonadaptive, they argue, arising from the constraints of constructional design rather than from function and adaptation. For example, they explained the tiny forelimbs of the dinosaur *Tyrannosaurus* as a "developmental correlate" that accompanies an increased size of the head and hind limbs, and they characterize adaptive explanations for this and other phenomena as the type of *Just So Stories* told by Rudyard Kipling.[b]

Although testing of adaptive explanations is often difficult (as is equally true for nonadaptive explanations), the fact that every organismal lineage has been subjected to selection makes it extremely likely that selection has affected most or all organismal characters, which therefore have or had some adaptive value. Transmission of a character between generations could escape selection because nonselective forces such as genetic drift, mutation, and migration influence its gene frequency, or it could be subject to stabilizing selection (see Chapter 22) so that even minor variants are eliminated. Even without stabilizing selection, the character, if it is more than just transitory, will sooner or later confront selection because of some effect it has, either by itself or together with other characters, on the relationship between the organism and its environment. How many persistent phenotypic characters exist that have never affected organism-environment interactions?

It is certainly reasonable to claim that nonselective causes that affect the gene frequencies involved in producing such a character do not continue to operate *ad infinitum* without selection intervening at some point. Thus, although we often may not know the adaptive explanation for a particular character, the search for such explanations in organisms whose survival depends on adapting to persistent (or intermittent) selective environmental pressures is a reasonable enterprise in evolutionary biology. To extend Gould and Lewontin's dinosaur example, why did *Tyrannosaurus* have tiny forelimbs but the 12 meter-long theropod dinosaur, *Spinosaurus*, the largest of the carnivorous dinosaurs, and also of the late Cretaceous, have relatively large forelimbs? The adaptationist position is basic to our understanding of biological evolution; a primary feature of all organic life is its subjection to selection. Selection for foreleg reduction in bipedal carnivorous dinosaurs may well have resulted in adaptations that improved balance and speed.

[a] Gould and Lewontin's claim that the spandrels of San Marco are "nonadaptive" has been countered by the demonstration that spandrels help support the overlying dome and are functional improvements over less functional "squinches."

[b] You can find further discussions of Gould and Lewontin's position on adaptationist explanations in Mayr (1983), Hall (2007b) and in the book edited by Dupré (1987).

but the adaptation can be revealed using large-scale temporal and spatial studies. Such changes can occur over long periods of a species' existence, as illustrated by Elisabeth Vrba's 1995 studies on patterns of change among African antelopes as they tracked climate and environmental changes during 0.5 My during the last ice age some 2.5 Mya.

Behaviors may be innate, needing no prior learning experience, or **learned**, improving over time through trial and error. **Innate behaviors** include *tropisms,* the directionally oriented growth patterns found in plants, fungi and sessile animals, and *taxes,* directionally oriented locomotion among animals. Examples of basic behaviors (i.e., fundamental to the success of the organisms that show them) are movements toward or away from light, gravity, environmental nutrients and chemicals. Some behaviors can be prompted by chemicals produced within organisms themselves, such as pheromones, which are molecular substances that many animals use to attract mates, lay down trails and warn off competitors. In the nematode *Caenorhabditis elegans,* the distinction between solitary and social feeding behavior can change because of a single amino acid substitution in a cellular receptor sensitive to secreted neuropeptide signals (de Bono and Bargmann, 1998).

Instincts

Relatively complex innate behaviors, often called **instincts**, may involve many behavioral components, as demonstrated in the courtship patterns of most animals and "dancing" patterns used by honeybees to communicate the direction and distance of food sources. Simple or complex, innate behaviors are often uniform within a species and, therefore, like other species-specific traits, seem to be entirely or almost entirely under intrinsic genetic control. For example, laboratory experiments show that various genes and associated neuroanatomical locations are directly involved in *Drosophila* courtship behavior (**Fig. 24-7**), and that male-female sexual orientation can be reversed genetically.

In natural populations of *Drosophila melanogaster,* allele differences in a foraging gene that produces a protein kinase used in signal transduction pathways (see Chapter

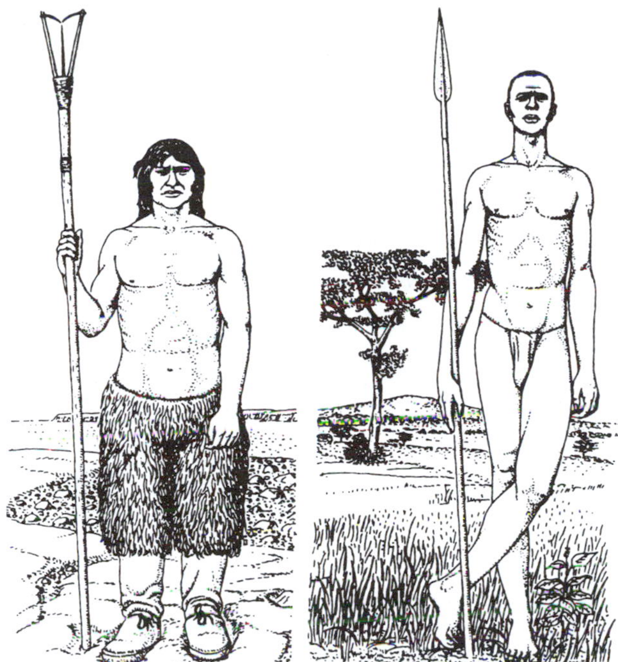

FIGURE 24-5 Difference in body proportions between a man from a group of arctic Inuit (Eskimo, *left*) and a man from a tribe in the Sudanese Nile (*right*), indicating differences in their adaptation to their prevailing climates. The proportionately greater bulk and smaller body surface of the Inuit (approximately 39 kg/m^2) helps to conserve heat, and the proportionately greater body surface of the Sudanese (approximately 34 kg/m^2) helps to dissipate heat. A similar explanation may account for differences in body form between Neanderthals and Cro-Magnon groups (see Fig. 20-29), each originating in different climates; the former more cold adapted, and the latter more equatorial. (From "The Distribution of Man," by William W. Howells, *Scientific American*, September 1960. © 1960 Eric Mose. Reprinted with permission.)

13) affect feeding behavior; "rovers" move longer distances than "sitters." These differences in foraging activity are selected by differences in population density; rovers are adaptive at high density where larvae must travel longer distances to obtain food; sitters are adaptive at lower densities where food is more available.[1]

Because an individual can modify its behavior to suit different environmental or social circumstances, **learned behavior** is more flexible and often more complex than innate behavior. Also, because these qualities are based on accommodating to a transitional and often unpredictable environment, learned behavior is difficult or impossible to program entirely genetically. Instead, learned behavior derives largely from practice (e.g., play and observation), allowing individuals to modify their behavior on the basis of their own or others' past experiences. Inexperienced mammalian juveniles must often be protected until they can use learned behaviors to deal with environmental and social problems or hazards. Toolmaking by Caledonian crows *Corvus moneduloides* on the island of New Caledonia is a fascinating nonmammalian example. Adults fashion barbed and hooked tools from leaves of pandanus palms. Adults in different parts of the island make different tools. The ability to make tools is thought to be passed on to the young by a social learning process. Indeed, these crows are no less sophisticated tool-makers than are chimpanzees (Hunt and Gray, 2003).

Because many neurological, muscular and sensory structures involved in the ability to learn and practice must be programmed into individuals, even learned behavior must have genetic components, and these have been widely investigated. For example, genes in different breeds of dogs influence tameness, playfulness and aggressiveness. In mice, a mutation called *disheveled* has no observable effect other than on social behavior, causing reduced social interaction and deficient nest building. In humans, we have considerable evidence for genes involved in nervous system disorders such as schizophrenia and manic depression, which alter human behavior. Such genetic influence and variation indicates that the capacities for most behavioral traits, like so many other adaptations, probably result from evolutionary selective forces. You may find it surprising, but behavior can be inferred from the fossil record using morphological features, the preservation of a fossil in a "living" position or situation, or evidence, such as a track-way left behind due to a behavior.[2]

Social Relationships as Evolutionarily Stable Strategies

Social relationships in which members of a group or species interact for breeding, feeding and defense, are areas of behavioral evolution given much attention. Chapter 20 discussed some behaviors involved in social interactions (**sociobiology**) as they relate to primates, but many behaviors, such as cooperation and dominance relations, apply to other social groups.[3]

In evaluating social interactions between individuals, various models that deal with the advantages of cooperation as opposed to the advantages of individual action have been proposed. For example, when two individuals confront each other in a social group, there may be conflicting interests between cooperation, in which each individual gains some advantage, and individual action, in which one individual gains greater immediate advantage than it could by cooperation. Such interactions are often referred to as **evolution-**

[1] See Osborne et al. (1997) for feeding behavior, and Sokolowski et al. (1997) for foraging.

[2] See Wahlstrom (1998) for genes and human behavioral disorders, and see H. W. Greene (1994) and Plavcan et al. (2002) for inferring behavior from fossils.

[3] See the texts by E.O. Wilson (1975, 1978) and the articles reprinted in Part VI of Appleman (2001).

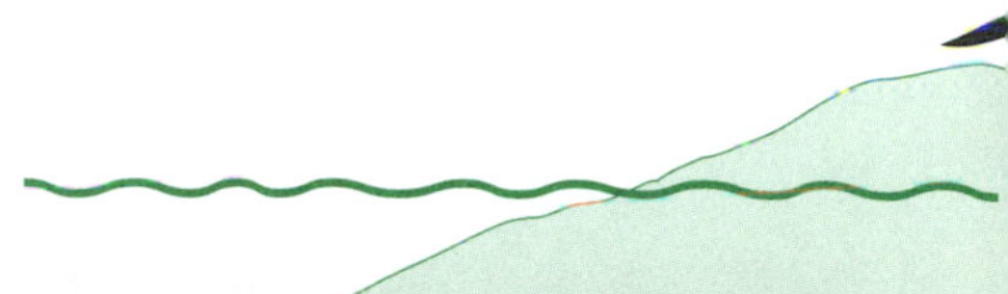

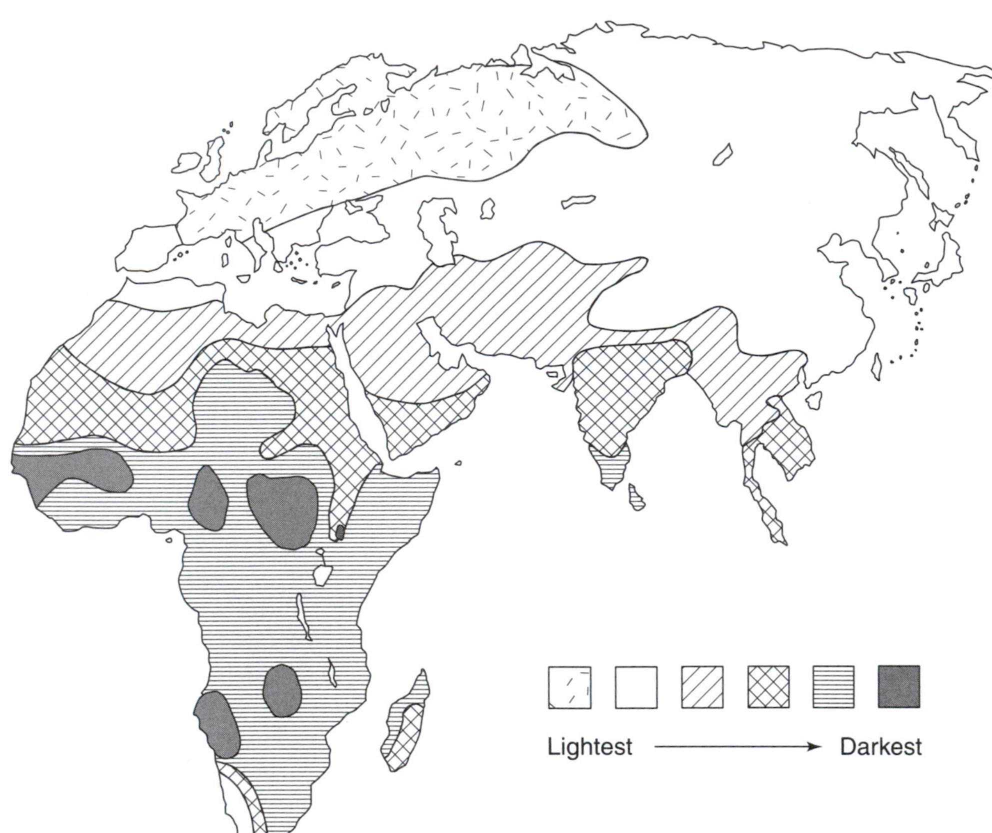

FIGURE 24-6 Distribution of skin color in human populations in Africa, Asia and Europe before A.D. 1400. (Adapted from Williams, B. J., 1979. *Evolution and Human Origins.* Harper & Row, New York.)

arily stable strategies. Cooperation can become an especially effective strategy, even between strangers (Nowak and Sigmund, 1998). Theorists have hotly evaluated the relationship of sociobiological determinants to human behavior and culture, as discussed in Chapter 25.

Among evolutionarily stable strategies is one called "tit for tat," in which an individual behaves cooperatively in a game's first move or interaction, and then repeats its opponent's previous move. Thus, an opponent acting selfishly is punished by a selfish response, while cooperative behavior is rewarded by a cooperative response. As Sigmund (1993) pointed out, "the advantage of Tit for Tat lies in it being quick to retaliate and quick to forgive."[4] A danger arises in tit for tat strategy; however, incorrectly evaluating an opponent's response as selfish can cause a cycle of retaliatory moves until a random correction occurs that reestablishes cooperation.

One solution is called "generous tit for tat," in which the probability of correcting such mistakes is greater than chance, because opponents occasionally overlook selfish responses. It has also been argued that "tit for tat" is highly susceptible to any errors in implementing the strategy: one mistake by either participant and the strategy fails. For such reasons, Nowak and Sigmund suggested that although "tit for tat" begins the process of successfully eliminating selfish exploiters, it will later be replaced by more cooperative strategies such as generous tit for tat, or tit for two tats, and so on (see Nowak and Sigmund (2005) and Nowak (2006) for up-to-date discussions).

When these concepts are extended to repetitive interactions among more than one pair of individuals, further strategies may develop, such as **reciprocal altruism**, in which individuals cooperate only with those who also cooperate. Individuals who do not cooperate can be identified early on, and may be banished or suffer retaliation and punishment.

Mating Systems: Sexual Competition and Selection

Mating systems are among the most important influences on social behavior. Although both sexes benefit if their offspring survive, males and females often differ in the cost of reproduction; females begin their reproductive careers by investing more resources in producing eggs than males do in producing sperm. Thus, a female's genes benefit if she discriminates in her choice of mates (**female choice**) to protect as much as possible her relatively expensive gametic

[4] May (1987) discussed some experiments that support the tit for tat model, while Maynard Smith (1988b) and others have elaborated other aspects of the application of game theory to social interactions.

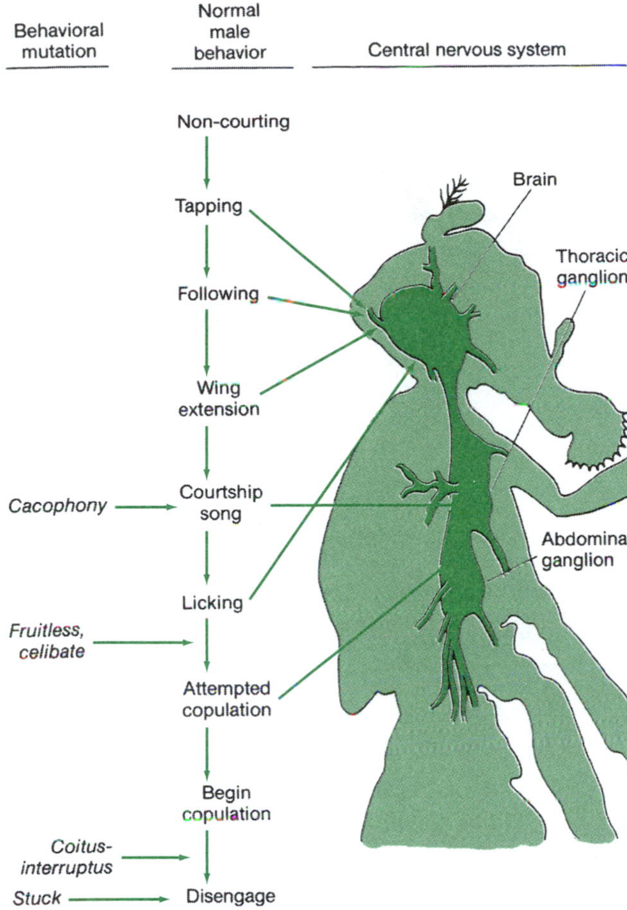

FIGURE 24-7 Sequence of the normal male courtship pattern in *Drosophila melanogaster* and localization of some of these stages to particular sections of the central nervous system. Some mutations with a primary effect on certain stages in the courtship pattern are shown in the *left column*. (Adapted from Strickberger, M. W., 1985. *Genetics*, 3rd ed. Macmillan, New York.)

output. For this same reason, there is an advantage for females to seek increased male **parental investment** in their offspring, a strategy especially notable among mammals and birds, where reproductive success can depend on relatively long-term commitment to their progeny.

Males, in contrast, can be more extravagant in disposing of their relatively inexpensive and more plentiful gametes. Genes carried by a male benefit when he fertilizes as many females as possible, often with relatively little discrimination. This conflict of interest between the sexes leads to a variety of mating patterns depending on a variety of factors, including the degree of parental care necessary for egg or infant survival, and which sex (or whether both sexes) provides such care. Even when males are normally involved in helping provide parental care, they may reduce such care if they have been "cuckolded" and the paternity of family offspring derives from another male.

In groups where females are primarily responsible for parental care — including many vertebrate species — males are likely to compete with each other for success in mating. As a result, selection can occur for traits that increase combative abilities of males (*intrasexual selection*) and/or traits that increase their attraction to females (*intersexual* or *epigamic selection*). As Darwin put it:

> Sexual selection depends on the success of certain individuals over others of the same sex, in relation to the propagation of the species; whilst natural selection depends on the success of both sexes, at all ages, in relation to the general conditions of life. The sexual struggle is of two kinds; in the one it is between the individuals of the same sex, generally the males, in order to drive away or kill their rivals, the females remaining passive; whilst in the other, the struggle is likewise between the individuals of the same sex, in order to excite or charm those of the opposite sex, generally the females, which no longer remain passive, but select the more agreeable partners.

Certainly, much evidence exists for intrasexual competition between males in polygynous species, where one male mates with many females. In such species, males may have special competitive armaments, such as horns and antlers, in order to gain access to females (**Fig. 24-8a**). Not surprisingly, these specialized masculine traits can lead to considerable sexual dimorphism (see Figs 19-17 and 19-18). For example, males are generally larger than females in such groups; male elephant seals (*Mirounga leonina*) are eight times larger than females. Longer horns in male dung beetles, *Onthophagus taurus,* is another excellent example; this case, as are many others is under hormonal control.[5]

Because of female choice, male ornaments can become quite conspicuous, as seen in the dramatic plumage of peacocks, birds of paradise and hummingbirds (Fig. 24-8b). In some cases, we can reasonably claim that although male decorative traits may enhance their breeding success, traits of this kind can cause increased susceptibility to predation; we can show mathematically that male decorative traits can become so exaggerated that their relative fitness in enhancing male fitness appears, at best, secondary to their value for sexual attraction. Such features can track environmental changes by rapid change. In barn swallows (*Hirundo rustica*), which are monogamous, the outermost tail feathers are 20 percent longer in males than in females. Over a 20-year period (1984–2003) and in response to changing climatic conditions at sites on the spring migration route, the length of these feathers in males increased by 11 mm but in females only by 3 mm, indicating that characters present in both

[5] For sexual dimorphism, especially of beetle horns, see Emlen (2000) and Emlen et al. (2005).

FIGURE 24-8 Two examples of sexual dimorphism (female on the left; male on the right. (a) Mature males of the extinct giant deer of Europe, *Megaloceros* (also called the giant Irish elk) had antlers more than 3.3 m wide and weighing about 45 kg. (From *Mammalian Evolution,* 1985 by Savage and Long. Reprinted by permission.) (b) Males (right) of the South American hummingbird, *Spathura underwoodi,* have long symmetrical tail feathers not found in the female (left). (From Darwin, C., 1871. *The Descent of Man and Selection in Relation to Sex.* Murray, London.)

sexes can be differentially responsive to selection in one sex (Møller and Szép, 2005).

As Fisher commented, this situation results because females continually mate with individuals whose attractiveness in such populations is passed on to their sons, and because the daughters of these females inherit their mother's preference for such phenotypically exaggerated males. What starts out as "fashion" in sexual attraction can escalate to extreme limits because of a selective cycle primarily devoted to mating success fed continually by genes that produce more exaggerated phenotypes and more exaggerated mating preferences (*runaway selection*). At a behavioral extreme of sexual selection are red back spider males, which place themselves within reach of the females' jaws to be cannibalized by the female with which they are mating. M. C. B. Andrade (1996) showed that this suicidal trait has adaptive features in that it allows such males to copulate longer and fertilize more eggs than their noncannibalized male competitors.

Nevertheless, female choice for nonadaptive traits in males is uncommon, and some reports demonstrate that female *Drosophila* fruit flies and *Colias* butterflies mate with those sexual partners that confer greater fitness on their offspring. In houseflies, developmental stability, as marked by symmetry in wing and leg lengths, enhances mating success and confers resistance to predation and fungal infection (Møller, 1996).

Among passerine birds, species with brightly colored males are associated with resistance to parasite infection. The offspring of female tree frogs choosing males with long mating calls gain advantages in growth and survival. Zahavi and Zahavi (1997) propose that such exaggerated traits, like the peacock's tail, are adaptive and selected because they signal to others that the carrier is in sufficiently good condition to expend the extra energy to produce and maintain the trait.

As for other adaptations, enough variations in ecological, genetic and evolutionary factors arise to prevent strict adherence to any widely applicable rules of sexual behavior. For example, male features may be selected on intra- and intersexual levels when females actively mate with males who are more successful in intrasexual combat. Female choice can even extend beyond coitus into sperm utilization, where male competition can occur through differences in genitalia and hormonal secretions for direct access to eggs. For example, the reproductive tract in female *Drosophila* is now known to be a coevolving battleground between the sexes. Males increase their sperm fertilization power by affecting female fertility, longevity and behavior through "sexually antagonistic" genes, and females attempt to respond with gene products that counter such effects. On the whole, differences in sexual behavior between groups may evolve quite rapidly and lead easily to barriers in gene exchange, as briefly discussed later.[6]

[6] See Clutton-Brock (1983) for variation and sexual behavior, Eberhard (1996) for access to eggs and A. G. Clark et al. (1999) for sexually antagonistic genes.

Origin of Species

When local adaptations occur within a species, whether through changes in morphology, physiology, behavior or all three, the subpopulations may still be able to combine and interbreed, forming a single population again. Thus, extensive migratory activity (gene flow) among individuals of a species may impede formation of subpopulations. Migratory species of birds average less than half the number of varieties of nonmigratory species: the greater the gene flow, the fewer the differences.

As a rule, therefore, barriers that reduce gene exchange between populations accelerate the formation of distinctive groups. Initially such barriers are primarily geographical and occur when populations bud off from one another and occupy different areas or environmental habitats. The potential for gene exchange, however, allows us to view all these different populations as members of a single species. Only when populations differ enough to inhibit gene exchange at all do we commonly view them as separate species.

Biological Species Revisited

To biologists, the *biological species concept* (see Chapter 11) — a species as an interbreeding group distinct from other such groups — is based on the knowledge that groups of sexually reproducing organisms exist in nature and that in many instances they are separated by "bridgeless gaps" across which interbreeding does not occur. The existence of species is supported by evidence that humans and other forms of life for whom such discrimination is essential recognize species as distinct groups. Predators of all kinds, for example, learn early to discriminate among varieties of prey and to select those that are palatable and can be used for food.

In groups that can verbalize species recognition, including early human societies, the distinctions made, in many cases, are strikingly similar to species classifications based on more sophisticated biological criteria. A tribe of New Guinea islanders uses distinct names for 137 species of birds found in this region, amazingly close to the 138 species recognized by ornithologists. Molecularly, taxonomic species designations in vertebrates also correlate, to a reasonable degree, with distinctions in mitochondrial DNA sequences. Because mitochondria are transmitted vertically without recombination, their DNA represents a lineage of historical events that can reflect species differences. Whether measured morphologically or molecularly, a qualitative change accompanied by reproductive separation or isolation usually marks the transition of population differences to species differences (**speciation**).[7]

What mechanisms prevent gene exchange between populations, and how do such mechanisms originate? Whatever the mechanisms, all must initiate or result in isolation of individuals or a group(s) from within an existing population.

[7] See (Mayr 1969b) for the New Guinea birds and Avise and Walker (1999) for the molecular analysis.

Isolating Mechanisms

Factors that prevent gene exchange among populations are known as **isolating mechanisms**. Terms such as *isolating barriers* or *reproductive barriers* might be more fitting, but isolating mechanisms, the term Dobzhansky proposed, remains the most common term for the concept of impediments to gene exchange. Most population geneticists consider all barriers, whether reproductive (biological), geographical, spatial or temporal, as isolating mechanisms. Speciation events among such geographically separated — **allopatric populations** — have been described in various ways, related to the kind and degree of separation and the opportunity for gene exchange.

The most common allopatric concept (also called vicariance) is of a formerly unified population that splits because of a natural physical barrier or because intervening geographical populations become extinct. **Peripatric speciation** describes the budding off of a small completely isolated "founder" colony from its larger more widespread parental population. **Parapatric** is used for a population at the periphery of a species that adapts to different environments but remains contiguous with its parent so that gene flow is possible between them (**Fig. 24-9**).

Where populations are sufficiently separated to prevent gene exchange, evolutionary biologists have considered whether, given the opportunity, many such populations would remain reproductively isolated. Some population biologists propose that we restrict the term *isolating mechanisms* to those that prevent gene exchange among populations in the same geographic locality; that is, to mechanisms that isolate **sympatric populations.** Mayr (1963) classified sympatric isolating mechanisms into two broad categories; those that operate before fertilization (prezygotic), and those that operate afterward (postzygotic).

Prezygotic Isolating Mechanisms

Among **prezygotic isolating mechanisms** are the three outlined below:

1. **Seasonal or Habitat Isolation.** Potential mates do not meet because they flourish in different seasons or in different habitats. For example, some plant species, such as the spiderworts *Tradescantia canaliculata* and *T. subaspera,* are sympatric throughout their geographical distribution and yet remain isolated because their flowers bloom at different seasons. Also, one species grows in sunlight and the other in deep shade.
2. **Behavioral or Sexual Isolation.** The sexes of two species of animals may appear together in the same locality, but their courtship patterns are sufficiently different to prevent mating. The distinctive songs of many birds, the special mating calls of certain frogs and, indeed, the

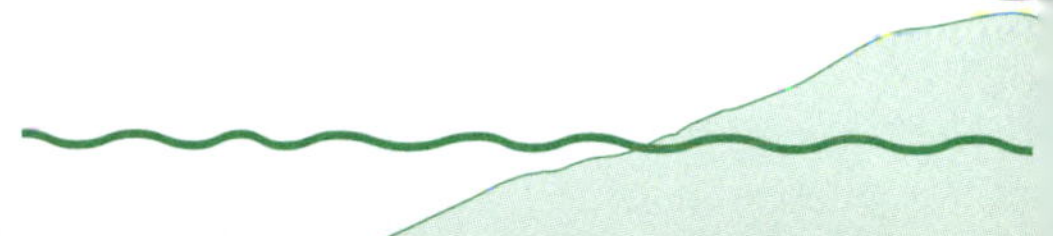

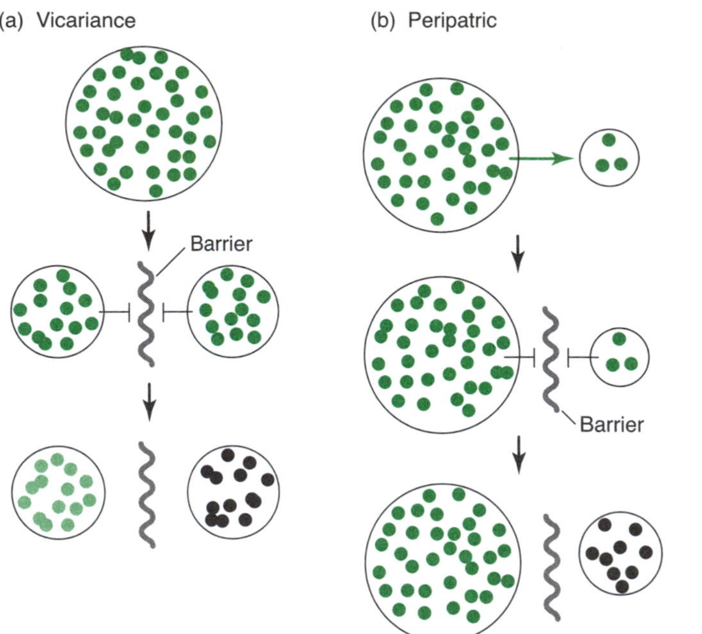

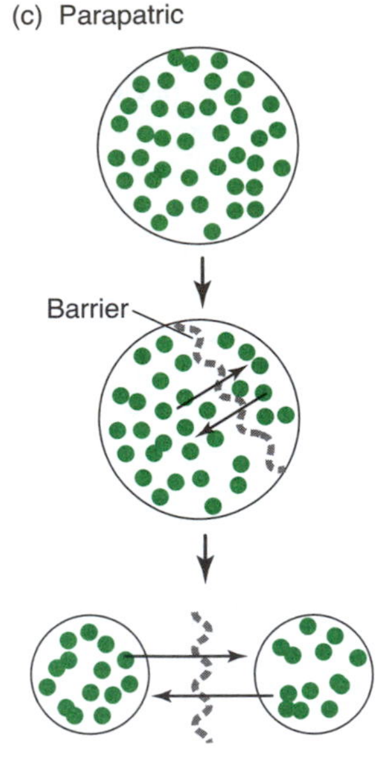

FIGURE 24-9 Comparison between three patterns of speciation: (a) Allopatric, where a barrier separates an original population into two, preventing gene flow (⊣); (b) Peripatric, where a small population is isolated from the original population and forms a founder population, again with no gene flow (⊣) between the populations; (c) Parapatric, where a barrier isolates a subpopulation but not sufficiently to prevent gene flow between the two populations.

sexual displays of most animals are attractive only to mates of the same species. Many plants have floral displays that discriminate between insect and bird pollinators, or attract only certain insect pollinators (see Box 14-1, Coevolution of Plants and Insects). Even where the morphological differences between two species are minimal, behavioral differences may prevent cross-fertilization. Thus, *D. melanogaster* and *D. simulans,* designated as sibling species because of their morphological similarity, normally do not mate with each other even when kept together in a single population cage. Male courtship in this group depends on their attraction to specific hydrocarbons in the female cuticle, and sexual isolation can be caused by only a few genetic differences.

3. **Mechanical Isolation.** Individuals attempt to mate, but cannot achieve fertilization because of difficulty in fitting together male-female genitalia. This type of incompatibility, long thought to be a primary isolating mechanism in animals, is no longer considered important. There is little evidence that mating in which the genitalia differ markedly is ever seriously attempted, although some exceptions exist among damselfly species and some other groups of species.

Postzygotic Isolating Mechanisms

Among **postzygotic isolating mechanisms** that prevent a successful interpopulation cross, even though mating has taken place, are the three outlined below:

1. **Gamete Mortality.** In this mechanism, the interspecific cross destroys either sperm or egg. Pollen grains in plants, for example, may be unable to grow pollen tubes in the styles of foreign species. In some *Drosophila* crosses, an insemination reaction in the vagina of the female causes swelling and prevents successful fertilization of the egg.
2. **Zygote Mortality** and **Hybrid Inviability.** The egg is fertilized, but the zygote either does not develop, or develops into an organism with reduced viability. Many such instances of incompatibility are known in plants and animals. For example, J. A. Moore (1949) made crosses among 12 frog species of the genus *Rana* and found a wide range of inviability. In some crosses, no egg cleavage occurred. In others, the cleavage and blastula stages were normal but gastrulation failed. In still others, early development was normal but later stages failed to develop.
3. **Hybrid Sterility.** The hybrid has normal viability but is reproductively deficient or sterile, a mechanism exemplified in the mule (progeny of a male donkey and female horse) and many other hybrids. Dobzhansky (1944, 1947) used hybrid sterility to provide a simple genetic answer to the question of how reproductive isolation can arise between two formerly interfertile populations. Assuming both populations begin with a two-locus genotype *aabb,* different adaptive mutations can be selected in each now separated population, one becoming *AAbb,* and the other *aaBB.* Each population retains its fertility, but epistatic interaction between the *A* and *B* alleles in the *A-B*-hybrid causes sterility and/or inviability. As time goes on, and more loci are differentially selected in each population, opportunity for such epistatic interactions increases. The observation that many genes are involved in *D. simulans-mauritania* hybrid male sterility indicates that

such epistatic interactions may be a leading cause for speciation.

In general, the barriers separating species are not confined to a single mechanism. The *Drosophila* sibling species *D. pseudoobscura* and *D. persimilis* are isolated from each other by habitat (*persimilis* usually lives in cooler regions and at higher elevations), courtship period (*persimilis* is usually more active in the morning, *pseudoobscura* in the evening) and mating behavior (the females prefer males of their own species). Although the distribution ranges of these two species overlap throughout large areas of the western United States, these isolating mechanisms are enough to keep the two species apart. To date, only a few cross-fertilized females have been found in nature among many thousands of flies examined. Even when cross-fertilization occurs, however, gene exchange is impeded, because the F_1 hybrid male is completely sterile and the progenies of fertile F_1 females backcrossed to males of either species show markedly lower viabilities than the parental stocks (**hybrid breakdown**).

Haldane's rule also calls for postzygotic isolation combining lethality and sterility. Haldane (1922) noted that the heterogametic sex is most commonly lethal or sterile in the F_1 of a cross between two groups or species. Many observations now support this rule (Coyne and Orr 1989).

Modes of Speciation

In 1889, Alfred Wallace proposed that natural selection might favor the establishment of mating barriers among populations if the hybrids were adaptively inferior; genotypes that did not mate to produce inferior hybrids would be selected over genotypes that did. According to this hypothesis, which Dobzhansky and others supported:

1. Selection for sexual isolation arises because most groups and species are strongly adapted to specific environments.
2. Maintenance of speciation enables populations to preserve their adaptive advantages from the disruption of gene flow from nonadapted groups.

Hybrids between two such highly adapted populations represent a genetic dilution of their parental gene complexes that can be of great disadvantage in the original environments. Genotypes that incorporate prezygotic isolating mechanisms would have the advantage of not wasting their gametes in producing deleterious offspring.

Full use of this mode of speciation demands that the different populations producing deleterious hybrids be exposed to each other in the same locality; only then could the more sexually isolated genotypes be specifically selected. Speciation, therefore, should occur in the following sequence:

1. genetic differentiation arises between allopatric populations;
2. overlap of these differentiated populations in a sympatric area; and
3. subsequent selection (also called *reinforcement*) for intensified sexual isolating mechanisms (**Fig. 24-10,** left column).

Sequential Allopatry and Sympatry

Evolutionary biologists have tried to demonstrate the above sequence among natural populations by comparing the degree of sexual isolation among different sympatric and allopatric populations. Sexual isolation should be strongest among sympatric populations of related species, because they are close enough to produce deleterious hybrids, and should be weakest among allopatric populations of species that are too distantly separated to produce such hybrids.

One such experiment tested the degree of sexual isolation between the sibling species *D. arizonensis* and *D. mojavensis* by attempting crosses in which the species strains derived from allopatric and sympatric origins. When the species strains came from sympatric origins the interspecific cross *arizonensis* × *mojavensis* occurred more rarely (4%; 14 out of 377 total matings) than when the strains came from allopatric origins (25%; 119 out of 473 total matings). In plants, V. Grant (1985) reported that of nine species in the annual herb *Gilia,* the most difficult to cross are the sympatric ones. The allopatric species, by contrast, show no barriers against intercrossing although all F_1 hybrids produced are sterile.

To these observations we can add the examination of hundreds of moth species, which showed that male scent-emitting organs used to attract females are significantly more common among species associated with the same host plant than among species associated with different host plants. Because these organs produce species-specific courtship pheromones, we can view them as sexual-isolating mechanisms that are more frequent in sympatric species (same host plants) than in allopatric species (different host plants).

In an experiment environmentally manipulating mating behavior, the normally isolated sibling species *D. pseudoobscura* and *D. persimilis* were used to demonstrate that prezygotic isolating mechanisms can be increased in sympatric populations. Although sexual isolation exists between these two species in nature and at normal temperatures in the laboratory, cold temperatures can cause a significant increase in interspecific mating. By identifying each of the two species with different homozygous recessive alleles, hybrids formed under these low-temperature conditions could be recognized and removed from interspecific population cages.

When this operation was performed each generation, fewer and fewer hybrids appeared. For example, after five

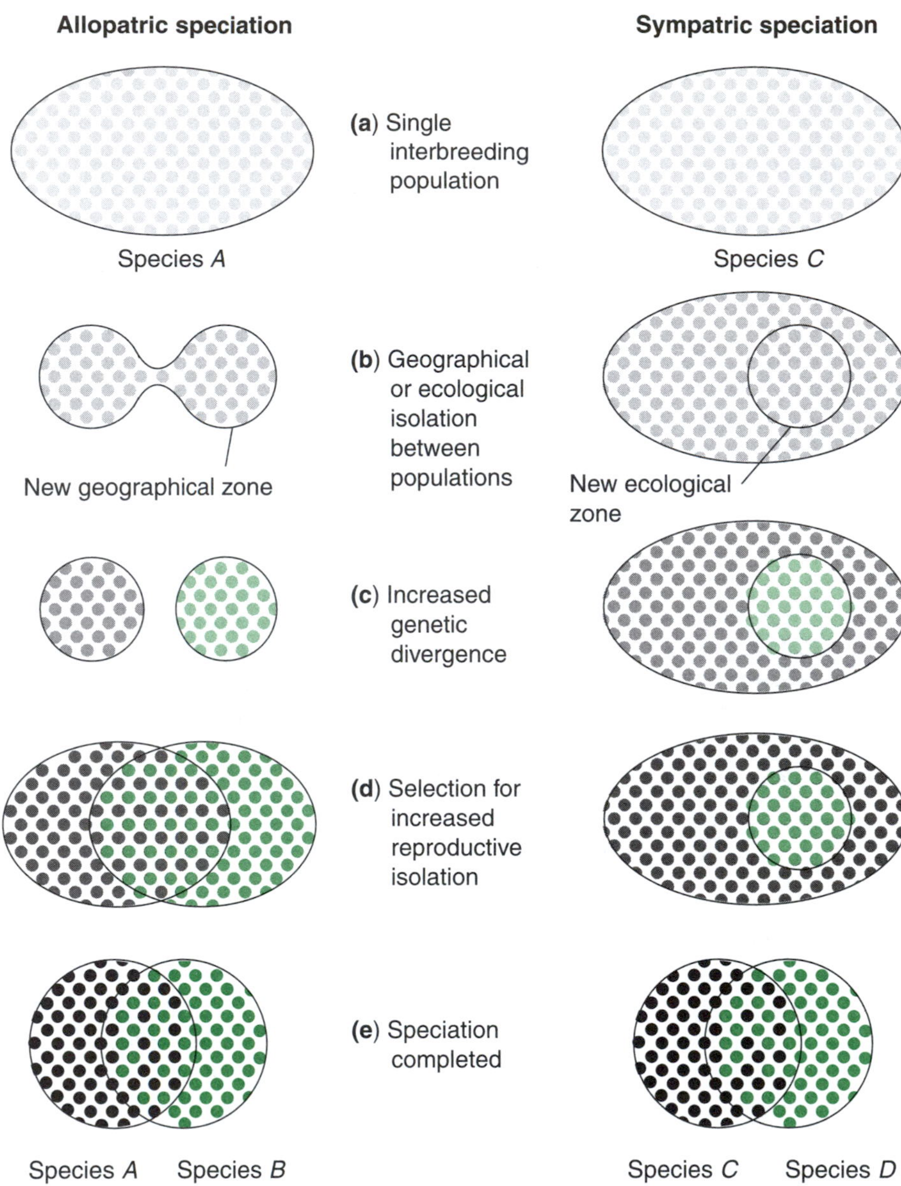

FIGURE 24-10 A simplified diagram of allopatric and sympatric speciation, showing different modes of divergence. In **allopatric speciation** (left column), a population (a) splits, or buds, into one or more new geographical zones (b) that allow genetic differentiation to occur among different geographical groups by means of genetic drift and selection (c). When each group has differentiated into a uniquely adapted genetic identity, geographical mixture of the groups can result in selection for increased reproductive isolation mechanisms among them. (d) Speciation is complete (e) when gene flow between the groups can no longer occur, even when they occupy the same locality. In **sympatric speciation** (right column), a population (a) splits into one or more groups that occupy different ecological zones, such as special habitats or food sources, within a single geographical locality (b). Increased genetic differentiation between the groups (c) permits selection for reproductive isolation mechanisms (d) that eventually lead to complete speciation (e). The difference between these models is the extent of physical separation involved in the initial genetic divergence between the groups. M. J. D. White (1973, 1978) discusses many examples and variations of these models, and Table 1 of Barton and Charlesworth compares a variety of speciation models. (From *Genetics,* Third Edition by Monroe W. Strickberger. Copyright ©1985 by Monroe W. Strickberger. Reprinted by permission of Prentice Hall, Inc., Upper Saddle River, NJ.)

generations, the frequency of hybrids in the mixed populations had fallen to 5 percent, from values that initially were as high as 50 percent. This was striking evidence that selection against hybrids caused rapid selection for sexual isolation that reduced hybrid formation. A somewhat similar experiment involved planting a mixture of yellow sweet and white flint strains of corn. By eliminating plants that produced the greatest proportion of heterozygotes, intercrossing was reduced from about 40 percent to less than 5 percent in five generations.

According to Butlin (1989) and others, it is debatable whether selection for prezygotic isolation mechanisms can occur between groups that have not already speciated. Butlin therefore suggests that because the experiments were conducted between species that were already reproductively isolated, we cannot view increased sexual isolation between *D. pseudoobscura* and *D. persimilis* as a cause for new speciation. According to Butlin, each species was selected for increased mate recognition (*reproductive character displacement*) and not for prezygotic isolation.

Nevertheless, the fact that we can experimentally increase prezygotic isolation by selecting against hybrids indicates that this isolating mechanism may well function in other cases where hybrid fitness declines. Mating tests between *D. pseudoobscura* females and *D. persimilis* males taken from natural populations showed that sexual isolation is increased in areas where populations overlap compared to areas where *D. persimilis* is absent (Noor, 1995). Similarly, when threespine sticklebacks (*Gasterosteus aculeatus*) are confronted with possible mates from different populations, they discriminate more between sympatric males than they do between allopatric males (Rundle and Schluter, 1998).

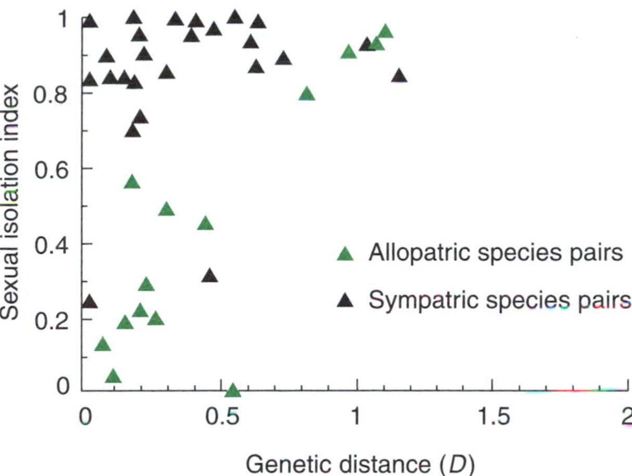

FIGURE 24-11 Measurements of the degree of sexual isolation for pairs of allopatric and sympatric *Drosophila* species where each pair has also been evaluated for Nei's *D* genetic distance. The sexual isolation index is based on comparing the frequency of heterospecific matings (matings between individuals from different species in the pair) and homospecific matings (matings between individuals from the same species in the pair) according to the following formula: Isolation = 1 – (heterospecific mating frequency)/(homospecific mating frequency). As Coyne and Orr (1997) point out and which is obvious in this illustration, when the species in a pair are closely related (genetic distance between them is small, for example, 0.5), they are more isolated from each other when they are sympatric than when they are allopatric. Also, for allopatric populations to achieve reproductive isolation requires a much greater genetic distance (*D* = .54) than reproductively isolated sympatric populations (*D* = 0.04). As a rough estimate, they propose that "speciation requires approximately 200,000 years among taxa that became sympatric and approximately 2.7 My among taxa that remain allopatric." (Adapted from Coyne, J. A., N. H. Barton, and M. Turelli, 1997. Perspective: A critique of Sewall Wright's shifting balance theory of evolution. *Evolution*, **51**, 643–671.)

Also supporting this view is an extensive survey showing that sexual isolation between pairs of *Drosophila* species of similar age is greater for sympatric than for allopatric species (**Fig. 24-11**).

One may argue that, given sufficient time, even allopatric populations will accumulate enough genetic differences to show sexual isolation when they come together in the same locality (**allopatric speciation**). In the *virilis* group of *Drosophila* species, the European *D. littoralis* is much more isolated from the American populations of *americana, texana* and *novamexicana* than are American species in the same group.

Further evidence for such allopatric differentiation comes from experiments in which a single population was separated into two or more groups for a considerable period and then tested for reproductive isolation. In one example, two replicate populations of *D. melanogaster,* raised in the laboratory under different conditions of temperature and humidity for six years, developed sexual isolation and hybrid sterility.

Bottlenecks and the Founder Effect

Whatever the speciation process between geographically separated populations, a number of evolutionary biologists suggest that it may proceed quite rapidly under some circumstances. They emphasize **founder effects,** or **bottlenecks** (see Chapter 22), in which a small, isolated peripatric population is subject to forces such as genetic drift, increased homozygosity caused by inbreeding and changes in the adaptive landscape (see Chapter 23), followed by radical changes in selection pressure. The combined effect of such forces may result in novel, **coadapted gene combinations** affecting behavioral, morphological and physiological traits that lead to reproductive isolation from neighboring and ancestral populations.

On the island of Hawaii, for example, Carson (1986, 1992) uses such concepts to help explain the origin of the 26 species of "picture-winged" Drosophilidae (see Fig. 22-11) in what may have been less than 500,000 years. Some of the founding events may have caused radical changes in male courtship behavior so that less discriminating females in these small, isolated populations were selected to respond to such changes; more discriminating females in ancestral or neighboring populations were unresponsive to such modified males. Courtship changes and ornamentation differences, however they arise, are powerful agents in initiating speciation. A high ratio of nonsynonymous (amino acid-changing) to synonymous ("silent") nucleotide substitutions in genes for mating behavior and other sex-related functions indicates that such genes play an important selective role in speciation (Civetta and Singh 1998a).

In opposition to these views on bottlenecks is the suggestion that the concept of speciation caused by single founder events has little theoretical support, because such events usually do not produce an immediately significant change in an isolated population. For example, it may take many generations for genetic drift to effectively modify gene frequencies. Barton and Charlesworth (1984) state that, "it is impossible to separate the effect of isolation, environmental differences, and continuous change by genetic drift [in moderately sized populations] from the impact of population bottlenecks [in small founder populations]." Even the impact of bottlenecks considered separately remains unresolved as some experiments do not support it (Moya et al., 1995), and phenotypic variation may even increase rather than decrease in populations that pass through bottlenecks.

Barton and Charlesworth find it difficult to accept that fixation of common alleles from a parental population would cause a significant developmental change and founder effect in a new population. By contrast, fixation of rare alleles might well cause greater developmental changes, but chances for becoming founders and for subsequent fixation are much less probable. Nevertheless, the founder effect model still gathers proponents. A peripheral population genetically isolated from its parent can evolve rapidly in new directions under strong selective conditions, sufficient to cause speciation. The relative importance of each suggested mode of speciation remains unclear and open to different interpretations.

Genes and Speciation

In a broad survey of pertinent laboratory experiments, Rice and Hostert (1993) claimed that reproductive barriers develop as secondary effects resulting from pleiotropy or by hitchhiking when disruptive selection between populations occurs for other genetic differences. Male hybrid sterility, caused by epistatic interaction between genes that are otherwise adaptive or neutral within a population (see also Turelli, 1998) may be an example of such speciation events. What can we say about the genes responsible for speciation?

For many, the search for "speciation genes" leads to sexual traits, because these are often basic in erecting the barriers that isolate species. Following such an approach, Civetta and Singh (1998b) measured variation between sexual traits (testis length) and nonsexual traits (femur length) for a group of *Drosophila* species. These comparisons showed that sexual traits exhibit greater variation between species and less variation within species, signifying that selection acts differently on sexual traits at different times; that is, sexual traits undergo greater selection for differences between populations during speciation, and greater selection for uniformity within species after attaining speciation. Civetta and Singh therefore propose that *changes in sexual traits correlate with early speciation events.* Studies pointing to similar genetic roles for other sexual traits in various organisms include investigations of pheromonal differences between *Drosophila* species, mating preferences in *Heliconius* passion-vine butterflies, evolution of mating type genes in the unicellular green alga *Chlamydomonas* and sperm-egg fertilization interaction in animals.

From what we now know, speciation events can occur in various ways and at various rates. In some groups, such as the Hawaiian Drosophilidae, speciation has been dramatically rapid and may well have involved fewer genes with greater phenotypic effects than in the slower speciation events in some other groups of *Drosophila* species. In plants, relatively few mutations seem to account for the rapid transition from teosinte to maize (see Fig. 21-1), and from bee-pollinated to hummingbird-pollinated species in the annual, *Mimulus.* In other groups allopatric speciation in the absence of bottlenecks may have been more common. Such allopatric speciation modes, however, do not exclude selection for sexual isolation between sympatric populations because of hybrid sterility or inviability, although as discussed earlier, that too remains unresolved.

Hybridization

Where species barriers break down to produce viable and fertile hybrids — and such instances do arise, especially in plants — *zones of hybridization* or *hybrid swarms* can develop in which genotypes and phenotypes differ from both parental species.

If a habitat exists to which the hybrids are better adapted than the parents, the new population may eventually become isolated from its parental populations. This mode of speciation is supported by detailed demonstrations of changes in chromosome number (ploidy levels) in plants (V. Grant, 1985) and animals (Bullini, 1994), although animal hybridization occurs more rarely because such chromosomal changes have greater impact on fitness.

One well-investigated example of plant species hybridization occurs in sunflowers of the genus, *Helianthus.* Three western United States species show rapid evolution of a hybrid *Helianthus anomalus* initially formed from a cross between *H. annuus* and *H. petiolaris* less than 60 generations ago (Ungerer et al., 1998). Interestingly, synthetic hybrids made experimentally by crossing the two parental sunflower species, and performing successive crosses and backcrosses, acquire genomes similar to the natural *H. anomalus* hybrid, incorporating similar parental genes while excluding others. Once an initial hybrid is formed, selection becomes an important factor in choosing genes that further develop its genetic architecture. Rieseberg and coworkers (1996) conclude that, "although the majority of interspecific gene interactions are indeed unfavorable or neutral, a small percentage of alien genes do appear to interact favorably in hybrids."

In some cases, fertile hybrids can act as intermediaries, introducing genes from one species into the other, thereby enhancing a species' ecological range and evolutionary flexibility; a phenomenon termed **introgressive hybridization.** Claims of introgression based on finding identical alleles in hybridizing species should exclude shared polymorphisms inherited from a common ancestor.

Studies by Rosemary and Peter Grant on two of Darwin's finches — *Geospiza fortis,* the medium ground finch and *Geospiza scandens,* the cactus finch (see Fig. 2-5) — on Daphne Island in the Galapagos, demonstrate that species-specific song, which is a learned trait that is culturally transmitted, can function as a reproductive barrier between species. Song in Darwin's finches is learned from the male parent during a short, sensitive period between 10 and 40 days after hatching. This mode of learning has a high fidelity in birds. Only rarely is a song misimprinted and then, not because of an inability of the young to learn the song but rather because it learns from an individual of another species rather than from its own species during its short sensitive period. For example, this has occurred when an egg is left in a nest that is taken over by another species or when a father dies leaving the mother (which does not sing) to raise the offspring. In the first case, the nestling learns its foster father's song; in the second, if its neighbor happens to be a male of another species, it will learn that song.

Because mating is based on recognition of song type, this cross-species learning can lead to hybridization. Hybridization is rare on Daphne Island, occurring in less than one percent of breeding pairs. Hybrids have beak sizes intermediate between their parental species and will only survive when seeds of the appropriate size are present. After a year of exceptionally heavy rains, the ecological condition of the island changed, an abundance of small soft seeds were produced and, under these altered conditions, hybrids with their intermediate beak size capable of eating these seeds survived. Hybrids did not mate with each other, probably because there are so few of them, but they backcrossed to one or other of their parental species according to their imprinted song type. In this way genes flowed from one species into another: *introgressive hybridization.*[8]

After this episode of introgressive hybridization, even with such a low rate of initial hybridization, over 30 percent of individual *Geospiza scandens* had some heterospecific genes due to high survival of several backcross generations. Introgression significantly increased the genetic and phenotypic variation of the *G. scandens* population. As the evolutionary consequence of episodic introgressive hybridization, it increases the genetic variation on which selection can later act. If the environment changes, rapid evolutionary change may ensue (see Box 2-1, Darwin's Finches). Episodic genetic introgression between closely related species is widespread in nature; examples are found in insects, cichlid fish (below), and sticklebacks and many species of plants, and could be an important contributor to rapid adaptive radiation.

Of course, hybrid sterility normally is a barrier to further evolution, but even then, specifically in plants, polyploidy may arise in a vegetatively propagating hybrid, enabling it to produce fertile gametes (allopolyploids; see Fig. 10-2). Because these gametes are diploid relative to the haploid gametes of the parental species, a new species is born at one stroke, fertile with itself or other such polyploid hybrids but sterile in crosses with either parental species.

Frequency of Hybridization

It has been difficult to document how often new hybrid species occur (Rieseberg, 1997). According to Ellstrand and coworkers (1996), the frequency of hybridization in vascular plants varies between families. Some have hardly any hybrid species while others have 50 percent or more. Of the 250,000 described plant species, they estimate only about 10 percent are hybrids.

The impact of hybrids on evolution may be more significant than their frequency; some plant hybrids may be the ancestors of entire lineages comprising many species and occupying many habitats. Dowling and Secor (1997) made similar claims for the significance of hybrids in animals. Arnold (1997), discussing plant and animal hybrids, argued that hybrid fitness is quite heterogeneous, and that some hybrids may be superior even in the parental environments. Although once considered uncommon, hybridization is beginning to receive more attention.

Can Species Differences Originate Sympatrically?

The sequence of evolutionary events in speciation seems to begin with the formation of distinctive populations — which could consist of only two individuals (one in hermaphrodites) — and end with reproductive isolation.

In this sequence, evolutionary geneticists disagree over the degree to which **geographical isolation** between populations is necessary to accumulate the initial genetic differences that lead to speciation. Many maintain that populations can only accumulate genetic differences when they are spatially separated enough to prevent gene exchange that might eradicate these differences. They propose that the speciation process takes hold only *after* this important early period of geographical separation, either by the accidental occurrence of isolating mechanisms or by later selection of isolating mechanisms because of defective hybrids.

Others propose that a population in a single locality selected for adaptation to different habitats within that locality could produce an increase in genetic variability (see disruptive selection) that would lead to *polymorphism.* Examples include polymorphism in the British peppered moth *Biston betularia,* mimicry in the butterfly *Papilio dardanus* and the evolution of distinct populations or varieties via mimicry in the butterfly *Heliconius erato* (Brower, 1996).

Under some circumstances, isolation between two or more selected groups might occur in the same locality, especially if the selected forms can exist independently of each other. Thoday and Gibson (1962) first presented evidence in selection experiments on bristle number in *D. melano-*

TABLE 24-4 Results of tests for mating preferences among *D. melanogaster* flies selected for high (H) bristle number and low (L) bristle number and in which males and females are given a free choice of mates

Generation of Selection	H × H	H × L	L × H	L × L
7	12	3	4	12
8	14	2	6	10
9	10	4	6	7
10	8	4	3	13
19	27	2	8	20
Total	71	15	27	62

Source: From J. M. Thoday, 1972. Disruptive selection, *Proc. Roy. Soc. Lond.* (B) **182,** 109–143. Reprinted by permission.

[8] See the studies in Grant and Grant (1994, 2002) and P.R. Grant et al. (2004).

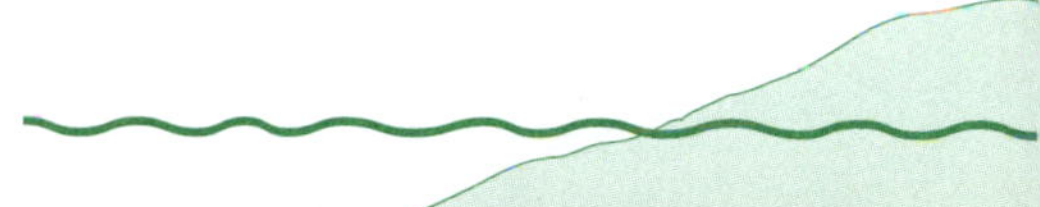

gaster. They selected flies each generation for high (H) and low (L) bristle number and found that, although mating was random, mating preferences of these flies went rapidly in the direction of positive assortative mating, H × H and L × L, with relatively few H × L and L × H matings (**Table 24-4**). Other experiments have since achieved increased isolation by disruptive selection between populations in the same locality.

Despite many attempts, however, population biologists have not replicated some findings of disruptive selection, and have asked whether any single locality in nature could consistently maintain divergent selective conditions long enough to produce sympatric speciation (Mayr, 1963). The primary issue revolves around selection's power and direction in an ecological isolate. Is selection strong enough to produce adaptive changes within a group that cause hybrid inviability and/or sterility (reproductive isolation) while the group continues to face gene flow from the surrounding population? Can reproductive isolation arise as a direct or indirect consequence of ecological adaptation?

However this matter is resolved, G. L. Bush (1994) and others used sympatric speciation (Fig. 24-10, right column) to explain the likelihood that various groups of insects speciated within a single geographical range by adapting to different kinds of host plants (apples and hawthorns) as food sources. Differences in the timing of fruit maturation between the plants cause these parasitic insects to emerge as adults at different times, providing a barrier preventing gene exchange between them and selecting genotypes that help restrict insects to their host plants. Ecological heterogeneity can certainly account for genetic diversity and contribute toward reproductive isolation, as shown in a *D. melanogaster* population selected for radically different experimental habitats. It seems that some or even many sympatric speciation events may have occurred (G. L. Bush, 1994).

Because of intensive studies from several investigators cichlid fishes have become a prominent example of recent and rapid speciation (see Fig. 12-18). Cichlids have undergone sympatric speciation, especially in small crater lakes where they can diversify ecologically and behaviorally. The combination of:

- aggressive behavior,
- territoriality,
- ability to consume specialized diets (to such an extent that one clade of seven species of cichlids in Lake Tanganyika has asymmetrical jaws, which are used to scrape scales from either the left or the right side of prey fish),
- phenotypic plasticity (see Box 17-3, Pharyngeal Jaws in Cichlid Fishes), and
- the availability of many microhabitats in the lakes,

all compensate for any lack of geographical isolation (T. Goldschmidt, 1996).

Phenotypic plasticity emerges as an important mode of incipient speciation, especially in situations in which changes in appearance (body form, pigmentation and so forth) are accompanied by changes in behavior, both of which can serve as mechanisms isolating individuals or groups within a population (West-Eberhard, 2003). Cichlid fishes show considerable evidence of both. In a recent study of cichlid species in a small (5 km diameter), shallow (200 m) and geologically recent (< 23,000 years old) volcanic crater lake, Lake Apoyo in Nicaragua, Barluenga and colleagues (2006) used a combined molecular, morphometric and ecological approach to demonstrate that:

- the lake has been populated only once by the Midas cichlid *Amphilophis citrinellus,* a benthic, robust, deep-bodied species that now is the most common species in the area;
- within less than 10,000 years a second species, the arrow cichlid (*A. zaliosus*) had evolved from the ancestral species, a species that is limnetic rather than benthic, elongate rather than high-bodied, and both ecologically and reproductively isolated from *A. citrinellus.*

Rates of Speciation

Mechanisms

Although we are just beginning to discern the underlying mechanisms of speciation, the presence of so many fossil and extant species enables some estimates of **evolutionary rates** from geological and paleontological data (see Chapters 6 and 13 through 20) or from biochemical changes and molecular clocks (see Chapter 12).

Rate determination in both cases is beset with problems. Should we measure rates in geological duration and periods (chronological time) or in generations (biological time)? What morphological or molecular features should we use to measure rates, and how do we determine the numbers and kinds of genes involved? No agreement has been reached on solutions to these problems.

On the paleontological level, taxonomic difficulties also intrude. "Lumpers" and "splitters" combine or split groups of organisms into different taxonomic categories (see Chapter 11). This might lead, for example, to different numbers of genera for the same lineages and thus to different generic evolutionary rates. Incompleteness of the fossil record and the frequent absence of evolutionarily intermediate groups also cloud many paleontological rate determinations. Nevertheless, as **Figure 24-12** shows, evolutionary rates among known fossil taxa of various mammalian groups

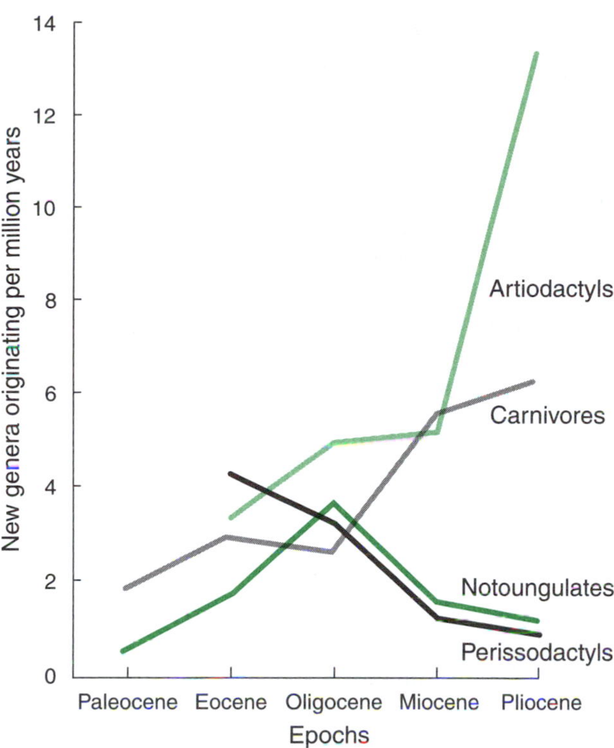

FIGURE 24-12 Evolutionary rates measured in terms of new genera originating per My in four clades (orders) of mammals during the Tertiary. (Adapted from Simpson, G. G., 1949. *The Meaning of Evolution*. Yale University Press, New Haven, CT.)

differ significantly, a finding that seems true of many other lineages for which we have fossil information.

Because rates often seem to vary over time *within any particular group*, one paleontological method has been to classify rates in terms of whether they are *slower or faster than the most typical rates in the fossil record*. As you might expect, "living fossils" such as *Latimeria* and *Neopilina* (see Chapter 3) lie at the slower end of evolutionary rates in their respective groups, while artiodactyls lie at the faster end of mammalian rates. G. G. Simpson gave the name **quantum evolution** to rapid evolutionary change, which is often marked by expansion into new adaptive zones and the origin of new taxa. As noted in Chapter 12, the term **macroevolution** is used to distinguish evolutionary changes at the species level and above. Evolution within the populations of a species is known as **microevolution.** Most population geneticists propose that the speciation process adds new directions to evolution, but that its mechanisms are similar to those used in nonspeciation changes.

Rates of Evolution

Rates of evolution have been approached in various ways, ranging from calibration of the fossil record to calculation of rates of change at the molecular level — the molecular clock or clocks (see Chapter 12) — to whether speciation occurs at a constant rate, is fast or slow, continuous or discontinuous, gradual or punctuated. Finding a metric that can compare rates of evolution/speciation is an important issue.

Darwin Units

Morphologically, a common proposal for evaluating evolutionary rate is the amount of change in a character divided by elapsed time, measured in **Darwin units (*D*)**, following a proposal made by Haldane. The value of such rates depends on completeness of the fossil record in providing valid phylogenetic information and accurate chronology.

Given such information, a rate in Darwins is calculated as the difference in a character's average dimension ($\bar{X}$) from time t_1 to time t_2, in natural logarithms (base e = 2.718):

$$\text{Rate} = (\ln \bar{X}_{t2} - \ln \bar{X}_{t1})/(t_2 - t_1)$$

Such logarithmic calculations offer the advantage of allowing evolutionary rate comparisons among organisms of different sizes. For example, a small femur in a rodent lineage that increased by .1 Darwin evolved twice as fast as a much larger elephant femur, which changed by only .05 Darwin, although the measured increase in size of the elephant's femur was larger. Illustrating both the power of sustained natural selection and taxon-specific rates of evolution, Reznick and coworkers (1997) reported that when transplanted from a high predator to a low predator environment for only several years, the guppy *Poecilia reticulata* evolved at a rate of thousands of Darwins.

Fenster and Sorhannus (1991) review these and other methods for obtaining numerical rates of morphological evolution. Interestingly, such rates show a contrast between slow changes in the fossil record measured in single Darwins or less, and more rapid evolutionary changes in extant populations, measured in thousands of Darwins, as in color changes in British peppered moths. Such marked differences reside in the numbers, kinds and durations of selective and environmental interactions experienced by fossil and contemporary populations.

Although the question of what accounts for evolutionary rate differences, whether on micro- or macroevolutionary levels, fossil or current, has not yet been satisfactorily answered, many biologists agree that such differences rely on various factors, three of which are outlined below:

1. **The structure of populations,** their size, genetic variability, distribution, and so on. Sewall Wright proposed that evolution can proceed more rapidly in a population subdivided into demes than in one more interconnected and homogeneous. Similarly, Ernst Mayr proposed the

rapid evolution of new taxa in small "founder" groups that break off from the peripheries of large populations (also called peripatric speciation).

2. **Adaptive and developmental constraints** that limit or dictate the structures and functions that organisms are able to achieve. Humans and most extant mammals, for example, cannot develop wings to escape predators. Such constraints do not mean that selection is ineffective, only that evolution proceeds along one path rather than another. Constraints at the morphological level may not reflect constraints or changes at other developmental levels (see Chapter 13). Thus, frogs with adult stages that have remained morphologically similar for about 200 My — showing a low rate of evolution — nevertheless have undergone considerable larval changes (including loss of the larval stage on multiple occasions) to produce the more than 3,000 extant species — showing a high rate of evolution.
3. **Changes in the direction and intensity of selection at different times in different groups**, the historical-environmental contingencies that all populations experience. Changes in the biological and physical environments that define the ecological niche (e.g., different habitats, food supplies, competitors, predators or parasites) are the stimuli for adaptation.

Supporting the importance of selection among these factors is the finding that rate differences do not necessarily correlate with generation time or available genetic variability, measured as electrophoretic protein differences (see Fig. 10-25). Some mammals with short generation times, such as opossums, evolved much more slowly than other mammals with much longer generation times, such as elephants. The slowly evolving horseshoe crab *Limulus* (see Fig. 11-3) shows as much electrophoretic genetic variability as more rapidly evolving invertebrates.

Punctuated or Gradual?

An important question still unresolved to the satisfaction of all is whether speciation involves mechanisms that differ from the mechanisms that cause changes within lineages.

In contrast to the view that the mechanisms of micro- and macroevolution are similar, proponents of **punctuated equilibria** such as Gould (1994, 2002) and Stanley (1973, 1979) argue that the mode of origin of new taxa is qualitatively different, as evidenced by the rapidity of macroevolutionary events in the fossil record. Punctuationists see the fossil record as long intervals of microevolutionary stasis, or equilibrium, during which relatively little change occurred, punctuated by rapid macroevolutionary periods during which new taxa arose. Some go further and propose that those new taxa arose through entirely new causes and mechanisms, but it is important to *separate the two issues of rates of change and modes of change*. Eldredge and Gould (1972), the initiators of the punctuated equilibrium theory, have moderated some of their early views, declaring that their dispute with neo-Darwinism is more a matter of emphasis on "species sorting" and "stasis" than on differences in speciation mechanisms. One can only agree with Gould (1994), that "macro- and microevolution should not be viewed as opposed, but as truly complementary."

Macroevolutionary changes are compatible with population genetics. From both neo-Darwinian and punctuationist views, the rate of evolution in a new population offshoot may be rapid compared to changes in its parental species, which remains tied to its more traditional ecological niche. The fact that paleontology can show evolutionary stasis in one or more branches of a group and rapid evolution in the group's other branches does not contradict the neo-Darwinian concept of evolution, but rather reflects the time scale over which paleontologists can view speciation events. Fossil data used to support punctuationism can be countered by data supporting gradualism and by data that indicate some speciation events are punctuationist and others are gradualist (Geary, 1990). In one of the last pieces he wrote, Ernst Mayr encapsulated the essence of the "problem" with **stasis** when he linked the issue to the interplay between variation and selection: "Periods of high phenotypic variation (as in the Pre-Cambrian, Cambrian) have alternated at other geological periods with periods of relative stasis. As natural selection is active at all periods, it would seem that the evolutionary change depends more on the availability of variation than that of selection" (2005, p. xviii).

Continuous or Discontinuous?

No neo-Darwinian rule prescribes that the rate of speciation be uniform, cannot be variable or even change abruptly. Evolutionary change is expressed in different patterns in different lineages, especially over different time scales. Two of many examples are as follows. In many plants, **allopolyploidy** — the cause of reproductive isolation in chromosomally doubled hybrids — is a common source of rapid speciation. Organisms experienced environmental interactions in the past different from the present; we have also long recognized that catastrophic events can account for considerable evolutionary changes, as can other unique historical phenomena. Such influences neither detract from the neo-Darwinian synthesis nor do they provide cause to make an explainable process unexplainable (see the introduction to Part 1).

Thus, although macroevolution may at times be associated with large environmental changes and microevolution with smaller ones, mechanisms used in one mode need not be excluded in the other. Regulatory mutations, bottlenecks and new directions and intensities of selection, as well as other presumed macroevolutionary mechanisms, also help explain microevolutionary events. Lenski and Travisano's

Born: March 12, 1934

Birthplace: Madrid, Spain

Undergraduate degree: B.S., University of Madrid, Spain

Graduate degrees: M.A., Columbia University, New York, 1963; Ph.D., Columbia University, New York, 1963

Postdoctoral training: Rockefeller University, New York, 1964–1965

Present position: Donald Bren Professor of Biological Sciences Department of Ecology and Evolutionary Biology University of California at Irvine[9]

Francisco J. Ayala

What prompted your initial interest in evolution?

I was born in Spain and went to school there. However, my early interests in evolution developed primarily outside of school in trying to understand why the living world was so diverse and especially the origin of humans. These interests were fed by reading Spanish translations of various books including such twentieth-century classics as *Genetics and the Origin of Species* by Theodosius Dobzhansky, and *Evolution: The Modern Synthesis* by Julian Huxley. Once started, I also read books by Richard Goldschmidt and C. H. Waddington. The effect of this exposure was to emphasize the relevance of genetics to the study of evolution, and I undertook experimental work with *Drosophila* at the University of Salamanca. My teachers there encouraged me to develop my interests further, and in 1961 I came to the United States to study for a Ph.D. under Theodosius Dobzhansky at Columbia University.

What do you think has been most valuable or interesting among the discoveries you have made in science?

I have been primarily involved in the study of genetic variation and the role it plays in evolution. Using techniques such as electrophoresis of enzymes, many of my experiments were oriented towards discovering the genetic differences that account for speciation in *Drosophila*, such as genetic distinctions between subspecies and between sibling species.

What areas of research are you (or your laboratory) presently engaged in?

More recently, I have also become interested in measuring genetic variation in parasitic protozoa such as trypanosomes, in attempting to understand the mechanisms by which these parasites rapidly adapt themselves to changes in host immune systems. In a sense, these genetic changes represent small, isolated, capsules of evolution. My publications also extend to various areas in the philosophy of biology such as teleology, reductionism, and the biological foundations of ethics. I have also written on the use of testimony by scientists in courts of law. A basic concern to which I have addressed considerable effort is the necessity for science education in schools, especially the teaching of evolution.

In which directions do you think future work in your field needs to be done?

The tremendous power of molecular biology offers the opportunity to answer long-standing evolutionary questions such as how extensive are the genetic differences between species, of what kinds are they, and at what rates do they occur. I intend to continue my recent research on the evolutionary history and population structure of *Plasmodium,* which causes malaria, a disease affecting several hundred million people each year.

What advice would you offer to students who are interested in a career in your field of evolution?

From my own experience, I believe it is extremely important for students to identify conceptual problems in their area of interest by reading the "masters," and by discussion and exchange with others in the field. Science is a community enterprise! Also, preparing for a career in science necessitates getting the best "tools" one can; in evolution these can include one or more disciplines such as biochemistry, systematics, mathematics and statistics.

[9] For a biography, visit http://www.faculty.uci.edu/profile.cfm?faculty_id=2134&term_list=ayala&name=Francisco%20 J.%20Ayala. Accessed May 22, 2007. Recent publication: Benton, M. J., and F. J. Ayala, 2003. Dating the tree of life. *Science,* **300**, 1698–1700.

EVOLUTION ON THE WEB biology.jbpub.com/book/evolution

(1994) study of laboratory bacterial populations undergoing many thousands of generations of selection shows that evolutionary changes that may seem "punctuational" are explainable as rapid increases in frequency of new mutations followed by static periods until further such mutations appear. When the opportunity for diversity ("adaptive radiation") presents itself by exposure to novel environments, bacteria adapt through simple mutations. Such "stop and go" events support the *"streetcar theory of evolution"* (Hammerstein, 1996), in which evolutionary advances pause until temporary genetic constraints are overcome by the entry of new genetic "passengers" (mutations) and replacement of old ones. Furthermore, what seems static when regarded at one level can be dynamic at another. Morphological differences among early mammalian Mesozoic fossils may seem relatively minor and inconspicuous, but we know that physiological changes, involving temperature regulation, lactation and viviparous reproduction, were dynamic in effect and highly significant evolutionarily (see Chapter 19).

KEY TERMS

Allen's rule
allopatric populations
allopatric speciation
behavior
behavioral isolation
Bergmann's rule
bottlenecks
coadapted gene combinations
ecogeographical rules
evolutionarily stable strategies
evolutionary rates
female choice
founder effects
genetic distance
groups
habitat isolation
hybrid breakdown
hybrid inviability
hybrid sterility
innate behaviors
instincts
introgressive hybridization
isolating mechanisms
learned behaviors
macroevolution
microevolution
parapatric
parental investment
peripatric speciation
postzygotic isolating mechanisms
prezygotic isolating mechanisms
punctuated equilibria
quantum evolution
Rapoport's rule
reciprocal altruism
seasonal isolation
sexual isolation
sociobiology
speciation
stasis
sympatric populations
sympatric speciation
zones of hybridization
zygotic mortality

DISCUSSION QUESTIONS

1. Human groups.
    a. How would you define a group?
    b. Do "pure" groups exist?
    c. What factors are involved in increasing and decreasing the number of groups in a species?
    d. How can we measure genetic distances among groups?
    e. What ecogeographical rules can we apply to groups?
2. Behavioral adaptations.
    a. How do innate and learned behaviors differ?
    b. Do learned behaviors have a genetic basis?
    c. What model of behavioral strategy can evolve when both cooperative and selfish responses can occur in interactions between individuals?
3. What are the conflicting interests between the two sexes of polygynous species in how they "choose" mating partners? How can such conflicting interests lead to exaggerated male–female dimorphism?
4. However species are defined, would you support the concept that species represent natural groupings of organisms? Why or why not?
5. What response would you offer to the statement, "All isolating mechanisms are equally efficient"?
6. What conditions would promote selection for prezygotic isolating mechanisms? Would you say that such selection could occur between groups of the same species, or that such selection could only occur between groups that have already speciated?
7. How would you support the concept that speciation can occur rapidly because of founding accidents or bottlenecks? What views oppose this concept?
8. How does sympatric speciation differ from allopatric speciation? What support is there for each of these modes of speciation?
9. Would you distinguish microevolution from macroevolution and if so, how?
10. What evidence can you offer for and against the punctuated equilibrium hypothesis?

EVOLUTION ON THE WEB

Explore evolution on the Internet! Visit the accompanying Web site for *Strickberger's Evolution*, Fourth Edition at http://www.biology.jbpub.com/book/evolution for Web exercises and links relating to topics covered in this chapter.

RECOMMENDED READING

Cavalli-Sforza, L. L., 1994. *The History and Geography of Human Genes.* Princeton University Press, Princeton, NJ.

Coyne, J. A., and H. A. Orr, 2004. *Speciation.* Sinauer and Associates, Sunderland, MA.

Eberhard, W. C., 1996. *Female Control: Sexual Selection by Cryptic Female Choice.* Princeton University Press, Princeton, NJ.

Emlen, D. J., 2000. Integrating development with evolution: A case study with beetle horns. *BioScience,* **50**, 403–418.

Hall, B. K., and W. M. Olson (eds.), 2003. *Keywords & Concepts in Evolutionary Developmental Biology.* Harvard University Press, Cambridge MA.

Hey, J., W. M. Fitch, and F. J. Ayala (eds.), 2005. *Systematics and the Origin of Species on Ernst Mayr's 100th Anniversary.* The National Academies Press, Washington, DC.

Levinton, J., 2001. *Genetics, Paleontology, and Macroevolution,* 2nd ed. Cambridge University Press, Cambridge, UK.

Maynard Smith, J., 1982. *Evolution and the Theory of Games.* Cambridge University Press, Cambridge, UK.

Maynard Smith, J. and E. Szathmáry, 1995. *The Major Transitions in Evolution.* Freeman, New York.

Nowak, M. A., 2006. *Evolutionary Dynamics.* Harvard University Press, Cambridge, MA.

Pagel, M. (ed. in chief), 2002. *Encyclopedia of Evolution.* 2 Volumes, Oxford University Press, New York.

Plavcan, J. M., R. F. Kay, W. J. Jungers, and C. P. van Schaik, 2002. *Reconstructing Behavior in the Primate Fossil Record.* Kluwer Academic/Plenum Publishers, New York.

Wilson, E. O., 1975. *Sociobiology: The New Synthesis.* Harvard University Press, Cambridge, MA.

Wilson, R. A. (ed.), 1999. *Species. New Interdisciplinary Essays.* MIT Press, Cambridge, MA.

Manipulating your environment is an ability associated with humans, indeed, often assumed to be restricted to humans. One aspect of environmental manipulation is the making of tools as an aid to obtaining food. Similar tool making from generation to generation is evidence of the ability to transfer knowledge of how to make tools through social interaction and cultural evolution and from generation to generation.

Tool making (and, therefore, cultural evolution) is not restricted to humans but is highly developed in chimpanzee (*Pan troglodytes*), as seen in the figure of the chimp using a stick as a tool. Tool making in mammals either predated the separation of chimps and hominins 6 Mya or arose by parallel evolution in the two groups after the split.

Both tool making and cultural transmission of the ability to make tools are seen in the New Caledonian crow (*Corvus moneduloides*), a species endemic to New Caledonia and the nearby Loyalty islands of Lifou, Maré and Ouvéa. The figure shows two crows, one using a section of a leaf to probe for insects, the other waiting in the wings. What the figure does not show is that the crows fashioned the tools by reducing the leaves to narrow strips. Nor does it show that crows can fashion leaves to strips of different sizes appropriate to the size of the hole being probed, that tools are stored for subsequent use, or that these abilities are learned over a year or more by imitating older individuals. (chimpanzee, © Norma Cornes/ShutterStock, Inc.; crows, © Behavioural Ecology Research Group, University of Oxford.)

PART

Evolution and Society: Past, Present and Future

Society and Religion

In Chapter 25, we turn to our own species to discuss cultural evolution and the influence that we have on the evolution of other species.

Cultural evolution appears in its most developed form among humans. Furthermore, unlike almost all other animals, humans transfer information from generation to generation through both genes and culture. In the nineteenth century, proponents of social Darwinism maintained that cultural differences evolved primarily by natural selection, as embodied in the concept "the survival of the fittest." This belief, which "justified" many social inequities, was based on the erroneous supposition that society, which often incorporates nonbiological goals and value systems, is governed by the same laws that govern biological evolution. Although biological factors are involved in most social patterns, they are often modified by cultural context. We discuss ways in which we have tried to influence our own evolution — social Darwinism, sociobiology, eugenics, genetic engineering, cloning — but we must remember that any desire to control our evolution is restricted by the potential incompatibility between cultural and biological fitness. We might wish to lengthen life span, but natural selection rarely acts on postreproductive individuals. We might wish to lower fertility, but biologically, high fertility has selective value, although socially it can lead to population and ecological crises.

Also discussed in Chapter 25 is the way we have influenced and continue to influence the genomes and therefore the evolution of other species through domestication, cloning, and manipulation of Earth, its climate and atmosphere. Starting in the Paleolithic era, when hunting and the widespread use of fire may have led to the extinction of larger animals, we have altered both the species mix and environment of almost the entire globe.

In Chapter 26 we discuss how, for many centuries, the religious rationale for social, cultural and biological systems was that the universe followed a designed order established by an intelligent deity. Until Copernicus and Galileo in the sixteenth century, no one had seriously challenged the idea of a powerful deity controlling the physical universe. In the new worldview they and others brought about, God appeared as an initial creator rather than as an incessant manipulator of the solar system. Darwinian evolution had a profound impact on biology and on many other fields, Western religion in particular. Its extraordinary influence fitted into the social, economic and technological developments that helped overthrow the old social order of feudalism and monarchy and brought about the rise of capitalism. The traditional religious rationale for social and biological systems was that the universe followed a designed order established by an intelligent deity. Evolutionary theory, by omitting a supreme being from the equation, was perceived as a dire threat to religious interests.

The advent of Darwinism posed threats to society, especially to religion, by suggesting that biological relationships, including the origin of humans and of all species, could be explained by natural selection without the intervention of a god. Many believed that evolutionary randomness and uncertainty had replaced a deity having conscious, purposeful, human characteristics. The Darwinian view that evolution is a historical process — extant organisms were not created spontaneously but formed in a succession of selective events that occurred in the past — contradicted the common religious view that there could be no design, biological or otherwise, without an intelligent designer, creationism and the intelligent design movement being manifestations of these beliefs. These issues are discussed in Chapter 26.

As is clear from this these final chapters, indeed, as should be clear from the entire book, many intellectual threads led to the modern theory of evolution, a theory that requires recognition that Earth is ancient, that there is a common inheritance within a biological group, and that natural events are explained by discoverable natural laws. It took a long time to weave these threads into an evolutionary tapestry. We hope that this presentation of the evidence for and the fact of evolution will enable you to appreciate more clearly the richness and beauty of that evolutionary tapestry.

25 Cultural and Human-Directed Evolution

Chapter Summary

Unlike most other animals, humans transfer information from generation to generation through both genes and culture. The tempo of cultural evolution has been so much more rapid than human biological evolution that we each gather new experience at a rate many times that of our ancestors.

In the nineteenth century, proponents of Social Darwinism maintained that cultural differences evolved primarily by natural selection, as embodied in the concept of "survival of the fittest." This belief, which "justified" many social inequities, was based on the erroneous supposition that society, which often incorporates nonbiological goals and value systems, is governed by the same laws as biological evolution.

A more sophisticated approach is sociobiology, the advocates of which propose that there is a biological basis for much of human culture and that genotypes predisposed toward cultural development accumulate through natural selection. Although biological factors are involved in most social patterns, they are often modified by cultural context. Cultural change may occur without biological input and cannot be explained simply by biological laws.

Any desire to control our evolution is restricted by the potential incompatibility between cultural and biological fitness. We might wish to lengthen life span, but natural selection rarely acts on post-reproductive individuals. We might wish to lower fertility, but biologically high fertility has had selective value, although socially it has led to population and ecological crises. A large proportion of the human population is or will be affected by deleterious alleles that persist because of forces that maintain genetic variability or because through the intervention of modern medicine individuals with genetic disorders are able to live longer and thus reproduce. Three approaches to controlling our own evolution — eugenics, genetic engineering and cloning — are discussed, as are human efforts to engineer the genomes of other species, all of which raise ethical concerns.

We have long pondered the state and future of our own species. Alone among living creatures, we are able to formulate and attempt to answer such questions as: How close are the ties between biology and culture? In which directions are humans evolving? Are human biological endowments satisfactory for human needs? What are the prospects for controlling our evolution and the evolution of other species? Here we begin our discussion by considering some important features of *Homo sapiens*.

Learning, Society, and Culture

Perhaps the most distinctive feature of our species is our mental capacity — what we call **intelligence** — and our consequent **ability to learn from our own experience** or from the experiences of others.

However measured, intelligence provides us with flexible adaptive behaviors that are far more complex than those attained by any other species. We have the capacity to analyze our environmental experiences, incorporate the lessons learned into our behavior and then create new environments over which we have considerable control.

Much human learning follows a Lamarckian pattern in the conscious acquisition and transmission of those behavioral responses that answer the needs of specific situations. Although some other organisms have the capacity to learn (see Chapter 24), they still primarily rely for survival on automatic responses genetically built into their nervous systems. Humans, in contrast, can flourish in a wide variety of environments, learning to acquire food, defend themselves, find shelter and perform various tasks in many specialized ways.

Most important, more than any other animal, humans can acquire and transmit such practices and behaviors — **culture** — through social exchanges involving language, teaching and imitation, both among individuals and between generations. Cultural transmission of learned behavior eliminates the hazards encountered when an individual must learn independently, by trial and error, to cope with environmental variables. Cultural transmission allows more successful imitative learning of adaptive practices, often developed over more than a single lifetime, into our social and cultural heritage.

Some writers argue that we should *broaden the term culture* to include any form of socially transmitted learned behavior even if the transmission is by imitation alone. This definition extends culture to organisms lacking speech. Chimpanzees, for example, are able to incorporate, through imitation, variations in learned behaviors into a community, resulting in a variety of social histories ("traditions") from group to group (see Chapter 20). Whiten and coworkers (1999) support this view by demonstrating that 39 "behaviour patterns, including tool usage, grooming and courtship behaviours, are customary or habitual in some [chimpanzee] communities," and by showing that differences in these patterns are unrelated to genetic differences between communities.

Because of socially mediated transmission, cultural changes — unlike changes in genetic makeup — are not restricted to passage from one generation to another, but may be proposed, accepted and used during most stages in the human life cycle in interactions between both related and unrelated individuals. This means that the cultural "parents" of individuals need not be their biological parents, nor need cultural parents derive from the same geographical area as their cultural offspring.

The kinds of isolation barriers that inhibit genetic exchange between biological species do not exist in the transmission of culture between human groups. Humans thus have two hereditary systems:

1. the **genetic system**, which transfers biological information from biological parent to offspring in the form of genes and chromosomes, and
2. the **extragenetic system**, which transfers cultural information from speaker to listener, from writer to reader, from performer to spectator and forms our cultural heritage.

Both systems are informational, producing their effects by instruction, the biological system through the information embodied in DNA via the coding properties of those macromolecules, the cultural system through social interactions coded in language and custom and embodied in records and traditions.

Relative Rates of Cultural and Biological Evolution

The changes produced by cultural heredity over the last 10,000 years have been dramatic. Somewhere during the Neolithic Age (see Fig. 20-24), the long-prevailing lifestyle of hunting–gathering–fishing began to give way to the cultivation of food using domesticated plants and animals.

TABLE 25-1 Major human expansions from the Neolithic age onward

Center of Origin	Area of Expansion	Time, Years Ago	Technologies	Crop/Product
Middle East	Europe, North Africa and Southwest Asia	10,000 to 5,000	Farming and domestication	Wheat, barley, goats, sheep, cattle
North China	North China	9,000 to 2,000	Farming and domestication	Millet, pigs
South China	Southeast Asia	8,000 to 3,000	Farming and domestication	Rice, pigs, water buffalo
Central America and North Andes	Americas	9,000 to 2,000	Farming	Corn, squash, beans
West Africa	sub-Saharan Africa	4,000 to 300	Farming	Millet, sorghum, cowpea, gourd
Eurasian steppes	Eurasia	5,000 to 300	Pastoral nomadism	horses, warfare
Southeast Asia or Philippines	Polynesia, Madagascar	5,000 to 1,000	Oceanic navigation	
Greek colonization	Mediterranean	4,000 to 2,400	Navigation and trade	

Source: From Cavalli-Sforza et al., 1993

Energies formerly expended in finding food could now be directed into the more productive methods of agriculture. Although originally developed in the Middle East, China and Central America, such changes spread rapidly. Within 1,000 years or so, many contiguous areas had begun some form of agriculture. Within 5,000 years, agriculture and the technologies it stimulated extended widely (**Table 25-1**).

Perhaps the most immediate and far-reaching effect of agriculture was to increase the food supply many-fold, resulting in larger populations and a greater population density in agricultural communities. What matters here is the change in human lifestyle. From bands of food gatherers, hunters and fishers mainly concerned with satisfying hunger, we moved to complex urban societies in which many if not most people could afford to pursue other occupations. Skills, abilities and behaviors that were used only occasionally if at all in the past became important adaptations for new technologies and lifestyles. Notwithstanding the uncertainties about whether sedentary communities preceded or followed agriculture, and whether agriculture was stimulated by climatic changes or by increased Neolithic social complexities, it did not take many centuries for villages to grow into towns, and towns into cities. These and other social effects of agriculture, which have been much discussed, have profoundly affected all areas of human interaction and creativity, from economics and politics to art, technology and science.[1]

One measure of how change continues to affect us is the time it takes to double our collective knowledge, a process that once took many thousands of years but now happens in a mere handful.[2] The rate with which purposeful modification of society takes place and the consequent rate of cultural change are limited by many factors, but — theoretically at least, and in the long run — primarily by human inventiveness. The generation time for cultural evolution is as rapid as communication methods can make it. We can now move from place to place faster than the speed of sound, and transfer ideas at electronic speed. Yet in contrast to the rapid changes associated with cultural and technological heredity, changes in human biological heredity since we adopted agriculture seem relatively small, if detectable at all. Indeed, the most distinguished possession of *Homo sapiens*, the human brain, shows no change in size over the last 100,000 years (see Chapter 20), nor is there any clear indication that any change in functioning occurred during this period. Why this difference in speed between cultural and biological evolution?

Oversimplifying (although not seriously), this contrast can be ascribed to differences between *two distinct types of evolution:* the mode of inheritance of acquired characters associated with cultural evolution, and the mode of inheritance through natural selection involved in biological evolution. The **Lamarckian mode of cultural evolution** is an extension of the method by which humans learn. It depends on conscious agents — humans with brains — who can modify inherited cultural information in a direction that offers them greater adaptiveness or utility. Furthermore, transmission occurs from mind to mind rather than through DNA. Thus, the information we receive from ancestors and

[1] See Bronowski (1974), Clutton-Brock (1999), Diamond (1997, 2005) and Mindell (2006) for broad-based and integrative approaches to these topics.

[2] A widely cited estimate, made by D. J. da Solla Price in 1963, is that of all the scientists who have ever lived, more than 90 percent were alive then, a figure echoed by Peter Gruss (President of the Max Planck Society for the Advancement of Science) in 2005.

contemporaries can be purposely changed to provide improved utility for ourselves, our offspring and others.

The high rate of cultural evolution is in striking contrast to the slow progress of natural selection, for reasons that should be apparent from earlier parts of the book. Organic evolution occurs through a process of selection (among other forces) acting on random genetic changes (mutation and recombination) to furnish an array of genotypes among which the environment selects only some for survival. Genetic evolution has been regarded as slow because it must await fortuitous changes in DNA sequences, their organization into the existing genome and selection of the resulting phenotype, before it can proceed. Each change may take many generations before it is incorporated into the population, although as we have seen, processes ranging from horizontal gene transfer in prokaryotes (see Chapter 10) to phenotypic plasticity in eukaryotes (see Chapters 13 and 24) can greatly speed the rate(s) of biological evolution.

This slow process of biological evolution is vastly different from the rapid, conscious selective process humans use to choose among behavioral alternatives, although the biological equipment needed to transmit and use cultural information (memory, perception, language ability, and so on) connects them both. Human minds have become agents of a novel selection mechanism by consciously choosing among alternatives because of their consequences. We say *minds* and not *mind* because cultural evolution relies on communication among individuals and on group interaction. Cultural evolution is vastly more than the sum of its parts, in that the products of a socially coordinated group of individuals — a city, a daily newspaper, an automobile factory, a cathedral, a film — are quantitatively and qualitatively more than such individuals could create alone.

Social Darwinism

Nevertheless, the facts that (1) human culture has at its source a biological foundation, and (2) both culture and biology arise from informational systems that evolve over time prompted various writers in the late nineteenth and early twentieth centuries to suggest that *biological concepts can be extended to society,* and that nature and culture share similar evolutionary mechanisms, especially natural selection. These ideas were developed into a body of thought, later called **Social Darwinism** (see Chapter 3), the chief concepts of which were:

- Differences among human individuals and groups arise through natural selection.
- Natural selection is the mechanism that leads to social class structures and to national differences with respect to economic, military and social power.

In particular, early writers on the subject saw parallels in the role of competition in the two spheres. Slogans such as "struggle for existence" and "survival of the fittest," when applied to social traits, enabled English Social Darwinists, especially Herbert Spencer (1820–1903), to suggest that cultural evolution was proceeding inevitably toward social and moral perfection, and approaching its culmination in English Victorian society.

Spencer's writings,[3] which ranged widely across biology, economics, philosophy and sociology, continued to exert their influence well into the twentieth century. Interestingly, although Spencer coined the term "survival of the fittest," implying the action of natural selection, he remained a Lamarckian in respect to biological evolution until relatively late in his life. Whatever its mechanisms, Spencer conceived evolution as a powerful mystical force that governed all spheres of existence and therefore justified social and economic policies that supported those who were most "morally fit." For many Protestant intellectuals, Spencer's belief in such an evolutionary "cosmic" power helped reconcile science to their religion and made his writings extremely popular.

In its harsher forms, the Spencerian approach became popular in various circles in the United States (where more than half a million copies of his books were sold), especially through the teachings of William Graham Sumner, the best-known American Social Darwinist, who in 1883 wrote: "We cannot go outside of this alternative: liberty, inequality, survival of the fittest; not-liberty, equality, survival of the unfittest. The former carries society forward and favors all its best members; the latter carries society downwards and favors all its worst members." Not surprisingly, many wealthy capitalists found such views to their liking. John D. Rockefeller, Jr., for example, whose father forged the gigantic Standard Oil Trust by destroying many smaller enterprises, justified such behavior with the observation that, "The growth of a large business is merely a survival of the fittest . . ." (quoted in Ghent, 1902, *Our Benevolent Feudalism*).

Along with objections raised by others, including T. H. Huxley in his *Evolution and Ethics,*[4] the major problem with Social Darwinism was its assumption that society (economics, politics, and so on) operates through the same laws as biology and for the same goals. As repeatedly pointed out,

[3] See for example, *The Factors of Organic Evolution* (Spencer, 1887) and *First Principles* (Spencer, 1897).

[4] A wonderful opportunity to evaluate Thomas Huxley's development of evolutionary ethics is available. Huxley's 1894 book *Evolution and Ethics, and Other Essays,* contains the text of the Romanes Lectures (*Evolution and Genetics*) he delivered at the University of Oxford in 1893. Fifty years later his grandson, Julian, delivered the Romanes Lectures on *Evolutionary Ethics,* and published both sets of lectures with an extensive introduction and commentary (T. H. and J. S. Huxley, 1947). Both grandfather and grandson affirm their belief in evolution as a means to enhance our moral progress; essays by both are reprinted in Appleman (2001).

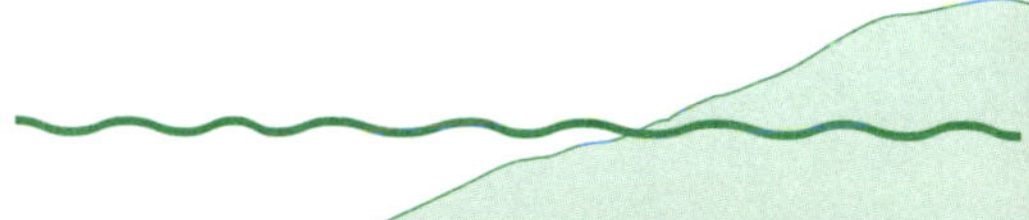

however, this assumption is false because no evidence exists that what is in biology, *is* or *ought to be* in society. The laws of the inheritance of wealth and power in society are entirely man-made; the laws of biological inheritance are not. Further, social rewards bestowed on individuals or groups may be unrelated to biological merit or even to presumed social merit. Because we can consciously select them, we can direct social goals toward almost any objective we choose: poverty or wealth, obedience or revolution, and so on.

Although Social Darwinism has no valid scientific basis — the socially "fit" are not necessarily the biologically "fit" — the philosophy attracted (and perhaps still attracts) individuals and groups who aspired to occupy superior positions over other individuals or groups, and has been used to justify or reinforce racism, genocide and social and national oppression. An extreme example is the role played during the 1930s and early 1940s by Adolf Hitler in the "racial health" movement. It was an important ideological element in the purposeful destruction of millions of people because they were considered members of "inferior" racial groups. Even in the United States, with its more democratic social heritage, laws were passed during the 1920s restricting immigration from eastern and southern Europe because of their "inferior" or "undesirable" races, while immigration of virtually all Asians was halted in 1882 by the Chinese Exclusion Acts, which were not repealed until 1943. (It must be added, however, that economic and political considerations played a large role in the push for immigration restrictions.) The United States was far from alone. For example, Australia enshrined what came to be known colloquially as the "White Australia Policy" in the second bill passed after confederation in 1901.

Sociobiology

Despite the corruption and failings of Social Darwinism, we can hardly deny that biology exerts influence on social interactions between people. The close mother-infant relationship is just one of many behaviors common to all human groups. But to what extent are social behaviors genetically determined? How much is nature and how much nurture?

E. O. Wilson's 1975 book, *Sociobiology: The New Synthesis,* launched a new field of science, but also evoked a firestorm of controversy. Defining **sociobiology** as "the systematic study of the biological basis of all social behavior," Wilson argued convincingly that animal behavior, like morphology, is shaped by natural selection: Behavioral traits that maximize an individual's reproductive success are more likely to be carried into the next generation than those that don't. Even such social behaviors as altruism — a difficult problem for evolutionary biologists because an organism sacrificing itself for the greater good may not leave any descendants — can be explained on the basis of genetic mechanisms, selection in this case acting not only on the fitness of an individual carrier of a favorable behavioral "gene," but also on that of the genetic relatives of such individuals ("kin selection" or "inclusive fitness"; see Chapter 23).

Drawing on methods used in population genetics, ethology, ecology and other disciplines, sociobiology provides a rationale for gauging how (and perhaps to what extent) biological causes account for social behaviors. In the years since Wilson's book, studies have been carried out on social behavior in a wide range of animals (see Chapter 24). For instance, why does a new dominant male in a pride of lions kill off his rival's offspring, and why will a female lion nurse not only her own cubs, but those of her close genetic relatives? (The actions help to ensure survival of, respectively, his own and her close relatives' genes into the next generation.)

Less successful has been the application of sociobiology to human behavior. Indeed, the major source of controversy in Wilson's landmark book was the last chapter, which extended the discussion to humans, as did a sequel, *On Human Nature,* published in 1978. Because of interactions between genes and environment, Wilson said, "there is no reason to regard most forms of human social behavior as qualitatively different from physiological and non-social psychological traits." According to this view, human nature is determined as much by heredity as by culture, and there are limits to how much we can change our behavior, (e.g., an individual's aggressive tendencies).

Although sociobiology deals less with the immediate causes of a particular human behavior than with its "ultimate" underlying evolutionary function, to its critics the discipline leads inevitably to the conclusion that most observed human social behaviors are biologically caused (with the potential although certainly not inevitable corollary that the status quo, including social inequities, is justified), a concept close to the views held by the Social Darwinists. But to the charge of determinism — to the implication that we are slaves to biological destiny — Wilson counters that he addressed this concern soon after publication of *Sociobiology:*

> The moment has come to stress that there is a dangerous trap in sociobiology, one which can be avoided only by constant vigilance. The trap is the naturalistic fallacy of ethics, which uncritically concludes that what is, should be. The 'what is' in human nature is to a large extent the heritage of a Pleistocene hunter-gatherer existence. When any genetic bias is demonstrated, it cannot be used to justify a continuing practice in present and future societies. ("Human Decency is Animal." *New York Times Magazine,* Oct. 12, 1975)

A second criticism of sociobiology — that we are treated as just part of nature rather than standing above it — questions whether Darwinian evolution is even an appropriate framework for discussing human behavior. Are humans not radically different from other animals? Do not our intelligence and culture transcend our animal nature and put us in a category of our own? As with so many areas of life, however, such a stark either/or choice is surely unnecessary; the challenge for sociobiology is to tease out the mix of biological and cultural evolution, an exceedingly difficult task given that we know next to nothing about our early (pre-civilization) history and that evolution proceeds over such vast time scales.

Perhaps the most important lessons to be learned are that biology and social behavior are tied to each other by many strands, some strong, others weak or imperceptible, and that social–behavioral responses, like other complex biological traits, are influenced by a wide spectrum of genetic variation, rather than being fixed in any group or population. Understanding social change, on the other hand, is quite a different task from understanding biological change: Societies and cultures cannot be explained as biological behaviors any more than biological behaviors can be understood as atomic interactions. The mechanisms, sequence, and ethics of cultural change involve different kinds of interactions and require explanations at a variety of levels of complexity (see Box 8-3, Complexity).

In summary, although the nature versus nurture debate has a long history and extends far beyond sociobiology, it is in this discipline, where the evolutionary history of our sociality is emphasized, that disagreements have been greatest. The field has been considerably discussed, commented upon, disputed and even vilified, not only in the years immediately following the publication of *Sociobiology* in 1975, but right up to the present day. In general, though, it has already provided us with valuable insights into the roots of our sociality.

Biology Versus Culture

The fact that cultural considerations can transcend biological considerations becomes apparent when we consider the topic of *human control over evolution*. In which direction are we to guide our evolution and the evolution of other species? What goals are we to set? These questions arise not from any underlying biological laws, but from the conscious cultural realization that we would like to improve ourselves and the world in which we live, and from the social technology that allows us to achieve such goals. This quest for human improvement, however, often comes face to face with our **biological limitations,** a few of which we now consider.

Although our lives have changed immeasurably as a result of our **advanced technology,** our genetic makeup has not. For example, we have become sedentary in occupation, but our bodies still require exercise. Many who live in surplus societies, such as the United States, tend to eat poorly, put on extra weight and suffer from the accompanying ills, including heart disease and diabetes. The stress of many aspects of modern life also finds much of the human species biologically unprepared, and we suffer from socially aggravated illnesses. In addition, pollution of various kinds caused by industry, automobiles, pesticides and tobacco leads to a variety of diseases, ranging from induced cancer to emphysema and silicosis.[5]

Perhaps one of the most important contrasts between what we are and what we would like to be lies in the difference between **biological and cultural maturity.** *Biologically,* our *efficiency* begins slowly to decline once we enter our 20s and 30s, ages that in the past coincided with reproduction or immediate post-reproduction. As measured by the contributions we make to various professions, however, our *cultural efficiency* often peaks a decade or more later. In effect, our cultural development is limited by our biological decline: Our biological heritage stresses early reproductive success, soon after which we begin to deteriorate physically, while our cultural development requires continued plasticity and longevity.

Modern societies also enjoy a long period of **post-reproductive longevity,** a feature that tends to be short or nonexistent in most organisms that attain reproductive maturity relatively early in their potential life span (Rose, 1994). Indeed, early in our own evolution, only about half the human population passed the age of 20 and probably not more than one in 10 lived beyond 40. These low longevity values extend into the period of the early Greeks, and even longer among some groups. Life expectancy remained between 20 and 30 years until the Middle Ages, then rose somewhat and has risen sharply among Europeans, Americans, Japanese and some others in the last century and a half, from about 40 years in 1850 to the present high 70s or low 80s (even now, very few people reach 100 years). Less advantaged countries also have seen dramatic increases, the major exception being those countries in Africa where the AIDS epidemic has reduced already low life expectancies to as little as the low 30s.[6]

These statistics are important because they indicate that as a result of improvements in sanitation, diet and medical practice over the last century and a half, natural selection now exerts relatively little influence on our "fitness," on the number

[5] See Gluckman and Hanson (2005) for evolutionary approaches to human health. A DVD, "Evolution — Why Bother?", produced by the American Institute of Biological Sciences and the U. S. Biological Sciences Curriculum Study (available at http://www.aibs.org/bookstore, accessed May 23, 2007) contains an up-to-date evaluation of how evolution impacts on agriculture, human health, the development of new drugs and other issues of the application of technology.

[6] For a table of rankings, see *The World Factbook,* a publication of the CIA available online at https://www.cia.gov/library/publications/the-world-factbook/index.html. Swaziland, with an estimated 2007 life expectancy of 32.23, occupied the lowest ranking of any country.

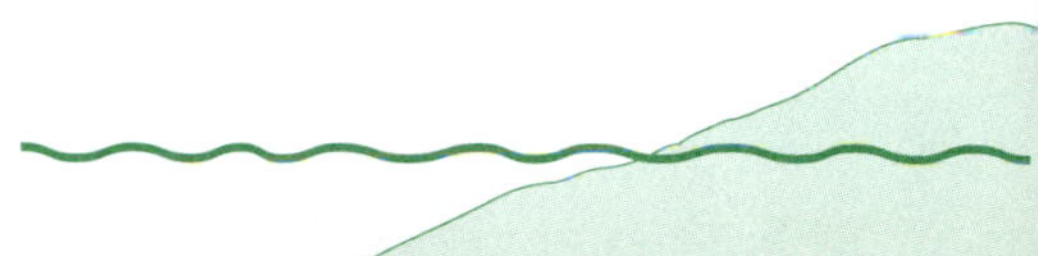

BOX 25-1 Population Explosion

In most countries, infant mortality rates have declined markedly over the past century (**Fig. 25-1**). From 15 or more percent of all births in 1900, infant mortality rates have fallen to less than one percent in more than 70 countries, a result of improved sanitation, nutrition and medical care, including control of infectious disease.[a] In most countries, however, it was (and still is) usual for birth rates to remain high for many years, even decades, following such a decrease. When combined with increased longevity, the result has been an exponential growth of the human population, an explosion that at its height in the 1960s saw a doubling every 35 years, a rate many thousands of times greater than that experienced in Paleolithic societies. Since 1900, the number of people has risen from a billion to the current 6.5 billion. Although the overall growth rate has dropped (the doubling time is now 61 years), the absolute numbers continue to climb. Today about 203,800 people are added to the world population every 24 hours, by far the majority of them in developing nations. The U.S. Census Bureau predicts a world population of 9.4 billion by the year 2050.[b]

This **population explosion** has grave implications not only for human beings, but also for the planet and all its other inhabitants. The more people there are, the more food, water, space, energy and other resources they need. The more resources are consumed, the more waste is produced. Our total impact on the environment, however, depends largely on how many resources each of us uses, and on how much waste each of us produces — our per capita **ecological footprint**,[c] which varies immensely between countries. A person living in the United States today consumes far more resources, produces far more garbage and emits far more carbon dioxide and other pollutants than does an inhabitant of, say, Nepal. Furthermore, per capita resource use in industrialized countries, which already far exceeds sustainable levels, continues to rise. At the same time,

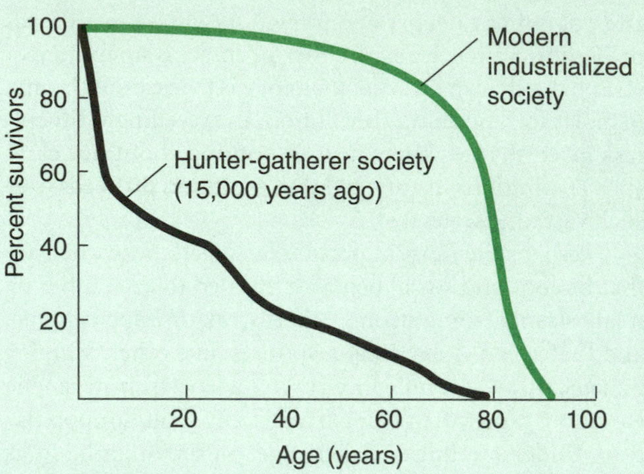

FIGURE 25-1 Survival curves for a population of hunter-gatherers who lived 15,000 years ago on the Mediterranean coast (based on skeletal remains) compared to a present-day population living in an industrialized society. (Adapted from May, R. M., 1987. More evolution of cooperation. *Nature*, **327** 15–17.)

of offspring we leave behind, which more and more is a matter of personal preference. Further, most of the ills resulting from our changed lives — diabetes, heart disease, cancer — affect us mostly in our later, post-reproductive years, so that there is little or no chance for natural selection to weed out susceptibility to these ills. An individual who has produced three children and at the age of 50 develops cancer or other diseases with genetic components is no less reproductively successful than an individual of the same age who has produced three children but does not suffer from such diseases.[7]

Our increasing longevity has had an even more dramatic impact, an impact that has grave implications not only for humanity but also for the world as a whole, and is discussed in **Box 25-1.**

[7] One may, of course, argue that children with healthy grandmothers are more fit than children with ill or absent grandmothers because they get more attention and care, although in present social situations, in the developed world at least, such caretaking functions can be assigned to other individuals and it is unlikely that grandparental attention adds to reproductive fitness.

Deleterious Alleles

Despite improvements in medical care, alleles that have obvious **deleterious effects** still exert an influence on population structure and result in human suffering. Here we consider changes in the frequencies of such alleles.

The number of children born with marked physical and/or mental abnormalities is conservatively estimated at about 20 to 25 in 1,000 births in the United States. The mortality rate ascribed to congenital malformations is around 15 percent of all infant deaths. Many other defects, not immediately noted at birth, become apparent during childhood years and are more widely prevalent than imagined. In various studies, about 30 percent or more of hospital admissions for children and 50 percent of all childhood deaths are ascribed to birth defects or to complications that such defects may have caused. Although not all birth defects are genetic, the proportion of those that are is high.

Table 25-2 lists estimates of the frequencies of some genetic disorders, which in total occur in approximately

hundreds of millions of people in developing countries are responding to improved economic conditions by adopting the patterns of resource use characteristic of wealthier regions.

Here we examine the complexities of just one concern associated with population growth: **access to food.** If current production, though rising, appears barely able to keep up with present requirements, what then of the future? This turns out to be a topic with many facets, some of them political. The list below is far from complete.

- The amount of arable land is finite, and the amount cultivated is at or approaching capacity in most countries.
- The average amount of arable land per person worldwide continues to shrink as populations expand and as land is taken out of production as a result of environmental problems and urban sprawl. (In 1950, there were 0.23 hectares per person, in 2000, 0.11 ha, a decline of more than 50 percent.)
- Increasing amounts of land are used to grow food for livestock, a practice that ultimately results in production of far fewer food calories for humans per unit of land.
- Substantial amounts of food are lost through faulty distribution and storage systems, and through pests and disease.
- Extreme poverty means that a sizable fraction of the world's population cannot afford to buy nutritious food. At the same time, large surpluses exist elsewhere.
- Unwise subsidies and policies as well as increasing globalization and domination of agriculture by multinational corporations distort production and markets almost everywhere.

Other threats to the food supply locally as well as globally include soil erosion, fertilizer and pesticide contamination, water supply depletion, global warming and extreme climatic events.

In terms of future food production, considerable amounts of arable land remain unused in several countries, and there is great potential to increase both yields and the nutritional value of crops, but as is evident from the list above, unless we can overcome the political, environmental and other problems that plague agriculture, it is unlikely that we will be able to feed the billions more who will be added to the world's population in the years to come.

In the meantime, famine remains a recurrent problem in some developing countries, especially in sub-Saharan Africa. Indeed, United Nations statistics show that 824 million people were suffering from chronic hunger in 2003. Improving the quality of life for an increasing population and ensuring that no one lives in abject poverty are surely among the most critical issues we face.

[a] In a recent analysis, F. Thomas et al. (2004) demonstrated that a significant amount of the variation in human birth weight from region to region reflects adaptation to local selection pressures, especially parasitism. See Gluckman and Hanson (2005) for causes of infant mortality.

[b] The infant mortality rates are taken from estimates in the 2006 *CIA World Factbook*, (https://www.cia.gov/library/publications/the-world-factbook/index.html).

[c] Ecological footprint is a measure defined as the amount of land and water required to produce the resources consumed and to absorb the waste of an individual, population or lifestyle.

3 percent of births. If we include genetic defects appearing later in life, such as muscular dystrophy, this frequency probably doubles. If in addition we include less obvious defects that nevertheless have strong genetic components — impaired resistance to stress and infection, and other physical and psychological weaknesses — the effects of harmful genes touch a majority of our population. Indeed, as of June 2, 2007, the Human Gene Mutation Database[8] lists 53,186 mutations in 2,056 genes, although most of these mutations are exceedingly rare.

The ubiquity of deleterious alleles with lethal effect was dramatically demonstrated in studies made of the offspring of cousin marriages. These studies, which use techniques of detecting and partitioning the genetic load caused by inbreeding, show that outwardly normal individuals in our society carry a genetic load equivalent to that of approximately one to eight deleterious **lethal alleles** (**lethal equivalents**) that, if homozygous, would cause early death. Two important questions arise:

- What accounts for the prevalence of these harmful alleles?
- What, if anything, can we do to rid ourselves of them?

The reasons for the high frequency of deleterious alleles are not fully agreed on, although there is little question that they originally arise through mutation. In one theory, held by the late Theodosius Dobzhansky and others, such alleles, although deleterious in homozygous condition, may offer considerable advantage to their heterozygous carriers by producing some sort of hybrid vigor. According to this theory, a gene will be maintained in the population although the homozygote produced by this gene is relatively inferior in fitness.

Another school, formerly headed by the late Hermann Muller, was based on the theory that such genes produce no advantage of any kind and that their frequency is high because the usual effect of natural selection has been artifi-

[8] The Human Mutation Database at the Institute of Medical Genetics in Cardiff, © 2007 Cardiff University. Available at http://www.hgmd.cf.ac.uk. Accessed June 2, 2007.

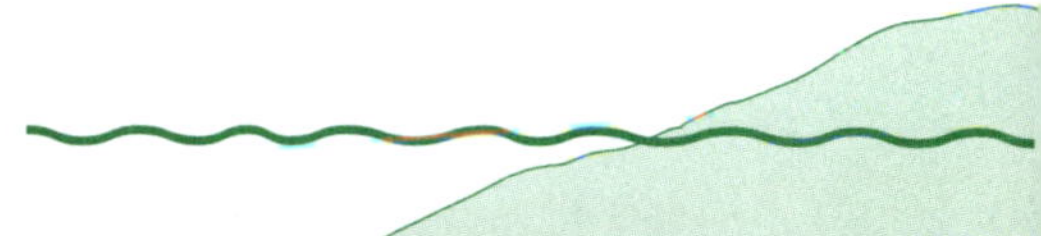

TABLE 25-2 Estimated frequencies (in percent) of some human genetic disorders

Disorder	Percent	Disorder	Percent
Single Gene Disorders		**Chromosomal Disorders**	
Autosomal recessive		Autosomal	
Mental retardation, severe	0.08	Trisomy 21	0.13
Cystic fibrosis	0.05	Trisomy 18	0.03
Deafness, severe (several forms)	0.05	Trisomy 13	0.02
Blindness, severe (several forms)	0.02	Other	0.02
Adrenogenital syndrome	0.01		0.20
Albinism	0.01	Sex chromosome	
Phenylketonuria	0.01	XO and X deletions	0.02
Other aminoacidurias	0.01	Other "severe" defects	0.01
Mucopolysaccharidoses (all forms)	0.005	XXY	0.1
Tay-Sachs disease	0.001	XXX	0.1
Galactosemia	0.0005	Others	0.015
	0.25		0.35
X-linked			
Duchenne muscular dystrophy	0.02	Total chromosomal disorders	0.55
Hemophilias A and B	0.01		
Others	0.02	**Multifactorial Disorders**	
	0.05	Congenital malformations	
Autosomal dominant		Spina bifida and anencephaly	0.45
Blindness (several forms)	0.01	Congenital heart defects	0.4
Deafness (several forms)	0.01	Pyloric stenosis	0.3
Marfan syndrome	0.005	Clubfoot	0.3
Achondroplasia	0.005	Cleft lip and palate	0.1
Neurofibromatosis	0.005	Dislocated hips	0.1
Myotonic dystrophy	0.005		1.65
Tuberous sclerosis	0.005		
All others	0.015	Total multifactorial disorders	> 1.65
	0.06		
Total single-gene disorders	0.36	**Total frequency of listed genetic disorders**	> 2.56

Source: From *American Scientist* **65:** 703–711, 1977 by C.J. Epstein and M.S. Golbus. Reprinted by permission.

cially reduced. According to Muller, genotypes that were formerly defective and would have been eliminated in earlier times are now kept alive by medical techniques that enable those with these defective alleles to pass them on to their offspring. As we know (see Chapter 21), a decrease in the selection coefficient against a particular gene causes an increase in the equilibrium frequency of the gene ($q = \sqrt{u/s}$ for a recessive allele, $p = u/s$ for a dominant allele). Thus, if deleterious alleles are not eliminated by selection, they will gradually increase in frequency in accordance with their mutation rate. Because the mutation rate is usually low, the frequency of any particular allele will increase rather slowly, but as there are many possible deleterious alleles, the allelic load will increase significantly. Nearsightedness is a trait whose frequency has most likely increased in recent periods, but can be corrected quite simply by an optometrist.

That natural selection no longer operates to eliminate many genotypes is not necessarily an undesirable feature of modern life. Few individuals would argue today that fire and clothes should be abolished because they are artificial devices that circumvent natural selection by permitting hairless phenotypes to survive in cold climates. It would also be difficult for us to return to the "good old" prevaccination, presanitation days of smallpox, diphtheria, typhus, cholera and plague.

What *is* an undesirable feature of modern life is that despite advanced medical care, a great many deleterious alleles still exert an effect. Beginning in the early twentieth century, and detailed below, there have been a variety of attempts to improve human genetic quality. Eugenics, the earliest of these attempts, had a much broader focus than the individual allele, perhaps understandable given the state of genetics at the time.

Eugenics

In his dialogue, *The Republic*, Plato suggested that humans could be improved through selective breeding. In his ideal philosopher-state, only the most physically and mentally fit individuals would be mated: any offspring would be raised by the state. Inferior types would be prevented from mating or their offspring destroyed, and the governing class would be selected only "from the most superior."

In its modern form, suggestions for improving the human gene pool come under the name **eugenics**, a term coined by Francis Galton in 1883 from the Greek words *eu* (good) and *gen* (birth). After reading the works of Darwin, his first cousin, Galton became concerned with the heredity of quantitative characteristics, especially intelligence. From 1865 on, he promoted the idea that the evolution of human traits through natural selection could be substituted by their evolution through selective breeding. ("What nature does blindly, slowly, and ruthlessly, man may do providently, quickly, and kindly.") Coming from a brilliant family himself, Galton was impressed by the way in which intellectual and personality traits tend to run in families. Convinced that such traits were inherited, and drawing on his knowledge of animal breeding, he concluded that, "judicious marriages over several generations" could "produce a highly gifted race of men," and thus thwart the "reversion to mediocrity" he believed to be threatening society as a result of excessive breeding by those who were not superior.

Galton's ideas were taken up and developed by many prominent thinkers. Like the Social Darwinists, however, early eugenicists reflected their own personal and cultural biases as to which traits they deemed undesirable, and strong racist and class-based biases played a part from the beginning. C. B. Davenport, an influential leader of the early twentieth century eugenics movement, exemplified this racist approach by using New Englanders as the standard of comparison for all American social groups irrespective of their country of origin. According to Davenport and other members of his Eugenics Record Office (which from 1910 to 1944 collected an enormous database of information about American families), particular groups could be characterized by social characteristics with identifiable hereditary components: Italian violence, Jewish mercantilism, Irish pauperism, and so on.

The early eugenicists, of course, knew next to nothing about genetics, and even after the rediscovery of Mendel's work in 1900 their comprehension remained of necessity limited. But as we now know, no evidence supports the biological superiority of any particular group in respect to social characteristics or intelligence. We also know that many characteristics we consider desirable, such as high intelligence, esthetic sensitivity, longevity and good physical health, are not caused by single genes that are easily identified but by complexes of many genes acting together and responsive to different environments. As we have learned from animal breeding, selective breeding to develop beneficial gene complexes is fraught with difficulty. The methods involve complicated schemes based on selection of parents along with testing of progeny under controlled environmental conditions. As I. M. Lerner (1954) and others discussed, the results of these experiments have, in general, enhanced certain complex characters to some degree but have usually caused the deterioration of others.

Although general intelligence does have a large genetic component, much of it is remarkably plastic, influenced as it is by such factors as prenatal diet, cultural surroundings, socio-economic environment, maternal uterine environment (Devlin et al., 1997) and even (in terms of measurement) the color of the interviewer administering an IQ test. Studies have shown that as socio-economic conditions improve, average intelligence scores also improve, although genetic differences between individuals remain.

Eugenics Policies

Unlike Social Darwinists, eugenicists believed in active intervention in breeding. Given that some early proponents were visionaries, there was at least a possibility that the movement could have developed into a serious attempt to diminish human suffering and improve the human gene pool, if only it had been stripped of racism and provincial prejudice. It was, however, working in a society where those attitudes were commonplace. Two avenues of action were followed:

- **positive eugenics** — attempts to increase the frequency of beneficial genes, and
- **negative eugenics** — attempts to decrease the frequency of harmful genes.

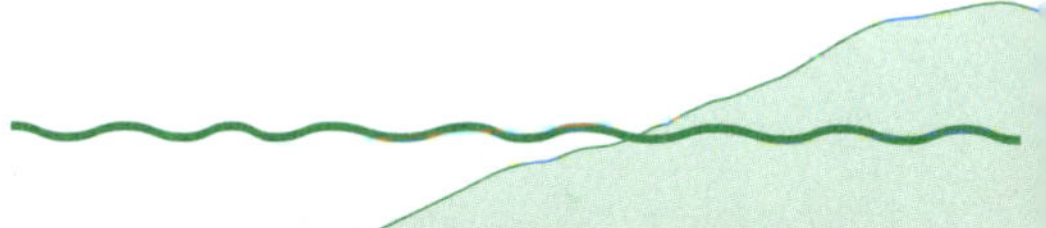

Francis Galton's approach of encouraging particular people to marry is one example of positive eugenics. Singapore's campaign in the 1990s, in which young graduates were offered inducements to produce children, is another. Unfortunately, however, negative policies dominated the vast majority of government programs instituted while eugenics held sway, roughly from the 1890s to the 1940s. So strong was the influence of the movement that almost every non-Catholic western country was involved. Other areas included the southernmost Latin American states (where policies favored whiter complexions) and, since the early 1990s, China.

Viewed today through the prism of human rights, we can only be appalled at the legislation — some of it draconian — enacted and enforced in the name of eugenics. Methods employed included restrictions on immigration and marriage[9]; racial segregation, including bans in the United States on marriage between whites and African Americans, overturned by the Supreme Court only in 1967; compulsory sterilization of the "feebleminded," certain criminals and others deemed unfit; forced abortions; and, finally, in Germany under the Nazis, genocide of those (especially Jews) regarded as racially inferior and thus a threat to the "purity" of the Aryan race. In the United States, home to the second largest eugenics movement (after Germany), marriage prohibitions were enacted in many states during the early decades of the twentieth century, and tens of thousands of individuals were sterilized, the last of them in the early 1960s.

Backing the movement was a large body of research, much of it involving statistics. But even from the early years, some geneticists and members of the general public were sharply critical of the methods employed and of conclusions drawn from the findings. Nevertheless, it took the excesses of the Nazi era (including some 225,000 sterilized between 1934 and 1937 alone; Kevles, 1995) before eugenics became anathema and ceased to exert an open influence. The desire to improve the human condition, however, continues.

Reproductive Technology and Eugenics

The search by today's parents for healthy children has many eugenic overtones. Genetic counseling before conception, artificial insemination, testing the fetus for inherited disorders, selective abortion of fetuses deemed defective, ultrasound examination and *in vitro* fertilization, including pre-implantation screening of embryos, are all in effect eugenic practices, albeit voluntary and on an individual basis. In countries where such techniques are widely used, the incidence of some congenital disorders has sharply declined in recent years. In Taiwan, for instance, the incidence of thalassemia among newborns dropped from 5.6 to 1.21 per 100,000 over eight years.

Because prenatal testing is expensive and often invasive, however, only ultrasound is carried out as a matter of routine; other tests are usually performed only when the family history (or mother's age, in the case of Down syndrome, in which there are three copies of chromosome 21) indicates a clear risk that a particular mutation or chromosomal abnormality may be present. The likely result is that the overall load of deleterious alleles in the newborn population has not declined significantly. Armand Leroi (2006) joins some other geneticists in asking whether we should be screening all fetuses for deleterious mutations. Armed with knowledge gained from sequencing the human genome (see below), he says, we may soon be able to design screening tools that test for all known disease-causing mutations, although — given the cost, inconvenience and risk to the fetus of prenatal screening — the most likely use of such tools will be in pre-implantation testing of embryos conceived via *in vitro* fertilization (IVF). About 1 percent of Americans and 4 percent of Danes currently begin life by IVF (Leroi).

Reproductive technologies, including genetic screening, evoke a host of ethical and social concerns, a few of which we list below:[10]

- What constitutes a disease? If I bear a gene that increases my susceptibility to a condition later in life, do I have a disease? Now that the human genome has been sequenced, we may have narrowed even further the number of people we consider normal.
- How do we (potential parents) make an informed choice among our options?
- Can I live a fulfilling life if I inherit a particular condition? If yes, should I (an embryo) be denied that right?
- Who has the right to know about my genetic makeup — my school? my employer? my insurance company? the government? In fact, who owns this information?

Despite the many ethical concerns, it is likely that the use of modern reproductive methods will continue to increase, and that over time they will have a larger and larger (though still limited) impact on our gene pool, at least in wealthy countries and well-off segments of others. (Oh brave new world!) But for now, and probably in the future, too, a far greater influence is likely to be modern medicine in general. In a recent study, for instance, Marelli et al. (2007) document that the incidence of severe congenital heart disease among adults in Quebec climbed steeply between 1985 and 2000. Before open-heart surgery, few if any of these patients would have survived childhood. Less benignly, and despite laws banning the practice, the widespread use in

[9] Eugenics was only one of many issues in immigration, marriage and segregation debates.

[10] For a longer list of questions, see the Human Genome Website, under Ethical Issues: http://www.ornl.gov/sci/techresources/Human_Genome/elsi/elsi.shtml. Accessed May 23, 2007.

India and China of ultrasound followed by abortion to ensure the birth of a male child is leading to significant imbalances in the male/female ratio of those societies.

So, if almost everyone now survives to have children, and if we can protect our children from natural selection so that they too go on to have children, will natural selection continue to operate on humans? The argument could be made that because only a handful of children now die of infectious disease in countries with robust health care systems, we may be opening the door to a population with weaker average immune systems. But again, does it matter? If, as expected, uncontrollable pandemics appear, the answer could well be yes.

Genetic Engineering

Humans have been changing the genetic makeup of organisms for a very long time. The **domestication** of plants and animals involved selection to promote some traits and eliminate others. In some species, the results have been dramatic, as we saw in the example of wheat (See Chapters 10 and 24). In the thousands of years since, we have created a world of agriculture, horticulture and plantation forestry by manipulating genes and genomes (**Table 25-3**), most commonly through selection.

The early 1970s, however, marked the beginning of a new era, the era of **genetic engineering**. Following the discovery that restriction enzymes can be used to cut sections of DNA at particular sequences, molecular biologists soon realized that they could splice together DNA fragments from different sources cut by such enzymes to form entirely new **recombinant DNA,** which could be used to introduce new traits to an organism (**Fig. 25-2**).

A basic technique of genetic engineering is to insert a new or modified nucleic acid sequence into a **vector,** such as a virus, plasmid or yeast artificial chromosome (YAC), that has the ability to carry these novel sequences into host cells and then host genomes where they can be propagated and amplified many times over. Briefly, we can divide gene manipulation into four major steps, diagrammed in (**Fig. 25-3**). The technology allows active intervention into the genetic material of any organism. Genomes of many viruses, bacteria, protists, animals and plants have been manipulated either by modifying their existing genes or by introducing new genetic material. The aim is to induce physical and physiological changes in a cell or organism that in turn affect such traits as protein production, resistance to infective agents, agricultural yield, nutritional value, toxic susceptibility, environmental stamina, tumor resistance, and so forth. Such manipulation of genetic traits allows us to move organismal evolution from the age-old province of random mutation and natural selection to *human-directed evolution.*

Genetic engineering is now big business, especially when it comes to agriculture. The first genetically modified (GM) organism, a tomato, was put on the U. S. market in 1994. Little more than a decade later, the area covered by just the four major GM crops in the United States (corn, soybeans, cotton and canola) had grown to 519,000 km^2 (128.3 million acres).[11] A large proportion of food on North American supermarket shelves now contains at least some GM content. Yet although by 2005 GM crops were being grown in more than 20 countries, they remained rare in Europe, where a huge public outcry against the technology on ethical, environmental and food safety grounds continued to have an impact on government policy.

With their early emphasis on increased resistance to insect pests, diseases and herbicides, the most common GM crops in use today were largely developed with financial profit in mind, but the potential is enormous. Improved nutrition (e.g., golden rice is rich in β carotene, provitamin A) is just one promised benefit.

GM technology has been applied in many other fields. Animals modified include fish (for enhanced growth, cold tolerance and human insulin production), cows (to produce human growth hormone, casein-enriched milk, and so on), pigs (for lower fat content) and mice (to produce fish oils). Although most of this work has yet to be approved for commercial application, in 2006 approval for release of ATryn, the first pharmaceutical drug created in a mammalian animal (a goat), was granted in Europe.[12] Indeed, the pharmaceutical industry is investing heavily in research into this method of drug production, an extension of the use of GM microorganisms and mammalian cells for the same purpose since the 1980s. Microorganisms have been used to produce a wide array of human and other proteins: human growth hormone, factor VIII for hemophilia sufferers, and so on.

In summary, the major benefits conferred by GM technology are that in contrast to more traditional breeding techniques, which are time-consuming and inexact, GM is more efficient, more precise in its choice of genes and allows genes from a far wider range of organisms to be introduced into the recipient organism's genome. When combined with cloning, see below, its scope may be further enhanced.

Concerns over GM technology

Perhaps the most fundamental charge leveled against genetic engineering is that we are interfering with nature. It is clear, however, that the physical/biological environment we consider "natural" is neither constant nor sacred. Organisms have regularly affected and changed the environment, and continue to do so, while humans have long engaged in a succession of interventions into "nature" that have changed its form, function, and substance in most localities and for many organisms.

[11] 2006 U. S. Department of Agriculture statistic.

[12] ATryn, an anti-clotting agent, is designed to be used during surgery on people with antithrombin deficiency.

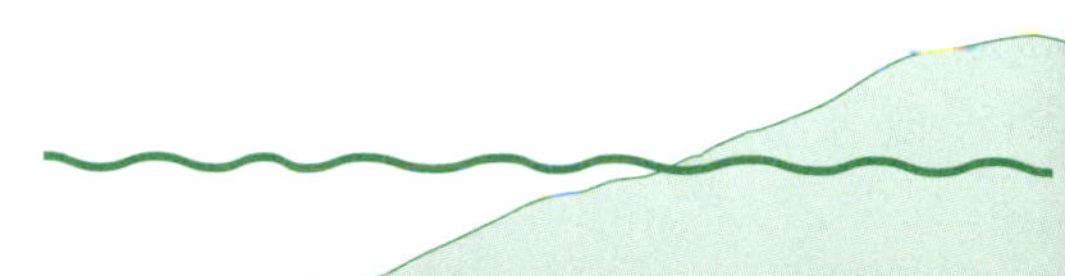

TABLE 25-3 Major discoveries and achievements over the past 150 years beginning with Mendel's experiments in 1856 and leading to sequencing of organismal genomes and genetic engineering

Year(s)	Advance
1856	Gregor Mendel's first crossbreeding experiments with the garden pea
1859	Publication of Charles Darwin's *On the Origin of Species*
1865	Presentation by Mendel of his results at the monthly meetings, 8 February and 8 March, of the *Naturforschenden Vereins* [The Natural Science Society] in Brno
1866	Publication of Mendel's paper *Versuche über Pflanzen-hybriden* (*Experiments on Plant Hybridization*) in the journal of the Natural Science Society; but his work attracts little attention
1871	Isolation by Friedrich Miescher of nuclein from nuclei, marking the first research with what we now know to be nucleic acids
1887	Proposal by August Weisman that during meiosis each chromosome carries many "determinants" of hereditary characters
1900	Mendel's 1866 experiments rediscovered independently by three researchers and publicized by William Bateson
1902/03	Chromosomal theory of inheritance proposed independently by William Sutton and Theodor Boveri
1902–09	Terms genetics, homozygote, heterozygote, allelomorph (allele) coined by William Bateson
1909	Demonstration by Archibald Garrod that some human diseases are inborn errors of metabolism, inherited as Mendelian recessive characters
1910–11	X-chromosomes, sex-linkage, mutant gene for eye color in *Drosophila* discovered by Thomas Hunt Morgan, who proposes exchange of chromosome segments by crossing over
1913	First genetic maps produced by Alfred Henry Sturtevant; recombination discovered; chromosomes shown to contain genes in a linear array
1918	Ronald Fisher's *On the Correlation Between Relatives on the Supposition of Mendelian Inheritance* launches population approach to genetics
1927	Discovery by Hermann Muller that X rays induce physical changes (mutations) in *Drosophila* genes
1941	Discovery that genes code for enzymes in the mold *Neurospora;* proposal by George Beadle and Edward Tatum of the "one gene–one enzyme" theory
1943–44	Evidence that bacteria contain genes; isolation of DNA and demonstration that it is genetic material
1949	Demonstration by Erwin Chargaff that the proportions of four nucleotides in DNA are the same in all cells within an individual but vary from species to species
1950s	François Jacob and Jacques Monod establish control functions located on chromosomes that turn the expression of genes on or off
1952	Proteins eliminated as basis of genes
1950	Transposons discovered in maize by Barbara McClintock
1953	Proposal by James Watson and Francis Crick of the three-dimensional double helical molecular structure of DNA, based in part on unpublished x-ray crystallographic data obtained by Rosalind Franklin and by Maurice Wilkins
1956	Humans shown to have 46 pairs of chromosomes
1957–58	Proposal by Francis Crick of the Central Dogma of molecular biology: genetic information passes in only one direction, from nucleic acid to protein
1958	Isolation of the first DNA replicating enzyme
1959	Recognition of the Down syndrome chromosome abnormality
1960	Discovery of messenger RNA, in bacterial cells
1961	Determination by Jacob, Crick, Sydney Brenner and others of the triplet nature of the genetic code, transcription of information in DNA into messenger RNA, and translation of mRNA into protein
1961	Nucleic acids hybridized
1962	Use of synthetic RNA to unravel the genetic code; discovery of repressor and transcriptional control of genes; cloning of differentiated (adult) toad cells
1964	Crick's Central Dogma shown not to hold for viruses
1966	Complete genetic code used in translating codons into amino acids established

(Continued)

TABLE 25-3 (continued)	
Year(s)	Advance
1970s	Isolation from the bacterium *Haemophilus influenzae* of a restriction enzyme that cuts DNA at specific sites, and development of method to recombine the fragments, launching recombinant DNA (genetic engineering); discovery of the retrovirus replication cycle 1972 First recombinant DNA molecules produced
1973	Introduction of recombinant plasmid vectors into bacteria for cloning
1975	DNA sequencing technology invented; use of radioactive probes and autoradiography to identify particular DNA fragments; cloning of hemoglobin DNA copied from messenger RNA (cDNA)
1976	Molecular diversity in immune system immunoglobulins elucidated; first clinical use of recombinant DNA (prenatal diagnosis of α thalassemia)
1977	Laboratory techniques for sequencing DNA invented; sequencing of bacteriophage DNA
1978	Organization and function of homeotic gene complex in *Drosophila* proposed; synthesis of the proinsulin peptide using recombinant DNA; mapping of the sickle cell mutation using its linkage to DNA ("RFLP") markers
1983	Discovery of homeobox as basic element of homeotic genes; direct detection of the nucleotide sequence of a specific mutant allele (sickle cell); identification of the HIV retrovirus in patients with AIDS; use of retrovirus as a vector to replace deficient human genes in cell culture; location on a chromosome and isolation for study of the gene defect that causes Huntington's disease; discovery of polymerase chain reaction (PCR)
1984	Targeting of mutations to specific genes in cell culture
1985	Polymerase chain reaction (PCR) used to amplify nucleotide sequences
1986	Reverse genetics developed: characterization of a gene with an unknown protein product (chronic granulomatous disease) by finding its chromosomal location
1987	Production of a vaccine (anti-hepatitis B) based on recombinant DNA technology; homologous recombination achieved in murine embryonic stem cells; mitochondrial DNA used to study human origins
1988	Incorporation and expression of genes inserted into mouse cells using retroviruses
1989	First human gene sequenced, and defect in the gene product shown to 'cause' cystic fibrosis
1990	Gene for defective adenine deaminase replaced in somatic cells of human patients
1992	First intensive linkage map of the entire human genome completed using 400 gene markers
1993	Marketing in the United States of FlavrSavr tomatoes, genetically engineered for longer shelf life
1995	DNA cloning of a major component in human chromosome end caps (telomeres) that can be used to produce an artificial human chromosome
1995–96	Bacterial and yeast genomes (*Haemophilus influenzae, Escherichia coli, Mycoplasma genitalium*) sequenced
1996	Birth of a lamb named Dolly, first animal to result from cloning an adult somatic cell
1997	Creation of the first artificial human chromosome possessing a centromere and telomeres
1998	Human embryonic stem cells cultured; first sequencing of a complete eukaryote genome (the nematode *C. elegans*)
1999	Sequencing of a human chromosome (22), and of complete fruit fly (*Drosophila melanogaster*) genome
2000	Sequencing of the genomes of *Saccharomyces cerevisiae,* and mustard cress (*Arabidopsis thaliana*); birth of first genetically modified primate, a rhesus monkey containing a fluorescent marker gene from a jellyfish
2001	First draft sequences of human genome completed independently by the Human Genome Project and Celera Genomics; discovery of tumor suppessor gene for prostate, breast and other cancers
2002	Mouse genome sequenced; birth of first genetically modified cat
2003	Sequencing of 99 percent of the human genome completed to 99.9 percent accuracy; marketing of GloFish, a pet fish to which a fluorescence gene was added
2005	Apparently successful work on human cloning carried out in South Korea revealed to be fraudulent
2006	Sequencing of the human genome completed; publication of a global map of p53 transcription-factor binding sites in the human genome, and of the first map of DNA methylation of a genome (*Arabidopsis*)
2007	Genomes of the purple sea urchin (*Stronglocentrotus drobachiensis*) and of the gray short-tailed possum (*Monodelphis domestica*) sequenced

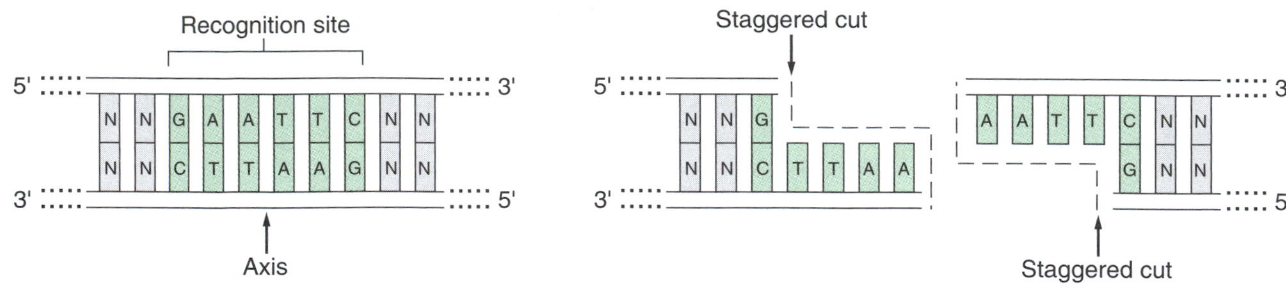

(a) Target sequence in vector DNA recognized by *Eco*RI restriction enzyme

(b) Staggered cuts at a single target sequence of vector DNA

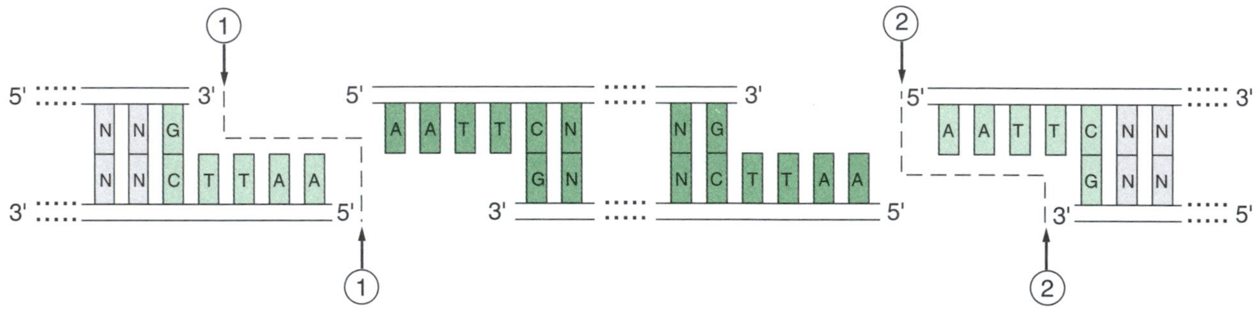

(c) Linear sequence of "foreign" DNA (N...N/N...N) extracted from cuts at two target sites (1) and (2) by *Eco*RI

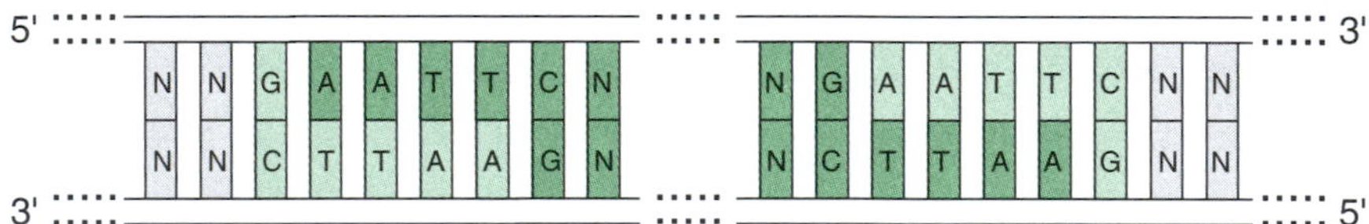

(d) Insertion of extracted foreign DNA sequence into cut target sequence of vector DNA

FIGURE 25-2 As discussed in Chapter 12, restriction endonucleases are enzymes that recognize specific DNA nucleotide sequences, and cleave the DNA double helix at or near these specific sites. The kinds of restriction enzymes most used in gene manipulation are those that recognize short "palindromic" sequences that have an axis of symmetry. In this illustration, the palindrome recognized by the bacterial *Escherichia coli* enzyme *Eco*RI is the double-strand hexanucleotide sequence GAATTC/CTTAAG, which reads the same from each 5'–3' or 3'–5' direction toward the central axis (N represents unspecified nucleotides and ... represents further extensions of the nucleotide sequence). In (a), this hexanucleotide target sequence is located in a vector, such as a virus or plasmid, which can later be inserted into a host cell. (b) Breakage of the target sequence is asymmetric, and the staggered cuts (*indicated by arrows*) produce two protruding but complementary single-strand ends, four nucleotides long, that can be considered as "sticky" or "cohesive." (c) DNA from a different ("foreign") source bearing the same restriction recognition sites can then be treated in the same way to generate one or more linear sequences, also with single-strand sticky ends. (d) Because of the cohesive nature of these ends, the foreign DNA (*shaded sequence*) can anneal by complementary base pairing with the vector DNA in which it is being inserted. The terminal nucleotides of the inserted foreign DNA can then be covalently joined to the adjacent nucleotides by a special DNA ligase enzyme, thus forming a recombinant DNA molecule derived from two (or more) sources. (Adapted from Strickberger, M. W., 1985. *Genetics,* 3d ed. Macmillan, New York.)

An important feature of GM technology is that the genes employed usually come from a species other than the receiving organism. To many, this poses a substantial ethical problem. But as shown in Chapter 10, gene flow (horizontal gene transfer) between organisms is a common event across the whole Tree of Life. Species barriers are continually being crossed by viruses, transposons (mobile self-replicating DNA sequences) and plasmids carrying genes from one species to another. For instance, we ourselves readily incorporate genes introduced by viruses and bacteria.

In agriculture, it has been standard practice for more than a century to develop new varieties through hybridization between different species. At first, the species concerned were closely related. By the middle of the twentieth century, however, methods had been developed that allowed crosses between more distantly related species. For instance, triticale, a widely grown cereal, is a hybrid between wheat and rye. Protests mounted against the *green revolution,* a major achievement of this type of breeding, were on the basis of environmental and social concerns, rather than the ethics and safety of gene transfer between species. With genetic engineering, however, the genes transferred have often come from a different phylum or even kingdom, which was not the case with earlier breeding practices.

Environmental concerns about GM crops are mostly focused on the perceived threats to local biodiversity and ecosystems, in particular the possibility that pollen from GM plants will fertilize closely related weeds (a problem if the crop is modified to resist pests or herbicides), closely related native plants (whose very existence might then be threatened), or non-GM crops of the same species. The long-term impact of such crosses will depend, of course, on the relative fitness of the gene(s) in question. Again, this is not a new thing; gene flow from crops altered by non-GM methods has occurred for many decades. Still, the concern remains, and great care must be taken before introducing any new GM crop. After all, our experience with the technology is still relatively limited, and long-term effects can be difficult to envisage.

The safety of food produced through GM technology is another concern. The potential for allergic reactions to newly created substances in food, for instance, is one area deserving special attention. The World Health Organization is just one of many bodies working to improve nutrition and safety guidelines in this and other areas for GM food. In **Box 25-2,** we address the nature of experimental risk.

Human Applications

Sequencing of the human genome, completed in 2006,[13] has provided a tremendous boost to both human genetics and the hope that we may someday be able to control inherited diseases. Two concurrent efforts were involved in this achievement, one carried out in laboratories around the world under the umbrella of the Human Genome Project

[13] Various drafts were announced over several years (see Table 25-3).

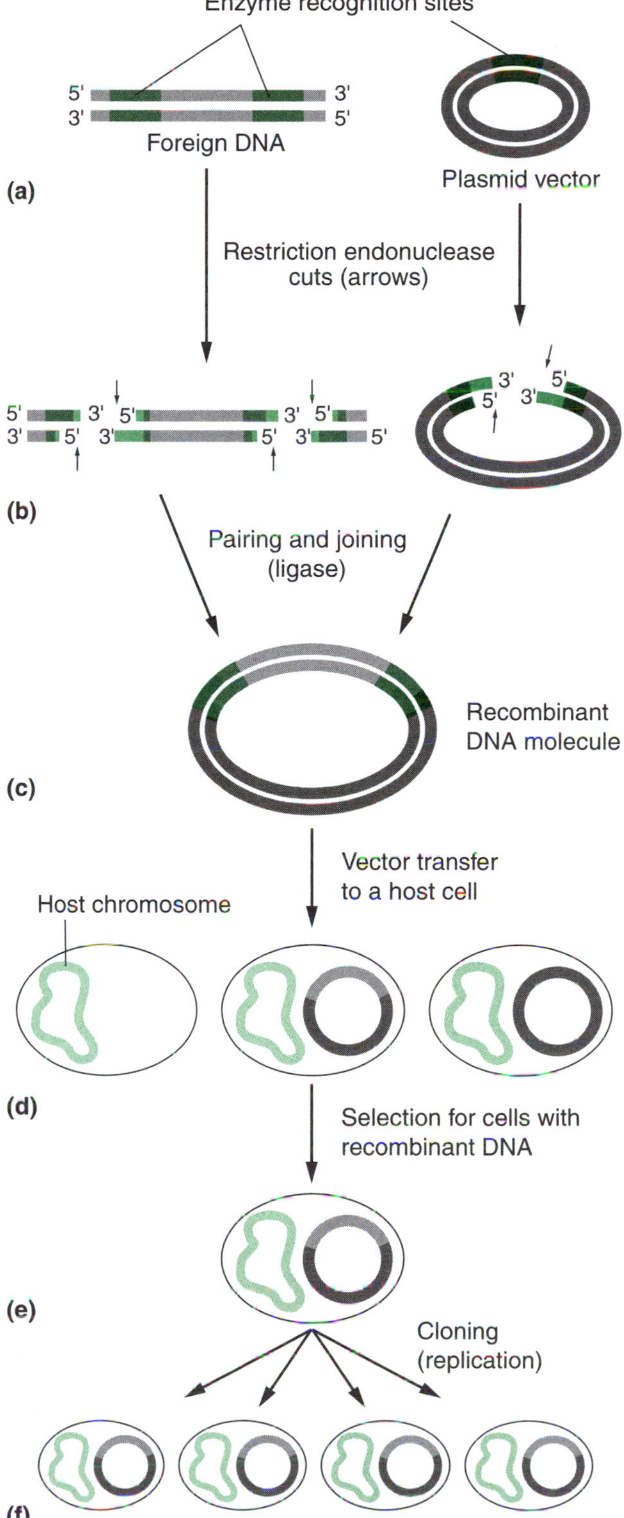

FIGURE 25-3 A general scheme for constructing a clone of recombinant DNA molecules using plasmids and restriction endonuclease enzymes. (a) A "foreign" DNA molecule and a plasmid vector are selected, both carrying recognition sites that can be cleaved by the same restriction endonuclease. (b) Cleavage produces one or more fragments of the foreign DNA, and opens the plasmid vector. (c) The cuts produced by the restriction enzyme are staggered so that complementary base pairing can occur between single-strand ("sticky") ends of the foreign DNA fragments and the opened plasmid DNA. A DNA ligase enzyme covalently bonds the two DNAs into a recombinant DNA molecule. (d) The plasmid carrying the foreign DNA is inserted into a host cell. (e) A selective process isolates cells carrying the recombinant DNA molecule. (f) Each isolate is further propagated, thereby generating a clone carrying the particular foreign DNA fragment that was incorporated in step (c). (Adapted from Strickberger, M. W., 1985. *Genetics,* 3d ed. Macmillan, New York.)

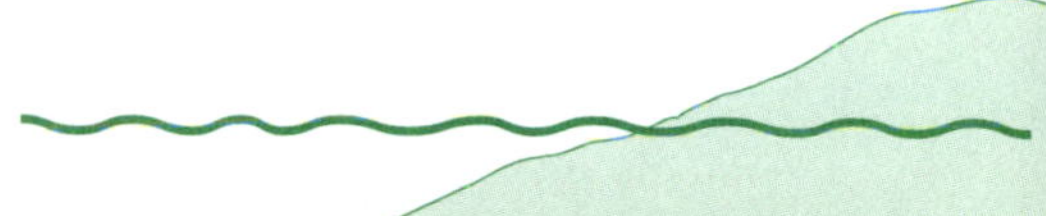

BOX 25-2 Regulating Risk and Safety

FROM THE FIRST DAYS of genetic modification technology, many have expressed concern over the potential dangers of recombinant DNA applications. It is a legitimate worry. How can we be sure that any type of genetic manipulation — indeed, of any type of experimentation — does not incur cause unforeseen, detrimental consequences? How do we regulate risk and safety? Difficult as it may be to ensure against all risk, we can evaluate potential difficulties and exercise precautions:

1. **It is important to realize that there are risks in almost any experimental procedure.** For example, a drug that reduces the effects of colds may turn out to be a potent carcinogen. Or an experimental strain of wheat may have the potential to expand limitlessly, devastating other crops. At the same time, we should keep in mind that "natural" catastrophes have, so far, exceeded those brought about by experimental procedures. (This assumes, of course, that those carrying out and monitoring experiments have the best interests of humanity at heart, an assumption that is not necessarily guaranteed.)
2. **There is no way of proving the absence of risk in an experimental procedure.** The ultimate effect of *any* activity, however trivial, is unpredictable; there is no way we can follow all its future interactions. Shall we then ban all beneficial technologies because there may be potential misuse or a possible unexpected deleterious effect in the future? In fact, had experiments in genetic engineering been banned, it would have excluded even our moderate success in treating cancer, AIDS and other serious diseases. Moreover, it seems foolhardy to restrict the benefits of agricultural improvements, practically all of which have involved gene manipulation.
3. **When we do evaluate risks, it seems logical to base these risks on more immediate and foreseeable consequences.** Again, the medical model seems appropriate: restrict experiments to tissue culture, animal models, clinical trials, and/or enclosed agricultural plots. Such methods allow main as well as side effects to be closely observed before expanding the use of experimental treatments to the general public. One point cannot be emphasized enough: the importance of running trials for sufficient time so as to allow long-term side effects to become apparent.
4. **For the most part, risk evaluations have functioned successfully.** For example, most pharmaceutical drugs on the market are effective for the purposes for which they were designed, and their side effects can be taken into account during treatment. Keep in mind that strict regulation by politically independent governmental agencies has been the primary agent in averting pharmaceutical disasters. Because these new genetic technologies concern matters that affect everyone, decisions should be publicly based.

Widening the argument, many would agree that we can justify intervention in "nature" if we follow the basic principle, "Above all, do no harm," that is, make sure our actions are beneficial or, at least, benign. Trade-offs are often necessary, because intervention for beneficial goals is not always benign or risk-free. In medicine, for example, some interventions involve surgery, radiation, chemotherapy and other potentially damaging procedures. General agreement prevails, however, that medical intervention should take into account the risk of damage and the quality of life: in other words, to evaluate the means and not just the ends. The same is true for genetic modification in all its guises.

(HGP), the other led by Craig Venter of Celera Genomics.[14] Among the goals of the HGP were identification of what turned out to be about 25,000 genes, determination of the sequences of the 3 billion chemical base pairs involved, and analysis of the ethical, legal and social issues arising from the project. A wide range of applications, covering many fields of human endeavor, is expected to flow from this work and from genomics in general in the coming decades.

An area of research evoking great interest is **gene therapy**, here defined as introduction of genetic material to treat or cure a disease or abnormal medical condition. Human trials, which began in 1990, have targeted several diseases and used a variety of methods and vectors. Although there have been a few successes among the hundreds of trials, a great many problems remain. As of the end of 2006, no gene therapy product had been approved for sale in the United States. In May, 2007, however, Epeius Biotechnologies Corp.'s Rexin-G™ was approved by the United States Federal Drug Administration as an orphan drug to treat pancreatic cancer.

Problems include the difficulty of getting the new genes into enough of the patient's cells, and then getting them to work; the short-term nature of the therapy; and reactions to (or infection by) the vector. In 1999, a patient in an American gene therapy trial died, most likely as a result of a severe immune response to the viral vector. A further major setback occurred in French trials carried out on young children with X-linked severe combined immune deficiency ("bubble boy" syndrome). Although nine of the ten children in this trial were successfully treated, three subsequently developed cancer. In a study by Woods et al. (2006), one third of mice administered the same gene as was used in the latter trial developed lymphoma later in life.

[14] In December 2006, Venter announced the sequencing of his own genome, the first for an individual person, and the first targeting all 46 chromosomes; the two human genome projects mapped only one of each pair. In October of the same year, a $10 million prize was offered to the first group to cheaply map the genomes of 100 people in 10 days.

Gene therapy is discussed here because it is so closely tied to genetic engineering. Its impact on the gene pool and thus its ability to alter the path of evolution is, however, limited to the fact that people who might otherwise die may now go on to produce offspring bearing the deleterious gene. The techniques have to date involved only somatic tissues; there is a strong consensus among scientists and indeed the general public that so extraordinary are the risks that there be no attempt as yet to modify human germ plasm. It is also considered unethical to alter genes that would change or "enhance" nondisease characters such as an individual's appearance or height. Other ethical concerns include, as with reproductive technologies, the definition of what constitutes a disability or disease (how do we define normal?), and whether, given the risks involved, treatment is warranted.

Clones and Cloning

A **clone** is *an organism descended from and genetically identical to another organism.* Under this definition, all offspring produced by asexual means — a method of reproduction used by bacteria and even many eukaryotes, especially plants — are clones. Clones provide genetic uniformity across the generations, an advantage when organisms face the same conditions for a long time, but lack genetic diversity. Sexual reproduction, on the other hand, has selective advantage for organisms facing changeable and unpredictable conditions, but is a comparatively expensive genetic gamble. The advantages and disadvantages of asexual versus sexual reproduction are discussed in Chapter 9.

Cloning, defined here as *the production of clones by artificial means,* has a long history in agriculture, horticulture and forestry. Methods such as taking plant cuttings, grafting and layering have been employed for centuries if not thousands of years; familiar plants reproduced in this way include potatoes, pineapples, many horticultural and forestry species and most trees bearing fruit or nuts. Plant clones also can be started in the laboratory in cell culture and transplanted into soil when they reach sufficient size.

When it comes to cloning animals, two modern laboratory methods are employed. In *embryo splitting,* an early-stage embryo with only a few cells is split into equal parts, each of which retains its ability to develop into a normal embryo (and is, therefore, totipotent). These cloned embryos are then implanted into the potential mother or mothers.

The other laboratory method of cloning, *somatic cell nuclear transfer,* has been the subject of enormous controversy and media attention in recent years and is what is usually referred to in discussions of cloning. In this method, a nucleus is removed from a diploid cell and inserted into an enucleated egg cell, and the now diploid egg develops into an embryo. The first such technique, invented in the 1950s, had some success when donor nuclei from frog embryos were transferred to enucleated eggs, but attempts to transfer post-embryonic nuclei showed that the nuclei were no longer totipotent. By the 1960s and 1970s, however, techniques had advanced to such an extent that eggs containing tadpole nuclei would develop into tadpoles and then metamorphose into adults (Gurdon, 1962).

By the 1990s, success had been achieved with a variety of mammal species, but always using nuclei from very early embryos. Nuclei from later embryos, and certainly from adults, appeared ineffective. The news in 1997 that Ian Wilmut and colleagues had succeeded in producing a sheep (named "Dolly") by transferring the nucleus of a mammary cell from a six-year-old ewe into one of its eggs, burst on the world like a thunderclap. If it could be done for sheep, why not for humans? Indeed, Wilmut told a British parliamentary committee that he expected a similar technique could be used to clone humans "within two years."

The major innovation of this work was the "reprogramming" of adult nuclei to assume the functions of a zygote nucleus and thus direct the development of a new individual. Since then, nuclei from a diverse range of adult tissues and a variety of animals — mouse, cow, pig, cat, goat, fish and more — have been used to produce clones. But these successes have been far from unqualified. For instance, the process is both expensive and highly inefficient — very few of the adult cell nuclei used result in living offspring — and abnormalities are common in those offspring that do survive to birth. Still, the implications were unparalleled. Would it now be possible for someone to clone a champion bull, a favorite pet or even him- or herself?

Reaction to the possibility of human cloning — most of it decidedly negative — was immediate. The media, representatives of medical, religious and other bodies, and members of the general public all voiced concerns about the ethics of the technology. Politicians in many jurisdictions acted swiftly. In the United States, for example, President Clinton banned all federally funded human cloning research, asked for a moratorium on nonfederally funded research and ordered the National Bioethics Advisory Commission to conduct hearings. Nevertheless, some individuals embarked on research in the field.

Little progress was made until 2004, when a South Korean biologist announced that he and his team had produced a human embryo by cloning; other advances in his lab were announced in quick succession. The aim of this work was not to generate an exact copy of a human being, but rather to produce a source of **stem cells**, thus opening the door to potential cures for a range of diseases. In late 2005 and early 2006, however, it became apparent that most of his results were fraudulent. Now regarded as one of the worst cases of fraud in the history of science, this has dealt a blow to stem

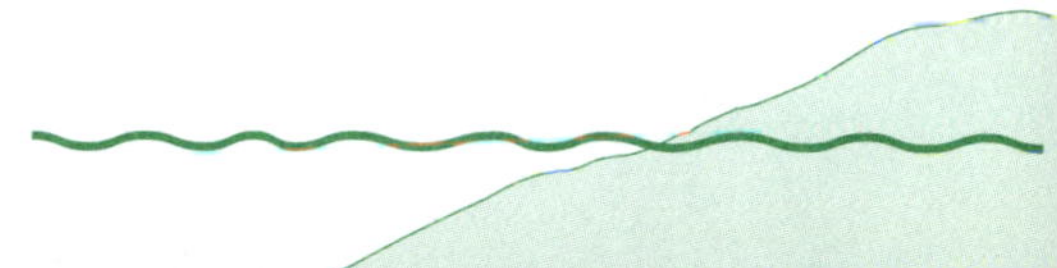

cell research. For now, the many ethical and experimental challenges posed by human cloning are moot.[15]

Stem cell research continues on cell lines obtained from frozen surplus embryos left over from *in vitro* fertilization (though in the United States there is currently a ban on creating new lines), and from other potential sources such as amniotic fluid and umbilical-cord blood. As with gene therapy, the hoped-for benefits are huge, but we have as yet little idea whether this line of research will yield fruitful results, let alone provide us with the tools to heal a large range of conditions. Work remains at a very early stage.

Other Human-Induced Evolutionary Changes

Until now, this chapter has concentrated on the role of culture as a force in human evolution, and on human efforts to direct the evolution of our own and other species. Here we briefly address the extraordinary impacts we humans have had throughout history on the distribution and gene pools of organisms other than those whose genomes we have tried directly to change, impacts that have mostly been inadvertent. Starting in the Paleolithic, when human actions such as hunting or the widespread use of fire may have led to the extinction of the megafaunas (very large animals) of Australia, the Americas and other parts of the world, we have altered both the species mix and environment of almost the entire globe. A notable example today is the plight of the ocean's fish stocks, all of which are under threat from overfishing. Populations of the northern cod, which lives off Newfoundland and Labrador, have not recovered at all from a disastrous decline in the years around 1990.

Wherever humans have moved, especially since the development of agriculture, they have brought with them a suite of other organisms, both domesticated (cattle, goats, etc.) and opportunistic (rats, weeds, disease-causing microbes, and so on). The changes wrought on local environments took a quantum leap when Europeans began venturing across oceans, a story told by Alfred Crosby in his 1986 book, *Ecological Imperialism: The Biological Expansion of Europe, 900–1900,* and more recently (1997) and comprehensively by Jared Diamond in *Guns, Germs, and Steel.* Among the effects have been mass extinctions, especially on islands (Hawaii is a prime example), displacement of native species by weedy ones (the list is very long) and a general homogenization and simplification of ecosystems. Fueled by globalization, such impacts continue today, so that the agents of destruction can arrive from any quarter (e.g., invasive species carried in the ballast water of ships; insect pests carried on lumber). Entire ecosystems are threatened or in some cases have disappeared.

The impacts on native humans in newly "discovered" regions were also calamitous. In the decades following the arrival of Europeans in the Americas and the Pacific, infectious diseases such as smallpox, measles and tuberculosis decimated populations that up until then had been isolated. A similar fate befell a variety of animal and plant species, especially trees (e.g., the American chestnut).

In *Collapse: How Societies Choose to Fail or Succeed,* Diamond (2005) details the consequences of exceeding one's resource base. Using a number of case studies (the Maya of Central America, the Norse settlement in Greenland, and more), he shows how human societies in different times and different places have collapsed as a result of changes in their environment, changes in large part precipitated through their own actions, including faulty agricultural practices, deforestation and water depletion, to name a few. Today our resource base is the whole world, and our impacts are accordingly global, extending even to climate change.[16]

What Lies Ahead?

Before us lie two very different worlds: A world where our ever-increasing numbers, our extravagant use of resources, and the growing divide between rich and poor threaten our very survival, and a world where dazzling new discoveries in the realms of medicine and genetics hold out the possibility that we may one day be able to alter the path of our inheritance.

Standing as we are at the threshold of the genetic revolution, we can only speculate on where the new and as yet unimagined technologies of the future will take us. The experimental challenges may now appear daunting, but then so did the quest to fly, perform open-heart surgery or walk on the moon. Perhaps the most pertinent question we can ask of this new adventure is: What are its goals? Even if we assign it the most moral of motives — the good of humankind — it remains to be determined whether this "good" can be known.

Major genetic transformations, however, remain far in the future. More relevant today is the need for us to overcome damaging social behaviors (including our militaristic tendencies) so we live within our resources and protect the other organisms that inhabit our planet. Reversing the current trends will be a daunting, perhaps impossible task. As Diamond (2005) demonstrates, a host of attitudinal and intellectual problems underlie our approach to the environment, atti-

---

[15] For a discussion of the many scientific, political and ethical questions involved, see Jane Maienschein's 2003 book, *Whose View of Life? Embryos, Cloning, and Stem Cells.*

[16] A large literature exists on global environmental problems. The Further Readings section of this chapter lists two books by Diamond that provide a convenient entry into the topic.

tudes that are echoed in other areas of our lives: failure to anticipate a problem, creeping normalcy (change occurs so slowly that we fail to notice it), denial once the problem becomes apparent, selfishness, and so on. The prime example today is global warming, which by the middle of the first decade of the twenty-first century could no longer be denied, but about which almost nothing significant was being done.

Ideally, our intellects and technologies should provide us with the tools to overcome these two linked quandaries. When added to our growing knowledge about and mastery over our biological makeup, we could imagine a new role for humanity: beneficial coauthors of our own evolution. Our lives — and the planet we inhabit — should be sacred to us, not because evolution has conferred such status, but because we ourselves have done so.

We turn in the next and last chapter to discussion of a very specific aspect of human culture — the ways in which humans have related and continue to relate their experience of nature to systems of belief, religion and evolution. Chapter 26 takes us back to historical perspectives introduced in Part I. The traditional religious rationale for social and biological systems was that the universe followed a designed order established by an intelligent deity or deities. By omitting God from the equation, evolutionary theory challenged that long-accepted worldview. Tension between religion and evolution remains, especially now that we can engineer other life forms, modify the environments of countless other species and create changes that affect the entire globe.

KEY TERMS

biological limitations
cloning
culture
ecological footprint
eugenics
gene therapy
genetic engineering
lethal alleles
longevity
negative eugenics
population explosion
recombinant DNA
Social Darwinism
sociobiology
vector

DISCUSSION QUESTIONS

1. Why is biological evolution considered Darwinian, and cultural evolution considered Lamarckian? Is such a division justified?
2. Social Darwinism.
   a. Can human social structures and social changes be explained by the Darwinian laws that govern biological evolution?
   b. Why is Social Darwinism no longer an acceptable concept?
3. Sociobiology.
   a. To what extent can biologically based social behaviors be culturally modified?
   b. Would you consider culture as merely an expression of underlying biology? Explain.
4. What explanations have geneticists offered to account for the frequency of traits caused by deleterious alleles? Can all such traits be eliminated? Why or why not?
5. Does alteration of the human genome through genetic technology have the potential to improve human health?
6. Discuss the implications of attempts to alter the human gene pool at a population level.
7. What are the potential impacts, both positive and negative, of human-directed alteration of the genomes of other species? Could such alterations affect the evolution of these other species?
8. What are the medium- and long-term implications of increased human life span and use of resources on our own and other species, and on the environment?

EVOLUTION ON THE WEB

Explore evolution on the Internet! Visit the accompanying Web site for *Strickberger's Evolution*, Fourth Edition, at http://www.biology.jbpub.com/book/evolution for Web exercises and links relating to topics covered in this chapter.

RECOMMENDED READING

Alcock, J., 2001. *The Triumph of Sociobiology.* Oxford University Press, New York.

Alcock, J., 2005. *Animal Behavior: An Evolutionary Approach.* Sinauer Associates, Sunderland, MA.

Crosby, A. 1986. *Ecological Imperialism: The Biological Expansion of Europe, 900–1900.* Cambridge University Press, Cambridge, UK.

Diamond, J. M., 1997. *Guns, Germs, and Steel: The Fates of Human Societies.* Norton, New York.

Diamond, J. M., 2005. *Collapse: How Societies Choose to Fail or Succeed.* Viking Books, New York.

Hawkins, M., 1997. *Social Darwinism in European and American Thought: 1860–1945.* Cambridge University Press, Cambridge, UK.

Jones, S., R. D. Martin, and D. R. Pilbeam (eds.), 1996. *The Cambridge Encyclopedia of Human Evolution.* Cambridge University Press, Cambridge, UK.

Kevles, D. J. 1995. *In the Name of Eugenics: Genetics and the Uses of Human Heredity.* Harvard University Press, Cambridge, MA.

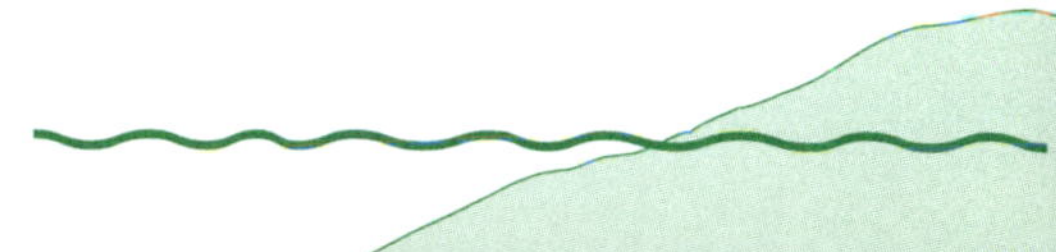

Leland, K. N., and G. Brown. 2002. *Sense and Nonsense: Evolutionary Perspectives on Human Behaviour.* Oxford University Press, Oxford, UK.

Maienschein, J., 2003. *Whose View of Life? Embryos, Cloning, and Stem Cells.* Harvard University Press, Cambridge, MA.

Milner, R., 1990. *The Encyclopedia of Evolution. Humanity's Search for Its Origins.* Facts On File, New York.

Mindell, D. P., 2006. *The Evolving World: Evolution in Everyday Life.* Harvard University Press, Cambridge, MA.

Ridley, M. 2000. *Genome: The Autobiography of a Species in 23 Chapters.* Harper Perennial, New York.

Wilson, E. O., 1975. *Sociobiology: The New Synthesis.* Harvard University Press, Cambridge, MA.

Wilson, E. O., 1977. Biology and the social sciences. *Daedalus,* **106,** 127–140.

Wilson, E. O., 1978. *On Human Nature.* Harvard University Press, Cambridge, MA.

26 Belief, Religion and Evolution

Chapter Summary

The roots of religious belief lie in human attempts to appeal to and control the forces of nature. From such roots arose the concepts of God and soul, both of which were thought to be eternal and intangible. The traditional religious rationale for social, cultural and biological systems was that the universe followed a designed order established by an intelligent deity. Along with belief in a god who created and maintained the universe, religion provided emotional solace, a set of ethical and moral values and support for the established social system.

Until Copernicus and Galileo in the sixteenth century, no one had seriously challenged the idea of a powerful deity controlling the physical universe. In the new worldview they and others ushered in, however, God appeared as an initial creator rather than as an incessant manipulator of the universe. The advent of Darwinism posed further threats to Western religion by suggesting that biological relationships, including the origin of humans and of all species, could be explained by natural selection without the intervention of a god. The Darwinian view that evolution is a historical process — extant organisms were not created spontaneously but rather formed in a succession of past events as they adapted to changing environments — contradicted the common religious view of design by an intelligent designer. Given the great age of Earth and the power of natural selection, complexities that seem unlikely as singular spontaneous events become evolutionarily probable events. To many, evolutionary randomness and uncertainty replaced a deity having conscious, purposeful, human characteristics.

Darwin's theory of evolution had a profound impact not only on biology, but also on virtually all spheres of human society and culture. Its extraordinary influence owes much to the era in which it appeared, a time of social, economic and technological developments, the overthrow of old social orders and the rise of capitalism. Since then, many Judeo-Christian denominations have evaded interpretation of evolutionary biological events or attempted compromises between traditional religious explanations and scientific ones; for example, God set evolution in progress by providing laws of nature that account for all subsequent events. Acceptance by individuals of religious explanations has been eroded further as we discover more and more natural explanations for the origin and modification of Earth and its inhabitants, recognize that ethics and morality can differ between different human societies, and that changes in such values need not depend on religious beliefs. Our ability to modify the genomes and the environments of other species and to create changes that affect the entire globe means that tension between religion and evolution continues.

Especially in the United States but also increasingly in Europe, fundamentalist religious groups who oppose evolution have attempted to prevent its teaching and to reduce or eliminate the subject of evolution in many biology textbooks. The "creation science" movement (which, despite its name, does not use the scientific method) and more recently "intelligent design" have both wielded extraordinary influence.

In the present chapter, we continue the themes of cultural and social evolution discussed in Chapter 25 by exploring how the theory of evolution impacted on a very specific aspect of human culture — systems of belief — especially those associated with established Judeo-Christian religions. In part, this chapter seeks to explain the roots of religion by taking us back to historical perspectives introduced in Part 1. It also evaluates contemporary issues such as why religion has persisted, and how to explain consciousness and the concept of the soul.

Science, society and culture are intimately interwoven. From a social point of view, the development of evolutionary theory in the late nineteenth century coincided with an all-pervasive political and economic revolution in social behavior. The economic challenges posed by capitalism and its new moneyed classes allowed — even encouraged — ideological challenges to the prevailing religious and philosophical systems that had supported the old social order, allowing European science to flourish. The influence of Social Darwinism and the rise of eugenics are changes discussed in the previous chapter.

As discussed in Chapter 2, the impact of Darwin's theory on all aspects of Victorian and subsequent societies was profound. Darwinian evolution offered a vast historical framework in which to understand biological change. By proposing that the form and function of living organisms did not arise by creation but rather by natural processes, Darwin's theory made it clear that species fixity was not natural. These radical ideas, which revolutionized biology, also affected sociology, anthropology, economics, politics, women's rights, fiction, poetry, linguistics, philosophy and psychology. Herbert Spencer, Karl Marx, Joseph Conrad, Thomas Hardy, Alfred Tennyson, George Eliot, George Bernard Shaw, Henri Bergson and Sigmund Freud are just a few of those who incorporated evolution into their studies, writings, politics and world views.[1]

Science and Religion

Our views of the world are influenced strongly by the culture in which we live. Different cultures place different emphases on how individuals perceive various events and relationships, and on how they explain these perceptions.

What we consider science is also culturally dependent in the sense that large differences can exist between cultures as to whether or how we should apply scientific concepts to events and experiences. Explanations that many of us accept as scientific — analyses based on the application to known and emerging knowledge of scientific principles and laws — others do not accept, or accept only to varying degrees.

In Western European culture, *nonliving phenomena* generally have been considered more amenable to scientific analyses than have matters that touch on *life itself or on human life in particular* (e.g., human behavior and interactions). Physics and chemistry were well established as sciences by the nineteenth century, whereas biology, especially development and evolution, continued to be subject to vitalistic interpretations (**vitalism**), according to which all living organisms are imbued with a vital force. Consequently, gaining freedom from cultural constraints has been more

[1] For Darwin's influence on novelists, see Levine (1988), Beer (1992) and the essays reprinted in Part IX (pp. 631–682) of Appleman (2001).

difficult for biology (especially for evolution) than for physics or chemistry.

Basis of Religious Belief

The various levels of religious development found in different cultures provide clues to the **evolution of religion** itself.

Religion first develops in a culture when, in an attempt to deal with aspects of experience they can neither control nor understand, people endow the forces of nature with the spirits of animals and humans. Ritual develops when people repeat ceremonies to help ensure their efficacy. Ritualized behavior seems to have become especially important in the transition from hunting to agricultural societies in which crops have to be planted and harvested at appropriate seasons each year and where individual efforts could be either rewarded or damned by external and uncontrollable forces — droughts, floods, volcanic eruptions, plagues of insects — that remained mysterious and arbitrary.

Underlying many human anxieties about the world, and often at the root of the difficulty in treating nature as an independent object of study, was the belief that nature reflected a supernatural evaluation of human affairs by imposing punishment and reward, calamity and prosperity. Because people saw the forces of nature as humanlike, religion sustained and encouraged the hope that those appeals that humans understood — gifts, sacrifice, obedience and loyalty — could appease nature's judgment and recrimination.

Religion's universal attractiveness is supported by the universality of such human emotions and behaviors as fear, devotion, insecurity, honor, dependency, care for others, aggression and guilt. Religion provided and preserved **belief systems** that helped explain and guide peoples' relationships to the world about them and provided answers to such questions as, "Where did we and our society come from, and why?" To these ends, the belief system of each culture usually offered comprehensive accounts of how and for what purpose the world was created (**Table 26-1**), integrating those accounts with views of human society, history, and behavior, more often than not made relevant to personal human experience. So, you can see how a theory of evolution by natural selection would disrupt much more than a belief in origins and design. Indeed, it disrupted an entire worldview and code of existence.

TABLE 26-1 Ten Creation Myths from Different Cultures

1. God arose from the depths of the ocean, created dry land, and then created all creatures on the hill at Eliopoulos at the center of the universe (Egypt).
2. God made sky and earth by splitting the powers of evil in half, and then produced humans for purposes of worship (Mesopotamia).
3. God creates all that is good and struggles with an evil being that creates all that is bad. Each struggle lasts about 3,000 years and will continue until evil is vanquished, at which time creation will be complete and perfect (Iran).
4. God, a female, divided the sky from the sea, and produced a serpent with whom she copulated. She then laid a giant egg out of which came the earth, its creatures, and all the heavenly bodies, as well as the subsidiary powers to rule these various entities (Greece).
5. God created himself from a golden egg, and from the various parts of his body everything was born. After a time, life is destroyed and the cycle begins again (India).
6. God created the universe in 6 days ending with the creation of humans, according to Genesis 1; or, God first created Adam in the Garden of Eden and then created animals and birds and eventually Eve, according to Genesis 2 (Israel).
7. God was a woman who produced twins — the sun and the moon. During various eclipses, the twins came together to create the various gods and spirits of earth and sky that rule over humans (Benin, Africa).
8. God created the world in four distinct periods, each separated by a flood (Yucatán).
9. God created the earth and its creatures from mud gathered in the webbed feet of ducks that swam on a primeval ocean (American Crow Indians).
10. The universe was originally in the shape of a hen's egg, out of which God emerged and chiseled its main physical features. After 18,000 years God died and the remainder of the world was derived from his body: the dome of the sky from his skull, rocks from his bones, soil from his flesh, rain from his sweat, plant life from his hair, and humans from his fleas (China).

Soul, God, and Consciousness

In support of a supernatural view of events, religion has relied on two basic concepts that probably arose early in history, the *soul* and *God.* Most religions consider the **soul,** or the spirit, or what many would call *the personality* or *self,* to be a conscious entity without physical attachments or properties.

According to various psychologists and anthropologists, the idea that the soul is separate from the body probably originated in the separation of mind from reality in dreams. The restrictions of space, time and even death vanish in dreams, and we can suppose that one aspect of human life, the soul, is immune to such restrictions. One of the nineteenth-century founders of anthropology, E. B. Tylor

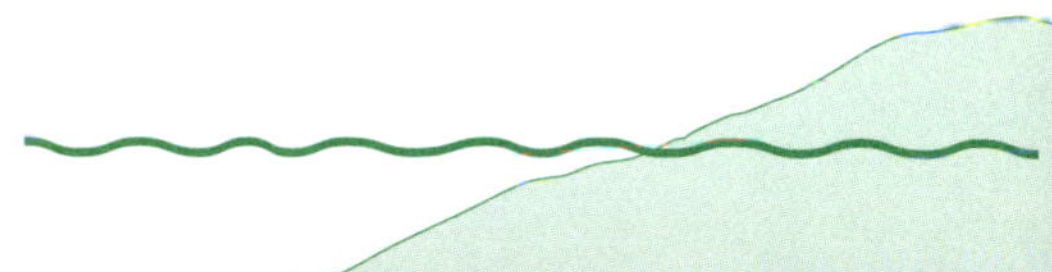

(1881) explained the soul concept among "primitive" tribes as follows:

> As it is well known by experience that men's bodies do not go on these excursions, the natural explanation is that every man's living self or soul is his phantom or image, which can go out of his body and see and be seen itself in dreams. Even waking men in broad daylight sometimes see these human phantoms, in what are called visions or hallucinations. They are further led to believe that the soul does not die with the body, but lives on after quitting it, for, although a man may be dead and buried, his phantom-figure continues to appear to the survivors in dreams and visions.[2]

This attitude is reinforced by the *difficulty of conceiving that the self has only a limited existence.* It is a formidable task to realize that the sensations and feelings that we use to perceive our relationships to the environment, and through which we integrate, evaluate and interact with the world around us, have a beginning and an end. Although one can intellectually conceive of death in relation to other creatures or phenomena or even to one's own body, it is impossible to "feel" death as one may feel and anticipate other events and sensations.[3] This awareness of self seems to be eternally present, because there is usually no self-knowledge of how it arrived or how it will end. Today, such concerns are reflected in research whose goal is to understand **consciousness,** which embodies the awareness of a self that is aware of itself, looks out at a world that it perceives and responds to it.

Theologians have often attempted to provide a "scientific" argument for the nonmaterial nature of the soul or the personality, claiming that intellectual processes cannot have a material origin. According to their argument, freedom of choice, ethics and all of the many complex ideas about which humans can think must be considered the fruits of nonbiological matter, because they are so obviously separate from purely physical processes. The biological view is quite different. Although we do not yet know the precise relationship between the matter of the brain (neurons, synapses, and so on) and the thoughts and feelings it produces, such a relationship exists and we have every reason to believe (as discussed in Box 20-5, Evolution of the Human Brain) that the complexities of thinking and feeling have evolved like any other trait.[4]

[2] For more recent analyses, see Richards (1987), Ruse (2001), Dennett (2006) and Mithen (2006).

[3] In one culture, death is a recycling of a person's spirit into another organismal form. Another believes death is a state of reward and punishment for an individual's behavior. Still another regards death as the end of a person's existence.

[4] See the recent books by Ruse (2005) and Dennett (2006) for how a psychologist and a philosopher of biology approach these issues.

As with the concept of the soul, the concept of God has many qualities that reflect human experience. Each of the various gods personifies, often in human form, forces or tasks that seem beyond our capabilities. How better to alleviate anxiety and fears about harvests, thunder, fertility, woods and rivers, than to believe that these elements of nature embody extensions of human behavior? Different religions endow dependent feelings with structures derived from their own societies, often with considerable imagination; reality could probably never produce pleasures and torments as great and enduring as those provided by the imaginations of believers in heaven and hell.

Darwinism and Religion

Darwinism had an especially dramatic impact on religion. To many of Darwin's religious contemporaries and to others since, *On the Origin of Species* (1859) and *The Descent of Man* (1871) raised issues of momentous importance. As pointed out by Ellegård (1958), "To the general public, Darwinism was at least as much a religious as a scientific question."

The stressful relationship between evolution and religion stemmed, not only from the vulnerability of religion, but also from the recognition that evolutionary concepts were not impregnable — as we saw in Chapter 3 when discussing scientific objections raised against evolution. Darwin pointed out that at least two discoveries could refute his theory:

1. An inexplicable reversion in the evolutionary sequence, such as evidence of humans in the Paleozoic or Mesozoic.
2. The finding of exactly the same species in two separated geographical locations whose presence was not caused by migration between these areas.

If such discoveries were made, Darwin recognized that he would have to abandon his view of evolution, whose major alternative seemed only a religious supernatural doctrine, in Darwin's words, "the common view of actual creation." No such evidence has been found in the century and a half since Darwin's theory was published in 1859.[5] Nevertheless, this dichotomy of views — evolution and religion — remains controversial.

To comprehend the role that evolution plays in modern life, we need to understand how religion and evolution interacted and interact. An early confrontation between the two views was a debate at Oxford soon after the publication of *The Origin.* Bishop Samuel Wilberforce (1805–1873) of the Church of England (**Fig. 26-1**) attacked Darwinian theory as incompatible with the Bible. Coached by the reli-

[5] 2009 will be the 150th anniversary of the publication of *On the Origin of Species,* so keep an eye out for celebrations on a grand scale.

FIGURE 26-1 Caricatures of Bishop Samuel Wilberforce (left) and Thomas Huxley (right), which appeared in the British magazine *Vanity Fair* — Wilberforce on 24 July, 1869; Huxley on 28 January, 1871. (Courtesy of *Vanity Fair*.)

gious former student of Cuvier, Richard Owen, the Bishop attempted to destroy Darwin's theory through scientific arguments. Wilberforce aimed his final point, not to the official debaters, but directly at Thomas Huxley, a member of the audience, when he asked whether it was through Huxley's grandfather or grandmother that Huxley claimed descent from a monkey. The wit of Huxley's response, recounted a few months later in a letter to his friend Frederick Dyster, has often been quoted:

> If, then, said I, the question is put to me would I rather have a miserable ape for a grandfather, or a man highly endowed by nature and possessed of great means and influence, and yet who employs these faculties and that influence for the mere purpose of introducing ridicule into a grave scientific discussion, I unhesitatingly affirm my preference for the ape.[6]

Unimpressed by scientific arguments, however, many nineteenth- and early twentieth-century theologians hammered away at the heresy of evolution. Wilberforce, accused Darwin of "a tendency to limit God's glory in creation." Cardinal Manning, a leader of English Catholicism, called Darwinism "a brutal philosophy — to wit, there is no God, and the ape is our Adam." Such religious attacks were worldwide, frequent, harsh and almost always focused on the same points: Religious opponents accused Darwinists of seeking "to do away with all ideas of God," "to produce in their readers a disbelief of the Bible," "to displace God by the unerring action of vagary," and to "destroy humanity's unique status."

In the most sensitive area of all — life itself — Darwinian evolution offered different answers to religion's claims of why life's important events occur. Darwin's works made clear that people no longer needed to believe that only the actions of a supernatural creator could explain biological relationships. Darwin presented the concept that nature entails continual change, unpredictable chance events, an unrelenting struggle for survival among living creatures and no obvious guidance. To religious believers natural selection substituted waste for economy by treating life as a continually expendable commodity. Darwin replaced what for many was an understandable view of nature — the creativity of a human-like God — by randomness and uncertainty.

How could such a heretical doctrine as evolution by natural selection have developed in a religious European country and become acceptable to so many of Darwin's learned compatriots? Among possible answers is one that demonstrates how the struggle between evolution and religion was part of the historical framework of the time.

Looking at evolution historically, we can see that by threatening basic religious concepts of a fixed universal order, evolution seemed to have exceeded the "game plan" for permissible ideological challenges. Nevertheless, there was no sustained political attempt to suppress evolutionary ideas, although they were attacked far and wide and even outlawed in some American states by the late 1920s. One reason for the relative freedom that evolution enjoyed elsewhere is probably its close ties with all other aspects of science. Science and technology were, after all, the mainstays of economic expansion. Many scientists were or became convinced by evolution, and most social leaders of the time must have felt that evolution was a scientific theory that would have to be tolerated.

They had considerable cause for concern. To understand the cause of that concern, we review those aspects of religion most directly threatened by evolution, although it is important always to remember that religious beliefs cannot be evaluated on the same scientific basis as scientific theories. Logically then, and as discussed in Chapter 25, biologists must be careful not to apply scientific explanations to areas such as ethics or moral judgments that lie outside the bounds of objective science.

[6] The suggestion that humans were related to the orangutan was raised in Chapter 1. Gorillas only became known after 1854 when the first bones were shipped from Africa, and in the early 1860s when a stuffed gorilla skin was taken along as a prop on lecture tours undertaken throughout England by an African explorer. "Victorians were horrified to think that these reputedly violent animals — distorted men in shape and size, representing the brutish, dark side of humanity — were possible ancestors (Browne, 2006, p. 95).

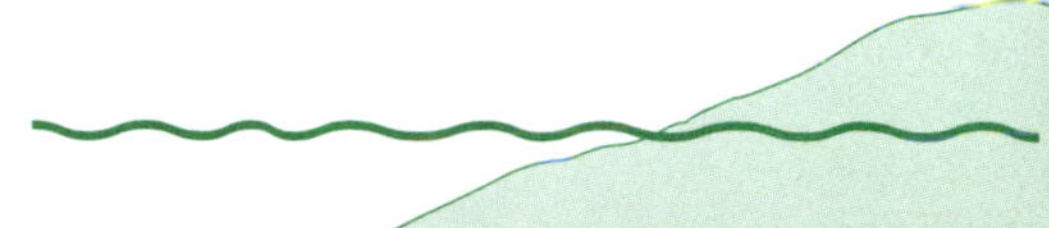

Persistence of Religion

The fear that Darwinism was an attempt to displace God in the sphere of creation was quite justified.[7] Consequently, for many evolution was a deep, abiding threat.

Hypotheses concerning evolution, whether conforming or dissenting, were evaluated, as in all science, by evidence; religious beliefs were asserted by faith or reference to a higher being. Indeed, evolution cannot answer questions that are central to religion such as, "Is there a divine purpose for the creation of humans?" or "Is there a divine purpose for the creation of any living species?" Science cannot even ask such questions. Why? Because, as outlined in the introductory essay to Part I, they cannot be answered using the scientific method.

Surprisingly, the conflict between seemingly irreconcilable points of view on the same subjects — the origin of species and the origin of humans — did not result in either the defeat of religion or the defeat of evolution. Evolution suffered almost no loss at all. In fact, tied to the scientific method and supported by it, evolution became the main unifying principle in biology and led to an expansion of research in almost every area, with the evolutionary view as strongly entrenched in biology as the atomic, molecular and gravitational theories are in the physical sciences. None of these principles can be observed, yet all are working models indispensable to a consistent and scientific comprehension of events.

One might think that, through the evidence and closely reasoned arguments offered by biologists, evolution had successfully undermined one of religion's prime justifications for itself — the special creation of humans. The demise of religion should therefore have been just a matter of a short time. However, it required an audacious leap to question the rule of nature by the wisdom of God.

First of all, the transition from regarding nature as operating in terms of anthropomorphic wisdom to operating in terms of the evolutionary opportunism expressed by the survival of the fittest was difficult. For many centuries, in the Judeo-Christian tradition it was, after all, easy to follow a pattern of simple parallels between society and nature: the father rules, the Lord rules, the King rules, the Pope rules, God rules. Even for those who questioned the legitimacy of kingly rule, it must have seemed far too audacious a leap to question the rule of nature by the wisdom of God. Even the philosopher Kant, who had an evolutionary approach to cosmology and to the origin of the solar system, found it at times abhorrent to admit that species could evolve, describing such notions as "ideas so monstrous that the reason shrinks before them." Darwin also is reported to have felt quite uncomfortable about his role in proposing the evolution of species and wrote in an 1844 letter to Joseph Hooker that "it is like confessing a murder."

One source of ambivalence was the established belief that humans were created in the image of God and endowed to rule over other biological and social groups. Kinship to those below, whether ape or servant, was a repugnant idea. An oft-reported example of this repugnance is the response of the wife of the Bishop of Worcester when informed that Huxley had announced that man was descended from apes: "Descended from apes! My dear let us hope that it is not true, but if it is, let us pray that it will not become generally known." Reinforcing these attitudes was the reaction of many of the European middle classes to the upheavals of the French Revolution and to the atheism of many of its leaders, reactions that helped strengthen the religious climate among the middle classes in nineteenth-century Europe, and that led, in various groups, at least for a time, to a more literal interpretation of the Bible.

Essential to the preservation of religion in the midst of the evolutionary bombardment was that religion answers a series of strong emotional needs such as a purpose in life, and offers answers to the important personal mysteries of birth and death, including the question of identity, "*Who am I?*" Religion also served as the main repository for the ethics and morals of society (what is right and wrong) and often served to maintain confidence in the social order and to unify nationalistic and provincial sentiments ("God is on our side"). In articulating his sociobiology in 1975, E. O. Wilson claimed that religion serves an important role in helping provide a common social identity to groups of individuals, and by increasing group power, confers "biological advantage" (see Chapter 25). Evolution, in contrast, deals with many basic questions of life that are of concern to religion, but as a science it did little to meet emotional needs.

Society therefore has held on to both evolution and religion with a sort of armed truce. With the exception of some fundamentalists, whose views are discussed in the following section, religion essentially withdrew from the domain of biological evolution, leaving both the origin of species and the origin of humans to evolutionary biologists and anthropologists. To help make evolution acceptable among some of the main religious groups, various theologians placed more emphasis on reinterpreting the Judeo-Christian Bible by either ignoring the creation story in Genesis or by describing it as allegorical or mythical. This enabled scientists and intellectuals who maintained religious affiliations ("theistic evolutionists") to insist that one could believe in both evolution and religion. However, such reinterpretations evoked considerable discomfort in orthodox believers because, like any other religious document, the Bible is an attempt to explain the unknown in a religious framework. To concede

---

[7] See Ruse (1988, 2005), Bleckmann (2006) and Dennett (2006) for critical analyses of these issues.

that parts of the Bible can be understood outside the religious structure opens the door to further loss of religious credibility, which generated considerable dissension among theologians, for many of whom, evolution was a deep, abiding threat.

The position of the Roman Catholic Church as enunciated by several recent Popes illustrates one type of accommodation between religion and evolution, namely acceptance of scientific findings when the evidence is incontrovertible. Although no supporter of evolution, Pope Pius XII (1876–1958) accepted the evidence that the universe was millions of years old. In 1993, in declaring that Galileo had not committed heresy in proposing the theory that the solar system revolved around the sun, Pope John Paul II (1920–2006) used Galileo's achievements as an example of how scientific knowledge could lead the church to reinterpret the writings in the Bible. Three years later, he declared that scientific findings show that evolution is more than a hypothesis it is "an effectively proven fact." Pope John Paul's acceptance was rational, predicated as it was on the basis that "truth cannot contradict truth." Such rationality makes the beliefs and arguments of creationists more difficult to comprehend.

The Question of Design

The need to believe that active intervention by a divine power was necessary to explain most or all observed events, and to allow the continued maintenance of the universe, did not require a search for natural laws to account for changes in the present, past or future.

The first significant cracks in the theological armor of continued divine intervention in nature were made in the **discoveries of natural laws regulating the motion of the solar system,** by Copernicus, Galileo and Kepler. Their genius was that all three provided simpler and more universal explanations of the workings of the physical universe than previously available. The elegance of their explanations made inevitable the rise of science and of the scientific method, for:

> A momentous change had come about when what scientists did came to be taken for granted, even by those who understood little or nothing of it. The crucial change in the making of the modern mind was the widespread acceptance of the idea that the world is essentially rational and explicable, though very wonderful and complicated (Roberts, 1985, p. 242).

These openings were considerably widened by the mechanistic explanations offered by Isaac Newton (1642–1727) on the motion of the solar system through the force of gravity, and the postulate of an unbounded infinite universe in which our world was proportionately smaller than even a grain of sand. Later, geologists such as Charles Lyell (1797–1895) extended this mechanistic approach by proposing how natural forces could mold Earth's surface. Although these scholars were not atheists, their findings about natural processes made through sciences such as mechanics, optics and chemistry, helped reduce God from a continually active, intervening agent to a prime force more like a master artisan who has designed logically contrived, self-functioning machines, and so led to a closer examination of the nature of God. If God functioned only as prime initiator of the universe, what was the need for God (and therefore for religion) at the self-functioning levels that followed the world's origin? Moreover, if the universe is logical, what was the logical purpose in creating it? Was the motivation to create perfection? If the world was not created perfect, what moral good could there have been in its creation?

An example of how essential it was to believe that each **design** had a creative purpose, is reflected in the statement by William Paley (1743–1805):

> There cannot be design without a designer; contrivance without a contriver; order without choice; arrangement without anything capable of arranging . . . Arrangement, disposition of parts, subservience of means to an end, relation of instruments to a use, imply the presence of intelligence and mind.

Simple common sense seemed to support supernatural design. Organismal design presupposes a designer, who by definition (but not based on any scientific evidence) is an intelligent supernatural being, a chain of reasoning that leads back to creation and a creator. So, it comes as no surprise to find that every culture/society has a creation story in its history or mythology (Table 26-1). Can there be a watch without a watchmaker, and by extension, can there be a person without a person-maker, laws without a lawmaker? To evolutionary theory, the essential challenge that religion poses has always been, "How from the disorder of random variability can nature achieve the beauty of adaptation without intelligent intervention?" Darwin's contribution was to answer this question by means of a phenomenon that no one had thoroughly explored before: natural selection.

However much these philosophical questions of design and designer, watch and watchmaker challenged religion, they did not pose as big a danger as did Darwin and his theory. In the most sensitive area of all — life itself — Darwinian evolution offered different answers to religion's claims of why life's important events occur. Darwin's works made clear that people no longer needed to believe that only the actions of a supernatural creator could explain biological relationships. God was neither "The Great Speciator" nor "The Great Sculptor of Nature." Instead, Darwin presented the concept that nature entails continual change, unpredictable chance events, an unrelenting struggle for survival among

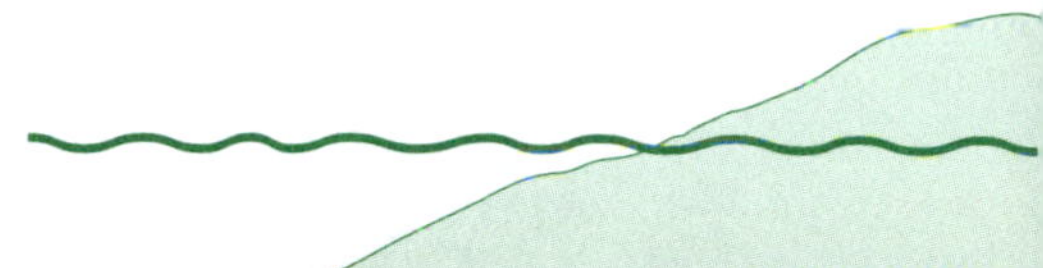

living creatures and no obvious guidance. *By viewing life as a continually expendable commodity rather than a divinely premeditated and consecrated goal,* Darwin replaced what many had seen as an understandable view of nature — the creativity of a human-like God — with the most heretical concepts of randomness and uncertainty and the fear that no one could understand the source and purpose of any natural event or design.

Nevertheless, difficulties remained for those nineteenth-century biologists who saw the strength of the evidence in support of evolution but who did not want to give up their faith. Some, such as Richard Owen (1804–1892), suggested that although natural selection may have caused some adaptations, many basic generalized patterns such as the vertebrate archetype and the parallel appearance of similar organs (such as eyes) in different groups, could only have arisen through design. Essentially, Owen moved the hand of the "designer" from specifying minor adaptations for species to devising major plans for higher taxonomic groups. Among other compromises attempted was that by Asa Gray (1810–1888), the American evolutionist, who proposed that the variability on which natural selection acts during evolution was itself specially created. To this Darwin replied that because not all variations are useful, it is inconceivable that evolution could be "designed" by such means.

Creationism and Intelligent Design

Fundamentalist religious groups have generally not accepted the uneasy truce between evolution and religious institutions in the Western world. In the United States, many people who believe in the Judeo-Christian account of creation in Genesis have formed political pressure groups to insist that their beliefs be treated as a scientific alternative to evolution in public education. According to a 1997 Gallup poll, 44 percent of Americans agreed with the statement that "God created humankind in its present form almost 10,000 years ago." A poll taken by the *New York Times* in November 2004 produced a statistic of 55 percent.

The origins of such groups date back at least to the early 1900s with strong roots among economically-threatened tenant farmers and small landholders, especially in the U.S. South and Southwest. Sociologists have suggested that believing in the literal truth of the Bible, revivalism and other aspects of fundamentalist religion helped many of these rural groups defend their way of life against domination and control by the more intellectual but exploitative northern and eastern social and economic establishments.

Whatever their initial motivations, fundamentalist groups were occasionally successful in pursuing anti-evolution goals in the South and Southwest during the first few decades of the last century. By the end of the 1920s, fundamentalists had introduced anti-evolution bills into a majority of U.S. state legislatures, and had passed some in various southern states. Probably the most famous confrontation between evolution and biblical creationism during that period was the 1925 trial of a schoolteacher, John Scopes, who was convicted of ignoring the ban against teaching evolution in Tennessee schools.

Although many biologists felt that the Scopes trial essentially defeated the intellectual validity of the creationist position, creationists apparently lost little ground in these regions and managed to have an impact on public education far beyond the South and Southwest. As Nelkin (1982) points out, by influencing textbook adoption procedures in various local and state school boards in the United States, creationists successfully minimized evolutionary explanations in secondary school textbooks for a long time; by 1942, more than 50 percent of high school biology teachers throughout the U.S. excluded any discussion of evolution from their courses.

The impetus for an increase of evolutionary teaching in U.S. secondary schools was the result of a movement to reform the science curriculum in the late 1950s and early 1960s, when it was realized that science education lagged behind that of other countries, specifically the Soviet Union, which in 1957 had launched the first space satellite, Sputnik. Among these innovations were new high school textbooks in both biological and social sciences (*Biological Science Curriculum Study; Man as a Course of Study*) that discussed evolution and analyzed changes in human social relationships. By the end of the 1960s anti-evolution laws were either repealed or declared unconstitutional.

Within the last decade, a number of societies and institutes established by fundamentalists for the propagation of creation science/intelligent design have entered the fray with the aim of including **creationism** in the science curriculum. Despite the name, there is little (if any) recognizable science in creation science. Although considerable literature deals with creationist attacks on evolution, the refusal of fundamentalist creationists to accept the scientific evidence shows no promise of ever being resolved.[8] By contrast, views in evolution, whether conforming or dissenting, are evaluated, as in all science, by evidence and not by faith or unverifiable decree.

Despite the overwhelming scientific evidence for evolution as a natural process, some religious groups adhering to creation have developed the notion of *intelligent design* as a purported scientific alternative to evolution. "Intelligent

[8] See Futuyma (1983), Pigliucci (2002), Scott (2005) and Brockman (2006) for analyses by scientists dealing with creationism. See http://www.hhmi.org/biointeractive/evolution/lectures.html#evodiscussion. (Accessed May 24, 2007) a Web site maintained by the Howard Hughes Medical Institute for discussions of religion and science.

design" is latter-day creationism. Indeed, in a landmark December 2005 decision in a U.S. federal district court trial (*Kitzmiller v. Dover* in Dover, PA), intelligent design was ruled a form of religion and not science, and as such "cannot be taught alongside evolution in sciences classes in U.S. public schools" (Scott, 2006).[9] Because intelligent design relies on supernatural explanations rather than natural causes, it is not a science. Religious arguments have explanatory power with respect to belief systems, but they are not scientific explanations and should not be confused with, or regarded as, scientific explanations.

Responses from evolutionary biologists to some of the more common creationist arguments are outlined in **Box 26-1.** Much of the discussion in this box relates to events, the evidence for which is provided and discussed in earlier chapters, of which perhaps the most important are:

- Analyses and explanations of historical events can be accepted by scientists, even without the ability to provide experimental verification (see Chapter 2).
- With respect to the origin of organic molecules, the existence of abiotic synthesis of amino acids and other basic biological molecules indicates that chemical reactions are biased to produce such molecules, and that natural selection can subsequently operate to increase their complexity and organization (see Chapter 7).
- The basic elements of organismal biology — distinctions between organisms, relationships of organisms to one another and how those distinctions and relationships arose — can all be understood and explained by evolution (see Chapters 18 and 24).
- Large numbers of different kinds of hominin fossils are known, from the 4.4 My old *Ardipithecus ramidus* to *Homo sapiens*. Humans are thus not a distinct creation (see Chapter 20).
- The theory of evolution (genetic changes over time) explains the historical course of biology in terms of natural processes, such as mutation, selection, genetic drift and migration, providing explanations consistent with all observations to date and demonstrating the fact of evolution (see Chapter 22).
- Mutations, even those not immediately adaptive, persist in populations through various mechanisms to provide the genetic variation allowing populations to evolve and improve fitness when environmental conditions change (see Chapter 23).
- Where genes and developmental processes associated with speciation have been identified they vary in different groups and circumstances (see Chapter 24).

[9] For the Web site containing the full decision in *Kitzmiller v. Dover* see: http://en.wikipedia.org/wiki/Kitzmiller_v._Dover_Area_School_District. Accessed May 24, 2007.

Specific examples discussed in Box 26-1 and often raised in contrary terms by creationists include:

- Geological strata are multilayered — some many kilometers thick — comprising sediments and fossils deposited over very long time periods (see Fig. 5-5). There is no geological or paleontological support for a biblical one-year worldwide flood.
- Fossils are not concentrated in any particular geological stratum and are found even in Precambrian deposits (see Figs. 5-5 and 8-2).
- Even when older geological strata have been elevated over younger strata (especially seen in association with mountain-building), the strata can be recognized as "older" by the kinds of fossils they contain, and by structural features that show continuity with nonoverturned deposits elsewhere (see Fig. 5-15).
- The demonstration of selection of resistance to insecticides in insects (see Fig. 10-26) and for rapid adaptation of cichlid fishes to microhabitats in African lakes (see Fig. 12-18) provides direct evidence for evolution.
- Selection for sight, hearing and vertebral support include steps that are adaptive (see Figs. 3-1, 18-11, 19-7), that is, "intermediate stages" are adaptive.
- Cultivated corn evolved from wild grass in about 7,000 years because it was continuously selected for food by Central American Indians (see Fig. 21-1), demonstrating the impact of humans on the evolution of other species.

Final Words

Many intellectual threads led to the modern theory of evolution, a theory that requires recognition that Earth is ancient, that there is a common inheritance within a biological group and that natural events can be explained by discoverable natural laws. Essential to our understanding of evolution is that:

- groups of organisms are bound together by their common inheritance;
- the past has been long enough for inherited changes to accumulate; and perhaps most essential of all,
- discoverable biological processes and natural relationships among organisms explain the reality of evolution.

It took a long time to weave these threads into an evolutionary tapestry. It is our hope that this presentation of the evidence for and the fact of evolution will enable you to appreciate more clearly the richness and beauty of that evolutionary tapestry.

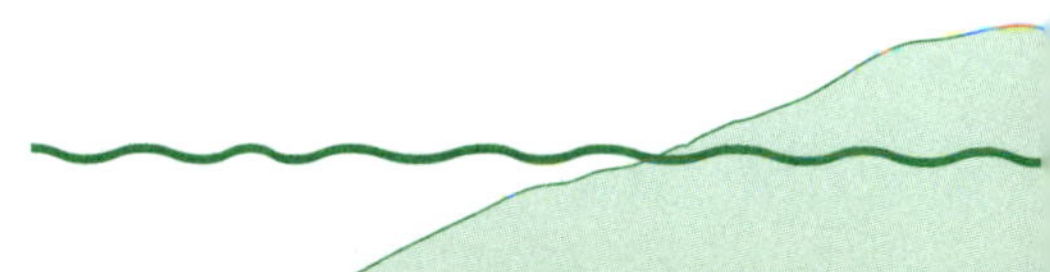

BOX 26-1 Responses to Creationist Arguments

THE FOLLOWING ARE responses by evolutionary biologists to some issues raised by creationists opposed to evolution.[a]

Creationist claim: *Evolution is a controversial subject even among evolutionary biologists, because they often engage in major disputes.*

Evolutionary biologist response (Stephen J. Gould): "Scientists regard debates on fundamental issues of theory as a sign of intellectual health and a source of excitement. Science — and how else can I say it — is most fun when it plays with interesting ideas, examines their implications, and recognizes that old information might be explained in surprisingly new ways. Evolutionary theory is now enjoying this uncommon vigor. Yet, amidst all this turmoil, no biologist has been led to doubt the fact that evolution occurred; we are debating how it happened. We are all trying to explain the same thing: the tree of evolutionary descent linking all organisms by ties of genealogy. Creationists pervert and caricature this debate by conveniently neglecting the common conviction that underlies it, and by falsely suggesting that evolutionists now doubt the very phenomenon we are struggling to understand."

Creationist claim: *Evolutionist belief in evolution is as much a matter of blind faith as belief in the Bible is to creationists.*

Evolutionist response (Kenneth E. Boulding): "The world scientific community has both many similarities to a world religion, but also important differences. It has, in the first place, a distinctive ethic of its own, which I am almost tempted to call "the four-fold way," as it has four essential components. The first is a high value on curiosity, which not all cultures possess. The second is a high value on veracity — that is, on not telling lies — which many other cultures also do not possess. The one thing that can get a scientist excommunicated from the scientific community is to be caught deliberately falsifying his [or her] results — that is, in telling lies. Error is often pardonable, but lying is the sin that cannot be forgiven. The third ethical principle is the high value on the testing of images of the world against the external world that they are supposed to map. Mere internal consistency is not enough, for there may be views of the world which are internally consistent, but which are, nevertheless, not true, in the sense that the real world does not conform to them. There are many methods of testing. Experiment is an important method where it is appropriate, though only perhaps a third to a half of the testing activities of science consist of experiment. Careful observation and recording, coupled with systematic analysis, is another important method, such as we have in celestial mechanics and in national economic statistics. Comparative studies of systems which are alike in many respects but differ in others is another important method; for instance, in medical research and the social sciences. Underlying all these, however, is a profound belief that the real world will speak for itself if it is asked the right questions. The fourth principle of scientific ethics is abstention from threat, embodied in the principle that people should be persuaded only by evidence and never by threat. This, of course, is in striking opposition to the ethics of many religious organizations and of all political organizations."

To Boulding's response we can emphasize the concept, discussed at the end of Chapter 2, that analyses and explanations of historical events can be accepted by scientists, even without experimental recapitulation. We regard geology, astronomy and evolutionary biology as historical sciences because their methodology has become sufficiently precise to allow explanations of past events that are "consistent with observations" of other past events and with present events.

Creationist claim: *Because no one has really seen all the events evolutionists claim, the most they can say is that evolution is a "theory" and not a "fact."*

Evolutionist response: As with previous points, this creationist argument shows a deep misunderstanding of science. There are many "theories" in science, from the atomic theory, to the molecular theory, to the theories of relativity and gravity, which are all based on explanations that account for observable events ("facts"). For example, the atomic theory was accepted by scientists because it explained chemical reactions, although the atoms themselves were invisible. Similarly, the theory of evolution accounts for the historical sequence of past and present organisms ("facts") by explaining their existence in terms of factors that caused changes in the genetic inheritance of organisms over time ("theory"). The facts of evolution come from the anatomical similarities and differences among organisms, the places where they live, the metabolic pathways they use, the embryological stages by which they develop, the fossil forms they leave behind, and the genetic, chromosomal, and molecular features that relate and separate them. These observable phenomena — all subjects of this book — show conclusively that "nothing in biology makes sense except in the light of evolution" (Dobzhansky, 1973). The theory of evolution (genetic changes over time) explains the historical course of biology (facts) in terms of natural processes, such as mutation, selection, genetic drift, and migration (see Chapter 22). These explanations are consistent with all observations so far. Evolution occurred — it is a "fact." Evolutionists also present the "theory" that explains this fact by the rules of science. Such an approach is not different from the "fact" that an apple dropped on Newton's head (or thereabouts), and he explained it with the "theory" of gravity.

Creationist claim: *"Creation science" is as much a biological science as evolution.*

Evolutionist response: An important part of biological science is the search for understandable explanations for the history of organisms: to seek answers to the questions, What happened and why did it happen this way? "Creation science" cannot be considered science, because creationists do not believe this search should be pursued scientifically, and instead offer biblical or religious explanations for the history of organisms. The following is a quotation from *Evolution? — The Fossils Say No!* by Duane T. Gish, then Associate Director of the Institute for Creation Research, the leading creationist organization:

[a] An extensive Internet source is the TalkOrigins Archive at http://www.talkorigins.org. Accessed May 24, 2007. Two further sites are http://www.nap.edu/catalog/6024.html#toc (*Science and Creationism: A View from the National Academy of Sciences,* 2d ed. The National Academies Press, Washington, DC) and http://www.hhmi.org/biointeractive/evolution/lectures.html#evodiscussion (Accessed May 24, 2007), BioInteractive, a Web site maintained by the Howard Hughes Medical Institute with discussions of religion and science.

> By creation we mean the bringing into being of the basic kinds of plants and animals by the process of sudden, or fiat, creation described in the first two chapters of Genesis. Here we find the creation by God of the plants and animals, each commanded to reproduce after its own kind using processes, which were essentially instantaneous. We do not know how God created, what processes He used, for God used processes which are not now operating anywhere in the natural universe. This is why we refer to divine creation as special creation. We cannot discover by scientific investigations anything about the creative processes used by God.

As the historian of evolution Michael Ruse (1988) points out:

> In science an explanation must explain more than that for which it was invented. Saying God created the eye of a dog, for example, does not explain anything about why it is structured the way it is, why it works the way it does, and why it is similar to the eye of a cat in some respects and different in others.

Creationist claim: *Because the second law of thermodynamics holds that entropy (disorder and disorganization) increases, organized life is thermodynamically impossible without the intervention of a creator who can circumvent this law.*

Evolutionist response: The second law of thermodynamics applies to closed systems isolated from external energy. It does not apply to open systems such as organisms that obtain energy from external sources, which they convert into negative entropy: organized forms. Many complex natural phenomena, such as snowflakes, tornadoes, and stalactites, are produced by the conversion of external energy into organized structures. Furthermore, as Hugo Franzen points out:

> If thermodynamics requires the intervention of a supernatural agent to originate life, it requires the continued intervention of that agent to sustain life. There would then be a sustained and repeated violation of the laws of thermodynamics in life processes [to produce all necessary catalytic molecules]. It is, of course, possible to accept this notion. It is not possible to accept it and thermodynamics as well.

Unlike creation, evolution is a natural event in perfect accord with all physical and chemical laws.

Creationist claim: *If evolution were true, we would find fossils of all transitional organisms, such as between fish and amphibians, reptiles and birds, apes and humans. According to creationist authority Gary Parker (1980), "Famous paleontologists at Harvard, the American Museum, and even the British Museum say that we have not a single example of evolutionary transition at all."*

Evolutionist response: Paleontologists have firmly identified many fossil intermediates, such as between fish and amphibians, reptiles and birds, reptiles and mammals, and apes and hominids. For example, Roger J. Cuffey (1972) provides long lists of paleontological sources describing fossil intermediates between species of the same genus as well as transitional forms between different genera, orders, and classes. These fossil intermediates are found in algae, ginkos, flowering plants, foraminiferans, corals, bryozoans, brachiopods, gastropods, bivalve mollusks, ammonites, trilobites, crustaceans, echinoderms, graptolites, conodonts, fishes, fish-tetrapods, amphibians, amphibian-reptiles, reptiles, reptile-birds, reptile-mammals, mammals, and hominids.

That paleontologists have not found all transitional fossils does not mean such organisms were absent: dead organisms are rarely preserved and transitional forms even less so because they are often in small populations that survived for relatively short time periods. Even when preserved, most fossils remain imbedded in rocks and inaccessible. The admission that we have not discovered everything does not mean we know nothing. Also, absence of evidence is not evidence of absence — that the sun is not visible at night does not imply that it vanishes in the evening and is recreated in the morning.

With respect to the obviously false claim by creationist Gary Parker, Niles Eldredge (1982), a paleontologist at the American Museum of Natural History writes:

> A prominent creationist interviewed a number of paleontologists at those institutions and elsewhere (actually, he never did get to Harvard). I was one of them. Some of us candidly admitted that there are some procedural difficulties in recognizing ancestors and that, yes, the fossil record is rather full of gaps. Nothing new there. This creationist then wrote letters to various newspapers, and even testified at hearings that the paleontologists he interviewed admitted that there are no intermediates in the fossil record. Thus, the lie has been perpetuated by Parker. All of the paleontologists interviewed have told me that they did cite examples of intermediates to the interviewer. The statement is an outright distortion of the willing admission by paleontologists concerned with accuracy that, to be sure, there are gaps in the fossil record. Such is creationist "scholarship."

Eldredge's complaint can be extended to many other such matters, from creationist literature to their practices in debates. Creationists consistently misrepresent evolutionary findings and often quote evolutionary biologists completely out of context. They repeat this tactic in debates: by the time the evolutionist has dealt with one distortion or misquotation, the creationist has introduced others. Their fundamental strategy is to deny any event, such as a transitional fossil, that questions creationist doctrine.

Thus, because creationists insist new species cannot appear after the initial creation period, new types are not novel species to them but only differences in "kind." That is, australopithecines who walked upright were just a "kind" of ape, and dog-sized four-toed leaf-eating eohippus (*Hyracotherium*) of the Eocene epoch was just a "kind" of horse. Transitional forms are therefore deemed not transitional at all. For example, *Archaeopteryx* with its reptilian teeth and skeleton was just a "bird," and *Ichthyostega* with its fishlike shape, vertebrae, and fin-rayed tail was just an "amphibian." Such distorted concepts obviously contribute nothing to how organismal distinctions and relationships are determined and how they can be understood and explained (see Chapters 18 and 24) — basic elements of biology.

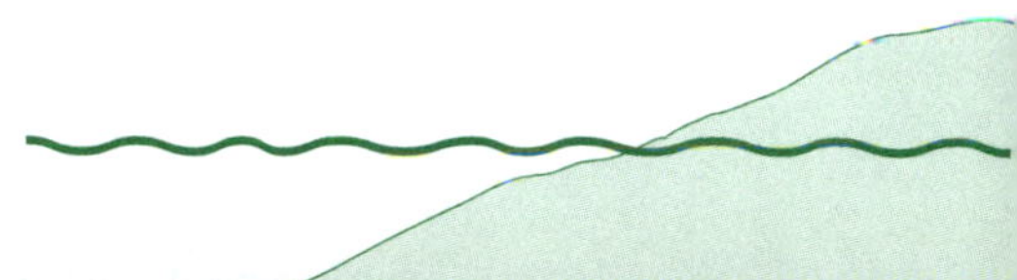

BOX 26-1 Responses to Creationist Arguments (*cont.*)

Creationist claim: *The earth is really young — perhaps 10,000 years old — and evolutionists' proposals for an older earth measured by decay rates of radioactive elements and other techniques are faulty and inaccurate.*

Evolutionist response (Joel Cracraft, 1983b): "Geologists have established, virtually beyond scientific doubt, that the earth is approximately 4.5 to 4.6 BY old. That the stratigraphic record of sediments can be sequentially dated by radiometric decay rates is not now a matter of question among geologists who study dating techniques. Each radioactive element decays at a specific and constant rate, and these rates are not influenced by external factors such as extremes of temperature or pre ssure. The creationists simply assert that these rates are not constant. Yet, at the [1981] Arkansas creation trial, every one of the creationists' geological witnesses — including Robert Gentry, their chief expert on radioactivity — testified that no scientific evidence exists which questions the constancy of these decay rates.

The creationists sometimes invoke a "singularity" at about 6,000 years corresponding to their suggested time for the Noachian "Flood"; at this singularity, the decay rates slowed down significantly — more or less to their present level. Prior to that time, the rates were much higher, giving the appearance that the earth is billions of years old, when it is only thousands. James Hopson of the University of Chicago has suggested to me a simple response to this supposition of a supernatural event: if the creationists are correct in believing that the earth is only thousands of years old, and that the decay rates at one time accelerated, then the amount of heat released from that amount of radioactive decay would have been sufficient — by a large margin — to have vaporized the earth.

By their own admission, the creationists cannot provide any scientific evidence for such dramatic changes in radioactive decay rates or their assertion that these rates do not measure the true age of the earth. The creationists turn to a supernatural "singularity," which is a belief derived from religion, not from the evidence of science."

Another evolutionary biologist's response (Niles Eldredge, 1982): "Perhaps the most dramatic demonstration of the validity and accuracy of modern geologic dating comes from the deep-sea cores stored by the thousands in various oceanographic institutions. The direct sequence is preserved in these drill cores, of course, and the microscopic fossils in them allow the usual "this-is-older-than-that" sort of relative dating to be done. We can also trace the pattern of changes in the orientation of the earth's magnetic field: as you go up a core, portions are positively charged, while others are negative. Major magnetic events, reflecting a flipping of the earth's magnetic poles, are recognizable, and the sequence of fossils, the same from core to core, always matches up with the magnetic history in the same fashion from core to core. *Then, when we obtain absolute dates from the cores (usually by using oxygen isotopes), we always* find that the date of the base of the "Jaramillo event" — one of the pole-switching episodes — yields a date of about 980,000 years ago. The dates are *always* the same (again, with a minor plus-or-minus factor). They are *always* in the right order. They are *always* in the tens or hundreds of thousands of years for the most recent dates, and in the millions of years further down the cores."

Creationist claim: *Evolutionists say they have found transitional forms between apes and humans, but these have turned out to be misclassified apes or misclassified humans. "The number of known human fossils would barely fill an average-sized coffin."*

Evolutionist response: Although primate fossil hunting has been a difficult pursuit, large numbers of different kinds of hominid fossils are known, from *Ardipithecus* to *Homo* (see Chapter 20). Kenneth E. Nahigian notes that even creationists who do not regard such fossils as evolutionary intermediates are impressed by the size of these findings:

> Creationist Michael J. Ord, in his book review of *Bones of Contention,* had this admission: "I was surprised to find that instead of enough fossils barely to fit into a coffin, there were over 4,000 hominid fossils as of 1976. Over 200 specimens have been classified as Neanderthal and about one hundred as Homo erectus. More of these fossils have been found since 1976." Marvin L. Lubenow, the author of *Bones of Contention,* wrote to the editor of the same creationist journal: "The current figures are even more impressive: over 220 *Homo erectus* fossil individuals to date and well over 300 Neanderthal fossil individuals discovered to date [1994]." Over 400 fossil individuals are now (2007) known.

Creationist claim: *Evolutionists claim that a complex living structure arose purely by chance — an event with such low probability as to be impossible in our universe.*

Evolutionist response: It is important to recognize that random chance events are not the same as evolutionary events based on natural selection. Chance can be defined as events whose causes are independent of each other, so that a succession of such events need bear no mutual relationship, and therefore often leads to disorganization ("chaos"). Evolution by natural selection, in contrast, marks events whose occurrence depends on previous events, so that a succession of such events can lead to organized structures and increased complexity. The non-organismal (abiotic) synthesis of amino acids and other basic biological molecules (see Chapter 7) also indicates that chemical reactions are biased to produce such molecules, and natural selection can subsequently operate to increase their complexity and organization. Although the exact steps to the origin of life are not yet known, this does not detract from our understanding of many basic mechanisms that enabled the present diversity of life to evolve, nor affect the evidence that such evolution occurred. Moreover, because the origin of life is a molecular biology question, and we are just learning molecular biology techniques, it is not surprising that we don't yet know exactly how life on Earth originated. Even so, we are learning how enzymatic reactions evolve, how metabolic pathways evolve, how developmental controls evolve, and how genetic systems evolve. That's a lot of learning about early life questions in a very short time.

Always missing in creationist arguments is an understanding of selection — that the alternative to design by a "creator" is not random chance, but selection. Selection is a sequential process that ties individual chance events into a "creative" sequence because particular steps in the sequence are "adaptive" and allowed to persist.

An adaptation is not a sudden event but the result of a succession of selective events, each with reasonable probability.

Thus, selection for sight leads to improved visual apparatus (see Fig. 3-1), selection for vertebral support leads to improved spinal apparatus (see Fig. 18-11), selection for hearing leads to improved auditory apparatus (see Fig. 19-7), and so forth — steps along the way are distinctively adaptive.

Creationist claim: *Natural selection can only eliminate misfits or produce minor changes, but cannot produce new adaptations. If selection was potent and evolution continuous, we should be bombarded with new species all the time. Where are they?*

Evolutionist response: First, selection, as explained above, is a "creative" force in that it leads to adaptations by a succession of selective organismal responses that improve fitness. Second, the answer to where new species can be found is simple: they are always about. New species have continually evolved over the last 3.5 BY and have continually replaced older species. Most organisms we see before us today are different from ancestors who may have lived only hundreds or thousands of generations ago, and even more different from ancestors who lived many hundreds of thousands of generations ago. We do not have film to show these changes, but we do have fossils and genetic techniques that indicate how different ancestral species were, and how novel extant species are.

The answer to why we don't commonly observe new species emerge is also simple: speciation takes time. In many groups, species formation is generally recognized when reproductive isolation occurs between two populations so that each can embark on its own distinctive evolutionary path. Because they involve many genetic and selective (adaptive) changes, such events can take many thousands of years. We can nevertheless document that intense selective differences have caused rapid genetic differentiation between populations, as have also certain large chromosomal mutations. For example, we know that cultivated corn evolved in only about 7,000 years from the wild grass, teosinte, because it was continuously selected for food by Central American Indians (see Fig. 21-1). Other selective effects, observed in both laboratory experiments and nature, include changes in temperature, food sources, habitats, and environmental toxins. Among such events investigated by population geneticists are fly populations that became resistant to insecticides (see Fig. 10-26), plants that developed new characters, *Drosophila* strains that developed specific sexual behaviors, cichlid fishes that rapidly adapted to different conditions in African lakes (see Fig. 12-18), and parasitic insects that adapted to different plant hosts. Although we have not yet discovered the exact genes necessary to produce speciation, we have identified some, and know they vary in different groups and circumstances (see Chapter 24).

We also know that certain chromosomal changes in structure and number can be more radical in their effects, and lead to spontaneous speciation events. Among such examples are translocations responsible for new species of evening primrose, *Oenothera* and doubled chromosome numbers responsible for new hybrid species of tobacco, *Nicotiana* (see Fig. 10-14).

Creationist claim: *Evolutionists cannot account for anomalies that contradict evolutionary concepts, such as:*

1. *The sun has much less angular (rotational) momentum than would be expected for its large mass.*
2. *There are deviations from the evolutionary concept that more recent geological strata should always be positioned over older strata.*
3. *Human and dinosaur footprints have been found together in Cretaceous rocks in Texas.*
4. *The finding of sudden large deposits of fossils is more in accord with catastrophic events such as the biblical Noachian flood than with slow evolutionary sedimentation processes.*
5. *Because mutations are deleterious and not beneficial, evolutionary selection can never lead to adaptation but only go from bad to worse.*

Evolutionist response: Scientists have demonstrated many times that these and other so-called anomalies are really "pseudoanomalies," yet they nevertheless keep reappearing in creationist literature.

1. The angular momentum of the sun was transferred to the planets by solar wind during condensation of the solar system.
2. Older geological strata can be boosted over younger strata in some places by "overthrusting," especially in mountain-building areas (see Fig. 5-15). They are nevertheless recognized as "older" by the kinds of fossils they contain, and by structural features that show continuity with non-overturned deposits elsewhere.
3. The Paluxy River anomalies (human footprints in Cretaceous deposits) have been demonstrated to be dinosaurian and the human footprints were recently carved by humans. (The earliest known fossil hominid footprints were made by australopithecines and date back about 3.7 My.)
4. Organismal fossils are not concentrated in any particular geological stratum, but are found even in Precambrian deposits (see Figs. 5-5 and 8-2). Geological strata are multilayered, some many kilometers thick, bearing sediments and fossils deposited over very long time periods. There is no geological or paleontological support for a biblical one-year worldwide flood.
5. Although most mutations are deleterious — organisms are generally so well adapted that a random genetic change will cause malfunction — some mutations, such as resistance to toxins or parasites, can and do improve fitness. Other mutations, even those not immediately adaptive, persist in populations through various mechanisms (see Chapter 23), and provide the genetic variation allowing populations to evolve and improve fitness when environmental conditions change.

Creationist claim: *In fairness to both sides of the argument, we need laws that creationism be taught along with evolution.*

Evolutionist response (Richard D. Alexander, 1987): "When a creationist, Darwinist, Marxist, or supporter of any other theory defends his or her views publicly, he or she does everyone a service. But when anyone attempts to establish laws or rules requiring that certain theories be taught or not be taught, he or she invites us to take a step toward totalitarianism. Whether a law is to prevent the teaching of a theory or to require it is immaterial. No laws were ever passed saying that evolution had to be taught in biology courses. The prestige of evolutionary theory has been built by its impact

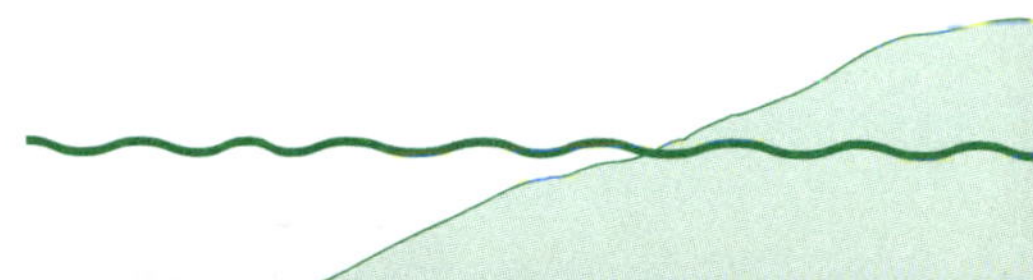

BOX 26-1 Responses to Creationist Arguments (*cont.*)

on the thousands of biologists who have learned its power and usefulness in the study of living things. No laws need to be passed for creationists to do the same thing. When creation theorists strive to introduce creation into the classroom as an alternative biological theory to evolution they must recognize that they are required to give creation the status of a falsifiable idea — that is, an idea that loses any special exemption from scrutiny, that is accepted as conceivably being false, and that must be continually tested until the question is settled. A science classroom is not the place for an idea that is revered as holy. The greatest threat to society and to our children is not whether students are exposed to wrong ideas — after all, many high school biology students are legally adults with voting privileges, and all high school biology students have already been exposed to many wrong ideas. What is important is whether each has been taught how and given the freedom to test new ideas, evaluate them, and respond appropriately."

To Alexander's statement, we can add the following: If "fairness" is a matter of teaching proposed alternate versions of science in schools, then other "believers" can rightfully claim that their concepts should also be given "equal time": astrology with astronomy, alchemy with chemistry, flat Earth with spherical Earth, geocentric solar system with heliocentric, phrenology (study of personality in cranial bumps) with psychology, Christian Science disease theory with germ theory, and so forth. Obviously, subjects to be taught in schools as science should be based on scientific concepts and judged by scientific rules of evidence rather than include nonscientific or pseudoscientific matter mandated by religious and special interest groups. This is why intelligent design had evoked such a strong reaction from the scientific community; it is creationism masquerading as science and so seeks a place alongside the teaching of evolution, as if the tenets of intelligent design can be approached scientifically, which they clearly cannot.[b]

Creationist claim: *By insisting that life and biological events are* ***not*** *the result of purposeful Godlike design, evolutionists attack the basis of all religious belief.*

Evolutionist Response: Among the different concepts of God that exist, the most developed form is a universal God who provided the laws of nature that account for all subsequent events — laws and events that can be investigated by science. The religion that creationists foster is of a more primitive God who fabricates unexplainable mysterious events and miraculous creations that violate natural laws. Evolutionists can oppose this creationist concept of God, yet still accept God concepts that accommodate belief in natural laws and evolutionary events.

[b] See Forrest and Gross (2004), Young and Edis (2004) and Mindell (2006) for the scientific arguments and case against intelligent design.

KEY TERMS

belief systems
consciousness
creationism
Darwinism
design
evolution of religion
God
religion
science
soul
vitalism

DISCUSSION QUESTIONS

1. How and why did the concepts of soul and god become established in religious institutions?
2. Why can creation myths not be approached using the scientific method?
3. The question of design.
   a. Why do some religious groups want to believe that a designer creates natural events?
   b. How does Darwinism, using the concept of natural selection, explain the design of organisms?
4. Why have religion and Darwinism existed side by side in Western society for more than 100 years despite conflicting explanations for natural events?
5. Select what you consider to be the best piece of evidence for organismal evolution and discuss it to show how the application of the scientific method demonstrates the reality of evolution.

EVOLUTION ON THE WEB

Explore evolution on the Internet! Visit the accompanying Web site for *Strickberger's Evolution*, Fourth Edition, at http://www.biology.jbpub.com/book/evolution for Web exercises and links relating to topics covered in this chapter.

RECOMMENDED READING

Alters, B. J., 2001. *Defending Evolution: A Guide to the Evolution/Creation Controversy.* Jones and Bartlett Publishers, Sudbury, MA.

Appleman, P., (ed.), 2001. *A Norton Critical Edition. Darwin: Texts, Commentary,* 3rd ed. W. W. Norton & Company, New York.

Brockman, J., 2006. *Intelligent Thought: Science Versus the Intelligent Design Movement.* Vintage Books, New York. [See the review, J. T. Bonner, 2006, *Nature,* **442,** 355–356.]

Cracraft, J., and R. W. Bybee (eds.), 2005. *Evolutionary Science and Society: Educating a New Generation.* Biological Sciences Curriculum Study, Washington, DC.

Dennett, D. C., 2006. *Breaking the Spell: Religion as a Natural Phenomenon.* Allen Lane, London, UK.

Forrest, B., and P. R. Gross, 2004. *Creationism's Trojan Horse: The Wedge of Intelligent Design.* Oxford University Press, New York.

La Barre, W., 1970. *The Ghost Dance: Origins of Religion.* Doubleday, New York.

Mindell, D. P., 2006. *The Evolving World: Evolution in Everyday Life.* Harvard University Press, Cambridge, MA.

Pigliucci, M., 2002. *Denying Evolution: Creationism, Scientism and the Nature of Science.* Sinauer Associates Inc., Sunderland, MA.

Ruse, M., 2001. *The Evolution Wars: A Guide to the Debates.* Rutgers University Press, New Jersey and London.

Ruse, M., 2005. *The Evolution-Creation Struggle.* Harvard University Press, Cambridge, MA.

Scott, E. C., 2005. *Evolution vs. Creationism: An Introduction.* University of California Press, Berkeley, CA.

Young, M., and T. Edis (eds.), 2004. *Why Intelligent Design Fails: A Scientific Critique of the New Creationism.* Rutgers University Press, New Brunswick, NJ.

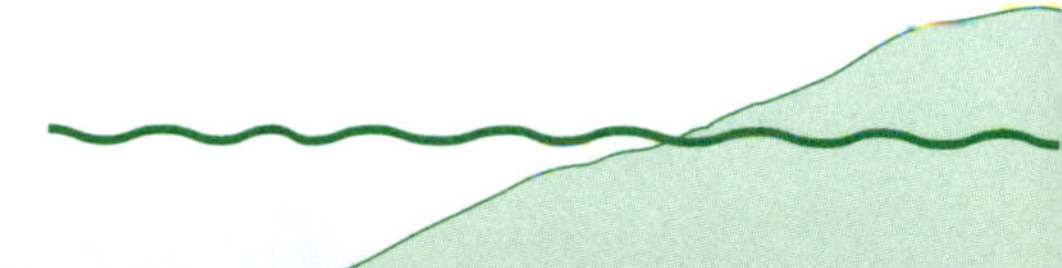

Literature Cited

A

Abel, T., G. L. Bryan, and M. L. Norman, 2002. The formation of the first star in the universe. *Science,* **295,** 93–98.

Abouheif, E., 2004. A framework for studying the evolution of gene networks underlying polyphenism: Insights from winged and wingless ant castes. In *Environment, Development, and Evolution.* B. K. Hall, R. D. Pearson, and G. B. Müller (eds.). MIT Press, Cambridge, MA, pp. 125–137.

Abouheif, E., M. Akam, W. J. Dickinson, P. W. H. Holland, et al., 1997. Homology and developmental genes. *Trends Genet.,* **13,** 432–433.

Abzhanov, A., M. Protas, B. R. Grant, P. R. Grant, and C. J. Tabin, 2004. Bmp4 and morphological variation of beaks in Darwin's finches. *Science,* **305,** 1462–1465.

Adams M. D., S. E. Celniker, R. A. Holt, C. A. Evans, et al., 2000. The genome sequence of *Drosophila melanogaster. Science,* **287,** 2185–2195.

Adelmann, H. B., 1966. *Marcello Malpighi and the Evolution of Embryology* (5 volumes). Cornell University Press, Ithaca, NY.

Adl, S. M., A. G. B. Simpson, M. A. Farmer, R. A. Andersen, et al., 2005. The new higher level classification of Eukaryotes with emphasis on the taxonomy of protists. *J. Eukaryot. Microbiol.,* **52,** 399–451.

Adman, E. T., L. C. Sieker, and L. H. Jensen, 1973. The structure of a bacterial ferredoxin. *J. Biol. Chem.,* **248,** 3987–3996.

Adoutte, A., G. Belavoine, N. Lartillot, and R. de Rosa, 1999. Animal evolution: The end of the intermediate taxa? *Trends Genet.,* **15,** 104–108.

Aguinaldo, A. M. A., J. M. Turbeville, L. S. Linford, M. D. Rivera, J. R. Garey, R. A. Raff, and J. A. Lake, 1997. Evidence for a clade of nematodes, arthropods and other moulting animals. *Nature,* **387,** 489–493.

Aiello, L., and C. Dean, 1990. *An Introduction to Human Evolutionary Anatomy.* Academic Press, London.

Akam, M., 1995. Hox genes and the evolution of diverse body plans. *Phil. Trans. R. Soc. Lond. B,* **349,** 313–319.

Akam, M., 1998. Hox genes: From master genes to micromanagers. *Curr. Biol.,* **8,** R676–R678.

Albertson, R. C., J. T. Streelman, T. D. Kocher, and P. C. Yelick, 2005. Integration and evolution of the cichlid mandible: The molecular basis of alternate feeding strategies. *Proc. Natl. Acad. Sci. USA,* **102,** 16287–16292.

Albertson, R. C., J. T. Streelman, and T. D. Kocher, 2000. The beak of the fish: genetic basis of adaptive shape differences among Lake Malawi cichlid fishes. Assessing morphological differences in an adaptive trait: a landmark-based morphometric approach. *J. Exp. Zool.,* **289,** 385–403.

Alcock, J., 2001. *The Triumph of Sociobiology.* Oxford University Press, New York.

Alcock, J., 2005. *Animal Behavior: An Evolutionary Approach.* Sinauer Associates, Sunderland, MA.

Alcock, J., 2006. *An Enthusiasm for Orchids: Sex and Deception in Plant Evolution.* Oxford University Press, Oxford, England.

Aldridge, R. J., D. E. G. Briggs, I. J. Sansom, and M. P. Smith, 1994. The latest vertebrates are the earliest. *Geol. Today,* **10,** 141–145.

Alexander, R. D., 1987. *The Biology of Moral Systems.* Aldine de Gruyter, New York.

Alexander, R. M., 1991. Apparent adaptation and actual performance. *Evol. Biol.,* **25,** 357–373.

Alleaume-Benharia, M., I. R. Pen, and O. Ronce, 2006. Geographical patterns of adaptation within a species' range: interactions between drift and gene flow. *J. Evol. Biol.,* **19,** 203–215.

Allwood, A. C., M. R. Walter, B. S. Kamber, P. Marshall, and I. W. Burch, 2006. Stromatolite reef from the Early Archaean era of Australia. *Nature,* **441,** 714–718.

Alters, B. J., 2001. *Defending Evolution: A Guide to the Evolution/Creation Controversy.* Jones and Bartlett, Sudbury, MA.

Alvarez, L. W., 1983. Experimental evidence that an asteroid impact led to the extinction of many species 65 million years ago. *Proc. Natl. Acad. Sci. USA,* **80,** 627–642.

Alvarez, W., and F. Asaro, 1992. The extinction of the dinosaurs. In *Understanding Catastrophe.* J. Bourriau (ed.). Cambridge University Press, Cambridge, England, pp. 28–56.

Amundson, R. (ed.), 2007. *On the Nature of Limbs* by Richard Owen, The University of Chicago Press, Chicago.

Anderson, W. F., 1992. Human gene therapy. *Science,* **256,** 808–813.

Anderson, W. F., 1998. Human gene therapy. *Nature,* **392** (Suppl.), 25–30.

Andersson, D. I., and D. Hughes, 1996. Muller's ratchet decreases fitness of a DNA-based microbe. *Proc. Natl. Acad. Sci. USA,* **93,** 906–907.

Andrade, M. C. B., 1996. Sexual selection for male sacrifice in the Australian red back spider. *Science,* **271,** 70–72.

Andrews, H. N. Jr., 1961. *Studies in Paleobotany.* Wiley, New York.

Andrews, P., 1987. Aspects of hominoid phylogeny. In *Molecules and Morphology in Evolution: Conflict or Compromise?* C. Patterson (ed.). Cambridge University Press, Cambridge, England, pp. 23–53.

Aparicio S., J. Chapman, E. Stupka, N. Putnam, et al., 2002. Whole-genome shotgun assembly and analysis of the genome of *Fugu rubripes. Science,* **297,** 1301–1310.

Appleman, O. (ed.), 2001. *A Norton Critical Edition. Darwin: Texts, Commentary,* 3rd ed. W. W. Norton & Company, New York.

Arabidopsis Genome Initiative, 2000. Analysis of the genome sequence of the flowering plant *Arabidopsis thaliana. Nature,* **408,** 796–815.

Aranzana, M. J., S. Kim, K. Zhao, E. Bakker, et al., 2005. Genome wide association mapping in Arabidopsis identifies previously known flowering time and pathogen resistance genes. *PloS Genet.,* **1,** 531–539.

Archer, M., and G. Clayton (eds.), 1984. *Vertebrate Zoogeography and Evolution in Australia.* Hesperian Press, Carlisle, Australia.

Archibald, J. D., 1996. *Dinosaur Extinction and the End of an Era.* Columbia University Press, New York.

Archibald, J. M., and P. J. Keeling, 2002. Recycled plastids: A "green movement" in eukaryotic evolution. *Trends Genet.,* **18,** 577–584.

Arenas-Mena, C., P. Martinez, R. A. Cameron, and E. H. Davidson, 1998. Expression of the Hox gene complex in the indirect development of a sea urchin. *Proc. Natl. Acad. Sci. USA,* **95,** 13062–13067.

Arendt, D., 2003. Evolution of eyes and photoreceptor cell types. *Int. J. Dev. Biol.,* **47,** 563–571.

Arendt, D., and K. Nübler-Jung, 1994. Inversion of dorsoventral axis? *Nature,* **371,** 26.

Arendt, D., and K. Nübler-Jung, 1999. Comparison of early nerve cord development in insects and vertebrates. *Development,* **126,** 2309–2325.

Arensburg, B., A. M. Tillier, B. Vandermeersch, H. Duday, L. A. Schepartz, and Y. Rak, 1989. A middle Paleolithic human hyoid bone. *Nature,* **338,** 758–760.

Arnold, M. L., 1997. *Natural Hybridization and Evolution.* Oxford University Press, Oxford, England.

Asfaw, B., T. White, O. Lovejoy, B. Latimer, S. Simpson, and G. Suwa, 1999. *Australopitchecus garhi:* A new species of early hominin from Ethiopia. *Science,* **284,** 629–635.

Ason, B., D. K. Darnell, B. Wittbrodt, E. Berezikov, et al., 2006. Differences in vertebrate microRNA expression. *Proc. Natl. Acad. Sci. USA,* **103,** 14385–14389.

Atchley, W. R., and B. K. Hall, 2001. A model for development and evolution of complex morphological structures. *Biol. Rev. Camb. Philos. Soc.,* **66,** 101–157.

Austin, C. R., and R. V. Short, 1976. *The Evolution of Reproduction.* Cambridge University Press, Cambridge, England.

Averof, M., and S. M. Cohen, 1997. Evolutionary origin of insect wings from ancestral gills. *Nature,* **385,** 627–630.

Averof, M., and N. H. Patel, 1997. Crustacean appendage evolution associated with changes in Hox gene expression. *Nature,* **388,** 682–686.

Avise, J. C., 1994. *Molecular Markers, Natural History and Evolution.* Chapman & Hall, London.

Avise, J., and D. Walker, 1999. Species realities and numbers in sexual vertebrates: Perspectives from an asexually transmitted genome. *Proc. Natl. Acad. Sci. USA,* **96,** 992–995.

Ax, P., 1987. *The Phylogenetic System: The Systematization of Organisms on the Basis of Their Phylogenesis.* Wiley, Chichester, England.

Ayala, F. J., A. Escalante, C. O'Huigin, and J. Klein, 1994. Molecular genetics of speciation and human origins. *Proc. Natl. Acad. Sci. USA,* **91,** 6787–6794.

Ayala, F. J., A. Rzhetsky, and F. J. Ayala, 1998. Origin of the metazoan phyla: Molecular clocks confirm paleontological estimates. *Proc. Natl. Acad. Sci. USA,* **95,** 606–611.

B

Bakker, R., 1986. *The Dinosaur Heresies.* Longman, Harlow, England.

Baldauf, S. L., and J. D. Palmer, 1993. Animals and fungi are each other's closest relative: Congruent evidence from multiple proteins. *Proc. Natl. Acad. Sci. USA,* **90,** 11558–11562.

Ballard, J. W., G. J. Olsen, D. P. Faith, W. A. Odgers, et al., 1992. Evidence from 12S ribosomal RNA sequences that onychophorans are modified arthropods. *Science,* **258,** 1345–1348.

Barber, J., and B. Andersson, 1994. Revealing the blueprint of photosynthesis. *Nature,* **370,** 31–34.

Barger, A. 2005. The midlife crisis of the cosmos. *Sci. Am.* **292,** 46–53.

Barluenga, M., K. N. Stölting, W. Salzburger, M. Muschick, and A. Meyer, 2006. Sympatric speciation in Nicaraguan crater lake cichlid fish. *Nature,* **439,** 719–723.

Barnabas, J., R. M. Schwartz, and M. O. Dayhoff, 1982. Evolution of major metabolic innovations in the Precambrian. *Origins of Life,* **12,** 81–91.

Barnes, R. D., 1980. *Invertebrate Zoology,* 4th ed. Saunders, Philadelphia, PA.

Barrow, J. D., 1994. *The Origin of the Universe.* Basic Books, New York.

Barton, N. H., and B. Charlesworth, 1984. Genetic revolutions, founder effects, and speciation. *Ann. Rev. Ecol. Syst.,* **15,** 133–164.

Basolo, A. L., 1994. The dynamics of Fisherian sex-ratio evolution: Theoretical and experimental investigations. *Amer. Nat.,* **144,** 473–490.

Bateson, W., 1886. The ancestry of the Chordata. *Q. J. Microsc. Sci.,* **26,** 535–571.

Batzer, M. A., M. Stoneking, M. Alegria-Hartman, H. Bazan, et al., 1994. African origin of human-specific polymorphic Alu insertions. *Proc. Natl. Acad. Sci. USA,* **91,** 12288–12292.

Beadle, G. W., and B. Ephrussi, 1937. Development of eye colors in *Drosophila:* Diffusable substances and their interrelations. *Genetics,* **22,** 76–86.

Beck, C. B., (ed.), 1976. *Origin and Early Evolution of Angiosperms.* Columbia University Press, New York.

Beck, F., D. B. Moffatt, and J. B. Lloyd, 1973. *Human Embryology and Genetics.* Blackwell, Oxford, England.

Beer, G., 1992. *Darwin's Plots. Evolutionary Narrative in Darwin, George Eliot and Nineteenth Century Fiction.* The University of Chicago Press, Chicago.

Begon, M., and M. Mortimer, 1986. *Population Ecology.* Second Edition. Blackwell, Oxford, England.

Begon, M., M. Mortimer, and D. J. Thompson, 1996. *Population Ecology: A Unified Study of Animals and Plants,* 3rd ed. Blackwell Science Ltd., Oxford, England.

Beiko, R. G., T. J. Harlow, and M. A. Ragan, 2005. Highways of gene sharing in prokaryotes. *Proc. Natl. Acad. Sci. USA,* **102,** 14332–14337.

Bejder, L., and B. K. Hall, 2002. Limbs in whales and limblessness in other vertebrates: mechanisms of evolutionary and developmental transformation and loss. *Evol. Dev.,* **4,** 445–458.

Bell, G., 1989. *Sex and Death in Protozoa: The History of Obsession.* Cambridge University Press, Cambridge, England.

Bell, G., and V. Koufopanou, 1991. The architecture of the life cycle in small organisms. *Phil. Trans. R. Soc. Lond. B.,* **332,** 81–89.

Bell, G., and A. O. Mooers, 1997. Size and complexity among multicellular organisms. *Biol. J. Linn. Soc.,* **60,** 345–363.

Bell, M. A., W. E. Aguirre, and N. J. Buck, 2004. Twelve years of contemporary armor evolution in a threespine stickleback population. *Evolution,* **58,** 814–824.

Belting, H.-G., C. S. Shashikant, and F. H. Ruddle, 1998. Modification of expression and cis-regulation of *Hoxc8* in the evolution of diverged axial morphology. *Proc. Natl. Acad. Sci. USA,* **95,** 2355–2360.

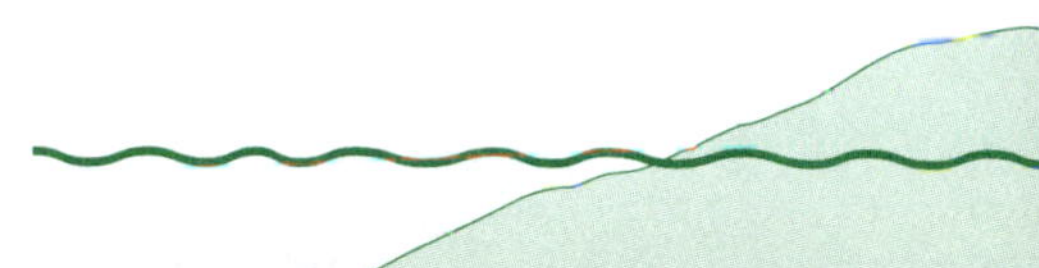

Benefit, B. R., and M. L. McCrossin, 1997. Earliest known Old World monkey skull. *Nature,* **388,** 368–371.

Bengtson, S., and Z. Yue, 1997. Fossilized metazoan embryos from the earliest Cambrian. *Science,* **277,** 1645–1648.

Benito-Gutiérrez, E., 2006. A gene catalogue of the amphioxus nervous system. *Int. J. Biol. Sci.,* **2,** 149–160.

Benner, S. A., M. A. Cohen, G. H. Gonnet, D. B. Berkowitz, and K. P. Johnsson, 1993. Reading the palimpsest: Contemporary biochemical data and the RNA world. In *The RNA World,* R. F. Gesteland and J. F. Atkins (eds.). Cold Spring Harbor Laboratory Press, Cold Spring Harbor, New York, pp. 27–70.

Bennett, M. D., and I. J. Leitch, 2005. Plant genome size research — a field in focus. *Ann. Bot.,* **95,** 1–6.

Bennett, M. D., I. J. Leitch, H. J. Price, and J. S. Johnston, 2003. Comparisons with *Caenorhabditis* (100 Mb) and *Drosophila* (175 Mb) using flow cytometry show genome size in *Arabidopsis* to be 157 Mb and thus 25% larger than the *Arabidopsis* Genome Initiative estimate of 125 Mb. *Ann. Bot.,* **91,** 547–557.

Benton, M. J., 1988. The relationships of the major group of mammals: New approaches. *Trends Ecol. Evol.,* **3,** 40–45.

Benton, M. J., 1991. *The Rise of the Mammals.* Apple Press, London.

Benton, M. J., 1995. Diversification and extinction in the history of life. *Science,* **268,** 52–58.

Benton, M. J., 1997. *Vertebrate Paleontology,* 2nd ed. Harper Collins, London.

Benton, M. J., 2003. *When Life Nearly Died: The Greatest Mass Extinction of All Time.* Thames and Hudson, London.

Benton, M. J., 2004. *Vertebrate Palaeontology,* 3rd ed. Blackwell Publishers, Oxford, England.

Benton, M. J., and F. J. Ayala, 2003. Dating the tree of life. *Science,* **300,** 1698–1700.

Berezikov, E., F. Thuemmler, L. W. van Laake, I. Kondova, et al., 2006. Diversity of microRNAs in human and chimpanzee brain. *Nat. Genet.,* **38,** 1375–1377.

Bergquist, P. R., 1985. Poriferan relationships. In *The Origin and Relationships of Lower Invertebrates.* S. Conway Morris, J. D. George, R. Gibson, and H. M. Platt (eds.). Clarendon Press, Oxford, England, pp. 14–27.

Bharathan, G., B.-J. Janssen, E. A. Kellogg, and N. Sinha, 1997. Did homeodomain proteins duplicate before the origin of angiosperms, fungi, and metazoa? *Proc. Natl. Acad. Sci. USA,* **94,** 13749–13753.

Bhattacharyya, M. K., A. M. Smith, T. H. N. Ellis, C. Hedley, and C. Martin, 1990. The wrinkled-seed character of pea described by Mendel is caused by a transposon-like insertion in a gene encoding starch-branching enzyme. *Cell,* **60,** 115–122.

Bird, A. P., 1995. Gene number, noise reduction and biological complexity. *Trends Genet.,* **11,** 94–100.

Birdsell, J., and C. Wills, 1996. Significant competitive advantage by meiosis and syngamy in the yeast *Saccharomyces cerevisiae. Proc. Natl. Acad. Sci. USA,* **93,** 908–912.

Biro, C., B. Benyó, C. Sansom, Á. Szlávecz, G. Fördös, T. Micsik, and Z. Benyó, 2003. A common periodic table of codons and amino acids. *Biochem. Biophys. Res. Commun.,* **306,** 408–415.

Bleckmann, C. A., 2006. Evolution and creationism in science: 1880–2000. *BioScience,* **56,** 151–158.

Blencowe, B. J., 2006. Alternative splicing: New insights from global analyses. *Cell,* **126,** 37–47.

Blomberg, S. P., and T. Garland, Jr., 2002. Tempo and mode in evolution: phylogenetic inertia, adaptation, and comparative methods. *J. Evol. Biol.,* **15,** 899–910.

Blondel, J., D. W. Thomas, A. Charmantier, P. Perret, et al., 2006. A thirty-year study of phenotypic and genetic variation of blue tits in Mediterranean habitat mosaics. *BioScience,* **56,** 661–673.

Bode, H., 2001. The role of Hox genes in axial patterning in *Hydra. Amer. Zool.,* **41,** 621–628.

Boesch, C., 1993. Aspects of transmission of tool-use in wild chimpanzees. In *Tools, Language and Cognition in Human Evolution,* K. R. Gibson and T. Ingold (eds.). Cambridge University Press, Cambridge, England, pp. 171–183.

Boesch-Achermann, H., and C. Boesch, 1994. Hominization in the rainforest: The chimpanzee's piece of the puzzle. *Evol. Anthropol.,* **3,** 9–16.

Bogdanova, M. I., R. G. Nager, and P. Monaghan, 2007. Age of the incubating parents affects nestling survival: an experimental study of the herring gull *Larus argentatus. J. Avian Biol.,* **38,** 83–93.

Boguski, M. S., 1998. Bioinformatics — A new era. In *Trends Guide in Bioinformatics.* Elsevier Trends Journals, Haywards Heath, West Sussex, England, pp. 1–3.

Bold, H. C., C. J. Alexopoulos, and T. Delevoryas, 1980. *Morphology of Plants and Fungi,* 4th ed. Harper & Row, New York.

Bonner, J. T., 1988. *The Evolution of Complexity by Means of Natural Selection.* Princeton University Press, Princeton, NJ.

Bonner, J. T., 2006. *Why Size Matters: From Bacteria to Blue Whales.* Princeton University Press, Princeton, NJ.

Borchiellini, C., N. Boury-Esnault, J. Vacelet, and Y. Le Parco, 1998. Phylogenetic analysis of the Hsp70 sequences reveals the monophyly of metazoa and specific phylogenetic relationships between animals and fungi. *Mol. Biol. Evol.,* **15,** 647–655.

Borradale, L. A., F. A. Potts, L. E. S. Eastham, and J. T. Saunders, 1959. *The Invertebrata,* 3rd ed., revised by G. A. Kerkut, Cambridge University Press, Cambridge, England.

Boury-Esnault, N., A. Ereskovsky, C. Bezac, and D. Tokina, 2003. Larval development in the Homoscleromorpha (Porifera, Demospongiae). *Invert. Biol.,* **122,** 187–202.

Boute, N., J. Y. Exposito, N. Boury-Esnault, J. Vacelet, et al., 1996. Type IV collagen in sponges, the missing link in basement membrane ubiquity. *Biol. Cellulaire,* **88,** 37–44.

Bouwens, R. J. and G. D. Illingworth, 2006. Rapid evolution of the most luminous galaxies during the first 900 million years. *Nature,* **443,** 189–192.

Bowler, P. J., 1996. *Life's Splendid Drama. Evolutionary Biology and the Reconstruction of Life's Ancestry 1860–1940.* The University of Chicago Press, Chicago.

Bowler, P. J., 2003. *Evolution: The History of an Idea,* 3rd ed. University of California Press, Berkeley, CA.

Bowler, P. J., 2005. Variation from Darwin to the modern synthesis. In *Variation: A Central Concept in Biology,* B. Hallgrímsson and B. K. Hall (eds.). Elsevier/Academic Press, Burlington, MA, pp. 9–27.

Bowring, S. A., D. H. Erwin, Y. G. Jin, M. W. Martin, K. Davidek, and W. Wang, 1998. U/Pb zircon geochronology and tempo of the End-Permian mass extinction. *Science,* **280,** 1039–1045.

Boyd, W. C., 1950. *Genetics and the Races of Man.* Little, Brown, Boston.

Bradshaw, H. D., Jr., S. M. Wilbert, K. G. Otto, and D. W. Shemske, 1995. Genetic mapping of floral traits associated with reproductive isolation in monkeyflowers (*Mimulus*). *Nature,* **376,** 762–765.

Brakefield, P. M., 1998. Receding black moths. *Trends Ecol. Evol.,* **13,** 376.

Bramble D. M., and D. E. Lieberman, 2004. Endurance running and the evolution of Homo. *Nature,* **432,** 345–352.

Brandes, J. A., N. Z. Boctor, G. D. Cody, B. A. Cooper, R. Hazen, and H. S. Yoder, Jr., 1998. Abiotic nitrogen reduction on early Earth. *Nature,* **395,** 365–367.

Brasier, M., and J. Antcliffe, 2004. Decoding the Ediacaran enigma. *Science,* **305,** 1115–1117.

Breaker, R. R., and G. F. Joyce, 1994. Emergence of a replicating species from an in vitro RNA evolution reaction. *Proc. Natl. Acad. Sci. USA,* **91,** 6093–6097.

Briggs, D. E. G., D. H. Erwin, and F. J. Collier, with photographs by Chip Clark, 1994. *The Fossils of the Burgess Shale.* Smithsonian Institution Press, Washington and London.

Britten, R. J., 1986. Rates of DNA sequence evolution differ between taxonomic groups. *Science,* **231,** 1393–1398.

Britten, R. J., 1998. Underlying assumptions of developmental models. *Proc. Natl. Acad. Sci. USA,* **95,** 9372–9377.

Britten, R. J., and E. H. Davidson, 1971. Repetitive and non-repetitive DNA sequences and a speculation on the origins of evolutionary novelty. *Q. Rev. Biol.,* **46,** 111–138.

Britten, R. J., and D. E. Kohne, 1968. Repeated sequences in DNA. *Science,* **161,** 259–540.

Brockman, J., 2006. *Intelligent Thought: Science Versus the Intelligent Design Movement.* Vintage Books, New York. [See the review by J. T. Bonner in *Nature,* 2006, **442,** 355–356.]

Bromham, L., A. Rambaut, R. Fortey, A. Cooper, and D. Penny, 1998. Testing the Cambrian explosion hypothesis by using a molecular dating technique. *Proc. Natl. Acad. Sci. USA,* **95,** 12386–12389.

Bronowski, J., 1974. *The Ascent of Man.* Little, Brown, Boston.

Brooks, D. R., and D. A. McLennan, 1991. *Phylogeny, Ecology, and Behavior: A Research Program in Comparative Biology.* The University of Chicago Press, Chicago.

Brower, A. V. Z., 1996. Parallel race formation and the evolution of mimicry in *Heliconius* butterflies: A phylogenetic hypothesis from mitochondrial DNA sequences. *Evolution,* **50,** 195–221.

Brower, L. P., (ed.), 1988. *Mimicry and the Evolutionary Process.* University of Chicago Press, Chicago.

Brown, M. D., S. H. Hosseini, A. Torroni, H.-J. Bandelt, et al., 1998. MtDNA haplogroup X: An ancient link between Europe/Western Asia and North America? *Amer. J. Hum. Genet.,* **63,** 1852–1861.

Brown, P., T. Sutikna, M. J. Morwood, R. P. Soejono, et al., 2004. A new small-bodied hominin from the Late Pleistocene of Flores, Indonesia. *Nature* **431,** 1055–1061.

Brown, T. A., and K. A. Brown, 1994. Ancient DNA: Using molecular biology to explore the past. *BioEssays,* **16,** 719–726.

Browne, E. J., 1995. *Charles Darwin: Voyaging.* Alfred A. Knopf, New York.

Browne, E. J., 2002. *Charles Darwin: The Power of Place. Volume II of a Biography.* Alfred A. Knopf, New York.

Browne, J., 2006. *Darwin's* Origin of Species. *A Biography.* Douglas & McIntyre, Vancouver, Canada.

Brunet, M., F. Guy, D. Pilbeam, H. T. Mackaye, et al., 2002. A new hominid from the Upper Miocene of Chad, Central Africa. *Nature* **418,** 145–151.

Brusca, R. C., and G. J. Brusca, 2003. *Invertebrates,* 2nd ed. Sinauer Associates, Sunderland, MA.

Budd, G. E., 2003. Animal phyla. In *Keywords and Concepts in Evolutionary Developmental Biology,* B. K. Hall and W. M. Olson (eds.). Harvard University Press, Cambridge, MA, pp. 1–10.

Buffetaut, E., and J.-M. Mazin (eds.), 2003. *Evolution and Paleobiology of Pterosaurs.* Geological Society of London Special Publication 217. Geological Society, London.

Buhl, D., 1974. Galactic clouds of organic molecules. *Origins of Life,* **5,** 29–40.

Bull, J. J., 1983. *Evolution of Sex Determining Mechanisms.* Benjamin/Cummings, Menlo Park, CA.

Bullini, L., 1994. Origin and evolution of animal hybrid species. *Trends Ecol. Evol.,* **9,** 422–426.

Bult, C. J., O. White, G. J. Olsen, L. Zhou, et al., 1996. Complete genome sequence of the methanogenic archaeon, *Methanococcus jannaschii. Science,* **273,** 1058–1073.

Bumpus, H. C., 1899. The elimination of the unfit as illustrated by the introduced sparrow. *Biol. Lect. Woods Hole,* pp. 209–226.

Burger, G., M. W., Gray, and B. F. Lang, 2004. Mitochondrial genomes: Anything goes. *Trends Genet.,* **19,** 709–716.

Burgers, P., and L. M. Chiappe, 1999. The wing of *Archaeopteryx* as a primary thrust generator. Nature, **399,** 60–62.

Buri, P., 1956. Gene frequency in small populations of mutant *Drosophila. Evolution,* **10,** 367–402.

Bush, G. L., 1994. Sympatric speciation in animals: New wine in old bottles. *Trends Ecol. Evol.,* **9,** 285–288.

Butlin, R., 1989. Reinforcement of premating isolation. In *Speciation and Its Consequences,* D. Otte and J. A. Endler (eds.). Sinauer Associates, Sunderland, MA, pp. 158–179.

Butterfield, N. J., 1994. Burgess Shale-type fossils from a Lower Cambrian shallow-shelf sequence in northwestern Canada. *Nature,* **369,** 477–479.

Byrne, J. A., S. Simonsson, P. S. Western, and J. B. Gurdon, 2003. Nuclei of adult mammalian somatic cells are directly reprogrammed to oct-4 stem cell gene expression by amphibian oocytes. *Curr. Biol.,* **13,** 1206–1213.

C

Cairns-Smith, A. G., 1982. *Genetic Takeover and the Mineral Origins of Life.* Cambridge University Press, Cambridge, England.

Calvin, M., 1969. *Chemical Evolution.* Oxford University Press, Oxford, England.

Campbell, B. G. 1998. *Human Evolution: An Introduction to Man's Adaptations.* Aldine de Gruyter, New York.

Canfield, M. A., M. A. Honein, N. Yuskiv, J. Xing, et al., 2006. National estimates and race/ethnicity-specific variation of selected birth defects in the United States, 1999–2001. *Birth Defects Res. (A) Clin. Mol. Teratol.,* **76,** 747–756.

Cann, R. L., M. Stoneking, and A. C. Wilson, 1987. Mitochondrial DNA and human evolution. *Nature,* **325,** 31–36.

Canup, R. M., 2004. Dynamics of lunar formation. *Annu. Rev. Astron. Astrophys.,* **42,** 441–475.

Carlson, T., 1913. Über Geschwindigkeit und grösze der Hefevermehrung in Würze. *Biochemische Zeitschrift,* **57,** 313–334.

Carpenter, K., (ed.), 2001. *The Armored Dinosaurs.* Indiana University Press, Bloomington, IN.

Carroll, R. L., 1988. *Vertebrate Paleontology and Evolution.* Freeman, New York.

Carroll, R. L., 1991. The origin of reptiles. In *Origins of the Higher Groups of Tetrapods: Controversy and Consensus,* H.-P. Schultze

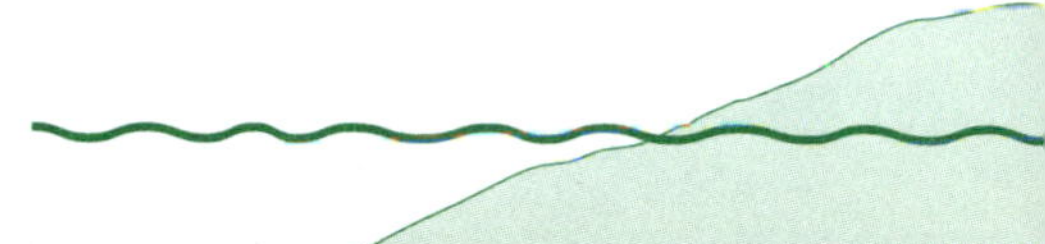

and L. Trueb (eds.). Cornell University Press, Ithaca, New York, pp. 331–353.

Carroll, R. L., 1995. Between fish and amphibian. *Nature,* **373,** 389–390.

Carroll, R. L., 1997. *Patterns and Processes of Vertebrate Evolution.* Cambridge University Press, Cambridge, England.

Carroll, R. L., and R. B. Holmes, 2007. Evolution of the appendicular skeleton of amphibians. In *Fins into Limbs: Evolution, Development and Transformation,* B. K. Hall (ed.). The University of Chicago Press, Chicago, pp. 186–224.

Carroll, R. L., A. Kuntz, and K. Albright, 1999. Vertebral development and amphibian evolution. *Evol. Devel.,* **1,** 36–48.

Carroll, S. B., 2005a. Evolution at two levels: on genes and form. *PLoS Biol* **3,** e245.

Carroll, S. B., 2005b. *Endless Forms Most Beautiful.* W. W. Norton & Co., New York.

Carroll, S. B., J. K. Grenier, and S. D. Weatherbee, 2005. *From DNA to Diversity: Molecular genetics and the Evolution of Animal Design,* 2nd ed. Blackwell Publishing, Malden, MA.

Carson, H. L., 1986. Sexual selection and speciation. In *Evolutionary Processes and Theory,* S. Karlin and E. Nevo (eds.). Academic Press, Orlando, FL, pp. 391–409.

Carson, H. L., 1992. Inversions in Hawaiian *Drosophila.* In Drosophila *Inversion Polymorphism,* C. B. Krimbas and J. R. Powell (eds.). CRC Press, Boca Raton, FL, pp. 407–439.

Cartmill, M. 1974. Rethinking primate origins. *Science,* **184,** 436–443.

Carvalho, A. B., M. C. Sampaio, F. R. Varandas, and L. B. Klaczko, 1998. An experimental demonstration of Fisher's principle: Evolution of sexual proportion by natural selection. *Genetics,* **148,** 719–731.

Casares, F., and R. S. Mann, 1998. Control of antennal versus leg development in *Drosophila. Nature,* **392,** 723–726.

Caskey, C. T., 1992. DNA-based medicine: Prevention and therapy. In *The Code of Codes: Scientific and Social Issues in the Human Genome Project,* P. J. Kevles and L. Hood (eds.). Harvard University Press, Cambridge, MA, pp. 112–135.

Castle, W. E., 1932. *Genetics and Eugenics,* 4th ed. Harvard University Press, Cambridge, MA.

Castle, W. E., and J. C. Phillips, 1914. Piebald Rats and Selection. Carnegie Institute of Washington Publication No. **195,** Washington, DC.

Cavalier-Smith, T., (ed.) 1985. *The Evolution of Genome Size.* Wiley, Chichester.

Cavalier-Smith, T., 1991. The evolution of the cells. In *Evolution of Life: Fossils, Molecules, and Culture,* S. Osawa and T. Honjo (eds.). Springer-Verlag, Tokyo, pp. 271–304.

Cavalier-Smith, T., 1993. Kingdom protozoa and its 18 phyla. *Microbiol. Rev.,* **57,** 953–994.

Cavalier-Smith, T., 2002. The phagotrophic origin of eukaryotes and phylogenetic classification of Protozoa. *Int. J. Syst. Evol. Microbiol.,* **52,** 297–354.

Cavalier-Smith, T., 2003. Protist phylogeny and the high-level classification of Protozoa. *Eur. J. Protistol.,* **39,** 338–348.

Cavalier-Smith, T., 2004. Only six kingdoms of life. *Proc. R. Soc. Lond. B,* **271,** 1251–1262.

Cavalli-Sforza, L. L., 1994. *The History and Geography of Human Genes.* Princeton University Press, Princeton, NJ.

Cavalli-Sforza, L. L., E. Minch, and J. L. Mountain, 1992. Coevolution of genes and languages revisited. *Proc. Natl. Acad. Sci. USA,* **89,** 5620–5624.

Cavalli-Sforza, L. L., P. Menozzi, and A. Piazza, 1993. Demic expansions and human evolution. *Science,* **259,** 639–646.

Chain, P. S., P. Hu, S. A. Malfatti, L. Radnedge, F. Larimer, et al., 2006. Complete genome sequence of *Yersinia pestis* strains Antiqua and Nepal516: evidence of gene reduction in an emerging pathogen. *J Bacteriol.,* **188,** 4453–4463.

Chang, M.-M., J. Zhang, and D. Miao, 2006. A lamprey from the Cretaceous Jehol biota of China. *Nature,* **441,** 972–974.

Charlesworth, B., 1996. The evolution of chromosomal sex determination and dosage compensation. *Curr. Biol.,* **6,** 149–162.

Charlesworth, B., P. Sniegowski, and W. D. Stephan, 1994. The evolutionary dynamics of repetitive DNA in eukaryotes. *Nature,* **371,** 215–220.

Charnov, E. L., 1982. *The Theory of Sex Allocation.* Princeton University Press, Princeton, NJ.

Chela-Flores, J., G. Lemarchand, and J. Oró, (eds.), 2002. *Astrobiology: Origins from the Big-Bang to Civilization.* Kluwer Academic Press, Dordrecht, The Netherlands.

Chen, I. A., and S. W. Szostak, 2004. Membrane growth can generate a transmembrane pH gradient in fatty acid vesicles. *Proc. Natl. Acad. Sci. USA,* **101,** 7965–7970.

Chen, I. A., K. Salehi-Ashtiani, and S. W. Szostak, 2005. RNA catalysis in model protocell vesicles. *J. Am. Chem. Soc.,* **127,** 13213–13219.

Chen, J.-Y., J. Dzik., G. D., Edgecombe, L. Ramskold, and G. Q. Zhou, 1995. A possible Early Cambrian chordate. *Nature,* **377,** 720–722.

Chen, K., and N. Rajewsky, 2006. Natural selection on human microRNA binding sites inferred from SNP data. *Nature Genet.,* **38,** 1452–1456.

Chern, J. P., K. H. Lin, Y. N. Su, M. Y. Lu, et al., 2006. Impact of a national β-thalassemia carrier screening program on the birth rate of thalassemia major. *Pediatr. Blood Cancer,* **46:** 72–76.

Chevalier-Skolnikoff, S. 1973. Facial expression of emotion in nonhuman primates. In *Darwin and Facial Expression.* P. Ekman (ed.). Academic Press, New York.

Cheverud, J. M., 2004. Modular pleiotropic effects of quantitative trait loci on morphological traits. In *Modularity in Development and Evolution.* G. Schlosser and G. P. Wagner (eds.). The University of Chicago Press, Chicago, pp. 132–153.

Chiappe, L. M., and L. Dingus, 2001*: Walking on Eggs: Discovering the Astonishing Secrets of the World of Dinosaurs.* Little Brown & Co., Boston.

Chinsamy, A., and W. J. Hillenius, 2004. Physiology of nonavian dinosaurs. In *The Dinosauria,* 2nd ed, D. B. Weishampel, P. Dodson, and H. Osmólska (eds.). University of California Press, Berkeley, CA, pp. 643–659.

Chourrout, D., F. Delsuc, P. Chourrout, R. B. Edvardsen, et al., 2006. Minimal ProtoHox cluster inferred from bilaterian and cnidarian Hox complements. *Nature,* **442,** 684–687.

Christiansen, F. B., 1984. The definition and measurement of fitness. In *Evolutionary Ecology,* B. Shorrocks (ed.). Blackwell, Oxford, England, pp. 65–79.

Chyba, C. F., T. C. Owen, and W.-H. Ip, 1995. Impact delivery of volatiles and organic molecules to Earth. In *Hazards Due to*

Comets and Asteroids, T. Gehrels (ed.). University of Arizona Press, Tucson, pp. 9–58.

Civetta, A., and R. S. Singh, 1998a. Sex-related genes, directional sexual selection, and speciation. *Mol. Biol. Evol.,* **15,** 901–909.

Civetta, A., and R. S. Singh, 1998b. Sex and speciation: Genetic architecture and evolutionary potential of sexual versus non-sexual traits in the sibling species of the *Drosophila melanogaster* complex. *Evolution,* **52,** 1080–1092.

Civeyrel, L., and D. Simberloff, 1996. A tale of two snails: is the cure worse than the disease? *Biodivers. Conserv.,* **5,** 1231–1252.

Clack, J. A., 2002. *Gaining Ground: The Origin and Evolution of Tetrapods.* Indiana University Press, Bloomington, IN.

Clapham, M. E., G. M. Narbonne, and J. G. Gehling, 2003. Paleoecology of the oldest-known animal communities: Ediacaran assemblages at Mistaken Point, Newfoundland. *Paleobiology,* **29,** 527–544.

Clark, A. G., D. J. Begun, and T. Prout, 1999. Female × male interactions in *Drosophila* sperm competition. *Science,* **283,** 217–220.

Clark, J. M., J. A. Hopson, R. Hernández, D. E. Fastovsky, and M. Montellano, 1998. Foot posture in a primitive pterosaur. *Nature,* **391,** 886–889.

Clark, R. B., 1964. *Dynamics in Metazoan Evolution.* Clarendon Press, Oxford, England.

Clark, R. B., 1979. Radiation of the metazoa. In *The Origin of the Major Invertebrate Groups,* M. R. House (ed.). Academic Press, London, pp. 55–102.

Clarke, D. K., E. A. Duarte, S. F. Elena, A. Moya, E. Domingo, and J. Holland, 1994. The Red Queen reigns in the kingdom of RNA viruses. *Proc. Natl. Acad. Sci. USA,* **91,** 4821–4824.

Clarke, K. U., 1973. *The Biology of Arthropods.* Arnold, London.

Clausen, J., D. D. Keck, and W. M. Hiesey, 1948. Experimental studies on the nature of species. III. Environmental responses of climatic races of *Achillea.* Carnegie Institute of Washington Publication No. **581,** 1–129.

Clayton, R. A., O. White, K. A. Ketchum, and J. C. Venter, 1997. The first genome from the third domain of life. *Nature,* **387,** 459–462.

Clemens, W. A. 1974. Purgatorius, an early paromomyid primate (Mammalia). *Science,* **184,** 903–905.

Cline, T. W., 1993. The *Drosophila* sex determination signal: How do flies count to two? *Trends Genet.,* **9,** 385–390.

Cline, T. W., and B. J. Meyer, 1996. Vive la différence: Males vs females in flies vs worms *Ann. Rev. Genet.,* **30,** 637–702.

Clodd, E., 1988. *Story of Creation.* Longmans Green, London.

Cloud, P., 1974. Evolution of ecosystems. *Am. Sci.,* **62,** 54–66.

Clutton-Brock, T. H., 1983. Selection in relation to sex. In *Evolution from Molecules to Men,* D. S. Bendall (ed.). Cambridge University Press, Cambridge, England, pp. 457–481.

Clutton-Brock, T. H., 1999. *A Natural History of Domesticated Mammals.* Second Edition. Cambridge University Press, Cambridge, England.

Coates, M. I., and J. A. Clack, 1990. Polydactyly in the earliest known tetrapod limbs. *Nature,* **347,** 66–69.

Coates, M. I., and J. A. Clack, 1991. Fish-like gills and breathing in the earliest known tetrapod. *Nature,* **352,** 234–236.

Coates, M. I., and M. Ruta, 2007. Skeletal changes in the transition from fins to limbs. In *Fins into Limbs: Evolution, Development and Transformation,* B. K. Hall (ed). The University of Chicago Press, Chicago, pp. 15–38.

Colbert, C. H., 1973. *Wandering Lands and Animals.* Hutchinson, London.

Colbert, E. H., and M. Morales, 1991. *Evolution of the Vertebrates: A History of the Backboned Animals Through Time,* 3rd ed. Wiley, New York.

Cole, A. G., and B. K. Hall, 2004. The nature and significance of invertebrate cartilages revisited: Distribution and histology of cartilage and cartilage-like tissues within the Metazoa. *Zoology* **107,** 261–274.

Collins, A. G., 1998. Evaluating multiple alternative hypotheses for the origin of Bilateria: An analysis of 18*S* rRNA molecular evidence. *Proc. Natl. Acad. Sci. USA,* **95,** 15458–15463.

Condie, K. C., 1997. *Plate Tectonics and Crustal Evolution,* 4th ed. Pergamon, New York.

Condie, K. C., and R. E. Sloan, 1998. *Origin and Evolution of Earth: Principles of Historical Geology.* Prentice Hall, Upper Saddle River, NJ.

Conn, H. W., 1900. *The Method of Evolution.* Putnam's, New York.

Conroy, G. C., G. B. Weber, H. Seidler, P. V. Tobias, A Kane, and B. Brunsden, 1998. Endocranial capacity in an early hominid cranium from Sterkfontein, South Africa. *Science,* **280,** 1730–1731.

Conway Morris, S., 1989. Burgess shale faunas and the Cambrian explosion. *Science,* **246,** 339–346.

Conway Morris, S., 1993. The fossil record and the early evolution of the metazoa. *Nature,* **361,** 219–225.

Conway Morris, S., 1998a. *The Crucible of Creation: The Burgess Shale and the Rise of Animals.* Oxford University Press, Oxford, England.

Conway Morris, S., 1998b. Metazoan phylogenies: Falling into place or falling to pieces? A paleontological perspective. *Curr. Opinion Genet. Devel.,* **8,** 662–667.

Conway Morris, S., 1998c. Eggs and embryos from the Cambrian. *BioEssays,* **20,** 676–682.

Conway Morris, S., J. D. George, R. Gibson, and H. M. Platt (eds.), 1985. *The Origins and Relationships of Lower Invertebrates.* Clarendon Press, Oxford, England.

Cook, T. A., 1914. *The Curves of Life.* Constable and Co., London.

Cooper, A., C. Mourer-Chauviré, G. K. Chambers, A. von Haesler, A. C. Wilson, and S. Pääbo, 1992. Independent origins of New Zealand moas and kiwis. *Proc. Natl. Acad. Sci. USA,* **89,** 8741–8744.

Cox, C. B., and P. D. Moore, 2005. *Biogeography: An Ecological and Evolutionary Approach,* 7th ed. Blackwell Publishing, Oxford, England.

Coyne, J. A., N. H. Barton, and M. Turelli, 1997. Perspective: A critique of Sewall Wright's shifting balance theory of evolution. *Evolution,* **51,** 643–671.

Coyne, J. A., and H. A. Orr, 1989. Two rules of speciation. In *Speciation and Its Consequences,* D. Otte and J. A. Endler (eds.). Sinauer Associates, Sunderland, MA, pp. 180–207.

Coyne, J. A., and H. A. Orr, 1997. "Patterns of speciation in *Drosophila*" revisited. *Evolution,* **51,** 295–303.

Coyne, J. A., and H. A. Orr, 2004. *Speciation.* Sinauer and Associates, Sunderland, MA.

Cracraft, J., 1983a. Species concepts and speciation analysis. *Curr. Ornithol.,* **1,** 159–187.

Cracraft, J., 1983b. The scientific response to creationism. In *Creationism, Science, and the Law: The Arkansas Case,* M. C. La Follette (ed.). MIT Press, Cambridge, MA, pp. 138–149.

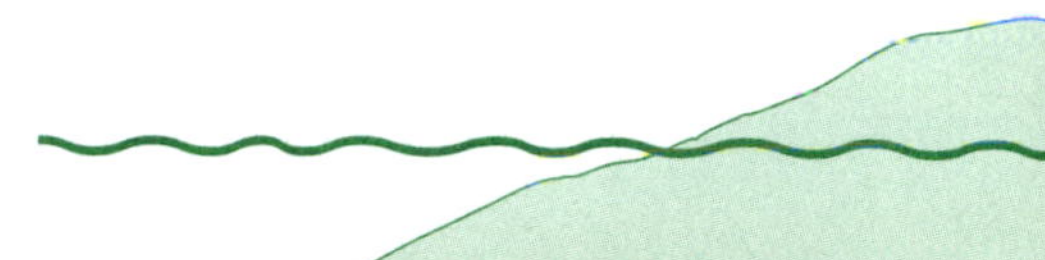

Cracraft, J., F. K. Barker, M. J. Braun, J. Harshman, et al., 2004. Phylogenetic relationships among modern birds (Neornithes): Toward an avian Tree of Life. In *Assembling the Tree of Life.* J. Cracraft and M. J. Donoghue (eds.). Oxford University Press, New York, pp. 468–489.

Cracraft, J., and R. W. Bybee (eds.), 2005. *Evolutionary Science and Society: Educating a New Generation.* Biological Sciences Curriculum Study, Washington, DC.

Cracraft, J., and M. J. Donoghue (eds.), 2004. *Assembling the Tree of Life.* Oxford University Press, Oxford, England.

Crepet, W. L., 2000. Progress in understanding angiosperm history, success, and relationships: Darwin's abominably "perplexing phenomenon." *Proc. Natl. Acad. Sci. USA,* **97,** 12939–12941.

Creuzet, S. E., S. Martinez, and N. M. Le Douarin, 2006. The cephalic neural crest exerts a critical effect on forebrain and midbrain development. *Proc. Natl. Acad. Sci. USA,* **97,** 4520–4524.

Crickmore, M. A., and R. S. Mann, 2006. Hox control of organ size by regulation of morphogen production and mobility. *Science,* **313,** 63–68.

Crompton, A. W., and F. A. Jenkins, Jr., 1979. Origin of mammals. In *Mesozoic Mammals: The First Two-Thirds of Mammalian History,* J. A. Lillegraven, Z. Kielan-Jaworowska, and W. A. Clemens (eds.). University of California Press, Berkeley, pp. 59–73.

Crompton, A. W., C. R. Taylor, and J. A. Jagger, 1978. Evolution of homeothermy in mammals. *Nature,* **272,** 333–336.

Crosby, A. 1986. *Ecological Imperialism: The Biological Expansion of Europe, 900–1900.* Cambridge University Press, Cambridge, UK.

Crow, J. F., 1957. Genetics of insect resistance to chemicals. *Ann. Rev. Entomol.,* **2,** 227–246.

Crow, J. F., 1988a. Eighty years ago: The beginnings of population genetics. *Genetics,* **119,** 473–476.

Crow, J. F., 1988b. The importance of recombination. In *The Evolution of Sex,* R. E. Michod and B. R. Levin (eds.). Sinauer Associates, Sunderland, MA, pp. 56–73.

Crow, J. F., 1992. Genetic load. In *Keywords in Evolutionary Biology,* E. F. Keller and E. A. Lloyd (eds.). Harvard University Press, Cambridge, MA, pp. 132–136.

Crow, J. F., 1993. Mutation, mean fitness, and genetic load. *Oxford Surv. Evol. Biol.,* **9,** 3–42.

Crow, J. F., 1997. The high spontaneous mutation rate: Is it a health risk? *Proc. Natl. Acad. Sci. USA,* **94,** 8380–8386.

Crow, J. F., 1998. 90 years ago: The beginning of hybrid maize. *Genetics,* **148,** 923–928.

Crow, J. F., and M. Kimura, 1970. *An Introduction to Population Genetics Theory.* Harper & Row, New York.

Cruickshank, R. H., and A. M. Paterson, 2006. The great escape: do parasites break Dollo's law? *Trends Parasitol.,* **22,** 509–515.

Cuffey, R. J., 1972. Paleontological evidence and organic evolution. In *Science and Creationism,* A. Montagu (ed.). Oxford University Press, Oxford, England, pp. 255–281.

Cummings, M. P., 1994. Transmission patterns of eukaryotic transposable elements: Arguments for and against horizontal transfer. *Trends Ecol. Evol.,* **9,** 141–145.

Currie, P. J., E. B. Koppelhus, M. A. Shugar, and J. L. Wright (eds.), 2004. *Feathered Dragons: Studies on the Transition from Dinosaurs to Birds.* Indiana University Press, Bloomington, IN.

Curry Rogers, K. A., and J. A. Wilson (eds.), 2005. *The Sauropods: Evolution and Paleobiology.* University of California Press, Berkeley, CA.

D

Dagan, T., and W. Martin, 2007. Ancestral genome sizes specify the minimum rate of lateral gene transfer during prokaryote evolution. *Proc. Natl. Acad. Sci. USA,* **104,** 870–875.

Dahn, R. D., M. C. Davis, W. N. Pappano, and N. H. Shubin, 2007. Sonic hedgehog function in chondrichthyans fins and the evolution of appendage patterning. *Nature,* **445,** 311–314.

Damasio, H., T. Grabowski, R. Frank, A. M. Galaburda, and A. R. Damasio, 1994. The return of Phineas Gage: Clues about the brain from the skull of a famous patient. *Science,* **264,** 1102–1105.

Damen, W. G. M., M. Hausdorf, E.-A. Seyfarth, and D. Tautz, 1998. A conserved mode of head segmentation in arthropods revealed by the expression pattern of Hox genes in a spider. *Proc. Natl. Acad. Sci. USA,* **95,** 10665–10670.

Danforth, B. N., S. Sipes, J. Fang, and S. G. Brady, 2006. The history of early bee diversification based on five genes plus morphology. *Proc. Natl. Acad. Sci. USA,* **103,** 15118–15123.

Daniels, S. B., K. R. Peterson, L. D. Strausbaugh, M. G. Kidwell, and A. Chovnick, 1990. Evidence for horizontal transmission of the P transposable element between *Drosophila* species. *Genetics,* **124,** 339–355.

Danilova, N., 2006. The evolution of immune mechanisms. *J. Exp. Biol. (Mol. Dev. Evol.),* **306B,** 496–520.

Darling, J. A., A. R. Reitzel, P. M. Burton, M. E. Mazza, J. F. Ryan, J. C. Sullivan, and J. R. Finnerty, 2005. Rising starlet: the starlet sea anemone, *Nematostella vectensis. BioEssays,* **27,** 211–221.

Dart, R., 1925. *Australopithecus africanus:* The man-ape of South Africa. *Nature,* **115,** 195–199.

Darwin, C., 1845. *Journal of Researches into the Natural History and Geology of the Countries Visited During the Voyage of H. M. S. Beagle Round the World, Under the Command of Capt. FitzRoy, R.N.* D. Appleton and Compant, New York.

Darwin, C., 1859. *On the Origin of Species by Means of Natural Selection or the Preservation of Favoured Races in the Struggle for Life.* Murray, London.

Darwin, C., 1862. *On the Various Contrivances by Which British and Foreign Orchids Are Fertilized by Insects, and on the Good Effects of Intercrossing.* Murray, London.

Darwin, C., 1871. *The Descent of Man, and Selection in Relation to Sex.* Murray, London.

Darwin, C., 1964. *On the Origin of Species.* Facsimile of the First Edition with an introduction by E. Mayr. Harvard University Press, Cambridge, MA. [paperback version, 2005].

Darwin, F., 1887. *The Life and Letters of Charles Darwin.* Appleton, New York.

Davidson, E. H., 2006. *The Regulatory Genome: Gene Regulatory Networks in Development and Evolution.* Elsevier/Academic Press, Burlington, MA.

Davidson, E. H., and R. J. Britten, 1973. Organization, transcription, and regulation in the animal genome. *Q. Rev. Biol.,* **48,** 565–613.

Davies, J., 1994. Inactivation of antibiotics and the dissemination of resistance genes. *Science,* **264,** 375–382.

Davis, A. W., and C-I. Wu, 1996. The broom of the sorcerer's apprentice: The fine structure of a chromosomal region causing reproductive isolation between two sibling species of *Drosophila. Genetics,* **143,** 1287–1298.

Davis, P. H., and V. H. Heywood, 1963. *Principles of Angiosperm Taxonomy.* Van Nostrand, Princeton, NJ.

Davison, A., 2006. The ovotestis: an undeveloped organ of evolution. *BioEssays*, **28**, 642–650.

Dawkins, R., 1986. *The Blind Watchmaker.* Longmans, Harlow, Essex, England.

Dawkins, R., 1995. *River Out of Eden: A Darwinian View of Life.* Basic Books, New York.

Day, M. H., 1986a. *Guide to Fossil Man,* 4th ed. Cassell, London.

Day, M. H., 1986b. Bipedalism: Pressures, origins and modes. In *Major Topics in Primate and Human Evolution,* B. Wood, L. Martin, and P. Andrews (eds.). Cambridge University Press, Cambridge, England, pp. 188–202.

Deacon, T. W., 1990. Rethinking mammalian brain evolution. *Am. Zool.,* **30,** 629–705.

Deamer, D. W., 1986. Role of amphiphilic compounds in the evolution of membrane structure on the early Earth. *Origins of Life,* **17,** 3–25.

Deamer, D. W., 2003. A giant step towards artificial life? *Trends Biotechnol.,* **23,** 336–338.

Debat, V, and P. David, 2001. Mapping phenotypes: canalization, plasticity and developmental stability. *Trends Ecol. Evol.,* **16,** 555–561.

de Beer, G. R., 1964. *Atlas of Evolution.* Nelson, London.

de Bono, M., and C. I. Bargmann, 1998. Natural variation in a neuropeptide Y receptor homolog modifies social behavior and food response in *C. elegans. Cell,* **94,** 679–689.

Deeming, D. C., 2006. Ultrastructural and functional morphology of eggshells supports the idea that dinosaur eggs were incubated buried in a substrate. *Palaeontology,* **49,** 171–185.

Dehal P., Y. Satou, R. K. Campbell, J. Chapman, et al., 2002. The draft genome of *Ciona intestinalis:* insights into chordate and vertebrate origins. *Science,* **298,** 2157–2167.

de Heinzelin, J., J. D. Clark, T. White, W. Hart, et al., 1999. Environment and behavior of 2.5 million-year-old Bouri hominids. *Science,* **284,** 625–629.

De Jong, W. W., 1998. Molecules remodel the mammalian tree. *Trends Ecol. Evol.,* **13,** 270–275.

Delevoryas, T., 1977. *Plant Diversification,* 2nd ed. Holt, Rinehart and Winston, New York.

Delsuc, F., H. Brinkmann, D. Chourrout, and H. Philippe, 2006. Tunicates and not cephalochordates are the closest living relatives of vertebrates. *Nature,* **439,** 965–968.

Delwiche, C. F., L. E. Graham, and N. Thomson, 1989. Lignin-like compounds and sporopollenin in Coleochaete, an algal model for plant ancestry. *Science,* **245,** 399–401.

Delwiche, C. F., M. Kuhsel, and J. D. Palmer, 1995. Phylogenetic analysis of *tufA* sequences indicates a cyanobacterial origin of all plastids. *Mol. Phylogenet. Evol.,* **4,** 110–128.

Delwiche, C. F., K. G. Karol, M. T. Cimino, and K. J. Sytsma, 2002. Phylogeny of the genus *Coleochaete* (Coleochaetales, Charophyta) and related taxa inferred by analysis of the chloroplast *gene rbcl. J. Phycol.,* **38,** *394*–403.

Dene, H. T., M. Goodman, and W. Prychodko, 1976. Immunodiffusion evidence on the phylogeny of primates. In *Molecular Anthropology,* M. Goodman and R. E. Tashian (eds.). Plenum Press, New York, pp. 171–195.

Dennett, D. C., 2006. *Breaking the Spell: Religion as a Natural Phenomenon.* Allen Lane, London.

DeRobertis, E. M., and Y. Sasai, 1996. A common plan for dorsoventral patterning in Bilateria. *Nature,* **380,** 37–40.

De Rosa, R., J.-L. Grenier, T. Andreevas, C. E. Cook, A. Adoutte, M. Akam, S. B. Carroll, and G. Balavoine, 1999. *Hox* genes in brachiopods and priapulids and protostome evolution. *Nature,* **399,** 772–776.

Desmond, A., and J. Moore, 1991. *Darwin.* Warner Books, New York.

de Solla Price, D. J., 1963. *Little Science, Big Science.* Columbia University Press, New York.

De Souza, S. J., M. Long, R. J. Klein, S. Roy, and W. Gilbert, 1998. Towards a resolution of the introns early/late debate: Only phase zero introns are correlated with the structure of ancient proteins. *Proc. Natl. Acad. Sci. USA,* **95,** 5094–5099.

Devlin, B., M. Daniels, and K. Roeder, 1997. The heritability of IQ. *Nature,* **388,** 468–471.

Dial, K. P., 2003. Wing-assisted incline running and the evolution of flight. *Science,* **299,** 402–404.

Dial, K. P., R. J. Randall, and T. R. Dial, 2006. What use is half a wing in the ecology and evolution of birds? *BioScience,* **56,** 437–445.

Diamond, J. M., 1995. The evolution of human creativity. In *Creative Evolution?!,* J. H. Campbell and J. W. Schopf (eds.). Jones and Bartlett, Boston, pp. 75–84.

Diamond, J. M., 1997. *Guns, Germs, and Steel: The Fates of Human Societies.* Norton, New York.

Diamond, J. M., 2005. *Collapse: How Societies Choose to Fail or Succeed.* Viking Books, New York.

Diao, X., M. Freeling, and D. Lisch, 2006. Horizontal transfer of a plant transposon. *Public Lib. Sci. Biol.,* **4,** 119–128.

Dingus, L., and T. Rowe, 1998. *The Mistaken Extinction: Dinosaur Evolution and the Origin of Birds.* Freeman, New York.

Dobzhansky, Th., 1937, 1941, 1951. *Genetics and the Origin of Species* (three eds.). Columbia University Press, New York.

Dobzhansky, Th., 1981. *Dobzhansky's Genetics of Natural Populations I-XLIII,* R. C. Lewontin, J. A. Moore, W. B. Provine, and B. Wallace (eds.). Columbia University Press, New York.

Dobzhansky, Th., 1944. Chromosomal races in *Drosophila pseudoobscura* and *D. persimilis.* Carnegie Inst. Wash. Publ. No. 554, Washington, DC, pp. 47–144.

Dobzhansky, Th., 1947. A directional change in the genetic constitution of a natural population of *Drosophila pseudoobscura. Heredity,* **1,** 53–64.

Dobzhansky, Th., 1973. Nothing in biology makes sense except in the light of evolution. *Amer. Biol. Teacher,* **35,** 125–129.

Dobzhansky, Th., and O. Pavlovsky, 1953. Indeterminate outcome of certain experiments on *Drosophila* populations. *Evolution,* 7, 198–210.

Doebley, J., A. Stec, and C. Gustus, 1995. *Teosinte branched1* and the origin of maize: Evidence for epistasis and the evolution of dominance. *Genetics,* **141,** 333–346.

Dong, X.-P., P. C. J. Donoghue, H. Cheng, and J.-B. Liu, 2004. Fossil embryos from the middle and late Cambrian period of Hunan, South China. *Nature,* **427,** 237–240.

Donoghue, M. J., 1992. Homology. In *Keywords in Evolutionary Biology,* E. F. Keller and E. A. Lloyd (eds.). Harvard University Press, Cambridge, MA, pp. 170–179.

Donoghue, P. C. J., 2002. Evolution of development of the vertebrate dermal and oral skeletons: unraveling concepts, regulatory theories, and homologies. *Paleobiology* **28,** 474–507.

Donoghue, P. C. J., P. L. Forey, and R. J. Aldridge, 2000. Conodont affinity and chordate phylogeny. *Biol. Rev. Camb. Philos. Soc.,* **75,** 191–251.

Doolittle, R. F., 1998. Microbial genomes opened up. *Nature,* **392,** 339–342.

Doolittle, W. F., 1978. Genes-in-pieces: Were they ever together? *Nature,* **272,** 581.

Doolittle, W. F., 1996. At the core of the Archaea. *Proc. Natl. Acad. Sci. USA,* **93,** 8797–8799.

Doolittle, W. F., 1998. You are what you eat: A gene transfer ratchet could account for bacterial genes in eukaryotic nuclear genomes. *Trends Genet.,* **14,** 307–311.

Doolittle, W. F., 1999. Lateral genomics. *Trends Cell Biol.,* **9,** M5–M8.

Doolittle, W. F., 2000. Uprooting the tree of life. *Sci. Am.* **282,** 90–95.

Doolittle, W. F., Y. Boucher, C. L. Nesbo, C. J. Douady, J. O. Andersson, and A. J. Roger, 2003. How big is the iceberg of which organelle genes in nuclear genomes are but the tip? *Phil. Trans. R. Soc. Lond. B. Biol. Sci.,* **358,** 39–57.

Doolittle, W. F., and C. Sapienza, 1980. Selfish genes, the phenotype paradigm, and genome evolution. *Nature,* **284,** 601–603.

Douglas, S. E., and S. Turner, 1991. Molecular evidence for the origin of plastids from a cyanobacterium-like ancestor. *J. Mol. Evol.,* **33,** 267–273.

Dowling, T. E., and C. L. Secor, 1997. The role of hybridization and introgression in the diversification of animals. *Ann. Rev. Ecol. Syst.,* **28,** 593–619.

Doyle, J. A., and L. J. Hickey, 1976. Pollen and leaves from the mid-Cretaceous Potomac Group and their bearing on early angiosperm evolution. In *Origin and Early Evolution of Angiosperms,* C. B. Beck (ed.). Columbia University Press, New York, pp. 139–206.

Drake, J. W., 1970. *The Molecular Basis of Mutation.* Holden-Day, San Francisco.

Duboule, D., and A. S. Wilkins, 1998. The evolution of 'bricolage.' *Trends Genet.,* **4,** 54–59.

Dudareva, N., and E. Pichersky (eds.), 2006. *Biology of Floral Scent.* CRC Press, Atlanta, GA.

Dudley, J. W., 1977. Seventy-six generations of selection for oil and protein percentages in maize. In *Proceedings of the International Conference on Quantitative Genetics,* E. Pollak, O. Kempthorne, and T. B. Bailey, Jr., (eds.). Iowa State University Press, Ames, pp. 459–473.

Duellman, W. E., and L. Trueb, 1986. *Biology of Amphibians.* McGraw-Hill, New York.

Dufaÿ, M., and M.-C. Anstett, 2004. Cheating is not always punished: killer female plants and pollination by deceit in the dwarf palm, *Chamaerops humilis. J. Evol. Biol.,* **17,** 862–868.

Dufour, H. D., Z. Chettouh, C. Deyts, R. de Rosa, et al., 2006. Precraniate origin of cranial motoneurons. *Proc. Natl. Acad. Sci. USA,* **103,** 8727–8732.

Dunn, L. C., 1964. Abnormalities associated with a chromosome region in the mouse. *Science,* **144,** 260–263.

Dupré, J., (ed.), 1987. *The Latest on the Best: Essays on Evolution and Optimality.* MIT Press, Cambridge, MA.

Dyke, G. J., R. L. Nudds, and J. M. V. Rayner, 2006. Limb disparity and wing shape in pterosaurs. *J. Evol. Biol.,* **19,** 1339–1342.

E

Eanes, W. F., M. Kirchner, and J. Yoon, 1993. Evidence for adaptive evolution of the *G6pd* gene in *Drosophila melanogaster* and *Drosophila simulans* lineages. *Proc. Natl. Acad. Sci. USA,* **90,** 7475–7479.

Eberhard, W. C., 1996. *Female Control: Sexual Selection by Cryptic Female Choice.* Princeton University Press, Princeton, NJ.

Eck, R. V., and M. O. Dayhoff, 1966. Evolution of the structure of ferredoxin based on living relics of primitive amino acid sequences. *Science,* **152,** 363–366.

Edwards, M. R., 1998. From a soup or a seed? Pyritic metabolic complexes in the origin of life. *Trends Ecol. Evol.,* **13,** 178–181.

Edwards, Y. J. K., K. Walter, G. McEwen, T. Vavouri, et al, 2005. Characterisation of conserved non-coding sequences in vertebrate genomes using bioinformatics, statistics and functional studies. *Comp. Biochem. Physiol., Part D* **1,** 3–15.

Eigen, M., 1992. *Steps Towards Life: A Perspective on Evolution.* Oxford University Press, Oxford, England.

Eigen, M., W. Gardiner, P. Schuster, and R. Winkler-Oswatitsch, 1981. The origin of genetic information. *Sci. Am.,* **244,** 88–118.

Eldredge, N., 1982. *The Monkey Business: A Scientist Looks at Creationism.* Washington Square Press, New York.

Eldredge, N., and S. J. Gould, 1972. Punctuated equilibria: An alternative to phyletic gradualism. In *Models in Paleobiology,* T. J. M. Schopf (ed.). Freeman, Cooper, San Francisco, pp. 82–115.

Ellegård, A., 1958. *Darwin and the General Reader.* Göteborgs Universitets Årsskrift, Gothenburg, Sweden. (Republished 1990 by The University of Chicago Press, Chicago.)

Elliott, P., 2003. Erasmus Darwin, Herbert Spencer, and the origins of the evolutionary worldview in British provincial scientific culture, 1770–1850. *Isis,* **94,** 1–29.

Ellstrand, N. C., R. Whitkus, and L. H. Rieseberg, 1996. Distribution of spontaneous plant hybrids. *Proc. Natl. Acad. Sci. USA,* **93,** 5090–5093.

Emlen, D. J., 2000. Integrating development with evolution: A case study with beetle horns. *BioScience,* **50,** 403–418.

Emlen, D. J., J. Hunt, and L. W. Simmons, 2005. Evolution of male- and sexual-dimorphism in the expression of beetle horns: Phylogenetic evidence for modularity, evolutionary lability, and constraint. *Am. Nat.,* **166,** S42–S68.

Endler, J. A., 1986. *Natural Selection in the Wild.* Princeton University Press, Princeton, NJ.

Endler, J. A., 1989. Conceptual and other problems in speciation. *In Speciation and Its Consequences,* D. Otte and J. A. Endler (eds.). Sinauer Associates, Sunderland, MA, pp. 625–648.

Engel, S. R., K. M. Hogan, J. F. Taylor, and S. K. Davis, 1998. Molecular systematics and paleobiogeography of the South American sigmodontine rodents. *Mol. Biol. Evol.,* **15,** 35–49.

Ephrussi, A., and R. Lehmann, 1992. Induction of germ cell formation by *oskar. Nature,* **358,** 387–392.

Epperson, B. K., 1995. Spotted distributions of genotypes under isolation by distance. *Genetics,* **140,** 1431–1440.

Erwin, D. H., 1993. *The Great Paleozoic Crisis: Life and Death in the Permian.* Columbia University Press, New York.

Erwin, D. H., 2006. *Extinction: How Life on Earth Nearly Ended 250 Million Years Ago.* Princeton University Press, Princeton, NJ, 296 pp.

Erwin, D. H. and E. H. Davidson, 2002. The last common bilaterian ancestor. *Development,* **129,** 3021–3032.

Erwin, D., J. Valentine, and D. Jablonski, 1997. The origin of animal body plans. *Am. Sci.,* **85,** 126–137.

Evans, D. A., N. J. Beukes, and J. L. Kirschvink. 1997. Low-latitude glaciation in the Palaeoproterozoic era. *Nature,* **386,** 262–266.

Evolution, 2006. A Scientific American Reader. The University of Chicago Press, Chicago.

Ewald, P. W., 1994. *Evolution of Infectious Disease.* Oxford University Press, Oxford, England.

Extavour, C. G., and M. Akam, 2003. Mechanisms of germ cell specification across the metazoans: epigenesis and preformation. *Development,* **130,** 5869–5884.

Eyre-Walker, A., and P. D. Keightley, 1999. High genomic deleterious mutation rates in hominids. *Nature,* **397,** 344–347.

F

Falconer, D. S., and T. F. C. Mackay, 1996. *Introduction to Quantitative Genetics,* 4th ed. Longman, Harlow, Essex, England.

Farley, J., 1977. *The Spontaneous Generation Controversy from Descartes to Oparin.* The Johns Hopkins University Press, Baltimore, MD.

Farmer, C., 1997. Did lungs and the intracardiac shunt evolve to oxygenate the heart in vertebrates? *Paleobiology,* **23,** 358–372.

Farmer, C. G., and D. R. Carrier, 2000. Pelvic aspiration in the American alligator, *Alligator mississippiensis. J. Exp. Biol.* **203,** 1769–1687.

Fedonkin, M. A., 1994. Vendian body fossils and trace fossils. In *Early Life on Earth,* S. Bengtson (ed.). Columbia University Press, New York, pp. 370–388.

Fedoroff, N. V., and J. E. Cohen, 1999. Plants and population: Is there time? *Proc. Natl. Acad. Sci. USA,* **96,** 5903–5907.

Feduccia, A., 1980. *The Age of Birds.* Harvard University Press, Cambridge, MA.

Feduccia, A., 1985. On why dinosaurs lacked feathers. In *The Beginnings of Birds,* M. K. Hecht, J. H. Ostrom, G. Viohl, and P. Wellnhofer (eds.). Freunde des Jura-Museums Eichstätt, Willibaldsburg, Eichstätt, Germany, pp. 75–79.

Feduccia, A., 1995. Explosive evolution in tertiary birds and mammals. *Science,* **267,** 637–638.

Feduccia, A., 1999a. 1,2,3 = 2,3,4: Accommodating the cladogram. *Proc. Natl. Acad. Sci. USA,* **96,** 4740–4742.

Feduccia, A., 1999b. *The Origin and Evolution of Birds.* Second Edition. Yale University Press, New Haven, CT.

Feelisch, M., and J. F. Martin, 1995. The early role of nitric oxide in evolution. *Trends Ecol. Evol.,* **10,** 496–499.

Feinsinger, P., 1983. Coevolution and pollination. In *Coevolution,* D. J. Futuyma and M. Slatkin (eds.). Sinauer Associates, Sunderland, MA, pp. 282–310.

Felsenstein, J., 1988. Phylogenies from molecular sequences: Inference and reliability. *Ann. Rev. Genet.,* **22,** 521–565.

Felsenstein, J., 2004. *Inferring Phylogenies.* Sinauer Associates, Sunderland, MA.

Fenchel, T., and B. J. Finlay, 1994. The evolution of life without oxygen. *Am. Sci.,* **82,** 22–29.

Feng, D.-F., G. Cho, and R. F. Doolittle, 1997. Determining divergence times with a protein clock: Update and reevaluation. *Proc. Natl. Acad. Sci. USA,* **94,** 13028–13033.

Fenner, F., and B. Fantini, 1999. *Biological Control of Vertebrate Pests: The History of Myxomatosis — An Experiment in Evolution.* CABI Publishing, Oxford, England.

Fenster, C. B., L. F. Galloway, and L. Chao, 1997. Epistasis and its consequences for the evolution of natural populations. *Trends Ecol. Evol.,* **12,** 282–286.

Fenster, E. J., and U. Sorhannus, 1991. On the measurement of morphological rates of evolution. *Evol. Biol.,* **25,** 375–410.

Ferguson, E. L., 1996. Conservation of dorsal-ventral patterning in arthropods and chordates. *Curr. Opin. Genet. Dev.,* **6,** 424–431.

Ferracin, A., 1981. A neutral theory of biogenesis. *Orig. Life Evol. Biosph.,* **11,** 369–385.

Ferris, J. P., 1992. Marine hydrothermal systems and the origin of life: Chemical markers of prebiotic chemistry in hydrothermal systems. *Orig. Life Evol. Biosph.,* **22,** 109–134.

Ferris, J. P., 2005a. Catalysis and prebiotic synthesis. In *Molecular Geomicrobiology,* J. F. Banfield, J. Cervini, and K. M. Nealson (eds.). Mineralogical Society of America, Chantilly, VA, pp. 187–210.

Ferris, J. P., 2005b. Mineral catalysis and prebiotic synthesis: Montmorillonite-catalyzed formation of RNA. *Elements,* **1,** 145–149.

Ferris, J. P., and P. C. Joshi, 1978. Chemical evolution from hydrogen cyanide: Photochemical decarboxylation of orotic acid and orotate derivatives. *Science,* **201,** 361–362.

Ferris, S. D., A. C. Wilson, and W. M. Brown, 1981. Evolutionary tree for apes and humans based on cleavage maps of mitochondrial DNA. *Proc. Natl. Acad. Sci. USA,* **78,** 2432–2436.

Fidalgo, M., R. R. Barrales, J. I. Ibeas, and J. Jimenez, 2006. Adaptive evolution by mutations in the FLO11 gene. *Proc. Natl. Acad. Sci. USA,* **103,** 11228–11233.

Field, K. G., G. J. Olsen, D. J. Lane, S. J. Giovannoni, et al., 1988. Molecular phylogeny of the animal kingdom. *Science,* **239,** 748–753.

Finlayson, C., F. G. Pacheco, J. Rodriguez-Vidal, D. A. Fa, et al., 2006. Late survival of Neanderthals at the southernmost extreme of Europe. *Nature,* **443,** 850–853.

Finnerty, J. R., 2001. Cnidarians reveal intermediate stages in the evolution of Hox clusters and axial complexity. *Am. Zool.,* **41,** 608–620.

Finnerty, J. R., and M. Q. Martindale, 1998. The evolution of the Hox cluster: Insights from outgroups. *Curr. Opin. Genet. Dev.,* **8,** 681–687.

Fisher, R. A., 1930. *The Genetical Theory of Natural Selection.* Clarendon, Oxford, England.

Fisher, R. A., 1958. *The Genetical Theory of Natural Selection,* 2d ed. Dover, New York.

Fitch, W. M., 1976. Molecular evolutionary clocks. In *Molecular Evolution,* F. J. Ayala (ed.). Sinauer Associates, Sunderland, MA, pp. 160–178.

Fitch, W. M., and C. J. Langley, 1976. Protein evolution and the molecular clock. *Fed. Proc.,* **35,** 2092–2097.

Fitch W. M., and E. Margoliash, 1967. Construction of phylogenetic trees. *Science,* **155,** 279–284.

Fitch, W. T., 2005. The evolution of music in comparative perspective. *Ann. N.Y. Acad. Sci.* **1060,** 29–49.

Flatt, T., 2005. The evolutionary genetics of canalization. *Q. Rev. Biol.,* **80,** 287–316.

Flatt, T., M.-P. Tu, and M. Tatar, 2005. Hormonal pleiotropy and the juvenile hormone regulation of *Drosophila* development and life history. *BioEssays,* **27,** 999–1010.

Flynn, J. J., and A. R. Wyss, 1998. Recent advances in South American mammalian paleontology. *Trends Ecol. Evol.,* **13,** 449–454.

Foley, R., 1987. *Another Unique Species: Patterns in Human Evolutionary Ecology.* Longman, Harlow, Great Britain.

Foley, R., 1995. *Humans Before Humanity.* Blackwell Publishers, Oxford, England.

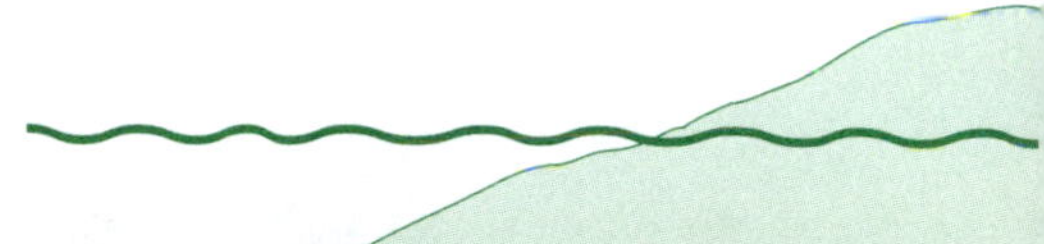

Foley, R., 1996. The adaptive legacy of human evolution: A search for the environment of evolutionary adaptedness. *Evol. Anthropol.*, **4**, 194–203.

Foote, M., J. P. Hunter, C. M. Janis, and J. J. Sepkoski, Jr., 1999. Evolutionary and preservational constraints on origins of biologic groups: Divergence times of eutherian mammals. *Science*, **283**, 1310–1314.

Ford, E. B., 1940. Genetic research in the Lepidoptera. *Ann. Eugen.*, **10**, 227–252.

Forey, P. L., C. J. Humphries, I. L. Kitching, R. W. Scotland, et al., 1992. *Cladistics: A Practical Course in Systematics.* Oxford University Press, Oxford, England.

Forrest, B., and P. R. Gross, 2004. *Creationism's Trojan Horse: The Wedge of Intelligent Design.* Oxford University Press, New York.

Fortey, R., 2002. *Fossils: The Key to the Past.* Natural History Museum, London.

Foster, A. S., and E. M. Gifford, Jr., 1974. *Comparative Morphology of Vascular Plants*, 2nd ed. Freeman, San Francisco.

Fox, C. W., D. A. Roff, and D. J. Fairbairn (eds.), 2001. *Evolutionary Ecology: Concepts and Case Studies.* Oxford University Press, New York.

Fox, R. M., and J. W. Fox, 1964. *Introduction to Comparative Entomology.* Reinhold, New York.

Fox, S. W., 1978. The origin and nature of protolife. In *The Nature of Life*, W. H. Heidecamp (ed.). University Park Press, Baltimore, pp. 23–92.

Fox, S. W., 1984. Proteinoid experiments and evolutionary theory. In *Beyond Neo-Darwinism*, M.-W. Ho and P. T. Saunders (eds.). Academic Press, London, pp. 15–60.

Fox, S. W., and K. Dose, 1972. *Molecular Evolution and the Origin of Life.* Freeman, San Francisco.

Fox, W. T., 1987. Harmonic analysis of periodic extinctions. *Paleobiology*, **13**, 257–271.

Franzen, H. F., 1983. Thermodynamics: The Red Herring, chapter 9 in *Did the Devil Make Darwin Do It?* D. B. Wilson (ed.). Iowa State Press, Ames, Iowa.

Franz-Odendaal, T. A., and B. K. Hall, 2006. Modularity and sense organs in the blind cavefish, *Astyanax mexicanus. Evol. Devel.*, **8**, 94–100.

Fraser, C. M., J. D. Gocayne, O. White, M. D. Adams, et al., 1995. The minimal gene complement of *Mycoplasma genitalium. Science*, **270**, 397–403.

Freedman, W. L., 1998. Measuring cosmological parameters. *Proc. Natl. Acad. Sci. USA*, **95**, 2–7.

Freeman, G. and Q. Martindale, 2002. Intracellular fate mapping and the origin of mesoderm in Phoronids. *Dev. Biol.*, **252**, 301–311.

Freeman, M., 1998. Complexity of EGF receptor signaling revealed in *Drosophila. Curr. Opin. Genet. Dev.*, **8**, 407–411.

Freitas, R., G. Zhang, and M. J. Cohn, 2006. Evidence that mechanisms of fin development evolved in the midline of early vertebrates. *Nature*, **442**, 1033–1037.

Futuyma, D. J., 1983. *Science on Trial: The Case for Evolution.* Pantheon Books, New York.

G

Gajewski, W. 1959. Evolution in the Genus *Geum, Evolution*, **13**, 378–388.

Galis, F., 1999. Why do almost all mammals have seven cervical vertebrae? Developmental constraints, *Hox* genes and cancer. *J. Exp. Zool. (Mol. Dev. Evol.)*, **285**, 19–26.

Galis, F., and J. A. J. Metz, 1998. Why are there so many cichlid species? *Trends Ecol. Evol.*, **13**, 1–2.

Galis, F., J. J. M. van Alphen, and J. A. J. Metz, 2001. Why five fingers? Evolutionary constraints on digit numbers. *Trends Ecol. Evol.*, **16**, 637–646.

Galton, F., 1871a. Pangenesis. *Nature*, **4**, 5–6.

Galton, F., 1871b. Experiments in pangenesis, by breeding from rabbits of a pure variety, into whose circulation blood taken from other varieties had previously been largely transfused. *Ann. Mag. Nat. Hist.*, **7**, 372–388.

Garcia-Fernàndez, J., and P. W. H. Holland, 1994. Archetypal organization of the amphioxus Hox gene cluster. *Nature*, **370**, 563–566.

Garcia-Ruiz, J. M., S. T. Hyde, A. M. Carnerup, A. G. Christy, et al., 2003. Self-assembled silica-carbonate structures and detection of ancient microfossils. *Science*, **302**, 1194–1197.

Gardner, R. A., and B. T. Gardner, 1969. Teaching sign language to a chimpanzee. *Science*, **165**, 664–672.

Garrigan, D., and M. F. Hammer, 2006. Reconstructing human origins in the genomic era. *Nature Rev. Genet.*, **7**, 669–680.

Garstang, W., 1928. The morphology of the Tunicata, and its bearings on the phylogeny of the Chordata. *Q. J. Microsc. Sci.*, **72**, 51–187.

Garstang, W., 1951. *Larval Forms and other Zoological Verses.* Blackwell Publishers, Oxford, England.

Gass, G. L., and J. A. Bolker, 2003. Modularity. In *Keywords and Concepts in Evolutionary Developmental Biology*, B. K. Hall and W. M. Olson (eds.). Harvard University Press, Cambridge, MA, pp. 260–267.

Gatesy, S. M., and K. M. Middleton, 2007. Skeletal adaptations for flight. In *Fins into Limbs: Evolution, Development and Transformation*, B. K. Hall (ed.). The University of Chicago Press, Chicago, pp. 270–283.

Gause, G. F., 1934. *The Struggle for Existence.* Williams & Wilkins, Baltimore.

Geary, D. H., 1990. Patterns of evolutionary tempo and mode in the radiation of *Melanopsis* (Gastropoda; Melanopsidae). *Paleobiology*, **16**, 492–511.

Gee, H., (ed.), 2001. *Rise of the Dragon: Readings from Nature on the Chinese Fossil Record.* The University of Chicago Press, Chicago.

Gelman, A., J. B. Carlin, H. S. Stern, and D. B. Rubin, 2003. *Bayesian Data Analysis*, 2nd ed. Chapman and Hall/CRC, Boca Raton, FL.

Gensel, P. G., and D. Edwards, (eds.), 2001. *Plants Invade the Land: Evolutionary and Environmental Perspectives.* Columbia University Press, New York.

Gerhart, J., 2000. Inversion of the chordate body axis: Are there alternatives? *Proc. Natl. Acad. Sci. USA*, **97**, 4445–4448.

Gerhart, J., and M. Kirschner, 1997. *Cells, Embryos, and Evolution: Toward a Cellular and Developmental Understanding of Phenotypic Variation and Evolutionary Adaptability.* Blackwell Science, Malden, MA.

Geritz, S. A. H., É. Kisdi, G. Meszéna, and J. A. J. Metz, 1998. Evolutionarily singular strategies and the adaptive growth and branching of the evolutionary tree. *Evol. Ecol.*, **12**, 37–57.

Gess, R. W., M. I. Coates, and B. S. Rubidge, 2006. A lamprey from the Devonian period of South Africa. *Nature,* **443,** 981–984.

Gest, H., and J. W. Schopf, 1983. Biochemical evolution of anaerobic energy conversion: The transition from fermentation to anoxygenic photosynthesis. In *Earth's Earliest Biosphere: Its Origin and Evolution,* J. W. Schopf (ed.). Princeton University Press, Princeton, NJ, pp. 135–148.

Ghimire, S. K., Y. Akashi, C. Maitani, M. Nakanishi, and K. Kato, 2005. Genetic Diversity and Geographical Differentiation in Asian Common Wheat (*Triticum aestivum* L.), Revealed by the Analysis of Peroxidase and Esterase Isozymes. *Breeding Sci.,* **55,** 75–185.

Ghiselin, M. T., 1984. Narrow approaches to phylogeny: A review of nine books on cladism. In *Oxford Surveys in Evolutionary Biology,* R. Dawkins and M. Ridley (eds.). Oxford University Press, Oxford, England, pp. 209–222.

Ghiselin, M. T., 1987. Species concepts, individuality, and objectivity. *Biol. Philos.,* **2,** 127–143.

Gibbs, H. L., and P. R. Grant, 1987. Oscillating selection on Darwin's finches. *Nature,* **327,** 511–513.

Gibbs, P. M., and A. Dugaiczyk, 1994. Reading the molecular clock from the decay of internal symmetry of a gene. *Proc. Natl. Acad. Sci. USA,* **91,** 3413–3417.

Gibbs, R. A., G. M. N. Weinstock, M. L. Metzker, D. M. Muzny, et al., 2004. Genome sequence of the Brown Norway rat yields insights into mammalian evolution. *Nature,* **428,** 493–521.

Gibson, G., and D. S. Hogness, 1996. Effect of polymorphism in the *Drosophila* regulatory gene *Ultrabithorax* on homeotic stability. *Science,* **271,** 200–203.

Gibson, G., and G. P. Wagner, 2000. Canalization in evolutionary genetics: a stabilizing theory? *BioEssays,* **22,** 372–380.

Gilad, Y., A. Oshlack, G. K. Smyth, T. P. Speed, and K. P. White, 2006. Expression profiling in primates reveals a rapid evolution of human transcription factors. Nature, **440,** 242–245.

Gillespie, J. H., 2004. *Population Genetics: A Concise Guide,* 2nd ed. Johns Hopkins University Press, Baltimore, MD.

Gillham, N. W., 1994. *Organelle Genes and Genomes.* Oxford University Press, Oxford, England.

Gingerich, P. D., 1998. Vertebrates and evolution. *Evolution,* **52,** 289–291.

Gish, D. T., 1972. *Evolution? — The Fossils Say No!* Creation-Life Publishers, San Diego, CA.

Givnish, T. J., 1997. Adaptive radiation and molecular systematics: Issues and approaches. In *Molecular Evolution and Adaptive Radiation,* T. J. Givnish and K. J. Sytsma (eds.). Cambridge University Press, Cambridge, England, pp. 1–54.

Glaessner, M. F., 1984. *The Dawn of Animal Life.* Cambridge University Press, Cambridge, England.

Glass, H. B., and C. C. Li, 1953. The dynamics of racial admixture: An analysis based on the American Negro. *Am. J. Hum. Genet.,* **5,** 1–20.

Glazier, D. S., 2006. The 3/4-power law is not universal: Evolution of isometric, ontogenetic metabolic scaling in pelagic animals. *BioScience,* **56,** 325–332.

Gluckman, P., and M. Hanson, 2005. *The Fetal Matrix. Evolution, Development and Disease.* Cambridge University Press, Cambridge, England.

Golding, G. B., and A. M. Dean, 1998. The structural basis of molecular adaptation. *Mol. Biol. and Evol.,* **15,** 355–369.

Goldman, N., 1998. Effects of sequence alignment procedures on estimates of phylogeny. *BioEssays,* **20,** 287–290.

Goldschmidt, R. B., 1940. *The Material Basis of Evolution.* Yale University Press, New Haven, CT.

Goldschmidt, T., 1996. *Darwin's Dreampond. Drama in Lake Victoria.* MIT Press, Cambridge, MA.

Golubic, S., and A. H. Knoll, 1993. Prokaryotes. In *Fossil Prokaryotes and Protists,* J. H. Lipps (ed.). Blackwell Scientific, Boston, pp. 51–76.

Goodall, J. 1986. *The Chimpanzees of Gombe: Patterns of Behavior.* Belknap Press of Harvard University Press, Harvard, MA.

Goode, J., (ed.), 2007. *Tinkering: The Microevolution of Development.* Novartis Foundation Symposium No. 284. John Wiley and Sons, Chichester, England.

Goodman, M., 1976. Toward a genealogical description of the primates. In *Molecular Anthropology,* M. Goodman and R. E. Tashian (eds.). Plenum Press, New York, pp. 321–353.

Goodman, M., A. E. Romero-Herrera, H. Dene, J. Czelusniak, and R. E. Tashian, 1982. Amino acid sequence evidence on the phylogeny of primates and other eutherians. In *Macromolecular Sequences in Systematic and Evolutionary Biology,* M. Goodman (ed.). Plenum Press, New York, pp. 115–191.

Goss, R. J., 1983. *Deer Antlers: Regeneration, Function and Evolution.* Academic Press, New York.

Gould, S. J., 1977. *Ontogeny and Phylogeny.* Harvard University Press, Cambridge, MA.

Gould, S. J., 1989. *Wonderful Life. The Burgess Shale and the Nature of History.* W. W. Norton & Co., New York.

Gould, S. J., 1994. Tempo and mode in the macroevolutionary reconstruction of Darwinism. *Proc. Natl. Acad. Sci. USA,* **91,** 9413–9417.

Gould, S. J., 2002. *The Structure of Evolutionary Theory.* The Belknap Press of Harvard University Press, Cambridge, MA.

Gould, S. J., and N. Eldredge, 1993. Punctuated equilibrium comes of age. *Nature,* **366,** 223–227.

Gould, S. J., and R. C. Lewontin, 1979. The spandrels of San Marco and the Panglossian paradigm: A critique of the adaptationist program. *Proc. Roy. Soc. Lond.,* **205,** 581–598.

Gould, S. J., and E. S. Vrba, 1982. Exaptation — a missing term in the science of form. *Paleobiology,* **8,** 4–15.

Gouy, M., and W-H. Li, 1989. Molecular phylogeny of the kingdoms Animalia, Plantae and fungi. *Mol. Biol. Evol.,* **6,** 109–122.

Gradstein, F. M., J. G. Ogg, A. G. Smith, F. P. Agterberg, et al., 2004. *A Geological Time Scale 2004.* Cambridge University Press, Cambridge, England.

Graham, J., and E. A. Thompson, 1998. Disequilibrium likelihoods for fine-scale mapping of a rare allele. *Am. J. Hum. Genet.* **63,** 1517–1530.

Graham, J. B., 1994. An evolutionary perspective for bimodal respiration: A biological synthesis of fish air breathing. *Am. Zool.,* **34,** 229–237.

Graham, L. E., 1993. *Origin of Land Plants.* Wiley, New York.

Grande, L., 2004. Categorizing various classes of morphological variation, and the importance of this to vertebrate paleontology. In *Mesozoic Fishes 3 — Systematics, Paleoenvironments and Biodiversity,* G. Arratia and A. Tintori (eds.). Verlag Dr. Friedrich Pfeil, München, Germany, pp.123–136.

Grant, P. R., 1986. *Ecology and Evolution of Darwin's Finches.* Princeton University Press, Princeton, NJ. [Reissued in 1999 with a Foreword by J. Weiner.]

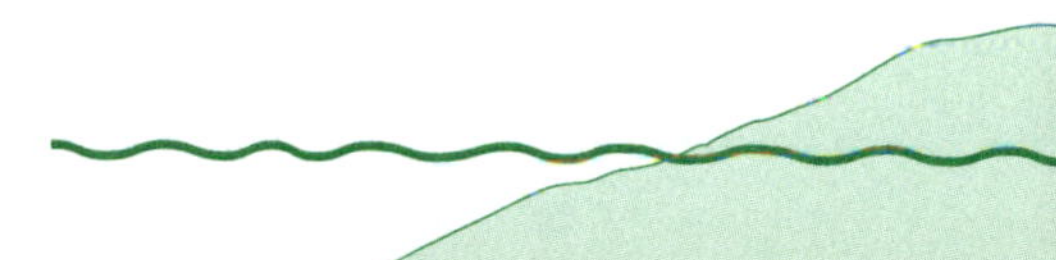

Grant, P. R., and B. R. Grant, 1994. Phenotypic and genetic effects of hybridization in Darwin's finches. *Evolution,* **48,** 297–316.

Grant, P. R., and B. R. Grant, 2002. Unpredictable evolution in a 30-year study of Darwin's finches. *Science,* **296,** 707–711.

Grant, P. R., and B. R. Grant, 2006. Evolution of character displacement in Darwin's finches. *Science,* **313,** 224–226.

Grant, P. R., B. R. Grant, L.F. Keller, J.A. Markert, and K. Petren, 2004. Convergent evolution of Darwin's finches caused by introgressive hybridization and selection. *Evolution* **58,** 1588–1599.

Grant P. R., B. R. Grant, J. N. Smith, I. J. Abbott, and L. K. Abbott, 1976. Darwin's finches: population variation and natural selection. *Proc. Natl. Acad. Sci. U.S.A* **73,** 257–261.

Grant, V., 1985. *The Evolutionary Process.* Columbia University Press, New York.

Grant, V., and K. A. Grant, 1965. *Flower Pollination in the Phlox Family.* Columbia University Press, New York.

Graves, J. A. M., 2006. Sex chromosome specialization and degeneration in mammals. *Cell,* **124,** 901–914.

Gray, M. W., 1993. Origin and evolution of organelle genomes. *Curr. Opin. Genet. Dev.* **3,** 884–890.

Gray, M. W., G. Burger, and B. F. Lang, 1999. Mitochondrial evolution. *Science,* **283,** 1476–1481.

Gray, M. W., G. Burger, and B. F. Lang, 2001. The origin and early evolution of mitochondria. *Genome Biol.,* **2,** 1–5.

Gray, M. W., R. Cedergren, Y. Abel and D. Sankoff, 1989. On the evolutionary origin of the plant mitochondrion and its genome. *Proc. Natl. Acad. Sci. USA,* **86,** 2267–2271.

Gray, M. W., D. Sankoff, and R. J. Cedergren, 1984. On the evolutionary descent of organisms and organelles: A global phylogeny based on a highly conserved structural core in small subunit ribosomal RNA. *Nucl. Acids Res.,* **12,** 5837–5852.

Greene, H. W., 1994. Homology and behavioral repertoires. In *Homology: The Hierarchical Basis of Comparative Biology,* B. K. Hall (ed.). Academic Press, San Diego, CA, pp. 370–391.

Gregory, T. R., 2002. A bird's eye view of the C-value enigma: genome size, cell size, and metabolic rate in the class Aves. *Evolution,* **56,** 121–130.

Gregory, W. K., 1929. *Our Face from Fish to Man. A Portrait Gallery of our Ancient Ancestors and Kinfolk Together with a Concise History of Our Best Features.* G. P. Putnam's Sons, London.

Gregory, W. K. 1951. *Evolution Emerging. A Survey of Changing Patterns from Primeval Life to Man,* 2 vol. The Macmillan Company, New York.

Greilhuber, J., J. Doleel, M. A. Lysák, and M. D. Bennett, 2005. The Origin, Evolution and Proposed Stabilization of the Terms 'Genome Size' and 'C-Value' to Describe Nuclear DNA Contents. *Ann. Bot.,* **95,** 255–260.

Griffiths, M., 1978. *The Biology of the Monotremes.* Academic Press, New York.

Grimaldi, D., and M. S. Engel, 2005. *Evolution of the Insects.* Cambridge University Press, New York.

Grützner, F., W. Rens, E. Tsend-Ayush, N. El-Mogharbel, et al., 2004. In the platypus a meiotic chain of ten sex chromosomes shares genes with the bird Z and mammal X chromosomes. *Nature,* **432,** 913–917.

Gulick, A., 1932. *Evolutionist and Missionary. John Thomas Gulick, Portrayed through Documents and Discussions.* The University of Chicago Press, Chicago.

Gupta, R. S., K. Aitken, M. Falah, and B. Singh, 1994.Cloning of *Giardia lamblia* heat shock protein HSP70 homologs: Implications regarding origin of eukaryotic cells and of endoplasmic reticulum. *Proc. Natl. Acad. Sci. USA,* **91,** 289–2899.

Gurdon, J. B., 1962. The developmental capacity of nuclei taken from intestinal epithelium cells of feeding tadpoles. *J. Embryol. Exp. Morphol.* **10,** 622–640.

H

Hadzi, J., 1963. *The Evolution of the Metazoa.* Macmillan, New York.

Haeckel, E., 1866. *Naturliche Schöpfungsgeschichte.* Reimer, Berlin.

Haeckel, E., 1874. The gastraea-theory, the phylogenetic classification of the animal kingdom and the homology of the germ-lamellae. *Q. J. Microsc. Sci.,* **14,** 142–165, 223–247.

Haeckel, E., 1904. *Kunstformen der Natur.* Bibliographisches Institut, Leipzig and Vienna.

Haeckel, E., 1905. *The Evolution of Man.* Translated from the 5th German ed. by J. McCabe, Watts, London.

Haile-Selassie, Y., 2001. Late Miocene hominids from the Middle Awash, Ethiopia. *Nature,* **412,** 178–181.

Halanych, K.M., 2004. The new animal phylogeny. *Ann. Rev. Ecol. Evol. Syst.,* **35,** 229–256.

Haldane, J. B. S., 1922. Sex ratio and unisexual sterility in hybrid animals. *J. Genet.,* **12,** 101–109.

Haldane, J. B. S., 1932. *The Causes of Evolution.* Harper & Row, New York. [Reprinted 1966, Cornell University Press, Ithaca, NY.]

Haldane, J. B. S., 1939. The theory of the evolution of dominance. *J. Genet.,* **37,** 365–374.

Hall, B. K. (ed.) 1994. *Homology: The Hierarchical Basis of Comparative Biology.* Academic Press, San Diego, CA. [paperback issued 2001]

Hall, B. K., 1995. Homology and embryonic development. *Evol. Biol.,* **28,** 1–37.

Hall, B. K., 1999a. *Evolutionary Developmental Biology,* 2nd ed. Kluwer Academic Publishers, Dordrecht, The Netherlands.

Hall, B. K., 1999b. *The Neural Crest in Development and Evolution.* Springer-Verlag, New York.

Hall, B. K., 2000a. The Evolution of the Neural Crest in Vertebrates. In *Regulatory Processes in Development,* Wenner-Gren International Series Volume 76, L. Olsson and C.-O. Jacobson (eds). *The Portland Press,* London, pp. 101–113.

Hall, B. K., 2000b. The neural crest as a fourth germ layer and vertebrates as quadroblastic not triploblastic. *Evol. Devel.,* **2,** 1–3.

Hall, B. K., 2001. Development of the clavicles in birds and mammals. *J. Exp. Zool.,* **289,** 153–161.

Hall, B. K., 2002. Palaeontology and evolutionary developmental biology: A science of the 19th and 21st centuries. *Palaeontology,* **45,** 647–669.

Hall, B. K., 2003a. Descent with modification: the unity underlying homology and homoplasy as seen through an analysis of development and evolution. *Biol. Rev. Camb. Philos. Soc.* **78,** 409–433.

Hall, B. K., 2003b. Unlocking the black box between genotype and phenotype: Cell condensations as morphogenetic (modular) units. *Biol. Philos.,* **18,** 219–247.

Hall, B. K., 2004. Evolution as the control of development by ecology. In *Environment, Development, and Evolution: Toward a Synthesis,* B. K. Hall, R. D. Pearson, and G. B. Müller (eds.). MIT Press, Cambridge, MA, pp. ix–xxiii.

Hall, B. K., 2005a. *Bones and Cartilage: Developmental and Evolutionary Skeletal Biology.* Elsevier Academic Press, London.

Hall, B. K., 2005b. Fifty years later: I. Michael Lerner's *Genetic Homeostasis* (1954): A valiant attempt to integrate genes, organisms and environment. *J. Exp. Zool. (Mol. Dev. Evol.),* **304B,** 187–197.

Hall, B. K., 2005c. Consideration of the neural crest and its skeletal derivatives in the context of novelty/innovation. *J. Exp. Zool. (Mol. Dev. Evol.),* **304B,** 548–557.

Hall, B. K., 2006a. Evolutionist and missionary: The Reverend John Thomas Gulick (1832–1923). Part I. Cumulative segregation — geographical isolation. *J. Exp. Zool. (Mol. Dev. Evol.),* **306B,** 407–418.

Hall, B. K., 2006b. Evolutionist and missionary: The Reverend John Thomas Gulick (1832–1923). Part 2. Cumulative segregation — geographical isolation. *J. Exp. Zool. (Mol. Dev. Evol.),* **306B,** 485–495.

Hall, B. K., 2006c. Homology and homoplasy. In: *Handbook of the Philosophy of Biology,* D. Gabbay, P. Thagard, and J. Woods (eds.), 2006, Volume 3 (*Philosophy of Biology),* M. Matthen and C. Stephens (eds.). Elsevier BV, Amsterdam, pp. 441–465.

Hall, B. K., 2006d. Tapping many sources: The adventitious roots of evo-devo in the 19th century. In *From Embryology to Evo-Devo: A History of Embryology in the 20th Century* (M. D. Laubichler and J. Maienschein), Symposium of the Dibner Institute for the History of Science, MIT. MIT Press, Cambridge, MA, pp. 467–497.

Hall, B. K., (ed.), 2007a. *Fins into Limbs. Development, Evolution and Transformation.* The University of Chicago Press, Chicago.

Hall, B. K., 2007b. 'Spandrels': Metaphor for morphological residue or entrée into evolutionary developmental mechanisms? In: *The Spandrels of San Marco 25 Years Later,* D. Walsh (ed.). Oxford University Press, Oxford, England (in preparation).

Hall, B. K., and W. M. Olson (eds.), 2003. *Keywords & Concepts in Evolutionary Developmental Biology.* Harvard University Press, Cambridge, MA.

Hall, B. K., R. D. Pearson, and G. B. Müller (eds.), 2004. *Environment, Development, and Evolution: Toward a Synthesis.* MIT Press, Cambridge, MA.

Hall, D. O., J. Lumsden, and E. Tel-Or, 1977. Iron-sulfur proteins and superoxide dismutases in the evolution of photosynthetic bacteria and algae. In *Chemical Evolution of the Early Precambrian,* C. Ponnamperuma (ed.). Academic Press, New York, pp. 191–210.

Hallam, A., 2005. *Catastrophes and Lesser Calamities.* Oxford University Press, Oxford, England.

Hallgrímsson, B., and B. K. Hall (eds.), 2005. *Variation: A Central Concept in Biology.* Elsevier/Academic Press, Burlington, MA.

Hallgrimsson B., K. Willmore, and B. K. Hall, 2002. Canalization, developmental stability, and morphological integration in primate limbs. *Am. J. Phys. Anthropol. (Yearbook)* **S45,** 131–158.

Halvorson, H., and J. Szulmajster, 1973. Differentiation: Sporogenesis and germination. In *Biochemistry of Bacterial Growth,* J. Mandelstam and K. McQuillen (eds.). John Wiley, New York, pp. 494–516.

Hamilton, A. J., and D. C. Baulcombe, 1999. A species of small antisense RNA in posttranscriptional gene silencing in plants. *Science,* **286,** 950–952.

Hamilton, W. D. 1964a. The genetical evolution of social behaviour. I. *J. Theor. Biol.* **7,** 1–16.

Hamilton, W. D. 1964b. The genetical evolution of social behaviour. II. *J. Theor. Biol.* **7,** 17–52.

Hamilton, W. D., R. Axelrod, and R. Tanese, 1990. Sexual reproduction as an adaptation to resist parasites (a review). *Proc. Natl. Acad. Sci. USA,* **87,** 3566–3573.

Hammer, M. F., T. Karafet, A. Rasanayagam, E. T. Wood, et al., 1998. Out of Africa and back again: Nested cladistic analysis of human Y chromosome variation. *Mol. Biol. Evol.,* **15,** 427–441.

Hammerstein, P., 1996. Darwinian adaptation, population genetics and the streetcar theory of evolution. *J. Math. Biol.,* **34,** 511–532.

Han, T. M., and B. Runnegar, 1992. Megascopic algae from the 2.1 billion-year-old Negaunee Iron Formation, Michigan. *Science,* **257,** 232–235.

Hanczyc, M. M., S. M. Fujikawa, and S. W. Szostak, 2003. Experimental models of primitive cellular compartments: encapsulation, growth, and division. *Science,* **302,** 618–622.

Harding, R. S. O., and G. Teleki (eds.), 1981. *Omnivorous Primates: Gathering and Hunting in Human Evolution.* Columbia University Press, New York.

Hardison, R. C., 1996. A brief history of hemoglobins: Plant, animal, protist, and bacteria. *Proc. Natl. Acad. Sci. USA,* **93,** 5675–5679.

Harland, W. B., 1964. Critical evidence for a great infra-Cambrian glaciation. *Int. J. Earth Sci.,* **54,** 45–61.

Harland, W. B., and M. J. S. Rudwick, 1964. The great infra-Cambrian glaciation. *Sci. Am.,* **211,** 28–36.

Harris, E. E., and J. Hey, 1999. X chromosome evidence for ancient human histories. *Proc. Natl. Acad. Sci. USA,* **96,** 3320–3324.

Harris, H., 1966. Enzyme polymorphisms in man. *Proc. Roy. Soc. Lond. (B),* **164,** 298–310.

Hartl, D. L., and A. G. Clark, 1997. *Principles of Population Genetics,* 3rd ed. Sinauer Associates, Sunderland, MA.

Hartl, D. L., and E. W. Jones, 1998. *Genetics: Principles and Analysis.* Jones and Bartlett, Sudbury, MA.

Hartmann, W. K., 1997. A brief history of the moon. *The Planetary Report,* **17,** 4–11.

Hartmann, W. K., and D. R. Davis, 1975. Satellite sized planetesimals and lunar origins. *Icarus,* **24,** 504–515.

Hartsoeker, N., 1694. *Essai de Dioptrique.* Anisson, Paris.

Hartwig, W. C., (ed.), 2002. *The Primate Fossil Record.* Cambridge University Press, Cambridge, England.

Harvati, K., and T. Harrison (eds.), 2006. *Neanderthals Revisited. New Approaches and Perspectives.* Springer, Netherlands.

Harvati, K., and T. D. Weaver, 2006. Human cranial anatomy and the differential preservation of population history and climate signatures. *Anat. Rec., Part A,* **288A,** 1225–1233.

Hästbacka, J., A. de la Chapelle, I. Kairila, P. Sistonen, et al., 1992. Linkage disequilibrium mapping in isolated founder populations: diastrophic dysplasia in Finland. *Nat. Genet.,* **2,** 204–211.

Hawkes, K., J. F. O'Connell, N. G. Blurton Jones, H. Alvarez, and E. L. Charnov, 1998. Grandmothering, menopause, and the evolution of human life histories. *Proc. Natl. Acad. Sci. USA,* **95,** 1336–1339.

Hawkesworth, C. J., and A. I. A. Kemp, 2006. Evolution of the continental crust. *Nature,* **443,** 811–817.

Hawking, S. W., 1988. *A Brief History of Time: From the Big Bang to Black Holes.* Bantam Books, Toronto.

Hawking, S. W., 2001. *The Universe in a Nutshell.* Bantam Books, Toronto.

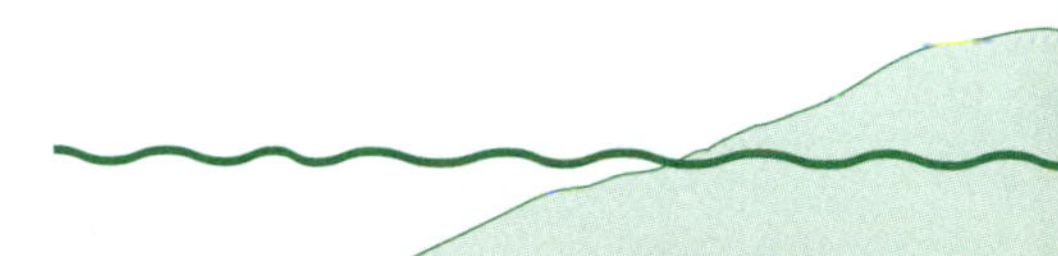

Hawkins, M., 1997. *Social Darwinism in European and American Thought: 1860–1945.* Cambridge University Press, Cambridge, England.

Hay, A., M. Barkoulas, and M. Tsiantis, 2006. Asymmetric leaves and auxin activities converge to repress Brevipedicellus expression and promote leaf development in *Arabidopsis. Development,* **133,** 3955–3961.

Hayes, K. J., and C. Hayes, 1951. The intellectual development of a home-raised chimpanzee. *Proc. Am. Philos. Soc.,* **95,** 105–109.

Hazen, R. M., 2005. *Genesis: The Scientific Quest for Life's Origins.* Joseph Henry Press, Washington, DC.

Hedges, S. B., 1994. Molecular evidence for the origin of birds. *Proc. Natl. Acad. Sci. USA,* **91,** 2621–2624.

Hedrick, P. W., 2000. *Genetics of Populations.* Second Edition. Jones and Bartlett, Boston, MA.

Heilmann, G., 1927. *The Origin of Birds.* Appleton, New York. (Reprinted 1972, Dover Books, New York.

Heinen, W., and A. M. Lauwers, 1996. Organic sulfur compounds resulting from FeS, H_2S, or HCl and CO_2. *Orig. Life Evol. Biosph.,* **26,** 131–150.

Hendrix, R. W., M. C. M. Smith, R. N. Burns, M. E. Ford, and G. F. Hatfull, 1999. Evolutionary relationships among diverse bacteriophages and prophages: All the world's a phage. *Proc. Natl. Acad. Sci. USA,* **96,** 2192–2197.

Henke, W., and I. Tattersall (eds.), 2007. *Handbook of Paleoanthropology. Vol. 1: Principles, Methods and Approaches, Vol 2: Primate Evolution and Human Origins, and Vol 3: Phylogeny of Hominines.* Springer-Verlag, New York.

Hennig, W., 1966. *Phylogenetic Systematics.* University of Illinois Press, Urbana, IL.

Hennig, W., 1975. "Cladistic analysis or cladistic classification?": A reply to Ernst Mayr. *Syst. Zool.,* **24,** 244–256.

Henze, K., A. Badr, M. Wettern, R. Cerff, and W. Martin, 1995. A nuclear gene of eubacterial origin in *Euglena gracilis* reflects cryptic endosymbioses during protist evolution. *Proc. Natl. Acad. Sci. USA,* **92,** 9122–9126.

Herrmann, B., and S. Hummel (eds.), 1994. *Ancient DNA.* Springer-Verlag, New York.

Hervé, P., N. Lartillot, and H. Brinkmann, 2005. Multigene analyses of bilaterian animals corroborate the monophyly of Ecdysozoa, Lophotrochozoa, and Protostomia. *Mol. Biol. Evol.,* **22,** 1246–1253.

Hey, J., 1999. The neutralist, the fly, and the selectionist. *Trends Ecol. Evol.,* **14,** 35–38.

Hey, J., W. M. Fitch, and F. J. Ayala (eds.), 2005. *Systematics and the Origin of Species on Ernst Mayr's 100th Anniversary.* The National Academies Press, Washington, DC.

Hibino, T., A. Nishino, and S. Amemiya, 2006. Phylogenetic correspondence of the body axes in bilaterians is revealed by the right-sided expression of *Pitx* genes in echinoderm larvae. *Dev. Growth Differ.,* **48,** 587–595.

Higuchi, R., B. Bowman, M. Freiberger, O. A. Ryder, and A. C. Wilson, 1984. DNA sequences from the quagga, an extinct member of the horse family. *Nature,* **312,** 282–284.

Hill, R. L., K. Brew, T. C. Vanaman, I. P. Trayer, and J. P. Mattock, 1969. The structure, function, and evolution of a-lactalbumin. *Brookhaven Symp. Biol.,* **21,** 139–152.

Hillenius, W. J., 1994. Turbinates in therapsids: Evidence for Late Permian origins of mammalian endothermy. *Evolution,* **48,** 207–229.

Hillis, D. M., 1994. Homology in molecular biology. In *Homology: The Hierarchical Basis of Comparative Biology,* B. K. Hall (ed.). Academic Press, San Diego, CA, pp. 339–368.

Hinkle, G., J. K. Wetterer, T. R. Schultz, and M. L. Sogin, 1994. Phylogeny of the attine ant fungi based on analysis of small subunit ribosomal RNA gene sequences. *Science,* **266,** 1695–1697.

Hino, O., J. R. Testa, K. H. Buetow, T. Taguchi, et al., 1993. Universal mapping probes and the origin of human chromosome 3. *Proc. Natl. Acad. Sci. USA,* **90,** 730–734.

Hirano, H.-Y., M. Eiguchi, and Y. Sano, 1998. A single base change altered the regulation of the Waxy gene at the posttranscriptional level during the domestication of rice. *Mol. Biol. Evol.,* **15,** 978–987.

Hodgkin, J., 1992. Genetic sex determination mechanisms and evolution. *BioEssays,* **14,** 253–261.

Hoffman, P. F., A. J. Kaufman, G. P. Halverson, and D. P. Schrag, 1998. A Neoproterozoic snowball Earth. *Science,* **281,** 1342–1346.

Hoffman, P. F., and D. P. Schrag. 2000. Snowball Earth. *Sci. Am.,* **282**(1), 68–75.

Holland, H. D., 1997. Evidence for life on Earth more than 3850 million years ago. *Science,* **275,** 38–39.

Holland, L. Z., and N. D. Holland, 1998. Developmental gene expression in amphioxus: New insights into the evolutionary origin of vertebrate brain regions, neural crest, and rostrocaudal segmentation. *Am. Zool.,* **38,** 647–658.

Holland, L. Z., V. Laudet, and M. Schubert, 2004. The chordate amphioxus: an emerging model organism for developmental biology. *Cell Mol. Life Sci.,* **61,** 2290–2308.

Holland, P. W. H., and J. Garcia-Fernàndez, 1996. Hox genes and chordate evolution. *Dev. Biol.,* **173,** 382–395.

Hölldobler, B., and E. O. Wilson, 1990. *The Ants.* Harvard University Press, Cambridge, MA.

Holmquist, G. P., and J. Filipski, 1994. Organization of mutations along the genome: A prime determinant of genome evolution. *Trends Ecol. Evol.,* **9,** 65–69.

Holt R. A., G. M. Subramanian, A. Halpern, G. G. Sutton, et al., 2002. The genome sequence of the malaria mosquito *Anopheles gambiae. Science,* **298,** 129–149.

Hone, D. W. E., T. M. Keesey, D. Pisani, and A. Purvis, 2005. Macroevolutionary trends in the Dinosauria: Cope's rule. *J. Evol. Biol.,* **18,** 587–595.

Hooper, J., 2002. *Of Moths and Men: Intrigue, Tragedy and the Peppered Moth.* Fourth Estate, London.

Horai, S., K. Hayasaka, R. Kondo, K. Tsugane, and N. Takahata, 1995. Recent African origin of modern humans revealed by complete sequences of hominoid mitochondrial DNAs. *Proc. Natl. Acad. Sci. USA,* **92,** 532–536.

Hori, H., and S. Osawa, 1979. Evolutionary change in RNA secondary structure and a phylogenetic tree of 54 5S RNA species. *Proc. Natl. Acad. Sci. USA,* **76,** 381–385.

Horowitz, N. H., 1945. On the evolution of biochemical synthesis. *Proc. Natl. Acad. Sci. USA,* **31,** 153–157.

Hou, X-G., R. J. Aldridge, J. Bergström, D. J. Siveter, and X.-H. Feng, 2004. *The Cambrian Fossils of Chengjiang, China: The Flowering of Early Animal Life.* Blackwell Science, Oxford, England.

Huber, C., and G. Wächtershäuser. 1998. Peptides by activation of amino acids with CO on (Ni, Fe)S surfaces: Implications for the origin of life. *Science,* **281,** 670.

Hughes, C., and R. Eastwood, 2006. Island radiation on a continental scale: Exceptional rates of plant diversification after uplift of the Andes. *Proc. Natl. Acad. Sci. USA,* **103,** 10334–10339.

Hughes, K. A., 1995. The evolutionary genetics of male life-history characters in *Drosophila melanogaster. Evolution,* **49,** 521–537.

Hull, D. L., 1988. *Science As a Process: An Evolutionary Account of the Social and Conceptual Development of Science.* University of Chicago Press, Chicago.

Hunt, G., and K. Roy, 2006. Climate change, body size evolution, and Cope's Rule in deep-sea ostracods. *Proc. Natl. Acad. Sci. USA,* **103,** 1347–1352.

Hunt, G. R., and R. D. Gray, 2003. Diversification and cumulative evolution in New Caledonian crow tool manufacture. *Proc. R. Soc. Lond. B.,* **270,** 867–874.

Hunter, C. P., and C. Kenyon, 1995. Specification of anteroposterior cell fates in *Caenorhabditis elegans* by *Drosophila* Hox proteins. *Nature,* **377,** 229–232.

Hunter, J. P., and J. Jernvall, 1995. The hypocone as a key innovation in mammalian evolution. *Proc. Natl. Acad. Sci. USA,* **92,** 10718–10722.

Hutchinson, G. E., 1978. *An Introduction to Population Ecology.* Yale University Press, New Haven, CT.

Huxley, J., 1942. *Evolution: The Modern Synthesis.* Allen & Unwin, London.

Huxley, J., 1947. The vindication of Darwinism. In *Touchstone for Ethics 1894–1943,* T. H. Huxley and J. Huxley (eds.). Harper & Brothers Publishers, New York and London, pp. 167–192.

Huxley, T. H., 1894. *Evolution and Ethics, and Other Essays.* Macmillan, London.

Huxley, T. H., 1910. *Lectures and Lay Sermons.* With an Introduction by Sir. Oliver Lodge. J. M. Dent & Sons, London.

Huxley, T. H., and Julian Huxley, 1947. *Touchstone for Ethics 1894–1943.* Harper & Brothers, New York and London.

Huysseune, A., J.-Y. Sire, and F. J. Meunier, 1994. Comparative study of lower pharyngeal jaw structure in two phenotypes of *Astatoreochromis alluaudi* (Teleostei: Cichlidae). *J. Morphol.,* **221,** 25–43.

Hyman, L., 1940. *The Invertebrates: Protozoa Through Ctenophora.* McGraw-Hill, New York.

Hyman, L., 1951. *The Invertebrates: Platyhelminthes and Rhynchocoela, the Acoelomate Bilateria.* McGraw-Hill, New York.

I

Imai, K. S., M. Levine, N. Satoh, and Y. Satou, 2006. Regulatory blueprint for a chordate embryo. *Science,* **312,** 1183–1187.

J

Jablonski, D., 1986. Background and mass extinctions: The alternation of macroevolutionary regimes. *Science,* **231,** 129–133.

Jablonski, D., 2000. Micro- and macroevolution: scale and hierarchy in evolutionary biology and paleobiology. In *Deep Time: Paleobiology's Perspective,* D. H. Erwin and S. L. Wing (eds.). *Paleobiology* 26 (Supplement to No. 4), The Paleontological Society, Lawrence, KS, pp. 15–22.

Jablonski, D., 2002. Survival without recovery after mass extinctions. *Proc. Natl. Acad. Sci. USA,* **99,** 8139–8144.

Jablonski, N. G., 2006. *Skin. A Natural History.* University of California Press, Berkeley, CA.

Jacob, F., 1977. Evolution and tinkering. *Science,* **196,** 1161–1166.

Jacob, F., and J. Monod, 1961. Genetic regulatory mechanisms in the synthesis of proteins. *J. Mol. Biol.* **3,** 318–356.

Jacob, T., E. Indriati, R. P. Soejono, K. Hsü, et al., 2006. Pygmoid Australomelanesian *Homo sapiens* skeletal remains from Liang Bua, Flores; population affinities and pathological anomalies. *Proc. Natl. Acad. Sci. USA,* **103,** 13421–13426.

Jaillon, O., J. M. Aury, F. Brunet, J. L. Petit, et al., 2004. Genome duplication in the teleost fish *Tetraodon nigroviridis* reveals the early vertebrate proto-karyotype. *Nature,* **431,** 946–957.

James, T. Y., F. Kauff, C. L. Schoch, P. B. Matheny, et al., 2006. Reconstructing the early evolution of fungi using a six-gene phylogeny. *Nature,* **443,** 818–822.

Janis, C. M., 1993. Tertiary mammal evolution in the context of changing climates, vegetation, and tectonic events. *Ann. Rev. Ecol. Syst.,* **24,** 467–500.

Janke, A., X. Xu, and U. Arnason, 1997. The complete mitochondrial genome of the wallaroo (*Macropis robustus*) and the phylogenetic relationship among Monotremata, Marsupialia, and Eutheria. *Proc. Natl. Acad. Sci. USA,* **94,** 1276–1281.

Januszkiewicz, A. J., and B. W. Robinson, 2007. Divergent walleye (*Sander vitreus*)-mediated inducible defenses in the centrarchid pumpkinseed sunfish (*Lepomis gibbobus*). *Biol. J. Linn. Soc.,* **90,** 25–36.

Jastrow, R., and M. H. Thompson, 1972. *Astronomy: Fundamentals and Frontiers.* Wiley, New York.

Jay, D. G., and W. Gilbert, 1987. Basic protein enhances the incorporation of DNA into lipid vesicles: Model for the formation of primordial cells. *Proc. Natl. Acad. Sci. USA,* **84,** 1978–1980.

Jeffery, W. Y., A. G. Strickler, S. Guiney, D. G. Heyser, and S. I. Tomarev, 2000. *Prox 1* in eye degeneration and sensory organ compensation during development and evolution of the cavefish *Astyanax. Devel. Genes Evol.,* **210,** 223–230.

Jenner, R., 2005. Foiling vertebrate inversion with the humble nemertean. *Palaeont. Assoc. Newsl.,* **58,** 32–39.

Jenner, R., 2006. Challenging received wisdoms: Some contributions of the new microscopy to the new animal phylogeny. *Integr. Comp. Biol.,* **46,** 93–103 (and see the other papers in this issue).

Jeong, S., A. Rokas, and S. B. Carroll, 2006. Regulation of body pigmentation by the abdominal-B Hox protein and its gain and loss in *Drosophila* evolution. *Cell,* **125,** 1387–1399.

Jerison, H. J., 1973. *Evolution of the Brain and Intelligence.* Academic Press, New York.

Jernvall, J. J., P. Hunter, and M. Fortelius, 1996. Molar tooth diversity, disparity, and ecology in Cenozoic ungulate radiations. *Science,* **274,** 1489–1492.

Ji, Q., Z. Luo, and S.-A. Ji, 1999. A Chinese triconodont mammal and mosaic evolution of the mammalian skeleton. *Nature,* **398,** 326–330.

Johannsen, W., 1903. *Über Erblichkeit in Populationen und in reinen Linien.* Fischer, Jena, Germany.

Johanson, D., and M. Edey, 1981. *Lucy: The Beginnings of Mankind.* Simon and Schuster, New York.

Jones, S., R. D. Martin, and D. R. Pilbeam (eds.), 1996. *The Cambridge Encyclopedia of Human Evolution.* Cambridge University Press, Cambridge, England.

Jorde, L. B., M. Bamshad, and A. R. Rogers, 1998. Using mitochondrial and nuclear DNA markers to reconstruct human evolution. *BioEssays,* **20,** 126–136.

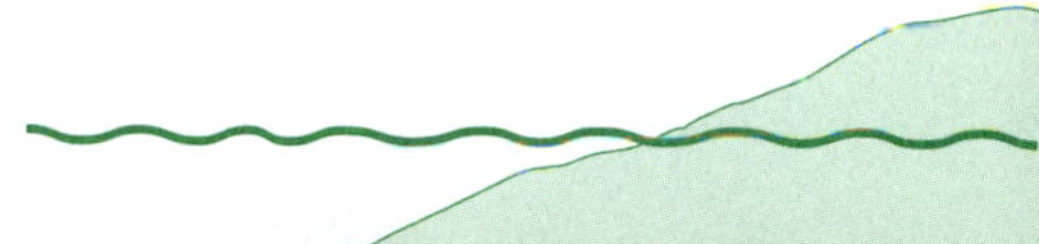

Joyce, G. F., 1989. RNA evolution and the origins of life. *Nature,* **338,** 217–224.

Joyce, G. F., and L. E. Orgel, 1993. Prospects for understanding the origin of the RNA world. In *The RNA World,* R. F. Gesteland and J. F. Atkins (eds.). Cold Spring Harbor Laboratory Press, Cold Spring Harbor, New York, pp. 1–25.

Joyce, G. F., A. W. Schwartz, S. L. Miller, and L. E. Orgel, 1987. The case for an ancestral genetic system involving simple analogues of the nucleotides. *Proc. Natl. Acad. Sci. USA,* **84,** 4398–4402.

Jukes, T. H., 1996. How did the molecular revolution start? What makes evolution happen? In *Evolution and the Molecular Revolution,* C. R. Marshall and J. W. Schopf (eds.). Jones and Bartlett, Sudbury, MA, pp. 31–52.

Jukes, T. H., and S. Osawa, 1991. Recent evidence for evolution of the genetic code. In *Evolution of Life: Fossils, Molecules, and Culture,* S. Osawa and T. Honjo (eds.). Springer, Tokyo, pp. 79–95.

K

Kaler, J. B. 2006. *The Cambridge Encyclopedia of Stars.* Cambridge University Press, Cambridge.

Kappen, C., and F. H. Ruddle, 1993. Evolution of a regulatory gene family: HOM/HOX genes. *Curr. Opin. Genet. Devel.,* **3,** 931–938.

Kappen, C., K. Schughart, and F. H. Ruddle, 1993. Early evolutionary origin of major homeodomain sequence classes. *Genomics,* **18,** 54–70.

Kardong, K. V., 2006. *Vertebrates: Comparative Anatomy, Function, Evolution,* 4th ed. WCB/McGraw-Hill, Boston.

Karn, M. N., and L. S. Penrose, 1951. Birth weight and gestation time in relation to maternal age, parity, and infant survival. *Ann. Eugen.,* **161,** 147–164.

Kasting, J. F., and D. Catling, 2003. Evolution of a heritable planet. *Annu. Rev. Astron. Astrophys.,* **41,** 429–463.

Katz, L. A., 1998. Changing perspectives on the origin of eukaryotes. *Trends Ecol. Evol.,* **13,** 493–497.

Kauffman, S. A., 1993. *The Origins of Order: Self-organization and Selection in Evolution.* Oxford University Press, New York.

Keeling, P. J., G. Burger, D. G. Durnford, B. F. Lang, et al., 2005. The tree of eukaryotes. *Trends Ecol. Evol.,* **20,** 670–676.

Keller, E. F. 1983. *A Feeling for the Organism.* W. H. Freeman and Company, New York.

Keller, E. F., and E. A. Lloyd, 1992. *Keywords in Evolutionary Biology.* Harvard University Press, Cambridge, MA.

Kemp, T. S., 2005. *The Origin and Evolution of Mammals.* Oxford University Press, Oxford and New York.

Kemp, T. S., 2006. The origin and early radiation of the therapsid mammal-like reptiles: a palaeobiological hypothesis. *J. Evol. Biol.,* **19,** 1231–1247.

Kemp, T. S., 2007. The origin of higher taxa: macroevolutionary processes, and the case of the mammals. *Acta Zool. (Stockh.),* **88,** 3–22.

Kenrick, P., and P. Davis, 2004. *Fossil Plants.* Natural History Museum, London.

Kenyon, D. H., and G. Steinman, 1969. *Biochemical Predestination.* McGraw-Hill, New York.

Kerkut, G. A., 1960. *Implications of Evolution.* Pergamon Press, Oxford, England.

Kermack, D. M., and K. A. Kermack, 1984. *The Evolution of Mammalian Characters.* Croom Helm, London.

Kermack, K. A., and F. Mussett, 1983. The ear in mammal-like reptiles and early mammals. *Acta Palaeontol. Pol.,* **28,** 147–158.

Kerr, R. A., 1997. Life goes to extremes in the deep earth — and elsewhere? *Science,* **276,** 703–704.

Kettlewell, H. B. D., 1973. *The Evolution of Melanism.* Clarendon Press, Oxford, England.

Kevles, D. J. 1995. *In the Name of Eugenics: Genetics and the Uses of Human Heredity.* Harvard University Press, Cambridge, MA.

Kidwell, M. G., 1994. The evolutionary history of the P family of transposable elements. *J. Hered.,* **85,** 339–346.

Kielan-Jaworowska, Z., R. L. Cifelli, and Z-.X. Luo, 2004. *Mammals from the Age of Dinosaurs. Origins, Evolution, and Structure.* Columbia University Press, New York.

Kimura, M., 1979. The neutral theory of molecular evolution. *Sci. Am.,* **241,** 94–104.

Kimura, M., 1983. *The Neutral Theory of Molecular Evolution.* Cambridge University Press, Cambridge, England.

Kimura, M., and J. F. Crow, 1964. The number of alleles that can be maintained in a finite population. *Genetics,* **49,** 725–738.

Kimura, M., and T. Ohta, 1972. Population genetics, molecular biometry, and evolution. In *Proceedings of the Sixth Berkeley Symposium on Mathematical Statistics and Probability,* vol. 5. University of California Press, Berkeley, pp. 43–68.

King, J. L., and T. H. Jukes, 1969. Non-Darwinian evolution. *Science,* **164,** 788–798.

King, M., 1993. *Species Evolution: The Role of Chromosome Change.* Cambridge University Press, Cambridge, England.

King, M. C., and A. C. Wilson, 1975. Evolution at two levels: Molecular similarities and biological differences between humans and chimpanzees. *Science,* **188,** 107–116.

Kingdon, J., 2004. *Lowly Origin: Where, When, and Why Our Ancestors First Stood Up.* Princeton University Press, Princeton, NJ.

Kingsland, S. E., 1995. *Modeling nature: Episodes in the History of Population Ecology,* 2nd ed. The University of Chicago Press, Chicago.

Kinzey, W. G., (ed.), 1987. *The Evolution of Human Behavior: Primate Models.* SUNY Press, Albany, NY.

Kirschvink, J. L., 1992. Late Proterozoic low-latitude global glaciation: the snowball Earth. In *The Proterozoic Biosphere,* J. W. Schopf and C. Klein (eds.). Cambridge University Press, Cambridge, England, pp. 51–52.

Kirschvink, J. L., E. J. Gaidos, L. E. Bertani, N. J. Beukes, et al., 2000. Paleoproterozoic Snowball Earth: Extreme climatic and geochemical global change and its biological consequences. *Proc. Natl. Acad. Sci. USA,* **97,** 1400–1405.

Kley, N. J., and M. Kearney, 2007. Adaptations for digging and burrowing. In *Fins into Limbs: Evolution, Development and Transformation,* B. K. Hall (ed.). The University of Chicago Press, Chicago, pp. 284–309.

Knoll, A. H., 1992. The early evolution of eukaryotes: A geological perspective. Science, **256,** 622–627.

Knoll, A. H., 2003. *Life on a Young Planet: The First Three Billion years of Evolution on Earth.* Princeton University Press, Princeton, NJ.

Knowles, L. L., 2004. The burgeoning field of statistical phylogeography. *J. Evol. Biol.,* **17,** 1–10.

Kocher, T. D., J. A. Conroy, K. R. McKaye, and J. R. Stauffer, 1993. Similar morphologies of cichlid fish in Lake Tanganyika and

Lake Malawi are due to convergence. *Mol. Phylog. Evol.,* **2,** 158–165.

Komdeur, J., S. Daan, J. Tinbergen, and C. Mateman, 1997. Extreme adaptive modification in sex ratio of the Seychelles warbler's eggs. *Nature,* **385,** 522–525.

Kondrashov, A. S., 1994. The asexual ploidy cycle and the origin of sex. *Nature,* **370,** 213–216.

Koonin, E. V., A. R. Mushegian, and P. Bork, 1996. Non-orthologous gene displacement. *Trends Genet.,* **12,** 334–336.

Kopp, R.E., J. L. Kirschvink, I. A. Hilburn, and C. Z. Nash, 2005. Was the Paleoproterozoic Snowball Earth a biologically-triggered climate disaster? *Proc. Natl. Acad. Sci. USA,* **102,** 11131–11136.

Kornegay, J. R., J. W. Schilling, and A. C. Wilson, 1994. Molecular adaptation of a leaf-eating bird: Stomach lysozyme of the hoatzin. *Mol. Biol. Evol.,* **11,** 921–928.

Kraft, P. G., C. E. Franklin, and M. W. Blows, 2006. Predator-induced phenotypic plasticity in tadpoles: extension or innovation. *J. Evol. Biol.,* **19,** 450–458.

Kragh, H., 1996. *Cosmology and Controversy: The Historical Development of Two Theories of the Universe.* Princeton University Press, Princeton, NJ.

Krebs, C. J., 2001. *Ecology: The Experimental Analysis of Distribution and Abundance,* 5th ed. Benjamin Cummings, San Francisco, CA.

Krebs, C. J., S. Boutin, R. Boonstra, A. R. E. Sinclair, et al., 1995. Impact of food and predation on the snowshoe hare cycle. *Science,* **269,** 1112–1115.

Krings, M., A. Stone, R. W. Schmitz, H. Kainitzki, et al., 1997. Neanderthal DNA sequences and the origin of modern humans. *Cell,* **90,** 19–30.

Kugelberg, E., E. Kofoid, A. B. Reams, D. I. Andersson, and J. R. Roth, 2006. Multiple pathways of selection gene amplification during adaptive mutation. *Proc. Natl. Acad. Sci. USA,* **103,** 17319–17324.

Kumar, S., and B. Hedges, 1998. A molecular timescale for vertebrate evolution. *Nature,* **392,** 917–920.

Kuo A. D., J. M. Donelan, and A. Ruina, 2005. Energetic consequences of walking like an inverted pendulum: Step-to-step transitions. *Exerc. Sport Sci. Rev.,* **33,** 88–97.

Kuratani, S., 2004. Evolution of the vertebrate jaw: Comparative embryology and molecular developmental biology reveal the factors behind evolutionary novelty. *J. Anat.,* **205,** 335–347.

Kuratani, S., Y. Nobusada, N. Horigome, and Y. Shigetani, 2001. Embryology of the lamprey and evolution of the vertebrate jaw: insights from molecular and developmental perspectives. *Phil. Trans. R. Soc. Lond. B. Biol. Sci.,* **356,** 15–32.

L

Lacalli, T. C., 2004. Evolutionary biology; light on ancient photoreceptors. *Nature,* **432,** 454–455.

Lacalli, T. C., 2006. Prospective protochordate homologs of vertebrate midbrain and MHB, with some thoughts on MHB origins. *Int. J. Biol. Sci.,* **2,** 104–109.

Lack, D., 1947. *Darwin's Finches: An Essay on the General Biological Theory of Evolution.* Cambridge University Press, Cambridge, England.

Lai, L., S. Nevo, and A. G. Steinberg, 1964. Acid phosphatases of human red cells: Predicted phenotype conforms to a genetic hypothesis. *Science,* **145,** 1187–1188.

Lake, J. A., and M. C. Rivera, 1994. Was the nucleus the first endosymbiont? *Proc. Natl. Acad. Sci. USA,* **91,** 2880–2881.

Land, M. F., and D.-E. Nilsson, 2002. *Animal Eyes.* Oxford University Press, Oxford, England.

Lander E. S., L. M. Linton, B. Birren, and C. Nusbaum, 2001. Initial sequencing and analysis of the human genome. *Nature,* **409,** 860–921.

Lane, N., 2005. *Power, Sex, Suicide. Mitochondria and the Meaning of Life.* Oxford University Press, Oxford, England.

Lankester, E. R., 1870. On the use of the term homology in modern zoology, and the distinction between homogenetic and homoplastic agreements. *Ann. Mag. Nat. Hist.,* **6,** Ser. 4, 34–43.

Larralde, R., M. P. Robertson, and S. L. Miller, 1995. Rates of decomposition of ribose and other sugars: Implications for chemical evolution. *Proc. Natl. Acad. Sci. USA,* **92,** 8158–8160.

Larsen, E. W., 1992. Tissue strategies as developmental constraints: Implications for animal evolution. *Trends Ecol. Evol.,* **7,** 414–417.

Laurin, M., M. Girondot, and M.-M. Loth, 2003. On the origin of and phylogenetic relationships among living amphibians. *Proc. Natl. Acad. Sci. USA,* **98,** 7380–7383.

Lawn, R. M., K. Schwartz, and L. Patthy, 1997. Convergent evolution of apolipoprotein (a) in primates and hedgehog. *Proc. Natl. Acad. Sci. USA,* **94,** 11992–11997.

Lawrence, J. G., and H. Ochman, 1998. Molecular archaeology of the *Escherichia coli* genome. *Proc. Natl. Acad. Sci. USA,* **95,** 9413–9417.

LeClerc, J. E., B. Li, W. L. Payne, and T. A. Cebula, 1996. High mutation frequency among *Escherichia coli* and *Salmonella* pathogens. *Science,* **274,** 1208–1211.

LeConte, A., 1888. *Evolution, Its Nature, Its Evidences, and Its Relation to Religious Thought.* Appleton, New York.

Lederberg, J., and E. M. Lederberg, 1952. Replica plating and indirect selection of bacterial mutants. *J. Bact.,* **63,** 399–406.

Lee, D. H., J. R. Granja, J. A. Martinez, K. Severin, and M. R. Ghadri, 1996. A self-replicating peptide. *Nature,* **382,** 525–528.

Lee, M. S. Y., 1999. Molecular phylogenies become functional. *Trends Ecol. Evol.,* **14,** 177–178.

Lee, M. S. Y., 2003. Species concepts and species reality: salvaging a Linnean rank. *J. Evol. Biol.,* **16,** 179–188.

Le Gros Clark, W. E., 1978. *The Fossil Evidence for Human Evolution,* 3rd ed. University of Chicago Press, Chicago.

Lehman, N., and G. F. Joyce, 1993. Evolution in vitro: Analysis of a lineage of ribozymes. *Curr. Biol.,* **3,** 723–734.

Leland, K. N., and G. Brown, 2002. *Sense and Nonsense: Evolutionary Perspectives on Human Behaviour.* Oxford University Press, London.

Lenski, R. E., and M. Travisano, 1994. Dynamics of adaptation and diversification: A 10,000-generation experiment with bacterial populations. *Proc. Natl. Acad. Sci. USA,* **91,** 6808–6814.

Lerner, I. M., 1954. *Genetic Homeostasis.* Oliver and Boyd, Edinburgh, Scotland [Reprinted 1970, Dover Publications, New York.]

Leroi, A. M., 2006. The future of neo-eugenics. Now that many people approve the elimination of certain genetically defective fetuses, is society closer to screening all fetuses for all known mutations? *EMBO Rep.,* **7,** 1184–1187.

Levin, D. A., (ed.), 1979. *Hybridization: An Evolutionary Perspective.* Dowden, Hutchinson and Ross, Stroudsburg, PA.

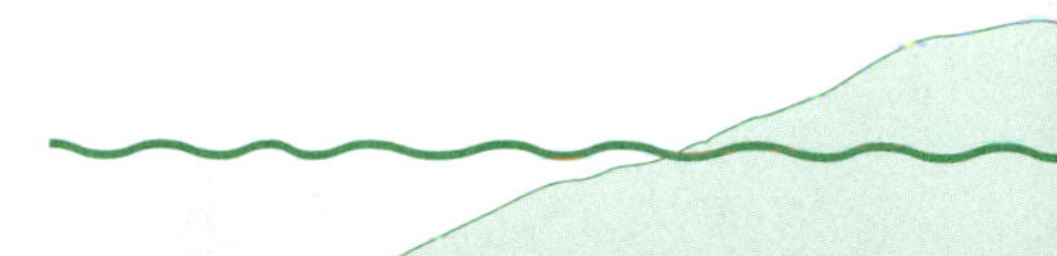

Levine, G., 1988. *Darwin and the Novelists; Patterns of Science in Victorian Fiction.* The University of Chicago Press, Chicago.

Levinton, J., 2001. *Genetics, Paleontology, and Macroevolution,* 2nd ed. Cambridge University Press, Cambridge, England.

Lewis, H., 1973. The origin of diploid neospecies in *Clarkia. Am. Nat.,* **107,** 161–170.

Lewontin, R. C., 1997. *Dobzhansky's Genetics* and *the Origin of Species: Is it still relevant? Genetics,* **147,** 351–355.

Lewontin, R. C., 2000. *The Triple Helix: Gene, Organism, and Environment.* Harvard University Press, Cambridge, MA.

Lewontin, R. C., and J. L. Hubby, 1966. A molecular approach to the study of genic heterozygosity in natural populations. II. Amount of variation and degree of heterozygosity in natural populations of *Drosophila pseudoobscura. Genetics,* **54,** 595–609.

Li, W.-H., 1993. So what about the molecular clock hypothesis? *Curr. Opin. Genet. Dev.,* **3,** 896–901.

Li, W.-H., 1997. *Molecular Evolution.* Sinauer Associates, Sunderland, MA.

Liang, Y. J., and S. C. Zhang, 2006. Demonstration of plasminogen-like protein in amphioxus with implications for the origin of vertebrate liver. *Acta Zool. (Stockholm),* **87,** 141–145.

Lieberman, P., 1991. *Uniquely Human: The Evolution of Speech, Thought, and Selfless Behavior.* Harvard University Press, Cambridge, MA.

Liem, K. F., 1974. Evolutionary strategies and morphological innovations: Cichlid pharyngeal jaws. *Syst. Zool.,* **22,** 425–441.

Liem, K. F., W. E. Bemis, W. F. Walker Jr., and L. Grande, 2001. *Functional Anatomy of the Vertebrates. An Evolutionary Perspective,* 3rd ed. Harcourt College Publishers, Fort Worth TX.

Lineweaver, C. H., and T. M. Davis, 2005. Misconceptions about the Big Bang. *Sci. Am.,* **292,** 24–33.

Lipmann, F., 1971. Attempts to map a process evolution of peptide biosynthesis. *Science,* **173,** 875–884.

Liu, K. W., Z.-J. Liu, L.-Q. Huang, L-Q. Li, L-J. Chen, and G.-D. Tang, 2006. Pollination: Self-fertilization strategy in an orchid. *Nature,* **441,** 945–946.

Loeb, A., 2006. The dark ages of the universe. *Sci. Am.,* **295,** 46–53.

Logsdon, J. M., Jr., A. Stoltzfus, and W. F. Doolittle, 1998. Molecular evolution: Recent cases of spliceosomal intron gain? *Current Biol.,* **8,** R560–R563.

Long, J. A., 1995. *The Rise of Fishes.* Johns Hopkins University Press, Baltimore, MD.

Long, J. A., B. K. Hall, K. J. McNamara, and M. M. Smith, 2006. The phylogenetic origin of jaws in vertebrates: developmental plasticity and heterochrony. *Kirtlandia,* Cleveland Museum of Natural History, Cleveland, OH.

Longair, M. S. 2006. *The Cosmic Century.* Cambridge University Press, Cambridge, United Kingdom.

Loomis, W. F., 1988. *Four Billion Years: An Essay on the Evolution of Genes and Organisms.* Sinauer Associates, Sunderland, MA.

Lopez-Vaamonde, C., N. Wikström, C. Labandeira, H. C. J. Godfray, et al., 2006. Fossil-calibrated molecular phylogenies reveal that leaf-mining moths radiated millions of years after their host plants. *J. Evol. Biol.,* **19,** 1314–1326.

Losos, J. B., T. R. Jackman, A. Larson, K. de Queiroz, and L. Rodriguez-Schettino, 1998. Contingency and determinism in replicated adaptive radiations of island lizards. *Science,* **279,** 2115–2118.

Louis, L. A., and R. M. McCourt, 2004. Green algae and the origin of land plants. *Am. J. Bot.,* **9,** 1535–1556.

Lovejoy, A. O., 1936. *The Great Chain of Being.* Harvard University Press, Cambridge, MA.

Lovejoy, C. O., 1981. The origin of man. *Science,* **211,** 341–350.

Lovejoy, C. O., 1988. The evolution of human walking. *Sci. Am.,* **259,** 118–125.

Lucchesi, J. C., 1994. The evolution of heteromorphic sex chromosomes. *BioEssays,* **16,** 81–83.

Lucchesi, J. C., 1998. Dosage compensation in flies and worms: The ups and downs of X-chromosome regulation. *Curr. Opin. Genet. Dev.,* **8,** 179–184.

Lutz, P. E., 1986. *Invertebrate Zoology.* Addison-Wesley, Reading, MA.

Lyell, C., 1830. *Principles of Geology, Being an Attempt to Explain the Former Changes of the Earth's Surface by References to Causes Now in Operation.* J. Murray, London. [Many subsequent editions, for example (i), 1990, with an introduction by M. J. S. Rudwick, University of Chicago Press, 3 volumes; (ii), 1997, edited with an introduction by J. A. Secord, Penguin Books].

Lynch, M., and B. Walsh, 1998. *Genetics and Analysis of Quantitative Traits.* Sinauer Associates, Sunderland, MA.

Lynch, V. J., and G. P. Wagner, 2006. The birth of the uterus. *Nat. Hist.,* **114,** 36–41.

M

Mable, B. K., and S. P. Otto, 1998. The evolution of lifecycles with haploid and diploid phases. *BioEssays,* **20,** 453–462.

MacFadden, B. J., 1992. *Fossil Horses: Systematics, Paleobiology, and Evolution of the Family Equidae.* Cambridge University Press, Cambridge, England.

MacFadden, B. J., N. Solounias, and T. E. Cerling, 1999. Ancient diets, ecology and extinction of 5-million-year-old horses from Florida. *Science,* **283,** 824–827.

Mackay, T. F. C., 1996. The nature of quantitative genetic variation revisited: Lessons from *Drosophila* bristles. *BioEssays,* **18,** 113–121.

Mackay, T. F. C., 2004. The genetic architecture of quantitative traits: lessons from *Drosophila. Curr. Opin. Genet. Dev.,* **14,** 253–257.

Mackay, T. F. C., and R. F. Lyman, 2005. *Drosophila* bristles and the nature of quantitative genetic variation. *Proc. Roy. Soc. Lond. (B),* **360,** 1513–1527.

Macleod, N., 2004. The extinction of all life and the sublime attraction of neocatastrophism. *Palaeont. Assoc. Newsl.,* **54,** 49–64.

Maienschein, J., 2003. *Whose View of Life? Embryos, Cloning, and Stem Cells,* Harvard University Press, Cambridge, MA.

Maisey, J. G., 1996. *Discovering Fossil Fish.* Henry Holt and Co., New York.

Maizels, N., and A. M. Weiner, 1994. Phylogeny from function: Evidence from the molecular fossil record that tRNA originated in replication, not translation. *Proc. Natl. Acad. Sci. USA,* **91,** 6729–6734.

Majerus, M. E. N., 1998. *Melanism: Evolution in Action.* Oxford University Press, Oxford, England.

Maley, L. E., and C. R. Marshall, 1998. The coming age of molecular systematics. *Science,* **279,** 505–506.

Mallatt, J., J.-Y. Chen, and N. D. Holland, 2003. Comment on "A new species of *Yunnanozoan* with implications for deuterostome evolution." *Science,* **300,** 1372c.

Malthus, T. R., 1798. An Essay on the Principle of Population. Reprinted in *A Norton Critical Edition. Darwin: Texts, Commentary,* 3rd ed, Selected and Edited by Philip Appleman, 2001. W. W. Norton & Company, New York, pp. 39–40.

Mann, W. M. 1938. Monkey folk. *Nat. Geograph.,* **73,** 615–655.

Manton, S. M., 1977. *The Arthropoda: Habits, Functional Morphology, and Evolution.* Clarendon Press, Oxford, England.

Marden, J. H., and M. G. Kramer, 1994. Surface-skimming stoneflies: A possible intermediate stage in insect flight evolution. *Science,* **266,** 427–430.

Marelli, A. J., A. A. Mackie, R. Ionescu-Ittu, E. Rahme, and L. Pilote, 2007. Congenital heart disease in the general population: Changing prevalence and age distribution. *Circulation,* **115,** 163–172.

Margulis, L., 1993. *Symbiosis in Cell Evolution: Microbial Communities in the Archean and Proterozoic Eons,* 2nd ed. Freeman, New York.

Margulis, L., 1998. *Symbiotic Planet: A New Look at Evolution.* Basic Books, New York.

Margulis, L., M. Chapman, R. Guerrero, and J. Hall, 2006. The last eukaryotic common ancestor (LECA): Acquisition of cytoskeletal motility from aerotolerant spirochetes in the Proterozoic Eon. *Proc. Natl. Acad. Sci. USA,* **103,** 13080–13085.

Margulis, L., and K. V. Schwartz, 1998. *Five Kingdoms. An Illustrated Guide to the Phyla of Life on Earth,* 3rd ed. W. H. Freeman and Co., New York. [1st edition, 1982].

Martin, R. D., 1990. *Primate Origins and Evolution: A Phylogenetic Reconstruction.* Chapman & Hall, London.

Martin, R. D., A. M. MacLarnon, J. L. Phillips, and W. B. Dobyns, 2006. Flores hominid: new species of microcephalic dwarf? *Anat. Rec. Part A* **288A,** 1123–1145.

Martin, T. E., 1996. Fitness costs of resource overlap among coexisting bird species. *Nature,* **380,** 338–340.

Martin, T. E., P. R. Martin, C. R. Olson, B. J. Heidinger, and J. J. Fontaine, 2000. Parental care and clutch sizes in North and South American birds. *Science,* **287,** 1482–1485.

Martin, W., H. Brinkmann, C. Savonna, and R. Cerff, 1993. Evidence for a chimeric nature of nuclear genomes: Eubacterial origin of eukaryotic glyceraldehyde-3-phosphate dehydrogenase genes. *Proc. Natl. Acad. Sci. USA,* **90,** 8692–8696.

Martin, W., and M. Müller, 1998. The hydrogen hypothesis for the first eukaryote. *Nature,* **392,** 37–41.

Martindale, M. Q., J. R. Finnerty, and J. Q. Henry, 2002. The Radiata and the evolutionary origins of the bilaterian body plan. *Mol. Phylogenet. Evol.,* **24,** 358–365.

Martindale, M. Q., and J. Q. Henry, 1999. Intracellular fate mapping in a basal metazoan, the ctenophore *Mnemiopsis leidyi,* reveals the origins of mesoderm and the existence of indeterminate cell lineages. *Dev. Biol.,* **214,** 243–257.

Martindale, M. Q., and M. J. Kourakis, 1999. Size doesn't matter. *Nature,* **399,** 730–731.

Martindale, M. Q., K. Pang, and J. R. Finnerty, 2004. Investigating the origins of triploblasty: 'mesodermal' gene expression in a diploblastic animal, the sea anemone *Nematostella vectensis* (phylum, Cnidaria; class, Anthozoa). *Development,* **131,** 2463–2474.

Martinez, P., J. P. Rast, C. Arena-Mena, and E. H. Davidson, 1999. Organization of an echinoderm Hox gene cluster. *Proc. Natl. Acad. Sci. USA,* **96,** 1469–1474.

Matsunaga, E., and S. Itoh, 1958. Blood groups and fertility in a Japanese population, with special reference to intrauterine selection due to maternal-fetal incompatibility. *Ann. Hum. Genet.,* **22,** 111–131.

Mattick, J., 1994. Intron evolution and function. *Curr. Opin. Genet. Dev.,* **4,** 823–831.

Matus, D. Q., K. Pang, H. Marlow, C. W. Dunn, et al., 2006. Molecular evidence for deep evolutionary roots of bilaterality in animal development. *Proc. Natl. Acad. Sci. USA,* **103,** 11195–11200.

Mauseth, J. D., 1998. *Botany: An Introduction to Plant Biology,* 2nd ed. Jones and Bartlett, Sudbury, MA.

May, R. M., 1987. More evolution of cooperation. *Nature,* **327,** 15–17.

Maynard Smith, J., 1978. *The Evolution of Sex.* Cambridge University Press, Cambridge, England.

Maynard Smith, J., 1982. *Evolution and the Theory of Games.* Cambridge University Press, Cambridge, England.

Maynard Smith, J., 1983. Game theory and the evolution of cooperation. In *Evolution from Molecules to Man,* D. S. Bendall (ed.). Cambridge University Press, Cambridge, England, pp. 445–456.

Maynard Smith, J., 1988a. The evolution of recombination. In *The Evolution of Sex,* R. E. Michod and B. R. Levin (eds.). Sinauer Associates, Sunderland, MA, pp. 106–125.

Maynard Smith, J., 1988b. *Toward a New Philosophy of Biology. Observations of an Evolutionist.* The Belknap Press of Harvard University Press, Cambridge, MA.

Maynard Smith, J., 1998. *Evolutionary Genetics.* Second edition. Oxford University Press, Oxford, England.

Maynard Smith, J., and E. Szathmáry, 1995. *The Major Transitions in Evolution.* Freeman, Oxford, England.

Mayr, E., 1942. *Systematics and the Origin of Species.* Columbia University Press, New York.

Mayr, E., 1963. *Animal Species and Evolution.* Harvard University Press, Cambridge, MA.

Mayr, E., 1969a. *Principles of Systematic Zoology.* McGraw-Hill, New York.

Mayr, E., 1969b. The biological meaning of species. *Biol. J. Linn. Soc.,* **1,** 311–320.

Mayr, E., 1981. Biological classification: Toward a synthesis of methodologies. *Science,* **214,** 510–516.

Mayr, E., 1982. *The Growth of Biological Thought. Diversity, Evolution, and Inheritance.* Harvard University Press, Cambridge, MA.

Mayr, E., 1983. How to carry out the adaptationist program. *Am. Nat.,* **121,** 324–334.

Mayr, E., 1987. The ethological status of species: Scientific progress and philosophical terminology. *Biol. Philos.,* **2,** 145–166.

Mayr, E., 1988. *Toward a New Philosophy of Biology. Observations of an Evolutionist.* The Belknap Press of Harvard University Press, Cambridge, MA.

Mayr, E., 1995. Systems of ordering data. *Biol. Philos.,* **10,** 419–434.

Mayr, E., 2001. *What Evolution Is.* With a Foreword by Jared Diamond. Basic Books, New York.

Mayr, E., 2005. Foreword. In *Variation: A central Concept in Biology,* B. Hallgrímsson and B. K. Hall (eds.). Elsevier/Academic Press, Burlington, MA, pp. xvii.

Mazet, F., and S. M. Shimeld, 2005. Molecular evidence from ascidians for the evolutionary origin of vertebrate cranial sensory placodes. *J. Exp. Zool. (Mol. Dev. Evol.),* **304B,** 340–346.

McCarthy, B. J., and M. N. Farquhar, 1972. The rate of change of DNA in evolution. *Brookhaven Symp. Biol.,* **23,** 1–41.

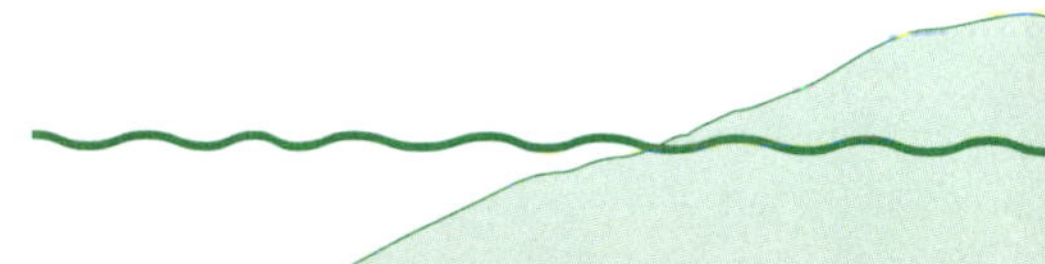

McCauley, D. W., and M. Bronner-Fraser, 2006. Importance of SoxE in neural crest development and the evolution of the pharynx. *Nature,* **441,** 750–752.

McClintock, B., 1950. The origin and behavior of mutable loci in maize. *Proc. Natl. Acad. Sci. USA,* **36,** 344–355.

McClintock, B., 1961. Some parallels between gene control systems in maize and in bacteria. *Am. Nat.* **95,** 265–277.

McClure, M. A., T. K. Vasi, and W. M. Fitch, 1994. Comparative analysis of multiple protein-sequence alignment methods. *Mol. Biol. Evol.,* **11,** 571–592.

McCollum, M. A., 1999. The robust australopithecine face: A morphogenetic perspective. *Science,* **284,** 301–305.

McGinnis, W., M. S. Levine, E. Hafen, A. Kuroiwa, and W. J. Gehring, 1984. A conserved DNA sequence in homeotic genes of the *Drosophila antennapedia* and *bithorax* complexes. *Nature,* **308,** 428–433.

McHenry, H. M., 1994. Tempo and mode in human evolution. *Proc. Natl. Acad. Sci. USA,* **91,** 6780–6786.

McKinney, M. L., and K. J. McNamara, 1991. *Heterochrony: The Evolution of Ontogeny.* Plenum Press, New York.

McNab, B. K., 2002. *The Physiological Ecology of Vertebrates: A View from Energetics.* Comstock Publishing Associates of Cornell University Press, Ithaca, NY.

McOuat, G., 1996. Species, Rules, and Meaning: The politics of language and the ends of definitions in nineteenth-century natural history. *Stud. Hist. Philos. Sci.,* **27,** 473–519.

McOuat, G., 2001. From cutting nature at its joints to measuring it: New kinds and new kinds of people in biology. *Stud. Hist. Philos. Sci.,* **32,** 613–645.

McShea, D. W., 2005. A universal generative tendency toward increased organismal complexity. In *Variation. A Central Concept in Biology,* B. Hallgrímsson and B. K. Hall (eds.). Elsevier/Academic Press, New York, pp. 435–453.

Medina, M., 2005. Genomes, phylogeny, and evolutionary systems biology. *Proc. Natl. Acad. Sci. USA,* **102** (Suppl 1), 6630–6635.

Mendel, G., 1866. *Versuch über Pflanzen-Hybriden.* (This is Mendel's classic paper, originally published in the *Proceedings of the Brünn Natural History Society.* It has been translated into English and reprinted under the title *Experiments in Plant Hybridization.*)

Mendelson, C. V., 1993. Acritarchs and prasinophytes. In *Fossil Prokaryotes and Protists,* J. H. Lipps (ed.). Blackwell Scientific, Boston, pp. 77–104.

Metschnikoff, E., 1884. Researches on the intracellular digestion of invertebrates. *Q. J. Microscop. Sci.,* **24,** 89–111.

Meyer, A., 1990. Ecological and evolutionary consequences of the trophic polymorphism in *Cichlasoma citrinellum* (Pisces: Cichlidae). *Biol. J. Linn. Soc.,* **39,** 279–299.

Meyer, A., T. D. Kocher, P. Basasibwaki, and A. C. Wilson, 1990. Monophyletic origin of Lake Victoria cichlid fishes suggested by mitochondrial DNA sequences. *Nature,* **347,** 550–553.

Miao, D., 1991. On the origins of mammals. In *Origins of the Higher Groups of Tetrapods,* H.-P. Schultze and L. Trueb (eds.). Cornell University Press, Ithaca, New York, pp. 579–597.

Michener, C. D., 1970. Diverse approaches to systematics. *Evol. Biol.,* **4,** 1–38.

Michener, C. D., 1977. Discordant evolution and the classification of allodapine bees. *Syst. Zool.,* **26,** 32–56.

Michod, R. E., and B. R. Levin (eds.), 1988. *The Evolution of Sex.* Sinauer Associates, Sunderland, MA.

Miele, E. A., D. R. Mills, and F. R. Kramer, 1983. Autocatalytic replication of a recombinant RNA. *J. Mol. Biol.,* **171,** 281–295.

Milkman, R. D., 1973. Electrophoretic variation in *Escherichia coli* from natural sources. *Science,* **182,** 1024–1026.

Millar, R., 1998. *The Piltdown Mystery: The Story Behind the World's Greatest Archaeological Hoax.* S. B. Publications, Seaford, East Sussex.

Miller, E. R., G. F. Gunnell, and R. D. Martin, 2005. Deep time and the search for anthropoid origins. *Am. J. Phys. Anthropol. Suppl.,* **41,** 60–95.

Miller, J. R., 1990. *X-Linked Traits: A Catalog of Loci in Nonhuman Animals.* Cambridge University Press, Cambridge, England.

Miller, S. L., 1953. A production of amino acids under possible primitive Earth conditions. *Science,* **117,** 528–529.

Miller, S. L., 1992. The prebiotic synthesis of organic compounds as a step toward the origin of life. In *Major Events in the History of Life,* J. W. Schopf (ed.). Jones and Bartlett, Boston, pp. 1–28.

Miller, S. L., and L. E. Orgel, 1974. *The Origins of Life on the Earth.* Prentice Hall, Englewood Cliffs, NJ.

Mills, D. R., F. R. Kramer, and S. Spiegelman, 1973. Complete nucleotide sequence of a replicating RNA molecule. *Science,* **180,** 916–927.

Milner, R., 1990. *The Encyclopedia of Evolution: Humanity's Search for Its Origins.* Facts On File, New York.

Milsom, C., and S. Rigby, 2003. *Fossils at a Glance.* Blackwell Science, Oxford, England.

Mindell, D. P., 2006. *The Evolving World: Evolution in Everyday Life.* Harvard University Press, Cambridge, MA.

Minelli, A., 1993. *Biological Systematics: The State of the Art.* Chapman & Hall, London.

Mirsky, A. E., and H. Ris, 1951. The deoxyribonucleic acid content of animal cells and its evolutionary significance. *J. Gen. Physiol.,* **34,** 451–462.

Mita K., M. Kasahara, S. Sasaki, Y. Nagayasu, 2004. The genome sequence of silkworm, *Bombyx mori. DNA Res.,* **11,** 27–35.

Mitchell, S., 2003. *Biological Complexity and Integrative Pluralism.* Cambridge University Press, Cambridge, England.

Mithen, S., 2006. *The Singing Neanderthals: The Origins of Music, Language, Mind, and Body.* Harvard University Press, Cambridge, MA.

Mivart, St. G. J., 1871. *On The Genesis of Species.* Appleton, New York.

Mizuuchi, K., 1992. Transpositional recombination: Mechanistic insights from studies of Mm and other elements. *Ann. Rev. Biochem.,* **61,** 1011–1051.

Moen, D. S., 2006. Cope's rule in cryptodiran turtles: do the body sizes of extant species reflect a trend of phyletic size increase? *J. Evol. Biol.,* **19,** 1210–1221.

Møller, A. P., 1996. Sexual selection, viability selection and developmental stability in the domestic fly *Musca domestica. Evolution,* **50,** 746–752.

Møller, A. P., and T. Szép, 2005. Rapid evolutionary change in a secondary sexual character linked to climate change. *J. Evol. Biol.,* **18,** 481–495.

Monnard, P.-A, C. L. Apel, A. Kanavarioti, and D. W. Deamer, 2002. Influence of ionic solutes on self-assembly and polymerization processes related to early forms of life: Implications for a prebiotic aqueous medium. *Astrobiology* **2,** 213–219.

Monod, J., 1971. *Chance and Necessity.* Knopf, New York.

Moore, J. A., 1949. Patterns of evolution in the genus *Rana.* In *Genetics, Paleontology, and Evolution,* G. L. Jepsen, E. Mayr, and

G. G. Simpson (eds.). Princeton University Press, Princeton, NJ, pp. 315–355.
Moore, K. L., and T. V. N. Persaud, 2003. *The Developing Human,* 7th ed. *Clinically Oriented Embryology.* Saunders/Elsevier, New York.
Morden, C. W., and A. R. Sherwood, 2002. Continued evolutionary surprises among dinoflagellates. *Proc. Natl. Acad. Sci. USA,* **99,** 11558–11560.
Moriyama, E. N., and J. R. Powell, 1996. Intraspecific nuclear DNA variation in *Drosophila. Mol. Biol. Evol.,* **13,** 261–277.
Morowitz, H. J., 1992. *Beginning of Cellular Life: Metabolism Recapitulates Biogenesis.* Yale University Press, New Haven, CT.
Morowitz, H. J., 2002. *The Emergence of Everything: How the World Became Complex.* Oxford University Press, New York.
Morton, L. E., 2005. Linkage disequilibrium maps and association mapping. *J. Clin. Invest.* **115,** 1425–1430.
Mountain, J. L., 1998. Molecular evolution and modern human origins. *Evol. Anthropol.,* **7,** 21–37.
Mountain, J. L., A. A. Lin, A. M. Bowcock, and L. L. Cavalli-Sforza, 1992. Evolution of modern humans: Evidence from nuclear polymorphisms. *Phil. Trans. Roy. Soc. London (Biol.),* **337,** 159–165.
Moya, A., A. Galiana, and F. J. Ayala, 1995. Founder-effect speciation theory: Failure of experimental corroboration. *Proc. Natl. Acad. Sci. USA,* **92,** 3983–3986.
Muchhala, N., 2006. Nectar bat stows huge tongue in its rib cage. *Nature,* **444,** 701–702.
Mueller, L. D., P. Guo, and F. J. Ayala, 1991. Density-dependent natural selection and trade-offs in life history traits. *Science,* **253,** 433–435.
Mulkidjanian, A. Y., E. V. Koonin, K. S. Makarova, S. L. Mekhedov, et al., 2006. The cyanobacterial genome core and the origin of photosynthesis. *Proc. Natl. Acad. Sci. USA,* **103,** 13126–13131.
Müller, G. B., J. Streicher, and R. J. Müller, 1996. Homeotic duplication of the pelvic body segment in regenerating tadpole tails induced by retinoic acid. *Devel. Genet. Evol.,* **206,** 344–348.
Muller, H. J., 1932. Some genetic aspects of sex. *Amer. Nat.,* **66,** 118–138.
Müller, J., and R. R. Reisz, 2005. Four well-constrained calibration points from the vertebrate fossil record for molecular clock estimates. *BioEssays,* **27,** 1069–1075.
Müller-Wille, S., 2003. Nature as a marketplace: The political economy of Linnaean botany. *Hist. Polit. Econ.* **35** (Suppl.), 154–172.
Mullis, K. B., F. Ferré, and R. A. Gibbs (eds.), 1994. *PCR: The Polymerase Chain Reaction.* Birkhäuser, Boston.
Murdin, P., (ed.), 2001. *Encyclopedia of Astronomy and Astrophysics* (Vols. 1–4). Institute of Physics, Bristol; Nature, London.
Murray, P. F., and P. Vickers-Rich, 2004. *Magnificent Mihirungs. The Colossal Flightless Birds of the Australian Dreamtime.* Indiana University Press, Bloomington, IN.

N

Nadeau, J. H., and D. Sankoff, 1997. Comparable rates of gene loss and functional divergence after genome duplications early in vertebrate evolution. *Genetics,* **147,** 1259–1266.
Nahigian, K. E., 1997. Impressions: An evening with Dr. Hugh Ross. *NCSE Reports,* **17,** 27–29.
Narbonne, G. M., 2004. Modular construction of early Ediacaran complex life forms. *Science,* **305,** 1141–1144.
Narbonne, G. M. and J. G. Gehling, 2003. Life after Snowball: the oldest complex Ediacaran fossils. *Geology,* **31,** 27–30.
Narlikar, J. V., 2002. *Introduction to Cosmology,* 3rd ed. Cambridge University Press, New York.
Neal, H. V., and H. W. Rand, 1939. *Comparative Anatomy.* Blakiston, Philadelphia, PA.
Nedelcu, A. M., T. Borza, and R. W. Lee, 2006. A land plant-specific multigene family in the unicellular *Mesostigma* argues for its close relationship to Streptophyta. *Mol. Biol. Evol.,* **23,** 1011–1015.
Nei, M., 1987. *Molecular Evolutionary Genetics.* Columbia University Press, New York.
Nei, M., 2005. Selectionism and neutralism in molecular evolution. *Mol. Biol. Evol.,* **22,** 2318–2342.
Nei, M., and A. K. Roychoudhury, 1993. Evolutionary relationships of human populations on a global scale. *Mol. Biol. Evol.,* **10,** 927–943.
Neidert, A. H., V. Virupannavar, G. W. Hooker, and J. A. Langeland, 2001. Lamprey *Dlx* genes and early vertebrate evolution. *Proc. Natl. Acad. Sci. USA,* **98,** 1665–1670.
Neisser, U., (ed.), 1998. *The Rising Curve: Long-Term Gains in IQ and Related Measures.* American Psychological Association, Washington, DC.
Nelkin, D., 1982. *The Creation Controversy; Science or Scripture in the Schools.* Norton, New York.
Nevo, E., 1978. Genetic variation in natural populations: Patterns and theory. *Theor. Popul. Biol.,* **13,** 121–177.
Nevo, E., 1983. Population genetics and ecology. In *Evolution from Molecules to Men,* D. S. Bendall (ed.). Cambridge University Press, Cambridge, England, pp. 287–321.
Nevo, E., M. Filippucci, C. Redi, A. Korol, and A. Beiles, 1994. Chromosomal speciation and adaptive radiation of mole rats in Asia Minor correlated with increased ecological stress. *Proc. Natl. Acad. Sci. USA,* **91,** 8160–8164.
Newell, N. D., 1959. The nature of fossil record. *Proc. Am. Phil. Soc.,* **103,** 264–285.
Newman, S. A., 2003. Hierarchy. In *Keywords and Concepts in Evolutionary Developmental Biology,* B. K. Hall and W. M. Olson (eds.). Harvard University Press, Cambridge, MA, pp. 169–174.
Ngai, J., M. M. Dowling, L. Buck, R. Axel, and A. Chess, 1993. The family of genes encoding odorant receptors in channel catfish. *Cell,* **72,** 657–666.
Nichols, S. A., W. Dirks, J. S. Pearse, and N. King, 2006. Early evolution of animal cell signalling and adhesion genes. *Proc. Natl. Acad. Sci. USA,* **103,** 12451–12456.
Nielsen, C., 2001. *Animal Evolution: Interrelationships of the Living Phyla,* 2nd ed. Oxford University Press, Oxford, England.
Nielsen, C., 2003. Proposing a solution to the Articulata-Ecdysozoa controversy. *Zool. Scripta,* **32,** 475–482.
Nijhout, H. F., and D. J. Emlen, 1998. Competition among body parts in the development and evolution of insect morphology. *Proc. Natl. Acad. Sci. USA,* **95,** 3685–3689.
Niklas, K. J., 1997. *The Evolutionary Biology of Plants.* University of Chicago Press, Chicago.
Niklas, K. J., 2003. Evolution of plant body plans and allometry. In *Keywords and Concepts in Evolutionary Developmental Biology,* B. K. Hall and W. M. Olson (eds.). Harvard University Press, Cambridge, MA. pp. 124–132.

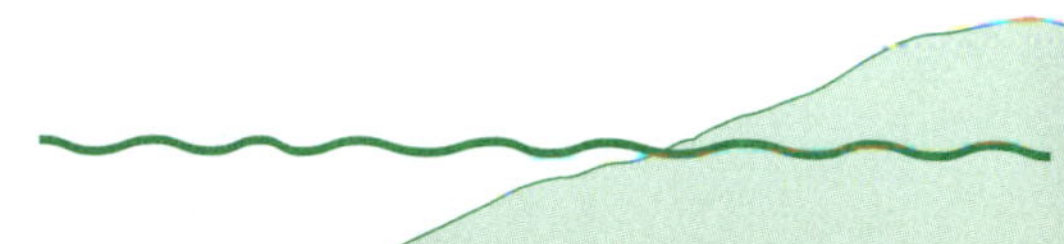

Niklas, K. J., 2004. The cell walls that bind the tree of life. *BioScience,* **54,** 831–841.

Nilsson, D.-E., 1996. Eye ancestry: Old genes for new eyes. *Curr. Biol.,* **6,** 39–42.

Nilsson, D.-E., and S. Pelger, 1994. A pessimistic estimate of the time required for an eye to evolve. *Proc. R. Soc. Lond. (B),* **256,** 53–58.

Nilsson, L. A., 1998. Deep flowers for long tongues. *Trends Ecol. Evol.,* **13,** 259–260.

Nomura, M., 1973. Assembly of bacterial ribosomes. *Science,* **179,** 864–873.

Noor, M. A., 1995. Speciation driven by natural selection in *Drosophila. Nature,* **375,** 674–675.

Northcutt, R. G., 2001. Changing views on brain evolution. *Brain Res. Bull.,* **55,** 663–674.

Northcutt, R. G., 2004. Taste buds: Development and evolution. *Brain Behav. Evol.,* **64,** 198–206.

Northcutt, R. G., 2005. The new head hypothesis revisited. *J. Exp. Zool. (Mol. Devel. Evol.),* **304B,** 274–297.

Northcutt, R. G., and C. Gans, 1983. The genesis of neural crest and epidermal placodes: A reinterpretation of vertebrate origins. *Q. Rev. Biol.,* **58,** 1–28.

Northcutt, R. G., and M. S. Northcutt, (eds.), 2004. *The Development of Vertebrate Sensory Organs.* Special issue of *Brain, Behaviour and Evolution* **66**(3); S. Karger AG, Basel, Switzerland.

Norton, O. R., 2002. *The Cambridge Encyclopedia of Meteorites.* Cambridge University Press, Cambridge, England.

Novacek, M. J., A. R. Wyss, and M. C. McKenna, 1988. The major groups of eutherian mammals. In *The Phylogeny and Classification of the Tetrapods,* vol. 2, M. J. Benton (ed.). Oxford University Press, Oxford, England, pp. 31–71.

Nowak, M. A., 2006. *Evolutionary Dynamics.* Harvard University Press, Cambridge, MA.

Nowak, M. A., and K. Sigmund, 1998. Evolution of indirect reciprocity by image scoring. *Nature,* **393,** 573–577.

Nowak, M. A., and K. Sigmund, 2005. Evolution of indirect reciprocity. *Nature,* **437,** 1291–1298.

Nowak, R., 1994. Mining treasures from junk DNA. *Science,* **263,** 608–610.

Nuismer, S. L., J. N. Thompson, and R. Gomulkiewicz, 2003. Coevolution between hosts and parasites with partially overlapping geographic ranges. *J. Evol. Biol.,* **16,** 1337–1345.

Nyhart, L. K., 1995. *Biology Takes Form. Animal Morphology and the German Universities, 1800–1900.* The University of Chicago Press, Chicago.

O

Odling-Smee, F. J., K. N. Laland, and M. W. Feldman, 2003. *Niche Construction. The Neglected Process in Evolution.* Princeton University Press, Princeton, NJ.

Ogasawara, M., H. Wada, H. Peters, and N. Satoh, 1999. Developmental expression of *Pax1/9* genes in urochordate and hemichordate gills: Insight into function and evolution of the pharyngeal epithelium. *Development,* **126,** 2539–2550.

Ohno, S., 1970, 1979. *Major Sex Determining Genes.* Springer-Verlag, Berlin.

Ohno, S., 1996. The notion of the Cambrian pananimalia genome. *Proc. Natl. Acad. Sci. USA,* **93,** 8475–8478.

Oldroyd, D. R., 1980. *Darwinian Impacts: An Introduction to the Darwinian Revolution.* Open University Press, Milton Keynes, Oxford, England.

Omland, K. E., 1997. Correlated rates of molecular and morphological evolution. *Evolution,* **51,** 1381–1393.

Oparin, A. I., 1924. *Proiskhozhdenie Zhizny* (The Origin of Life). Moscovsky Robotschii, Moscow. (Original Russian edition of Oparin's theory; a revised edition was published in English in 1938, and reprinted 1953 by Dover Publications.)

Orgel, L. E., 2004. Prebiotic chemistry and the origin of the RNA world. *Crit. Rev. Biochem. Mol. Biol.,* **39,** 99–123.

Orgel, L. E., and F. H. C. Crick, 1980. Selfish DNA: The ultimate parasite. *Nature,* **284,** 604–607.

O'Riain, M. J., J. U. M. Jarvis, R. Alexander, R. Buffenstein, and C. Peeters, 2000. Morphological castes in a vertebrate. *Proc. Natl. Acad. Sci. USA,* **97,** 13194–13197.

Orme, C. D. L., D. L. J. Quicke, J. M. Cook, and A. Purvis, 2002. Body size does not predict species richness among the metazoan phyla. *J. Evol. Biol.,* **15,** 235–247.

Orr, H. A., 1997. Haldane's rule. *Ann. Rev. Ecol. Syst.,* **28,** 195–218.

Orr, H. A., and J. A. Coyne, 1992. The genetics of adaptation: A reassessment. *Am. Nat.,* **140,** 725–742.

Orzack, S. H., and E. Sober (eds.), 2001. *Adaptation and Optimality.* Cambridge University Press, Cambridge.

Osawa, S., 1995. *Evolution of the Genetic Code.* Oxford University Press, Oxford, England.

Osborne, K. A., A. Robichon, and E. Burgess, 1997. Natural behavior polymorphism due to a cGMP-dependent protein kinase of *Drosophila. Science,* **277,** 834–836.

Ospovat, D., 1981. *The Development of Darwin's Theory: Natural History, Natural Theology, and Natural Selection, 1838–1859.* Cambridge University Press, Cambridge, England.

Ostrom, J. H., 1974. *Archaeopteryx* and the origin of flight. *Q. Rev. Biol.,* **49,** 27–47.

Ostrom, J. H., 1991. The question of the origin of birds. In *Origins of the Higher Groups of Tetrapods: Controversy and Consensus.* H.-P. Schultze and L. Trueb (eds.). Cornell University Press, Ithaca, New York, pp. 467–484.

Owen, R., 1848. *On the Archetype and Homologies of the Vertebrate Skeleton.* Voorst, London.

Owen, R., 1849. *On the Nature of Limbs.* J. van Voorst, London. [To be reissued Fall 2007, ed. Ronald Amundson, with introductory essays and Foreword by B. K. Hall. The University of Chicago Press, Chicago.]

P

Pääbo, S., 1993. Ancient DNA. *Sci. Am.,* **269,** 86–92.

Pääbo, S., 2003. The mosaic that is our genome. *Nature,* **421,** 409–412.

Pace, N, 2006. Time for a change. *Nature,* **441,** 289.

Padian, K., 1985. The origins and aerodynamics of flight in extinct vertebrates. *Paleontology,* **28,** 413–433.

Padian, K., and L. M. Chiappe, 1998. The origin of birds and their flight. *Sci. Am.,* **278,** 38–47.

Pagel, M., (ed. in chief), 2002. *Encyclopedia of Evolution,* 2 vols. Oxford University Press, New York.

Paland, S., and M. Lynch, 2006. Transitions to asexuality result in excess amino acid substitutions. *Science,* **311,** 990–992.

Palmer, J. D., 2003. The symbiotic birth and spread of plasmids: how many times and whodunit? *J. Phycol.,* **39,** 4–11.

Panchen, A. L., 1992. *Classification, Evolution, and the Nature of Biology.* Cambridge University Press, Cambridge, England.

Panganiban, G., S. M. Irvine, C. Lowe, H. Roehl, et al., 1997. The origin and evolution of animal appendages. *Proc. Natl. Acad. Sci. USA,* **94,** 5162–5166.

Parker, G., 1980. *Creation: The Facts of Life.* Creation-Life Publishers, San Diego, CA.

Parker, S. T., 1990. Why big brains are so rare: Energy costs of intelligence and brain size in anthropoid primates. In *"Language" and Intelligence in Monkeys and Apes,* S. T. Parker and K. R. Gibson (eds.). Cambridge University Press, Cambridge, England, pp. 129–154.

Parker, W. K. 1890. On the skull of *Tarsipes rostratus. Stud. Mus. Zool. Univ. Coll. Dundee,* 79–83 + plate.

Parkhurst, S. M., and P. M. Meneely, 1994. Sex determination and dosage compensation: Lessons from flies and worms. *Science,* **264,** 924–932.

Parra, E., A. Marcini, J. Akey, J., J. Martinson, et al., 1998. Estimating African-American admixture proportions by use of population-specific alleles. *Am. J. Hum. Genet.* **63,** 1839–1851.

Paterson, H. E. H., 1985. The recognition concept of species. In *Species and Speciation,* E. Vrba (ed.). Transvaal Museum Monograph 4. Pretoria, South Africa, pp. 21–29.

Patterson, N., D. J. Richter, S. Gnerre, E. S. Lander, and D. Reich, 2006. Genetic evidence for complex speciation of humans and chimpanzees. *Nature,* **441,** 1103–1108.

Pawlowski, J., J.-I. Montoya-Burgos, J. F. Fahrni, J. Wüest, and L. Zaninetti, 1996. Origin of the Mesozoa inferred from 18S rRNA gene sequences. *Mol. Biol. Evol.,* **13,** 1128–1132.

Peck, J. R., 1994. A ruby in the rubbish: Beneficial mutations, deleterious mutations and the evolution of sex. *Genetics,* **137,** 597–606.

Peichel, C. L., K. S. Nereng, K. A. Ohgi, B. L. E. Cole, et al., 2001. The genetic architecture of divergence between threespine stickleback species. *Nature* **414,** 901–905.

Peleg, G., G. Katzir, O. Peleg, M. Kamara, et al., 2006. Heredity family signature of facial expression. *Proc. Natl. Acad. Sci. USA,* **103,** 15921–15926.

Pendleton, J. W., B. K. Nagai, M. T. Murtha, and F. H. Ruddle, 1993. Expansion of the Hox gene family and the evolution of chordates. *Proc. Natl. Acad. Sci. USA,* **90,** 6300–6304.

Pennisi, E., 1998. Genome data shake the tree of life. *Science,* **280,** 672–674.

Penny, D., 2005. An interpretive review of the origin of life research. *Biol. Philos.* **20,** 633–671.

Phillips, J., 1860. *Life on the Earth: Its Origin and Succession.* MacMillan and Co., Cambridge and London. [Reprint edition, 1980, Arno Press, New York.

Phillips, M. L., A. W. Young, C. Senior, M. Brammer, et al., 1997. A specific neural substrate for perceiving facial expressions of disgust. *Nature,* **389,** 495–498.

Pianka, E. R., 1988. *Evolutionary Ecology,* 4th ed. Harper & Row, New York.

Piatagorsky, J., and G. J. Wistow, 1989. Enzyme/crystallins: Gene sharing as an evolutionary strategy. *Cell,* **57,** 197–199.

Pickstone, J. V., 2001. *Ways of Knowing. A New History of Science, Technology and Medicine.* The University of Chicago Press, Chicago.

Pigliucci, M., 2001. *Phenotypic Plasticity: Beyond Nature and Nurture.* The Johns Hopkins University Press, Baltimore and London.

Pigliucci, M., 2002. *Denying Evolution: Creationism, Scientism and the Nature of Science.* Sinauer Associates Inc., Sunderland, MA.

Pigliucci, M., and K. Preston (eds.), 2004. *Phenotypic Integration: Studying the Ecology and Evolution of Complex Phenotypes.* Oxford University Press, Oxford, England.

Pimm, S. L., G. J. Russell, J. L. Gittleman, and T. M. Brooks, 1995. The future of biodiversity. *Science,* **269,** 347–350.

Plavcan, J. M., R. F. Kay, W. J. Jungers, and C. P. van Schaik, 2002. *Reconstructing Behavior in the Primate Fossil Record.* Kluwer Academic/Plenum Publishers, New York.

Pollard, K. S., S. R. Salama, N. Lambert, M. A. Lambot, et al., 2006. An RNA gene expressed during cortical development evolved rapidly in humans *Nature,* **443,** 167–172.

Polly, P. D., 2007. Limbs in mammalian evolution. In *Fins into Limbs: Evolution, Development and Transformation,* B. K. Hall (ed.). The University of Chicago Press, Chicago, pp. 245–268.

Portmann, A., 1990. *A Biologist Looks at Humankind.* (Translated from an earlier German edition by J. Schaefer.) Columbia University Press, New York.

Potts, R., 1992. The hominid way of life. In *The Cambridge Encyclopedia of Human Evolution,* S. Jones, R. Martin, and D. Pilbeam (eds.). Cambridge University Press, Cambridge, England, pp. 325–334.

Pough, F. H., C. M. Janis, and J. B. Heiser, 2005. *Vertebrate Life.* 7th ed. Prentice Hall, Upper Saddle River, NJ.

Poulin, R., 2006. *Evolutionary Ecology of Parasites.* Second Edition. Princeton University Press, Princeton, NJ.

Prakash, S., R. C. Lewontin, and J. L. Hubby, 1969. A molecular approach to the study of genic heterozygosity in natural populations. IV. Patterns of genic variation in central, marginal and isolated populations of *Drosophila pseudoobscura. Genetics,* **61,** 841–858.

Premack, D., 1971. Language in the chimpanzee? *Science,* **172,** 808–822.

Provine, W. B., 1986. *Sewall Wright and Evolutionary Biology.* The University of Chicago Press, Chicago.

Prusiner, S. B., 1995. The prion diseases. *Sci. Am.,* **272,** 48–57.

Ptashne, M., 1992. *A Genetic Switch: Phage λ and Higher Organisms.* Second Edition. Blackwell Scientific, Cambridge, MA.

Purugganan, M. D., and J. I. Suddith, 1998. Molecular population genetics of the *Arabidopsis* Cauliflower regulatory gene: Nonneutral evolution and naturally occurring variation in floral homeotic function. *Proc. Natl. Acad. Sci. USA,* **95,** 8130–8134.

Q

Qiu, Y.-L., L. Li, B. Wang, Z. Chen, et al., 2006. The deepest divergence in land plants inferred from phylogenetic evidence. *Proc. Natl. Acad. Sci. USA,* **103,**15511–15516.

Queller, D. C., J. E. Strassmann, and C. R. Bridges, 1993. Microsatellites and kinship. *Trends Ecol. Evol.,* **8,** 285–288.

Qumsiyeh, M. B., 1994. Evolution of number and morphology of mammalian chromosomes. *Heredity,* **85,** 455–465.

R

Raff, R. A., 1992. Direct-developing sea urchins and the evolutionary reorganization of early development. *BioEssays,* **14,** 211–218.

Raff, R. A., 1996. *The Shape of Life: Genes, Development, and the Evolution of Animal Form.* University of Chicago Press, Chicago.

Raff, R. A., G. A. Wray, and J. J. Henry, 1991. Implications of radical evolutionary changes in early development for concepts of developmental constraint. In *New Perspectives on Evolution,* L. Warren and H. Koprowski (eds.). Wiley, New York, pp. 189–207.

Ramsköld, L., and X.-G. Hou, 1991. New early Cambrian animal and onychophoran affinities of enigmatic metazoans. *Nature,* **351,** 225–228.

Rasmussen, A.-S., and U. Arnason, 1999. Molecular studies suggest that cartilaginous fishes have a terminal position in the piscine tree. *Proc. Natl. Acad. Sci. USA,* **96,** 2177–2182.

Ratnieks, F. L. W., and P. K. Visscher, 1989. Worker policing in the honeybee. *Nature,* **342,** 796–797.

Raup, D. M., 1966. Geometric analysis of shell coiling: General problems. *J. Paleontol.,* **40,** 1178–1190.

Raup, D. M., 1991. *Extinction: Bad Genes or Bad Luck?* Norton, New York.

Raup, D. M., 1994. The role of extinction in evolution. *Proc. Natl. Acad. Sci. USA,* **91,** 6758–6763.

Raup, D. M., and J. J. Sepkoski, Jr., 1986. Periodic extinction of families and genera. *Science,* **231,** 833–835.

Ravasz, E., A. L. Somera, D. A. Mongru, Z. N. Oltvai, and A.-L. Barabási, 2002. Hierarchical organization of modularity in metabolic networks. *Science,* **297,** 1551–1555.

Ready, J. S., I. Sampaio, H. Schneider, C. Vinson, et al., 2006. Color forms of Amazonian cichlid fish represent reproductively isolated species. *J. Evol. Biol.,* **19,** 1139–1148.

Reeck, G. R., C. de Haen, D. C. Teller, R. F. Doolittle, et al., 1987. "Homology" in proteins and nucleic acids: a terminology muddle and a way out of it. *Cell,* **60,** 667.

Rees, M. J., 2000. *New Perspectives in Astrophysical Cosmology,* 2nd ed. Cambridge University Press, Cambridge, England.

Rees, M. J. 2001a. *Our Cosmic Habitat.* Princeton University Press, Princeton, NJ.

Rees, M. J. 2001b. *Before the Beginning.* Addison-Wesley, Reading, MA.

Regier, J. C., and J. W. Shultz, 1997. Molecular phylogeny of the major arthropod groups indicates polyphyly of crustaceans and a new hypothesis for the origin of hexapods. *Mol. Biol. Evol.,* **14,** 902–913.

Ren, D., 1998. Flower-associated *Brachycera* flies as fossil evidence for Jurassic angiosperm origins. *Science,* **280,** 85–88.

Rendel, J. M., 1967. *Canalization and Gene Control.* Logos Press, London.

Retallack, G. J., 1994. Were the Ediacaran fossils lichens? *Paleobiology,* **20,** 523–544.

Rey, M., S. Ohno, J. A. Pinter-Toro, A. Llobell, and T. Bentez, 1998. Unexpected homology between inducible cell wall protein QID74 of filamentous fungi and BR3 salivary protein of the insect *Chironomus. Proc. Natl. Acad. Sci. USA,* **95,** 6212–6216.

Reznick, D. N., F. H. Shaw, F. H. Rodd, and R. G. Shaw, 1997. Evaluation of the rate of evolution in natural populations of guppies (*Poecilia reticulata*). *Science,* **275,** 1934–1937.

Rice, W. R., 1994. Degeneration of a nonrecombining chromosome. *Science,* **263,** 230–232.

Rice, W. R., 1998. Male fitness increases when females are eliminated from gene pool: Implications for the Y chromosome. *Proc. Natl. Acad. Sci. USA,* **95,** 6217–6221.

Rice, W. R., and E. E. Hostert, 1993. Perspective: Laboratory experiments on speciation: What have we learned in forty years? *Evolution,* **47,** 1637–1653.

Richards, G. D, 2006. Genetic, physiologic and ecogeographic factors contributing to variation in *Homo sapiens: Homo floresiensis* reconsidered. *J. Evol. Biol.,* **19,** 1744–1767.

Richards, R. J., 1987. *Darwin and the Emergence of Evolutionary Theories of Mind and Behavior.* The University of Chicago Press, Chicago.

Richards, R. J., 1992. *The Meaning of Evolution: The Morphological Construction and Ideological Reconstruction of Darwin's Theory.* The University of Chicago Press, Chicago.

Ricklefs, R. E., and G. L. Miller, 2000. *Ecology,* 4th ed. W.H. Freeman, New York.

Ridley, M., 1996. *The Evolution of Virtue: Human Instincts and the Evolution of Cooperation.* Viking, New York.

Ridley, M., 2006. *Genome: The Autobiography of a Species in 23 Chapters.* Harper Perennial, New York.

Rieppel, O.,1984. Atomism, transformism, and the fossil record. *Zool. J. Linn. Soc.,* **82,** 17–32.

Rieseberg, L. H., 1997. Hybrid origins of plant species. *Ann. Rev. Ecol. Syst.,* **28,** 359–389.

Rieseberg, L. H., B. Sinervo, C. R. Linder, M. C. Ungerer, and D. M. Arias, 1996. Role of gene interactions in hybrid speciation: Evidence from ancient and experimental hybrids. *Science,* **272,** 741–745.

Riley, M. A., 1989. Nucleotide sequence of the Xdh region in *Drosophila pseudoobscura* and an analysis of the evolution of synonymous codons. *Mol. Biol. Evol.,* **6,** 33–52.

Riordan, M., and W. A. Zajc, 2006. The first few microseconds. *Sci. Am.,* **294,** 34A–41.

Roberts, J. M., 1985. *The Triumph of the West.* British Broadcasting Corporation, London.

Robinson, G. E., 1998. From society to genes with the honey bee. *Am. Sci.,* **86,** 456–462.

Robinson, J. T., 1972. *Early Hominid Posture and Locomotion.* University of Chicago Press, Chicago.

Rockwood, L. L., 2006. *Introduction to Population Ecology.* Blackwell, Malden, MA.

Roff, D. A., 1997. *Evolutionary Quantitative Genetics.* Chapman and Hall, New York.

Roger, A. J., S. G. Svärd, J. Tovar, C. G. Clark, et al., 1998. A mitochondrial-like chaperonin 60 gene in *Giardia lamblia*: Evidence that diplomonads once harbored an endosymbiont related to the progenitor of mitochondria. *Proc. Natl. Acad. Sci. USA,* **95,** 229–234.

Rokas, A., B. L. Williams, N. King, and S. B. Carroll, 2003. Genome-scale approaches to resolving incongruence in molecular phylogeny. *Science,* **425,** 789–804.

Rokas, A., D. Krüger, and S. B. Carroll, 2005. Animal evolution and the molecular signature of radiations compressed in time. *Science,* **310,** 1933–1938.

Romanes, G. J., 1910. *Darwin, and After Darwin.* Vols I–III. Open Court, Chicago.

Romer, A. S., 1968. *The Procession of Life.* World Publishing, Cleveland.

Romer, A. S., and T. S. Parsons, 1977. *The Vertebrate Body,* 5th ed. Saunders, Philadelphia, PA.

Ronshaugen, M., N. McGinnis, and W. McGinnis, 2002. Hox protein mutation and macroevolution of the insect body plan. *Nature,* **415,** 914–917.

Rosa-Milinar, E., and M. B. Pritz (eds), 2004. *Hindbrain*

Evolution, Development, and Organization. Special issue of *Brain, Behaviour and Evolution,* **66**(4). S. Karger AG, Basel, Switzerland.

Rose, M. R., 1994. *Evolutionary Biology of Aging,* revised ed. Oxford University Press, New York.

Rose, M. R., and L. D. Mueller, 2006. *Evolution and Ecology of the Organism.* Prentice-Hall, Upper Saddle River, NJ.

Rosenberg, K., and W. Trevathan, 1996. Bipedalism and human birth dilemma revisited. *Evol. Anthropol.,* **4,** 161–168.

Roughgarden, J., 1996. *Theory of Population Genetics and Evolutionary Ecology: An Introduction.* Prentice Hall, New York.

Rouse, G. W., S. K. Goffredi, and R. C. Vrijenhoek, 2005. *Osedax:* Bone-eating marine worms with dwarf males. *Science,* **305,** 668–671.

Roush, R., and D. R. McKenzie, 1987. Ecological genetics of insecticide and acaricide resistance. *Ann. Rev. Entomol.,* **32,** 361–380.

Rowlands, T., P. Baumann, and S. P. Jackson, 1994. The TATA-binding protein: A general transcription factor in eukaryotes and archaebacteria. *Science,* **264,** 1326–1329.

Rudwick, M. J. S., 1985. *The Meaning of Fossils. Episodes in the History of Palaeontology,* 2nd ed. The University of Chicago Press, Chicago.

Rudwick, M. J. S., 1995. *Scenes From Deep Time. Early Pictorial Representations of the Prehistoric World.* The University of Chicago Press, Chicago.

Rudwick, M. J. S., 2005. *Bursting the Limits of Time. The Reconstruction of Geohistory in the Age of Revolution.* The University of Chicago Press, Chicago.

Ruff, C. B., E. Trinkaus, and T. W. Holliday, 1997. Body mass and encephalization in Pleistocene *Homo. Nature,* **387,** 173–176.

Ruiz-Trillo, I., M. Riutort, D. T. J. Littlewood, E. A. Herniou, and J. Baguñà, 1999. Acoel flatworms: Earliest extant bilaterian metazoans, not members of Platyhelminthes. *Science,* **283,** 1919–1923.

Rumbaugh, D. M., 1977. *Language Learning by a Chimpanzee: The Lana Project.* Academic Press, New York.

Rundle, H. D., and D. Schluter, 1998. Reinforcement of stickleback mate preferences: Sympatry breeds contempt. *Evolution,* **52,** 200–208.

Ruppert, E. E., R. S. Fox, and R. D. Barnes, 2004. *Invertebrate Zoology: A Functional Evolutionary Approach,* 7th ed. Brooks Cole, Thomson Learning, Belmont, CA.

Ruse, M., 1979a. *The Darwinian Revolution.* The University of Chicago Press, Chicago.

Ruse, M., 1979b. *Sociobiology: Sense or Nonsense?* Reidel, Dordrecht, Netherlands.

Ruse, M., (ed.), 1987. *Biol. Philos.,* **2,** 127–225. (An issue devoted to species concepts.)

Ruse, M., 1988. *But Is It Science? The Philosophical Question in the Creation/Evolution Controversy.* Prometheus Books, Buffalo, NY.

Ruse, M., 1996. *From Monad to Man: The Concept of Progress in Evolutionary Biology.* Harvard University Press, Cambridge, MA.

Ruse, M., 2001. *The Evolution Wars: A Guide to the Debates.* Rutgers University Press, New Jersey and London.

Ruse, M., 2005. *The Evolution-Creation Struggle.* Harvard University Press, Cambridge, MA.

Russell, M., 2003. *Piltdown Man: The Secret Life of Charles Dawson and the World's Greatest Archaeological Hoax.* Tempus Publishing, Gloucestershire, England.

Russo, C. A. M., N. Takezaki, and M. Nei, 1996. Efficiencies of different genes and different tree-building methods in recovering a known vertebrate phylogeny. *Mol. Biol. Evol.,* **13,** 525–536.

Rutherford, S. L., and S. Lindquist, 1998. Hsp90 as a capacitor for morphological evolution. *Nature,* **396,** 336–342.

Ruvolo, M., S. Zehr, M. von Dornum, D. Pan, B. Chang, and J. Lin, 1993. Mitochondrial COII sequences and modern human origins. *Mol. Biol. Evol.,* **10,** 1115–1135.

S

Saccheri, I., M. Kuussaari, M. Kankare, P. Vikman, et al., 1998. Inbreeding and extinction in a butterfly metapopulation. *Nature,* **392,** 491–494.

Sáenz-de-Miera, L. E., and F. J. Ayala, 2004. Complex evolution of orthologous and paralogous decarboxylase genes. *J. Evol. Biol.,* **17,** 55–66.

Sander, P. M., O. Mateus, T. Laven, and N. Knötschke, 2006. Bone histology indicates insular dwarfism in a new Late Jurassic sauropod dinosaur. *Nature,* **441,** 739–741.

Sanjúan, R., and S. F. Elena, 2006. Epistasis correlates to genomic complexity. *Proc. Natl. Acad. Sci. USA,* **103,** 14402–14405.

Sargent, T. D., C. D. Millar, and D. M. Lambert, 1998. The "classical" explanation of industrial melanism. *Evol. Biol.,* **30,** 299–322.

Sarich, V. M., and A. C. Wilson, 1966. Quantitative immunochemistry and the evolution of primate albumins: Microcomplement fixation. *Science,* **154,** 1563–1566.

Sassanfar, M., and J. W. Szostak, 1993. An RNA motif that binds ATP. *Nature,* **364,** 550–553.

Sato, A., C. O'Huigin, F. Figueroa, P. R. Grant, et al., 1999. Phylogeny of Darwin's finches as revealed by mtDNA sequences. *Proc. Natl. Acad. Sci. USA,* **96,** 5101–5106.

Satoh, N., 1994. *Developmental Biology of Ascidians.* Cambridge University Press, Cambridge. England.

Savage, R. J. G., and M. R. Long, 1986. *Mammalian Evolution.* British Museum, London.

Savage-Rumbaugh, E. S., and D. M. Rumbaugh, 1993. The emergence of language. In *Tools, Language and Cognition in Human Evolution,* K. R. Gibson and T. Ingold (eds.). Cambridge University Press, Cambridge, England, pp. 86–108.

Savard, L., P. Li, S. H. Strauss, M. W. Chase, M. Michaud, and J. Bousquet, 1994. Chloroplast and nuclear gene sequences indicate Late Pennsylvanian time for the last common ancestor of extant seed plants. *Proc. Natl. Acad. Sci. USA,* **91,** 5163–5167.

Scheiner, S. M., 2002. Selection experiments and the study of phenotypic plasticity. *J. Evol. Biol.,* **15,** 889–898.

Scherer, G., and M. Schmid, (eds.), 2001. *Genes and Mechanisms in Vertebrate Sex Determination.* Birkhäuser Verlag, Basel.

Schierwater, B., and K. Kuhn, 1998. Homology of Hox genes and the zootype concept in early metazoan evolution. *Mol. Phylogenet. Evol.,* **9,** 375–381.

Schlichting, C. D., and M. Pigliucci, 1998. *Phenotypic Evolution: A Reaction Norm Perspective.* Sinauer Associates, Sunderland, MA.

Schlosser, G., and G. P. Wagner (eds.). 2004. *Modularity in Development and Evolution.* The University of Chicago Press, Chicago.

Schluter, D., 2000. *The Ecology of Adaptive Radiation.* Oxford University Press, Oxford.

Schopf, J. W. 1978. The evolution of the earliest cells. *Sci. Am.* **239,** 110–134.

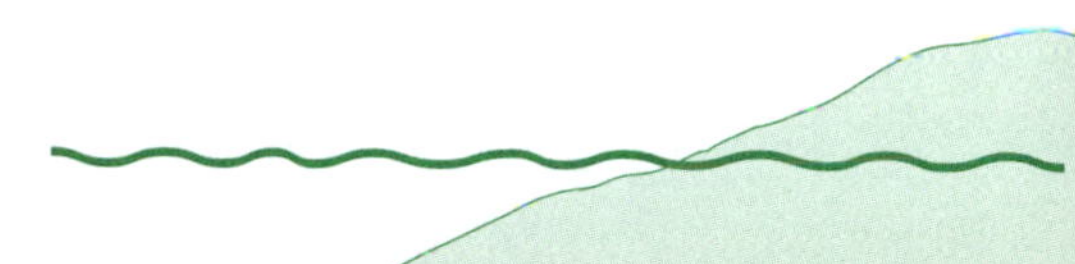

Schopf, J. W., 1996. Metabolic memories of Earth's earliest biosphere. In *Evolution and the Molecular Revolution,* C. R. Marshall and J. W. Schopf (eds.). Jones and Bartlett, Sudbury, MA, pp. 73–107.

Schopf, J. W., J. M. Hayes, and M. R. Walter, 1983. Evolution of earth's earliest ecosystems: Recent progress and unsolved problems. In *Earth's Earliest Biosphere,* J. W. Schopf (ed.). Princeton University Press, Princeton, NJ, pp. 361–384.

Schopf, J. W., and M. R. Walter, 1983. Archean microfossils: New evidence of ancient microbes. In *Earth's Earliest Biosphere: Its Origin and Evolution,* J. W. Schopf (ed.). Princeton University Press, Princeton, NJ, pp. 214–239.

Schoustra, S. E., M. Slakhorst, A. J. M. Debets, and R. F. Hoekstra, 2005. Comparing artificial and natural selection in rate of adaptation to genetic stress in *Aspergillus nidulans. J. Evol. Biol.,* **18,** 771–778.

Schulter, D., 2000. *The Ecology of Adaptive Radiation.* Oxford University Press, Oxford, England.

Schultz, A. H. 1969. *The Life of Primates.* Universe Books, New York.

Schwenk, K., 1995. A utilitarian approach to evolutionary constraint. *Zoology,* **98,** 251–262.

Schwenk, K., and G. P. Wagner, 2003. Constraint. In *Keywords and Concepts in Evolutionary Developmental Biology,* B. K. Hall and W. M. Olson (eds.). Harvard University Press, Cambridge, MA, pp. 52–61.

Scott, E. C., 2000. Not (just) in Kansas anymore. *Science,* **288,** 813–815.

Scott, E. C., 2005. *Evolution vs. Creationism: An Introduction.* University of California Press, Berkeley, CA.

Scott, E. C., 2006. Creationism and evolution: It's the American way. *Cell,* **124,** 449–450.

Seger, J., 1985. Intraspecific resource competition as a cause of sympatric speciation. In *Evolution: Essays in Honour of John Maynard Smith,* P. J. Greenwood, P. H. Harvey, and M. Slatkin (eds.). Cambridge University Press, Cambridge, England, pp. 43–53.

Seger, J., 1996. Exoskeletons out of the closet. *Science,* **274,** 941.

Seidel, K., and V. Schmid, 2006. Mesodermal anatomies in cnidarian polyps and medusae. *Int. J. Devel. Biol.* **50,** 589–599.

Seilacher, A., 1989. Vendozoa: Organismic construction in the Proterozoic biosphere. *Lethaia,* **22,** 229–239.

Seilacher, A., 1992. Vendobionta and Psammocorallia: Lost constructions of Precambrian evolution. *J. Geol. Soc. Lond.,* **149,** 607–613.

Seilacher, A., P. K. Bose, and F. Pflüger, 1998.Triploblastic animals more than 1 billion years ago: Trace fossils from India. *Science,* **282,** 80–83.

Selden, P., and J. Nudds, 2004. *Evolution of Fossil Ecosystems.* Manson Publishing Ltd., London.

Semaw, S., P. Renne, J. W. K. Harris, C. S. Feibel, et al., 1997. 2.5-million-year-old stone tools from Gona, Ethiopia. *Nature,* **385,** 333–336.

Sempere, L. F., C. N. Cole, M. A. McPeek, and K. J. Petersen, 2006. The phylogenetic distribution of Metazoan microRNAs; Insight into evolutionary complexity and constraint. *J. Exp. Zool. (Mol. Dev. Evol.),* **306B,** 575–588.

Senter, P., 2007. Necks for sex: sexual selection as an explanation for sauropod dinosaur neck elongation. *J. Zool.,* **271,** 45–53.

Senut, B., M. Pickford, D. Gommery, P. Mein, et al., 2001. First hominid from the Miocene. (Lukeino formation, Kenya). *C. R. Acad. Sci. Paris, Sciences de la Terre et des planètes (Earth and Planetary Sciences),* 1–9.

Seymour, R. S., C. L. Bennett-Stamper, S. D. Johnston., D. R. Carrier, and G. C. Grigg, 2004. Evidence for endothermic ancestors of crocodiles at the stem of archosaur evolution. *Physiol. Biochem. Zool.,* **77,** 1051–1067.

Shadan, F. F., and L. P. Villarreal, 1993. Coevolution of persistently infecting small DNA viruses and their hosts linked to host-interactive regulatory domains. *Proc. Natl. Acad. Sci. USA,* **90,** 4117–4121.

Shanahan, T., 2004. *The Evolution of Darwinism: Selection, Adaptation and Progress in Evolutionary Biology.* Cambridge University Press, New York.

Shapiro, M. D., H. Shubin, and J. P. Downs, 2007. Limb diversity and digit reduction in reptilian evolution. In *Fins into Limbs: Evolution, Development and Transformation,* B. K. Hall (ed.). The University of Chicago Press, Chicago, pp. 225–244.

Sharpton, V. L., K. Burke, A. Camargo-Zanoguera, S. A. Hall, et al., 1993. Chicxulub multiring impact basin: Size and other characteristics derived from gravity analysis. *Science,* **261,** 1564–1567.

Shipman, P., 2001. *The Man Who Found the Missing Link: The Extraordinary Life of Eugène Dubois.* Simon & Schuster, New York.

Shipman, P., and P. Storm, 2002. Missing links: Eugène Dubois and the origins of Paleoanthropology. *Evol. Anthropol.,* **11,** 108–116.

Shu, D., S. C. Morris, Z. F. Zhang, J. N. Liu, et al., 2003a. A new species of *Yunnanozoan* with implications for deuterostome evolution. *Science,* **299,** 1380–1384.

Shu, D., S. C. Morris, J. Han, Z. F. Zhang, et al., 2003b. Head and backbone of the Early Cambrian vertebrate *Haikouichthys. Nature,* **421,** 526–529.

Shu, D., X. Zhang, and L. Chen, 1996. Reinterpretation of *Yunnanozoan* as the earliest known hemichordate. *Nature,* **380,** 428–430.

Shubin, N., C. Tabin, and S. Carroll, 1997. Fossils, genes, and the evolution of animal limbs. *Nature,* **388,** 639–648.

Sibley, C. G., and J. E. Ahlquist, 1984. The phylogeny of primates as indicated by DNA-DNA hybridization. *J. Mol. Evol.,* **20,** 2–15.

Sibley, C. G., and J. E. Ahlquist, 1987. Avian phylogeny reconstructed from comparisons of the genetic material, DNA. In *Molecules and Morphology in Evolution: Conflict or Compromise?* C. Patterson (ed.). Cambridge University Press, Cambridge, England, pp. 95–121.

Sibley, C. G., and J. E. Ahlquist, 1990. *Phylogeny and Classification of Birds.* Yale University Press, New Haven.

Sidow, A., and W. K. Thomas, 1994. A molecular evolutionary framework for eukaryotic model organisms. *Curr. Biol.,* **4,** 596–603.

Sigmund, K., 1993. *Games of Life: Explorations in Ecology, Evolution, and Behaviour.* Oxford University Press, Oxford, England.

Sigmundsson, F., 2006. Magma does the splits. *Nature,* **442,** 251–252.

Silberbauer, G., 1981. Hunter/gatherers of the Central Kalahari. In *Omnivorous Primates: Gathering and Hunting in Human Evolution,* R. S. O. Harding and G. Teleki (eds.). Columbia University Press, New York, pp. 455–498.

Silva, J. C., and M. G. Kidwell, 2000. Horizontal transfer and selection in the evolution of P elements. *Mol. Biol. Evol.,* **17,** 1542–1557.

Silver, L. M., 1993. The peculiar journey of a selfish chromosome: Mouse t haplotypes and meiotic drive. *Trends Genet.,* **9,** 250–254.

Simon, H. A., 1962. The architecture of complexity. *Proc. Am. Phil. Soc.,* **106,** 467–482.

Simons, A. M., 2002. The continuity of microevolution and macroevolution. *J. Evol. Biol.,* **15,** 688–701.

Simpson, A. G. B., 2003. Cytoskeletal organization, phylogenetic affinities and systematics in the contentious taxon Excavata (Eukaryota). *Int. J. Syst. Evol. Microbiol.,* **53,** 1759–1777.

Simpson, A. G. B., and A. J. Roger, 2004. The real 'kingdoms' of eukaryotes. *Curr. Biol,* **14,** R693–R696.

Simpson, G.G., 1945. *The Meaning of Evolution.* Yale University Press, New Haven, CT.

Simpson, G. G., 1961. *Principles of Animal Taxonomy.* Columbia University Press, New York.

Simpson, G. G., 1980. *Why and How: Some Problems and Methods in Historical Biology.* Pergamon Press, Oxford, England.

Singh, N., K. W. Barbour, and F. G. Berger, 1998. Evolution of transcriptional regulatory elements within the promoter of a mammalian gene. *Mol. Biol. and Evol.,* **15,** 312–325.

Skelton, P., (ed.), 2003. *The Cretaceous World.* The Open University, Milton Keynes, England.

Skulan, J., 2000. Has the importance of the amniote egg been overstated? *Zool. J. Linn. Soc.,* **130,** 235–261.

Slack, J. M. W., P. W. H. Holland, and C. E. Graham, 1993. The zootype and the phylotypic stage. *Nature,* **361,** 490–492.

Slack, K. E., C. M. Jones, T. Ando, G. L. Harrison, et al., 2006. Early penguin fossils, plus mitochondrial genomes, calibrate avian evolution. *Mol. Biol. Evol.* **23,**1144–1155.

Sleigh, M. A., 1979. Radiation of the eukaryote Protista. In *The Origin of Major Invertebrate Groups,* M. R. House (ed.). Academic Press, London, pp. 23–54.

Smith, F. A., J. L. Betancourt, and J. H. Brown, 1995. Evolution of body size in the wood rat over the past 25,000 years of climate change. *Science,* **270,** 2012–2014.

Smith, J. E., (ed.), 1971. *The Invertebrate Panorama.* Universe Books, New York.

Smith, J. V., 1998. Biochemical evolution. I. Polymerization on internal organophilic silica surfaces of dealuminated zeolites and feldspars. *Proc. Natl. Acad. Sci. USA,* **95,** 3370–3375.

Smith, M. M., and B. K. Hall, 1990. Developmental and evolutionary origins of vertebrate skeletogenic and odontogenic tissues. *Biol. Rev. Camb. Philos. Soc.,* **65,** 277–374.

Smith, M. M., and B. K. Hall, 1993. A developmental model for evolution of the vertebrate exoskeleton and teeth: The role of cranial and trunk neural crest. *Evol. Biol.,* **27,** 387–448.

Smith, T. B., 1993. Disruptive selection and the genetic basis of bill size polymorphism in the African finch *Pyrenestes. Nature,* **363,** 618–620.

Smouse, P. E., and W.-H. Li, 1987. Likelihood analysis of mitochondrial restriction-cleavage patterns for the human-chimpanzee-gorilla trichotomy. *Evolution,* **41,** 1162–1176.

Sober, E., and D. S. Wilson, 1998. *Unto Others: The Evolution and Psychology of Unselfish Behavior.* Harvard University Press, Cambridge, MA.

Sogin, M. L., 1991. Early evolution and the origin of eukaryotes. *Curr. Opin. Genet. Dev.,* **1,** 457–463.

Sokolowski, M. B., H. S. Pereira, and K. Hughes, 1997. Evolution of foraging behavior in *Drosophila* by density-dependent selection. *Proc. Natl. Acad. Sci. USA,* **94,** 7373–7377.

Solem, G. A., 1974. *The Shell Makers.* Wiley, New York.

Soltis, D. E., and P. S. Soltis, 1995. The dynamic nature of polyploid genomes. *Proc. Natl. Acad. Sci. USA,* **92,** 8089–8091.

Soltis, P. S., S. E. Soltis, V. Savolainen, P. R. Crane, and T. G. Barraclough, 2002. Integration of molecular and fossil evidence for molecular living fossils. *Proc. Natl. Acad. Sci. USA,* **99,** 4430–4435.

Sparke, L. S., and J. S. Gallagher, 2000. *Galaxies in the Universe.* Cambridge University Press, Cambridge, England.

Spencer, F., 1990. *Piltdown: A Scientific Forgery.* Oxford University Press, Oxford, England.

Spencer, H., 1887. *The Factors of Organic Evolution.* D. Appleton and Co., New York.

Spencer, H., 1897. *First Principles.* D. Appleton and Co., New York.

Spoor, F., B. Wood, and F. Zonneveld, 1994. Implications of early hominid labyrinthine morphology for evolution of human bipedal locomotion. *Nature,* **369,** 645–648.

Stahl, B. J., 1974. *Vertebrate History: Problems in Evolution.* McGraw-Hill, New York.

Stanhope, M. J., V. G. Waddell, O. Madsen, W. de Jong, et al., 1998. Molecular evidence for multiple origins of Insectivora and for a new order of endemic African insectivore mammals. *Proc. Natl. Acad. Sci. USA,* **95,** 9967–9972.

Stanier, R. Y., 1970. Some aspects of the biology of cells and their possible evolutionary significance. *Symp. Soc. Gen. Microbiol.,* **20,** 1–38.

Stanier, R. Y., and C. B. van Neil, 1941. The main outlines of bacterial classification. *J. Bacteriol.,* **42,** 437–466.

Stanier, R. Y., and C. B. van Neil, 1962. The concept of a bacterium. *Arch. Mikrobiol.* **42,** 17–35.

Stanley, S. M., 1973. An ecological theory for the sudden origin of multicellular life in the late Precambrian. *Proc. Natl. Acad. Sci. USA,* **70,** 1486–1489.

Stanley, S. M., 1979. *Macroevolution: Process and Product.* Freeman, San Francisco.

Stearns, S. C. (ed.), 1987. *The Evolution of Sex and Its Consequences.* Birkhaüser Verlag, Basel, Switzerland.

Stearns, S. C. 1989. The evolutionary significance of phenotypic plasticity. *BioScience,* **37,** 436–445.

Stearns, S. C., C. M. Kaiser, and T. J. Kawecki, 1995. The differential genetic and environmental canalization of fitness components in *Drosophila melanogaster. J. Evol. Biol.,* **8,** 539–557.

Stebbins, G. L., 1974. *Flowering Plants: Evolution Above the Species Level.* Harvard University Press, Cambridge, MA.

Steel, R., and A. P. Harvey, 1979. *The Encyclopaedia of Prehistoric Life.* Mitchell-Beazley, London.

Stein, W. E., and J. S. Boyer, 2006. Evolution of land plant architecture: beyond the telome theory. *Paleobiology,* **32,** 450–482.

Stemmer, W. P. C., 1994. Rapid evolution of a protein by DNA shuffling. *Nature,* **370,** 389–391.

Stern, C., 1973. *Principles of Human Genetics.* Third Edition. Freeman, San Francisco.

Stevens, K. A., and J. M. Parrish, 1999. Neck posture and feeding habits of two Jurassic sauropod dinosaurs. *Science,* **284,** 798–800.

Stevens, L., C. J. Goodnight, and S. Kalisz, 1995. Multilevel selection in natural populations of *Impatiens capensis. Am. Nat.,* **145,** 513–526.

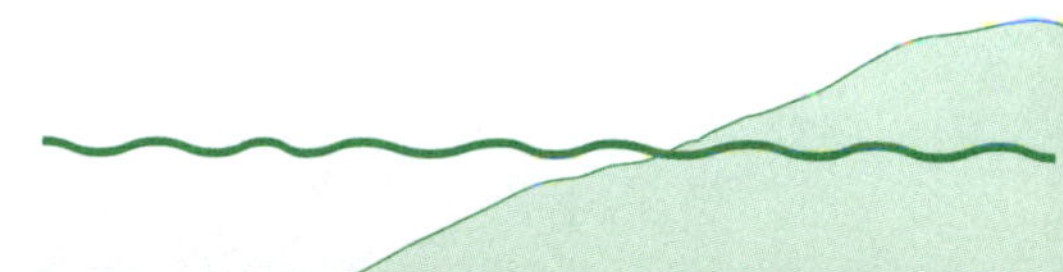

Stevens, P. F., 1994. *The Development of Biological Classification: Antoine-Laurent de Jussieu, Nature, and the Natural System.* Columbia University Press, New York.

Stewart, C. B., and A. C. Wilson, 1987. Sequence conversion and functional adaptation of stomach lysozymes from foregut fermenters. *Cold Spring Harbor Symp. Quant. Biol.*, **52**, 891–899.

Stewart, W. N., and G. Rothwell, 1993. *Paleobotany and the Evolution of Plants.* Second Edition. Cambridge University Press, Cambridge, England.

Stokes, H. W., and B. G. Hall, 1985. Sequence of the ebgR gene of *Escherichia coli:* Evidence that the EBG and LAC operons are descended from a common ancestor. *Mol. Biol. Evol.*, **2,** 478–483.

Stokes, M. D., and N. D. Holland, 1998. The lancelet. *Am. Sci.*, **86,** 552–560.

Stone, J. R., and B. K. Hall, 2004. Latent homologues for the neural crest as an evolutionary novelty. *Evol. Devel.*, **6,** 123–129.

Stone, J. R., and B. K. Hall, 2006. A system for analyzing features in studies integrating ecology, development and evolution. *Biol. & Philos.* **21,** 25–40.

Stone, W. S., 1962. The dominance of natural selection and the reality of superspecies (species groups) in the evolution of *Drosophila. Univ. of Texas Publ.*, **6205,** 507–537.

Stoneking, M., 1993. DNA and recent human evolution. *Evol. Anthropol.*, **2,** 60–71.

Straus, N. A., 1976. Repeated DNA in eukaryotes. In *Handbook of Genetics,* vol. 5, R. C. King (ed.). Plenum Press, New York, pp. 3–29.

Strauss, E., 1999. Can mitochondrial clocks keep time? *Science*, **283,** 1435–1438.

Strickberger, M. W., 1985. *Genetics,* 3d ed. Macmillan, New York.

Stringer, C. B., and C. Gamble, 1993. *In Search of the Neanderthals: Solving the Puzzle of Human Origins.* Thames and Hudson, London.

Sturmbauer, C., J. S. Levinton, and J. Christy, 1996. Molecular phylogeny analysis of fiddler crabs: Test of the hypothesis of increasing behavioral complexity in evolution. *Proc. Natl. Acad. Sci. USA,* **93,** 10855–10857.

Sultan, S. E., and S. C. Stearns, 2005. Environmentally contingent variation: Phenotypic plasticity and norms of reaction. In *Variation. A Central Concept in Biology,* B. Hallgrímsson and B. K. Hall (eds). Elsevier/Academic Press, New York, pp. 303–332.

Sumner, W. G., 1883. *What Social Classes Owe to Each Other.* Harper & Brothers, New York.

Suzuki, Y., and H. D. Nijhout, 2006. Evolution of a polyphenism by genetic accommodation. *Science,* **311,** 650–652.

Swofford, D. L., G. J. Olsen, P. J. Waddell, and D. M. Hillis, 1996. Phylogenetic inference. In *Molecular Systematics,* 2nd ed. D. M. Hillis, C. Moritz, and B. K. Mable (eds.). Sinauer Associates, Sunderland, MA, pp. 407–514.

Sword, G. A., 1999. Density dependent warning coloration. *Nature,* **397,** 217.

Szathmáry, E., 1989. The emergence, maintenance and transitions of the earliest evolutionary units. *Oxford SUV. Evol. Biol.*, **6,** 169–205.

Szathmáry, E., 1991. Four letters in the genetic alphabet: A frozen evolutionary optimum? *Proc. R. Soc. Lond. (B),* **245,** 91–99.

Szathmáry, E., and J. Maynard Smith, 1995. The major evolutionary transition. *Nature,* **374,** 227–232.

T

Takahata, N., 1996. Neutral theory of molecular evolution. *Curr. Opin. Genet. Dev.*, **6,** 767–772.

Talbot, S. L., and G. F. Shields, 1996. Phylogeography of brown bears (*Ursus arctos*) of Alaska and paraphyly within the Ursidae. *Mol. Phylogenet. Evol.*, **5,** 477–494.

Tanke, D. H., and K. Carpenter, (eds.), 2001. *Mesozoic Vertebrate Life.* Indiana University Press, Bloomington, IN. [A *Festschrift* to celebrate 25 years of paleontological research by Philip J. Currie.]

Tanner, N. M., 1987. The chimpanzee model revisited and the gathering hypothesis. In *The Evolution of Human Behavior: Primate Models,* W. G. Kinzey (ed.). SUNY Press, Albany, NY, pp. 3–27.

Tarbuck, E. J., and F. K. Lutgens, 2006. *Earth Science,* 9th ed. Prentice Hall, Englewood Cliffs, NJ.

Tarsitano, S. F., A. P. Russell, F. Horne, C. Plummer, and K. Millerchip, 2000. On the evolution of feathers from an aerodynamic and constructional view point. *Am. Zool.*, **40,** 676–686.

Taylor, A. L., 1970. Current linkage map of *Escherichia coli. Bacteriol. Rev.*, **34,** 155–175.

Taylor, D. W., and L. J. Hickey, 1990. An Aptian plant with attached leaves and flowers: Implications for angiosperm origin. *Science,* **247,** 702–704.

Taylor, S. R., 1998. *Destiny or Chance: Our Solar System Evolution and Its Place in the Cosmos.* Cambridge University Press, Cambridge, England.

Taylor, T. N., and E. L. Taylor, 1993. *The Biology and Evolution of Fossil Plants.* Prentice Hall, Englewood Cliffs, NJ.

Tchernavin, V., 1937. Skulls of salmon and trout. A brief study of their differences and breeding changes. *Salmon & Trout Mag.*, **88,** 235–242.

Templeton, A. R., 1986. Relations of humans to African apes: A statistical appraisal of diverse types of data. In *Evolutionary Processes and Theory,* S. Karlin and E. Nevo (eds.). Academic Press, Orlando, FL, pp. 365–388.

Templeton, A. R., 1989. The meaning of species and speciation: A genetic perspective. In *Speciation and Its Consequences,* D. Otte and J. A. Endler (eds.). Sinauer Associates, Sunderland, MA, pp. 3–27.

Templeton, A. R., 1997. Out of Africa? What do genes tell us? *Curr. Opin. Genet. Dev.*, **7,** 841–847.

Teotónio, H., and M. R. Rose, 2001. Reverse evolution. *Evolution,* **55,** 653–660.

The *C. elegans* Sequencing Consortium, 1998. Genome sequence of the nematode *C. elegans*: a platform for investigating biology. *Science,* **282,** 2012–2018.

The Chimpanzee Sequencing and Analysis Consortium, 2005. Initial sequence of the chimpanzee genome and comparison with the human genome. *Nature,* **437,** 69–87.

The Honeybee Genome Sequencing Consortium, 2006. *Nature,* **443,** 931–949.

The Sea Urchin Genome Sequencing Project Consortium, 2006. The genome of the sea urchin *Strongylocentrotus purpuratus. Science,* **314,** 941–952.

Theunissen, B., 1981. *Eugène Dubois and the Ape-Man from Java.* Kluwer Academic Publishers, Dordrecht, The Netherlands.

The yeast genome directory, 1997. *Nature,* **387,** 5.

Thoday, J. M., and J. B. Gibson, 1962. Isolation by disruptive selection. *Nature,* **193,** 1164–1166.

Thomas, B. A., and R. A. Spicer, 1987. *The Evolution and Palaeobiology of Land Plants.* Croom Helm, London.

Thomas, F., A. T. Teriokhin, E. V. Budilova, S. P. Brown, et al., 2004. Human birthweight evolution across contrasting environments. *J. Evol. Biol.,* **17,** 542–553.

Thompson, D'A. W., 1942. *On Growth and Form,* 2d ed. Cambridge University Press, Cambridge, England.

Thompson, J. N., 2005. *The Geographic Mosaic of Coevolution.* The University of Chicago Press, Chicago.

Thompson, K. S., 1995. *HMS Beagle: The Story of Darwin's Ship.* W. W. Norton, New York.

Thorpe, J. P., 1982. The Molecular Clock Hypothesis: Biochemical evolution, genetic differentiation and systematics. *Annu. Rev. Ecol. Syst.,* **13,** 139–168.

Till-Bottraud, I., D. Joly, D. Lachaise, and R. N. Snook, 2005. Pollen and sperm heteromorphism: convergence across kingdoms. *J. Evol. Biol.,* **18,** 1–18.

TIME, 2006. Nature's Extremes. Inside the Great Natural Disasters that Shape Life on Earth. Time, Inc., New York.

Tooby, J., and I. DeVore, 1987. The reconstruction of hominid behavioral evolution through strategic modeling. In *The Evolution of Human Behavior: Primate Models,* W. G. Kinzey (ed.). SUNY Press, Albany, NY, pp. 183–237.

Trimble, V. 2006. Early photons from the early universe. *New Astron. Revs.,* **50,** 844–849.

Trivers, R., 1985. *Social Evolution.* Benjamin/Cummings, Menlo Park, CA.

True, J. R., and S. B. Carroll, 2002. Gene co-option in physiological and morphological evolution. *Annu. Rev. Cell Dev. Biol.,* **18,** 53–80.

Turelli, M., 1998. The causes of Haldane's rule. *Science,* **282,** 889–891.

Tuttle, R. H. 1972. *Functional and Evolutionary Biology of Hylobatid Hands and Feet.* Gibbon and Siamang: Karger, Basel.

Tylor, E. B., 1881. *Anthropology: An Introduction to the Study of Man and Civilization.* Appleton, New York.

Tyndale-Biscoe, H., 2005. *Life of Marsupials.* CSIRO, Canberra, Australia.

Tyson, E., and M. v. d. Gucht, 1699. *Orang-outang, sive, Homo sylvestris, or, The anatomy of a pygmie compared with that of a monkey, an ape, and a man to which is added, A philological essay concerning the pygmies, the cynocephali, the satyrs and sphinges of the ancients: wherein it will appear that they are all either apes or monkeys, and not men, as formerly pretended.* London, Printed for Thomas Bennet . . . and Daniel Brown . . . and are to be had of Mr. Hunt.

U

Ueno, Y., Y. Isozaki, H. Yurimoto, and S. Maruyama, 2001. Carbon isotope signatures of individual Archean microfossils (?) from Western Australia. *Int. Geol. Rev.,* **43,** 186–212.

Ueno, Y., K. Yamada, N. Yoshida, S. Maruyama, and Y. Isozaki, 2006. Evidence from fluid inclusions for microbial methanogenesis in the early Archaean era. *Nature,* **440,** 516–519.

Ungerer, M. C., S. J. Baird, J. Pan, and L. H. Rieseberg, 1998. Rapid hybrid speciation in wild sunflowers. *Proc. Natl. Acad. Sci. USA,* **95,** 11757–11762.

Unrau, P. J., and D. P. Bartel, 1998. RNA-catalysed nucleotide synthesis. *Nature,* **395,** 260–263.

Unwin, D. M., and N. N. Bakhurina, 1994. *Sordes pilosus* and the nature of the pterosaur flight apparatus. *Nature,* **371,** 62–64.

V

Valentine, J. W., 1994. Late Precambrian bilaterians: Grades and clades. *Proc. Natl. Acad. Sci. USA,* **91,** 6751–6757.

Valentine, J. W., 1997. Cleavage patterns and the topology of the metazoan tree of life. *Proc. Natl. Acad. Sci. USA,* **94,** 8001–8005.

Valentine, J. W., 2003. Cell types, numbers, and body plan complexity. In *Keywords and Concepts in Evolutionary Developmental Biology,* B. K. Hall and W. M. Olson (eds.). Harvard University Press, Cambridge, MA, pp. 35–43.

Valentine, J. W., 2004. *On the Origin of Phyla.* The University of Chicago Press, Chicago.

Valentine, J. W., S. M. Awramik, P. W. Signor, and P. M. Sadler, 1991. The biological explosion at the Precambrian-Cambrian boundary. *Evol. Biol.,* **25,** 279–356.

Valentine, J. W., A. G. Collins, and C. P. Meyer, 1994. Morphological complexity increase in metazoans. *Paleobiology,* **20,** 131–142.

Valley, J. W., 2005. A cool early Earth? *Sci. Am.* **291,** 58–65.

Van Laere, A. S., M. Nguyen, M. Braunschweig, C. Nezer, et al., 2003. A regulatory mutation in IGF2 causes a major QTL effect on muscle growth in the pig. *Nature,* **425,** 832–836.

Van Valen, L., 1976. Ecological species, multispecies, and oaks. *Taxon,* **25,** 233–239.

Venter J. C., M. D. Adams, E. W, Myers, P. W. Li, et al., 2001. The sequence of the human genome. *Science,* **291,** 1304–1351.

Vetter, R. D., 1991. Symbiosis and the evolution of novel trophic strategies: Thiotrophic organisms at hydrothermal vents. In *Symbiosis As a Source of Evolutionary Innovation,* L. Margulis and R. Fester (eds.). MIT Press, Cambridge, MA, pp. 219–245.

Vickaryous, M., and B. K. Hall, 2006a. Human cell type diversity, evolution development classification with special reference to cells derived from the neural crest. *Biol. Rev. Camb. Philos. Soc.,* **81,** 425–455.

Vickaryous, M., and B. K. Hall, 2006b. Homology of the reptilian coracoid and a reappraisal of the evolution and development of the amniote pectoral apparatus. *J. Anat.* **208**: 263–285.

Vila, C., J. A. Leonard, A. Götherström, S. Marklund, et al., 2001. Widespread origins of domestic horse lineages. *Science,* **291,** 474–477.

Vrba, E. S., 1989. Levels of selection and sorting with special reference to the species level. *Oxford Surv. Evol. Biol.,* **6,** 111–168.

Vrba, E. S., 1995. The fossil record of African antelopes (Mammalia, Bovidae) in relation to human evolution and paleoclimate. In *Paleoclimate and Evolution with Emphasis on Human Origins,* E. S. Vrba, G. H. Denton, T. C. Partridge, and L. H. Burckle (eds.). Yale University Press, New Haven, CT, pp. 385–414.

Vrba, E. S., 2004. Ecology, development, and evolution; Perspectives from the fossil record. In *Environment, Development and Evolution,* B. K. Hall, R. D. Pearson, and G. B Müller (eds.). MIT Press, Cambridge, MA, pp. 85–105.

W

Wada, H., and N. Satoh, 1994. Details of the evolutionary history from invertebrates to vertebrates, as deduced from

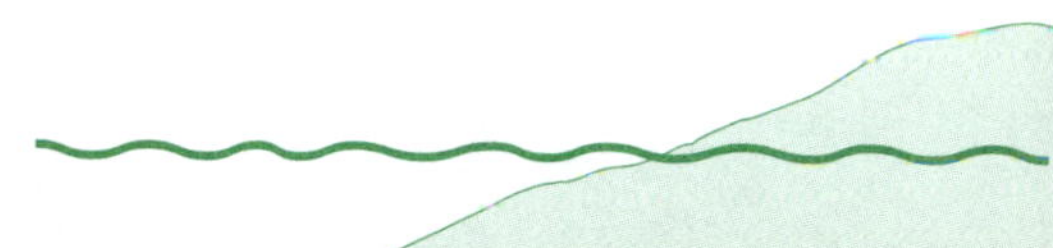

the sequences of 18*S* rDNA. *Proc. Natl. Acad. Sci. USA,* **91,** 1801–1804.

Wada, H., P. W. H. Holland, and N. Satoh, 1996. Origin of patterning in neural tubes. *Nature,* **384,** 123.

Wada, H., H. Saiga, N. Satoh, and P. W. Holland, 1998. Tripartite organization of the ancestral chordate brain and the antiquity of placodes: Insights from ascidian *Pax-2/5/8, Hox* and *Otx* genes. *Development,* **125,** 1113–1122.

Waddington, C. H., 1942. Canalization of development and the inheritance of acquired characters. *Nature,* **150,** 563.

Waddington, C. H., 1957. *The Strategy of the Genes: A Discussion of Some Aspects of Theoretical Biology.* Allen & Unwin, London.

Waddington, C. H., 1962. *New Patterns in Genetics and Development.* Columbia University Press, New York.

Waddington, C. H., and E. Robertson, 1966. Selection for developmental canalization. *Genet. Res. Camb.,* **7,** 303–312.

Wade, M. J., and C. J. Goodnight, 1998. Perspective: The theories of Fisher and Wright in the context of metapopulations: When nature does small experiments. *Evolution,* **52,** 1537–1553.

Wagner, G. P., G. Booth, and H. Bagheri-Chaichian, 1997. A population genetic theory of canalization. *Evolution,* **51,** 329–347.

Wahlstrom, J., 1998. *Genetics and Psychiatric Disorders.* Elsevier Science, New York.

Wainright, P. O., G. Hinkle, M. L. Sogin, and S. K. Stickel, 1993. Monophyletic origins of the metazoa: An evolutionary link with fungi. *Science,* **260,** 340–342.

Wake, M. H., 1990. The evolution of integration of biological systems: An evolutionary perspective through studies of cells, tissues, and organs. *Amer. Zool.,* **30,** 897–906.

Wake, M. H., 1995. The current status of the *Diversitas* program and its implementation. *Biol. Intern.,* **31,** 7–18.

Wald, G., 1974. Fitness in the universe: Choices and necessities. In *Cosmochemical Evolution and the Origins of Life,* J. Oró, S. L. Miller, C. Ponnamperuma, and R. S. Young (eds.). D. Reidel, Dordrecht, Netherlands, pp. 7–27.

Walker, A., and R. Leakey (eds.), 1993. *The Nariokotome* Homo erectus *Skeleton.* Harvard University Press, Cambridge, MA.

Wallace, A. R., 1889. *Darwinism: An Exposition of the Theory of Natural Selection with Some of Its Applications.* Macmillan, London.

Wallace, B., 1970. *Genetic Load: Its Biological and Conceptual Aspects.* Prentice Hall, Englewood Cliffs, NJ.

Wallis, G. P., 1999. Do animal mitochondrial genes recombine? *Trends Genet.,* **14,** 209–210.

Wang, R.-X., 2003. Differential strength of sex-biased hybrid inferiority in impeding gene flow may be a cause of Haldane's rule. *J. Evol. Biol.,* **16,** 353–361.

Ward, P., C. Labandeira, M. Laurin, and R. A. Berner, 2006. Confirmation of Romer's Gap as a low oxygen interval constraining the timing of initial arthropod and vertebrate terrestrialization. *Proc. Natl. Acad. Sci. USA,* **103,** 16818–16822.

Waser, N. M., and J. Ollerton, 2006. *Plant-Pollinator Interactions: From Specialization to Generalization.* The University of Chicago Press, Chicago.

Waterman, A. J., B. E Frye, K. Johansen, A. G. Kluge, et al., 1971. *Chordate Structure and Function.* Macmillan, New York.

Waterston R. H., K. Lindblad-Toh, E. Birney, J. Rogers, et al., 2002. Initial sequencing and comparative analysis of the mouse genome. *Nature,* **420,** 520–562.

Watson, J. D., M. Gilman, J. Witkowski, and M. Zoller, 1992. *Recombinant DNA,* 2nd ed. Scientific American Books, New York.

Waxman, D., and S. Gravilets, 2005. 20 questions on adaptive dynamics. *J. Evol. Biol.,* **18,** 1139–1154.

Webster, B. L., R. A. Copley, R. A. Jenner, J. A. Mackenzie-Dodds, et al., 2006. Mitogenomics and phylogenomics reveal priapulid worms as extant models of the ancestral Ecdysozoan. *Evol. Dev.,* **8,** 502–510.

Webster, J. A., J. L. Gittleman, and A. Purvis, 2004. The life history legacy of evolutionary body size change in carnivores. *J. Evol. Biol.,* **17,** 369–407.

Wedel, M. J., 2003. Vertebral pneumaticity, air sacs, and the physiology of sauropod dinosaurs. *Paleobiology,* **29,** 243–255.

Wedmann, S., S. Bradler, and J. Rust, 2007. The first fossil leaf insect: 47 million years of specialized cryptic morphology and behavior. *Proc. Natl. Acad. Sci. USA,* **104,** 565–569.

Weiner, J. S., 1955. *The Piltdown Forgery.* Oxford University Press, Oxford, England (reprinted 2003).

Weishampel, D. B., P. Dodson, and H. Osmólska (eds.), 2004. *The Dinosauria.* Second Edition. University of California Press, Berkeley, CA.

Weishampel, D. B., and N. M. White (eds.), 2003. *The Dinosaur Papers 1676–1906.* Smithsonian Books, Washington, DC.

Weiss, K. M., 2002. Biology's theoretical kudzu: The irrepressible illusion of teleology. *Evol. Anthrop.,* **11,** 4–8.

Weiss, K. M., and A. V. Buchanan, 2004. *Genetics and the Logic of Evolution.* Wiley-Liss, Hoboken, NJ.

Weitzman, P. D. J., 1985. Evolution in the citric acid cycle. In *Evolution of Prokaryotes,* K. H. Schleifer and E. Stackebrandt (eds.). Academic Press, London, pp. 253–275.

West-Eberhard, M. J., 2003. *Developmental Plasticity and Evolution.* Oxford University Press, Oxford, England.

Wetherill, G. W., 1995. How special is Jupiter? *Nature,* **373,** 470.

Wheelis, M. L., O. Kandler, and C. R. Woese, 1992. On the nature of global classification. *Proc. Natl. Acad. Sci. USA,* **89,** 2930–2934.

White, M. J. D., 1973. *Animal Cytology and Evolution,* 3d ed. Cambridge University Press, Cambridge, England.

White, M. J. D., 1978. *Modes of Speciation.* Freeman, San Francisco.

White, T. D., B. Asfaw, D. DeGusta, H. Gilbert, et al., 2003. Pleistocene *Homo sapiens* from Middle Awash, Ethiopia. *Nature,* **423,** 742–747.

White, T. D., G. Suwa, and B. Asfaw, 1995. *Australopithecus ramidus,* a new species of early hominid from Aramis, Ethiopia. *Nature,* **375,** 88.

Whiten, A., J. Goodall, M. C. McGrew, T. Nishida, et al., 1999. Culture in chimpanzees. *Nature,* **399,** 682–685.

Whiting, M. F., and W. C. Wheeler, 1994. Insect homeotic transformation. *Nature,* **368,** 696.

Whittaker, R. H., 1959. On the broad classification of organisms. *Quart. Rev. Biol.,* **34,** 210–226.

Whittle, C.-A., and M. O. Johnston, 2003. Male-biased transmission of deletions mutations to the progeny in *Arabidopsis thaliana. Proc. Natl. Acad. Sci. USA,* **100,** 4055–4059.

Wicht H, and T. C. Lacalli, 2005. The nervous system of amphioxus: structure, development, and evolutionary significance. *Can. J. Zool.,* **83,** 122–150.

Wicken, M., and K. Takayama, 1994. Deviants — or emissaries. *Nature,* **367,** 17–18.

Wiley, E. O., 1978. The evolutionary species concept reconsidered. *Syst. Zool.,* **27,** 17–26.

Wiley, E. O., 1981. *Phylogenetics: Theory and Practice of Phylogenetic Systematics.* Wiley, New York.

Wilkins, A. S., 2002. *The Evolution of Developmental Pathways.* Sinauer Assoc., Sunderland, MA.

Wilkins, A. S., 2003. Canalization and genetic assimilation. In *Keywords and Concepts in Evolutionary Developmental Biology.* B. K. Hall and W. M. Olson (eds.). Harvard University Press, Cambridge, MA, pp. 23–30.

Williams, B. J., 1979. *Evolution and Human Origins.* Harper & Row, New York.

Williams, G. C., 1966. *Adaptation and Natural Selection: A Critique of Some Current Evolutionary Thought.* Princeton University Press, Princeton, NJ.

Williams, G. C., 1992. *Natural Selection: Domains, Levels and Challenges.* Oxford University Press, New York.

Willmer, P., 1990. *Invertebrate Relationships: Patterns in Animal Evolution.* Cambridge University Press, Cambridge, England.

Wills, C., 1981. *Genetic Variability.* Clarendon Press, Oxford, England.

Wilmut, I., A. E. Schnieke, J. McWhir, A. J. Kind, and K. H. Campbell, 1997. Viable offspring derived from fetal and adult mammalian cells. *Nature,* **385,** 810–813.

Wilner, E., 2006. Darwin's artificial selection as an experiment. *Stud. Hist. Philos. Sci, Part C: Stud. Hist. Philos. Biol. Biomed. Sci.,* **37,** 26–40.

Wilson, A. C., 1975. Evolutionary importance of gene regulation. *Stadler Symp.,* **7,** 117–133.

Wilson, A. C., L. R. Maxson, and V. M. Sarich, 1974. Two types of molecular evolution: Evidence from studies of interspecific hybridization. *Proc. Natl. Acad. Sci. USA,* **71,** 2843–2847.

Wilson, A. C., H. Ochman, and E. M. Prager, 1987. Molecular time scale for evolution. *Trends in Genet.,* **3,** 241–247.

Wilson, D. S., 1992. Group selection. In *Keywords in Evolutionary Biology,* E. Fox Keller and E. A. Lloyd (eds.). Harvard University Press, Cambridge, MA, pp. 145–148.

Wilson, D. S., and J. W. Szostak, 1999. *In vitro* selection of functional nucleic acids. *Annu. Rev. Biochem.,* **68,** 611–647.

Wilson, E. O., 1975. *Sociobiology: The New Synthesis.* Harvard University Press, Cambridge, MA.

Wilson, E. O., 1978. *On Human Nature.* Harvard University Press, Cambridge, MA.

Wilson, E. O., 1992. *The Diversity of Life.* Harvard University Press, Cambridge, MA.

Wilson, R. A. (ed.), 1999. *Species. New Interdisciplinary Essays.* MIT Press, Cambridge, MA.

Wilson, R. A., 2005. *Genes and the Agents of Life. The Individual in the Fragile Sciences.* Cambridge University Press, Cambridge, England.

Wimsatt, W. C., 1986. Developmental constraints, generative entrenchment, and the innate-acquired distinction. In *Integrating Scientific Disciplines,* W. Bechte (ed.), Martinus Nijhoff, Dordrecht, Holland, pp. 185–208.

Winchester, S., 2001. *The Map that Changed the World.* Viking Press, New York.

Winnepenninckx, B., T. Backeljau, L. Y. Mackey, J. M. Brooks, et al., 1995. 18*S* rRNA data indicate that Aschelminthes are polyphyletic in origin and consist of at least three distinct clades. *Mol. Biol. Evol.,* **12,** 1132–1137.

Winnepenninckx, B., T. Backeljau, and R. De Wachter, 1996. Investigation of molluscan phylogeny on the basis of 18*S* rRNA sequences. *Mol. Biol. Evol.,* **13,** 1306–1317.

Winsor, M. P., 2003. Non-essentialist methods in pre-Darwinian taxonomy. *Biol. Philos.,* **18,** 387–400.

Winsor, M. P., 2006. Linnaeus' biology was not essentialist. *Ann. Missouri Bot. Gard.,* **93,** 2–7.

Wise, S. B., and D. W. Stock, 2006. Conservation and divergence of Bmp21, Bmp2b, and Bmp4 expression patterns within and between dentitions of teleost fish. *Evol. Dev.,* **8,** 511–523.

Wistow, G. J., 1993. Identification of lens crystallin: A model system for gene recruitment. *Methods Enzymol.,* **224,** 563–575.

Woese, C. R., 1981. Archaebacteria. *Sci. Am.,* **244,** 92–122.

Woese, C. R., 1983. The primary lines of descent and the universal ancestor. In *Evolution from Molecules to Men,* D. S. Bendall (ed.). Cambridge University Press, Cambridge, England, pp. 209–233.

Woese, C. R., 1985. Why study evolutionary relationships among bacteria? In *Evolution of Prokaryotes,* K. H. Schleifer and E. Stackebrandt (eds.). Academic Press, London, pp. 1–30.

Woese, C. R., 1987. Bacterial evolution. *Microbiol. Rev.,* **51,** 221–271.

Woese, C. R., 1998a. The universal ancestor. *Proc. Natl. Acad. Sci. USA,* **95,** 6854–6859.

Woese, C. R., 1998b. Default taxonomy: Ernst Mayr's view of the microbial world. *Proc. Natl. Acad. Sci. USA,* **95,** 11043–11046.

Woese, C. R., B. A. Debrunner-Vossbrinck, H. Oyaizu, E. Stackbrandt, and W. Ludwig, 1985. Gram-positive bacteria: Possible photosynthetic ancestry. *Science,* **229,** 762–765.

Woese, C. R., O. Kandler, and M. L. Whellis, 1990. Toward a natural system of organisms: Proposal for the domains Archaea, Bacteria and Eucarya. *Proc. Natl. Acad. Sci. USA,* **87,** 4576–4579.

Wolfe, S. L., 1981. *Biology of the Cell,* 2nd ed. Wadsworth, Belmont, CA.

Wolfenden, R. V., P. M. Collis, and C. C. F. Southgate, 1979. Water, protein folding and the genetic code. *Science,* **206,** 575–577.

Wolpoff, M. H., 1989. Multiregional evolution: The fossil alternative to Eden. In *The Human Revolution: Behavioural and Biological Perspectives on the Origin of Modern Humans,* P. Mellars and C. Stringer (eds.). Edinburgh University Press, Edinburgh, pp. 62–108.

Wolpoff, M. H., and R. Caspari, 1997. *Race and Human Evolution.* Simon and Schuster, New York.

Womack, J. E., and Y. D. Moll, 1986. Gene map of the cow: Conservation of linkage with mouse and man. *J. Hered.,* **77,** 2–7.

Wood, B., 1997. The oldest whodunnit in the world. *Nature,* **385,** 292–293.

Wood, B., 2005. *Human Evolution. A Very Short Introduction.* Oxford University Press, Oxford, England.

Wood, B., and M. Collard, 1999. The human genus. *Science,* **284,** 65–71.

Wood, W. B., 1980. Bacteriophage T4 morphogenesis as a model for assembly of subcellular structure. *Q. Rev. Biol.,* **55,** 353–367.

Woods, N. B., V. Bottero, M. Schmidt, C. von Kalle and I. M. Verma, 2006. Gene therapy: therapeutic gene causing lymphoma. *Nature,* **440,**1123.

Worthy, T. H., A. J. D. Tennyson, M. Archer, A. M. Musser, et al., 2006. Miocene mammal reveals a Mesozoic ghost lineage on insular New Zealand, southwest Pacific. *Proc. Natl. Acad. Sci. USA,* **103,** 19419–19423.

Wray, G. A., and E. Abouheif, 1998. When is homology not homology? *Curr. Opin. Genet. Dev.,* **8,** 675–680.

Wright, M. C., and G. F. Joyce, 1997. Continuous *in vitro* evolution of catalytic function. *Science,* **276,** 614–617.

Wright, S., 1921. Systems of mating. *Genetics,* **6,** 111–178.

Wright, S., 1931. Evolution in Mendelian populations. *Genetics* **16,** 97–159.

Wright, S., 1949. Adaptation and selection. In *Genetics, Paleontology, and Evolution,* G. L. Jepson, G. G. Simpson, and E. Mayr (eds.). Princeton University Press, Princeton, NJ, pp. 365–389.

Wright, S., 1951. The genetic structure of populations. *Ann. Eugen,* **15,** 323–354.

Wright, S., 1963. Genic interaction. In *Methodology in Mammalian Genetics,* W. J. Burdette (ed.). Holden-Day, San Francisco, pp. 159–192.

Wright, S., 1978. *Evolution and the Genetics of Populations: Vol. 4. Variability Within and Among Natural Populations.* University of Chicago Press, Chicago.

Wright, S., 1988. Surfaces of selective value revisited. *Am. Nat.,* **131,** 115–123.

Wu, C-I., N. A. Johnson, and M. F. Palopoli, 1996. Haldane's rule and its legacy: Why are there so many sterile males? *Trends Ecol. Evol.,* **11,** 281–284.

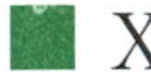

X

Xu, X, Z. Zhou, and X. Wang, 2000. The smallest known non-avian theropod dinosaur. *Nature,* **408,** 705–708.

Y

Yamamoto, Y., L. Espinasa, D. W. Stock, and W. R. Jeffery, 2003. Development and evolution of craniofacial patterning in mediated by eye-dependent and -independent processes in the cavefish *Astyanax. Evol. Dev.* **5,** 435–446.

Yokoyama, R., and S. Yokoyama, 1990. Convergent evolution of the red- and green-like pigment in fish, *Astyanax fasciatus,* and human. *Proc. Natl. Acad. Sci. USA,* **87,** 9315–9318.

You, H.-I., M. C. Lamanna, J. D. Harris, L. M. Chiappe, et al., 2006. A nearly modern amphibious bird from the early cretaceous of northwestern China. *Science,* **312,** 1640–1643.

Young, M., and T. Edis (eds.) 2004. *Why Intelligent Design Fails: A Scientific Critique of the New Creationism.* Rutgers University Press, New Brunswick, NJ.

Young, N. M., and L. MacLatchy, 2004. The phylogenetic position of *Morotopithecus. J. Hum. Evol.,* **46,** 163–184.

Yu J., S. Hu, J. Wang, G. K. Wong, et al., 2002. A draft sequence of the rice genome (*Oryza sativa* L. ssp. *indica). Science,* **296,** 79–92.

Yule, G. U., 1902. Mendel's laws and their probable relations to intra-racial heredity. *New Phytol.,* **1,** 193–207, 222–238.

Yunis, J. J., and O. Prakash, 1982. The origin of man: A chromosomal pictorial legacy. *Science,* **215,** 1525–1529.

Z

Zahavi, A, and A. Zahavi, 1997. The *Handicap Principle: A Missing Piece of Darwin's Puzzle.* Oxford University Press, New York.

Zardoya, R., Y. Cao, M. Hasegawa, and A. Meyer, 1998. Searching for the closest living relative(s) of tetrapods through evolutionary analysis of mitochondrial and nuclear data. *Mol. Biol. Evol.,* **15,** 506–517.

Zardoya, R., and A. Meyer, 1998. Complete mitochondrial genome suggests diapsid affinities of turtles. *Proc. Natl. Acad. Sci. USA,* **95,** 14226–14231.

Zeilik, M., 2002. *Astronomy: The Evolving Universe* (9th ed.). Cambridge University Press, Cambridge, England.

Zelditch, M. L., (ed.), 2001. *Beyond Heterochrony. The Evolution of Development.* Wiley-Liss, New York.

Zhang, B., and T. R. Cech, 1997. Peptide bond formation by *in vitro* selected ribozymes. *Nature,* **390,** 96–100.

Zhang, J., and M. Nei, 1996. Evolution of *Antennapedia*-class homeobox genes. *Genetics,* **142,** 295–303.

Zhang, X., J. Yazaki, A. Sundaresan, S. Cokus et al., 2006. Genome-wide high-resolution mapping and functional analysis of DNA methylation in *Arabidopsis. Cell,* **126,** 1189–1201.

Zhang, X. H.-F., and L. Chasin, 2006. Comparison of multiple vertebrate genomes reveals the birth and evolution of human exons. *Proc. Natl. Acad. Sci. USA,* **103,** 13427–13432.

Zhou, Z., and F. Zhang, (eds.), 2002. *Proceedings of the 5th Symposium of the Society of Avian Paleontology and Evolution, Beijing, 1–4 June, 2000.* Science Press, Beijing, China.

Ziegler, A. M., C. R. Scotese, W. S. McKerrow, M. E. Johnson, and R. K. Bombach, 1979. Paleozoic paleogeography. *Ann. Rev. Earth Planet. Sci.,* **7,** 473–502.

Zielinski, W. S., and L. E. Orgel, 1987. Autocatalytic synthesis of a tetra nucleotide analogue. *Nature,* **327,** 346–347.

Zimmermann, W., 1952. Main results of the "telome theory." *Paleobotanist,* **1,** 456–470.

Zubay, G., 1988. *Biochemistry,* 2nd ed. Macmillan, New York.

Glossary[1]

A

Abiotic Substances of nonbiological origin; environments characterized by the absence of organisms.

Acidic A compound that produces an excess of hydrogen ions (H^+) when dissolved in water and so has a pH value less than 7.0. (*See also* **pH scale.**)

Acoelomate An animal that lacks a coelom (internal body cavity).

Acrocentric Chromosomes whose centromeres are near one end between metacentric and telocentric locations.

Active transport Biochemical transport requiring the input of energy, for example, hydrolysis of adenosine-5'-triphosphate (ATP).

Adaptation The relationship between structure and/or function and an organism's environment that makes the structure or function suitable for ("adapted" to) life in that environment.

Adaptive landscape A model originally devised by Sewall Wright that describes a topography in which populations with high fitness occupy peaks and those with low fitness occupy valleys.

Adaptive peak *See* **Adaptive landscape.**

Adaptive radiation The diversification of a single species or group of related species into new ecological or geographical zones to produce a larger number of species or groups.

Adenosine triphosphate (ATP) An organic compound commonly involved in the transfer of phosphate bond energy, composed of adenosine (an adenine base + a D-ribose sugar) and three phosphate groups. (*See also* **Active transport.**)

Aerobic metabolism The use of oxygen for reactions that provide energy for metabolism from the oxidative breakdown of food molecules; living in the presence of oxygen. (*See also* **Anaerobic metabolism.**)

Aerobic respiration An electron transport system in which oxygen serves as the terminal electron acceptor.

Algae Photosynthetic eukaryotes now divided among several super kingdoms of life.

Allantois An extraembryonic membrane that functions in gas exchange and excretion of waste products in amniote embryos. (*See also* **Chorioallantoic membrane.**)

Allele One of the alternative forms of a single gene; a nucleotide sequence occurring at a given locus on a chromosome.

[1]The particular science in which some terms are used is included in parentheses, especially when similar terms are used in more than one science.

Allele frequency The proportion of different alleles in a population. (*See also* **Gene pool.**)

Allen's rule The generalization that mammals tend to have shorter extremities in colder than in warmer climates.

Allometry The proportionate relationship between body parts and the size of the organism.

Allopatric Species or populations in separate geographical locations. (*See also* **Allopatric speciation.**)

Allopatric speciation Speciation between populations that are geographically separated. (*See also* **Allopatric.**)

Allopolyploid An organism or species that has more than two sets of chromosomes because of hybridization and chromosome doubling.

Allotetraploid An individual or species with double the chromosome number because of hybridization

Allozyme The particular form (amino acid sequence) of an enzyme produced by a particular allele at a gene locus when there are different possible forms of the enzyme, each produced by a different allele.

Alternate splicing The mechanism by which single genes produce different functional proteins by rearrangement of exons.

Alternation of generations A life cycle in which a multicellular haploid stage alternates with a multicellular diploid stage.

Altricial Born or hatched in a state requiring parental care; unable to survive on its own.

Altruism (Animal behavior) Behavior that benefits the reproductive success of other individuals because of an actual or potential sacrifice of reproductive success by an individual (the altruist).

Amino acids Organic molecules of the general formula R–CH(NH_2)–COOH, possessing both basic (NH_2) and acidic (COOH) groups, as well as a side group (R) specific for each type of amino acid.

Amino group An $-NH_2$ group.

Amniota (amniotes) The clade including all extant and extinct vertebrates with embryos encased in extraembryonic membranes.

Amniotic egg The type of egg produced by reptiles, birds and mammals (Amniota), in which the embryo is enveloped in a series of membranes (amnion, allantois, chorion) that help sustain its development, usually outside the body of the female.

Anaerobic glycolysis *See* **Glycolysis.**

Anaerobic metabolism Energy for metabolic processes obtained from the oxidative breakdown of food molecules in the absence of molecular oxygen; living in the absence of oxygen. (*See also* **Aerobic metabolism.**)

Anaerobic respiration An electron transport system in which substances other than oxygen (e.g., sulfates, nitrates, methane), serve as the terminal electron acceptor.

Anagenesis Evolution within a single lineage (usually a species) as opposed to cladogenesis where a group (species) diverges into two or more branches. (*See also* **Cladogenesis, Phyletic evolution.**)

Analogous *See* **Analogy**

Analogy The possession of a similar character by two or more species caused by factors other than common genetic ancestry. (*See also* **Convergence, Homology.**)

Ancestral state The characteristic(s) of an organism that gave rise to related groups of organisms.

Aneuploidy The gain or loss of chromosomes leading to a number that is not an exact multiple of the basic haploid chromosome set.

Angiosperms Flowering plants with floral reproductive structures and encapsulated seeds.

Angstrom (Å) A length one ten billionth (10^{-10}) of a meter.

Angular gyrus A region of the parietal lobe of the brain that performs a number of important functions related to cognition and language.

Antibody A protein produced by the immune system that binds to a substance (antigen) typically foreign to the organism. (*See also* **Antigen.**)

Anticodon A sequence of three nucleotides (a triplet) on transfer RNA that is complementary to the codon on messenger RNA, and that specifies placement of a particular amino acid in a polypeptide during translation. (*See also* **Codon.**)

Antigen A substance, typically foreign to an organism, which initiates antibody formation and is bound by the activated antibody. (*See also* **Antibody.**)

Apocrine Exocrine glands that secrete by producing vesicles (small membranous sacs) that bud off from the plasma membrane. (*See also* **Exocrine.**)

Apomixis (apomictic) In plants, asexual reproduction from seeds. (*See also* **Parthenogenesis.**)

Apomorphy A character derived from, yet different from the ancestral condition. (*See also* **Synapomorphy.**)

Apoptosis Cell death that is genetically programmed.

A-P axis The major axis of symmetry of bilateral animals where the head is at the anterior and the tail is posterior.

Arboreal Living predominantly in trees.

Archaebacteria Sulfur-dependent, methane-producing and halophilic bacteria with cell walls and a single chromosome. Differ from eubacteria in ribosomal structures, and in the possession of introns by some forms. Originated early in the history of life.

Archaeopteryx Perhaps the most famous, the earliest and the least derived fossil bird, which lived in the Jurassic Period (150-155 Mya), and provides a link between nonavian reptiles and birds.

Archenteron The cavity that forms during gastrulation of animal embryos and that later forms the primitive gut. (*See also* **Endoderm, Gastrula.**)

Archetype A body plan ("*Bauplan*") characteristic of clades of organisms, usually phyla, but can characterize lower taxonomic levels.

Arthropods The clade of animals (phylum) that includes insects, spiders and crustaceans with hard exoskeletons, a segmented body plan and jointed appendages.

Articular A bone in the lower jaw in nonmammalian vertebrates that contributes to the jaw joint by articulating with the quadrate and that is modified as an ear ossicle in mammals. (*See also* **Quadrate.**)

Artificial selection The process of selection of organisms by humans to change one or more traits. (*See* **Selection.**)

Asexual reproduction Offspring produced by one parent in the absence of sexual fertilization or in the absence of gamete formation.

Assortative mating Mating among individuals on the basis of their phenotypic or genotypic similarities (positive assortative) or differences (negative assortative), rather than mating on a random basis.

Atmosphere The gaseous envelope surrounding a celestial body such as Earth.

Autocatalytic reaction Chemical reactions in which the agent that promotes (catalyzes) the reaction is a product of the reaction.

Autopolyploid A species or organism that has more than two sets of chromosomes (polyploid) derived from one or more duplications in a single ancestral source.

Autosome A chromosome other than a sex chromosome.

Autotroph An organism capable of synthesizing complex organic compounds needed for growth from simple inorganic environmental substrates.

B

Bacteriophage A virus (phage) that parasitizes bacteria.

Balanced genetic load The decrease in overall fitness of a population caused by genotypes (e.g., homozygotes for deleterious recessives) whose alleles persist in the population because they confer selective advantages in other genotypic combinations (e.g., heterozygote advantage).

Balanced polymorphism The persistence of two or more different genetic forms through selection (e.g., heterozygote advantage) rather than because of mutation or other evolutionary forces.

Banded iron formations Sedimentary rocks, some as old as 3.76 Mya, comprised of thin layers of iron oxides alternating with thin iron-poor layers of chert and shale. The iron oxide layers are believed to result from the photosynthetic activity of cyanobacteria.

Baryon (Greek, *heavy*) The collective name for those subatomic particles composed of three quarks; includes protons and neutrons that together form atomic nuclei.

Baryonic matter Matter composed largely of baryons, including all atoms and thus almost all matter we experience in everyday life.

Basal character A feature found in ancestral taxa of a lineage. (*See also* **Derived character.**)

Basalt A fine-grained igneous rock found in oceanic crust and produced in lava flows.

Base (nucleotide) The nitrogenous component of the nucleotide unit in nucleic acids, consisting of either a purine (adenine, A, or guanine, G) or pyrimidine (thymine, T, or cytosine, C, in DNA; uracil, U, or cytosine, C, in RNA). (*See also* **Purine, Pyrimidine.**)

Base pairs *See* **Complementary base pairs.**

Base substitutions Nucleotide changes that substitute one base for another in DNA.

Basic (alkaline) A compound that produces an excess of hydroxyl (OH^-) ions when dissolved in water and has a pH value greater than 7.0.

Batesian mimicry The similarity in appearance of a harmless species (the mimic) to a species that is harmful or distasteful to predators (the model), maintained because of selective advantage to the relatively rare mimic.

Bauplan (plural, **Baupläne**) The structural body plan that characterizes a group of organisms such as a phylum. (*See also* **Archetype.**)

Bayesian inference A method likelihood analysis based on a given probability.

Behavior (behavioral adaptation) *See* **Altruism, Culture, Dominance, Imprinting, Instinct, Learning.**

Belief systems Codes of conduct, often with a moral or religious component, used by humans to guide their relationships with one another, with the world around them, and, in the case of religious systems of belief, with a supernatural being.

Benthic Refers to the floor of a body of water and to organisms that live in on or near it.

Bergmann's rule The generalization that animals living in colder climates tend to be larger than those of the same group living in warmer climates.

Big Bang theory A theory that the universe (time and space) expanded rapidly from an extremely hot, dense state about 13.7 Bya and that in the years since it has continued to cool and expand. (*See also* **Evolutionary universe.**)

Bilateral symmetry The condition in which the left and right halves of an organism are equivalent so that the organism can be divided in half along a single plane.

Binary fission Replication of an organism by its division into two mostly equal parts, the common form of asexual reproduction in prokaryotes and protistan eukaryotes.

Binomial expansion The binomial $(a + b)$ raised to a power n $[(a + b)^n]$ where a and b represent alternative states whose sum equals the probability of 1.

Binomial nomenclature The Linnaean principle of designating a species by two names: the genus name followed by the species name (for example *Homo sapiens*).

Biogenetic law The theory proposed by Ernst Haeckel that stages in the development of an individual (ontogeny) recapitulate the evolutionary history (phylogeny) of the clades to which the organism belongs.

Biogeographical realm A spatially limited region characterized by a distinctive biota.

Biogeography The study of the spatial (geographical) distributions of organisms.

Biological species concept The thesis that the primary criterion for separating one species from another is reproductive isolation; the formation of species following reproductive isolation.

Biosphere That part of Earth containing all living organisms.

Biota All the fauna and flora of a given region or time period.

Biotic Relating to or produced by biological organisms.

Bipedalism The exclusive use of the hind limbs for locomotion. (*See also* **Brachiation.**)

Blastocoele The cavity that forms within the blastula during animal embryonic development.

Blastopore The opening formed by the invagination of cells in the embryonic gastrula, connecting its cavity (archenteron) to the outside. In protostome phyla, the blastopore is the site of the future mouth. In deuterostomes the blastopore becomes the anus and the mouth forms elsewhere.

Blastula A hollow sphere enclosed within a single layer of cells, occurring at an early stage of animal development animals.

Blending inheritance The now-abandoned theory that offspring inherit a dilution, or blend of parental traits, rather than particles (genes) that determine those traits.

Body plan See *Bauplan.*

Bottleneck effect A form of genetic drift that occurs when a population is reduced in size (population crash) and later expands in numbers (population flush). The enlarged population that results may have gene frequencies distinctly different than before the bottleneck. (*See also* **Founder effect.**)

Box (Boxes) In molecular genetics, a DNA sequence specifying a protein that binds to DNA to act as a transcription factor activating or repressing other sequences of DNA. Families of such transcription factor genes share the same box.

Brachiation A mode of locomotion in which an animal suspends itself from tree branches and moves about the canopy by swinging between branches or along branches (*See also* **Bipedalism.**)

Bryophytes Mosses and liverworts.

C

Calorie The amount of heat necessary to raise the temperature of 1 gram of water by 1 degree centigrade at a pressure of 1 atmosphere.

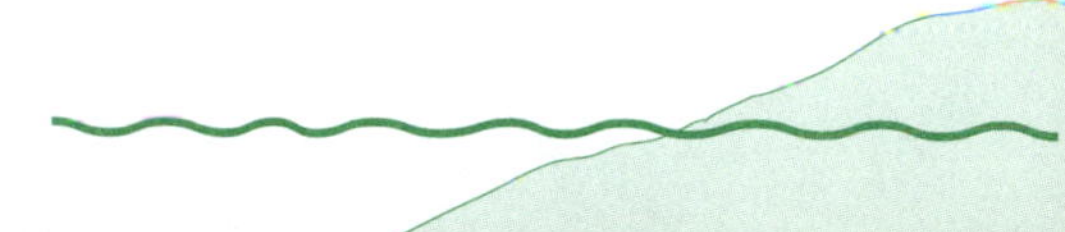

Calvin cycle A cyclic series of light-independent reactions that accompany photosynthesis to reduce carbon dioxide to carbohydrate in the presence of water.

Cambrian explosion The appearance of fossils of many different forms of animals in the Cambrian Period, interpreted as a rapid diversification (explosion) of body plans.

Cambrian Period The interval between about 545 and 495 My before the present, marking the appearance of many fossilized organisms, many with hardened skeletons (the Cambrian explosion). It is considered the beginning of the Phanerozoic time scale (eon) and is the first period in the Paleozoic era.

Canalization Restriction of phenotypic variation by intrinsic genetic or developmental mechanisms.

Canalizing selection Artificial or natural selection that results in reduced variation in expression of a phenotypic trait.

Carbohydrate A compound in which the hydrogen and oxygen atoms bonded to carbons are commonly in a ratio of 2/1 [e.g., glucose ($C_6H_{12}O_6$), starch $(C_6H_{12}O_6)_n$ and cellulose $(C_6H_{10}O_5)_n$].

Carbonaceous Composed of or containing carbon.

Carbonaceous meteorites Meteorites containing carbon compounds; also known as chondrites.

Carboxylic acid An organic compound with an acidic group consisting of a carbon with a double-bond attachment to an oxygen atom and a single-bond attachment to a hydroxyl group (O=C–OH).

Carnivores Flesh eaters; organisms (almost entirely animal, but some plants) that feed on animals.

Carrying capacity (K) The maximum number of organisms in a population that can be sustained in a given environment.

Catalyst A substance that lowers the energy necessary to activate a reaction but is not itself consumed or altered in the reaction.

Catastrophism (earth sciences) The eighteenth- and nineteenth-century theory that fossilized organisms and changes in geological strata were produced by periodic, violent and widespread catastrophic events rather than by naturally explainable events based on laws that act uniformly through time. (*See also* **Uniformitarianism.**)

Cell wall The rigid or semi rigid extracellular envelope (outside the plasma membrane) that gives shape to plant, algal, fungal and bacterial cells.

Celsius scale (°C) A scale of temperature in which the melting point of ice is taken as 0° and the boiling point of water as 100°, measured at 1 atmosphere of pressure.

Cenozoic (Caenozoic) era The period from 65 Mya to the present, marked by reduction in dinosaur diversity and radiation of mammals. The third and most recent era of the Phanerozoic eon, divided into two major periods, the Tertiary and Quaternary.

Central dogma (molecular biology) The tenet that the direction of flow of genetic information is from DNA → RNA → proteins.

Centromere The chromosome region in eukaryotic nuclei to which spindle fibers attach during cell division.

Centrosome A cellular organelle in eukaryotic cells that functions as the place of origin of the microtubules for the mitotic spindle during cell division.

Character A feature, trait or property of an organism or population. (*See also* **Basal character, Derived character.**)

Character displacement Divergence in the appearance or measurement of a character between two species when their geographical distributions overlap, compared to the similarity of the character in the two species when they are geographically separated.

Character state The designation of a character into (usually) binary states used to code characters in phylogenetic analysis.

Chert A sedimentary rock composed largely of tiny quartz crystals (SiO_2) precipitated from aqueous solutions.

Chiasma The place where two homologous chromosomes establish close contact and where exchange of homologous parts takes place between nonsister chromatids during meiosis.

Chimera An organism (or part of an organism) made from (composed of cells from) two different organisms, either from the same or from different species.

Chlorophyll A green pigment found in the chloroplasts of plants that collects light used in photosynthesis.

Chloroplast A chlorophyll-containing, membrane-bound organelle that is the site of photosynthesis in the cells of plants and some protistans. Chloroplasts contain their own genetic material (circular DNA without histones) and are believed to be descendants of cyanobacteria that entered eukaryotic cells via endosymbiosis.

Chondrites *See* **Carbonaceous meteorites.**

Chordates Members of the Phylum Chordata (which includes the vertebrates), united by the presence of a notochord and dorsal nerve cord.

Chorioallantoic membrane An extraembryonic membrane that forms from the fusion of the chorion and the allantois to form the outer extraembryonic membrane of amniotes. (*See also* **Allantois.**)

Chromatid One of the two sister products of chromosome replication in eukaryotes, marked by an attachment between sister chromatids at the centromere.

Chromosome aberration A change in the gene sequence of a chromosome caused by deletion, duplication, inversion or translocation.

Chromosomes Elongate, threadlike structures in the nuclei of eukaryotes containing the nuclear genes. (*See also* **Polytene chromosomes.**)

***Cis*-regulation** The process by which an element of DNA that is adjacent to a gene interacts with the promoter of that gene to control binding sites for transcriptional activators or repressors.

Citric acid cycle *See* **Krebs cycle.**

Clade A natural grouping of taxa (all the descendants) derived from a single common ancestor.

Cladistics A mode of classification in which taxa are principally grouped on the basis of their shared possession of similar ("derived") characters that differ from the ancestral condition.

Cladogenesis Branching evolution involving the splitting and divergence of a lineage into two or more lineages. (*See also* **Anagenesis, Phyletic evolution.**)

Cladogram A tree diagram representing phylogenetic relationships among taxa.

Class A taxonomic rank that stands between phylum and order. A phylum may include one or more classes, and a class may include one or more orders.

Classification The grouping of organisms into a hierarchy of categories commonly ranging from species through genera, families, orders, classes, phyla and kingdom, each category reflecting one or more significant features.

Cleavage The process of cellular division of a zygote immediately after fertilization to form a multicellular embryo, the blastula.

Cline A gradient of phenotypic or genotypic change in a population or species correlated with the direction or orientation of some environmental feature, such as a river, mountain range, north–south transect or altitude.

Clone A group of organisms derived by asexual reproduction from a single ancestral individual; a group of cells that are identical because they arose from a single cell.

Cloning (genetics) Techniques for producing identical copies of a section of genetic material by inserting a DNA sequence into a cell, such as a bacterium, where it can be replicated.

Cnidaria Metazoans, commonly known as corals, jellyfish and hydrozoans.

Coacervate An aggregation of colloidal particles in liquid phase that persists for a period of time as suspended membranous droplets. (*See also* **Microspheres, Protocells.**)

Coadaptation The action of selection in producing adaptive combinations of alleles at two or more different gene loci.

Codominance The independent phenotypic expression of two different alleles in a heterozygote.

Codon The triplet of adjacent nucleotides in messenger RNA that codes for a specific amino acid carried by a specific transfer RNA or that codes for termination of translation (STOP codons). Placement of the amino acid is based on complementary pairing between the anticodon on tRNA and the codon on mRNA. (*See also* **Anticodon.**)

Coelom (Coelome) The internal body cavity of eucoelomate animals (true coelomates) formed within mesoderm. (*See also* **Hemocoel.**)

Coenzymes Non-protein enzyme-associated organic molecules (e.g., NAD, FAD and coenzyme A) that participate in enzymatic reactions by acting as intermediate carriers of electrons, atoms, or groups of atoms.

Coevolution Evolutionary changes in one or more genes, developmental processes, organs or species in response to changes in genes, developmental processes, organs or species.

Cofactor A small molecule that may be organic (i.e., a coenzyme) or inorganic (i.e., a metal ion), which is required by an enzyme in order to function.

Cohort Individuals of a population all of which are of the same age.

Colinearity Applied to Hox genes to signify a correspondence between the positions of the genes on a chromosome and the antero-posterior parts of the body whose development they control.

Collision theory (Cosmology) Proposed in 1749 by Georges Louis Leclerc, Comte de Buffon, the idea that a comet (or star) struck the sun and broke off fragments that formed the planets.

Commensalism An association between organisms of different species in which one species benefits from the relationship and the other species is not affected significantly.

Competition Relationship between organismal units (e.g., individuals, groups, species) attempting to exploit a limited common resource in which each unit inhibits, to varying degrees, the survival or reproduction of another unit by means other than predation.

Complementary base pairs Nucleotides on one strand of a nucleic acid that hydrogen bond with nucleotides on another strand according to the rule that pairing between purine and pyrimidine bases is restricted to certain combinations: A pairs with T in DNA, A pairs with U in RNA, and G pairs with C in both DNA and RNA.

Complexity A state of intricate organization caused by arrangement or interaction among different component parts or processes: the greater the number of interacting parts, the greater the complexity. (*See also* **Progress.**)

Concerted evolution The process by which a series of nucleotide sequences or different members of a gene family remain similar or identical through time.

Condensation (by dehydration) The formation of a covalent bond between two molecules by removal of H_2O.

Condensation theory (Cosmology) The theory that the planets arose by the collision and accretion of planetesimals.

Consciousness The ability of an organisms to have an awareness of itself, to perceive the world around it and to respond to it.

Constraint (Biology) Describes the factors that limit character variation or evolutionary direction; genotypic or phenotypic channeling and evolutionary trends caused mostly by processes involving adaptation.

Continental drift Theory proposed in 1912 by Alfred Wegener, according to which large landmasses move relative to each other across Earth's surface. (*See also* **Paleomagnetism, Sea-floor spreading, Tectonic plates.**)

Continuous variation Character variations (such as height in humans), the distribution of which follows a series of small, nondiscrete quantitative steps, from one extreme to the other. (*See also* **Quantitative character.**)

Convergence (Convergent evolution) The evolution of similar characters (genes or morphologies) in genetically unrelated

or distantly related species (homoplasy), mostly because they have been subjected to similar environmental selective pressures.

Cope's rule The generalization (not always confirmed) that body size tends to increase in an animal lineage during its evolution.

Core (Earth sciences) The innermost layer of Earth, composed mostly of iron with some nickel, and consisting of a solid inner core and an outer, liquid core

Correlation The degree to which two measured characters tend to vary in the same quantitative direction (positive correlation) or in opposite directions (negative correlation).

Cosmic microwave background (CMB) A form of electromagnetic radiation that fills the entire universe and is a relic of the Big Bang. Tiny differences in the temperature of the CMB, which averages 2.7 K, reflect minute fluctuations in the matter of the universe and provide a snapshot of the early universe.

Cosmological constant The energy density of the smooth vacuum, introduced by Albert Einstein in his theory of general relativity to represent the inbuilt tendency of space-time to expand; a constant required to explain the universe as stationary.

Cosmological redshift Reflects the fact that light (photons) from distant galaxies is proportional to their distance from the observer and has been stretched so that the wavelength moves toward the red end of the spectrum; hence redshifted. (*See also* **Redshift.**)

Cosmology Study of the structure and evolution of the universe.

Covalent bond A strong chemical bond that results from the sharing of electrons between two atoms.

Crepuscular A lifestyle characterized by greatest activity around the morning and evening hours (around sunrise and sunset).

Creationism The belief that each different kind of organism was individually created by a supernatural force.

Critical density (Cosmology) The density of atoms (10^{-29}g/cm^3 or six atoms of hydrogen/m^3) that determines the geometry of the universe as flat. A higher density of atoms and the universe would collapse on itself; a lower density and the universe would continue to expand forever.

Cropping (Ecology) The recognition that predators feed on the most abundant prey species, thereby reducing their numbers and allowing other species to use resources formerly monopolized by the dominant prey.

Crossing Over The process in which the chromatids of two homologous chromosomes exchange genetic material. (*See also* **Recombination.**)

Crown Group The last common ancestor of the extant members of a clade and all its descendants. (*See also* **Stem group.**)

Crust (Earth sciences) The exterior portion of the earth floating on the mantle — from which it is separated by the Mohorovicic discontinuity — divided into continental (older) and oceanic (younger) crust, and composed of igneous, sedimentary and metamorphic rocks.

Cryptic A feature that resembles (mimics) the background and so is less visible.

Cryptic (Sibling) species Two or more species that are so similar that they cannot be distinguished morphologically.

Cryptogams A term of convenience, no longer valid taxonomically, that covers all non-seed-bearing plants and other organisms that reproduce by spores, including ferns, mosses, lichen and fungi.

Ctenophora Metazoans commonly known as comb jellies.

Culture The learned behaviors and practices common to a social group.

Cursorial Adapted for running on land.

Cyanobacteria Photosynthetic prokaryotes possessing chlorophyll a but not chlorophyll b. Many are photosynthetic aerobes (requiring oxygen) some are anaerobes (not requiring oxygen). Formerly called blue-green algae, their color is caused by a bluish pigment masking the chlorophyll.

Cytochromes Proteins containing iron-porphyrin (heme) complexes that function as hydrogen or electron carriers in respiration and photosynthesis.

Cytology The study of cells — their structures, functions, components and life histories.

Cytoplasm All cellular material within the plasma membrane, excluding the nucleus and nuclear membrane.

Cytoplasmic inheritance *See* **Maternal inheritance.**

Cytoskeleton Tubular or membrane-bound elements in the cytoplasm, the major ones of which are microtubules and microfilaments.

D

Dark ages (Cosmology) The time during the early life of the universe before the evolution of stars, when ordinary matter in the universe consisted of neutral hydrogen and helium, with a little lithium, and the universe was extremely dense, hot and dark.

Dark energy (Cosmology) A force that does not interact with light but is detectable through its gravitational effects, which accounts for 74% of the density of the universe and which is thought to cause the acceleration of the expansion of the universe.

Dark matter That portion of the universe (22 percent) that neither emits nor absorbs light, which is known only from its gravitational effects on itself and on baryons, and whose composition is unknown.

Darwinism The theory, proposed by Charles Darwin, that biological evolution has led to the many different highly adapted species through natural selection acting on hereditary variations in populations to give rise to descent with modification.

Darwin units (*D*) The amount of change in a character divided by time over which the change occurred.

Deciduous (teeth) Teeth that are replaced during development by permanent teeth.

Degenerate (redundant) code Part of the genetic code for which there is more than one triplet codon for a particular amino acid but where a specific codon cannot code for more than one amino acid. (*See also* **Genetic code.**)

Dehydrogenase An enzyme that catalyzes the removal of hydrogen from a molecule (oxidation).

Deleterious allele An allele whose effect reduces the adaptive value of its carrier, either when present in homozygous condition (recessive allele) or in heterozygous condition (dominant or partially dominant allele).

Deletion A genetic change in which a section of DNA or chromosome has been lost.

Deme A local population of a species; in sexual forms, a local interbreeding group.

Density dependent The dependence of population growth and size on factors directly related to the numbers of individuals in a particular locality (e.g., competition for food, accumulation of waste products).

Density independent The dependence of population growth on factors (climatic changes, meteorite impacts, and so on) unrelated to the numbers of individuals in a particular locality.

Dentary A dermal bone that forms around Meckel's cartilage (the cartilaginous core of the mandible).

Derived character A character whose structure or form differs from the ancestral (basal) character. (*See also* **Basal character.**)

Design (Religion) The concept that the complexity of organisms requires belief in a supernatural power (designer) to explain the diversity of life.

Deuterium An isotope of hydrogen containing one proton and one neutron, giving it twice the mass of an ordinary hydrogen atom.

Deuterostomes Coelomate animal phyla in which the embryonic blastopore becomes the anus.

Development The progression from egg to adult in multicellular organisms.

Developmental control gene A regulatory gene high in a gene network that controls other (often many other) genes.

Diaphragm A sheet-like muscle that separates the abdominal from the thoracic cavity in mammals.

Diapsida (diapsids) The clade of all extant and extinct amniotes with a post-orbital double fenestration of the skull.

Dicotyledons Flowering plants (angiosperms) in which the embryo bears two seed leaves (cotyledons).

Differentiation Changes that occur in the structure and/or function of cells and/or tissues.

Dimorphism Presence in a population or species of two morphologically distinctive types of individuals, for example, differences between males and females, pigmented and nonpigmented forms.

Dinosaurs Extinct terrestrial carnivorous or herbivorous reptiles that existed during the Mesozoic Era; members of the clades Saurischia and Ornithischia

Dioecious Organisms in which the male and female sex organs are in separate individuals.

Diploblastic An animal whose embryos are comprised only of two major germ layers: ectoderm and endoderm (e.g., Cnidaria).

Diploid An organism whose somatic cell nuclei possess two sets of chromosomes (2n), usually one from the male and one from the female parent, providing two different (heterozygous) or similar (homozygous) alleles for each gene.

Directional selection Selection that causes the phenotype of a character to shift in one direction.

Discontinuous variation Character variations that are sufficiently different from each other that they fall into non-overlapping classes.

Displacement (Language) The ability of language to refer to things that are separated from the speaker in time or space.

Disruptive selection Selection that tends to favor the survival of organisms in a population that are at opposite phenotypic extremes for a particular character and eliminates individuals with intermediate values.

Diurnal A lifestyle characterized by activity during the day rather than at night (nocturnal).

Divergent evolution Change leading to differences between lineages.

DNA (deoxyribonucleic acid) A nucleic acid that serves as the genetic material of all cells and many viruses; composed of nucleotides that are usually polymerized into long double-stranded chains, each nucleotide characterized by the presence of a deoxyribose sugar.

DNA ligase An enzyme that joins sections of DNA together.

Domain In molecular biology, an amino acid sequence within a polypeptide chain that performs a particular function. The term has also been used in systematics to provide a tripartite division of organisms — Archaea, Bacteria, Eucarya — as a substitute for the rank of superkingdom, which commonly designates prokaryotes and eukaryotes.

Domestication The modification of plants or animals to suite human needs, often effected through selection.

Dominance (social) The result of behavioral interactions between individuals in a group in which one or more individuals, sustained by aggression or other behaviors, rank higher than others in controlling the conduct of group members.

Dominant allele The allele that determines the phenotype in heterozygotes. (*See also* **Allele.**)

Dorsal The back side or upper surface of an animal; opposite of ventral. (In vertebrates, the surface defined by the location of the spinal column.)

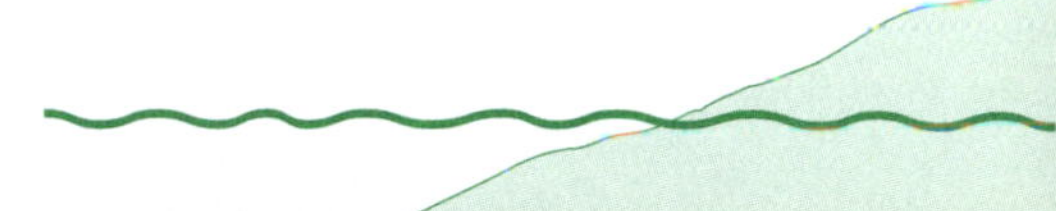

Dosage compensation A mechanism that compensates for difference in the number of X (or Z) chromosomes between males and females, with the consequence that the metabolic activities (gene expression) of their X-linked genes are equalized.

Double fertilization A distinctive feature of angiosperm plants in which two nuclei from a male pollen tube fertilize the female gametophyte, one producing a diploid embryo and the other producing polyploid (usually triploid) nutritional endosperm.

Downstream gene A gene(s) controlled by (downstream of) another gene, usually a regulatory gene. (*See also* **Upstream gene.**)

Drift *See* **Genetic drift.**

Duplication Instances in which a particular section of DNA, chromosome segment, chromosome or entire genome occurs more than once.

Dwarfism (*See* **Insular dwarfism.**)

Dwarf planets A class of planets in our solar system based on size, established in 2006 to accommodate Pluto, Ceres, and UB313. Unlike planets, dwarf planets do not dominate their neighborhoods.

E

Ear ossicles *See* **Middle ear ossicles**

Ecdysoza One of two groups (the other being Lophotrochozoa) of protostomes, Ecdysoza being animals such as arthropods and nematodes with a molt in the life cycle. (*See also* **Lophotrochoza.**)

Ecogeographical rules Generalizations that correlate adaptational tendencies of species with environmental factors such as climate or altitude. (*See also* **Allen's rule, Bergmann's rule, Gloger's rule.**)

Ecological niche The environmental habitat of a population or species, including the resources it uses and its interactions with other organisms.

Ecology The study of the relations between organisms and their environment, in terms of their numbers, distributions and life cycles.

Ecomorph A phenotypic subset of a population of a species associated with a particular environment. (*See also* **Group.**)

Ectoderm The outermost layer of cells that covers early animal embryos, from which nerve tissues and outermost epidermal tissues are derived.

Ectothermic An organism having a variable body temperature primarily determined by the ambient (environmental) temperature.

Ediacaran fauna Soft-bodied fossils found in South Australia and other places, dating to a Precambrian Period lasting 60 or more My and with unknown relationships to prokaryotes or eukaryotes.

Electron carrier In oxidation–reduction reactions, a molecule that acts alternatively as an electron donor (becomes oxidized) and as an electron acceptor (becomes reduced).

Embden-Meyerhof glycolytic pathway The biochemical pathway leading to pyruvic acid and providing a net yield of two high-energy phosphate bonds in ATP used as chemical energy by cells.

Embryology The study of the development of organisms from their inception to birth/hatching, and often into later life history stages such as larvae.

Embryophyte Green (land) plants that use photosynthesis to obtain energy and that reproduce on land.

Enation theory The earliest leaves to evolve, thin microphylls, evolved from extensions of tissues along the stem and not from small branches. (*See also* **Telome theory.**)

Encephalization quotient (EQ) The relative brain size as the ratio of actual brain weight to the brain weight expected for an animal of the same body size.

Endemic A species or population that is specific (indigenous) to a particular geographic region.

Endocranial cast A cast of the inside of the cranial vault.

Endocrine gland A gland that secretes directly into the bloodstream. (*See also* **Exocrine gland.**)

Endocytosis Cellular engulfment of outside material, followed by its transfer into the cellular interior encapsulated in a membrane.

Endoderm The layer of cells that lines the embryonic gut (archenteron) during the early stages of development in animals, and which later forms the epithelial lining of the intestinal tract and internal organs such as the liver, lung and urinary bladder. (*See also* **Archenteron.**)

Endonucleases Enzymes that fragment DNA chains. (*See also* **Restriction enzymes.**)

Endosymbiosis A relationship between two different organisms, in which one (the endosymbiont) lives within the tissues or cell of the other, benefiting either or both. Eukaryotic organelles, such as mitochondria and chloroplasts, had an endosymbiotic prokaryotic origin.

Endothermic A body temperature maintained by internal physiological mechanisms at a level independent of the ambient (environmental) temperature.

Enhancer A region of the DNA (nucleotide sequence) of a gene that binds to another nucleotide sequence within the gene so that the promoter sequence can be transcribed. Individual genes may contain more than one enhancer.

Entropy The measure of disorder of a physical system. In a closed system, to which energy is not added, the second law of thermodynamics essentially states that entropy, or energy unavailable for work, will remain constant or increase but never decrease. Living systems, however, are open systems, to which energy is added from sunlight and other sources, and order can therefore arise from disorder in such systems, that is, energy available for work can increase and entropy can decrease.

Environment The complex of external conditions, abiotic and biotic, that affects and interacts with organisms or populations to provide the facilities and resources that enable hereditary data (genotypes) to produce organismal features (phenotypes).

Enzyme A protein that catalyzes chemical reactions.

Eon A major division of the geological time scale, often divided into two eons beginning from the origin of Earth 4.5 Bya: the Precambrian or Cryptozoic (rarity of life forms) and the Phanerozoic (abundance of life forms).

Epigenesis The theory that tissues and organs are formed by interaction between cells and substances that appear during development, rather than being preformed in the zygote. (*See also* **Preformationism.**)

Epigenetic The sum of the genetic and nongenetic factors that influence gene action.

Epistasis Interaction between nonallelic genes.

Epistatic interactions *See* **Epistasis.**

Epithelium (plural, **Epithelia**) One of the two fundamental types of cellular organization in animals, consisting of a sheet of laterally connected and interacting cells resting on an extracellular matrix — the basement membrane — that is produced by the epithelium. (*See also* **Mesenchyme.**)

Epoch One of the categories into which geological time is divided; a subdivision of a geological period. Periods are often divided into three epochs: Early, Middle and Late; for example, Early Cambrian.

Epoch of recombination (Cosmology) The time 380,000 years after the Big Bang when the temperature had cooled to 3000 K, atoms began to form and the universe became transparent.

Equilibrium (Genetics) The persistence of the same allelic frequencies over a series of generations. Equilibria may be stable or unstable. In a stable equilibrium (e.g., when the heterozygote is superior in fitness to the homozygotes), the population returns to a particular equilibrium value when the allelic frequencies have been disturbed. In an unstable equilibrium (e.g., when the heterozygote is inferior in fitness to the homozygotes), such disturbances are not followed by a return to equilibrium frequencies.

Era A division of geological time that stands between the eon and the period. The Phanerozoic eon is divided into Paleozoic, Mesozoic and Cenozoic eras; and each era is divided into two or more periods.

Estrus (Oestrous) The interval during which female mammals exhibit maximum sexual receptivity, usually coinciding with the release of eggs from the ovary.

Eubacteria Prokaryotes, other than archaebacteria, marked by sensitivity to particular antibiotics, and by the incorporation of muramic acid into their cell walls.

Euchromatin Normally staining chromosomal regions that possesses most of the active genes. (*See also* **Heterochromatin.**)

Eucoelomates *See* **Coelom.**

Eugenics The belief that humanity can be improved by selective breeding.

Eukaryotes Organisms whose cells contain nuclear membranes, mitochondrial organelles and other characteristics that distinguish them from prokaryotes. Eukaryotes may be unicellular or multicellular and include protistans, fungi, plants and animals.

Euploidy State of an organism or cell whose chromosome number is an exact multiple of n, the single (haploid) set of chromosomes (e.g., 2n, 3n, 4n, 5n).

Eutheria *See* **Placentals.**

Eutrophication The process in which an aquatic ecosystem becomes overloaded with nutrients, thereby increasing its organic productivity and causing an accumulation of debris and severe reduction in water quality.

Evo-devo *See* **Evolutionary developmental biology.**

Evolution Genetic changes in populations of organisms through time that lead to differences among them.

Evolutionary developmental biology The field in biology in which developmental processes are studied for their insights into evolutionary processes.

Evolutionarily stable strategies The result of a balance between cooperation and individual action among interacting individuals in a social group.

Evolutionary trees Arrangements of the evolutionary history of one or more groups of organisms presented in a tree-like branching pattern.

Evolutionary universe (Cosmology) A more formal title for the Big Bang hypothesis to describe the origin and evolution of the universe. (*See also* **Big Bang theory.**)

Evolvability The concept that the genotypes of some organisms evolve more rapidly/readily than do the genotypes of other organisms.

Exaptation A character that was adaptive under a prior set of conditions and later provides the initial stage (is "co-opted") for the evolution of a new adaptation under a different set of conditions. (*See also* **Preadaptation.**)

Exocrine gland A gland that secretes into a duct or directly on to an external surface. (*See also* **Endocrine gland.**)

Exon An expressed nucleotide sequence in a gene that is transcribed into messenger RNA and spliced together with the transcribed sequences of other exons from the same gene. Exons are separated from one another by intervening nontranslated sequences (*see* **Intron**). Intron–exon split genes are commonly found in eukaryotes but are almost entirely absent in prokaryotes.

Exon shuffling The recombination or exclusion of exons such that they remain active as sources of genetic information.

Expanding universe The theory that the space-time between galaxies is expanding so that the universe is continuing to increase.

Extant Currently in existence.

Extinction The disappearance of a species or higher taxon.

Extracellular matrix Material secreted by mesenchymal and epithelial cells that surrounds mesenchymal cells and forms a basement membrane for epithelial cells.

Extranuclear inheritance Patterns of heredity in which the transmission is not via nuclear genes, for example, transmission by mitochondrial genes.

F

F (inbreeding coefficient) *See* **Inbreeding coefficient.**

Family A taxonomic category that stands between order and genus; an order may comprise a number of families, each of which contains a number of genera.

Fauna All animals of a particular region or time period.

Fecundity A measure of the potential production of offspring.

Feedback When the products of a process affect its own function.

Fermentation The anaerobic degradation of glucose (glycolysis) or related molecules, yielding energy and organic end products.

Ferns (Pterophyta) Spore-bearing plants that carry their sporangia on the fronds.

Fertility A trait measured by the number of viable offspring produced.

Fertilization The fusion of a sperm with an egg that allows sperm and egg nucleus to fuse as a zygote nucleus and development to be initiated.

Filter feeder An animal that obtains its food by filtering suspended food particles from water.

Fitness Central to evolutionary theory evaluating genotypes and populations, fitness has had many definitions, ranging from comparing growth rates to comparing long-term survival rates. The basic fitness concept that population geneticists commonly use is relative reproductive success, as governed by selection in a particular environment.

Fixation Achievement of a frequency of 100 percent (monomorphism) by an allele or genotype that begins in a population at a lesser frequency (polymorphism).

Fixity of species A theory held by Linnaeus and others that members of a species could only produce progeny like themselves, and therefore each species was fixed in its particular form(s) at the time of its creation.

Flora All plants of a particular region or time period.

Flower (Botany) The reproductive structure of many seed-producing plants; often brightly colored and elaborate in form.

Forebrain The rostral-most portion of the brain in vertebrates divided into two vesicles (diencephalon and telencephalon) that give rise to a majority of the brain in mammals. The diencephalon gives rise to the thalamus, hypothalamus, subthalamus, prethalamus and pretectum. The telencephalon gives rise to the cerebral cortex.

Fossils The geological remains, impressions or traces of organisms that existed in the past.

Founder effect The effect caused when a few individuals ("founders") derived from a large population begin a new colony. Since these founders carry only a small fraction of the parental population's genetic variability, radically different gene frequencies can become established in the new colony. (*See also* **Bottleneck effect.**)

Frequency-dependent selection Instances where the effect of selection on a phenotype or genotype depends on its frequency (e.g., a genotype that is rare may have a higher adaptive value than when it is common).

Frozen accident The theory that an accidental event in the distant past was responsible for the presence of a universal feature in living organisms. Such events may include an accident in which the present genetic code was used by a group of early organisms that managed to survive some population bottleneck, thereby conferring this particular code on later organisms.

G

Galaxy A system of numerous stars, held together by mutual gravitational effects, and often spiral or elliptical in shape. Galaxies, in turn, are grouped into clusters and superclusters. Our own supercluster, centered on Virgo, contains many thousands of galaxies and is more than 100 million light-years across. Large galaxies typically have old, metal-poor stars in their outskirts, and old, metal-rich stars at their center. (*See also* **Milky Way.**)

Gamete A germ cell (eggs in females, sperm in males) that is usually haploid and that fuses with a germ cell of the opposite sex to form a zygote (usually diploid) at fertilization.

Gametophyte The haploid life cycle phase in plants.

Gamma ray A high-frequency, highly penetrating radiation emitted in nuclear reactions.

Gastrula A (typically) cuplike embryonic stage in animal development that follows the blastula stage. Its hollow cavity (archenteron) is lined with endoderm and opens to the outside through a blastopore. (*See also* **Archenteron, Haeckel's gastrula hypothesis.**)

Gene A unit of hereditary genetic material composed of a segment of DNA (sequence of nucleotides), usually with a specific function such as coding for a protein or sometimes coding for RNA.

Genealogy A record of familial ties and ancestral connections among members of a group.

Gene duplication *See* **Duplication.**

Gene family Two or more gene loci in an organism whose similarities in nucleotide sequences indicate they have been derived by duplication from a common ancestral gene.

Gene flow The migration of genes into a population from other populations by interbreeding.

Gene frequency The proportion of a particular allele among all alleles at a gene locus. (Also called allele or allelic frequency.)

Gene locus The chromosomal position (nucleotide sequence) occupied by a particular gene.

Gene mutation *See* **Mutation.**

Gene pair The two alleles present in a diploid organism at a specific gene locus on two homologous chromosomes.

Gene pool All the genes present in the gametes of individuals in a sexually reproducing population. (*See also* **Allele frequency.**)

Gene regulation The processes by which a gene is turned on or off or by which the level of activity of a gene is controlled. (*See also* **cis-regulation.**)

Gene therapy Human-directed repair or replacement of genes that cause inherited diseases. When confined to somatic (body) cells rather than to sex cells (sperm or eggs), such gene repairs are not passed on to future generations.

General theory of relativity Expanding on his initial 1905 theory in which (1) the speed of light is identical for all observers no matter where they are in relation to the light source and (2) observers moving at a constant speed see the same physical laws, Einstein theorized that the relativity of space and time is equal ($E = mc^2$; E is energy, m is mass, and c is the speed of light), meaning that the acceleration of gravity can be calculated by the curvature of space and time.

Genetic code The sequences of nucleotide triplets (codons) on messenger RNA that specify each of the different kinds of amino acids positioned on polypeptides during the translation process. With few exceptions, the genetic code used by all organisms is identical: the 20 amino acids are each specified by the same codons (total: 61 codons), and the same three triplet codons are used to terminate polypeptide synthesis. (*See also* **Degenerate redundant code, Universal genetic code.**)

Genetic crossing over *See* **Crossovers (chromosomes), Recombination.**

Genetic death The inability of a genotype to reproduce itself because of selection.

Genetic distance A measure of the divergence among populations based on their differences in frequencies of given alleles.

Genetic drift The random change in frequency of alleles in a population. These can be caused by sampling errors that are of greater magnitude in small populations or by bottlenecks (founder effects), when population size is suddenly reduced to a few individuals. As population size increases, random drift becomes less important and selection more important in causing gene frequency changes.

Genetic engineering Manipulation of genetic material from different sources to produce new combinations that are then introduced into organisms in which such genetic material does not normally occur.

Genetic load The loss in average fitness of individuals in a population because the population carries deleterious alleles or genotypes. (*See also* **Balanced genetic load, Mutational load.**)

Genetic polymorphism The presence of two or more alleles at a gene locus over a succession of generations. (Called balanced polymorphism when the persistence of the different alleles cannot be accounted for by mutation alone.)

Genome The complete genetic constitution of a cell or individual.

Genotype The genetic constitution of an individual.

Genus (plural, **Genera**) A taxonomic category that stands between family and species. A family may comprise a number of genera, each of which contains a number of species that are presumably related to each other by descent from a common ancestor. In taxonomic binomial nomenclature, the genus is used as the first of two words in naming a species; for example, *Homo* (genus) *sapiens* (species).

Geographic isolation The separation between populations caused by geographic distance or geographic barriers.

Geographic speciation *See* **Allopatric speciation.**

Geological dating The determination of the ages of rocks and strata using a variety of methods. (*See also* **Radiometric dating.**)

Geological strata A series of discrete layers laid down on top of each other as rock by natural forces such as lava flows, siltation, infilling of a marsh, and so on.

Geological time scale The correlation between rocks (or the fossils contained in them) and time periods of the past.

Germ layers *See* **Ectoderm, Endoderm, Mesoderm.**

Germ line *See* **Germ Plasm.**

Germ plasm Cells in animals that are exclusively devoted to transmitting hereditary information to offspring, in contrast to somatic cells, which comprise all other tissues of the body. (*See also* **Germ plasm theory.**)

Germ plasm theory The theory that, in some animals, gametes form from a special part of the egg (the germ plasm) and so cannot be influenced by environmental influences acting on the somatic tissues. (*See also* **Germ plasm.**)

Ghost lineage Extension of the range of an extinct taxon beyond its oldest known occurrence in the fossil record (often noted in phylogenetic trees as a dotted line extending toward the present).

Gloger's rule The generalization that warm-blooded (endothermic) animals tend to have more pigmentation in warm, humid areas than in cool, dry areas.

Glycolysis The energy-producing conversion of glucose to pyruvate under anaerobic conditions (fermentation). Subsequent steps may yield lactic acid or ethanol.

Gondwana The supercontinent in the Southern Hemisphere formed from the breakup of the larger Pangaea landmass about 180 Mya. Gondwana was composed of what are now South America, Africa, Antarctica, Australia and India.

Grade A level of phenotypic organization or adaptation reached by one or more species. Distantly related or

unrelated species that reach the same grade are considered to have undergone parallel or convergent evolution.

Granite A coarse-grained igneous rock commonly intruded into continental crust.

Great Chain of Being The eighteenth century theory that instead of a static universe, there is a continuous progression of stages leading to a superior supernatural being; the transformation of the "Ladder of Nature" into a succession of moving platforms.

Group A population(s) in a species that shares a geographically and/or ecologically identifiable origin and has gene frequencies and phenotypic characters that distinguish it from other groups. Because of the large amount of genetic and phenotypic variability in most species, the number of groups that can be identified is often arbitrary.

Group selection Selection acting on the attributes of a group of related individuals in competition with other groups rather than only on the attributes of an individual in competition with other individuals. (*See also* **Selection.**)

Gymnosperms A group of vascular plants with seeds unenclosed in an ovary (naked); mainly cone-bearing trees.

H

Habitat The place and conditions in which an organism normally lives. (*See also* **Environment.**)

Haeckel's Gastraea theory The theory that metazoans developed from swimming hollow-balled colonies of flagellated protozoans that evolved an anterior-posterior orientation in searching for food. The anterior cells, specialized for digestion, invaginated through a circular blastopore to form a digestive archenteron. This bilayered cup, called a gastrula or Gastraea, was, according to Ernst Haeckel, the progenitor of the gastrula developmental stage found in some present-day metazoans.

Haldane's rule The rule that if one sex is absent, rare or sterile in a cross between two species with sex-determining chromosomes, that sex is the heterogametic one.

Half-life (radioactivity) The time required for the decay of one-half the original amount of a radioactive isotope. Each radioactive isotope has a distinctive and constant half-life.

Haplodiploidy A reproductive system found in some animals, such as bees and wasps, in which males develop from unfertilized eggs and are therefore haploid, while females develop from fertilized eggs and are therefore diploid.

Haploid Cells or organisms that have only one set (1n) of chromosomes.

Haplotype A sequence of nucleotides, restriction sites or marker genes inherited as a linked unit from one parent. Since more than a single genetic locus may be involved, a haplotype may be composed of a string of alleles.

Hardy-Weinberg equilibrium (principle) The conservation of gene (allelic) and genotype frequencies in large populations under conditions of random mating and in the absence of evolutionary forces, such as selection, migration and genetic drift, which act to change gene frequencies.

Hemizygous Genes, such as those on the X chromosome in a male mammal (hemizygote), which are unpaired in a diploid cell.

Hemocoel (Haemocoel) An internal animal body cavity that arises from the internal cavity (blastocoel) of the embryonic blastula. (*See also* **Coelom.**)

Hemoglobin A protein with as basic unit of an iron-containing porphyrin (heme) that reversibly binds oxygen attached to a globin polypeptide chain usually no less than 140 amino acids long.

Herbivores Animals that feed mainly on plants.

Heritability In a general sense, the degree to which variations in the phenotype of a character are caused by genetic differences. Traits with high heritabilities can be more easily modified by selection than traits with low heritabilities. A measure of an organism's potential to respond to selection.

Hermaphrodite An individual possessing both male and female sexual reproductive systems. (*See also* **Monoecious.**)

Heterocercal Fish tail in which the vertebral axis is curved (usually upward).

Heterochromatin A region of the eukaryotic chromosome that stains differently from euchromatin because of its tightly compacted structure. Compared to euchromatin, heterochromatin is characterized by few active genes and many more repetitive DNA sequences. Heterochromatin constitutes about 15 percent of the human genome and about 30 percent of the *Drosophila* genome, much of it located on either side of chromosome centromeres. In *Drosophila*, heterochromatin also constitutes almost the entire Y chromosome.

Heterochrony A term Haeckel proposed to describe changes in timing of an organ's development during evolution. Such changes were used to explain departures from the "recapitulation" of phylogeny expected during ontogeny of descendant species. (*See* **Biogenetic law.**) Its present usage varies but still hinges on a phylogenetic change in developmental timing, whether of one organ relative to other organs, or of one organ relative to the same ancestral organ. Among the consequences of heterochrony are shifts in relative development of reproductive and nonreproductive tissue. (*See also* **Paedomorphosis.**) Such changes can cause an organism (a) to appear more juvenile because its non-reproductive tissues develop more slowly (neoteny), or (b) reach sexual maturity earlier because its reproductive tissues develop more rapidly (progenesis). (*See also* **Heterotopy.**)

Heterodont dentition An animal with structural and functional differences among its teeth, such as incisors and molars.

Heterogametic The sex that produces two kinds of gametes for sex determination in offspring, one kind for males and the other for females. The heterogametic sex is the male in mammals and the female in birds. (*See also* **Sex chromosomes.**)

Heterosis (Hybrid vigor) The increase in vigor and performance that can result when two different, often inbred strains are crossed. Because each inbred parental strain may be homozygous for different deleterious recessive alleles, heterosis has been ascribed by some geneticists to the superiority of heterozygotes. (*See* **Heterozygote advantage.**)

Heterotopy A term Haeckel proposed to describe changes in the position of an organ during evolution. (*See also* **Heterochrony.**)

Heterotroph An organism that cannot use inorganic materials to synthesize the organic compounds needed for growth but obtains them by feeding on other organisms or their products, such as a carnivore, herbivore, parasite, scavenger or saprophyte.

Heterozygote The situation in a genotype or an individual in which the two copies of a gene are different; having different alleles at a particular gene locus on homologous chromosomes (for example, *Aa* in a diploid).

Heterozygote advantage (superiority) The superior fitness of some heterozygotes relative to homozygotes. (*See also* **Heterosis, Overdominance.**)

Hierarchy A term used to designate an ordered grouping of the items within a system, often associated with increasing levels of complexity or organization.

Histones A family of small acid-soluble (basic) proteins that are tightly bound to eukaryotic nuclear DNA molecules and help fold DNA into thick chromosome filaments.

Hitchhiking When a gene persists in a population, not because of selection, but because of close linkage to one or more selected genes. (*See also* **Linkage disequilibrium.**)

Homeobox A transcriptional sequence of 180 base pairs that unites genes known as homeotic, homeobox or Hox genes. Homologous homeobox sequences are found throughout metazoan phyla, and the chromosomal organization of their homeobox-containing genes often follows the linkage order noted in *Drosophila.* The overall conservation of homeobox sequences, of the genes containing them, and of their linkage orders, indicate common developmental functions in different phyla preserved for many hundreds of millions of years, extending back to Precambrian times. (*See also* **Homeodomain, Homeotic genes, Homeotic mutations.**)

Homeodomain A specific protein domain that is shared among a family of transcription factors (homeotic genes) and that is transcribed from a conserved DNA sequence known as the homeobox. (*See also* **Homeobox, Homeotic mutations.**)

Homeostasis A term classically defined by W. B. Cannon to denote the tendency of a physiological system to react to an external disturbance so that the system is not displaced from normal values. Probably its most common use applies to traits that measure or perform at constant values in the face of disturbing forces. An example is the persistence of a specific phenotype although confronted with genetic or environmental differences (e.g., canalization).

Homeotic genes Genes that share the homeobox of 180 base pairs. In *Drosophila,* homeotic genes are organized into two chromosomally separate groups, the antennapedia and bithorax complexes, in which each group contains several independently functioning homeotic genes. Each of these homeotic genes contains a nucleotide sequence (homeobox) coding for a DNA-binding polypeptide (homeodomain) involved in embryonic development along the animal's anterior-posterior axis.

Homeotic mutations (homeosis) Homeosis was defined by William Bateson as "something [that] has been changed into the likeness of something else." In modern genetic usage, homeotic mutations cause the development of tissue in an inappropriate position; for example, the *bithorax* mutations in *Drosophila* produce an extra set of wings. (*See also* **Homeotic genes.**)

Hominid A member of the family Hominidae, which includes humans, whose earliest fossils can now be dated to about 4 Mya (genus *Australopithecus*). Previously used exclusively for humans and the fossil species most closely related to humans. (*See also* **Hominin.**)

Hominin All taxa on the human lineage after separation from the common ancestor with chimpanzee; members of the tribe *hominini.* In recent publications hominin has the same meaning as hominid in older literature.

Hominoids A group (superfamily Hominoidea) that includes hominids (Hominidae), gibbons (Hylobatidae), and apes (Pongidae).

Homogametic The sex that produces only one kind of gamete for sex determination in offspring, thus causing sex differences among offspring to depend on the kind of gamete contributed by the heterogametic sex. The homogametic sex is the female in mammals and the male in birds. (*See also* **Sex chromosomes.**)

Homologous *See* **Homology.**

Homologous chromosomes Chromosomes that pair during meiosis, each pair usually possessing a similar sequence of genes.

Homology The similarity of biological features (genes, structures, behaviors) in different species or groups because of their descent from a common ancestor. Homologous genes do not necessarily produce the same features, because an ancestral gene may be recruited for different functions in different lineages. Also, as is obvious from convergence, functional, morphological and developmental similarities lineages can be produced by nonhomologous genetic elements of independent evolutionary origin.

Homoplasy Character similarity that arose independently in different groups through parallelism or convergence. (*See also* **Convergence, Parallelism.**)

Homozygote The situation in a genotype or an individual in which the two copies of a gene are the same; having the same alleles at a particular gene locus on homologous chromosomes (for example, *AA*).

Horizontal (Lateral) transmission Transfer of genes from one organism to another without reproduction.

Hox gene A homeobox gene in vertebrates, taking the gene name from the first two letters of the orthologous gene in *Drosophila* and adding an x.

Hunter-gatherer societies A form of human subsistence based on hunting animals for meat as well as foraging for other foods such as plants, insects and scavenged meat.

Hybrid breakdown, inviability, sterility Hybrids that suffer from loss of fitness and reproductive failure.

Hybridization (Genetics) Reproduction between individuals from two species to form a fertile offspring (the hybrid).

Hybridization (Molecular biology) The formation of double-stranded molecules of DNA or RNA from a complementary single strand.

Hybrids (hybridization) Offspring of a cross between genetically different parents or groups.

Hybrid vigor *See* **Heterosis.**

Hydrogen bond A weak, noncovalent bond between a hydrogen atom and an electronegative atom such as oxygen.

Hydrogen ion A proton (H^+) that in aqueous solution exists only in hydrated form (H_3O^+, hydronium ion).

Hydrolysis Splitting of a molecule by the addition of the three atoms from a water molecule (H_2O).

Hydrophilic A compound (e.g., charged molecule) or part of a compound (e.g., polar group) that has an affinity for water molecules.

Hydrophobic Compounds, such as lipids, which do not readily interact with water but tend to dissolve in organic solvents.

Hydrostatic pressure The pressure exerted by a liquid. When the liquid is in an elastic, muscularly controlled container (e.g., the coelom of a worm), changes in shape of the container can be effected by muscularly generated hydrostatic pressure.

Hypoxia The reduction of oxygen supply to tissues.

I

Idealism The philosophy that all objects, including the universe, have no independent existence apart from the minds of those perceiving them; perceiving objects as ideal forms.

Igneous rock A rock such as basalt (fine-grained) and granite (coarse-grained), formed by the cooling of molten material from Earth's interior.

Imaginal disks Clusters of cells set aside in the larvae of insects such as *Drosophlia* from which the structures of the adult fly are formed.

Implantation (Mammals) The attachment and invasion of the embryo into the uterine wall.

Imprinting (Behavior) The learning process by which newborn/hatched organisms associate with an object (usually a parent or an adult of the same species, or an environment) and orient their behavior toward that object. Fish imprinting on a home river, newly hatched chicks imprinting on a parent are two well-studied examples.

Imprinting (Genetics) The silencing of genes in germ cells by methylation of DNA. As a state, imprinting is usually sex-specific and is removed after development of a new individual begins from the imprinted gamete(s).

Inbreeding Mating between genetically related individuals, often resulting in increased homozygosity in their offspring.

Inbreeding coefficient (F) The probability that the two alleles of a gene in a diploid organism are identical because they originated from a single allele in a common ancestor.

Inbreeding depression Decrease in the average value of a character, or in growth, vigor, fertility and survival, as a result of inbreeding.

Inclusive fitness The fitness of an allele or genotype measured not only by its effect on an individual but also by its effect on related individuals that also possess it (kin selection).

Incus Meaning "anvil," the incus is the middle of the sequence of three ossicles in the mammalian middle ear. (*See also* **Malleus, Stapes.**)

Independent assortment A basic principle of Mendelian genetics — that a gamete will contain a random assortment of alleles from different chromosomes because chromosome pairs orient randomly toward opposite poles during meiosis.

Industrial melanism The effect of soot and other dark-colored pollution in industrial areas in increasing the frequency of darkly pigmented (melanic) forms perhaps because of selection by predators against nonpigmented or lightly pigmented forms.

Ingroup Taxa more closely related to one another (monophyletic) than they are to any other group.

Inheritance The transmission of information from generation to generation. (*See also* **Blending inheritance, Extranuclear inheritance, Maternal inheritance.**)

Inheritance of acquired characters The theory used by Lamarck to explain evolutionary adaptations — that phenotypic characters acquired by interaction with the environment during the lifetime of an individual are transmitted to its offspring.

Innovation A feature in a descendant not found in its ancestors.

Insectivore An animal (usually a mammal) that feeds primarily on insects.

Instinct An inherited (innate), relatively inflexible behavior pattern that is often activated by one or several environmental factors (releasers).

Insular dwarfism The process of the reduction of body size in organisms isolated on islands.

Intercentrum (plural, **intercentra**) The anterior element, which with the posterior element or pleurocentrum, forms individual vertebrae. (*See also* **Pleurocentrum.**)

Intrinsic rate of natural increase The potential rate at which a population can increase in an environment free of limiting factors.

Introgressive hybridization The incorporation of genes from one species into the gene pool of another because some fertile hybrids are produced from crosses between the two species.

Intron A nucleotide sequence (region of DNA) within a gene that is transcribed to produce mRNA but the mRNA is not translated into protein.

Intrusion (Geology) Igneous rock inserted within or between geological strata rather than being deposited on Earth's surface.

Inversion An aberration in which a section of DNA or chromosome has been inverted 180 degrees, so that the sequence of nucleotides or genes within the inversion is now reversed with respect to its original order in the DNA or chromosome.

Ion An atom or group of atoms carrying a positive or negative electrostatic charge.

Isolating mechanisms Biological mechanisms that act as barriers to gene exchange or interbreeding between populations.

Isomerase An enzyme that catalyzes the rearrangement of atoms within a molecule.

Isotope One of several forms of an element, with a distinctive mass based on the number of neutrons in the atomic nucleus. Radioactive isotopes decay at a rate that is constant for each isotope and release ionizing radiation as they decay. (*See* also **Half-life, Radiometric dating.**)

Iteroparous Reproducing on several distinct occasions during a lifetime. (*See also* **Semelparous.**)

K

Karyotype The characteristic chromosome complement of a cell, individual or species.

Kelvin scale (K) A temperature scale in which absolute zero (the point at which molecules oscillate at their lowest possible frequency, 273°C) is designated as 0 K, and the boiling point of water as 373 K.

Kilocalorie (kcal) Unit comprising 1,000 calories. (*See also* **Calorie.**)

Kingdom The highest inclusive category of taxonomic classification. Each kingdom includes phyla or subkingdoms. (*See* also **Domain, Supergroups**)

Kin selection Selection effects (e.g., altruism) that influence the survival and reproductive success of genetically related individuals (kin). This contrasts with selection confined solely to an individual and its own offspring. (*See also* **Inclusive fitness.**)

Knuckle-walking Quadrupedal gait of chimpanzees and gorillas, performed by curling the fingers toward the palm of the hand and using the backs (dorsal surfaces) of the knuckles to support the weight of the front part of the body.

Krebs cycle The cyclic series of reactions in the mitochondrion in which pyruvate is degraded to carbon dioxide and hydrogen protons and electrons. The latter are then passed into the oxidative phosphorylation pathway to generate ATP.

K-selection Selection based on a population being maintained at or near the limit of its carrying capacity; selection is theoretically for improved competitive ability rather than for rapid numerical increase.

K value *See* **Carrying capacity.**

L

Lactation Formation and secretion of milk in maternal mammary glands for nursing offspring; a distinctive characteristic of mammals.

Ladder of Nature A concept based on Aristotle's view (the Scale of Nature) that nature can be represented as a succession of stages or ranks that leads from inanimate matter through plants, lower animals, higher animals and finally to the level of humans. (*See also* **Great Chain of Being.**)

Lamarckian inheritance The concept that the phenotype of an organism is itself hereditary: that characters acquired or lost during life experience, as well as characters that organisms acquire in order to meet environmental needs, can be transmitted to offspring. Lamarck proposed that it is through such means that changes in organisms (evolution) takes place. (*See also* **Inheritance of acquired characters, Use and disuse.**)

Language A structured system of communication among individuals using vocal, visual or tactile signs to describe thoughts, feelings, concepts and observations. Rather than communication, some writers emphasize the representational nature of language, defining it as a symbolic system used to store and retrieve information about experiences and concepts.

Larva (plural, **larvae**) A sexually immature stage in various animal groups, often with a form and diet distinct from those of the adult.

Larynx A structure in tetrapods formed at the junction of the trachea (airway) and pharynx. Comprised of a cartilaginous skeleton, it contains the vocal folds and serves to separate the respiratory tract from the gastrointestinal tract.

Lateral transmission. *See* **Horizontal transmission.**

Laurasia The supercontinent in the Northern Hemisphere (comprising what is now North America, Greenland, Europe and parts of Asia) formed from the breakup of Pangaea about 180 Mya.

Learning Acquisition of a behavior through experience.

Lethal allele An allele whose effect prevents its carrier from reaching sexual maturity when present in homozygous condition for a recessive lethal or in either heterozygous or homozygous condition for a dominant lethal. (*See also* **Allele.**)

Life The capability of performing various organismal functions such as metabolism, growth and reproduction of genetic material.

Ligand A signaling molecule that by binding to a receptor molecule in the cell membrane initiates the next step in a gene pathway.

Life cycle The series of stages that takes place between the formation of zygotes in one generation of a species and the formation of zygotes in the next generation. (Also life history: the series of stages experienced by an individual of a species, from birth to death.)

Life-history trait Features associated with the survival and reproduction of an individual such as number of offspring produced, age at reproduction, growth rate.

Light-year The distance traveled by light, moving at 186,000 miles a second, in a solar year; approximately 6×10^{12} miles or 9.5×10^{12} kilometers.

Lineage An evolutionary sequence, arranged in linear order from an ancestral (stem) group or species to a descendant (crown) group or species (or *vice versa*).

Linkage (Genetics) The occurrence of two or more gene loci on the same chromosome.

Linkage disequilibrium The absence of linkage equilibrium (i.e., the presence of nonrandom associations between alleles at different loci). (*See also* **Hitchhiking.**)

Linkage equilibrium The attainment of genotypic frequencies in a population indicating that recombination between two or more gene loci has reached the point at which their alleles are now found in random genotypic combinations.

Linkage map The linear sequence of known genes on a chromosome obtained from recombination data.

Lipids Organic compounds such as fats, waxes and steroids that tend to be more soluble in organic solvents of low polarity (e.g., ether, chloroform) than in more polar solvents (e.g., water).

Lissamphibia The clade that includes all recent amphibians (frogs, toads, newts, salamanders, caecilians).

Lithosphere (Earth sciences) The term for the crust plus the uppermost portion of the mantle of Earth.

Living fossil An existing species whose similarity to a fossil taxon indicates that very few morphological changes have occurred over a long period of geological time.

Locus (plural, **loci**) The site (nucleotide sequence) on a chromosome occupied by a specific gene. Some researchers use it more broadly as a synonym of gene.

Logistic growth curve Population growth that follows a sigmoid (S-shaped) curve in which numbers increase slowly at first, then rapidly, and finally level off as the population reaches its maximum size or carrying capacity for a particular environment.

Longevity The average life span of individuals in a population.

Lophotrochozoa One of two groups (the other being Ecdysoza) of protostomes, Lophotrochozoa being animals such as mollusks, annelids, brachiopods with tentacles. (*See also* **Ecdysoza.**)

Lumbar The region of the trunk pertaining to those vertebrae that are caudal or inferior to the last rib-bearing vertebra and the sacrum (pelvis).

M

Macroevolution The pattern of evolution at and above the level of the species. Fossils are our chief evidence for macroevolution. (*See also* **Macroevolution, Punctuated equilibrium.**)

Macromolecules Very large polymeric molecules such as proteins, nucleic acids and polysaccharides.

Macromutation A theory that explains the origin of a new species or higher taxonomic category by a single large mutation rather than by selection acting on many mutations. Although most geneticists agree that mutations can produce major as well as minor developmental changes, no single mutation is yet known that can cause an instantaneous speciation event, probably because such a sudden large radical change would dislocate normal genetic and developmental processes.

Magnetosphere A magnetic "bubble" that envelops Earth, shielding it from incoming radiation, cosmic rays, ionized particles and from particles carried by solar winds.

Malleus Small, hammer-shaped bone (ossicles) in the inner ear that connects the eardrum with the incus and transmits sound. (*See also* **Incus, Stapes.**)

Mammals Homeothermic vertebrates that suckle their offspring with milk produced in the mammary glands, have hair, three middle ear ossicles and a neocortex region in the brain.

Mammary glands One or more pairs of ventrally placed glands used by mammalian females for nursing offspring. (*See also* **Lactation.**)

Mandible The lower jaw of vertebrates

Mantle (Earth sciences) A layer of partly plastic or ductile rock 2900 km thick, that has experienced repeated melting and crystallization, and which constituting four-fifths of Earth's volume.

Marsupials Mammals of the infraclass Metatheria possessing, among other characters, a reproductive process in which tiny live young at an early stage of development are born and then nursed in a female pouch (marsupium).

Maternal inheritance Transmission of heredity information from the female parent by deposition of gene products into the egg cytoplasm.

Maximum likelihood estimates A statistical method based on probability distribution that finds the maximum estimate of the likely distribution for a set or sets of data.

Meiosis The two eukaryotic cell (maturation) divisions that produce haploid gametes (animals) or spores (plants) from a diploid cell. One is a reduction division that ensures that each gamete or spore contains one representative of each pair of homologous chromosomes in the parental cell.

Meiotic drive *See* **Segregation distortion.**

Mendelian population A group of interbreeding, diploid individuals that exchange genes through sexual reproduction.

Mendel's laws *See* **Independent assortment, Segregation.**

Mesenchyme One of the two fundamental cell types in animals, consisting of loosely connected cells in an extracellular matrix that may be "solid" as in bone or fluid as in blood. (*See also* **Epithelium.**)

Mesoderm The embryonic tissue layer between ectoderm and endoderm in triploblastic animals that gives rise to muscle tissue, kidneys, blood, internal cavity linings, and so on.

Mesoglea The gelatinous layer located between the ectoderm and endoderm in cnidarians and ctenophorans.

Mesozoa A clade (phylum) of some 50 species, all parasitic on marine invertebrates, considered by some to be degenerate, but usually regarded as a separate phylum. (*See also* **Placozoa.**)

Mesozoic era The middle era of the Phanerozoic eon, covering an approximately 220-million-year interval between the Paleozoic (ending about 248 Mya) and Cenozoic (beginning about 65 Mya). It is marked by the origin of mammals in the earliest period of the era (Triassic), the dominance of dinosaurs throughout the last two periods of the era (Jurassic and Cretaceous), and the origin of angiosperms.

Messenger RNA (mRNA) An RNA molecule produced by transcription from a DNA template, bearing a sequence of triplet codons used to specify the sequence of amino acids in a polypeptide.

Metabolic pathway A sequence of enzyme-catalyzed reactions that convert a precursor substance to one or more end products.

Metabolism A network of enzyme-catalyzed reactions used by living organisms to maintain themselves.

Metacentric A chromosome whose centromere is at or near the center.

Metamerism Division of an animal body or a major portion of the body into a series of similar segments along the anterior-posterior axis. (*See also* **Segmentation.**)

Metamorphic rock Rock that has been subjected to high but non-melting temperatures and pressures, causing chemical and physical changes.

Metamorphosis (Zoology) The transition from one form into another during the life cycle, for example, a larva into an adult.

Metaphyta The Kingdom containing the green plants; also known as Embryophyta, Plantae).

Metatheria *See* **Marsupials.**

Metazoa Multicellular animals.

Microevolution Evolutionary changes within populations of a species. (*See also* **Macroevolution.**)

Micro RNA (miRNA) Short (18 to 25 nucleotide) sequences of noncoding single-stranded RNA that regulate the translation of proteins in plants and animals by binding to matching target mRNAs leading to destruction of the mRNA.

Microsatellites Tandem repeats of short di-, tri-, and tetra-nucleotide sequences such as cytosine-adenine-cytosine-adenine-cytosine-adenine, and so on. Such loci are abundant (humans are estimated to possess at least 35,000 loci for repeats of the C–A sequence) and mutate (change in sequence number) at a relatively high rate.

Microspheres Microscopic membrane-bound spheres formed when proteinoids are boiled in water and allowed to cool. Some cell-like properties, such as osmosis, growth in size and selective absorption of chemicals, have been ascribed to them. (*See also* **Coacervates, Protocells.**)

Middle ear ossicles Small bones in the middle ear of tetrapods that transmit sound from the eardrum to the oval window of the inner ear. (*See also* **Incus, Malleus, Stapes.**)

Migration (Ecology) Movement of a population to a different geographical area or its periodic passage from one region to another.

Migration (Genetics) The transfer of genes from one population into another by interbreeding (gene flow).

Milky Way (Cosmology) A large spiral galaxy containing more than 200 billion stars and their planets, including our own solar system and surrounded by more than a dozen much smaller galaxies. (*See also* **Galaxy.**)

Mimicry Resemblance of individuals in one species (mimics) to individuals in another (models) as a result of selection. (*See also* **Batesian mimicry, Müllerian mimicry.**)

Missing link A specimen or species of an extinct organism thought to be intermediate between two clades and so to provide a link between them.

Mitochondria Organelle in eukaryotic cells that use an oxygen-requiring electron transport system to transfer chemical energy derived from the breakdown of food molecules to ATP. Mitochondria have their own genetic material (circular DNA without histones) and generate some mitochondrial proteins by using their own protein-synthesizing apparatus. (Most mitochondrial proteins are coded by nuclear DNA and produced on cytoplasmic ribosomes.)

Mitosis The mode of eukaryotic cell division that produces two daughter cells possessing the same chromosome complement as the parent cell.

Modern synthesis (Evolutionary theory) *See* **Neo-Darwinism.**

Modifier gene A gene whose effect alters the phenotypic expression of one or more genes at loci other than its own.

Modularity The concept that units of life, such as gene networks, aggregations of cells and organ primordia, develop and evolve as units (modules) that interact with other modules.

Molecular clock The rate at which nucleotides are substituted over evolutionary time.

Monocotyledons Flowering plants (angiosperms) in which the embryo bears one seed leaf (cotyledon).

Monoecious An individual bearing both male and female organs. (*See also* **Hermaphrodite.**)

Monomers The subunits linked together to form a polymer (e.g., nucleotides in nucleic acids, amino acids in proteins, sugars in polysaccharides).

Monophyletic A taxonomic group united by having arisen from a single ancestral lineage.

Monotremes Egg-laying mammals, presently restricted to Australasia; echidnas (*Tachyglossus, Zaglossus*) and the platypus (*Ornithorhynchus*).

Morganucodonts The earliest mammals with a dentary-squamosal joint, which is a mammalian synapomorphy.

Morphogenesis Development of the form (morphology) of an organism or part of an organism.

Morphological species Assemblages of individuals with shared structural features that allow them to be separated from other assemblages.

Morphology The study (science) of the anatomical form and structure of organisms.

Morphospace A three-dimensional representation of the morphological features of an organisms or groups of organisms used to show how much of the possible range of morphologies is expressed.

Müllerian mimicry Sharing of a common warning coloration or pattern among a number of species that are all dangerous or toxic to predators; resemblances maintained because of common selective advantage.

Muller's ratchet The generalization that because of sampling errors, populations more easily lose that class of individuals bearing the fewest harmful mutations, so that classes with increasing numbers of such mutations tend to increase with time.

Multigene family *See* **Gene family.**

Mutagenesis Production of mutations by chemical treatment or radiation.

Mutation A change in gene structure and often function; change in the nucleotide sequence of genetic material whether by substitution, duplication, insertion, deletion or inversion.

Mutational load That portion of the genetic load caused by production of deleterious genes through recurrent mutation.

Mutualism A relationship among different species in which the participants benefit.

Mya Million years ago.

N

Natural scientist (natural historian) The term used before the twentieth century when referring to individuals we would now refer to as biologists, botanists, zoologists, physicists, and so forth.

Natural selection Differential reproduction or survival of replicating organisms caused by agencies other than humans. Because such differential selective effects are widely prevalent, and often act on hereditary (genetic) variations, natural selection is a common major cause for a change in the gene frequencies of a population that leads to a new distinctive genetic constitution (evolution). (*See also* **Adaptation, Artificial selection, Fitness, Selection.**)

Nebular hypothesis Originally proposed by Emanuel Kant and Laplace in the eighteenth century to explain the origin of the universe; a cloud of gas and dust (a nebula) collapses as gravitational force overcome the pressure of the gas.

Negative eugenics Proposals to eliminate deleterious genes from the human gene pool by identifying their carriers and restraining or discouraging their reproduction.

Neocortex The outer layer of the cerebral hemispheres of the mammalian brain; the part of the brain associated with higher cognitive functions such as spatial reasoning and (in humans) language.

Neo-Darwinism The theory (also called the Modern synthesis) of evolution as a change in the frequencies of genes introduced by mutation, with natural selection as the most important, although not the only, cause for such changes.

Neoteny The retention of juvenile morphological traits in the sexually mature adult. (*See also* **Heterochrony, Paedomorphosis, Progenesis.**)

Neural crest A fourth germ layer found in vertebrates and from which arise skeletal tissues, cells that form the dentine of teeth, pigment cells, peripheral nerves and ganglia, some hormone-synthesizing cells, valves and septa of the heart and other cell types.

Neutral mutation A mutation that does not affect the fitness of an organism in a particular environment.

Neutral theory of molecular evolution The theory that most mutations that contribute to genetic variability (genetic polymorphism on the molecular level) consist of alleles that are neutral in respect to the fitness of the organism and that their frequencies can be explained in terms of mutation rate and genetic drift.

Neutron stars (pulsars) The collapsed core of an exploding star or supernova Type II.

Niche *See* **Ecological niche.**

Nocturnal A lifestyle characterized by nighttime activity.

Nondisjunction The failure of homologous chromosomes (or sister chromatids) to separate from each other during one of the two meiotic anaphase stages and go to opposite poles, leading to one daughter cell having both homologues (or sister chromatids) and the other daughter cell none.

Nonrandom mating *See* **Assortative mating.**

Nonsense mutation A mutation that produces a codon that terminates the translation of a polypeptide prematurely. Such codons were previously called "nonsense" codons but are now generally called "stop codons" or "chain termination codons."

Notochord The dorsal axial supporting rod found in all chordates.

Nucleic acid An organic acid polymer, such as DNA or RNA, composed of a sequence of nucleotides.

Nucleotide A molecular unit consisting of a purine or pyrimidine base, a ribose (RNA) or deoxyribose (DNA) sugar and one or more phosphate groups.

Nucleus A membrane-enclosed eukaryotic organelle that contains all the histone-bound DNA in the cell (i.e., practically all the cellular genetic material).

Numerical taxonomy A statistical method for classifying organisms by comparing them on the basis of measurable phenotypic characters and giving each character equal weight. The degree of overall similarity between individuals or groups is then calculated, and a decision is made as to their classification.

O

Occipital condyle Articular facets on the base of the skull that articulate with the first cervical vertebra.

Omnivores Animals that feed on both plants and animals.

Ontogeny The development of an individual from zygote to maturity.

Operon A cluster of coordinately regulated structural genes. In prokaryotes and a few eukaryotes, this cluster is transcribed as a unit into a single long ("polycistronic") messenger RNA molecule that is then translated into a sequence of individual gene products that often function together.

Order A taxonomic category between class and family. A class may contain a number of orders, each of which contains a number of families.

Organelles Functional intracellular membrane-enclosed bodies such as nuclei, mitochondria and chloroplasts.

Organic Carbon-containing compounds. Also refers to features or products characteristic of organisms or life.

Organic catalyst *See* **Catalyst.**

Organism A living entity. (*See* **Life.**)

Orthogenesis The concept that evolution proceeds in a particular direction because of internal or vitalistic causes.

Orthologous genes Gene loci in different species that are sufficiently similar in their nucleotide sequences (or amino acid sequences of their protein products) to suggest they originated from a common ancestral gene.

Outgroup A taxon that diverged from an ingroup before the members of the ingroup diverged from one another. (*See also* **Ingroup.**)

Overdominance Instances when the phenotypic expression of a heterozygote (for example, A_1A_2) is more extreme than that of either homozygote (e.g., A_1A_1 or A_2A_2). Overdominance has been considered a cause for hybrid vigor. (*See* **Heterosis.**)

Oviparous Females that lay eggs that develop outside the body.

Oxidation-reduction Reactions in which electrons are transferred from one atom or molecule (the reducing agent, which is oxidized by losing electrons) to another (the oxidizing agent, which is reduced by gaining electrons).

Oxidative phosphorylation A process that produces ATP by transferring electrons to oxygen.

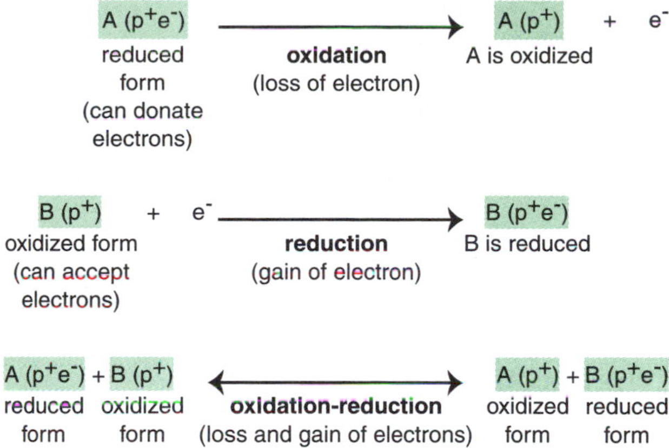

Oxidizing (Chemistry) A chemical reaction carried out by agents that have the ability to combine with oxygen to form other substances. (*See also* **Reducing.**)

Ozone (O_3) A molecule consisting of three atoms of oxygen.

P

Paleobiology The study of the biology of extinct organisms and their ecosystems.

Paleomagnetism The magnetic fields of ferrous (iron-containing) materials in ancient rocks. Among other applications, paleomagnetism provides information on the position of landmasses and continents relative to Earth's magnetic poles at the time that the rocks were formed and thus can be used to describe the historical movement of continents relative to each other (continental drift).

Paleontology The study of extinct fossil organisms or traces of organisms.

Paleozoic (Palaeozoic) era The first era of the Phanerozoic eon extending from 545 to about 248 Mya.

Pangaea A very large supercontinent formed about 250 Mya comprising most or all of the present continental landmasses. (*See also* **Gondwana, Laurasia.**)

Pangenesis The theory of heredity, held by Darwin and others, that small, particulate "gemmules," or "pangenes" are produced by each of the various tissues of an organism and sent to the gonads where they are incorporated into gametes.

Panin (Anthropology) The clade of primates including hominins and chimpanzees.

Panmixis (panmictic) *See* **Random mating.**

Panspermia The theory that life was introduced on Earth from elsewhere in the universe.

Parallel evolution The evolution of similar characters in related lineages that do not share a recent common ancestor. (*See also* **Convergence.**)

Paralogous genes Two or more different gene loci in the same organism that are sufficiently similar in their nucleotide sequences (or in the amino acid sequences of their protein

products) to indicate they originated from one or more duplications of a common ancestral gene. (*See also* **Gene family.**)

Parapatric Geographically adjacent, nonoverlapping species or populations that at the zone of contact do not interbreed.

Paraphyletic A taxonomic grouping which includes some descendants of a single common ancestor, but not all.

Parasitism An association between species in which individuals of one (the parasite) obtain their nutrients by living on or in the tissues of the other species (the host), often with harmful effects to the host.

Parental investment Parental provision of resources to offspring with the effect of increasing the offspring's reproductive success.

Parsimony method Choice of a phylogenetic tree that minimizes the number of evolutionary changes necessary to explain species divergence.

Parthenogenesis Development of an egg without fertilization.

Partial (incomplete) dominance Instances where two different alleles of a gene in a heterozygote produce a phenotypic effect intermediate between the effects produced by the two homozygotes.

Pedomorphosis The incorporation of adult sexual features into immature developmental stages. Causes for such effects are changes in developmental speed of sexual tissues relative to nonsexual tissues. In progenesis, sexual development is so rapid that the sexually mature form may remain quite small in size. In neoteny, sexual maturation proceeds normally, but somatic development slows down, producing a full-sized sexual adult with juvenile appearance. A classic neotenous form is the Mexican axolotl, an amphibian salamander that can retain its gills even when sexually mature. (*See also* **Heterochrony.**)

Pelagic Refers to an entire body of water and the organisms within it, excluding the bottom (benthic) zone and the area close to land.

Pentadactyly (Pentadactyl) Literally, "five fingers," referring to the theory that the digits of tetrapods are built on an ancient five-digit plan, although we now know that the first tetrapods had more than five digits.

Peptide (polypeptide) An organic molecule composed of a sequence of amino acids covalently linked by peptide bonds (a bond formed between the amino group of one amino acid and the carboxyl group of another through the elimination of a water molecule).

Period (Geological) A major subdivision of an era of geological time distinguished by a particular system of rocks and associated fossils.

Peripatric Populations that bud off from the geographic periphery of a parental population and become genetically isolated.

Phagocytic Cellular engulfment of external material. (*See also* **Endocytosis.**)

Phanerozoic eon A major division of the geological time scale marked by the relatively abundant appearance of fossilized skeletons of multicellular organisms, dating from about 545 Mya to the present.

Pharyngeal jaws An additional set of jaws that develop from posterior pharyngeal arches in some teleost fish.

Pharyngula *See* **Phylotypic stage.**

Phenetics *See* **Cladistics.**

Phenotype The characters that constitute the structural, functional and behavioral; properties of an organism.

Phenotypic plasticity Variation in the phenotype expressed in response to environmental changes and indicative of underlying genotypic plasticity.

Philtrum A vertical groove or depression at the midline of the face between the nose and the upper lip.

Phloem The conducting tissue of green land plants that takes nutrient to all parts of the plant. (*See also* **Xylem.**)

Phosphorylation The addition of one or more phosphate groups (HPO_4^{2-}) to a compound (e.g., the phosphorylation of ADP to ATP).

Photoautotroph *See* **Autotroph.**

Photon The quantum of electromagnetic energy, a particle with zero mass, no electric charge and an indefinitely long lifetime.

Photosynthesis The synthesis of organic compounds from carbon dioxide and water through a process that begins with the capture of light energy by chlorophyll.

pH scale [The negative logarithm of the hydrogen ion (H^-) concentration in an aqueous solution.] A scale used for measuring acidity (pH less than 7) and alkalinity (pH greater than 7), given that pure water has a neutral pH of 7.

Phyletic evolution Evolutionary changes within a single nonbranching lineage. Although new species are produced by this lineage over time (chronospecies) there is no increase in the number of species existing at any one time. (*See also* **Anagenesis.**)

Phylogenetic branching *See* **Cladogenesis, Phylogenetic evolution.**

Phylogenetic evolution (Also called branching evolution.) Evolutionary changes that produce two or more lineages that diverge from a single ancestral lineage. (*See also* **Cladogenesis.**)

Phylogenetic systematics *See* **Cladistics**

Phylogenetic tree A branching diagram showing the relationships and evolutionary lineages of one or more groups of organisms. (*See also* **Tree of life.**).

Phylogeny The evolutionary history of a species or group of species in terms of their derivations and connections. A phylogenetic tree is a schematic diagram designed to represent that evolution — ideally, a portrait of genetic relationships.

Phylogeography The study of the evolutionary processes regulating the geographic distributions of lineages/groups by reconstructing genealogies of individual genes, groups of genes or populations.

Phylotypic stage A proposed stage in embryonic development that characterizes basic features in the body plan of a phylum, such as the pharyngula of vertebrates.

Phylum (plural, **phyla**) The major taxonomic category below the level of kingdom, used to include classes of organisms that may be phenotypically quite different but share some general features or body plan.

Placenta A mammalian organ formed by union between the female uterine lining and embryonic membranes that provides nutrition to the embryo, allows exchange of gases, and aids elimination of embryonic waste products.

Placentals Mammals of the infraclass Eutheria, possessing, among other features, a reproductive process that uses a placenta to nourish their young until a relatively advanced stage of development compared to other mammalian groups (monotremes and marsupials).

Placozoa A clade (phylum) comprised of the single species *Trichoplax adhaerens,* with uncertain affinities to metazoans. (*See also* **Mesozoa.**)

Planetesimals (Cosmology) Aggregations of dust and gas that form once aggregates reach one kilometer in diameter, allowing gravity to influence their formation. As they expand planetesimals can become protoplanets and finally planets.

Planula hypothesis The theory that metazoans evolved from small organisms, each consisting of a solid balls of cells (planula) similar to embryonic stages of sponges and Cnidaria.

Plasma membrane The boundary membrane consisting of phospholipids and proteins surrounding the cytoplasm of a cell.

Plasmid A self-replicating, circular DNA element that can exist outside the host chromosome. There are various kinds, some maintaining more than one copy per cell.

Plasticity. *See* **Phenotypic plasticity.**

Plate tectonics The geological theory that Earth's crust is comprised of moving plates. (*See also* **Tectonic plates.**)

Pleiotropy Phenotypic effects of a single gene on more than one character.

Plesiomorphy Instances when a species character is similar to that character in an ancestral species.

Pleurocentrum (plural, **pleurocentra**) The posterior element, which with the anterior element or interocentrum, forms individual vertebrae. (*See also* **Interocentrum.**)

Pollen (Botany) Small grains composed of protein, produced by the male organs (anthers) of seed plants (flowers, trees, grasses, weeds) and containing the male DNA.

Polygene A gene that interacts with other genes to produce an aspect of the phenotype.

Polymer A molecule composed of many repeating subunits (monomers) linked together by covalent bonds.

Polymerase An enzyme that catalyzes the synthesis of a polymer by linking its component monomers together.

Polymerase chain reaction (PCR) A laboratory technique that can replicate a sequence of DNA nucleotides into millions of copies in a very short time.

Polymorphism The presence of two or more genetic or phenotypic variants in a population. Usually refers to genetic variations where the frequency of the rarest type is not maintained by mutation alone. (*See also* **Balanced polymorphism; Seasonal polymorphism.**)

Polypeptide *See* **Peptide.**

Polyphyletic The presumed derivation of a single taxonomic group from two or more different ancestral lineages through convergent or parallel evolution.

Polyploidy When the number of chromosome sets (n) is greater than the diploid number ($2n$).

Polytene chromosomes Giant chromosomes consisting of many chromatids arranged in parallel resulting in bands that reveal chromosome structure.

Polytypic species A species consisting of individuals of two or more forms which may be varieties, races or subspecies.

Population A group of conspecific organisms occupying a more or less well-defined geographical region and exhibiting reproductive continuity from generation to generation. (*See also* **Deme, Mendelian population.**)

Population explosion A term used, usually for human populations, for rapid increases in numbers of individuals, with the implication that the rate of increase is unsustainable

Position effect The effect of a gene on the phenotype because of a change in the position of the gene on the chromosome.

Postadaptive mutations The concept that appropriate mutations arise only after bacteria have encountered a selective agent.

Preadaptation A character that was adaptive under a prior set of conditions and later provides the initial stage (is "co-opted") for the evolution of a new adaptation under a different set of conditions. (*See also* **Exaptation.**)

Precambrian eon A major division of the geological time scale that includes all eras from Earth's origin about 4.5 Bya to the beginning of the Phanerozoic eon, about 545 Mya. The Precambrian (also known as the Cryptozoic) is marked biologically by the appearance of prokaryotes about 3.5 Bya and small, non-skeletonized multicellular organisms in the Ediacarian period about 50 or 60 My before the Phanerozoic.

Predation The killing and consumption of one living organism — the prey — by another, the predator.

Preformationism The theory that an organism is preformed at conception in the form of a miniature adult and that embryonic development consists of enlargement of preformed structures.

Prions The smallest agent of infection, composed of a hydrophobic protein but neither DNA nor RNA.

Progenesis When sexual maturity of an animal occurs earlier in the life cycle than in related taxa because the reproductive

tissues develop more rapidly than do somatic tissues. (*See also* **Heterochrony, Neoteny.**)

Progenote The hypothetical ancestral cellular form that gave rise to archaebacteria, eubacteria and eukaryotes.

Progress (Evolution) A controversial concept in evolutionary biology that evolution has been accompanied by change reflected in increasing complexity (*See also* **Complexity.**)

Prokaryotes Single-celled organisms (including Eubacteria and Archaea) that lack histone-bound DNA, endoplasmic reticulum, a membrane-enclosed nucleus and other cellular organelles found in eukaryotes.

Promoter A DNA nucleotide sequence that enables transcription (RNA synthesis) by serving as the starting point for transcription and as the site for binding the enzyme RNA polymerase. (*See also* **Enhancer.**)

Protein A macromolecule composed of one or more polypeptide chains of amino acids, coiled and folded into specific shapes based on its amino acid sequences.

Proteinoids Synthetic polymers produced by heating a mixture of amino acids. Some show protein-like properties in respect to enzyme activity, color test reactions, hormonal activity, and so on.

Protista One of the four eukaryotic kingdoms; includes protozoa, algae, slime molds and some other groups. (Called "Protoctista" by some biologists.)

Protocell A membrane-bounded system containing molecules considered an early step in the origin of cells. (*See also* **Coacervates, Microspheres.**)

Protogalaxy (Cosmology) A galaxy in the process of formation.

Protoplanet (Cosmology) A forming planet that arises from the aggregation of planetesimals in a protoplanetary disc.

Protoplasm Cellular material within the plasma membrane but outside the nucleus.

Protostomes Coelomate phyla in which the embryonic blastopore becomes the mouth.

Prototheria *See* **Monotremes.**

Prototissue The term for the precursors of tissues as seen in sponges.

Pseudocoelomates Animals with a coelom derived from a persistent embryonic blastocoel, largely unlined with mesodermal tissue.

Pulsars *See* **Neutron stars.**

Punctuated equilibrium The view that evolution of a lineage follows a pattern of long intervals in which there is relatively little change (stasis or equilibrium), punctuated by short bursts of speciation during which new taxa arise.

Purine A nitrogenous base composed of two joined ring structures, one five-member and one six-member, commonly present in nucleotides as adenine (A) or guanine (G).

Pyrimidine A nitrogenous base composed of a single six-member ring, commonly present in nucleotides as thymine (T), cytosine (C) or uracil (U).

Q

Quadrate A bone in nonmammalian vertebrates that contributes to the jaw joint by articulating with the articular, and that is modified as an ear ossicle in mammals. (*See also* **Articular.**)

Quadrupedal Tetrapod locomotion using all four limbs.

Quantitative character A character whose phenotype can be numerically measured or evaluated; a character displaying continuous variation.

Quantitative trait loci (QTLs) Regions of a chromosome containing genes (alleles) that influence a quantitative trait such as height or weight.

Quantum evolution A rapid increase in the rate of evolution over a relatively short period of time.

R

Racemic mixture A mixture of two kinds of molecules whose structures are similar but which are mirror images of each other (one kind cannot be superimposed on the other). Each of the two molecular forms rotates the plane of polarized light in a particular direction, but the racemic mixture is optically inactive.

Radiation (phylogenetic) *See* **Adaptive radiation.**

Radioactivity Emission of radiation by certain elements as their atomic nuclei undergo changes.

Radiometric dating (Earth sciences) The dating of rocks by measuring the proportions present of a radioactive isotope and the stable products of its decay. Since the rate at which a particular radioactive isotope decays is constant, these proportions provide an estimate of the age of the rock, which can often be confirmed by dating with other radioactive elements.

Random genetic drift *See* **Genetic drift.**

Random mating Mating within a population regardless of the phenotype or genotype of the sexual partner (panmixis).

Range The geographical limits of the region habitually traversed by an individual or occupied by a population or species.

Reaction norm A measure of the responsiveness of a feature of the phenotype to a range of levels or concentrations of an environmental factor, temperature or predator abundance for example. A reaction norm is plotted as the phenotypic response (size, shape, number of elements, and so forth) against the environmental parameter (temperatures, predator abundance, and so forth).

Receptor A protein in the cell membrane that binds to a ligand to allow gene action to continue.

Recessive allele An allele without phenotypic effect in a heterozygote.

Recessive lethal An allele whose presence in homozygous condition causes lethality. (*See also* **Lethal allele.**)

Reciprocal altruism A mutually beneficial exchange of altruistic behavioral acts between individuals. (*See also* **Altruism.**)

Recombinant DNA A DNA molecule composed of nucleotide sequences from different sources.

Recombination (Genetics) A chromosomal exchange process (*see* **Crossovers**) that produces offspring that have gene combinations different from those of their parents. (Also used by some geneticists to describe the results of independent assortment.)

Red giant (Cosmology) A star that has exhausted its core hydrogen fuel so that only the center is hot enough to burn and in which hydrogen fusion continues in a shell around the core.

Red Queen hypothesis The theory that adaptive evolution in one species of a community causes a deterioration of the environment of other species. As a consequence, each species must evolve as fast as it can in order "to stay in the same place" (to survive).

Redshift (Cosmology) The degree to which the photons reaching Earth have been stretched so that their wavelength moves toward the red end of the spectrum.

Reducing (Chemistry) A chemical reaction carried out by reducing agents that brings about the reduction of another chemical substance at the same time as being and oxidized. (*See also* **Oxidizing.**)

Reduction-oxidation *See* **Oxidation-reduction.**

Reductionism The theory that explanations for events at one level of complexity can or should be reduced to explanations at a more basic level. For example, that all biological events should be explained in the form of chemical reactions.

Regulation (developmental biology) The ability of an embryo or part of a developing embryo to compensate for the loss of parts.

Regulator gene A gene that controls other genes, either by turning them on or off or by regulating their rate.

Relative fitness The relative reproductive success of an allele or genotype as compared to other alleles or genotypes.

Repetitive DNA DNA nucleotide sequences that are repeated many times in the genome.

Replication Doubling of DNA.

Repressor A regulator gene that produces a repressor (usually a protein) that binds to a particular nucleotide sequence and prevents transcription.

Repressor protein *See* **Repressor.**

Reproductive isolation The absence of gene exchange between populations. (*See* **Isolating mechanisms.**)

Reproductive success The proportion of reproductively fertile offspring produced by a genotype relative to other genotypes. (*See also* **Fitness.**)

Restriction enzymes Enzymes that recognize particular nucleotide sequences and cut DNA molecules at or near those sequences. (*See also* **Endonucleases.**)

Restriction fragment length polymorphisms (RFLPs) Differences between individuals in the size of DNA fragments for a particular DNA section cut by restriction enzymes. These are inherited in Mendelian fashion, and furnish a basis for estimating genetic variation. They also provide linkage markers used to track mutant genes between generations.

Retinal Tapetum *See* ***Tapetum lucidum***

Ribosomal RNA (rRNA) RNA sequences that are incorporated into the structure of ribosomes.

Ribosomes Intracellular particles composed of ribosomal RNA and proteins that furnish the site at which messenger RNA molecules are translated into polypeptides.

Ribozymes Sequences of RNA nucleotides that can perform catalytic roles.

RNA (ribonucleic acid) A typically single-strand nucleic acid, characterized by the presence of a ribose sugar in each nucleotide, whose sequences serve either as messenger RNA, ribosomal RNA or transfer RNA in cells, or as genetic material in some viruses. In contrast to the base composition of DNA, RNA usually bears uracil instead of thymine.

RNA editing Information changes in RNA molecules by the addition, deletion, or transformation of ribonucleotide bases after these molecules have been transcribed from their DNA templates.

RNA interference (RNAi) The process of regulating gene transcription using small interference RNA (siRNA). A class of short (18 to 25 nucleotide) sequences of non-coding RNA that repress the translation of proteins by binding to matching target mRNAs leading to degradation of the mRNA.

RNA splicing The joining of exons by the excision of introns.

RNA world The theory that RNA nucleotide sequences possessing catalytic and self-replicating capabilities predated catalytic protein systems in pre-biological times.

***r*-selection** Selection in populations subject to rapidly changing environments with highly fluctuating food resources. Theoretically, selection in such populations emphasizes adaptations for rapid population growth rather than for the competitive ability experienced in K-selected populations.

S

Saltation The theory that new species or higher taxa originate abruptly.

Sampling error (gene frequencies) Variability in gene frequencies caused by the fact that not all samples taken from a population have the same gene frequency as the population itself.

Saprophyte An organism that feeds on decomposing organic material.

Sauropsids A clade of amniotes that includes all recent and all or almost all extinct reptiles (excluding the Synapsida), and birds.

Scavenger An organism that habitually feeds on animals that died naturally or accidentally, or that were killed by another carnivore.

Sea-floor spreading Expansion of oceanic crust through the deposition of mantle material along oceanic ridges. (*See also* **Continental drift, Tectonic plates.**)

Seasonal polymorphism The presence of different morphological types of a species in different seasons, for example, dry and wet season forms of a butterfly. (*See also* **Polymorphism.**)

Sedimentary rock Rock formed by the hardening of accumulated particles (sediments) that had been transported by agents such as wind and water. Sedimentary rocks are the prime source of fossils. (*See also* **Geological strata.**)

Seed (Botany) A complex structure of plants containing the embryo along with parental diploid and haploid tissues. (*See also* **Pollen.**)

Segmentation The repetition of body structures along an animal's anterior-posterior axis, as found generally in annelids, arthropods and chordates. (*See also* **Metamerism.**)

Segregation The Mendelian principle that the two different alleles of a gene pair in a heterozygote segregate from each other during meiosis to produce two kinds of gametes in equal ratios, each with a different allele.

Segregational genetic load *See* **Balanced genetic load.**

Segregation distortion (meiotic drive) Aberrant segregation ratios among the gametes produced by heterozygotes because of the presence of certain alleles (segregation distorters).

Selection A composite of all the forces that cause differential survival and differential reproduction among genetic variants. When the selective agencies are primarily those of human choice, the process is called artificial selection; when the selective agencies are not those of human choice, it is called natural selection. Although evolutionary biologists recognize other factors that contribute to genetic change, and, therefore, to evolution (e.g., Mutation, Genetic drift), selection remains the most commonly accepted cause to account for organismal adaptive features. However, selection does not have the foresight nor can development supply the means to enable a single population to face every eventuality; that is, although selection is a cause for evolutionary change, the amount and direction of change is limited by an organism's past history. (*See also* **Fitness, Relative fitness.**)

Selection coefficient (symbol *s*) A relative measure of the effect of selection, usually in terms of the loss of fitness endured by a genotype, given that the genotype with greatest fitness has a value of 1.

Self-assembly The spontaneous aggregation of macromolecules into biological configurations that can have functional value.

Selfish DNA The theory that the persistence of DNA sequences with no discernible cellular function (e.g., various repetitive DNA sequences) arises from the likelihood that, once present in the genome, they are impossible to remove without the death of the organism — that is, they act as "selfish," or "junk" DNA, which the cell has no choice but to replicate along with functional DNA.

Semelparous organisms that breed only once during their lifetimes (*See also* **Iteroparous.**)

Semipermeable membrane A membrane that selectively permits transmission of certain molecules but not others.

Senescence The process of deterioration with aging.

Septum (plural, **septa**) A dividing wall or partition between sections of an organism.

Serial ("iterative") homology Similarities between parts of the same organism, such as the vertebrae of a vertebrate or the different kinds of hemoglobin molecules produced by a mammal. The genetic basis for such homology can often be ascribed to gene duplications that have diverged over time but still produce somewhat similar effects. (*See* also **Gene family, Paralogous genes.**)

Sessile Attached to a substrate. An organism whose behavior is mostly nonmotile.

Sex chromosomes Chromosomes associated with determining the difference in sex. These chromosomes are alike in the homogametic sex (for example, XX) but differ in the heterogametic sex (for example, XY).

Sex linkage Genes linked on a sex chromosome.

Sex ratio The relative proportions of males and females in a population.

Sexual dimorphism When males and females of a species have distinctive phenotypes.

Sexual reproduction Zygotes produced by the union of genetic material from different sexes through gametic fertilization.

Sexual selection Selection that acts directly on mating success through direct competition between members of one sex for mates (intrasexual selection), or through choices made between them by the opposite sex (epigamic selection) or through a combination of both selective modes. In any of these cases, sexual selection may cause exaggerated phenotypes to appear in the sex on which it is acting (large antlers, striking colors, and so on).

Short tandem repeat polymorphisms (STRPs) Sequences or numbers of repeating nucleotide strings distributed throughout the genome.

Sibling species Species so similar to each other morphologically that they are difficult to distinguish but that are reproductively isolated. (Sometimes called "cryptic species.")

Sickle cell anemia Destruction of red blood cells in humans homozygous for the autosomal sickle cell gene, H^S; clinically presents as a usually fatal form of hemolytic anemia.

Signal transduction The process whereby a receptor is activated inside a cell to regulate gene activity.

Sister group (sister taxon) A group (taxon) that is the closest relative of another group (taxon). Derives from the concept that each significant evolutionary step marks a dichotomous split that produces two sister taxa equal to each other in rank.

Small interference RNA (siRNA) Twenty to twenty-five-nucleotide RNA sequences that assemble into RNA-induced silencing complexes (RISCs) that they guide to complementary RNA sequences, which they cleave and destroy.

Snowball Earth The theory that for 10 My about 600 Mya when the average temperature was around -40°C, Earth was enveloped in a blanket of ice as much as 1 km thick, and that four such episodes may have occurred between 750 and 600 Mya.

Social Darwinism The theory that social and cultural differences in human societies (political, economic, military, religious, and so on) arise through processes of natural selection, similar to those that account for biological differences among populations and species.

Sociobiology The study of the biological basis of social behavior.

Somatic cells (or tissues) All body cells (also known as soma) other than those that produce sperm or eggs. (*See* also **Germ plasm.**)

Space-time (Cosmology) The four-dimensional construct proposed to contain all celestial bodies (stars, planets, galaxies, black holes). Because it continues to inflate, space-time is responsible for the ongoing expansion of the universe.

Speciation The splitting of one species into two or more new species (*See* **Cladogenesis, Phylogenetic evolution.**) or the transformation of one species into a new species over time. (*See also* **Anagenesis, Phyletic evolution.**)

Species A basic taxonomic category for which there are various definitions. Among these are an interbreeding or potentially interbreeding group of populations reproductively isolated from other groups (the biological species concept) and a lineage evolving separately from others with its own unitary evolutionary role and tendencies (Simpson's evolutionary species concept). Employing the terms of population genetics, some definitions can be combined into the concept that a species is a population of individuals bearing distinctive genes and gene frequencies, separated from other species by biological barriers preventing gene exchange.

Split gene A gene whose nucleotide sequence is divided into exons and introns.

Spontaneous generation The theory that complex organisms can appear spontaneously from inert materials without biological parentage; life without parents.

Sporangia The organ in plants and fungi that produces or contains spores.

Sporophyte The diploid spore-producing stage of plants.

Squamosal A cranial bone that forms much of the temporal region and part of the mammalian zygomatic (malar) arch.

Stabilizing selection Selection that favors the survival of organisms in a population that are at an intermediate phenotypic value for a particular character, thus eliminating extreme phenotypes. (Also called "normalizing selection.")

Stapes Meaning "stirrup," the incus is the innermost of the sequence of three ossicles in the mammalian middle ear. It transmits sound vibrations to the oval window of the inner ear. (*See also* **Incus, Malleus.**)

Stasis A period of time (usually geological) without evident evolutionary change. (*See also* **Stabilizing selection.**)

Stem cells Cells of multicellular organisms capable of producing more than one different cell type.

Stem group The group of extinct organisms considered closest to and more basal than the most basal members of a clade. (*See also* **Crown group.**)

Stop codon One of the three messenger RNA codons (UAA, UAG, UGA) that terminates the translation of a polypeptide. (Also called "chain-termination codon" or "nonsense codon.")

Stone tools Stones that have been modified with the intention of use.

Strata *See* **Geological strata.**

Stromatolites Laminated rocks produced by layered accretions of benthic microorganisms (mainly filamentous cyanobacteria) that trap or precipitate sediments.

Structural gene A DNA nucleotide sequence that codes for RNA or protein. Some definitions restrict this term to a protein-coding gene.

Subduction (Earth sciences) The process by which a tectonic plate descends (is subducted) beneath the edge of another plate into the mantle, often giving rise to earthquakes and/or volcanic activity.

Subspecies A taxonomic subdivision of a species often distinguished by special phenotypic characters and by its origin or localization in a given geographical region. Like other species subdivisions (*see* **Group**), a subspecies can still interbreed successfully with the remainder of the species. However, in some cases, interbreeding capabilities are unknown, and subspecies designations (e.g., *Homo sapiens neanderthalensis* and *Homo sapiens sapiens*) are based entirely on phenotype.

Supergroups Major groupings of eukaryotes at levels that replace kingdoms in some classifications. (*See also* **Kingdoms.**)

Supernova *See* **Type Ia supernova, Type IIa supernova.**

Survivorship The proportion of individuals born at a given time (cohort) who survive to a given age.

Symbiont A participant in the interactive association (symbiosis) between two individuals or two species. This term is often restricted to mutually beneficial associations (mutualism).

Sympatric Species or populations whose geographical distributions coincide or overlap.

Sympatric speciation Speciation that occurs between populations occupying the same geographic range.

Symplesiomorphy A trait (feature, character) shared between two or more taxa and also shared with other taxa with which those two or more taxa share a last common ancestor; a shared or ancestral state of a character.

Synapomorphy The possession by two or more related lineages of the same phenotypic character derived from a different but homologous character in the ancestral lineage.

Synapsids All extant and extinct amniotes with a single fenestration in the temporal region of the skull (all 4,500 species of extant mammals and their fossil relatives).

Syntax The rules that underlie the structure of language.

Synteny The retention of homologous genes on the same chromosome in different species, irrespective of their linkage order.

Synthetic theory of evolution *See* **Neo-Darwinism.**

Systematics Although defined by G. G. Simpson as the study of the diversity of organisms and all their comparative and evolutionary relationships, it is often used interchangeably with the terms classification and taxonomy.

T

Tactile communication Communication by physical touch (e.g., grooming in primates).

Tapetum lucidum A retinal layer that reflects incoming light back through the retina.

Taxon (plural, **taxa**) A taxonomic unit at any level of classification.

Taxonomy The principles and procedures used in classifying organisms.

Tectonic plates The fairly rigid plates composing Earth's crust whose boundaries are marked by earthquake belts and volcanic chains. In oceanic regions, accretions to these plates occur at mid oceanic ridges (sea-floor spreading), and they are subducted under other plates at the deep oceanic trenches. Continental masses ride on some of these plates, accounting for continental drift and such processes as the mountain building that occurs when these plates collide.

Teleology The theory that natural processes such as development or evolution are guided by their final stage (*telos*) or for some particular purpose, for example, "the reason plants engage in photosynthesis and animals seek food is for survival, and the ultimate purpose of survival is for reproductive success."

Telocentric Chromosomes whose centromeres are located at one end.

Telome theory Thin branches (*telomes*) evolved toward greater complexity and vascularization to produce the leaves and branches of ferns and vascular plants, or regressed toward a single unbranched form, producing bryophytes. (*See also* **Enation theory.**)

Terrestrial On the ground; on or of the planet Earth.

Tetrapod Literal meaning, "four-footed." Commonly used to specify a member of the land-evolved vertebrates: amphibians, reptiles and mammals.

Thallophyta A division of plants that includes fungi (thought to be a form of algae) and algae.

Therapsids Previously known as "mammal-like reptiles," composed mainly of fairly large herbivorous and carnivorous forms, which were dominant reptilian stocks during the Permian and Triassic periods. Mammals are descended from the cynodont therapsids.

Therian A subclass of Mammalia consisting of marsupial and placental mammals. This group is defined by the possession of tribosphenic molars.

Thoracic Pertaining to the chest region; extent defined by those vertebrae bearing ribs.

Tissue A group of cells that all perform a similar function in a multicellular organism.

Tooth replacement The development and eruption of two or more sets of teeth during the life of an individual vertebrate.

Trait *See* **Character.**

Transcription The process by which the synthesis of an RNA molecule (e.g., messenger RNA) is initiated and completed on a DNA template by RNA polymerase enzyme.

Transcription factors Gene products that bind to specific (regulatory) sequences of other genes to control their activity.

Transfer RNA (tRNA) Relatively small RNA molecules (about 80 nucleotides long) that carry specific amino acids to the ribosome for polypeptide synthesis. Each kind of tRNA has a unique anticodon complementary to messenger RNA codons that specify the placement of particular amino acids in the polypeptide chain.

Transgenic An organism (sometimes a cell) containing a gene from another organism (or cell) that has been incorporated into its genome.

Translation The protein-synthesizing process that takes place on the ribosome, linking together a particular sequence of amino acids (polypeptide) on the basis of information received from a particular sequence of codons on messenger RNA.

Translocation The movement during division of a sequence of nucleotides to a different position in the genome.

Transposable elements *See* **Transposons.**

Transposons (transposable elements) Nucleotide sequences that produce enzymes to promote their own movement from one chromosomal site to another and that may carry additional genes such as those for antibiotic resistance.

Tree of Life A phylogenetic tree of all the organisms on Earth today. Based on the proposal by Charles Darwin that the similarities and differences between organisms reflected a single, branched and hierarchical tree of nature.

Triploblastic An animal that produces all three major types of cell layers during development — ectoderm, endoderm and mesoderm.

Tritium An isotope of hydrogen that has three times the mass of an ordinary hydrogen atom.

Tympanic membrane The ear drum or tympanic membrane consists of a thin skin-like tissue that separates the middle and outer ears of tetrapods.

Typology The study of organic diversity based on the principle that all members of a taxonomic group conform to a basic plan, and variation among them is of little or no significance. (*See also* **Archetype.**)

U

Ultraviolet radiation Electromagnetic radiation at wavelengths between about 4 and 400 nanometers, shorter than visible light but longer than X-rays. UV radiation is absorbed by purine and pyrimidine ring structures and is therefore quite damaging to nucleic acid genetic material.

Unequal crossing over The result of improper pairing between chromatids, causing their crossover products to differ from each other in the amounts of genetic material.

Ungulate A hoofed mammal.

Uniformitarianism The theory in earth sciences, popularized by Lyell, that none of the forces active in past Earth history were different from those active today.

Universal genetic code The use of the same genetic code in all living organisms. (A few codons differ from the universal code in mitochondria, mycoplasmas and some ciliated protozoa.)

Upstream gene A regulatory gene(s), that controls (is upstream of) another gene or genes. (*See also* **Downstream gene.**)

Use and disuse A theory used by Lamarck to explain evolution as resulting from the transmission of characters that became enhanced or diminished because of their use or disuse, respectively, during the life experience of individuals. (*See also* **Lamarckian inheritance.**)

Uterus An organ formed as a specialization of the reproductive tract in mammals in which the early embryo implants. In placental mammals, an elaborate placenta develops within the uterus around the embryo.

V

Variability The propensity of genotypes or phenotypes to vary.

Variation A term commonly used to indicate differences in the qualitative or quantitative values of a character among individual members of a population, whether molecules, cells or organisms.

Vascular plants Land plants that have special water- and food-conducting vessels and tissues (xylem and phloem).

Vector A vehicle (e.g., plasmid or virus) used to carry a genetically engineered DNA sequence into a cell.

Vertebrae *See* **Intercentrum, Pleurocentrum.**

Vertical transmission Transmission of heredity from parent to offspring.

Vestigial organs Organs or structures that appear to be small and functionless but can be shown to be homologous with ancestral organs and structures that were larger and functional.

Virus A small intracellular infectious agent, often parasitic, often composed of DNA or RNA in a protein coat, and that depends on the host cell to replicate its genetic material and to synthesize its proteins.

Visual-Spatial Reasoning The ability to use visual and spatial information for problem solving.

Vitalism The concept that the activities of living organisms cannot be explained by any underlying physical or chemical principles but arise from unknowable internal or supernatural causes.

Viviparity (viviparous reproduction) Mode of reproduction in which eggs develop into live young before leaving the maternal body.

W

Wernicke's area A region of the brain, located where the temporal and parietal lobes meet, that is critical to the comprehension of speech and the formulation of coherent and meaningful speech.

White dwarf (Cosmology) The name for the final dying stage in the life of a planet, visible as a faint object in the sky.

Wild type The most commonly observed phenotype or genotype for a particular character. Variations from wild type are considered mutants.

X

X chromosome The name given in various groups to a sex chromosome usually present twice in the homogametic sex (XX) and only once in the heterogametic sex (XY or XO).

X-linked genes Genes present on the X chromosome. (*See* **Sex linkage.**)

Xylem The conducting tissue of green land plants that takes water to all parts of the plant. (*See also* **Phloem.**)

Y

Y chromosome A sex chromosome present only in the heterogametic (XY) sex.

Z

Zootype A proposed stage in development characterized by the expression of a particular set of genes that governs spatial development in animals.

Zygote The cell formed by the union of male and female gametes.

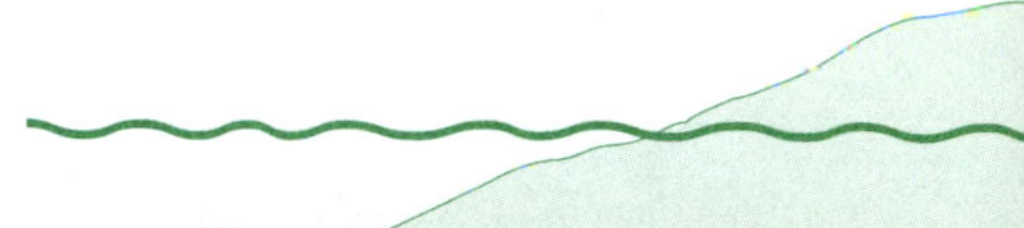

Index

Note: Pages containing a box, figure, or table with a list of data or summary information are indicated by *b*, *f* or *t* before the page number. For information on particular topics such as groups of organisms (e.g., plants, vertebrates, metazoans), molecules (e.g., amino acids) or evolutionary concepts and processes (e.g., adaptive landscape, alleles, selection) check under the appropriate entry. For information on origins, evolution, character definitions, extinctions and relationships see the entries for individual topics such as those above and see the entries for "Origin of," "Evolution," "Characters defining major groups," Extinctions" and "Phylogenetic relationships," respectively.

A

D

E

F

H

I

P

Q

R

S

V

W

X

Y

Z

Photo Credits

All Part Opener Images: © Photos.com

Chapter Opener Images

Chapter 1: © Trout55/ShutterStock, Inc.; **Chapter 2:** © Photos.com; **Chapter 3:** © Canoneer/ShutterStock, Inc.; **Chapter 4:** Courtesy of NASA/JPL-Caltech and The Hubble Heritage Team (STScI/AURA); **Chapter 5:** © Jim Lopes/ShutterStock, Inc.; **Chapter 6:** © Trout55/ShutterStock, Inc.; **Chapter 7:** © Artsilensecome/ShutterStock, Inc.; **Chapter 8:** Courtesy of Monte Later/Yellowstone National Park/NPS; **Chapter 9:** © AbleStock/Alamy Images; **Chapter 10:** © Sebastian Kaulitzki/ShutterStock, Inc.; **Chapter 11:** © Ken Schulze/ShutterStock, Inc.; **Chapter 12:** © Photodisc; **Chapter 13:** © Melba Photo Agency/Alamy Images; **Chapter 14:** © Richard Robinson/ShutterStock, Inc.; **Chapter 15:** Photo by Melissa Mott, courtesy of Nicole King, Ph.D., University of California, Berkeley; **Chapter 16:** © Thierry Maffeis/ShutterStock, Inc.; **Chapter 17:** © emin kuliyev/ShutterStock, Inc.; **Chapter 18:** © Lynsey Allan/ShutterStock, Inc.; **Chapter 19:** © Chris Hill/ShutterStock, Inc.; **Chapter 20:** © Laurie L. Snidow/ShutterStock, Inc.; **Chapter 21:** © Robert Adrian Hillman/ShutterStock, Inc.; **Chapter 22:** © Anita Huszti/ShutterStock, Inc.; **Chapter 23:** © EcoPrint/ShutterStock, Inc.; **Chapter 24:** © Ian Cartwright/StockByte/age fotostock; **Chapter 25:** © Antonio Petrone/ShutterStock, Inc.; **Chapter 26:** © Valenta/ShutterStock, Inc.

Profile Images

Chapter 1: © Kris Snibbe/Harvard University News Office; **Chapter 4:** Courtesy of Doctor Virginia Trimble, University of California, Irvine; **Chapter 6:** Courtesy of Doctor William Schopf, Professor of Paleobiology and Director of IGPP CSEOL, University of California; **Chapter 8:** Courtesy of Lynn Margulis, University of Massachusetts (photo by Jerry Bauer); **Chapter 12:** Courtesy of Doctor Walter M. Fitch, Professor of Ecology and Evolutionary Biology at the University of California, Irvine; **Chapter 15:** Photo courtesy of Doctor James W. Valentine, University of California, Berkeley; **Chapter 18:** Courtesy of Doctor Michael Benton, Professor of Vertebrate Paleontology at the University of Bristol; **Chapter 23:** Courtesy of Doctor B. Rosemary Grant, Lecturer in Ecology and Senior Research Biology, Princeton University; **Chapter 24:** Courtesy of Doctor Francisco J. Ayala, Professor of Philosophy at the University of California.

Unnumbered Images

Page 313 (shell): © Photos.com; **Page 313** (bread): © Jones and Bartlett Publishers. Photographed by Kimberly Potvin; **Page 313** (sunflower): © Photos.com; **Page 313** (DNA): © Photos.com; **Page 595** (fly): © Frank B. Yuwono/ShutterStock, Inc.; **Page 635** (ape): © Norma Cornes/ShutterStock, Inc.; **Page 635** (crows): © Behavioral Ecology Research Group, University of Oxford.

Metric and US Conversions and Abbreviations for Units of Measure

Length/Distance

1 millimeter (mm) = 0.039 inches (in)

1 inch (in) = 25.4 mm

1 meter (m) = 3.2808 feet (ft)

1 foot = 0.3048 m

1 kilometer (km) = 0.6214 miles

1 mile =1.6093 km

Weight

1 gram (g) = 0.0352 ounces (oz)

1 ounce (oz) = 28.350 grams (g)

16 oz = 1 pound (lb)

1 lb = 454 g

Volume

1 liter (L) = 0.264 US gallons (gal)

1 US gallon (gal) = 3.785 liter (L)

Temperature Scale Conversions

°Celsius → °Fahrenheit	°C = (°F – 32)/1.8
°Fahrenheit → °Celsius	°F = (1.8 × °C) + 32
Kelvin (K) → °Celsius	K = °C – 273.15
°Celsius → Kelvin	°C = K + 273.15

Thus, 3 K is –270 °C (–454 °F).
The freezing point of water is 273 K (0 °C = 32 °F).
The boiling point of water is 373 K (100 °C = 212 °F).